MOHSEN MOSTAJABI, PE
ATLANTA, GA.

Contents—2004 HVAC SYSTEMS AND

Contents—2003 HVAC APPLICATIONS

2006 ASHRAE® HANDBOOK

REFRIGERATION

Inch-Pound Edition

American Society of Heating, Refrigerating and Air-Conditioning Engineers, Inc.

1791 Tullie Circle, N.E., Atlanta, GA 30329

(404) 636-8400 http://www.ashrae.org

DEDICATED TO THE ADVANCEMENT OF

THE PROFESSION AND ITS ALLIED INDUSTRIES

Volunteer members of ASHRAE Technical Committees and others compiled the information in this handbook, and it is generally reviewed and updated every four years. Comments, criticisms, and suggestions regarding the subject matter are invited. Any errors or omissions in the data should be brought to the attention of the Editor. Additions and corrections to Handbook volumes in print will be published in the Handbook published the year following their verification and, as soon as verified, on the ASHRAE Internet Web site.

DISCLAIMER

ASHRAE has compiled this publication with care, but ASHRAE has not investigated, and ASHRAE expressly disclaims any duty to investigate, any product, service, process, procedure, design, or the like that may be described herein. The appearance of any technical data or editorial material in this publication does not constitute endorsement, warranty, or guaranty by ASHRAE of any product, service, process, procedure, design, or the like. ASHRAE does not warrant that the information in this publication is free of errors. The entire risk of the use of any information in this publication is assumed by the user.

ISBN 1-931862-86-9
ISSN 1930-7195

The paper for this book is acid free and was manufactured with post-consumer pulp from sources using sustainable forestry practices.

CONTENTS

DISTRIBUTION OF CHILLED AND FROZEN FOOD

INDUSTRIAL APPLICATIONS

LOW-TEMPERATURE APPLICATIONS

REFRIGERATION EQUIPMENT

UNITARY REFRIGERATION EQUIPMENT

GENERAL

Additions and Corrections

Index

Composite index to the 2003 HVAC Applications, 2004 HVAC Systems and Equipment, 2005 Fundamentals, and 2006 Refrigeration volumes

Comment Pages

CONTRIBUTORS

In addition to the Technical Committees, the following individuals contributed significantly to this volume. The appropriate chapter numbers follow each contributor's name.

David F. Ward (1)
HEC Energy Services

Thomas K. O'Donnell (1, 36)
Gallo Winery

Robert A. Jones (2)
Sporlan Valve

Gene Troy (2)
IIAR

Ronald A. Cole (3)
R.A. Cole and Associates, Inc.

Piotr Domanski (3)
NIST

Martin Timm (4)
Praxair, Inc.

George C. Briley (4, 26)
Technicold Services, Inc.

Robert Doerr (5)
Trane Co.

Tom Werkema (5)
Arkema, Inc.

David Wilson (5)
Honeywell International, Inc.

Alan Cohen (6)
UOP LLC

Jay E. Field (6)
Trane Co.

Cary Haramoto (6)
Parker-Hannifin Corp.

Fred F. Polley, Jr. (6)
Zeochem

Robert W. Woods (6)
GRACE Davison

David R. Bellas (7)
CPI Engineering Services, Inc.

Richard L. Hall (7)
Battelle

Thomas J. Leck (7)
DuPont Fluoroproducts

Steven Pohlman (7)
Shrieve Chemical Products

Ngoc Dung Rohatgi (7)
Spauschus Associates, Inc.

Chris Seeton (7)
University of Illinois

Liwen Wei (7)
Shrieve Chemical Products

Ward D. Wells (7)
DuPont

Warren L. Beeton (8)
Copeland Corporation

Denis F. Clodic (8)
Amines CEP

William Dietrich (8)
York International

Daniel J. Miles (8)
Vacuum Technology Inc.

Avraham Shitzer (9)
Israel Institute of Technology

Bryan R. Becker (9, 10, 15, 16, 28)
University of Missouri, Kansas City

Brian A. Fricke (9, 10, 15, 16, 28)
University of Missouri, Kansas City

Donald J. Cleland (10, 11, 14)
Massey University

Gordon Follette (11, 12, 14, 15, 16, 22, 24, 27)
Follette Engineering, Inc.

Dennis Halsey (11, 14)
FES Systems, Inc.

Daniel Dettmers (11, 14, 20, 25)
University of Wisconsin, Madison

K.N. Bramadathan (12)
Christian Medical College

Godan Nambudiripad (12)
General Mills

William Sperber (12)
Cargill, Inc.

Don Fenton (13)
Kansas State University

S.A. Sherif (13)
University of Florida

John Topliss (13, 34)
Refrigeration Components RCC Canada, Ltd.

James F. Thompson (15, 22, 24)
University of California, Davis

Patricia A. Curtis (17, 18)
Auburn University

Gideon Zeidler (17, 18, 21)
University of California

Ed Fuhrmann (19)

Lloyd Wickett (19)
Flemming & Wickett

Todd Jekel (20)
University of Wisconsin, Madison

Scott Rankin (20)
University of Wisconsin, Madison

Kenneth E. Anderson (21)
North Carolina State University

Mo Samimi (21)
Food and Drug Administration

Giustino Mastro (22, 23)
University of Vermont

Thomas Ressler (23)
McCormack Manufacturing, Inc.

Burt Campbell (25)
Citrus Maintenance and Welding

Rod Cole (25)
Florida's Natural Growers

George Johnston (25)
Florida's Natural Growers

Allan Redd (25)
Fru-Con Engineering

Lenzie Kenyon (26)
Industrial Control Concepts, Inc.

Evans Lizardos (26)
Lizardos Engineering

Carla Panetta (26)
Lizardos Engineering

Bob Streets (26)
Benham Companies, Inc.

David Phelps (27)
Covert Engineers, Inc.

Jeff Rootring (28)
American Institute of Baking

Ronald H. Zelch (28)
American Ingredients Company

Nemat Lofti (29)
Hershey Foods Corporation

Peter D. Lord (29)
Hershey Foods Corporation

H. Douglas Souder (29)
Hershey Foods Corporation

Jeff Berge (30)
Ingersoll-Rand Company

James J. Bushnell (30, 31, 32)
HVAC Consulting Services

Roy H. Bleiberg (31)
ABS America

Chris Spunar (31)
Carrier Marine Systems

Dave Waugh (31)
Carrier Marine Systems

Richard A. Johnson (32)
Boeing

Paul Hough (33)
Armstrong World Industries, Inc.

Kelly Huang (34)
Trico Refrigeration Ltd.

Ronald H. Strong (34, 35)
R.H. Strong & Associates, Inc.

Eric Bradley (35)
R.H. Strong & Associates, Inc.

Claude Dumas (35)
City of Montreal Building Services

Reinhold Kittler (35)
Hudson Industrial Consulting

Ted Martin (35)
Wrights Mesa

Brian Simkins (35)
SPX Cooling Technologies

Bill Wladyka (35)
W.L. Wladyka Systems Consulting

Jim Shepherd (37)
Toromont Process Systems

Sanford A. Klein (38)
University of Wisconsin, Madison

William A. Little (38)
MMR Technologies, Inc.

Gregory F. Nellis (38)
University of Wisconsin, Madison

John M. Pfotenhauer (38)
University of Wisconsin, Madison

Eric B. Ratts (38)
University of Michigan, Dearborn

Kathleen Posteraro (39)
Pittsburgh Corning Corporation

John Bischof (40)
University of Minnesota

Ken Diller (40)
University of Texas, Austin

Bumsoo Han (40)
University of Texas, Arlington

Donald C. Erickson (41)
Energy Concepts Co.

Jay Kohler (41)
York International

Uwe Rockenfeller (41)
Rocky Research Corporation

Roland Ares (42)
Ares Corporation

Michel Lecompte (42)
Refplus, Inc.

Gary Price (42)
York International

Donald K. Miller (43)
MDK Engineering Corporation

Allan N. Podhorodeski (43)
P-B Engineering, Inc.

Dennis Littwin (44)
Fujikoki America

Ernest W. Schumacher (44)

John Sluga (44)
Hansen Technologies Corporation

Nick Zupp (44)
Hays Fluid Controls

Alexander Leyderman (45)
Fairchild Controls

John O'Brien (45)
UTC Carrier

Sami Zendah (45)
Copeland Corporation

David Hinde (46)
Hill PHOENIX

David Menninger (46)
The Kroger Co.

Scott Mitchell (46)
Southern California Edison

John Murray (46)
Sporlan Valve Co.

Ramin Faramarzi (46, 47)
Southern California Edison

Bruce Heirlmeier (46, 47)
Zero Zone, Inc.

Carl Roberts (46, 47)
Zero Zone, Inc.

Van Baxter (47)
Oak Ridge National Laboratory

David Cowen (47)
Food Service Technology Center

Brian Krafthefer (47)
Honeywell Laboratories

Lester Nakata (47)
Oahu Sales, Inc.

Pradeep Bansal (48)
University of Auckland

John Dieckmann (48)
TIAX LLC

Lindsey Roke (48)
Fisher & Paykel

Lawrence R. Wethje (48)
AHAM

ASHRAE HANDBOOK COMMITTEE

William S. Fleming, Chair

2006 Refrigeration Volume Subcommittee: **Norm Maxwell,** Chair

Richard A. Evans **Brian C. Krafthefer** **William J. McCartney** **Florentino J. Mendez** **Benjamin P. Sun**

ASHRAE HANDBOOK STAFF

W. Stephen Comstock,
Publisher
Director of Communications and Education

Mark S. Owen, Editor

Heather E. Kennedy, Associate Editor

Nancy F. Thysell, Typographer/Page Designer

David Soltis, Manager and **Jayne E. Jackson**
Publishing Services

ASHRAE TECHNICAL COMMITTEES, TASK GROUPS, AND TECHNICAL RESOURCE GROUPS

SECTION 1.0—FUNDAMENTALS AND GENERAL

1.1 Thermodynamics and Psychrometrics
1.2 Instruments and Measurement
1.3 Heat Transfer and Fluid Flow
1.4 Control Theory and Application
1.5 Computer Applications
1.6 Terminology
1.7 Business, Management, and General Legal Education
1.8 Mechanical Systems Insulation
1.9 Electrical Systems
1.10 Cogeneration Systems
1.11 Electric Motors and Motor Control
1.12 Moisture Management in Buildings
TG1.EASB Exergy Analysis for Sustainable Buildings

SECTION 2.0—ENVIRONMENTAL QUALITY

2.1 Physiology and Human Environment
2.2 Plant and Animal Environment
2.3 Gaseous Air Contaminants and Gas Contaminant Removal Equipment
2.4 Particulate Air Contaminants and Particulate Contaminant Removal Equipment
2.5 Global Climate Change
2.6 Sound and Vibration Control
2.7 Seismic and Wind Restraint Design
2.8 Building Environmental Impacts and Sustainability
TG2.UVAS Ultraviolet Air and Surface Treatment
TRG Blast, Chemical and Biological Remediation

SECTION 3.0—MATERIALS AND PROCESSES

3.1 Refrigerants and Secondary Coolants
3.2 Refrigerant System Chemistry
3.3 Refrigerant Contaminant Control
3.4 Lubrication
3.6 Water Treatment
3.8 Refrigerant Containment

SECTION 4.0—LOAD CALCULATIONS AND ENERGY REQUIREMENTS

4.1 Load Calculation Data and Procedures
4.2 Climate Information
4.3 Ventilation Requirements and Infiltration
4.4 Building Materials and Building Envelope Performance
4.5 Fenestration
4.7 Energy Calculations
4.10 Indoor Environmental Modeling

SECTION 5.0—VENTILATION AND AIR DISTRIBUTION

5.1 Fans
5.2 Duct Design
5.3 Room Air Distribution
5.4 Industrial Process Air Cleaning (Air Pollution Control)
5.5 Air-to-Air Energy Recovery
5.6 Control of Fire and Smoke
5.7 Evaporative Cooling
5.8 Industrial Ventilation Systems
5.9 Enclosed Vehicular Facilities
5.10 Kitchen Ventilation
5.11 Humidifying Equipment

SECTION 6.0—HEATING EQUIPMENT, HEATING AND COOLING SYSTEMS AND APPLICATIONS

6.1 Hydronic and Steam Equipment and Systems
6.2 District Energy
6.3 Central Forced-Air Heating and Cooling Systems
6.5 Radiant and Convective Space Heating and Cooling
6.6 Service Water Heating
6.7 Solar Energy Utilization
6.8 Geothermal Energy Utilization
6.9 Thermal Storage
6.10 Fuels and Combustion

SECTION 7.0—BUILDING PERFORMANCE

7.1 Integrated Building Design
7.3 Operation and Maintenance Management
7.4 Building Operation Dynamics
7.5 Smart Building Systems
7.6 Systems Energy Utilization
7.7 Testing and Balancing
7.8 Owning and Operating Costs
7.9 Building Commissioning

SECTION 8.0—AIR-CONDITIONING AND REFRIGERATION SYSTEM COMPONENTS

8.1 Positive Displacement Compressors
8.2 Centrifugal Machines
8.3 Absorption and Heat-Operated Machines
8.4 Air-to-Refrigerant Heat Transfer Equipment
8.5 Liquid-to-Refrigerant Heat Exchangers
8.6 Cooling Towers and Evaporative Condensers
8.7 Combustion Gas Turbine Inlet Air Cooling Systems
8.8 Refrigerant System Controls and Accessories
8.9 Residential Refrigerators and Food Freezers
8.10 Mechanical Dehumidification Equipment and Heat Pipes
8.11 Unitary and Room Air Conditioners and Heat Pumps
8.12 Desiccant Dehumidification Equipment and Components

SECTION 9.0—BUILDING APPLICATIONS

9.1 Large-Building Air-Conditioning Systems
9.2 Industrial Air Conditioning
9.3 Transportation Air Conditioning
9.4 Applied Heat Pump/Heat Recovery Systems
9.5 Residential and Small-Building Applications
9.6 Healthcare Facilities
9.7 Educational Facilities
9.8 Large-Building Air-Conditioning Applications
9.9 Mission Critical Facilities, Technology Spaces and Electronic Equipment
9.10 Laboratory Systems
9.11 Clean Space
9.12 Tall Buildings
TG9.JF Justice Facilities

SECTION 10.0—REFRIGERATION SYSTEMS

10.1 Custom-Engineered Refrigeration Systems
10.2 Automatic Icemaking Plants and Skating Rinks
10.3 Refrigerant Piping
10.4 Ultralow-Temperature Systems and Cryogenics
10.5 Refrigerated Distribution and Storage Facilities
10.6 Transport Refrigeration
10.7 Commercial Food and Beverage Cooling Display and Storage
10.8 Refrigeration Load Calculations
10.9 Refrigeration Application for Foods and Beverages
10.10 Management of Lubricant in Circulation

ASHRAE Research: Improving the Quality of Life

The American Society of Heating, Refrigerating and Air-Conditioning Engineers is the world's foremost technical society in the fields of heating, ventilation, air conditioning, and refrigeration. Its members worldwide are individuals who share ideas, identify needs, support research, and write the industry's standards for testing and practice. The result is that engineers are better able to keep indoor environments safe and productive while protecting and preserving the outdoors for generations to come.

One of the ways that ASHRAE supports its members' and industry's need for information is through ASHRAE Research. Thousands of individuals and companies support ASHRAE Research annually, enabling ASHRAE to report new data about material properties and building physics and to promote the application of innovative technologies.

Chapters in the ASHRAE Handbook are updated through the experience of members of ASHRAE Technical Committees and through results of ASHRAE Research reported at ASHRAE meetings and published in ASHRAE special publications and in *ASHRAE Transactions*.

For information about ASHRAE Research or to become a member, contact ASHRAE, 1791 Tullie Circle, Atlanta, GA 30329; telephone: 404-636-8400; www.ashrae.org.

Preface

The 2006 *ASHRAE Handbook—Refrigeration* covers the refrigeration equipment and systems for applications other than human comfort. This book includes information on cooling, freezing, and storing food; industrial applications of refrigeration; and low-temperature refrigeration. Primarily a reference for the practicing engineer, this volume is also useful for anyone involved in cooling and storage of food products.

This edition includes a new chapter (8), Refrigerant Containment, Recovery, Recycling, and Reclamation, reflecting this topic's importance. An accompanying CD-ROM contains all the volume's chapters (in both I-P and SI units) in searchable electronic format.

Some of the other revisions and additions are as follows:

- Chapter 2, System Practices for Halocarbon Refrigerants, has added information on safety considerations in using copper tubing for refrigerant lines.
- Chapter 3, System Practices for Ammonia and Carbon Dioxide Refrigerants, has a new title to reflect the addition of a new section on carbon dioxide as a refrigerant.
- Chapter 5, Refrigerant System Chemistry, has been reorganized and contains updated information on environmental acceptability.
- Chapter 7, Lubricants in Refrigerant Systems, has been thoroughly updated, with several new figures on various refrigerant/lubricant combinations, plus new information on considerations for carbon dioxide and CFC conversions.
- Chapter 9, Thermal Properties of Foods, contains updated surface heat transfer coefficients for food products.
- Chapter 10, Cooling and Freezing Times of Foods, has been extensively updated, with new geometric shape factors, and a new section comparing freezing time estimating methods.
- Chapter 11, Commodity Storage Requirements, has updated requirements for vegetables, fresh fruits, and melons.
- Chapter 13, Refrigeration Load, has new material on heat gain from cooler floors and coil defrosting.
- Chapter 14, Refrigerated Facility Design, has new discussion of design considerations for interstitial spaces.
- Chapter 15, Methods of Precooling Fruits, Vegetables, and Cut Flowers, contains new information on product requirements, loads, and time estimation methods, with new figures and tables on cooling methods.
- Chapter 16, Industrial Food-Freezing Systems, updated throughout, has added discussion of dehydration losses.
- Chapter 18, Poultry Products, has been updated, particularly on freezing effects on product quality, and control of texture and tenderness.
- Chapter 22, Deciduous Tree and Vine Fruit, contains new information on controlled-atmosphere storage and transport of some fruits.
- Chapter 27, Processed, Precooked, and Prepared Foods, has a revised section on potato products, and updates for regulatory changes.
- Chapter 38, Cryogenics, now includes low-temperature and integrated thermal and transport properties of cryogens and cryogenic materials; revised text on using refrigerant mixtures in cryocoolers, including discussion of the Kleemenko cycle; and updates on pulse tube cryocoolers.
- Chapter 44, Refrigerant-Control Devices, has updates on float switches, condenser-pressure-regulating valves, and pressure-relief devices, and a new section on liquid level sensors.
- Chapter 45, Factory Dehydrating, Charging, and Testing, has a substantially revised section on testing with refrigerant.
- Chapter 46, Retail Food Store Refrigeration and Equipment, has been reorganized for clarity, and has new discussions on self-contained versus remote systems, energy efficiency opportunities, refrigerated storage rooms, and interaction with supermarket air-conditioning systems.
- Chapter 47, Food Service and General Commercial Refrigeration Equipment, largely revised, has an expanded section on refrigerated cabinets, and added discussion on refrigeration systems, vending machines, ice machines, preparation tables, and energy efficiency opportunities.
- Chapter 48, Household Refrigerators and Freezers, has been substantially updated and reorganized, including adding a new table showing energy consumption testing standards from around the world.

This volume is published, both as a bound print volume and in electronic format on a CD-ROM, in two editions: one using inch-pound (I-P) units of measurement, the other using the International System of Units (SI).

Corrections to the 2003, 2004, and 2005 Handbook volumes can be found on the ASHRAE Web site at http://www.ashrae.org and in the Additions and Corrections section of this volume. Corrections for this volume will be listed in subsequent volumes and on the ASHRAE Web site.

Reader comments are enthusiastically invited. To suggest improvements for a chapter, **please comment using the form on the ASHRAE Web site** or, using the cutout comment pages at the end of this volume's index, write to Handbook Editor, ASHRAE, 1791 Tullie Circle, Atlanta, GA 30329, or fax 678-539-2187, or e-mail mowen@ashrae.org.

Mark S. Owen
Editor

CHAPTER 1

LIQUID OVERFEED SYSTEMS

OVERFEED systems force excess liquid, either mechanically or by gas pressure, through organized-flow evaporators, separate it from the vapor, and return it to the evaporators.

Terminology

Low-pressure receiver. Sometimes referred to as an **accumulator**, this vessel acts as the separator for the mixture of vapor and liquid returning from the evaporators. A constant refrigerant level is usually maintained by conventional control devices.

Pumping unit. One or more mechanical pumps or gas-operated liquid circulators are arranged to pump overfeed liquid to the evaporators. The pumping unit is located below the low-pressure receiver.

Wet returns. These are connections between the evaporator outlets and low-pressure receiver through which the mixture of vapor and overfeed liquid is drawn.

Liquid feeds. These are connections between the pumping unit outlet and evaporator inlets.

Flow control regulators. These devices regulate overfeed flow into the evaporators. They may be needle valves, fixed orifices, calibrated manual regulating valves, or automatic valves designed to provide a fixed liquid rate.

Advantages and Disadvantages

The main advantages of liquid overfeed systems are high system efficiency and reduced operating expenses. These systems have lower energy cost and fewer operating hours because

- The evaporator surface is used efficiently through good refrigerant distribution and completely wetted internal tube surfaces.
- The compressors are protected. Liquid slugs resulting from fluctuating loads or malfunctioning controls are separated from suction gas in the low-pressure receiver.
- Low-suction superheats are achieved where suction lines between the low-pressure receiver and the compressors are short. This minimizes discharge temperature, preventing lubrication breakdown and minimizing condenser fouling.
- With simple controls, evaporators can be hot-gas defrosted with little disturbance to the system.
- Refrigerant feed to evaporators is unaffected by fluctuating ambient and condensing conditions. Flow control regulators do not need to be adjusted after initial setting because overfeed rates are not generally critical.
- Flash gas resulting from refrigerant throttling losses is removed at the low-pressure receiver before entering the evaporators. This gas is drawn directly to the compressors and eliminated as a factor in system low-side design. It does not contribute to increased pressure drops in the evaporators or overfeed lines.
- Refrigerant level controls, level indicators, refrigerant pumps, and oil drains are generally located in equipment rooms, which are under operator surveillance or computer monitoring.
- Because of ideal entering suction gas conditions, compressors last longer. There is less maintenance and fewer breakdowns. The oil circulation rate to the evaporators is reduced as a result of the low compressor discharge superheat and separation at the low-pressure receiver (Scotland 1963).
- Automatic operation is convenient.

The following are possible disadvantages:

- In some cases, refrigerant charges are greater than those used in other systems.
- Higher refrigerant flow rates to and from evaporators cause liquid feed and wet return lines to be larger in diameter than high-pressure liquid and suction lines for other systems.
- Piping insulation, which is costly, is generally required on all feed and return lines to prevent condensation, frosting, or heat gain.
- Installed cost may be greater, particularly for small systems or those with fewer than three evaporators.
- Operation of the pumping unit requires added expenses that are offset by the increased efficiency of the overall system.
- Pumping units may require maintenance.
- Pumps sometimes have cavitation problems caused by low available net positive suction head.

Generally, the more evaporators used, the more favorable the initial costs for liquid overfeed compared to a gravity recirculated or flooded system (Scotland 1970). Liquid overfeed systems compare favorably with thermostatic valve feed systems for the same reason. For small systems, the initial cost for liquid overfeed may be higher than for direct expansion.

Ammonia Systems. Easy operation and lower maintenance are attractive features for even small ammonia systems. However, for ammonia systems operating below 0°F evaporating temperature, some manufacturers do not supply direct-expansion evaporators because of unsatisfactory refrigerant distribution and control problems.

OVERFEED SYSTEM OPERATION

Mechanical Pump

Figure 1 shows a simplified pumped overfeed system in which a constant liquid level is maintained in a low-pressure receiver. A mechanical pump circulates liquid through the evaporator(s). The two-phase return mixture is separated in the low-pressure receiver. Vapor is directed to the compressor(s). Makeup refrigerant enters the low-pressure receiver by means of a refrigerant metering device.

Figure 2 shows a horizontal low-pressure receiver with a minimum pump pressure, two service valves in place, and a strainer on the suction side of the pump. Valves from the low-pressure receiver to the pump should be selected for minimal pressure drop. The strainer protects hermetic pumps when oil is miscible with the

The preparation of this chapter is assigned to TC 10.1, Custom Engineered Refrigeration Systems.

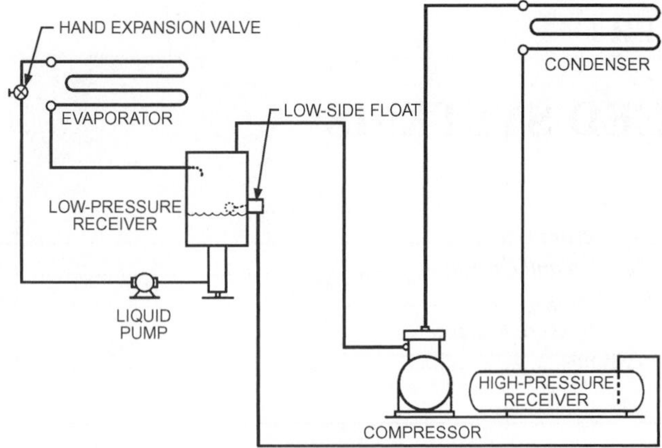

Fig. 1 Liquid Overfeed with Mechanical Pump

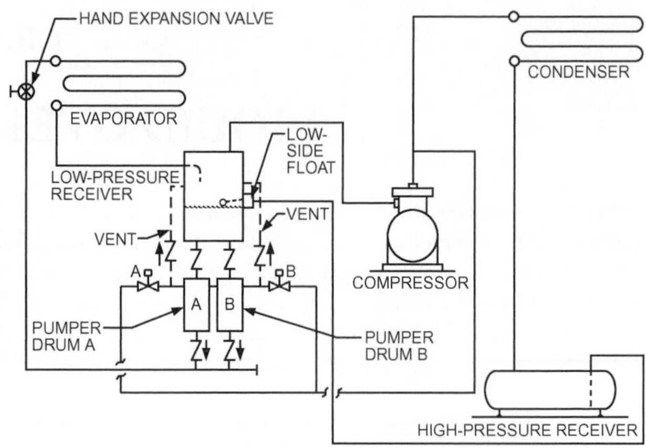

Fig. 3 Double-Pumper-Drum System

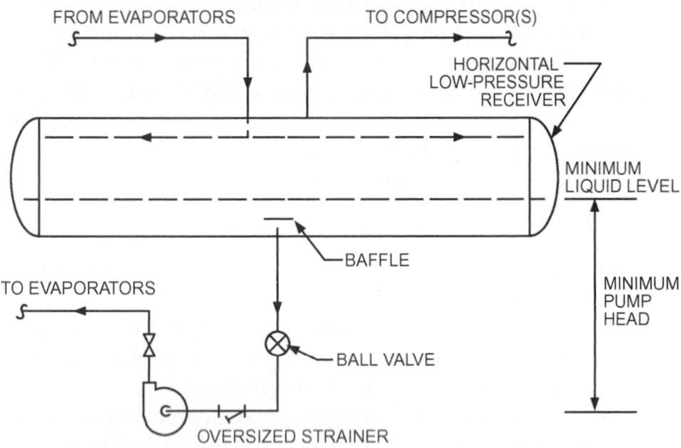

Fig. 2 Pump Circulation, Horizontal Separator

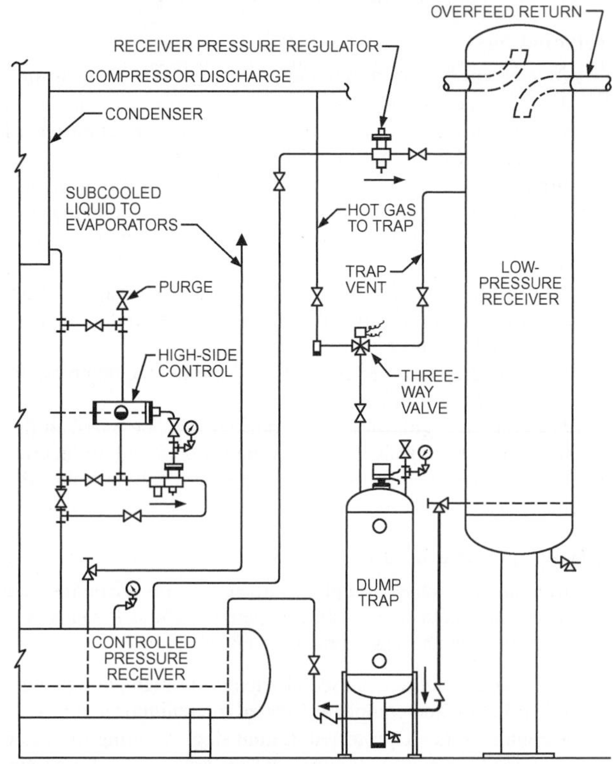

Fig. 4 Constant-Pressure Liquid Overfeed System

refrigerant. It should have a free area twice the transverse cross-sectional area of the line in which it is installed. With ammonia, consider using a suction strainer. Open-drive pumps do not require strainers. If no strainer is used, a dirt leg should be used to reduce the risk of solids getting into the pump.

Generally, minimum pump pressure should be at least double the net positive suction pressure to avoid cavitation. Liquid velocity to the pump should not exceed 3 fps. Net positive suction head and flow requirements vary with pump type and design; consult the pump manufacturer for specific requirements. The pump should be evaluated over the full range of operation at low and high flow. Centrifugal pumps have a flat curve and have difficulty with systems in which discharge pressure fluctuates.

Gas Pump

Figure 3 shows a basic gas-pumped liquid overfeed system, with pumping power supplied by gas at condenser pressure. In this system, a level control maintains the liquid level in the low-pressure receiver. There are two pumper drums; one is filled by the low-pressure receiver, and the other is drained as hot gas pushes liquid from the pumper drum to the evaporator. Pumper drum B drains when hot gas enters the drum through valve B. To function properly, the pumper drums must be correctly vented so they can fill during the fill cycle.

Another common arrangement is shown in Figure 4. In this system, high-pressure liquid is flashed into a controlled-pressure receiver that maintains constant liquid pressure at the evaporator inlets, resulting in continuous liquid feed at constant pressure. Flash gas is drawn into the low-pressure receiver through a receiver pressure regulator. Excess liquid drains into a liquid dump trap from the low-pressure receiver. Check valves and a three-way equalizing valve transfer liquid into the controlled-pressure receiver during the dump cycle. Refined versions of this arrangement are used for multistage systems.

REFRIGERANT DISTRIBUTION

To prevent underfeeding and excessive overfeeding of refrigerants, metering devices regulate the liquid feed to each evaporator and/or evaporator circuit. An automatic regulating device continuously controls refrigerant feed to the design value. Other common devices are hand expansion valves, calibrated regulating valves, orifices, and distributors.

It is time-consuming to adjust hand expansion valves to achieve ideal flow conditions. However, they have been used with some success in many installations before more sophisticated controls were available. One factor to consider is that standard hand expansion valves are designed to regulate flows caused by the relatively high pressure differences between condensing and evaporating pressure. In overfeed systems, large differences do not exist, so valves with larger orifices may be needed to cope with the combination of increased refrigerant quantity and relatively small pressure differences. Caution is necessary when using larger orifices because controllability decreases as orifice size increases.

Calibrated, manually operated regulating valves reduce some of the uncertainties involved in using conventional hand expansion valves. To be effective, the valves should be adjusted to the manufacturer's recommendations. Because refrigerant in the liquid feed lines is above saturation pressure, the lines should not contain flash gas. However, liquid flashing can occur if excessive heat gains by the refrigerant and/or high pressure drops build up in feed lines.

Orifices should be carefully designed and selected; once installed, they cannot be adjusted. They are generally used only for top- and horizontal-feed multicircuit evaporators. Foreign matter and congealed oil globules can restrict flow; a minimum orifice of 0.1 in. is recommended. With ammonia, the circulation rate may have to be increased beyond that needed for the minimum orifice size because of the small liquid volume normally circulated. Pumps and feed and return lines larger than minimum may be needed. This does not apply to halocarbons because of the greater liquid volume circulated as a result of fluid characteristics.

Conventional multiple-outlet distributors with capillary tubes of the type usually paired with thermostatic expansion valves have been used successfully in liquid overfeed systems. Capillary tubes may be installed downstream of a distributor with oversized orifices to achieve the required pressure reduction and efficient distribution.

Existing gravity-flooded evaporators with accumulators can be connected to liquid overfeed systems. Changes may be needed only for the feed to the accumulator, with suction lines from the accumulator connected to the system wet return lines. An acceptable arrangement is shown in Figure 5. Generally, gravity-flooded evaporators have different circuiting arrangements from overfeed evaporators. In many cases, the circulating rates developed by thermosiphon action are greater than those used in conventional overfeed systems.

Example 1. Find the orifice diameter of an ammonia overfeed system with a refrigeration load per circuit of 1.27 tons and a circulating rate of 7. Evaporating temperature is –30°F, pressure drop across the orifice is 8 psi, and the coefficient of discharge for the orifice is 0.61. The circulation per circuit is 0.528 gpm.

Solution: Orifice diameter may be calculated as follows:

$$d = \left[\frac{Q}{aC_d\sqrt{p/S}}\right]^{0.5} \tag{1}$$

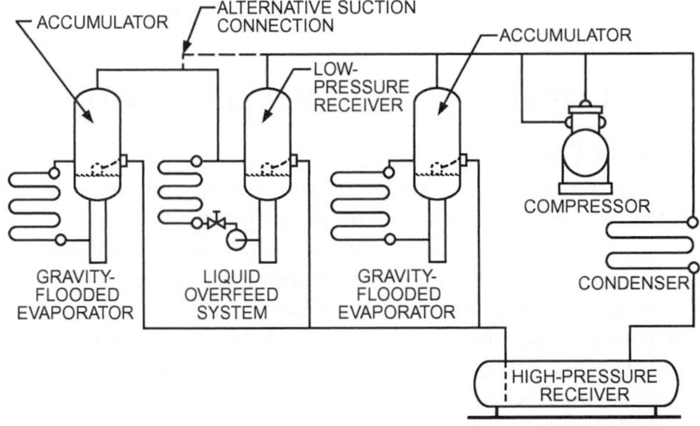

Fig. 5 Liquid Overfeed System Connected on Common System with Gravity-Flooded Evaporators

where

d = orifice diameter, in.
a = units conversion, 29.81
Q = discharge through orifice, gpm
p = pressure drop through orifice, psi
S = specific gravity of fluid relative to water at –30°F = 5.701/8.336 = 0.6839
C_d = coefficient of discharge for orifice

$$d = \left[\frac{0.528}{29.81 \times 0.61\sqrt{8/0.6839}}\right]^{0.5} = 0.092 \text{ in.}$$

Note: As noted in the text, use a 0.1 in. diameter orifice to avoid clogging.

OIL IN SYSTEM

Despite reasonably efficient compressor discharge oil separators, oil finds its way into the system low-pressure sides. In ammonia overfeed systems, most of this oil can be drained from low-pressure receivers with suitable oil drainage facilities. In low-temperature systems, a separate valved and pressure-protected, noninsulated oil drain pot can be placed in a warm space at the accumulator. The oil/ammonia mixture flows into the pot, and the refrigerant evaporates. This arrangement is shown in Figure 6. At subatmospheric pressures, high-pressure vapor must be piped into the oil pot to force oil out. Because of oil's low solubility in liquid ammonia, thick oil globules circulate with the liquid and can restrict flow through strainers, orifices, and regulators. To maintain high efficiency, oil should be removed from the system by regular draining.

Except at low temperatures, halocarbons are miscible with oil. Therefore, positive oil return to the compressor must be ensured. There are many methods, including oil stills using both electric heat and heat exchange from high-pressure liquid or vapor. Some arrangements are discussed in Chapter 2. At low temperatures, oil skimmers must be used because oil migrates to the top of the low-pressure receiver.

Build-up of excessive oil in evaporators must not be allowed because it rapidly decreases efficiency. This is particularly critical in evaporators with high heat transfer rates associated with low volumes, such as flake ice makers, ice cream freezers, and scraped-surface heat exchangers. Because refrigerant flow rate is high, excessive oil can accumulate and rapidly reduce efficiency.

CIRCULATING RATE

In a liquid overfeed system, the **circulating number** or **rate** is the mass ratio of liquid pumped to amount of vaporized liquid. The amount of liquid vaporized is based on the latent heat for the refrigerant at the evaporator temperature. The **overfeed rate** is the ratio of liquid to vapor returning to the low-pressure receiver. When vapor leaves an evaporator at saturated vapor conditions with no excess liquid, the circulating rate is 1 and the overfeed rate is 0. With a

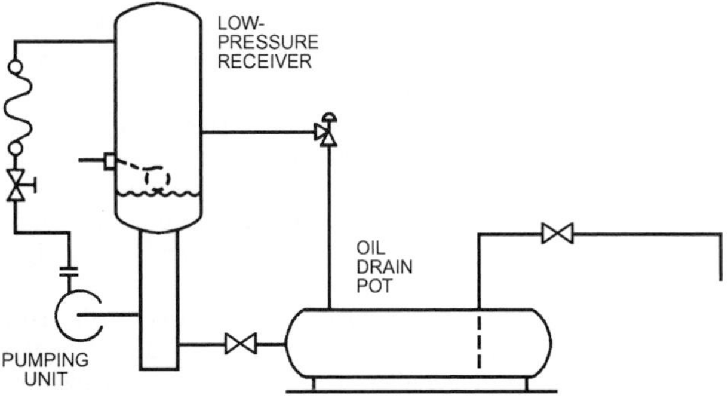

Fig. 6 Oil Drain Pot Connected to Low-Pressure Receiver

Table 1 Recommended Minimum Circulating Rate

Refrigerant	Circulating Rate*
Ammonia (R-717)	
Downfeed (large-diameter tubes)	6 to 7
Upfeed (small-diameter tubes)	2 to 4
R-22, upfeed	3
R-134a	2

*Circulating rate of 1 equals evaporating rate.

circulating rate of 4, the overfeed rate at full load is 3; at no load, it is 4. Most systems are designed for steady flow. With few exceptions, load conditions may vary, causing fluctuating temperatures outside and within the evaporator. Evaporator capacities vary considerably; with constant refrigerant flow to the evaporator, the overfeed rate fluctuates.

For each evaporator, there is an ideal circulating rate for every loading condition that gives the minimum temperature difference and best evaporator efficiency (Lorentzen 1968; Lorentzen and Gronnerud 1967). With few exceptions, it is impossible to predict ideal circulating rates or to design a plant for automatic adjustment of the rates to suit fluctuating loads. The optimum rate can vary with heat load, pipe diameter, circuit length, and number of parallel circuits to achieve the best performance. High circulating rates can cause excessively high pressure drops through evaporators and wet return lines. Return line sizing (see the section on Line Sizing) can affect the ideal rates. Many evaporator manufacturers specify recommended circulating rates for their equipment. Rates in Table 1 agree with these recommendations.

Because of distribution considerations, higher circulating rates are common with top-feed evaporators. In multicircuit systems, refrigerant distribution must be adjusted to provide the best possible results. Incorrect distribution can cause excessive overfeed or starvation in some circuits. Manual or automatic regulating valves can control flow for the optimum or design value.

Halocarbon densities are about twice that of ammonia. If halocarbons R-22, R-134a, and R-502 are circulated at the same rate as ammonia, they require 6 to 8.3 times more energy for pumping to the same height than the less-dense ammonia. Because pumping energy must be added to the system load, halocarbon circulating rates are usually lower than those for ammonia. Ammonia has a relatively high latent heat of vaporization, so for equal heat removal, much less ammonia mass must be circulated compared to halocarbons.

Although halocarbons circulate at lower rates than ammonia, the wetting process in the evaporators is still efficient because of the liquid and vapor volume ratios. For example, at –40°F evaporating temperature, with constant flow conditions in the wet return connections, similar ratios of liquid and vapor are experienced with a circulating rate of 4 for ammonia and 2.5 for R-22, R-502, and R-134a. With halocarbons, some additional wetting is also experienced because of the solubility of the oil in these refrigerants.

When bottom feed is used for multicircuit coils, a minimum feed rate per circuit is not necessary because orifices or other distribution devices are not required. The circulating rate for top-feed and horizontal feed coils may be determined by the minimum rates from the orifices or other distributors in use.

Figure 7 provides a method for determining the liquid refrigerant flow (Niederer 1964). The charts indicate the amount of refrigerant vaporized in a 1 ton system with circulated operation having no flash gas in the liquid feed line. The value obtained from the chart may be multiplied by the desired circulating rate and total refrigeration to determine total flow.

Pressure drop through flow control regulators is usually 10 to 50% of the available feed pressure. Pressure at the outlet of the flow regulators must be higher than the vapor pressure at the low-pressure receiver by an amount equal to the total pressure drop of the two-phase mixture through the evaporator, any evaporator pressure regulator, and wet return lines. Pressure loss could be 5 psi in a typical system. When using recommended liquid feed sizing practices, assuming a single-story building, the frictional pressure drop from pump discharge to evaporators is about 10 psi. Therefore, a pump for 20 to 25 psi should be satisfactory in this case, depending on the lengths and sizes of feed lines, quantity and types of fittings, and vertical lift involved.

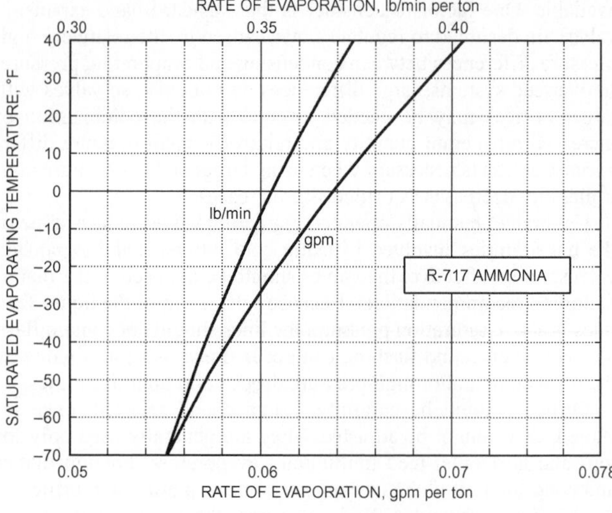

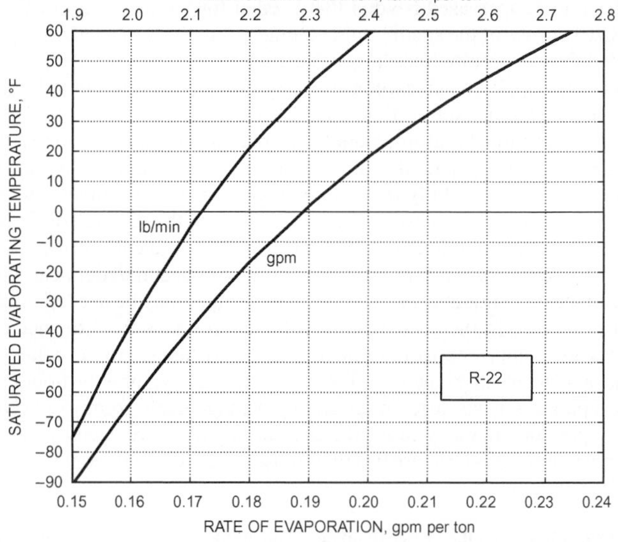

Fig. 7 Charts for Determining Rate of Refrigerant Feed (No Flash Gas)

PUMP SELECTION AND INSTALLATION

Types of Pumps

Mechanical pumps, gas pressure pumping systems, and injector systems are available for liquid overfeed systems.

Types of mechanical pump drives include open, semihermetic, magnetic clutch, and hermetic. Rotor arrangements include positive rotary, centrifugal, and turbine vane. Positive rotary and gear pumps are generally operated at slow speeds up to 900 rpm. Whatever type of pump is used, take care to prevent flashing at the pump suction and/or within the pump itself.

Centrifugal pumps are typically used for larger volumes, whereas semihermetic pumps are best suited for halocarbons at or below atmospheric refrigerant saturated pressure. Regenerative turbines are used with relatively high pressure and large swings in discharge pressure.

Open pumps are fitted with a wide variety of packing or seals. For continuous duty, a mechanical seal with an oil reservoir or a liquid refrigerant supply to cool, wash, and lubricate the seals is commonly used. Experience with the particular application or the recommendations of an experienced pump supplier are the best guide for selecting the packing or seal. The motor and pump can be magnetically coupled instead of shaft coupled to eliminate shaft seals. A small immersion electric heater in the oil reservoir can be used with low-temperature systems to ensure that the oil remains fluid. Motors should have a service factor that compensates for drag on the pump if the oil is cold or stiff.

Considerations should include ambient temperatures, heat leakage, fluctuating system pressures from compressor cycling, internal bypass of liquid to pump suction, friction heat, motor heat conduction, dynamic conditions, cycling of automatic evaporator liquid and suction stop valves, action of regulators, gas entrance with liquid, and loss of subcooling by pressure drop. Another factor to consider is the time lag caused by the heat capacity of pump suction, cavitation, and net positive suction head factors (Lorentzen 1963).

The motor and stator of hermetic pumps are separated from the refrigerant by a thin nonmagnetic membrane. The metal membrane should be strong enough to withstand system design pressures. Normally, the motors are cooled and the bearings lubricated by liquid refrigerant bypassed from the pump discharge. It is good practice to use two pumps, one operating and one standby.

Installing and Connecting Mechanical Pumps

Because of the sensitive suction conditions of mechanical pumps on overfeed systems, the manufacturer's application and installation specifications must be followed closely. Suction connections should be as short as possible, without restrictions, valves, or elbows. Angle or full-flow ball valves should be used. Using valves with horizontal valve spindles eliminates possible traps. Gas binding is more likely with high evaporating pressures.

Installing discharge check valves prevents backflow. Relief valves should be used, particularly for positive-displacement pumps. Strainers are not usually installed in ammonia pump suction lines because they plug with oil. Strainers, although a poor substitute for a clean installation, protect halocarbon pumps from damage by dirt or pipe scale.

Pump suction connections to liquid legs (vertical drop legs from low-pressure receivers) should be made above the bottom of the legs to allow collection space for solids and sludge. Consider using vortex eliminators, particularly when submersion of the suction inlet is insufficient to prevent the intake of gas bubbles. Lorentzen (1963, 1965) gives more complete information.

Sizing the pump suction line is important. The general velocity should be about 3 fps. Small lines cause restrictions; oversized lines can cause bubble formation during evaporator temperature decrease because of the heat capacity of the liquid and piping. Oversized lines also increase heat gain from ambient spaces. Oil heaters for the seal lubrication system keep the oil fluid, particularly during subzero operation. Thermally insulating all cold surfaces of pumps, lines, and receivers increases efficiency.

CONTROLS

The liquid level in the low-pressure receiver can be controlled by conventional devices such as low-pressure float valves, combinations of float switch and solenoid valve with manual regulator, thermostatic level controls, electronic level sensors, or other proven automatic devices. High-level float switches are useful in stopping compressors and/or operating alarms; they are mandatory in some areas. Solenoid valves should be installed on liquid lines (minimum sized) feeding low-pressure receivers so that positive shutoff is automatically achieved with system shutdown. This prevents excessive refrigerant from collecting in low-pressure receivers, which can cause carryover at start-up.

To prevent pumps from operating without liquid, low-level float switches can be fitted on liquid legs. An alternative device, a differential pressure switch connected across pump discharge and suction connections, stops the pump without interrupting liquid flow. Cavitation can also cause this control to operate. When hand expansion valves are used to control the circulation rate to evaporators, the orifice should be sized for operation between system high and low pressures. Occasionally, with reduced inlet pressure, these valves can starve the circuit. Calibrated, manually adjusted regulators are available to meter the flow according to the design conditions. An automatic flow-regulating valve specifically for overfeed systems is available.

Liquid and suction solenoid valves must be selected for refrigerant flow rates by mass or volume, not by refrigeration ratings from capacity tables. Evaporator pressure regulators should be sized according to the manufacturer's ratings for overfeed systems. Notify the manufacturer that valves being ordered are for overfeed application, because slight modifications may be required. When evaporator pressure regulators are used on overfeed systems for controlling air defrosting of cooling units (particularly when fed with very-low-temperature liquid), refrigerant heat gain may be achieved by sensible, not latent, effect. In such cases, other defrosting methods should be investigated. The possibility of connecting the units directly to high-pressure liquid should be considered, especially if the loads are minor.

When a check valve and a solenoid valve are paired on an overfeed system liquid line, the check valve should be downstream from the solenoid valve. When the solenoid valve is closed, dangerous hydraulic pressure can build up from expansion of the trapped liquid as it is heated. When evaporator pressure regulators are used, entering liquid pressure should be high enough to cause flow into the evaporator.

Multicircuit systems must have a bypass relief valve in the pump discharge. The relief valve's pressure should be set considering the back pressure on the valve from the low-pressure receiver. For example, if the low-pressure receiver is set at 50 psi and the maximum discharge pressure from the pump is 150 psi, the relief valve should be set at 100 psi. When some circuits are closed, excess liquid is bypassed into the low-pressure receiver rather than forced through the evaporators still in operation. This prevents higher evaporating temperatures from pressurizing evaporators and reducing capacities of operating units. Where low-temperature liquid feeds can be isolated manually or automatically, relief valves can be installed to prevent damage from excessive hydraulic pressure.

EVAPORATOR DESIGN

Considerations

There is an ideal refrigerant feed and flow system for each evaporator design and arrangement. An evaporator designed for gravity-flooded operation cannot always be converted to an overfeed arrangement, and vice versa; neither can systems always be designed to circulate the optimum flow rate. When top feed is used to ensure good distribution, a minimum quantity per circuit must be circulated, generally about 0.5 gpm. In bottom-feed evaporators, distribution is less critical than in top or horizontal feed because each circuit fills with liquid to equal the pressure loss in other parallel circuits.

Circuit length in evaporators is determined by allowable pressure drop, load per circuit, tubing diameter, overfeed rate, type of refrigerant, and heat transfer coefficients. The most efficient circuiting is determined in most cases through laboratory tests conducted by the evaporator manufacturers. Their recommendations should be followed when designing systems.

Top Feed Versus Bottom Feed

System design must determine whether evaporators are to be top fed or bottom fed, although both feed types can be installed in a single system. Each feed type has advantages; no arrangement is best for all systems.

Advantages of **top feed** include

- Smaller refrigerant charge
- Possibly smaller low-pressure receiver
- Possible absence of static pressure penalty
- Better oil return
- Quicker, simpler defrost arrangements

For halocarbon systems with greater fluid densities, the refrigerant charge, oil return, and static pressure are very important.

Bottom feed is advantageous in that

- Distribution considerations are less critical
- Relative locations of evaporators and low-pressure receivers are less important
- System design and layout are simpler

The top-feed system is limited by the relative location of components. Because this system sometimes requires more refrigerant circulation than bottom-feed systems, it has greater pumping load, possibly larger feed and return lines, and increased line pressure drop penalties. In bottom-feed evaporators, multiple headers with individual inlets and outlets can be installed to reduce static pressure penalties. For high lift of return overfeed lines from the evaporators, dual suction risers eliminate static pressure penalties (Miller 1974, 1979).

Distribution must be considered when using a vertical refrigerant feed, because of static pressure variations in the feed and return header circuits. For example, for equal circuit loadings in a horizontal-airflow unit cooler, using gradually smaller orifices for bottom-feed circuits than for upper circuits can compensate for pressure differences.

When the top-feed free-draining arrangement is used for air-cooling units, liquid solenoid control valves can be used during the defrost cycle. This applies in particular to air, water, or electric defrost units. Any liquid remaining in the coils rapidly evaporates or drains to the low-pressure receiver. Defrost is faster than in bottom-feed evaporators.

REFRIGERANT CHARGE

Overfeed systems need more refrigerant than dry expansion systems. Top-feed arrangements have smaller charges than bottom-feed systems. The amount of charge depends on evaporator volume, circulating rate, sizes of flow and return lines, operating temperature differences, and heat transfer coefficients. Generally, top-feed evaporators operate with the refrigerant charge occupying about 25 to 40% of the evaporator volume. The refrigerant charge for the bottom-feed arrangement occupies about 60 to 75% of the evaporator volume, with corresponding variations in the wet returns. Under some no-load conditions in up-feed evaporators, the charge may occupy 100% of the evaporator volume. In this case, the liquid surge volume from full load to no load must be considered in sizing the low-pressure receiver (Miller 1971, 1974).

Evaporators with high heat transfer rates, such as flake ice makers and scraped-surface heat exchangers, have small charges because of small evaporator volumes. The amount of refrigerant in the low side has a major effect on the size of the low-pressure receiver, especially in horizontal vessels. The cross-sectional area for vapor flow in horizontal vessels is reduced with increasing liquid level. It is important to ascertain the evaporator refrigerant charge with fluctuating loads for correct vessel design, particularly for a low-pressure receiver that does not have a constant level control but is fed through a high-pressure control.

START-UP AND OPERATION

All control devices should be checked before start-up. If mechanical pumps are used, the direction of operation must be correct. System evacuation and charging procedures are similar to those for other systems. The system must be operating under normal conditions to determine the total required refrigerant charge. Liquid height is established by liquid level indicators in the low-pressure receivers.

Calibrated, manually operated regulators should be set for the design conditions and adjusted for better performance when necessary. When hand expansion valves are used, the system should be started by opening the valves about one-quarter to one-half turn. When balancing is necessary, the regulators should be cut back on circuits not starved of liquid, to force the liquid through underfed circuits. The outlet temperature of the return line from each evaporator should be the same as the main return line's saturation temperature, allowing for pressure drops. Starved circuits are indicated by temperatures higher than those for adequately fed circuits. Excessive feed to a circuit increases evaporator temperature because of excessive pressure drop.

The relief bypass from the liquid line to the low-pressure receiver should be adjusted and checked to ensure that it is functioning. During operation, the pump manufacturer's recommendations for lubrication and maintenance should be followed. Regular oil draining procedures should be established for ammonia systems; the quantities of oil added to and drained from each system should be compared, to determine whether oil is accumulating. Oil should not be drained in halocarbon systems. Because of oil's miscibility with halocarbons at high temperatures, it may be necessary to add oil to the system until an operating balance is achieved (Soling 1971; Stoecker 1960).

Operating Costs and Efficiency

Operating costs for overfeed systems are generally lower than for other systems (though not always, because of the various inefficiencies that exist from system to system and from plant to plant). For existing dry expansion plants converted to liquid overfeed, the operating hours, power, and maintenance costs are reduced. Efficiency of the early gas pump systems has been improved by using high-side pressure to circulate overfeed liquid. This type of system is indicated in the controlled-pressure system shown in Figure 4. Refinements of the double-pumper-drum arrangement (see Figure 3) have also been developed.

Gas-pumped systems, which use refrigerant gas to pump liquid to the evaporators or to the controlled-pressure receiver, require additional compressor volume, from which no useful refrigeration is obtained. These systems consume 4 to 10% or more of the compressor power to maintain refrigerant flow.

If condensing pressure is reduced as much as 10 psi, the compressor power per unit of refrigeration drops by about 7%. Where outdoor dry- and wet-bulb conditions allow, a mechanical pump can be used to pump gas with no effect on evaporator performance. Gas-operated systems must, however, maintain the condensing pressure within a much smaller range to pump the liquid and maintain the required overfeed rate.

LINE SIZING

The liquid feed line to the evaporator and wet return line to the low-pressure receiver cannot be sized by the method described in Chapter 36 of the 2005 *ASHRAE Handbook—Fundamentals*. Figure 7 can be used to size liquid feed lines. The circulating rate from Table 1 is multiplied by the evaporating rate. For example, an evaporator with a circulating rate of 4 that forms vapor at a rate of 5 lb/min needs a feed line sized for $4 \times 5 = 20$ lb/min.

Alternative methods that may be used to design wet returns include the following:

- Use one pipe size larger than calculated for vapor flow alone.
- Use a velocity selected for dry expansion reduced by the factor $\sqrt{1/\text{Circulating Rate}}$. This method suggests that the wet-return velocity for a circulating rate of 4 is $\sqrt{1/4} = 0.5$, or half that of the acceptable dry-vapor velocity.
- Use the design method described by Chaddock et al. (1972). The report includes tables of flow capacities at 2°F drop per 100 ft of horizontal lines for R-717 (ammonia), R-12, R-22, and R-502.

When sizing refrigerant lines, the following design precautions should be taken:

- Carefully size overfeed return lines with vertical risers because more liquid is held in risers than in horizontal pipe. This holdup increases with reduced vapor flow and increases pressure loss because of gravity and two-phase pressure drop.
- Use double risers with halocarbons to maintain velocity at partial loads and to reduce liquid static pressure loss (Miller 1979).
- Add the equivalent of a 100% liquid static height penalty to the pressure drop allowance to compensate for liquid holdup in ammonia systems that have unavoidable vertical risers.
- As alternatives in severe cases, provide traps and a means of pumping liquids, or use dual-pipe risers.
- Install low-pressure drop valves so the stems are horizontal or nearly so (Chisholm 1971).

LOW-PRESSURE RECEIVER SIZING

Low-pressure receivers are also called liquid separators, suction traps, accumulators, liquid/vapor separators, flash coolers, gas and liquid coolers, surge drums, knockout drums, slop tanks, or low-side pressure vessels, depending on their function and user preference.

Low-pressure receiver sizing is determined by the required liquid holdup volume and allowable gas velocity. The volume must accommodate fluctuations of liquid in the evaporators and overfeed return lines as a result of load changes and defrost periods. It must also handle swelling and foaming of the liquid charge in the receiver, which is caused by boiling during temperature increase or pressure reduction. At the same time, a liquid seal must be maintained on the supply line for continuous circulation devices. A separating space must be provided for gas velocity low enough to cause a minimum entrainment of liquid drops into the suction outlet. Space limitations and design requirements result in a wide variety of configurations (Lorentzen 1966; Miller 1971; Niemeyer 1961; Scheiman 1963, 1964; Sonders and Brown 1934; Stoecker 1960; Younger 1955).

In selecting a gas-and-liquid separator, adequate volume for the liquid supply and a vapor space above the minimum liquid height for liquid surge must be provided. This requires analysis of operating load variations. This, in turn, determines the **maximum operating liquid level**. Figures 8 and 9 identify these levels and the important parameters of vertical and horizontal gravity separators.

Vertical separators maintain the same separating area with level variations, whereas separating areas in horizontal separators change with level variations. **Horizontal separators** should have inlets and outlets separated horizontally by at least the vertical separating distance. A useful arrangement in horizontal separators distributes the inlet flow into two or more connections to reduce turbulence and horizontal velocity without reducing the residence time of the gas flow within the shell (Miller 1971).

In horizontal separators, as the horizontal separating distance increases beyond the vertical separating distance, the residence time of vapor passing through increases so that higher velocities than allowed in vertical separators can be tolerated. As the separating distance reduces, the amount of liquid entrainment from gravity separators increases. Table 2 shows the gravity separation velocities. For surging loads or pulsating flow associated with large step changes in capacity, the maximum steady-flow velocity should be reduced to a value achieved by a suitable multiplier such as 0.75.

The gas-and-liquid separator may be designed with baffles or eliminators to separate liquid from the suction gas returning from the top of the shell to the compressor. More often, enough separation space is allowed above the liquid level for this purpose. Such a design is usually of the vertical type, with a separation height above the liquid level of 24 to 36 in. The shell diameter is sized to keep suction gas velocity low enough to allow liquid droplets to separate and not be entrained with the returning suction gas off the top of the shell.

Although separators are made with length-to-diameter (L/D) ratios of 1/1 increasing to 10/1, the least expensive separators usually have L/D ratios between 3/1 and 5/1. Vertical separators are normally used for systems with reciprocating compressors. Horizontal separators may be preferable where vertical height is critical and/or where large volume space for liquid is required. The procedures for designing vertical and horizontal separators are different.

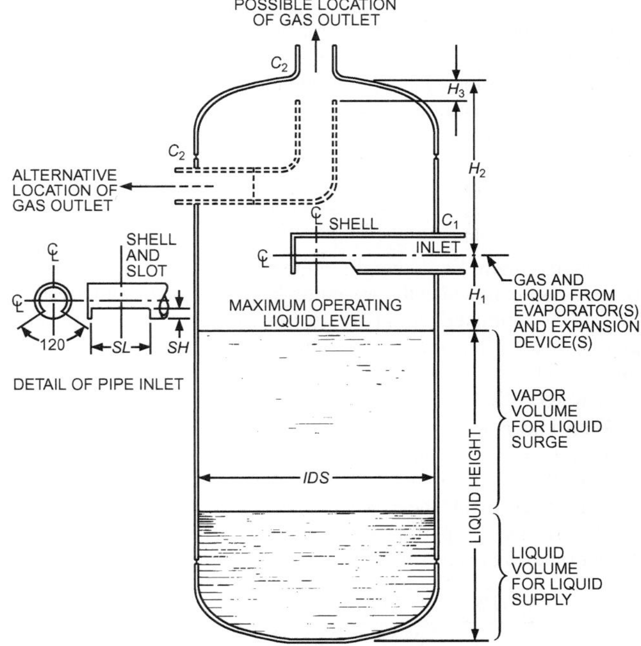

C_1 = inlet pipe diameter, OD, in.
C_2 = outlet pipe diameter, OD, in.
D = vertical separation distance, in. (actual)
H_1 = height of C_1 above maximum liquid level, in.; for pseudo D = 24 in.; $\sqrt{7.33\,Q/V}$ (at D = 24 in.)
H_2 = location of C_1 from inside top of shell, in.; $D + 0.5 \times$ depth of curved portion of head or 2 in.
H_3 = location of gas exit point for alternative location of C_2 measured from inside top of shell, in.; $0.5 \times$ depth of curved portion of shell or 2 in., whichever is greater
IDS = internal diameter of shell; $\sqrt{183\,Q/V}$, in.
Q = maximum gas flow in shell at maximum sustained operating conditions, cfm
SH = slot height = $C_1/4$, in.
SL = slot length = $3C_1$, in.
V = separation velocity, fpm

Fig. 9 Basic Vertical Gravity Gas and Liquid Separator

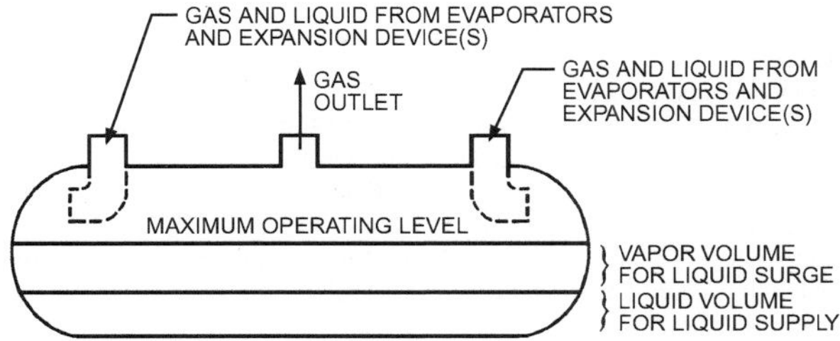

Fig. 8 Basic Horizontal Gas-and-Liquid Separator

Table 2 Maximum Effective Separation Velocities for R-717, R-22, R-12, and R-502, with Steady Flow Conditions

Temp., °F	Vertical Separation Distance, in.	Maximum Steady Flow Velocity, fpm			
		R-717	R-22	R-12	R-502
+50	10	29	13	16	11
	24	125	62	70	50
	36	139	77	85	62
+20	10	42	19	22	15
	24	172	86	96	69
	36	195	102	115	83
–10	10	61	27	32	22
	24	253	120	135	97
	36	281	141	159	116
–40	10	95	41	47	33
	24	392	173	198	140
	36	428	205	230	165
–70	10	158	65	72	50
	24	649	267	303	212
	36	697	310	351	247

Source: Adapted from Miller (1971).

A vertical gas-and-liquid separator is shown in Figure 9. The end of the inlet pipe C_1 is capped so that flow dispersion is directed down toward the liquid level. The suggested opening is four times the transverse internal area of the pipe. Height H_1 with a 120° dispersion of the flow reaches approximately 70% of the internal diameter of the shell.

An alternative inlet pipe with a downturned elbow or mitered bend can be used. However, the jet effect of entering fluid must be considered to avoid undue splashing. The pipe outlet must be a minimum distance of *IDS*/5 above the maximum liquid level in the shell. H_2 is measured from the outlet to the inside top of the shell. It equals D + 0.5 times the depth of the curved portion of the head.

For the alternative location of C_2, determine *IDS* from the following equation:

$$IDS = \sqrt{\frac{183Q}{V} + C_2^2} \tag{2}$$

The maximum liquid height in the separator is a function of the type of system in which the separator is being used. In some systems this can be estimated, but in others, previous experience is the only guide for selecting the proper liquid height. Accumulated liquid must be returned to the system by a suitable means at a rate comparable to its collection rate.

With a horizontal separator, the vertical separation distance used is an average value. The top part of the horizontal shell restricts gas flow so that the maximum vertical separation distance cannot be used. If H_t represents the maximum vertical distance from the liquid level to the inside top of the shell, the average separation distance as a fraction of *IDS* is as follows:

H_t/IDS	D/IDS	H_t/IDS	D/IDS
0.1	0.068	0.6	0.492
0.2	0.140	0.7	0.592
0.3	0.215	0.8	0.693
0.4	0.298	0.9	0.793
0.5	0.392	1.0	0.893

The suction connection(s) for refrigerant gas leaving the horizontal shell must be at or above the location established by the average distance for separation. The maximum cross-flow velocity of gas establishes residence time for the gas and any entrained liquid droplets in the shell. The most effective removal of entrainment occurs when residence time is the maximum practical. Regardless of the number of gas outlet connections for uniform distribution of gas flow, the cross-sectional area of the gas space is

$$A_x = \frac{288DQ}{VL} \tag{3}$$

where

A_x = minimum transverse net cross-sectional area or gas space, in²
D = average vertical separation distance, in.
Q = total quantity of gas leaving vessel, cfm
L = inside length of shell, in.
V = separation velocity for separation distance used, fpm

For nonuniform distribution of gas flow in the horizontal shell, determine the minimum horizontal distance for gas flow from point of entry to point of exit as follows:

$$RTL = \frac{144QD}{VA_x} \tag{4}$$

where

RTL = residence time length, in.
Q = maximum flow for that portion of the shell, cfm

All connections must be sized for the flow rates and pressure drops permissible and must be positioned to minimize liquid splashing. Internal baffles or mist eliminators can reduce vessel diameter; however, test correlations are necessary for a given configuration and placement of these devices.

An alternative formula for determining separation velocities that can be applied to separators is

$$v = k\sqrt{\frac{\rho_l - \rho_v}{\rho_v}} \tag{5}$$

where

v = velocity of vapor, fps
ρ_l = density of liquid, lb/ft³
ρ_v = density of vapor, lb/ft³
k = factor based on experience without regard to vertical separation distance and surface tension for gravity separators

In gravity liquid/vapor separators that must separate heavy entrainment from vapors, use a k of 0.1. This gives velocities equivalent to those used for 12 to 14 in. vertical separation distance for R-717 and 14 to 16 in. vertical separation distance for halocarbons. In knockout drums that separate light entrainment, use a k of 0.2. This gives velocities equivalent to those used for 36 in. vertical separation distance for R-717 and for halocarbons.

REFERENCES

Chaddock, J.B., D.P. Werner, and C.G. Papachristou. 1972. Pressure drop in the suction lines of refrigerant circulation systems. *ASHRAE Transactions* 78(2):114-123.

Chisholm, D. 1971. Prediction of pressure drop at pipe fittings during two-phase flow. *Proceedings of the IIR Conference*, Washington, D.C.

Lorentzen, G. 1963. Conditions of cavitation in liquid pumps for refrigerant circulation. *Progress Refrigeration Science Technology* I:497.

Lorentzen, G. 1965. How to design piping for liquid recirculation. *Heating, Piping & Air Conditioning* (June):139.

Lorentzen, G. 1966. On the dimensioning of liquid separators for refrigeration systems. *Kältetechnik* 18:89.

Lorentzen, G. 1968. Evaporator design and liquid feed regulation. *Journal of Refrigeration* (November-December):160.

Lorentzen, G. and R. Gronnerud. 1967. On the design of recirculation type evaporators. *Kulde* 21(4):55.

Miller, D.K. 1971. Recent methods for sizing liquid overfeed piping and suction accumulator-receivers. *Proceedings of the IIR Conference*, Washington, D.C.

Miller, D.K. 1974. Refrigeration problems of a VCM carrying tanker. *ASHRAE Journal* 11.

Miller, D.K. 1979. Sizing dual suction risers in liquid overfeed refrigeration systems. *Chemical Engineering* 9.

Niederer, D.H. 1964. Liquid recirculation systems—What rate of feed is recommended. *The Air Conditioning & Refrigeration Business* (December).

Niemeyer, E.R. 1961. Check these points when designing knockout drums. *Hydrocarbon Processing and Petroleum Refiner* (June).

Scheiman, A.D. 1963. Size vapor-liquid separators quicker by nomograph. *Hydrocarbon Processing and Petroleum Refiner* (October).

Scheiman, A.D. 1964. Horizontal vapor-liquid separators. *Hydrocarbon Processing and Petroleum Refiner* (May).

Scotland, W.B. 1963. Discharge temperature considerations with multicylinder ammonia compressors. *Modern Refrigeration* (February).

Scotland, W.B. 1970. Advantages, disadvantages and economics of liquid overfeed systems. *ASHRAE Symposium Bulletin* KC-70-3: *Liquid overfeed systems*.

Soling, S.P. 1971. Oil recovery from low temperature pump recirculating hydrocarbon systems. *ASHRAE Symposium Bulletin* PH-71-2: *Effect of oil on the refrigeration syst*em.

Sonders, M. and G.G. Brown. 1934. Design of fractionating columns, entrainment and capacity. *Industrial & Engineering Chemistry* (January).

Stoecker, W.F. 1960. How to design and operate flooded evaporators for cooling air and liquids. *Heating, Piping & Air Conditioning* (December).

Younger, A.H. 1955. How to size future process vessels. *Chemical Engineering* (May).

BIBLIOGRAPHY

Chaddock, J.B. 1976. Two-phase pressure drop in refrigerant liquid overfeed systems—Design tables. *ASHRAE Transactions* 82(2):107-133.

Chaddock, J.B., H. Lau, and E. Skuchas. 1976. Two-phase pressure drop in refrigerant liquid overfeed systems—Experimental measurements. *ASHRAE Transactions* 82(2):134-150.

Geltz, R.W. 1967. Pump overfeed evaporator refrigeration systems. *Air Conditioning, Heating & Refrigeration News* (January 30, February 6, March 6, March 13, March 20, March 27).

Lorentzen, G. and A.O. Baglo. 1969. An investigation of a gas pump recirculation system. *Proceedings of the Xth International Congress of Refrigeration*, p. 215. International Institute of Refrigeration, Paris.

Richards, W.V. 1959. Liquid ammonia recirculation systems. *Industrial Refrigeration* (June):139.

Richards, W.V. 1970. Pumps and piping in liquid overfeed systems. *ASHRAE Symposium Bulletin* KC-70-3: *Liquid overfeed systems*.

Slipcevic, B. 1964. The calculation of the refrigerant charge in refrigerating systems with circulation pumps. *Kältetechnik* 4:111.

Thompson, R.B. 1970. Control of evaporators in liquid overfeed systems. *ASHRAE Symposium Bulletin* KC-70-3: *Liquid overfeed systems*.

Watkins, J.E. 1956. Improving refrigeration systems by applying established principles. *Industrial Refrigeration* (June).

CHAPTER 2

SYSTEM PRACTICES FOR HALOCARBON REFRIGERANTS

REFRIGERATION is the process of moving heat from one location to another by use of refrigerant in a closed cycle. Oil management; gas and liquid separation; subcooling, superheating, and piping of refrigerant liquid and gas; and two-phase flow are all part of refrigeration. Applications include air conditioning, commercial refrigeration, and industrial refrigeration.

Desired characteristics of a refrigeration system may include

- Year-round operation, regardless of outdoor ambient conditions
- Possible wide load variations (0 to 100% capacity) during short periods without serious disruption of the required temperature levels
- Frost control for continuous-performance applications
- Oil management for different refrigerants under varying load and temperature conditions
- A wide choice of heat exchange methods (e.g., dry expansion, liquid overfeed, or flooded feed of the refrigerants) and use of secondary coolants such as salt brine, alcohol, and glycol
- System efficiency, maintainability, and operating simplicity
- Operating pressures and pressure ratios that might require multistaging, cascading, and so forth

A successful refrigeration system depends on good piping design and an understanding of the required accessories. This chapter covers the fundamentals of piping and accessories in halocarbon refrigerant systems. Hydrocarbon refrigerant pipe friction data can be found in petroleum industry handbooks. Use the refrigerant properties and information in Chapters 2, 19, and 20 of the 2005 *ASHRAE Handbook—Fundamentals* to calculate friction losses.

For information on refrigeration load, see Chapter 12. For R-502 information, refer to the 1998 *ASHRAE Handbook—Refrigeration.*

Piping Basic Principles

The design and operation of refrigerant piping systems should (1) ensure proper refrigerant feed to evaporators; (2) provide practical refrigerant line sizes without excessive pressure drop; (3) prevent excessive amounts of lubricating oil from being trapped in any part of the system; (4) protect the compressor at all times from loss of lubricating oil; (5) prevent liquid refrigerant or oil slugs from entering the compressor during operating and idle time; and (6) maintain a clean and dry system.

REFRIGERANT FLOW

Refrigerant Line Velocities

Economics, pressure drop, noise, and oil entrainment establish feasible design velocities in refrigerant lines (Table 1).

The preparation of this chapter is assigned to TC 10.3, Refrigerant Piping.

Table 1 Recommended Gas Line Velocities

Suction line	900 to 4000 fpm
Discharge line	2000 to 3500 fpm

Higher gas velocities are sometimes found in relatively short suction lines on comfort air-conditioning or other applications where the operating time is only 2000 to 4000 h per year and where low initial cost of the system may be more significant than low operating cost. Industrial or commercial refrigeration applications, where equipment runs almost continuously, should be designed with low refrigerant velocities for most efficient compressor performance and low equipment operating costs. An owning and operating cost analysis will reveal the best choice of line sizes. (See Chapter 36 of the 2003 *ASHRAE Handbook—HVAC Applications* for information on owning and operating costs.) Liquid lines from condensers to receivers should be sized for 100 fpm or less to ensure positive gravity flow without incurring backup of liquid flow. Liquid lines from receiver to evaporator should be sized to maintain velocities below 300 fpm, thus minimizing or preventing liquid hammer when solenoids or other electrically operated valves are used.

Refrigerant Flow Rates

Refrigerant flow rates for R-22 and R-134a are indicated in Figures 1 and 2. To obtain total system flow rate, select the proper rate value and multiply by system capacity. Enter curves using saturated refrigerant temperature at the evaporator outlet and actual liquid temperature entering the liquid feed device (including subcooling in condensers and liquid-suction interchanger, if used).

Because Figures 1 and 2 are based on a saturated evaporator temperature, they may indicate slightly higher refrigerant flow rates than are actually in effect when suction vapor is superheated in excess of the conditions mentioned. Refrigerant flow rates may be reduced approximately 3% for each 10°F increase in superheat in the evaporator.

Suction-line superheating downstream of the evaporator from line heat gain from external sources should not be used to reduce evaluated mass flow, because it increases volumetric flow rate and line velocity per unit of evaporator capacity, but not mass flow rate. It should be considered when evaluating suction-line size for satisfactory oil return up risers.

Suction gas superheating from use of a liquid-suction heat exchanger has an effect on oil return similar to that of suction-line superheating. The liquid cooling that results from the heat exchange reduces mass flow rate per ton of refrigeration. This can be seen in Figures 1 and 2 because the reduced temperature of the liquid supplied to the evaporator feed valve has been taken into account.

Superheat caused by heat in a space not intended to be cooled is always detrimental because the volumetric flow rate increases with no compensating gain in refrigerating effect.

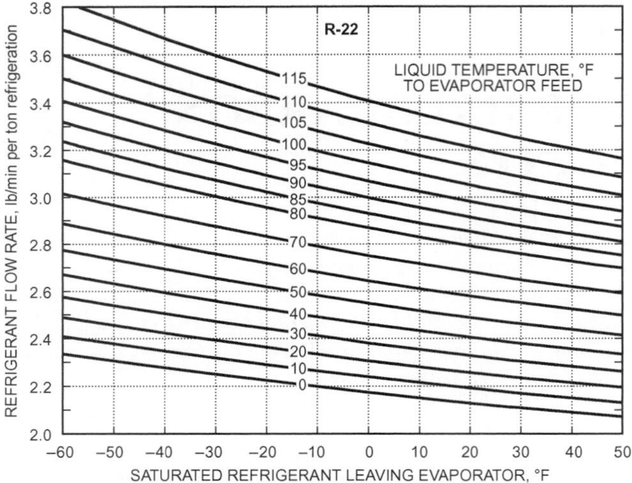

Fig. 1 Flow Rate per Ton of Refrigeration for Refrigerant 22

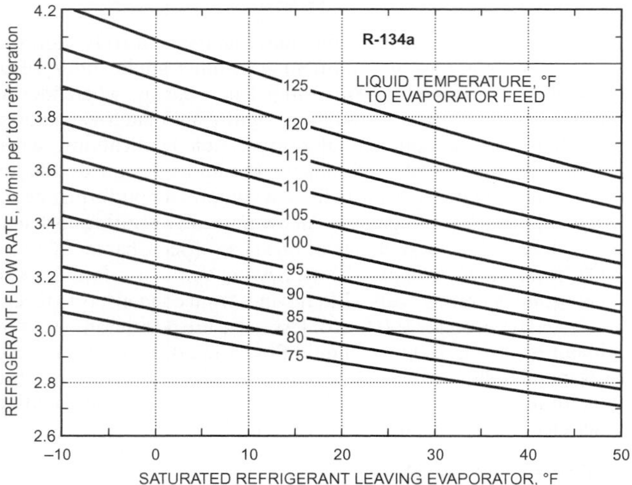

Fig. 2 Flow Rate per Ton of Refrigeration for Refrigerant 134a

REFRIGERANT LINE SIZING

In sizing refrigerant lines, cost considerations favor minimizing line sizes. However, suction and discharge line pressure drops cause loss of compressor capacity and increased power usage. Excessive liquid line pressure drops can cause liquid refrigerant to flash, resulting in faulty expansion valve operation. Refrigeration systems are designed so that friction pressure losses do not exceed a pressure differential equivalent to a corresponding change in the saturation boiling temperature. The primary measure for determining pressure drops is a given change in saturation temperature.

Pressure Drop Considerations

Pressure drop in refrigerant lines reduces system efficiency. Correct sizing must be based on minimizing cost and maximizing efficiency. Table 2 indicates the approximate effect of refrigerant pressure drop on an R-22 system operating at a 40°F saturated evaporator temperature with a 100°F saturated condensing temperature.

Pressure drop calculations are determined as normal pressure loss associated with a change in saturation temperature of the refrigerant. Typically, the refrigeration system is sized for pressure losses of 2°F or less for each segment of the discharge, suction, and liquid lines.

Table 2 Approximate Effect of Gas Line Pressure Drops on R-22 Compressor Capacity and Power[a]

Line Loss, °F	Capacity, %	Energy, %[b]
Suction Line		
0	100	100
2	96.4	104.8
4	92.9	108.1
Discharge Line		
0	100	100
2	99.1	103.0
4	98.2	106.3

[a]For system operating at 40°F saturated evaporator temperature and 100°F saturated condensing temperature.
[b]Energy percentage rated at hp/ton.

Liquid Lines. Pressure drop should not be so large as to cause gas formation in the liquid line, insufficient liquid pressure at the liquid feed device, or both. Systems are normally designed so that pressure drop in the liquid line from friction is not greater than that corresponding to about a 1 to 2°F change in saturation temperature. See Tables 3 to 9 for liquid-line sizing information.

Liquid subcooling is the only method of overcoming liquid line pressure loss to guarantee liquid at the expansion device in the evaporator. If subcooling is insufficient, flashing occurs in the liquid line and degrades system efficiency.

Friction pressure drops in the liquid line are caused by accessories such as solenoid valves, filter-driers, and hand valves, as well as by the actual pipe and fittings between the receiver outlet and the refrigerant feed device at the evaporator.

Liquid-line risers are a source of pressure loss and add to the total loss of the liquid line. Loss caused by risers is approximately 0.5 psi per foot of liquid lift. Total loss is the sum of all friction losses plus pressure loss from liquid risers.

Example 1 illustrates the process of determining liquid-line size and checking for total subcooling required.

Example 1. An R-22 refrigeration system using copper pipe operates at 40°F evaporator and 105°F condensing. Capacity is 5 tons, and the liquid line is 100 ft equivalent length with a riser of 20 ft. Determine the liquid-line size and total required subcooling.

Solution: From Table 3, the size of the liquid line at 1°F drop is 5/8 in. OD. Use the equation in Note 3 of Table 3 to compute actual temperature drop. At 5 tons,

Actual temperature drop	=	$1.0(5.0/6.7)^{1.8}$	=	0.59°F
Estimated friction loss	=	0.59 × 3.05	=	1.8 psi
Loss for the riser	=	20 × 0.5	=	10 psi
Total pressure losses	=	10.0 + 1.8	=	11.8 psi
R-22 saturation pressure at 105°F condensing (see R-22 properties in Chapter 20, 2005 *ASHRAE Handbook—Fundamentals*)				210.8 psig
Initial pressure at beginning of liquid line				210.8 psig
Total liquid line losses			−	11.8 psi
Net pressure at expansion device			=	199 psig

The saturation temperature at 199 psig is 101.1°F.

Required subcooling to overcome the liquid losses = (105.0 − 101.1) or 3.9°F

Refrigeration systems that have no liquid risers and have the evaporator below the condenser/receiver benefit from a gain in pressure caused by liquid weight and can tolerate larger friction losses without flashing. Regardless of the liquid-line routing when flashing occurs, overall efficiency is reduced, and the system may malfunction.

The velocity of liquid leaving a partially filled vessel (such as a receiver or shell-and-tube condenser) is limited by the height of the liquid above the point at which the liquid line leaves the vessel, whether or not the liquid at the surface is subcooled. Because liquid

Table 3 Suction, Discharge, and Liquid Line Capacities in Tons for Refrigerant 22 (Single- or High-Stage Applications)

Line Size		Suction Lines (Δt = 2°F)					Discharge Lines (Δt = 1°F, Δp = 3.05 psi)		Line Size		Liquid Lines	
		Saturated Suction Temperature, °F					Saturated Suction Temperature, °F				See notes a and b	
Type L Copper, OD		−40	−20	0	20	40	−40	40	Type L Copper, OD		Vel. = 100 fpm	Δt = 1°F Δp = 3.05
		Corresponding Δp, psi/100 ft: 0.79	1.15	1.6	2.22	2.91						
1/2		—	—	—	0.40	0.6	0.75	0.85	1/2		2.3	3.6
5/8		—	0.32	0.51	0.76	1.1	1.4	1.6	5/8		3.7	6.7
7/8		0.52	0.86	1.3	2.0	2.9	3.7	4.2	7/8		7.8	18.2
1 1/8		1.1	1.7	2.7	4.0	5.8	7.5	8.5	1 1/8		13.2	37.0
1 3/8		1.9	3.1	4.7	7.0	10.1	13.1	14.8	1 3/8		20.2	64.7
1 5/8		3.0	4.8	7.5	11.1	16.0	20.7	23.4	1 5/8		28.5	102.5
2 1/8		6.2	10.0	15.6	23.1	33.1	42.8	48.5	2 1/8		49.6	213.0
2 5/8		10.9	17.8	27.5	40.8	58.3	75.4	85.4	2 5/8		76.5	376.9
3 1/8		17.5	28.4	44.0	65.0	92.9	120.2	136.2	3 1/8		109.2	601.5
3 5/8		26.0	42.3	65.4	96.6	137.8	178.4	202.1	3 5/8		147.8	895.7
4 1/8		36.8	59.6	92.2	136.3	194.3	251.1	284.4	4 1/8		192.1	1263.2
Steel									Steel			
IPS	SCH								IPS	SCH		
1/2	40	—	0.38	0.58	0.85	1.2	1.5	1.7	1/2	80	3.8	5.7
3/4	40	0.50	0.8	1.2	1.8	2.5	3.3	3.7	3/4	80	6.9	12.8
1	40	0.95	1.5	2.3	3.4	4.8	6.1	6.9	1	80	11.5	25.2
1 1/4	40	2.0	3.2	4.8	7.0	9.9	12.6	14.3	1 1/4	80	20.6	54.1
1 1/2	40	3.0	4.7	7.2	10.5	14.8	19.0	21.5	1 1/2	80	28.3	82.6
2	40	5.7	9.1	13.9	20.2	28.5	36.6	41.4	2	40	53.8	192.0
2 1/2	40	9.2	14.6	22.1	32.2	45.4	58.1	65.9	2 1/2	40	76.7	305.8
3	40	16.2	25.7	39.0	56.8	80.1	102.8	116.4	3	40	118.5	540.3
4	40	33.1	52.5	79.5	115.9	163.2	209.5	237.3	4	40	204.2	1101.2

Notes:

1. Table capacities are in tons of refrigeration.
   Δp = pressure drop from line friction, psi per 100 ft of equivalent line length
   Δt = corresponding change in saturation temperature, °F per 100 ft
2. Line capacity for other saturation temperatures Δt and equivalent lengths L_e

$$\text{Line capacity} = \text{Table capacity}\left(\frac{\text{Table } L_e}{\text{Actual } L_e} \times \frac{\text{Actual } \Delta t}{\text{Table } \Delta t}\right)^{0.55}$$

3. Saturation temperature Δt for other capacities and equivalent lengths L_e

$$\Delta t = \text{Table } \Delta t \left(\frac{\text{Actual } L_e}{\text{Table } L_e}\right)\left(\frac{\text{Actual capacity}}{\text{Table capacity}}\right)^{1.8}$$

4. Values based on 105°F condensing temperature. Multiply table capacities by the following factors for other condensing temperatures.

Condensing Temperature, °F	Suction Line	Discharge Line
80	1.11	0.79
90	1.07	0.88
100	1.03	0.95
110	0.97	1.04
120	0.90	1.10
130	0.86	1.18
140	0.80	1.26

[a]Sizing shown is recommended where any gas generated in receiver must return up condensate line to condenser without restricting condensate flow. Water-cooled condensers, where receiver ambient temperature may be higher than refrigerant condensing temperature, fall into this category.

[b]Line pressure drop Δp is conservative; if subcooling is substantial or line is short, a smaller size line may be used. Applications with very little subcooling or very long lines may require a larger line.

Table 4 Suction, Discharge, and Liquid Line Capacities in Tons for Refrigerant 22 (Intermediate- or Low-Stage Duty)

Line Size	Suction Lines (Δt = 2°F)*							Discharge Lines (Δt = 2°F)*	Liquid Lines
	Saturated Suction Temperature, °F								
Type L Copper, OD	−90	−80	−70	−60	−50	−40	−30		
5/8								0.7	See Table 3
7/8	0.18	0.25	0.34	0.46	0.61	0.79	1.0	1.9	
1 1/8	0.36	0.51	0.70	0.94	1.2	1.6	2.1	3.8	
1 3/8	0.6	0.9	1.2	1.6	2.2	2.8	3.6	6.6	
1 5/8	1.0	1.4	1.9	2.6	3.4	4.5	5.7	10.5	
2 1/8	2.1	3.0	4.1	5.5	7.2	9.3	11.9	21.7	
2 5/8	3.8	5.3	7.2	9.7	12.7	16.5	21.1	38.4	
3 1/8	6.1	8.5	11.6	15.5	20.4	26.4	33.8	61.4	
3 5/8	9.1	12.7	17.3	23.1	30.4	39.4	50.2	91.2	
4 1/8	12.9	18.0	24.5	32.7	43.0	55.6	70.9	128.6	
5 1/8	23.2	32.3	43.9	58.7	77.1	99.8	126.9	229.5	
6 1/8	37.5	52.1	71.0	94.6	124.2	160.5	204.2	369.4	

Notes:

1. Table capacities are in tons of refrigeration.
   Δp = pressure drop from line friction, psi per 100 ft of equivalent line length
   Δt = corresponding change in saturation temperature, °F per 100 ft
2. Line capacity for other saturation temperatures Δt and equivalent lengths L_e

$$\text{Line capacity} = \text{Table capacity}\left(\frac{\text{Table } L_e}{\text{Actual } L_e} \times \frac{\text{Actual } \Delta t}{\text{Table } \Delta t}\right)^{0.55}$$

3. Saturation temperature Δt for other capacities and equivalent lengths L_e

$$\Delta t = \text{Table } \Delta t \left(\frac{\text{Actual } L_e}{\text{Table } L_e}\right)\left(\frac{\text{Actual capacity}}{\text{Table capacity}}\right)^{1.8}$$

4. Refer to refrigerant thermodynamic property tables (Chapter 20 of the 2005 *ASHRAE Handbook—Fundamentals*) for pressure drop corresponding to Δt.
5. Values based on 0°F condensing temperature. Multiply table capacities by the following factors for other condensing temperatures. Flow rates for discharge lines are based on −50°F evaporating temperature.

Condensing Temperature, °F	Suction Line	Discharge Line
−30	1.09	0.58
−20	1.06	0.71
−10	1.03	0.85
0	1.00	1.00
10	0.97	1.20
20	0.94	1.45
30	0.90	1.80

*See section on Pressure Drop Considerations.

in the vessel has a very low (or zero) velocity, the velocity V in the liquid line (usually at the vena contracta) is $V^2 = 2gh$, where h is the liquid height in the vessel. Gas pressure does not add to the velocity unless gas is flowing in the same direction. As a result, both gas and liquid flow through the line, limiting the rate of liquid flow. If this factor is not considered, excess operating charges in receivers and flooding of shell-and-tube condensers may result.

No specific data are available to precisely size a line leaving a vessel. If the height of liquid above the vena contracta produces the desired velocity, liquid leaves the vessel at the expected rate. Thus, if the level in the vessel falls to one pipe diameter above the bottom of the vessel from which the liquid line leaves, the capacity of copper lines for R-22 at 3 lb/min per ton of refrigeration is approximately as follows:

OD, in.	Tons
1 1/8	14
1 3/8	25
1 5/8	40
2 1/8	80
2 5/8	130
3 1/8	195
4 1/8	410

The whole liquid line need not be as large as the leaving connection. After the vena contracta, the velocity is about 40% less. If the line continues down from the receiver, the value of h increases. For a 200 ton capacity with R-22, the line from the bottom of the receiver should be about 3 1/8 in. After a drop of 1 ft, a reduction to 2 5/8 in. is satisfactory.

Suction Lines. Suction lines are more critical than liquid and discharge lines from a design and construction standpoint. Refrigerant lines should be sized to (1) provide a minimum pressure drop at full load, (2) return oil from the evaporator to the compressor under minimum load conditions, and (3) prevent oil from draining from an active evaporator into an idle one. A pressure drop in the suction line reduces a system's capacity because it forces the compressor to operate at a lower suction pressure to maintain a desired evaporating temperature in the coil. The suction line is normally sized to have a pressure drop from friction no greater than the equivalent of about a 2°F change in saturation temperature. See Tables 3 to 15 for suction line sizing information.

At suction temperatures lower than 40°F, the pressure drop equivalent to a given temperature change decreases. For example, at –40°F suction with R-22, the pressure drop equivalent to a 2°F change in saturation temperature is about 0.8 psi. Therefore,

Table 5 Suction, Discharge, and Liquid Line Capacities in Tons for Refrigerant 134a (Single- or High-Stage Applications)

Line Size		Suction Lines (Δt = 2°F)					Discharge Lines (Δt = 1°F, Δp = 2.2 psi/100 ft)			Line Size		Liquid Lines See notes a and b	
		Saturated Suction Temperature, °F					Saturated Suction Temperature, °F						
		0	10	20	30	40							
Type L Copper, OD		Corresponding Δp, psi/100 ft								Type L Copper, OD			
		1.00	1.19	1.41	1.66	1.93	0	20	40			Velocity = 100 fpm	Δt = 1°F Δp = 2.2
1/2		0.14	0.18	0.23	0.29	0.35	0.54	0.57	0.59	1/2		2.13	2.79
5/8		0.27	0.34	0.43	0.54	0.66	1.01	1.07	1.12	5/8		3.42	5.27
7/8		0.71	0.91	1.14	1.42	1.75	2.67	2.81	2.94	7/8		7.09	14.00
1 1/8		1.45	1.84	2.32	2.88	3.54	5.40	5.68	5.95	1 1/8		12.10	28.40
1 3/8		2.53	3.22	4.04	5.02	6.17	9.42	9.91	10.40	1 3/8		18.40	50.00
1 5/8		4.02	5.10	6.39	7.94	9.77	14.90	15.70	16.40	1 5/8		26.10	78.60
2 1/8		8.34	10.60	13.30	16.50	20.20	30.80	32.40	34.00	2 1/8		45.30	163.00
2 5/8		14.80	18.80	23.50	29.10	35.80	54.40	57.20	59.90	2 5/8		69.90	290.00
3 1/8		23.70	30.00	37.50	46.40	57.10	86.70	91.20	95.50	3 1/8		100.00	462.00
3 5/8		35.10	44.60	55.80	69.10	84.80	129.00	135.00	142.00	3 5/8		135.00	688.00
4 1/8		49.60	62.90	78.70	97.40	119.43	181.00	191.00	200.00	4 1/8		175.00	971.00
5 1/8		88.90	113.00	141.00	174.00	213.00	323.00	340.00	356.00	—		—	—
6 1/8		143.00	181.00	226.00	280.00	342.00	518.00	545.00	571.00	—		—	—
Steel										Steel			
IPS	SCH									IPS	SCH		
1/2	80	0.22	0.28	0.35	0.43	0.53	0.79	0.84	0.88	1/2	80	3.43	4.38
3/4	80	0.51	0.64	0.79	0.98	1.19	1.79	1.88	1.97	3/4	80	6.34	9.91
1	80	1.00	1.25	1.56	1.92	2.33	3.51	3.69	3.86	1	80	10.50	19.50
1 1/4	40	2.62	3.30	4.09	5.03	6.12	9.20	9.68	10.10	1 1/4	80	18.80	41.80
1 1/2	40	3.94	4.95	6.14	7.54	9.18	13.80	14.50	15.20	1 1/2	80	25.90	63.70
2	40	7.60	9.56	11.90	14.60	17.70	26.60	28.00	29.30	2	40	49.20	148.00
2 1/2	40	12.10	15.20	18.90	23.10	28.20	42.40	44.60	46.70	2 1/2	40	70.10	236.00
3	40	21.40	26.90	33.40	41.00	49.80	75.00	78.80	82.50	3	40	108.00	419.00
4	40	43.80	54.90	68.00	83.50	101.60	153.00	160.00	168.00	4	40	187.00	853.00

Notes:

1. Table capacities are in tons of refrigeration.
   Δp = pressure drop from line friction, psi per 100 ft of equivalent line length
   Δt = corresponding change in saturation temperature, °F per 100 ft
2. Line capacity for other saturation temperatures Δt and equivalent lengths L_e

$$\text{Line capacity} = \text{Table capacity}\left(\frac{\text{Table } L_e}{\text{Actual } L_e} \times \frac{\text{Actual } \Delta t}{\text{Table } \Delta t}\right)^{0.55}$$

3. Saturation temperature Δt for other capacities and equivalent lengths L_e

$$\Delta t = \text{Table } \Delta t \left(\frac{\text{Actual } L_e}{\text{Table } L_e}\right)\left(\frac{\text{Actual capacity}}{\text{Table capacity}}\right)^{1.8}$$

4. Values based on 105°F condensing temperature. Multiply table capacities by the following factors for other condensing temperatures.

Condensing Temperature, °F	Suction Line	Discharge Line
80	1.158	0.804
90	1.095	0.882
100	1.032	0.961
110	0.968	1.026
120	0.902	1.078
130	0.834	1.156

[a]Sizing shown is recommended where any gas generated in receiver must return up condensate line to the condenser without restricting condensate flow. Water-cooled condensers, where receiver ambient temperature may be higher than refrigerant condensing temperature, fall into this category.

[b]Line pressure drop Δp is conservative; if subcooling is substantial or line is short, a smaller size line may be used. Applications with very little subcooling or very long lines may require a larger line.

low-temperature lines must be sized for a very low pressure drop, or higher equivalent temperature losses, with resultant loss in equipment capacity, must be accepted. For very low pressure drops, any suction or hot-gas risers must be sized properly to ensure oil entrainment up the riser so that oil is always returned to the compressor.

Where pipe size must be reduced to provide sufficient gas velocity to entrain oil up vertical risers at partial loads, greater pressure drops are imposed at full load. These can usually be compensated for by oversizing the horizontal and down run lines and components.

Discharge Lines. Pressure loss in hot-gas lines increases the required compressor power per unit of refrigeration and decreases compressor capacity. Table 2 illustrates power losses for an R-22 system at 40°F evaporator and 100°F condensing temperature. Pressure drop is minimized by generously sizing lines for low friction losses, but still maintaining refrigerant line velocities to entrain and carry oil along at all loading conditions. Pressure drop is normally designed not to exceed the equivalent of a 2°F change in saturation temperature. Recommended sizing tables are based on a 1°F change in saturation temperature.

Location and Arrangement of Piping

Refrigerant lines should be as short and direct as possible to minimize tubing and refrigerant requirements and pressure drops. Plan piping for a minimum number of joints using as few elbows and other fittings as possible, but provide sufficient flexibility to absorb compressor vibration and stresses caused by thermal expansion and contraction.

Arrange refrigerant piping so that normal inspection and servicing of the compressor and other equipment is not hindered. Do not obstruct the view of the oil-level sight glass or run piping so that it interferes with removing compressor cylinder heads, end bells, access plates, or any internal parts. Suction-line piping to the compressor should be arranged so that it will not interfere with removal of the compressor for servicing.

Provide adequate clearance between pipe and adjacent walls and hangers or between pipes for insulation installation. Use sleeves that are sized to permit installation of both pipe and insulation through floors, walls, or ceilings. Set these sleeves prior to pouring of concrete or erection of brickwork.

Run piping so that it does not interfere with passages or obstruct headroom, windows, and doors. Refer to ASHRAE *Standard* 15 and other governing local codes for restrictions that may apply.

Protection Against Damage to Piping

Protection against damage is necessary, particularly for small lines, which have a false appearance of strength. Where traffic is heavy, provide protection against impact from carelessly handled hand trucks, overhanging loads, ladders, and fork trucks.

Piping Insulation

All piping joints and fittings should be thoroughly leak-tested before insulation is sealed. Suction lines should be insulated to prevent sweating and heat gain. Insulation covering lines on which moisture can condense or lines subjected to outside conditions must be vapor-sealed to prevent any moisture travel through the insulation or condensation in the insulation. Many commercially available types are provided with an integral waterproof jacket for this purpose. Although the liquid line ordinarily does not require insulation, suction and liquid lines can be insulated as a unit on installations where the two lines are clamped together. When it passes through a warmer area, the liquid line should be insulated to minimize heat gain. Hot-gas discharge lines usually are not insulated; however, they should be insulated if the heat dissipated is objectionable or to prevent injury from high-temperature surfaces. In the latter case, it is not essential to provide insulation with a tight vapor seal because moisture condensation is not a problem unless the line is located outside. Hot-gas defrost lines are customarily insulated to minimize heat loss and condensation of gas inside the piping.

All joints and fittings should be covered, but it is not advisable to do so until the system has been thoroughly leak-tested. See Chapter 33 for additional information.

Vibration and Noise in Piping

Vibration transmitted through or generated in refrigerant piping and the resulting objectionable noise can be eliminated or minimized by proper piping design and support.

Two undesirable effects of vibration of refrigerant piping are (1) physical damage to the piping, which can break brazed joints and, consequently, lose charge; and (2) transmission of noise through the piping itself and through building construction that may come into direct contact with the piping.

In refrigeration applications, piping vibration can be caused by rigid connection of the refrigerant piping to a reciprocating compressor. Vibration effects are evident in all lines directly connected to the compressor or condensing unit. It is thus impossible to eliminate vibration in piping; it is only possible to mitigate its effects.

Flexible metal hose is sometimes used to absorb vibration transmission along smaller pipe sizes. For maximum effectiveness, it should be installed parallel to the crankshaft. In some cases, two isolators may be required, one in the horizontal line and the other in the vertical line at the compressor. A rigid brace on the end of the flexible hose away from the compressor is required to prevent vibration of the hot-gas line beyond the hose.

Flexible metal hose is not as efficient in absorbing vibration on larger pipes because it is not actually flexible unless the ratio of length to diameter is relatively great. In practice, the length is often limited, so flexibility is reduced in larger sizes. This problem is best solved by using flexible piping and isolation hangers where the piping is secured to the structure.

When piping passes through walls, through floors, or inside furring, it must not touch any part of the building and must be supported only by the hangers (provided to avoid transmitting vibration to the building); this eliminates the possibility of walls or ceilings acting as sounding boards or diaphragms. When piping is erected where access is difficult after installation, it should be supported by isolation hangers.

Vibration and noise from a piping system can also be caused by gas pulsations from the compressor operation or from turbulence in the gas, which increases at high velocities. It is usually more apparent in the discharge line than in other parts of the system.

When gas pulsations caused by the compressor create vibration and noise, they have a characteristic frequency that is a function of the number of gas discharges by the compressor on each revolution. This frequency is not necessarily equal to the number of cylinders, because on some compressors two pistons operate together. It is also varied by the angular displacement of the cylinders, such as in V-type compressors. Noise resulting from gas pulsations is usually objectionable only when the piping system amplifies the pulsation by resonance. On single-compressor systems, resonance can be reduced by changing the size or length of the resonating line or by installing a properly sized hot-gas muffler in the discharge line immediately after the compressor discharge valve. On a paralleled compressor system, a harmonic frequency from the different speeds of multiple compressors may be apparent. This noise can sometimes be reduced by installing mufflers.

When noise is caused by turbulence and isolating the line is not effective enough, installing a larger-diameter pipe to reduce gas velocity is sometimes helpful. Also, changing to a line of heavier wall or from copper to steel to change the pipe natural frequency may help.

Table 6 Suction, Discharge, and Liquid Line Capacities in Tons for Refrigerant 404a (Single- or High-Stage Applications)

Line Size		Suction Lines ($\Delta t = 2$°F)						Discharge Lines ($\Delta t = 1$°F, $\Delta p = 3.55$ psi)						Liquid Lines		
		Saturated Suction Temperature, °F						Saturated Suction Temperature, °F						See note a		
Type L Copper, OD		−60	−40	−20	0	20	40	−60	−40	−20	0	20	40	Velocity = 100 fpm	$\Delta t = 1$°F Drop	$\Delta t = 5$°F Drop
		Corresponding Δp, psi/100 ft						Corresponding Δp, psi/100 ft								
		0.64	0.97	1.41	1.96	2.62	3.44	3.55	3.55	3.55	3.55	3.55	3.55		$\Delta p = 3.6$	$\Delta p = 17.4$
1/2		0.05	0.09	0.15	0.24	0.36	0.53	0.56	0.61	0.65	0.70	0.75	0.79	1.3	2.6	6.09
5/8		0.09	0.16	0.28	0.44	0.68	1.00	1.04	1.14	1.23	1.31	1.40	1.48	2.1	4.9	11.39
3/4		0.15	0.28	0.47	0.76	1.15	1.70	1.77	1.93	2.09	2.23	2.38	2.51	3.1	8.1	18.87
7/8		0.24	0.43	0.73	1.17	1.78	2.63	2.73	2.98	3.22	3.44	3.66	3.87	4.4	12.8	29.81
1 1/8		0.49	0.88	1.49	2.37	3.61	5.31	5.52	6.01	6.49	6.96	7.40	7.81	7.5	25.9	60.17
1 3/8		0.86	1.54	2.59	4.13	6.28	9.23	9.60	10.46	11.29	12.10	12.87	13.58	11.4	45.2	104.41
1 5/8		1.36	2.44	4.10	6.53	9.92	14.57	15.14	16.49	17.80	19.07	20.28	21.41	16.1	71.4	164.68
2 1/8		2.83	5.07	8.52	13.53	20.51	30.06	31.29	34.08	36.80	39.43	41.93	44.26	28.0	147.9	339.46
2 5/8		5.03	8.97	15.07	23.88	36.16	52.96	55.04	59.95	64.74	69.36	73.76	77.85	43.2	261.2	597.42
3 1/8		8.05	14.34	24.02	38.05	57.56	84.33	87.66	95.48	103.11	110.47	117.48	124.00	61.7	416.2	950.09
3 5/8		11.98	21.31	35.73	56.53	85.39	125.18	129.88	141.46	152.76	163.67	174.05	183.71	83.5	618.4	1407.96
4 1/8		16.93	30.09	50.32	79.66	120.39	176.20	182.83	199.13	215.05	230.40	245.01	258.61	108.5	871.6	1982.40
5 1/8		30.35	53.85	89.97	142.32	214.82	313.91	325.75	354.81	383.16	410.51	436.55	460.78	169.1	1554.2	3525.99
6 1/8		48.89	86.74	144.47	228.50	344.70	502.77	521.74	568.28	613.69	657.49	699.20	738.00	243.1	2497.7	5648.67
8 1/8		101.60	179.88	299.39	472.46	710.75	1037.34	1076.62	1172.66	1266.36	1356.75	1442.81	1522.89	424.6	5159.7	11660.71
Steel																
IPS	SCH															
3/8	80	0.04	0.07	0.11	0.18	0.27	0.39	0.40	0.44	0.47	0.51	0.54	0.57	1.3	1.9	4.3
1/2	80	0.08	0.14	0.22	0.35	0.53	0.76	0.79	0.86	0.93	0.99	1.06	1.12	2.1	3.8	8.5
3/4	80	0.18	0.31	0.51	0.79	1.18	1.71	1.78	1.93	2.09	2.24	2.38	2.51	3.9	8.6	19.2
1	80	0.35	0.60	0.99	1.55	2.32	3.36	3.48	3.79	4.09	4.38	4.66	4.92	6.5	16.9	37.5
1 1/4	80	0.75	1.30	2.13	3.33	4.97	7.20	7.45	8.12	8.77	9.39	9.99	10.54	11.6	36.3	80.3
1 1/2	80	1.14	1.98	3.26	5.08	7.57	10.96	11.35	12.37	13.35	14.31	15.21	16.06	16.0	55.3	122.3
2	40	2.65	4.61	7.55	11.78	17.57	25.45	26.36	28.71	31.01	33.22	35.33	37.29	30.4	128.4	283.5
2 1/2	40	4.23	7.34	12.04	18.74	27.94	40.49	41.93	45.67	49.32	52.84	56.19	59.31	43.3	204.7	450.9
3	40	7.48	12.98	21.26	33.11	49.37	71.55	74.10	80.71	87.16	93.38	99.31	104.82	66.9	361.6	796.8
4	40	15.30	26.47	43.34	67.50	100.66	145.57	150.75	164.20	177.32	189.98	202.03	213.24	115.3	735.6	1623.0
5	40	27.58	47.78	78.24	121.87	181.32	262.52	272.21	296.49	320.19	343.04	364.80	385.05	181.1	1328.2	2927.2
6	40	44.58	77.26	126.52	197.09	293.24	424.04	439.72	478.94	517.21	554.13	589.28	621.99	261.7	2148.0	4728.3
8	40	91.40	158.09	258.81	402.66	599.91	867.50	898.42	978.56	1056.75	1132.18	1203.99	1270.82	453.2	4394.4	9674.1
10	40	165.52	286.19	468.14	728.40	1083.73	1569.40	1625.34	1770.31	1911.78	2048.23	2178.15	2299.05	714.4	7938.5	17,477.4
12	ID[b]	264.36	457.37	748.94	1163.62	1733.87	2507.30	2600.54	2832.50	3058.84	3277.16	3485.04	3678.47	1024.6	12,681.8	27,963.7
14	30	342.81	592.13	968.21	1506.59	2244.98	3246.34	3362.07	3661.96	3954.59	4236.83	4505.59	4755.67	1249.2	16,419.6	36,152.5
16	30	493.87	852.84	1395.24	2171.13	3230.27	4678.48	4845.26	5277.44	5699.16	6105.92	6493.24	6853.65	1654.7	23,662.2	52,101.2

[a] Sizing shown is recommended where any gas generated in receiver must return up condensate line to condenser without restricting condensate flow. Water-cooled condensers, where receiver ambient temperature may be higher than refrigerant condensing temperature, fall into this category.

[b] Pipe inside diameter is same as nominal pipe size.

Notes: 1. Table capacities are in tons of refrigeration.

Δp = pressure drop from line friction, psi per 100 ft of equivalent line length

Δt = corresponding change in saturation temperature, °F per 100 ft

2. Line capacity for other saturation temperatures Δt and equivalent lengths L_e

$$\text{Line capacity} = \text{Table capacity}\left(\frac{\text{Table } L_e}{\text{Actual } L_e} \times \frac{\text{Actual } \Delta t}{\text{Table } \Delta t}\right)^{0.55}$$

3. Saturation temperature Δt for other capacities and equivalent lengths L_e

$$\Delta t = \text{Table } \Delta t \left(\frac{\text{Actual } L_e}{\text{Table } L_e}\right)\left(\frac{\text{Actual capacity}}{\text{Table capacity}}\right)^{1.8}$$

4. Tons based on standard refrigerant cycle of 105°F liquid and saturated evaporator outlet temperature. Liquid tons based on 20°F evaporator temperature.

5. Thermophysical properties and viscosity data based on calculations from NIST REFPROP program Version 6.01.

6. For brazed Type L copper tubing larger than 1 1/8 in. OD for discharge or liquid service, see Safety Requirements section.

7. Values based on 105°F condensing temperature. Multiply table capacities by the following factors for other condensing temperatures.

Cond. Temp., °F	Suction Line	Discharge Line
80	1.246	0.870
90	1.150	0.922
100	1.051	0.974
110	0.948	1.009
120	0.840	1.026
130	0.723	1.043

Table 7 Suction, Discharge, and Liquid Line Capacities in Tons for Refrigerant 507 (Single- or High-Stage Applications)

Line Size		Suction Lines ($\Delta t = 2°F$)						Discharge Lines ($\Delta t = 1°F$, $\Delta p = 3.65$ psi)						Liquid Lines		
		Saturated Suction Temperature, °F						Saturated Suction Temperature, °F						See note a		
Type L Copper, OD		−60	−40	−20	0	20	40	−60	−40	−20	0	20	40	Velocity = 100 fpm	$\Delta t = 1°F$ Drop $\Delta p = 3.65$	$\Delta t = 5°F$ Drop $\Delta p = 17.8$
		Corresponding Δp, psi/100 ft						Corresponding Δp, psi/100 ft								
		0.67	1.01	1.46	2.02	2.71	3.6	3.65	3.65	3.65	3.65	3.65	3.65			
1/2		0.05	0.09	0.15	0.24	0.37	0.55	0.55	0.60	0.65	0.70	0.75	0.79	1.3	2.5	5.96
5/8		0.09	0.17	0.28	0.45	0.69	1.02	1.04	1.13	1.22	1.31	1.40	1.48	2.0	4.7	11.13
3/4		0.16	0.28	0.48	0.77	1.17	1.74	1.76	1.92	2.08	2.24	2.38	2.52	3.0	7.9	18.45
7/8		0.25	0.44	0.74	1.18	1.81	2.68	2.72	2.97	3.22	3.45	3.68	3.89	4.2	12.5	29.14
1 1/8		0.50	0.90	1.51	2.40	3.66	5.41	5.48	5.99	6.49	6.96	7.41	7.84	7.2	25.2	58.74
1 3/8		0.88	1.57	2.63	4.18	6.35	9.41	9.54	10.42	11.28	12.11	12.90	13.63	11.0	44.0	102.09
1 5/8		1.39	2.48	4.17	6.61	10.04	14.84	15.04	16.43	17.79	19.09	20.34	21.50	15.6	69.5	161.04
2 1/8		2.91	5.17	8.65	13.70	20.76	30.66	31.03	33.90	36.70	39.40	41.96	44.36	27.1	144.0	331.97
2 5/8		5.15	9.14	15.27	24.19	36.62	54.04	54.69	59.74	64.68	69.43	73.96	78.18	41.8	254.3	584.28
3 1/8		8.24	14.61	24.40	38.55	58.29	85.90	86.95	94.98	102.84	110.39	117.58	124.29	59.6	405.2	929.27
3 5/8		12.27	21.75	36.22	57.15	86.47	127.52	129.07	140.99	152.66	163.87	174.54	184.50	80.6	601.0	1377.19
4 1/8		17.34	30.66	51.13	80.55	121.93	179.33	181.70	198.48	214.91	230.69	245.71	259.74	104.8	847.0	1935.27
5 1/8		31.09	54.88	91.25	143.93	217.14	319.89	323.48	353.35	382.60	410.70	437.44	462.40	163.3	1513.6	3449.44
6 1/8		49.99	88.20	146.87	230.77	348.36	512.29	518.62	566.52	613.40	658.45	701.32	741.34	234.8	2427.4	5526.55
8 1/8		103.91	182.97	303.62	477.80	720.09	1057.14	1070.49	1169.35	1266.13	1359.11	1447.60	1530.21	410.1	5019.4	11,383.18
Steel																
IPS	**SCH**															
3/8	80	0.04	0.07	0.12	0.18	0.27	0.39	0.40	0.43	0.47	0.51	0.54	0.57	1.2	1.9	4.2
1/2	80	0.08	0.14	0.23	0.35	0.53	0.77	0.78	0.86	0.93	0.99	1.06	1.12	2.1	3.7	8.3
3/4	80	0.18	0.31	0.51	0.80	1.20	1.74	1.76	1.93	2.09	2.24	2.39	2.52	3.8	8.4	18.7
1	80	0.35	0.61	1.01	1.57	2.34	3.41	3.45	3.77	4.08	4.38	4.67	4.94	6.3	16.4	36.6
1 1/4	80	0.76	1.32	2.16	3.36	5.02	7.32	7.39	8.08	8.74	9.39	10.00	10.57	11.2	35.2	78.4
1 1/2	80	1.16	2.01	3.29	5.12	7.65	11.15	11.26	12.30	13.32	14.30	15.23	16.10	15.5	53.8	119.4
2	40	2.70	4.68	7.65	11.89	17.76	25.88	26.15	28.56	30.93	33.20	35.36	37.38	29.4	124.8	276.7
2 1/2	40	4.31	7.45	12.18	18.93	28.24	41.17	41.59	45.43	49.19	52.80	56.24	59.45	41.9	198.9	440.6
3	40	7.63	13.19	21.54	33.45	49.90	72.75	73.50	80.29	86.93	93.32	99.39	105.06	64.6	351.5	777.9
4	40	15.57	26.88	43.92	68.12	101.75	148.00	149.53	163.33	176.85	189.84	202.20	213.74	111.4	714.9	1586.3
5	40	28.10	48.52	79.19	122.99	183.27	266.91	270.00	294.93	319.34	342.79	365.11	385.94	174.9	1290.8	2857.5
6	40	45.48	78.45	128.06	198.91	296.40	431.69	436.14	476.41	515.85	553.73	589.78	623.44	252.8	2087.5	4622.0
8	40	93.13	160.66	261.94	406.93	606.38	882.01	891.10	973.39	1053.96	1131.36	1205.02	1273.79	437.7	4270.8	9443.9
10	40	168.64	290.60	473.82	735.12	1095.44	1595.65	1612.10	1760.97	1906.72	2046.75	2180.00	2304.41	690.0	7715.1	17,086.7
12	ID[b]	269.75	464.87	758.01	1174.36	1752.56	2553.03	2579.36	2817.55	3050.75	3274.79	3488.00	3687.06	989.6	12,324.9	27,298.3
14	30	349.22	601.87	979.92	1520.49	2269.19	3300.65	3334.69	3642.64	3944.13	4233.77	4509.42	4766.76	1206.5	15,957.5	35,292.2
16	30	503.20	866.37	1414.32	2191.17	3265.09	4756.74	4805.79	5249.60	5684.09	6101.51	6498.76	6869.63	1598.2	22,996.2	50,861.5

[a]Sizing shown is recommended where any gas generated in receiver must return up condensate line to condenser without restricting condensate flow. Water-cooled condensers, where receiver ambient temperature may be higher than refrigerant condensing temperature, fall into this category.

[b]Pipe inside diameter is same as nominal pipe size.

Notes: 1. Table capacities are in tons of refrigeration.

Δp = pressure drop from line friction, psi per 100 ft of equivalent line length

Δt = corresponding change in saturation temperature, °F per 100 ft

2. Line capacity for other saturation temperatures Δt and equivalent lengths L_e

$$\text{Line capacity} = \text{Table capacity}\left(\frac{\text{Table } L_e}{\text{Actual } L_e} \times \frac{\text{Actual } \Delta t}{\text{Table } \Delta t}\right)^{0.55}$$

3. Saturation temperature Δt for other capacities and equivalent lengths L_e

$$\Delta t = \text{Table } \Delta t \left(\frac{\text{Actual } L_e}{\text{Table } L_e}\right)\left(\frac{\text{Actual capacity}}{\text{Table capacity}}\right)^{1.8}$$

4. Tons based on standard refrigerant cycle of 105°F liquid and saturated evaporator outlet temperature. Liquid tons based on 20°F evaporator temperature.

5. Thermophysical properties and viscosity data based on calculations from NIST REFPROP program Version 6.01.

6. For brazed Type L copper tubing larger than 1 1/8 in. OD for discharge or liquid service, see Safety Requirements section.

7. Values based on 105°F condensing temperature. Multiply table capacities by the following factors for other condensing temperatures.

Cond. Temp., °F	Suction Line	Discharge Line
80	1.267	0.873
90	1.163	0.924
100	1.055	0.975
110	0.944	1.005
120	0.826	1.014
130	0.701	1.024

Table 8 Suction, Discharge, and Liquid Line Capacities in Tons for Refrigerant 410a (Single- or High-Stage Applications)

Line Size		Suction Lines ($\Delta t = 2°F$)						Discharge Lines ($\Delta t = 1°F$, $\Delta p = 4.75$ psi)						Liquid Lines		
		Saturated Suction Temperature, °F						Saturated Suction Temperature, °F						See note a		
Type L Copper, OD		−60	−40	−20	0	20	40	−60	−40	−20	0	20	40	Velocity = 100 fpm	$\Delta t = 1°F$ Drop $\Delta p = 4.75$	$\Delta t = 5°F$ Drop $\Delta p = 23.3$
		Corresponding Δp, psi/100 ft						Corresponding Δp, psi/100 ft								
		0.84	1.27	1.85	2.57	3.46	4.5	4.75	4.75	4.75	4.75	4.75	4.75			
1/2		0.10	0.17	0.27	0.42	0.62	0.89	1.13	1.17	1.22	1.26	1.30	1.33	2.0	4.6	10.81
5/8		0.18	0.31	0.51	0.79	1.17	1.67	2.11	2.20	2.29	2.36	2.43	2.49	3.2	8.6	20.24
3/4		0.31	0.53	0.87	1.35	2.00	2.84	3.59	3.74	3.88	4.02	4.14	4.23	4.7	14.3	33.53
7/8		0.48	0.83	1.35	2.08	3.08	4.39	5.53	5.76	5.99	6.19	6.38	6.52	6.7	22.6	52.92
1 1/8		0.98	1.69	2.74	4.22	6.23	8.86	11.16	11.64	12.09	12.50	12.88	13.17	11.4	45.8	106.59
1 3/8		1.72	2.95	4.78	7.34	10.85	15.41	19.39	20.21	21.00	21.72	22.37	22.88	17.4	79.7	185.04
1 5/8		2.73	4.67	7.56	11.61	17.14	24.28	30.63	31.92	33.16	34.30	35.33	36.14	24.6	125.9	291.48
2 1/8		5.69	9.71	15.71	24.05	35.45	50.19	63.20	65.88	68.44	70.78	72.90	74.57	42.8	260.7	601.13
2 5/8		10.09	17.17	27.74	42.45	62.53	88.43	111.20	115.90	120.41	124.53	128.25	131.20	66.0	459.7	1056.39
3 1/8		16.15	27.44	44.24	67.77	99.53	140.83	177.12	184.62	191.80	198.36	204.29	208.98	94.2	733.0	1680.52
3 5/8		24.06	40.84	65.81	100.50	147.66	208.65	262.44	273.54	284.19	293.90	302.70	309.64	127.4	1087.5	2491.00
4 1/8		33.98	57.58	92.66	141.61	208.22	293.70	369.45	385.08	400.07	413.75	426.13	435.90	165.7	1530.2	3500.91
5 1/8		60.95	103.03	165.73	253.05	370.82	523.21	658.32	686.18	712.88	737.26	759.31	776.72	258.2	2729.8	6228.40
6 1/8		98.05	166.00	266.14	405.75	594.85	839.82	1054.47	1099.10	1141.87	1180.91	1216.24	1244.13	371.1	4383.7	9980.43
8 1/8		203.77	344.31	551.73	840.04	1229.69	1733.02	2176.50	2268.62	2356.89	2437.49	2510.41	2567.98	648.3	9049.5	20,561.73
Steel																
IPS	**SCH**															
3/8	80	0.08	0.13	0.21	0.32	0.46	0.65	0.81	0.84	0.88	0.91	0.93	0.95	1.9	3.4	7.6
1/2	80	0.16	0.26	0.41	0.62	0.91	1.27	1.59	1.66	1.73	1.78	1.84	1.88	3.2	6.7	15.0
3/4	80	0.35	0.59	0.93	1.41	2.04	2.86	3.59	3.74	3.88	4.02	4.14	4.23	6.0	15.1	33.6
1	80	0.69	1.15	1.83	2.75	4.00	5.59	7.02	7.32	7.60	7.86	8.10	8.28	10.0	29.5	65.8
1 1/4	80	1.49	2.48	3.92	5.90	8.58	12.00	15.03	15.67	16.28	16.83	17.34	17.74	17.7	63.3	140.9
1 1/2	80	2.28	3.79	5.98	9.01	13.06	18.27	22.89	23.86	24.79	25.64	26.41	27.01	24.4	96.6	214.7
2	40	5.30	8.80	13.89	20.91	30.32	42.43	53.16	55.41	57.57	59.54	61.32	62.73	46.4	224.2	498.0
2 1/2	40	8.46	14.02	22.13	33.29	48.23	67.48	84.56	88.14	91.57	94.70	97.53	99.77	66.2	356.5	793.0
3	40	14.98	24.81	39.10	58.81	85.22	119.26	149.44	155.76	161.82	167.36	172.37	176.32	102.2	630.0	1398.4
4	40	30.58	50.56	79.68	119.77	173.76	242.63	304.02	316.88	329.21	340.47	350.66	358.70	176.1	1284.6	2851.7
5	40	55.19	91.27	143.84	216.23	312.97	437.56	548.97	572.20	594.46	614.79	633.19	647.71	276.5	2313.7	5137.0
6	40	89.34	147.57	232.61	349.71	506.16	707.69	886.76	924.29	960.25	993.09	1022.80	1046.26	399.6	3741.9	8308.9
8	40	182.90	301.82	475.80	715.45	1035.51	1445.92	1811.80	1888.48	1961.96	2029.05	2089.76	2137.68	692.0	7655.3	16,977.6
10	40	331.22	546.64	860.67	1292.44	1870.67	2615.83	3277.74	3416.46	3549.40	3670.77	3780.59	3867.29	1090.7	13,829.2	30,716.4
12	ID[b]	529.89	873.19	1376.89	2064.68	2992.85	4185.32	5244.38	5466.33	5679.03	5873.23	6048.94	6187.65	1564.3	22,125.4	49,074.9
14	30	685.86	1130.48	1779.99	2673.23	3875.08	5410.92	6780.14	7067.08	7342.06	7593.13	7820.29	7999.63	1907.2	28,647.5	63,445.8
16	30	988.28	1628.96	2569.05	3852.37	5575.79	7797.98	9771.20	10,184.73	10,581.02	10,942.85	11,270.23	11,528.68	2526.4	41,220.5	91,435.1

[a] Sizing shown is recommended where any gas generated in receiver must return up condensate line to condenser without restricting condensate flow. Water-cooled condensers, where receiver ambient temperature may be higher than refrigerant condensing temperature, fall into this category.

[b] Pipe inside diameter is same as nominal pipe size.

Notes:

1. Table capacities are in tons of refrigeration.
   Δp = pressure drop from line friction, psi per 100 ft of equivalent line length
   Δt = corresponding change in saturation temperature, °F per 100 ft
2. Line capacity for other saturation temperatures Δt and equivalent lengths L_e

$$\text{Line capacity} = \text{Table capacity}\left(\frac{\text{Table } L_e}{\text{Actual } L_e} \times \frac{\text{Actual } \Delta t}{\text{Table } \Delta t}\right)^{0.55}$$

3. Saturation temperature Δt for other capacities and equivalent lengths L_e

$$\Delta t = \text{Table } \Delta t \left(\frac{\text{Actual } L_e}{\text{Table } L_e}\right)\left(\frac{\text{Actual capacity}}{\text{Table capacity}}\right)^{1.8}$$

4. Tons based on standard refrigerant cycle of 105°F liquid and saturated evaporator outlet temperature. Liquid tons based on 20°F evaporator temperature.
5. Thermophysical properties and viscosity data based on calculations from NIST REFPROP program Version 6.01.
6. For brazed Type L copper tubing larger than 5/8 in. OD for discharge or liquid service, see Safety Requirements section.
7. Values based on 105°F condensing temperature. Multiply table capacities by the following factors for other condensing temperatures.

Cond. Temp., °F	Suction Line	Discharge Line
80	1.170	0.815
90	1.104	0.889
100	1.035	0.963
110	0.964	1.032
120	0.889	1.096
130	0.808	1.160

Table 9 Suction, Discharge, and Liquid Line Capacities in Tons for Refrigerant 407c (Single- or High-Stage Applications)

Line Size		Suction Lines (Δt = 2°F)						Discharge Lines (Δt = 1°F, Δp = 3.3 psi)						Liquid Lines		
		Saturated Suction Temperature, °F						Saturated Suction Temperature, °F						See note a		
Type L Copper, OD		−60	−40	−20	0	20	40	−60	−40	−20	0	20	40	Velocity = 100 fpm	Δt = 1°F Drop Δp = 3.5	Δt = 5°F Drop Δp = 16.9
		Corresponding Δp, psi/100 ft						Corresponding Δp, psi/100 ft								
		0.435	0.7	1.06	1.55	2.16	2.92	3.3	3.3	3.3	3.3	3.3	3.3			
1/2		0.04	0.08	0.14	0.23	0.36	0.54	0.71	0.75	0.78	0.82	0.86	0.89	2.1	3.8	8.90
5/8		0.08	0.15	0.26	0.43	0.68	1.02	1.33	1.40	1.47	1.54	1.61	1.67	3.4	7.1	16.68
3/4		0.14	0.26	0.45	0.74	1.16	1.74	2.26	2.38	2.50	2.62	2.73	2.84	4.9	11.8	27.66
7/8		0.21	0.40	0.70	1.15	1.79	2.68	3.48	3.67	3.86	4.05	4.22	4.38	6.9	18.7	43.73
1 1/8		0.44	0.82	1.42	2.33	3.63	5.42	7.05	7.43	7.82	8.19	8.53	8.86	11.8	37.9	88.21
1 3/8		0.77	1.43	2.48	4.07	6.33	9.45	12.25	12.92	13.59	14.23	14.83	15.40	18.0	66.2	153.45
1 5/8		1.23	2.27	3.93	6.44	10.00	14.93	19.33	20.39	21.44	22.46	23.40	24.30	25.5	104.7	241.93
2 1/8		2.56	4.74	8.18	13.37	20.72	30.90	39.99	42.17	44.35	46.45	48.40	50.27	44.4	217.1	499.23
2 5/8		4.55	8.42	14.49	23.64	36.62	54.50	70.56	74.41	78.25	81.96	85.40	88.70	68.5	383.7	879.85
3 1/8		7.30	13.47	23.15	37.76	58.34	86.88	112.34	118.47	124.59	130.50	135.97	141.22	97.7	611.3	1401.50
3 5/8		10.90	20.08	34.44	56.15	86.64	128.89	166.39	175.47	184.54	193.29	201.39	209.17	132.2	907.9	2076.59
4 1/8		15.42	28.37	48.62	79.21	122.10	181.34	234.63	247.42	260.22	272.56	283.98	294.95	171.8	1281.5	2923.40
5 1/8		27.70	50.85	86.97	141.60	218.05	323.50	417.91	440.69	463.48	485.46	505.80	525.33	267.8	2288.8	5209.13
6 1/8		44.70	81.91	140.04	227.86	350.42	519.62	670.58	707.15	743.71	778.97	811.62	842.96	385.0	3676.9	8344.10
8 1/8		92.98	170.14	290.93	471.55	725.11	1072.54	1383.29	1458.72	1534.15	1606.88	1674.23	1738.88	672.4	7599.4	17,220.64
Steel																
IPS	**SCH**															
3/8	80	0.04	0.07	0.11	0.18	0.27	0.40	0.52	0.55	0.57	0.60	0.63	0.65	2.0	2.9	6.4
1/2	80	0.07	0.13	0.22	0.35	0.54	0.79	1.02	1.07	1.13	1.18	1.23	1.28	3.4	5.7	12.6
3/4	80	0.16	0.30	0.50	0.80	1.22	1.79	2.29	2.42	2.54	2.66	2.78	2.88	6.2	12.8	28.4
1	80	0.32	0.58	0.98	1.57	2.38	3.50	4.50	4.74	4.99	5.22	5.44	5.65	10.3	25.1	55.6
1 1/4	80	0.69	1.25	2.10	3.37	5.12	7.50	9.63	10.15	10.68	11.18	11.65	12.10	18.4	53.7	118.9
1 1/2	80	1.06	1.91	3.21	5.13	7.79	11.44	14.66	15.46	16.26	17.03	17.74	18.43	25.4	82.0	181.1
2	40	2.49	4.46	7.47	11.93	18.13	26.57	34.04	35.89	37.75	39.54	41.20	42.79	48.1	190.3	420.6
2 1/2	40	3.97	7.11	11.90	19.01	28.83	42.25	54.25	57.21	60.16	63.02	65.66	68.19	68.6	303.2	669.0
3	40	7.04	12.59	21.05	33.59	50.94	74.66	95.76	100.99	106.21	111.24	115.90	120.38	106.0	535.7	1182.3
4	40	14.38	25.70	42.97	68.47	103.84	152.24	195.04	205.68	216.31	226.57	236.06	245.18	182.6	1092.0	2405.3
5	40	26.00	46.36	77.55	123.61	187.25	274.21	351.31	370.46	389.62	408.09	425.19	441.61	286.8	1969.0	4343.2
6	40	42.13	75.15	125.49	199.88	302.82	443.47	568.16	599.14	630.12	659.99	687.65	714.21	414.5	3184.3	7015.7
8	40	86.32	153.84	256.66	408.86	619.47	907.26	1162.36	1225.74	1289.12	1350.24	1406.83	1461.15	717.7	6514.5	14,334.3
10	40	156.54	278.57	464.86	739.58	1120.60	1638.95	2102.83	2217.49	2332.15	2442.72	2545.10	2643.38	1131.3	11,784.6	25,932.3
12	ID[b]	250.23	445.65	742.54	1183.19	1790.17	2622.17	3359.45	3542.64	3725.82	3902.46	4066.02	4223.03	1622.5	18,826.0	41,491.5
14	30	324.38	576.93	961.33	1529.58	2317.81	3395.13	4349.77	4586.95	4824.14	5052.85	5264.62	5467.92	1978.2	24,374.8	53,641.7
16	30	468.29	831.27	1385.24	2204.17	3340.17	4885.19	6258.81	6600.09	6941.37	7270.46	7575.17	7867.69	2620.4	35,126.4	77,305.8

[a]Sizing shown is recommended where any gas generated in receiver must return up condensate line to condenser without restricting condensate flow. Water-cooled condensers, where receiver ambient temperature may be higher than refrigerant condensing temperature, fall into this category.

[b]Pipe inside diameter is same as nominal pipe size.

Notes: 1. Table capacities are in tons of refrigeration.

Δp = pressure drop from line friction, psi per 100 ft of equivalent line length

Δt = corresponding change in saturation temperature, °F per 100 ft

2. Line capacity for other saturation temperatures Δt and equivalent lengths L_e

$$\text{Line capacity} = \text{Table capacity}\left(\frac{\text{Table } L_e}{\text{Actual } L_e} \times \frac{\text{Actual } \Delta t}{\text{Table } \Delta t}\right)^{0.55}$$

3. Saturation temperature Δt for other capacities and equivalent lengths L_e

$$\Delta t = \text{Table } \Delta t \left(\frac{\text{Actual } L_e}{\text{Table } L_e}\right)\left(\frac{\text{Actual capacity}}{\text{Table capacity}}\right)^{1.8}$$

4. Tons based on standard refrigerant cycle of 105°F liquid and saturated evaporator outlet temperature. Liquid tons based on 20°F evaporator temperature.

5. Thermophysical properties and viscosity data based on calculations from NIST REFPROP program Version 6.01.

6. For brazed Type L copper tubing larger than 2 1/8 in. OD for discharge or liquid service, see Safety Requirements section.

7. Values based on 105°F condensing temperature. Multiply table capacities by the following factors for other condensing temperatures.

Cond. Temp., °F	Suction Line	Discharge Line
80	1.163	0.787
90	1.099	0.872
100	1.033	0.957
110	0.966	1.036
120	0.896	1.109
130	0.824	1.182

Refrigerant Line Capacity Tables

Tables 3 to 9 show line capacities in tons of refrigeration for R-22, R-134a, R-404a, R-507, R-410a, and R-407c. Capacities in the tables are based on the refrigerant flow that develops a friction loss, per 100 ft of equivalent pipe length, corresponding to a 2°F change in the saturation temperature (Δt) in the suction line, and a 1°F change in the discharge line. The capacities shown for liquid lines are for pressure losses corresponding to 1 and 5°F change in saturation temperature and also for velocity corresponding to 100 fpm. Tables 10 to 15 show capacities for the same refrigerants based on reduced suction line pressure loss corresponding to 1.0 and 0.5°F per 100 ft equivalent length of pipe. These tables may be used when designing system piping to minimize suction line pressure drop.

The refrigerant line sizing capacity tables are based on the Darcy-Weisbach relation and friction factors as computed by the Colebrook function (Colebrook 1938, 1939). Tubing roughness height is 0.000005 ft for copper and 0.00015 ft for steel pipe. Viscosity extrapolations and adjustments for pressures other than 1 atm were based on correlation techniques as presented by Keating and Matula (1969). Discharge gas superheat was 80°F for R-134a and 105°F for R-22.

The refrigerant cycle for determining capacity is based on saturated gas leaving the evaporator. The calculations neglect the presence of oil and assume nonpulsating flow.

For additional charts and discussion of line sizing refer to Atwood (1990), Timm (1991), and Wile (1977).

Equivalent Lengths of Valves and Fittings

Refrigerant line capacity tables are based on unit pressure drop per 100 ft length of straight pipe, or per combination of straight pipe, fittings, and valves with friction drop equivalent to a 100 ft length of straight pipe.

Generally, pressure drop through valves and fittings is determined by establishing the equivalent straight length of pipe of the same size with the same friction drop. Line sizing tables can then be used directly. Tables 16 to 18 give equivalent lengths of straight pipe for various fittings and valves, based on nominal pipe sizes.

The following example illustrates the use of various tables and charts to size refrigerant lines.

Example 2. Determine the line size and pressure drop equivalent (in degrees) for the suction line of a 30 ton R-22 system, operating at 40°F suction and 100°F condensing temperatures. Suction line is copper tubing, with 50 ft of straight pipe and six long-radius elbows.

Solution: Add 50% to the straight length of pipe to establish a trial equivalent length. Trial equivalent length is 50 × 1.5 = 75 ft. From Table 3 (for 40°F suction, 105°F condensing), 33.1 tons capacity in 2 1/8 in. OD results in a 2°F loss per 100 ft equivalent length. Referring to Note 4, Table 3, capacity at 40°F evaporator and 100°F condensing temperature is 1.03 × 33.1 = 34.1 ton. This trial size is used to evaluate actual equivalent length.

Straight pipe length	=	50.0 ft
Six 2 in. long-radius elbows at 3 ft each (Table 16)	=	19.8 ft
Total equivalent length	=	69.8 ft

$$\Delta t = 2(69.8/100)(30/34.1)^{1.8} = 1.1\text{°F or } 1.6 \text{ psi}$$

Oil Management in Refrigerant Lines

Oil Circulation. All compressors lose some lubricating oil during normal operation. Because oil inevitably leaves the compressor with the discharge gas, systems using halocarbon refrigerants must return this oil at the same rate at which it leaves (Cooper 1971).

Oil that leaves the compressor or oil separator reaches the condenser and dissolves in the liquid refrigerant, enabling it to pass readily through the liquid line to the evaporator. In the evaporator, the refrigerant evaporates, and the liquid phase becomes enriched in oil. The concentration of refrigerant in the oil depends on the evaporator temperature and types of refrigerant and oil used. The viscosity of the oil/refrigerant solution is determined by the system parameters. Oil separated in the evaporator is returned to the compressor by gravity or by drag forces of the returning gas. Oil's effect on pressure drop is large, increasing the pressure drop by as much as a factor of 10 (Alofs et al. 1990).

One of the most difficult problems in low-temperature refrigeration systems using halocarbon refrigerants is returning lubrication oil from the evaporator to the compressors. Except for most centrifugal compressors and rarely used nonlubricated compressors, refrigerant continuously carries oil into the discharge line from the compressor. Most of this oil can be removed from the stream by an oil separator and returned to the compressor. Coalescing oil separators are far better than separators using only mist pads or baffles; however, they are not 100% effective. Oil that finds its way into the system must be managed.

Oil mixes well with halocarbon refrigerants at higher temperatures. As temperature decreases, miscibility is reduced, and some oil separates to form an oil-rich layer near the top of the liquid level in a flooded evaporator. If the temperature is very low, the oil becomes a gummy mass that prevents refrigerant controls from functioning, blocks flow passages, and fouls heat transfer surfaces. Proper oil management is often key to a properly functioning system.

In general, direct-expansion and liquid overfeed system evaporators have fewer oil return problems than do flooded system evaporators because refrigerant flows continuously at velocities high enough to sweep oil from the evaporator. Low-temperature systems using hot-gas defrost can also be designed to sweep oil out of the circuit each time the system defrosts. This reduces the possibility of oil coating the evaporator surface and hindering heat transfer.

Flooded evaporators can promote oil contamination of the evaporator charge because they may only return dry refrigerant vapor back to the system. Skimming systems must sample the oil-rich layer floating in the drum, a heat source must distill the refrigerant, and the oil must be returned to the compressor. Because flooded halocarbon systems can be elaborate, some designers avoid them.

System Capacity Reduction. Using automatic capacity control on compressors requires careful analysis and design. The compressor can load and unload as it modulates with system load requirements through a considerable range of capacity. A single compressor can unload down to 25% of full-load capacity, and multiple compressors connected in parallel can unload to a system capacity of 12.5% or lower. System piping must be designed to return oil at the lowest loading, yet not impose excessive pressure drops in the piping and equipment at full load.

Oil Return up Suction Risers. Many refrigeration piping systems contain a suction riser because the evaporator is at a lower level than the compressor. Oil circulating in the system can return up gas risers only by being transported by returning gas or by auxiliary means such as a trap and pump. The minimum conditions for oil transport correlate with buoyancy forces (i.e., density difference between liquid and vapor, and momentum flux of vapor) (Jacobs et al. 1976).

The principal criteria determining the transport of oil are gas velocity, gas density, and pipe inside diameter. Density of the oil/refrigerant mixture plays a somewhat lesser role because it is almost constant over a wide range. In addition, at temperatures somewhat lower than –40°F, oil viscosity may be significant. Greater gas velocities are required as temperature drops and the gas becomes less dense. Higher velocities are also necessary if the pipe diameter increases. Table 19 translates these criteria to minimum refrigeration capacity requirements for oil transport. Suction risers must be sized for minimum system capacity. Oil must be returned to the compressor at the operating condition corresponding to the minimum displacement and minimum suction temperature at which the

Table 10 Suction Line Capacities in Tons for Refrigerant 22 (Single- or High-Stage Applications)

Line Size		Saturated Suction Temperature, °F									
Type L Copper, OD		−40		−20		0		20		40	
		$\Delta t = 1°F$	$\Delta t = 0.5°F$	$\Delta t = 1°F$	$\Delta t = 0.5°F$	$\Delta t = 1°F$	$\Delta t = 0.5°F$	$\Delta t = 1°F$	$\Delta t = 0.5°F$	$\Delta t = 1°F$	$\Delta t = 0.5°F$
		$\Delta p = 0.393$	$\Delta p = 0.197$	$\Delta p = 0.577$	$\Delta p = 0.289$	$\Delta p = 0.813$	$\Delta p = 0.406$	$\Delta p = 1.104$	$\Delta p = 0.552$	$\Delta p = 1.455$	$\Delta p = 0.727$
1/2		0.07	0.05	0.12	0.08	0.18	0.12	0.27	0.19	0.40	0.27
5/8		0.13	0.09	0.22	0.15	0.34	0.23	0.52	0.35	0.75	0.51
3/4		0.22	0.15	0.37	0.25	0.58	0.39	0.86	0.59	1.24	0.85
7/8		0.35	0.24	0.58	0.40	0.91	0.62	1.37	0.93	1.97	1.35
1 1/8		0.72	0.49	1.19	0.81	1.86	1.27	2.77	1.90	3.99	2.74
1 3/8		1.27	0.86	2.09	1.42	3.25	2.22	4.84	3.32	6.96	4.78
1 5/8		2.02	1.38	3.31	2.26	5.16	3.53	7.67	5.26	11.00	7.57
2 1/8		4.21	2.88	6.90	4.73	10.71	7.35	15.92	10.96	22.81	15.73
2 5/8		7.48	5.13	12.23	8.39	18.97	13.04	28.19	19.40	40.38	27.84
3 1/8		11.99	8.22	19.55	13.43	30.31	20.85	44.93	31.00	64.30	44.44
3 5/8		17.89	12.26	29.13	20.00	45.09	31.03	66.81	46.11	95.68	66.09
4 1/8		25.29	17.36	41.17	28.26	63.71	43.85	94.25	65.12	134.81	93.22
Steel											
IPS	SCH										
3/8	80	0.06	0.04	0.10	0.07	0.15	0.10	0.21	0.15	0.30	0.21
1/2	80	0.12	0.08	0.19	0.13	0.29	0.20	0.42	0.30	0.60	0.42
3/4	80	0.27	0.18	0.43	0.30	0.65	0.46	0.95	0.67	1.35	0.95
1	80	0.52	0.36	0.84	0.59	1.28	0.89	1.87	1.31	2.64	1.86
1 1/4	40	1.38	0.96	2.21	1.55	3.37	2.36	4.91	3.45	6.93	4.88
1 1/2	40	2.08	1.45	3.32	2.33	5.05	3.55	7.38	5.19	10.42	7.33
2	40	4.03	2.81	6.41	4.51	9.74	6.85	14.22	10.01	20.07	14.14
2 1/2	40	6.43	4.49	10.23	7.19	15.56	10.93	22.65	15.95	31.99	22.53
3	40	11.38	7.97	18.11	12.74	27.47	19.34	40.10	28.23	56.52	39.79
4	40	23.24	16.30	36.98	26.02	56.12	39.49	81.73	57.53	115.24	81.21
5	40	42.04	29.50	66.73	47.05	101.16	71.27	147.36	103.82	207.59	146.38
6	40	68.04	47.86	108.14	76.15	163.77	115.21	238.29	168.07	335.71	236.70
8	40	139.48	98.06	221.17	155.78	334.94	236.21	488.05	344.19	686.71	484.74
10	40	252.38	177.75	400.53	282.05	606.74	427.75	881.59	622.51	1243.64	876.79
12	ID*	403.63	284.69	639.74	451.09	969.02	683.22	1410.30	995.80	1987.29	1402.63

Δp = pressure drop from line friction, psi per 100 ft equivalent line length
Δt = change in saturation temperature corresponding to pressure drop, °F per 100 ft

*Pipe inside diameter is same as nominal pipe size.

Table 11 Suction Line Capacities in Tons for Refrigerant 134a (Single- or High-Stage Applications)

Line Size		Saturated Suction Temperature, °F									
Type L Copper, OD		0		10		20		30		40	
		$\Delta t = 1°F$	$\Delta t = 0.5°F$	$\Delta t = 1°F$	$\Delta t = 0.5°F$	$\Delta t = 1°F$	$\Delta t = 0.5°F$	$\Delta t = 1°F$	$\Delta t = 0.5°F$	$\Delta t = 1°F$	$\Delta t = 0.5°F$
		$\Delta p = 0.50$	$\Delta p = 0.25$	$\Delta p = 0.60$	$\Delta p = 0.30$	$\Delta p = 0.71$	$\Delta p = 0.35$	$\Delta p = 0.83$	$\Delta p = 0.42$	$\Delta p = 0.97$	$\Delta p = 0.48$
1/2		0.10	0.07	0.12	0.08	0.16	0.11	0.19	0.13	0.24	0.16
5/8		0.18	0.12	0.23	0.16	0.29	0.20	0.37	0.25	0.45	0.31
7/8		0.48	0.33	0.62	0.42	0.78	0.53	0.97	0.66	1.20	0.82
1 1/8		0.99	0.67	1.26	0.86	1.59	1.08	1.97	1.35	2.43	1.66
1 3/8		1.73	1.18	2.21	1.51	2.77	1.89	3.45	2.36	4.25	2.91
1 5/8		2.75	1.88	3.50	2.40	4.40	3.01	5.46	3.75	6.72	4.61
2 1/8		5.73	3.92	7.29	5.00	9.14	6.27	11.40	7.79	14.00	9.59
2 5/8		10.20	6.97	12.90	8.87	16.20	11.10	20.00	13.80	24.70	17.00
3 1/8		16.20	11.10	20.60	14.20	25.90	17.80	32.10	22.10	39.40	27.20
3 5/8		24.20	16.60	30.80	21.20	38.50	26.50	47.70	32.90	58.70	40.40
4 1/8		34.20	23.50	43.40	29.90	54.30	37.40	67.30	46.50	82.60	57.10
5 1/8		61.30	42.20	77.70	53.60	97.20	67.10	121.00	83.20	148.00	102.00
6 1/8		98.80	68.00	125.00	86.30	157.00	108.00	194.00	134.00	237.00	165.00
Steel											
IPS	SCH										
1/2	80	0.16	0.11	0.20	0.14	0.25	0.17	0.30	0.21	0.37	0.26
3/4	80	0.36	0.25	0.45	0.31	0.56	0.39	0.69	0.48	0.84	0.59
1	80	0.70	0.49	0.88	0.61	1.09	0.77	1.34	0.94	1.64	1.15
1 1/4	40	1.84	1.29	2.31	1.62	2.87	2.02	3.54	2.48	4.31	3.03
1 1/2	40	2.77	1.94	3.48	2.44	4.32	3.03	5.30	3.73	6.47	4.55
2	40	5.35	3.75	6.72	4.72	8.33	5.86	10.30	7.20	12.50	8.78
2 1/2	40	8.53	5.99	10.70	7.53	13.30	9.35	16.30	11.50	19.90	14.00
3	40	15.10	10.60	18.90	13.30	23.50	16.50	28.90	20.30	35.20	24.80
4	40	30.80	21.70	38.70	27.20	48.00	33.80	58.80	41.50	71.60	50.50
5	40	55.60	39.20	69.80	49.10	86.50	60.93	106.00	74.95	129.00	91.00
6	40	89.90	63.40	113.00	79.60	140.00	98.50	172.00	121.00	209.00	148.00

Δp = pressure drop from line friction, psi per 100 ft equivalent line length
Δt = change in saturation temperature corresponding to pressure drop, °F per 100 ft

Table 12 Suction Line Capacities in Tons for Refrigerant 404a (Single- or High-Stage Applications)

Line Size		Saturated Suction Temperature, °F											
Type L Copper, OD		−60		−40		−20		0		20		40	
		Δt = 1°F Δp = 0.32	Δt = 0.5°F Δp = 0.16	Δt = 1°F Δp = 0.485	Δt = 0.5°F Δp = 0.243	Δt = 1°F Δp = 0.705	Δt = 0.5°F Δp = 0.353	Δt = 1°F Δp = 0.98	Δt = 0.5°F Δp = 0.49	Δt = 1°F Δp = 1.31	Δt = 0.5°F Δp = 0.655	Δt = 1°F Δp = 1.72	Δt = 0.5°F Δp = 0.86
1/2		0.03	0.02	0.06	0.04	0.10	0.07	0.16	0.11	0.25	0.17	0.37	0.25
5/8		0.06	0.04	0.11	0.08	0.19	0.13	0.30	0.21	0.46	0.32	0.69	0.47
3/4		0.10	0.07	0.19	0.13	0.32	0.22	0.52	0.35	0.79	0.54	1.17	0.80
7/8		0.16	0.11	0.29	0.20	0.50	0.34	0.80	0.55	1.22	0.84	1.81	1.24
1 1/8		0.33	0.23	0.60	0.41	1.02	0.70	1.63	1.12	2.48	1.70	3.66	2.52
1 3/8		0.59	0.40	1.05	0.72	1.78	1.22	2.84	1.95	4.33	2.98	6.38	4.40
1 5/8		0.93	0.63	1.67	1.14	2.82	1.93	4.50	3.09	6.84	4.71	10.08	6.95
2 1/8		1.94	1.33	3.48	2.38	5.86	4.02	9.33	6.42	14.19	9.78	20.86	14.42
2 5/8		3.45	2.36	6.17	4.23	10.38	7.13	16.50	11.37	25.04	17.30	36.79	25.48
3 1/8		5.53	3.78	9.87	6.78	16.57	11.40	26.36	18.17	39.90	27.63	58.65	40.65
3 5/8		8.24	5.64	14.70	10.09	24.66	16.98	39.19	27.05	59.27	41.08	86.99	60.38
4 1/8		11.66	7.99	20.74	14.27	34.82	24.00	55.29	38.19	83.67	57.95	122.65	85.08
5 1/8		20.91	14.35	37.20	25.58	62.32	43.03	98.68	68.35	149.15	103.62	218.80	151.93
6 1/8		33.70	23.17	59.82	41.25	100.16	69.25	158.78	109.86	239.61	166.38	350.99	244.04
8 1/8		70.16	48.33	124.35	85.75	207.70	143.94	329.02	228.24	496.00	344.71	725.34	504.94
Steel													
IPS	SCH												
3/8	80	0.03	0.02	0.05	0.03	0.08	0.06	0.13	0.09	0.19	0.13	0.27	0.19
1/2	80	0.05	0.04	0.09	0.07	0.16	0.11	0.25	0.17	0.37	0.26	0.54	0.38
3/4	80	0.12	0.08	0.21	0.15	0.36	0.25	0.56	0.39	0.83	0.59	1.21	0.85
1	80	0.24	0.17	0.42	0.29	0.70	0.49	1.09	0.77	1.63	1.15	2.37	1.67
1 1/4	40	0.52	0.36	0.91	0.63	1.50	1.05	2.34	1.65	3.50	2.46	5.07	3.57
1 1/2	40	0.80	0.55	1.39	0.97	2.29	1.60	3.57	2.51	5.33	3.76	7.74	5.45
2	40	1.86	1.30	3.24	2.26	5.32	3.74	8.30	5.85	12.40	8.73	17.96	12.66
2 1/2	40	2.96	2.07	5.16	3.61	8.48	5.96	13.23	9.32	19.71	13.92	28.57	20.17
3	40	5.25	3.68	9.13	6.41	15.01	10.54	23.37	16.47	34.83	24.59	50.48	35.63
4	40	10.75	7.53	18.64	13.06	30.59	21.53	47.64	33.61	71.01	50.12	102.93	72.64
5	40	19.42	13.61	33.64	23.67	55.22	38.85	86.00	60.66	128.09	90.47	185.40	130.81
6	40	31.37	22.07	54.45	38.36	89.29	62.97	139.08	98.09	207.08	146.31	299.84	211.53
8	40	64.28	45.29	111.50	78.62	182.58	128.75	284.48	200.61	423.62	299.27	613.41	433.35
10	40	116.63	82.09	201.92	142.37	330.75	233.20	514.60	363.34	766.32	541.35	1108.13	783.91
12	ID*	186.39	131.47	322.98	227.70	528.22	373.02	823.24	580.40	1224.19	866.05	1772.90	1252.32
14	30	241.28	170.14	418.14	294.77	683.87	482.92	1064.28	751.41	1585.02	1119.62	2295.51	1621.44
16	30	348.15	245.48	602.49	424.62	985.62	695.84	1533.35	1082.76	2284.15	1613.40	3302.98	2336.63

Notes:

1. Δt = change in saturation temperature corresponding to pressure drop, °F per 100 ft.
2. Tons based on standard refrigerant cycle of 105°F liquid and saturated evaporator outlet temperature. Liquid tons based on 20°F evaporator temperature.
3. Thermophysical properties and viscosity data based on calculations from NIST REFPROP program Version 6.01.
4. Values based on 105°F condensing temperature. Multiply table capacities by the following factors for other condensing temperatures.

*Pipe inside diameter is same as nominal pipe size.

Condensing Temperature, °F	Suction Line
80	1.246
90	1.150
100	1.051
110	0.948
120	0.840
130	0.723

Table 13 Suction Line Capacities in Tons for Refrigerant 507 (Single- or High-Stage Applications)

Line Size		Saturated Suction Temperature, °F											
		−60		−40		−20		0		20		40	
Type L Copper, OD		Δt = 1°F Δp = 0.335	Δt = 0.5°F Δp = 0.168	Δt = 1°F Δp = 0.505	Δt = 0.5°F Δp = 0.253	Δt = 1°F Δp =0.73	Δt = 0.5°F Δp = 0.365	Δt = 1°F Δp = 1.01	Δt = 0.5°F Δp = 0.505	Δt = 1°F Δp = 1.355	Δt = 0.5°F Δp = 0.678	Δt = 1°F Δp = 1.8	Δt = 0.5°F Δp = 0.9
1/2		0.03	0.02	0.06	0.04	0.10	0.07	0.16	0.11	0.25	0.17	0.37	0.26
5/8		0.06	0.04	0.11	0.08	0.19	0.13	0.31	0.21	0.47	0.32	0.70	0.48
3/4		0.11	0.07	0.19	0.13	0.33	0.22	0.52	0.36	0.80	0.55	1.20	0.82
7/8		0.17	0.11	0.30	0.20	0.51	0.35	0.81	0.56	1.24	0.85	1.85	1.27
1 1/8		0.34	0.23	0.61	0.42	1.03	0.71	1.65	1.13	2.52	1.73	3.74	2.57
1 3/8		0.60	0.41	1.07	0.73	1.81	1.24	2.88	1.97	4.39	3.02	6.51	4.49
1 5/8		0.95	0.65	1.70	1.16	2.87	1.96	4.56	3.13	6.94	4.78	10.28	7.09
2 1/8		1.99	1.36	3.55	2.43	5.95	4.09	9.44	6.51	14.37	9.92	21.28	14.70
2 5/8		3.53	2.42	6.29	4.31	10.54	7.24	16.70	11.51	25.40	17.54	37.53	25.99
3 1/8		5.66	3.88	10.06	6.90	16.82	11.57	26.69	18.40	40.48	27.98	59.74	41.41
3 5/8		8.43	5.78	14.98	10.29	25.04	17.24	39.62	27.38	60.13	41.61	88.62	61.51
4 1/8		11.92	8.18	21.16	14.53	35.37	24.37	55.91	38.62	84.73	58.69	124.94	86.66
5 1/8		21.40	14.71	37.90	26.11	63.19	43.71	99.99	69.10	151.06	104.94	222.92	154.78
6 1/8		34.51	23.73	60.98	42.03	101.79	70.33	160.57	111.29	242.71	168.52	357.63	248.63
8 1/8		71.74	49.40	126.80	87.41	210.91	145.98	332.73	230.83	502.46	349.13	739.16	514.43
Steel													
IPS	**SCH**												
3/8	80	0.03	0.02	0.05	0.03	0.08	0.06	0.13	0.09	0.19	0.13	0.28	0.19
1/2	80	0.06	0.04	0.10	0.07	0.16	0.11	0.25	0.17	0.37	0.26	0.55	0.38
3/4	80	0.13	0.09	0.22	0.15	0.36	0.25	0.56	0.39	0.84	0.59	1.23	0.87
1	80	0.25	0.17	0.43	0.30	0.71	0.50	1.10	0.77	1.65	1.16	2.41	1.70
1 1/4	40	0.53	0.37	0.93	0.65	1.52	1.07	2.37	1.66	3.54	2.49	5.16	3.63
1 1/2	40	0.81	0.57	1.41	0.99	2.32	1.63	3.61	2.54	5.39	3.80	7.87	5.54
2	40	1.90	1.33	3.29	2.31	5.39	3.79	8.38	5.90	12.53	8.83	18.26	12.86
2 1/2	40	3.03	2.12	5.25	3.68	8.58	6.04	13.35	9.40	19.93	14.07	29.05	20.51
3	40	5.37	3.76	9.29	6.52	15.19	10.69	23.58	16.64	35.22	24.85	51.33	36.23
4	40	10.95	7.69	18.93	13.32	30.96	21.79	48.07	33.92	71.78	50.66	104.65	73.86
5	40	19.77	13.90	34.20	23.98	55.89	39.37	86.80	61.22	129.59	91.45	188.50	133.18
6	40	32.06	22.52	55.36	38.95	90.37	63.73	140.36	98.99	209.38	147.89	304.85	215.38
8	40	65.72	46.21	113.19	79.83	185.07	130.51	287.10	202.42	428.18	302.50	623.68	440.60
10	40	119.01	83.76	205.02	144.56	334.75	236.03	519.34	366.70	774.58	547.19	1126.66	797.04
12	ID*	190.34	134.16	327.88	231.16	535.50	377.46	830.83	585.64	1237.39	875.38	1802.55	1273.31
14	30	246.21	173.66	424.56	299.30	693.31	488.76	1074.09	758.20	1602.11	1131.69	2333.91	1648.57
16	30	354.73	250.14	611.65	431.90	997.35	704.12	1550.19	1092.75	2308.78	1630.79	3358.23	2375.74

Notes:

1. Δt = change in saturation temperature corresponding to pressure drop, °F per 100 ft.
2. Tons based on standard refrigerant cycle of 105°F liquid and saturated evaporator outlet temperature. Liquid tons based on 20°F evaporator temperature.
3. Thermophysical properties and viscosity data based on calculations from NIST REFPROP program Version 6.01.
4. Values based on 105°F condensing temperature. Multiply table capacities by the following factors for other condensing temperatures.

*Pipe inside diameter is same as nominal pipe size.

Condensing Temperature, °F	Suction Line
80	1.267
90	1.163
100	1.055
110	0.944
120	0.826
130	0.701

Table 14 Suction Line Capacities in Tons for Refrigerant 410a (Single- or High-Stage Applications)

Line Size		Saturated Suction Temperature, °F											
		−60		−40		−20		0		20		40	
Type L Copper, OD		Δt = 1°F Δp = 0.42	Δt = 0.5°F Δp = 0.21	Δt = 1°F Δp = 0.635	Δt = 0.5°F Δp = 0.318	Δt = 1°F Δp = 0.925	Δt = 0.5°F Δp = 0.463	Δt = 1°F Δp = 1.285	Δt = 0.5°F Δp = 0.643	Δt = 1°F Δp = 1.73	Δt = 0.5°F Δp = 0.865	Δt = 1°F Δp = 2.25	Δt = 0.5°F Δp = 1.125
1/2		0.06	0.04	0.11	0.08	0.18	0.13	0.29	0.20	0.43	0.29	0.61	0.42
5/8		0.12	0.08	0.21	0.14	0.35	0.24	0.54	0.37	0.80	0.55	1.15	0.79
3/4		0.21	0.14	0.36	0.25	0.60	0.41	0.92	0.63	1.37	0.94	1.96	1.34
7/8		0.33	0.22	0.57	0.38	0.92	0.63	1.43	0.98	2.12	1.45	3.02	2.08
1 1/8		0.67	0.46	1.15	0.79	1.88	1.28	2.90	1.99	4.29	2.95	6.12	4.22
1 3/8		1.18	0.80	2.02	1.38	3.28	2.25	5.06	3.47	7.49	5.15	10.65	7.34
1 5/8		1.87	1.27	3.20	2.19	5.20	3.56	8.00	5.50	11.84	8.16	16.82	11.62
2 1/8		3.90	2.66	6.66	4.57	10.80	7.42	16.60	11.43	24.53	16.94	34.82	24.06
2 5/8		6.92	4.74	11.81	8.11	19.15	13.16	29.37	20.24	43.30	29.96	61.42	42.54
3 1/8		11.10	7.59	18.88	12.98	30.56	21.03	46.84	32.36	69.12	47.78	97.93	67.88
3 5/8		16.54	11.32	28.12	19.33	45.48	31.32	69.66	48.14	102.68	71.03	145.29	100.82
4 1/8		23.37	16.04	39.75	27.34	64.13	44.26	98.29	67.89	144.70	100.22	204.80	142.08
5 1/8		41.90	28.80	71.16	49.04	114.79	79.27	175.44	121.50	257.95	179.21	365.02	253.76
6 1/8		67.56	46.54	114.71	79.08	184.50	127.75	282.30	195.66	414.50	287.76	586.12	407.59
8 1/8		140.71	96.90	238.00	164.42	382.64	265.15	583.63	405.01	858.05	596.10	1208.61	843.44
Steel													
IPS	SCH												
3/8	80	0.05	0.04	0.09	0.06	0.15	0.10	0.22	0.16	0.32	0.23	0.45	0.32
1/2	80	0.11	0.07	0.18	0.13	0.29	0.20	0.44	0.31	0.64	0.45	0.89	0.63
3/4	80	0.25	0.17	0.41	0.29	0.65	0.46	0.99	0.69	1.44	1.01	2.01	1.42
1	80	0.48	0.34	0.81	0.57	1.28	0.90	1.94	1.36	2.81	1.98	3.94	2.78
1 1/4	40	1.04	0.73	1.74	1.22	2.76	1.94	4.16	2.92	6.04	4.25	8.45	5.96
1 1/2	40	1.60	1.11	2.66	1.86	4.21	2.96	6.35	4.46	9.20	6.48	12.90	9.09
2	40	3.73	2.60	6.19	4.34	9.79	6.88	14.72	10.38	21.40	15.08	29.94	21.09
2 1/2	40	5.94	4.16	9.85	6.93	15.59	10.98	23.46	16.53	34.03	24.02	47.62	33.61
3	40	10.52	7.37	17.43	12.25	27.60	19.43	41.47	29.26	60.13	42.44	84.14	59.39
4	40	21.48	15.08	35.60	25.06	56.24	39.58	84.52	59.63	122.57	86.51	171.56	121.08
5	40	38.84	27.30	64.25	45.21	101.52	71.51	152.52	107.63	221.30	156.17	309.01	218.33
6	40	62.85	44.23	104.14	73.26	164.15	115.77	246.64	174.04	357.45	252.55	499.76	353.09
8	40	128.81	90.62	212.93	150.18	336.18	236.70	504.51	355.89	731.21	516.58	1022.43	722.30
10	40	233.22	164.52	385.68	271.93	608.06	428.73	912.58	644.70	1322.74	934.44	1847.00	1306.62
12	ID*	372.99	263.04	616.79	434.92	972.73	685.64	1459.96	1029.64	2113.09	1494.90	2955.02	2087.38
14	30	483.55	340.47	798.65	563.02	1259.39	887.82	1887.38	1333.03	2735.91	1932.59	3826.11	2702.56
16	30	696.69	491.23	1150.59	812.45	1811.67	1279.02	2724.04	1921.21	3942.69	2784.92	5505.32	3894.62

Notes:

1. Δt = change in saturation temperature corresponding to pressure drop, °F per 100 ft.
2. Tons based on standard refrigerant cycle of 105°F liquid and saturated evaporator outlet temperature. Liquid tons based on 20°F evaporator temperature.
3. Thermophysical properties and viscosity data based on calculations from NIST REFPROP program Version 6.01.
4. Values based on 105°F condensing temperature. Multiply table capacities by the following factors for other condensing temperatures.

*Pipe inside diameter is same as nominal pipe size.

Condensing Temperature, °F	Suction Line
80	1.170
90	1.104
100	1.035
110	0.964
120	0.889
130	0.808

Table 15 Suction Line Capacities in Tons for Refrigerant 407c (Single- or High-Stage Applications)

Line Size		Saturated Suction Temperature, °F											
		−60		−40		−20		0		20		40	
Type L Copper, OD		Δt = 1°F Δp = 0.218	Δt = 0.5°F Δp = 0.109	Δt = 1°F Δp = 0.35	Δt = 0.5°F Δp = 0.175	Δt = 1°F Δp = 0.53	Δt = 0.5°F Δp = 0.265	Δt = 1°F Δp = 0.775	Δt = 0.5°F Δp = 0.388	Δt = 1°F Δp = 1.08	Δt = 0.5°F Δp = 0.54	Δt = 1°F Δp = 1.46	Δt = 0.5°F Δp = 0.73
1/2		0.03	0.02	0.05	0.04	0.09	0.06	0.16	0.11	0.25	0.17	0.37	0.25
5/8		0.05	0.04	0.10	0.07	0.18	0.12	0.30	0.20	0.46	0.32	0.70	0.48
3/4		0.09	0.06	0.17	0.12	0.31	0.21	0.51	0.34	0.79	0.54	1.19	0.81
7/8		0.15	0.10	0.27	0.18	0.48	0.32	0.78	0.53	1.23	0.84	1.84	1.26
1 1/8		0.30	0.20	0.56	0.38	0.97	0.66	1.60	1.09	2.49	1.71	3.74	2.56
1 3/8		0.52	0.36	0.98	0.66	1.70	1.16	2.79	1.91	4.35	2.98	6.52	4.48
1 5/8		0.83	0.57	1.56	1.06	2.69	1.84	4.43	3.03	6.89	4.72	10.30	7.09
2 1/8		1.75	1.19	3.24	2.22	5.61	3.84	9.20	6.32	14.31	9.84	21.36	14.73
2 5/8		3.11	2.12	5.77	3.94	9.94	6.82	16.30	11.20	25.30	17.42	37.75	26.08
3 1/8		5.00	3.41	9.24	6.32	15.91	10.92	26.05	17.90	40.34	27.82	60.23	41.58
3 5/8		7.45	5.09	13.77	9.44	23.72	16.29	38.75	26.70	59.97	41.40	89.47	61.81
4 1/8		10.57	7.22	19.47	13.36	33.52	23.03	54.73	37.71	84.60	58.46	126.06	87.32
5 1/8		19.00	13.00	35.00	24.01	60.09	41.35	97.90	67.62	151.22	104.74	225.14	156.10
6 1/8		30.67	21.04	56.41	38.77	96.82	66.66	157.64	108.96	243.24	168.35	361.69	251.08
8 1/8		63.98	43.94	117.40	80.79	201.22	138.52	326.82	226.06	503.94	349.69	748.45	520.10
Steel													
IPS	SCH												
3/8	80	0.02	0.02	0.05	0.03	0.08	0.05	0.13	0.09	0.19	0.13	0.28	0.20
1/2	80	0.05	0.03	0.09	0.06	0.15	0.11	0.25	0.17	0.38	0.27	0.56	0.39
3/4	80	0.11	0.08	0.21	0.14	0.35	0.24	0.56	0.39	0.86	0.60	1.26	0.88
1	80	0.22	0.15	0.40	0.28	0.68	0.48	1.10	0.77	1.68	1.18	2.47	1.73
1 1/4	40	0.48	0.33	0.87	0.61	1.48	1.03	2.36	1.66	3.60	2.53	5.28	3.72
1 1/2	40	0.74	0.51	1.34	0.93	2.25	1.57	3.61	2.53	5.49	3.86	8.06	5.67
2	40	1.73	1.20	3.12	2.18	5.25	3.68	8.39	5.90	12.77	8.98	18.71	13.18
2 1/2	40	2.77	1.92	4.98	3.49	8.36	5.85	13.39	9.41	20.35	14.30	29.82	21.00
3	40	4.92	3.42	8.82	6.16	14.81	10.39	23.66	16.66	35.96	25.31	52.68	37.18
4	40	10.07	7.04	18.05	12.62	30.24	21.26	48.33	34.01	73.25	51.63	107.39	75.79
5	40	18.24	12.74	32.63	22.88	54.64	38.40	87.11	61.38	132.31	93.20	193.65	136.64
6	40	29.56	20.67	52.87	37.10	88.32	62.14	141.05	99.22	213.85	150.90	313.17	221.26
8	40	60.72	42.53	108.35	76.09	181.10	127.34	288.49	203.43	437.43	308.63	640.64	452.60
10	40	110.03	77.14	196.18	137.91	327.97	230.56	522.51	368.41	791.25	558.98	1158.92	817.38
12	ID*	176.17	123.83	314.18	220.86	523.74	369.31	834.64	588.33	1265.85	892.90	1851.38	1307.58
14	30	228.34	160.22	406.04	286.29	678.86	477.98	1080.58	762.91	1636.44	1155.99	2397.05	1693.24
16	30	329.42	231.48	585.97	413.05	978.61	689.83	1557.07	1099.28	2358.16	1665.74	3454.36	2440.00

Notes:

1. Δt = change in saturation temperature corresponding to pressure drop, °F per 100 ft.
2. Tons based on standard refrigerant cycle of 105°F liquid and saturated evaporator outlet temperature. Liquid tons based on 20°F evaporator temperature.
3. Thermophysical properties and viscosity data based on calculations from NIST REFPROP program Version 6.01.
4. Values based on 105°F condensing temperature. Multiply table capacities by the following factors for other condensing temperatures.

*Pipe inside diameter is same as nominal pipe size.

Condensing Temperature, °F	Suction Line
80	1.163
90	1.099
100	1.033
110	0.966
120	0.896
130	0.824

Table 16 Fitting Losses in Equivalent Feet of Pipe
(Screwed, Welded, Flanged, Flared, and Brazed Connections)

Nominal Pipe or Tube Size, in.	Smooth Bend Elbows						Smooth Bend Tees			
	90° Std[a]	90° Long-Radius[b]	90° Street[a]	45° Std[a]	45° Street[a]	180° Std[a]	Flow Through Branch	Straight-Through Flow: No Reduction	Straight-Through Flow: Reduced 1/4	Straight-Through Flow: Reduced 1/2
3/8	1.4	0.9	2.3	0.7	1.1	2.3	2.7	0.9	1.2	1.4
1/2	1.6	1.0	2.5	0.8	1.3	2.5	3.0	1.0	1.4	1.6
3/4	2.0	1.4	3.2	0.9	1.6	3.2	4.0	1.4	1.9	2.0
1	2.6	1.7	4.1	1.3	2.1	4.1	5.0	1.7	2.2	2.6
1 1/4	3.3	2.3	5.6	1.7	3.0	5.6	7.0	2.3	3.1	3.3
1 1/2	4.0	2.6	6.3	2.1	3.4	6.3	8.0	2.6	3.7	4.0
2	5.0	3.3	8.2	2.6	4.5	8.2	10.0	3.3	4.7	5.0
2 1/2	6.0	4.1	10.0	3.2	5.2	10.0	12.0	4.1	5.6	6.0
3	7.5	5.0	12.0	4.0	6.4	12.0	15.0	5.0	7.0	7.5
3 1/2	9.0	5.9	15.0	4.7	7.3	15.0	18.0	5.9	8.0	9.0
4	10.0	6.7	17.0	5.2	8.5	17.0	21.0	6.7	9.0	10.0
5	13.0	8.2	21.0	6.5	11.0	21.0	25.0	8.2	12.0	13.0
6	16.0	10.0	25.0	7.9	13.0	25.0	30.0	10.0	14.0	16.0
8	20.0	13.0	—	10.0	—	33.0	40.0	13.0	18.0	20.0
10	25.0	16.0	—	13.0	—	42.0	50.0	16.0	23.0	25.0
12	30.0	19.0	—	16.0	—	50.0	60.0	19.0	26.0	30.0
14	34.0	23.0	—	18.0	—	55.0	68.0	23.0	30.0	34.0
16	38.0	26.0	—	20.0	—	62.0	78.0	26.0	35.0	38.0
18	42.0	29.0	—	23.0	—	70.0	85.0	29.0	40.0	42.0
20	50.0	33.0	—	26.0	—	81.0	100.0	33.0	44.0	50.0
24	60.0	40.0	—	30.0	—	94.0	115.0	40.0	50.0	60.0

[a]*R/D* approximately equal to 1. [b]*R/D* approximately equal to 1.5.

Table 17 Special Fitting Losses in Equivalent Feet of Pipe

Nominal Pipe or Tube Size, in.	Sudden Enlargement, *d/D*			Sudden Contraction, *d/D*			Sharp Edge		Pipe Projection	
	1/4	1/2	3/4	1/4	1/2	3/4	Entrance	Exit	Entrance	Exit
3/8	1.4	0.8	0.3	0.7	0.5	0.3	1.5	0.8	1.5	1.1
1/2	1.8	1.1	0.4	0.9	0.7	0.4	1.8	1.0	1.8	1.5
3/4	2.5	1.5	0.5	1.2	1.0	0.5	2.8	1.4	2.8	2.2
1	3.2	2.0	0.7	1.6	1.2	0.7	3.7	1.8	3.7	2.7
1 1/4	4.7	3.0	1.0	2.3	1.8	1.0	5.3	2.6	5.3	4.2
1 1/2	5.8	3.6	1.2	2.9	2.2	1.2	6.6	3.3	6.6	5.0
2	8.0	4.8	1.6	4.0	3.0	1.6	9.0	4.4	9.0	6.8
2 1/2	10.0	6.1	2.0	5.0	3.8	2.0	12.0	5.6	12.0	8.7
3	13.0	8.0	2.6	6.5	4.9	2.6	14.0	7.2	14.0	11.0
3 1/2	15.0	9.2	3.0	7.7	6.0	3.0	17.0	8.5	17.0	13.0
4	17.0	11.0	3.8	9.0	6.8	3.8	20.0	10.0	20.0	16.0
5	24.0	15.0	5.0	12.0	9.0	5.0	27.0	14.0	27.0	20.0
6	29.0	22.0	6.0	15.0	11.0	6.0	33.0	19.0	33.0	25.0
8	—	25.0	8.5	—	15.0	8.5	47.0	24.0	47.0	35.0
10	—	32.0	11.0	—	20.0	11.0	60.0	29.0	60.0	46.0
12	—	41.0	13.0	—	25.0	13.0	73.0	37.0	73.0	57.0
14	—	—	16.0	—	—	16.0	86.0	45.0	86.0	66.0
16	—	—	18.0	—	—	18.0	96.0	50.0	96.0	77.0
18	—	—	20.0	—	—	20.0	115.0	58.0	115.0	90.0
20	—	—	—	—	—	—	142.0	70.0	142.0	108.0
24	—	—	—	—	—	—	163.0	83.0	163.0	130.0

Note: Enter table for losses at smallest diameter *d*.

compressor will operate. When suction or evaporator pressure regulators are used, suction risers must be sized for actual gas conditions in the riser.

For a single compressor with capacity control, the minimum capacity is the lowest capacity at which the unit can operate. For multiple compressors with capacity control, the minimum capacity is the lowest at which the last operating compressor can run.

Riser Sizing. The following example demonstrates the use of Table 19 in establishing maximum riser sizes for satisfactory oil transport down to minimum partial loading.

Example 3. Determine the maximum size suction riser that will transport oil at minimum loading, using R-22 with a 40 ton compressor with capacity in steps of 25, 50, 75, and 100%. Assume the minimum system loading is 10 tons at 40°F suction and 105°F condensing temperatures with 15°F superheat.

Solution: From Table 19, a 2 1/8 in. OD pipe at 40°F suction and 90°F liquid temperature has a minimum capacity of 7.5 tons. When corrected to 105°F liquid temperature using the chart at the bottom of Table 19, minimum capacity becomes 7.2 tons. Therefore, 2 1/8 in. OD pipe is suitable.

Based on Table 19, the next smaller line size should be used for marginal suction risers. When vertical riser sizes are reduced to provide satisfactory minimum gas velocities, pressure drop at full load increases considerably; horizontal lines should be sized to keep total pressure drop within practical limits. As long as horizontal lines are level or pitched in the direction of the compressor, oil can be transported with normal design velocities.

Because most compressors have multiple capacity-reduction features, gas velocities required to return oil up through vertical suction risers under all load conditions are difficult to maintain. When the suction riser is sized to allow oil return at the minimum operating capacity of the system, pressure drop in this portion of the line may be too great when operating at full load. If a correctly sized suction riser imposes too great a pressure drop at full load, a double suction riser should be used.

Oil Return up Suction Risers: Multistage Systems. Oil movement in the suction lines of multistage systems requires the same design approach as that for single-stage systems. For oil to flow up along a pipe wall, a certain minimum drag of gas flow is required. Drag can be represented by the friction gradient. The following sizing data may be used for ensuring oil return up vertical suction lines for refrigerants other than those listed in Tables 19 and 20. The line size selected should provide a pressure drop equal to or greater than that shown in the chart.

Saturation Temperature, °F	Line Size	
	2 in. or less	Above 2 in.
0	0.35 psi/100 ft	0.20 psi/100 ft
−50	0.45 psi/100 ft	0.25 psi/100 ft

Double Suction Risers. Figure 3 shows two methods of double suction riser construction. Oil return in this arrangement is accomplished at minimum loads, but it does not cause excessive pressure drops at full load. Sizing and operation of a double suction riser are as follows:

1. Riser A is sized to return oil at minimum load possible.
2. Riser B is sized for satisfactory pressure drop through both risers at full load. The usual method is to size riser B so that the combined cross-sectional area of A and B is equal to or slightly greater than the cross-sectional area of a single pipe sized for acceptable pressure drop at full load without regard for oil return at minimum load. The combined cross-sectional area, however, should not be greater than the cross-sectional area of a single pipe that would return oil in an upflow riser under maximum load.
3. A trap is introduced between the two risers, as shown in both methods. During part-load operation, gas velocity is not sufficient to return oil through both risers, and the trap gradually fills up with oil until riser B is sealed off. The gas then travels up riser A only with enough velocity to carry oil along with it back into the horizontal suction main.

The oil holding capacity of the trap is limited to a minimum by close-coupling the fittings at the bottom of the risers. If this is not done, the trap can accumulate enough oil during part-load operation to lower the compressor crankcase oil level. Note in Figure 3 that riser lines A and B form an inverted loop and enter the horizontal suction line from the top. This prevents oil drainage into the risers, which may be idle during part-load operation. The same purpose can be served by running risers horizontally into the main, provided that the main is larger in diameter than either riser.

Often, double suction risers are essential on low-temperature systems that can tolerate very little pressure drop. Any system using

Table 18 Valve Losses in Equivalent Feet of Pipe

Nominal Pipe or Tube Size, in.	Globe[a]	60° Wye	45° Wye	Angle[a]	Gate[b]	Swing Check[c]	Lift Check
3/8	17	8	6	6	0.6	5	Globe and vertical lift same as globe valve[d]
1/2	18	9	7	7	0.7	6	
3/4	22	11	9	9	0.9	8	
1	29	15	12	12	1.0	10	
1 1/4	38	20	15	15	1.5	14	
1 1/2	43	24	18	18	1.8	16	
2	55	30	24	24	2.3	20	
2 1/2	69	35	29	29	2.8	25	
3	84	43	35	35	3.2	30	
3 1/2	100	50	41	41	4.0	35	
4	120	58	47	47	4.5	40	
5	140	71	58	58	6.0	50	
6	170	88	70	70	7.0	60	
8	220	115	85	85	9.0	80	Angle lift same as angle valve
10	280	145	105	105	12.0	100	
12	320	165	130	130	13.0	120	
14	360	185	155	155	15.0	135	
16	410	210	180	180	17.0	150	
18	460	240	200	200	19.0	165	
20	520	275	235	235	22.0	200	
24	610	320	265	265	25.0	240	

Note: Losses are for valves in fully open position and with screwed, welded, flanged, or flared connections.

[a]These losses do not apply to valves with needlepoint seats.

[b]Regular and short pattern plug cock valves, when fully open, have same loss as gate valve. For valve losses of short pattern plug cocks above 6 in., check with manufacturer.

[c]Losses also apply to inline, ball check valve.

[d]For *Y* pattern globe lift check valve with seat approximately equal to nominal pipe diameter, use values of 60° wye valve for loss.

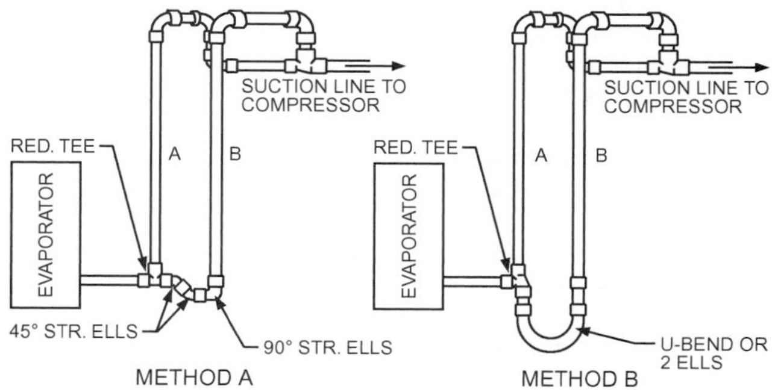

Fig. 3 Double-Suction Riser Construction

Table 19 Minimum Refrigeration Capacity in Tons for Oil Entrainment up Suction Risers (Type L Copper Tubing)

Refrigerant	Saturated Suction Temp., °F	Suction Gas Temp., °F	Pipe OD, in. 1/2	5/8	3/4	7/8	1 1/8	1 3/8	1 5/8	2 1/8	2 5/8	3 1/8	3 5/8	4 1/8
		Area, in²	0.146	0.233	0.348	0.484	0.825	1.256	1.780	3.094	4.770	6.812	9.213	11.970
22	–40.0	–30.0	0.067	0.119	0.197	0.298	0.580	0.981	1.52	3.03	5.20	8.12	11.8	16.4
		–10.0	0.065	0.117	0.194	0.292	0.570	0.963	1.49	2.97	5.11	7.97	11.6	16.1
		10.0	0.066	0.118	0.195	0.295	0.575	0.972	1.50	3.00	5.15	8.04	11.7	16.3
	–20.0	–10.0	0.087	0.156	0.258	0.389	0.758	1.28	1.98	3.96	6.80	10.6	15.5	21.5
		10.0	0.085	0.153	0.253	0.362	0.744	1.26	1.95	3.88	6.67	10.4	15.2	21.1
		30.0	0.086	0.154	0.254	0.383	0.747	1.26	1.95	3.90	6.69	10.4	15.2	21.1
	0.0	10.0	0.111	0.199	0.328	0.496	0.986	1.63	2.53	5.04	8.66	13.5	19.7	27.4
		30.0	0.108	0.194	0.320	0.484	0.942	1.59	2.46	4.92	8.45	13.2	19.2	26.7
		50.0	0.109	0.195	0.322	0.486	0.946	1.60	2.47	4.94	8.48	13.2	19.3	26.8
	20.0	30.0	0.136	0.244	0.403	0.608	1.18	2.00	3.10	6.18	10.6	16.6	24.2	33.5
		50.0	0.135	0.242	0.399	0.603	1.17	1.99	3.07	6.13	10.5	16.4	24.0	33.3
		70.0	0.135	0.242	0.400	0.605	1.18	1.99	3.08	6.15	10.6	16.5	24.0	33.3
	40.0	50.0	0.167	0.300	0.495	0.748	1.46	2.46	3.81	7.60	13.1	20.4	29.7	41.3
		70.0	0.165	0.296	0.488	0.737	1.44	2.43	3.75	7.49	12.9	20.1	29.3	40.7
		90.0	0.165	0.296	0.488	0.738	1.44	2.43	3.76	7.50	12.9	20.1	29.3	40.7
134a	0.0	10.0	0.089	0.161	0.259	0.400	0.78	1.32	2.03	4.06	7.0	10.9	15.9	22.1
		30.0	0.075	0.135	0.218	0.336	0.66	1.11	1.71	3.42	5.9	9.2	13.4	18.5
		50.0	0.072	0.130	0.209	0.323	0.63	1.07	1.64	3.28	5.6	8.8	12.8	17.8
	10.0	20.0	0.101	0.182	0.294	0.453	0.88	1.49	2.31	4.61	7.9	12.4	18.0	25.0
		40.0	0.084	0.152	0.246	0.379	0.74	1.25	1.93	3.86	6.6	10.3	15.1	20.9
		60.0	0.081	0.147	0.237	0.366	0.71	1.21	1.87	3.73	6.4	10.0	14.6	20.2
	20.0	30.0	0.113	0.205	0.331	0.510	0.99	1.68	2.60	5.19	8.9	13.9	20.3	28.2
		50.0	0.095	0.172	0.277	0.427	0.83	1.41	2.17	4.34	7.5	11.6	17.0	23.6
		70.0	0.092	0.166	0.268	0.413	0.81	1.36	2.10	4.20	7.2	11.3	16.4	22.8
	30.0	40.0	0.115	0.207	0.335	0.517	1.01	1.70	2.63	5.25	9.0	14.1	20.5	28.5
		60.0	0.107	0.193	0.311	0.480	0.94	1.58	2.44	4.88	8.4	13.1	19.1	26.5
		80.0	0.103	0.187	0.301	0.465	0.91	1.53	2.37	4.72	8.1	12.7	18.5	25.6
	40.0	50.0	0.128	0.232	0.374	0.577	1.12	1.90	2.94	5.87	10.1	15.7	22.9	31.8
		70.0	0.117	0.212	0.342	0.528	1.03	1.74	2.69	5.37	9.2	14.4	21.0	29.1
		90.0	0.114	0.206	0.332	0.512	1.00	1.69	2.61	5.21	8.9	14.0	20.4	28.3

Notes:

1. Refrigeration capacity in tons is based on 90°F liquid temperature and superheat as indicated by listed temperature. For other liquid line temperatures, use correction factors in table at right.
2. Values computed using ISO 32 mineral oil for R-22. R-134a computed using ISO 32 ester-based oil.

Refrigerant	Liquid Temperature, °F 50	60	70	80	100	110	120	130	140
22	1.17	1.14	1.10	1.06	0.98	0.94	0.89	0.85	0.80
134a	1.26	1.20	1.13	1.07	0.94	0.87	0.80	0.74	0.67

these risers should include a suction trap (accumulator) and a means of returning oil gradually.

For systems operating at higher suction temperatures, such as for comfort air conditioning, single suction risers can be sized for oil return at minimum load. Where single compressors are used with capacity control, minimum capacity is usually 25 or 33% of maximum displacement. With this low ratio, pressure drop in single suction risers designed for oil return at minimum load is rarely serious at full load.

When multiple compressors are used, one or more may shut down while another continues to operate, and the maximum-to-minimum ratio becomes much larger. This may make a double suction riser necessary.

The remaining suction line portions are sized to allow a practical pressure drop between the evaporators and compressors because oil is carried along in horizontal lines at relatively low gas velocities. It is good practice to give some pitch to these lines toward the compressor. Traps should be avoided, but when that is impossible, the risers from them are treated the same as those leading from the evaporators.

Preventing Oil Trapping in Idle Evaporators. Suction lines should be designed so that oil from an active evaporator does not drain into an idle one. Figure 4A shows multiple evaporators on different floor levels with the compressor above. Each suction line is brought upward and looped into the top of the common suction line to prevent oil from draining into inactive coils.

Figure 4B shows multiple evaporators stacked on the same level, with the compressor above. Oil cannot drain into the lowest evaporator because the common suction line drops below the outlet of the lowest evaporator before entering the suction riser.

Figure 4C shows multiple evaporators on the same level, with the compressor located below. The suction line from each evaporator drops down into the common suction line so that oil cannot drain into an idle evaporator. An alternative arrangement is shown in Figure 4D for cases where the compressor is above the evaporators.

Figure 5 illustrates typical piping for evaporators above and below a common suction line. All horizontal runs should be level or pitched toward the compressor to ensure oil return.

Traps shown in the suction lines after the evaporator suction outlet are recommended by thermal expansion valve manufacturers to prevent erratic operation of the thermal expansion valve. Expansion valve bulbs are located on the suction lines between the evaporator and these traps. The traps serve as drains and help prevent liquid from accumulating under the expansion valve bulbs during compressor off cycles. They are useful only where straight runs or risers are encountered in the suction line leaving the evaporator outlet.

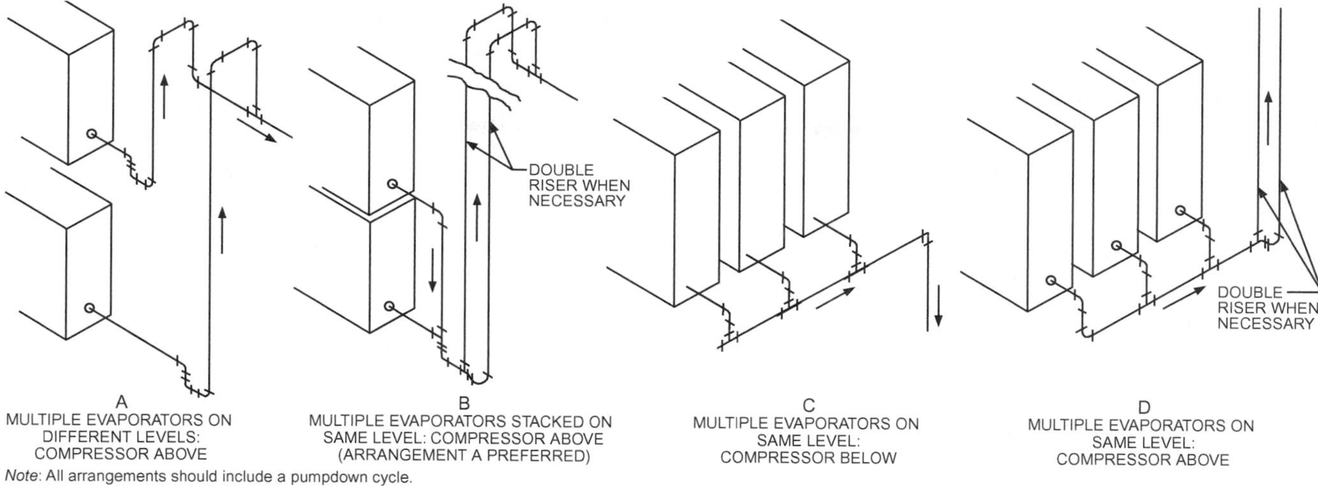

Fig. 4 Suction Line Piping at Evaporator Coils

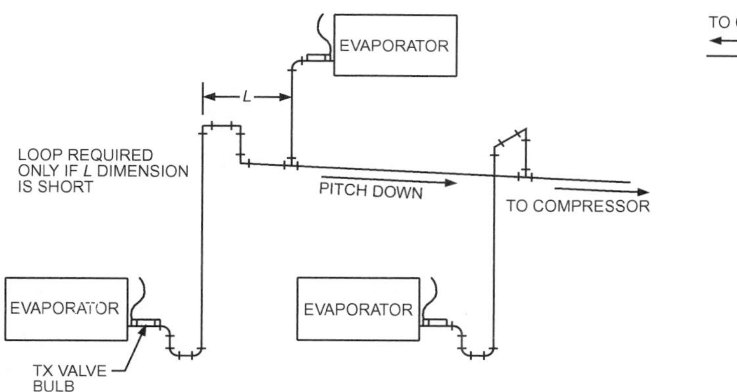

Fig. 5 Typical Piping from Evaporators Located above and below Common Suction Line

DISCHARGE (HOT-GAS) LINES

Hot-gas lines should be designed to

- Avoid trapping oil at part-load operation
- Prevent condensed refrigerant and oil in the line from draining back to the head of the compressor
- Have carefully selected connections from a common line to multiple compressors
- Avoid developing excessive noise or vibration from hot-gas pulsations, compressor vibration, or both

Oil Transport up Risers at Normal Loads. Although a low pressure drop is desired, oversized hot-gas lines can reduce gas velocities to a point where the refrigerant will not transport oil. Therefore, when using multiple compressors with capacity control, hot-gas risers must transport oil at all possible loadings.

Minimum Gas Velocities for Oil Transport in Risers. Minimum capacities for oil entrainment in hot-gas line risers are shown in Table 20. On multiple-compressor installations, the lowest possible system loading should be calculated and a riser size selected to give at least the minimum capacity indicated in the table for successful oil transport.

In some installations with multiple compressors and with capacity control, a vertical hot-gas line, sized to transport oil at minimum load, has excessive pressure drop at maximum load. When this problem exists, either a double riser or a single riser with an oil separator can be used.

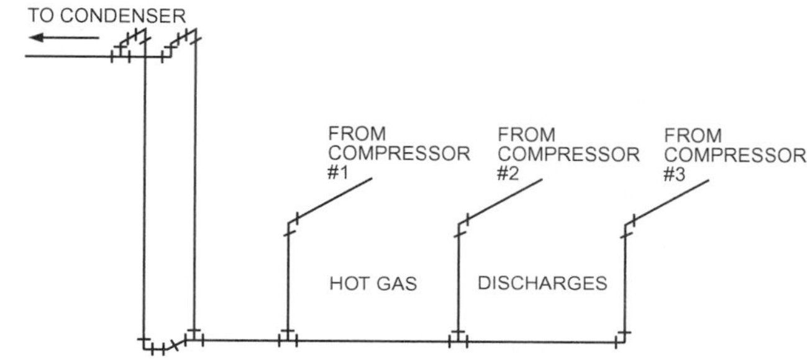

Fig. 6 Double Hot-Gas Riser

Double Hot-Gas Risers. A double hot-gas riser can be used the same way it is used in a suction line. Figure 6 shows the double riser principle applied to a hot-gas line. Its operating principle and sizing technique are described in the section on Double Suction Risers.

Single Riser and Oil Separator. As an alternative, an oil separator in the discharge line just before the riser allows sizing the riser for a low pressure drop. Any oil draining back down the riser accumulates in the oil separator. With large multiple compressors, separator capacity may dictate the use of individual units for each compressor located between the discharge line and the main discharge header. Horizontal lines should be level or pitched downward in the direction of gas flow to facilitate travel of oil through the system and back to the compressor.

Piping to Prevent Liquid and Oil from Draining to Compressor Head. Whenever the condenser is located above the compressor, the hot-gas line should be trapped near the compressor before rising to the condenser, especially if the hot-gas riser is long. This minimizes the possibility of refrigerant, condensed in the line during off cycles, draining back to the head of the compressor. Also, any oil traveling up the pipe wall will not drain back to the compressor head.

The loop in the hot-gas line (Figure 7) serves as a reservoir and traps liquid resulting from condensation in the line during shutdown, thus preventing gravity drainage of liquid and oil back to the compressor head. A small high-pressure float drainer should be installed at the bottom of the trap to drain any significant amount of refrigerant condensate to a low-side component such as a suction accumulator or low-pressure receiver. This float prevents excessive build-up of liquid in the trap and possible liquid hammer when the compressor is restarted.

Table 20 Minimum Refrigeration Capacity in Tons for Oil Entrainment up Hot-Gas Risers (Type L Copper Tubing)

Refrig-erant	Saturated Temp., °F	Discharge Gas Temp., °F	Pipe OD, in. 1/2	5/8	3/4	7/8	1 1/8	1 3/8	1 5/8	2 1/8	2 5/8	3 1/8	3 5/8	4 1/8
		Area, in^2	0.146	0.233	0.348	0.484	0.825	1.256	1.780	3.094	4.770	6.812	9.213	11.970
22	80.0	110.0	0.235	0.421	0.695	1.05	2.03	3.46	5.35	10.7	18.3	28.6	41.8	57.9
		140.0	0.223	0.399	0.659	0.996	1.94	3.28	5.07	10.1	17.4	27.1	39.6	54.9
		170.0	0.215	0.385	0.635	0.960	1.87	3.16	4.89	9.76	16.8	26.2	38.2	52.9
	90.0	120.0	0.242	0.433	0.716	1.06	2.11	3.56	5.50	11.0	18.9	29.5	43.0	59.6
		150.0	0.226	0.406	0.671	1.01	1.97	3.34	5.16	10.3	17.7	27.6	40.3	55.9
		180.0	0.216	0.387	0.540	0.956	1.88	3.18	4.92	9.82	16.9	26.3	38.4	53.3
	100.0	130.0	0.247	0.442	0.730	1.10	2.15	3.83	5.62	11.2	19.3	30.1	43.9	60.8
		160.0	0.231	0.414	0.884	1.03	2.01	3.40	5.26	10.5	18.0	28.2	41.1	57.0
		190.0	0.220	0.394	0.650	0.982	1.91	3.24	3.00	9.96	17.2	26.8	39.1	54.2
	110.0	140.0	0.251	0.451	0.744	1.12	2.19	3.70	5.73	11.4	19.6	30.6	44.7	62.0
		170.0	0.235	0.421	0.693	1.05	2.05	3.46	3.35	10.7	18.3	28.6	41.8	57.9
		200.0	0.222	0.399	0.658	0.994	1.94	3.28	5.06	10.1	17.4	27.1	39.5	54.8
	120.0	150.0	0.257	0.460	0.760	1.15	2.24	3.78	5.85	11.7	20.0	31.3	45.7	63.3
		180.0	0.239	0.428	0.707	1.07	2.08	3.51	5.44	10.8	18.6	29.1	42.4	58.9
		210.0	0.225	0.404	0.666	1.01	1.96	3.31	5.12	10.2	17.6	27.4	40.0	55.5
134a	80.0	110.0	0.199	0.360	0.581	0.897	1.75	2.96	4.56	9.12	15.7	24.4	35.7	49.5
		140.0	0.183	0.331	0.535	0.825	1.61	2.72	4.20	8.39	14.4	22.5	32.8	45.6
		170.0	0.176	0.318	0.512	0.791	1.54	2.61	4.02	8.04	13.8	21.6	31.4	43.6
	90.0	120.0	0.201	0.364	0.587	0.906	1.76	2.99	4.61	9.21	15.8	24.7	36.0	50.0
		150.0	0.184	0.333	0.538	0.830	1.62	2.74	4.22	8.44	14.5	22.6	33.0	45.8
		180.0	0.177	0.320	0.516	0.796	1.55	2.62	4.05	8.09	13.9	21.7	31.6	43.9
	100.0	130.0	0.206	0.372	0.600	0.926	1.80	3.05	4.71	9.42	16.2	25.2	36.8	51.1
		160.0	0.188	0.340	0.549	0.848	1.65	2.79	4.31	8.62	14.8	23.1	33.7	46.8
		190.0	0.180	0.326	0.526	0.811	1.58	2.67	4.13	8.25	14.2	22.1	32.2	44.8
	110.0	140.0	0.209	0.378	0.610	0.942	1.83	3.10	4.79	9.57	16.5	25.7	37.4	52.0
		170.0	0.191	0.346	0.558	0.861	1.68	2.84	4.38	8.76	15.0	23.5	34.2	47.5
		200.0	0.183	0.331	0.534	0.824	1.61	2.72	4.19	8.38	14.4	22.5	32.8	45.5
	120.0	150.0	0.212	0.383	0.618	0.953	1.86	3.14	4.85	9.69	16.7	26.0	37.9	52.6
		180.0	0.194	0.351	0.566	0.873	1.70	2.88	4.44	8.88	15.3	23.8	34.7	48.2
		210.0	0.184	0.334	0.538	0.830	1.62	2.74	4.23	8.44	14.5	22.6	33.0	45.8

Notes:
1. Refrigeration capacity in tons based on saturated suction temperature of 20°F with 15°F superheat at indicated saturated condensing temperature with 15°F subcooling. For other saturated suction temperatures with 15°F superheat, use correction factors in the table at right.
2. Table computed using ISO 32 mineral oil for R-22, and ISO 32 ester-based oil for R-134a.

Refrigerant	Saturated Suction Temperature, °F −40	−20	0	+40
22	0.92	0.95	0.97	1.02
134a	—	—	0.96	1.04

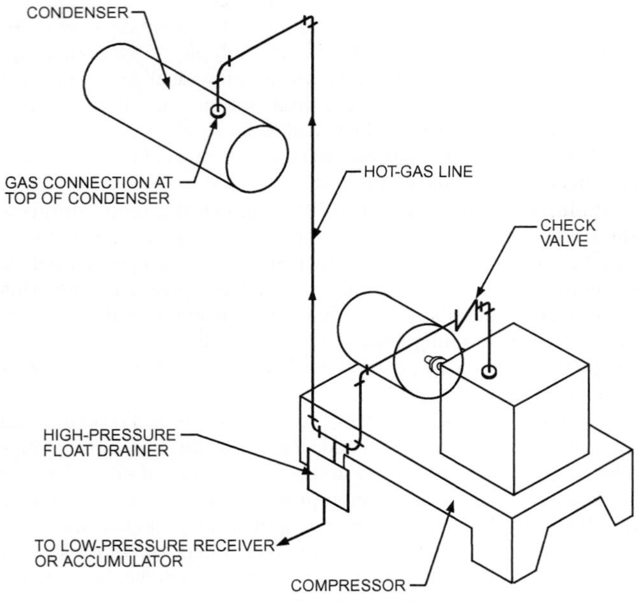

Fig. 7 Hot-Gas Loop

For multiple-compressor arrangements, each discharge line should have a check valve to prevent gas from active compressors from condensing on heads of idle compressors.

For single-compressor applications, a tightly closing check valve should be installed in the hot-gas line of the compressor whenever the condenser and the receiver ambient temperature are higher than that of the compressor. The check valve prevents refrigerant from boiling off in the condenser or receiver and condensing on the compressor heads during off cycles.

This check valve should be a piston type, which will close by gravity when the compressor stops running. The use of a spring-loaded check may incur chatter (vibration), particularly on slow-speed reciprocating compressors.

For compressors equipped with water-cooled oil coolers, a water solenoid and water-regulating valve should be installed in the water line so that the regulating valve maintains adequate cooling during operation, and the solenoid stops flow during the off cycle to prevent localized condensing of the refrigerant.

Hot-Gas (Discharge) Mufflers. Mufflers can be installed in hot-gas lines to dampen discharge gas pulsations, reducing vibration and noise. Mufflers should be installed in a horizontal or downflow portion of the hot-gas line immediately after it leaves the compressor.

Because gas velocity through the muffler is substantially lower than that through the hot-gas line, the muffler may form an oil trap.

The muffler should be installed to allow oil to flow through it and not be trapped.

DEFROST GAS SUPPLY LINES

Sizing refrigeration lines to supply defrost gas to one or more evaporators is not an exact science. The parameters associated with sizing the defrost gas line are related to allowable pressure drop and refrigerant flow rate during defrost.

Engineers use an estimated two times the evaporator load for effective refrigerant flow rate to determine line sizing requirements. Pressure drop is not as critical during the defrost cycle, and many engineers use velocity as the criterion for determining line size. The effective condensing temperature and average temperature of the gas must be determined. The velocity determined at saturated conditions gives a conservative line size.

Some controlled testing (Stoecker 1984) has shown that, in small coils with R-22, the defrost flow rate tends to be higher as the condensing temperature is increased. The flow rate is on the order of two to three times the normal evaporator flow rate, which supports the estimated two times used by practicing engineers.

Table 21 provides guidance on selecting defrost gas supply lines based on velocity at a saturated condensing temperature of 70°F. It is recommended that initial sizing be based on twice the evaporator flow rate and that velocities from 1000 to 2000 fpm be used for determining the defrost gas supply line size.

Gas defrost lines must be designed to continuously drain any condensed liquid.

RECEIVERS

Refrigerant receivers are vessels used to store excess refrigerant circulated throughout the system. Their purpose is to

- Provide pumpdown storage capacity when another part of the system must be serviced or the system must be shut down for an extended time. In some water-cooled condenser systems, the condenser also serves as a receiver if the total refrigerant charge does not exceed its storage capacity.
- Handle the excess refrigerant charge that occurs with air-cooled condensers using flooding condensing pressure control (see the section on Head Pressure Control for Refrigerant Condensers).
- Accommodate a fluctuating charge in the low side and drain the condenser of liquid to maintain an adequate effective condensing surface on systems where the operating charge in the evaporator and/or condenser varies for different loading conditions. When an evaporator is fed with a thermal expansion valve, hand expansion valve, or low-pressure float, the operating charge in the evaporator varies considerably depending on the loading. During low load, the evaporator requires a larger charge because boiling is not as intense. When load increases, the operating charge in the evaporator decreases, and the receiver must store excess refrigerant.
- Hold the full charge of the idle circuit on systems with multicircuit evaporators that shut off liquid supply to one or more circuits during reduced load and pump out the idle circuit.

Connections for Through-Type Receiver. When a through-type receiver is used, liquid must always flow from condenser to receiver. Pressure in the receiver must be lower than that in the condenser outlet. The receiver and its associated piping provide free flow of liquid from the condenser to the receiver by equalizing pressures between the two so that the receiver cannot build up a higher pressure than the condenser.

If a vent is not used, piping between condenser and receiver (condensate line) is sized so that liquid flows in one direction and gas flows in the opposite direction. Sizing the condensate line for 100 fpm liquid velocity is usually adequate to attain this flow. Piping should slope at least 0.25 in/ft and eliminate any natural liquid traps. Figure 8 illustrates this configuration.

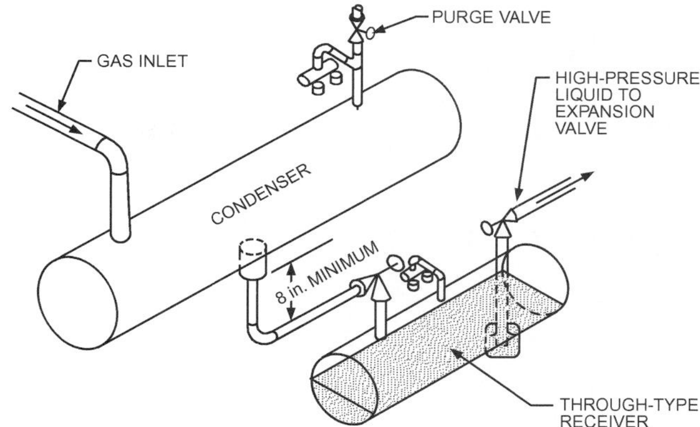

Fig. 8 Shell-and-Tube Condenser to Receiver Piping (Through-Type Receiver)

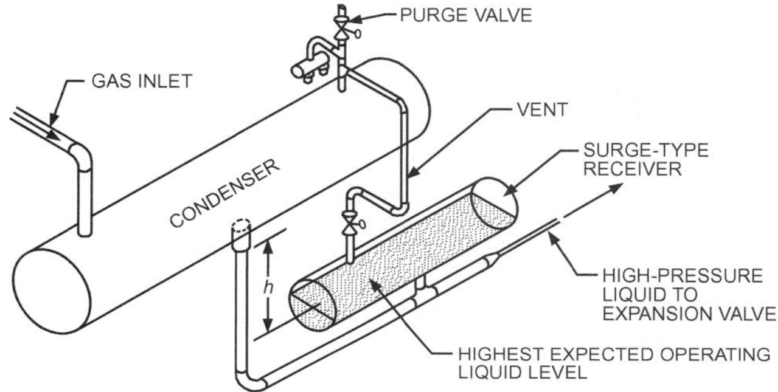

Fig. 9 Shell-and-Tube Condenser to Receiver Piping (Surge-Type Receiver)

Piping between the condenser and the receiver can be equipped with a separate vent (equalizer) line to allow receiver and condenser pressures to equalize. This external vent line can be piped either with or without a check valve in the vent line (see Figures 10 and 11). If there is no check valve, prevent discharge gas from discharging directly into the vent line; this should prevent a gas velocity pressure component from being introduced on top of the liquid in the receiver. When the piping configuration is unknown, install a check valve in the vent with flow in the direction of the condenser. The check valve should be selected for minimum opening pressure (i.e., approximately 0.5 psi). When determining condensate drop leg height, allowance must be made to overcome both the pressure drop across this check valve and the refrigerant pressure drop through the condenser. This ensures that there will be no liquid backup into an operating condenser on a multiple-condenser application when one or more of the condensers is idle. The condensate line should be sized so that velocity does not exceed 150 fpm.

The vent line flow is from receiver to condenser when receiver temperature is higher than condensing temperature. Flow is from condenser to receiver when air temperature around the receiver is below condensing temperature. Flow rate depends on this temperature difference as well as on the receiver surface area. Vent size can be calculated from this flow rate.

Connections for Surge-Type Receiver. The purpose of a surge-type receiver is to allow liquid to flow to the expansion valve without exposure to refrigerant in the receiver, so that it can remain subcooled. The receiver volume is available for liquid that is to be removed from the system. Figure 9 shows an example of connections for a surge-type receiver. Height h must be adequate for a liquid pressure at least as large as the pressure loss through the condenser, liquid line, and vent line at the maximum temperature difference between

Table 21 Refrigerant Flow Capacity Data For Defrost Lines

Pipe Size		R-22 Mass Flow Data, lb/h			R-134a Mass Flow Data, lb/h			R-404a Mass Flow Data, lb/h			R-507 Mass Flow Data, lb/h			R-410a Mass Flow Data, lb/h			R-407c Mass Flow Data, lb/h		
		Velocity, fpm			Velocity, fpm			Velocity, fpm			Velocity, fpm			Velocity, fpm			Velocity, fpm		
Copper[a]		1000	2000	3000	1000	2000	3000	1000	2000	3000	1000	2000	3000	1000	2000	3000	1000	2000	3000
1/2		150	300	450	110	220	330	220	440	660	233	465	698	221	442	662	147	294	441
5/8		240	480	720	170	350	520	354	707	1061	374	747	1121	355	709	1064	236	472	708
3/4		350	710	1060	260	510	770	528	1056	1584	558	1116	1674	530	1059	1589	352	705	1057
7/8		500	1000	1500	360	720	1090	734	1467	2201	775	1550	2325	736	1471	2207	490	979	1469
1 1/8		850	1700	2550	620	1230	1850	1251	2502	3752	1321	2643	3964	1254	2509	3763	835	1670	2504
1 3/8		1300	2590	3890	940	1880	2820	1905	3810	5715	2013	4025	6037	1910	3821	5731	1272	2543	3814
1 5/8		1840	3670	5510	1330	2660	3990	2697	5393	8090	2849	5697	8546	2704	5408	8112	1800	3599	5399
2 1/8		3190	6390	9580	2310	4630	6940	4691	9382	14,073	4955	9911	14,866	4704	9408	14,112	3131	6262	9392
2 5/8		4930	9850	14,800	3570	7140	10,700	7234	14,468	21,702	7642	15,283	22,925	7254	14,508	21,762	4828	9656	14,484
3 1/8		7030	14,100	21,100	5100	10,200	15,300	10,326	20,651	30,977	10,907	21,815	32,722	10,354	20,708	31,062	6891	13,783	20,674
3 5/8		9510	19,000	28,500	6900	13,800	20,700	13,966	27,932	41,897	14,753	29,505	44,258	14,004	28,008	42,012	9321	18,641	27,962
4 1/8		12,400	24,700	37,100	9000	17,900	26,900	18,155	36,309	54,464	19,178	38,355	57,533	18,204	36,409	54,613	12,116	24,233	36,349
5 1/8		19,300	38,500	57,800	14,000	27,900	41,900	28,294	56,588	84,882	29,888	59,776	89,665	28,372	56,743	85,115	18,883	37,767	56,650
6 1/8		27,700	55,400	83,100	20,100	40,100	60,200	40,674	81,347	122,021	42,965	85,931	128,896	40,785	81,571	122,356	27,146	54,291	81,436
8 1/8		48,400	96,700	145,100	35,100	70,100	105,200	71,046	142,092	213,138	75,049	150,099	225,148	71,241	142,483	213,724	47,416	94,832	142,248
Steel																			
IPS	**SCH**																		
3/8	80	150	290	440	110	210	320	213	426	639	225	450	675	214	427	641	142	284	427
1/2	80	240	480	720	180	350	530	355	710	1,065	375	750	1125	356	712	1,068	237	474	711
3/4	80	450	890	1340	320	650	970	656	1,311	1,966	692	1385	2077	657	1,315	1,972	438	875	1312
1	80	740	1480	2230	540	1080	1610	1090	2181	3271	1152	2304	3455	1093	2187	3280	728	1455	2183
1 1/4	80	1540	3090	4630	1120	2240	3360	1945	3889	5833	2054	4108	6162	1950	3900	5850	1298	2596	3893
1 1/2	80	2100	4200	6300	1520	3050	4570	2679	5357	8036	2830	5659	8489	2686	5372	8058	1788	3576	5363
2	40	3460	6930	10,400	2510	5020	7530	5087	10,173	15,260	5373	10,746	16,120	5101	10,201	15,302	3395	6790	10,184
2 1/2	40	4940	9870	14,800	3580	7160	10,700	7252	14,503	21,755	7660	15,320	22,981	7272	14,543	21,815	4840	9679	14,519
3	40	7620	15,200	22,900	5530	11,100	16,600	11,199	22,398	33,596	11,830	23,660	35,489	11,230	22,459	33,689	7474	14,948	22,422
4	40	13,100	26,300	39,400	9520	19,000	28,600	19,297	38,594	57,891	20,384	40,769	61,153	19,350	38,700	58,050	12,879	25,758	38,636
5	40	20,600	41,300	61,900	15,000	29,900	44,900	30,302	60,603	90,905	32,009	64,018	96,027	30,385	60,770	91,155	20,223	40,447	60,670
6	40	29,800	59,600	89,400	21,600	43,200	64,800	43,793	87,586	131,379	46,261	92,521	138,782	43,913	87,827	131,740	29,227	58,455	87,682
8	40	51,600	103,300	154,900	37,400	74,800	112,300	75,833	151,666	227,498	80,106	160,212	240,318	76,041	152,083	228,124	50,611	101,222	151,832
10	40	81,400	162,800	244,100	59,000	118,00	176,900	119,530	239,061	358,591	126,266	252,532	378,797	119,859	239,718	359,576	79,775	159,549	239,323
12	ID[b]	116,700	233,400	350,200	84,600	169,200	253,800	171,437	342,874	514,311	181,098	362,195	543,293	171,908	343,817	515,725	114,417	228,834	343,251
14	30	—	—	—	—	—	—	209,013	418,027	627,040	220,791	441,582	662,374	209,588	419,176	628,764	139,496	278,991	418,486
16	30	—	—	—	—	—	—	276,874	553,748	830,622	292,476	584,951	877,427	277,635	555,270	832,905	184,786	369,571	554,357

Note: Refrigerant flow data based on saturated condensing temperature of 70°F.

[a]For brazed Type L copper tubing for defrost service, see Safety Requirements section.

[b]Pipe inside diameter is same as nominal pipe size.

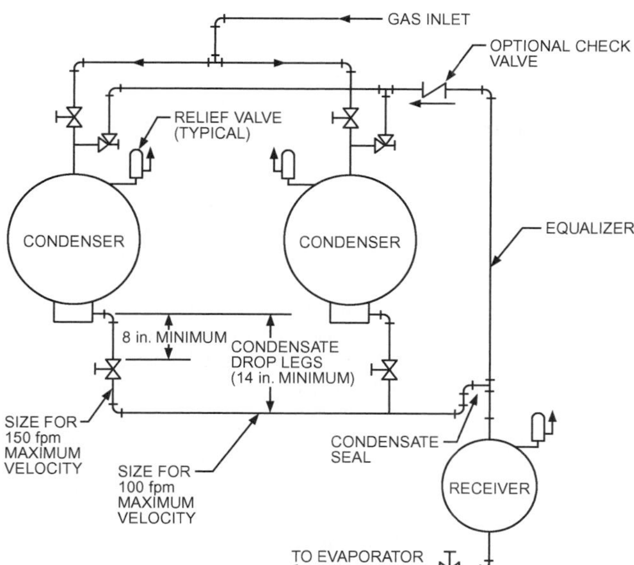

Fig. 10 Parallel Condensers with Through-Type Receiver

the receiver ambient and the condensing temperature. Condenser pressure drop at the greatest expected heat rejection should be obtained from the manufacturer. The minimum value of h can then be calculated and a decision made as to whether the available height will permit the surge-type receiver.

Multiple Condensers. Two or more condensers connected in series or in parallel can be used in a single refrigeration system. If connected in series, the pressure losses through each condenser must be added. Condensers are more often arranged in parallel. Pressure loss through any one of the parallel circuits is always equal to that through any of the others, even if it results in filling much of one circuit with liquid while gas passes through another.

Figure 10 shows a basic arrangement for parallel condensers with a through-type receiver. Condensate drop legs must be long enough to allow liquid levels in them to adjust to equalize pressure losses between condensers at all operating conditions. Drop legs should be 6 to 12 in. higher than calculated to ensure that liquid outlets remain free-draining. This height provides a liquid pressure to offset the largest condenser pressure loss. The liquid seal prevents gas blow-by between condensers.

Large single condensers with multiple coil circuits should be piped as though the independent circuits were parallel condensers. For example, if the left condenser in Figure 10 has 2 psi more pressure drop than the right condenser, the liquid level on the left is about 4 ft higher than that on the right. If the condensate lines do not have enough vertical height for this level difference, liquid will back up into the condenser until pressure drop is the same through both circuits. Enough surface may be covered to reduce condenser capacity significantly.

Condensate drop legs should be sized based on 150 fpm velocity. The main condensate lines should be based on 100 fpm. Depending on prevailing local and/or national safety codes, a relief device may have to be installed in the discharge piping.

Figure 11 shows a piping arrangement for parallel condensers with a surge-type receiver. When the system is operating at reduced load, flow paths through the circuits may not be symmetrical. Small pressure differences are not unusual; therefore, the liquid line junction should be about 2 or 3 ft below the bottom of the condensers. The exact amount can be calculated from pressure loss through each path at all possible operating conditions.

When condensers are water-cooled, a single automatic water valve for the condensers in one refrigeration system should be used.

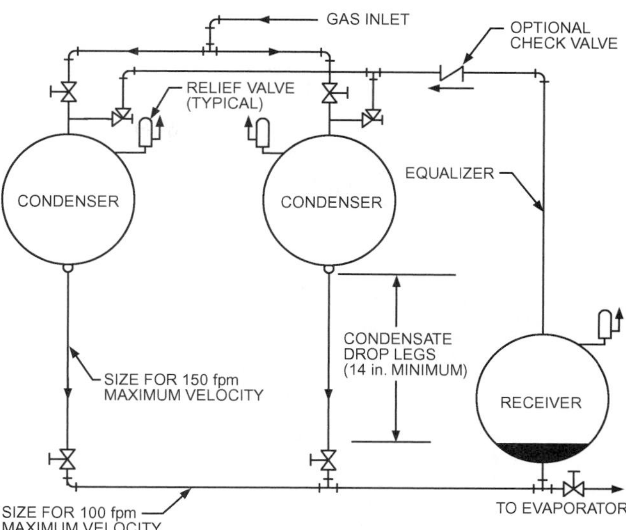

Fig. 11 Parallel Condensers with Surge-Type Receiver

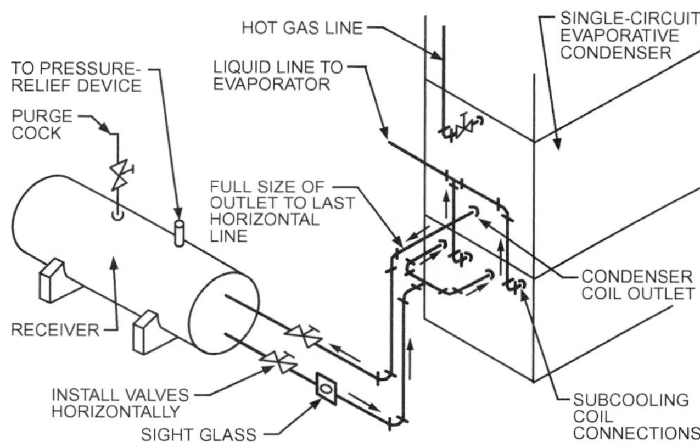

Fig. 12 Single-Circuit Evaporative Condenser with Receiver and Liquid Subcooling Coil

Individual valves for each condenser in a single system cannot maintain the same pressure and corresponding pressure drops.

With evaporative condensers (Figure 12), pressure loss may be high. If parallel condensers are alike and all are operated, the differences may be small, and condenser outlets need not be more than 2 or 3 ft above the liquid line junction. If fans on one condenser are not operated while the fans on another condenser are, then the liquid level in the one condenser must be high enough to compensate for the pressure drop through the operating condenser.

When the available level difference between condenser outlets and the liquid-line junction is sufficient, the receiver may be vented to the condenser inlets (Figure 13). In this case, the surge-type receiver can be used. The level difference must then be at least equal to the greatest loss through any condenser circuit plus the greatest vent line loss when the receiver ambient is greater than the condensing temperature.

AIR-COOLED CONDENSERS

Refrigerant pressure drop through air-cooled condensers must be obtained from the supplier for the particular unit at the specified load. If refrigerant pressure drop is low enough and the arrangement is practical, parallel condensers can be connected to allow for

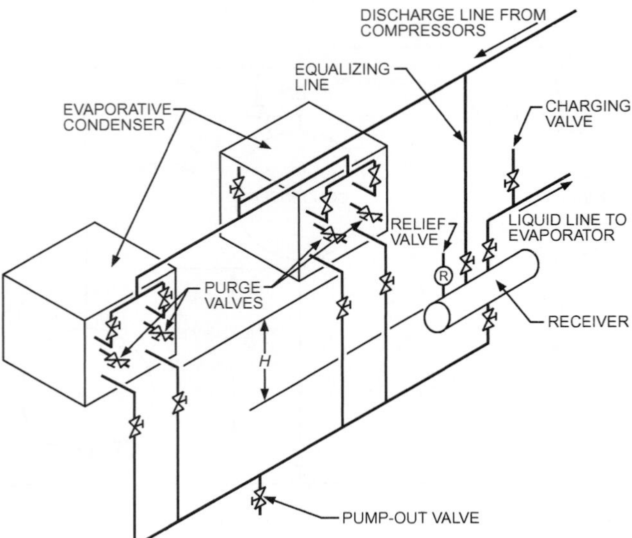

Fig. 13 Multiple Evaporative Condensers with Equalization to Condenser Inlets

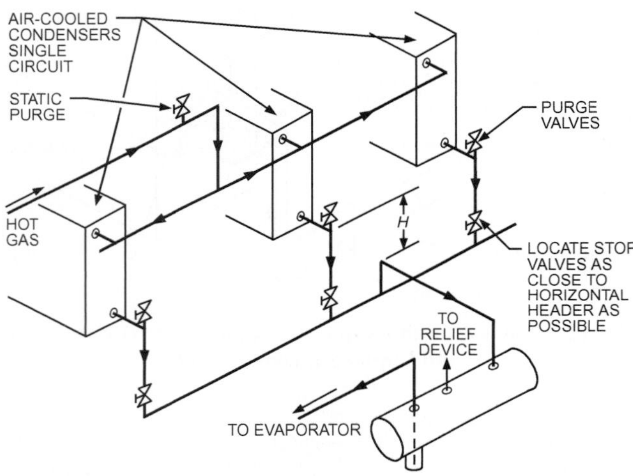

Fig. 14 Multiple Air-Cooled Condensers

capacity reduction to zero on one condenser without causing liquid backup in active condensers (Figure 14). Multiple condensers with high pressure drops can be connected as shown in Figure 14, provided that (1) the ambient at the receiver is equal to or lower than the inlet air temperature to the condenser; (2) capacity control affects all units equally; (3) all units operate when one operates, unless valved off at both inlet and outlet; and (4) all units are of equal size.

A single condenser with any pressure drop can be connected to a receiver without an equalizer and without trapping height if the condenser outlet and the line from it to the receiver can be sized for sewer flow without a trap or restriction, using a maximum velocity of 100 fpm. A single condenser can also be connected with an equalizer line to the hot-gas inlet if the vertical drop leg is sufficient to balance refrigerant pressure drop through the condenser and liquid line to the receiver.

If unit sizes are unequal, additional liquid height *H*, equivalent to the difference in full-load pressure drop, is required. Usually, condensers of equal size are used in parallel applications.

If the receiver cannot be located in an ambient temperature below the inlet air temperature for all operating conditions, sufficient extra height of drop leg *H* is required to overcome the equivalent differences in saturation pressure of the receiver and the condenser. Subcooling by the liquid leg tends to condense vapor in the receiver to reach a balance between rate of condensation, at an intermediate saturation pressure, and heat gain from ambient to the receiver. A relatively large liquid leg is required to balance a small temperature difference; therefore, this method is probably limited to marginal cases. Liquid leaving the receiver is nonetheless saturated, and any subcooling to prevent flashing in the liquid line must be obtained downstream of the receiver. If the temperature of the receiver ambient is above the condensing pressure only at part-load conditions, it may be acceptable to back liquid into the condensing surface, sacrificing the operating economy of lower part-load head pressure for a lower liquid leg requirement. The receiver must be adequately sized to contain a minimum of the backed-up liquid so that the condenser can be fully drained when full load is required. If a low-ambient control system of backing liquid into the condenser is used, consult the system supplier for proper piping.

PIPING AT MULTIPLE COMPRESSORS

Multiple compressors operating in parallel must be carefully piped to ensure proper operation.

Suction Piping

Suction piping should be designed so that all compressors run at the same suction pressure and so that oil is returned in equal proportions. All suction lines should be brought into a common suction header to return oil to each crankcase as uniformly as possible. Depending on the type and size of compressors, oil may be returned by designing the piping in one or more of the following schemes:

- Oil returned with the suction gas to each compressor
- Oil contained with a suction trap (accumulator) and returned to the compressors through a controlled means
- Oil trapped in a discharge line separator and returned to the compressors through a controlled means (see the section on Discharge Piping)

The suction header is a means of distributing suction gas equally to each compressor. Header design can be to freely pass the suction gas and oil mixture or to provide a suction trap for the oil. The header should be run above the level of the compressor suction inlets so oil can drain into the compressors by gravity.

Figure 15 shows a pyramidal or yoke-type suction header to maximize pressure and flow equalization at each of three compressor suction inlets piped in parallel. This type of construction is recommended for applications of three or more compressors in parallel. For two compressors in parallel, a single feed between the two compressor takeoffs is acceptable. Although not as good for equalizing flow and pressure drops to all compressors, one alternative is to have the suction line from evaporators enter at one end of the header instead of using the yoke arrangement. Then the suction header may have to be enlarged to minimize pressure drop and flow turbulence.

Suction headers designed to freely pass the gas/oil mixture should have branch suction lines to compressors connected to the side of the header. Return mains from the evaporators should not be connected into the suction header to form crosses with the branch suction lines to the compressors. The header should be full size based on the largest mass flow of the suction line returning to the compressors. The takeoffs to the compressors should either be the

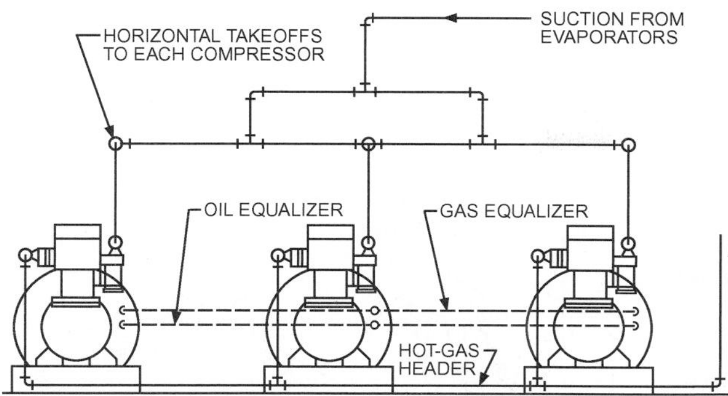

Note: Gas equalizer must be large enough to approximate identical crankcase pressure in all compressors with any combination of idle and operating compressors (any pressure difference is reflected by difference in oil level).

Fig. 15 Suction and Hot-Gas Headers for Multiple Compressors

same size as the suction header or be constructed so that the oil will not trap within the suction header. The branch suction lines to the compressors should not be reduced until the vertical drop is reached.

Suction traps are recommended wherever (1) parallel compressors, (2) flooded evaporators, (3) double suction risers, (4) long suction lines, (5) multiple expansion valves, (6) hot-gas defrost, (7) reverse-cycle operation, or (8) suction-pressure regulators are used.

Depending on system size, the suction header may be designed to function as a suction trap. The suction header should be large enough to provide a low-velocity region in the header to allow suction gas and oil to separate. See the section on Low-Pressure Receiver Sizing in Chapter 1 to arrive at recommended velocities for separation. Suction gas flow for individual compressors should be taken off the top of the suction header. Oil can be returned to the compressor directly or through a vessel equipped with a heater to boil off refrigerant and then allow oil to drain to the compressors or other devices used to feed oil to the compressors.

The suction trap must be sized for effective gas and liquid separation. Adequate liquid volume and a means of disposing of it must be provided. A liquid transfer pump or heater may be used. Chapter 1 has further information on separation and liquid transfer pumps.

An oil receiver equipped with a heater effectively evaporates liquid refrigerant accumulated in the suction trap. It also assumes that each compressor receives its share of oil. Either crankcase float valves or external float switches and solenoid valves can be used to control the oil flow to each compressor.

A gravity-feed oil receiver should be elevated to overcome the pressure drop between it and the crankcase. The oil receiver should be sized so that a malfunction of the oil control mechanism cannot overfill an idle compressor.

Figure 16 shows a recommended hookup of multiple compressors, suction trap (accumulator), oil receiver, and discharge line oil separators. The oil receiver also provides a reserve supply of oil for compressors where oil in the system outside the compressor varies with system loading. The heater mechanism should always be submerged.

Discharge Piping

The piping arrangement in Figure 15 is suggested for discharge piping. The piping must be arranged to prevent refrigerant liquid and oil from draining back into the heads of idle compressors. A check valve in the discharge line may be necessary to prevent refrigerant and oil from entering the compressor heads by migration. It is recommended that, after leaving the compressor head, the piping be routed to a lower elevation so that a trap is formed to allow for drainback of refrigerant and oil from the discharge line when flow rates

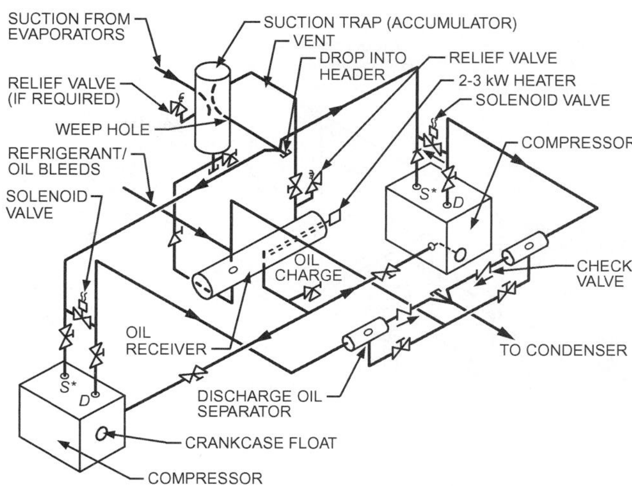

**Note:* Solenoid valve open when compressor is idle to avoid check valve leakage from condensing on heads. Not needed if oil separator has bleeder to outlet.

Fig. 16 Parallel Compressors with Gravity Oil Flow

are reduced or the compressors are off. If an oil separator is used in the discharge line, it may suffice as the trap for drainback for the discharge line.

A bullheaded tee at the junction of two compressor branches and the main discharge header should be avoided because it causes increased turbulence, increased pressure drop, and possible hammering in the line.

When an oil separator is used on multiple-compressor arrangements, oil must be piped to return to the compressors. This can be done in various ways, depending on the oil management system design. Oil may be returned to an oil receiver that is the supply for control devices feeding oil back to the compressors.

Interconnection of Crankcases

When two or more compressors are to be interconnected, a method must be provided to equalize the crankcases. Some compressor designs do not operate correctly with simple equalization of the crankcases. For these systems, it may be necessary to design a positive oil float control system for each compressor crankcase. A typical system allows oil to collect in a receiver that, in turn, supplies oil to a device that meters it back into the compressor crankcase to maintain a proper oil level (Figure 16).

Compressor systems that can be equalized should be placed on foundations so that all oil equalizer tapping locations are exactly level. If crankcase floats (as in Figure 16) are not used, an oil equalization line should connect all crankcases to maintain uniform oil levels. The oil equalizer may be run level with the tapping, or, for convenient access to compressors, it may be run at the floor (Figure 17). It should never be run at a level higher than that of the tapping.

For the oil equalizer line to work properly, equalize the crankcase pressures by installing a gas equalizer line above the oil level. This line may be run to provide head room (Figure 17) or run level with tapping on the compressors. It should be piped so that oil or liquid refrigerant will not be trapped.

Both lines should be the same size as the tapping on the largest compressor and should be valved so that any one machine can be taken out for repair. The piping should be arranged to absorb vibration.

PIPING AT VARIOUS SYSTEM COMPONENTS

Flooded Fluid Coolers

For a description of flooded fluid coolers, see Chapter 37 of the 2004 *ASHRAE Handbook—HVAC Systems and Equipment.*

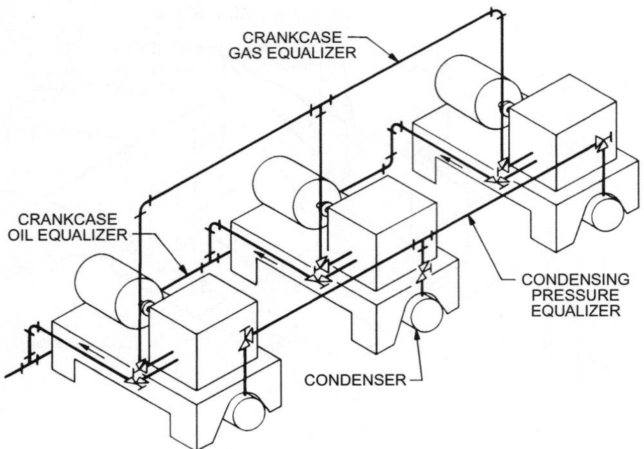

Fig. 17 Interconnecting Piping for Multiple Condensing Units

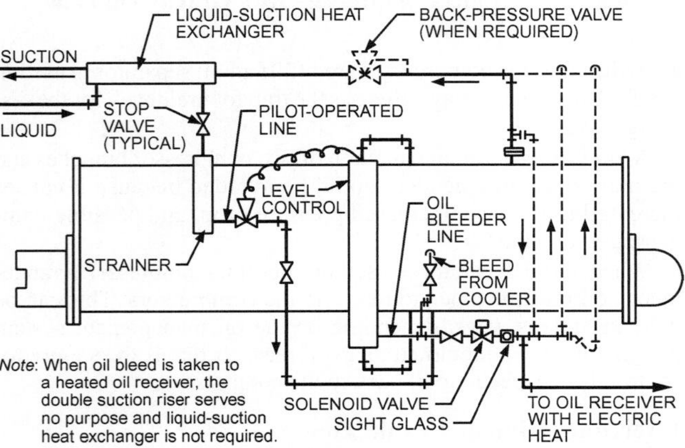

Fig. 18 Typical Piping at Flooded Fluid Cooler

Shell-and-tube flooded coolers designed to minimize liquid entrainment in the suction gas require a continuous liquid bleed line (Figure 18) installed at some point in the cooler shell below the liquid level to remove trapped oil. This continuous bleed of refrigerant liquid and oil prevents the oil concentration in the cooler from getting too high. The location of the liquid bleed connection on the shell depends on the refrigerant and oil used. For refrigerants that are highly miscible with the oil, the connection can be anywhere below the liquid level.

Refrigerant 22 can have a separate oil-rich phase floating on a refrigerant-rich layer. This becomes more pronounced as evaporating temperature drops. When R-22 is used with mineral oil, the bleed line is usually taken off the shell just slightly below the liquid level, or there may be more than one valved bleed connection at slightly different levels so that the optimum point can be selected during operation. With alkyl benzene lubricants, oil/refrigerant miscibility may be high enough that the oil bleed connection can be anywhere below the liquid level. The solubility charts in Chapter 7 give specific information.

Where the flooded cooler design requires an external surge drum to separate liquid carryover from suction gas off the tube bundle, the richest oil concentration may or may not be in the cooler. In some cases, the surge drum has the highest concentration of oil. Here, the refrigerant and oil bleed connection is taken from the surge drum. The refrigerant and oil bleed from the cooler by gravity. The bleed sometimes drains into the suction line so oil can be returned to the compressor with the suction gas after the accompanying liquid refrigerant is vaporized in a liquid-suction heat interchanger. A better method is to drain the refrigerant/oil bleed into a heated receiver that boils refrigerant off to the suction line and drains oil back to the compressor.

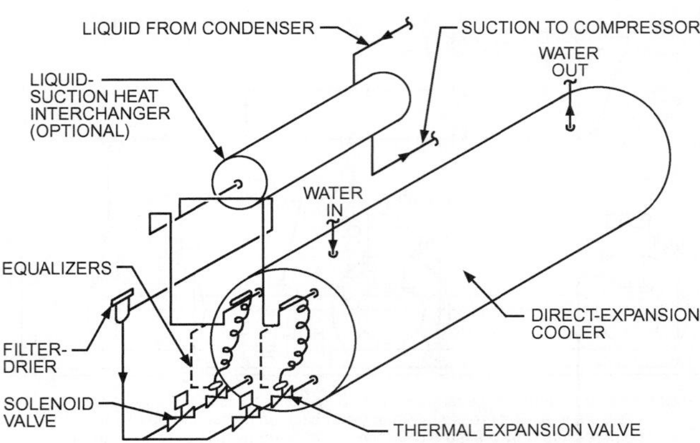

Fig. 19 Two-Circuit Direct-Expansion Cooler Connections (for Single-Compressor System)

Refrigerant Feed Devices

For further information on refrigerant feed devices, see Chapter 44. The pilot-operated low-side float control (Figure 18) is sometimes selected for flooded systems using halocarbon refrigerants. Except for small capacities, direct-acting low-side float valves are impractical for these refrigerants. The displacer float controlling a pneumatic valve works well for low-side liquid level control; it allows the cooler level to be adjusted within the instrument without disturbing the piping.

High-side float valves are practical only in single-evaporator systems, because distribution problems result when multiple evaporators are used.

Float chambers should be located as near the liquid connection on the cooler as possible because a long length of liquid line, even if insulated, can pick up room heat and give an artificial liquid level in the float chamber. Equalizer lines to the float chamber must be amply sized to minimize the effect of heat transmission. The float chamber and its equalizing lines must be insulated.

Each flooded cooler system must have a way of keeping oil concentration in the evaporator low, both to minimize the bleedoff needed to keep oil concentration in the cooler low and to reduce system losses from large stills. A highly efficient discharge gas/oil separator can be used for this purpose.

At low temperatures, periodic warm-up of the evaporator allows recovery of oil accumulation in the chiller. If continuous operation is required, dual chillers may be needed to deoil an oil-laden evaporator, or an oil-free compressor may be used.

Direct-Expansion Fluid Chillers

For further information on these chillers, see Chapter 38 in the 2004 *ASHRAE Handbook—HVAC Systems and Equipment*. Figure 19 shows typical piping connections for a multicircuit direct-expansion chiller. Each circuit contains its own thermostatic expansion and solenoid valves. One solenoid valve can be wired to close at reduced system capacity. The thermostatic expansion valve bulbs should be located between the cooler and the liquid-suction interchanger, if used. Locating the bulb downstream from the interchanger can cause excessive cycling of the thermostatic expansion valve because the flow of high-pressure liquid through the interchanger ceases when the thermostatic expansion valve closes; consequently, no heat is available from the high-pressure liquid, and the

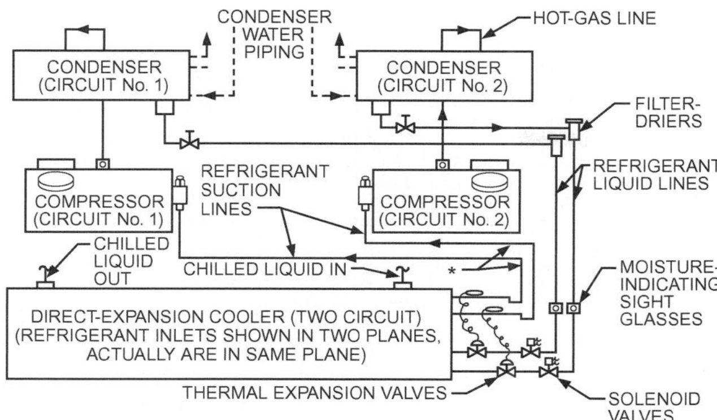

Fig. 20 Typical Refrigerant Piping in Liquid Chilling Package with Two Completely Separate Circuits

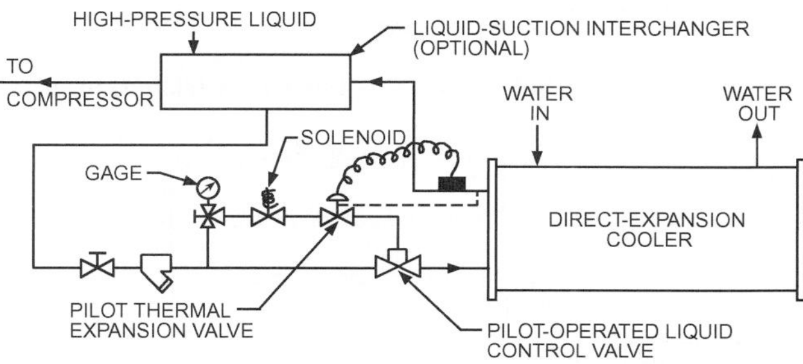

Fig. 21 Direct-Expansion Cooler with Pilot-Operated Control Valve

cooler must starve itself to obtain the superheat necessary to open the valve. When the valve does open, excessive superheat causes it to overfeed until the bulb senses liquid downstream from the interchanger. Therefore, the remote bulb should be positioned between the cooler and the interchanger.

Figure 20 shows a typical piping arrangement that has been successful in packaged water chillers having direct-expansion coolers. With this arrangement, automatic recycling pumpdown is needed on the lag compressor to prevent leakage through compressor valves, allowing migration to the cold evaporator circuit. It also prevents liquid from slugging the compressor at start-up.

On larger systems, the limited size of thermostatic expansion valves may require use of a pilot-operated liquid valve controlled by a small thermostatic expansion valve (Figure 21). The small thermostatic expansion valve pilots the main liquid control valve. The equalizing connection and bulb of the pilot thermostatic expansion valve should be treated as a direct-acting thermal expansion valve. A small solenoid valve in the pilot line shuts off the high side from the low during shutdown. However, the main liquid valve does not open and close instantaneously.

Direct-Expansion Air Coils

For further information on these coils, see Chapter 21 of the 2004 *ASHRAE Handbook—HVAC Systems and Equipment*. The most common ways of arranging direct-expansion coils are shown in Figures 22 and 23. The method shown in Figure 23 provides the superheat needed to operate the thermostatic expansion valve and is effective for heat transfer because leaving air contacts the coldest evaporator surface. This arrangement is advantageous on low-temperature applications, where the coil pressure drop represents an appreciable change in evaporating temperature.

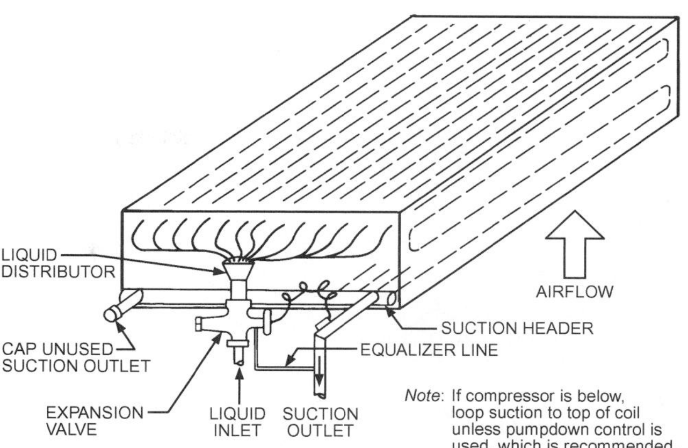

Fig. 22 Direct-Expansion Evaporator (Top-Feed, Free-Draining)

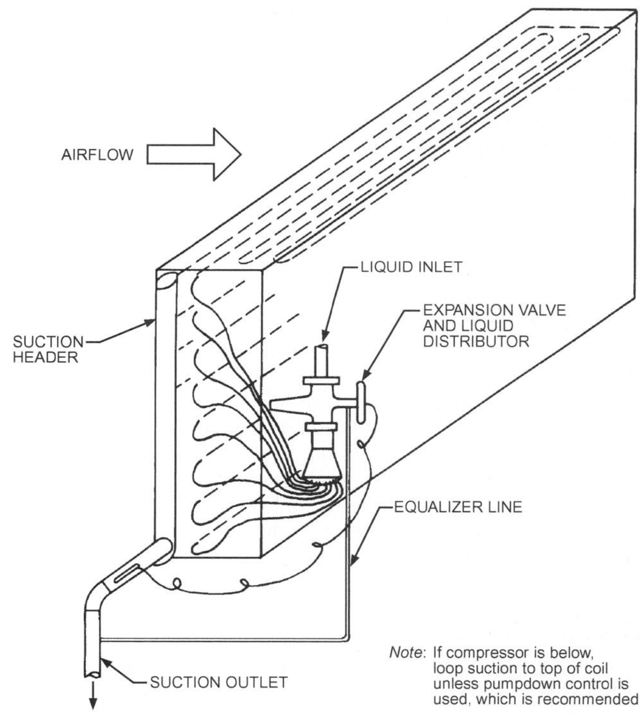

Fig. 23 Direct-Expansion Evaporator (Horizontal Airflow)

Direct-expansion air coils can be located in any position as long as proper refrigerant distribution and continuous oil removal facilities are provided.

Figure 22 shows top-feed, free-draining piping with a vertical up-airflow coil. In Figure 23, which illustrates a horizontal-airflow coil, suction is taken off the bottom header connection, providing free oil draining. Many coils are supplied with connections at each end of the suction header so that a free-draining connection can be used regardless of which side of the coil is up; the other end is then capped.

In Figure 24, a refrigerant upfeed coil is used with a vertical downflow air arrangement. Here, the coil design must provide sufficient gas velocity to entrain oil at lowest loadings and to carry it into the suction line.

Pumpdown compressor control is desirable on all systems using downfeed or upfeed evaporators, to protect the compressor against

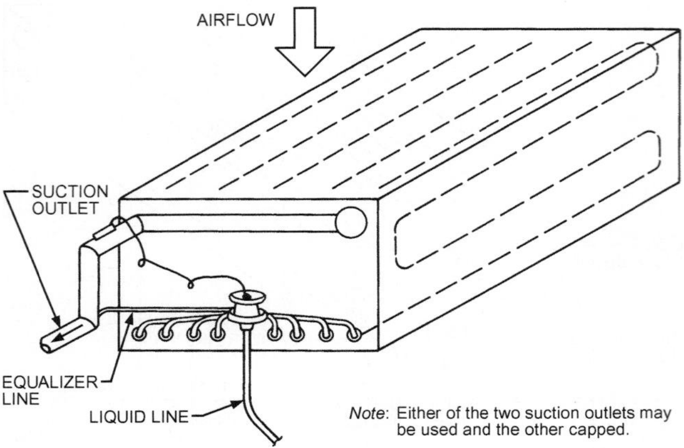

Fig. 24 Direct-Expansion Evaporator (Bottom-Feed)

a liquid slugback in cases where liquid can accumulate in the suction header and/or the coil on system off cycles. Pumpdown compressor control is described in the section on Keeping Liquid from Crankcase During Off Cycles.

Thermostatic expansion valve operation and application are described in Chapter 44. Thermostatic expansion valves should be sized carefully to avoid undersizing at full load and oversizing at partial load. The refrigerant pressure drops through the system (distributor, coil, condenser, and refrigerant lines, including liquid lifts) must be properly evaluated to determine the correct pressure drop available across the valve on which to base the selection. Variations in condensing pressure greatly affect the pressure available across the valve, and hence its capacity.

Oversized thermostatic expansion valves result in cycling that alternates flooding and starving the coil. This occurs because the valve attempts to throttle at a capacity below its capability, which causes periodic flooding of the liquid back to the compressor and wide temperature variations in the air leaving the coil. Reduced compressor capacity further aggravates this problem. Systems having multiple coils can use solenoid valves located in the liquid line feeding each evaporator or group of evaporators to close them off individually as compressor capacity is reduced.

For information on defrosting, see Chapter 42.

Flooded Evaporators

Flooded evaporators may be desirable when a small temperature differential is required between the refrigerant and the medium being cooled. A small temperature differential is advantageous in low-temperature applications.

In a flooded evaporator, the coil is kept full of refrigerant when cooling is required. The refrigerant level is generally controlled through a high- or low-side float control. Figure 25 represents a typical arrangement showing a low-side float control, oil return line, and heat interchanger.

Circulation of refrigerant through the evaporator depends on gravity and a thermosiphon effect. A mixture of liquid refrigerant and vapor returns to the surge tank, and the vapor flows into the suction line. A baffle installed in the surge tank helps prevent foam and liquid from entering the suction line. A liquid refrigerant circulating pump (Figure 26) provides a more positive way of obtaining a high circulation rate.

Taking the suction line off the top of the surge tank causes difficulties if no special provisions are made for oil return. For this reason, the oil return lines in Figure 25 should be installed. These lines are connected near the bottom of the float chamber and also just below the liquid level in the surge tank (where an oil-rich liquid refrigerant exists). They extend to a lower point on the suction line

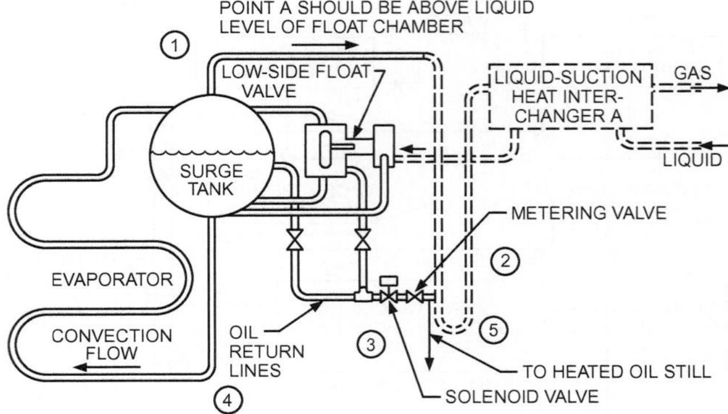

Notes: 1. Looped suction line is not needed when oil bleed goes to still. Heat exchanger is optional.
2. Trap allows gravity return of oil-rich liquid refrigerant to suction line ahead of interchanger.
3. Solenoid valve normally closed except when compressor is operating and heater is off.
4. Do not use pilot-operated solenoid valve.
5. To heated oil still is a better arrangement.

Fig. 25 Flooded Evaporator (Gravity Circulation)

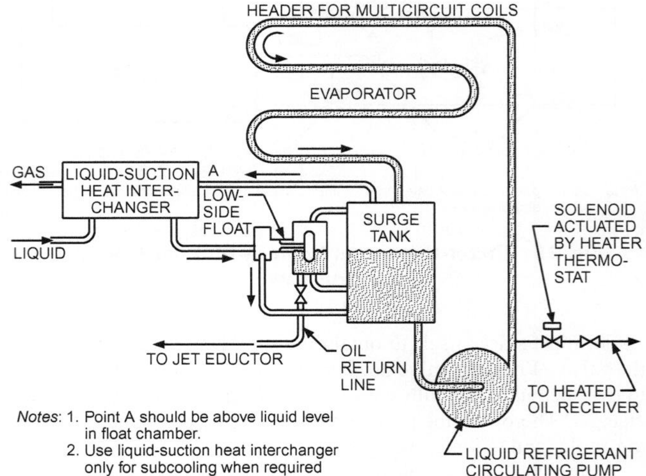

Notes: 1. Point A should be above liquid level in float chamber.
2. Use liquid-suction heat interchanger only for subcooling when required

Fig. 26 Flooded Evaporator (Forced Circulation)

to allow gravity flow. Included in this oil return line is (1) a solenoid valve that is open only while the compressor is running and (2) a metering valve that is adjusted to allow a constant but small-volume return to the suction line. A liquid-line sight glass may be installed downstream from the metering valve to serve as a convenient check on liquid being returned.

Oil can be returned satisfactorily by taking a bleed of refrigerant and oil from the pump discharge (Figure 26) and feeding it to the heated oil receiver. If a low-side float is used, a jet ejector can be used to remove oil from the quiescent float chamber.

REFRIGERATION ACCESSORIES

Liquid-Suction Heat Exchangers

Generally, liquid-suction heat exchangers subcool liquid refrigerant and superheat suction gas. They are used for one or more of the following functions:

- *Increasing efficiency of the refrigeration cycle.* Efficiency of the thermodynamic cycle of certain halocarbon refrigerants can be increased when the suction gas is superheated by removing heat from the liquid. This increased efficiency must be evaluated against the effect of pressure drop through the suction side of the exchanger, which forces the compressor to operate at a lower suction pressure. Liquid-suction heat exchangers are most beneficial at low suction temperatures. The increase in cycle efficiency for systems operating in the air-conditioning range (down to about 30°F evaporating temperature) usually does not justify their use. The heat exchanger can be located wherever convenient.

- *Subcooling liquid refrigerant to prevent flash gas at the expansion valve.* The heat exchanger should be located near the condenser or receiver to achieve subcooling before pressure drop occurs.
- *Evaporating small amounts of expected liquid refrigerant returning from evaporators in certain applications.* Many heat pumps incorporating reversals of the refrigerant cycle include a suction-line accumulator and liquid-suction heat exchanger arrangement to trap liquid floodbacks and vaporize them slowly between cycle reversals.

If an evaporator design makes a deliberate slight overfeed of refrigerant necessary, either to improve evaporator performance or to return oil out of the evaporator, a liquid-suction heat exchanger is needed to evaporate the refrigerant.

A flooded water cooler usually incorporates an oil-rich liquid bleed from the shell into the suction line for returning oil. The liquid-suction heat exchanger boils liquid refrigerant out of the mixture in the suction line. Exchangers used for this purpose should be placed in a horizontal run near the evaporator. Several types of liquid-suction heat exchangers are used.

Liquid and Suction Line Soldered Together. The simplest form of heat exchanger is obtained by strapping or soldering the suction and liquid lines together to obtain counterflow and then insulating the lines as a unit. To maximize capacity, the liquid line should always be on the bottom of the suction line, because liquid in a suction line runs along the bottom (Figure 27). This arrangement is limited by the amount of suction line available.

Shell-and-Coil or Shell-and-Tube Heat Exchangers (Figure 28**).** These units are usually installed so that the suction outlet drains the shell. When the units are used to evaporate liquid refrigerant returning in the suction line, the free-draining arrangement is not recommended. Liquid refrigerant can run along the bottom of the heat exchanger shell, having little contact with the warm liquid coil, and drain into the compressor. By installing the heat exchanger at a slight angle to the horizontal (Figure 29) with gas entering at the bottom and leaving at the top, any liquid returning in the line is trapped in the shell and held in contact with the warm liquid coil, where most of it is vaporized. An oil return line, with a metering valve and solenoid valve (open only when the compressor is running), is required to return oil that collects in the trapped shell.

Concentric Tube-in-Tube Heat Exchangers. The tube-in-tube heat exchanger is not as efficient as the shell-and-finned-coil type. It is, however, quite suitable for cleaning up small amounts of excessive liquid refrigerant returning in the suction line. Figure 30 shows typical construction with available pipe and fittings.

Plate Heat Exchangers. Plate heat exchangers provide high-efficiency heat transfer. They are very compact, have low pressure drop, and are lightweight devices. They are good for use as liquid subcoolers.

For air-conditioning applications, heat exchangers are recommended for liquid subcooling or for clearing up excess liquid in the suction line. For refrigeration applications, heat exchangers are recommended to increase cycle efficiency, as well as for liquid subcooling and removing small amounts of excess liquid in the suction line. Excessive superheating of the suction gas should be avoided.

Two-Stage Subcoolers

To take full advantage of the two-stage system, the refrigerant liquid should be cooled to near the interstage temperature to reduce the amount of flash gas handled by the low-stage compressor. The net result is a reduction in total system power requirements. The amount of gain from cooling to near interstage conditions varies among refrigerants.

Figure 31 illustrates an open or flash-type cooler. This is the simplest and least costly type, which has the advantage of cooling liquid to the saturation temperature of the interstage pressure. One disadvantage is that the pressure of cooled liquid is reduced to interstage pressure, leaving less pressure available for liquid transport. Although the liquid temperature is reduced, the pressure drops correspondingly, and the expansion device controlling flow to the cooler must be large enough to pass all the liquid refrigerant flow. Failure of this valve could allow a large flow of liquid to the upper-stage compressor suction, which could seriously damage the compressor.

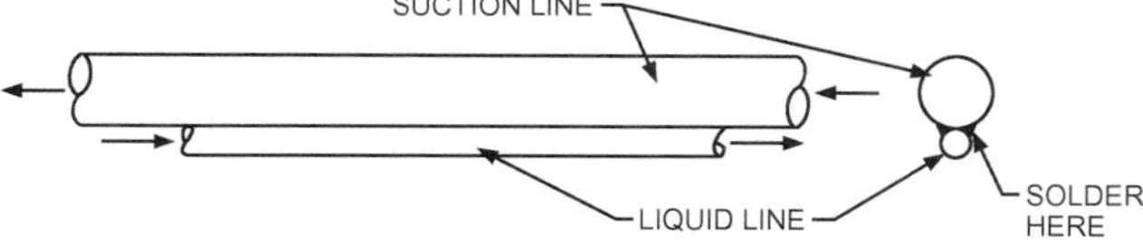

Fig. 27 Soldered Tube Heat Exchanger

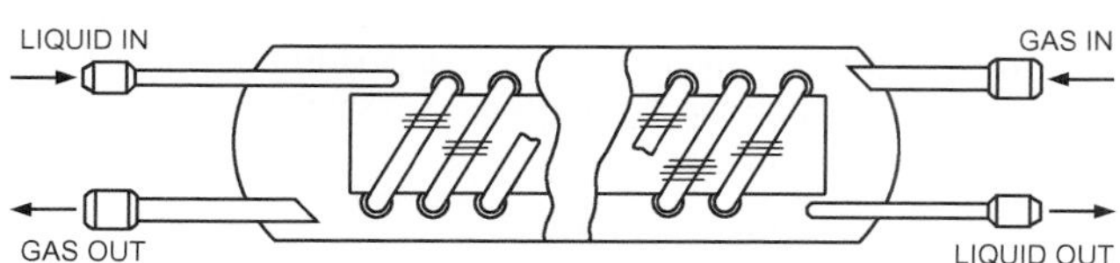

Fig. 28 Shell-and-Finned-Coil Heat Exchanger

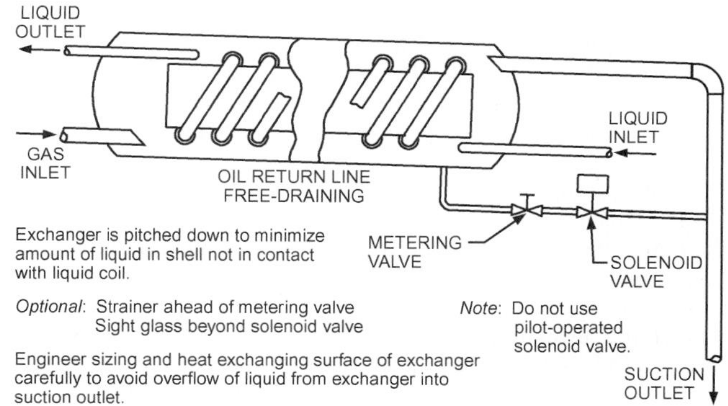

Fig. 29 Shell-and-Finned-Coil Exchanger Installed to Prevent Liquid Floodback

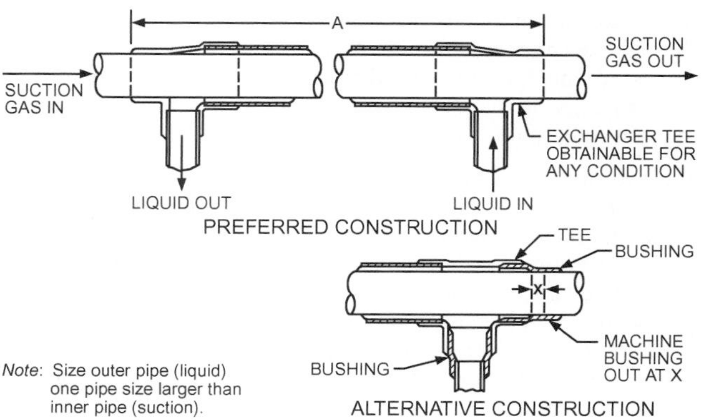

Fig. 30 Tube-in-Tube Heat Exchanger

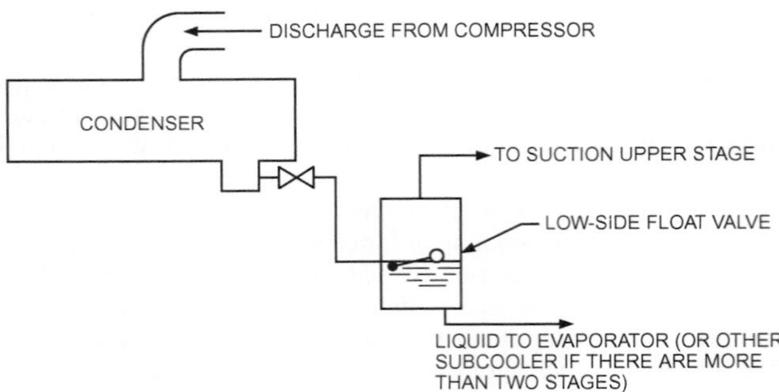

Fig. 31 Flash-Type Cooler

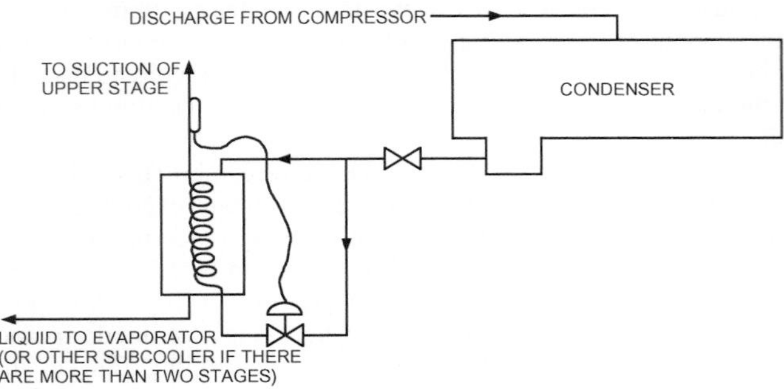

Fig. 32 Closed-Type Subcooler

Liquid from a flash cooler is saturated, and liquid from a cascade condenser usually has little subcooling. In both cases, the liquid temperature is usually lower than the temperature of the surroundings. Thus, it is important to avoid heat input and pressure losses that would cause flash gas to form in the liquid line to the expansion device or to recirculating pumps. Cold liquid lines should be insulated, because expansion devices are usually designed to feed liquid, not vapor.

Figure 32 shows the closed or heat exchanger type of subcooler. It should have sufficient heat transfer surface to transfer heat from the liquid to the evaporating refrigerant with a small final temperature difference. Pressure drop should be small, so that full pressure is available for feeding liquid to the expansion device at the low-temperature evaporator. The subcooler liquid control valve should be sized to supply only the quantity of refrigerant required for the subcooling. This prevents a tremendous quantity of liquid from flowing to the upper-stage suction in the event of a valve failure.

Discharge Line Oil Separators

Oil is always in circulation in systems using halocarbon refrigerants. Refrigerant piping is designed to ensure that this oil passes through the entire system and returns to the compressor as fast as it leaves. Although well-designed piping systems can handle the oil in most cases, a discharge-line oil separator can have certain advantages in some applications (see Chapter 44), such as

- In systems where it is impossible to prevent substantial absorption of refrigerant in the crankcase oil during shutdown periods. When the compressor starts up with a violent foaming action, oil is thrown out at an accelerated rate, and the separator immediately returns a large portion of this oil to the crankcase. Normally, the system should be designed with pumpdown control or crankcase heaters to minimize liquid absorption in the crankcase.
- In systems using flooded evaporators, where refrigerant bleedoff is necessary to remove oil from the evaporator. Oil separators reduce the amount of bleedoff from the flooded cooler needed for operation.
- In direct-expansion systems using coils or tube bundles that require bottom feed for good liquid distribution and where refrigerant carryover from the top of the evaporator is essential for proper oil removal.
- In low-temperature systems, where it is advantageous to have as little oil as possible going through the low side.
- In screw-type compressor systems, where an oil separator is necessary for proper operation. The oil separator is usually supplied with the compressor unit assembly directly from the compressor manufacturer.
- In multiple compressors operating in parallel. The oil separator can be an integral part of the total system oil management system.

In applying oil separators in refrigeration systems, the following potential hazards must be considered:

- Oil separators are not 100% efficient, and they do not eliminate the need to design the complete system for oil return to the compressor.
- Oil separators tend to condense out liquid refrigerant during compressor off cycles and on compressor start-up. This is true if the condenser is in a warm location, such as on a roof. During the off cycle, the oil separator cools down and acts as a condenser for refrigerant that evaporates in warmer parts of the system. A cool oil separator may condense discharge gas and, on compressor start-up, automatically drain it into the compressor crankcase. To minimize this possibility, the drain connection from the oil separator can be connected into the suction line. This line should be equipped with a shutoff valve, a fine filter, hand throttling and solenoid valves, and a sight glass. The throttling valve should be adjusted so that flow through this line is only a little greater than would normally be expected to return oil through the suction line.
- The float valve is a mechanical device that may stick open or closed. If it sticks open, hot gas will be continuously bypassed to the compressor crankcase. If the valve sticks closed, no oil is returned to the compressor. To minimize this problem, the separator can be supplied without an internal float valve. A separate external float trap can then be located in the oil drain line from the separator preceded by a filter. Shutoff valves should isolate the filter and trap. The filter and traps are also easy to service without stopping the system.

The discharge line pipe size into and out of the oil separator should be the full size determined for the discharge line. For separators that have internal oil float mechanisms, allow enough room to remove the oil float assembly for servicing.

Depending on system design, the oil return line from the separator may feed to one of the following locations:

- Directly to the compressor crankcase
- Directly into the suction line ahead of the compressor
- Into an oil reservoir or device used to collect oil, used for a specifically designed oil management system

When a solenoid valve is used in the oil return line, the valve should be wired so that it is open when the compressor is running. To minimize entrance of condensed refrigerant from the low side, a thermostat may be installed and wired to control the solenoid in the oil return line from the separator. The thermostat sensing element should be located on the oil separator shell below the oil level and set high enough so that the solenoid valve will not open until the separator temperature is higher than the condensing temperature. A superheat-controlled expansion valve can perform the same function. If a discharge line check valve is used, it should be downstream of the oil separator.

Surge Drums or Accumulators

A surge drum is required on the suction side of almost all flooded evaporators to prevent liquid slopover to the compressor. Exceptions include shell-and-tube coolers and similar shell-type evaporators, which provide ample surge space above the liquid level or contain eliminators to separate gas and liquid. A horizontal surge drum is sometimes used where headroom is limited.

The drum can be designed with baffles or eliminators to separate liquid from the suction gas. More often, sufficient separation space is allowed above the liquid level for this purpose. Usually, the design is vertical, with a separation height above the liquid level of 24 to 30 in. and with the shell diameter sized to keep suction gas velocity low enough to allow liquid droplets to separate. Because these vessels are also oil traps, it is necessary to provide oil bleed.

Although separators may be fabricated with length-to-diameter (*L*/*D*) ratios of 1/1 up to 10/1, the lowest-cost separators are usually for *L*/*D* ratios between 3/1 and 5/1.

Compressor Floodback Protection

Certain systems periodically flood the compressor with excessive amounts of liquid refrigerant. When periodic floodback through the suction line cannot be controlled, the compressor must be protected against it.

The most satisfactory method appears to be a trap arrangement that catches liquid floodback and (1) meters it slowly into the suction line, where the floodback is cleared up with a liquid-suction heat interchanger; (2) evaporates the liquid 100% in the trap itself by using a liquid coil or electric heater, and then automatically returns oil to the suction line; or (3) returns it to the receiver or to one of the evaporators. Figure 29 illustrates an arrangement that handles moderate liquid floodback, disposing of liquid by a combination of boiling off in the exchanger and limited bleedoff into the suction line. This device, however, does not have sufficient trapping volume for most heat pump applications or hot-gas defrost systems using reversal of the refrigerant cycle.

For heavier floodback, a larger volume is required in the trap. The arrangement shown in Figure 33 has been applied successfully in reverse-cycle heat pump applications using halocarbon refrigerants. It consists of a suction-line accumulator with enough volume to hold the maximum expected floodback and a large enough diameter to separate liquid from suction gas. Trapped liquid is slowly bled off through a properly sized and controlled drain line into the suction line, where it is boiled off in a liquid-suction heat exchanger between cycle reversals.

With the alternative arrangement shown, the liquid/oil mixture is heated to evaporate the refrigerant, and the remaining oil is drained into the crankcase or suction line.

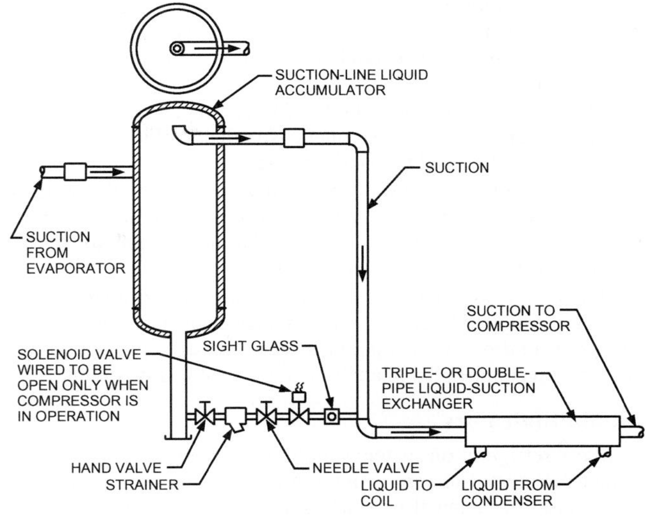

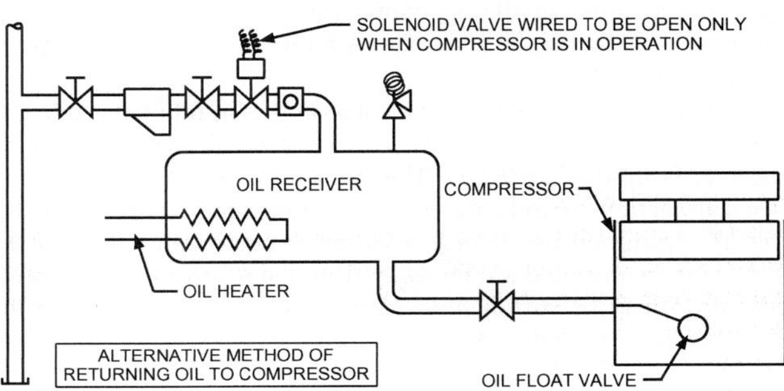

Fig. 33 Compressor Floodback Protection Using Accumulator with Controlled Bleed

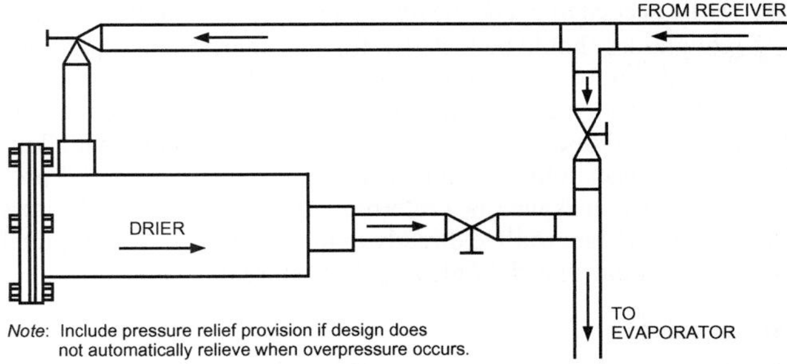

Note: Include pressure relief provision if design does not automatically relieve when overpressure occurs.

Fig. 34 Drier with Piping Connections

Refrigerant Driers and Moisture Indicators

The effect of moisture in refrigeration systems is discussed in Chapters 5 and 6. Using a permanent refrigerant drier is recommended on all systems and with all refrigerants. It is especially important on low-temperature systems to prevent ice from forming at expansion devices. A **full-flow drier** is always recommended in hermetic compressor systems to keep the system dry and prevent decomposition products from getting into the evaporator in the event of a motor burnout.

Replaceable-element filter-driers are preferred for large systems because the drying element can be replaced without breaking any refrigerant connections. The drier is usually located in the liquid line near the liquid receiver. It may be mounted horizontally or vertically with the flange at the bottom, but it should never be mounted vertically with the flange on top because any loose material would then fall into the line when the drying element was removed.

A three-valve bypass is usually used, as shown in Figure 34, to provide a way to isolate the drier for servicing. The refrigerant charging connection should be located between the receiver outlet valve and liquid-line drier so that all refrigerant added to the system passes through the drier.

Reliable moisture indicators can be installed in refrigerant liquid lines to provide a positive indication of when the drier cartridge should be replaced.

Strainers

Strainers should be used in both liquid and suction lines to protect automatic valves and the compressor from foreign material, such as pipe welding scale, rust, and metal chips. The strainer should be mounted in a horizontal line, oriented so that the screen can be replaced without loose particles falling into the system.

A liquid-line strainer should be installed before each automatic valve to prevent particles from lodging on the valve seats. Where multiple expansion valves with internal strainers are used at one location, a single main liquid-line strainer will protect all of these. The liquid-line strainer can be located anywhere in the line between the condenser (or receiver) and the automatic valves, preferably near the valves for maximum protection. Strainers should trap the particle size that could affect valve operation. With pilot-operated valves, a very fine strainer should be installed in the pilot line ahead of the valve.

Filter-driers dry the refrigerant and filter out particles far smaller than those trapped by mesh strainers. No other strainer is needed in the liquid line if a good filter-drier is used.

Refrigeration compressors are usually equipped with a built-in suction strainer, which is adequate for the usual system with copper piping. The suction line should be piped at the compressor so that the built-in strainer is accessible for servicing.

Both liquid- and suction-line strainers should be adequately sized to ensure sufficient foreign material storage capacity without excessive pressure drop. In steel piping systems, an external suction-line strainer is recommended in addition to the compressor strainer.

Liquid Indicators

Every refrigeration system should have a way to check for sufficient refrigerant charge. Common devices used are liquid-line sight glass, mechanical or electronic indicators, and an external gage glass with equalizing connections and shutoff valves. A properly installed sight glass shows bubbling when the charge is insufficient.

Liquid indicators should be located in the liquid line as close as possible to the receiver outlet, or to the condenser outlet if no receiver is used (Figure 35). The sight glass is best installed in a vertical section of line, far enough downstream from any valve that the resulting disturbance does not appear in the glass. If the sight glass is installed too far away from the receiver, the line pressure drop may be sufficient to cause flashing and bubbles in the glass, even if the charge is sufficient for a liquid seal at the receiver outlet.

When sight glasses are installed near the evaporator, often no amount of system overcharging will give a solid liquid condition at the sight glass because of pressure drop in the liquid line or lift. Subcooling is required here. An additional sight glass near the evaporator may be needed to check the refrigerant condition at that point.

Sight glasses should be installed full size in the main liquid line. In very large liquid lines, this may not be possible; the glass can then be installed in a bypass or saddle mount that is arranged so that any gas in the liquid line will tend to move to it. A sight glass with double ports (for back lighting) and seal caps, which provide added protection against leakage, is preferred. Moisture-liquid indicators large enough to be installed directly in the liquid line serve the dual purpose of liquid-line sight glass and moisture indicator.

Oil Receivers

Oil receivers serve as reservoirs for replenishing crankcase oil pumped by the compressors and provide the means to remove refrigerant dissolved in the oil. They are selected for systems having any of the following components:

- Flooded or semiflooded evaporators with large refrigerant charges
- Two or more compressors operated in parallel
- Long suction and discharge lines
- Double suction line risers

A typical hookup is shown in Figure 33. Outlets are arranged to prevent oil from draining below the heater level to avoid heater burnout and to prevent scale and dirt from being returned to the compressor.

Purge Units

Noncondensable gas separation using a purge unit is useful on most large refrigeration systems where suction pressure may fall below atmospheric pressure (see Figure 11 of Chapter 3).

HEAD PRESSURE CONTROL FOR REFRIGERANT CONDENSERS

For more information on head pressure control, see Chapter 35 of the 2004 *ASHRAE Handbook—HVAC Systems and Equipment.*

Water-Cooled Condensers

With water-cooled condensers, head pressure controls are used both to maintain condensing pressure and to conserve water. On cooling tower applications, they are used only where it is necessary to maintain condensing temperatures.

Condenser-Water-Regulating Valves

The shutoff pressure of the valve must be set slightly higher than the saturation pressure of the refrigerant at the highest ambient temperature expected when the system is not in operation. This ensures that the valve will not pass water during off cycles. These valves are usually sized to pass the design quantity of water at about a 25 to 30 psi difference between design condensing pressure and valve shutoff pressure. Chapter 44 has further information.

Water Bypass

In cooling tower applications, a simple bypass with a manual or automatic valve responsive to head pressure change can also be used to maintain condensing pressure. Figure 36 shows an automatic three-way valve arrangement. The valve divides water flow between the condenser and the bypass line to maintain the desired condensing pressure. This maintains a balanced flow of water on the tower and pump.

Evaporative Condensers

Among the methods used for condensing pressure control with evaporative condensers are (1) cycling the spray pump motor; (2) cycling both fan and spray pump motors; (3) throttling the spray water; (4) bypassing air around duct and dampers; (5) throttling air

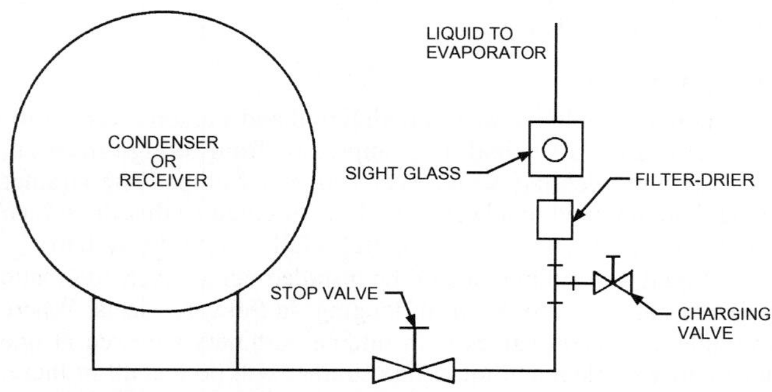

Fig. 35 Sight Glass and Charging Valve Locations

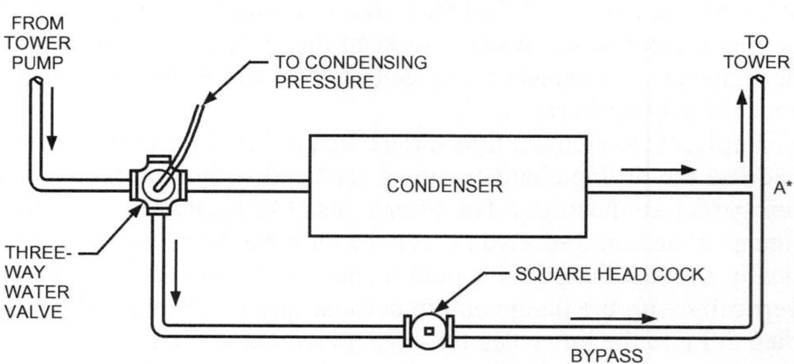

*A is alternative location for three-way valve, depending on valve design. Check manufacturer's recommendations.

Fig. 36 Head Pressure Control for Condensers Used with Cooling Towers (Water Bypass Modulation)

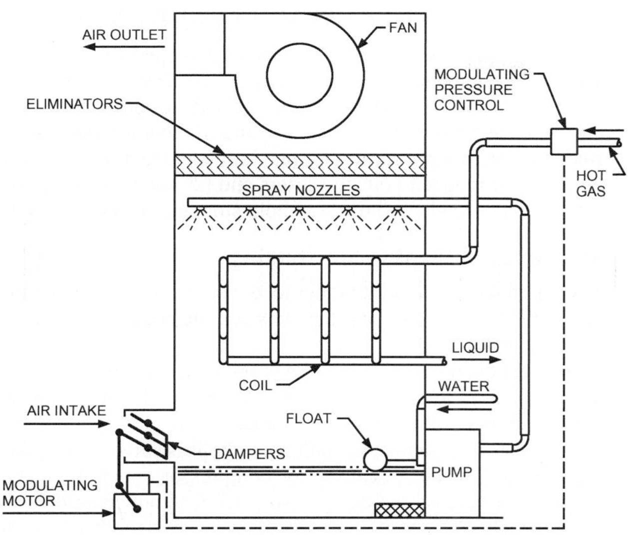

Fig. 37 Head Pressure Control for Evaporative Condenser (Air Intake Modulation)

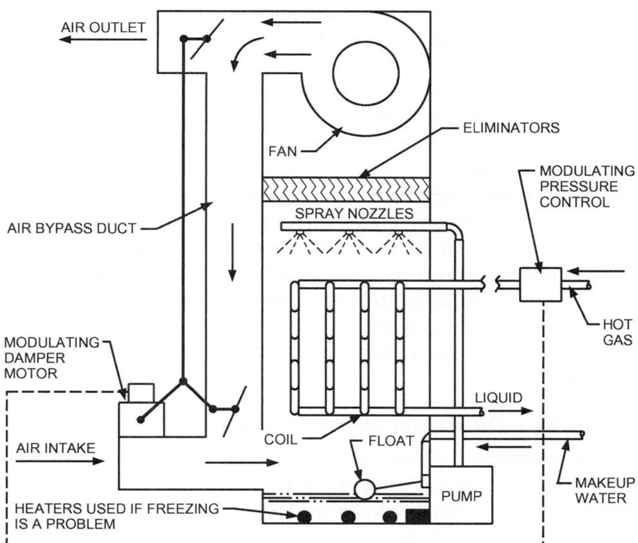

Fig. 38 Head Pressure Control for Evaporative Condenser (Air Bypass Modulation)

via dampers, on either inlet or discharge; and (6) combinations of these methods. For further information, see Chapter 35 of the 2004 *ASHRAE Handbook—HVAC Systems and Equipment.*

In water pump cycling, a pressure control at the gas inlet starts and stops the pump in response to head pressure changes. The pump sprays water over the condenser coils. As head pressure drops, the pump stops and the unit becomes an air-cooled condenser.

Constant pressure is difficult to maintain with coils of prime surface tubing because as soon as the pump stops, the pressure goes up and the pump starts again. This occurs because these coils have insufficient capacity when operating as an air-cooled condenser. The problem is not as acute with extended-surface coils. Short-cycling results in excessive deposits of mineral and scale on the tubes, decreasing the life of the water pump.

One method of controlling head pressure is using cycle fans and pumps. This minimizes water-side scaling. In colder climates, an indoor water sump with a remote spray pump(s) is required. The fan cycling sequence is as follows:

Upon dropping head pressure

- Stop fans.
- If pressure continues to fall, stop pumps.

Upon rising head pressure

- Start fans.
- If pressure continues to rise, start pumps.

Damper control (Figure 37) may be incorporated in systems requiring more constant head pressures (e.g., some systems using thermostatic expansion valves). One drawback of dampers is formation of ice on dampers and linkages.

Figure 38 incorporates an air bypass arrangement for controlling head pressure. A modulating motor, acting in response to a modulating pressure control, positions dampers so that the mixture of recirculated and cold inlet air maintains the desired pressure. In extremely cold weather, most of the air is recirculated.

Air-Cooled Condensers

Methods for condensing pressure control with air-cooled condensers include (1) cycling fan motor, (2) air throttling or bypassing, (3) coil flooding, and (4) fan motor speed control. The first two methods are described in the section on Evaporative Condensers.

The third method holds condensing pressure up by backing liquid refrigerant up in the coil to cut down on effective condensing surface. When head pressure drops below the setting of the modulating control valve, it opens, allowing discharge gas to enter the liquid drain line. This restricts liquid refrigerant drainage and causes the condenser to flood enough to maintain the condenser and receiver pressure at the control valve setting. A pressure difference must be available across the valve to open it. Although the condenser would impose sufficient pressure drop at full load, pressure drop may practically disappear at partial loading. Therefore, a positive restriction must be placed parallel with the condenser and the control valve. Systems using this type of control require extra refrigerant charge.

In multiple-fan air-cooled condensers, it is common to cycle fans off down to one fan and then to apply air throttling to that section or modulate the fan motor speed. Consult the manufacturer before using this method, because not all condensers are properly circuited for it.

Using ambient temperature change (rather than condensing pressure) to modulate air-cooled condenser capacity prevents rapid cycling of condenser capacity. A disadvantage of this method is that the condensing pressure is not closely controlled.

KEEPING LIQUID FROM CRANKCASE DURING OFF CYCLES

Control of reciprocating compressors should prevent excessive accumulation of liquid refrigerant in the crankcase during off cycles. Any one of the following control methods accomplishes this.

Automatic Pumpdown Control (Direct-Expansion Air-Cooling Systems)

The most effective way to keep liquid out of the crankcase during system shutdown is to operate the compressor on automatic pumpdown control. The recommended arrangement involves the following devices and provisions:

- A liquid-line solenoid valve in the main liquid line or in the branch to each evaporator.
- Compressor operation through a low-pressure cutout providing for pumpdown whenever this device closes, regardless of whether the balance of the system is operating.

- Electrical interlock of the liquid solenoid valve with the evaporator fan, so refrigerant flow stops when the fan is out of operation.
- Electrical interlock of refrigerant solenoid valve with safety devices (e.g., high-pressure cutout, oil safety switch, and motor overloads), so that the refrigerant solenoid valve closes when the compressor stops.
- Low-pressure control settings such that the cut-in point corresponds to a saturated refrigerant temperature lower than any expected compressor ambient air temperature. If the cut-in setting is any higher, liquid refrigerant can accumulate and condense in the crankcase at a pressure corresponding to the ambient temperature. Then, the crankcase pressure would not rise high enough to reach the cut-in point, and effective automatic pumpdown would not be obtained.

Crankcase Oil Heater (Direct-Expansion Systems)

A crankcase oil heater with or without single (nonrecycling) pumpout at the end of each operating cycle does not keep liquid refrigerant out of the crankcase as effectively as automatic pumpdown control, but many compressors equalize too quickly after stopping automatic pumpdown control. Crankcase oil heaters maintain the crankcase oil at a temperature higher than that of other parts of the system, minimizing the absorption of the refrigerant by the oil.

Operation with the single pumpout arrangement is as follows. Whenever the temperature control device opens the circuit, or the manual control switch is opened for shutdown purposes, the crankcase heater is energized, and the compressor keeps running until it cuts off on the low-pressure switch. Because the crankcase heater remains energized during the complete off cycle, it is important that a continuous live circuit be available to the heater during the off time. The compressor cannot start again until the temperature control device or manual control switch closes, regardless of the position of the low-pressure switch.

This control method requires

- A liquid-line solenoid valve in the main liquid line or in the branch to each evaporator
- Use of a relay or the maintained contact of the compressor motor auxiliary switch to obtain a single pumpout operation before stopping the compressor
- A relay or auxiliary starter contact to energize the crankcase heater during the compressor off cycle and deenergize it during the compressor on cycle
- Electrical interlock of the refrigerant solenoid valve with the evaporator fan, so that refrigerant flow is stopped when the fan is out of operation
- Electrical interlock of refrigerant solenoid valve with safety devices (e.g., high-pressure cutout, oil safety switch, and motor overloads), so that the refrigerant flow valve closes when the compressor stops

Control for Direct-Expansion Water Chillers

Automatic pumpdown control is undesirable for direct-expansion water chillers because freezing is possible if excessive cycling occurs. A crankcase heater is the best solution, with a solenoid valve in the liquid line that closes when the compressor stops.

Effect of Short Operating Cycle

With reciprocating compressors, oil leaves the crankcase at an accelerated rate immediately after starting. Therefore, each start should be followed by a long enough operating period to allow the oil level to recover. Controllers used for compressors should not produce short-cycling of the compressor. Refer to the compressor manufacturer's literature for guidelines on maximum or minimum cycles for a specified period.

HOT-GAS BYPASS ARRANGEMENTS

Most large reciprocating compressors are equipped with unloaders that allow the compressor to start with most of its cylinders unloaded. However, it may be necessary to further unload the compressor to (1) reduce starting torque requirements so that the compressor can be started both with low-starting-torque prime movers and on low-current taps of reduced voltage starters and (2) allow capacity control down to 0% load conditions without stopping the compressor.

Full (100%) Unloading for Starting

Starting the compressor without load can be done with a manual or automatic valve in a bypass line between the hot-gas and suction lines at the compressor.

To prevent overheating, this valve is open only during the starting period and closed after the compressor is up to full speed and full voltage is applied to the motor terminals.

In the control sequence, the unloading bypass valve is energized on demand of the control calling for compressor operation, equalizing pressures across the compressor. After an adequate delay, a timing relay closes a pair of normally open contacts to start the compressor. After a further time delay, a pair of normally closed timing relay contacts opens, deenergizing the bypass valve.

Full (100%) Unloading for Capacity Control

Where full unloading is required for capacity control, hot-gas bypass arrangements can be used in ways that will not overheat the compressor. In using these arrangements, hot gas should not be bypassed until after the last unloading step.

Hot-gas bypass should (1) give acceptable regulation throughout the range of loads, (2) not cause excessive superheating of the suction gas, (3) not cause any refrigerant overfeed to the compressor, and (4) maintain an oil return to the compressor.

Hot-gas bypass for capacity control is an artificial loading device that maintains a minimum evaporating pressure during continuous compressor operation, regardless of evaporator load. This is usually done by an automatic or manual pressure-reducing valve that establishes a constant pressure on the downstream side.

Four common methods of using hot-gas bypass are shown in Figure 39. Figure 39A illustrates the simplest type; it will dangerously overheat the compressor if used for protracted periods of time. Figure 39B shows the use of hot-gas bypass to the exit of the evaporator. The expansion valve bulb should be placed at least 5 ft downstream from the bypass point of entrance, and preferably further, to ensure good mixing.

In Figure 39D, the hot-gas bypass enters after the evaporator thermostatic expansion valve bulb. Another thermostatic expansion valve supplies liquid directly to the bypass line for desuperheating. It is always important to install the hot-gas bypass far enough back in the system to maintain sufficient gas velocities in suction risers and other components to ensure oil return at any evaporator loading.

Figure 39C shows the most satisfactory hot-gas bypass arrangement. Here, the bypass is connected into the low side between the expansion valve and entrance to the evaporator. If a distributor is used, gas enters between the expansion valve and distributor. Refrigerant distributors are commercially available with side inlet connections that can be used for hot-gas bypass duty to a certain extent. Pressure drop through the distributor tubes must be evaluated to determine how much gas can be bypassed. This arrangement provides good oil return.

Solenoid valves should be placed before the constant-pressure bypass valve and before the thermal expansion valve used for liquid injection desuperheating, so that these devices cannot function until they are required.

Control valves for hot gas should be close to the main discharge line because the line preceding the valve usually fills with liquid when closed.

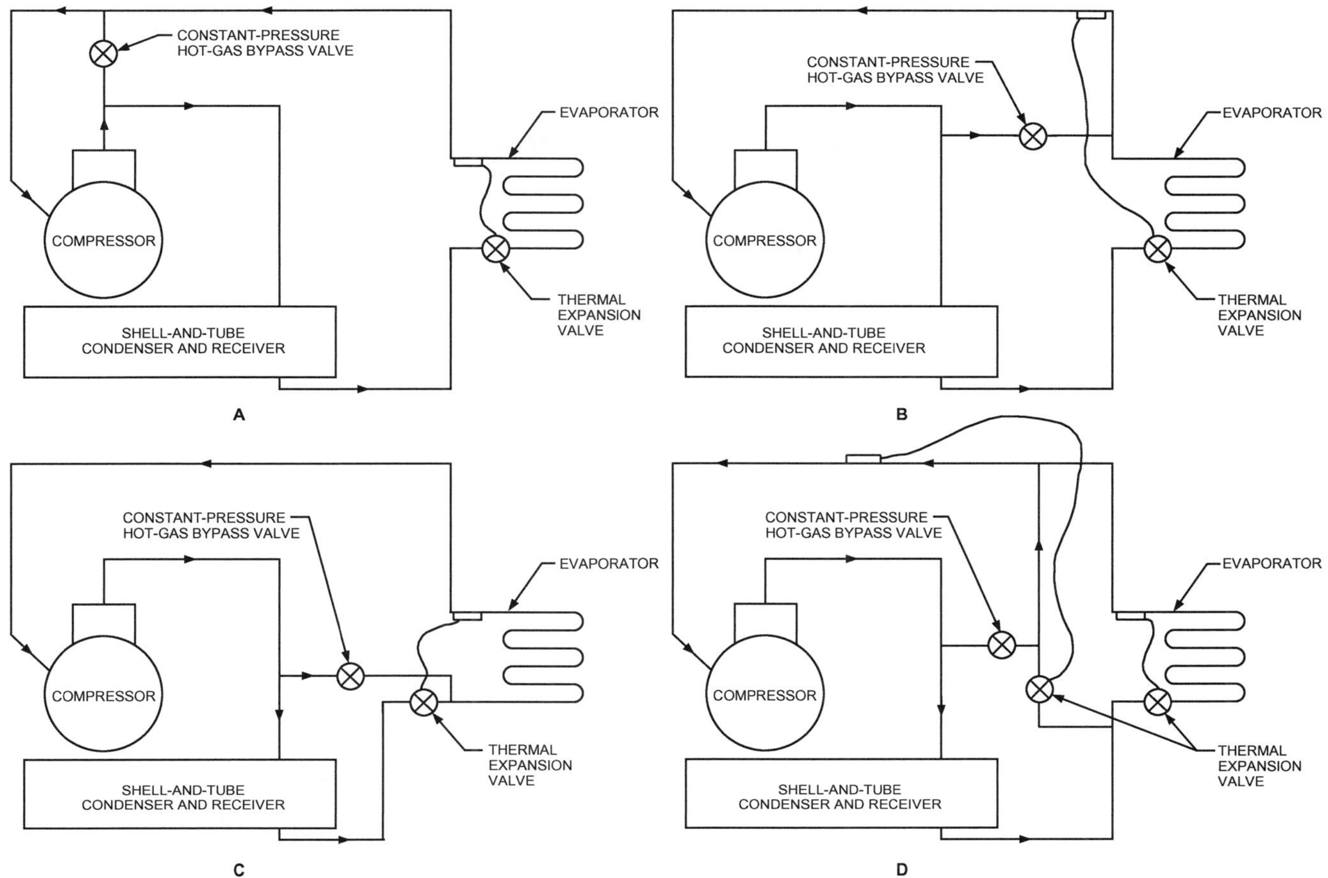

Fig. 39 Hot-Gas Bypass Arrangements

The hot-gas bypass line should be sized so that its pressure loss is only a small percentage of the pressure drop across the valve. Usually, it is the same size as the valve connections. When sizing the valve, consult a control valve manufacturer to determine the minimum compressor capacity that must be offset, refrigerant used, condensing pressure, and suction pressure.

When unloading (Figure 39C), head pressure control requirements increase considerably because the only heat delivered to the condenser is that caused by the motor power delivered to the compressor. Discharge pressure should be kept high enough that the hot-gas bypass valve can deliver gas at the required rate. The condenser head pressure control must be capable of meeting this condition.

Safety Requirements

ASHRAE *Standard* 15 and ASME *Standard* B31.5 should be used as guides for safe practice because they are the basis of most municipal and state codes. However, some ordinances require heavier piping and other features. The designer should know the specific requirements of the installation site. Only A106 Grade A or B or A53 Grade A or B should be considered for steel refrigerant piping.

The designer should know that the rated internal working pressure for Type L copper tubing decreases with (1) increasing metal operating temperature, (2) increasing tubing size (OD), and (3) increasing temperature of joining method. Hot methods used to join drawn pipe (e.g., brazing or welding) produce joints as strong as surrounding pipe, but reduce the strength of the heated pipe material to that of annealed material. Particular attention should be paid when specifying use of copper in conjunction with newer, high-pressure refrigerants (e.g., R-404a, R-507, R-410a, R-407c) because some of these refrigerants can achieve operating pressures as high as 500 psia and operating temperatures as high as 300°F at a typical saturated condensing condition of 130°F.

REFERENCES

Alofs, D.J., M.M. Hasan, and H.J. Sauer, Jr. 1990. Influence of oil on pressure drop in refrigerant compressor suction lines. *ASHRAE Transactions* 96:1.

ASHRAE. 2004. Safety standard for refrigeration systems. ANSI/ASHRAE *Standard* 15-2004.

ASME. 2001. Refrigeration piping and heat transfer components. ANSI/ASME *Standard* B31.5-2001. American Society of Mechanical Engineers, New York.

Atwood, T. 1990. Pipe sizing and pressure drop calculations for HFC-134a. *ASHRAE Journal* 32(4):62-66.

Colebrook, D.F. 1938, 1939. Turbulent flow in pipes. *Journal of the Institute of Engineers* 11.

Cooper, W.D. 1971. Influence of oil-refrigerant relationships on oil return. *ASHRAE Symposium Bulletin* PH71(2):6-10.

Jacobs, M.L., F.C. Scheideman, F.C. Kazem, and N.A. Macken. 1976. Oil transport by refrigerant vapor. *ASHRAE Transactions* 81(2):318-329.

Keating, E.L. and R.A. Matula. 1969. Correlation and prediction of viscosity and thermal conductivity of vapor refrigerants. *ASHRAE Transactions* 75(1).

Stoecker, W.F. 1984. Selecting the size of pipes carrying hot gas to defrosted evaporators. *International Journal of Refrigeration* 7(4):225-228.

Timm, M.L. 1991. An improved method for calculating refrigerant line pressure drops. *ASHRAE Transactions* 97(1):194-203.

Wile, D.D. 1977. *Refrigerant line sizing*. ASHRAE.

CHAPTER 3

SYSTEM PRACTICES FOR AMMONIA AND CARBON DIOXIDE REFRIGERANTS

AMMONIA

CUSTOM-ENGINEERED ammonia (R-717) refrigeration systems often have design conditions that span a wide range of evaporating and condensing temperatures. Examples are (1) a food freezing plant operating from +50 to –50°F; (2) a candy storage requiring 60°F db with precise humidity control; (3) a beef chill room at 28 to 30°F with high humidity; (4) a distribution warehouse requiring multiple temperatures for storing ice cream, frozen food, meat, and produce and for docks; and (5) a chemical process requiring multiple temperatures ranging from +60 to –60°F. Ammonia is the refrigerant of choice for many industrial refrigeration systems.

The figures in this chapter are for illustrative purposes only, and may not show all the required elements (e.g., valves). For safety and minimum design criteria for ammonia systems, refer to ASHRAE *Standard* 15, IIAR *Bulletin* 109, IIAR *Standard* 2, and applicable state and local codes.

See Chapter 13 for information on refrigeration load calculations.

Ammonia Refrigerant for HVAC Systems

Using ammonia for HVAC systems has received renewed interest, in part because of the scheduled phaseout and increasing costs of chlorofluorocarbon (CFC) and hydrochlorofluorocarbon (HCFC) refrigerants. Ammonia secondary systems that circulate chilled water or another secondary refrigerant are a viable alternative to halocarbon systems, although ammonia is inappropriate for direct refrigeration systems (ammonia in the air unit coils) for HVAC applications. Ammonia packaged chilling units are available for HVAC applications. As with the installation of any air-conditioning unit, all applicable codes, standards, and insurance requirements must be followed.

SYSTEM SELECTION

In selecting an engineered ammonia refrigeration system, several design decisions must be considered, including whether to use (1) single-stage compression, (2) economized compression, (3) multistage compression, (4) direct-expansion feed, (5) flooded feed, (6) liquid recirculation feed, and (7) secondary coolants.

Single-Stage Systems

The basic single-stage system consists of evaporator(s), a compressor, a condenser, a refrigerant receiver (if used), and a refrigerant control device (expansion valve, float, etc.). Chapter 1 of the 2005 *ASHRAE Handbook—Fundamentals* discusses the compression refrigeration cycle.

The preparation of this chapter is assigned to TC 10.3, Refrigerant Piping.

Economized Systems

Economized systems are frequently used with rotary screw compressors. Figure 1 shows an arrangement of the basic components. Subcooling the liquid refrigerant before it reaches the evaporator reduces its enthalpy, resulting in a higher net refrigerating effect. Economizing is beneficial because the vapor generated during subcooling is injected into the compressor partway through its compression cycle and must be compressed only from the economizer port pressure (which is higher than suction pressure) to the discharge pressure. This produces additional refrigerating capacity with less increase in unit energy input. Economizing is most beneficial at high pressure ratios. Under most conditions, economizing can provide operating efficiencies that approach that of two-stage systems, but with much less complexity and simpler maintenance.

Economized systems for variable loads should be selected carefully. At approximately 75% capacity, most screw compressors revert to single-stage performance as the slide valve moves such that the economizer port is open to the compressor suction area.

A flash economizer, which is somewhat more efficient, may often be used instead of the shell-and-coil economizer (Figure 1). However, ammonia liquid delivery pressure is reduced to economizer pressure.

Multistage Systems

Multistage systems compress gas from the evaporator to the condenser in several stages. They are used to produce temperatures of –15°F and below. This is not economical with single-stage compression.

Single-stage reciprocating compression systems are generally limited to between 5 and 10 psig suction pressure. With lubricant-injected economized rotary screw compressors, where the discharge

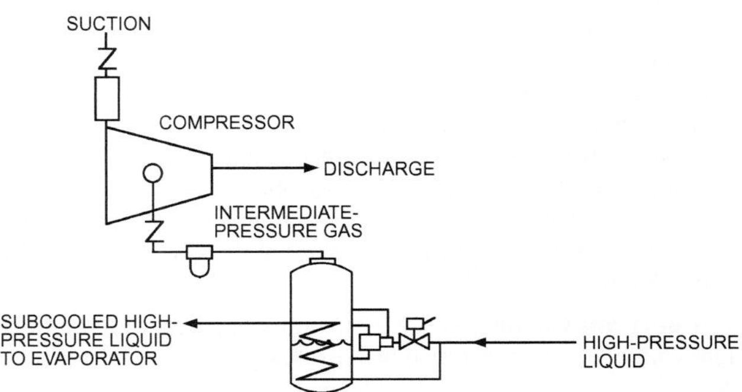

Fig. 1 Shell-and-Coil Economizer Arrangement

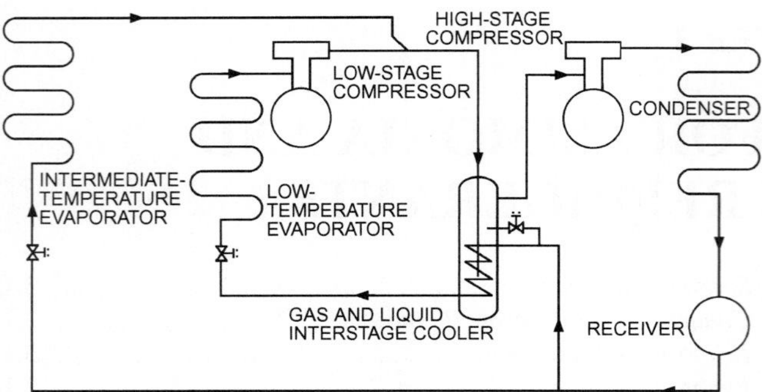

Fig. 2 Two-Stage System with High- and Low-Temperature Loads

temperatures are lower because of the lubricant cooling, the low-suction temperature limit is about –40°F, but efficiency is very low. Two-stage systems are used down to about –70 or –80°F evaporator temperatures. Below this temperature, three-stage systems should be considered.

Two-stage systems consist of one or more compressors that operate at low suction pressure and discharge at intermediate pressure and have one or more compressors that operate at intermediate pressure and discharge to the condenser (Figure 2).

Where either single- or two-stage compression systems can be used, two-stage systems require less power and have lower operating costs, but they can have a higher initial equipment cost.

EQUIPMENT

Compressors

Compressors available for single- and multistage applications include the following:

- Reciprocating
  - Single-stage (low-stage or high-stage)
  - Internally compounded
- Rotary vane
- Rotary screw (low-stage or high-stage, with or without economizing)

The reciprocating compressor is the most common compressor used in small, 100 hp or less, single-stage or multistage systems. The screw compressor is the predominant compressor above 100 hp, in both single- and multistage systems. Various combinations of compressors may be used in multistage systems. Rotary vane and screw compressors are frequently used for the low-pressure stage, where large volumes of gas must be moved. The high-pressure stage may be a reciprocating or screw compressor.

When selecting a compressor, consider the following:

- System size and capacity requirements.
- Location, such as indoor or outdoor installation at ground level or on the roof.
- Equipment noise.
- Part- or full-load operation.
- Winter and summer operation.
- Pulldown time required to reduce the temperature to desired conditions for either initial or normal operation. The temperature must be pulled down frequently for some applications for a process load, whereas a large cold-storage warehouse may require pulldown only once in its lifetime.

Lubricant Cooling. When a reciprocating compressor requires lubricant cooling, an external heat exchanger using a refrigerant or secondary cooling is usually added. Screw compressor lubricant cooling is covered in detail in the section on Screw Compressors.

Compressor Drives. The correct electric motor size(s) for a multistage system is determined by pulldown load. When the final low-stage operating level is –100°F, the pulldown load can be three times the operating load. Positive-displacement reciprocating compressor motors are usually selected for about 150% of operating power requirements for 100% load. The compressor's unloading mechanism can be used to prevent motor overload. Electric motors should not be overloaded, even when a service factor is indicated. For screw compressor applications, motors should be sized by adding 10% to the operating power. Screw compressors have built-in unloading mechanisms to prevent motor overload. The motor should not be oversized, because an oversized motor has a lower power factor and lower efficiency at design and reduced loads.

Steam turbines or gasoline, natural gas, propane, or diesel internal combustion engines are used when electricity is unavailable, or if the selected energy source is cheaper. Sometimes they are used in combination with electricity to reduce peak demands. The power output of a given engine size can vary as much as 15% depending on the fuel selected.

Steam turbine drives for refrigerant compressors are usually limited to very large installations where steam is already available at moderate to high pressure. In all cases, torsional analysis is required to determine what coupling must be used to dampen out any pulsations transmitted from the compressor. For optimum efficiency, a turbine should operate at a high speed that must be geared down for reciprocating and possibly screw compressors. Neither the gear reducer nor the turbine can tolerate a pulsating backlash from the driven end, so torsional analysis and special couplings are essential.

Advantages of turbines include variable speed for capacity control and low operating and maintenance costs. Disadvantages include higher initial costs and possible high noise levels. The turbine must be started manually to bring the turbine housing up to temperature slowly and to prevent excess condensate from entering the turbine.

The standard power rating of an engine is the absolute maximum, not the recommended power available for continuous use. Also, torque characteristics of internal combustion engines and electric motors differ greatly. The proper engine selection is at 75% of its maximum power rating. For longer life, the full-load speed should be at least 10% below maximum engine speed.

Internal combustion engines, in some cases, can reduce operating cost below that for electric motors. Disadvantages include (1) higher initial cost of the engine, (2) additional safety and starting controls, (3) higher noise levels, (4) larger space requirements, (5) air pollution, (6) requirement for heat dissipation, (7) higher maintenance costs, and (8) higher levels of vibration than with electric motors. A torsional analysis must be made to determine the proper coupling if engine drives are chosen.

Condensers

Condensers should be selected on the basis of total heat rejection at maximum load. Often, the heat rejected at the start of pulldown is several times the amount rejected at normal, low-temperature operating conditions. Some means, such as compressor unloading, can be used to limit the maximum amount of heat rejected during pulldown. If the condenser is not sized for pulldown conditions, and compressor capacity cannot be limited during this period, condensing pressure might increase enough to shut down the system.

Evaporators

Several types of evaporators are used in ammonia refrigeration systems. Fan-coil, direct-expansion evaporators can be used, but they are not generally recommended unless the suction temperature is 0°F or higher. This is due in part to the relative inefficiency of the direct-expansion coil, but more importantly, the low mass flow rate of ammonia is difficult to feed uniformly as a liquid to the coil. Instead, ammonia fan-coil units designed for recirculation (overfeed) systems are preferred. Typically, in this type of system, high-pressure ammonia from the system high stage flashes into a large vessel at the evaporator pressure, from which it is pumped to the evaporators at an overfeed rate of 2.5 to 1 to 4 to 1. This type of system is standard and very efficient. See Chapter 1 for more details.

Flooded shell-and-tube evaporators are often used in ammonia systems in which indirect or secondary cooling fluids such as water, brine, or glycol must be cooled.

Some problems that can become more acute at low temperatures include changes in lubricant transport properties, loss of capacity caused by static head from the depth of the pool of liquid refrigerant in the evaporator, deterioration of refrigerant boiling heat transfer coefficients caused by lubricant logging, and higher specific volumes for the vapor.

The effect of pressure losses in the evaporator and suction piping is more acute in low-temperature systems because of the large change in saturation temperatures and specific volume in relation to pressure changes at these conditions. Systems that operate near zero absolute pressure are particularly affected by pressure loss.

The depth of the pool of boiling refrigerant in a flooded evaporator exerts a liquid pressure on the lower part of the heat transfer surface. Therefore, the saturation temperature at this surface is higher than that in the suction line, which is not affected by the liquid pressure. This temperature gradient must be considered when designing the evaporator.

Spray shell-and-tube evaporators, though not commonly used, offer certain advantages. In this design, the evaporator's liquid depth penalty can be eliminated because the pool of liquid is below the heat transfer surface. A refrigerant pump sprays liquid over the surface. Pump energy is an additional heat load to the system, and more refrigerant must be used to provide the net positive suction head (NPSH) required by the pump. The pump is also an additional item that must be maintained. This evaporator design also reduces the refrigerant charge requirement compared to a flooded design (see Chapter 1).

Vessels

High-Pressure Receivers. Industrial systems generally incorporate a central high-pressure refrigerant receiver, which serves as the primary refrigerant storage location in the system. It handles refrigerant volume variations between the condenser and the system's low side during operation and pumpdowns for repairs or defrost. Ideally, the receiver should be large enough to hold the entire system charge, but this is not generally economical. The system should be analyzed to determine the optimum receiver size. Receivers are commonly equalized to the condenser inlet and operate at the same pressure as the condenser. In some systems, the receiver is operated at a pressure between the condensing pressure and the highest suction pressure to allow for variations in condensing pressure without affecting the system's feed pressure.

If additional receiver capacity is needed for normal operation, use extreme caution in the design. Designers usually remove the inadequate receiver and replace it with a larger one rather than install an additional receiver in parallel. This procedure is best because even slight differences in piping pressure or temperature can cause the refrigerant to migrate to one receiver and not to the other.

Smaller auxiliary receivers can be incorporated to serve as sources of high-pressure liquid for compressor injection or thermosiphon, lubricant cooling, high-temperature evaporators, and so forth.

Intercoolers (Gas and Liquid). An intercooler (subcooler/desuperheater) is the intermediate vessel between the high and low stages in a multistage system. One purpose is to cool discharge gas of the low-stage compressor to prevent overheating the high-stage compressor. This can be done by bubbling discharge gas from the low-stage compressor through a bath of liquid refrigerant or by mixing liquid normally entering the intermediate vessel with the discharge gas as it enters above the liquid level. Heat removed from the discharge gas is absorbed by the evaporation of part of the liquid and eventually passes through the high-stage compressor to the condenser. Disbursing the discharge gas below a level of liquid refrigerant separates out any lubricant carryover from the low-stage compressor. If liquid in the intercooler is to be used for other purposes, such as liquid makeup or feed to the low stage, periodic lubricant removal is important.

Another purpose of the intercooler is to lower the temperature of liquid used in the system low stage. Lowering refrigerant temperature increases the refrigeration effect and reduces the low-stage compressor's required displacement, thus reducing its operating cost.

Intercoolers for two-stage compression systems can be shell-and-coil or flash. Figure 3 depicts a shell-and-coil intercooler incorporating an internal pipe coil for subcooling high-pressure liquid before it is fed to the low stage of the system. Typically, the coil subcools liquid to within 10°F of the intermediate temperature.

Vertical **shell-and-coil intercoolers** with float valve feed perform well in many applications using ammonia refrigerant systems. The vessel must be sized properly to separate liquid from vapor that is returning to the high-stage compressor. The superheated gas inlet pipe should extend below the liquid level and have perforations or slots to distribute the gas evenly in small bubbles. Adding a perforated baffle across the area of the vessel slightly below the liquid level protects against violent surging. A float switch that shuts down the high-stage compressor when the liquid level gets too high should always be used with this type of intercooler in case the feed valve fails to control properly.

The **flash intercooler** is similar in design to the shell-and-coil intercooler, except for the coil. The high-pressure liquid is flash-cooled to the intermediate temperature. Use caution in selecting a flash intercooler because all the high-pressure liquid is flashed to intermediate pressure. Though colder than that of the shell-and-coil intercooler, liquid in the flash intercooler is not subcooled and is susceptible to flashing from system pressure drop. Two-phase liquid feed to control valves may cause premature failure because of the wire-drawing effect of the liquid/vapor mixture.

Figure 4 shows a vertical shell-and-coil intercooler as piped into the system. The liquid level is maintained in the intercooler by a float that controls the solenoid valve feeding liquid into the shell side of the intercooler. Gas from the first-stage compressor enters the lower section of the intercooler, is distributed by a perforated

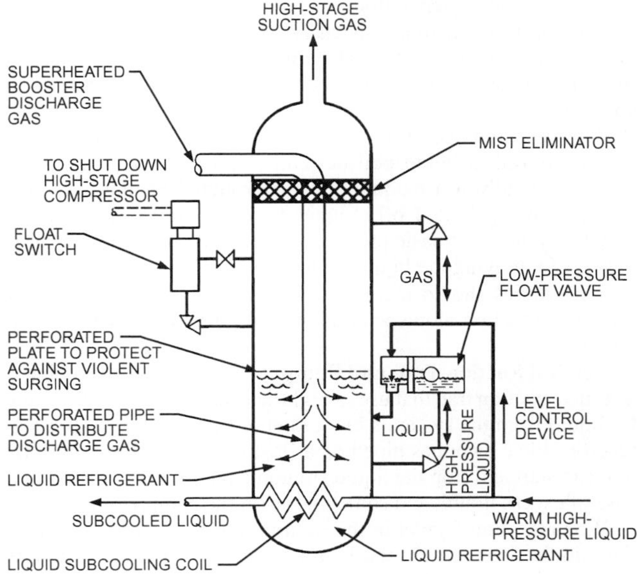

Fig. 3 Intercooler

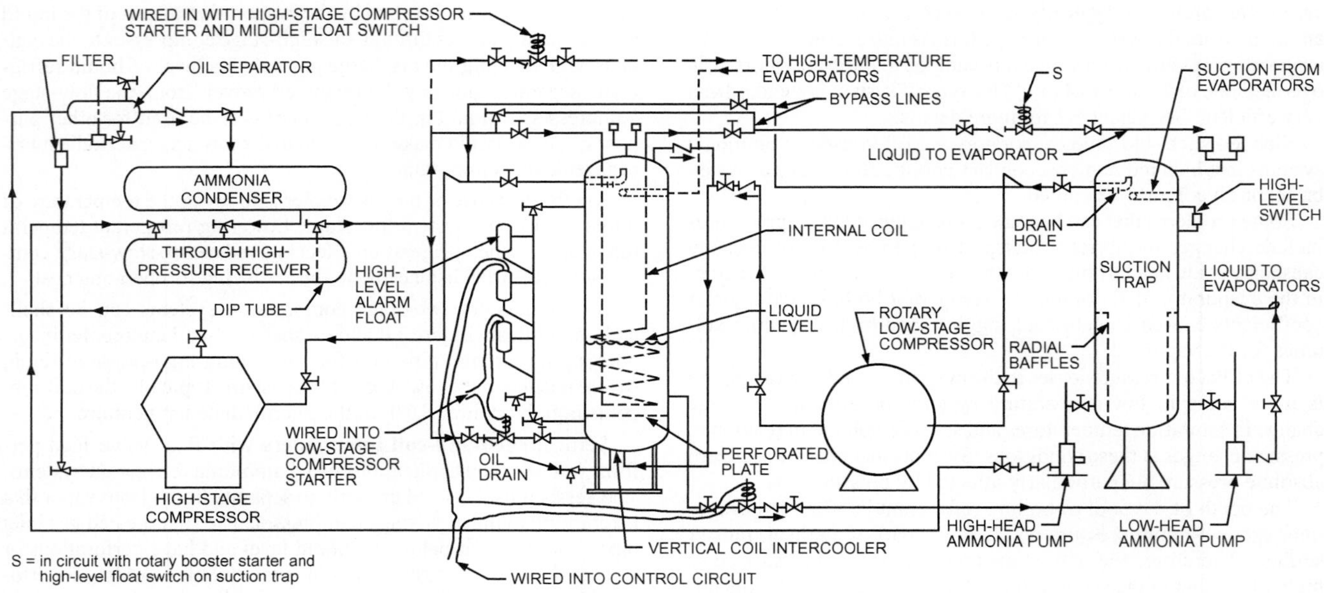

Fig. 4 Arrangement for Compound System with Vertical Intercooler and Suction Trap

plate, and is then cooled to the saturation temperature corresponding to intermediate pressure.

When sizing any intercooler, the designer must consider (1) low-stage compressor capacity; (2) vapor desuperheating, liquid make-up requirements for the subcooling coil load, or vapor cooling load associated with the flash intercooler; and (3) any high-stage side loading. The volume required for normal liquid levels, liquid surging from high-stage evaporators, feed valve malfunctions, and liquid/vapor must also be analyzed.

Necessary accessories are the liquid level control device and high-level float switch. Though not absolutely necessary, an auxiliary oil pot should also be considered.

Suction Accumulator. A suction accumulator (also known as a knockout drum, suction trap, pump receiver, recirculator, etc.) prevents liquid from entering the suction of the compressor, whether on the high stage or low stage of the system. Both vertical and horizontal vessels can be incorporated. Baffling and mist eliminator pads can enhance liquid separation.

Suction accumulators, especially those not intentionally maintaining a level of liquid, should have a means to remove any build-up of ammonia liquid. Gas boil-out coils or electric heating elements are costly and inefficient.

Although it is one of the more common and simplest means of liquid removal, a liquid boil-out coil (Figure 5) has some drawbacks. Generally, warm liquid flowing through the coil is the source of liquid being boiled off. Liquid transfer pumps, gas-powered transfer systems, or basic pressure differentials are a more positive means of removing the liquid (Figures 6 and 7).

Accessories should include a high-level float switch for compressor protection along with additional pump or transfer system controls.

Vertical Suction Trap and Pump. Figure 8 shows the piping of a vertical suction trap that uses a high-head ammonia pump to transfer liquid from the system's low-pressure side to the high-pressure receiver. Float switches piped on a float column on the side of the trap can start and stop the liquid ammonia pump, sound an alarm in case of excess liquid, and sometimes stop the compressors.

When the liquid level in the suction trap reaches the setting of the middle float switch, the liquid ammonia pump starts and reduces the liquid level to the setting of the lower float switch, which stops the liquid ammonia pump. A check valve in the discharge line

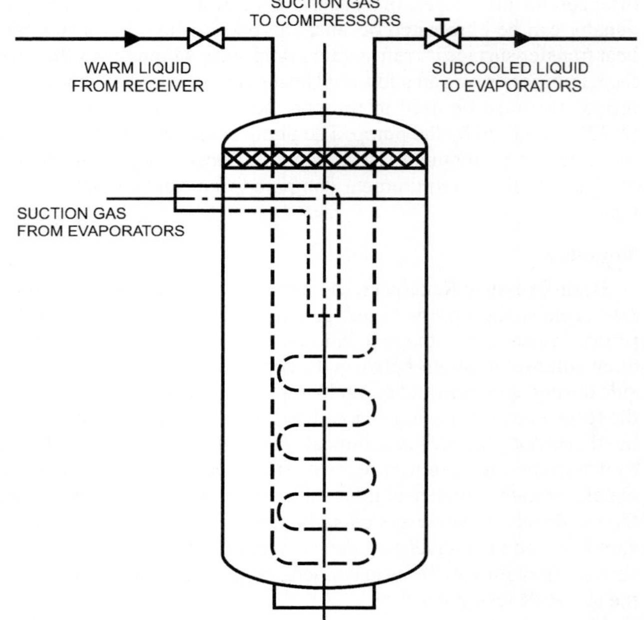

Fig. 5 Suction Accumulator with Warm Liquid Coil

of the ammonia pump prevents gas and liquid from flowing backward through the pump when it is not in operation. Depending on the type of check valve used, some installations have two valves in a series as an extra precaution against pump backspin.

Compressor controls adequately designed for starting, stopping, and capacity reduction result in minimal agitation, which helps separate vapor and liquid in the suction trap. Increasing compressor capacity slowly and in small increments reduces liquid boiling in the trap, which is caused by the refrigeration load of cooling the refrigerant and metal mass of the trap. If another compressor is started when plant suction pressure increases, it should be brought on line slowly to prevent a sudden pressure change in the suction trap.

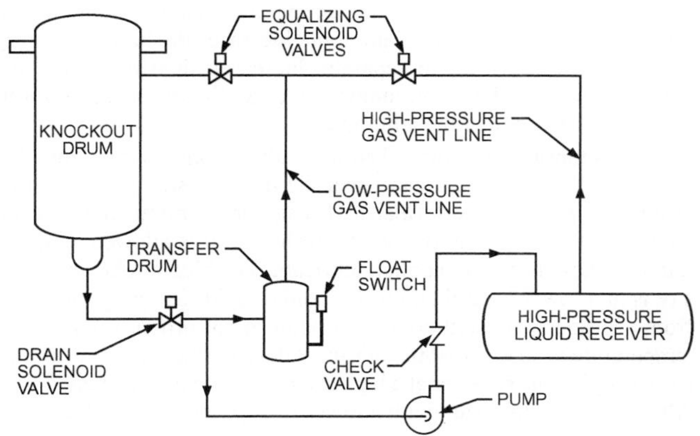

Fig. 6 Equalized Pressure Pump Transfer System

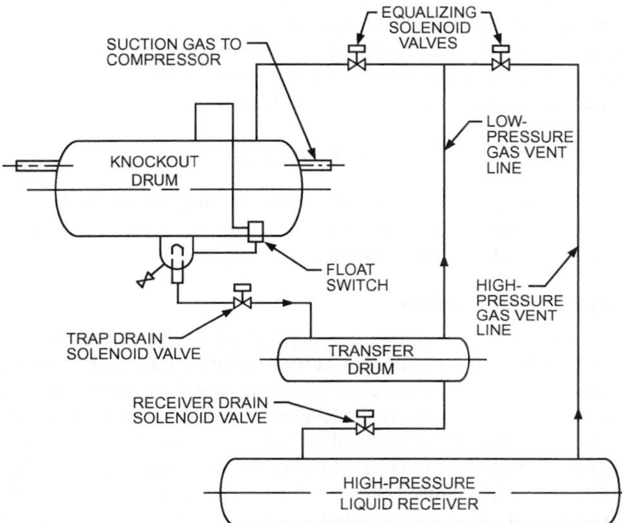

Fig. 7 Gravity Transfer System

A high level of liquid in a suction trap should activate an alarm or stop the compressors. Although eliminating the cause is the most effective way to reduce a high level of excess surging liquid, a more immediate solution is to stop part of the compression system and raise plant suction pressure slightly. Continuing high levels indicate insufficient pump capacity or suction trap volume.

Liquid Level Indicators. Liquid level can be indicated by visual indicators, electronic sensors, or a combination of the two. Visual indicators include individual circular reflex level indicators (bull's-eyes) mounted on a pipe column or stand-alone linear reflex glass assemblies (Figure 9). For operation at temperatures below the frost point, transparent plastic frost shields covering the reflex surfaces are necessary. Also, the pipe column must be insulated, especially when control devices are attached.

Electronic level sensors can continuously monitor liquid level. Digital or graphic displays of liquid level can be locally or remotely monitored (Figure 10).

Level indicators should have adequate isolation valves. High-temperature glass tube indicators should incorporate stop check or excess-flow valves for isolation and safety.

Purge Units. A noncondensable gas separator (purge unit) is useful in most plants, especially when suction pressure is below atmospheric pressure. Purge units on ammonia systems are piped to carry noncondensables (air) from the receiver and condenser to the

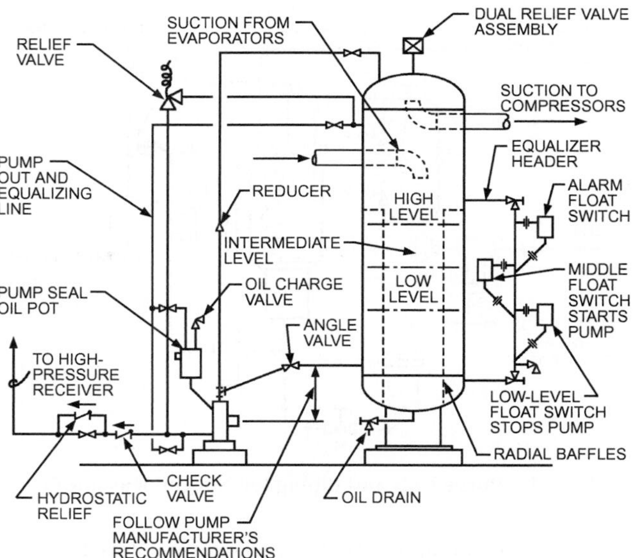

Fig. 8 Piping for Vertical Suction Trap and High-Head Pump

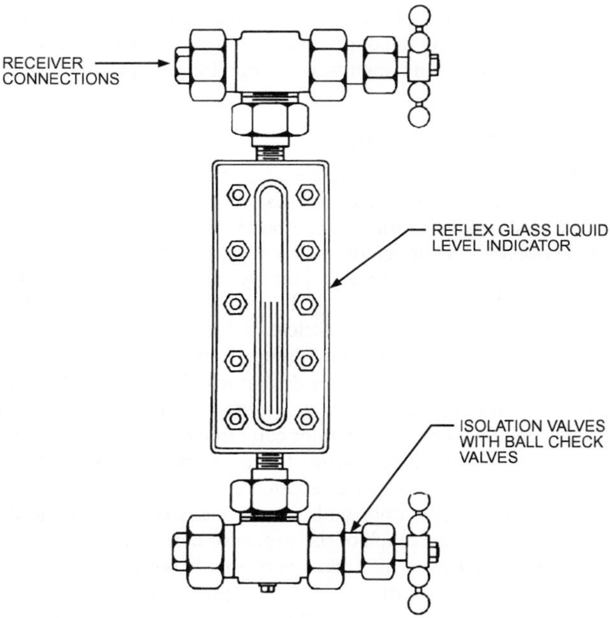

Fig. 9 Gage Glass Assembly for Ammonia

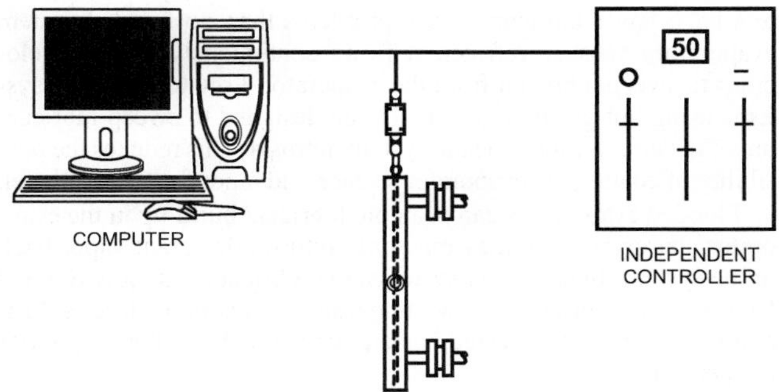

Fig. 10 Electronic Liquid Level Control

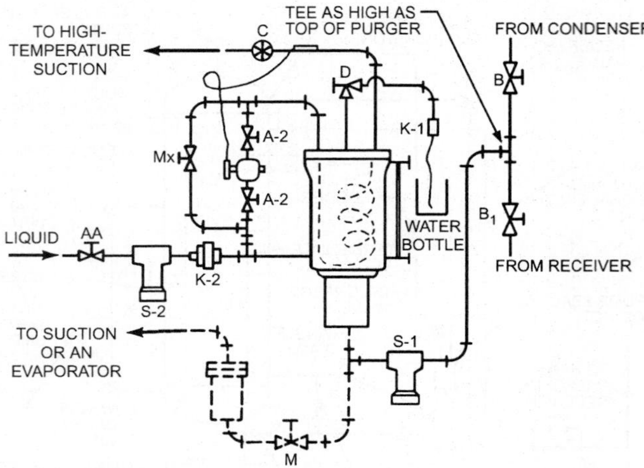

Fig. 11 Purge Unit and Piping for Noncondensable Gas

purger, as shown in Figure 11. High-pressure liquid expands through a coil in the purge unit, providing a cold spot in the purge drum. The suction from the coil should be taken to one of the low-temperature suction mains. Ammonia vapor and noncondensable gas are drawn into the purge drum, and the ammonia condenses on the cold surface. When the drum fills with air and other noncondensables, a float valve in the purger opens and allows them to leave the drum and pass into the open water bottle. Purge units are available for automatic operation.

Lubricant Management

Most lubricants are immiscible in ammonia and separate out of the liquid easily when flow velocity is low or when temperatures are lowered. Normally, lubricants can be easily drained from the system. However, if the temperature is very low and the lubricant is not properly selected, it becomes a gummy mass that prevents refrigerant controls from functioning, blocks flow passages, and fouls heat transfer surfaces. Proper lubricant selection and management is often the key to a properly functioning system.

In two-stage systems, proper design usually calls for lubricant separators on both the high- and low-stage compressors. A properly designed coalescing separator can remove almost all the lubricant that is in droplet or aerosol form. Lubricant that reaches its saturation vapor pressure and becomes a vapor cannot be removed by a separator. Separators with some means of cooling the discharge gas condense much of the vapor for consequent separation. Selection of lubricants that have very low vapor pressures below 180°F can minimize carryover to 2 or 3 ppm. Care must be taken, however, to ensure that refrigerant is not condensed and fed back into the compressor or separator, where it can lower lubricity and cause compressor damage.

In general, direct-expansion and liquid overfeed system evaporators have fewer lubricant return problems than do flooded system evaporators because refrigerant flows continuously at good velocities to sweep lubricant from the evaporator. Low-temperature systems using hot-gas defrost can also be designed to sweep lubricant out of the circuit each time the system defrosts. This reduces the possibility of coating the evaporator surface and hindering heat transfer.

Flooded evaporators can promote lubricant build-up in the evaporator charge because they may only return refrigerant vapor back to the system. In ammonia systems, the lubricant is simply drained from the surge drum. At low temperatures, this procedure is difficult if the lubricant selected has a pour point above the evaporator temperature.

Lubricant Removal from Ammonia Systems. Most lubricants are miscible with liquid ammonia only in very small proportions. The proportion decreases with the temperature, causing lubricant to separate. The evaporation of ammonia increases the lubricant ratio, causing more lubricant to separate. Increased density causes the lubricant (saturated with ammonia at the existing pressure) to form a separate layer below the ammonia liquid.

Unless lubricant is removed periodically or continuously from the point where it collects, it can cover the heat transfer surface in the evaporator, reducing performance. If gage lines or branches to level controls are taken from low points (or lubricant is allowed to accumulate), these lines will contain lubricant. The higher lubricant density is at a lower level than the ammonia liquid. Draining lubricant from a properly located collection point is not difficult unless the temperature is so low that the lubricant does not flow readily. In this case, keeping the receiver at a higher temperature may be beneficial. Alternatively, a lubricant with a lower pour point can be selected.

Lubricant in the system is saturated with ammonia at the existing pressure. When the pressure is reduced, ammonia vapor separates, causing foaming.

Draining lubricant from ammonia systems requires special care. Ammonia in lubricant foam normally starts to evaporate and produces a smell. Operators should be made aware of this. On systems where lubricant is drained from a still, a spring-loaded drain valve, which closes if the valve handle is released, should be installed.

CONTROLS

Refrigerant flow controls are discussed in Chapter 44. The following precautions are necessary in the application of certain controls in low-temperature systems.

Liquid Feed Control

Many controls available for single-stage, high-temperature systems may be used with some discretion on low-temperature systems. If the liquid level is controlled by a low-side float valve (with the float in the chamber where the level is controlled), low pressure and temperature have no appreciable effect on operation. External float chambers, however, must be thoroughly insulated to prevent heat influx that might cause boiling and an unstable level, affecting the float response. Equalizing lines to external float chambers, particularly the upper line, must be sized generously so that liquid can reach the float chamber, and gas resulting from any evaporation may be returned to the vessel without appreciable pressure loss.

The superheat-controlled (thermostatic) expansion valve is generally used in direct-expansion evaporators. This valve operates on the difference between bulb pressure, which is responsive to suction temperature, and pressure below the diaphragm, which is the actual suction pressure.

The thermostatic expansion valve is designed to maintain a preset superheat in suction gas. Although the pressure-sensing part of the system responds almost immediately to a change in conditions, the temperature-sensing bulb must overcome thermal inertia before its effect is felt on the power element of the valve. Thus, when compressor capacity decreases suddenly, the expansion valve may overfeed before the bulb senses the presence of liquid in the suction line and reduces the feed. Therefore, a suction accumulator should be installed on direct-expansion low-temperature systems with multiple expansion valves.

Controlling Load During Pulldown

System transients during pulldown can be managed by controlling compressor capacity. Proper load control reduces compressor capacity so that energy requirements stay within the motor and condenser capacities. On larger systems using screw compressors, a current-sensing device reads motor amperage and adjusts the capacity control device appropriately. Cylinders on reciprocating compressors can be unloaded for similar control.

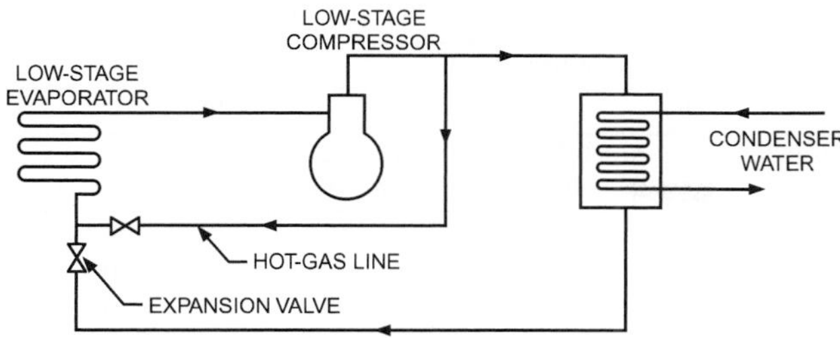

Fig. 12 Hot-Gas Injection Evaporator for Operations at Low Load

Alternatively, a downstream, outlet, or crankcase pressure regulator can be installed in the suction line to throttle suction flow if the pressure exceeds a preset limit. This regulator limits the compressor's suction pressure during pulldown. The disadvantage of this device is the extra pressure drop it causes when the system is at the desired operating conditions. To overcome some of this, the designer can use external forces to drive the valve, causing it to be held fully open when the pressure is below the maximum allowable. Systems using downstream pressure regulators and compressor unloading must be carefully designed so that the two controls complement each other.

Operation at Varying Loads and Temperatures

Compressor and evaporator capacity controls are similar for multi- and single-stage systems. Control methods include compressor capacity control, hot-gas bypass, or evaporator pressure regulators. Low pressure can affect control systems by significantly increasing the specific volume of the refrigerant gas and the pressure drop. A small pressure reduction can cause a large percentage capacity reduction.

System load usually cannot be reduced to near zero, because this results in little or no flow of gas through the compressor and consequent overheating. Additionally, high pressure ratios are detrimental to the compressor if it is required to run at very low loads. If the compressor cannot be allowed to cycle off during low load, an acceptable alternative is a **hot-gas bypass**. High-pressure gas is fed to the low-pressure side of the system through a downstream pressure regulator. The gas should be desuperheated by injecting it at a point in the system where it is in contact with expanding liquid, such as immediately downstream of the liquid feed to the evaporator. Otherwise, extremely high compressor discharge temperatures can result. The artificial load supplied by high-pressure gas can fill the gap between the actual load and the lowest stable compressor operating capacity. Figure 12 shows such an arrangement.

Electronic Control

Microprocessor- and computer-based control systems are becoming the norm for control systems on individual compressors as well as for entire system control. Almost all screw compressors use microprocessor control systems to monitor all safety functions and operating conditions. These machines are frequently linked together with a programmable controller or computer for sequencing multiple compressors so that they load and unload in response to system fluctuations in the most economical manner. Programmable controllers are also used to replace multiple defrost time clocks on larger systems for more accurate and economical defrosting. Communications and data logging allow systems to operate at optimum conditions under transient load conditions even when operators are not in attendance.

PIPING

Local codes or ordinances governing ammonia mains should be followed, in addition to the recommendations here.

Recommended Material

Because copper and copper-bearing materials are attacked by ammonia, they are not used in ammonia piping systems. Steel piping, fittings, and valves of the proper pressure rating are suitable for ammonia gas and liquid.

Ammonia piping should conform to ASME *Standard* B31.5, and to IIAR *Standard* 2, which states the following:

1. Liquid lines 1.5 in. and smaller shall be not less than Schedule 80 carbon steel pipe.
2. Liquid lines 2 through 6 in. shall be not less than Schedule 40 carbon steel pipe.
3. Liquid lines 8 through 12 in. shall be not less than Schedule 20 carbon steel pipe.
4. Vapor lines 6 in. and smaller shall be not less than Schedule 40 carbon steel pipe.
5. Vapor lines 8 through 12 in. shall be not less than Schedule 20 carbon steel pipe.
6. Vapor lines 14 in. and larger shall be not less than Schedule 10 carbon steel pipe.
7. All threaded pipe shall be Schedule 80.
8. Carbon steel pipe shall be ASTM *Standard* A53 Grade A or B, Type E (electric resistance welded) or Type S (seamless); or ASTM *Standard* A106 (seamless), except where temperature-pressure criteria mandate a higher specification material. *Standard* A53 Type F is not permitted for ammonia piping.

Fittings

Couplings, elbows, and tees for threaded pipe are for a minimum of 3000 psi design pressure and constructed of forged steel. Fittings for welded pipe should match the type of pipe used (i.e., standard fittings for standard pipe and extra-heavy fittings for extra-heavy pipe).

Tongue-and-groove or ANSI flanges should be used in ammonia piping. Welded flanges for low-side piping can have a minimum 150 psi design pressure rating. On systems located in high ambients, low-side piping and vessels should be designed for 200 to 225 psig. The high side should be 250 psig if the system uses water-cooled or evaporative cooled condensing. Use 300 psig minimum for air-cooled designs.

Pipe Joints

Joints between lengths of pipe or between pipe and fittings can be threaded if the pipe size is 1.25 in. or smaller. Pipe 1.5 in. or larger should be welded. An all-welded piping system is superior.

Threaded Joints. Many sealants and compounds are available for sealing threaded joints. The manufacturer's instructions cover compatibility and application method. Do not use excessive amounts or apply on female threads because any excess can contaminate the system.

Welded Joints. Pipe should be cut and beveled before welding. Use pipe alignment guides and provide a proper gap between pipe ends so that a full-penetration weld is obtained. The weld should be made by a qualified welder, using proper procedures such as the *Welding Procedure Specifications*, prepared by the National Certified Pipe Welding Bureau (NCPWB).

Gasketed Joints. A compatible fiber gasket should be used with flanges. Before tightening flange bolts to valves, controls, or flange unions, properly align pipe and bolt holes. When flanges are used to straighten pipe, they put stress on adjacent valves, compressors, and controls, causing the operating mechanism to bind. To prevent leaks, flange bolts are drawn up evenly when connecting the flanges. Flanges at compressors and other system components must not move or indicate stress when all bolts are loosened.

Union Joints. Steel (3000 psi) ground joint unions are used for gage and pressure control lines with screwed valves and for joints up to 0.75 in. When tightening this type of joint, the two pipes must be axially aligned. To be effective, the two parts of the union

Table 1 Suction Line Capacities in Tons for Ammonia with Pressure Drops of 0.25 and 0.50°F per 100 ft Equivalent

Steel Line Size		Saturated Suction Temperature, °F					
		−60		−40		−20	
IPS	SCH	Δt = 0.25°F Δp = 0.046	Δt = 0.50°F Δp = 0.092	Δt = 0.25°F Δp = 0.077	Δt = 0.50°F Δp = 0.155	Δt = 0.25°F Δp = 0.123	Δt = 0.50°F Δp = 0.245
3/8	80	0.03	0.05	0.06	0.09	0.11	0.16
1/2	80	0.06	0.10	0.12	0.18	0.22	0.32
3/4	80	0.15	0.22	0.28	0.42	0.50	0.73
1	80	0.30	0.45	0.57	0.84	0.99	1.44
1 1/4	40	0.82	1.21	1.53	2.24	2.65	3.84
1 1/2	40	1.25	1.83	2.32	3.38	4.00	5.80
2	40	2.43	3.57	4.54	6.59	7.79	11.26
2 1/2	40	3.94	5.78	7.23	10.56	12.50	18.03
3	40	7.10	10.30	13.00	18.81	22.23	32.09
4	40	14.77	21.21	26.81	38.62	45.66	65.81
5	40	26.66	38.65	48.68	70.07	82.70	119.60
6	40	43.48	62.83	79.18	114.26	134.37	193.44
8	40	90.07	129.79	163.48	235.38	277.80	397.55
10	40	164.26	236.39	297.51	427.71	504.98	721.08
12	ID*	264.07	379.88	477.55	686.10	808.93	1157.59
Steel Line Size		Saturated Suction Temperature, °F					
		0		20		40	
IPS	SCH	Δt = 0.25°F Δp = 0.184	Δt = 0.50°F Δp = 0.368	Δt = 0.25°F Δp = 0.265	Δt = 0.50°F Δp = 0.530	Δt = 0.25°F Δp = 0.366	Δt = 0.50°F Δp = 0.582
3/8	80	0.18	0.26	0.28	0.40	0.41	0.53
1/2	80	0.36	0.52	0.55	0.80	0.82	1.05
3/4	80	0.82	1.18	1.26	1.83	1.87	2.38
1	40	1.62	2.34	2.50	3.60	3.68	4.69
1 1/4	40	4.30	6.21	6.63	9.52	9.76	12.42
1 1/2	40	6.49	9.34	9.98	14.34	14.68	18.64
2	40	12.57	18.12	19.35	27.74	28.45	36.08
2 1/2	40	20.19	28.94	30.98	44.30	45.37	57.51
3	40	35.87	51.35	54.98	78.50	80.40	101.93
4	40	73.56	105.17	112.34	160.57	164.44	208.34
5	40	133.12	190.55	203.53	289.97	296.88	376.18
6	40	216.05	308.62	329.59	469.07	480.96	609.57
8	40	444.56	633.82	676.99	962.47	985.55	1250.34
10	40	806.47	1148.72	1226.96	1744.84	1786.55	2263.99
12	ID*	1290.92	1839.28	1964.56	2790.37	2862.23	3613.23

Note: Capacities are in tons of refrigeration resulting in a line friction loss (Δp in psi per 100 ft equivalent pipe length), with corresponding change (Δt in °F per 100 ft) in saturation temperature.

*The inside diameter of the pipe is the same as the nominal pipe size.

must match perfectly. Ground joint unions should be avoided if at all possible.

Pipe Location

Piping should be at least 7.5 ft above the floor. Locate pipes carefully in relation to other piping and structural members, especially when lines are to be insulated. The distance between insulated lines should be at least three times the thickness of the insulation for screwed fittings, and four times for flange fittings. The space between the pipe and adjacent surfaces should be three-fourths of these amounts.

Hangers located close to the vertical risers to and from compressors keep the piping weight off the compressor. Pipe hangers should be placed no more than 8 to 10 ft apart and within 2 ft of a change in direction of the piping. Hangers should be designed to bear on the outside of insulated lines. Sheet metal sleeves on the lower half of the insulation are usually sufficient. Where piping penetrates a wall, a sleeve should be installed, and where the pipe penetrating the wall is insulated, it must be adequately sealed.

Piping to and from compressors and to other components must provide for expansion and contraction. Sufficient flange or union joints should be located in the piping so components can be assembled easily during installation and also disassembled for servicing.

Pipe Sizing

Table 1 presents practical suction line sizing data based on 0.25°F and 0.50°F differential pressure drop equivalent per 100 ft total equivalent length of pipe. For data on equivalent lengths of valves and fittings, refer to Tables 10, 11, and 12 in Chapter 2. Table 2 lists data for sizing suction and discharge lines at 1°F differential pressure drop equivalent per 100 ft equivalent length of pipe, and for sizing liquid lines at 100 fpm. Charts prepared by Wile (1977) present pressure drops in saturation temperature equivalents. For a complete discussion of the basis of these line sizing charts, see Timm (1991). Table 3 presents line sizing information for pumped liquid lines, high-pressure liquid lines, hot-gas defrost lines, equalizing lines, and thermosiphon lubricant cooling ammonia lines.

Valves

Stop Valves. These valves should be placed in the inlet and outlet lines to all condensers, vessels, evaporators, and long lengths of

Table 2 Suction, Discharge, and Liquid Line Capacities in Tons for Ammonia (Single- or High-Stage Applications)

Steel Line Size		Suction Lines (Δt = 1°F) Saturated Suction Temperature, °F					Discharge Lines	Steel Line Size		Liquid Lines	
IPS	SCH	−40 Δp = 0.31	−20 Δp = 0.49	0 Δp = 0.73	20 Δp = 1.06	40 Δp = 1.46	Δt = 1°F Δp = 2.95	IPS	SCH	Velocity = 100 fpm	Δp =2.0 psi Δt = 0.7°F
3/8	80	—	—	—	—	—	—	3/8	80	8.6	12.1
1/2	80	—	—	—	—	—	3.1	1/2	80	14.2	24.0
3/4	80	—	—	—	2.6	3.8	7.1	3/4	80	26.3	54.2
1	80	—	2.1	3.4	5.2	7.6	13.9	1	80	43.8	106.4
1 1/4	40	3.2	5.6	8.9	13.6	19.9	36.5	1 1/4	80	78.1	228.6
1 1/2	40	4.9	8.4	13.4	20.5	29.9	54.8	1 1/2	80	107.5	349.2
2	40	9.5	16.2	26.0	39.6	57.8	105.7	2	40	204.2	811.4
2 1/2	40	15.3	25.9	41.5	63.2	92.1	168.5	2 1/2	40	291.1	1292.6
3	40	27.1	46.1	73.5	111.9	163.0	297.6	3	40	449.6	2287.8
4	40	55.7	94.2	150.1	228.7	333.0	606.2	4	40	774.7	4662.1
5	40	101.1	170.4	271.1	412.4	600.9	1095.2	5	40	—	—
6	40	164.0	276.4	439.2	667.5	971.6	1771.2	6	40	—	—
8	40	337.2	566.8	901.1	1366.6	1989.4	3623.0	8	40	—	—
10	40	611.6	1027.2	1634.3	2474.5	3598.0	—	10	40	—	—
12	ID*	981.6	1644.5	2612.4	3963.5	5764.6	—	12	ID*	—	—

Notes:

1. Table capacities are in tons of refrigeration.
   Δp = pressure drop due to line friction, psi per 100 ft of equivalent line length
   Δt = corresponding change in saturation temperature, °F per 100 ft
2. Line capacity for other saturation temperatures Δt and equivalent lengths L_e

$$\text{Line capacity} = \text{Table capacity}\left(\frac{\text{Table } L_e}{\text{Actual } L_e} \times \frac{\text{Actual } \Delta t}{\text{Table } \Delta t}\right)^{0.55}$$

3. Saturation temperature Δt for other capacities and equivalent lengths L_e

$$\Delta t = \text{Table } \Delta t \left(\frac{\text{Actual } L_e}{\text{Table } L_e}\right)\left(\frac{\text{Actual capacity}}{\text{Table capacity}}\right)^{1.8}$$

4. Values based on 90°F condensing temperature. Multiply table capacities by the following factors for other condensing temperatures:

Condensing Temperature, °F	Suction Lines	Discharge Lines
70	1.05	0.78
80	1.02	0.89
90	1.00	1.00
100	0.98	1.11

5. Discharge and liquid line capacities based on 20°F suction. Evaporator temperature is 0°F. The capacity is affected less than 3% when applied from −40 to +40°F extremes.

*The inside diameter of the pipe is the same as the nominal pipe size.

Table 3 Liquid Ammonia Line Capacities

(Capacity in tons of refrigeration, except as noted)

Nominal Size, in.	Pumped Liquid Overfeed Ratio 3:1	4:1	5:1	High-Pressure Liquid at 3 psi[a]	Hot-Gas Defrost[a]	Equalizer High Side[b]	Thermosiphon Lubricant Cooling Lines Gravity Flow,[c] 1000 Btu/h Supply	Return	Vent
1/2	10	7.5	6	30	—	—	—	—	—
3/4	22	16.5	13	69	4	50	—	—	—
1	43	32.5	26	134	8	100	—	—	—
1 1/4	93.5	70	56	286	20	150	—	—	—
1 1/2	146	110	87.5	439	30	225	200	120	203
2	334	250	200	1016	50	300	470	300	362
2 1/2	533	400	320	1616	92	500	850	530	638
3	768	576	461	2886	162	1000	1312	870	1102
4	1365	1024	819	—	328	2000	2261	1410	2000
5	—	—	—	—	594	—	3550	2214	3624
6	—	—	—	—	970	—	5130	3200	6378
8	—	—	—	—	—	—	8874	5533	11596

Source: Wile (1977).

[a]Hot-gas line sizes are based on 1.5 psi pressure drop per 100 ft of equivalent length at 100 psig discharge pressure and three times evaporator refrigeration capacity.

[b]Line sizes based on experience using total system evaporator tons.

[c]From Frick Co. (1995). Values for line sizes above 4 in. are extrapolated.

pipe so they can be isolated in case of leaks and to facilitate pumping out by evacuation. Sections of liquid piping that can be valved off and isolated must be protected with a relief device.

Installing globe-type stop valves with the valve stems horizontal lessens the chance (1) for dirt or scale to lodge on the valve seat or disk and cause it to leak or (2) for liquid or lubricant to pocket in the area below the seat. Wet suction return lines (recirculation system) should use angle valves to reduce the possibility of liquid pockets and reduce pressure drop.

Welded flanged or weld-in-line valves are desirable for all line sizes; however, screwed valves may be used for 1 1/4 in. and smaller lines. Ammonia globe and angle valves should have the following features:

- Soft seating surfaces for positive shutoff (no copper or copper alloy)
- Back seating to permit repacking the valve stem while in service
- Arrangement that allows packing to be tightened easily

- All-steel construction (preferable)
- Bolted bonnets above 1 in., threaded bonnets for 1 in. and smaller

Consider seal cap valves in refrigerated areas and for all ammonia piping. To keep pressure drop to a minimum, consider angle valves (as opposed to globe valves).

Control Valves. Pressure regulators, solenoid valves, and thermostatic expansion valves should be flanged for easy assembly and removal. Valves 1.5 in. and larger should have welded companion flanges. Smaller valves can have threaded companion flanges.

A strainer should be used in front of self-contained control valves to protect them from pipe construction material and dirt. A ceramic filter installed in the pilot line to the power piston protects the close tolerances from foreign material when pilot-operated control valves are used.

Solenoid Valves. Solenoid valve stems should be upright, with their coils protected from moisture. They should have flexible conduit connections, where allowed by codes, and an electric pilot light wired in parallel to indicate when the coil is energized. A manual opening stem is useful for emergencies.

Solenoid valves for high-pressure liquid feed to evaporators should have soft seats for positive shutoff. Solenoid valves for other applications, such as in suction, hot-gas, or gravity feed lines, should be selected for the pressure and temperature of the fluid flowing and for the pressure drop available.

Relief Valves. Safety valves must be provided in conformance with ASHRAE *Standard* 15 and Section VIII, Division 1, of the ASME *Boiler and Pressure Vessel Code*. For ammonia systems, IIAR *Bulletin* 109 also addresses the subject of safety valves.

Dual relief valve arrangements allow testing of the relief valves (Figure 13). The three-way stop valve is constructed so that it is always open to one of the relief valves if the other is removed to be checked or repaired.

Isolated Line Sections

Sections of piping that can be isolated between hand valves or check valves can be subjected to extreme hydraulic pressures if cold liquid refrigerant is trapped in them and subsequently warmed. Additional safety valves for such piping must be provided.

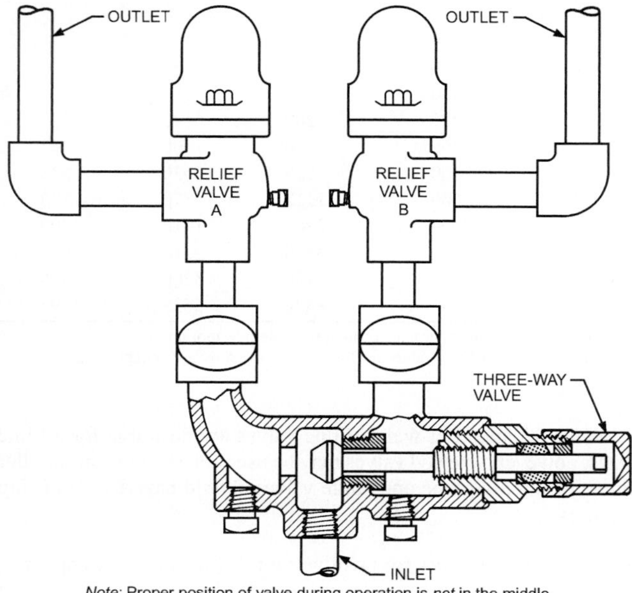

Fig. 13 Dual Relief Valve Fitting for Ammonia

Insulation and Vapor Retarders

Chapter 33 covers insulation and vapor retarders. Insulation and effective vapor retarders on low-temperature systems are very important. At low temperatures, the smallest leak in the vapor retarder can allow ice to form inside the insulation, which can totally destroy the integrity of the entire insulation system. The result can significantly increase load and power usage.

RECIPROCATING COMPRESSORS

Piping

Figure 14 shows a typical piping arrangement for two compressors operating in parallel off the same suction main. Suction mains should be laid out with the objective of returning only clean, dry gas to the compressor. This usually requires a suction trap sized adequately for gravity gas and liquid separation based on permissible gas velocities for specific temperatures. A dead-end trap can usually trap only scale and lubricant. As an alternative, a shell-and-coil accumulator with a warm liquid coil may be considered. Suction mains running to and from the suction trap or accumulator should be pitched toward the trap at 1/8 in. per foot for liquid drainage.

In sizing suction mains and takeoffs from mains to compressors, consider how the pressure drop in the selected piping affects the compressor size required. First costs and operating costs for compressor and piping selections should be optimized.

Good suction line systems have a total friction drop of 1 to 3°F pressure drop equivalent. Practical suction line friction losses should not exceed 0.5°F equivalent per 100 ft equivalent length.

A well-designed discharge main has a total friction loss of 1 to 2 psi. Generally, a slightly oversized discharge line is desirable to hold down discharge pressure and, consequently, discharge temperature and energy costs. Where possible, discharge mains should be pitched (1/8 in/ft) toward the condenser, without creating a liquid trap; otherwise, pitch should be toward the discharge line separator.

High- and low-pressure cutouts and gages and lubricant pressure failure cutout are installed on the compressor side of the stop valves to protect the compressor.

Lubricant Separators. Lubricant separators are located in the discharge line of each compressor (Figure 14A). A high-pressure float valve drains lubricant back into the compressor crankcase or lubricant receiver. The separator should be placed as far from the compressor as possible, so the extra pipe length can be used to cool the discharge gas before it enters the separator. This reduces the temperature of the ammonia vapor and makes the separator more effective.

Liquid ammonia must not reach the crankcase. Often, a valve (preferably automatic) is installed in the drain from the lubricant separator, open only when the temperature at the bottom of the separator is higher than the condensing temperature. Some manufacturers install a small electric heater at the bottom of a vertical lubricant trap instead. The heater is actuated when the compressor is not operating. Separators installed in cold conditions must be insulated to prevent ammonia condensation.

A filter is recommended in the drain line on the downstream side of the high-pressure float valve.

Lubricant Receivers. Figure 14B illustrates two compressors on the same suction line with one discharge-line lubricant separator. The separator float drains into a lubricant receiver, which maintains a reserve supply of lubricant for the compressors. Compressors should be equipped with crankcase floats to regulate lubricant flow to the crankcase.

Discharge Check Valves and Discharge Lines. Discharge check valves on the downstream side of each lubricant separator prevent high-pressure gas from flowing into an inactive compressor and causing condensation (Figure 14A).

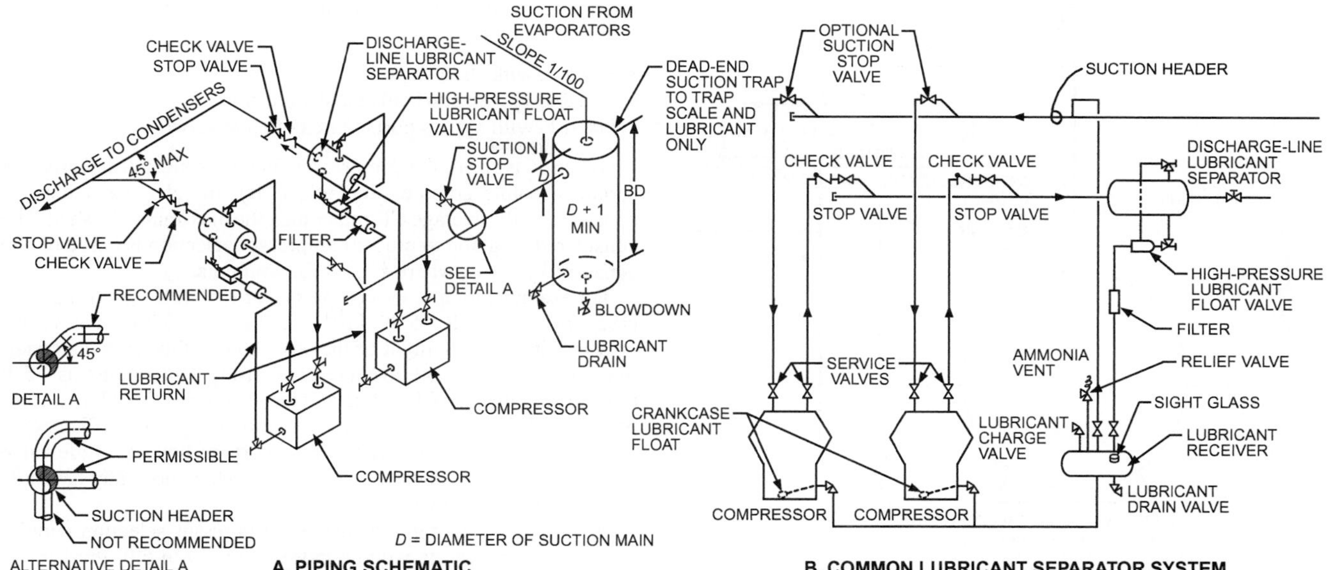

Fig. 14 Schematic of Reciprocating Compressors Operating in Parallel

The discharge line from each compressor should enter the discharge main at a 45° maximum angle in the horizontal plane so the gas flows smoothly.

Unloaded Starting. Unloaded starting is frequently needed to stay within the torque or current limitations of the motor. Most compressors are unloaded either by holding the suction valve open or by external bypassing. Control can be manual or automatic.

Suction Gas Conditioning. Suction main piping should be insulated, complete with vapor retarder to minimize thermal losses, to prevent sweating and/or ice build-up on the piping, and to limit superheat at the compressor. Additional superheat results in increased discharge temperatures and reduces compressor capacity. Low discharge temperatures in ammonia plants are important to reduce lubricant carryover and because compressor lubricant can carbonize at higher temperatures, which can cause cylinder wall scoring and lubricant sludge throughout the system. Discharge temperatures above 250°F should be avoided at all times. Lubricants should have flash-point temperatures above the maximum expected compressor discharge temperature.

Cooling

Generally, ammonia compressors are constructed with internally cast cooling passages along the cylinders and/or in the top heads. These passages provide space for circulating a heat transfer medium, which minimizes heat conduction from the hot discharge gas to the incoming suction gas and lubricant in the compressor's crankcase. An external lubricant cooler is supplied on most reciprocating ammonia compressors. Water is usually the medium circulated through these passages (**water jackets**) and the lubricant cooler at a rate of about 0.1 gpm per ton of refrigeration. Lubricant in the crankcase (depending on type of construction) is about 120°F. Temperatures above this level reduce the lubricant's lubricating properties.

For compressors operating in ambients above 32°F, water flow is sometimes controlled entirely by hand valves, although a solenoid valve in the inlet line is desirable to automate the system. When the compressor stops, water flow must be stopped to keep residual gas from condensing and to conserve water. A water-regulating valve, installed in the water supply line with the sensing bulb in the water return line, is also recommended. This type of cooling is shown in Figure 15.

The thermostat in the water line leaving the jacket serves as a safety cutout to stop the compressor if the temperature becomes too high.

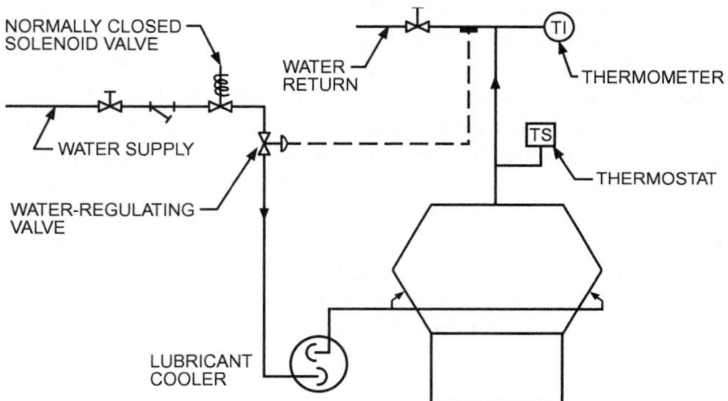

Fig. 15 Jacket Water Cooling for Ambient Temperatures Above Freezing

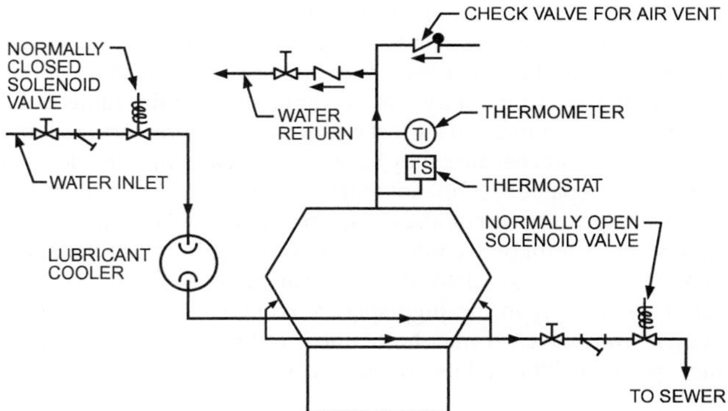

Fig. 16 Jacket Water Cooling for Ambient Temperatures Below Freezing

For compressors where ambient temperatures may be below 32°F, a means for draining the jacket on shutdown to prevent freeze-up must be provided. One method is shown in Figure 16. Water flow is through the normally closed solenoid valve, which is energized when the compressor starts. Water then circulates through the lubricant cooler and the jacket, and out through the water return line. When the compressor stops, the solenoid valve in the water inlet line

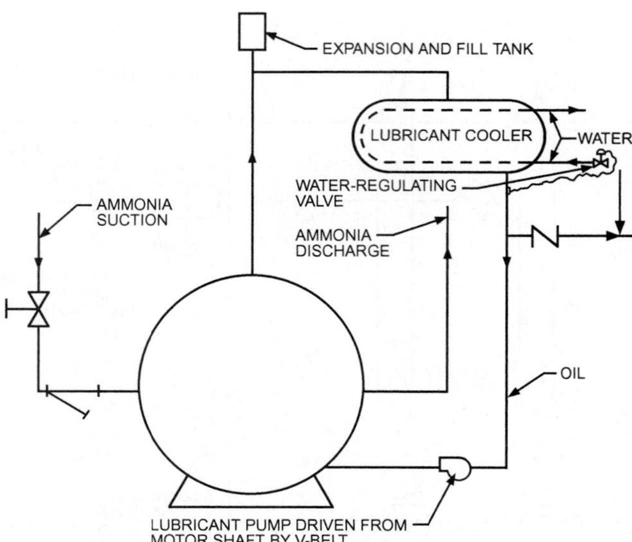

Fig. 17 Rotary Vane Booster Compressor Cooling with Lubricant

is deenergized and stops water flow to the compressor. At the same time, the solenoid valve opens to drain the water out of the low point to wastewater treatment. The check valves in the air vent lines open when pressure is relieved and allow the jacket and cooler to be drained. Each flapper check valve is installed so that water pressure closes it, but absence of water pressure allows it to swing open.

For compressors in spaces below 32°F or where water quality is very poor, cooling is best handled by using an inhibited glycol solution or other suitable fluid in the jackets and lubricant cooler and cooling with a secondary heat exchanger. This method for cooling reciprocating ammonia compressors eliminates fouling of the lubricant cooler and jacket normally associated with city water or cooling tower water.

ROTARY VANE, LOW-STAGE COMPRESSORS

Piping

Rotary vane compressors have been used extensively as low-stage compressors in ammonia refrigeration systems. Now, however, the screw compressor has largely replaced the rotary vane compressor for ammonia low-stage compressor applications. Piping requirements for rotary vane compressors are the same as for reciprocating compressors. Most rotary vane compressors are lubricated by injectors because they have no crankcase. In some designs, a lubricant separator, lubricant receiver, and cooler are required on the discharge of these compressors; a pump recirculates lubricant to the compressor for both cooling and lubrication. In other rotary vane compressor designs, a discharge lubricant separator is not used, and lubricant collects in the high-stage suction accumulator or intercooler, from which it may be drained. Lubricant for the injectors must periodically be added to a reservoir.

Cooling

The compressor jacket is cooled by circulating a cooling fluid, such as water or lubricant. Lubricant is recommended, because it will not freeze and can serve both purposes (Figure 17).

SCREW COMPRESSORS

Piping

Helical screw compressors are the choice for most industrial refrigeration systems. All helical screw compressors have a constant-volume (displacement) design. The volume index V_i refers to the internal volume ratio of the compressor. There are three types of screw compressors:

- Fixed V_i with slide valve
- Variable V_i with slide valve and slide stop
- Fixed V_i with bypass ports in lieu of slide valve

When V_i is fixed, the compressor functions most efficiently at a certain absolute compression ratio (CR). In selecting a fixed-V_i compressor, the average CR rather than the maximum CR should be considered. A guide to proper compressor selection is based on the equation V_i^k = CR, where k = 1.4 for ammonia.

For example, for a screw compressor at 10°F (38.5 psia) and 95°F (195.8 psia) with CR = 5.09, $V_i^{1.4}$ = 5.09 and V_i = 3.20. Thus, a compressor with V_i = 3.6 might be the best choice. If the ambient conditions are such that the average condensing temperature is 75°F (140.5 psia), then the CR is 3.65 and the ideal V_i is 2.52. Thus, a compressor with V_i = 2.4 is the proper selection to optimize efficiency.

Fixed-V_i compressors with bypass ports in lieu of a slide valve are often applied as booster compressors, which normally have a V_i requirement of less than 2.9.

A variable-V_i compressor makes compressor selection simpler because it can vary its volume index from 2.0 to 5.0; thus, it can automatically match the internal pressure ratio in the compressor with the external pressure ratio.

Typical flow diagrams for screw compressor packages are shown in Figures 18 (for indirect cooling) and 19 (for direct cooling with refrigerant liquid injection). Figure 20 illustrates a variable-V_i compressor that does not require a full-time lube pump but rather a pump to prelube the bearings. Full-time lube pumps are required when fixed- or variable-V_i compressors are used as low-stage compressors. Lubrication systems require at least a 75 psi pressure differential for proper operation.

Lubricant Cooling

Lubricant in screw compressors may be cooled three ways:

- Liquid refrigerant injection
- Indirect cooling with glycol or water in a heat exchanger
- Indirect cooling with boiling high-pressure refrigerant used as the coolant in a thermosiphon process

Refrigerant injection cooling is shown schematically in Figures 19 and 21. Depending on the application, this cooling method usually decreases compressor efficiency and capacity but lowers equipment cost. Most screw compressor manufacturers publish a derating curve for this type of cooling. Injection cooling for low-stage compression has little or no penalty on compressor efficiency or capacity. However, efficiency can be increased by using an indirectly cooled lubricant cooler. With this configuration, heat from the lubricant cooler is removed by the evaporative condenser or cooling tower and is not transmitted to the high-stage compressors.

Refrigerant liquid for liquid-injection oil cooling must come from a dedicated supply. The source may be the system receiver or a separate receiver; a 5 min uninterrupted supply of refrigerant liquid is usually adequate.

Indirect or thermosiphon lubricant cooling for low-stage screw compressors rejects the lubricant cooling load to the condenser or auxiliary cooling system; this load is not transferred to the high-stage compressor, which improves system efficiency. Indirect lubricant cooling systems using glycol or water reject the lubricant cooling load to a section of an evaporative condenser, a separate evaporative cooler, or a cooling tower. A three-way lubricant control valve should be used to control lubricant temperature.

Thermosiphon lubricant cooling is the industry standard. In this system, high-pressure refrigerant liquid from the condenser, which boils at condensing temperature/pressure (usually 90 to 95°F design), cools lubricant in a tubular heat exchanger. Typical thermosiphon

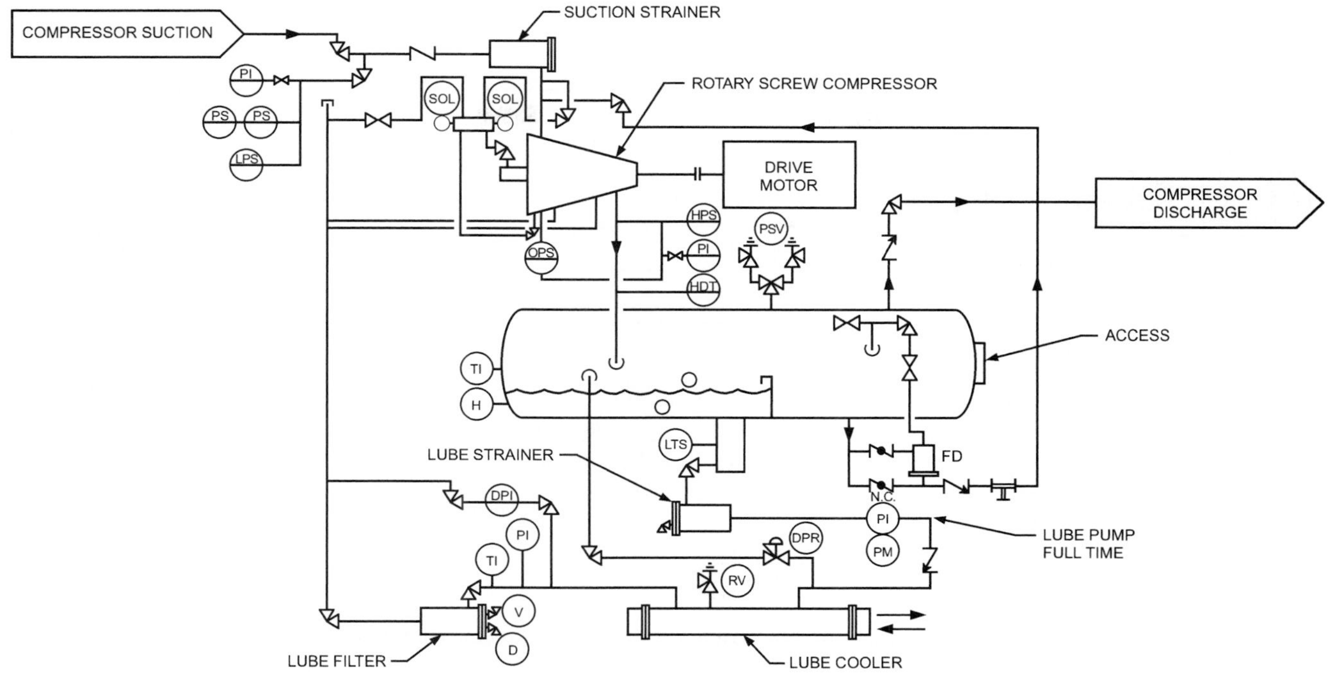

Fig. 18 Fixed-V_i Screw Compressor Flow Diagram with Indirect Lubricant Cooling

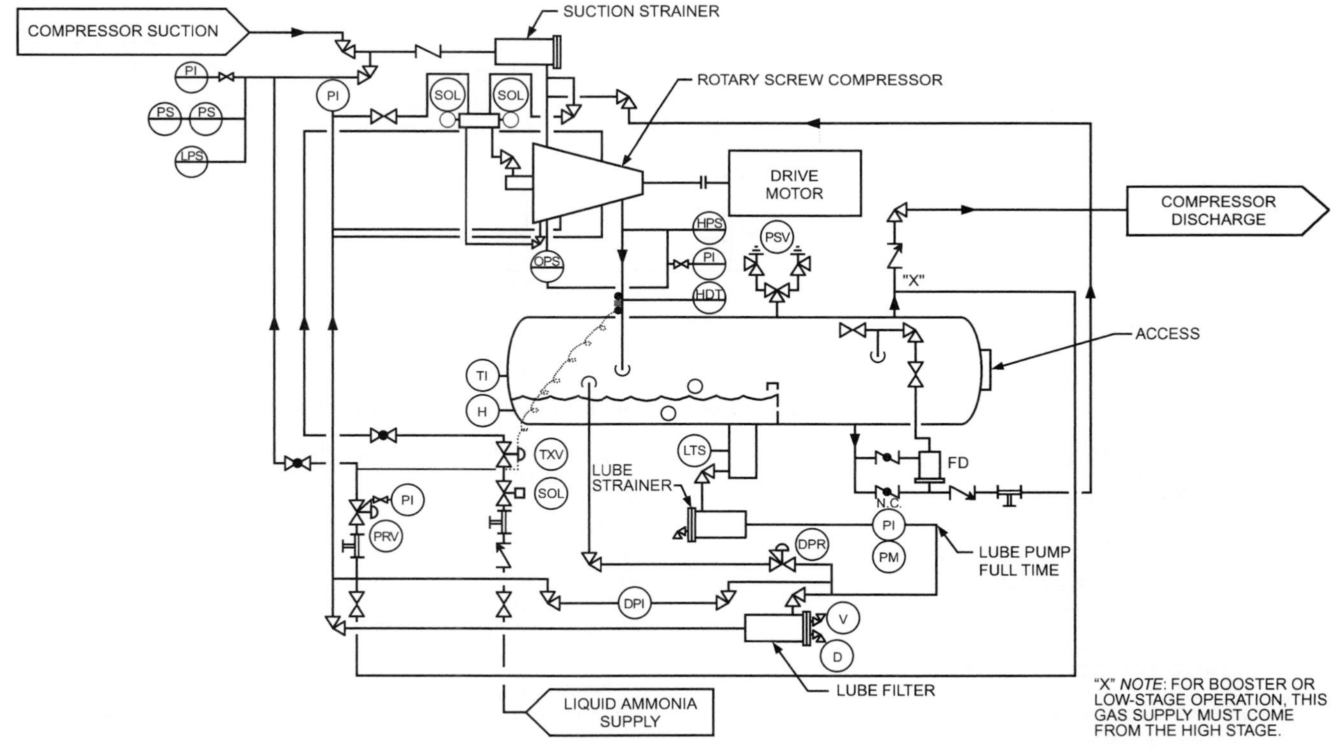

Fig. 19 Fixed-V_i Screw Compressor Flow Diagram with Liquid Injection Cooling

lubricant cooling arrangements are shown in Figures 18, 20, 22, 23, and 24. Note on all figures that the refrigerant liquid supply to the lubricant cooler receives priority over the feed to the system low side. It is important that the gas equalizing line (vent) off the top of the thermosiphon receiver be adequately sized to match the lubricant cooler load to prevent the thermosiphon receiver from becoming gas-bound.

Figure 25 shows a typical capacity control system for a fixed-V_i screw compressor. The four-way valve controls the slide valve position and thus the compressor capacity from typically 100 to 10% with a signal from an electric, electronic, or microprocessor controller. The slide valve unloads the compressor by bypassing vapor back to the suction of the compressor.

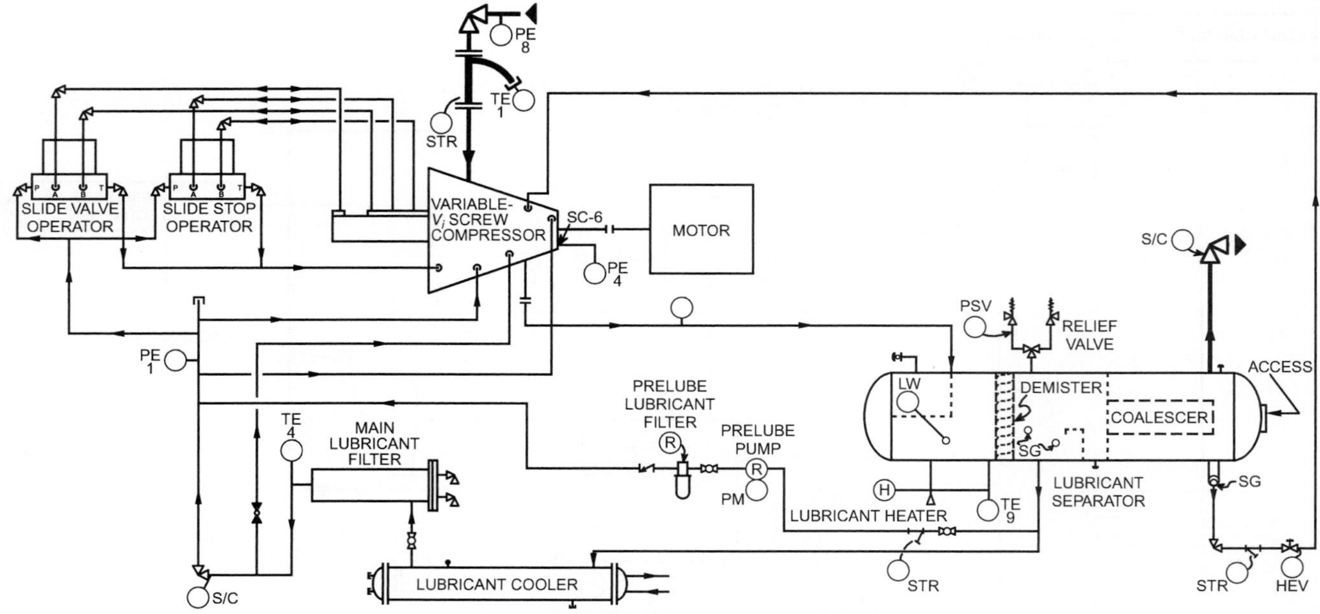

Fig. 20 Flow Diagram for Variable-V_i Screw Compressor High-Stage Only

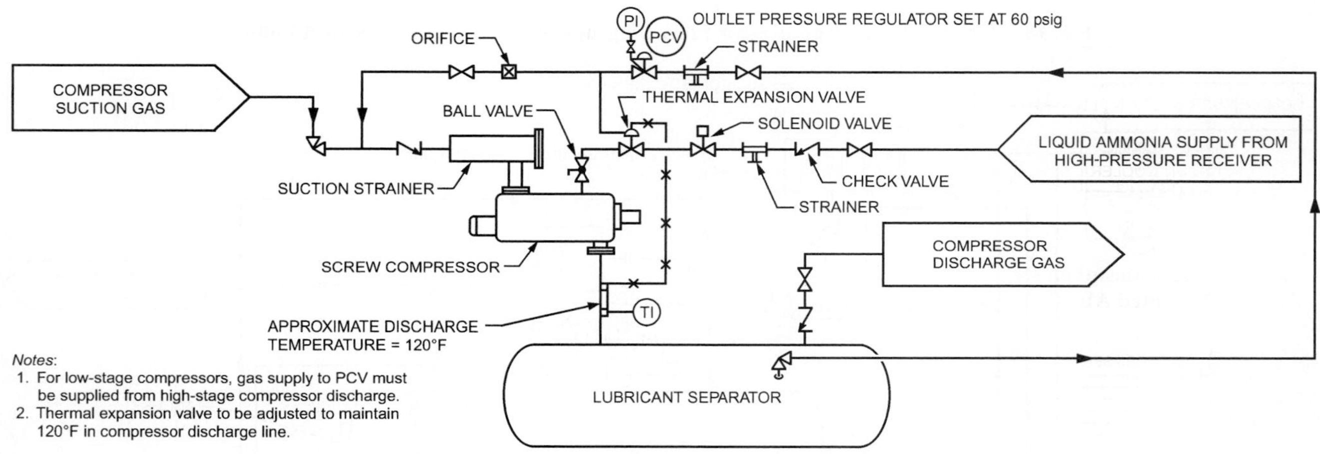

Fig. 21 Flow Diagram for Screw Compressors with Refrigerant Injection Cooling

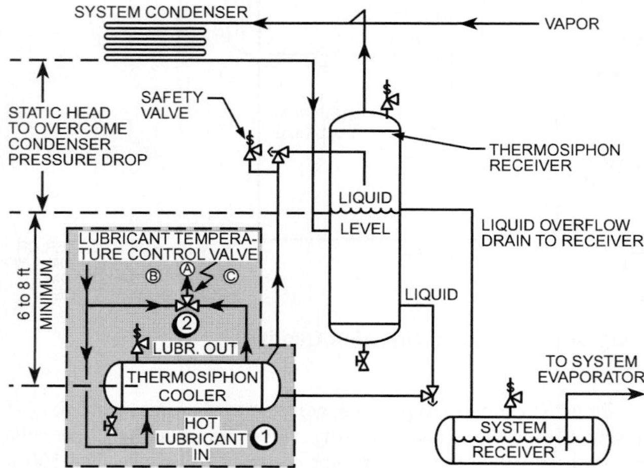

Fig. 22 Typical Thermosiphon Lubricant Cooling System with Thermosiphon Accumulator

Figure 26 shows a typical capacity and volume index control system in which two four-way control valves take their signals from a computer controller. One four-way valve controls capacity by positioning the slide valve in accordance with the load, and the other positions the slide stop to adjust the compressor internal pressure ratio to match system suction and discharge pressure. The slide valve works the same as that on fixed-V_i compressors. Volume index is varied by adjusting the slide stop on the discharge end of the compressor.

Screw compressor piping should generally be installed in the same manner as for reciprocating compressors. Although screw compressors can ingest some liquid refrigerant, they should be protected against liquid carryover. Screw compressors are furnished with both suction and discharge check valves.

CONDENSER AND RECEIVER PIPING

Properly designed piping around the condensers and receivers keeps the condensing surface at its highest efficiency by draining liquid ammonia out of the condenser as soon as it condenses and keeping air and other noncondensables purged.

Horizontal Shell-and-Tube Condenser and Through-Type Receiver

Figure 27 shows a horizontal water-cooled condenser draining into a through (top inlet) receiver. Ammonia plants do not always require controlled water flow to maintain pressure. Usually, pressure is adequate to force the ammonia to the various evaporators without water regulation. Each situation should be evaluated by comparing water costs with input power cost savings at lower condenser pressures.

Water piping should be arranged so that condenser tubes are always filled with water. Air vents should be provided on condenser heads and should have hand valves for manual purging.

Receivers must be below the condenser so that the condensing surface is not flooded with ammonia. The piping should provide (1) free drainage from the condenser and (2) static height of ammonia above the first valve out of the condenser greater than the pressure drop through the valve.

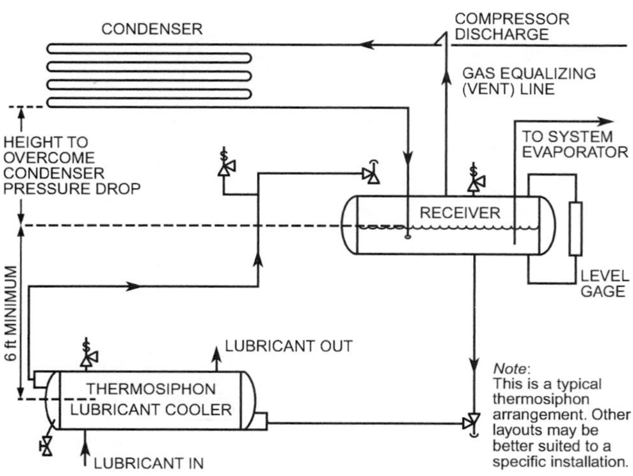

Fig. 23 Thermosiphon Lubricant Cooling System with Receiver Mounted Above Thermosiphon Lubricant Cooler

The drain line from condenser to receiver is designed on the basis of 100 fpm maximum velocity to allow gas equalization between condenser and receiver. Refer to Table 2 for sizing criteria.

Parallel Horizontal Shell-and-Tube Condensers

Figure 28 shows two condensers operating in parallel with one through-type (top inlet) receiver. The length of horizontal liquid drain lines to the receiver should be minimized, with no traps permitted. Equalization between the shells is achieved by keeping liquid velocity in the drain line less than 100 fpm. The drain line can be sized from Table 2.

EVAPORATIVE CONDENSERS

Evaporative condensers are selected based on the wet-bulb temperature in which they operate. The 1% design wet bulb is that wet-bulb temperature that will be equalled or exceeded 1% of the months of June through September, or 29.3 h. Thus, for the majority of industrial plants that operate at least at part load all year, the wet-bulb temperature is below design 99.6% of the operating time. The resultant condensing pressure will only equal or exceed the design condition during 0.4% of the time if the design wet-bulb temperature and peak design refrigeration load occur coincidentally. This peak condition is more a function of how the load is calculated, what load diversity factor exists or is used in the calculation, and what safety factor is used in the calculations, than of the size of the condenser.

Location

If an evaporative condenser is located with insufficient space for air movement, the effect is the same as that imposed by an inlet damper, and the fan may not deliver enough air. In addition, evaporative condenser discharge air may recirculate, which adds to the problem. The high inlet velocity causes a low-pressure region to develop around the fan inlet, inducing flow of discharge air into that region. If the obstruction is from a second condenser, the problem can be even more severe because discharge air from the second condenser flows into the air intake of the first.

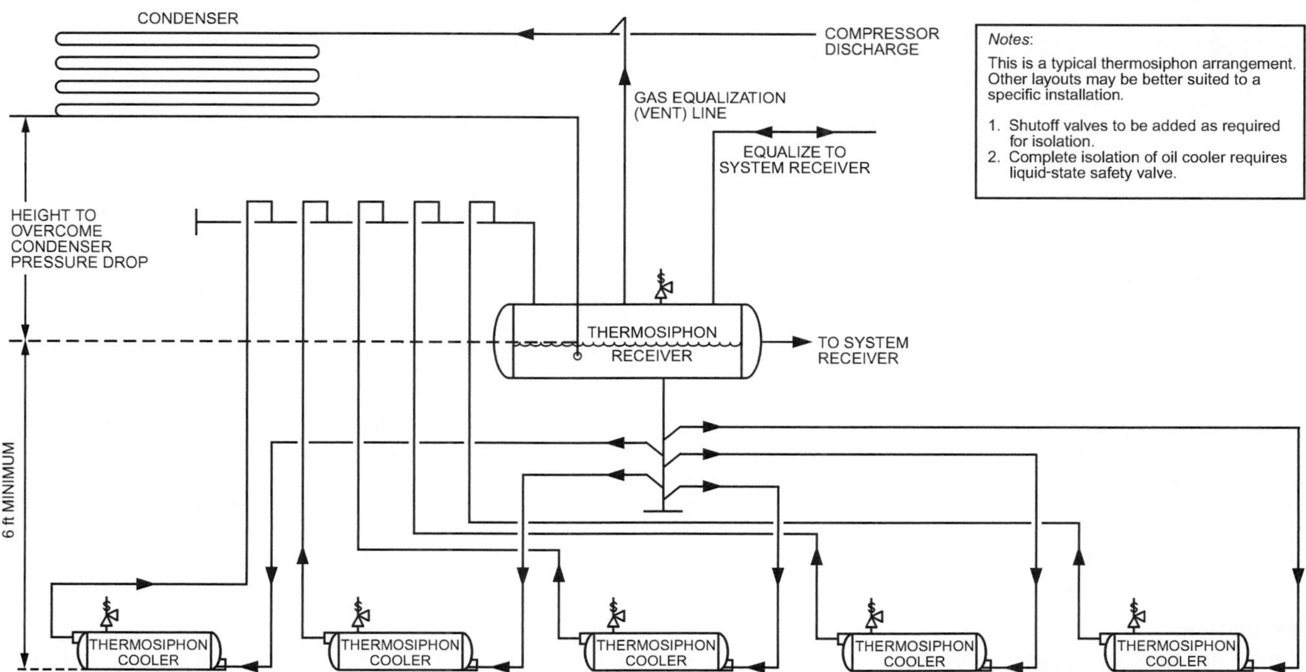

Fig. 24 Typical Thermosiphon System with Multiple Oil Coolers

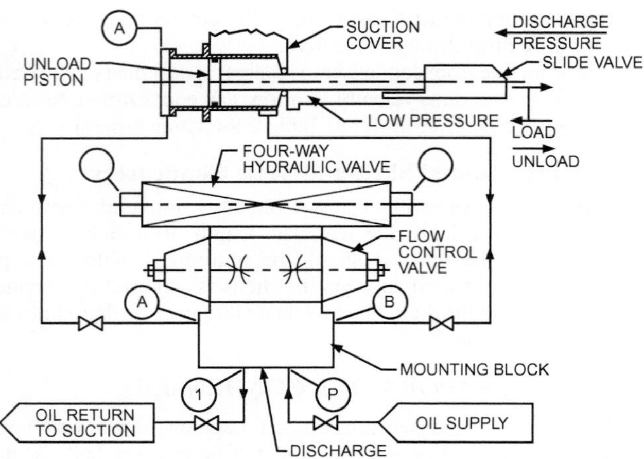

Fig. 25 Typical Hydraulic System for Slide Valve Capacity Control for Screw Compressor with Fixed V_i

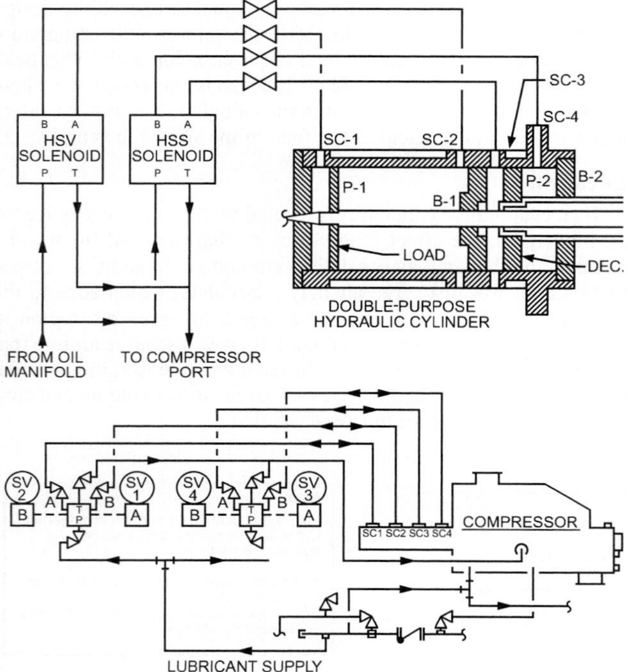

Fig. 26 Typical Positioning System for Slide Valve and Slide Stop for Variable-V_i Screw Compressor

Prevailing winds can also contribute to recirculation. In many areas, the winds shift with the seasons; wind direction during the peak high-humidity season is the most important consideration.

The tops of condensers should always be higher than any adjacent structure to eliminate downdrafts that might induce recirculation. Where this is impractical, discharge hoods can be used to discharge air far enough away from the fan intakes to avoid recirculation. However, the additional static pressure imposed by a discharge hood must be added to the fan system. Fan speed can be increased slightly to obtain proper air volume.

Installation

A single evaporative condenser used with a through-type (top inlet) receiver can be connected as shown in Figure 29. The receiver must always be at a lower pressure than the condensing

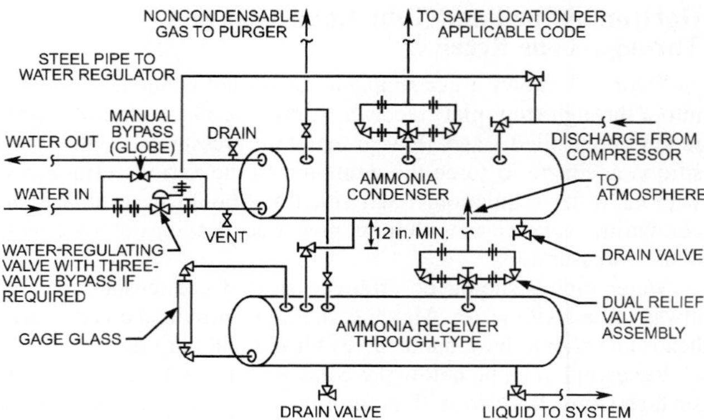

Fig. 27 Horizontal Condenser and Top Inlet Receiver Piping

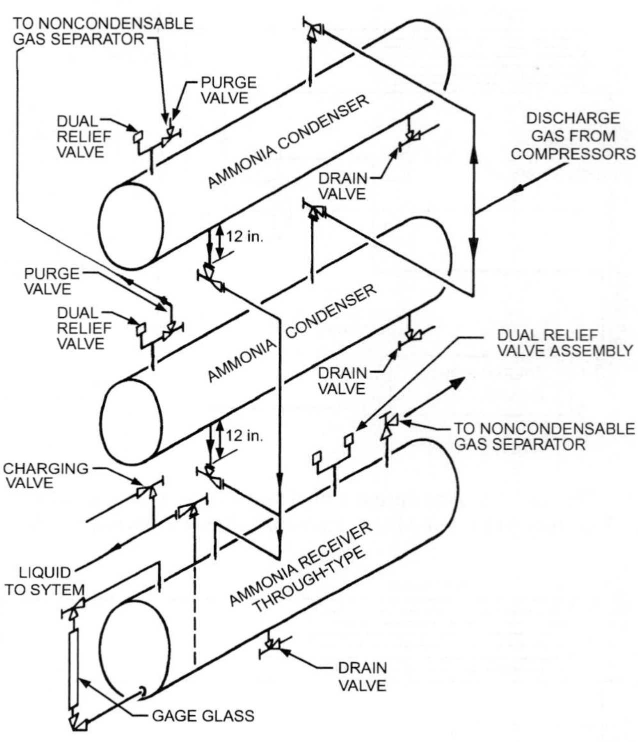

Fig. 28 Parallel Condensers with Top Inlet Receiver

pressure. Design ensures that the receiver is cooler than the condensing temperature.

Installation in Freezing Areas. In areas having ambient temperatures below 32°F, water in the evaporative condenser drain pan and water circuit must be kept from freezing at light plant loads. When the temperature is at freezing, the evaporative condenser can operate as a dry-coil unit, and the water pump(s) and piping can be drained and secured for the season.

Another method of keeping water from freezing is to place the water tank inside and install it as illustrated in Figure 30. When outdoor temperature drops, the condensing pressure drops, and a pressure switch with its sensing element in the discharge pressure line stops the water pump; the water is then drained into the tank. An alternative is to use a thermostat that senses water or outdoor ambient temperature and stops the pump at low temperatures. Exposed piping and any trapped water headers in the evaporative condenser should be drained into the indoor water tank.

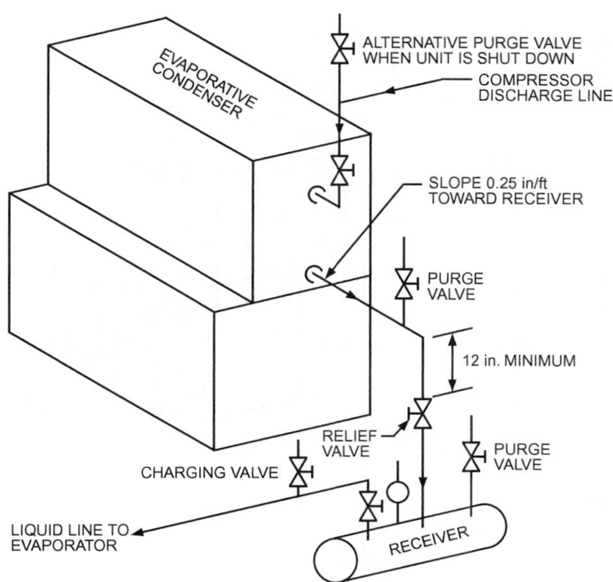

Fig. 29 Single Evaporative Condenser with Top Inlet Receiver

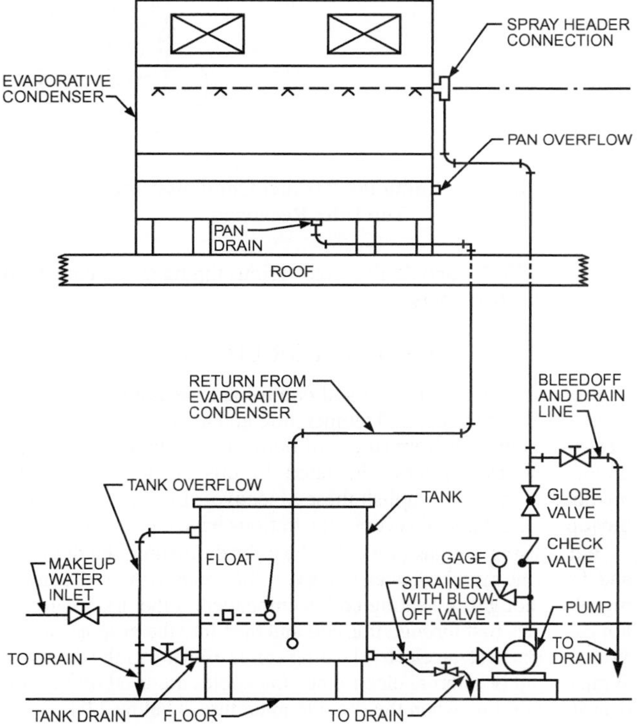

Fig. 30 Evaporative Condenser with Inside Water Tank

Air volume capacity control methods include inlet, outlet, or bypass dampers; two-speed fan motors; or fan cycling in response to pressure controls.

Liquid Traps. Because all evaporative condensers have substantial pressure drop in the ammonia circuit, liquid traps are needed at the outlets when two or more condensers or condenser coils are installed (Figure 31). Also, an equalizer line is necessary to maintain stable pressure in the receiver to ensure free drainage from condensers. For example, assume a 1 psi pressure drop in the operating condenser in Figure 31, which is producing a lower pressure

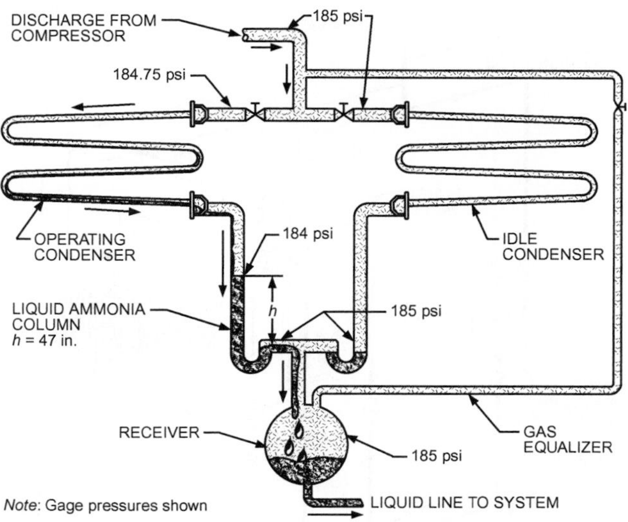

Fig. 31 Two Evaporative Condensers with Trapped Piping to Receiver

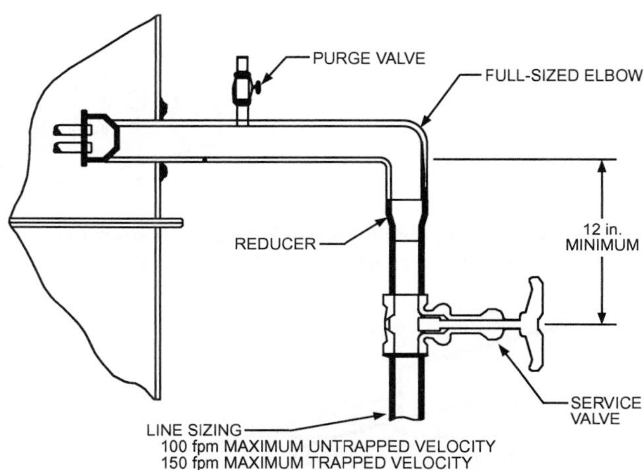

Fig. 32 Method of Reducing Condenser Outlet Sizes

(184 psig) at its outlet compared to the idle condenser (185 psig) and the receiver (185 psig). The trap creates a liquid seal so that a liquid height h of 47 in. (equivalent to 1 psi) builds up in the vertical drop leg and not in the condenser coil.

The trap must have enough height above the vertical liquid leg to accommodate a liquid height equal to the maximum pressure drop encountered in the condenser. The example illustrates the extreme case of one unit on and one off; however, the same phenomenon occurs to a lesser degree with two condensers of differing pressure drops when both are in full operation. Substantial differences in pressure drop can also occur between two different brands of the same size condenser or even different models produced by the same manufacturer.

The minimum recommended height of the vertical leg is 5 ft for ammonia. This vertical dimension h is shown in all evaporative condenser piping diagrams. This height is satisfactory for operation within reasonable ranges around normal design conditions and is based on the maximum condensing pressure drop of the coil. If service valves are installed at the coil inlets and/or outlets, the pressure drops imposed by these valves must be accounted for by increasing the minimum 5 ft drop-leg height by an amount equal to the valve pressure drop in height of liquid refrigerant (Figure 32).

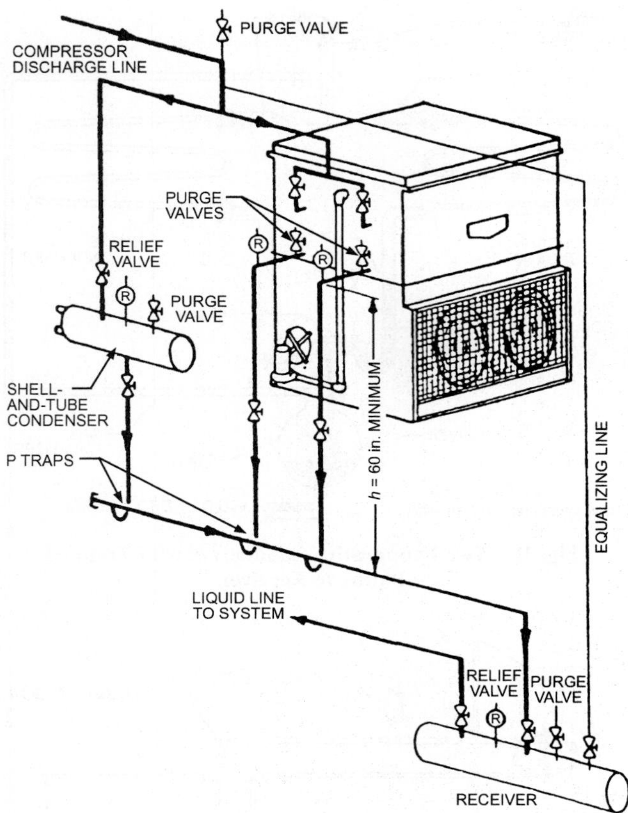

Fig. 33 Piping for Shell-and-Tube and Evaporative Condensers with Top Inlet Receiver

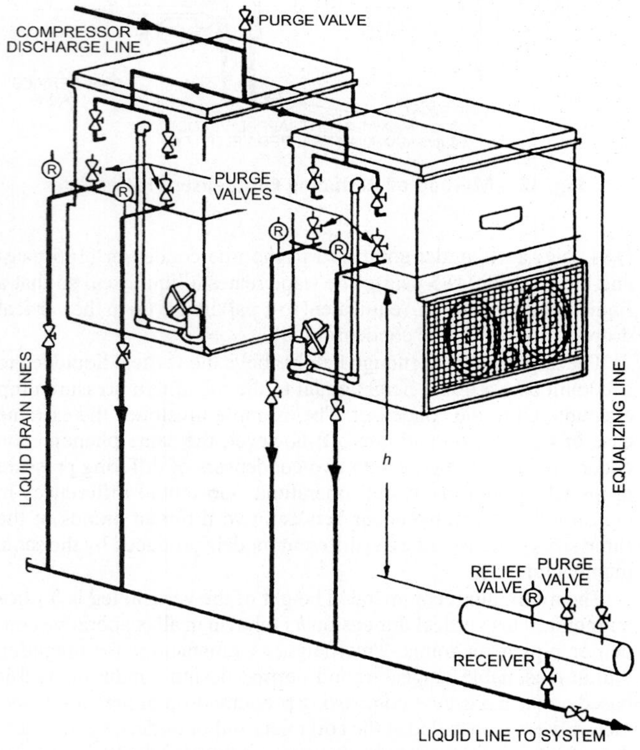

Fig. 34 Piping for Parallel Condensers with Surge-Type Receiver

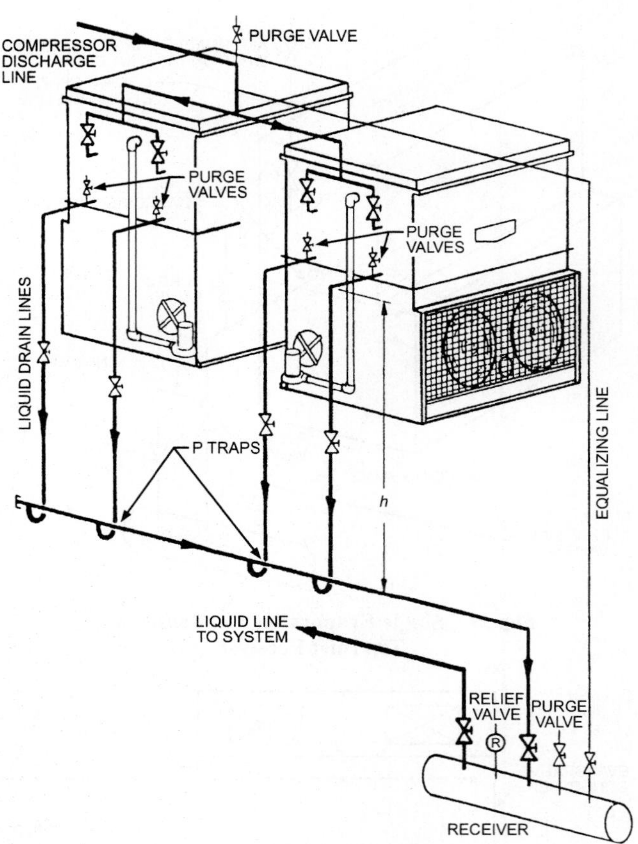

Fig. 35 Piping for Parallel Condensers with Top Inlet Receiver

Figures 33, 34, and 35 illustrate various piping arrangements for evaporative condensers.

EVAPORATOR PIPING

Proper evaporator piping and control are necessary to keep the cooled space at the desired temperature and also to adequately protect the compressor from surges of liquid ammonia out of the evaporator. The evaporators illustrated in this section show some methods used to accomplish these objectives. In some cases, combinations of details on several illustrations have been used.

When using hot gas or electric heat for defrosting, the drain pan and drain line must be heated to prevent the condensate from refreezing. With hot gas, a heating coil is embedded in the drain pan. The hot gas flows first through this coil and then into the evaporator coil. With electric heat, an electric heating coil is used under the drain pan. Wraparound or internal electric heating cables are used on the condensate drain line when the room temperature is below 32°F.

Figure 36 illustrates a thermostatic expansion valve on a unit cooler using hot gas for automatic defrosting. Because this is an automatic defrosting arrangement, hot gas must always be available at the hot-gas solenoid valve near the unit. The system must contain multiple evaporators so the compressor is running when the evaporator to be defrosted is shut down. The hot-gas header must be kept in a space where ammonia does not condense in the pipe. Otherwise, the coil receives liquid ammonia at the start of defrosting and is unable to take full advantage of the latent heat of hot-gas condensation entering the coil. This can also lead to severe hydraulic shock loads. If the header must be in a cold space, the insulated hot-gas main must be drained to the suction line by a high-pressure float.

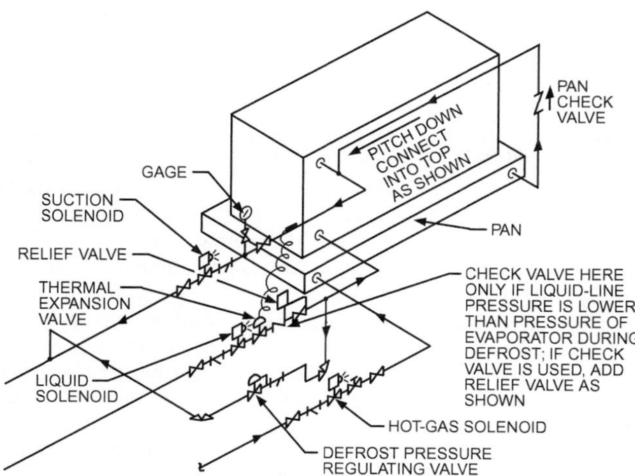

Fig. 36 Piping for Thermostatic Expansion Valve Application for Automatic Defrost on Unit Cooler

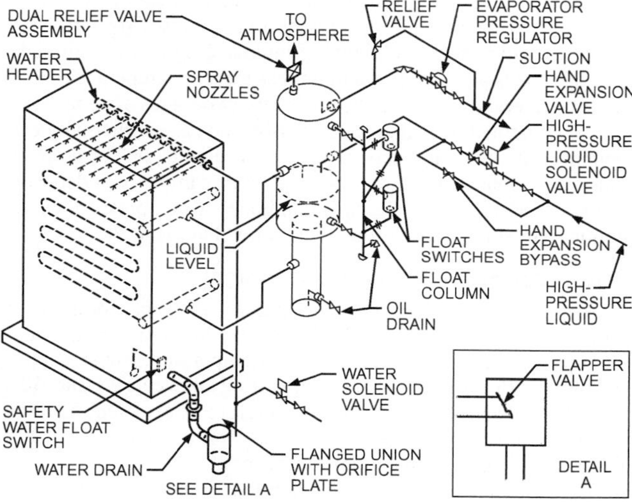

Fig. 37 Arrangement for Automatic Defrost of Air Blower with Flooded Coil

The liquid- and suction-line solenoid valves are open during normal operation only and are closed during the defrost cycle. When defrost starts, the hot-gas solenoid valve is opened. Refer to IIAR *Bulletin* 116 for information on possible hydraulic shock when the hot-gas defrost valve is opened after a defrost.

A defrost pressure regulator maintains a gage pressure of about 70 to 80 psi in the coil.

Unit Cooler—Flooded Operation

Figure 37 shows a flooded evaporator with a close-coupled low-pressure vessel for feeding ammonia into the coil and automatic water defrost.

The lower float switch on the float column at the vessel controls opening and closing of the liquid-line solenoid valve, regulating ammonia feed into the unit to maintain a liquid level. The hand expansion valve downstream of the solenoid valve should be adjusted so that it does not feed ammonia into the vessel more quickly than the vessel can accommodate while raising the suction pressure of gas from the vessel no more than 1 or 2 psi.

The static height of liquid in the vessel should be sufficient to flood the coil with liquid under normal loads. The higher float switch should be wired into an alarm circuit and possibly a compressor shutdown circuit for when the liquid level in the vessel is too high. With flooded coils having horizontal headers, distribution between the multiple circuits is accomplished without distributing orifices.

A combination evaporator pressure regulator and stop valve is used in the suction line from the vessel. During operation, the regulator maintains a nearly constant back pressure in the vessel. A solenoid coil in the regulator mechanism closes it during the defrost cycle. The liquid solenoid valve should also be closed at this time. One of the best means of controlling room temperature is a room thermostat that controls the effective setting of the evaporator pressure regulator.

A spring-loaded relief valve is used around the suction pressure regulator and is set so that the vessel is kept below 125 psig.

A solenoid valve unaffected by downstream pressure is used in the water line to the defrost header. The defrost header is constructed so that it drains at the end of the defrost cycle and the downstream side of the solenoid valve drains through a fixed orifice.

Unless the room is maintained above 32°F, the drain line from the unit should be wrapped with a heater cable or provided with another heat source and then insulated to prevent defrost water from refreezing in the line.

Water line length in the space leading up to the header and the length of the drain line in the cooled space should be kept to a minimum. A flapper or pipe trap on the end of the drain line prevents warm air from flowing up the drain pipe and into the unit.

An air outlet damper may be closed during defrosting to prevent thermal circulation of air through the unit, which affects the temperature of the cooled space. The fan is stopped during defrost.

This type of defrosting requires a drain pan float switch for safety control. If the drain pan fills with water, the switch overrides the time clock to stop flow into the unit by closing the water solenoid valve.

There should be a 5 min delay at the end of the water spray part of the defrosting cycle so water can drain from the coil and pan. This limits ice build-up in the drain pan and on the coils after the cycle is completed.

On completion of the cycle, the low-pressure vessel may be at about 75 psig. When the unit is opened to the much-lower-pressure suction main, some liquid surges out into the main; therefore, it may be necessary to gradually bleed off this pressure before fully opening the suction valve in order to prevent thermal shock. Generally, a suction trap in the engine room removes this liquid before the gas stream enters the compressors.

The type of refrigerant control shown in Figure 37 can be used on brine spray units where brine is sprayed over the coil at all times to pick up the condensed water vapor from the airstream. The brine is reconcentrated continually to remove water absorbed from the airstream.

High-Side Float Control

When a system has only one evaporator, a high-pressure float control can be used to keep the condenser drained and to provide a liquid seal between the high and low sides. Figure 38 illustrates a brine or water cooler with this type of control. The high-side float should be located near the evaporator to avoid insulating the liquid line.

The amount of ammonia in this type of system is critical because the charge must be limited so that liquid will not surge into the suction line under the highest loading in the evaporator. Some type of suction trap should be used. One method is to place a horizontal shell above the cooler, with suction gas piped into the bottom and out the top. The reduction of gas velocity in this shell causes liquid to separate from the gas and draw back into the chiller.

Coolers should include a liquid indicator. A reflex glass lens with a large liquid chamber and vapor connections for boiling liquids and a plastic frost shield to determine the actual level should be used. A refrigeration thermostat measuring chilled-fluid temperature as it exits the cooler should be wired into the compressor starting circuit to prevent freezing.

A flow switch or differential pressure switch should prove flow before the compressor starts. The fluid to be cooled should be piped into the lower portion of the tube bundle and out of the top portion.

Low-Side Float Control

For multiple evaporator systems, low-side float valves are used to control the refrigerant level in flooded evaporators. The low-pressure float in Figure 39 has an equalizer line from the top of the float chamber to the space above the tube bundle and an equalizer line out of the lower side of the float chamber to the lower side of the tube bundle.

For positive shutoff of liquid feed when the system stops, a solenoid valve in the liquid line is wired so that it is only energized when the brine or water pump motor is operating and the compressor is running.

A reflex glass lens with large liquid chamber and vapor connections for boiling liquids should be used with a plastic frost shield to determine the actual level, and with front extensions as required.

Usually a high-level float switch is installed above the operating level of the float to shut the liquid solenoid valve if the float should overfeed.

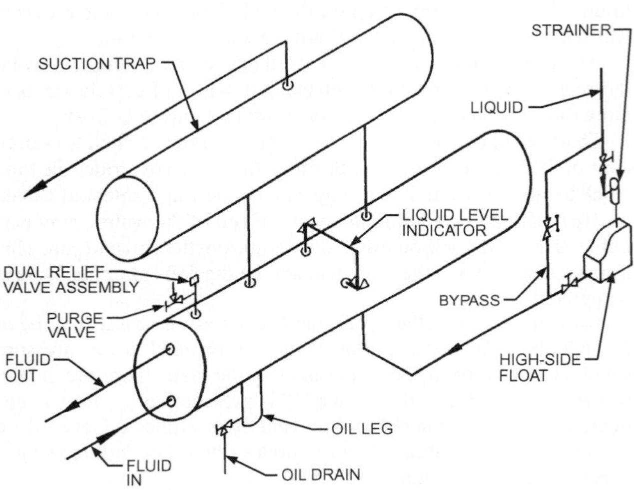

Fig. 38 Arrangement for Horizontal Liquid Cooler and High-Side Float

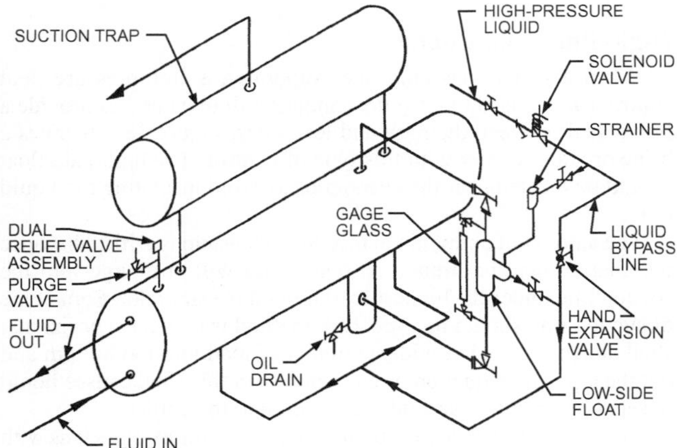

Fig. 39 Piping for Evaporator and Low-Side Float with Horizontal Liquid Cooler

MULTISTAGE SYSTEMS

As pressure ratios increase, single-stage ammonia systems encounter problems such as (1) high discharge temperatures on reciprocating compressors causing lubricant to deteriorate, (2) loss of volumetric efficiency as high pressure leaks back to the low-pressure side through compressor clearances, and (3) excessive stresses on compressor moving parts. Thus, manufacturers usually limit the maximum pressure ratios for multicylinder reciprocating machines to approximately 7 to 9. For screw compressors, which incorporate cooling, compression ratio is not a limitation, but efficiency deteriorates at high ratios.

When the overall system pressure ratio (absolute discharge pressure divided by absolute suction pressure) begins to exceed these limits, the pressure ratio across the compressor must be reduced. This is usually done by using a multistage system. A properly designed two-stage system exposes each of the two compressors to a pressure ratio approximately equal to the square root of the overall pressure ratio. In a three-stage system, each compressor is exposed to a pressure ratio approximately equal to the cube root of the overall ratio. When screw compressors are used, this calculation does not always guarantee the most efficient system.

Another advantage to multistaging is that successively subcooling liquid at each stage of compression increases overall system operating efficiency. Additionally, multistaging can accommodate multiple loads at different suction pressures and temperatures in the same refrigeration system. In some cases, two stages of compression can be contained in a single compressor, such as an internally compounded reciprocating compressor. In these units, one or more cylinders are isolated from the others so they can act as independent stages of compression. Internally compounded compressors are economical for small systems that require low temperature.

Two-Stage Screw Compressor System

A typical two-stage, two-temperature screw compressor system provides refrigeration for high- and low-temperature loads (Figure 40). For example, the high-temperature stage supplies refrigerant to all process areas operating between 28 and 50°F. An 18°F intermediate suction temperature is selected. The low-temperature stage requires a –35°F suction temperature for blast freezers and continuous or spiral freezers.

The system uses a flash intercooler that doubles as a recirculator for the 18°F load. It is the most efficient system available if the screw compressor uses indirect lubricant cooling. If refrigerant injection cooling is used, system efficiency decreases. This system is efficient for several reasons:

1. Approximately 50% of the booster (low-stage) motor heat is removed from the high-stage compressor load by the thermosiphon lubricant cooler.

   Note: In any system, thermosiphon lubricant cooling for booster and high-stage compressors is about 10% more efficient than injection cooling. Also, plants with a piggyback, two-stage screw compressor system without intercooling or injection cooling can be converted to a multistage system with indirect cooling to increase system efficiency approximately 15%.
2. Flash intercoolers are more efficient than shell-and-coil intercoolers by several percent.
3. Thermosiphon lubricant cooling of the high-stage screw compressor provides the highest efficiency available. Installing indirect cooling in plants with liquid injection cooling of screw compressors can increase compressor efficiency by 3 to 4%.
4. Thermosiphon cooling saves 20 to 30% in electric energy during the low-temperature months. When outside air temperature is low, the condensing pressure can be decreased to 90 to 100 psig in most ammonia systems. With liquid injection cooling, the condensing pressure can only be reduced to approximately 125 to 130 psig.

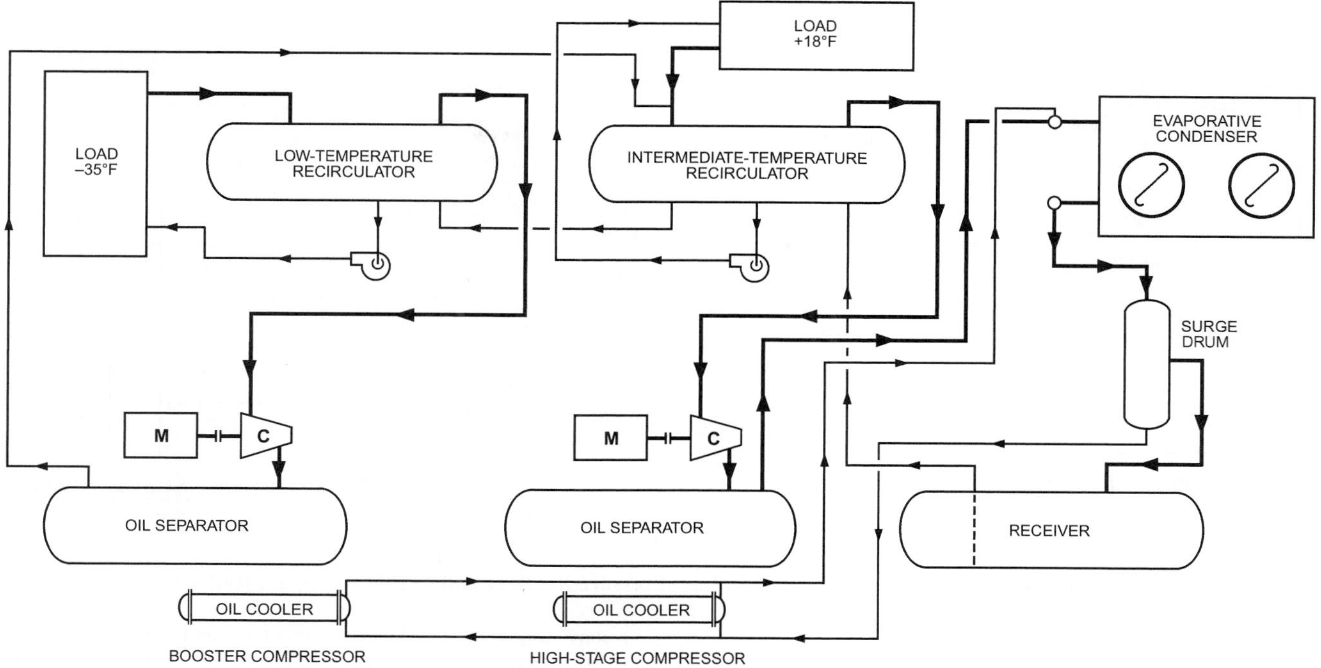

Fig. 40 Compound Ammonia System with Screw Compressor Thermosiphon Cooled

5. Variable-V_i compressors with microprocessor control require less total energy when used as high-stage compressors. The controller tracks compressor operating conditions to take advantage of ambient conditions as well as variations in load.

Converting Single-Stage into Two-Stage Systems

When plant refrigeration capacity must be increased and the system is operating below about 10 psig suction pressure, it is usually more economical to increase capacity by adding a compressor to operate as the low-stage compressor of a two-stage system than to implement a general capacity increase. The existing single-stage compressor then becomes the high-stage compressor of the two-stage system. When converting, consider the following:

- The motor on the existing single-stage compressor may have to be increased in size when used at a higher suction pressure.
- The suction trap should be checked for sizing at the increased gas flow rate.
- An intercooler should be added to cool the low-stage compressor discharge gas and to cool high-pressure liquid.
- A condenser may need to be added to handle the increased condensing load.
- A means of purging air should be added if plant suction gage pressure is below zero.
- A means of automatically reducing compressor capacity should be added so that the system will operate satisfactorily at reduced system capacity points.

LIQUID RECIRCULATION SYSTEMS

The following discussion gives an overview of liquid recirculation (liquid overfeed) systems. See Chapter 1 for more complete information. For additional engineering details on liquid overfeed systems, refer to Stoecker (1988).

In a liquid ammonia recirculation system, a pump circulates ammonia from a low-pressure receiver to the evaporators. The low-pressure receiver is a shell for storing refrigerant at low pressure and is used to supply evaporators with refrigerant, either by gravity or by a low-head pump. It also takes suction from the evaporators and separates gas from the liquid. Because the amount of liquid fed into the evaporator is usually several times the amount that actually evaporates there, liquid is always present in the suction return to the low-pressure receiver. Frequently, three times the evaporated amount is circulated through the evaporator (see Chapter 1).

Generally, the liquid ammonia pump is sized by the flow rate required and a pressure differential of about 25 psi. This is satisfactory for most single-story installations. If there is a static lift on the pump discharge, the differential is increased accordingly.

The low-pressure receiver should be sized by the cross-sectional area required to separate liquid and gas and by the volume between the normal and alarm liquid levels in the low-pressure receiver. This volume should be sufficient to contain the maximum fluctuation in liquid from the various load conditions (see Chapter 1).

Liquid at the pump discharge is in the subcooled region. A total pressure drop of about 5 psi in the piping can be tolerated.

The remaining pressure is expended through the control valve and coil. Pressure drop and heat pickup in the liquid supply line should be low enough to prevent flashing in the liquid supply line.

Provisions for liquid relief from the liquid main back to the low-pressure receiver are required, so when liquid-line solenoid valves at the various evaporators are closed, either for defrosting or for temperature control, the excess liquid can be relieved back to the receiver. Generally, relief valves used for this purpose are set at about 40 psi differential when positive-displacement pumps are used. When centrifugal pumps are used, a hand expansion valve or a minimum flow orifice is acceptable to ensure that the pump is not dead-headed.

The suction header between evaporators and low-pressure receiver should be pitched 1% to allow excess liquid flow back to the low-pressure receiver. The header should be designed to avoid traps.

Liquid Recirculation in Single-Stage System. Figure 41 shows the piping of a typical single-stage system with a low-pressure receiver and liquid ammonia recirculation feed.

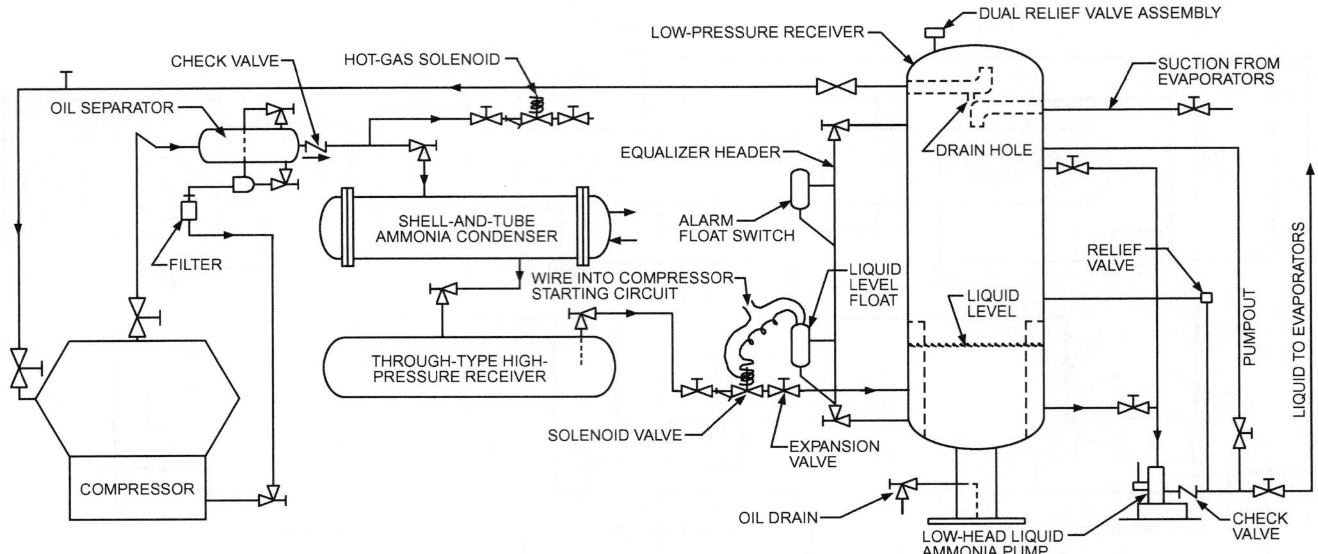

Fig. 41 Piping for Single-Stage System with Low-Pressure Receiver and Liquid Ammonia Recirculation

Hot-Gas Defrost

This section was taken from a technical paper by Briley and Lyons (1992). Several methods are used for defrosting coils in areas below 35°F room temperature:

- Hot refrigerant gas (the predominant method)
- Water
- Air
- Combinations of hot gas, water, and air

The evaporator (air unit) in a liquid recirculation system is circuited so that the refrigerant flow provides maximum cooling efficiency. The evaporator can also work as a condenser if the necessary piping and flow modifications are made. When the evaporator operates as a condenser and the fans are shut down, hot refrigerant vapor raises the surface temperature of the coil enough to melt any ice and/or frost on the surface so that it drains off. Although this method is effective, it can be troublesome and inefficient if the piping system is not properly designed.

Even when fans are not operating, 50% or more of the heat given up by the refrigerant vapor may be lost to the space. Because the heat transfer rate varies with the temperature difference between coil surface and room air, the temperature/pressure of the refrigerant during defrost should be minimized.

Another reason to maintain the lowest possible defrost temperature/pressure, particularly in freezers, is to keep the coil from steaming. Steam increases refrigeration load, and the resulting icicle or frost formation must be dealt with. Icicles increase maintenance during cleanup; ice formed during defrost tends to collect at the fan rings, which sometimes restricts fan operation.

Defrosting takes slightly longer at lower defrost pressures. The shorter the time heat is added to the space, the more efficient the defrost. However, with slightly extended defrost times at lower temperature, overall defrosting efficiency is much greater than at higher temperature/pressure because refrigeration requirements are reduced.

Another loss during defrost can occur when hot, or uncondensed, gas blows through the coil and relief regulator and vents back to the compressor. Some of this gas load cannot be contained and must be vented to the compressor through the wet return line. It is most energy-efficient to vent this hot gas to the highest suction possible; an evaporator defrost relief should be vented to the intermediate or high-stage compressor if the system is two-stage. Figure 42 shows a conventional hot-gas defrost system for evaporator coils of 15 tons of refrigeration and below. Note that the wet return is above the evaporator and that a single riser is used.

Defrost Control. Because defrosting efficiency is low, frequency and duration of defrosting should be kept to the minimum necessary to keep the coils clean. Less defrosting is required during winter than during hotter, more humid periods. An effective energy-saving measure is to reset defrost schedules in the winter.

Several methods are used to initiate the defrost cycle. **Demand defrost**, actuated by a pressure device that measures air pressure drop across the coil, is a good way of minimizing total daily defrost time. The coil is defrosted automatically only when necessary. Demand initiation, together with a float drainer to dump the liquid formed during defrost to an intermediate vessel, is the most efficient defrost system available (Figure 43).

The most common defrost control method, however, is **time-initiated, time-terminated**; it includes adjustable defrost duration and an adjustable number of defrost cycles per 24 h period. This control is commonly provided by a defrost timer.

Estimates indicate that the load placed on a refrigeration system by a coil during defrost is up to three times the operating design load. Thus, it is important to properly engineer hot-gas defrost systems.

Designing Hot-Gas Defrost Systems. Several approaches are followed in designing hot-gas defrost systems. Figure 43 shows a typical demand defrost system for both upfeed and downfeed coils. This design returns defrost liquid to the system's intermediate pressure. An alternative is to direct defrost liquid into the wet suction. A float drainer or thermostatic trap with a hot-gas regulator installed at the hot-gas inlet to the coil is much better than the relief regulator (see Figure 43).

Most defrost systems installed today (Figure 42) use a time clock to initiate defrost; the demand defrost system shown in Figure 43 uses a low differential pressure switch to sense the air pressure drop across the coil and actuate the defrost. A thermostat terminates the defrost cycle. A timer is used as a back-up to ensure the defrost terminates.

Sizing and Designing Hot-Gas Piping. Hot gas is supplied to the evaporators in two ways:

1. The preferred method is to install a pressure regulator set at approximately 100 psig in the equipment room at the hot-gas takeoff and size the piping accordingly.

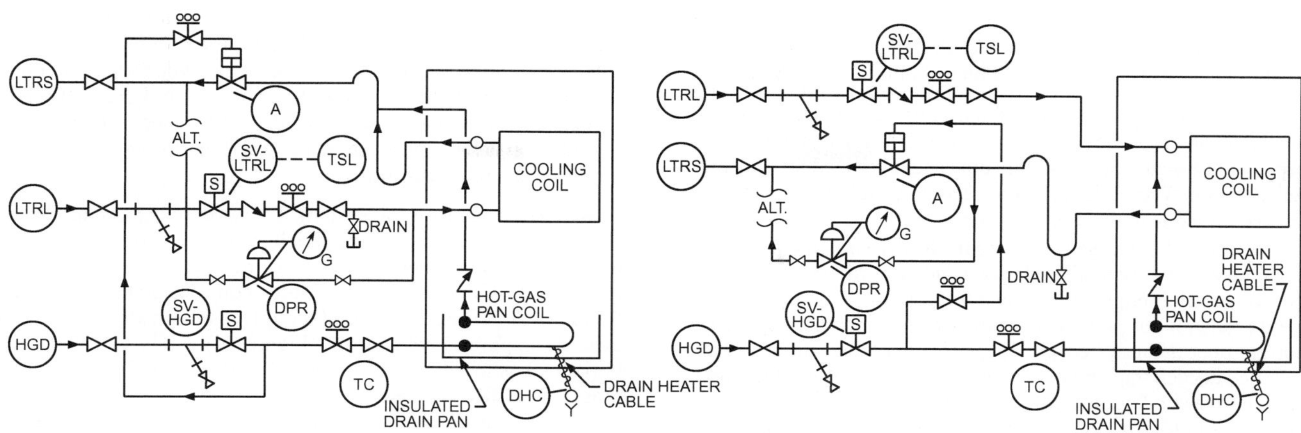

UPFEED COILS

DOWNFEED OR CROSSFEED COILS

LEGEND	
A	= Gas-powered automatic shutoff valve
DHC	= Defrost heater cable
DPR	= Differential pressure regulator (set at ±90 psig)
G	= 0 to 150 psi, 3 1/2 in. gage with valve
HGD	= Hot-gas defrost supply (185 to 100 psig)
LTRL	= Low-temperature recirculated liquid
LTRS	= Low-temperature recirculated suction
SV-HGD	= Hot-gas defrost solenoid valve
SV-LTRL	= Solenoid valve, low-temperature recirculated liquid
TC	= Time clock
TL	= Thermostat: operates SV-LTRL on demand
TSL	= Room thermostat: operates SV-LTRLF on demand

DEFROST CYCLE ACTUATED BY TIME CLOCK

SEQUENCE OF OPERATION

A. Deenergize SV-LTRL and stop fans.
Delay 2 to 3 min for downfeed and crossfeed coils and 4 to 8 min for upfeed coils.
Energize SV-HGD, which closes valve A.

B. Coil defrost period terminated by Time Clock. (Keep defrost time to a minimum: just enough to clean and drain pan.)
TC deenergizes SV-HGD, which opens valve A.

C. Energize SV-LTRL.
Delay 2 min to freeze any remaining water on coil.

D. Start fans.

Fig. 42 Conventional Hot-Gas Defrost Cycle
(For coils with 15 tons refrigeration capacity and below)

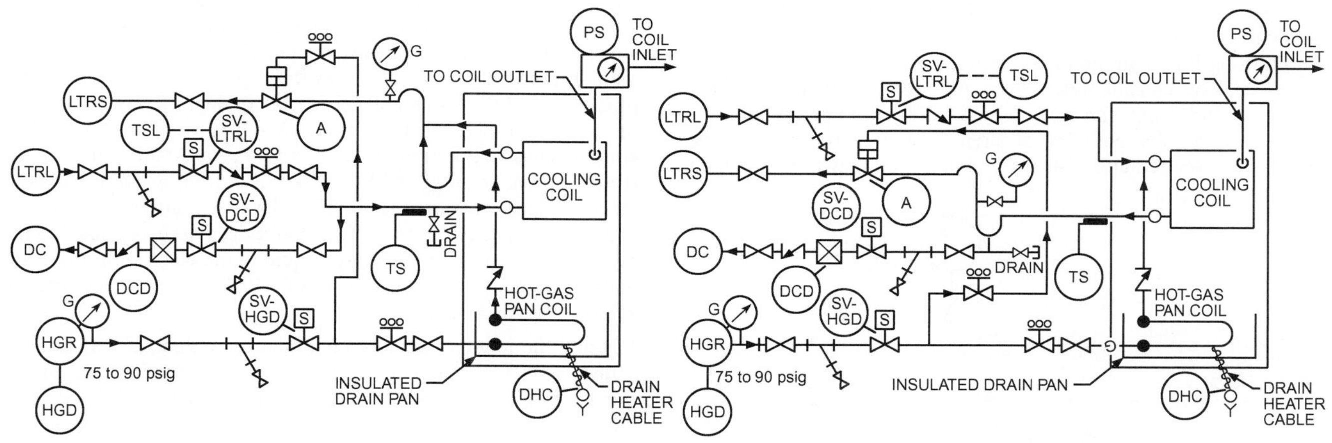

UPFEED COILS

DEFROST CYCLE ACTUATED BY PRESSURE SWITCH AT PREDETERMINED SETTING (approximately 1 in. water gage)

DOWNFEED OR CROSSFEED COILS

LEGEND			
A	= Gas-powered automatic shutoff valve	LTRS	= Low-temperature recirculated suction
DC	= Return liquid defrost condensate	PS	= Air-side pressure switch
DCD	= Defrost condensate return drainer	SV-DCD	= Solenoid valve, defrost condensate return drainer
DHC	= Defrost heater cable	SV-HGD	= Hot-gas defrost solenoid valve
G	= 0 to 150 psi, 3 1/2 in. gage with valve	SV-LTRL	= Recirculated liquid-line solenoid valve
HGD	= Hot-gas defrost supply	TS	= Thermostat: terminates hot-gas defrost cycle
HGR	= Hot-gas regulator valve	TSL	= Room thermostat: operates SV-LTRL on demand
LTRL	= Low-temperature recirculated liquid		

SEQUENCE OF OPERATION

A. Deenergize SV-LTRL and stop fans. Delay energize SV-HGD closing A. Coil will defrost.

B. Coil defrost period terminated by TS set at 40°F. Use override timer after 30 min to force SV-HGD to deenergize if TS malfunctions.

C. Deenergize SV-HGD via thermostat TS, opening valve A.

D. Energize SV-LTRL. Delay 2 to 4 min.

E. Start fans.

F. To save energy, return liquid to intercooler of intermediate-temperature recirculator.
Alternative: Return liquid to wet suction downstream of stop valve.

Note: This defrost control method assumes a hot-gas outlet pressure regulator in equipment room. If high-pressure hot gas is used, an outlet pressure regulator with electric shutoff should be substituted for SV-HGD and set at 75 to 90 psig.

Fig. 43 Demand Defrost Cycle
(For coils with 15 tons refrigeration capacity and below)

2. The alternative is to install a pressure regulator at each evaporator or group of evaporators and size the piping for minimum design condensing pressure, which should be 75 to 85 psig.

A maximum of one-third of the coils in a system should be defrosted at one time. If a system has 300 tons of refrigeration capacity, the main hot-gas supply pipe could be sized for 100 tons of refrigeration. The outlet pressure-regulating valve should be sized in accordance with the manufacturer's data.

Reducing defrost hot-gas pressure in the equipment room has advantages, notably that less liquid condenses in the hot-gas line as the condensing temperature drops to 52 to 64°F. A typical equipment room hot-gas pressure control system is shown in Figure 44. If hot-gas lines in the system are trapped, a condensate drainer must be installed at each trap and at the low point in the hot-gas line (Figure 45). Defrost condensate liquid return piping from coils where a float or thermostatic valve is used should be one size larger than the liquid feed piping to the coil.

Demand Defrost. The following are advantages and features of demand defrost:

- It uses the least energy for defrost.
- It increases total system efficiency because coils are off-line for a minimum amount of time.
- It imposes less stress on the piping system because there are fewer defrost cycles.
- Regulating hot gas to approximately 100 psig in the equipment room gives the gas less chance of condensing in supply piping. Liquid in hot-gas systems may cause problems because of the hydraulic shock created when the liquid is forced into an evaporator (coil). Coils in hot-gas pans may rupture as a result.
- Draining liquid formed during defrost with a float or thermostatic drainer eliminates hot-gas blow-by normally associated with pressure-regulating valves installed around the wet suction return line pilot check valve.
- Returning liquid ammonia to the intercooler or high-stage recirculator saves considerable energy. A 20 ton refrigeration coil defrosting for 12 min can condense up to 24 lb/min of ammonia, or 288 lb total. The enthalpy difference between returning to the low-stage recirculator (–40°F) and the intermediate recirculator (+20°F) is 64 Btu/lb, for 18,432 Btu total or 7.68 tons of refrigeration removed from the –40°F booster for 12 min. This assumes that only liquid is drained and is the saving when liquid is drained to the intermediate point, not the total cost to defrost. If a pressure-reducing valve is used around the pilot check valve, this rate could double or triple because hot gas flows through these valves in unmeasurable quantities.

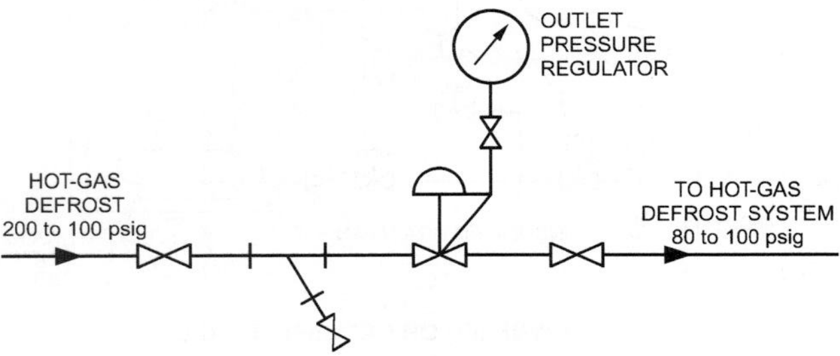

Fig. 44 Equipment Room Hot-Gas Pressure Control System

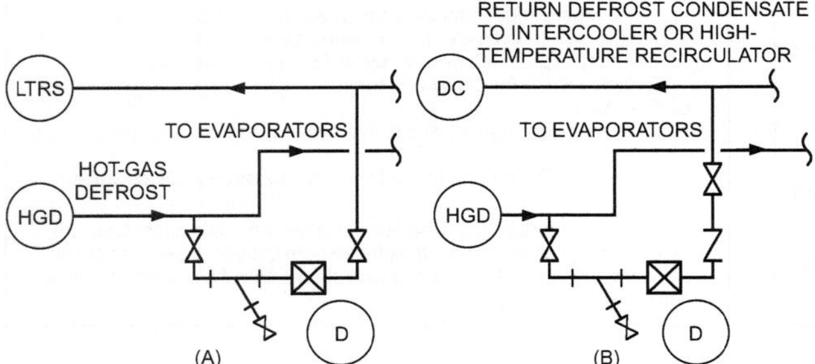

Fig. 45 Hot-Gas Condensate Return Drainer

Soft Hot-Gas Defrost System. This system is particularly well suited to large evaporators and should be used on all coils of 15 tons of refrigeration or over. It eliminates the valve clatter, pipe movements, and noise associated with large coils during hot-gas defrost. Soft hot-gas defrost can be used for upfeed or downfeed coils; however, the piping systems differ (Figure 46). Coils operated in the horizontal plane must be orificed. Vertical coils that usually are crossfed are also orificed.

Soft hot-gas defrost is designed to increase coil pressure gradually as defrost begins. This is accomplished by a small hot-gas feed having a capacity of about 25 to 30% of the estimated duty with a solenoid and a hand expansion valve adjusted to bring the pressure up to about 40 psig in 3 to 5 min. (See Sequence of Operation in Figure 46.) After defrost, a small suction-line solenoid is opened so that the coil can be brought down to operation pressure gradually before liquid is introduced and the fans started. The system can be initiated by a pressure switch; however, for large coils in spiral or individual quick freezing systems, manual initiation is preferred.

This system eliminates check valve chatter and most, if not all, liquid hammer (i.e., hydraulic problems in the piping). In addition, the last three features listed in the section on Demand Defrost apply to soft hot-gas defrost.

Double Riser Designs for Large Evaporator Coils

Static pressure penalty is the pressure/temperature loss associated with a refrigerant vapor stream bubbling through a liquid bath. If speed in the riser is high enough, it will carry over a certain amount of liquid, thus reducing the penalty. For example, at –40°F ammonia has a density of 43.07 lb/ft^3, which is equivalent to a pressure of 43.07/144 = 0.30 psi per foot of depth. Thus, a 16 ft riser has a column of liquid that exerts 16 × 0.30 = 4.8 psi. At –40°F, ammonia has a saturation pressure of 10.4 psia. At the bottom of the riser then, the pressure is 4.8 + 10.4 = 15.2 psia, which is the saturation pressure of ammonia at –27°F. This 13°F difference amounts to a 0.81°F penalty per foot of riser. If a riser were oversized to the point that the vapor did not carry liquid to the wet return, the evaporator would be at –27°F instead of –40°F. This problem can be solved in several ways:

- Install the low-temperature recirculated suction (LTRS) line below the evaporator. This method is very effective for downfeed evaporators. Suction from the coil should not be trapped. This arrangement also ensures lubricant return to the recirculator.
- Where the LTRS is above the evaporator, install a liquid return system below the evaporator (Figure 47). This arrangement eliminates static penalty, which is particularly advantageous for plate, individual quick freeze, and spiral freezers.
- Use double risers from the evaporator to the LTRS (Figure 48).

If a single riser is sized for minimum pressure drop at full load, the static pressure penalty is excessive at part load, and lubricant return could be a problem. If the single riser is sized for minimum load, then riser pressure drop is excessive and counterproductive.

Double risers solve these problems (Miller 1979). Figure 48 shows that, when maximum load occurs, both risers return vapor and liquid to the wet suction. At minimum load, the large riser is sealed by liquid ammonia in the large trap, and refrigerant vapor flows through the small riser. A small trap on the small riser ensures that some lubricant and liquid return to the wet suction.

Risers should be sized so that pressure drop, calculated on a dry-gas basis, is at least 0.3 psi per 100 ft. The larger riser is designed for

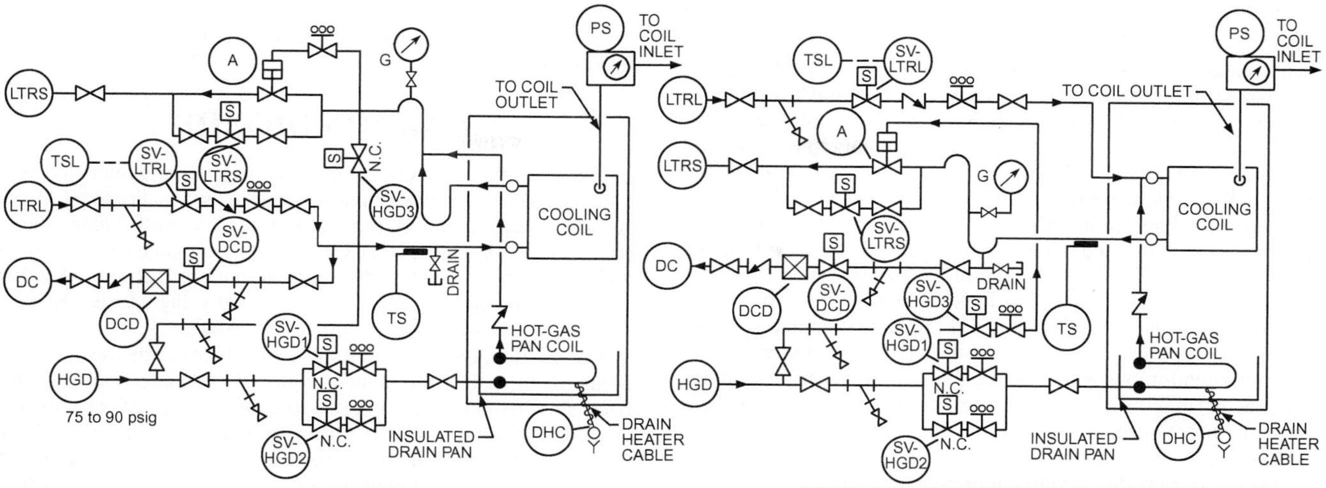

UPFEED COILS

HOT-GAS DEFROST CYCLE ACTUATED BY AIR-SIDE PRESSURE SWITCH (PS) AT PREDETERMINED SETTING (approximately 1 in. water gage)

LEGEND

A	= Gas-powered automatic shutoff valve LTRS	SV-HGD1	= Hot-gas solenoid valve (main hot-gas defrost supply)
DC	= Defrost condensate return liquid	SV-HGD2	= Hot-gas solenoid valve (25% hot-gas defrost supply)
DCD	= Defrost condensate drainer	SV-HGD3	= Hot-gas solenoid valve (to operate valve A)
DHC	= Defrost heater cable	SV-LTRL	= Liquid-line solenoid valve
G	= 0 to 150 psi, 3 1/2 in. gage with valve	SV-LTRS	= Solenoid valve bypass to LTRS
HGD	= Hot-gas defrost supply 75 to 90 psig	TS	= Thermostat: terminates hot-gas defrost cycle
LTRL	= Low-temperature recirculated liquid	TSL	= Thermostat: operates SV-LTRL on demand
LTRS	= Low-temperature recirculated suction		
PS	= Air-side pressure switch		
PS2	= Pressure switch to operate SV-HGD1		
SV-DCD	= Solenoid valve, defrost condensate return drainer		

DOWNFEED OR CROSSFEED COILS (SINGLE RISER)

SEQUENCE OF OPERATION

A. Deenergize SV-LTRL and stop fans. Energize SV-HGD3, which closes valve A.
B. After 60 s, energize SV-HGD2 to gradually increase coil internal pressure.
C. When coil pressure reaches about 40 psig, pressure switch PS2 energizes SV-HGD1 and defrost continues.
D. When liquid reaches about 40°F, temperature switch (TS) closes SV-HGD1 and SV-HGD2.
E. Wait 1 min and open small solenoid valve SV-LTRS. Then wait 2 to 4 min.
F. Deenergize SV-HGD3, which opens valve A and closes SV-LTRS.
G. Energize SV-LTRL and wait 2 to 4 min.
H. Start fans.
I. To save energy, return defrost condensate (DC) to intercooler or intermediate-temperature recirculator.
Alternative: Return liquid to LTRS downstream of stop valve.

Note 1: This defrost method assumes an outlet pressure regulator in equipment room. If high-pressure hot gas is used, an outlet pressure regulator is installed at the HGD inlet.
Note 2: Soft hot-gas defrost also may be manually initiated.

Fig. 46 Soft Hot-Gas Defrost Cycle
(For coils with 15 tons refrigeration capacity or above)

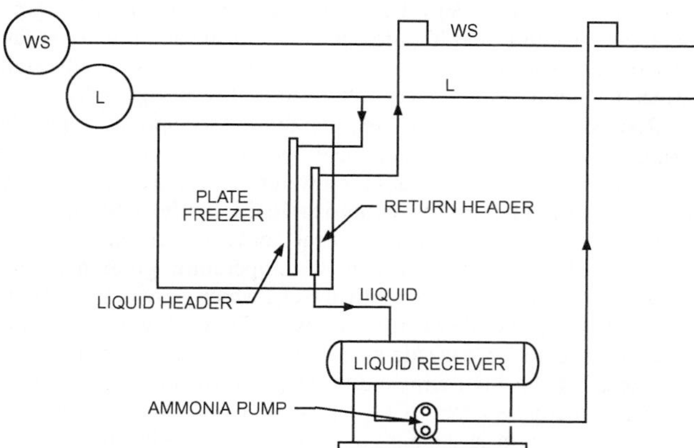

Fig. 47 Recirculated Liquid Return System

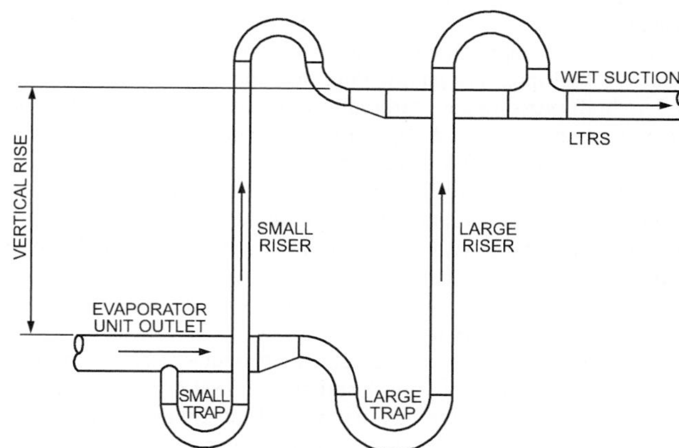

Fig. 48 Double Low-Temperature Suction Risers

approximately 65 to 75% of the flow and the small one for the remainder. This design results in a velocity of approximately 5000 fpm or higher. Some coils may require three risers (large, medium, and small).

Over the years, freezer capacity has grown. As they became larger, so did the evaporators (coils). Where these freezers are in line and the product to be frozen is wet, the defrost cycle can be every 4 or 8 h. Many production lines limit defrost duration to 30 min. If coils are large (some coils have a refrigeration capacity of 200 to 300 tons), it is difficult to design a hot-gas defrost system that can complete a safe defrost in 30 min. Sequential defrost systems, where coils are defrosted alternately during production, are feasible but require special treatment.

SAFETY CONSIDERATIONS

Ammonia is an economical choice for industrial systems. Although ammonia has superior thermodynamic properties, it is considered toxic at low concentration levels of 35 to 50 ppm. Large quantities of ammonia should not be vented to enclosed areas near open flames or heavy sparks. Ammonia at 16 to 25% by volume burns and can explode in air in the presence of an open flame.

The importance of ammonia piping is sometimes minimized when the main emphasis is on selecting major equipment pieces. Mains should be sized carefully to provide low pressure drop and avoid capacity or power penalties caused by inadequate piping.

Rusting pipes and vessels in older systems containing ammonia can create a safety hazard. Oblique x-ray photographs of welded pipe joints and ultrasonic inspection of vessels may be used to disclose defects. Only vendor-certified parts for pipe, valving, and pressure-containing components according to designated assembly drawings should be used to reduce hazards. Cold liquid refrigerant should not be confined between closed valves in a pipe where the liquid can warm and expand to burst piping components. Rapid multiple pulsations of ammonia liquid in piping components (e.g., those developed by cavitation forces or hydraulic hammering from compressor pulsations with massive slugs of liquid carryover to the compressor) must be avoided to prevent equipment and piping damage and injury to personnel.

Hydraulic shock, also known as **hydraulic hammering** (or, in steam and water systems, **water hammer**) can be particularly hazardous. Symptoms in an operating system are loud clattering or banging or pipes moving suddenly and erratically.

The hazard arises out of great, potentially destructive forces that are suddenly generated inside the piping by hydraulic shocks (Lloyko 1992; Shelton and Jacobi 1997a, 1997b) The fundamental flow-phenomenon descriptions of the transients that can cause these shocks are **vapor-propelled**, **liquid slugs**, and **condensation-induced shock**. The phenomenon typically arises when high-pressure warm vapor and cold liquid mix in the same pipe or other refrigerating system element. This most often can occur within, and in the piping around, air-cooling units before, during, and after hot-gas defrosting and in parts of gas-pressure liquid circulating systems, but other parts of the system can also be vulnerable (Glennon and Cole 1998).

Conditions that are most conducive to development of hydraulic transients are pressures lower than approximately 20 psia, with saturated or subcooled liquid present, which are then exposed to high-pressure vapors at pressures equal to or greater than those commonly used for hot-gas defrosting (85 psia). Standing liquid in a horizontal pipe can be excited sufficiently to cause hydraulic shocks when exposed to high-pressure gas flowing over the top of it. In situations where high-pressure gas is introduced regularly, it is necessary that steps be taken as part of the operating protocol to rid the pipe or other system element of liquid before introducing hot gas.

To that end, all valves in horizontal lines should be installed with their stems horizontal, and air-cooling units should be subjected to a complete pumpdown before introducing hot gas for defrosting. Failures occur most often at the pipe or header end closure. These pipe sections, where unavoidable, should be kept very short, less than 24 pipe diameters in length.

Most service problems are caused by inadequate precautions during design, construction, and installation (ASHRAE *Standard* 15; IIAR *Standard* 2). Ammonia is a powerful solvent that removes dirt, scale, sand, or moisture remaining in the pipes, valves, and fittings during installation. These substances are swept along with the suction gas to the compressor, where they are a menace to the bearings, pistons, cylinder walls, valves, and lubricant. Most compressors are equipped with suction strainers and/or additional disposable strainer liners for the large quantity of debris that can be present at initial start-up.

Moving parts are often scored when a compressor is run for the first time. Damage starts with minor scratches, which increase progressively until they seriously affect compressor operation or render it inoperative.

A system that has been carefully and properly installed with no foreign matter or liquid entering the compressor will operate satisfactorily for a long time. As piping is installed, it should be power rotary wire brushed and blown out with compressed air. The piping system should be blown out again with compressed air or nitrogen before evacuation and charging. See ASHRAE *Standard* 15 for system piping test pressure.

CARBON DIOXIDE

Because of its good environmental properties and relative safety, there is renewed and widespread interest in carbon dioxide (CO_2, or R-744) as a refrigerant. Because of its low critical-point temperature (87.8°F) and high pressure, CO_2 presents some unusual technological requirements compared to conventional refrigerants. Another constraint in applying CO_2 is its relatively high triple point at –69.8°F and coincident pressure of 75.1 psia.

Carbon dioxide was used in the early stages of the refrigeration industry, but it lost the competition with halocarbon refrigerants because of its high operating pressure and the loss of capacity and coefficient of performance (COP) when rejecting heat near or above the critical point. In an application with comparable components and heat rejection to ambient air, a CO_2 system's COP is about 40% lower than that of a system using conventional refrigerants. New heat exchanger technology and system components allow CO_2 to reach competitive efficiency levels. The CO_2 efficiency deficit becomes less of a problem in a system when heat is rejected far below the critical point (e.g., in a low-temperature stage of a commercial refrigeration cascade cycle). In this application, the advantageous transport properties of CO_2 and the CO_2-geared design may outweigh CO_2's slight thermodynamic disadvantage.

Table 4 presents selected thermophysical properties of refrigerants used in refrigeration. The high operating pressures (e.g., 490.8 psia at a saturation temperature of 30°F, or 969.6 psia at 80°F) present some unique challenges for containment and safety. However, CO_2 has a number of attractive thermophysical properties and other characteristics. Compared to counterpart halocarbon refrigerants, it has low viscosity, high thermal conductivity, and high vapor density. It is nontoxic, nonflammable, readily available, and of low cost; it has no ozone depletion potential (ODP), and negligible direct global warming potential (GWP). Drawbacks include high operating pressures for medium- and high-temperature refrigeration, and the fact that, during a catastrophic release from the system, CO_2 can adversely affect humans at lower concentrations than HFC refrigerants.

Recently, CO_2 has been intensely studied for application as the primary refrigerant in transcritical mobile air conditioners and vending machines. CO_2 heat pump water heaters are already commercially available in a few countries. In this application, transcritical operation (i.e., rejection of heat above the critical point) is beneficial because it allows good temperature glide matching between the water and supercritical CO_2, which provides COP benefit. In large industrial systems, CO_2 is used as the low-temperature-stage refrigerant in cascade-system arrangements, typically with ammonia as the high-temperature-stage refrigerant. In medium-sized commercial systems, CO_2 is also used as the low-temperature-stage refrigerant in cascade-system arrangements with HFCs and, in

Table 4 Thermophysical Properties for Selected Refrigerants

	CO_2	NH_3	R-134a	R-22	R-404A	R-507A
T_{crit}, °F	87.8	270.1	213.9	205.1	161.6	159.1
P_{crit}, psia	1070.0	1643.7	588.8	723.7	540.8	537.4
P_{sat}, psia	490.8	59.7	40.8	69.7	85.4	87.4
h_{fg},[a] Btu/lb$_m$	100.8	544.7	85.8	88.6	71.9	70.2
ρ_f,[a] lb$_m$/ft^3	58.3	40.0	81.1	80.2	72.1	72.5
ρ_g,[b] lb$_m$/ft^3	5.88	0.21	0.87	1.28	1.84	1.94
c_p,[a] Btu/lb$_m$·°F	0.60	1.10	0.32	0.28	0.33	0.33
k,[a] Btu/h·ft·°F	0.065	0.325	0.053	0.055	0.043	0.043
μ,[a] lb$_m$/ft·s	6.8×10^{-5}	1.2×10^{-4}	8.0×10^{-4}	1.5×10^{-4}	1.2×10^{-4}	1.2×10^{-4}

Source: Lemmon et al. (2002) [a]For saturated liquid at 30°F [b]For saturated vapor at 30°F

a few cases, ammonia or hydrocarbons as the high-temperature-stage refrigerant.

In addition, carbon dioxide is used in low-temperature systems as a two-phase working fluid in a typical secondary coolant loop. The primary refrigerant is ammonia, R-22, or R-404A in a two-stage configuration, or R-507 in a single-stage configuration. In one arrangement, CO_2 is circulated through the process cooler as a secondary fluid, without changing phase, thereby experiencing a temperature rise in the process cooler. The process cooler must be sized and circuited to accommodate the nonisothermal flow of the coolant. The necessity of a larger process cooler and the high pressure rating of the secondary loop is offset by CO_2's superior thermophysical properties and low pumping cost. The CO_2 is cooled in a chiller and pumped to the process cooler, just as in a conventional secondary coolant loop.

In another arrangement, the secondary loop is configured as a low-temperature liquid recirculation system with liquid CO_2 pumped to the process cooler. In this case, the CO_2 evaporates in the process, typically in an overfeed arrangement, with theoretically isothermal flow in the process cooler. This may result in a smaller process cooler than in the non-phase-change flow arrangement. The liquid/vapor mixture is returned to the pump receiver/liquid separator, and the vapor is condensed in a chiller/condenser. In this case, the primary refrigerating plant is as described previously and the CO_2 system circulates in a phase-change manner, but operates at a single pressure. Power to move the CO_2 is imparted by the pump and by the vapor pressure difference that results from the temperature difference between the CO_2 in the receiver and the refrigerant in the primary system chiller/CO_2 condenser.

This arrangement is also used in a conventional cascade system arrangement, where the low-temperature-stage compressor operates on CO_2 rather than on the primary refrigerant. Whether this arrangement or the low-stage compressor with the primary refrigerant is used typically depends on the size and nature of the low-temperature load. The size and operating cost of the low-stage compressor dictate whether it should compress CO_2 or the primary refrigerant.

REFERENCES

ASHRAE. 2004. Safety standard for refrigeration systems. ANSI/ASHRAE *Standard* 15-2004.

ASME. 2004. Rules for construction of pressure vessels. *Boiler and pressure vessel code*, Section VIII, Division 1. American Society of Mechanical Engineers, New York.

ASME. 2001. Refrigeration piping and heat transfer components. ANSI/ASME *Standard* B31.5-2001. American Society of Mechanical Engineers, New York.

ASTM. 2005. Specification for pipe, steel, black and hot-dipped, zinc-coated, welded and seamless. ANSI/ASTM *Standard* A53/A53M-05. American Society for Testing and Materials, West Conshohocken, PA.

ASTM. 2004. Specification for seamless carbon steel pipe for high-temperature service. ANSI/ASTM *Standard* A106/A106M-04b. American Society for Testing and Materials, West Conshohocken, PA.

Briley, G.C. and T.A. Lyons. 1992. Hot gas defrost systems for large evaporators in ammonia liquid overfeed systems. *IIAR Technical Paper* 163. International Institute of Ammonia Refrigeration, Arlington, VA.

Frick Co. 1995. Thermosyphon oil cooling. *Bulletin* E70-900Z (August). Frick Company, Waynesboro, PA.

Glennon, C. and R.A. Cole. 1998. Case study of hydraulic shock events in an ammonia refrigerating system. *IIAR Technical Paper.* International Institute of Ammonia Refrigeration, Arlington, VA.

IIAR. 1998. Minimum safety criteria for a safe ammonia refrigeration system. *Bulletin* 109. International Institute of Ammonia Refrigeration, Arlington, VA.

IIAR. 1999. Equipment, design, and installation of ammonia mechanical refrigeration systems. ANSI/IIAR *Standard* 2-1999. International Institute of Ammonia Refrigeration, Arlington, VA.

IIAR. 1992. Avoiding component failure in industrial refrigeration systems caused by abnormal pressure or shock. *Bulletin* 116. International Institute of Ammonia Refrigeration, Arlington, VA.

Lemmon, E.W., M.O. McLinden, and M.L. Huber. 2002. NIST reference fluids thermodynamic and transport properties—REFPROP, v. 7.0. *NIST Standard Reference Database* 23. National Institute of Standards and Technology, Gaithersburg, MD.

Lloyko, L. 1992. Condensation induced hydraulic shock. *IIAR Technical Paper.* International Institute of Ammonia Refrigeration, Arlington, VA.

Miller, D.K. 1979. Sizing dual-suction risers in liquid overfeed refrigeration systems. *Chemical Engineering* (September 24).

NCPWB. *Welding procedure specifications*. National Certified Pipe Welding Bureau, Rockville, MD.

Shelton, J.C. and A.M. Jacobi. 1997a. A fundamental study of refrigerant line transients: Part 1—Description of the problem and survey of relevant literature. *ASHRAE Transactions* 103(1):65-87.

Shelton, J.C. and A.M. Jacobi. 1997b. A fundamental study of refrigerant line transients: Part 2—Pressure excursion estimates and initiation mechanisms. *ASHRAE Transactions* 103(2):32-41.

Stoecker, W.F. 1988. Chapters 8 and 9 in *Industrial refrigeration.* Business News, Troy, MI.

Timm, M.L. 1991. An improved method for calculating refrigerant line pressure drops. *ASHRAE Transactions* 97(1):194-203.

Wile, D.D. 1977. Refrigerant line sizing. *Final Report*, ASHRAE Research Project RP-185.

BIBLIOGRAPHY

BAC. 1983. *Evaporative condenser engineering manual.* Baltimore Aircoil Company, Baltimore, MD.

Bradley, W.E. 1984. Piping evaporative condensers. In *Proceedings of IIAR Meeting*, Chicago. International Institute of Ammonia Refrigeration, Arlington, VA.

Cole, R.A. 1986. Avoiding refrigeration condenser problems. *Heating/Piping/Air-Conditioning*, Parts I and II, 58(7, 8).

IIR. 2000. Carbon dioxide as a refrigerant. *15th Informatory Note on Refrigerants*. International Institute of Refrigeration, Paris. http://www.iifiir.org/2enpubnotes.php.

Gillies, A. 2005. Design consideration when using carbon dioxide in industrial refrigeration systems. *Proceedings of the Natural Working Fluids 2004, 6th IIR-Gustav Lorentzen Conference*. International Institute of Refrigeration, Paris.

Lloyko, L. 1989. Hydraulic shock in ammonia systems. *IIAR Technical Paper* T-125. International Institute of Ammonia Refrigeration, Arlington, VA.

Nuckolls, A.H. The comparative life, fire, and explosion hazards of common refrigerants. *Miscellaneous Hazard* 2375. Underwriters Laboratory, Northbrook, IL.

Schiesaro, P. and H. Kruse. 2002. Development of a two stage CO_2 supermarket system. *Proceedings of New Technologies in Commercial Refrigeration*. International Institute of Refrigeration, Paris.

Strong, A.P. 1984. Hot gas defrost—A-one-a-more-a-time. *IIAR Technical Paper* T-53. International Institute of Ammonia Refrigeration, Arlington, VA.

Swalha, S. 2005. Using CO_2 in supermarket refrigeration. *ASHRAE Journal* 47(8):26-30.

CHAPTER 4

SECONDARY COOLANTS IN REFRIGERATION SYSTEMS

SECONDARY coolants are liquids used as heat transfer fluids that change temperature as they gain or lose heat energy without changing into another phase. For lower refrigeration temperatures, this requires a coolant with a freezing point below that of water. This chapter discusses design considerations for components, system performance requirements, and applications for secondary coolants. Related information can be found in Chapters 2, 3, 20, 21, and 36 of the 2005 *ASHRAE Handbook—Fundamentals*.

COOLANT SELECTION

A secondary coolant must be compatible with other materials in the system at the pressures and temperatures encountered for maximum component reliability and operating life. The coolant should also be compatible with the environment and the applicable safety regulations, and should be economical to use and replace.

The coolant should have a minimum freezing point of 5°F below and preferably 15°F below the lowest temperature to which it will be exposed. When subjected to the lowest temperature in the system, coolant viscosity should be low enough to allow satisfactory heat transfer and reasonable pressure drop.

Coolant vapor pressure should not exceed that allowed at the maximum temperature encountered. To avoid a vacuum in a low-vapor-pressure secondary coolant system, the coolant can be pressurized with pressure-regulated dry nitrogen in the expansion tank. However, some special secondary coolants such as those used for computer circuit cooling have a high solubility for nitrogen and must therefore be isolated from the nitrogen with a suitable diaphragm.

Load Versus Flow Rate

The secondary coolant pump is usually in the return line upstream of the chiller. Therefore, the pumping rate is based on the density at the return temperature. The mass flow rate for a given heat load is based on the desired temperature range and required coefficient of heat transfer at the average bulk temperature.

To determine heat transfer and pressure drop, the specific gravity, specific heat, viscosity, and thermal conductivity are based on the average bulk temperature of coolant in the heat exchanger, noting that film temperature corrections are based on the average film temperature. Trial solutions of the secondary coolant-side coefficient compared to the overall coefficient and total log mean temperature difference (LMTD) determine the average film temperature. Where the secondary coolant is cooled, the more viscous film reduces the heat transfer rate and raises the pressure drop compared to what can be expected at the bulk temperature. Where the secondary coolant is heated, the less viscous film approaches the heat transfer rate and pressure drop expected at the bulk temperature.

The more turbulence and mixing of the bulk and film, the better the heat transfer and higher the pressure drop. Where secondary coolant velocity in the tubes of a heat transfer device results in laminar flow, heat transfer can be improved by inserting spiral tapes or spring turbulators that promote mixing the bulk and film. This usually increases pressure drop. The inside surface can also be spirally grooved or augmented by other devices. Because the state of the art of heat transfer is constantly improving, use the most cost-effective heat exchanger to provide optimum heat transfer and pressure drop. Energy costs for pumping secondary coolant must be considered when selecting the fluid to be used and the heat exchangers to be installed.

Pumping Cost

Pumping costs are a function of the secondary coolant selected, load and temperature range where energy is transferred, pump pressure required by the system pressure drop (including that of the chiller), mechanical efficiencies of the pump and driver, and electrical efficiency and power factor where the driver is an electric motor. Small centrifugal pumps, operating in the range of approximately 50 gpm at 80 ft of head to 150 gpm at 70 ft of head, for 60 Hz applications, typically have 45 to 65% efficiency, respectively. Larger pumps, operating in the range of 500 gpm at 80 ft of head to 1500 gpm at 70 ft of head, for 60 Hz applications, typically have 75 to 85% efficiency, respectively.

A pump should operate near its peak operating efficiency for the flow rate and pressure that usually exist. Secondary coolant temperature increases slightly from energy expended at the pump shaft. If a semihermetic electric motor is used as the driver, motor inefficiency is added as heat to the secondary coolant, and the total kilowatt input to the motor must be considered in establishing load and temperatures.

Performance Comparisons

Assuming that the total refrigeration load at the evaporator includes the pump motor input and brine line insulation heat gains, as well as the delivered beneficial cooling, tabulating typical secondary coolant performance values assists in the coolant selection. A 1.06 in. ID smooth steel tube evaluated for pressure drop and internal heat transfer coefficient at the average bulk temperature of 20°F and a temperature range of 10°F for 7 fps tube-side velocity provides comparative data (Table 1) for some typical coolants. Table 2 ranks the same coolants comparatively, using data from Table 1.

For a given evaporator configuration, load, and temperature range, select a secondary coolant that gives satisfactory velocities, heat transfer, and pressure drop. At the 20°F level, hydrocarbon and halocarbon secondary coolants must be pumped at a rate of 2.3 to 3.0 times the rate of water-based secondary coolants for the same temperature range.

Higher pumping rates require larger coolant lines to keep the pump's pressure and brake horsepower requirement within reasonable limits. Table 3 lists approximate ratios of pump power for secondary coolants. Heat transferred by a given secondary coolant affects the cost and perhaps the configuration and pressure drop of a chiller and other heat exchangers in the system; therefore, Tables 2 and 3 are only guides of the relative merits of each coolant.

The preparation of this chapter is assigned to TC 10.1, Custom Engineered Refrigeration Systems.

Table 1 Secondary Coolant Performance Comparisons

Secondary Coolant	Concentration (by Weight), %	Freeze Point, °F	gpm/ton[a]	Pressure Drop,[b] psi	Heat Transfer Coefficient[c] h_i, Btu/h·ft^2·°F
Propylene glycol	39	–5.1	2.56	2.91	205
Ethylene glycol	38	–6.9	2.76	2.38	406
Methanol	26	–5.3	2.61	2.05	473
Sodium chloride	23	–5.1	2.56	2.30	558
Calcium chloride	22	–7.8	2.79	2.42	566
Aqua ammonia	14	–7.0	2.48	2.44	541
Trichloroethylene	100	–123	7.44	2.11	432
d-Limonene	100	–142	6.47	1.48	321
Methylene chloride	100	–142	6.39	1.86	585
R-11	100	–168	7.61	2.08	428

[a]Based on inlet secondary coolant temperature at pump of 25°F.
[b]Based on one length of 16 ft tube with 1.06 in. ID and use of Moody Chart (1944) for an average velocity of 7 fps. Input/output losses equal one Vel. $H_D(V^2\rho/2g)$ for 7 fps velocity. Evaluations are at a bulk temperature of 20°F and a temperature range of 10°F.
[c]Based on curve fit equation for Kern's (1950) adaptation of Sieder and Tate's (1936) heat transfer equation using 16 ft tube for L/D = 181 and film temperature of 5°F lower than average bulk temperature with 7 fps velocity.

Table 2 Comparative Ranking of Heat Transfer Factors at 7 fps*

Secondary Coolant	Heat Transfer Factor
Propylene glycol	1.000
d-Limonene	1.566
Ethylene glycol	1.981
R-11	2.088
Trichloroethylene	2.107
Methanol	2.307
Aqua ammonia	2.639
Sodium chloride	2.722
Calcium chloride	2.761
Methylene chloride	2.854

*Based on Table 1 values using 1.06 in. ID tube 16 ft long. Actual ID and length vary according to specific loading and refrigerant applied with each secondary coolant, tube material, and surface augmentation.

Table 3 Relative Pumping Energy Required*

Secondary Coolant	Energy Factor
Aqua ammonia	1.000
Methanol	1.078
Propylene glycol	1.142
Ethylene glycol	1.250
Sodium chloride	1.295
Calcium chloride	1.447
d-Limonene	2.406
Methylene chloride	3.735
Trichloroethylene	4.787
R-11	5.022

*Based on same pump pressure, refrigeration load, 20°F average temperature, 10°F range, and freezing point (for water-based secondary coolants) 20 to 23°F below lowest secondary coolant temperature.

Other Considerations

Corrosion must be considered when selecting coolant, inhibitor, and system components. The effect of secondary coolant and inhibitor toxicity on the health and safety of plant personnel or consumers of food and beverages must be considered. The flash point and explosive limits of secondary coolant vapors must also be evaluated.

Examine the secondary coolant stability for anticipated moisture, air, and contaminants at the temperature limits of materials used in the system. Skin temperatures of the hottest elements determine secondary coolant stability.

If defoaming additives are necessary, their effect on thermal stability and coolant toxicity must be considered for the application.

DESIGN CONSIDERATIONS

Secondary coolant vapor pressure at the lowest operating temperature determines whether a vacuum could exist in the secondary coolant system. To keep air and moisture out of the system, pressure-controlled dry nitrogen can be applied to the top level of secondary coolant (e.g., in the expansion tank or a storage tank). Gas pressure over the coolant plus the pressure created at the lowest point in the system by the maximum vertical height of coolant determine the minimum internal pressure for design purposes. The coincident highest pressure and lowest secondary coolant temperature dictate the design working pressure (DWP) and material specifications for the components.

To select proper relief valve(s) with settings based on the system DWP, the highest temperatures to which the secondary coolant could be subjected should be considered. This temperature occurs in case of heat radiation from a fire in the area or the normal warming of the valved-off sections. Normally, a valved-off section is relieved to an unconstrained portion of the system and the secondary coolant can expand freely without loss to the environment.

Safety considerations for the system are found in ASHRAE *Standard* 15. Design standards for pressure piping can be found in ASME *Standard* B31.5, and design standards for pressure vessels in Section VIII of the ASME *Boiler and Pressure Vessel Code*.

Piping and Control Valves

Piping should be sized for reasonable pressure drop using the calculation methods in Chapters 2 and 36 of the 2005 *ASHRAE Handbook—Fundamentals.* Balancing valves or orifices in each of the multiple feed lines help distribute the secondary coolant. A reverse-return piping arrangement balances flow. Control valves that vary flow are sized for 20 to 80% of the total friction pressure drop through the system for proper response and stable operation. Valves sized for pressure drops smaller than 20% may respond too slowly to a control signal for a flow change. Valves sized for pressure drops over 80% can be too sensitive, causing control cycling and instability.

Storage Tanks

Storage tanks can shave peak loads for brief periods, limit the size of refrigeration equipment, and reduce energy costs. In off-peak hours, a relatively small refrigeration plant cools a secondary coolant stored for later use. A separate circulating pump sized for the maximum flow needed by the peak load is started to satisfy peak load. Energy cost savings are enhanced if the refrigeration equipment is used to cool secondary coolant at night, when the cooling medium for heat rejection is generally at the lowest temperature.

The load profile over 24 h and the temperature range of the secondary coolant determine the minimum net capacity required for the refrigeration plant, pump sizes, and minimum amount of secondary coolant to be stored. For maximum use of the storage tank volume at the expected temperatures, choose inlet velocities and locate connections and tank for maximum stratification. Note, however, that maximum use will probably never exceed 90% and, in some cases, may equal only 75% of the tank volume.

Example 1. Figure 1 depicts the load profile and Figure 2 shows the arrangement of a refrigeration plant with storage of a 23% (by weight) sodium chloride secondary coolant at a nominal 20°F. During the peak load of 50 tons, a range of 8°F is required. At an average temperature of 24°F, with a range of 8°F, the coolant's specific heat c_p is 0.791 Btu/lb·°F. At 28°F, the weight per unit volume of coolant ρ_L at the pump = [1.183(62.4 lb/ft³)]/(7.48 gal/ft³); at 20°F, ρ_L = [1.185(62.4 lb/ft³)]/(7.48 gal/ft³).

Determine the minimum size storage tank for 90% use, minimum capacity required for the chiller, and sizes of the two pumps. The chiller and the chiller pump run continuously. The secondary coolant storage pump runs only during the peak load. A control valve to the load source diverts all coolant to the storage tank during a zero-load condition, so that the initial temperature of 20°F is restored in the tank. During low load, only the required flow rate for a range of 8°F at the load source is used; the balance returns to the tank and restores the temperature to 20°F.

Solution: If x is the minimum capacity of the chiller, determine the energy balance in each segment by subtracting the load in each segment from x. Then multiply the result by the time length of the respective segments, and add as follows:

$$\begin{aligned} 6(x-0)+4(x-50)+14(x-9) &= 0 \\ 6x+4x-200+14x-126 &= 0 \\ 24x &= 326 \\ x &= 13.58 \text{ tons} \end{aligned}$$

Calculate the secondary coolant flow rate W at peak load:

$$W = (50 \times 200)/(0.791 \times 8) = 1580.3 \text{ lb/min}$$

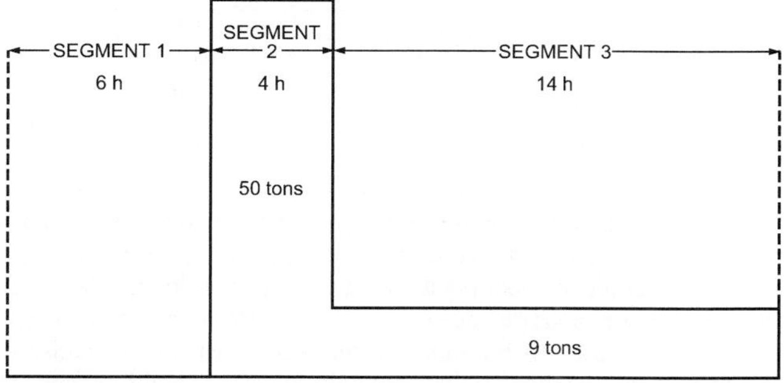

Fig. 1 Load Profile of Refrigeration Plant Where Secondary Coolant Storage Can Save Energy

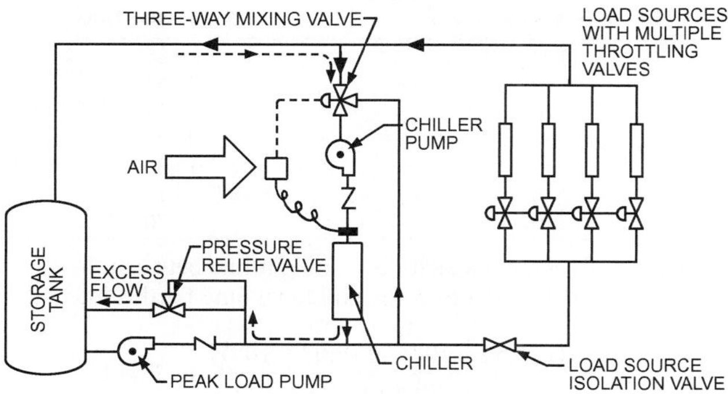

Fig. 2 Arrangement of System with Secondary Coolant Storage

For the chiller at 15 tons, the secondary coolant flow rate is

$$W = (15 \times 200)/(0.791 \times 8) = 474.1 \text{ lb/min}$$

Therefore, the coolant flow rate to the storage tank pump is 1580.3 – 474.1 = 1106.2 lb/min. Chiller pump size is determined by

$$474.1/[(1.183 \times 62.4)/7.48] = 48 \text{ gpm}$$

Calculate the storage tank pump size as follows:

$$1106.2/[(1.185 \times 62.4)/7.48] = 112 \text{ gpm}$$

Using the concept of stratification in the storage tank, the interface between warm return and cold stored secondary coolant falls at the rate pumped from the tank. Because the time segments fix the total amount pumped and the storage tank pump operates only in segment 2 (see Figure 1), the minimum tank volume V at 90% use is determined as follows:

$$\text{Total mass} = [(1106.2 \text{ lb/min})(60 \text{ min/h})(4 \text{ h})]/0.90 = 295{,}000 \text{ lb}$$

and

$$V = 295{,}000/[(1.185 \times 62.4)/7.48] = 29{,}840 \text{ gal}$$

A larger tank (e.g., 50,000 gal) provides flexibility for longer segments at peak load and accommodates potential mixing. It may be desirable to insulate and limit heat gains to 8000 Btu/h for the tank and lines. Energy use for pumping can be limited by designing for 46 ft head. With the smaller pump operating at 51% efficiency and the larger pump at 52.5% efficiency, pump heat added to the secondary coolant is 3300 and 7478 Btu/h, respectively.

For cases with various time segments and their respective loads, the maximum load for segment 1 or 3 with the smaller pump operating cannot exceed the net capacity of the chiller minus insulation and pump heat gain to the secondary coolant. For various combinations of segment time lengths and cooling loads, the recovery or restoration rate of the storage tank to the lowest temperature required for satisfactory operation should be considered.

As load source circuits shut off, excess flow is bypassed back to the storage tank (Figure 2). The temperature setting of the three-way valve is the normal return temperature for full flow through the load sources.

When only the storage tank requires cooling, flow is as shown by the dashed lines with the load source isolation valve closed. When storage tank temperature is at the desired level, the load isolation valve can be opened to allow cooling of the piping loops to and from the load sources for full restoration of storage cooling capacity.

Expansion Tanks

Figure 3 shows a typical closed secondary coolant system without a storage tank; it also illustrates different control strategies. The reverse-return piping assists flow balance. Figure 4 shows a secondary coolant strengthening unit for salt brines. Secondary coolant expansion tank volume is determined by considering the total coolant inventory and differences in coolant density at the lowest temperature t_1 of coolant pumped to the load location and the maximum temperature. The expansion tank is sized to accommodate a residual volume with the system coolant at t_1, plus an expansion volume and vapor space above the coolant. A vapor space equal to 20% of the expansion tank volume should be adequate. A level indicator, used to prevent overcharging, is calibrated at the residual volume level versus lowest system secondary coolant temperature.

Example 2. Assume a 50,000 gal charge of 23% sodium chloride secondary coolant at t_1 of 20°F in the system. If 100°F is the maximum temperature, determine the size of the expansion tank required. Assume that the residual volume is 10% of the total tank volume and that the vapor space at the highest temperature is 20% of the total tank volume.

$$\text{ETV} = \frac{V_S[(SG_1/SG_2)-1]}{1-(R_F+V_F)}$$

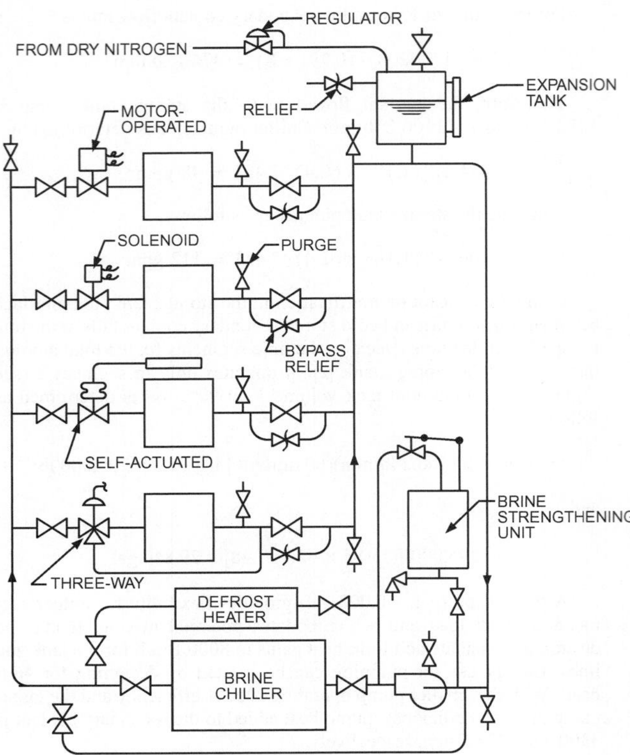

Fig. 3 Typical Closed Salt Brine System

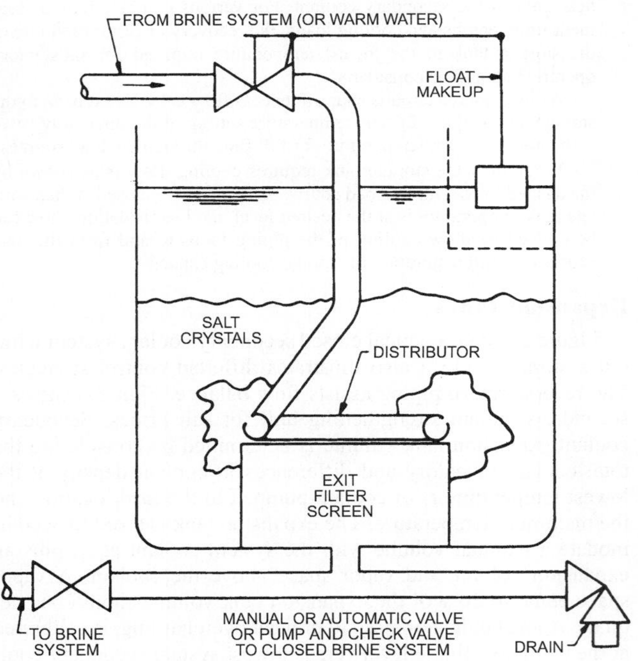

Fig. 4 Brine Strengthening Unit for Salt Brines Used as Secondary Coolants

where

ETV = expansion tank volume

V_S = system secondary coolant volume at temperature t_1

SG_1 = specific gravity at t_1

SG_2 = specific gravity at maximum temperature

R_F = residual volume of tank liquid (low level) at t_1, expressed as a fraction

V_F = volume of vapor space at highest temperature, expressed as a fraction

If the specific gravity of the secondary coolant is 1.185 at 20°F and 1.155 at 100°F, the tank volume is

$$\text{ETV} = \frac{50{,}000[(1.185/1.155) - 1]}{1 - (0.10 + 0.20)} = 1855 \text{ gal}$$

Pulldown Time

Example 1 is based on a static situation of secondary coolant temperature at two different loads: normal and peak. The length of time for pulldown from 100°F to the final 20°F may need to be calculated. For graphical solution, required heat extraction versus secondary coolant temperature is plotted. Then, by iteration, pulldown time is solved by finding the net refrigeration capacity for each increment of coolant temperature change. A mathematical method may also be used.

The 15 ton system in the examples has a 30.03 ton capacity at a maximum of 50°F saturated suction temperature (STP). For pulldown, a compressor suction pressure regulator (holdback valve) is sometimes used. The maximum secondary coolant temperature must be determined when the holdback valve is wide open and the STP is at 50°F. For Example 1, this is at 70°F coolant temperature. As coolant temperature is further reduced with a constant 48.1 gpm, refrigeration system capacity gradually reduces until a 15 ton capacity is reached with 26°F coolant in the tank. Further cooling to 20°F is at reduced capacity.

Temperatures of the secondary coolant mass, storage tanks, piping, cooler, pump, and insulation must all be reduced. In Example 1, as the coolant drops from 100 to 20°F, the total heat removed from these items is as follows:

Brine Temperature, °F	Total Heat Removed, Million Btu
100	31.54
80	23.62
70	19.67
60	15.73
40	7.85
20	0

From a secondary coolant temperature of 100 to 70°F, the refrigeration system capacity is fixed at 30.03 tons, and the time for pulldown is essentially linear (system net tons for pulldown is less than the compressor capacity because of heat gain through insulation and added pump heat). In Example 1, pump heat was not considered. When recognizing the variable heat gain for a 95°F ambient, and the pump heat as the secondary coolant temperature is reduced, the following net capacity is available for pulldown at various secondary coolant temperatures:

Brine Temperature, °F	Net Capacity, Tons
100	29.86
80	29.58
70	29.44
60	25.28
40	17.80
26	14.10
20	12.70

A curve fit shows capacity is a straight line between the values for 100 and 70°F. Therefore, the pulldown time for this interval is

$$\theta = \frac{[(31.54 \times 10^6) - (19.67 \times 10^6)]}{12{,}000[(29.86 + 29.44)/2]} = 33.4 \text{ h}$$

From 70 to 20°F, the capacity curve fits a second-degree polynomial equation as follows:

$$q = 9.514809086 + 0.1089883647t + 0.002524039t^2$$

where

t = secondary coolant temperature, °F
q = capacity for pulldown, tons

Using the arithmetic average pulldown net capacity from 70 to 20°F, the time interval would be

$$\theta = \frac{19.67 \times 10^6}{12{,}000[(29.44 + 12.7)/2]} = 77.8 \text{ h}$$

If the logarithmic (base e) mean average net capacity for this temperature interval is used, the time is

$$\theta = \frac{19.67 \times 10^6}{(19.91 \times 12{,}000)} = 82.3 \text{ h}$$

This is a difference of 4.5 h, and neither solution is correct. A more exact calculation uses a graphical analysis or calculus. One mathematical approach determines the heat removed per degree of secondary coolant temperature change per ton of capacity. Because the coolant's heat capacity and heat leakage change as the temperature drops, the amount of heat removed is best determined by first fitting a curve to the data for total heat removed versus secondary coolant temperature. Then a series of iterations for secondary coolant temperature ±1°F is made as the temperature is reduced. The polynomial equations may be solved by computer or calculator with a suitable program or spreadsheet. The time for pulldown is less if supplemental refrigeration is available for pulldown or if less secondary coolant is stored.

The correct answer is 88.1 h, which is 7% greater than the logarithmic mean average capacity and 13% greater than the arithmetic average capacity over the temperature range.

Therefore, total time for temperature pulldown from 100 to 20°F is

$$\theta = 33.4 + 88.1 = 121.5 \text{ h}$$

System Costs

Various alternatives may be evaluated to justify a new project or system modification. Means (updated annually) lists the installed cost of various projects. NBS (1978) and Park and Jackson (1984) discuss engineering and life-cycle cost analysis. Using various time-value-of-money formulas, payback for storage tank handling of peak loads compared to large refrigeration equipment and higher energy costs can be evaluated. Trade-offs in these costs (initial, maintenance, insurance, increased secondary coolant, loss of space, and energy escalation) all must be considered.

Corrosion Prevention

Corrosion prevention requires choosing proper materials and inhibitors, routine testing for pH, and eliminating contaminants. Because potentially corrosive calcium chloride and sodium chloride salt brine secondary coolant systems are widely used, test and adjust the brine solution monthly. To replenish salt brines in a system, a concentrated solution may be better than a crystalline form, because it is easier to handle and mix.

A brine should not be allowed to change from alkaline to acidic. Acids rapidly corrode the metals ordinarily used in refrigeration and ice-making systems. Calcium chloride usually contains sufficient alkali to render the freshly prepared brine slightly alkaline. When any brine is exposed to air, it gradually absorbs carbon dioxide and oxygen, which eventually make the brine slightly acid. Dilute brines dissolve oxygen more readily and generally are more corrosive than concentrated brines. One of the best preventive measures is to make a closed rather than open system, using a regulated inert gas over the surface of a closed expansion tank (see Figure 2). However, many systems, such as ice-making tanks, brine-spray unit coolers, and brine-spray carcass chill rooms, cannot be closed.

A brine pH of 7.5 for a sodium or calcium chloride system is ideal, because it is safer to have a slightly alkaline rather than a slightly acid brine. Operators should check pH regularly.

If a brine is acid, the pH can be raised by adding caustic soda dissolved in warm water. If a brine is alkaline (indicating ammonia leakage into the brine), carbonic gas or chromic, acetic, or hydrochloric acid should be added. Ammonia leakage must be stopped immediately so that the brine can be neutralized.

In addition to controlling pH, an inhibitor should be used. Generally, sodium dichromate is the most effective and economical for salt brine systems. The granular dichromate is bright orange and readily dissolves in warm water. Because it dissolves very slowly in cold brine, it should be dissolved in warm water and added to the brine far enough ahead of the pump so that only a dilute solution reaches the pump. Recommended quantities are 125 lb/1000 ft^3 of calcium chloride brine, and 200 lb/1000 ft^3 of sodium chloride brine.

Adding sodium dichromate to the salt brine does not make it noncorrosive immediately. The process is affected by many factors, including water quality, specific gravity of the brine, amount of surface and kind of material exposed in the system, age, and temperature. Corrosion stops only when protective chromate film has built up on the surface of the zinc and other electrically positive metals exposed to the brine. No simple test is available to determine chromate concentration. Because the protection afforded by sodium dichromate treatment depends greatly on maintaining the proper chromate concentration in the brine, brine samples should be analyzed annually. The proper concentration for calcium chloride brine is 7.58 gr/gal (as $Na_2Cr_2O_7 \cdot 2H_2O$); for sodium chloride brine, it is 12.128 gr/gal (as $Na_2Cr_2O_7 \cdot 2H_2O$).

Crystals and concentrated solutions of sodium dichromate can cause severe skin rash, so avoid contact. If contact does occur, wash the skin immediately. *Warning: sodium dichromate should not be used for brine spray decks, spray units, or immersion tanks where food or personnel may come in contact with the spray mist or the brine itself.*

Polyphosphate/silicate and orthophosphate/boron mixtures in water-treating compounds are useful for sodium chloride brines in open systems. However, where the rate of spray loss and dilution is very high, any treatment other than density and pH control is not economical. For the best protection of spray unit coolers, housings and fans should be of a high quality, hot-dipped galvanized construction. Stainless steel fan shafts and wheels, scrolls, and eliminators are desirable.

Although nonsalt secondary coolants described in this chapter are generally noncorrosive when used in systems for long periods, recommended inhibitors should be used, and pH should be checked occasionally.

Steel, iron, or copper piping should not be used to carry salt brines. Use copper nickel or suitable plastic. Use all-steel and iron tanks if the pH is not ideal. Similarly, calcium chloride systems usually have all-iron and steel pumps and valves to prevent electrolysis in the presence of acidity. Sodium chloride systems usually have all-iron or all-bronze pumps. When pH can be controlled in a system, brass valves and bronze fitted pumps may be satisfactory. A stainless steel pump shaft is desirable. Consider salt brine composition and temperature to select the proper rotary seal or, for dirtier systems, the proper stuffing box.

APPLICATIONS

Applications for secondary coolant systems are extensive (see Chapters 11 to 37). A glycol coolant prevents freezing in solar collectors and outdoor piping. Secondary coolants heated by solar collectors or by other means can be used to heat absorption cooling equipment, to melt a product such as ice or snow, or to heat a building. Process heat exchangers can use a number of secondary coolants to transfer heat between locations at various temperature levels. Using secondary coolant storage tanks increases the availability of cooling and heating and reduces peak demands for energy.

Each supplier of refrigeration equipment that uses secondary coolant flow has specific ratings. Flooded and direct-expansion coolers, dairy plate heat exchangers, food processing, and other air, liquid, and solid chilling devices come in various shapes and sizes. Refrigerated secondary coolant spray wetted-surface cooling and humidity control equipment has an open system that absorbs moisture while cooling and then continuously regenerates the secondary coolant with a concentrator. Although this assists cooling, dehumidifying, and defrosting, it is not strictly a secondary coolant flow application for refrigeration, unless the secondary coolant also is used in the coil. Heat transfer coefficients can be determined from vendor rating data or by methods described in Chapter 3 of the 2005 *ASHRAE Handbook—Fundamentals* and appropriate texts.

A primary refrigerant may be used as a secondary coolant in a system by being pumped at a flow rate and pressure high enough that the primary heat exchange occurs without evaporation. The refrigerant is then subsequently flashed at low pressure, with the resulting flash gas drawn off to a compressor in the conventional manner.

REFERENCES

ASHRAE. 2004. Safety standard for refrigeration systems. ANSI/ASHRAE *Standard* 15-2004.

ASME. 2001. Refrigeration piping and heat transfer components. ANSI/ASME *Standard* B31.5-2001. American Society of Mechanical Engineers, New York.

ASME. 2004. Rules for construction of pressure vessels. *Boiler and pressure vessel code*, Section VIII-2004. American Society of Mechanical Engineers, New York.

Kern, D.Q. 1950. *Process heat transfer*, p. 134. McGraw-Hill, New York.

Means. Updated annually. *Means mechanical cost data*. RSMeans, Kingston, MA.

Moody, L.F. 1944. Frictional factors for pipe flow. *ASME Transactions* (November):672-673.

NBS. 1978. Life cycle costing. *National Bureau of Standards Building Science Series* 113. SD Catalog Stock No. 003-003-01980-1, U.S. Government Printing Office, Washington, D.C.

Park, W.R. and D.E. Jackson. 1984. *Cost engineering analysis*, 2nd ed. John Wiley & Sons, New York.

Sieder, E.N. and G.E. Tate. 1936. Heat transfer and pressure drop of liquids in tubes. *Industrial and Engineering Chemistry* 28(12):1429.

CHAPTER 5

REFRIGERANT SYSTEM CHEMISTRY

GOOD understanding of the chemical interactions between refrigerant, lubricant, and materials in a refrigeration system is necessary for designing reliable systems that have a long service life. This chapter covers the chemical aspects of both historical refrigerants and newer refrigerants and blends. Physical aspects such as measurement and contaminant control (including moisture) are discussed in Chapter 6. Physical properties of lubricants are discussed in Chapter 7.

REFRIGERANTS

Environmental Acceptability

Common chlorine-containing refrigerants contribute to depletion of the ozone layer. A material's **ozone depletion potential (ODP)** is a measure of its ability, compared to CFC-11, to destroy stratospheric ozone.

Halocarbon refrigerants also can contribute to global warming and are considered greenhouse gases. The **global warming potential (GWP)** of a greenhouse gas is an index describing its ability, compared to CO_2 (which has a very long atmospheric lifespan), to trap radiant energy. The GWP, therefore, is connected to a particular time scale (e.g., 100 or 500 years). For regulatory purposes, the convention is to use the **100-year integrated time horizon (ITH)**.

Appliances using a given refrigerant also consume energy, which indirectly produces CO_2 emissions that contribute to global warming; this indirect effect is frequently much larger than the refrigerant's direct effect. An appliance's **total equivalent warming impact (TEWI)** is based on the refrigerant's direct warming potential and indirect effect of the appliance's energy use The **life cycle climate performance (LCCP)**, which includes the TEWI as well as cradle-to-grave considerations such as the climate change effect of manufacturing the refrigerant, transportation-related energy, and end-of-life disposal, is becoming more prevalent.

Environmentally preferred refrigerants (1) have low or zero ODP, (2) provide good system efficiency, and (3) have low GWP or TEWI values. Hydrogen-containing compounds such as the hydrochlorofluorocarbon HCFC-22 or the hydrofluorocarbon HFC-134a have shorter atmospheric lifetimes than chlorofluorocarbons (CFCs) because they are largely destroyed in the lower atmosphere by reactions with OH radicals, resulting in lower ODP and GWP values.

Tables 1 and 2 show boiling points, atmospheric lifetimes, ODPs, GWPs, and flammabilities of new refrigerants and the refrigerants being replaced. ODP values were established through the Montreal Protocol and are unlikely to change. ODP values calculated using the latest scientific information are sometimes lower but are not used for regulatory purposes. Because HFCs do not contain chlorine atoms, their ODP values are essentially zero (Ravishankara et al. 1994).

GWP values were established as a reference point using Intergovernmental Panel on Climate Change (IPCC 1995) assessment values, as shown in Table 1, and are the official numbers used for reporting and compliance purposes to meet requirements of the United Nations Framework Convention on Climate Change (UNFCCC) and Kyoto Protocol. However, lifetimes and GWPs have since been reviewed (IPCC 2001) and are shown in Table 2, representing the most recent published values based on an updated assessment of the science. These values are subject to review and may change with future reassessments, but are currently not used for regulatory compliance purposes. Table 3 shows bubble points and calculated ODPs and GWPs for refrigerant blends, using the latest scientific assessment values.

Compositional Groups

Chlorofluorocarbons. CFC refrigerants such as R-12, R-11, R-114, and R-115 have been used extensively in the air-conditioning and refrigeration industries. Because of their chlorine content, these materials have significant ODP values. The Montreal Protocol, which governs the elimination of ozone-depleting substances, was strengthened at the London meeting in 1990 and confirmed at the Copenhagen meeting in 1992. In accordance with this international agreement, production of CFCs in industrialized countries was totally phased out as of January 1, 1996. Production in developing countries will be phased out in 2010, although many have already made considerable phaseout progress.

Hydrochlorofluorocarbons. HCFC refrigerants such as R-22 and R-123 have shorter atmospheric lifetimes (and lower ODP values) than CFCs. Nevertheless, the Montreal Protocol limited developed-country consumption of HCFCs beginning January 1, 1996, using a cap equal to 2.8% of the 1989 ODP weighted consumption of CFCs plus the 1989 ODP-weighted consumption of HCFCs. The CAP was reduced by 35% by January 1, 2004, and will be reduced by 65% on January 1, 2010; 90% by January 1, 2015; 99.5% by January 1, 2020; and total phaseout by January 1, 2030. From 2020 to 2030, HCFCs may only be used to service existing equipment. Developing countries must freeze HCFC ODP consumption at 2015 levels in 2016, and completely phase out by January 1, 2040.

In addition to the requirements of the Montreal Protocol, several countries have established their own regulations on HCFC phaseout of HCFCs. The United States has met the Montreal Protocol's requirements by banning consumption of R-141b (primarily used as a foam-blowing agent) on January 1, 2003, and phasing out HCFC-142b (primarily foams) and HCFC-22 for original equipment manufacturers (OEMs) beginning January 1, 2010. Production for service needs is allowed to continue. Production and consumption of all other HCFCs will be frozen on January 1, 2015. On January 1, 2020, production and consumption of R-22 and R-142b will be banned, followed by a ban on production and consumption of all other HCFCs on January 1, 2030. As required by the Montreal Protocol, from 2020 to 2030, virgin HCFCs may only be used to service existing equipment.

The preparation of this chapter is assigned to TC 3.2, Refrigerant System Chemistry.

Table 1 Refrigerant Properties: Regulatory Compliance Values Used by Governments for UNFCCC Reporting and Kyoto Protocol Compliance

Refrigerant	Structure	Boiling Point,[a] °F	Atmospheric Lifetime,[b] Years	ODP[c]	GWP, ITH 100-Year	Flammable?
E125	$CHF_2OC F_3$	–43.6	165[a]		15,300[a]	No
E143	CHF_2OCH_2F	85.8[d]				Yes
E143a	CF_3OCH_3	–11.4	5.7[a]		5400[a]	Yes
11	CCl_3F	74.7	50	1	4600[a]	No
12	CCl_2F_2	–21.6	102	1	10,600[a]	No
22	$CHClF_2$	–41.4	12.1	0.055	1900[a]	No
23	CHF_3	–115.8	264		11,700	No
32	CH_2F_2	–61.1	5.6		650	Yes
113	CCl_2FCClF_2	117.7	85	0.8	6000[a]	No
114	$CClF_2CClF_2$	38.5	300	1	9800[a]	No
115	$CClF_2CF_3$	–38.0	1700	0.6	10,300[a]	No
116	CF_3CF_3	–108.8	10,000		11,400[a]	No
123	$CHCl_2CF_3$	82.0	1.4	0.02	120[a]	No
124	$CHClFCF_3$	10.4	6.1	0.022	620[a]	No
125	CHF_2CF_3	–54.6	32.6		2800	No
134a	CH_2FCF_3	–16.0	14.6		1300	No
142b	$CClF_2CH_3$	15.8	18.4	0.065	2300[a]	Yes
143	CH_2FCHF_2	41.0	3.8		300	Yes
143a	CF_3CH_3	–53.0	48.3		3800	Yes
152a	CHF_2CH_3	–11.2	1.5		140	Yes
218	$CF_3CF_2CF_3$	–33.9	2600[a]		8600[a]	No
227ea	CF_3CHFCF_3	3.9	36.5		2900	No
236ea	$CF_3CHFCHF_2$	43.7[d]	10[d]		9400[a]	No
236fa	$CF_3CH_2CF_3$	29.5	209		6300	No
245ca	$CHF_2CF_2CH_2F$	–13.2	6.6		560	Yes
245fa	$CF_3CH_2CHF_2$	59.2	8.8[a]		820[a]	No

[a]Data from Calm and Hourahan (1999).
[b]Data from IPCC (1995).
[c]Data from Montreal Protocol (2003).
[d]Data from Chapter 5 of the 2002 *ASHRAE Handbook—Refrigeration*.

Table 2 Refrigerant Properties: Current IPCC Scientific Assessment Values

Refrigerant	Structure	Boiling Point, °F	Atmospheric Lifetime, Years	ODP	GWP, ITH[a] 100-Year	Flammable?[b]
E125	CHF_2OCF_3	–43.6	165[c]		14,900	No
E143	CHF_2OCH_2F	85.1[b]			57	Yes
E143a	CF_3OCH_3	–11.4	5.7[c]		750	Yes
11	CHI_3F	74.7	50	1	4600	No
12	CCl_2F_2	–21.6	102	1	10,600	No
22	$CHClF_2$	–41.4	12.1	0.055	1700	No
23	CHF_3	–115.8	264		12,000	No
32	CH_2F_2	–61.1	5.6		550	Yes
113	CCl_2FCF_2Cl	117.7	85	0.8	6000	No
114	$CClF_2CClF_2$	38.5	300	1	9800	No
115	ClF_2CF_3	–38.0	1700	0.6	7200	No
116	CF_3CF_3	–108.8	10,000		11,900[c]	No
123	$CHCl_2CF_3$	82.0	1.4	0.02	120	No
124	$CHClFCF_3$	10.4	6.1	0.022	620	No
125	CHF_2CF_3	–54.6	32.6		3400	No
134a	CH_2FCF_3	–15.0	14.6		1300	No
142b	CH_3CClF_2	15.8	18.4	0.065	2400	Yes
143	CH_2FCHF_2	41.0	3.8		330	Yes
143a	CH_3CF_3	–53.0	48.3		4300	Yes
152a	CH_3CHF_2	–11.2	1.5		120	Yes
218	$CF_3CF_2CF_3$	–33.9	2600[c]		8600[c]	No
227ea	CF_3CHFCF_3	3.9	36.5		3500	No
236ea	$CF_3CHFCHF_2$	43.7[b]	10[b]		1200	No
236fa	$CF_3CH_2CF_3$	29.5	209		9400	No
245ca	$CHF_2CF_2CH_2F$	77.2	6.6		640	Yes
245fa	$CF_3CH_2CHF_2$	59.2	8.8[c]		950	No

[a]Data from IPCC (2001). [b]Data from ASHRAE *Standard* 34. [c]Data from Calm and Hourahan (1999).

Table 3 Properties of Refrigerant Blends[a]

Refrig-erant	Composition	Bubble Point,[b] °F	ODP[c]	GWP,[d] 100-Year ITH
401A	(22/152a/124)/(53/13/34)	–27.9	0.027	1100
401B	(22/152a/124)/(61/11/28)	–30.8	0.028	1200
401C	(22/152a/124)/(33/15/52)	–19.1	0.025	900
402A	(125/C_3H_8/22)/(60/2/38)	–56.2	0.013	2700
402B	(125/C_3H_8/22)/(38/2/60)	–52.6	0.020	2300
403A	(C_2H_6/22/218)/(5/75/20)	–54.0	0.026	3000
403B	(C_2H_6/22/218)/(5/56/39)	–56.6	0.019	4300
404A	(125/143a/134a)/(44/52/4)	–51.2	0	3800
405A	(22/152a/142b/C318)/(45/7/5.5/42.5)	–27.2	0.018	5200
406A	(22/600a/142b)/(55/4/41)	–26.9	0.036	1900
407A	32/125/134a)/(20/40/40)	–49.5	0	2000
407B	(32/125/134a)/(10/70/20)	–52.2	0	2700
407C	(32/125/134a)/(23/25/52)	–46.5	0	1700
407D	(32/125/134a)/(15/15/70)	–39.1	0	1500
407E	(32/125/134a)/(25/15/60)	–45.2	0	1400
408A	(125/143a/22)/(7/46/47)	–48.3	0.016	3000
409A	(22/124/142b)/(60/25/15)	–30.5	0.039	1500
409B	(22/124/142b)/(65/25/10)	–32.1		
410A	(32/125)/(50/50)	–60.5	0	2000
411A	(R-1270/22/152a)/(1.5/87.5/11.0)	–39.1	0.030	1500
411B	(1270/22/152a)/(3/94/3)	–42.9	0.032	1600
412A	(22/218/142b)/(70/5/25)	–36.4	0.035	2200
413A	(218/134a/600a)/(9/88/3)	–23.1	0	1900
414A	(22/124/600a/142b)/(51/28.5/4/16.5)	–29.2	0.032	1400
414B	(22/124/600a/142b)/(50/39/1.5/9.5)	–27.2	0.031	1300
415A	(22/152a)/(82/18)	–35.5	0.028	1400
415B	(22/152a)/(25/75)	–17.9	0.009	500
416A	(134a/124/600)/(59/39.5/1.5)	–10.1	0.010	1000
417A	(125/134a/600)/(46.6/50/3.4)	–36.4	0.000	2200
418A	(290/22/152a)/(1.5/96/2.5)	–42.2	0.33	1600
500	(12/152a)/(73.8/26.2)	–28.5	0.605	7900
502	(22/115)/(48.8/51.2)	–49.4	0.221	4500
503	23/13/(40.1/59.9)	–127.8	0.599	13,000
507A	(125/143a)/(50/50)	–52.1	0	3900
508A	(23/116)/(39/61)	–125.3	0	12,000
508B	(23/116)/(46/54)	–124.6	0	12,000
509A	(22/218)/(44/56)	–57.6	0.015	5600

[a]Data from IPCC (2001).
[b]Data from ARI *Standard* 700.
[c]Data from Calm (2001).
[d]GWPs are weight fraction average for GWP values of individual components.

The European Union has already reduced the consumption cap on HCFCs and accelerated the phase-out schedule. E.U. consumption of HCFCs was reduced 15% on January 1, 2002, by 55% on January 1, 2003, and by 70% on January 1, 2004; future reductions are to be by 75% on January 1, 2008, and total phaseout on January 1, 2010. They also implemented several use restrictions on HCFCs in air-conditioning and refrigeration equipment.

U.S. and E.U. phaseout schedules allow continued, limited manufacture for developing-country needs or for export to other countries where HCFCs are still legally used.

Atmospheric studies (Calm et al. 1999; Wuebbles and Calm 1997) suggest that phaseout of HCFC refrigerants, with low atmospheric lives, low ozone depletion potentials, low global warming potentials, low emissions, and high thermodynamic efficiencies, will result in an increase in global warming, but have a negligible effect on ozone depletion.

HCFC-22 is the most widely used hydrochlorofluorocarbon. R-410A is now the leading alternative for HCFC-22 for new equipment. R-407C is another HCFC-22 replacement and can be used in retrofits as well as in new equipment. HCFC-123 is used commercially in large chillers.

Hydrofluorocarbons. These refrigerants contain no chlorine atoms, so their ODP is zero. HFC methanes, ethanes, and propanes have been extensively considered for use in air conditioning and refrigeration.

Fluoromethanes. Mixtures that include R-32 (difluoromethane, CH_2F_2) are being promoted as a replacement for R-22 and R-502. For very-low-temperature applications, R-23 (trifluoromethane, CHF_3) has been used as a replacement for R-13 and R-503 (Atwood and Zheng 1991).

Fluoroethanes. Refrigerant 134a (CF_3CH_2F) of the fluoroethane series is used extensively as a direct replacement for R-12 and as a replacement for R-22 in higher-temperature applications. R-125 and R-143a are used in azeotropes or zeotropic blends with R-32 and/or R-134a as replacements for R-22 or R-502. R-152a is flammable and less efficient than R-134a in applications using suction-line heat exchangers (Sandvordenker 1992), but it is still being considered for R-12 replacement. R-152a is also being considered as a component, with R-22 and R-124, in zeotropic blends (Bateman et al. 1990; Bivens et al. 1989) that can be R-12 and R-500 alternatives.

Fluoropropanes. Desmarteau et al. (1991) identified a number of fluoropropanes as potential refrigerants. R-245ca is being considered as a chlorine-free replacement for R-11. Evaluation by Doerr et al. (1992) showed that R-245ca is stable and compatible with key components of the hermetic system. However, Smith et al. (1993) demonstrated that R-245ca is slightly flammable in humid air at room temperature. Keuper et al. (1996) investigated R-245ca performance in a centrifugal chiller; they found that the refrigerant might be useful in new equipment but posed some problems when used as a retrofit for R-11 and R-123 machines. R-245fa is used as a chlorine-free replacement for R-11 and R-141b in foams, and is being considered as a refrigerant and commercialized in organic Rankine-cycle and waste-heat-recovery systems. R-236fa has been commercialized as a replacement for R-114 in naval centrifugal chillers.

Fluoroethers. Booth (1937), Eiseman (1968), Kopko (1989), O'Neill (1992), O'Neill and Holdsworth (1990), and Wang et al. (1991) proposed these compounds as refrigerants. Fluoroethers are usually more physiologically and chemically reactive than fluorinated hydrocarbons. Fluorinated ethers have been used as anesthetics and convulsants (Krantz and Rudo 1966; Terrell et al. 1971a, 1971b). Reactivity with glass is characteristic of some fluoroethers (Doerr et al. 1993; Gross 1990; Simons et al. 1977). Misaki and Sekiya (1995, 1996) investigated 1-methoxyperfluoropropane (boiling point 93.6°F) and 2-methoxyperfluoropropane (boiling point 84.9°F) as potential low-pressure refrigerants. Bivens and Minor (1997) reviewed the status of fluoroethers currently under consideration and concluded that none appear to have a balance of refrigerant fluid requirements to challenge the HFCs.

Hydrocarbons. Hydrocarbons such as propane, *n*-butane (R-600), isobutane (R-600a), and blends of these are being used as refrigerants. Hydrocarbons have zero ODP and low GWP. However, they are very flammable, which is a serious obstacle to their widespread use as refrigerants. Hydrocarbons are commonly used in small proportions in mixtures with nonflammable halogenated refrigerants and in small equipment requiring low refrigerant charges. Hydrocarbons are currently used in air-conditioning and refrigeration equipment in Europe and China (Lohbeck 1996; Mianmiam 1996; Powell 1996).

Ammonia. Used extensively in large, open-type compressors for industrial and commercial applications, ammonia (R-717) has high refrigerating capacity per unit displacement, low pressure losses in connecting piping, and low reactivity with refrigeration lubricants (mineral oils). See Chapter 3 for detailed information.

The toxicity and flammability of ammonia offset its advantages. Ammonia is such a strong irritant to the human nose (detectable below 5 ppm) that people automatically avoid exposure to it. Ammonia is considered toxic at 35 to 50 ppm. Ammonia/air mixtures are flammable, but only within a narrow range of 15.2 to 27.4% by volume. These mixtures can explode but are difficult to ignite because they require an ignition source of at least 1200°F.

Carbon Dioxide. Some governments are promoting use of CO_2 in refrigeration and air-conditioning cycles. Trial cascade systems are being used in Europe, and some countries in the European Union are promoting transcritical carbon dioxide systems to replace HFC-134a in automotive air-conditioning systems. Higher costs are expected because of the higher pressures and transcritical cycle.

Refrigerant Analysis

With the introduction of many new pure refrigerants and refrigerant mixtures, interest in refrigerant analysis has increased. Refrigerant analysis is addressed in ARI *Standards* 700 and 700c. Gas chromatographic methods are available to determine purity determination of R-134a and R-141b (Gehring et al. 1992a, 1992b). Gehring (1995) discusses measurement of water in refrigerants Bruno and Caciari (1994) and Bruno et al. (1995) have done extensive work developing chromatographic methods for analysis of refrigerants using a graphitized carbon black column with a coating of hexafluoropropene. Bruno et al. (1994) also published refractive indices for some alternative refrigerants. There is interest in developing methods for field analysis of refrigerant systems. Systems for field analysis of both oils and refrigerants are commercially available. Rohatgi et al. (2001) compared ion chromatography to other analytical methods for determining chloride, fluoride, and acids in refrigerants. They also investigated sample vessel surfaces and liners for absorption of hydrochloric and oleic acids.

Flammability and Combustibility

Refrigerant flammability testing is defined in UL *Standard* 2182, Section 7. For many refrigerants, flammability is enhanced by increased temperature and humidity. These factors must be controlled accurately to obtain reproducible, reliable data.

Fedorko et al. (1987) studied the flammability envelope of R-22/air as a function of pressure (up to 200 psia) and fuel (R-22)-to-oxygen ratio. They found that R-22 was nonflammable under 75 psia. In addition, the flammable compositions between 30 and 45% generated maximum heats of reaction. Their results were in general agreement with those of Sand and Andrjeski (1982), who found that pressurized mixtures of R-22 and at least 50% air are combustible. R-11 and R-12 did not ignite under similar conditions.

Lindley (1992) and Reed and Rizzo (1991), using different experimental arrangements, studied R-134a's combustibility at high temperature and pressure. Lindley notes that the results depend on the equipment used. Reed and Rizzo showed that R-134a is combustible above 15 psig at room temperature and air concentrations greater than 80% by volume. At 350°F, combustibility was observed at pressures above 5 psig and air concentrations above 60% by volume. Lindley found flammability limits of 8 to 22% by volume in air at 340°F and 100 psia. Both researchers found R-134a to be nonflammable at ambient conditions and under the likely operating conditions of air-conditioning and refrigeration equipment. Blends of R-22/152a/114 combusted above 180°F at atmospheric pressure and above, with air concentrations above 80% by volume (Reed and Rizzo 1991).

Richard and Shankland (1991) followed ASTM *Standard* E681's method to study flammability of R-32, R-141b, R-142b, R-152a, R-152, R-143, R-161, methylene chloride, 1,1,1-trichloroethane, propane, pentane, dimethyl ether, and ammonia. They used several ignition methods, including the electrically activated match ignition source specified in ASHRAE *Standard* 34. They also reported on the critical flammability ratio of mixtures such as R-32/125, R-143a/134a, R-152a/125, propane/R-125, R-152a/22, R-152a/124, and R-152a/134a. The critical flammability ratio is the maximum amount of flammable component that a mixture can contain and still be nonflammable, regardless of the amount of air. These data are important because mixtures containing flammable components are being considered as refrigerants.

Zhigang et al. (1992) published data on flammability of R-152a/22 mixtures. Their measured lower flammability limit in air of R-152a is 11.4% by volume, though values reported in the literature range from 4.7 to 16.8% by volume. Richard and Shankland (1991) reported an average flammable range of 4.1 to 20.2% by mass for R-152a. Zhigang et al. (1992) also provide data on flame length as a function of R-22 concentration. They found that the flame no longer existed somewhere between 17 and 40% R-22 by mass in the mixture. This is in apparent disagreement with Richard and Shankland's (1991) data, which showed a critical flammability ratio of 57.1% R-22 by mass. Comparison is difficult because results depend on the apparatus and methods used. Grob (1991), reporting on flammabilities of R-152a, R-141b, and R-142b, describes R-152a as having "the lowest flammable mixture percentage, highest explosive pressure and highest potential for ignition of the refrigerants studied." Womeldorf and Grosshandler (1995) used an opposed-flow burner to evaluate flammability limits of refrigerants.

CHEMICAL REACTIONS

Halocarbons

Thermal Stability in the Presence of Metals. All common halocarbon refrigerants have excellent thermal stability, as shown in Table 4. Bier et al. (1990) studied R-12, R-134a, and R-152a. For R-134a in contact with metals, traces of hydrogen fluoride (HF) were detected after 10 days at 392°F. This decomposition did not increase much with time. R-152a showed traces of HF at 356°F after five days in a steel container. Bier et al. suggested that vinyl fluoride forms during thermal decomposition of R-152a, and can then react with water to form acetaldehyde. Hansen and Finsen (1992) conducted lifetime tests on small hermetic compressors with a ternary mixture of R-22/152a/124 and an alkyl benzene lubricant. In agree-

Table 4 Inherent Thermal Stability of Halocarbon Refrigerants

Refrigerant	Formula	Decomposition Rated at 400°F in Steel, % per yr[a]	Temperature at Which Decomposition Readily Observed in Laboratory,[b] °F	Temperature at Which 1%/Year Decomposes in Absence of Active Materials, °F	Major Gaseous Decomposition Products[c]
22	$CHClF_2$	—	800	480	CF_2CF_2,[d] HCl
11	CCl_3F	2	1100	570[e]	R-12, Cl_2
114	$CClF_2CClF_2$	1	1100	710	R-12
115	$CClF_2CF_3$	—	1160	740	R-13
12	CCl_2F_2	Less than 1	1400	930	R-13, Cl_2
13	CClF3	—	1550	1000[f]	R-14, Cl_2, R-116

Sources: Borchardt (1975), DuPont (1959, 1969), and Norton (1957).
[a]Data from UL *Standard* 207.
[b]Decomposition rate is about 1% per min.
[c]Data from Borchardt (1975).
[d]Various side products are also produced, here and with the other refrigerants, some of which may be quite toxic.
[e]Conditions were not found where this reaction proceeds homogeneously.
[f]Rate behavior too complex to permit extrapolation to 1% per year.

Table 5 Rate of Hydrolysis in Water (Grams per Litre of Water per Year)

Refrig-erant	Formula	14.7 psi at 86°F		Saturation Pressure at 122°F
		Water Alone	With Steel	with Steel
113	CCl_2FCClF_2	<0.005	50	40
11	CCl_3F	<0.005	10	28
12	CCl_2F_2	<0.005	1	10
21	$CHCl_2F$	<0.01	5	9
114	$CClF_2$-$CClF$	<0.005	1	3
22	$CHClF_2$	<0.01	0.1	—

Source: DuPont (1959, 1969).

ment with Bier et al., they found that vinyl fluoride and acetaldehyde formed in the compressor. Aluminum, copper, and brass and solder joints lower the temperature at which decomposition begins. Decomposition also increases with time.

Under extreme conditions, such as above red heat or with molten metal temperatures, refrigerants react exothermically to produce metal halides and carbon. Extreme temperatures may occur in devices such as centrifugal compressors if the impeller rubs against the housing when the system malfunctions. Using R-12 as the test refrigerant, Eiseman (1963) found that aluminum was most reactive, followed by iron and stainless steel. Copper is relatively unreactive. Using aluminum as the reactive metal, Eiseman reported that R-14 causes the most vigorous reaction, followed by R-22, R-12, R-114, R-11, and R-113. Dekleva et al. (1993) studied the reaction of various CFCs, HCFCs, and HFCs in vapor tubes at very high temperatures in the presence of various catalysts and measured the onset temperature of decomposition. These data also showed HFCs to be more thermally stable than CFCs and HCFCs, and that, when molten aluminum is in contact with R-134a, a layer of unreactive aluminum fluoride forms and inhibits further reaction.

Hydrolysis. Halogenated refrigerants are susceptible to reaction with water (hydrolysis), but the rates of reaction are so slow that they are negligible (Table 5). Desiccants (see Chapter 6) are used to keep refrigeration systems dry. Cohen (1993) investigated compatibilities of desiccants with R-134a and refrigerant blends.

Ammonia

Reactions involving ammonia, oxygen, oil degradation acids, and moisture are common factors in the formation of ammonia compressor deposits. Sedgwick (1966) suggested that ammonia or ammonium hydroxide reacts with organic acids produced by oxidation of the compressor oil to form ammonium salts (soaps), which can decompose further to form amides (sludge) and water. The reaction is as follows:

$$NH_3 + RCOOH \Leftrightarrow RCOONH_4 \Leftrightarrow RCONH_2 + H_2O$$

Water may be consumed or released during the reaction, depending on system temperature, metallic catalysts, and pH (acidic or basic). Compressor deposits can be minimized by keeping the system clean and dry, preventing entry of air, and maintaining proper compressor temperatures. Ensure that ester lubricants and ammonia are not used together, because large quantities of soaps and sludges would be produced.

At atmospheric pressure, ammonia starts to dissociate into nitrogen and hydrogen at about 570°F in the presence of active catalysts such as nickel and iron. However, because these high temperatures are unlikely to occur in open-type compression systems, thermal stability is not a problem. Ammonia attacks copper in the presence of even small amounts of moisture; therefore, except for some specialty bronzes, copper-bearing materials and copper plating are excluded in ammonia systems. (See the section on Copper Plating for more information.)

Table 6 Influence of Type of Alcohol on Ester Viscosity

Type of Alcohol	Ester Viscosity at 104°F, centistokes
Neopentyl glycol (NPG)	13.3
Glycerin (GLY)	31.9
Trimethylolpropane (TMP)	51.7
Pentaerythritol (PER)	115

Note: Ester derived using the same carboxylic acid.

Table 7 R-134a Miscibility and Viscosity of Several Pentaerythritol-Based Esters

Acid Used	R-134a Miscibility at 20% Ester, °F	Ester Viscosity at 104°F, centistokes
5 carbon, linear	<–94	15.6
6 carbon, linear	–53	18.5
7 carbon, linear	34	21.2
8 carbon, linear	>149	26.7
9 carbon, linear	>149	31.0
5 carbon, branched	<–94	25.2
8 carbon, branched	5	44.4
9 carbon, branched	17	112.9

Source: Jolley (1997).

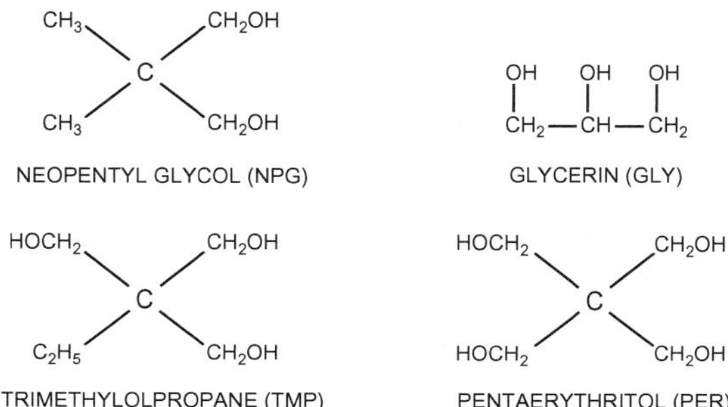

Fig. 1 Types of Alcohols Used for Ester Synthesis

Lubricants

Lubricants now in use and under consideration for new refrigerants are mineral oils, alkyl benzenes, polyol esters, polyalkylene glycols, modified polyalkylene glycols, and polyvinyl ethers. Gunderson and Hart (1962) give an excellent introduction to synthetic lubricants, including polyglycols and esters.

Polyol Esters. Commercial esters (Jolley 1991) are manufactured from four types of alcohols; (1) neopentyl glycol (NPG), with two OH reaction sites; (2) glycerin (GLY), with three OH sites; (3) trimethylolpropane (TMP) with three OH sites; and (4) pentaerythritol (PER), with four OH sites. Formulas for the four alcohol types are shown in Figure 1. Viscosities of the esters formed by reaction of a given acid with each of the four alcohol types are given in Table 6.

Polyol esters are widely used as lubricants in HCFC refrigerant systems, mainly because of their physical properties. Because they are made from a wide variety of materials, polyol esters can be designed to optimize desired physical characteristics. The system chemistry of the lubricant can be significantly influenced by the type and chain length of the carboxylic acid used to prepare the ester.

Table 7 gives R-134a miscibility and viscosity data for several esters based on pentaerythritol. Clearly, polyol ester lubricants rapidly lose refrigerant miscibility when linear carbon chain lengths exceed six carbons. Using branched chain acids to prepare these lubricants can greatly enhance refrigerant miscibility. Chain branching also enables preparation of higher-viscosity esters, which are needed in some industrial refrigeration applications.

The thermal stability of polyol esters is well known. Esters made from polyols that possess a central neo structure, which consists of a carbon atom attached to four other carbon atoms (i.e., structures corresponding to NPG, TMP, and PER in Figure 1), have outstanding thermal stability. Gunderson and Hart (1962) reviewed research measuring the thermal stability of various polyol esters and dibasic acid esters at 500°F by heating them in evacuated tubes for up to 250 h. These tests demonstrated the increased thermal stability expected from neo ester structures, with dibasic acid esters decomposing three times faster than the polyol esters.

Hydrolysis of Esters. An alcohol and an organic acid react to produce an organic ester and water; this reaction is called esterification, and it is reversible. The reverse reaction of an ester and water to produce an alcohol and an organic acid is called hydrolysis:

$$\underset{\text{Ester}}{\mathrm{RCOOR'}} + \underset{\text{Water}}{\mathrm{HOH}} \Leftrightarrow \underset{\text{Acid}}{\mathrm{RCOOH}} + \underset{\text{Alcohol}}{\mathrm{R'OH}}$$

Hydrolysis may be the most important chemical stability issue associated with esters. The degree to which esters are subject to hydrolysis is related to their processing parameters [particularly total acid number (TAN), degree of esterification, nature of the catalyst used during production, and catalyst level remaining in the polyol ester after processing] and their structure. Dick et al. (1996) demonstrated that (1) using polyol esters prepared with acids known as α-branched acids significantly reduces ester hydrolysis and (2) using α-branched esters with certain additives can eliminate hydrolysis.

Hydrolysis is undesirable in refrigeration systems because free carboxylic acid can react with and corrode metal surfaces. Metal carboxylate soaps that may be produced by hydrolysis can also block capillary tubes. Davis et al. (1996) reported that polyol ester hydrolysis proceeds through autocatalytic reaction, and determined reaction rate constants for hydrolysis using sealed-tube tests. Jolley et al. (1996) and others used compressor testing, along with variations of the ASHRAE *Standard* 97 sealed-tube test, to examine the potential for lubricant hydrolysis in operating systems. Compressor tests run with lubricant saturated with water (2000 ppm) have gone 2000 h with no significant capillary tube blockage, indicating that under normal, much drier operating conditions, little or no detrimental ester hydrolysis occurs with use of polyol ester lubricants. Hansen and Snitkjær (1991) demonstrated ester hydrolysis in compressor life tests run without desiccants and in sealed tubes. They detected hydrolysis by measuring the total acid number and showed that desiccants can reduce the extent of hydrolysis in a compressor. They concluded that, with filter-driers, refrigeration systems using esters and R-134a can be very reliable.

Greig (1992) ran the thermal and oxidation stability test (TOST) by heating an oil/water emulsion to 203°F and bubbling oxygen through it in the presence of steel and copper. Appropriate additives can suppress hydrolysis of esters. Although agreeing that esters can be used in refrigeration, Jolley et al. (1996) point out that some additives are themselves subject to hydrolysis. Cottington and Ravner (1969) and Jones et al. (1969) studied the effect of tricresyl phosphate, a common antiwear agent, on ester decomposition.

Field and Henderson (1998) studied the effect of elevated levels of organic acids and moisture on corrosion of metals in the presence of R-134a and POE lubricant. Copper, brass, and aluminum showed little corrosion, but cast iron and steel were severely corroded. At 392°F, iron caused the POE lubricant to break down, even in the absence of additional acid and moisture. Similar chemistry was reported by Klauss et al. (1970), who found that high-temperature (600°F) decomposition of POE was catalyzed by iron. Naidu et al. (1988) showed that this POE/iron reaction did not occur at a measurable rate at 365°F. Cottington and Ravner (1969) reported that the presence of TCP inhibits the POE/iron reaction, which Lilje (2000) concluded is a high-energy process and does not occur in properly operating refrigeration systems. Field lubricant analysis data, after 5 years of operation, support this conclusion: no lubricant degradation was observed (Riemer and Hansen 1996).

Polyalkylene Glycols (PAGs). Polyalkylene glycols are of the general formula RO—[CH_2—CHR′—O]—R′. They are used as lubricants in automotive applications that use R-134a. Linear PAGs can have one or two terminal hydroxyl groups. Modified PAG molecules have both ends capped by various groups. Sundaresan and Finkenstadt (1990) discuss the use of PAGs and modified PAGs in refrigeration compressors. Short and Cavestri (1992) present data on PAGs.

These lubricants and their additive packages may (1) oxidize, (2) degrade thermally, (3) react with system contaminants such as water, and/or (4) react with refrigerant or system materials such as polyester films.

Oxidation is usually not a problem in hermetic systems using hydrocarbon oils, because no oxygen is available to react with the lubricant. However, if a system is not adequately evacuated or if air is allowed to leak into the system, organic acids and sludges can be formed. Clark et al. (1985) and Lockwood and Klaus (1981) found that iron and copper catalyze the oxidative degradation of esters. These reaction products are detrimental to the refrigeration system and can cause failure. Komatsuzaki et al. (1991) have suggested that the oxidative breakdown products of PAG lubricants and perhaps of esters are volatile, whereas those of mineral oils are more likely to include sludges.

Sanvordenker (1991) studied thermal stability of PAG and ester lubricants and found that, above 400°F, water is one of the decomposition products of esters (in the presence of steel) and of PAG lubricants. He recommends that polyol esters be used with metal passivators to enhance their stability when in contact with metallic bearing surfaces, which can experience 400°F temperatures. Sanvordenker presents data on the kinetics of the thermal decomposition of polyol esters and PAGs. These reactions are catalyzed by metal surfaces in the following order: low carbon steel > aluminum > copper (Naidu et al. 1988).

Lubricant Additives

Additives are often used to improve lubricant performance in refrigeration systems, and have become more important as use of HFC refrigerants has increased. Chlorine in CFC refrigerants acted as an antiwear agent, so mineral-oil lubricants needed minimal or no additives to provide wear protection. HFC refrigerants such as R-134a do not contain chlorine and thus do not provide this antiwear benefit. Additives such as antioxidants, detergents, dispersants, rust inhibitors, etc., are not normally used because the conditions they treat are absent from most refrigeration systems. Many HFC/polyol ester refrigeration systems function well without lubricant additives. However, some systems that have aluminum wear surfaces require an additive to supplement wear protection. Antiwear protection is likely to be necessary in future systems with lower-viscosity lubricants to improve energy efficiency, especially if branched-acid polyol esters are used. Randles et al. (1996) discuss the advantages and disadvantages of using additives in polyol ester lubricants for refrigeration systems.

The active ingredient in antiwear additives is typically phosphorous, sulfur, or both. Organic phosphates, phosphites, and phosphonates are typical phosphorous-containing antiwear agents. Tricresylphosphate (TCP) is the best known of these. Sulfurized olefins and disulfides are typical of sulfur-containing additives for wear protection. Zinc dithiophosphates are the best examples of mixed additives. Vinci and Dick (1995) showed that additives containing phosphorous can perform well as antiwear agents, and that sulfur-containing additives are not thermally stable as determined by the ASHRAE *Standard* 97 sealed-tube stability test.

Other additives used in HFC/polyol ester combinations are foam-producing agents (compressor start-up noise reduction) and hydrolysis inhibitors. Vinci and Dick (1995) show that a combination of antiwear additive and hydrolysis inhibitor can produce exceptional performance in both wear and capillary tube blockage in bench testing and long-term compressor endurance tests. Sanvordenker (1991) has shown that iron surfaces can catalyze the decomposition of esters at 400°F. He proposed using a metal passivator additive to minimize this effect in systems where high temperatures are possible. Schmitz (1996) describes the use of a siloxane ester foaming agent for noise reduction. Swallow et al. (1995, 1996) suggested using additives to control the release of refrigerant vapor from polyol ester lubricants.

System Reactions

Average strengths of carbon/chlorine, carbon/hydrogen, and carbon/fluorine bonds are 78, 93, and 100 kcal/mole, respectively (Pauling 1960). The relative stabilities of refrigerants that contain chlorine, hydrogen, and fluorine bonded to carbon can be understood by considering these bond strengths. The CFCs have characteristic reactions that depend largely on the presence of the C—Cl bond. Spauschus and Doderer (1961) concluded that R-12 can react with a hydrocarbon oil by exchanging a chlorine for a hydrogen. In this reaction, characteristic of chlorine-containing refrigerants, R-12 forms the reduction product R-22, R-22 forms R-32 (Spauschus and Doderer 1964), and R-115 forms R-125 (Parmelee 1965). For R-123, Carrier (1989) demonstrated that the reduction product R-133a is formed at high temperatures.

Factor and Miranda (1991) studied the reaction between R-12, steel, and oil sludge. They concluded that it can proceed by a predominantly Friedel-Crafts mechanism in which Fe^{3+} compounds are key catalysts. They also concluded that oil sludge can be formed by a pathway that does not generate R-22. They suggest that, except for the initial formation of Fe^{3+} salts, the free-radical mechanism plays only a minor role. Further work is needed to clarify this mechanism.

Huttenlocher (1992) tested 23 refrigerant-lubricant combinations for stability in sealed glass tubes. HFC refrigerants were shown to be very stable even at temperatures much higher than normal operating temperatures. HCFC-124 and HCFC-142b were slightly more reactive than the HFCs, but less reactive than CFC-12. HCFC-123 was less reactive than CFC-11 by a factor of approximately 10.

Fluoroethers were studied as alternative refrigerants. Sealed-glass-tube and Parr bomb stability tests with E-245 (CF_3—CH_2—O—CHF_2) showed evidence of an autocatalytic reaction with glass that proceeds until either the glass or the fluoroether is consumed (Doerr et al. 1993). High pressures (about 2000 psi) usually cause the sealed glass tubes to explode.

Breakdown of CFCs and HCFCs can usually be tracked by observing the concentration of reaction products formed. Alternatively, the amount of fluoride and chloride formed in the system can be observed. For HFCs, no chloride will be formed, and reaction products are highly unlikely because the C—F bond is strong. Decomposition of HFCs is usually tracked by measuring the fluoride ion concentration in the system (Spauschus 1991; Thomas and Pham 1989; Thomas et al. 1993); according to this test, R-125, R-32, R-143a, R-152a, and R-134a are quite stable.

The possibility that hydrogen fluoride released by the breakdown of the refrigerants being studied will react with glass of the sealed tube is a concern. Sanvordenker (1985) confirmed this possibility with R-12. Spauschus et al. (1992) found no evidence of fluoride on the glass surface of sealed tubes with R-134a.

Figures 2 and 3 show sealed-tube test data for reaction rates of R-22 and R-12 with oil in the presence of copper and mild steel. Formation of chloride ion was taken as a measure of decomposition. These figures show the extent to which temperature accelerates reactions, and that R-22 is much less reactive than R-12. The data only illustrate the chemical reactivities involved and do not represent actual rates in refrigeration systems.

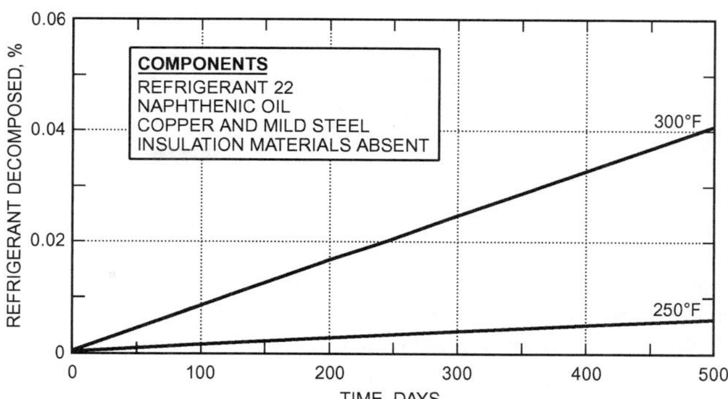

Fig. 2 Stability of Refrigerant 22 Control System
(Kvalnes and Parmelee 1957)

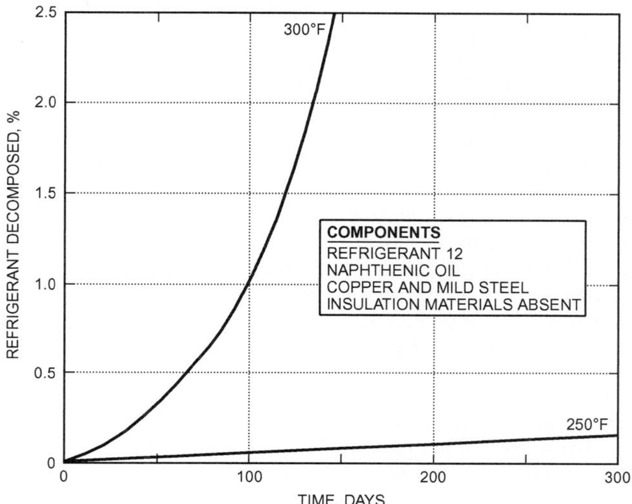

Fig. 3 Stability of Refrigerant 12 Control System
(Kvalnes and Parmelee 1957)

The chemistry in CFC systems retrofitted to use HFC refrigerants and their lubricants is an area of growing interest. Corr et al. (1992) point out that a major problem is the effect of chlorinated residues in the new system. Komatsuzaki et al. (1991) showed that R-12 and R-113 degrade PAG lubricants. Powers and Rosen (1992) performed sealed-tube tests and concluded that the threshold of reactivity for R-12 in R-134a and PAG lubricant is between 1 and 3%.

Copper Plating

Copper plating is the formation of a copper film on steel surfaces in refrigeration and air-conditioning compressors. A blush of copper is often discernible on compressor bearing and valve surfaces when machines are cut apart. After several hours of exposure to air, this thin film becomes invisible, probably because metallic copper is converted to copper oxide. In severe cases, the copper deposit can build up to substantial thickness and interfere with compressor operation. Extreme copper plating can cause compressor failure.

Although the exact mechanism of copper plating is not completely understood, early work by Spauschus (1963), Steinle and Bosch (1955), and Steinle and Seeman (1951, 1953) demonstrated that three distinct steps must occur: (1) copper oxidation, (2) solubilization and transport of copper ions, and (3) deposition of copper onto iron or steel.

In step 1, copper oxidizes from the metallic (0 valent) state to either the +1 or +2 oxidation state. Under normal operating conditions, this chemical process does not occur with a lubricant, and is unlikely to occur with carboxylic acids. The most likely source of oxidizing agents is system contaminants, such as air (oxygen), chlorine-containing species (CFC refrigerants or cleaning solvents, solder fluxes), or strong acids.

Step 2 is dissolution of the copper ions. Spauschus postulated that an organic complex of the copper and olefins is the soluble species in mineral oils. Oxygen-containing lubricants are much more likely to solubilize metal ions and/or complexes via coordination with the oxygen atoms. Once soluble, the copper can move throughout the refrigeration system.

Step 3 is deposition of the copper onto iron surfaces, an electrochemical process in which electrons transfer from iron to copper, resulting in copper metal (0 valent) plating on the surface of the iron and the concomitant generation of iron ions. This is more likely to occur on hot, clean iron surfaces and is often seen on bearing surfaces.

Thomas and Pham (1989) compared copper plating in R-12/mineral oil and R-134a/PAG systems. They showed that R-134a/PAG systems produced much less total copper (in solution and as precipitate) than R-12/mineral oil systems, and that water did not significantly affect the amount of copper produced. In the R-134a/PAG system, copper was largely precipitated. In the R-12/mineral oil system, the copper was found in solution when dry and precipitated when wet. Walker et al. (1960) found that water below the saturation level had no significant effect on copper plating for R-12/mineral oil systems. Spauschus (1963) observed that copper plating in sealed glass tubes was more prevalent with medium-refined naphthenic pale oil than with a highly refined white oil. He concluded that the refrigerant/lubricant reaction was an essential precursor for gross copper plating. The excess acid produced by refrigerant decomposition had little effect on copper solubility, but facilitated plating. Herbe and Lundqvist (1996, 1997) examined a large number of systems retrofitted from R-12 to R-134a for contaminants and copper plating. They reported that copper plating did not occur in retrofitted systems where the level of contaminants was low.

Contaminant Generation by High Temperature

Hermetic motors can overheat well beyond design levels under adverse conditions such as line voltage fluctuations, brownouts, or inadequate airflow over the condenser coils. Under these conditions, motor winding temperatures can exceed 300°F. Prolonged exposure to these thermal excursions can damage motor insulation, depending on the insulation materials' thermal stability and reactivity with the refrigerant and lubricant, and the temperature levels encountered.

Another potential for high temperatures is in the bearings. Oil-film temperatures in hydrodynamically lubricated journal bearings are usually not much higher than the bulk oil temperature; however, in elastohydrodynamic films in bearings with a high slide/roll ratio, the temperature can be several hundred degrees above the bulk oil temperature (Keping and Shizhu 1991). Local hot spots in boundary lubrication can reach very high temperatures, but fortunately, the amount of material exposed to these temperatures is usually very small. The appearance of methane or other small hydrocarbon molecules in the refrigerant indicates lubricant cracking by high bearing temperatures.

Thermal decomposition of organic insulation materials and some types of lubricants produces noncondensable gases such as carbon dioxide and carbon monoxide. These gases circulate with the refrigerant, increasing the discharge pressure and lowering unit efficiency. At the same time, compressor temperature and deterioration rate of the insulation or lubricant increase. Liquid decomposition products circulate with the lubricating oil either in solution or as colloidal suspensions. Dissolved and suspended decomposition products circulate throughout the refrigeration system, where they clog oil passages; interfere with operation of expansion, suction, and discharge valves; or plug capillary tubes.

Appropriate control mechanisms in the refrigeration system minimize exposure to high temperatures. Identifying potential reactions, performing adequate laboratory tests to qualify materials before field use, and finding means to remove contaminants generated by high-temperature excursions are equally important (see Chapter 6).

COMPATIBILITY OF MATERIALS

Electrical Insulation

Insulation on electric motors is affected by the refrigerant and/or the lubricant in two main ways: extraction of insulation polymer into the refrigerant or absorption of refrigerant by the polymer.

Extraction of insulation material causes embrittlement, delamination, and general degradation of the material. In addition, extracted material can separate from solution, deposit out, and cause components to stick or passages (e.g., capillary tubes) to clog.

Refrigerant absorption can change the material's dielectric strength or physical integrity through softening or swelling. Rapid desorption (off-gassing) of refrigerant caused by internal heating can be more serious, because it results in high internal pressures that cause blistering or voids within the insulation, decreasing its dielectric or physical strength.

In compatibility studies of 10 refrigerants and 7 lubricants with 24 motor materials in various combinations, Doerr and Kujak (1993) showed that R-123 was absorbed to the greatest extent, but R-22 caused more damage because of more rapid desorption and higher internal pressures. They also observed insulation damage after desorption of R-32, R-134, and R-152a in a 300°F oven, but not as much as with R-22.

Compatibility studies of motor materials were also conducted under retrofit conditions in which materials were exposed to the original refrigerant/mineral oil followed by exposure to the alternative refrigerant/polyolester lubricant (Doerr and Waite 1995, 1996a). Alternative refrigerants included R-134a, R-407C, R-404A, and R-123. Most motor materials were unaffected, except for increased brittleness in polyethylene terephthalate (PET) caused by moisture and blistering between layers of sheet insulation from the adhesive. Many of the same materials were completely destroyed when exposed to ammonia; the magnet wire enamel was degraded, and the PET sheet insulation completely disappeared, having been converted to a terephthalic acid diamide precipitate (Doerr and Waite 1996b).

Ratanaphruks et al. (1996) determined the compatibility of metals, desiccants, motor materials, plastics, and elastomers with the HFCs R-245ca, R-245fa, R-236ea, and R-236fa, and HFE-125. Most metals and desiccants were compatible. Plastics and elastomers were compatible except for excessive absorption of refrigerant or lubricant (resulting in unacceptable swelling) observed with fluoropolymers, hydrogenated nitrile butyl rubber, and natural rubber. Corr et al. (1994) tested compatibility with R-22 and R-502 replacements. Kujak and Waite (1994) studied the effect on motor materials of HFC refrigerants with polyol ester lubricants containing elevated levels of moisture and organic acids. They concluded that a 500 ppm moisture level in polyol ester lubricant had a greater effect on the motor materials than an organic acid level of 2 mg KOH/g. Exposure to R-134a/polyol ester with a high moisture level had less effect than exposure to R-22/mineral oil with a low moisture level.

Ellis et al. (1996) developed an accelerated test to determine the life of motor materials in alternative refrigerants using a simulated stator unit. Hawley-Fedder (1996) studied breakdown products of a simulated motor burnout in HFC refrigerant atmospheres.

Magnet Wire Insulation. Magnet wire is coated with heat-cured enamels. The most common insulation is a polyester base coat followed by a polyamide imide top coat; a polyester imide base coat is also used. Acrylic and polyvinyl formal enamels are found on older motors. An enameled wire with an outer layer of polyester-glass is used in larger hermetic motors for greater wire separation and thermal stability.

Magnet wire insulation is the primary source of electrical insulation and the most critical in compatibility with refrigerants. Most electrical tests (NEMA *Standard* MW 1000) are conducted in air and may not be valid for hermetic motors. For example, wire enamels absorb R-22 up to 15 to 30% by mass (Hurtgen 1971) and at different rates, depending on their chemical structure, degree of cure, and conditions of exposure to the refrigerant. Refrigerant permeation is shown by changes in electrical, mechanical, and physical properties of the wire enamels. Fellows et al. (1991) measured dielectric strength, Paschen curve minimum, dielectric constant, conductivity, and resistivity for 19 HFCs in order to predict electrical properties in the presence of these refrigerants.

Wire enamels in refrigerant vapor typically exhibit dielectric loss with increasing temperature, as shown in Figure 4. Depending on the atmosphere and degree of cure, each wire enamel or enamel/varnish combination exhibits a characteristic temperature t_{max}, above which dielectric losses increase sharply. Table 8 shows values of t_{max} for several hermetic enamels. Continued heating above t_{max} causes aging, shown by the irreversible alteration of dielectric properties and increased conductance of the insulating material.

Spauschus and Sellers (1969) showed that the change rate in conductance is a quantitative measure of aging in a refrigerant environment. They proposed aging rates for varnished and unvarnished enamels at two levels of R-22 pressure, typical of high- and low-side hermetic motor operation.

Apart from the effects on long-term aging, R-22 can also affect the short-term insulating properties of some wire enamels. Beacham and Divers (1955) demonstrated that polyvinyl formal's resistance drops drastically when it is submerged in liquid R-22. A parallel experiment using R-12 showed a much smaller drop, followed by quick recovery to the original resistance. The relatively rapid permeation of R-22 into polyvinyl formal, coupled with R-22's low volume resistivity and other electrical properties of the two refrigerants, explains the phenomenon.

With certain combinations of coatings and refrigerants, wire coatings can soften, which can cause the insulation to fail. Table 9 shows data on softening measured in terms of abrasion resistance for a number of wire enamels exposed to R-22. At the end of the shortest soaking period, the urethane-modified polyvinyl formal had lost all its abrasion resistance. All the other insulations, except polyimide, lost abrasion resistance more slowly, approaching, over three months, the rate of the urethane-polyvinyl formal. The polyimide showed only a minimal effect, although its abrasion resistance was originally among the lowest.

Because of the time dependency of softening, which is related to the rate of R-22 permeation into the enamel, Sanvordenker and Larime (1971) proposed that comparative tests on magnet wire be made only after the enamel is completely saturated with refrigerant, so that the effect on enamel properties of long-term exposure to R-22 can be evaluated.

The second consequence of R-22 permeation is blistering, caused by the rapid change in pressure and temperature after a wire enamel is exposed to R-22. Heating greatly increases the internal pressure as the dissolved R-22 expands; because the polymer film has already been softened, portions of the enamel lift up in the form of blisters. Although blistered wire has a poor appearance, field experience indicates that mild blistering is not cause for concern, as long as the blisters do not break and the enamel film remains flexible. Modern wire enamels have the characteristics mentioned previously and maintain dielectric strength even after blistering. However, hermetic wire enamel with strong resistance is preferred.

Varnishes. After the stator of an electric motor is wound, it is usually treated with a varnish by a vacuum-and-pressure impregnation process for form-wound, high-voltage motors or a dip-and-bake process for low-voltage, random wound motors. The varnished motor is cured in a 275 to 350°F oven. The varnish holds the windings together in the magnetic field and acts as a secondary source of electrical insulation. The windings have a tendency to move, and independent movement of the wires abrades and wears the insulation. High-voltage motors contain form-wound coils wrapped with a porous fiberglass, which is saturated with varnish and cured as an additional layer.

Many different chemicals are used as motor varnishes. The most common are epoxies, polyesters, phenolics, and modified polyimides. Characteristics important to a varnish are good adhesion and bond strength to the wire enamel; flexibility and strength under both heat and cold; thermal stability; good dielectric properties; and chemical compatibility with wire enamel, sheet insulation, and refrigerant/lubricant mixture.

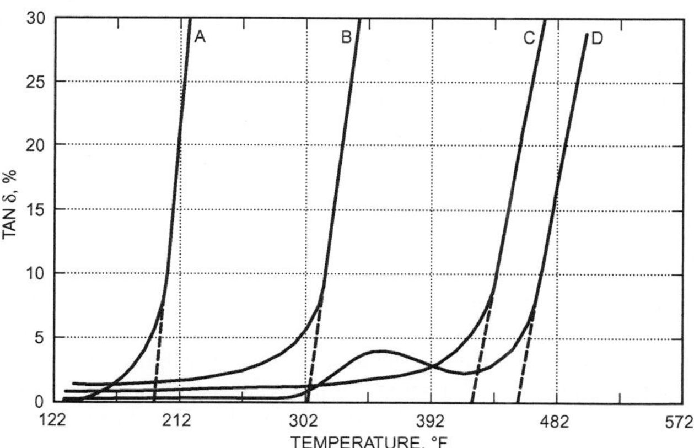

Fig. 4 Loss Curves of Various Insulating Materials
(Spauschus and Sellers 1969)

Table 8 Maximum Temperature t_{max} for Hermetic Wire Enamels in R-22 at 65 psia

Enamel Type	t_{max}, °F
Acrylic	226
Polyvinyl formal	277
Isocyanate-modified polyvinyl formal	304
Polyamide imide	361
Polyester imide	419
Polyimide	450

Table 9 Effect of Liquid R-22 on Abrasion Resistance

		After Time in Liquid R-22		
Magnet Wire Insulation	As Received	7 to 10 Days	One Month	Three Months
Urethane/polyvinyl formal batch 1	40	3	2	2
Urethane/polyvinyl formal batch 2	42	2	2	7
Polyester imide batch 1	44	15	18	6
Polyester imide batch 2	24	10	5	6
Dual-coat amide/imide top coat, polyester base	79	35	23	11
Dual-coat, polyester	35	5	5	9
Polyimide	26	25	23	21

Source: Sanvordenker and Larime (1971).

Varnish compatibility is determined by exposing the cured varnish (in the form of a section of a thin disk and varnished magnet wire in single strands, helical coils, and twisted pairs) to a refrigerant at elevated temperatures. The varnish's properties are then compared to samples not exposed to refrigerant and to other exposed samples placed in a hot oven to rapidly remove absorbed refrigerant. Disk sections are evaluated for absorption, extraction, degradation, and changes in flexibility. The single strands are wound around a mandrel, and the varnish is examined for flexibility and effect on the wire enamel. In many cases, the varnish does not flex as well as the enamel; if bound tightly to the enamel, the varnish removes it from the copper wire. The helical coils are evaluated for bond strength (ASTM *Standard* D2519) before and after exposure to a refrigerant/lubricant mixture. The twisted pair is tested for dielectric breakdown voltage, or burnout time, while subjected to resistance heating (ASTM *Standard* D1676).

During compatibility testing of motor materials with alternative refrigerants, researchers observed that varnish can absorb considerable amounts of refrigerant, especially R-123. Doerr (1992) studied the effects of time and temperature on absorption and desorption rates of R-123 and R-11 by epoxy motor varnishes. Absorption was faster at higher temperatures. Desorption was slow at temperatures as high as 250°F. The equilibrium absorption value for R-123 was linearly dependent on temperature, with higher absorption at lower temperatures. Absorption of R-11 remained the same at all test temperatures.

Ground Insulation. Sheet insulation material is used in slot liners, phase insulation, and wedges in hermetic motors. The sheet material is usually a PET film or an aramid (aromatic polyamide) mat, used singly or laminated together. PET or aramid films have excellent dielectric properties and good chemical resistance to refrigerants and oils.

The PET film selected must contain little of the low-relative-molecular-mass polymers that exhibit temperature-dependent solubility in mineral lubricants and tend to precipitate as noncohesive granules at temperatures lower than those of the motor. Another limitation is that, like most polyesters, this film is susceptible to degradation by hydrolysis; however, the amount of water required is more than that generally found in refrigerant systems. Sundaresan and Finkenstadt (1991) discuss the effect of synthetic lubricants on PET films. Dick and Malone (1996) reported that low-viscosity POEs tend to extract more low-oligomeric PET components than higher-viscosity esters.

Elastomers

Refrigerants, oils, or mixtures of both can, at times, extract enough filler or plasticizer from an elastomer to change its physical or chemical properties. This extracted material can harm the refrigeration system by increasing its chemical reactivity or by clogging screens and expansion devices. Many elastomers are unsuitable for use with refrigerants because of excessive swelling or shrinkage (e.g., some neoprenes tend to shrink in HFC refrigerants, and nitriles swell in R-123). Hamed and Seiple (1993a, 1993b) determined swell data on 95 elastomers in 10 refrigerants and seven lubricants. Compatibility data on general classifications of elastomers such as neoprenes or nitriles should be used with caution because results depend on the particular formulation. Users should be aware that elastomeric behavior is strongly affected by the elastomer's specific formulation as well as by its general type.

Plastics

The effect of refrigerants on plastics usually decreases as the amount of fluorine in the molecule increases. For example, R-12 has less effect than R-11, whereas R-13 is almost entirely inert. Cavestri (1993) studied the compatibility of 23 engineering plastics with alternative refrigerants and lubricants.

Each type of plastic material should be tested for compatibility with the refrigerant before use. Two samples of the same type of plastic might be affected differently by the refrigerant because of differences in polymer structure, relative molecular mass, and plasticizer.

CHEMICAL EVALUATION TECHNIQUES

Chemical problems can often be attributed to inadequate testing of a new material, improper application of a previously tested material, or inadvertent introduction of contaminants into the system. Three techniques are used to chemically evaluate materials: (1) sealed-tube material tests, (2) component tests, and (3) system tests.

Sealed-Tube Material Tests

The glass sealed-tube test, described in ASHRAE *Standard* 97, is widely used to assess stability of refrigerant system materials and to identify chemical reactions that are likely to occur in operating units.

Generally, glass tubes are charged with refrigerant, oil, metal strips, and other materials to be tested, and then sealed and aged at elevated temperatures for a specified time. The tubes are inspected for color and appearance and compared to control tubes that are processed identically to the specimen tubes, but might contain a reference material rather than the test material. Contents can be analyzed for changes by gas, ion, or liquid chromatography; infrared spectroscopy; specific ion electrode; or wet methods, such as total acid number analysis.

The sealed-tube test was originally designed to compare lubricants, but it is also effective in testing other materials. For example, Huttenlocher (1972) evaluated zinc die castings, Guy et al. (1992) reported on compatibilities of motor insulation materials and elastomers, and Mays (1962) studied R-22 decomposition in the presence of 4A-type molecular sieve desiccants.

Although the sealed tube is very useful, it has some disadvantages. Because chemical reactions likely to occur in a refrigeration system are greatly magnified, results can be misinterpreted. Also, reactions in which mechanical energy plays a role (e.g., in a failing bearing) are not easily studied in a static sealed tube.

Despite its proven utility, the sealed-tube test is only a screening tool and not a full simulation of a refrigeration system. Sealed-tube tests alone should not be used to predict field behavior. Material selection for refrigerant systems requires follow-up with component or system tests or both.

Component Tests

Component tests carry material evaluations a step beyond sealed-tube tests: materials are tested not only in the proper environment, but also under dynamic conditions. Motorette (enameled wire, ground insulation, and other motor materials assembled into a simulated motor) tests used to evaluate hermetic motor insulation, as described in Underwriters Laboratories (UL) *Standard* 984, are a good example of this type of test. Component tests are conducted in large pressure vessels or autoclaves in the presence of a lubricant and a refrigerant. Unlike sealed-tube tests, in which temperature and pressure are the only means of accelerating aging, autoclave tests can include external stresses (e.g., mechanical vibration, on/off electrical voltages, liquid refrigerant floodback) that may accelerate phenomena likely to occur in an operating system.

System Tests

System tests can be divided into two major categories:

- Testing a sufficient number of systems under a broad spectrum of operating conditions to obtain a good, statistical reference base. Failure rates of units containing new materials can be compared to those of units containing proven materials.
- Testing under well-controlled conditions. Temperatures, pressures, and other operating conditions are continuously monitored.

Refrigerant and lubricant are chemically analyzed before, during, and after the test.

In most cases, tests are conducted under severe operating conditions to obtain results quickly. Analyzing lubricant and refrigerant samples during the test and inspecting the components after teardown can yield information on the (1) nature and rate of chemical reactions taking place in the system, (2) products formed by these reactions, and (3) possible effects on system life and performance. Accurate interpretation of these data determines system operating limits that keep chemical reactions at an acceptable level.

REFRIGERANT DATABASE AND ARTI/MCLR RESEARCH PROJECTS

The U.S. Department of Energy (DOE) funded research through the Air-Conditioning and Refrigeration Technology Institute to accelerate phaseout of CFCs and conversion to alternative refrigerants and lubricants. The funds supported about 40 materials compatibility and lubricants research (MCLR) projects and a refrigerants database (Calm 2001).

The database covers over 5000 references with abstracts or summaries of presentations and technical papers. Performance, physical property, toxicity, compatibility, and flammability data for refrigerants are also included.

REFERENCES

ARI. 2004. Specifications for fluorocarbon refrigerants. *Standard* 700-2004. Air-Conditioning and Refrigeration Institute, Arlington, VA.

ARI. 1999. Appendix C to ARI *Standard* 700—Analytical procedures for ARI *Standard* 700-99. *Standard* 700c-99. Air-Conditioning and Refrigeration Institute, Arlington, VA.

ASHRAE. 2004. Number designation and safety classification of refrigerants. ANSI/ASHRAE *Standard* 34-2004.

ASHRAE. 1999. Sealed glass tube method to test the chemical stability of material for use within refrigerant systems. ANSI/ASHRAE *Standard* 97-1999 (RA2003).

ASTM. 2003. Test methods for film-insulated magnet wire. ANSI/ASTM *Standard* D1676-03. American Society for Testing and Materials, West Conshohocken, PA.

ASTM. 2002. Test method for bond strength of electrical insulating varnishes by the helical coil test. ANSI/ASTM *Standard* D2519-02. American Society for Testing and Materials, West Conshohocken, PA.

ASTM. 2004. Test method for concentration limits of flammability of chemicals (vapors and gases). *Standard* E681-04. American Society for Testing and Materials, West Conshohocken, PA.

Atwood, T. and J. Zheng. 1991. Cascade refrigeration systems: The HFC-23 solution. *International CFC and Halon Alternatives Conference*, Baltimore, MD, sponsored by The Alliance for Responsible CFC Policy, Arlington, VA, pp. 442-450.

Bateman, D.J., D.B. Bivens, R.A. Gorski, W.D. Wells, R.A. Lindstrom, R.A. Morse, and R.L. Shimon. 1990. Refrigerant blends for the automotive air conditioning aftermarket. *SAE Technical Paper Series* 900216. Society of Automotive Engineers, Warrendale, PA.

Beacham, E.A. and R.T. Divers. 1955. Some practical aspects of the dielectric properties of refrigerants. *Refrigerating Engineering* (July):33.

Bier, K., M. Crone, M. Tuerk, W. Leuckel, M. Christill, and B. Leisenheimer. 1990. Studies of the thermal stability and ignition behavior and combustion properties of the refrigerants R-152a and R-134a. *DKV-Tagungsbericht* 17:169-191.

Bivens, D.B. and B.H. Minor. 1997. Fluoroethers and other next-generation fluids. *Proceedings of Refrigerants for the 21st Century*, Gaithersburg, MD.

Bivens, D.B., R.A. Gorski, W.D. Wells, A. Yokozeki, R.A. Lindstrom, and R.L. Shimon. 1989. Evaluation of fluorocarbon blends as automotive air conditioning refrigerants. *SAE Technical Paper Series* 890306. Society of Automotive Engineers, Warrendale, PA.

Booth, H.S. 1937. *Halogenated methyl ethers*. U.S. Patent 2,066,905.

Borchardt, H.J. 1975. *Du Pont innovation* G(2).

Bruno, T.J. and M. Caciari. 1994. Retention of halocarbons on a hexafluoropropylene epoxide modified graphitized carbon black, Part 1: Methane based fluids. *Journal of Chromatography* A 672:149-158.

Bruno, T.J., M. Wood, and B.N. Hansen. 1994. Refractive indices of alternative refrigerants. *Journal of Research of the National Institute of Standards and Technology* 99(3):263-266.

Bruno, T.J., M. Caciari, and K.H. Wertz. 1995. Retention of halocarbons on a hexafluoropropylene epoxide modified graphitized carbon black, part 4: Propane based fluids. *Journal of Chromatography A* 708:293-302.

Calm, J.M. 2001. *ARTI refrigerant database*. Air-Conditioning and Refrigeration Technology Institute, Arlington, VA.

Calm, J.M. and G.C. Hourahan. 1999. Physical, safety, and environmental data for refrigerants. *HPAC Engineering* (August).

Calm, J.M., D.J. Wuebbles, and A.K. Jain. 1999. Impacts of global ozone and climate from use and emissions of 2,2-dichloro-1,1,1-trifluoroethane (HCFC-123). *Journal of Climatic Change* 42:439-474.

Carrier. 1989. Decomposition rates of R-11 and R-123. Carrier Corporation, Syracuse, NY. Available from *ARTI refrigerant database*. Air-Conditioning and Refrigeration Technology Institute, Arlington, VA.

Cavestri, R.C. 1993. Compatibility of refrigerants and lubricants with plastics. *Final Report* DOE/CE 23810-15. *ARTI refrigerant database* (December). Air-Conditioning and Refrigeration Technology Institute, Arlington, VA.

Clark, D.B., E.E. Klaus, and S.M. Hsu. 1985. The role of iron and copper in the degradation of lubricating oils. *Journal of ASLE* (May):80-87. Society of Tribologists and Lubrication Engineers, Park Ridge, IL.

Cohen, A.P. 1993. Test methods for the compatibility of desiccants with alternative refrigerants. *ASHRAE Transactions* 99(1):408-412.

Corr, S., R.D. Gregson, G. Tompsett, A.L. Savage, and J.A. Schukraft. 1992. Retrofitting large refrigeration systems with R-134a. *Proceedings of the International Refrigeration Conference—Energy Efficiency and New Refrigerants* 1:221-230. Purdue University, West Lafayette, IN.

Corr, S., P. Dowdle, G. Tompsett, R. Yost, T. Dekleva, J. Allison, and R. Brutsch. 1994. Compatibility of non-metallic motor components with RH22 and R-502 replacements. ASHRAE Winter Meeting, New Orleans, LA. *ARTI refrigerant database* 4216. Air-Conditioning and Refrigeration Technology Institute, Arlington, VA.

Cottington, R.L. and H. Ravner. 1969. Interactions in neopentyl polyol ester-tricresyl phosphate-iron systems at 500°F. *ASLE Transactions* 12:280-286. Society of Tribologists and Lubrication Engineers, Park Ridge, IL.

Davis, K.E., S.T. Jolley, and J.R. Shanklin. 1996. Hydrolytic stability of polyolester refrigeration lubricants. *ARTI refrigerant database*. Air-Conditioning and Refrigeration Technology Institute, Arlington, VA.

Dekleva, T.D., A.A. Lindley, and P. Powell. 1993. Flammability and reactivity of select HFCs and mixtures. *ASHRAE Journal* (12):40-47.

Desmarteau, D., A. Beyerlein, S. Hwang, Y. Shen, S. Li, R. Mendonca, K. Naik, N.D. Smith, and P. Joyner. 1991. Selection and synthesis of fluorinated propanes and butanes as CFC and HCFC alternatives. *International CFC and Halon Alternatives Conference*, sponsored by The Alliance for Responsible CFC Policy, Arlington, VA, pp. 396-405.

Dick, D.L., and G.R. Malone. 1996. Compatibility of polymeric materials with low-viscosity refrigeration lubricants. *Proceedings of the 1996 International Refrigeration Conference*, Purdue University, West Lafayette, IN, pp. 391-396.

Dick, D.L., J.N. Vinci, K.E. Davis, G.R. Malone, and S.T. Jolley. 1996. *ARTI refrigerant database*. Air-Conditioning and Refrigeration Technology Institute, Arlington, VA.

Doerr, R.G. 1992. Absorption of HCFC-123 and CFC-11 by epoxy motor varnish. *ASHRAE Transactions* 98(2):227-234.

Doerr, R.G. and S.A. Kujak. 1993. Compatibility of refrigerants and lubricants with motor materials. *ASHRAE Journal* 35(8):42-47.

Doerr, R.G., D. Lambert, R. Schafer, and D. Steinke. 1992. Stability and compatibility studies of R-245ca, CHF_2—CF_2—CH_2F, a potential low-pressure refrigerant. *International CFC and Halon Alternatives Conference*, Washington, D.C., pp. 147-152.

Doerr, R.G. and T.D. Waite. 1995. Compatibility of refrigerants and lubricants with motor materials under retrofit conditions. *Proceedings of the International CFC and Halon Alternatives Conference*, Washington, D.C., pp. 159-168.

Doerr, R.G. and T.D. Waite. 1996a. Compatibility of ammonia with motor materials. ASHRAE Annual Meeting, February, Atlanta, GA. *ARTI refrigerant database*. Air-Conditioning and Refrigeration Technology Institute, Arlington, VA.

Doerr, R.G. and T.D. Waite. 1996b. Compatibility of refrigerants and lubricants with motor materials under retrofit conditions. *Final Report* DOE/CE 23810-63. *ARTI refrigerant database*. Air-Conditioning and Refrigeration Technology Institute, Arlington, VA.

Doerr, R.G., D. Lambert, R. Schafer, and D. Steinke. 1993. Stability studies of E-245 fluoroether, CF_3—CH_2—O—CHF_2. *ASHRAE Transactions* 99(2):1137-1140.

DuPont. 1959. Properties and application of the "Freon" fluorinated hydrocarbons. *Technical Bulletin* B-2. Freon Products Division, E.I. du Pont de Nemours and Co.

DuPont. 1969. Stability of several "Freon" compounds at high temperatures. *Technical Bulletin* XIA, Freon Products Division, E.I. du Pont de Nemours and Co.

Eiseman, B.J. 1968. *Chemical processes*. U.S. Patent 3,362,180.

Eiseman, B.J., Jr. 1963. Reactions of chlorofluoro-hydrocarbons with metals. *ASHRAE Journal* 5(5):63.

Ellis, P.F., A.F. Ferguson, and K.T. Fuentes. 1996. Accelerated test methods for predicting the life of motor materials exposed to refrigerant-lubricant mixtures. *Report* DOE/CE 23810-69. *ARTI refrigerant database*. Air-Conditioning and Refrigeration Technology Institute, Arlington, VA.

Factor, A. and P.M. Miranda. 1991. An investigation of the mechanism of the R-12-oil-steel reaction. *Wear* 150:41-58.

Fedorko, G., G. Fredrick, and J.G. Hansel. 1987. Flammability characteristics of chlorodifluoromethane (R-22)-oxygen-nitrogen mixtures. *ASHRAE Transactions* 93(2):716-724.

Fellows, B.R., R.C. Richard, and I.R. Shankland. 1991. Electrical characterization of alternative refrigerants. *Proceedings of the XVIIIth International Congress of Refrigeration*, Montreal, vol. 45, p. 398.

Field, J.E. and D.R. Henderson. 1998. Corrosion of metals in contact with new refrigerants/lubricants at various moisture and organic acid levels. *ASHRAE Transactions* 104(1).

Gehring, D.G. 1995. How to determine concentration of water in system refrigerants. *ASHRAE Journal* 37(9):52-55.

Gehring, D.G., D.J. Barsotti, and H.E. Gibbon. 1992a. Chlorofluorocarbons alternatives analysis, part I: The determination of HFC-134a purity by gas chromatography. *Journal of Chromatographic Science* 30:280.

Gehring, D.G., D.J. Barsotti, and H.E. Gibbon. 1992b. Chlorofluorocarbons alternatives analysis, part II: The determination of HCFC-141b purity by gas chromatography. *Journal of Chromatographic Science* 30:301.

Greig, B.D. 1992. Formulated polyol ester lubricants for use with HFC-134a. The role of additives and conversion of existing CFC-12 plant to HFC-134a. *Proceedings of the International CFC and Halon Alternatives Conference*, Washington, D.C., pp 135-145.

Grob, D.P. 1991. Summary of flammability characteristics of R-152a, R-141b, R-142b and analysis of effects in potential applications: Household refrigerators. 42nd Annual International Appliance Technical Conference, Batavia, IL.

Gross, T.P. 1990. Sealed tube tests—Grace ether (E-134). *ARTI refrigerant database* RDB0904. Air-Conditioning and Refrigeration Technology Institute, Arlington, VA.

Gunderson, R.C. and A.W. Hart. 1962. *Synthetic lubricants*. Reinhold Publishing, New York.

Guy, P.D., G. Tompsett, and T.W. Dekleva. 1992. Compatibilities of nonmetallic materials with R-134a and alternative lubricants in refrigeration systems. *ASHRAE Transactions* 98(1):804-816.

Hamed, G.R. and R.H. Seiple. 1993a. Compatibility of elastomers with refrigerant/lubricant mixtures. *ASHRAE Journal* 35(8):173-176.

Hamed, G.R. and R.H. Seiple. 1993b. Compatibility of refrigerants and lubricants with elastomers. *Final Report* DOE/CE 23810-14. *ARTI refrigerant database*. Air-Conditioning and Refrigeration Technology Institute, Arlington, VA.

Hansen, P.E. and L. Finsen. 1992. Lifetime and reliability of small hermetic compressors using a ternary blend HCFC-22/HFC-152a/HCFC-124. *International Refrigeration Conference—Energy Efficiency and New Refrigerants*, D. Tree, ed. Purdue University, West Lafayette, IN.

Hansen, P.E. and L. Snitkjær. 1991. Development of small hermetic compressors for R-134a. *Proceedings of the XVIIIth International Congress of Refrigeration*, Montreal, vol. 223, p. 1146.

Hawley-Fedder, R. 1996. Products of motor burnout. *Report* DOE/CE 23810-74. *ARTI refrigerant database*. Air-Conditioning and Refrigeration Technology Institute, Arlington, VA.

Herbe, L. and P. Lundqvist. 1996. Refrigerant retrofit in Sweden—Field and laboratory studies. *Proceedings of the International Refrigeration Conference*, Purdue University, West Lafayette, IN, pp. 71-76.

Herbe, L. and P. Lundqvist. 1997. CFC and HCFC refrigerants retrofit—Experiences and results. *International Journal of Refrigeration* 20(1): 49-54.

Hurtgen, J.R. 1971. R-22 blister testing of magnet wire. *Proceedings of the 10th Electrical Insulation Conference*, Chicago, pp. 183-185.

Huttenlocher, D.F. 1972. Accelerated sealed-tube test procedure for Refrigerant 22 reactions. *Proceedings of the 1972 Purdue Compressor Technology Conference*, Purdue University, West Lafayette, IN.

Huttenlocher, D.F. 1992. Chemical and thermal stability of refrigerant-lubricant mixtures with metals. *Final Report* DOE/CE 23810-5. *ARTI refrigerant database*. Air-Conditioning and Refrigeration Technology Institute, Arlington, VA.

IPCC. 1995. *Climate change 1995: The science of climate change*. Contribution of Working Group I to the Second Assessment of the Intergovernmental Panel on Climate Change. Cambridge University Press, U.K.

IPCC. 2001. *Climate change 2001: The scientific basis*. Contribution of Working Group I to the Third Assessment Report of the Intergovernmental Panel on Climate Change. Cambridge University Press, U.K.

Jolley, S.T. 1991. The performance of synthetic ester lubricants in mobile air conditioning systems. U.S. DOE Grant No. DE-F-G02-91 CE 23810. Paper presented at the International CFC and Halon Alternatives Conference, Baltimore, MD, sponsored by The Alliance for Responsible CFC Policy, Arlington, VA. *ARTI refrigerant database* RDB2C07, Air-Conditioning and Refrigeration Technology Institute, Arlington, VA.

Jolley, S.T. 1997. Polyolester lubricants for use in environmentally friendly refrigeration applications. American Chemical Society National Meeting, San Francisco, CA. *Preprints of Papers, Division of Petroleum Chemistry* 42(1):238-241.

Jolley, S.T., K.E. Davis, and G.R. Malone. 1996. The effect of desiccants in HFC refrigeration systems using ester lubricants. *ARTI refrigerant database*. Air-Conditioning and Refrigeration Technology Institute, Arlington, VA.

Jones, R.L., H.L. Ravner, and R.L. Cottington. 1969. Inhibition of iron-catalyzed neopentyl polyol ester thermal degradation through passivation of the active metal surface by tricresyl phosphate. ASLE/ASME Lubrication Conference, Houston, TX.

Keping, H. and W. Shizhu. 1991. Analysis of maximum temperature for thermo elastohydrodynamic lubrication in point contacts. *Wear* 150: 1-10.

Keuper, E.F., F.B. Hamm, and P.R. Glamm. 1996. Evaluation of R-245ca for commercial use in low pressure chiller. *Final Report* DOE/CE 23810-67. *ARTI refrigerant database* (March). Air-Conditioning and Refrigeration Technology Institute, Arlington, VA.

Klaus, E.E., E.J. Tewksbury, and S.S Fietelson. 1970. Thermal characteristics of some organic esters. *ASLE Transactions* 13(1):11-20.

Ko, M., R.L. Shia, and N.D. Sze. 1997. *Report on calculations of global warming potentials*. Prepared for AFEAS, Contract P97-134, July.

Komatsuzaki, S., Y. Homma, K. Kawashima, and Y. Itoh. 1991. Polyalkylene glycol as lubricant for HCFC-134a compressors. *Lubrication Engineering* (Dec.):1018-1025.

Kopko, W.L. 1989. Extending the search for new refrigerants. *Proceedings of CFC Technology Conference*, Gaithersburg, MD.

Krantz, J.C. and F.G. Rudo. 1966. The fluorinated anesthetics. In *Handbook of experimental pharmacology*, O. Eichler, H. Herken, and A.D. Welch, eds., vol. 20, part 1, pp. 501-564. Springer-Verlag, Berlin.

Kujak, S.A. and T.D. Waite. 1994. Compatibility of motor materials with polyolester lubricants: Effect of moisture and weak acids. *Proceedings of the International Refrigeration Conference*, Purdue University, West Lafayette, IN, pp. 425-429.

Kvalnes, D.E. and H.M. Parmelee. 1957. Behavior of Freon-12 and Freon-22 in sealed-tube tests. *Refrigerating Engineering* (November):40.

Lilje, K. 2000. The impact of chemistry on the use of polyol ester lubricants in refrigeration. *ASHRAE Transactions* 106(2):661-667.

Lindley, A.A. 1992. KLEA 134a flammability characteristics at high-temperatures and pressures. *ARTI refrigerant database*. Air-Conditioning and Refrigeration Technology Institute, Arlington, VA.

Lockwood, F. and E.E. Klaus. 1981. Ester oxidation—The effect of an iron surface. *ASLE Transactions* 25(2):236-244. Society of Tribologists and Lubrication Engineers, Park Ridge, IL.

Lohbeck, W. 1996. Development and state of conversion to hydrocarbon technology. *Proceedings of the International Conference on Ozone Protection Technologies*, Washington, D.C., pp. 247-251.

Mays, R.L. 1962. Molecular sieve and gel-type desiccants for refrigerants. *ASHRAE Transactions* 68:330.

Mianmiam, Y. 1996. *Proceedings of the International Conference on Ozone Protection Technologies*, Washington, D.C., pp. 260-266.

Misaki, S. and A. Sekiya. 1995. Development of a new refrigerant. *Proceedings of the International CFC and Halon Alternatives Conference*, Washington, D.C., pp. 278-285.

Misaki, S. and A. Sekiya. 1996. Update on fluorinated ethers as alternatives to CFC refrigerants. *Proceedings of the International Conference on Ozone Protection Technologies*, Washington, D.C., pp. 65-70.

Montreal Protocol. 2003. *Montreal Protocol handbook for the international treaties for the protection of the ozone layer*, 6th ed., Annexes A, B, and C. Secretariat for the Vienna Convention for the Protection of the Ozone Layer and the Montreal Protocol on Substances That Deplete the Ozone Layer, United Nations Environment Programme, Nairobi.

Naidu, S.K., B.E. Klaus, and J.L. Duda. 1988. Thermal stability of esters under simulated boundary lubrication conditions. *Wear* 121:211-222.

NEMA. 2003. Magnet wire. *Standard* MW 1000-03. National Electrical Manufacturer's Association, Rosslyn, VA.

Norton, F.J. 1957. Rates of thermal decomposition of $CHClF_2$ and Cl_2F_2. *Refrigerating Engineering* (September):33.

O'Neill, G.J. 1992. *Synthesis of fluorinated dimethyl ethers*. U.K. Patent Application 2,248,617.

O'Neill, G.J. and R.S. Holdsworth. 1990. *Bis(difluoromethyl) ether refrigerant*. U.S. Patent 4,961,321.

Parmelee, H.M. 1965. Sealed-tube stability tests on refrigerant materials. *ASHRAE Transactions* 71(1):154.

Pauling, L. 1960. The nature of the chemical bond and the structure of molecules and crystals. In *An introduction to modern structural chemistry*. Cornell University, Ithaca, NY.

Powell, L. 1996. Field experience of HC's in commercial applications. *Proceedings of International Conference on Ozone Protection Technologies*, Washington, D.C., pp. 237-246.

Powers, S. and S. Rosen. 1992. Compatibility testing of various percentages of R-12 in R-134a and PAG lubricant. International CFC and Halon Alternatives Conference, Washington, D.C.

Randles, S.J., P.J. Tayler, S.H. Colmery, R.W. Yost, A.J. Whittaker, and S. Corr. 1996. The advantages and disadvantages of additives in polyol esters: An overview. *ARTI refrigerant database*. Air-Conditioning and Refrigeration Technology Institute, Arlington, VA.

Ratanaphruks, K., M.W. Tufts, A.S. Ng, and N.D. Smith. 1996. Material compatibility evaluations of HFC-245ca, HFC-245fa, HFE-125, HFC-236ea, and HFC-236fa. *Proceedings of the International Conference on Ozone Protection Technologies*, Washington, D.C., pp. 113-122.

Ravishankara, A.R., A.A. Turnipseed, N.R. Jensen, and R.F. Warren. 1994. Do hydrofluorocarbons destroy stratospheric ozone? *Science* 248:1217-1219.

Reed, P.R. and J.J. Rizzo. 1991. Combustibility and stability studies of CFC substitutes with simulated motor failure in hermetic refrigeration equipment. *Proceedings of the XVIIIth International Congress of Refrigeration*, Montreal, vol. 2, pp. 888-891.

Richard, R.G. and I.R. Shankland. 1991. Flammability of alternative refrigerants. *Proceedings of the XVIIIth Congress of Refrigeration*, Montreal, p. 384.

Riemer, A. and P.E. Hansen. 1996. Analysis of R-134a cabinets from the first series production in 1990. *Proceedings of the 1996 International Refrigeration Conference*, Purdue University, West Lafayette, IN, pp. 501-505.

Rohatgi, N.D.T., T.T. Whitmire, and R.W. Clark. 2001. Chlorine, fluoride and acidity measurements in refrigerants. *ASHRAE Transactions* 107(1): 141-146.

Sand, J.R. and D.L. Andrjeski. 1982. Combustibility of chlorodifluoromethane. *ASHRAE Journal* 24(5):38-40.

Sanvordenker, K.S. 1985. Mechanism of oil-R12 reactions—The role of iron catalyst in glass sealed tubes. *ASHRAE Transactions* 91(1A):356-363.

Sanvordenker, K.S. 1991. Durability of HFC-134a compressors—The role of the lubricant. *Proceedings of the 42nd Annual International Appliance Technical Conference*, University of Wisconsin, Madison.

Sanvordenker, K.S. and M.W. Larime. 1971. Screening tests for hermetic magnet wire insulation. *Proceedings of the 10th Electrical Insulation Conference*, Institute of Electrical and Electronics Engineers, Piscataway, NJ, pp. 122-136.

Schmitz, R. 1996. *Method of making foam in an energy efficient compressor*. U.S. Patent 549,908.

Sedgwick, N.V. 1966. *The organic chemistry of nitrogen*, 3rd ed. Clarendon Press, Oxford, U.K.

Short, G.D. and R.C. Cavestri. 1992. High viscosity ester lubricants for alternative refrigerants. *ASHRAE Transactions* 98(1):789-795.

Simons, G.W., G.J. O'Neill, and J.A. Gribens. 1977. *New aerosol propellants for personal products*. U.S. Patent 4,041,148.

Smith, N.D., K. Ratanaphruks, M.W. Tufts, and A.S. Ng. 1993. R-245ca: A potential far-term alternative for R-11. *ASHRAE Journal* 35(2):19-23.

Spauschus, H.O. 1963. Copper transfer in refrigeration oil solutions. *ASHRAE Journal* 5:89.

Spauschus, H.O. 1991. Stability requirements of lubricants for alternative refrigerants. *Paper* 148. XVIIIth International Congress of Refrigeration, Montreal.

Spauschus, H.O. and G.C. Doderer. 1961. Reaction of Refrigerant 12 with petroleum oils. *ASHRAE Journal* 3(2):65.

Spauschus, H.O. and G.C. Doderer. 1964. Chemical reactions of Refrigerant 22. *ASHRAE Journal* 6(10).

Spauschus, H.O. and R.A. Sellers. 1969. Aging of hermetic motor insulation. *IEEE Transactions* E1-4(4):90.

Spauschus, H.O., G. Freeman, and T.L. Starr. 1992. Surface analysis of glass from sealed tubes after aging with HFC-134a. *ARTI refrigerant database* RDB2729. Air-Conditioning and Refrigeration Technology Institute, Arlington, VA.

Steinle, H. and R. Bosch. 1955. Versuche über Kupferplattierung an Kaltemaschinen. *Kaltetechnik* 7(4):101-104.

Steinle, H. and W. Seemann. 1951. Über die Kupferplattierung in Kaltemaschinen. *Kaltetechnik*, 3(8):194-197.

Steinle, H. and W. Seemann. 1953. Ursache der Kupferplattierung in Kaltemaschinen. *Kaltetechnik* 5(4):90-94.

Sundaresan, S.G. and W.R. Finkenstadt. 1990. Status report on polyalkylene glycol lubricants for use with HFC-134a in refrigeration compressors. *Proceedings of the 1990 USNC/IIR-Purdue Conference*, ASHRAE-Purdue Conference, pp. 138-144.

Sundaresan, S.G. and W.R. Finkenstadt. 1991. Degradation of polyethylene terephthalate films in the presence of lubricants for HFC-134a: A critical issue for hermetic motor insulation systems. *International Journal of Refrigeration* 14:317.

Swallow, A., A. Smith, and B. Greig. 1995. Control of refrigerant vapor release from polyol ester/halocarbon working fluids. *ASHRAE Transactions* 101(2):929-934.

Swallow, A.P., A.M. Smith, and D.G.V. Jones. 1996. Control of refrigerant vapor release from polyolester/halocarbon working fluids. *Proceedings of the International Conference on Ozone Protection Technologies*, Washington, D.C., pp. 123-132.

Terrell, R.C., L. Spears, A.J. Szur, J. Treadwell, and T.R. Ucciardi. 1971a. General anesthetics. 1. Halogenated methyl ethers as anesthetics agents. *Journal of Medical Chemistry* 14:517.

Terrell, R.C., L. Spears, T. Szur, T. Ucciardi, and J.F. Vitcha. 1971b. General anesthetics. 3. Fluorinated methyl ethyl ethers as anesthetic agents. *Journal of Medical Chemistry* 14:604.

Thomas, R.H. and H.T. Pham. 1989. Evaluation of environmentally acceptable refrigerant/lubricant mixtures for refrigeration and air conditioning. SAE Passenger Car Meeting and Exposition. Society of Automotive Engineers, Warrendale, PA.

Thomas, R.H., W.T. Wu, and R.H. Chen. 1993. The stability of R-32/125 and R-125/143a. *ASHRAE Transactions* 99(2).

UL. 2001. Refrigerant-containing components and accessories, nonelectrical, 7th ed. ANSI/UL *Standard* 207-01. Underwriters Laboratories, Northbrook, IL.

UL. 1996. Hermetic refrigerant motor-compressors, 7th ed. *Standard* 984-96. Underwriters Laboratories, Northbrook, IL.

UL. 1994. Refrigerants. ANSI/UL *Standard* 2182-94. Underwriters Laboratories, Northbrook, IL.

Vinci, J.N. and D.L. Dick. 1995. Polyol ester lubricants for HFC refrigerants: A systematic protocol for additive selection. *International CFC and Halon Alternatives Conference*, Baltimore, MD, sponsored by The Alliance for Responsible CFC Policy, Arlington, VA.

Walker, W.O., S. Rosen, and S.L. Levy. 1960. A study of the factors influencing the stability of mixtures of Refrigerant 22 and refrigerating oils. *ASHRAE Transactions* 66:445.

Wang, B., J.L. Adcock, S.B. Mathur, and W.A. Van Hook. 1991. Vapor pressures, liquid molar volumes, vapor non-idealities and critical properties of some fluorinated ethers: CF_3—O—CF_2—O—CF_3, CF_3—O—CF_2—CF_2H, C—CF_2—CF_2—CF_2—O—, CF_3—O—CF_2H, and CF_3—O—CH_3 and of CCl_3F and CF_2ClH. *Journal of Chemical Thermodynamics* 23:699-710.

Womeldorf, C. and W. Grosshandler. 1995. Lean flammability limits as a fundamental refrigerant property. *Final Report* DOE/CE 23810-68. *ARTI refrigerant database*. Air-Conditioning and Refrigeration Technology Institute, Arlington, VA.

Wuebbles, D.J. and J.M. Calm. 1997. An environmental rationale for retention of endangered chemicals. *Science* 278:1090-1091.

Zhigang, L., L. Xianding, Y. Jianmin, T. Xhoufang, J. Pingkun, C. Zhehua, L. Dairu, R. Mingzhi, Z. Fan, and W. Hong. 1992. Application of HFC-152a/HCFC-22 blends in domestic refrigerators. *ARTI refrigerant database* RDB2514. Air-Conditioning and Refrigeration Technology Institute, Arlington, VA.

BIBLIOGRAPHY

ASTM. 2004. Test method for acid number of petroleum products by potentiometric titration. ANSI/ASTM *Standard* D664-04. American Society for Testing and Materials, West Conshohocken, PA.

Calm, J.M. and G.C. Hourahan. 2001. Refrigerant data summary. *Engineered Systems* (November):74-88.

DeMore, W.B., S.P. Sander, D.M. Golden, R.F. Hampson, M.J. Kurylo, C.J. Howard, A.R. Ravishankara, C.E. Kolb, and M.J. Molina. 1997. Chemical kinetics and photochemical data for use in stratospheric modeling. *Evaluation* 12, January 15. NASA and JPL, California Institute of Technology.

Gierczak, T., R.K. Talukdar, J.B. Burkholder, R.W. Portman, J.S. Daniel, S. Solomon, and A.R. Ravishankara. 1996. Atmospheric fate and greenhouse warming potentials of HFC 236fa and HFC 236ea. *Journal of Geophysical Research* 101(D8):12,905-12,911.

Houghton, J.T., L.G. Filho, B.A. Callander, N. Harris, A. Kattenberg, and K. Maskell. 1995. *Climate change 1994: The science of climate change*. Contribution of WGI to First Assessment Report of the International Panel on Climate Change. Cambridge University Press, U.K.

Houghton, J.T., L.G. Filho, B.A. Callander, N. Harris, A. Kattenberg, and K. Maskell. 1996. *Climate change 1995: The science of climate change*. Contribution of WGI to Second Assessment Report of the International Panel on Climate Change. Cambridge University Press, U.K.

Keuper, E.F. 1996. Performance characteristics of R-11, R-123 and R-245ca in direct drive low pressure chillers. *Proceedings of the International Compressor Conference*, Purdue University, West Lafayette, IN, pp. 749-754.

Reed, P.R. and H.O. Spauschus. 1991. HCFC-124: Applications, properties and comparison with CFC-114. *ASHRAE Journal* 40(2).

Sanvordenker, K.S. 1992. R-152a versus R-134a in a domestic refrigerator-freezer, energy advantage or energy penalty. *Proceedings of the International Refrigeration Conference—Energy Efficiency and New Refrigerants*, Purdue University, West Lafayette, IN.

CHAPTER 6

CONTROL OF MOISTURE AND OTHER CONTAMINANTS IN REFRIGERANT SYSTEMS

MOISTURE

MOISTURE (water) is an important and universal contaminant in refrigeration systems. The amount of moisture in a refrigerant system must be kept below an allowable maximum for satisfactory operation. Moisture must be removed from components during manufacture and assembly to minimize the amount of moisture in the completed assembly. Any moisture that enters during installation or servicing should be removed promptly.

Sources of Moisture

Moisture in a refrigerant system results from

- Inadequate equipment drying in factories and service operations
- Introduction during installation or service operations in the field
- Low-side leaks, resulting in entrance of moisture-laden air
- Leakage of water-cooled heat exchangers
- Oxidation of some hydrocarbon lubricants that produce moisture
- Wet lubricant, refrigerant, or desiccant
- Moisture entering a nonhermetic refrigerant system through hoses and seals

Drying equipment in the factory is discussed in Chapter 43. Proper installation and service procedures as given in ASHRAE *Standard* 147 minimize the second, third, and fourth sources. Lubricants are discussed in Chapter 7. If purchased refrigerants and lubricants meet specifications and are properly handled, the moisture content generally remains satisfactory. See the section on Electrical Insulation under Compatibility of Materials in Chapter 5 and the section on Motor Burnouts in this chapter.

Effects of Moisture

Excess moisture in a refrigerating system can cause one or all of the following undesirable effects:

- Ice formation in expansion valves, capillary tubes, or evaporators
- Corrosion of metals
- Copper plating
- Chemical damage to motor insulation in hermetic compressors or other system materials
- Hydrolysis of lubricants and other materials
- Sludge formation

Ice or solid hydrate separates from refrigerants if the water concentration is high enough and the temperature low enough. Solid hydrate, a complex molecule of refrigerant and water, can form at temperatures higher than those required to separate ice. Liquid water forms at temperatures above those required to separate ice or solid hydrate. Ice forms during refrigerant evaporation when the relative saturation of vapor reaches 100% at temperatures of 32°F or below.

The separation of water as ice or liquid also is related to the solubility of water in a refrigerant. This solubility varies for different refrigerants and with temperature (Table 1). Various investigators have obtained different results on water solubility in R-134a and R-123. The data presented here are the best available. The greater the solubility of water in a refrigerant, the less the possibility that ice or liquid water will separate in a refrigerating system. The solubility of water in ammonia, carbon dioxide, and sulfur dioxide is so high that ice or liquid water separation does not occur.

The concentration of water by mass at equilibrium is greater in the gas phase than in the liquid phase of R-12 (Elsey and Flowers

Table 1 Solubility of Water in Liquid Phase of Certain Refrigerants

Temp., °F	Solubility, ppm (by weight)								
	R-11	R-12	R-13	R-22	R-113	R-114	R-123	R-134a	R-502
160	460	700	—	4100	460	450	2600	4200	1780
150	400	560	—	3600	400	380	2300	3600	1580
140	340	440	—	3150	344	320	2000	3200	1400
130	290	350	—	2750	290	270	1800	2800	1220
120	240	270	—	2400	240	220	1600	2400	1080
110	200	210	—	2100	200	180	1400	2000	930
100	168	165	—	1800	168	148	1200	1800	810
90	140	128	—	1580	140	120	1000	1500	690
80	113	98	—	1350	113	95	900	1300	580
70	90	76	—	1140	90	74	770	1100	490
60	70	58	44	970	70	57	660	880	400
50	55	44	—	830	55	44	560	730	335
40	44	32	26	690	44	33	470	600	278
30	34	23.3	—	573	34	25	400	490	225
20	26	16.6	14	472	26	18	330	390	180
10	20	11.8	—	384	20	13	270	320	146
0	15	8.3	7	308	15	10	220	250	115
–10	11	5.7	—	244	11	7	180	200	90
–20	8	3.8	3	195	8	5	140	150	69
–30	6	2.5	—	152	6	3	110	120	53
–40	4	1.7	1	120	—	2	90	89	40
–50	3	1.1	—	91	—	1.5	70	66	30
–60	2	0.7	—	68	—	1	53	49	22
–70	1	0.4	—	50	—	0.6	40	35	16
–80	0.8	0.3	—	37	—	0.4	30	25	11
–90	0.5	0.1	—	27	—	0.2	22	18	8
–100	0.3	0.1	—	19	—	0.1	16	12	5

Data on R-134a adapted from Thrasher et al. (1993) and Allied-Signal Corporation. Data on R-123 adapted from Thrasher et al. (1993) and E.I. DuPont de Nemours & Company. Remaining data adapted from E.I. DuPont de Nemours & Company and Allied-Signal Corporation.

The preparation of this chapter is assigned to TC 3.3, Refrigerant Contaminant Control.

Table 2 Distribution of Water Between Vapor and Liquid Phases of Certain Refrigerants

Temperature, °F	Water in Vapor/Water in Liquid, mass %/mass %						
	R-11	R-12	R-22	R-114	R-123	R-134a	R-502
100	30.1	5.5	0.400	13.3	4.25	0.542	0.63
90	—	6.1	0.397	14.4	4.57	0.571	0.64
80	34.8	6.3	0.405	16.0	4.53	0.573	0.65
70	—	7.5	0.404	17.6	4.69	0.585	0.66
60	43.1	8.2	0.401	19.7	4.82	0.625	0.66
50	—	9.0	0.391	21.7	4.97	0.637	0.66
40	50.5	9.9	0.390	24.7	5.16	0.650	0.66
30	—	11.2	0.378	26.9	5.17	0.654	0.65
20	57.9	11.9	0.351	29.5	5.09	0.639	0.61
0	62.4	13.1	0.301	32.1	4.91	0.584	0.53
–20	71.0	15.3	0.251	37.4	4.78	0.543	0.47
–40	—	17.1	0.203	52.2	4.43	0.484	0.40

Data adapted from E.I. DuPont de Nemours & Company, Inc.

1949). The opposite is true for Refrigerants 22 and 502. The ratio of mass concentrations differs for each refrigerant; it also varies with temperature. Table 2 shows the distribution ratios of water in the vapor phase to water in the liquid phase for common refrigerants. It can be used to calculate the equilibrium water concentration of the liquid-phase refrigerant if the gas phase concentration is known, and vice versa. The water content in the vapor phase is determined by

$$W = \left[\frac{P_w}{(P_w)^0}\right]\left[\frac{(d_w)^0}{(d_R)^0}\right] \tag{1}$$

where

W = mass water/mass refrigerant
P_w = partial pressure of water vapor
$(P_w)^0$ = partial pressure of water vapor at saturation
$(d_w)^0$ = density of water vapor at saturation
$(d_R)^0$ = density of refrigerant vapor at saturation

Freezing at expansion valves or capillary tubes can occur when excessive moisture is present in a refrigerating system. Formation of ice or hydrate in evaporators can partially insulate the evaporator and reduce efficiency or cause system failure. Excess moisture can cause corrosion and enhance copper plating (Walker et al. 1962). Other factors affecting copper plating are discussed in Chapter 5.

The moisture required for freeze-up is a function of the amount of refrigerant vapor formed during expansion and the distribution of water between the liquid and gas phases downstream of the expansion device. For example, in an R-12 system with a 110°F liquid temperature and a –20°F evaporator temperature, refrigerant after expansion is 41.3% vapor and 58.7% liquid (by mass). The percentage of vapor formed is determined by

$$\%\text{ Vapor} = 100\frac{h_{L(liquid)} - h_{L(evap)}}{h_{fg(evap)}} \tag{2}$$

where

$h_{L(liquid)}$ = saturated liquid enthalpy for refrigerant at liquid temperature
$h_{L(evap)}$ = saturated liquid enthalpy for refrigerant at evaporating temperature
$h_{fg(evap)}$ = latent heat of vaporization of refrigerant at evaporating temperature

Table 1 lists the saturated water content of the R-12 liquid phase at –20°F as 3.8 ppm. Table 2 is used to determine the saturated vapor phase water content as

$$3.8\text{ ppm} \times 15.3 = 58\text{ ppm}$$

When the vapor contains more than the saturation quantity (100% rh), free water will be present as a third phase. If the temperature is below 32°F, ice will form. Using the saturated moisture values and the liquid-vapor ratios, the critical water content of the circulating refrigerant can be calculated as

$$\begin{aligned} 3.8 \times 0.587 &= 2.2\text{ ppm} \\ 58.0 \times 0.413 &= \underline{24.0\text{ ppm}} \\ & 26.2\text{ ppm} \end{aligned}$$

Maintaining moisture levels below critical value keeps free water from the low side of the system.

The previous analysis can be applied to all refrigerants and applications. An R-22 system with 110°F liquid and –20°F evaporating temperatures reaches saturation when the moisture circulating is 139 ppm. Note that this value is less than the liquid solubility, 195 ppm at –20°F.

Excess moisture causes paper or polyester motor insulation to become brittle, which can cause premature motor failure. However, not all motor insulations are affected adversely by moisture. The amount of water in a refrigerant system must be small enough to avoid ice separation, corrosion, and insulation breakdown.

Polyol ester lubricants (POEs), which are used largely with hydrofluorocarbons (HFCs), absorb substantially more moisture than do mineral oils, and do so very rapidly on exposure to the atmosphere. Once present, the moisture is difficult to remove. Hydrolysis of POEs can lead to formation of acids and alcohols that, in turn, can negatively affect system durability and performance (Griffith 1993). For these reasons, POEs should not be exposed to ambient air except for very brief periods required for compressor installation. Also, adequate driers are particularly important elements for equipment containing POEs.

Exact experimental data on the maximum permissible moisture level in refrigerant systems are not known because so many factors are involved.

Drying Methods

Equipment in the field is dried by decontamination, evacuation, and driers. Before opening equipment for service, refrigerant must be isolated or recovered into an external storage container (see Chapter 8). After installation or service, noncondensable gases (air) should be removed with a vacuum pump connected preferably to both suction and discharge service ports. The absolute pressure should be reduced to 1 mm of mercury or less, which is below the vapor pressure of water at ambient temperature. External or internal heat may be required to vaporize water in the system. Take care not to overheat the equipment. Even with these procedures, small amounts of moisture trapped under a lubricant film, adsorbed by the motor windings, or located far from the vacuum pump are difficult to remove. Evacuation will not remove any significant amount of water from polyol ester lubricants used in HFC systems.

It is good practice to install a drier. Larger systems frequently use a drier with a replaceable core. The core may need to be changed several times before the proper degree of dryness is obtained. A moisture indicator in the liquid line can indicate when the system has been dried satisfactorily.

Special techniques are required to remove free water in a refrigeration or air-conditioning system from a burst tube or water chiller leak. Refrigerant should be transferred to a pumpdown receiver or recovered in a separate storage tank. Parts of the system may have to be disassembled and the water drained from system low points. In some large systems, the semihermetic or open-drive compressor may need to be cleaned by disassembling and hand-wiping the various parts. Decontamination work should be performed before reinstallation of compressors, particularly hermetic units. After reassembly, the compressor should be dried further by passing dry nitrogen through the system and by heating and evacuation. Using internal heat, by circulating warmed water on the water side of water-cooled equipment, is preferred. Drying may take an extended period and require frequent changes of the vacuum pump lubricant. Liquid-line driers should be replaced and temporary suction line driers installed. During initial operation, driers need to be changed often. Decontamination procedures use large temporary driers. Properly performed decontamination eliminates the need for frequent on-board liquid-line drier changes.

If refrigerant in the pumpdown receiver is to be reused, it must be thoroughly dried before being reintroduced into the system. One method begins by drawing a liquid refrigerant sample and recording the ambient temperature. If chemical analysis of the sample by a qualified laboratory reveals a moisture content at or near the water solubility in Table 1 at the recorded temperature, then free water is probably present. In that case, a recovery unit with a suction filter-drier and/or a moisture/lubricant trap must be used to transfer the bulk of the refrigerant from the receiver liquid port to a separate tank. When the free water reaches the tank liquid port, most of the remaining refrigerant can be recovered through the receiver vapor port. The water can then be drained from the pumpdown receiver.

Moisture Indicators

Moisture-sensitive elements that change color according to moisture content can gage the system's moisture level; the color changes at a low enough level to be safe. Manufacturers' instructions must be followed because the color change point is also affected by liquid-line temperature and the refrigerant used.

Moisture Measurement

Techniques for measuring the amount of moisture in a compressor, or in an entire system, are discussed in Chapter 45. The following methods are used to measure the moisture content of various halocarbon refrigerants. The moisture content to be measured is generally in the parts-per-million range, and the procedures require special laboratory equipment and techniques.

The **Karl Fischer method** is suitable for measuring the moisture content of a refrigerant, even if it contains mineral oil. Although different firms have slightly different ways of performing this test and get somewhat varying results, the method remains the common industry practice for determining moisture content in refrigerants. The refrigerant sample is bubbled through predried methyl alcohol in a special sealed glass flask; any water present remains with the alcohol. In **volumetric titration**, Karl Fischer reagent is added, and the solution is immediately titrated to a "dead stop" electrometric end point. The reagent reacts with any moisture present so that the amount of water in the sample can be calculated from a previous calibration of the Karl Fischer reagent.

In **coulometric titration** (ARI *Standard* 700c), water is titrated with iodine that is generated electrochemically. The instrument measures the quantity of electric charge used to produce the iodine and titrate the water and calculates the amount of water present.

These titration methods, considered among the most accurate, are also suitable for measuring the moisture content of unused lubricant or other liquids. Special instruments designed for this particular analysis are available from laboratory supply companies. Haagen-Smit et al. (1970) describe improvements in the equipment and technique that significantly reduce analysis time.

The **gravimetric method** for measuring moisture content of refrigerants is described in ASHRAE *Standards* 35 and 63.1. It is not widely used in the industry. In this method, a measured amount of refrigerant vapor is passed through two tubes in series, each containing phosphorous pentoxide (P_2O_5). Moisture present in the refrigerant reacts chemically with the P_2O_5 and appears as an increase in mass in the first tube. The second tube is used as a tare. This method is satisfactory when the refrigerant is pure, but the presence of lubricant produces inaccurate results, because the lubricant is weighed as moisture. Approximately 200 g of refrigerant is required for accurate results. Because the refrigerant must pass slowly through the tube, analysis requires many hours.

DeGeiso and Stalzer (1969) discuss the **electrolytic moisture analyzer**, which is suitable for high-purity refrigerants. Other electronic hygrometers are available that sense moisture by the adsorption of water on an anodized aluminum strip with a gold foil overlay (Dunne and Clancy 1984). Calibration is critical to obtain maximum accuracy. These hygrometers give a continuous moisture reading and respond rapidly enough to monitor changes. Data showing dry-down rates can be gathered with these instruments (Cohen 1994). Brisken (1955) used this method in a study of moisture migration in hermetic equipment.

Thrasher et al. (1993) used nuclear magnetic resonance spectroscopy to determine the moisture solubilities in R-134a and R-123. Another method, infrared spectroscopy, is used for moisture analysis, but requires a large sample for precise results and is subject to interference if lubricant is present in the refrigerant.

Desiccants

Desiccants used in refrigeration systems adsorb or react chemically with the moisture contained in a liquid or gaseous refrigerant/lubricant mixture. Solid desiccants, used widely as dehydrating agents in refrigerant systems, remove moisture from both new and field-installed equipment. The desiccant is contained in a device called a drier (also spelled dryer) or filter-drier and can be installed in either the liquid or the suction line of a refrigeration system.

Desiccants must remove most of the moisture and not react unfavorably with any other materials in the system. Activated alumina, silica gel, and molecular sieves are the most widely used desiccants acceptable for refrigerant drying. Water is physically adsorbed on the internal surfaces of these highly porous desiccant materials.

Activated alumina and silica gel have a wide range of pore sizes, which are large enough to adsorb refrigerant, lubricant, additives, and water molecules. Pore sizes of molecular sieves, however, are uniform, with an aperture of approximately 0.3 nm for a type 3A molecular sieve or 0.4 nm for a type 4A molecular sieve. The uniform openings exclude lubricant molecules from the adsorption surfaces. Molecular sieves can be selected to exclude refrigerant molecules, as well. This property gives the molecular sieve the advantage of increasing water capacity and improving chemical compatibility between refrigerant and desiccant (Cohen 1993, 1994; Cohen and Blackwell 1995). The drier or desiccant manufacturer can provide information about which desiccant adsorbs or excludes a particular refrigerant.

Drier manufacturers offer combinations of desiccants that can be used in a single drier and may have advantages over a single desiccant because they can adsorb a greater variety of refrigeration contaminants. Two combinations are activated alumina with molecular sieves and silica gel with molecular sieves. Activated carbon is also used in some mixtures.

Desiccants are available in granular, bead, and block forms. Solid core desiccants, or block forms, consist of desiccant beads, granules, or both held together by a binder (Walker 1963). The binder is usually a nondesiccant material. Suitable filtration, adequate contact between desiccant and refrigerant, and low pressure drop are obtained by properly sizing the desiccant particles used to make up the core, and by the proper geometry of the core with respect to the flowing refrigerant. Beaded molecular sieve desiccants have higher water capacity per unit mass than solid-core desiccants. The composition and form of the desiccant is varied by drier manufacturers to achieve the desired properties.

Desiccants that take up water by chemical reaction are not recommended. Calcium chloride reacts with water to form a corrosive liquid. Barium oxide is known to cause explosions. Magnesium perchlorate and barium perchlorate are powerful oxidizing agents, which are

Table 3 Reactivation of Desiccants

Desiccant	Temperature, °F
Activated alumina	400 to 600
Silica gel	350 to 600
Molecular sieves	500 to 660

potential explosion hazards in the presence of lubricant. Phosphorous pentoxide is an excellent desiccant, but its fine powdery form makes it difficult to handle and produces a high resistance to gas and liquid flow. A mixture of calcium oxide and sodium hydroxide, which has limited use as an acid scavenger, should not be used as a desiccant.

Desiccants readily adsorb moisture and must be protected against it until ready for use. If a desiccant has picked up moisture, it can be reactivated under laboratory conditions by heating for about 4 h at a suitable temperature, preferably with a dry-air purge or in a vacuum oven (Table 3). Only adsorbed water is driven off at the temperatures listed, and the desiccant is returned to its initial activated state. Avoid repeated reactivation and excessive temperatures during reactivation, which may damage the desiccant. Desiccant in a refrigerating equipment drier should not be reactivated for reuse, because of lubricant and other contaminants in the drier as well as possible damage caused by overheating the drier shell.

Equilibrium Conditions of Desiccants. Desiccants in refrigeration and air-conditioning systems function on the equilibrium principle. If an activated desiccant contacts a moisture-laden refrigerant, the water is adsorbed from the refrigerant/water mixture onto the desiccant surface until the vapor pressures of the adsorbed water (i.e., at the desiccant surface) and the water remaining in the refrigerant are equal. Conversely, if the vapor pressure of water on the desiccant surface is higher than that in the refrigerant, water is released into the refrigerant/water mixture, and equilibrium is reestablished.

Adsorbent desiccants function by holding (adsorbing) moisture on their internal surfaces. The amount of water adsorbed from a refrigerant by an adsorbent at equilibrium is influenced by (1) pore volume, pore size, and surface characteristics of the adsorbent; (2) temperature and moisture content of the refrigerant; and (3) solubility of water in the refrigerant.

Figures 1 to 3 are equilibrium curves (known as **adsorption isotherms**) for various adsorbent desiccants with R-12 and R-22. These curves are representative of commercially available materials. The adsorption isotherms are based on the technique developed by Gully et al. (1954), as modified by ASHRAE *Standard* 35. ASHRAE *Standards* 35 and 63.1 define the moisture content of the refrigerant as equilibrium point dryness (EPD), and the moisture held by the desiccant as water capacity. The curves show that for any specified amount of water in a particular refrigerant, the desiccant holds a corresponding specific quantity of water.

Figures 1 and 2 show moisture equilibrium curves for three common adsorbent desiccants in drying R-12 and R-22 at 75°F. As shown, desiccant capacity can vary widely for different refrigerants when the same EPD is required. Generally, a refrigerant in which moisture is more soluble requires more desiccant for adequate drying than one that has less solubility.

Figure 3 shows the effect of temperature on moisture equilibrium capacities of activated alumina and R-12. Much higher water capacities are obtained at lower temperatures, demonstrating the advantage of locating alumina driers at relatively cool spots in the system. The effect of temperature on molecular sieves' water capacity is much smaller. ARI *Standard* 710 requires determining the water capacity for R-12 at an EPD of 15 ppm, and for R-22 at 60 ppm. Each determination must be made at 75°F (see Figures 1 and 2) and 125°F.

Figure 4 shows water capacity of a molecular sieve in liquid R-134a at 125°F. These data were obtained using the Karl Fischer method similar to that described in Dunne and Clancy (1984). Cavestri and Schafer (1999) determined water capacities for three

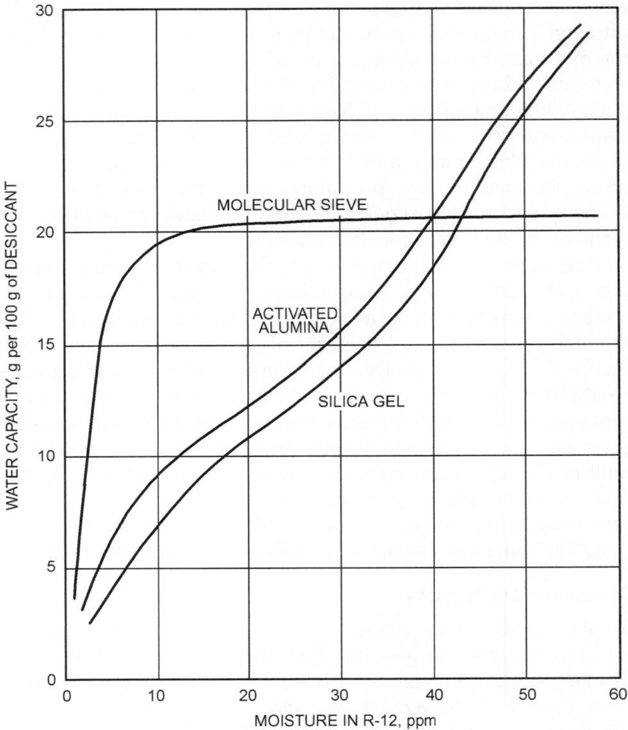

Fig. 1 Moisture Equilibrium Curves for R-12 and Three Common Desiccants at 75°F

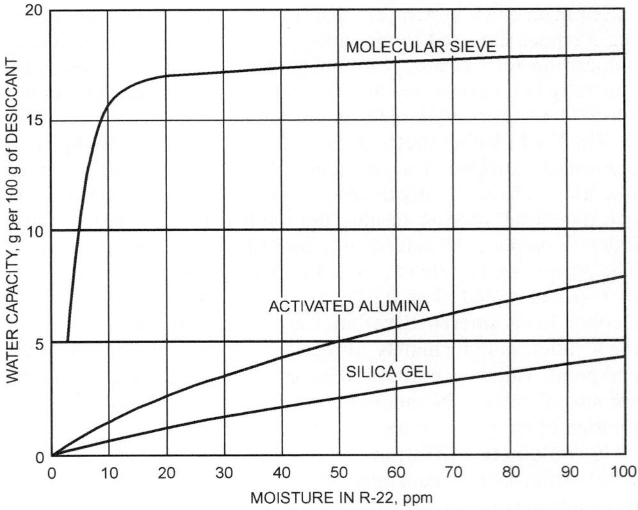

Fig. 2 Moisture Equilibrium Curves for R-22 and Three Common Desiccants at 75°F

common desiccants in R-134a when POE lubricant was added to the refrigerant. Figures 5 and 6 show water capacity for type 3A molecular sieves, activated alumina beads, and bonded activated alumina cores in R-134a and 2% POE lubricant at 75°F and 125°F.

Although the figures show that molecular sieves have greater water capacities than activated alumina or silica gel at the indicated EPD, all three desiccants are suitable if sufficient quantities are used. Cost, operating temperature, other contaminants present, and equilibrium capacity at the desired EPD must be considered when choosing a desiccant for refrigerant drying. Consult the desiccant manufacturer for information and equilibrium curves for specific desiccant/refrigerant systems.

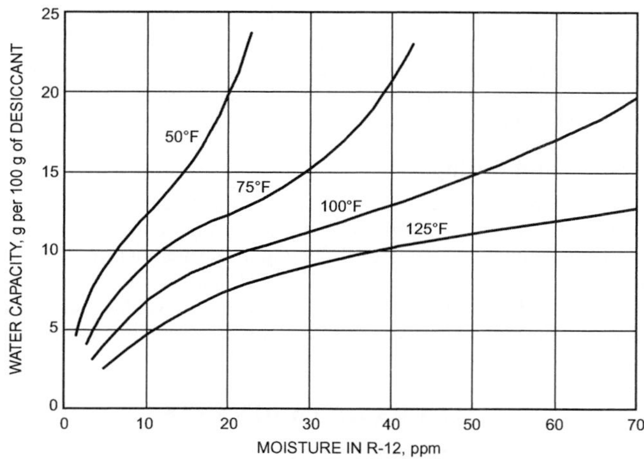

Fig. 3 Moisture Equilibrium Curves for Activated Alumina at Various Temperatures in R-12

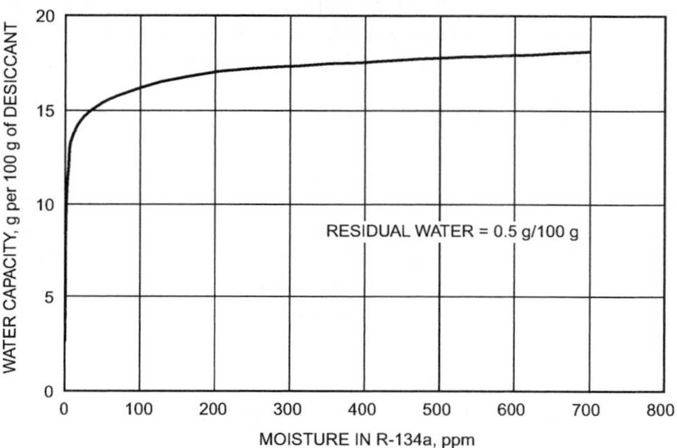

Fig. 4 Moisture Equilibrium Curve for Molecular Sieve in R-134a at 125°F

(Courtesy UOP, Reprinted with permission.)

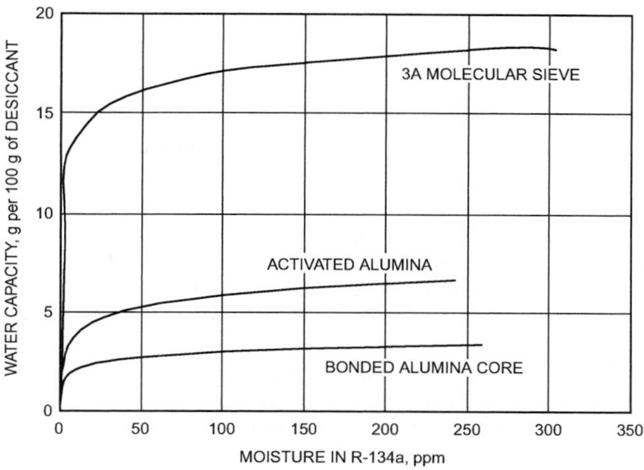

Fig. 5 Moisture Equilibrium Curves for Three Common Desiccants in R-134a and 2% POE Lubricant at 75°F

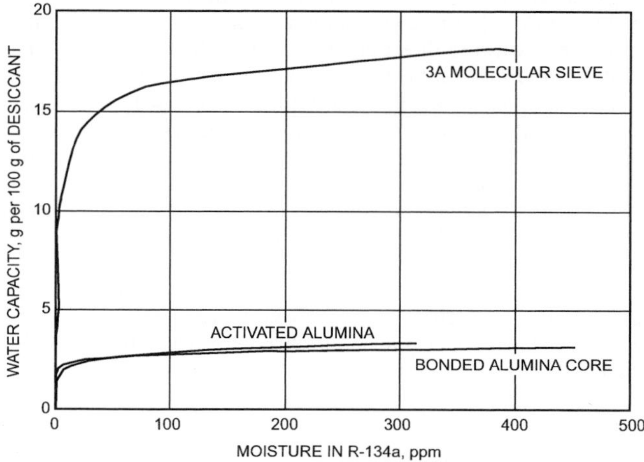

Fig. 6 Moisture Equilibrium Curves for Three Common Desiccants in R-134a and 2% POE Lubricant at 125°F

Activated carbon technically is not a desiccant, but it is often used in filter-driers to scavenge waxes and insoluble resins. The other common desiccants do not remove these contaminants, which can plug expansion devices and reduce system capacity and efficiency. Activated carbon is typically incorporated into bound desiccant blocks along with molecular sieve and activated alumina.

Desiccant Applications

In addition to removing water, desiccants may adsorb or react with acids, dyes, chemical additives, and refrigerant lubricant reaction products.

Acids. Generally, acids can harm refrigerant systems. The amount of acid a refrigerant system tolerates depends in part on the size, mechanical design and construction of the system, type of motor insulation, type of acid, and amount of water in the system.

Desiccants' acid removal capacity is difficult to determine because the environment is complex. Hoffman and Lange (1962) and Mays (1962) showed that desiccants remove acids from refrigerants and lubricants by adsorption and/or chemical reaction. Hoffman and Lange also showed that the loading of water on the desiccant, type of desiccant, and type of acid play major roles in a desiccant's ability to remove acids from refrigerant systems. In addition, acids formed in these systems can be inorganic, such as HCl and HF, or a mixture of organic acids. All of these factors must be considered to establish acid capacities of desiccants.

Cavestri and Schooley (1998) determined the inorganic acid capacity of desiccants. Both molecular sieve and alumina desiccants remove inorganic acids such as HCl and HF from refrigerant systems. Molecular sieves remove these acids through irreversible chemisorption: the acids form their respective salts with the molecular sieve's sodium and potassium cations. Alumina removes such acids principally by reversible physical adsorption.

Colors. Colored materials frequently are adsorbed by activated alumina and silica gel and occasionally by calcium sulfate and molecular sieves. Leak detector dyes may lose their effectiveness in systems containing desiccants. The interaction of the dye and drier should be evaluated before putting a dye in the system.

Lubricant Deterioration Products. Lubricants can react chemically to produce substances that are adsorbed by desiccants. Some of these are hydrophobic and, when adsorbed by the desiccant, may reduce the rate at which it can adsorb liquid water. However, the rate and capacity of the desiccant to remove water dissolved in the refrigerant are not significantly impaired (Walker et al. 1955). Often, reaction products are sludges or powders that can be filtered out mechanically by the drier.

Chemicals. Refrigerants that can be adsorbed by desiccants cause the drier temperature to rise considerably when the refrigerant is first admitted. This temperature rise is not the result of moisture in the refrigerant, but the adsorption heat of the refrigerant. Lubricant additives may be adsorbed by silica gel and activated alumina. Because of small pore size, molecular sieves generally do not adsorb additives or lubricant.

Driers

A drier is a device containing a desiccant. It collects and holds moisture, but also acts as a filter and adsorber of acids and other contaminants.

To prevent moisture from freezing in the expansion valve or capillary tube, a drier is installed in the liquid line close to these devices. Hot locations should be avoided. Driers can function on the low-pressure side of expansion devices, but this is not the preferred location (Jones 1969).

Moisture is reduced as liquid refrigerant passes through a drier. However, Krause et al. (1960) showed that considerable time is required to reach moisture equilibrium in a refrigeration unit. The moisture is usually distributed throughout the entire system, and time is required for the circulating refrigerant/lubricant mixture to carry the moisture to the drier. Cohen (1994) and Cohen and Dunne (1987) discuss the kinetics of drying refrigerants in circulating systems.

Loose-filled driers should be mounted vertically, with downward refrigerant flow. In this configuration, both gravity and drag forces act in the downward direction on the beads. Settling of the beads creates a void space at the top, which is not a problem.

Vertical orientation with upward flow, where gravity and drag act in opposite directions, should be avoided because the flow will likely fluidize the desiccant beads, causing the beads to move against each another. This promotes attrition or abrasion of the beads, producing fine particles that can contaminate the system. Settling creates a void space between the retention screens, promoting fluidization.

Horizontal mounting should also be avoided with a loose-filled drier because bead settling creates a void space that promotes fluidization, and may also produce a channel around the beads that reduces drying effectiveness.

Driers are also used effectively to clean systems severely contaminated by hermetic motor burnouts and mechanical failures (see the section on System Cleanup Procedure after Hermetic Motor Burnout).

Drier Selection

The drier manufacturer's selection chart lists the amount of desiccants, flow capacity, filter area, water capacity, and a specific recommendation on the type and refrigeration capacity of the drier for various applications.

The equipment manufacturer must consider the following factors when selecting a drier:

1. The **desiccant** is the heart of the drier and its selection is most important. The section on Desiccants has further information.
2. The drier's **water capacity** is measured as described in ARI *Standard* 710. Reference points are set arbitrarily to prevent confusion arising from determinations made at other points. The specific refrigerant, amount of desiccant, and effect of temperature are all considered in the statement of water capacity.
3. The **liquid-line** flow capacity is listed at 1 psi pressure drop across the drier by the official procedures of ARI *Standard* 710 and ANSI/ASHRAE *Standard* 63.1. Rosen et al. (1965) described a closed-loop method for evaluating filtration and flow characteristics of liquid-line refrigerant driers. The flow capacity of suction-line filters and filter-driers is determined according to ARI *Standard* 730 and ANSI/ASHRAE *Standard* 78. ARI *Standard* 730 gives recommended pressure drops for selecting suction-line filter-driers for permanent and temporary installations. Flow capacity may be reduced quickly when critical quantities of solids and semisolids are filtered out by the drier. Whenever flow capacity drops below the machine's requirements, the drier should be replaced.
4. Although limits for particle size vary with refrigerant system size and design, and with the geometry and hardness of the particles, manufacturers publish **filtration capabilities** for comparison.

Testing and Rating

Desiccants and driers are tested according to the procedures of ASHRAE *Standards* 35 and 63. Driers are rated under ARI *Standard* 710. Minimum standards for listing of refrigerant driers can be found in UL *Standard* 207. ANSI/ASHRAE *Standard* 63.2 specifies a test method for filtration testing of filter-driers. No ARI Standard has been developed to give rating conditions for publication of filtration capacity.

OTHER CONTAMINANTS

Refrigerant filter-driers are the principal devices used to remove contaminants from refrigeration systems. The filter-drier is not a substitute for good workmanship or design, but a maintenance tool necessary for continued and proper system performance. Contaminants removed by filter-driers include moisture, acids, hydrocarbons with a high molecular weight, oil decomposition products, and insoluble material, such as metallic particles and copper oxide.

Metallic Contaminants and Dirt

Small contaminant particles frequently left in refrigerating systems during manufacture or servicing include chips of copper, steel, or aluminum; copper or iron oxide; copper or iron chloride; welding scale; brazing or soldering flux; sand; and other dirt. Some of these contaminants, such as copper chloride, develop from normal wear or chemical breakdown during system operation. Solid contaminants vary widely in size, shape, and density. Solid contaminants create problems by

- Scoring cylinder walls and bearings
- Lodging in the motor insulation of a hermetic system, where they act as conductors between individual motor windings or abrade the wire coating when flexing of the windings occurs
- Depositing on terminal blocks and serving as a conductor
- Plugging expansion valve screen or capillary tubing
- Depositing on suction or discharge valve seats, significantly reducing compressor efficiency
- Plugging oil holes in compressor parts, leading to improper lubrication
- Increasing the rate of chemical breakdown [e.g., at elevated temperatures, R-22 decomposes more readily when in contact with iron powder, iron oxide, or copper oxide (Norton 1957)]
- Plugging driers

Liquid-line filter-driers, suction filters, and strainers isolate contaminants from the compressor and expansion valve. Filters minimize return of particulate matter to the compressor and expansion valve, but the capacity of permanently installed liquid and/or suction filters must accommodate this particulate matter without causing excessive, energy-consuming pressure losses. Equipment manufacturers should consider the following procedures to ensure proper operation during the design life:

1. Develop cleanliness specifications that include a reasonable value for maximum residual matter. Some manufacturers specify allowable quantities in terms of internal surface area. ASTM *Standard* B280 allows a maximum of 0.0035 g of contaminants per square foot of internal surface.
2. Multiply the factory contaminant level by a factor of five to allow for solid contaminants added during installation. This factor depends on the type of system and the previous experience of the installers, among other considerations.

3. Determine maximum pressure drop to be incurred by the suction or liquid filter when loaded with the quantity of solid matter calculated in Step 2.
4. Conduct pressure drop tests according to ASHRAE *Standard* 63.2.
5. Select driers for each system according to its capacity requirements and test data. In addition to contaminant removal capacity, tests can evaluate filter efficiency, maximum escaped particle size, and average escaped particle size.

Very small particles passing through filters tend to accumulate in the crankcase. Most compressors tolerate a small quantity of these particles without allowing them into the oil pump inlet, where they can damage running surfaces.

Organic Contaminants: Sludge, Wax, and Tars

Organic contaminants in a refrigerating system with a mineral oil lubricant can appear when organic materials such as oil, insulation, varnish, gaskets, and adhesives decompose. As opposed to inorganic contaminants, these materials are mostly carbon, hydrogen, and oxygen. Organic materials may be partially soluble in the refrigerant/lubricant mixture or may become so when heated. They then circulate in the refrigerating system and can plug small orifices. Organic contaminants in a refrigerating system using a synthetic polyol ester lubricant may also generate sludge. The following contaminants should be avoided:

- Paraffin (typically found in mineral oil lubricants)
- Silicone (found in some machine lubricants)
- Phthalate (found in some machine lubricants)

Whether mineral oil or synthetic lubricants are used, some organic contaminants remain in a new refrigerating system during manufacture or assembly. For example, excessive brazing paste introduces a waxlike contaminant into the refrigerant stream. Certain cutting lubricants, corrosion inhibitors, or drawing compounds frequently contain paraffin-based compounds. These lubricants can leave a layer of paraffin on a component that may be removed by the refrigerant/lubricant combination and generate insoluble material in the refrigerant stream. Organic contamination also results during the normal method of fabricating return bends. The die used during forming is lubricated with these organic materials, and afterwards the return bend is brazed to the tubes to form the evaporator and/or condenser. During brazing, residual lubricant inside the tubing and bends can be baked to a resinous deposit.

If organic materials are handled improperly, certain contaminants remain. Resins used in varnishes, wire coating, or casting sealers may not be cured properly and can dissolve in the refrigerant/lubricant mixture. Solvents used in washing stators may be adsorbed by the wire film and later, during compressor operation, carry chemically reactive organic extractables. Chips of varnish, insulation, or fibers can detach and circulate in the system. Portions of improperly selected or cured rubber parts or gaskets can dissolve in the refrigerant.

Refrigeration-grade mineral oil decomposes under adverse conditions to form a resinous liquid or a solid frequently found on refrigeration filter-driers. These mineral oils decompose noticeably when exposed for as little as 2 h to temperatures as low as 250°F in an atmosphere of air or oxygen. The compressor manufacturer should perform all high-temperature dehydrating operations on the machines before adding the lubricant charge. In addition, equipment manufacturers should not expose compressors to processes requiring high temperatures unless the compressors contain refrigerant or inert gas.

The result of organic contamination is frequently noticed at the expansion device. Materials dissolved in the refrigerant/lubricant mixture, under liquid line conditions, may precipitate at the lower temperature in the expansion device, resulting in restricted or plugged capillary tubes or sticky expansion valves. A few milligrams of these contaminants can render a system inoperative. These materials have physical properties that range from a fluffy powder to a solid resin entraining inorganic debris. If the contaminant is dissolved in the refrigerant/lubricant mixture in the liquid line, it will not be removed by a filter-drier.

Chemical identification of these organic contaminants is very difficult. Infrared spectroscopy can characterize the type of organic groups present in contaminants. Materials found in actual systems vary from waxlike aliphatic hydrocarbons to resinlike materials containing double bonds, carbonyl groups, and carboxyl groups. In some cases, organic compounds of copper and/or iron have been identified.

These contaminants can be eliminated by carefully selecting materials and strictly controlling cleanliness during manufacture and assembly of the components as well as the final system. Because heat degrades most organic materials and enhances chemical reactions, operating conditions with excessively high discharge or bearing surface temperatures must be avoided to prevent formation of degradation products.

Residual Cleaning Agents

Mineral Oil Systems. Solvents used to clean compressor parts are likely contaminants if left in refrigerating equipment. Solvents in this category are considered pure liquids without additives. If additives are present, they are reactive materials and should not be in a refrigerating system. Some solvents are relatively harmless to the chemical stability of the refrigerating system, whereas others initiate or accelerate degradation reactions. For example, the common mineral spirits solvents are considered harmless. Other common compounds react rapidly with hydrocarbon lubricants (Elsey et al. 1952).

Polyol Ester Lubricated Systems. Typical solvents used in cleaning mineral oil systems are not compatible with polyol ester lubricants. Several chemicals must be avoided to reduce or eliminate possible contamination and sludge generation. In addition to paraffin, silicone, and phthalate contaminants, a small amount of the following contaminants can cause system failure:

- Chlorides (typically found in chlorinated solvents)
- Acid or alkali (found in some water-based cleaning fluids)
- Water (component of water-based cleaning fluids)

Noncondensable Gases

Gases, other than the refrigerant, are another contaminant frequently found in refrigerating systems. These gases result (1) from incomplete evacuation, (2) when functional materials release sorbed gases or decompose to form gases at an elevated temperature during system operation, (3) through low-side leaks, and (4) from chemical reactions during system operation. Chemically reactive gases, such as hydrogen chloride, attack other components, and, in extreme cases, the refrigerating unit fails.

Chemically inert gases, which do not liquefy in the condenser, reduce cooling efficiency. The quantity of inert, noncondensable gas that is harmful depends on the design and size of the refrigerating unit and on the nature of the refrigerant. Its presence contributes to higher-than-normal head pressures and resultant higher discharge temperatures, which speed up undesirable chemical reactions.

Gases found in hermetic refrigeration units include nitrogen, oxygen, carbon dioxide, carbon monoxide, methane, and hydrogen. The first three gases originate from incomplete air evacuation or a low-side leak. Carbon dioxide and carbon monoxide usually form when organic insulation is overheated. Hydrogen has been detected when a compressor is experiencing serious bearing wear. These gases are also found where a significant refrigerant/lubricant reaction has occurred. Only trace amounts of these gases are present in well-designed, properly functioning equipment.

Doderer and Spauschus (1966), Gustafsson (1977), and Spauschus and Olsen (1959) developed sampling and analytical techniques for establishing the quantities of contaminant gases present in refrigerating systems. Kvalnes (1965), Parmelee (1965), and Spauschus and Doderer (1961, 1964) applied gas analysis techniques to sealed tube tests to yield information on stability limitations of refrigerants, in conjunction with other materials used in hermetic systems.

Motor Burnouts

Motor burnout is the final result of hermetic motor insulation failure. During burnout, high temperatures and arc discharges can severely deteriorate the insulation, producing large amounts of carbonaceous sludge, acid, water, and other contaminants. In addition, a burnout can chemically alter the compressor lubricant, and/or thermally decompose refrigerant in the vicinity of the burn. Products of burnout escape into the system, causing severe cleanup problems. If decomposition products are not removed, replacement motors fail with increasing frequency.

Although the Refrigeration Service Engineers Society (RSES 1988) differentiates between mild and severe burnouts, many compressor manufacturers' service bulletins treat all burnouts alike. A rapid burn from a spot failure in the motor winding results in a mild burnout with little lubricant discoloration and no carbon deposits. A severe burnout occurs when the compressor remains online and burns over a longer period, resulting in highly discolored lubricant, carbon deposits, and acid formation.

Because the condition of the lubricant can be used to indicate the amount of contamination, the lubricant should be examined during the cleanup process. Wojtkowski (1964) stated that acid in R-22/mineral oil systems should not exceed 0.05 total acid number (mg KOH per g oil). Commercial acid test kits can be used for this analysis. An acceptable acid number for other lubricants has not been established.

Various methods are recommended for cleaning a system after hermetic motor burnout (RSES 1988). However, the suction-line filter-drier method is commonly used (see the section on System Cleanup Procedure after Hermetic Motor Burnout).

Field Assembly

Proper field assembly and maintenance are essential for contaminant control in refrigerating systems and to prevent undesirable refrigerant emissions to the atmosphere. Driers may be too small or carelessly handled so that drying capacity is lost. Improper tube-joint soldering is a major source of water, flux, and oxide scale contamination. Copper oxide scale from improper brazing is one of the most frequently observed contaminants. Careless tube cutting and handling can introduce excessive quantities of dirt and metal chips. Take care to minimize these sources of internal contamination. In addition, because an assembled system cannot be dehydrated easily, oversized driers should be installed. Even if components are delivered sealed and dry, weather and the amount of time the unit is open during assembly can introduce large amounts of moisture.

In addition to internal sources, external factors can cause a unit to fail. Too small or too large transport tubing, mismatched or misapplied components, fouled air condensers, scaled heat exchangers, inaccurate control settings, failed controls, and improper evacuation are some of these factors.

SYSTEM CLEANUP PROCEDURE AFTER HERMETIC MOTOR BURNOUT

This procedure is limited to positive-displacement hermetic compressors. Centrifugal compressor systems are highly specialized and are frequently designed for a particular application. A centrifugal system should be cleaned according to the manufacturer's recommendations. All or part of the procedure can be used, depending on factors such as severity of burnout and size of the refrigeration system.

After a hermetic motor burnout, the system must be cleaned thoroughly to remove all contaminants. Otherwise, a repeat burnout will *likely* occur. Failure to follow these minimum cleanup recommendations as quickly as possible increases the potential for repeat burnout.

Procedure

A. **Make sure a burnout has occurred.** Although a motor that will not start appears to be a motor failure, the problem may be improper voltage, starter malfunction, or a compressor mechanical fault (RSES 1988). Investigation should include the following steps:

   1. Check for proper voltage.
   2. Check that the compressor is cool to the touch. An open internal overload could prevent the compressor from starting.
   3. Check the compressor motor for improper grounding using a megohmmeter or a precision ohmmeter.
   4. Check the external leads and starter components.
   5. Obtain a small sample of oil from the compressor, examine it for discoloration, and analyze it for acidity.

B. **Safety.** In addition to electrical hazards, service personnel should be aware of the hazard of acid burns. If the lubricant or sludge in a burned-out compressor must be touched, wear rubber gloves to avoid a possible acid burn.

C. **Cleanup after a burnout.** Just as proper installation and service procedures are essential to prevent compressor and system failures, proper system cleanup and installation procedures when installing the replacement compressor are also essential to prevent repeat failures. Key elements of the recommended procedures are as follows:

   1. U.S. federal regulations require that the refrigerant be isolated in the system or recovered into an external storage container to avoid discharge into the atmosphere. Before opening any portion of the system for inspection or repairs, refrigerant should be recovered from that portion until the vapor pressure reduces to less than 12.7 psia (4 in. Hg vacuum) for R-22 or 9.8 psia (10 in. Hg vacuum) for CFC or other HCFC systems.
   2. Remove the burned-out compressor and install the replacement. Save a sample of the new compressor lubricant that has not been exposed to refrigerant and store in a sealed glass bottle. This will be used later for comparison.
   3. Inspect all system controls such as expansion valves, solenoid valves, check valves, etc. Clean or replace if necessary.
   4. Install an oversized drier in the suction line to protect the replacement compressor from any contaminants remaining in the system. Install a pressure tap upstream of the filter-drier, to allow measuring the pressure drop from tap to service valve during the first hours of operation to determine whether the suction line drier needs to be replaced.
   5. Remove the old liquid-line drier, if one exists, and install a replacement drier of the next larger capacity than is normal for this system. Install a moisture indicator in the liquid line if the system does not have one.
   6. Evacuate and leak-check the system or portion opened to the atmosphere according to the manufacturer's recommendations.
   7. Recharge the system and begin operations according to the manufacturer's start-up instructions, typically as follows:

      a. Observe pressure drop across the suction line drier for the first 4 h. Follow the manufacturer's guide; otherwise, compare to pressure drop curve in Figure 7 and replace driers as required.

b. After 24 to 48 h, check pressure drop and replace driers as required. Take a lubricant sample and check with an acid test kit. Compare the lubricant sample to the initial sample saved at the time the replacement compressor was installed. Cautiously smell the lubricant sample. Replace lubricant if acidity persists or if color or odor indicates.

c. After 7 to 10 days or as required, repeat step b.

D. **Additional suggestions**

1. If sludge or carbon has backed up into the suction line, swab it out or replace that section of the line.
2. If a change in the suction line drier is required, change the lubricant in the compressor each time the cores are changed, if compressor design permits.
3. Remove the suction line drier after several weeks of system operation to avoid excessive pressure drop in the suction line. This problem is particularly significant on commercial refrigeration systems.
4. Noncondensable gases may be produced during burnout. With the system off, compare the head pressure to the saturation pressure after stabilization at ambient temperature. Adequate time must be allowed to ensure stabilization. If required, purge the charge by recycling it or submit the purged material for reclamation.

Special System Characteristics and Procedures

Because of unique system characteristics, the procedures described here may require adaptations.

A. If a lubricant sample cannot be obtained from the new compressor, find another way to get a sample from the system.

1. Install a tee and a trap in the suction line. An access valve at the bottom of the trap permits easy lubricant drainage. Only 0.5 oz of lubricant is required for an acid analysis. Be certain the lubricant sample represents lubricant circulating in the system. It may be necessary to drain the trap and discard the first amount of lubricant collected, before collecting the sample to be analyzed.

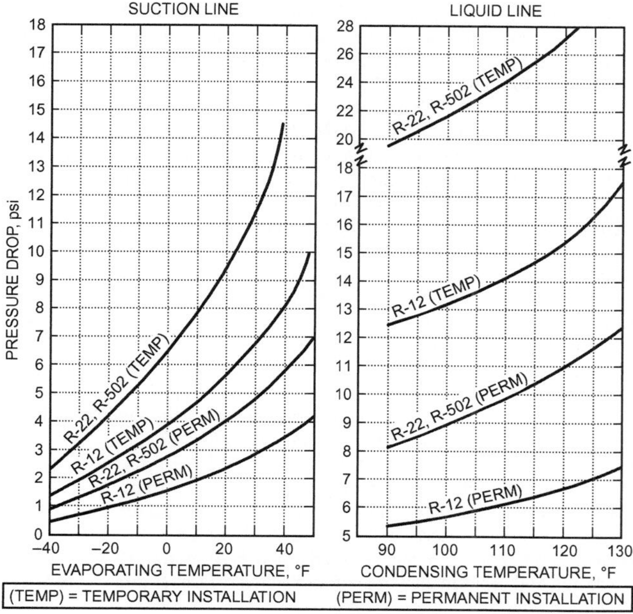

Fig. 7 Maximum Recommended Filter-Drier Pressure Drop

2. Make a trap from 1 3/8 in. copper tubing and valves. Attach this trap to the suction and discharge gage port connections with a charging hose. By blowing discharge gas through the trap and into the suction valve, enough lubricant will be collected in the trap for analysis. This trap becomes a tool that can be used repeatedly on any system that has suction and discharge service valves. Be sure to clean the trap after every use to avoid cross contamination.

B. On semihermetic compressors, remove the cylinder head to determine the severity of burnout. Dismantle the compressor for solvent cleaning and hand wiping to remove contaminants. Consult the manufacturer's recommendations on compressor rebuilding and motor replacement.

C. In rare instances on a close-coupled system, where it is not feasible to install a suction line drier, the system can be cleaned by repeated changes of the cores in the liquid line drier and repeated lubricant changes.

D. On heat pumps, the four-way valve and compressor should be carefully inspected after a burnout. In cleaning a heat pump after a motor burnout, it is essential to remove any drier originally installed in the liquid line. These driers may be replaced for cleanup, or a biflow drier may be installed in the common reversing liquid line.

E. Systems with a critical charge require a particular effort for proper operation after cleanup. If an oversized liquid-line drier is installed, an additional charge must be added. Check with the drier manufacturer for specifications. However, no additional charge is required for the suction line drier that may be added.

F. The new compressor should not be used to pull a vacuum. Refer to the manufacturer's recommendations for evacuation. Normally, the following method is used, after determining that there are no refrigerant leaks in the system:

a. Pull a high vacuum to an absolute pressure of less than 500 μm Hg for several hours.

b. Allow the system to stand several hours to be sure the vacuum is maintained. This requires a good vacuum pump and an accurate high-vacuum gage.

CONTAMINANT CONTROL DURING RETROFIT

Because of the phaseout of CFCs, existing refrigeration and air-conditioning systems are commonly retrofitted to alternative refrigerants. The term "refrigerant" in this section refers to a fluorocarbon working fluid offered as a possible replacement for a CFC, whether that replacement consists of one chemical, an azeotrope of two chemicals, or a blend of two or more chemicals. The terms "retrofitting" and "conversion" are used interchangeably to mean the modification of an existing refrigeration or air-conditioning system designed to operate on a CFC so that it can safely and effectively operate on an HCFC or HFC refrigerant. This section only covers the contaminant control aspects of such conversions. Equipment manufacturers should be consulted for guidance regarding the specifics of actual conversion. Industry standards and manufacturers' literature are also available that contain supporting information (e.g., UL *Standards* 2170, 2171, and 2172).

Contaminant control concerns for retrofitting a CFC system to an alternative refrigerant fall into the following categories:

- **Cross-contamination of old and new refrigerants.** This should be avoided even though there are usually no chemical compatibility problems between the CFCs and their replacement refrigerants. One problem with mixing refrigerants is that it is difficult to determine system performance after retrofit. Pressure/temperature relationships are different for a blend of two refrigerants than for each refrigerant individually. A second concern with mixing refrigerants is that if the new refrigerant charge must be removed in the future, the mixture may not be reclaimable (DuPont 1992).

- **Cross-contamination of old and new lubricant.** Equipment manufacturers generally specify that the existing lubricant be replaced with the lubricant they consider suitable for use with a given HFC refrigerant. In some cases, the new lubricant is incompatible with the old one or with chlorinated residues present. In other cases, the old lubricant is insoluble with the new refrigerant and tends to collect in the evaporator, interfering with heat transfer. For example, when mineral oil is replaced by a polyol ester lubricant during retrofit to an HFC refrigerant, a typical recommendation is to reduce the old oil content to 5% or less of the nominal oil charge (Castrol 1992). Some retrofit recommendations specify lower levels of acceptable contamination for polyol ester lubricant/HFC retrofits, so original equipment manufacturers recommendations should be obtained before attempting a conversion.
- **Chemical compatibility of old system components with new fluids.** One of the preparatory steps in a retrofit is to confirm that either the existing materials in the system are acceptable or that replacement materials are on hand to be installed in the system during the retrofit. Fluorocarbon refrigerants generally have solvent properties, and some are very aggressive. This characteristic can lead to swelling and extrusion of polymer O rings, undermining their sealing capabilities. Material can also be extracted from polymers, varnishes, and resins used in hermetic motor windings. These extracts can then collect in expansion devices, interfering with system operation. Residual manufacturing fluids such as those used to draw wire for compressor motors can be extracted from components and deposited in areas where they can interfere with operation. Suitable materials of construction have been identified by equipment manufacturers for use with HFC refrigerant systems.

Drier media must also be chemically compatible with the new refrigerant and effective in removing moisture, acid, and particulate in the presence of the new refrigerant. Drier media commonly used with CFC refrigerants tend to accept small HFC refrigerant molecules and lose moisture retention capability (Cohen and Blackwell 1995). Drier media have been developed that minimize this tendency.

CHILLER DECONTAMINATION

Chiller decontamination is used to clean reciprocating, rotary screw, and centrifugal machines. Large volumes of refrigerant are circulated through a contaminated chiller while continuously being reclaimed. It has been used successfully to restore many chillers to operating specifications. Some chillers have been saved from early retirement by decontamination procedures. Variations of the procedure are myriad and have been used for burnouts, water-flooded barrels, particulate incursions, chemical contamination events, brine leaks, and oil strips. One frequently used technique is to perform numerous batch cycles, thus increasing the velocity-based cleansing component. Excess oil is stripped out to improve chiller heat transfer efficiency. The full oil charge can be removed in preparation for refrigerant conversion.

Low-pressure units require different machinery than high-pressure units. It is best to integrate decontamination and mechanical services early into one overall procedure. On machines that require compressor rebuild, it is best to perform decontamination work while the compressor is removed or before it is rebuilt, particularly for reciprocating units. Larger-diameter or relocated access ports may be requested. The oil sump will be drained. For chillers that cannot be shut down, special online techniques have been developed using reclamation. The overall plan is coordinated with operations personnel to prevent service interruptions. For some decontamination projects, it is advantageous to have the water boxes open; in other cases, closed. Intercoolers offer special challenges.

REFERENCES

ARI. 1999. Appendix C to ARI *Standard* 700—Analytical procedures for ARI *Standard* 700. *Standard* 700c-1999. Air-Conditioning and Refrigeration Institute, Arlington, VA.

ARI. 2004. Liquid-line driers. *Standard* 710-04. Air-Conditioning and Refrigeration Institute, Arlington, VA.

ARI. 2001. Flow-capacity rating and application of suction-line filters and filter-driers. *Standard* 730-01. Air-Conditioning and Refrigeration Institute, Arlington, VA.

ASHRAE. 1992. Method of testing desiccants for refrigerant drying. *Standard* 35-1992.

ASHRAE. 1995. Method of testing liquid line refrigerant driers. ANSI/ASHRAE *Standard* 63.1-1995 (RA 01).

ASHRAE. 1996. Method of testing liquid line filter-drier filtration capability. ANSI/ASHRAE *Standard* 63.2-1996.

ASHRAE. 1985. Method of testing flow capacity of suction line filter driers. ANSI/ASHRAE *Standard* 78-1985 (RA 03).

ASHRAE. 2002. Reducing the release of halogenated refrigerants from refrigerating and air-conditioning equipment and systems. ANSI/ASHRAE *Standard* 147-2002.

ASTM. 2003. Standard specification for seamless copper tube for air conditioning and refrigeration field service. *Standard* B280-03. American Society for Testing and Materials, West Conshohocken, PA.

Brisken, W.R. 1955. Moisture migration in hermetic refrigeration systems as measured under various operating conditions. *Refrigerating Engineering* (July):42.

Castrol. 1992. *Technical Bulletin* 2. Castrol Industrial North America, Specialty Products Division, Irvine, CA.

Cavestri, R.C. and W.R. Schafer. 1999. Equilibrium water capacity of desiccants in mixtures of HFC refrigerants and appropriate lubricants. *ASHRAE Transactions* 104(2):60-65.

Cavestri, R.C. and D.L. Schooley. 1998. Test methods for inorganic acid removal capacity in desiccants used in liquid line driers. *ASHRAE Transactions* 104(1B):1335-1340.

Cohen, A.P. 1993. Test methods for the compatibility of desiccants with alternative refrigerants. *ASHRAE Transactions* 99(1):408-412.

Cohen, A.P. 1994. Compatibility and performance of molecular sieve desiccants with alternative refrigerants. *Proceedings of the International Conference: CFCs, The Day After.* International Institute of Refrigeration, Paris.

Cohen, A.P. and C.S. Blackwell. 1995. Inorganic fluoride uptake as a measure of relative compatibility of molecular sieve desiccants with fluorocarbon refrigerants. *ASHRAE Transactions* 101(2):341-347.

Cohen, A.P. and S.R. Dunne. 1987. Review of automotive air-conditioning drydown rate studies—The kinetics of drying Refrigerant 12. *ASHRAE Transactions* 93(2):725-735.

DeGeiso, R.C. and R.F. Stalzer. 1969. Comparison of methods of moisture determination in refrigerants. *ASHRAE Journal* (April).

Doderer, G.C. and H.O. Spauschus. 1966. A sealed tube-gas chromatograph method for measuring reaction of Refrigerant 12 with oil. *ASHRAE Transactions* 72(2):IV.4.1-IV.4.4.

Dunne, S.R. and T.J. Clancy. 1984. Methods of testing desiccant for refrigeration drying. *ASHRAE Transactions* 90(1A):164-178.

DuPont. 1992. Acceptance specification for used refrigerants. *Bulletin* H-31790-1. E.I. DuPont de Nemours and Company, Wilmington, DE.

Elsey, H.M. and L.C. Flowers. 1949. Equilibria in Freon-12—Water systems. *Refrigerating Engineering* (February):153.

Elsey, H.M., L.C. Flowers, and J.B. Kelley. 1952. A method of evaluating refrigerator oils. *Refrigerating Engineering* (July):737.

Griffith, R. 1993. Polyolesters are expensive, but probably a universal fit. *Air Conditioning, Heating & Refrigeration News* (May 3):38.

Gully, A.J., H.A. Tooke, and L.H. Bartlett. 1954. Desiccant-refrigerant moisture equilibria. *Refrigerating Engineering* (April):62.

Gustafsson, V. 1977. Determining the air content in small refrigeration systems. Purdue Compressor Technology Conference.

Haagen-Smit, I.W., P. King, T. Johns, and E.A. Berry. 1970. Chemical design and performance of an improved Karl Fischer titrator. *American Laboratory* (December).

Hoffman, J.E. and B.L. Lange. 1962. Acid removal by various desiccants. *ASHRAE Journal* (February):61.

Jones, E. 1969. Liquid or suction line drying? *Air Conditioning and Refrigeration Business* (September).

Krause, W.O., A.B. Guise, and E.A. Beacham. 1960. Time factors in the removal of moisture from refrigerating systems with desiccant type driers. *ASHRAE Transactions* 66:465-476.

Kvalnes, D.E. 1965. The sealed tube test for refrigeration oils. *ASHRAE Transactions* 71(1):138-142.

Mays, R.L. 1962. Molecular sieve and gel-type desiccants for Refrigerants 12 and 22. *ASHRAE Journal* (August):73.

Norton, F.J. 1957. Rates of thermal decomposition of $CHClF_2$ and CF_2Cl_2. *Refrigerating Engineering* (September):33.

Parmelee, H.M. 1965. Sealed tube stability tests on refrigeration materials. *ASHRAE Transactions* 71(1):154-161.

Rosen, S., A.A. Sakhnovsky, R.B. Tilney, and W.O. Walker. 1965. A method of evaluating filtration and flow characteristics of liquid line driers. *ASHRAE Transactions* 71(1):200-205.

RSES. 1988. *Standard procedure for replacement of components in a sealed refrigerant system (compressor motor burnout)*. Refrigeration Service Engineers Society, Des Plaines, IL.

Spauschus, H.O. and G.C. Doderer. 1961. Reaction of Refrigerant 12 with petroleum oils. *ASHRAE Journal* (February):65.

Spauschus, H.O. and G.C. Doderer. 1964. Chemical reactions of Refrigerant 22. *ASHRAE Journal* (October):54.

Spauschus, H.O. and R.S. Olsen. 1959. Gas analysis—A new tool for determining the chemical stability of hermetic systems. *Refrigerating Engineering* (February):25.

Thrasher, J.S., R. Timkovich, H.P.S. Kumar, and S.L. Hathcock. 1993. Moisture solubility in Refrigerant 123 and Refrigerant 134a. *ASHRAE Transactions* 100(1):346-357.

UL. 2001. Standard for refrigerant-containing components and accessories, nonelectrical. ANSI/UL *Standard* 207-01. Underwriters Laboratories, Northbrook, IL.

UL. 1993. Field conversion/retrofit of products to change to an alternative refrigerant—Construction and operation. *Standard* 2170-93. Underwriters Laboratories, Northbrook, IL.

UL. 1993. Field conversion/retrofit of products to change to an alternative refrigerant—Insulating material and refrigerant compatibility. *Standard* 2171-93. Underwriters Laboratories, Northbrook, IL.

UL. 1993. Field conversion/retrofit of products to change to an alternative refrigerant—Procedures and methods. *Standard* 2172-93. Underwriters Laboratories, Northbrook, IL.

Walker, W.O. 1963. Latest ideas in use of desiccants and driers. *Refrigerating Service & Contracting* (August):24.

Walker, W.O., J.M. Malcolm, and H.C. Lynn. 1955. Hydrophobic behavior of certain desiccants. *Refrigerating Engineering* (April):50.

Walker, W.O., S. Rosen, and S.L. Levy. 1962. Stability of mixtures of refrigerants and refrigerating oils. *ASHRAE Journal* (August):59.

Wojtkowski, E.F. 1964. System contamination and cleanup. *ASHRAE Journal* (June):49.

BIBLIOGRAPHY

Boing, J. 1973. Desiccants and driers. *RSES Service Manual*, Section 5, 620-16B. Refrigeration Service Engineers Society, Des Plaines, IL.

Burgel, J., N. Knaup, and H. Lotz. 1988. Reduction of CFC-12 emission from refrigerators in the FRG. *International Journal of Refrigeration* 11(4).

Byrne, J.J., M. Shows, and M.W. Abel. 1996. *Investigation of flushing and clean-out methods for refrigeration equipment to ensure system compatibility.* Air-Conditioning and Refrigeration Technology Institute, Arlington, VA. DOE/CE/23810-73.

DuPont. 1976. *Mutual solubilities of water with fluorocarbons and fluorocarbon-hydrate formation.* DuPont de Nemours and Company, Wilmington, DE.

Guy, P.D., G. Tompsett, and T.W. Dekleva. 1992. Compatibilities of nonmetallic materials with R-134a and alternative lubricants in refrigeration systems. *ASHRAE Transactions* 98(1):804-816.

Jones, E. 1964. Determining pressure drop and refrigerant flow capacities of liquid line driers. *ASHRAE Journal* (February):70.

Kauffman, R.E. 1992. Sealed tube tests of refrigerants from field systems before and after recycling. *ASHRAE Transactions* 99(2):414-424.

Kitamura, K., T. Ohara, S. Honda, and H. SakaKibara. 1993. A new refrigerant-drying method in the automotive air conditioning system using HFC-134a. *ASHRAE Transactions* 99(1):361-367.

Manz, K.W. 1988. Recovery of CFC refrigerants during service and recycling by the filtration method. *ASHRAE Transactions* 94(2):2145-2151.

McCain, C.A. 1991. Refrigerant reclamation protects HVAC equipment investment. *ASHRAE Journal* 33(4).

Sundaresan, S.G. 1989. Standards for acceptable levels of contaminants in refrigerants. In *CFCs—Time of Transition*, pp. 220-223. ASHRAE.

Walker, W.O. 1960. Contaminating gases in refrigerating systems. In *RSES Service Manual*, Section 5, 620-15. Refrigeration Service Engineers Society, Des Plaines, IL.

Walker, W.O. 1985. Methyl alcohol in refrigeration. In *RSES Service Manual*, Section 5, 620-17A. Refrigeration Service Engineers Society, Des Plaines, IL.

Zahorsky, L.A. 1967. Field and laboratory studies of wax-like contaminants in commercial refrigeration equipment. *ASHRAE Transactions* 73(1):II.1.1-II.1.9.

Zhukoborshy, S.L. 1984. Application of natural zeolites in refrigeration industry. *Proceedings of the International Symposium on Zeolites*, Portoroz, Yugoslavia (September).

CHAPTER 7

LUBRICANTS IN REFRIGERANT SYSTEMS

THE primary function of a lubricant is to reduce friction and minimize wear. It achieves this by interposing a film between moving surfaces that reduces direct solid-to-solid contact or lowers the coefficient of friction.

Understanding the role of a lubricant requires analysis of the surfaces to be lubricated. Although bearing surfaces and other machined parts may appear and feel smooth, close examination reveals microscopic peaks (asperities) and valleys. Lubricant, in sufficient amounts, creates a layer thicker than the maximum height of the mating asperities, so that moving parts ride on a lubricant cushion.

These dual conditions are not always easily attained. For example, when the shaft of a horizontal journal bearing is at rest, static loads squeeze out the lubricant, producing a discontinuous film with metal-to-metal contact at the bottom of the shaft. When the shaft begins to turn, there is no layer of liquid lubricant separating the surfaces. As the shaft picks up speed, lubricating fluid is drawn into the converging clearance between the bearing and the shaft, generating a hydrodynamic pressure that eventually can support the load on an uninterrupted fluid film (Fuller 1984).

Various regimes or conditions of lubrication can exist when surfaces are in motion with respect to one another. Regimes of lubrication are defined as follows:

- *Full fluid film or hydrodynamic*. Mating surfaces are completely separated by the lubricant film.
- *Mixed fluid film or quasi-hydrodynamic*. Occasional or random surface contact occurs.
- *Boundary*. Gross surface-to-surface contact occurs because the bulk lubricant film is too thin to separate the mating surfaces.

Various lubricating oils are used to separate and lubricate contacting surfaces. Separation can be maintained by a boundary layer on a metal surface, a fluid film, or a combination of both.

In addition to preventing surface contact and reducing friction, lubricants also remove heat, provide a seal to keep out contaminants or to retain pressures, inhibit corrosion, and remove debris created by wear. Lubricating oils are best suited to meet these various requirements.

Viscosity is the most important property to consider in choosing a lubricant under full fluid film conditions. Under boundary conditions, the asperities are the contact points that take much, if not all, of the load. The resulting contact pressures are usually enough to cause welding and surface deformation. However, even under these conditions, wear can be controlled effectively with nonfluid, multimolecular films formed on the surface. These films must be strong enough to resist rupturing, yet have acceptable frictional and shear characteristics to reduce surface fatigue, adhesion, abrasion, and corrosion, which are the four major sources (either singularly or together) of rapid wear under boundary conditions.

Additives have also been developed to improve lubrication under boundary and mixed lubrication conditions. These materials are characterized by terms such as oiliness agents, lubricity improvers, and antiwear additives. They form a film on the metal surface through polar (physical) attraction and/or chemical action. These films or coatings result in lower coefficients of friction under loads. In chemical action, the temperature increase from friction-generated heat causes a reaction between the additive and the metal surface. Films such as iron sulfide and iron phosphate can form, depending on the additives and energy available for the reaction. In some instances, organic phosphates and phosphites are used in refrigeration oils to improve boundary and mixed lubrication. The nature and condition of the metal surfaces are important. Refrigeration compressor designers often treat ferrous pistons, shafts, and wrist pins with phosphating processes that impart a crystalline, soft, and smooth film of metal phosphate to the surface. This film helps provide the lubrication needed during break-in. Additives are often the synthesized components in lubricating oils. The slightly active nonhydrocarbon components left in commercially refined mineral oils give them their natural film-forming properties.

TESTS FOR BOUNDARY AND MIXED LUBRICATION

Film strength or *load-carrying ability* often describe lubricant lubricity characteristics under boundary conditions. Both mixed and boundary lubrication are evaluated by the same tests, but test conditions are usually less severe for mixed. Laboratory tests to evaluate lubricants measure the degree of scoring, welding, or wear. However, bench tests cannot be expected to accurately simulate actual field performance in a given compressor and are, therefore, merely screening devices. Some tests have been standardized by ASTM and other organizations.

In the **four-ball extreme-pressure method** (ASTM *Standard* D2783), the antiwear property is determined from the average scar diameter on the stationary balls and is stated in terms of a load-wear index. The smaller the scar, the better the load-wear index. The maximum load-carrying capability is defined in terms of a weld point (i.e., the load at which welding by frictional heat occurs).

The **Falex** method (ASTM *Standard* D2670) allows wear measurement during the test itself, and scar width on the V-blocks and/or mass loss of the pin is used to measure antiwear properties. Load-carrying capability is determined from a failure, which can be caused by excess wear or extreme frictional resistance. The **Timken** method (ASTM *Standard* D2782) determines the load at which rupture of the lubricant film occurs, and the **Alpha LFW-1** machine (ASTM *Standard* D2714) measures frictional force and wear.

The preparation of this chapter is assigned to TC 3.4, Lubrication.

The FZG gear test facility can provide useful information on how a lubricant performs in a gear box. Specific applications include gear-driven centrifugal compressors in which lubricant dilution by refrigerant is expected to be quite low.

However, because all these machines operate in air, available data may not apply to a refrigerant environment. Divers (1958) questioned the validity of tests in air, because several oils that performed poorly in Falex testing have been used successfully in refrigerant systems. Murray et al. (1956) suggest that halocarbon refrigerants can aid in boundary lubrication. Refrigerant 12, for example, when run hot in the absence of oil, reacted with steel surfaces to form a lubricating film. These studies emphasize the need for laboratory testing in a simulated refrigerant environment.

In Huttenlocher's (1969) simulation method, refrigerant vapor is bubbled through the lubricant reservoir before the test to displace the dissolved air. Refrigerant is bubbled continually during the test to maintain a blanket of refrigerant on the lubricant surface. Using the Falex tester, Huttenlocher showed the beneficial effect of R-22 on the load-carrying capability of the same lubricant compared with air or nitrogen. Sanvordenker and Gram (1974) describe a further modification of the Falex test using a sealed sample system.

Both R-12 (a CFC) and R-22 (an HCFC) atmospheres improved a lubricant's boundary lubrication characteristics when compared with tests in air. HFC refrigerants, which are chlorine-free, contribute to increased wear, compared to a chlorinated refrigerant with the same lubricant.

Komatsuzaki and Homma (1991) used a modified four-ball tester to determine antiseizure and antiwear properties of R-12 and R-22 in mineral oil and R-134a in a propylene glycol.

Test parameters must simulate as closely as possible the system conditions (base material from which test specimens are made, their surface condition, processing methods, and operating temperature). There are several bearings or rubbing surfaces in a refrigerant compressor, each of which may use different materials and may operate under different conditions. A different test may be required for each bearing. Moreover, bearings in hermetic compressors have very small clearances. Permissible bearing wear is minimal because wear debris remains in the system and can cause other problems even if clearances stay within working limits. Compressor system mechanics must be understood to perform and interpret simulated tests.

Some aspects of compressor lubrication are not suitable for laboratory simulation; for instance, return of liquid refrigerant to the compressor can cause lubricant to dilute or wash away from the bearings, creating conditions of boundary lubrication. Tests using operating refrigerant compressors have also been considered, and one such wear test has been proposed as a German Standard (DIN *Standard* 8978). The test is functional for a given compressor system and may allow comparison of lubricants within that class of compressors. However, it is not designed to be a generalized test for the boundary lubricating capability of a lubricant. Other tests using radioactive tracers in refrigerant systems have given useful results (Rembold and Lo 1966).

Although most boundary lubrication testing is performed at or near atmospheric pressure, testing some refrigerants at atmospheric pressures yields less meaningful results. Atmospheric or low-pressure sealed operation with refrigerant bubbled through the lubricant during the test has yielded positive results for refrigerants with a normal evaporation pressure within 145 psi of the testing pressure under the normal compressor operating temperature range. Refrigerants that operate at high pressure, such as CO_2, and zeotropic refrigerant blends, such as R-410A, require testing at near-operation elevated test pressures.

REFRIGERATION LUBRICANT REQUIREMENTS

Regardless of size or system application, refrigerant compressors are classified as either positive-displacement or dynamic. Both function to increase the pressure of the refrigerant vapor. Positive-displacement compressors increase refrigerant pressure by reducing the volume of a compression chamber through work applied to the mechanism (scroll, reciprocating, rotary, and screw). In contrast, dynamic compressors increase refrigerant pressure by a continuous transfer of angular momentum from the rotating member. As the gas decelerates, the imparted momentum is converted into a pressure rise. Centrifugal compressors function based on these principles.

Refrigerant compressors require lubricant to do more than simply lubricate bearings and mechanism elements. Oil delivered to the mechanism serves as a barrier that separates gas on the discharge side from gas on the suction sides. Oil also acts as a coolant, transferring heat from the bearings and mechanism elements to the crankcase sump, which, in turn, transfers heat to the surroundings. Moreover, oil helps reduce noise generated by moving parts inside the compressor. Generally, the higher the lubricant's viscosity, the better the sealing and noise reduction capabilities.

A hermetic system, in which the motor is exposed to the lubricant, requires a lubricant with electrical insulating properties. Refrigerant gas normally carries some lubricant with it as it flows through the condenser, flow-control device, and evaporator. This lubricant must return to the compressor in a reasonable time and must have adequate fluidity at low temperatures. It must also be free of suspended matter or components such as wax that might clog the flow control device or deposit in the evaporator and adversely affect heat transfer. In a hermetic system, the lubricant is charged only once, so it must function for the compressor's lifetime. The chemical stability required of the lubricant in the presence of refrigerants, metals, motor insulation, and extraneous contaminants is perhaps the most important characteristic distinguishing refrigeration lubricants from those used for all other applications (see Chapter 5).

Although compression components of centrifugal compressors require no internal lubrication, rotating shaft bearings, seals, and couplings must be adequately lubricated. Turbine or other types of lubricants can be used when the lubricant is not in contact or circulated with the refrigerant.

An ideal lubricant does not exist; a compromise must be made to balance the requirements. A high-viscosity lubricant seals gas pressure best, but may offer more frictional resistance. Slight foaming can reduce noise, but excessive foaming can carry too much lubricant into the cylinder and cause structural damage. Lubricants that are most stable chemically are not necessarily good lubricants. The lubricant should not be considered alone, because it functions as a lubricant/refrigerant mixture.

The precise relationship between composition and performance is not well defined. Standard ASTM bench tests can provide information such as (1) viscosity, (2) viscosity index, (3) color, (4) density, (5) refractive index, (6) pour point, (7) aniline point, (8) oxidation resistance, (9) dielectric breakdown voltage, (10) foaming tendency in air, (11) moisture content, (12) wax separation, and (13) volatility. Other properties, particularly those involving interactions with a refrigerant, must be determined by special tests described in the refrigeration literature. Nonstandard properties include (1) mutual solubility with various refrigerants; (2) chemical stability in the presence of refrigerants and metals; (3) chemical effects of contaminants or additives that may be in the oils; (4) boundary film-forming ability; (5) solubility of air; and (6) viscosity, vapor pressure, and density of oil-refrigerant mixtures.

MINERAL OIL COMPOSITION AND COMPONENT CHARACTERISTICS

For typical applications, the numerous compounds in refrigeration oils of mineral origin can be grouped into the following structures: (1) paraffins, (2) naphthenes (cycloparaffins), (3) aromatics, and (4) nonhydrocarbons. **Paraffins** consist of all straight-chain and branched-carbon-chain saturated hydrocarbons. Isopentane and *n*-pentane are examples of paraffinic hydrocarbons. **Naphthenes** are also completely saturated but consist of cyclic or ring structures; cyclopentane is a typical example. **Aromatics** are unsaturated cyclic hydrocarbons containing one or more rings characterized by alternate double bonds; benzene is a typical example. **Nonhydrocarbon** molecules contain atoms such as sulfur, nitrogen, or oxygen in addition to carbon and hydrogen.

The preceding structural components do not necessarily exist in pure states. In fact, a paraffinic chain frequently is attached to a naphthenic or aromatic structure. Similarly, a naphthenic ring to which a paraffinic chain is attached may in turn be attached to an aromatic molecule. Because of these complications, mineral oil composition is usually described by carbon type and molecular analysis.

In **carbon type analysis**, the number of carbon atoms on the paraffinic chains, naphthenic structures, and aromatic rings is determined and represented as a percentage of the total. Thus, % C_P, the percentage of carbon atoms having a paraffinic configuration, includes not only free paraffins but also paraffinic chains attached to naphthenic or to aromatic rings.

Similarly, % C_N includes carbon atoms on free naphthenes as well as those on naphthenic rings attached to aromatic rings, and % C_A represents carbon on aromatic rings. Carbon analysis describes a lubricant in its fundamental structure, and correlates and predicts many physical properties of the lubricant. However, direct methods of determining carbon composition are laborious. Therefore, common practice uses a correlative method, such as the one based on the refractive index-density-relative molecular mass (n-d-m) (Van Nes and Weston 1951) or one standardized by ASTM *Standard* D2140 or D3288. Other methods include ASTM *Standard* D2008, which uses ultraviolet absorbency, and a rapid method using infrared spectrophotometry and calibration from known oils.

Molecular analysis is based on methods of separating structural molecules. For refrigeration oils, important structural molecules are (1) saturates or nonaromatics, (2) aromatics, and (3) nonhydrocarbons. All free paraffins and naphthenes (cycloparaffins), as well as mixed molecules of paraffins and naphthenes, are included in the saturates. However, any paraffinic and naphthenic molecules attached to an aromatic ring are classified as aromatics. This representation of lubricant composition is less fundamental than carbon analysis. However, many properties of the lubricant relevant to refrigeration can be explained with this analysis, and the chromatographic methods of analysis are fairly simple (ASTM *Standards* D2007 and D2549; Mosle and Wolf 1963; Sanvordenker 1968).

Traditional classification of oils as paraffinic or naphthenic refers to the number of paraffinic or naphthenic molecules in the refined lubricant. Paraffinic crudes contain a higher proportion of paraffin wax, and thus have a higher viscosity index and pour point than to naphthenic crudes.

Component Characteristics

Saturates have excellent chemical stability, but poor solubility with polar refrigerants such as R-22; they are also poor boundary lubricants. Aromatics are somewhat more reactive but have very good solubility with refrigerants and good boundary lubricating properties. Nonhydrocarbons are the most reactive but are beneficial for boundary lubrication, although the amounts needed for that purpose are small. A lubricant's reactivity, solubility, and boundary lubricating properties are affected by the relative amounts of these components in the lubricant.

The saturate and aromatic components separated from a lubricant do not have the same viscosity as the parent lubricant. For the same boiling point range, saturates are much less viscous, and aromatics are much more viscous, than the parent lubricant. For the same viscosity, aromatics have higher volatility than saturates. Also, saturates have lower density and a lower refractive index, but a higher viscosity index and molecular mass than the aromatic component of the same lubricant.

Among the saturates, straight-chain paraffins are undesirable for refrigeration applications because they precipitate as wax crystals when the lubricant cools to its pour point, and tend to form flocs in some refrigerant solutions (see the section on Wax Separation). Branched-chain paraffins and naphthenes are less viscous at low temperatures and have extremely low pour points.

Nonhydrocarbons are mostly removed during refining of refrigeration oils. Those that remain are expected to have little effect on the lubricant's physical properties, except perhaps on its color, stability, and lubricity. Because not all the nonhydrocarbons (e.g., sulfur compounds) are dark, even a colorless lubricant does not necessarily guarantee the absence of nonhydrocarbons. Kartzmark et al. (1967) and Mills and Melchoire (1967) found indications that nitrogen-bearing compounds cause or act as catalysts toward oil deterioration. The sulfur and oxygen compounds are thought to be less reactive, with some types considered to be natural inhibitors and lubricity enhancers.

Solvent refining, hydrofinishing, or acid treatment followed by a separation of the acid tar formed are often used to remove more thermally unstable aromatic and unsaturated compounds from the base stock. These methods also produce refrigeration oils that are free from carcinogenic materials sometimes found in crude oil stocks.

The properties of the components naturally are reflected in the parent oil. An oil with a very high saturate content, as is frequently the case with paraffinic oils, also has a high viscosity index, low specific gravity, high relative molecular mass, low refractive index, and low volatility. In addition, it would have a high aniline point and would be less miscible with polar refrigerants. The reverse is true of naphthenic oils. Table 1 lists typical properties of several mineral-based refrigeration oils.

SYNTHETIC LUBRICANTS

The limited solubility of mineral oils with R-22 and R-502 originally led to the investigation of synthetic lubricants for refrigeration use. In more recent times, the lack of solubility of mineral oils in nonchlorinated fluorocarbon refrigerants, such as R-134a and R-32, has led to the commercial use of some synthetic lubricants. Gunderson and Hart (1962) describe a number of commercially available synthetic lubricants, such as synthetic paraffins, polyglycols, dibasic acid esters, neopentyl esters, silicones, silicate esters, and fluorinated compounds. Sanvordenker and Larime (1972) describe the properties of synthetic lubricants, alkylbenzenes, and phosphate esters in regard to refrigeration applications using chlorinated fluorocarbon refrigerants. Phosphate esters are unsuitable for refrigeration use because of their poor thermal stability. Although very stable and compatible with refrigerants, fluorocarbon lubricants are expensive. Among the others, only synthetic paraffins have poor miscibility relations with R-22. Dibasic acid esters, neopentyl esters, silicate esters, and polyglycols all have excellent viscosity temperature relations and remain miscible with R-22 and R-502 to very low temperatures. At this time, the three most commonly used synthetic lubricants are alkylbenzene, for R-22 and R-502 service, and polyglycols and polyol esters, for use with R-134a and refrigerant blends using R-32.

There are two basic types of alkylbenzenes: branched and linear. The products are synthesized by reacting an olefin or chlorinated paraffin with benzene in the presence of a catalyst. Catalysts commonly used for this reaction are aluminum chloride and hydrofluoric acid. After the catalyst is removed, the product is distilled into fractions. The relative size of these fractions can be changed by adjusting the relative molecular mass of the side chain (olefin or chlorinated paraffin) and by changing other variables. The quality of alkylbenzene refrigeration lubricant varies, depending on the type (branched or linear) and manufacturing scheme. In addition to good

Table 1 Typical Properties of Refrigerant Lubricants

		Mineral Lubricants				Synthetic Lubricants				
Property	**Source**	**Naphthenic**			**Paraffinic**	**Alkyl-benzene**	**Ester**		**Glycol**	
Viscosity, cSt (SSU) at 100°F	ASTM D445	33.1 (155)	61.9 (287)	68.6 (318)	34.2 (160)	31.7 (149)	30 (142)	100 (463)	29.9 (141)	90 (417)
Viscosity index	ASTM D2270	0	0	46	95	27	111	98	210	235
Specific gravity	ASTM D1298	0.913	0.917	0.9	0.862	0.872	0.995	0.972	0.99	1.007
Color	ASTM D1500	0.5	1	1	0.5					
Refractive index	ASTM D1747	1.5015	1.5057	1.4918	1.4752					
Molecular weight	ASTM D2503	300	321	345	378	320	570	840	750	1200
Pour point, °F	ASTM D97	–45	–40	–35	0	–50	–54	–22	–51	–40
Floc point, °F	ASHRAE 86	–68	–60	–60	–31	–100				
Flash point, °F	ASTM D92	340	360	400	395	350	453	496	399	334
Fire point, °F	ASTM D92	390	400	450	450	365				
Composition							Branched-acid penta-erythritol	Branched-acid penta-erythritol	Monol mono-functional poly-propylene glycol	Diol di-functional poly-propylene glycol
Carbon-type										
% C_A	Van Nes and	14	16	7	3	24				
% C_N	Weston (1951)	43	42	46	32	None				
% C_P		43	42	47	65	76				
Molecular composition	ASTM D2549									
% Saturates		62	59	78	87	None				
% Aromatics		38	41	22	13	100				
Aniline point, °F	ASTM D611	160	165	197	220	125				
Critical solution temp. with R-22, °F	—	25	35	74	81	–100				

solubility with refrigerants, such as R-22 and R-502, these lubricants have better high-temperature and oxidation stability than mineral oil-based refrigeration oils. Typical properties for a branched alkylbenzene are shown in Table 1.

Polyalkylene glycols (PAGs) derive from ethylene oxide or propylene oxide. Polymerization is usually initiated either with an alcohol, such as butyl alcohol, or by water. Initiation by an alcohol results in a monol (mono-end-capped); initiation by water results in a diol (uncapped). Another type is the double-end-capped PAG, a monocapped PAG that is further reacted with alkylating agents. PAGs are common lubricants in automotive air-conditioning systems using R-134a. PAGs have excellent lubricity, low pour points, good low-temperature fluidity, and good compatibility with most elastomers. Major concerns are that these oils are somewhat hygroscopic, are immiscible with mineral oils, and require additives for good chemical and thermal stability (Short 1990).

Polyalphaolefins (PAOs) are normally manufactured from linear α-olefins. The first step in manufacture is synthesizing a mixture of oligomers in the presence of a $BF_3 \cdot ROH$ catalyst. Several parameters (e.g., temperature, type of promoters) can be varied to control the distribution of the oligomers formed. The second step involves a hydrogenation processing of the unsaturated oligomers in the presence of a metal catalyst (Shubkin 1993). PAOs have good miscibility with R-12 and R-114. Some R-22 applications have been tried but are limited by the low miscibility of the fluid in R-22. PAOs are immiscible in R-134a (Short 1990), and are mainly used as an immiscible oil in ammonia systems.

Neopentyl esters (polyol esters) are derived from a reaction between an alcohol (usually pentaerythritol, trimethylolpropane, or neopentyl glycol) and a normal or branched carboxylic acid. For higher viscosities, a dipentaerythritol is often used. Acids are usually selected to give the correct viscosity and fluidity at low temperatures matched to the miscibility requirements of the refrigerant. Complex neopentyl esters are derived by a sequential reaction of the polyol with a dibasic acid followed by reaction with mixed monoacids (Short 1990). This results in a lubricant with a higher relative molecular mass, high viscosity indices, and higher ISO viscosity grades. Polyol ester lubricants are used commercially with HFC refrigerants in all types of compressors.

LUBRICANT ADDITIVES

Additives are used to enhance certain lubricant properties or impart new characteristics. They generally fall into three groups: polar compounds, polymers, and compounds containing active elements such as sulfur or phosphorous. Additive types include (1) pour-point depressants for mineral oil, (2) floc-point depressants for mineral oil, (3) viscosity index improvers for mineral oil, (4) thermal stability improvers, (5) extreme pressure and antiwear additives, (6) rust inhibitors, (7) antifoam agents, (8) metal deactivators, (9) dispersants, and (10) oxidation inhibitors.

Some additives offer performance advantages in one area but are detrimental in another. For example, antiwear additives can reduce wear on compressor components, but because of the chemical reactivity of these materials, the additives can reduce the lubricant's overall stability. Some additives work best when combined with other additives. They must be compatible with materials in the system (including the refrigerant) and be present in the optimum concentration; too little may be ineffective, whereas too much can be detrimental or offer no incremental improvement.

In general, additives are not required to lubricate a refrigerant compressor. However, additive-containing lubricants give highly satisfactory service, and some (e.g., those with antiwear additives) offer performance advantages over straight mineral oils. Their use is justified as long as the user knows of their presence, and provided the additives do not significantly degrade with use. Additives can often be used with synthetic lubricants to reduce wear because, unlike mineral oil, they do not contain nonhydrocarbon components such as sulfur.

An additive is only used after thorough testing to determine whether it is (1) removed by system dryers, (2) inert to system components, (3) soluble in refrigerants at low temperatures so as not to cause deposits in capillary tubes or expansion valves, and (4) stable at high temperatures to avoid adverse chemical reactions such as harmful deposits. This can best be done by sealed-tube and compressor testing using the actual additive/base lubricant combination intended for field use.

LUBRICANT PROPERTIES

Viscosity and Viscosity Grades

Viscosity defines a fluid's resistance to flow. It can be expressed as absolute or dynamic viscosity (centipoises, cP), kinematic viscosity (centistokes, cSt), or Saybolt Seconds Universal viscosity (abbreviated SSU or SUS). ASTM *Standard* D2161 contains tables to convert SSU to kinematic viscosity. The density must be known to convert kinematic viscosity to absolute viscosity; that is, absolute or dynamic viscosity (cP) equals density (g/cm^3) times kinematic viscosity (cSt). Refrigeration oils are sold in viscosity grades, and ASTM *Standard* D2422 describes a system of standardized viscosity grades.

In selecting the proper viscosity grade, the environment to which the lubricant will be exposed must be considered. Lubricant viscosity decreases if temperatures rise or if the refrigerant dissolves appreciably in the lubricant, and directly affects refrigeration compressor and system performance.

A large reduction in the viscosity of the lubricating fluid may affect the lubricant's lubricity and, more likely, its sealing function, depending on the nature of the machinery. The design of some hermetically sealed units (e.g., single-vane rotary) requires lubricating fluid to act as an efficient sealing agent. In reciprocating compressors, the lubricant film is spread over the entire area of contact between the piston and cylinder wall, providing a very large area to resist leakage from the high- to the low-pressure side. In a single-vane rotary type, however, the critical sealing area is a line contact between the vane and a roller. In this case, viscosity reduction is serious, and using sufficiently high-viscosity-grade materials is essential to ensure proper sealing.

Another consideration is the viscosity effect of lubricants on power consumption. Generally, the lowest safe viscosity-grade material that meets all requirements is chosen for a given refrigeration application. A practical method for determining the minimum safe viscosity is to calculate the total volumetric efficiency of a given compressor using several lubricants of widely varying viscosities. The lowest-viscosity lubricant that gives satisfactory volumetric efficiency should be selected. Tests should be run at several ambient temperatures (e.g., 70, 90, and 110°F). As a guideline, Table 2 lists recommended viscosity ranges for various refrigeration systems.

The International Organization for Standardization (ISO) established a series of viscosity levels as a standard for specifying or selecting lubricant for industrial applications. This system, covered in the United States by ASTM *Standard* D2422, is designed to eliminate intermediate or unnecessary viscosity grades while providing enough grades for operating equipment. The system reference point is kinematic viscosity at 40°C, and each viscosity grade with suitable tolerances is identified by the kinematic viscosity at this temperature. Therefore, an ISO VG 32 grade lubricant identifies a lubricant with a viscosity of 32 mm^2/s at 40°C. Table 3 lists standardized viscosity grades of lubricants.

Viscosity Index

Lubricant viscosity decreases as temperature increases and increases as temperature decreases. The relationship between

Table 2 Recommended Viscosity Ranges

Small and Commercial Systems

Refrigerant	Type of Compressor	Lubricant Viscosities at 100°F SSU	cSt
Ammonia	Screw	280 to 300	60 to 65
	Reciprocating	150 to 300	32 to 65
Carbon dioxide	Reciprocating	280 to 300[a]	60 to 65[a]
R-11	Centrifugal	280 to 300	60 to 65
R-12	Centrifugal	280 to 300	60 to 65
	Reciprocating	150 to 300	32 to 65
	Rotary	280 to 300	60 to 65
R-123	Centrifugal	280 to 300	60 to 65
R-22	Centrifugal	280 to 400	60 to 86
	Reciprocating	150 to 300	32 to 65
	Scroll	280 to 300	60 to 65
	Screw	280 to 800	60 to 173
R-134a	Scroll	100 to 300	22 to 68
	Screw	150 to 500	32 to 100
	Centrifugal	280 to 300	60 to 65
R-407C	Scroll	100 to 300	22 to 68
	Reciprocating	150 to 300	32 to 68
R-410A	Scroll	100 to 300	22 to 68
Halogenated	Screw	150 to 4000	32 to 800

Industrial Refrigeration[b]

Type of Compressor	Lubricant Viscosities at 100°F SSU	cSt
Where lubricant may enter refrigeration system or compressor cylinders	150 to 300	32 to 65
Where lubricant is prevented from entering system or cylinders:		
In force-feed or gravity systems	500 to 600	108 to 129
In splash systems	150 to 160	32 to 34
Steam-driven compressor cylinders when condensate is reclaimed for ice-making	High-viscosity lubricant [30 to 35 cSt (140 to 165 SSU) viscosity at 210°F]	

[a]Some applications may require lighter lubricants of 75 to 85 SSU (14 to 17 cSt); others, heavier lubricants of 500 to 600 SSU (108 to 129 cSt).

[b]Ammonia and carbon dioxide compressors with splash, force-feed, or gravity circulating systems.

temperature and kinematic viscosity is represented by the following equation (ASTM *Standard* D341):

$$\log \log[\nu + 0.7 + f(\nu)] = A + B \log T \quad (1)$$

where

ν = kinematic viscosity, cSt
$f(\nu)$ = additive function of kinematic viscosity, only used below 2 cSt
T = thermodynamic temperature, K or °R
A, B = constants for each lubricant

This relationship is the basis for the viscosity/temperature charts published by ASTM and allows a straight-line plot of viscosity over a wide temperature range. Figure 1 shows a plot for a naphthenic mineral oil (LVI) and a synthetic lubricant (HVI). This plot is applicable over the temperature range in which the oils are homogenous liquids.

The slope of the viscosity/temperature lines is different for different lubricants. The viscosity/temperature relationship of a lubricant is described by an empirical number called the **viscosity index (VI)** (ASTM *Standard* D2270). A lubricant with a high viscosity index (HVI) shows less change in viscosity over a given temperature range than a lubricant with a low viscosity index (LVI). In the example shown in Figure 1, both oils possess equal viscosities (32 cSt or

Table 3 Viscosity System for Industrial Fluid Lubricants (ASTM *Standard* D2422)

Viscosity System Grade Identification	Midpoint Viscosity, cSt at 40°C	Kinematic Viscosity Limits cSt at 40°C		Approximate Equivalents, SSU Units
		Minimum	Maximum	
ISO VG 2	2.2	1.98	2.42	32
ISO VG 3	3.2	2.88	3.52	*
ISO VG 5	4.6	4.14	5.06	40
ISO VG 7	6.8	6.12	7.48	*
ISO VG 10	10	9.00	11.00	60
ISO VG 15	15	13.50	16.50	75
ISO VG 22	22	19.80	24.20	105
ISO VG 32	32	28.80	35.20	150
ISO VG 46	46	41.40	50.60	215
ISO VG 68	68	61.20	74.80	315
ISO VG 100	100	90	110	465
ISO VG 150	150	135	165	700
ISO VG 220	220	198	242	1000
ISO VG 320	320	288	352	1500
ISO VG 460	460	414	506	2150
ISO VG 680	680	612	748	3150
ISO VG 1000	1000	900	1100	4650
ISO VG 1500	1500	1350	1650	7000

*The 36 and 50 SSU grades are not standard grades in the United States.

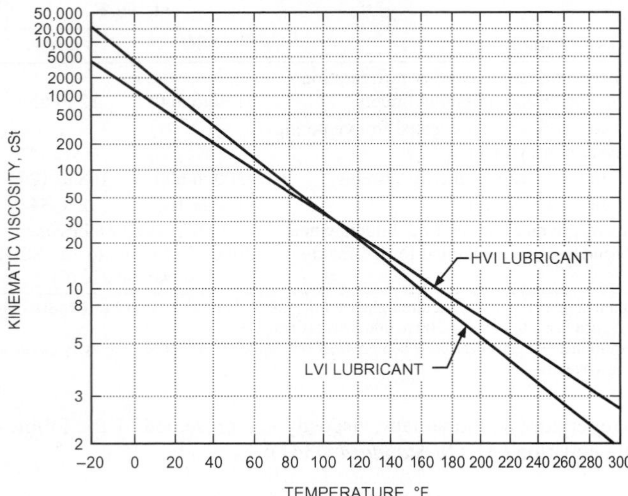

Fig. 1 Viscosity/Temperature Chart for 108 cSt (500 SSU) HVI and LVI Lubricants

151 SSU) at 104°F. However, the viscosity of the LVI lubricant, increases to 520 cSt (2400 SSU) at 32°F, whereas the HVI lubricant's viscosity increases only to 280 cSt (1280 SSU).

The viscosity index is related to the mineral oil's composition. Generally, an increase in cyclic structure (aromatic and naphthenic) decreases VI. Paraffinic oils usually have a high viscosity index and low aromatic content. Naphthenic oils, on the other hand, have a lower viscosity index and are usually higher in aromatics. For the same base lubricant, VI decreases as aromatic content increases. Generally, among common synthetic lubricants, polyalphaolefins, polyalkylene glycols, and polyol esters have high viscosity indices. As shown in Table 1, alkylbenzenes have lower viscosity indices.

Generally, for the same type of fluids with similar refrigerant solubility characteristics, higher-VI oils means better full-film fluid lubrication at elevated compressor temperature. At lower evaporator temperatures, however, fluids with lower VI and lower viscosity and fluidity characteristics can provide better oil return and less viscosity drag across the overall temperature range.

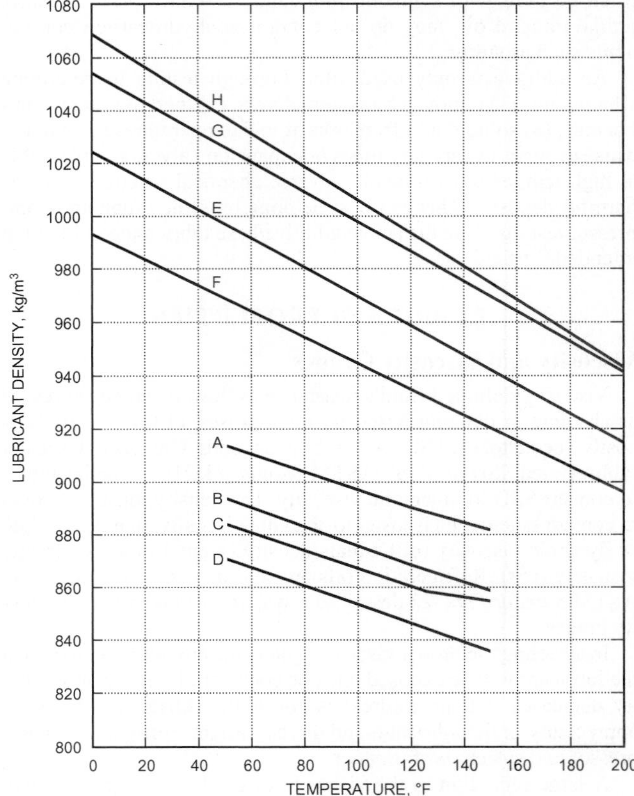

Lubricants		Viscosity at 100°F		Ref.
		cSt	SSU	
A	Naphthene	64.7	300	1
B	Naphthene	15.7	80	1
C	Paraffin	64.7	300	1
D	Paraffin	32.0	150	1
E	Branched-acid POE	32		2
F	Branched-acid POE	100		2
G	Polypropylene glycol mono butyl ether	32		2
H	Polyoxypropylene diol	80		2

References: 1. Albright and Lawyer (1959) 2. Cavestri (1993)

Fig. 2 Variation of Refrigeration Lubricant Density with Temperature

Density

Figure 2 shows published values for pure lubricant densities over a range of temperatures. These density/temperature curves all have approximately the same slope and appear merely to be displaced from one another. If the density of a particular lubricant is known at one temperature but not over a range of temperatures, a reasonable estimate at other temperatures can be obtained by drawing a line paralleling those in Figure 2.

Density indicates the composition of a lubricant for a given viscosity. As shown in Figure 2, naphthenic oils are usually denser than paraffinic oils, and synthetic lubricants are generally denser than mineral oils. Also, the higher the aromatic content, the higher the density. For equivalent compositions, higher-viscosity oils have higher densities, but the change in density with aromatic content is greater than it is with viscosity.

Relative Molecular Mass

In refrigeration applications, the relative molecular mass of a lubricant is often needed. Albright and Lawyer (1959) showed that, on a molar basis, Refrigerants 22, 115, 13, and 13B1 have about the same viscosity-reducing effects on a paraffinic lubricant.

For most mineral oils, a reasonable estimate of the average molecular mass can be obtained by a standard test (ASTM *Standard* D2502), based on kinematic viscosities at 100°F and 210°F; or from viscosity/gravity correlations of Mills et al. (1946). Direct methods (ASTM *Standard* D2503) can also be used when greater precision is needed or when the correlative methods are not applicable.

Pour Point

Any lubricant intended for low-temperature service should be able to flow at the lowest temperature that it will encounter. This requirement is usually met by specifying a suitably low pour point. The pour point of a lubricant is defined as the lowest temperature at which it will pour or flow, when tested according to the standard method prescribed in ASTM *Standard* D97.

Loss of fluidity at the pour point may manifest in two ways. Naphthenic oils and synthetic lubricants usually approach the pour point by a steady increase in viscosity. Paraffinic oils, unless heavily dewaxed, tend to separate out a rigid network of wax crystals, which may prevent flow while still retaining unfrozen liquid in the interstices. Pour points can be lowered by adding pour-point depressants, which are believed to modify the wax structure, possibly by depositing a film on the surface of each wax crystal, so that the crystals no longer adhere to form a matrix and do not interfere with the lubricant's ability to flow. Pour-point depressants are not suitable for use with halogenated refrigerants.

Standard pour test values are significant in selection of oils for ammonia and carbon dioxide systems using alkylbenezene or mineral oils, and any other system in which refrigerant and lubricant are almost totally immiscible. In such a system, any lubricant that gets into the low side is essentially refrigerant-free; therefore, the pour point of the lubricant itself determines whether loss of fluidity, congealment, or wax deposition occurs at low-side temperatures.

Because oil in the low-pressure side of halogenated refrigerant systems contains significant amounts of dissolved refrigerant, the pour-point test, which is conducted on pure oils and in air, is of little significance. Viscosity of lubricant/refrigerant solutions at low-side conditions and wax separation (or floc test) are important considerations.

A lubricant's pour point should not be confused with its freezing point. Pour point is determined by exposing the lubricant to a low temperature for a short time. Refrigeration lubricants will solidify after long-term exposure to low temperature, even if the temperature is higher than the pour point.

Some refrigerant lubricants stored at low temperatures become unstable. Typically, components of the lubricant separate and form crystals or deposits on the container.

Volatility: Flash and Fire Points

Because boiling ranges and vapor pressure data on lubricants are not readily available, an indication of a lubricant's volatility is obtained from the flash and fire points (ASTM *Standard* D92). These properties are normally not significant in refrigeration equipment. However, some refrigerants, such as sulfur dioxide, ammonia, and methyl chloride, have a high ratio of specific heats (c_p/c_v) and consequently have a high adiabatic compression temperature. These refrigerants frequently carbonize oils with low flash and fire points when operating in high ambient temperatures. Lubricant can also carbonize in some applications that use halogenated refrigerants and require high compression ratios (such as domestic refrigerator-freezers operating in high ambient temperatures). Because such carbonization or coking of the valves is not necessarily accompanied by general lubricant deterioration, the tendency of a lubricant to carbonize is referred to as **thermal instability**, as opposed to chemical instability. Some manufacturers circumvent these problems by using paraffinic oils, which in comparison to naphthenic oils have higher flash and fire points. Others prevent them through appropriate design.

Table 4 Increase in Vapor Pressure and Temperature

Temperature, °F	Vapor Pressure 32 cSt (150 SSU) Oil	
	Alkylbenzene, mm Hg	Naphthene Base, mm Hg
300	0.72	0.93
325	1.58	1.92
350	3.36	3.78
375	6.70	7.15
400	13.0	13.1
425	24.3	23.0
450	43.8	39.4

Vapor Pressure

Vapor pressure is the pressure at which the vapor phase of a substance is in equilibrium with the liquid phase at a specified temperature. The composition of the vapor and liquid phases (when not pure) influences equilibrium pressure. With refrigeration lubricants, the type, boiling range, and viscosity also affect vapor pressure; naphthenic oils of a specific viscosity grade generally show higher vapor pressures than paraffinic oils.

Vapor pressure of a lubricant increases with increasing temperature, as shown in Table 4. In practice, the vapor pressure of a refrigeration lubricant at an elevated temperature is negligible compared with that of the refrigerant at that temperature. The vapor pressure of narrow-boiling petroleum fractions can be plotted as straight-line functions. If the lubricant's boiling range and type are known, standard tables may be used to determine the lubricant's vapor pressure up to 760 mm Hg at any given temperature (API 1970).

Aniline Point

Aniline, an aromatic compound, is more soluble in oils containing a greater quantity of similar compounds. The temperature at which a lubricant and aniline are mutually soluble is the lubricant's aniline point (ASTM *Standard* D611). Therefore, the relative aromaticity of a mineral oil can be determined by its solubility in aniline. In comparing mineral oils, lower aniline points correspond to the presence of more naphthenic and/or aromatic molecules.

Aniline point can also predict a mineral oil's effect on elastomer seal materials. Generally, a highly naphthenic lubricant swells a specific elastomer material more than a paraffinic lubricant, because the aromatic and naphthenic compounds in a naphthenic lubricant are more soluble. However, aniline point gives only a general indication of lubricant/elastomer compatibility. Within a given class of elastomer material, lubricant resistance varies widely because of differences in compounding practiced by the elastomer manufacturer. Finally, in some retrofit applications, a high-aniline-point mineral oil may cause elastomer shrinkage and possible seal leakage.

Elastomers behave differently in synthetic lubricants, such as alkylbenzenes, polyalkylene glycols, and polyol esters, than in mineral oils. For example, an alkylbenzene has an aniline point lower than that of a mineral oil of the same viscosity grade. However, the amount of swell in a chloroneoprene O ring is generally less than that found with mineral oil. For these reasons, lubricant-elastomer compatibility needs to be tested under conditions anticipated in actual service.

Solubility of Refrigerants in Oils

All gases are soluble to some extent in lubricants, and many refrigerant gases are highly soluble. For instance, chlorinated refrigerants are miscible with most oils at any temperature likely to be encountered. Nonchlorinated refrigerants, however, are often limited

Table 5 Absorption of Low-Solubility Refrigerant Gases in Oil

Ammonia[a] (Percent by Mass)					
Absolute Pressure, psi	Temperature, °F 32	68	149	212	302
14.2	0.246	0.180	0.105	0.072	0.054
28.4	0.500	0.360	0.198	0.144	0.108
42.7	0.800	0.540	0.304	0.228	0.166
57.0	—	0.720	0.398	0.300	0.222
142.0	—	—	1.050	0.720	0.545

Carbon Dioxide[b] (Percent by Mass)				
Absolute Pressure, psi	Temperature, °F 32	68	149	212
14.7	0.26	0.19	0.13	0.072

[a]Type of oil: Not given (Steinle 1950)
[b]Type of oil: HVI oil, 34.8 cSt (163 SSU) at 100°F (Baldwin and Daniel 1953)

to the polar synthetic lubricants such as polyol ester or PAG oils. The amount dissolved depends on gas pressure and lubricant temperature, and on their natures. Because refrigerants are much less viscous than lubricants, any appreciable amount in solution markedly reduces viscosity.

Two refrigerants usually regarded as poorly soluble in mineral oil are ammonia and carbon dioxide. Data showing the slight absorption of these gases by mineral oil are given in Table 5. The amount absorbed increases with increasing pressure and decreases with increasing temperature. In ammonia systems, where pressures are moderate, the 1% or less refrigerant that dissolves in the lubricant should have little, if any, effect on lubricant viscosity. However, operating pressures in CO_2 systems tend to be much higher (not shown in Table 5), and the quantity of gas dissolved in the lubricant may be enough to substantially reduce viscosity. At 390 psig, for example, Beerbower and Greene (1961) observed a 69% reduction when a 32 cSt (150 SSU) lubricant (HVI) was tested under CO_2 pressure at 80°F.

LUBRICANT/REFRIGERANT SOLUTIONS

The behavior of lubricant/refrigerant solutions is determined by their mutual solubility in the relevant temperature and pressure ranges. For instance, chlorinated refrigerants such as R-22 and R-114 may show limited solubilities with some lubricants at evaporator temperatures (exhibited in the form of phase separation) and unlimited solubilities in the higher temperature regions of a refrigerant system. In some systems using HFC refrigerants, a second, distinct two-phase region may occur at high temperatures. For such refrigerants, solubility studies must therefore be carried out over an extended temperature range.

Because halogenated refrigerants have such high solubilities, the lubricating fluid can no longer be treated as a pure lubricant, but rather as a lubricant/refrigerant solution whose properties are markedly different from those of pure lubricant. The amount of refrigerant dissolved in a lubricant depends on the pressure and temperature. Therefore, lubricating fluid composition is different in different sections/stages of a refrigeration system, and changes from the time of start-up until the system attains the steady state. The most pronounced effect is on viscosity.

For example, refrigerant and lubricant in a compressor crankcase are assumed to be in equilibrium, and the viscosity is as shown in Figure 44. If lubricant in the crankcase at start-up is 75°F, viscosity of pure 32 ISO VG branched-acid polyol ester is about 60 cSt. Under operating conditions, lubricant in the crankcase is typically about 125°F. At this temperature, viscosity of the pure lubricant is about 20 cSt. If R-134a is the refrigerant and the pressure in the crankcase is 51 psia, viscosity of the lubricant/refrigerant mixture at start-up is about 10 cSt and decreases to 9 cSt at 125°F.

Thus, if only lubricant properties are considered, an erroneous picture of the system is obtained. As another example, when lubricant returns from the evaporator to the compressor, the highest viscosity does not occur at the lowest temperature, because the lubricant contains a large amount of dissolved refrigerant. As temperature increases, the lubricant loses some of the refrigerant and the viscosity peaks at a point away from the coldest spot in the system.

Similarly, properties of the working fluid (a high-refrigerant-concentration solution) are also affected. The vapor pressure of a lubricant/refrigerant solution is markedly lower than that of the pure refrigerant. Consequently, the evaporator temperature is higher than if the refrigerant is pure. Another result is what is sometimes called flooded start-up. When the crankcase and evaporator are at about the same temperature, fluid in the evaporator (which is mostly refrigerant) has a higher vapor pressure than fluid in the crankcase (which is mostly lubricant). This difference in vapor pressures drives refrigerant to the crankcase, where it is absorbed in the lubricant until the pressures equalize. At times, moving parts in the crankcase may be completely immersed in this lubricant/refrigerant solution. At start-up, the change in pressure and turbulence can cause excessive amounts of liquid to enter the cylinders, causing damage to the valves and starving the crankcase of lubricant. Use of crankcase heaters to prevent such problems caused by highly soluble refrigerants is discussed in Chapter 2 and by Neubauer (1958). Problems associated with rapid outgassing from the lubricant are more pronounced with synthetic oils than with mineral oils. Synthetic oils release absorbed refrigerant more quickly and have a lower surface tension, which results in a lack of the stable foam found with mineral oils (Swallow et al. 1995).

Density

When estimating the density of a lubricant/refrigerant solution, the solution is assumed ideal so that the specific volumes of the components are additive. The formula for calculating the ideal density ρ_{id} is

$$\rho_{id} = \frac{\rho_o}{1 + W(\rho_o/\rho_R - 1)} \tag{2}$$

where

ρ_o = density of pure lubricant at solution temperature
ρ_R = density of refrigerant liquid at solution temperature
W = mass fraction of refrigerant in solution

For some combinations, the actual density of a lubricant/refrigerant solution may deviate from the ideal by as much as 8%. The solutions are usually more dense than calculated, but sometimes they are less. For example, R-11 forms ideal solutions with oils, whereas R-12 and R-22 show significant deviations. Density correction factors for R-12 and R-22 solutions are depicted in Figure 3. The corrected densities can be obtained from the relation

$$\text{Mixture density} = \rho_m = \frac{\rho_{id}}{A} \tag{3}$$

where A is the density correction factor read from Figure 3 at the desired temperature and refrigerant concentration.

Van Gaalen et al. (1990, 1991a, 1991b) provide values of density for four refrigerant/lubricant pairs: R-22/mineral oil, R-22/alkylbenzene, R-502/mineral oil, and R-502/alkylbenzene. Figures 4 to 7 provide data on the variation of density with temperature for R-134a in combination with a 32 ISO VG polyol ester, a 100 ISO VG polyol ester, a 32 ISO VG polyalkylene glycol, and an 80 ISO VG polyalkylene glycol, respectively (Cavestri 1993).

Cavestri (1993) provides density data as a function of temperature and pressure for R-134A/polyol ester oils as shown in Figures 4 to 7. Additionally, Cavestri and Schafer (2000) provide comparable

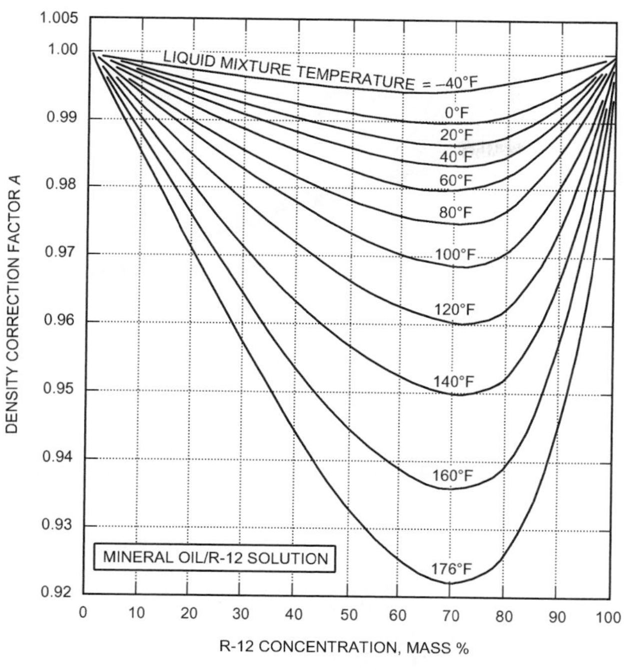

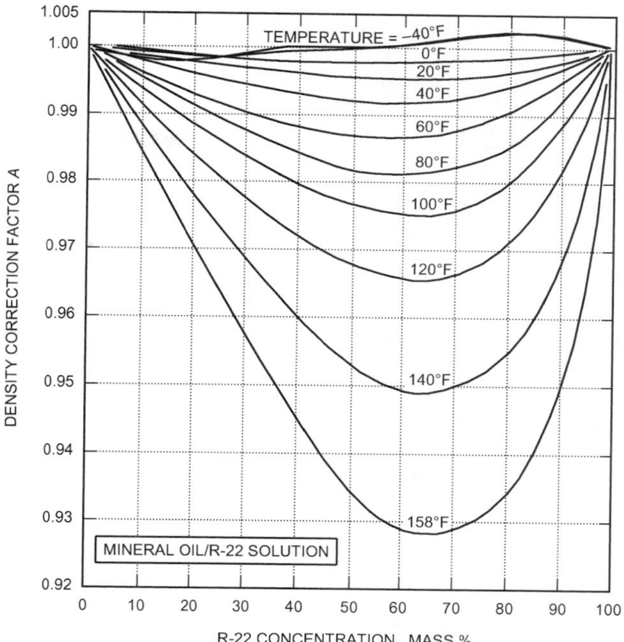

Fig. 3 Density Correction Factors
(Loffler 1959)

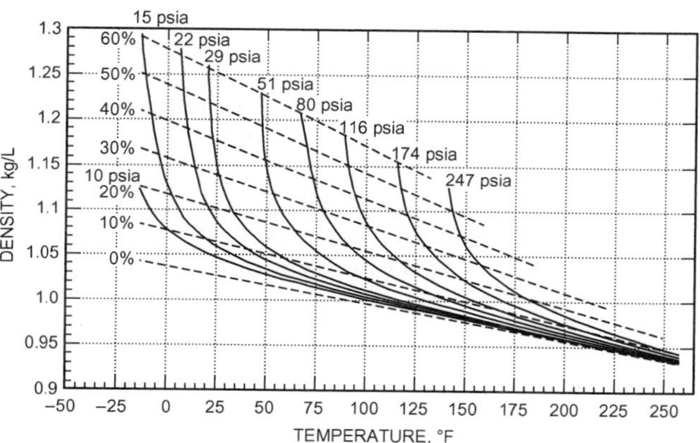

Fig. 4 Density as Function of Temperature and Pressure for Mixture of R-134a and 32 ISO VG Branched-Acid Polyol Ester Lubricant

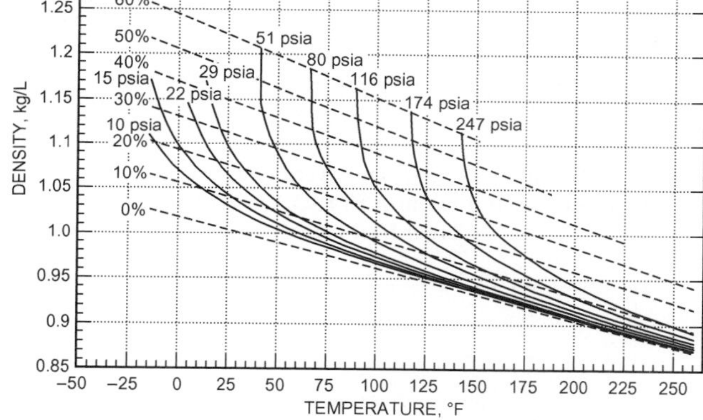

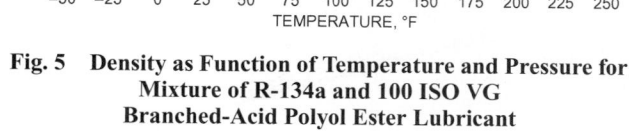

Fig. 5 Density as Function of Temperature and Pressure for Mixture of R-134a and 100 ISO VG Branched-Acid Polyol Ester Lubricant

density data for R-410A/polyol ester oils, as shown in Figures 8 to 11, and Cavestri (1993) provide comparable density data for R-507A/polyol ester and polyether lubricants in Figure 12 to 14.

Thermodynamics and Transport Phenomena

Dissolving lubricant in liquid refrigerant affects the working fluid's thermodynamic properties. Vapor pressures of refrigerant/lubricant solutions at a given temperature are always less than the vapor pressure of pure refrigerant at that temperature. Therefore, dissolved lubricant in an evaporator leads to lower suction pressures and higher evaporator temperatures than those expected from pure refrigerant tables. Bambach (1955) gives an enthalpy diagram for R-12/lubricant solutions over the range of compositions from 0 to 100% lubricant and temperatures from –40 to 240°F. Spauschus (1963) developed general equations for calculating thermodynamic functions of refrigerant/lubricant solutions and applied them to the special case of R-12/mineral oil solutions.

Pressure/Temperature/Solubility Relations

When a refrigerant is in equilibrium with a lubricant, a fixed amount of refrigerant is present in the lubricant at a given temperature and pressure. This is evident if the Gibbs phase rule is applied to basically a two-phase, two-component mixture. The lubricant, although a mixture of several compounds, may be considered one component, and the refrigerant the other. The two phases are the liquid phase and the vapor phase. The phase rule defines this mixture as having two degrees of freedom. Normally, the variables involved are pressure, temperature, and compositions of the liquid and vapor. Because the vapor pressure of the lubricant is negligible compared with that of the refrigerant, the vapor phase is essentially pure refrigerant, and only liquid-phase composition needs to be considered. If the pressure and temperature are defined, the system is invariant (i.e., the liquid phase can have only one composition). This is a different but more precise way of stating that a lubricant/

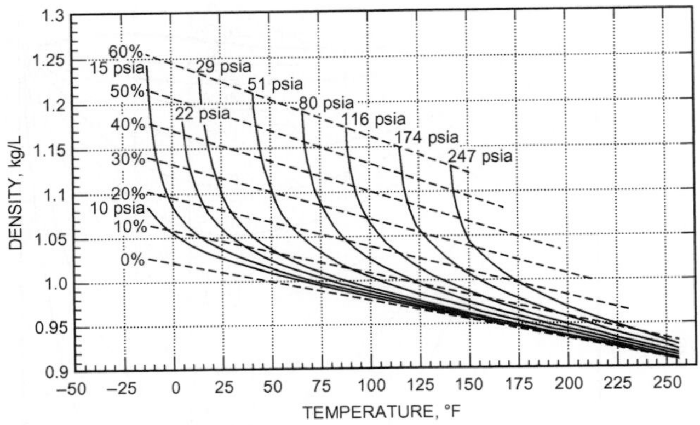

Fig. 6 Density as Function of Temperature and Pressure for Mixture of R-134a and 32 ISO VG Polypropylene Glycol Butyl Ether Lubricant

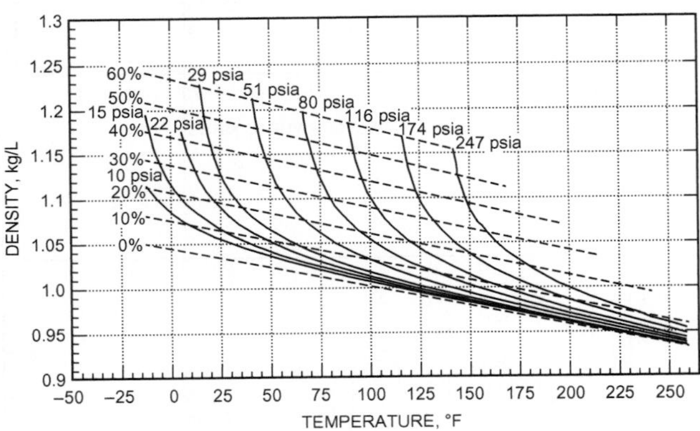

Fig. 7 Density as Function of Temperature and Pressure for Mixture of R-134a and 80 ISO VG Polyoxypropylene Glycol Diol Lubricant

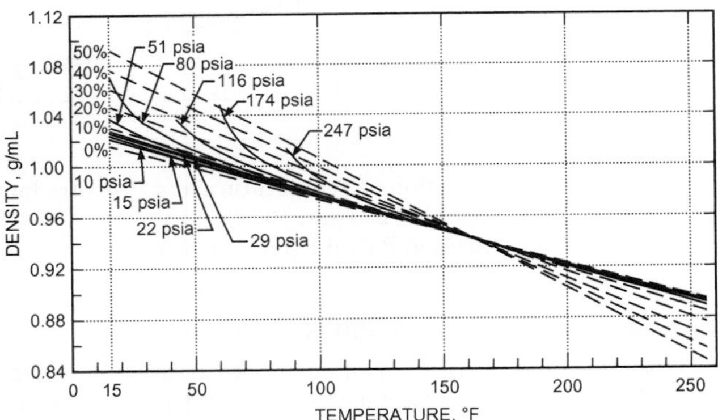

Fig. 8 Density as Function of Temperature and Pressure for Mixture of R-410A and 32 ISO VG Branched-Acid Polyol Ester Lubricant

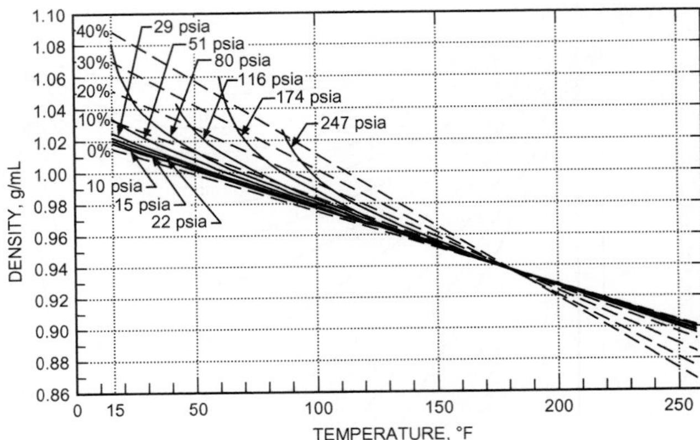

Fig. 9 Density as Function of Temperature and Pressure for Mixture of R-410A and 68 ISO VG Branched-Acid Polyol Ester Lubricant

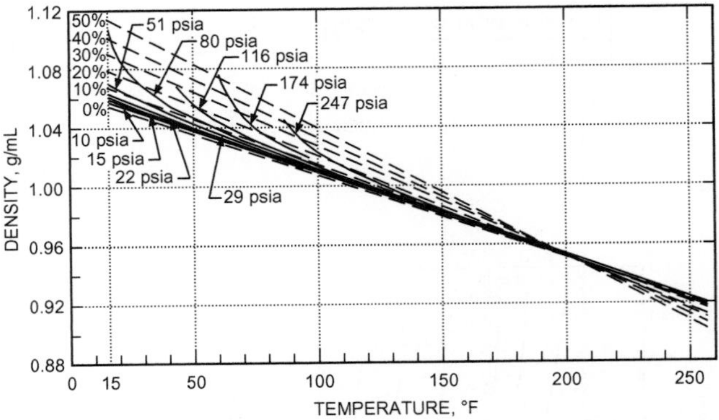

Fig. 10 Density as Function of Temperature and Pressure for Mixture of R-410A and 32 ISO VG Mixed-Acid Polyol Ester Lubricant

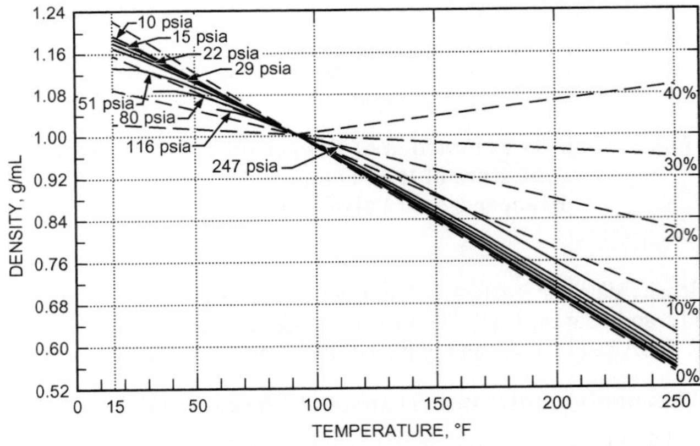

Fig. 11 Density as Function of Temperature and Pressure for Mixture of R-410A and 68 ISO VG Mixed-Acid Polyol Ester Lubricant

refrigerant mixture of a known composition exerts a certain vapor pressure at a certain temperature. If the temperature changes, the vapor pressure also changes.

Pressure/temperature/solubility relations are usually presented in the form shown in Figure 15. On this graph, $P_1°$ and $P_2°$ represent the saturation pressures of the pure refrigerant at temperatures t_1 and t_2, respectively. Point E_1 represents an equilibrium condition where one and only one composition of the liquid, represented by W_1, is possible at pressure P_1. If system temperature increases to t_2, some liquid refrigerant evaporates and the equilibrium point shifts to E_2, corresponding to a new pressure and composition. In either case, the lubricant/ refrigerant solution exerts a vapor pressure less than that of the pure refrigerant at the same temperature.

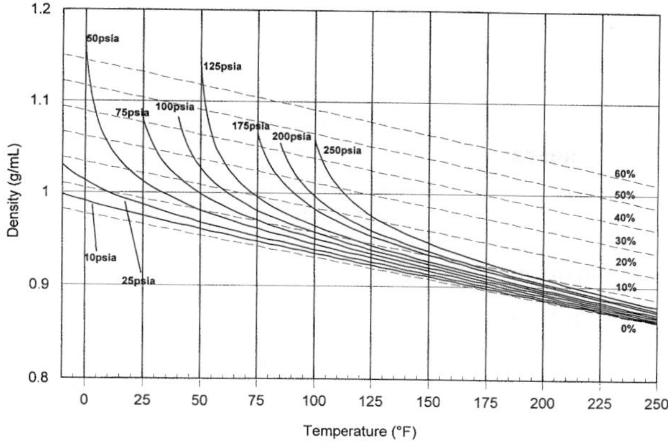

Fig. 12 Density as Function of Temperature and Pressure for Mixture of R-507A and 32 ISO VG Branched-Acid Polyol Ester Lubricant

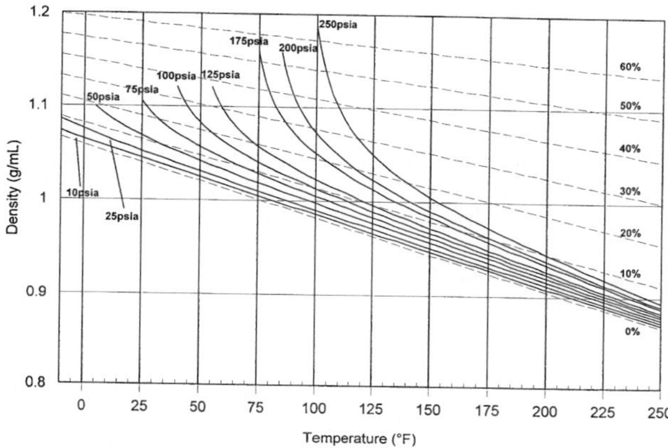

Fig. 13 Density as Function of Temperature and Pressure for Mixture of R-507A and 68 ISO VG Branched-Acid Polyol Ester Lubricant

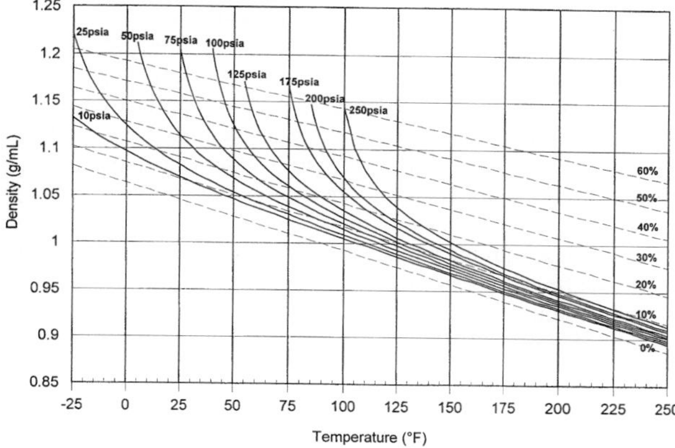

Fig. 14 Density as Function of Temperature and Pressure for Mixture of R-507A and 68 ISO VG Tetrahydrofural Alcohol-Initiated, Methoxy-Terminated, Propylene Oxide Polyether Lubricant

Mutual Solubility

In a compressor, the lubricating fluid is a solution of refrigerant dissolved in lubricant. In other parts of the refrigerant system, the solution is a lubricant in liquid refrigerant. In both instances, either lubricant or refrigerant could exist alone as a liquid if the other were not present; therefore, any distinction between the dissolving and dissolved components merely reflects a point of view. Either liquid can be considered as dissolving the other (mutual solubility).

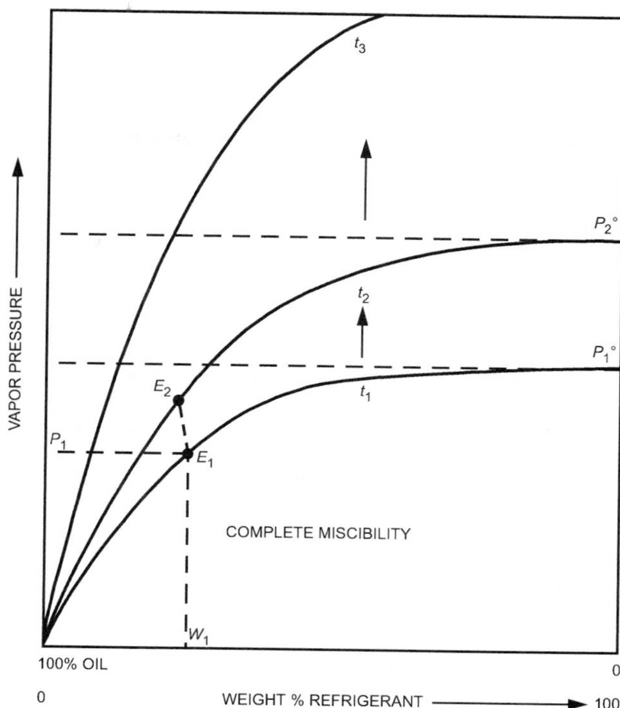

Fig. 15 P-T-S Diagram for Completely Miscible Refrigerant/Lubricant Solutions

Table 6 Mutual Solubility of Refrigerants and Mineral Oil

	Partially Miscible			
Completely Miscible	High Miscibility	Intermediate Miscibility	Low Miscibility	Immiscible
R-11	R-123	R-22	R-13	Ammonia
R-12		R-114	R-14	CO_2
R-113			R-115	R-134a
			R-152a	R-407C
			R-C318	R-410A
			R-502	

Refrigerants are classified as completely miscible, partially miscible, or immiscible, according to their mutual solubility relations with mineral oils. Because several commercially important refrigerants are partially miscible, further designation as having high, intermediate, or low miscibility is shown in Table 6.

Completely miscible refrigerants and lubricants are mutually soluble in all proportions at any temperature encountered in a refrigeration or air-conditioning system. This type of mixture always forms a single liquid phase under equilibrium conditions, no matter how much refrigerant or lubricant is present.

Partially miscible refrigerant/lubricant solutions are mutually soluble to a limited extent. Above the critical solution temperature (CST) or consolute temperature, many refrigerant/lubricant mixtures in this class are completely miscible, and their behavior is identical to that just described. R-134a and some synthetic lubricants exhibit a region of immiscibility at higher temperatures.

Below the critical solution temperature, however, the liquid may separate into two phases. This does not mean that the lubricant and refrigerant are insoluble in each other. Each liquid phase is a solution; one is lubricant-rich and the other refrigerant-rich. Each phase may contain substantial amounts of the leaner component, and these two solutions are themselves immiscible with each other.

The importance of this concept is best illustrated by R-502, which is considered a low-miscibility refrigerant with a high CST as well as a broad immiscibility range. However, even at –4°F, the lubricant-rich phase contains about 20 mass % of dissolved refrigerant (see Figure 18). Other examples of partially miscible systems are R-22, R-114, and R-13 with mineral oils.

The basic properties of the immiscible region can be recognized by applying the phase rule. With three phases (two liquid and one vapor) and two components, there can be only one degree of freedom. Therefore, either temperature or pressure automatically determines the composition of both liquid phases. If system pressure changes, the temperature of the system changes and the two liquid phases assume somewhat different compositions determined by the new equilibrium conditions.

Figure 16 illustrates the behavior of partially miscible mixtures. Point C on the graph represents the critical solution temperature t_3. There are three separate regions below this temperature on the diagram. Reading from left to right, a family of the smooth solid curves represents a region of completely miscible lubricant-rich solutions. These curves are followed by a wide break representing a region of partial miscibility, in which there are two immiscible liquid phases. On the right side, the partially miscible region disappears into a second completely miscible region of refrigerant-rich solutions. A dome-shaped envelope (broken-line curve OCR) encloses the partially miscible region; everywhere outside this dome the refrigerant and lubricant are completely miscible. In a sense, Figure 16 is a variant of Figure 15 in which the partial miscibility dome (OCR) blots out a substantial portion of the continuous solubility curves.

Under the dome (i.e., in the immiscible region), points E_1 and E_2 on the temperature line t_1 represent the two phases coexisting in equilibrium. These two phases differ considerably in composition (W_1 and W_2) but have the same refrigerant pressure P_1. The solution pressure P_1 lies not far below the saturation pressure of pure refrigerant $P_1°$. Commonly, refrigerant/lubricant solutions near the partial miscibility limit show less reduction in refrigerant pressure than is observed at the same lubricant concentration with completely miscible refrigerants.

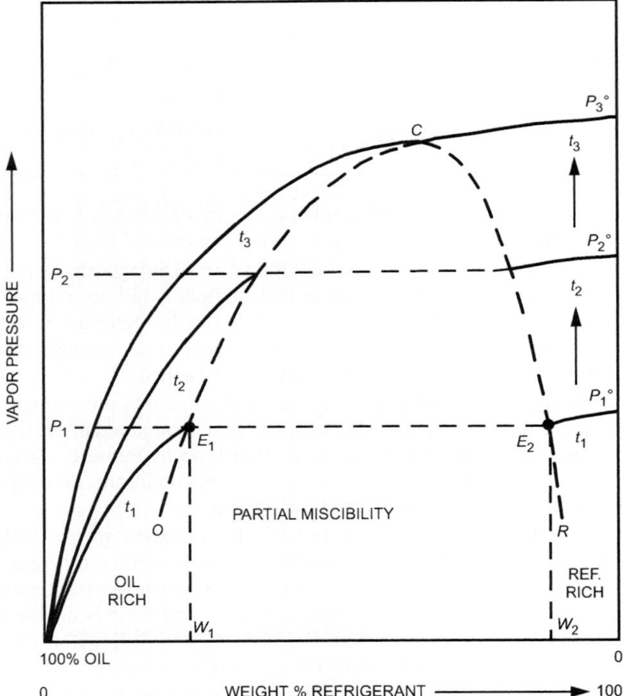

Fig. 16 P-T-S Diagram for Partially Miscible Refrigerant/Oil Solutions

Totally immiscible lubricant/refrigerant solutions are defined in this chapter as only very slightly miscible. In such mixtures, the immiscible range is so broad that mutual solubility effects can be ignored. Critical solution temperatures are seldom found in totally immiscible mixtures. Examples are ammonia and lubricant, and carbon dioxide and mineral oil.

Effects of Partial Miscibility in Refrigerant Systems

Evaporator. The evaporator is the coldest part of the system, and the most likely location for immiscibility or phase separation to occur. If evaporator temperature is below the critical solution temperature, phase separation is likely in some part of the evaporator. Fluid entering the evaporator is mostly liquid refrigerant containing a small fraction of lubricant, whereas liquid leaving the evaporator is mostly lubricant, because the refrigerant is in vapor form. No matter how little lubricant the entering refrigerant carries, the liquid phase, as it progresses through the evaporator, passes through the critical composition (usually 15 to 20% lubricant in the total liquid phase).

Phase separation in the evaporator can sometimes cause problems. In a dry-type evaporator, there is usually enough turbulence for the phases to emulsify. In this case, the heat transfer characteristics of the evaporator may not be significantly affected. In flooded-type evaporators, however, the working fluid may separate into layers, and the lubricant-rich phase may float on top of the boiling liquid. In addition to the heat transfer, partial miscibility may also affect the lubricant's return from the evaporator to the crankcase. Usually the lubricant is moved by high-velocity suction gas transferring momentum to the droplets of lubricant on the return line walls.

If a lubricant-rich layer separates at the evaporator temperatures, this viscous, nonvolatile liquid can migrate and collect in pockets or blind passages not easily reached by the high-velocity suction gas. The lubricant return problem may be magnified and, in some cases, oil logging can occur. System design should take into account all these possibilities, and evaporators should be designed to promote entrainment (see Chapter 2). Oil separators are frequently required in the discharge line to minimize lubricant circulation when refrigerants of poor solvent power are used or in systems involving very low evaporator temperatures (Soling 1971).

Crankcase. With some refrigerant and lubricant pairs, such as R-502 and mineral oil, or even with R-22 in applications such as heat pumps, phase separation sometimes occurs in the crankcase when the system is shut down. When this happens, the refrigerant-rich layer settles to the bottom, often completely immersing the pistons, bearings, and other moving parts. At start-up, the fluid that lubricates these moving parts is mostly refrigerant with little lubricity, and bearings may be severely damaged. Turbulence at start-up may cause liquid refrigerant to enter the cylinders, carrying large amounts of lubricant with it. Precautions in design prevent such problems in partially miscible systems.

Condenser. Partial miscibility is not a problem in the condenser, because the liquid flow lies in the turbulent region and the temperatures are relatively high. Even if phase separation occurs, there is little danger of layer separation, the main obstacle to efficient heat transfer.

Solubility Curves and Miscibility Diagrams

Figure 17 shows mutual solubility relations of partially miscible refrigerant/lubricant mixtures. More than one curve of this type can be plotted on a miscibility diagram. Each single dome then represents the immiscible ranges for one lubricant and one refrigerant. Miscibility curves for R-13, R-13B1, R-502 (Parmelee 1964), R-22, and mixtures of R-12 and R-22 (Walker et al. 1957) are shown in Figure 18. Miscibility curves for R-13, R-22, R-502, and R-503 in

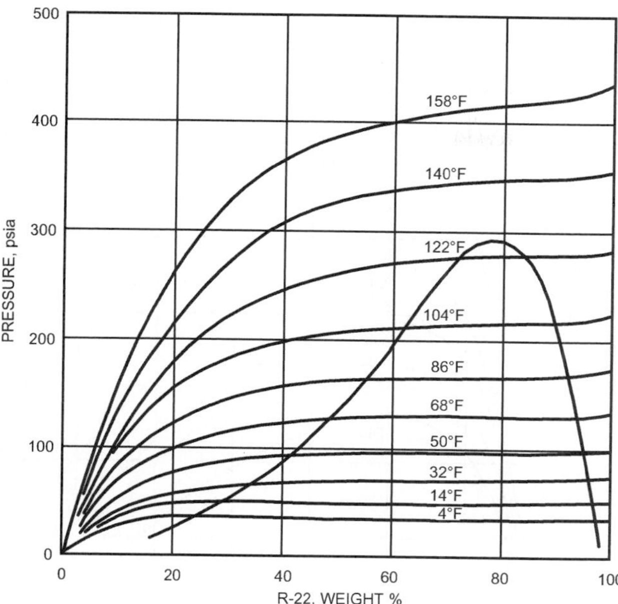

Fig. 17 P-T-S Relations of R-22 with 43 cSt (200 SSU) White Oil (0% C_A, 55% C_N, 45% C_P)
(Spauschus 1964)

an alkylbenzene refrigeration lubricant are shown in Figure 19. Comparison with Figure 18 illustrates the greater solubility of refrigerants in this type of lubricant.

Effect of Lubricant Type on Solubility and Miscibility

On a mass basis, low-viscosity oils absorb more refrigerant than high-viscosity oils do. Also, naphthenic oils absorb more than paraffinic oils. However, when compared on a mole basis, some confusion arises. Paraffinic oils absorb more refrigerant than naphthenic oils (i.e., reversal of the mass basis), and there is little difference between a 15.7 cSt (80 SSU) and a 64.7 cSt (300 SSU) naphthenic lubricant (Albright and Lawyer 1959; Albright and Mandelbaum 1956). The differences on either basis are small (i.e., within 20% of each other). Comparisons of oils by carbon type analyses are not available, but in view of the data on naphthenic and paraffinic types, differences between oils with different carbon type analyses, except perhaps for extreme compositions, are unlikely.

The effect of lubricant type and composition on miscibility is better defined than solubility. When the critical solution temperature (CST) is used as the criterion of miscibility, oils with higher aromatic contents show a lower CST. Higher-viscosity-grade oils show a higher CST than lower-viscosity-grade oils, and paraffinic oils show a higher CST than naphthenic oils (see Figures 20 and 32). When the entire dome of immiscibility is considered, a similar result is noticeable. Oils with a lower CST usually show a narrowed immiscibility range (i.e., the mutual solubility is greater at any given temperature).

Miscibility of R-22 with Lubricants

Parmelee (1964) showed that polybutyl silicate improves miscibility with R-22 (and also R-13) at low temperatures. Alkylbenzenes, by themselves or mixed with mineral oils, also have better miscibility with R-22 than do mineral oils alone (Seeman and Shellard 1963). Polyol esters, which are HFC miscible, are completely miscible with R-22 irrespective of viscosity grade.

For mineral oils, Walker et al. (1962) provide detailed miscibility diagrams of 12 brand-name oils commonly used for refrigeration

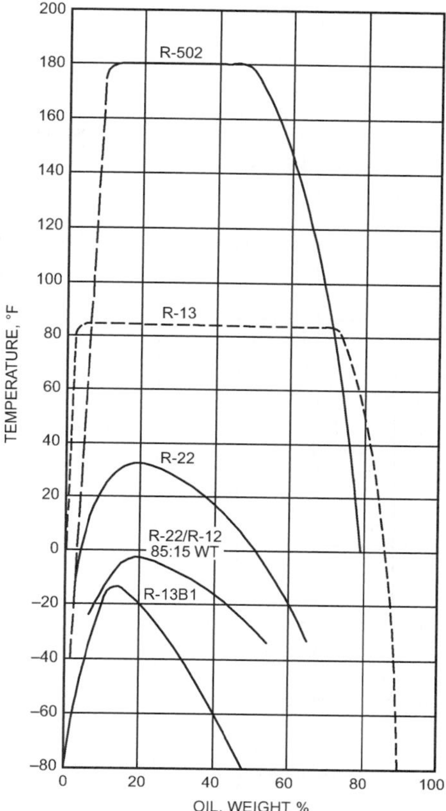

Fig. 18 Critical Solubilities of Refrigerants with 32 cSt (150 SSU) Naphthenic Lubricant (C_A 12%, C_N 44%, C_P 44%)

systems. Walker's data show that, in every case, higher-viscosity lubricant of the same base and type has a higher critical solution temperature.

Loffler (1957) provides complete miscibility diagrams of R-22 and 18 oils. Some properties of the oils used and the critical solution temperatures are summarized in Table 7. Although precise correlations are not evident in the table, certain trends are clear. For the same viscosity grade and base, the effect of aromatic carbon content is seen in oils 2, 3, 7, and 8 and between oils 4 and 6. Similarly, for the same viscosity grades, the effect of paraffinic structure (with essentially the same % C_A) is noticeable between oils 6 and 17 and between oils 8 and 18.

According to Loffler, the most pronounced effect on the critical solution temperature is exerted by the lubricant's aromatic content; the table indicates that the paraffinic structure reduces miscibility compared with naphthenic structures. Sanvordenker (1968) reported miscibility relations of saturated and aromatic fractions of mineral oils as a function of their physical properties. The critical solution temperatures with R-22 increase with increasing viscosities for the saturates, as well as for the aromatics. For equivalent viscosities, aromatic fractions with naphthenic linkages show lower critical solution temperatures than aromatics with only paraffinic linkages.

Pate et al. (1993) developed miscibility data for 10 refrigerants and 14 lubricants. Table 8 lists lower and upper critical solution temperatures for several of the refrigerant/lubricant pairs studied.

Solubilities and Viscosities of Lubricant/Refrigerant Solutions

Although the differences are small on a mass basis, naphthenic oils are better solvents than paraffinic oils. When considering the viscosity of lubricant/refrigerant mixtures, naphthenic oils show

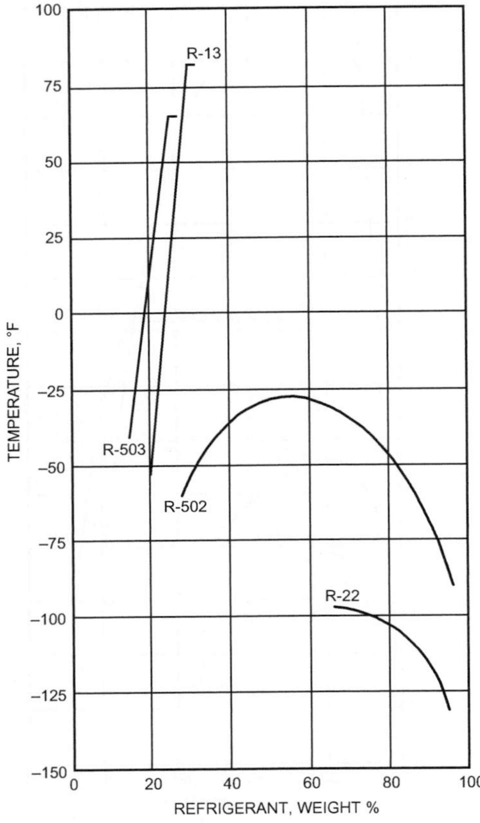

Fig. 19 Critical Solubilities of Refrigerants with 32 cSt (150 SSU) Alkylbenzene Lubricant

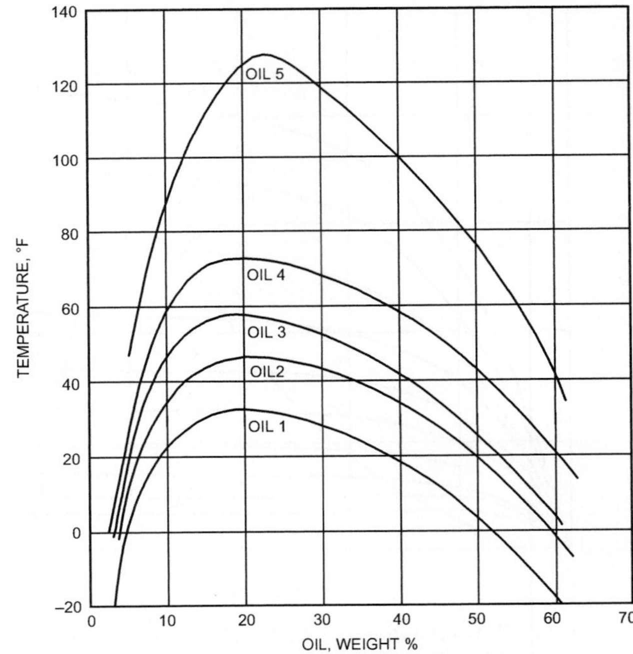

Oil No.	Viscosity at 100°F cSt	SSU	Compositions, % C_A	C_N	C_P	Ref.
1	34.0	159	12	44	44*	1
2	33.5	157	7	46	47*	1
3	63.0	292	12	44	44*	1
4	67.7	314	7	46	47*	1
5	41.3	192	0	55	45	2

References: 1. Walker et al. (1957) 2. Spauschus (1964)
*Estimated composition, not in original reference.

Fig. 20 Effect of Oil Properties on Miscibility with R-22

Table 7 Critical Miscibility Values of R-22 with Different Oils

Oil No.	Oil Base Type[a]	Approximate Viscosity Grade SSU	cSt	Viscosity at 122°F Converted to SSU	to cSt	Carbon-Type Composition %C_A	%C_N	%C_P	Critical Solution Temperature, °F
2	N	75	15	63	11.2	23	34	43	−35
3	N	75	15	60	10.2	2.5	48.5	49	21
1	N	150	32	92	18.5	13	43	44	37
8	N	200	46	118	24.6	0.6	45	55	75
7	N	200	46	127	26.7	2.8	44	54	68
5	N	200+	46+	132	27.9	22	30	47	3
4	N	250	46	135	28.6	26	28	46	−4
6	N	250	46	140	29.7	4	45	51	63
13	N	500	100	253	54.5	1.9	41	56	None[b]
12	N	500	100	282	60.7	4	41	55	None[b]
11	N	700	150	320	69.1	7	40	53	None[b]
10	N	1000	220	434	93.2	21	27	52	61
9	N	1200	220	502	109.0	27	24	50	48
18	P	200+	46+	138	29.3	0.5	33	67	None[b]
17	P	250	46	148	31.6	3.5	34	63	None[b]
16	P	300	68	164	35.2	6.4	30	63	None[b]
15	P	350	68	210	45.2	14.3	25	61	None[b]
14	P	400	100	232	50.0	18.1	22	60	111[c]

[a]P = Paraffinic, N = Naphthenic
[b]Never completely miscible at any temperature
[c]A second (inverted) miscibility dome was observed above 136°F. Above this temperature, the oil/R-22 mixture again separated into two immiscible solutions.

Table 8 Critical Solution Temperatures for Selected Refrigerant/Lubricant Pairs

Refrig-erant	Lubricant	Critical Solution Temperature, °F Lower	Upper
R-22	ISO 32 Naphthenic mineral oil	23	>140
	ISO 32 Modified polyglycol	10	>140
	ISO 68 Naphthenic mineral oil	59	>140
R-123	ISO 68 Naphthenic mineral oil	−38	>140
	ISO 58 Polypropylene glycol butyl monoether	−58	57
R-134a	ISO 58 Polypropylene glycol butyl monoether	−58	133
	ISO 32 Modified polyglycol	50	>194
	ISO 22 Pentaerythritol, mixed-acid ester	−44	>194
	ISO 58 Polypropylene glycol butyl monoether	−51	43
	ISO 100 Polypropylene glycol diol	−50	52
	ISO 100 Pentaerythritol, mixed-acid ester	−31	>90
	ISO 100 Pentaerythritol, branched-acid ester	−51	54

Source: Pate et al. 1993.

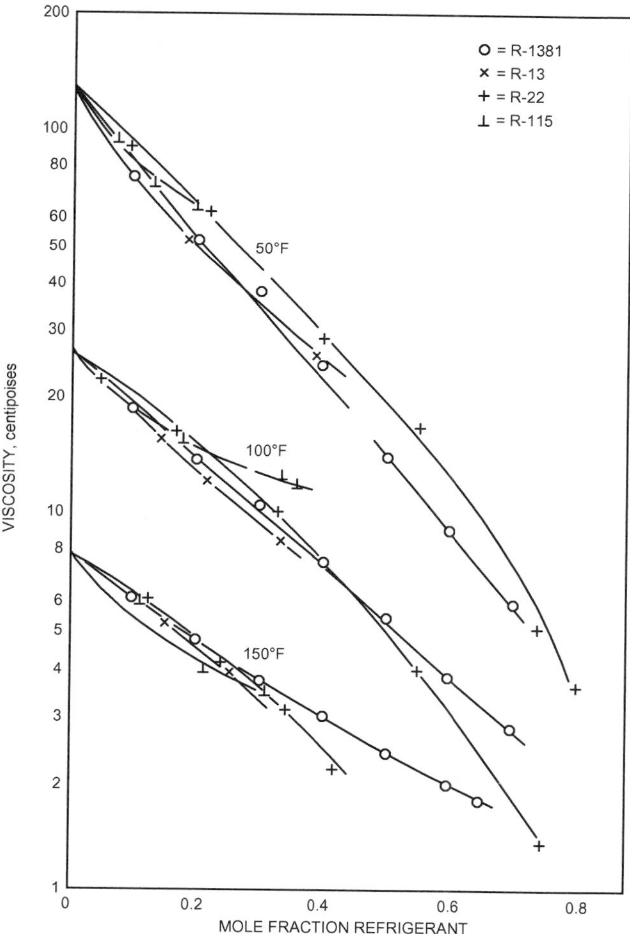

Fig. 21 Viscosity of Mixtures of Various Refrigerants and 32 cSt (150 SSU) Paraffinic Oil
(Albright and Lawyer 1959)

greater viscosity reduction than paraffinic oils for the same mass percent of dissolved refrigerant. When the two effects are compounded, under the same conditions of temperature and pressure, a naphthenic lubricant in equilibrium with a given refrigerant shows a significantly lower viscosity than a paraffinic lubricant.

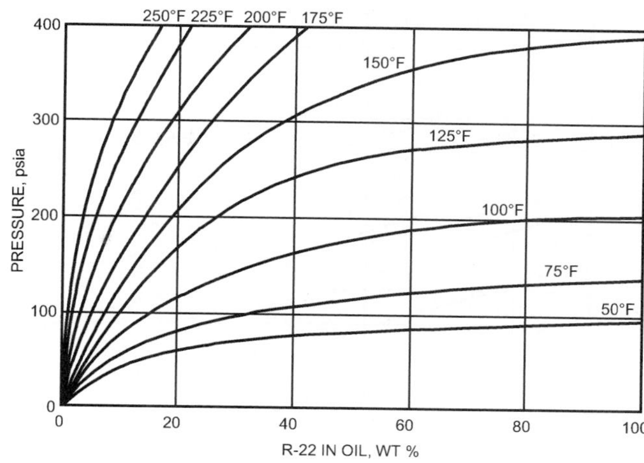

Fig. 22 Solubility of R-22 in 32 cSt (150 SSU) Naphthenic Oil

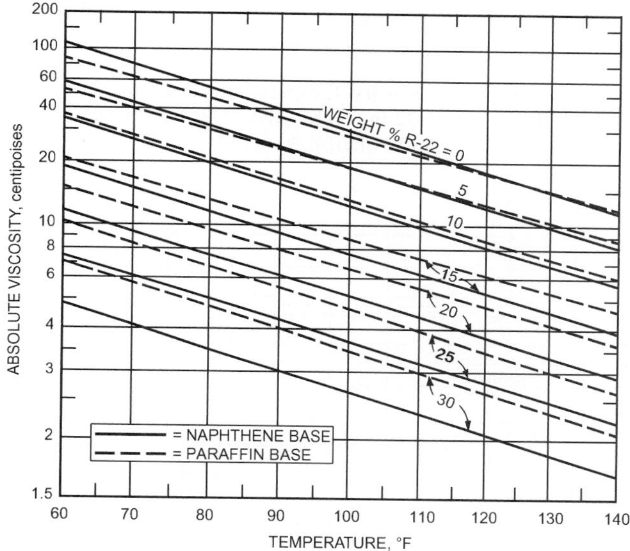

Fig. 23 Viscosity/Temperature Chart for Solutions of R-22 in 32 cSt (150 SSU) Naphthene and Paraffin Base Oils

Refrigerants also differ in their viscosity-reducing effects when the solution concentration is measured in mass percent. However, when the solubility is plotted in terms of mole percent, the reduction in viscosity is approximately the same, at least for Refrigerants 13, 13B1, 22, and 115 (Figure 21).

Spauschus (1964) reports numerical vapor pressure data on a R-22/white oil system; solubility/viscosity graphs on naphthenic and paraffinic oils have been published by Albright and Mandelbaum (1956), Little (1952), and Loffler (1960). Some discrepancies, particularly at high R-22 contents, have been found in data on viscosities that apparently could not be attributed to the properties of the lubricant and remain unexplained. However, general plots reported by these authors are satisfactory for engineering and design purposes.

Spauschus and Speaker (1987) compiled references of solubility and viscosity data. Selected solubility/viscosity data are summarized in Figure 17 and Figures 22 to 34.

Where possible, solubilities have been converted to mass percent to provide consistency among the various charts. Figure 17 and Figures 22 through 26 contain data on R-22 and oils, Figure 27 on R-502, Figures 28 and 29 on R-11, Figures 30 and 31 on R-12,

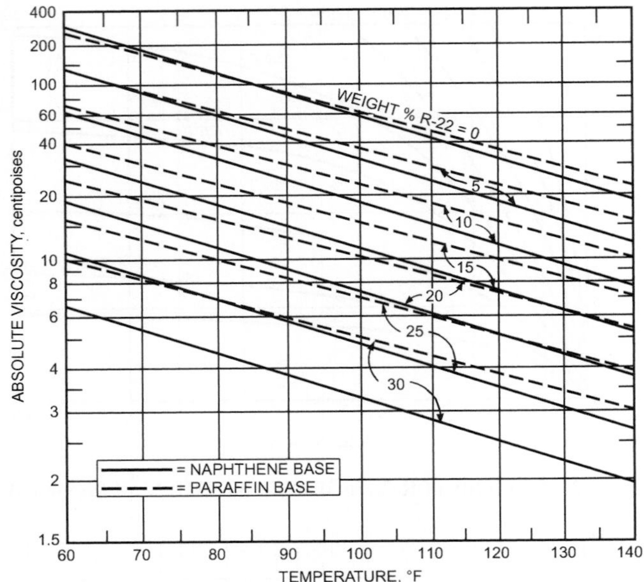

Fig. 24 Viscosity/Temperature Chart for Solutions of R-22 in 65 cSt (300 SSU) Naphthene and Paraffin Base Oils

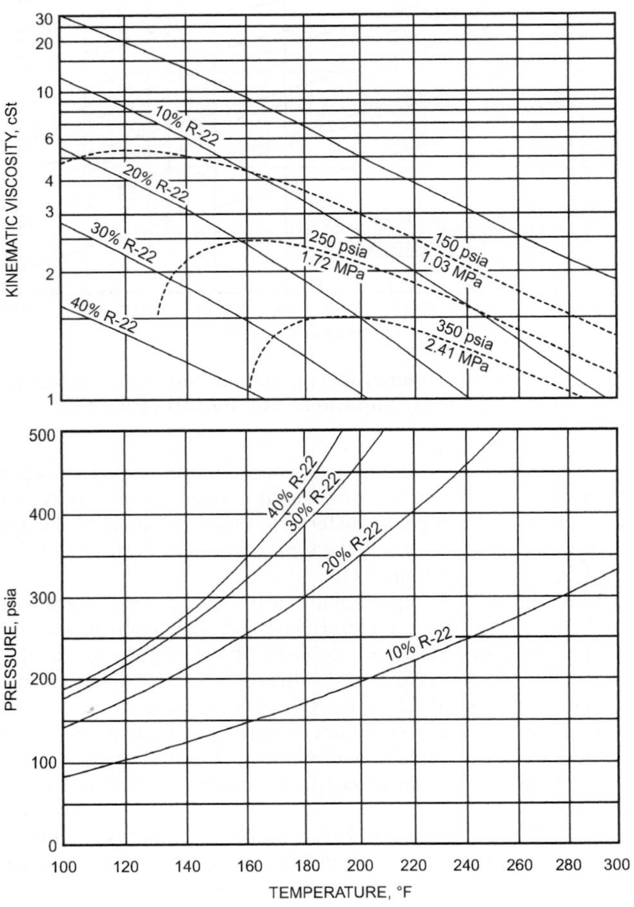

Fig. 25 Viscosity/Temperature Chart for Solutions of R-22 in 32 cSt (150 SSU) Naphthenic Oil
(Van Gaalen et al. 1990, 1991a)

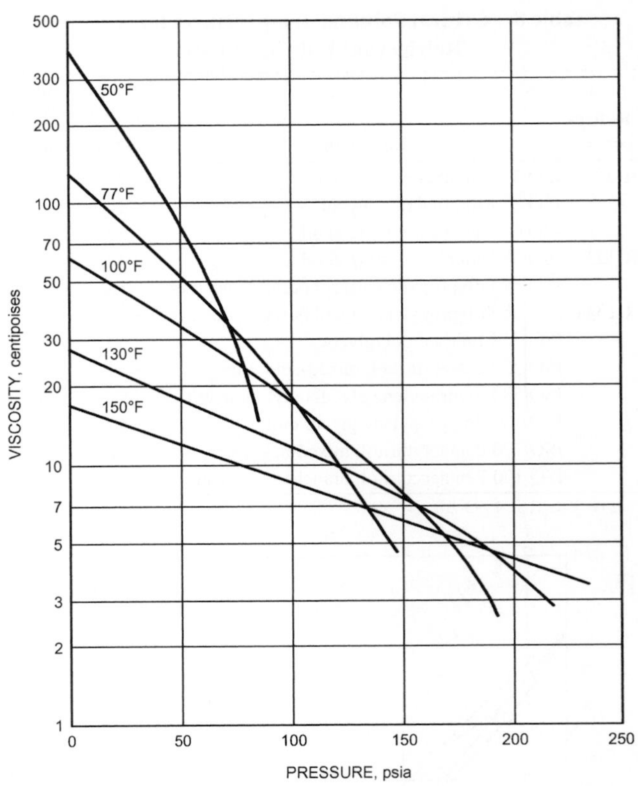

Fig. 26 Viscosity of Mixtures of 65 cSt (300 SSU) Paraffin Base Oil and R-22
(Albright and Mandelbaum 1956)

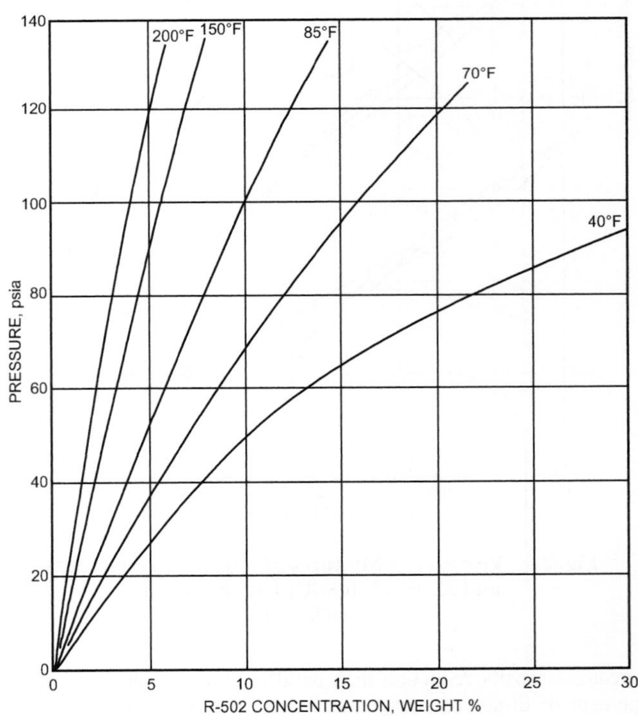

Fig. 27 Solubility of R-502 in 32 cSt (150 SSU) Naphthenic Oil (C_A 12%, C_N 44%, C_P 44%)

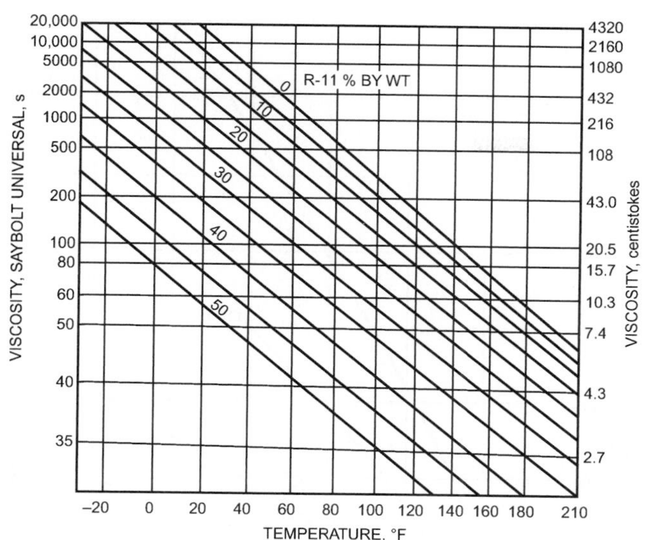

Fig. 28 Viscosity/Temperature Curves for Solutions of R-11 in 65 cSt (300 SSU) Naphthene Base Oil

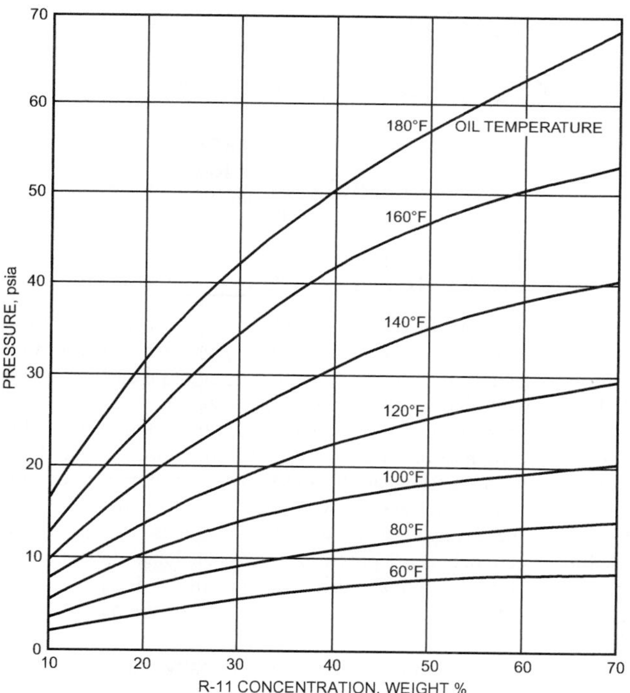

Fig. 29 Solubility of R-11 in 65 cSt (300 SSU) Oil

and Figures 32 and 33 on R-114. Figure 34 contains data on the solubility of various refrigerants in alkylbenzene lubricant. Viscosity/solubility characteristics of mixtures of R-13B1 and lubricating oils were investigated by Albright and Lawyer (1959). Similar studies on R-13 and R-115 are covered by Albright and Mandelbaum (1956).

The solubility of refrigerants in oils, in particular of HFC refrigerants in ester oils, is usually determined experimentally. Wahlstrom and Vamling (2000) developed a predictive scheme based on group contributions for the solubilities of pentaerythritol

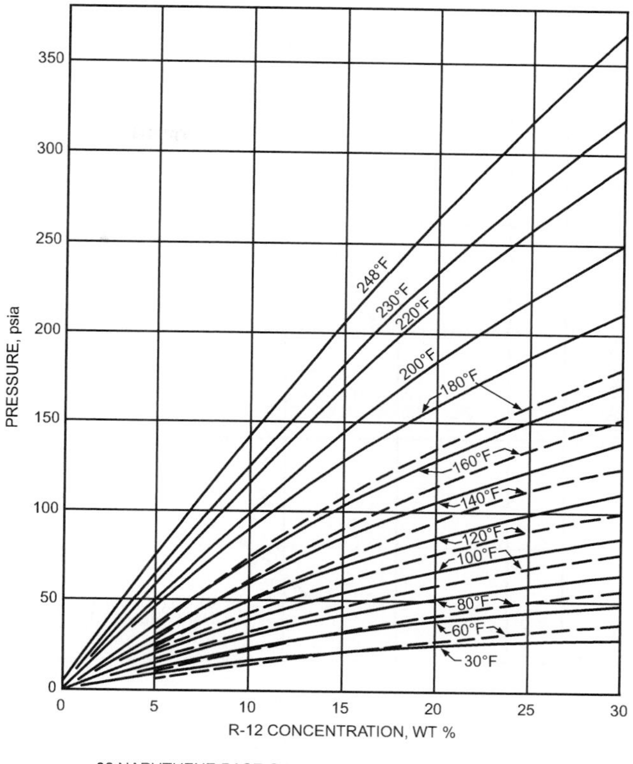

Fig. 30 Solubility of R-12 in Refrigerant Oils

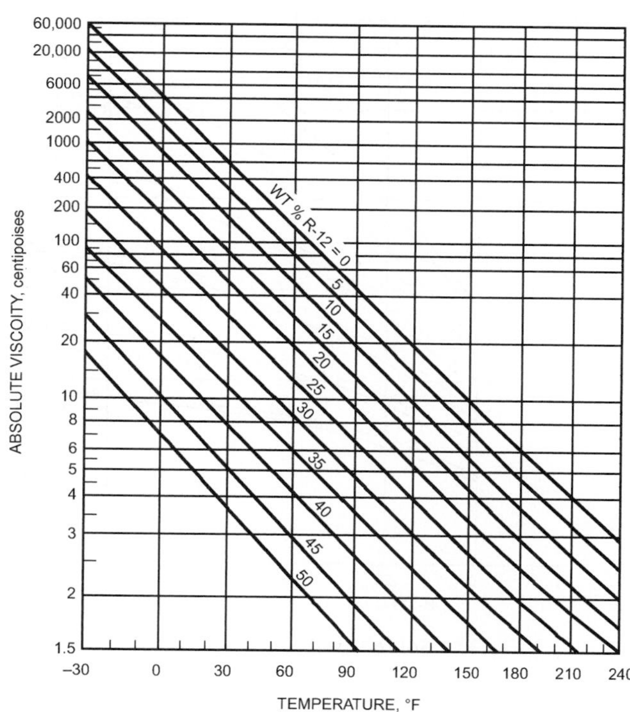

Fig. 31 Viscosity/Temperature Chart for Solutions of R-12 in 32 cSt (150 SSU) Naphthene Base Oil

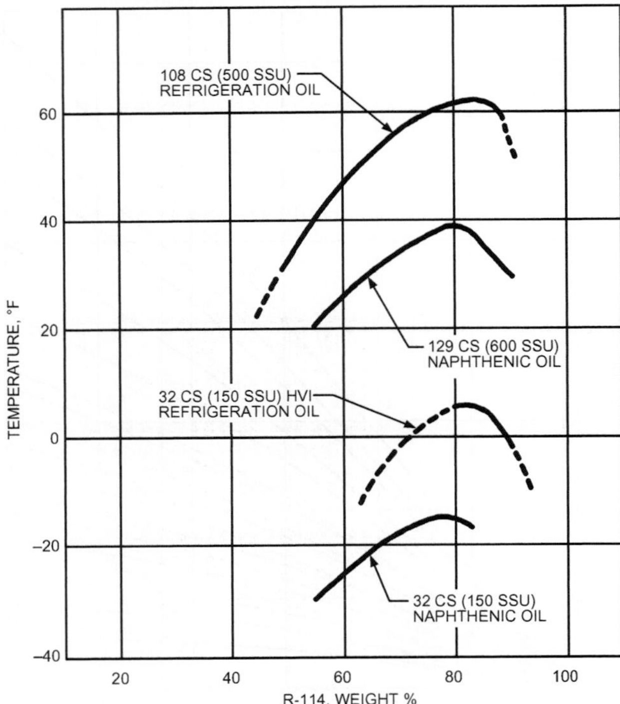

Fig. 32 Critical Solution Temperatures of R-114/Oil Mixtures

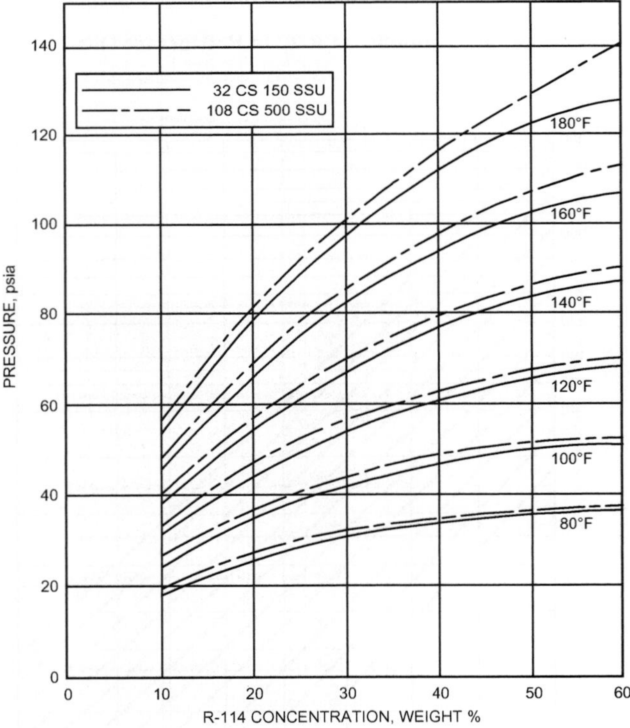

Fig. 33 Solubility of R-114 in HVI Oils

esters and five HFCs (HFC-125, HFC-134a, HFC-143a, HFC-152a, and HFC-32). The scheme uses a modified Flory-Huggins model and a Unifac model. With these schemes, knowing only the structure of the pentaerythritol and the HFC refrigerant, the solubility can be predicted.

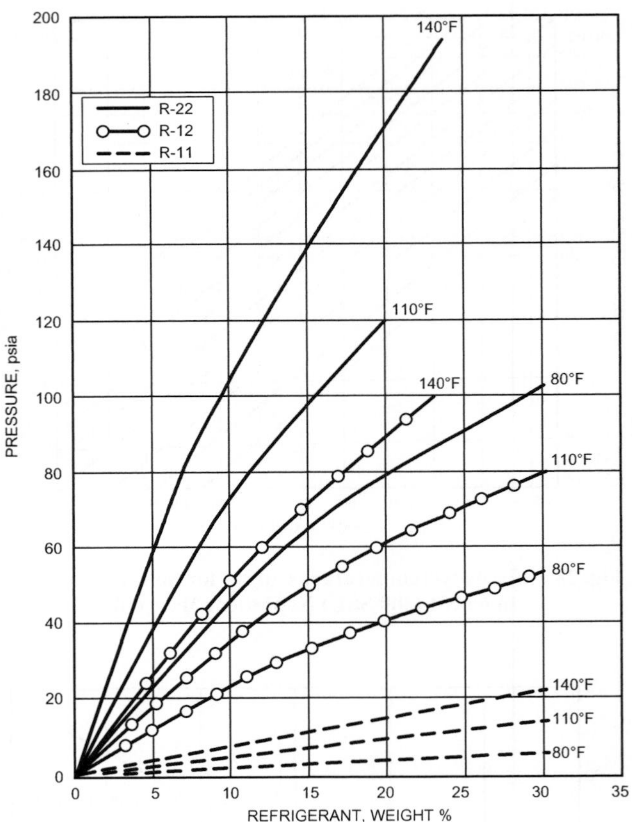

Fig. 34 Solubility of Refrigerants in 32 cSt (150 SSU) Alkylbenzene Oil

LUBRICANT RETURN FROM EVAPORATORS

Regardless of a lubricant's miscibility relations with refrigerants, for a refrigeration system to function properly, the lubricant must return adequately from the evaporator to the crankcase. Parmelee (1964) showed that lubricant viscosity, saturated with refrigerant under low pressure and low temperature, is important in providing good lubricant return. Viscosity of the lubricant-rich liquid that accompanies the suction gas changes with rising temperatures on its way back to the compressor. Two opposing factors then come into play. First, increasing temperature tends to decrease the viscosity of the fluid. Second, because pressure remains unchanged, the increasing temperature also tends to drive off some of the dissolved refrigerant from the solution, thereby increasing its viscosity (Loffler 1960).

Figures 35 to 37 show variation in viscosity with temperature and pressure for three lubricant/refrigerant solutions ranging from –40 to 70°F. In all cases, viscosities of the solutions passed through maximum values as temperature changed at constant pressure, a finding that was also consistent with previous data obtained by Bambach (1955) and Loffler (1960). According to Parmelee, the existence of a viscosity maximum is significant, because the lubricant-rich solution becomes most viscous not in the coldest regions in the evaporator, but at some intermediate point where much of the refrigerant has escaped from the lubricant. This condition is possibly in the suction line. Velocity of the return vapor, which may be high enough to move the lubricant/refrigerant solution in the colder part of the evaporator, may be too low to achieve the same result at the point of maximum viscosity. The designer must consider this factor to minimize any lubricant return problems. Chapters 2 and 3 have further information on velocities in return lines.

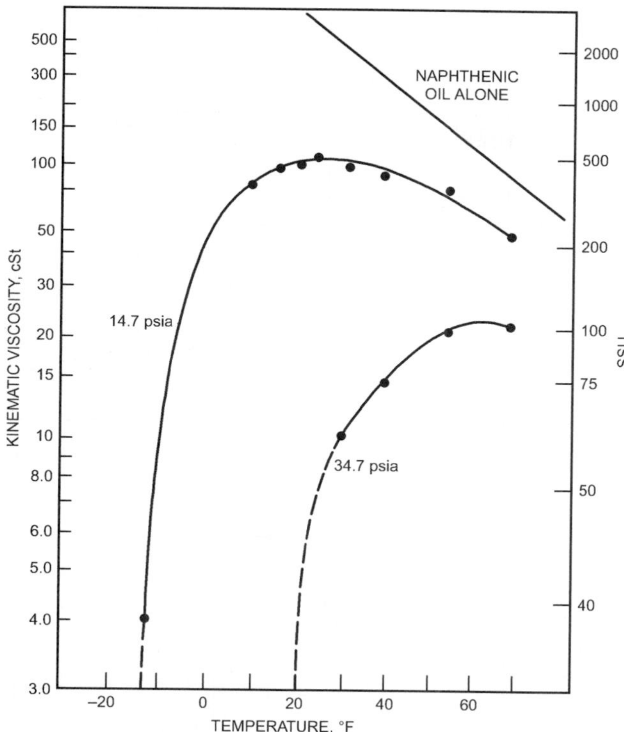

Fig. 35 Viscosity of R-12/Oil Solutions at Low-Side Conditions
(Parmelee 1964)

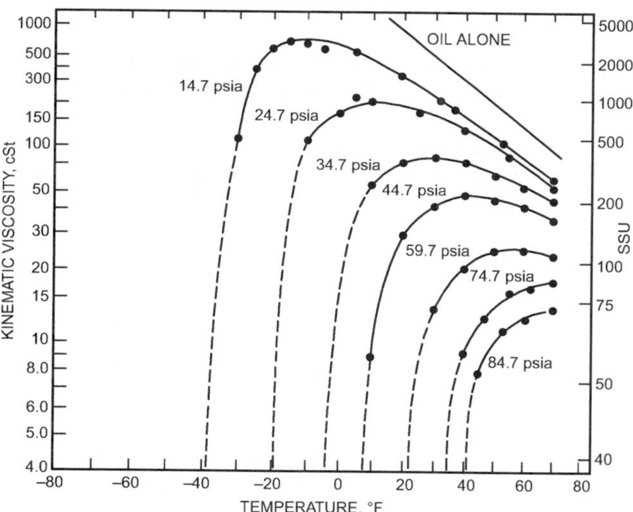

Fig. 36 Viscosity of R-22/Naphthenic Oil Solutions at Low-Side Conditions
(Paarmelee 1964)

Another aspect of viscosity data at the evaporator conditions is shown in Figure 38, which compares a synthetic alkylbenzene lubricant with a naphthenic mineral oil. The two oils are the same viscosity grade, but the highly aromatic alkylbenzene lubricant has a much lower viscosity index in the pure state and shows a higher viscosity at low temperatures. However, at 19.7 psia or approximately −40°F evaporator temperature, the viscosity of the lubricant/R-502 mixture is considerably lower for alkylbenzene than for naphthenic lubricant. In spite of the lower viscosity index, alkylbenzene returns more easily than naphthenic lubricant.

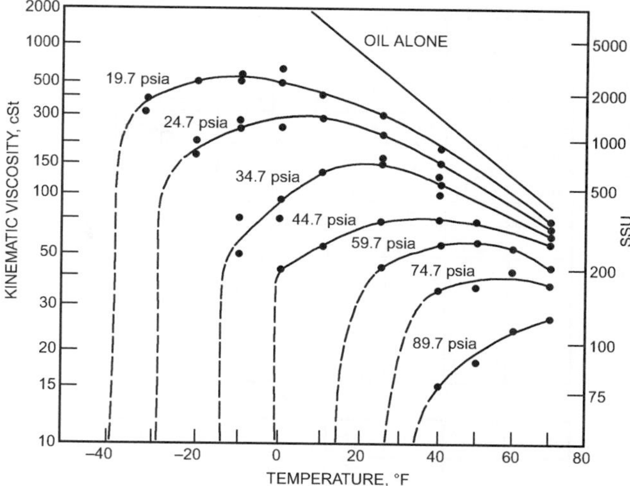

Fig. 37 Viscosity of R-502/Naphthenic Oil Solutions at Low-Side Conditions

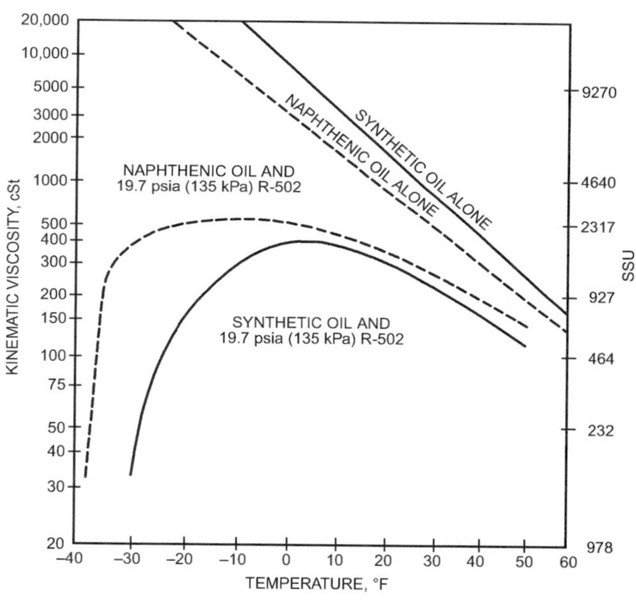

Fig. 38 Viscosities of Solutions of R-502 with 32 cSt (150 SSU) Naphthenic Oil (C_A 12%, C_N 44%, C_P 44%) and 32 cSt (150 SSU) Synthetic Alkylbenzene Oil

Estimated viscosity/temperature/pressure relationships for a naphthenic lubricant with R-502 are shown in Figure 39. Figures 40 and 41 show viscosity/temperature/pressure plots of alkylbenzene and R-22 and R-502, respectively, based on experimental data from Van Gaalen et al. (1991a, 1991b). Figures 42 and 43 show viscosity/temperature/pressure data for mixtures of R-134a and a 32 ISO VG polyalkylene glycol and an 80 ISO VG polyalkylene glycol, respectively. Figures 44 and 45 show similar data for R-134a and a 32 ISO VG polyol ester and a 100 ISO VG polyol ester, respectively (Cavestri 1993). Cavestri and Schafer (2000) provide viscosity data as a function of temperature and pressure for R-410A/polyol ester oils, as shown in Figures 46 to 49. Viscosity and pressure data at constant concentrations are given in Figures 50 to 53. Comparable viscosity/temperature/pressure data for R-507A/polyol ester and polyether lubricants are shown in Figures 54 to 56; and viscosity/pressure data

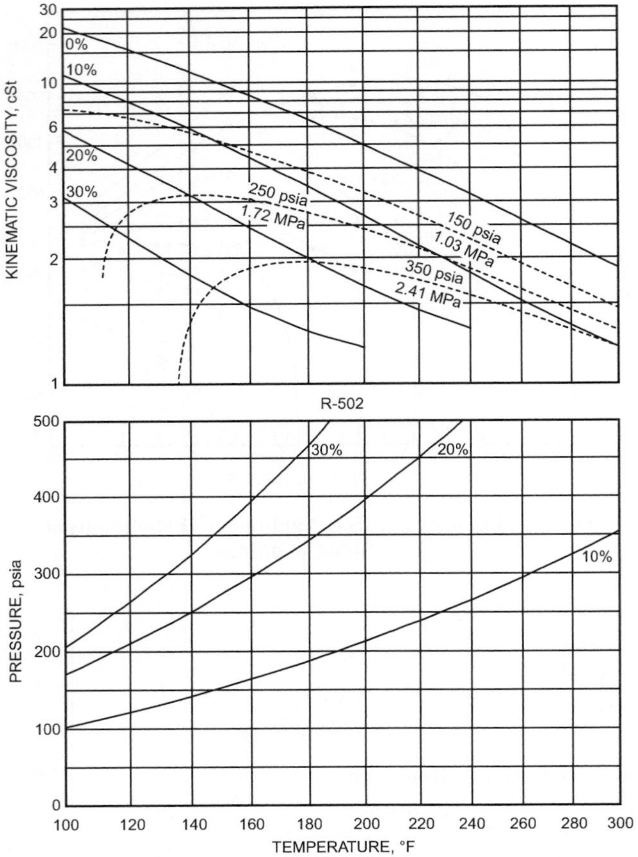

Fig. 39 Viscosity/Temperature/Pressure Chart for Solutions of R-502 in 32 Naphthenic Oil

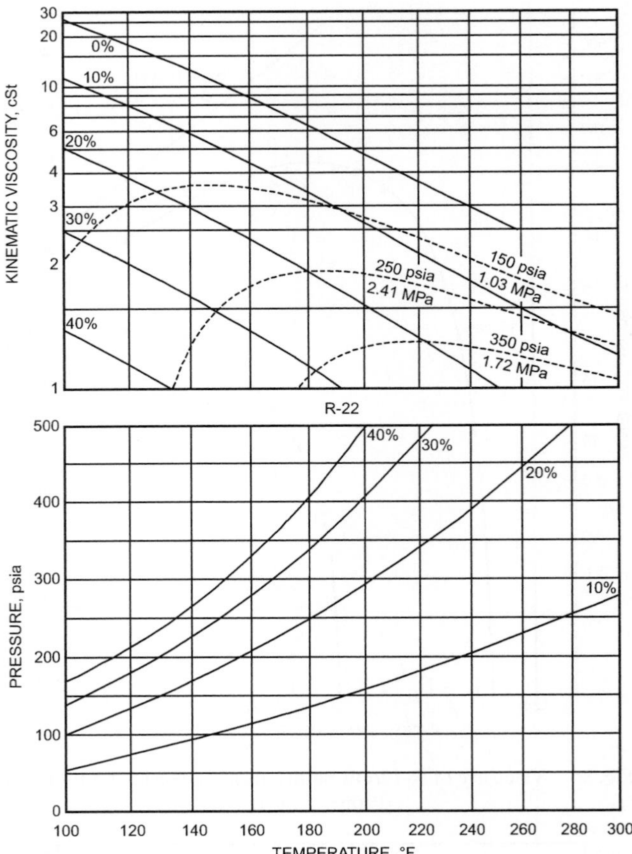

Fig. 40 Viscosity/Temperature/Pressure Chart for Solutions of R-22 in 32 cSt (150 SSU) Alkylbenzene Oil

at constant concentrations are given in Figures 57 to 59, respectively (Cavestri et al. 2005).

Sundaresan and Radermacher (1996) observed oil return in a small air-to-air heat pump. Three refrigerant lubricant pairs (R-22/mineral oil, R-407C/mineral oil, and R-407C/polyol ester) were studied under four conditions (steady-state cooling, steady-state heating, cyclic operation, and a simulated lubricant pumpout situation). The lubricant returned rapidly to the compressor in the R-22/mineral oil and R-407C/polyol ester tests, but oil return was unreliable in the R-407C/mineral oil test.

Kesim et al. (2000) developed general relationships for calculating the required refrigerant speed to carry lubricant oil up vertical sections of refrigerant lines. They assumed the thickness of the oil film to be 2% of the inner pipe diameter. They converted these minimum speeds to the corresponding refrigeration load or capacities for R-134a and copper suction and discharge risers.

WAX SEPARATION (FLOC TESTS)

Wax separation properties are of little importance with synthetic lubricants because they do not contain wax or waxlike molecules. However, petroleum-derived lubricating oils are mixtures of large numbers of chemically distinct hydrocarbon molecules. At low temperatures in the low-pressure side of refrigeration units, some of the larger molecules separate from the bulk of the lubricant, forming waxlike deposits. This wax can clog capillary tubes and cause expansion valves to stick, which is undesirable in refrigeration systems. Bosworth (1952) describes other wax separation problems.

In selecting a lubricant to use with completely miscible refrigerants, the wax-forming tendency of the lubricant can be determined by the floc test. The **floc point** is the highest temperature at which waxlike materials or other solid substances precipitate when a mixture of 10% lubricant and 90% R-12 is cooled under specific conditions. Because different refrigerant and lubricant concentrations are encountered in actual equipment, test results cannot be used directly to predict performance. The lubricant concentration in the expansion devices of most refrigeration and air-conditioning systems is considerably less than 10%, resulting in significantly lower temperatures at which wax separates from lubricant/refrigerant mixture. ASHRAE *Standard* 86 describes a standard method of determining floc characteristics of refrigeration oils in the presence of R-12.

Attempts to develop a test for the floc point of partially miscible refrigerants with R-22 have not been successful. The solutions being cooled often separate into two liquid phases. Once phase separation occurs, the components of the lubricant distribute themselves into lubricant-rich and refrigerant-rich phases in such a way that the highly soluble aromatics concentrate into the refrigerant phase, and the less soluble saturates concentrate into the lubricant phase. Waxy materials stay dissolved in the refrigerant-rich phase only to the extent of their solubility limit. On further cooling, any wax that separates from the refrigerant-rich phase migrates into the lubricant-rich phase. Therefore, a significant floc point cannot be obtained with partially miscible refrigerants once phase separation has occurred. However, lack of flocculation does not mean lack of wax separation. Wax may separate in the lubricant-rich phase, causing it to congeal. Parmelee (1964) reported such phenomena with a paraffinic lubricant and R-22.

Floc point might not be reliable when applied to used oils. Part of the original wax may already have been deposited, and the used

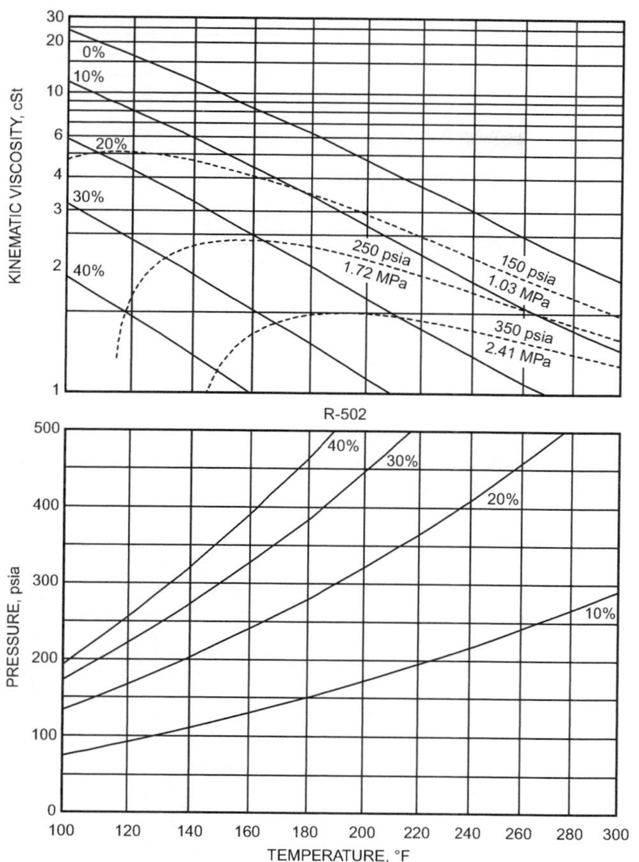

Fig. 41 Viscosity/Temperature/Pressure Chart for Solutions of R-502 in 32 cSt (150 SSU) Alkylbenzene Oil

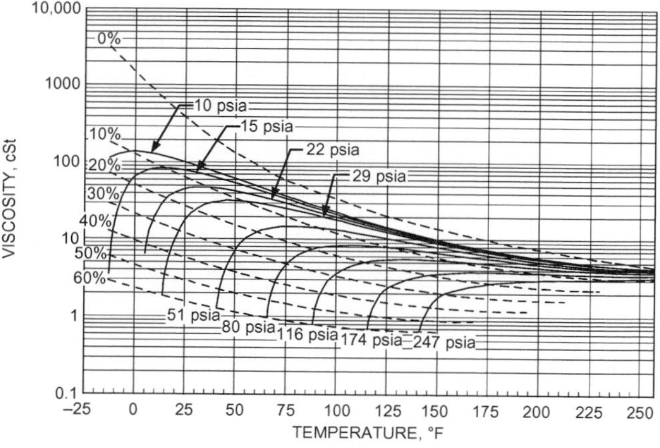

Fig. 42 Viscosity/Temperature/Pressure Plot for 32 ISO VG Polypropylene Glycol Butyl Mono Ether with R-134a

lubricant may contain extraneous material from the operating equipment.

Good design practice suggests selecting oils that do not deposit wax on the low-pressure side of a refrigeration system, regardless of single-phase or two-phase refrigerant/lubricant solutions. Mechanical design affects how susceptible equipment is to wax deposition. Wax deposits at sharp bends, and suspended wax particles build up on the tubing walls by impingement. Careful design avoids bends and materially reduces the tendency to deposit wax.

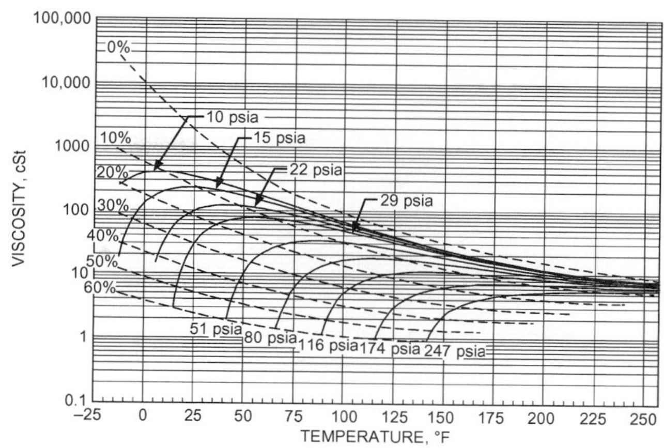

Fig. 43 Viscosity/Temperature/Pressure Plot for 80 ISO VG Polyoxypropylene Diol with R-134a

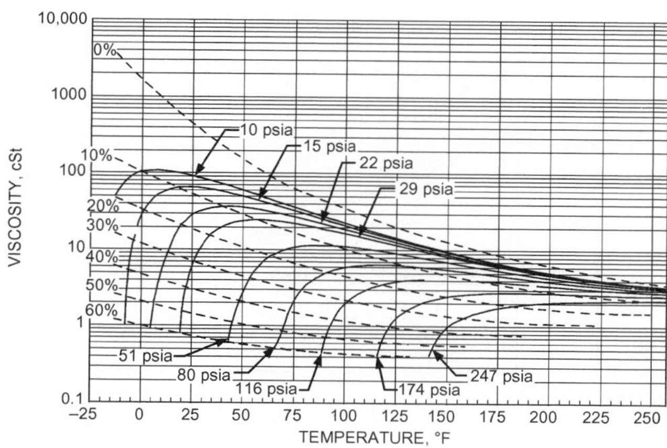

Fig. 44 Viscosity/Temperature/Pressure Plot for 32 ISO VG Branched-Acid Polyol Ester with R-134a
(Cavestri 1993)

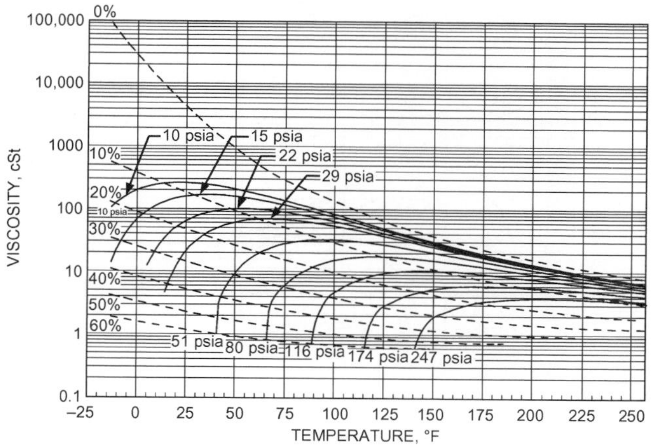

Fig. 45 Viscosity/Temperature/Pressure Plot for 100 ISO VG Branched-Acid Polyol Ester with R-134a
(Cavestri 1993)

SOLUBILITY OF HYDROCARBON GASES

Hydrocarbon gases such as propane (R-290) and ethylene (R-1150) are fully miscible with most compressor lubricating oils

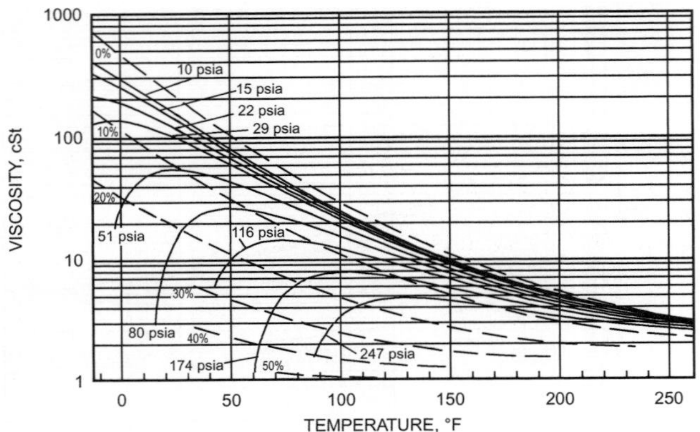

Fig. 46 Viscosity/Temperature/Pressure Plot for Mixture of R-410A and 32 ISO VG Mixed-Acid Polyol Ester Lubricant
(Cavestri and Schafer 2000)

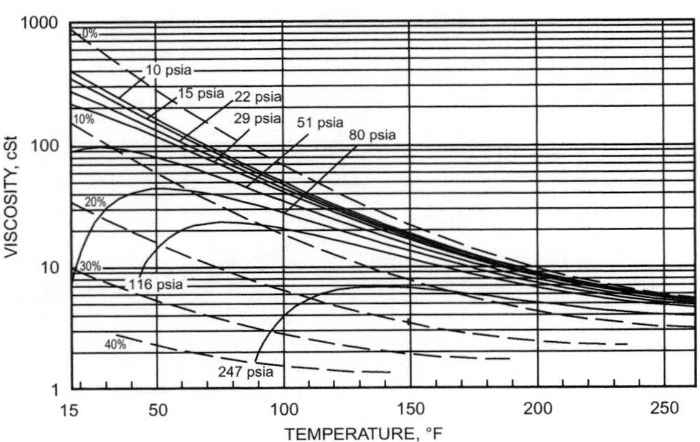

Fig. 47 Viscosity/Temperature/Pressure Plot for Mixture of R-410A and 68 ISO VG Mixed-Acid Polyol Ester Lubricant
(Cavestri and Schafer 2000)

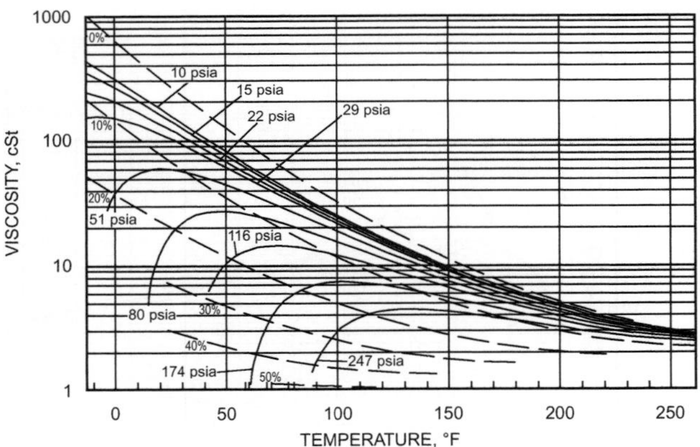

Fig. 48 Viscosity/Temperature/Pressure Plot for Mixture of R-410A and 32 ISO VG Branched-Acid Polyol Ester Lubricant
(Cavestri and Schafer 2000)

and are absorbed by the lubricant, except for some synthetic lubricants. The lower the boiling point or critical temperature, the less soluble the gas, all other values being equal. Gas solubility increases with decreasing temperature and increasing pressure (see Figures 60, 61, and 65). As with other lubricant-miscible refrigerants, absorption of the hydrocarbon gas reduces lubricant viscosity.

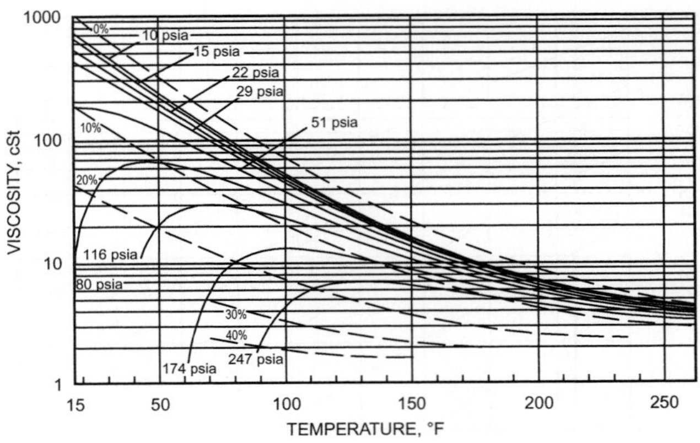

Fig. 49 Viscosity/Temperature/Pressure Plot for Mixture of R-410A and 68 ISO VG Branched-Acid Polyol Ester Lubricant
(Cavestri and Schafer 2000)

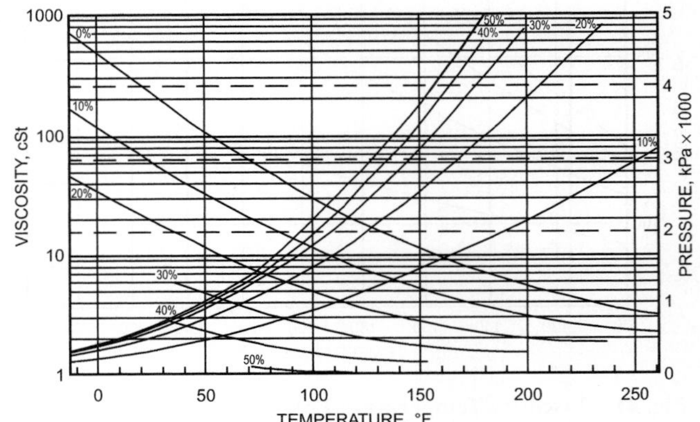

Fig. 50 Viscosity as Function of Temperature and Pressure at Constant Concentrations for Mixture of R-410A and 32 ISO VG Mixed-Acid Polyol Ester Lubricant
(Cavestri and Schafer 2000)

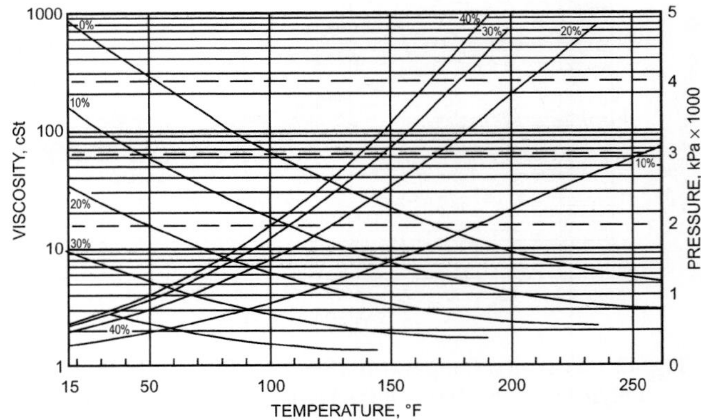

Fig. 51 Viscosity as Function of Temperature and Pressure at Constant Concentrations for Mixture of R-410A and 68 ISO VG Mixed-Acid Polyol Ester Lubricant

LUBRICANTS FOR CARBON DIOXIDE

There is renewed interest in using carbon dioxide as a refrigerant in air-conditioning, heat pump, industrial refrigeration, and some

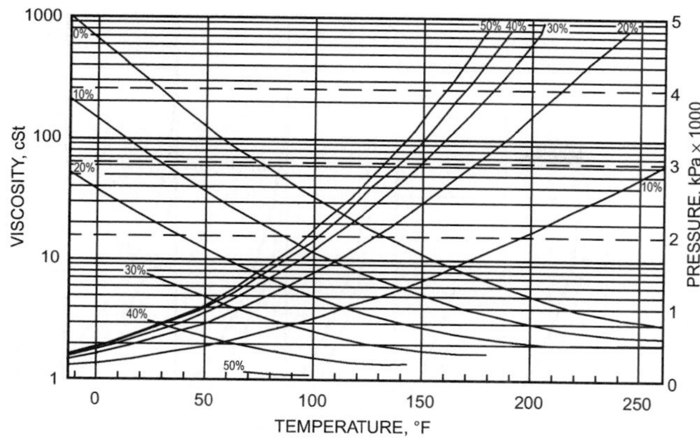

Fig. 52 Viscosity as Function of Temperature and Pressure at Constant Concentrations for Mixture of R-410A and 32 ISO VG Branched-Acid Polyol Ester Lubricant

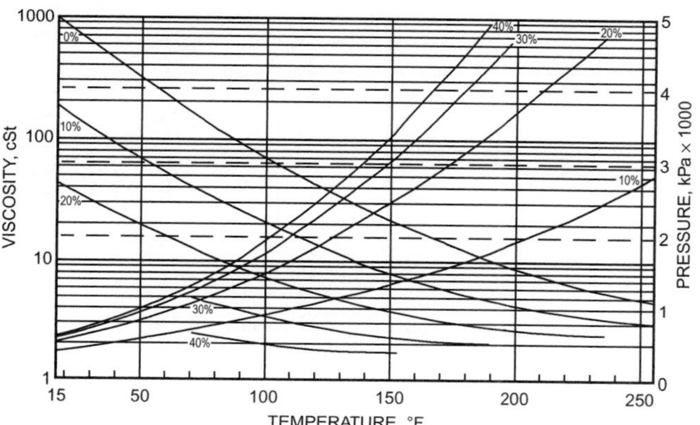

Fig. 53 Viscosity as Function of Temperature and Pressure at Constant Concentrations for Mixture of R-410A and 68 ISO VG Branched-Acid Polyol Ester Lubricant

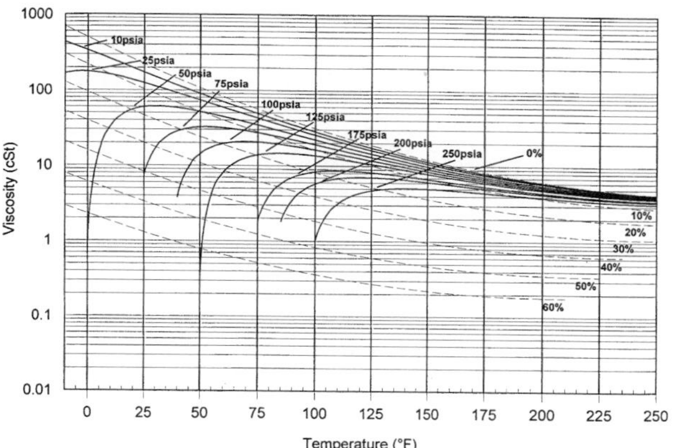

Fig. 54 Viscosity/Temperature/Pressure Plot for Mixture of R-507A and 32 ISO VG Branched-Acid Polyol Ester Lubricant

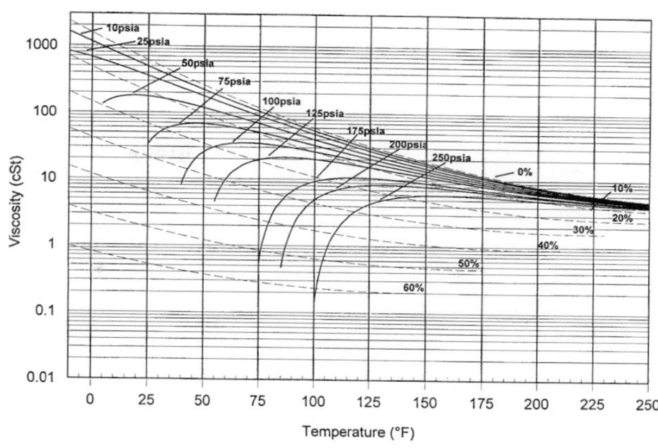

Fig. 55 Viscosity/Temperature/Pressure Plot for Mixture of R-507A and 68 ISO VG Branched-Acid Polyol Ester Lubricant

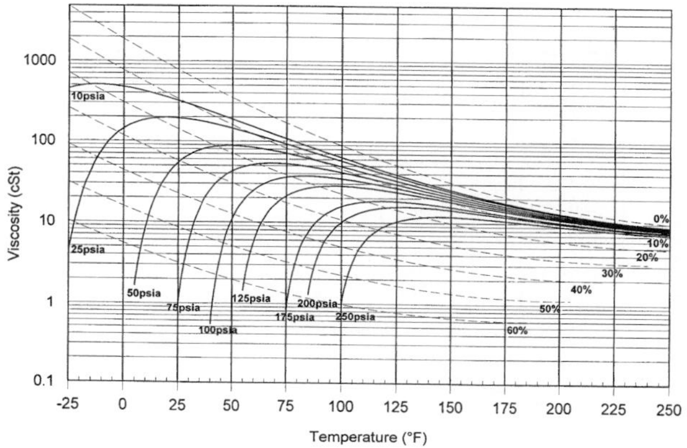

Fig. 56 Viscosity/Temperature/Pressure Plot for Mixture of R-507A and 68 ISO VG Tetrahydrofural Alcohol-Initiated, Methoxy-Terminated, Propylene Oxide Polyether Lubricant

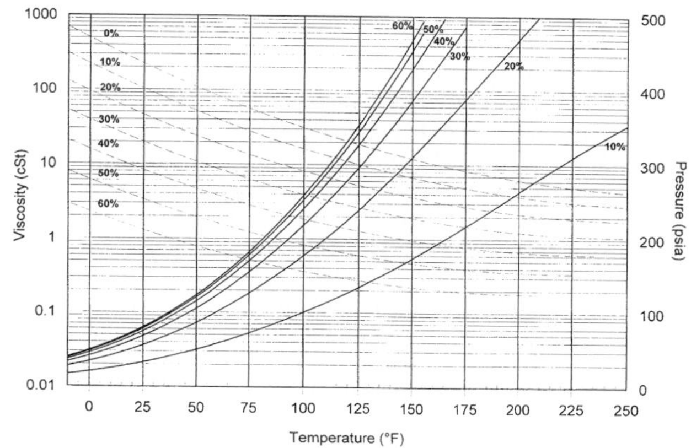

Fig. 57 Viscosity as Function of Temperature and Pressure at Constant Concentrations for Mixture of R-507A and 32 ISO VG Branched-Acid Polyol Ester Lubricant

high-temperature drying applications. Proper lubricant selection depends on the operation of the proposed system. In the 1920s and 1930s, when CO_2 was initially used, lubricant selection was relatively easy because only nonmiscible mineral oils were available. A wide selection of synthetic lubricants is now available, but different types of lubricants are better for different systems. CO_2 systems can be divided into two basic cycles: cascade and transcritical. In cascade systems, carbon dioxide is used as the low-temperature refrigerant and circulates from a machine room out into the plant for cooling. Because its low critical temperature (87.76°F) limits air-sourced heat rejection, CO_2 is also used in a transcritical system: the condenser does not condense carbon dioxide to the liquid phase, but

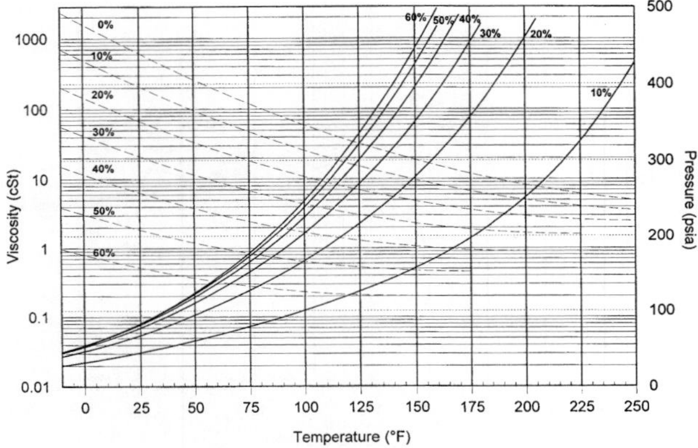

Fig. 58 Viscosity as Function of Temperature and Pressure at Constant Concentrations for Mixture of R-507A and 68 ISO VG Branched-Acid Polyol Ester Lubricant

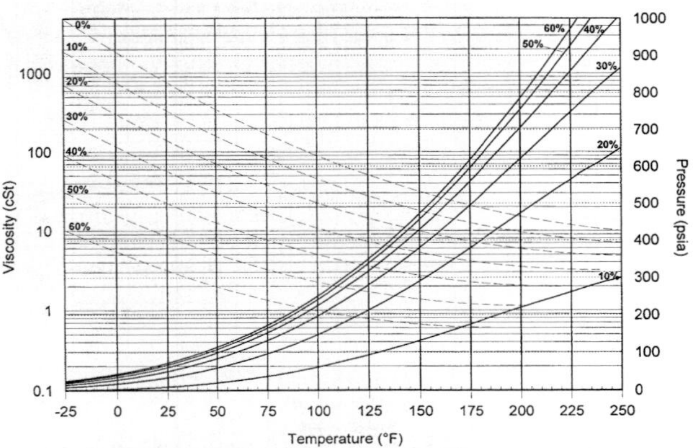

Fig. 59 Viscosity as Function of Temperature and Pressure at Constant Concentrations for Mixture of R-507A and 68 ISO VG Tetrahydrofural Alcohol-Initiated, Methoxy-Terminated, Propylene Oxide Polyether Lubricant

only cools it as a supercritical fluid. Lubricants in CO_2 systems are either completely immiscible or only partially miscible. Figure 62 shows that mineral oil (MO), alkylbenzene (AB), and polyalphaolefins (PAO) are considered completely immiscible, although they do dissolve some carbon dioxide; polyalkylene glycols (PAGs) are partially miscible, and polyol esters (POE) only have a small miscibility gap. Polyvinyl ether (PVE) lubricants behave much like POE lubricants and have only a small immiscibility region.

In low-temperature industrial ammonia/CO_2 cascade systems, PAO oils are generally used with very large oil separators on the compressor discharge. Although POE lubricants are generally preferred in low-temperature applications, it is generally felt that the consequences of a mistake of charging POE into an ammonia system far exceeds the cost of the additional oil separation components. PAO lubricants, such as mineral oil and alkylbenzene, are considered completely immiscible with CO_2, and if lubricant is carried over to the evaporators, it is likely to collect and foul heat exchange surfaces and block refrigerant flow.

For transcritical systems, PAGs are currently the lubricants of choice. PAG lubricants allow for lower-quality, "wet" CO_2 to be used in the system because it does not form the acids experienced in POE systems. Ikeda et al. (2004) found that the electrical resistivity of PAGs can be acceptable in semihermetic and hermetic systems. POE lubricants can also be used in transcritical systems as long as

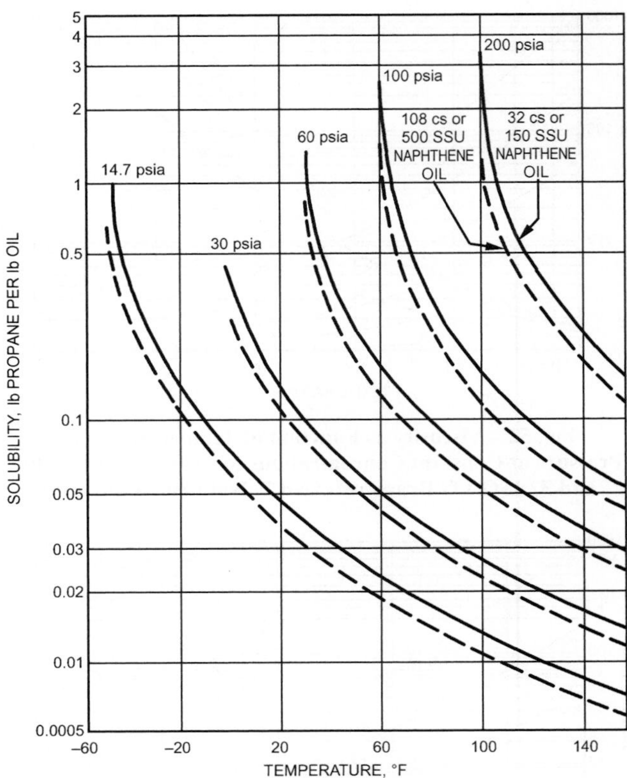

Fig. 60 Solubility of Propane in Oil
(Witco)

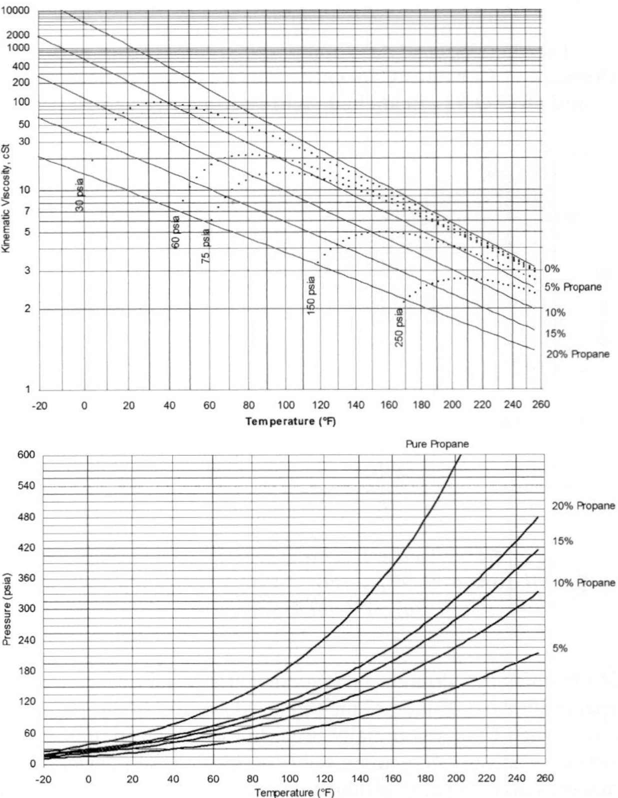

Fig. 61 Viscosity/Temperature/Pressure Chart for Propane and 32 ISO Mineral Oil

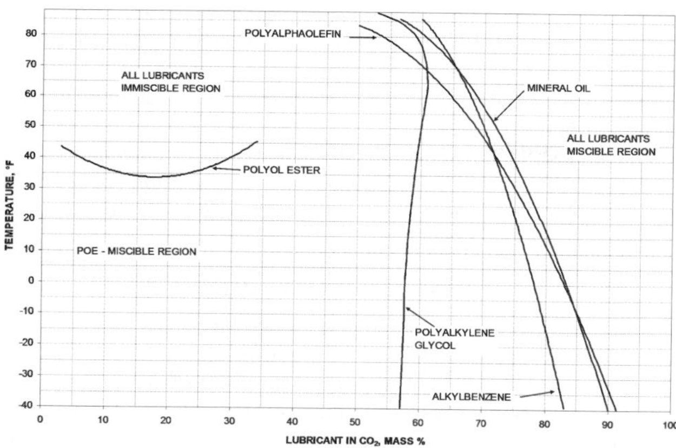

Fig. 62 Miscibility Limits of 220 ISO Lubricants with Carbon Dioxide
(Seeton et al. 2000)

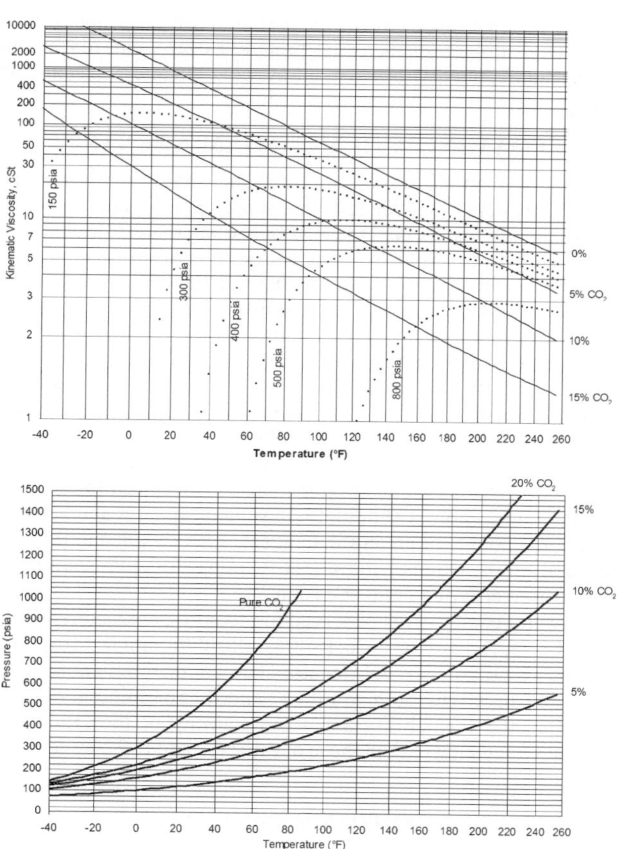

Fig. 63 Viscosity/Temperature/Pressure Chart for CO_2 and 55 ISO Polyol Ester

the significant viscosity reduction of the mixture is taken into account in design, and dry carbon dioxide is used. Figure 63 shows a viscosity chart for a 55 ISO POE with carbon dioxide.

As in the section on Lubricant/Refrigerant Solutions, a compressor crankcase can be used as an example of the significant viscosity reduction in CO_2/lubricant mixtures. If lubricant in the crankcase at start-up is 74°F, the viscosity of pure 54 ISO POE in Figure 64 is about 100 cSt. Under operating conditions, lubricant in the crankcase is typically about 126°F. At this temperature, the viscosity of the pure lubricant is about 35 cSt. In a carbon dioxide system operating with an evaporator pressure of 32°F, crankcase pressure is approximately 510 psi, and the viscosity of the lubricant/refrigerant mixture at start-up is about 2 cSt and climbs to 6 cSt at 126°F as CO_2 boils from solution.

Densities of CO_2/lubricant solutions deviate far from the ideal, and the approximation in the section on Lubricant Properties will not give meaningful results, as shown in Figure 64.

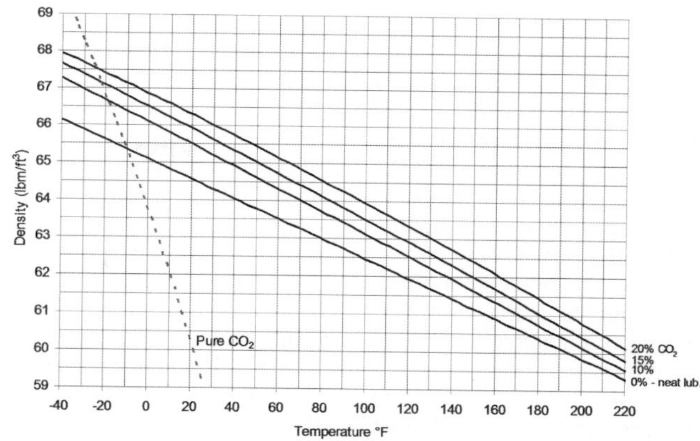

Fig. 64 Density Chart for CO_2 and 55 ISO Polyol Ester

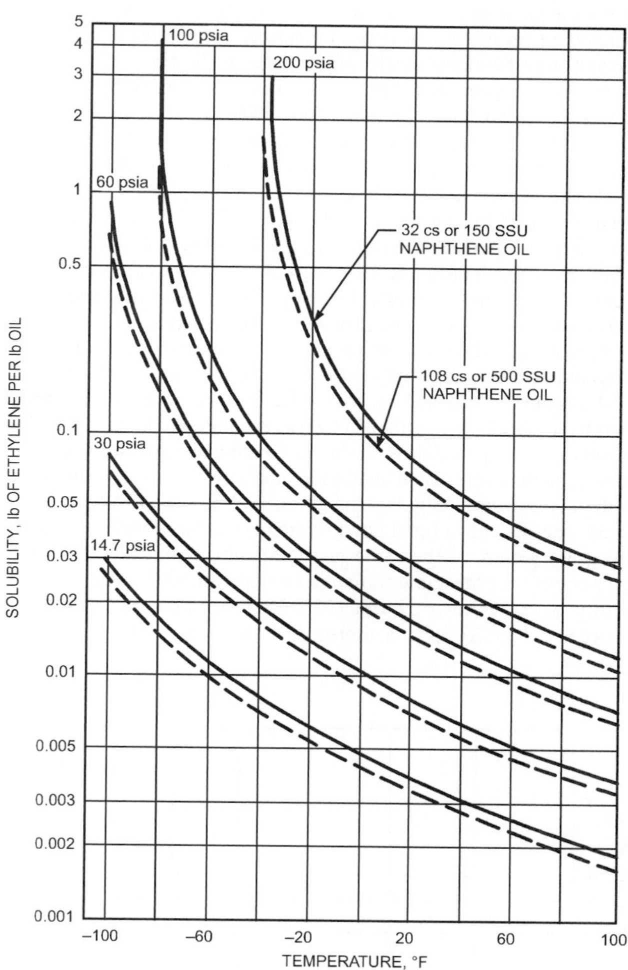

Fig. 65 Solubility of Ethylene in Oil
(Witco)

SOLUBILITY OF WATER IN LUBRICANTS

Refrigerant systems must be dry internally because high moisture content can cause ice to form in the expansion valve or capillary tube, corrosion of bearings, reactions that affect lubricant/refrigerant stability, or other operational problems.

As with other components, the refrigeration lubricant must be as dry as practical. Normal manufacturing and refinery handling practices result in moisture content of about 30 ppm for almost all hydrocarbon-based lubricants. Polyalkylene glycols generally contain several hundred ppm of water. Polyol esters usually contain 50 to 100 ppm moisture. However, this amount may increase between the time of shipment from the refinery and the time of actual use, unless proper preventive measures are taken. Small containers are usually sealed. Tank cars are not normally pressure-sealed or nitrogen-blanketed except when shipping synthetic polyol ester and polyalkylene lubricants, which are quite hygroscopic.

During transit, changes in ambient temperatures cause the lubricant to expand and contract and draw in humid air from outside. Depending on the extent of such cycling, the lubricant's moisture content may be significantly higher than at the time of shipment. Users of large quantities of refrigeration oils frequently dry the lubricant before use. Chapter 38 in the 2004 *ASHRAE Handbook—HVAC Systems and Equipment* discusses methods of drying lubricants. Normally, removing any moisture present also deaerates the lubricant.

Because POE and PAG lubricants are quite hygroscopic, when they are in a refrigeration system they should circulate through a filter-drier designed for liquids. A filter-drier can be installed in the line carrying liquid refrigerant or in a line returning lubricant to the compressor. The material in the filter-drier must be compatible with the lubricant. Also, desiccants can remove some additives in the lubricant.

Spot checks show that water solubility data for transformer oils obtained by Clark (1940) also apply to refrigeration oils (Figure 66).

A simple method, previously used in industry to detect free water in refrigeration oils, is the dielectric breakdown voltage (ASTM *Standard* D877), which is designed to control moisture and other contaminants in electrical insulating oils. The method does not work with polyester and polyalkylene glycol oils, however. According to Clark, the dielectric breakdown voltage decreases with increasing moisture content at the same test temperature and increases with temperature for the same moisture content. At 80°F, when the solubility of water in a 32 cSt (150 SSU) naphthenic lubricant is between 50 to 70 ppm, a dielectric breakdown voltage of about 25 kV indicates that no free water is present in the lubricant. However, the lubricant may contain dissolved water up to the solubility limit. Therefore, a dielectric breakdown voltage of 35 kV is commonly specified to indicate that the moisture content is well below saturation.

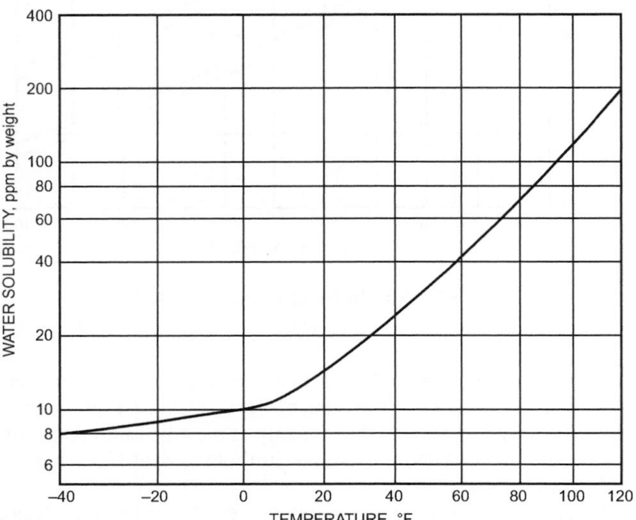

Fig. 66 Solubility of Water in Mineral Oil

The ASTM *Standard* D877 test is not sensitive below about 60% saturation. Current practice is to measure total moisture content directly by procedures such as the Karl Fischer (ASTM *Standard* D1533) method.

SOLUBILITY OF AIR IN LUBRICANTS

Refrigerant systems should not contain excessive amounts of air or other noncondensable gases. Oxygen in air can react with the lubricant to form oxidation products. More importantly, nitrogen in the air (which does not react with lubricant) is a noncondensable gas that can interfere with performance. In some systems, the tolerable volume of noncondensables is very low. Therefore, if the lubricant is added after the system is evacuated, it must not contain an excessive amount of dissolved air or other noncondensable gas. Dissolved air is removed when a vacuum is used to dry the lubricant. However, if the deaerated lubricant is stored under pressure in dry air, it will reabsorb air in proportion to the pressure (Baldwin and Daniel 1953). Dry nitrogen blankets are preferred over using dry air for keeping lubricants dry, because introducing air into a system can cause problems with unintended oxidation.

FOAMING AND ANTIFOAM AGENTS

Excessive foaming of the lubricant is undesirable in refrigeration systems. Brewer (1951) suggests that abnormal refrigerant foaming reduces the lubricant's effectiveness in cooling the motor windings and removing heat from the compressor. Too much foaming also can cause too much lubricant to pass through the pump and enter the low-pressure side. Foaming in a pressure oiling system can result in starved lubrication under some conditions.

However, moderate foaming is beneficial in refrigeration systems, particularly for noise suppression. A foamy layer on top of the lubricant level dampens the noise created by the moving parts of the compressor. Moderate foam also lubricates effectively, yet it is pumpable, which minimizes the risk of vapor lock of the oil pump at start-up. There is no general agreement on what constitutes excessive foaming or how it should be prevented. Some manufacturers add small amounts of an antifoam agent, such as silicone fluid, to refrigerator oils. Others believe that foaming difficulties are more easily corrected by equipment design.

Goswami et al. (1997) observed the foaming characteristics of R-32, R-125, R-134a, R-143a, R-404A, R-407C, and R-410A with two ISO 68 polyol ester lubricants. They compared them to R-12 and R-22 paired with both an ISO 32 and ISO 68 mineral oil, and found that the foamability and foam stability of the HFC/POE pairs were much lower than those of the R-12 and R-22/mineral oil pairs.

OXIDATION RESISTANCE

Refrigeration oils are seldom exposed to oxidizing conditions in hermetic systems. Once a system is sealed against air and moisture, a lubricant's oxidation resistance is not significant unless it reflects the chemical stability. Handling and manufacturing practices include elaborate care to protect lubricants against air, moisture, or any other contaminant. Oxidation resistance by itself is rarely included in refrigeration lubricant specifications.

Nevertheless, oxidation tests are justified, because oxidation reactions are chemically similar to the reactions between oils and refrigerants. An oxygen test, using power factor as the measure, correlates with established sealed-tube tests. However, oxidation resistance tests are not used as primary criteria of chemical reactivity, but rather to support the claims of chemical stability determined by sealed-tube and other tests.

Oxidation resistance may become a prime requirement during manufacture. The small amount of lubricant used during compressor assembly and testing is not always completely removed before the system is dehydrated. If subsequent dehydration is done in a stream of hot, dry air, as is frequently the case, the hot oxidizing conditions can make the residual lubricant gummy, leading to stuck bearings, overheated motors, and other operating difficulties. Oxidation of polyglycol lubricants at 302°F produces degradation products that remove zinc from brass surfaces, leaving behind a layer of soft, porous copper. Compressors can fail prematurely if this layer wears off excessively in loaded sling contacts (Tseregounis 1993). For these purposes, the lubricant should have high oxidation resistance. However, lubricant used under such extreme conditions should be classed as a specialty process lubricant rather than a refrigeration lubricant.

CHEMICAL STABILITY

Refrigeration lubricants must have excellent chemical stability. In the enclosed refrigeration environment, the lubricant must resist chemical attack by the refrigerant in the presence of all the materials encountered, including various metals, motor insulation, and any unavoidable contaminants trapped in the system. The presence of air and water is the most common cause of problems with chemical stability of lubricants in refrigeration and air-conditioning systems. This is true for all lubricants, especially for polyol esters and, to some extent, for polyalkylene glycols. Water may also react with CO_2 refrigerant to form carbonic acid, leading to lubricant instability and copper plating issues (Randles et al. 2003). Lubricant might react with a chlorine-containing refrigerant at elevated temperatures, and the reaction can be catalyzed by metals. Polyol esters react chemically with ammonia (Briley 2004; Randles et al. 2003). Methods to evaluate chemical stability of lubricant/refrigerant mixtures are covered in Chapter 5.

Various phenomena in an operating system (e.g., sludge formation, carbon deposits on valves, gumming, copper plating of bearing surfaces) have been attributed to lubricant decomposition in the presence of refrigerant. In addition to direct reactions of the lubricant and refrigerant, the lubricant may also act as a medium for reactions between the refrigerant and motor insulation, particularly when the refrigerant extracts lighter components of the insulation. Factors affecting the stability of various components such as wire insulation materials in hermetic systems are also covered in Chapter 5. In addition, the presence of residual process chemicals (e.g., brazing fluxes, cleaners, degreasers, cooling lubricants, metalworking fluids, corrosion inhibitors, rust preventives, sealants) may lead to insoluble material restricting or plugging capillary tubes (Cavestri and Schooley 1996; Dekleva et al. 1992) or chemical reactions in POE/HFC systems (Lilje 2000; Rohatgi 2003).

Effect of Lubricant Type

Mineral oils differ in their ability to withstand chemical attack by a given refrigerant. In an extensive laboratory sealed-tube test program, Walker et al. (1960, 1962) showed that darkening, corrosion of metals, deposits, and copper plating occur less in paraffinic oils than in naphthenic oils. Using gas analysis, Doderer and Spauschus (1965) and Spauschus and Doderer (1961) show that a white oil containing only saturates and no aromatics is considerably more stable in the presence of R-12 and R-22 than a medium-refined lubricant is. Steinle (1950) reported the effect of oleoresin (nonhydrocarbons) and sulfur content on the reactivity of the lubricant, using the Philipp test. A decrease in oleoresin content, accompanied by a decrease in sulfur and aromatic content, showed improved chemical stability with R-12, but the oil's lubricating properties became poorer. Schwing's (1968) study on a synthetic polyisobutyl benzene lubricant reports that it is not only chemically stable but also has good lubricating properties.

HFC refrigerants are chemically very stable and show very little tendency to degrade under conditions found in refrigeration and air-conditioning systems. HFC refrigerants are therefore not a factor in degradation of lubricants that might be used with them. Hygroscopic synthetic POE and PAG lubricants are less chemically stable with chlorinated refrigerants than mineral oil because of the interaction of moisture with the refrigerant at high temperatures.

CONVERSION FROM CFC REFRIGERANTS TO OTHER REFRIGERANTS

Choice of Refrigerant Lubricants

The most common conversion from a CFC refrigerant to another refrigerant is retrofitting to use HCFC or HFC refrigerants. Once a refrigerant is identified, in addition to the system and design changes needed to accommodate the new refrigerant chemistry, a suitable lubricant must be selected. Adequate refrigerant miscibility, long-term stability, low hygroscopicity, minimum safe viscosity grades, high lubricity, and low-temperature characteristics (e.g., pour point) are some of the criteria used to identify an acceptable replacement.

In addition to common HCFCs and HFCs such as R-134a, R-404A, R-407C, R-410A, and R-507A, alternative refrigerants such as hydrocarbon gases (e.g., propane), carbon dioxide (CO_2), and ammonia (NH_3) are gaining popularity. Generally, neopentyl polyol esters and polyalkylene glycols are commonly used as miscible lubricants with HFC refrigerants; polyalphaolefins (immiscible) and polyol esters (miscible) may be used with CO_2, depending on system requirements. Ammonia systems may also be designed to handle either miscible (polyalkylene glycols) or immiscible (mineral oils or polyalphaolefins) lubricants.

Because both R-134a/PAG and R-134a/POE systems are used in automotive air conditioning, close attention must be paid to which PAG or POE lubricant package is specified by the system's or vehicle's manufacturer. Mixing lubricants can cause serious air-conditioning system problems. Overcharging with lubricant can make the system oil-logged and less efficient, and possibly result in premature compressor failure (Scaringe 1998).

Flushing

Often, flushing is the only way to remove old lubricant. The flushing medium may be liquid refrigerant, an intermediate fluid, or the lubricant that will be charged with the alternative refrigerant. Liquid CFC refrigerants may be circulated through the entire system, although other refrigerants or commercially available flush solvents may be used. The refrigerant is recovered with equipment modified or specially designed for this use.

The refrigeration equipment must be operated during the flush process if intermediate fluids and lubricants are used for flushing. The system is charged with the flushing material and CFC refrigerant and operated long enough to allow the refrigerant to pass multiple times through the system. The time required varies with operating temperatures and system complexity, but a common recommendation is to flush for at least eight hours. After operation, the lubricant charge is drained from the compressor. This process is repeated until the lubricant in the drained material is reduced to a specified level. Chemical test kits or portable refractometers are available to determine the amount of old lubricant that is mixed with the recovered flush material. The system designer or manufacturer may be able to offer guidance on acceptable levels of residual previous lubricant. Many contractors simply operate the system and closely monitor performance to determine whether additional flushing is necessary. Excessive amounts of residual old oil may increase energy consumption or make the system unable to reach the desired temperature.

Finally, in any refrigerant conversion, as when any major service is done on a system, it is important to check for refrigerant leaks around gaskets, valves, and elastomeric seals or O rings. The change in oil or refrigerant type may affect the gaskets' ability to continue to maintain proper seals. This is especially true if the gaskets or seals are embrittled by age or have been exposed to less than optimum operating conditions, such as excessive heat.

REFERENCES

Albright, L.F. and J.D. Lawyer. 1959. Viscosity-solubility characteristics of mixtures of Refrigerant 13B1 and lubricating oils. *ASHRAE Journal* (April):67.

Albright, L.F. and A.S. Mandelbaum. 1956. Solubility and viscosity characteristics of mixtures of lubricating oils and "Freon-13 or -115." *Refrigerating Engineering* (October):37.

API. 1999. *Technical data book—Petroleum refining*, 6th ed. American Petroleum Institute, Washington, D.C.

ASHRAE. 1994. Methods of testing the floc point of refrigeration grade oils. ANSI/ASHRAE *Standard* 86-1994 (RA01).

ASTM. 2005. Test method for flash and fire points by Cleveland open cup tester. ANSI/ASTM *Standard* D92-05a. American Society for Testing and Materials, West Conshohocken, PA.

ASTM. 2005. Test method for pour point of petroleum products. ANSI/ASTM *Standard* D97-05a. American Society for Testing and Materials, West Conshohocken, PA.

ASTM. 2003. Viscosity-temperature charts for liquid petroleum products. ANSI/ASTM *Standard* D341-03. American Society for Testing and Materials, West Conshohocken, PA.

ASTM. 2004. Test method for kinematic viscosity of transparent and opaque liquids (and the calculation of dynamic viscosity). ANSI/ASTM *Standard* D445-04e2. American Society for Testing and Materials, West Conshohocken, PA.

ASTM. 2004. Test methods for aniline point and mixed aniline point of petroleum products and hydrocarbon solvents. ANSI/ASTM *Standard* D611-04. American Society for Testing and Materials, West Conshohocken, PA.

ASTM. 2002. Test method for dielectric breakdown voltage of insulating liquids using disk electrodes. *Standard* D877-02e1. American Society for Testing and Materials, West Conshohocken, PA.

ASTM. 1999. Test method for density, relative density (specific gravity), or API gravity of crude petroleum and liquid petroleum products by hydrometer method. ANSI/ASTM *Standard* D1298-99(2005). American Society for Testing and Materials, West Conshohocken, PA.

ASTM. 2004. Test method for ASTM color of petroleum products (ASTM color scale). ANSI/ASTM *Standard* D1500-04a. American Society for Testing and Materials, West Conshohocken, PA.

ASTM. 2000. Test method for water in insulating liquids by coulometric Karl Fischer titration. *Standard* D1533-00(2005). American Society for Testing and Materials, West Conshohocken, PA.

ASTM. 1999. Test method for refractive index of viscous materials. ANSI/ASTM *Standard* D1747-99(2004)e1. American Society for Testing and Materials, West Conshohocken, PA.

ASTM. 2003. Test method for characteristic groups in rubber extender and processing oils and other petroleum-derived oils by the clay-gel absorption chromatographic method. ANSI/ASTM *Standard* D2007-03. American Society for Testing and Materials, West Conshohocken, PA.

ASTM. 1991. Test method for ultraviolet absorbance and absorptivity of petroleum products. ANSI/ASTM *Standard* D2008-91(2001). American Society for Testing and Materials, West Conshohocken, PA.

ASTM. Test method for carbon-type composition of insulating oils of petroleum origin. *Standard* D2140-03. American Society for Testing and Materials, West Conshohocken, PA.

ASTM. 2005. Practice for conversion of kinematic viscosity to Saybolt universal viscosity or to Saybolt furol viscosity. ANSI/ASTM *Standard* D2161-05. American Society for Testing and Materials, West Conshohocken, PA.

ASTM. 2004. Practice for calculating viscosity index from kinematic viscosity at 40 and 100°C. ANSI/ASTM *Standard* D2270-04. American Society for Testing and Materials, West Conshohocken, PA.

ASTM. 1997. Classification of industrial fluid lubricants by viscosity system. ANSI/ASTM *Standard* D2422-97(2002). American Society for Testing and Materials, West Conshohocken, PA.

ASTM. 2004. Test method for estimation of molecular weight (relative molecular mass) of petroleum oils from viscosity measurements. ANSI/ASTM *Standard* D2502-04. American Society for Testing and Materials, West Conshohocken, PA.

ASTM. 1992. Test method for relative molecular mass (molecular weight) of hydrocarbons by thermoelectric measurement of vapor pressure. ANSI/ASTM *Standard* D2503-92(2002)e1. American Society for Testing and Materials, West Conshohocken, PA.

ASTM. 2002. Test method for separation of representative aromatics and nonaromatics fractions of high-boiling oils by elution chromatography. ANSI/ASTM *Standard* D2549-02. American Society for Testing and Materials, West Conshohocken, PA.

ASTM. 1995. Test method for measuring wear properties of fluid lubricants (Falex pin and vee block method). ANSI/ASTM *Standard* D2670-95 (2004). American Society for Testing and Materials, West Conshohocken, PA.

ASTM. 1994. Test method for calibration and operation of the falex block-on-ring friction and wear testing machine. ANSI/ASTM *Standard* D2714-94(2003). American Society for Testing and Materials, West Conshohocken, PA.

ASTM. 2002. Test method for measurement of extreme-pressure properties of lubricating fluids (Timken method). ANSI/ASTM *Standard* D2782-02. American Society for Testing and Materials, West Conshohocken, PA.

ASTM. 2003. Test method for measurement of extreme-pressure properties of lubricating fluids (four-ball method). ANSI/ASTM *Standard* D2783-03. American Society for Testing and Materials, West Conshohocken, PA.

ASTM. 2003. Test methods for magnet-wire enamels. ANSI/ASTM *Standard* D3288-03. American Society for Testing and Materials, West Conshohocken, PA.

Baldwin, R.R. and S.G. Daniel. 1953. *Journal of the Institute of Petroleum* 39:105.

Bambach, G. 1955. The behavior of mineral oil-F12 mixtures in refrigerating machines. *Abhandlungen des Deutschen Kältetechnischen Vereins*, No. 9. (Translated by Carl Demrick.) Also see abridgement in *Kältetechnik* 7(7):187.

Beerbower, A. and D.F. Greene. 1961. The behavior of lubricating oils in inert gas atmospheres. *ASLE Transactions* 4(1):87.

Bosworth, C.M. 1952. Predicting the behavior of oils in refrigeration systems. *Refrigerating Engineering* (June):617.

Brewer, A.F. 1951. Good compressor performance demands the right lubricating oil. *Refrigerating Engineering* (October):965.

Briley, G.C. 2004. Selecting lubricant for the ammonia refrigeration system. *ASHRAE Journal* 46(8):66

Cavestri, R.C. 1993. Measurement of the solubility, viscosity and density of synthetic lubricants with HFC-134a. ASHRAE Research Project RP-716, *Final Report*.

Cavestri, R. and D. Schooley. 1996. Compatibility of manufacturing process fluids with R-134a and polyolester lubricants. DOE/CE/23810-55, *Final Report*.

Cavestri, R.C. and W.R. Schafer. 2000. Measurement of solubility, viscosity, and density of R-410A refrigerant/lubricant mixtures. *ASHRAE Transactions* 106(1):277.

Cavestri, R.C, J.R. Thuermer, and D. Seeger-Clevenger. 1993. Measurement of solubility, viscosity, and density of R-507A (R-125/R-143a; 50:50) refrigerant mixtures. ASHRAE Research Project RP-1253, *Final Report*.

Clark, F.M. 1940. Water solution in high-voltage dielectric liquids. *Electrical Engineering Transactions* 59(8):433.

Dekleva, T.W., R. Yost, S. Corr, R.D. Gregson, G. Tompsett, T. Nishizawa, and Y. Obata. 1992. Investigations into the potential effects of process chemicals and materials on the long-term performance of home appliances. *Proceedings of the International CFC and Halon Alternatives Conference*, Washington D.C.

DIN. 1973. Wear test for refrigerant compressors. *Standard* 8978. Deutsches Institut für Normung, Berlin.

Divers, R.T. 1958. Better standards are needed for refrigeration lubricants. *Refrigeration Engineering* (October):40.

Doderer, G.C. and H.O. Spauschus. 1965. Chemical reactions of R-22. *ASHRAE Transactions* 71(I):162.

Fuller, D.D. 1984. *Theory and practice of lubrication for engineers*. John Wiley & Sons, New York.

Goswami, D.Y., D.O. Shah, C.K. Jotshi, S. Bhagwat, M. Leung, and A.S. Gregory. 1997. Foaming characteristics of HFC refrigerants. *ASHRAE Journal* 39(6):39-44.

Gunderson, R.C. and A.W. Hart. 1962. *Synthetic lubricants*. Reinhold, New York.

Huttenlocher, D.F. 1969. A bench scale test procedure for hermetic compressor lubricants. *ASHRAE Journal* (June):85.

Ideka, H., J. Yagi, and K. Yagaguchi. 2004. Evaluation of various compressor lubricants for a carbon dioxide heat pump system. *Proceedings of the 6th IIR-Gustav Lorentzen Conference of Natural Working Fluids*. Glasgow.

Kartzmark, R., J.B. Gilbert, and L.W. Sproule. 1967. Hydrogen processing of lube stocks. *Journal of the Institute of Petroleum* 53:317.

Kesim, S.C., K. Albayrak, and A. Ileri. 2000. Oil entrainment in vertical refrigerant piping. *International Journal of Refrigeration* 23(2000): 626-631.

Komatsuzaki, S. and Y. Homma. 1991. Antiseizure and antiwear properties of lubricating oils under refrigerant gas environments. *Lubrication Engineering* 47(3):193.

Lilje, K.C. 2000. Impact of chemistry on the use of polyol ester lubricants in refrigeration. *ASHRAE Transactions* 106(2):661-667.

Little, J.L. 1952. Viscosity of lubricating oil-Freon-22 mixtures. *Refrigerating Engineering* (Nov.):1191.

Loffler, H.J. 1957. The effect of physical properties of mineral oils on their miscibility with R-22. *Kältetechnik* 9(9):282.

Loffler, H.J. 1959. Density of oil-refrigerant mixtures. *Kältetechnik*. 11(3):70.

Loffler, H.J. 1960. Viscosity of oil-refrigerant mixtures. *Kältetechnik*. 12(3):71.

Mills, I.W., A.E. Hirschler, and S.S. Kurtz, Jr. 1946. Molecular weight-physical property correlations for petroleum fractions. *Industrial and Engineering Chemistry* 38:442-450.

Mills, I.W. and J.J. Melchoire. 1967. Effect of aromatics and selected additives on oxidation stability of transformer oils. *Industrial and Engineering Chemistry* (Product Research and Development) 6:40.

Mosle, H. and W. Wolf. 1963. *Kältetechnik* 15:11.

Murray, S.F., R.L. Johnson, and M.A. Swikert. 1956. Difluoro-dichloromethane as a boundary lubricant for steel and other metals. *Mechanical Engineering* 78(3):233.

Neubauer, E.T. 1958. Compressor crankcase heaters reduce oil foaming. *Refrigerating Engineering* (June):52.

Parmelee, H.M. 1964. Viscosity of refrigerant-oil mixtures at evaporator conditions. *ASHRAE Transactions* 70:173.

Pate, M.B., S.C. Zoz, and L.J. Berkenbosch. 1993. Miscibility of lubricants with refrigerants. *Report* DOE/CE/23810-18. U.S. Department of Energy, Washington, D.C.

Randles, S.J., S. Pasquin, and P.T. Gibb. 2003. A critical assessment of synthetic lubricant technologies for alternative refrigerants. *X European Conference on Technological Innovations in Air Conditioning and Refrigeration Industry with Particular Reference to New Refrigerants, New European Regulations, New Plants—The Cold Chain*, Milan.

Rembold, U. and R.K. Lo. 1966. Determination of wear of rotary compressors using the isotope tracer technique. *ASHRAE Transactions* 72:VI.1.1.

Rohatgi, N.D.T. 2003. Effects of system materials towards the breakdown of POE lubricants and HFC refrigerants. ASHRAE Research Project RP-1158, *Final Report.*

Sanvordenker, K.S. 1968. Separation of refrigeration oil into structural components and their miscibility with R-22. *ASHRAE Transactions* 74(I): III.2.1.

Sanvordenker, K.S. and W.J. Gram. 1974. Laboratory testing under controlled environment using a Falex machine. Compressor Technology Conference, Purdue University.

Sanvordenker, K.S. and M.W. Larime. 1972. A review of synthetic oils for refrigeration use. ASHRAE Symposium, Lubricants, Refrigerants and Systems—Some Interactions.

Scaringe, R.P. 1998. *Environmentally safe refrigerant service techniques for motor vehicle air conditioning technicians—A self study course for EPA 609 motor vehicle A/C certification in the proper use of refrigerants, including recovery, recycling, and reclamation*. Mainstream Engineering Corporation, Rockledge, FL. http://www.epatest.com/manual609.html.

Schwing, R.C. 1968. Polyisobutyl benzenes and refrigeration lubricants. *ASHRAE Transactions* 74(1):III.1.1.

Seeman, W.P. and A.D. Shellard. 1963. Lubrication of Refrigerant 22 machines. IX International Congress of Refrigeration, Paper III-7, Munich.

Seeton, C.J., J. Fahl, and D.R. Henderson. 2000. Solubility, viscosity, boundary lubrication and miscibility of CO_2 and synthetic lubricants. *Proceedings of the 4th IIR-Gustav Lorentzen Conference of Natural Working Fluids*, pp. 417-424, Purdue University, West Lafayette, IN.

Short, G.D. 1990. Synthetic lubricants and their refrigeration applications. *Lubrication Engineering* 46(4):239.

Shubkin, R.L. 1993. Polyalphaolefins. In *Synthetic lubricants and high performance functional fluids*. Marcel Dekker, New York.

Soling, S.P. 1971. Oil recovery from low temperature pump recirculating halocarbon systems. ASHRAE Symposium PH-71-2.

Spauschus, H.O. 1963. Thermodynamic properties of refrigerant-oil solutions. *ASHRAE Journal* (April):47; (October):63.

Spauschus, H.O. 1964. Vapor pressures, volumes and miscibility limits of R-22-oil solutions. *ASHRAE Transactions* 70:306.

Spauschus, H.O. and G.C. Doderer. 1961. Reaction of Refrigerant 12 with petroleum oils. *ASHRAE Journal* (February):65.

Spauschus, H.O. and L.M. Speaker. 1987. A review of viscosity data for oil-refrigerant solutions. *ASHRAE Transactions* 93(2):667.

Steinle, H. 1950. *Kaltemaschinenole*. Springer-Verlag, Berlin, 81.

Sundaresan, S.G. and R. Radermacher. 1996. Oil return characteristics of refrigerant oils in split heat pump system. *ASHRAE Journal* 38(8):57.

Swallow, A., A. Smith, and B. Greig. 1995. Control of refrigerant vapor release from polyol ester/halocarbon working fluids. *ASHRAE Transactions* 101(2):929.

Tseregounis, S.I. 1993. Chemical effects of a polyglycol on brass surfaces as determined by XPS/depth profiling. *Applied Surface Science* 64(2): 147-165.

Van Gaalen, N.A., M.B. Pate, and S.C. Zoz. 1990. The measurement of solubility and viscosity of oil/refrigerant mixtures at high pressures and temperatures: Test facility and initial results for R-22/naphthenic oil mixtures. *ASHRAE Transactions* 96(2):183.

Van Gaalen, N.A., S.C. Zoz, and M.B. Pate. 1991a. The solubility and viscosity of solutions of HCFC-22 in naphthenic oil and in alkylbenzene at high pressures and temperatures. *ASHRAE Transactions* 97(1):100.

Van Gaalen, N.A., S.C. Zoz, and M.B. Pate. 1991b. The solubility and viscosity of solutions of R-502 in naphthenic oil and in an alkylbenzene at high pressures and temperatures. *ASHRAE Transactions* 97(2): 285.

Van Nes, K. and H.A. Weston. 1951. *Aspects of the constitution of mineral oils*. Elsevier, New York.

Walker, W.O., A.A. Sakhanovsky, and S. Rosen. 1957. Behavior of refrigerant oils and Genetron-141. *Refrigerating Engineering* (March):38.

Walker, W.O., S. Rosen, and S.L. Levy. 1960. A study of the factors influencing the stability of the mixtures of Refrigerant 22 and refrigerating oils. *ASHRAE Transactions* 66:445.

Walker, W.O., S. Rosen, and S.L. Levy. 1962. Stability of mixtures of refrigerants and refrigerating oils. *ASHRAE Transactions* 68:360.

Wahlstrom, A. and L. Vamling. 2000. Development of models for prediction of solubility for HFC working fluids in pentaerythritol ester compressor oils. *International Journal of Refrigeration* 23(2000):597-608.

Witco. Sonneborn Division, *Bulletin* 8846.

CHAPTER 8

REFRIGERANT CONTAINMENT, RECOVERY, RECYCLING, AND RECLAMATION

CONTAINMENT of refrigerant is an important consideration during service and maintenance of refrigeration systems. The potential environmental effect of chlorofluorocarbon (CFC) and hydrochlorofluorocarbon (HCFC) refrigerants on ozone depletion, and of these and hydrofluorocarbon (HFC) refrigerants on global warming, make it imperative that refrigerants are confined to closed systems and recovered during service and at the end of life. Containment must be considered in all phases of a system's life, including (1) design and construction of leaktight and easily serviced systems, (2) leak detection and repair, (3) recovery during service, and (4) recovery system disposal. Additional reference can be found in ASHRAE *Standard* 147-2002.

EMISSIONS TYPES

Refrigerant emissions to the atmosphere are often generically called *losses*, without distinguishing the causes. However, emission types are very different, and their causes must be identified before they can be controlled. Clodic (1997) identified six types:

- Sources of **fugitive emissions** cannot be precisely located.
- **Tightness degradation** is caused by temperature variations, pressure cycling, and vibrations that can lead to unexpected and significant increases of refrigerant emission rates.
- **Component failures** mostly originate from poor construction or faulty assembly.
- **Losses during refrigerant handling** occur mainly when charging the system, and opening the system without previously recovering the refrigerant.
- **Accidental losses** are unpredictable and are caused by fires, explosions, sabotage, theft, etc.
- **Losses at equipment disposal** are caused by intentionally venting, rather than recovering, refrigerant at the end of system life.

DESIGN

The potential for leakage is first affected by system design. Every attempt must be made to design systems that are leaktight for the length of their useful service lives, and reliable, to minimize the need for service. Selection of materials, joining techniques, and design for easy installation and service access are critical factors in designing leaktight systems.

For example, leaktight service valves should be installed to allow removal of replaceable components from the cooling system, and located for efficient liquid refrigerant recovery.

The preparation of this chapter is assigned to TC 3.8, Refrigerant Containment.

Design should minimize charge, to reduce the amount of released refrigerant in case of catastrophic loss. There are many opportunities for refrigerant charge reduction in initial design. Heat exchangers, piping, and components should be selected to reduce the amount of refrigerant in the system (but not at the expense of energy efficiency).

INSTALLATION

Proper installation is critical to proper operation and containment during the useful life of refrigerating systems. Tight joints and proper piping materials are required. Later service requirements are minimized by proper cleaning of joints before brazing, purging the system with an inert gas (e.g., nitrogen) during brazing, and evacuation to remove noncondensables. Use an inert gas purge to prevent oxides, which can contaminate the system. Proper charging and careful system performance and leak checks should be performed. At installation, systems should be carefully charged per design specifications to prevent overcharging, which can potentially lead to a serious release of excess refrigerant, and make it impossible to transfer the entire charge into the receiver for service. The installer also has the opportunity to find manufacturing defects before the system begins operation.

SERVICING AND DECOMMISSIONING

Proper service is critical in reducing emissions. Refrigerating systems must be monitored to ensure that they are well sealed, properly charged, and operating properly. The service technician must study service records to determine any history of leakage or malfunction. The equipment should be checked to detect leaks before significant charge is lost. During system maintenance, refrigerant should not be released; instead, it should be isolated in the system or recovered by equipment capable of handling the specific refrigerant.

A maintenance document allows the user to monitor additions and removals of refrigerant, and whether recharging operations are actually associated with repairs of leaks. Maintenance documents are mandatory in a number of countries, because they enable authorities to check the actual consumption of refrigerants. In a retrofit, the new refrigerant must be noted in the service record and clearly marked on equipment. Technicians must follow manufacturers' retrofit procedures, because some system components may be incompatible with different refrigerants. Failure to perform proper retrofits may result in system failure and subsequent loss of refrigerant.

When a system is decommissioned, recover the refrigerant for recycling, reuse, or disposal (usually by incineration). Special care is required to properly clean or reclaim used refrigerants to industry-recognized standards (see ARI *Standard* 700).

TRAINING

Technician training is essential for proper handling and containment of refrigerants. Training must provide a basic understanding of the environmental effects of refrigerants; recovery, recycling, and reclamation of refrigerants; leak checks and repairs; and introduction to new refrigerants. The service operator requires continuous training to understand new designs, new refrigerants and their compatibility with lubricants, new low-emission purge units, retrofitting requirements, and service practices.

Some countries have mandatory training and certification for refrigerant technicians. In other countries, private groups provide voluntary certification and training programs for service technicians.

LEAK DETECTION

Leak detection is a basic element for manufacturing, installing, and servicing systems, because it makes it possible to measure and improve containment of refrigerant. Leak detection must be performed as the final step after the system is completed in the factory or in the field. It is good practice (and mandatory in some countries) to regularly leak test the equipment.

There are three general types of leak detection: global, local, and automated performance monitoring.

Global Detection

These methods indicate whether a leak exists, but do not identify its location. They are useful at the end of construction, and when the system is opened for repair or retrofit.

System Checking. These approaches are applicable to a system that has been emptied of its charge.

- Pressurize the system with a tracer gas and isolate it. A pressure drop within a specified time indicates leakage.
- Evacuate the system and measure the vacuum level over a specified time. A pressure rise indicates leakage.
- Place the system in a chamber and charge with a tracer gas. Then evacuate the chamber and monitor it for leaks with a mass spectrometer or residual gas analyzer.
- Evacuate the system and place it in an atmosphere with a tracer gas. Monitor for leaks with a mass spectrometer or residual gas analyzer.

Many of these tests use a tracer gas, often nitrogen or helium. It is not good practice to use a refrigerant as the tracer gas.

Continuous Monitoring During Operation. Electronic leak detectors in machinery rooms may be efficient if (1) they are sensitive enough to refrigerant dilution in the air, and (2) air is circulated properly in the room.

Local Detection

These methods pinpoint locations of leaks, and are usually used during servicing. Sensitivity varies widely; it is usually stated as ppm/volume but, for clarity, mass flow rates (oz/year) are often used.

- **Visual checks** locate large leaks (≥2 oz/year) by seeking telltale traces of oil at joints.
- **Soapy water detection (bubble testing)** is simple and inexpensive; a trained operator can pinpoint leaks of 1 oz/year or more.
- **Tracer color** added to the oil/refrigerant mixture shows the leak's location. The tracer must be compatible with the various materials used in the refrigeration circuit.
- **Electronic detectors** of different techniques can detect leaks as low as 0.25 to 2 oz/year, according to their sensitivity. They must be used with proper care and training.
- **Ultrasonic detectors** register noise generated by the flow of gas exiting through the leak, and are less sensitive than electronic detectors; they are easily disturbed by air circulation.
- **Helium and HFC mass spectrometers** with probes or hoods can detect leaks at very low levels (less than 0.1 oz/year).

Automated Performance Monitoring Systems

Monitoring parameters such as temperatures and pressures helps identify any change in the equipment. It also provides data useful for performing diagnostics on the condition of heat exchanger surfaces, proper refrigerant pumping, and shortage of refrigerant charge. Automated diagnostic programs are now being developed to produce pre-alarm messages as soon as a drift is observed. These developments are in their early stages, but their general adoption would give better control over refrigerant leaks. Equipment room monitors are currently used. On low-pressure systems, it is also possible to monitor equipment tightness by monitoring purge unit run time, which can indicate leaks.

RECOVERY, RECYCLING, AND RECLAMATION

The procedures involved in removing contaminants when recycling refrigerants are similar to those discussed in Chapter 6. Service techniques, proper handling and storage, and possible mixing or cross contamination of refrigerants are of concern. Building owners, equipment manufacturers, and contractors are concerned about reintroducing refrigerants with unacceptable levels of contaminants into refrigeration equipment. Contaminated refrigerant can negatively affect system performance, and may lead to equipment failure and release of refrigerant into the atmosphere.

Installation and Service Practices

Proper installation and service procedures, including proper evacuation and leak checking, are essential to minimize major repairs. Service lines should be made of low-permeability hose material and should include shutoff valves. Larger systems should include isolation valves and pumpdown receivers. ASHRAE *Standard* 147 describes equipment, installation, and service requirements.

Recovering refrigerant to an external storage container and then returning the refrigerant for cleanup inside the refrigeration system is similar to the procedure described in Chapter 6. Some additional air and moisture contamination may be introduced in the service procedure. In general, because contaminants are distributed throughout the system, the refrigeration system must be cleaned regardless of whether the refrigerant is isolated in the receiver, recovered into a storage container, recycled, reclaimed, or replaced with new refrigerant. The advantage of new, reclaimed, or recycled refrigerant is that a properly cleaned system is not recontaminated by impure refrigerant.

Contaminants

Contaminants found in recovered refrigerants are discussed in detail in Chapter 6. The main contaminants are moisture, acid, noncondensables, particulates, high-boiling residue (lubricant and sludge), and other condensable gases (Manz 1995).

Moisture is normally dissolved in the refrigerant or lubricant, but sometimes free water is present. Moisture is removed by passing the refrigerant through a filter-drier. Some moisture is also removed by lubricant separation.

Acid consists of organic and inorganic types. Organic acids are normally contained in the lubricant and are removed in the oil separator and in the filter-drier. Inorganic acids, such as hydrochloric acid, are removed by noncondensable purging, reaction with metal surfaces, and the filter-drier.

Noncondensable gases consist primarily of air. These gases can come from refrigeration equipment or can be introduced during servicing. Control consists of minimizing infiltration through proper equipment construction and installation (ASHRAE *Standard* 147). Proper service equipment construction, connection techniques, and maintenance procedures (e.g., during filter-drier change) also reduce air contamination. Typically, a vapor purge is used to remove air.

Suction filters, oil separators, and filter-driers remove **particulates**.

High-boiling residues consist primarily of refrigerant lubricant and sludge. Because different refrigeration systems use different lubricants and because it is a collection point for other contaminants, the lubricant is considered a contaminant. High-boiling residues are removed by separators designed to extract lubricant from vapor-phase refrigerant, or by distillation.

Other **condensable gases** consist mainly of other refrigerants. They can be generated in small quantities by high-temperature operation or during a burnout. In rare cases, refrigerants may be mixed intentionally for performance or to top off with substitutes. To maintain purity of the used refrigerant supply as well as the performance and durability of the particular system, refrigerants should not be mixed. In general, separation of other condensable gases, if possible, can only be done at a fully equipped reclamation center.

Mixed refrigerants are a special case of other condensable gases in that the refrigerant would not meet product specifications even if all moisture, acids, particulates, lubricant, and noncondensables were removed. Inadvertent mixing may occur because of a failure to

- Dedicate and clearly mark containers for specific refrigerants
- Clear hoses or recovery equipment before switching to a different refrigerant
- Test suspect refrigerant before consolidating it into large batches
- Use proper retrofit procedures

Recovery

To **recover** means to remove refrigerant in any condition from a system and to store it in an external container. Recovery reduces refrigerant emissions to the atmosphere and is a necessary first or concurrent step to either recycling or reclamation. The largest potential for service-related emissions of refrigerant occurs during recovery. These emissions consist of refrigerant left in the system (recovery efficiency) and losses through service connections (Manz 1995).

The key to reducing emissions is proper recovery equipment and techniques. The recovery equipment manufacturer and technician must share this responsibility to minimize refrigerant loss to the atmosphere. Training in handling halocarbon refrigerants is required to learn the proper techniques (RSES 1991).

Important: *Recover refrigerants into an approved container and keep containers for different refrigerants separate. Do not fill containers over 80% of capacity, because liquid expansion with rising temperature could cause loss of refrigerant through the pressure-relief valve, or even rupture of the container.*

Medium- and high-pressure refrigerants are commonly recovered using a compressor-based recovery unit to pump the refrigerant directly into a storage container (Manz 1995). Such a system is shown in Figure 1. Minimum functions include evaporation, compression, condensation, storage, and control. Where possible, the recovery unit should be connected to both the high- and low-side ports to hasten the process. Removal of the refrigerant as a liquid, especially where the refrigerant is to be reclaimed, greatly speeds the process (Clodic and Sauer 1994). As a variation, a refrigeration unit may be used to cool the storage container to transfer the refrigerant directly. For low-pressure refrigerants (e.g., R-11), a compressor or vacuum pump may be used to lower pressure in the storage container and raise pressure in the vapor space of the refrigeration system so that the liquid refrigerant will flow without evaporation. An alternative is to use a liquid pump to transfer the refrigerant (Manz 1995). A pumpdown unit (e.g., a condensing unit) is required to remove vaporized refrigerant remaining after liquid removal is complete. Recovery systems for use at a factory for charging or leak-testing operations are likely to be larger and of specialized construction to meet the manufacturer's specific needs (Parker 1988).

Components, such as an accumulator, in which liquid could be trapped may need to be gently heated with a thermostatically controlled heating blanket or a warm-air gun to remove all the refrigerant. Good practice requires watching for a pressure rise after recovery is completed to determine whether the recovery unit needs to be restarted to remove all refrigerant. Where visual inspection is possible, these components can be identified by frosting or condensation on external surfaces to the level of the liquid refrigerant inside.

For fast refrigerant transfer, the entire liquid phase must be recovered without evaporating it, or evaporating only a very small fraction. Depending on the particular refrigeration circuit, special methods needs to be developed, access may have to be created, and components may need to be modified; these modifications must be simple and fast (Clodic and Sauer 1994). Lubricant separation is essential in systems where used refrigerant is to be introduced without reclaiming. It may take longer to pump out vapor and separate lubricant, but clean recovery units, storage containers, and refrigeration systems are usually worth the extra time (Manz 1995).

Recycling

To **recycle** means to reduce contaminants in used refrigerants by separating lubricant, removing noncondensables, and using devices such as filter-driers to reduce moisture, acidity, and particulate matter. The term usually applies to procedures implemented at the field job site or at a local service shop. Industry guidelines (ARI 1994) and federal regulations (EPA 1996) specify maximum contaminant levels in recycled refrigerant for certified recycling equipment under ARI *Standard* 740.

Recycling conserves limited supplies of regulated refrigerants (e.g., R-12). A single-pass recycling schematic is shown in Figure 2 (Manz 1995). In the single-pass recycling unit, refrigerant is processed by oil separation and filter-drying in the recovery path. Typically, air and noncondensables are not removed during recovery, and are handled at a later time.

In a multiple-pass recycling unit (Figure 3) the refrigerant is typically processed through an oil separator during recovery. The filter-drier may be placed in the compressor suction line, a bypass recycling loop, or both. During a continuous recycling loop, refrigerant is withdrawn from the storage tank, processed through filter-

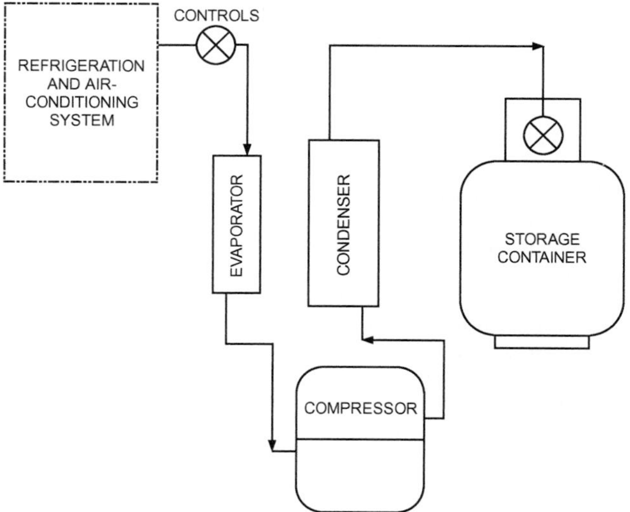

Fig. 1 Recovery Functions

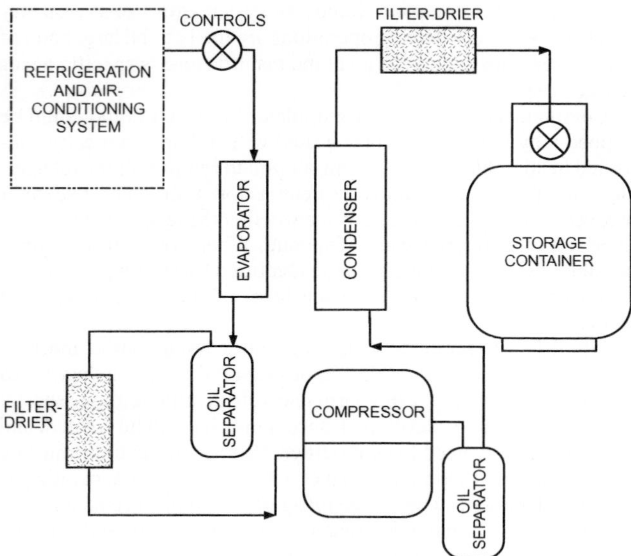

Fig. 2 Single-Pass Recycling

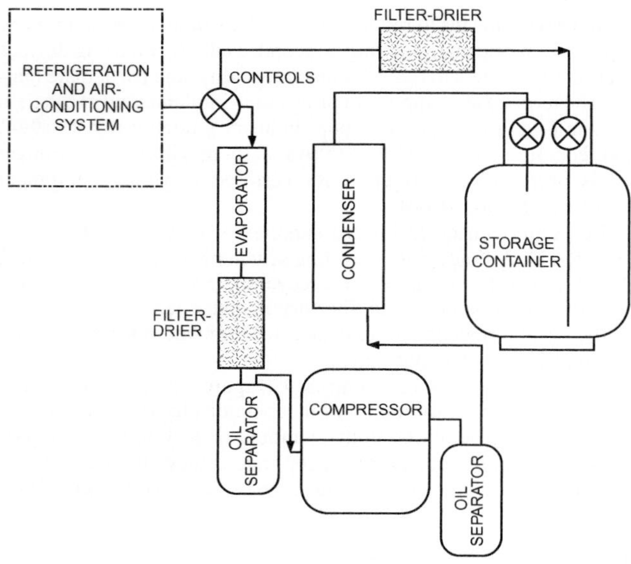

Fig. 3 Multiple-Pass Recycling

driers, and returned to the storage container. Noncondensable purge is accomplished during this recycling loop.

The primary function of the filter-drier is to remove moisture, and its the secondary function is to remove acid, particulate, and sludge (Manz 1995). The ability of the filter-drier to remove moisture and acid from used refrigerants is improved if the lubricant is separated before passing through the filter-drier (Kauffman 1992). Moisture indicators are typically used to indicate when a filter-drier change is required. For some refrigerants, these devices cannot indicate a moisture level as low as the purity level required by ARI *Standard* 700. Devices such as an in-situ mass flow meter can be used to accurately determine when to change the filter-drier and still meet purity requirements (Manz 1995).

Important: *The service technician must change recovery/recycling unit filter-driers frequently, as directed by the indicators and manufacturer's instructions.*

The primary advantage of recycling is that this operation can be performed at the job site or at a local service shop, thus avoiding transportation costs. The chance of mixing refrigerants is reduced if recycling is done at the service shop instead of consolidating refrigerant batches for shipment to a reclamation facility. Recycling equipment cannot separate mixed refrigerants to bring them back to product specifications.

A preliminary investigation of recycling refrigerants R-404A, R-410A, and R-507 showed that the refrigerant blend compositions changed less than 1% after 23 repetitions (Manz 1996). The study showed similar moisture removal capabilities as for R-22 under the test conditions. The most difficult contaminants to remove were noncondensable gases (air).

Industry Guidelines. ARI's (1994) *Industry Recycling Guide* 2 (IRG-2), Handling and Reuse of Refrigerants in the United States, includes a flowchart that outlines the following options:

Option 1: Return refrigerant to the system without recycling.

Option 2: Recycle refrigerant and return it to the original system or one with the same owner.

Option 3: Recycle the refrigerant and test to verify conformance to ARI *Standard* 700 before reuse in a different owner's equipment, provided that the refrigerant remains constantly in the contractor's custody and control from recovery through recycling to reuse.

Option 4: Send refrigerant to a certified reclaimer.

IRG-2 states, "Used refrigerants shall not be sold, or used in a different owner's equipment, unless the refrigerant has been analyzed and found to meet requirements of ARI *Standard* 700."

IRG-2 provides maximum contaminant levels of recycled refrigerants in the same owner's equipment, and lists the following reasons for concern over mixed refrigerant:

- Effect on performance and operating characteristics that may affect equipment capacity and efficiency
- Effect on materials compatibility, lubrication, equipment life, and warranty costs
- Increased service and repair requirements and higher operating costs
- High cost or inability to separate refrigerants
- High cost of disposal and loss of refrigerant for future service

Equipment Standards

Recovery and recycling equipment comes in various sizes, shapes, and functions. ARI *Standard* 740 establishes methods of testing for rating performance by type of equipment, designated refrigerants, liquid or vapor recovery rate, final recovery vacuum, recycle rate, and trapped refrigerant. The Standard requires that refrigerant emissions caused by lubricant draining, noncondensable purging, and clearing between refrigerant types not exceed 3% by mass.

In the test method, recycled refrigerant (sometimes called "dirty cocktail") is processed to determine its level of contamination. Measurements include moisture content, chloride ions, acidity, high-boiling residue, particulates/solids, and noncondensables. Each refrigerant is sampled when the first filter-drier is changed, when levels are expected to be highest. U.S. regulations (EPA 1996) require that recycling equipment meet the maximum contaminant levels in IRG-2 for recycled refrigerant (Option 2).

The basic distinction between recycling and reclamation is best illustrated by associating recycling equipment with certification (ARI *Standard* 740) and reclamation with analysis (ARI *Standard* 700). ARI *Standard* 740 covers certification testing of recycling equipment using a "standard contaminated refrigerant sample" in lieu of chemical analysis of each batch. It allows comparison of equipment performance under controlled conditions. In contrast, ARI *Standard* 700 is based on chemical analysis of a refrigerant sample from each batch after contaminant removal. ARI *Standard* 700 allows refrigerant analysis and comparison of contaminant levels to product specifications.

ARI *Standard* 740 applies to single-refrigerant systems and their normal contaminants. It does not apply to refrigerant systems or storage containers with mixed refrigerants, and does not attempt to rate equipment's ability to remove different refrigerants and other condensable gases from recovered refrigerant. Responsibility is placed on the equipment operator to identify those situations and to treat them accordingly. One uncertainty associated with recycled refrigerants is describing purity levels when offering the refrigerant for resale; appropriate field measurement techniques do not exist for all contaminants listed in ARI *Standard* 700. IRG-2 discusses possible options.

The Society of Automotive Engineers (SAE) has Standards for refrigerant recovery equipment used to service motor vehicle air-conditioners.

Special Considerations and Equipment for Handling Multiple Refrigerants

Different refrigerants must be kept separate. Storage containers should meet applicable standards for transportation and use with that refrigerant, as specified in ARI *Guideline* K. Disposable cylinders are not recommended (RSES 1991); if they are used, the remaining refrigerant heels should be recovered before cylinder disposal.

Containers should be filled per ARI *Guideline* K and marked with the refrigerant type. Container colors for recovered refrigerants and for new and reclaimed refrigerants are specified in ARI *Guidelines* K and N, respectively.

Recovery/recycling (R/R) equipment capable of handling more than one refrigerant is readily available and often preferred. Equipment should only be used for labeled refrigerants. When switching refrigerants, a significant amount of the previous refrigerant remains in the R/R equipment, particularly in the condenser section (Manz 1991), and must be removed, preferably by connecting the condenser section to the compressor suction with isolation/bypass valves and connecting the compressor discharge directly to the storage container (Manz 1995). After the bulk of the refrigerant has been removed, the system should be evacuated before changing to the appropriate storage container for the new refrigerant. This procedure should also include all lines and connecting hoses and may include replacement of the filter-driers.

The need for purging noncondensables is determined by comparing the refrigerant pressure to the saturation pressure of pure refrigerant at the same temperature. Circulation to achieve thermal equilibrium may be required to eliminate the effect of the temperature difference on pressures. A sealed bulb is often used to determine the saturation pressure for a single-refrigerant system. When purging air from multiple-refrigerant R/R equipment, the difference in saturation pressures between refrigerants far exceeds any allowable partial pressure caused by noncondensables. Special equipment and/or techniques are required (Manz 1991, 1995).

Refrigerant to be recovered may be vapor or liquid. To optimize recovery, R/R equipment must be able to handle each of these states. For some equipment, this may involve one hookup or piece of R/R equipment for liquid and a separate one for vapor. In general, a single hookup is desired. When handling multiple refrigerants, traditional liquid-flow-control devices such as capillary tubes or expansion valves either compromise performance or simply do not work. Possible solutions include (1) the operator watching a sight glass for liquid flow and switching a valve; (2) multiple flow-control devices with a refrigerant selection switch; and (3) a two-bulb expansion valve, which controls temperature differential across the evaporator (Manz 1991).

Reclamation

To **reclaim** means to process used refrigerant to new product specifications. It usually implies use of processes or procedures available only at a reprocessing or manufacturing facility. Chemical analysis is required to determine that appropriate product specifications have been met. U.S. regulations (EPA 1996) require that refrigerants must meet ARI *Standard* 700 contaminant levels in order to be sold, using option 4 discussed in IRG-2 (ARI 1994) (see Industry Guidelines in the section on Refrigerant Recycling). The EPA (1996) requires use of certified reclaimers for option 4 (reclamation), based on ARI *Standard* 700.

Some equipment warranties, especially those for smaller consumer appliances, may not allow use of refrigerants reclaimed to purity levels specified in ARI *Standard* 700. For small appliances (e.g., refrigerators and freezers), consult the manufacturer's literature before charging with reclaimed refrigerants.

Reclamation has traditionally been used for systems containing more than 100 lb refrigerant (O'Meara 1988). The reclaimer often provides help in furnishing shipping containers and labeling instructions. Many reclaimers use air-conditioning and refrigeration wholesalers as collection points for refrigerant. Refrigerant mixing at the consolidation points is possible. If the refrigerant is contaminated beyond limits, the price paid for the refrigerant may be reduced or the shipment may be refused. One advantage of reclaimed refrigerants is in the availability of purity level data when offering the refrigerant for resale; this information is generally not available for recycled refrigerant.

Purity Standards

ARI *Standard* 700 discusses halocarbon refrigerants, regardless of source, and defines acceptable levels of contaminants, which are the same as the *Federal Specifications for Fluorocarbon Refrigerants* BB-F-1421B. It specifies laboratory analysis methods for each contaminant. Only fully equipped laboratories with trained personnel are presently capable of performing the analysis.

Because ARI *Standard* 700 is based on chemical analysis of a sample from each batch after removal, it is not concerned with the level of contaminants before removal (Manz 1995). This difference from the "standard contaminated refrigerant sample" required in ARI *Standard* 740 is the basic distinction between analysis/reclamation and certification/recycling.

SAE *Standards* J1991 and J2099 list recycled refrigerant purity levels for mobile air-conditioning systems using R-12 and R-134a, respectively.

REFERENCES

ARI. 1994. Handling and reuse of refrigerants in the United States. *Industry Recycling Guide* 2 (IRG-2). Air-Conditioning and Refrigeration Institute, Arlington, VA.

ARI. 2004. Specifications for fluorocarbon refrigerants. *Standard* 700. Air-Conditioning and Refrigeration Institute, Arlington, VA.

ARI. 1998. Refrigerant recovery/recycling equipment. *Standard* 740. Air-Conditioning and Refrigeration Institute, Arlington, VA.

ARI. 2004. Containers for recovered non-flammable fluorocarbon refrigerants. *Guideline* K. Air-Conditioning and Refrigeration Institute, Arlington, VA.

Byrne, J.J., M. Shows, and M.W. Abel. 1996. Investigation of flushing and clean-out methods for refrigeration equipment to ensure system compatibility. *Report* 23810-73. U.S. Department of Energy, Washington, D.C.

Clodic, D. 1997. *Zero leaks*. ASHRAE.

Clodic, D. and F. Sauer. 1994. *The refrigerant recovery book*. ASHRAE.

EPA. 1996. Regulations governing sale of refrigerant. 61FR68505. *U.S. Federal Register*, U.S. Government Printing Office, Washington, D.C.

Kauffman, R.E. 1992. Chemical analysis and recycling of used refrigerant from field systems. *ASHRAE Transactions* 98(1).

Manz. K.W. 1988. Recovery of CFC refrigerants during service and recycling by the filtration method. *ASHRAE Transactions* 94(2).

Manz, K.W. 1991. How to handle multiple refrigerants in recovery and recycling equipment. *ASHRAE Journal* 33(4).

Manz, K.W. 1995. *The challenge of recycling refrigerants*. Business News Publishing, Troy, MI.

Manz, K.W. 1996. Recycling alternate refrigerants R-404A, R-410A, and R-507. *Proceedings of the International Conference on Ozone Protection Technologies*, Frederick, MD, pp. 411-419.

O'Meara, D.R. 1988. Operating experiences of a refrigerant recovery services company. *ASHRAE Transactions* 94(2).

Parker, R.W. 1988. Reclaiming refrigerant in OEM plants. *ASHRAE Transactions* 94(2).

RSES. 1991. *Refrigerant service for the 90's*, 1st edition. Refrigeration Service Engineers Society, Des Plaines, IL.

BIBLIOGRAPHY

ARI. 2002. Assignment of refrigerant color containers. *Guideline* N. Air-Conditioning and Refrigeration Institute, Arlington, VA.

ASHRAE. 2002. Reducing the release of halogenated refrigerants from refrigerating and air-conditioning equipment and systems. ANSI/ASHRAE *Standard* 147-2002.

Kauffman, R.E. 1992. Sealed tube tests of refrigerants from field systems before and after recycling. *ASHRAE Transactions* 99(2).

McCain, C.A. 1991. Refrigerant reclamation protects HVAC equipment investment. *ASHRAE Journal* 33(4).

SAE. 1999. Standard of purity for use in mobile air-conditioning systems. *Standard* J1991. Society of Automotive Engineers, Warrendale, PA.

SAE. 1999. Standard of purity for recycled HFC-134a (R-134a) for use in mobile air-conditioning systems. *Standard* J2099. Society of Automotive Engineers, Warrendale, PA.

CHAPTER 9

THERMAL PROPERTIES OF FOODS

THERMAL properties of foods and beverages must be known to perform the various heat transfer calculations involved in designing storage and refrigeration equipment and estimating process times for refrigerating, freezing, heating, or drying of foods and beverages. Because the thermal properties of foods and beverages strongly depend on chemical composition and temperature, and because many types of food are available, it is nearly impossible to experimentally determine and tabulate the thermal properties of foods and beverages for all possible conditions and compositions. However, composition data for foods and beverages are readily available from sources such as Holland et al. (1991) and USDA (1975). These data consist of the mass fractions of the major components found in foods. Thermal properties of foods can be predicted by using these composition data in conjunction with temperature-dependent mathematical models of thermal properties of the individual food constituents.

Thermophysical properties often required for heat transfer calculations include density, specific heat, enthalpy, thermal conductivity, and thermal diffusivity. In addition, if the food is a living organism, such as a fresh fruit or vegetable, it generates heat through respiration and loses moisture through transpiration. Both of these processes should be included in heat transfer calculations. This chapter summarizes prediction methods for estimating these thermophysical properties and includes examples on the use of these prediction methods. Tables of measured thermophysical property data for various foods and beverages are also provided.

THERMAL PROPERTIES OF FOOD CONSTITUENTS

Constituents commonly found in foods include water, protein, fat, carbohydrate, fiber, and ash. Choi and Okos (1986) developed mathematical models for predicting the thermal properties of these components as functions of temperature in the range of –40 to 300°F (Table 1); they also developed models for predicting the thermal properties of water and ice (Table 2). Table 3 lists the composition of various foods, including the mass percentage of moisture, protein, fat, carbohydrate, fiber, and ash (USDA 1996).

THERMAL PROPERTIES OF FOODS

In general, thermophysical properties of a food or beverage are well behaved when its temperature is above its initial freezing point. However, below the initial freezing point, the thermophysical properties vary greatly because of the complex processes involved during freezing.

Table 1 Thermal Property Models for Food Components ($-40 \le t \le 300$°F)

Thermal Property	Food Component	Thermal Property Model
Thermal conductivity, Btu/(h·ft·°F)	Protein	$k = 9.0535 \times 10^{-2} + 4.1486 \times 10^{-4}t - 4.8467 \times 10^{-7}t^2$
	Fat	$k = 1.0722 \times 10^{-1} - 8.6581 \times 10^{-5}t - 3.1652 \times 10^{-8}t^2$
	Carbohydrate	$k = 1.0133 \times 10^{-1} + 4.9478 \times 10^{-4}t - 7.7238 \times 10^{-7}t^2$
	Fiber	$k = 9.2499 \times 10^{-2} + 4.3731 \times 10^{-4}t - 5.6500 \times 10^{-7}t^2$
	Ash	$k = 1.7553 \times 10^{-1} + 4.8292 \times 10^{-4}t - 5.1839 \times 10^{-7}t^2$
Thermal diffusivity, ft²/h	Protein	$\alpha = 2.3170 \times 10^{-3} + 1.1364 \times 10^{-5}t - 1.7516 \times 10^{-8}t^2$
	Fat	$\alpha = 3.8358 \times 10^{-3} - 2.4128 \times 10^{-7}t - 4.5790 \times 10^{-10}t^2$
	Carbohydrate	$\alpha = 2.7387 \times 10^{-3} + 1.3198 \times 10^{-5}t - 2.7769 \times 10^{-8}t^2$
	Fiber	$\alpha = 2.4818 \times 10^{-3} + 1.2873 \times 10^{-5}t - 2.6553 \times 10^{-8}t^2$
	Ash	$\alpha = 4.5565 \times 10^{-3} + 8.9716 \times 10^{-6}t - 1.4644 \times 10^{-8}t^2$
Density, lb/ft³	Protein	$\rho = 8.3599 \times 10^{1} - 1.7979 \times 10^{-2}t$
	Fat	$\rho = 5.8246 \times 10^{1} - 1.4482 \times 10^{-2}t$
	Carbohydrate	$\rho = 1.0017 \times 10^{2} - 1.0767 \times 10^{-2}t$
	Fiber	$\rho = 8.2280 \times 10^{1} - 1.2690 \times 10^{-2}t$
	Ash	$\rho = 1.5162 \times 10^{2} - 9.7329 \times 10^{-3}t$
Specific heat, Btu/(lb·°F)	Protein	$c_p = 4.7442 \times 10^{-1} + 1.6661 \times 10^{-4}t - 9.6784 \times 10^{-8}t^2$
	Fat	$c_p = 4.6730 \times 10^{-1} + 2.1815 \times 10^{-4}t - 3.5391 \times 10^{-7}t^2$
	Carbohydrate	$c_p = 3.6114 \times 10^{-1} + 2.8843 \times 10^{-4}t - 4.3788 \times 10^{-7}t^2$
	Fiber	$c_p = 4.3276 \times 10^{-1} + 2.6485 \times 10^{-4}t - 3.4285 \times 10^{-7}t^2$
	Ash	$c_p = 2.5266 \times 10^{-1} + 2.6810 \times 10^{-4}t - 2.7141 \times 10^{-7}t^2$

Source: Choi and Okos (1986)

The preparation of this chapter is assigned to TC 10.9, Refrigeration Application for Foods and Beverages.

Table 2 Thermal Pr[illegible]ty Models for Water and Ice ($-40 \le t \le 300$°F)

	Thermal Property	Thermal Property Model
Water	Thermal conductivity, Btu/(h·ft·[illegible])	$k_w = 3.1064 \times 10^{-1} + 6.4226 \times 10^{-4}t - 1.1955 \times 10^{-6}t^2$
	Thermal diffusivity, ft²/h	$\alpha_w = 4.6428 \times 10^{-3} + 1.5289 \times 10^{-5}t - 2.8730 \times 10^{-8}t^2$
	Density, lb/ft³	$\rho_w = 6.2174 \times 10^{1} + 4.7425 \times 10^{-3}t - 7.2397 \times 10^{-8}t^2$
	Specific heat, Btu/(lb·°F) [illegible]r temp[illegible]ature range of –40 to 32°F)	$c_w = 1.0725 - 5.3992 \times 10^{-3}t + 7.3361 \times 10^{-5}t^2$
	Specific heat, Btu/(l[illegible] (For te[illegible]perature range of 32 to 300°F)	$c_w = 9.9827 \times 10^{-1} - 3.7879 \times 10^{-5}t + 4.0347 \times 10^{-7}t^2$
Ice	Thermal condu[illegible]tu/(h·[illegible]°F)	$k_{ice} = 1.3652 - 3.1648 \times 10^{-3}t + 1.8108 \times 10^{-5}t^2$
	Thermal diff[illegible]t²/h	$\alpha_{ice} = 5.0909 \times 10^{-2} - 2.0371 \times 10^{-4}t + 1.1366 \times 10^{-6}t^2$
	Density[illegible]	$\rho_{ice} = 5.7385 \times 10^{1} - 4.5333 \times 10^{-3}t$
	Speci[illegible]tu/(lb·°F)	$c_{ice} = 4.6677 \times 10^{-1} + 8.0636 \times 10^{-4}t$

Source: Choi [illegible]86)

[illegible]reezing point of a food is somewhat lower than the [illegible]t of pure water because of dissolved substances in the T[illegible] the food. At the initial freezing point, some of the water f[illegible]od crystallizes, and the remaining solution becomes more [illegible]trated. Thus, the freezing point of the unfrozen portion of the [illegible] is further reduced. The temperature continues to decrease as [illegible]aration of ice crystals increases the concentration of solutes in [illegible]olution and depresses the freezing point further. Thus, the ice and water fractions in the frozen food depend on temperature. Because the thermophysical properties of ice and water are quite different, thermophysical properties of frozen foods vary dramatically with temperature. In addition, the thermophysical properties of the food above and below the freezing point are drastically different.

WATER CONTENT

Because water is the predominant constituent in most foods, water content significantly influences the thermophysical properties of foods. Average values of moisture content (percent by mass) are given in Table 3. For fruits and vegetables, water content varies with the cultivar as well as with the stage of development or maturity when harvested, growing conditions, and amount of moisture lost after harvest. In general, values given in Table 3 apply to mature products shortly after harvest. For fresh meat, the water content values in Table 3 are at the time of slaughter or after the usual aging period. For cured or processed products, the water content depends on the particular process or product.

INITIAL FREEZING POINT

Foods and beverages do not freeze completely at a single temperature, but rather over a range of temperatures. In fact, foods high in sugar content or packed in high syrup concentrations may never be completely frozen, even at typical frozen food storage temperatures. Thus, there is not a distinct freezing point for foods and beverages, but an initial freezing point at which crystallization begins.

The initial freezing point of a food or beverage is important not only for determining the food's proper storage conditions, but also for calculating thermophysical properties. During storage of fresh fruits and vegetables, for example, the commodity temperature must be kept above its initial freezing point to avoid freezing damage. In addition, because there are drastic changes in the thermophysical properties of foods as they freeze, a food's initial freezing point must be known to model its thermophysical properties accurately. Experimentally determined values of the initial freezing point of foods and beverages are given in Table 3.

ICE FRACTION

To predict the thermophysical properties of frozen foods, which depend strongly on the fraction of ice in the food, the mass fraction of water that has crystallized must be determined. Below the initial freezing point, the mass fraction of water that has crystallized in a food is a function of temperature.

In general, foods consist of water, dissolved solids, and undissolved solids. During freezing, as some of the liquid water crystallizes, the solids dissolved in the remaining liquid water become increasingly more concentrated, thus lowering the freezing temperature. This unfrozen solution can be assumed to obey the freezing point depression equation given by Raoult's law (Pham 1987). Thus, based on Raoult's law, Chen (1985) proposed the following model for predicting the mass fraction of ice x_{ice}:

$$x_{ice} = \frac{x_s R T_o^2 (t_f - t)}{M_s L_o (t_f - 32)(t - 32)} \tag{1}$$

where

x_s = mass fraction of solids in food
M_s = relative molecular mass of soluble solids, lb_m/mol
R = universal gas constant = 1.986 Btu/lb mol·°R
T_o = freezing point of water = 491.7°R
L_o = latent heat of fusion of water at 491.7°R = 143.4 Btu/lb
t_f = initial freezing point of food, °F
t = food temperature, °F

The relative molecular mass of the soluble solids in the food may be estimated as follows:

$$M_s = \frac{x_s R T_o^2}{-L_o (x_{wo} - x_b)(t_f - 32)} \tag{2}$$

where x_{wo} is the mass fraction of water in the unfrozen food and x_b is the mass fraction of bound water in the food (Schwartzberg 1976). Bound water is the portion of water in a food that is bound to solids in the food, and thus is unavailable for freezing.

The mass fraction of bound water may be estimated as follows:

$$x_b = 0.4 x_p \tag{3}$$

where x_p is the mass fraction of protein in the food.

Substituting Equation (2) into Equation (1) yields a simple way to predict the ice fraction (Miles 1974):

$$x_{ice} = (x_{wo} - x_b)\left(1 - \frac{t_f - 32}{t - 32}\right) \tag{4}$$

Because Equation (4) underestimates the ice fraction at temperatures near the initial freezing point and overestimates the ice fraction at lower temperatures, Tchigeov (1979) proposed an empirical relationship to estimate the mass fraction of ice:

$$x_{ice} = \frac{1.105 x_{wo}}{1 + \dfrac{0.7138}{\ln[1 + (t_f - t)/1.8]}} \tag{5}$$

Fikiin (1996) notes that Equation (5) applies to a wide variety of foods and provides satisfactory accuracy.

Table 3 Unfrozen Composition Data, Initial Freezing Point, and Specific Heats of Foods*

Food Item	Moisture Content, % x_{wo}	Protein, % x_p	Fat, % x_f	Carbohydrate Total, % x_c	Carbohydrate Fiber, % x_{fb}	Ash, % x_a	Initial Freezing Point, °F	Specific Heat Above Freezing, Btu/lb·°F	Specific Heat Below Freezing Btu/lb·°F	Latent Heat of Fusion, Btu/lb
Vegetables										
Artichokes, globe	84.94	3.27	0.15	10.51	5.40	1.13	29.8	0.93	0.48	122
Jerusalem	78.01	2.00	0.01	17.44	1.60	2.54	27.5	0.87	0.54	112
Asparagus	92.40	2.28	0.20	4.54	2.10	0.57	30.9	0.96	0.43	133
Beans, snap	90.27	1.82	0.12	7.14	3.40	0.66	30.7	0.95	0.44	130
lima	70.24	6.84	0.86	20.16	4.90	1.89	30.9	0.84	0.49	101
Beets	87.58	1.61	0.17	9.56	2.80	1.08	30.0	0.93	0.46	126
Broccoli	90.69	2.98	0.35	5.24	3.00	0.92	30.9	0.96	0.43	130
Brussels sprouts	86.00	3.38	0.30	8.96	3.80	1.37	30.6	0.93	0.46	123
Cabbage	92.15	1.44	0.27	5.43	2.30	0.71	30.4	0.96	0.44	132
Carrots	87.79	1.03	0.19	10.14	3.00	0.87	29.5	0.94	0.48	126
Cauliflower	91.91	1.98	0.21	5.20	2.50	0.71	30.6	0.96	0.44	132
Celeriac	88.00	1.50	0.30	9.20	1.80	1.00	30.4	0.93	0.45	126
Celery	94.64	0.75	0.14	3.65	1.70	0.82	31.1	0.97	0.42	136
Collards	90.55	1.57	0.22	7.11	3.60	0.55	30.6	0.96	0.44	130
Corn, sweet, yellow	75.96	3.22	1.18	19.02	2.70	0.62	30.9	0.86	0.47	109
Cucumbers	96.01	0.69	0.13	2.76	0.80	0.41	31.1	0.98	0.41	138
Eggplant	92.03	1.02	0.18	6.07	2.50	0.71	30.6	0.96	0.44	132
Endive	93.79	1.25	0.20	3.35	3.10	1.41	31.8	0.97	0.40	135
Garlic	58.58	6.36	0.50	33.07	2.10	1.50	30.6	0.76	0.52	84
Ginger, root	81.67	1.74	0.73	15.09	2.00	0.77	—	0.90	0.46	117
Horseradish	78.66	9.40	1.40	8.28	2.00	2.26	28.8	0.88	0.51	113
Kale	84.46	3.30	0.70	10.01	2.00	1.53	31.1	0.91	0.44	121
Kohlrabi	91.00	1.70	0.10	6.20	3.60	1.00	30.2	0.96	0.45	131
Leeks	83.00	1.50	0.30	14.15	1.80	1.05	30.7	0.90	0.46	119
Lettuce, iceberg	95.89	1.01	0.19	2.09	1.40	0.48	31.6	0.98	0.39	138
Mushrooms	91.81	2.09	0.42	4.65	1.20	0.89	30.4	0.95	0.44	132
Okra	89.58	2.00	0.10	7.63	3.20	0.70	28.8	0.95	0.49	129
Onions	89.68	1.16	0.16	8.63	1.80	0.37	30.4	0.94	0.45	129
dehydrated flakes	3.93	8.95	0.46	83.28	9.20	3.38	—	—	—	6
Parsley	87.71	2.97	0.79	6.33	3.30	2.20	30.0	0.94	0.46	126
Parsnips	79.53	1.20	0.30	17.99	4.90	0.98	30.4	0.89	0.48	114
Peas, green	78.86	5.42	0.40	14.46	5.10	0.87	30.9	0.90	0.47	113
Peppers, freeze-dried	2.00	17.90	3.00	68.70	21.30	8.40	—	—	—	3
sweet, green	92.19	0.89	0.19	6.43	1.80	0.30	30.7	0.96	0.43	132
Potatoes, main crop	78.96	2.07	0.10	17.98	1.60	0.89	30.9	0.88	0.46	113
sweet	72.84	1.65	0.30	24.28	3.00	0.95	29.7	0.83	0.50	104
Pumpkins	91.60	1.00	0.10	6.50	0.50	0.80	30.6	0.95	0.43	132
Radishes	94.84	0.60	0.54	3.59	1.60	0.54	30.7	0.97	0.42	136
Rhubarb	93.61	0.90	0.20	4.54	1.80	0.76	30.4	0.97	0.44	135
Rutabaga	89.66	1.20	0.20	8.13	2.50	0.81	30.0	0.94	0.46	129
Salsify (vegetable oyster)	77.00	3.30	0.20	18.60	3.30	0.90	30.0	0.87	0.49	110
Spinach	91.58	2.86	0.35	3.50	2.70	1.72	31.5	0.96	0.42	132
Squash, summer	94.20	0.94	0.24	4.04	1.90	0.58	31.1	0.97	0.42	135
winter	87.78	0.80	0.10	10.42	1.50	0.90	30.6	0.93	0.45	126
Tomatoes, mature green	93.00	1.20	0.20	5.10	1.10	0.50	30.9	0.96	0.42	134
ripe	93.76	0.85	0.33	4.64	1.10	0.42	31.1	0.97	0.43	135
Turnip	91.87	0.90	0.10	6.23	1.80	0.70	30.0	0.96	0.45	132
greens	91.07	1.50	0.30	5.73	3.20	1.40	31.6	0.96	0.42	131
Watercress	95.11	2.30	0.10	1.29	1.50	1.20	31.5	0.97	0.40	137
Yams	69.60	1.53	0.17	27.89	4.10	0.82	—	0.83	0.49	100
Fruits										
Apples, fresh	83.93	0.19	0.36	15.25	2.70	0.26	30.0	0.91	0.47	120
dried	31.76	0.93	0.32	65.89	8.70	1.10	—	0.61	0.68	46
Apricots	86.35	1.40	0.39	11.12	2.40	0.75	30.0	0.92	0.47	124
Avocados	74.27	1.98	15.32	7.39	5.00	1.04	31.5	0.88	0.47	107
Bananas	74.26	1.03	0.48	23.43	2.40	0.80	30.6	0.85	0.48	107
Blackberries	85.64	0.72	0.39	12.76	5.30	0.48	30.6	0.93	0.46	123
Blueberries	84.61	0.67	0.38	14.13	2.70	0.21	29.1	0.91	0.49	122
Cantaloupes	89.78	0.88	0.28	8.36	0.80	0.71	29.8	0.94	0.46	129
Cherries, sour	86.13	1.00	0.30	12.18	1.60	0.40	28.9	0.92	0.49	124
sweet	80.76	1.20	0.96	16.55	2.30	0.53	28.8	0.89	0.51	116
Cranberries	86.54	0.39	0.20	12.68	4.20	0.19	30.4	0.93	0.46	124

Table 3 Unfrozen Composition Data, Initial Freezing Point, and Specific Heats of Foods* (*Continued*)

Food Item	Moisture Content, % x_{wo}	Protein, % x_p	Fat, % x_f	Carbohydrate Total, % x_c	Carbohydrate Fiber, % x_{fb}	Ash, % x_a	Initial Freezing Point, °F	Specific Heat Above Freezing, Btu/lb·°F	Specific Heat Below Freezing Btu/lb·°F	Latent Heat of Fusion, Btu/lb
Currants, European black	81.96	1.40	0.41	15.38	0.00	0.86	30.2	0.89	0.47	118
red and white	83.95	1.40	0.20	13.80	4.30	0.66	30.2	0.92	0.47	120
Dates, cured	22.50	1.97	0.45	73.51	7.50	1.58	3.7	0.55	0.55	32
Figs, fresh	79.11	0.75	0.30	19.18	3.30	0.66	27.7	0.88	0.54	113
dried	28.43	3.05	1.17	65.35	9.30	2.01	—	0.60	0.98	41
Gooseberries	87.87	0.88	0.58	10.18	4.30	0.49	30.0	0.94	0.47	126
Grapefruit	90.89	0.63	0.10	8.08	1.10	0.31	30.0	0.95	0.45	131
Grapes, American	81.30	0.63	0.35	17.15	1.00	0.57	29.1	0.89	0.49	117
European type	80.56	0.66	0.58	17.77	1.00	0.44	28.2	0.88	0.52	116
Lemons	87.40	1.20	0.30	10.70	4.70	0.40	29.5	0.94	0.48	126
Limes	88.26	0.70	0.20	10.54	2.80	0.30	29.1	0.94	0.48	127
Mangos	81.71	0.51	0.27	17.00	1.80	0.50	30.4	0.89	0.47	117
Melons, casaba	92.00	0.90	0.10	6.20	0.80	0.80	30.0	0.95	0.45	132
honeydew	89.66	0.46	0.10	9.18	0.60	0.60	30.4	0.94	0.44	129
watermelon	91.51	0.62	0.43	7.18	0.50	0.26	31.3	0.95	0.42	132
Nectarines	86.28	0.94	0.46	11.78	1.60	0.54	30.4	0.92	0.45	124
Olives	79.99	0.84	10.68	6.26	3.20	2.23	29.5	0.90	0.49	115
Oranges	82.30	1.30	0.30	15.50	4.50	0.60	30.6	0.91	0.47	118
Peaches, fresh	87.66	0.70	0.90	11.10	2.00	0.46	30.4	0.93	0.45	126
dried	31.80	3.61	0.76	61.33	8.20	2.50	—	0.61	0.83	46
Pears	83.81	0.39	0.40	15.11	2.40	0.28	29.1	0.91	0.49	120
Persimmons	64.40	0.80	0.40	33.50	0.00	0.90	28.0	0.78	0.55	92
Pineapples	86.50	0.39	0.43	12.39	1.20	0.29	30.2	0.92	0.46	124
Plums	85.20	0.79	0.62	13.01	1.50	0.39	30.6	0.91	0.45	123
Pomegranates	80.97	0.95	0.30	17.17	0.60	0.61	26.6	0.88	0.55	116
Prunes, dried	32.39	2.61	0.52	62.73	7.10	1.76	—	0.61	0.84	46
Quinces	83.80	0.40	0.10	15.30	1.90	0.40	28.4	0.91	0.51	120
Raisins, seedless	15.42	3.22	0.46	79.13	4.00	1.77	—	0.49	0.49	22
Raspberries	86.57	0.91	0.55	11.57	6.80	0.40	30.9	0.95	0.46	124
Strawberries	91.57	0.61	0.37	7.02	2.30	0.43	30.6	0.96	0.44	132
Tangerines	87.60	0.63	0.19	11.19	2.30	0.39	30.0	0.93	0.46	126
Whole Fish										
Cod	81.22	17.81	0.67	0.0	0.0	1.16	28.0	0.90	0.51	117
Haddock	79.92	18.91	0.72	0.0	0.0	1.21	28.0	0.90	0.51	115
Halibut	77.92	20.81	2.29	0.0	0.0	1.36	28.0	0.89	0.52	112
Herring, kippered	59.70	24.58	12.37	0.0	0.0	1.94	28.0	0.78	0.54	86
Mackerel, Atlantic	63.55	18.60	13.89	0.0	0.0	1.35	28.0	0.80	0.53	91
Perch	78.70	18.62	1.63	0.0	0.0	1.20	28.0	0.89	0.51	113
Pollock, Atlantic	78.18	19.44	0.98	0.0	0.0	1.41	28.0	0.88	0.51	112
Salmon, pink	76.35	19.94	3.45	0.0	0.0	1.22	28.0	0.88	0.52	110
Tuna, bluefin	68.09	23.33	4.90	0.0	0.0	1.18	28.0	0.82	0.52	98
Whiting	80.27	18.31	1.31	0.0	0.0	1.30	28.0	0.90	0.51	115
Shellfish										
Clams	81.82	12.77	0.97	2.57	0.0	1.87	28.0	0.90	0.51	117
Lobster, American	76.76	18.80	0.90	0.50	0.0	2.20	28.0	0.87	0.51	110
Oysters	85.16	7.05	2.46	3.91	0.0	1.42	28.0	0.91	0.51	122
Scallop, meat	78.57	16.78	0.76	2.36	0.0	1.53	28.0	0.89	0.51	113
Shrimp	75.86	20.31	1.73	0.91	0.0	1.20	28.0	0.87	0.52	109
Beef										
Brisket	55.18	16.94	26.54	0.0	0.0	0.80	—	0.76	0.56	79
Carcass, choice	57.26	17.32	24.05	0.0	0.0	0.81	28.0	0.77	0.55	82
select	58.21	17.48	22.55	0.0	0.0	0.82	28.9	0.78	0.54	83
Liver	68.99	20.00	3.85	5.82	0.0	1.34	28.9	0.83	0.52	99
Ribs, whole (ribs 6-12)	54.54	16.37	26.98	0.0	0.0	0.77	—	0.75	0.55	78
Round, full cut, lean and fat	64.75	20.37	12.81	0.0	0.0	0.97	—	0.81	0.52	93
full cut, lean	70.83	22.03	4.89	0.0	0.0	1.07	—	0.84	0.51	102
Sirloin, lean	71.70	21.24	4.40	0.0	0.0	1.08	28.9	0.84	0.50	103
Short loin, porterhouse steak, lean	69.59	20.27	8.17	0.0	0.0	1.01	—	0.83	0.51	100
T-bone steak, lean	69.71	20.78	7.27	0.0	0.0	1.27	—	0.83	0.51	100
Tenderloin, lean	68.40	20.78	7.90	0.0	0.0	1.04	—	0.82	0.51	98
Veal, lean	75.91	20.20	2.87	0.0	0.0	1.08	—	0.87	0.50	109

Table 3 Unfrozen Composition Data, Initial Freezing Point, and Specific Heats of Foods* (*Continued*)

Food Item	Moisture Content, % x_{wo}	Protein, % x_p	Fat, % x_f	Carbohydrate Total, % x_c	Carbohydrate Fiber, % x_{fb}	Ash, % x_a	Initial Freezing Point, °F	Specific Heat Above Freezing, Btu/lb·°F	Specific Heat Below Freezing Btu/lb·°F	Latent Heat of Fusion, Btu/lb
Pork										
Backfat	7.69	2.92	88.69	0.0	0.0	0.70	—	0.52	0.71	11
Bacon	31.58	8.66	57.54	0.09	0.0	2.13	—	0.64	0.64	45
Belly	36.74	9.34	53.01	0.0	0.0	0.49	—	0.67	0.80	53
Carcass	49.83	13.91	35.07	0.0	0.0	0.72	—	0.74	0.74	71
Ham, cured, whole, lean	68.26	22.32	5.71	0.05	0.0	3.66	—	0.83	0.53	98
country cured, lean	55.93	27.80	8.32	0.30	0.0	7.65	—	0.75	0.55	80
Shoulder, whole, lean	72.63	19.55	7.14	0.0	0.0	1.02	28.0	0.86	0.53	104
Sausage										
Braunschweiger	48.01	13.50	32.09	3.13	0.0	3.27	—	0.72	0.57	69
Frankfurter	53.87	11.28	29.15	2.55	0.0	3.15	28.9	0.75	0.55	77
Italian	51.08	14.25	31.33	0.65	0.0	2.70	—	0.74	0.57	74
Polish	53.15	14.10	28.72	1.63	0.0	2.40	—	0.75	0.56	77
Pork	44.52	11.69	40.29	1.02	0.0	2.49	—	0.70	0.58	64
Smoked links	39.30	22.20	31.70	2.10	0.0	4.70	—	0.67	0.59	56
Poultry Products										
Chicken	65.99	18.60	15.06	0.0	0.0	0.79	27.0	1.04	0.79	95
Duck	48.50	11.49	39.34	0.0	0.0	0.68	—	0.73	0.59	70
Turkey	70.40	20.42	8.02	0.0	0.0	0.88	—	0.84	0.54	101
Egg										
White	87.81	10.52	0.0	1.03	0.0	0.64	30.9	0.93	0.43	126
dried	14.62	76.92	0.04	4.17	0.0	4.25	—	0.55	0.50	21
Whole	75.33	12.49	10.02	1.22	0.0	0.94	30.9	0.87	0.47	108
dried	3.10	47.35	40.95	4.95	0.0	3.65	—	0.49	0.48	4
Yolk	48.81	16.76	30.87	1.78	0.0	1.77	30.9	0.73	0.54	70
salted	50.80	14.00	23.00	1.60	0.0	10.60	1.0	0.72	0.91	73
sugared	51.25	13.80	22.75	10.80	0.0	1.40	25.0	0.73	0.61	74
Lamb										
Composite of cuts, lean	73.42	20.29	5.25	0.0	0.0	1.06	28.6	0.86	0.51	105
Leg, whole, lean	74.11	20.56	4.51	0.0	0.0	1.07	—	0.86	0.51	107
Dairy Products										
Butter	17.94	0.85	81.11	0.06	0.0	0.04	—	0.57	0.63	26
Cheese										
Camembert	51.80	19.80	24.26	0.46	0.0	3.68	—	0.74	0.80	74
Cheddar	36.75	24.90	33.14	1.28	0.0	3.93	8.8	0.66	0.73	53
Cottage, uncreamed	79.77	17.27	0.42	1.85	0.0	0.69	29.8	0.89	0.48	114
Cream	53.75	7.55	34.87	2.66	0.0	1.17	—	0.75	0.70	77
Gouda	41.46	24.94	27.44	2.22	0.0	3.94	—	0.69	0.66	59
Limburger	48.42	20.05	27.25	0.49	0.0	3.79	18.7	0.72	0.67	70
Mozzarella	54.14	19.42	21.60	2.22	0.0	2.62	—	0.75	0.59	78
Parmesan, hard	29.16	35.75	25.83	3.22	0.0	6.04	—	0.62	0.70	42
Processed American	39.16	22.15	31.25	1.30	0.0	5.84	19.6	0.67	0.66	56
Roquefort	39.38	21.54	30.64	2.00	0.0	6.44	2.7	0.67	0.80	57
Swiss	37.21	28.43	27.45	3.38	0.0	3.53	14.0	0.66	0.69	53
Cream										
Half and half	80.57	2.96	11.50	4.30	0.0	0.67	—	0.89	0.52	116
Table	73.75	2.70	19.31	3.66	0.0	0.58	28.0	0.86	0.53	106
Heavy whipping	57.71	2.05	37.00	2.79	0.0	0.45	—	0.78	0.55	83
Ice Cream										
Chocolate	55.70	3.80	11.0	28.20	1.20	1.00	21.9	0.74	0.66	80
Strawberry	60.00	3.20	8.40	27.60	0.30	0.70	21.9	0.76	0.65	86
Vanilla	61.00	3.50	11.00	23.60	0.0	0.90	21.9	0.77	0.65	88
Milk										
Canned, condensed, sweetened	27.16	7.91	8.70	54.40	0.0	1.83	5.0	0.56	—	39
Evaporated	74.04	6.81	7.56	10.04	0.0	1.55	29.5	0.85	0.50	106
Skim	90.80	3.41	0.18	4.85	0.0	0.76	—	0.94	0.43	130
dried	3.16	36.16	0.77	51.98	0.0	7.93	—	0.43	—	5
Whole	87.69	3.28	3.66	4.65	0.0	0.72	30.9	0.93	0.43	126
dried	2.47	26.32	26.71	38.42	0.0	6.08	—	0.44	—	3
Whey, acid, dried	3.51	11.73	0.54	73.45	0.0	10.77	—	0.40	—	5
sweet, dried	3.19	12.93	1.07	74.46	0.0	8.35	—	0.40	—	5

Table 3 Unfrozen Composition Data, Initial Freezing Point, and Specific Heats of Foods* (*Continued*)

Food Item	Moisture Content, % x_{wo}	Protein, % x_p	Fat, % x_f	Carbohydrate Total, % x_c	Carbohydrate Fiber, % x_{fb}	Ash, % x_a	Initial Freezing Point, °F	Specific Heat Above Freezing, Btu/lb·°F	Specific Heat Below Freezing Btu/lb·°F	Latent Heat of Fusion, Btu/lb
Nuts, Shelled										
Almonds	4.42	19.95	52.21	20.40	10.90	3.03	—	0.53	—	6
Filberts	5.42	13.04	62.64	15.30	6.10	3.61	—	0.50	—	8
Peanuts, raw	6.50	25.80	49.24	16.14	8.50	2.33	—	0.53	—	9
dry roasted with salt	1.55	23.68	49.66	21.51	8.00	3.60	—	0.50	—	2
Pecans	4.82	7.75	67.64	18.24	7.60	1.56	—	0.52	—	7
Walnuts, English	3.65	14.29	61.87	18.34	4.80	1.86	—	0.50	—	5
Candy										
Fudge, vanilla	10.90	1.10	5.40	82.30	0.0	0.40	—	0.45	—	15
Marshmallows	16.40	1.80	0.20	81.30	0.10	0.30	—	0.48	—	24
Milk chocolate	1.30	6.90	30.70	59.20	3.40	1.50	—	0.44	—	2
Peanut brittle	1.80	7.50	19.10	69.30	2.00	1.50	—	0.42	—	3
Juice and Beverages										
Apple juice, unsweetened	87.93	0.06	0.11	11.68	0.10	0.22	—	0.92	0.43	126
Grapefruit juice, sweetened	87.38	0.58	0.09	11.13	0.10	0.82	—	0.92	0.43	126
Grape juice, unsweetened	84.12	0.56	0.08	14.96	0.10	0.29	—	0.90	0.43	121
Lemon juice	92.46	0.40	0.29	6.48	0.40	0.36	—	0.95	0.41	133
Lime juice, unsweetened	92.52	0.25	0.23	6.69	0.40	0.31	—	0.95	0.41	133
Orange juice	89.01	0.59	0.14	9.85	0.20	0.41	31.3	0.93	0.42	128
Pineapple juice, unsweetened	85.53	0.32	0.08	13.78	0.20	0.30	—	0.91	0.43	123
Prune juice	81.24	0.61	0.03	17.45	1.00	0.68	—	0.89	0.45	117
Tomato juice	93.90	0.76	0.06	4.23	0.40	1.05	—	0.96	0.41	135
Cranberry-apple juice drink	82.80	0.10	0.0	17.10	0.10	0.0	—	0.89	0.44	119
Cranberry-grape juice drink	85.60	0.20	0.10	14.00	0.10	0.10	—	0.91	0.43	123
Fruit punch drink	88.00	0.0	0.0	11.90	0.10	0.10	—	0.92	0.43	126
Club soda	99.90	0.0	0.0	0.0	0.0	0.10	—	1.00	0.39	144
Cola	89.40	0.0	0.0	10.40	0.0	0.10	—	0.93	0.42	129
Cream soda	86.70	0.0	0.0	13.30	0.0	0.10	—	0.91	0.43	125
Ginger ale	91.20	0.0	0.0	8.70	0.0	0.0	—	0.94	0.41	131
Grape soda	88.80	0.0	0.0	11.20	0.0	0.10	—	0.93	0.42	128
Lemon-lime soda	89.50	0.0	0.0	10.40	0.0	0.10	—	0.93	0.42	129
Orange soda	87.60	0.0	0.0	12.30	0.0	0.10	—	0.92	0.43	126
Root beer	89.30	0.0	0.0	10.60	0.0	0.10	—	0.93	0.42	128
Chocolate milk, 2% fat	83.58	3.21	2.00	10.40	0.50	0.81	—	0.90	0.44	120
Miscellaneous										
Honey	17.10	0.30	0.0	82.40	0.20	0.20	—	0.48	—	25
Maple syrup	32.00	0.00	0.20	67.20	0.0	0.60	—	0.58	—	46
Popcorn, air-popped	4.10	12.00	4.20	77.90	15.10	1.80	—	0.49	—	6
oil-popped	2.80	9.00	28.10	57.20	10.00	2.90	—	0.48	—	4
Yeast, baker's, compressed	69.00	8.40	1.90	18.10	8.10	1.80	—	0.85	0.52	100

*Composition data from USDA (1996). Initial freezing point data from Table 1 in Chapter 30 of the 1993 *ASHRAE Handbook—Fundamentals* and USDA (1968). Specific heats calculated from equations in this chapter. Latent heat of fusion obtained by multiplying water content expressed in decimal form by 144 Btu/lb, the heat of fusion of water (Table 1 in Chapter 30 of the 1993 *ASHRAE Handbook—Fundamentals*).

Example 1. A 300 lb beef carcass is to be frozen to 0°F. What are the masses of the frozen and unfrozen water at 0°F?

Solution:

From Table 3, the mass fraction of water in the beef carcass is 0.58 and the initial freezing point for the beef carcass is 28.9°F. Using Equation (5), the mass fraction of ice is

$$x_{ice} = \frac{1.105 \times 0.58}{1 + \dfrac{0.7138}{\ln[1 + (28.9 - 0)/1.8]}} = 0.51$$

The mass fraction of unfrozen water is

$$x_u = x_{wo} - x_{ice} = 0.58 - 0.51 = 0.07$$

The mass of frozen water at 0°F is

$$x_{ice} \times 300 \text{ lb} = 0.51 \times 300 = 153 \text{ lb}$$

The mass of unfrozen water at 0°F is

$$x_u \times 300 \text{ lb} = 0.07 \times 300 = 21 \text{ lb}$$

DENSITY

Modeling the density of foods and beverages requires knowledge of the food porosity, as well as the mass fraction and density of the food components. The density ρ of foods and beverages can be calculated accordingly:

$$\rho = \frac{(1 - \varepsilon)}{\sum x_i / \rho_i} \tag{6}$$

where ε is the porosity, x_i is the mass fraction of the food constituents, and ρ_i is the density of the food constituents. The porosity ε is required to model the density of granular foods stored in bulk, such as grains and rice. For other foods, the porosity is zero.

SPECIFIC HEAT

Specific heat is a measure of the energy required to change the temperature of a food by one degree. Therefore, the specific heat of foods or beverages can be used to calculate the heat load imposed on the refrigeration equipment by the cooling or freezing of foods and beverages. In unfrozen foods, specific heat becomes slightly lower as the temperature rises from 32°F to 68°F. For frozen foods, there is a large decrease in specific heat as the temperature decreases. Table 3 lists experimentally determined values of the specific heats for various foods above and below freezing.

Unfrozen Food

The specific heat of a food, at temperatures above its initial freezing point, can be obtained from the mass average of the specific heats of the food components. Thus, the specific heat of an unfrozen food c_u may be determined as follows:

$$c_u = \sum c_i x_i \tag{7}$$

where c_i is the specific heat of the individual food components and x_i is the mass fraction of the food components.

A simpler model for the specific heat of an unfrozen food is presented by Chen (1985). If detailed composition data are not available, the following expression for specific heat of an unfrozen food can be used:

$$c_u = 1.0 - 0.55x_s - 0.15x_s^3 \tag{8}$$

where c_u is the specific heat of the unfrozen food in Btu/lb·°F and x_s is the mass fraction of the solids in the food.

Frozen Food

Below the food's freezing point, the sensible heat from temperature change and the latent heat from the fusion of water must be considered. Because latent heat is not released at a constant temperature, but rather over a range of temperatures, an apparent specific heat must be used to account for both the sensible and latent heat effects. A common method to predict the apparent specific heat of foods is (Schwartzberg 1976)

$$c_a = c_u + (x_b - x_{wo})\Delta c + Ex_s\left[\frac{RT_o^2}{M_w(t-32)^2} - 0.8\Delta c\right] \tag{9}$$

where

c_a = apparent specific heat
c_u = specific heat of food above initial freezing point
x_b = mass fraction of bound water
x_{wo} = mass fraction of water above initial freezing point
0.8 = constant
Δc = difference between specific heats of water and ice = $c_w - c_{ice}$
E = ratio of relative molecular masses of water M_w and food solids M_s ($E = M_w/M_s$)
R = universal gas constant = 1.986 Btu/lb mol·°R
T_o = freezing point of water = 491.7°R
M_w = relative molecular weight, lb_m/mol
t = food temperature, °F

The specific heat of the food above the freezing point may be estimated with Equation (7) or (8).

Schwartzberg (1981) developed an alternative method for determining the apparent specific heat of a food below the initial freezing point, as follows:

$$c_a = c_f + (x_{wo} - x_b)\left[\frac{L_o(t_o - t_f)}{t_o - t}\right] \tag{10}$$

where

c_f = specific heat of fully frozen food (typically at –40°F)
t_o = freezing point of water = 32°F
t_f = initial freezing point of food, °F
t = food temperature, °F
L_o = latent heat of fusion of water = 143.4 Btu/lb

Experimentally determined values of the specific heat of fully frozen foods are given in Table 3.

A slightly simpler apparent specific heat model, which is similar in form to that of Schwartzberg (1976), was developed by Chen (1985). Chen's model is an expansion of Siebel's equation (Siebel 1892) for specific heat and has the following form:

$$c_a = 0.37 + 0.30x_s + \frac{x_s RT_o^2}{M_s(t-32)^2} \tag{11}$$

where

c_a = apparent specific heat, Btu/lb·°F
x_s = mass fraction of solids
R = universal gas constant
T_o = freezing point of water = 491.7°R
M_s = relative molecular mass of soluble solids in food
t = food temperature, °F

If the relative molecular mass of the soluble solids is unknown, Equation (2) may be used to estimate the molecular mass. Substituting Equation (2) into Equation (11) yields

$$c_a = 0.37 + 0.30x_s - \frac{L_o(x_{wo} - x_b)(t_f - 32)}{(t-32)^2} \tag{12}$$

Example 2. Three hundred pounds of lamb meat is to be cooled from 50°F to 32°F. Using the specific heat, determine the amount of heat that must be removed from the lamb.

Solution:

From Table 3, the composition of lamb is given as follows:

x_{wo} = 0.7342 x_f = 0.0525
x_p = 0.2029 x_a = 0.0106

Evaluate the specific heat of lamb at an average temperature of (50 + 32)/2 = 41°F. From Tables 1 and 2, the specific heat of the food constituents may be determined as follows:

$c_w = 9.9827 \times 10^{-1} - 3.7879 \times 10^{-5}(41) + 4.0347 \times 10^{-7}(41)^2$
$= 0.9974$ Btu/lb · °F

$c_p = 4.7442 \times 10^{-1} + 1.6661 \times 10^{-4}(41) - 9.6784 \times 10^{-8}(41)^2$
$= 0.4811$ Btu/lb · °F

$c_f = 4.6730 \times 10^{-1} + 2.1815 \times 10^{-4}(41) - 3.5391 \times 10^{-7}(41)^2$
$= 0.4756$ Btu/lb · °F

$c_a = 2.5266 \times 10^{-1} + 2.6810 \times 10^{-4}(41) - 2.7141 \times 10^{-7}(41)^2$
$= 0.2632$ Btu/lb · °F

The specific heat of lamb can be calculated with Equation (7):

$c = \sum c_i x_i = (0.9974)(0.7342) + (0.4811)(0.2029)$
$+ (0.4756)(0.0525) + (0.2632)(0.0106)$

$c = 0.858$ Btu/lb·°F

The heat to be removed from the lamb is thus

$Q = mc\Delta T = 300 \times 0.858(50 - 32) = 4630$ Btu

ENTHALPY

The change in a food's enthalpy can be used to estimate the energy that must be added or removed to effect a temperature change. Above the freezing point, enthalpy consists of sensible energy; below the freezing point, enthalpy consists of both sensible and latent energy. Enthalpy may be obtained from the definition of constant-pressure specific heat:

$$c_p = \left(\frac{\partial H}{\partial T}\right)_p \tag{13}$$

where c_p is constant pressure specific heat, H is enthalpy, and T is temperature. Mathematical models for enthalpy may be obtained by integrating expressions of specific heat with respect to temperature.

Unfrozen Food

For foods at temperatures above their initial freezing point, enthalpy may be obtained by integrating the corresponding expression for specific heat above the freezing point. Thus, the enthalpy H of an unfrozen food may be determined by integrating Equation (7) as follows:

$$H = \sum H_i x_i = \sum \int c_i x_i \, dT \tag{14}$$

where H_i is the enthalpy of the individual food components and x_i is the mass fraction of the food components.

In Chen's (1985) method, the enthalpy of an unfrozen food may be obtained by integrating Equation (8):

$$H = H_f + (t - t_f)(1.0 - 0.55x_s - 0.15x_s^3) \tag{15}$$

where

H = enthalpy of food, Btu/lb
H_f = enthalpy of food at initial freezing temperature, Btu/lb
t = temperature of food, °F
t_f = initial freezing temperature of food, °F
x_s = mass fraction of food solids

The enthalpy at initial freezing point H_f may be estimated by evaluating either Equation (17) or (18) at the initial freezing temperature of the food, as discussed in the following section.

Frozen Foods

For foods below the initial freezing point, mathematical expressions for enthalpy may be obtained by integrating the apparent specific heat models. Integration of Equation (9) between a reference temperature T_r and food temperature T leads to the following expression for the enthalpy of a food (Schwartzberg 1976):

$$H = (T - T_r) \times \left\{ c_u + (x_b - x_{wo})\Delta c + Ex_s\left[\frac{RT_o^2}{18(T_o - T_r)(T_o - T)} - 0.8\,\Delta c\right]\right\} \tag{16}$$

Generally, the reference temperature T_r is taken to be 419.7°R (–40°F), at which point the enthalpy is defined to be zero.

By integrating Equation (11) between reference temperature T_r and food temperature T, Chen (1985) obtained the following expression for enthalpy below the initial freezing point:

$$H = (t - t_r)\left[0.37 + 0.30x_s + \frac{x_s RT_o^2}{M_s(t - 32)(t_r - 32)}\right] \tag{17}$$

where

H = enthalpy of food
R = universal gas constant
T_o = freezing point of water = 491.7°R

Substituting Equation (2) for the relative molecular mass of the soluble solids M_s simplifies Chen's method as follows:

$$H = (t - t_r)\left[0.37 + 0.30x_s - \frac{(x_{wo} - x_b)L_o(t_f - 32)}{(t_r - 32)(t - 32)}\right] \tag{18}$$

As an alternative to the enthalpy models developed by integration of specific heat equations, Chang and Tao (1981) developed empirical correlations for the enthalpy of foods. Their enthalpy correlations are given as functions of water content, initial and final temperatures, and food type (meat, juice, or fruit/vegetable). The correlations at a reference temperature of –50°F have the following form:

$$H = H_f\left[y\bar{T} + (1 - y)\bar{T}^z\right] \tag{19}$$

where

H = enthalpy of food, Btu/lb
H_f = enthalpy of food at initial freezing temperature, Btu/lb
$\bar{T}$ = reduced temperature, $\bar{T} = (T - T_r)/(T_f - T_r)$
T_r = reference temperature (zero enthalpy) = 409.7°R (–50°F)
y, z = correlation parameters

By performing regression analysis on experimental data available in the literature, Chang and Tao (1981) developed the following correlation parameters y and z used in Equation (19):

Meat Group:

$$y = 0.316 - 0.247(x_{wo} - 0.73) - 0.688(x_{wo} - 0.73)^2$$
$$z = 22.95 + 54.68(y - 0.28) - 5589.03(y - 0.28)^2 \tag{20}$$

Fruit, Vegetable, and Juice Group:

$$y = 0.362 + 0.0498(x_{wo} - 0.73) - 3.465(x_{wo} - 0.73)^2$$
$$z = 27.2 - 129.04(y - 0.23) - 481.46(y - 0.23)^2 \tag{21}$$

They also developed correlations to estimate the initial freezing temperature T_f for use in Equation (19). These correlations give T_f as a function of water content:

Meat Group:

$$T_f = 488.12 + 2.65x_{wo} \tag{22}$$

Fruit/Vegetable Group:

$$T_f = 517.61 - 88.54x_{wo} + 66.73x_{wo}^2 \tag{23}$$

Juice Group:

$$T_f = 216.85 + 589.23x_{wo} - 317.68x_{wo}^2 \tag{24}$$

In addition, the enthalpy of the food at its initial freezing point is required in Equation (19). Chang and Tao (1981) suggest the following correlation for determining the food's enthalpy at its initial freezing point H_f:

$$H_f = 4.21 + 0.17416x_{wo} \tag{25}$$

Table 4 presents experimentally determined values for the enthalpy of some frozen foods at a reference temperature of –40°F as well as the percentage of unfrozen water in these foods.

Example 3. A 300 lb beef carcass is to be frozen to a temperature of 0°F. The initial temperature of the beef carcass is 50°F. How much heat must be removed from the beef carcass during this process?

Solution:

From Table 3, the mass fraction of water in the beef carcass is 0.5821, the mass fraction of protein in the beef carcass is 0.1748, and the initial freezing point of the beef carcass is 28.9°F. The mass fraction of solids in the beef carcass is

$$x_s = 1 - x_{wo} = 1 - 0.5821 = 0.4179$$

The mass fraction of bound water is given by Equation (3):

$$x_b = 0.4x_p = 0.4 \times 0.1748 = 0.0699$$

The enthalpy of the beef carcass at 0°F is given by Equation (18) for frozen foods:

$$H_0 = \left[0-(-40)\right]\left[0.37 + 0.30 \times 0.4179 - \frac{(0.5821-0.0699)143.4(28.9-32)}{(-40-32)(0-32)}\right] = 23.77 \text{ Btu/lb}$$

The enthalpy of the beef carcass at the initial freezing point is determined by evaluating Equation (18) at the initial freezing point:

$$H_f = \left[28.9-(-40)\right]\left[0.37 + 0.30 \times 0.4179 - \frac{(0.5821-0.0699)143.4(28.9-32)}{(-40-32)(28.9-32)}\right] = 104.42 \text{ Btu/lb}$$

The enthalpy of the beef carcass at 50°F is given by Equation (15) for unfrozen foods:

$$H_{50} = 104.42 + (50-28.9) \times [1 - 0.55(0.4179) - 0.15(0.4179)^3]$$
$$= 120.44 \text{ Btu/lb}$$

Thus, the amount of heat removed during the freezing process is

$$Q = m\Delta H = m(H_{50} - H_0)$$
$$= 300(120.44 - 23.77) = 29{,}000 \text{ Btu}$$

THERMAL CONDUCTIVITY

Thermal conductivity relates the conduction heat transfer rate to the temperature gradient. A food's thermal conductivity depends on factors such as composition, structure, and temperature. Early work in the modeling of thermal conductivity of foods and beverages includes Eucken's adaption of Maxwell's equation (Eucken 1940). This model is based on the thermal conductivity of dilute dispersions of small spheres in a continuous phase:

$$k = k_c \frac{1 - [1 - a(k_d/k_c)]b}{1 + (a-1)b} \tag{26}$$

where

k = conductivity of mixture
k_c = conductivity of continuous phase
k_d = conductivity of dispersed phase
a = $3k_c/(2k_c + k_d)$
b = $V_d/(V_c + V_d)$
V_d = volume of dispersed phase
V_c = volume of continuous phase

In an effort to account for the different structural features of foods, Kopelman (1966) developed thermal conductivity models for homogeneous and fibrous foods. Differences in thermal conductivity parallel and perpendicular to the food fibers are accounted for in Kopelman's fibrous food thermal conductivity models.

For an isotropic, two-component system composed of continuous and discontinuous phases, in which thermal conductivity is independent of direction of heat flow, Kopelman (1966) developed the following expression for thermal conductivity k:

$$k = k_c\left[\frac{1-L^2}{1-L^2(1-L)}\right] \tag{27}$$

where k_c is the thermal conductivity of the continuous phase and L^3 is the volume fraction of the discontinuous phase. In Equation (27), thermal conductivity of the continuous phase is assumed to be much larger than that of the discontinuous phase. However, if the opposite if true, the following expression is used to calculate the thermal conductivity of the isotropic mixture:

$$k = k_c\left[\frac{1-M}{1-M(1-L)}\right] \tag{28}$$

where $M = L^2(1 - k_d/k_c)$ and k_d is the thermal conductivity of the discontinuous phase.

For an anisotropic, two-component system in which thermal conductivity depends on the direction of heat flow, such as in fibrous food materials, Kopelman (1966) developed two expressions for thermal conductivity. For heat flow parallel to food fibers, thermal conductivity $k_=$ is

$$k_= = k_c\left[1 - N^2\left(1 - \frac{k_d}{k_c}\right)\right] \tag{29}$$

where N^2 is the volume fraction of the discontinuous phase. If the heat flow is perpendicular to the food fibers, then thermal conductivity $k_\perp$ is

$$k_\perp = k_c\left[\frac{1-P}{1-P(1-N)}\right] \tag{30}$$

where $P = N(1 - k_d/k_c)$.

Levy (1981) introduced a modified version of the Maxwell-Eucken equation. Levy's expression for the thermal conductivity of a two-component system is as follows:

$$k = \frac{k_2[(2+\Lambda) + 2(\Lambda-1)F_1]}{(2+\Lambda) - (\Lambda-1)F_1} \tag{31}$$

where Λ is the thermal conductivity ratio ($\Lambda = k_1/k_2$), and k_1 and k_2 are the thermal conductivities of components 1 and 2, respectively. The parameter F_1 introduced by Levy is given as follows:

$$F_1 = 0.5\left\{\left(\frac{2}{\sigma} - 1 + 2R_1\right) - \left[\left(\frac{2}{\sigma} - 1 + 2R_1\right)^2 - \frac{8R_1}{\sigma}\right]^{0.5}\right\} \tag{32}$$

where

$$\sigma = \frac{(\Lambda-1)^2}{(\Lambda+1)^2 + (\Lambda/2)} \tag{33}$$

and R_1 is the volume fraction of component 1, or

$$R_1 = \left[1 + \left(\frac{1}{x_1} - 1\right)\left(\frac{\rho_1}{\rho_2}\right)\right]^{-1} \tag{34}$$

Here, x_1 is the mass fraction of component 1, ρ_1 is the density of component 1, and ρ_2 is the density of component 2.

To use Levy's method, follow these steps:

1. Calculate thermal conductivity ratio Λ
2. Determine volume fraction of constituent 1 using Equation (34)
3. Evaluate σ using Equation (33)
4. Determine F_1 using Equation (32)
5. Evaluate thermal conductivity of two-component system using Equation (31)

When foods consist of more than two distinct phases, the previously mentioned methods for the prediction of thermal conductivity must be applied successively to obtain the thermal conductivity of

Table 4 Enthalpy of Frozen Foods

Food	Water Content, % by mass		Temperature, °F															
			−40	−20	−10	−5	0	5	10	15	18	20	22	24	26	28	30	32
Fruits and Vegetables																		
Applesauce	82.8	Enthalpy, Btu/lb	0	11	17	21	25	30	36	43	49	56	61	71	84	114	145	147
		% water unfrozen	—	5	7	9	11	14	17	20	25	28	33	41	52	76	100	—
Asparagus, peeled	92.6	Enthalpy, Btu/lb	0	8	14	16	19	22	26	30	34	37	40	44	51	63	101	162
		% water unfrozen	—	—	—	4	5	6	7	9	10	12	16	20	28	55	100	—
Bilberries	85.1	Enthalpy, Btu/lb	0	10	15	18	22	25	30	37	41	45	50	56	67	87	149	151
		% water unfrozen	—	—	5	6	7	9	11	14	17	19	22	27	35	50	100	—
Carrots	87.5	Enthalpy, Btu/lb	0	10	15	18	22	26	31	37	41	45	50	57	68	88	152	154
		% water unfrozen	—	—	5	6	7	9	11	14	17	19	22	27	35	50	100	—
Cucumbers	95.4	Enthalpy, Btu/lb	0	8	13	16	18	21	24	27	30	32	35	38	43	52	78	167
		% water unfrozen	—	—	—	—	—	—	—	—	6	7	8	9	12	18	36	100
Onions	85.5	Enthalpy, Btu/lb	0	10	16	20	24	28	34	40	46	52	57	66	79	105	149	151
		% water unfrozen	—	5	7	8	9	12	15	18	21	24	28	35	45	65	100	—
Peaches, without stones	85.1	Enthalpy, Btu/lb	0	10	16	20	24	28	34	42	47	53	59	67	81	108	148	150
		% water unfrozen	—	5	7	8	10	12	15	18	22	26	30	37	48	69	100	—
Pears, Bartlett	83.8	Enthalpy, Btu/lb	0	10	17	21	25	29	35	42	47	53	59	69	83	111	146	148
		% water unfrozen	—	6	8	9	10	12	15	19	23	27	31	38	49	72	100	—
Plums, without stones	80.3	Enthalpy, Btu/lb	0	12	19	24	28	33	40	50	57	64	73	85	113	139	141	143
		% water unfrozen	—	8	11	13	16	18	22	28	34	38	46	55	71	100	—	—
Raspberries	82.7	Enthalpy, Btu/lb	0	10	16	19	22	26	31	38	42	46	52	59	71	92	146	148
		% water unfrozen	—	4	6	7	8	9	12	15	18	21	24	30	39	56	100	—
Spinach	90.2	Enthalpy, Btu/lb	0	8	14	16	19	22	26	29	32	35	38	42	48	59	93	158
		% water unfrozen	—	—	—	—	—	—	5	7	9	10	11	14	18	25	50	100
Strawberries	89.3	Enthalpy, Btu/lb	0	9	15	18	21	25	29	34	39	41	45	51	60	77	127	158
		% water unfrozen	—	—	—	5	6	7	8	10	13	15	18	21	28	40	79	100
Sweet cherries, without stones	77.0	Enthalpy, Btu/lb	0	12	20	24	29	35	42	51	59	67	76	89	110	134	136	138
		% water unfrozen	—	9	12	14	17	20	25	32	38	43	50	62	80	100	—	—
Tall peas	75.8	Enthalpy, Btu/lb	0	10	17	21	25	30	36	43	49	54	61	70	86	114	137	139
		% water unfrozen	—	6	8	10	12	15	18	22	27	30	37	44	57	82	100	—
Tomato pulp	92.9	Enthalpy, Btu/lb	0	10	14	17	20	23	27	32	36	39	42	47	54	68	112	163
		% water unfrozen	—	—	—	—	—	5	6	8	10	12	14	18	22	31	62	100
Fish and Meat																		
Cod	80.3	Enthalpy, Btu/lb	0	10	15	18	21	24	28	33	36	39	43	48	56	73	123	139
		% water unfrozen	10	10	10	11	12	13	14	16	18	20	22	26	32	45	88	100
Haddock	83.6	Enthalpy, Btu/lb	0	9	15	18	21	24	28	33	36	39	43	48	56	73	127	145
		% water unfrozen	8	8	9	9	10	11	12	14	15	17	19	23	29	42	86	100
Perch	79.1	Enthalpy, Btu/lb	0	9	14	17	20	23	27	32	35	38	42	46	53	68	117	137
		% water unfrozen	10	10	11	11	12	13	14	16	17	19	21	24	30	41	83	100
Beef, lean, fresh[a]	74.5	Enthalpy, Btu/lb	0	9	15	18	21	24	27	32	35	38	42	48	57	74	119	131
		% water unfrozen	10	10	11	12	12	13	15	18	20	22	24	28	37	48	92	100
lean, dried	26.1	Enthalpy, Btu/lb	0	9	14	17	20	24	28	31	—	33	—	36	—	38	—	40
		% water unfrozen	96	96	96	97	98	99	100	—	—	—	—	—	—	—	—	—
Eggs																		
White	86.5	Enthalpy, Btu/lb	0	9	14	16	19	22	25	29	31	33	36	40	45	55	87	151
		% water unfrozen	—	—	—	—	—	—	—	—	10	12	13	14	17	22	48	100
Yolk	50.0	Enthalpy, Btu/lb	0	9	14	16	19	22	25	29	31	33	35	38	42	47	65	98
		% water unfrozen	—	—	—	—	—	—	—	—	—	—	20	23	27	32	66	100
	40.0	Enthalpy, Btu/lb	0	9	14	17	20	23	26	31	33	35	38	41	46	53	76	82
		% water unfrozen	20	—	—	—	24	—	27	—	30	—	34	38	43	54	89	100
Whole, with shell[b]	66.4	Enthalpy, Btu/lb	0	9	13	15	18	20	23	27	29	31	34	37	41	49	73	121
Bread																		
White	37.3	Enthalpy, Btu/lb	0	9	13	15	18	21	26	34	40	45	51	55	56	57	58	59
Whole wheat	42.4	Enthalpy, Btu/lb	0	9	13	15	18	22	27	36	43	48	55	62	67	68	69	70

Source: Adapted from Dickerson (1968) and Riedel (1951, 1956, 1957a, 1957b, 1959).
[a]Data for chicken, veal, and venison nearly matched data for beef of same water content (Riedel 1957a, 1957b).
[b]Calculated for mass composition of 58% white (86.5% water) and 32% yolk (50% water).

the food product. For example, in the case of frozen food, the thermal conductivity of the ice and liquid water mix is calculated first by using one of the earlier methods mentioned. The resulting thermal conductivity of the ice/water mix is then combined successively with the thermal conductivity of each remaining food constituent to determine the thermal conductivity of the food product.

Numerous researchers have proposed using parallel and perpendicular (or series) thermal conductivity models based on analogies with electrical resistance (Murakami and Okos 1989). The parallel model is the sum of the thermal conductivities of the food constituents multiplied by their volume fractions:

$$k = \sum x_i^v k_i \tag{35}$$

where x_i^v is the volume fraction of constituent i. The volume fraction of constituent i can be found from the following equation:

$$x_i^v = \frac{x_i/\rho_i}{\sum(x_i/\rho_i)} \tag{36}$$

The perpendicular model is the reciprocal of the sum of the volume fractions divided by their thermal conductivities:

$$k = \frac{1}{\sum(x_i^v/k_i)} \tag{37}$$

These two models have been found to predict the upper and lower bounds of the thermal conductivity of most foods.

Tables 5 and 6 list the thermal conductivities for many foods (Qashou et al. 1972). Data in these tables have been averaged, interpolated, extrapolated, selected, or rounded off from the original research data. Tables 5 and 6 also include ASHRAE research data on foods of low and intermediate moisture content (Sweat 1985).

Example 4. Determine the thermal conductivity and density of lean pork shoulder meat at –40°F. Use both the parallel and perpendicular thermal conductivity models.

Solution:

From Table 3, the composition of lean pork shoulder meat is:

$$x_{wo} = 0.7263 \qquad x_f = 0.0714$$
$$x_p = 0.1955 \qquad x_a = 0.0102$$

In addition, the initial freezing point of lean pork shoulder meat is 28°F. Because the pork's temperature is below the initial freezing point, the fraction of ice in the pork must be determined. Using Equation (4), the ice fraction becomes

$$x_{ice} = (x_{wo} - x_b)\left(1 - \frac{t_f - 32}{t - 32}\right) = (x_{wo} - 0.4x_p)\left(1 - \frac{t_f - 32}{t - 32}\right)$$

$$= [0.7263 - (0.4)(0.1955)]\left(1 - \frac{28 - 32}{-40 - 32}\right) = 0.6121$$

The mass fraction of unfrozen water is then

$$x_w = x_{wo} - x_{ice} = 0.7263 - 0.6121 = 0.1142$$

Using the equations in Tables 1 and 2, the density and thermal conductivity of the food constituents are calculated at the given temperature –40°F:

$$\rho_w = 6.2174 \times 10^1 + 4.7425 \times 10^{-3}(-40) - 7.2397 \times 10^{-5}(-40)^2$$
$$= 61.868 \text{ lb/ft}^3$$

$$\rho_{ice} = 5.7385 \times 10^1 - 4.5333 \times 10^{-3}(-40)$$
$$= 57.566 \text{ lb/ft}^3$$

$$\rho_p = 8.3599 \times 10^1 - 1.7979 \times 10^{-2}(-40)$$
$$= 84.318 \text{ lb/ft}^3$$

$$\rho_f = 5.8246 \times 10^1 - 1.4482 \times 10^{-2}(-40)$$
$$= 58.825 \text{ lb/ft}^3$$

$$\rho_a = 1.5162 \times 10^2 - 9.7329 \times 10^{-3}(-40)$$
$$= 152.01 \text{ lb/ft}^3$$

$$k_w = 3.1064 \times 10^{-1} + 6.4226 \times 10^{-4}(-40) - 1.1955 \times 10^{-6}(-40)^2$$
$$= 0.2830 \text{ Btu/h·ft·°F}$$

$$k_{ice} = 1.3652 - 3.1648 \times 10^{-3}(-40) + 1.8108 \times 10^{-5}(-40)^2$$
$$= 1.521 \text{ Btu/h·ft·°F}$$

$$k_p = 9.0535 \times 10^{-2} + 4.1486 \times 10^{-4}(-40) - 4.8467 \times 10^{-7}(-40)^2$$
$$= 0.07317 \text{ Btu/h·ft·°F}$$

$$k_f = 1.3273 \times 10^{-1} - 8.8405 \times 10^{-4}(-40) - 3.1652 \times 10^{-8}(-40)^2$$
$$= 0.1680 \text{ Btu/h·ft·°F}$$

$$k_a = 1.7553 \times 10^{-1} + 4.8292 \times 10^{-4}(-40) - 5.1839 \times 10^{-7}(-40)^2$$
$$= 0.1554 \text{ Btu/h·ft·°F}$$

Using Equation (6), the density of lean pork shoulder meat at –40°F can be determined:

$$\sum\frac{x_i}{\rho_i} = \frac{0.6121}{57.566} + \frac{0.1142}{61.868} + \frac{0.1955}{84.318} + \frac{0.0714}{58.825} + \frac{0.0102}{152.01}$$
$$= 1.6078 \times 10^{-2}$$

$$\rho = \frac{1 - \varepsilon}{\sum x_i/\rho_i} = \frac{1 - 0}{1.6078 \times 10^{-2}} = 62.2 \text{ lb/ft}^3$$

Using Equation (36), the volume fractions of the constituents can be found:

$$x_{ice}^v = \frac{x_{ice}/\rho_{ice}}{\sum x_i/\rho_i} = \frac{0.6121/57.566}{1.6078 \times 10^{-2}} = 0.6613$$

$$x_w^v = \frac{x_w/\rho_w}{\sum x_i/\rho_i} = \frac{0.1142/61.868}{1.6078 \times 10^{-2}} = 0.1148$$

$$x_p^v = \frac{x_p/\rho_p}{\sum x_i/\rho_i} = \frac{0.1955/84.318}{1.6078 \times 10^{-2}} = 0.1442$$

$$x_f^v = \frac{x_f/\rho_f}{\sum x_i/\rho_i} = \frac{0.0714/58.825}{1.6078 \times 10^{-2}} = 0.0755$$

$$x_f^a = \frac{x_a/\rho_a}{\sum x_i/\rho_i} = \frac{0.0102/152.01}{1.6078 \times 10^{-2}} = 0.0042$$

Using the parallel model, Equation (35), the thermal conductivity becomes

$$k = \sum x_i^v k_i = (0.6613)(1.521) + (0.1148)(0.2830)$$
$$+ (0.1442)(0.0731) + (0.0755)(0.1680) + (0.0042)(0.1554)$$
$$k = 1.06 \text{ Btu/h·ft·°F}$$

Using the perpendicular model, Equation (37), the thermal conductivity becomes

$$k = \frac{1}{\sum x_i^v/k_i} = \left(\frac{0.6613}{1.521} + \frac{0.1148}{0.2830} + \frac{0.1442}{0.07317} + \frac{0.0755}{0.1680} + \frac{0.0042}{0.1554}\right)^{-1}$$
$$k = 0.304 \text{ Btu/h·ft·°F}$$

Table 5 Thermal Conductivity of Foods

Food [a]	Thermal Conductivity Btu/h·ft·°F	Temperature, °F	Water Content, % by mass	Reference[b]	Remarks
Fruits, Vegetables					
Apples	0.242	46.4	—	Gane (1936)	Tasmanian French crabapple, whole fruit; 0.3 lb
dried	0.127	73.4	41.6	Sweat (1985)	Density = 54 lb/ft^3
Apple juice	0.323	68	87	Riedel (1949)	Refractive index at 68°F = 1.35
	0.365	176	87		
	0.291	68	70		Refractive index at 68°F = 1.38
	0.326	176	70		
	0.225	68	36		Refractive index at 68°F = 1.45
	0.251	176	36		
Applesauce	0.317	84.2	—	Sweat (1974)	
Apricots, dried	0.217	73.4	43.6	Sweat (1985)	Density = 82 lb/ft^3
Beans, runner	0.230	48.2	—	Smith et al. (1952)	Density = 47 lb/ft^3; machine sliced, scalded, packed in slab
Beets	0.347	82.4	87.6	Sweat (1974)	
Broccoli	0.222	21.2	—	Smith et al. (1952)	Density = 35 lb/ft^3; heads cut and scalded
Carrots	0.387	3.2	—	Smith et al. (1952)	Density = 37 lb/ft^3; scraped, sliced and scalded
pureed	0.728	17.6	—	Smith et al. (1952)	Density = 56 lb/ft^3; slab
Currants, black	0.179	1.4	—	Smith et al. (1952)	Density = 40 lb/ft^3
Dates	0.195	73.4	34.5	Sweat (1985)	Density = 82 lb/ft^3
Figs	0.179	73.4	40.4	Sweat (1985)	Density = 77 lb/ft^3
Gooseberries	0.159	5	—	Smith et al. (1952)	Density = 36 lb/ft^3; mixed sizes
Grapefruit juice vesicle	0.267	86	—	Bennett et al. (1964)	Marsh, seedless
Grapefruit rind	0.137	82	—	Bennett et al. (1964)	Marsh, seedless
Grape, green, juice	0.328	68	89	Riedel (1949)	Refractive index at 68°F = 1.35
	0.369	176	89		
	0.287	68	68		Refractive index at 68°F = 1.38
	0.320	176	68		
	0.229	68	37		Refractive index at 20°C = 1.45
	0.254	176	37		
	0.254	77	—	Turrell and Perry (1957)	Eureka
Grape jelly	0.226	68	42.0	Sweat (1985)	Density = 82 lb/ft^3
Nectarines	0.338	47.5	82.9	Sweat (1974)	
Onions	0.332	47.5	—	Saravacos (1965)	
Orange juice vesicle	0.251	86	—	Bennett et al. (1964)	Valencia
Orange rind	0.103	86	—	Bennett et al. (1964)	Valencia
Peas	0.277	8.6	—	Smith et al. (1952)	Density = 44 lb/ft^3; shelled and scalded
	0.228	26.6	—		
	0.182	44.6	—		
Peaches, dried	0.209	73.4	43.4	Sweat (1985)	Density = 79 lb/ft^3
Pears	0.344	47.7	—	Sweat (1974)	
Pear juice	0.318	68	85	Riedel (1949)	Refractive index at 68°F = 1.36
	0.363	176	85		
	0.274	68	60		Refractive index at 68°F = 1.40
	0.307	176	60		
	0.232	68	39		Refractive index at 68°F = 1.44
	0.258	176	39		
Plums	0.143	3.2	—	Smith et al. (1952)	Density = 38 lb/ft^3; 1.57 in. dia.; 2.0 in. long
Potatoes, mashed	0.630	8.6	—	Smith et al. (1952)	Density = 61 lb/ft^3; tightly packed slab
Potato salad	0.277	35.6	—	Dickerson and Read (1968)	Density = 63 lb/ft^3
Prunes	0.217	73.4	42.9	Sweat (1985)	Density = 76 lb/ft^3
Raisins	0.194	73.4	32.2	Sweat (1985)	Density = 86 lb/ft^3
Strawberries	0.636	6.8	—	Smith et al. (1952)	Mixed sizes, density = 50 lb/ft^3, slab
	0.555	5	—		Mixed sizes in 57% sucrose syrup, slab
Strawberry jam	0.195	68	41.0	Sweat (1985)	Density = 82 lb/ft^3
Squash	0.290	46.4	—	Gane (1936)	
Meat and Animal By-Products					
Beef, lean =[a]	0.292	37.4	75	Lentz (1961)	Sirloin; 0.9% fat
	0.820	5	75		
	0.248	68	79	Hill et al. (1967)	1.4% fat
	0.826	5	79		
	0.231	42.8	76.5	Hill (1966), Hill et al. (1967)	2.4% fat
	0.786	5	76.5		
⊥[a]	0.277	68	79	Hill et al. (1967)	Inside round; 0.8% fat
	0.780	5	79		
	0.237	42.8	76	Hill (1966), Hill et al. (1967)	3% fat
	0.659	5	76		
	0.272	37.4	74	Lentz (1961)	Flank; 3 to 4% fat
	0.647	5	74		
ground	0.235	42.8	67	Qashou et al. (1970)	12.3% fat; density = 59 lb/ft^3
	0.237	39.2	62		16.8% fat; density = 61 lb/ft^3
	0.203	42.8	55		18% fat; density = 58 lb/ft^3

Table 5 Thermal Conductivity of Foods (*Continued*)

Food[a]	Thermal Conductivity Btu/h·ft·°F	Temperature, °F	Water Content, % by mass	Reference[b]	Remarks
Beef, ground (*continued*)	0.210	37.4	53		22% fat; density = 59 lb/ft³
Beef brain	0.287	95	77.7	Poppendiek et al. (1965-1966)	12% fat; 10.3% protein; density = 63 lb/ft³
Beef fat	0.110	95	0.0	Poppendiek et al. (1965-1966)	Melted 100% fat; density = 51 lb/ft³
	0.133	95	20		Density = 54 lb/ft³
⊥[a]	0.125	35.6	9	Lentz (1961)	89% fat
	0.166	15.8	9		
Beef kidney	0.303	95	76.4	Poppendiek et al. (1965-1966)	8.3% fat, 15.3% protein; density = 64 lb/ft³
Beef liver	0.282	95	72	Poppendiek et al. (1965-1966)	7.2% fat, 20.6% protein
Beefstick	0.172	68	36.6	Sweat (1985)	Density = 66 lb/ft³
Bologna	0.243	68	64.7	Sweat (1985)	Density = 62 lb/ft³
Dog food	0.184	73.4	30.6	Sweat (1985)	Density = 77 lb/ft³
Cat food	0.188	73.4	39.7	Sweat (1985)	Density = 71 lb/ft³
Ham, country	0.277	68	71.8	Sweat (1985)	Density = 64 lb/ft³
Horse meat ⊥[a]	0.266	86	70	Griffiths and Cole (1948)	Lean
Lamb ⊥[a]	0.263	68	72	Hill et al. (1967)	8.7% fat
	0.647	5	72		
=[a]	0.231	68	71	Hill et al. (1967)	9.6% fat
	0.734	5	71		
Pepperoni	0.148	68	32.0	Sweat (1985)	Density = 66 lb/ft³
Pork fat	0.124	37.4	6	Lentz (1961)	93% fat
	0.126	5	6		
Pork, lean =[a]	0.262	68	76	Hill et al. (1967)	6.7% fat
	0.820	8.6	76		
⊥[a]	0.292	68	76	Hill et al. (1967)	6.7% fat
	0.751	6.8	76		
lean flank	0.266	36.0	—	Lentz (1961)	3.4% fat
	0.705	5	—		
lean leg =[a]	0.276	39.2	72	Lentz (1961)	6.1% fat
	0.861	5	72		
⊥[a]	0.263	39.2	72	Lentz (1961)	6.1% fat
	0.745	5	72		
Salami	0.180	68	35.6	Sweat (1985)	Density = 60 lb/ft³
Sausage	0.247	77	68	Nowrey and Woodams (1968),	Mixture of beef and pork; 16.1% fat, 12.2% protein
	0.222	77	62	Woodams (1965)	Mixture of beef and pork; 24.1% fat, 10.3% protein
Veal ⊥[a]	0.272	68	75	Hill et al. (1967)	2.1% fat
	0.797	5	75		
=[a]	0.257	82.4	75	Hill et al. (1967)	2.1% fat
	0.844	5	75		
Poultry and Eggs					
Chicken breast ⊥[a]	0.238	68	69−75	Walters and May (1963)	0.6% fat
with skin	0.211	68	58−74	Walters and May (1963)	0−30% fat
Turkey, breast ⊥[a]	0.287	37.4	74	Lentz (1961)	2.1% fat
	0.797	5	74		
leg ⊥[a]	0.287	39.2	74	Lentz (1961)	3.4% fat
	0.711	5	74		
breast = ⊥[a]	0.290	37.4	74	Lentz (1961)	2.1% fat
	0.884	5	74		
Egg, white	0.322	96.8	88	Spells (1958, 1960-1961)	
whole	0.555	17.6	—	Smith et al. (1952)	Density = 61 lb/ft³
yolk	0.243	87.8	50.6	Poppendiek et al. (1965-1966)	32.7% fat; 16.7% protein, density = 64 lb/ft³
Fish and Sea Products					
Fish, cod	0.324	33.8	—	Jason and Long (1955), Long (1955)	
	0.976	5	—	Long (1955)	
⊥[a]	0.309	37.4	83	Lentz (1961)	0.1% fat
	0.844	5	83		
Fish, herring	0.462	-2.2	—	Smith et al. (1952)	Density = 57 lb/ft³; whole and gutted
Fish, salmon ⊥[a]	0.307	37.4	67	Lentz (1961)	12% fat; *Salmo salar* from Gaspe peninsula
	0.716	5	67		
	0.288	41	73	Lentz (1961)	5.4% fat; *Oncorhynchus tchawytscha* from British Columbia
	0.653	5	73		
Seal blubber ⊥[a]	0.114	41	4.3	Lentz (1961)	95% fat
Whale blubber ⊥[a]	0.121	64.4	—	Griffiths and Cole (1948)	Density = 65 lb/ft³
Whale meat	0.375	89.6	—	Griffiths and Hickman (1951)	Density = 67 lb/ft³
	0.832	15.8	—		
	0.740	10.4	—	Smith et al. (1952)	0.51% fat; density = 62 lb/ft³
Dairy Products					
Butterfat	0.100	42.8	0.6	Lentz (1961)	
	0.103	5	0.6		

Table 5 Thermal Conductivity of Foods (*Continued*)

Food[a]	Thermal Conductivity Btu/h·ft·°F	Temperature, °F	Water Content, % by mass	Reference[b]	Remarks
Butter	0.114	39.2	—	Hooper and Chang (1952)	
Buttermilk	0.329	68	89	Riedel (1949)	0.35% fat
Milk, whole	0.335	82.4	90	Leidenfrost (1959)	3% fat
	0.302	35.6	83	Riedel (1949)	3.6% fat
	0.318	68	83		
	0.339	122	83		
	0.355	176	83		
skimmed	0.311	35.6	90	Riedel (1949)	0.1% fat
	0.327	68	90		
	0.350	122	90		
	0.367	176	90		
evaporated	0.281	35.6	72	Riedel (1949)	4.8% fat
	0.291	68	72		
	0.313	122	72		
	0.326	176	72		
	0.263	35.6	62	Riedel (1949)	6.4% fat
	0.273	68	62		
	0.295	122	62		
	0.307	176	62		
	0.273	73.4	67	Leidenfrost (1959)	10% fat
	0.291	105.8	67		
	0.298	140	67		
	0.304	174.2	67		
	0.187	78.8	50	Leidenfrost (1959)	15% fat
	0.196	104	50		
	0.206	138.2	50		
	0.210	174.2	50		
Whey	0.312	35.6	90	Riedel (1949)	No fat
	0.328	68	90		
	0.364	122	90		
	0.370	176	90		
Sugar, Starch, Bakery Products, and Derivatives					
Sugar beet juice	0.318	77	79	Khelemskii and Zhadan (1964)	
	0.329	77	82		
Sucrose solution	0.309	32	90	Riedel (1949)	Cane or beet sugar solution
	0.327	68	90		
	0.351	122	90		
	0.367	176	90		
	0.291	32	80		
	0.309	68	80		
	0.330	122	80		
	0.347	176	80		
	0.273	32	70		
	0.289	68	70		
	0.310	122	70		
	0.325	176	70		
	0.256	32	60		
	0.272	68	60		
	0.290	122	60		
	0.303	176	60		
	0.239	32	50		
	0.252	68	50		
	0.270	122	93 to 80		
	0.283	176	93 to 80		
	0.221	32	40		
	0.233	68	40		
	0.251	122	40		
	0.262	176	40		
Glucose solution	0.311	35.6	89	Riedel (1949)	
	0.327	68	89		
	0.347	122	89		
	0.369	176	89		
	0.294	35.6	80		
	0.309	68	80		
	0.330	122	80		

Table 5 Thermal Conductivity of Foods (*Continued*)

Food [a]	Thermal Conductivity Btu/h·ft·°F	Temperature, °F	Water Content, % by mass	Reference[b]	Remarks
Glucose solution (*continued*)					
	0.346	176	80		
	0.276	35.6	70		
	0.291	68	70		
	0.311	122	70		
	0.326	176	70		
	0.258	35.6	60		
	0.272	68	60		
	0.289	122	60		
	0.306	176	60		
Corn syrup	0.325	77	—	Metzner and Friend (1959)	Density = 72 lb/ft^3
	0.280	77	—		Density = 82 lb/ft^3
	0.270	77	—		Density = 84 lb/ft^3
Honey	0.290	35.6	80	Reidy (1968)	
	0.240	156.2	80		
Molasses syrup	0.200	86	23	Popov and Terentiev (1966)	
Cake, angel food	0.057	73.4	36.1	Sweat (1985)	Density = 9.4 lb/ft^3, porosity: 88%
applesauce	0.046	73.4	23.7	Sweat (1985)	Density = 19 lb/ft^3, porosity: 78%
carrot	0.049	73.4	21.6	Sweat (1985)	Density = 20 lb/ft^3, porosity: 75%
chocolate	0.061	73.4	31.9	Sweat (1985)	Density = 21 lb/ft^3, porosity: 74%
pound	0.076	73.4	22.7	Sweat (1985)	Density = 30 lb/ft^3, porosity: 58%
yellow	0.064	73.4	25.1	Sweat (1985)	Density = 19 lb/ft^3, porosity: 78%
white	0.047	73.4	32.3	Sweat (1985)	Density = 28 lb/ft^3, porosity: 62%
Grains, Cereals, and Seeds					
Corn, yellow	0.081	89.6	0.9	Kazarian (1962)	Density = 47 lb/ft^3
	0.092	89.6	14.7		Density = 47 lb/ft^3
	0.099	89.6	30.2		Density = 42 lb/ft^3
Flaxseed	0.066	89.6	—	Griffiths and Hickman (1951)	Density = 41 lb/ft^3
Oats, white English	0.075	80.6	12.7	Oxley (1944)	
Sorghum	0.076	41	13	Miller (1963)	Hybrid Rs610 grain
	0.087		22		
Wheat, No. 1, northern hard spring	0.078	93.2	2	Moote (1953)	Values taken from plot of series of values given by authors
	0.086	—	7	Babbitt (1945)	
	0.090	—	10		
	0.097	—	14		
	0.000	32			
Wheat, soft white winter	0.070	87.8	5	Kazarian (1962)	Values taken from plot of series of values given by author; Density = 49 lb/ft^3
	0.075	87.8	10		
	0.079	87.8	15		
Fats, Oils, Gums, and Extracts					
Gelatin gel	0.302	41	94–80	Lentz (1961)	Conductivity did not vary with concentration in range tested (6, 12, 20%)
	1.236	5	94		6% gelatin concentration
	1.121	5	88		12% gelatin concentration
	0.815	5	80		20% gelatin concentration
Margarine	0.135	41	—	Hooper and Chang (1952)	Density = 62 lb/ft^3
Oil, almond	0.102	39.2	—	Wachsmuth (1892)	Density = 57 lb/ft^3
cod liver	0.098	95	—	Spells (1958, 1960-1961)	
lemon	0.090	42.8	—	Weber (1880)	Density = 51 lb/ft^3
mustard	0.098	77	—	Weber (1886)	Density = 64 lb/ft^3
nutmeg	0.090	39.2	—	Wachsmuth (1892)	Density = 59 lb/ft^3
olive	0.101	44.6	—	Weber (1880)	Density = 57 lb/ft^3
	0.097	89.6	—	Kaye and Higgins (1928)	Density = 57 lb/ft^3
	0.096	149	—		
	0.092	304	—		
	0.090	365	—		
peanut	0.097	39.2	—	Wachsmuth (1892)	Density = 57 lb/ft^3
	0.098	77	—	Woodams (1965)	
rapeseed	0.092	68	—	Kondrat'ev (1950)	Density = 57 lb/ft^3
sesame	0.102	39.2	—	Wachsmuth (1892)	Density = 57 lb/ft^3

[a] ⊥ indicates heat flow perpendicular to grain structure, and = indicates heat flow parallel to grain structure.

[b] References quoted are those on which given data are based, although actual values in this table may have been averaged, interpolated, extrapolated, selected, or rounded off.

Table 6 Thermal Conductivity of Freeze-Dried Foods

Food	Thermal Conductivity, Btu/h·ft·°F	Temperature, °F	Pressure, psia	Reference[b]	Remarks
Apple	0.0090	95	0.000386	Harper (1960, 1962)	Delicious; 88% porosity; 5.1 tortuosity factor; measured in air
	0.0107	95	0.00305		
	0.0163	95	0.0271		
	0.0234	95	0.418		
Peach	0.0095	95	0.000870	Harper (1960, 1962)	Clingstone; 91% porosity; 4.1 tortuosity factor; measured in air
	0.0107	95	0.00312		
	0.0161	95	0.0271		
	0.0237	95	0.387		
	0.0249	95	7.40		
Pears	0.0107	95	0.000309	Harper (1960, 1962)	97% porosity; measured in nitrogen
	0.0120	95	0.00283		
	0.0177	95	0.0271		
	0.0242	95	0.312		
	0.0261	95	10.0		
Beef =[a]	0.0221	95	0.000212	Harper (1960, 1962)	Lean; 64% porosity; 4.4 tortuosity factor; measured in air
	0.0238	95	0.00329		
	0.0307	95	0.0345		
	0.0358	95	0.392		
	0.0377	95	14.7		
Egg albumin gel	0.0227	106	14.7	Saravacos and Pilsworth (1965)	2% water content; measured in air
	0.0075	106	0.00064	Saravacos and Pilsworth (1965)	Measured in air
Turkey =[a]	0.0166	—	0.000773	Triebes and King (1966)	Cooked white meat; 68 to 72% porosity; measured in air
	0.0256	—	0.00218		
	0.0408	—	0.0677		
	0.0497	—	0.309		
	0.0536	—	14.3		
⊥[a]	0.0098	—	0.000812	Triebes and King (1966)	Cooked white meat; 68 to 72% porosity; measured in air
	0.0101	—	0.00274		
	0.0128	—	0.0193		
	0.0241	—	0.181		
	0.0339	—	12.7		
Potato starch gel	0.0053	—	0.000624	Saravacos and Pilsworth (1965)	Measured in air
	0.0083	—	0.0262		
	0.0168	—	0.320		
	0.0227	—	14.9		

[a]⊥ indicates heat flow perpendicular to grain structure, and = indicates heat flow parallel to grain structure.
[b]References quoted are those on which given data are based, although actual values in this table may have been averaged, interpolated, extrapolated, selected, or rounded off.

Example 5. Determine the thermal conductivity and density of lean pork shoulder meat at a temperature of –40°F. Use the isotropic model developed by Kopelman (1966).

Solution:

From Table 3, the composition of lean pork shoulder meat is

$$x_{wo} = 0.7263 \qquad x_f = 0.0714$$
$$x_p = 0.1955 \qquad x_a = 0.0102$$

In addition, the initial freezing point of lean pork shoulder is 28°F. Because the pork's temperature is below the initial freezing point, the fraction of ice within the pork must be determined. From Example 4, the ice fraction was found to be

$$x_{ice} = 0.6121$$

The mass fraction of unfrozen water is then

$$x_w = x_{wo} - x_{ice} = 0.7263 - 0.6121 = 0.1142$$

Using the equations in Tables 1 and 2, the density and thermal conductivity of the food constituents are c1alculated at the given temperature, –40°F (refer to Example 4):

$$\rho_w = 61.868 \text{ lb/ft}^3 \qquad k_w = 0.2830 \text{ Btu/h·ft·°F}$$
$$\rho_{ice} = 57.566 \text{ lb/ft}^3 \qquad k_{ice} = 1.521 \text{ Btu/h·ft·°F}$$
$$\rho_p = 84.318 \text{ lb/ft}^3 \qquad k_p = 0.07317 \text{ Btu/h·ft·°F}$$
$$\rho_f = 58.825 \text{ lb/ft}^3 \qquad k_f = 0.1680 \text{ Btu/h·ft·°F}$$
$$\rho_a = 152.01 \text{ lb/ft}^3 \qquad k_a = 0.1554 \text{ Btu/h·ft·°F}$$

Now, determine the thermal conductivity of the ice/water mixture. This requires the volume fractions of the ice and water:

$$x_w^v = \frac{x_w/\rho_w}{\sum \frac{x_i}{\rho_i}} = \frac{0.1142/61.868}{\frac{0.1142}{61.868} + \frac{0.6121}{57.566}} = 0.1479$$

$$x_{ice}^v = \frac{x_{ice}/\rho_{ice}}{\sum \frac{x_i}{\rho_i}} = \frac{0.6121/57.566}{\frac{0.1142}{61.868} + \frac{0.6121}{57.566}} = 0.8521$$

Note that the volume fractions calculated for the two-component ice/water mixture are different from those calculated in Example 4 for lean pork shoulder meat. Because the ice has the largest volume fraction in the two-component ice/water mixture, consider the ice to be the "continuous" phase. Then, L from Equation (27) becomes

$$L^3 = x_w^v = 0.1479$$
$$L^2 = 0.2797$$
$$L = 0.5288$$

Because $k_{ice} > k_w$ and the ice is the continuous phase, the thermal conductivity of the ice/water mixture is calculated using Equation (27):

$$k_{ice/water} = k_{ice}\left[\frac{1 - L^2}{1 - L^2(1 - L)}\right]$$

$$= 1.521\left[\frac{1 - 0.2797}{1 - 0.2797(1 - 0.5288)}\right] = 1.2619 \text{ Btu/h·ft·°F}$$

The density of the ice/water mixture then becomes

$$\rho_{ice/water} = x_w^v \rho_w + x_{ice}^v \rho_{ice}$$
$$= (0.1479)(61.868) + (0.8521)(57.566)$$
$$= 58.202 \text{ lb/ft}^3$$

Next, find the thermal conductivity of the ice/water/protein mixture. This requires the volume fractions of the ice/water and the protein:

$$x_p^v = \frac{x_p/\rho_p}{\sum \frac{x_i}{\rho_i}} = \frac{0.1955/84.318}{\frac{0.1955}{84.318} + \frac{0.7263}{58.202}} = 0.1567$$

$$x_{ice/water}^v = \frac{x_{ice/water}/\rho_{ice/water}}{\sum \frac{x_i}{\rho_i}} = \frac{0.7263/58.202}{\frac{0.1955}{84.318} + \frac{0.7263}{58.202}} = 0.8433$$

Note that these volume fractions are calculated based on a two-component system composed of ice/water as one constituent and protein as the other. Because protein has the smaller volume fraction, consider it to be the discontinuous phase.

$$L^3 = x_p^v = 0.1567$$
$$L^2 = 0.2907$$
$$L = 0.5391$$

Thus, the thermal conductivity of the ice/water/protein mixture becomes

$$k_{ice/water/protein} = k_{ice/water}\left[\frac{1 - L^2}{1 - L^2(1 - L)}\right]$$

$$= 1.2619\left[\frac{1 - 0.2907}{1 - 0.2907(1 - 0.5391)}\right]$$

$$= 1.0335 \text{ Btu/h·ft·°F}$$

The density of the ice/water/protein mixture then becomes

$$\rho_{ice/water/protein} = x_{ice/water}^v \rho_{ice/water} + x_p^v \rho_p$$
$$= (0.8433)(58.202) + (0.1567)(84.318)$$
$$= 62.294 \text{ lb/ft}^3$$

Next, find the thermal conductivity of the ice/water/protein/fat mixture. This requires the volume fractions of the ice/water/protein and the fat:

$$x_f^v = \frac{x_f/\rho_f}{\sum \frac{x_i}{\rho_i}} = \frac{0.0714/58.825}{\frac{0.0714}{58.825} + \frac{0.9218}{62.294}} = 0.0758$$

$$x_{i/w/p}^v = \frac{x_{i/w/p}/\rho_{i/w/p}}{\sum \frac{x_i}{\rho_i}} = \frac{0.9218/62.294}{\frac{0.0714}{58.825} + \frac{0.9218}{62.294}} = 0.9242$$

$$L^3 = x_f^v = 0.0758$$
$$L^2 = 0.1791$$
$$L = 0.4232$$

Thus, the thermal conductivity of the ice/water/protein/fat mixture becomes

$$k_{i/w/p/f} = k_{i/w/p}\left[\frac{1 - L^2}{1 - L^2(1 - L)}\right]$$

$$= 1.0335\left[\frac{1 - 0.1791}{1 - 0.1791(1 - 0.4232)}\right]$$

$$= 0.9461 \text{ Btu/h·ft·°F}$$

The density of the ice/water/protein/fat mixture then becomes

$$\rho_{i/w/p/f} = x_{i/w/p}^v \rho_{i/w/p} + x_f^v \rho_f$$
$$= (0.9242)(62.294) + (0.0758)(58.825)$$
$$= 62.031 \text{ lb/ft}^3$$

Finally, the thermal conductivity of the lean pork shoulder meat can be found. This requires the volume fractions of the ice/water/protein/fat and the ash:

$$x_a^v = \frac{x_a/\rho_a}{\sum \frac{x_i}{\rho_i}} = \frac{0.0102/152.01}{\frac{0.0102}{152.01} + \frac{0.9932}{62.031}} = 0.0042$$

$$x_{i/w/p/f}^v = \frac{\frac{x_{i/w/p/f}}{\rho_{i/w/p/f}}}{\sum \frac{x_i}{\rho_i}} = \frac{\frac{0.9932}{62.031}}{\frac{0.0102}{152.01} + \frac{0.9932}{62.031}} = 0.9958$$

$$L^3 = x_a^v = 0.0042$$
$$L^2 = 0.0260$$
$$L = 0.1613$$

Thus, the thermal conductivity of the lean pork shoulder meat becomes

$$k_{pork} = k_{i/w/p/f}\left[\frac{1 - L^2}{1 - L^2(1 - L)}\right]$$

$$= 0.9461\left[\frac{1 - 0.0260}{1 - 0.0260(1 - 0.1613)}\right]$$

$$= 0.942 \text{ Btu/h·ft·°F}$$

The density of the lean pork shoulder meat then becomes

$$\rho_{pork} = x_{i/w/p/f}^v \rho_{i/w/p/f} + x_a^v \rho_a$$
$$= (0.9958)(62.031) + (0.0042)(152.01)$$
$$= 62.4 \text{ lb/ft}^3$$

THERMAL DIFFUSIVITY

For transient heat transfer, the important thermophysical property is thermal diffusivity α, which appears in the Fourier equation:

$$\frac{\partial T}{\partial \theta} = \alpha\left[\frac{\partial^2 T}{\partial x^2} + \frac{\partial^2 T}{\partial y^2} + \frac{\partial^2 T}{\partial z^2}\right] \tag{38}$$

where x, y, z are rectangular coordinates, T is temperature, and θ is time. Thermal diffusivity can be defined as follows:

$$\alpha = \frac{k}{\rho c} \tag{39}$$

where α is thermal diffusivity, k is thermal conductivity, ρ is density, and c is specific heat.

Experimentally determined values of food's thermal diffusivity are scarce. However, thermal diffusivity can be calculated using Equation (39), with appropriate values of thermal conductivity, specific heat, and density. A few experimental values are given in Table 7.

Table 7 Thermal Diffusivity of Foods

Food	Thermal Diffusivity, Centistokes	Water Content, % by mass	Fat Content, % by mass	Apparent Density, lb/ft³	Temperature, °F	Reference
Fruits and Vegetables						
Apple, Red Delicious, whole[a]	0.14	85	—	52.4	32 to 86	Bennett et al. (1969)
dried	0.096	42	—	53.4	73	Sweat (1985)
Applesauce	0.11	37	—	—	41	Riedel (1969)
	0.11	37	—	—	149	Riedel (1969)
	0.12	80	—	—	41	Riedel (1969)
	0.14	80	—	—	149	Riedel (1969)
Apricots, dried	0.11	44	—	82.6	73	Sweat (1985)
Bananas, flesh	0.12	76	—	—	41	Riedel (1969)
	0.14	76	—	—	149	Riedel (1969)
Cherries, flesh[b]	0.13	—	—	65.5	32 to 86	Parker and Stout (1967)
Dates	0.10	35	—	82.3	73	Sweat (1985)
Figs	0.096	40	—	77.4	73	Sweat (1985)
Jam, strawberry	0.12	41	—	81.7	68	Sweat (1985)
Jelly, grape	0.12	42	—	82.4	68	Sweat (1985)
Peaches[b]	0.14	—	—	59.9	36 to 90	Bennett (1963)
dried	0.12	43	—	78.6	73	Sweat (1985)
Potatoes, whole	0.13	—	—	65 to 67	32 to 158	Mathews and Hall (1968), Minh et al. (1969)
mashed, cooked	0.12	78	—	—	41	Riedel (1969)
	0.15	78	—	—	149	Riedel (1969)
Prunes	0.12	43	—	76.1	73	Sweat (1985)
Raisins	0.11	32	—	86.1	73	Sweat (1985)
Strawberries, flesh	0.13	92	—	—	41	Riedel (1969)
Sugar beets	0.13	—	—	—	32 to 140	Slavicek et al. (1962)
Meats						
Codfish	0.12	81	—	—	41	Riedel (1969)
	0.14	81	—	—	149	Riedel (1969)
Halibut[c]	0.15	76	1	66.8	104 to 149	Dickerson and Read (1975)
Beef, chuck[d]	0.12	66	16	66.2	104 to 149	Dickerson and Read (1975)
round[d]	0.13	71	4	68.0	104 to 149	Dickerson and Read (1975)
tongue[d]	0.13	68	13	66.2	104 to 149	Dickerson and Read (1975)
Beefstick	0.11	37	—	65.5	68	Sweat (1985)
Bologna	0.13	65	—	62.4	68	Sweat (1985)
Corned beef	0.11	65	—	—	41	Riedel (1969)
	0.13	65	—	—	149	Riedel (1969)
Ham, country	0.14	72	—	64.3	68	Sweat (1985)
smoked[d]	0.12	64	—	—	41	Riedel (1969)
	0.13	64	14	68.0	104 to 149	Dickerson and Read (1975)
Pepperoni	0.093	32	—	66.1	68	Sweat (1985)
Salami	0.13	36	—	59.9	68	Sweat (1985)
Cakes						
Angel food	0.26	36	—	9.2	73	Sweat (1985)
Applesauce	0.12	24	—	18.7	73	Sweat (1985)
Carrot	0.12	22	—	20.0	73	Sweat (1985)
Chocolate	0.12	32	—	21.2	73	Sweat (1985)
Pound	0.12	23	—	30.0	73	Sweat (1985)
Yellow	0.12	25	—	18.7	73	Sweat (1985)
White	0.10	32	—	27.8	73	Sweat (1985)

[a]Data apply only to raw whole apple.
[b]Freshly harvested.
[c]Stored frozen and thawed before test.
[d]Data apply only where juices exuded during heating remain in food samples.

HEAT OF RESPIRATION

All living foods respire. During respiration, sugar and oxygen combine to form CO_2, H_2O, and heat as follows:

$$C_6H_{12}O_6 + 6O_2 \rightarrow 6CO_2 + 6H_2O + 2528 \text{ Btu} \quad (40)$$

In most stored plant products, little cell development takes place, and the greater part of respiration energy is released as heat, which must be taken into account when cooling and storing these living commodities (Becker et al. 1996a). The rate at which this chemical reaction takes place varies with the type and temperature of the commodity.

Becker et al. (1996b) developed correlations that relate a commodity's rate of carbon dioxide production to its temperature. The carbon dioxide production rate can then be related to the commodity's heat generation rate from respiration. The resulting correlation gives the commodity's respiratory heat generation rate W in Btu/h·lb as a function of temperature t in °F:

$$W = 0.00460 f(t)^g \quad (41)$$

The respiration coefficients f and g for various commodities are given in Table 8.

Table 8 Commodity Respiration Coefficients

Commodity	Respiration Coefficients f	Respiration Coefficients g	Commodity	Respiration Coefficients f	Respiration Coefficients g
Apples	5.6871×10^{-4}	2.5977	Onions	3.668×10^{-4}	2.538
Blueberries	7.2520×10^{-5}	3.2584	Oranges	2.8050×10^{-4}	2.6840
Brussels sprouts	0.0027238	2.5728	Peaches	1.2996×10^{-5}	3.6417
Cabbage	6.0803×10^{-4}	2.6183	Pears	6.3614×10^{-5}	3.2037
Carrots	0.050018	1.7926	Plums	8.608×10^{-5}	2.972
Grapefruit	0.0035828	1.9982	Potatoes	0.01709	1.769
Grapes	7.056×10^{-5}	3.033	Rutabagas (swedes)	1.6524×10^{-4}	2.9039
Green peppers	3.5104×10^{-4}	2.7414	Snap beans	0.0032828	2.5077
Lemons	0.011192	1.7740	Sugar beets	8.5913×10^{-3}	1.8880
Lima beans	9.1051×10^{-4}	2.8480	Strawberries	3.6683×10^{-4}	3.0330
Limes	2.9834×10^{-8}	4.7329	Tomatoes	2.0074×10^{-4}	2.8350

Source: Becker et al. (1996b).

Fruits, vegetables, flowers, bulbs, florists' greens, and nursery stock are storage commodities with significant heats of respiration. Dry plant products, such as seeds and nuts, have very low respiration rates. Young, actively growing tissues, such as asparagus, broccoli, and spinach, have high rates of respiration, as do immature seeds such as green peas and sweet corn. Fast-developing fruits, such as strawberries, raspberries, and blackberries, have much higher respiration rates than do fruits that are slow to develop, such as apples, grapes, and citrus fruits.

In general, most vegetables, other than root crops, have a high initial respiration rate for the first one or two days after harvest. Within a few days, the respiration rate quickly lowers to the equilibrium rate (Ryall and Lipton 1972).

Fruits that do not ripen during storage, such as citrus fruits and grapes, have fairly constant rates of respiration. Those that ripen in storage, such as apples, peaches, and avocados, increase in respiration rate. At low storage temperatures, around 32°F, the rate of respiration rarely increases because no ripening takes place. However, if fruits are stored at higher temperatures (50 to 60°F), the respiration rate increases because of ripening and then decreases. Soft fruits, such as blueberries, figs, and strawberries, decrease in respiration with time at 32°F. If they become infected with decay organisms, however, respiration increases.

Table 9 lists the heats of respiration as a function of temperature for a variety of commodities, and Table 10 shows the change in respiration rate with time. Most commodities in Table 9 have a low and a high value for heat of respiration at each temperature. When no range is given, the value is an average for the specified temperature and may be an average of the respiration rates for many days.

When using Table 9, select the lower value for estimating the heat of respiration at equilibrium storage, and use the higher value for calculating the heat load for the first day or two after harvest, including precooling and short-distance transport. In storage of fruits between 32 and 40°F, the increase in respiration rate caused by ripening is slight. However, for fruits such as mangoes, avocados, or bananas, significant ripening occurs at temperatures above 50°F and the higher rates listed in Table 9 should be used. Vegetables such as onions, garlic, and cabbage can increase heat production after a long storage period.

TRANSPIRATION OF FRESH FRUITS AND VEGETABLES

The most abundant constituent in fresh fruits and vegetables is water, which exists as a continuous liquid phase in the fruit or vegetable. Some of this water is lost through transpiration, which involves the transport of moisture through the skin, evaporation, and convective mass transport of the moisture to the surroundings (Becker et al. 1996b).

The rate of transpiration in fresh fruits and vegetables affects product quality. Moisture transpires continuously from commodities during handling and storage. Some moisture loss is inevitable and can be tolerated. However, under many conditions, enough moisture may be lost to cause shriveling. The resulting loss in mass not only affects appearance, texture, and flavor of the commodity, but also reduces the salable mass (Becker et al. 1996a).

Many factors affect the rate of transpiration from fresh fruits and vegetables. Moisture loss is driven by a difference in water vapor pressure between the product surface and the environment. Becker and Fricke (1996a) state that the product surface may be assumed to be saturated, and thus the water vapor pressure at the commodity surface is equal to the water vapor saturation pressure evaluated at the product's surface temperature. However, they also report that dissolved substances in the moisture of the commodity tend to lower the vapor pressure at the evaporating surface slightly.

Evaporation at the product surface is an endothermic process that cools the surface, thus lowering the vapor pressure at the surface and reducing transpiration. Respiration within the fruit or vegetable, on the other hand, tends to increase the product's temperature, thus raising the vapor pressure at the surface and increasing transpiration. Furthermore, the respiration rate is itself a function of the commodity's temperature (Gaffney et al. 1985). In addition, factors such as surface structure, skin permeability, and airflow also effect the transpiration rate (Sastry et al. 1978).

Becker et al. (1996c) performed a numerical, parametric study to investigate the influence of bulk mass, airflow rate, skin mass transfer coefficient, and relative humidity on the cooling time and moisture loss of a bulk load of apples. They found that relative humidity and skin mass transfer coefficient had little effect on cooling time, whereas bulk mass and airflow rate were of primary importance. Moisture loss varied appreciably with relative humidity, airflow rate, and skin mass transfer coefficient; bulk mass had little effect. Increased airflow resulted in a decrease in moisture loss; increased airflow reduces cooling time, which quickly reduces the vapor pressure deficit, thus lowering the transpiration rate.

The driving force for transpiration is a difference in water vapor pressure between the surface of a commodity and the surrounding air. Thus, the basic form of the transpiration model is as follows:

$$\dot{m} = k_t(p_s - p_a) \tag{42}$$

where $\dot{m}$ is the transpiration rate expressed as the mass of moisture transpired per unit area of commodity surface per unit time. This rate may also be expressed per unit mass of commodity rather than per unit area of commodity surface. The transpiration coefficient k_t is the mass of moisture transpired per unit area of commodity, per unit water vapor pressure deficit, per unit time. It may also be expressed per unit mass of commodity rather than per unit area of commodity

Table 9 Heat of Respiration of Fresh Fruits and Vegetables Held at Various Temperatures

	Heat of Respiration, Btu/day per Ton of Produce						
Commodity	**32°F**	**41°F**	**50°F**	**59°F**	**68°F**	**77°F**	**Reference**
Apples							
Yellow, transparent	1513	2665	—	7889	12,392	—	Wright et al. (1954)
Delicious	757	1117	—	—	—	—	Lutz and Hardenburg (1968)
Golden Delicious	793	1189	—	—	—	—	Lutz and Hardenburg (1968)
Jonathan	865	1295	—	—	—	—	Lutz and Hardenburg (1968)
McIntosh	793	1189	—	—	—	—	Lutz and Hardenburg (1968)
Early cultivars	720-1369	1153-2342	3062-4503	3962-6844	4323-9005	—	IIR (1967)
Late cultivars	396-793	1008-1549	1513-2306	2053-4323	3242-5403	—	IIR (1967)
Average of many cultivars	505-901	1117-1585	—	2990-6808	3711-7709	—	Lutz and Hardenburg (1968)
Apricots	1153-1261	1405-1982	2449-4143	4683-7565	6484-11,527	—	Lutz and Hardenburg (1968)
Artichokes, globe	5007-9907	7025-13,220	1203-21,649	1704-31,951	3004-51,403	—	Rappaport and Watada (1958), Sastry et al. (1978)
Asparagus	6015-17,651	12,032-30,043	23,630-67,146	35,086-72,152	60,121-10,228	—	Lipton (1957), Sastry et al. (1978)
Avocados	*b	*b	—	13,616-34,581	16,246-76,439	—	Biale (1960), Lutz and Hardenburg (1968)
Bananas							
Green	*b	*b	†b	4431-7626	6484-11,527	—	IIR (1967)
Ripening	*b	*b	†b	6484-9726	7204-18,011	—	IIR (1967)
Beans							
Lima, unshelled	2306-6628	4323-7925	—	22,046-27,449	29,250-39,480	—	Lutz and Hardenburg (1968), Tewfik and Scott (1954)
shelled	3890-7709	6412-13,436	—	—	46,577-59,509	—	Lutz and Hardenburg (1968), Tewfik and Scott (1954)
Snap	*b	7529-7709	12,032-12,824	18,731-20,533	26,044-28,673	—	Ryall and Lipton (1972), Watada and Morris (1966)
Beets, red, roots	1189-1585	2017-2089	2594-2990	3711-5115	—	—	Ryall and Lipton (1972), Smith (1957)
Berries							
Blackberries	3458-5043	6304-10,086	11,527-20,893	15,489-32,060	28,818-43,227	—	IIR (1967)
Blueberries	505-2306	2017-2702	—	7529-13,616	11,419-19,236	—	Lutz and Hardenburg (1968)
Cranberries	*b	901-1008	—	—	2413-3999	—	Anderson et al. (1963), Lutz and Hardenburg (1968)
Gooseberries	1513-1909	2702-2990	—	4791-7096	—	—	Lutz and Hardenburg (1968), Smith (1966)
Raspberries	3890-5512	6808-8501	6124-12,248	18,119-22,334	25,215-54,033	—	Haller et al. (1941), IIR (1967), Lutz and Hardenburg (1968)
Strawberries	2702-3890	3602-7313	10,807-20,893	15,634-20,317	22,514-43,154	37,247-46,468	IIR (1967), Lutz and Hardenburg (1968), Maxie et al. (1959)
Broccoli, sprouting	4107-4719	7601-35,226	—	38,256-74,890	61,274-75,106	85,805-23,376	Morris (1947), Lutz and Hardenburg (1968), Scholz et al. (1963)
Brussels sprouts	3386-5295	7096-10,698	13,904-18,623	21,037-23,523	19,848-41,894	—	Sastry et al. (1978), Smith (1957)
Cabbage							
Penn State[c]	865	2089-2234	—	4935-6988	—	—	Van den Berg and Lentz (1972)
White, winter	1081-1801	1621-3062	2702-3962	4323-5944	7925-9006	—	IIR (1967)
spring	2089-2990	3890-4719	6412-7313	11,815-12,609	—	—	Sastry et al. (1978), Smith (1957)
Red, early	1693-2161	3423-3783	5224-61,238	8105-9366	12,248-12,608	—	IIR (1967)
Savoy	3422-4683	5584-6484	11,527-13,509	19,272-21,794	28,818-32,420	—	IIR (1967)
Carrots, roots							
Imperator, Texas	3386	4323	6916	8718	15,526	—	Scholz et al. (1963)
Main crop, United Kingdom	757-1513	1296-2666	2161-3423	6448-14,589 at 65°F	—	—	Smith(1957)
Nantes, Canada[d]	684	1477	—	4755-6232	—	—	Van den Berg and Lentz (1972)
Cauliflower							
Texas	3926	4503	7456	10,158	17,687	—	Scholz et al. (1963)
United Kingdom	1693-5295	4323-6015	9006-10,734	14,841-18,047	—	—	Smith (1957)
Celery							
New York, white	1585	2413	—	8215	14,229	—	Lutz and Hardenburg (1968)
United Kingdom	1117-1585	2017-2810	4323-6015	8609-9221 at 65°F	—	—	Smith(1957)
Utah, Canada[e]	1117	1982	—	6556	—	—	Van den Berg and Lentz (1972)
Cherries							
Sour	296-2918	2810-2918	—	6015-11,022	8609-11,022	11,708-15,634	Hawkins (1929), Lutz and Hardenburg (1968)

Table 9 Heat of Respiration of Fresh Fruits and Vegetables Held at Various Temperatures (*Continued*)

Commodity	Heat of Respiration, Btu/day per Ton of Produce						Reference
	32°F	41°F	50°F	59°F	68°F	77°F	
Sweet	901-1189	2089-3098	—	5512-9907	6196-7025	—	Gerhardt et al. (1942), Lutz and Hardenburg (1968), Micke et al. (1965)
Corn, sweet with husk, Texas	9366	17,111	24,676	35,878	63,543	89,695	Scholz et al. (1963)
Cucumbers, California	*b	*b	5079-6376	5295-7313	6844-10,591	—	Eaks and Morris (1956)
Figs, mission	—	2413-2918	4863-5079	10,807-13,940	12,536-20,929	18,731-20,929	Claypool and Ozbek (1952), Lutz and Hardenburg (1968)
Garlic	648-2413	1296-2125	2017-2125	2413-6015	2197-3999	—	Mann and Lewis (1956), Sastry et al. (1978)
Grapes							
Labrusca, Concord	612	1189	—	3494	7204	8501	Lutz (1938), Lutz and Hardenburg (1968)
Vinifera, Emperor	288-505	684-1296	1801	2197-2594	—	5512-6628	Lutz and Hardenburg (1968), Pentzer et al. (1933)
Thompson seedless	432	1045	1693	—	—	—	Wright et al. (1954)
Ohanez	288	720	2	—	—	—	Wright et al. (1954)
Grapefruit							
California Marsh	*b	*b	*b	2594	3890	4791	Haller et al. (1945)
Florida	*b	*b	*b	2810	3494	4214	Haller et al. (1945)
Horseradish	1801	2377	5800	7204	9834	—	Sastry et al. (1978)
Kiwifruit	616	1455	2889	—	3858-4254	—	Saravacos and Pilsworth (1965)
Kohlrabi	2197	3602	6916	10,807	—	—	Sastry et al. (1978)
Leeks	2089-3062	4323-6412	11,815-15,021	18,227-25,756	—	—	Sastry et al. (1978), Smith (1957)
Lemons, California, Eureka	*b	*b	*b	3494	5007	5727	Haller et al. (1945)
Lettuce							
Head, California	2017-3711	2918-4395	6015-8826	8501-9006	13,220	—	Sastry et al. (1978)
Texas	2306	2918	4791	7925	12,536	181 at 180°F	Lutz and Hardenburg, (1968), Watt and Merrill (1963)
Leaf, Texas	5079	6448	8681	13,869	22,118	32,275	Scholz et al. (1963)
Romaine, Texas	—	4575	7817	9762	15,093	23,883	Scholz et al. (1963)
Limes, Persian	*b	*b	576-1261	1296-2306	1513-4107	3314-10,014	Lutz and Hardenburg (1968)
Mangoes	*b	*b	—	9907	16,534-33,356	26,441	Gore (1911), Karmarkar and Joshe (1941b), Lutz and Hardenburg (1968)
Melons							
Cantaloupes	*b	1909-2197	3423	7420-8501	9834-14,229	13,725-15,741	Lutz and Hardenburg (1968), Sastry et al. (1978), Scholz et al. (1963)
Honeydew	—	*b	1765	2594-3494	4395-5259	5800-7601	Lutz and Hardenburg (1968), Pratt and Morris (1958), Scholz (1963)
Watermelon	*b	*b	1657	—	3818-5512	—	Lutz and Hardenburg (1968), Scholz et al. (1963)
Mint[l]	1769-3306	6614	16,754-20,061	23,148-29,981	36,595-50,041	56,655-69,883	Hruschka and Want (1979)
Mushrooms	6196-9618	15,634	—	—	58,104-69,738	—	Lutz and Hardenburg (1968), Smith (1964)
Nuts (kind not specified)	181	360	720	720	1081	—	IIR (1967)
Okra, Clemson	*b	76,043	19,236	32,132	57,527	76,040 at 85°F	Scholz et al. (1963)
Olives, Manzanillo	*b	*b		4791-8609	8501-10,807	9006-13,436	Maxie et al. (1959)
Onions							
Dry, Autumn Spice[f]	505-684	793-1477	—	2089-5548	—	—	Van den Berg and Lentz (1972)
White Bermuda	648	757	1585	2449	3711	6196 at 80°F	Scholz et al. (1963)
Green, New Jersey	2306-4899	3819-15,021	7961-12,968	14,553-21,434	17,205-34,225	21,541-46,217	Lutz and Hardenburg (1968)
Oranges							
Florida	684	1405	2702	4611	6628	7817 at 80°F	Haller et al. (1945)
California, w. navel	*b	1405	2990	5007	6015	7997	Haller et al. (1945)
Valencia	*b	1008	2594	2810	3890	4611	Haller et al. (1945)
Papayas	*b	*b	2485	3314-4791	—	8609-21,613	Jones (1942), Pantastico (1974)
Parsley[l]	7277-10,140	14,549-18,738	28,879-36,155	31,746-49,163	43,208-56,216	67,902-75,174	Hruschka and Want (1979)

Table 9 Heat of Respiration of Fresh Fruits and Vegetables Held at Various Temperatures (*Continued*)

Commodity	Heat of Respiration, Btu/day per Ton of Produce						Reference
	32°F	41°F	50°F	59°F	68°F	77°F	
Parsnips							
United Kingdom	2558-3423	1946-3854	4503-5800	7096-9438	—	—	Smith (1957)
Canada, Hollow Crown[g]	793-1801	1369-3386	—	4755-10,195	—	—	Van den Berg and Lentz (1972)
Peaches							
Elberta	829	1441	3458	7565	13,509	19,812 at 80°F	Haller et al. (1932)
Several cultivars	901-1405	1405-2017	—	7313-9330	13,040-22,549	17,939-26,837	Lutz and Hardenburg (1968)
Peanuts							
Cured[h]	3 at 85°F					51 at 85°F	Thompson et al. (1951)
Not cured, Virginia Bunch[i]						3120 at 85°F	Schenk (1959, 1961)
Dixie Spanish						1823 at 85°F	Schenk (1959, 1961)
Pears							
Bartlett	684-1513	1117-2197	—	3314-13,220	6628-15,417	—	Lutz and Hardenburg (1968)
Late ripening	576-793	1296-3062	1729-4143	6124-9366	7204-16,210	—	IIR (1967)
Early ripening	576-1081	1621-3423	2161-4683	7565-11,887	8645-19,812	—	IIR (1967)
Peas							
Green-in-pod	6700-10,302	12,139-16,822	—	39,372-44,595	54,105-79,645	75,646-83,067	Lutz and Hardenburg (1968), Tewfik and Scott (1954)
Shelled	10,410-16,642	17,435-21,444	—	—	76,871-10,893	—	Lutz and Hardenburg (1968), Tewfik and Scott (1954)
Peppers, sweet	*[b]	*[b]	3170	5043	9654	—	Scholz et al. (1963)
Persimmons		1296		2594-3098	4395-5295	6412-8826	Gore (1911), Lutz and Hardenburg (1968)
Pineapple							
Mature green	*[b]	*[b]	1225	2846	5331	7817 at 80°F	Scholz et al. (1963)
Ripening	*[b]	*[b]	1657	3999	8790	13,797	Scholz et al. (1963)
Plums, Wickson	432-648	865-1982	1981-2522	2630-2737	3962-5727	6160-15,634	Claypool and Allen (1951)
Potatoes							
California white, rose							
immature	*[b]	2594	3098-4611	3098-6808	3999-9932		Sastry et al. (1978)
mature	*[b]	1296-1513	1467-2197	1467-2594	1467-3494		Sastry et al. (1978)
very mature	*[b]	1117-1513	1513	1513-2197	2017-2630		Sastry et al. (1978)
Katahdin, Canada[j]	*[b]	865-936		1729-2234			Van den Berg and Lentz (1972)
Kennebec	*[b]	793-936		936-1982			Van den Berg and Lentz (1972)
Radishes							
With tops	3206-3818	4214-4611	6808-8105	15,417-17,146	27,341-30,043	34,869-42,470	Lutz and Hardenburg (1968)
Topped	1189-1296	1693-1801	3314-3494	6124-7204	10,519-10,807	14,841-16,751	Lutz and Hardenburg (1968)
Rhubarb, topped	1801-2918	2413-3999		6808-10,014	8826-12,536		Hruschka (1966)
Rutabaga, Laurentian, Canada[k]	432-612	1045-1124		2342-3458			Van den Berg and Lentz (1972)
Spinach							
Texas		10,122	24,387	39,409	50,683		Scholz et al. (1963)
United Kingdom, summer	2558-4719	6015-7096	12,896-16,534		40,777-47,657 at 65°F		Smith (1957)
winter	3854-5584	6448-13,869	15,021-22,766		42,938-53,673 at 65°F		Smith (1957)
Squash							
Summer, yellow, straight-neck	†[b]	†[b]	7709-8105	16,534-20,028	18,731-21,434		Lutz and Hardenburg (1968)
Winter butternut	*[b]	*[b]				16,318-26,908	Lutz and Hardenburg (1968)
Sweet Potatoes							
Cured, Puerto Rico	*[b]	*[b]	†[b]	3530-4863			Lewis and Morris (1956)
Yellow Jersey	*[b]	*[b]	†[b]	4863-5079			Lewis and Morris (1956)
Noncured	*[b]	*[b]	*[b]	6304		11,923-16,138	Lutz and Hardenburg (1968)
Tomatoes							
Texas, mature green	*[b]	*[b]	*[b]	4503	7637	9402 at 80°F	Scholz et al. (1963)
ripening	*[b]	*[b]	*[b]	5872	8933	10,627 at 80°F	Scholz et al. (1963)
California mature green	*[b]	*[b]	*[b]		5295-7709	6592-10,591	Workman and Pratt (1957)

Table 9 Heat of Respiration of Fresh Fruits and Vegetables Held at Various Temperatures (*Continued*)

Commodity	Heat of Respiration, Btu/day per Ton of Produce						Reference
	32°F	41°F	50°F	59°F	68°F	77°F	
Turnip, roots	1909	2089-2197		4719-5295	5295-5512		Lutz and Hardenburg (1968)
Watercress[l]	3306	9920	20,061-26,674	29,981-43,208	66,576-76,719	76,720-96,561	Hruschka and Want (1979)

[a]Column headings indicate temperatures at which respiration rates were determined, within 2°F, except where the actual temperatures are given.

[b]The symbol * denotes a chilling temperature. The symbol † denotes the temperature is borderline, not damaging to some cultivars if exposure is short.

[c]Rates are for 30 to 60 days and 60 to 120 days storage, the longer storage having the higher rate, except at 32°F, where they were the same.

[d]Rates are for 30 to 60 days and 120 to 180 days storage, respiration increasing with time only at 59°F.

[e]Rates are for 30 to 60 days storage.

[f]Rates are for 30 to 60 days and 120 to 180 days storage; rates increased with time at all temperatures as dormancy was lost.

[g]Rates are for 30 to 60 days and 120 to 180 days; rates increased with time at all temperatures.

[h]Shelled peanuts with about 7% moisture. Respiration after 60 h curing was almost negligible, even at 85°F.

[i]Respiration for freshly dug peanuts, not cured, with about 35 to 40% moisture. During curing, peanuts in the shell were dried to about 5 to 6% moisture, and in roasting are dried further to about 2% moisture.

[j]Rates are for 30 to 60 days and 120 to 180 days with rate declining with time at 41°F but increasing at 59°F as sprouting started.

[k]Rates are for 30 to 60 days and 120 to 180 days; rates increased with time, especially at 59°F where sprouting occurred.

[l]Rates are for 1 day after harvest.

Table 10 Change in Respiration Rates with Time

Commodity	Days in Storage	Heat of Respiration, Btu/day per Ton of Produce		Reference
		32°F	41°F	
Apples, Grimes	7	648	2882 at 50°F	Harding (1929)
	30	648	3854	
	80	648	2413	
Artichokes, globe	1	9907	13,220	Rappaport and Watada (1958)
	4	5512	7709	
	16	3314	5727	
Asparagus, Martha Washington	1	17,652	2316	Lipton (1957)
	3	8682	14,337	
	16	6160	6629	
Beans, lima, in pod	2	6593	7925	Tewfik and Scott (1954)
	4	4431	6376	
	6	3890	5836	
Blueberries, Blue Crop	1	1585	—	
	2	584	—	
		1261	—	
Broccoli, Waltham 29	1	—	16,102	
	4	—	9690	
	8	—	7277	
Corn, sweet, in husk	1	11,312	—	Scholz et al. (1963)
	2	8106	—	
	4	6772	—	
Figs, Mission	1	2882	—	Claypool and Ozbek (1952)
	2	2630	—	
	12	2630	—	
Garlic	10	865	1982	Mann and Lewis (1956)
	30	1333	3314	
	180	3098	7277	
Lettuce, Great Lakes	1	3747	4395	Pratt et al. (1954)
	5	1982	33	
	10	1765	3314	
Olives, Manzanillo	1	—	8610 at 60°F	Maxie et al. (1960)
	5	—	6376	
	10	—	4864	
Onions, red	1	360	—	Karmarkar and Joshe (1941a)
	30	541	—	
	120	720	—	
Plums, Wickson	2	432	865	Claypool and Allen (1951)
	6	432	1549	
	18	648	1982	
Potatoes	2	—	1333	Morris (1959)
	6	—	1765	
	10	—	1549	
Strawberries, Shasta	1	3873	6305	Maxie et al. (1959)
	2	2918	6772	
	5	2918	7277	
Tomatoes, Pearson, mature green	5	—	706 at 70°F	Workman and Pratt (1957)
	15	—	6160	
	20	—	5295	

surface. The quantity $(p_s - p_a)$ is the water vapor pressure deficit. The water vapor pressure at the commodity surface p_s is the water vapor saturation pressure evaluated at the commodity surface temperature; the water vapor pressure in the surrounding air p_a is a function of the relative humidity of the air.

In its simplest form, the transpiration coefficient k_t is considered to be constant for a particular commodity. Table 11 lists values for the transpiration coefficients k_t of various fruits and vegetables (Sastry et al. 1978). Because of the many factors that influence transpiration rate, not all the values in Table 11 are reliable. They are to be used primarily as a guide or as a comparative indication of various commodity transpiration rates obtained from the literature.

Fockens and Meffert (1972) modified the simple transpiration coefficient to model variable skin permeability and to account for airflow rate. Their modified transpiration coefficient takes the following form:

$$k_t = \frac{1}{\dfrac{1}{k_a} + \dfrac{1}{k_s}} \qquad (43)$$

where k_a is the air film mass transfer coefficient and k_s is the skin mass transfer coefficient. The variable k_a describes the convective mass transfer that occurs at the surface of the commodity and is a function of airflow rate. The variable k_s describes the skin's diffusional resistance to moisture migration.

The air film mass transfer coefficient k_a can be estimated by using the Sherwood-Reynolds-Schmidt correlations (Becker et al. 1996b). The Sherwood number is defined as follows:

Table 11 Transpiration Coefficients of Certain Fruits and Vegetables

Commodity and Variety	Transpiration Coefficient, ppm/h·in. Hg	Commodity and Variety	Transpiration Coefficient, ppm/h·in. Hg	Commodity and Variety	Transpiration Coefficient, ppm/h·in. Hg
Apples		**Leeks**		**Pears**	
Jonathan	430	Musselburgh	12,600	Passe Crassane	974
Golden Delicious	710	*Average for all varieties*	**9600**	Beurre Clairgeau	986
Bramley's Seedling	510	**Lemons**		*Average for all varieties*	**840**
Average for all varieties	**510**	Eureka			
Brussels Sprouts		dark green	2760	**Plums**	
Unspecified	40,100	yellow	1700	Victoria	
Average for all varieties	**75,000**	*Average for all varieties*	**2270**	unripe	2410
Cabbage		**Lettuce**		ripe	1400
Penn State ballhead		Unrivalled	106,000	Wickson	1510
trimmed	3300	*Average for all varieties*	**90,200**	*Average for all varieties*	**1660**
untrimmed	4920	**Onions**			
Mammoth		Autumn Spice		**Potatoes**	
trimmed	2920	uncured	1170	Manona	
Average for all varieties	**2720**	cured	535	mature	304
Carrots		Sweet White Spanish		Kennebec	
Nantes	20,000	cured	1500	uncured	2080
Chantenay	21,500	*Average for all varieties*	**730**	cured	730
Average for all varieties	**14,700**	**Oranges**		Sebago	
Celery		Valencia	710	uncured	1920
Unspecified varieties	25,400	Navel	1270	cured	462
Average for all varieties	**21,500**	*Average for all varieties*	**1430**	*Average for all varieties*	**540**
Grapefruit		**Parsnips**			
Unspecified varieties	380	Hollow Crown	23,500	**Rutabagas**	
Marsh	670			Laurentian	5710
Average for all varieties	**990**	**Peaches**			
Grapes		Redhaven			
Emperor	960	hard mature	11,200	**Tomatoes**	
Cardinal	1220	soft mature	12,400	Marglobe	864
Thompson	2480	Elberta	3330	Eurocross BB	1410
Average for all varieties	**1500**	*Average for all varieties*	**6970**	*Average for all varieties*	**1710**

Note: Sastry et al. (1978) gathered these data as part of a literature review. Averages reported are the average of all published data found by Sastry et al. for each commodity. Specific varietal data were selected because they considered them highly reliable.

$$\text{Sh} = \frac{k'_a d}{\delta} \tag{44}$$

where k'_a is the air film mass transfer coefficient, d is the commodity's diameter, and δ is the coefficient of diffusion of water vapor in air. For convective mass transfer from a spherical fruit or vegetable, Becker and Fricke (1996b) recommend using the following Sherwood-Reynolds-Schmidt correlation, which was taken from Geankoplis (1978):

$$\text{Sh} = 2.0 + 0.552\,\text{Re}^{0.53}\,\text{Sc}^{0.33} \tag{45}$$

Re is the Reynolds number ($\text{Re} = u_\infty d/\nu$) and Sc is the Schmidt number ($\text{Sc} = \nu/\delta$), where u_∞ is the free stream air velocity and ν is the kinematic viscosity of air. The driving force for k'_a is concentration. However, the driving force in the transpiration model is vapor pressure. Thus, the following conversion from concentration to vapor pressure is required:

$$k_a = \frac{1}{R_{wv}T}k'_a \tag{46}$$

where R_{wv} is the gas constant for water vapor and T is the absolute mean temperature of the boundary layer.

The skin mass transfer coefficient k_s, which describes the resistance to moisture migration through the skin of a commodity, is based on the fraction of the product surface covered by pores. Although it is difficult to theoretically determine the skin mass transfer coefficient, experimental determination has been performed by Chau et al. (1987) and Gan and Woods (1989). These experimental values of k_s are given in Table 12, along with estimated values of k_s for grapes,

Table 12 Commodity Skin Mass Transfer Coefficient

	Skin Mass Transfer Coefficient, k_s, lb/ft²·h·in. Hg			
Commodity	**Low**	**Mean**	**High**	**Standard Deviation**
Apples	2.77×10^{-4}	4.17×10^{-4}	5.67×10^{-4}	7.49×10^{-5}
Blueberries	2.38×10^{-3}	5.47×10^{-3}	8.46×10^{-3}	1.60×10^{-3}
Brussels sprouts	2.41×10^{-2}	3.32×10^{-2}	4.64×10^{-2}	6.09×10^{-3}
Cabbage	6.24×10^{-3}	1.68×10^{-2}	3.25×10^{-2}	7.09×10^{-3}
Carrots	7.94×10^{-2}	3.90×10^{-1}	9.01×10^{-1}	1.90×10^{-1}
Grapefruit	2.72×10^{-3}	4.19×10^{-3}	5.54×10^{-3}	8.24×10^{-4}
Grapes	—	1.00×10^{-3}	—	—
Green peppers	1.36×10^{-3}	5.39×10^{-3}	1.09×10^{-2}	1.77×10^{-3}
Lemons	2.72×10^{-3}	5.19×10^{-3}	8.74×10^{-3}	1.60×10^{-3}
Lima beans	8.16×10^{-3}	1.08×10^{-2}	1.43×10^{-2}	1.47×10^{-3}
Limes	2.60×10^{-3}	5.54×10^{-3}	8.69×10^{-3}	1.40×10^{-3}
Onions	—	2.22×10^{-3}	—	—
Oranges	3.45×10^{-3}	4.29×10^{-3}	5.34×10^{-3}	5.24×10^{-4}
Peaches	3.40×10^{-3}	3.55×10^{-2}	1.15×10^{-1}	1.30×10^{-2}
Pears	1.31×10^{-3}	1.71×10^{-3}	3.00×10^{-3}	3.72×10^{-4}
Plums	—	3.44×10^{-3}	—	—
Potatoes	—	1.59×10^{-3}	—	—
Rutabagas (swedes)	—	2.91×10^{-1}	—	—
Snap beans	8.64×10^{-3}	1.41×10^{-2}	2.50×10^{-2}	4.42×10^{-3}
Sugar beets	2.27×10^{-2}	8.39×10^{-2}	2.18×10^{-1}	5.02×10^{-2}
Strawberries	9.86×10^{-3}	3.40×10^{-2}	6.62×10^{-2}	1.20×10^{-2}
Tomatoes	5.42×10^{-4}	2.75×10^{-3}	6.07×10^{-3}	1.67×10^{-3}

Source: Becker and Fricke (1996a)

onions, plums, potatoes, and rutabagas. Note that three values of skin mass transfer coefficient are tabulated for most commodities. These values correspond to the spread of the experimental data.

SURFACE HEAT TRANSFER COEFFICIENT

Although the surface heat transfer coefficient is not a thermal property of a food or beverage, it is needed to design heat transfer equipment for processing foods and beverages where convection is involved. Newton's law of cooling defines the surface heat transfer coefficient h as follows:

$$q = hA(t_s - t_\infty) \quad (47)$$

where q is the heat transfer rate, t_s is the surface temperature of the food, t_∞ is the surrounding fluid temperature, and A is the surface area of the food through which the heat transfer occurs.

The surface heat transfer coefficient h depends on the velocity of the surrounding fluid, product geometry, orientation, surface roughness, and packaging, as well as other factors. Therefore, for most applications h must be determined experimentally. Researchers have generally reported their findings as correlations, which give the Nusselt number as a function of the Reynolds number and the Prandtl number.

Experimentally determined values of the surface heat transfer coefficient are given in Table 13. The following guidelines are important for using the table:

- Use a Nusselt-Reynolds-Prandtl correlation or a value of the surface heat transfer coefficient that applies to the Reynolds number called for in the design.
- Avoid extrapolations.
- Use data for the same heat transfer medium, including temperature and temperature difference, that are similar to the design conditions. The proper characteristic length and fluid velocity, either free stream or interstitial, should be used in calculating the Reynolds and Nusselt numbers.

Evaluation of Thermophysical Property Models

Numerous composition-based thermophysical property models have been developed, and selecting appropriate ones from those available can be challenging. Becker and Fricke (1999) and Fricke and Becker (2001, 2002) quantitatively evaluated selected thermophysical property models by comparison to a comprehensive experimental thermophysical property data set compiled from the literature. They found that for ice fraction prediction, the equation by Chen (1985) performed best, followed closely by that of Tchigeov (1979). For apparent specific heat capacity, the model of Schwartzberg (1976) performed best, and for specific enthalpy prediction, the Chen (1985) equation gave the best results. Finally, for thermal conductivity, the model by Levy (1981) performed best.

Table 13 Surface Heat Transfer Coefficients for Food Products

1	2	3	4	5	6	7	8	9	10
Product	**Shape and Length, in.**[a]	**Transfer Medium**	**Δt and/or Temp. t of Medium, °F**	**Velocity of Medium, ft/s**	**Reynolds Number Range**[b]	**h, Btu/h·ft²·°F**	**Nu-Re-Pr Correlation**[c]	**Reference**	**Comments**
Apple	Spherical	Air	$t = 81$	0	N/A	2.0	N/A	Kopelman et al. (1966)	N/A indicates that data were not reported in original article
Jonathan	2.0			1.3		3.0			
				3.0		4.8			
				6.7		8.0			
				17.0		9.4			
	2.3			0		2.0			
				1.3		3.0			
				3.0		4.9			
				6.7		7.9			
				17.0		9.6			
	2.4			0		2.0			
				1.3		2.8			
				3.0		4.6			
				6.7		6.9			
				17.0		8.9			
Red Delicious	2.5	Air	$\Delta t = 41$, $t = 31$	4.9	N/A	4.8	N/A	Nicholas et al. (1964)	Thermocouples at center of fruit
				15.0		10.0			
	2.8			4.9		2.5			
				15.0		6.5			
	3.0			0		1.8			
				4.9		4.0			
				9.8		5.8			
				15.0		6.1			
	2.2	Water	$\Delta t = 46$	0.90		16.0			
	2.8		$t = 32$			14.0			
	3.0					9.8			
Beef carcass	142 lb*	Air	$t = -3$	5.9	N/A	3.8	N/A	Fedorov et al. (1972)	*For size indication
	187 lb*			1.0		1.8			
patties	Slab	Air	$t = -26$ to -18	9.2 to 20	2000 to 7500	N/A	$Nu = 1.37 Re^{0.282} Pr^{0.3}$	Becker and Fricke (2004)	Unpackaged patties. Characteristic dimension is patty thickness. 7 points in correlation.
Cake	Cylinder or brick	Air	$t = -40$ to 32	6.9 to 9.8	4000 to 80,000	N/A	$Nu = 0.00156 Re^{0.960} Pr^{0.3}$	Becker and Fricke (2004)	Packaged and unpackaged. Characteristic dimension is cake height. 29 points in correlation.

Table 13 Surface Heat Transfer Coefficients for Food Products (*Continued*)

1	2	3	4	5	6	7	8	9	10
Product	**Shape and Length, in.**[a]	**Transfer Medium**	**Δt and/or Temp. t of Medium, °F**	**Velocity of Medium, ft/s**	**Reynolds Number Range**[b]	**h, Btu/ $h \cdot ft^2 \cdot °F$**	**Nu-Re-Pr Correlation**[c]	**Reference**	**Comments**
Cheese	Brick	Air	$t = -29$ to 36	9.8	6000 to 30,000	N/A	$Nu = 0.0987 Re^{0.560} Pr^{0.3}$	Becker and Fricke (2004)	Packaged and unpackaged. Characteristic dimension is minimum dimension. 7 points in correlation.
Cucumbers	Cylinder 1.5	Air	$t = 39$	3.28 4.10 4.92 5.74 6.56	N/A	3.2 305 3.8 4.1 4.7	$Nu = 0.291 Re^{0.592} Pr^{0.333}$	Dincer (1994)	Diameter = 38 mm Length = 160 mm
Eggs, Jifujitori	1.3	Air	$\Delta t = 81$	6.6 to 26	6000 to 15,000	N/A	$Nu = 0.46 Re^{0.56}$ $\pm 1.0\%$	Chuma et al. (1970)	5 points in correlation
Leghorn	1.7	Air	$\Delta t = 81$	6.6 to 26	8000 to 25,000	N/A	$Nu = 0.71 Re^{0.55}$ $\pm 1.0\%$	Chuma et al. (1970)	5 points in correlation
Entrees	Brick	Air	$t = -36$ to 32	9.2 to 16	5000 to 20,000	N/A	$Nu = 1.31 Re^{0.280} Pr^{0.3}$	Becker and Fricke (2004)	Packaged. Characteristic dimension is minimum dimension. 42 points in correlation.
Figs	Spherical 1.85	Air	$t = 39$	3.61 4.92 5.74 8.20	N/A	4.2 4.6 4.8 5.8	$Nu = 1.560 Re^{0.426} Pr^{0.333}$	Dincer (1994)	
Fish, Pike, perch, sheatfish	N/A	Air	N/A	3.2 to 22	5000 to 35,000	N/A	$Nu = 4.5 Re^{0.28} \pm 10\%$	Khatchaturov (1958)	32 points in correlation
Fillets	N/A	Air	$t = -40$ to -18	8.9 to 23	1000 to 25,000	N/A	$Nu = 0.0154 Re^{0.818} Pr^{0.3}$	Becker and Fricke (2004)	Packaged and unpackaged. Characteristic dimension is minimum dimension. 28 points in correlation.
Grapes	Cylinder 0.43	Air	$t = 39$	3.28 4.10 4.92 5.74 6.56	N/A	5.4 6.0 6.7 7.2 7.4	$Nu = 0.291 Re^{0.592} Pr^{0.333}$	Dincer (1994)	Diameter = 11 mm Length = 22 mm
Hams Boneless Processed	$G^* =$ 0.4 to 0.45 $^*G$ = Geometrical factor for shrink-fitted plastic bag	Air	$\Delta t = 132$ $t = 150$	N/A	1000 to 86,000	N/A	$Nu = 0.329 Re^{0.564}$	Clary et al. (1968) $G = 1/4 + 3/(8A^2) + 3/(8B^2)$ $A = a/Z$, $B = b/Z$ A = characteristic length = 0.5 min. dist. ⊥ to airflow a = minor axis b = major axis Correlation on 18 points Recalc. with min. distance ⊥ to airflow Calculated Nu with 1/2 char. length	
	N/A	Air	$t = -10$ $t = -55$ $t = -60$ $t = -70$ $t = -80$	2.0	N/A	3.6 3.6 3.5 3.5 3.2	N/A	Van den Berg and Lentz (1957)	38 points total Values are averages
Meat	Slabs 0.91 thick	Air	$t = 32$	1.8 4.6 12.0	N/A	1.9 3.5 6.2	N/A	Radford et al. (1976)	
Oranges, grapefruit, tangelos, bulk packed	Spheroids 2.3 3.1 2.1	Air	$\Delta t = 70$ to 56 $t = 16$	0.36–1.1	35,000 to 135,000	11.7*	$Nu = 5.05 Re^{0.333}$	Bennett et al. (1966) Bins 42 × 42 × 16 in. 36 points in correlation. Random packaging. Interstitial velocity. *Average for oranges	
	Spheroids 3.0 4.2	Air	$\Delta t = 91$ $t = 32$	0.17 to 6.7	180 to 18,000	N/A	$Nu = 1.17 Re^{0.529}$	Baird and Gaffney (1976)	20 points in correlation Bed depth: 26 in.
Peas Fluidized bed	Spherical N/A	Air	$t = -15$ to -35	4.9 to 2.4 ± 1.0	1000 to 4000	N/A	$Nu = 3.5 \times 10^{-4} Re^{1.5}$	Kelly (1965)	Bed depth: 2 in.
Bulk packed	Spherical N/A	Air	$t = -15$ to -35	4.9 to 2.4 ± 1.0	1000 to 6000	N/A	$Nu = 0.016 Re^{0.95}$	Kelly (1965)	

Table 13 Surface Heat Transfer Coefficients for Food Products (*Continued*)

1	2	3	4	5	6	7	8	9	10
Product	**Shape and Length, in.**[a]	**Transfer Medium**	**Δt and/or Temp. t of Medium, °F**	**Velocity of Medium, ft/s**	**Reynolds Number Range**[b]	**h, Btu/ $h \cdot ft^2 \cdot °F$**	**Nu-Re-Pr Correlation**[c]	**Reference**	**Comments**
Pears	Spherical 2.36	Air	$t = 39$	3.28 4.10 4.92 5.74 6.56	N/A	2.2 2.5 2.8 2.8 3.4	$Nu = 1.560 Re^{0.426} Pr^{0.333}$	Dincer (1994)	
Pizza	Slab	Air	$t = -29$ to -15	9.8 to 12	3000 to 12,000	N/A	$Nu = 0.00517 Re^{0.891} Pr^{0.3}$	Fricke and Becker (2004)	Packaged and unpackaged. Characteristic dimension is pizza thickness. 12 points in correlation.
Potatoes Pungo, bulk packed	Ellipsoid N/A N/A	Air	$t = 40$	2.2 4.0 4.5 5.7	3000 to 9000	2.5* 3.4 3.6 4.3	$Nu = 0.364 Re^{0.558} Pr^{1/3}$ (at top of bin)	Minh et al. (1969)	Use interstitial velocity to calculate Re Bin is 30 × 20 × 9 in. *Each h value is average of 3 reps with airflow from top to bottom
Patties, fried	Slab	Air	$t = -26$ to -18	7.5 to 11	1000 to 6000	N/A	$Nu = 0.00313 Re^{1.06} Pr^{0.3}$	Becker and Fricke (2004)	Unpackaged. Characteristic dimension is patty thickness. 8 points in correlation.
Poultry Chickens, turkeys	2.6 to 20.8 lb*	**	$\Delta t = 32$	***	N/A	74 to 83	N/A	Lentz (1969)	Vacuum packaged *To give indications of size. **$CaCl_2$ Brine, 26% by mass ***Moderately agitated Chickens 2.4 to 6.4 lb Turkeys 11.9 to 21 lb
Chicken breast	N/A	Air	$t = -29$ to 28	3.3 to 9.8	1000 to 11,000	N/A	$Nu = 0.0378 Re^{0.837} Pr^{0.3}$	Becker and Fricke (2004)	Unpackaged. Characteristic dimension is minimum dimension. 22 points in correlation.
Sausage	Cylinder	Air	$t = -40$ to 8.6	8.9 to 9.8	4500 to 25,000	N/A	$Nu = 7.14 Re^{0.170} Pr^{0.3}$	Becker and Fricke (2004)	Unpackaged. Characteristic dimension is sausage diameter. 14 points in correlation.
Soybeans	Spherical 2.6	Air	N/A	22	1200 to 4600	N/A	$Nu = 1.07 Re^{0.64}$	Otten (1974)	8 points in correlation Bed depth: 1.3 in.
Squash	Cylinder 1.8	Water	1.64 3.28 4.92	0.16	N/A	47.9 36.1 29.2	N/A	Dincer (1993)	Diameter = 1.8 in. Length = 6.1 in.
Tomatoes	Spherical 2.75	Air	$t = 39$	3.28 4.10 4.92 5.74 6.56	N/A	1.9 2.3 2.4 2.6 3.0	$Nu = 1.560 Re^{0.426} Pr^{0.333}$	Dincer (1994)	
Karlsruhe substance	Slab 3.0	Air	$\Delta t = 96$ $t = 100$	N/A	N/A	2.9	N/A	Cleland and Earle (1976)	Packed in aluminum foil and brown paper
Milk Container	Cylinder 2.8 × 3.9 2.8 × 5.9 2.8 × 9.8	Air	$\Delta t = 9.5$	N/A	$Gr = 10^6$ to 5×10^7	N/A	$Nu = 0.754 Gr^{0.264}$	Leichter et al. (1976)	Emissivity = 0.7 300 points in correlation L = characteristic length All cylinders 2.8 in. dia.
Acrylic	Ellipsoid 3.0 (minor axis) G = 0.297 to 1.0	Air	$\Delta t = 80$	6.9 to 26	12,000 to 50,000	N/A	$Nu = aRe^b$ $a = 0.32 - 0.22G$ $b = 0.44 + 0.23G$	Smith et al. (1971)	$G = 1/4 + 3/(8A^2) + 3/(8B^2)$ A = minor length/char. length B = major length/char. length Char. length = 0.5 × minor axis Use twice char. length to calculate Re
	Spherical 3.0	Air	$t = 24$	2.17 4.04 4.46 5.68	3700 to 10,000	2.6* 2.5 3.9 3.8	$Nu = 2.58 Re^{0.303} Pr^{1/3}$	Minh et al. (1969)	Random packed. Interstitial velocity used to calculate Re Bin dimensions: 30 × 18 × 24 in. *Values for top of bin

[a]Characteristic length is used in Reynolds number and illustrated in the Comments column (10) where appropriate.
[b]Characteristic length is given in column 2; free stream velocity is used, unless specified otherwise in the Comments column (10).
[c]Nu = Nusselt number, Re = Reynolds number, Gr = Grashof number, Pr = Prandtl number.

SYMBOLS

a = parameter in Equation (26): $a = 3k_c/(2k_c + k_d)$
A = surface area
b = parameter in Equation (26): $b = V_d/(V_c + V_d)$
c = specific heat
c_a = apparent specific heat
c_f = specific heat of fully frozen food
c_i = specific heat of ith food component
c_p = constant-pressure specific heat
c_u = specific heat of unfrozen food
d = commodity diameter
E = ratio of relative molecular masses of water and solids: $E = M_w/M_s$
f = respiration coefficient given in Table 8
F_1 = parameter given by Equation (32)
g = respiration coefficient given in Table 8
Gr = Grashof number
h = surface heat transfer coefficient
H = enthalpy
H_f = enthalpy at initial freezing temperature
H_i = enthalpy of ith food component
k = thermal conductivity
k_1 = thermal conductivity of component 1
k_2 = thermal conductivity of component 2
k'_a = air film mass transfer coefficient (driving force: vapor pressure)
k_a = air film mass transfer coefficient (driving force: concentration)
k_c = thermal conductivity of continuous phase
k_d = thermal conductivity of discontinuous phase
k_i = thermal conductivity of the ith component
k_s = skin mass transfer coefficient
k_t = transpiration coefficient
$k_=$ = thermal conductivity parallel to food fibers
$k_\perp$ = thermal conductivity perpendicular to food fibers
L^3 = volume fraction of discontinuous phase
L_o = latent heat of fusion of water at 32°F = 144 Btu/lb
m = mass
$\dot{m}$ = transpiration rate
M = parameter in Equation (28) = $L^2(1 - k_d/k_c)$
M_s = relative molecular mass of soluble solids
M_w = relative molecular mass of water
Nu = Nusselt number
N^2 = volume fraction of discontinuous phase
P = parameter in Equation (30) = $N(1 - k_d/k_c)$
Pr = Prandtl number
p_a = water vapor pressure in air
p_s = water vapor pressure at commodity surface
q = heat transfer rate
Q = heat transfer
R = universal gas constant = 1.986 Btu/lb mol·°R
R_1 = volume fraction of component 1
Re = Reynolds number
R_{wv} = universal gas constant for water vapor
Sc = Schmidt number
Sh = Sherwood number
t = food temperature, °F
t_f = initial freezing temperature of food, °F
t_r = reference temperature = –40°F
t_s = surface temperature, °F
t_∞ = ambient temperature, °F
T = food temperature, °R
T_f = initial freezing point of food, °R
T_o = freezing point of water; T_o = 491.7°R
T_r = reference temperature = 419.7°R (–40°F)
$\overline{T}$ = reduced temperature
u_∞ = free stream air velocity
V_c = volume of continuous phase
V_d = volume of discontinuous phase
W = rate of heat generation from respiration, Btu/h·lb
x_1 = mass fraction of component 1
x_a = mass fraction of ash
x_b = mass fraction of bound water
x_c = mass fraction of carbohydrate
x_f = mass fraction of fat
x_{fb} = mass fraction of fiber
x_i = mass fraction of ith food component
x_{ice} = mass fraction of ice
x_p = mass fraction of protein
x_s = mass fraction of solids
x_{wo} = mass fraction of water in unfrozen food
x_i^v = volume fraction of ith food component
y = correlation parameter in Equation (19)
z = correlation parameter in Equation (19)

Greek

α = thermal diffusivity
δ = diffusion coefficient of water vapor in air
Δc = difference in specific heats of water and ice = $c_{water} - c_{ice}$
ΔH = enthalpy difference
Δt = temperature difference
ε = porosity
θ = time
Λ = thermal conductivity ratio = k_1/k_2
ν = kinematic viscosity
ρ = density of food
ρ_1 = density of component 1
ρ_2 = density of component 2
ρ_i = density of ith food component
σ = parameter given by Equation (33)

REFERENCES

Acre, J.A. and V.E. Sweat. 1980. Survey of published heat transfer coefficients encountered in food processes. *ASHRAE Transactions* 86(2):235-260.

Anderson, R.E., R.E. Hardenburg, and H.C. Baught. 1963. Controlled atmosphere storage studies with cranberries. *Journal of the American Society for Horticultural Science* 83:416.

Babbitt, J.D. 1945. The thermal properties of wheat in bulk. *Canadian Journal of Research* 23F:338.

Baird, C.D. and J.J. Gaffney. 1976. A numerical procedure for calculating heat transfer in bulk loads of fruits or vegetables. *ASHRAE Transactions* 82:525-535.

Becker, B.R. and B.A. Fricke. 1996a. Transpiration and respiration of fruits and vegetables. In *New Developments in Refrigeration for Food Safety and Quality*, pp. 110-121. International Institute of Refrigeration, Paris, and American Society of Agricultural Engineers, St. Joseph, MI.

Becker, B.R. and B.A. Fricke. 1996b. Simulation of moisture loss and heat loads in refrigerated storage of fruits and vegetables. In *New Developments in Refrigeration for Food Safety and Quality*, pp. 210-221. International Institute of Refrigeration, Paris, and American Society of Agricultural Engineers, St. Joseph, MI.

Becker, B.R. and B.A. Fricke. 1999. Food thermophysical property models. *International Communications in Heat & Mass Transfer* 26(5):627-636.

Becker, B.R. and B.A. Fricke. 2004. Heat transfer coefficients for forced-air cooling and freezing of selected foods. *International Journal of Refrigeration* 27(5):540-551.

Becker, B.R., A. Misra, and B.A. Fricke. 1996a. A numerical model of moisture loss and heat loads in refrigerated storage of fruits and vegetables. Frigair '96 Congress and Exhibition, Johannesburg.

Becker, B.R., A. Misra, and B.A. Fricke. 1996b. Bulk refrigeration of fruits and vegetables, Part I: Theoretical considerations of heat and mass transfer. *International Journal of HVAC&R Research* (now *HVAC&R Research*) 2(2):122-134.

Becker, B.R., A. Misra, and B.A. Fricke. 1996c. Bulk refrigeration of fruits and vegetables, Part II: Computer algorithm for heat loads and moisture loss. *International Journal of HVAC&R Research* (now *HVAC&R Research*) 2(3):215-230.

Bennett, A.H. 1963. Thermal characteristics of peaches as related to hydrocooling. *Technical Bulletin* 1292. U.S. Department of Agriculture, Washington, D.C.

Bennett, A.H., W.G. Chace, and R.H. Cubbedge. 1964. Thermal conductivity of Valencia orange and Marsh grapefruit rind and juice vesicles. *ASHRAE Transactions* 70:256-259.

Bennett, A.H., J. Soule, and G.E. Yost. 1966. Temperature response of Florida citrus to forced-air precooling. *ASHRAE Journal* 8(4):48-54.

Bennett, A.H., W.G. Chace, and R.H. Cubbedge. 1969. Heat transfer properties and characteristics of Appalachian area, Red Delicious apples. *ASHRAE Transactions* 75(2):133.

Bennett, A.H., W.G. Chace, and R.H. Cubbedge. 1970. Thermal properties and heat transfer characteristics of Marsh grapefruit. *Technical Bulletin* 1413. U.S. Department of Agriculture, Washington, D.C.

Biale, J.B. 1960. Respiration of fruits. *Encyclopedia of Plant Physiology* 12:536.

Chang, H.D. and L.C. Tao. 1981. Correlations of enthalpies of food systems. *Journal of Food Science* 46:1493.

Chau, K.V., R.A. Romero, C.D. Baird, and J.J. Gaffney. 1987. Transpiration coefficients of fruits and vegetables in refrigerated storage. ASHRAE Research Project RP-370, *Final Report*.

Chen, C.S. 1985. Thermodynamic analysis of the freezing and thawing of foods: Enthalpy and apparent specific heat. *Journal of Food Science* 50:1158.

Choi, Y. and M.R. Okos. 1986. Effects of temperature and composition on the thermal properties of foods. In *Food Engineering and Process Applications*, vol. 1, pp. 93-101. M. LeMaguer and P. Jelen, eds. Elsevier Applied Science, London.

Chuma, Y., S. Murata, and S. Uchita. 1970. Determination of heat transfer coefficients of farm products by transient method using lead model. *Journal of the Society of Agricultural Machinery* 31(4):298-302.

Clary, B.L., G.L. Nelson, and R.E. Smith. 1968. Heat transfer from hams during freezing by low temperature air. *Transactions of the ASAE* 11:496-499.

Claypool, L.L. and F.W. Allen. 1951. The influence of temperature and oxygen level on the respiration and ripening of Wickson plums. *Hilgardea* 21:129.

Claypool, L.L. and S. Ozbek. 1952. Some influences of temperature and carbon dioxide on the respiration and storage life of the Mission fig. *Proceedings of the American Society for Horticultural Science*, vol. 60, p. 266.

Cleland, A.C. and R.L. Earle. 1976. A new method for prediction of surface heat transfer coefficients in freezing. *Bulletin de L'Institut International du Froid* Annexe 1976-1:361-368.

Dickerson, R.W., Jr. 1968. Thermal properties of food. In *The Freezing Preservation of Foods*, 4th ed., vol. 2. D.K. Tressler, W.B. Van Arsdel, and M.T. Copley, eds. AVI., Westport, CT.

Dickerson R.W., Jr. and R.B. Read, Jr. 1968. Calculation and measurement of heat transfer in foods. *Food Technology* 22:37.

Dickerson, R.W. and R.B. Read. 1975. Thermal diffusivity of meats. *ASHRAE Transactions* 81(1):356.

Dincer, I. 1993. Heat-transfer coefficients in hydrocooling of spherical and cylindrical food products. *Energy* 18(4):335-340.

Dincer, I. 1994. Development of new effective Nusselt-Reynolds correlations for air-cooling of spherical and cylindrical products. *International Journal of Heat and Mass Transfer* 37(17):2781-2787.

Eaks, J.L. and L.L. Morris. 1956. Respiration of cucumber fruits associated with physiological injury at chilling temperatures. *Plant Physiology* 31:308.

Eucken, A. 1940. Allgemeine Gesetzmassigkeiten für das Warmeleitvermogen verschiedener Stoffarten und Aggregatzustande. *Forschung auf dem Gebiete des Ingenieurwesens, Ausgabe A* 11(1):6.

Fedorov, V.G., D.N. Il'Inskiy, O.A. Gerashchenko, and L.D. Andreyeva. 1972. Heat transfer accompanying the cooling and freezing of meat carcasses. *Heat Transfer—Soviet Research* 4:55-59.

Fikiin, K.A. 1996. Ice content prediction methods during food freezing: A Survey of the Eastern European Literature. In *New Developments in Refrigeration for Food Safety and Quality*, pp. 90-97. International Institute of Refrigeration, Paris, and American Society of Agricultural Engineers, St. Joseph, MI.

Fockens, F.H. and H.F.T. Meffert. 1972. Biophysical properties of horticultural products as related to loss of moisture during cooling down. *Journal of Science of Food and Agriculture* 23:285-298.

Fricke, B.A. and B.R. Becker. 2001. Evaluation of thermophysical property models for foods. *International Journal of HVAC&R Research* (now *HVAC&R Research*) 7(4):311-330.

Fricke, B.A. and B.R. Becker. 2002. Evaluation of thermophysical property models for foods (RP-888). *Technical Paper* 4519, presented at the 2002 ASHRAE Winter Meeting, January 12-16, Atlantic City.

Fricke, B.A. and B.R. Becker. 2004. Calculation of food freezing times and heat transfer coefficients (RP-1123). *ASHRAE Transactions* 110(2): 145-157.

Gaffney, J.J., C.D. Baird, and K.V. Chau. 1985. Influence of airflow rate, respiration, evaporative cooling, and other factors affecting weight loss calculations for fruits and vegetables. *ASHRAE Transactions* 91(1B): 690-707.

Gan, G. and J.L. Woods. 1989. A deep bed simulation of vegetable cooling. In *Agricultural Engineering*, pp. 2301-2308. V.A. Dodd and P.M. Grace, eds. A.A. Balkema, Rotterdam.

Gane, R. 1936. The thermal conductivity of the tissue of fruits. *Annual Report*, p. 211. Food Investigation Board, U.K.

Geankoplis, C.J. 1978. *Transport processes and unit operations*. Allyn & Bacon, Boston.

Gerhardt, F., H. English, and E. Smith. 1942. Respiration, internal atmosphere, and moisture studies of sweet cherries during storage. *Proceedings of the American Society for Horticultural Science*, vol. 41, p. 119.

Gore, H.C. 1911. Studies on fruit respiration. *USDA Bureau Chemistry Bulletin* 142.

Griffiths, E. and D.H. Cole. 1948. Thermal properties of meat. *Society of Chemical Industry Journal* 67:33.

Griffiths, E. and M.J. Hickman. 1951. *The thermal conductivity of some nonmetallic materials*, p. 289. Institute of Mechanical Engineers, London.

Haller, M.H., P.L. Harding, J.M. Lutz, and D.H. Rose. 1932. The respiration of some fruits in relation to temperature. *Proceedings of the American Society for Horticultural Science*, vol. 28, p. 583.

Haller, M.H., D.H. Rose, and P.L. Harding. 1941. Studies on the respiration of strawberry and raspberry fruits. *USDA Circular* 613.

Haller, M.H., et al. 1945. Respiration of citrus fruits after harvest. *Journal of Agricultural Research* 71(8):327.

Harding, P.L. 1929. Respiration studies of grimes apples under various controlled temperatures. *Proceedings of the American Society for Horticultural Science*, vol. 26, p. 319.

Harper, J.C. 1960. Microwave spectra and physical characteristics of fruit and animal products relative to freeze-dehydration. *Report* 6, Army Quartermaster Food and Container Institute for the Armed Forces, ASTIA AD 255 818, 16.

Harper, J.C. 1962. Transport properties of gases in porous media at reduced pressures with reference to freeze-drying. *American Institute of Chemical Engineering Journal* 8(3):298.

Hawkins, L.A. 1929. Governing factors in transportation of perishable commodities. *Refrigerating Engineering* 18:130.

Hill, J.E. 1966. *The thermal conductivity of beef*, p. 49. Georgia Institute of Technology, Atlanta.

Hill, J.E., J.D. Leitman, and J.E. Sunderland. 1967. Thermal conductivity of various meats. *Food Technology* 21(8):91.

Holland, B., A.A. Welch, I.D. Unwin, D.H. Buss, A.A. Paul, and D.A.T. Southgate. 1991. *McCance and Widdowson's—The composition of foods*. Royal Society of Chemistry and Ministry of Agriculture, Fisheries and Food, Cambridge, U.K.

Hooper, F.C. and S.C. Chang. 1952. Development of the thermal conductivity probe. *Heating, Piping and Air Conditioning* 24(10):125.

Hruschka, H.W. 1966. Storage and shelf life of packaged rhubarb. *USDA Marketing Research Report*, p. 771.

Hruschka, H.W. and C.Y. Want. 1979. Storage and shelf life of packaged watercress, parsley, and mint. *USDA Marketing Research Report*, p. 1102.

IIR. 1967. *Recommended conditions for the cold storage of perishable produce*, 2nd ed., International Institute of Refrigeration, Paris.

Jason, A.C., and R.A.K. Long. 1955. The specific heat and thermal conductivity of fish muscle. *Proceedings of the 9th International Congress of Refrigeration*, Paris, 1:2160.

Jones, W.W. 1942. Respiration and chemical changes of papaya fruit in relation to temperature. *Plant Physiology* 17:481.

Karmarkar, D.V. and B.M. Joshe. 1941a. Respiration of onions. *Indian Journal of Agricultural Science* 11:82.

Karmarkar, D.V. and B.M. Joshe. 1941b. Respiration studies on the Alphonse mango. *Indian Journal of Agricultural Science* 11:993.

Kaye, G.W.C. and W.F. Higgins. 1928. The thermal conductivities of certain liquids. *Proceedings of the Royal Society of London* A117:459.

Kazarian, E.A. 1962. *Thermal properties of grain*, p. 74. Michigan State University, East Lansing.

Kelly, M.J. 1965. Heat transfer in fluidized beds. *Dechema Monographien* 56:119.

Khatchaturov, A.B. 1958. Thermal processes during air-blast freezing of fish. *Bulletin of the IIR*, Annexe 1958-2:365-378.

Khelemskii, M.Z. and V.Z. Zhadan. 1964. Thermal conductivity of normal beet juice. *Sakharnaya Promyshlennost* 10:11.

Kondrat'ev, G.M. 1950. Application of the theory of regular cooling of a two-component sphere to the determination of heat conductivity of poor heat conductors (method, sphere in a sphere). *Otdelenie Tekhnicheskikh Nauk, Isvestiya Akademii Nauk* 4(April):536.

Kopelman, I.J. 1966. *Transient heat transfer and thermal properties in food systems*. Ph.D. dissertation, Michigan State University, East Lansing.

Kopelman, I., J.L. Blaisdell, and I.J. Pflug. 1966. Influence of fruit size and coolant velocity on the cooling of Jonathan apples in water and air. *ASHRAE Transactions* 72(1):209-216.

Leichter, S., S. Mizrahi, and I.J. Kopelman. 1976. Effect of vapor condensation on rate of warming up of refrigerated products exposed to humid atmosphere: Application to the prediction of fluid milk shelf life. *Journal of Food Science* 41:1214-1218.

Leidenfrost, W. 1959. Measurements on the thermal conductivity of milk. *ASME Symposium on Thermophysical Properties*, p. 291. Purdue University, IN.

Lentz, C.P. 1961. Thermal conductivity of meats, fats, gelatin gels, and ice. *Food Technology* 15(5):243.

Lentz, C.P. 1969. Calorimetric study of immersion freezing of poultry. *Journal of the Canadian Institute of Food Technology* 2(3):132-136.

Levy, F.L. 1981. A modified Maxwell-Eucken equation for calculating the thermal conductivity of two-component solutions or mixtures. *International Journal of Refrigeration* 4:223-225.

Lewis, D.A. and L.L. Morris. 1956. Effects of chilling storage on respiration and deterioration of several sweet potato varieties. *Proceedings of the American Society for Horticultural Science* 68:421.

Lipton, W.J. 1957. *Physiological changes in harvested asparagus* (Asparagus officinalis) *as related to temperature*. University of California, Davis.

Long, R.A.K. 1955. Some thermodynamic properties of fish and their effect on the rate of freezing. *Journal of the Science of Food and Agriculture* 6:621.

Lutz, J.M. 1938. Factors influencing the quality of american grapes in storage. *USDA Technical Bulletin* 606.

Lutz, J.M. and R.E. Hardenburg. 1968. The commercial storage of fruits, vegetables, and florists and nursery stocks. *USDA Handbook* 66.

Mann, L.K. and D.A. Lewis. 1956. Rest and dormancy in garlic. *Hilgardia* 26:161.

Mathews, F.W., Jr. and C.W. Hall. 1968. Method of finite differences used to relate changes in thermal and physical properties of potatoes. *ASAE Transactions* 11(4):558.

Maxie, E.C., F.G. Mitchell, and A. Greathead. 1959. Studies on strawberry quality. *California Agriculture* 13(2):11, 16.

Maxie, E.C., P.B. Catlin, and H.T. Hartmann. 1960. Respiration and ripening of olive fruits. *Proceedings of the American Society for Horticultural Science* 75:275.

Metzner, A.B. and P.S. Friend. 1959. Heat transfer to turbulent non-Newtonian fluids. *Industrial and Engineering Chemistry* 51:879.

Micke, W.C., F.G. Mitchell, and E.C. Maxie. 1965. Handling sweet cherries for fresh shipment. *California Agriculture* 19(4):12.

Miles, C.A. 1974. Meat freezing—Why and how? *Proceedings of the Meat Research Institute*, Symposium No. 3, Bristol, 15.1-15.7.

Miller, C.F. 1963. *Thermal conductivity and specific heat of sorghum grain*, p. 79. Texas Agricultural and Mechanical College, College Station.

Minh, T.V., J.S. Perry, and A.H. Bennett. 1969. Forced-air precooling of white potatoes in bulk. *ASHRAE Transactions* 75(2):148-150.

Moote, I. 1953. The effect of moisture on the thermal properties of wheat. *Canadian Journal of Technology* 31(2/3):57.

Morris, L.L. 1947. A study of broccoli deterioration. *Ice and Refrigeration* 113(5):41.

Murakami, E.G., and M.R. Okos. 1989. Measurement and prediction of thermal properties of foods. In *Food Properties and Computer-Aided Engineering of Food Processing Systems,* pp. 3-48. R.P. Singh and A.G. Medina, eds. Kluwer Academic, Dordrecht.

Nicholas, R.C., K.E.H. Motawi, and J.L. Blaisdell. 1964. Cooling rate of individual fruit in air and in water. *Quarterly Bulletin*, Michigan State University Agricultural Experiment Station 47(1):51-64.

Nowrey, J.E. and E.E. Woodams. 1968. Thermal conductivity of a vegetable oil-in-water emulsion. *Journal of Chemical and Engineering Data* 13(3): 297.

Otten, L. 1974. Thermal parameters of agricultural materials and food products. *Bulletin of the IIR* Annexe 1974-3:191-199.

Oxley, T.A. 1944. The properties of grain in bulk; III—The thermal conductivity of wheat, maize and oats. *Society of Chemical Industry Journal* 63:53.

Pantastico, E.B. 1974. Handling and utilization of tropical and subtropical fruits and vegetables. In *Postharvest Physiology.* AVI Publishing, Westport, CT.

Parker, R.E. and B.A. Stout. 1967. Thermal properties of tart cherries. *Transactions of the ASAE* 10(4):489-491, 496.

Pentzer, W.T., C.E. Asbury, and K.C. Hamner. 1933. The effect of sulfur dioxide fumigation on the respiration of Emperor grapes. *Proceedings of the American Society for Horticultural Science* 30:258.

Pham, Q.T. 1987. Calculation of bound water in frozen food. *Journal of Food Science* 52(1):210-212.

Polley, S.L., O.P. Snyder, and P. Kotnour. 1980. A compilation of thermal properties of foods. *Food Technology* 34(11):76-94.

Popov, V.D. and Y.A. Terentiev. 1966. Thermal properties of highly viscous fluids and coarsely dispersed media. *Teplofizicheskie Svoistva Veshchestv, Akademiya Nauk, Ukrainskoi SSSR, Respublikanskii Sbornik* 18:76.

Poppendiek, H.F., N.D. Greene, P.M. Morehouse, R. Randall, J.R. Murphy, and W.A. Morton. 1965-1966. Annual report on thermal and electrical conductivities of biological fluids and tissues. *ONR Contract* 4094 (00), A-2, GLR-43 Geoscience Ltd., 39.

Pratt, H.K. and L.L. Morris. 1958. Some physiological aspects of vegetable and fruit handling. *Food Technology in Australia* 10:407.

Pratt, H.K., L.L. Morris, and C.L. Tucker. 1954. Temperature and lettuce deterioration. *Proceedings of the Conference on Transportation of Perishables*, p. 77. University of California, Davis.

Qashou, MS., G. Nix, R.I. Vachon, and G.W. Lowery. 1970. Thermal conductivity values for ground beef and chuck. *Food Technology* 23(4):189.

Qashou, M.S., R.I. Vachon, and Y.S. Touloukian. 1972. Thermal conductivity of foods. *ASHRAE Transactions* 78(1):165-183.

Radford, R.D., L.S. Herbert, and D.A. Lorett. 1976. Chilling of meat—A mathematical model for heat and mass transfer. *Bulletin de L'Institut International du Froid*, Annexe 1976(1):323-330.

Rappaport, L. and A.E. Watada. 1958. Effects of temperature on artichoke quality. *Proceedings of the Conference on Transportation of Perishables*, p. 142. University of California, Davis.

Riedel, L. 1949. Thermal conductivity measurements on sugar solutions, fruit juices and milk. *Chemie-Ingenieur-Technik* 21(17):340-341.

Riedel, L. 1951. The refrigeration effect required to freeze fruits and vegetables. *Refrigeration Engineering* 59:670.

Riedel, L. 1956. Calorimetric investigation of the freezing of fish meat. *Kaltetechnik* 8:374-377.

Riedel, L. 1957a. Calorimetric investigation of the meat freezing process. *Kaltetechnik* 9(2):38-40.

Riedel, L. 1957b. Calorimetric investigation of the freezing of egg white and yolk. *Kaltetechnik* 9:342.

Riedel, L. 1959. Calorimetric investigations of the freezing of white bread and other flour products. *Kaltetechnik* 11(2):41.

Riedel, L. 1969. Measurements of thermal diffusivity on foodstuffs rich in water. *Kaltetechnik* 21(11):315-316.

Reidy, G.A. 1968. *Values for thermal properties of foods gathered from the literature*. Ph.D. dissertation, Michigan State University, East Lansing.

Ryall. A.L. and W.J. Lipton. 1972. Vegetables as living products. Respiration and heat production. In *Transportation and Storage of Fruits and Vegetables*, vol. 1. AVI Publishing, Westport, CT.

Saravacos, G.D. 1965. Freeze-drying rates and water sorption of model food gels. *Food Technology* 19(4):193.

Saravacos, G.D. and M.N. Pilsworth. 1965. Thermal conductivity of freeze-dried model food gels. *Journal of Food Science* 30:773.

Sastry, S.K., C.D. Baird, and D.E. Buffington. 1978. Transpiration rates of certain fruits and vegetables. *ASHRAE Transactions* 84(1).

Sastry, S.K. and D.E. Buffington. 1982. Transpiration rates of stored perishable commodities: A mathematical model and experiments on tomatoes. *ASHRAE Transactions* 88(1):159-184.

Schenk, R.U. 1959. Respiration of peanut fruit during curing. *Proceedings of the Association of Southern Agricultural Workers* 56:228.

Schenk, R.U. 1961. Development of the peanut fruit. *Georgia Agricultural Experiment Station Bulletin N.S.*, vol. 22.

Scholz, E.W., H.B. Johnson, and W.R. Buford. 1963. Heat evolution rates of some Texas-grown fruits and vegetables. *Rio Grande Valley Horticultural Society Journal* 17:170.

Schwartzberg, H.G. 1976. Effective heat capacities for the freezing and thawing of food. *Journal of Food Science* 41(1):152-156.

Schwartzberg, H.G. 1981. Mathematical analysis of the freezing and thawing of foods. Tutorial presented at the AIChE Summer Meeting, Detroit, MI.

Siebel, J.E. 1892. Specific heat of various products. *Ice and Refrigeration* 256.

Slavicek, E., K. Handa, and M. Kminek. 1962. Measurements of the thermal diffusivity of sugar beets. *Cukrovarnicke Listy* 78:116.

Smith, F.G., A.J. Ede, and R. Gane. 1952. The thermal conductivity of frozen foodstuffs. *Modern Refrigeration* 55:254.

Smith, R.E., A.H. Bennett, and A.A. Vacinek. 1971. Convection film coefficients related to geometry for anomalous shapes. *ASAE Transactions* 14(1):44-47.

Smith, R.E., G.L. Nelson, and R.L. Henrickson. 1976. Analyses on transient heat transfer from anomalous shapes. *ASAE Transactions* 10(2):236.

Smith, W.H. 1957. The production of carbon dioxide and metabolic heat by horticultural produce. *Modern Refrigeration* 60:493.

Smith, W.H. 1964. The storage of mushrooms. *Ditton and Covent Garden Laboratories Annual Report*, p. 18. Great Britain Agricultural Research Council.

Smith, W.H. 1966. The storage of gooseberries. *Ditton and Covent Garden Laboratories Annual Report*, p. 13. Great Britain Agricultural Research Council.

Spells, K.E. 1958. *The thermal conductivities of some biological fluids.* Flying Personnel Research Committee, Institute of Aviation Medicine, Royal Air Force, Farnborough, England, FPRC-1071 AD 229 167, 8.

Spells, K.E. 1960-1961. The thermal conductivities of some biological fluids. *Physics in Medicine and Biology* 5:139.

Sweat, V.E. 1974. Experimental values of thermal conductivity of selected fruits and vegetables. *Journal of Food Science* 39:1080.

Sweat, V.E. 1985. Thermal properties of low- and intermediate-moisture food. *ASHRAE Transactions* 91(2):369-389.

Tchigeov, G. 1979. *Thermophysical processes in food refrigeration technology.* Food Industry, Moscow.

Tewfik, S. and L.E. Scott. 1954. Respiration of vegetables as affected by postharvest treatment. *Journal of Agricultural and Food Chemistry* 2:415.

Thompson, H., S.R. Cecil, and J.G. Woodroof. 1951. Storage of edible peanuts. *Georgia Agricultural Experiment Station Bulletin*, vol. 268.

Triebes, T.A. and C.J. King. 1966. Factors influencing the rate of heat conduction in freeze-drying. *I and EC Process Design and Development* 5(4):430.

Turrell, F.M. and R.L. Perry. 1957. Specific heat and heat conductivity of citrus fruit. *Proceedings of the American Society for Horticultural Science* 70:261.

USDA. 1968. Egg pasteurization manual. ARS *Publication* 74-48. U.S. Department of Agriculture, Agricultural Research Service, Washington, D.C.

USDA. 1975. Composition of foods. *Agricultural Handbook* 8. U.S. Department of Agriculture, Washington, D.C.

USDA. 1996. *Nutrient database for standard reference,* release 11. U.S. Department of Agriculture, Washington, D.C.

Van den Berg, L. and C.P. Lentz. 1957. Factors affecting freezing rates of poultry immersed in liquid. *Food Technology* 11(7):377-380.

Van den Berg, L. and C.P. Lentz. 1972. Respiratory heat production of vegetables during refrigerated storage. *Journal of the American Society for Horticultural Science* 97:431.

Wachsmuth. R. 1892. Untersuchungen auf dem Gebiet der inneren Warmeleitung. *Annalen der Physik* 3(48):158.

Walters, R.E. and K.N. May. 1963. Thermal conductivity and density of chicken breast muscle and skin. *Food Technology* 17(June):130.

Watada, A.E. and L.L. Morris. 1966. Effect of chilling and nonchilling temperatures on snap bean fruits. *Proceedings of the American Society for Horticultural Science* 89:368.

Watt, B.K. and A.L. Merrill. 1963. Composition of foods. *USDA Handbook* 8.

Weber, H.F. VII. 1880. Untersuchungen über die Warmeleitung in Flussigkeiten. *Annael der Physik* 10(3):304.

Weber, H.F. 1886. The thermal conductivity of drop forming liquids. *Exner's Reportorium* 22:116.

Woodams, E.E. 1965. *Thermal conductivity of fluid foods,* p. 95. Cornell University, Ithaca, NY.

Workman, M. and H.K. Pratt. 1957. Studies on the physiology of tomato fruits; II, Ethylene production at 20°C as related to respiration, ripening and date of harvest. *Plant Physiology* 32:330.

Wright, R.C., D.H. Rose, and T.H. Whiteman. 1954. The commercial storage of fruits, vegetables, and florists and nursery stocks. *USDA Handbook* 66.

CHAPTER 10

COOLING AND FREEZING TIMES OF FOODS

PRESERVATION of food is one of the most significant applications of refrigeration. Cooling and freezing food effectively reduces the activity of microorganisms and enzymes, thus retarding deterioration. In addition, crystallization of water reduces the amount of liquid water in food and inhibits microbial growth (Heldman 1975).

Most commercial food and beverage cooling and freezing operations use air-blast convection heat transfer; only a limited number of products are cooled or frozen by conduction heat transfer in plate freezers. Thus, this chapter focuses on convective heat transfer.

For air-blast convective cooling and freezing operations to be cost-effective, refrigeration equipment should fit the specific requirements of the particular cooling or freezing application. The design of such refrigeration equipment requires estimation of the cooling and freezing times of foods and beverages, as well as the corresponding refrigeration loads.

Numerous methods for predicting the cooling and freezing times of foods and beverages have been proposed, based on numerical, analytical, and empirical analysis. Selecting an appropriate estimation method from the many available methods can be challenging. This chapter reviews selected procedures available for estimating the air-blast convective cooling and freezing times of foods and beverages, and presents examples of these procedures. These procedures use the thermal properties of foods, discussed in Chapter 9.

THERMODYNAMICS OF COOLING AND FREEZING

Cooling and freezing food is a complex process. Before freezing, sensible heat must be removed from the food to decrease its temperature to the initial freezing point of the food. This initial freezing point is somewhat lower than the freezing point of pure water because of dissolved substances in the moisture within the food. At the initial freezing point, a portion of the water within the food crystallizes and the remaining solution becomes more concentrated, reducing the freezing point of the unfrozen portion of the food further. As the temperature decreases, ice crystal formation increases the concentration of the solutes in solution and depresses the freezing point further. Thus, the ice and water fractions in the frozen food, and consequently the food's thermophysical properties, depend on temperature.

Because most foods are irregularly shaped and have temperature-dependent thermophysical properties, exact analytical solutions for their cooling and freezing times cannot be derived. Most research has focused on developing semianalytical/empirical cooling and freezing time prediction methods that use simplifying assumptions.

COOLING TIMES OF FOODS AND BEVERAGES

Before a food can be frozen, its temperature must be reduced to its initial freezing point. This cooling process, also known as precooling or chilling, removes only sensible heat and, thus, no phase change occurs.

Air-blast convective cooling of foods and beverages is influenced by the ratio of the external heat transfer resistance to the internal heat transfer resistance. This ratio (the Biot number) is

$$\mathrm{Bi} = hL/k \qquad (1)$$

where h is the convective heat transfer coefficient, L is the characteristic dimension of the food, and k is the thermal conductivity of the food (see Chapter 9). In cooling time calculations, the characteristic dimension L is taken to be the shortest distance from the thermal center of the food to its surface. Thus, in cooling time calculations, L is half the thickness of a slab, or the radius of a cylinder or a sphere.

When the Biot number approaches zero ($\mathrm{Bi} < 0.1$), internal resistance to heat transfer is much less than external resistance, and the lumped-parameter approach can be used to determine a food's cooling time (Heldman 1975). When the Biot number is very large ($\mathrm{Bi} > 40$), internal resistance to heat transfer is much greater than external resistance, and the food's surface temperature can be assumed to equal the temperature of the cooling medium. For this latter situation, series solutions of the Fourier heat conduction equation are available for simple geometric shapes. When $0.1 < \mathrm{Bi} < 40$, both the internal resistance to heat transfer and the convective heat transfer coefficient must be considered. In this case, series solutions, which incorporate transcendental functions to account for the influence of the Biot number, are available for simple geometric shapes.

Simplified methods for predicting the cooling times of foods and beverages are available for regularly and irregularly shaped foods over a wide range of Biot numbers. In this chapter, these simplified methods are grouped into two main categories: (1) those based on f and j factors, and (2) those based on equivalent heat transfer dimensionality. Furthermore, the methods based on f and j factors are divided into two subgroups: (1) those for regular shapes, and (2) those for irregular shapes.

Cooling Time Estimation Methods Based on *f* and *j* Factors

All cooling processes exhibit similar behavior. After an initial lag, the temperature at the thermal center of the food decreases exponentially (Cleland 1990). As shown in Figure 1, a cooling curve depicting this behavior can be obtained by plotting, on semilogarithmic axes, the fractional unaccomplished temperature difference Y versus time. Y is defined as follows:

$$Y = \frac{T_m - T}{T_m - T_i} = \frac{T - T_m}{T_i - T_m} \qquad (2)$$

where T_m is the cooling medium temperature, T is the product temperature, and T_i is the initial temperature of the product.

This semilogarithmic temperature history curve consists of an initial curvilinear portion, followed by a linear portion. Empirical

The preparation of this chapter is assigned to TC 10.9, Refrigeration Application for Foods and Beverages.

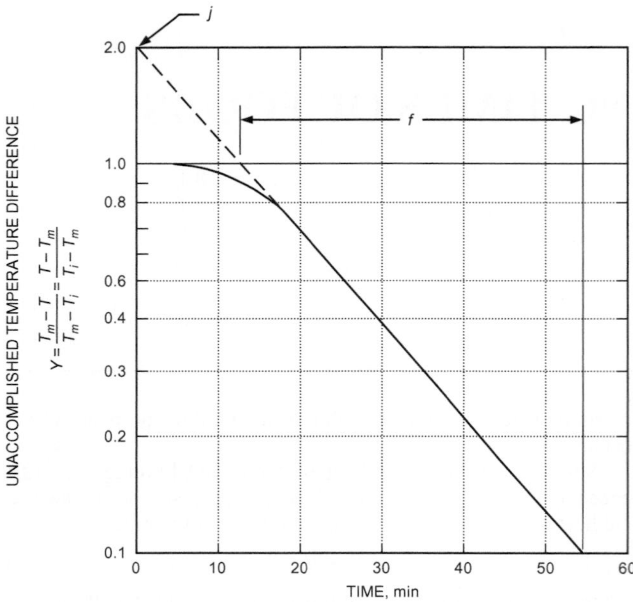

Fig. 1 Typical Cooling Curve

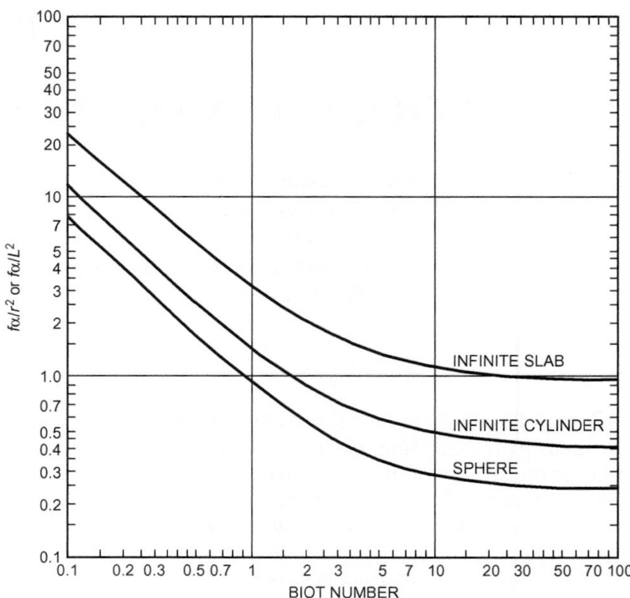

Fig. 2 Relationship Between $f\alpha/r^2$ and Biot Number for Infinite Slab, Infinite Cylinder, and Sphere

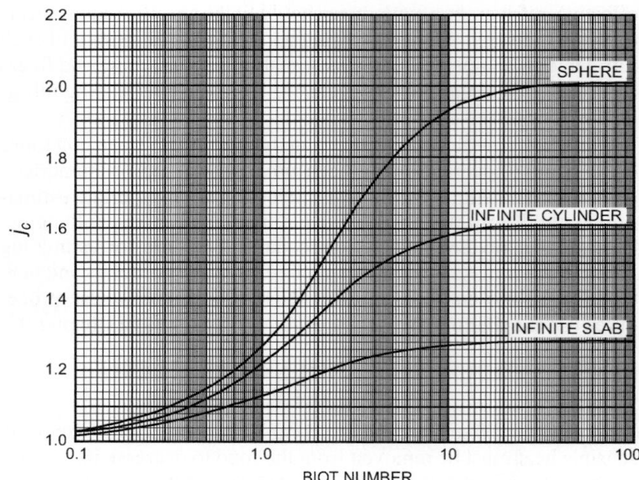

Fig. 3 Relationship Between j_c Value for Thermal Center Temperature and Biot Number for Various Shapes

formulas that model this cooling behavior incorporate two factors, *f* and *j*, which represent the slope and intercept, respectively, of the temperature history curve. The *j* factor is a measure of lag between the onset of cooling and the exponential decrease in the temperature of the food. The *f* factor represents the time required for a 90% reduction in the nondimensional temperature difference. Graphically, the *f* factor corresponds to the time required for the linear portion of the temperature history curve to pass through one log cycle. The *f* factor is a function of the Biot number, and the *j* factor is a function of the Biot number and the location within the food.

The general form of the cooling time model is

$$Y = \frac{T_m - T}{T_m - T_i} = je^{-2.303\theta/f} \tag{3}$$

where θ is the cooling time. This equation can be rearranged to give cooling time explicitly as

$$\theta = \frac{-f}{2.303}\ln\left(\frac{Y}{j}\right) \tag{4}$$

Determination of *f* and *j* Factors for Slabs, Cylinders, and Spheres

From analytical solutions, Pflug et al. (1965) developed charts for determining *f* and *j* factors for foods shaped either as infinite slabs, infinite cylinders, or spheres. They assumed uniform initial temperature distribution in the food, constant surrounding medium temperature, convective heat exchange at the surface, and constant thermophysical properties. Figure 2 can be used to determine *f* values and Figures 3 to 5 can be used to determine *j* values. Because the *j* factor is a function of location within the food, Pflug et al. presented charts for determining *j* factors for center, mass average, and surface temperatures.

As an alternative to Figures 2 to 5, Lacroix and Castaigne (1987a) presented expressions for estimating *f* and j_c factors for the thermal center temperature of infinite slabs, infinite cylinders, and spheres. These expressions, which depend on geometry and Biot number, are summarized in Tables 1 to 3. In these expressions, α is the thermal diffusivity of the food (see Chapter 9) and *L* is the characteristic dimension, defined as the shortest distance from the thermal center of the food to its surface. For an infinite slab, *L* is the half thickness. For an infinite cylinder or a sphere, *L* is the radius.

By using various combinations of infinite slabs and infinite cylinders, the *f* and *j* factors for infinite rectangular rods, finite cylinders, and rectangular bricks may be estimated. Each of these shapes can be generated by intersecting infinite slabs and infinite cylinders: two infinite slabs of proper thickness for the infinite rectangular rod, one infinite slab and one infinite cylinder for the finite cylinder, or three infinite orthogonal slabs of proper thickness for the rectangular brick. The *f* and *j* factors of these composite bodies can be estimated by

$$\frac{1}{f_{comp}} = \sum_i \left(\frac{1}{f_i}\right) \tag{5}$$

$$j_{comp} = \prod_i j_i \tag{6}$$

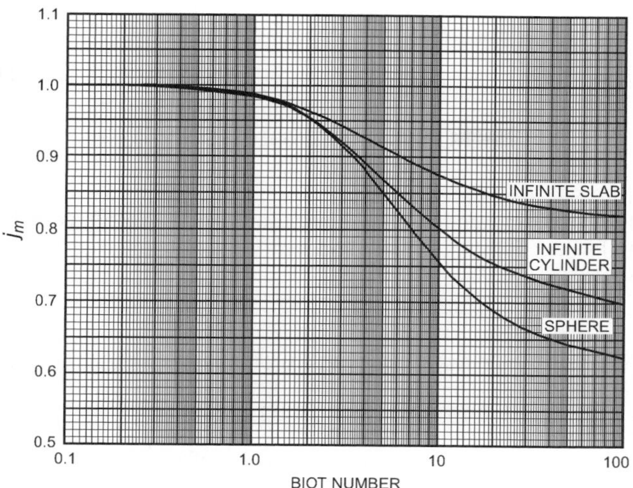

Fig. 4 Relationship Between j_m Value for Mass Average Temperature and Biot Number for Various Shapes

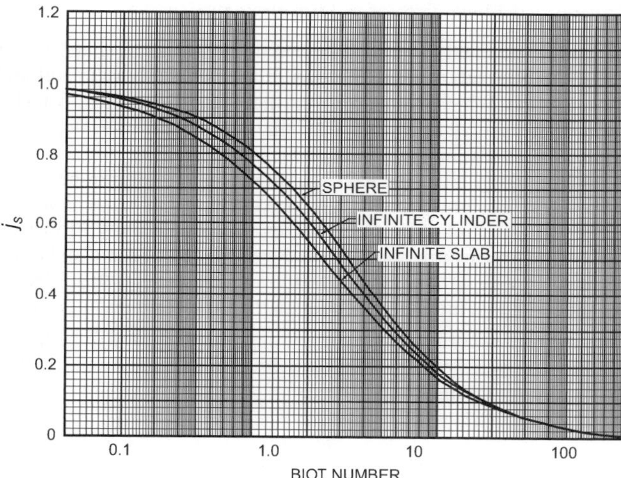

Fig. 5 Relationship Between j_s Value for Surface Temperature and Biot Number for Various Shapes

where the subscript i represents the appropriate infinite slab(s) or infinite cylinder. To evaluate the f_i and j_i of Equations (5) and (6), the Biot number must be defined, corresponding to the appropriate infinite slab(s) or infinite cylinder.

Table 1 Expressions for Estimating f and j_c Factors for Thermal Center Temperature of Infinite Slabs

Biot Number Range	Equations for f and j factors
$\mathrm{Bi} \le 0.1$	$\frac{f\alpha}{L^2} = \frac{\ln 10}{\mathrm{Bi}}$ $j_c = 1.0$
$0.1 < \mathrm{Bi} \le 100$	$\frac{f\alpha}{L^2} = \frac{\ln 10}{u^2}$ $j_c = \frac{2\sin u}{u + \sin u \cos u}$ *where* $u = 0.860972 + 0.312133\ln(\mathrm{Bi}) + 0.007986[\ln(\mathrm{Bi})]^2 - 0.016192[\ln(\mathrm{Bi})]^3 - 0.001190[\ln(\mathrm{Bi})]^4 + 0.000581[\ln(\mathrm{Bi})]^5$
$\mathrm{Bi} > 100$	$\frac{f\alpha}{L^2} = 0.9332$ $j_c = 1.273$

Source: Lacroix and Castaigne (1987a)

Table 2 Expressions for Estimating f and j_c Factors for Thermal Center Temperature of Infinite Cylinders

Biot Number Range	Equations for f and j factors
$\mathrm{Bi} \le 0.1$	$\frac{f\alpha}{L^2} = \frac{\ln 10}{2\,\mathrm{Bi}}$ $j_c = 1.0$
$0.1 < \mathrm{Bi} \le 100$	$\frac{f\alpha}{L^2} = \frac{\ln 10}{v^2}$ $j_c = \frac{2J_1(v)}{v[J_0^2(v) - J_1^2(v)]}$ *where* $v = 1.257493 + 0.487941\ln(\mathrm{Bi}) + 0.025322[\ln(\mathrm{Bi})]^2 - 0.026568[\ln(\mathrm{Bi})]^3 - 0.002888[\ln(\mathrm{Bi})]^4 + 0.001078[\ln(\mathrm{Bi})]^5$ and $J_0(v)$ and $J_1(v)$ are zero and first-order Bessel functions, respectively.
$\mathrm{Bi} > 100$	$\frac{f\alpha}{L^2} = 0.3982$ $j_c = 1.6015$

Source: Lacroix and Castaigne (1987a)

Determination of f and j Factors for Irregular Shapes

Smith et al. (1968) developed, for the case of irregularly shaped foods and Biot number approaching infinity, a shape factor called the geometry index G, which is obtained as follows:

$$G = 0.25 + \frac{3}{8B_1^2} + \frac{3}{8B_2^2} \tag{7}$$

where B_1 and B_2 are related to the cross-sectional areas of the food:

$$B_1 = \frac{A_1}{\pi L^2} \qquad B_2 = \frac{A_2}{\pi L^2} \tag{8}$$

where L is the shortest distance between the thermal center of the food and its surface, A_1 is the minimum cross-sectional area containing L, and A_2 is the cross-sectional area containing L that is orthogonal to A_1.

G is used in conjunction with the inverse of the Biot number m and a nomograph (shown in Figure 6) to obtain the characteristic value M_1^2. Smith et al. showed that the characteristic value M_1^2 can be related to the f factor by

$$f = \frac{2.303 L^2}{M_1^2 \alpha} \tag{9}$$

where α is the thermal diffusivity of the food. In addition, an expression for estimating a j_m factor used to determine the mass average temperature is given as

$$j_m = 0.892 e^{-0.0388 M_1^2} \tag{10}$$

Table 3 Expressions for Estimating f and j_c Factors for Thermal Center Temperature of Spheres

Biot Number Range	Equations for f and j factors
$Bi \le 0.1$	$\frac{f\alpha}{L^2} = \frac{\ln 10}{3\,Bi}$ $j_c = 1.0$
$0.1 < Bi \le 100$	$\frac{f\alpha}{L^2} = \frac{\ln 10}{w^2}$ $j_c = \frac{2(\sin w - w\cos w)}{w - \sin w \cos w}$ where $w = 1.573729 + 0.642906\ln(Bi) + 0.047859[\ln(Bi)]^2 - 0.03553[\ln(Bi)]^3 - 0.004907[\ln(Bi)]^4 + 0.001563[\ln(Bi)]^5$
$Bi > 100$	$\frac{f\alpha}{L^2} = 0.2333$ $j_c = 2.0$

Source: Lacroix and Castaigne (1987a)

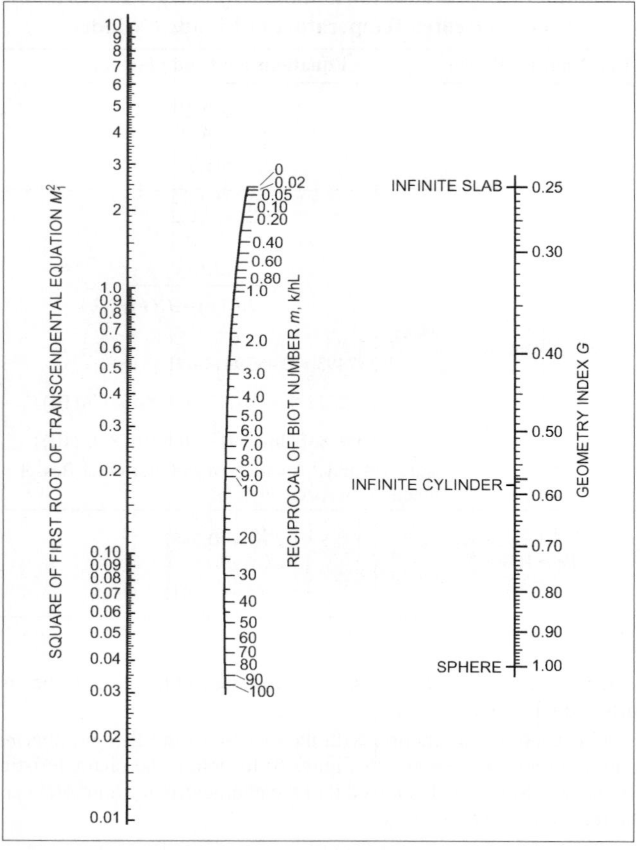

Fig. 6 Nomograph for Estimating Value of M_1^2 from Reciprocal of Biot Number and Smith's (1966) Geometry Index

As an alternative to estimating $M_1{}^2$ from the nomograph developed by Smith et al. (1968), Hayakawa and Villalobos (1989) obtained regression formulas for estimating $M_1{}^2$. For Biot numbers approaching infinity, their regression formula is

$$\ln(M_1^2) = 2.2893825 + 0.35330539X_g - 3.8044156X_g^2 - 9.6821811X_g^3 - 12.0321827X_g^4 - 7.1542411X_g^5 - 1.6301018X_g^6 \quad (11)$$

where $X_g = \ln(G)$. Equation (11) is applicable for $0.25 \le G \le 1.0$. For finite Biot numbers, Hayakawa and Villalobos (1989) gave the following:

$$\ln(M_1^2) = 0.92083090 + 0.83409615X_g - 0.78765739X_b - 0.04821784X_gX_b - 0.04088987X_g^2 - 0.10045526X_b^2 + 0.01521388X_g^3 + 0.00119941X_gX_b^3 + 0.00129982X_b^4 \quad (12)$$

where $X_g = \ln(G)$ and $X_b = \ln(1/Bi)$. Equation (12) is applicable for $0.25 \le G \le 1.0$ and $0.01 \le 1/Bi \le 100$.

Cooling Time Estimation Methods Based on Equivalent Heat Transfer Dimensionality

Product geometry can also be considered using a shape factor called the **equivalent heat transfer dimensionality** (Cleland and Earle 1982a), which compares total heat transfer to heat transfer through the shortest dimension. Cleland and Earle developed an expression for estimating the equivalent heat transfer dimensionality of irregularly shaped foods as a function of Biot number. This overcomes the limitation of the geometry index G, which was derived for the case of Biot number approaching infinity. However, the cooling time estimation method developed by Cleland and Earle requires the use of a nomograph. Lin et al. (1993, 1996a, 1996b) expanded on this method to eliminate the need for a nomograph.

In the method of Lin et al., the cooling time of a food or beverage is estimated by a first term approximation to the analytical solution for convective cooling of a sphere:

$$\theta = \frac{3\rho c L^2}{\omega^2 k E}\ln\left(\frac{j}{Y}\right) \quad (13)$$

Equation (13) is applicable for center temperature if $Y_c < 0.7$ and for mass average temperature if $Y_m < 0.55$, where Y_c is the fractional unaccomplished temperature difference based on final center temperature and Y_m is the fractional unaccomplished temperature difference based on final mass average temperature. In Equation (13), θ is cooling time, ρ is the food's density, c is the food's specific heat, L is the food's radius or half-thickness, k is the food's thermal conductivity, j is the lag factor, E is the equivalent heat transfer dimensionality, and ω is the first root (in radians) of the following transcendental function:

$$\omega \cot\omega + Bi - 1 = 0 \quad (14)$$

In Equation (13), the equivalent heat transfer dimensionality E is given as a function of Biot number:

$$E = \frac{Bi^{4/3} + 1.85}{\frac{Bi^{4/3}}{E_\infty} + \frac{1.85}{E_0}} \quad (15)$$

E_0 and E are the equivalent heat transfer dimensionalities for the limiting cases of $Bi = 0$ and $Bi \to \infty$, respectively. The definitions of E_0 and E use the dimensional ratios β_1 and β_2:

$$\beta_1 = \frac{\text{Second shortest dimension of food}}{\text{Shortest dimension of food}} \quad (16)$$

$$\beta_2 = \frac{\text{Longest dimension of food}}{\text{Shortest dimension of food}} \quad (17)$$

For two-dimensional, irregularly shaped foods, E_0 (the equivalent heat transfer dimensionality for Bi = 0) is given by

$$E_0 = \left(1 + \frac{1}{\beta_1}\right)\left[1 + \left(\frac{\beta_1 - 1}{2\beta_1 + 2}\right)^2\right] \quad (18)$$

For three-dimensional, irregularly shaped foods, E_0 is

$$E_0 = 1.5\frac{\beta_1 + \beta_2 + \beta_1^2(1 + \beta_2) + \beta_2^2(1 + \beta_1)}{\beta_1\beta_2(1 + \beta_1 + \beta_2)} - \frac{[(\beta_1 - \beta_2)^2]^{0.4}}{15} \quad (19)$$

For finite cylinders, bricks, and infinite rectangular rods, E_0 may be determined as follows:

$$E_0 = 1 + \frac{1}{\beta_1} + \frac{1}{\beta_2} \quad (20)$$

For spheres, infinite cylinders, and infinite slabs, E_0 = 3, 2, and 1, respectively.

For both two-dimensional and three-dimensional food items, the general form for E at Bi $\rightarrow \infty$ is given as

$$E_\infty = 0.75 + p_1 f(\beta_1) + p_2 f(\beta_2) \quad (21)$$

where

$$f(\beta) = \frac{1}{\beta^2} + 0.01 p_3 \exp\left[\beta - \frac{\beta^2}{6}\right] \quad (22)$$

with β_1 and β_2 as previously defined. The geometric parameters p_1, p_2, and p_3 are given in Table 4 for various geometries.

Lin et al. (1993, 1996a, 1996b) also developed an expression for the lag factor j_c applicable to the thermal center of a food as

$$j_c = \frac{\text{Bi}^{1.35} + \dfrac{1}{\lambda}}{\dfrac{\text{Bi}^{1.35}}{j_\infty} + \dfrac{1}{\lambda}} \quad (23)$$

where j_∞ is as follows:

$$j_\infty = 1.271 + 0.305\exp(0.172\gamma_1 - 0.115\gamma_1^2) + 0.425\exp(0.09\gamma_2 - 0.128\gamma_2^2) \quad (24)$$

and the geometric parameters λ, γ_1, and γ_2 are given in Table 4.

For the mass average temperature, Lin et al. gave the lag factor j_m as follows:

$$j_m = \mu j_c \quad (25)$$

where

$$\mu = \left(\frac{1.5 + 0.69\,\text{Bi}}{1.5 + \text{Bi}}\right)^N \quad (26)$$

and N is the number of dimensions of a food in which heat transfer is significant (see Table 4).

Algorithms for Estimating Cooling Time

The following suggested algorithm for estimating cooling time of foods and beverages is based on the equivalent heat transfer dimensionality method by Lin et al. (1993, 1996a, 1996b).

Table 4 Geometric Parameters

Shape	N	p_1	p_2	p_3	γ_1	γ_2	λ
Infinite slab ($\beta_1 = \beta_2 = \infty$)	1	0	0	0	∞	∞	1
Infinite rectangular rod ($\beta_1 \geq 1$, $\beta_2 = \infty$)	2	0.75	0	−1	$4\beta_1/\pi$	∞	γ_1
Brick ($\beta_1 \geq 1$, $\beta_2 \geq \beta_1$)	3	0.75	0.75	−1	$4\beta_1/\pi$	$1.5\beta_2$	γ_1
Infinite cylinder ($\beta_1 = 1$, $\beta_2 = \infty$)	2	1.01	0	0	1	∞	1
Infinite ellipse ($\beta_1 > 1$, $\beta_2 = \infty$)	2	1.01	0	1	β_1	∞	γ_1
Squat cylinder ($\beta_1 = \beta_2$, $\beta_1 \geq 1$)	3	1.01	0.75	−1	$1.225\beta_1$	$1.225\beta_2$	γ_1
Short cylinder ($\beta_1 = 1$, $\beta_2 \geq 1$)	3	1.01	0.75	−1	β_1	$1.5\beta_2$	γ_1
Sphere ($\beta_1 = \beta_2 = 1$)	3	1.01	1.24	0	1	1	1
Ellipsoid ($\beta_1 \geq 1$, $\beta_2 \geq \beta_1$)	3	1.01	1.24	1	β_1	β_2	γ_1

Source: Lin et al. (1996b)

1. Determine thermal properties of the food (see Chapter 9).
2. Determine surface heat transfer coefficient for cooling (see Chapter 9).
3. Determine characteristic dimension L and dimensional ratios β_1 and β_2 using Equations (16) and (17).
4. Calculate Biot number using Equation (1).
5. Calculate equivalent heat transfer dimensionality E for food geometry using Equation (15). This calculation requires evaluation of E_0 and E using Equations (18) to (22).
6. Calculate lag factor corresponding to thermal center and/or mass average of food using Equations (23) to (26).
7. Calculate root of transcendental equation given in Equation (14).
8. Calculate cooling time using Equation (13).

The following alternative algorithm for estimating the cooling time of foods and beverages is based on the use of f and j factors.

1. Determine thermal properties of food (see Chapter 9).
2. Determine surface heat transfer coefficient for cooling process (see Chapter 9).
3. Determine characteristic dimension L of food.
4. Calculate Biot number using Equation (1).
5. Calculate f and j factors by one of the following methods:
   (a) Method of Pflug et al. (1965): Figures 2 to 5.
   (b) Method of Lacroix and Castaigne (1987a): Tables 1, 2, and 3.
   (c) Method of Smith et al. (1968): Equations (7) to (10) and Figure 6.
   (d) Method of Hayakawa and Villalobos (1989): Equations (11) and (12) in conjunction with Equations (7) to (10).
6. Calculate cooling time using Equation (4).

SAMPLE PROBLEMS FOR ESTIMATING COOLING TIME

Example 1. A piece of ham, initially at 160°F, is to be cooled in a blast freezer. The air temperature within the freezer is 30°F and the surface heat transfer coefficient is estimated to be 8.5 Btu/h·ft²·°F. The overall dimension of the ham is 4 by 6.5 by 11 in. Estimate the time required for the mass average temperature of the ham to reach 50°F. Thermophysical properties for ham are given as follows:

c = 0.89 Btu/lb·°F
k = 0.22 Btu/h·ft·°F
ρ = 67.5 lb/ft³

Solution: Use the algorithm based on the method of Lin et al. (1993, 1996a, 1996b).

Step 1: Determine the ham's thermal properties (c, k, ρ).
These were given in the problem statement.

Step 2: Determine the heat transfer coefficient h.
The heat transfer coefficient is given as $h = 8.5$ Btu/h·ft²·°F.

Step 3: Determine the characteristic dimension L and dimensional ratios β_1 and β_2.

For cooling time problems, the characteristic dimension is the shortest distance from the thermal center of a food to its surface. Assuming that the thermal center of the ham coincides with its geometric center, the characteristic dimension becomes

$$L = (4/12 \text{ ft})/2 = 0.1667 \text{ ft}$$

The dimensional ratios then become [Equations (16) and (17)]

$$\beta_1 = \frac{6.5}{4} = 1.625$$

$$\beta_2 = \frac{11}{4} = 2.75$$

Step 4: Calculate the Biot number.

$$\text{Bi} = hL/k = (8.5)(0.1667)/0.22 = 6.44$$

Step 5: Calculate the heat transfer dimensionality.
Using Equation (19), E_0 becomes

$$E_0 = 1.5\frac{1.625 + 2.75 + 1.625^2(1 + 2.75) + 2.75^2(1 + 1.625)}{(1.625)(2.75)(1 + 1.625 + 2.75)} - \frac{[(1.625 - 2.75)^2]^{0.4}}{15} = 2.06$$

Assuming the ham to be ellipsoidal, the geometric factors can be obtained from Table 4:

$$p_1 = 1.01 \qquad p_2 = 1.24 \qquad p_3 = 1$$

From Equation (22),

$$f(\beta_1) = \frac{1}{1.625^2} + (0.01)(1)\exp\left(1.625 - \frac{1.625^2}{6}\right) = 0.4114$$

$$f(\beta_2) = \frac{1}{2.75^2} + (0.01)(1)\exp\left(2.75 - \frac{2.75^2}{6}\right) = 0.1766$$

From Equation (21),

$$E_\infty = 0.75 + 1.01 \times 0.4114 + 1.24 \times 0.1766 = 1.38$$

Thus, using Equation (15), the equivalent heat transfer dimensionality becomes

$$E = \frac{6.44^{4/3} + 1.85}{\dfrac{6.44^{4/3}}{1.38} + \dfrac{1.85}{2.06}} = 1.44$$

Step 6: Calculate the lag factor applicable to the mass average temperature.

From Table 4, $\lambda = \beta_1$, $\gamma_1 = \beta_1$, and $\gamma_2 = \beta_2$. Using Equation (24), j_∞ becomes

$$j_\infty = 1.271 + 0.305\exp[(0.172)(1.625) - (0.115)(1.625)^2] + 0.425\exp[(0.09)(2.75) - (0.128)(2.75)^2] = 1.78$$

Using Equation (23), the lag factor applicable to the center temperature becomes

$$j_c = \frac{6.44^{1.35} + \dfrac{1}{1.625}}{\dfrac{6.44^{1.35}}{1.78} + \dfrac{1}{1.625}} = 1.72$$

Using Equations (25) and (26), the lag factor for the mass average temperature becomes

$$j_m = \left[\frac{1.5 + (0.69)(6.44)}{1.5 + 6.44}\right]^3 (1.72) = 0.721$$

Step 7: Find the root of transcendental Equation (14):

$$\omega \cot \omega + \text{Bi} - 1 = 0$$

$$\omega \cot \omega + 6.44 - 1 = 0$$

$$\omega = 2.68$$

Step 8: Calculate cooling time.
The unaccomplished temperature difference is

$$Y = \frac{T_m - T}{T_m - T_i} = \frac{30 - 50}{30 - 160} = 0.1538$$

Using Equation (13), the cooling time becomes

$$\theta = \frac{3 \times 67.5 \times 0.89(0.1667)^2}{(2.68)^2 \times 0.22 \times 1.44}\ln\left(\frac{0.721}{0.1538}\right) = 3.40 \text{ h}$$

Example 2. Repeat the cooling time calculation of Example 1, but use Hayakawa and Villalobos' (1989) estimation algorithm based on the use of f and j factors.

Solution.

Step 1: Determine the thermal properties of the ham.
The thermal properties of ham are given in Example 1.

Step 2: Determine the heat transfer coefficient.
From Example 1, $h = 8.5$ Btu/h·ft²·°F.

Step 3: Determine the characteristic dimension L and the dimensional ratios β_1 and β_2.
From Example 1, $L = 0.1667$ ft, $\beta_1 = 1.625$, $\beta_2 = 2.75$.

Step 4: Calculate the Biot number.
From Example 1, Bi = 6.44.

Step 5: Calculate the f and j factors using the method of Hayakawa and Villalobos (1989).

For simplicity, assume the cross sections of the ham to be ellipsoidal. The area of an ellipse is the product of π times half the minor axis times half the major axis, or

$$A_1 = \pi L^2 \beta_1 \qquad A_2 = \pi L^2 \beta_2$$

Using Equations (7) and (8), calculate the geometry index G:

$$B_1 = \frac{A_1}{\pi L^2} = \frac{\pi L^2 \beta_1}{\pi L^2} = \beta_1 = 1.625$$

$$B_2 = \frac{A_2}{\pi L^2} = \frac{\pi L^2 \beta_2}{\pi L^2} = \beta_2 = 2.75$$

$$G = 0.25 + \frac{3}{(8)(1.625)^2} + \frac{3}{(8)(2.75)^2} = 0.442$$

Using Equation (12), determine the characteristic value M_1^2:

$$X_g = \ln(G) = \ln(0.442) = -0.816$$

$$X_b = \ln(1/\text{Bi}) = \ln(1/6.44) = -1.86$$

$$\begin{aligned}\ln(M_1^2) &= 0.92083090 + 0.83409615(-0.816) - 0.78765739(-1.86)\\ &\quad - 0.04821784(-0.816)(-1.86) - 0.04088987(-0.816)^2\\ &\quad - 0.10045526(-1.86)^2 + 0.01521388(-0.816)^3\\ &\quad + 0.00119941(-0.816)(-1.86)^3 + 0.00129982(-1.86)^4\\ &= 1.27\end{aligned}$$

$$M_1^2 = 3.56$$

From Equation (9), the f factor becomes

$$f = \frac{2.303L^2}{M_1^2\alpha} = \frac{2.303L^2\rho c}{M_1^2 k}$$

$$f = \frac{(2.303)(0.1667)^2(67.5)(0.89)}{(3.56)(0.22)} = 4.91 \text{ h}$$

From Equation (10), the j factor becomes

$$j_m = 0.892e^{(-0.0388)(3.56)} = 0.777$$

Step 6: Calculate cooling time.

From Example 1, the unaccomplished temperature difference was found to be $Y = 0.1538$. Using Equation (4), the cooling time becomes

$$\theta = -\frac{4.91}{2.303}\ln\left(\frac{0.1538}{0.777}\right) = 3.45 \text{ h}$$

FREEZING TIMES OF FOODS AND BEVERAGES

As discussed at the beginning of this chapter, freezing of foods and beverages is not an isothermal process but rather occurs over a range of temperatures. This section discusses Plank's basic freezing time estimation method and its modifications; methods that calculate freezing time as the sum of the precooling, phase change, and subcooling times; and methods for irregularly shaped foods. These methods are divided into three subgroups: (1) equivalent heat transfer dimensionality, (2) mean conducting path, and (3) equivalent sphere diameter. All of these freezing time estimation methods use thermal properties of foods (Chapter 9).

Plank's Equation

One of the most widely known simple methods for estimating freezing times of foods and beverages was developed by Plank (1913, 1941). Convective heat transfer is assumed to occur between the food and the surrounding cooling medium. The temperature of the food is assumed to be at its initial freezing temperature, which is constant throughout the freezing process. Furthermore, constant thermal conductivity for the frozen region is assumed. Plank's freezing time estimation is as follows:

$$\theta = \frac{L_f}{T_f - T_m}\left(\frac{PD}{h} + \frac{RD^2}{k_s}\right) \tag{27}$$

where L_f is the volumetric latent heat of fusion (see Chapter 9), T_f is the initial freezing temperature of the food, T_m is the freezing medium temperature, D is the thickness of the slab or diameter of the sphere or infinite cylinder, h is the convective heat transfer coefficient, k_s is the thermal conductivity of the fully frozen food, and P and R are geometric factors. For an infinite slab, $P = 1/2$ and $R = 1/8$. For a sphere, $P = 1/6$ and $R = 1/24$; for an infinite cylinder, $P = 1/4$ and $R = 1/16$.

Plank's geometric factors indicate that an infinite slab of thickness D, an infinite cylinder of diameter D, and a sphere of diameter D, if exposed to the same conditions, would have freezing times in the ratio of 6:3:2. Hence, a cylinder freezes in half the time of a slab and a sphere freezes in one-third the time of a slab.

Modifications to Plank's Equation

Various researchers have noted that Plank's method does not accurately predict freezing times of foods and beverages. This is because, in part, Plank's method assumes that foods freeze at a constant temperature, and not over a range of temperatures as is the case in actual food freezing processes. In addition, the frozen food's thermal conductivity is assumed to be constant; in reality, thermal conductivity varies greatly during freezing. Another limitation of Plank's equation is that it neglects precooling and subcooling, the removal of sensible heat above and below the freezing point. Consequently, researchers have developed improved semianalytical/empirical cooling and freezing time estimation methods that account for these factors.

Cleland and Earle (1977, 1979a, 1979b) incorporated corrections to account for removal of sensible heat both above and below the food's initial freezing point as well as temperature variation during freezing. Regression equations were developed to estimate the geometric parameters P and R for infinite slabs, infinite cylinders, spheres, and rectangular bricks. In these regression equations, the effects of surface heat transfer, precooling, and final subcooling are accounted for by the Biot, Plank, and Stefan numbers, respectively.

In this section, the Biot number is defined as

$$\text{Bi} = \frac{hD}{k_s} \tag{28}$$

where h is the convective heat transfer coefficient, D is the characteristic dimension, and k_s is the thermal conductivity of the fully frozen food. In freezing time calculations, the characteristic dimension D is defined to be twice the shortest distance from the thermal center of a food to its surface: the thickness of a slab or the diameter of a cylinder or a sphere.

In general, the Plank number is defined as follows:

$$\text{Pk} = \frac{C_l(T_i - T_f)}{\Delta H} \tag{29}$$

where C_l is the volumetric specific heat of the unfrozen phase and ΔH is the food's volumetric enthalpy change between T_f and the final food temperature (see Chapter 9). The Stefan number is similarly defined as

$$\text{Ste} = \frac{C_s(T_f - T_m)}{\Delta H} \tag{30}$$

where C_s is the volumetric specific heat of the frozen phase.

In Cleland and Earle's method, Plank's original geometric factors P and R are replaced with the modified values given in Table 5, and the latent heat L_f is replaced with the volumetric enthalpy change of the food ΔH_{14} between the freezing temperature T_f and the final center temperature, assumed to be 14°F. As shown in Table 5, P and R are functions of the Plank and Stefan numbers. Both parameters should be evaluated using the enthalpy change ΔH_{14}. Thus, the modified Plank equation takes the form

$$\theta = \frac{\Delta H_{14}}{T_f - T_m}\left(\frac{PD}{h} + \frac{RD^2}{k_s}\right) \tag{31}$$

where k_s is the thermal conductivity of the fully frozen food.

Equation (31) is based on curve-fitting of experimental data in which the product final center temperature was 14°F. Cleland and Earle (1984) noted that this prediction formula does not perform as well in situations with final center temperatures other than 14°F. Cleland and Earle proposed the following modified form of Equation (31) to account for different final center temperatures:

$$\theta = \frac{\Delta H_{14}}{T_f - T_m}\left(\frac{PD}{h} + \frac{RD^2}{k_s}\right)\left[1 - \frac{1.65\ \text{Ste}}{k_s}\ln\left(\frac{T_c - T_m}{T_{ref} - T_m}\right)\right] \tag{32}$$

where T_{ref} is 14°F, T_c is the final product center temperature, and ΔH_{14} is the volumetric enthalpy difference between the initial freezing temperature T_f and 14°F. The values of P, R, Pk, and Ste should be evaluated using ΔH_{14}, as previously discussed.

Hung and Thompson (1983) also improved on Plank's equation to develop an alternative freezing time estimation method for infinite slabs. Their equation incorporates the volumetric change in enthalpy ΔH_0 for freezing as well as a weighted average temperature difference between the food's initial temperature and the freezing medium temperature. This weighted average temperature difference ΔT is given as follows:

$$\Delta T = (T_f - T_m) + \frac{\left(T_i - T_f\right)^2 \frac{C_l}{2} - \left(T_f - T_c\right)^2 \frac{C_s}{2}}{\Delta H_0} \tag{33}$$

where T_c is the food's final center temperature and ΔH_0 is its enthalpy change between initial and final center temperatures; the latter is assumed to be 0°F. Empirical equations were developed to estimate P and R for infinite slabs as follows:

$$P = 0.7306 - 1.083\,\text{Pk} + \text{Ste}\left(15.40U - 15.43 + 0.01329\frac{\text{Ste}}{\text{Bi}}\right) \tag{34}$$

$$R = 0.2079 - 0.2656U(\text{Ste}) \tag{35}$$

where $U = \Delta T/(T_f - T_m)$. In these expressions, Pk and Ste should be evaluated using the enthalpy change ΔH_0. The freezing time prediction model is

$$\theta = \frac{\Delta H_0}{\Delta T}\left(\frac{PD}{h} + \frac{RD^2}{k_s}\right) \tag{36}$$

Cleland and Earle (1984) applied a correction factor to the Hung and Thompson model [Equation (36)] and improved the prediction accuracy of the model for final temperatures other than 0°F. The correction to Equation (36) is as follows:

$$\theta = \frac{\Delta H_0}{\Delta T}\left(\frac{PD}{h} + \frac{RD^2}{k_s}\right)\left[1 - \frac{1.65\,\text{Ste}}{k_s}\ln\left(\frac{T_c - T_m}{T_{ref} - T_m}\right)\right] \tag{37}$$

where T_{ref} is 0°F, T_c is the product final center temperature and ΔH_0 is the volumetric enthalpy change between the initial temperature T_i and 0°F. The weighted average temperature difference ΔT, Pk, and Ste should be evaluated using ΔH_0.

Precooling, Phase Change, and Subcooling Time Calculations

Total freezing time θ is as follows:

$$\theta = \theta_1 + \theta_2 + \theta_3 \tag{38}$$

where θ_1, θ_2, and θ_3 are the precooling, phase change, and subcooling times, respectively.

DeMichelis and Calvelo (1983) suggested using Cleland and Earle's (1982a) equivalent heat transfer dimensionality method, discussed in the Cooling Times of Foods and Beverages section of this chapter, to estimate precooling and subcooling times. They also suggested that the phase change time be calculated with Plank's equation, but with the thermal conductivity of the frozen food evaluated at temperature $(T_f + T_m)/2$, where T_f is the food's initial freezing temperature and T_m is the temperature of the cooling medium.

Lacroix and Castaigne (1987a, 1987b, 1988) suggested the use of f and j factors to determine precooling and subcooling times of foods and beverages. They presented equations (see Tables 1 to 3) for estimating the values of f and j for infinite slabs, infinite cylinders, and spheres. Note that Lacroix and Castaigne based the Biot number on the shortest distance between the thermal center of the food and its surface, not twice that distance.

Lacroix and Castaigne (1987a, 1987b, 1988) gave the following expression for estimating precooling time θ_1:

$$\theta_1 = f_1 \log\left(j_1 \frac{T_m - T_i}{T_m - T_f}\right) \tag{39}$$

where T_m is the coolant temperature, T_i is the food's initial temperature, and T_f is the initial freezing point of the food. The f_1 and j_1 factors are determined from a Biot number calculated using an average thermal conductivity, which is based on the frozen and unfrozen food's thermal conductivity evaluated at $(T_f + T_m)/2$. See Chapter 9 for the evaluation of food thermal properties.

The expression for estimating subcooling time θ_3 is

$$\theta_3 = f_3 \log\left(j_3 \frac{T_m - T_f}{T_m - T_c}\right) \tag{40}$$

where T_c is the final temperature at the center of the food. The f_3 and j_3 factors are determined from a Biot number calculated using the thermal conductivity of the frozen food evaluated at the temperature $(T_f + T_m)/2$.

Lacroix and Castaigne model the phase change time θ_2 with Plank's equation:

$$\theta_2 = \frac{L_f D^2}{(T_f - T_m)k_c}\left(\frac{P}{2\text{Bi}_c} + R\right) \tag{41}$$

where L_f is the food's volumetric latent heat of fusion, P and R are the original Plank geometric shape factors, k_c is the frozen food's thermal conductivity at $(T_f + T_m)/2$, and Bi_c is the Biot number for the subcooling period ($\text{Bi}_c = hL/k_c$).

Lacroix and Castaigne (1987a, 1987b) adjusted P and R to obtain better agreement between predicted freezing times and experimental data. Using regression analysis, Lacroix and Castaigne suggested the following geometric factors:

For **infinite slabs**

$$P = 0.51233 \tag{42}$$

$$R = 0.15396 \tag{43}$$

For **infinite cylinders**

$$P = 0.27553 \tag{44}$$

$$R = 0.07212 \tag{45}$$

For **spheres**

$$P = 0.19665 \tag{46}$$

$$R = 0.03939 \tag{47}$$

For **rectangular bricks**

$$P = P'\left(-0.02175\frac{1}{\text{Bi}_c} - 0.01956\frac{1}{\text{Ste}} - 1.69657\right) \tag{48}$$

$$R = R'\left(5.57519\frac{1}{\text{Bi}_c} + 0.02932\frac{1}{\text{Ste}} + 1.58247\right) \tag{49}$$

For rectangular bricks, P' and R' are calculated using the expressions given in Table 5 for the P and R of bricks.

Pham (1984) also devised a freezing time estimation method, similar to Plank's equation, in which sensible heat effects were considered by calculating precooling, phase-change, and subcooling times separately. In addition, Pham suggested using a mean freezing point, assumed to be 3°F below the initial freezing point of the food, to account for freezing that takes place over a range of temperatures. Pham's freezing time estimation method is stated in terms of the

Table 5 Expressions for P and R

Shape	P and R Expressions	Applicability
Infinite slab	$P = 0.5072 + 0.2018\,\text{Pk} + \text{Ste}\left(0.3224\,\text{Pk} + \frac{0.0105}{\text{Bi}} + 0.0681\right)$ $R = 0.1684 + \text{Ste}(0.2740\,\text{Pk} - 0.0135)$	$2 \le h \le 88$ Btu/h·ft²·°F $0 \le D \le 4.7$ in. $T_i \le 104$°F $-49 \le T_m \le 5$°F
Infinite cylinder	$P = 0.3751 + 0.0999\,\text{Pk} + \text{Ste}\left(0.4008\,\text{Pk} + \frac{0.0710}{\text{Bi}} - 0.5865\right)$ $R = 0.0133 + \text{Ste}(0.0415\,\text{Pk} + 0.3957)$	$0.155 \le \text{Ste} \le 0.345$ $0.5 \le \text{Bi} \le 4.5$ $0 \le \text{Pk} \le 0.55$
Sphere	$P = 0.1084 + 0.0924\,\text{Pk} + \text{Ste}\left(0.231\,\text{Pk} - \frac{0.3114}{\text{Bi}} + 0.6739\right)$ $R = 0.0784 + \text{Ste}(0.0386\,\text{Pk} - 0.1694)$	$0.155 \le \text{Ste} \le 0.345$ $0.5 \le \text{Bi} \le 4.5$ $0 \le \text{Pk} \le 0.55$
Brick	$P = P_2 + P_1[0.1136 + \text{Ste}(5.766P_1 - 1.242)]$ $R = R_2 + R_1[0.7344 + \text{Ste}(49.89R_1 - 2.900)]$ where $P_2 = P_1\left[1.026 + 0.5808\,\text{Pk} + \text{Ste}\left(0.2296\,\text{Pk} + \frac{0.0182}{\text{Bi}} + 0.1050\right)\right]$ $R_2 = R_1[1.202 + \text{Ste}(3.410\,\text{Pk} + 0.7336)]$ and $P_1 = \frac{\beta_1\beta_2}{2(\beta_1\beta_2 + \beta_1 + \beta_2)}$ $R_1 = \frac{Q}{2}\left[(r-1)(\beta_1 - r)(\beta_2 - r)\ln\left(\frac{r}{r-1}\right) - (s-1)(\beta_1 - s)(\beta_2 - s)\ln\left(\frac{s}{s-1}\right)\right] + \frac{1}{72}(2\beta_1 + 2\beta_2 - 1)$ in which $\frac{1}{Q} = 4\left[(\beta_1 - \beta_2)(\beta_1 - 1) + (\beta_2 - 1)^2\right]^{1/2}$ $r = \frac{1}{3}\left\{\beta_1 + \beta_2 + 1 + [(\beta_1 - \beta_2)(\beta_1 - 1) + (\beta_2 - 1)^2]^{1/2}\right\}$ $s = \frac{1}{3}\left\{\beta_1 + \beta_2 + 1 - [(\beta_1 - \beta_2)(\beta_1 - 1) + (\beta_2 - 1)^2]^{1/2}\right\}$ and $\beta_1 = \frac{\text{Second shortest dimension of food}}{\text{Shortest dimension of food}}$ $\beta_2 = \frac{\text{Longest dimension of food}}{\text{Shortest dimension of food}}$	$0.155 \le \text{Ste} \le 0.345$ $0 \le \text{Pk} \le 0.55$ $0 \le \text{Bi} \le 22$ $1 \le \beta_1 \le 4$ $1 \le \beta_2 \le 4$

Source: Cleland and Earle (1977, 1979a, 1979b)

volume and surface area of the food and is, therefore, applicable to foods of any shape. This method is given as

$$\theta_i = \frac{Q_i}{hA_s\Delta T_{mi}}\left(1 + \frac{\text{Bi}_i}{k_i}\right) \qquad i = 1, 2, 3 \tag{50}$$

where θ_1 is the precooling time, θ_2 is the phase change time, θ_3 is the subcooling time, and the remaining variables are defined as shown in Table 6.

Pham (1986a) significantly simplified the previous freezing time estimation method to yield

$$\theta = \frac{V}{hA_s}\left(\frac{\Delta H_1}{\Delta T_1} + \frac{\Delta H_2}{\Delta T_2}\right)\left(1 + \frac{\text{Bi}_s}{4}\right) \tag{51}$$

in which

$$\Delta H_1 = C_l(T_i - T_{fm}) \qquad \Delta H_2 = L_f + C_s(T_{fm} - T_c) \tag{52}$$

$$\Delta T_1 = \frac{T_i + T_{fm}}{2} - T_m \qquad \Delta T_2 = T_{fm} - T_m \tag{53}$$

where C_l and C_s are volumetric specific heats above and below freezing, respectively, T_i is the initial food temperature, L_f is the volumetric latent heat of freezing, and V is the volume of the food.

Pham suggested that the mean freezing temperature T_{fm} used in Equations (52) and (53) mainly depended on the cooling medium temperature T_m and product center temperature T_c. By curve fitting to existing experimental data, Pham (1986a) proposed the following equation to determine the mean freezing temperature for use in Equations (52) and (53):

$$T_{fm} = 23.46 + 0.263T_c + 0.105T_m \tag{54}$$

where all temperatures are in °F.

Geometric Considerations

Equivalent Heat Transfer Dimensionality. Similar to their work involving cooling times of foods, Cleland and Earle (1982b) also introduced a geometric correction factor, called the **equivalent heat transfer dimensionality *E***, to calculate the freezing times of irregularly shaped foods. The freezing time of an irregularly shaped object θ_{shape} was related to the freezing time of an infinite slab θ_{slab} using the equivalent heat transfer dimensionality:

$$\theta_{shape} = \theta_{slab}/E \tag{55}$$

Table 6 Definition of Variables for Freezing Time Estimation Method

Process	Variables
Precooling	$i = 1$ $k_1 = 6$ $Q_1 = C_l(T_i - T_{fm})V$ $\mathrm{Bi}_1 = (\mathrm{Bi}_l + \mathrm{Bi}_s)/2$ $\Delta T_{m1} = \dfrac{(T_i - T_m) - (T_{fm} - T_m)}{\ln\left(\dfrac{T_i - T_m}{T_{fm} - T_m}\right)}$
Phase change	$i = 2$ $k_2 = 4$ $Q_2 = L_f V$ $\mathrm{Bi}_2 = \mathrm{Bi}_s$ $\Delta T_{m2} = T_{fm} - T_m$
Subcooling	$i = 3$ $k_3 = 6$ $Q_3 = C_s(T_{fm} - T_c)V$ $\mathrm{Bi}_3 = \mathrm{Bi}_s$ $\Delta T_{m3} = \dfrac{(T_{fm} - T_m) - (T_o - T_m)}{\ln\left(\dfrac{T_{fm} - T_m}{T_o - T_m}\right)}$

Source: Pham (1984)

Notes: A_s = area through which heat is transferred
Bi_l = Biot number for unfrozen phase
Bi_s = Biot number for frozen phase
Q_1, Q_2, Q_3 = heats of precooling, phase change, and subcooling, respectively
$\Delta T_{m1}, \Delta T_{m2}, \Delta T_{m3}$ = corresponding log-mean temperature driving forces
T_c = final thermal center temperature
T_{fm} = mean freezing point, assumed 3°F below initial freezing point
T_o = mean final temperature
V = volume of food

Freezing time of the infinite slab is then calculated from one of the many suitable freezing time estimation methods.

Using data collected from a large number of freezing experiments, Cleland and Earle (1982b) developed empirical correlations for the equivalent heat transfer dimensionality applicable to rectangular bricks and finite cylinders. For rectangular brick shapes with dimensions D by $\beta_1 D$ by $\beta_2 D$, the equivalent heat transfer dimensionality was given as follows:

$$E = 1 + W_1 + W_2 \tag{56}$$

where

$$W_1 = \left(\frac{\mathrm{Bi}}{\mathrm{Bi}+2}\right)\frac{5}{8\beta_1^3} + \left(\frac{2}{\mathrm{Bi}+2}\right)\frac{2}{\beta_1(\beta_1+1)} \tag{57}$$

and

$$W_2 = \left(\frac{\mathrm{Bi}}{\mathrm{Bi}+2}\right)\frac{5}{8\beta_2^3} + \left(\frac{2}{\mathrm{Bi}+2}\right)\frac{2}{\beta_2(\beta_2+1)} \tag{58}$$

For finite cylinders where the diameter is smaller than the height, the equivalent heat transfer dimensionality was given as

$$E = 2.0 + W_2 \tag{59}$$

In addition, Cleland et al. (1987a, 1987b) developed expressions for determining the equivalent heat transfer dimensionality of infinite slabs, infinite and finite cylinders, rectangular bricks, spheres, and two- and three-dimensional irregular shapes. Numerical methods were used to calculate the freezing or thawing times for these shapes. A nonlinear regression analysis of the resulting numerical data yielded the following form for the equivalent heat transfer dimensionality:

Table 7 Geometric Constants

Shape	G_1	G_2	G_3
Infinite slab	1	0	0
Infinite cylinder	2	0	0
Sphere	3	0	0
Finite cylinder (diameter > height)	1	2	0
Finite cylinder (height > diameter)	2	0	1
Infinite rod	1	1	0
Rectangular brick	1	1	1
Two-dimensional irregular shape	1	1	0
Three-dimensional irregular shape	1	1	1

Source: Cleland et al. (1987a)

$$E = G_1 + G_2E_1 + G_3E_2 \tag{60}$$

where

$$E_1 = X\left(2.32/\beta_1^{1.77}\right)\frac{1}{\beta_1} + \left[1 - X\left(2.32/\beta_1^{1.77}\right)\right]\frac{0.73}{\beta_1^{2.50}} \tag{61}$$

$$E_2 = X\left(2.32/\beta_1^{1.77}\right)\frac{1}{\beta_2} + \left[1 - X\left(2.32/\beta_2^{1.77}\right)\right]\frac{0.50}{\beta_2^{3.69}} \tag{62}$$

and G_1, G_2, and G_3 are given in Table 7. In Equations (61) and (62), the function X with argument ϕ is defined as

$$X(\phi) = \phi/\left(\mathrm{Bi}^{1.34} + \phi\right) \tag{63}$$

Using the freezing time prediction methods for infinite slabs and various multidimensional shapes developed by McNabb et al. (1990), Hossain et al. (1992a) derived infinite series expressions for E of infinite rectangular rods, finite cylinders, and rectangular bricks. For most practical freezing situations, only the first term of these series expressions is significant. The resulting expressions for E are given in Table 8.

Hossain et al. (1992b) also presented a semianalytically derived expression for the equivalent heat transfer dimensionality of two-dimensional, irregularly shaped foods. An equivalent "pseudoelliptical" infinite cylinder was used to replace the actual two-dimensional, irregular shape in the calculations. A pseudoellipse is a shape that depends on the Biot number. As the Biot number approaches infinity, the shape closely resembles an ellipse. As the Biot number approaches zero, the pseudoelliptical infinite cylinder approaches an infinite rectangular rod. Hossain et al. (1992b) stated that, for practical Biot numbers, the pseudoellipse is very similar to a true ellipse. This model pseudoelliptical infinite cylinder has the same volume per unit length and characteristic dimension as the actual food. The resulting expression for E is as follows:

$$E = 1 + \frac{1 + \dfrac{2}{\mathrm{Bi}}}{\beta_2^2 + \dfrac{2\beta_2}{\mathrm{Bi}}} \tag{64}$$

In Equation (64), the Biot number is based on the shortest distance from the thermal center to the food's surface, not twice that distance. Using this expression for E, the freezing time θ_{shape} of two-dimensional, irregularly shaped foods can be calculated with Equation (55).

Hossain et al. (1992c) extended this analysis to predicting freezing times of three-dimensional, irregularly shaped foods. In this work, the irregularly shaped food was replaced with a model ellipsoid shape having the same volume, characteristic dimension, and smallest cross-sectional area orthogonal to the characteristic dimension, as the actual food item. An expression was presented for E of a pseudoellipsoid as follows:

Table 8 Expressions for Equivalent Heat Transfer Dimensionality

Shape	Expressions for Equivalent Heat Transfer Dimensionality E
Infinite rectangular rod ($2L$ by $2\beta_1 L$)	$$E = \left(1+\frac{2}{\mathrm{Bi}}\right)\left\{\left(1+\frac{2}{\mathrm{Bi}}\right) - 4\sum_{n=1}^{\infty}\left[\frac{(\sin z_n)}{z_n^3\left(1+\frac{\sin^2 z_n}{\mathrm{Bi}}\right)\left(\frac{z_n}{\mathrm{Bi}}\sinh(z_n\beta_1)+\cosh(z_n\beta_1)\right)}\right]\right\}^{-1}$$ where z_n are roots of $\mathrm{Bi} = z_n \tan(z_n)$ and $\mathrm{Bi} = hL/k$, where L is the shortest distance from the center of the rectangular rod to the surface.
Finite cylinder, height exceeds diameter (radius L and height $2\beta_1 L$)	$$E = \left(2+\frac{4}{\mathrm{Bi}}\right)\left\{\left(1+\frac{2}{\mathrm{Bi}}\right) - 8\sum_{n=1}^{\infty}\left[y_n^3 J_1(y_n)\left(1+\frac{y_n^2}{\mathrm{Bi}^2}\right)\left(\cosh(\beta_1 y_n)+\frac{y_n}{\mathrm{Bi}}\sinh(\beta_1 y_n)\right)\right]^{-1}\right\}^{-1}$$ where y_n are roots of $y_n J_1(y_n) - \mathrm{Bi} J_0(y_n) = 0$; J_0 and J_1 are Bessel functions of the first kind, order zero and one, respectively; and $\mathrm{Bi} = hL/k$, where L is the radius of the cylinder.
Finite cylinder, diameter exceeds height (radius $\beta_1 L$ and height $2L$)	$$E = \left(1+\frac{2}{\mathrm{Bi}}\right)\left\{\left(1+\frac{2}{\mathrm{Bi}}\right) - 4\sum_{n=1}^{\infty}\frac{\sin z_n}{z_n^2(z_n+\cos z_n \sin z_n)\left(I_0(z_n\beta_1)+\frac{z_n}{\mathrm{Bi}}I_1(z_n\beta_1)\right)}\right\}^{-1}$$ where z_n are roots of $\mathrm{Bi} = z_n \tan(z_n)$; I_0 and I_1 are Bessel function of the second kind, order zero and one, respectively; and $\mathrm{Bi} = hL/k$, where L is the radius of the cylinder.
Rectangular brick ($2L$ by $2\beta_1 L$ by $2\beta_2 L$)	$$E = \left(1+\frac{2}{\mathrm{Bi}}\right)\left\{\left(1+\frac{2}{\mathrm{Bi}}\right) - 4\sum_{n=1}^{\infty}\left[\frac{\sin z_n}{z_n^3\left(1+\frac{\sin^2 z_n}{\mathrm{Bi}}\right)\left[\frac{z_n}{\mathrm{Bi}}\sinh(z_n\beta_1)+\cosh(z_n\beta_1)\right]}\right] - 8\beta_2^2\sum_{n=1}^{\infty}\sum_{m=1}^{\infty}\left[\sin z_n \sin z_m\left[\left(\cosh(z_{nm})+\frac{z_{nm}}{\mathrm{Bi}\beta_2}\sinh(z_{nm})\right) z_n z_m z_{nm}^2\left(1+\frac{1}{\mathrm{Bi}}\sin^2 z_n\right)\left(1+\frac{1}{\mathrm{Bi}\beta_1}\sin^2 z_m\right)\right]^{-1}\right]\right\}^{-1}$$ where z_n are roots of $\mathrm{Bi} = z_n \tan(z_n)$; z_m are the roots of $\mathrm{Bi}\beta_1 = z_m \tan(z_m)$; $\mathrm{Bi} = hL/k$, where L is the shortest distance from the thermal center of the rectangular brick to the surface; and z_{nm} is given as $$z_{nm}^2 = z_n^2\beta_2^2 + z_m^2\left(\frac{\beta_2}{\beta_1}\right)^2$$

Source: Hossain et al. (1992a)

Table 9 Summary of Methods for Determining Equivalent Heat Transfer Dimensionality

Slab		Cleland et al. (1987a, 1987b) Equations (60) to (63)	
Infinite cylinder		Cleland et al. (1987a, 1987b) Equations (60) to (63)	
Sphere		Cleland et al. (1987a, 1987b) Equations (60) to (63)	
Finite cylinder (diameter > height)		Cleland et al. (1987a, 1987b) Equations (60) to (63)	Hossain et al. (1992a) Table 8
Finite cylinder (height > diameter)	Cleland and Earle (1982a, 1982b) Equations (58) and (59)	Cleland et al. (1987a, 1987b) Equations (60) to (63)	Hossain et al. (1992a) Table 8
Infinite rod		Cleland et al. (1987a, 1987b) Equations (60) to (63)	Hossain et al. (1992a) Table 8
Rectangular brick	Cleland and Earle (1982a, 1982b) Equations (56) to (58)	Cleland et al. (1987a, 1987b) Equations (60) to (63)	Hossain et al. (1992a) Table 8
2-D irregular shape (infinite ellipse)		Cleland et al. (1987a, 1987b) Equations (60) to (63)	Hossain et al. (1992b) Equation (64)
3-D irregular shape (ellipsoid)		Cleland et al. (1987a, 1987b) Equations (60) to (63)	Hossain et al. (1992b) Equation (65)

$$E = 1 + \frac{1 + \frac{2}{\mathrm{Bi}}}{\beta_1^2 + \frac{2\beta_1}{\mathrm{Bi}}} + \frac{1 + \frac{2}{\mathrm{Bi}}}{\beta_2^2 + \frac{2\beta_2}{\mathrm{Bi}}} \tag{65}$$

In Equation (65), the Biot number is based on the shortest distance from the thermal center to the surface of the food, not twice that distance. With this expression for E, freezing times θ_{shape} of three-dimensional, irregularly shaped foods may be calculated using Equation (55).

Table 9 summarizes the methods that have been discussed for determining the equivalent heat transfer dimensionality of various geometries. These methods can be used with Equation (55) to calculate freezing times.

Mean Conducting Path. Pham's freezing time formulas, given in Equations (50) and (51), require knowledge of the Biot number. To calculate the Biot number of a food, its characteristic dimension must be known. Because it is difficult to determine the characteristic dimension of an irregularly shaped food, Pham (1985) introduced the concept of the **mean conducting path**, which is the mean heat transfer length from the surface of the food to its thermal center, or $D_m/2$. Thus, the Biot number becomes

$$\mathrm{Bi} = \frac{hD_m}{k} \tag{66}$$

where D_m is twice the mean conducting path.

For rectangular blocks of food, Pham (1985) found that the mean conducting path was proportional to the geometric mean of the block's two shorter dimensions. Based on this result, Pham (1985) presented an equation to calculate the Biot number for rectangular blocks of food:

$$\frac{\mathrm{Bi}}{\mathrm{Bi}_o} = 1 + \left\{\left[1.5\sqrt{\beta_1} - 1\right]^{-4} + \left[\left(\frac{1}{\beta_1} + \frac{1}{\beta_2}\right)\left(1 + \frac{4}{\mathrm{Bi}_o}\right)\right]^{-4}\right\}^{-0.25} \tag{67}$$

where Bi_o is the Biot number based on the shortest dimension of the block D_1, or $\mathrm{Bi}_o = hD_1/k$. The Biot number can then be substituted into a freezing time estimation method to calculate the freezing time for rectangular blocks.

Pham (1985) noted that, for squat-shaped foods, the mean conducting path $D_m/2$ could be reasonably estimated as the arithmetic mean of the longest and shortest distances from the surface of the food to its thermal center.

Equivalent Sphere Diameter. Ilicali and Engez (1990) and Ilicali and Hocalar (1990) introduced the **equivalent sphere diameter** concept to calculate the freezing time of irregularly shaped foods. In this method, a sphere diameter is calculated based on the volume and the volume-to-surface-area ratio of the irregularly shaped food. This equivalent sphere is then used to calculate the freezing time of the food item.

Considering an irregularly shaped food item where the shortest and longest distances from the surface to the thermal center were designated as D_1 and D_2, respectively, Ilicali and Engez (1990) and Ilicali and Hocalar (1990) defined the volume-surface diameter D_{vs} as the diameter of a sphere having the same volume-to-surface-area ratio as the irregular shape:

$$D_{vs} = 6V/A_s \tag{68}$$

where V is the volume of the irregular shape and A_s is its surface area. In addition, the volume diameter D_v is defined as the diameter of a sphere having the same volume as the irregular shape:

$$D_v = (6V/\pi)^{1/3} \tag{69}$$

Because a sphere is the solid geometry with minimum surface area per unit volume, the equivalent sphere diameter $D_{eq,s}$ must be greater than D_{vs} and smaller than D_v. In addition, the contribution of the volume diameter D_v has to decrease as the ratio of the longest to the shortest dimensions D_2/D_1 increases, because the object will be essentially two-dimensional if $D_2/D_1 \gg 1$. Therefore, the equivalent sphere diameter $D_{eq,s}$ is defined as follows:

$$D_{eq,s} = \frac{1}{\beta_2 + 1}D_v + \frac{\beta_2}{\beta_2 + 1}D_{vs} \tag{70}$$

Thus, predicting the freezing time of the irregularly shaped food is reduced to predicting the freezing time of a spherical food with diameter $D_{eq,s}$. Any of the previously discussed freezing time methods for spheres may then be used to calculate this freezing time.

Evaluation of Freezing Time Estimation Methods

As noted previously, selecting an appropriate estimation method from the plethora of available methods can be challenging for the designer. Thus, Becker and Fricke (1999a, 1999b, 1999c, 2000a, 2000b) quantitatively evaluated selected semianalytical/empirical food freezing time estimation methods for regularly and irregularly shaped foods. Each method's performance was quantified by comparing its numerical results to a comprehensive experimental freezing time data set compiled from the literature. The best-performing methods for each shape are listed in Table 10.

Algorithms for Freezing Time Estimation

The following suggested algorithm for estimating the freezing time of foods and beverages is based on the modified Plank equation presented by Cleland and Earle (1977, 1979a, 1979b). This algorithm is applicable to simple food geometries, including infinite slabs, infinite cylinders, spheres, and three-dimensional rectangular bricks.

1. Determine thermal properties of food (see Chapter 9).
2. Determine surface heat transfer coefficient for the freezing process (see Chapter 9).
3. Determine characteristic dimension D and dimensional ratios β_1 and β_2 using Equations (16) and (17).
4. Calculate Biot, Plank, and Stefan numbers using Equations (28), (29), and (30), respectively.
5. Determine geometric parameters P and R given in Table 5.
6. Calculate freezing time using Equation (31) or (32), depending on the final temperature of the frozen food.

The following algorithm for estimating freezing times of foods and beverages is based on the method of equivalent heat transfer dimensionality. It is applicable to many food geometries, including infinite rectangular rods, finite cylinders, three-dimensional rectangular bricks, and two- and three-dimensional irregular shapes.

1. Determine thermal properties of the food (see Chapter 9).

Table 10 Estimation Methods of Freezing Time of Regularly and Irregularly Shaped Foods

Shape	Methods
Infinite slab	Cleland and Earle (1977), Hung and Thompson (1983), Pham (1984, 1986a)
Infinite cylinder	Cleland and Earle (1979a), Pham (1986a), Lacroix and Castaigne (1987a)
Short cylinder	Cleland et al. (1987a, 1987b), Hossain et al. (1992a), equivalent sphere diameter technique
Rectangular brick	Cleland and Earle (1982b), Cleland et al. (1987a, 1987b), Hossain et al. (1992a)
Two-dimensional irregular shape	Hossain et al. (1992b)
Three-dimensional irregular shape	Hossain et al. (1992c), equivalent sphere diameter technique

2. Determine surface heat transfer coefficient for the freezing process (see Chapter 9).
3. Determine characteristic dimension D and dimensional ratios β_1 and β_2 using Equations (16) and (17).
4. Calculate Biot, Plank, and Stefan numbers using Equations (28), (29), and (30), respectively.
5. Calculate freezing time of an infinite slab using a suitable method. Suitable methods include
   (a) Equation (31) or (32) in conjunction with the geometric parameters P and R given in Table 5.
   (b) Equation (36) or (37) in conjunction with Equations (33), (34), and (35).
6. Calculate the food's equivalent heat transfer dimensionality. Refer to Table 9 to determine which equivalent heat transfer dimensionality method is applicable to the particular food geometry.
7. Calculate the freezing time of the food using Equation (55).

SAMPLE PROBLEMS FOR ESTIMATING FREEZING TIME

Example 3. A rectangular brick-shaped package of beef (lean sirloin) measuring 1.5 by 4.5 by 6 in. is to be frozen in a blast freezer. The beef's initial temperature is 50°F, and the freezer air temperature is −22°F. The surface heat transfer coefficient is estimated to be 7.4 Btu/h·ft²·°F. Calculate the time required for the thermal center of the beef to reach 14°F.

Solution: Because the food is a rectangular brick, the algorithm based on the modified Plank equation by Cleland and Earle (1977, 1979a, 1979b) is used.

Step 1: Determine the thermal properties of lean sirloin.

As described in Chapter 9, the thermal properties can be calculated as follows:

Property	At −40°F (Fully Frozen)	At 14°F (Final Temp.)	At 28.9°F (Initial Freezing Point)	At 50°F (Initial Temp.)
Density, lb/ft³	$\rho_s = 63.6$	$\rho_s = 63.6$	$\rho_l = 67.2$	$\rho_l = 67.2$
Enthalpy, Btu/lb	—	$H_s = 35.81$	$H_l = 117.8$	—
Specific heat, Btu/lb·°F	$c_s = 0.504$	—	—	$c_l = 0.840$
Thermal conductivity, Btu/(h·ft·°F)	$k_s = 0.96$	—	—	—

Volumetric enthalpy difference between the initial freezing point and 14°F:

$$\Delta H_{14} = \rho_l H_l - \rho_s H_s$$

$$\Delta H_{14} = (67.2)(117.8) - (63.6)(35.8) = 5640 \text{ Btu/ft}^3$$

Volumetric specific heats:

$$C_s = \rho_s c_s = (63.6)(0.504) = 32.05 \text{ Btu/(ft}^3\cdot°\text{F)}$$

$$C_l = \rho_l c_l = (67.2)(0.84) = 56.45 \text{ Btu/(ft}^3\cdot°\text{F)}$$

Step 2: Determine the surface heat transfer coefficient.

The surface heat transfer coefficient is estimated to be 7.4 Btu/h·ft²·°F.

Step 3: Determine the characteristic dimension D and the dimensional ratios β_1 and β_2.

For freezing time problems, the characteristic dimension D is twice the shortest distance from the thermal center of the food to its surface. For this example, $D = 1.5/12 = 0.125$ ft.

Using Equations (16) and (17), the dimensional ratios then become

$$\beta_1 = 4.5/1.5 = 3$$

$$\beta_2 = 6.0/1.5 = 4$$

Step 4: Using Equations (28) to (30), calculate the Biot, Plank, and Stefan numbers.

$$\text{Bi} = \frac{hD}{k_s} = \frac{(7.4)(0.125)}{0.96} = 0.964$$

$$\text{Pk} = \frac{C_l(T_i - T_f)}{\Delta H_{10}} = \frac{56.45(50 - 28.9)}{5640} = 0.211$$

$$\text{Ste} = \frac{C_s(T_f - T_m)}{\Delta H_{10}} = \frac{32.05[28.9 - (-22)]}{5640} = 0.289$$

Step 5: Determine the geometric parameters P and R for the rectangular brick.

Determine P from Table 5.

$$P_1 = \frac{(3)(4)}{2[(3)(4) + 3 + 4]} = 0.316$$

$$P_2 = 0.316\left\{1.026 + (0.5808)(0.211) + 0.289\left[(0.2296)(0.211) + \frac{0.0182}{0.964} + 0.1050\right]\right\} = 0.379$$

$$P = 0.379 + 0.316\{0.1136 + 0.289[(5.766)(0.316) - 1.242]\} = 0.468$$

Determine R from Table 5.

$$\frac{1}{Q} = 4[(3-4)(3-1) + (4-1)^2]^{1/2} = 10.6$$

$$r = \frac{1}{3}\left\{3 + 4 + 1 + \left[(3-4)(3-1) + (4-1)^2\right]^{1/2}\right\} = 3.55$$

$$s = \frac{1}{3}\left\{3 + 4 + 1 - \left[(3-4)(3-1) + (4-1)^2\right]^{1/2}\right\} = 1.78$$

$$R_1 = \frac{1}{(10.6)(2)}\left\{(3.55-1)(3-3.55)(4-3.55)\ln\left[\frac{3.55}{3.55-1}\right] - (1.78-1)(3-1.78)(4-1.78)\ln\left[\frac{1.78}{1.78-1}\right]\right\} + \frac{1}{72}[(2)(3) + (2)(4) - 1] = 0.0885$$

$$R_2 = 0.0885\{1.202 + 0.289[(3.410)(0.211) + 0.7336]\} = 0.144$$

$$R = 0.144 + 0.0885\{0.7344 + 0.289[(49.89)(0.0885) - 2.900]\} = 0.248$$

Step 6: Calculate the beef's freezing time.

Because the final temperature at the thermal center of the beef is given to be 14°F, use Equation (31) to calculate the freezing time:

$$\theta = \frac{5640}{28.9 - (-22)}\left[\frac{(0.468)(0.125)}{7.4} + \frac{(0.248)(0.125)^2}{0.96}\right] = 1.32 \text{ h}$$

Example 4. Orange juice in a cylindrical container, 1.0 ft diameter by 1.5 ft tall, is to be frozen in a blast freezer. The initial temperature of the juice is 41°F and the freezer air temperature is −31°F. The surface heat transfer coefficient is estimated to be 5.3 Btu/h·ft²·°F. Calculate the time required for the thermal center of the juice to reach 0°F.

Solution: Because the food is a finite cylinder, the algorithm based on the method of equivalent heat transfer dimensionality (Cleland et al. 1987a, 1987b) is used. This method requires calculation of the freezing time of an infinite slab, which is determined using the method of Hung and Thompson (1983).

Step 1: Determine the thermal properties of orange juice.

Using the methods described in Chapter 9, the thermal properties of orange juice are calculated as follows:

Property	At −40°F (Fully Frozen)	At 0°F (Final Temp.)	At 41°F (Initial Temp.)
Density, lb/ft³	$\rho_s = 60.6$	$\rho_s = 60.6$	$\rho_l = 64.9$
Enthalpy, Btu/lb	—	$H_s = 17.5$	$H_l = 164$
Specific heat, Btu/lb · °F	$c_s = 0.420$	—	$c_l = 0.933$
Thermal cond., Btu/lb · ft · °F	$k_s = 1.29$	—	—
Initial freezing temperature: $T_f = 31.3$°F			

Volumetric enthalpy difference between $T_i = 41$°F, and 0°F:

$$\Delta H_0 = \rho_l H_l - \rho_s H_s$$

$$\Delta H_0 = (64.9)(164.0) - (60.6)(17.5) = 9580 \text{ Btu/ft}^3$$

Volumetric specific heats:

$$C_s = \rho_s c_s = (60.6)(0.420) = 25.45 \text{ Btu/ft}^3\cdot°\text{F}$$

$$C_l = \rho_l c_l = (64.9)(0.933) = 60.55 \text{ Btu/ft}^3\cdot°\text{F}$$

Step 2: Determine the surface heat transfer coefficient.

The surface heat transfer coefficient is estimated to be 5.3 Btu/h·ft²·°F.

Step 3: Determine the characteristic dimension D and the dimensional ratios β_1 and β_2.

For freezing time problems, the characteristic dimension is twice the shortest distance from the thermal center of the food item to its surface. For the cylindrical sample of orange juice, the characteristic dimension is equal to the diameter of the cylinder:

$$D = 1.0 \text{ ft}$$

Using Equations (16) and (17), the dimensional ratios then become

$$\beta_1 = \beta_2 = \frac{1.5}{1.0} = 1.5$$

Step 4: Using Equations (28) to (30), calculate the Biot, Plank, and Stefan numbers.

$$\text{Bi} = \frac{hD}{k_s} = \frac{(5.3)(1.0)}{1.29} = 4.11$$

$$\text{Pk} = \frac{C_l(T_i - T_f)}{\Delta H_0} = \frac{(60.55)(41 - 31.3)}{9580} = 0.0613$$

$$\text{Ste} = \frac{C_s(T_f - T_m)}{\Delta H_0} = \frac{(25.45)[31.3 - (-31)]}{9580} = 0.166$$

Step 5: Calculate the freezing time of an infinite slab.

Use the method of Hung and Thompson (1983). First, find the weighted average temperature difference given by Equation (33).

$$\Delta T = [31.3 - (-31)] + \frac{(41 - 31.3)^2(60.55/2) - (31.3 - 0)^2(25.45/2)}{9580} = 61.3 \text{ °F}$$

Determine the parameter U:

$$U = \frac{61.3}{31.3 - (-31)} = 0.984$$

Determine the geometric parameters P and R for an infinite slab using Equations (34) and (35):

$$P = 0.7306 - (1.083)(0.0613) + (0.166)\left[(15.40)(0.984) - 15.43 + \frac{(0.01329)(0.166)}{4.11}\right] = 0.616$$

$$R = 0.2079 - (0.2656)(0.984)(0.166) = 0.165$$

Determine the freezing time of the slab using Equation (36):

$$\theta = \frac{9580}{61.3}\left[\frac{(0.616)(1.0)}{5.3} + \frac{0.165(1.0)^2}{1.29}\right] = 38.2 \text{ h}$$

Step 6: Calculate the equivalent heat transfer dimensionality for a finite cylinder.

Use the method presented by Cleland et al. (1987a, 1987b), Equations (60) to (63), to calculate the equivalent heat transfer dimensionality. From Table 7, the geometric constants for a cylinder are

$$G_1 = 2 \qquad G_2 = 0 \qquad G_3 = 1$$

Calculate E_2:

$$\phi = \frac{2.32}{\beta_2^{1.77}} = \frac{2.32}{1.5^{1.77}} = 1.132$$

$$X(1.132) = \frac{1.132}{4.11^{1.34} + 1.132} = 0.146$$

$$E_2 = \frac{0.146}{1.5} + (1 - 0.146)\frac{0.50}{1.5^{3.69}} = 0.193$$

Thus, the equivalent heat transfer dimensionality E becomes

$$E = G_1 + G_2E_1 + G_3E_2$$

$$E = 2 + (0)(E_1) + (1)(0.193) = 2.193$$

Step 7: Calculate freezing time of the orange juice using Equation (55):

$$\theta_{shape} = \theta_{slab}/E = 38.2/2.193 = 17.4 \text{ h}$$

SYMBOLS

A_1 = cross-sectional area in Equation (8), ft²
A_2 = cross-sectional area in Equation (8), ft²
A_s = surface area of food, ft²
B_1 = parameter in Equation (7)
B_2 = parameter in Equation (7)
Bi = Biot number
Bi_1 = Biot number for precooling = $(\text{Bi}_l + \text{Bi}_s)/2$
Bi_2 = Biot number for phase change = Bi_s
Bi_3 = Biot number for subcooling = Bi_s
Bi_c = Biot number evaluated at $k_c = hD/k_c$
Bi_l = Biot number for unfrozen food = hD/k_l
Bi_o = Biot number based on shortest dimension = hD_1/k
Bi_s = Biot number for fully frozen food = hD/k_s
c = specific heat of food, Btu/lb·°F
C_l = volumetric specific heat of unfrozen food, Btu/ft³·°F
C_s = volumetric specific heat of fully frozen food, Btu/ft³·°F
D = slab thickness or cylinder/sphere diameter, ft
D_1 = shortest dimension, ft
D_2 = longest dimension, ft
$D_{eq,s}$ = equivalent sphere diameter, ft
D_m = twice the mean conducting path, ft
D_v = volume diameter, ft
D_{vs} = volume-surface diameter, ft
E = equivalent heat transfer dimensionality
E_0 = equivalent heat transfer dimensionality at Bi = 0
E_1 = parameter given by Equation (61)
E_2 = parameter given by Equation (62)
E_∞ = equivalent heat transfer dimensionality at Bi → ∞
f = cooling time parameter
f_1 = cooling time parameter for precooling
f_3 = cooling time parameter for subcooling
f_{comp} = cooling parameter for a composite shape
G = geometry index
G_1 = geometric constant in Equation (60)
G_2 = geometric constant in Equation (60)
G_3 = geometric constant in Equation (60)
h = heat transfer coefficient, Btu/h·ft²·°F
$I_0(x)$ = Bessel function of second kind, order zero
$I_1(x)$ = Bessel function of second kind, order one
j = cooling time parameter

j_1 = cooling time parameter for precooling
j_3 = cooling time parameter for subcooling
j_c = cooling time parameter applicable to thermal center
j_{comp} = cooling time parameter for a composite shape
j_m = cooling time parameter applicable to mass average
j_s = cooling time parameter applicable to surface temperature
$J_0(x)$ = Bessel function of first kind, order zero
$J_1(x)$ = Bessel function of first kind, order one
j_∞ = lag factor parameter given by Equation (24)
k = thermal conductivity of food, Btu/h·ft·°F
k_c = thermal conductivity of food evaluated at $(T_f + T_m)/2$, Btu/h·ft·°F
k_l = thermal conductivity of unfrozen food, Btu/h·ft·°F
k_s = thermal conductivity of fully frozen food, Btu/h·ft·°F
L = half thickness of slab or radius of cylinder/sphere, ft
L_f = volumetric latent heat of fusion, Btu/ft^3
m = inverse of Biot number
M_1^2 = characteristic value of Smith et al. (1968)
N = number of dimensions
p_1 = geometric parameter given in Table 4
p_2 = geometric parameter given in Table 4
p_3 = geometric parameter given in Table 4
P = Plank's geometry factor
P' = geometric factor for rectangular bricks calculated using method in Table 5
P_1 = intermediate value of Plank's geometric factor
P_2 = intermediate value of Plank's geometric factor
Pk = Plank number = $C_l(T_i - T_f)/\Delta H$
Q = parameter in Table 5
Q_1 = volumetric heat of precooling, Btu/ft^3
Q_2 = volumetric heat of phase change, Btu/ft^3
Q_3 = volumetric heat of subcooling, Btu/ft^3
r = parameter given in Table 5
R = Plank's geometry factor
R' = geometric factor for rectangular bricks calculated using method in Table 5
R_1 = intermediate value of Plank's geometric factor
R_2 = intermediate value of Plank's geometric factor
s = parameter given in Table 5
Ste = Stefan number = $C_s(T_f - T_m)/\Delta H$
T = product temperature, °F
T_c = final center temperature of food, °F
T_f = initial freezing temperature of food, °F
T_{fm} = mean freezing temperature, °F
T_i = initial temperature of food, °F
T_m = cooling or freezing medium temperature, °F
T_o = mean final temperature, °F
T_{ref} = reference temperature for freezing time correction factor, °F
u = parameter given in Table 1
U = parameter in Equations (34) and (35) = $\Delta T/(T_f - T_m)$
v = parameter given in Table 2
V = volume of food, ft^3
w = parameter given in Table 3
W_1 = parameter given by Equation (57)
W_2 = parameter given by Equation (58)
x = coordinate direction
$X(\phi)$ = function given by Equation (63)
X_b = parameter in Equation (12)
X_g = parameter in Equations (11) and (12)
y = coordinate direction
Y = fractional unaccomplished temperature difference
Y_c = fractional unaccomplished temperature difference based on final center temperature
Y_m = fractional unaccomplished temperature difference based on final mass average temperature
y_n = roots of transcendental equation; $y_n J_1(y_n) - \text{Bi}\, J_0(y_n) = 0$
z = coordinate direction
z_m = roots of transcendental equation; $\text{Bi}\beta_1 = z_m \tan(z_m)$
z_n = roots of transcendental equation; $\text{Bi} = z_n \tan(z_n)$
z_{nm} = parameter given in Table 8

Greek

α = thermal diffusivity of food, ft^2/h
β_1 = ratio of second shortest dimension to shortest dimension, Equation (16)
β_2 = ratio of longest dimension to shortest dimension, Equation (17)
γ_1 = geometric parameter from Lin et al. (1996b)
γ_2 = geometric parameter from Lin et al. (1996b)
ΔH = volumetric enthalpy difference, Btu/ft^3
ΔH_1 = volumetric enthalpy difference = $C_l(T_i - T_{fm})$, Btu/ft^3
ΔH_2 = volumetric enthalpy difference = $L_f + C_s(T_{fm} - T_c)$, Btu/ft^3
ΔH_{14} = volumetric enthalpy difference between initial freezing temperature T_f and 14°F, Btu/ft^3
ΔH_0 = volumetric enthalpy difference between initial temperature T_i and 0°F, Btu/ft^3
ΔT = weighted average temperature difference in Equation (33), °F
ΔT_1 = temperature difference = $(T_i + T_{fm})/2 - T_m$, °F
ΔT_2 = temperature difference = $T_{fm} - T_m$, °F
ΔT_{m1} = temperature difference for precooling, °F
ΔT_{m2} = temperature difference for phase change, °F
ΔT_{m3} = temperature difference for subcooling, °F
θ = cooling or freezing time, h
θ_1 = precooling time, h
θ_2 = phase change time, h
θ_3 = tempering time, h
θ_{shape} = freezing time of an irregularly shaped food, h
θ_{slab} = freezing time of an infinite slab-shaped food, h
λ = geometric parameter from Lin et al. (1996b)
μ = parameter given by Equation (26)
ρ = density of food, lb/ft^3
ϕ = argument of function X, Equation (63)
ω = first root of Equation (14)

REFERENCES

Becker, B.R. and B.A. Fricke. 1999a. Evaluation of semi-analytical/empirical freezing time estimation methods, part I: Regularly shaped food items. *International Journal of HVAC&R Research* (now *HVAC&R Research*) 5(2):151-169.

Becker, B.R. and B.A. Fricke. 1999b. Evaluation of semi-analytical/empirical freezing time estimation methods, part II: Irregularly shaped food items. *International Journal of HVAC&R Research* (now *HVAC&R Research*) 5(2):171-187.

Becker, B.R. and B.A. Fricke. 1999c. Freezing times of regularly shaped food items. *International Communications in Heat and Mass Transfer* 26(5):617-626.

Becker, B.R. and B.A. Fricke. 2000a. Evaluation of semi-analytical/empirical freezing time estimation methods, part I: Regularly shaped food items (RP-888). *Technical Paper* 4352, presented at the ASHRAE Winter Meeting, Dallas.

Becker, B.R. and B.A. Fricke. 2000b. Evaluation of semi-analytical/empirical freezing time estimation methods, part II: Irregularly shaped food items (RP-888). *Technical Paper* 4353, presented at the ASHRAE Winter Meeting, Dallas.

Cleland, A.C. 1990. *Food refrigeration processes: Analysis, design and simulation*. Elsevier Science, London.

Cleland, A.C. and R.L. Earle. 1977. A comparison of analytical and numerical methods of predicting the freezing times of foods. *Journal of Food Science* 42(5):1390-1395.

Cleland, A.C. and R.L. Earle. 1979a. A comparison of methods for predicting the freezing times of cylindrical and spherical foodstuffs. *Journal of Food Science* 44(4):958-963, 970.

Cleland, A.C. and R.L. Earle. 1979b. Prediction of freezing times for foods in rectangular packages. *Journal of Food Science* 44(4):964-970.

Cleland, A.C. and R.L. Earle. 1982a. A simple method for prediction of heating and cooling rates in solids of various shapes. *International Journal of Refrigeration* 5(2):98-106.

Cleland, A.C. and R.L. Earle. 1982b. Freezing time prediction for foods—A simplified procedure. *International Journal of Refrigeration* 5(3): 134-140.

Cleland, A.C. and R.L. Earle. 1984. Freezing time predictions for different final product temperatures. *Journal of Food Science* 49(4):1230-1232.

Cleland, D.J., A.C. Cleland, and R.L. Earle. 1987a. Prediction of freezing and thawing times for multi-dimensional shapes by simple formulae—Part 1: Regular shapes. *International Journal of Refrigeration* 10(3): 156-164.

Cleland, D.J., A.C. Cleland, and R.L. Earle. 1987b. Prediction of freezing and thawing times for multi-dimensional shapes by simple formulae—Part 2: Irregular shapes. *International Journal of Refrigeration* 10(4): 234-240.

DeMichelis, A. and A. Calvelo. 1983. Freezing time predictions for brick and cylindrical-shaped foods. *Journal of Food Science* 48: 909-913, 934.

Hayakawa, K. and G. Villalobos. 1989. Formulas for estimating Smith et al. parameters to determine the mass average temperature of irregularly shaped bodies. *Journal of Food Process Engineering* 11(4):237-256.

Heldman, D.R. 1975. *Food process engineering*. AVI, Westport, CT.

Hossain, M.M., D.J. Cleland, and A.C. Cleland. 1992a. Prediction of freezing and thawing times for foods of regular multi-dimensional shape by using an analytically derived geometric factor. *International Journal of Refrigeration* 15(4):227-234.

Hossain, M.M., D.J. Cleland, and A.C. Cleland. 1992b. Prediction of freezing and thawing times for foods of two-dimensional irregular shape by using a semi-analytical geometric factor. *International Journal of Refrigeration* 15(4):235-240.

Hossain, M.M., D.J. Cleland, and A.C. Cleland. 1992c. Prediction of freezing and thawing times for foods of three-dimensional irregular shape by using a semi-analytical geometric factor. *International Journal of Refrigeration* 15(4):241-246.

Hung, Y.C. and D.R. Thompson. 1983. Freezing time prediction for slab shape foodstuffs by an improved analytical method. *Journal of Food Science* 48(2):555-560.

Ilicali, C. and S.T. Engez. 1990. A simplified approach for predicting the freezing or thawing times of foods having brick or finite cylinder shape. In *Engineering and food*, vol. 2, pp. 442-456. W.E.L. Speiss and H. Schubert, eds. Elsevier Applied Science, London.

Ilicali, C. and M. Hocalar. 1990. A simplified approach for predicting the freezing times of foodstuffs of anomalous shape. In *Engineering and food*, vol. 2, pp. 418-425. W.E.L. Speiss and H. Schubert, eds. Elsevier Applied Science, London.

Lacroix, C. and F. Castaigne. 1987a. Simple method for freezing time calculations for infinite flat slabs, infinite cylinders and spheres. *Canadian Institute of Food Science and Technology Journal* 20(4):252-259.

Lacroix, C. and F. Castaigne. 1987b. Simple method for freezing time calculations for brick and cylindrical shaped food products. *Canadian Institute of Food Science and Technology Journal* 20(5):342-349.

Lacroix, C. and F. Castaigne. 1988. Freezing time calculation for products with simple geometrical shapes. *Journal of Food Process Engineering* 10(2):81-104.

Lin, Z., A.C. Cleland, G.F. Serrallach, and D.J. Cleland. 1993. Prediction of chilling times for objects of regular multi-dimensional shapes using a general geometric factor. *Refrigeration Science and Technology* 1993-3:259-267.

Lin, Z., A.C. Cleland, D.J. Cleland, and G.F. Serrallach. 1996a. A simple method for prediction of chilling times for objects of two-dimensional irregular shape. *International Journal of Refrigeration* 19(2):95-106.

Lin, Z., A.C. Cleland, D.J. Cleland, and G.F. Serrallach. 1996b. A simple method for prediction of chilling times: Extension to three-dimensional irregular shaped. *International Journal of Refrigeration* 19(2):107-114.

McNabb, A., G.C. Wake, and M.M. Hossain. 1990. Transition times between steady states for heat conduction: Part I—General theory and some exact results. *Occasional Publications in Mathematics and Statistics* 20, Massey University, New Zealand.

Pflug, I.J., J.L. Blaisdell, and J. Kopelman. 1965. Developing temperature-time curves for objects that can be approximated by a sphere, infinite plate, or infinite cylinder. *ASHRAE Transactions* 71(1):238-248.

Pham, Q.T. 1984. An extension to Plank's equation for predicting freezing times for foodstuffs of simple shapes. *International Journal of Refrigeration* 7:377-383.

Pham, Q.T. 1985. Analytical method for predicting freezing times of rectangular blocks of foodstuffs. *International Journal of Refrigeration* 8(1): 43-47.

Pham, Q.T. 1986a. Simplified equation for predicting the freezing time of foodstuffs. *Journal of Food Technology* 21(2):209-219.

Pham, Q.T. 1986b. Freezing of foodstuffs with variations in environmental conditions. *International Journal of Refrigeration* 9(5):290-295.

Pham, Q.T. 1987. A converging-front model for the asymmetric freezing of slab-shaped food. *Journal of Food Science* 52(3):795-800.

Pham, Q.T. 1991. Shape factors for the freezing time of ellipses and ellipsoids. *Journal of Food Engineering* 13:159-170.

Plank, R. 1913. Die Gefrierdauer von Eisblocken. *Zeitschrift für die gesamte Kälte Industrie* 20(6):109-114.

Plank, R. 1941. Beitrage zur Berechnung und Bewertung der Gefriergeschwindigkeit von Lebensmitteln. *Zeitschrift für die gesamte Kälte Industrie* 3(10):1-24.

Smith, R.E. 1966. *Analysis of transient heat transfer from anomalous shape with heterogeneous properties*. Ph.D. dissertation, Oklahoma State University, Stillwater.

Smith, R.E., G.L. Nelson, and R.L. Henrickson. 1968. Applications of geometry analysis of anomalous shapes to problems in transient heat transfer. *Transactions of the ASAE* 11(2):296-302.

CHAPTER 11

COMMODITY STORAGE REQUIREMENTS

THIS chapter presents information on storage requirements of many perishable foods that enter the market on a commercial scale. Also included is a short discussion on the storage of furs and fabrics. The data are based on the storage of fresh, high-quality commodities that have been properly harvested, handled, and cooled.

Tables 1 and 2 present recommended storage requirements for various products. Some products require a curing period before storage. Other products require different storage conditions, depending on their intended use.

The recommended temperatures are optimum for long storage and are commodity temperatures, not air temperatures. For short storage, higher temperatures are often acceptable. Conversely, products subject to chilling injury can sometimes be held at a lower temperature for a short time without injury. Exceptions include bananas, cranberries, cucumbers, eggplant, melons, okra, pumpkins, squash, white potatoes, sweet potatoes, and tomatoes. The minimum recommended temperature for these products should be strictly followed.

The listed storage lives are based on typical commercial practice. Special treatments can, in certain instances, extend storage life significantly.

Thermal properties of many of these products, including water content, freezing point, specific heat, and latent heat of fusion, are listed in Chapter 9. Also, because fresh fruits and vegetables are living products, they generate heat that should be included as part of the storage refrigeration load. The approximate heat of respiration for various fruits and vegetables is also listed in Chapter 9.

REFRIGERATED STORAGE

Cooling

Because products deteriorate much faster at warm than at low temperatures, rapid removal of field heat by cooling to the storage temperature substantially increases the market life of the product. Chapter 15 describes various cooling methods.

Deterioration

The environment in which harvested produce is placed may greatly influence not only the respiration rate but also other changes and products formed in related chemical reactions. In fruits, these changes are described as ripening. In many fruits, such as bananas and pears, the process of ripening is required to develop the maximum edible quality. However, as ripening continues, deterioration begins and the fruit softens, loses flavor, and eventually undergoes tissue breakdown.

In addition to deterioration after harvest by biochemical changes within the product, desiccation and diseases caused by microorganisms are also important.

Deterioration rate is greatly influenced by temperature and is generally reduced as temperature is lowered. The specific relationships between temperature and deterioration rate vary considerably among commodities and diseases. A generalization, assuming a nominal deterioration rate of 1 for a fruit at 30°F, is as follows:

Approximate Deterioration Rate of Fresh Produce

Temperature, °F	Relative Deterioration Rate
68	8 to 10
50	4 to 5
41	3
37	2
32	1.25
30	1

For example, fruit that remains marketable for 12 days when stored at 30°F may last only 12/3 = 4 days when stored at 41°F. The best temperature to slow down deterioration is often the lowest temperature that can safely be maintained without freezing the commodity, which is 1 to 2°F above the freezing point of the fruit or vegetable.

Some produce will not tolerate low storage temperatures. Severe physiological disorders that develop because of exposure to low but not freezing temperatures are classed as **chilling injury**. The banana is a classic example of a fruit displaying chilling injury symptoms, and storage temperatures must be elevated accordingly. Certain apple varieties exhibit this characteristic, and prolonged storage must be at a temperature well above that usually recommended. An apple variety's degree of susceptibility to chilling may vary with climatic and growing factors. Products susceptible to chilling injury, its symptoms, and the lowest safe temperature are discussed in Chapters 9, 22, 23, and 24.

Desiccation

Water loss, which causes a product to shrivel, is a physical factor related to the evaporative potential of air, and can be expressed as follows:

$$p_D = \frac{p(100 - \phi)}{100}$$

where

p_D = vapor pressure deficit, indicating combined influence of temperature and relative humidity on evaporative potential of air
p = vapor pressure of water at given temperature
ϕ = relative humidity, percent

For example, comparing the evaporative potential of air in storage rooms at 32°F and 50°F db, with 90% rh in each room, the vapor pressure deficit at 32°F is 0.018 in. Hg, whereas at 50°F it is 0.036 in. Hg. Thus, if all other factors are equal, commodities tend to lose water twice as fast at 50°F db as at 32°F at the same relative humidity. For equal water loss at the two temperatures, the rh has to be maintained at 95% at 50°F in comparison to 90% at 32°F. These comparisons are not precise because the water in fruits and vegetables contains a sufficient quantity of dissolved sugars and other chemicals to cause the water to be in equilibrium with water vapor in the air at 98 to 99% rh instead of 100% rh. This property is described by the water activity a_w of the product. Lowering the vapor pressure deficit by lowering the air temperature is an excellent means of reducing water loss during storage.

The preparation of this chapter is assigned to TC 10.5, Refrigerated Distribution and Storage Facilities.

Table 1 Storage Requirements of Vegetables, Fresh Fruits, and Melons

Common Name (Other Common Name)	Scientific Name	Storage Temp., °F	Relative Humidity, %	Highest Freezing Temp., °F	Ethylene Production Rate[a]	Ethylene Sensitivity[b]	Respiration Rate[c]	Approximate Postharvest Life	Observations and Beneficial CA[d] Conditions
Acerola (Barbados cherry)	*Malpighia glabra*	32	85 to 90	29.5				6 to 8 weeks	
African horned melon (kiwano)	*Cucumis africanus*	55 to 59	90		Low	Moderate		3 to 6 months	
Amaranth (pigweed)	*Amaranthus* spp.	32 to 36	95 to 100		Very low	Moderate		10 to 14 days	
Anise (fennel)	*Foeniculum vulgare*	32 to 36	90 to 95	30.0				2 to 3 weeks	
Apple									
Not chilling sensitive	*Malus pumila*	30	90 to 95	29.3	Very high	High	Low	3 to 6 months	2 to 3% O_2 1 to 2% CO_2
Chilling sensitive	*Malus pumila* cv. Yellow Newton, Grimes golden, McIntosh	40	90 to 95	29.3	Very high	High	Low	1 to 2 months	2 to 3% O_2 1 to 2% CO_2
Apricot	*Prunus armeniaca*	31 to 32	90 to 95	30.0	Moderate	Moderate	Low	1 to 3 weeks	2 to 3% O_2 2 to 3% CO_2
Artichokes									
Chinese	*Stachys affinia*	32	90 to 95		Very low	Very Low		1 to 2 weeks	
Globe	*Cynara acolymus*	32	95 to 100	29.8	Very low	Low	High	2 to 3 weeks	2 to 3% O_2 3 to 5% CO_2
Jerusalem	*Helianthus tuberosus*	31 to 32	90 to 95	27.5	Very low	Low	Low	4 months	
Arugula	*Eruca vesicaria* var. *sativa*	32	95 to 100		Very low	High	Moderate	7 to 10 days	
Asian pear (nashi)	*Pyrus serotina P. pyrifolia*	34	90 to 95	29.1	High	High	Low	4 to 6 months	
Asparagus, green or white	*Asparagus officinalis*	36.5	95 to 100	31.0	Very low	Moderate	Very high	2 to 3 weeks	5 to 12% CO_2
Atemoya	*Annona squamosa* x *A. cherimola*	55	85 to 90		High	High		2 to 4 weeks	3 to 5% O_2 5 to 10% CO_2
Avocado									
Fuchs, Pollock	*Persea americana* cv. Fuchs, Pollock	55	85 to 90	30.4	High	High	Moderate	2 weeks	
Fuerte, Hass	*Persea americana* cv. Fuerte, Hass	37 to 45	85 to 90	29.1	High	High	Moderate	2 to 4 weeks	2 to 5% O_2 3 to 10% CO_2
Lula, Booth	*Persea americana* cv. Lula, Booth	40	90 to 95	30.4	High	High	Moderate	4 to 8 weeks	
Babaco (mountain papaya)	*Carica candamarcensis*	45	85 to 90					1 to 3 weeks	
Banana	*Musa paradisiaca* var. *sapientum*	55 to 59	90 to 95	30.5	Moderate	High	Low	1 to 4 weeks	2 to 5% O_2 2 to 5% CO_2
Barbados cherry	see Acerola								
Beans									
Fava (broad)	*Vicia faba*	32	90 to 95					1 to 2 weeks	
Lima	*Phaseolous lunatus*	41 to 43	95	31.0	Low	Moderate	Moderate	5 to 7 days	
Long (yard-long)	*Vigna sesquipedalis*	40 to 45	90 to 95		Low	Moderate		7 to 10 days	
Snap (wax, green)	*Phaseolus vulgaris*	40 to 45	95	30.7	Low	Moderate	Moderate	7 to 10 days	2 to 3% O_2 4 to 7% CO_2
Winged	*Psophocarpus tetragonolobus*	50	90					4 weeks	
Beet									
Bunched	*Beta vulgaris*	32	98 to 100	31.3	Very low	Low	Low	10 to 14 days	
Topped	*Beta vulgaris*	32	98 to 100	30.4	Very low	Low	Low	4 months	
Berries									
Blackberry	*Rubus* spp.	31 to 32	90 to 95	30.5	Low	Low	Moderate	3 to 6 days	5 to 10% O_2 15 to 20% CO_2

Table 1 Storage Requirements of Vegetables, Fresh Fruits, and Melons (*Continued*)

Common Name (Other Common Name)	Scientific Name	Storage Temp., °F	Relative Humidity, %	Highest Freezing Temp., °F	Ethylene Production Rate[a]	Ethylene Sensitivity[b]	Respiration Rate[c]	Approximate Postharvest Life	Observations and Beneficial CA[d] Conditions
Blueberry	*Vaccinium corymbosum*	31 to 32	90 to 95	29.7	Low	Low	Low	10 to 18 days	2 to 5% O_2 12 to 20% CO_2
Cranberry	*Vaccinium macrocarpon*	36 to 41	90 to 95	30.4	Low	Low	Low	8 to 16 weeks	1 to 2% O_2 0 to 5% CO_2
Dewberry	*Rubus* spp.	31 to 32	90 to 95	29.7	Low	Low		2 to 3 days	
Elderberry	*Rubus* spp.	31 to 32	90 to 95	30.4	Low	Low		5 to 14 days	
Loganberry	*Rubus* spp.	31 to 32	90 to 95	29.7	Low	Low		2 to 3 days	
Raspberry	*Rubus idaeus*	31 to 32	90 to 95	30.4	Low	Low	Moderate	3 to 6 days	5 to 10% O_2 15 to 20% CO_2
Strawberry	*Fragaria* spp.	32	90 to 95	30.5	Low	Low	Low	7 to 10 days	5 to 10% O_2 15 to 20% CO_2
Bittermelon (bitter gourd)	*Momordica*	50 to 54	85 to 90		Low	Moderate	Moderate	2 to 3 weeks	2 to 3% O_2 5% CO_2
Black salsify (scorzonera)	*Scorzonera hispanica*	32 to 34	95 to 98		Very low	Low		6 months	
Bok choy	*Brassica chinensis*	32	95 to 100		Very low	High		3 weeks	
Breadfruit	*Artocarpus altilis*	55 to 59	85 to 90					2 to 4 weeks	
Broccoli	*Brassica oleracea* var. *Italica*	32	95 to 100	31.0	Very low	High	Moderate	10 to 14 days	1 to 2% O_2 5 to 10% CO_2
Brussels sprouts	*Brassica oleracea* var. *Gemnifera*	32	95 to 100	30.5	Very low	High	Moderate	3 to 5 weeks	1 to 2% O_2 5 to 7% CO_2
Cabbage									
Chinese (Napa)	*Brassica campestris* var. *Pekinensis*	32	95 to 100	30.4	Very low	High	Low	2 to 3 months	1 to 2% O_2 0 to 6% CO_2
Common, early crop	*Brassica oleracea* var. *Capitata*	32	98 to 100	30.4	Very low	High	Low	3 to 6 weeks	
Common, late crop	*Brassica oleracea* var. *Capitata*	32	95 to 100	30.4	Very low	High	Low	5 to 6 months	3 to 5% O_2 3 to 7% CO_2
Cactus leaves (nopalitos)	*Opuntia* spp.	41 to 50	90 to 95		Very low	Moderate		2 to 3 weeks	
Cactus fruit (prickly pear fruit)	*Opuntia* spp.	41	85 to 90	28.7	Very low	Moderate		2 to 6 weeks	
Caimito	see Sapotes								
Calamondin	see Citrus								
Canistel	see Sapotes								
Carambola (starfruit)	*Averrhoa carambola*	48 to 50	85 to 90	29.8			Low	3 to 4 weeks	
Carrot									
Topped	*Daucus carota*	32	98 to 100	29.5	Very low	High	Low	3 to 6 months	No CA benefit
Bunched, immature	*Daucus carota*	32	98 to 100	29.5	Very low	High	Moderate	10 to 14 days	Ethylene causes bitterness
Cashew, apple	*Anacardium occidentale*	32 to 36	85 to 90					5 weeks	
Cassava (yucca, manioc)	*Manihot esculenta*	32 to 41	85 to 90		Very low	Low	Low	1 to 2 months	No CA benefit
Cauliflower	*Brassica oleracea* var. *Botrytis*	32	95 to 98	30.5	Very low	High	Moderate	3 to 4 weeks	2 to 5% O_2 2 to 5% CO_2
Celeriac	*Apium graveolens* var. *Rapaceum*	32	98 to 100	30.4	Very low	Low	Low	6 to 8 months	2 to 4% O_2 2 to 3% CO_2
Celery	*Apium graveolens* var. *Dulce*	32	98 to 100	31.1	Very low	Moderate	Low	1 to 2 months	1 to 4% O_2 3 to 5% CO_2
Chard	*Beta vulgaris* var. *Cida*	32	95 to 100		Very low	High		10 to 14 days	
Chayote	*Sechium edule*	45	85 to 90				Low	4 to 6 weeks	
Cherimoya (custard apple)	*Annona cherimola*	55	90 to 95	28.0	High	High	Very high	2 to 4 weeks	3 to 5% O_2 5 to 10% CO_2
Cherries									
Sour	*Prunus cerasus*	32	90 to 95	28.9			Low	3 to 7 days	3 to 10% O_2 10 to 12% CO_2
Sweet	*Prunus avium*	30 to 32	90 to 95	28.2			Low	2 to 3 weeks	10 to 20% O_2 20 to 25% CO_2
Chicory	see Endive								

Table 1 Storage Requirements of Vegetables, Fresh Fruits, and Melons (*Continued*)

Common Name (Other Common Name)	Scientific Name	Storage Temp., °F	Relative Humid-ity, %	Highest Freezing Temp., °F	Ethylene Production Rate[a]	Ethylene Sensitivity[b]	Respi-ration Rate[c]	Approximate Postharvest Life	Observations and Beneficial CA[d] Conditions
Chiles	see Peppers								
Chinese broccoli (gailan)	*Brassica alboglabra*	32	95 to 100		Very low	High		10 to 14 days	
Chives	*Allium schoenoprasum*	32	95 to 100		Very low	High		2 to 3 weeks	
Cilantro (Chinese parsley)	*Coriandrum sativum*	32 to 36	95 to 100		Very low	High	High	2 weeks	
Citrus									
Calamondin orange	*Citrus reticulta* x. *Fortunella* spp.	48 to 50	90	28.4			Low	2 weeks	
Grapefruit									
CA, AZ, dry areas	*Citrus paradisi*	58 to 59	85 to 90	30.0	Very low	Moderate	Low	6 to 8 weeks	3 to 10% O_2 5 to 10% CO_2
FL, humid areas	*Citrus paradisi*	50 to 59	85 to 90	30.0	Very low	Moderate	Low	6 to 8 weeks	3 to 10% O_2 5 to 10% CO_2
Kumquat	*Fortunella japponica*	40	90 to 95				Low	2 to 4 weeks	
Lemon	*Citrus limon*	50 to 55	85 to 90	29.5			Low	1 to 6 months	5 to 10% O_2 0 to 10% CO_2 Store at 32 to 40°F for <1 mo.
Lime (Mexican, Tahitian or Persian)	*Citrus aurantifolia; C. latifolia*	48 to 50	85 to 90	29.1			Low	6 to 8 weeks	5 to 10% O_2 0 to 10% CO_2
Orange									
CA, AZ, dry areas	*Citrus sinensis*	37 to 48	85 to 90	30.5	Very low	Moderate	Low	3 to 8 weeks	5 to 10% O_2 0 to 5% CO_2
FL, humid areas	*Citrus sinensis*	32 to 36	85 to 90	30.5	Very low	Moderate	Low	8 to 12 weeks	5 to 10% O_2 0 to 5% CO_2
Blood orange	*Citrus sinensis*	40 to 45	90 to 95	30.5			Low	3 to 8 weeks	5 to 10% O_2 0 to 5% CO_2
Seville (sour)	*Citrus aurantium*	50	85 to 90	30.5	Low		Low	12 weeks	
Pomelo	*Citrus grandis*	45 to 48	85 to 90	29.1			Low	12 weeks	
Tangelo (minneola)	*Citrus reticulata* x *paradisi*	45 to 50	85 to 95	30.4			Low		
Tangerine (mandarin)	*Citrus reticulata*	40 to 45	90 to 95	30.0	Very low	Moderate	Low	2 to 4 weeks	
Coconut	*Cocos nucifera*	32 to 36	89 to 85	30.4				1 to 2 months	
Collards and kale	*Brassica oleracea* var. *Acephala*	32	95 to 100	31.1	Very low	High	High	10 to 14 days	
Corn, sweet and baby	*Zea mays*	32	95 to 98	31.0	Very low	Low	High	5 to 8 days	2 to 4% O_2 5 to 10% CO_2
Cucumber	*Cucumis sativus*	50 to 54	85 to 90	31.1	Low	High	Low	10 to 14 days	3 to 5% O_2 0 to 5% CO_2
Cucumber, pickling	*Cucumis sativus*	40	95 to 100		Low	High		7 days	3 to 5% O_2 3 to 5% CO_2
Currants	*Ribes sativum; R. nigrum; R. rubrum*	31 to 32	90 to 95	30.2	Low	Low		1 to 4 weeks	
Custard apple	see Cherimoya								
Daikon (Oriental radish)	*Raphanus sativus*	32 to 34	95 to 100		Very low	Low		4 months	
Dasheen	see Taro								
Date	*Phoenix dactylifera*	0 to 32	75	3.7	Very low	Low	Low	6 to 12 months	
Dill	see Herbs								
Durian	*Durio zibethinus*	40 to 43	85 to 90					6 to 8 weeks	3 to 5% O_2 5 to 15% CO_2
Eggplant	*Solanum melongena*	50 to 54	90 to 95	30.5	Low	Moderate	Low	1 to 2 weeks	3 to 5% O_2 0% CO_2
Endive (escarole)	*Cichorium endivia*	32	95 to 100	31.6	Very low	Moderate	High	2 to 4 weeks	

Table 1 Storage Requirements of Vegetables, Fresh Fruits, and Melons (*Continued*)

Common Name (Other Common Name)	Scientific Name	Storage Temp., °F	Relative Humidity, %	Highest Freezing Temp., °F	Ethylene Production Rate[a]	Ethylene Sensitivity[b]	Respiration Rate[c]	Approximate Postharvest Life	Observations and Beneficial CA[d] Conditions
Belgian endive (Witloof chicory)	*Cichorium intybus*	36 to 37	95 to 98		Very low	Moderate		2 to 4 weeks	Light causes greening 3 to 4% O_2 4 to 5% C_2
Feijoa (pineapple guava)	*Feijoa selloiana*	41 to 50	90		Moderate	Low		2 to 3 weeks	
Fennel	see Anise								
Fig, fresh	*Ficus carica*	31 to 32	85 to 90	27.7	Moderate	Low	Low	7 to 10 days	5 to 10% O_2 15 to 20% CO_2
Garlic	*Allium sativum*	32	65 to 70	30.5	Very low	Low	Low	6 to 7 months	0.5% O_2 5 to 10% CO_2
Ginger	*Zingiber officinale*	55	65		Very low	Low		6 months	No CA benefit
Gooseberry	*Ribes grossularia*	31 to 32	90 to 95	30.0	Low	Low	Low	3 to 4 weeks	
Granadilla	see Passionfruit								
Grape[e]									
Table grape	*Vitis vinifera*	31 to 32	90 to 95	27.1	Very low	Low	Low	1 to 6 months	2 to 5% O_2 1 to 3% CO_2 to 4 weeks: 5 to 10% O_2 10 to 15% CO_2
American grape	*Vitis labrusca*	30 to 31	90 to 95	29.5	Very low	Low	Low	2 to 8 weeks	
Grapefruit	see Citrus						Low		
Guava	*Psidium guajava*	41 to 50	90		Low	Moderate	Moderate	2 to 3 weeks	
Herbs, fresh culinary									5 to 10% O_2 5 to 10% CO_2
Basil	*Ocimum basilicum*	50	90		Very low	High		7 days	
Chives	*Allium schoenorasum*	32	95 to 100	30.4	Low	Moderate			
Dill	*Anethum graveolens*	32	95 to 100	30.7	Very low	High		1 to 2 weeks	
Epazote	*Chenopodium ambrosioides*	32 to 41	90 to 95		Very low	Moderate		1 to 2 weeks	
Mint	*Mentha* spp.	32	95 to 100		Very low	High		2 to 3 weeks	
Oregano	*Origanum vulgare*	32 to 41	90 to 95		Very low	Moderate		1 to 2 weeks	
Parsley	*Petroselinum crispum*	32	95 to 100	30.0	Very low	High	Very high	1 to 2 months	
Perilla (shiso)	*Perilla frutescens*	50	95		Very low	Moderate		7 days	
Sage	*Salvia officinalis*	32	90 to 95					2 to 3 weeks	
Thyme	*Thymus vulgaris*	32	90 to 95					2 to 3 weeks	
Horseradish	*Amoracia rusticana*	30 to 32	98 to 100	28.7	Very low	Low		10 to 12 months	
Husk tomato	see Tomatillo								
Jaboticaba	*Myrciaria cauliflora* = *Eugenia cauliflora*	55 to 59	90 to 95					2 to 3 days	
Jackfruit	*Artocarpus heterophyllus*	55	85 to 90		Moderate	Moderate		2 to 4 weeks	
Jerusalem artichoke	see Artichoke								
Jicama (yambean)	*Pachyrrhizus erosus*	55 to 65	85 to 90		Very low	Low	Low	1 to 2 months	
Jujube (Chinese date)	*Ziziphus jujuba*	36.5 to 50	85 to 90	29.1	Low	Moderate		1 month	
Kaki	see Persimmon								
Kale	see Collards and kale								
Kiwano	see African horned melon								
Kiwifruit (Chinese gooseberry)	*Actinidia chinensis*	32	90 to 95	30.4	Low	High	Low	3 to 5 months	1 to 2% O_2 3 to 5% CO_2
Kohlrabi	*Brassica oleracea* var. *Gongylodes*	32	98 to 100	30.2	Very low	Low	Low	2 to 3 months	
Lo Bok	see Daikon								
Langsat (lanzone)	*Aglaia* sp.; *Lansium* sp.	52 to 58	85 to 90					2 weeks	

Table 1 Storage Requirements of Vegetables, Fresh Fruits, and Melons (*Continued*)

Common Name (Other Common Name)	Scientific Name	Storage Temp., °F	Relative Humidity, %	Highest Freezing Temp., °F	Ethylene Production Rate[a]	Ethylene Sensitivity[b]	Respiration Rate[c]	Approximate Postharvest Life	Observations and Beneficial CA[d] Conditions
Leafy greens									
Cool-season	various	32	95 to 100	31.0	Very low	High		10 to 14 days	
Warm-season	various	45 to 50	95 to 100	31.0	Very low	High		5 to 7 days	
Leek	*Allium porrum*	32	95 to 100	30.7	Very low	Moderate	Moderate	2 months	1 to 2% O_2 2 to 5% CO_2
Lemon	see Citrus								
Lettuce	*Lactuca sativa*	32	98 to 100	31.7	Very low	High	Low	2 to 3 weeks	2 to 5% O_2 0% CO_2
Lime	see Citrus								
Longan	*Dimocarpus longan = Euphoria longan*	34 to 36	90 to 95	27.7				2 to 4 weeks	
Loquat	*Eriobotrya japonica*	32	90	28.6				3 weeks	
Luffa (Chinese okra)	*Luffa* spp.	50 to 54	90 to 95		Low	Moderate		1 to 2 weeks	
Lychee (litchi)	*Litchi chinensis*	34 to 36	90 to 95		Moderate	Moderate	Low	3 to 5 weeks	3 to 5% O_2 3 to 5% CO_2
Malanga (tania, new cocoyam)	*Xanthosoma sagittifolium*	45	70 to 80		Very low	Low		3 months	
Mamey	see Sapotes								
Mandarin	see Citrus								
Mango	*Mangifera indica*	55	85 to 90	29.5	Moderate	Moderate	Moderate	2 to 3 weeks	3 to 5% O_2 5 to 10% CO_2
Mangosteen	*Garcinia mangostana*	55	85 to 90		Moderate	High		2 to 4 weeks	
Melons									
Cantaloupes and other netted melons	*Cucurbita melo* var. *reticulatus*	36 to 41	95	29.8	High	Moderate	Low	2 to 3 weeks	3 to 5% O_2 10 to 15% CO_2
Casaba	*Cucurbita melo*	45 to 50	85 to 90	30.2	Low	Low		3 to 4 weeks	3 to 5% O_2 5 to 10% CO_2
Crenshaw	*Cucurbita melo*	45 to 50	85 to 90	30.0	Moderate	High		2 to 3 weeks	3 to 5% O_2 5 to 10% CO_2
Honeydew, orange-flesh	*Cucurbita melo*	41 to 50	85 to 90	30.0	Moderate	High	Low	3 to 4 weeks	3 to 5% O_2 5 to 10% CO_2
Persian	*Cucurbita melo*	45 to 50	85 to 90	30.5	Moderate	High		2 to 3 weeks	3 to 5% O_2 5 to 10% CO_2
Mint	see Herbs								
Mombin	see Spondias								
Mushrooms	*Agaricus*, other genera	32	90	30.4	Very low	Moderate	High	7 to 14 days	3 to 21% O_2 15 to 15% CO_2
Mustard greens	*Brassica juncea*	32	90 to 95		Very low	High		7 to 14 days	
Nashi	see Asian pear								
Nectarine	*Prunus persica*	31 to 32	90 to 95	30.4	Moderate	Moderate	Low	2 to 4 weeks	1 to 2% O_2 3 to 5% CO_2 Internal breakdown at 37 to 50°F
Okra	*Abelmoschus esculentus*	45 to 50	90 to 95	28.7	Low	Moderate	High	7 to 10 days	Air 4 to 10% CO_2
Olives, fresh	*Olea europea*	41 to 50	85 to 90	29.5	Low	Moderate	Low	4 to 6 weeks	2 to 3% O_2 0 to 1% CO_2
Onion									
Mature bulbs, dry	*Allium cepa*	32	65 to 70	30.5	Very low	Low	Low	1 to 8 months	1 to 3% O_2 5 to 10% CO_2
Green	*Allium cepa*	32	95 to 100	30.4	Low	High	Moderate	3 weeks	2 to 4% O_2 10 to 20% CO_2
Orange	see Citrus								
Papaya	*Carica papaya*	45 to 55	85 to 90	30.4			Low	1 to 3 weeks	2 to 5% O_2 5 to 8% CO_2
Parsley	see Herbs								

Table 1 Storage Requirements of Vegetables, Fresh Fruits, and Melons (*Continued*)

Common Name (Other Common Name)	Scientific Name	Storage Temp., °F	Relative Humidity, %	Highest Freezing Temp., °F	Ethylene Production Rate[a]	Ethylene Sensitivity[b]	Respiration Rate[c]	Approximate Postharvest Life	Observations and Beneficial CA[d] Conditions
Parsnips	*Pastinaca sativa*	32	95 to 100	30.4	Very low	High	Low	4 to 6 months	Ethylene causes bitterness
Passionfruit	*Passiflora* spp.	50	85 to 90		Very high	Moderate	Very high	3 to 4 weeks	
Peach	*Prunus persica*	31 to 32	90 to 95	30.4	High	Moderate	Low	2 to 4 weeks	1 to 2% O_2 3 to 5% CO_2 Internal breakdown at 37 to 50°F
Pear, American[e]	*Pyrus communis*	29 to 31	90 to 95	28.9	High	High	Low	2 to 7 months	Cultivar variations 1 to 3% O_2 0 to 5% CO_2
Peas									
In pods (snow, snap, and sugar peas)	*Pisum sativum*	32 to 34	90 to 98	31.0	Very low	Moderate	Very high	1 to 2 weeks	2 to 3% O_2 2 to 3% CO_2
Southern peas (cowpeas)	*Vigna sinensis* = *V. unguiculata*	40 to 41	95					6 to 8 days	
Pepino (melon pear)	*Solanum muricatum*	41 to 50	95		Low	Moderate		4 weeks	
Peppers									
Bell pepper or paprika	*Capsicum annuum*	45 to 50	95 to 98	30.7	Low	Low	Low	2 to 3 weeks	2 to 5% O_2 2 to 5% CO_2
Hot peppers (chiles)	*Capsicum annuum* and *C. frutescens*	41 to 50	85 to 95	30.7	Low	Moderate		2 to 3 weeks	3 to 5% O_2 5 to 10% CO_2
Persimmon (kaki)	*Dispyros kaki*								3 to 5% O_2 5 to 8% CO_2
Fuyu	*Dispyros kaki* var. *Fuyu*	32	90 to 95	28.0	Low	High	Low	1 to 3 months	
Hachiya	*Dispyros kaki* var. *Hachiya*	32	90 to 95	28.0	Low	High	Low	2 to 3 months	
Pineapple	*Ananas comosus*	45 to 55	85 to 90	30.0	Low	Low	Low	2 to 4 weeks	2 to 5% O_2 5 to 10% CO_2
Plantain	*Musa paradisiaca* var. *paradisiaca*	55 to 59	90 to 95	30.5	Low	High		1 to 5 weeks	
Plums and prunes	*Prunus domestica*	31 to 32	90 to 95	30.5	Moderate	Moderate	Low	2 to 5 weeks	1 to 2% O_2 0 to 5% CO_2
Pomegranate	*Punica granatum*	41	90 to 95	26.6			Low	2 to 3 months	3 to 5% O_2 5 to 10% CO_2
Potato									
Early crop	*Solanum tuberosum*	50 to 59	90 to 95	30.5	Very low	Moderate	Low	10 to 14 days	No CA benefit
Late crop	*Solanum tuberosum*	40 to 54	95 to 98	30.5	Very low	Moderate	Low	5 to 10 months	No CA benefit
Pumpkin	*Cucurbita maxima*	54 to 59	50 to 70	30.5	Very low	Moderate	Low	2 to 3 months	
Quince	*Cydonia oblonga*	31 to 32	90	28.4	Low	High		2 to 3 months	
Raddichio	*Cichorium intybus*	32 to 34	95 to 100					4 to 8 weeks	
Radish	*Raphanus sativus*	32	95 to 100	30.7	Very low	Low	Low	1 to 2 months	1 to 2% O_2 2 to 3% CO_2
Rambutan	*Nephelium lappaceum*	54	90 to 95		High	High		1 to 3 weeks	3 to 5% O_2 7 to 12% CO_2
Rhubarb	*Rheum rhaponticum*	32	95 to 100	30.4	Very low	Low	Low	2 to 4 weeks	
Rutabaga	*Brassica napus* var. *Napobrassica*	32	98 to 100	30.0	Very low	Low	Low	4 to 6 months	
Sage	see Herbs								
Salsify (vegetable oyster)	*Trapopogon porrifolius*	32	95 to 98	30.0	Very low	Low	Low	2 to 4 months	
Sapotes									
Black sapote	*Diospyros ebenaster*	55 to 59	85 to 90	27.8				2 to 3 weeks	
Caimito (star apple)	*Chrysophyllum cainito*	37	90	29.8				3 weeks	
Canistel (eggfruit)	*Pouteria campechiana*	55 to 59	85 to 90	28.7				3 weeks	

Table 1 Storage Requirements of Vegetables, Fresh Fruits, and Melons (*Continued*)

Common Name (Other Common Name)	Scientific Name	Storage Temp., °F	Relative Humidity, %	Highest Freezing Temp., °F	Ethylene Production Rate[a]	Ethylene Sensitivity[b]	Respiration Rate[c]	Approximate Postharvest Life	Observations and Beneficial CA[d] Conditions
Mamey sapote	*Calocarpum mammosum*	55 to 59	90 to 95		High	High		2 to 3 weeks	
Sapodilla (chicosapote)	*Achras sapota*	59 to 68	85 to 90		High	High		2 weeks	
White sapote	*Casimiroa edulis*	68	85 to 90	28.4				2 to 3 weeks	
Scorzonera	see Black salsify								
Shallot	*Allium cepa* var. *ascalonicum*	32 to 36	65 to70	30.7	Low	Low			
Soursop	*Annona muricata*	55	85 to 90					1 to 2 weeks	
Spinach	*Spinacia oleracea*	32	95 to 100	31.5	Very low	High	Low	10 to 14 days	5 to 10% O_2 5 to 10% CO_2
Spondias (mombin, wi apple, jobo, hogplum)	*Spondias* spp.	55	85 to 90					1 to 2 weeks	
Sprouts from seeds		32	95 to 100					5 to 9 days	
Alfalfa sprouts	*Medicago sativa*	32	95 to 100					7 days	
Bean sprouts	*Phaseolus* sp.	32	95 to 100					7 to 9 days	
Radish sprouts	*Raphanus* sp.	32	95 to 100					5 to 7 days	
Squash									
Summer, soft rind (courgette)	*Cucurbita pepo*	45 to 50	95	31.1	Low	Moderate	Low	1 to 2 weeks	3 to 5% O_2 5 to 10% CO_2
Winter, hard rind (calabash)	*Cucurbita moschata; C. maxima*	54 to 59	50 to 70	30.5	Low	Moderate	Low	2 to 3 months	Large differences among varieties
Star apple	see Sapotes								
Starfruit	see Carambola								
Sweet potato or yam	*Ipomea batatas*	55 to 59	85 to 95	29.7	Very low	Low	Low	4 to 7 months	
Sweetsop (sugar apple, custard apple)	*Annona squamosa; Annona* spp.	45	85 to 90		High	High		4 weeks	3 to 5% O_2 5 to 10% CO_2
Tamarillo (tree tomato)	*Cyphomandra betacea*	37 to 40	85 to 95		Low	Moderate		10 weeks	
Tamarind	*Tamarindus indica*	36 to 45	90 to 95	25.4	Very low	Very Low		3 to 4 weeks	
Taro (cocoyam, eddoe, dasheen)	*Colocasia esculenta*	45 to 50	85 to 90	30.4			Low	4 months	No CA benefit
Thyme	see Herbs								
Tomatillo (husk tomato)	*Physalis ixocarpa*	45 to 55	85 to 90		Very low	Moderate	Low	3 weeks	
Tomato									
Mature, green	*Lycopersicon esculentum*	50 to 55	90 to 95	31.1	Very low	High	Low	2 to 5 weeks	3 to 5% O_2 2 to 3% CO_2
Firm, ripe	*Lycopersicon esculentum*	46 to 50	85 to 90	31.1	High	Low	Low	1 to 3 weeks	3 to 5% O_2 3 to 5% CO_2
Turnip root	*Brassica campetris* var. *Rapifera*	32	95	30.2	Very low	Low	Low	4 to 5 months	
Water chestnut	*Eleocharis dulcis*	32 to 36	85 to 90					2 to 4 months	
Watercress (garden cress)	*Lepidium sativum; Nasturtium officinales*	32	95 to 100	31.5	Very low	High	High	2 to 3 weeks	
Watermelon	*Citrullus vulgaris*	50 to 59	90	31.3	Very low	High	Low	2 to 3 weeks	No CA benefit
Yam	*Dioscorea* spp.	59	70 to 80	30.0	Very low	Low		2 to 7 months	
Yucca	see Cassava								

Note: Recommendations in this table are general guidelines. Recommended storage conditions and expected postharvest life for a specific produce item may be different from those listed here because of variations in growing conditions and postharvest care. Also, new cultivars (varieties) of a particular item may require different conditions and have a very different expected postharvest life from that listed in the table. Empty cells indicate that no data are available. For updates on guidelines, refer to the University of California Web site at http://postharvest.ucdavis.edu.

[a]Very low = <0.1 μL/(kg·h) at 68°F
Low = 0.1 to 1.0 μL/(kg·h)
Moderate = 1.0 to 10.0 μL/(kg·h)
High = 10 to 100 μL/(kg·h)
Very high = >100 μL/(kg·h)

[b]Detrimental effects include yellowing, softening, increased decay, abscission, and browning.

[c]At recommended storage temperature.
Low = <20 mg CO_2/(kg·h)
Moderate = <40 mg CO_2/(kg·h)
High = <60 mg CO_2/(kg·h)
Very high = >60 mg CO_2/(kg·h)

[d]CA = controlled atmosphere.

[e]For a more complete listing of grapes and pears, see International Institute of Refrigeration (IIR 2000).

Source: Appendix B, Thompson et al. (2000). Copyright University of California Board of Regents. Used by permission.

Table 2 Storage Requirements of Other Perishable Products

Product	Storage Temp., °F	Relative Humidity, %	Approximate Storage Life[a]
Fish			
Haddock, cod, perch	31 to 34	95 to 100	12 days
Hake, whiting	32 to 34	95 to 100	10 days
Halibut	31 to 34	95 to 100	18 days
Herring, kippered	32 to 36	80 to 90	10 days
smoked	32 to 36	80 to 90	10 days
Mackerel	32 to 34	95 to 100	6 to 8 days
Menhaden	34 to 41	95 to 100	4 to 5 days
Salmon	31 to 34	95 to 100	18 days
Tuna	32 to 36	95 to 100	14 days
Frozen fish	–20 to –4	90 to 95	6 to 12 months
Shellfish[a]			
Scallop meat	32 to 34	95 to 100	12 days
Shrimp	31 to 34	95 to 100	12 to 14 days
Lobster, American	41 to 50	In sea water	Indefinitely
Oysters, clams			
(meat and liquid)	32 to 36	100	5 to 8 days
in shell	41 to 50	95 to 100	5 days
Frozen shellfish	–30 to –4	90 to 95	3 to 8 months
Beef			
Beef, fresh, average	28 to 34	88 to 95	1 week
Beef carcass			
choice, 60% lean	32 to 39	85 to 90	1 to 3 weeks
prime, 54% lean	32 to 34	85	1 to 3 weeks
sirloin cut (choice)	32 to 34	85	1 to 3 weeks
round cut (choice)	32 to 34	85	1 to 3 weeks
dried, chipped	50 to 59	15	6 to 8 weeks
Liver	32	90	5 days
Veal, lean	28 to 34	85 to 95	3 weeks
Beef, frozen	–10 to 0	90 to 95	6 to 12 months
Pork			
Pork, fresh, average	32 to 34	85 to 90	3 to 7 days
carcass, 47% lean	32 to 34	85 to 90	3 to 5 days
bellies, 35% lean	32 to 34	85	3 to 5 days
fatback, 100% fat	32 to 34	85	3 to 7 days
shoulder, 67% lean	32 to 34	85	3 to 5 days
frozen	–10 to 0	90 to 95	4 to 8 months
Ham, 74% lean	32 to 34	80 to 85	3 to 5 days
light cure	37 to 41	80 to 85	1 to 2 weeks
country cure	50 to 59	65 to 70	3 to 5 months
frozen	–10 to 0	90 to 95	6 to 8 months
Bacon, medium fat class	37 to 41	80 to 85	2 to 3 weeks
cured, farm style	61 to 64	85	4 to 6 months
cured, packer style	34 to 39	85	2 to 6 weeks
frozen	–10 to 0	90 to 95	2 to 4 months
Sausage, links or bulk	32 to 34	85	1 to 7 days
country, smoked	32	85	1 to 3 weeks
Frankfurters, average	32	85	1 to 3 weeks
Polish style	32	85	1 to 3 weeks
Lamb			
Fresh, average	28 to 34	85 to 90	3 to 4 weeks
Choice, lean	32	85	5 to 12 days
Leg, choice, 83% lean	32	85	5 to 12 days
Frozen	–10 to 0	90 to 95	8 to 12 months
Poultry			
Poultry, fresh, average	28 to 32	95 to 100	1 to 3 weeks
Chicken, all classes	28 to 32	95 to 100	1 to 4 weeks
Turkey, all classes	28 to 32	95 to 100	1 to 4 weeks
breast roll	–4 to –1		6 to 12 months
frankfurters	0 to 15		6 to 16 months
Duck	28 to 32	95 to 100	1 to 4 weeks
Poultry, frozen	–10 to 0	90 to 95	12 months
Meat (Miscellaneous)			
Rabbits, fresh	32 to 34	90 to 95	1 to 5 days
Dairy Products			
Butter	32	75 to 85	2 to 4 weeks
Butter, frozen	–10	70 to 85	12 to 20 months
Cheese, cheddar			
long storage	32 to 34	65	12 months
short storage	40	65	6 months
processed	40	65	12 months
grated	40	65	12 months
Ice cream, 10% fat	–20 to –15	90 to 95	3 to 23 months
premium	–30 to –40	90 to 95	3 to 23 months
Milk			
fluid, pasteurized	39 to 43		7 days
grade A (3.7% fat)	32 to 34		2 to 4 months
raw	32 to 39		2 days
dried, whole	70	Low	6 to 9 months
nonfat	45 to 70	Low	16 months
evaporated	40		24 months
evaporated, unsweetened	70		12 months
condensed, sweetened	40		15 months
Whey, dried	70	Low	12 months
Eggs			
Shell	29 to 32[b]	80 to 90	5 to 6 months
farm cooler	50 to 55	70 to 75	2 to 3 weeks
Frozen			
Whole	0		1 year plus
yolk	0		1 year plus
white	0		1 year plus
whole egg solids	35 to 40	Low	6 to 12 months
Yolk solids	35 to 40	Low	6 to 12 months
Flake albumen solids	Room	Low	1 year plus
Dry spray albumen solids	Room	Low	1 year plus
Candy			
Milk chocolate	0 to 34	40	6 to 12 months
Peanut brittle	0 to 34	40	1.5 to 6 months
Fudge	0 to 34	65	5 to 12 months
Marshmallows	0 to 34	65	3 to 9 months
Miscellaneous			
Alfalfa meal	0	70 to 75	1 year plus
Beer, keg	35 to 40		3 to 8 weeks
bottles and cans	35 to 40	65 or below	3 to 6 months
Bread	0		3 to 13 weeks
Canned goods	32 to 60	70 or lower	1 year
Cocoa	32 to 40	50 to 70	1 year plus
Coffee, green	35 to 37	80 to 85	2 to 4 months
Fur and fabrics	34 to 40	45 to 55	Several years
Honey	50		1 year plus
Hops	28 to 32	50 to 60	Several months
Lard (without	45	90 to 95	4 to 8 months
antioxidant)	0	90 to 95	12 to 14 months
Nuts	32 to 50	65 to 75	8 to 12 months
Oil, vegetable, salad	70		1 year plus
Oleomargarine	35	60 to 70	1 year plus
Orange juice	30 to 35		3 to 6 weeks
Popcorn, unpopped	32 to 40	85	4 to 6 weeks
Yeast, baker's compressed	31 to 32		
Tobacco, hogshead	50 to 65	50 to 65	1 year
bales	35 to 40	70 to 85	1 to 2 years
cigarettes	35 to 46	50 to 55	6 months
cigars	35 to 50	60 to 65	2 months

Note: The text in this chapter or the appropriate commodity chapter gives additional information on many of the commodities listed. For a complete listing of frozen food practical storage life, see IIR (1986).

[a]Storage life is not based on maintaining nutritional value.
[b]Eggs with weak albumen freeze just below 30°F.

Other important factors in desiccation include product size, surface-to-volume ratio, the kind of protective surface on the product, and air movement. Of these, the storage operator can control only the last, and this control is greatly influenced by the container, kind of pack, and stacking arrangement (i.e., the ability of the air to move past individual fruits and vegetables).

As a rule, shrivelling does not become a serious market problem until fruits lose about 5% of their weight, but any loss reduces the salable amount. Moisture losses of 3 to 6% are enough to cause a marked loss of quality for many kinds of produce. A few kinds may lose 10% moisture and still be marketable, although some trimming may be necessary, as for stored cabbage.

The vapor pressure deficit cannot be kept at a zero level, but it should be maintained as low as possible. A maximum of about 0.018 in. Hg, which corresponds to 90% rh at 32°F, is recommended. Some compromise is possible for short storage periods. In many instances, the refrigerated storage operator may find it desirable to add moisture, or, in special cases, the owner of the produce may find it desirable to use moisture barriers such as film liners.

REFRIGERATED STORAGE PLANT OPERATION

Checking Temperatures and Humidity

To maintain top product quality, temperature in the cold storage room must be accurately maintained. Variations of 2 to 3°F in the product temperature above or below the desired temperature are too large in most cases. Storage rooms should be equipped either with accurate thermostats or with manual controls that receive frequent attention.

In refrigerated storage rooms, thermometers are usually placed at a height of about 5 ft for convenient reading. Temperatures should be monitored where they might be undesirably high or low; one or two aisle temperatures is not enough. A record of both product and air temperatures is necessary to determine performance of the storage plant. A thermometer or recording device of good quality that is periodically calibrated is essential.

Temperature in less accessible storage locations, such as the middle of stacks, can be obtained conveniently with distant-reading equipment such as thermocouples or electrical-resistance thermometers.

Storage instructions or recommendations usually specify a relative humidity within 3 to 5% of the desired levels. An ordinary sling psychrometer at temperatures of 32°F or lower cannot be read that closely. An error of 0.5°F in reading either the wet- or the dry-bulb thermometer will cause an error of 5% rh. Carefully calibrated thermometers graduated to 0.1°F with a range of 25 to 40°F are best adapted for this purpose in fruit storage. A convenient device for measuring humidity consists of a pair of these thermometers, mounted in a short length of metal casing attached to a spring or motor-operated fan that draws air past the thermometers at a speed of 3 ft/s or faster. Thermometers should be placed so that they will not be heated by the fan motor, and they should be read quickly to prevent warming (these are often referred to as aspirated psychrometers). The advantage of this instrument over the sling psychrometer is that it can be left in the room long enough to get a true wet-bulb reading. This may require 15 min or more if ice is formed on the wet bulb. Under these conditions, a thin coating of ice is preferable to a thick one in getting accurate readings. Hair hygrometers are satisfactory if not subjected to sudden large changes in humidity and temperature and if checked regularly with a psychrometer. There are now a number of electronic relative humidity probes based on capacitance and resistance of water-absorbing membranes; they can be accurate to within 3% for a wide range of temperatures and relative humidity if they are regularly calibrated. Many are sensitive to condensation, so this should be avoided (e.g., taking a probe from a cold area into the ambient air).

Perhaps a more accurate method of determining relative humidity is to electrically record the dew-point temperature of the air and to use a resistance thermometer of suitable sensitivity to record the ambient or dry-bulb temperature. From these temperature records, relative humidity can be calculated.

Air Circulation

Air must be circulated to keep refrigerated storage rooms at a uniform temperature throughout. Commodity temperatures in a storage room may vary because air temperature rises as air passes through the room and absorbs heat from the commodity; also, heat leakage may vary in different parts of the storage. In a duct system, air near the return ducts is warmer than the air near delivery ducts. In many storages, refrigeration units are installed over the center aisle. Air circulates from the center of rooms outward to the walls, down through the rows of produce, and back up through the center of the room.

Rapid air circulation is needed most during removal of field heat. Sometimes this is best done in separate cooling rooms that have more refrigerating and air-moving capacity than regular refrigerated storage rooms (see Chapter 15). After field heat is removed, a high air velocity is usually undesirable. Air movement is needed only to remove respiratory heat, if any, and heat entering the room through exterior surfaces and doorways. Excessive air movement can increase moisture loss from a product, with a resulting loss of both weight and quality. However, some air, circulated by fans or blowers, must be uniformly distributed through all parts of the room. Also, if air circulation or refrigeration is turned off, some precaution is needed to ensure that the stored product does not become too warm, particularly for respiring products.

The type of container and the manner of stacking are important factors that influence cooling performance. An elaborate system for air distribution is useless if poor stacking prevents airflow. If spacing is irregular, wider spaces will get a greater volume of air than narrower ones. If some spaces are partially blocked, dead air zones will occur, with resultant higher temperatures.

Sanitation and Air Purification

The refrigerated storage and product containers must be clean. Fruits and vegetables coming into the storage are generally contaminated with mold spores, which can enter through punctures or breaks in the skin of the product. Contaminated commodities can foster rapid development of the spores, which are carried by air currents throughout the storage. Removing decaying raw material from storage and sanitizing the product container reduces the problem. If mold contamination is excessive, the storage develops a musty or moldy odor that is quickly picked up by the fruits and vegetables, a fault of many apple varieties and other products held for several months in refrigerated storage. This problem can best be controlled by means of special cleaners, sanitizers, and deodorizers (see Chapters 10, 22, 23, and 24 for details on product diseases).

During several months of storage, even at 31°F, molds may grow on the surface of packages and on the walls and ceilings of rooms under high relative humidity. These surface molds generally will not rot fruits and vegetables. However, because surface molds are unsightly, storage warehouses should have a thorough cleaning at least once a year. Good air circulation alone is of considerable value in minimizing growth of surface molds. If floors and walls become moldy, they can be scrubbed with a cleaner containing sodium hypochlorite or trisodium phosphate, then rinsed and aired. Field boxes and equipment can be cleaned with 0.25% calcium hypochlorite solutions or by exposing to steam for 2 min.

All inspected plants and warehouses operate under regulations with sanitation requirements clearly set forth in inspection service orders. Plants should be constructed to prevent the entrance of insects and rodents. This involves ratproof building construction and adequate screening. For doors frequently opened, special measures must be taken to prevent the entrance of insects.

The frequency of cleanup and the detergents and sanitizing agents used should be specified by the quality control leader. A representative from quality control should inspect all areas after cleanup and determine whether a suitable job has been done. Waste and miscellaneous trash accumulation areas must receive special attention around warehouses, because they become breeding places for rodents and insects. In any food warehouse, the successful quality control and sanitation program depends on the cooperation and vigilance of management.

Air may be purified in storage rooms where odors or volatiles may contribute to off-flavors and hasten deterioration. Air may be cleaned with trays or canisters containing 6 to 14 mesh activated carbon. Pinewood volatiles are removed by activated carbon air-purifying units. Some produce volatiles are also removed, but ethylene, a ripening gas, is not removed by activated carbon alone.

Air washing with water to remove volatiles does not retard fruit ripening, but it may increase the relative humidity and thus help maintain good fruit appearance by reducing weight loss.

Removal of Produce from Storage

When produce is removed from storage, undue warming and condensation of moisture, which promote decay and deterioration, must be prevented. Because many storages are built on railroad sidings, canvas tunnels should be installed between the car and the storage through which the produce will be conveyed to minimize warming and condensation of moisture.

When produce is removed from storage for distribution to wholesale and retail markets, the storage operator can do little to prevent undesirable condensation. Warming the packages until they are above the dew point of the air would prevent it, but this takes time and space and is seldom practical. Deterioration in flavor and condition proceeds rapidly after long storage periods. Therefore, the produce should be moved to consumers as rapidly as possible.

STORAGE OF FROZEN FOODS

Frozen foods deteriorate during the period between production and consumption. The extent of deterioration depends mainly on storage temperature and time, although other factors, such as protection provided by the package, are important. Bacteria in frozen foods may be killed during freezing and frozen storage, but all the bacteria present are never completely destroyed. When defrosting, frozen foods are still subject to bacterial decomposition.

OTHER PRODUCTS

Beer

Because beer in bottles or cans is either pasteurized or filtered to destroy or remove the living yeast cells, it does not require as low a storage temperature as keg beer. Bottled beer may be stored at an ordinary room temperature of 70 to 75°F, but for convenience, it is often stored with keg beer at a lower temperature of 35 to 40°F. The bottled product should be protected from strong light, especially direct sunlight. Storage life varies from 3 to 6 months, depending largely on the method of processing and packaging. Keg beer, stored at 35 to 40°F, usually has a storage life of 3 to 6 weeks.

Canned Foods

Canned foods that are heat-processed in hermetically sealed containers do not benefit from refrigerated storage if they are to be stored for no longer than 2 or 3 months at temperatures that rarely exceed 75°F. However, seasonal commodities produced to provide an inventory for an entire year or longer do benefit from reduced-temperature/humidity storage, which delays the onset of considerable color, texture, and flavor changes, loss of nutrients, and container corrosion. Notable examples are canned asparagus, cherries, and catsup. Reduced temperature and humidity storage for canned goods is essential in environments where ambient temperatures and humidities regularly exceed 90°F and 70% rh.

Dried Foods

Dried foods and feeds, particularly those expected to supply a high level of protein, such as dehydrated milk or alfalfa meal, should be protected against high temperature and humidity. Good packaging, such as canning in vacuum, can maintain dried food nutritive value and quality for a year or longer if ambient temperatures do not exceed 90°F regularly. When stored in bulk or in bags that are not good water vapor barriers, storage at 40% rh and 70°F should be limited to less than one year, and at 90°F to less than 6 months. At 60% rh and 70°F, storage should be limited to 6 months, and at 90°F to not more than 3 to 4 months. The only process by which dried foods can maintain quality and nutritive values for a year and longer at elevated temperatures is by packaging in zero oxygen. Similar or better results can be obtained with 0°F storage regardless of adequacy of the package or relative humidity.

Furs and Fabrics

Refrigerated storage effectively protects furs, floor coverings, garments, and other materials containing wool against insect damage. Commonly used refrigerated storage temperatures do not kill the insects, but inactivate them, preventing insect damage while the susceptible items are in storage (Table 3). However, if insects are present, the article is susceptible to damage as soon as it is removed from refrigerated storage.

Articles should be free of any possible infestation before placement in refrigerated storage. Those items that can be cleaned should be so treated. Others can be either fumigated or mothproofed. Food should not be stored with fur garments.

Furs and garments should be stored at 34 to 40°F. A temperature of 40°F is most widely used commercially. The low temperature not only inactivates fabric insects but also preserves the integrity and luster of furs and the tensile strength of fabrics.

Continuous storage below 34 to 40°F is a wasteful expense as far as protection from insect damage is concerned. Some storage firms maintain constant temperatures in their fur vaults between 14 and 32°F and claim excellent results. However, no research indicates that temperatures in this range are required for storing dressed furs or fabrics. Cured raw furs (but not processed) should be stored at –10 to 10°F with 45 to 60% rh and will keep up to 2 years.

Honey

Both extracted (liquid) and comb honey can be held satisfactorily in common dry storage for about a year. The slow darkening and flavor deterioration at ordinary room temperatures becomes objectionable after this time. Although cold storage is not necessary, temperatures below 50°F maintain original quality for several years and retard or prevent fermentation. The range between 50 and 65°F

Table 3 Temperature and Time Requirements for Killing Moths in Stored Clothing

Storage Temperature, °F	All Eggs Dead After, Days	All Larvae Dead After, Days	All Adults Dead After, Days
–0.4 to 5	1	2	1
5 to 10	2	21[a]	1
10 to 15	4	—	1
15 to 19	—	—	1
20 to 25	21	67	4
25 to 30	21	125[b]	7
30 to 35	—	283[c]	—

Table adapted from USDA *Publication* AMS-57 (1955).
[a]50 to 95% of larvae may be killed in 2 days.
[b]A few larvae survived this period.
[c]Larvae survived this period.

should be avoided, because it promotes granulation, which increases the probability of fermentation of raw (unheated) honey.

As storage temperature increases in the 80 to 100°F range, deterioration is accelerated; temperatures constantly above 85°F are unsuitable, and above 90°F, quite damaging.

Honey for export is best kept in cold storage, because the half-life for honey diastase at 77°F is about 17 months.

Raw honey of greater than 20% moisture is always in danger of fermentation; the likelihood is much less at or below 18.6% moisture. Below 17% moisture, raw honey will not ordinarily ferment. Granulation increases the possibility of fermentation of raw honey by increasing the moisture content of the liquid portion. Properly pasteurized honey will not ferment at any moisture content. Granulated honey can be reliquefied by warming to 120 to 140°F.

Comb honey should not be stored above 60% rh to avoid moisture absorption through the wax, which leads to fermentation.

Finely granulated honey (honey spread, Dyce process honey, and honey cream) must not be stored above about 75°F. Higher temperatures eventually cause partial liquefaction and destroy the texture. Any subsequent regranulation by lower temperatures produces an undesirable coarse texture. For holding more than 4 months, cold storage is required.

Maple Syrup

Maple syrup keeps indefinitely at room temperatures without darkening or losing flavor, if packed hot (at or within a few degrees of its boiling point) and in clean containers, which are promptly closed airtight and laid on their sides or inverted to self-sterilize the closure, and then cooled. Refrigerated storage is not necessary. However, once opened, syrup in a bottle, can, or drum may become contaminated by organisms in the air. Mold or yeast spores, which may be present in improperly pasteurized syrup, though unable to germinate in full-density syrup, may grow in the thin syrup on the surface caused by condensed water. Small packages not completely sterile, containing spores, can be kept free of vegetative growth by periodically inverting the containers to redisperse any thin syrup on the surface caused by condensed water. Maple syrup should never be packaged at temperatures below 180°F. After pasteurizing, cool the syrup as quickly as possible to prevent stack burn, which darkens the syrup and causes a lowering of its grade.

Nursery Stock and Cut Flowers

Low temperature (31 to 33°F) and dry packaging prevent, or at least greatly retard, flower disintegration and extend storage life. The temperature and storage life given in Table 4 for cut flowers allow for a reasonable shelf life after removal from storage; however, the storage period may be extended beyond that listed. These conditions, though not widely used commercially, are recommended. Proper dry packing requires a moisture/vaporproof container in which flowers can be sealed. No free water is added because the package prevents almost all water loss.

Flowers held in water should not be crowded in the containers and should be arranged on shelves or racks to allow good air circulation. Forced air circulation should be provided, but flowers must be kept out of a direct draft. Use clean water and clean containers.

Many kinds of nursery stock can also be stored at temperatures of 31 to 35°F. Open packages and harden flowers before marketing if the blooms have been stored for long periods. Flowers conditioned at about 50°F following storage regain full turgidity most rapidly. Cut or crush stem ends and then place in water or a food solution at 80 to 100°F for 6 to 8 h.

Many kinds of cut flowers and greens are injured if stored in the same room with certain fruits, principally apples and pears, which give off gases (such as ethylene) during ripening. These gases cause premature aging of blooms and may defoliate greens. Greens should not be stored in the same room with cut flowers because the greens, acting in the same way as fruit, can hasten bloom deterioration.

Table 4 Storage Conditions for Cut Flowers and Nursery Stock

Commodity	Storage Temp., °F	Relative Humidity, %	Approx. Storage Life	Method of Holding	Highest Freezing Point, °F
Cut Flowers					
Calla Lily	40	90 to 95	1 week	Dry pack	
Camellia	45	90 to 95	3 to 6 days	Dry pack	30.6
Carnation	31 to 32	90 to 95	3 to 4 weeks	Dry pack	30.8
Chrysanthemum	31 to 32	90 to 95	3 to 4 weeks	Dry pack	30.5
Daffodil (Narcissus)	32 to 33	90 to 95	1 to 3 weeks	Dry pack	31.8
Dahlia	40	90 to 95	3 to 5 days	Dry pack	
Gardenia	32 to 34	90 to 95	2 weeks	Dry pack	31.0
Gladiolus	36 to 42	90 to 95	1 week	Dry pack	31.4
Iris, tight buds	31 to 32	90 to 95	2 weeks	Dry pack	30.6
Lily, Easter	32 to 35	90 to 95	2 to 3 weeks	Dry pack	31.1
Lily-of-the-valley	31 to 32	90 to 95	2 to 3 weeks	Dry pack	
Orchid	45 to 55	90 to 95	2 weeks	Water	31.4
Peony, tight buds	32 to 35	90 to 95	4 to 6 weeks	Dry pack	30.1
Rose, tight buds	32	90 to 95	2 weeks	Dry pack	31.2
Snapdragon	40 to 42	90 to 95	1 to 2 weeks	Dry pack	30.4
Sweet Peas	31 to 32	90 to 95	2 weeks	Dry pack	30.4
Tulips	31 to 32	90 to 95	2 to 3 weeks	Dry pack	
Greens					
Asparagus (plumosus)	35 to 40	90 to 95	2 to 3 weeks	Polylined cases	26.0
Fern, dagger and wood	30 to 32	90 to 95	2 to 3 months	Dry pack	28.9
Fern, leatherleaf	34 to 40	90 to 95	1 to 2 months	Dry pack	
Holly	32	90 to 95	4 to 5 weeks	Dry pack	27.0
Huckleberry	32	90 to 95	1 to 4 weeks	Dry pack	26.7
Laurel	32	90 to 95	2 to 4 weeks	Dry pack	27.6
Magnolia	35 to 40	90 to 95	2 to 4 weeks	Dry pack	27.0
Rhododendron	32	90 to 95	2 to 4 weeks	Dry pack	27.6
Salal	32	90 to 95	2 to 3 weeks	Dry pack	26.8
Bulbs					
Amaryllis	38 to 45	70 to 75	5 months	Dry	30.8
Caladium	70	70 to 75	2 to 4 months		29.7
Crocus	48 to 63		2 to 3 months		29.7
Dahlia	40 to 48	70 to 75	5 months	Dry	28.7
Gladiolus	45 to 50	70 to 75	5 to 8 months	Dry	28.2
Hyacinth	63 to 68		2 to 5 months		29.3
Iris, Dutch, Spanish	68 to 77	70 to 75	4 months	Dry	29.3
Gloriosa	50 to 63	70 to 75	3 to 4 months	Polyliner	
Candidum	31 to 33	70 to 75	1 to 6 months	Polyliner and peat	
Croft	31 to 33	70 to 75	1 to 6 months	Polyliner and peat	
Longiflorum	31 to 33	70 to 75	1 to 10 months	Polyliner and peat	28.9
Speciosum	31 to 33	70 to 75	1 to 6 months	Polyliner and peat	
Peony	33 to 35	70 to 75	5 months	Dry	
Tuberose	40 to 45	70 to 75	4 to 12 months	Dry	
Tulip	63	70 to 75	2 to 6 months	Dry	27.6
Nursery Stock					
Trees and shrubs	32 to 36	95	4 to 5 months	*	
Rose bushes	31 to 36	85 to 95	4 to 5 months	Bare rooted w/polyliner	
Strawberry plants	30 to 32	80 to 85	8 to 10 months	Bare rooted w/polyliner	29.9
Rooted cuttings	31 to 36	85 to 95		Polywrap	
Herbaceous perennials	27 to 28	80 to 85	4 to 8 months	*	
	31 to 35	80 to 85	3 to 7 months		
Christmas trees	22 to 32	80 to 85	6 to 7 weeks		

Data from USDA *Agricultural Handbook* 66.
*For details for various trees, shrubs, and perennials, see ANSI/ANLA *Standard* Z60.1.

Greens, bulbs, and certain nursery stock are usually packaged or crated when stored. Some bulbs and nursery stock are packed in damp moss or similar material, and low temperatures are required to keep them dormant. Polyethylene wraps or box liners are very effective for maintaining quality of strawberry plants, bare-root rose bushes, and certain cuttings and other nursery stock in storage. Strawberry plants can be stored up to 10 months in polyethylene-lined crates at 30 to 32°F.

Popcorn

Store popcorn at 32 to 40°F and about 85% rh. This relative humidity yields the optimum popping condition and the desired moisture content of about 13.5%.

Vegetable Seeds

Seeds generally benefit from low temperatures and low-humidity storage. High temperatures and high humidity favor loss of viability. Most vegetable seeds undergo no significant decrease in germination during one season when stored at 50°F and 50% rh. Full viability is retained far longer than one year as temperature and humidity are reduced. A temperature of 20°F and 15 to 25% rh are considered ideal but rarely necessary unless seed viability must be maintained for many years. Aster, pepper, tomato, and lettuce seeds stored under these conditions had equal or better viability after 13 years than did the fresh seed.

Low moisture content of the seed is important for germination. Hemp seed containing 9.5% moisture was mostly dead after 12 years storage at 50°F, but lost only 12% viability when moisture content was 5.7%. If stored at about 0°F, moisture percentage should not exceed 10% for many species. Seed with higher moisture content, when stored at 0°F, eventually equilibrates at a lower moisture level but may initially suffer frost damage. At higher temperatures, if it is impossible to keep humidity low enough, seeds must be stored in moistureproof containers.

REFERENCES

Thompson, J.F., P. Brecht, R.T. Hinsch, and A.A. Kader. 2000. Marine container transport of chilled perishable produce. *Publication* 21595. University of California, Division of Agriculture and Natural Resources.

USDA. 1955. Protecting stored furs from insects. *Publication* AMS-57. U.S. Department of Agriculture, Washington, D.C.

USDA. 2004. The commercial storage of fruits, vegetables, and florist and nursery stocks. *Agricultural Handbook* 66 (draft). U.S. Department of Agriculture, Agricultural Research Service, Washington, D.C. (http://www.ba.ars.usda.gov/hb66)

BIBLIOGRAPHY

ANLA. 2004. American standard for nursery stock. ANSI/ANLA *Standard* Z60.1-2004. American Nursery & Landscape Association, Washington, D.C.

Hunt Ashby, B. 1987. Protecting perishable foods during transport by truck. USDA *Handbook* 669. U.S. Department of Agriculture, Washington, D.C.

IIR. 1986. *General principles for the freezing, storage and thawing of foodstuffs: Recommendations for the processing and handling of frozen foods*, 3rd ed. International Institute of Refrigeration, Paris.

IIR. 1990. *Principles of refrigerated preservation of perishable foodstuffs: Manual of refrigerated storage in the warmer developing countries.* International Institute of Refrigeration, Paris.

IIR. 1993. *Cold stores guide.* International Institute of Refrigeration, Paris.

IIR. 2000. *Recommendations for chilled storage of perishable produce.* International Institute of Refrigeration, Paris.

ISO. 1974. Avocados—Guide for storage and transport. *Standard* 2295. International Organization for Standardization, Geneva.

Kader, A.A. 1986. Biochemical basis for effect of controlled and modified atmospheres on fruits and vegetables. *Food Technology* 40(5):99-100 and 102-104.

Scott, V.N. 1989. Interaction of factors to control microbial spoilage of refrigerated foods. *Journal of Food Protection* 52(6):431-435.

Schlimme, D.V., M.A. Smith, and L.M. Ali. 1991. Influence of freezing rate, storage temperature, and storage duration on the quality of cooked turkey breast roll. *ASHRAE Transactions* 97(1):214-220.

Schlimme, D.V., M.A. Smith, and L.M. Ali. 1991. Influence of freezing rate, storage temperature, and storage duration on the quality of turkey frankfurters. *ASHRAE Transactions* 97(1):221-227.

Tressler, D.K. and C.F. Evers. 1957. *The freezing preservation of foods*, 3rd ed. AVI Publishing, Westport, CT.

Valenzuela Segura, G., D.D. Delgado, and D.R. Ramirez. 1972. *Handling, storage and transport systems for exported refrigerated perishable foods.* Revista del Instituto de Investigaciones Technologicas, Bogotá, Columbia.

Webb, B.H., A.H. Johnson, and J.A. Alford. 1973. *Fundamentals of dairy chemistry.* AVI Publishing, Westport, CT.

CHAPTER 12

FOOD MICROBIOLOGY AND REFRIGERATION

REFRIGERATION'S largest overall application is the prevention or retardation of microbial, physiological, and chemical changes in foods. Even at temperatures near the freezing point, foods may deteriorate through growth of microorganisms, changes caused by enzymes, or chemical reactions. Holding foods at low temperatures merely reduces the rate at which these changes take place. A few spoilage organisms can grow at or below temperatures at which food begins to freeze.

Refrigeration also plays a major role in maintaining a safe food supply. Overall, the leading factor causing foodborne illness is improper food-holding temperatures. Another important factor is improperly sanitized equipment. Engineering directly affects the safety and stability of the food supply in design of cleanable equipment and facilities, as well as maintenance of environmental conditions that inhibit microbial growth. This chapter briefly discusses the microbiology of foods and the effect of design decisions on the production of safe and wholesome foods. Methods of applying refrigeration to specific foods are discussed in Chapters 17 to 29.

BASIC MICROBIOLOGY

Microorganisms play several roles in a food production facility. They can contribute to food spoilage, producing off-odors and flavors, or altering product texture or appearance through slime production and pigment formation. Some organisms cause disease; others are beneficial and are required to produce foods such as cheese, wine, and sauerkraut through fermentation.

Microorganisms fall into four categories: bacteria, yeasts, molds, and viruses. Bacteria are the most common foodborne pathogens. Bacterial growth rates, under optimum conditions, are generally faster than those of yeasts and molds, making bacteria a prime cause of spoilage, especially in refrigerated, moist foods. Bacteria have many shapes, including spheres (cocci), rods (bacilli), or spirals (spirochetes), and are usually between 0.3 and 5 to 10 μm in size. Bacteria can grow in a wide range of environments. Some, notably *Clostridium* and *Bacillus* spp., form endospores (i.e., resting states with extensive temperature, desiccation, and chemical resistance).

Yeasts and molds become important in situations that restrict the growth of bacteria, such as in acidic or dry products. Yeasts can cause gas formation in juices and slime formation on fermented products. Mildew (black mold) on humid surfaces and mold formation on spoiled foods are also common. Some molds produce very powerful toxins (mycotoxins) that, if consumed, may be fatal.

Viruses are obligate intracellular parasites that are specific to an individual host. All viruses, including human viruses (e.g., hepatitis A), cannot multiply outside living cells or tissue. Refrigeration design features must include facilities for good employee handwashing and sanitation practices to minimize potential for product contamination. Bacterial viruses (phages), however, may contribute to starter culture failure in bacterial fermentations if proper isolation, ventilation, and sanitation procedures are not followed. The use of commercial concentrated cultures, selected for phage resistance, has greatly reduced this problem.

Sources of Microorganisms

Bacteria, yeasts, and molds are widely distributed in water, soil, air, plant materials, and the skin and intestinal tracts of humans and animals. Practically all unprocessed foods are contaminated with a variety of spoilage and, sometimes, pathogenic microorganisms because foods act as excellent media for bacterial multiplication. Food processing environments that contain residual food material will naturally select for the microorganisms that are most likely to spoil the particular product.

Microbial Growth

Changes in microbial populations follow a generalized growth curve (Figure 1). An initial lag phase occurs as organisms adapt to new environmental conditions and start to grow. The lag phase is very important because the maximum extension of shelf life and length of production runs are directly related to the length of the lag phase. After adaptation, the culture enters into the maximum (logarithmic) growth rate, and control of microbial growth is not possible without major sanitation or other drastic measures. Numbers can double as fast as every 20 to 30 min under optimum conditions.

Toxin production and spore maturation, if possible, usually occur at the end of the exponential phase as the culture enters a stationary phase. At this time, essential nutrients are depleted and/or inhibitory by-products are accumulated. Eventually, culture viability declines; the rate depends on the organism, medium, and other environmental characteristics. Although refrigeration prolongs generation time and reduces enzyme activity and toxin production, in most cases, it will not restore lost product quality or safety.

CRITICAL MICROBIAL GROWTH REQUIREMENTS

Factors that influence microbial growth can be divided into two categories: (1) intrinsic factors that are a function of the food itself

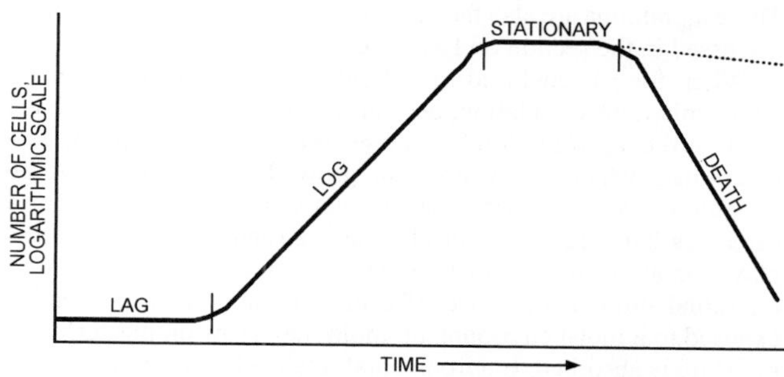

Fig. 1 Typical Microbial Growth Curve

The preparation of this chapter is assigned to TC 10.9, Refrigeration Application for Foods and Beverages.

Table 1 Approximate Minimum Water Activity for Growth of Microorganisms

Organism	a_w	Foods
Pseudomonads	0.98	Fresh fruits, vegetables, meats
Salmonella spp., *E. coli*	0.95	Many processed foods
Listeria monocytogenes	0.93	
Bacillus cereus	0.92	Salted butter
Staphylococcus aureus	0.86	Fermented sausage
Molds	0.84	Soft, moist pet food
	0.80	Pancake syrup, jam
	0.70	Corn syrup
Xerotrophic molds	0.65	Caramels
Osmophilic yeasts	0.62	
Limit of microbial growth	0.60	Wheat flour
	0.40	Nonfat dry milk

and (2) extrinsic factors that are a function of the environment in which a food is held.

Intrinsic Factors

Intrinsic factors affecting microbial growth include nutrients, inhibitors, biological characteristics, water activity, pH, and presence of competing microorganisms in a food. Although engineering practices have little effect on these parameters, an understanding of how intrinsic factors influence growth is useful in predicting the types of microorganisms that may be present.

Nutrients. Like other living organisms, microorganisms require food to grow. Carbon and energy sources are usually sugars and starches. Nitrogen requirements are met by the presence of protein. Vitamins and minerals are also necessary. Lactic acid bacteria have rather exacting nutritional requirements, but many aerobic spore formers have tremendous enzymatic capabilities that allow growth on a wide variety of substrates. Cleanable systems facilitate removal of residual food material and deprive microorganisms of the nutrients required for growth, thus preventing a buildup of organisms in the environment.

Inhibitors. Either naturally occurring or added as preservatives, inhibitors may be present in food. Preservatives are not substitutes for hygienic practices and, with time, microorganisms may develop resistance. A cleanable processing system is still essential in preventing development of a resistant population.

Competing Microorganisms. The presence of one type of microorganism affects other organisms in foods. Some organisms produce inhibiting compounds or grow faster; others are better able to use the available nutrients in a food matrix.

Water Activity. All life-forms require water for growth. Water activity a_w refers to the availability of water in a food system and is defined at a given temperature as

$$a_w = \frac{\text{Vapor pressure of solution (food)}}{\text{Vapor pressure of solute (water)}}$$

The minimum water activities for growth of a variety of microorganisms, along with representative foods, are listed in Table 1. These a_w minima are also factors in environmental humidity control discussed in the section on Extrinsic Factors.

When food is enclosed in airtight packaging or in a chamber with limited air circulation, an equilibrium a_w is achieved that is equal to the a_w of the food. In these situations, the a_w of the food determines which organisms can grow. If the same foods are exposed to reduced environmental relative humidity, such as meat carcasses hanging in a controlled aging room or vegetables displayed in an open case, surface dehydration acts as an inhibitor to microbial growth. Likewise, if a dry product, such as bread, is exposed to a moist environment, mold may grow on the surface as moisture is absorbed. Environmental relative humidity thus significantly affects product shelf life.

Table 2 Minimum Growth Temperatures for Some Bacteria in Foods

Organism	Possible Significance	Approximate Minimum Growth Temperature, °F
Staphylococcus aureus	Foodborne disease	50
Salmonella spp.	Foodborne disease	42
Clostridium botulinum,	Foodborne disease	
proteolytic		50
nonproteolytic		38
Lactobacillus and *Leuconostoc*	Spoilage of fresh and cured meats	32
Listeria monocytogenes	Foodborne disease	34
Acinetobacter spp.	Spoilage of precooked foods	31
Pseudomonads	Spoilage of raw fish, meats, poultry, and dairy products	31

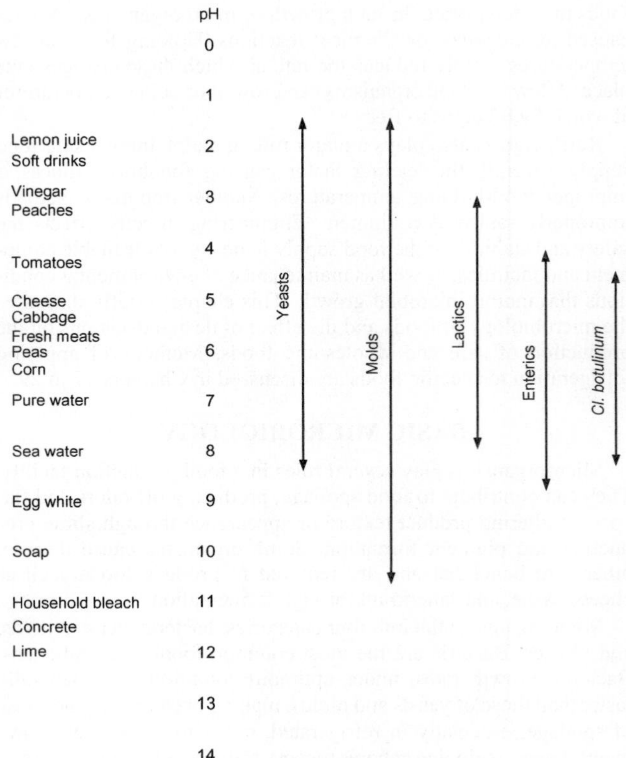

Fig. 2 pH Ranges for Microbial Growth and Representative Examples

pH. For most microorganisms, optimal growth occurs at neutral pH (7.0). Few organisms grow under alkaline conditions, but some, such as yeasts, molds, and lactic acid bacteria, are acid tolerant. Figure 2 depicts pH values of a variety of foods and limiting pH values for microorganisms.

Extrinsic Factors

Extrinsic factors that influence the growth of microorganisms include temperature, environmental relative humidity, and oxygen levels. Refrigeration and ventilation systems play a major role in controlling these factors.

Temperature. Microorganisms can grow over a wide range of temperatures. Minimum growth temperatures for a variety of spoilage and pathogenic bacteria of significance in foods are summarized in Table 2. Previously, 45°F was thought to be sufficient to control growth of pathogenic organisms. However, the emergence of psychrophilic pathogens, such as *Listeria monocytogenes*, has demonstrated the need for lower temperatures. In the United States, 41°F is now recognized as the upper limit for safe refrigeration temperature, although in some cases 34°F or lower may be more appropriate. Foods that can support growth of pathogenic microorganisms should not be held between 41 and 140°F for more than 2 h.

Temperature is used to categorize microorganisms. Those capable of growth above 113°F, with optimum growth at 130 to 150°F, are **thermophiles**. Thermophilic growth can be extremely rapid, with generation times of 10 to 20 min. Thermophiles can become a problem in blanchers and other equipment that maintain food at elevated temperatures for extended periods. These organisms die or do not grow at refrigeration temperatures.

Mesophiles grow best between 68 and 113°F. Most pathogens are in this group, with optimum growth temperatures around 98.6°F (i.e., body temperature). Mesophiles also include a number of spoilage organisms. Growth of mesophiles is quite rapid, with typical generation times of 20 to 30 min. Because mesophiles grow so rapidly, perishable foods must be cooled as fast as possible to prevent spoilage or potential unsafe conditions. Also, slower cooling rates cause mesophiles to adapt and grow at lower temperatures. With mild temperature abuse, prolific growth can occur, leading to spoilage or a potential health hazard.

Psychrophiles can grow at 41°F; some are able to grow at temperatures as low as 23°F and are a primary cause of spoilage of perishable foods. Psychrophilic growth is slow compared to mesophilic and thermophilic growth, with maximum growth rates of 1 to 2 h or longer. However, control of psychrophilic growth is a major requirement in products with extended shelf life. Because many psychrophiles have optimum temperatures in the mesophilic range, what may seem to be an insignificant increase in temperature can have a major effect on the growth rate of spoilage organisms. Growth roughly doubles with each 5°F increase in temperature. In practice, shelf life of fresh meat is maximized at 29°F and is reduced 50% by holding at 36°F. Meat freezes at 28°F.

For all critical growth factors, the range over which growth can occur is characteristic for a given organism. The range for growth is narrower than that for survival. For example, the maximum temperature for growth is slightly above the optimum, and death usually occurs just slightly above the maximum. This is not the case at the lower end of the temperature range. Survival of psychrophilic and most mesophilic microorganisms is enhanced by low storage temperatures. Freezing is not an effective lethal process; some organisms, notably gram-negative bacteria, are damaged by freezing and may die slowly, but others are extremely resistant. In fact, freezing is used as an effective means of preserving microbial cultures at extremely low temperatures (e.g., –110°F).

Environmental Relative Humidity. Water, previously discussed as an essential intrinsic growth factor, is also a major extrinsic factor. Environmental water acts as a vector for transmission of microorganisms from one location to another through foot traffic or aerosols. Refrigeration drain pans and drip coils have been identified as significant contributors of *L. monocytogenes* contamination in food processing environments. Aerosols have also transmitted the agent that causes Legionnaires' disease. High relative humidity in cold rooms is a particular problem and leads to black mold build-up on walls and ceilings as well as growth of organisms in drains and other reservoirs of water. Condensation on ceilings supports microbial growth and can drip onto product contact surfaces. Inadequately drained equipment collects stagnant water and supports microbial growth that is easily transported throughout a production facility when people walk through puddles. It is extremely important to control environmental relative humidity in food production environments. Control measures are discussed further in the section on Regulations and Standards.

Oxygen. Microorganisms are frequently classified by their oxygen requirement. Strictly aerobic microorganisms, such as molds and pseudomonads, require oxygen for growth. Conversely, strict anaerobes, such as *Clostridium* spp., cannot grow in the presence of oxygen. Facultatively anaerobic microorganisms (e.g., coliforms) grow with or without oxygen present, and microaerophiles, such as lactobacilli, grow best in conditions with reduced oxygen levels. Controlled-atmosphere (CA) chambers for fruit storage use lower oxygen levels to prolong storage life by retarding growth of spoilage organisms as well as influencing ripening. Vacuum packing also uses this extrinsic growth factor by inhibiting the growth of strict aerobes.

Biological Diversity

Most bacteria prefer specific environmental habitats, but some are more broadly adaptable because of their enzyme systems. Some grow well at low temperatures with or without oxygen (e.g., *Listeria monocytogenes*) or survive high temperatures (e.g., *Bacillus cereus*) and still grow at body temperature and cause illness.

DESIGN FOR CONTROL OF MICROORGANISMS

Microorganisms can be controlled by one of three mechanisms: contamination prevention, growth prevention, or destruction of the organisms themselves. Design of refrigeration and ventilation systems can affect all these areas.

Contamination Prevention

To prevent the entry of microorganisms into food production areas, ventilation systems must provide adequately clean air. Because bacteria are generally transported through air on dust particles, 95% filters (as defined in ASHRAE *Standard* 52.1) are sufficient to remove most microorganisms. High-efficiency particulate air (HEPA) filters provide sterile air and are used for cleanrooms.

Air-filtering materials must remain dry. Wet filters in ventilation systems support microbial growth, and organisms are transported throughout the production facility in the air. All ventilation systems must also be protected from water and condensate formation to prevent mold growth. This may require increased airflow or dehumidification systems. Positive pressure in the production environment prevents the entry of airborne contamination from sources other than ventilation ducts. Air intakes for production areas must not face areas that are prone to contamination, such as puddles on roofs or nesting sites for birds. Microbial contamination from airborne sources has been recorded, but quantitative assessments are not available to specify design parameters.

Refrigeration drip pans are a significant source of *L. monocytogenes* contamination. Condensation drip pans should be plumbed directly to drain to prevent contamination of floors and subsequent transport of organisms throughout a production facility. Drip pans must be easily accessible to allow scheduled cleaning, thus preventing organism growth. Air defrost should be avoided in critical areas. Continuous glycol-sprayed evaporative coils offer advantages, because glycol has been found to trap and kill microorganisms. Being hygroscopic, glycol depresses the air's dew point, providing a drier environment.

Traffic flow through a production facility should be planned to minimize contact between raw and cooked products, as mandated in USDA regulations for plants that cook meat products. Straight-line flow of a raw product from one end of a facility to the other prevents cross contamination. Walls separating raw product from cooked (or dirty from clean), with positive pressure in the cooked area, should be considered, because this provides the best protection. Provide adequate storage facilities to allow separate storage of raw ingredients from processed products, especially in facilities that handle meat products, which are a significant source of *Salmonella.* Raw meat must not be stored with cooked meats and/or vegetables or dairy products.

Growth Prevention

Water control is one of the most effective and most frequently overlooked means of inhibiting microbial growth. All ventilation systems, piping, equipment, and floors must be designed to drain completely. Residual standing water supports rapid microbial growth, and foot and forklift traffic transports organisms from puddles throughout the facility.

Condensation on ceilings and chilled pipes also supports microbial growth and may drip onto product contact surfaces that are not adequately protected. Preventing condensation is essential to prevent contamination. Insulation of pipe and/or dehumidifying systems may be necessary, particularly in chilled rooms. Increased airflow may also be useful in removing residual moisture. Maintaining 70% rh prevents the growth of all but the most resistant microorganisms; less than 60% rh prevents all microbial growth on facility surfaces (see Table 1).

Sanitation procedures use much water and leave the facility and surfaces wet. Adequate dehumidification should be provided to remove moisture during and after sanitation.

Limiting relative humidity is not always possible. For example, aging meat carcasses requires relative humidities of 90 to 95% to prevent excessive drying of the tissue. In these cases, a temperature of 29°F, just above product freezing point, should be used to inhibit microbial deterioration. Temperatures below 41°F inhibit the most common organisms that cause foodborne illness; however, 34°F is required to inhibit *L. monocytogenes*. Airflow, relative humidity, and temperature must be finely balanced to achieve maximum shelf life with limited deterioration of quality.

Freezing is also an effective means of microbial control. Limited death may occur during freezing, especially during slow freezing of gram-negative bacteria. However, freezing is not a reliable way to kill microorganisms. Because no microbial growth occurs in frozen foods, as long as a product remains well below its freezing point, microbial safety issues are nonexistent. Frozen foods must be stored below 0°F for legal and quality reasons.

Destruction of Organisms

High temperature is an effective means of inactivating microorganisms and is used extensively in blanching, pasteurization, and canning. Moist heat is far more effective than dry heat. High temperatures (170°F) may also be used for sanitation when chemicals are not used. Although hot-water sanitation is effective against vegetative forms of bacteria, spores are not affected.

In addition to heat, high pressure, pulsed electric fields, high-energy white light, irradiation, ultraviolet light, hydrogen peroxide, ozone, and sanitation chemicals are effective in destroying microorganisms.

THE ROLE OF HACCP

Many of the procedures for control of microorganisms are managed by the Hazard Analysis and Critical Control Point (HACCP) system of food safety. Developed in the food industry since the 1960s, HACCP is a preventive system that builds safety control features into the food product's design and production. The HACCP system is used to manage physical, chemical, and biological hazards. The approach is described in the seven principles developed by the National Advisory Committee on Microbiological Criteria for Foods (NACMCF 1998):

1. Conduct a hazard analysis and identify control measures.
2. Identify critical control points.
3. Establish critical limits.
4. Establish monitoring procedures.
5. Establish corrective actions.
6. Establish verification procedures.
7. Establish record-keeping and documentation procedures.

Each food manufacturing site should have an HACCP team to develop and implement its HACCP plan. The team is multidisciplinary, with members experienced in plant operations, product development, food microbiology, etc. Because of their knowledge of the manufacturing facility and the process equipment, engineers are important members of the HACCP team. They help identify potential hazards and respective control measures, implement the HACCP plan, and verify its effectiveness.

SANITATION

Cleaning and sanitation are key elements that incorporate all three strategies for control of microorganisms. Cleaning controls microbial growth by removing the residual food material required for proliferation. Sanitizing kills most of the bacteria that remain on surfaces, preventing subsequent contamination of food being produced. Most microbial issues that occur in food processing environments are caused by unclean equipment, sometimes because of design. Therefore, equipment and facilities designed with cleaning and sanitizing in mind maximize the effectiveness of control.

Products that are frozen before packaging are particularly vulnerable to contamination. Many freezing tunnels in food processing facilities are difficult or impossible to clean because of limited access and poor drainage. Although freezing temperatures control microbial growth, proliferation of organisms does occur during downtime, such as on weekends. The following points should be considered during design to minimize potential problems:

- Provide good access for the cleaning crew to facilitate cleaning and adequate lighting (50 fc) to allow inspection of all surfaces.
- Eliminate inaccessible parts and features that allow product accumulation.
- Design equipment that is easy to dismantle using few tools, especially for areas that are difficult to clean. Design air-handling ducts for easy cleaning. Provide removable spools or access doors. Washable fabric ducts designed and approved for such use are another option.
- Use smooth and nonporous construction materials to prevent product accumulation. Materials must tolerate common cleaning and sanitizing chemicals listed in Table 3. Consult sanitation personnel to identify chemicals likely to be used. Give special attention to insulation materials, many of which are porous. Insulation must be protected from water to avoid saturation and resultant microbial growth. An effective method is a well-sealed PVC or stainless steel cover. Avoid using fiberglass batts in food processing plants.
- All equipment must drain completely.
- Consult references and regulations on sanitary design principles.

Innovation is needed in the area of drying after cleaning is complete. Providing adequately sloped surfaces and sufficient drains to handle water is important. Dehumidification systems and/or increased airflow in new and existing systems could greatly reduce problems associated with water, especially in cleaning chilled production environments.

Standard water washing procedures are not appropriate for some food production facilities, such as dry mix, chocolate, peanut butter, or flour milling operations. Refrigeration or ventilation systems for these plants must be made to facilitate dry cleaning, reduce condensation, and restrict water to a very confined area if it is absolutely necessary.

Table 3 Common Cleaning and Sanitizing Chemicals

Cleaning Compounds	Sanitizers
Caustic	Chlorine
Chlorinated alkaline detergents	Iodophors
Acid cleaners	Quaternary ammonium compounds
	Acid sanitizers

REGULATIONS AND STANDARDS

Facilities and equipment should be designed and installed to minimize microbial growth and maximize ease of sanitation. Take care to use material that can withstand moisture and chemicals.

The food industry has many equipment materials, fabrication, and installation standards. In the United States, examples include those by the International Association for Food Protection (IAFP), which publishes the *3-A Dairy Standards*; the Baking Industry Sanitation Standards Committee (BISSC); and the U.S. Department of Agriculture's (USDA) Food Safety and Inspection Service (FSIS), which ensures that the meat, poultry, and egg product food supply is safe, wholesome, and properly labeled.

Chapter VII, Section 701(A) of the Federal Food, Drug and Cosmetic Act as amended establishes current Good Manufacturing Practices (GMPs) in manufacturing, processing, packaging, or holding human food. These GMPs are listed in 21CFR110.

BIBLIOGRAPHY

ASHRAE. 1992. Gravimetric and dust-spot procedures for testing air-cleaning devices used in general ventilation for removing particulate matter. *Standard* 52.1-1992.

BACS. 1989. *The control of* Legionellae *by safe and effective operation of cooling systems. A code of practice*. British Association of Chemical Specialties, Lancaster.

Bibek, R. 1996. *Fundamental food microbiology*. CRC Press, Boca Raton, FL.

CFR. 2005. Current good manufacturing practice in manufacturing, packing, or holding human food. 21CFR110. *Code of Federal Regulations*, Government Printing Office, Washington, D.C.

CIBSE. 1987. Minimising the risk of Legionnaires' disease. *Technical Memorandum* TM13. Chartered Institute of Building Services Engineers, London.

FDA. 2005. *Food Code*. U.S. Food and Drug Administration, Washington, D.C.

Graham, D.J. 1991-1992. Sanitary design—A mind set. A nine-part serial in *Dairy, Food and Environmental Sanitation*.

IAMFES. *3-A Sanitary Standards*. International Association of Milk, Food, and Environmental Sanitarians, Ames, IA.

IIR. 1986. *Recommendations for the processing and handling of frozen foods*, 3rd ed. International Institute of Refrigeration, Paris.

IIR. 2000. *Recommendations for chilled storage of perishable produce*, 4th ed. International Institute of Refrigeration, Paris.

Imholte, T.J. as revised by T.K. Imholte-Tauscher. 1999. *Engineering for food safety and sanitation*. Technical Institute for Food Safety, Woodinville, WA.

Jay, J.M. 2000. *Modern food microbiology*, 6th ed. Springer-Verlag, New York.

Marriott, N.G. 1989. *Principles of food sanitation*, 2nd ed. Plenum, New York.

NACMCF. 1998. Hazard analysis and critical control point system. *Journal of Food Protection* 61:1246-1259.

Todd, E. 1990. Epidemiology of food-borne illness: North America. *The Lancet* 336:788-793.

UL. 2000. Meat and poultry plant equipment. ANSI/UL *Standard* 2128-2000. Underwriters Laboratories, Northbrook, IL.

CHAPTER 13

REFRIGERATION LOAD

TOTAL refrigeration load includes (1) transmission load, which is heat transferred into the refrigerated space through its surface; (2) product load, which is heat removed from and produced by products brought into and kept in the refrigerated space; (3) internal load, which is heat produced by internal sources (e.g., lights, electric motors, and people working in the space); (4) infiltration air load, which is heat gain associated with air entering the refrigerated space; and (5) equipment-related load.

The first four segments of load constitute the net heat load for which a refrigeration system is to be provided; the fifth segment consists of all heat gains created by the refrigerating equipment. Thus, net heat load plus equipment heat load is the total refrigeration load for which a compressor must be selected.

This chapter contains load calculating procedures and data for the first four segments and load determination recommendations for the fifth segment. Information needed for refrigeration of specific foods can be found in Chapters 15 and 17 to 29.

TRANSMISSION LOAD

Sensible heat gain through walls, floor, and ceiling is calculated at steady state as

$$q = UA\,\Delta t \qquad (1)$$

where

q = heat gain, Btu/h
A = outside area of section, ft²
Δt = difference between outside air temperature and air temperature of the refrigerated space, °F

The overall coefficient of heat transfer U of the wall, floor, or ceiling can be calculated by the following equation:

$$U = \frac{1}{1/h_i + x/k + 1/h_o} \qquad (2)$$

where

U = overall heat transfer coefficient, Btu/h·ft²·°F
x = wall thickness, in.
k = thermal conductivity of wall material, Btu·in/h·ft²·°F
h_i = inside surface conductance, Btu/h·ft²·°F
h_o = outside surface conductance, Btu/h·ft²·°F

A value of 1.6 Btu/h·ft²·°F for h_i and h_o is frequently used for still air. If the outer surface is exposed to 15 mph wind, h_o is increased to 6 Btu/h·ft²·°F.

With thick walls and low conductivity, the resistance x/k makes U so small that $1/h_i$ and $1/h_o$ have little effect and can be omitted from the calculation. Walls are usually made of more than one material; therefore, the value x/k represents the composite resistance of the materials. The U-factor for a wall with flat parallel surfaces of materials 1, 2, and 3 is given by the following equation:

$$U = \frac{1}{x_1/k_1 + x_2/k_2 + x_3/k_3} \qquad (3)$$

The thermal conductivity of several cold storage insulations are listed in Table 1. These values increase with age due to factors discussed in Chapter 23 of the 2005 *ASHRAE Handbook—Fundamentals*. Chapter 25 of the 2005 *ASHRAE Handbook—Fundamentals* includes more complete tables listing the thermal properties of various building and insulation materials.

Table 2 lists minimum insulation thicknesses of expanded polyisocyanurate board recommended by the refrigeration industry. These thicknesses may need to be increased to offset heat gain caused by building components such as wood and metal studs, webs in concrete masonry, and metal ties that bridge across the insulation and reduce the thermal resistance of the wall or roof. Chapter 25 of the 2005 *ASHRAE Handbook—Fundamentals* describes how to calculate heat gain through walls and roofs with thermal bridges. Metal surfaces of prefabricated or insulated panels have a negligible effect on thermal performance and need not be considered in calculating the U-factor.

In most cases, the temperature difference Δt can be adjusted to compensate for solar effect on heat load. Values in Table 3 apply over a 24 h period and are added to the ambient temperature when calculating wall heat gain.

Table 1 Thermal Conductivity of Cold Storage Insulation

Insulation	Thermal Conductivity[a] k, Btu·in/h·ft²·°F
Polyurethane board (R-11 expanded)	0.16 to 0.18
Polyisocyanurate, cellular (R-141b expanded)	0.19
Polystyrene, extruded (R-142b)	0.24
Polystyrene, expanded (R-142b)	0.26
Corkboard[b]	0.30
Foam glass[c]	0.31

[a]Values are for a mean temperature of 75°F, and insulation is aged 180 days.
[b]Seldom-used insulation. Data are only for reference.
[c]Virtually no effects from aging.

Table 2 Minimum Insulation Thickness

Storage Temperature, °F	Expanded Polyisocyanurate Thickness	
	Northern U.S., in.	Southern U.S., in.
50 to 60	2	2
40 to 50	2	2
25 to 40	2	3
15 to 25	3	3
0 to 15	3	4
–15 to 0	4	4
–40 to –15	5	5

The preparation of this chapter is assigned to TC 10.8, Refrigeration Load Calculations.

Table 3 Allowance for Sun Effect

Typical Surface Types	East Wall, °F	South Wall, °F	West Wall, °F	Flat Roof, °F
Dark-colored surfaces				
Slate roofing	8	5	8	20
Tar roofing				
Black paint				
Medium-colored surfaces				
Unpainted wood	6	4	6	15
Brick				
Red tile				
Dark cement				
Red, gray, or green paint				
Light-colored surfaces				
White stone	4	2	4	9
Light colored cement				
White paint				

Note: Add to the normal temperature difference for heat leakage calculations to compensate for sun effect. Do not use for air-conditioning design.

In most cases, the temperature difference Δt can be adjusted to compensate for solar effect on heat load. Values given in Table 3 apply over a 24 h period and are added to the ambient temperature when calculating wall heat gain.

Latent heat gain due to moisture transmission through walls, floors, and ceilings of modern refrigerated facilities is negligible. Data in Chapter 25 of the 2005 *ASHRAE Handbook—Fundamentals* may be used to calculate this load if moisture permeable materials are used.

Chapter 28 of the 2005 *ASHRAE Handbook—Fundamentals* gives outdoor design temperatures for major cities; values for 0.4% should be used.

Additional information on thermal insulation may be found in Chapters 23 and 24 of the 2005 *ASHRAE Handbook—Fundamentals.* Chapter 30 of the 2005 *ASHRAE Handbook—Fundamentals* discusses load calculation procedures in greater detail.

Heat Gain from Cooler Floors

Heat gain through cooler concrete slab floors is predicted using procedures developed by Chuangchid and Krarti (2000), who developed a simplified correlation of the total slab heat gain for coolers based on analytical results reported earlier by Chuangchid and Krarti (1999). Parameters in the solution include slab size and thermal resistance, insulation thermal resistance, soil thermal conductivity, water table depth and temperature, and indoor and outdoor air temperatures. The design procedure accommodates four slab insulation configurations: no insulation, uniform horizontal insulation, partial-horizontal-perimeter insulation, and partial-vertical-perimeter insulation. The slab size characteristic is expressed as the ratio of slab area A to exposed slab perimeter P. The result is an estimate of the annual mean and amplitude cooler floor heat gain that, when combined, gives the instantaneous floor heat at a specific time of year. The time-variation of the ground-coupled heat gain $q(t)$ for cooler slabs is

$$q(t) = q_m - q_a \cos[\omega(t - \phi)] \quad (4)$$

where

q_m = annual mean slab floor heat gain, Btu/h
q_a = amplitude of annual variation slab floor heat gain, Btu/h
ϕ = phase lag between cooler floor heat gain and outdoor air temperature variation, days
t = time, days
ω = constant for annual angular frequency, 0.0172 radians/day

These quantities are functions of input parameters such as building dimensions, soil properties, and insulation thermal resistance. Note that the time of origin ($t = 0$) in Equation (4) corresponds to the time when the soil surface temperature is minimum (typically January 15 in most U.S. locations). Details of the calculation method are given by Chuangchid and Krarti (2000).

Table 4 Example Input Data Required to Estimate Cooler Floor Heat Gain

Required Information	Example Values
Soil thermal conductivity k_s	10.47 Btu·in/h·ft²·°F
Soil thermal diffusivity α_s	7.66×10^{-6} ft²/s
Total floor area A	32 ft × 50 ft = 1600 ft²
Exposed perimeter P	164 ft
Slab thickness y	4 in.
Slab thermal resistance R_f	3.33 ft²·h·°F/Btu
Insulation thermal resistance R_i (partial uniform, partial, vertical)	20 ft²·h·°F/Btu
Partial insulation length	3 ft
Cooler inside temperature T_r	35°F
Annual mean T_m	40°F
Annual amplitude T_a	60°F
Water table depth b	Ignored (if depth is greater than 7 ft, water table effect can be ignored)

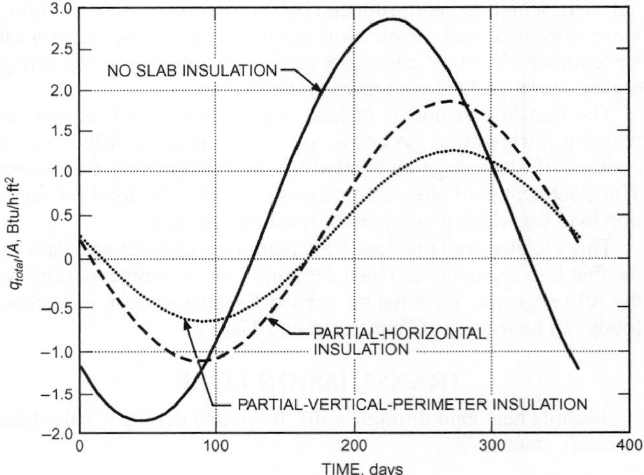

Fig. 1 Variation of Cooler Floor Heat Gain over One Year for Conditions in Table 4

For the conditions in Table 4, results for cooler slab floor heat gain are computed based on the method. As shown in Figure 1, heat gain through the cooler floor significantly varies and may even be negative at certain times of the year. The influence of partial-horizontal and partial-vertical-perimeter insulation on cooler floor heat gain for Table 4 conditions can also be seen in Figure 1.

Figure 2 shows the results for maximum heat transfer per unit area q_{max}/A by calculation where typical cold and warm climate temperatures are applied. These outdoor temperatures are given in Table 5. q_{max}/A is the sum of the mean and amplitude of heat gain per unit area and is, therefore, the maximum heat transfer rate that occurs at some time during the year. The influence of floor slab width and length is captured by the A/P parameter. Widths of both 10 and 30 m were used to generate Figure 2. The fact that the q_{max}/A curves in Figure 2 coalesce for different slab areas while still having the same A/P ratio and the same indoor, annual mean outdoor, and annual amplitude outdoor temperatures suggests that slab area does not affect the cooler floor slab heat gain rate per unit area with floor areas greater than about 100 m².

Table 5 Typical Annual and Annual Amplitude Outdoor Temperatures for Warm and Cold Climates

Temperature	Typical Cold Climate, °F	Typical Warm Climate, °F
Annual mean T_m	40	70
Annual amplitude T_a	60	30

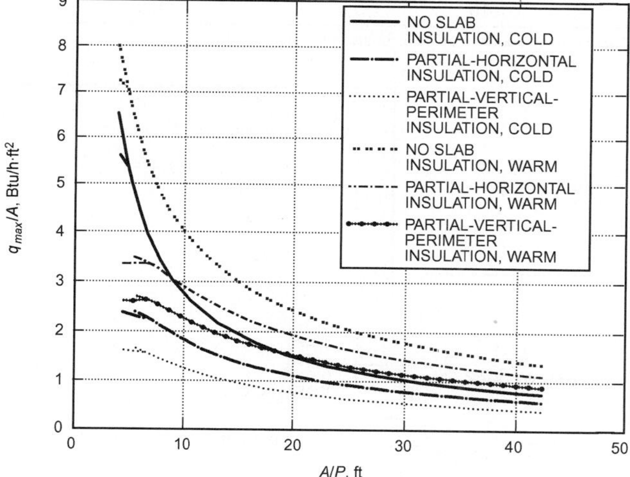

Fig. 2 Variation of q_{max}/A with A/P Using Conditions from Tables 4 and 5

PRODUCT LOAD

The primary refrigeration loads from products brought into and kept in the refrigerated space are (1) heat that must be removed to bring products to storage temperature and (2) heat generated by products (mainly fruits and vegetables) in storage. The quantity of heat to be removed can be calculated as follows:

1. Heat removed to cool from initial temperature to some lower temperature above freezing:

$$Q_1 = mc_1(t_1 - t_2) \tag{5}$$

2. Heat removed to cool from initial temperature to freezing point of product:

$$Q_2 = mc_1(t_1 - t_f) \tag{6}$$

3. Heat removed to freeze product:

$$Q_3 = mh_{if} \tag{7}$$

4. Heat removed to cool from freezing point to final temperature below freezing point:

$$Q_4 = mc_2(t_f - t_3) \tag{8}$$

where

Q_1, Q_2, Q_3, Q_4 = heat removed, Btu
m = weight of product, lb
c_1 = specific heat of product above freezing, Btu/lb·°F
t_1 = initial temperature of product above freezing, °F
t_2 = lower temperature of product above freezing, °F
t_f = freezing temperature of product, °F
h_{if} = latent heat of fusion of product, Btu/lb
c_2 = specific heat of product below freezing, Btu/lb·°F
t_3 = final temperature of product below freezing, °F

The refrigeration capacity required for products brought into storage is determined from the time allotted for heat removal and assumes that product is properly exposed to remove the heat in that time. The calculation is

$$q = \frac{Q_2 + Q_3 + Q_4}{n} \tag{9}$$

where

q = average cooling load, Btu/h
n = allotted time, h

Equation (9) only applies to uniform entry of product into storage. The refrigeration load created by nonuniform loading of a warm product may be much greater over a short period. See Chapter 15 for information on calculating the cooling load of warm product.

Specific heats above and below freezing for many products are given in Table 3 of Chapter 9. A product's latent heat of fusion may be estimated by multiplying the water content of the product (expressed as a decimal) by the latent heat of fusion of water, which is 144 Btu/lb. Most foods freeze between 26 and 31°F. When the exact freezing temperature is not known, assume that it is 28°F.

Example 1. 200 lb of lean beef is to be cooled from 65 to 40°F, then frozen and cooled to 0°F. The moisture content is 69.5%, so the latent heat is estimated as 144 × 0.695 = 100 Btu/lb. Estimate the cooling load.

Solution:

Specific heat of beef before freezing is listed in Table 3, Chapter 9, as 0.84 Btu/lb·°F; after freezing, 0.51 Btu/lb·°F.

To cool from 65 to 40°F in a chilled room:

$$200 \times 0.84(65 - 40) = 4200 \text{ Btu}$$

To cool from 40°F to freezing point in freezer:

$$200 \times 0.84(40 - 28) = 2016 \text{ Btu}$$

To freeze: $200 \times 100 = 20{,}000$ Btu

To cool from freezing to storage temperature:

$$200 \times 0.51(28 - 0) = 2856 \text{ Btu}$$

Total: $4200 + 2016 + 20{,}000 + 2856 = 29{,}072$ Btu

(Example 3 in Chapter 9 shows an alternative calculation method.)

Fresh fruits and vegetables respire and release heat during storage. This respiration heat varies by product and its temperature; the colder the product, the less heat of respiration. Table 9 in Chapter 9 gives heat of respiration rates for various products.

Calculations in Example 1 do not cover heat gained from product containers brought into the refrigerated space. When pallets, boxes, or other packing materials are a significant portion of the total mass introduced, this heat load should be calculated.

Equations (5) to (9) are used to calculate the total heat gain. Any moisture removed appears as latent heat gain. The amount of moisture involved is usually provided by the end-user as a percentage of product mass; so, with such information, the latent heat component of the total heat gain may be determined. Subtracting the latent heat component from the total heat gain determines the sensible heat component.

INTERNAL LOAD

Electrical Equipment. All electrical energy dissipated in the refrigerated space (from lights, motors, heaters, and other equipment) must be included in the internal heat load. Heat equivalents of electric motors are listed in Table 6.

Forklifts. Forklifts in some facilities can be a large and variable contributor to the load. Although many forklifts may be in a space at one time, they do not all operate at the same energy level. For example, the energy used by a forklift while it is elevating or lowering forks is different than when it is moving.

Processing Equipment. Grinding, mixing, or cooking equipment may be in refrigerated areas of food processing plants. Other

Table 6 Heat Gain from Typical Electric Motors

Motor Name-plate or Rated Horse-power	Motor Type	Nominal rpm	Full Load Motor Effi-ciency, %	A: Motor in, Driven Equip-ment in, Btu/h	B: Motor out, Driven Equip-ment in, Btu/h	C: Motor in, Driven Equip-ment out, Btu/h
				Location of Motor and Driven Equipment with Respect to Conditioned Space or Airstream		
0.05	Shaded pole	1500	35	360	130	240
0.08				580	200	380
0.125				900	320	590
0.16				1160	400	760
0.25	Split phase	1750	54	1180	640	540
0.33			56	1500	840	660
0.50			60	2120	1270	850
0.75	3-Phase	1750	72	2650	1900	740
1			75	3390	2550	850
1.5			77	4960	3820	1140
2			79	6440	5090	1350
3			81	9430	7640	1790
5			82	15,500	12,700	2790
7.5			84	22,700	19,100	3640
10			85	29,900	24,500	4490
15			86	44,400	38,200	6210
20			87	58,500	50,900	7610
25			88	72,300	63,600	8680
30			89	85,700	76,300	9440
40				114,000	102,000	12,600
50				143,000	127,000	15,700
60				172,000	153,000	18,900
75			90	212,000	191,000	21,200
100				283,000	255,000	28,300
125				353,000	318,000	35,300
150			91	420,000	382,000	37,800
200				569,000	509,000	50,300
250				699,000	636,000	62,900

heat sources include equipment for packaging, glue melting, or shrink wrapping. Another possible load is the makeup air for equipment that exhausts air from a refrigerated space.

People. People add to the heat load, in amounts depending on factors such as room temperature, type of work being done, type of clothing worn, and size of the person. Heat load from a person q_p may be estimated as

$$q_p = 1295 - 11.5t \qquad (10)$$

where t is the temperature of the refrigerated space in °F. Table 7 shows the average load from people in a refrigerated space as calculated from Equation (10).

When people first enter a storage they bring in additional surface heat. Thus, when many people enter and leave every few minutes, the load is greater than that listed in Table 7 and must be adjusted. A conservative adjustment is to multiply the values calculated in Equation (10) by 1.25.

Latent Load. The latent heat component of the internal load is usually very small compared to the total refrigeration load and is customarily regarded as all sensible heat in the total load summary. However, the latent heat component should be calculated where water is involved in processing or cleaning.

INFILTRATION AIR LOAD

Heat gain from infiltration air and associated equipment loads can amount to more than half the total refrigeration load of distribution warehouses and similar applications.

Table 7 Heat Equivalent of Occupancy

Refrigerated Space Temperature, °F	Heat Equivalent/Person, Btu/h
50	720
40	840
30	950
20	1050
10	1200
0	1300
–10	1400

Note: Heat equivalent may be estimated by Equation (10).

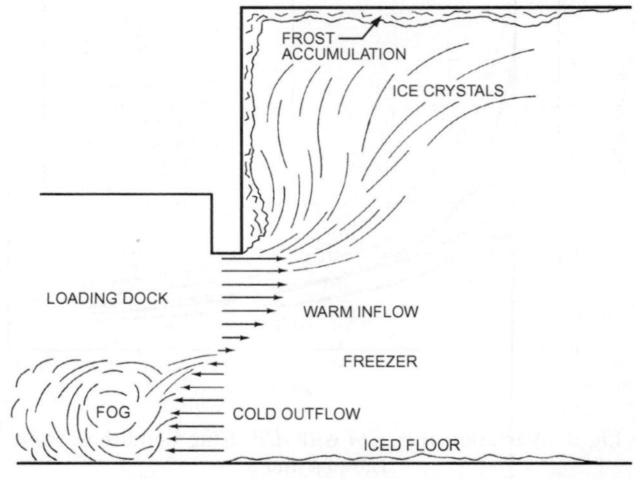

Fig. 3 Flowing Cold and Warm Air Masses for Typical Open Freezer Doors

Infiltration by Air Exchange

Infiltration most commonly occurs because of air density differences between rooms (Figures 3 and 4). For a typical case where the air mass flowing in equals the air mass flowing out minus any condensed moisture, the room must be sealed except at the opening in question. If the cold room is not sealed, air may flow directly through the door (discussed in the following section).

Heat gain through doorways from air exchange is as follows:

$$q_t = qD_tD_f(1 - E) \qquad (11)$$

where

q_t = average heat gain for the 24 h or other period, Btu/h
q = sensible and latent refrigeration load for fully established flow, Btu/h
D_t = doorway open-time factor
D_f = doorway flow factor
E = effectiveness of doorway protective device

Gosney and Olama (1975) developed the following air exchange equation for fully established flow:

$$q = 795.6A(h_i - h_r)\rho_r(1 - \rho_i/\rho_r)^{0.5}(gH)^{0.5}F_m \qquad (12)$$

where

q = sensible and latent refrigeration load, Btu/h
A = doorway area, ft²
h_i = enthalpy of infiltration air, Btu/lb
h_r = enthalpy of refrigerated air, Btu/lb
ρ_i = density of infiltration air, lb/ft³
ρ_r = density of refrigerated air, lb/ft³
g = gravitational constant = 32.174 ft/s²
H = doorway height, ft
F_m = density factor

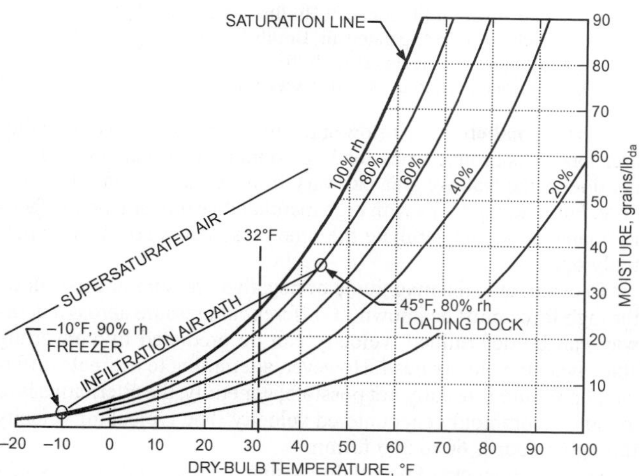

Fig. 4 Psychrometric Depiction of Air Exchange for Typical Freezer Doorway

Table 8 Sensible Heat Ratio R_s for Infiltration from Outdoors to Refrigerated Spaces

Outdoors Temp. °F	rh, %	Cold Space at 90% rh Dry-Bulb Temperature, °F −30	−20	−10	0	10	20	30	40	50	60
100	50	0.59	0.57	0.55	0.53	0.51	0.49	0.48	0.46	0.45	0.45
	40	0.65	0.63	0.61	0.59	0.57	0.56	0.54	0.53	0.54	0.57
	30	0.71	0.69	0.68	0.66	0.65	0.64	0.63	0.64	0.66	0.76
	20	0.79	0.77	0.76	0.75	0.74	0.74	0.75	0.78	0.87	—
95	60	0.58	0.56	0.54	0.52	0.49	0.47	0.45	0.43	0.42	0.41
	50	0.62	0.60	0.58	0.56	0.54	0.52	0.51	0.49	0.48	0.50
	40	0.67	0.66	0.64	0.62	0.60	0.59	0.57	0.57	0.58	0.64
	30	0.73	0.72	0.70	0.69	0.67	0.66	0.66	0.68	0.72	0.89
90	60	0.61	0.59	0.57	0.55	0.52	0.50	0.48	0.46	0.45	0.45
	50	0.65	0.63	0.61	0.59	0.57	0.55	0.54	0.52	0.52	0.56
	40	0.70	0.68	0.67	0.65	0.63	0.61	0.61	0.61	0.63	0.74
	30	0.76	0.74	0.73	0.71	0.70	0.69	0.70	0.72	0.80	—

$$F_m = \left[\frac{2}{1 + (\rho_r/\rho_i)^{1/3}}\right]^{1.5} \tag{13}$$

(Chapter 6 of the 2005 *ASHRAE Handbook—Fundamentals* and the appropriate ASHRAE psychrometric chart list air enthalpy and density values.)

Equation (14), when used with Figure 5, is a simplification of Equation (12):

$$q = 3790WH^{1.5}\left(\frac{Q_s}{A}\right)\left(\frac{1}{R_s}\right) \tag{14}$$

where

- q = sensible and latent refrigeration load, Btu/h
- Q_s/A = sensible heat load of infiltration air per square foot of doorway opening as read from Figure 5, ton/ft^2
- W = doorway width, ft
- R_s = sensible heat ratio of the infiltration air heat gain, from Table 7 or 8 (or from a psychrometric chart)

The values of R_s in Tables 8 and 9 are based on 90% rh in the cold room. A small error occurs where these values are used for 80 or 100% rh. This error, with loss of accuracy due to simplification, results in a maximum error for Equation (14) of approximately 4%.

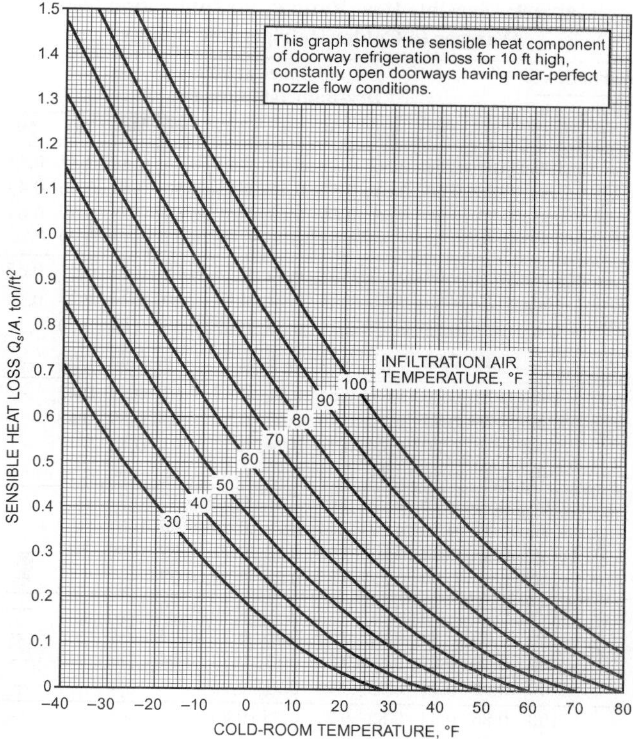

Fig. 5 Sensible Heat Gain by Air Exchange for Continuously Open Door with Fully Established Flow

For cyclical, irregular, and constant door usage, alone or in combination, the doorway open-time factor can be calculated as

$$D_t = \frac{(P\theta_p + 60\theta_o)}{3600\theta_d} \tag{15}$$

where

- D_t = decimal portion of time doorway is open
- P = number of doorway passages
- θ_p = door open-close time, seconds per passage
- θ_o = time door simply stands open, min
- θ_d = daily (or other) time period, h

The typical time θ_p for conventional pull-cord-operated doors ranges from 15 to 25 s per passage. The time for high-speed doors ranges from 5 to 10 s, although it can be as low as 3 s. The time for θ_o and θ_d should be provided by the user. Hendrix et al. (1989) found that steady-state flow becomes established 3 s after the cold-room door is opened. This fact may be used as a basis to reduce θ_p in Equation (15), particularly for high-speed doors, which may significantly reduce infiltration.

The doorway flow factor D_f is the ratio of actual air exchange to fully established flow. Fully established flow occurs only in the unusual case of an unused doorway standing open to a large room or to the outdoors, and where cold outflow is not impeded by obstructions (e.g., stacked pallets in or adjacent to the flow path in or outside the cold room). Under these conditions, D_f is 1.0.

Hendrix et al. (1989) found that a flow factor D_f of 0.8 is conservative for a 28°F temperature difference when traffic flow equals one entry and exit per minute through fast-operating doors. Downing and Meffert (1993) found a flow factor of 1.1 at temperature differences of 12 and 18°F. Based on these results, the recommended flow factor for cyclically operated doors with temperature differentials less than 20°F is 1.1, and the recommended flow factor for higher differentials is 0.8.

Table 9 Sensible Heat Ratio R_s for Infiltration from Warmer to Colder Refrigerated Spaces

Warm Space Temp. °F	Warm Space rh, %	Cold Space at 90% rh Dry-Bulb Temperature, °F −40	−30	−20	−10	0	10	20	30	40	50
70	100	0.60	0.58	0.56	0.53	0.50	0.47	0.44	0.41	0.37	0.34
	80	0.66	0.64	0.61	0.59	0.56	0.53	0.50	0.48	0.46	0.44
	60	0.72	0.70	0.68	0.66	0.63	0.61	0.59	0.58	0.59	0.64
	40	0.79	0.78	0.76	0.75	0.73	0.72	0.71	0.73	0.80	—
60	100	0.66	0.64	0.62	0.59	0.56	0.52	0.49	0.45	0.41	0.35
	80	0.71	0.69	0.67	0.64	0.62	0.59	0.56	0.53	0.52	0.53
	60	0.77	0.75	0.73	0.71	0.69	0.67	0.65	0.65	0.70	—
	40	0.83	0.82	0.81	0.79	0.78	0.77	0.78	0.83	—	—
50	100	0.72	0.70	0.67	0.64	0.61	0.57	0.53	0.49	0.43	—
	80	0.76	0.74	0.72	0.70	0.67	0.64	0.61	0.59	0.62	—
	60	0.81	0.80	0.78	0.76	0.74	0.72	0.71	0.75	—	—
	40	0.87	0.86	0.84	0.83	0.82	0.82	0.85	—	—	—
40	100	0.77	0.75	0.72	0.69	0.66	0.62	0.57	0.51	—	—
	80	0.81	0.79	0.77	0.74	0.72	0.69	0.66	0.67	—	—
	60	0.85	0.84	0.82	0.80	0.78	0.77	0.79	0.99	—	—
	40	0.90	0.89	0.88	0.87	0.86	0.88	0.97	—	—	—
30	100	0.82	0.80	0.77	0.74	0.70	0.66	0.59	—	—	—
	80	0.85	0.83	0.81	0.79	0.76	0.73	0.73	—	—	—
	60	0.88	0.87	0.86	0.84	0.83	0.83	0.94	—	—	—
	40	0.92	0.91	0.90	0.90	0.91	0.96	—	—	—	—
20	100	0.86	0.84	0.82	0.79	0.75	0.69	—	—	—	—
	80	0.89	0.87	0.85	0.83	0.81	0.80	—	—	—	—
	60	0.91	0.90	0.89	0.88	0.88	0.95	—	—	—	—
	40	0.94	0.94	0.93	0.94	0.97	—	—	—	—	—
10	100	0.90	0.88	0.86	0.83	0.78	—	—	—	—	—
	80	0.92	0.90	0.89	0.87	0.86	—	—	—	—	—
	60	0.94	0.93	0.92	0.92	0.96	—	—	—	—	—
	40	0.96	0.96	0.96	0.98	—	—	—	—	—	—
0	100	0.92	0.91	0.89	0.85	—	—	—	—	—	—
	80	0.94	0.93	0.92	0.91	—	—	—	—	—	—
	60	0.96	0.95	0.95	0.97	—	—	—	—	—	—
	40	0.97	0.97	0.98	—	—	—	—	—	—	—

The effectiveness E of open-doorway protective devices is 0.95 or higher for newly installed strip, fast-fold, and other non-tight-closing doors. However, depending on the traffic level and door maintenance, E may quickly drop to 0.8 on freezer doorways and to about 0.85 for other doorways. Airlock vestibules with strip or push-through doors have an effectiveness ranging between 0.95 and 0.85 for freezers and between 0.95 and 0.90 for other doorways. The effectiveness of air curtains ranges from very poor to more than 0.7. For a wide-open door with no devices, $E = 0$ in Equation (11).

Infiltration by Direct Flow Through Doorways

Negative pressure elsewhere in the building, created by mechanical air exhaust without mechanical air replenishment, is a common cause of heat gain from infiltration of warm air. In refrigerated spaces with constantly or frequently open doorways or other through-the-room passageways, this air flows directly through the doorway. The effect is identical to that of open doorways exposed to wind, and heat gain may be very large. Equation (16) for heat gain from infiltration by direct inflow provides the basis for either correcting negative pressure or adding to refrigeration capacity:

$$q_t = 60VA(h_i - h_r)\rho_r D_t \qquad (16)$$

where

q_t = average refrigeration load, Btu/h
V = average air velocity, ft/min
A = opening area, ft^2
h_i = enthalpy of infiltration air, Btu/lb
h_r = enthalpy of refrigerated air, Btu/lb
ρ_r = density of refrigerated air, lb/ft^3
D_t = decimal portion of time doorway is open

A is the smaller of the inflow and outflow openings. If the smaller area has leaks around truck loading doors in well-maintained loading docks, the leakage area can vary from 0.3 ft^2 to over 1.0 ft^2 per door. For loading docks with high merchandise movement, the facility manager should estimate the time these doors are fully or partially open.

To evaluate V, the magnitude of negative pressure or other flow-through force must be known. If differential pressure across a doorway can be determined, velocity can be predicted by converting static head to velocity head. However, attempting to estimate differential pressure is usually not possible; generally, the alternative is to assume a commonly encountered velocity. The typical air velocity through a door is 60 to 300 ft/min.

The effectiveness of non-tight-closing devices on doorways subject to infiltration by direct airflow cannot be readily determined. Depending on the pressure differential, its tendency to vary, and the ratio of inflow area to outflow area, the effectiveness of these devices can be very low.

Sensible and Latent Heat Components

When calculating q_t for infiltration air, the sensible and latent heat components may be obtained by plotting the infiltration air path on the appropriate ASHRAE psychrometric chart, determining the air sensible heat ratio R_s from the chart, and calculating as follows:

$$\text{Sensible heat: } q_s = q_t R_s \qquad (17)$$

$$\text{Latent heat: } q_l = q_t(1 - R_s) \qquad (18)$$

where $R_s = \Delta h_s / \Delta h_t$.

EQUIPMENT RELATED LOAD

Heat gains associated with refrigeration equipment operation consist essentially of the following:

- Fan motor heat where forced-air circulation is used
- Reheat where humidity control is part of the cooling
- Heat from defrosting where the refrigeration coil operates at a temperature below freezing and must be defrosted periodically, regardless of the room temperature

Fan motor heat must be computed based on the actual electrical energy consumed during operation. Propeller fan motors are mounted in the airstream on many cooling units because the cold air extends the power range of the motor. For example, a standard motor in a −10°F freezer operates satisfactorily at a 25% overload to the rated (nameplate) power. Heat gain from fan motors should be based on the actual run time. Generally, fans on cooling units operate continuously except during the defrost period. However, fans may be cycled on and off to control temperature and save energy.

Cole (1989) characterized and quantified the heat load associated with defrosting using hot gas. Other common defrost methods use electricity or water. Generally, heat gain from a cooling unit with electric defrost is greater than the same unit with hot-gas defrost; heat gain from a unit with water defrost is even less. Moisture that evaporates into the space during defrost must also be added to the refrigeration load.

Some heat from defrosting is added only to the refrigerant, and the rest is added to the space. To accurately select refrigeration equipment, a distinction should be made between equipment heat loads that are in the refrigerated space and those that are introduced directly to the refrigerating fluid.

Equipment heat gain is usually small at space temperatures above approximately 30°F. Where reheat or other artificial loads are not imposed, total equipment heat gain is about 5% or less of the total load. However, equipment heat gain becomes a major portion of the total load at freezer temperatures. For example, at –20°F the theoretical contribution to total refrigeration load from fan power and coil defrosting alone can exceed, for many cases, 15% of the total load (assuming proper control of defrosting so that the space is not heated excessively).

ASHRAE research project RP-1094 (Sherif et al. 2002) found that, where excessive moisture persists during normal freezer operation and/or freezer air becomes supersaturated for an extended period of time, significantly more heat gain from coil defrosting should be included in calculating refrigeration load. Excessive moisture and supersaturated air result if care is not taken to prevent air infiltration into the freezer and/or if a high coil temperature difference (TD) is imposed. Coil TD is the difference between the temperatures of air entering the coil and refrigerant inside the coil. Excessive coil TDs typically exist if entering air temperature rises, refrigerant temperature drops, or both. Coil refrigerant temperatures typically drop as an intuitive response by freezer operators to the rising air temperature inside the freezer. Air temperatures typically rise after excessive coil frost build-up, resulting in coil heat transfer performance degradation. Excessive frost build-up typically occurs if supersaturated air is allowed inside the freezer. Sherif et al. (2002) suggested guidelines to prevent this chain reaction of events by using the saturation curve of the psychrometric chart as a guide to select the proper combination of coil entering air temperature, refrigerant temperature, and sensible heat ratio for a given freezer temperature. The study imposed upper limits on the allowable coil TD for prescribed values of these variables. In cases where excessive moisture and/or supersaturated air are allowed for an extended period of time during the refrigeration portion of the refrigeration/defrost (R/D) cycle, the study recommends that the fan/defrost contribution be raised from 15% to as high as 30%, depending on the degree of supersaturation and how long it prevailed in the freezer.

SAFETY FACTOR

Generally, the calculated load is increased by a factor of 10% to allow for possible discrepancies between design criteria and actual operation. This factor should be selected in consultation with the facility user and should be applied individually to the first four heat load segments.

A separate factor should be added to the coil-defrosting portion of the equipment load for freezer applications that use dry-surface refrigerating coils. However, few data are available to predict heat gain from coil defrosting. For this reason, the experience of similar facilities should be sought to obtain an appropriate defrosting safety factor. Similar facilities should have similar room sensible heat ratios.

The nature of frost accumulation on cooling coils also affects cooling unit performance and, therefore, refrigeration load. A very-low-density frost forms under certain conditions, particularly where the room sensible heat ratio is more than a few points below 1.0. This type of frost is difficult to remove and tends to block airflow through the cooling coils more readily. Removing this frost requires more frequent and longer periods of defrosting of cooling units, which increases the refrigeration load.

Example 2. Calculate the total refrigeration load for a freezer storage with design criteria as follows:

Design Criteria
- Summer: 92°F db, 80°F wb
- Comments: 0.4% summer and winter conditions
- Room dimensions: 133 by 222 by 30 ft
- Floor area: 29,526 ft²
- Pallet positions: 4800
- Turns per yr: 20
- Use factor: 90%

Ambient Design Conditions

Design Room Temperature, –10°F

		Sun Effect, °F	Surface Temperature, °F
Roof		10	102
Floor		0	60
Wall,	east	0	92
	north	0	92
	west[a]	0	28
	south[b]	0	45

[a]Adjacent to refrigerated meat room held at 28°F
[b]Adjacent to refrigerated truck dock held at 44.5°F

Insulation Thickness

		in.	k, Btu·in/h·ft²·°F	U, Btu/h·ft²·°F	R, ft²·h·°F/Btu
Roof		6	0.142	0.02367	42.25
Floor		6	0.188	0.03133	31.91
Wall,	east	4	0.121	0.03025	33.06
	north	4	0.121	0.03025	33.06
	west	4	0.121	0.03025	33.06
	south	4	0.121	0.03025	33.06

k = thermal conductivity $U = k$/in. thickness R = thermal resistance

Solution:

Heat Transmission

	Length, ft	Width, ft	Height, ft	Adj. Temp.	U, Btu/h·ft²·°F	Area A, ft²	Δt, °F	Load, tons*
Roof	133	222	0	102	0.02367	29,526	112	6.52
Floor	133	222	0	60	0.03133	29,526	70	5.40
Wall,								
east	222	0	30	92	0.03025	6660	102	1.71
north	133	0	30	92	0.03025	3990	102	1.03
west	222	0	30	28	0.03025	6660	38	0.64
south	133	0	30	45	0.03025	3990	55	0.55
							Safety, 20%	3.17
							Total Transmission Load, tons	19.02

*$UA\Delta t$/12,000 U = Heat transfer coefficient

Product (see Chapter 9)
- Pallets per 24 h: 420
- Pounds per pallet: 2500
- Mass flow: 43,750 lb/h
- Temperature in: 5°F
- Temperature out: –10°F
- Specific heat: 0.45 Btu/lb·°F

Product Load = (43,750)(15)(0.45)/12,000 = 24.61 tons

Motors (Other than air unit fan motors): None

Infiltration

Door Openings	**Door Type 1**	**Door Type 2**	**Door Type 3**
From	**Dock**	**Dry**	**Meat**
To	**Freezer**	**Freezer**	**Freezer**
Door width, ft	8	8	8
Door height, ft	10	10	10
Enthalpy h_i	14.4	43.8	9.5
Enthalpy h_r	–2	–2	–2
Density ρ_r	0.0883	0.0883	0.0883
Density ρ_i	0.0782	0.0697	0.081
Doorway flow factor D_f	0.7	0.7	0.7
Doorway time factor D_t	0.14583	0.0219	0.0417
Effectiveness device E_f	0	0	0
Number of doors	3	1	1
Load per door, tons	4.61	2.55	0.79
Total load, tons	13.84	2.55	0.79
		Infiltration Load, tons	17.19

$$\text{Infiltration Load} = \frac{795.6A(h_i - h_r)\rho_r(1 - \rho_i/\rho_r)^{0.5}(gH)^{0.5}F_m[D_f D_t(1 - E_f)]}{12{,}000}$$

A = doorway area, ft^2
h_f = enthalpy incoming air through doorway from adjacent area, Btu/lb
h_r = enthalpy room air, Btu/lb
ρ_r = density room air, lb/ft^3
ρ_i = density incoming air, lb/ft^3
g = gravitational acceleration, 32.2 ft/s^2
H = doorway height, ft
F_m = density factor = $[2/(1 + (\rho_r/\rho_i)^{1/3}]^{1.5}$
D_f = doorway flow factor
D_t = percentage time period doorway is open during 1 h period, average, expressed as a decimal
E_f = effectiveness factor for open-doorway protective device such as air curtain or plastic strip curtain

Lights
Lighting level: 1.0 W/ft^2
Floor area: 29,526 ft^2

Lighting Load = (1.0)(29,526)(3.413)/12,000 = 8.40 tons

People
Number of persons: 3
Room temperature: –10°F

People Load = (3)[1295 – 11.5(–10)]/12,000 = 0.35 tons

Trucks
Number of trucks: 3
Horsepower per truck: 7.5

Truck Load = (3)(7.5)(2545)/12,000 = 4.77 tons

Fans
Number of fans: 15
Horsepower per fan: 1.50 nominal (4960 Btu/h heat gain per Table 6)

Fan Load = (15)(4960)/12,000 = 6.20 tons

Load Summary	**Load, tons**
Transmission	19.02
Product	24.61
Motor	0.00
Infiltration	17.19
Lighting	8.40
People	0.35
Trucks	4.77
Fan motors	6.20
Subtotal	80.54
Safety 10%	8.05
Total Load, tons	88.59

LOAD DIVERSITY

When computing refrigeration load, the most conservative approach is to calculate each part at its expected **peak value**. The combined result can overstate the actual total load by as much as 20 to 50%.

The reason for such overestimates is that, typically, all the loads do not occur at the same time of day. Furthermore, many of them are not always at their maximum value when they do occur. The consequence is that oversized refrigerating equipment is often installed for a plant. Some of it may not ever run, or if a single piece of equipment runs, it may do so inefficiently.

There are two ways that this mismatch of estimated load to actual load is addressed: (1) a rigorous computational method, or (2) a "rule-of-thumb" adjustment to the final estimate, determined by judgment.

The rigorous approach is to use the **hour-by-hour load calculation** method. In the past, when most calculations were done manually, this procedure was tedious and time consuming and therefore, not often used. The advent of computer software to perform such calculations eases the task. It is necessary, however, to have comprehensive operating data for the facility under consideration so complete and precise information regarding load magnitude and time of occurrence can be input to the computer. Ballard (1992) describes the procedure.

The other method, used by experienced engineers, is to use a **diversity factor**. Based on analysis of the load data and an understanding of how, or more importantly, when or how often each load element will occur, the designer will often apply a factor ranging from 0.7 to 0.85 to the calculated final total load. That result is the load on which the selection of equipment is based.

REFERENCES

Ballard, R.N. 1992. Calculating refrigeration loads on an hour-by-hour basis: Part I—Building envelope and Part II—Infiltration and internal heat sources. *ASHRAE Transactions* 98(2):658-663, 664-669.

Chuangchid, P. and M. Krarti. 1999. Validation of analytical models for foundation heat transfer from slab-on-grade floors. Presented at ASME Solar Engineering Conference.

Chuangchid, P. and M. Krarti. 2000. Parametric analysis and development of a design tool for foundation heat gain for refrigerated warehouses. *ASHRAE Transactions* 106(2):240-250.

Cole, R.A. 1989. Refrigeration loads in a freezer due to hot gas defrost, and their associated costs. *ASHRAE Transactions* 95(2):1149-1154.

Downing, C.C. and W.A. Meffert. 1993. Effectiveness of cold-storage door infiltration protective devices. *ASHRAE Transactions* 99(2):356-366.

Gosney, W.B. and H.A.L. Olama. 1975. Heat and enthalpy gains through cold room doorways. *Proceedings of the Institute of Refrigeration*, vol. 72, pp. 31-41.

Hendrix, W.A., D.R. Henderson, and H.Z. Jackson. 1989. Infiltration heat gains through cold storage room doorways. *ASHRAE Transactions* 95(2).

Sherif, S.A., P.J. Mago, and R.S. Theen. 2002. A study to determine heat loads due to coil defrosting—Phase II. ASHRAE Research Project RP-1094, *Final Report*. Solar Energy and Energy Conversion Laboratory, Department of Mechanical Engineering, University of Florida, Gainesville.

BIBLIOGRAPHY

Cole, R.A. 1984. Infiltration: A load calculation guide. *Proceedings of the International Institute of Ammonia Refrigeration, 6th Annual Meeting*, San Francisco (February).

Cole, R.A. 1987. Infiltration load calculations for refrigerated warehouses. *Heating/Piping/Air Conditioning* (April).

Dickerson, R.W. 1972. Computing heating and cooling rates of foods. *ASHRAE Symposium Bulletin* 72-03.

Fisher, D.V. 1960. Cooling rates of apples packed in different bushel containers, stacked at different spacings in cold storage. *ASHRAE Transactions* 66.

Hamilton, J.J., D.C. Pearce, and N.B. Hutcheon. 1959. What frost action did to a cold storage plant. *ASHRAE Journal* 1(4).

Haugh, C.G., W.J. Standelman, and V.E. Sweat. 1972. Prediction of cooling/freezing times for food products. *ASHRAE Symposium Bulletin* 72-03.

Hovanesian, J.D., H.F. Pfost, and C.W. Hall. 1960. An analysis of the necessity to insulate floors of cold storage rooms at 35°F. *ASHRAE Transactions* 66.

Jones, B.W., B.T. Beck, and J.P. Steele. 1983. Latent loads in low humidity rooms due to moisture. *ASHRAE Transactions* 89(1).

Kayan, C.F. and J.A. McCague. 1959. Transient refrigeration loads as related to energy-flow concepts. *ASHRAE Journal* 1(3).

Meyer, C.S. 1964. "Inside-out" design developed for low-temperature buildings. *ASHRAE Journal* 5(4).

Pedersen, C.O., D.E. Fisher, J.D. Spitler, and R.J. Liesen. 1998. *Cooling and heating load calculation principles*. ASHRAE.

Pham, O.T. and D.W. Oliver. 1983. Infiltration of air into cold stores. Meat Industry Research Institute of New Zealand. Presented at IIF-IIR 16th International Congress of Refrigeration, Paris.

Pichel, W. 1966. Soil freezing below refrigerated warehouses. *ASHRAE Journal* 8(10).

Powell, R.M. 1970. Public refrigerated warehouses. *ASHRAE Journal* 12(8).

CHAPTER 14

REFRIGERATED-FACILITY DESIGN

REFRIGERATED facilities are any buildings or sections of a building that achieve controlled storage conditions using refrigeration. Two basic storage facilities are (1) coolers that protect commodities at temperatures usually above 32°F and (2) low-temperature rooms (freezers) operating under 32°F to prevent spoilage or to maintain or extend product life.

The conditions within a closed refrigerated chamber must be maintained to preserve the stored product. This refers particularly to seasonal, shelf life, and long-term storage. Specific items for consideration include

- Uniform temperatures
- Length of airflow pathway and impingement on stored product
- Effect of relative humidity
- Effect of air movement on employees
- Controlled ventilation, if necessary
- Product entering temperature
- Expected duration of storage
- Required product outlet temperature
- Traffic in and out of storage area

In the United States, the U.S. Public Health Service Food and Drug Administration developed the Food Code (FDA 1997), which provides model requirements for safeguarding public health and ensuring that food is unadulterated. The code is a guide for establishing standards for all phases of handling refrigerated foods. It treats receiving, handling, storing, and transporting refrigerated foods and calls for sanitary as well as temperature requirements. These standards must be recognized in the design and operation of refrigerated storage facilities.

Regulations of the Occupational Safety and Health Administration (OSHA), Environmental Protection Agency (EPA), U.S. Department of Agriculture (USDA), and other standards must also be incorporated in warehouse facility and procedures.

Refrigerated facilities may be operated for or by a private company for storage or warehousing of their own products, as a public facility where storage services are offered to many concerns, or both. Important locations for refrigerated facilities, public or private, are (1) point of processing, (2) intermediate points for general or long-term storage, and (3) final distributor or distribution point.

The five categories for the classification of refrigerated storage for preservation of food quality are

- Controlled atmosphere for long-term fruit and vegetable storage
- Coolers at temperatures of 32°F and above
- High-temperature freezers at 27 to 28°F
- Low-temperature storage rooms for general frozen products, usually maintained at –5 to –20°F
- Low-temperature storages at –5 to –20°F, with a surplus of refrigeration for freezing products received at above 0°F

The preparation of this chapter is assigned to TC 10.5, Refrigerated Distribution and Storage Facilities.

Note that, because of ongoing research, the trend is toward lower temperatures for frozen foods. Refer to Chapters 23 and 24 of the 2005 *ASHRAE Handbook—Fundamentals* and Chapter 11 of this volume for further information.

INITIAL BUILDING CONSIDERATIONS

Location

Private refrigerated space is usually adjacent to or in the same building with the owner's other operations.

Public space should be located to serve a producing area, a transit storage point, a large consuming area, or various combinations of these to develop a good average occupancy. It should also have the following:

- Convenient location for producers, shippers, and distributors, considering the present tendency toward decentralization and avoidance of congested areas
- Good railroad switching facilities and service with minimum switching charges from all trunk lines to plant tracks if a railhead is necessary to operate the business profitably
- Easy access from main highway truck routes as well as local trucking, but avoiding location on congested streets
- Ample land for trucks, truck movement, and plant utility space plus future expansion
- Location with a reasonable land cost
- Adequate power and water supply
- Provisions for surface, waste, and sanitary water disposal
- Consideration of zoning limitations and fire protection
- Location away from residential areas, where noise of outside operating equipment (i.e., fans and engine-driven equipment on refrigerated vehicles) would be objectionable
- External appearance that is not objectionable to the community
- Minimal tax and insurance burden
- Plant security
- Favorable undersoil bearing conditions and good surface drainage

Plants are often located away from congested areas or even outside city limits where the cost of increased trucking distance is offset by better plant layout possibilities, a better road network, better or lower-priced labor supply, or other economies of operation.

Configuration and Size Determination

Building configuration and size of a cold-storage facility are determined by the following factors:

- Is receipt and shipment of goods to be primarily by rail or by truck? Shipping practices affect the platform areas and internal traffic pattern.
- What relative percentages of merchandise are for cooler and for freezer storage? Products requiring specially controlled conditions, such as fresh fruits and vegetables, may justify or demand several individual rooms. Seafood, butter, and nuts also require special treatment. Where overall occupancy may be reduced

because of seasonal conditions, consider providing multiple-use spaces.

- What percentage is anticipated for long-term storage? Products that are stored long term can usually be stacked more densely.
- Will the product be primarily in small or large lots? The drive-through rack system or a combination of pallet racks and a mezzanine have proved effective in achieving efficient operation and effective use of space. Mobile or moving rack systems are also valid options.
- How will the product be palletized? Dense products such as meat, tinned fruit, drums of concentrate, and cases of canned goods can be stacked very efficiently. Palletized containers and special pallet baskets or boxes effectively hold meat, fish, and other loose products. The slip sheet system, which requires no pallets, eliminates the waste space of the pallet and can be used effectively for some products. Pallet stacking racks make it feasible to use the full height of the storage and palletize any closed or boxed merchandise.
- Will rental space be provided for tenants? Rental space usually requires special personnel and office facilities. An isolated area for tenant operations is also desirable. These areas are usually leased on a unit area basis, and plans are worked into the main building layout.

The owner of a prospective refrigerated facility may want to obtain advice from specialists in product storage, handling, and movement systems.

Stacking Arrangement

Typically, the height of refrigerated spaces is at least 28 to 35 ft or more clear space between the floor and structural steel to allow forklift operation. Pallet rack systems use the greater height. The practical height for stacking pallets without racks is 15 to 18 ft. Clear space above the pallet stacks is used for air units, air distribution, lighting, and sprinkler lines. Generally, 6 to 10 ft minimum clear height is required from top of product to bottom of support structure to ensure there is no interference with drain pan and drain lines of air units. Greater clear heights are usually required if automated or mechanized equipment is used. Overhead space is inexpensive, and because the refrigeration requirement for extra height is not significant in the overall plant cost, a minimum of 20 ft clear height is desirable. Greater heights are valuable if automated or mechanized material handling equipment is contemplated. The effect of high stacking arrangements on insurance rates should be investigated.

Floor area in a facility where diverse merchandise is to be stored can be calculated on the basis of 8 to 10 lb per gross cubic foot, to allow about 40% for aisles and space above the pallet stacks. In special-purpose or production facilities, products can be stacked with less aisle and open space, with an allowance factor of about 20%.

Building Design

Most refrigerated facilities are single-story structures. Small columns on wide centers allow palletized storage with minimal lost space. This type of building usually provides additional highway truck unloading space. The following characteristics of single-story design must be considered: (1) horizontal traffic distances, which to some extent offset the vertical travel required in a multistory building; (2) difficulty of using the stacking height with many commodities or with small-lot storage and movement of goods; (3) necessity for treatment of the floor below freezers to give economical protection against possible ground heaving; and (4) high land cost for building capacity. A one-story facility with moderate or low stacking heights has a high cost per unit area because of the high ratio of construction costs and added land cost to product storage capacity. However, first cost and operating cost are usually lower than for a multistory facility.

One-Story Configuration

Figure 1 shows the layout of a one-story −10°F freezer that complies with current practices. The following essential items and functions are considered:

- Refrigeration machinery room
- Refrigerated shipping docks with seal-cushion closures on the doors
- Automatic doors
- Batten doors or strip curtains
- Low-temperature storage held at −10°F or lower
- Pallet-rack systems to facilitate handling of small lots and to comply with first-in, first-out inventory, which is required for some products
- Blast freezer or separate sharp freezer room for isolation of products being frozen
- Cooler or convertible space
- Space for brokerage offices
- Space for empty pallet storage and repair
- Space for shop and battery charging
- Automatic sprinklers in accordance with National Fire Protection Association (NFPA) regulations
- Trucker/employee break area
- Valve stations for underfloor heating
- Evaporative condenser(s) location

Other areas that must be in a complete operable facility are

- Electrical area
- Shipping office
- Administration office
- Personnel welfare facilities

A modified one-story design is sometimes used to reduce horizontal traffic distances and land costs. An alternative is to locate nonproductive services (including offices and the machinery room) on a second-floor level, usually over the truck platform work area, to allow full use of the ground floor for production work and storage. However, potential vibration of the second floor from equipment below must be considered.

One-story design or modification thereof gives the maximum capacity per unit of investment with a minimum overall operating

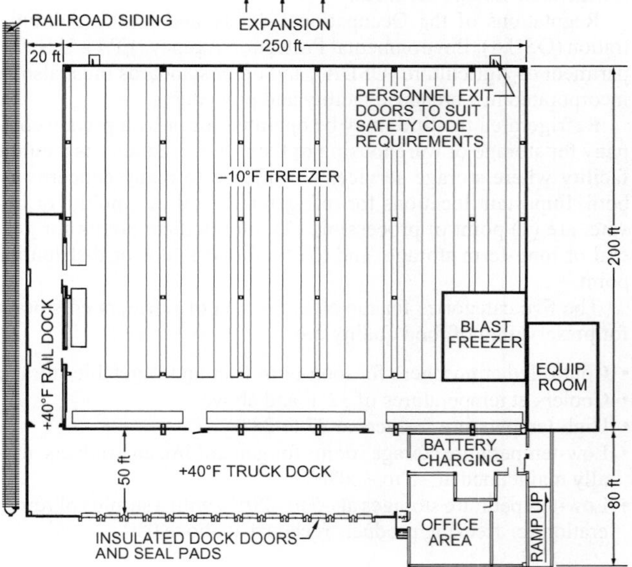

Fig. 1 Typical Plan for One-Story Refrigerated Facility

expense, including amortization, refrigeration, and labor. Mechanization must be considered, as well. In areas where land availability or cost is a concern, a high-rise refrigerated storage building may be a viable option.

Designs that provide minimum overall costs restrict office facilities and utility areas to a minimum. They also include ample dock area to ensure efficiency in loading and unloading merchandise.

Shipping and Receiving Docks

Temperature control regulations for all steps of product handling have led to designing the trucking dock as a refrigerated anteroom to the cold storage area. Dock refrigeration is an absolute necessity in humid and warm climates. Typically, loading and unloading transport vehicles is handled by separate work crews. One crew moves the product in and out of the vehicles, and a warehouse crew moves the product in and out of the refrigerated storage. This procedure may allow merchandise to accumulate on the shipping dock. Maintaining the dock at 35 to 45°F offers the following advantages:

- Refrigeration load in the low-temperature storage area, where energy demand per unit capacity of refrigeration is higher, is reduced.
- Less ice or frost forms in the low-temperature storage because less humid and warm air infiltrates the area.
- Refrigerated products held on the dock maintain a more favorable temperature, thus maintaining product quality.
- Packaging remains in good condition because it stays drier. Facility personnel are more comfortable because temperature differences are smaller.
- Less maintenance on forklifts and other equipment is required because condensation is reduced.
- Need for anterooms or vestibules to the freezer space is reduced or eliminated.
- Floor areas stay drier, particularly in front of freeze door areas. This assists in housekeeping and improves safety.

Utility Space

Space for a general office, locker room, and machinery room is needed. A superintendent's office and a warehouse records office should be located near the center of operations, and a checker's office should be in view of the dock and traffic arrangement. Rented space should be isolated from warehouse operations.

The machinery room should include ample space for refrigeration equipment and maintenance, adequate ventilation, standby capacity for emergency ventilation, and adequate segregation from other areas. Separate exits are required by most building codes. A maintenance shop and space for parking, charging, and servicing warehouse equipment should be located adjacent to the machinery room. Electrically operated material-handling equipment is used to eliminate inherent safety hazards of combustion-type equipment. Battery-charging areas should be designed with high roofs and must be ventilated, because of the potential for combustible fumes from charging.

Specialized Storage Facilities

Material handling methods and storage requirements often dictate design of specialized storage facilities. Automated material handling within the storage, particularly for high stack piling, may be an integral part of the structure or require special structural treatments. Controlled-atmosphere and minimal-air-circulation rooms require special building designs and mechanical equipment to achieve design requirements. Drive-in and/or drive-through rack systems can improve product inventory control and can be used in combination with stacker cranes, narrow-aisle high stacker cranes, and automatic conveyors. Mobile racking systems may be considered where space is at a premium.

In general, specialized storage facilities may be classified as follows:

- Public refrigerated facility with several chambers designed to handle all commodities. Storage temperatures range from 35 to 60°F (with humidity control) and to –20°F (without humidity control).
- Refrigerated facility area for case and break-up distribution, automated to varying degrees. The area may incorporate racks with pallet spaces to facilitate distribution.
- Facility designed for a processing operation with bulk storage for frozen ingredients and rack storage for palletized outshipment of processed merchandise. An efficient adaptation frequently seen is to adjoin the refrigerated facility to the processing plant.
- Public refrigerated facility serving several production manufacturers for storing and inventorying products in lots and assembling outshipments.
- Mechanized refrigerated facility with stacker cranes, racks, infeed and outfeed conveyors, and conveyor vestibules. Such a facility may have an interior ceiling 60 to 100 ft high. Evaporators should be mounted in the highest internal area to help remove moisture from outside air infiltration. A penthouse to house the evaporators can be accessed through doorways on the roof for maintenance, providing a means to control condensate drip, and allowing added rack storage space in the freezer area.

Controlled-Atmosphere Storage Rooms

Controlled-atmosphere (CA) storage rooms may be required for storing some commodities, particularly fresh fruits and vegetables that respire, consuming oxygen (O_2) and producing carbon dioxide (CO_2) in the process. The storage life of such products may be greatly lengthened by a properly controlled environment, which includes control of temperature, humidity, and concentration of noncondensable gases (O_2, CO_2, and nitrogen). Hermetically sealing the room to provide such an atmosphere is challenging, often requiring special gastight seals. Although information is available for some commodities, the desired atmosphere usually must be determined experimentally for the commodity as produced in the specific geographic location that the storage room is to serve.

Commercial application of controlled-atmosphere storage has historically been limited to fresh fruits and vegetables that respire. Storage spaces may be classified as having either (1) product-generated atmospheres, in which the room is sufficiently well sealed that the natural oxygen consumption and CO_2 generation by the fruit balance infiltration of O_2 into the space and exhaust of CO_2 from the space; or (2) externally generated atmospheres, in which nitrogen generators, CO_2 scrubbers, or O_2 consumers supplement normal respiration of the fruit to create the desired atmospheric composition. The second type of system can cope with a poorly sealed room, but the cost of operation may be high; even with the external gas generator system, a hermetically sealed room is desired.

In most cases, a CO_2 scrubber is required, unless the total desired O_2 and CO_2 content is 21%, which is the normal balance between O_2 and CO_2 during respiration. Carbon dioxide may be removed by (1) passing room air over dry lime that is replaced periodically; (2) passing air through wet caustic solutions in which the caustic (typically sodium hydroxide) is periodically replaced; (3) using water scrubbers, in which CO_2 is absorbed from the room air by a water spray and then desorbed from the water by outdoor air passed through the water in a separate compartment; (4) using monoethanolamine scrubbers, in which the solution is regenerated periodically by a manual process or continuously by automatic equipment; or (5) using dry adsorbents automatically regenerated on a cyclic basis.

Systems of room sealing to prevent outside air infiltration include (1) galvanized steel lining the walls and ceiling of the room and interfaced into a floor sealing system; (2) plywood with an impervious sealing system applied to the inside face; and (3) carefully applied sprayed urethane finished with mastic, which also serves as a fire retardant.

A room is considered sufficiently sealed if, under uniform temperature and barometric conditions, 1 h after the room is pressurized to 1 in. of water (gage), 0.1 to 0.2 in. remains. A room with external gas generation is considered satisfactorily sealed if it loses pressure at double the above rate, and the test prescribed for a room with product-generated atmosphere is about one air change of the empty room in a 30 day period.

Extreme care in all details of construction is required to obtain a seal that passes these tests. Doors are well sealed and have sills that can be bolted down; electrical conduits and special seals around pipe and hanger penetrations must allow some movement while keeping the hermetic seal intact. Structural penetrations through the seal must be avoided, and the structure must be stable. Controlled-atmosphere rooms in multifloor buildings, where the structure deflects appreciably under load, are extremely difficult to seal.

Gasket seals are normally applied at the cold side of the insulation, so that they may be easily maintained and points of leakage can be detected. However, this placement causes some moisture entrapment, and the insulation materials must be carefully selected so that this moisture causes minimal damage. In some installations, cold air with a dew point lower than the inside surface temperature is circulated through the space between the gas seal and the insulation to provide drying of this area. Chapters 22 and 24 have additional information on conditions required for storage of various commodities in controlled-atmosphere storage rooms.

Automated Warehouses

Automated warehouses usually contain tall, fixed rack arrangements with stacker cranes under fully automatic, semiautomatic, or manual control. The control systems can be tied into a computer system to retain a complete inventory of product and location.

The following are some of the advantages of automation:

- First-in, first-out inventory can be maintained.
- Enclosure structure is high, requiring a minimum of floor space and providing favorable cost per cubic foot.
- Product damage and pilfering are minimized.
- Direct material handling costs are minimized.

The following are some of the disadvantages:

- First cost of the racking system and building are very high compared to conventional designs.
- Access may be slower, depending on product flow and locations.
- Cooling equipment may be difficult to access for maintenance, unless installed in a penthouse.
- Air distribution must be carefully evaluated.

Refrigerated Rooms

Refrigerated rooms may be appropriate for long-term storage at temperatures other than the temperature of the main facility, for bin storage, for controlled-atmosphere storage, or for products that deteriorate with active air movement. Mechanically cooled walls, floors, and ceilings may be economical options for controlling the temperature. Embedded pipes or air spaces through which refrigerated air is recirculated can provide the cooling; with this method, heat leakage is absorbed into the walls and prevented from entering the refrigerated space.

The following must be considered in the initial design of the room:

- Initial cooldown of the product, which can impose short-term peak loads
- Service loads when storing and removing product
- Odor contamination from products that deteriorate over long periods
- Product heat of respiration

Supplementary refrigeration or air-conditioning units in the refrigerated room that operate only as required can usually alleviate such problems.

Construction Methods

Cold storage, more than most construction, requires correct design, high-quality materials, good workmanship, and close supervision. Design should ensure that proper installation can be accomplished under various adverse job site conditions. Materials must be compatible with each other. Installation must be done by careful workers directed by an experienced, well-trained superintendent. Close cooperation between the general, roofing, insulation, and other contractors increases the likelihood of a successful installation.

Enclosure construction methods can be classified as (1) insulated structural panel, (2) mechanically applied insulation, or (3) adhesive or spray-applied foam systems. These construction techniques seal the insulation within an airtight, moisturetight envelope that must not be violated by major structural components.

Three methods are used to achieve an uninterrupted vapor retarder/insulation envelope. The first and simplest is total encapsulation of the structural system by an exterior vapor retarder/insulation system under the floor, on the outside of the walls, and over the roof deck (Figure 2). This method offers the least number of penetrations through the vapor retarder, as well as the lowest cost.

The second method is an entirely interior system in which the vapor retarder envelope is placed within the room, and insulation is added to the walls, floors, and suspended ceiling (Figure 3). As with an exterior system, the moisture barrier is best applied to the outside of the enclosures. This technique is used where walls and ceilings must be washed, where an existing structure is converted

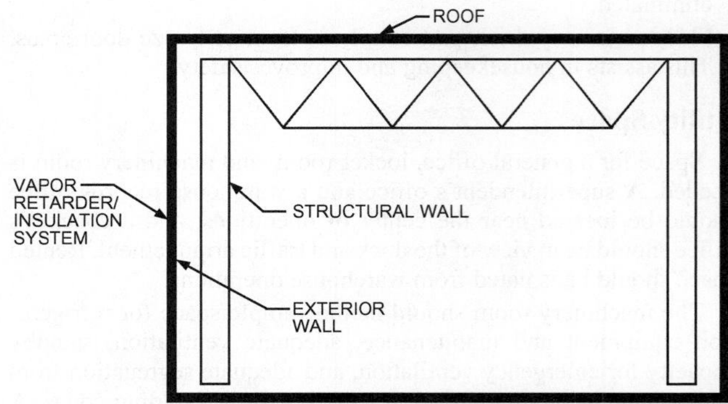

Fig. 2 Total Exterior Vapor Retarder/Insulation System

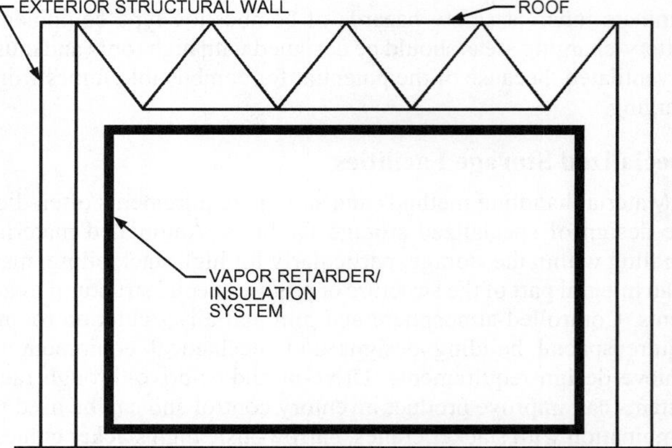

Fig. 3 Entirely Interior Vapor Retarder/Insulation System

to refrigerated space, or for smaller rooms that are located within large coolers or unrefrigerated facilities or are part of a food-processing facility. Special-purpose rooms require separate analysis to determine proper moisture barrier location.

The third method is interior/exterior construction (Figure 4), which involves an exterior curtain wall of masonry or similar material tied to an interior structural system. Adequate space allows the vapor retarder/insulation system to turn up over a roof deck and be incorporated into a roofing system, which serves as the vapor retarder. This method is a viable alternative, although it allows more interruptions in the vapor retarder than the exterior system.

The total exterior vapor retarder system (see Figure 2) is best because it has the fewest penetrations and the lowest cost. Areas of widely varying temperature should be divided into separate envelopes to retard heat and moisture flow between them (Figure 5).

Space Adjacent to Envelope

Condensation at the envelope is usually caused by high humidity and inadequate ventilation. Poor ventilation occurs most often within a dead air space such as a ceiling plenum, hollow masonry unit, through-metal structure, or beam cavity. All closed air spaces should be eliminated, except those large enough to be ventilated adequately. Ceiling plenums, for instance, are best ventilated by mechanical vents that move air above the envelope, or with mechanical dehumidification. See the section on Suspended Ceilings and Other Interstitial Spaces for more information.

If possible, the insulation envelope and vapor retarder should not be penetrated. All steel beams, columns, and large pipes that project through the insulation should be vapor-sealed and insulated with a 4 ft wrap of insulation. Conduit, small pipes, and rods should be insulated at four times the regular wall insulation thickness. In both cases, the thickness of insulation on the projection should be half that on the regular wall or ceiling. Voids within metal projections should be filled. Where practicable, the wrap insulation should be located, and the metal projection sealed, on the warm side. Vapor sealing on the inside of conduits prevents moisture flow through the conduits.

Other considerations include the following:

- Vapor retardants should be placed on the warm side of insulation systems.
- Prefabricated, self-locking wall panels also serve as vapor barriers.
- Roof vapor membranes are often used; these large rubber sheets are laid over roof insulation, overlapped, sealed, and attached to the roof with nonpenetrating fasteners or covered with small stone ballast, which is light-colored to help reflect solar heat.

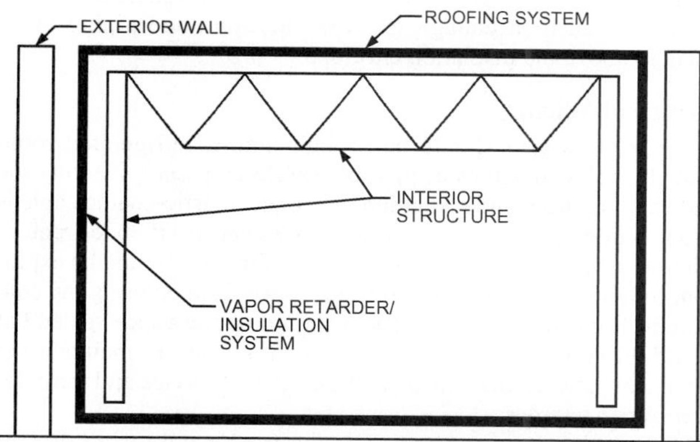

Fig. 4 Interior/Exterior Vapor Retarder/Insulation System

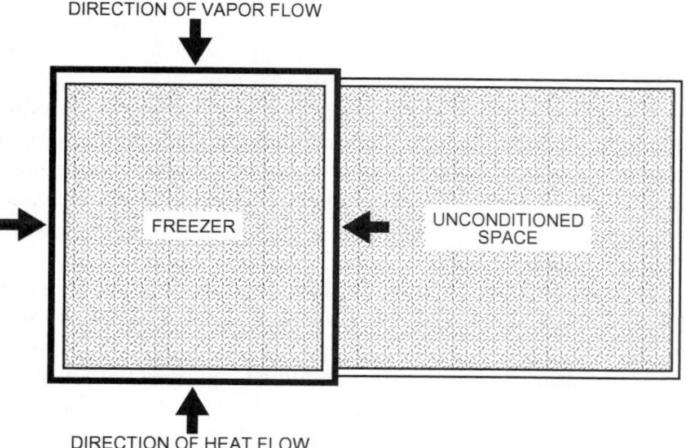

Fig. 5 Separate Exterior Vapor Retarder Systems for Each Area of Significantly Different Temperature

Air/Vapor Treatment at Junctions

Air and vapor leakage at wall/roof junctions is perhaps the predominant construction problem in cold storage facilities.

When a cold room of interior/exterior design (see Figure 4) is lowered to operating temperature, the structural elements (roof deck and insulation) contract and can pull the roof away from the wall. Negative pressure in the space of the wall/roof junction can cause warm, moist air to leak into the room and form frost and ice. Therefore, proper design and construction of the air/vapor seal is critical.

An air/vapor flashing sheet system (a transition from the roof vapor retarder to the exterior wall vapor retarder) is best for preventing leakage. A good corner flashing sheet must be flexible, tough, airtight, and vaportight. Proper use of flexible insulation at overlaps, mastic adhesive, and a good mastic sealer ensure leak-free performance. To remain airtight and vaportight during the life of the facility, a properly constructed vapor retarder should

- Be flexible enough to withstand building movements that may occur at operating temperatures
- Allow for thermal contraction of the insulation as the room is pulled down to operating temperature
- Be constructed with a minimum of penetrations that might cause leaks (wall ties and structural steel that extend through the corner flashing sheet may eventually leak no matter how well sealed during construction; minimize these, and make them accessible for maintenance)
- Have corner flashing sheet properly lapped and sealed with adhesive and mechanically fastened to the wall vapor retarder
- Have corner flashing sealed to roof without openings
- Have floor to exterior vapor retarders that are totally sealed

The interior/exterior design (see Figure 4) is likely to be unsuccessful at the wall/roof junction because of extreme difficulty in maintaining an airtight or vaportight environment.

The practices outlined for the wall/roof junction apply for other insulation junctions. The insulation manufacturer and designer must coordinate details of the corner flashing design.

Poor design and shoddy installation cause moist air leakage into the facility, resulting in frost and ice formation, energy loss, poor appearance, loss of useful storage space, and, eventually, expensive repairs.

Floor Construction

Refrigerated facilities held above freezing need no special underfloor treatment. A below-the-floor vapor retarder is needed in facilities held below freezing, however. Without underfloor heating, the subsoil eventually freezes; any moisture in this soil also freezes and causes floor frost heaving. In warmer climates, underfloor tubes vented to ambient air may be sufficient to prevent heaving. Artificial heating, either by air circulated through underfloor ducts or by glycol circulated through plastic pipe, is the preferred method to prevent frost heaving. Electric heating cables installed under the floor can also be used to prevent frost formation. The choice of heating method depends on energy cost, reliability, and maintenance requirements. Air duct systems should be screened to keep rodents out and sloped for drainage to remove condensation.

Future facility expansion must be considered when underfloor heating systems are being designed. Therefore, a system including artificial heating methods that do not require building exterior access is preferred. Wearing-slab heating under the dock area in front of the freezer doors helps eliminate moisture at the door and floor joints.

Surface Preparation

When an adhesive is used, the surface against which the insulating material is to be applied should be smooth and dust-free. Where room temperatures are to be below freezing, masonry walls should be leveled and sealed with cement back plaster. Smooth poured concrete surfaces may not require back plastering.

No special surface preparation is needed for a mechanical fastening system, assuming that the surface is reasonably smooth and in good repair.

The surface must be warm and dry for a sprayed-foam system. Any cracks or construction joints must be prepared to prevent projection through the sprayed insulation envelope. All loose grout and dust must be removed to ensure a good bond between the sprayed foam insulation and the surface. Very smooth surfaces may require special bonding agents.

No special surface preparation is needed for insulated panels used as a building lining, assuming the surfaces are sound and reasonably smooth. Grade beams and floors should be true and level where panels serve as the primary walls.

Finishes

Insulated structural panels with metal exteriors and metal or reinforced plastic interior faces are prevalent for both coolers and freezers. They keep moisture from the insulation, leaving only the joints between panels as potential areas of moisture penetration. They are also available with surface finishes that meet government requirements.

For sanitary washdowns, a scrubbable finish is sometimes required. Such finishes generally have low permeance; when one is applied on the inside surface of the insulation, a lower-permeance treatment is required on the outside of the insulation.

All insulated walls and ceilings should have an interior finish. The finish should be impervious to moisture vapor and should not serve as a vapor retarder, except for panel construction. The permeance of the in-place interior finish should be significantly greater than the permeance of the in-place vapor retarder.

To select an interior finish to meet the installation's in-use requirements, consider the following factors: (1) fire resistance, (2) washdown requirements, (3) mechanical damage, (4) moisture and gas permeance, and (5) government requirements. All interior walls of insulated spaces should be protected by bumpers and curbs wherever there is a possibility of damage to the finish.

Suspended Ceilings and Other Interstitial Spaces

It is not uncommon to have interstitial spaces above or adjacent to cold spaces in refrigerated facilities. The reason for the space may be design (e.g., an older facility with an air space used as insulation, a drop ceiling, or a production space that requires a cleanable ceiling surface), or facility expansion (e.g., adding freezer space next to an existing freezer that cannot structurally support what would be the common wall). Regardless, if air in the space has no ventilation or conditioning, moisture in the air will condense onto the cold surface, and can lead to structural failure of the envelope through corrosion, frost penetration in the cold space, or other forms of deterioration if not kept in check.

Several methods are available to deal with the moisture. The space could be sealed airtight, with a dehumidifier inside to maintain a low dew point in the space. This method is preferable in warm and humid climates. The sealed space could also be heated to ensure the cold surface is always above the air's dew point. This is uncommon because of the heat load it transmits to the refrigerated space.

The most common method used to prevent condensation is continuously ventilating outside air into the interstitial space. This keeps the insulated surface temperature above the dew point of the interstitial space, thus preventing moisture condensing from the air on the surface. Roof-mounted exhaust fans and uniformly spaced vents around the perimeter of the plenum are typically used to ventilate suspended ceilings; similar arrangements can be used for other spaces. Be certain, though, that fan and inlet louvers are placed to provide good air distribution across the entire cold surface. The cold surface should also be covered with a vapor retarder attached with flashing to the wall insulation on the top or warm side. Finally, beware of overventilating the space: this only reduces the insulating effect of the dead air space.

Suspended ceilings are often designed for light foot traffic for inspection and maintenance of piping and electrical wiring. Fastening systems for ceiling panels include spline, U channel, and camlock. To minimize problems with ceiling penetrations during both installation and ongoing maintenance, wall panel penetrations may be preferable when possible.

Floor Drains

Floor drains should be avoided if possible, particularly in freezers. If they are necessary, they should have short, squat dimensions and be placed high enough to allow the drain and piping to be installed above the insulation envelope.

Electrical Wiring

Electrical wiring should be brought into a refrigerated room through as few locations as possible (preferably one), piercing the wall vapor retarder and insulation only once. Plastic-coated cable is recommended for this service where codes permit. If codes require conduit, the last fitting on the warm side of the run should be explosionproof and sealed to prevent water vapor from entering the cold conduit. Light fixtures in the room should not be vapor-sealed but should allow free passage of moisture. Take care to maintain the vapor seal between the outside of the electrical service and the cold-room vapor retarder.

Heat tracing is suggested inside the freezer only from the air-handling unit drain outlet panel to the insulated wall panel. Heat tracing within the wall could be a possible fire hazard and also cannot be serviced. Drain tracing can continue external to the freezer on a separate electrical circuit.

Tracking

Cold-room product suspension tracking, wherever possible, should be erected and supported within the insulated structure, entirely independent of the building itself. This eliminates flexure of the roof structure or overhead members, and simplifies maintenance.

Cold-Storage Doors

Doors should be strong yet light enough for easy opening and closing. Hardware should be of good quality, so that it can be set to compress the gasket uniformly against the casing. All doors to rooms operating below freezing should be equipped with heaters.

In-fitting doors are not recommended for rooms operating below freezing unless they are provided with heaters, and they should not be used at temperatures below 0°F with or without heaters.

See the subsection on Doors in the section on Applying Insulation for more information.

Hardware

All metal hardware, whether within the construction or exposed to conditions that will rust or corrode the base metal, should be heavily galvanized, plated, or otherwise protected. It is best to choose materials not subject to corrosion or rust from exposure to vapor condensation and cleaning agents used in the facility.

Refrigerated Docks

The purpose of the refrigerated facility (e.g., distribution, in-transit storage, or seasonal storage) dictates the loading dock requirements. Shipping docks and corridors should provide liberal space for (1) movement of goods to and from storage, (2) storage of pallets and idle equipment, (3) sorting, and (4) inspecting. The dock should be at least 30 ft wide. Commercial-use facilities usually require more truck dock space than specialized storage facilities because of the variety of products handled.

Floor heights of refrigerated vehicles vary widely but are often greater than those of unrefrigerated vehicles. Rail dock heights and building clearances should be verified by the railroad serving the plant. A dock height of 54 in. above the rail is typical for refrigerated rail cars. Three to five railroad car spots per million cubic feet of storage should be planned.

Truck dock heights must comply with the requirements of fleet owners and clients, as well as the requirements of local delivery trucks. Trucks generally require a 54 in. height above the pavement, although local delivery trucks may be much lower. Some reefer trucks are up to 58 in. above grade. Adjustable ramps at some truck spots will partly compensate for height variations. If dimensions permit, seven to ten truck spots per million cubic feet should be provided in a public refrigerated facility.

Refrigerated docks maintained at temperatures of 35 to 45°F require about 5 tons of refrigeration per 1000 ft^2 of floor area; however, actual load calculation should be done per ASHRAE methodology (see Chapter 13). Cushion-closure seals around the truck doorways reduce infiltration of outside air. Be sure to avoid gaps, particularly beneath the leveling plate between the truck and the dock. An inflatable or telescoping enclosure can be extended to seal the space between a railcar and the dock. Insulated doors for docks must be mounted on the inside walls. The relatively high costs of doors, cushion closures, and refrigeration influence dock size and number of doors.

Schneider System

The Schneider system, and modifications thereof, is a cold-storage construction and insulation method primarily used in the western United States, with most of the installations in the Pacific Northwest. It is an interior/exterior vapor retarder system, as illustrated in Figure 4. The structure uses concrete tilt-up walls and either glue-laminated wood beams or bowstring trusses for the roof. Fiberglass batts coupled with highly efficient vapor retarders and a support framework are used to insulate the walls and roof. The floor slab construction, insulation, and underfloor heat are conventional for refrigerated facilities.

The key to success for the Schneider system is an excellent vapor retarder system that is professionally designed and applied, with special emphasis on the wall/roof junction. Fiberglass has a high permeability rating and loses its insulating value when wet. It is therefore absolutely essential that the vapor retarder system perform at high efficiency. Typical wall vapor retarder materials include aluminum B foil and heavy-gage polyethylene, generously overlapped and adhered to the wall with a full coating of mastic. The roofing materials act as a vapor retarder for the roof. The vapor retarder at the wall/roof junction is usually a special aluminum foil assembly installed to perform efficiently in all weather conditions.

Fiberglass insulation applied to the wall is usually 10 to 12 in. thick for freezers and 6 to 8 in. for coolers. It is retained by offset wood or fabricated fiberglass-aluminum sheathed studs on 24 in. centers. Horizontal girts are used at intervals for bracing. The inside finish is 1 in. thick perforated higher-density fiberglass panels that can breathe to allow any moisture that passes through the vapor retarder to be deposited as frost on the evaporator coils.

Fiberglass insulation applied to the roof structure is usually 12 to 14 in. thick for freezers and 8 to 10 in. for coolers, and is applied between 2 by 12 in. or 2 by 14 in. joists that span the glue-laminated wood beams, purlins, or trusses. The exterior finish is the same as described for walls. Battens attached to the underside of the joists hold the finish panels and insulation in place.

Advantages of the Schneider system over insulated panels, assuming equal effectiveness of the vapor retarder/insulation envelopes over time, include lower first cost for structures over 40,000 ft^2, lower operating cost, and fewer interior columns. Disadvantages include a less clean appearance, unsuitability where washdown is required, impracticality where a number of smaller rooms are required, and a smaller number of capable practitioners (i.e., architects, engineers, contractors) available.

REFRIGERATION SYSTEMS

Types of Refrigeration Systems

Refrigeration systems can be broadly classified as unitary or applied. In this context, *unitary* systems are designed by manufacturers, assembled in factories, and installed in a refrigerated space as prepackaged units. Heat rejection and compression equipment is either within the same housing as the low-temperature air-cooling coils or separated from the cooling section. Such units normally use hydrochlorofluorocarbon (HCFC) and hydrofluorocarbon (HFC) refrigerants.

Applied units denote field-engineered and -erected systems and form the vast majority of large refrigerated (below-freezing) facility systems. Installations generally have a central machinery room or series of machinery rooms convenient to electrical distribution services, outside service entrance, etc., located as close to the refrigerated space as possible to reduce piping losses (pressure drop), piping costs, refrigerant charge, and thermal losses. Essentially made to order, applied systems are generally designed and built from standard components obtained from one or more suppliers. Key components include compressors, motors, fan-coil units, receivers, pump circulation systems, controls, refrigerant condensers (evaporative and shell-and-tube), and other pressure vessels.

The refrigeration system for a refrigerated facility should be selected in the early stages of planning. If the facility is a single-purpose, low-temperature storage building, most types of systems can be used. However, if commodities to be stored require different temperatures and humidities, a system must be selected that can meet the demands using isolated rooms at different conditions.

Using factory-built packaged unitary equipment may have merit for the smallest structures and for a multiroom facility that requires a variety of storage conditions. Conversely, the central compressor room has been the accepted standard for larger installations, especially where energy conservation is important.

Multiple centrally located, single-zone condensing units have been used successfully in Japan and other markets where high-rise refrigerated structures are used or where local codes drive system selection.

Direct refrigeration, either a flooded or pumped recirculation system serving fan-coil units, is a dependable choice for a central compressor room. Refrigeration compressors, programmable logic controllers, and microprocessor controls complement the central engine room refrigeration equipment.

Choice of Refrigerant

The choice of refrigerant is a very important decision in the design of refrigerated facilities. Typically, ammonia has been used, particularly in the food and beverage industries, but R-22 has been and is used, as well. Some low-temperature facilities now also use R-507A or R-404A, which are replacements of choice for R-502 and R-22. Factors to consider when choosing refrigerants include

- Cost
- Safety code issues, (e.g., code requirements regarding the use of refrigerant in certain types of occupied spaces)
- System refrigerant charge requirements [e.g., charges above 10,000 lb of NH_3 may require government-mandated process safety management (PSM) and risk management plan (RMP)]
- State and local codes, which may require full- or part-time operators with a specific level of expertise
- Effects on global warming and ozone depletion (ammonia has no effect on either)

Load Determination

Loads for refrigerated facilities of the same capacity vary widely. Many factors, including building design, indoor and outdoor temperatures, and especially the type and flow of goods expected and the daily freezing capacity, contribute to the load. Therefore, no simple design rules apply. Experience from comparable buildings and operations is valuable, but any projected operation should be analyzed. Compressor and room cooling equipment should be designed for maximum daily requirements, which will be well above any monthly average.

Load factors to be considered include

- Heat transmission through insulated enclosures
- Heat and vapor infiltration load from warm air passing into refrigerated space and improper air balance
- Heat from pumps or fans circulating refrigerant or air, power equipment, personnel working in refrigerated space, product-moving equipment, and lights
- Heat removed from goods in lowering their temperatures from receiving to storage temperatures
- Heat removed in freezing goods received unfrozen
- Heat produced by goods in storage
- Other loads, such as office air conditioning, car precooling, or special operations inside the building
- Refrigerated shipping docks
- Heat released from automatic defrost units by fan motors and defrosting, which increases overall refrigerant system requirements
- Blast freezing or process freezing

High humidity, warm temperatures, or manual product handling may dramatically affect design, particularly that of the refrigeration system.

A summation of the average proportional effect of the load factors is shown in Table 1 as a percentage of total load for a facility in the southern United States. Both the size and the effect of the load factors are influenced by the facility design, usage, and location.

Heat leakage or transmission load can be calculated using the known overall heat transfer coefficient of various portions of the insulating envelope, the area of each portion, and the temperature difference between the lowest cold-room design temperature and highest average air temperature for three to five consecutive days at the building location. For freezer storage floors on ground, the average yearly ground temperature should be used.

Heat infiltration load varies greatly with the size of room, number of openings to warm areas, protection on openings, traffic through openings, and cold and warm air temperatures and humidities. Calculation should be based on experience, remembering that most of the load usually occurs during daytime operations. Chapter 13 presents a complete analysis of refrigeration load calculation.

Heat from goods received for storage can be approximated from the quantity expected daily and the source. Generally, 10 to 20°F of temperature reduction can be expected, but for some newly processed items and for fruits and vegetables direct from harvesting, 60°F or more temperature reduction may be required. For general public cold storage, the load may range from 4 to 8 tons of cooling capacity per million cubic feet to allow for items received direct from harvest in a producing area.

Table 1 Refrigeration Design Load Factors for Typical 100,000 ft² Single-Floor Freezer*

Refrigeration Load Factors	Long-Term Storage Cooling Capacity		Short-Term Storage Cooling Capacity		Distribution Operation Cooling Capacity	
	Tons	%	Tons	%	Tons	%
Transmission losses	98	49	98	43	98	36
Infiltration	10	5	20	9	40	15
Internal operation loads	50	25	56	24	62	22
Cooling of goods received	7	3	15	6	30	11
Other factors	35	18	41	18	45	16
Total design capacity	200	100	230	100	275	100

Note: Based on a facility located in the southern United States using a refrigerated loading dock, automatic doors, and forklift material handling.
*See Chapter 13.

The freezing load varies from zero for the pure distribution facility, where the product is received already frozen, to the majority of the total for a warehouse near a producing area. The freezing load depends on the commodity, the temperature at which it is received, and method of freezing. More refrigeration is required for blast freezing than for still freezing without forced-air circulation.

Heat is produced by many commodities in cooler storage, principally fruits and vegetables. Heat of respiration is a sizable factor, even at 32°F, and is a continuing load throughout the storage period. Refrigeration loads should be calculated for maximum expected occupancy of such commodities.

Manual handling of product may add 30 to 50% more load to a facility in tropical areas due to constant interruption of the cold barriers at doors and on loading docks.

Unit Cooler Selection

Fan-Coil Units. These units may have direct-expansion, flooded, or recirculating liquid evaporators with either primary or finned-coil surfaces or a brine spray coolant. Storage temperature, packaging method, type of product, etc., must be considered when selecting a unit. Coil surface area, temperature difference between the refrigerant coil and the return air, and volumetric airflow depend on the application. Brine spray systems circulate a chemical mixture and water over the coil by spraying onto the coil upstream on the air side of the coil, to prevent frost formation on the coil. Filtration and other brine conditioning equipment are located outside of controlled-temperature areas. The sprayed brine is not a salt-water brine but rather a water-based glycol solution. Manufacturers claim these units can reduce microbial levels to help protect product from contamination. The units work well if they are maintained, but can be more expensive to purchase and operate and require additional room (for the regeneration equipment). They do not add defrost heat to the room and can often be placed above doorways to remove moisture in troublesome facilities to keep infiltration down to tolerable levels. Failure to maintain units can lead to contamination from dust, odor, and biological pollutants.

Fans are normally of the axial propeller type, but may be centrifugal if a high static discharge loss is expected. In refrigerated facilities, fan-coil units are usually draw-through (i.e., room air is drawn through the coil and discharged through the fan). Blow-through units are used in special applications, such as fruit storages, where refrigerant and air temperatures must be close. Heat from the motor is absorbed immediately by the coil on a blow-through unit and does not enter the room. Motor heat must be

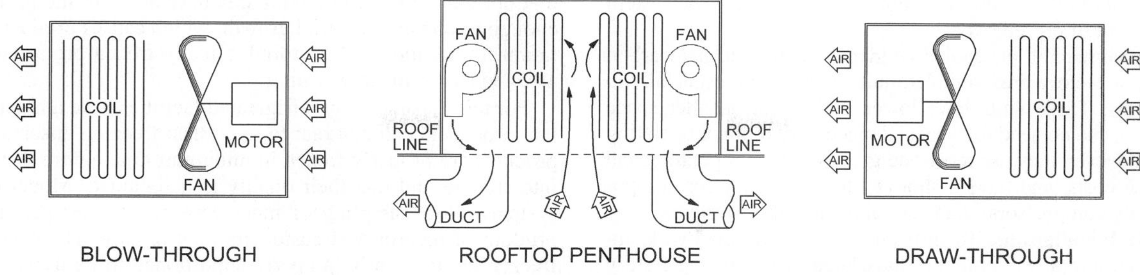

Fig. 6 Typical Fan-Coil Unit Configurations for Refrigerated Facilities

added to the room load with both draw-through and blow-through units. Figure 6 illustrates fan-coil units commonly used in refrigerated facility construction.

When selecting fan-coil units, consider the throw, or distance air must travel to cool the farthest area. Failure to properly consider throw and unit location can result in areas of stagnant air and hot spots in the refrigerated space (Crawford et al. 1992). Consult manufacturers' recommendations in all cases. Do not rely on guesses or rules of thumb to select units with proper airflow. Units vary widely in fan type, design of the diffuser leaving the fan-coil, and coil air pressure drop.

Defrosting. All fan-coils normally operate below room dew-point conditions. Fan coils operating below approximately 38°F will require some defrosting. Common methods of defrost in rooms 36°F and above include

- Air defrost
- Hot-gas defrost
- Electric defrost
- Water defrost

Rooms colder than 36°F normally use

- Hot-gas defrost
- Electric defrost

Units located above entrances to a refrigerated space tend to draw in warm, moist air from adjacent spaces and frost the coil quickly. If this occurs, more frequent defrosting is required to maintain the efficiency of the cooling coil. When the coil approach line crosses into the supersaturated region, a particularly unfavorable frost almost immediately clogs the coil, very rapidly decreasing performance (Sherif et al. 2001). Cleland and O'Hagan (2003) developed criteria to estimate when this will occur, providing a way to avoid this problem through redesign of the coil and/or the facility (e.g., so the load has a higher sensible heat ratio).

A properly engineered and installed system can be automatically defrosted successfully with hot gas, desiccant dehumidifier, water, electric heat, or continuously sprayed brine. The sprayed-brine system has the advantage of producing the full refrigeration capacity at all times; however, it does require a supply and return pipe system with a means of boiling off the absorbed condensed moisture, and can be subject to contamination with odors, biological pollution, or airborne dust.

Condensate Drains. When coils defrost, condensate that has formed as ice or frost on the coils melts. This new condensate collects in a pan beneath the coil and flows into collection drains outside of the freezer space. Because the space is cold, condensate pans are connected to the hot-gas defrost system or otherwise heated to prevent ice formation. Likewise, all condensate drain lines must be wrapped in heat-tracing tape and trapped outside of the refrigerated space to ensure that condensate can drain unrestricted.

Valve Selection. Refer to Chapter 44 and manufacturers' literature for specific information on control valve type and selection (sizing).

Valving Arrangements. Proper refrigerant feed valve, block valve, and defrosting valve arrangements are critical to the performance of all fan-coil units.

Various valve piping schemes are used. See Chapter 3 for typical piping arrangements.

Valve Location. Good valve location ensures convenient maintenance of control and service block valves. The owner/designer has some options in most plants. If penthouse units are used, all valves are generally located outside the penthouse and are accessible from the roof. Fan-coil units mounted in the refrigerated space are generally hung from the ceiling and must be accessed via personnel lift cage on a forklift or other service vehicle. It is recommended that valve stations be located outside the freezer storage area if possible to ensure that refrigerant leaks do not enter storage areas and also to facilitate maintenance.

System Considerations. For refrigerated temperatures below −25°F, two-stage compression is generally used. Compound compressors having capacity control on each stage may be used. For variable loads, separate high- and low-stage (or booster) compressors, each with capacity control or of different capacities, may provide better operation. Depending on the degree of capacity redundancy desired, two or more compressors can be selected at each suction temperature level. This also permits shutting one or more compressors down during colder months when load is reduced. Redundancy can also be provided on many systems by cross-connecting the piping such that a nonoperating high-stage compressor can also be run as a temporary low-stage single-stage compressor in case a booster compressor is down. Other combinations of cross connection are possible. If blast freezers are included, pipe connections should be arranged so that sufficient booster capacity for the blast freezers can be provided by the low-stage suction pressure compressor, while the other booster is at higher suction pressure for the freezer room load. Interstage pressure and temperatures are usually selected to provide refrigeration for loading dock cooling and for rooms above 32°F.

In a two-stage system, liquid refrigerant should be precooled at the high-stage suction pressure (interstage) to reduce the low-stage load. An automatic purger to remove air and other noncondensable gases is essential. Almost all compressors used in the refrigeration industry for facility designs use oil for lubrication. All these compressors lose a certain amount of oil from the compressor unit into the condenser and the low side of the system. Both halocarbon and ammonia plants should have means of recovering oil from all low-side vessels and heat exchangers where oil tends to accumulate. This includes low-pressure receivers, suction accumulators, pumper drums, shell and tube evaporators, surge tanks on gravity recirculation systems, intercoolers, subcoolers, and economizers. The compressor should have a good discharge oil separator. The means of recovering oil are different for halocarbons and ammonia. Oil is usually recovered from ammonia systems manually and then discarded, whereas oil can be recovered manually or automatically from halocarbon systems and is usually reused in the system. Refer to Chapters 1 to 7 in this volume for more information. Oil logs should be kept to record both the amount of oil added to the system and the amount of oil removed.

Use of commercial, air-cooled condenser, packaged halocarbon refrigerant, or factory preassembled units is common, especially in smaller plants. These units have lower initial cost, smaller space requirements, and no need for a special machinery room or operating engineer. However, they use more energy, have higher operating and maintenance costs, and have a shorter life expectancy for components (usually compressors) than central refrigeration systems.

Multiple Installations. To distribute air without ductwork, installations of multiple fan-coil units have been used. For single-story buildings, air-handling units installed in penthouses with ducted or nonducted air distribution arrangements have been used to make full use of floor space in the storage area (Figure 7). Either prefabricated or field-erected refrigeration systems or cooling units connected to a central plant can be incorporated in penthouse design.

Unitary cooling units are located in a penthouse, with distributing ductwork projected through the penthouse floor and under the insulated ceiling below. Return air passes up through the penthouse floor grille. This system avoids the interference of fan-coil units hung below the ceiling in the refrigerated chamber and facilitates maintenance access.

Condensate drain piping passes through the penthouse insulated walls and onto the main storage roof. Refrigerant mains and electrical conduit can be run over the roof on suitable supports to the central compressor room or to packaged refrigeration units on the adjacent roof. Thermostats and electrical equipment can be housed in the penthouse.

A personnel access door to the penthouse is required for convenient equipment service. The inside insulated penthouse walls and ceiling must be vaportight to keep condensation from deteriorating the insulation and to maintain the integrity of the building vapor retarder. Some primary advantages of penthouses are

- Cooling units, catwalks, and piping do not interfere with product storage space and are not subject to physical damage from stacking truck operations.
- Service to all cooling equipment and controls can be handled by one individual from a grated floor or roof deck location.
- Maintenance and service costs are minimized.
- Main piping, control devices, and block valves are located outside the refrigerated space.
- If control and block valves are located outside the penthouse, any refrigerant leaks will occur outside the refrigerated space.

Freezers

Freezers within refrigerated facilities are generally used to freeze products or to chill products from some higher temperature to storage temperature. Failure to properly cool the incoming product transfers the product cooldown load to the facility, greatly increasing facility operating costs. Of perhaps greater concern is that dormant storage in a cold area may not cool the product fast enough to prevent bacterial growth, which causes product deterioration. In addition, other stored, already frozen products may be affected by localized warming.

For this reason, many refrigerated facilities have a **blast freezer** that producers can contract to use. Blast freezing ensures that the products are properly frozen in minimum time before they are put into storage and that their quality is maintained. Modern control systems allow sampling of inner core product temperatures and printout of records that customers may require. The cost of blast freezer service can be properly apportioned to its users, allowing higher efficiency and lower cost for other cold-storage customers.

Although there are many types of freezers, including belt, tray, contact plate, spiral, and other packaged types, the most common arrangement used in refrigerated facilities is designed to accept pallets of products from a forklift. The freezing area is large and free from obstructions, and has large doors. See Chapter 16 for more information on freezing systems.

Figure 8 illustrates a typical blast freezer used in a refrigerated facility. Air temperatures are normally about −30°F, but may be higher or lower, depending upon the product being frozen. Once the room is filled to design capacity, it is sealed and the system is started. The refrigeration process time can be controlled by a time clock, by manual termination, or by measuring internal product temperature and stopping the process once the control temperature is reached. The last method gives optimum performance. Once the product is frozen, the pallets are transferred to general refrigerated storage areas.

Because the blast freezer normally operates intermittently, freezer owners should try to operate it during the times when energy cost is lowest. Unfortunately, food products must be frozen as quickly as possible, and products are usually delivered during times of peak electrical rates. Alternative power sources, such as natural gas engines or diesel drives, should be considered. Although these normally have first cost and maintenance cost premiums, they are not subject to time-varying energy rates and may offer savings.

Defrost techniques for blast freezers are similar to normal defrost methods for refrigerated facility fan-coil units. Coils can often be defrosted after the product cooling cycle is completed or while the freezer is being emptied for the next load.

Pumped refrigerant recycling systems and flooded surge drum coils have both been used with success. Direct-expansion coils may be used, but the designer should be careful with expansion valve systems to address coil circuitry, refrigerant liquid overfeed, oil return,

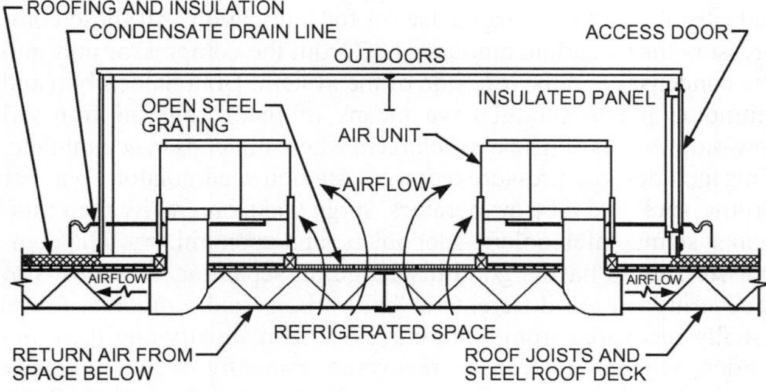

Fig. 7 Penthouse Cooling Units

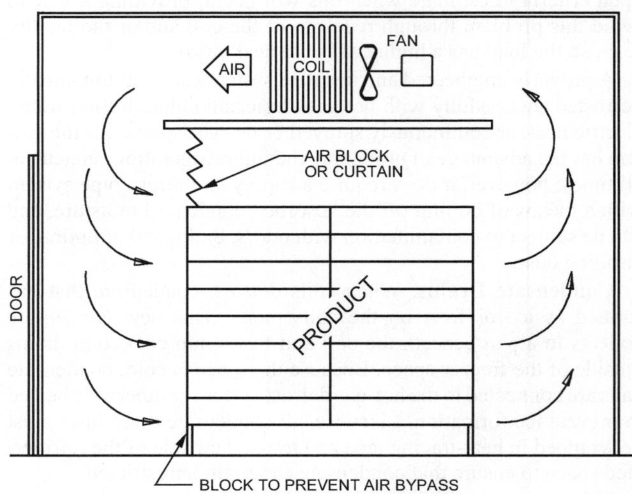

Fig. 8 Typical Blast Freezer

defrost, shutdown liquid inventory management, and so forth. Conventional oil removal devices should be supplied on flooded coil and pumped systems, because the blast freezer is normally the lowest-temperature system in the facility and may accumulate oil over time. Materials of construction for systems operating below –20°F and subject to ASME code conformance should comply with the latest ASME *Standard* B31.5. See Chapter 40 for further information on low-temperature materials. Floor heating may be convenient if products are damp or wet during loading.

Most blast freezers are accessed from a refrigerated space, so that products can be moved directly from the freezer to storage racks. Also, blast freezers can be used for storage when not operating.

Controls

The term *controls* refers to any mechanism or device used to start, stop, adjust, protect, or monitor the operation of a moving or functional piece of equipment. Controls for any system can be as simple as electromechanical devices such as pressure switches and timer relays or as complex as a complete digital control system with analog sensors and a high-speed communications network connected to a supervisory computer station. Because controls are required in every industry, there is a wide variety from which to choose. In recent years, the industrial refrigeration industry has moved away from the use of electromechanical devices and toward the use of specialized microprocessors, programmable logic controllers, and computers for unit and system control.

Electromechanical devices for control will continue to be used for some time and may never be replaced entirely for certain control functions (e.g., use of relays for electrical high current isolation and float switches for refrigerant high level shutdowns). See Chapter 44 for more information.

Microprocessor controls and electronic sensors generally offer the following advantages over electromechanical control:

- More accurate readings and therefore more accurate control
- Easier operation
- Greater flexibility through adjustable set points and operating parameters
- More information concerning operating conditions, alarms, failures, and troubleshooting
- Capability for interfacing with remote operator stations

There are four main areas of control in all refrigerated facility systems with a central compressor room:

- **Compressor package control.** Minimum requirements: orderly start-up, orderly shutdown, capacity control to maintain suction pressure, alarm monitoring, and safety shutdown.
- **Condenser control.** Minimum requirements: fan and water pump start and stop to maintain a reasonable constant or floating refrigerant discharge pressure.
- **Evaporator control.** Minimum requirements: control of air unit fans and refrigerant liquid feed to maintain room air temperatures and staging of air unit refrigerant valve stations to provide automatic coil defrosts.
- **Refrigerant flow management.** Minimum requirements: maintenance of desired refrigerant levels in vessels, control of valves and pumps to transfer refrigerant as needed between vessels and air units in the system, and proper shutdowns in the event of refrigerant overfeed or underfeed.

Other areas of control, such as refrigerant leak detection and alarm, sequencing of multiple compressors for energy efficiency, and underfloor warming system control, may be desired.

Because of the wide variety and fast-changing capabilities of control components and systems available, it is impossible to define or recommend an absolute component list. However, it is possible to provide guidelines for the design and layout of the overall control system, regardless of the components used or the functions to be controlled. This design or general layout can be termed a control system architecture.

All control systems consist of four main building blocks:

- **Controller(s):** Microprocessor with control software
- **Input/outputs (I/Os):** Means of connecting devices or measurements to the controller
- **Operator interface(s):** Means of conveying information contained in the controller to a human being
- **Interconnecting media:** Means of transferring information between controllers, I/Os, and operator interfaces

The control system architecture defines the quantity, location, and function of these basic components. The architecture determines the reliability, expandability, operator interface opportunities, component costs, and installation costs of a control system. Therefore, the architecture should be designed before any controls component manufacturers or vendors are selected.

The following are the basic steps in designing a refrigerated facility control system:

Step 1. Define the control tasks. This step should provide a complete and detailed I/O listing, including quantity and type. With this list and a little experience and knowledge of available hardware, the type, quantity, and processing power of the necessary controllers can be determined.

Step 2. Determine physical locations of controlled devices and measurements to be taken. If remote I/Os or multiple controllers are located close to the devices and sensors, field wiring installation costs can be reduced. To avoid extra costs or impracticalities, the environments of the various locations must be compared with the environmental specifications of the hardware to be placed in them. Maintenance requirements can also affect the selection of physical location of the I/Os and controller.

Step 3. Determine control task integration requirements. Control tasks that require and share the same information (such as a discharge pressure reading for starting both a condenser fan and a condenser water pump) must be accomplished either with the same controller or with multiple controllers that share information via interconnecting media. Tasks that do not share information can be performed by separate controllers. Using multiple controllers minimizes the chance of catastrophic control failures. With multiple controllers that share information, the interconnecting media must be robust, with minimal chance of failure for critical tasks. In particular, the speed of data transfer between controllers must be suitable to maintain the control accuracy required.

Step 4. Determine operator interface requirements. This includes noting which controllers must have a local or remote interface, how many remote interface stations are required, and defining the hardware and software requirements of the interfaces.

Step 5. Select the interconnecting media between controllers and their remote I/Os, between different controllers, and between controllers and operator interfaces. The interconnecting medium to remote I/Os is typically defined by the controller manufacturer; it must be robust and high-speed, because controllers' decisions depend on real-time data. The interconnecting medium between controllers themselves is also typically defined by the controller manufacturer; speed requirements depend on the tasks being performed with the shared information. For media connecting controllers and operator interfaces, speed is typically not as critical because the control continues even if the connection fails. For the operator interface connection, speed of accessing a controller's data is not as critical as having access to all the available data from the controller.

Step 6. Evaluate the architecture for technical merit. The first five steps should produce a list of controllers, their locations, their operator interfaces, and their control tasks. Once the list is complete, the selected controllers should be evaluated for both processor memory available for programming and processor I/O capacity available for current and future requirements. The selected interconnecting media should be evaluated for distance and speed limitations. If any weaknesses are found, a different model, type, or even manufacturer of the component should be selected.

Step 7. Evaluate the architecture for software availability. The best microprocessor is of little good if no software exists to make it operate. It must be ascertained that software exists or can easily be written to provide (1) information transfer between controllers and operator interfaces; (2) the programming functions needed to perform the control tasks; and (3) desired operator interface capabilities such as graphics, historical data, reports, and alarm management. Untested or proprietary software should be avoided.

Step 8. Evaluate the architecture for failure conditions. Determine how the system will operate with a failure of each controller. If the failure of a particular controller would be catastrophic, more controllers can be used to further distribute the control tasks, or electromechanical components can be added to allow manual completion of the tasks. For complex tasks that are impossible to control manually, it is essential that spare or backup control hardware be in stock and that operators be trained in the troubleshooting and reinstallation of control hardware and software.

Step 9. Evaluate the proposed architecture for cost, including field wiring, components, start-up, training, downtime, and maintenance costs. All these costs must be considered together for a fair and proper evaluation. If budgets are exceeded, then Steps 1 through 8 must be repeated, removing any nonessential control tasks and reducing the quantity of controllers, I/Os, and operator interfaces.

Once the control system architecture is designed, specifics of software operation should be determined. This includes items such as set points necessary for a control task, control algorithms and calculations used to determine output responses, graphic screen layouts, report layouts, alarm message wording, and so forth. More detail is necessary, but excessive time spent determining the details of software operation may be better applied to further definition and refinement of the system architecture. If the system architecture is solid, the software can always be modified as needed. With improper architecture, functional additions or corrections can be costly, time consuming, and sometimes impossible.

For more information on controls and their design and application, see Chapter 15 of the 2005 *ASHRAE Handbook—Fundamentals* and Chapters 41 and 46 of the 2003 *ASHRAE Handbook—HVAC Applications*.

INSULATION TECHNIQUES

The two main functions of an insulation envelope are to reduce the refrigeration requirements for the refrigerated space and to prevent condensation. See Chapter 33 for further information.

Vapor Retarder System

The primary concern in the design of a low-temperature facility is the vapor retarder system, which should be as close to 100% effective as is practical. The success or failure of an insulation envelope is due entirely to the effectiveness of the vapor retarder system in preventing water vapor transmission into and through the insulation.

The driving force behind water vapor transmission is the difference in vapor pressure across the vapor retarder. Once water vapor passes a vapor retarder, a series of detrimental events begins. Water migrating into the insulation may condense or solidify, which diminishes the thermal resistance of the insulation and eventually destroys the envelope. Ice formation inside the envelope system usually grows and physically forces the building elements apart to the point of failure.

Another practical function of the vapor retarder is to stop air infiltration, which can be driven by atmospheric pressure or ventilation.

After condensing or freezing, some water vapor in the insulation revaporizes or sublimes and is eventually drawn to the refrigeration coil and disposed of by the condensate drain, but the amount removed is usually not sufficient to dry out the insulation unless the vapor retarder break is located and corrected.

The vapor retarder must be located on the warm side of the insulation. Each building element inside the prime retarder must be more permeable than the last to permit moisture to move through it, or it becomes a site of condensation or ice. This precept is abandoned for the sake of sanitation at the inside faces of coolers. However, the inside faces of freezers are usually permitted to breathe by leaving the joints uncaulked in panel construction, or by using less permeable surfaces for other forms of construction. Factory-assembled insulation panels endure this double vapor retarder problem better than other types of construction.

In walls with insufficient insulation, the temperature at the inside wall surface may, during certain periods, reach the dew point of the migrating water vapor, causing condensation and freezing. This can also happen to a wall that originally had adequate insulation but, through condensation or ice formation in the insulation, lost part of its insulating value. In either case, ice deposited on the wall gradually pushes the insulation and protective covering away from the wall until the insulation structure collapses.

It is extremely important to properly install vapor retarders and seal joints in the vapor retarder material to ensure continuity from one surface to another (i.e., wall to roof, wall to floor, or wall to ceiling). Failure of vapor retarder systems for refrigerated facilities is almost always caused by poor installation. The contractor must be experienced in installation of vapor retarder systems to be able to execute a vaportight system.

Condensation on the inside of the cooler is unacceptable because (1) the wet surface provides a culture base for bacterial growth, and (2) any dripping onto the product gives cause for condemnation of the product in part or in whole.

Stagnant or dead air spots behind beams or inside metal roof decks can allow localized condensation. This moisture can be from within the cooler or freezer (i.e., not necessarily from a vapor retarder leak).

No vapor retarder system is 100% effective. A system is successful when the rate of moisture infiltration equals the rate of moisture removal by refrigeration, with no detectable condensation.

Types of Insulation

Rigid Insulation. Insulation materials, such as polystyrene, polyisocyanurate, polyurethane, and phenolic material, have proven satisfactory when installed with the proper vapor retarder and finished with materials that provide fire protection and a sanitary surface. Selection of the proper insulation material should be based primarily on the economics of the installed insulation, including the finish, sanitation, and fire protection.

Panel Insulation. Use of prefabricated insulated panels for insulated wall and roof construction is widely accepted. These panels can be assembled around the building structural frame or against masonry or precast walls. Panels can be insulated at the factory with either polystyrene or urethane. Other insulation materials do not lend themselves to panelized construction.

The basic advantage, besides economics, of using insulated panel construction is that repair and maintenance are simplified because the outer skin also serves as the vapor retarder and is accessible. This is of great benefit if the structure is to be enlarged in the future. Proper vapor retarder tie-ins then become practical.

Foam-in-Place Insulation. This application method has gained acceptance as a result of developments in polyurethane insulation and equipment for installation. Portable blending machines with a spray or frothing nozzles feed insulation into the wall, floor, or ceiling cavities to fill without joints the space provided for monolithic insulation

Table 2 Recommended Insulation R-Values

Type of Facility	Temperature Range, °F	Thermal Resistance R, °F·ft²·h/Btu		
		Floors	Walls/Suspended Ceilings	Roofs
Coolers[a]	40 to 50	Perimeter insulation only[c]	25	30 to 35
Chill coolers[a]	25 to 35	20	24 to 32	35 to 40
Holding freezer	−10 to −20	27 to 32	35 to 40	45 to 50
Blast freezers[b]	−40 to −50	30 to 40	45 to 50	50 to 60

Note: Because of wide variation in cost of energy and insulation materials based on thermal performance, a recommended R-value is given as a guide in each area of construction. For more exact values, consult a designer and/or insulation supplier.

[a]If a cooler may be converted to a freezer in the future, the owner should consider insulating the facility with higher R-values from the freezer section.

[b]R-values shown are for a blast freezer built within an unconditioned space. If the blast freezer is built within a cooler or freezer, consult a designer and/or insulation supplier.

[c]If high room relative humidity is desired, then floor insulation at least equal to that in the walls is recommended.

construction. This material does not provide significant vapor resistance; its application in floor construction should be limited.

Precast Concrete Insulation Panels. This specialized form of construction has been successful when proper vapor retarder and other specialized elements are incorporated. As always, vapor retarder continuity is the key to a successful installation.

Insulation Thickness

The R-value of insulation required varies with the temperature held in the refrigerated space and the conditions surrounding the room. Table 2 shows recommended R-values for different types of facilities. The range in R-values is due to variations in energy cost, insulation materials, and climatic conditions. For more exact values, consult a designer and/or insulation supplier. Insulation with R-values lower than those shown should not be used.

APPLYING INSULATION

The method and materials used to insulate roofs, ceilings, walls, floors, and doors need careful consideration.

Roofs

The suspended ceiling method of construction is preferred for attaining a complete thermal and vapor envelope. Insulating materials may be placed on the roof or floor above the refrigerated space rather than adhered to the structural ceiling. If this type of construction is not feasible, and the insulation must be installed under a concrete or other ceiling, then the vapor retarder, insulation, and finish materials should be mechanically supported from the structure above rather than relying on adhesive application only. Suspending a wood or metal deck from the roof structure and applying insulation and a vapor retarder to the top of the deck is another method of hanging ceiling insulation. Skill of application and attention to positive air and vapor seals are essential to continued effectiveness.

Suspended insulated ceilings, whether built-up or prefabricated, should be adequately ventilated to maintain near-ambient conditions in the plenum space; this minimizes both condensation and deterioration of vapor retarder materials (see the section on Suspended Ceilings). Permanent sealing is needed around insulating hanger rods, columns, conduit, and other penetrations.

The structural designer usually includes roofing expansion joints when installing insulation on top of metal decking or concrete structural slabs for a building larger than 100 by 100 ft. Because the refrigerated space is not normally subject to temperature variations, structural framing is usually designed without expansion or contraction joints if it is entirely enclosed within the insulation envelope. Board insulation laid on metal decking should be installed in two or more layers with the seams staggered. An examination of the coefficients of linear expansion for typical roof construction materials illustrates the need for careful attention to this phase of the building design.

Although asphalt built-up roofs have been used, loosely laid membrane roofing has become popular and requires little maintenance.

Walls

Wall construction must be designed so that as few structural members as possible penetrate the insulation envelope. Insulated panels applied to the outside of the structural frame prevent conduction through the framing. Where masonry or concrete wall construction is used, structural framing must be independent of the exterior wall. The exterior wall cannot be used as a bearing wall unless a suspended insulated ceiling is used.

Where interior insulated partitions are required, a double-column arrangement at the partition prevents structural members from penetrating the wall insulation. For satisfactory operation and long life of the insulation structure, envelope construction should be used wherever possible.

Governing codes for fire prevention and sanitation must be followed in selecting a finish or panel. For conventional insulation materials other than prefabricated panels, a vapor retarder system should be selected.

Abrasion-resistant membranes, such as 10 mil thick black polyethylene film with a minimum of joints, are suitable vapor retarders. Rigid insulation can then be installed dry and finished with plaster or sheet finishes, as the specific facility requires. In refrigerated facilities, contraction of the interior finish is of more concern than expansion because temperatures are usually held far below installation ambient temperatures.

Floors

Freezer buildings have been constructed without floor insulation, and some operate without difficulty. However, the possibility of failure is so great that this practice is seldom recommended.

Underfloor ice formation, which causes heaving of floors and columns, can be prevented by heating the soil or fill under the insulation. Heating can be by air ducts, electric heating elements, or pipes through which a liquid is recirculated (see the section on Floor Construction).

The air duct system works well for smaller storages. For a larger storage, it should be supplemented with fans and a source of heat if the pipe is more than 100 ft long. The end openings should be screened to keep out rodents, insects, and any material that might close off the air passages. The ducts must be sloped for drainage to remove condensed moisture. Perforated pipes should not be used.

The electrical system is simple to install and maintain if the heating elements are run in conduit or pipe so they can be replaced; however, operating costs may be very high. Adequate insulation should be used because it directly influences energy consumption.

The pipe grid system, shown in Figure 9, is usually best because it can be designed and installed to warm where needed and can later be regulated to suit varying conditions. Extensions of this system can be placed in vestibules and corridors to reduce ice and wetness on floors. The underfloor pipe grid also facilitates future expansion. A heat exchanger in the refrigeration system, steam, or gas engine exhaust can provide a source of heat for this system. The temperature of the recirculated fluid is controlled at 50 to 70°F, depending on design requirements. Almost universally, the pipes are made of plastic.

The pipe grid system is usually placed in the base concrete slab directly under the insulation. If the pipe is metal, a vapor retarder should be placed below the pipe to prevent corrosion. The fluid should be an antifreeze solution such as propylene glycol with the proper inhibitor.

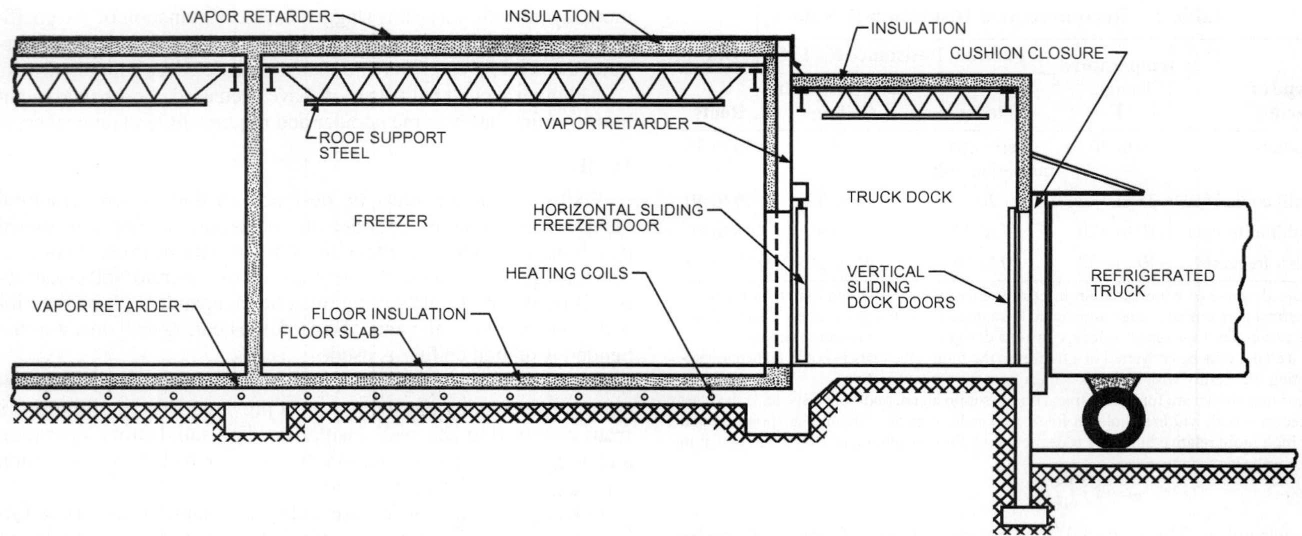

Fig. 9 Typical One-Story Construction with Underfloor Warming Pipes

The amount of warming for any system can be calculated and is about the same for medium-sized and large refrigerated spaces regardless of ambient conditions. The calculated heat input requirement is the floor insulation leakage based on the temperature difference between the 40°F underfloor earth and room temperature (e.g., 50°F temperature difference for a −10°F storage room). The flow of heat from the earth, about 1.3 Btu/h per square foot of floor, serves as a safety factor.

Freezer Doorways

An important factor in warehouse productivity is maintaining safe working conditions at doorways in high-usage freezers. At doorways, infiltration air mixes with air inside the freezer, forming airborne ice crystals. These crystals can accumulate on walls, ceilings, and nearby appurtenances, and can cause icy conditions on the floor. Consequences include danger to pedestrians, damage from skidding vehicles, premature frost clogging of nearby evaporators, and decreased productivity.

A **freezer vestibule** is any small room or airlock device with properly designed air curtains that impose little restriction on traffic flow but still counter adverse effects by reducing outside air infiltration.

Electrically heated traffic doors effectively eliminate doorway frost and ice.

Whether freezer vestibules or electrically heated doors are used, to calculate loads properly, see the section on Infiltration Air Load in Chapter 13, for door-open time per doorway pass-through and for time required to reach fully established flow upon each door opening.

Doors

The selection and application of cold-storage doors are a fundamental part of cold-storage facility design and have a strong bearing on the overall economy of facility operation. The trend is to have fewer and better doors. Manufacturers offer many types of doors supplied with the proper thickness of insulation for the intended use. Four basic types of doors are swinging, horizontal sliding, vertical sliding, and double-acting. Door manufacturers' catalogs give detailed illustrations of each. Doors used only for personnel cause few problems. In general, a standard swinging personnel door, 3 ft wide by 6.5 ft high and designed for the temperature and humidity involved, is adequate.

The proper door for heavy traffic areas should provide maximum traffic capacity with minimum loss of refrigeration and require minimum maintenance.

The following are factors to consider when selecting cold storage doors:

- Automatic doors are a primary requirement with forklift and automatic conveyor material-handling systems.
- Careless forklift operators are a hazard to door operation and effectiveness. Guards can be installed but are effective only when the door is open. Photoelectric and ultrasonic beams across the doorway or proximity loop control on both sides of the doorway can provide additional protection by monitoring objects in the door openings or approaches. These systems can also control door opening and closing.
- Selection of automatic door systems to suit traffic requirements and building structure may require experienced technical guidance.
- To ensure continuous door performance, the work area near the doors must be supervised, and the doors must have planned maintenance.
- Cooled or refrigerated shipping platforms increase door efficiency and reduce door maintenance, because the humidity and temperature difference across the doorway is lower. Icing of the door is lessened, and fogging in traffic ways is reduced.

Biparting and Other Doors. Air curtains, plastic or rubber strip curtains, and biparting doors give varied effectiveness. Strip curtains are not accepted by USDA standards if open product moves through the doorway. Often, the curtain seems to the forklift driver to be a substitute for the door, so the door is left open with a concurrent loss of refrigerated air. Quick-operating powered doors of fabric or rigid plastic are beneficial for draft control.

Swinging and Sliding Doors. A door with hinges on the right edge (when observed from the side on which the operating hardware is mounted) is called a right-hand swing. A door that slides to the right to open (when observed from the side of the wall on which it is mounted) is called a right-slide door.

Vertical Sliding Doors. These doors, which are hand- or motor-operated with counterbalanced springs or weights, are used on truck receiving and shipping docks.

Refrigerated-Room Doors. Doors for pallet material handling are usually automatic horizontal sliding doors, either single-slide or biparting.

Metal or Plastic Cladding. Light metal cladding or a reinforced plastic skin protects most doors. Areas of abuse must be further protected by heavy metal, either partial or full-height.

Heat. To prevent ice formation and resultant faulty door operation, doors are available with automatic electric heat, not only in the sides, head, and sill of the door or door frame, but also in switches and cover hoods of power-operated units. Such heating elements are necessary on all four edges of double-acting doors in low-temperature rooms. Safe devices that meet electrical codes must be used.

Bumpers and Guard Posts. Power-operated doors require protection from abuse. Bumpers embedded in the floor on both sides of the wall and on each side of the passageway help preserve the life of the door. Correctly placed guard posts protect sliding doors from traffic damage.

Buck and Anchorage. Effective door operation is impossible without good buck and anchorage provisions. Recommendations of the door manufacturers should be coordinated with wall construction.

Door Location. Doors should be located to accommodate safe and economical material handling. Irregular aisles and blind spots in trafficways near doors should be avoided.

Door Size. A hinged insulated door opening should provide at least 1 ft clearance on both sides of a pallet. Thus, 6 ft should be the minimum door width for a 4 ft wide pallet load. Double-acting doors should be 8 ft wide. Specific conditions at a particular doorway can require variations from this recommendation. A standard height of 10 ft accommodates all high-stacking forklifts.

Sill. A concrete sill minimizes the rise at the door sill. A thermal break should be provided in the floor slab at or near the plane of the front of the wall.

Power Doors. Horizontal sliding doors are standard when electric operation is provided. The two-leaf biparting unit keeps opening and closing time to a minimum, and the door is out of the way and protected from damage when open. Also, because leading edges of both leaves have safety edges, personnel, doors, trucks, and product are protected. A pull cord is used for opening, and a time-delay relay, proximity-loop control, or photoelectric cell controls closing. Potential for major door damage may be reduced by proper location of pull-cord switches. Doors must be protected from moisture and frost with heat or baffles. Preferably, low-moisture air should be introduced near door areas. Automatic doors should have a preventive maintenance program to check gaskets, door alignment, electrical switches, safety edges, and heating circuits. Safety releases on locking devices are necessary to prevent entrapment of personnel.

Fire-Rated Doors. Available in both swinging and sliding types, fire-rated doors are also insulated. Refrigerated buildings have increased in size, and their contents have increased in value, so insurance companies and fire authorities are requiring fire walls and doors.

Large Door Openings. Door openings that can accommodate forklifts with high masts, two-pallet-high loads, and tractor-drawn trailers are large enough to cause appreciable loss of refrigeration. Infiltration of moisture is objectionable because it forms as condensate or frost on stacked merchandise and within the building structure. Door heights up to 10 or 12 ft are frequently required, especially where drive-through racks are used. Refrigeration loss and infiltration of moisture can be particularly serious when doors are located in opposite walls of a refrigerated space and cross flow of air is possible. It is important to reduce infiltration with enclosed refrigerated loading docks and, in some instances, with one-way traffic vestibules.

OTHER CONSIDERATIONS

Temperature Pulldown

Because of the low temperatures in freezer facilities, contraction of structural members in these spaces will be substantially greater than in any surrounding ambient or cooler facilities. Therefore, contraction joints must be properly designed to prevent structural damage during facility pulldown.

The first stage of temperature reduction should be from ambient down to 35°F at whatever rate of reduction the refrigeration system can achieve.

The room should then be held at that temperature until it is dry. Finishes are especially subject to damage when temperatures are lowered too rapidly. Portland cement plaster should be fully cured before the room is refrigerated.

If there is a possibility that the room is airtight (most likely for small rooms, 20 by 20 ft maximum), swinging doors should be partially open during pulldown to relieve the internal vacuum caused by the cooling of the air, or vents should be provided. Permanent air relief vents are needed for continual operation of defrosts in small rooms with only swinging doors. Both conditions of possible air heating during defrost and cooling should be considered in design of air vents and reliefs.

The concrete slab will contract during pulldown, causing slab/wall joints, contraction joints, and other construction joints to open. At the end of the holding period (i.e., at 35°F), any necessary caulking should be done.

An average time for drying is 72 h. However, there are indicators that may be used, such as watching the rate of frost formation on the coils or measuring the rate of moisture removal by capturing the condensation during defrost.

After the refrigerated room is dry, the temperature can then be reduced again at whatever rate the refrigeration equipment can achieve until the operating temperature is reached. Rates of 10°F per day have been used in the past, but if care has been taken to remove all the construction moisture in the previous steps, faster rates are possible without damage.

Material-Handling Equipment

Both private and public refrigerated facilities can house high-volume, year-round operations with fast-moving order pick areas backed by in-transit bulk storage. Distribution facilities may carry 300 to 3000 items or as many as 30,000 lots. Palletized loads stored either in bulk or on racks are transported by forklifts or high-rise storage/retrieval machines in a 0 to −20°F environment. Standard battery-driven forklifts that can lift up to 25 ft can service one-deep, two-deep reach-in, drive-in, drive-through, or gravity flow storage racks. Special forklifts can lift up to 60 ft.

Automated storage/retrieval machines make better use of storage volume, require fewer personnel, and reduce the refrigeration load because the facility requires less roof and floor area. This equipment operates in a height range of 23 to 100 ft to service one-deep, two-deep reach-in, two- to twelve-deep roll pin, or gravity flow pallet storage racks. Computers and bar code identification allow a system to automatically control the retrieval, transfer, and delivery of products. In addition, these systems can record product location and inventory and load several delivery trucks simultaneously from one order pick conveyor and sorting device.

A refrigerated plant may have two or more material-handling systems if the storage area contains fast- and slow-moving reserve storage, plus slow-moving order pick. Fast-moving items may be delivered and order-picked by a conventional forklift pallet operation with up to 30 ft stacking heights. In the fast-moving order pick section, the storage room internal height is raised to accommodate storage/retrieval machines; reserve pallet storage; order pick slots; multilevel palletizing; and the infeed, discharge, and order pick conveyors. Mezzanines may be considered to provide maximum access to the order pick slots. Intermediate-level fire protection sprinklers may be required in the high rack or mezzanine areas above 14 ft high.

Fire Protection

Ordinary wet sprinkler systems can be applied to refrigerated spaces above freezing. In rooms below freezing, entering water freezes if a sprinkler head malfunctions or is mechanically damaged.

If this occurs, the affected piping must be removed. In lower-temperature spaces, a dry air or nitrogen-charged system should be selected.

Designing a dry sprinkler system operating in areas below 32°F requires special knowledge and should not be undertaken without expert guidance. Freezer storage with rack storages 30 ft high or higher may require special design, and the initial design should be shown to the insuring company.

Local regulations may require ceiling isolation smoke curtains and smoke vents near the roof in large refrigerated chambers. These features allow smoke to escape and help firefighters locate the fire. If the building does not have a sprinkler system, central reporting or warning systems are available for hazardous areas.

Inspection and Maintenance

Buildings dimensions can change because of settling, temperature change, and other factors; thus, cold-storage facilities should be inspected regularly to spot problems early, so that preventive maintenance can be performed in time to avert serious damage.

Inspection and maintenance procedures fall into two areas: basic system (floor, wall, and roof/ceiling systems); and apertures (doors, frames, and other access to cold storage rooms).

Basic System

- Stack pallets at a sufficient distance (18 in.) from walls or ceiling to permit air circulation.
- Examine walls and ceiling at random every month for frost buildup. If build-up persists, locate the break in the vapor retarder.
- For insulated ceilings below a plenum, inspect the plenum areas for possible roof leaks or condensation.
- If condensation or leaks are detected, make repairs immediately.

Apertures

- Remind personnel to close doors quickly to reduce frosting in rooms.
- Check the rollers and door travel periodically to ensure that the seal at the door edge is effective. If leaks are detected, adjust the door to restore a moisture- and airtight condition.
- Check doors and door edges to detect damage from forklifts or other traffic. Repair any damage immediately to prevent door icing or motor overload due to excessive friction.
- Lubricate doors according to the maintenance schedule from the door manufacturer to ensure free movement and complete closure.
- Periodically check seals around openings for ducts, piping, and wiring in the walls and ceiling.

REFERENCES

Cleland, D.J. and A.N. O'Hagan. 2003. Performance of an air cooling coil under frosting conditions. *ASHRAE Transactions* 109(1):243-250.

Crawford, R.R., J.P. Mavec, and R.A. Cole. 1992. Literature survey on recommended procedures for the selection, placement, and type of evaporators for refrigerated warehouses. *ASHRAE Transactions* 98(1): 500-513.

FDA. 2005. *Food Code*. U.S. Food and Drug Administration, Department of Health and Human Services, College Park, MD. (http://www.cfsan.fda.gov/~dms/fc05-toc.html)

Sherif, S.A., P. Mago, N.K. Al-Mutawa, R.S. Theen, and K. Bilen. 2001. Psychrometrics in the supersaturated frost zone (RP-1094). *ASHRAE Transactions* 107(2):753-767.

BIBLIOGRAPHY

Aldrich, D.F. and R.H. Bond. 1985. Thermal performance of rigid cellular foam insulation at subfreezing temperatures. In *Thermal performance of the exterior envelopes of buildings III*, pp. 500-509. ASHRAE.

ASME. 2001. Refrigeration piping and heat transfer components. *Standard* B31.5-2001. American Society of Mechanical Engineers, New York.

Baird, C.D., J.J. Gaffney, and M.T. Talbot. 1988. Design criteria for efficient and cost effective forced air cooling systems for fruits and vegetables. *ASHRAE Transactions* 94(1):1434-1454.

Ballou, D.F. 1981. A case history of a frost heaved freezer floor. *ASHRAE Transactions* 87(2):1099-1105.

Beatty, K.O., E.B. Birch, and E.M. Schoenborn. 1951. Heat transfer from humid air to metal under frosting conditions. *Journal of the ASRE* (December):1203-1207.

Cole, R.A. 1989. Refrigeration load in a freezer due to hot gas defrost and their associated costs. *ASHRAE Transactions* 95(2):1149-1154.

Coleman, R.V. 1983. Doors for high rise refrigerated storage. *ASHRAE Transactions* 89(1B):762-765.

Corradi, G. 1973. Air cooling units for the refrigerating industry and new equipment. *Revue generale du froid* 1(January):45-51.

Courville, G.E., J.P. Sanders, and P.W. Childs. 1985. Dynamic thermal performance of insulated metal deck roof systems. In *Thermal performance of the exterior envelopes of buildings III*, pp. 53-63. ASHRAE.

D'Artagnan, S. 1985. The rate of temperature pulldown. *ASHRAE Journal* 27(9):36.

Downing, C.G. and W.A. Meefert. 1993. Effectiveness of cold storage infiltration protection devices. *ASHRAE Transactions* 99(2):356-366.

Duminil, M., A. Ionov, B. Gazinski, and G. Cano-Munoz. 2002. *Insulation and airtightness of cold rooms*. International Institute of Refrigeration, Paris.

Hampson, G.R. 1981. Energy conservation opportunities in cold storage warehouses. *ASHRAE Transactions* 87(2):845-849.

Hendrix, W.A., D.R. Henderson, and H.Z. Jackson. 1989. Infiltration heat gains through cold storage room doorways. *ASHRAE Transactions* 95(2):1155-1168.

Holske, C.F. 1953. Commercial and industrial defrosting: General principles. *Refrigerating Engineering* 61(3):261-262.

Kerschbaumer, H.G. 1971. Analysis of the influence of frost formation on evaporators and of the defrost cycles on performance and power consumption of refrigerating systems. *Proceedings of the 13th International Congress of Refrigeration*, pp. 1-12.

Kurilev, E.S. and M.Z. Pechatnikov. 1966. Patterns of airflow distribution in cold storage rooms. *Bulletin of the International Institute of Refrigeration*, Annex 1966-1, pp. 573-579.

Lehman, D.C. and J.E. Ferguson. 1982. A modified jacketed cold storage design. *ASHRAE Transactions* 88(2):228-334.

Lotz, H. 1967. Heat and mass transfer and pressure drop in frosting finned coils: Progress in refrigeration science and technology. *Proceedings of the 12th International Congress of Refrigeration*, Madrid, vol. 2.

Niederer, D.H. 1976. Frosting and defrosting effects on coil heat transfer. *ASHRAE Transactions* 82(1):467-473.

Paulson, B.A. 1988. Air distribution in freezer areas. *International Institute of Ammonia Refrigeration 10th Annual Meeting*, pp. 261-271.

Powers, G.L. 1981. Ambient gravity air flow stops freezer floor heaving. *ASHRAE Transactions* 87(2):1107-1116.

Sainsbury, G.F. 1985. Reducing shrinkage through improved design and operation in refrigerated facilities. *ASHRAE Transactions* 91(1B): 726-734.

Sastry, S.K. 1985. Factors affecting shrinkage of foods in refrigerated storage. *ASHRAE Transactions* 91(1B):683-689.

Shaffer, J.A. 1983. Foundations and superstructure systems for stacker crane high-rise freezers and coolers. *ASHRAE Transactions* 89(1):753-756.

Sherman, M.H. and D.T. Grimsrud. 1980. Infiltration-pressurization correlation: Simplified physical modeling. *ASHRAE Transactions* 86(2): 778-807.

Soling, S.P. 1983. High rise refrigerated storage. *ASHRAE Transactions* 89(1B):737-761.

Stoecker, W.F. 1960. Frost formations on refrigeration coils. *ASHRAE Transactions* 66:91-103.

Stoecker, W.F. 1988. *Industrial refrigeration technical principles and practices text book*. Industrial Refrigeration, Business News Publishing.

Stoecker, W.F., J.J. Lux, and R.J. Kooy. 1983. Energy considerations in hot-gas defrosting of industrial refrigeration coils. *ASHRAE Transactions* 89(2A):549-568.

Treschel, H.R., P.R. Achenbach, and J.R. Ebbets. 1985. Effect of an exterior air infiltration barrier on moisture condensation and accumulation within insulated frame wall cavities. *ASHRAE Transactions* 91(2A):545-559.

Tye, R.P., J.P. Silvers, D.C. Brownell, and S.E. Smith. 1985. New materials and concepts to reduce energy losses through structural thermal bridges. In *Thermal performance of the exterior envelopes of buildings III*, pp. 739-750. ASHRAE.

Voelker, J.T. 1983. Insulating considerations for stacker crane high rise freezers and coolers. *ASHRAE Transactions* 89(1B):766-768.

Wang, I.H. and Touber. 1987. Prediction of airflow pattern in cold stores based on temperature measurements. *Proceedings of Commission D, International Congress of Refrigeration*, Vienna, pp. 52-60.

CHAPTER 15

METHODS OF PRECOOLING FRUITS, VEGETABLES, AND CUT FLOWERS

PRECOOLING is the rapid removal of field heat from freshly harvested fruits and vegetables before shipping, storage, or processing. Prompt precooling inhibits growth of microorganisms that cause decay, reduces enzymatic and respiratory activity, and reduces moisture loss. Thus, proper precooling reduces spoilage and retards loss of preharvest freshness and quality (Becker and Fricke 2002).

Precooling requires greater refrigeration capacity and cooling medium movement than do storage rooms, which hold commodities at a constant temperature. Thus, precooling is typically a separate operation from refrigerated storage and requires specially designed equipment (Fricke and Becker 2003). Precooling can be done by various methods, including hydrocooling, vacuum cooling, air cooling, and contact icing. These methods rapidly transfer heat from the commodity to a cooling medium such as water, air, or ice. Cooling times from several minutes to over 24 hours may be required.

PRODUCT REQUIREMENTS

During postharvest handling and storage, fresh fruits and vegetables lose moisture through their skins through transpiration. Commodity deterioration, such as shriveling or impaired flavor, may result if moisture loss is high. To minimize losses through transpiration and increase market quality and shelf life, commodities must be stored in a low-temperature, high-humidity environment. Various skin coatings and moisture-proof films can also be used during packaging to significantly reduce transpiration and extend storage life (Becker and Fricke 1996a).

Metabolic activity in fresh fruits and vegetables continues for a short period after harvest. The energy required to sustain this activity comes from respiration, which involves oxidation of sugars to produce carbon dioxide, water, and heat. A commodity's storage life is influenced by its respiratory activity. By storing a commodity at low temperature, respiration is reduced and senescence is delayed, thus extending storage life. Proper control of oxygen and carbon dioxide concentrations surrounding a commodity is also effective in reducing the respiration rate (Becker and Fricke 1996a).

Product physiology, in relation to harvest maturity and ambient temperature at harvest time, largely determines precooling requirements and methods. Some products are highly perishable and must begin cooling as soon as possible after harvest; examples include asparagus, snap beans, broccoli, cauliflower, sweet corn, cantaloupes, summer squash, vine-ripened tomatoes, leafy vegetables, globe artichokes, brussels sprouts, cabbage, celery, carrots, snow peas, and radishes. Less perishable produce, such as white potatoes, sweet potatoes, winter squash, pumpkins, and mature green tomatoes, may need to be cured at a higher temperature. Cooling of these products is not as important; however, some cooling is necessary if ambient temperature is high during harvest.

The preparation of this chapter is assigned to TC 10.9, Refrigeration Application for Food and Beverages.

Commercially important fruits that need immediate precooling include apricots; avocados; all berries except cranberries; tart cherries; peaches and nectarines; plums and prunes; and tropical and subtropical fruits such as guavas, mangos, papayas, and pineapples. Tropical and subtropical fruits of this group are susceptible to chilling injury and thus need to be cooled according to individual temperature requirements. Sweet cherries, grapes, pears, and citrus fruit have a longer postharvest life, but prompt cooling is essential to maintain high quality during holding. Bananas require special ripening treatment and therefore are not precooled. Chapter 11 lists recommended storage temperatures for many products.

CALCULATION METHODS

Heat Load

The refrigeration capacity needed for precooling is much greater than that for holding a product at a constant temperature or for slow cooling. Although it is imperative to have enough refrigeration for effective precooling, it is uneconomical to have more than is normally needed. Therefore, heat load on a precooling system should be determined as accurately as possible.

Total heat load comes from product, surroundings, air infiltration, containers, and heat-producing devices such as motors, lights, fans, and pumps. Product heat accounts for the major portion of total heat load, and depends on product temperature, cooling rate, amount of product cooled in a given time, and specific heat of the product. Heat from respiration is part of the product heat load, but it is generally small. Chapter 13 discusses how to calculate the refrigeration load in more detail.

Product temperature must be determined accurately to calculate heat load accurately. During rapid heat transfer, a temperature gradient develops in the product, with faster cooling causing larger gradients. This gradient is a function of product properties, surface heat transfer parameters, and cooling rate. Initially, for example, hydrocooling rapidly reduces the temperature of the exterior of a product, but may not change the center temperature at all. Most of the product mass is in the outer portion. Thus, calculations based on center temperature would show little heat removal, though, in fact, substantial heat has been extracted. For this reason, the product mass-average temperature must be used for product heat load calculations (Smith and Bennett 1965).

The product cooling load can then be calculated as

$$Q = mc_p(t_i - t_{ma}) \qquad (1)$$

where m is the mass of product being cooled, c_p is the product's specific heat, t_i is the product's initial temperature, and t_{ma} is the product's final mass average temperature. Specific heats of various fruits and vegetables can be found in Chapter 9.

Precooling Time Estimation Methods

Efficient precooler operation involves (1) proper sizing of refrigeration equipment to maintain a constant cooling medium temperature, (2) adequate flow of the cooling medium, and (3) proper product residence time in the cooling medium. Thus, to properly design a precooler, it is necessary to estimate the time required to cool the commodities from their initial temperature (usually the ambient temperature at harvest) to the final temperature, just before shipping and/or storage. For a specified cooling medium temperature and flow rate, this cooling time dictates the residence time in the precooler that is required for proper cooling (Fricke and Becker 2003).

Accurate estimations of precooling times can be obtained by using finite-element or finite-difference computer programs, but the effort required makes this impractical for the design or process engineer. In addition, two- and three-dimensional simulations require time-consuming data preparation and significant computing time. Most research to date has been in the development of semianalytical/ empirical precooling time estimation methods that use simplifying assumptions, but nevertheless produce accurate results.

Fractional Unaccomplished Temperature Difference

All cooling processes exhibit similar behavior. After an initial lag, the temperature at the food's thermal center decreases exponentially (see Chapter 10). As shown in Figure 1, a cooling curve depicting this behavior can be obtained by plotting, on semilogarithmic axes, the fractional unaccomplished temperature difference Y [Equation (2)] versus time (Fricke and Becker 2004).

$$Y = \frac{t_m - t}{t_m - t_i} = \frac{t - t_m}{t_i - t_m} \tag{2}$$

where t_m is the cooling medium temperature, t_i is the initial commodity temperature, and t is the commodity final mass average temperature. This semilogarithmic temperature history curve consists of an initial curvilinear portion, followed by a linear portion. Simple empirical formulas that model this cooling behavior, such as half-cooling time and cooling coefficient, have been proposed for estimating the cooling time of fruits and vegetables.

Half-Cooling Time

A common concept used to characterize the cooling process is the half-cooling time, which is the time required to reduce the temperature difference between the commodity and the cooling medium by half (Becker and Fricke 2002). This is also equivalent to the time required to reduce the fractional unaccomplished temperature difference Y by half.

The half-cooling time is independent of initial temperature and remains constant throughout the cooling period as long as the cooling medium temperature remains constant (Becker and Fricke 2002). Therefore, once the half-cooling time has been determined for a given commodity, cooling time can be predicted regardless of the commodity's initial temperature or cooling medium temperature.

Product-specific nomographs have been developed, which, when used in conjunction with half-cooling times, can provide estimates of cooling times for fruits and vegetables (Stewart and Couey 1963). In addition, a general nomograph (Figure 2) was constructed to calculate hydrocooling times of commodities based on their half-cooling times (Stewart and Couey 1963). In Figure 2, product temperature is plotted along the vertical axis versus time measured in half-cooling periods along the horizontal axis. At zero time, the product temperature is the initial commodity temperature; at infinite time, product temperature equals water temperature. To use Figure 2, draw a straight line from the initial commodity temperature at zero time (left axis) to the commodity temperature at infinite time [i.e., the water temperature (right axis)]. Then draw a horizontal line at the final commodity temperature (left and right axes). The intersection of these two lines determines the number of half-cooling periods required (bottom axis). Multiply the half-cooling time for the particular commodity by the number of half-cooling periods to obtain the hydrocooling time.

The following example illustrates the use of the general nomograph for determining hydrocooling time.

Example 1. Assume that topped radishes with a half-cooling time of 2.2 min are to be hydrocooled using 32°F water. How long would it take to hydrocool the radishes from 80° to 50°F?

Solution. Using the general nomograph in Figure 2, draw a straight line from 80° on the left to 32°F on the right. Then draw a horizontal line at the final commodity temperature, 50°F. These lines intersect at 1.4 half-cooling periods. Multiply this by the half-cooling time (2.2 min) to obtain the total hydrocooling time of 3.1 min.

Using nomographs can be time consuming and cumbersome, however. Cooling time θ of fruits and vegetables may be determined without the use of nomographs by using the half-cooling time Z:

$$\theta = \frac{-Z \ln(Y)}{\ln(2)} \tag{3}$$

Values of half-cooling times for the hydrocooling of numerous commodities have been reported (Bennett 1963; Dincer 1995; Dincer and Genceli 1994, 1995; Guillou 1958; Nicholas et al. 1964; O'Brien and Gentry 1967; Stewart and Couey 1963). Tables 1 to 3 summarize half-cooling time data for a variety of commodities.

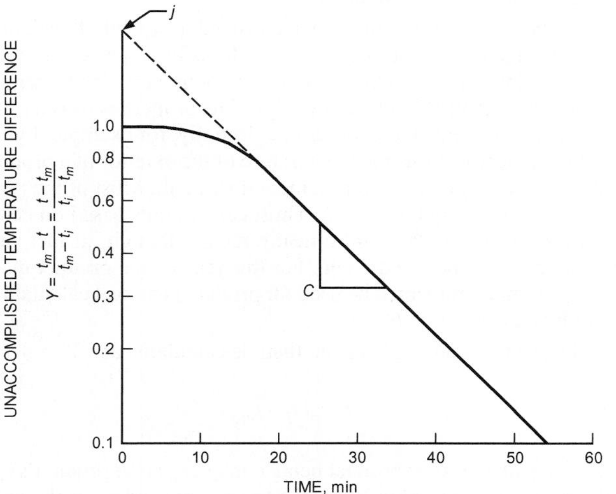

Fig. 1 Typical Cooling Curve

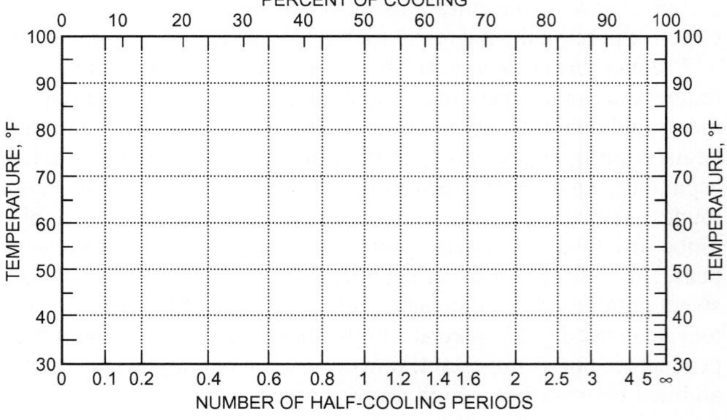

Fig. 2 General Nomograph to Determine Half-Cooling Periods (Stewart and Couey 1963)

Table 1 Half-Cooling Times for Hydrocooling of Various Commodities

Commodity	Commodity Size	Container	Half-Cooling Time, min.
Artichoke		None (completely exposed)	8
		Crate, lid off, paper liner	12
Asparagus	Medium	Completely exposed	1.1
		Lidded pyramid crate, spears upright	2.2
Broccoli		Completely exposed	2.1
		Crate with paper liner, lid off	2.2
		Crate without liner, lid off	3.1
Brussels sprouts		Completely exposed	4.4
		Carton, lid open	4.8
		Jumble stack (9 in. deep)	6.0
Cabbage		Completely exposed	69
		Carton, lid open	81
		Jumble stack (four layers)	81
Carrots, topped	Large	Completely exposed	3.2
		50 lb mesh bag	4.4
Cauliflower, trimmed		Completely exposed	7.2
Celery	2 Dozen	Completely exposed	5.8
		Crate, lidded, paper liner	9.1
Sweet corn, in husks	5 Dozen	Completely exposed	20
		Wirebound corn crate, lidded	28
Peas, in pod		Completely exposed (flood)	1.9
		1 bushel basket, lid off (flood)	2.8
		1 bushel basket, lidded (submersion)	3.5
Potatoes		Completely exposed	11
		Jumble stack (five layers, 9 in. deep)	11
Radishes		Completely exposed	1.1
		Crate, lid off, three layers of bunches, 9 in. deep	1.9
		Carton, lid open, three layers of bunches, 9 in. deep	1.4
topped		Completely exposed	1.6
		Jumble stack (9 in. deep)	2.2
Tomatoes		Completely exposed	10
		Jumble stack, five layers, 10 in. deep	11

Source: Stewart and Couey (1963).

Cooling Coefficient

Cooling time may also be predicted using the cooling coefficient C. As shown in Figure 1, the cooling coefficient is minus the slope of the $\ln(Y)$ versus time curve, constructed on a semilogarithmic axis from experimental observations of time and temperature (Becker and Fricke 2002). The cooling coefficient indicates the change in the fractional unaccomplished temperature difference per unit cooling time (Dincer and Genceli 1994). The cooling coefficient depends on the commodity's specific heat and thermal conductance to the surroundings (Guillou 1958). Using the cooling coefficient for a particular cooling process, cooling time θ may be estimated as

$$\theta = -\frac{1}{C}\ln\left(\frac{Y}{j}\right) \tag{4}$$

The lag factor j is a measure of the time between the onset of cooling and the point at which the slope of the $\ln(Y)$ versus θ curve becomes constant [i.e., the time required for the $\ln(Y)$ versus θ curve to become linear]. The lag factor j can be found by extending the linear portion of the semilogarithmic cooling curve to the $\ln(Y)$ axis; the intersection is the lag factor j.

By substituting $Y = 0.5$ into Equation (4), which corresponds to the half-cooling time, cooling coefficient C can be related to half-cooling time Z as follows:

$$Z = \frac{\ln(2j)}{C} \tag{5}$$

Cooling coefficients have been reported by Dincer (1995, 1996), Dincer and Genceli (1994, 1995), Henry and Bennett (1973), and Henry et al. (1976) for hydrocooling and hydraircooling (see the Cooling Methods section for discussion of these methods) various commodities, as summarized in Tables 2 to 4.

Other Semianalytical/Empirical Precooling Time Estimation Methods

Chapter 10 discusses various semianalytical/empirical methods for predicting cooling times of regularly and irregularly shaped foods. These cooling time estimation methods are grouped into two main categories: those based on (1) f and j factors (for either regular or irregular shapes), and (2) equivalent heat transfer dimensionality.

Numerical Techniques

Becker and Fricke (1996b, 2001) and Becker et al. (1996a, 1996b) developed a numerical technique for determining cooling rates as well as latent and sensible heat loads caused by bulk refrigeration of fruits and vegetables. This computer model can predict commodity moisture loss during refrigerated storage and the temperature distribution within the refrigerated commodity, using a porous media approach to simulate the combined phenomena of transpiration, respiration, airflow, and convective heat and mass transfer. Using this numerical model, Becker et al. (1996b) found that increased airflow decreases moisture loss by reducing cooling time, which quickly reduces the vapor pressure deficit between the commodity and surrounding air, thus lowering the transpiration rate. They also found that bulk mass and airflow rate were of primary importance to cooling time, whereas relative humidity had little effect on cooling time.

COOLING METHODS

The principal methods of precooling are hydrocooling, forced-air cooling, forced-air evaporative cooling, package icing, and vacuum cooling. Precooling may be done in the field, in central cooling facilities, or at the packinghouse.

HYDROCOOLING

In hydrocooling, commodities are sprayed with chilled water, or immersed in an agitated bath of chilled water. Hydrocooling is effective and economical; however, it tends to produce physiological and pathological effects on certain commodities; therefore, its use is limited (Bennett 1970). In addition, proper sanitation of the hydrocooling water is necessary to prevent bacterial infection of commodities. Commodities that are often hydrocooled include asparagus, snap beans, carrots, sweet corn, cantaloupes, celery, snow peas, radishes, tart cherries, and peaches. Cucumbers, peppers, melons, and early crop potatoes are sometimes hydrocooled. Apples and citrus fruits are rarely hydrocooled. Hydrocooling is not popular for citrus fruits because of their long marketing season; good postharvest holding ability; and susceptibility to increased peel injury, decay, and loss of quality and vitality after hydrocooling.

Hydrocooling is rapid because the cold water flowing around the commodities causes the commodity surface temperature to essentially equal that of the water (Ryall and Lipton 1979). Thus, the resistance to heat transfer at the commodity surface is negligible. The rate of internal cooling of the commodity is limited by the rate

Table 2 Lag Factors, Cooling Coefficients, and Half-Cooling Times for Hydrocooling Various Fruits and Vegetables

Commodity and Size	Temperature, °F			Water Flow Rate, ft/min	Crate Load, lb	Lag Factor j	Cooling Coefficient C, s^{-1}	Half-Cooling Time Z, s	Reference
	Initial	Final	Water						
Cucumbers l = 6.3 in. d = 1.5 in.	72	39		9.8	11	1.291	0.001601	546.6	Dincer and Genceli 1994
					22	1.177	0.001567	592.3	
					33	1.210	0.001385	638.2	
					44	1.251	0.001243	737.6	
			33	9.8	11	1.037	0.001684	432.9	Dincer 1995
					22	1.228	0.001675	536.4	
					33	1.222	0.001629	548.5	
					44	1.237	0.001480	612.1	
Eggplant l = 5.6 in. d = 1.8 in.	71			9.8	11	1.077	0.000822	933.9	Dincer 1995
					22	1.109	0.000794	1003	
					33	1.195	0.000870	1011	
					44	1.206	0.000770	1143	
Peaches d = 2.2 in.	70	39		9.8	11	1.067	0.001585		Dincer 1996
					44	1.113	0.001201		
Pears d = 2.4 in.	73	39	34	9.8	11	1.119	0.001434	561.6	Dincer and Genceli 1995
					22	1.157	0.001419	591.0	
					33	1.078	0.001296	592.8	
					44	1.366	0.001151	873.1	
		36		9.8	11	1.076	0.001352		Dincer 1996
					44	1.366	0.001151		
Plums d = 1.5 in.	72	36		9.8	11	1.122	0.003017		Dincer 1996
					44	1.171	0.002279		
Squash l = 6.1 in. d = 1.8 in.	71		33	9.8	11	1.172	0.001272	669.6	Dincer 1995
					22	1.202	0.001186	739.8	
					33	1.193	0.001087	799.9	
					44	1.227	0.001036	866.6	
Tomatoes d = 2.8 in.	70		33	9.8	11	1.209	0.001020	865.4	Dincer 1995
					22	1.310	0.000907	1062	
					33	1.330	0.000800	1222	
					44	1.322	0.000728	1336	
		39		9.8	11	1.266	0.000953		Dincer 1996
					44	1.335	0.000710		

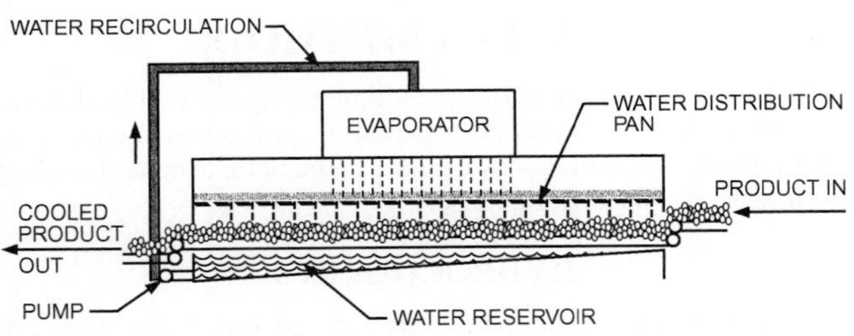

Fig. 3 Schematic of Shower Hydrocooler
(USDA 2004)

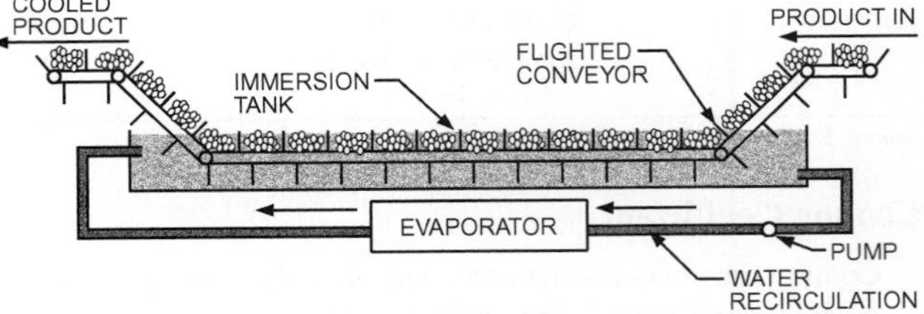

Fig. 4 Schematic of Immersion Hydrocooler
(USDA 2004)

of heat transfer from the interior to the surface, and depends on the commodity's volume in relation to its surface area, as well as its thermal properties. For example, Stewart and Lipton (1960) showed a substantial difference in half-cooling time for sizes 36 and 45 cantaloupes. A weighted average of temperatures taken at different depths showed that 20 min was required to half-cool size 36 melons and only 10 min for size 45.

Hydrocooling also has the advantage of causing no commodity moisture loss. In fact, it may even rehydrate slightly wilted product (USDA 2004). Thus, from a consumer standpoint, the quality of hydrocooled commodities is high; from the producer's standpoint, the salable weight is high. In contrast, other precooling methods such as vacuum or air cooling may lead to significant commodity moisture loss and wilting, thus reducing product quality and salable weight.

Commodities may be hydrocooled either loose or in packaging (which must allow for adequate water flow within and must tolerate contact with water without losing strength). Plastic or wood containers are well suited for use in hydrocoolers. Corrugated fiberboard containers can be used in hydrocoolers, if they are wax-dipped to withstand water contact (USDA 2004).

Types of Hydrocoolers

Hydrocooler designs can generally be divided into two categories: shower-type and immersion. In a **shower hydrocooler**, the commodities pass under a shower of chilled water (Figure 3), which is typically achieved by flooding a perforated pan with chilled

Table 3 Cooling Coefficients and Half-Cooling Times for Hydraircooling Sweet Corn and Celery

Commodity	Crate Type	Spray Nozzle Type	Water Flow Rate, ft³/min	Airflow Rate, ft³/min	Cooling Coefficient C, s^{-1}	Half-Cooling Time, s	Reference
Sweet corn	Wirebound	Coarse	12	0	0.000347		Henry and Bennett 1973
			12	0	0.000444		
			7.4	0	0.000642		
			13	0	0.000336		
		Medium	11	0	0.000406		
			6.7	0	0.000406		
			6.7	—	0.000414		
			13	0	0.000492		
			13	—	0.000542		
			13	1000	0.000447		
			13	1600	0.000486		
			13	2760	0.000564		
		Flood pan	33	0	0.000464		
			53	0	0.000567		
		Coarse	13	0		2170	Henry et al. 1976
		Medium	11	0		1730	
			13	1000		1570	
			13	1590		1440	
			13	2770		1220	
		Flood pan	5.3	0		1290	
Celery	Vacuum-cooling		6.1	2000		3710	Henry et al. 1976
			6.1	4200		2360	
			6.1	6470		2310	
	Hydrocooling		6.1	1800		1890	
			6.1	3500		1790	
			6.1	5000		1390	
	Well-ventilated		6.1	1800		2170	
			6.1	4000		1490	
			6.1	5100		1050	

Table 4 Cooling Coefficients for Hydrocooling Peaches

Hydrocooling Method	Water Flow	Water Temp., °F	Fruit Temp., °F Initial	Final	Cooling Coefficient, s^{-1}
Flood, peaches in 3/4 bushel baskets	0.667 ft³/min·ft²	35	88	47	0.00105
	1.33 ft³/min·ft²	35	85	44	0.00111
		40	82	49	0.000941
		45	82	49	0.00144
	2.00 ft³/min·ft²	35	91	39	0.00183
		45	89	51	0.00174
		55	88	58	0.00139
Immersion	2.67 ft³/min	35	85	44	0.00123
	5.35 ft³/min	35	85	42	0.00137
	2.67 ft³/min	45	88	49	0.00168
	5.35 ft³/min	45	86	49	0.00172
	8.00 ft³/min	45	86	51	0.00130

Source: Bennett (1963).

water. Gravity forces the water through the perforated pan and over the commodities. Shower hydrocoolers may have conveyors for continuous product flow, or may be operated in batch mode. Water flow rates typically range from 0.17 to 0.33 gal/s per square foot of cooling area (Bennett et al. 1965; Boyette et al. 1992; Ryall and Lipton 1979). **Immersion hydrocoolers** (Figure 4) consist of large, shallow tanks that contain agitated, chilled water. Crates or boxes of commodities are loaded onto a conveyor at one end of the tank, travel submerged along the length of the tank, and are removed at the opposite end. For immersion hydrocooling, a water velocity of 3 to 4 in/s is suggested (Bennett 1963; Bennett et al. 1965).

In large packing facilities, flooded ammonia refrigeration systems are often used to chill hydrocooling water. Cooling coils are placed directly in a tank through which water is rapidly circulated. Refrigerant temperature inside the cooling coils is typically 28°F, producing a chilled-water temperature of about 34°F. Because of the high cost of acquiring and operating mechanical refrigeration units, they are typically limited to providing chilled water for medium- to high-volume hydrocooling operations.

Smaller operations may use crushed ice rather than mechanical refrigeration to produce chilled water. Typically, large blocks of ice are transported from an ice plant to the hydrocooler, and then crushed and added to the hydrocooler's water reservoir. The initial cost of an ice-cooled hydrocooler is much less than that of one using mechanical refrigeration. However, for an ice-cooled hydrocooler to be economically viable, a reliable source of ice must be available at a reasonable cost (Boyette et al. 1992).

Variations on Hydrocooling

Henry and Bennett (1973) and Henry et al. (1976) describe **hydraircooling**, in which a combination of chilled water and chilled air is circulated over commodities. Hydraircooling requires less water for cooling than conventional hydrocooling, and also reduces the maintenance required to keep the cooling water clean. Cooling rates equal to, and in some cases better than, those obtained in conventional unit load hydrocoolers are possible.

Robertson et al. (1976) describe a process in which vegetables are frozen by **direct contact with aqueous freezing media**. The aqueous freezing media consists of a 23% NaCl solution. Freezing times of less than one minute were reported for peas, diced carrots, snow peas, and cut green beans, and a cost analysis indicated that freezing with aqueous freezing media was competitive to air-blast freezing.

Lucas and Raoult-Wack (1998) note that immersion chilling and freezing using aqueous refrigerating media have the advantage of shorter process times, energy savings, and better food quality compared to air-blast chilling or freezing. The main disadvantage is absorption of solutes from the aqueous solution by food. Immersion chilling or freezing with aqueous refrigerating media can be applied to a broad range of foods, including pork, fish, poultry, peppers, beans, tomatoes, peas, and berries.

As an alternative to producing chilled water with mechanical refrigeration or ice, **well water** can be used, provided that the water temperature is at least 10°F lower than that of the product to be cooled. However, the well water must not contain chemicals and biological pollutants that could render the product unsuitable for human consumption (Gast and Flores 1991).

Hydrocooler Efficiency

Hydrocooling efficiency is reduced by heat gain to the water from surrounding air. Other heat sources that reduce effectiveness include solar loads, radiation from hot surfaces, and conduction from the surroundings. Protection from these sources enhances efficiency. Energy can also be lost if a hydrocooler operates at less than full capacity or intermittently, or if more water than necessary is used (Boyette et al. 1992).

To increase hydrocooler energy efficiency, the following factors should be considered during design and operation (Boyette et al. 1992):

- Insulate all refrigerated surfaces and protect the hydrocooler from wind and direct sunlight.
- Use plastic strip curtains on both the inlet and outlet of conveyor hydrocoolers to reduce infiltration heat gain.
- Operate the hydrocooler at maximum capacity.
- Consider using thermal storage, in which chilled water or ice is produced and stored during periods of low energy demand and is subsequently used along with mechanical refrigeration to chill hydrocooling water during periods of peak energy demand. Thermal storage reduces the size of the required refrigeration equipment and may decrease energy costs.
- Use an appropriately sized water reservoir. Because energy is wasted when hydrocooling water is discarded after operation, this waste can be minimized by not using an oversized water reservoir. On the other hand, it may be difficult to maintain consistent hydrocooling water temperature and flow rate with an undersized water reservoir.

Hydrocooling Water Treatment

The surface of wet commodities provides an excellent site for diseases to thrive. In addition, because hydrocooling water is recirculated, decay-producing organisms can accumulate in the hydrocooling water and can easily spread to other commodities being hydrocooled. Thus, to reduce the spread of disease, hydrocooling water must be treated with mild disinfectants.

Typically, hydrocooling water is treated with chlorine to minimize the levels of decay-producing organisms (USDA 2004). Chlorine (gaseous, or in the form of hypochlorous acid from sodium hypochlorite) is added to the hydrocooling water, typically at the level of 50 to 100 ppm. However, chlorination only provides a surface treatment of the commodities; it is not effective at neutralizing an infection below the commodity's surface.

The chlorine level in the hydrocooling water must be checked at regular intervals to ensure that the proper concentration is maintained. Chlorine is volatile and disperses into the air at a rate that increases with increasing temperature (Boyette et al. 1992). Furthermore, if ice cooling is used, melting in the hydrocooling water dilutes the chlorine in solution.

The effectiveness of the chlorine in the hydrocooling water strongly depends on the pH of the hydrocooling water, which should be maintained at 7.0 for maximum effectiveness (Boyette et al. 1992).

To minimize debris accumulation in the hydrocooling water, it may be necessary to wash commodities before hydrocooling. Nevertheless, hydrocooling water should be replaced daily, or more often if necessary. Take special care when disposing of hydrocooling water, because it often contains high concentrations of sediment, pesticides, and other suspended matter. Depending on the municipality, hydrocooling water may be considered an industrial wastewater and, thus, a hydrocooler owner may be required to obtain a wastewater discharge permit (Boyette et al. 1992). In addition to daily replacement of hydrocooling water, shower pans and/or debris screens should be cleaned daily, or more often if necessary, for maximum efficiency.

FORCED-AIR COOLING

Theoretically, air cooling rates can be comparable to hydrocooling under certain conditions of product exposure and air temperature. In air cooling, the optimum value of the surface heat transfer coefficient is considerably smaller than in cooling with water. However, Pflug et al. (1965) showed that apples moving through a cooling tunnel on a conveyer belt cool faster with air at 20°F approaching the fruit at 600 fpm than they would in a water spray at 35°F. For this condition, they calculated an average film coefficient of heat transfer of 7.3 Btu/h·ft^2·°F. They noted that the advantage of air is its lower temperature and that, if water were reduced to 34°F, the time for water cooling would be less. Note, however, that air temperatures could be more difficult to manage without specifically fine control below 34°F.

In tests to evaluate film coefficients of heat transfer for anomalous shapes, Smith et al. (1970) obtained an experimental value of 6.66 Btu/h·ft^2·°F for a single Red Delicious apple in a cooling tunnel with air approaching at 1570 fpm. At this airflow rate, the logarithmic mean surface temperature of a single apple cooled for 0.5 h in air at 20°F is approximately 35°F. The average temperature difference across the surface boundary layer is, therefore, 15°F and the rate of heat transfer per square foot of surface area is

$$q/A = 6.66 \times 15 = 100 \text{ Btu/h·ft}^2$$

For these conditions, the cooling rate compares favorably with that obtained in ideal hydrocooling. However, these coefficients are based on single specimens isolated from surrounding fruit. Had the fruit been in a packed bed at equivalent flow rates, the values would have been less because less surface area would have been exposed to the cooling fluid. Also, the evaporation rate from the product surface significantly affects the cooling rate.

Because of physical characteristics, mostly geometry, various fruit and vegetable products respond differently to similar treatments of airflow and air temperature. For example, in a packed bed under similar conditions of airflow and air temperature, peaches cool faster than potatoes.

Surface coefficients of heat transfer are sensitive to the physical conditions involved among objects and their surroundings. Soule et al. (1966) obtained experimental surface coefficients ranging from 9 to 12 Btu/h·ft^2·°F for bulk lots of Hamlin oranges and Orlando tangelos with air approaching at 225 to 350 fpm. Bulk bins containing 1000 lb of 2.85 in. diameter Hamlin oranges were cooled from 80°F to a final mass-average temperature of 46.5°F in 1 h with air at 330 fpm (Bennett et al. 1966). Surface heat transfer coefficients for these tests averaged slightly above 11 Btu/h·ft^2·°F. On the basis of a log mean air temperature of 44°F, the calculated half-cooling time was 0.27 h.

By correlating data from experiments on cooling 2.8 in. diameter oranges in bulk lots with results of a mathematical model, Baird and Gaffney (1976) found surface heat transfer coefficients of 1.5 and 9 Btu/h·ft^2·°F for approach velocities of 11 and 412 fpm, respectively. A Nusselt-Reynolds heat transfer correlation representing data from six experiments on air cooling of 2.8 in. diameter oranges and seven experiments on 4.2 in. diameter grapefruit, with approach air velocities ranging from 5 to 412 fpm, gave the relationship $Nu = 1.17 Re^{0.529}$, with a correlation coefficient of 0.996.

Ishibashi et al. (1969) constructed a staged forced-air cooler that exposed bulk fruit to air at a progressively declining temperature

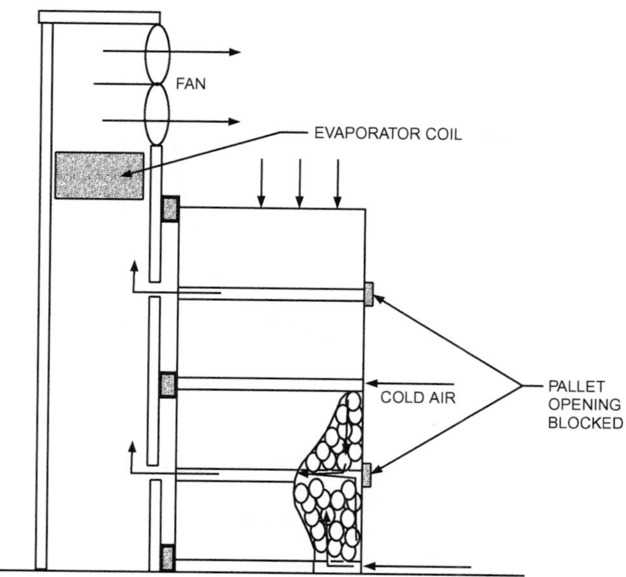

Fig. 5 Serpentine Forced-Air Cooler

(50, 32, and 14°F) as the fruit was conveyed through the cooling tunnel. Air approached at 700 fpm. With this system, 2.5 in. diameter citrus fruit cooled from 77 to 41°F in 1 h. Their half-cooling time of 0.32 h compares favorably with a half-cooling time of 0.30 h for similarly cooled Delicious apples at an approach air velocity of 400 fpm (Bennett et al. 1969). Perry and Perkins (1968) obtained a half-cooling time of 0.5 h for potatoes in a bulk bin with air approaching at 250 fpm, compared to 0.4 h for similarly treated peaches and 0.38 h for apples. Optimum approach velocity for this type of cooling is in the range of 300 to 400 fpm, depending on conditions and circumstances.

Commercial Methods

Produce can be satisfactorily cooled (1) with air circulated in refrigerated rooms adapted for that purpose, (2) in rail cars using special portable cooling equipment that cools the load before it is transported, (3) with air forced through the voids of bulk products moving through a cooling tunnel on continuous conveyors, (4) on continuous conveyors in wind tunnels, or (5) by the forced-air method of passing air through the containers by pressure differential. Each of these methods is used commercially, and each is suitable for certain commodities when properly applied. Figure 5 shows a schematic of a serpentine forced-air cooler.

In circumstances where air cannot be forced directly through the voids of products in bulk, using a container type and load pattern that permit air to circulate through the container and reach a substantial part of the product surface is beneficial. Examples of this are (1) small products such as grapes and strawberries that offer appreciable resistance to airflow through voids in bulk lots, (2) delicate products that cannot be handled in bulk, and (3) products that are packed in shipping containers before precooling.

Forced-air or pressure cooling involves definite stacking patterns and baffling of stacks so that cooling air is forced through, rather than around, individual containers. Success requires a container with vent holes in the direction air will move and a minimum of packaging materials that would interfere with free air movement through the containers. Under these conditions, a relatively small pressure differential between the two sides of the containers results in good air movement and excellent heat transfer. Differential pressures in use are about 0.25 to 3 in. of water, with airflows ranging from 1 to 3 cfm/lb of product.

Because cooling air comes in direct contact with the product being cooled, cooling is much faster than with conventional room cooling. This gives the advantage of rapid product movement through the cooling plant, and the size of the plant is one-third to one-fourth that of an equivalent cold room type of plant.

Mitchell et al. (1972) noted that forced-air cooling usually cools in one-fourth to one-tenth the time needed for conventional room cooling, but it still takes two to three times longer than hydrocooling or vacuum cooling.

A proprietary direct-contact heat exchanger cools air and maintains high humidities using chilled water as a secondary coolant and a continuously wound polypropylene monofilament packing. It contains about 2000 linear feet of filament per cubic foot of packing section. Air is forced up through the unit while chilled water flows downward. The dew-point temperature of air leaving the unit equals the entering water temperature. Chilled water can be supplied from coils submerged in a tank. Build-up of ice on the coils provides an extra cooling effect during peak loads. This design also allows an operator to add commercial ice during long periods of mechanical equipment outage.

In one portable, forced-air method, refrigeration components are mounted on flat bed trailers and the warm, packaged produce is cooled in refrigerated transport trailers. Usually the refrigeration equipment is mounted on two trailers; one holds the forced-air evaporators and the other holds compressors, air-cooling condensers, a high-pressure receiver, and electrical gear. The loaded produce trailers are moved to the evaporator trailer and the product is cooled. After cooling, the trailer is transported to its destination.

Effects of Containers and Stacking Patterns

Accessibility of the product to the cooling medium, essential to rapid cooling, may involve both access to the product in the container and to the individual container in a stack. This effect is evident in the cooling rate data of various commodities in various types of containers reported by Mitchell et al. (1972). Parsons et al. (1972) developed a corrugated paperboard container venting pattern for palletized unit loads that produced cooling rates equal to those from conventional register stacked patterns. Fisher (1960) demonstrated that spacing apple containers on pallets reduced cooling time by 50% compared to pallet loads stacked solidly. A minimum of 5% sidewall venting is recommended.

Palletization is essential for shipment of many products, and pallet stability is improved if cartons are packed closely together. Thus, cartons and packages should be designed to allow ample airflow though the stacked products. Amos et al. (1993) and Parsons et al. (1972) showed the importance of vent sizes and location to obtain good cooling in palletized loads without reducing container strength. Some operations wrap palletized products in polyethylene to increase stability. In this case, the product may need to be cooled before it is palletized.

Moisture Loss in Forced-Air Cooling

The information in this section is drawn from Thompson et al. (2002).

Moisture loss in forced-air cooling ranges from very little to amounts significant enough to damage produce. Factors that affect moisture loss include product initial temperature and transpiration coefficient, humidity, exposure to airflow after cooling, and whether waxes or moisture-resistant packaging is used.

High initial temperature results in high moisture loss; this can be minimized by harvesting at cooler times of day (i.e., early morning or night), and cooling (or at least shading) products immediately after harvest. Keep reheat during packing to a minimum.

The primary advantage of high humidity during cooling is that product packaging can absorb moisture, which reduces the packaging's absorption of moisture from the product itself.

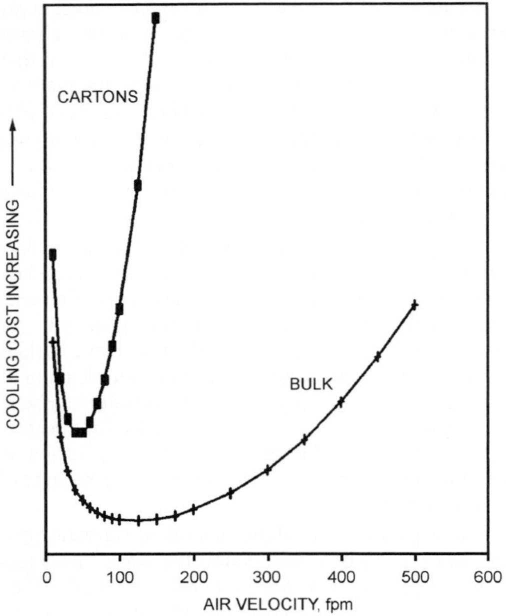

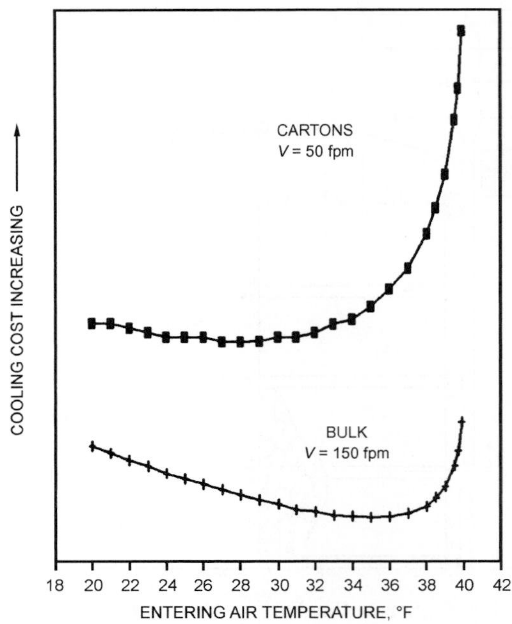

Fig. 6 Engineering-Economic Model Output for a Forced-Air Cooler

High transpiration coefficients also increase moisture loss. For example, carrots, with a high transpiration rate, can lose 0.6 to 1.8% of their original, uncooled weight during cooling. Polyethylene packaging has reduced moisture loss in carrots to 0.08%, although cooling times are about five times longer. Film box liners, sometimes used for packing products with low transpiration coefficients (e.g., apples, pears, kiwifruit, and grapes), are also useful in reducing moisture loss, but they also increase the time required to cool products. Some film box liners are perforated to reduce condensation; liners used to package grapes must also include an SO_2-generating pad to reduce decay.

To prevent exposing product to unnecessary airflow, forced-air coolers should reduce or stop airflow as soon as the target product temperature is reached. Otherwise, moisture loss will continue unless the surrounding air is close to saturation. One method is to link cooler fan control to return air plenum temperature, slowing fan speeds as the temperature of the return air approaches that of the supply air.

Computer Solution

Baird et al. (1988) developed an engineering economic model for designing forced-air cooling systems. Figure 6 shows the type of information that can be obtained from the model. By selecting a set of input conditions (which varies with each application) and varying approach air velocity, entering air temperature, or some other variable, the optimum (minimum-cost) value can be determined. The curves in Figure 6 show that selection of air velocity for containers is critical, whereas selection of entering air temperature is not as critical until the desired final product temperature of 40°F is approached. The results shown are for four cartons deep with a 4% vent area in the direction of airflow, and they would be quite different if the carton vent area was changed. Other design parameters that can be optimized using this program are the depth of product in direction of airflow and the size of evaporators and condensers.

FORCED-AIR EVAPORATIVE COOLING

This approach cools air with an evaporative cooler, passing air through a wet pad before it comes into contact with product and packaging, instead of using mechanical refrigeration. A correctly designed and operated evaporative cooler produces air a few degrees above the outside wet-bulb temperature, at high humidity (about 90% rh), and is more energy-efficient than mechanical refrigeration (Kader 2002). In most of California, for instance, product temperatures of 60 to 70°F can be achieved. This method is suited for products that are best held at moderate temperatures, such as tomatoes, or for those that are marketed soon after harvest.

For more information on evaporative cooling equipment and applications, see Chapter 51 of the 2003 *ASHRAE Handbook—HVAC Applications*, and Chapter 19 of the 2004 *ASHRAE Handbook—HVAC Systems and Equipment*.

PACKAGE ICING

Finely crushed ice placed in shipping containers can effectively cool products that are not harmed by contact with ice. Spinach, collards, kale, brussels sprouts, broccoli, radishes, carrots, and green onions are commonly packaged with ice (Hardenburg et al. 1986). Cooling a product from 95 to 35°F requires melting ice equal to 38% of the product's mass. Additional ice must melt to remove heat leaking into the packages and to remove heat from the container. In addition to removing field heat, package ice can keep the product cool during transit.

Pumping **slush ice** or liquid ice into the shipping container through a hose and special nozzle that connect to the package is used for cooling some products. Some systems can ice an entire pallet at one time.

Top icing, or placing ice on top of packed containers, is used occasionally to supplement another cooling method. Because corrugated containers have largely replaced wooden crates, use of top ice has decreased in favor of forced-air and hydrocooling. Wax-impregnated corrugated containers, however, allow icing and hydrocooling of products after packaging.

Flaked or crushed ice can be manufactured on site and stored in an ice bunker for later use; for short-season cooling requirements with low ice demands (e.g., a few tons a day), it may be more economical to buy block ice and crush it on site. Another option is to rent liquid ice equipment for on-site production.

The cooling capacity of ice is 144 Btu/lb; 1 lb of ice will reduce the temperature of 3 lb of produce by approximately 50°F. However, commercial ice-injection systems require significantly more ice beyond that needed for produce cooling. For example, 20 lb of broccoli requires about 32 lb of manufactured ice (losses occur in product cooling, transport, and equipment heat gain; also, a remainder of ice is required in the box on delivery to the customer). The high ice requirement makes liquid icing energy-inefficient and expensive (Thompson et al. 2002). Other disadvantages of ice cooling include (1) weight of the ice, which decreases the net product weight in a vehicle; and (2) the need for water-resistant packaging to prevent water damage to other products; and (3) safety hazards during storage. These disadvantages can be minimized if ice is used for temperature maintenance in transit rather than for cooling, or by using gel-pack ice (often used for flowers), which is sealed in a leakproof bag.

VACUUM COOLING

Vacuum cooling of fresh produce by rapid evaporation of water from the product works best with vegetables having a high ratio of surface area to volume and a high transpiration coefficient. In vacuum refrigeration, water, as the primary refrigerant, vaporizes in a flash chamber under low pressure. Pressure in the chamber is lowered to the saturation point corresponding to the lowest required temperature of the water.

Vacuum cooling is a batch process. The product to be cooled is loaded into the flash chamber, the system is put into operation, and the product is cooled by reducing the pressure to the corresponding saturation temperature desired. The system is then shut down, the product removed, and the process repeated. Because the product is normally at ambient temperature before it is cooled, vacuum cooling can be thought of as a series of intermittent operations of a vacuum refrigeration system in which water in the flash chamber is allowed to come to ambient temperature before each start. The functional relationships for determining refrigerating capacity are the same in each case.

Cooling is achieved by boiling water, mostly off the surface of the product to be cooled. The heat of vaporization required to boil the water is furnished by the product, which is cooled accordingly. As pressure is further reduced, cooling continues to the desired temperature level. The saturation pressure for water at 212°F is 760 mm Hg; at 32°F, it is 4.58 mm Hg. Commercial vacuum coolers normally operate in this range.

Although the cooling rate of lettuce could be increased without danger of freezing by reducing the pressure to 3.8 mm Hg, corresponding to a saturation temperature of 27°F, most operators do not reduce the pressure below that which freezes water because of the extra work involved and the freezing potential.

Pressure, Volume, and Temperature

In vacuum cooling, the thermodynamic process is assumed to take place in two phases. In the first phase, the product is assumed to be loaded into the flash chamber at ambient temperature, and the temperature in the flash chamber remains constant until saturation pressure is reached. At the onset of boiling, the small remaining amount of air in the chamber is replaced by the water vapor, the first phase ends, and the second phase begins simultaneously. The second phase continues at saturation until the product has cooled to the desired temperature.

If the ideal gas law is applied for an approximate solution in a commercial vacuum cooler, the pressure/volume relationships are

$$\text{Phase 1} \quad pv = 29{,}318 \text{ ft}\cdot\text{lb/lb}$$
$$\text{Phase 2} \quad pv^{1.056} = 66{,}370 \text{ ft}\cdot\text{lb/lb}$$

where p is absolute pressure and v is specific volume.

The pressure/temperature relationship is determined by the value of ambient and product temperature. Based on 90°F for this value, the temperature in the flash chamber theoretically remains constant at 90°F as the pressure reduces from atmospheric to saturation, after which it declines progressively along the saturation line. These relationships are illustrated in Figure 7. Product temperature responds similarly, but varies depending on where temperature is measured in

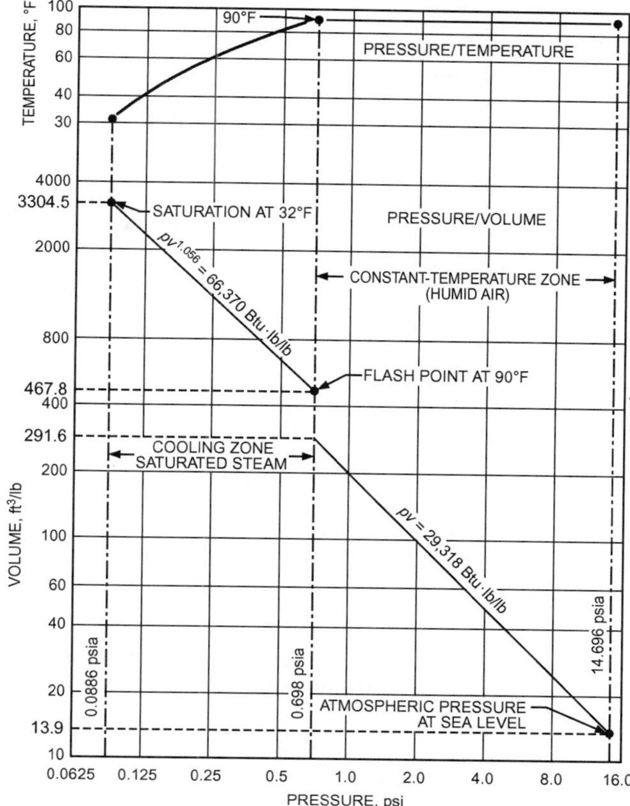

Fig. 7 Pressure, Volume, and Temperature in a Vacuum Cooler Cooling Product from 90 to 32°F

the product, physical characteristics of the product, and amount of product surface water available. Although it is possible for some vaporization to occur in intercellular spaces beneath the product surface, most water is vaporized off the surface. The heat required to vaporize this water is also taken off the product surface, where it flows by conduction under the thermal gradient produced. Thus, the rate of cooling depends on the relation of surface area to volume of product and the rate at which the vacuum is drawn in the flash chamber.

Because water is the sole refrigerant, the amount of heat removed from the product depends on the mass of water vaporized m_v and its latent heat of vaporization L. Assuming an ideal condition, with no heat gain from surroundings, total heat Q removed from the product is

$$Q = m_v L \tag{6}$$

The amount of moisture removed from the product during vacuum cooling, then, is directly related to the product's specific heat and the amount of temperature reduction accomplished. A product with a specific heat capacity of 0.95 Btu/lb·°F theoretically loses 1% moisture for each 11°F reduction in temperature. In a study of vacuum cooling of 16 different vegetables, Barger (1963) showed that cooling of all products was proportional to the amount of moisture evaporated from the product. Temperature reductions averaged 9 to 10°F for each 1% of weight loss, regardless of the product cooled. This weight loss may reduce the amount of money the grower receives as well as the turgor and crispness of the product. Some vegetables are sprayed with water before or during cooling to reduce this loss.

Commercial Systems

The four types of vacuum refrigeration systems that use water as the refrigerant are (1) steam ejector, (2) centrifugal, (3) rotary, and

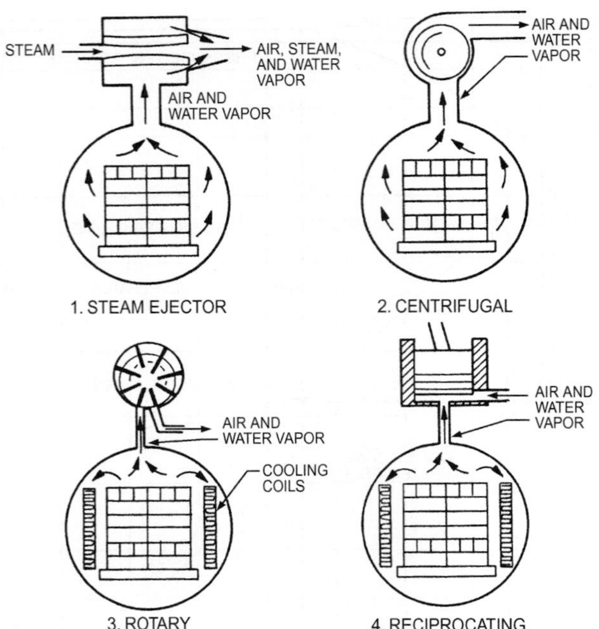

Fig. 8 Schematic Cross Sections of Vacuum-Producing Mechanisms

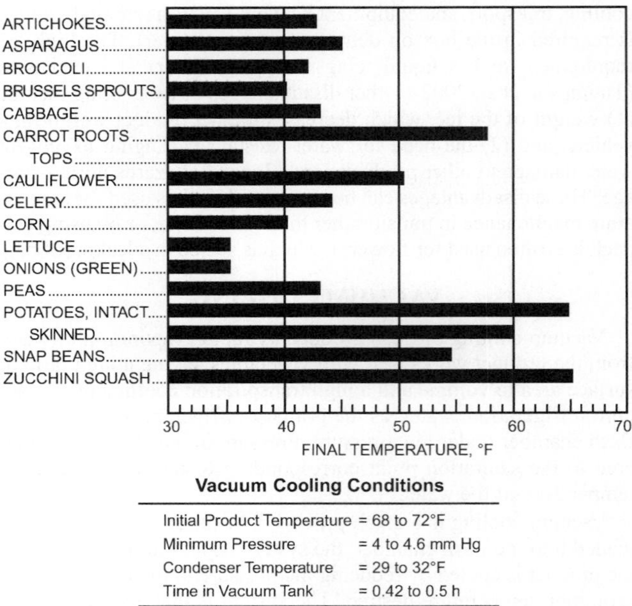

Vacuum Cooling Conditions

Initial Product Temperature	= 68 to 72°F
Minimum Pressure	= 4 to 4.6 mm Hg
Condenser Temperature	= 29 to 32°F
Time in Vacuum Tank	= 0.42 to 0.5 h

Fig. 9 Comparative Cooling of Vegetables Under Similar Vacuum Conditions

(4) reciprocating. A schematic of the vacuum-producing mechanism of each is illustrated in Figure 8.

Of these, the steam ejector type is best suited for displacing the extremely high volumes of water vapor encountered at the low pressures needed in vacuum cooling. It also has the advantage of having few moving parts, thus requiring no compressor to condense the water vapor. High-pressure steam is expanded through a series of jets or ejectors arranged in series and condensed in barometric condensers mounted below the ejectors. Cooling water for condensing is accomplished by means of an induced-draft cooling tower. In spite of these advantages, few steam ejector vacuum coolers are used today, because of the inconvenience of using steam and the lack of portability. Instead, vacuum coolers are mounted on semitrailers to follow seasonal crops.

The centrifugal compressor is also a high-volume pump and can be adapted to water vapor refrigeration. However, its use in vacuum cooling is limited because of inherent mechanical difficulties at the high rotative speeds required to produce the low pressures needed.

Both rotary and reciprocal vacuum pumps can produce the low pressures needed, and they also have the advantage of portability. Being positive-displacement pumps, however, they have low volumetric capacity; therefore, vacuum coolers using rotary or reciprocating pumps have separate refrigeration systems to condense much of the water vapor that evaporates off the product, thus substantially reducing the volume of water vapor passing through the pump. Ideally, when it can be assumed that all water vapor is condensed, the required refrigeration capacity equals the amount of heat removed from the product during cooling.

The condenser must contain adequate surface to condense the large amount of vapor removed from the produce in a few minutes. Refrigeration is furnished from cold brine or a direct-expansion system. A very large peak load occurs from rapid condensing of so much vapor. Best results are obtained if the refrigeration plant is equipped with a large brine or icemaking tank having enough stored refrigeration to smooth out the load. A standard three-tube plant, with capacity to handle three cars per hour, has a peak refrigeration load of at least 250 tons.

To increase cooling effectiveness and reduce product moisture loss, the product is sometimes wetted before cooling begins. However, iceberg lettuce is rarely prewetted. A modification of vacuum cooling circulates chilled water over the product throughout the cooling process. Among the chief advantages are increased cooling rates and residual refrigeration that is stored in the chilled water after each vacuum process. It also prevents water loss from products that show objectionable wilting after conventional vacuum cooling.

Applications

Because vacuum cooling is generally more expensive, particularly in capital cost, than other cooling methods, its use is primarily restricted to products for which vacuum cooling is much faster or more convenient. Lettuce is ideally adapted to vacuum cooling. The numerous individual leaves provide a large surface area and the tissues release moisture readily. It is possible to freeze lettuce in a vacuum chamber if pressure and condenser temperatures are not carefully controlled. However, even lettuce does not cool entirely uniformly. The fleshy core, or butt, releases moisture more slowly than the leaves. Temperatures as high as 43°F have been recorded in core tissue when leaf temperatures were down to 33°F (Barger 1961).

Other leafy vegetables such as spinach, endive, escarole, and parsley are also suitable for vacuum cooling. Vegetables that are less suitable but adaptable by wetting are asparagus, snap beans, broccoli, brussels sprouts, cabbage, cauliflower, celery, green peas, sweet corn, leeks, and mushrooms. Of these vegetables, only cauliflower, celery, cabbage, and mushrooms are commercially vacuum cooled in California. Fruits are generally not suitable, except some berries. Cucumbers, cantaloupes, tomatoes, dry onions, and potatoes cool very little because of their low surface-to-mass ratio and relatively impervious surface. The final temperatures of various vegetables when vacuum cooled under similar conditions are illustrated in Figure 9.

The rate of cooling and final temperature attained by vacuum cooling are largely affected by the commodity's ratio of surface area to its mass and the ease with which it gives up water from its tissues. Consequently, the adaptability of fruits and vegetables varies tremendously for this method of precooling. For products that have a low surface-to-mass ratio, high temperature gradients occur. To prevent the surface from freezing before the desired mass-average temperature is reached, the vacuum pump is switched off and on ("bounced") to keep the saturation temperature above freezing.

Mechanical vacuum coolers have been designed in several sizes. Most installations use cylindrical or rectangular retorts. For portability, some vacuum coolers and associated refrigeration equipment have been placed on flat-bed trailers.

SELECTING A COOLING METHOD

Packing house size and operating procedures, response of product to the cooling method, and market demands largely dictate the cooling method used. Other factors include whether the product is packaged in the field or in a packing house, product mix, length of cooling season, and comparative costs of dry versus water-resistant cartons. In some cases, there is little question about the type of cooling to be used. For example, vacuum cooling is most effective on lettuce and other similar vegetables. Peach packers in the southeastern United States and some vegetable and citrus packers are satisfied with hydrocooling. Air (room) cooling is used for apples, pears, and citrus fruit. In other cases, choice of cooling method is not so clearly defined. Celery and sweet corn are usually hydrocooled, but they may be vacuum cooled as effectively. Cantaloupes may be satisfactorily cooled by several methods. *Note*: sweet cherries are often hydrocooled in packing houses but are air cooled if orchard packed.

When more than one method can be used, cost becomes a major consideration. Although rapid forced-air cooling is more costly than hydrocooling, if the product does not require rapid cooling, a forced-air system can operate almost as economically as hydrocooling. In a study to evaluate costs of hypothetical precooling systems for citrus fruit, Gaffney and Bowman (1970) found that the cost for forced-air cooling in bulk lots was 20% more than that for hydrocooling in bulk and that forced-air cooling in cartons costs 45% more than hydrocooling in bulk.

Table 5 summaries precooling and cooling methods suggested for various commodities.

COOLING CUT FLOWERS

Because of their high rates of respiration and low tolerance for heat, deterioration in cut flowers is rapid at field temperatures. Refrigerated highway vans do not have the capacity to remove the field heat in sufficient time to prevent some deterioration from occurring (Farnham et al. 1979). Forced-air cooling is commonly used by the flower industry. As with most fruits and vegetables, the cooling rate of cut flowers varies substantially among the various types. Rij et al. (1979) found that the half-cooling time for packed boxes of gypsophila was about 3 min compared to about 20 min for chrysanthemums at airflows ranging from 80 to 260 cfm per box. Within this range, cooling time was proportional to the reciprocal of airflow but varied less with airflow than with flower type.

SYMBOLS

A = product surface area, ft^2
c_p = specific heat of product, Btu/lb·°F
C = cooling coefficient, reciprocal of hours
j = lag factor
L = heat of vaporization, Btu/lb
m = mass of product, lb
m_v = mass of water vaporized, lb
p = pressure, lb/ft^2
q = cooling load or rate of heat transfer, Btu/h
Q = total heat, Btu
t = temperature of any point in product, °F
t_i = initial uniform product temperature, °F
t_m = temperature of cooling medium, °F
t_o = surrounding temperature, °F
t_{ma} = mass-average temperature, °F
v = specific volume of water vapor, ft^3/lb

Table 5 Cooling Methods Suggested for Horticultural Commodities

Commodity	Size of Operation: Large	Small
Tree fruits		
Citrus	R	R
Deciduous [a]	FA, R, HC	FA
Subtropical	FA, R	FA
Tropical	FA, R	FA
Berries	FA	FA
Grapes [b]	FA	FA
Leafy vegetables		
Cabbage	VC, FA	FA
Iceberg lettuce	VC	FA
Kale, collards	VC, R, WV	FA
Leaf lettuces, spinach, endive, escarole, Chinese cabbage, bok choy, romaine	VC, FA, WV, HC	FA
Root vegetables		
With tops [c]	HC, PI, FA	HC, FA
Topped	HC, PI	HC, PI, FA
Irish potatoes, sweet potatoes [d]	R w/evap coolers, HC	R
Stem and flower vegetables		
Artichokes	HC, PI	FA, PI
Asparagus	HC	HC
Broccoli, Brussels sprouts	HC, FA, PI	FA, PI
Cauliflower	FA, VC	FA
Celery, rhubarb	HC, WV, VC	HC, FA
Green onions, leeks	PI, HC	PI
Mushrooms	FA, VC	FA
Pod vegetables		
Beans	HC, FA	FA
Peas	FA, PI, VC	FA, PI
Bulb vegetables		
Dry onions [e]	R	R, FA
Garlic	R	
Fruit-type vegetables [f]		
Cucumbers, eggplant	R, FA, FA-EC	FA, FA-EC
Melons		
Cantaloupes, muskmelons, honeydew, casaba	HC, FA, PI	FA, FA-EC
Crenshaw	FA, R	FA, FA-EC
Watermelons	FA, HC	FA, R
Peppers	R, FA, FA-EC, VC	FA, FA-EC
Summer squashes, okra	R, FA, FA-EC	FA, FA-EC
Sweet corn	HV, VC, PI	HC, FA, PI
Tomatillos	R, FA, FA-EC	FA, FA-EC
Tomatoes	R, FA, FA-EC	
Winter squashes	R	R
Fresh herbs		
Not packaged [g]	HC, FA	FA, R
Packaged	FA	FA, R
Cactus		
Leaves (nopalitos)	R	FA
Fruit (tunas or prickly pears)	R	FA
Ornamentals		
Cut flowers [h]	FA, R	FA
Potted plants	R	R

R = Room cooling
HC = Hydrocooling
FA = Forced-air cooling
VC = Vacuum cooling
WV = Water spray vacuum cooling
PI = Package icing
FA-EC = Forced-air evaporative cooling

[a]Apricots cannot be hydrocooled.
[b]Grapes require rapid cooling facilities adaptable to sulfur dioxide fumigation.
[c]Carrots can be vacuum cooled.
[d]With evaporative coolers, facilities for potatoes should be adapted to curing.
[e]Facilities should be adapted to curing onions.
[f]Fruit-type vegetables are sensitive to chilling but at varying temperatures.
[g]Fresh herbs can be easily damaged by water beating in hydrocooler.
[h]When cut flowers are packaged, only use forced-air cooling.

Reprinted with permission from A.A. Kader (2001).

V = air velocity, fpm
Y = temperature ratio $(t - t_o)/(t_i - t_o)$
Z = half-cooling time, h
θ = cooling time, h

REFERENCES

Amos, N.D., D.J. Cleland, and N.H. Banks. 1993. Effect of pallet stacking arrangement on fruit cooling rates within forced-air pre-coolers. *Refrigeration Science and Technology* 3:232-241.

Baird, C.D. and J.J. Gaffney. 1976. A numerical procedure for calculating heat transfer in bulk loads of fruits or vegetables. *ASHRAE Transactions* 82(2):525.

Baird, C.D., J.J. Gaffney, and M.T. Talbot. 1988. Design criteria for efficient and cost effective forced air cooling systems for fruits and vegetables. *ASHRAE Transactions* 94(1):1434.

Barger, W.R. 1961. Factors affecting temperature reduction and weight loss of vacuum-cooled lettuce. *USDA Marketing Research Report* 469.

Barger, W.R. 1963. Vacuum precooling—A comparison of cooling of different vegetables. *USDA Marketing Research Report* 600.

Becker, B.R. and B.A. Fricke. 1996a. Transpiration and respiration of fruits and vegetables. In *New Developments in Refrigeration for Food Safety and Quality*, pp. 110-121. International Institute of Refrigeration, Paris.

Becker, B.R. and B.A. Fricke. 1996b. Simulation of moisture loss and heat loads in refrigerated storage of fruits and vegetables. In *New Developments in Refrigeration for Food Safety and Quality*, pp. 210-221. International Institute of Refrigeration, Paris.

Becker, B.R. and B.A. Fricke. 2001. A numerical model of commodity moisture loss and temperature distribution during refrigerated storage. In *Applications of Modelling as an Innovative Technology in the Agri-Food Chain*, M.L.A.T.M. Hertog and B.R. MacKay, eds., pp. 431-436. International Society for Horticultural Science, Leuven, Belgium.

Becker, B.R. and B.A. Fricke. 2002. Hydrocooling time estimation methods. *International Communications in Heat and Mass Transfer* 29(2): 165-174.

Becker, B.R., A. Misra, and B.A. Fricke. 1996a. Bulk refrigeration of fruits and vegetables, part I: Theoretical considerations of heat and mass transfer. *International Journal of HVAC&R Research* (now *HVAC&R Research*) 2(2):122-134.

Becker, B.R., A. Misra, and B.A. Fricke. 1996b. Bulk refrigeration of fruits and vegetables, part II: Computer algorithm for heat loads and moisture loss. *International Journal of HVAC&R Research* (now *HVAC&R Research*) 2(3):215-230.

Bennett, A.H. 1963. Thermal characteristics of peaches as related to hydrocooling. U.S. Department of Agriculture, *Technical Bulletin* 1292.

Bennett, A.H. 1970. Principles and equipment for precooling fruits and vegetables. *ASHRAE Symposium Bulletin* SF-4-70. Symposium on Precooling of Fruits and Vegetables, San Francisco.

Bennett, A.H., R.E. Smith, and J.C. Fortson. 1965. Hydrocooling peaches—A practical guide for determining cooling requirements and cooling times. *USDA Agriculture Information Bulletin* 298 (June).

Bennett, A.H., J. Soule, and G.E. Yost. 1966. Temperature response of citrus to forced-air precooling. *ASHRAE Journal* 8(4):48.

Bennett, A.H., J. Soule, and G.E. Yost. 1969. *Forced-air precooling for Red Delicious apples*. USDA, Agricultural Research Service ARS 52-41.

Boyette, M.D., E.A. Estes, and A.R. Rubin. 1992. Hydrocooling. *Postharvest Technology Series* AG-414-4. North Carolina Cooperative Extension Service, Raleigh.

Dincer, I. 1995. An effective method for analysing precooling process parameters. *International Journal of Energy Research* 19(2):95-102.

Dincer, I. 1996. Convective heat transfer coefficient model for spherical products subject to hydrocooling. *Energy Sources* 18(6):735-742.

Dincer, I. and O.F. Genceli. 1994. Cooling process and heat transfer parameters of cylindrical products cooled both in water and in air. *International Journal of Heat & Mass Transfer* 37(4):625-633.

Dincer, I. and O.F. Genceli. 1995. Cooling of spherical products: Part I—Effective process parameters. *International Journal of Energy Research* 19(3):205-218.

Farnham, D.S., F.J. Marousky, D. Durkin, R. Rij, J.F. Thompson, and A.M. Kofranek. 1979. Comparison of conditioning, precooling, transit method, and use of a floral preservative on cut flower quality. *Proceedings, Journal of American Society of Horticultural Science* 104(4):483.

Fisher, D.V. 1960. Cooling rates of apples packed in different bushel containers and stacked at different spacing in cold storage. *ASHRAE Journal* (July):53.

Fricke, B.A. 2006. Precooling fruits and vegetables using hydrocooling. *ASHRAE Journal* 48(2):20-28.

Fricke, B.A. and B.R. Becker. 2003. Comparison of hydrocooling time estimation methods. *Proceedings of the 21st IIR International Congress of Refrigeration: Serving the Needs of Mankind*, August 17-22, 2003, Washington, D.C. *Paper* ICR0432.

Gaffney, J.J. and E.K. Bowman. 1970. An economic evaluation of different concepts for precooling citrus fruits. *ASHRAE Symposium Bulletin* SF-4-70. Symposium on Precooling Fruits and Vegetables, San Francisco (January).

Gast, K.L.B. and R.A. Flores. 1991. Precooling produce: Fruits and vegetables. *Postharvest management of commercial horticultural crops*, MF1002. Kansas State University Cooperative Extension Service, Manhattan.

Guillou, R. 1958. Some engineering aspects of cooling fruits and vegetables. *Transactions of the ASAE* 1(1):38, 39, 42.

Hardenburg, R.E., A.E. Watada, and C.Y. Wang. 1986. The commercial storage of fruits, vegetables, and florist and nursery stocks. *USDA Agricultural Handbook* 66.

Henry, F.E. and A.H. Bennett. 1973. "Hydraircooling" vegetable products in unit loads. *Transactions of the ASAE* 16(4):731-733.

Henry, F.E. A.H. Bennett, and R.H. Segall. 1976. Hydraircooling—A new concept for precooling pallet loads of vegetables. *ASHRAE Transactions* 82(2):541.

Ishibashi, S., R. Kojima, and T. Kaneko. 1969. Studies on the forced-air cooler. *Journal of the Japanese Society of Agricultural Machinery* 31(2).

Kader, A.A. 2001. *Post Harvest Technology of Horticultural Crops*. University of California, Division of Agriculture and Natural Resources.

Lucas, T. and A.L. Raoult-Wack. 1998. Immersion chilling and freezing in aqueous refrigerating media: Review and future trends. *International Journal of Refrigeration* 21(6):419-429.

Mitchell, F.G., R. Guillou, and R.A. Parsons. 1972. Commercial cooling of fruits and vegetables. *Manual* 43. University of California, Division of Agricultural and Natural Resources.

Nicholas, R.C., K.E.H. Motawi, and J.L. Blaisdell. 1964. Cooling rates of individual fruit in air and in water. *Michigan State University Agricultural Experiment Station Quarterly Bulletin* 47:51-64.

O'Brien, M. and J.P. Gentry. 1967. Effect of cooling methods on cooling rates and accompanying desiccation of fruits. *Transactions of the ASAE* 10(5):603-606.

Parsons, R.A., F.G. Mitchell, and G. Mayer. 1972. Forced-air cooling of palletized fresh fruit. *Transactions of the ASAE* 15(4):729.

Perry, R.L. and R.M. Perkins. 1968. Hydrocooling sweet corn. *Paper* 68-800. American Society of Agricultural Engineering, St. Joseph, MI.

Pflug, I.J., J.L. Blaisdell, and I.J. Kopelman. 1965. Developing temperature-time curves for objects that can be approximated by a sphere, infinite plate, or infinite cylinder. *ASHRAE Transactions* 71(1):238.

Rij, R.E., J.F. Thompson, and D.S. Farnham. 1979. Handling, precooling, and temperature management of cut flower crops for truck transportation. USDA-SEA *Western Series* 5 (June).

Robertson, G.H., J.C. Cipolletti, D.F. Farkas, and G.E. Secor. 1976. Methodology for direct contact freezing of vegetables in aqueous freezing media. *Journal of Food Science* 41(4):845-851.

Ryall, A.L. and W.J. Lipton. 1979. *Handling, transportation and storage of fruits and vegetables*. AVI Publishing, Westport, CT.

Smith, R.E. and A.H. Bennett. 1965. Mass-average temperature of fruits and vegetables during transient cooling. *Transactions of the ASAE* 8(2):249.

Smith, R.E., A.H. Bennett, and A.A. Vacinek. 1970. Convection film coefficients related to geometry for anomalous shapes. *Transactions of the ASAE* 13(2).

Soule, J., G.E. Yost, and A.H. Bennett. 1966. Certain heat characteristics of oranges, grapefruit and tangelos during forced-air precooling. *Transactions of the ASAE* 9(3):355.

Stewart, J.K. and H.M. Couey. 1963. Hydrocooling vegetables—A practical guide to predicting final temperatures and cooling times. *USDA Marketing Research Report* 637.

Stewart, J.K. and W.J. Lipton. 1960. Factors influencing heat loss in cantaloupes during hydrocooling. *USDA Marketing Research Report* 421.

Thompson, J.F., T.R. Rumsey, R.F. Kasmir, and C.H. Crisosto. 2002. Commercial cooling of fruits, vegetables, and flowers. *Publication* 21567. University of California, Division of Agricultural and Natural Resources.

USDA. 2004. *The commercial storage of fruits, vegetables, and florist and nursery stocks*. Agricultural Research Service, U.S. Department of Agriculture, Washington, D.C.

BIBLIOGRAPHY

Ansari, F.A. and A. Afaq. 1986. Precooling of cylindrical food products. *International Journal of Refrigeration* 9(3):161-163.

Arifin, B.B. and K.V. Chau. 1988. Cooling of strawberries in cartons with new vent hole designs. *ASHRAE Transactions* 94(1):1415-1426.

Bennett, A.H. 1962. Thermal characteristics of peaches as related to hydrocooling. *USDA Technical Bulletin* 1292.

Bennett, A.H., W.G. Chace, Jr., and R.H. Cubbedge. 1969. Heat transfer properties and characteristics of Appalachian area Red Delicious apples. *ASHRAE Transactions* 75(2):133.

Bennett, A.H., W.G. Chace, Jr., and R.H. Cubbedge. 1970. Thermal properties and heat transfer characteristics of marsh grapefruit. *USDA Technical Bulletin* 1413.

Beukema, K.J., S. Bruin, and J. Schenk. 1982. Heat and mass transfer during cooling and storage of agricultural products. *Chemical Engineering Science* 37(2):291-298.

Burton, K.S., C.E. Frost, and P.T. Atkey. 1987. Effect of vacuum cooling on mushroom browning. *International Journal of Food Science & Technology* 22(6):599-606.

Chau, K.V. 1994. Time-temperature-humidity relations for the storage of fresh commodities. *ASHRAE Transactions* 100(2):348-353.

Chuntranuluck, S., C.M. Wells, and A.C. Cleland. 1998. Prediction of chilling times of foods in situations where evaporative cooling is significant—Part 1: Method development. *Journal of Food Engineering* 37: 111-125.

Chuntranuluck, S., C.M. Wells, and A.C. Cleland. 1998. Prediction of chilling times of foods in situations where evaporative cooling is significant—Part 2: Experimental testing. *Journal of Food Engineering* 37: 127-141.

Chuntranuluck, S., C.M. Wells, and A.C. Cleland. 1998. Prediction of chilling times of foods in situations where evaporative cooling is significant—Part 3: Applications. *Journal of Food Engineering* 37:143-157.

Flockens, I.H. and H.F.T. Meffert. 1972. Biophysical properties of horticultural products as related to loss of moisture during cooling down. *Journal of the Science of Food and Agriculture* 23:285-298.

Gan, G. and J.L. Woods. 1989. A deep bed simulation of vegetable cooling. *Land and Water Use: Proceedings of the 11th International Congress on Agricultural Engineering*, Dublin, pp. 2301-2308.

Gariepy, Y., G.S.V. Raghavan, and R. Theriault. 1987. Cooling characteristics of cabbage. *Canadian Agricultural Engineering* 29(1):45-50.

Grizell, W.G. and A.H. Bennett. 1966. *Hydrocooling stacked crates of celery and sweet corn*. USDA, Agricultural Research Service, ARS 52-12.

Hackert, J.M., R.V. Morey, and D.R. Thompson. 1987. Precooling of fresh market broccoli. *Transactions of the ASAE* 30(5):1489-1493.

Harvey, J.M. 1963. Improved techniques for vacuum cooling vegetables. *ASHRAE Journal* 5(1):41-44.

Hayakawa, K. 1978. Computerized simulation for heat transfer and moisture loss from an idealized fresh produce. *Transactions of the ASAE* 21(5):1015-1024.

Hayakawa, K. and J. Succar. 1982. Heat transfer and moisture loss of spherical fresh produce. *Journal of Food Science* 47(2):596-605.

Isenberg, F.M.R., R.F. Kasmire, and J.E. Parson. Vacuum cooling vegetables. *Information Bulletin* 186. Cornell University Cooperative Extensive Service, Ithaca, NY.

Kader, A.A., R.F. Kasmire, and J.F. Thompson. 1992. Cooling horticultural commodities. *Publication* 3311. University of California, Division of Agricultural and Natural Resources.

Rohrbach, R.P., R. Ferrell, E.O. Beasley, J.R. Fowler. 1984. Precooling blueberries and muscadine grapes with liquid carbon dioxide. *Transactions of the ASAE* 27(6):1950-1955.

Smith, W.L. and W.H. Redit. 1968. Postharvest decay of peaches as affected by hot-water treatments, cooling methods, and sanitation. *USDA Marketing Research Report* 807.

Smith, R.E., G.L. Nelson, and R.L. Henrickson. 1967. Analyses on transient heat transfer from anomalous shapes. *Transactions of the ASAE* 10(2):236.

Smith, R.E., G.L. Nelson, and R.L. Henrickson. 1968. Applications of geometry analysis of anomalous shapes to problems in transient heat transfer. *Transactions of the ASAE* 11(2):296.

Thompson, J.F., Y.L. Chen, and T.R. Rumsey. 1987. Energy use in vacuum coolers for fresh market vegetables. *Applied Engineering in Agriculture* 3(2):196-199.

Woods, J.L. 1990. Moisture loss from fruits and vegetables. *Postharvest News and Information* 1(3):195-199.

CHAPTER 16

INDUSTRIAL FOOD-FREEZING SYSTEMS

FREEZING is a method of food preservation that slows the physical changes and chemical and microbiological activity that cause deterioration in foods. Reducing temperature slows molecular and microbial activity in food, thus extending useful storage life. Although every product has an individual ideal storage temperature, most frozen food products are stored at 0 to –30°F. Chapter 11 lists frozen storage temperatures for specific products.

Freezing reduces the temperature of a product from ambient to storage level and changes most of the water in the product to ice. Figure 1 shows the three phases of freezing: (1) cooling, which removes sensible heat, reducing the temperature of the product to the freezing point; (2) removal of the product's latent heat of fusion, changing the water to ice crystals; and (3) continued cooling below the freezing point, which removes more sensible heat, reducing the temperature of the product to the desired or optimum frozen storage temperature. Values for specific heats, freezing points, and latent heats of fusion for various products are given in Chapter 9.

The longest part of the freezing process is removing the latent heat of fusion as water turns to ice. Many food products are sensitive to freezing rate, which affects yield (dehydration), quality, nutritional value, and sensory properties. The freezing method and system selected can thus have substantial economic impact.

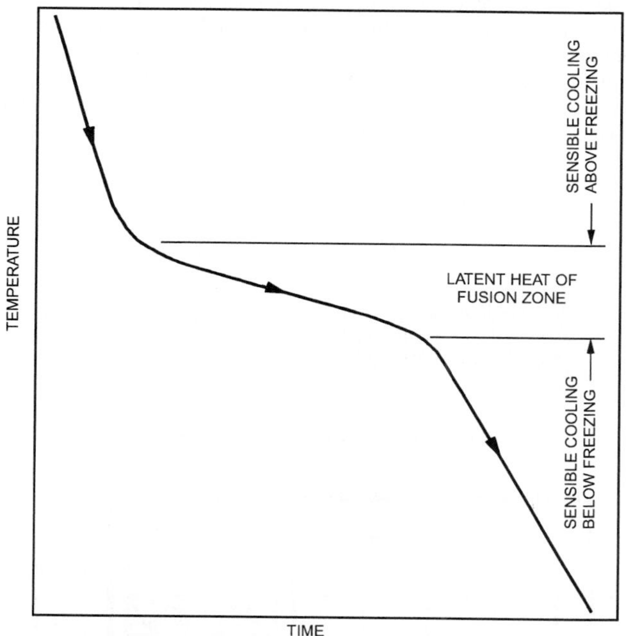

Fig. 1 Typical Freezing Curve

When selecting freezing methods and systems for specific products, consider special handling requirements, capacity, freezing times, quality, yield, appearance, first cost, operating costs, automation, space availability, and upstream/downstream processes.

This chapter covers general freezing methods and systems. Additional information on freezing specific products is covered in Chapters 14, 17 to 20, and 25 to 29. Related information can be obtained in Chapters 9 and 10, which cover thermal properties of foods as well as their cooling and freezing times. Information on refrigeration system practices is given in Chapters 1 to 3.

FREEZING METHODS

Freezing systems can be grouped by their basic method of extracting heat from food products:

Blast freezing (convection). Cold air is circulated over the product at high velocity. The air removes heat from the product and releases it to an air/refrigerant heat exchanger before being recirculated.

Contact freezing (conduction). Food, packaged or unpackaged, is placed on or between cold metal surfaces. Heat is extracted by direct conduction through the surfaces, which are directly cooled by a circulating refrigerated medium.

Cryogenic freezing (convection and/or conduction). Food is exposed to an environment below –76°F by spraying liquid nitrogen or liquid carbon dioxide into the freezing chamber.

Cryomechanical freezing (convection and/or conduction). Food is first exposed to cryogenic freezing and then finish-frozen through mechanical refrigeration.

Special freezing methods, such as liquid immersion (e.g., brines for packaged products), are covered under the specific product chapters.

BLAST FREEZERS

Blast freezers use air as the heat transfer medium and depend on contact between the product and the air. Sophistication in airflow control and conveying techniques varies from crude blast-freezing chambers to carefully controlled impingement freezers.

The earliest blast freezers consisted of cold storage rooms with extra fans and a surplus of refrigeration. Improved airflow control and mechanization of conveying techniques have made heat transfer more efficient and product flow less labor-intensive.

Although **batch freezing** is still widely used, more sophisticated freezers integrate freezing into a continuous production line. This **process-line freezing** has become essential for large-volume,

The preparation of this chapter is assigned to TC 10.9, Refrigeration Application for Foods and Beverages.

high-quality, cost-effective operations. A wide range of blast freezer systems are available, including

Batch	**Continuous/Process-Line**
• Cold storage rooms	• Straight belts (two-stage, multipass)
• Stationary blast cells	• Fluidized beds
• Push-through trolleys	• Fluidized belts
	• Spiral belts
	• Carton (carrier)

Cold Storage Rooms

Although a cold storage room is not considered a freezing system, it is sometimes used for this purpose. Because a storage room is not designed to be a freezer, it should only be used for freezing in exceptional cases. Freezing is generally so slow that the quality of most products suffers. The quality of the already frozen products stored in the room is jeopardized because the excess refrigeration load may raise the temperature of the frozen products considerably. Also, flavors from warm products may be transferred.

Stationary Blast Cell Freezing Tunnels

The stationary blast cell (Figure 2) is the simplest freezer that can be expected to produce satisfactory results for most products. It is an insulated enclosure equipped with refrigeration coils and axial or centrifugal fans that circulate air over the products in a controlled way. Products are usually placed on trays, which are then placed into racks so that an air space is left between adjacent layers of trays. The racks are moved in and out of the tunnel manually using a pallet mover. It is important that the racks be placed so that air bypass is minimized. The stationary blast cell is a universal freezer, because almost all products can be frozen in a blast cell. Vegetables and other products (e.g., bakery items, meat patties, fish fillets, prepared foods) may be frozen either in cartons or unpacked and spread in a layer on trays. However, product losses from spillage, damage, and dehydration can be greater, and product quality can be reduced. In some instances, this type of freezer is also used to reduce to 0°F or below the temperature of palletized, cased products that have previously been frozen through the latent heat of fusion zone by other means. The flexibility of a blast cell makes it suitable for small quantities of varied products; however, labor requirement is relatively high and product movement is slow.

Push-Through Trolley Freezers

The push-through trolley freezer (Figure 3), in which the racks are fitted with wheels, incorporates a moderate degree of mechanization. Racks are usually moved on rails by a pushing mechanism, which can be hydraulically or electrically powered. This type of freezer is similar to the stationary blast cell, except that labor costs and product handling time are decreased. This system is widely used to **crust-freeze** (quick-chill) wrapped packages of raw poultry and for irregularly shaped products. Another version uses a chain drive to move the trolleys through the freezer.

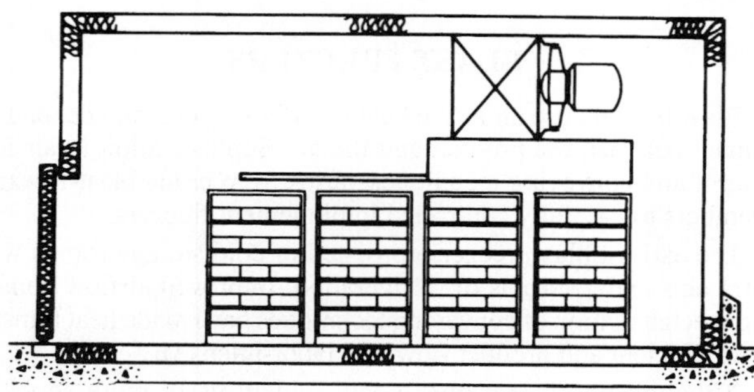

Fig. 2 Stationary Blast Cell

Straight Belt Freezers

The first mechanized blast freezers consisted of a wire mesh belt conveyor in a blast room, which satisfied the need for continuous product flow. A disadvantage to these early systems was the poorly controlled airflow and resulting inefficient heat transfer. Current versions use controlled vertical airflow, which forces cold air up through the product layer, thereby creating good contact with the product particles. Straight belt freezers are generally used with fruits, vegetables, French fried potatoes, cooked meat toppings (e.g., diced chicken), and cooked shrimp.

The principal design is the **two-stage belt freezer** (Figure 4), which consists of two mesh conveyor belts in series. The first belt initially precools or crust-freezes an outer layer or crust to condition the product before transferring it to the second belt for freezing to 0°F or below. Transfer between belts helps to redistribute the product on the belt and prevents product adhesion to the belt. To ensure uniform cold air contact and effective freezing, products should be distributed uniformly over the entire belt. Two-stage freezers are generally operated at 15 to 25°F refrigerant temperatures in the precool section and –25 to –40°F in the freezing section. Capacities range from 1 to 50 tons of product per hour, with freezing times from 3 to 50 min.

When products to be frozen are hot (e.g., French fries from the fryer at 180 to 200°F), another cooling section is added ahead of the normal precool section. This section supplies either refrigerated air at approximately 50°F or filtered ambient air to cool the product and congeal the fat. Refrigerated air is preferred because filtered ambient air has greater temperature variations and may contaminate the product.

Multipass Straight Belt Freezers

For larger products with longer freezing times (up to 60 min) and higher capacity requirements (0.5 to 6 ton/h), a single straight belt freezer would require a very large floor space. Required floor space can be reduced by stacking belts above each other to form either (1) a single-feed/single-discharge multipass system (usually three passes) or (2) multiple single-pass systems (multiple infeeds and discharges) stacked one on top of the other. The multipass (triple-pass) arrangement (Figure 5) provides another benefit in that the product, after being surface frozen on the first (top) belt,

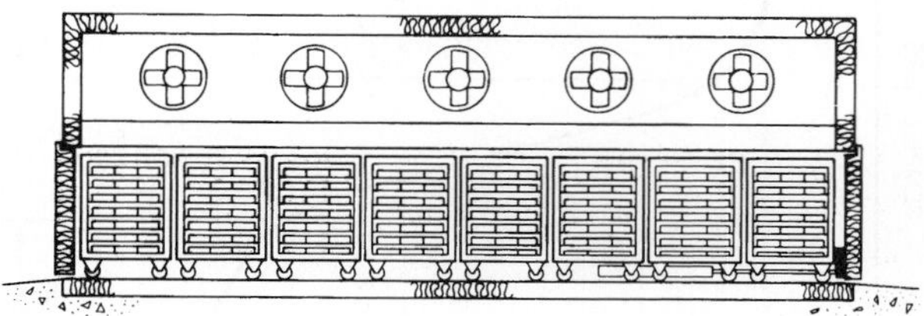

Fig. 3 Push-Through Trolley Freezer

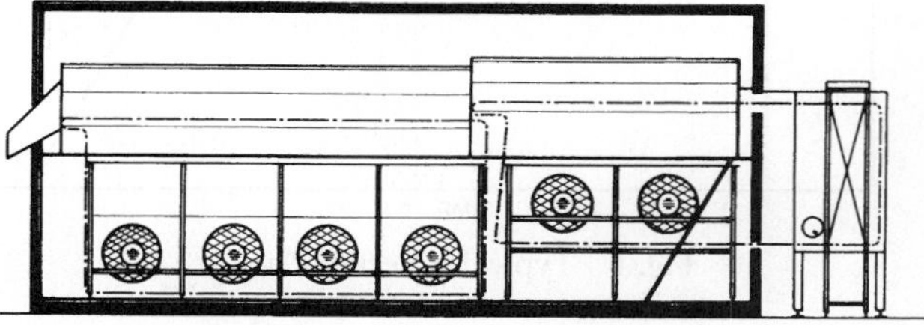

Fig. 4 Two-Stage Belt Freezer

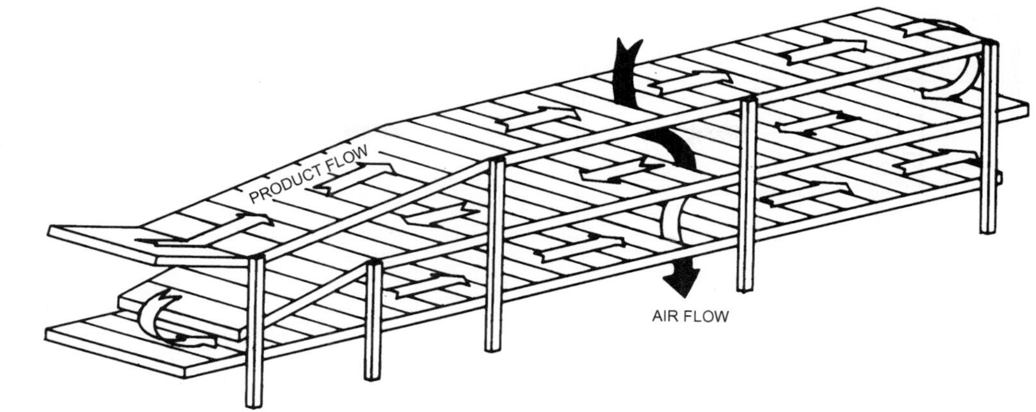

Fig. 5 Multipass, Straight Belt Freezer

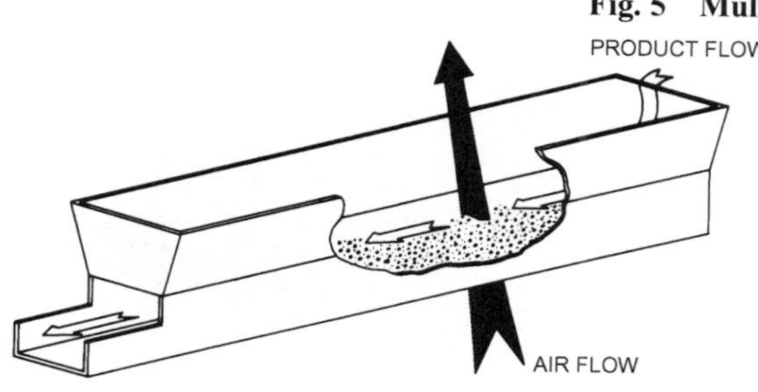

Fig. 6 Fluidized Bed Freezer

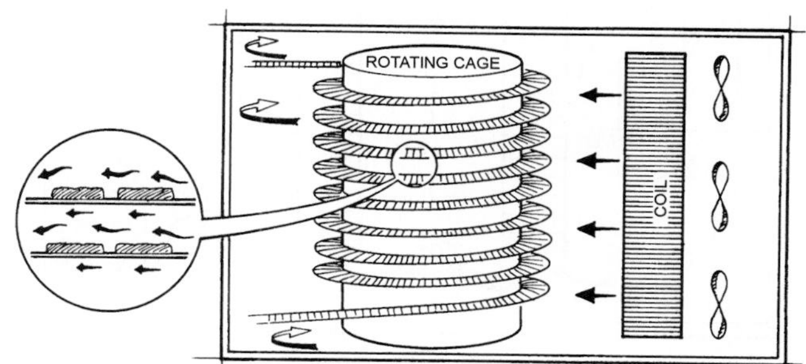

Fig. 7 Horizontal Airflow Spiral Freezer

may be stacked more deeply on the lower belts. Thus, the total belt area required is reduced, as is the overall size of the freezer. However, this system has a potential for product damage and product jams at the belt transfers.

Fluidized Bed Freezers

This freezer uses air both as the medium of heat transfer and for transport; the product flows through the freezer on a cushion of upward-flowing cold air (Figure 6). This design is well suited for small, uniform-sized particulate products such as peas, diced vegetables, and small fruit.

The high degree of fluidization improves the heat transfer rate and allows good use of floor space. The technique is limited to well-dewatered products of uniform size that can be readily fluidized and transported through the freezing zone. Because the principle depends on rapid crust-freezing of the product, the operating refrigerant temperature must be −40°F or lower, giving air temperatures of −20°F or lower. Fluidized bed freezers are normally manufactured as packaged, factory-assembled units with capacity ranges of 1 to 10 ton/h. Particulate products generally have a freezing time of 3 to 15 min.

Fluidized Belt Freezers

A hybrid of the two-stage belt freezer and the fluidized bed freezer, the fluidized belt freezer has a fluidizing section in the first belt stage. An increased air resistance is designed under the first belt to provide fluidizing conditions for wet incoming product, but the belt is there to help transport heavier, less uniform products that do not fluidize fully. Once crust-frozen, the product can be loaded deeper for greater efficiency on the second belt. Two-stage fluidized belt freezers operate at −30 to −35°F refrigerant temperature and in capacity ranges from 1 to 50 ton/h. A good order-of-magnitude estimate of total refrigeration load for individually quick-frozen (IQF) freezers is 40 tons of refrigeration per ton of product per hour. Small freezers require about 10 to 15% more capacity per ton of product per hour.

Spiral Belt Freezers

This freezer is generally used for products with long freezing times (generally 10 min to 3 h), and for products that require gentle handling during freezing. An endless conveyor belt that can be bent laterally is wrapped cylindrically, one tier below the last; this configuration requires minimal floor space for a relatively long belt. The original spiral belt principle uses a spiraling rail system to carry the belt, although more recent designs use a proprietary self-stacking belt requiring less overhead clearance. The number of tiers in the spiral can be varied to accommodate different capacities. In addition, two or more spiral towers can be used in series for products with long freezing times. Spiral freezers are available in a range of belt widths and are manufactured as packaged, modular, and field-erected models to accommodate various upstream processes and capacity requirements.

Airflow varies from open, unbaffled spiral conveyors to flow through extensive baffling and high-pressure fans. Horizontal airflow is applied to spiral freezers (Figure 7) by axial fans mounted along one side. The fans blow air horizontally across the spiral conveyor with minimal baffling limited to two portions of the spiral circumference. The rotation of the cage and belt produces a rotisserie effect, with product moving past the high-velocity cold air near the discharge, aiding in uniform freezing.

Several proprietary designs are available to control airflow. One design (Figure 8) has a mezzanine floor that separates the freezer into two pressure zones. Baffles around the outside and inside of the belt form an air duct so that air flows up or down around the product as the conveyor moves the product. The controlled airflow reduces freezing time for some products.

Another design (Figure 9) splits the airflow so that the coldest air contacts the product both as it enters and as it leaves the freezer. The coldest air introduced on the incoming, warm product may increase surface heat transfer and freeze the surface more rapidly, which may reduce product dehydration.

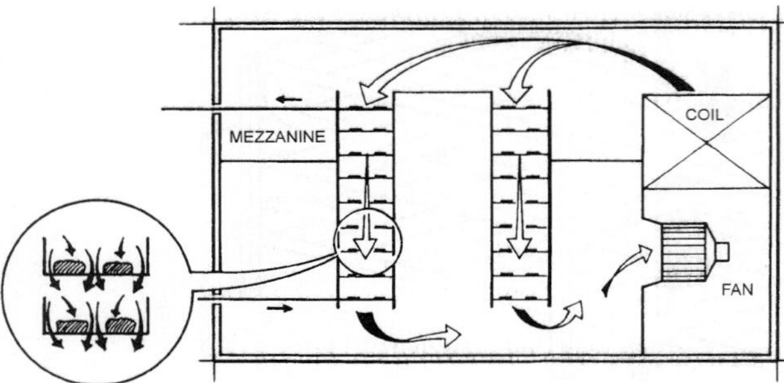

Fig. 8 Vertical Airflow Spiral Freezer

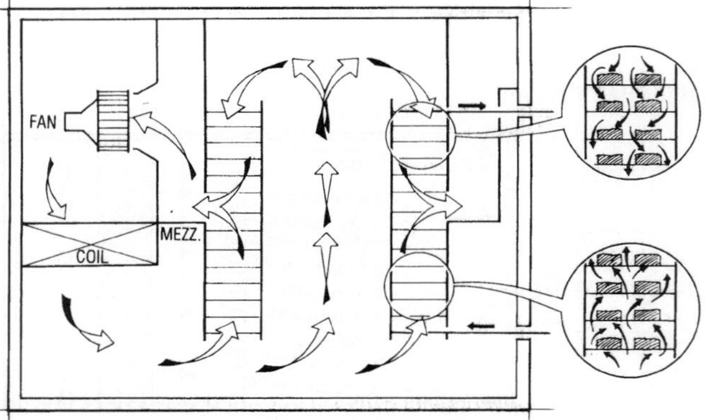

Fig. 9 Split Airflow Spiral Freezer

Typical products frozen in spiral belt freezers include raw and cooked meat patties, fish fillets, chicken portions, pizza, and a variety of packaged products. Spiral freezers are available in a wide range of capacities, from 0.5 to 10 ton/h. They dominate today's frozen food industry and account for the majority of unpackaged nonparticulate frozen food production, as well as many packaged products.

Impingement Freezers

In this design (Figure 10), cold air flows perpendicular to the product's largest surfaces at a relatively high velocity. Air nozzles with corresponding return ducts are mounted above and below the conveyors. The airflow constantly interrupts the boundary layer that surrounds the product, enhancing the surface heat transfer rate. The technique may therefore reduce freezing time of products with large surface-to-mass ratios (thin hamburger patties, for example). Impingement freezers are designed with single-pass or multipass straight belts. Freezing times are 1 to 10 min. Cost-effective application is limited to thin food products (less than 1 in. thick).

Carton Freezers

The carton (or carrier) freezer (Figure 11) is a very-high-capacity freezer (5 to 20 ton/h) for medium to large cartons of products such as red meat, poultry, and ice cream. These units are also used as chillers for meat products and block cheese.

In the top section of the freezer, a row of loaded product carriers is pushed toward the rear of the freezer, while on the lower section they are returned to the front. Elevating mechanisms are located at both ends. A carrier is similar to a bookcase with shelves. When it is indexed in the loading/discharge end of the freezer, the already-frozen product is pushed off each shelf one row at a time onto a discharge conveyor. When the carrier is indexed up, this shelf aligns with the loading station, where new products are continuously pushed onto the carrier before it is moved once again to the rear of the freezer. Refrigerated air is circulated over the cartons by forced

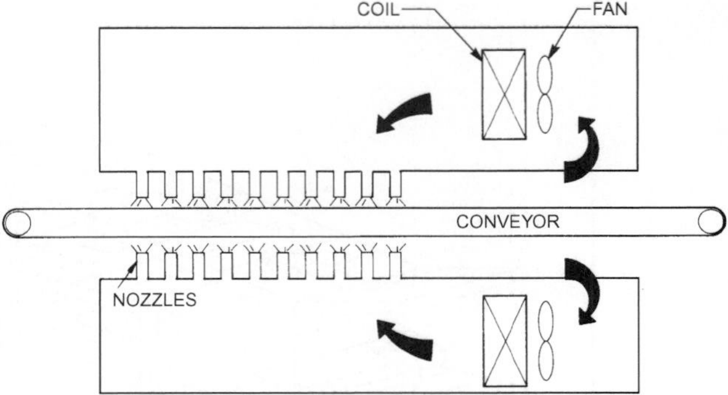

Fig. 10 Impingement Freezer

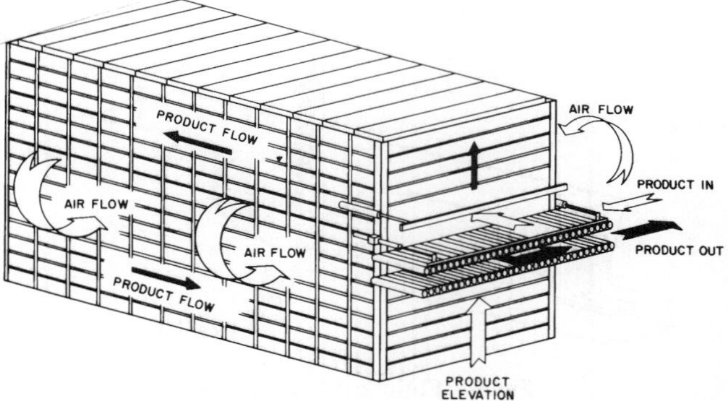

Fig. 11 Carton (Carrier) Freezer

convection. Generally, the air and product are arranged in cross flow, but some designs have air flowing either with or counter to the product (i.e., along the length of the freezer).

Computerized systems are now available to control shelf loading, unloading, and movement to freeze and/or cool products with different retention times in the same unit simultaneously. This increased flexibility is particularly useful and cost-effective where different sizes and cuts are prevalent (e.g., red meat and poultry products).

CONTACT FREEZERS

A contact freezer's primary means of heat transfer is conduction; the product or package is placed in direct contact with a refrigerated surface. Contact freezers can be categorized as follows:

Batch
- Manual horizontal plate
- Manual vertical plate

Process-Line
- Automatic plate
- Contact belt (solid stainless steel)
- Specialized design

The most common type of contact freezer is the **contact plate freezer**, in which the product is pressed between metal plates. Refrigerant is circulated inside channels in the plates, which ensures efficient heat transfer and results in short freezing times, provided that the product is a good conductor of heat, as in the case of fish fillets, chopped spinach, or meat offal. However, packages or cavities should be well filled, and if metal trays are used, they should not be distorted.

Manual and Automatic Plate Freezers

Contact plate freezers (Figure 12) are available in horizontal or vertical arrangements with manual loading/unloading. Horizontal plate freezers are also available in an automatic loading/unloading

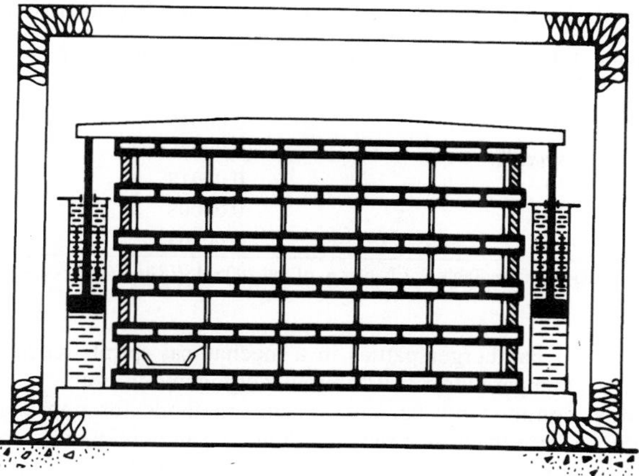

Fig. 12 Plate Freezer

version, which generally accommodates higher capacities and continuous operation. The advantage of good heat transfer in contact plate freezers is gradually reduced with increasing product thickness. For this reason, thickness is often limited to 2 to 3 in. Contact plate freezers operate efficiently because they require no fans, they are very compact, and there is no extra heat transfer between the refrigerant and a heat transfer medium. An advantage with packaged products is that pressure from the plates minimizes any bulging that may occur. Thus, packages are even and square within close tolerances. Automatic plate freezers accommodate up to 200 packages per minute, with freezing times from 10 to 150 min. When greater capacities are required, freezers are placed in series with associated conveyor systems to handle loading and unloading packages.

Specialized Contact Freezers

A combination of air and contact freezing is used for wet fish fillets and similar soft, wet products with relatively large, flat surfaces. The continuous, solid stainless steel belt is typically 4 to 6 ft wide and may be 100 ft long. Product is loaded onto the belt at one end of the freezer and then travels in a fixed position through the freezing zone to the discharge end. Freezing is usually accomplished both by conduction through the belt to a cooling medium below the belt and by convection through controlled airflow above the belt, or by convection only through high-velocity air above and below the belt. This freezer design produces attractive product, but a drawback is the physical size of the freezer. Capacities for typical products are generally limited to 1 to 2.5 ton/h, with a freezing time of less than 30 min.

Another specialized contact freezer conveys food products on a continuous plastic film over a low-temperature (–40°F) refrigerated plate. Contact with the film freezes approximately the bottom 0.04 in. of products in about 1 min. This equipment is used to eliminate product deformation or wire mesh belt markings on products that are flat, wet and sticky, soft, or in need of hand shaping before entering an air blast freezer. Another benefit of contact prefreezing is that it reduces dehydration losses in the subsequent freezing step. Examples of products suitable for contact prefreezing are marinated, boneless chicken breasts and thin fish fillets.

CRYOGENIC FREEZERS

Cryogenic (or gas) freezing is often an alternative for (1) small-scale production, (2) new products, (3) overload situations, or (4) seasonal products. Cryogenic freezers use liquid nitrogen or liquid carbon dioxide (CO_2) as the refrigeration medium, and the freezers may be batch cabinets, straight belt freezers, spiral conveyors, or liquid immersion freezers.

Liquid Nitrogen Freezers

The most common type of liquid nitrogen freezer is a straight-through, single-belt, process-line tunnel. Liquid nitrogen at –320°F is introduced at the outfeed end of the freezer directly onto the product; as the liquid nitrogen vaporizes, those cold vapors are circulated toward the infeed end, where they are used for precooling and initial freezing of the product. The "warmed" vapors (typically –50°F) are then discharged to the atmosphere. The low temperature of the liquid and vaporous nitrogen provides rapid freezing, which can improve quality and reduce dehydration for some products. However, the freezing cost is high because of the cryogen's cost, and the surface of high-water-content products may crack if precautions are not taken.

Consumption of liquid nitrogen is in the range of 0.9 to 2.0 lb of nitrogen per pound of product, depending on the water content and temperature of the product. Although this translates into relatively high operating costs, the small initial investment makes liquid nitrogen freezers cost-effective for some applications.

Carbon Dioxide Freezers

A similar cryogenic freezing method places boiling (subliming) CO_2 in direct contact with foods frozen in a straight belt or spiral freezer. Carbon dioxide boils at approximately –110°F, and the system operates like a liquid nitrogen freezing system, consuming cryogenic liquid as it freezes product. Applications for CO_2 freezing include producing individual quick-frozen (IQF) diced poultry, pizza toppings, and seafood.

CRYOMECHANICAL FREEZERS

Although the technique is not new, cryomechanical freezing (combination of cryogenic and blast freezing) applications are increasing. High-value, sticky products, such as IQF shrimp, and wet, delicate products, such as IQF strawberries and IQF cane berries, are common applications for these systems.

A typical cryomechanical freezer has an initial immersion step in which the product flows through a bath of liquid nitrogen to set the product surface. This rapid surface freezing reduces dehydration and improves the handling characteristics of the product, thus minimizing sticking and clumping. The cryogenically crust-frozen product is then transferred directly into a mechanical freezer, where the remainder of the heat is removed and the product temperature is reduced to 0°F or lower. The cryogenic step is sometimes retrofitted to existing mechanical freezers to increase their capacity. The mechanical freezing step makes operating costs lower than for cryogenic-only freezing.

OTHER FREEZER SELECTION CRITERIA

Reliability

Because of the harsh operating conditions, the freezing system is probably the most vulnerable equipment in a process line. A process line usually incorporates only one freezer, which makes reliability a major concern.

To achieve normal equipment life expectancy, freezing systems must be designed and constructed with adequate safety factors for electrical/mechanical components and with materials that can withstand harsh environments and rugged usage.

Hygiene

Cleanability and sanitary design are as important as reliability. Freezing systems should (1) have a minimum number of locations where the product can hang up, (2) be constructed of noncorrosive, safe materials, and (3) be equipped with manual and/or automatic sanitation systems for washdown and cleanup. If the equipment is not or cannot be properly cleaned and sanitized, product contamination can result. These features are particularly important for chilled, partially cooked, and fully cooked products that may not be fully reheated or properly prepared before consumption.

Quality

The quality of processed food products is affected by physical changes and by rates of microbiological activity and chemical reactions, each of which is influenced by the rate of temperature change. See Chapter 12 for more information on food microbiology. The freezing process physically changes the food product; the rate of physical change or freezing time determines the size of ice crystals produced.

At a slow freezing rate, initially formed ice crystals can grow to a relatively large size; fast freezing forces more crystals to be seeded with a smaller average size. However, different-sized ice crystals are formed because the product's surface freezes faster than its inner parts. Large ice crystals may damage cell walls of the product, usually increasing loss of juices during thawing. For some products, this can greatly affect the texture and flavor of the remaining product tissue.

The influence of freezing time is more apparent in some products than in others. For strawberries, a shorter freezing time can significantly reduce drip loss. Drip loss is 20% for strawberries frozen in 12 h but only 8% for strawberries frozen in 15 min. Cryogenic systems perform the same freezing function in 8 min or less, reducing drip loss to less than 5%.

A well-applied mechanical, cryogenic, or cryomechanical freezer can crust-freeze products rapidly, minimizing loss of natural juices, aromatics, and flavor essences and maintaining higher, more marketable product quality. Also, lower storage temperature and fewer, less severe temperature fluctuations tend to help preserve quality. However, long-term storage can diminish any benefits of more rapid freezing.

Economics

Ironically, freezing equipment is considered both the most expensive and the least expensive link in the modern processing chain. Although the freezer frequently represents the single largest investment in a line, its operating costs are usually only 3 to 5% of the total. Packaging costs may vary widely but generally are several times greater than total freezing cost.

One essential factor to consider when choosing freezing equipment is the loss in product mass that occurs during freezing. The cost of this loss may be about the same as the operating cost of the freezer for inexpensive products such as peas; the loss is even more significant for expensive products such as seafood.

Loss of product mass during freezing may be caused by mechanical losses, downgrading, and dehydration (shrink). **Mechanical losses** occur from products dropping to the floor or sticking to conveyor belts, and are specific to each processing plant. A modern freezer should produce minimal losses in this category. **Downgrading losses** refer to product damage, breakage, and other occurrences that render the product unsalable at the top-quality price. For most products, a modern freezing system should incur minimal losses from damage and breakage.

Dehydration losses occur in any freezing system. Evaporation of water vapor from unpackaged products during freezing becomes evident as frost builds up on evaporator surfaces. Frost is also caused by excessive infiltration of warm, moist air into the freezer. Still air inside a diffusiontight carton often creates larger dehydration losses than individual quick freezing of unpackaged products. Heat transfer is poor because no circulation of air occurs within the package. The resulting evaporation of moisture can be significant; however, the frost stays inside the carton.

Unpackaged food products cannot be chilled or frozen without losing at least some moisture. The most important factor affecting food dehydration is the time it takes for the product to freeze: the faster the freezing (within practical and cost limitations), the less dehydration shrinkage and the higher the quality. Most mechanical in-line freezers have speeds comparable to that of cryogenics: 12 to 15 min for hamburger patties in a mechanical spiral freezer, for instance, compared to 8 to 10 min in a cryogenic freezer.

Table 1 Moisture-Carrying Capacity of Air (Saturated)

Temperature, °F	Ratio of Water to Dry Air, by Weight
0	0.0008
−10	0.00046
−20	0.00026
−30	0.00015
−40	0.00008
−50	0.00004

Source: Adapted from Table 2, Chapter 6, of the 2005 *ASHRAE Handbook—Fundamentals*.

Most common mechanical freezing methods for unpackaged foods use air as the heat transfer medium; the amount of moisture air absorbs depends on temperature and pressure, as shown in Table 1. For more specific values, see Table 2 in Chapter 6 of the 2005 *ASHRAE Handbook—Fundamentals*.

From Table 1, 0°F air can, if it becomes saturated, absorb 10 times the amount of moisture from product that −40°F air would.

In air, moisture flows away from high-vapor-pressure or high-temperature areas (e.g., food products) toward areas of lower pressure or temperature (e.g., air). Higher air temperatures increase the likelihood of product moisture loss, but only during the initial freezing, before the outer crust freezes or is sealed with ice.

Many manufacturers use high evaporator temperatures to reduce energy costs; when coupled with high air volumes and an air temperature close to the refrigerant temperature, this method has the added benefit of reducing food dehydration. Lower evaporator temperatures tend to remove more moisture from air, which in turn leads to greater moisture being robbed from the product. Using the smallest achievable air temperature rise is helpful; for mechanical systems, try to keep the temperature difference between coil air and refrigerant to 10 to 12°F.

Increasing air quantities also helps reduce shrinkage. Airflow around the product should be evenly distributed, to allow good heat transfer.

Drip loss varies according to initial product quality, amount of surface or product moisture available, and time to freeze. Quicker freezing times reduce drip loss by sealing the surface.

The rate of moisture diffusion, and how readily a product releases moisture to air, also affects shrinkage. Different products have different moisture diffusion rates, which can be difficult to estimate or measure.

A poorly designed freezing system for unpackaged products operates with dehydration losses of easily 3 to 4%, but well-designed mechanical or cryogenic freezing systems can be built to operate with losses near 0.5%. Liquid nitrogen tunnels normally operate with a dehydration loss of about 0.4 to 1.25%, which occurs when the nitrogen gas is circulated over the product at the infeed end of the freezer. Infeed circulation is sometimes needed to temper the product and to use the nitrogen's heat capacity most efficiently. Nitrogen immersion freezers have lower dehydration losses but use more liquid nitrogen. A CO_2 freezer using jet impingement operates with a dehydration loss of about 0.5 to 1.25%.

In general, the faster the product surface temperature is reduced, the lower the dehydration rate. Although air relative humidity affects dehydration during frozen storage, it has little effect on dehydration rates during freezing.

REFRIGERATION SYSTEMS

Most mechanical food freezers use ammonia as the refrigerant and are equipped with either liquid overfeed or gravity-flooded evaporators. The choice of evaporator system depends on freezer size and configuration, space limitations, freezer location, existing plant systems (where applicable), relative cost, and end user preference. For systems with three or more evaporators, a liquid overfeed system is usually less costly to install and operate.

Evaporators may be defrosted with water, hot gas, or a combination of both. Defrost systems can be manual, manual start/automatic run, or fully automatic. Coil defrost can take place at a shift change or be sequential, so that the freezer remains in continuous operation for long periods. Selection of a defrost system depends on plant and product requirements, water supply and disposal situation, sanitation regulations, and end user preference.

With a liquid overfeed system, carefully consider refrigeration line sizing and potential static head penalties if the liquid overfeed recirculator is remote from the freezer. Locating the liquid recirculation equipment adjacent to and below the evaporators provides for the most efficient and productive operation of the freezing equipment. In particular, vertical risers in wet suction lines can result in liquid logging (retention), leading to excessive pressure drops. See Chapter 1 for design considerations.

A design evaporator temperature for the freezer should be selected to achieve the lowest overall capital and operating cost possible for the freezer and the other high- and low-side refrigeration components, while remaining consistent with product requirements and other plant operating conditions.

If there is significant air infiltration into the freezer, then special air-cooling coil designs using large fin spacing (or staggered fin space) may be necessary to avoid excessive deterioration of performance by frosting.

In a mechanical system, using a –50°F evaporator temperature instead of –40°F increases the utility bill by about 15%. A system with lower evaporator temperature may have slightly higher first cost, but will cause significantly less shrinkage. For temperatures below –40°F, a CO_2/NH_3 cascade refrigeration system should be considered, and may be less expensive than a two-stage NH_3 system.

Operation

Modern conveyor freezers are equipped with programmable logic controls (PLCs) and/or computer control systems that can monitor and control key elements of freezer operation to maximize productivity, product quality, and safety. Items to be monitored and controlled include belt speeds, air and refrigerant temperatures, air and refrigerant pressures, evaporator defrost cycles, belt washers and dryers, amperage for electric motors, safety and alarm functions, and other variables specific to the products being frozen.

The presence of electronic controls alone does not guarantee freezer performance; human operators are still needed. The number and specialty of operators required depends on the size of the plant and the quantity of freezers. A small plant may have a combination operator covering belt production and refrigeration. In larger plants, a freezer operator may oversee production while a specialist attends to the refrigeration cycle. Long-term success requires well-trained, knowledgeable operators who make the proper adjustments as changes occur in the process.

Maintenance

Freezing systems operate in a harsh environment in which some of the components are hidden from view by the enclosure and product. Many freezers operate 5000 to 7000 h per year. The best freezers are ruggedly constructed and well designed for easy maintenance. Nevertheless, a well-run maintenance program is essential to productivity and safety.

Freezer manufacturers supply operation and maintenance manuals with key instructions, information on components, parts lists, and suggestions regarding maintenance and safety inspections and tasks on a planned-frequency basis. Manufacturers also provide training programs for maintenance technicians.

It is important for plants to have a sufficient number of properly trained maintenance technicians to maintain all systems. Duties include prescribed inspections, routine maintenance tasks, troubleshooting, and required maintenance during nonproduction periods. Depending on plant size, technicians may be individual mechanics, electricians, and refrigeration specialists or combinations of the three.

It is suggested that plants hire contract services when they are not able to cover any or all maintenance and operation functions adequately with their own personnel.

BIBLIOGRAPHY

Becker, B.R and B.A. Fricke. 1999. Evaluation of semi-analytical/empirical freezing time estimation methods. Part I: Regularly shaped food items. *International Journal of HVAC&R Research* (now *HVAC&R Research*) 5(2):151-169.

Becker, B.R. and B.A. Fricke. 1999. Evaluation of semi-analytical/empirical freezing time estimation methods. Part II: Irregularly shaped food items. *International Journal of HVAC&R Research* (now *HVAC&R Research*) 5(2):171-187.

Becker, B.R. and B.A. Fricke. 2004. Heat transfer coefficients for forced-air cooling and freezing of selected foods. *International Journal of Refrigeration* 27(5):540-551.

Briley, G.C. 2002. Moisture loss during freezing. *ASHRAE Journal* 44(11):68.

Bustabad, O.M. 1999. Weight loss during freezing and the storage of frozen meat. *Journal of Food Engineering* 41(1):1-11.

Campanone, L.A., V.O. Salvadori, and R.H. Mascheroni. 2001. Weight loss during freezing and storage of unpackaged foods. *Journal of Food Engineering* 47(2):69-79.

Campanone, L.A., V.O. Salvadori, and R.H. Mascheroni. 2005. Food freezing with simultaneous surface dehydration: Approximate prediction of freezing time. *International Journal of Heat & Mass Transfer* 48(6): 1205-1213.

Campanone, L.A., V.O. Salvadori, and R.H. Mascheroni. 2005. Food freezing with simultaneous surface dehydration: Approximate prediction of weight loss during freezing and storage. *International Journal of Heat & Mass Transfer* 48(6):1195-1204.

Cleland, A.C. and S. Ozilgen. 1998. Thermal design calculations for food freezing equipment—Past, present and future. *International Journal of Refrigeration* 21(5):359-371.

Delgado, A.E. and D.-W. Sun. 2001. Heat and mass transfer models for predicting freezing processes—A review. *Journal of Food Engineering* 47(3):157-174.

Fricke, B.A. and B.R. Becker. 2002. Calculation of heat transfer coefficients for foods. *International Communications in Heat & Mass Transfer* 29 (6):731-740.

Fricke, B.A. and B.R. Becker. 2004. Calculation of food freezing times and heat transfer coefficients. *ASHRAE Transactions* 110(2):145-157.

Gruda, Z. 1979. Increasing of freezers efficiency through the improvement of the heat transfer conditions between air and product—New inventions in fluidized bed freezers regarding soft products freezing. *Proceedings of the 15th IIR International Congress of Refrigeration*, Padova, Italy, pp. 1199-1208.

Ionov, A.G., O.K. Bogljubsky, and B.H. Erlikhman. 1979. Analysis of cooling systems of plate freezers. *Proceedings of the 15th IIR International Congress of Refrigeration*, Padova, Italy, pp. 461-468.

Norwig, J.F. and D.R. Thompson. 1984. Review of dehydration during freezing. *Transactions of the ASAE* 27(5):1619-1624.

Salvadori, V.O. and R.H. Mascheroni. 2002. Analysis of impingement freezers performance. *Journal of Food Engineering* 54(2):133-140.

Sheen, S. and L.F. Whitney. 1990. Modelling heat transfer in fluidized beds of large particles and its applications in the freezing of large food items. *Journal of Food Engineering* 12(4):249-265.

Stoecker, W.F. 1998. *Industrial refrigeration handbook*. McGraw-Hill, New York.

CHAPTER 17

MEAT PRODUCTS

AROUND the world about 4 to 5 million (0.4 million in the United States) four-legged animals such as hogs, cattle, calves, buffalo, water buffalo, lambs, sheep, goats, and deer are slaughtered each day to supply the demand for red meats and their products. The majority of these animals are slaughtered in commercial slaughterhouses (abattoirs) under supervision, although a small portion (0.08% in the United States) are still killed on the farm. The slaughter process from live animals to packaged meat products is illustrated in Figure 1.

SANITATION

Sound sanitary practices should be applied at all stages of food processing, not only to protect the public but to meet aesthetic requirements. In this respect, meat processing plants are no different from other food plants. The same principles apply regarding sanitation of buildings and equipment; provision of sanitary water supplies and wash facilities; disposal of waste materials; insect and pest control; and proper use of sanitizers, germicides, and fungicides. All U.S. meat plants operate under regulations set forth in inspection service orders. For detailed sanitation guidelines to be followed in all plants producing meat under federal inspection, refer to the U.S. Department of Agriculture's *Agriculture Handbook* 570, the Food Safety and Inspection Service (FSIS), and Marriott (1994).

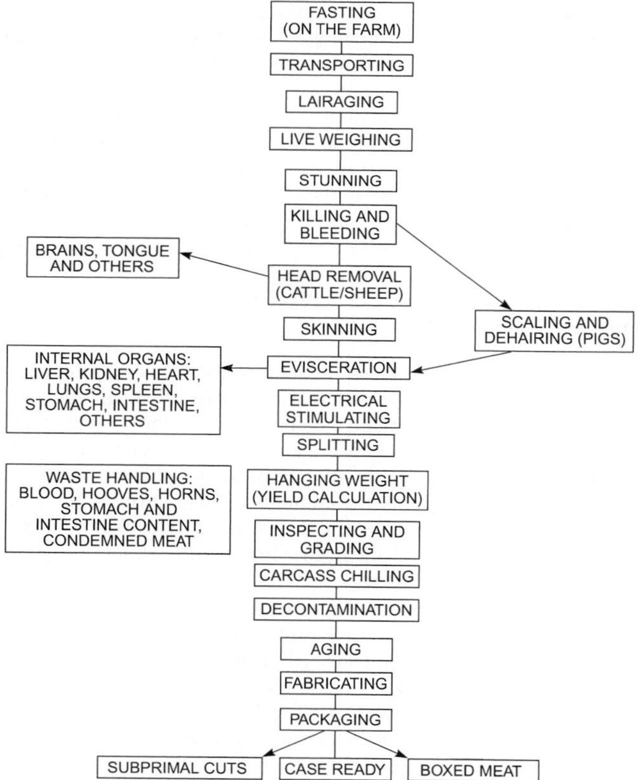

Fig. 1 Steps of Meat Processing

Proper safeguards and good manufacturing practices should minimize bacterial contamination and growth. This involves using clean raw materials, clean water and air, sanitary handling throughout, good temperature control (particularly in coolers and freezers), and scrupulous between-shift cleaning of all surfaces in contact with the product.

Precooked products present additional problems because conditions are favorable for bacterial growth after the product cools to below 130°F. In addition, potential pathogen growth may be enhanced because their competitor organisms were destroyed during cooking. Any delay in processing at this stage allows surviving microorganisms to multiply, especially when cooked and cooled meat is handled and packed into containers before processing and freezing. Creamed products offer especially favorable conditions for bacterial growth. Filled packages should be removed immediately and quickly chilled, which not only reduces the time for growth, but can also reduce the number of bacteria.

It is even more important during processing to avoid any opportunity for growth of pathogenic bacteria that may have entered the product. Although these organisms do not grow as quickly at temperatures below 40°F, they can survive freezing and prolonged frozen storage.

Storage at about 25°F allows growth of psychrophilic spoilage bacteria, but at 14°F these, as well as all other bacteria, are dormant. Even though some cells of all bacteria types die during storage, activity of the survivors is quickly renewed with rising temperature. The processor should recommend safe preparation practices to the consumer. The best procedure is to provide instructions for cooking the food without preliminary thawing. In the freezer, sanitation is confined to keeping physical cleanliness and order, preventing access of foreign odors, and maintaining the desired temperatures.

Role of HACCP

Many of the procedures for the control of microorganisms are managed by the Hazard Analysis and Critical Control Point (HACCP) system of food safety, which is described in Chapter 12. It is a logical process of preventive measures that can control food safety problems. HACCP plans are required by the USDA in all plants. One aspect of the plan recommends that red meat carcasses and variety meats be chilled to 40°F within 24 h, and that this temperature be maintained during storage, shipping, and product display.

The preparation of this chapter is assigned to TC 10.9, Refrigeration Application for Foods and Beverages.

CARCASS CHILLING AND HOLDING

A hot-carcass cooler removes live animal heat as rapidly as possible. Side effects such as cold shortening, which can reduce tenderness, must be considered. Electrical stimulation can minimize cold shortening. Rapid temperature reduction is important in reducing the growth rate of microorganisms that may exist on carcass surfaces. Conditions of temperature, humidity, and air motion must be considered to attain desired meat temperatures within the time limit and to prevent excessive shrinkage, bone taint, sour rounds, surface slime, mold, or discoloration. The carcass must be delivered with a bright, fresh appearance.

Spray Chilling Beef

Spraying cold water intermittently on beef carcasses for 3 to 8 h during chilling is currently the normal procedure in commercial beef slaughter plants (Johnson et al. 1988). Basically, this practice reduces evaporative losses and speeds chilling. Regulations do not allow the chilled carcass to exceed the prewashed hot-carcass weight. The carcass is chilled to a large extent by evaporative cooling. As the carcass surface tissue dries, moisture migrates toward the surface, where it evaporates. Eventually, an equilibrium is reached when the temperature differential narrows and reduces evaporative loss.

When carcasses were shrouded, once a common method for reducing weight loss (shrink), typical evaporative losses ranged from 0.75 to 2.0% for an overnight chill (Kastner 1981). Allen et al. (1987) found that spray-chilled beef sides lost 0.3% compared with 1.5% for non-spray-chilled sides. Although variation in carcass shrink of spray-chilled sides was influenced by carcass spacing, other factors, especially those affecting the dynamics of surface tissue moisture, may be involved. Carcass washing, length of spray cycle, and carcass fatness also affect shrinkage. With enough care, however, carcass cooler shrink can be nearly eliminated.

Loin eye muscle color and shear force are not affected by spray chilling, but fat color can be lighter in spray-chilled compared to non-spray-chilled sides. Over a 4 day period, color changes and drip losses in retail packs for rib steaks and round roasts were not related to spray chilling (Jones and Robertson 1989). Spray chilling could provide a moderate reduction in carcass shrinkage during cooling without having a detrimental influence on muscle quality.

Vacuum-packaged inside rounds from spray-chilled sides had significantly more purge (i.e., air removed) (0.9 lb or 0.26%) than those from conventionally chilled sides. Spacing treatments where foreshanks were aligned in opposite directions and where they were aligned in the same direction but with 6 in. between sides both result in less shrink during a 24 h spray-chill period than the treatment where foreshanks were aligned in the same direction but with all sides tightly crowded together (Allen et al. 1987). Some studies with beef (Hamby et al. 1987) and pork (Greer and Dilts 1988) indicated that bacterial populations of conventionally and spray-chilled carcasses were not affected by chilling method (Dickson 1991). However, Acuff (1991) and others showed that using a sanitizer (chlorine, 200 ppm, or organic acid, 1 to 3%) significantly reduces carcass bacterial counts.

Chilling Time

Although certain basic principles are identical, beef and hog carcass chilling differs substantially. The massive beef carcass is only partially chilled (although shippable) at the end of the standard overnight period. The average hog carcass may be fully chilled (but not ready for cutting) in 8 to 12 h; the balance of the period accomplishes only temperature equalization.

The beef carcass surface retains a large amount of wash water, which provides much evaporative cooling in addition to that derived from actual shrinkage, but evaporative cooling of the hog carcass, which retains little wash water, occurs only through actual shrinkage. A beef carcass, without skin and destined largely for sale as fresh cuts, must be chilled in air temperatures high enough to avoid freezing and damage to appearance. Although it must subsequently be well tempered for cutting and scheduled for in-plant processing, a hog carcass, including the skin, can tolerate a certain amount of surface freezing. Beef carcasses can be chilled with an overnight shrinkage of 0.5%, whereas equally good practice on hog carcasses results in 1.25 to 2% shrinkage.

The bulk (16 to 20 h) of beef chilling is done overnight in high-humidity chilling rooms with a large refrigeration and air circulation capacity. The rest of the chilling and temperature equalization occurs during a subsequent holding or storage period that averages 1 day, but can extend to 2 or 3 days, usually in a separate holding room with a low refrigeration and air circulation capacity.

Some packers load for shipment the day after slaughter, because some refrigerated transport vehicles have ample capacity to remove the remaining internal heat in round or chuck beef during the first two days in transit. This practice is most important in rapid delivery of fresh meat to the marketplace. Carcass beef that is not shipped the day after slaughter should be kept in a beef-holding cooler at temperatures of 34 to 36°F with minimum air circulation to avoid excessive color change and weight loss.

Refrigeration Systems for Coolers

Refrigeration systems commonly used in carcass chilling and holding rooms are operated with ammonia as the primary refrigerant and are of three general types: dry coils, chilled brine spray, and sprayed coil.

Dry-Coil Refrigeration. Dry-coil systems comprise most chilling and holding room installations. Dry-coil systems usually include unit coolers equipped with coils, defrosting equipment, and fans for air/vapor circulation. Because the coils operate without continuous brine spray, eliminators are not required. Coils are usually finned, with fins limited to 3 or 4 per inch or with variable fin spacing to avoid icing difficulties. Units may be mounted on the floor, overhead on the rail beams, or overhead on converted brine spray decks.

Dry-coil systems operated at surface temperatures below 32°F build up a coating of frost or ice, which ultimately reduces airflow and cooling capacity. Coils must therefore be defrosted periodically, normally every 4 to 24 h for coils with 3 or 4 fins per inch, to maintain capacity. The rate of build-up, and hence defrosting frequency, decreases with large coil capacity and high evaporating pressure.

Defrosting may be done either manually or automatically by the following methods:

- **Hot-gas defrost** introduces hot gas directly from the system compressors into the evaporator coils, with fans off. Evaporator suction is throttled to maintain a coil pressure of about 60 to 75 psig (at approximately 40 to 50°F). The coils then act as condensers and supply heat for melting the ice coating. Other evaporators in the system must supply the compressor load during this period. Hot-gas defrost is rapid, normally requiring 10 to 30 min for completion. See Chapter 3 for further information about hot-gas defrost piping and control.
- **Coil spray defrost** is accomplished (with fans turned off) by spraying the coil surfaces with water, which supplies heat to melt the ice coating. Suction and feed lines are closed, with pressure relief from the coil to the suction line to minimize the refrigeration effect. Enough water at 50 to 75°F must be used to avoid freezing on the coils, and care must be taken to ensure that drain lines do not freeze. Sprayed water tends to produce some fog in the refrigerated space. Coil spray defrost may be more rapid than hot-gas defrost.
- **Room air defrost** (for rooms 35°F or higher) is done with fans running while suction and feed lines are closed (with pressure relief from coil to suction line), to allow build-up of coil pressure

and melting the ice coating by transfer of heat out of the air flowing across the coils. Refrigeration therefore continues during defrosting, but at a drastically reduced rate. Room air defrost is slow; the time required may vary from 30 min to several hours if the coils are undersized for dry-coil operation.

- **Electric defrost** uses electric heaters with fans either on or off. During defrost, refrigerant flow is interrupted.

Unit coolers may be defrosted by any one or combinations of the first three methods. All methods involve reduced chilling capacity, which varies with time loss and heat input. Hot-gas and coil spray defrost interrupt chilling only for short periods, but they introduce some heat into the space. Room air defrost severely reduces the chilling rate for long periods, but the heat required to vaporize ice is obtained entirely from the room air.

Evaporator controls customarily used in carcass chilling and holding rooms include refrigerant feed controls, evaporator pressure controls, and air circulation control.

Refrigerant feed controls are designed to maintain, under varying loads, as high a liquid level in the coil as can be carried without excessive liquid spillover into the suction line. This is done by using an expansion valve that throttles liquid from supply pressure (typically 150 psig) to evaporating pressure (usually 20 psig or higher). Throttling flashes some of the liquid to gas, which chills the remaining liquid to saturation temperature at the lower pressure. If it does not bypass the coil, flashed gas tends to reduce flooding of the interior coil surface, thus lowering coil efficiency.

The valve used may be a hand-controlled expansion valve supervised by operator judgment alone, a thermal expansion valve governed by the degree of suction gas superheat, or a float valve (or solenoid valve operated by a float switch) governed by the level of feed liquid in a surge drum placed in the coil suction line. This surge drum suction trap allows ammonia flashed to gas during throttling to flow directly to the suction line, bypassing the coil. The trap may be small and placed just high enough so that its level governs that in the coils by gravity transfer. Or, as in ammonia recirculation, it may be placed below coil level so that the liquid is pumped mechanically through the coils in much greater quantity than is required for evaporation. In the latter case, the trap is large enough to carry its normal operating level plus all the liquid flowing through the coils, thus effectively preventing liquid spillover to the compressors. Nevertheless, it is necessary in all cases to provide further protection at the compressors' liquid return.

Present practice strongly favors liquid ammonia recirculation, mainly because of the greater coil heat transfer rates with the resultant greater refrigerating capacity over other systems (see Chapter 3). Some have coils mounted above the rail beams with 4 to 6 ft of ceiling head space. Air is forced through the coils, sometimes using two-speed fans.

Manual and thermal expansion valves do not provide good coil flooding under varying loads and do not bypass flashed feed gas around the coils. As a result, evaporators so controlled are usually rated 15 to 25% less in capacity than those controlled by float valve or ammonia recirculation.

Evaporator pressure controls regulate coil temperature, and thereby the rate of refrigeration, by varying evaporating pressure in the coil by using a throttling valve in the evaporator suction line downstream from the surge drum suction trap. All such valves impose a definite loss on the refrigeration system; the amount varies directly with pressure drop through the valve. This increases the work of compression for a given refrigeration effect.

The valve used to control evaporating pressure may be a manual suction valve set solely by operator judgment, or a back-pressure valve actuated by coil pressure or temperature or by a temperature-sensing element somewhere in the room. Manual suction valves require excessive attention when loads fluctuate. The coil-controlled back-pressure valve seeks to hold a constant coil temperature but does not control room temperature unless the load is constant. Only the room-controlled compensated back-pressure valve responds to room temperature.

Air circulation control is frequently used when an evaporator must handle separate load conditions differing greatly in magnitude, such as the load in chilling rooms that are also used as holding rooms or for the negligible load on weekends. The use of two-speed fan motors (operated at reduced speed during the periods of light load) or turning the fans off and on can control air circulation to a degree.

Chilled Brine Spray Systems. These are generally being abandoned in favor of other systems, because of their large required building space, inherent low capacity, brine carryover tendencies, and difficulty of control.

Sprayed Coil Systems. These consist of unit coolers equipped with coils, brine spray banks, eliminators to prevent brine carryover, and fans for air/vapor circulation. The units are usually mounted (without ductwork) either on the floor or overhead on converted brine spray decks. Refrigeration is supplied by the primary refrigerant in the coils. Chilled or nonchilled recirculated brine is continually sprayed over the coils, eliminating ice formation and the need for periodic defrosting.

The predominant brine used is sodium chloride, with caustic soda or another additive for controlling pH. Because sodium chloride brine is corrosive, bare-pipe coils (without fins) are generally used. The brine is also highly corrosive to the rail system and other cooler equipment.

Propylene glycol with added inhibitor complexes is another coil spray solution used in place of sodium chloride. As with sodium chloride brine, propylene glycol is constantly diluted by moisture condensed out of the spaces being refrigerated and must be concentrated by evaporating water from it. Reconcentration requires special equipment designed to minimize glycol losses. Sludge that accumulates in the concentrator may become an operating problem; to avoid it, additives must be selected and pH closely controlled. Finned coils are usually used with propylene glycol.

Because it is noncorrosive compared to sodium chloride, propylene glycol greatly reduces the cost of unit cooler construction as well as maintenance of space equipment.

Other Systems. Considerable attention is directed to system designs that reduce the amount of evaporative cooling at the time of entrance into the cooler and eliminate ceiling rail and beam condensation and drip. Good results have been achieved by using low-temperature blast chill tunnels before entrance into the chill room. The volume of ceiling condensate is reduced because the rate of evaporative cooling is reduced in proportion to the degree of surface cooling. Room condensation has been reduced by adding heat above carcasses (out of the main air stream), fans, minimized hot-water usage during cleanup, better dry cleanup, timing of cleanup, and using wood rail supports.

Grade and yield sorting, with its simultaneous filling of several rooms, has shortened the chilling time available if refrigeration is kept off during the filling cycle. Its effect has to be offset by more chill rooms and more installed refrigeration capacity. If full refrigeration is kept at the start of filling, peak load is reduced to the rooms being filled. Hot-carcass cutting has been started with only a short chilling time. Cryogenic chilling has also been tested for hot-carcass chilling.

Beef Cooler Layout and Capacity

Carcass halves or sides are supported by hooks suspended from one-wheel trolleys running on overhead rails. The trolleys are generally pushed from the dressing floor to the chilling room by powered conveyor chains equipped with fingers that engage the trolleys, which are then distributed manually over the chilling and holding room rail system. Chilling and holding room rails are commonly placed on 3 to 4 ft centers in the holding rooms, with pullout or sorting rails between them. Rails must be placed a minimum of 2 ft from the nearest obstruction, such as a wall or building column, and the tops of the rails must be at least 11 ft above the floor. Supporting beams should be a minimum of 6 ft below the ceiling for optimum air distribution. Regulations for some of these dimensions and applicable to new construction in plants engaged in interstate commerce are issued by the Meat Inspection Division of the FSIS.

To ensure effective air circulation, carcass sides should be placed on rails in both chilling and holding coolers so that they do not touch each other. Required spacing varies with the size of the carcass and averages 2.5 ft per two sides of beef. In practice, however, sides are often more crowded.

A chilling room should be of such size that the last carcass loaded into it does not materially retard the chilling of the first carcass. Although size is not as critical as in the case of the hog carcass chill room (because of the slower chill), to better control shrinkage and condensation, it is desirable to limit chill cooler size to hold not more than 4 h of the daily kill. Holding coolers may be as large as desired because they can maintain more uniform temperature and humidity.

Overall plant chilling and holding room capacities vary widely, but chilling coolers generally require a capacity equal to the daily kill; holding coolers require 1 to 2 times the daily kill.

Beef Carcasses. Dressed beef carcasses, each split into two sides, range in weight from approximately 300 to 1000 lb, averaging about 700 lb per head. Specific heats of carcass components range from 0.50 Btu/lb·°F for fat to 0.8 Btu/lb·°F or more for lean muscle, averaging about 0.75 Btu/lb·°F for the carcass as a whole.

An animal's body temperature at slaughter is about 102°F. After slaughter, physiological changes generate heat and tend to increase carcass temperature, while heat loss from the surface tends to lower it.

The largest part of the carcass is the round, and at any given stage of the chilling cycle its center has the highest temperature of all carcass parts. This **deep round temperature** (about 105°F when the carcass enters the chilling cooler) is therefore universally used as a measure of chilling progress. If it is to be significant, the temperature must be taken accurately. Incorrect techniques give results as much as 10°F lower than actual deep round temperature. An accurate technique that yields consistent results is shown in Figure 2: a fast-reacting, easily read stem dial thermometer, calibrated before and after tests, is inserted upward to the full depth through the hole in the aitch bone.

At the time of slaughter, the water content of beef muscle is approximately 75% of the total weight. Thereafter, gradual surface drying occurs, resulting in weight loss or shrinkage. Shrinkage and its measurement are greatly affected by the final dressing operations: weighing and washing. Weighing must be done before washing if weights are to reflect actual product shrinkage.

A beef carcass retains large amounts of wash water on its surface, which it carries into the cooler. Loss of this water, occurring in the form of vapor, does not constitute actual product loss. However, it must be considered when estimating system capacity because the vapor must be condensed on the coils, thus constituting an important part of the refrigeration load.

The amount of wash water retained by the carcass depends on its condition and on washing techniques. A carcass typically retains 8 lb, part of which is lost by drip and part by evaporation. Water pressures used in washing vary from 50 to 300 psig, and temperatures from 60 to 115°F.

To minimize spoilage, a carcass should be reduced to a uniform temperature of about 35°F as rapidly as possible. In practice, deep round temperatures of 60°F (measured as in Figure 2, with surface temperatures of 35 to 45°F, are common at the end of the first day's chill period.

To prevent surface slime formation, most carcasses are cut, vacuum packaged, and boxed within 24 to 72 h. Otherwise, a carcass surface requires a certain dryness during storage. Exposed beef muscle chilled to an actual temperature of 36°F will not slime readily if dried at the surface to a water content of 90% of dry weight (47.4% of total weight). Such a surface is in vapor-pressure equilibrium with a surrounding atmosphere at the same temperature (36°F) and 96% rh. In practice, a room at 32 to 34°F and approximately 90% rh will maintain a well-chilled carcass in good condition without slime (Thatcher and Clark 1968).

Chilling-Drying. Curves of carcass temperature in a chilling-holding cycle are shown in Figure 3. Note that some heat loss occurs before a carcass enters the chilling cooler. Evaporative cooling of surface water dominates the initial stages; as chilling progresses, the rate of losses by evaporative surface cooling diminishes and sensible transfer of heat from the carcass surface increases. Note that the time/temperature rates of change are subject to variations between summer and winter ambient conditions, which influence system capacity.

Transfer rate is increased both by more rapid circulation of air and lower air temperature, but these are limited by the necessity of avoiding surface freezing.

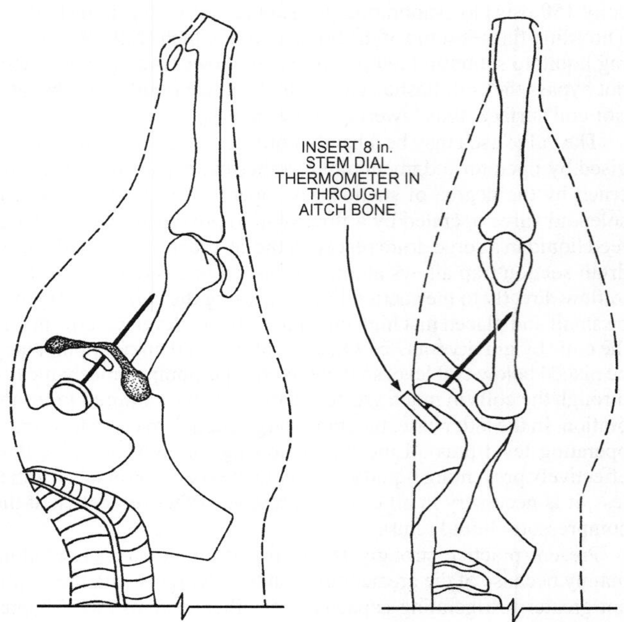

Fig. 2 Deep Round Temperature Measurement in Beef Carcass

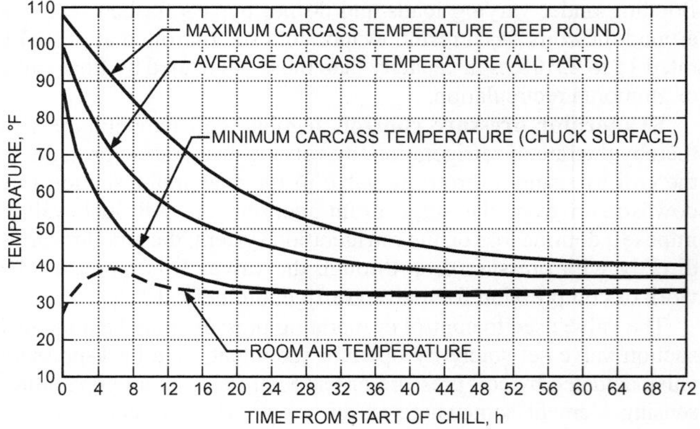

Fig. 3 Beef Carcass Chill Curves

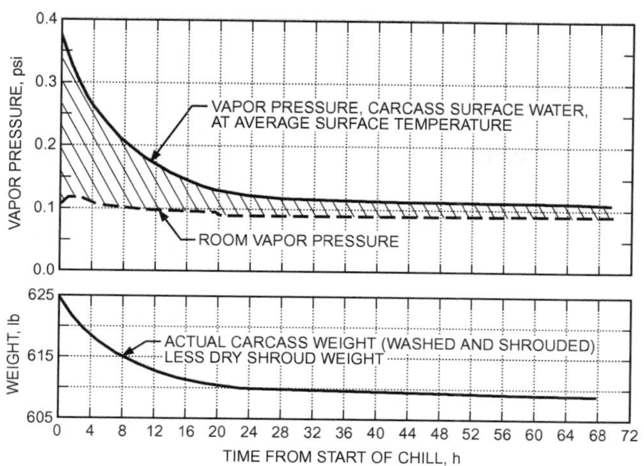

Fig. 4 Beef Carcass Shrinkage Rate Curves

Estimated differences in vapor pressure between surface water (at average surface temperature) and atmospheric vapor during a typical chilling-holding cycle, and the corresponding shrinkage curve for an average carcass, are shown in Figure 4. Note the tremendous vapor-pressure differences during the early chill cycle, when the carcass is warm. Evaporative loss could be reduced by beginning the chill with room temperature high, then lowering it slowly to minimize the pressure difference between carcass surface water and room vapor at all times. However, this slows the chill and prolongs the period of rapid evaporation. Quick chilling is favored, but the cold shortening effect and bacterial growth must be considered in carcass quality and keeping time.

Evaporation from the warm carcass in cool air is nearly independent of room relative humidity because the warm carcass surface generates a much higher vapor pressure than the cooler vapor surrounding the carcass. If the space surrounding a warm carcass is saturated, evaporation forms fog, which can be observed at the beginning of any chill.

Evaporation from the well-chilled carcass with surface temperature at or near room temperature is different. The spread between surface and room vapor pressures approaches zero when room air is near saturation. Evaporation proceeds slowly, without forming fog. Evaporation does not cease when a room is saturated; it ceases only if the carcass is chilled through to room temperature, and no heat transfer is taking place.

The ultimate disposition of water condensed on the coils depends on the coil's surface temperature and its method of operation. In continuous defrost (sprayed-coil) operation, condensed and trapped water dilute the solution sprayed over the coil. In nonfrosting dry-coil operation, condensed water falls to the evaporator pan and drains to the sewer. Water frozen on the coil is lost to the sewer if removed by hot-gas or coil spray defrost. Periodic room air defrost, however, vaporizes part of the ice and returns it to room atmosphere, losing the remainder to drain. This defrost method is not normally used in beef chill or holding coolers because temperatures are not suitable, and thus the defrost period is excessive, resulting in abnormal room temperature variations. Most chill and holding evaporators are automatically defrosted with water or hot gas on selected time cycles. The weight changes that take place in an average beef carcass are given in Table 1.

Chilling the beef carcass is not completed in the chill cooler but continues at a reduced rate in the holding cooler. A well-chilled carcass entering the holding cooler shows minimum holding shrinkage; a poorly chilled one shows high holding shrinkage.

If shrinkage values are to have any significance, they must be carefully derived. Actual product loss must be determined by first weighing the dry carcass before washing and then weighing it out of the cooler. In-motion weights are not sufficiently precise; carcasses must be weighed at rest. Scales must be accurate, and, if possible, the same scale should be used for both weighings. If shrinkage is to have any comparison value, it must be measured on carcasses chilled to the same temperature, because chilling occurs largely by evaporative weight loss.

Table 1 Weight Changes in Beef Carcass

Chilling Cooler	Weight, lb
Initial dry weight	615
Wash water pickup	8
Initial wet weight	623
Spray chill water use	16
Drip (not evaporated)	10
Weight at maximum (8 h postmortem)	633
Weight loss 8 to 24 h postmortem	18
Net weight loss (ideal)	0
Holding Cooler	
Weight loss/day	3
Final weight (48 h postmortem)	602

Design Conditions and Refrigeration Load. Equipment selection should be based on conditions at peak load, when product loss is greatest. Room losses, equipment heat, and carcass heat add up to a total load that varies greatly, not only in magnitude but in proportion of sensible to total heat (sensible heat ratio), throughout the chill. As the chill progresses, vapor load decreases and sensible load becomes more predominant.

Under peak chilling load, excess moisture condenses into fog, enough to warm the air/vapor/fog mixture to the sensible heat ratio of the heat removal process. The heat removal process of the coil therefore underestimates the actual rate of water removal by the amount of vapor condensed to fog (Table 2).

Fog does not generally form under later chilling-room loads and all holding-room loads, although it may form locally and then vaporize. Sensible heat ratios of air/vapor heat gain and air/vapor heat removal are then equal (Table 3).

Beef chilling rooms generally have enough evaporator capacity to hold room temperature under load approximately as shown in Figure 3. This increases room temperature to 35 to 40°F, with a gradual reduction to 32 to 34°F. However, many installations provide greater capacity, particularly dry-coil systems, which thereby avoid excessive coil frosting. In batch-loaded coolers, room temperature may be as low as 25°F under peak load, provided it is raised to 30°F as the chill progresses, without surface freezing of the beef. The shrinkage improvement effected by these lower temperatures, however, tends to be less than expected (in beef chilling) because of the relatively small part played by sensible heat transfer.

Standard holding room practice calls for providing enough evaporator capacity to keep the room temperature at 32 to 34°F at all times. Holding room coils sized at peak load, low air/vapor circulation rate, and a coil temperature 10°F below room temperature tend to maintain the approximately 90% rh that avoids excessive shrinkage and prevents surface sliming.

From the average temperature curve shown in Figure 3 and the shrinkage curve in Figure 4, certain generalizations useful in calculating carcass chilling load may be made. In the chilling cooler, the average carcass temperature is reduced approximately 55°F, from about 102°F to about 47°F, in 20 h. Simultaneously, about 14.3 lb of water is vaporized for each 625 lb carcass entering the chill; only 4.3 lb of this is actual shrinkage. The losses of sensible heat and water occur at about the same rate. In the sample load calculations, this is calculated at an average of 10% for the first 4 h of chill for sensible heat and 13% for the evaporation of moisture, which roughly

Table 2 Load Calculations for Beef Chilling

Cooler size, ft:	62 × 74 × 18.5
Cooler capacity:	476 carcasses
Average carcass weight:	625 lb
Assumed chill rate:	50°F in 20 h
Assumed air circulation:	167,400 cfm
Loading time:	3.3 h maximum
Assumed air to coil:	33°F, 100% rh
Assumed fan motive power:	36 hp
Specific heat of beef:	0.75 Btu/(lb·°F)
Air density:	0.08 lb/ft³

	Loads, Btu/h		
Heat Gain—Room Load	**Sensible**	**Latent**	**Total**
1. Transmission, infiltration, personnel, fan motor, lights, and equipment heat	162,000	4100	166,100
2. Product heat (average, first 4 h):			
a. 476 × 625 × 0.75 × 50 × 0.1	1,115,700		
b. 476 × 14.3 × 0.13 × 1070[a]	–947,000	947,000	1,115,700
3. Total heat gain (room load), kW (Items 1 + 2a + 2b)	330,700	951,100	1,281,800
Heat Removal—Coil Load			
4. Air circulation, lb/h dry air 167,400 × 0.08 × 60 = 803,500	—	—	—
5. Heat removed per lb of dry air, total heat (Item 3)/(Item 4) = 1.6 Btu/lb	—	—	—
6. Air/vapor enthalpy, Btu/lb dry air:			
a. Air to coil, 33°F, 100% rh	7.927	4.242	12.17
b. Btu removed, temperature drop 3.7°F	–0.927	–0.672	–1.60
c. Air from coil, 29.3°F, 100% rh	7.000	3.570	10.57
7. Cool air/vapor heat removal, Btu (Item 4) × (Item 6b)	744,000	539,000	1,285,000
8. Room vapor condensed to fog (Item 7) – (Item 3)	411,300	–411,300	
9. Water (ice) removed by coil 476 × 14.3 × 0.13 × 144[b]		127,000	127,000
10. Total heat removal (coil load), Btu/h (Items 3 + 8 + 9)	742,000	666,800	1,408,800

[a]Heat of vaporization [b]Heat of fusion

Table 3 Load Calculations for Beef Holding

Cooler size, ft:	100 × 136 × 18.5
Cooler capacity, one day's kill:	1120 carcasses
Average carcass weight:	610 lb
Assumed chill rate:	7.5°F in 24 h
Assumed air circulation:	91,200 cfm
Assumed air to coil:	34°F, 96% rh
Assumed fan motive power:	24 hp
Specific heat of beef:	0.75 Btu/(lb·°F)
Air density:	0.08 lb/ft³

	Loads, Btu/h		
Heat Gain—Room Load	**Sensible**	**Latent**	**Total**
1. Transmission, infiltration, personnel, fan motor, lights, and equipment heat	245,000	10,000	255,000
2. Product heat (average, first 4 h):			
a. 1120 × 610 × 0.75 × 7.5 × 0.05	192,000		
b. 1120 × 1.8 × 0.06 × 1070[a]	–130,000	130,000	192,000
3. Total heat gain (room load), Btu/h (Items 1 + 2a + 2b)	307,000	140,000	447,000
Heat Removal—Coil Load			
4. Air circulation, lb/h dry air 91,200 × 0.08 × 60 = 437,800	—	—	—
5. Heat removed per lb of dry air, Btu (Item 3)/(Item 4) = 1.021	—	—	—
6. Air/vapor enthalpy, Btu/lb dry air:			
a. Air to coil, 34°F, 96% rh	8.167	4.227	12.394
b. Btu removed, temperature drop 2.9°F	–0.700	–0.323	–1.027
c. Air from coil, 31.1°F, 100% rh	7.467	3.904	11.367
7. Coil air/vapor heat removal, Btu/h	307,000	140,000	447,000
8. Room vapor condensed to fog	—	—	—
9. Water (ice) removed by coil 1120 × 1.8 × 0.06 × 144[b]	—	17,400	17,400
10. Total heat removal (coil load), Btu/h (Items 3 + 8 + 9)	307,000	157,400	464,400

[a]Heat of vaporization [b]Heat of fusion

agrees with the curves of Figures 3 and 4. This is the maximum rate of chill and is used for sizing refrigeration equipment and piping.

In the holding cooler, the average carcass temperature is reduced from 47 to 39.5°F in 24 h. Simultaneously, about 1.8 lb of water is vaporized per carcass. With spray chilling, this shrinkage approaches zero. Here also, the losses of sensible heat and water occur at about the same rate. The sample load calculations are figured at a 5% average for the first 4 h for sensible heat and 6% for latent heat.

Under peak chilling and holding room conditions, water trapped and condensed out by the coils imposes a further latent load on the evaporators: heat extracted to freeze condensed water into ice, or to chill the returning warmed and strengthened spray solution. In the absence of a more complex evaluation, this load may be considered equal to the latent heat of fusion (144 Btu/lb) of the water removed.

Based on the data just mentioned, cooler loads may be calculated as shown in Tables 2 and 3. Transmission, infiltration, personnel, and equipment loads are estimated by standard methods such as those discussed in Chapter 13.

The complete calculation is made to illustrate the heat removal process associated with chilling-drying the carcass. In particular, it illustrates that the sensible heat ratio of heat transfer in the coil cannot be used to measure the amount of water removed from the space when fog is involved.

Evaporator Selection. Evaporator selection is a procedure of approximation only, because of the inaccuracies of load determination on the one hand and of predicting sustained field performance of coils on the other. Furthermore, there is rarely complete freedom of specification; for example, the air/vapor circulation rate for a given coil may be limited to avoid spray solution carryover or excessive fan power.

Sprayed- and dry-coil systems perform equally well with respect to shrinkage, if compressor capacity is adequate and the evaporators are correct for the system selected. Evaporator requirements vary widely by system type. Comparative evaporator data on a typical, successful flooded coil installation in the chilling cooler are presented in Table 4.

The coil overall heat transfer coefficient and airflows shown describe sustained field performance under actual chilling conditions and loads; they should not be confused with clean coil test ratings. Heat transfer coefficient varies greatly with the character of the coil and its operation and is influenced by variables such as (1) the ratio of extended-to-prime surface, which may range from 7-to-1 to 21-to-1 in standard dry coils; (2) coil depth, which typically ranges from 8 to 12 rows in sprayed coils and from 4 to 10

Table 4 Sample Evaporator Installations for Beef Chilling[a]

Cooler size, ft:	62 × 74 × 18.5
Cooler capacity:	476 carcasses
Deep-round chill:	to 50°F in 20 h
Design load:	1,412,000 Btu/h
Coil operation:	liquid recirculation
Loading time:	3.3 h
Average carcass weight:	625 lb
Assumed air to coil:	33°F, 100% rh
Sensible heat ratio:	53%

	Dry Coil
Coil Description:	
Type of coil	Finned
Fin spacing, fins/in.	4
Coil depth, number of pipe rows	8
Coil face area	20.4 ft^2
Coil surface area, total	2238 ft^2
Fan Description, Airflow:	
Type of fan	Centrifugal
Flow through coil	9300 cfm
Flow, coil face area	455 cfm/ft^3
Fan motive power	2 hp
Unit Rating[b] (Total Heat):	
TD for capacity rating, °F[c]	10
Chilling capacity, Btu/h	81,000
Temperature drop, air through coil, °F	3.7
Equipment for 520 Carcasses:	
Number units required	18
Total motive power, fans and pumps, hp	36
Coil surface per carcass, ft^2	88
Airflow per carcass, cfm	350

[a]Data describe actual installations, but other successful installations may be different.
[b]Ratings shown are estimated from performance of actual systems. Dry-coil ratings are at average frost conditions, with airflow reduced by frost obstruction. This example describes actual installations; it is not to be interpreted as an accepted standard. Other installations, using both more and less equipment, are also successful.
[c]TD is temperature difference between refrigerant and air.

rows in dry coils; (3) fin spacing, which may be 3 or 4 per inch in typical dry coils; (4) condition of the surface, either continuously defrosted or generally coated with frost; and (5) airflow, which may vary from 250 to 750 cfm per square foot of coil face area.

Greater temperature differences (TDs) than those shown are sometimes used, but a higher TD is valid only at high room temperatures. The lower TD (10°F) shown for dry coils is desirable to limit frosting. Many dry-coil evaporators have higher ratios of extended-to-prime surface and higher airflows per unit face area than shown.

Because of difficulties in obtaining accurate shrinkage figures on carcasses chilled to a specified degree, opinions vary widely on the coil capacity required for good chilling. Although data describe actual successful installations, other successful installations may differ.

Boxed Beef

The majority of beef slaughterhouse output is in the form of sections of the carcass, vacuum packed in plastic bags and shipped in corrugated boxes. Standard cuts can be sold at cost savings to the market. The shipping density is much greater, with easier material handling, and bones and fat are removed where their value as a by-product is greatest. Customers purchase only the sections they need, and the trim loss at final processing to primal cuts is minimized.

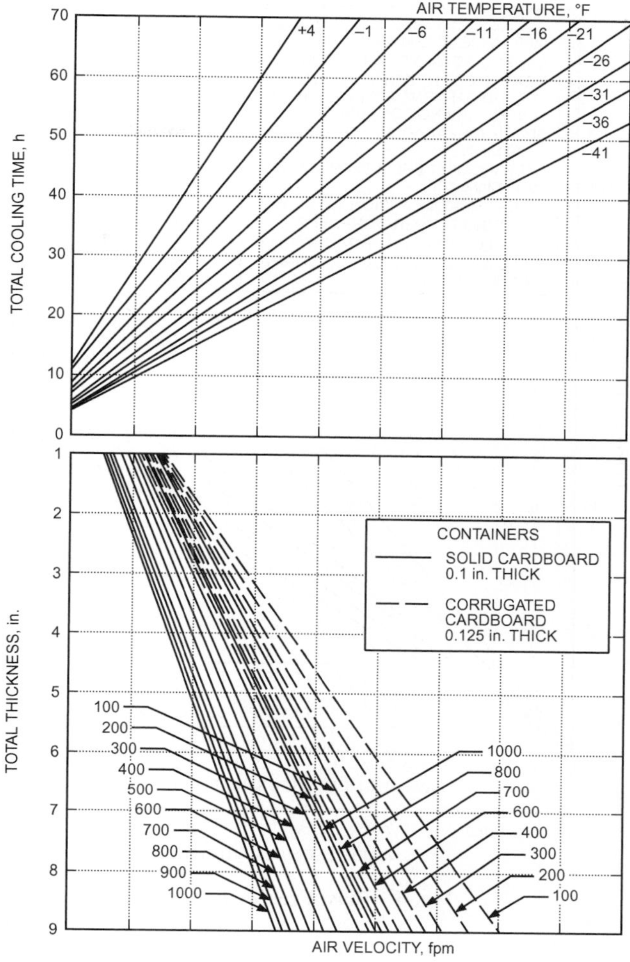

Fig. 5 Freezing Times of Boneless Meat

Vacuum packaging with added carbon dioxide, nitrogen, or a combination of gases has the following advantages:

- Creates anaerobic conditions, preventing growth of mold (which is aerobic and requires the presence of oxygen for growth)
- Provides more sanitary conditions for carcass breaking
- Retains moisture, retards shrinkage
- Excludes bacteria entry and extends shelf life
- Retards bloom until opened

After normal chilling, a carcass is broken into primal cuts, vacuum packed, and boxed for shipment. Temperatures are usually held at 28°F to prevent development of pathogenic organisms. Aging of the beef continues after vacuum packaging and during shipment, because exclusion of oxygen or addition of gases does not slow enzymatic action in the muscle.

Freezing Times of Boneless Meat. Cooling boneless meat from 50 to 10°F requires the removal of about 133 Btu/lb of lean meat (74% water), most of which is latent heat liberated when liquid water in the meat changes to ice. Most of the time needed to freeze meat is spent cooling from 30 to 25°F.

For boneless meat in cartons, freezing rate depends on the temperature and velocity of surrounding air and on the thickness and thermal properties of the carton and the meat itself.

Figure 5 shows the effects of the first three factors on cooling times for lean meat in two carton types. For example, the chart shows a total cooling time of about 24 h for solid cardboard cartons

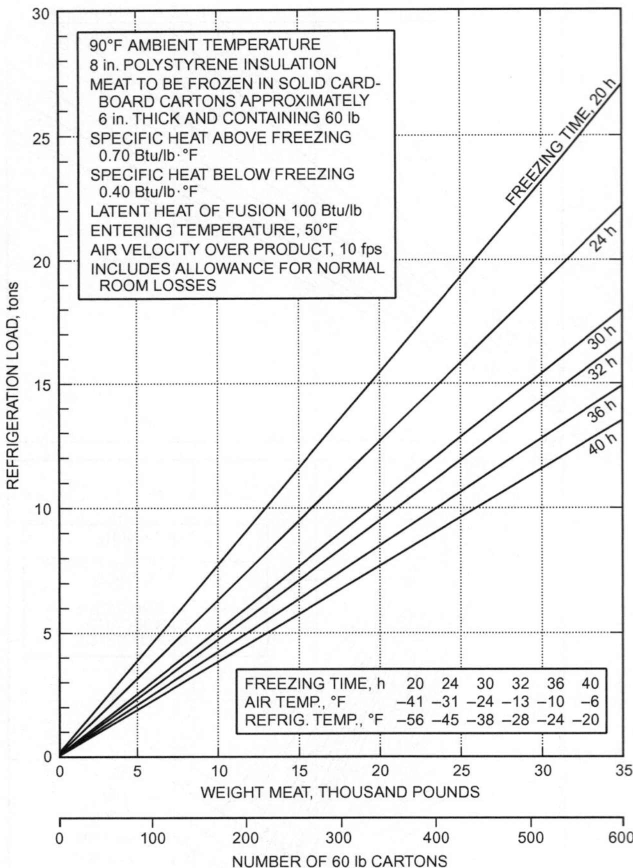

Fig. 6 Blast Freezer Loads

5 in. thick at an air temperature of –26°F and a velocity of 400 fpm. The corresponding air temperatures and cooling times may be found for any specified thickness of carton and air velocity. Conversely, the chart can be used to find combinations of air velocity and temperature needed to freeze cartons of a particular thickness in a specified time (Figure 6).

Accuracy of the estimated freezing times is about 3% for air velocities greater than 400 fpm. Calculations are based on Plank's equation, as modified by Earle. A latent heat of 107 Btu/lb and an average freezing point of 28°F are assumed for lean meat.

Increasing the fat content of meat reduces the water content and hence the latent heat load. Thermal conductivity of the meat is reduced at the same time, but the overall effect is for freezing times to drop as the percentage of fat rises. Actual cooling times for mixtures of lean and fatty tissue should therefore be somewhat less than the times obtained from the chart. For meat with 15% fat, the reduction is about 17%.

Hog Chilling and Tempering

The internal temperature of hog carcasses entering the chill coolers from the killing floor varies from 100 to 106°F. The specific heat shown in Chapter 9 is 0.74 Btu/lb·°F, but in practice 0.7 to 0.75 Btu/lb·°F is used because changed feeding techniques have created leaner hogs. Dressed weight varies from 90 to 450 lb approximately; the average is near 180 lb. Present practice requires dressed hogs to be chilled and tempered to an internal ham temperature of 37 to 39°F overnight. This limits the chilling and tempering time to 12 to 18 h.

Cooler and refrigeration equipment must be designed to chill the hogs thoroughly with no frozen parts at the time the carcasses are moved to the cutting floor. Carcass crowding, reducing exposure to circulated chilled air, and excessively high peak temperatures are all detrimental to proper chilling.

The following hog cooler design details provide

- Sufficiently quick chilling to retard bacterial development and prevent deterioration
- Cooler shrinkage from 0.1 to 0.2%
- Firm carcasses that are dry and bright without frozen surface or internal frost, suitable for efficient cutting

Hog Cooler Design. The capacity of hog coolers is set by the dressing rate of hogs and the planned hours of operation. However, on a one-shift basis, it is economically sound to provide cooler hanging capacity for 10 h dressing to properly handle chilling large sows that require more than 24 h exposure in the chill room; handle increased dressing volumes when market conditions warrant overtime operation; and have some flexibility in unloading and loading the cooler during normal operations. On a two-shift basis, extra cooler capacity for overtime operation is not necessary.

Rail height should be 9 ft to provide both good air circulation and adequate clearance between the floor and the largest dressed hog. (In the United States, this is a requirement of the FSIS and most state regulations.) Rails should be spaced at a minimum of 30 in. on centers to provide sufficient clearance for hanging hogs and to prevent contact between carcasses.

The spacing of hogs on the rail varies according to the size of the hog. There should be at least 1.5 to 2 in. between the flank of one carcass and the back of the carcass immediately in front of it. Rail spacing of 13 in. on centers is normal for 180 lb dressed hogs.

Many meat-packing refrigeration engineers maintain that several hog chill coolers with a capacity of 2 h loading for 300 to 600 kill/h or 4 h loading for lower killing capacities is more economical than one large chill cooler.

The hog cooler should be designed on the following basis:

- Total amount of hanging rail should be equal to
  - 10 h × Rate of kill × 12 ft/13 hogs (one-shift operation)
  - 16 h × Rate of kill × 12 ft/13 hogs (two-shift operation)
  - For combination carcass loading, hogs and cattle, calves, or sheep, the figures should be modified accordingly.
- Rail height should be 9 ft. It may be 11 ft for combination beef and hog coolers.
- Rail spacing should be a minimum of 30 in.
- Inside building height varies, depending on the type of refrigeration equipment installed. Clear height of 6 ft above the rail support is adequate for space to install units, piping, and controls; it provides sufficient plenum over the rails to ensure even air distribution over carcasses.

Preliminary, intensive batch, inline chilling is practiced in some plants, resulting in smaller shrinkages with the variations in chilling systems.

Selecting Refrigeration Equipment. Both floor units and units installed above the rail supports are used. Floor units, with a top discharge outlet equipped with a short section of duct to discharge air in a space over the rail supports, are used by a few pork processors. A few brine spray units equipped with ammonia coils and water or hot-gas defrost units are being used.

Two types of units are available for installation above the rail supports. One uses a blower fan to force air below and through a horizontally placed coil with the air discharging horizontally from the front or top of the unit. The other consists of a vertical coil with axial fans or blowers to force air through the unit. Both units are designed for various types of liquid feed control.

Horizontal units are equipped for both hot-gas and water defrosting. All have finned surfaces designed for hog chill cooler operation. Evaporator controls should be provided as for beef carcass coolers.

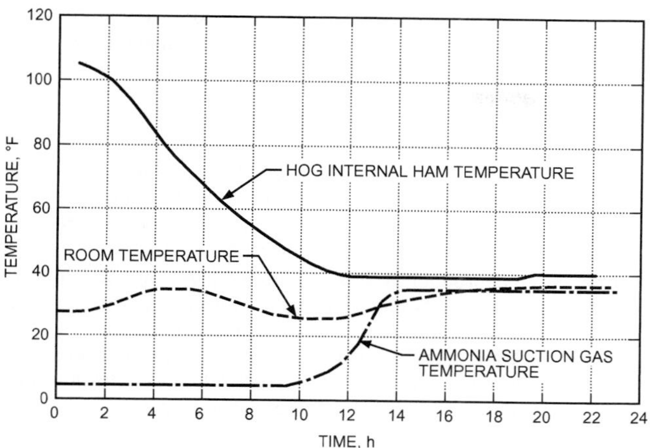

Fig. 7 Composite Hog Chilling Time/Temperature Curves

Careful selection of units and use of automatic controls, including liquid recirculation, provides an air circulation, temperature, and humidity balance that chills hogs with minimum shrinkage in the quickest time.

Temperature control should be set to provide an opening room temperature of 26 to 28°F. As the cooler is loaded, the suction temperature decreases to provide the additional refrigeration effect required to handle increased refrigeration load and maintain room temperature at 30 to 32°F. Ample compressor and unit capacity is required to achieve these results.

Dry coils selected to maintain a 10 to 12°F temperature difference (refrigerant to air) at peak operation provides adequate coil surface and a TD of 1 to 5°F prior to opening and about 10 h after closing the cooler. This low temperature difference results in economical high-humidity conditions during the entire chilling cycle. Cutting practices typically use a high initial chilling TD and a lower TD at the end of the chilling cycle.

Sample Calculation. The hog chilling time/temperature curves (Figure 7) are composite curves developed from several operation tests. The relation of room temperature and ammonia suction gas temperature curves show that the refrigeration load decreased about 9 h after closing the cooler. After about 9 h, the room temperature is increased and the hog is tempered to an internal ham temperature of 37 to 39°F.

Table 5 was prepared using empirical calculations to coordinate product and unit refrigeration loads. A shortcut method for determining hog chill cooler refrigeration loads is presented in Table 6. The latent heat of the product has been neglected, since the latent heat of evaporation is equal to the reduction of sensible heat load of the product. Total sensible heat was used in all calculations.

Example 1: Select cooling units for 600 hogs/h at 180 lb average dressed weight using 2 h loading time cooler.

Four coolers minimum requirement (five desirable)
Each cooler

Capacity 1200 hogs	=	11.3 tons (Table 6)
Product peak load = 6 × 14	=	<u>84.0 tons</u> (Table 5)
Total	=	95.3 tons

Select 18 units of 5.3 tons at 10.3°F temperature difference per cooler
Approximately 198,000 cfm
Air changes per minute = 198,000/68,000 = 2.9

Refrigeration of the hog cutting room, where the carcass is cut up into its primal parts, is an important factor in maintaining product quality. A maximum dry-bulb temperature of 50°F should be maintained. This level is low enough to prevent excessive rise in product temperature during its relatively short stay in the cutting room and also complies with USDA-FSIS requirements.

Chilled carcasses entering the room may have surface temperatures as low as 30°F. Unless the dew point of the air in the cutting room is kept below 30°F, moisture will condense on the surface of the product, providing an excellent medium for bacterial growth.

Floor, walls, and all machinery on the cutting floor must be thoroughly cleaned at the end of each day's operation. Cleaning releases a large amount of vapor in the room that, unless quickly removed, condenses on walls, ceiling, floor, and machinery surfaces. When outside dew-point temperature is less than the room temperature, vapor can be removed by installing fans to continuously exhaust the room during cleanup.

These fans should operate only during cleanup and as long as required to remove vapor produced by cleaning. An exhaust capability of five air changes per hour should be satisfactory. When the room temperature is lower than the outside dew-point temperature, this method cannot be used. Water vapor must then be condensed out on the evaporator surfaces.

Many people work in the hog cutting room. Attention should be given to the sensible heat from personnel, normally 1000 Btu/h per person. Heat from electric motors must also be included in the refrigeration load.

Special consideration should also be given to latent heat load from knife boxes, wash water, workers, and infiltration. If the sensible heat load is not sufficient, this high latent heat load must be offset by reheat at the refrigeration units in order to maintain the desired low relative humidity.

The quantity of air circulated is influenced by the amount of sensible heat to be picked up and the relative humidity to be maintained, but is usually between 7 and 12 complete air changes each hour. The

Table 5 Product Refrigeration Load, Tons

Hours	Cooler Loading Time, h			
	1	2	4	8
1	7.20*	7.20	7.20	7.20
2	6.80	14.00*	14.00	14.00
3	6.49	13.29	20.49	20.49
4	6.19	12.68	26.88*	26.68
5	5.94	12.13	25.42	32.62
6	5.71	11.65	24.33	38.33
7	5.56	11.27	23.40	43.89
8	5.44	11.00	22.65	49.33*
9	5.38	10.82	22.09	47.51
10	—	5.38	16.38	39.71
11	—	—	10.82	34.22
12	—	—	5.38	28.03
13	—	—	—	22.09
14	—	—	—	16.38
15	—	—	—	10.82
16	—	—	—	5.38

*Values are for peak load: 100 hogs/h at 180 lb dressed weight average, chilled from 102 to 38°F in 16 h. Based on operating test, not laboratory standards.

Table 6 Average Chill Cooler Loads Exclusive of Product

Cooler Capacity	Room Dimensions, ft	Floor Area, ft^2	Room Volume, ft^3	Refrigeration tons*
1200 Hogs	40 × 100 × 17	4000	68,000	11.3
2400 Hogs	80 × 100 × 17	8000	136,000	22.6
3600 Hogs	120 × 100 × 17	12,000	204,000	33.9
4800 Hogs	100 × 160 × 17	16,000	272,000	45.2
6000 Hogs	100 × 200 × 17	20,000	340,000	56.5

*Based on 6000 ft^3 room volume/ton (or use detailed calculations for building heat gains, infiltration, people, lights, and average unit cooler motors from Chapter 30 of the 2005 *ASHRAE Handbook—Fundamentals*).

air distribution pattern requires careful attention to prevent drafts on the workers.

Forced-air units are satisfactory for refrigerating cutting floors. Ceiling height must be sufficient to accommodate the units. A wide selection of forced-air units may be applied to these rooms. They can be floor or ceiling mounted, with either dry-coil or wetted-surface units arranged for flooded, recirculated, or direct-expansion refrigerant systems.

Suction pressure regulators should be provided for both flooded and direct-expansion units. Automatic dry- and wet-bulb controls are essential for best operation.

Pork Trimmings

Pork trimmings come from the chilled hog carcass, principally from the primal cuts: belly, plate, back fat, shoulder, and ham.

Trimmings per hog average 4 to 8 lb. Only trimmings used in sausage or canning operations are discussed here.

In the cutting or trimming room, trimmings are usually between 38 and 45°F; an engineer must design for the higher temperature. The product requires only moderate chilling to be in proper condition for grinding, if it is to be used locally in sausage or canning operations. If it is to be stored or shipped elsewhere, hard chilling is required. Satisfactory final temperature for local processing is 28°F. This is the average temperature after tempering and should not be confused with surface temperature immediately after chilling. Trimmings may be much cooler on the surface than on the interior immediately after chilling, especially if they have been quick-chilled.

Good operating practice requires rapid chilling of pork trimmings as soon as possible after removal from the primal cuts. This retards enzymatic action and microbial growth, which are responsible for poor flavor, rancidity, loss of color, and excessive shrinkage.

The choice of chilling method depends largely on local conditions and consists of a variation of air temperature, air velocity, and method of achieving contact between air and meat. Continuous belt equipment using low-temperature air or fluids to obtain lower shrinkage is available.

Truck Chilling. Economic conditions may require existing chilling or freezer rooms to be used. Some may require an overnight chill, others less than an hour. The following methods make use of existing facilities:

- Trimmings are often chilled with CO_2 snow before grinding to prevent excessive temperature rise during grinding and blending. Additionally, finished products, such as sausage or hamburger in chubs, may be crusted or frozen in a glycol/brine liquid contact chiller.
- Trimmings are put on truck pans to a depth of 2 to 4 in. and held in a suitable cooler kept at 30°F. This method requires a short chill time and results in a near-uniform temperature (30°F) of trimmings.
- Trimmings are spread 4 or 5 in. deep on truck pans in a 0°F freezer and held for 5 to 6 h, or until the meat is well stiffened with frost. Using temperatures below 0°F (with or without fans) expedites chilling if time is limited. After trimmings are hard-chilled, they are removed from the metal pans and tightly packed into suitable containers. They are held in a 26 to 28°F room until shipped or used. Average shrinkage using this system is 0.5% up to the time they are put in the containers.
- Trimmings from the cutting or trimming room are put in a meat truck and held in a cooler at 28 to 30°F. This method usually requires an overnight chill and is not likely to reach a temperature of 32°F in the center of the load.
- If trimmings are not to be used within one week, they should be frozen immediately and held at –10°F or lower.

Fresh Pork Holding

Fresh pork cuts are usually packed on the cutting floor. If they are not shipped the same day that they are cut and packed, they should be held in a cooler with a temperature of 20 to 28°F.

Forced-air cooling units are frequently used for holding room service because they provide better air circulation and more uniform temperatures throughout the room, minimize ceiling condensation caused by air entering doorways from adjacent warmer areas (because of traffic), and eliminate the necessity of coil scraping or drip troughs if hot-gas defrost is used.

Cooling units may be the dry type with hot-gas defrost, or wetted surface with brine spray. Units should have air diffusers to prevent direct air blast on the products. Unless the room shape is very odd, discharge ductwork should not be necessary.

Because the product is boxed and wrapped and the holding period is short, humidity control is not too important. Various methods of automatic control may be used. CO_2 has been used in boxes of pork cuts. Care must be taken to maintain the ratio of pounds of CO_2 to pounds of meat for the retention period. The enclosures must be relieved and ventilated in the interest of life safety.

Calf and Lamb Chilling

Dry coils (either the between-the-rail type, the suspended type above the rail, or floor units) are typically used for calf and lamb chilling. The same type of refrigerating units used for pork may be used for lamb, with some modifications. For example, in chilling lambs and calves, it is desirable to reduce air changes over the carcass by using two-speed motors, using the higher speed for the initial chill and reducing the rate of air circulation when carcass temperatures are reduced, approximately 4 to 6 h after the cooler is loaded.

Lambs usually weigh 40 to 80 lb, with an approximate average dressed carcass weight of 50 lb. Sheep weigh up to an average of approximately 125 lb and readily take refrigeration. Adequate coil surface should be installed to maintain a room temperature below 30°F and 90 to 95% rh in the loading period. The evaporating capacity should be based on an average 10°F temperature differential between refrigerant and room air temperature, with an opening room temperature of 32°F.

Compensating back-pressure-regulating valves, which vary the evaporator pressure as room temperature changes, should be used. As room and carcass temperatures drop, the temperature differential is reduced, thus holding a high relative humidity (40 to 45%). At the end of a 4 to 6 h chill period, air over the carcass may be reduced to help keep product bloom and color.

Carcasses should not touch each other. They enter the cooler at 98 to 102°F, with the carcass temperature taken at the center of the heavy section of the rear leg. The specific heat of a carcass is 0.7 Btu/lb·°F. Air circulation for the first 4 to 6 h should be approximately 50 to 60 changes per hour, reduced to 10 to 12 changes per hour. The carcass should reach 34 to 36°F internally in about 12 to 14 h and should be held at that point with 85 to 90% rh room air until shipped or otherwise processed. This gives the least possible shrinkage and prevents excess surface moisture.

In calf-chilling coolers, approximately the same procedure is acceptable, with carcasses hung on 12 to 15 in. centers. The dressed weight varies at different locations, with an approximate 85 to 90 lb average in dairy country and a 200 to 350 lb (sometimes heavier) average in beef-producing localities. The same time and temperature relationship and air velocities for chilling lambs are used for chilling calves, except when calves are chilled with the hide on. Also, the time may be extended for air circulation. Air circulation need not be curtailed in hide-on chilling because rapid cooling gives better color to these carcasses after they are skinned.

Refrigerating capacity for lamb- and calf-chill coolers is calculated the same as for other coolers, but additional capacity should be added to allow reduced air circulation and maintain close temperature differential between room air and refrigerant.

Chilling and Freezing Variety Meats

The temperature of variety meats must be lowered rapidly to 28 to 30°F to reduce spoilage. Large boxes are particularly difficult to cool. For example, a 5 in. deep box containing 70 lb of hot variety meat can still have a core temperature as high as 60°F 24 h after it enters a –20°F freezer. Variety meats in boxes or packages more than 3 in. thick may be chilled very effectively during freezing by adding dry ice to the center of the box.

For design calculations, variety meat has an initial temperature of about 100°F. Specific heats of variety meats vary with the percentage of fat and moisture in each. For design purposes, a specific heat of 0.75 Btu/lb·°F should be used.

Quick Chilling. A better and more widely used method consists of quick chilling at lower temperatures and higher air velocities, using the same type of truck equipment as in the overnight chilling method. This method is also used for chilling trimmings. Careful design of the quick-chilling cabinet or room is needed to provide for the refrigeration load imposed by the hot product. One industry survey shows that approximately 50% of large establishments use quick-chill in their variety meat operations.

The quick-chilling cabinet or room should be designed to operate at an air temperature of approximately –20 to –40°F, with air velocities over the product of 500 to 1000 fpm. During initial loading, the air temperature may rise to 0°F. In quick-chilling unit design, refrigeration coils are used with axial-flow fans for air circulation.

The recommended defrosting method is with water and/or hot gas, except where units with continuous defrost are used. The product is chilled to the point where the outside is frosted or frozen and a temperature of 28 to 30°F is obtained when the product later reaches an even temperature throughout in the packing or tempering room. The time required to chill the product by this method depends on the depth of product in the pans, size of individual pieces, air temperature, and velocity. Normally, 0.5 to 4 h is a satisfactory chill period to attain the required 28 to 30°F temperature. In addition to the obvious savings in time and space, an important advantage of this method is the low total shrinkage, averaging only 0.5 to 1%. These values were obtained in the same survey as those in Table 7.

Packaging Before Chilling. Another method of handling variety meats involves packaging the product before chilling, as near as possible to the killing floor. Packed containers are placed on platforms and frozen in a freezer. Separators should allow air circulation between packages.

This method is used in preparing products for frozen shipment or freezer storage. The internal temperature of the product should reach 25°F for prompt transfer to a storage freezer. For immediate shipment, the internal temperature of the product must be reduced to 0°F; this may be done by longer retention in the quick freezer. Here package material and size, particularly package thickness, largely determine the rate of freezing. For example, a 5 in. thick box takes at least 16 h to freeze, depending on the type of product, package material, size, and loading method.

The dry-bulb air temperature in these freezers is kept at –40 to –20°F, with air velocities over the product at 500 to 1000 fpm.

Table 7 Storage Life of Meat Products

	Months			
	Temperature, °F			
Product	**10**	**0**	**–10**	**–20**
Beef	4 to 12	6 to 18	12 to 24	12+
Lamb	3 to 8	6 to 16	12 to 18	12+
Veal	3 to 4	4 to 14	8	12
Pork	2 to 6	4 to 12	8 to 15	10
Chopped beef	3 to 4	4 to 6	8	10
Pork sausage	1 to 2	2 to 6	3	4
Smoked ham and bacon	1 to 3	2 to 4	3	4
Uncured ham and bacon	2	4	6	6
Beef liver	2 to 3	2 to 4		
Cooked foods	2 to 3	2 to 4		

The time required to reach the desired internal temperature depends on refrigeration capacity, size of largest package, insulating properties of package material, and so forth. A generous safety factor should be used in sizing evaporator coils. These freezers are best incorporated in refrigerated rooms. Defrosting is by water or hot gas, except where units with continuous defrost are used. Shrinkage varies in the range of only 0.5 to 1%.

Initial freezing equipment cost and design load can be reduced if carbon dioxide is included in packaging as part of the operational plan. Another efficient cooling method uses plate freezers to form blocks of product that can be loaded on pallets with minimal packaging.

Packaging and Storage

Packages for variety meats do not have standard sizes or dimensions. Present requirements are a package that will stand shipping, with sizes to suit individual establishments. The package's importance becomes more apparent with the hot-pack freezing method. Standardized sizes and package materials promote faster chilling and more economical handling.

Storage of variety meats depends on its end use. For short storage (under one week) and local use, 28 to 30°F is considered a good internal product temperature. If stored for shipping, the internal temperature of the product should be kept at 0°F or below. Recommended length of storage is controversial; type of package, freezer temperature and relative humidity, amount of moisture removed in original chill, and variations of the products themselves all affect storage life.

Packers' storage time recommendations vary from 2 to 6 months and longer, because variety meats pick up rancidity on the surface and soft muscle tissue dehydrates while freezing. More rapid freezing and vaporproof packaging are important in increasing storage life.

Packaged Fresh Cuts

In packaging fresh cuts of meat intended for direct placement into retail display cases, sanitation of the processing room is particularly important. The same environmental concerns also apply to some processors of precooked, ready-to-eat products.

Uncooked fresh cuts are packaged in sealed packages with an atmosphere of sterile nitrogen/oxygen/carbon dioxide mixture to control pathogens and organic activity. Shelf life is extended from days to weeks.

It is important to prepare and package this product in an environment free of harmful bacteria and other pathogens, and to transport these products at a continuously controlled temperature to the market display case. Techniques to accomplish this include

- Processing room temperature of 36 to 38°F
- A semi-cleanroom environment with positive air pressure created by highly filtered, refrigerated outdoor air
- Keeping only packaging film and pouches in the room (e.g., no boxes or cardboard)
- A program of follow-through with temperature-monitoring devices shipped with the product, and returned

Cleanroom techniques include an isolated workcrew entering through a sanitation anteroom, changing outer garments, wearing hair nets, using footbath sanitation, and handwashing with disinfection. Facilities for frequent microbiological testing should be provided.

Refrigeration Load Computations

The average evaporator refrigerating load for a typical chilling process above freezing may be computed as follows:

$$q_r = m_m c_m (t_1 - t_2) + m_t c_t (t_1 - t_2) + q_w + q_i + q_m \quad (1)$$

where

q_r = refrigeration load, Btu/h
m_m = weight of meat, lb/h
m_t = weight of trucks, lb/h
c_m = specific heat of meat, Btu/lb·°F
c_t = specific heat of truck, containers, or platforms (0.12 for steel), Btu/lb·°F
t_1 = average initial temperature, °F
t_2 = average final temperature, °F
q_w = heat gain through room surfaces, Btu/h
q_i = heat gain from infiltration, Btu/h
q_m = heat gain from equipment and lighting, Btu/h

The following example illustrates the method of computing the refrigeration load for a quick-chill operation.

Example 2. Find the refrigeration load for chilling six trucks of offal from a maximum temperature of 100 to 34°F in 2 h. Each truck weighs 400 lb empty and holds 720 lb of offal. The specific heat is 0.12 Btu/lb·°F for the truck and 0.75 Btu/lb·°F for the offal. The room temperature is to be held at 0°F, with an outdoor temperature of 40°F and 70% rh. The walls, ceiling, and floor gain 72 Btu/ft^2 and have an area of 947 ft^2. The room volume is 1881 ft^3 and 12 air changes in 24 h are assumed.

Solution: Values for substitution in Equation (1) are as follows:

m_m = 6 × 720/2 = 2160 lb/h
c_m = 0.75 Btu/lb·°F
$t_1 - t_2$ = 100 – 34 = 66°F
m_t = 6 × 400/2 = 1200 lb/h
c_t = 0.12 Btu/lb·°F
q_w = 72 × 947/24 = 2841 Btu/h
q_i = (1.12 × 1881 × 12)/24 = 1053 Btu/h where
1.12 = heat removed per ft^3 of air entering, Btu
q_m = (10 × 2545) + (200 × 3.4) = 26,130 Btu/h
10 = assumed horsepower
2545 = Btu per horsepower hour
200 = lights, W
3.4 = Btu/W·h

Substituting in Equation (1),

$$q_r = (2160 \times 0.75 \times 66) + (1200 \times 0.12 \times 66) + 2841 + 1053 + 26{,}140 = 146{,}448 \text{ Btu/h} = 12.2 \text{ tons}$$

Good practice is to add 10 to 25% to the computed refrigeration load.

PROCESSED MEATS

Prompt chilling, handling, and storage under controlled temperatures help in production of mild and rapidly cured and smoked meats. The product is usually transferred directly from the smokehouse to a refrigerated room, but sometimes a drop of 10 to 30°F can occur if the transfer time is appreciable.

Because the day's production is not usually removed from the smokehouses at one time, the refrigeration load is spread over nearly 24 h. Table 8 outlines temperatures, relative humidities, and time required in refrigerated rooms used in handling smoked meats.

Prechilling smoked meat reduces drips of moisture and fat, thus increasing yield. Meats can be chilled at higher temperatures, with air velocities of up to 500 fpm (Table 8). At lower temperatures, air velocities of 1000 fpm and higher are used. Chilling in the hanging or wrapping and packaging rooms results in slow chilling and high temperatures when packing. Slow chilling is not desirable for a product that is to be stored or shipped a considerable distance.

Meats handled through smoke and into refrigerated rooms are hung or racked on cages that are moved on an overhead track or mounted on wheels. Sometimes the product is transferred from suspended cages to wheel-mounted cages between smoking and subsequent handling.

Table 8 Room Temperatures and Relative Humidities for Smoking Meats

	Room Conditions			
	°F Dry-Bulb	% Relative Humidity	Final Product Temp., °F	Time, h
Prechill method				
Hams, picnics, etc.				
High temperature	38 to 40	80	60	8 to 10
Low temperature	26 to 28	80	60	2 to 3
Derind bacon				
Normal	26 to 28	80	28	8 to 10
Blast	0 to 10	80	26	2 to 3
Hanging or tempering				
Ham, picnics, etc.	45 to 50	70	50 to 55	
Derind bacon	26 to 28	70	26 to 28	
Wrapping or packaging				
Hams, picnics, etc.	45 to 50	70		
Storage	28 to 45	70		

Smoked hams and picnic meats must be chilled as rapidly as possible through the incubation temperature range of 105 to 50°F. A product requiring cooking before eating is brought to a minimum internal temperature of 140°F to destroy possible live trichinae, whereas one not requiring cooking before eating is brought to a minimum internal temperature of 155°F.

Maximum storage room temperature should be 40°F db when delivery from the plant to retail outlets is made within a short time. A room dry-bulb temperature of 28 to 32°F is desirable when delivery is to points distant from the plant and transfer is made through controlled low-humidity rooms, docks, cars, or trucks, keeping the dew point below that of the product.

Bacon usually reaches a maximum temperature of 125°F in the smokehouse. Because most sliced smoked bacon is packaged, it may be transferred directly to the chill room if it has been skinned before smoking. If bacon is to be skinned after smoking, it is usually allowed to hang in the smokehouse vestibule for 2 to 4 h, until it drops to 90°F before skinning.

Bacon is usually molded and sliced at temperatures just below 28°F. Chill rooms are usually designed to reduce the bacon's internal temperature to 26°F in 24 h or less, requiring a room dry-bulb temperature of 18 to 20°F. A tempering room (which also serves as storage for stock reserve), held at the exact temperature at which bacon is sliced, is often used.

Bacon can be molded either after tempering, in which case it is moved directly to the slicing machines, or after the initial hardening, and then be transferred to the tempering room. In the latter case, care should be taken that none of the slabs is below 24°F so that the product will not crack during molding. Bacon cured by the pickle injection process generally shows fewer pickle pockets if it is molded after hardening, placed no more than eight slabs high on pallets, and held in the tempering room.

In any of these rooms, air distribution must be uniform. To minimize shrinkage, the air supply from floor-mounted unit coolers should be delivered through slotted ducts or by closed ducts supplying properly spaced diffusers directed so that no high-velocity airstreams impinge on the product itself. The exception is in blast chill rooms, which need high air velocities but subject the product to the condition for only a short time.

Refrigeration may be supplied by floor or ceiling-mounted dry- or wet-coil units. If the latter are selected, water, hot-gas, or electric defrost must also be used.

Many processors use three methods of chilling smoked meats: rapid blast, direct-contact spraying of brine, and cryogenic. Direct-contact spraying is especially emphasized, because it minimizes shrinkage, increases shelf life, and provides more uniform chilling. This method is usually carried out in special enclosures designed to combat the detrimental effects of salt brine. Color and salt taste may need close monitoring in contact spraying.

The product should enter the slicing room chilled to a uniform internal temperature not to exceed 50°F for beef rounds and 40 to 45°F for other fresh carcass parts, depending on the individual packer's temperature standard. Internal temperatures below 26°F tend to cause shattering of products such as bacon during slicing and slow processing. For that type of product, temperatures above 32°F cause improper shingling from the slicing machine.

The slicing and packaging room temperature and air movement are usually the result of a compromise between the physical comfort demands of operating personnel and the product's requirements. The design room dry-bulb temperature should be below 50°F, according to USDA-FSIS regulations.

An objectionable amount of condensation on the product may occur. To guard against this, the coil temperature should be maintained below the temperature of bacon entering the room, thus keeping the room air's dew point below the product temperature. Product should be exposed to room air for the shortest possible time.

Bacon Slicing and Packaging Room

Exhaust ventilation should remove smoke and fumes from the sealing and packaging equipment and comply with OSHA occupancy regulations. Again, consider using heat exchangers to reduce the resulting increased refrigeration load.

Refrigeration for this room may be supplied by forced-air units (floor or ceiling mounted, dry or wet coil) or finned-tube ceiling coils. Dry coils should have defrost facilities if coil temperatures are to be kept at more than several degrees below freezing. Air discharge and return should be evenly distributed, using ductwork if necessary. To avoid drafts on personnel, air velocities in the occupied zone should be in the range of 25 to 35 fpm. The temperature differential between primary air and room air should not exceed 10°F, to ensure personnel comfort. For optimum comfort and dew-point control, reheat coils are necessary.

Where ceiling heights are adequate, multiple ceiling units can be used to minimize the amount of ductwork. Automatic temperature and humidity controls are desirable in cooling units.

To provide draft-free conditions, drip troughs with suitable drainage should be added to finned-tube ceiling coils. However, it is difficult, if not impossible, to maintain a room air dew point low enough to approach the product temperature. Some installations operating with relative humidities of 60 to 70% do not have product condensation problems.

One control method consists of individual coil banks connected to common liquid and suction headers. Each bank is equipped with a thermal expansion valve. The suction header has an automatically operated back-pressure valve. The thermostatically controlled dual back-pressure regulator and liquid-header solenoid are both controlled by a single thermostat. This arrangement provides a simple automatic defrosting cycle.

Another system uses fin coils with glycol sprays. Humidity is controlled by varying the concentration strength of the glycol and the refrigerant temperature.

Sausage Dry Rooms

Refrigeration or air conditioning is integral to year-round sausage dry rooms. The purpose of these systems is to produce and control air conditions for proper moisture removal from the sausage.

Various dry sausages are manufactured, for the most part uncooked. This sausage is generally of two distinct types: smoked and unsmoked. Keeping qualities depend on curing ingredients, spices, and removal of moisture from the product by drying.

FSIS regulates the minimum temperature and amount of time that the product must be held after stuffing and before release, depending on the method of production. The dry room temperature should not be lower than 45°F, and the length of time product is held in the dry room depends on the sausage's diameter after stuffing and preparation method used.

After stuffing, sausages are held at a temperature of 60 to 75°F and 75 to 95% rh in the sausage greenroom to develop the cure. Sausages are suspended from sticks at the time of stuffing and may be held on the trucks or railed cages or be transferred to racks in the greenroom. Sausages in 3.5 to 4 in. diameter casings are generally spaced about 6 in. on centers on the sausage sticks. The length of time sausages are held in the greenroom depends on the preparation method, type and dimensions of the sausage, the operator, and the sausage maker's judgment about proper flavor, pH, and other characteristics.

Varieties that are not smoked are then transferred to the sausage dry room; those that are to be smoked are transferred from the greenroom to the smokehouse and then to the dry room.

In the dry room, approximately 30% of the moisture is removed from the sausage, to a point at which the sausage will keep for a long time, virtually without refrigeration. The drying period required depends on the amount of moisture to be removed to suit trade demand, type of sausage, and type of casing. Moisture transmission characteristics of synthetic casings vary widely and greatly influence the rate of drying. Sausage diameter is probably the most important factor influencing the drying rate.

Small-diameter sausages, such as pepperoni, have more surface in proportion to the weight of material than do large-diameter sausages. Furthermore, moisture from the interior has to travel a much shorter distance to reach the surface, where it can evaporate. Thus, drying time for small-diameter sausages is much shorter than for large-diameter sausages.

Typical conditions in the dry room are approximately 45 to 55°F and 60 to 75% rh. Some sausage makers favor the lower range of temperatures for unsmoked varieties of dry sausage and the higher range for smoked varieties.

In processing dry sausage, moisture should only be removed from the product at the rate at which the moisture comes to the casing surface. Any attempt to hasten drying rate results in overdrying the sausage surface, a condition known as case hardening. This condition is identified by a dark ring inside the casing, close to the surface of the sausage, which precludes any further attempt to remove moisture from the interior of the sausage. On the other hand, if sausage is dried too slowly, excessive mold occurs on the casing surface, usually leading to an unsatisfactory appearance. (An exception is the Hungarian salami, which requires a high humidity so that prolific mold growth can occur and flourish.)

As with any other cool or refrigerated space, sausage dry rooms should be properly insulated to prevent temperatures in adjoining spaces from influencing the temperature in the room. Ample insulation is especially important for dry rooms located adjacent to rooms of much lower temperature or rooms on the top floor, where the ceiling may be exposed to relatively high temperatures in summer and low temperatures in winter.

Insulation should be adequate to prevent inner surfaces of the walls, floor, and ceiling of the dry room from differing more than a degree or two from the average temperature in the room. Otherwise, condensation is possible because of the high relative humidity in these rooms, which leads to mold growth on the surfaces themselves and, in some cases, on the sausages as well.

Sticks of sausage are generally supported on permanent racks built into the dry room. In the past, these were frequently made of wood; however, sanitary requirements have virtually outlawed the use of wood for this purpose in new construction. The uprights and rails for the racks are now made of either galvanized pipe, hot-dip galvanized steel, or stainless steel. Rails for supporting sausage sticks should be spaced vertically at a distance that leaves ample room for air circulation below the bottom row and between the top row and the ceiling. Spacing between rails (usually not less than 1 or 2 ft) depends on the length of sausage stick used by the individual manufacturer.

Horizontal spacing between sausages should be such that they do not touch at any point, to prevent mold formation or improper development of color. Generally, with large 4 in. diameter sausages, spacing of 6 in. on centers is adequate.

Dry-Room Equipment. In general, two types of refrigeration equipment are used to attain the required conditions in a dry room. The most common is a refrigeration-reheat system, in which room air is circulated either through a brine spray or over a refrigerated coil and sufficiently cooled to reduce the dew point to the temperature required in the room. The other type involves spraying a hygroscopic liquid over a refrigerating coil in the dehumidifier, thus condensing moisture from the air without the severe overcooling usually required by refrigeration-reheat systems. The chief advantage of this arrangement is that refrigeration and heating loads are greatly reduced.

Use of any type of liquid, brine or hygroscopic, requires periodic tests and adjusting the pH to minimize equipment corrosion. Although most systems depend on a type of liquid spray to prevent frost build-up on the refrigerating coils, some successful rooms use dry coils with hot-gas or water defrost.

Air for conditioning the dry room is normally drawn through the refrigerating and dehumidifying systems by a suitable blower fan (or fans) and discharged into the distribution ductwork.

Rooms used exclusively for small-diameter products with a rapid drying rate may actually have air leaving the room to return to the conditioning unit at a lower dry-bulb temperature and greater density than at which it is introduced. A dry-room designer needs to know what the room will be used for to determine the natural circulation of the room air. Supply and return ducts can then be arranged to take advantage of and accelerate this natural circulation to provide thorough mixing of incoming dry air with air in the room.

Regardless of the location of supply and return ducts, care should be taken to prevent strong drafts or high-velocity airstreams from impinging on the product, which leads to local overdrying and unsatisfactory products.

Study of air circulation within the product racks shows that, as air passes over the sausages and moisture evaporates from them, this air becomes cooler and heavier, and thus tends to drop toward the bottom of the room, creating a vertical downward air movement in the sausage racks. This natural tendency must be considered in designing duct installation if uniform conditions are to be achieved.

An example of the calculation involved in designing a sausage dry room follows. These calculations apply to a room used for assorted sausages, with an average drying time of approximately 30 days. They would not be directly applicable to a room used primarily for very large salami (which has a much longer drying period) or small-diameter sausage.

In the latter case, using the air-circulating rate shown in this example allows the air to absorb so much moisture in passing through the room that it is difficult to obtain uniform conditions throughout the space. Furthermore, the amount of refrigeration required to lower the air temperature enough to produce the required low inlet-air dew point becomes excessive. An air circulation rate of 12 air changes per hour should therefore be considered average for use in average rooms. The actual circulating rate should be adjusted to obtain the best compromise of refrigeration load and air uniformity for the particular type of product handled.

Example 3. Air conditioning for sausage drying room

Room Dimensions:

40 ft, 2 in. by 33 ft, 6 in. by 11 ft, 6 in.
Floor space: 1350 ft^2
Volume: 15,600 ft^3
Outdoor wall area: 980 ft^2
Partition wall area: 770 ft^2

Hanging Capacity:

Number of racks: 12
Length of racks: 27 ft
Number of rails high: 5
Spacing of sticks: 2 per foot of rail
Number of pieces of sausage per stick: 7
Average weight per sausage: 4 lb
Total weight: 12 × 27 × 5 × 2 × 7 × 4 = 90,720 lb of product
Assume 90,000 lb weight green hanging capacity
Loading per day: 1500 lb

Assumed Outdoor Conditions (Summer):

95°F db; 74.5°F wb; 39% rh; h = 37.8 Btu/lb; 66°F dp; 96 gr/lb

Dry-Room Conditions Desired:

55°F db; 50°F wb; 70% rh; h = 20.2 Btu/lb; 46°F dp; 46 gr/lb

Sensible Heat Calculations:

Walls (2 in. insulation):		
980(95 – 55) × 0.10	=	3920 Btu/h
Partition (4 in. insulation):		
770(95 – 55) × 0.067	=	2060 Btu/h
Floor and ceiling:		
2700(55 – 55) × 0.10	=	none
Infiltration: 0.5 × 15,600(95 – 55) × 0.243 × 0.075	=	5700 Btu/h
Lights: 600 W × 3.415	=	2050 Btu/h
Motors: 5 × 1 hp × 2546	=	12,730 Btu/h
Product: 1500 × 0.8(95 – 55)/24	=	2000 Btu/h
Total sensible heat	=	28,460 Btu/h

Latent Heat Calculations:

Product		
90,000 × 0.30 × 1000/(60 × 24)	=	18,750 Btu/h
(18,750 × 7000)/(60 × 1000)	=	2188 gr/min
Infiltration		
0.5 × 15,600 × 0.075(37.8 – 20.2)	=	10,300 Btu/h
(10,300 – 5700)7000/(60 × 1000)	=	537 gr/min
Total grains of moisture = 2188 + 537	=	2725 gr/min

Assume 12 air changes per hour, with an empty room volume of 15,600 ft^3 = (15,600)(12/60)(0.075) = 234 lb of air per minute. Then each pound of air must absorb 2725/234 = 11.6 gr of moisture. Because air at the desired room condition carries 46 gr/lb, the air must enter the room with only 46 – 11.6 = 34.4 gr/lb, corresponding to 41.8°F db and 40°F wb (h = 15.3 Btu/lb).

Temperature rise from sensible heat gain (air specific heat = 0.243 Btu/lb·°F):

28,460/(234 × 0.243 × 60) = 8.3°F db

Temperature drop caused by evaporative cooling from latent heat of product only:

18,750/(234 × 0.243 × 60) = 5.5°F

Net temperature rise in the room = 8.3 – 5.5 = 2.8°F, or air entering the room must be 55 – 2.8 = 52.2°F db at 38.5°F dp (34.5 gr/lb).

Refrigerating load = 234(20.2 – 15.3) × 60 = 68,800 Btu/h

Reheat load = 234(17.8 – 15.3) × 60 = 35,100 Btu/h

Room load = 234(20.2 – 17.8) × 60 = 33,700 Btu/h

Lard Chilling

In federally inspected plants, the USDA-FSIS designates the types of pork fats that, when rendered, are classified as lard. Other pork fats, when rendered, are designated as rendered pork fats. The following data for refrigeration requirements may be used for either product type. Rendering requires considerable heat, and the subsequent temperature of the lard at which refrigeration is to be applied may be as high as 120°F.

The fundamental requirement of the FSIS is good sanitation through all phases of handling. Avoid using copper or copper-bearing alloys that come in contact with lard, because minute traces of copper lower product stability.

Lard has the following properties:

Specific gravity at	0°F	= 0.99
	70°F	= 0.93
	160°F	= 0.88
Heat of solidification		= 48 Btu/lb
Melting begins at –32 to –40°F.		
Melting ends at 110 to 115°F.		
Point of half fusion is around 40°F.		
Specific heat in solid state	–110°F	= 0.28 Btu/lb·°F
	–40°F	= 0.34 Btu/lb·°F
Specific heat in liquid state	110°F	= 0.50 Btu/lb·°F
	212°F	= 0.52 Btu/lb·°F

In lard production, refrigeration is applied so that the final product has enough texture and a firm consistency. The finest possible crystal structure is desired.

Calculations for chilling 1000 lb of lard per hour are

Initial temperature:	120°F
Final temperature:	80°F
Heat of solidification:	48 Btu/lb
Specific heat:	0.50 Btu/lb·°F

$$S_f = 100\frac{t_e - t_f}{t_e - t_b} = 100\frac{115 - 80}{115 - (-40)} = 22.6\%$$

where

S_f = percent solidification at final temperature
t_e = temperature at which melting ends
t_f = final temperature
t_b = temperature at which melting begins

Latent heat of solidification:

$$48(22.6/100) = 10.8 \text{ Btu/lb}$$

Sensible heat removed:

$$0.50(120 - 80) = 20 \text{ Btu/lb}$$

Total heat removed:

$$1000(10.8 + 20) = 30{,}800 \text{ Btu/h} = 2.57 \text{ tons refrigeration}$$

Assuming a 15% loss because of radiation, for example, in the process, the required refrigeration to chill 1000 lb of lard per hour is 1.15 × 2.57 = 2.96 tons.

Filtered lard at 120°F can be chilled and plasticized in compact internal swept-surface chilling units, which use either ammonia or halogenated hydrocarbons. A refrigerating capacity of about 36,000 Btu/h per 1000 lb of lard handled per hour for the product only should be provided. Additional refrigeration for the requirements of heat equivalent to the work done by the internal swept-surface chilling equipment is needed.

When operating this type of equipment, it is essential to keep the refrigerant free of oil and other impurities so that the heat transfer surface does not form a film of oil to act as insulation and reduce the unit's capacity. Some installations have oil traps connected to the liquid refrigerant leg on the floor below to provide an oil accumulation drainage space.

Safety requirements for this type of chilling equipment are described in ASHRAE *Standard* 15. Note that these units are pressure vessels and, as such, require properly installed and maintained safety valves.

The recommended storage temperature for packaged refined lard is 31 to 33°F. The storage temperature required for prime steam lard in metal containers is 40°F or below for up to a 6 month storage period. Lard stored for a year or more should be kept at 0°F.

Blast and Storage Freezers

The standard method of sharp-freezing a product destined for storage freezers is to freeze the product directly from the cutting floor in a blast freezer until its internal temperature reaches the holding room temperature. The product is then transferred to holding or storage freezers.

Product to be sharp-frozen may be bagged, wrapped, or boxed in cartons. Individual loads are usually placed on pallets, dead skids, or in wire basket containers. In general, the larger the ratio of surface exposed to blast air to the volume of either the individual piece or the product's container, the greater the rate of freezing. Product loads should be placed in a blast freezer to ensure that each load is well exposed to the blast air and to minimize possible short-circuiting of the airflow. Each layer on a load should be separated by 2 in. spacers to give the individual pieces as much exposure to the blast air as possible.

The most popular types of blast equipment are self-contained air-handling or cooling units that consist of a fan, evaporator, and other elements in one package. They are usually used in multiples and placed in the blast freezer to provide optimum blast air coverage. Unit fans should be capable of high air velocity and volumetric flow; two air changes per minute is the accepted minimum.

The coils of the evaporator may have either a wet or a dry surface. See Chapter 42 for information on defrosting.

Blast chill design temperatures vary throughout the industry. Most designs are within –20 to –40°F. For low temperatures, booster compressors that discharge through a desuperheater into the general plant suction system are used.

Blast freezers require sufficient insulation and good vapor barriers. If possible, a blast freezer should be located so that temperature differentials between it and adjacent areas are minimized, to decrease insulation costs and refrigeration losses.

Blast freezer entrance doors should be power operated. Suitable vestibules should also be provided as air locks to decrease infiltration of outside air.

Besides normal losses, heat calculations for a blast freezer should include loads imposed by material handling equipment (e.g., electric trucks, skids, spacers) and packaging materials for the product. Some portion of any heat added under the floor to prevent frost heaving must also be added to the room load.

Storage freezers are usually maintained at 0 to –15°F. If the plant operates with several high and low suction pressures, the evaporators can be tied to a suitable plant suction system. The evaporators can also be tied to a booster compressor system; if the booster system is operated intermittently, provisions must be made to switch to a suitable plant suction system when the booster system is down. Storage freezer coils can be defrosted by hot gas, electricity, or water. Emphasis should be placed on not defrosting too quickly with hot gas (because of pipe expansion) and on providing well-insulated, sloped, heat-traced drains and drain pans to prevent freeze-ups.

Direct-Contact Meat Chilling

Continuous processes for smoked and cooked wieners use direct sodium chloride brine tanks or deluge tunnels to chill the meat as soon as it comes out of the cooker. Every day, the brine is prepared fresh in 2 to 13% solutions, depending on chilling temperature and salt content of the meat.

Cooling is usually done on sanitary stainless steel surface coolers, which are either refrigerated coils or plates in cabinets. Using this type of unit allows coolant temperatures near the freezing point without damaging the cooler; damage may occur when brine is confined in a tubular cooler. Brine quantities should be enough to fully wet the surface cooler and fill the distribution troughs of the deluge.

Another type of continuous process uses a conveyor belt to move wieners through the cooking and smoking process, and then drops them into a brine tank. Pumped brine moves the product to the end of the tank, where it is removed by hand and inserted into peeling and packaging lines.

FROZEN MEAT PRODUCTS

Handling and selling consumer portions of frozen meats have many potential advantages compared with merchandising fresh meat. Preparation and packaging can be done at the packinghouse, allowing economies of mass production, by-product savings, lower transportation costs, and flexibility in meeting market demands. At the retail level, frozen meat products reduce space and investment requirements and labor costs.

Freezing Quality of Meat

After an animal is slaughtered, physiological and biochemical reactions continue in the muscle until the complex system supplying energy for work has run down and the muscle goes into rigor. These changes continue for up to 32 h postmortem in major beef muscles. Hot boning with electrical stimulation renders meat tender on a continuous basis without conventional chilling. Freezing meat or cutting carcasses for freezing before these changes complete causes cold- and thaw-shortening, which render meat tough. The best time to freeze meat is either after rigor has passed or later, when natural tenderization is more or less complete. Natural tenderization is completed during 7 days of aging in most major beef muscles. Where flavor is concerned, freezing as soon as tenderization is complete is desirable.

For frozen pork, the age of the meat before freezing is even more critical than it is for beef. Pork loins aged 7 days before freezing deteriorate more rapidly in frozen storage than loins aged 1 to 3 days. In tests, a difference could be detected between 1 and 3 day old loins, favoring those only 1 day old. With frozen pork loin roasts from carcasses chilled for 1 to 7 days, the flavor of lean and fat in the roasts was progressively poorer with longer holding time after slaughter.

Effect of Freezing on Quality

Freezing affects the quality (including color, tenderness, and amount of drip) of meat.

Color. The color of frozen meat depends on the rate of freezing. Tests in which prepackaged, steak-size cuts of beef were frozen by immersion in liquid or exposure to an air blast at between –20 and –40°F revealed that airblast freezing at –20°F produced a color most similar to that of the unfrozen product. An initial meat temperature of 32°F was necessary for best results (Lentz 1971).

Flavor and Tenderness. Flavor does not appear to be affected by freezing per se, but tenderness may be affected, depending on the condition of the meat and the rate and end temperature of freezing. Faster freezing to lower temperatures was found to increase tenderness; however, consensus on this effect has not been reached.

Drip. The rate of freezing generally affects the amount of drip, and meat nutrients, such as vitamins, that are lost from cut surfaces after thawing. Faster freezing tends to reduce the amount of drip, although many other factors, such as the pH of meat, also have an effect on drip.

Changes in Fat. Pork fat changes significantly in 112 days at –5°F, whereas beef fat shows no change in 260 days at this temperature. At –20 and –30°F, no measurable change occurs in either meat in one year.

The relationship of fat rancidity and oxidation flavor has not been clearly established for frozen meat, and the usefulness of antioxidants in reducing flavor changes during frozen storage is doubtful.

Storage and Handling

Pork remains acceptable for a shorter storage period than beef, lamb, and veal because of differences in fatty acid chain length and saturation in the different species. Storage life is also related to storage temperature. Because animals within a species vary greatly in nutritional and physiological backgrounds, their tissues differ in susceptibility to change when stored. Because of differences between meat animals, packaging methods, and acceptability criteria, a wide range of storage periods is reported for each type of meat (see Table 7).

Lentz (1971) found that color and flavor of frozen beef change perceptibly at storage temperatures down to –40°F in 1 to 90 days (depending on temperature) for samples held in the dark. Changes were much more rapid (1 to 7 days) for samples exposed to light. Color changes were less pronounced after thawing than when frozen.

Reports on the effect of different storage temperatures on fat oxidation and palatability of frozen meats indicate that a temperature of 0°F or lower is desirable. Cuts of pork back fat held at 20, 10, 0, and –10°F show increases in peroxide value; free fatty acid is most pronounced at the two higher temperatures. For storage of 48 weeks, 0°F or lower is essential to avoid fat changes. Pork rib roasts of 0°F showed little or no flavor change up to 8 months, whereas at 10°F, fat was in the early stages of rancidity in 4 months. Ground beef and ground pork patties stored at 10, 0, and –10°F indicate that meats must be stored at 0°F or lower to retain good quality for 5 to 8 months. For longer storage, –20°F is desirable.

The desirable flavor in pork loin roasts stored at –6 to –8°F, with maximum fluctuations of 5 to 8°F, decreased slightly, apparently without significant difference between treatments. Fluctuations from 0 to 10°F did not harm quality.

Storage temperature is perhaps more critical with meat in frozen meals because of the differing stability of the various individual dishes included. Frozen meals show marked deterioration of most of the foods after 3 months at 13 to 15°F.

Storage and Handling Practices. Surveys of practices in the industry indicate why some product reaches the consumer in poor condition. One unpublished survey indicated that 10% of frozen foods may be at 6°F or higher in warehouses, 16°F or higher in assembly rooms, 21°F or higher during delivery, and 17°F or higher in display cases. All these temperatures should be maintained at 0°F for complete protection of the product.

Packaging

At the time of freezing, a package or packaging material serves to hold the product and prevent it from losing moisture. Other functions of the wrapper or box become important as soon as the storage period begins. Ideal packaging material in direct contact with meat should have low moisture vapor transmission rate; low gas transmission rate; high wet strength; grease resistance; flexibility over a temperature range including subfreezing; freedom from odor, flavor, and any toxic substance; easy handling and application characteristics adaptable to hand or machine use; and reasonable price. Individually or collectively, these properties are desired for good appearance of the package, protection against handling, preventing dehydration (which is unsightly and damages the product), and keeping oxygen out of the package.

Desiccation through use of unsuitable packaging material is one of the major problems with frozen foods. Another problem is that of distorted or damaged containers caused either by lack of expansion space for the product in freezing or by selection of low-strength box material.

Whenever free space is present in a container of frozen food, ice sublimes and condenses on the film or package. Temperature fluctuation increases the severity of frost deposition.

SHIPPING DOCKS

A refrigerated shipping dock can eliminate the need for assembling orders on the nonrefrigerated dock or other area, or using a more valuable storage space for this purpose. This is especially true for freezer operations. Some businesses do not really need a refrigerated order assembly area. One example is a packing plant that ships out whole carcasses or sides in bulk quantities and does not need a large area in which to assemble orders. Many are constructed without any dock at all, simply having the load-out doors lead directly into the carcass-holding cooler, requiring increased refrigerating capacity around the shipping doors to prevent undue temperature rise in the coolers during shipping.

A refrigerated shipping dock can perform a second function of reducing the refrigeration load, which is most important in the case of freezers but serves almost as valuable a function with coolers. Even with cooler operations, installation of a refrigerated dock greatly reduces the load on the cooler's refrigerating units and ensures a more stable temperature within the cooler. At the same time, it is possible to only provide refrigeration to maintain dock temperatures on the order of 40 to 45°F, so that the refrigerating units can be designed to operate with a wet coil. In this way, frost build-up on the units is avoided and the capacity of the units themselves substantially increased, making it unnecessary to install as many or as large units in this area.

For freezers, units should be designed and selected to maintain a dock temperature slightly above freezing, usually about 35°F. With this dock temperature, orders may be assembled and held before shipment without the risk of defrosting the frozen product, and workers can assemble orders in a much more comfortable space than the freezer. The design temperature should be low enough that the dew point of the dock atmosphere is below the product temperature. Condensation on product surfaces is one step in developing off-condition product.

With a dock temperature of 35°F, the temperature difference between the freezer itself and outdoor summer conditions is split roughly in half. Because airflow through loading doors or other openings is proportional to the square root of the temperature difference, this results in an approximate 30% reduction in airflow through the doors (both those into the dock itself and those from the dock into the freezer). At the same time, by cooling outdoor air to approximately 35°F, in most cases about 50% of the total heat in the outdoor air is removed by the refrigerating units on the dock.

Because using a refrigerated dock reduces airflow through the door into the freezer by approximately 30%, and 50% of the heat in air that does pass through this door is removed, the net effect is to reduce the infiltration load on units in the freezer itself by about 65%. This is not a net gain; because an equal number of these units operate at a much higher temperature, the power required to remove heat on the dock is substantially lower than it would be if heat were allowed to enter the freezer.

The infiltration load from the shipping door, whether it opens directly into a cooler or freezer or into a refrigerated dock, is extremely high. Even with well-maintained foam or inflatable door seals, a great deal of warm air leaks through the doors whenever they are open. This air infiltration may be calculated approximately by

$$V = CHW\theta(H)^{0.5}(t_1 - t_2)^{0.5} \qquad (2)$$

where

V = air volume, ft³/h at higher-temperature condition
C = 1.4 = empirical constant selected to convert airflow into ft³/h and to account for contraction of airstream as it passes through door and for obstruction created by truck parked at door with only nominal sealing
H = door height, ft
W = door width, ft
θ = time door is open, min/h
t_1 = outdoor air temperature or air at higher temperature, °F
t_2 = temperature of air in dock or cooler, °F

Time θ is estimated, based on the time the door is assumed to be obstructed or partially obstructed. If doors have good, well-maintained seals that will tightly seal the average truck to the building, this time is assumed as only the time necessary to spot the truck at the door and complete the air seal.

The unit cooler providing refrigeration for the dock area should be ceiling-suspended with a horizontal air discharge. Each unit should be aimed toward the outer wall and above each of the truck loading doors, if possible, so that cold air strikes the wall and is deflected downward across the door. This downward airflow just inside the door tends to oppose the natural airflow of entering warm air, thus helping reduce the total amount of infiltration.

In general, a between-the-rails unit cooler has proved most successful for this purpose, because it distributes air over a fairly wide area and at low outlet velocity. This airflow pattern does not create severe drafts in the working area and is more acceptable to employees working in the refrigerated space. The preceding comments and equation for determining air infiltration also apply to shipping doors that open directly into storage or shipping coolers.

ENERGY CONSERVATION

Water, a utility previously considered free, frequently has the most rapid rate increases. Coupled with high sewer rates, it is the largest single-cost item in some plants. If fuel charges are added to the hot-water portion of water usage, water is definitely the most costly utility. Costs can be reduced by better dry cleanup, use of heat exchangers, use of filters and/or settling basins to collect solids and greases, use of towers and/or evaporative condensers, not using water for product transport, and an active conservation program.

Air is needed for combustion in steam generators, sewage aeration, air coolers or evaporative condensers, and blowing product through lines. Used properly in conjunction with heat exchangers, air can reduce other utility costs (fuel, sewage, water, and electricity). Nearly all plants need close monitoring of valves either leaking through or left open in product conveying. Low-pressure blowers are frequently used in place of high-pressure air, reducing initial investment and operating costs of driving equipment.

Steam generation is a source of large savings through efficient boiler operation (fuel and water sides). Reduced use of hot water and sterilizer boxes, and proper use of equipment in plants with electric and steam drives, should be promoted. Sizable reductions can be made by scavenging heat from process-side steam and hot water and by systematically checking steam traps. In some plants, excess hot water and low-energy heat can be recovered using heat exchangers and better heat balances.

Electrical energy needs can be reduced by

- Properly sizing, spacing, and selecting light fixtures and an energy program of keeping lights off (lights comprise 25 to 33% of an electric bill)
- Monitoring and controlling the demand portion of electricity use
- Checking and sizing motors to their actual loads for operation within the more efficient ranges of their curves
- Adjusting the power factor to reduce initial costs in transformers, switchgear, and wiring
- Lubricating properly to cut power demands

Although refrigeration is not a direct utility, it involves all or some of the factors just mentioned. Energy use in refrigeration systems can be reduced by

- Operating with lower condenser and higher compressor suction pressures
- Properly removing oil from the system

- Purging noncondensable gases from the system
- Adequately insulating floors, ceilings, walls, and hot and cold lines
- Using energy exchangers on exhaust and air makeup
- Keeping doors closed to cut humidity or prevent an infusion of warmer air
- Installing high-efficiency motors
- Maintaining compressors at peak efficiency
- Keeping condensers free of scale and dirt
- Using proper water treatment in the condensing system
- Operating with a microprocessor-based management system

Utility savings are also possible when use is considered with product line flows and storage space. A strong energy conservation program not only saves total energy but frequently results in greater product yields and product quality improvements, and thus increased profits. Prerigor or hot processing of pork and beef products greatly reduces the energy required for postmortem chilling. Removing waste fat and bone before chilling reduces the amount of chilling space by 30 to 35% per beef carcass.

REFERENCES

Acuff, G.R. 1991. Acid decontamination of beef carcasses for increased shelf life and microbiological safety. *Proceedings of the Meat Industry Resources Conference*, Chicago.

Allen, D.M., M.C. Hunt, A.L. Filho, R.J. Danler, and S.J. Goll. 1987. Effects of spray chilling and carcass spacing on beef carcass cooler shrink and grade factors. *Journal of Animal Science* 64:165.

Dickson, J.S. 1991. Control of *Salmonella typhimurium, Listeria monocytogenes*, and *Escherichia coli* O157:H7 on beef in a model spray chilling system. *Journal of Food Science* 56:191.

Earle, R.L. Physical aspects of the freezing of cartoned meat. *Bulletin* 2, Meat Industry Research Institute of New Zealand.

Greer, G.G. and B.D. Dilts. 1988. Bacteriology and retail case life of spray-chilled pork. *Canadian Institute of Food Science Technology Journal* 21:295.

Hamby, P.L., J.W. Savell, G.R. Acuff, C. Vanderzant, and H.R. Cross. 1987. Spray-chilling and carcass decontamination systems using lactic and acetic acid. *Meat Science* 21:1.

Johnson, R.D., M.C. Hunt, D.M. Allen, C.L. Kastner, R.J. Danler, and C.C. Schrock. 1988. Moisture uptake during washing and spray chilling of Holstein and beef-type carcasses. *Journal of Animal Science* 66:2180.

Jones, S.M. and W.M. Robertson. 1989. The effects of spray-chilling carcasses on the shrinkage and quality of beef. *Meat Science* 24:177-188.

Kastner, C.L. 1981. Livestock and meat: Carcasses, primal and subprimals. In *CRC handbook of transportation and marketing in agriculture*, pp. 239-258. E.E. Finney, Jr., ed. CRC Press, Boca Raton, FL.

Lentz, C.P. 1971. Effect of light and temperature on color and flavor of prepackaged frozen beef. *Canadian Institute of Food Technology Journal* 4:166.

Marriott, N.G. 1994. *Principles of food sanitation*, 3rd ed. Chapman & Hall, New York.

USDA-FSIS. U.S. inspected meats and poultry packing plants—A guide to construction and layout. *Agriculture Handbook* 570. U.S. Department of Agriculture.

BIBLIOGRAPHY

ASHRAE. 2004. Safety standard for refrigeration systems. ANSI/ASHRAE *Standard* 15-2004.

Heitter, E.F. 1975. Chlor-chill. *Proceedings of the Meat Industry Resources Conference AMIF*, pp. 31-32. Arlington, VA.

CHAPTER 18

POULTRY PRODUCTS

POULTRY, and broilers in particular, are the most widely grown farm animal on earth. Two major challenges face the poultry industry: (1) keeping food safe from human pathogens carried by poultry in small numbers that could multiply, sometimes to dangerous levels, during processing, handling, and meal preparation; and (2) developing environmentally sound, economical waste management facilities. Innovative engineering and refrigeration are a part of the solutions for these issues.

PROCESSING

Processing is composed of three major segments:

- **Dressing**, where the birds are placed on moving line, killed, and defeathered.
- **Eviscerating**, where the viscera are removed, the carcass is chilled, and the birds are inspected and graded.
- **Further processing**, where the largest portion of the carcasses are cut up, deboned, and processed into various products. The products are packaged and stored chilled or frozen.

A schematic processing flowsheet is described in Figure 1; equipment layout for the dressing area is given in Figure 2 and for the eviscerating area in Figure 3. The space needed in the production area for the various activities is shown in Figure 4. A modern, highly automated poultry processing plant processes 1 to 3 million birds per week. In the 1970s, a standard U.S. plant was processing 1500 birds per hour (2 shifts, 5 days), or close to 120,000 birds per week. Barbut (2000) describes processing in detail.

CHILLING

Poultry products in the United States may be chilled to 26°F or frozen to lower than 26°F. Means of refrigeration include ice, mechanically cooled water or air, dry ice (carbon dioxide sprays), and liquid nitrogen sprays. Continuous chilling and freezing systems, with various means for conveying the product, are common. According to USDA regulations (1990), poultry carcasses weighing less than 4 lb should be chilled to 40°F or below in less than 4 h, carcasses of 4 to 8 lb in less than 6 h, and carcasses of more than 8 lb in less than 8 h. In air-chilling ready-to-cook poultry, the carcasses' internal temperature should reach 40°F or less within 16 hours (9CFR381.66).

Slow air chilling was considered adequate for semiscalded, uneviscerated poultry in the past. But with the transformation to eviscerated, ready-to-cook, sometimes subscalded, poultry, air chilling was replaced by chilling in tanks of slush ice. Immersion chilling is more rapid than air chilling, prevents dehydration, and effects a net absorption of water of 4 to 12%. Per U.S. regulations (9CFR441.10), water retention in raw carcasses and parts must be shown to be an unavoidable consequence of processing, to the specifications of the Food Safety and Inspection Service (FSIS). Additionally, water-retaining poultry must carry a label stating the maximum percentage of water retained. Objections to this weight gain from external water, a concern that water chillers can be recontamination points, and the high cost of disposing of waste water in an environmentally sound manner have encouraged some operators to consider returning to air chillers.

Continuous-immersion slush ice chillers, which are fed automatically from the end of the evisceration conveyer line, have replaced slush ice tank chilling, a batch process. In general, tanks are only used to hold iced, chilled carcasses before cutting up, or to age before freezing.

The following types of continuous chillers are used:

- **Continuous drag chillers.** Suspended carcasses are pulled through troughs containing agitated cool water and ice slush.

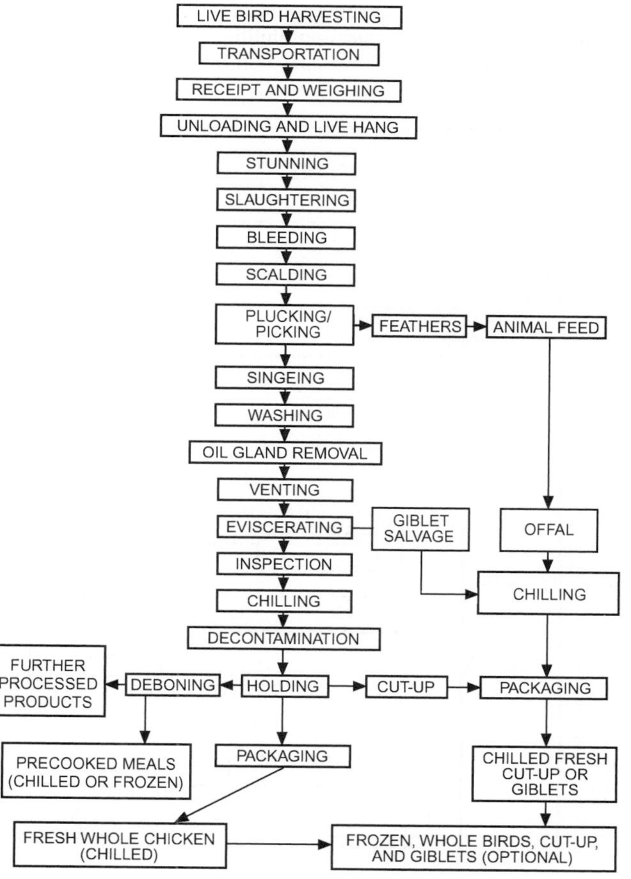

Fig. 1 Processing Sequence of Fresh Poultry

The preparation of this chapter is assigned to TC 10.9, Refrigeration Application for Foods and Beverages.

- **Slush ice chillers.** Carcasses are pushed by a continuous series of power-driven rakes.
- **Concurrent tumble systems.** Free-floating carcasses pass through horizontally rotating drums suspended in tanks of, successively, cool water and ice slush. Movement of the carcasses is regulated by the flow rate of recirculated water in each tank.
- **Counterflow tumble chillers.** Carcasses are carried through tanks of cool water and ice slush by horizontally rotating drums with helical flights on the inner surface of the drums
- **Rocker vat systems.** Carcasses are conveyed by the recirculating water flow and agitated by an oscillating, longitudinally oriented paddle. Carcasses are removed automatically from the tanks by continuous elevators.

These chillers can reduce the internal temperature of broilers from 90 to 40°F in 20 to 40 min, at processing speeds of 5000 to 10,000 birds/h (Figure 5). Chillers must meet food safety requirements (see, e.g., 9CFR381.66) and the facility's Hazard Analysis of Critical Control Points (HACCP) plan (see Chapter 12).

Adjuncts and replacements for continuous-immersion chilling should be used, if available, because immersion chilling is believed to be a major cause of bacterial contamination. Water spray chilling, air blast chilling, carbon dioxide snow, or liquid nitrogen spray are alternatives, but with the following limitations:

- Liquid water has a much higher heat transfer coefficient than any gas at the same temperature of cooling medium, so water immersion chilling is more rapid and efficient than gas chilling.
- Water spray chilling, without recirculation, requires much greater amounts of water than immersion chilling.
- Product appearance should be equivalent for water immersion or spray chilling, but inferior for air blast, carbon dioxide, or nitrogen chilling, because of surface dehydration.
- Air chilling without packaging could cause a 1 to 2% loss of moisture, whereas water immersion chilling allows from 4 to 15% moisture uptake, and water spray chilling up to 4% moisture uptake. Salt-brine chilling is the fastest chilling medium, but has little use in fresh poultry chilling.

Coolant temperature and degree of contact between coolant and product are most important in transferring heat from the carcass surface to the cooling water. The heat transfer coefficient between the carcass and the water can be as high as 630 Btu/h·ft²·°F. Mechanical agitation, injection of air, or both can improve the heat transfer rate (Veerkamp 1995). Veerkamp and Hofmans (1974) expressed heat removed from poultry carcasses by the following empirical relationship.

$$\frac{\Delta Q}{\Delta Q_i} = (-0.009 \log h + 0.73)\log \theta - (0.194 \log h - 0.187)\log m + 0.564 \log h - 2.219 \quad (1)$$

where

h = apparent heat transfer coefficient, Btu/h·ft²·°F
m = mass of the carcass, lb
θ = cooling time, s
ΔQ_i = maximum heat removal, Btu

Figure 5 shows time-temperature curves in a commercial counterflow chiller and compares calculated and measured values.

With adequately washed carcasses and adequate chiller overflow in counterflow to the carcasses, the bacterial count on carcasses should be reduced by continuous water-immersion chilling. However, incidence of a particular low-level contaminant, such as *Salmonella*, may increase during continuous water-immersion chilling; this can be controlled by chlorinating the chill water. However, for chlorine to be effective, the water's pH should be <7.0.

Spray chilling without recirculation has reduced bacterial surface counts 85 to 90% (Peric et al. 1971). Microbe transfer by spray chilling is unlikely. Chilling with air, carbon dioxide, or nitrogen presents no obvious microbiological hazards, although good sanitary practices

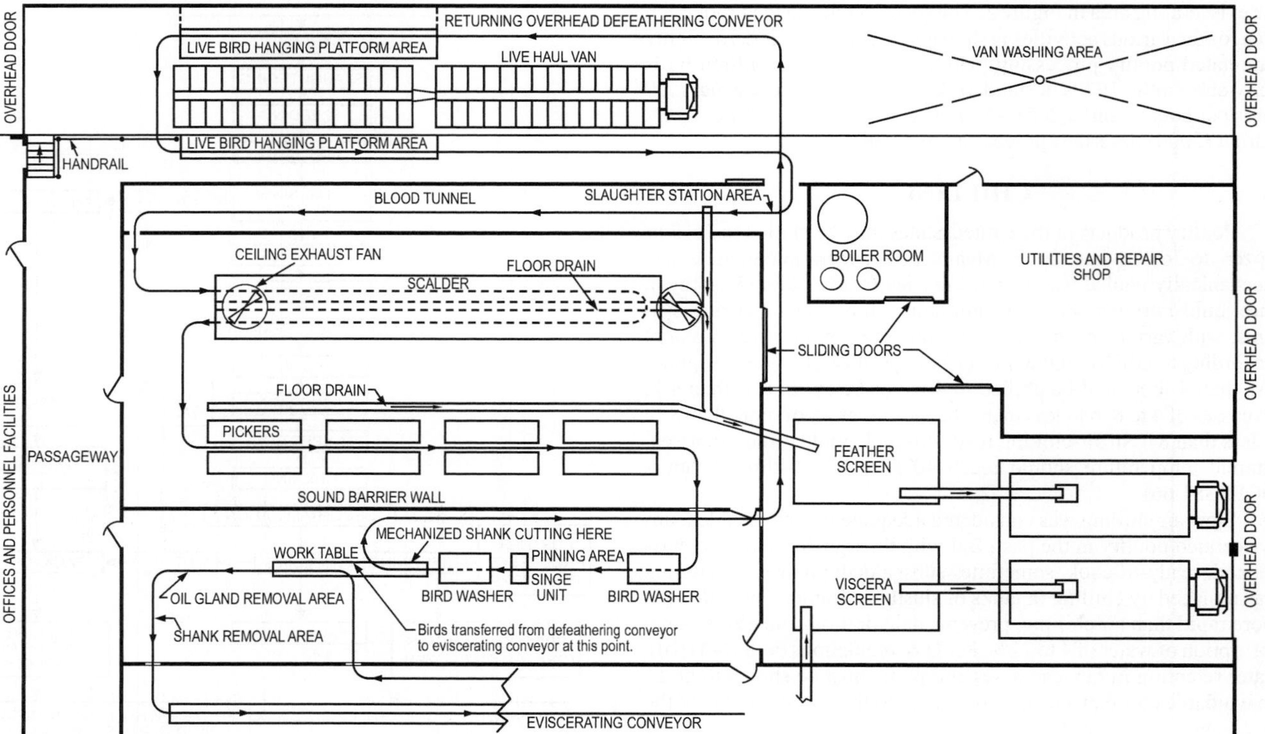

Fig. 2 Typical Equipment Layout for Live Bird Receiving, Slaughtering, and Defeathering Areas

are essential. If the surface of the carcass freezes as a part of the chilling process, the bacterial load may be reduced as much as 90%.

Air or **gas chilling** is commonly used in Europe. In air-blast and evaporative chilling, heat is conducted partly by the air-to-carcass contact and partly by evaporation of moisture from the carcass surface. The amount of water removed by evaporation depends on the carcass temperature, but even at 14°F it is about 1%. The apparent heat transfer coefficient ranges from 16 to 63 Btu/h·ft^2·°F. Major disadvantages of air chilling are slow cooling, dripping from one bird to another in multitiered chillers, and weight loss during chilling. A diagram of a one-tiered evaporative air chiller is given in Figure 6. To reduce contamination, it is very important that birds do not touch or drip on each other if multiple layers are used.

Cryogenic gases are generally used in long insulated tunnels through which the product is conveyed on an endless belt. Some freezing of the outer layer (crust freezing) usually occurs, and the temperature is allowed to equilibrate to the final, intended chill temperature. Some plants use a combination of continuous water immersion chilling to reach 35 to 40°F and a cryogenic gas tunnel to reach 28°F. The water-chilled poultry, either whole or cut up, is generally packaged before gas chilling to prevent dehydration.

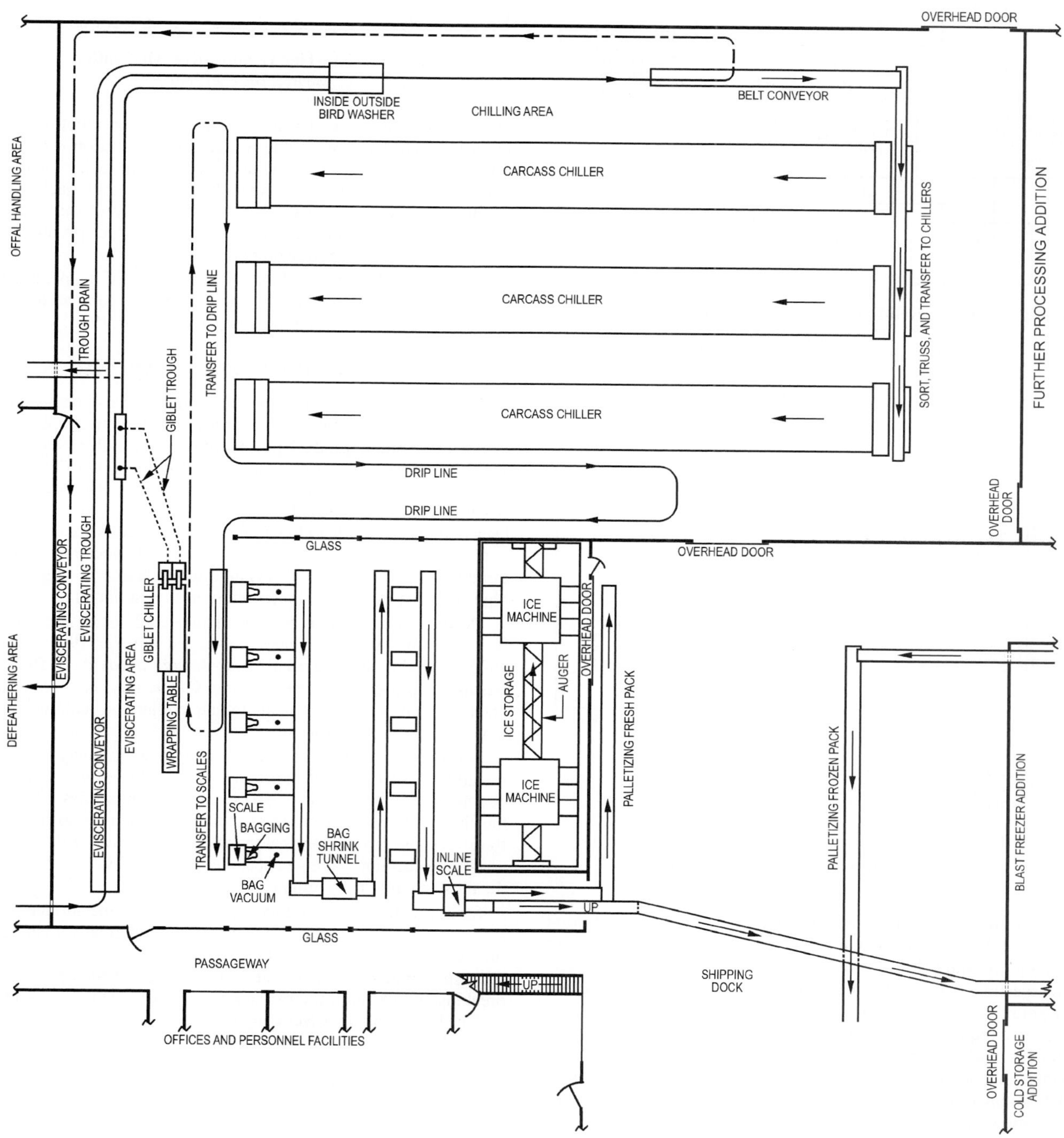

Fig. 3 Typical Equipment Layout for Eviscerating, Chilling, and Packaging Areas

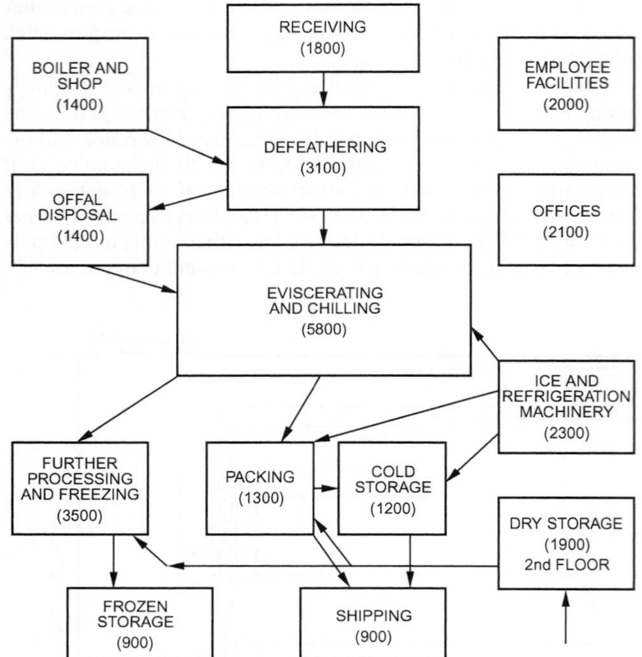

Fig. 4 Space-Relationship-Flow Diagram for Poultry Processing Plant
(Square feet of floor space needed)

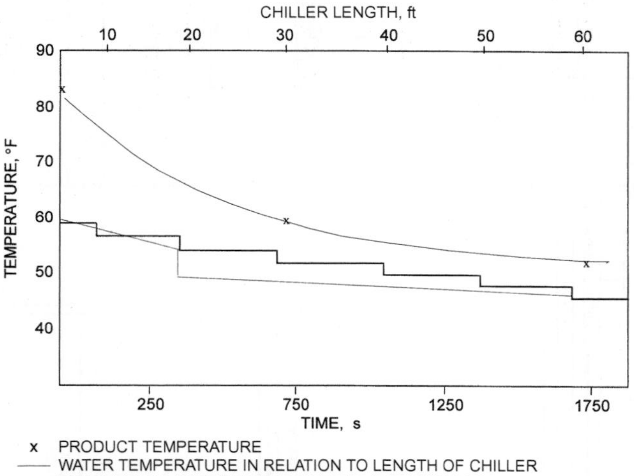

Fig. 5 Broiler and Coolant Temperatures in Countercurrent Immersion Chiller

Ice requirements per bird for continuous immersion chilling depend on entering carcass temperatures and weight, entering water temperature, and exit water and carcass temperature. For a counterflow system, 60°F entering water and 65°F exit water, 0.25 lb of ice per pound of carcass is a reasonable estimate. This may be compared to a requirement of 0.5 to 1 lb of ice per pound of poultry for static ice slush chilling in tanks. For continuous counterflow water-immersion chillers, if plant water temperature is considerably above 65°F, it may be economical to use a heat exchanger between incoming plant water and exiting (overflow) chill water.

Ice production for chilling is usually a complete in-plant operation, with large piping and pumps to convey small crystalline ice or

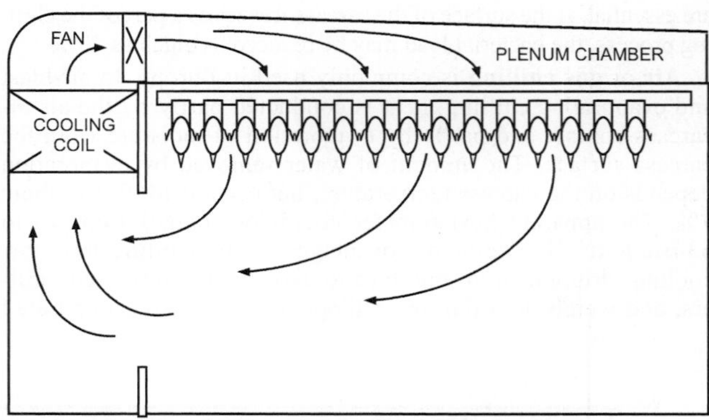

Fig. 6 One-Tier Evaporative Air Chiller
(Source: Bishop 1980)

ice slush to the point of use. To reduce ice consumption, some immersion chillers are double-walled and depend on circulating refrigerant to chill the water in the chiller. The chiller has an ammonia or refrigerant lubricant between the outer and inner jacket, with the inner jacket serving as the heat transfer medium. Agitation or a defrost cycle must be provided during periods of slack production to prevent the chiller from freezing up.

Chilling and holding to about 28°F, the point of incipient freezing, gives the product a much longer shelf life compared with a product held at ice-pack temperatures (Stadelman 1970).

DECONTAMINATION OF CARCASSES

Contamination of poultry meat by foodborne pathogens during processing can be potentially dangerous if microbes multiply to critical numbers and/or produce poisonous toxins (Zeidler 1996, 1997). The Hazard Analysis of Critical Control Points (HACCP) system (see Chapter 12 and the section on HACCP Systems in Poultry Processing) was specifically developed for each food to eliminate or keep pathogen levels very low so food-related illnesses cannot break out. Appropriate refrigeration and strict temperature control throughout the food channel is vital to suppress microbial growth in high-moisture perishable foods and meats in particular.

Decontamination steps are now being added just before chilling. Numerous methods have been developed (Bolder 1997; Mulder 1995), including **lactic acid** (1%), **hydrogen peroxide** (0.5%), and **trisodium phosphate (TSP) sprays**. **Ozone** (O_3) is a strong oxidizer and can be used to decontaminate chiller and scalding water; however, it is very corrosive.

Gamma irradiation of poultry is approved in many countries, including the United States; products are available for sale in a few outlets. The public's fear of this technique limits sales. However, the threat of food poisoning is reducing objections to irradiated foods because irradiation is very effective, and can kill 95.5% of non-spore-forming pathogens (Stone 1995). A dose of 250 krad is the most suitable for poultry.

Steam under vacuum effectively kills 99% of the surface bacteria on beef and pork carcasses and is used commercially. In this continuous system, the carcass is carried on a rail to a chamber. A vacuum is pulled and steam at 290°F is applied for 25 ms. Upon breaking the vacuum, the carcass surface is cooled to prevent the surface from cooking. USDA engineers developed steam equipment for poultry in 1996.

FURTHER PROCESSING

Most chickens and turkeys, for both chilled and frozen distribution, are cut up in the processing plant. More than 90% of the broilers in the United States are sold as cut-up products produced at the processing plant. The cutting procedure is almost fully automatic.

Backs and necks are often mechanically deboned, giving a comminuted slurry that is frozen in rectangular flat cartons containing about 60 lb. Turkey breasts, legs, and drumsticks are available as separate film-packaged parts, and turkey thigh meat is marketed as a ground product resembling hamburger. Partial cooking and breading and battering of broiler parts is done in poultry processing plants.

Unit Operations

The following types of equipment used for further processing of poultry products are also used in red meat facilities.

Size Reduction and Mixing Machines. Several types of size-reduction and mixing equipment are available.

- In **grinding**, meat is conveyed by an auger and forced through a grinding plate.
- **Flaking** is done by cutting blades locked at a specific angle on a rotating drum. Flaking does not extensively break muscle cells, as in grinding, and moisture loss and dripping are limited. Product texture resembles muscle texture.
- **Chopping** is generally conducted with silent cutter equipment. Meat is placed in a rotating bowl with ice, which is used to keep the temperature low, and vertical rotating blades chop the moving meat. The length of chopping time determines the particle size. The end product is used in hot dogs and sausages.
- **Mixing, tumbling, and injecting machines** produce a uniform product out of various meats and nonmeat ingredients such as salt, sugar, dairy or egg proteins, spices, and flavorings. Together with salt, mixing also helps extract myosin, which acts likes a glue in holding the product together.
- **Injection machines** insert an accurate and repeatable volume of liquid that contains salt and flavorings into large chunks of muscle meats such as turkey breasts or whole turkeys. The procedure disperses these ingredients better and faster than soaking in brine and marinade. It also protects the meat from drying during cooking, especially at home.
- **Automated** systems consist of conveyor belts that pass meat into a channel where a cross head assembly of needles is lowered into the product. The hollow needles pierce the meat and marinade is pumped in through a small orifice in the side of each needle. Each needle is independently suspended so bones are not penetrated (Smith and Acton 2001). Production line speeds are fast, averaging up to 10,000 lb/h or greater.
- **Tumblers** shaped like concrete mixers tumble injected large meat chunks mostly under vacuum. The tumbling helps distribute injected brine and spices throughout the meat. Tumbling is a widespread method of commercial marination.

Shaping Forms and Dimension. These machines establish the form, size, and desired weight of size-reduced meats.

- **Stuffing machines** make hot dogs and sausages by stuffing meat emulsion into the casing. Modern stuffing machines operate under vacuum to eliminate bubbles and other textural defects. Dough products or muscle meats are also stuffed with other meats, fruit or vegetable pieces, etc., using equipment that was originally designed to stuff doughnuts with jelly.
- **Forming machines** make hamburgers and nuggets. They are basically presses that force the meat through a plate with holes of various sizes and shapes.
- **Metal molds.** Many products such as turkey rolls and luncheon meats are made from meat chunks, which are placed into metal molds and cooked to produce a restructured log. The meat is chilled in the molds before being released.
- **Coating.** Batter and breading give the product a uniform shape as well as higher palatability and weight. Products are carried on belts through ingredients that coat the products, which are fried immediately after.

Cooking Techniques. Many meat products are produced as ready-to-eat meals that need warming only or are eaten cold. These products are fully cooked in the plant by various methods. Other products are produced as ready-to-cook and skip the cooking step.

- **Smoking/cooking** is a popular method, in which smoke from slow-burning wood outside the cooking chamber flows over the hanging product. To eliminate some smoke carcinogenic compounds and to accelerate the process, liquid smoke is used to treat the product before cooking (Lazar 1997). Smoking is done best on a dry, uncooked surface, which better absorbs the smoke ingredients. Smokehouses are generally the bottleneck of the process, and their high capital cost and large size limits the number of units in the plant. Every product is cooked to a specific internal temperature, commonly between 145 and 175°F, followed by immediate chilling by water showers from sprinklers located in the cooking chamber.
- **Continuous hot-air ovens** cook hamburgers and chicken breast products. These ovens accelerate cooking and reduce labor compared to batch-type equipment. Wireless, solid-state temperature monitoring devices that travel with the product optimize and record the cooking process. Indirect heat sources are used to prevent pink or red discoloration of some poultry products exposed to gases from the direct-heat gas jet (Smith and Acton 2001).
- **Cooking in water bath** is a fast and low-cost way to cook meats because of better heat transfer than in air cooking. Product is protected from the water by waterproof plastic packaging. Most operations are batch-type.
- **Frying** provides higher palatability at the cost of increasing fat content. Frying provides crispness as the hot oil above 212°F replaces the water in the skin, batter, and breading. Frying is a fast method of cooking because of oil's high heat transfer capacity. Oil quality is critical to good product quality; oil problems translate into poor appearance, flavor, and odor of finished product.

There are three basic types of poultry meat products:

- **Whole-muscle products**, such as nuggets, rolls, Buffalo wings, and schnitzels
- **Coarsely ground products**, such as ground poultry meat, loaves, and meatballs
- **Emulsified products**, such as hot dogs and bologna

Figure 7 gives a flow chart for preparing these product groups; batch and continuous heat processing (i.e., cooking and chilling) are illustrated in Figures 8 and 9.

FREEZING

Effect on Product Quality

Generally, lower temperature and protection from atmospheric oxygen reduces oxidation rancidity and extends storage life. At ≤50°F, most microbial growth and enzymatic activity drop to almost zero because most of the cellular water molecules are fixed in a crystalline structure, but reactions may continue slowly down to –80°F. Most commercial holding freezers range from –4 to –20°F, whereas air-blast individual quick freeze (IQF) freezers use high air velocity (2500 ft/min at ≤–20°F) to rapidly remove heat. Powdered carbon dioxide (CO_2 "snow") may be added to product before closing the box container to accelerate freezing. In any freezing application, raw or finished products must be packaged to exclude air and protect the surface from excessive drying (freezer burn). Poultry muscle that is frozen and held at –4 to –20°F should retain its quality for 6 to 10 months. The least desirable temperature range for holding products is –12 to –14°F, at which the phase transition between intercellular crystalline ice and a combination of ice and water occurs. Frequent

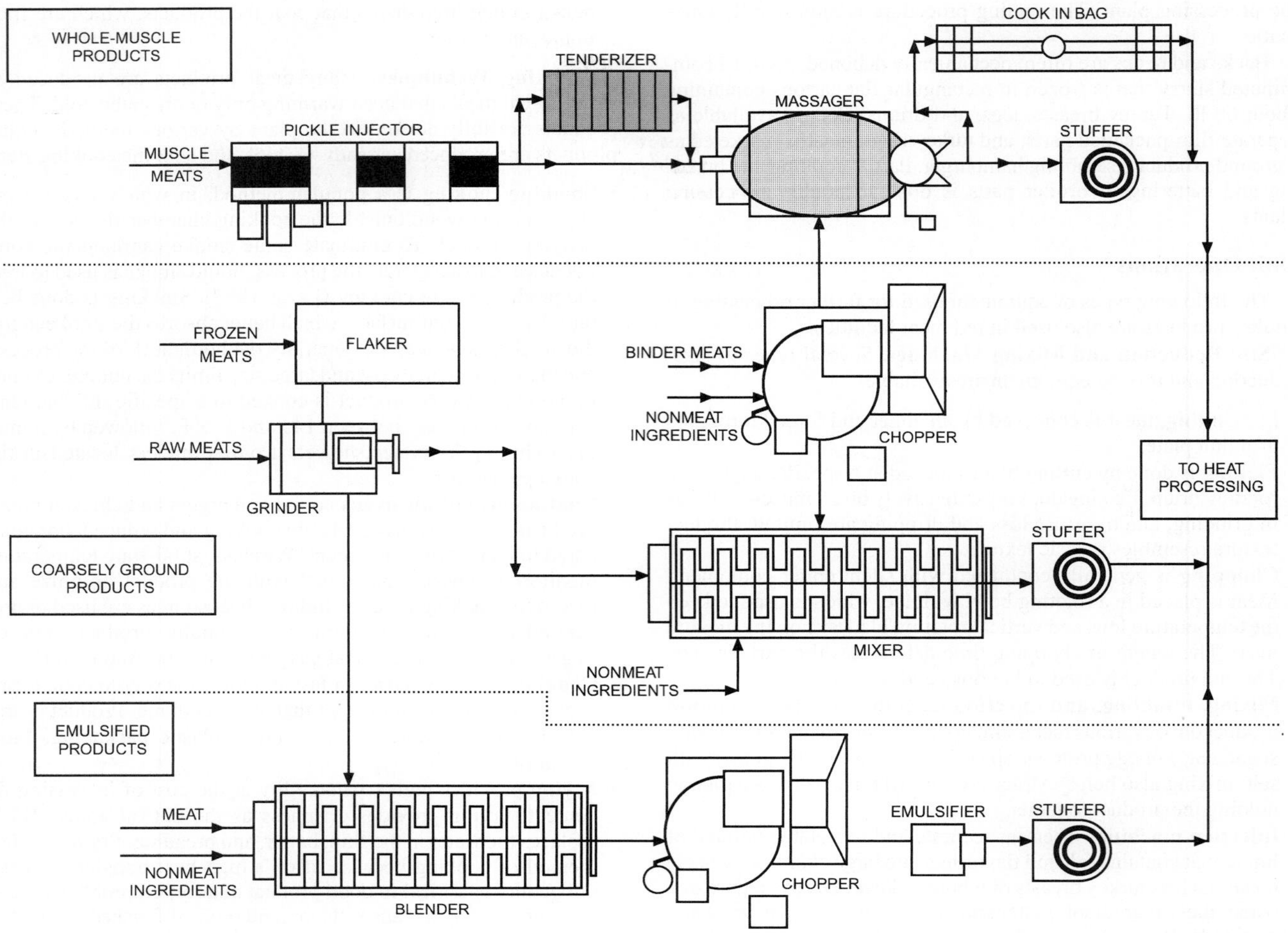

Fig. 7 Meat Products Processing Flow Chart

cycling of the refrigeration system through this temperature zone causes large ice crystal formation in muscle cells and excessive purge (water loss) when thawed (Keeton 2001).

USDA regulations define frozen poultry as cooled to 26°F or lower. This rule prevents the practice of cooling meat to above 0°F, thawing it in destination, and selling it as fresh. Poultry that is frozen to less than 0°F is now called deep frozen.

The freezing rate of diced cooked chicken meat does affect the quality of the frozen meat. Hamre and Stadelman (1967a) reported that cryogenic freezing procedures were desirable because the resulting color was lighter, but too rapid a freezing rate resulted in the meat cubes shattering. The freeze-drying rates for rapidly frozen material were slower than for products frozen by slower methods. Hamre and Stadelman (1967b) indicated that tenderness of freeze-dried chicken after rehydration was affected by freezing rate prior to drying. Liquid nitrogen spray or carbon dioxide snow freezing were selected as preferred methods for overall quality of diced cooked chicken meat to be freeze-dried.

Freezing Methods

Air Blast Tunnel Freezers. Air blast tunnel freezers use air temperatures of –20°F and air velocities of 2500 ft/min. To obtain high air velocity over the product, the blast tunnel should be completely loaded across its cross section, with product units properly spaced to ensure airflow around all sides and no large openings that might allow bypassing of the airstream.

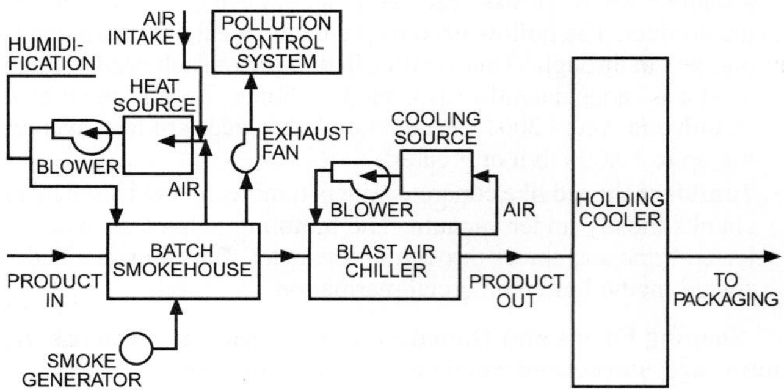

Fig. 8 Heat Processing of Meat Products by Batch Smoker/Cooker

Individual Quick Frozen (IQF) Products. This method creates a crust on the bottom of the product, which moves on thin, disposable plastic sheets. IQF works well for marinated bones, chicken breast, and chicken tenders because they are moist and softer than other parts and tend to stick to freezer belts. The plastic sheet keeps the product from sticking.

Freezer Conveyors. Automated units may be designed to handle packages, cartons, or unwrapped pieces of chicken or turkey. The product may be transported through the freezing chamber on belts or trays. One such system adapts to all sizes of whole birds.

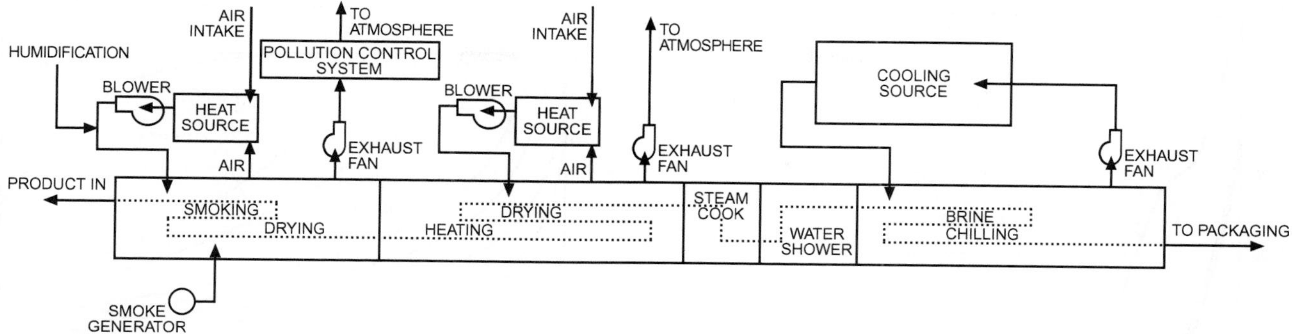

Fig. 9 Heat Processing of Meat Products by Continuous Smoker/Cooker

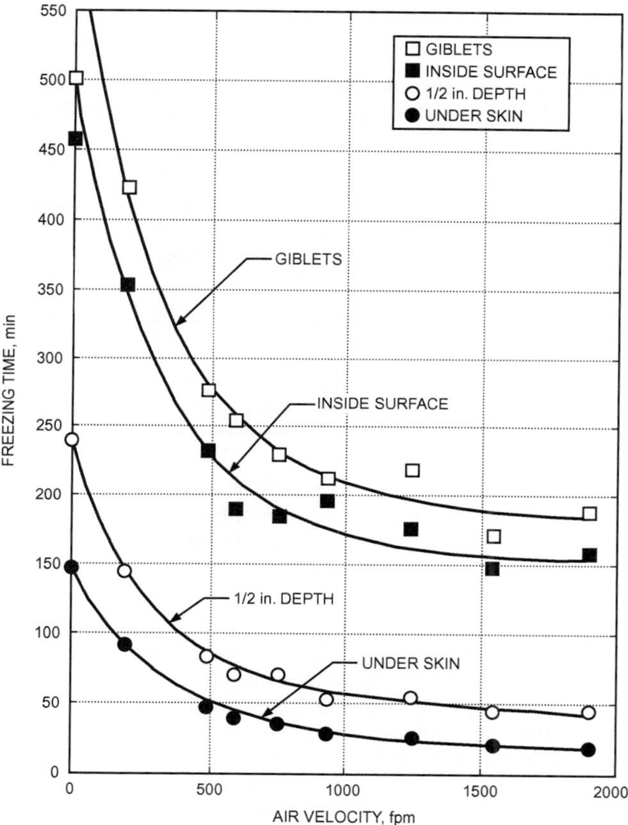

Fig. 10 Relation Between Freezing Time and Air Velocity
(van den Berg and Lentz 1958)

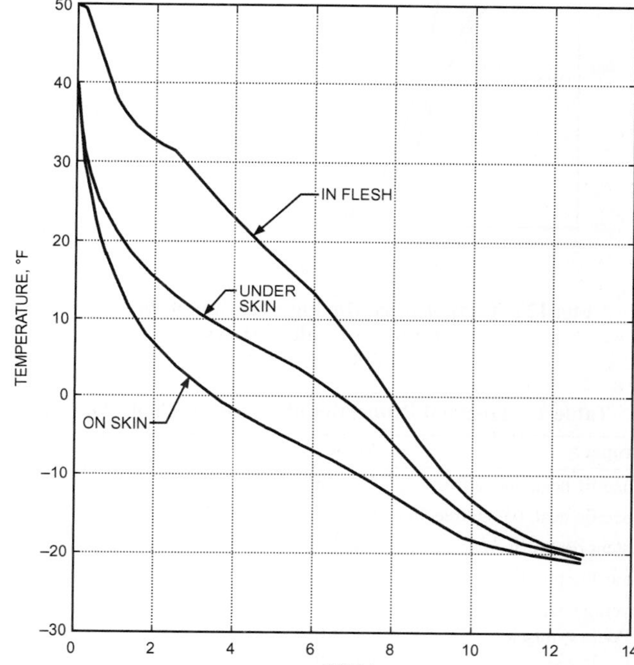

Note: For 21 lb, bronze tom turkeys on shelves in air blast.

Fig. 11 Temperature During Freezing of Packaged, Ready-to-Cook Turkeys
(Klose et al. 1955)

Predicting Freezing or Thawing Times

The following equation can be used to predict freezing and thawing time with an accuracy of about 10% (Calvelo 1981; Cleland and Earle 1984; Cleland et al. 1982).

$$\theta_f = \rho \frac{\Delta H}{\Delta t}\left(\frac{d}{6h} + \frac{d^2}{24k}\right) \quad (2)$$

where

θ_f = freezing time, h
ρ = product density, lb/ft³
d = equivalent diameter of product, ft
ΔH = enthalpy difference, Btu/lb
Δt = temperature difference between air and mean freezing temperature, °F
h = heat transfer coefficient, Btu/h·ft²·°F
k = thermal conductivity at mean freezing temperature, Btu·ft/h·ft²·°F

PACKAGING

Most packaged poultry is now tray-packed, either for frozen or chilled distribution. All-plastic packages and automated packaging lines using plastic film have been engineered. Changes in packaging methods and materials are so rapid that the best sources of information on this subject are manufacturers and distributors of films and packages. They are listed in the most recent Encyclopedia Issue of *Modern Packaging*.

Packages for frozen, whole, and ready-to-cook poultry consist principally of plastic film bags that are tough and reasonably impermeable to moisture vapor and air. The commonly used polyvinylidene chloride, polyethylene, and polyester films are sufficient barriers to water vapor and air to give adequate protection for normal commercial times and temperatures. Turkeys, ducks, and geese are packaged mostly in the whole, ready-to-cook form; frozen chickens appear whole and in packaged, cut-up form.

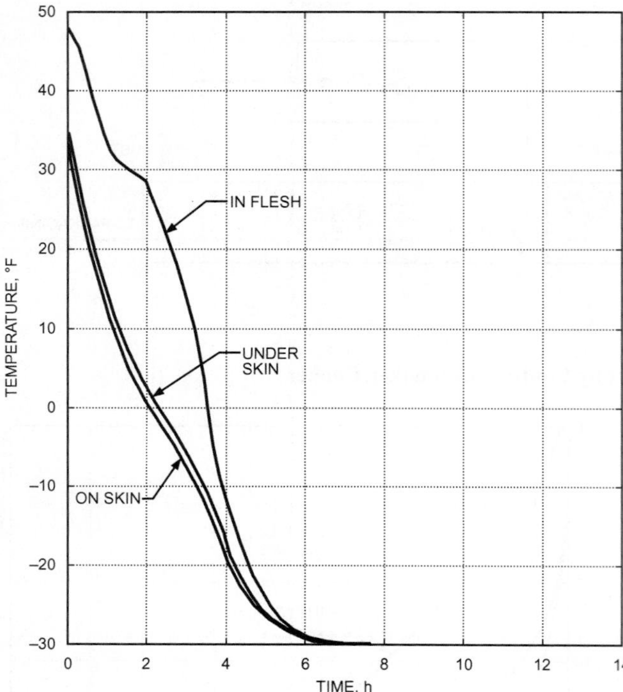

Fig. 12 Temperature During Freezing of Packaged, Ready-to-Cook Turkeys
(Klose et al. 1955)

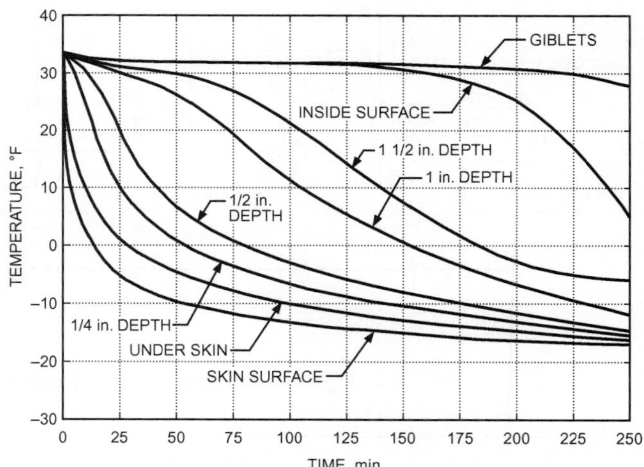

Fig. 13 Temperatures at Various Depths in Breast of 15 lb Turkeys During Immersion Freezing at –20°F
(Lentz and van den Berg 1957)

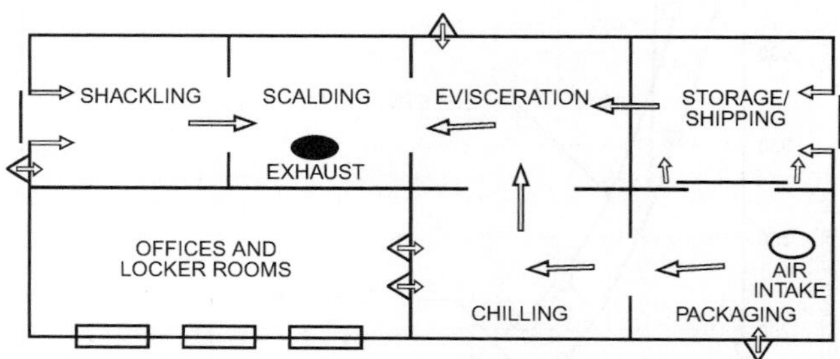

Fig. 14 Air Movement Pattern in Positively Pressurized Poultry Processing Plant
(Further processing is not included)
(*Source*: Keener 2000)

Table 1 Thermal Properties of Ready-to-Cook Poultry

Property	Value	Reference
Specific heat, above freezing	0.70 Btu/lb·°F	Pflug (1957)
Specific heat, below freezing	0.37 Btu/lb·°F	Pflug (1957)
Latent heat of fusion	106 Btu/lb	Pflug (1957)
Freezing point	27°F	Pflug (1957)
Average density		
Poultry muscle	67 lb/ft³	
Poultry skin	64 lb/ft³	
Thermal conductivity, Btu/h·ft·°F		
Broiler breast muscle =	0.24 at 80°F	Walters and May (1963)
Broiler breast muscle ⊥	0.29 at 68°F	Sweat et al. (1973)
Broiler breast muscle ⊥	0.80 at –4°F	Sweat et al. (1973)
Broiler breast muscle ⊥	0.87 at –40°F	Sweat et al. (1973)
Broiler dark muscle ⊥	0.90 at –40°F	Sweat et al. (1973)
Turkey breast muscle ⊥	0.73 at –4°F	Sweat et al. (1973)
Turkey breast muscle =	0.93 at –4°F	Sweat et al. (1973)
Turkey leg muscle ⊥	0.83 at –4°F	Lentz (1961)

⊥ indicates heat flow perpendicular to the muscle fibers.
= indicates heat flow parallel to the muscle fibers.

Large fiberboard cartons or containers for holding and shipping from 2 to 12 individually packaged birds should be rectangular to facilitate palletizing, and should be strong enough to support 16 ft high stacked loads common in refrigerated warehouses. If rapid freezing is necessary for contents (e.g., fryer turkeys), holes or cutaway sections in the sides and ends are needed to permit rapid airflow across the poultry surfaces in the air-blast freezer.

AIRFLOW SYSTEMS IN POULTRY PROCESSING PLANTS

Appropriate air-handling systems in poultry processing plants are vital for maintaining product quality and safety as well as for employees' health and comfort. Moisture, dust, and microorganisms, some of which are hazardous to human health, become airborne at the beginning of the slaughtering process in the unloading, shackling, killing, scalding, and defeathering areas. This aerosol must be treated to protect finished products and workers from contact. Specific work on airflow systems in poultry processing plants and aerosol handling were conducted by Heber et al. (1997) and Keener (2000). Reviews of articles on airflow systems appear in ACGIH (1995) and Burfoot et al. (2001). A typical arrangement of the airflow system in a poultry processing facility is shown in Figure 14. There, air moves from the cleanest cold-storage and packaging areas to the dirtiest parts (shackling and killing) of the plant. Unfortunately, in many poultry processing plants, airflow systems have had a low priority, and renovations often ignore correcting airflow system deficiencies or adjusting the system to the renovated plant.

Historically, many processing plants were ventilated using negative-pressure systems in which uncontrolled fresh air entered the plant through doors, windows, and exhaust hoods. Currently, positive-pressure ventilation systems are used, because they better control internal airflow and incoming fresh air. An air pressure gradient prevents contaminated air produced at the beginning of the process from reaching the finished product areas, while exhausting it from already-dirty areas. Air enters the plant through doors and openings in the unloading and shackling sections and through shipping areas. An air intake is also located in the packaging area, and the exhausting outlets are located in the scalding area. Fans are routinely installed in the chilling area to better recirculate the moist air to prevent condensation. Airflow balance within a room depends on the location of openings in the rooms and their size. In a positive-pressure ventilated system, the packaging area (the cleanest area in the processing plant) has the greatest static pressure, and the defeathering and scalding areas are neutral. As a result, air moves away from the finished product area, where incoming air is filtered and controlled.

The demand for poultry meat has dramatically increased since the mid-1970s and is still growing. To accommodate this growth, processing plants are often being renovated and expanded, but frequently, these projects were designed without sufficient consideration for their effect on the plant ventilation system. Often, moist and dusty air migrates from the slaughter area into the further processing area, and condensation on ceilings and structures results in moisture dripping onto the processing lines, floors, and employees.

This type of air movement can recontaminate in-process and finished products, reducing quality and shelf life and creating a potential health hazard to plant workers and consumers. Airborne microorganisms, including several pathogens, are attached to dust and tiny feather particles, which become airborne in the shackling and slaughtering areas and can remain suspended for a long time. For example, one of the most dangerous pathogens in poultry processing plants is *Listeria monocytogenes*, which is well adapted to grow in low temperatures and can survive long periods in evaporators' drip pans, creating a secondary contamination source. Because many cooked poultry products are eaten cold or warm, pathogens such as *Salmonella*, *Campylobacter*, and *Listeria* in recontaminated products are not destroyed before consumption and could result in serious illnesses and fatalities. Outbreaks with fatalities have been recorded in countries around the world, with severe economic losses by the processing companies and growers. The presence of *Listeria* in cooked poultry could result in immediate product recall. In contrast, raw poultry products have lower risk because they are fully cooked before consumption, destroying all pathogens in the process.

Airflow System Consideration During Renovation

During structural changes, such as providing new doors or wall openings or increasing or altering processing capacity, airflow pattern will probably be affected. Therefore, before renovations take place, the ideal and practical parameters of the airflow system should be reestablished. The evaluation should be conducted by qualified HVAC practitioners and consider all areas of the plant, not just the renovation area. Parameters should include airflow patterns, static pressures, air speed, air temperature, and relative humidity. A follow-up evaluation should be conducted to determine the deviation from the ideal pattern to minimize changes in airflow patterns and production of stagnant areas, and to prevent movement of contaminated air into the finished product areas. In addition, serious attention should be paid to moisture-producing parameters: for example, processing an additional 100,000 chickens per day adds about 150 to 160 lb of water vapor per hour, adding 10 employees generates 3 to 10 lb of water vapor per hour, and sanitation with hot water increases plant humidity. Proper consideration and evaluation of these parameters can help provide safe products and a healthy atmosphere for workers.

PLANT SANITATION

Poultry meat is highly perishable because it composed of nutrients that are ideal for microbial growth. During processing, excessive amounts of meat and drippings soil equipment and floors. If not thoroughly cleaned and sanitized, it becomes a source of bacterial growth that can recontaminate incoming new meats. Therefore, specific cleaning teams clean the plant at the end of the working day using steam, soap, and sanitizing agents. In many instances, work is stopped and certain equipment is cleaned every few hours.

In January 1997, the rules for meat inspection changed dramatically (USDA/FSIS 1996). Processing plants are required to (1) inspect their own processes by writing and implementing their own sanitation standard operation procedures (SSOP), (2) monitor the processes, and (3) take corrective action when necessary. Precise records should be kept in a format ready for instant review by purchasers.

Proper sanitation should be addressed when the structure, processing equipment, and refrigeration systems are designed. The plant structure should be designed to prevent pests such as mice, rats, cockroaches, and birds from entering the facility and finding places to hide that cannot be reached. This includes drainage, sewage, windows, vents, etc. Equipment should be designed for easy cleaning and easy assembly and disassembly. It should not have any areas on which product particles can accumulate. Refrigeration systems should be designed to restrict airflow from raw to cooked meat areas and to eliminate possible condensation and dripping into the product or into drip pans that cannot be reached for easy cleaning.

Clearly written procedures, constant training of employees, and adequate numbers of employees are essential for successful implementation of the program. Also, constant management commitment is vital.

HACCP Systems in Poultry Processing

Hazard Analysis of Critical Control Points (HACCP) is a logical process of preventative measures that can control food safety problems. HACCP is a process control system designed to identify and prevent and microbial and other hazards in food production. It is designed to prevent problems before they occur and to correct deviations as soon as they are detected. This method of control emphasizes a preventative approach rather than a reactive approach, which can reduce the dependence on final product testing. The fundamentals of HACCP are described in Chapter 12.

HACCP systems are used in poultry processing to improve the safety of fresh meats and their products. HACCP programs are required by the USDA in all plants.

Poultry is associated with numerous microbial pathogens that occur naturally in wild birds, rats, mice, and cockroaches. Poultry is contaminated by feed containing feces of these pests. They are potentially transferred to the meat during processing from unclean equipment, processing water, air, and human hands, hair, or clothing. Strict temperature control throughout the system will strongly suppress microbial growth, keeping pathogen levels too low to generate foodborne illness outbreaks. **In most outbreaks, temperature control breakdown or temperature abuse is involved** (Zeidler 1996).

The major pathogens associated with raw poultry are various types of *Salmonella* and *Campylobacter jejuni*, which recently became the leading pathogen in poultry meat. HACCP programs cover production farms, processing plant, and shipping trucks. Water baths (as in chilling and scalding areas) could easily spread pathogens, and the circulating water must be treated. The aerosol, places where condensation may accumulate, backup of sewage, and used processing water are also potential contamination risk areas. Reducing human touch, bird-to-bird contact, and dripping from bird to bird during air chilling, as well as increased automation, help reduce contamination. Appropriate temperature control throughout the system is vital as foodborne disease outbreaks always involve temperature abuse.

TENDERNESS CONTROL

Texture is considered the most important characteristic of poultry meat and is most affected by the bird's age and by processing procedures.

Tenderness in cooked poultry meat is a prerequisite to acceptability. Relative tenderness decreases as birds mature, and this toughness has always been considered in the recommendations for cooking birds of various ages. However, another type of toughness depends primarily on the length of time that the carcass is held unfrozen before cooking. Birds cooked before they have time to pass through rigor are very tough. Normal tenderization after slaughter is arrested by freezing. For birds held at 40°F, complete tenderization occurs for all muscles within 24 h and for many muscles in a much shorter time.

Other factors that interfere with normal tenderization are immersion in 140°F water and cutting into the muscle. Formerly, birds were held unfrozen for enough time in the normal channels of processing and use to allow adequate tenderization. Shorter chilling periods, more rapid freezing, and cooking without a preliminary thawing period have shortened the period during which tenderization can occur to such an extent that toughness has become a potential consumer complaint.

Hanson et al. (1942) observed a rapid increase in tenderness within the first 3 h of holding and a gradual increase thereafter. Shannon et al. (1957), working with hand-picked stewing hens, found increased toughness because of increased scalding temperature or time, in the ranges of 120 to 195°F and 5 to 160 s. However, the differences in toughness that occurred within the limits of temperature and time, necessary or practical in commercial plants, were quite small.

Tenderness is also increased by reducing the extent of beating received by the birds during picking operations. Turkey fryers should be held at least 12 h above freezing to develop optimum tenderness. Holding fryers at 0°F for 6 months and longer has no tenderizing effect, but holding in a thawed state (35°F) after frozen storage has as much tenderizing effect as an equal period of chilling before freezing. Turkeys frozen 1 h after slaughter are adequately tenderized by holding for 3 days at 28°F, a temperature at which the carcass is firm and no important quality loss occurs for the period involved. Behnke et al. (1973) confirmed this effect for Leghorn hens.

Overall processing efficiency is improved by cutting up the carcass directly from the end of the eviscerating line, packaging the parts, and then chilling the still-warm packaged product in a low-temperature air blast or cryogenic gas tunnel. Webb and Brunson (1972) reported that cutting the breast muscle and removing a wing at the shoulder joint before chilling significantly decreased tenderness of treated muscles, though cut carcasses were aged in ice slush before cooking. Klose et al. (1972) found that, under commercial plant conditions, making an eight-piece hot-cut before chilling and aging significantly reduced tenderness of breast and thigh muscles, compared to cutting after chilling. Smith et al. (1966) indicated that too-rapid chilling of poultry might have a toughening effect, similar to cold shortening observed in red meats.

Post-mortem electrical stimulation can prevent some toughness while providing some tenderization. In electrical stimulation (which is very different from preslaughter stunning), electricity is pulsed through a recently bled carcass still on the shackles. The electricity enters the head from a charged plate and exits the carcass where the feet contact the metal shackle. The electrical characteristics and timing cause two effects: the pulses excite the muscle and speed onset of rigor mortis, and cause such forceful contractions that the filaments are torn, reducing the integrity of the protein network responsible for toughness (Sams 2001).

DISTRIBUTION AND RETAIL HOLDING REFRIGERATION

Chilled poultry, handled under proper conditions, is an excellent product. However, there are limitations in its marketability because of the relatively short shelf life caused by bacterial deterioration. Bacterial growth on poultry flesh, as on other meats, has a high temperature coefficient. Studies based on total bacterial counts have shown that birds held at 36°F for 14 days are equivalent to those held at 50°F for 5 days or 75°F for 1 day. Spencer and Stadelman (1955) found that birds at 31°F had 8 days of additional shelf life over those at 38°F.

The generation time of psychrophilic organisms isolated from chickens was 10 to 35 h at 32°F, depending on the species studied (Ingraham 1958). Raising the temperature to 36°F reduced generation time to 8 to 14 h, again depending on the species.

Frequent cleaning of processing equipment, as well as thorough washing of the eviscerated carcasses, is essential. Goresline et al. (1951) reported a substantial decrease in bacterial contamination and an increase in shelf life by the use of 20 ppm of chlorine in processing and chilling water. Water is routinely chlorinated in the United States, but chlorine is not allowed to touch poultry meat in some European countries.

Because shelf life is limited considerably by bacterial growth (slime formation) on the skin layer, it is reasonable to assume that drastic changes in the skin surface, such as removal of the epidermal layer by high-temperature scalding, might appreciably affect shelf life. Ziegler and Stadelman (1955) reported approximately 1 day more chilled shelf life for 128°F scalded birds than for 140°F scalded ones.

Chickens, principally broilers, are sold as whole, ready-to-cook; cut-up, ready-to-cook; or boneless, skinless ready-to-cook. Poultry may be shipped in wax-coated corrugated containers, but most is consumer-packaged at the processing plant. A number of precooked poultry meat products are sold in wholesale and retail markets as refrigerated, nonfrozen products. Such items are usually vacuum-packaged or packaged in either a carbon dioxide or nitrogen gas atmosphere. The desired temperature for such products is also 28 to 30°F.

PRESERVING QUALITY IN STORAGE AND MARKETING

Important qualities of frozen poultry include appearance, flavor, and tenderness. Optimum quality requires care in every phase of the marketing sequence, from the frozen storage warehouse, through transportation facilities, wholesaler, retailer, and finally to the frozen food case or refrigerator in the home.

Tissue Darkening. Darkening of the bones occurs in immature chickens and has become more prevalent as broilers are marketed at younger and younger ages. During chilled storage or during freezing and defrosting, some of the pigment normally contained inside the bones of particularly young chickens leaches out and discolors adjacent tissues. This discoloration does not affect the palatability of the product. Brant and Stewart (1950) found that development of dark bones was greatly reduced by a combination of freezing and storage at –30°F and immediate cooking after rapid thawing. Aside from this combination, freezing rate, temperature and length of storage, and temperature fluctuations during storage were not found to have a significant effect.

Further research suggested that freezing and thawing not only liberated hemoglobin from the bone marrow cells but modified the bone structure to allow penetration by the released pigment. Roasting pieces of chicken 0.5 h prior to freezing reduced discoloration of the bone. Ellis and Woodroof (1959) found that heating legs and thighs to 180°F before freezing effectively controlled meat darkening. Methods of preheating, in order of preference, include microwave oven, steam, radiant heat oven, and deep fat frying.

Dehydration. During storage, poultry may become dehydrated, causing a condition known as **freezer burn**. Dehydration can be controlled by humidification, lowering storage temperatures, or packaging the product adequately (Smith et al. 1990).

Rancidity. Poultry fat becomes rancid during very long storage periods or at extremely high storage temperatures. Rancidity in frozen, eviscerated whole poultry stored for 12 months is not a serious problem if the bird is packaged in essentially impermeable film and held at 0°F or below. Danger of rancidification is greatly increased when poultry is cut up before freezing and storage, because of the increased surface exposed to atmospheric oxygen.

Length of Storage. Klose et al. (1959) studied quality losses in frozen, packaged, and cut-up frying chickens over temperatures of –30 to 20°F and storage periods from 1 month to 2 years. All commercial-type samples examined were acceptable after storage at 0°F of at least 6 months, and some were stable for more than a year. In a comparison of a superior (moisture/vaporproof) commercial package with a fair commercial package, increased adequacy of packaging resulted in as much extension in storage life as a decrease in storage temperature of about 20°F. The results indicate that no statement on storage life can have general value unless the packaging condition is accurately specified.

Frozen storage tests by Klose et al. (1960) on commercial packs of ready-to-cook ducklings and ready-to-cook geese established that these products have frozen storage lives similar to other commercial forms of poultry. Ducks and geese should be stored at 0°F or below to maintain their original high quality for 8 to 12 months.

Incorporation of polyphosphates into poultry meat by adding it to the chilling water has been shown to increase shelf life in frozen or refrigerated storage and to control loss of moisture in refrigerated storage and during thawing and cooking.

Storage of Precooked Poultry. Studies on frozen fried chicken indicated that precooking produces a product much less stable than a raw product. Rancidity development is the limiting factor, and is detected in the meat slightly sooner than in the skin and fatty coating of the fried product. The marked beneficial effect of oxygen (air)-free packaging was demonstrated in tests in which detectable off-flavors were observed at 0°F in air-packed samples after 2 months, whereas nitrogen-packed samples developed no off-flavors for periods exceeding 12 months.

Cooling precooked parts in ice water before breading was found to reduce TBA (thiobarbituric acid, a measure of rancidity from fat oxidation) values of precooked parts (Webb and Goodwin 1970). In this study, no difference in rancidity was noted for chicken stored 6, 8, or 10 months. By removing the skin from precooked broilers, TBA values were lower, but yield and tenderness were reduced. No difference was detected in the TBA values of thighs frozen in liquid refrigerant with or without skin. Chicken parts that were blast-frozen without skin were less rancid than those frozen with skin. Precooked frozen chicken parts browned for 120 s at 400°F were less rancid than those parts browned at 300°F (Love and Goodwin 1974).

In contrast to a loosely packed product such as frozen fried chicken, Hanson and Fletcher (1958) reported that a solid-pack product such as chicken and turkey pot pies, in which cooked poultry is surrounded by sauce or gravy, with consequent exclusion of air, had a storage life at 0°F of at least 1 year. As is the case with raw poultry, turkey products have less fat stability than chicken products, but stability can be increased by substituting more stable fats in the sauces or by using antioxidants. A quality defect found in precooked frozen products containing a sauce or gravy is a liquid separation and curdled appearance of the sauce or gravy when thawed for use. This separation is extremely sensitive to storage temperature. Sauces can be stored at least five times as long at 0°F as at 10°F before separation takes place. Hanson et al. (1951) established that flour in the sauce was the cause of the separation, and found, among a large number of alternative thickening agents, that waxy rice flour produced superior stability. Sauces and gravies prepared with waxy rice flour are completely stable for about a year at 0°F.

Because precooked frozen foods are not apt to be sterilized in the reheating process in the home, the processor has an added responsibility to keep bacterial counts in the product well below hazardous levels. Extra precautions should be taken in general plant sanitation, in rapid chilling and freezing of cooked products, and in seeing that products do not reach a temperature that will permit bacterial growth at any time during storage or distribution.

THAWING

Under ordinary conditions, poultry should be kept frozen until shortly before its consumption. The general procedure is to defrost in air or in water. No significant difference has been found in palatability between thawing in oven, refrigerator, room, or water.

For turkeys that have been scalded at high temperatures and fast-frozen to give a light appearance, the temperature in retail storage and display must be kept as low as possible (0°F is reasonable) to prevent darkening. Thawing in the package will minimize darkening.

The safest procedure for thawing poultry to hold the bird in the refrigerator (35 to 40°F) for 2 to 4 days, depending on the size of the bird.

REFERENCES

ACGIH. 1995. Ventilation aspects of indoor air quality. In *Industrial Ventilation: A Manual of Recommended Practice*, 22nd ed. American Conference of Governmental Industrial Hygienists, Cincinnati, OH.

Barbut, S. 2000. Poultry processing and product technology. In *Encyclopedia of food science and technology*, pp. 1563-1973. Francis, J.F., ed. John Wiley & Sons, New York.

Behnke, J.R., O. Fennema, and R.W. Haller. 1973. Quality changes in prerigor poultry at –3°C. *Journal of Food Science* (38):275.

Bolder, N.M. 1997. Decontamination of meat and poultry carcasses. *Trends in Food Science and Technology*. 8:221-227.

Brant, A.W. and G.F. Stewart. 1950. Bone darkening in frozen poultry. *Food Technology* (4):168.

Burfoot, D., K. Brown, Y. Xu, S.V. Reavell, and K. Hall. 2001. Localized air delivery system in the food industry. *Trends in Food Science and Technology* 11:410-418.

Calvelo, B. 1981. Recent studies on meat freezing. In *Development in Meat Sciences*, vol. 2, pp. 125-158. R. Laurie, ed. Applied Science Publishing, London.

CFR. 2003. Poultry products inspection regulations—Temperatures and chilling and freezing procedures. *Code of Federal Regulations* 9CFR381.66. U.S. Government Printing Office, Washington, D.C.

CFR. 2005. Consumer protection standards. Raw products—Retained water. *Code of Federal Regulations* 9CFR441.10. U.S. Government Printing Office, Washington, D.C.

Cleland, A.C. and R.L. Earle. 1984. Assessment of freezing time prediction formula. *Journal of Food Science* 49:1034-1042.

Cleland, A.C., R.L. Earle, and D.J. Cleland. 1982. The effect of freezing rate on the accuracy of numerical freezing calculations. *International Journal of Refrigeration* 5:294-301.

Ellis, C. and J.G. Woodroof. 1959. Prevention of darkening in frozen broilers. *Food Technology* 13:533.

Goresline, H.E., M.A. Howe, E.R. Baush, and M.F. Gunderson. 1951. In plant chlorination does a 3-way job. *U.S. Egg and Poultry Magazine* 4:12.

Hamre, M.L. and W.J. Stadelman. 1967a. Effect of various freezing methods on frozen diced chicken. *Quick Frozen Foods* 29(4):78.

Hamre, M.L. and W.J. Stadelman. 1967b. The effect of the freezing method on tenderness of frozen and freeze dried chicken meat. *Quick Frozen Foods* 30(8):50.

Hanson, H.L. and L.R. Fletcher. 1958. Time-temperature tolerance of frozen foods. Part XII, Turkey dinners and turkey pies. *Food Technology* 12:40.

Hanson, H.L., A. Campbell, and H. Lineweaver. 1951. Preparation of stable frozen sauces and gravies. *Food Technology* 5:432.

Hanson, H.L., G.F. Stewart, and B. Lowe. 1942. Palatability and histological changes occurring in New York dressed broilers held at 1.7°C (35°F). *Food Research* 7:148.

Heber, J.H., M.W. Peugh, R.H. Linton, N.J. Zimmerman, and K. Lutgring. 1997. *The effect of processing and airflow parameters on microbial aerosol dispersion in poultry plants*. ASHRAE Research Project RP-834, Final Report.

Ingraham, J.L. 1958. Growth of psychrophilic bacteria. *Journal of Bacteriology* 6:75.

Keener, K.M. 2000. *Air quality intervention strategies in the processing plant: A system approach*. North Carolina Cooperative Extension Service Publication.

Keeton, J.T. 2001. Formed and emulsion products. In *Poultry Meat Processing*. CRC Press, New York.

Klose, A.A., A.A. Campbell, and H.L. Hanson. 1960. Stability of frozen ready-to-cook ducks and geese. *Poultry Science* 39:1136.

Klose, A.A., M.F. Pool, and H. Lineweaver. 1955. Effect of fluctuating temperatures on frozen turkeys. *Food Technology* 9:372.

Klose, A.A., M.F. Pool, M.B. Wiele, H.L. Hanson, and H. Lineweaver. 1959. Time-temperature tolerance of frozen foods: Ready-to-cook cut-up chicken. *Food Technology* 13:477.

Klose, A.A., R.N. Sayre, D. deFrenery, and M.F. Pool. 1972. Effect of hot cutting and related factors in commercial boiler processing on tenderness. *Poultry Science* 51:634.

Lazar, V. 1997. Natural vs. liquid smoke. *Meat Processing* 36(9):28-31.

Lentz, C.P. 1961. Thermal conductivity of meats, fats, gelatin, gels, and ice. *Food Technology* 15:243.

Lentz, C.P. and L. van den Berg. 1957. Liquid immersion freezing of poultry. *Food Technology* 11:247.

Love, B.E. and T.L. Goodwin. 1974. Effects of cooking methods and browning temperatures on yields of poultry parts. *Poultry Science* 53:1391.

Mulder, R.W.A.W. 1995. Decontamination of broiler carcasses. *Misset World Poultry* 11(3):39-43.

Peric, M., E. Rossmanith, and L. Leistner. 1971. Verbesserung der microbiologischen Qualität von Schlachthänchen durch die Sprühkühlung. *Die Fleischwirtschaft* April:574.

Pflug, I.J. 1957. Immersion freezing found to improve poultry appearance. *Frosted Food Field* June:17.

Sams, A.R. 2001. *Poultry meat processing*. CRC Press, New York.

Shannon, W.G., W.W. Marion, and W.J. Stadelman. 1957. Effect of temperature and time of scalding on the tenderness of breast meat of chicken. *Food Technology* 11:284.

Smith, D.P. and F.C. Acton. 2001. Marination, cooking, and curing of poultry products. In *Poultry Meat Processing*. CRC Press, New York.

Smith, J.P., H.S. Ramaswami, and K. Simpson. 1990. Development in food packaging technology, part II: Storage aspects. *Trends in Food Science Technology* 1(5):111-118.

Smith, M.C., Jr., M.D. Judge, and W.J. Stadelman. 1966. A cold shortening effect in avian muscle. *Journal of Food Science* 31:450.

Spencer, J.V. and W.J. Stadelman. 1955. Effect of certain holding conditions on shelf life of fresh poultry meat. *Food Technology* 9:358.

Stadelman, W.J. 1970. 28 to 32°F temperature is ideal for preservation, storage and transportation of poultry. ASHRAE *Journal* 12(3):61.

Stone, D.R. 1995. Can irradiation zap consumer resistance? *Poultry Marketing and Technology* 3(2):20.

Sweat, V.E., C.G. Haugh, and W.J. Stadelman. 1973. Thermal conductivity of chicken meat at temperatures between –75 and 20°C. *Journal of Food Science* 38:158.

USDA/FSIS. 1990. *Poultry products inspection regulations*. Chapter 3, Sub-Chapter C, Part 381. Washington, D.C.

USDA/FSIS. 1996. *Sanitation standard operation procedures (SSOP) reference guide.*

van den Berg, L. and C.P. Lentz. 1958. Factors affecting freezing rate and appearance of eviscerated poultry frozen in air. *Food Technology* 12:183.

Veerkamp, C.H. 1995. Chilling, freezing and thawing. In *Processing of Poultry,* pp. 103-125. G.C. Mead, ed. Chapman & Hall, London.

Veerkamp, C.H. and G. J. P. Hofmans. 1974. Factors influencing cooling of poultry carcasses. *Journal of Food Science* 39:980-984.

Walters, R.E. and K.N. May. 1963. Thermal conductivity and density of chicken breast, muscle and skin. *Food Technology* 17:808.

Webb, J.E. and C.C. Brunson. 1972. Effects of eviscerating line trimming on tenderness of broiler breast meat. *Poultry Science* 51:200.

Webb, J.E., R.L. Dake, and R.E. Wolfe. 1989. Method of eliminating aging step in poultry processing. U.S. Patent No. 4,860,403. August 29.

Webb, J.E. and T.L. Goodwin. 1970. Precooked chicken: Effect of cooking methods and batter formula on yields and storage conditions on 2-thiobarbituric acid values. *British Poultry Science* 11:171.

Zeidler, G. 1996. How can food-borne microorganisms make you ill. *Misset World Poultry* 12(X).

Zeidler, G. 1997. New light on foodborne and waterborne diseases. *Misset World Poultry* 13(9):10-12.

Ziegler, F. and W.J. Stadelman. 1955. The effect of different scald water temperatures on the shelf life of fresh, non-frozen fryers. *Poultry Science* 34:237.

BIBLIOGRAPHY

Babbot, S. 2001. *Poultry product processing*. Technomic, Lancaster, PA.

Bowers, P. 1997. In-plant irradiation emerges. *Poultry Marketing and Technology* 5(4):18.

Bowers, P. 1997. Hot off the bone. *Poultry Marketing and Technology.* 5(4):14.

Brant, A.W., J.W. Goble, J.A. Hamann, C.J. Wabeck, and R.E. Walters. 1982. Guidelines for establishing and operating broiler processing plants. *USDA Agricultural Handbook* No. 581.

Clatfelter, K.A. and J.E. Webb. 1987. Method of eliminating aging step in poultry processing. U.S. Patent No. 4,675,947, June 30.

Hoggins, J. 1986. Chilling broiler chicken: An overview. In *Proceedings of Recent Advances and Development in the Refrigeration of Meat by Chilling*, pp. 133-147. International Institute of Refrigeration, Paris.

Elliott, R.P. and R.P. Straka. 1964. Rate of microbial deterioration of chicken meat at 2°C after freezing and thawing. *Poultry Science* 43:81.

Evans, T. 1997. Watt poultry statistical yearbook. *Poultry International* 36(9).

Herwill, J. 1986. What to do before meat hits your boning line. *Broiler Industry* 19(1):124-128.

Kotula, A.W., J.E. Thomson, and J.A. Kinner. 1960. Water absorption by eviscerated broilers during washing and chilling. USDA, Agricultural Marketing Service, *Marketing Research Report* No. 438 (October).

Mogens, J. 1986. Chilling broiler chicken: An overview. In *Proceedings of Recent Advances and Development in the Refrigeration of Meat by Chilling*, pp. 133-141. International Institute of Refrigeration, Paris.

Mountney, G.J. 1976. Plant layout. In *Poultry Products Technology*, 2nd ed. pp. 116-131. AVI Pub. Westport, CT.

Poulson, B.A. 1990. *Food plant air quality management*. King, Owatonn, MN.

Schlimme, D.V., M.A. Smith, and L.M. Ali. 1991. Influence of freezing rate, storage temperature, and storage duration of the quality of turkey frankfurters. *ASHRAE Transactions* 97(1):214-220.

Schlimme, D.V., M.A. Smith, and L.M. Ali. 1991. Influence of freezing rate, storage temperature, and storage duration of the quality of cooked turkey breast roll. *ASHRAE Transactions* 97(1):221-227.

Stadelman, W.J., V.M. Olson, G.A. Shemwell, and S. Pasch. 1988. Scalding. In *Egg and Poultry Meat Processing*, pp. 127-128. Ellis Horwood, Chichester, England.

Todd, E.C.D. 1980. Poultry associated foodborne diseases—Its occurrence, cost, source and prevention. *Journal of Food Protection* 43:129-139.

USDA and U.C. Davis. 1975. Guidelines for turkey processing plant layout. *USDA Marketing Research Report Number 1036*. Washington, D.C.

USDA and University of Georgia. 1970. Guidelines for poultry processing plant layout. *USDA Marketing Research Report No. 878*. Washington, D.C.

Wells, F.E., J.V. Spencer, and W.J. Stadelman. 1958. Effect of packaging materials and techniques on shelf life of fresh poultry meat. *Food Technology* 12:425.

Willis, R., B. Lowe, and G.F. Stewart. 1948. Poultry storage at subfreezing temperatures—Comparisons at –10 and +10°F. *Refrigerating Engineering* 56:237.

Zeidler, G. 1997. Changes in consumer behaviors and in economic and demographic trends in the US as reflected in successful new poultry product introductions. In *Poultry Meat Quality*, pp. 43-14-32. J. Kijowski and J. Piskell, eds.

CHAPTER 19

FISHERY PRODUCTS

THE major types of fish and shellfish harvested from North American waters and used for food include the following:

- Groundfish (haddock, cod, whiting, flounder, and ocean perch), lobster, clams, scallops, snow crab, shrimp, capelin, herring, and sardines from New England and Atlantic Canada
- Oysters, clams, scallops, striped bass, and blue crab from the Middle and South Atlantic
- Shrimp, oysters, red snapper, clams, and mullet from the Gulf Coast
- Lake herring, chubs, carp, buffalofish, catfish, yellow perch, and yellow pike from the Mississippi Valley and Great Lakes
- Alaska pollock, Pacific pollock, tuna, halibut, salmon, Pacific cod, various species of flatfish, king and snow crab (*Chinoecetes opelio*; about 200,000,000 lb annually), dungeness crab, scallops, shrimp, and oysters from the Pacific Coast and Alaska
- Catfish, salmon, trout, oysters, and mussels from aquaculture operations in various locations

Fish harvested from tropical waters are reported to have a substantially longer shelf life than fish harvested from cold waters, possibly because of the bacterial flora naturally associated with the fish. Bacteria associated with fish from tropical waters are mainly gram-negative mesophiles, whereas those that cause spoilage of fish during refrigerated storage are usually gram-negative psychrophiles. The time required for this bacterial population shift (from mesophiles to psychrophiles) after refrigeration may account for the increased shelf life.

The major industrial fish used for fish meal and oil is menhaden from the Atlantic and Gulf coasts. Also, fish parts not used for human consumption are often used to manufacture fish meal and oil.

Fish meal and oil are the principal components of feed used in the aquaculture of trout and salmon, and is a dietary component for poultry and pigs. Fish oil is used in margarine, in paints, and in the tanning industry. It is also refined for pharmaceutical purposes.

This chapter covers preservation and processing of fresh and frozen fishery products; handling of fresh fish aboard vessels and ashore; the technology of freezing fish; and present commercial trends in freezing, frozen storage, and distribution of seafood.

See Chapter 27 for additional information regarding fishery products for precooked and prepared foods, and Chapter 31 for more on marine refrigeration.

HACCP System. Many procedures for control of microorganisms are managed by the Hazard Analysis and Critical Control Point (HACCP) system of food safety. Each food manufacturing site should have a HACCP team to develop and implement its HACCP plan. See Chapter 12 for additional information on sanitation.

The preparation of this chapter is assigned to TC 10.9, Refrigeration Application for Foods and Beverages.

FRESH FISHERY PRODUCTS

CARE ABOARD VESSELS

After fish are brought aboard a vessel, they must be promptly and properly handled to ensure maximum quality. Trawl-caught fish on the New England and Canadian Atlantic coasts, such as haddock and cod, are usually eviscerated, washed, and then iced down in the pens of the vessel's hold. Canadian (offshore), Icelandic, U.K., and other European fleets ice fish in boxes for optimum quality. Because of their small size, other groundfish (e.g., ocean perch, whiting, flounder) are not eviscerated and are not always washed. Instead, they are iced down directly in the hold of the vessel.

Crustaceans, such as lobsters and many species of crabs, are usually kept alive on the vessel without refrigeration. Warm-water shrimp are beheaded, washed, and stored in ice in the hold; on some vessels, however, the catch is frozen either in refrigerated brine or in plate freezers. Cold-water shrimp are stored whole in ice or in chilled sea water, or they may be cooked in brine, chilled, and stored in containers surrounded with ice.

Freshwater fish in the Great Lakes and Mississippi River areas are caught in trap nets, haul seines, or gill nets. They are sorted according to species into 50 or 100 lb boxes, which are kept on the deck of the vessel. In most cases, fishing vessels carry ice aboard, and fish are landed the day they are caught.

Freshwater fish in Canadian lakes are iced down in the summertime and stored at collecting stations on the lakes, where they are picked up by a collecting boat with a refrigerated hold. Winter-caught Canadian freshwater and Arctic saltwater fish are usually weather-frozen on the ice immediately after catching and are marketed as frozen fish.

Line-caught fish of the Pacific Northwest, such as halibut caught largely by bottom long-line gear and salmon caught by trolling gear, are eviscerated, washed, and iced in the pens of the vessel. Pacific salmon caught by seines and gill nets for cannery use are usually stored whole for several days, either aboard vessels or ashore in tanks of seawater refrigerated to 30°F. A small but significant volume of halibut is held similarly in refrigerated seawater aboard vessels. Tuna caught offshore by seiners or clipper vessels are usually brine-frozen at sea. However, tuna caught inshore by smaller trollers or seiners are often iced in the round or refrigerated with a brine spray.

Fish raised by aquaculture farms are usually harvested and sold as required by the fresh fish market. They are usually shipped in containers in which they are surrounded by ice.

Icing

Fish lose quality because of bacterial or enzymatic activity or both. Reducing storage temperature retards these activities significantly, thus delaying spoilage and autolytic deterioration.

Low temperatures are particularly effective in delaying growth of psychrophilic bacteria, which are primarily responsible for spoilage of nonfatty fish. The shelf life of species such as haddock and cod is doubled for each 7 to 10°F decrease in storage temperature within the range of 60 to 30°F.

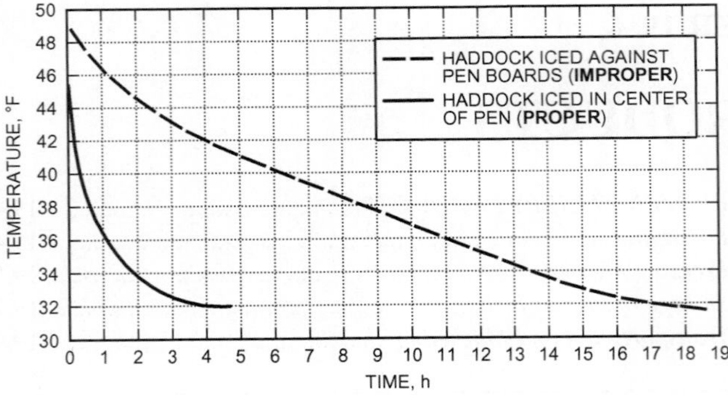

Fig. 1 Cooling Rate of Properly and Improperly Iced Haddock

To be effective, ice must be clean when used. Bacteriological tests on ice in the hold of a fishing vessel showed bacterial counts as high as 5 billion per gram of ice. These results indicate that (1) chlorinated or potable water should be used to make the ice at the plant, (2) ice should be stored under sanitary conditions, and (3) unused ice should be discarded at the end of each trip.

Both flake and crushed block ice are used aboard fishing vessels, although flake ice is more common because it is cheaper to produce and easier to handle mechanically.

The amount of ice used aboard vessels varies with the particular fishery and vessel; however, it is essential to provide enough ice around the fish to obtain a proper cooling rate (Figure 1). A common ratio of ice to fish used in bulk icing on New England trawlers is one part ice to three parts fish. Experiments on British trawlers in boxing fish at sea with one part ice to two parts fish demonstrated improved quality in the landed fish, and, as ice has become more plentiful and less costly relative to the value of fish, the ratio of ice to fish continues to increase. Some vessels use mechanical refrigeration to retard ice melting en route to the fishing grounds; however, the hold temperature must be controlled after fish are taken to allow the ice to melt for effective cooling of the fish.

Saltwater Icing

Iced fish storage temperatures must be maintained close to the freezing point of fish. To obtain lower ice temperatures, the freezing point may be depressed by adding salt to the water from which ice is made. Adequate amounts of ice made from a 3% solution of sodium chloride brine maintain a storage environment of about 30°F. Tests conducted on haddock storage in saltwater ice aboard a fishing vessel showed that, under parallel conditions, fish iced with saltwater ice cooled more quickly and to a lower temperature than fish iced with plain ice. However, the saltwater ice melted more quickly because of its lower latent heat and greater temperature differential. Therefore, once the saltwater ice melted, fish stored in this ice rose to a higher temperature than those in plain ice. Because it is not always possible to replenish ice on fish at sea, sufficient quantities of saltwater ice must be used initially to make up for its faster melting rate.

In making ice from water containing a preservative, rapid freezing and/or using a stabilizing dispersant is essential to prevent migration of the additive to the center of the ice block. This problem is not encountered in flake ice because flake ice machines freeze water rapidly into thin layers of ice, thus fixing additives within the flakes. Chapter 34 describes the manufacture of flake ice in more detail.

Use of Preservatives

In the United States and Canada, the use of antibiotics in ice or in dips for treatment of whole or gutted fish, shucked scallops, and unpeeled shrimp is prohibited by regulation.

Storage of Fish in Refrigerated Seawater

Refrigerated seawater (RSW) is used commercially for preserving fish. On the Pacific Coast, substantial quantities of net-caught salmon are stored in RSW aboard barges and cannery tenders for delivery to the canneries. Most salmon seiners now use RSW systems. It is often a condition of sale. On the East and Gulf coasts, RSW installations on fishing vessels are used for chilling and holding menhaden and industrial species needed for production of meal, oil, and pet food. On the east and west coasts of Canada, RSW installations are used for chilling and holding herring and capelin, which are processed on shore for their roe. Other, more limited applications of RSW include holding Pacific halibut and Gulf shrimp aboard a vessel; chilling and holding Maine sardines in shore tanks for canning, and short-term holding of Pacific groundfish in shore tanks for later filleting.

With groundfish and shrimp, RSW works well for short-term storage (2 to 4 days), but is not suitable for longer periods because excessive salt uptake, accelerated rancidity, poorer texture, and increased bacterial spoilage may result. These problems can be partially overcome by introducing carbon dioxide (CO_2) gas into the RSW; holding in RSW saturated with CO_2 can increase the storage life of some species of fish by about 1 week. Additionally, RSW reduces (1) handling that results from bulk storage of the fish and (2) pressure on the fish as a result of buoyancy, faster cooling, and lower storage temperature.

In many RSW systems, refrigeration is provided by ammonia flowing through external chillers (which gives the best results) or pipe coils in the tanks.

Boxing at Sea

There are many advantages to using containers or boxes instead of bulk storage aboard fishing vessels. Using containers reduces pressure on fish stowed in a vessel's hold. Because significant reductions in handling during and after unloading are possible, mechanical damage and product temperature rise may be virtually eliminated, and handling costs may be reduced. Fish can be sorted into boxes by size and species as soon as they are caught. Boxed fish lend themselves more readily to mechanized handling, such as machine filleting, because they are generally firmer and of more uniform shape; fillet yields are generally better than they are with bulk-stored fish.

Boxing at sea is not generally practiced in the United States, except by some inshore vessels. The principal problems with converting a fishing vessel from bulk to boxed storage are increased labor required by the crew for handling the boxes, reduced hold capacity, and relatively large investment for boxes. Many fisheries have difficulties working out the logistics for ensuring prompt return of properly cleaned boxes to the vessel. Most of these problems have been solved in European, Canadian (offshore), and South American (hake) fleets. Using nonreturnable containers for boxing at sea simplifies logistics and reduces initial capital outlay; it has proved justifiable in some U.S. fisheries.

Reusable containers for boxing at sea are usually made of plastic. Careful icing is necessary to minimize the surface area of fish in contact with the box. Plastic provides more heat transfer resistance than aluminum in vessels with uninsulated fish holds and for in-plant storage prior to processing.

All fish boxes must be equipped with drains, preferably directed outside the boxes on the bottom of a stack.

SHORE PLANT PROCEDURE AND MARKETING

Proper use of ice and adherence to good sanitary practices ensure maintenance of iced fish freshness during unloading from the vessel, at the shore plant, during processing, and throughout the distribution chain. Fish landed in good quality spoil rapidly if these practices are not carried out.

Table 1 Organoleptic Quality Criteria for Fish

Factor	Good Quality	Poor Quality
Eyes	Bright, transparent, often protruding	Cloudy, often pink, sunken
Odor	Sweet, fishy, similar to seaweed	Stale, sour, presence of sulfides, amines
Color	Bright, characteristic of species, sometimes pearlescent at correct light angles	Faded, dull
Texture	Firm, may be in rigor, elastic to finger pressure	Soft, flabby, little resilience, presence of fluid
Belly	Walls intact, vent pink, normal shape	Often ruptured, bloated, vent brown, protruding
Organs (including gills)	Intact, bright, easily recognizable	Soft to liquid, gray homogeneous mass
Muscle tissue	White or characteristic of species and type	White flesh pink to gray, spreading of blood color around backbone

Fish unloaded from the vessel are usually graded by the buyer for species, size, and minimum quality specification. A price is based in part on the quality in relation to market requirements. Fish also may be inspected by local and federal regulatory agencies for wholesomeness and sanitary condition. Organoleptic criteria are most important for evaluating quality; however, there is a growing acceptance, particularly in Canada and some European countries, of objective chemical and physical tests as indexes of quality loss or spoilage. Organoleptic (sensory) quality criteria vary somewhat among species, but the information in Table 1 can be used as a general guide in judging the quality of whole fish.

In New England and the Canadian Atlantic provinces, groundfish may be placed in boxes and trucked to the shore plant or conveyed directly from the hold or deck to the shore plant. Single- or double-wall insulated boxes are normally used; wooden boxes are rarely used because they are a source of microbiological contamination. Ice should be applied generously to each box of fish, even if the period before processing is only a few hours. Fish awaiting processing for more than a few hours should be iced heavily and stored in insulated containers or in single-wall boxes in a chill room refrigerated to 35°F. If refrigerated facilities are not available, boxes of fish should be kept in a cool section of the plant that is clean and sanitary and has adequate drainage.

Large boxes of resin-coated plywood or reinforced fiberglass that hold up to 1000 lb of fish and ice are used by some plants in preference to icing fish overnight on the floor. These **tote boxes** are moved and stacked by forklift, can be used for trucking fish to other plants, and make better use of plant floor space. Generally, fish awaiting processing should not be kept longer than overnight.

Fresh fish are marketed in different forms: fillets, whole fish, dressed-head on, dressed-headed (head removed), and, in some instances, steaks. The method of preparing fish for marketing depends largely on the species of fish and on consumer preference. For example, groundfish such as cod and haddock are usually marketed as fillets or as dressed-headed fish. Freshwater fish such as catfish and bullheads are usually dressed and skinned; lake trout are not skinned, but are merely dressed; and lake herring are marketed in dressed, round, or filleted form.

PACKAGING FRESH FISH

Most fresh fish is packaged in institutional containers of 5 to 35 lb capacity at the point of processing. Polyethylene trays, steel cans, aluminum trays, plastic-coated solid boxes, wax-impregnated corrugated fiberboard boxes, foamed polystyrene boxes, and polyethylene bags are used.

Fresh fish is often packaged while it still contains process heat from wash water. In these cases, it is advantageous to use a packaging material that is a good heat conductor. The fresh fish industry makes little use of controlled prechilling equipment in packaging. As a result, product temperatures may never reach the optimum level after packaging. Traditionally, institutional fresh fish travels packed in wet ice; in this case, it may cool to the proper level in transit even if process heat is initially present. However, there is a trend toward using leaktight shipping containers for fresh fish because modern transportation equipment is not designed to handle wet shipments. Also, some customers want to avoid the cost of transporting ice yet demand a product that is uniformly chilled to 32 to 36°F when it reaches their door. Shippers who use leaktight shipping containers have to upgrade their product temperature control systems to ensure that the fish reaches ice temperature before packaging. Rapid prechilling systems that result in crust freezing can be applied to some fresh seafood products, but this practice must be used with discretion because partial freezing harms quality.

Some general requirements for institutional containers that hold products such as fillets, steaks, and shucked shellfish are (1) sufficient rigidity to prevent pressure on the product, even when containers are stacked or heavily covered with ice; and (2) measures to prevent ice-melt water from contaminating the product. Some containers have drains to allow drip from the fish itself to run off. Others are sealed and may be gastight, which increases shelf life. One problem associated with sealed containers is a strong odor when the package is first opened. Although this odor may be foul, it soon dissipates and has no adverse effect on quality. Dressed or whole fish may be placed in direct contact with ice in a gastight container.

Leaktight shipping containers are used with nonrefrigerated transportation systems, such as air freight, and consequently require insulation. Foamed polystyrene is particularly suitable. For typical air freight shipments, the most economical thickness of insulation is between 1 and 2 in. To maintain product temperature in transit, shippers use either dry ice, packaged wet ice, packaged gel refrigerant, or wet ice with absorbent padding in the bottom of the container. Foamed polystyrene containers may be of molded construction or of the composite type, in which foam inserts and a plastic liner are used with a corrugated fiberboard box.

At the retail level, fresh fish may be handled in two ways. Stores with service counters display fish in unpackaged form. However, markets without service counters sometimes package fish before displaying for sale. Both types of outlets receive product in institutional containers. If fish is prepackaged at the market, labor and packaging costs may be high, and product temperature is likely to rise. Often, relatively warm fish is placed in a foam tray, wrapped, and displayed in a meat case at 40°F or more. This drastically reduces shelf life of the fish. Centralized prepackaging at the point of initial processing appears to have many important advantages over the present system. A number of retail chains have suppliers prepackage product under controlled temperature and sanitary conditions.

FRESH FISH STORAGE

The maximum storage life of fish varies with the species. In general, the storage life of East and West Coast fish, properly iced and stored in refrigerated rooms at 35°F, is 10 to 15 days, depending on its condition when unloaded from the boat. Generally, freshwater fish properly iced in boxes and stored in refrigerated rooms may be held for only 7 days. Both of these time limitations refer to the period between landing/processing and consumption.

Table 2 Optimal Radiation Dose Levels and Shelf Life at 33°F for Some Species of Fish and Shellfish

Species	Optimal Radiation Dose, Rads Air Packed	Shelf Life, Weeks
Oysters, shucked, raw	200,000	3 to 4
Shrimp	150,000	4
Smoked chub	100,000	6
Yellow perch	300,000	4
Petrale sole	200,000	2 to 3 (4 to 5 when vac pac)
Pacific halibut	200,000	2 (4 when vac pac)
King crabmeat	200,000	4 to 6
Dungeness crabmeat	200,000	3 to 6
English sole	200,000 to 300,000	4 to 5
Soft-shell clam meat	450,000	4
Haddock	150,000 to 250,000	3 to 4
Pollock	150,000	4
Cod	150,000	4 to 5
Ocean perch	250,000	4
Mackerel	250,000	4 to 5
Lobster meat	150,000	4

Cold-storage facilities for fresh fish should be maintained at about 35°F with over 90% rh. Air velocity should be limited to control ice loss. Temperatures less than 32°F retard ice melting and can result in excessive fish temperatures. This is particularly important when storing round fish such as herring, which generate heat from autolytic processes.

Floors should have adequate drainage with ample slopes toward drains. All inside surfaces of a cold storage room should be easy to clean and able to withstand corrosive effects of frequent washings with antimicrobial compounds.

Irradiation of Fresh Seafood

Ionizing radiation can double or triple the normal shelf life of refrigerated, unfrozen fish and shellfish stored at 33°F (Table 2). No off-odors, adverse nutritional effects, or other changes are imparted to the product by the radiation treatment. However, irradiation of fish is still not common and is not permitted in some jurisdictions.

Modified-Atmosphere (MA) Packaging

A product environment with modified levels of nitrogen, CO_2, and oxygen can curtail bacterial growth and extend shelf life of fresh fish. For example, whole haddock stored in a 25% CO_2 atmosphere from the time it is caught keeps about twice as long as it would in air. However, a modified atmosphere does not inhibit all microbes, and spoilage bacteria, because of their great number, usually restrict growth of the few pathogenic bacteria present. Traditionally, the obvious signs of spoilage serve as the safeguard against eating fish that may have dangerous levels of pathogenic bacteria.

Because modified-atmosphere packaging can be a safety hazard, it is being introduced slowly in several countries under close monitoring by regulatory agencies. This type of packaging requires complete knowledge of regulations and a good control system that maintains proper temperature and sanitation levels.

FROZEN FISHERY PRODUCTS

The production of frozen fishery products varies with geographical location and includes primarily the production of groundfish fillets, scallops, breaded precooked fish sticks, breaded raw fish portions, fish roe, and bait and animal food in northeastern states and in Atlantic Canada; round or dressed halibut and salmon, halibut and salmon steaks, groundfish fillets, surimi, herring roe, and bait and animal food in northwestern states and in British Columbia; halibut, groundfish fillets, crab, salmon, and surimi in Alaska (salmon roe in Alaska is called "ikura"); shrimp, oysters, crabs, and other shellfish and crustaceans in the Gulf of Mexico and southern Atlantic states; and round or dressed fish in the areas bordering on the Great Lakes.

Fish from these areas differ considerably in both physical and chemical composition. For example, cod or haddock are readily adaptable to freezing and have a comparatively long storage life, but other fatty species, such as mackerel, tend to become rancid during frozen storage and therefore have a relatively short storage life. The differences in composition and marketing requirements of many species of fish require consideration of the specific product's quality maintenance and methods of packaging, freezing, cold storage, and handling.

Temperature is the most important factor limiting the storage life of frozen fish. Below freezing, bacterial activity as a cause of spoilage is limited. However, even fish frozen within a few hours of catching and stored at –20°F very slowly deteriorates until it becomes unattractive and unpleasant to eat.

Fish proteins are permanently altered during freezing and cold storage. This denaturation occurs quickly at temperatures not far below freezing; even at 0°F, fish deteriorates rapidly. Badly stored fish is easily recognized: the thawed product is opaque, white, and dull, and juice is easily squeezed from it. Although properly stored product is firm and elastic, poorly stored fish is spongy, and in very bad cases, the flesh breaks up. Instead of the succulent curdiness of cooked fresh fish, cooked denatured samples have a wet and sloppy consistency at first and, on further chewing, become dry and fibrous.

Other factors that determine how quickly quality deteriorates in cold storage are initial quality and composition of the fish, protection of the fish from dehydration, freezing method, and environment during storage and transport. These factors are reflected in four principal phases of frozen fish production and handling: packaging, freezing, cold storage, and transportation.

Today, many species are brought from warm and tropical waters where parasites and toxins could infect them. In addition, food dishes that use raw seafood, such as sushi and sashimi, have gained wide popularity, making them a potential health risk. Parasites are not life-threatening but can cause pain and inconvenience. They are easily destroyed by cooking or by deep freezing (–40°F). Marine toxins could be deadly and are not affected by temperature. Susceptible species should not be eaten during periods when toxins could be developed.

PACKAGING

Materials for packaging frozen fish are similar to those for other frozen foods. A package should (1) be attractive and appeal to the consumer, (2) protect the product, (3) allow rapid, efficient freezing and easy handling, and (4) be cost-effective.

Package Considerations in Freezing

Refrigeration equipment and packaging materials are frequently purchased without considering the effect of package size on freezing rate and efficiency. For example, a thin consumer package has a faster rate of product freezing, lower total freezing cost, higher handling cost, and higher packaging material cost; a thicker institutional-type package has the opposite qualities.

Tests indicate that the time required to freeze packaged fish fillets in a plate freezer is directly proportional to the square of the package thickness. Thus, if it takes 3 h to freeze packaged fish fillets 2 in. thick, it takes about 4.7 h to freeze packaged fish fillets 2.5 in. thick. Insulating effects of packaging material, fit of the product in the package, and total package surface area must be considered. A packing material with low moisture-vapor permeability has an insulating effect, which increases freezing time and cost.

The rate of heat transfer through packaging is inversely proportional to its thickness; therefore, packaging material should be (1) thin enough to produce rapid freezing and an adequate moisture-vapor barrier in frozen storage and (2) thick enough to withstand heavy abuse. Aluminum foil cartons and packages offer an advantage in this regard.

Proper fit of package to product is essential; otherwise, the insulating effect of the air space formed reduces the product's freezing rate and increases freezing cost. The surface area of the package is also important because of its relation to the size of the freezer shelves or plates. Maximum use of freezer space can be obtained by designing the package so that it fits the freezer properly. Often, however, these factors cannot be changed and still meet customer requirements for a specific package.

Package Considerations for Frozen Storage

Fish products lose considerable moisture and become tough and fibrous during frozen storage unless a package with low moisture-vapor permeability is specified. The package in contact with the product must also be resistant to oils or moisture exuded from the product, or the oils will go rancid and the package material will soften. The package must fit the product tightly to minimize air spaces and thereby reduce moisture migration from the product to the inside surfaces of the package.

Unless temperatures are very low or special packaging is used, fish oils oxidize in frozen storage, producing an off-flavor. One effective approach is to replace the air surrounding the frozen fish with pure nitrogen and seal the fish in a leak-proof bag made of an oxygen-impervious material.

Types of Packages

Packaging consists of either paperboard cartons coated with various waterproofing materials or cartons laminated with moisture-vapor-resistant films and heat-sealable overwrapping materials with a low moisture-vapor permeability. Paperboard cartons are usually made of a bleached kraft stock, coated with a suitable fortified wax, polyethylene, or other plastic material.

Overwrapping materials should be highly resistant to moisture transmission, inexpensive, heat sealable, adaptable to machinery application, and attractive in appearance. Various types of hot-melt-coated waxed paper, cellophane, polyethylene, and aluminum foil are available in different forms and laminate combinations to best suit each product.

Consumer Packages. These usually hold less than 1 lb and are generally printed, bleached paperboard coated with wax or polyethylene and closed with adhesive. Fish sticks and portions, shrimp, scallops, crabmeat, and precooked dinners and entrees are packaged in this way. For dinners and entrees, rigid plastic, pressboard, or aluminum trays are used inside the printed paperboard package. Rigid plastic or pressboard packages are more common because they are better for microwave cooking. Packaging these products is normally mechanized.

Materials such as polyethylene combined with cellophane, polyvinylidene chloride, or polyester and combinations of other plastic materials are used with high-speed automatic packaging machines to package shrimp, dressed fish, fish fillets, fish portions, and fish steaks before freezing. In some instances, wrapping material has been torn by fins protruding from the fish, but otherwise, this method of packaging is satisfactory and offers considerable protection against dehydration and rancidity at a comparatively low cost. This packaging method has also created new markets for merchandising frozen fish products. Boil-in-bag pouches made of polyester-polyethylene and combinations of foil, polyethylene, and paper are used for packaging shrimp, fish fillets, and entrees. These packages are also suitable for microwave cooking.

Institutional Packages. The 5 lb and larger cartons used in the institutional trade are commonly constructed of bleached paperboard that has been waxed or polyethylene coated. Folding cartons with self-locking covers, full-telescoping covers, or glued closures are used. Often, cartons are packaged inside a corrugated master carton or are shrink-wrapped in polyethylene film.

Products such as fish fillets and steaks are individually wrapped in cellophane or another moisture-vapor-resistant film and then packed in the carton. Fish, such as headed and dressed whiting and scallop meats, are packed into the carton and covered with a sheet of cellophane. The cover is then put in place and the package is frozen upside down in the freezer. Raw, unbreaded products, such as shrimp, scallops, fillets, and steaks, are sometimes individually quick frozen (IQF) before packaging. IQF products can be glazed to enhance moisture retention. This method is preferred over freezing after packaging because it leads to a product that is more convenient to handle and sometimes obviates the need to thaw the fish before cooking.

For institutional frozen fish, the trend is toward printed paperboard folding cartons coated with moisture-vapor-resistant materials instead of waxed paper or cellophane overwrap, though "shatter pack" bulk is also common. Some frozen fish products and seafood entrees for institutional markets are packaged in aluminum or rigid plastic trays so they may be heated within the package.

FREEZING METHODS

Product characteristics, such as size and shape, freezing method, and rate of freezing, affect quality, appearance, and production cost.

Quick freezing offers the following advantages:

- Chills the product rapidly, preventing bacterial spoilage
- Facilitates rapid handling of large quantities of product
- Makes use of conveyors and automatic devices practical, thus materially reducing handling costs
- Promotes maximum use of the space occupied by the freezer
- Produces a packaged product of uniform appearance, with a minimum of voids or bulges

For further information, see Chapters 9, 10, and 16.

Blast Freezing

Blast freezers for fishery products are generally small rooms or tunnels in which cold air is circulated by one or more fans over an evaporator and around the product to be frozen, which is on racks or shelves. A refrigerant such as ammonia, a halocarbon, or brine flowing through a pipe coil evaporator furnishes the necessary refrigeration effect.

Static pressure in these rooms is considerable, and air velocities average between 500 and 1500 fpm, with 1200 fpm being common. Air velocities between 500 and 1000 fpm give the most economical freezing. Lower air velocities slow down product freezing, and higher velocities increase unit freezing costs considerably.

Some factories have blast freezers in which conveyors move fish continuously through a blast room or tunnel. These freezers are built in several configurations, including (1) a single pass through the tunnel, (2) multiple passes, (3) spiral belts, and (4) moving trays or carpets. The configuration and type of conveyor belt or freezing surface depend on the type and quantity of product to be frozen, space available to install the equipment, and capital and operating costs of the freezer.

Batch-loaded blast freezers are used for freezing shrimp, fish fillets, steaks, scallops, and breaded precooked products in institutional packages; round, dressed, and panned fish; and shrimp, clams, oysters, and salmon roe (ikura) packed in metal cans.

Conveyor blast freezers are widely used to freeze products before packaging. These products include all types of breaded, precooked seafoods; IQF fillets, loins, tails, steaks, scallops, and shrimp; and raw, breaded fish portions. In the case of portions, which are sliced or sawed from blocks, the function of the blast freezer is to harden the batter and breading before packaging and lower the temperature of the frozen fish for storage if it has been tempered for slicing.

Dehydration of product (freezer burn) may occur in freezing unpackaged whole or dressed fish in blast freezers unless the air velocity is kept to about 500 fpm and the period of exposure to the air is controlled. Consumer packages of fish fillets or fish-fillet blocks requiring close dimensional tolerances bulge and distort during freezing unless restrained. In blast rooms or tunnels, product can be frozen on specially designed trucks, enabling distribution of pressure on the surfaces of the package and remedying this condition. It is difficult to control product expansion on conveyor installations.

Freezing times for various sizes of packaged fishery products are shown in Figure 2.

Plate Freezing

In the multiplate freezer, refrigerant flows through connected passageways in horizontal movable plates stacked vertically in an insulated cabinet or room. The plate freezer is used extensively in freezing fishery products in consumer cartons and in 5 and 10 lb institutional cartons. Fish to be plate frozen should be properly packaged to minimize air spaces. Spacers should be used between the plates during freezing to prevent crushing or bulging of the package. For most products, spacer thickness should be about 0.03 to 0.06 in. less than that of the package.

Where very close package tolerances are required, as in the manufacture of fish fillet blocks, a metal frame or tray is used to hold packages during freezing. The frame or tray is generally the same width as the package and the length of one or two blocks. It must be rigid enough to prevent bulging and to hold the fish block's dimensions. This is sometimes done with rigid spacers that limit the tray's weight and cost.

Fish blocks are available in two common sizes: 16.5 lb (19 by 10 by 2.5 in.) and 18.5 lb (19 by 11.5 by 2.5 in.). Other blocks are sized for special applications. Fish can be packed in the block with the long dimension of the fillets along the length of the frame (long-pack) or along the width of the frame (cross-pack). The orientation depends on the eventual cutting pattern and type of cutting used to convert the block into a finished product.

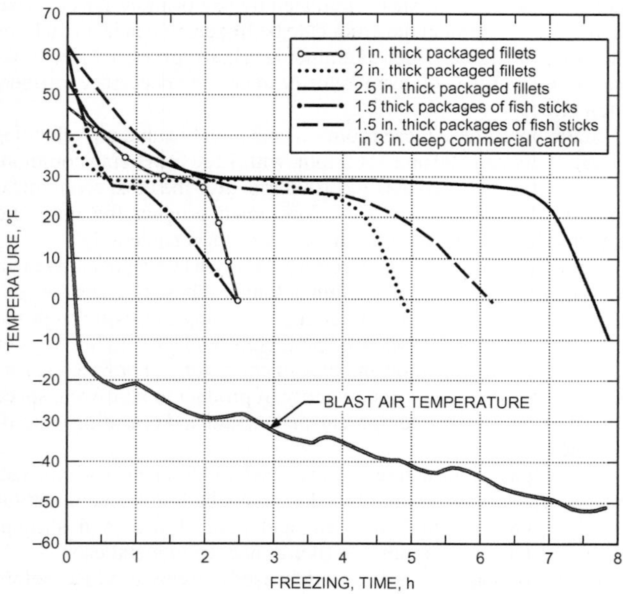

Fig. 2 Freezing Time of Fish Fillets and Fish Sticks in Tunnel Blast Freezer
Air Velocity 500 to 1000 fpm

A tray is not necessary for other packaged seafoods, such as shrimp, fillets, fish sticks, or scallops, where close package tolerances are not as essential. Therefore, an automatic continuous plate freezer with properly sized spacers is satisfactory for these products.

Plate freezers provide rapid and efficient freezing of packaged fish products. The freezing time and energy required for freezing packaged fish sticks is greater than that for fish fillets because heat transfer is slowed by the air space within the package. Energy required to freeze a unit mass of product increases with thickness. The freezing times of consumer and institutional size packages of fish fillets and fish sticks are shown in Figure 3.

Immersion Freezing

Immersion in low-temperature brine was one of the first methods used for quick-freezing fishery products. Numerous direct-immersion freezing machines were developed for whole or panned fish. These machines were generally unsuitable for packaged fish products, which make up the bulk of frozen fish production, and have been replaced by methods using air cooling, contact with refrigerated plates or shelves, and combinations of these methods.

Immersion freezing is used primarily for freezing tuna at sea and, to a lesser extent, for shrimp, salmon, and Dungeness crab, as well as king crab and Alaska snow crab (*C. opelio*). Extensive research has been conducted on brine freezing of groundfish aboard vessels, but this method is not in commercial use.

An important consideration is selection of a suitable freezing medium. The medium should be nontoxic, acceptable to public health regulatory agencies, easy to renew, and inexpensive; it should also have a low freezing temperature and viscosity. It is difficult to obtain a freezing medium that meets all these requirements. Sodium chloride brine and a mixture of glucose and salt in water are acceptable media. The glucose reduces salt penetration into the fish and provides a protective glaze.

Liquid nitrogen spray and CO_2 are coming into wider use for IQF seafood products such as shrimp. Although the cost per unit mass is high, fish frozen by these methods is of good quality, there is virtually no weight loss from dehydration, and there are space and equipment

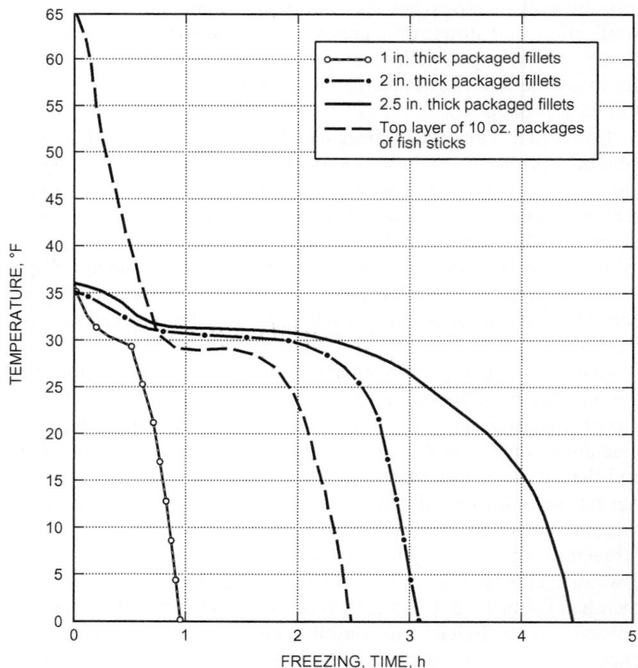

Fig. 3 Freezing Time of Fish Fillets and Fish Sticks in Plate Freezer

savings. Fish should not be directly immersed in the liquid nitrogen, because this will cause the flesh to shatter and rupture.

Immersion Freezing of Tuna. Most tuna harvested by the U.S. fleet is brine-frozen aboard the fishing vessel. Freezing at sea enables the vessel to make extended voyages and return to port with a full payload of high-quality fish.

Tuna are frozen in brine wells, which are lined with galvanized pipe coils on the inside. Direct expansion of ammonia into the evaporator coils provides the refrigeration effect. Wells are designed so that tuna can be precooled and washed with refrigerated seawater and then frozen in an added sodium chloride brine. After the fish are frozen, the brine is pumped overboard, and the tuna are kept in 10°F dry storage. Before unloading, the fish are thawed in 30°F brine. In some cases, the fish are thawed in tanks at the cannery. If the fish are thawed ashore, thawing on the vessel is not required beyond the stage needed to separate those fused together in the vessel's wells.

Sometimes tuna are held in the wells for a long time before freezing or are frozen very slowly because of high well temperatures caused by overloading, insufficient refrigeration capacity, or inadequate brine circulation. These practices have a detrimental effect on product quality, especially for smaller fish, which are more subject to salt penetration and quality changes. Tuna that are not promptly and properly frozen may undergo excessive changes, absorb excessive quantities of salt, and possibly be bacteriologically spoiled when landed. Some freezing times for tuna of various sizes are shown in Figure 4.

Specialized Contact Freezers. Fish frozen by this method are placed on a solid stainless steel belt that slowly moves the fillets through a tunnel, where they are frozen not only by air blast but also by direct contact between the conveyor belt and a thin layer of glycol pumped through the plates that support the belt. A refrigerant, such as ammonia or a halocarbon, also flows through separate channels in the plates. This provides the refrigeration effect with minimal temperature difference between the evaporating refrigerant and the product.

Freezing Fish at Sea

Freezing fish at sea has found increasing commercial application in leading fishery nations such as Japan, Russia, the United Kingdom, Norway, Spain, Portugal, Poland, Iceland, and the United States. Including freezer trawlers, factory ships, and refrigerated transports in fisheries, hundreds of large freezer vessels operate throughout the world. U.S. factory-freezer trawlers, factory surimi trawlers, and floating factory ships supplied by catcher vessels operate off Alaska, mainly processing Alaskan pollock, cod, and flounder.

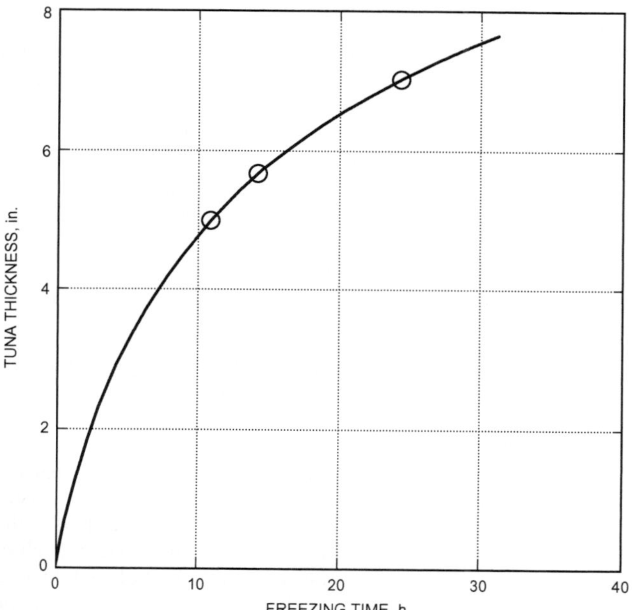

Fig. 4 Freezing Time for Tuna Immersed in Brine

Freezing groundfish at sea is uncommon in the northeastern United States, largely because fresh fish commands a better price than frozen fish. For the same reason, East Coast U.S. producers avoid putting their product into frozen packs if they can sell it fresh. Hence, much of the frozen fish used in the United States (except Alaskan fish) is imported.

Factory vessels are equipped to catch, process, and freeze fish at sea and to use the waste material to manufacture fish meal and oil. A large European factory vessel measures 280 ft in length, displaces 3700 tons, and is equipped to stay at sea for about 80 days without being refueled. About 65 to 100 people are required to operate the vessel and to process and handle the fish. Most vessels of this type use contact-plate freezers. The freezers can freeze about 30 tons of fish per day, and the total capacity of the frozen fish hold may be as high as 750 tons.

Because the factory trawler stays at sea for long periods, it can fully use its space for storing fish. However, because of limited available labor, frozen packs are generally of the less labor-intensive types.

The freezer trawler was designed to resolve the disadvantages associated with factory freezer vessels. It is smaller and equipped to freeze fish in bulk for later thawing and processing ashore. Freezer trawlers use vertical plate freezers to freeze dressed fish in blocks of about 100 lb.

Some countries use freezer trawlers to supply raw material to shore-based processing plants producing frozen fish products. This allows the trawlers to fill their holds in distant waters and transport the fish to home base, where it becomes frozen raw material that is held in storage until required for processing. In some cases, trawlers have been designed as dual fisheries, fishing and freezing groundfish blocks during part of the year and catching, processing, and freezing Northern shrimp for the rest of the year.

STORAGE OF FROZEN FISH

Fishery products may undergo undesirable changes in flavor, odor, appearance, and texture during frozen storage. These changes are attributable to dehydration (moisture loss) of the fish, oxidation of oils or pigments, and enzyme activity in the flesh. The rate at which these changes occur depends on the (1) composition of the species of fish, (2) level and constancy of storage room temperature and humidity, and (3) protection offered by suitable packaging materials and glazing compounds.

Composition

The composition of a particular species of fish affects its frozen storage life considerably. Fish with high oil content, such as some species of salmon, tuna, mackerel, and herring, have a comparatively short frozen storage life because of rancidity that results from oxidation of oils and pigments in the flesh. Certain fish, such as sablefish, are quite resistant to oxidative deterioration in frozen storage, despite their high oil content. Rancidity development is less pronounced in fish with a low oil content. Therefore, lean fish such as haddock and cod, if handled properly, can be kept in frozen storage for many months without serious loss of quality. The relative susceptibility of various species of fish to oxidative changes during frozen storage is shown in Table 3.

Storage Conditions

Temperature. Quality loss of frozen fish in storage depends primarily on temperature and duration of storage. Fish stored at –20°F has a shelf life of more than a year. In Canada, the Department of Fisheries recommends a storage temperature of –15°F or lower. Storage above –10°F, even for a short period, results in rapid loss of quality. Time/temperature tolerance studies show that frozen seafoods have memory; that is, each time they are subjected to high temperatures or poor handling practices, the loss in quality is recorded. When the product is finally thawed, the total effect of each mistreatment is reflected in product quality at the consumer level. Continuous storage at temperatures lower than –15°F reduces oxidation, dehydration, and enzymatic changes, resulting in longer product shelf life. From the time they are frozen until they reach the consumer, frozen seafoods should be kept as close to –15°F as possible. The shelf life of frozen fish products stored at different temperatures is given in Table 4. Note the increase in shelf life at the lowest temperatures.

For many years, it was thought too costly to operate refrigerated warehouses below –10°F. However, improvements in design and operation of refrigeration equipment have made such temperatures economically possible. Surimi production by West Coast-based factory ships has led to construction of ultracold storage rooms. Japanese standards call for this product to be kept at –22°F.

Table 3 Relative Susceptibility of Representative Species of Fish to Oxidative Changes in Frozen Storage

Severe	Moderate	Minor	Very Slight
Pink salmon	Chum salmon	Cod	Yellow pike
Rockfish	Coho salmon	Haddock	Yellow perch
Lake chub	King salmon	Flounder	Crab
Whiting	Halibut	Sole	Lobster
Red salmon	Ocean perch	Sablefish	
	Herring	Oysters	
	Mackerel		
	Tuna		
	Lake herring		
	Sheepshead		
	Lake trout		

Table 4 Effect of Storage Temperature on Shelf Life of Frozen Fishery Products

Product	Temperature, °F	Shelf Life, Months
Packaged haddock fillets	10	4 to 5
	0	11 to 12
	–20	Longer than 12
Packaged cod fillets	10	5
	0	6
	–10	10 to 11
Packaged pollock fillets*	20	1
	10	2
	0	8
	–10	11
	–20	24
Packaged ocean perch fillets	15	1.5 to 2
	10	3.5 to 4
	0	6 to 8
	–10	9 to 10
Packaged striped bass fillets	15	4
	0	9
Glazed whole halibut	10	3
	0	6
	–10	9
	–20	12
Whole bluefin tuna	10	4
	0 to –5	8
	–20	12
Glazed whole herring	0	6
	–17	9
Packaged mackerel fillets	15	2
	0	3
	–10	3 to 5

*Prepared from 1 day old iced fish.

Humidity. High relative humidity in cold-storage rooms tends to reduce evaporation of moisture from the product. The relative humidity of air in the refrigerated room is directly affected by the temperature difference between room cooling coils and room temperature. A large temperature difference decreases relative humidity and accelerates the rate of moisture withdrawal from the frozen product; a small temperature difference has the opposite effect.

Relative humidity in commercial cold storages is 10 to 20% higher than that of an empty cold storage because of constant evaporation of moisture from the product. In a cold storage operating at 0°F, with 70% rh and pipe coil temperature of –10°F, the moisture-vapor pressure of air in the package (in direct contact with the frozen fish) is 0.0185 psia; air in the cold storage is at a vapor pressure of 0.0132 psia, and the moisture-vapor pressure at the coils is 0.0108 psia. These differences in moisture-vapor pressure result in considerable product moisture loss unless it is adequately protected by suitable packaging materials or glazing compounds. The evaporator coils in the freezer should be sized properly so that the desired high relative humidities can be obtained. However, because of material costs and space limitations, a temperature difference of 10°F between evaporator coils and room air is the most practical.

Packaging and Glazing

Adequate packaging of fishery products is important in preventing product dehydration and consequent quality loss, as discussed in the Packaging section under Frozen Fishery Products. Individual fish, whether frozen in the round or dressed, cannot usually be suitably packaged; therefore, they must be protected by a glazing compound.

A glaze acts as a protective coating against the two main causes of deterioration during storage: dehydration and oxidation. It protects against dehydration by preventing moisture from leaving the product and against oxidation by mechanically preventing air contact with the product. It may also minimize these changes chemically with an antioxidant.

Storage life of fishery products can be maximized by using the following procedures:

- Select only high-quality fish for freezing.
- Use moisture-vapor-resistant packaging materials and fit package tightly around product, or use a modified atmosphere and oxygen-barrier package.
- Freeze fish immediately after processing or packaging.
- Glaze frozen fish before packaging.
- Glaze round, unpackaged fish before cold storage.
- Put fish in frozen storage immediately after freezing and glazing, if required.
- Store frozen fish at –15°F or lower.
- Renew glaze on round, unpackaged fish as required during frozen storage.

The recommended protection and expected storage life for various species of fish at 0°F are shown in Table 5.

Space Requirements

Packaged products such as fillets and steaks are usually packed in cardboard master cartons for storage and shipment. These master cartons are stacked on pallets and transferred to various areas of the cold-storage room by forklift. Master cartons are strong enough to support one or two pallet loads placed on the shelf of each rack in cold storage. In cold storages without racks, cartons should be stacked to a height that does not crush the bottom cartons. Cartons for products in packages that contain a lot of air, such as IQF fillets,

Table 5 Storage Conditions and Storage Life of Frozen Fish

Fish	Recommended Protection*	Storage Life (0°F), Months
Chub, pink salmon	Ice glazing and packaging	4 to 6
Mackerel, sea herring, pollock, chub, smelts	Ice glazing and packaging	5 to 9
Pacific sardines, tuna	Packaging	4 to 6
Buffalofish, flounder, halibut, ocean perch, rockfish, sablefish, red, sockeye, silver or coho salmon, whiting, shrimp	Packaging	7 to 12
Haddock, blue pike, cod, hake, lingcod	Packaging	Over 12

*All packaging should be with moisture-resistant films.

Table 6 Space Requirements for Frozen Fishery Products

Commodity	Product Package	Container for Storage	Space Required, lb/ft^3
Fish sticks, breaded shrimp, breaded scallops	8 or 10 oz	Corrugated master containers	25 to 30
Fillets, steaks, small dressed fish	1, 5, or 10 lb	Corrugated master containers	50 to 60
Shrimp	2.5 and 5 lb	Corrugated master containers	35
Panned, frozen fish (mackerel, herring, chub)	None	Wooden or fiberboard boxes	35
Round halibut	None	Wooden box	30 to 35
		Stacked loose	38
Round groundfish (cod, etc.)	None	Stacked loose	32
Round salmon	None	Stacked loose	33 to 35

must be stronger than those for solid packages of fish to resist crushing during storage.

Whole or dressed fish frozen in blocks in metal pans, such as mackerel, chub, or whiting, are removed from the pans after freezing, glazed, and then packaged in wooden boxes lined with wax-impregnated paper or in cardboard cartons.

Round fish stored in wooden boxes can be easily reglazed periodically during frozen storage. Space requirements for storing fishery products are shown in Table 6.

Thawing Frozen Fish. Frozen fishery products can be thawed by circulating air or water. Thawing fish should not be allowed to rise above refrigerated temperatures; otherwise, rapid deterioration may occur. Thawing is slower and more difficult than freezing when done to ensure quality maintenance. Each application should be carefully designed.

TRANSPORTATION AND MARKETING

Temperature and humidity conditions recommended for frozen storage should also be applied during transportation and marketing to minimize product quality loss. Shipment in nonrefrigerated or improperly refrigerated carriers, exposure to high ambient temperatures during transfer from one environment to another, improper loading of common carriers or display cases, equipment failure, and other poor practices lead to increased product temperature and, consequently, to quality loss.

Frozen fish is transported under mechanical refrigeration in trucks, railroad cars, or ships. Most of these vehicles can maintain temperatures of 0°F or lower. Additional information on equipment used in the transportation and marketing of frozen fish and other foods is given in Chapters 11, 13, 14, 30, 31, 32, and 46 of this volume and in Chapter 28 of the 1994 *ASHRAE Handbook—Refrigeration*.

To minimize quality loss during transportation and marketing, use the following procedures:

1. Transport frozen fish in refrigerated carriers (mechanical or dry ice systems) with ample capacity to maintain 0°F over long distances.
2. Precool refrigerated carriers to at least 10°F before loading.
3. Remove frozen products from the warehouse only when the carrier is ready to be loaded. Load directly into the refrigerated carrier; do not allow product to sit on the dock.
4. Check fish temperature with a thermometer before loading.
5. Do not stack frozen fish directly against floors or walls of the carrier. Provide floor and wall racks or strips to allow air circulation around the entire load.
6. Continuously record the refrigerated carrier temperature during transit. Use an alarm to warn of equipment failure.
7. Measure product temperature when it is removed from the common carrier at its destination.
8. If products are shipped in an insulated container, apply sufficient dry ice to maintain temperatures of 0°F or lower for the duration of the trip.
9. Maintain food delivery or breakup rooms at 0 to 10°F. Do not hold products in breakup rooms any longer than necessary.
10. When received at the retail store, place the product in a 0°F storage room immediately.
11. Hold display cases in retail stores at 0°F or lower.
12. Do not overload display cases, especially above the frost line.
13. Record display case temperature. Provide an alarm to warn of an excessive rise in temperature.
14. Because of the accelerated deterioration of frozen fish products in the distribution and retail chain, hold products in these areas for as short a period as possible.

BIBLIOGRAPHY

Barnett, H.J, R.W. Nelson, P.J. Hunter, S. Bauer, and H. Groninger. 1971. Studies on the use of carbon dioxide dissolved in refrigerated brine for the preservation of whole fish. *Fishery Bulletin* 69(2).

Bibek, R. 1996. *Fundamental food microbiology*. CRC, Boca Raton, FL.

Charm, S.E. and P. Moody. 1966. Bound water in haddock muscle *ASHRAE Journal* 8(4):39.

Dassow, J.A. and D.T. Miyauchi. 1965. Radiation preservation of fish and shellfish of the Northeast Pacific and Gulf of Mexico. *Radiation preservation of foods*, L.J. Ronsivalli, M.A. Steinberg, and H.L. Seagran, eds. National Academy of Science *Publication* 1273. Washington, D.C.

Feiger, E.A. and C.W. du Bois. 1952. Conditions affecting the quality of frozen shrimp. *Refrigerating Engineering* (September):225.

Holston, J. and S.R. Pottinger. 1954. Some factors affecting the sodium chloride content of haddock during brine freezing and water thawing. *Food Technology* 8(9):409.

Kader, A.A., ed. 1992. *Postharvest technology of horticultural crops*, 2nd ed. University of California, Division of Agriculture and Natural Resources.

NACMCF. 1992. Hazard Analysis and Critical Control Point System. National Advisory Committee on Microbiological Criteria for Foods. *International Journal of Food Microbiology* 16:1-23.

Nelson, R.W. 1963. Storage life of individually frozen Pacific oyster meats glazed with plain water or with solutions of ascorbic acid or corn syrup solids. *Commercial Fisheries Review* 25(4):1.

Peters, J.A. 1964. Time-temperature tolerance of frozen seafood. *ASHRAE Journal* 6(8):72.

Peters, J.A., E.H. Cohen, and F.J. King. 1963. Effect of chilled storage on the frozen storage life of whiting. *Food Technology* 17(6):109.

Peters, J.A. and J.W. Slavin. 1958. Comparative keeping quality, cooling rates, and storage temperatures of haddock held in fresh water ice and salt water ice. *Commercial Fisheries Review* 20(1):6.

Ronsivalli, L.J. and J.W. Slavin. 1965. Pasteurization of fishery products with gamma rays from a cobalt 60 source. *Commercial Fisheries Review* 27(10):1.

Stansby, M.E., ed. 1976. *Industrial fishery technology*, 2nd ed. Robert E. Krieger Publishing, Huntington, NY.

Tressler, D.K, W.B. van Arsdel, and M.J. Copley, eds. 1968. *The freezing preservation of foods*, 4th ed. AVI Publishing, Westport, CT.

Wagner, R.L, A.F. Bezanson, and J.A. Peters. 1969. Fresh fish shipments in the BCF insulated leakproof container. *Commercial Fisheries Review* 31(8 and 9):41.

CHAPTER 20

DAIRY PRODUCTS

RAW milk is either processed for beverage milks, creams, and related milk products for marketing, or is used for the manufacture of dairy products. Milk is defined in the U.S. *Code of Federal Regulations* and the *Grade A Pasteurized Milk Ordinance* (PMO). Milk products are defined in 21CFR131 to 135. Public Law 519 defines butter. Note that there are many nonstandard dairy-based products that may be processed and manufactured by the equipment described in this chapter. Dairy plant operations include receiving raw milk; purchase of equipment, supplies, and services; processing milk and milk products; manufacture of frozen dairy desserts, butter, cheeses, and cultured products; packaging; maintenance of equipment and other facilities; quality control; sales and distribution; engineering; and research.

Farm cooling tanks and most dairy processing equipment manufactured in the United States meet the requirements of the *3-A Sanitary Standards* (IAMFES). These standards set forth the minimum design criteria acceptable for composition and surface finishes of materials in contact with the product; construction features such as minimum inside radii; accessibility for inspection and manual cleaning; criteria for mechanical and chemical cleaning or sanitizing in place (CIP and SIP); insulation of nonrefrigerated holding and transport tanks; and other factors that may adversely affect product quality and safety or the ease of cleaning and sanitizing equipment. Also available is *3-A Accepted Practices*, which deals with construction, installation, operation, and testing of certain systems rather than individual items of equipment.

The *3-A Sanitary Standards* and *Accepted Practices* are developed by the 3-A Standards Committees, which are composed of conferees representing state and local sanitarians, the U.S. Public Health Service, dairy processors, and equipment manufacturers. Compliance with the *3-A Sanitary Standards* is voluntary, but manufacturers who comply and have authorization from the 3-A Symbol Council may affix to their equipment a plate bearing the 3-A Symbol, which indicates to regulatory inspectors and purchasers that the equipment meets the pertinent sanitary standards.

MILK PRODUCTION AND PROCESSING

Handling Milk at the Dairy

Most dairy farms have bulk tanks to receive, cool, and hold milk. Tank capacity ranges from 200 to 5000 gal, with a few larger tanks. As cows are mechanically milked, the milk flows through sanitary pipelines to an insulated stainless steel bulk tank. An electric-motor-driven mechanical agitator stirs the milk, and mechanical refrigeration begins to cool it even during milking.

The *Pasteurized Milk Ordinance* (PMO) requires a tank to have sufficient refrigerated surface at the first milking to cool to 50°F or less within 4 h of the start of the first milking and to 45°F or less within 2 h after completion of milking. During subsequent milkings, there must be enough refrigerating capacity to prevent the temperature of the blended milk from rising above 50°F. The nameplate must state the maximum rate at which milk may be added and still meet the cooling requirements of the *3-A Sanitary Standards*.

Automatic controls maintain the desired temperature within a preset range in conjunction with agitation. Some dairies continuously record temperatures in the tank, a practice required by the PMO for bulk milk tanks manufactured after January 1, 2000. Because milk is picked up from the farm tank daily or every other day, milk from the additional milkings generally flows into the reservoir cooled from the previous one. Some large dairy farms may use a plate or tubular heat exchanger for rapid cooling. Cooled milk may be stored in an insulated silo tank (a vertical cylinder 10 ft or more in height).

Milk in the farm tank is pumped into a stainless steel tank on a truck for delivery to the dairy plant or receiving station. The tanks are well insulated to alleviate the need for refrigeration during transportation. Temperature rise when testing the tank full of water should not be more than 2°F in 18 h, when the average temperature difference between the water and the atmosphere surrounding the tank is 30°F.

The most common grades of raw milk are Grade A and Manufacturing Grade. Grade A raw milk is used for market milk and related products such as cream. Surplus Grade A milk is used for ice cream or manufactured products. To produce Grade A milk, the dairy farmer must meet state and federal standards; a few municipal governments also have raw milk regulations.

For raw milk produced under the provisions of the Grade A PMO recommended by the U.S. Public Health Service, the dairy farmer must have healthy cows and adequate facilities (barn, milkhouse, and equipment), maintain satisfactory sanitation of these facilities, and have milk with a bacteria count of less than 100,000 per mL for individual producers. Commingled raw milk cannot have more than 300,000 counts per mL. The milk should not contain pesticides, antibiotics, sanitizers, and so forth. However, current methods detect even minute traces of these prohibited substances, and total purity is difficult. Current regulators require no positive results on drug residue. Milk should be free of objectionable flavors and odors.

Receiving and Storing Milk

A milk processing plant receives, standardizes, processes, packages, and merchandises milk products that are safe and nutritious for human consumption. Most dairy plants either receive raw milk in bulk from a producer or arrange for pickup directly from dairy farms. The milk level in a farm tank is measured with a dipstick or a direct-reading gage, and the volume is converted to weight. Fat test and weight are common measures used to base payment to the farmer. A few organizations and the state of California include the percent of nonfat solids and protein content.

Plants can determine the amount of milk received by (1) weighing the tanker, (2) metering milk while pumping from the tanker to

The preparation of this chapter is assigned to TC 10.9, Refrigeration Application for Foods and Beverages.

a storage tank, or (3) using load cells on the storage tank or other methods associated with the amount in the storage tank.

Milk is generally received more rapidly than it is processed, so ample storage capacity is needed. A holdover supply of raw milk at the plant may be needed for start-up before arrival of the first tankers in the morning. Storage may also be required for nonprocessing days and emergencies. Storage tanks vary in size from 1000 to 60,000 gal. The tanks have a stainless steel lining and are well insulated.

The *3-A Sanitary Standards* for silo-type storage tanks specify that the insulating material should be of a nature and an amount sufficient to prevent freezing during winter in colder climates, or an average 18 h temperature change of no more than 3°F in the tank filled with water when the average temperature differential between the water and the surrounding air is 30°F. Inside tanks should have a minimum insulation R-value of 8, whereas partially or wholly outside tanks have a minimum R-value of 12. R-value units are $ft^2 \cdot °F \cdot h/Btu$. For horizontal storage tanks, the allowable temperature change under the same conditions is 2°F.

Agitation is essential to maintain uniform milkfat distribution. Milk held in large tanks, such as the silo type, is continuously agitated with a slow-speed propeller driven by a gearhead electric motor or with filtered compressed air. The tank may or may not have refrigeration, depending on the temperature of the milk flowing into it and the maximum holding time.

If refrigeration is provided for milk in a storage tank, it may be by use of a refrigerated jacket around the interior lining of the silo or tank. This cooling surface may be an annular space from a plate welded to the outside of the lining for direct refrigerant cooling or circulation of chilled water or a water/propylene glycol solution. Another system provides a distributing pipe at the top for chilled liquid to flow down the lining and drain from the bottom. Some plants pass milk through a plate cooler (heat exchanger) to keep all milk directed into the storage tanks at 40°F or less. Direct refrigerant cooling must be carefully applied to prevent milk from freezing on the lining. This limits the evaporator temperature to approximately 25 to 28°F.

Separation and Clarification

Before pasteurizing, milk and cream are standardized and blended to control the milkfat content within legal and practical limits. Nonfat solids may also need to be adjusted for some products; some states require added nonfat solids, especially for lowfat milk such as 2% (fat) milk. Table 1 shows the approximate legal milkfat and nonfat solids requirements for milks and creams in the United States.

One means of obtaining the desired fat standard is by separating a portion of the milk. The required amount of cream or skim milk is returned to the milk to control the final desired fat content. Milk with excessive fat content may be processed through a standardizer-clarifier that removes fat to a predetermined percentage (0.1 to 2.0%) and clarifies it at the same time. To increase the nonfat solids, condensed skim milk or low-heat nonfat dry milk may be added.

Milk separators are enclosed and fed with a pump. Separators designed to separate cold milk, usually not below 40°F, have increased capacity and efficiency as milk temperature increases. Capacity of a separator is doubled as milk temperature rises from 40 to 90°F. The efficiency of fat removal with a cold milk separator decreases as temperature decreases below 40°F. The maximum efficiency for fat removal is attained at approximately 45 to 50°F or above. Milk is usually separated at 70 to 90°F, but not above 100°F in warm milk separators. If raw, warmed milk or cream is to be held for more than 20 min before pasteurizing, it should be immediately recooled to 40°F or below after separation.

The pump supplying milk to the separator should be adjusted to supply milk at the desired rate without causing a partial churning action.

An automated process uses a meter-based system that controls the separation, fat and/or nonfat solids content, and ingredient addition for a variety of common products. If the initial fat tests fed into the computer are correct, the accuracy of the fat content of the standardized product is ±0.01%.

At an early stage between receiving and before pasteurizing, the milk or resulting skim milk and cream should be filtered or clarified, optimally during the transfer from the pickup tanker into the plant equipment. A clarifier removes extraneous matter and leucocytes, thus improving the appearance of homogenized milks.

Table 1 U.S. Requirements for Milkfat and Nonfat Solids in Milks and Creams

	Legal Minimum					
	Milkfat, %			Nonfat Solids, %		
Product	Federal	Range	Most Often	Federal	Range	Most Often
Whole milk	3.25	3.0 to 3.8	3.25	8.25	8.0 to 8.7	8.25
Lowfat milk	0.5	0.5 to 2.0	2.0	8.25	8.25 to 10.0	8.25
Skim milk	0.5*	0.1 to 0.5	0.5*	8.25	8.25 to 9.0	8.25
Flavored milk	—	2.8 to 3.8	3.25	8.25	7.5 to 10.0	8.25
Half-and-half	10.5	10.0 to 18.0*	10.5	—	—	—
Light (coffee) cream	18.0	16.0 to 30.0*	18.0	—	—	—
Light whipping cream	30.0	30.0 to 36.0*	30.0	—	—	—
Heavy cream	36.0	36.0 to 40.0	36.0	—	—	—
Sour cream	18.0	14.4 to 20.0	18.0	—	—	—

*Maximum

Pasteurization and Homogenization

There are two systems of pasteurization: batch and continuous. The minimum feasible processing rate for continuous systems is about 2000 lb/h. Therefore, batch pasteurization is used for relatively small quantities of liquid milk products. The product is heated in a stainless steel-lined vat to not less than 145°F and held at that temperature or above for not less than 30 min. The Grade A PMO requires that batch or vat pasteurizers keep the vapor space above liquid product at a temperature at least 5°F higher than the minimum required temperature of pasteurization during the holding period. Pasteurizing vats are heated with hot water or steam vapor in contact with the outer surface of the lining. One heating method consists of spraying heated water around the top of the lining. It flows to the bottom, where it drains into a sump, is reheated by steam injection, and returns to the spray distributor. Steam-regulating valves control the hot-water temperature. The maximum temperature difference between the milk or milk product throughout the vat during its holding period must not exceed 1°F. Therefore, the vat must have adequate agitation throughout the holding period. Whole and lowfat milk, half-and-half, and coffee cream are cooled, usually in the vat, to 130°F and then homogenized. Cooling is continued in a heat exchanger (e.g., a plate or tubular unit) to 40°F or lower and then packaged.

Plate coolers may have two sections, one using plant water and the second using chilled water or propylene glycol. The temperature of the product leaving the cooler depends on the flow rates and temperature of the cooling medium.

Most pasteurizing vats are constructed and installed so that the plant's cold water is used for initial product cooling after pasteurization. For final vat cooling, refrigerated water or propylene glycol is recirculated through the jacket of the vat to attain a product temperature of 40°F or less. Cooling time to 40°F should be less than 1 h.

High-temperature short-time (HTST) pasteurization is a continuous process in which milk is heated to at least 161°F and held at this temperature for at least 15 s. The complete pasteurizing system usually consists of a series of heat exchanger plates contained in a press, a milk balance tank, one or more milk pumps, a holding tube, flow diversion valve, automatic controls, and sources of hot water or steam and chilled water or propylene glycol for heating and cooling the milk, respectively. Homogenizers are used in many HTST systems as timing pumps used to process Grade A products. The heat exchanger plates are arranged so that milk to be heated or cooled flows between two plates, and the heat exchange medium flows in the opposite direction between alternate pairs of plates.

Ports in the plates are arranged to direct the flow where desired, and gaskets are arranged so that any leakage will be from the product to the heating or cooling media, to minimize potential for product contamination. Terminal plates are inserted to divide the press into three sections (heating, regenerating, and cooling) and arranged with ports for inlet and outlet of milk, hot water, or steam for heating, and chilled water or propylene glycol for cooling. To provide a sufficient heat-exchange surface for the temperature change desired in a section, milk flow is arranged for several passes through each section. The capacity of the pasteurizer can be increased by arranging several streams for each pass made by the milk. The capacity range of a complete HTST pasteurizer is 100 to about 100,000 lb/h. A few shell-and-tube and triple-tube HTST units are in use, but the plate type is by far the most prevalent.

Figure 1 shows one example of a flow diagram for an HTST plate pasteurizing system. Raw product is first introduced into a constant-level (or balance) tank from a storage tank or receiving line by either gravity or a pump. A uniform level is maintained in this tank by a float-operated valve or similar device. A booster pump is often used to direct flow through the regeneration section. The product may be clarified and/or homogenized or directly pumped to the heating section by a timing pump. From the heating section, the product continues through a holding tube to the flow diversion valve. If the product is at or above the preset temperature, it passes back through the opposite sides of the plates in the regeneration section and then through the final cooling section. The flow diversion valve is set at 161°F or above; if the product is below this minimum temperature, it is diverted back into the balance tank for repasteurization. Heat exchange in the regeneration section causes cold raw milk to be heated by hot pasteurized milk going downstream from the heater section and flow diversion valve. According to the PMO, the pasteurized milk pressure must be maintained at least 1 psi above the raw. The flow rate and temperature change are about the same for both products.

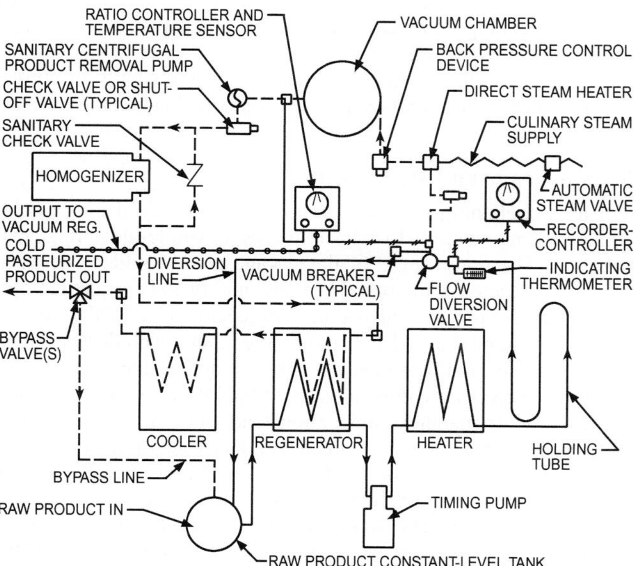

Fig. 1 Flow Diagram of Plate HTST Pasteurizer with Vacuum Chamber

Most HTST heat exchangers achieve 80 to 90% regeneration. The cost of additional equipment to obtain more than 90% regeneration should be compared with savings in the increased regeneration to determine feasibility. The percentage of regeneration may be calculated as follows for equal mass flow rates on either side of the regenerator:

$$\frac{138°\text{F (regeneration)} - 40°\text{F (raw product)}}{161°\text{F (pasteurization)} - 40°\text{F (raw product)}} = \frac{98}{121} = 81\%$$

The temperature of a product going into the cooling section can be calculated if the percent regeneration is known and the raw product and pasteurizing temperatures are determined. If they are 80%, 45°F, and 161°F, respectively,

$$(161 - 45) \times 0.80 = 92.8°\text{F}$$
$$161 - 92.8 = 68.2°\text{F}$$

The product should be cooled to at least 40°F, preferably lower, to compensate for the heat gain while in the sanitary pipelines and during the packaging process (including filling, sealing, casing, and transfer into cold storage). Average temperature increases of milk between discharge from the HTST unit's cooling section and arrival at the cold storage in various containers are as follows: glass bottles, 8°F; preformed paperboard cartons, 6°F; formed paperboard, 5°F; and semirigid plastic, 4°F.

Some plate pasteurizing systems are equipped with a cooling section using propylene glycol solution to cool the milk or milk product to temperatures lower than are practical by circulating only chilled water. This requires an additional section in the plate heat exchanger, a glycol chiller, a pump for circulating the glycol solution, and a product-temperature-actuated control to regulate the flow of glycol solution and prevent product freezing.

Some plants use propylene glycol exclusively for cooling, thus avoiding the use of chilled water and the requirement for two separate cooling sections. Milk is usually cooled with propylene glycol to approximately 34°F, then packaged. The lower temperature allows the milk to absorb heat from the containers and still maintain a low enough temperature for excellent shelf life. Milk should not be cooled to less than 33.5°F because of the tendency toward increased foaming in this range. Propylene glycol is usually chilled to approximately 28 to 30°F for circulation through the milk-cooling section.

Product flow rate through the pasteurizer may be more or less than the filling rate of the packaging equipment. Pasteurized product storage tanks are generally used to hold the product until it is packaged.

The number of plates in the pasteurizing unit is determined by the volume of product needed per unit of time, desired percentage of regeneration, and temperature differentials between the product and heating and cooling media. The heating section usually has ample surface so that the temperature of hot water entering the section is no more than 2 to 6°F higher than the pasteurizing, or outlet, temperature of the product. This temperature difference is often called the **approach** of the heat exchanger section.

On larger units, steam may be used for the heater section instead of hot water. The cooling section is usually sized so that the temperature of pasteurized product leaving the section is about 4 to 5°F higher than the entering temperature of chilled water or propylene glycol.

The holding tube size and length are selected so that not less than 15 s will elapse for the product to flow from one end of the tube to the other. An automatic, power-actuated, flow diversion valve, controlled by a temperature recorder-controller, is located at the outlet end of the holding tube and diverts flow back to the raw product constant-level tank as long as the product is below the minimum set pasteurizing temperature. The product timing pump is a variable-speed, positive-displacement, rotary type that can be sealed by the local government milk plant inspector at a maximum speed and volume. This ensures a product dwell time of not less than 15 s in the holding tube.

To reduce undesirable flavors and odors in milk (usually caused by specific types of dairy cattle feed), some plants use a vacuum process in addition to the usual pasteurization. Milk from the flow diversion valve passes through a direct steam injector or steam infusion chamber and is heated with culinary steam to 180 to 200°F. The milk is then immediately sprayed into a vacuum chamber, where it cools by evaporation to the pasteurizing temperature and is promptly pumped to the regeneration section of the pasteurizing unit. The vacuum in the evaporating chamber is automatically controlled so that the same amount of moisture is removed as was added by steam condensate. Noncondensable gases are removed by the vacuum pump, and vapor from the vacuum chamber is condensed in a heat exchanger cooled by the plant water.

The vacuum chamber can be installed with any type of HTST pasteurizer. In some plants, after preheating in the HTST system, the product is further heated by direct steam infusion or injection. It then is deaerated in the vacuum chamber. The product is pumped from the chamber by a timing pump through final heating, holding, flow diversion valve, and regenerative and cooling sections. Homogenization may occur either immediately after preheating for pasteurization or after the product passes through the flow diversion valve. Preferred practice is to homogenize after deaeration if the product is heated by direct steam injection and deaerated.

Where volatile weed and feed taints in the milk are mild, some processors use only a vacuum treatment to reduce off-flavor. The main objection to vacuum treatment alone is that, to be effective, the vacuum must be low enough to cause some evaporation, and the moisture so removed constitutes a loss of product. The vacuum chamber may be installed immediately after preheating, where it effectively deaerates the milk before heating, or immediately after the flow diversion valve, where it is more effective in removing volatile taints.

Nearly all milk processed in the United States is homogenized to improve stability of the milkfat emulsion, thus preventing creaming (concentration of the buoyant milkfat at the top of containerized milk) during normal shelf life. The homogenizer is a high-pressure reciprocating pump with three to seven pistons, fitted with a special homogenizing valve. Several types of homogenizing valves are used, all of which subject fat globules in the milk stream to enough shear to divide into several smaller globules. Homogenizing valves may either be single or two in series.

For effective homogenization of whole milk, fat globules should be 2 μm or less in diameter. The usual temperature range is from 130 to 180°F, and the higher the temperature within this range, the lower the pressure required for satisfactory homogenization. The homogenizing pressure for a single-stage homogenizing valve ranges from about 1200 to 2500 psi for milk; for a two-stage valve, from 1200 to 2000 psi on the first stage plus 300 to 700 psi on the second, depending on the design of the valve and the product temperature and composition. To conserve energy, use the lowest homogenizing pressure consistent with satisfactory homogenization: the higher the pressure, the greater the power requirements.

Packaging Milk Products

Cold product from the pasteurizer cooling section flows to the packaging machine and/or a surge tank 1000 to 10,000 gal or larger. These tanks are stainless steel, well insulated, and have agitation and usually refrigeration.

Milk and related products are packaged for distribution in paperboard, plastic, or glass containers in various sizes. Fillers vary in design. Gravity flow is used, but positive piston displacement is used on paper machines. Filling speeds range from roughly 16 to 250 units/min, but vary with container size. Some fillers handle only one size, whereas others may be adjusted to automatically fill and seal several size containers. Paperboard cartons are usually formed on the line ahead of filling, but may be preformed before delivery to the plant. Semirigid plastic containers may be blow-molded on the line ahead of the filler or preformed. Plastic pouches (called bags) arrive at the plant ready for filling and sealing. Filling dispenser cans and bags is a semimanual operation.

The paperboard milk carton consists of a 16 mil thick kraft paperboard from virgin paper with a 1 mil polyethylene film laminated onto the inside and a 0.75 mil film onto the outside. Gas or electric heaters supply heat for sealing while pressure is applied.

Blow-molded plastic milk containers are fabricated from high-density polyethylene resin. The resin temperature for blow-forming varies from 340 to 425°F. The molded gallon weighs approximately 60 to 70 g, and the one-half gallon, about 45 g. Contact the blow-molding equipment manufacturer for refrigeration requirements of a specific machine. The refrigeration demand to cool the mold head and clutch is large enough to require consideration in planning a plastic blow-molded operation. Blow-molding equipment may use stand-alone direct-expansion water chillers, or combine blow-molding refrigeration with the central refrigeration system to achieve better overall efficiency.

Packages containing the product may be placed into cases mechanically. Stackers place cases five or six high, and conveyors transfer stacks into the cold storage area.

Equipment Cleaning

Several automatic clean-in-place (CIP) systems are used in milk processing plants. These may involve holding and reusing the detergent solution or the preparation of a fresh solution (single-use) each day. Programming automatic control of each cleaning and sanitizing step also varies. Tanks, vats, and other large equipment can be cleaned by using spray balls and similar devices that ensure complete coverage of soiled surfaces. Tubing, HTST units, and equipment with relatively low volume may be cleaned by the full-flood system. Solutions should have a velocity of not less than 5 fps and must be in contact with all soiled surfaces. Surfaces used for heating milk products, such as in batch or HTST pasteurization, are more difficult to clean than other equipment surfaces. Other surfaces difficult to clean are those in contact with products that are high in fat, contain added solids and/or sweeteners, or are highly viscous. The usual cleaning steps for this equipment are a warm-water rinse, hot-acid-solution wash, rinse, hot-alkali-solution wash, and rinse. Time, temperature, concentration, and velocity may need to be adjusted for effective cleaning. Just before use, surfaces in contact with product should be sanitized with chemical solution, hot water, or steam. During CIP, the cooling section is isolated from the supply of chilled water or propylene glycol to minimize parasitic load on the refrigeration system.

Milk Storage and Distribution

Cases containing packaged products are conveyed into a cold-storage room or directly to delivery trucks for wholesale or retail distribution. The temperature of the storage area should be between 33 and 40°F, and for improved keeping quality, the product temperature in the container on arrival in storage should be 40°F or less.

The refrigeration load for cold-storage areas includes transmission through the building envelope, product and packaging materials temperature reduction, internally generated loads (e.g., lights, equipment motors, personnel), infiltration load from air exchange with other spaces and the environment, and refrigeration equipment-related load (e.g., fan motors, defrost). See Chapter 13 for a more detailed discussion of refrigeration load calculations.

Moisture load in these storage areas is generally high, which can lead to high humidity or wet conditions if evaporators are not selected properly. These applications usually require higher temperature differences between refrigerant and refrigerated-space set-point temperatures to achieve lower humidity. In addition, supply air temperatures should be controlled to prevent product freezing. Using reheat coils to provide humidity control is not recommended, because bacteriological growth on these surfaces could be rapid. Evaporators for these applications should have automatic coil defrost to remove the rapidly forming frost as required. Defrost cycles add to the refrigeration load and should be considered in the design.

A proprietary system used in some plants sprays coils continuously with an aqueous glycol solution to prevent frost from forming on the coil. These fan-coil units eliminate defrosting, can control humidity to an acceptable level with less danger of product freezing, and reduce bacteriological contamination. The glycol absorbs the water, which is continuously reconcentrated in a separate apparatus with the addition of heat to evaporate the water absorbed at the coil. A separate load calculation and analysis is required for these systems.

The floor space required for cold storage depends on product volume, height of stacked cases, packaging type (glass requires more space than paperboard), handling (mechanized or manual), and number of processing days per week. A 5 day processing week requires a capacity for holding product supply for 2 days. A very general estimate is that 100 lb of milk product in paperboard cartons can be stored per square foot of area. Approximately one-third more area should be allowed for aisles. Some automated, racked storages are used for milk products, and can be more economical than manually operated storages.

Milk product may be transferred by conveyor from storage room to dock for loading onto delivery trucks. In-floor drag-chain conveyors are commonly used, especially for retail trucks. Refrigeration losses are reduced if the load-out doorway has an air seal to contact the doorway frame of the truck as it is backed to the dock.

Distribution trucks need refrigeration to protect quality and extend storage life of milk products. Refrigeration capacity must be sufficient to maintain Grade A products at 45°F or less. Many plants use insulated truck trailer bodies with integral refrigerating systems powered by an engine or that can be plugged into a remote electric power source when it is parked. In some facilities, cold plates in the truck body are connected to a coolant source in the parking space. These refrigerated trucks can also be loaded when convenient and held over at the connecting station until the next morning.

Half-and-Half and Cream

Half-and-half is standardized at 10.5 to 12% milkfat and, in most areas of the United States, to about the same percent nonfat milk solids. Coffee cream should be standardized at 18 to 20% milkfat. Both are pasteurized, homogenized, cooled, and packaged similarly to milk. Milkfat content of whipping cream is adjusted to 30 to 35%. Take care during processing to preserve the whipping properties; this includes the omission of the homogenization step.

Buttermilk, Sour Cream, and Yogurt

Retail buttermilk is not from the butter churn but is instead a cultured product. To reduce microorganisms to a low level and improve the body of the resulting buttermilk, skim milk is pasteurized at 180°F or higher for 0.5 to 1 h and cooled to 70 to 72°F. One percent of a lactic acid culture (starter) specifically for buttermilk is added and the mixture incubated until firmly coagulated by the correct lactic acid production (pH 4.5). The product is cooled to 40°F or less with gentle agitation to inhibit serum separation after packaging and distribution. Salt and/or milkfat (0.5 to 1.0%) in the form of cream or small fat granules may be added. Packaging equipment and containers are the same as for milk. Pasteurizing, setting, incubating, and cooling are usually accomplished in the same vat. Rapid cooling is necessary, so chilled water is used. If a 500 gal vat is used, as much as 25 to 30 tons of refrigeration may be needed. Some plants have been able to cool buttermilk with a plate heat exchanger without causing a serum separation problem (wheying off).

Cultured half-and-half and cultured sour cream are also manufactured this way. Rennet may be added at a rate of 0.5 mL (diluted in water) per 10 gal cream. Take care to use an active lactic culture and to prevent postpasteurization contamination by bacteriophage, bacteria, yeast, or molds. An alternative method consists of packaging the inoculated cream, incubating, and then cooling by placing packages in a refrigerated room.

For yogurt, skim milk may be used, or milkfat standardized to 1 to 5%, and a 0.1 to 0.2% stabilizer may be added. Either vat pasteurization at 150 to 200°F for 0.5 to 1 h or HTST at 185 to 285°F for 15 to 30 s can be used. For optimum body, milk homogenization is at 130 to 150°F and 500 to 2000 psi. After cooling to between 100 and 110°F, the product is inoculated with a yogurt culture. Incubation for 1.5 to 2.0 h is necessary; the product is then cooled to about 90°F, packaged, incubated 2 to 3 h (acidity 0.80 to 0.85%), and chilled to 40°F or below in the package. Varying yogurt cultures and manufacturing procedures should be selected on the basis of consumer preferences. Numerous flavorings are used (fruit is quite common), and sugar is usually added. The flavoring material may be added at the same time as the culture, after incubation, or ahead of packaging. In some dairy plants, a fruit (or sauce) is placed into the package before filling with yogurt.

Refrigeration

The refrigerant of choice for production plants is usually ammonia (R-717). Some small plants may use halocarbon refrigerants; in large plants, halocarbons may be used with a centralized ammonia refrigeration system for special, small applications. The halocarbon refrigerant of choice is currently R-22; however, the Montreal Protocol outlines a phaseout schedule for the use of R-22 and other hydrochlorofluorocarbon (HCFC) refrigerants. Currently, no consensus alternative for R-22 has been identified. Two HFC blends, R-507 and R-404a, are currently favored for refrigeration applications.

Product plants use single-stage compression, and new applications are equipped with rotary screw compressors with microprocessors and automatic control. Older plants may be equipped with reciprocating compressors, but added capacity is generally with rotary screw compressors.

Most refrigerant condensing is accomplished with evaporative condensers. Freeze protection is required in cold climates, and materials of construction are an important consideration in subtropical climates. Water treatment is required.

Evaporators or cooling units for milk storage areas use either direct ammonia (direct-expansion, flooded or liquid overfeed), chilled water, or propylene glycol. In choosing new systems, evaluation should involve capital requirements, operating costs, ammonia charges, and plant safety.

Direct use of ammonia has the potential for the lowest operating cost because the refrigeration system does not have the increased losses associated with exchanging heat with a secondary cooling medium (chilled water or propylene glycol). However, direct use of ammonia requires larger system charges and more ammonia in production areas.

To limit ammonia charges in production areas, many plants use a secondary cooling system that circulates chilled water or propylene glycol where needed. If chilled water is used, it must be supplied at 33 to 34°F to cool milk products below 40°F. Chilled water is often used in combination with falling-film water chillers and ice-building chillers to cool water so close to its freezing point. Ice-building and falling-film chillers should be compared for each application, considering both initial capital and operating costs. Sizing ice builders to build ice during periods when chilled water is not required allows installation of a refrigeration system with considerably less capacity than is required for the peak cooling load. When chilled water is required, melting ice adds cooling capacity to that supplied by the refrigeration system. Additional information on ice thermal storage is found in Chapter 34 of the 2003 *ASHRAE Handbook—HVAC Applications*. The advantage of this system is a lower ammonia charge compared to the direct use of ammonia.

Other plants use propylene glycol at 28 to 30°F for process cooling requirements. This system cools propylene glycol in a welded-plate or shell-and-tube heat exchanger. The ammonia feed system is either gravity-flooded or liquid-overfeed. Advantages to this system are a reduced ammonia charge compared to direct use of ammonia (especially with a plate heat exchanger) and a lower cooling fluid temperature to achieve lower milk product temperatures. This system may have a higher operating cost, because there is no stored refrigeration, and possibly higher pumping requirements compared to chilled water. Commercially available propylene glycol packages for closed cooling systems include biological growth and corrosion inhibitors. The concentration of propylene glycol necessary in the system should be determined by consulting the glycol manufacturer to ensure adequate freeze protection as well as protection against biological growth and corrosion.

In addition, there are combination systems in which chilled water is used for most of the process requirements and a separate, smaller propylene glycol system is used in final cooling sections to provide lower milk product temperatures.

Other plant refrigeration loads, such as air conditioning of process areas, may be met with the central ammonia refrigeration system. The choice between chilled water and propylene glycol may also depend on the plant winter climate conditions and location of piping serving the loads.

Most new or expanded plants rely on automated operation and computer controls for operating and monitoring the refrigeration systems. There also is a trend to use welded-plate heat exchangers for water and propylene glycol cooling in milk product plants and to reduce or eliminate direct ammonia refrigeration in plant process areas. This approach may add somewhat to the capital and operating costs, but it can substantially reduce the ammonia charge in the system and confines ammonia to the refrigeration machine room area.

BUTTER MANUFACTURE

Much of the butter production is in combination butter-powder plants. These plants get the excess milk production after current market needs are met for milk products, frozen dairy desserts, and, to some extent, cheeses. Consequently, seasonal variation in the volume of butter manufactured is large; spring is the period of highest volume, fall the lowest.

Separation and Pasteurization

After separation, cream with 30 to 40% fat content is either pumped to the pasteurizer or cooled to 45°F and held for later pasteurization. Cream from cold milk separation does not need to be recooled except for extended storage. Cream is received, weighed, sampled, and, in some plants, graded according to flavor and acidity. It is pumped to a refrigerated storage vat and cooled to 45°F if held for a short period or overnight. Cream with developed acidity is warmed to 80 to 90°F, and neutralized to 0.12 to 0.15% titratable acidity just before pasteurization. If acidity is above 0.40%, it is neutralized with a soda-type compound in aqueous solution to about 0.30% and then to the final acidity with aqueous lime solution. Sodium neutralizers include $NaHCO_3$, Na_2CO_3, and NaOH. Limes are $Ca(OH)_2$, MgO, and CaO.

Batch pasteurization is usually at 155 to 175°F for 0.5 h, depending on intended storage temperature and time. HTST continuous pasteurization is at 185 to 250°F for at least 15 s. HTST systems may be plate or tubular. After pasteurization, the cream is immediately cooled. The temperature range is 40 to 55°F, depending on the time that the cream will be held before churning, whether it is ripened, season (higher in winter because of fat composition), and churning method. Ripening consists of adding a flavor-producing lactic starter to tempered cream and holding until acidity has developed to 0.25 to 0.30%. The cream is cooled to prevent further acid development and warmed to the churning temperature just before churning. First, tap water is used to reduce the temperature to between 80 and 100°F. Refrigerated water or brine is then used to reduce the temperature to the desired level. The cream may be cooled by passing the cooling medium through a revolving coil in the vat or through the vat jacket, or by using a plate or tubular cooler. Ripening cream is not common in the United States, but is customary in some European countries such as Denmark.

If the temperature of 1000 lb of cream is to be reduced by refrigerated water from 104 to 39°F, and the specific heat is 0.85 Btu/lb·°F, the heat to be removed is

$$1000(104 - 39)0.85 = 55{,}250 \text{ Btu}$$

This heat can be removed by 55,250/144 = 384 lb of ice at 32°F plus 10% for mechanical loss.

The temperature of refrigerated water commonly used for cooling cream is 33 to 34°F. The ice-builder system is efficient for this purpose. Brine or glycol is not currently used. About 265 gal of cream can be cooled from 100 to 40°F in a vat using refrigerated water in an hour.

After a vat of cream has cooled to the desired temperature, the temperature increases during the following 3 h because heat is liberated when fat changes from liquid to a crystal form. It may increase several degrees, depending on the rapidity with which the cream was cooled, the temperature to which it was cooled, the richness of the cream, and the properties of the fat.

Rishoi (1951) presented data in Figure 2 that show the thermal behavior of cream heated to 167°F followed by rapid cooling to 86°F and to 50.7°F, as compared with cream heated to 122°F and cooled rapidly to 88.5°F and to 53.6°F. The curves indicate that when cream is cooled to a temperature at which the fat remains liquid, the cooling rate is normal, but when the cream is cooled to a temperature at which some fractions of the fat have crystallized, a spontaneous temperature rise takes place after cooling.

Rishoi also determined the amount of heat liberated by the part of the milkfat that crystallizes in the temperature range of 85 to 33°F. The results are shown in Figure 3 and Table 2.

Table 2 shows that, at a temperature below 50°F, about one-half of the liberated heat evolved in less than 15 s. The heat liberated during fat crystallization constitutes a considerable portion of the refrigeration load required to cool fat-rich cream. Rishoi states,

> If we assume an operation of cooling cream containing 40% fat from about 150 to 40°F, heat of crystallization evolved represents about 14% of the total heat to be removed. In plastic cream containing 80% fat it represents about 30% and in pure milkfat oil about 40%.

Churning

To maintain the yellow color of butter from cream that came from cows on green pasture in spring and early summer, yellow coloring can be added to the cream to match the color obtained naturally during other periods of the year. After cooling, pasteurized cream should be held a minimum of 2 h and preferably overnight. It is tempered to the desired batch churning temperature, which varies with the season and feed of the cows but ranges from 45°F in early summer to 56°F in winter, to maintain a churning time 0.5 to 0.75 h. Lower churning time results in soft butter that is more difficult (or impossible) to work into a uniform composition.

Most butter is churned by continuous churns, but some batch units remain in use, especially in smaller butter factories. Batch churns are usually made of stainless steel, although a few aluminum ones are still in use. They are cylinder, cube, cone, or double cone in shape. The inside surface of metal churns is sandblasted during fabrication to reduce or prevent butter from sticking to the surface.

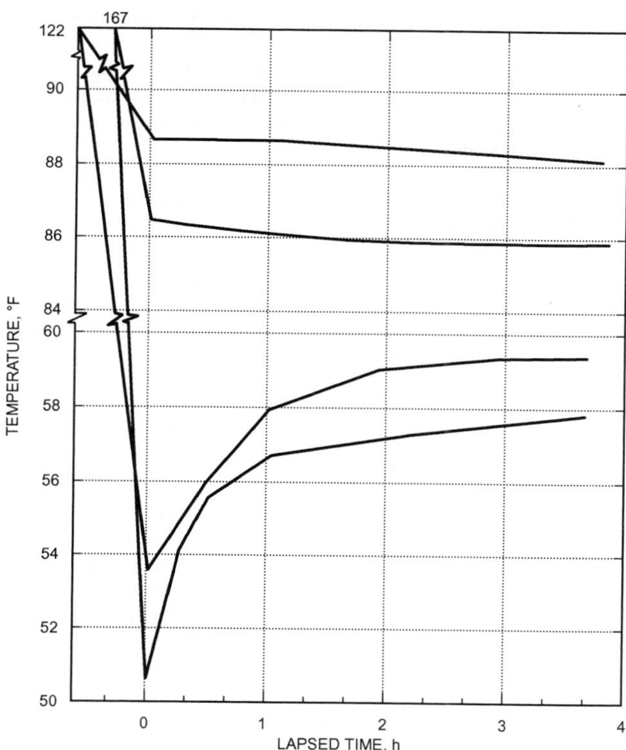

Fig. 2 Thermal Behavior of Cream Heated to 167°F Followed by Rapid Cooling to 86°F and to 50.7°F; Comparison with Cream Heated to 122°F, then Rapid Cooling to 88.5°F and to 53.6°F

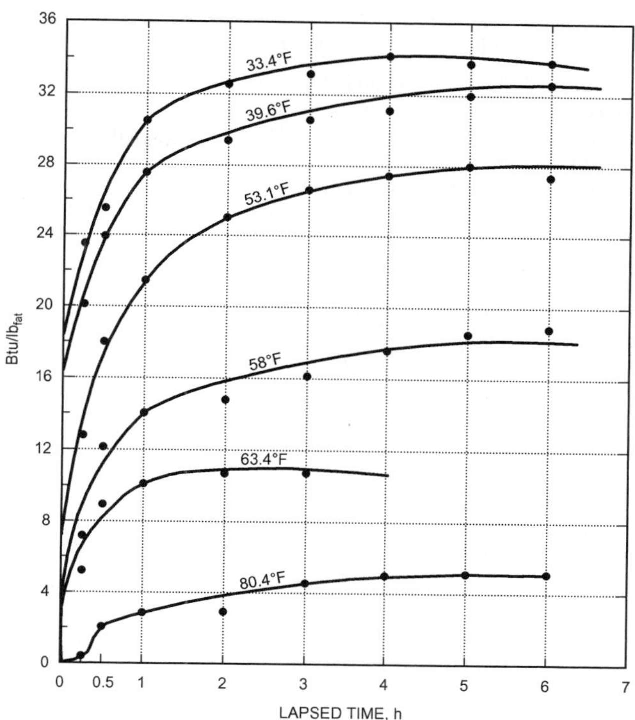

Fig. 3 Heat Liberated from Fat in Cream Cooled Rapidly from Approximately 86°F to Various Temperatures
(Rishoi 1951)

Table 2 Heat Liberated from Fat in Cream Cooled Rapidly from about 86°F to Various Temperatures

Calculated temperature for zero time, °F:	33.4	39.6	53.1	58.0	63.4	80.4	85.6*
First observed temperature:	36.2	43.7	54.5	58.6	63.8	80.4	85.6
Final equilibrium temperature:	39.3	46.1	58.0	60.7	65.3	82.6	85.6
Lapsed time, min	**Btu Liberated per Pound**						
0.25	18.3	16.7	7.75	2.9	2.3	0	0
15	23.6	20.0	12.8	7.2	5.2	0.35	0
30	25.5	23.9	18.0	12.2	9.0	2.1	0
60	30.4	27.5	21.4	14.0	10.2	2.7	0
120	32.4	29.3	25.0	14.7	10.6	2.9	0
180	32.9	30.2	26.6	16.0	10.6	4.3	0
240	34.0	30.9	27.2	17.5	10.6	4.9	
300	33.6	31.8	27.9	18.4		5.0	
360	33.6	32.5	27.2	18.8		5.0	

Percent heat liberated at zero time compared with that at equilibrium: 54.5, 51.3, 27.7, 20.7, 21.7.
Percent total heat liberated compared with that liberated at about 32°F: 100.0, 95.7, 82.0, 55.0, 31.0, 12.5, 0.
Iodine values of three samples of butter produced while these tests were in progress were 28.00, 28.55, and 28.24.
*Cooled in an ice-water bath.

Metal churns may have accessories to draw a partial vacuum or introduce an inert gas (e.g., nitrogen) under pressure. Working under a partial vacuum reduces air in the butter. Churns have two or more speeds, with the faster rate for churning. The higher speed should provide maximum agitation of the cream, usually between 0.25 to 0.5 rev/s.

When churning, temperature is adjusted and the churn is filled to 40 to 48% of capacity. The churn is revolved until the granules break out and attain a diameter of 5 mm or slightly larger. The buttermilk, which should have no more than 1% milkfat, is drained. The butter may or may not be washed. The purpose of washing is to remove buttermilk and temper the butter granules if they are too soft for adequate working. Wash water temperature is adjusted to 0 to 10°F below churning temperature. The preferred procedure is to spray wash water over granules until it appears clear from the churn drain vent. The vent is then closed, and water is added to the churn until the volume of butter and water is approximately equal to the former amount of the cream. The churn revolves slowly 12 to 15 times and drained or held for an additional 5 to 15 min for tempering so granules will work into a mass of butter without becoming greasy.

The butter is worked at a slow speed until free moisture is no longer extruded. Free water is drained, and the butter is analyzed for moisture content. The amount of water needed to obtain the desired content (usually 16.0 to 18.0%) is calculated and added. Salt may be added to the butter. The salt content is standardized between 1.0 and 2.5% according to customer demand.

Dry salt may be added either to a trench formed in the butter or spread over the top of the butter. It also may be added in moistened form, using the water required for standardizing the composition to not less than 80.0% fat. Working continues until the granules are completely compacted and the salt and moisture droplets are uniformly incorporated. Moisture droplets should become invisible to normal vision with adequate working. Most churns have ribs or vanes, which tumble and fold the butter as the churn revolves. The butter passes between the narrow slit of shelves attached to the shell and the roll. A leaky butter is inadequately worked, possibly leading to economic losses because of weight reduction and shorter keeping quality. The average composition of U.S. butter on the market has these ranges:

Fat	80.0 to 81.2%	Moisture	16.0 to 18.0%
Salt	1.0 to 2.5%	Curd, etc.	0.5 to 1.5%

Cultured skim milk is added to unsalted butter as part of the moisture and thoroughly mixed in during working. On rare occasions, cultured skim milk may be used to increase acid flavor and the diacetyl content associated with butter flavor.

Butter may be removed manually from small churns, but it is usually emptied mechanically. One method is to dump butter from the churn directly into a stainless steel boat on casters or a tray that has been pushed under the churn with the door removed. Butter in boats may be augered to the hopper for printing (forming the butter into retail sizes) or pumping into cartons 60 to 68 lb in size. The bulk cartons are held cold before printing or shipment. Butter may be stored in the boats or trays and tempered until printing. A hydraulic lift may be used for hoisting the trays and dumping the butter into the hopper. Cone-shaped churns with a special pump can be emptied by pumping butter from churn to hopper.

Continuous Churning

The basic steps in two of the continuous buttermaking processes developed in the United States are as follows:

1. Fat emulsion in the cream is destabilized and the serum separated from the milkfat.
2. The butter mix is prepared by thoroughly blending the correct amount of milkfat, water, salt, and cultured skim milk (if necessary).
3. This mixture is worked and chilled at the same time.
4. Butter is extruded at 38 to 50°F with a smooth body and texture.

Some European continuous churns consist of a single machine that directly converts cream to butter granules, drains off the buttermilk, and washes and works the butter, incorporating the salt in continuous flow. Each brand of continuous churn may vary in equipment design and specific operation details for obtaining the optimum composition and quality control of the finished product. Figure 4 shows a flow diagram of a continuous churn.

In one such system, milk is heated to 110°F and separated to cream with 35 to 50% fat and skim milk. The cream is pasteurized at 203°F for 16 s, cooled to a churning temperature of 46 to 55°F, and held for 6 h. It then enters the balance tank and is pumped to the churning cylinder, where it is converted to granules and serum in less than 2 s by vigorous agitation. Buttermilk is drained off and the granules are sprayed with tempered wash water while being agitated.

Next, salt, in the form of 50% brine prepared from microcrystalline sodium chloride, is fed into the product cylinder by a proportioning pump. If needed, yellow coloring may be added to the brine. High-speed agitators work the salt and moisture into the butter in the texturizer section and then extrude it to the hopper for packaging into bulk cartons or retail packages. The cylinders on some designs have a cooling system to maintain the desired temperature of the butter from churning to extrusion. The butterfat content is adjusted by fat test of the cream, churning temperature of the cream, and flow rate of product.

Continuous churns are designed for CIP. The system may be automated or the cream tank may be used to prepare the detergent solution before circulation through the churn after the initial rinsing.

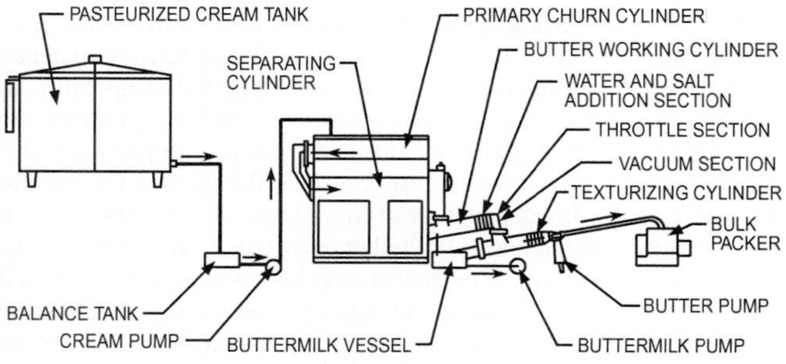

Fig. 4 Flow Diagram of Continuous Butter Manufacture

Packaging Butter

Printing is the process of forming (or cutting) butter into retail sizes. Each print is then wrapped with parchment or parchment-coated foil. The wrapped prints may be inserted in paperboard cartons or overwrapped in cellophane, glassine, and so forth, and heat-sealed. For institutional uses, butter may be extruded into slabs. These are cut into patties, embossed, and each slab of patties wrapped in parchment paper. Most common numbers of patties are 48 to 72 per lb.

Butter keeps better if stored in bulk. If the butter is intended to be stored for several months, the temperature should not be above 0°F, and preferably below –20°F. For short periods, 32 to 40°F is satisfactory for bulk or printed butter. Butter should be well protected to prevent absorption of off-odors during storage and weight loss from evaporation, and to minimize surface oxidation of fat.

The specific heat of butter and other dairy products at temperatures varying from 32 to 140°F is given in Table 3. The butter temperature when removed from the churn ranges from 56 to 62°F. Assuming a temperature of 60°F of packed butter, the heat that must be removed from 1000 lb to reduce the temperature to 32°F is

$$1000(60 - 32)0.52 = 14,560 \text{ Btu}$$

It is assumed that the average specific heat at the given range of temperatures is 0.52 Btu/lb·°F. Heat to be removed from butter containers and packaging material should be added.

Deterioration of Butter in Storage

Undesirable flavor in butter may develop during storage because of (1) growth of microorganisms (proteolytic organisms causing putrid and bitter off-flavors); (2) absorption of odors from the atmosphere; (3) fat oxidation; (4) catalytic action by metallic salts; (5) activity of enzymes, principally from microorganisms; and (6) low pH (high acid) of salted butter.

Normally, microorganisms do not grow below 32°F; if salt-tolerant bacteria are present, their growth will be slow below 32°F. Microorganisms do not grow at 0°F or below, but some may survive in butter held at this temperature. It is important to store butter in a room free of atmospheric odors. Butter readily absorbs odors from the atmosphere or from odoriferous materials with which it comes into contact.

Oxidation causes a stale, tallowy flavor. Chemical changes take place slowly in butter held in cold storage, but are hastened by the presence of metals or metallic oxides.

Table 3 Specific Heats of Milk and Milk Derivatives, Btu/lb·°F

	32°F	59°F	104°F	140°F
Whey	0.978	0.976	0.974	0.972
Skim milk	0.940	0.943	0.952	0.963
Whole milk	0.920	0.938	0.930	0.918
15% cream	0.750	0.923	0.899	0.900
20% cream	0.723	0.940	0.880	0.886
30% cream	0.673	0.983	0.852	0.860
45% cream	0.606	1.016	0.787	0.793
60% cream	0.560	1.053	0.721	0.737
Butter	(0.512)*	(0.527)*	0.556	0.580
Milkfat	(0.445)*	(0.467)*	0.500	0.530

*For butter and milkfat, values in parentheses were obtained by extrapolation, assuming that the specific heat is about the same in the solid and liquid states.

With almost 100% replacement of tinned copper equipment with stainless steel equipment, a tallowy flavor is not as common as in the past. Factors that favor oxidation are light, high acid, high pH, and metal.

Enzymes present in raw cream are inactivated by current pasteurization temperatures and holding times. The only enzymes that may cause butter deterioration are those produced by microorganisms that gain entrance to the pasteurized cream and butter or survive pasteurization. The chemical changes caused by enzymes present in butter are retarded by lowering the storage temperature.

A fishy flavor may develop in salted butter during cold storage. Development of the defect is favored by high acidity (low pH) of the cream at the time of churning and by metallic salts. With the use of stainless steel equipment and proper control of the butter's pH, this defect now occurs very rarely. For salted butter to be stored for several months, even at –10°F, it is advisable to use good-quality cream; avoid exposing the milk or cream to strong light, copper, or iron; and adjust any acidity developed in the cream so that the butter serum has a pH of 6.8 to 7.0.

Total Refrigeration Load

Some dairy plants that manufacture butter also process and manufacture other products such as ice cream, fluid milk, and cottage cheese. A single central refrigeration system is used to provide refrigeration to all of these loads. The method of determining the refrigeration load is illustrated by the following example.

Example 1. Determine the product refrigeration load for a plant manufacturing butter from 12,600 lb of 30% cream per day in three churnings.

Solution: Assume that refrigeration is accomplished with chilled water from an ice builder. See Figure 5 for a workflow diagram.

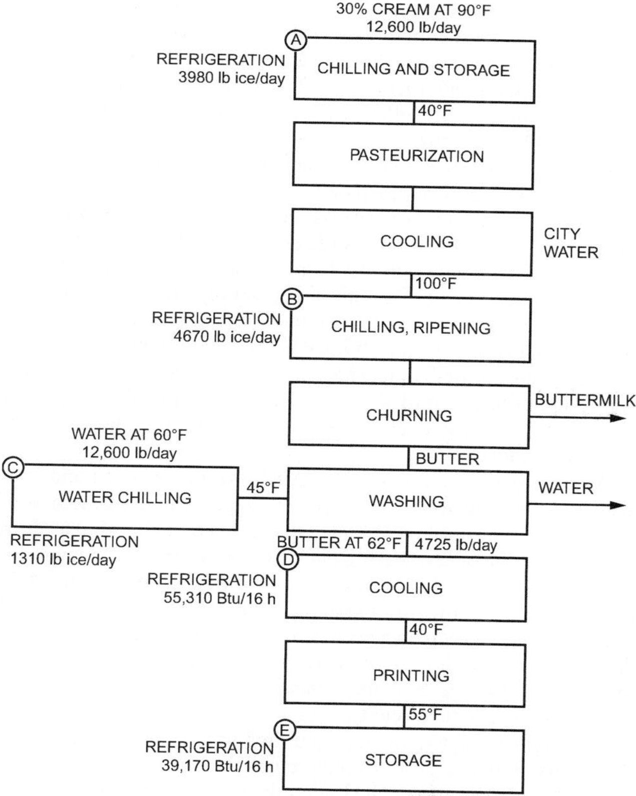

Fig. 5 Butter Flow Diagram

The cream is cooled in steps A and B. The butter is then cooled through steps C, D, and E. Refrigerated water is normally used as a cooling medium in steps A, B, and C. The ice builder system is used to produce 34°F water, and the load should be expressed in pounds of ice that must be melted to handle steps A, B, and C. This load is added to the refrigerated water load from the various other products such as milk, cottage cheese, and so forth, in sizing the ice builder.

A. If cream is separated in the plant rather than on the farm, it must be cooled from 90°F separating temperature to 40°F for holding until it is processed.

$$\frac{12{,}600(90-40)0.91}{144} = 3980\ \text{lb}_{ice}/\text{day}$$

B. After pasteurization, the temperature of the cream is reduced to approximately 100°F with city water, then down to 40°F with refrigeration.

$$\frac{12{,}600(100-40)0.89}{144} = 4670\ \text{lb}_{ice}/\text{day}$$

C. After churning, the 60°F butter wash water (city water) is usually cooled to 45°F, then used to wash the butter granules. A mass of water equal to the mass of cream churned may be used.

$$\frac{12{,}600(60-45)1.00}{144} = 1310\ \text{lb}_{ice}/\text{day}$$

Total ice load	9960 lb_{ice}/day
Plus 10% mechanical loss	1000 lb_{ice}/day
Total ice required	10,960 lb_{ice}/day

D. Approximately 4725 lb of butter is obtained. (12,600 lb cream × 30% fat = 3780 lb of fat. If butter contains approximately 80% fat, 3780 lb divided by 80% equals approximately 4725 lb of butter.) The butter temperature going into the refrigerated storage room is usually about 62°F and must be cooled to 40°F in the following 16 h. (For long-term storage, butter is held at –10 to 0°F.) The average specific heat for butter over this range is 0.52 Btu/lb·°F.

4725(62 – 40)0.52 =	54,050 Btu
300 lb (metal container) × (75 – 40)0.12 =	1,260 Btu
Total/24 h	55,310 Btu

E. After 24 h or longer, the butter is removed from the cooler to be cut and wrapped in 1 lb or smaller units. During this process, the butter temperature rises to approximately 55°F, which constitutes another product load in the cooler when it goes back for storage.

4725(55 – 40)0.52 =	36,860 Btu
200 lb (paper container) × (75 – 40)0.33 =	2,310 Btu
Total/24 h	39,170 Btu

Total of Steps D and E, Product Load in Cooler:

$$\frac{39{,}170 + 55{,}310}{16\ \text{h}} = 5910\ \text{Btu/h}$$

Whipped Butter

To whip butter by the batch method, the butter is tempered to 62 to 70°F, depending on factors such as the season and type of whipper. The butter is cut into slabs for placing into the whipping bowl. The whipping mechanism is activated, and air is incorporated until the desired overrun (volume increase) is obtained, usually between 50 and 100%. Whipped butter is packaged mechanically or manually into semirigid plastic containers.

With one continuous system, butter directly from cooler storage is cut into pieces and augered until soft. However, it can be tempered and the augering step omitted. The butter is then pumped into a cylindrical continuous whipper that uses the same principles as those for incorporating air in ice cream. Air or nitrogen is incorporated until the desired overrun is obtained. Another continuous method (used less commercially) is to melt butter or standardize butter oil to the composition of butter with moisture and salt. The fluid product is pumped through a chiller/whipper. Metered air or nitrogen provides overrun control. Soft whipped butter is pumped to the hopper of the filler and packaged in rigid or semirigid containers, such as plastic. It is chilled and held in storage at 32 to 40°F.

CHEESE MANUFACTURE

Approximately 800 cheeses have been named, but there are only 18 distinctly different types. A few of the more popular types in the United States are cheddar, cottage, roquefort or blue, cream, ricotta, mozzarella, Swiss, edam, and provolone. Such details of manufacture as setting (starter organisms, enzyme, milk or milk product, temperature, and time), cutting, heating (cooking), stirring, draining, pressing, salting, and curing (including temperature and humidity control) are varied to produce a characteristic variety and its optimum quality.

Production of cheddar cheese in the United States currently exceeds the other cured varieties; however, mozzarella production is a close second and is gaining fast. For uncured cheese varieties, cottage cheese production is much greater than that of the others. Another trend in the cheese industry is large factories. These plants may have sufficient curing facilities for the total production. If not, the cheese is shipped to central curing plants.

The physical shape of cured cheese varies considerably. Barrel cheese is common; it is cured in a metal barrel or similar impervious container in units of approximately 500 lb. Cheese may also be cured in rectangular metal containers holding 2000 lb.

The microbiological flora of cured cheese are important in the development of flavor and body. Heating the milk for cheese is the general practice. The milk may be pasteurized at the minimum HTST conditions or be given a subpasteurization treatment that results in a positive phosphatase test, which checks for inactivated enzymes indicating the presence of raw milk. Subpasteurization is possible with good-quality milk (low level of spoilage microorganisms and pathogens). Such milk treatments give the cheese some characteristics of raw-milk cheese in curing, such as production of higher flavors, in a shorter time. Pasteurization to produce phosphatase-negative milk is used in making soft, unripened varieties of cheese and some of the more perishable of the ripened types such as camembert, limburger, and munster.

The standards and definitions of the Food and Drug Administration (FDA) and of most state regulatory agencies require that cheese that is not pasteurized must be cured for not less than 60 days at not less than 35°F. Raw-milk cheese contains not only lactic-acid-producing organisms such as *Lactococcus lactis*, which are added to the milk during cheesemaking, but also the heterogeneous mixture of microorganisms present in the raw milk, many of which may produce gas and off-flavors in the cheese. Pasteurization gives some control over the bacterial flora of the cheese.

Freshly manufactured cured cheese is rubbery in texture and has little flavor; perhaps the more characteristic flavor is slightly acid. The presence of definite flavor(s) in freshly made cheese indicates poor quality, probably resulting from off-flavored milk. On curing under proper conditions, however, the body of the cheese breaks down, and the nut-like, full-bodied flavor characteristic of aged cheese develops. These changes are accompanied by certain chemical and physical changes during curing. The calcium paracaseinate of cheese gradually changes into proteoses, amino acids, and ammonia. These changes are a part of ripening and may be controlled by time and temperature of storage. As cheese cures, varying degrees of lipolytic activity also occur. In the case of blue or roquefort cheese, this partial fat breakdown contributes substantially to the characteristic flavor.

During curing, microbiological development produces changes according to the species and strains present. It is possible to predict from the microorganism data some of the usual defects in cheddar. In some cheeses (e.g., Swiss), gas production accompanies the desirable flavor development.

Cheese quality is evaluated on the basis of a scorecard. Flavor and odor, body and texture, and color and finish are principal factors. They are influenced by milk quality, the skill of manufacturing (including starter preparation), and the control effectiveness of maintaining optimum curing conditions.

Cheddar Cheese

Manufacture. Raw or pasteurized whole milk is tempered to 86 to 88°F and pumped to a cheese vat, which typically holds approximately 40,000 lb of milk. It is set by adding 0.75 to 1.25% active cheese starter and possibly annatto yellow color, depending on market demand. After 15 to 30 min, when the milk has reached the proper acidity (0.05 to 0.1%), 99 mL of single-strength rennet per 1000 lb milk is diluted in water 1:40 and slowly added with agitation of milk in the vat. After a quiescent period of 25 to 30 min, the curd should have developed proper firmness. The curd is cut into 0.25 to 0.40 in. cubes. After 15 to 30 min of gentle agitation, cooking is begun by heating water in the vat jacket using steam or hot water for 30 to 40 min. The curd and whey should increase 2°F per 5 min, and a temperature of 100 to 102°F is maintained for approximately 45 min.

In batch systems, the whey is drained and curd is trenched along both sides of the vat, allowing a narrow area free of curd the length of the midsection of the vat. Slabs about 10 in. long are cut and inverted at 15 min periods during the cheddaring process (matting together of curd pieces). When acidity of the small whey drainage is at a pH of 5.3 to 5.2, the slabs are milled (cut into small pieces) and returned to the vat for salting and stirring, or the curd goes to a machine that automatically adds salt and uniformly incorporates it into the curd. Weighed curd goes into hoops, which are placed into a press, and 20 psi is applied. After 0.5 to 1 h, the hoops are taken out of the press, the bandage adjusted to remove wrinkles, and then the cheese is pressed overnight at 25 to 30 psi or higher. Cheese may be subjected to a vacuum treatment to improve body by reducing or eliminating air pockets. After the surface is dried, the cheese is coated by dipping into melted paraffin or wrapped with one of several plastic films, or oil with a plastic film, and sealed. Yield is about 10 lb per 100 lb of milk.

Faster and more mechanized methods of making cheddar cheese have evolved. The stirred curd method (which omits the cheddaring step) is being used by more cheesemakers. Deep circular or oblong cheese vats with special, reversible agitators and means for cutting the curd are becoming popular. Curd is pumped from these vats to draining and matting tables with sloped bottoms and low sides, then milled, salted, and hooped. In one method, curd (except for Odenburg cheddar) is carried and drained by a draining/matting conveyor with a porous plastic belt to a second belt for cheddaring and transport to the mill. The milled curd is then carried to a finishing table or conveyor, where it is salted, stirred, and moved out for hooping or to block formers.

Another system, imported from Australia, is used in a number of cheddar cheese factories. This system requires a short method of setting. After the curd is cut and cooked, it is transferred to a series of perforated stainless steel troughs traveling on a conveyor for draining and partial fusion. The slabs are then transferred into buckets of a forming conveyor, transferred again to transfer buckets, and finally to compression buckets where cheddaring takes place. Cheddared slabs are discharged to a slatted conveyor, which carries them to the mill and then to a final machine where the milled curd is salted, weighed, and hooped.

Curing. Curing temperature and time vary widely among cheddar plants. A temperature of 50°F cures the cheddar more rapidly than lower temperatures. The higher the temperature above 50°F to about 80°F, the more rapid the curing and the more likely that off-flavors will develop. At 50°F, 3 to 4 months are required for a mild to medium cheddar flavor. Six months or more are necessary for an aged (sharp) cheddar cheese. Relative humidity should be roughly 70%. Cheddar intended for processed cheese is cured in many plants at 70°F because of the economy of time. Some experts suggest that cheddar, after its coating or wrapping, should be held in cold storage at approximately 40°F for about 30 days, then transferred to the 50°F curing room. During cold storage, the curd particles knit together, forming a close-bodied cheese. The small amount of residual lactose is slowly converted to lactic acid, along with other changes in optimum curing.

The maximum legal moisture content of cheddar is 39% and the fat must be not less than 50% of total solids. The amount of moisture directly affects the curing rate to some extent within the normal range of 34 to 39%. Cheese with loose or crumbly body and a high acidity is less likely to cure properly. For best curing, the cheese should have a sodium chloride content of 1.5 to 2.0%. A lower percentage encourages off-flavors to develop, and higher amounts retard flavor development.

Moisture Losses. Weight loss of cheese during curing is largely attributed to moisture loss. Paraffined cheddar cheese going into cure averages approximately 37% moisture. After a 12-month cure at 40°F, paraffined cheese averages approximately 33% moisture. This is a real loss to the cheese manufacturer unless the cheese is sold on the basis of total solids. Control of humidity can have an important role in moisture loss. Figure 6 shows the loss from paraffined longhorns in boxes held at 38°F and 70% rh over 12 months. The conditions were well controlled, but the average loss was 7%. The high loss shown on the graph was influenced by the larger surface area in 12 lb longhorns, compared to 70 lb cheddars. Curing the cheese within a good-quality sealed wrapper having a low moisture transmission (but some oxygen and carbon dioxide) largely eliminates moisture loss.

Provolone and Mozzarella (Pasta Filata Types)

Provolone is an Italian plastic-curd cheese representative of a large group of pasta filata cheeses. These cheeses vary widely in size and composition, but they are all manufactured by a similar method. After the curd has been matted, like cheddar, it is cut into slabs, which are worked and stretched in hot water at 150 to 180°F. The curd is kneaded and stretched in the hot water until it reaches a temperature of about 135°F. The maker then takes the amount necessary for one cheese and folds, rolls, and kneads it by hand to give the cheese its characteristic shape and smooth, closed surface. Molding machines have been developed for large-scale operations to eliminate this hand labor. The warm curd of some varieties of pasta filata is placed in molds and submerged in or sprayed with 36°F cold water to harden into the desired shape. The hardened cheese is then salted in batch or continuous brine tanks for final cooling and salting, depending on the size and variety. Some pasta filata cheese, such as mozzarella for pizza, is packaged for shipment with wrappers to protect it for the period it is held before use. This cheese may be sealed under vacuum in plastic bags for prolonged holding.

Provolone is salted by submersion in 24% sodium chloride solution at 45°F for 1 to 3 days, depending on the size, and then hung to dry. If a smoked flavor is desired, it is then transferred to the smokehouse and exposed to hickory or other hardwood smoke for 1 to 3 days. The cheese is hung in a curing room for 3 weeks at 55°F and then for 2 to 10 months at 40°F. Size and shape vary, but the most common in the United States is 14 lb and pear-shaped. Moisture content ranges from 37 to 45% and the salt from 2 to 4%. Milkfat usually comprises 46 to 47% of the total solids. The yield is roughly 9.5 lb per 100 lb of milk.

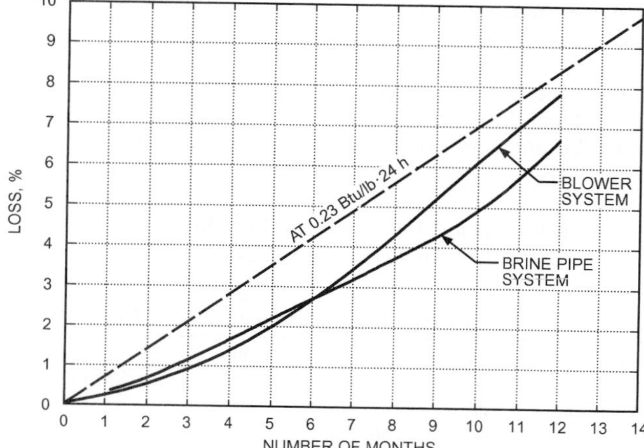

Fig. 6 Cheese Shrinkage in Storage

Swiss Cheese

One of the distinguishing characteristics of Swiss cheese is the eye formation during curing. These eyes result from the development of CO_2. Raw or heat-treated milk is tempered to 95°F and pumped to a large kettle or vat. One starter unit, consisting of 27 mL of *Propioni bacterium shermanii*, 165 mL of *Lactococcus thermophilus*, and 165 mL of *Lactobacillus bulgaricus*, is added per 1100 lb of milk. After mixing, 77 mL of rennet per 1100 lb is diluted 1:40 with water and slowly added with agitation of the milk. Curd is cut when firm (after 25 to 30 min) into very small granules. After 5 min, curd and whey are agitated for 40 min, and then the steam is released into the jacket without water. Curd is heated slowly to 122 to 130°F in 30 to 45 min. Without additional steam, the cooking is continued until curd is firm and has no tendency to stick when a group of particles is squeezed together (0.5 to 1 h and whey pH 6.3). Curd is dipped into hoops (160 lb) and pressed lightly for 6 h, redressing and turning the hoops every 2 h. Pressing is continued overnight. The next step is soaking the cheese in brine until it has about 1.5% salt. Table 4 shows temperature and time at which curing occurs. The minimum milkfat content is 43% by weight of solids and the maximum moisture content is 41% by weight (21CFR133).

Roquefort and Blue Cheese

Roquefort and blue cheese require a mold (*Penicillium roquefortii*) to develop the typical flavor. Roquefort is made from ewes' milk in France. Blue cheese in the United States is made from cow's milk. The equipment used for the manufacture and curing of blue cheese is the same as that used for cheddar, with a few exceptions. The hoops are 7.5 in. in diameter and 6 in. high. They have no top or bottom covers and are thoroughly perforated with small holes. A manually or pneumatically operated device with 50 needles, which are 6 in. long and 0.125 in. in diameter, is used to punch holes in the curd wheels. An apparatus is also needed to feed moisture into the curing room to maintain at least 95% rh without causing a drip onto the cheese.

Table 4 Swiss Cheese Manufacturing Conditions

Processing Step	Temperature, °F	Relative Humidity, %	Time
Setting	95	—	0.4 to 0.5 h
Cooking	122 to 130	—	1.0 to 1.5 h
Pressing	80 to 85	—	12 to 15 h
Salting (brine)	50 to 52	—	2 to 3 days
Cool room hold	50 to 60	90	10 to 14 days
Warm room hold	70 to 75	80 to 85	3 to 6 weeks
Cool room hold	40 to 45	80 to 85	4 to 10 mos

The milk may be raw or pasteurized and separated. The cream is bleached and may be homogenized at low pressure. Skim milk is added to the cream, and the milk is set with 2 to 3% active lactic starter. After 30 min, 3 to 4 oz of rennet per 1100 lb is diluted with water (1:35) and thoroughly mixed into the milk. When the curd is firm (after 30 min), it is cut into 5/8 in. cubes. Agitation begins 5 min later. After whey acidity is 0.14% (1 h), the temperature is raised to 92°F and held for 20 min. The whey is drained and trenched. Approximately 2 lb of coarse salt and 1 oz of *P. roquefortii* powder are mixed into each 100 lb of curd.

The curd is transferred to stainless steel perforated cylinders (hoops). These hoops are inverted each 15 min for 2 h on a drain cloth, and curd matting continues overnight. The hoops are removed and surfaces of the wheels covered with salt. The cheese is placed in a controlled room at 60°F and 85% rh and resalted daily for 4 more days (5 days total). Small holes are punched through the wheels of cheese from top to bottom of the flat surfaces to provide oxygen for mold growth. The cheese is placed in racks on its curved edge in the curing room and held at 50 to 55°F and not less than 95% humidity. At the end of the month, the cheese surfaces are cleaned; the cheese is wrapped in foil and placed in a 36 to 40°F cold room for 2 to 4 months (Table 5). The surfaces are again scraped clean, and the wheels are wrapped in new foil for distribution.

Originally, roquefort and blue cheese were cured in caves with high humidity and constant cool temperature. Refrigerating insulated blue cheese curing rooms to the optimum temperature is not difficult. However, maintaining a uniform relative humidity of not less than 95% without excessive expense seems to be an engineering challenge, at least in some plants.

Cottage Cheese

Cottage cheese is made from skim milk. It is a soft, unripened curd and generally has a cream dressing added to it. There are small and large curd types, and may have added fruits or vegetables. Plant equipment may consist of receiving apparatus, storage tanks, clarifier/separator, pasteurizer, cheese vats with mechanical agitation, curd pumps, drain drum, blender, filler, conveyors, and accessory items such as refrigerated trucks, laboratory testing facilities, and whey disposal equipment. The largest vats have a 45,000 lb capacity. The basic steps are separation, pasteurization, setting, cutting, cooking, draining and washing, creaming, packaging, and distribution.

Skim milk is pasteurized at the minimum temperature and time of 161°F for 15 s to avoid adversely affecting curd properties. If heat treatment is substantially higher, the manufacturing procedure must be altered to obtain good body and texture quality and reduce curd loss in the whey. Skim milk is cooled to the setting temperature, which is 86 to 90°F for the short set (5 to 6 h) and 70 to 72°F for the overnight set (12 to 15 h). A medium set is used in a few plants. For the short set, 5 to 8% of a good cultured skim milk (starter) and 1 to 1.5 mL of rennet diluted in water are added per 1000 lb of skim milk. For the long set, 0.25 to 1% starter and 0.5 to 1 mL of diluted rennet per 1000 lb are thoroughly mixed into the skim milk. The use of rennet is optional.The setting temperature is maintained until the curd is ready to cut. The whey acidity at cutting time depends on the total solids content of the skim milk (0.55% for 8.7% and 0.62% for 10.5%). The pH is typically 4.80, but it may be necessary to adjust for specific make procedures.

The curd is cut into 0.5 in. cubes for large curd and 0.25 in. for small curd cottage cheese. After the cut curd sets for 10 to 15 min, heat is applied to water in the vat jacket to maintain a temperature rise in the curd and whey of 2°F each 5 min. In very large vats, jacket heating is not practical, and superheated culinary steam in small jet streams is used directly in the vat; 20 to 30 min after cutting, very gentle agitation is applied. Heating rate may be increased to 3 or 4°F per 5 min as the curd firms enough to resist shattering. Cooking is completed when the cubes contain no whey pockets and have the desired firmness. The final temperature of curd and whey is usually 120 to 130°F, but some cheesemakers heat to 145°F when making the small curd.

After cooking, the hot water in both the jacket and the whey is drained. Wash water temperature is adjusted to about 70°F for the first washing and added gently to the vat to reduce curd temperature to 80 to 85°F. After gentle stirring and a brief hold, the water/whey mixture is drained. The temperature of the second wash is adjusted to reduce the curd temperature to 50 to 55°F, and to 40°F with a third wash. Water for the last wash may have 3 to 5 ppm of added chlorine. The curd is trenched for adequate drainage. The dressing is made from lowfat cream, salt, and usually 0.1 to 0.4% stabilizer based on cream weight. Salt averages 1%, and milkfat must be 4% or more in creamed cottage cheese or 2% in lowfat cottage cheese. The dressing is cooled to 40°F and blended into the curd.

A cheese vat can be reused sooner if the cheese pumps quickly convey curd and whey after cooking to a special tank for whey drainage, washing, and blending of dressing and curd. Creamed cottage cheese is transferred mechanically to an automatic packaging machine. One type of filler uses an oscillating cylinder that holds a specific volume. Another type has a piston in a cylinder that discharges a definite volume. Common retail containers are 32, 16, 12, and 8 oz sizes of semirigid plastic. Cottage cheese is perishable and must be stored at 40°F or lower to prolong the keeping quality to 2 or 3 weeks. A good yield is 15.5 lb of curd per 100 lb of skim milk with 9% total solids.

Other Cheeses

Table 6 presents data on a few additional common varieties of cheese in the United States. Except for soft ripened cheeses such as camembert and liederkranz, freezing cheese results in undesirable texture changes. This can be serious, as in the case of cream cheese, where a mealy, pebbly texture results. Other types, such as brick and limburger, undergo a slight roughening of texture, which is undesirable but which still might be acceptable to certain consumers. As a general rule, cheese should not be subjected to temperatures below 29°F.

When cured cheese is held above the melting point of milkfat, it becomes greasy because of oiling off. The oiling-off point of all types of cheese except processed cheese begins at 68 to 70°F. Consequently, storage should be substantially below the melting point (Table 7). Uncured cheese (i.e., cottage, cream) is highly perishable and thus should not be stored above 45°F and preferably at 35°F.

Table 5 Typical Blue Cheese Manufacturing Conditions

Processing Step	Temperature, °F	Relative Humidity, %	Time
Setting	85 to 86	—	1 h
Acid development	85 to 86	—	1 h
(after cutting)	92	—	120 s
Curd matting	70 to 75	80 to 90	18 to 24 h
Dry salting	60	85	5 days
Curing	50 to 55	>95	30 days
Additional curing	36 to 40	80	60 to 120 days

Table 6 Curing Temperature, Humidity, and Time of Some Cheese Varieties

Variety	Curing Temperature, °F	Relative Humidity, %	Curing Time
Brick	50 to 65	90	60 days
Romano	50 to 60	85	5 to 12 mos
Mozzarella	70	85	24 to 72 h
Edam	50 to 60	85	3 to 4 mos
Parmesan	55 to 60	85 to 90	14 mos
Limburger	50 to 60	90	2 to 3 mos

Table 7 Temperature Range of Storage for Common Types of Cheese

Cheese	Ideal Temperature, °F	Maximum Temperature, °F
Brick	30 to 34	50
Camembert	30 to 34	50
Cheddar	30 to 34	60
Cottage	30 to 34	45
Cream	32 to 34	45
Limburger	30 to 34	50
Neufchatel	32 to 34	45
Processed American	40 to 45	75
Processed brick	40 to 45	75
Processed limburger	40 to 45	75
Processed Swiss	40 to 45	75
Roquefort	30 to 34	50
Swiss	30 to 34	60
Cheese foods	40 to 45	55

Processing protects cheese from oiling off. By heating the bulk cheese to temperatures of 140 to 180°F, and incorporating emulsifying salts, a more stable emulsion is formed than in natural or nonprocessed cheese. Processed cheese will not oil off even at melting temperatures. Because of the temperatures used in processing, processed cheese is essentially a pasteurized product. Microorganisms causing changes in the body and flavor of the cheese during cure are largely destroyed; thus, there is practically no further flavor development. Consequently, the maximum permissible storage temperature for processed cheese is considerably higher than any of the other types. Table 7 shows the maximum temperatures of storage for cheese of various types.

Refrigerating Cheese Rooms

Cheeses that are to be dried before wrapping or waxing enter the cooler at approximately room temperature. Sufficient refrigerating capacity must be provided to reduce the cheeses to drying-room temperature. The product load may be taken as 1 ton for each 12,000 to 15,000 lb per day. Product load in a cheese-drying room is usually small compared to total room load. Extreme accuracy in calculating product load is not warranted.

When determining peak refrigerating load in a cheese-drying room, remember that peak cheese production may coincide with periods of high ambient temperature. In addition, these rooms normally open directly into the cheese-making room, where both temperature and humidity are quite high. Also, traffic in and out of the drying room may be heavy; therefore, ample allowance for door losses should be made. Two to three air changes per hour are quite possible during the flush season. See Chapter 13 for information on load calculations.

To maintain desired humidity, refrigerating units for the cheese-drying room should be sized to handle the peak summer load with not more than a 15°F difference between the return air temperature and evaporator temperature. Units operated from a central refrigerating system should be equipped with suction pressure regulators.

Temperature may be controlled through a room thermostat controlling a solenoid valve in the liquid supply to the unit or units, assuming a central refrigeration system is being used. Fans should be allowed to run continuously. A modulating suction-pressure regulator is not a satisfactory temperature control for a cheese-drying room because it causes undesirable variations in humidity.

Air circulation should only be enough to ensure uniform temperature and humidity throughout the room. Strong drafts or air currents should be avoided because they cause uneven drying and cracking of the cheeses. The most satisfactory refrigerating units are the ceiling-suspended between-the-rails type or the penthouse type. One unit for each 400 to 500 ft^2 of floor area usually ensures uniform conditions. One unit should be placed near the door to the room to cool warm, moist air before it can spread over the ceiling. Otherwise, condensation dripping from the ceiling and mold growth could result.

Humidity control during winter may present problems in cold climates. Because most of the peak-season refrigeration load is due to insulation losses and warm air entering through the door, refrigerating units may not operate enough during cold weather to remove moisture released by the cheese, resulting in excess humidity and improper drying. Within certain limits, the sensible load must be increased to meet the latent load. One way to do this is to run evaporators in a modified hot-gas defrost mode with fans energized to increase the sensible load on the space. If there are several units in the room, the refrigeration may be turned off on some while the evaporating pressure is lowered. Fans should be left running to ensure uniform conditions throughout the room. If these adjustments are not sufficient, or if automatic control of humidity is desired, it is necessary to use reheat coils (electric heaters, steam or hot-water coils, or hot gas from the refrigerating system) in the airstream leaving the units. A heating capacity of 15 to 20% of the refrigerating capacity of the units is usually sufficient to maintain humidity control.

A humidistat may be used to operate the heaters when humidity rises above the desired level. The heater should be wired in series, with a second room thermostat set to shut it off if room temperature becomes excessive. Because of variations in size and shape of drying rooms, it is impossible to generalize about air velocities and capacities. Airflow should be regulated so that the cheese feels moist for the first 24 h and then becomes progressively drier and firmer.

Calculating product refrigeration load for a cheese-curing room involves a simple computation of heat to be removed from the cheese at the incoming temperature to bring it to curing temperature, using 0.65 Btu/lb·°F as the specific heat of cheese. For most varieties, heat given off during curing is negligible.

Although fermentation of lactose to lactic acid is an exothermic reaction, this process is substantially completed in the first week after cheese is made; further heat given off during curing is of no significance. Assuming that average conditions for American cheese curing are approximately 45°F and 70% rh, if 30 to 35°F refrigerant is used in the cooling system, a humidity of about 70% will be maintained.

FROZEN DAIRY DESSERTS

Ice cream is the most common frozen dairy dessert. Legal guidelines for the composition of frozen dairy desserts generally follow federal standards. The amount of air incorporated during freezing is controlled for the prepackaged products by the standard specifying the minimum density, 4.5 lb/gal, and/or a minimum density of food solids, 1.6 solids/gal (21CFR135).

The basic dairy components of frozen dairy dessert are milk, cream, and condensed or nonfat dry milk. Some plants also use butter, butter oil, buttermilk (liquid or dry), and dry or concentrated sweet whey. The acid-type whey (e.g., from cottage cheese) can be used for sherbets.

Ice Cream

Milkfat content (called butterfat by some standards) is one of the principal factors in the legal standards for ice cream. Fats in other ingredients such as eggs, nuts, cocoa, or chocolate do not satisfy the legal minimum. Federal standards set the minimum milkfat content at 8% for bulky flavored ice cream mixes (e.g., chocolate) and 10% or above for the other flavors (e.g., vanilla). Manufacturers, however, usually make two or more grades of ice cream, one being competitively priced with the minimum legal fat content, and the others richer in fat, higher in total solids, and lower in overrun for a special trade. This ice cream may be made with a fat content of 16 or 18%, although most ice cream fat content ranges from 10 to 12%.

Serum solids content designates the nonfat solids from milk. The chief components of milk serum are lactose, milk proteins (casein, albumin, and globulin), and milk salts (sodium, potassium, calcium, and magnesium as chlorides, citrates, and phosphates). The following average composition for serum solids is useful for general calculations: lactose, 54.5%; milk proteins, 37.0%; and milk salts, 8.5%.

The serum solids in ice cream produce a smoother texture, better body, and better melting characteristics. Because serum solids are relatively inexpensive compared with fat, they are used liberally. The total solids content usually is kept below 40%.

The lower limit in the serum solids content, 6 to 7%, is found in a homemade type of ice cream, where the only dairy ingredients are milk and cream. Ice creams with an unusually high fat content are also kept near this serum solids content so that the total solids content will not be excessive. Most ice cream, however, is made with condensed or nonfat dry milk added to bring the serum solids content within the range of 10 to 11.5%. The upper extreme of 12 to 14% serum solids can avoid sandiness (gritty texture) only where rapid product sales turnover or other special means are used.

The sugar content of ice cream is of special interest because of its effect on the freezing point of the mix and its hardening behavior. The extreme range of sugar content encountered in ice cream is from 12 to 18%, with 16% being most representative of the industry. The chief sugar used is sucrose (cane or beet sugar), in either granulated or liquid form. Many manufacturers use dextrose and corn syrup solids to replace part of the sucrose. Some manufacturers prefer sucrose in liquid form, or in a mixture with syrup, because of lower cost and easier handling in tank car lots. In some instances, 50% of the sucrose content has been replaced by other sweetening agents. A more common practice is to replace one-fourth to one-third of the sucrose with dextrose or corn syrup solids, or a combination of the two.

Practically all ice cream is made with a stabilizer to help maintain a smooth texture, especially under the conditions that prevail in retail cabinets. Manufacturers who do not use stabilizers offset this omission by a combination of factors such as a high fat and solids content, the use of superheated condensed milk to help smooth the texture and impart body, and a sales program designed to provide rapid turnover. The most common stabilizing substances are carboxymethylcellulose (CMC) and sodium alginate, a product made from giant kelp gathered off the coast of California. Gelatin is used for some ice cream mixes that are to be batch pasteurized. Other stabilizers are locust or carob bean gum, gum arabic or acacia, gum tragacanth, gum Karaya, psyllium seed gum, and pectin. The amount of stabilizer commonly used in ice cream ranges from 0.20 to 0.35% of the mass of the mix.

Many plants now combine an emulsifier with the stabilizer to produce a smoother and richer product. The emulsifier reduces the surface tension between the water and fat phase to produce a drier-appearing product.

Egg solids in the form of fresh whole eggs, frozen eggs, or powdered whole eggs or yolks are used by some manufacturers. Flavor and color may motivate this choice, but the most common reason for selecting them is to aid the whipping qualities of the mix. The amount required is about 0.25% egg solids, with 0.50% being about the maximum content for this purpose. To obtain the desired result, the egg yolk should be in the mix at the time it is being homogenized.

In frozen custards or parfait ice cream, the presence of eggs in liberal amounts and the resulting yellow color are identifying characteristics. Federal standards specify a minimum 1.4% egg yolk solids content for these products.

Ice Milk

Ice milk commonly contains 3 to 4% fat (but not less than 2% or more than 7%) and 13 to 15% serum solids; formulations with respect to sugar and stabilizers are similar to those for ice cream. The sugar content in ice milk is somewhat higher, to build up the total solids content. The stabilizer content is also higher in proportion to the higher water content of ice milk. Overrun is approximately 70%.

Soft Ice Milk or Ice Cream

Machines that serve freshly frozen ice cream are common in roadside stands, retail ice cream stores, and restaurants. These establishments must meet sanitation requirements and have facilities for proper cleaning of the equipment, but very few blend and process the ice cream mix used. The mix is usually supplied either from a plant specializing in producing ice cream mix only or from an established ice cream plant. The mix should be cooled to about 35°F at the time of delivery, and the ice cream outlet should have ample refrigerated space to store the mix until it is frozen. To be served in a soft condition, this ice cream mix is usually frozen stiffer than would be customary for a regular plant operation with a 30 to 50% overrun. Some mixes are prepared only for soft serve. They are 1 to 2% greater in serum solids and have 0.5% stabilizer/emulsifier to aid in producing a smooth texture. Overrun is limited to 30 to 40% during freezing to the soft-serve condition.

Frozen Yogurt

Hard- or soft-serve frozen yogurt is similar to lowfat ice cream in composition and processing. The significant exception is the presence of a live culture in the yogurt.

Sherbets

Sherbets are fruit- (and mint-) flavored frozen desserts characterized by their high sugar content and tart flavor. They must weigh not less than 6 pounds to the gallon and contain between 1 and 2% milkfat and not more than 5% by mass of total milk solids (21CFR135). Although the milk solids can be supplied by milk, the general practice is to supply them by using ice cream mix. Typically, a solution of sugars and stabilizer in water is prepared as a base for sherbets of various flavors. To 70 lb of sherbet base, 20 lb of flavoring and 10 lb of ice cream mix are added. The sugar content of sherbets ranges from 25 to 35%, with 28 to 30% being most common. One example of a sherbet formula is milk solids, 5%, of which 1.5 is milkfat; sugar, 13%; corn syrup solids, 22%; stabilizer, 0.3%; and flavoring, acid, and water, 59.7%.

In sherbets, and even more so in ices, a high overrun is not desirable because the resulting product will appear foamy or spongy under serving conditions. Overrun should be kept within 25 to 40%. This fact and the problem of preventing bleeding (syrup leakage from the frozen product) emphasize the importance of the choice of stabilizers. If gelatin is selected as the stabilizer, the freezing conditions must be managed to avoid an excessive overrun. The gums added to ice cream are commonly used as the stabilizer in sherbets and ices.

Ices

Ices contain no milk solids, but closely resemble sherbets in other respects. To offset the lack of solids from milk, the sugar content of ices is usually slightly higher (30 to 32%) than in sherbets. A combination of sugars should be used to prevent crusty sugar crystallization, just as in the case of sherbets. The usual procedure is to make a solution of the sugars and stabilizer, from which different flavored ices may be prepared by adding the flavoring in the same general manner as mentioned for sherbets.

Ices contain few ingredients with lubricating qualities and often cause extensive wear on scraper blades in the freezer. Frequent resharpening of the blades is necessary. Where a number of freezers are available, and the main production is ice cream, it is desirable to confine freezing of ices and sherbets to a specific freezer or freezers, which should then receive special attention to resharpening.

Making Ice Cream Mix

The chosen composition for a typical ice cream would be

Fat	12.5%	Sugar	15.0%
Serum solids	10.5%	Stabilizer	0.3%

Mixing and Pasteurizing. Generally, the liquid dairy ingredients are placed in a vat equipped with suitable means of agitation to keep the sugar in suspension until it is dissolved. The dry ingredients are then added, with precautions to prevent lump formation of products such as stabilizers, nonfat dry milk solids, powdered eggs, and cocoa. Gelatin should be added while the temperature is still low to allow time for the gelatin to imbibe water before its dissolving is promoted by heat. Dry ingredients that tend to form lumps may be successfully added by first mixing them with some of the dry sugar so that moisture may penetrate freely. Where vat agitation is not fully adequate, sugar may be withheld until the liquid portion of the mix is partly heated so that promptness of solution avoids settling out.

The mix is pasteurized to destroy any pathogenic organisms, to lower the bacterial count to enhance the keeping quality of the mix and comply with bacterial count standards, to dissolve the dry ingredients, and to provide a temperature suitable for efficient homogenization. A pasteurizing treatment of 155°F maintained for 0.5 h is the minimum allowed. The mix should be homogenized at the pasteurizing temperature. Vat batches should be homogenized in 1 h and preferably less.

Practically all ice cream plants use continuous pasteurization using plate heat exchange equipment for heating and cooling the mix. If some solid ingredients are selected, such as skim milk powder and granulated sugar, a batch is made in a mixing tank at a temperature of 100 to 140°F. This preheated mix is then pumped through a heating section of the plate unit, where it is heated to a temperature of 175°F or higher, and held for 25 s while passing through a holding tube. The mix is then homogenized and pumped to the precooling plate section using city, well, or cooling tower water as the cooling medium. Final cooling may be done in an additional plate section, using chilled water as the cooling medium, or through a separate mix cooler. A propylene glycol medium is sometimes used for cooling to temperatures just above the freezing point.

Large plants generally use all liquid ingredients, especially if the production is automated and computerized. The ingredients are blended at 40 to 60°F. The mix passes through the product-to-product regeneration section of a plate heat exchanger with about 70% regeneration during preheating. The mix is HTST heated to not less than 175°F, homogenized, and held for 25 s. Greater heat treatment is common, and 220°F for 40 s is not unusual. The final heating may be accomplished with plate equipment, a swept-surface heat exchanger, or a direct steam injector or infusor.

Steam injection and infusion equipment may be followed by vacuum chamber treatment, in which the mix is flash-cooled to 180 to 190°F by a partial vacuum. It is further cooled through a regenerative plate section and additionally cooled indirectly to 40°F or less with chilled water. The chief advantage of the vacuum treatment is the flavor improvement of the mix if prepared from raw materials of questionable quality.

Homogenizing the Mix. Homogenization disperses the fat in a very finely divided condition so it will not churn out during freezing. Most of the fat in milk and cream is in globules <2 μm in diameter that form clumps 3 to 7 μm in diameter. Some of the clumps can be 12 μm or larger in diameter, especially if there has been some churning incidental to handling. In a properly homogenized mix, globules are seldom over 2 μm in diameter.

Cooling and Holding Mix. Methods of final cooling of ice cream mix after pasteurization depend on the equipment used and the final mix temperature desired. The mix should be as cold as possible, to about 30°F minimum for greater capacity and less refrigeration load on the ice cream freezers. Smaller plants generally use vat holding pasteurization with either a Baudelot (falling-film) surface cooler or a plate heat exchanger, both with precooling and final cooling sections. Precooling may be done with city, well, or cooling tower water, and mix leaving the precooling section is about 10°F warmer than the entering water temperature. The Baudelot cooler may be arranged for final cooling with chilled water, propylene glycol, or direct-expansion refrigerant. A final mix temperature of 30 to 33°F can be obtained over the surface cooler using propylene glycol or refrigerant. Final mix temperature when using chilled water is about 40°F.

For larger ice cream plants, where low mix temperature is desired and where plate pasteurizing equipment is installed, it may be desirable to use separate equipment for the final cooling. Where the mix is preheated to about 140°F, it will be precooled to about 10°F warmer than the entering precooling water temperature; final cooling can be done in a remote cooler. An ammonia-jacketed, scraped-surface chiller is often selected. Where cold liquid mix is used through a continuous pasteurizing, high-heat vacuum system with regeneration at about 70%, the temperature of the mix to the final cooling unit is 85°F, assuming 40°F original mix temperature and 190°F temperature of mix returning through the regenerating section.

Where plants have ample ice cream mix holding tank capacity (allowing mix to be held overnight), part of the final mix cooling may be accomplished by means of a refrigerated surface built into the tanks. Using refrigerated mix holding tanks, the average rate of cooling may be estimated at 1°F/h. Mixes with gelatin as a stabilizer should be aged 24 h to allow time for the gelatin to fully set. Mixes made with sodium alginate or other vegetable-type stabilizers develop maximum viscosity on being cooled, and can be used in the freezer immediately.

Freezing

The ice cream freezer freezes the mix to the desired consistency and whips in the desired amount of air in a finely divided condition. The aim is to conduct the freezing and later hardening to obtain the smoothest possible texture.

Freezing an ice cream mix means, of course, freezing a mixed solution. The solutes that determine the freezing point are the lactose and soluble salts contained in the serum solids and the sugars added as sweetening agents. Other constituents of the mix affect the freezing point only indirectly, by displacing water and affecting the in-water concentration of the solutes mentioned. Leighton (1927) developed a reliable method for computing the freezing points of ice cream mixes from their known composition. He added the lactose and sucrose content of the mix, expressed their concentration in terms of parts of sugar per 100 parts of water, and determined the freezing-point depression caused by the sugars by reference to published data for sucrose. This computation is justified because lactose and sucrose have the same molecular mass.

$$\text{\% Lactose in mix} = 0.545(\text{\% Serum solids})$$

$$\frac{(\text{\% Lactose} + \text{\% Sucrose})100}{\text{\% Water in mix}} = \begin{array}{l}\text{Parts lactose + Sucrose}\\ \text{per 100 parts water}\end{array}$$

To the freezing-point depression caused by these sugars, he added the depression caused by the soluble milk salts. The depression caused by the salts is computed as follows:

$$\text{Freezing-point depression caused by salt solids in °F} = \frac{4.27(\text{\% Serum solids})}{\text{\% Water in mix}}$$

Table 8 presents the freezing points of various ice creams and a typical sherbet and an ice, as computed by Leighton's method. The freezing point represents the temperature at which freezing

Table 8 Freezing Points of Typical Ice Creams, Sherbet, and Ice

	Composition of the Mix, %					
	Fat	Serum Solids	Sugar	Stabilizer	Water	Freezing Point, °F
Ice cream	8.5	11.5	15	0.4	64.6	27.59
	10.5	11.0	15	0.35	63.15	27.57
	12.5	10.5	15	0.30	61.7	27.55
	14.0	9.5	15	0.28	61.22	27.68
	16.0	8.5	15	0.25	60.25	27.79
	10.5	8.4	S 12, D 4	0.40	64.7	27.39
Sherbet	1.2	1.0	S 22, D 8	0.50	67.3	25.97
Ice	0	0	S 23, D 9	0.50	67.5	25.68

S = Sucrose D = Dextrose

Table 9 Freezing Behavior of Typical Ice Cream*

Water Frozen to Ice, %	Freezing Point of Unfrozen Portion, °F	Water Frozen to Ice, %	Freezing Point of Unfrozen Portion, °F
0	27.55	40	24.40
5	27.35	45	23.63
10	27.05	50	22.62
15	26.78	55	21.42
20	26.40	60	19.79
25	26.04	70	14.99
30	25.70	80	5.14
35	25.03	90	–22.29

*Composition, %: fat, 12.5; serum solids, 10.5; sugar, 15; stabilizer, 0.30; water, 61.7.

begins. As in the case of all solutions, the unfrozen portion becomes more concentrated as the freezing progresses, and the freezing temperature therefore decreases as freezing progresses. In a simple solution, containing only one solute, this trend progresses until the unfrozen portion represents a saturated solution of the solute, and thereafter the temperature remains constant until freezing has been completed. This temperature is known as the **cryohydric point** of the solute. In a mixed solution such as ice cream, which contains several sugars and a number of salts, no such point can be recognized.

Sugars remain in solution in a supersaturated state in the unfrozen portion of the product, because, by the time the saturation point has been reached, the temperature is so low and viscosity so high that essentially a glass state exists. In a mixed solution, however, the temperature required for complete freezing must be somewhat below the cryohydric point of the solute with the lowest cryohydric point. In ice cream, that solute is calcium chloride, contained as a component of serum solids. The cryohydric point of calcium chloride is –59.8°F. Therefore, ice cream ranges from 0 to 100% frozen within the approximate range of 27.5 to –67°F.

Therefore, the temperature to which ice cream has been frozen becomes a measure of the degree to which the water has been frozen, as illustrated by Table 9. In the table, the freezing points of the unfrozen portions of the third ice cream listed in Table 8 have been computed when 0 to 90% of the original water has been frozen.

Refrigeration Requirements. Exact calculation of refrigeration requirements is complicated by the number of factors involved. The specific heat of the mix varies with its composition. According to Zhadan (1940), the specific heat of food products may be computed by assuming the following specific heats in Btu/lb·°F for the chief components: carbohydrates, 0.34; proteins, 0.37; fats, 0.40; and water, 1.00. Salts are normally not included. Where they are present in significant amounts, as in ice cream (8.5% of the serum solids), a specific heat of 0.20 is accurate. The value given by Zhadan for fats is apparently for solid fats. For liquid milkfat, Hammer and Johnson (1913) found the specific heat to be 0.52. In addition, their data clearly show that the latent heat of fusion of fats becomes involved. From their data, the latent heat of fusion of milkfat is about 35 Btu/lb.

The change from liquid to solid fat occurs over a wide temperature range, approximately 80 to 40°F; in changing from solid to liquid fat, the range is approximately 50 to 105°F. This wide discrepancy between solidifying and melting behavior is apparently because milkfat is a mixture of glycerides, and mutual solubility of the glycerides is involved. In any case, the latent heat of fusion of fat is involved in cooling the mix from the pasteurizing and homogenizing temperature down to the aging temperature of 38 to 40°F. Instead of making detailed calculations, a specific heat of 0.80 Btu/lb·°F is assumed for ice cream mix, which is high for mixes ranging from 36 to 40% total solids.

In calculating the refrigeration required for freezing and hardening, a single value of a specific heat for frozen ice cream cannot be chosen. As shown in Table 9, any change in temperature in freezing and hardening involves some latent heat of fusion of the water, as well as the sensible heat of the unfrozen mix and the ice. Near the initial freezing point, much more latent heat of fusion is involved per degree temperature change than in well-hardened ice cream (e.g., at –10 to –11°F). For this reason, instead of using an overall value of specific heat, freezing load may be computed as follows:

1. First, determine the temperature to which the freezing is to occur; then determine (by calculations such as those used to develop Table 9) how much water will be converted to ice. The heat to be removed is the product of the heat of fusion of ice and the mass of water frozen.
2. Compute the sensible heat that must be removed in the desired temperature change, by treating the product as a mix; that is, use the specific heat for ice cream mix. The temperature change times the mass of product times 0.80 = sensible heat to be removed.

In such a calculation, the water present is treated as though it all remained in a liquid form until the desired temperature had been reached, although ice was forming progressively. Because ice has a specific heat of 0.492 Btu/lb·°F instead of 1.00 Btu/lb·°F as for water, this calculation errs in the direction of generous refrigeration. To offset this, the freezer agitation develops friction heat. Approximately 80% of the energy input in the motor of the freezer is converted to heat in the product. Where the product is frozen to a stiff consistency, power requirements increase, and should be added to the load calculation.

A gallon of ice cream mix weighs from 9 lb, for mixes with a high fat content, to 9.2 lb, for mixes with a low fat content and a high content of serum solids and sugar. The mass of a unit volume of ice cream varies with the mix and overrun (volume increase) according to the following relationship:

$$\text{Percentage overrun} = \frac{100(\text{Wt/gal of mix} - \text{Wt/gal of ice cream})}{\text{Wt/gal of ice cream}}$$

Freezing Ice Cream. Both batch and continuous ice cream freezers are in general use. Both are arranged with a freezer cylinder having either an annular space or coils around the cylinder, where cooling is accomplished by direct refrigerant cooling, either in a flooded arrangement with an accumulator or controlled by a thermostatic expansion valve. The freezer cylinder has a dasher, which revolves within the cylinder. Sharp metal blades on the dasher scrape the cylinder's inner surface to remove the frozen film of ice cream as it forms. Some batch freezers use plastic dashers and blades.

Batch freezers range in size from 2 to 40 quarts of ice cream per batch, the smaller sizes being used for retail or soft ice cream operations, and the 40 quart size used in small commercial ice cream plants or in large plants for running small special-order quantities. Batch freezers larger than 40 quarts have not been used extensively since the development of the continuous freezer.

In operation, a measured quantity of mix is placed in the freezer cylinder and the required flavor, fruit, or nuts are added as freezing of the mix progresses. Freezing is continued until the desired consistency is obtained in the operator's judgment or by the indication of a meter showing an increase in the current drawn by the motor as the partly frozen mix stiffens. At the desired point of freezing, the refrigeration is cut off from the freezer cylinder, usually by closing the refrigerant suction valve. The dasher continues operating until enough air has been taken into the mix by whipping action to produce the desired overrun, which is checked by taking and weighing a sample from the freezer. When the desired overrun is obtained, the entire batch is discharged from the freezer cylinder into cans or cartons, and the machine is then ready for a new batch of mix.

Output of a batch freezer varies with blade sharpness, refrigeration supplied, and overrun desired. The average maximum output for commercial batch freezers is 8 batches per hour. This schedule allows 3 to 4 min to freeze, 2 to 3 min to whip, and about 1 min to empty the ice cream and refill with mix. For this time schedule, it is assumed the ice cream is drawn from the freezer at not over 100% overrun, at a temperature of about 24°F and at a refrigerant temperature around the freezer cylinder of about –15°F.

Continuous ice cream freezers range in size from 40 to 2700 gal/h at 100% overrun, and they are used almost exclusively in commercial ice cream plants. Where large capacities are required, multiple units are installed with the ice cream discharge from several machines connected together to supply the requirements of automatic or semiautomatic packaging or filling machines. In operation, the ice cream mix is continuously pumped to the freezer cylinder by a positive displacement rotary pump. Air pressure within the cylinder is maintained from 20 psig to more than 100 psig, supplied by either a separate air compressor or drawn in with the mix through the mix pump. The mix entering the rear of the freezer cylinder becomes partly frozen and takes on the overrun because of air pressure and agitation of the dasher and freezer blades as it moves to the front of the cylinder and is discharged.

The output capacity of most continuous freezers can be varied from 50 to 100% rated capacity by regulating the variable-speed control supplied for the mix pump. Continuous freezers can be used for nearly every flavor of frozen dessert. Where flavors requiring nuts, whole fruits, or candy pellets are run, the base or unflavored mix is run through the continuous freezer and then passed through a fruit feeder, which automatically feeds and mixes the flavor particles into the ice cream. Ice cream can be discharged from continuous freezers at temperatures of 25°F, as required for ice cream bar (novelty) operations, up to a very stiff consistency at 20°F, as required for automatic filling of small packages.

Special low-temperature ice cream freezers are available to produce very stiff ice cream for extruded shapes, stickless bars, and sandwiches. Ice cream temperatures as low as 16°F can be drawn with some mixes. When ample refrigerating effect is supplied, ice cream discharge temperature can be varied by regulating the evaporator temperature around the freezer cylinder with a suction-pressure regulating valve. For filling cans and cartons, the average discharge temperature from the continuous freezer is about 22°F, when operating with ammonia in a flooded system at about –25°F.

To calculate accurately the refrigeration requirement for freezing the ice cream mix in the freezer, the weight of the mix per gallon and the amount of water should be known. This can be checked by weighing, knowing the percentage of water, or by calculating the weight from the mix formula, as in Example 2.

Example 2. Find the weight of mix for the following composition (by percent): milkfat, 12.0; serum solids, 10.5; sugar, 16.0; stabilizer, 0.25; egg solids, 0.25; and water, 61.0.

Solution: The specific gravity of the mix is

$$\frac{100}{\left(\frac{\%\ \text{Milkfat}}{0.93}+\frac{\%\ \text{Solids, not fat}}{1.58}+\frac{\%\ \text{Water}}{1.00}\right)}$$

$$=\frac{100}{(12/0.93)+(27/1.58)+(61/1.00)}=1.099$$

Wt per gal of mix = Wt of 1 gal of water × Specific gravity of mix

$$= 8.345 \times 1.099 = 9.17 \text{ lb}$$

The overrun in ice cream varies from 60 to 100%, which affects the required refrigeration. For a continuous freezer, the required refrigeration may be calculated as in Example 3.

Example 3. Assume a typical ice cream mix as listed in Example 2 with 100% overrun. The mix contains 61% water and goes to the freezer at a temperature of 40°F. Freezing would start in this mix at about 27°F, and 48% of the water would be frozen at 22°F.

The weight of mix required to produce 100 gal of ice cream is

$$\frac{100}{100+\%\ \text{Overrun}}\times 100\ \text{gal}\times \text{Wt mix per gal}$$

For the ice cream being considered, the weight of mix required for 100 gal would be

$$\frac{100}{100+100}\times 100\times 9.17 = 459\ \text{lb}$$

Calculations of capacity required to freeze 100 gal/h of ice cream are as follows:

Sensible heat of mix: 459 × (40 – 27) × 0.80 =	4,770 Btu/h
Latent heat: 459 × 0.61 × 0.48 × 144 =	19,350 Btu/h
Sensible heat of slush: 459 × (27 – 22) × 0.65 =	1,490 Btu/h
Heat from motors: 5.5 hp × 2545 =	14,000 Btu/h
Total =	39,610 Btu/h
5% losses from freezer and piping (estimated) =	1,980 Btu/h
Total refrigeration =	41,590 Btu/h
=	3.47 ton

Under these conditions, 3.5 tons of cooling capacity per 100 gal/h of 100% overrun ice cream is required.

In continuous freezer operations, heat gain from motors and losses from freezer and piping remains about the same at all levels of overrun, but the necessary refrigerating effect varies with the weight of mix required to produce 100 gal of ice cream, as shown in Table 10.

Hardening Ice Cream. After leaving the freezer, ice cream is in a semisolid state and must be further refrigerated to become solid enough for storage and distribution. The ideal serving temperature for ice cream is about 8°F; it is considered hard at 0°F. To retain a smooth texture in .hardened ice cream, the remaining water content

Table 10 Continuous Freezing Loads for Typical Ice Cream Mix

Overrun, %	Ammonia Refrigeration at 3 psig Suction Pressure, tons per 100 gal/h
60	4.04
70	3.88
80	3.74
90	3.61
100	3.50
110	3.39
120	3.30

must be frozen rapidly, so that the ice crystals formed will be small. For this reason, most hardening rooms are maintained at –20°F, and some as low as –30°F Most modern hardening rooms have forced-air circulation, usually from fan-coil evaporators. With the ice cream containers arranged to allow air circulation around them, the hardening time is about one-half that in rooms having overhead coils or coil shelves and gravity circulation. With forced-air circulation in the hardening room and average plant conditions, ice cream in 2.5 or 5 gal containers (or smaller packages in wire basket containers), all spaced to allow air circulation, hardens in about 10 h. Hardening rooms are usually sized to allow space for a minimum of three times the daily peak production and for a stock of all flavors, with the sizing based on 10 gal/ft^2 of floor area in a 9 ft high room when stacked loose, which includes aisles.

Some larger plants use ice cream hardening carton (carrier) freezers, which discharge into a low-temperature storage room. Because of the various size packages to be hardened, most tunnels are the air-blast type, operating at temperatures of –30 to –40°F and, in some cases, as low as –50°F. Containers under one-half gallon are usually hardened in these blast tunnels in about 4 h.

Contact-plate hardening machines are also used. They must continuously and automatically load and unload to introduce packages from the filler without delay. Compared to carton (carrier) freezers, contact-plate hardeners save space and power and eliminate package bulging. They are limited to packages of uniform thickness having parallel flat sides. These freezers are described in Chapter 16.

Temperature in the storage room is held at about –20°F. Space in storage rooms can be estimated at 25 gal/ft^2 when palletized and stacked solid 6 ft high, including space for aisles. Many storage rooms today use pallet storage and racking systems. These rooms may be 30 ft tall or more, some using stacker-crane automation. Freezer storages are described in Chapter 14.

Refrigeration required to harden ice cream varies with the temperature from the freezer and the overrun. The following example calculates the refrigeration required to harden a typical ice cream mix.

Example 4. Assume ice cream with 100% overrun enters the hardening room at a temperature of 25°F. At this temperature, approximately 30% of the water would be frozen in the ice cream freezer with the remainder to be frozen in the hardening room. The weight of one gallon of ice cream at 100% overrun, from a mix weighing 9.18 lb/gal, is 4.59 lb. The mix is assumed to contain 61% water, and the hardening room is at –20°F. Calculate the refrigeration required to harden the ice cream in Btu/gal.

Solution:

Latent heat of hardening: 4.59 × 0.61 × 0.70 × 144	=	282 Btu
Sensible heat: 4.59 × (25 + 20) × 0.50	=	103 Btu
Total	=	385 Btu
Loss due to heat of container and exposure to outside air, assumed 10%	=	40 Btu
Total Btu per gallon to harden	=	425 Btu

Percent overrun, when calculated on the basis of the quantity of ice cream delivered by the freezer or the quantity placed in the hardening room, would affect the refrigeration required, as shown in Table 11.

Table 11 Hardening Loads for Typical Ice Cream Mix

Overrun, %	Hardening Load, Btu/gal
60	532
70	500
80	470
90	447
100	425
110	405
120	386

Example 5. Calculate the refrigeration load in an ice cream hardening room, assuming 1000 gal of ice cream at 100% overrun are to be hardened in 10 h in a forced-air circulation room at a temperature of –20°F. The hardening room, for three times this daily output, should have 300 ft^2 of floor area measuring approximately 15 by 20 by 9 ft high. The total insulated surface of 1230 ft^2 requires 4 to 6 in. of urethane or equivalent. For this example, the heat conductance through the insulated surface is selected at 0.04 Btu/h·ft^2·°F. The average ambient temperature is assumed to be 90°F.

Solution:

Heat leakage: 1230 × 0.04 × (90 + 20)	=	5,410 Btu/h
Heat from fan motor: 2 hp × 2545/0.85	=	5,990 Btu/h
Heat from lights: 600 W × 3.412	=	2,050 Btu/h
Air infiltration and persons in room (approximately 20% leakage)	=	1,080 Btu/h
Hardening 1000 gal ice cream × 425 Btu/gal in 10 h	=	42,500 Btu/h
Total	=	57,030 Btu/h
	=	4.75 tons

Additional refrigeration load calculation information is located in Chapter 13.

Other products, such as sherbets, ices, ice milk, and novelties, usually represent a small percentage of the total output of the plant, but should be included in the total requirement of the hardening room.

Ice Cream Bars and Other Novelties

Ice cream plants may manufacture and merchandise a limited number of the many novelties. The most common are chocolate-coated ice cream bars, flavored ices, fudge pops, drumsticks, ice cream sundae cups, ice cream sandwiches, and so forth. Small plants freeze most of these products, especially those with sticks, in metal trays containing 24 molds, which are submerged in a special brine tank with a built-in evaporator surface and brine agitation. The product mix is prepared in a tank and cooled to 35 to 40°F. A controlled quantity of mix is poured into the tray molds or measured in with a dispenser. Tray molds are placed in the brine tank for complete freezing. Brine temperature is –30 to –36°F. The freezing rate should be rapid to result in small ice crystals, but it varies with the product and generally is 15 to 20 min. The frozen product is loosened from the molds by momentarily melting the outer layers of the product in a water bath. It is immediately removed from the molds; each is separately wrapped or put in a novelty bag and promptly placed in frozen storage for distribution.

Example 6. Show the refrigeration calculations to freeze 1200 flavored ices per h at 3 oz per pop, based on the mix containing 85% water.

Solution:

Estimated mass flow of mix: 1200 × 3/16 × 1.06 (sp gr)	=	239 lb/h
Cooling mix: 239 × (40 – 27) × 0.87	=	2,700 Btu/h
Freezing: 239 × 0.85 × 144	=	29,250 Btu/h
Subcooling: 239 × (27 + 30) × 0.5	=	6,810 Btu/h
Cooling trays (50/h) 50 × 8 lb × (60 + 30) × 0.12	=	4,320 Btu/h
Heat from agitator: 1 hp × 2545	=	2,550 Btu/h
Leakage through tank, 3 × 12 × 3 ft deep	=	750 Btu/h
Loss, top of tank and piping (assumed)	=	7,500 Btu/h
Total refrigeration load	=	53,880 Btu/h
	=	4.5 tons

In making ice cream, ice milk, and similar kinds of bars, the mix is processed through the freezer and is extruded in a viscous form at about 22°F. Using similar calculations, the estimated refrigeration load to freeze 100 dozen would be 2.2 tons for 3 oz ice cream bars with 100% overrun.

The equipment to make and package novelties in large plants is available in several designs and capacities. Some are limited to the manufacture of one or a few kinds of similar novelties. Other machines have more versatility; for example, they can be used to make novelties with or without sticks, coated or uncoated, and of numerous sizes, shapes, and flavor combinations. Some of these machines include packaging in a bag or wrap, plus placement and sealing in a carton in units of 6, 8, 12, 14, 18, 24, or 48. In other plants, a separate packaging unit may be required. Some units harden the product by air at a temperature within the range of –35 to –46°F. Brine, usually calcium chloride, with a specific gravity of 1.275 or more and a temperature of –28 to –38°F may be the hardening medium. Capacity varies with the shape and size of the specific product, but is commonly in the range of 3500 to 35,000 or more per hour. Novelty equipment in plants may be semiautomatic or automatic in performance of the necessary functions.

An example of a simple novelty machine is one that has two parallel conveyor chains on which the mold strips are fastened. The molds are conveyed through filling, stick inserting, freezing, and defrosting stages. The extractor conveyor removes the frozen product from the mold cups and carries it to packaging or through dipping; it is then discharged at packaging. In the meantime, the molds go through a wash and rinse and back to be filled. The novelty is either bagged or wrapped by machine, grouped and placed into cartons, and conveyed to cold storage.

Refrigeration Compressor Equipment Selection and Operation

Nearly all commercial ice cream plants, particularly larger ones, use ammonia multistage systems. Some smaller plants operate continuous ice cream freezers and refrigerate hardening rooms to acceptable temperatures with single-stage refrigerant compressors. In most cases, these smaller plants operate reciprocating compressors at conditions above the maximum compression ratio recommended by the manufacturer.

For economical operation, and to maintain reasonable limits of compression ratio, ice cream plants normally use multistage compression. For freezing ice cream, producing frozen novelties, and refrigerating an ice cream hardening room to –20°F, one or more booster compressors may be used at the same suction pressure, discharging into second-stage compressors, which also handle the mix cooling and ingredient cold storage room loads. If a carton freezer is used at a temperature of –40°F or below for ice cream hardening, two low-suction pressure systems should be used, the lower one for the carton freezer and the higher one for the ice cream freezers and storage room. Both low-suction pressure systems discharge into the high-stage compressor system. For plants with carton freezers arranged for large volumes, an analysis of operating costs may indicate savings in using three-stage compression with the low-temperature booster compressors used for the tunnel, discharging into the second-stage booster compressor system used for freezers and storage, and then the second-stage booster compressor system discharging into the third- or high-stage compressor system.

High-temperature loads in an ice cream plant usually consist of refrigeration for cooling and holding cream, cooling ice cream mix after pasteurization, cooling for mix holding tanks, refrigeration for the ingredient cold-storage room, and air conditioning for the production areas. If direct refrigerant cooling is used for these loads, then compressor selections for the high stage can be made at about 20°F saturated suction temperature and combined with the compressor capacity required to handle the booster discharge load. Approximately the same high-stage suction temperature can be estimated if ice cream mix and mix holding tanks are cooled by chilled water from a falling-film water chilling system. If an ice builder supplies chilled water for cooling pasteurized ice cream mix, it may be desirable to provide a separate compressor system to handle this refrigerating load rather than meeting all of the high-temperature loads at the reduced suction temperature required to make ice.

Refrigeration is a significant and important cost in an ice cream plant because of the relatively large refrigeration capacity required at low suction pressure (temperature). It is imperative to use efficient two- or three-stage compression systems at the highest suction pressures and lowest discharge pressures practical to achieve the desired product temperatures.

Effectiveness of the heat transfer surfaces is reduced by oil films, excessive ice and frost, scale, noncondensable gases, abnormal temperature differentials, clogged sprays, improper liquid circulation, poor airflow, and foreign materials in the system. Adequate operations and maintenance procedures for all components and systems should be used to ensure maximum performance and safety.

Process operation performance is also critical to the effectiveness of the refrigeration system. Items that adversely affect ice cream freezing rates include dull scraper blades, high mix inlet temperatures, low ice cream discharge temperatures, overrun below specifications, and incorrect mix composition and/or viscosity.

Rooms and storage areas should be well maintained to preserve insulation integrity. This includes doors and passageways, which may be a major source of air infiltration load.

New and updated ice cream plants should be equipped with microprocessor compressor controls and an overall computerized control system for operations and monitoring. When properly used, these controls help provide safe, efficient operation of the refrigeration system.

ULTRAHIGH-TEMPERATURE (UHT) STERILIZATION AND ASEPTIC PACKAGING (AP)

Ultrahigh-temperature sterilization of liquid dairy products destroys microorganisms with a minimum adverse effect on sensory and nutritional properties. **Aseptic packaging** containerizes the sterilized product without recontamination. Sterilization, in the true sense, is the destruction or elimination of all viable microorganisms. In industry, however, the term *sterilized* may refer to a product that does not deteriorate microbiologically, but in which viable organisms may have survived the sterilization process. In other words, heat treatment renders the product safe for consumption and imparts an extended shelf life microbiologically.

Sterilization Methods and Equipment

Retort sterilization of milk products has been a commercial practice for many years. It consists of sterilizing the product after hermetically sealing it in a metal or glass container. The heat treatment is sufficiently severe to cause a definite cooked off-flavor in milk and to decrease the heat-labile nutritional constituents of milk products. UHT-AP has the advantage of causing less cooked flavor, color change, and loss of vitamins while producing the same sterilization effect as the retort method.

UHT-AP has been applied to common fluid milk products (whole milk, 2% milk, skim milk, and half-and-half), various creams, flavored milks, evaporated milk, and such frozen dessert mixes as ice cream, ice milk, milk shakes, soft-serve, and sherbets. UHT-sterilized dairy foods include eggnog, salad dressings, sauces, infant preparations, puddings, custards, and nondairy coffee whiteners and toppings.

UHT sterilization is accomplished by rapidly heating the product to the sterilizing temperature, holding the temperature for a definite number of seconds, and then rapid cooling. The methods have been classified as direct steam or indirect heating. Advantages of direct methods include the following: (1) faster heating, (2) longer processing intervals between equipment cleanings, and (3) the flow rate is easier to change. Advantages of indirect methods include the following: (1) greater regeneration potential, (2) potable steam is not necessary, and (3) viscous products and those with small pieces of solids can be processed with the scraped-surface unit.

The direct steam method is subdivided into injection or infusion. In direct injection, steam is forced into the product, preferably in small streamlets, with enough turbulence to minimize localized overheating of the milk surfaces that the steam initially contacts. In infusion, the product is sprayed into a steam chamber. Advantages of infusion over injection are (1) slightly less steam pressure is required (with exceptions), (2) less localized overheating of a portion of the product, and (3) more flexibility for change of the product flow rate. Vacuum chambers are required for direct steam methods to remove the water added during heating.

The three important indirect systems are tubular, plate, and cylinder with mechanical agitation. In the tubular type, the tube diameter must be small and the velocity of flow high to maximize heat transfer into the product.

Essential components for direct steam injection are storage or balance tank, timing pump, preheater (tubular or plate), steam injector or infuser unit, holding unit, flow-diversion valve, vacuum chamber, aseptic pump, aseptic homogenizer, plate or tubular cooler, and control instruments. The minimum items of equipment for steam infusion are the same, except that the infuser is used to heat the product from the preheat to the sterilization temperature.

The necessary equipment for indirect systems is similar: storage or balance tank, timing pump, preheater (tubular or plate type, and preferably regenerative), homogenizer, final plate or tubular heater, holding tube or plate, flow-diversion valve, cooler (one to three stages), and control instruments. The mechanically agitated heat exchanger replaces the tubes or plates in the final heating stage. Otherwise, the same items of equipment are used for this system of sterilization.

In addition to the basic equipment, many combinations of essential and supplemental items of UHT equipment are available. For example, one variation on the indirect system is to use the pump portion of the homogenizer as a timing pump when it is installed after the balance tank. The first stage of homogenization may occur after preheating, and the second stage may occur after precooling. A vacuum chamber may be placed in the line after preheating, for precooling after sterilization, or installed in both locations. A condenser in the vacuum chamber allows the advantages of deaeration without moisture losses that otherwise would occur in the indirect system. In Europe, some indirect systems have a hold of several minutes after preheating, to reduce the rate of solids accumulation on the final heating surfaces of the tubes or plates. A bactofuge may be included in the line after preheating to reduce a high microbiological content, especially of bacterial spores.

Self-acting controls and other instrumentation are available to ensure automatic operation in nearly every respect. Particularly important is automatic control of temperatures for preheating, sterilizing, and precooling in the vacuum chamber, and to some extent, of the final temperature before packaging. This may include temperature-sensing elements to control heating and cooling and pressure-sensing elements for operating pneumatic valves. The cleaning cycle may be automated, beginning with a predetermined solids accumulation on specific heating surfaces. Timers regulate the various cleaning and rinsing steps.

In some systems, one or more aseptic surge tanks are installed between the UHT sterilizer and the AP equipment. Aseptic surge tanks allow either the sterilizer or AP equipment to continue operation if the other goes off-line. It also makes the use of two or more AP units easier than direct flow from the UHT sterilizer to the AP machines.

When aseptic surge tanks are used, they must be constructed to withstand the steam pressure required for equipment sterilization and be provided with a sterile air venting system. Aseptic surge tanks may be unloaded by applying sterile air to force product out to the AP equipment. The pressure for air unloading can be controlled at a constant value, making uniform filling possible even when one of several AP machines is removed from service.

Aseptic surge tanks make it possible to hold bulk product, even for several days, until it is convenient to package it.

Basic Steps. After the formula is prepared and the product standardized, the processing steps are (1) preheat to 150 to 170°F by a plate or tubular heat exchanger; (2) heat to a sterilization temperature of 285 to 300°F; (3) hold for 1 to 20 s at sterilizing temperature; and (4) cool to 40 to 100°F, depending on product keeping quality needs. Cooling may be by one to three stages; generally, two are used. The direct steam method requires at least two cooling stages. The first is flash cooling in a vacuum chamber to 150 to 170°F to remove moisture equal to the steam injected during sterilization. The second stage reduces the temperature to within 50 to 100°F. A third stage is required in most plants if the temperature is lowered to 35 to 50°F.

Products with fat are homogenized to increase stability of the fat emulsion. The direct method requires homogenization after sterilization and precooling. Homogenization may follow preheating or precooling, but usually follows preheating in the indirect method. Efficient homogenization is very important in delaying the formation of a cream layer during storage.

Sterilized plain milks (such as whole, 2%, and skim milk) are most vulnerable to having a cooked off-flavor. Consequently, the aim is to have low sterilization temperature and time consistent with satisfactory keeping quality. The total cumulative heat treatment is directly related to the intensity of the cooked off-flavor. The total processing time from preheating to cooling varies widely among systems. Most operations in the United States range from 30 to 200 s; in European UHT processes, it may be much longer.

Several factors influence the minimum sterilization temperature and time needed to control adverse effects on flavor and physical, chemical, and nutritional changes. Type of product, initial number of spores and their heat resistance, total solids of the product, and pH are the most important factors. Obviously, the relationship is direct for the number and heat resistance of the spores. Total solids also have a direct relationship, but for an acid pH, it is inverse.

Several terms are used to describe UHT's effect on the microbiological population. **Decimal reduction** refers to a reduction of 90% (e.g., 100 to 10, or one log cycle). An example of a three-decimal reduction is 10,000 to 10. **Decimal reduction time**, or ***D* value**, is the time required to obtain a 90% decrease. **Sterilizing effect**, or **bactericidal effect**, is the number of decimal reductions obtained and expressed as a logarithmic reduction ($\log_{10}$ initial count minus $\log_{10}$ final count). A sterilizing effect of six means one organism remaining from a million per mL (10^6), and seven would be one remaining in 10 mL (a final count of 10^{-1}).

The ***Z* value** is the temperature increase required to reduce the *D* value by one log cycle (90% reduction of microorganisms with the time held constant). The ***F* value** (thermal death time) is the time required to reduce the number of microorganisms by a stated amount or to a specific number. For example, assuming a *D* value of 36 s for *Bacillus substilis* spores at 250°F and a need to reduce the spores from 10^6 per mL to <1 per mL, the thermal treatment time would be 6×36 s = 216 s (*F* value).

Aseptic Packaging

Aseptic fillers are available for coated metal cans, glass bottles, plastic/paperboard/foil cartons, thermoformed plastic containers, blow-molded plastic containers, and plastic pouches. Aseptic can equipment includes a can conveyor and sterilizing compartment, filling chamber, lid sterilizing compartment, sealing unit, and instrument controls. The procedure sterilizes cans with steam at 550°F as they are conveyed, fills them by continuous flow, simultaneously sterilizes the lids, places the lids on the cans, and seals the lids onto the cans. Pressure control apparatus is not used for entry or exit of cans.

A similar system is used for glass bottles or jars. The jars are conveyed into a turret chamber; air is removed by vacuum; the jars are then sterilized for 2 s with wet steam at 60 psi and moved into the filler. The temperature of the glass equalizes to 120°F and the filling takes place. Next, the transfer is to the capper for placement of sterile caps, which are screwed onto the jars. The filling and capping space is maintained at 500°F.

Several aseptic blow-mold forming and filling systems have been developed. Each system is different, but the basic steps using molten plastic are (1) extruded into a parison, (2) extended to the bottom of the mold, (3) mold closed, (4) preblown with compressed air that inflates the plastic film into a bottle shape, (5) parison cut and the neck pinched, (6) final air application, (7) bottle filled and foam removed, (8) top sealed, and (9) mold opened and filled bottle ejected.

The basic steps in the manufacture of aseptic, thermoformed plastic containers are as follows: (1) a sheet of plastic (e.g., polystyrene) is drawn from a roll through the heating compartment and then multistamped into units, which constitute the containers; (2) these units are conveyed to the filler, which is located in a sterile atmosphere, and are filled; (3) a sheet of sterilized foil is heat sealed to the container tops; and (4) each container is separated by scoring and cutting.

One of the two aseptic systems for the plastic/paperboard/foil cartons draws the material from a roll through a concentrated hydrogen peroxide bath to destroy the microorganisms. The peroxide is removed by drawing the sheet between twin rolls, by exposure to ultraviolet light and hot air, or by superheated, sterilized air forced through small slits at high velocity. The packaging material is drawn downward in a vertical, sterile compartment for forming, filling by continuous flow, sealing, separation, and ejection.

In the other plastic/paperboard/foil aseptic system, the prepared, flat blanks are formed and the bottoms are heat sealed. In the next step, the inside surfaces are fogged with hydrogen peroxide. Sterilized hot air dissipates the peroxide. The cartons are conveyed into the aseptic filling and then into top-sealing compartments. Air forced into these two areas is rendered devoid of microorganisms by high-efficiency filters.

Operational Problems. Aseptic operational problems are reduced by careful installation of satisfactory equipment. The equipment should comply with *3-A Sanitary Standards*. Milk and milk products that are processed to be commercially sterile and aseptically packaged must also meet the *Grade A Pasteurized Milk Ordinance* and be processed in accordance with 21CFR113. Generally, the simplest system, with a minimum of equipment for product contact surfaces and processing time, is desirable. It is specifically important to have as few pumps and nonwelded unions as possible, particularly those with gaskets. The gaskets and O rings in unions, pumps, and valves are much more difficult to clean and sterilize than are the smooth surfaces of chambers and tubing. Automatic controls, rather than manual attention, is generally more satisfactory.

Complete cleaning and sterilizing of the processing and packaging equipment are essential. Milk solids accumulate rapidly on heated surfaces; therefore, cleaning may be necessary after 0.5 h of processing for tubular or plate UHT heat exchangers, although cleaning after 3 to 4 h is more common. Cleaning for the sterilizer, filler, and accessory equipment usually involves the CIP method for the rinse and alkali cleaning cycle, rinse, acid cleaning cycle, and rinse. Some plants only periodically acid-clean the storage tanks and packaging equipment (e.g., once or twice a week). Steam sterilization just before processing is customary. At 8 to 10 psig of wet steam, 1.5 to 2.0 h (or a shorter time at higher steam pressure) may be required. Water sterilized by steam injection or the indirect method can be used for rinsing and for the cleaning solution.

Survival of spores during UHT processing, or subsequent recontamination of the product before the container is sealed, is a constant threat. Inadequate sealing of the container also may be troublesome with certain types of containers. Another source of poststerilization contamination is airborne microorganisms, which may contact the product through inadequate sterilization of air that enters the storage vat for the processed product or through air leaks into the product upstream of the sterilized product pumps or homogenizer, if pressure is reduced. During packaging, air may contaminate the inside of the container or the product itself during filling and sealing.

Quality Control

Poor quality of raw materials must be avoided. The higher the spore count of the product before sterilization, the larger the spore survival number at a constant sterilization temperature and time. Poor quality can also contribute to other product defects (off-flavor, short keeping quality) because of sensory, physical, or chemical changes. Heat stability of the raw product must be considered.

A good-quality sterilized product has a pleasing, characteristic flavor and color that are similar to pasteurized samples. The cooked flavor should be slight or negligible, with no unpleasant aftertaste. The product should be free of microorganisms and adulterants such as insecticides, herbicides, and peroxide or other container residues. It should have good physical, sensory, and keeping quality.

Deterioration in storage may be evaluated by holding samples at 70, 89, 98, or 113°F for 1 or 2 weeks. The number of samples for storage testing should be selected statistically and should include samplings of the first and last of each product packaged during the processing day. To identify the source of microbiological spoilage, continuous aseptic sampling into standard-sized containers after sterilization and/or just ahead of packaging may be practiced. Sampling rate should be set to change containers each hour.

The rate of change in storage of sterilized milk products is directly related to the temperature. Commercial practice varies, with storage ranging from 35°F to room temperature, which may reach 95°F or higher. In plain milks, the cooked flavor may decrease the first few days, and then remain at its optimum for 2 to 3 weeks at 70°F before gradually declining. When milk is held at 70°F, a slight cream layer becomes noticeable in approximately 2 weeks and slowly continues until much of the fat has risen to the top. Thereafter, the cream layer becomes increasingly difficult to reincorporate or reemulsify.

Viscosity increases slightly the first few weeks at 70°F and then remains fairly stable for 4 to 5 months. Thereafter, gelation gradually occurs. However, milks vary in stability to gelation, depending on factors such as feeds, stage of lactation, preheat treatment, and homogenization pressure. Adding sodium tetraphosphate to some milks causes gelatin to develop more slowly.

Occasionally, some sterilized milk products develop a sediment on the bottom of the container because of crystallization of complex salt compounds or sugars. Browning can also occur during storage. Usually, off-flavors develop more rapidly and render the product unsalable before the off-color becomes objectionable.

Heat-Labile Nutrients

Results reported by researchers on the effects of UHT sterilization on heat-sensitive constituents of milk products lack consistency. The variability may be attributed to the analytical methods and to the difference in total heat treatment among various UHT systems, especially in Europe. In a review, Van Eeckelen and Heijne (1965) summarized the effect of UHT sterilization on milk as follows: slight or none for vitamins A, B_2, and D, carotene, pantothenic acid, nicotinic acid, biotin, and calcium; and no decrease in biological value of the proteins. The decreases were 3 to 10%, thiamine; 0 to 30%, B_6; 10 to 20%, B_{12}; 25 to 40%, C; 10%, folic acid; 2.4 to 66.7%, lysine; 34%, linoleic acid; and 13%, linolenic acid. Protein digestibility was decreased slightly. A substantial loss of vitamins C, B_6, and B_{12} occurred during a 90 day storage. Brookes (1968) reported that Puschel found that babies fed sterilized milk averaged a gain of 27 g per day, compared to 20 g for the control group.

EVAPORATED, SWEETENED CONDENSED, AND DRY MILK

Evaporated Milk

Raw milk intended for processing into evaporated milk should have a heat stability quality with little (preferably no) developed acidity. As milk is received, it should be filtered and held cold in a storage tank. The milkfat is standardized to nonfat solids at the ratio of 1:2.2785. It is then preheated to 200 to 205°F for 10 to 20 min or 240 to 260°F for 60 to 360 s to reduce product denaturation during sterilization. Moisture is removed by batch or (usually) continuous evaporation until the total solids have been concentrated to 2.25 times the original content.

Condensed product is pumped from the evaporator and, with or without additional heating, is homogenized at 2000 to 3000 psi and 120 to 140°F. It is cooled to 45°F and held in storage tanks for restandardization to not less than 7.9% milkfat and 25.9% total solids. The product is pumped to the packaging unit for filling cans made from tin-coated sheet steel. Filled cans are conveyed continuously through a retort, where the product is rapidly heated with hot water and steam to 245°F and held for 15 min to complete the sterilization. Rapid cooling with water to 80 to 90°F follows. The evaporated milk is agitated while in the retort by the can movement. Application of labels and placement of cans in shipping cartons are done automatically.

Storage at room temperature is common, but deterioration of flavor, body, and color is decreased by lowering the storage temperature to 50 to 60°F. Relative humidity should be less than 50% to reduce can and label deterioration. The recommended inversion of cases during storage to reduce fat separation is shown in Table 12.

Sweetened Condensed Milk

Sweetened condensed milk is manufactured similarly to evaporated milk in several aspects. One important difference, however, is that added sugar replaces heat sterilization to extend storage life. Filtered cold milk is held in tanks and standardized to 1:2.2942 (fat to nonfat solids). The milk is preheated to 145 to 160°F, homogenized at 2500 psi, and then heated to 180 to 200°F for 5 to 15 min or to 240 to 300°F for 30 s to 5 min. The milk is condensed in a vacuum pan to slightly higher than a 2:1 ratio. Liquid sugar (pasteurized) is added at the rate of 18 to 20 lb/100 lb of condensed milk.

As the mixture is pumped from the vacuum pan, it is cooled through a heat exchanger to 86°F and held in a vat with an agitator. Nuclei for proper lactose crystallization are provided by adding finely powdered lactose (200-mesh). The product is cooled slowly, taking an hour to reach 75°F with agitation. Then cooling is continued more rapidly to 60°F. Improper crystallization forms large crystals, which cause sandiness (gritty texture). The sweetened condensed milk is pumped to a packaging unit for filling into retail cans and sealing. Labeling cans and placement in cases is mechanized, similar to the process used for evaporated milk. The product is usually stored at room temperature, but the keeping quality is improved if stored below 70°F.

Condensing Equipment. Both batch and continuous equipment are used to reduce the moisture content of fluid milk products. The continuous types have single, double, triple, or more evaporating effects. The improvement in efficiency with multiple effects is shown in Table 13 by the reduction in steam required to evaporate 1 lb of water.

A simple evaporator is the horizontal tube. In this design, the tubes are in the lower section of a vertical chamber. During operation, water vapor is removed from the top and the product, from the bottom of the unit. For the vertical short-tube evaporator, the chamber design may be similar to the horizontal tube. The long-tube vertical unit may be designed to operate with a rising or falling film in the tubes; the latter is common. For the falling film, the product Reynolds number should be greater than 2000 for good heat transfer. Falling-film units may have a high *k*-factor at low temperature differentials, resulting in low steam requirements per mass of water evaporated per area of heating surface. Falling-film units have a rapid start-up and shutdown. Thermocompressing and mechanical compressing evaporators have the advantage of operating efficiently at lower temperatures, thus reducing the adverse effect on heat-sensitive constituents. Vapors removed from the product are compressed and used as a source of heat for additional evaporation.

Plate evaporators are also used. They are similar to plate heat exchangers used for pasteurization in that they have a frame and a number of plates gasketed to carry the product in a passage between two plates and the heating medium in adjacent passages. They differ in that, in addition to ports for product, they have large ports to carry vapor to a vapor separator. Vapors flow from the separator chamber to a condenser similar to those used for other types of evaporators. Plate evaporators require less head space for installation than other types, may be enlarged or decreased in capacity by a change in the number of plates, and offer a very efficient heat exchange surface.

Equipment Operation. Positive pumps of the reciprocating type are often used to obtain 24 in. Hg vacuum in the chamber. Steam jet ejectors may be used for 25 in. Hg vacuum, for one stage; two stages permit 28 in. Hg vacuum; and three stages, 29.8 in. Hg vacuum. Condensers between stages remove heat and may reduce the amount of vapor for the following stage. Either a centrifugal or reciprocating pump may be used to remove water from the condenser. A barometric leg may also be placed at the bottom of a 34 ft or longer condenser to remove water by gravity.

Dry Milk and Nonfat Dry Milk

There are two important methods of drying milk: spray and drum. Each has modifications, such as the foam spray and the vacuum drum drying methods. Spray drying exceeds by far the other methods for drying milk, and the largest volume of dried dairy product is skim milk.

In the manufacture of spray-dried nonfat dry milk (NDM), cold milk is preheated to 90°F and separated, and the skim milk for low-heat NDM is pasteurized at 161°F for 15 s or slightly higher and/or longer. It is condensed with caution to restrict total heat denaturation of the serum protein to less than 10%. This requires using a low-temperature evaporator or operating the first effect of a regular double-effect evaporator at a reduced temperature. After increasing the total solids to 40 to 45%, the condensed skim milk is continuously pumped from the evaporator through a heat exchanger to increase the temperature to 145°F. The concentrated skim milk is filtered and enters a positive pump operating at 3000 to 4000 psi,

Table 12 Inversion Times for Cases of Evaporated Milk in Storage

Storage Temperature, °F	Time
90	1 month initially and each 15 days
80	1 to 2 months
70	2 to 3 months
60	3 to 6 months

Table 13 Typical Steam Requirements for Evaporating Water from Milk

No. of Evaporating Effects	Steam Required, lb steam/lb water
Single	1.30 to 1.00
Double	0.60 to 0.50
Triple	0.40 to 0.35
Quadruple	0.30 to 0.25

which forces the product through a nozzle with a very small orifice, producing a mist-like spray in the drying chamber. Hot air of 290 to 400°F or higher dries the milk spray rapidly. Nonfat dry milk with 2.5 to 4.0% moisture is conveyed from the drier by pneumatic or mechanical means, then cooled, sifted, and packaged. Packages for industrial users are 50 or 100 lb bags.

High-heat nonfat dry milk is used principally in bread and other bakery products. The manufacturing procedure is the same as for low-heat NDM except that (1) the pasteurization temperature is well above the minimum (e.g., 175°F for 20 s or higher); (2) after pasteurization, the skim milk is heated to 185 to 195°F for 15 to 20 min, condensed; and (3) the concentrate is heated to 160 to 165°F before filtering and then is spray dried, similar to the process for low-heat NDM. Storage of low- or high-heat NDM is usually at room temperature.

Dry Whole Milk. Raw whole milk in storage tanks is standardized at a ratio of fat to nonfat solids of 1:2.769. The milk is preheated to 160°F, filtered or clarified, and homogenized at 160°F and 3000 psi on the first stage and 750 psi on the second stage. Heating continues to 200°F with a 180 s hold. The milk is drawn into the evaporator and the total solids are condensed to 45%. The product is continuously pumped from the evaporator, reheated in a heat exchanger to 160°F, and spray dried to 1.5 to 2.5% moisture. Dry whole milk (DWM) is cooled (not below dew point) and sifted through a 12-mesh screen. For industrial use within 2 or 3 months, the dry whole milk is packaged in 50 lb bags and held at room temperature or, preferably, well below 70°F.

To retard oxidation, the dry whole milk may be containerized in large metal drums or in retail-sized cans unsealed and subjected to 28 in. Hg vacuum. Less than 2% oxygen in the head space of the package after a week of storage is a common aim. Oxygen desorption from the entrapped content in lactose is slow, and two vacuum treatments may be necessary with a 7 to 10 day interval between them. Warm DWM directly from the drier desorbs oxygen more quickly than cooled DWM. Nitrogen is used to restore atmospheric pressure after each vacuum treatment. After the hold period for the first vacuum treatment, the DWM in the drums is dumped into a hopper, mechanically packaged into retailed-sized metal cans, and given the second vacuum treatment.

Foam spray drying allows the total solids to be increased to 50 to 60% in the evaporator before drying. Gas (compressed air or nitrogen) is distributed, by means of a small mixing device, into the condensed product between the high-pressure pump and the spray nozzle. A regulator and needle valve are used to adjust the gas flow into the product. Gas use is approximately 0.5 ft^3/gal of concentrated product. Otherwise, the procedure is the same as for regular drying. Foam spray-dried NDM has poor sinkability but good reconstitutability in water. The density is roughly half that of regular spray-dried NDM. The additional equipment for foam spray drying is limited to a compressor, storage drum, pressure regulator, and a few accessory items. The cost is relatively small, especially if compressed air is used.

Spray driers are made in various shapes and sizes, with one or many spray nozzles. Horizontal-spray driers may be box-shaped or a teardrop design. Vertical spray driers are usually cone- or silo-shaped.

Heat Transfer Calculations. The typical atomization in U.S. spray-drying plants is produced by a high-pressure pump that forces liquid through a small orifice in a nozzle designed to give a spreading effect as it emerges from the nozzle. Single-nozzle driers have an orifice opening diameter of 0.107 to 0.177 in. The diameter for multinozzle driers is 0.025 to 0.052 in. In Europe, the spinning disk is the most common means of atomizing in milk drying plants. Droplet sizes of 50 to 250 μm in diameter are usual. Droplet size has an inverse relationship to the rate of drying at a uniform hot-air temperature. Larger droplets require a higher air temperature and/or longer exposure than the smaller ones.

Other essential steps in spray drying are (1) moving, filtering, and heating the air; (2) incorporating the hot air with the product droplets; and (3) removing the moisture vapors and separating the moist air from the product particles. After passing through a rough or intermediate filter, the air is heated indirectly by steam coils or directly with a gas flame to 250 to 500°F. During the short drying exposure time, the air temperature drops to 160 to 200°F.

Thermal efficiency is the percentage of the total heat used to evaporate the water during the drying process. Efficiency is improved by heat recovery from exhaust air, decreased radiation loss, and high drying air temperature versus a low outlet air temperature. Roughly 2.2 to 3.2 lb of steam are needed to evaporate 1 lb of moisture in the drier.

$$\text{Thermal efficiency} = \frac{(1 - R/100)(t_1 - t_2)}{t_1 - t_0}$$

where

R = radiation loss, percentage of temperature decrease in drier
t_1 = inlet air temperature, °F
t_2 = outlet air temperature, °F
t_0 = ambient air temperature, °F

Most of the dried particles are separated from the drying air by gravity and fall to the bottom of the drier or the collectors. Fine particles are removed by directing the air/powder mixture through bag filters or a series of cyclone collectors. Air movement in the cyclone is designed to provide a centrifugal force to separate product particles. In general, several small-diameter cyclones with a fixed pressure drop will be more efficient for removal of fines than two large units.

The drier has sensing elements to continuously record the hot-air (inlet) temperature and moist-air (outlet) temperature. During drying, these temperatures are adjusted with a steam valve or gas inlet valve.

Drum Drying

Relatively little skim or whole milk is drum-dried. Drum-dried products, when reconstituted, have a cooked or scorched flavor compared to spray-dried products. Heat treatment during drying denatures the protein and results in a high insolubility index. In preparation for drying, skim milk is separated or whole milk is standardized to 1:2.769 [e.g., 3.2 fat and 8.86 solids-not-fat (SNF)]. The product is filtered or clarified, homogenized after preheating, and pasteurized. If the resulting dry product is intended for bakery purposes, the milk is heated to approximately 185°F for 10 min. The fluid product may be concentrated by moisture evaporation to not more than 2 to 1. The product is then dried on the drum(s): skim milk to not more than 4.0%, and whole milk to not more than 2.5% moisture. A blade pressed against the drum scrapes off the sheet of dried product. An auger conveys the dry material to the hammer mill, where it is pulverized and sifted through an 8-mesh screen. Drum-dried milks are usually packaged at the sifter into 50 to 100 lb kraft bags with a plastic liner.

A double-drum drier, with drums spaced 0.02 to 0.043 in. apart, is more common than a single drum for drying milk. Cast iron is used more often in drum construction than stainless steel or alloy steel and chrome plate steel. The knife metal must be softer than the drum. End plates on the drums create a reservoir into which the product, at 185°F, is sprayed the length of the drums. The steam-heated drums boil the product continuously as a thin film adheres to the revolving drums. After about 0.875 of one revolution, the film of product is dry and is scraped off. Drums normally revolve between 10 and 19 rpm. Steam pressure inside the drums is approximately 70 to 90 psi, as indicated by the pressure gage at the inlet of the condensate trap.

Steam pressure is adjusted for drying the product to the desired moisture content. Superheated steam scorches the product. Condensate inside the drums must be continuously removed, while the exterior vapors from the product are exhausted from the building with a hood and fan system. Capacity, dried product quality, and moisture content depend on many factors. Some important ones are steam pressure in drums, rotation speed of drums, total solids of product, smoothness of drum surface and sharpness of the knives, properly adjusted gap between the two drums, liquid level in drum reservoir, and product temperature as it enters the reservoir.

REFERENCES

Brookes, H. 1968. New developments in longlife milk and dairy products. *Dairy Industries* (May).

CFR. 2005a. Thermally processed low-acid foods packaged in hermetically sealed containers. 21CFR113. *Code of Federal Regulations*, U.S. Government Printing Office, Washington, D.C.

CFR. 2005b. Milk and cream. 21CFR131. *Code of Federal Regulations*, U.S. Government Printing Office, Washington, D.C.

CFR. 2005c. Cheeses and related cheese products. 21CFR133. *Code of Federal Regulations*, U.S. Government Printing Office, Washington, D.C.

CFR. 2005d. Frozen desserts. 21CFR135. *Code of Federal Regulations*, U.S. Government Printing Office, Washington, D.C.

FDA. 2001. *Grade "A" pasteurized milk ordinance*. U.S. Food and Drug Administration, Washington, D.C.

Hammer, B.W. and A. R. Johnson. 1913. The specific heat of milk and milk derivatives. *Research Bulletin* 14, Iowa Agricultural Experiment Station.

IAMFES. *3-A sanitary standards*. International Association of Milk, Food, and Environmental Sanitarians, Ames, IA.

Leighton, A. 1927. On the calculation of the freezing point of ice cream mixes and of the quantities of ice separated during the freezing process. *Journal of Dairy Science* 10:300.

Rishoi, A.H. 1951. Physical characteristics of free and globular milkfat. American Dairy Science Association, Annual Meetings (June).

Van Eeckelen, M. and J.J.I.G. Heijne. 1965. Nutritive value of sterilized milk. In *Milk sterilization*. Food and Agricultural Organization of the United Nations, Rome.

Zhadan, V.Z. 1940. Specific heat of foodstuffs in relation to temperature. *Kholod'naia Prom.* 18(4):32. (Russian) Cited from Stitt and Kennedy.

BIBLIOGRAPHY

Arbuckle, W.S. 1972. *Ice cream*, 2nd ed. AVI Publishing, Westport, CT.

Burdick, R. 1991. Salt brine cooling systems in the cheese industry. International Institute of Ammonia Refrigeration, *1991 Annual Meeting Technical Papers*.

Farrall, A.W. 1963. *Engineering for dairy and food products*. John Wiley & Sons, New York.

Griffin, R.C. and S. Sacharow. 1970. *Food packaging*. AVI Publishing, Westport, CT.

Hall, C.W. and T.I. Hedrick. 1971. *Drying of milk and milk products*. AVI Publishing, Westport, CT.

Henderson, F.L. 1971. *The fluid milk industry*. AVI Publishing, Westport, CT.

Judkins, H.F. and H.A. Keener. 1960. *Milk production and processing*. John Wiley & Sons, New York.

Kosikowski, F.V. 1966. *Cheese and fermented milk foods*. Published by author, Ithaca, NY.

Reed, G.H. 1970. *Refrigeration*. Hart Publishing, New York.

Sanders, G.P. Cheese varieties and descriptions. *Agriculture Handbook* 54. U.S. Department of Agriculture, U.S. Government Printing Office, Washington, D.C.

Webb, B.W. and E.A. Whittier. 1970. *Byproducts from milk*. AVI Publishing, Westport, CT.

Wilcox, G. 1971. *Milk, cream and butter technology*. Noyes Data Corporation, Park Ridge, NJ.

Wilster, G.H. 1964. *Practical cheesemaking*, 10th ed. Oregon State University Bookstore, Corvallis.

CHAPTER 21

EGGS AND EGG PRODUCTS

ABOUT 69% of the table eggs produced in the United States are sold as shell eggs. The remainder are further processed into liquid, frozen, or dehydrated egg products that are used in food service or as an ingredient in food products. Small amounts of further processed eggs are converted to retail egg products, mainly mayonnaise, salad dressings, and egg substitutes. Shell egg processing includes cleaning, washing, drying, candling for interior and exterior defects, sizing, and packaging. Further processed eggs require shell removal, filtering, blending, pasteurization, and possibly freezing or dehydration.

After processing, shell eggs intended for use within several weeks are stored at 39 to 45°F and relative humidities of 75 to 80%. These conditions reduce the evaporation of water from the egg, which would reduce the egg's weight and hasten breakdown of the albumen (an indicator of quality and grade). Shell eggs are also refrigerated during transportation, during short- and long-term storage, in retail outlets, and at the institutional and consumer levels.

Research has shown that microbial growth can be curtailed by holding eggs at less than 41°F. USDA regulations require eggs to be kept in an ambient temperature below 45°F until they reach the consumer, to prevent the growth of *Salmonella* (see October 27, 1992, United States Federal Register). Storage and display areas must be refrigerated and able to maintain ambient temperatures at 45°F.

SHELL EGGS

EGG STRUCTURE AND COMPOSITION

Physical Structure

The parts of an egg are shown in Figure 1, and physical properties of eggs are given in Table 1.

The **shell** is about 11% of the egg weight and is deposited on the exterior of the outer shell membrane. It consists of a mammillary layer and a spongy layer. The shell contains large numbers of pores (approximately 17,000) that allow water, gases, and small particles (e.g., microorganisms) to move through the shell. A thin, clear film (cuticle) on the exterior of the shell covers the pores. This material is thought to retard the passage of microbes through the shell and serves to prevent moisture loss from the egg's interior. The shape and structure of the shell provide enormous resistance to pressure stress, but very little resistance to breakage caused by impact.

Tough **fibrous shell membranes** surround the albumen. As the egg ages, cools, and loses moisture, an air cell develops on the large end of the egg between these two membranes. The size of the air cell is an indirect measure of the egg's age and is used to evaluate interior quality.

The preparation of this chapter is assigned to TC 10.9, Refrigeration Application for Foods and Beverages.

The **white** (albumen) constitutes about 58% of the egg weight. The white consists of a thin, inner chalaziferous layer of firm protein containing fibers that twist into chalazae on the polar ends of the yolk. These structures (Figure 1) anchor the yolk in the center of the egg, also known as the inner thick. The albumen consists of inner thin, outer thick, and outer thin layers.

The **yolk** constitutes approximately 31% of the egg weight. It consists of a yolk (vitelline) membrane and concentric rings of six yellow layers and narrow white layers (Figure 1). In the intact egg, these layers are not visible. Most of the egg's lipids and cholesterol are bounded into a lipoprotein complex that is found more in the white layers. The yolk contains the germinal disc, which consists of about 20,000 cells if the egg is fertile. However, eggs produced for human consumption are not fertile because the hens are raised without roosters.

Table 1 Physical Properties of Chicken Eggs

Property	Whole Egg	Albumen	Yolk
Solids, %	26.4	11.5	52.5
pH (fresh eggs)		7.6	6.0
Density, lb/ft^3	67.5	64.7	64.7
Surface tension, psi			6.38×10^{-4}
Freezing point, °F		31.2	31.0
Specific heat, Btu/lb·°F	0.772		
Viscosity, centipoise			
Thick white		164	
Thin white		4	
Electrical conductivity, mho/cm × 10^{-4}		8.25	0.07
Water activity, % relative humidity		97.8	98.1

Source: Burley and Vadehra (1989).

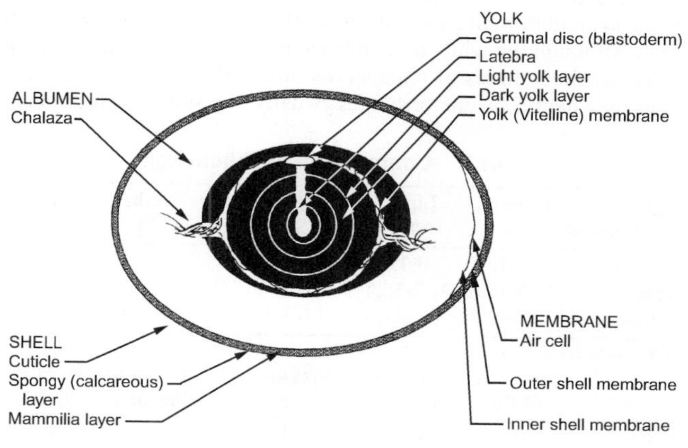

Fig. 1 Structure of an Egg

Chemical Composition

The weight of the chicken egg varies from 35 to 80 g or more. The main factors affecting weight and size are the bird's age, breed, and strain. Nutritional adequacy of the ration and ambient temperature of the laying house also influence egg size. Size affects the egg's composition, because the proportion of the parts changes as egg weight increases. For example, small eggs laid by young pullets just coming into production will have relatively more yolk and less albumen than eggs laid by older hens. Table 2 presents the general composition of a typical egg weighing 60 g.

The shell is low in water content and high in inorganic solids, mainly calcium carbonate as calcite crystals plus small amounts of phosphorus and magnesium and some trace minerals. Most of the shell's organic matter is protein. It is found in the matrix fibers closely associated with the calcite crystals and in the cuticle layer covering the shell surface. Protein fibers are also present in the pore canals extending through the shell structures to the cuticle, and in the two shell membranes. The membranes contain keratin, a protein that makes the membranes tough even though they are very thin.

Egg albumen, or egg white, is a gel-like substance consisting of ovomucin fibers and globular-type proteins in an aqueous solution. Ovalbumin is the most abundant protein in egg white. When heated to about 140°F, coagulation occurs and the albumen becomes firm. Several fractions of ovoglobulins have been identified by electrophoretic and chromatographic analyses. These proteins impart excellent foaming and beating qualities to egg white when making cakes, meringues, candies, etc. Ovomucin is partly responsible for the viscous characteristic of raw albumen and also has a stabilizing effect on egg-white foams, an important property in cakes and candy.

Egg white contains a small amount of carbohydrates. About half is present as free glucose and half as glycoproteins containing mannose and galactose units. In dried egg products, glucose interacts with other egg components to produce off-colors and off-flavors during storage; therefore, glucose is enzymatically digested before drying.

The yolk comprises one third of the edible portion of the egg. Its major components are water (48 to 52%), lipids (33%), and proteins (17%). The yolk contains all of the fatty material of the egg. The lipids are very closely associated with the proteins. These very complex lipoproteins give yolk special functional properties, such as emulsifying power in mayonnaise and foaming and coagulating powers in sponge cakes and doughnuts.

Nutritive Value

Eggs are a year-round staple in the diet of nearly every culture. The composition and nutritive value of eggs differ among the various avian species. However, only the chicken egg is considered here, as it is the most widely used for human foods.

Eggs contain high-quality protein, which supplies essential amino acids that cannot be produced by the body or that cannot be synthesized at a rate sufficient to meet the body's demands. Eggs are also an important source of minerals and vitamins in the human diet. Although the white and yolk are low in calcium, they contain substantial quantities of phosphorus, iron, and trace minerals. Except for vitamin C, one or two eggs daily can supply a significant portion of the recommended daily allowance for most vitamins, particularly the vitamins A and B_{12}. Eggs are second only to fish liver oils as a natural food source of vitamin D.

Fatty acids in the yolk are divided into saturated and unsaturated in a ratio of 1:1.8, with the latter further subdivided into mono- and polyunsaturated fatty acids in a ratio of 1:0.3. Eggs are a source of oleic acid, a monounsaturated fatty acid; they also contain polyunsaturated linoleic acid, an essential fatty acid. The fatty acid composition of eggs and the balance of saturated to unsaturated fatty acids can be changed by modifying the hen's diet. Several commercial egg products with modified lipids have been marketed.

Table 2 Composition of Whole Egg

Egg Component	Protein, %	Lipid, %	Carbohydrate, %	Ash, %	Water, %
Albumen	9.7-10.6	0.03	0.4-0.9	0.5-0.6	88.0
Yolk	15.7-16.6	31.8-35.5	0.2-1.0	1.1	51.1
Whole egg	12.8-13.4	10.5-11.8	0.3-1.0	0.8-1.0	75.5

Note: Shell is not included in above percentages.

	Percent of Egg	Calcium Carbonate	Magnesium Carbonate	Calcium Phosphate	Organic Matter
Shell	11	94.0	1.0	1.0	4.0

Source: Stadelman and Cotterill (1990).

EGG QUALITY AND SAFETY

Quality Grades and Weight Classes

In the United States, the Egg Products Inspection Act of 1970 requires that all eggs moving in interstate commerce be graded for size and quality. USDA standards for quality of individual shell eggs are shown in Table 3. The quality of shell eggs begins to decline immediately after the egg is laid. Aging of the egg thins the albumen and increases the size of the air cell. Carbon dioxide migration from the egg increases albumen pH and decreases vitelline membrane strength.

Classes for shell eggs are shown in Table 4. The average weight of shell eggs from commercial flocks varies with age, strain, diet, and environment. Practically all eggs produced on commercial poultry farms are processed mechanically. They are washed, candled, sized, then packed. Eggs are oiled at times to extend internal quality when they are to be transported long distances over a number of days. Although eggs are sold by units of 6, 12, 18, or 30 per package, the packaged eggs must maintain a minimum weight that relates to the egg size.

Table 3 United States Standards for Quality of Shell Eggs

Quality Factor	AA Quality	A Quality	B Quality
Shell	Clean Unbroken Practically normal	Clean Unbroken Practically normal	Clean to slightly stained[a] Unbroken Abnormal
Air cell	1/8 in. or less in depth Unlimited movement and free or bubbly	3/16 in. or less in depth Unlimited movement and free or bubbly	Over 3/16 in. in depth Unlimited movement and free or bubbly
White	Clear Firm	Clear Reasonably firm	Weak and watery Small blood and meat spots present[b]
Yolk	Outline slightly defined Practically free from defects	Outline fairly well defined Practically free from defects	Outline plainly visible Enlarged and flattened Clearly visible germ development but no blood Other serious defects

For eggs with dirty or broken shells, the standards of quality provide two additional qualities. These are:

Dirty	Check
Unbroken. Adhering dirt or foreign material, prominent stains, moderate stained areas in excess of B quality.	Broken or cracked shell but membranes intact, not leaking.[c]

[a]Moderately stained areas permitted (1/32 of surface if localized, or 1/16 if scattered).
[b]If they are small (aggregating not more than 1/8 in. in diameter).
[c]Leaker has broken or cracked shell and membranes, and contents are leaking or free to leak.
Source: *Federal Register*, 7CFR56, May 1, 1991. USDA *Agriculture Handbook* No. 75, p. 18.

Table 4 United States Egg Weight Classes for Consumer Grades

Size or Weight Class	Minimum Net Weight per Dozen, oz	Minimum Net Weight per 30-Dozen Case, lb	Minimum Weight for Individual Eggs, oz
Jumbo	30	56.0	2.42
Extra Large	27	50.5	2.17
Large	24	45.0	1.92
Medium	21	39.5	1.67
Small	18	34.0	1.42
Peewee	15	28.0	

Quality Factors

Besides legal requirements, egg quality encompasses all the characteristics that affect an egg's acceptability to a particular user. The specific meaning of quality may vary. To a producer, it might mean the number of cracked or loss eggs that cannot be sold, or the percentage of undergrades on the grade-out slip. Processors associate quality with prominence of yolk shadow under the candling light and resistance of the shell to damage on the automated grading and packing lines. The consumer looks critically at shell texture and cleanliness and the appearance of the broken-out egg and considers these factors in their relationship to a microbially safe product.

Shell Quality. Strength, texture, porosity, shape, cleanliness, soundness, and color are factors determining shell quality. Of these, shell soundness is the most important. It is estimated that about 10% of all eggs produced are cracked or broken between oviposition and retail sale. Eggs that have only shell damage can be salvaged only for their liquid content, but eggs that have both shell and shell membrane ruptured are regarded as a loss and cannot be used for human consumption. Shell strength is highly dependent on shell thickness and crystalline structure, which is affected by genetics, nutrition, length of continuous lay, disease, and environmental factors.

Eggs with smooth shells are preferred over those with a sandy texture or prominent nodules that detract from the egg's appearance. Eggs with rough or thin shells or other defects are often weaker than those with smooth shells. Although shell texture and thickness deteriorate as the laying cycle progresses, the exact causes of these changes are not fully understood. Some research suggests that debris in the oviduct collects on the shell membrane surface, resulting in rough texture formation (nodules).

The number and structure of pores are factors in microbial penetration and loss of carbon dioxide and water. Eggs without a cuticle or with a damaged cuticle are not as resistant to water loss, water penetration, and microbial growth as those with this outer proteinaceous covering. External oiling of the shell provides additional protection.

Eggs have an oval shape with shape indexes (breadth/length × 100) ranging from 70 to 74. Eggs that deviate excessively from this norm are considered less attractive and break more readily in packaging and in transit. Egg shape is changing to a more rounded shape, which is resulting in a stronger shell.

Shells with visible soil or deep stains are not allowed in a high-quality pack of eggs. Furthermore, soil usually contains a heavy load of microorganisms that may penetrate the shell, get into the contents, and cause spoilage.

Shell color is a breed characteristic. Brown shells owe their color to a reddish-brown pigment, ooporphyrin, which is derived from hemoglobin. The highest content of the pigment is near the surface of the shell. White shells contain a small amount of ooporphyrin, too, but it degrades soon after laying by exposure to light. Brown-shelled eggs tend to vary in color.

Albumen Quality. Egg white viscosity differs in various areas of the egg. A dense layer of albumen is centered in the middle and is most visible when the egg is broken out onto a flat surface. Raw albumen has a yellowish-green cast. In high-quality eggs, the white should stand up high around the yolk with minimum spreading of the outer thin layer of the albumen. The quality of thick albumen in the freshly laid egg is affected by genetics, duration of continuous production, and environmental factors. Albumen quality generally declines with age, especially in the last part of the laying cycle. Breakdown of thick white is a continuing process in eggs held for food marketing or consumption. The rate of quality loss depends on holding conditions and the length of time required to cool the egg.

Intensity of color is associated with the amount of riboflavin in the ration. The albumen of top-quality eggs should be free of any blood or meat spots. Incidence of non-meat spots such as blood spots and related problems has been reduced to such a low level by genetic selection that it is no longer a serious concern.

The chalazae may be very prominent in some eggs and can create a negative reaction from consumers who are unfamiliar with these structures (see Figure 1). The twisted, rope-like cords are merely extensions of the chalaziferous layer surrounding the yolk and are a normal part of the egg. The chalazae stabilize the yolk in the center of the egg.

Yolk Quality. Shape and color are the principal characteristics of yolk quality. In a freshly laid egg, the yolk is nearly spherical, and when the egg is broken out onto a flat surface, the yolk stands high with little change in shape. Shell and albumen tend to decline in quality as the hen ages. However, yolk quality, as measured by shape, remains relatively constant throughout the laying cycle.

Yolk shape depends on the strength of the vitelline membrane and the chalaziferous albumen layer surrounding the yolk. After oviposition, these structures gradually undergo physical and chemical changes that decrease their ability to keep the yolk's spherical shape. These changes alter the integrity of the vitelline membrane so that water passes from the white into the yolk, increasing the yolk's size and weakening the membrane.

Color as a quality factor of yolk depends on the desires of the user. Most consumers of table eggs favor a light to medium yellow color, but some prefer a deeper yellowish orange hue. Processors of liquid, frozen, and dried egg products generally desire a darker yolk color than users of table eggs because these products are used in making mayonnaise, doughnuts, noodles, pasta, and other foods that depend on eggs for their yellowish color. If laying hens are confined, yolk color is easily regulated by adjusting the number of carotenoid pigments supplied in the hen's diet. Birds with access to growing grasses and other plants usually produce deep-colored yolks of varying hues.

Yolk defects that detract from their quality include blood spots, embryonic development, and mottling. Blood on the yolk can be from (1) hemorrhages occurring in the follicle at the time of ovulation, or (2) embryonic development that has reached the blood-forming stage. The second source is a possibility only in breeding flocks where males are present.

Yolk surface mottling or discoloration can be present in the fresh egg or may develop during storage and marketing. Very light mottling, resulting from an uneven distribution of moisture under the surface of the vitelline membrane, can often be detected on close examination, but this slight defect usually passes unnoticed and is of little concern. Certain coccidiostats (nicarbazin) and wormers (piperazine citrate and dibutylin dilaurate) have been reported to cause mottled yolks and should not be used above recommended levels in layer rations. More serious are the olive-brown mottled yolks produced by rations containing cottonseed products with excessive amounts of free gossypol. This fat-soluble compound reacts with iron in the yolk to give the discoloration. Cottonseed meal may also have cyclopropanoid compounds that increase vitelline membrane permeability. When iron from the yolk passes through the membrane and reacts with the conalbumen of the white, a pink pigment is formed in the albumen. Cyclopropanoid compounds also cause yolks to have a higher proportion of saturated fats than normal, giving the yolks a pasty, custard-like consistency when they are cooled.

Flavors and Odors. When birds are confined and fed a standard ration, eggs have a uniform and mild flavor. Off-flavors can be caused by rations with poor-quality fish meal containing rancid oil or by birds having access to garlic, certain wild seeds, or other materials foreign to normal poultry rations. Off-flavors or odors from rations are frequently found in the yolk, because many compounds imparting off-flavors are fat-soluble. Once eggs acquire off-flavor during storage, their quality is unacceptable to consumers. Eggs have a great capacity to absorb odors from the surrounding atmosphere (Carter 1968). Storage should be free from odor sources such as apples, oranges, decaying vegetable matter, gasoline, and organic solvents (Stadelman and Cotterill 1990). If this cannot be avoided, odors can be controlled with charcoal absorbers or periodic ventilation.

Control and Preservation of Quality

Egg quality is evaluated by shell appearance, air-cell size, and the apparent thickness of the yolk and white. Some changes that occur during storage are caused by chemical reaction and temperature effects. As the egg ages, the pH of the white increases, the thick white thins, and the yolk membrane thins. Ultimately, the white becomes quite watery, although total protein content changes very little. Some coincidental loss in flavor usually occurs, although it develops more slowly. A low storage temperature and shell oiling slow down the escape of carbon dioxide and moisture and prevent shrinkage and thinning of the white. Clear white mineral oil sprayed on the shell after washing partially protects the egg, but its use in commercial operations is diminishing. Rapid cooling will also reduce moisture loss.

Egg quality loss is slowed by maintaining egg temperatures near the freezing point. Albumen freezes at 31.2°F, and the yolk freezes at 31°F. Stadelman et al. (1954) and Tarver (1964) found that eggs stored for 15 or 16 days at 45 to 50°F had significantly better quality than eggs stored at 57 to 61°F.

Stadelman and Cotterill (1990) recommend that storage humidity be maintained between 75 and 80%. As a rule, eggs lose about 1% of their weight per week in storage. When large amounts of eggs are palletized, humidity in the center of the pallet may be higher than that of the surrounding air. Therefore, airflow through the eggs is needed to remove excess humidity above 95% to prevent mold growth and decay.

Albumen quality loss is associated with carbon dioxide loss from the egg. Quality losses can be reduced by increasing carbon dioxide levels around the eggs. Controlled-atmosphere storage and modified-atmosphere packaging have been studied, but they are not used commercially because eggs typically do not need long-term storage. Oiling also helps retard carbon dioxide and moisture loss.

Egg Spoilage and Safety

Microbiological Spoilage. Shell eggs deteriorate in three distinct ways: (1) decomposition by bacteria and molds, (2) changes from chemical reactions, and (3) changes because of absorption of flavors and odors from the environment. Dirty or improperly cleaned eggs are the most common source of bacterial spoilage. Dirty eggs are contaminated with bacteria. Improper washing by immersing the egg in water colder than the eggs or water with high iron content increases the possibility of contamination, although it removes evidence of dirt. Most improperly cleaned eggs spoil during long-term storage. Therefore, extremely high sanitary standards are required when washing eggs that will go into long-term storage.

Eggs contaminated with certain microorganisms spoil quickly, resulting in black, red, or green rot, crusted yolks, mold, etc. However, eggs occasionally become heavily contaminated without any outward manifestations of spoilage. Clean, fresh eggs are seldom contaminated internally. It has been shown that egg sweating caused by fluctuations in environmental temperatures or humidity does not result in increased bacteria and/or mold spoilage (Ernst et al. 1998).

Preventing Microbial Spoilage. Egg quality can be severely jeopardized by invasion of microorganisms that cause off-odors and off-flavors. With frequent gathering, proper cleaning, and refrigeration, sound-shell eggs that move quickly through market channels have few spoilage problems.

Sound-shell eggs have a number of mechanical and chemical defenses against microbial attack. Although most of the shell pores are too large to impede bacterial movement, the cuticle layer, and possibly materials within the pores, offer some protection, especially if the shell surface remains dry. Bacteria that successfully penetrate the shell are next confronted by a second set of physical barriers, the shell membranes.

Microorganisms reaching the albumen find it unfavorable for growth. Movement is retarded by the egg white's viscosity. Also, most bacteria prefer a pH near neutral, but the pH of egg white, initially at 7.6 when newly laid, increases to 9.0 or more after several days, providing a deterring alkaline condition.

Conalbumen, which is believed to be the main microbial defense system of albumen, complexes with iron, zinc, and copper, thus making these elements unavailable to the bacteria and restricting their growth. The chelating potential increases with the rise in albumen pH.

Eggs can ward off a limited quantity of organisms, but should be handled in a manner that minimizes contamination. Egg washing must be done with care. Proper overflow, maintenance of a minimum water temperature of 90°F (as required by USDA regulations), and use of a sufficient quantity of approved detergent-sanitizer are important for effective cleaning. The wash water should be at least 20°F warmer than the internal temperature of the eggs to be washed. Likewise, the rinse water should be a few degrees higher than the wash water. Under these conditions, the contents of the eggs expand to create a positive pressure, which tends to repel penetrations of the shell by microorganisms.

Regular changes of the wash water, as well as thorough daily cleaning of the washing machine, are very important. When the wash water temperature exceeds the egg temperature by more than 50°F, an inordinate number of cracks in the shells, called thermal cracks, occur. Excessive shell damage also occurs if the washer and its brushes are not properly adjusted. Most egg processors use wash waters at temperatures of 110 to 125°F.

In-Shell Egg Pasteurization

In-shell egg pasteurization is a process of reducing the potential pathogenic organisms in intact shell eggs. These would be used in institutional settings where susceptible human populations want to eat eggs cooked in their intact state. This process is covered by the 1997 USDA/FDA joint published initial standards for the processing and labeling of pasteurized shell eggs. The FDA defined the target shell egg pasteurization criterion as a "5-log reduction in *Salmonella* count" per egg.

The supply of eggs for this process are USDA Grade AA eggs which contain 0% checks. These eggs must go through traditional egg processing before diversion to the pasteurization process. Typically, because of the increased costs of the process, only large and extra-large eggs are used. This process takes graded shell eggs through a series of baths that raise the internal temperature of the egg to destroy *Salmonella* and other potential pathogens. During heating, the eggs are agitated by air bubbles created by air injection at the bottom of the tanks. The eggs are then rapidly cooled in water baths to an internal temperature of 45°F. The chilling process stops the pasteurization process, after which a protective seal is applied to the shell surface to preserve the safety and quality of the egg.

HACCP Plan for Shell Eggs

Many of the procedures for the control of microorganisms are managed by the Hazard Analysis for Critical Control Points (HACCP), which is currently implemented in U.S. egg farms, egg packaging sites, egg processing facilities, and the distribution system. Information on the fundamentals of the HACCP system can be found in Chapter 12.

HACCP systems in the egg industry focus mainly on the prevention of *Salmonella* food poisoning. In the past, *S. typhimurium* was the leading strain in food poisoning related to eggs. However, since 1985 *S. enteritidis* has taken the leading role in egg-related salmonellosis illnesses (about 25%).

Salmonella is found naturally in the intestines of mice, rats, snakes, and wild birds and not in domesticated chickens. Chicken feed, which attracts rodents and birds, is the main source of chicken intestine contamination. Unfortunately, *S. enteritidis* can invade the hen ovaries and contaminate the developing yolks, thus being transferred into the egg interior. There it is unreachable by sanitizing agents. Pasteurization of eggs in the shell is one method of dealing with this internal contamination. Fortunately, only a very small portion of eggs are internally contaminated. Because the number of internal bacteria is very small, immediate cooling to 45°F and preferably to 41°F will suppress bacterial growth to below the hazard level until the egg is consumed, normally 10 to 30 days after being laid.

SHELL EGG PROCESSING

Off-Line and In-Line Processing

Poultry farms either send eggs to a processing plant or package them themselves. On commercial farms, the hens reside in cages with sloped floors. Eggs immediately roll onto a gathering tray or conveyor, where they are either gathered by hand, packed on flats, and stored for transport to an processing line (off-line); or the eggs are conveyed directly from the poultry house to a packing machine (in-line) operation. Machines can package both in-line and off-line eggs, thereby increasing the flexibility of the operation (Figure 2). Off-line operations have coolers both for incoming eggs and for outgoing finished product (Figure 3). An in-line operation has only one cooler for the outgoing finished product (Figure 4).

Figure 5 illustrates material flow during egg packaging in an off-line facility. Egg packaging machines wash the eggs by brushing with warm detergent solution followed by rinsing with warm water and sanitizing with an approved sanitizing agent. Sodium hydrochloride is most commonly used.

The eggs are then dried by air and moved by conveyor, which rotates the egg as they enter the candling booth. There, a strong light source under the conveyor illuminates the eggs' internal and shell defects. Two operators (candlers) remove defective eggs. The eggs are then weighed and sized automatically and the different sizes are packaged into cartons (12 eggs) or flats (20 or 30 eggs).

Automated candling can now detect and remove eggs with cracks, dirt, and internal defects, with little human intervention. This has raised the limit of 250 cases per hour (with manual candling) to 500 to 800 cases per hour. However, only very large facilities and egg-breaking operations tend to use automated candling; many others still operate at 250 to 300 cases an hour. In shell egg packaging, speed is limited by case and pallet packaging, which are not automated.

Kuney et al. (1992) demonstrated the high cost of good eggs overpulled in error by candlers. Machine speed was the major factor related to overpulling. Packaging is another area that could be automated because feeding packaging materials, packaging cartons or flats into cases, and palletizing are still largely manual operations.

EFFECT OF REFRIGERATION ON EGG QUALITY AND SAFETY

Refrigeration is the most effective and practical means for preserving quality of shell eggs. It is widely used in farm holding rooms, processing plants, and in marketing channels. Refrigeration conditions for shell eggs to prevent quality loss during short- and long-term storage are as follows:

Temperature, °F	Relative Humidity, %	Storage Period
45	75 to 80	2 to 3 weeks
39 to 45	75 to 80	2 to 4 weeks
29 to 31	85 to 92	5 to 6 months

A relative humidity of 75 to 80% in egg storage rooms must be maintained to prevent moisture loss with a subsequent loss of egg weight. Too high a relative humidity causes mold growth,

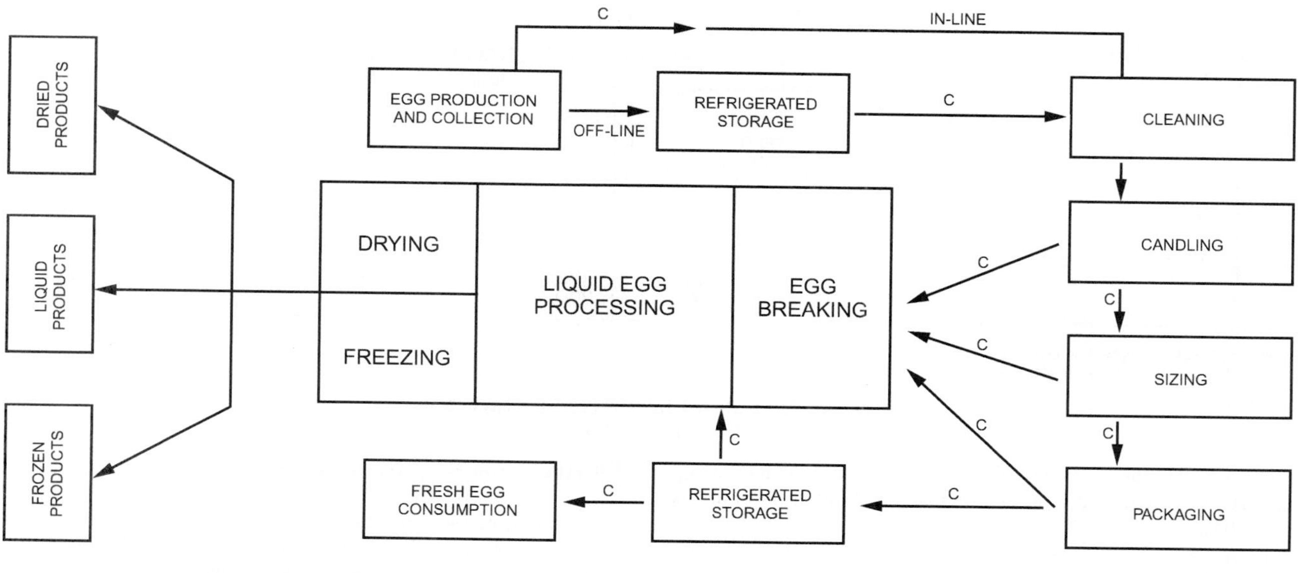

Fig. 2 Unit Operations in Off-Line and In-Line Egg Packaging

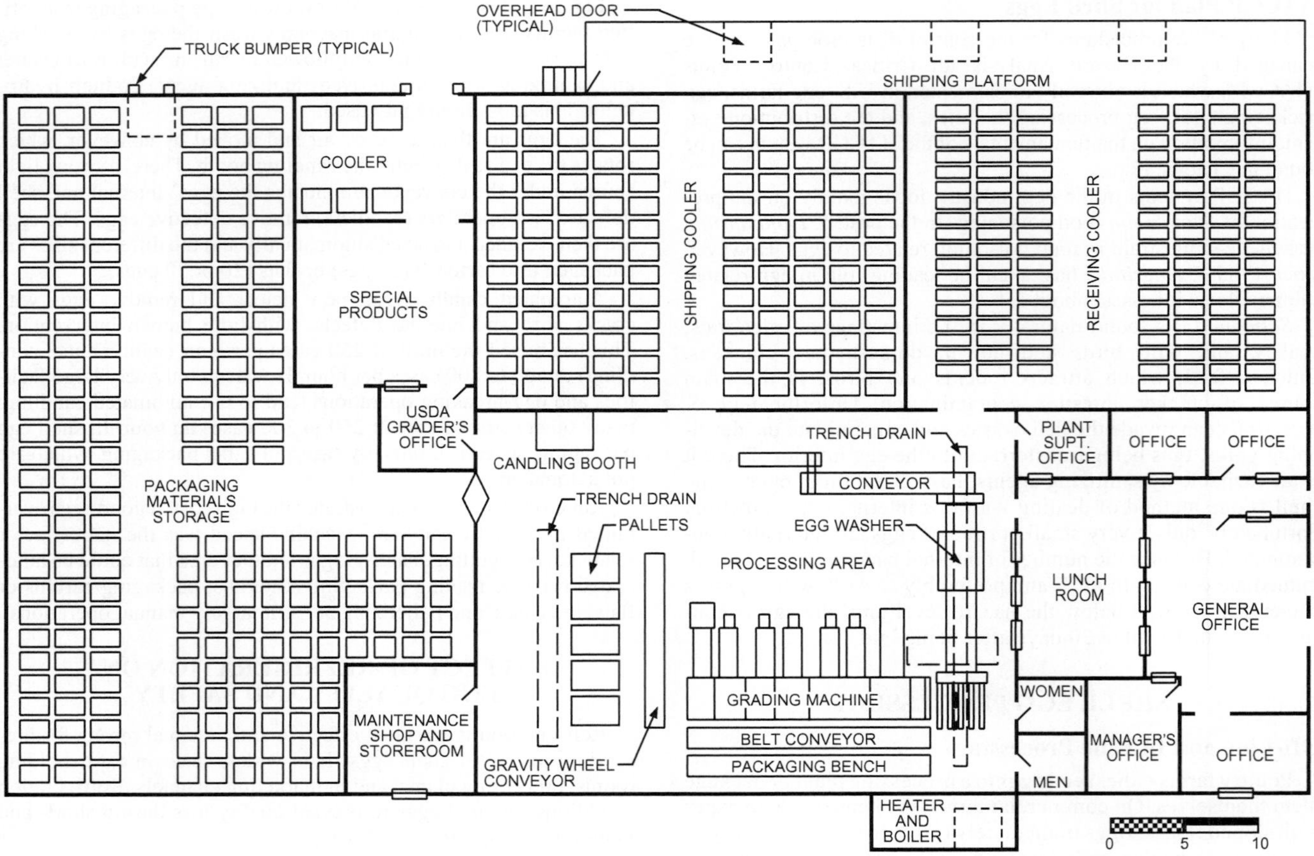

Fig. 3 Off-Line Egg Processing Operation
(Goble 1980)

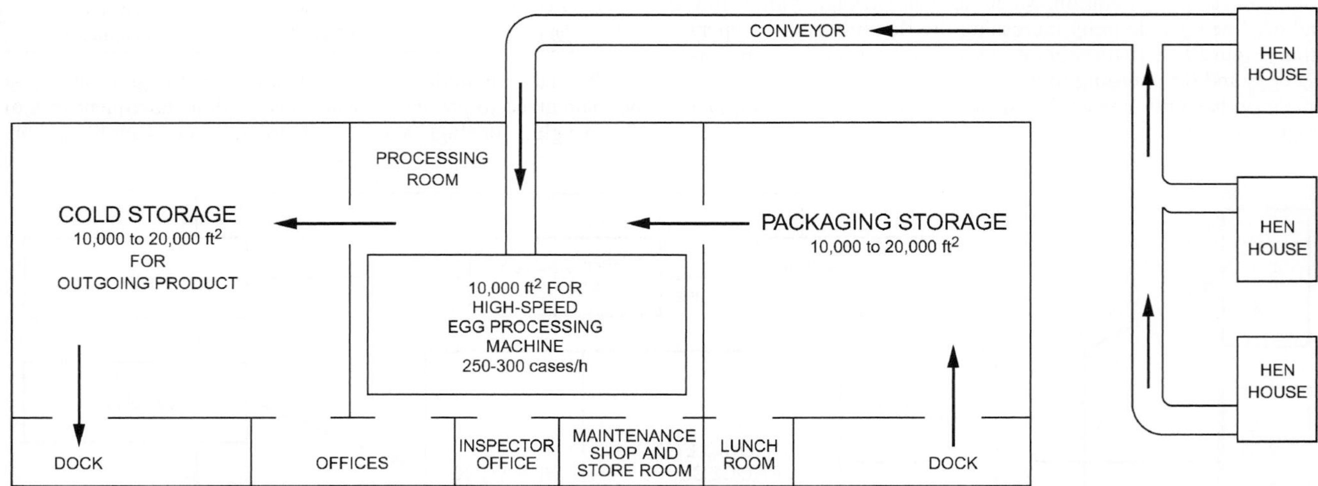

Building is 30,000 ft² equally divided to packaging storage, processing room, and cold storage with possible expansion from cold storage side to 50,000 ft².

Fig. 4 Typical In-Line Processing Operation
(Zeidler and Riley 1993)

which can penetrate the pores of the shell and contaminate the egg contents. Mold will grow on eggs when the relative humidity is above 90%.

For long-term storage, eggs should be kept just above their freezing point, 31°F. However, long-term storage is seldom used because most eggs are consumed within a short period. Low temperatures can cause sweating (i.e., condensation of moisture on the shell).

Refrigeration Requirement Issues

Temperature has a profound effect on *Salmonella enteritidis* on and in eggs. Research has shown that the growth rate of *S. enteritidis* in eggs is directly proportional to the temperature at which the eggs were stored. Holding eggs at 39 to 46°F reduced the heat resistance of *S. enteritidis* and suggested that not only does refrigeration

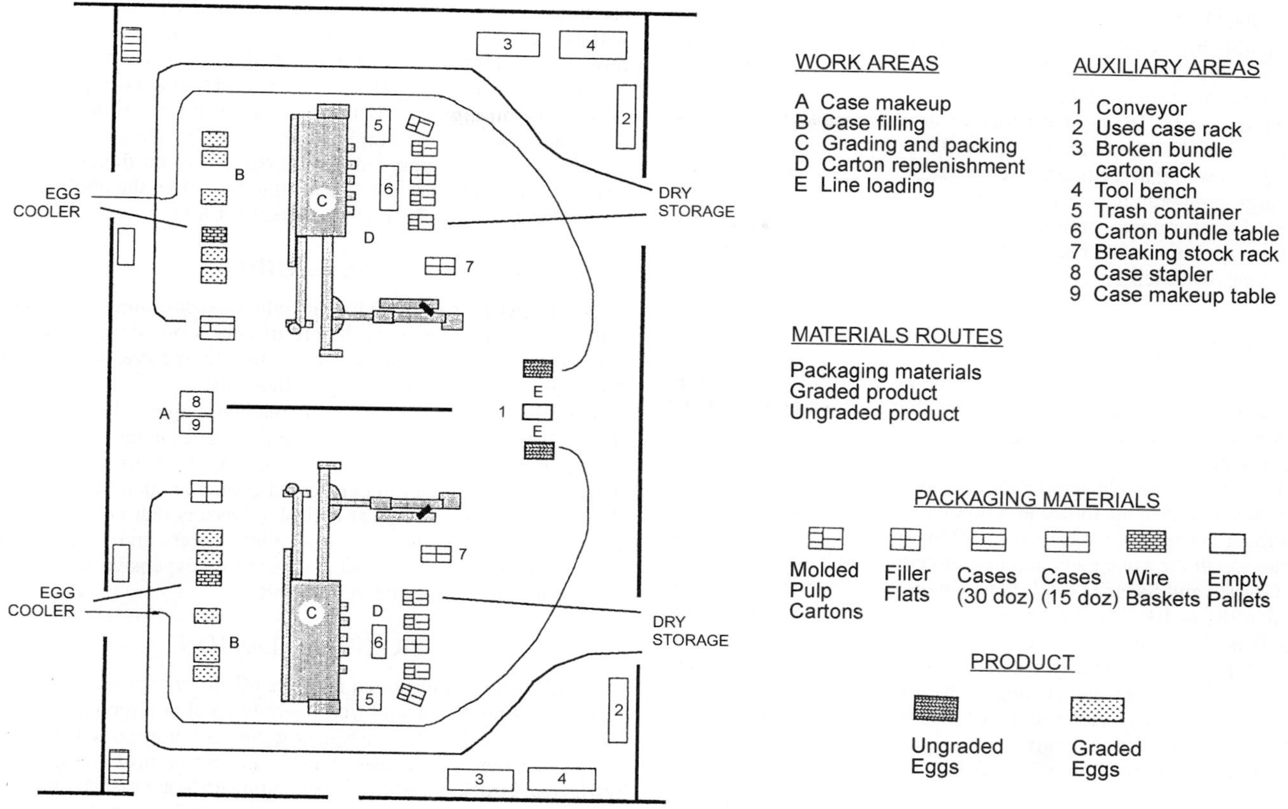

Fig. 5 Material Flow in Off-Line Operation
(Hamann et al. 1978)

reduce the level of microbial multiplication in shell eggs, but it lowers the temperature at which the organism is killed during cooking.

At present, most shell eggs in the United States are refrigerated to 45°F after processing. Commonly, they are transported in refrigerated trucks and displayed in refrigerated retail displays.

Condensation on Eggs

Moisture often condenses on the shell surface when cold eggs are moved from cool storage into hot and humid outside conditions or if the temperature varies widely inside the cooler. Sweating results in a wet egg, and the egg may adhere to the packaging material. The ability of any microbes present on the shell to penetrate the shell is not increased (Ernst et al. 1998). However, wet eggs are more likely to become stained when handled.

Plastic wrapped around the pallets to stabilize the load for shipping can also prevent moisture loss and increase humidity in the load, which can cause mold problems when eggs are held too long in this condition.

Condensation or sweating can be predicted from a psychrometric chart. Table 5 lists typical conditions in which sweating may occur.

Table 5 Ambient Conditions When Moisture Condenses on Cold Eggs

Outside Temperature, °F	Outside Relative Humidity, % Egg Temperature 45°F	55°F
55	70	—
60	60	85
65	50	70
70	40	60
75	35	50
80	30	43
85	25	36
90	21	31

Czarick and Savage (1992) reported that incoming egg temperature in an in-line system reached equilibrium with the layer house temperature. House temperatures are often maintained at 75 to 80°F; however, 90°F sometimes occurs.

Initial Egg Temperatures

Cooling requirements for shell eggs obviously vary with the weight of eggs to be cooled and their initial temperature. Anderson et al. (1992) showed that incoming egg temperature depends on the type of processing operation and time of year. In off-line plants, eggs typically arrive at the plant with internal temperatures ranging from 62 to 68°F. Before processing, the eggs are placed in a cooler, which is held between 50 and 60°F. In in-line plants, eggs are conveyed directly from poultry houses to the packing area. Anderson et al. measured incoming egg temperatures ranging from 88 to 96°F.

Egg Temperatures After Processing

Experience has shown that quality defects are more readily detected when eggs are allowed to age. Thus, in off-line processing, eggs from production units are usually stored overnight at 55 to 60°F before processing. With present cooling methods, eggs require about 48 h in cold storage to cool completely.

Cooling eggs before processing is limited by the temperature rise the shells can tolerate without cracking, which is about 62°F. Most processors wash eggs in warm water ranging from 115 to 125°F (Anderson et al. 1992). This wash temperature could cause shell cracking if eggs are initially cooled to the minimum temperature prescribed by USDA (41°F). Therefore, the lowest egg temperature acceptable before processing is about 60°F. In contrast, eggs in in-line operations are commonly processed while still warm from the house and are packaged warm.

Hillerman (1955) reported that wash water kept at 115°F increases internal egg temperature by 0.4°F per second. Anderson (1993) showed that the internal temperature of eggs rose because of the high temperatures during washing, resulting in an 8 to 12°F internal temperature increase above their starting temperature. As a result, egg temperature after washing and packing in in-line systems can typically reach 75 to 85°F, and in rare cases may reach nearly 100°F.

Cooling Rates

Henderson (1957) showed that air rates of 105 to 600 fpm flowing past an individual egg caused it to cool within one hour by 90% of the difference between the initial egg temperature and the temperature of the cooling air. Eggs packed in filler flats required 4 to 5 h to achieve 90% of total possible temperature drop. Bell and Curley (1966) reported that 55°F air forced around fiber flats in vented corrugated fiberboard boxes cooled eggs from 90 to 60°F in 2 to 5 h. Unvented cartons with the same pack required more than 30 h to cool.

Czarick and Savage (1992) placed eggs with an internal temperature of 81 to 100°F either on fiber flats and stacked six high or in egg flats placed in 6-flat (15 dozen) fiber cases. The eggs were then placed in a 50°F cooler. Eggs in the outermost cells of the cased flats cooled to 50°F in 9 h and all eggs in the fiber flats cooled to 50°F in 24 h. However, eggs at the center of the cases had not reached 50°F after 36 h. They found that it took more than 5 days for a pallet of eggs in cases to cool from 85 or 90°F to 45°F in a 45°F cold room.

Egg moisture loss is not increased by rapid cooling. Funk (1935) found that weight loss was the same for eggs in wire baskets cooled in 1 h with circulating air or 15 h with still air.

Cooling for Storage

With current handling practices, packed eggs require more than one week of storage before they reach the temperature of the storage room. This slow cooling results in egg temperatures in the optimal growth range for *S. enteritidis* from 24 to 72 h after processing. Packaging materials effectively insulate the eggs from the surrounding environment, especially in the center of the pallet. In addition, pallets are often stacked touching each other and may be wrapped in plastic, which further insulates the inner cases and reduces airflow. Also, most eggs are moved from storage within hours of processing, so they are barely cooled. But delaying shipment to allow the eggs to cool results in less-than-fresh products being delivered to the consumer, and interior quality suffers.

Adequate air flow through a box requires that the box be vented. In a study done for fruits and vegetables, Baird et al. (1988) showed that cooling cost increases rapidly when carton face vent area decreased below 4% of the total area. Other packing materials, such as liners, wraps, flats, or cartons, must not prevent air that enters the box from contacting the eggs. Also, cases must be stacked to allow air to circulate freely around the pallets.

Because of the inefficiency of cooling eggs in containers, it would seem best to cool them before packing. Eggs could be cooled between washing and packing just before being placed into cartons and then cases. A cooler has been developed specifically for in-line cooling to capitalize on the cooling rate of individual shell eggs. This would allow the use of current packaging. However, existing equipment is not designed to incorporate this procedure.

Accelerated Cooling Methods

Forced-Air Cooling. Henderson (1957) showed that forced ventilation of palletized eggs produced cooling times close to that of cooling individual eggs. Thompson et al. (2000) arranged a 30-case pallet of eggs so that a 1000 cfm fan drew 40°F air through openings in the cases. The eggs were cooled to less than 45°F within 1 to 3 h. This cooling method can be used in an existing refrigerated storage room with little additional investment.

Cryogenic. Curtis et al. (1995) showed that eggs exposed to a –60°F carbon dioxide environment for 3 min continued to cool after packaging and 15 min later were at 45°F. The process maintained egg quality and did not increase the incidence of shell cracking. This process has been refined to allow the cooling process to occur in an –70°F environment for 80 sec.

PACKAGING

Shell eggs are packaged for the individual consumer or the institutional user. Consumer packs are usually a one dozen carton or variations of it. The institutional user usually receives shell eggs in 30 dozen cases on twelve 30-egg filler-flats.

Consumer cartons are generally made of paper pulp, foam plastic, or clear plastic. Some cartons have openings in the top for viewing the eggs, which also facilitates cooling. Cartons are generally delivered to the retailer in corrugated containers that hold 15 to 30 dozen eggs, in wire or plastic display baskets that hold 15 dozen eggs, or on rolling display carts. Wire baskets and rolling racks allow more rapid cooling, but are also more expensive and take up more space in storage and in transport.

TRANSPORTATION

Shell eggs are transported from the off-line egg production site to egg processing plants, and from there to local or regional retail and food service outlets. Less frequently, eggs are transported from one state to another or overseas. Truck transport is most common and refrigerated trucks capable of maintaining 45°F are mandatory in the United States, with an exemption for small producer-packers with an annual egg production from 3000 or fewer hens.

Cases and baskets are generally stored and transported on pallets in 30-case lots (five cases high with six cases per layer). The common carrier for local and long-distance hauling is the refrigerated tractor/trailer combination. Trailers carry 24 to 26 30-case pallets of eggs, often of one size category. A typical load of 720 to 780 cases weighs about 44,000 lb. Some additional cases may be added when small or medium eggs are being transported. Eggs are not generally stacked above six cases high to allow the cold air to travel to the rear of the trailer and to minimize crushing of lower-level cases.

Interregional shipment of eggs is quite common, with production and consumption areas often 1500 miles apart. Such shipments usually require two to three days using team driving.

Local transportation of eggs may be with similar equipment, especially when delivered to retailer warehouses. Smaller trucks with capacities of 250 to 400 cases are often used when multiple deliveries are required. Local deliveries are commonly made directly to retail or institutional outlets. Individual store deliveries require a variety of egg sizes to be placed on single pallets. This assembly operation in the processing plant is very labor-intensive. Local delivery may involve multiple short stops and considerable opening and closing of the storage compartment, with resultant loss of cooling. Many patented truck designs are available to protect cargo from temperature extremes during local delivery, yet none has been adopted by the egg industry.

A 1993 USDA survey found that over 80% of the trucks used to deliver eggs were unsuitable to maintain 45°F. Damron et al. (1994), in a survey of three egg transport companies in Tampa and Dallas, found the average temperature of trailers during nonstop warehouse deliveries was 46.4°F. The front of the trailer averaged below 45°F 20 to 25% of the time while the back of the trailer was below 45°F 65% of the time. The loads were below 45°F 37% of the time while the reefer discharge was below 40°F.

Trailers used for store-door deliveries had temperatures averaging approximately 45°F at the start of the route; however, some areas only reached a low of 48.5°F. As the deliveries continued and the volume of eggs decreased, temperatures increased and temperature recovery never occurred.

EGG PRODUCTS

Egg products are classified into four groups according to the American Egg Board (www.aeb.org):

1. Refrigerated egg products
2. Frozen egg products
3. Dried egg products
4. Specialty egg products (including hard-cooked eggs, omelets, scrambled eggs, egg substitutes)

Most of these products are not seen at the retail level, but are used as further processed ingredients by the food processing industry for such products as mayonnaise, salad dressing, pasta, quiches, bakery products, and eggnog. Other egg products, such as deviled eggs, Scottish eggs, frozen omelets, egg patties, and scrambled eggs, are prepared for fast food and institutional food establishments, hotels, and restaurants. In recent years, several products such as egg substitutes (which are made from egg whites) and scrambled eggs have appeared. Yet to be developed are large-volume items such as aseptically filled, ultrapasteurized, chilled liquid egg and low-cholesterol chilled liquid eggs.

EGG BREAKING

Egg breaking transforms shell eggs into liquid products: whole egg, egg white, and yolk. Liquid egg products are chilled, frozen, or dried. These items can be used as is or are processed as an ingredient in food products. Only a few products, such as hard-cooked eggs, do not use the breaking operation system. Dried egg powder, which is the oldest processed egg product, lost ground as a proportion of total egg products, whereas chilled egg products are booming because of their superior flavor, aroma, pronounced egg characteristics, and convenience. Most liquid egg products (about 44% of all egg products) must be consumed in a relatively short time because of their short storage life. Frozen or dried egg products may be stored considerably longer.

Surplus, small, and cracked eggs are the major supply source for egg-breaking operations. Those eggs must be cleaned in the same manner as shell eggs. Washing and loading of eggs to be broken must be conducted in a separate room from the breaking operation (Figure 6). Eggs with broken shell membrane (leakers) or blood spots are not allowed to be broken for human consumption. Most breaking operations are close to production areas, and in many cases are merely a separate area of a shell egg processing and packaging facility. An egg-breaking operation usually receives its eggs from several processing plants in the area that do not have breaking equipment. Storage and transport of eggs, and especially of cracked eggs, reduces the quality of the end product.

Two types of egg-breaking equipment are available:

1. **Basket centrifuge.** Shell eggs are dumped into a centrifuge and a whole egg liquid is collected. Several states and some local health authorities ban this equipment for breaking eggs for human consumption because of the high risk of contamination. Similar centrifuges are used to extract liquid egg residue from the discarded egg shells. This inedible product is used mostly for pet food.
2. **Egg breaker and separator.** These machines can process up to 100 cases per hour (36,000 eggs), which is still slow compared to up to 500 cases per hour (180,000 eggs) handled by modern table egg packaging equipment.

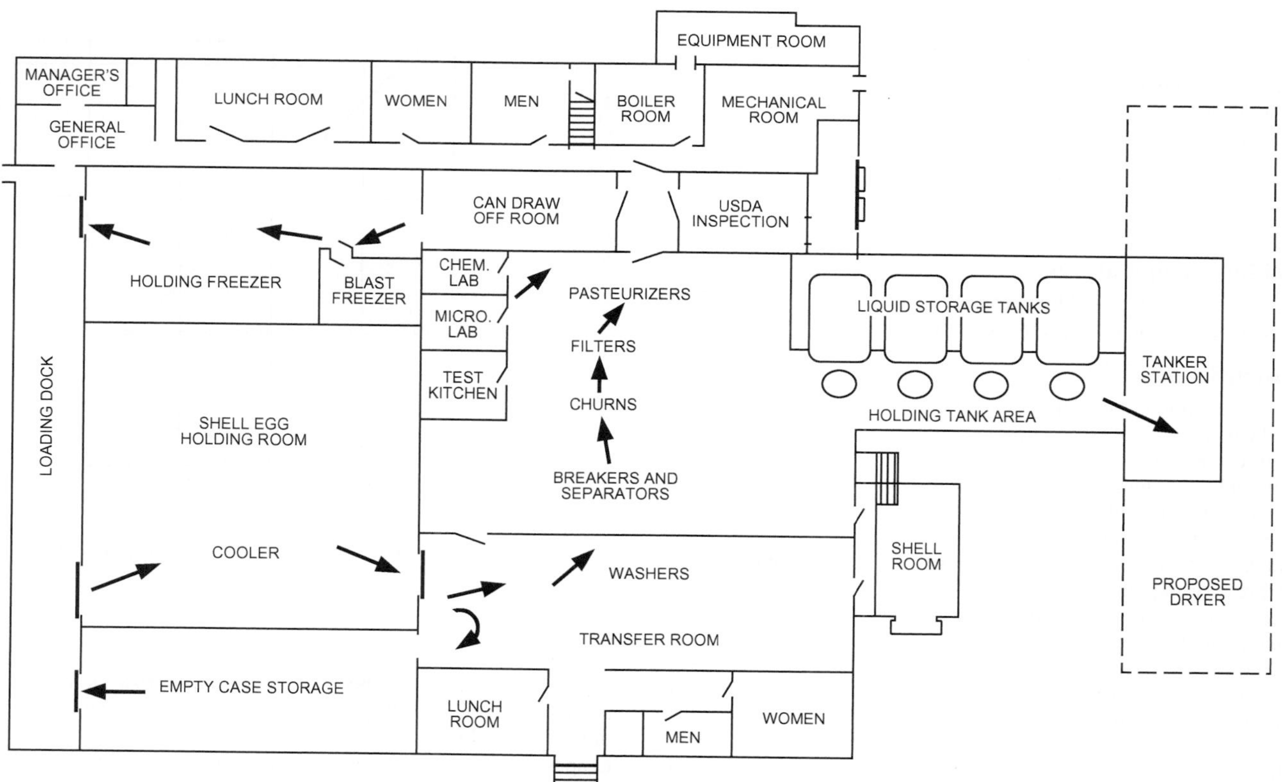

Fig. 6 Floor Plan and Material Flow in Large Egg-Breaking Plant
(Courtesy of Seymour Food)

Table 6 Minimum Cooling and Temperature Requirements for Liquid Egg Products
(Unpasteurized product temperature within 2 h from time of breaking)

Product	Liquid (Other Than Salt Product) Held 8 h or Less	Liquid (Other Than Salt Product) Held in Excess of 8 h	Liquid Salt Product	Temperatures Within 2 h after Pasteurization	Temperatures Within 3 h after Stabilization
Whites (not to be stabilized)	55°F or lower	45°F or lower	—	45°F or lower	—
Whites (to be stabilized)	70°F or lower	55°F or lower	—	55°F or lower	[a]
All other products (except product with 10% or more salt added)	45°F or lower	40°F or lower	If held 8 h or less, 45°F or lower. If held more than 8 h, 40°F or lower.		
Liquid egg product (10% or more salt added)	—	If held 30 h or less, 65°F or lower. If held in excess of 30 h, 45°F or lower.	—	65°F or lower[b]	—

Source: Inspection of eggs (7CFR57), January 1, 2005.
[a]Stabilized liquid whites should be dried as soon as possible after removal of glucose. Limit storage of stabilized liquid whites to that necessary for continuous operation.
[b]Cooling should be continued to ensure that any salt product held over 24 h is cooled and maintained at 45°F or lower.

Table 7 Pasteurization Requirements of Various Egg Products

Liquid Egg Products	Minimum Temperature, °F	Minimum Holding Time, minutes
Albumen (without use of chemicals)	134 132	3.5 6.2
Whole egg	140	3.5
Whole egg blends (less than 2% added non-egg ingredients)	142 140	3.5 6.2
Fortified whole eggs and blends (24 to 38% egg solids, 2 to 12% non-egg ingredients)	144 142	3.5 6.2
Salt whole egg (2% salt added)	146 144	3.5 6.2
Sugar whole egg (2 to 12% sugar added)	142 140	3.5 6.2
Plain yolk	142 140	3.5 6.2
Sugar yolk (2% or more sugar added)	146 144	3.5 6.2
Salt yolk (2 to 12% salt added)	146 144	3.5 6.2

Source: Regulations governing the inspection of eggs and egg products (9CFR590).

Holding Temperatures

Prepasteurization holding temperatures required by the USDA for out-of-shell liquid egg products are listed in Table 6.

Pasteurization

In the United States, the USDA requires all egg products made by the breaking process to be pasteurized and free of salmonella and requires all plants to be inspected. The minimum required temperatures and holding times for pasteurization of each type of egg product are listed in Table 7.

Plate heat exchangers, commonly used for pasteurization of milk and dairy products, are also commonly used for liquid egg products. Before entering the heat exchanger, the liquid egg is moved through a clarifier that removes solid particles such as vitelline (the yolk membrane) and shell pieces.

Egg white solids may be made *Salmonellae*-negative by heat treatments. Spray-dried albumen is heated in closed containers so that the temperature throughout the product is not less than 130°F for not less than 7 days, until it is free of *Salmonellae*. For pan-dried albumen, the requirement is 125°F for 5 days until it is free of *Salmonellae*. For the dried whites to be labeled pasteurized, the USDA requires that each lot be sampled, cultured, and found to contain no viable *Salmonellae*.

Table 8 Minimum Pasteurization Requirements in Various Countries

Country	Temperature, °F	Time, minutes
Great Britain	148	2.5
Poland	151-154	3
China (PRC)	146	2.5
Australia	144.5	2.5
Denmark	149-156.5	1.5-3
United States	140	3.5

Source: Stadelman et al. (1988).

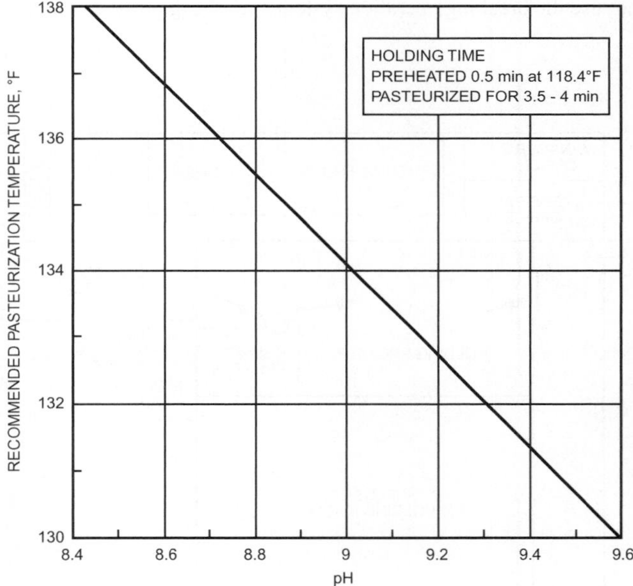

Fig. 7 Effect of pH on Pasteurization Temperature of Egg White

Temperature, time, and pH affect the pasteurization of liquid eggs. Various countries specify different pasteurization time, temperature, and pH, but all specifications provide the same pasteurization effects (Table 8). Higher pH requires lower pasteurization temperature, and pH 9.0 is most commonly used for egg whites (Figure 7). Various egg products demonstrate different destruction curves (Figure 8); therefore, different pasteurization conditions were set for these products (Table 8).

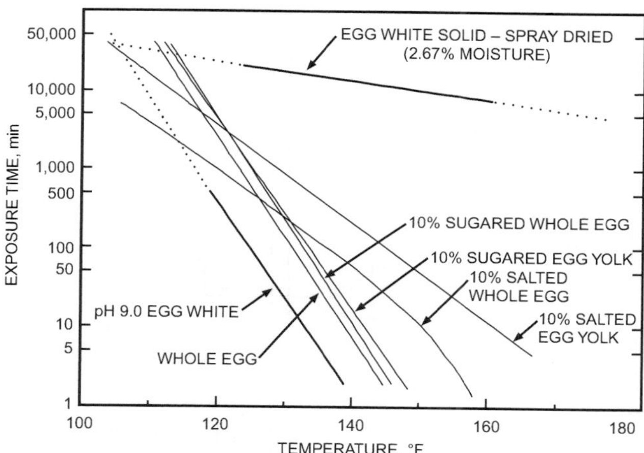

Fig. 8 Thermal Destruction Curves of Several Egg Products
(Stadelman and Cotterill 1990)

Table 9 Liquid and Solid Yields From Shell Eggs

Constituent	Liquid (% by weight) Mean	Range	Solids, %
Shell	10.5	7.8 to 13.6	99.0
Whites	58.5	53.1 to 68.9	11.5
Yolk	31.0	24.8 to 35.5	52.5
Edible whole egg	89.5	86.4 to 92.2	24.5

Source: Shenstone (1968).

Egg whites are more sensitive to higher temperatures than whole eggs or yolk, and therefore will coagulate. Thus, lactic acid is added to adjust the pH to 7.0 to allow the egg whites to withstand 141 to 143°F. Egg whites can be pasteurized at 125 to 127°F for 1.5 min if, after the heat treatment, 0.075 to 0.1% hydrogen peroxide is added for 2 min, followed by its elimination with the enzyme catalase. Liquid yolk, on the other hand, requires higher temperatures for pasteurization than liquid whole eggs (144°F for 3.5 min).

Yields

The ratios of white, yolk, and shell vary with the size of the egg. During the laying cycle, the hens lay small, medium, and large eggs, which have different proportions of yolk and white. Therefore, the distribution of egg sizes that the breaking plant receives during the year varies with season, breed, egg prices, and surplus sizes. As a result, processing yields of white, yolk, and shell vary accordingly (Table 9).

REFRIGERATED LIQUID EGG PRODUCTS

Liquid egg products are extremely perishable and should be cooled immediately after pasteurization to below 40°F and kept cool at 34 to 40°F during storage. Refrigerated liquid egg products are convenient to use, do not need defrosting, and can be delivered in bulk tank trucks, totes, or pails, which reduces packaging costs. However, shelf life at 34 to 30°F is about 2 to 3 weeks; therefore, this product is used mostly as an ingredient in further food processing and manufacturing.

Extending the shelf life of liquid egg products is difficult because egg proteins are much more heat-sensitive than dairy proteins. As a result, ultrapasteurized liquid eggs must be kept under refrigeration whereas ultrapasteurized milk can be kept at room temperature. Ball et al. (1987) used ultrapasteurization and aseptic packaging to extend the shelf life of refrigerated whole eggs to 24 weeks.

Chilled Egg Products

Chilled or Frozen Liquid. Whole egg, yolk, and whites are the major high-volume products.

Stabilized Egg Products. Additives in yolk products to be frozen prevent coagulation during thawing. Ten percent salt is added to yolks used in mayonnaise and salad dressings, and 10% sugar is added to yolks used in baking, ice cream, and confectionery manufacturing. Whole egg products are also fortified with salt or sugar according to finished product specifications. However, egg whites are not fortified, because they do not have gelation problems during defrosting.

UHT Products. High-temperature processing (UHT) was initially aimed at producing sterile milk with superior palatability and shelf life by replacing conventional sterilization at 250°F for about 12 to 20 min with 275°F for 2 to 5 s. UHT treatment of liquid eggs is more complicated, because egg proteins are more sensitive to heat treatment; therefore, UHT liquid eggs must be kept under refrigeration.

In one study, researchers applied aseptic processing and packaging technology to extend the shelf life of liquid egg products to several months under refrigerated (40°F) conditions. According to the USDA, the process condition for extended-shelf-life liquid whole egg is about 147°F for 3.5 min. Ultrapasteurized, aseptically filled, chilled, whole liquid egg product is now limited to institutional food establishments in the United States, although retail products are available in some European countries.

Egg Substitutes. Substitutes are made from egg whites, which do not contain cholesterol or fat. The yolk is replaced with vegetable oil, food coloring, gums, and nonfat dry milk. Recent formulations have reduced the fat content to almost zero. These products are packaged in cardboard containers and sold frozen or chilled in numerous formula variations. Aseptic packaging extends the shelf life of the refrigerated product.

Low-Cholesterol Eggs. Many techniques have been developed to remove cholesterol from eggs, yet no commercial product is currently available.

FROZEN EGG PRODUCTS

Egg products are usually frozen in cartons, plastic bags, 30-lb plastic cans, or 55-gal drums (for bulk shipment). Table 3 in Chapter 9 lists thermal properties involved in freezing egg products. Freezing is usually by air blasts at temperatures ranging from –10 to –40°F. Pasteurized products designated for freezing must be frozen solid or cooled to a temperature of at least 10°F within 60 h after pasteurization. Newer freezing techniques for products containing cooked white (e.g., deviled eggs, egg rolls) include individual quick freezing (IQF) at very low temperatures (–4 to –240°F).

Defrosting. Frozen eggs may be defrosted below 45°F in approved metal tanks in 40 to 48 h. If defrosted at higher temperatures (up to 50°F), the time cannot exceed 24 h. Running water can be used for defrosting. When the frozen mass is crushed by crushers, all sanitary precautions must be followed.

DEHYDRATED EGG PRODUCTS

Spray drying is the most common method for egg dehydration. However, other methods are used for specific products such as scrambled eggs, which are made by freeze drying, and egg white products, which are usually made by pan drying to produce a flake-like product. In spray drying (Figure 9), liquid is atomized by nozzles operating at 500 to 600 psi. The centrifugal atomizer, in which a spinning disc or rod rotates at 3500 to 50,000 rpm, creates a hollow cone pattern for the liquid, which enters the drying chamber. The atomized droplets meet a 250 to 450°F hot air cyclone, which is created and driven by a fan blowing in the opposite direction. Because the surface area of the atomized liquid is so large, moisture

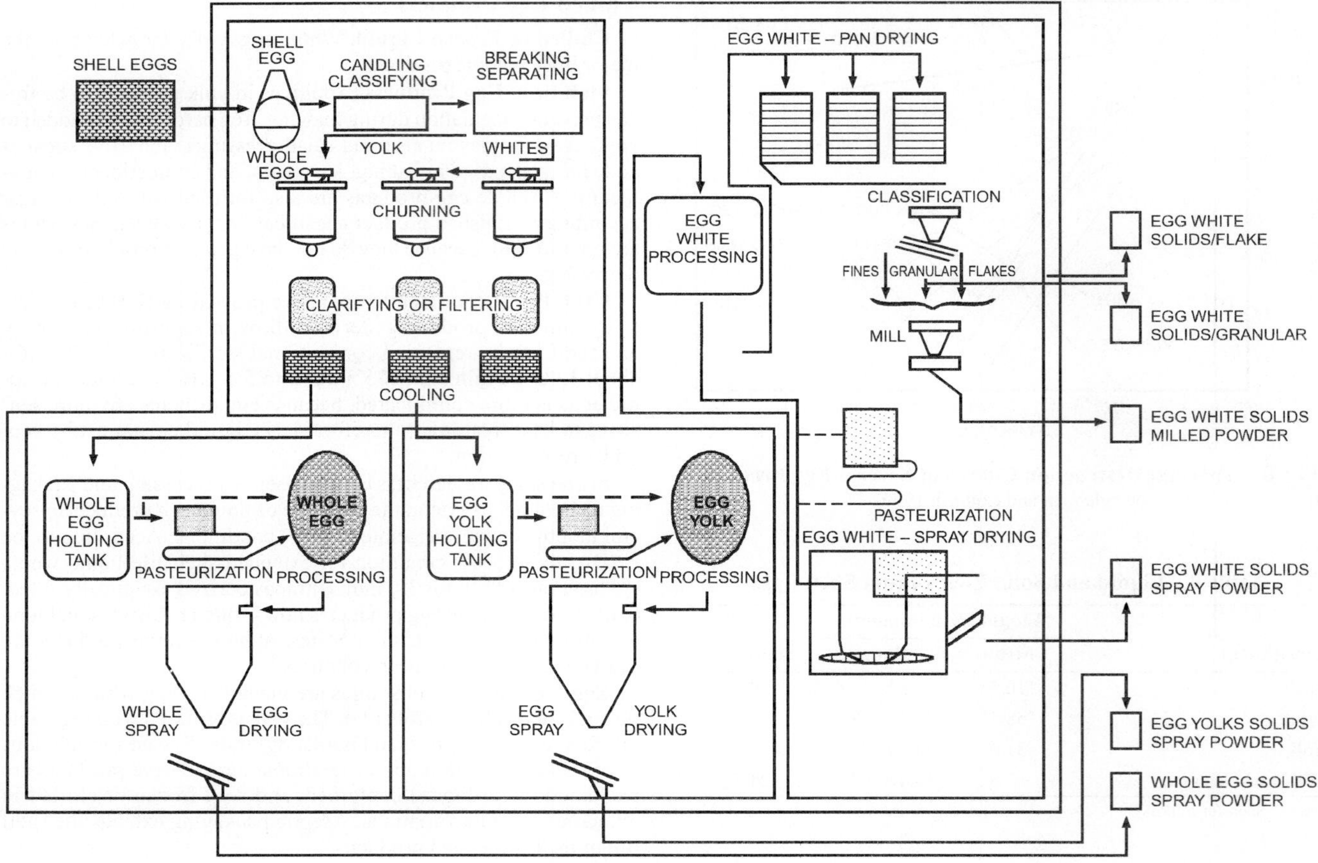

Fig. 9 Steps in Egg Product Drying

evaporates very rapidly. The dry product is separated from the air, cooled, and, in many cases, sifted before being packaged into fiber drums lined with vapor retarder liners. Military specifications usually call for gas-packaging in metal cans. Moisture level in dehydrated products is usually around 5%, whereas in pan dryer products it is around 2%.

Spray dryers are classified as vertical or horizontal. However, there are large variations in methods of atomizing, drying air movement, and powder separation.

Whole egg, egg white, and yolk products naturally contain reduced sugar. To extend shelf life and to prevent color change through browning (Maillard reaction), the glucose in the egg is removed by baker's yeast, which consumes the glucose in 2 to 3 h at 86°F. Many commercial firms replace the baker's yeast method with a glucose oxidase-catalase enzyme process because it is more controllable. The enzyme-treated liquid is then pasteurized in continuous heat exchangers at 142°F for 4 min and dried. Whole egg and yolk powder have excellent emulsifying, binding, and heat coagulating properties, whereas egg white possesses whipping capabilities.

Dry egg products are used in production of baked goods such as sponge cakes, layer cakes, pound cakes, doughnuts, and cookies. Numerous dry products exist because it is possible to dry eggs together with other ingredients such as milk, other dairy products, sucrose, corn syrup, and other carbohydrates.

Common Dried Products. Figure 9 shows processing steps for several dried products. Common dried products include

- Pan-dried egg whites, spray-dried egg white solids, whole egg solids, yolk solids
- Stabilized (desugared) whole egg, stabilized yolk
- Free-flowing (sodium silicoaluminate) whole egg solids, free-flowing yolk solids
- Dry blends (whole egg or yolk with carbohydrates, such as sucrose, corn syrup)
- Dry blends with dairy products, such as scrambled egg mix

EGG PRODUCT QUALITY

Criteria usually used in evaluating egg product quality are odor, yolk color, bacteria count, solids and fat content (for yolk and whole egg), yolk content (for whites), and performance. All users want a wholesome product with a normal odor that performs satisfactorily in the ways it will be used. For noodles, a high solids content and color are important. Bakers are particular about performance: whites do not perform well in angel food cake if excessive yolk is present. They test the foaming performance of whites based on the height and volume of angel food cake and meringue. Performance is also critical for candy (using whites). Salad dressing and mayonnaise are used to evaluate the performance of the yolk as an emulsifier, and emulsion stability is tested.

SANITARY STANDARDS AND PLANT SANITATION

In the United States, *Egg 3-A Sanitary Standards and Accepted Practices* are formulated by the cooperative efforts of the U.S. Public Health Service; the U.S. Department of Agriculture; the Poultry and Egg Institute of America; the Dairy Industry Committee; International Association of Milk, Food, and Environmental Sanitarians; and the Dairy Food Industries Supply Association. The Standards are published by the *Journal of Food Protection* (formerly the *Journal of Milk and Food Technology*).

Egg processing facilities and equipment require daily cleaning and sanitation. Plastic egg flats should be sanitized after each use to avoid microbial contamination of eggs. Chlorine or quaternary-based sanitizers are often used for egg washing and for cleaning equipment, egg flats, floor, walls, etc. Water for egg washing should have low iron content (below 2 ppm) to prevent bacterial growth.

Filters in forced-air egg drying equipment should be cleaned a minimum of once per week. Egg processing rooms should be well ventilated. Inlet air filters should be cleaned weekly. Egg cooling rooms should be kept clean and free from dust or molds.

HACCP Program for Egg Products

Food regulations in the USA require food companies to operate under Current Good Manufacturing Practices (CGMPs, CFR100). Egg products must also be produced under 9CFR590, Egg Products Inspection Act. Many egg companies have chosen to implement Hazard Analysis and Critical Control Point (HACCP) programs to further ensure the safety of their products. HACCP programs rely heavily on CGMPs and other programs (collectively called **prerequisite** programs). Some of these programs include

- Standard sanitation operating procedures (SSOPs)
- Pest control program
- Customer complaint and recall programs
- Maintenance program
- Training programs

When all of the prerequisite programs are properly implemented and satisfied, the HACCP program is used to monitor, control, verify, and record critical points in the process. Critical control points in egg products processing include pasteurization time and temperature, and prevention of post process contamination.

REFERENCES

Anderson, K.E. 1993. Refrigeration and removal of heat from eggs. *World Poultry* 11(9):40-43.

Anderson, K.E., P.A. Curtis, and F.T. Jones. 1992. Legislation ignores technology: Heat loss from commercially packed eggs in post-processing coolers. *Egg Industry* 98(5):11.

Baird, C.D., J.J. Gafney, and M.T. Talbot. 1988. Design criteria for efficient and cost effective forced air cooling systems for fruits and vegetables. *ASHRAE Transactions* 94(1):1434-1454.

Ball, H.R., Jr., M. Hamid-Samimi, P.M. Foegeding, and K.R. Swartzel. 1987. Functional and microbial stability of ultrapasteurized aseptically packaged refrigerated whole egg. *Journal of Food Science* 52:1212-1218.

Bell, D.D. and R.G. Curley. 1966. Egg cooling rates affected by containers. *California Agriculture* 20(6):2-3.

Burley, R.W. and D.V. Vadehra. 1989. *The avian egg: Chemistry and biology.* John Wiley & Sons, New York.

Carter, T.C., ed. 1968. *Egg quality: A study of the hen's egg.* Oliver and Boyd, Edinburgh.

CFR. 2005. Inspection of eggs (Egg Inspection Act). *Code of Federal Register* 7CFR57. U.S. Department of Agriculture, U.S. Government Printing Office, Washington, D.C.

Curtis, P.A., K.E. Anderson, and F.T. Jones. 1995. Cryogenic gas for rapid cooling of commercially processed shell eggs prior to packaging. *Journal of Food Protection* 58:389-394.

Czarick, M. and S. Savage. 1992. Egg cooling characteristics in commercial egg coolers. *Journal of Applied Poultry Research* 1:258-270.

Damron, B.L., C.R. Douglas, and R.D. Jacobs. 1994. Temperature patterns in commercial egg transport vehicles. *Journal of Applied Poultry Research* 3:193-198.

Ernst, R.A., L. Fuqua, H.P. Riemann, and S. Himathongkham. 1998. Effect of sweating on shell penetration of *Salmonella enteritidis. Journal of Applied Poultry Research* 7:81-84.

Funk, E.M. 1935. The cooling of eggs. *Missouri Agricultural Experiment Station Bulletin No. 350.*

Goble, J.W. 1980. Designing egg shell grading and packaging plants. *USDA Marketing Research Report No. 1105.*

Hamann, J.A., R.G. Walters, E.D. Rodda, G. Serpa, and E.W. Spangler. 1978. Shell egg processing plant design. *USDA/ARS Market Research Report No. 912.*

Henderson, S.M. 1957. On-the-farm egg processing: Cooling. *Agricultural Engineering* 38(8):598-601, 605.

Hillerman, J.P. 1955. Quick cooling for better eggs. *Pacific Poultryman*, pp. 18-20.

Kuney, D.R., S. Bokhari, G. Zeidler, R. Ernst, and D. Bell. 1992. Factors affecting candling errors. *Proceedings of the 1992 Egg Processing, Packaging and Marketing Seminar*, San Bernardino and Modesto, CA.

Shenstone, F.S. 1968. The gross composition, chemistry, and physico-chemical basis of organization of the yolk and white. In *Egg quality: A study of the hen's egg.* T.C. Carter, ed. Oliver and Boyd, Edinburgh.

Stadelman, W.J. and O.J. Cotterill, eds. 1990. *Egg science and technology*, 3rd ed. Food Products Press, Binghamton, NY.

Stadelman, W.J., E.L. Baum, J.G. Darroch, and H.G. Walkup. 1954. A comparison of quality in eggs marketing with and without refrigeration. *Food Technology* 8:89-102.

Stadelman, W.J., V.M. Olson, G.A. Shemwell, and S. Pasch. 1988. *Egg and poultry meat processing.* Ellis Horwood, Chichester, U.K.

Tarver, F.R. 1964. The influences of rapid cooling and storage conditions on shell egg quality. *Food Technology* 18(10):1604-1606.

Thompson, J.F., J. Knutson, R.A. Ernst, D. Kuney, H. Riemann, S. Himathongkham, and G. Zeidler. 2000. Rapid cooling of shell eggs. *Journal of Applied Poultry Research* 9:258-268.

USDA. 1990. Egg grading manual. *Agriculture Handbook* 75. U.S. Department of Agriculture, Agricultural Marketing Service.

Zeidler, G. and D. Riley. 1993. The role of humidity in egg refrigeration. *Proceedings, 1993 Egg Processing, Packaging, and Marketing Seminar.*

BIBLIOGRAPHY

Cotterill, O.J. 1981 (Revised in 1990). *A scientist speaks about egg products.* American Egg Board, Park Ridge, IL.

Dawson, L.E. and J.A. Davidson. 1951. Farm practices and egg quality: Part III. Egg-holding conditions as they affect decline in quality. *Quarterly Bulletin*, Michigan Agricultural Experiment Station 34(1):105-144.

Fraser, A.C., M.M. Bain, and S.E. Solomon. 1998. Organic matrix morphology and distribution in the palisade layer of egg shells sampled at selected periods during day. *British Poultry Science* 38:225-228.

Henderson, S.M. 1958. On-the-farm egg processing: Moisture loss. *Agricultural Engineering* 39(1):28-30, 34.

Lucore, L.A., F.T. Jones, K.E. Anderson, and P.A. Curtis. 1997. Internal and external bacterial counts from shells of eggs washed in a commercial-type processor at various wash-water temperatures. *Journal of Food Protection* 60(11):1324-1328.

Rhorer, A.R. 1991. What every producer should know about refrigeration. *Egg Industry* (May/June):16-25.

Schumang, J.D., B.W. Sheldon, I.M. Vandepopuliere, and H.R. Ball, Jr. 1997. Immersion heat treatments for inactivation of *Salmonella enteritidis* with intact eggs. *Journal of Applied Microbiology* 83:438-444.

Stadelman, W.J. 1992. Eggs and egg products. In *Encyclopedia of Food Science and Technology*, vol. 2. Y.H. Hui, ed. John Wiley & Sons, New York.

Tharrington, J.B., P.A. Curtis, K.E. Anderson, and F.T. Jones. 1999. Shell quality of eggs from historic strains of SCWL chickens and the relationship of egg shape to shell strength. *Proceedings of XIV European Symposium on the Quality of Eggs and Egg Products*, pp. 77-83.

USDA. 1991. *Criteria for shelf-life of refrigerated liquid egg products.* U.S. Department of Agriculture, Agricultural Marketing Service.

Van Rest, D.J. 1967. Operations research on egg management. *Transactions of the American Society of Agricultural Engineers*, St. Joseph, MI (Dec.): 752-755.

Wells, R.G. and C.G. Belyavin. 1984. *Egg quality: Current problems and recent advances.* Butterworth's, London.

CHAPTER 22

DECIDUOUS TREE AND VINE FRUIT

THE most obvious losses from marketing fruit crops are caused by mechanical injury, decay, and aging. Losses in moisture, vitamins, and sugars are less obvious, but they adversely affect quality and nutrition. Rough handling and holding at undesirably high or low temperatures increase loss. Loss can be substantially reduced by greater care in handling and by following recommended storage practices.

FRUIT STORAGE AND HANDLING CONSIDERATIONS

Quality and Maturity

Maximum storage life can be obtained only by storing high-quality commodities soon after harvest. Different lots of fruit may vary greatly in storage behavior because of variety, climate, soil and cultural conditions, maturity, and handling practices. When fruit is transported from a distance, is grown under unfavorable conditions, or is deteriorated, proper storage allowance should be made.

Fresh fruit for storage should be as free as possible from skin breaks, bruises, and decay. These defects reduce the value of the product and may cause rapid deterioration not only of the damaged fruit, but also of fruit stored nearby. Damaged fruit often produces more ethylene, which can cause rapid ripening of many types of climacteric fruit. For the same reason, it is unwise to store fruit or vegetables having different storage characteristics together; some may emit ethylene, causing a more sensitive crop to ripen prematurely. Natural cooling in well-ventilated storage slows down or halts these processes.

The amount of incipient decay infection, which influences storage potential of grapes and apples, can be predicted early. Only lots with good storage potential should be held for late-season marketing.

Fruit maturity at harvest time determines its refrigerated storage life and quality. For any given produce, there is a maturity best suited for refrigerated storage. Undermature produce will not ripen or develop good quality during or after refrigerated storage. For many crops, excessively overmature produce deteriorates quickly during storage, although there are some exceptions for late-harvested fruit (in particular, late-harvested kiwifruit). Determination of maturity can be a complex problem. A number of measurements are used, depending on the crop; these include penetromer firmness, color, degree-days since flowering or fruit set, soluble solids, or other physical, chemical, or biological tests. In critical cases, a combination of tests may be used.

Handling and Harvesting

Rising handling costs have encouraged the use of bulk handling and large storage bins for many kinds of fruit. Moving, loading, and stacking bins by forklift trucks must be done carefully to maintain proper ventilation and refrigeration of the product. Bins should not be so deep that excessive weight damages the produce near the bottom.

Mechanical harvesters for fruit frequently cause some bruising. This damage can materially reduce the quality of the produce.

Storage and Transportation

As in storage, losses from deterioration during distribution are affected by temperature, moisture, diseases, and mechanical damage. Gradual aging and deterioration are continuous after harvest. Time in transit may represent a large portion of postharvest life for some commodities, such as cherries and strawberries. Thus, the environment during this period largely determines produce salability when it reaches the consumer.

To prevent undue warming and condensation of moisture, which promote decay and deterioration, fruit-handling systems must be well-designed to minimize rewarming and moisture condensation on the product. For example, fruit should not be removed from cool storage and left unattended for significant periods of time before loading and transport in refrigerated vehicles. When the product is removed from cool storage, it should be consumed as quickly as possible or retained at low temperature.

Details on storage and handling of common fruit are given in the following sections. For more information on storage requirements and physical properties of specific commodities, see Table 1 in Chapter 11. Table 1 in this chapter shows recommended controlled-atmosphere (CA) and modified-atmosphere (MA) conditions for fruit other than apples and pears (Kader 2001). Also see Table 1 in Chapter 24 for guidelines on mixing produce in storage and transportation.

This chapter describes proper postharvest handling guidelines for selected fruits. Additional information on these and many other fruits can be found at postharvest.ucdavis.edu and www.ba.ars.usda.gov/hb66/contents.html.

APPLES

Apples are not only the most important fruit stored on a tonnage basis, but their average storage period is considerably greater than that of any other fruit. Storage may be short for early varieties and those going into processing, but cold storage is critical to proper handling and marketing.

Recommended storage temperature depends on the cultivar. For most varieties, cool storage at 32 to 34°F is recommended. Specific recommendations for each commercial cultivar are usually available from marketing organizations [or see Kader et al. (2002)].

Storage life depends on harvest maturity, elapsed time and temperature between harvest and storage, cooling rate in storage, and sometimes cultural factors. The best storage potential is usually in apples that are mature but have not yet attained their peak of respiration when harvested. However, the grower is inclined to sacrifice

The preparation of this chapter is assigned to TC 10.9, Refrigeration Application for Foods and Beverages.

storage quality for the better color often gained in red varieties by holding them longer on the tree. Even if harvesting begins at the proper time, fruit picked last may be at an advanced stage of maturity. Such late-harvested apples do not have good storage characteristics; neither do those harvested on the immature side, but this is seldom a problem with apples intended for storage before marketing. Harvest at proper maturity, careful handling, and prompt storage after harvest are conducive to long storage life.

Chilling injury is the term commonly applied to disorders that occur at low storage temperatures where freezing is not a factor. The exact mutual relationship of the many types of chilling injury is unknown. The principal disorders classed as chilling injuries in apples are (1) soft scald, (2) soggy breakdown, (3) brown core, and (4) internal browning. Varieties susceptible to one or more of these disorders are Rome Beauty, Braeburn, Jonathan, Golden Delicious, Empire, Grimes Golden, McIntosh, Rhode Island Greening, and Yellow Newtown. In addition to variable susceptibility by variety, there are also yearly variations related to climate, fruit size, and cultural factors.

The following practices affect the condition of apples held for both conventional and controlled atmosphere storage:

Maturity. Because there is no reliable maturity index, growers must use personal experience of the variety, area, or orchard to decide when the crop is mature. Availability of labor, size of operation and crop, weather, storage facilities, and intended length of storage also affect the time of harvest.

Handling to Storage. For optimum storage, apples should be cooled within one or two days of harvest because they can deteriorate as much during one day at field temperatures as during one week at proper storage temperature. If other factors prevent final packaging, fruit can be cooled and stored in field bins. In this case, no grading will have been done to remove substandard product. Subsequent grading may increase the level of bruising, especially if the fruit is still cold when handled.

Normally, apples are placed in storage and cooled by the room refrigeration equipment to about 32°F in 1 to 3 days. Hydrocooling is sometimes used, but requires careful disease control. It also interferes with scald inhibitors, which must be applied to warm fruit.

Table 1 Summary of Controlled Atmosphere Requirements and Recommendations for Fruits Other Than Apples and Pears

Commodity	Temperature Range,[a] °F	Controlled Atmosphere[b] % Oxygen	% Carbon Dioxide	Commercial Use as of June 2001
Apricot	32 to 41	2 to 3	2 to 3	
Asian pear	32 to 41	2 to 4	0 to 1	Limited use on some cultivars
Avocado	41 to 55	2 to 5	3 to 10	During marine transport
Banana	54 to 61	2 to 5	2 to 5	During marine transport
Blackberry	32 to 41	5 to 10	15 to 20	Within pallet covers during transport
Blueberry	32 to 41	2 to 5	12 to 20	Limited use during transport
Cactus pear	41 to 50	2 to 3	2 to 5	
Cherimoya and Atemoya	46 to 59	3 to 5	5 to 10	
Cherry, sweet	32 to 41	3 to 10	10 to 15	Within pallet covers or marine containers during transport
Cranberry	36 to 41	1 to 2	0 to 5	
Durian	54 to 68	3 to 5	5 to 15	
Fig	32 to 41	5 to 10	15 to 20	Limited use during transport
Grape	32 to 41	2 to 5	1 to 3	Incompatible with SO_2
	32 to 41	5 to 10	10 to 15	Can be used instead of SO_2 for decay control up to four weeks
Grapefruit	50 to 59	3 to 10	5 to 10	
Guava	41 to 59	2 to 5	0 to 1	
Kiwifruit	32 to 41	1 to 2	3 to 5	Expanding use during transport and storage; C_2H_4 must be maintained below 20 ppb
Lemon	50 to 59	5 to 10	0 to 10	
Lime	50 to 59	5 to 10	0 to 10	
Lychee (litchi)	41 to 54	3 to 5	3 to 5	
Mango	50 to 59	3 to 7	5 to 8	Increasing use during marine transport
Nectarine	32 to 41	1 to 2	3 to 5	Limited use during marine transport
	32 to 41	4 to 6	15 to 17	Reduces chilling injury (internal breakdown) of some cultivars
Olive	41 to 50	2 to 3	0 to 1	Limited use to extend processing season
Orange	41 to 50	5 to 10	0 to 5	
Papaya	50 to 59	2 to 5	5 to 8	
Peach, clingstone	32 to 41	1 to 2	3 to 5	Limited use to extend canning season
Peach, freestone	32 to 41	1 to 2	3 to 5	Limited use during marine transport
	32 to 41	4 to 6	15 to 17	Reduces incidence and severity of internal breakdown (chilling injury) of some cultivars
Persimmon	32 to 41	3 to 5	5 to 8	Limited use of modified atmosphere packaging
Pineapple	46 to 55	2 to 5	5 to 10	Waxing is used to create modified atmosphere and reduce endogenous brown spot
Plum	32 to 41	1 to 2	0 to 5	Limited use for long-term storage of some cultivars
Pomegranate	41 to 50	3 to 5	5 to 10	
Rambutan	46 to 59	3 to 5	7 to 12	
Raspberry	32 to 41	5 to 10	15 to 20	Within pallet covers during transport
Strawberry	32 to 41	5 to 10	15 to 20	Within pallet covers during transport
Sweetsop (custard apple)	54 to 68	3 to 5	5 to 10	

Source: Kader (2001).
[a]Usual or recommended range; 90 to 95% rh is recommended.
[b]Specific CA combination depends on cultivar, temperature, and duration of storage. Recommendations are for transport or storage beyond two weeks. Exposure to lower O_2 or higher CO_2 concentrations for shorter durations may be used to control some physiological disorders, pathogens, or insects.

Controlled-Atmosphere Storage

Controlled-atmosphere (CA) storage is important in extending the market life of certain apple varieties. Chilling injury is eliminated in some varieties by elevating the storage temperature to about 40°F and altering the composition of the atmosphere.

Only apples of good quality and high storage potential should be placed in CA storage. Harvest maturity and handling practices are crucial; only fruit harvested at proper maturity should be considered. In any one district, this limits the number of apples suitable for CA storage to only a few days' harvest. Immature apples or those retained on the tree to gain better color, as is often done with Delicious and McIntosh, are equally undesirable.

Table 2 lists optimum levels of O_2, CO_2, and temperature for CA storage of apples. It also indicates storage life and whether the specific variety is susceptible to storage scald. This information was obtained from a worldwide survey of postharvest scientists who work on pome fruits.

Chapter 14 discusses systems and methods for achieving specific CA conditions as well as construction techniques and details for the rooms and spaces.

Storage Diseases and Deterioration

Storage problems in apples may be caused either by invading microorganisms or by the fruit's own physiological processes. Physiological disorders, although sometimes resembling rots, are related to biochemical processes within the fruit. Susceptibility to such disorders is often a variety characteristic, but it may be influenced by cultural and climatic factors and storage temperature.

Alternaria Rot. Dark brown to black, firm, fairly dry to dry storage decay centering at wounds, in skin cracks, in core area, or in scald patches; one of the blackest of storage decays. *Control*: Cultural practices that produce apples of good finish and prevent skin diseases and injuries that open the way for infection.

Table 2 Optimum Levels for Controlled Atmosphere Storage of Apples

Cultivar	Country	Region	Optimum O_2, %	Optimum CO_2, %	Optimum Temperature, °F	Storage Life, months
Alwa	Poland	Skierniewice	1.5	1.5	32 to 37	7
Ampion	Poland	Skierniewice	1.5	1.5	32 to 37	7
Arlet	Poland	Skierniewice	3	5	32 to 37	8*
Bancroft	Poland	Skierniewice	1.5	1.5	32	9
Bellena Roma	Spain	Lleida	3	2.5	32	*
Blanquilla	Spain	Lleida	3	3	32	6 to 7
Bonza	Australia	Victoria	1.5 to 1.8	1	32	*
Boskoop	Belgium	Heverlee	2	0.7	37 to 38	6*
Braeburn	Australia	Victoria	1.5 to 1.8	0.8 to 1	34	*
	Belgium	Heverlee	2	1	34	6
	France	St. Remy	1.5	0.8 to 1.2	33 to 34	6*
	Italy	Milan	1	1	33 to 34	8 to 9*
	New Zealand	New Zealand	3	1	33	6
	South Africa	Stellenbosch	1.5	1.5	31	8 to 9
	United States	Washington	2	<0.5	35	*
Bramley's Seedling	United Kingdom	Kent	1	5	39	11*
Cortland	Canada	Nova Scotia	2.5	4.5	37	8 to 10*
	United States	New York	2 to 3	2 to 3% one month, then 5%	36	4 to 6*
Cox's Orange Pippin	Belgium	Heverlee	2	0.7	37 to 38	5
	New Zealand	New Zealand	2	2	37	4 to 5
	Netherlands	Wageningen	1.2 to 1.4	<1	39	6.5
	United Kingdom	Kent	1.3	4	38	7
Elstar	Belgium	Heverlee	2	1	34	7
	Canada	Nova Scotia	2.5	4.5	32 to 33	No data
	Netherlands	Wageningen	1.2	2.5	35	7
Empire	Canada	Nova Scotia	2.5	0.5 to 1	34 to 36	No data
	United States	Michigan	1.5	2.5	37	9
		New York	2 to 3	2 to 3	34 to 36	5 to 10
Fiesta	New Zealand	New Zealand	2	2	33	6
	Poland	Skierniewice	1.5	1.5	32 to 37	7
Firmgold	Australia	Victoria	1.5 to 1.8	2 to 2.5	32	No data
Fuji	Australia	S. Australia	2	1	32	8*
		Victoria	2 to 2.5	2	32	*
	Canada	British Columbia	1.2	1	32	9*
			0.7	2	32	9*
	France	St. Remy	2 to 2.5	1 to 2	32 to 34	7 to 8*
	Italy	Milan	1	1	32	*
	United States	Washington	1	1	34	9*
			1	1	34	11*
		California	1.5	<0.5	32 to 34	7 to 9*
Gala	Australia	Victoria	1.5 to 2	1	32	*
	Canada	British Columbia	1.2	1	32	6
	France	St. Remy	1.5	2	32 to 34	4 to 5
	Italy	Milan	3	2	34 to 36	6
	New Zealand	New Zealand	2	2	33	No data
	Poland	Skierniewice	1.5	1.5	37	7

*Indicates variety of cultivars is subject to storage scald.

Table 2 Optimum Levels for Controlled Atmosphere Storage of Apples (*Continued*)

Cultivar	Country	Region	Optimum O_2, %	Optimum CO_2, %	Optimum Temperature, °F	Storage Life, months
Gala (*continued*)	Spain	Lleida	2	2	36	2 to 9
	United States	Washington	1	1	34	4
	Netherlands	Wageningen	1.2	2	34	5.5
Gala-Mondiel	United Kingdom	Kent	1	5	35	7*
Galaxy	Australia	Victoria	1.5 to 2	1	32	No data
Gloster	Netherlands	Wageningen	1.2	3	34	7.5
	Austria	Austria	2.25	3	35	10
	Canada	Nova Scotia	2.5	4.5	32 to 33	10
	Poland	Skierniewice	1.5	2	32 to 37	7 to 8
Golden Delicious	Australia	South Australia	2	1	32	7
		Victoria	1.5 to 2	1	32	*
	Austria	Austria	2 to 3	3	35	9 to 10
	Belgium	Heverlee	2	2	33	10*
	Canada	Nova Scotia	2.5	3	32 to 35	10
		British Columbia	1	1.5	32	11
	France	St. Remy	1 to 1.5	2 to 3	34 to 36	7 to 9*
	Israel	Israel	1.2	2	32	No data
			1 to 1.5	2	32	No data
	Italy	Milan	1	1.5	32 to 34	8 to 9
	New Zealand	New Zealand	2	2	33	6
	South Africa	Stellenbosch	1.5	3	31	9*
	Spain	Lleida	2.5	2.5	33	9*
		Murcia	2 to 3	4	33	7 to 8*
	Netherlands	Wageningen	1.1	4	34	8
	United States	Washington	1	2 to 3	34	9
		West Virginia	1	0	32	7 to 9
		Michigan	1.5	3	32 to 34	9
		New York	2 to 2.5	2 to 3	32	8 to 10*
Granny Smith	Australia	South Australia	2	1	32	8*
		Victoria	1.5 to 1.8	1	34	*
	France	St. Remy	0.8 to 1.2	0.8 to 1	32 to 36	7 to 8*
	Israel	Israel	1.2	3.5	32	No data
			0.8 to 1.5	2 to 5	32	No data
	South Africa	Stellenbosch	1.5	0 to 1	31 to 33	No data
	Spain	Lleida	2.5	4	36	7 to 8*
	United States	Washington	1	1	34	11*
		California	1.2	1	33	8*
Gravenstein	Canada	Nova Scotia	2.5	2.5	37	4*
Idared	Austria	Austria	2	2 to 2.5	36 to 37	No data
	Belgium	Heverlee	2	1.5	34	7
	Canada	Nova Scotia	2.5	0.5 to 1.5	32 to 37	10*
	France	St. Remy	3	3	36 to 39	7 to 8
	United Kingdom	Kent	1.3	4	39	4*
	United States	Michigan	1.5	3	32	9
		New York	2 to 2.5	2 to 3	34	7 to 9*
Jester	Poland	Skierniewice	1.5	1.5	32 to 37	6*
Jonagold	Australia	Victoria	1.5 to 1.8	1	32	*
	Austria	Austria	2	2	36	10
	Belgium	Heverlee	1	2.5	33	9*
	Canada	Nova Scotia	2.5	4.5	32 to 33	10
		British Columbia	1.2	1.5	32	9
	Poland	Skierniewice	1.5	2	37	9
	Spain	Lleida	2	3	33	5 to 8*
	Netherlands	Wageningen	1 to 1.2	4	34	9*
	United Kingdom	Kent	1.3	4	35	6*
	United States	New York	2 to 2.5	2 to 3	32	5 to 7*
Jonagold Red	Australia	Victoria	1.5 to 1.8	1	32	*
Jonamac	United States	New York	2 to 3	2 to 3% one month, then 5%	36	3*
			2 to 3	2 to 3	32	3*
			2 to 3	2 to 3	32	3*
Jonathan	Australia	South Australia	2	1	32	8*
		Victoria	1.5 to 1.8	1	34	*
	Poland	Skierniewice	3	5	32	7
	United States	Michigan	1.5	3	32	6
Lady Williams	Australia	Victoria	1.5 to 1.8	1	34	*
		South Australia	2	1	32	8*

*Indicates variety of cultivars is subject to storage scald.

Table 2 Optimum Levels for Controlled Atmosphere Storage of Apples (*Continued*)

Cultivar	Country	Region	Optimum O_2, %	Optimum CO_2, %	Optimum Temperature, °F	Storage Life, months
Law Rome	United States	Michigan	1.5	3	32	9*
		New York	2 to 2.5	2	32	7 to 9*
Ligol	Poland	Skierniewice	1.5	1.5	32	7.5*
Lobo	Canada	Nova Scotia	2.5	4.5	37 to 38	No data
Lodel	Poland	Skierniewice	3	5	32	6
Macfree	Canada	Nova Scotia	2.5	4.5	32 to 33	4
McIntosh	Canada	Nova Scotia	1.5	1.2	37	8
			2.5	5	35 to 37	8
	Poland	Skierniewice	1.5	1	36	8
	United States	Michigan	1.5	3	37	6*
		New York	3	2 to 3% one month, then 5%	36	5 to 7*
McIntosh (not Marshall)	Canada	British Columbia	2.5	4.5	37	8 to 10
McIntosh-Marshall	Canada	British Columbia	2.5	4.5	37	8 to 10
	United States	New York	4 to 4.5	2 to 3% one month, then 5%	36	No data
Melrose	France	St. Remy	2 to 3	3 to 5	32 to 37	4 to 6*
	Poland	Skierniewice	3	5	32	8
Moira	Canada	Nova Scotia	2.5	4.5	32 to 33	<2
Mutsu	Australia	Victoria	1.5 to 1.8	1	34	*
	United States	Michigan	1.5	3	32	9*
		New York	2 to 2.5	2 to 3	32	6 to 8*
Nashi (Nijisseki)	Australia	Victoria	1.5 to 1.8	1	34	*
		South Australia	2	1	32	5
Northern Spy	Canada	Nova Scotia	2.5	2	32 to 33	No data
	United States	Michigan	1.5	3	32	12
Nova Easygro	Canada	Nova Scotia	2.5	4.5	32 to 33	4
Novamac	Canada	Nova Scotia	2.5	4.5	37 to 38	4
Novaspy	Canada	Nova Scotia	2.5	4.5	32 to 33	10
Pink Lady	Australia	South Australia	2	1	32	9
			1.5 to 1.8	1	32	*
Prima	Canada	Nova Scotia	2.5	4.5	32 to 33	<2
Priscilla	Canada	Nova Scotia	2.5	4.5	32 to 33	<2
Boskoop	Netherlands	Wageningen	1.2	0.7	40 to 41	5.5*
Red Delicious	Australia	South Australia	1.8 to 2	2 to 2.5	32	*
	Canada	Nova Scotia	2.5	4.5	32 to 33	10*
		British Columbia	1.2	1	32 to 34	11*
			0.7	1	32 to 34	11*
	Israel	Israel	1.5	2	32	6 to 9*
	Italy	Milan	0.8	1	32 to 34	8
	Japan	Aomori	2	1.5	32	9*
	New Zealand	New Zealand	1.5	1.5	33	6*
	United States	West Virginia	1	0	32	7 to 9*
		Michigan	1.5	3	32	9*
		New York	2 to 2.5	2	32	8 to 10*
		Washington	1	3	34	9*
Red Delicious-Early Red One	Spain	Lleida	2.5	3	32	8*
Red Delicious-Hi Early	Australia	Victoria	2	1	32	8*
Red Delicious-Starking	South Africa	Stellenbosch	1.5	1.5	31	9*
	Spain	Lleida	2.5	2	32	8*
Red Delicious-Top Red	France	St. Remy	1.5	1.8 to 2.2	32 to 34	6 to 7*
	South Africa	Stellenbosch	1	1	31	9*
Rome	United States	West Virginia	1	0	32	7 to 9*
Royal Gala	Australia	South Australia	2	1	32	5
	South Africa	Stellenbosch	1	1	31	8
	United States	California	1.5	1.5	32	6
Rubin	Poland	Skierniewice	1.5	1.5	32	7
Runkel	United States	Michigan	1.5	3	32	9
Spartan	Canada	Nova Scotia	2.5	2.5	32 to 33	10
		British Columbia	1.2	1.5	32	11
	United States	New York	2 to 2.5	2 to 3	32	6 to 8*
Splendor	Canada	Nova Scotia	2.5	4.5	32 to 33	No data
Sturmer Pippin	New Zealand	New Zealand	2	2	36	4 to 5
Sundowner	Australia	South Australia	2	1	32	9
		Victoria	1.5 to 2	1	32	*
York	United States	West Virginia	1	0	32	7 to 9*

Source: Mitcham (1997).

*Indicates variety of cultivars is subject to storage scald.

Ammonia Gas Discoloration. Circular spots centering at lenticels; dull green on unblushed side and brown to black on blushed side. Injury may disappear from slightly affected fruit. *Control*: Ventilate as soon as possible. Examine fruit for injury at various points in the room because some sections may escape damage.

Bitter Pit. Many small, sunken bruise-like spots, usually on the calyx half of the fruit. Masses of brown, spongy tissue occur adjacent to surface pits, or may be found deeper in the flesh. In storage, spongy tissue near surface loses moisture and tends to become hollow. New areas may appear and develop in storage. *Control*: Apply boron and calcium, as recommended, in the orchard. Follow cultural practices that promote regular bearing and stabilize moisture. Store fruit of proper maturity and cool promptly to 32°F. Keep humidity high enough to prevent moisture loss.

Blue Mold Rot (*Penicillium*). Spots of various sizes with decayed tissue that is soft and watery and can be readily scooped out of the surrounding healthy flesh. Rot is usually as deep as it is wide. Advanced stages have white tufts of mold that turn bluish-green, because spores are produced under moist conditions. Affected tissue has moldy or musty flavor and odor. This is the most prevalent type of storage decay of apples. *Control*: Handle carefully to prevent skin breaks. Cool promptly to 32°F. Use fungicides in wash treatments. Keep picking boxes, packing house, and storage room sanitary. Whitewash walls and ceiling.

Brown Core. No external symptoms. It first appears as slight browning or discoloration of core tissue between seed cavities. Later, part or all of the flesh between seed cavities and the core line may become brown. This is serious in McIntosh and other susceptible varieties stored for long periods at 30°F. *Control*: Pick at proper stage of maturity. Use CA storage at 36 to 38°F. A disorder with similar symptoms results from exposure to excessive concentrations of CO_2.

Freezing Injury. Water-soaked, rubbery condition of large areas or of entire apple. Vasculars (water-conducting strands) are brown. Bruised areas in frozen apples are large, with wrinkled gray to light brown surface. Moisture is lost rapidly from affected areas. In refrigerator cars, it is most prevalent on floor and at doorways; in storage rooms, most injury is in bottom layer boxes, near coils, or against walls next to freezer storage. *Control*: Heat car during freezing weather. Prevent cold pockets in storage rooms by adequate air circulation. Minimize handling of fruit while frozen. Thaw at 40 to 50°F. Move thawed fruit into trade channels promptly; do not allow it to become overripe.

Internal Breakdown. Mealy breakdown of internal tissue in overripe fruit. Flesh is soft. Surface is often duller and darker than normal. It is hastened by too high storage temperature, freezing, bruising, or presence of water core, which it often follows. *Control*: Pick before overmature. Cool promptly at temperatures as near 32°F as possible for varieties that tolerate that temperature. Watch ripening rate, particularly of fruit with water core.

Internal Browning. No abnormal skin appearance. Sometimes it appears only around core; the apple's outer fleshy portion remains normal in appearance. Occasionally, only outer flesh is involved, but is usually accompanied by browning around the core. Disease develops uniformly throughout tissue, and occurs in firm, sound apples. *Control*: Use CA at 38°F for Empire and other susceptible varieties.

Jonathan Spot. Slate-brown to black, entirely superficial or very slightly sunken, skin-deep spots in color-bearing cells of skin. In some varieties, spots center at lenticels. *Control*: Refrigerate promptly, because this disease is greatly aggravated by delayed storage. Use CA storage.

Lenticel Rots. Bullseye rot (*Neofabrabraea*): most common of group; important only in apples from Northwest; spots fairly firm, pale centers, decay mealy, may penetrate nearly as deep as wide. Fisheye rot (*Corticum*): tough leathery spot, often follows scab; decayed tissue stringy. Side rots (*Phialophora*): spots shallow with tender skin, decayed tissue wet, slippery. *Control*: Harvest at prime maturity; store and cool promptly; use forecasting technique for bullseye rot to determine potential keeping quality.

Scab (*Venturia*). Occasionally, active scab spots on fruit at time of storage enlarge. Fruit may be infected in orchard but show no disease at the time of storage. Disease may subsequently develop in storage as small brown or jet black spots in peel, often without breaking cuticle of fruit. *Control*: Follow recommended orchard spray schedule.

Scald. Diffuse browning and killing of skin of fruit stored for several months. Ordinarily most prevalent on immature fruit or on green portions of fruit. *Control*: Pick apples when well matured. Treat with scald-inhibiting chemicals. Scald develops less on fruit in controlled atmospheres.

Soft Scald. Sharply defined or slightly sunken ribbon-like areas in the skin. Affected tissue is shallow and rubbery. It is most severe on Jonathan, Golden Delicious, and Wealthy. *Control*: Store promptly. Use recommended controlled atmospheres, temperatures, and lengths in storage for each variety.

Soggy Breakdown. Light brown, moist, rubbery, definitely delimited areas in cortex of apple; not visible on surface. It is worst in Grimes Golden, Wealthy, and Golden Delicious. *Control*: Same as for soft scald.

Water Core. Hard, glassy, water-soaked regions in flesh of apple at core or under skin. Extent decreases during storage but predisposes fruit to internal breakdown. *Control*: Pick as soon as mature. Watch fruit in storage and move before it becomes overripe.

PEARS

Bartlett is the most important pear variety, exceeding the total of all others by a wide margin. Other Pacific Coast varieties are Hardy, Comice, Anjou, Bosc, and Winter Nelis. The eastern states have limited varieties because of the severe problem of fire blight, and primarily grow the Kieffer variety. Although most Bartlett and Hardy and many Winter Nelis pears are canned, cold storage before ripening for canning is the usual procedure. A 10 day to 2 week cold storage period for Bartlett pears is commonly used by canners because it improves ripening uniformity. Substantial quantities may also be stored for periods approaching maximum storage life of the variety to better use processing facilities.

Maturity at harvest strongly affects subsequent storage life, as it does for apples. However, unlike apples, pears do not ripen on the tree, nor do most varieties ripen at cold storage temperatures. If harvested too early, they are subject to excessive water loss in storage. If allowed to become overmature on the tree, their storage life is shortened, and they may be highly susceptible to scald and core breakdown. Flesh firmness, as measured by a pressure tester, is perhaps the best measure of potential storage life of pears from any single orchard. For the Bartlett variety, a firmness of 19 to 17 lb, measured with a Magnus-Taylor pressure tester or similar device using a 0.31 in. diameter plunger head, indicates best storage quality. If average firmness is as low as 15 lb, storage for any prolonged period is likely to produce poor-quality fruit. Pressure-test information for each lot of pears going into storage can be very helpful to both the fruit owner and the cold storage operator in determining the storage program.

Careful harvesting and handling are essential to good storage quality. Bruises and skin breaks are likely sites for infection by microorganisms. Varieties such as Winter Nelis and Bosc are highly susceptible to punctures caused by stems broken during harvesting. Comice is also easily damaged because of its very tender skin. Pears are harvested into pallet bins holding about 1000 lb of fruit. Care in dumping fruit from a picking container is important in keeping mechanical damage to a minimum.

For best storage quality, rapid cooling after harvest is essential. Pears ripen rapidly at elevated temperatures but do not soften or change color in the early ripening stages. Therefore, a considerable part of the storage life may be used up without a visible change in the fruit. If cold-storage rooms do not have adequate refrigeration and air circulation capacity for rapidly cooling fruit, consider precooling in special rooms (or hydrocooling) before placing in the storage room. When warm fruit is placed in a room with cold fruit, the loading arrangements should be such that the temperature of the cold fruit is not elevated.

Pears are very sensitive to temperature and should be stored at 30°F and 90 to 95% rh. Recommendations as low as 29°F have been made, but the risk of freezing injury is great unless the temperature in all parts of the room can be controlled precisely. Pears are not subject to chilling injury as some apple varieties are, so elevated storage temperatures are not required. The stacking arrangement recommended for apples in the cold storage room also applies for pears.

Because pears lose water more readily than most apple varieties, good humidity conditions in the storage room must be maintained. For long storage, 90 to 95% rh is recommended. Perforated film box liners are excellent for moisture loss control.

The approximate storage life of pears at 30°F is shown in Table 3. These values assume an additional time for transportation and marketing. If Bartlett pears for canning are harvested at the best stage of maturity and quickly cooled to 30°F, their safe storage life may be as long as 4 months, because marketing involves only ripening for processing. However, quality deteriorates during storage, particularly as the maximum storage life is approached.

After removal from storage, best dessert quality (for both canning and fresh use) is attained if pears are ripened in a controlled temperature range of about 60 to 70°F. For cannery fruit, ripening at 68 to 72°F is more practical than at lower temperatures because the shorter time involved reduces overhead costs with no measurable difference in quality.

Controlled-Atmosphere Storage

The storage life of Bartlett pears can be extended to 5 to 6 months at 30°F for fruit of desirable maturity (17 to 20 lb firmness) in controlled-atmosphere storage. Bartlett pears of advanced maturity are intolerant to elevated CO_2 and develop core and flesh browning within a few weeks. They are tolerant of low O_2, but have less storage potential than pears of desirable maturity. Chronological age and pressure test (under 16 lb firmness) are evidence of advanced maturity.

Commercial CA storage of pears has not been considered as necessary as it is in the apple industry. Because no low-temperature disorders have been recognized, there has not been a need for further extension of the storage life.

Table 3 lists optimum levels of O_2 and CO_2, storage time, and CA disorders for pear varieties at 30 to 32°F. This information was obtained through a worldwide survey of postharvest scientists who work on pome fruits.

Many pears from western states, when packed for storage before marketing, have perforated polyethylene liners in the container. Although these liners give excellent protection against water loss, there is no agreement about their value in modifying the atmosphere in the container.

Table 3 Commercial Controlled Atmosphere Conditions for Pear Varieties[a]

Variety	O_2, %	CO_2, %	Storage, months	CA Disorders[b]
Abate Fetel	1	1	5 to 6	IB
Alejandrina	3	2	4 to 5	IB
Anjou, d'Anjou	1 to 2.5	0 to 0.5	7 to 8	IB, PBC, Cav
Bartlett, (= William's Bon Chretien)	1 to 2	0 to 0.5	3 to 5	CF, PBC
Blanquilla, (= Blanca de Aranjuez)	3	3	6 to 7	
Bosc, Kaiser	1 to 2.5	0.5 to 1.5	4 to 8	PBC, Cav
Buena Luisa (= Buona Luisa)	3	2	6	IB, CF
Clapp's Favorite	2	<0.7	3 to 4	IB, PBC
Comice (= Doyenne du Comice, Comizio)	1.5 to 4	0.5 to 4	5 to 6	IB (overmature)
Conference	1 to 2.5	0.6 to 1.5	6 to 8	BH, IB, Cav
Coscia	1.5	2 to 3	6 to 7	CF
Flor d'Hivern (= Inverno)	3	3	4 to 5	IB
Forelle	1.5	0 to 1.5	6 to 7	
General Leclerc	2 to 3	2 to 3	3 to 5	
Grand Champion	3	2 to 2.5	4	
Hardy	2 to 3	3 to 5	4 to 6	
Josephine	1 to 2	1 to 2	8	
Krystalli	2	1 to 2	3 to 5	
Limonera, Llimonera	3	3	3 to 4	
Packham's Triumph	1.5 to 1.8	1.5 to 2.5	7 to 9	CB
Passe Crassane (= Passa Crassane)	3	4 to 5	5 to 8	IB
Rocha	2	2	8	
Spadona	1.5 to 2.5	1.5 to 3.5	8 to 9	IB
Nashi, Asian pears				
Chojuro	2	1 to 2	3 to 4	
Kosui	1 to 2	0 to 2	3 to 4	
Nijiseiki (= 20th Century)	0.5 to 3	0 to 1	5	
Tsu Li	1 to 2	0 to 3	3 to 5	IB
Ya Li[c]	4 to 5	0 to 4	3 to 4	IB, Cav, CI?

Source: Richardson and Kupferman (1997).
[a]Optimum storage temperature is 30 to 32°F, unless otherwise indicated.
[b]CA disorder abbreviations are as follows:
IB = internal breakdown or browning
BH = brown heart
PBC = pithy brown core
CF = core flush
Cav = cavity, usually lens-shaped
CI = chilling injury
[c]Ya Li may show a type of chill injury at <41°F

Storage Diseases and Deterioration

The principal storage disorders in pears are (1) core breakdown, (2) scald and failure to ripen, and (3) fungus rots.

Core Breakdown. Often accompanies scald. Soft, brown breakdown in core area has an acrid, disagreeable odor of acetaldehyde. *Control*: Do not allow pears to become overmature on tree. Cool promptly. Store at 30°F. Ripen between 65 and 75°F.

Core breakdown is associated with overmaturity at harvest. This problem has become more important because of growth regulator sprays used to keep pears from dropping during the harvest season. Pressure-test information on late-harvested fruit is helpful in locating lots susceptible to core breakdown. Records of the time lapse between harvest and storage are important, because pressure-test information may not be a true measure of relative storage quality when storage is delayed. Pear color, particularly in California Bartletts, is a very poor measure of potential storage life because of great variability among pears from different districts.

Scald. Brown to black softening of large areas of skin and tissues immediately beneath skin; often accompanies core breakdown. Affected areas slough off readily. Acetaldehyde odor and flavor are prominent. *Control*: Pick before overmature. Cool promptly. Store only for proper period. Oiled paper wraps do not control scald.

Pear scald is associated with overlong storage and lost ripening capacity. It is not related to apple scald and cannot be controlled by any supplemental treatments. The problem develops progressively earlier as the temperature is raised above 30°F. Yellowing of the fruit is the principal storage symptom; Bartlett and Bosc are the two most susceptible varieties. Anjou and Comice may not develop scald but do lose their capacity to ripen. Periodic inspection is desirable to ensure that green pear varieties are removed from storage before yellowing progresses to the danger point. Yellow pears may show no scald in storage but may develop scald on removal to a ripening temperature. If pear scald does show in storage, the pears have been kept too long and may be worthless.

Anjou Scald. Anjou pears are often affected with a surface browning more superficial than common scald and distinct from it, resembling apple scald. Anjou scald is controlled by oiled paper wraps and scald-inhibiting chemicals.

Gray Mold Rot (*Botrytis*). Extensive, firm, dull brown, water-soaked decay with bleached border. Dirty white to gray extensive mycelia form nests of decayed fruit. *Control*: Wrap fruit in copper-impregnated paper. Use fungicide in spray or wax on packing line. Cool promptly to 30°F.

Gray mold rot caused by *Botrytis cinerea* grows at cold storage temperatures and can be a serious threat to long-stored winter varieties such as Anjou and Winter Nelis. Without control measures, the disease may spread from one fruit to another by contact.

Alternaria Rot. Dark brown to black surface. Decayed tissue is gray to black, dry in center, gelatinous at edge, easily removable as core from surrounding flesh. It is found late in storage season, usually at punctures. *Control*: Prevent skin breaks. Remove from storage at first appearance of trouble.

Brown Core. Various degrees of pithy brown core and desiccated air pockets in Anjou and Bartlett pears stored in sealed, polyethylene-lined boxes with inadequate permeability. Prolonged storage in concentrations of 4% or more CO_2 often produces brown core, particularly when pears are harvested at advanced maturity or are cooled slowly after packing. *Control*: Harvest at proper maturity. Cool promptly. Store at 30°F. Use perforated film liners to maintain CO_2 level at 0 to 0.5%.

Freezing Injury. Glassy, water-soaked external appearance with tan pithy area around core in Bartlett and Anjou pears exposed for 4 to 6 weeks just below their freezing point. Pears frozen sharply may break down completely or show abruptly sunken large pits where slightly bruised while frozen. *Control*: Keep transit and storage temperature above 30°F.

GRAPES

Grapes are widely grown in the United States, but over 90% are grown in California, which produces the *Vitis vinifera* species almost exclusively. This species can withstand the rigors of handling, transport, and storage required of table grapes for wide distribution over a long marketing period. Almost all of this fruit is precooled and much of it stored for varying periods before consumption. On the other hand, for fresh use, fruit of the species *Vitis labrusca* (Eastern type) is largely limited to local market distribution.

Grapes grow relatively slowly and should be mature before harvest because all of their ripening occurs on the vine. *Mature* here means the stage of physiological development when the fruit appears pleasing to the eye and can be eaten with satisfaction. However, grapes should not be overripe, because this predisposes them to two serious postharvest disorders: (1) weakening of stem attachment in some varieties, such as Thompson seedless, which causes the berries to separate from the pedicel attachment; and (2) progressively greater susceptibility to invading decay organisms. Danger of fruit decay is increased with exposure to rain or excessively damp weather before harvest (conditions favorable for field infections by *Botrytis cinerea Pers*).

Cooling and Storage

Grapes are vulnerable to the drying effect of air because of their relatively large surface-to-volume ratio, especially that of the stems. Stem condition is an important quality factor and an excellent indicator of the past treatment of the fruit. Stems should be maintained in a fresh green condition not only for appearance but because they become brittle when dry and are apt to break. The stem of a grape cluster, unlike that of other fruit, is the handle by which the fruit is carried; if breakage (shatter) occurs, the fruit is lost for all practical purposes even though the shattered berries may still be in excellent condition. Therefore, careful attention should be paid to operations that minimize moisture loss.

The rate of water loss is especially high before and during precooling, because grapes are normally harvested under hot, dry conditions. Field heat should be removed promptly after the fruit is picked to minimize the grapes' exposure to low-vapor-pressure conditions. Volume and temperature of precooling air, its velocity past or through the containers (lugs), and accessibility of the fruit to this air are significant factors in the rate of heat removal. These factors are drastically influenced by the location and amount of venting of the containers, alignment of the containers (air channels), and packing materials such as curtains, cluster wraps, and pads.

Table grapes are initially cooled with a forced-air cooling system. A pressure gradient is set up so that there is a positive flow of cold air through the fruit from one vented side of the container to the other. The containers are arranged so that the air must pass *through* the containers before returning to the refrigeration surface. Precooling time is usually 3 to 12 h, depending on packaging system and airflow.

The recommended storage temperature for *Vitis vinifera* (European or California-type) grapes is 30°F. The relative humidity should be 90 to 95%. Although temperatures as low as 29°F have not been injurious to well-matured fruit of some varieties, other varieties with low sugar content have been damaged by exposure to 31°F. Grape storage plants in California should provide uniform air circulation in the rooms. Fruit should be forced-air-cooled to less than 39°F before storage. During initial storage, a well-distributed airflow of 100 cfm per ton of grapes is needed to finish the cooling. After the fruit has been precooled, air velocity should be reduced to a rate that maintains uniform temperatures throughout the room (no more than 10 to 20 fpm in the channels between the lugs).

The greatest change that occurs in stored grapes is water loss. The first noticeable effect is drying and browning of stems and pedicels. This becomes evident with a loss of only 1 to 2% of the mass of the fruit. When loss reaches 3 to 5%, the fruit loses its turgidity and softens.

Maintaining 90 to 95% rh in grape storage is often a problem, especially at the beginning of the storage season when the rooms are being filled with dry lugs. Each lug absorbs 0.33 to 0.67 lb of water over a month, and, unless moisture is supplied to the room, this water must come from the fruit. Spray humidification is effective at minimizing shrinkage. With proper balance of water and air pressure and the correct type of nozzle, a fine spray can be obtained that will vaporize readily even at 31°F.

Fumigation

Vinifera grapes must be fumigated with sulfur dioxide (SO_2) after they are packed to prevent or retard the spread of decay. The treatment surface-sterilizes the fruit, particularly wounds made during handling.

Fumigation with SO_2 in storage prevents new infections of the fruit but does not control infections that have already occurred in the vineyard. Frequently, these have not developed far enough to be detected at harvest and consequently are the primary cause of decay in storage. Harvey (1955, 1984) describes a method of measuring field infection to forecast decay during storage; the forecast indicates lots that are sound and can be safely stored, and those that are likely to decay and should be marketed early.

Common practice during initial cooling is to fumigate the fruit in the evening. In this way, precooling is not delayed, and fumigation can be done after most of the working crew has left. This initial treatment often becomes the responsibility of the refrigeration personnel.

Amount of Sulfur Dioxide. Other commodities should not be treated along with the grapes or even held where the fumigant can reach them, because most of them are very easily injured by the gas. Because grapes also can be injured, they should be exposed to the minimum quantity of SO_2 required, which depends on the following:

- Decay potential and condition of the fruit
- Amount of fruit to be treated
- Type of containers and packing materials
- Air velocity and uniformity of air distribution
- Size of the room
- Losses from leakage and sorption on walls

Much of the following information on sulfur dioxide fumigation methods of table grapes is from *Bulletin* 1932, University of California Division of Agriculture and Natural Resources, 1992.

The amount of SO_2 for effective control of *Botrytis* varies with the length of time that grapes are exposed to the gas. The dosage of SO_2 to kill *Botrytis* spores or mycelia during fumigation is 100 ppm·h.

Fumigation Methods. The following two methods are currently used for fumigation. Both give adequate decay control but differ in the quantity of SO_2 used and the application method.

Traditional fumigation may be used for initial fumigation when grapes are received at a facility, and for weekly fumigation during long-term storage. Relatively high SO_2 concentrations are added for 20 to 30 min, then the remaining gas is scrubbed or vented from the room.

Traditional initial fumigation can use either circulating-air or forced-air fumigation. Each is used either in combination with initial cooling or as a separate operation. In **circulating-air fumigation**, air flows past, but not through, palletized boxes. Penetration of SO_2 into the innermost boxes depends on the speed of air past the pallets and the types of box and packing materials. A minimum airspeed of 140 fpm is required for maximum penetration.

Airflow systems for **forced-air fumigation** are the same as systems used for forced-air cooling. Sulfur dioxide gas is introduced into the room air and forced through the boxes, resulting in rapid and thorough penetration, even for bagged or tissue-wrapped fruit.

Airflow for many forced-air fumigation rooms is typical of forced-air coolers (about 1 cfm per lb of product stored); however, good SO_2 penetration has been observed even with lower airflow (0.5 cfm/lb) and slower cooling times.

The maximum permitted SO_2 concentration for initial fumigation is 10,000 ppm. Although a few operators regularly use this level in small circulating-air fumigation chambers, many operators use 5000 ppm for initial fumigation. Levels used for a particular facility must be determined by using SO_2 dosimeter tubes to measure actual SO_2 penetration into boxes.

The amount of SO_2 needed for a traditional fumigation can be calculated by

$$W_s = \frac{AVC}{10^6}$$

where

W_s = quantity of SO_2 required, lb
A = 1.67 lb SO_2/ft^3 at 70°F and 1.82 lb SO_2/ft^3 at 32°F
V = room volume, ft^3
C = desired SO_2 concentration, ppm

Sulfur dioxide concentration may also be expressed in terms of the percentage of SO_2 in the room atmosphere (1% equals 10,000 ppm; 0.5% equals 5000 ppm).

New fumigation facilities may be restricted from releasing any SO_2 into the outside atmosphere. Water scrubbing can remove SO_2 from the room atmosphere without venting. The most effective systems route refrigeration return air through a water spray or pad assembly. Water can absorb SO_2 at a rate of 10 lb SO_2 per 1400 gal H_2O, if the water is at 32°F and becomes completely saturated with the fumigant. At 70°F, water will absorb only half as much SO_2.

Because absorption efficiency drops as water nears saturation with SO_2, the actual amount of water used in practice will be several times the theoretical amount. This water cannot be reused, and must be disposed of. Sodium hydroxide or potassium hydroxide can be added to the scrubber water to increase the amount of SO_2 it can absorb. For large storage rooms, portable scrubbers used by some operators often require long periods of operation to yield adequate fumigant level reductions. If not maintained, their efficiency can be reduced dramatically by sulfite salts (plugged nozzles are a common problem).

Traditional storage room fumigation is done weekly to prevent the spread of decay from *Botrytis*-infected fruit. In cold-storage rooms, it is similar to traditional circulating-air initial fumigation. The maximum permitted SO_2 concentration for storage fumigation is 5000 ppm (CFR 2005), although many operators use 2500 ppm to fumigate filled storage rooms, and lower levels for partially filled rooms.

Traditional storage fumigation has a number of disadvantages. Large rooms are required, and poorly designed airflow systems and/or nonuniform placement of grape pallets can cause even lower levels of SO_2 penetration and greater variations than in initial fumigation. The short fumigation time results in high SO_2 levels in room air at the end of fumigation; excess fumigant must be vented or scrubbed before the room is safe for reentry.

In **total utilization fumigation**, the amount of SO_2 applied is balanced with the amount absorbed by fruit, boxes, and the room itself. Because fumigation is prolonged and the quantity of fumigant is calculated so closely, nearly all of the SO_2 is absorbed by fruit, packaging materials, and room surfaces. At the end of fumigation, the SO_2 concentration in the room air is usually less than 2 ppm.

Total utilization initial fumigation can *only* be used with precooling. For complete gas absorption, the gas must be kept in contact with the product and room surfaces for at least a few hours. Without precooling, the grapes would be exposed to warm air, which would desiccate the stems. Waiting several hours for a drop in SO_2 concentration would also unnecessarily delay cooling. With simultaneous fumigation and precooling, the fruit is quickly cooled and effectively fumigated. Forced-air total utilization initial fumigation may use 75% less SO_2 than traditional fumigation, and consistently provides an in-box **CT** (the SO_2 concentration in ppm times exposure time in hours) in excess of 100 ppm·h in all package types. Airflow considerations for traditional initial fumigation also apply to total utilization initial fumigation with circulating air.

The quantity of SO_2 required for effective decay control can be calculated using the following equation and the factors listed in Table 4.

$$W_s = \frac{BQR}{10{,}000}$$

where

W_s = quantity of SO_2 required, lb
B = SO_2 factor from Table 4 (lb of SO_2 per 10,000 boxes)
Q = quantity of boxes in room
R = ratio of actual box count to maximum box count in room (R should equal 0.5 if box count drops below 50% of room capacity)

Each box stored occupies 2.5 to 3.5 ft^3 of room volume. Calculations are based on a box occupying 3 ft^3 of room volume (10,000 boxes in a 30,000 ft^3 room).

Because polystyrene does not absorb sulfur dioxide as readily as wood and fiberboard materials, expanded polystyrene (EPS) boxes have lower SO_2 factors than wood-paper laminated (TKV) boxes. Although there are no reliable industry data on which to base factors for fiberboard boxes, laboratory studies show that they absorb more

Table 4 Factors for Determining Amount of SO_2 Needed for Forced-Air Fumigation Using Total Utilization System

Box Type[b]	SO_2 Factor, lb/10,000 boxes[a]	
	Good SO_2 Penetration	Poor SO_2 Penetration
EPS	1.5	3.0
TKV	3.7	6.3

Source: University of California (1992)
[a]Factor is based on boxes that weigh 20 to 25 lb gross.
[b]Fiberboard boxes should probably be fumigated using TKV factors, although there are no industry data available to make a reliable recommendation.

Table 5 Factors for Determining Amount of SO_2 Needed for Storage Room Fumigation

Box Type[b]	SO_2 Factor, lb/10,000 boxes[a]	
	Good SO_2 Penetration	Poor SO_2 Penetration
EPS	3.0	7.5
TKV	6.3	14.0

Source: University of California (1992)
[a]Factor is based on boxes that weigh 20 to 25 lb gross.
[b]Fiberboard boxes should probably be fumigated using TKV factors, although there are no industry data available to make a reliable recommendation.

SO_2 and would have higher factors than TKV boxes (Harvey et al. 1988). The higher factors ("poor SO_2 penetration" in Table 4) should be used for boxes that have low SO_2 penetration rates because of poor venting or packing materials that reduce gas movement into the box. Grapes with a high potential for decay may also require high SO_2 levels.

As rooms are emptied of grapes, the minimum amount of SO_2 needed is influenced by the absorption by room surfaces and coils and the amount of air leakage. The amount of SO_2 used should not be less than that required for a half-full room. Because initial fumigation rooms may differ in construction and operation, each room must be calibrated to determine the SO_2 quantity needed for a CT of 100 ppm·h in packed boxes.

The procedure for total utilization storage-room fumigation is similar to that for total utilization initial forced-air fumigation: SO_2 flows past the outside surfaces of palletized boxes rather than being forced through the boxes. Lanes are usually stacked two to three pallets high with 4 to 6 in. between lanes.

The amount of SO_2 required for effective decay control is calculated by the same methods as for total utilization forced-air fumigation. However, SO_2 penetration into boxes can be much poorer in a storage room than in an initial fumigation room, so the SO_2 factors are greater (Table 5). The amount of SO_2 should never be below that required for a half-full room. Higher factors should be used for boxes with poor SO_2 penetration characteristics because of poor venting, overpacking with fruit, or packing with materials that reduce air movement through the box.

EPS and TKV boxes should be stored in separate rooms, as should wrapped, plain-pack, and bagged grapes. Even so, because of the great variability in SO_2 penetration among boxes of the same type, there is still potential for differences in decay control and residue levels in a room filled with similar boxes.

Fans should always be on high speed during the first 1 to 2 h of fumigation, and airspeeds past all pallets should be greater than 140 fpm. Pallets should be stacked neatly with a 4 to 6 in. gap between lanes so that airflow is not blocked.

Calibration. All storage rooms should be monitored to determine the effectiveness of the fumigation program by conducting CT, residue, and *Botrytis*-control tests as follows:

- **CT.** A CT of at least 100 ppm·h should be indicated by dosimeter tubes placed among the grapes in boxes located in the hardest-to-fumigate positions in the room (typically, center boxes in pallets in areas with the least airflow). Dosimeter tubes should be placed in the boxes immediately before fumigating and removed and read at completion. Excessive SO_2 use is indicated if all dosimeter tubes have a color change along their entire length.
- **Residues.** Residue analyses should be conducted on grapes removed from the easiest-to-penetrate positions. High-residue areas (typically top corner boxes in the highest-airflow areas) can be located by using dosimeter tubes. Because of the variability of residue analysis, residues over 3 ppm SO_2 should be viewed with concern.
- ***Botrytis* control.** Grapes stored for a prolonged period should be inspected for the number of *Botrytis*-infected berries. Initial fumigation kills spores of this fungus. Fumigated rooms are nearly sterile; living *Botrytis* spores are not present in the atmosphere in these rooms. Infected berries found later during storage result from field infections. Inadequate fumigation and temperatures above 35°F during cold storage after initial fumigation allow mycelial development, and berry-to-berry spread (nesting).

Frequency of Storage-Room Fumigation. Storage rooms should be fumigated regularly and frequently enough to control mycelia from infected berries before they can spread to adjacent berries. The speed of mycelium growth varies with temperature. Berry temperature during storage should be as close as possible to 31°F. Industry experience and test results indicate that a maximum fumigation interval of 7 days is adequate to prevent *Botrytis* spread. Lengthening the fumigation interval may lead to a greater amount or spread of decay.

Precautions. Sulfur dioxide has certain properties that demand care in its use as a fumigant in cold-storage plants. Concentrations recommended for grape fumigation in storage can cause respiratory spasms and death if the victim cannot escape the fumes. When working in even weak concentrations of SO_2, goggles to protect against eye injury and a gas mask fitted with canister for acid gases (not the usual canister for ammonia gas) should be worn. Concentrations as low as 30 to 40 ppm can be detected by smell. It requires several times these concentrations to cause discomfort.

Because a small segment of the population may experience severe allergic reactions to sulfites, the U.S. Environmental Protection Agency allows a 10 ppm tolerance for sulfite residues in table grapes (CFR 2005; Harvey et al. 1988). Fruit with residues exceeding the tolerance cannot be marketed.

Another precaution about SO_2 that cannot be overemphasized is its injurious effects on other produce. For this reason, care must be taken that only grapes are stored in the room that is to be fumigated and that there are no leaks through walls or halls to adjacent rooms storing other produce.

Periodic fruit inspection is recommended to check whether the SO_2 gas is reaching the center of the stacks or whether some grapes are being overtreated. If the pedicels and stems retain a yellow or green color and broken berries show no mold and appear to be dried or seared, the gas has reached the fruit in question and is having the desired effect. Serious bleaching on unbroken grapes means too high a concentration or too long an exposure; there should be better distribution of the gas, lower concentration, or shorter fumigation periods.

Diseases

Blue Mold Rot (*Penicillium*). Watery, mushy condition, with early production of typical bluish green spores on berries and stems. There is moldy odor and flavor. *Control*: Prevent deterioration by careful handling and prompt refrigeration, preferably to 32°F. Fumigate with SO_2 in storage.

Cladosporium Rot. Black, firm, shallow decay that produces an olive green surface mold. It is common on stored grapes harvested early in the season. Infections occur on small growth cracks at the blossom end and sides of the grape. *Control*: Precool and store grapes promptly at 32°F. After harvest, fumigate with SO_2 to reduce spread.

Gray Mold Rot (*Botrytis*). Early stage: slip skin with no mold growth. Later: nest of fairly firm decay covered with abundant fine gray mold and grayish brown, velvety spore masses. *Control*: Cull decay when packing. Fumigate grapes with SO_2. For storage, cool rapidly to 30°F. Use forecasting to determine safe storage periods. Use short storage for grapes harvested in rainy periods or after slight freezes.

Rhizopus Rot. Soft, mushy, leaky decay causing staining of lugs. Coarse, extensive mycelia and black sporangia develop under moist conditions. *Control*: Prevent skin breaks. Cool promptly to below 50°F.

Sulfur Dioxide Injury. Bleached sunken areas on berry at skin breaks or cap-stem attachment. Decolorized portions have disagreeable astringent flavor. Full severity does not appear until cool grapes are warmed. *Control*: Apply proper concentration and distribution of gas for recommended period.

Storage Life

The normal storage life of major varieties of California table grapes at 30°F is shown in Table 6. Under exceptional conditions, sound fruit will keep longer than indicated; for example, Emperor grapes have been held in good condition for 7 months, and Thompson seedless for 4 months. The EPA registration label limits the number of legal fumigations permitted for various cultivars. This limit may restrict permissible storage time.

The storage life of grapes is affected most by the attention given to selecting and preparing the fruit. Grapes should be picked at the best maturity for storage, especially Thompson seedless and Ohanez. Stems and pedicels should be well developed, and the fruit should be firm and mature. Soft and weak fruit should not be stored. The display lug is a satisfactory package for storage because it can be cooled and fumigated easily.

Cooling to 40 to 45°F is advised for grapes that are to be in transit a day or two before reaching storage. Take special care during transit so that decay does not start. It is not good practice to delay fumigation until the grapes reach a distant storage plant, because during picking and packing, many berries are injured enough for mold to begin unless the fruit is fumigated promptly.

For *labrusca* (Eastern type) grapes, a storage temperature of 32°F and humidity of 85% are recommended. Care in packing and handling the fruit, minimum delay before storage, and prompt cooling are important for best results with these varieties, as with *vinifera* grapes. Eastern varieties are not fumigated with SO_2 because of their susceptibility to injury from it. Storage life of important commercial varieties at 32°F is shown in Table 7.

Refrigeration System Materials and Practices

SO_2 is corrosive to many materials; the following materials and practices are recommended to provide satisfactory equipment durability and economical operation:

Table 6 Storage Life of California Table Grapes at 32°F

Variety	Storage Life, Months
Emperor, Ohanez, Ribier	3 to 5
Malaga, Red Malaga, Cornichon	2 to 3
Thompson seedless, Tokay	1 to 2.5
Muscat, Cardinal	1 to 1.5

Table 7 Storage Life of *Labrusca* Grapes at 32°F

Variety	Storage Life, Weeks
Catawba	5 to 8
Concord, Delaware	4 to 7
Niagara, Moore	3 to 6
Worden	3 to 5

Evaporators

- Coils: an aluminum alloy resistant to SO_2; also can be coated with a material to prevent chemical attack
- Housing: aluminum sheet metal with 304 stainless steel fasteners
- Fan blades: cast aluminum or mild steel coated with a baked-on food-grade powder coating
- Fan guards: mild steel with baked-on food-grade powder coating
- Fan motors: totally enclosed fan-cooled (TEFC) or totally enclosed air-over (TEAO) with cast-iron or cast-steel housing; specify a high-quality paint job
- Fan motor shafts: coat with nickel-based anti-seize material
- Coil connections: use dielectric connectors between coil connections and external piping

Piping to Evaporators

- Use aluminum or stainless steel pipe of correct weight or schedule for the refrigerant. Use stainless steel pipe hangers and hanger rods.
- Locate hand valves and control valves outside the cold-storage space to minimize corrosion and for easy maintenance.
- Locate evaporators as close to outside walls as practical to minimize length of piping between control valves and evaporators.
- Revert to piping materials normally used when the piping is outside the cold storage environment. Use dielectric connections between dissimilar materials.

Maintenance and Operation

A successful SO_2 fumigation program requires that cold-storage facilities be well operated and maintained. Leaky door seals, inoperative SO_2 input lines or fans, improperly functioning or unsafe refrigeration equipment, or poorly operating and maintained ductwork reduce system and program effectiveness. Safe, efficient operations require that equipment and buildings be kept in good working order.

PLUMS

Plums are not suitable for long storage. Commercial storage life varies from 1 to 8 weeks, depending on variety. The Italian prune can be held for no more than 2 weeks before marketing begins.

Plums intended for storage should be harvested at a high soluble solids level for the variety, although doing so may delay harvest beyond the normal picking date. Harvested fruit should be carefully graded to remove disease, defects, and injuries before packing in the shipping container.

Fruit should be thoroughly cooled before storage. Forced-air or hydrocooling may be done in the 900 to 1000 lb bulk bins that are used for harvest. Shipping containers are normally vented to aid cooling after packing. Although most fruit is air-cooled in conventional room coolers, some shippers use forced air to cool fruit quickly in bulk bins or shipping containers.

Many varieties can usually be stored for 1 month at 30 to 32°F with 90 to 95% rh. Results of storage-life tests have been varied, with some lots in certain seasons remaining in good condition even after 4 to 5 months in storage. Other lots in some seasons have not been held satisfactorily beyond 2 months. Fruit with the highest soluble solids has consistently shown the longest storage life, even when harvested several weeks after the completion of commercial harvest. Some plum varieties benefit from CA storage.

Storage Diseases and Deterioration

Plum deterioration appears as changes in appearance and flavor. A poststorage holding period should be used in judging the condition of stored fruit. Fruit that appears bright and flavorful in storage can show severe deterioration when removed to room temperature for 2 to 3 days.

Some flesh softening and a gradual loss of varietal flavor and tartness occur even at low storage temperatures. The first visual sign of deterioration is the development of translucence, first around the pit, then extending outward through the fruit, followed by progressively more severe flesh browning following the same pattern. The first noticeable loss in flavor is generally associated with the first symptoms of translucence. It is necessary to cut through the fruit to judge condition, because fruit held under good storage conditions may appear sound from the outside while being seriously deteriorated internally. See the section on Sweet Cherries for diseases. See the section on Diseases under Peaches and Nectarines for information on cold-storage and sulfur dioxide injuries.

SWEET CHERRIES

Harvesting Techniques

Sweet cherries for storage must be harvested with stems attached.

Cooling

Rapid cooling to 30°F is essential if the fruit is to be stored. Hydrocooling has been used successfully; wetting is tolerable as long as the fruit remains cold. Fungicidal postharvest sprays or dips are helpful in reducing decay during storage.

Forced-air or pressure cooling can be used to quickly cool the fruit without the problem of wetting. Moisture loss and stem drying can be minimized by rapid movement from the field to the cooler, rapid cooling, and maintaining low temperatures and high humidity during cooling and storage.

Storage

When sweet cherries are stored, they are normally held in shipping containers, often with polyethylene liners. These liners allow increased CO_2 gas around the fruit, which tends to reduce decay rates and increase storage time. Cherries should be stored at 30°F and may be held 2 weeks after harvest and still retain enough quality for shipment to market.

Controlled atmospheres with 10 to 15% CO_2 and 3 to 10% O_2 help maintain firmness and bright, full color during storage. Polyethylene liners can extend market life. The liner must be perforated when removed from storage. Modified-atmosphere bags are now used commercially.

Diseases

Alternaria and Cladosporium Rot. Light brown, dry, firm decay lining skin breaks that can be removed easily from surrounding healthy tissue. Mycelia on the area are fine and white above and dark green below. *Control*: Remove cherries with cracks and other skin breaks at packing. Use fungicide in spray or sizer on packing line.

Blue Mold Rot (*Penicillium*). Circular, flat spots covering conical, soft, mushy decay that can be scooped out cleanly from surrounding healthy flesh. White fungus tufts turning to bluish green develop on surface. Odor and flavor are musty. *Control*: Prevent skin breaks. Use fungicide in spray or sizer on the packing line. Market promptly. Refrigerate promptly to 32°F.

Brown Rot (*Monilinia*). See the section on Diseases under Peaches and Nectarines. *Control*: Follow recommended orchard spray practices. Use fungicide in spray or sizer on packing line. Refrigerate promptly to 32°F. Package cherries in polyethylene bags to reduce desiccation of stem and fruit, preserve color, and reduce decay development.

Gray Mold Rot (*Botrytis*). Light brown, fairly firm, watery decay covered with extensive delicate, dirty-white mycelia. On completely decayed cherries, grayish-brown velvety spores may be found. *Control*: Handle carefully. Use fungicide in spray or sizer on packing line. Refrigerate promptly to 32°F.

Rhizopus Rot. Extensive soft, leaking decay with little change from normal color. Coarse mycelia and black spore heads are prominent under moist conditions. More prevalent in upper-layer packages in refrigerator car. *Control*: Rhizopus develops very slowly at temperatures below 50°F, so storage at recommended temperature keeps decay in check.

PEACHES AND NECTARINES

This discussion relates primarily to peaches but also applies to nectarines in many respects.

Storage Varieties

Peaches do not adapt well to prolonged storage. However, if they are sound and well matured, most freestone varieties can be stored for up to 2 weeks (some freestone and most clingstone for up to 4 weeks) without noticeable deterioration in flavor, texture, or appearance. Storage life appears to depend on harvest season. Early varieties, particularly freestone peaches grown in Florida and early clingstones grown in the Southeast, have an extremely short storage life and should be used as soon as possible after harvest. However, some late-season varieties can be safely stored for up to 6 weeks. In the West, the Rio Oso Gem and several other varieties can be stored for 4 to 6 weeks before being marketed.

Harvest Techniques

Peaches for fresh consumption must be in a condition to survive a postharvest holding period of several days to several weeks. The fruit must be sound and bruise-free and must be handled delicately during harvesting and packing. With widespread use of bulk bins or pallet boxes, hand-picked fruit requires extra-careful handling. Hydrodumpers are generally used for dumping pallet bins. With proper care, pallet boxes cause less bruising than small field boxes.

Cooling

Cooling peaches to 40°F soon after harvesting is essential to retention of quality and control of decay. Peaches begin to soften and decay in a few hours without proper temperature management. All peaches shipped out of the Southeast are hydrocooled, originally in flood-type hydrocoolers as a final operation after being packed in containers. In the West, most fresh peaches are hydrocooled or forced-air cooled in pressure coolers before packing to remove field heat rapidly for postharvest holding. Forced-air cooling is used after packing to complete the cooling. With forced-air or pressure cooling, peaches or nectarines in two-layer plastic tray packs with 6% side-vented corrugated containers cool 80% in about 6 h with an airflow of 0.2 cfm per pound of fruit.

Storage

Peaches are normally stored in corrugated or tray pack shipping containers.

An environment of 31°F and 90 to 95% rh with very low air movement is best for peaches. Under these conditions, peaches can be held for 2 to 6 weeks, depending on variety.

The same storage conditions may be used for nectarines; however, they are somewhat more susceptible to shrivel than peaches are. Air velocity in the storage room should be as low as possible but still maintain proper storage temperatures. Frequently check fruit at the edge of alleyways, for example, to detect the first signs of shrivel.

Good experimental CA results have been obtained with peaches and nectarines held in 1 to 2% O_2 with 3 to 5% CO_2 at 32°F. Extended storage of 6 to 9 weeks is possible. Fruit ripens or softens with good flavor and is juicy on removal. Low-temperature breakdown, which is usually encountered with lengthy storage, is controlled by CA. Although CA reduces decay, it does not completely control it; thus, a fungicide is needed for extended storage.

Diseases

Brown Rot (*Monilinia*). Extensive firm, brown, unsunken areas turning dark brown to black in the center and generally covered with yellowish gray spore masses. Skin clings tightly to center of old lesions. *Control*: Follow recommended field and postharvest control measures involving use of heat treatments and fungicides. Refrigerate promptly to as near 32°F as feasible.

Cold Storage Injury. Flavor loss, dry and mealy texture. Breakdown starting around pit is grayish brown, water-soaked, or mealy. *Control*: Refrigerate promptly to 32°F. Breakdown appears earlier at 38°F. Store for only 2 to 4 weeks, depending on variety.

Pustular Spot (*Coryneum*). Common on peaches from the West, occasionally on Eastern fruit. Small purplish-red spots grow up to 0.5 in. in diameter, becoming brown, sunken, with white center. *Control*: Treat with orchard sprays. Cool harvested fruit to below 45°F.

Rhizopus Rot. Extensive, fairly firm, watery decay with uniformly brown surface color. Skin slips readily from center of lesions. Coarse mycelia and black spherical sporangia develop. *Control*: Store cannery peaches at 32°F before ripening. Prevent skin breaks. Follow recommended field and postharvest control measures. Refrigerate promptly to as near 32°F as feasible.

Sulfur Dioxide Injury. Bleached and pitted areas on fruit surface. After removal from refrigeration, injured areas of peaches are brown, dry, and collapsed. Skin may slough off. *Control*: Avoid SO_2 contact of peaches (and other stone fruit) in storage or in transit with grapes.

Sour Rot. An unfamiliar postharvest disease in peaches noticed in some packing sheds in southeastern states. First signs of the infection may be peaches that are easily skinned by brushes and belts on the packing line. Affected peaches then develop softened and sunken brown lesions that eventually become covered with a white or creamy exudation. Infected areas generally emit a vinegar-like, sour odor. *Control*: Chlorination of dump tank water, chlorination of hydrocooling water, and careful culling of all overripe, bruised, and damaged fruit. In short, good shed sanitation and quality control are the keys to eliminating sour rot.

APRICOTS

Apricots are not stored for a prolonged time but may be held for 2 or 3 weeks if they are picked while firm enough not to bruise. Unfortunately, this maturity does not yield good dessert-quality fruit. Care must be used in sizing and packing fruit going into storage, because small surface bruises can become infected with disease-producing organisms. Chapter 11 has further details.

Apricots for short-term storage are harvested in much the same way as freestone peaches, precooled, and placed in storage promptly. Storage temperature should be 32°F with 90 to 95% rh.

Diseases and Deterioration

See the section on Peaches and Nectarines.

BERRIES

Blackberries, raspberries, and related berries cannot be stored for more than 2 or 3 days even at 31°F with a relative humidity of 90 to 95%. An atmosphere with 20 to 40% CO_2 increases storage life by 3 or 4 days by inhibiting fungal rots.

As they come from the field, cranberries are stored in field boxes at 36 to 40°F and 90 to 95% rh. They are usually not stored longer than 2 months. Storage at 30 to 32°F causes chilling and physiological breakdown. Modified atmospheres have not extended the storage life of fresh cranberries beyond that attainable in conventional storage.

Diseases

Cladosporium Rot. Surfaces of berries covered with olive to olive green mold. In raspberries, the mold is most abundant on inside or cup of berry. *Control*: Avoid bruising; pack and ship promptly. Refrigerate to 32°F.

Gray Mold Rot. Causes soft, watery rot. Fruit may be covered with dense, dusty gray fungus, which spreads rapidly in package, forming nests. *Control*: Avoid bruising. Refrigerate to 32°F in transit and use modified-atmosphere packaging (MAP).

Anthracnose (*Gloeosporium* sp.). Berries may be completely rotted and show masses of spores glistening in salmon-colored droplets on fruit. *Control*: Refrigerate to 32°F.

Alternaria Rot. Affected berries remain firm and show gray-white woolly fungal growth from injured cap-stem areas. Nesting occurs in tight clusters scattered throughout containers. *Control*: Refrigerate to 32°F.

Chilling Injury. Berries held for 4 or more weeks become tough and rubbery; surfaces are dull in appearance, red throughout. *Control*: Hold fruit at 37°F.

Fungus Rots (Several Fungi). Limited portions or entire berries are brown, soft, or collapsed. Some berries turn into water bags. *Control*: Spray in field. Handle carefully. Reduce temperature to 38°F after harvest.

STRAWBERRIES

Diseases

Gray Mold Rot (*Botrytis*). Brown, fairly firm, fairly dry decay. Dirty-gray mold and grayish-brown velvety spore masses are present. Nesting is common. *Control*: Apply recommended fungicides in field. Handle carefully to prevent skin breaks. Cull out all diseased berries. Cool promptly to 40°F or below.

Leather Rot (*Phytophthora*). Large, slightly discolored tough areas with indefinite purplish margins. Vascular system is browned; flavor is bitter. *Control*: Mulch plants to keep berries from contact with infested soil. Cool promptly to 40°F or below.

Rhizoctonia Rot. Hard dark brown decay on one side of berry, usually small quantities of soil adhering. This develops shortly after harvest. *Control*: Mulch plants to keep berries from contact with infested soil. Cull thoroughly.

Rhizopus Rot. Mushy, leaky collapse of berries associated with coarse black mycelia and sporangia. Extensive red staining of containers from leaking juice. *Control*: Reduce temperature promptly to 40°F or below. Handle carefully to prevent skin breaks.

FIGS

Diseases

Alternaria Spot. White fungal growth on surfaces that soon darkens. As fungus spots enlarge, tissue beneath becomes slightly sunken. *Control*: Cool promptly after harvest, hold at 45°F in transit.

Black Mold Rot (*Aspergillus*). First appears as a dirty white to pink color of skin and pulp. White mold growth develops within fig. Cavities formed in fruit become lined with black spore masses. *Control*: Store fresh figs at 32°F and 85 to 90% rh.

SUPPLEMENTS TO REFRIGERATION

Antiseptic Washes

Many types of fruit are washed before packing to remove dirt and improve appearance. In some cases, hydrocoolers are used to remove field heat. If water is recirculated, it may become heavily contaminated with decay-producing bacteria and fungi. Chlorine can be added at 50 to 100 ppm to control build-up of these organisms. Other fungicides may also be used, but they must be legally registered for the specific application.

Protective Packaging

Proper packaging protects against bruising, moisture loss, and spread of disease. Packaging materials may also contain chemicals to control spoilage. Packages must have good stacking strength for palletizing and perform well under high humidity.

Selective Marketing

The potential storage life of grapes and apples can be predicted within a few weeks after they are stored. Thus, those with a short storage life may be marketed while still in good condition, and longer-lived products can be stored for late-season marketing. Samples taken from each lot placed in storage are kept for a few weeks at temperatures and relative humidities that favor rapid development of decay. Grapes that will not keep long can be detected in about 2 weeks, and apples in about 60 days. Because both kinds of fruit may be stored for several months, knowing their potential storage life can significantly reduce spoilage losses.

Heat Treatment

Heat treatments to reduce decay also kill insects on and microorganisms near the surface of the fruit without leaving a residue. For example, brown rot and rhizopus rot of peaches are reduced by exposing the fruit for 1.5 min in 130°F water or for 3 min in 120°F water.

Fungicides

Fungicides may be applied during cleaning, brushing, or waxing of some fruit. Only fungicides registered for the particular fruit and application may be used.

Irradiation

Gamma radiation effectively controls decay in some products. High dosages can cause discoloration, softening, or flavor loss. Commercial application of gamma radiation is limited because of the cost and size of equipment needed for the treatment and uncertainty about the acceptability of irradiated foods to the consumer (Hardenburg et al. 1986).

Ultraviolet lamps are sometimes used to control bacteria and mold in refrigerated storage. Although ultraviolet light kills bacteria and fungi that are sufficiently exposed to the direct rays, it does not reduce decay of packaged fruit in storage. Even ultraviolet light directed on fruit as it passed over a grader did not control decay.

REFERENCES

CFR. 2005. Tolerances and exemptions from tolerances for pesticide chemicals in food; Specific tolerances—Sulfur dioxide. 40CFR180.144. *Code of Federal Regulations,* U.S. Government Printing Office, Washington, D.C.

EPA. 1999. *EPA compendium of registered pesticides*, vols. I and II. U.S. Environmental Protection Agency. U.S. Government Printing Office, Washington, D.C.

EPA. 2000. *Pesticide tolerance index system (TIS) information retrieval*, v. 1.0. U.S. Environmental Protection Agency. Database available at http://www.epa.gov/pesticides/food/viewtols.htm.

Hardenburg, R.E., A.E. Watada, and C.Y. Wang. 1986. The commercial storage of fruit, vegetables, and florist and nursery stocks. *USDA Agriculture Handbook* 66. U.S. Department of Agriculture, Washington, D.C.

Harvey, J.M. 1955. A method of forecasting decay in California storage grapes. *Phytopathology* 45:229-232.

Harvey, J.M. 1984. Instructions for forecasting decay in table grapes for storage. *ARS-7*. U.S. Department of Agriculture, Washington, D.C.

Harvey, J.M., C.M. Harris, T.A. Hanke, and P.L. Hartsell. 1988. Sulfur dioxide fumigation of table grapes: Relative sorption of SO_2 by fruit and packages, SO_2 residues, decay, and bleaching. *American Journal of Enology and Viticulture* 39:132-136.

Kader, A.A. 2001. Optimal controlled atmospheres for horticultural perishables. *Postharvest Horticultural Series* 22A. Postharvest Technology Research and Information Center, University of California, Davis.

Kader, A.A., R.F. Kasmire, and J.F. Thompson. 2002. Cooling horticultural commodities: Selecting a cooling method. *Publication* 3311. University of California Division of Agriculture and Natural Resources.

Mitcham, E.J. 1997. Apples and pears. *Proceedings of the International Controlled Atmosphere Research Conference*, University of California, Davis, vol. 2.

Richardson, D.G. and E. Kupferman. 1997. Controlled atmosphere of pears. *Proceedings of the International Controlled Atmosphere Research Conference*, University of California, Davis, vol. 2.

University of California. 1992. Sulfur dioxide fumigation of table grapes. *Bulletin* 1932. University of California Division of Agriculture and Natural Resources.

BIBLIOGRAPHY

Chau, K.V., C.D. Baird, P.C. Talasila, and S.A. Sargent. 1992. Development of time-temperature-humidity relations for fresh fruits and vegetables. ASHRAE Research Project RP-678, *Final Report*.

Nelson, K.E. 1979. Harvesting and handling California table grapes for market. *Publication* 4095. University of California.

Ryall, A.L. and W.T. Pentzer. 1982. *Handling, transportation, and storage of fruits and vegetables*, 2nd ed., vol. 2, *Fruits and tree nuts*. AVI Publishing, Westport, CT.

Thompson, J.F., P. Brecht, R.T. Hinsch, and A.A. Kader. 2000. Marine container transport of chilled perishable produce. *Publication* 21595. University of California Division of Agriculture and Natural Resources.

CHAPTER 23

CITRUS FRUIT, BANANAS, AND SUBTROPICAL FRUIT

THIS chapter covers the harvesting, handling, processing, storage requirements, and possible disorders of fresh market citrus fruit grown in Florida, California, Texas, and Arizona; of bananas; and of subtropical fruit grown in California, Florida, Hawaii, and Puerto Rico.

CITRUS FRUIT

MATURITY AND QUALITY

The degree of citrus fruit ripeness at the time of harvest is the most important factor determining eating quality. Oranges and grapefruit do not improve in palatability after harvest. They contain practically no starch and do not undergo marked composition changes after they are picked (as do apples, pears, and bananas), and their sweetness comes from the natural sugars they contain when picked.

Citrus fruit ripeness increases slowly and is closely correlated with increases in diameter and weight. Citrus fruit must be of high quality when harvested to ensure quality during storage and shelf life.

Quality is often associated with the fruit rind's appearance, firmness, thickness, texture, freedom from blemishes, and color. However, quality determination should be based on flesh texture, juiciness, soluble solids (principally sugars), total acid, aromatic constituents, and vitamin and mineral content. Age is also important. Immature fruit is usually coarse and very acid or tart and has an internal texture that is ricey or coarse. Overripe fruit held on the tree too long may become insipid, develop off-flavors, and possess short transit, storage, and shelf life. The importance of having good-quality fruit at harvest cannot be overemphasized. The main objective thereafter is to maintain quality and freshness.

HARVESTING AND PACKING

Picking

Citrus fruit is harvested in the United States throughout the year, depending on the growing area and kind of fruit. Figure 1 shows the approximate commercial shipping seasons for Florida, California/Arizona, and Texas citrus. Trained crews from independent packing houses or large associations do the picking, which is scheduled to meet market demands. Fruit that is not handled through cooperatives is normally sold on the tree to shippers or processors and is picked at the latter's discretion.

Pickers carefully remove fruit from the trees, either by hand or with special clippers, and then place the fruit in picking bags that are emptied into field boxes. An increasing amount of fruit is handled in bulk, so pickers put fruit into pallet boxes or wheeled carts. In some

The preparation of this chapter is assigned to TC 10.9, Refrigeration Application for Foods and Beverages.

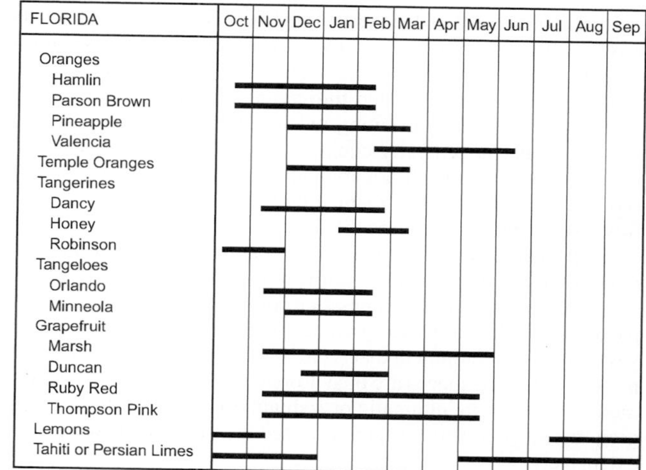

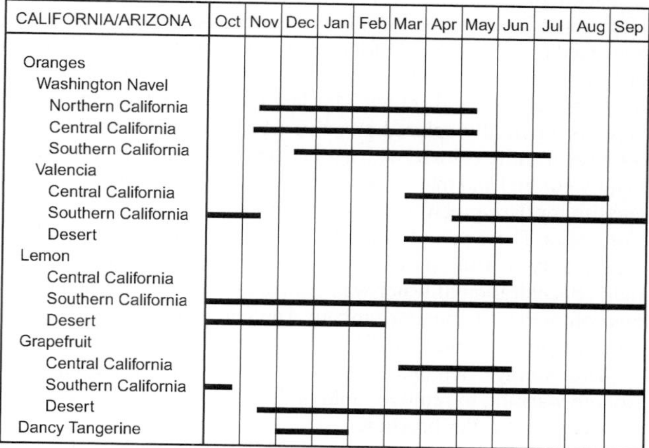

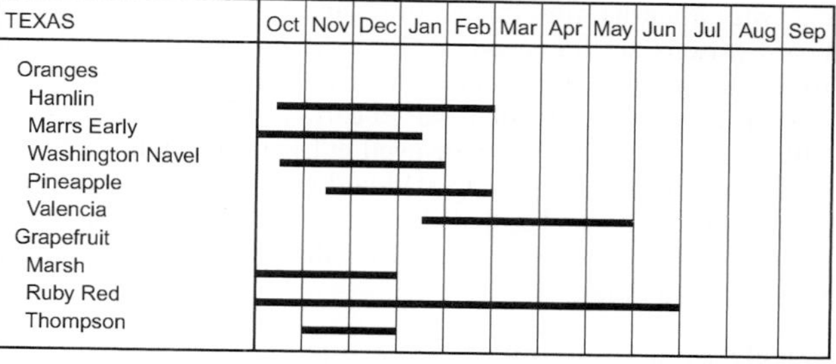

Fig. 1 Approximate Commercial Shipping Season for U.S. Citrus

cases, especially when fruit is picked for processing, it is loaded loose into open truck trailers. In Florida, over 90% of the oranges and slightly more than 50% of the grapefruit and specialty fruit are processed, while in California and Arizona less than 35% of all citrus fruit is processed.

At the beginning of the season, fruit is often spot-picked; only the riper, larger, or outside fruit is harvested. Later, the trees are picked clean. In California, lemons usually are picked for size with the aid of sizing rings.

Various labor-saving devices have been tested, including mechanical platforms and positioners, tree shakers with catch frames, and air blasts for fruit removal. Mechanical harvesting, however, is limited to a very small percentage of the total crop. Because of damage incurred, fruit intended only for processing is mechanically harvested. Preharvest sprays have also been developed to improve color and to loosen fruit to facilitate harvest.

Handling

After fruit is received at the packinghouse, it is removed from the boxes or bulk containers carefully to prevent damage to the fruit. It is then presized to remove oversized or undersized fruit. Before washing, fruit may be floated through a soak tank, which usually contains a detergent for cleaning and an antiseptic for decay control.

The washer is generally equipped with transverse brushes that revolve up to 120 rpm. If not applied at the soak tank, soap or antiseptic may be dribbled or foamed on the first series of brushes. A fresh water spray then rinses the fruit.

Fruit then passes under fans that circulate warm air through the moving pieces. When dried, fruit is polished and waxed (typically with a water wax), and then passes through a second drying. It then passes over roller conveyor grading tables. After grading, it is conveyed to sizing equipment that separates the pieces into the standard packing sizes; the pieces are then dropped at stations for hand packing or conveyed to automatic or semiautomatic box-filling or bagging machines. Electronic sizing, based on machine vision, is used extensively.

Packinghouse handling of California lemons for fresh market is interrupted by an extended storage period. After washing, fruit is conveyed to a sorting table for color separation by electronic means or by human eye. Usually, separation is into four colors: dark green, light green, silver, and yellow. The dark green is a full green; the light green is a partially colored green (a green with color well broken); the silver is fully colored with a green tip (stylar end); and the yellow is fully colored and mature with no green showing. The normal storage life for dark green fruit is 4 to 6 months; for yellow fruit, it is 3 to 4 weeks. These periods are approximate, because storage (or keeping) quality of fruit varies considerably with season and grove. A light concentration of water/wax emulsion is usually applied to lemons before they are put into storage. The section on Storage has more information on storing the different varieties of citrus fruit.

After storage, lemons are waxed, then sized and packed. Poststorage washing to remove mold is desirable but requires a washer with very soft roller brushes.

Shipping and storage containers vary considerably for the different types of citrus fruit. In California and Florida, 0.8 bushel fiberboard cartons are the standard. In addition, over 15% of Florida fresh fruit is consumer-packed in mesh and polyethylene bags that are normally shipped in 40 lb master cartons. After the packages are filled and closed, they are conveyed either to precooling rooms to await shipment or directly to standard refrigerator cars or trucks. Containers are stacked so that air distribution is uniform throughout the load. Either slipsheets or wooden pallets are commonly used for palletized handling.

Accelerated Coloring or Sweating

All varieties of citrus fruit must be mature before they are picked. Maturity standards are based on internal attributes of soluble solids, acid percentage, and juice percentage. Color is not always a criterion of maturity. The natural change of color in oranges from dark green to deep orange is a gradual process while the fruit remains on the tree. The fruit remains dark green from its formation until it is nearly full size and approaching maturity; then the color may change very rapidly. Color change is influenced greatly by temperature variations. A few cold nights followed by warm days may turn very green oranges to deep orange. Color changes in lemons and grapefruit similarly turn the fruit yellow. Unfavorable weather conditions may delay coloring even after maturity.

Up to a certain point, the natural color changes in Valencia oranges follow the trend described, but complete or nearly complete orange color generally develops some time before the fruit is mature. Some regreening of Valencias may occur after the fruit has reached its prime. Navel oranges in California, as well as the Florida varieties of Hamlin, Parson Brown, and Pineapple harvested in late fall and early winter, may be mature and of good eating quality even if the rind is green. Grapefruit, lemons, tangerines, tangelos, and other specialty fruit may also be mature enough for eating before they are fully colored. Because the consumer is accustomed to fruit of characteristic color, poorly colored fruit is put through a coloring or **degreening** process in special rooms, bulk bins, or trailer degreening equipment.

These units are equipped to maintain temperature and humidity at desired levels. Approximately 5 ppm ethylene is maintained in the air. The concentration of ethylene and duration of degreening depends on the variety of fruit and the amount of chlorophyll to be removed. During degreening, fresh air is introduced into the room, and a relative humidity of 90 to 96% is maintained. In Florida, temperatures of 82 to 85°F are recommended, whereas in California temperatures of 65 to 70°F are used. Lemons are usually degreened at 60°F. In California, the process is called **sweating** instead of coloring or degreening.

Oranges, grapefruit, and specialty citrus fruit requiring ethylene treatment are frequently degreened upon delivery to the packinghouse, but they may receive a fungicide drench before degreening. Lemons are washed and graded or color-separated before being degreened.

Color-Added Treatment

A high percentage of Florida's early and midseason varieties of oranges and tangelos receive color-added treatment with a certified food dye that causes the rind of pale fruit to take on a brighter and more uniform orange color. This is usually in addition to degreening with ethylene gas. In color-added treatment, the fruit is subjected for 2 to 3 min to the dye solution, which is maintained at 120°F. Treatment can be in an immersion tank filled with vegetable dye solution, or the dye can be flooded on the fruit as it passes on a roller conveyor. The immersion tank is between the washer and the wax applicator. Oranges with the desired color at harvest, as well as tangerines and grapefruit, bypass the dye tank; alternatively, the flow of dye may be cut off as the fruit passes over the equipment. Standards for maturity are slightly higher in Florida for oranges given color-added treatment. California oranges are not artificially colored.

Cooling

After the fruit is packed, it is cooled. Cooling room efficiency depends on

- Cooling air rate per railcar load (at least 3000 cfm)
- Relative humidity of supply air (95% or above)
- Temperature of supply air entering room (no more than 2°F below the selected cooling temperature)

The fruit may also be cooled in a refrigerated truck trailer or container after it has been loaded.

In California, air is used to cool oranges but not lemons or grapefruit. In Florida, specialty fruit such as Temple oranges, tangerines, and tangelos may be cooled. Chapter 15 discusses cooling practices and equipment used for various commodities in more detail.

TRANSPORTATION

Fruit packed in piggybacks, trucks, ship vans, or rail cars should be stowed in appropriate modifications of the spaced bonded block to ensure good air circulation, uniform temperature, and stable load. No dunnage is required. Such stowing provides continuous air channels through the interior of the load and improves the likelihood of sound arrival. Trailers and containers that circulate air from the bottom provide uniform temperatures throughout the load with a regular bonded-block stow.

In Florida, the present quarantine treatment for the Caribbean fruit fly, *Anastrepha suspensa*, is to subject an export load of citrus to specified temperatures for up to 24 days (Ismail et al. 1986). This treatment may be implemented in containers or in a ship's hold.

A uniform sample of 1500 fruit is withdrawn from a shipment before ship or container loading and is held at 80°F or higher. These fruit are then examined after a 10 day incubation period. If an infestation of *A. suspensa* is found, the entire load must undergo the long treatment process. Table 1 details temperature and time schedules.

STORAGE

Performance of any citrus storage facility depends on three conditions: (1) provision of sufficient capacity for peak loads; (2) an evaporator and secondary refrigerating surface sufficient to permit operation at high back pressures, which prevents low humidity and lowers operating costs; and (3) efficient air distribution, which ensures velocities high enough to effect rapid initial cooling and volumetric flows great enough to permit operation during storage with only a small temperature rise between delivery and return air. Chapters 11, 13, and 14 have further information on storage design.

Oranges

Valencia oranges grown Florida and Texas can be stored successfully for 8 to 12 weeks at 32 to 34°F with a relative humidity of 85 to 90%. The same requirements apply to Pope's Summer orange, a late-maturing Valencia-type orange. A temperature range of 40 to 44°F for 4 to 6 weeks is suggested for California oranges. Arizona Valencias harvested in March store best at 48°F, but fruit harvested in June store best at 38°F.

Oranges lose moisture rapidly, so high humidity should be maintained in the storage rooms. For storage longer than the usual transit and distribution periods, 85 to 90% relative humidity is recommended.

Table 1 Quarantine Treatment of Citrus Fruit for Caribbean Fruit Fly

	Temperature,[a] °F	Days
Short Treatment[b]	33	10
	34	12
	35	14
	36	17
Long Treatment[b]	33	14
	33.5	16
	34	17
	34.5	19
	35	20
	35.5	22
	36	24

[a]Required center pulp temperatures.

[b]To avoid chilling injury, a conditioning period of 7 days at 59°F is recommended before cold treatment.

Florida and Texas oranges are particularly susceptible to stem end rots. Citrus fruit from all producing areas are subject to blue and green mold rot. These decays develop in the packinghouse, in transit, in storage, and in the market, but they can be greatly reduced if fruit is properly treated. Proper temperature is effective in reducing decay. However, once storage fruit is removed to room temperature, decay develops rapidly.

Prolonged holding at relatively low temperatures may cause the development of physiological rind disorders (mainly aging, pitting, and watery breakdown) not ordinarily encountered at room temperature. This possibility often complicates orange storage. California and Arizona oranges are generally more susceptible to low-temperature rind disorders than Florida oranges.

Successful long storage of oranges requires harvest at proper maturity, careful handling, good packinghouse methods, fungicidal treatments, and prompt storage after harvest.

The rate of respiration of citrus fruit is usually much lower than that of most stone fruit and green vegetables and somewhat lower than that of apples. Navel oranges have the highest respiration rate, followed by Valencia oranges, grapefruit, and lemons. Heat from respiration is a relatively small part of the heat load. Table 2 shows heat generated through respiration.

Grapefruit

Florida and Texas grapefruit is frequently placed in storage for 4 to 6 weeks without serious loss from decay and rind breakdown. The recommended temperature is 50°F. A temperature range of 58 to 60°F is recommended for storing California and Arizona grapefruit.

A relative humidity of 85 to 90% is usually recommended for storage rooms containing grapefruit. Weight and water loss occur rapidly and can be avoided by maintaining the correct humidity and taking the additional precaution of a wax coating.

Long storage of grapefruit may cause decay and rind breakdown during or after storage. Proper prestorage treatments with fungicides greatly reduce these problems. Also, stored fruit should be inspected periodically for the least symptom of rind pitting or excessive decay so that storage can be terminated if necessary.

Export may require 10 days to 4 weeks of storage in a refrigerated hold and may present problems similar to those encountered in refrigerated storage. Marsh Seedless and Ruby Red grapefruit picked before January retain appearance best when stored at 60°F. With riper fruit, 50 to 55°F is better for export shipments. Very ripe fruit harvested in April and May, however, develops excessive decay following after at 50 to 60°F.

Lemons

Most of the lemon crop is picked during the period of least consumption and stored until consumer demand justifies shipment. Lemons are generally stored near producing areas rather than consuming areas.

Table 2 Heat of Respiration of Citrus Fruit

	Heat of Respiration, Btu per Ton of Fruit per Day					
	Oranges			Grapefruit		Lemons
Temp., °F	Florida	Calif. Navels	Calif. Valencias	Florida	Calif. Marsh	Calif. Eureka
32	700	900	400	500	500	700
40	1400	1400	1000	1100	800	1100
50	2700	3000	2600	1500	2000	2500
60	4600	5000	2800	2800	2600	3500
70	6600	6000	3900	3500	3900	5000
80	7800	8000	4600	4200	4800	5700

Source: Haller et al. (1945)

All lemons, except the relatively small percentage that are ripe when harvested, must be conditioned or cured and degreened before shipping. When lemons are stored prior to shipment, the curing and degreening processes occur during storage. These lemons are usually stored at 52 to 55°F and 86 to 88% rh. Local conditions may suggest slight modifications of these values.

Lemons picked green but intended for immediate marketing, (e.g., most lemons grown in the desert portions of Arizona and California) are degreened and cured for 6 to 10 days at 72 to 78°F and 88 to 90% rh. The thin-skinned Pryor strain of Lisbon lemons degreens in about 6 days, whereas the thick-skinned old-line Lisbon requires as long as 10 days.

Lemon storage rooms must have accurately controlled temperature and relative humidity; the air should be clean and uniformly circulated to all parts of the room. Ventilation should be sufficient to remove harmful metabolic products. Air-conditioning equipment is necessary to provide satisfactory storage conditions, because natural atmospheric conditions are not suitable for the necessary length of time.

A uniform storage temperature between 50 and 55°F is important. Fluctuating or low temperatures cause lemons to develop an undesirable color or bronzing of the rind. Temperatures 50°F and lower can stain or darken the membranes dividing the pulp segments and may affect flavor. Temperatures above 55°F shorten storage life and promote the growth of decay-producing organisms.

A relative humidity of 86 to 88% is generally considered satisfactory for lemon storage, although a slightly lower humidity may be desirable in some locations. Higher humidities prevent proper curing, encourage mold growth on walls and containers, and hasten decay; much lower humidities cause excessive shrinkage.

Stacking fruit containers properly in storage rooms is important to ensure uniform air circulation and temperature control. Stacks should be at least 2 in. apart, and rows should be 4 in. apart; trucking aisles at least 12 ft wide should be provided at intervals.

Specialty Citrus Fruit

In Florida, small amounts of various specialty citrus fruit are grown commercially. These types of fruit, which are usually channeled to fresh market, include tangerines, tangerine hybrids (Murcott Honey oranges, Temple oranges, tangelos), King oranges, and other mandarin-type fruit.

Careful handling during picking and packing is especially necessary for these types of fruit. Because of their perishable nature and limited shelf life, specialty citrus fruit should not be stored longer than required for orderly marketing (2 to 4 weeks). A temperature of 38 to 40°F at 90 to 95% rh is recommended. Adequate precooling and continuous refrigeration during transit are required.

Tahiti or Persian limes are also grown in southern Florida. This is the only citrus fruit marketed while it is green in color. The fully ripe (yellow) fruit lacks consumer appeal and is undesirable for fresh market. Limes should be picked while still green, but after the fruit has lost the dimpled appearance around the blossom end. Good-quality fruit may be stored satisfactorily for 6 to 8 weeks at 48 to 50°F. However, mature fruit gradually turns yellow at this temperature. Preventing desiccation is very important, as is a relative humidity above 85%. Pitting occurs below 45°F, and temperatures above those recommended allow stem end rot to develop.

Controlled-Atmosphere Storage

Tests of modified- or controlled-atmosphere (CA) storage for oranges, grapefruit, lemons, and limes have found minor benefits but have not shown that CA storage extends storage or market life. For this reason, CA storage is not generally recommended for citrus fruit. Atmospheres used for storage of apples and other deciduous fruit are unsatisfactory for citrus fruit and lead to rind injuries, off-flavors, and decay.

STORAGE DISORDERS AND CONTROL

Postharvest Diseases

Citrus fruit often carries incipient fungus infections when harvested. Decay organisms may also enter minor injuries caused during harvesting and handling. Major postharvest diseases (with symptoms) encountered in storage include the following:

Alternaria Rot. Usually a black, dry, deeply penetrating decay at stylar end of navel oranges; a slimy, leaden-brown storage decay of core starting at stem end in other citrus fruit. *Control*: Provide optimum growing conditions. Harvest oranges before they are overripe. Do not store tree-ripe lemons. Restrict storage period for other lots known to be weak. Green buttons indicate strong fruit.

Anthracnose (*Colletotrichum*). Leathery, dark brown, sunken spots or irregular areas. Internal affected tissues dark gray, fading through pink to normal color. Most serious with degreened early-season tangerines, tangerine hybrids, and long-stored oranges, and long-stored grapefruit. *Control*: Use recommended postharvest fungicide. Avoid long storage and move promptly.

Blue (and Green) Mold Rot (*Penicillium*). Soft, watery, decolorized lesions that, under moist conditions, become quickly covered with blue or olive-green powdery spores. *Control*: Prevent skin breaks. Use recommended fungicides in washes. Cool fruit to as near to 32°F as practicable.

Brown Rot (*Phytophthora*). Extensive firm, brown decay with a penetrating rancid odor. Chiefly on fruit from California and Arizona. *Control*: Orchard spraying and good sanitation. Submerge fruit at packing for 2 min in 114°F water.

Sour Rot (*Geotrichum*). Soft, watery rot with sour smell after peel injuries. Similar to early stages of mold rot, except that no powdery spores are formed. Most serious on lemons and mandarin-type fruit. *Control*: Avoid peel injuries at harvest; refrigerate at lowest practical temperature. Approved fungicides are of little or no value.

Stem End Rot (*Diplodia*; *Phomopsis*). Pliable, fairly firm, extensive brown decay starting at stem. Sour, pungent odor. Prevalent in Florida and found occasionally in Arizona and California fruit. *Control*: Treat harvested fruit promptly in recommended fungicides and cool promptly below 50°F.

The U.S. Environmental Protection Agency (EPA) approves several chemical fungicides for postharvest use on citrus fruit. These include thiabendazole (TBZ), orthophenylphenol (OPP or SOPP), and imazalil. These materials are applied after washing and before waxing or are incorporated in the wax coating. Under certain conditions, it is beneficial to use a combination of these materials because all are not equally effective against the same organism. Strains of the blue and green molds (*Penicillium*) that are resistant to certain fungicides have developed in citrus storage houses, so care must be taken in selecting the fungicide and the time of application.

Physiological Disturbances

Various physiological conditions can also cause defects. Using fruit at prime maturity and proper handling after harvest can eliminate these defects. Proper temperature and humidity levels are required during handling, storage, and transit. The following are the physiological disorders and symptoms:

Stem End Rind Breakdown. Small to large sunken, drying, discolored, firm areas in skin around stem button or on the upper part of fruit. *Control*: Pick before overmature. Avoid overheating in packinghouse treatments. Wax fruit. Store for limited period only in fairly high relative humidity (85 to 90%). Follow storage temperatures recommended for variety and growing area.

Freezing Injury. Field freezing is found scattered through boxes. Transit and storage freezing are worse in exposed fruit in bottom-layer boxes or those nearest cooling coils. Affected fruit may show water-soaked areas in rind. The internal tissue is disorganized, water-soaked, and milky and has rind flavor. Frozen fruit loses moisture, causing drying, separation of juice vesicles, and buckling of segment walls. The freezing point of citrus fruit is about 28.5°F.

Internal Decline. In lemons, core tissues near the stylar end break down and dry, becoming pink. *Control*: Maintain optimum moisture conditions in grove.

Pitting. Depressed areas of 0.1 to 0.8 in. diameter in the peel of citrus fruit. Affected tissues collapse and may appear bleached or brown. Pits occur anywhere on the fruit and may coalesce to form large irregular areas. The cause is not fully understood. In general, pitting is a low-temperature disorder. However, lack of immediate cold storage for Florida grapefruit has accentuated pitting. *Control*: Follow storage temperatures recommended for cultivar and growing area.

BANANAS

HARVESTING AND TRANSPORTATION

Bananas do not ripen satisfactorily on the plant; even if they did, deterioration of ripe fruit is too rapid to allow shipping from tropical growing areas to distant markets. Bananas are harvested when the fruit is mature but unripe, with dark green peels and hard, starchy, inedible pulps. Each banana plant produces a single stem of bananas that contains from 50 to 150 individual pieces of fruit (or fingers). The stem is cut from the plant as a unit with fingers attached and transported to nearby boxing stations.

Bananas are removed from the stem, washed, and cut into consumer-sized cluster units of four or more fingers. The clusters are packed in protective fiberboard cartons that contain 40 lb of fruit. The cartons move by rail from the tropical boxing stations to port and then are loaded into the holds of refrigerated ships. On the ship, fruit is cooled to the optimum carrying temperature, usually 56 to 58°F, depending on variety.

Bananas are unloaded still green and unripe at seaboard and transported under refrigeration at a holding temperature of 58°F to interior wholesale distribution centers by both truck and railcar. The objective is to maintain the product in an optimal environment and move it to its destination as quickly as possible to minimize postharvest deterioration.

DISEASES AND DETERIORATION

Bananas are subject to various diseases and physiological disorders. Proper temperature and moisture during storage and careful handling slow aging and development of decay.

Anthracnose (Ripe Rot) (*Gloeosporium*). Shallow black spots on stems of ripening fruit. Under moist conditions, pink spore masses cover center of spots. Dark discoloration of skin may extend from stem end over entire fruit. *Control*: Protect fruit from mechanical injury; damage is reduced if fruit is transported in corrugated boxes. Schedule ripening so that fruit can be marketed and consumed before appearance of defect.

Black Rot (*Ceratocystis*). Transmitted from wounds via fibrovascular system of plant; progresses into crowns and stem ends of fingers, and produces brownish-black areas in peel at fruit ends. As fruit ripens, skin becomes grayish-black and water-soaked. Pulp is rarely affected. *Control*: In the tropics, dip or spray freshly cut tips and bunches with fungicides before boxing. Avoid mechanical injury and maintain sanitation program from tropics to ripening room.

Chilling Injury. Dull gray skin color with increased tendency to darken on slight bruising. Latex in green fruit does not bleed freely and will be clear rather than cloudy. Subsurface peel tissue streaked with brown. Turning or ripe bananas are more susceptible to injury than green fruit. *Control*: Avoid temperatures below 55°F. Moving air accelerates chilling.

Fungus Rots (Several Fungi). Extensive soft rot of scarred, split, or broken fruit. Affected skin and flesh are moist and brown to black. Under high humidity, the surface is often covered with mold. *Control*: Handle fruit carefully to avoid bruising and mechanical injury. Cool stored fruit rapidly to 56°F.

EXPOSURE TO EXCESSIVE TEMPERATURES

Fruit pulp temperatures only a few degrees below optimum holding temperatures, although considerably above the actual freezing point of bananas, can cause chilling injury. Severity varies directly with the duration of exposure and indirectly with temperature. It is primarily a peel injury in which some surface cells of the banana peel are killed. The contents of the dead cells eventually darken because of oxidation and give the fruit a dull appearance. Both green and ripe bananas are susceptible to chilling injury; severely chilled green bananas never ripen properly. Fruit pulp temperatures only a few degrees above the optimum holding temperatures can cause fruit to ripen prematurely in transit.

Once bananas arrive at wholesale distribution centers, they are unloaded and placed in specially equipped processing rooms for controlled ripening. As soon as the bananas have ripened to an edible state, they are rushed to retail because ripening cannot be stopped. Even under ideal refrigeration, ripe bananas eventually progress to the point where they are too ripe to sell.

WHOLESALE PROCESSING FACILITIES

Wholesale banana processing facilities are distinguished from general wholesale produce storage facilities by special banana ripening rooms. The ripening room controls initiation and completion of fruit ripening; Figure 2 shows a typical banana room. The ability to ripen bananas properly is so critically linked to the design of the ripening rooms that major banana importers maintain technical staffs that specialize in banana room design. These technical staffs provide free, nonobligatory consultation to wholesalers, architects, engineers, contractors, and others involved in banana ripening facility design, construction, and operation.

A typical banana processing facility consists of a bank of five or more individual ripening rooms. For design purposes, one complete turnover per week is assumed; therefore, the combined capacity of all rooms should approximately equal total weekly volume, allowing for seasonal variations.

Each load is scheduled for optimum ripeness on a particular day. Fruit shipped from this load a day ahead of schedule will be underripe; fruit shipped a day late will be overripe. Therefore, shipping for several days out of one room is not practical. There should be at least as many rooms as there are retail shipping days per week.

Because bananas cannot be processed on a continuous flow basis, individual room capacities are multiples of carlots, usually one-half or one carlot. As the capacity of transportation equipment has increased in recent years, the design capacity of processing rooms has also increased. It is generally cheaper to build one or two

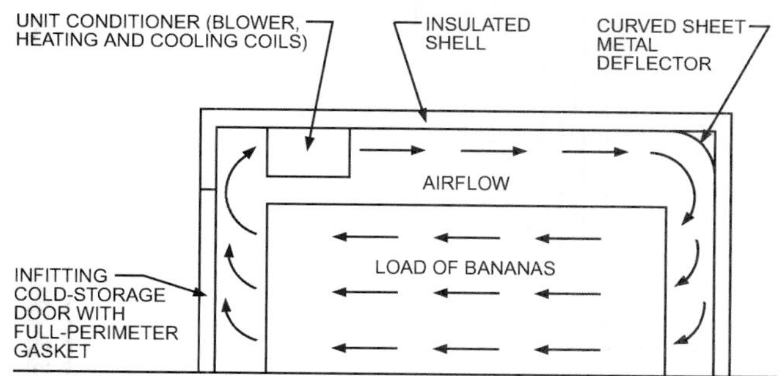

Fig. 2 Banana Room (Side View)

large rooms than several small rooms with equivalent total capacity. However, minimizing construction cost is not the pertinent consideration; having all bananas in each particular processing room reach optimum ripeness for shipment to retail at the same time is more important.

Airtightness

Exposing fruit to ethylene gas, introduced into the room from cylinders, initiates ripening. The dose is 1 ft^3 of ethylene gas per 1000 ft^3 of room air space. Ethylene is explosive in air at concentrations between 2.75 and 28.6%. Many ethylene systems gas the fruit automatically over a 24 h period.

To be effective, the gas must be confined to the ripening room for 24 h, so banana rooms must be airtight. Floor drains must be individually trapped to prevent gas leakage. Special care should be taken to seal all penetrations in room walls where refrigerant piping, plumbing lines, and the like enter rooms. Doors should have single-seal gaskets all around and sweep gaskets at the floor line.

Refrigeration

A direct-expansion halocarbon system is recommended. Because of ammonia's harmful effect on bananas, direct-expansion ammonia systems should not be used. Malfunctioning refrigeration equipment during processing could cause heavy product losses, so each ripening room should have a completely separate system despite high initial installation costs.

For maintenance-free operation in the high-humidity environment of processing rooms, evaporator coils should have a fin spacing of 4 fins/in. Coils should be amply sized and capacity-rated at a design temperature difference of 15°F with a refrigerant temperature of 40°F. Air temperatures used during processing range from 45 to 65°F. Because of the danger of banana chilling, refrigerant temperatures below 40°F are not recommended. With programmers, suction pressure control devices or hot-gas bypass systems must be installed.

Refrigeration Load Calculations

These calculations are based on the same methods used for other fresh fruit. A typical half-carlot-capacity banana room holds approximately 432 boxes of fruit. Approximate outside dimensions for a three-tier forklift-type room, shown in Figure 3, are 30 ft long by 6 ft wide by 22 ft high. Pallets are 48 by 40 in. and are stacked 6 pallets deep in each of 3 tiers, totaling 18 pallets per room. Boxes for use in banana rooms are approximately 10 in. high by 16 in. wide by 22 in. long and are stacked 4 boxes per each of 6 layers, totaling 24 boxes per pallet. With 18 pallets per room, 432 boxes of bananas can be stored (18 pallets by 24 boxes). Each box has a net weight of 42 lb and a gross weight of 47 lb.

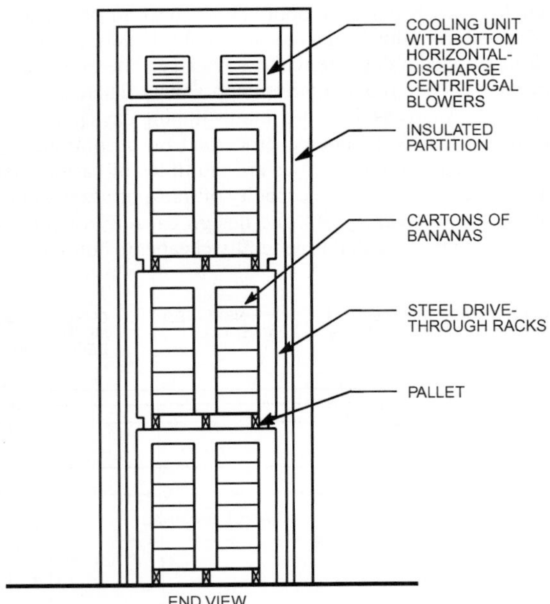

Fig. 3 Three-Tier Forklift Banana Room (End View)

Transmission load is calculated in the normal manner; the air change load is negligible. The electrical load is based on continuous operation of multikilowatt fan motor(s). The peak heat of respiration is 0.5 Btu/h per pound multiplied by the total net weight of bananas in the ripening room. For product cooling, the specific heat of bananas is 0.8 Btu/lb·°F multiplied by the total net weight of the bananas, plus the total tare weight of the cartons multiplied by 0.4 Btu/lb·°F, the specific heat of fiberboard. The total calculated load is thus approximately 60 Btu/h per box. A pulldown rate of 1°F/h is assumed. Total system design capacity is calculated by assuming simultaneous peak respiration and pulldown load.

Heating

Heat is not required during most ripening cycles. However, occasional loads may come in at temperatures below desired levels for treatment with ethylene gas, making heating necessary. Many banana room refrigeration units come with integrated electrical heating elements. If electrical heating strips are used, they should be enclosed in a corrosion-resistant sheath and have a surface temperature of not more than 800°F in dead still air; this temperature limitation is necessary because of the proximity of the heating strips to refrigerant coils and the inherent danger should leakage occur. Portable plug-in electric heaters are also used. Heating system capacity should be sufficient to raise load temperature at a rate of 1°F/h. Open-flame gas heaters should never be used in banana rooms for two reasons: (1) ethylene gas used during ripening is explosive at certain concentrations; and (2) the necessary room airtightness could easily result in the open-flame heaters' consuming the available oxygen within the space, thereby extinguishing the flame and permitting raw gas to enter the room.

Air Circulation

Table 3 shows the fruit pulp temperature schedules for 4 to 8 day ripening cycles. A temperature variation of only a few degrees considerably alters the rate of fruit ripening. For even ripening, fruit temperatures must be uniform throughout the room, so the volumetric airflow circulated throughout the entire load must be comparatively large. Centrifugal fans are necessary. They are installed for bottom horizontal discharge, so that the top boxes are not chilled immediately in front of the unit. Fan air output should be rated at 0.62 in. of water external static pressure. Because of heat of respiration, heat must be continually withdrawn from the product even when it is held at a constant temperature. Temperature variation in the load is therefore inevitable, with warmer fruit downstream relative to the circulated air. Unit conditioners at the front of the room over the door discharge toward the rear of the room. This arrangement leaves riper fruit near the door to be shipped first.

For improved air distribution, a sheet metal (or other suitable material) air deflector curved to a 90° arc is mounted full width on the back room wall. This deflector reduces turbulence and directs the air downward for return through the load.

Airflow Requirements

Air volumetric flow requirements are calculated on the basis of conditions required at the end of the pulldown period. Assume a maximum allowable fruit temperature variation of 2°F, an air temperature drop through the cooling unit of 2°F, and product temperature reduction proceeding at a rate of 0.2°F/h. During initial pulldown, the air quantity so calculated will give about a 5.5°F drop through the cooling unit. The general equations are

Table 3 Fruit Temperatures for Banana Ripening

Ripening Schedule	Temperature, °F							
	1st Day	2nd Day	3rd Day	4th Day	5th Day	6th Day	7th Day	8th Day
Four days	64	64	62	60				
Five days	62	62	62	62	60			
Six days	62	62	60	60	60	58		
Seven days	60	60	60	60	60	58	58	
Eight days	58	58	58	58	58	58	58	58

$$q_t = q_r + q_p \quad (1)$$

$$q_t = \dot{m} c_p \Delta t \quad (2)$$

where

q_t = total heat removed, Btu/h
q_r = heat of respiration, Btu/h
q_p = pulldown load, Btu/h
$\dot{m}$ = mass flow rate of air, lb/h
c_p = specific heat of air = 0.24 Btu/lb·°F
Δt = temperature change of air, °F

The values of q_r and q_p can be determined using the heat of respiration and specific heat values given in the section on Refrigeration Load Calculations. For calculation of q_p, assume a temperature reduction rate of 0.2°F/h is occurring at the end of the pulldown period.

$$q_r = 0.5 \text{ Btu/h·lb} \times 42 \text{ lb/box} = 21 \text{ Btu/h per box}$$

$$q_p = 0.2\text{°F/h}[(0.8 \text{ Btu/lb·°F} \times 42 \text{ lb/box}) + (0.4 \text{ Btu/lb·°F} \times 0.5 \text{ lb/box})] = 6.76 \text{ Btu/h per box}$$

$$q_t = 21 + 6.76 = 27.76 \text{ Btu/h per box}$$

At equilibrium, the air temperature Δt equals the fruit temperature Δt, and

$$m = q_t/c_p\Delta t = (27.76 \text{ Btu/h·box})/(0.24 \text{ Btu/lb·°F} \times 2\text{°F}) = 57.83 \text{ lb}_{air}\text{/h per box}$$

$$\text{Volumetric flow rate} = (57.83 \text{ lb/h·box})/(0.075 \text{ lb/ft}^3 \times 60 \text{ min/h}) = 12.85 \text{ cfm per box}$$

For a room with 432 boxes, the airflow should be 432 × 13.02 = 5600 cfm at 0.62 in. of water external static pressure.

Humidity

A high relative humidity around the fruit is important during banana ripening. Bananas ripened under low-humidity conditions are more susceptible to handling damage. When bananas were ripened on the stem, naked fruit was directly exposed to the moving airstream, and automatic room humidifiers were used to prevent excessive fruit dehydration. However, banana room humidifiers are not required with tropical boxing. The fiberboard carton shields the fruit from the moving airstream. In addition, ample sizing of evaporator coils keeps the temperature difference across the coil to 10 to 15°F, thereby limiting dehumidification. Both natural transpiration of the fruit and airtight room design also contribute to high room humidity.

Controls

Ripening room air temperatures are varied frequently during banana processing. Temperatures should be controlled by remote bulb-type thermostats, with bulbs for heating and cooling mounted in the return airstream within the ripening room to prevent short-cycling of equipment. Thermostats should be mounted on the exterior of the ripening room and have a range of 45 to 70°F, calibrated in 1°F increments, with no more than 2°F differential. The thermostats are best mounted on a control panel having a selector switch providing heating and cooling with continuous fan operation.

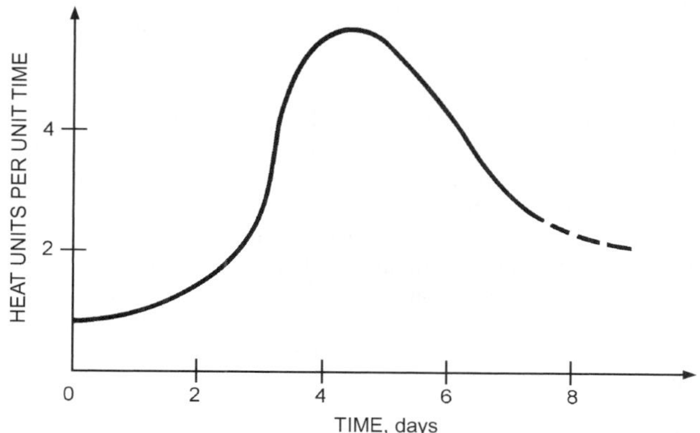

Fig. 4 Heat of Respiration During Banana Ripening

Automatic temperature controllers or programmers are installed in new facilities. Bananas produce heat continuously, but the rate of heat production varies considerably during the ripening cycle. Figure 4 shows a generalized heat-of-respiration curve. Although more complex, the removal of heat of respiration from the load can be viewed, for the purpose of analysis, as a simple conduction process. Applying the general conduction equation,

$$q = \frac{kA\Delta t}{L} \quad (3)$$

$$\frac{k}{L} = \frac{1}{R}$$

where

q = heat of respiration, Btu/h
k = thermal conductivity of packaging material, Btu·in/h·ft²·°F
A = surface area of packaging material, ft²
Δt = banana temperature minus air temperature, °F
L = thickness of packaging material, in.
R = thermal resistance of packaging material, ft²·°F·h/Btu

If kA/L is a constant, Δt must vary as q. Assuming the banana temperature is also constant, room air temperature must be lowered as q increases during ripening. However, the exact value of q at any particular point in the ripening cycle is unknown.

With the conventional, manually adjustable air-sensing thermostat-control system, fruit temperatures are taken manually with a pulp thermometer, and thermostat settings are continually adjusted to followed ripening schedules. This is essentially a trial-and-error procedure.

In contrast, the temperature programmer has a remote bulb, which is placed in a box of bananas in the load. Since the bulb senses fruit temperature directly, heat of respiration is compensated for during ripening. Fruit temperature is automatically adjusted to follow preset cycles.

SUBTROPICAL FRUIT

AVOCADOS

Avocado cultivars grown in California are not grown commercially in Florida, and vice versa. Florida cultivars tend to be larger-fruited than those from California. In California, the Fuerte variety accounts for 75% of the annual crop and is available from October through March. Hass (black skin) is available from April through September. In order of importance, Florida cultivars are Booth 8, Lula, Waldin, and Booth 7. Waldin appears on the market in August, followed by Booth 8 in September, and Booth 7 and Lula from October to February.

The best storage temperature for cold-tolerant Florida avocado cultivars such as Booth 8 and Lula is 40°F. All Florida summer avocado cultivars, such as Waldin, are cold-intolerant and store best at 54 to 55°F. A few cultivars, such as Fuerte, store best at 45°F. Cold-tolerant cultivars can be held in storage a month or longer, but storage of cold-intolerant cultivars is usually limited to 2 weeks because of their susceptibility to softening and chilling injury. The best ripening temperature for avocados is 60°F, but temperatures from 55 to 75°F are usually satisfactory. Temperatures above 79°F frequently cause off-flavor, skin discoloration, uneven ripening, and increased decay.

Storage Disorders

Anthracnose (*Colletotrichum*). Scattered black spots covering firm, decayed tissue that can be removed easily from surrounding flesh. Pink spore masses form on the spots under moist conditions. *Control*: Prevent blemishes and other breaks in the skin.

Chilling Injury. Typified by small to large sunken pits in the skin that turn brown or black and are often accompanied by general browning of the skin and light smoky streaks in the flesh, which develop independently. *Control*: Store at proper temperatures.

MANGOES

The important early cultivars Tommy Atkins and Irwin mature during June and July, followed by such midseason cultivars as the Kent and Palmer, which mature during July and August. The most important cultivar produced in Florida is the large-fruited Keitt, which is late-season and matures during August and September.

Optimum storage temperature for mangoes is 54 to 55°F for 2 to 3 weeks, although 50°F is adequate for some cultivars for shorter periods. Mangoes are subject to chilling injury at temperatures below 50°F. The best ripening temperatures for mangoes are from 70 to 75°F, but temperatures of 60 to 65°F are also satisfactory under certain conditions. At 60 to 65°F, the fruit develops a bright and most attractive skin color, but the flavor is usually tart and requires an additional 2 to 3 days at 70 to 75°F to attain sweetness. Mangoes ripened at 80°F and higher frequently have a strong flavor and mottled skin.

Storage Disorders

Anthracnose (*Colletotrichum*). Large scattered black spots in the skin of ripening fruit. Under moist conditions, pink spore masses develop in spots. *Control*: By regular spraying on the tree and by use of hot water treatment (131°F for 0.12 h) after harvest.

Chilling Injury. Pitting of the skin, which sometimes develops a gray cast. Fruit with chilling injury usually does not ripen uniformly. *Control*: Store at proper temperatures.

PINEAPPLES

Fresh pineapples are available throughout the year, but the supply is much larger from March through June. Only three pineapple cultivars are commercially important in the United States: the Smooth Cayenne from Hawaii and the Red Spanish and Smooth Cayenne from Puerto Rico.

Pineapples harvested at the half-ripe stage can be held for 2 weeks at 45 to 55°F and still have a shelf life of about 1 week. Continuous maintenance of storage temperature is as important as the specific storage temperature. Ripe fruit should be held at 45 to 47°F. Harvesting at the mature green stage is not recommended because some fruit would be too immature to ripen. Mature green fruit is especially susceptible to chilling injury at temperatures below 50°F.

Storage Disorders

Black Rot (*Ceratocystis*). Extensive damage to tissues, which become soft and leaky, ranging from normal to jet black in color. *Control*: Treat freshly cut stem parts with benzoic acid-talc dust, prevent bruising, and cool to 50°F.

Brown Rot (*Penicillium; Fusarium*). Brown, firm decay starting at eyes or cracks; it is common on overripe fruit. *Control*: Provide good growing conditions and market before the fruit is overripe.

Chilling Injury. Fruit takes on a dull hue, the flesh develops water soaking, and the core darkens. *Control*: Hold at recommended temperatures.

REFERENCES

Haller, M.H., D.H. Rose, J.M. Lutz, and P.L. Harding. 1945. Respiration of citrus fruits after harvest. *Journal of Agricultural Resources* 71:327.

Ismail, M.A., T.T. Hatton, D.J. Dezman, and W.R. Miller. 1986. In transit cold treatment of Florida grapefruit shipped to Japan in refrigerated van containers: Problems and recommendations. *Proceedings of the Florida State Horticultural Society* 99:117-121.

BIBLIOGRAPHY

Chace, W.G., Jr., J.J. Smoot, and R.H. Cubbedge. 1979. Storage and transportation of Florida citrus fruits. *Florida Citrus Industry* 51:16.

Chau, K.V., C.D. Baird, P.C. Talasila, and S.A. Sargent. 1992. Development of time-temperature-humidity relations for fresh fruits and vegetables. ASHRAE Research Project RP-678, *Final Report*.

Hardenburg, R.E., A.E. Watada, and C.Y. Wang. 1986. The commercial storage of fruits, vegetables, and florist and nursery stocks. *USDA Handbook* 66. U.S. Department of Agriculture, Washington, D.C.

McCornack, A.A., W.F. Wardowski, and G.E. Brown. 1976. Postharvest decay control recommendations for Florida citrus fruit. *Florida Cooperative Extension Service Circular* 359-A.

Smoot, J.J., L.G. Houck, and H.B. Johnson. 1971. Market diseases of citrus and other subtropical fruits. *USDA Handbook*, p. 398.

Wardowski, W.F., S. Nagy, and W. Grierson. 1986. *Fresh citrus fruits*. AVI Publishing, Westport, CT.

CHAPTER 24

VEGETABLES

LOSSES (shrinkage) in marketing fresh vegetables (harvesting, handling, packing, storing, and retailing) are caused, in part, by overly high temperatures during handling, storage, and transport, which increase ripening, decay, and the loss of edible quality and nutrient values. Some cases may involve freezing or chilling injury from overly low temperatures. Other serious losses are caused by mechanical injury from careless or rough handling and by shrinkage or wilting because of moisture loss. Shrinkage can be reduced substantially by following recommended handling, cooling, transport, and storage practices. Improved packaging, refrigerated transport, and awareness of refrigeration's role in maintaining quality throughout marketing have made it possible to transport vegetables in field-fresh condition to distant markets.

This chapter covers postharvest handling, cooling, packaging, in-transit preservation, and storage at destination locations for fresh vegetables. It also gives storage requirements for specific vegetables, including potential product deterioration due to improper handling and storage conditions. Vegetable precooling is covered in Chapter 15, and vegetable processing and freezing in Chapter 27. Chapter 11 also provides storage requirements for many types of vegetables.

PRODUCT SELECTION AND QUALITY MAINTENANCE

The principal hazards to quality retention during marketing include

- Metabolic changes (composition, texture, color) associated with respiration, ripening, and senescence (aging)
- Moisture loss with resultant wilting and shriveling
- Bruising and other mechanical injuries
- Parasitic diseases
- Physiological disorders
- Freezing and chilling injury
- Flavor and nutritional changes
- Growth (sprouting, rooting)
- Ethylene-caused injury

Fresh vegetables are living tissues and have a continuing need for O_2 for respiration. During respiration, stored food such as sugar is converted to heat energy, and the product loses quality and food value. In maintaining commodity temperatures during storage or transportation, some of the refrigeration load can be attributed to respiration. For example, a 20,000 lb load of asparagus cooled to 39°F can produce enough heat of respiration during a cross-country trip to melt 7900 lb of ice.

Vegetables that respire the fastest often have greater handling problems because they are the most perishable. Variations are caused by the type of plant part involved. For example, root crops such as carrots and radishes have lower respiration rates than fruit vegetables (cucumber, pepper) and sprouts (asparagus). Refrigeration is the best method of slowing respiration and other life processes. Chapters 9 and 11 give more information on the respiration rates of many vegetables.

Vegetables are usually covered with natural populations of microorganisms, which will cause decay under the right conditions. Deterioration from decay is probably the greatest source of spoilage during marketing. When mechanical injuries break the skin of the produce, decay organisms enter. If it is then exposed to warm (especially warm, humid) conditions, infection usually increases. Adequate refrigeration is the best method of controlling decay because low temperatures control growth of most microorganisms.

Many color changes associated with ripening and aging can be delayed by refrigeration. For example, broccoli may show yellowing in 1 day on a nonrefrigerated counter, but remain green at least 3 to 4 days in a refrigerated display.

Refrigeration can retard deterioration caused by chemical and biological reactions. Freshly harvested asparagus will lose 50% of its vitamin C content in 1 day at 68°F, whereas it takes 4 days at 50°F or 12 days at 32°F to lose this amount (Lipton 1968). Recommended conditions for long-term storage are listed in Table 1 of Chapter 11.

Loss of moisture with consequent wilting and shriveling is one of the obvious ways to lose freshness. **Transpiration** is the loss of water vapor from living tissues. Moisture losses of 3 to 6% are enough to cause a marked loss of quality for many kinds of vegetables. A few commodities may lose 10% or more in moisture and still be marketable, although some trimming may be necessary, such as for stored cabbage. For more on transpiration, see Chapter 9.

Postharvest Handling

After harvest, most highly perishable vegetables should be removed from the field as rapidly as possible and refrigerated, or they should be graded and packaged for marketing. Because aging and deterioration continue after harvest, marketable life depends greatly on temperature and care in physical handling.

The effects of rough handling are cumulative. Several small bruises on a tomato can produce an off-flavor. Bruising also stimulates the ripening rate of products such as tomatoes and thereby shortens potential storage and shelf life. Mechanical damage increases moisture loss; skinned potatoes may lose 3 to 4 times as much weight as nonskinned ones.

Use care in stacking bulk bins in storage, to maintain proper ventilation and refrigeration of the product. Bins should not be so deep that excessive weight damages product near the bottom.

Quality maintenance is further aided by

- Harvesting at optimum maturity or quality
- Handling carefully to avoid mechanical injury
- Handling rapidly to minimize deterioration
- Providing protective containers and packaging
- Using preservative chemical, heat, or modified-atmosphere treatments
- Enforcing good plant sanitation procedures while handling
- Precooling to remove field heat

The preparation of this chapter is assigned to TC 10.9, Refrigeration Application for Foods and Beverages.

- Providing high relative humidity to minimize moisture loss
- Providing proper refrigeration throughout marketing

Cooling

Rapid cooling of a commodity after harvest, before or after packaging or before it is stored or moved in transit, reduces deterioration of the more perishable vegetables. The faster field heat is removed after harvest, the longer produce can be maintained in good marketable condition. Cooling slows natural deterioration, including aging and ripening; slows growth of decay organisms (and thereby the development of rot); and reduces wilting, because water losses occur much more slowly at low temperatures than at high temperatures. After cooling, produce should be refrigerated continuously at recommended temperatures. If warming is allowed, much of the benefit of prompt precooling may be lost.

Types of cooling include hydrocooling, vacuum cooling, air cooling, and cooling with contact ice and top ice, which are discussed in detail in Chapter 15. The choice of cooling method depends on factors such as refrigeration sources and costs, volume of product shipped, and compatibility with the product.

Protective Packaging and Waxing

Vegetables for transit and destination storage should be packed in containers with adequate stacking strength and durability to protect against crushing under high humidity. Bulging crates should be stacked on their sides or stripped between layers to keep weight off the commodity. Many vegetables are stored or shipped in corrugated fiberboard containers, but fiberboard weakening by moisture absorption at high storage humidities is frequently a serious problem.

Fiberboard strength has improved, and its rate of moisture absorption slowed. Special fiberboard treatments allow use of some cartons in hydrocooling and with package and top ice. Cartons may be strengthened for stacking using dividers, wooden corner posts, and full telescoping covers.

Produce is often consumer-packaged at production locations using many types of trays, wraps, and film bags, which may present special transit and handling problems when master containers for these packages lack stacking strength.

Desiccation often can be minimized by using moisture-retentive plastic packaging materials. Polyethylene film box liners, pallet covers, and tarpaulins may be helpful in reducing moisture loss. Plastic films, if sealed or tightly tied, may restrict transfer of CO_2, O_2, and water vapor, leading to harmful concentrations of these respiratory gases; films also restrict heat transfer, which retards the rate of cooling (Hardenburg 1971).

Waxes are applied to rutabagas, cucumbers, mature green tomatoes, and cantaloupes, and to a lesser extent to peppers, turnips, sweet potatoes, and certain other crops. With products such as cucumbers and root crops, waxing reduces moisture loss and shriveling. With some products, an improved glossy appearance is the main advantage. Thin wax coatings may give little if any protection against moisture loss; coatings that are too heavy may increase decay and breakdown. Waxing alone does not control decay, but waxing combined with fungicides may be beneficial. Waxing is not recommended for potatoes either before or after storage (Hardenburg et al. 1959).

IN-TRANSIT PRESERVATION

Good equipment is available to transport perishable commodities to market under refrigeration by rail, trucks, piggyback trailers, and containers. High relative humidity (about 95% rh) is desirable for most vegetables to prevent moisture loss and wilting. Many vegetables benefit from 95 to 100% rh. Humidity in both iced and mechanically refrigerated cars and trailers is usually high.

Cooling Vehicle and Product

Vehicles used to ship vegetables that require low transit temperatures should have their interiors cooled before loading to prevent produce loaded near container walls from warming under hot ambient conditions or cooling too much under cold ambient conditions. In temperate climates with low humidity, the container should be cooled to the carrying temperature. If loading from an open dock in a humid environment, the container should be cooled to the dew point of the outside air. Temperatures below the dew point cause water to condense on walls, which may damage fiberboard packages. In all cases, refrigeration should be turned off when container doors are open, to prevent moisture from condensing on the evaporator coils.

Generally, vegetables that require a low temperature during shipping should be cooled before they are loaded into transport vehicles. Cooling produce in tightly loaded refrigerator cars or containers is slow, and the portion of the load exposed to the cold air discharge may be frozen when the interior of the load is still warm. Highway trailers do not have adequate airflow to remove field heat from perishable produce.

Packaging, Loading, and Handling

In this section, the term "boxes" is used for corrugated containers and "containers" is used for marine containers. Boxes must protect the commodity, allow heat exchange as necessary, and serve as an appropriate merchandising unit with sufficient strength to withstand normal handling. Freight container tariffs describe approved containers and loading procedures.

Boxes should be loaded to take advantage of their maximum strength and to allow adequate stripping or use of spacers to hold the load in alignment. Proper vertical alignment of containers is essential to obtain their maximum stacking strength. Previously, when boxes were hand-stacked, stacks were spaced to provide channels for air circulation. Air channels are no longer specified with the use of pallet loads.

When different types of boxes are used in the same load, stacks should be separated so that one type will not damage another. If separation of stacks is impossible, boxes made of lighter material, such as fiberboard, should always be loaded on top of heavier wood boxes.

Providing Refrigeration and Air Circulation

Safe transit temperatures for various vegetables are given in Table 1. For safety, the suggested thermostat settings for cool-season vegetables are usually 2 to 4°F above the freezing point.

Perishables are often shipped in loads with other commodities. When this occurs, the loads should be set up so that different types of produce have compatible temperature ranges and ethylene sensitivities. Use Table 2, which groups common fruits and vegetables by temperature range and ethylene sensitivity, to select compatible produce. Produce in the same column can be safely held at the same temperature range. Mixing produce from different temperature ranges can compromise produce quality, especially with longer transit times. The greater the difference in recommended temperatures, the greater the potential for quality loss.

Dry vegetables (row 1 of Table 2) should not be mixed with any other produce in the table, and should be held in a 50 to 70% rh environment to prevent decay. Most of the vegetables in the lowest temperature range (32 to 36°F) are sensitive to moisture loss and should be held at more than 90% rh or packaged to minimize water loss. Other vegetables and fruit should be held at 85 to 95% rh.

Ethylene-sensitive vegetables (row 2) should not be mixed with ethylene-producing fruits (row 5). If, for some reason, they must be mixed, damage may be reduced by using a fresh-air exchange rate of 45 cfm (Thompson et al. 2000) and/or ethylene scrubbers. In some cases, a controlled atmosphere (CA) will allow ethylene-sensitive produce to be shipped with ethylene-producing produce, but the

Table 1 Optimal Transit Temperatures for Various Vegetables

Vegetable	Desirable Transit Temperature, °F	Suggested Thermostat Setting,[a] °F	Highest Freezing Point,[b] °F
Artichokes	32	33	29.8
Asparagus	32 to 35	35	30.9
Beans, lima	37 to 41	37	31.0
Beans, snap	40 to 45	45	30.7
Beets, topped	32	34	30.4
Broccoli	32	34	30.9
Brussels sprouts	32	34	30.6
Cabbage	32	34	30.4
Cantaloupes	36 to 41	37	29.8
Carrots, topped	32	33	29.5
Cauliflower	32	34	30.6
Celery	32	34	31.1
Corn, sweet	32	34	30.9
Cucumbers	50 to 55	50	31.1
Eggplant	46 to 54	50	30.6
Endive and escarole	32	34	31.8
Greens, leafy	32	34	—
Honeydew melon	45 to 50	45	30.4
Lettuce	32	34	31.6
Onions, dry	32 to 39	35	30.6
Onions, green	32	34	30.4
Peas, green	32	34	30.9
Peppers, sweet	45 to 50	46	30.7
Potatoes			
Early crop	50 to 60	50	30.9
Late crop	39 to 50	40	30.9
For chipping			
Early crop	64 to 70	64 to 70	30.9
Late crop	50 to 60	50 to 60	30.9
Radishes	32	34	30.7
Spinach	32	34	31.5
Squash			
Summer	41 to 50	41	31.1
Winter	50 to 55	50	30.5
Sweet potatoes	55 to 61	55	29.7
Tomatoes			
Mature green	55 to 70	55	30.9
Pink	50	50	30.6
Watermelons	50 to 60	50	31.3

Sources: *USDA Agricultural Handbook* 195 (Redit 1969), *USDA Marketing Research Report* 196 (Whiteman 1957), and *USDA Agricultural Handbook* 66 (USDA 2004).
[a]For U.S. shipments of vegetables in mechanically refrigerated cars under Rule 710 and in trailers under Rule 800 of the Perishable Protective Tariff.
[b]Highest temperature at which freezing occurs.

acceptable produce combinations and atmosphere prescriptions are not well documented. Produce that neither is sensitive to nor produces ethylene (rows 3 and 4) can be mixed with produce above or below them in the same temperature column.

Some produce can exchange odors with other items. See the notes at the bottom of Table 2 for precautions.

Certain produce have a short postharvest life and are not suited for container shipments, especially if held at nonoptimal temperatures. For example, asparagus has a maximum postharvest life of 3 weeks at 36°F. If shipped at 32°F in a load of broccoli, it is subject to chilling injury after only 10 days. Modified-atmosphere (MA) packaging can sometimes increase produce life and allow shipping to destinations that require several weeks' transport time. If a CA environment is used, it should, at a minimum, not reduce the postharvest life of any of the mixed commodities.

Table 2 also applies to fruits covered in Chapter 22.

With the many kinds of refrigeration, heating, and ventilating services available, the shipper has only to specify the desired transport temperature. Generally, the shipper or the receiver is responsible for selecting the protective service for the commodity in transit. Protective services are described in detail in USDA (2000).

Protection from Cold

In winter, vegetables must be protected from freezing. Refrigerated transport vehicles, equipped to handle the full range of both fresh and frozen commodities, are also designed to provide heat for cold-weather protection. Heat is supplied by electric heating elements or by reverse-cycle operation of the refrigerating unit (hot gas from the compressor is circulated in the cooling coils). The change from cooling to heating is done automatically by thermostatic controls.

In-transit freezing is a major problem in moving late-crop potatoes from storage to market. Cars and trailers should be warmed before loading, protected during loading with canvas door shields or loading tunnels, and loaded properly.

Checking and Cleaning Equipment

If the thermostat is out of adjustment by a few degrees, products may be damaged by freezing or by insufficient refrigeration. Thermostats should be calibrated at regular intervals to ensure that the proper amount of refrigeration is furnished. Trailers and railcars should be cleaned carefully before loading. Debris from previous shipments may contaminate loads and should be swept from floors and floor racks.

Modified Atmospheres in Transit

Various systems provide controlled or modified atmospheres in railcars and marine containers. Although the gas concentration for each process may differ, all systems alter the levels of O_2, CO_2, and nitrogen surrounding the produce.

The O_2 concentration in air is usually lowered to 1 to 5% because this level tends to depress the respiration rate.

No single modified atmosphere can be expected to benefit more than a few commodities; crop requirements and tolerances are quite specific (see Table 1 in Chapter 11). Also, some vegetables may tolerate an atmosphere at one temperature but not at another, or they may tolerate a modified atmosphere for only a limited time.

The load compartment must be fairly tight to maintain desired atmospheres. MA equipment is predominantly installed in vehicles used for long-haul transport (over five days' transit time) of chilled perishables. Atmosphere systems used in transport either vaporize liquid nitrogen or generate nitrogen by passing heated compressed air through hollow fibers.

Temperature is critical in determining the benefits of modified atmospheres. If temperatures are higher than recommended, increased respiration lowers O_2 and raises CO_2 levels toward anaerobic levels. In addition, decay and other deterioration may increase rather than reduce. Modified atmospheres should never be used as a substitute for good temperature management.

Use of controlled atmospheres is discussed later in this chapter. For further information on refrigerated transport by truck, ship, and air, see Chapters 30 to 32.

PRESERVATION IN DESTINATION FACILITIES

Wholesale warehouse facilities usually do not have a wide range of controlled temperature rooms to provide optimum storage conditions for each kind of produce, which is not necessary for short holding. It is, however, necessary to separate products into groupings by temperature and humidity. Table 3 groups fruits and vegetables into three temperature groups, and the low-temperature group is subdivided into two humidity groups. This allows a convenient number of storage conditions for distribution centers, retail stores, and food service operations. Group 1, at 32 to 36°F, includes the majority of produce volume in most North American food-handling operations.

Table 2 Compatible Produce for Long-Distance Transport

Produce	Recommended Storage Temperatures						
	32 to 36°F			40 to 45°F	45 to 50°F		55 to 65°F
Dry vegetables	Dry onion[1,3,9]	Garlic					Ginger[5] Pumpkin Squash, winter
Ethylene-sensitive vegetables	Arugula[a] Asparagus Belgian endive Bok choy Broccoflower Broccoli[a] Brussels sprouts Cabbage[1] Carrot[1,3] Cauliflower Celery[1,3,9] Chard[a]	Chicory Chinese cabbage Collards[a] Cut vegetables Endive Escarole Green onion[7] Herbs (not basil) Kailon[a] Kale[a] Leek[8] Lettuce	Mint Mushroom[a,7] Mustard greens[a] Parsley Parsnip Snow pea[a] Spinach[a] Sweet pea[a] Turnip greens Watercress	Beans, snap, etc.[a,10] Cactus leaves Potato, late crop[1] Southern peas[a]	Basil[a] Chayote Cucumber[a] Eggplant[a,5] Kiwano Long bean Okra Pepper (chili) Squash, summer[a] Tomatillo Watermelon		Potato, early crop[a] Tomato, mature green
Vegetables (not ethylene sensitive)	Amaranth[a] Anise Artichoke Bean sprouts[a] Beet Celeriac Daikon Horseradish	Jerusalem artichoke Kohlrabi Lo bok Radicchio Radish Rhubarb[7] Rutabaga	Salsify Scorzonera Shallot Sweet corn[7] Swiss chard Turnip Water chestnut		Calabaza Haricot vert Pepper, bell[10] Winged bean Loofah (Luffa)[a,b]		Cassava Jicama Sweet potato (Boniato) Taro (malanga) Yam Tomato, ripe[a,b]
Fruits and melons (very low ethylene producing)	Barbados cherry Bitter melon Blackberry[a] Blueberry Caimito Cashew apple Cherry Coconut Currant Date Gooseberry Grape[6,7,8]	Longan Loquat Lychee Orange, FL[4,9] Persimmon Raspberry[a] Strawberry[a]		Blood orange[4,9] Cactus pear (tuna) Jujube Kumquat Mandarin[4,9] Olive Orange, CA, AZ[4,9] Pepino Pomegranate Tamarind Tangerine[4,9]	Babaco Calamondin[a] Carambola Casaba melon Cranberry Grapefruit[4,9] Juan Canary melon Lemon[4,9] Lime[4,9] Pineapple[2,10] Pummelo[4,9]	Tamarillo Tangelo[4,9] Ugli fruit	Breadfruit Canistel Grapefruit, CA, AZ[4,9] Jaboticaba[a]
Ethylene-producing fruits and melons	Apple[1,3,9] Apricot Avocado, ripe[2,10] Cantaloupe[c] Cut fruits Fig[1,7,8] Kiwifruit Nectarine Peach Pear, Asian[1,9] Pear, European[1,9]	Plum Prune Quince		Durian Feijoa Guava Honeydew melon Persian melon	Avocado, unripe Crenshaw melon Custard apple Passion fruit		Atemoya Banana Cherimoya Jackfruit Mamey Mango Mangosteen Papaya Plantain Rambutan Sapote Soursop[a]

Source: University of California (1997). Reprinted with permission.

Note: Produce in the same temperature column can be safely mixed. Ethylene-sensitive vegetables should not be mixed with ethylene-producing fruits and vegetables. Dry vegetables should not be mixed with other fruits and vegetables.

[a]Less than 14 day shelf life at recommended temperature and normal atmosphere conditions.

[b]Produces moderate amounts of ethylene and should be treated as an ethylene-producing fruit.

[c]Recommended storage temperatures = 36 to 39°F.

1. Odors from apples and pears are absorbed by cabbage, carrots, celery, figs, onions, and potatoes.
2. Avocado odor is absorbed by pineapple.
3. Celery absorbs odor from onions, apples, and carrots.
4. Citrus fruit absorbs odor from strongly scented fruits and vegetables.
5. Ginger odor is absorbed by eggplant.
6. Sulfur dioxide released from pads used with table grapes will damage other produce.
7. Green onion odor is absorbed by figs, grapes, mushrooms, rhubarb, and sweet corn.
8. Leek odor is absorbed by figs and grapes.
9. Onion odor is absorbed by apples, celery, pears, and citrus.
10. Pepper odor is absorbed by beans, pineapples, and avocados.

Table 3 Compatible Fresh Fruits and Vegetables During 7 Day Storage Wholesale and Retail Handling Operations

Produce	Group 1	Group 2	Group 3
	32 to 36°F and 90 to 98% rh	45 to 50°F and 85 to 95% rh	55 to 65°F and 85 to 95% rh
Vegetables	1A Alfalfa sprouts Amaranth Anise Artichoke Arugula* Asparagus* Beans; fava, lima Bean sprouts Beet Belgian endive* Bok choy* Broccoflower* Broccoli* Brussel sprouts* Cabbage* Carrot* Cauliflower* Celeriac Celery* Chard* Chinese cabbage* Chinese turnip Collards* Corn; sweet, baby Cut vegetables* Daikon Endive*/chicory* Escarole* Fennel Garlic Green onion* Herbs* (not basil) Horseradish Jerusalem artichoke Kailon* Kale* Kohlrabi Leek* Lettuce* Mint* Mushroom* Mustard greens* Parsley* Parsnip* Radicchio Radish Rutabaga Rhubarb Salsify Scorzonera Shallot Spinach* Snow pea* Sweet pea* Swiss chard Turnip Turnip greens* Water chestnut Watercress*	2 Basil* Beans;* snap, green, wax Cactus leaves (nopales)* Calabaza Chayote* Cowpea (Southern pea)* Cucumber* Eggplant* Kiwano (horned melon)* Long bean* Malanga Okra* Pepper;* bell, chili Squash; summer (soft rind)* Tomatillo* Winged bean	3 Bitter melon Cassava Dry onion Ginger Jicama Potato* Pumpkin Squash; winter (hard rind)* Sweet potato (Boniato)* Taro (dasheen) Tomato;* ripe, partially ripe, mature green Yam
Fruits and Melons	1B Apple Apricot Avocado, ripe Barbados cherry Blackberry Blueberry Boysenberry Caimito Cantaloupe[†] Cashew apple Cherry Coconut Currant Cut fruits Date Dewberry Elderberry Fig Gooseberry Grape Kiwifruit* Loganberry Longan Loquat Lychee Nectarine Peach Pear; Asian, European Persimmon* Plumcot Plum Pomegranate Prune Quince Raspberry Strawberry	2 Avocado, unripe Babaco Cactus pear, tuna Calamondin Carambola Cranberry Custard apple Durian Feijoa Granadilla Grapefruit* Guava Juan Canary melon Kumquat Lemon* Lime* Limequat Mandarin Olive Orange Passion fruit Pepino Pineapple Pummelo Sugar apple Tamarillo Tamarind Tangelo Tangerine Ugli fruit Watermelon*	3 Atemoya Banana Breadfruit Canistel Casaba melon Cherimoya Crenshaw melon Jaboticaba Jackfruit Honeydew melon Mamey Mango Mangosteen Papaya Persian melon Plantain Rambutan Sapodilla Sapote Soursop

Source: University of California (1997). Reprinted with permission.
Note: Ethylene level should be kept below 1 ppm in all storage areas.

*Products marked with an asterisk are sensitive to ethylene damage.
[†]Recommended storage temperatures = 36 to 39°F.

Group 1A is vegetables, many of which are wilting- and ethylene-sensitive, and 1B is low-temperature fruits that rarely need high humidity and mostly do not produce ethylene. Group 2, at 45 to 50°F, has citrus and subtropical fruits and many fruit-type vegetables. Group 3, at 55 to 65°F, contains some common root vegetables, winter (hard-rind) squashes, and most of the tropical fruits and melons. Group 3 conditions are also used for ripening products such as pears and stone fruit.

Distribution centers should have separate cold rooms for each of the four combinations of temperature and humidity. Groups 1A and 1B can be combined if only three rooms are available, but fruits and vegetables should be separated in the room. Retail stores rarely have three temperature-controlled rooms; Group 3 products can be held in an air-conditioned corridor or preparation room.

Ethylene must be kept below 1 ppm in all storage rooms, particularly in Group 1A, which holds most of the ethylene-sensitive produce. Ethylene can be minimized by using separate storage rooms, night outside ventilation, and ethylene-absorbing equipment. Operators should consider using air curtains, flap doors, or other infiltration reduction devices whenever doors to cold rooms must be opened often or for prolonged periods.

Mature green tomatoes (for ripening) usually need separate, temperature-controlled rooms where 55 to 70°F with 85 to 90% rh can be maintained to delay or speed ripening as desired.

REFRIGERATED STORAGE CONSIDERATIONS

The refrigeration requirement of any storage facility must be based on peak refrigeration load, which usually occurs when outside temperatures are high and warm produce is being moved into the plant for cooling and storage. Peak load depends on the amount of commodity received each day, its temperature when is placed under refrigeration, its specific heat, and final temperature attained. Information on general aspects of cold storage design and operation can be found in Chapters 11, 13, and 14.

Sprout Inhibitors

Sprout inhibitors are used when cold storage facilities are lacking or if low temperatures might injure the vegetable or affect its processing quality. In-storage sprouting of onions, potatoes, and carrots can be inhibited by spraying the plants a few weeks before harvest with a solution of maleic hydrazide. Potatoes are also sprayed or dipped in a solution that inhibits sprouting.

Gamma irradiation suppresses sprouting of onions, sweet potatoes, and white potatoes at dosages of 0.05 to 0.15 kGy. Dosages above 0.15 kGy cause breakdown and increased decay in white potatoes (Kader et al. 1984).

Controlled- and Modified-Atmosphere Storage

Refrigeration is most effective in retarding respiration and lengthening storage life. For some products, reducing the O_2 level in the storage air and/or increasing the CO_2 level as a supplement to refrigeration can extend storage life. Careful control of the concentration of O_2 and CO_2 is essential. If all of the O_2 is used, produce will suffocate and may develop an alcoholic off-flavor in a few days. CO_2 given off in respiration or from dry ice may accumulate to injurious levels.

CA storage usually uses external generators, nitrogen, or dry ice to create desired atmospheres rather than using product respiration in a gastight room. MA packaging, on the other hand, can obtain its gas concentrations either by initial flushing of the package (usually polyethylene) or by product respiration, which reaches equilibrium concentrations within a few days in cold storage. More rapid equilibration is accomplished by minimizing the initial air volume in the package, or by starting with warm product in the package before cold storage. Polyethylene packaging is useful because it transmits CO_2 about 5 times as readily as it transmits O_2, thus allowing low (5%) O_2 levels to prevail with low CO_2 levels (2 to 3%), preventing injury. If higher CO_2 levels are desirable, then MA packaging with pinholes should be used, because holes transfer both gases equally well. Pinholes, however, have no temperature forgiveness: they transmit at the same rate even if refrigeration is lost, which leads to anaerobic conditions (very low O_2, high CO_2) in the package. Polyethylene is more forgiving of temperature abuse because its gas transmission rate increases with temperature (doubles with a 22°F rise), although less than respiration does (doubles with a 9°F rise). MA packaging preserves product by lowering its respiration rate and, equally important, by retaining its moisture with 95 to 100% rh in the package. Again, MA packaging using pinholes can cause excessive in-package condensate problems because, although the holes can transmit adequate O_2, they transmit much less water vapor than large permeable polyethylene surfaces (Moyls et al. 1998).

Table 1 in Chapter 11 lists CA and MA requirements and recommendations.

Hypobaric storage, or storage at reduced atmospheric pressure, is another supplement to refrigeration that involves principles similar to those in controlled-atmosphere storage. At atmospheric pressures 0.1 or 0.2 of normal, several kinds of produce have an extended storage life. The benefits are attributed both to the low O_2 level maintained and to continuous removal of ethylene, CO_2, and possibly other metabolically active gases. Their rates of production under hypobaric ventilation are also lower.

Positive hypobaric ventilation is required to achieve the desired low ethylene concentration within and around produce to retard ripening. The continuous flow of water-saturated air at a low pressure flushes away emanated gases and prevents weight loss.

Despite its benefits, cost makes hypobaric storage not commercially viable.

Injury

Chilling injury is caused by exposure to low but nonfreezing temperatures, often between 32 to 50°F. At these temperatures, vegetables are weakened because they are unable to carry on normal metabolic processes. Often, chilled vegetables look sound when removed from low temperatures. However, symptoms (pitting or other skin blemishes, internal discoloration, or failure to ripen) become evident in a few days at warmer temperatures (Morris and Platenius 1938). Vegetables that have been chilled may be particularly susceptible to decay. ***Alternaria* rot** is often severe on tomatoes, squash, peppers, and cantaloupes that have been chilled. Tomatoes that have been severely chilled usually ripen slowly and rot rapidly.

Both time and temperature are involved in chilling injury. Damage may occur in a short time if temperatures are considerably below the safe storage temperatures, but a product may withstand a few degrees in the danger zone for a longer time. Also, the effects of chilling are cumulative. Low temperatures in the field before harvest, in transit, and in storage all contribute to the total effects of chilling. A list of vegetables susceptible to chilling injury together with the symptoms and the lowest safe temperatures are shown in Table 3.

Table 4 Vegetables Susceptible to Chilling Injury at Moderately Low but Nonfreezing Temperatures

Commodity	Approximate Lowest Safe Temperature, °F	Character of Injury Between 32°F and Safe Temperature[a]
Beans (snap)	45	Pitting and russeting
Cucumbers	50	Pitting; watersoaked spots, decay
Eggplants	45	Surface scald; *Alternaria* rot
Melons		
Cantaloupes	a	Pitting; surface decay
Honeydew	45 to 50	Pitting; failure to ripen
Casaba	45 to 50	Pitting; surface decay
Crenshaw and Persian	45 to 50	Pitting; surface decay
Watermelons	40	Pitting; objectionable flavor
Okra	45	Discoloration; watersoaked areas; pitting; decay
Peppers, sweet	45	Sheet pitting, *Alternaria* rot on pods and calyxes
Potatoes	37 to 39	Mahogany browning (Chippewa and Sebago); sweetening[b]
Pumpkins and hard-shell squash	50	Decay, especially *Alternaria* rot
Sweet potatoes	55	Decay; pitting; internal discoloration
Tomatoes		
Ripe	50[a]	Watersoaking and softening; decay
Mature green	55	Poor color when ripe; *Alternaria* rot

Source: *USDA Agricultural Handbook* 66 (USDA 2004).
[a]See text.
[b]Often appears only after removal to warm temperatures, as in marketing.

A vegetable's **freezing point** (see Table 1) is the highest temperature at which ice crystal formation in the tissues has been recorded experimentally. Most vegetables have a freezing point between 28 and 31°F (Whiteman 1957). Different vegetables vary widely in susceptibility to freezing injury.

Tissues injured by freezing generally appear water-soaked. Even though some vegetables are somewhat tolerant of freezing, it is desirable to avoid freezing temperatures because they shorten storage life (Parsons and Day 1970).

To minimize damage, fresh commodities should not be handled while frozen. Fast thawing damages tissues, but very slow thawing (32 to 34°F), allows ice to remain in the tissues too long. Thawing at 39°F is suggested (Lutz 1936).

STORAGE OF VARIOUS VEGETABLES

In the following sections, the temperatures and relative humidity recommendations (shown in parentheses) are the optimum for maximum storage in the fresh condition. For short storage, higher temperatures may be satisfactory for some commodities. Temperature requirements represent commodity temperatures that should be maintained. Much of this information is taken from *USDA Agricultural Handbook* 66 (USDA 2004).

The quality of each lot of produce should be determined at the time of storage, and regular inspections should be made during storage. Such vigilance will permit early detection of disease so that the affected commodity can be moved out of storage before serious loss occurs.

Artichokes, Globe (32°F and 95 to 100% rh)

Globe artichokes are seldom stored, but for temporary holding a temperature of about 32°F is recommended. A high relative humidity (at least 95%) will help prevent wilting. This product should keep for 2 weeks in storage if buds are uninjured and wilting is prevented. Perforated polyethylene liners with twenty-three 1/4 in. holes/ft^2 help retard moisture loss. Hydrocooling or room cooling to 39°F on the day of harvest reduces deterioration.

Gray Mold Rot (*Botrytis*). The most common decay of harvested artichokes. Reddish brown to dark brown firm rot (see Table 5, Note 2). *Control*: Practice sanitation in the field. Refrigerate promptly.

Asparagus (32 to 36°F and 95 to 100% rh)

Asparagus deteriorates very rapidly at temperatures above 36°F and especially at room temperature. It loses sweetness, tenderness, and flavor, and decay develops later. If the storage period is 10 days or less, 32°F is recommended; asparagus is subject to chilling injury if held longer at this temperature. Asparagus is not ordinarily stored except temporarily, but at 36°F with a high relative humidity, it can be kept in salable condition for 3 weeks. However, after a long haul to market, even under refrigeration, it cannot be expected to keep longer than 1 to 2 weeks.

Asparagus should be cooled immediately after cutting. Hydrocooling is the usual method. During transit or storage, the butts of asparagus should be placed on some moist absorbent material to prevent loss of moisture and to maintain freshness of the spears. Sometimes asparagus bunches are set in shallow pans of water in storage.

Bacterial Soft Rot. Mushy, soft, watersoaked areas on tips and cut ends of asparagus (see Table 5, Note 1). *Control*: Avoid excessive bruising of tips; cool to 39°F.

Fusarium Rot. Watersoaked areas changing through yellow to brown, chiefly on asparagus tips; white to pink delicate mold. *Control*: Cool and ship at temperatures of 39°F or lower; handle promptly. Keep tips dry on way to market.

Phytophthora Rot. Large, watersoaked, or brownish lesions at the side of cut asparagus stalks. Lesions later are extensively shriveled. *Control*: Cool to 39°F. Maintain low transit temperatures. Market promptly.

Beans, Green or Snap (40 to 45°F and 95% rh)

Green or snap beans are probably best stored at 39 to 45°F, where they may keep for about 1 week. Even these recommended temperatures cause some chilling injury, but are best for short storage. When stored at 40°F or below for 3 to 6 days, surface pitting and russet discoloration may appear in a day or two following removal for marketing. The russeting will be especially noticeable in the centers of the containers where condensed moisture remains. *Contact icing is not recommended.* To prevent wilting, the relative humidity should be maintained at 90 to 95%. Beans for processing can be stored up to 10 days at 40°F. Containers of beans should be stacked to allow abundant air circulation; otherwise, the temperature may rise from the heat of respiration. Beans stored too long or at too high a temperature are subject to such decays as watery soft rot, slimy soft rot, and rhizopus rot.

Anthracnose (*Colletotrichum*). Circular or oval sunken spots; reddish brown around border with tan centers that frequently bear pink spore mounds. *Control*: Use resistant varieties. Plant disease-free seed. Refrigerate harvested beans promptly to 45°F.

Bacterial Blight (*Pseudomonas*). Small, greasy-appearing, watersoaked spots in pod. Older spots show red at the center, with watersoaked area surrounding and penetrating to the seed. *Control*: Keep the field sanitary. Use disease-free seed. Reduce the transit temperature promptly to 45°F.

Bacterial Soft Rot. See Table 5, Note 1.

Cottony Leak (*Pythium*). Pods with large, watersoaked spots, accompanied by abundant white cottony mold. *Control*: Sort out diseased pods in packing. Reduce transit temperature promptly to 45°F.

Freezing Injury. Slight freezing results in watersoaked mottling on surface of exposed pods. Severely frozen beans become completely watersoaked, limp, and dry out rapidly. Snap beans freeze at about 31°F.

Table 5 Notes on Diseases of General Occurrence

Note 1. Bacterial Soft Rot

Occurs on various vegetables as dark green, greasy or watersoaked soft spots and areas in leaves and stems. Soft, mushy, yellowish spots or soupy areas on stems, roots, and tubers of vegetables. Frequently accompanied by repulsive odor from secondary invaders. *Control*: Use sanitation practices during picking and packing to reduce contamination of harvested product. Use bactericides in postharvest wash treatments. Where possible avoid bruising and injury. Shade harvested produce in the field and reduce temperature promptly to 39°F or lower for commodities that can withstand low temperatures.

Note 2. Gray Mold Rot (*Botrytis*)

Decayed tissues are fairly firm to semiwatery. Watersoaked grayish tan to brownish in color. Gray mold and grayish brown, with conspicuous velvety spore masses. *Control*: Use sanitation practices during harvesting and packing. Avoid wounds as much as possible. Use storage and transit temperatures as low as otherwise practicable because decay progresses even at 32°F.

Note 3. Rhizopus Rot

Decayed tissues are watersoaked, leaky, and softer than those with gray mold rot or watery soft rot. Coarse mycelium and black spore heads develop under moist conditions. Nesting is common. *Control*: Insofar as possible, avoid injury and bruising. Reduce temperature promptly and maintain below 50°F for commodities that can withstand low temperatures.

Note 4. Watery Soft Rot (*Sclerotinia*)

Decayed tissues are watersoaked, with slightly pinkish or brownish tan borders; very soft and watery in later stages, accompanied by development of fine white to dingy cottony mold and black to brown mustard-seed-like bodies called *sclerotia*. Nesting is common. *Control*: Use sanitary practices in harvesting and packing. Cull out specimens with discolored or dead portions. Maintain temperature as low as practicable because rot progresses even at 32°F. Do not store commodities known to have watery soft rot at harvest time.

Russeting. Chestnut brown or rusty, diffuse surface discoloration on both sides of pods. *Control*: Permit no surface moisture on warm beans. Cool promptly; avoid temperatures lower than 45°F.

Soil Rot (*Rhizoctonia*). Large, reddish brown, sunken, decayed spots on pods. Cream-colored or brown mycelium and irregular chocolate-colored sclerotia may develop. Nesting is common. *Control*: Maintain transit temperatures of 45 to 50°F.

Watery Soft Rot. Presence of large black sclerotia in white mold helps to separate this from cottony leak (see Table 5, Note 4).

Beans, Lima (37 to 41°F and 95% rh)

For best quality, lima beans should not be stored. Unshelled lima beans can be stored for about 1 week at 37 to 41°F. They should be used promptly after removal because the pods discolor rapidly at room temperature. Even with only 1 week of refrigerated storage, the pods may develop rusty brown or brown specks, spots, and larger discolored areas that reduce salability of the beans. The pod discoloration will increase sharply during an additional day at 70°F following storage.

Beets (32°F and 98 to 100% rh)

Topped beets are subject to wilting because of the rapid loss of water when the storage atmosphere is too dry. When stored at 32°F with at least 95% rh, they should keep for 4 to 6 months. Before beets are stored, they should be topped and sorted to remove all beets that are diseased and showing mechanical injury. Bunch beets under the same conditions will keep 1 to 2 weeks. Contact icing is recommended. The containers should be well ventilated and stacked to allow air circulation.

Bacterial Soft Rot. See Table 5, Note 1.

Broccoli (32°F and 95 to 100% rh)

Italian or sprouting broccoli is highly perishable and is usually stored for only a brief period as needed for orderly marketing. Good salable condition, fresh green color, and vitamin C content are best maintained at 32°F. If it is in good condition and is stored with adequate air circulation and spacing between containers to avoid heating, broccoli should keep satisfactorily 10 to 14 days at 32°F. Longer storage is undesirable because leaves discolor, buds may drop off, and tissues soften. The respiration rate of freshly harvested broccoli is high comparable to that of asparagus, beans, and sweet corn. This high rate of respiration should be considered when storing broccoli, especially if it is held without package ice.

Brussels Sprouts (32°F and 95 to 100% rh)

Brussels sprouts can be stored in good condition for a maximum of 3 to 5 weeks at the recommended temperature of 32°F. Deterioration, yellowing of the sprouts, and discoloration of the stem are rapid at temperatures of 50°F and above. Rate of deterioration is twice as fast at 40°F as at 32°F. Loss of moisture through transpiration is rather high even if the relative humidity is kept at the recommended level. Film packaging is useful in preventing moisture loss. As with broccoli, sufficient air circulation and spacing between packages is desirable to allow good cooling and to prevent yellowing and decay.

Cabbage (32°F and 98 to 100% rh)

A large percentage of the late crop of cabbage is stored and sold during the winter and early spring, or until the new crop from southern states appears on the market. If it is stored under proper conditions, late cabbage should keep for 3 to 4 months. The longest-keeping varieties belong to the Danish class. Early crop cabbage, especially southern grown, has a limited storage life of 3 to 6 weeks.

Cabbage is successfully held in common storage in northern states, where a fairly uniform inside temperature of 32 to 36°F is maintained. An increasing quantity of cabbage is now held in mechanically refrigerated storage, but in some seasons its value does not justify the expense. Use of controlled atmospheres to supplement refrigeration aids quality retention. An atmosphere with 2.5 to 5% O_2 and 2.5 to 5% CO_2 can extend the storage life of late cabbage. Cabbage should not be stored with fruits emitting ethylene. Concentrations of 10 to 100 ppm of ethylene cause leaf abscission and loss of green color within 5 weeks at 32°F.

Pallet bins are used as both field and storage containers, so the cabbage requires no handling from the time of harvest until preparation for shipment. Before the heads are stored, all loose leaves should be trimmed away; only 3 to 6 tight wrapper leaves should be left on the head. Loose leaves interfere with ventilation between heads, which is essential for successful storage. When removed from storage, the heads should be trimmed again to remove loose and damaged leaves.

Chinese cabbage can be stored for 2 to 3 months at 32°F with 95 to 100% rh.

***Alternaria* Leaf Spot.** Small to large spots bearing brown to black mold. This spotting opens the way for other decays. *Control*: Avoid injuries. Maintain 32 to 34°F temperature in transit and storage. Practice sanitation in storage rooms.

Bacterial Soft Rot. This slimy decay frequently starts in the pith of a cut stem or in leaf spots caused by other organisms (see Table 5, Note 1).

Black Leaf Speck. Small, sharply sunken, brown or black specks occurring anywhere on outer leaves or on leaves throughout the head. Occurs under refrigeration in transit and storage, and in association with sharp temperature drops. *Control*: No effective control measures are known.

Freezing Injury. Heads frozen slightly may thaw without apparent injury. Freezing injury first appears as brown streaks in the stem, then as light brown watersoaking of the heart leaves and stem. *Control*: Prevent any extended exposure to temperatures below 31°F.

Watery Soft Rot. See Table 5, Note 4.

Carrots (32°F and 98 to 100% rh)

Carrots are best stored at 32°F with a very high relative humidity. Like beets, they are subject to rapid wilting if the humidity is low. For long storage, carrots should be topped and free from cuts and bruises. If they are in good condition when stored and promptly cooled after harvest, mature carrots should keep 5 to 9 months. Carrots lose moisture readily and wilting results. Humidity should be kept high, but condensation or dripping on the carrots should be avoided, since this causes decay.

Most carrots for the fresh market are not fully mature. Immature carrots are prepackaged in polyethylene bags either at the shipping point or in terminal markets. They are usually moved into marketing channels soon after harvest, but they can be stored for a short period to avoid a market glut. If the carrots are cooled quickly and all traces of leaf growth are removed, they can be held 4 to 6 weeks at 32°F.

In Texas, immature carrots are often stored in clean 50 lb burlap sacks. The sacks of carrots should be stacked in such a manner that at least one surface of each sack is in contact with top ice at all times. Top ice provides some of the necessary refrigeration and prevents dehydration. Bunched carrots may be stored 10 to 14 days at 32°F. Contact icing is recommended.

Bitterness in carrots, which may develop in storage, is due to abnormal metabolism caused by the ethylene emitted by apples, pears, and some other fruits and vegetables. It can also be caused by other sources such as internal combustion engines. Bitterness can be prevented by storing carrots away from products that give off ethylene.

Bacterial Soft Rot. See Table 5, Note 1.

Black Rot (*Stemphylium*). Fairly firm black decay at the crown, on the side, or at the tips of harvested roots. *Control*: Avoid bruising. Store at 32°F.

Crater Rot (*Rhizoctonia*). Circular brown craters with white to cream-colored mold in the center. Develops under high humidity in cold storage. *Control*: Field sanitation measures. Avoid surface moisture on storage roots.

Freezing Injury. Roots are flabby and on cutting show radial cracks in the flesh of the central part and tangential cracks in the outer part. *Control*: Prevent exposure to temperatures below 30°F.

Gray Mold Rot. See Table 5, Note 2.

Rhizopus Rot. See Table 5, Note 3.

Watery Soft Rot. See Table 5, Note 4.

Cauliflower (32°F and 95% rh)

Cauliflower may be stored for 3 to 4 weeks at 32°F with about 95% rh. Successful storage depends on retarding the aging of the head or curd, preventing decay marked by spotting or watersoaking of the white curd, and preventing yellowing and dropping of the leaves. When it is necessary to hold cauliflower temporarily out of cold storage, packing in crushed ice will aid in keeping it fresh. Freezing causes a grayish brown discoloration, softening of the curd, and a watersoaked condition. Affected tissues are rapidly invaded by soft rot bacteria.

Much of the cauliflower now marketed has closely trimmed leaves and is prepackaged in perforated cellophane overwraps that are then packed in fiberboard containers. Vacuum cooling is a fairly efficient method of cooling prepackaged cauliflower. In general, use of controlled atmospheres with cauliflower has not been promising. Atmospheres containing 5% CO_2 or higher are injurious to cauliflower, although the damage may not be apparent until after cooking.

Bacterial Soft Rot. See Table 5, Note 1.

Brown Rot (*Alternaria*). Brown or black spotting of the curd. *Control*: Use seed treatment and field spraying. Keep the curds dry. Maintain low transit temperatures. Store at 32°F.

Celery (32°F and 98 to 100% rh)

Celery is a relatively perishable commodity, and for storage of 1 to 2 months, it is essential that a commodity temperature of 32°F be maintained. The rh should be high enough to prevent wilting (98 to 100%). Considerable heat is given off because of respiration, and for this reason the stacks of crates should be separated and dunnage used to allow cold air circulation under and over the crates and between the bottom crates and the floor. Forced-air circulation should be provided; otherwise there may be a 3 to 4°F temperature differential between the top and the bottom of the room.

Celery can be cooled by forced-air cooling, hydrocooling, or vacuum cooling. Hydrocooling is the most common cooling method; temperatures should be brought to as near 32°F as possible. In practice, temperatures are reduced to 40 to 45°F. Vacuum cooling is widely used for celery packed in corrugated cartons for long-distance shipment.

Bacterial Soft Rot. See Table 5, Note 1.

Black Heart. Brown or black discoloration of tips or all of the heart leaves. Affected celery should not be stored because of rapid development of bacterial soft rot. *Control*: Good cultural practices with special attention to available calcium. Harvest promptly after celery is mature.

Early Blight (*Cercospora*). Circular pale yellow spots on leaflets. In advanced stages, spots coalesce and become brown to ashen gray. No spots develop in storage, but affected lots lose moisture and their fresh appearance. *Control*: Control early blight in the field by spraying or dusting.

Freezing Injury. Characteristic loosening of the epidermis is best demonstrated by twisting the leaf stem. Severe freezing causes celery to become limp and to dry out rapidly. Freezing may cause watery soft rot and bacterial soft rot. The freezing point is about 31°F.

Late Blight (*Septoria*). Small (1/8 in. or less), yellowish, indefinite spots in the leaflet and elongated spots on the leaf stalk bearing black fruiting bodies of pinpoint size on the surface and surrounding green tissue. Development of blight in storage is probably negligible, but it opens the way for storage decays. *Control*: Control late blight in the field with sanitary measures and fungicides. Store infected lots for short periods only.

Mosaic. Leaflets are mottled; the stalk shows brownish sunken streaks. Badly affected stalks shrivel. *Control*: Eradicate weeds that carry the virus. Grade out all discolored stalks at packing time.

Watery Soft Rot (Pink Rot). This is the principal decay of celery. It is often severe on field-frozen stock and on celery harvested after prolonged cool, moist weather. In early stages, it often has a pink color. *Control*: Grade out all discolored stalks at packing time. Storage at 32°F will retard but not prevent the disease.

Corn, Sweet (32°F and 95 to 98% rh)

Sweet corn is highly perishable and is seldom stored except to temporarily protect an excess supply. Corn, as it usually arrives on the market, should not be expected to keep for more than 4 to 8 days even in 32°F storage.

The sugar content, which so largely determines quality in corn and decreases rapidly at ordinary temperatures, decreases less rapidly if the corn is kept at about 32°F. The loss of sugar is about 4 times as rapid at 50°F as it is at 32°F.

Sweet corn should be cooled promptly after harvest. Usually, corn is hydrocooled, but vacuum cooling is also satisfactory if the corn is prewetted and top-iced after cooling. Where precooling facilities are not available, corn can be cooled with package and top ice. Sweet corn should not be handled in bulk unless it is copiously iced because of its tendency to heat throughout the pile. Sweet corn ears should be trimmed to remove most shank material before shipment to minimize moisture loss and prevent kernel denting. A loss of 2% moisture from sweet corn can result in objectionable kernel denting.

Cucumbers (50 to 55°F and 95% rh)

Cucumbers can be held only for short periods of 10 to 14 days at 50 to 55°F with a relative humidity of 95%. Cucumbers held at 45°F or below for longer periods develop surface pitting or dark-colored watery areas. These blemishes indicate chilling injury. Such areas soon become infected and decay rapidly when the cucumbers are moved to warmer temperatures. Slight chilling may develop in 2 days at 32°F and severe chilling injury within 6 days at 32°F. The susceptibility of cucumbers to chilling injury does not preclude their exposure to temperatures below 50°F for short intervals as long as they are used immediately after removal from cold storage. Chilling symptoms develop rapidly only at higher temperatures. Thus, 2 days at 32°F or 4 days at 39°F are both harmless. Waxing is of some value in reducing weight loss and in giving a brighter appearance. Shrink wrapping with polyethylene film can also prevent the loss of turgidity.

At temperatures of 50°F and above, cucumbers ripen rather rapidly as the green color changes to yellow. Ripening is accelerated if they are stored in the same room with ethylene-producing crops for more than a few hours.

Anthracnose (*Colletotrichum*). Circular, sunken, watersoaked spots that soon produce pink spore masses in the center. Later, the spots turn black. *Control*: Use disease-free seed and fungicidal applications in the field.

Bacterial Soft Rot. See Table 5, Note 1.

Bacterial Spot (*Pseudomonas*). Small, circular, watersoaked spots, later chalky or moist with gummy exudate. *Control*: If possible, avoid shipping infected cucumbers. Pack them dry and maintain temperatures as near optimum as practicable.

Black Rot (*Mycosphaerella*). Irregular, brownish, watersoaked spots of varying size, later nearly black. Black fruiting bodies are sometimes present. *Control*: Exclude infected cucumbers from the pack, if possible. Reduce carrying temperatures to about 50°F.

Chilling Injury. Numerous sunken and slightly watersoaked areas in the skin of cucumbers after removal from storage, found on cucumbers stored for longer than a week at temperatures below 45°F. *Control*: Store cucumbers at temperatures between 50 and 55°F, for no longer than 2 weeks.

Cottony Leak (*Pythium*). Large, greenish, watersoaked lesions. Luxuriant, white, cottony mold over wet decay. *Control*: Exclude infected cucumbers from the pack, if possible. Reduce carrying temperatures to about 50°F.

Freezing Injury. Large areas in cucumbers that are soft, flabby, watersoaked, and wrinkled, especially toward the stem end. *Control*: Prevent exposure to temperatures below 31°F.

Watery Soft Rot. See Table 5, Note 4.

Eggplants (46 to 54°F and 90 to 95% rh)

Eggplants are not adapted to long storage. They cannot be kept satisfactorily even at the optimum temperatures of 46 to 54°F for over a week and still retain good condition during retailing. Eggplants are subject to chilling injury at temperatures below 45°F. Surface scald or bronzing and pitting after sand scarring are symptoms of chilling injury. Eggplants that have been chilled are subject to decay by *Alternaria* when they are removed from storage. Exposure to ethylene for 2 or more days hastens deterioration.

Cottony Leak (*Pythium*). Decayed areas are large, bleached, discolored (tan), wrinkled, moist, and soft; later they exhibit abundant cottony mold. *Control*: Reduce carrying temperatures to about 46°F.

Fruit Rot (*Phomopsis*). Numerous, somewhat circular brown spots that later coalesce over much of the fruit with pycnidia dotting the older lesions. This is a very common decay of eggplant. *Control*: Use fungicide sprays in the field. Reduce carrying temperature to about 46°F. Move the fruit promptly if decay is evident.

Endive and Escarole (32°F and 95 to 100% rh)

Endive and escarole are leafy vegetables not adapted to long storage. Even at 32°F, which is considered the best storage temperature, they will not keep satisfactorily for more than 2 or 3 weeks. They should keep somewhat longer if they are stored with cracked ice in or around the packages. Some desirable blanching usually occurs in endive held in storage.

Garlic, Dry (32°F and 65 to 70% rh)

If it is in good condition and is well cured when stored, garlic should keep at 32°F for 6 to 7 months. Garlic cloves sprout most rapidly at 40 to 64°F; therefore, prolonged storage at this temperature should be avoided. In California, it is frequently put in common storage, where it can be held 3 to 4 months or sometimes longer if the building can be kept cool, dry, and well ventilated.

Blue Mold Rot (*Penicillium*). Soft, spongy, or powdery dry decay of cloves. Affected cloves finally break down completely into gray or tan powdery masses. *Control*: Prevent bruising; keep garlic dry.

Waxy Breakdown. Yellow or amber waxy translucent breakdown of the outside cloves. *Control*: No control measures have been developed.

Greens, Leafy (32°F and 95 to 100% rh)

Leafy greens such as collards, chard, and beet and turnip greens are very perishable and should be held as close to 32°F as possible. At this temperature, they can be held 10 to 14 days. They are commonly shipped with package and top ice to maintain freshness and are handled like spinach. Kale packed with polyethylene crate liners should keep at least 3 weeks at 32°F or 1 week at 40°F. Vitamin content and quality are retained better when wilting is prevented.

Lettuce (32°F and 95 to 100% rh)

Lettuce is highly perishable. To minimize deterioration, it requires a temperature as close to its freezing point as possible without actually freezing it. Lettuce will keep about twice as long at 32°F as at 37°F. If it is in good condition when stored, lettuce should keep 2 to 3 weeks at 32°F with a high relative humidity. Most lettuce is packed in cartons and vacuum-cooled to about 34 to 36°F soon after harvest. It should then be immediately loaded into refrigerated cars or trailers for shipment or placed in cold storage rooms for holding prior to shipping.

An increasing quantity of lettuce is shipped in modified atmospheres to aid quality retention. Modified atmospheres are a supplement to proper transit refrigeration but are not a substitute for refrigeration. Head lettuce is not tolerant of CO_2 and is injured by concentrations of 2 to 3% or higher. Romaine is injured by 10% CO_2, but not by 5% at 32°F. Leaf lettuce and chopped romaine tolerate 10% CO_2, and chopped, shredded iceberg lettuce tolerates 15% CO_2 (Gorny 1997).

Excess wrapper leaves are usually trimmed off before sale or use, so it is suggested that lettuce be trimmed to two wrapper leaves before packaging (rather than the usual five or six) to save space and weight. The extra wrapper leaves are not needed to maintain quality.

Bacterial Soft Rot. The most common cause of spoilage in transit and storage. Often, it starts on bruised leaves. This decay normally is the controlling factor in determining the storage life of lettuce and is much less serious at 32°F than at higher temperatures (see Table 5, Note 1).

Brown Stain. Lesions that are typically tan, brown, or even black and about 1/4 in. wide and 1/2 in. long with distinct margins that are darker than the slightly sunken centers. The margins give a halo effect. The lesions develop on head leaves just under the cap leaves. The heart and wrapper leaves are not affected. Brown stain is caused by CO_2 accumulation from normal product respiration in railcars or trailers. *Control*: Ventilate to keep CO_2 below 2% in transport vehicles by keeping one water drain open. Enclose bags of hydrated lime (in vehicles shipped under a modified atmosphere) to absorb CO_2.

Gray Mold. See Table 5, Note 2.

Pink Rib. Characterized by diffuse pink discoloration near the bases of the midribs of the outer head leaves. In heads with severe symptoms, all but the youngest head leaves may be pink and discoloration may reach into large veins. The cause has not been identified but shipment in low O_2 atmospheres at undesirably high temperatures (50°F) can accentuate the disease. It is most common in hard to overmature lettuce. *Control*: Store and ship lettuce at recommended low temperatures.

Russett Spotting. This occasionally causes serious losses. Small tan or rust-colored pitlike spots appear mostly on the midrib but possibly develop on other parts of leaves. Exposure to ethylene and to storage or transport temperatures above 37°F are the main causes of this disorder. Hard lettuce is more susceptible to it than firm lettuce. *Control*: Avoid storing or shipping lettuce with apples, pears, or other products that give off ethylene. Precool lettuce adequately to 34 to 37°F and refrigerate it continuously. Shipment in a low O_2 atmosphere (1 to 8%) allows effective control.

Rusty Brown Discoloration. A serious market disorder of western head lettuce; a diffuse discoloration which tends to follow the veins but also spreads to adjacent tissue. The disorder starts on the outer head leaves but in severe cases may affect all leaves. *Control*: There is no known control method.

Tipburn. Dead brown areas along the edges and tips of inner leaves. This is considered to be of field origin, but occasionally the severity of the disease increases after harvest. *Control*: Keep the affected stock well cooled and market it promptly after unloading to avoid secondary bacterial rots.

Watery Soft Rot. See Table 5, Note 4.

Melons

Persians should keep at 45 to 50°F for up to 2 weeks; **honeydews** for 2 to 3 weeks; and **casabas** for 4 to 6 weeks. These melons will definitely be injured in 8 days at temperatures as low as 32°F. Honeydews are usually given an 18 to 24 h ethylene treatment (5000 ppm) to obtain uniform ripening. Pulp temperature should be 70°F or above during treatment. Honeydews must be mature when harvested; immature melons fail to ripen even if treated with ethylene.

Cantaloupes harvested at the hard-ripe stage (less than full slip) can be held about 15 days at 36 to 39°F. Lower temperatures may cause chilling injury. Full-slip hard-ripe cantaloupes can be held for a maximum of 10 to 14 days at 32 to 36°F. They are more resistant to chilling injury. Cantaloupes are precooled by hydrocooling or forced-air cooling before loading or by top-icing after loading in railcars or trucks.

Watermelons are best stored at 50 to 60°F and should keep for 2 to 3 weeks. Watermelons decay less at 32°F than at 40°F, but they tend to become pitted and have an objectionable flavor after 1 week at 32°F. At low temperatures, they are subject to various symptoms of chilling injury (loss of flavor and fading of red color). Watermelons should be consumed within 2 to 3 weeks after harvest, primarily because of the gradual loss of crispness.

***Alternaria* Rot.** Irregular, circular, brownish spots, sometimes with concentric rings, later covered with black mold. Often found on melons that have been chilled. *Control*: Avoid chilling temperatures. If cold melons are to be held at room temperature, they should be so stacked that condensed moisture will evaporate readily. Market melons promptly.

Anthracnose (*Colletotrichum*). Numerous greenish, elevated spots with yellow centers, later sunken and covered with moist pink spore masses. *Control*: Apply recommended field control measures.

Chilling Injury. Honeydew and honeyball melons stored for 2 weeks or longer at temperatures of 32 to 34°F sometimes show large, irregular, water soaked, sticky areas in the rind. *Control*: Store melons at 45 to 50°F.

Cladosporium Rot. Small black shallow spots later covered with velvety green mold. On cantaloupes, this rot is evident on extensive shallow areas at the stem ends or at points of contact between melons and it can be rubbed off easily. *Control*: Control measures are the same as for *Alternaria* rot.

Fusarium Rot. Brown areas on white melons; white or pink mold over indefinite spots on green melons. Affected tissue is spongy and soft with white or pink mold. *Control*: Avoid mechanical injuries; reduce carrying temperatures to 45°F.

Phytophthora Rot. Brown slightly sunken areas; later water-soaked and covered with a wet, appressed, whitish mold. *Control*: Cull out the affected fruits during packing. Reduce carrying temperatures to 45°F.

Rhizopus Rot. The affected melon is soft but not soupy and leaky as it is in similar decay on other vegetables. Coarse fungus strands may be demonstrated in decayed tissue (see Table 5, Note 3).

Stem End Rot (*Diplodia*). Fairly firm brown decay usually starting at the stem end and affecting a large part of the watermelon. Black fruiting bodies develop later. *Control*: At the time of loading in cars, recut the stems and treat them with Bordeaux paste or another recommended fungicide.

Mushrooms (32°F and 95% rh)

Mushrooms are usually sold in a retail market within 24 to 48 h after harvesting. They keep in good salable condition at 32°F for 5 days, at 39°F for 2 days, and at 50°F or above for about 1 day. A rh of 95% is recommended during storage. While being transported or displayed for retail sale, mushrooms should be refrigerated. Deterioration is marked by brown discoloration of the surfaces, elongation of the stalks, and opening of the veils. Black stems and open veils are correlated with dehydration.

Controlled atmosphere storage can prolong the shelf life of mushrooms held at 50°F, if the O_2 concentration in the atmosphere is 9% or the CO_2 concentration is 25 to 50%.

Moisture-retentive film overwraps of caps usually help in reducing moisture loss.

Okra (45 to 50°F and 90 to 95% rh)

Okra deteriorates rapidly and is normally stored only briefly before marketing or processing. It has a very high respiration rate at warm temperatures. Okra in good condition can be kept satisfactorily for 7 to 10 days at 45 to 50°F. A relative humidity of 90 to 95% is desirable to prevent wilting. At temperatures below 45°F, okra is subject to chilling injury which is shown by surface discoloration, pitting, and decay. Holding okra for 3 days at 32°F may cause pitting. Contact or top ice causes water spotting in 2 or 3 days at all temperatures.

Fresh okra bruises easily; the damaged areas blacken within a few hours. A bleaching injury may also develop when okra is held in hampers for more than 24 h without refrigeration.

Onions (32°F and 65 to 70% rh)

A comparatively low relative humidity is essential in the successful storage of dry onions. However, humidities as high as 85% and forced-air circulation also have given satisfactory results. At higher humidities, at which most other vegetables keep best, onions develop root growth and decay; at too high a temperature, sprouting occurs. Storage at 32°F with 65 to 70% rh is recommended to keep them dormant.

Onions should be adequately cured in the field, in open sheds, or by artificial means before, or in, storage. The most common method of curing in northern areas is by forced ventilation in storage. Onions are considered cured when the necks are tight and the outer scales are dried until they rustle. If they are not cured, onions are likely to decay in storage.

Onions are stored in 50 lb bags, in crates, in pallet boxes that hold about 1000 lb of loose onions, or in bulk bins. Bags of onions are frequently stored on pallets. Bagged onions should be stacked to allow proper air circulation.

In the northern onion-growing states, onions of the globe type are generally held in common storage because average winter temperatures are sufficiently low. They should not be held after early March unless they have been treated with maleic hydrazide in the field to reduce sprout growth.

Refrigerated storage is often used to hold onions for marketing in late spring. Onions to be held in cold storage should be placed there immediately after curing. A temperature of 32°F will keep onions dormant and reasonably free from decay, provided the onions are sound and well cured when stored. Sprout growth indicates too high a storage temperature, poorly cured bulbs, or immature bulbs. Root growth indicates the relative humidity is too high.

Globe onions can be held for 6 to 8 months at 32°F. Mild or Bermuda types can usually be held at 32°F for only 1 to 2 months. Spanish onions are often stored; if well matured, they can be held at 32°F at least until January or February. In California, onions of the sweet Spanish type are held at 32°F until April or May.

Onions are damaged by freezing, which appears as watersoaking of the scales when cut after thawing. If allowed to thaw slowly and without handling, onions that have been slightly frozen may recover with little perceptible injury. When onions are removed from storage in warm weather, they may sweat due to condensation of moisture. This may favor decay. Warming onions gradually (for example to 50°F over 2 to 3 days) with good air movement should avoid the difficulty. Onions should not be stored with other products that tend to absorb odors.

Onion sets require practically the same temperature and humidity conditions as onions, but because they are smaller in size they tend to pack more solidly. They are handled in approximately 25 lb bags and should be stacked to allow the maximum air circulation.

Green onions (scallions) and green shallots are usually marketed promptly after harvest. They can be stored 3 to 4 weeks at 32°F with 95% rh. Crushed ice spread over the onions aids in supplying moisture. Packaging in polyethylene film also aids in preventing moisture loss. Storage life of green onions at 32°F can be extended to 8 weeks by (1) packaging them in perforated polyethylene bags or in waxed cartons and (2) holding them in a controlled atmosphere of 1% O_2 with 5% CO_2.

Ammonia Gas Discoloration. Exposure of onions to 1% ammonia in air for 24 h causes the surface of yellow onions to turn brown, red onions to turn deep metallic black, and white onions to turn greenish yellow. *Control*: Ventilate storage rooms as soon as possible after exposure.

Bacterial Soft Rot. This decay often affects one or more scales in the interior of the bulb. Decayed tissue is more mushy than gray mold rot (see Table 5, Note 1).

Black Mold (*Aspergillus*). Black powdery spore masses on the outermost scale or between outer scales. *Control*: Store onions at just above 32°F and at 65% rh.

Freezing Injury. A watersoaked, grayish yellow appearance of all the outer fleshy scales results from a slight freezing injury. All scales are affected and become flabby with severe injury. Opaque areas appear in affected scales. *Control*: Prevent exposure to 30°F temperatures and lower. Thaw frozen onions at 40°F.

Fusarium Bulb Rot. Semiwatery to dry decay progressing up the scales from the base. Decay is usually covered with dense, low-lying white to pinkish mold. *Control*: Do not store badly affected lots. Pull out infected bulbs in slightly affected lots. Store onions at 32°F.

Gray Mold Rot (*Botrytis*). This is the most common type of onion decay; it usually starts at the neck, affecting all scales equally. Decay often is pinkish (see Table 5, Note 2). *Control*: Cure onions thoroughly. Protect them from rain. Store them at just above 32°F.

Smudge. Black blotches or aggregations of minute black or dark green dots on the outer drying scales of white onions. Under moist conditions sunken yellow spots develop on fleshy scales. *Control*: Protect onions from rain after harvest. Store them just above 32°F.

Translucent Scale. The outer 2 or 3 scales are gray and water-soaked, as in freezing. The entire scale may not be discolored; no opaque area is noticeable. Sometimes translucent scale is found in the field. *Control*: No control is known. Store onions at 32°F after curing.

Parsley (32°F and 95 to 100% rh)

Parsley should keep 1 to 2.5 months at 32°F and for a somewhat shorter period at 36 to 39°F. High humidity is essential to prevent desiccation. Package icing is often beneficial.

Bacterial Soft Rot. See Table 5, Note 1.

Watery Soft Rot. See Table 5, Note 4.

Parsnips (32°F and 98 to 100% rh)

Topped parsnips have similar storage requirements to topped carrots and should keep for 2 to 6 months at 32°F. Parsnips held at 32 to 34°F for 2 weeks after harvest attain a sweetness and high quality equal to that of roots subjected to frosts for 2 months in the field. Ventilated polyethylene box or basket liners can aid in preventing moisture loss. Parsnips are not injured by slight freezing while in storage, but they should be protected from hard freezing. They should be handled with great care while frozen. The main storage problems with parsnips are decay, surface browning, and their tendency to shrivel. Refrigeration and high relative humidity will retard deterioration.

Bacterial Soft Rot. See Table 5, Note 1.

Canker (*Itersonilia* sp.). An organism enters through fine rootlets and through injuries. The surface of the infected parsnip is first brown to reddish; later, it turns black where a depressed canker is formed. *Control*: Follow the recommended field spray program. Practice crop rotation.

Gray Mold Rot (*Botrytis*). See Table 5, Note 2.

Watery Soft Rot. See Table 5, Note 4.

Peas, Green (32°F and 95 to 98% rh)

Green peas lose part of their sugar rapidly if they are not refrigerated promptly after harvest. They should keep in salable condition for 1 to 2 weeks at 32°F. Top icing is beneficial in maintaining freshness. Peas keep better unshelled than shelled.

Bacterial Soft Rot. See Table 5, Note 1.

Gray Mold Rot (*Botrytis*). See Table 5, Note 2.

Watery Soft Rot. See Table 5, Note 4.

Peas, Southern (40 to 41°F and 95% rh)

Freshly harvested southern peas at the mature-green stage should have a storage life of 6 to 8 days at 40 to 41°F with high relative humidity. Without refrigeration, they remain edible for only about 2 days, the pods yellowing in 3 days and showing extensive decay in 4 to 6 days.

Peppers, Dry Chili or Hot

Chili peppers, after drying to a moisture content of 10 to 15%, are stored in nonrefrigerated warehouses for 6 to 9 months. The moisture content is usually low enough to prevent fungus growth. A relative humidity of 60 to 70% is desirable. Polyethylene-lined bags are recommended to prevent changes in moisture content. Manufacturers of chili pepper products hold part of their raw material supply in cold storage at 32 to 50°F, but they prefer to grind the peppers as soon as possible and to store them in the manufactured form in airtight containers.

Peppers, Sweet (45 to 55°F and 90 to 95% rh)

Sweet peppers can be stored for a maximum of 2 to 3 weeks at 45 to 55°F. They are subjected to chilling injury if they are stored at temperatures below 45°F. The symptoms of this injury are surface pitting and discoloration near the calyx, which develops a few hours after removal from storage. At temperatures of 32 to 36°F peppers usually develop pitting in a few days. When stored at temperatures above 55°F, ripening (red color) and decay develop rapidly. Rapid cooling of harvested sweet peppers is essential in reducing marketing losses. It can be done by forced-air cooling, hydrocooling, or vacuum cooling. Forced air cooling is the preferred method. Peppers are often waxed commercially, which reduces in-transit chafing and moisture loss.

Bacterial Soft Rot. See Table 5, Note 1.

Freezing Injury. The outer wall is soft, flabby, watersoaked, and dark green in color. The core and seeds turn brown with severe freezing. Sweet peppers freeze at about 31°F.

Gray Mold Rot (*Botrytis*). See Table 5, Note 2.

Rhizopus Rot. See Table 5, Note 3.

Potatoes (Temperature, see following; 90 to 95% rh)

The proper potato storage environment will promote the most rapid healing of bruises and cuts, reduce rot penetration to a minimum, allow the least weight and other storage losses to occur, and reduce to a minimum the deleterious quality changes that might occur during storage.

Early-crop potatoes are usually stored only during congested periods. They are more perishable and do not keep as well or as long as late-crop tubers. Refrigerated storage at 40°F, following a curing period of 4 or 5 days at 70°F, is recommended; they also can be stored for about 2 months at 50°F without curing. If early-crop potatoes are to be used for chipping or French frying, storage at 70°F is recommended. Holding these potatoes in cold storage (even at moderate temperatures of 50 to 55°F) for only a few days causes excessive accumulation of reducing sugars, which results in production of dark-colored chips.

Late-crop potatoes produced in the northern half of the United States are usually stored. The greater part of the crop is held in nonrefrigerated commercial and farm storages, but some potatoes are held in refrigerated storages. Potatoes in nonrefrigerated storages are usually held in bulk bins 8 to 20 ft deep. Shallower bins are used in milder climates. Some potatoes are stored in pallet boxes. In refrigerated warehouses, potatoes can be stored in sacks, pallet boxes, or bulk.

Late-crop potatoes should be cured immediately after harvest by being held at 50 to 61°F and high relative humidity for about 10 to 14 days to permit suberization and wound periderm formation (healing of cuts and bruises). If properly cured, they should keep in sound dormant condition at 38 to 40°F with 95% rh for 5 to 8 months. A lower temperature is not desirable, except in the case of seed stock for late planting. For this purpose, 37°F is best. At 37°F or below, Irish potatoes tend to become sweet. For ordinary table use, potatoes stored at 39°F are satisfactory, but they will probably be unsatisfactory for chipping or French frying unless desugared or conditioned at about 64 to 70°F for 1 to 3 weeks before use. However, conditioning may be costly and good results are often uncertain.

Potatoes will remain dormant at 50°F for 2 to 4 months; since tubers from this temperature are more desirable for both table use and processing than those from 40°F, late-crop potatoes intended for use within 4 months should be stored at 50°F and those for later use at 40°F. All potatoes should be stored in the dark to prevent greening.

A storage temperature of 50 to 55°F is recommended for most cultivars of potatoes intended for chip manufacture. At these temperatures, they usually remain in satisfactory condition if their reducing-sugar content is low enough when they are initially stored. Storage at 61 to 64°F is less desirable because shrinkage, internal and external sprout growth, and decay are greater at these temperatures than at 50 to 55°F. Russet Burbank potatoes for table stock or for chipping are stored at 45°F with 95% rh.

Potatoes usually do not sprout until 2 to 3 months after harvest, even at 50 to 61°F. However, after 2 to 3 months of storage, sprouting occurs in potatoes stored at temperatures above 39°F and particularly at temperatures around 61°F. Although limited sprouting does not affect potatoes for food purposes, badly sprouted stock shrivels and is difficult to handle and market.

Certain growth-regulating chemicals have been approved by the U.S. Food and Drug Administration to control or reduce sprouting on potatoes. Potatoes treated with chemical sprout inhibitors should not be stored in the same warehouse with seed potatoes. If 90% rh or slightly higher is maintained, potatoes will have the best quality and the least amount of shrinkage. Cunningham et al. (1971) recommend 95% rh or higher for late-crop potatoes. Condensation on the ceiling and resultant moisture drip is sometimes a problem when very high humidity is maintained.

Ventilation or air circulation in potato storage is needed to provide and maintain optimum temperature and relative humidity throughout storage and the tubers it contains. In northern states, where average outdoor temperatures during storage are low, little circulation or ventilation is needed. Shell or perimeter circulation is extensively used in these areas for seed and table stock potatoes. Forced circulation through the potatoes is required for the higher-temperature storage of processing potatoes and for table and seed stock in the warmer parts of the late-crop area. Rapid air circulation may lower the relative humidity of the air immediately surrounding the potatoes; it is conducive to drying and weight loss, which may be desirable if there are disease problems but undesirable with sound potatoes because of increased shrinkage. For late-crop Idaho potatoes, a uniform airflow, which does not have to be continuous, of 0.0065 cfm/lb is recommended. With this ventilation, Russet Burbank potatoes stored at 45°F with 95% rh should keep in good condition for 10 months or longer.

Potatoes can absorb objectionable flavors or odors from fruit, nuts, eggs, dairy products, and volatile chemicals; consequently they should not be kept in the same room.

Bacterial Ring Rot. Yellow, soft, cheesy decay of the thin layer of tissue in the vascular ring. The outer 0.25 in. of the tuber and the inner part may appear normal. *Control*: Use disease-free seed; store promptly at 40°F.

Bacterial Soft Rot. See Table 5, Note 1. This disease probably causes more loss in the early and intermediate crops than all other potato diseases combined.

Freezing Injury. If frozen solidly, tubers become soft and cream-colored and exude moisture. Slightly frozen tubers show darkening of the vascular ring and dull gray to black areas in the flesh. Potatoes freeze at about 29 to 31°F.

Fusarium Rot. Brown to black, spongy, and fairly dry; white or pink mold inside cavities of stored potatoes. *Control*: Avoid cutting and bruising during harvesting. After proper curing, maintain well-ventilated storage at 40°F.

Late Blight (*Phytophthora*). Reddish brown to black granular discoloration of the outer 0.12 to 0.25 in. of tuber. The affected tissue is firm to rock hard. *Control*: Apply recommended fungicides in the field. Kill vines prior to harvesting tubers or keep tubers away from blighted vines at harvest. Keep them dry; store at 40°F; market promptly.

Leak (*Pythium*). A large, gray to black, moist, decayed area starting at bruises or the stem end of the tuber. The internal tissue is granular and cream-colored at first, turning through reddish brown to inky black. *Control*: Prevent bruising. Refrigerate tubers to 40°F and keep them dry.

Mahogany Browning. Reddish brown patches or blotches in the flesh of tubers. Chippewa and Katahdin varieties are most susceptible. This differs from flesh discoloration caused by freezing in being reddish brown instead of gray. *Control*: Store at 40°F or above because lower temperatures cause the discoloration.

Net Necrosis. Dark brown vascular ring and vascular netting of the flesh, most prominent at the stem end, but extends well toward the bud end; increases during storage. *Control*: Reduce storage temperature promptly to 40°F; the infected tubers show symptoms earlier at higher temperatures.

Scald and Surface Discoloration. On early potatoes, this appears as sunken injured areas; later, it turns black and sticky and is followed by bacterial rots. *Control*: Move potatoes promptly to market; cool to 40°F.

Southern Bacterial Wilt. Moist, sticky exudation from the vascular ring when the tuber is cut. Sometimes there is advanced mushy decay in the center of the tuber. *Control*: Avoid shipping infected tubers; market promptly.

Stem End Browning. Dark brown to black vascular tissue, occurs in streaks that extend 0.4 to 1 in. into the flesh from the stem end; develops during storage. *Control*: After curing, reduce the storage temperature promptly to 40°F. Higher temperatures allow rapid development in susceptible lots.

Tuber Rot (*Alternaria*). Black to purplish, slightly sunken, shallow, irregularly shaped lesions, 0.25 to 1 in. in diameter, developing during storage. *Control*: Apply recommended fungicides in the field. Keep the tubers away from blighted vines as much as possible at harvest. If the tubers are damp, inspect them frequently during storage and use forced-air ventilation to dry up excess moisture.

Pumpkins and Squash

Hard-shell winter squash, such as the Hubbards, can be successfully stored for 6 months or longer at 50 to 55°F with a relative humidity of 60 to 75%. Dry storage is needed for quality retention. All specimens should be well-matured, carefully handled, and free from injury or decay when stored. Hubbard and other dark-green-skinned squashes should not be stored near apples, as the ethylene from apples may cause the squashes' skin to turn orange-yellow. Most varieties of **pumpkins** do not keep in storage for as long as the usual storage varieties of squash. Such varieties as Connecticut Field and Cushaw do not keep well and cannot be kept in good condition for more than 2 to 3 months. **Acorn squash** can be stored satisfactorily for 5 to 8 weeks at 50°F. **Butternut squash** should keep at least 2 to 3 months at 50°F with 50% rh.

Summer squash, such as yellow crookneck and giant straightneck, are harvested at an immature stage for best quality. The skin is tender and these varieties are easily wounded and perishable. The storage temperature range for summer squash is 41 to 50°F with 95% rh. A temperature of 41°F is best for zucchini squash stored up to 2 weeks, since they are sensitive to chilling injury.

Black Rot (*Mycosphaerella*). Hard, dry, black decay, dotted with minute black pimplelike fruiting bodies that occurs at stem ends or sides of the fruit. *Control*: Avoid skin breaks on the fruit; handle promptly.

Dry Rots (*Alternaria*; *Cladosporium*; *Fusarium*). Small, deep, dry, decayed areas. The decayed portion is easily lifted out of the surrounding healthy tissue. The surface mold is low-growing and either greenish black or pinkish white in color. *Control*: Prevent skin breaks. Do not store hard shell squashes below 50°F.

Rhizopus Rot. See Table 5, Note 3.

Radishes (32°F and 95 to 100% rh)

After harvest, topped spring radishes should be precooled quickly, often by hydrocooling, to 41°F or below. If they are then packaged in polyethylene bags, radishes can usually be held 3 to 4 weeks at 32°F and for a somewhat shorter time at 40°F.

Bunched radishes with tops are more perishable. They can be stored at 32°F and 95 to 100% rh for 1 to 2 weeks.

Rhubarb (32°F and 95% rh)

Fresh rhubarb stalks wilt and decay rapidly. Rhubarb in good condition can be stored 2 to 4 weeks at 32°F with a 95% rh or above. Moisture loss during holding or storage can be minimized by using nonsealed polyethylene crate liners or by film wrapping consumer size bunches. Removing and discarding leaf blades at harvest is desirable, because it not only reduces the possibility of decay and weight loss but also reduces shipping weight and package size by one-third. Rhubarb is usually marketed with about 3/8 in. of the leaf blade attached to the petiole. Splitting of the petiole is more serious if the entire leaf is removed.

Fresh rhubarb cut into 1 in. pieces and packaged in 1 lb perforated polyethylene bags can be held 2 to 3 weeks at 32°F with high relative humidity. Splitting of cut ends and curling of these pieces in film bags may be a problem if marketing is at warm temperatures.

Gray Mold Rot (*Botrytis*). Grayish, smoke-colored growths and grayish brown spore masses on stalks. *Control*: Refrigerate to 32°F.

Phytophthora Rots (*Phytophthora*). Watery, greenish brown, sunken lesions starting at the base of the leafstalk and causing brown decay. *Control*: Decay is retarded with transit and storage temperatures below 40°F.

Rutabagas (32°F and 98 to 100% rh)

Rutabagas lose moisture and shrivel readily if they are not stored under high humidity conditions. A hot paraffin wax coating, often given to rutabagas, is effective in preventing wilting and loss of weight; it also slightly improves appearance. Too heavy a wax coating may produce severe injury from internal breakdown caused by suboxidation. Rutabagas in good condition, when stored, should keep 4 to 6 months at 32°F.

Freezing Injury. Rare, because the commodity can stand slight freezing without injury. Severe freezing causes watersoaking and light browning of the flesh, a mustard odor, and fermentation. *Control*: Prevent repeated slight freezing or severe freezing.

Gray Mold Rot (*Botrytis*). See Table 5, Note 2.

Spinach (32°F and 95 to 98% rh)

Spinach is very perishable and can be stored for only short periods of 10 to 14 days at 32°F with 95 to 98% rh. It will deteriorate rapidly at higher temperatures. Spinach is commercially vacuum cooled and forced-air cooled. If it is thoroughly cooled, it can be held for 10 to 14 days at 32°F without the addition of any package ice prior to storage. When precooling facilities are not available, crushed ice should be placed in each package to provide rapid cooling and to take care of the heat of respiration. Top ice is also beneficial.

Bacterial Soft Rot. See Table 5, Note 1.

Downy Mildew (*Peronospora*). A field disease, commonly found at the marketing stage as pale yellow irregular areas in the leaves. Downy gray mold is present on the lower surface. *Control*: Control it in the field; market promptly.

White Rust (*Albugo*). Slight yellowing of areas in the leaf above white blisterlike pustules filled with white masses of spores. *Control*: Control it in the field.

Sweet Potatoes (55 to 60°F, 85 to 90% rh)

Most sweet potatoes are stored in nonrefrigerated commercial or farm storages. Preliminary curing at 84°F and 90 to 95% rh for 4 to 7 days is essential in the healing of injuries received in harvesting and handling and in preventing the entrance of decay organisms. After curing, the temperature should be reduced to 55 to 61°F, usually by ventilating the storage, and the relative humidity should be retained at 85 to 90%. Most varieties will keep satisfactorily for 4 to 7 months under these conditions. Weight loss of 2 to 6% can be expected during curing and about 2% a month during subsequent storage.

Usually, sweet potatoes will not keep satisfactorily if they have been subjected to excessively wet soil conditions just before harvest or chilled before or after harvest by exposure to temperatures of 50°F or below. Short periods at temperatures as low as 50°F need not cause alarm; but after a few days at lower temperatures, sweet potatoes may develop discoloration of the flesh, internal breakdown, increased susceptibility to decay, and off-flavors when cooked.

Temperatures above 61°F stimulate development of sprouting (especially at high humidities), pithiness, and internal cork (a virus disease). Refrigeration is frequently used in large sweet potato storages to extend the marketing season into warm weather when ventilation will not maintain low enough temperatures.

Sweet potatoes are usually stored in slatted crates or bushel baskets. Palletization of crates and use of pallet boxes facilitates handling. Sweet potatoes are usually washed and graded and are sometimes waxed before being shipped to market. They may be treated with a fungicide to reduce decay during marketing.

Black Rot (*Ceratocystis*). Greenish black decay, frequently fairly shallow, and sometimes circular in outline at the surface. *Control*: Follow recommended field and postharvest control measures. Heat treatment of seed roots at 106 to 109°F for 24 h will prevent development of black rot.

Chilling Injury. Brown tinged with black discolored areas scattered or associated with vasculars. The interior becomes pithy. Chilling injury is often produced by exposure to lower temperatures for only a few days. Uncured roots are more sensitive than cured ones. *Control*: Store sweet potatoes at 55 to 60°F.

Freezing Injury. Soft, leaky condition of the flesh. The outer layer of the potato is dark brown. *Control*: Do not subject potatoes to low temperatures; sweet potatoes may freeze at 30°F.

Rhizopus Rot. See Table 5, Note 3. *Control*: Cure potatoes for 4 to 7 days at 84°F before storage. Follow recommended field and postharvest control measures.

Tomatoes (Mature Green, 55 to 70°F; Ripe, 50°F; 90 to 95% rh)

Mature green tomatoes cannot be successfully stored at temperatures that greatly delay ripening, even at a temperature of 55°F, which is considered to be a nonchilling temperature. Tomatoes held for 2 weeks or longer at 55°F may develop an abnormal amount of decay and fail to reach as intense a red color as tomatoes ripened promptly at 64 to 70°F. Temperatures of 68 to 72°F, and a relative humidity of 90 to 95% are probably used most extensively in commercial ripening of mature green tomatoes. At temperatures above 70°F, decay is generally increased. A temperature range of 57 to 61°F is probably the most desirable for slowing ripening without increasing decay problems. At this temperature, the more mature fruit will ripen enough to be packaged for retailing in 7 to 14 days. Tomatoes should be kept out of cold, wet rooms because, in addition to potential chilling injury, extended refrigeration damages the ability of fruit to develop desirable fresh tomato flavor.

Ethylene gas is sometimes used to hasten and give more even ripening to mature green tomatoes. In ripening rooms, a concentration of one part ethylene per 5000 parts of air daily for 2 to 4 days will usually shorten the ripening period by about 2 days at 68 to 72°F. Some tomatoes are gassed with ethylene in loaded railcars prior to shipping. Adding ethylene has little or no effect on tomatoes just before or after they have started to turn pink. Tomatoes themselves give off considerable ethylene as they ripen. Interest is increasing in the commercial use of low O_2 atmospheres of 3 to 5% during storage or transport to retard ripening and decay.

Storage temperatures below 50°F are especially harmful to mature green tomatoes; these chilling temperatures make the fruit susceptible to *Alternaria* decay during subsequent ripening. Increased decay during ripening occurs following 6 days' exposure to 32°F, or 9 days at 39°F (see Table 2).

Firm ripe tomatoes may be held at 45 to 50°F with a relative humidity of 85 to 90% overnight or over a holiday or weekend. Tomatoes showing 50 to 75% of the surface colored (the usual ripeness when packed for retailing) cannot be successfully stored for more than 1 week and be expected to have a normal shelf life during retailing. Such fruits should also be held at 45 to 50°F and 85 to 90% rh. A storage temperature of 50 to 55°F is recommended for pink-red to firm red tomatoes raised in greenhouses.

When it is necessary to hold firm ripe tomatoes for the longest possible time, consistent with immediate consumption on removal from storage, such as on board a ship or for an overseas military base, they can be held at 32 to 36°F for up to 3 weeks, with some loss in quality. Mature green, turning, or pink tomatoes should be ripened before storage at this low temperature.

***Alternaria* Rot.** Decayed area is brown to black, with or without a definite margin. Lesions are firm; rot extends into the flesh of the fruit. Dense, velvety, olive green or black spore masses frequently grow over affected surfaces. *Control*: Avoid mechanical injuries at packing time. Avoid temperatures below 50°F in green fruit.

Bacterial Soft Rot. See Table 5, Note 1.

Cladosporium Rot. Thin, brownish blemishes or black shiny spots of shallow decay, later covered by green, velvety mold. *Control*: Take care in harvesting and packing. Ship high quality tomatoes free of field chilling injury under protective services that will provide temperatures of 55 to 68°F.

Late Blight Rot (*Phytophthora*). Greenish brown to brown, roughened areas with a rusty tan margin. *Control*: Apply recommended field control measures. Cull tomatoes carefully before packing.

Phoma Rot. Slightly sunken, moderately penetrating, black areas at the edge of the stem scar and elsewhere on the fruit. Black pimplelike fruiting bodies develop later. Decayed tissues are firm and brown to black in color. Phoma rot is found in eastern-grown tomatoes. *Control*: Apply field control measures. Exercise care in harvesting and packing. Avoid temperatures below 55°F.

Rhizopus Rot. See Table 5, Note 3.

Soil Rot (*Rhizoctonia*). Small circular brown spots, frequently with concentric ring markings; later, large, brown, and fairly firm lesions. In advanced stages, under warm conditions, cream-colored or brown mycelium and irregular sclerotia may develop. *Control*: Before packing, sort out tomatoes with early lesions if the disease is prevalent.

Turnips (32°F and 95% rh)

Turnips in good condition can be expected to keep 4 to 5 months at 32°F with 90 to 95% rh. At higher temperatures (41°F and above), decay will develop much more rapidly than at 32°F. Injured or bruised turnips should not be stored. Store turnips in slatted crates or bins and allow good circulation around containers.

Dipping turnips in hot melted paraffin wax gives them a glossy appearance and is of some value in reducing moisture loss during handling. However, waxing is primarily to aid in marketing and is not recommended prior to long-term storage.

Turnip greens are usually stored for only short periods (10 to 14 days). They should keep about as well as spinach at 32°F with crushed ice in the packages.

REFERENCES

Cunningham, H.H., M.V. Zaehringer, and W.C. Sparks. 1971. Storage temperature for maintenance of internal quality in Idaho Russet Burbank potatoes. *American Potato Journal* 48:320.

Gorny, J.R. 1997. Fresh-cut fruits and vegetables map. *Proceedings of the International Controlled Atmosphere Research Conference,* University of California, Davis, vol. 5.

Hardenburg, R.E. 1971. Effect of in-package environment on keeping quality of fruits and vegetables. *HortScience* 6(3):198.

Hardenburg, R.E., H. Findlen, and H.W. Hruschka. 1959. Waxing potatoes—Its effect on weight loss, shrivelling, decay, and appearance. *American Potato Journal* 36:434.

Harvey, J.M. 1965. Nitrogen—Its strategic role in produce freshness. *Produce Marketing* 8(7):17.

Isenberg, F.M. and R.M. Sayles. 1969. Modified atmosphere storage of Danish cabbage. *Journal of the American Society for Horticultural Science* 94(4):447.

Kader, A.A., W.J. Lipton, H.J. Reitz, D.W. Smith, E.W. Tilton, and W.M. Urbain. 1984. *Irradiation of plant products: Comments from CAST 1984-1*. Council of Agricultural Science and Technology, Ames, IA.

Lipton, W.J. 1968. Effect of temperature on asparagus quality. *Proceedings of the Conference on Transportation of Perishables*, Davis, CA, p. 147.

Lutz, J.M. 1936. The influences of rate of thawing on freezing injury of apples, potatoes and onions. *Proceedings of the American Society for Horticultural Science* 33:227.

Morris, L.L. and H. Platenius. 1938. Low temperature injury to certain vegetables after harvest. *Proceedings of the American Society for Horticultural Science* 36:609.

Moyls, A.L., D.-L. McKenzie, R.P. Hocking, P.M.A. Toivonon, P. Delaquis, B. Girard, and G. Mazza. 1998. Variability in O_2, CO_2 and H_2O transmission rates among commercial polyethlene films for modified atmosphere packaging. *Transactions of the American Society of Agricultural Engineers*, 41(5):1441-1446.

Parsons, C.S. and R.E. Anderson. 1970. Progress on controlled-atmosphere storage of tomatoes, peaches and nectarines. *United Fresh Fruit and Vegetables Association Yearbook* 175.

Parsons, C.S. and R.H. Day. 1970. Freezing injury of root crops—Beets, carrots, parsnips, radishes, and turnips. *USDA Marketing Research Report* 866. U.S. Department of Agriculture, Washington, D.C.

Redit, W.H. 1969. Protection of rail shipments of fruits and vegetables. *USDA Handbook* 195. U.S. Department of Agriculture, Washington, D.C.

Thompson, J., P. Brecht, R.T. Hinsch, and A.A. Kader. 2000. Marine container transport of chilled perishable produce. *Publication* 21595, University of California, Division of Agriculture and Natural Resources.

Whiteman, T.M. 1957. Freezing points of fruits, vegetables and florist stocks. *USDA Marketing Research Report* 196 (December).

University of California. 1997. Table on compatible fresh fruits and vegetables during 7 day storage. *Publication* 21560. Division of Agriculture and Natural Resources.

USDA. 2000. Protecting perishable foods during transport by truck. *USDA Handbook* 669. U.S. Department of Agriculture, Transportation and Marketing Division, Washington, D.C.

USDA. 2004. The commercial storage of fruits, vegetables, and florist and nursery stocks. *USDA Agriculture Handbook* 66 (April 2004 draft). U.S. Department of Agriculture, Washington, D.C. http://www.ba.ars.usda.gov/hb66.

BIBLIOGRAPHY

Appleman, C.O., and J.M. Arthur. 1919. Carbohydrate metabolism in green sweet corn. *Journal of Agricultural Research* 17:137.

Bogardus, R.K., and J.M. Lutz. 1961. Maintaining the fresh quality in produce in wholesale warehouses. *Agricultural Marketing* 6(12):8.

Dewey, D.H., R.C. Herner, and D.R. Dilley. 1969. Controlled atmospheres for the storage and transport of horticultural crops. *Horticultural Report* 9, Michigan State University (July).

Lipton, W.J. 1965. Post-harvest responses of asparagus spears to high carbon dioxide and low oxygen atmospheres. *Proceedings of the American Society for Horticultural Science* 86:347.

Stewart, J.K. and M.J. Ceponis. 1968. Effects of transit temperatures and modified atmospheres on market quality of lettuce shipped in nitrogen-refrigerated and mechanically refrigerated trailers. *USDA Marketing Research Report* 832 (December).

Stewart, J.K., M.J. Ceponis, and L. Beraha. 1970. Modified atmosphere effects on the market quality of lettuce shipped by rail. *USDA Marketing Research Report* 863.

USDA. [Annual]. *Agricultural Statistics*. U.S. Department of Agriculture, Washington, D.C. http://www.usda.gov./nass/pubs/agstats.htm.

Watada, A.E. and L.L. Morris. 1966. Effect of chilling and nonchilling temperatures on snap bean fruits. *Proceedings of the American Society for Horticultural Science*, vol. 89, p. 368.

CHAPTER 25

FRUIT JUICE CONCENTRATES AND CHILLED JUICE PRODUCTS

CITRUS products, especially orange juice, comprise the largest percentage of the total volume of juices sold in the United States. Much of the technology used in processing noncitrus juices was developed from citrus processing.

ORANGE JUICE

ORANGE CONCENTRATE

Processed orange juice is sold in four principal forms:

1. Frozen concentrate (3-plus-1 concentration, in which three volumes of water are added to one volume of concentrate for reconstitution) in a variety of package sizes. These are the familiar retail products.
2. Concentrate in bulk at 65° Brix. This is an intermediate product that is bought and sold daily as futures on the Commodity Exchange. Most of this product will ultimately be sold in one of the other forms.
3. Chilled orange juice, which is ready to drink when poured from the carton. It is either reconstituted concentrate or nonconcentrated juice. By law, these two products must be labeled "from concentrate" or "not from concentrate."
4. Institutional or restaurant concentrates in special packaging at 4-plus-1 or higher concentrations.

After processing, frozen citrus concentrates in retail (3-plus-1) packages must be stored at 0°F. Highly concentrated bulk juice (65° Brix) may be satisfactorily stored at about 15°F. Chilled single-strength juices are stored at about 30 to 32°F.

Figure 1 shows a schematic flow diagram of citrus processing. The **Brix scale** is a hydrometer scale that indicates the percentage by weight of sugar in a solution at a specified temperature.

Selecting, Handling, and Processing Fresh Fruit

Selection. Fruit is selected for proper quality and maturity. Some fruit that is blemished but sound in quality, referred to as packinghouse eliminations, is used. A major portion of the crop is taken directly from the grove to the processing plant. To be mature, the fruit must have the proper Brix-acid ratio and the juice content and Brix must be above specified values. Fruit should be handled without delay because no real maturing occurs after harvesting; instead, the temperature and condition of the fruit determine the rate of deterioration. Citrus fruit is sufficiently rugged to withstand mechanical handling on conveyors, elevators, and belts, provided that the fruit is processed within a day or two after picking. Samples are taken mechanically as fruit enters the bins, and records of chemical analyses are maintained. Usually fruit from two or more bins is used simultaneously to improve uniformity. The fruit passes over inspection tables both before and after temporary storage in bins, and damaged or deteriorated fruit is removed.

Washing. Before juice extraction, the fruit is wetted by sprays. The wetting agent is dispensed onto the fruit as it travels over rotating brushes. Water sprays near the end of the washer unit rinse the fruit. A sanitizing solution may be used to sanitize conveyors and elevators.

Juice Extraction. Individual high-speed mechanical juice extractors handle from 300 to 700 pieces of fruit per minute. Some machines halve the fruit and ream or squeeze the juice from the half. Other machines insert a tube through the middle of the fruit and squeeze the juice through fine holes into the tube, at the same time sieving away the seeds and large pieces of membrane. After the juice has been extracted, it passes to finishers that remove the remaining seeds, pieces of peel, and excess cell or fruit membrane. In the past, this was a comparatively simple process involving one or two stages, but it has become complicated in recent years and varies extensively from plant to plant. Usually, one or two stages of screw-type finishers separate most of the pulp from the juice.

Pulp washing, in which soluble solids in separated segment and cell walls are recovered by countercurrent extraction with water, is permitted in Florida, provided that the resultant extract is not used in frozen orange concentrates. It may be used in other formulated products permitted by the Federal Standard of Identity for frozen concentrated orange juice.

The juice or pulp wash liquor from the finishers may require treatment in high-speed desludging centrifuges that remove suspended matter before transferring the juice to the evaporator. These centrifuges have peripheral discharges that open and close at intervals to discharge a thick suspension of pulp cells. This operation decreases the viscosity of juice in the evaporator, improves the efficiency of evaporation, and improves the appearance of the final product. Special means are used to classify orange pulp for inclusion in products with a high pulp content.

Heat Treatment. When frozen concentrated orange juice was first developed, minimal heat treatment was used to maintain optimum flavor. Such concentrate, if prepared from good, sound fruit, remains stable for a considerable time at 0°F and for nearly a year at 5°F. However, with large-scale production, it is not possible to ensure storage below 5°F. Concentrates originally of good quality tend to gel or clarify rapidly during storage. Heat treatment inactivates enzymes responsible for the development of these defects during improper storage. Earlier methods used steam-heated plate pasteurizers, but heat treatment is now almost universally included as an integral part of heat conservation in the evaporation process.

The preparation of this chapter is assigned to TC 10.9, Refrigeration Application for Foods and Beverages.

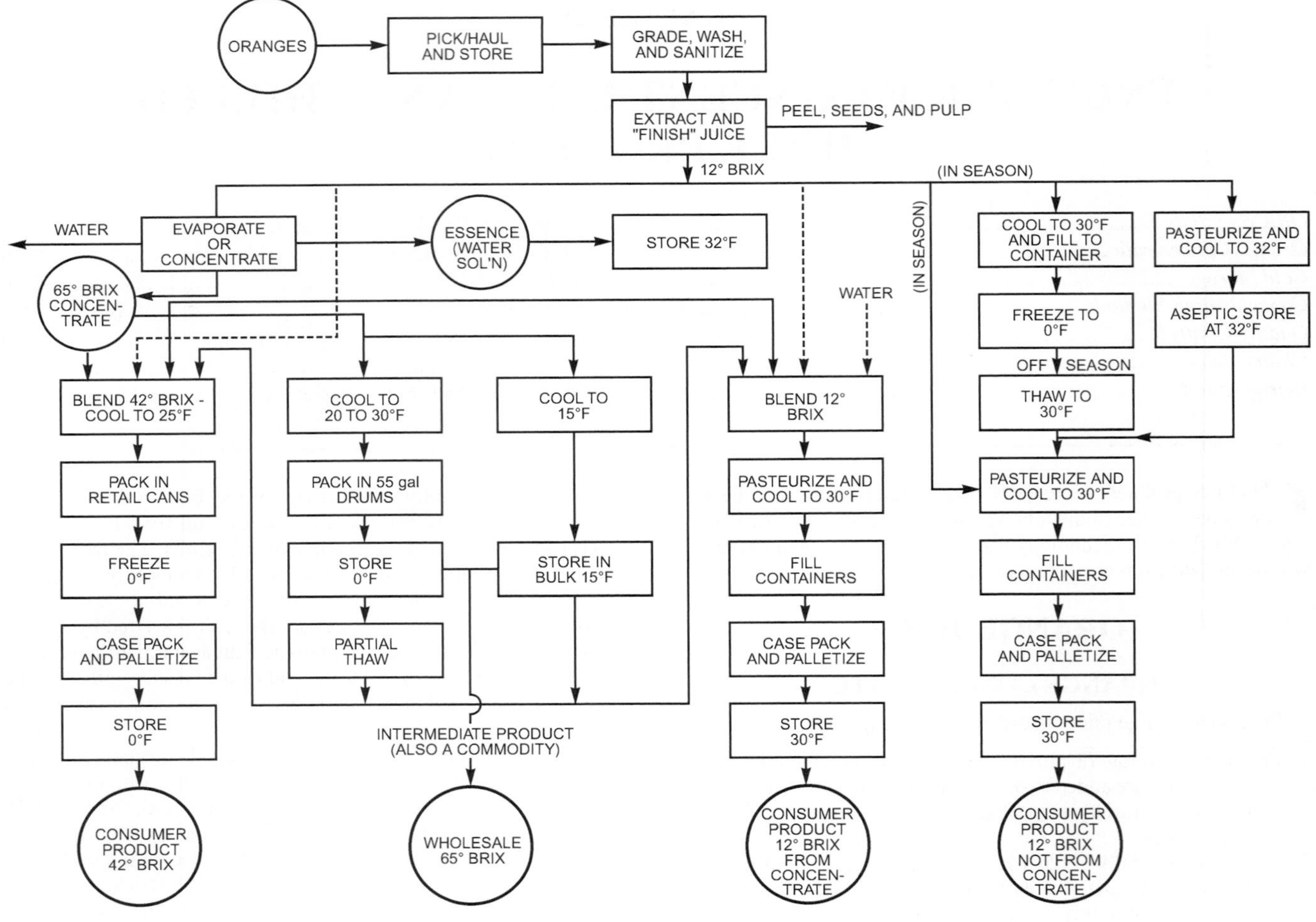

Fig. 1 Citrus Processing Schematic

Evaporation-Water Removal. Removal of water follows juice extraction and preparation. See the section on Concentration Methods for details.

Flavor Fortification. Originally, frozen concentrated orange juice was overconcentrated, and fresh juice (**cutback juice**) was added to reduce the concentration to the desired level and provide fresh flavor in the final product. This process is used extensively in frozen concentrates, but flavor levels cannot be standardized by this method alone.

Essential oil from orange peel does not by itself supply completely balanced fresh flavor, but it is now used extensively to control flavor intensity. Some **peel oil** is found in the cutback juice, but little remains in juice from the evaporator. Adding peel oil to the finished concentrate, at approximately 0.014% by volume in the reconstituted juice, has become standard practice.

Several variations have been introduced to supplement the use of peel oil and cutback fresh juice for flavor fortification. In most cases, vapor from the first stage of concentration is used to produce an **essence** that is restored to the final concentrate to enhance flavor. Although cutback fresh juice and cold-pressed peel oil are commonly added, most operations depend heavily on essence recovery and its incorporation into the final product.

Blending, Packaging, and Freezing. The final step in processing is the blending of concentrate with flavor-enhancing components such as cold-pressed orange oil and liquid essence. The 65° Brix concentrate is then reduced to about 42° Brix, the requirement for 3-plus-1 product, which reconstitutes to about 12° Brix when used by the consumer. The availability of cold-pressed oil and essence when no fresh fruit is available makes blending and packaging possible year-round.

Most retail product is packaged in fiber-foil cans, although aluminum and tinned steel cans may be used. Sizes range from 6 to 64 fluid ounces. Gable-top milk cartons and specialized large disposable plastic reservoirs for dispensers are also used.

Before filling, the product is cooled to about 25 to 35°F. This is usually done in the swept-surface cold-wall tank in which the concentrate is blended, but it may also be done in a bladed heat exchanger or, preferably, in a plate-type unit. Alternatively, the 65° Brix concentrate from the evaporator may be precooled (unblended) and packed in 55 gal open-top drums or delivered to large bulk storage tanks.

The filled cans pass through blast freezers, where their temperature is reduced to 0°F or below in 45 to 90 min. Cans are transported on link belts, and air handlers are arranged so that air at about –25°F is forced down through the loaded belt.

Thermal Properties. In calculating cooling and freezing requirements, concentrates may be considered to approximate sucrose solutions of the same concentration. In the range of 60 to 65° Brix, the specific heat is about 0.68 Btu/lb·°F; for 42° Brix, the value is about 0.74 Btu/lb·°F. Thermal conductivity in the liquid state is in the range of 0.17 to 0.18 Btu·ft/h·ft^2·°F. Regarding the heat of fusion for freezing tunnel design, the value should be 70 Btu/lb in the range of 30 to 0°F; however, experience has shown that a value of about 100 Btu/lb should be used. This allows for extraneous heat gains, coil defrosting, and so forth.

COLD STORAGE

Cold storage facilities for citrus processing can be divided into three categories according to temperature requirements of 0, 15, and 30°F.

Finished goods for retail and institutional markets are stored at 0°F in insulated, refrigerated buildings. Bulk 65° Brix product packed in drums is also stored at 0°F. The refrigerated buildings range from a few thousand square feet to a few acres. Other than the usual insulation requirements, two factors are critical to the design: (1) the vapor retarder outside the insulation must be as close to hermetic as possible, and (2) irrespective of the insulation in the floor, a heat supply must be installed beneath the insulation to maintain the temperature below the floor at about 32°F. Otherwise, the floor will ultimately heave from ice formation below the floor.

The 15°F buildings are used for bulk storage of 65° Brix concentrate. In a typical installation, a 15°F building would house several large stainless steel tanks, ranging from a few thousand to 200,000 gal each. At the stated temperature, the product is barely pumpable, requiring sanitary positive-displacement pumps. Because the temperature is virtually impossible to change after the product is in the tank, the product must be cooled to storage temperature before it is introduced to the tanks. Cooling usually occurs in a plate heat exchanger.

Finally, 30°F storage rooms are used largely for chilled single-strength juice in retail packages or for not-from-concentrate bulk tank systems. This product is discussed in the sections on Chilled Juice and Refrigeration.

CONCENTRATION METHODS

The three major methods for producing concentrates are (1) high-temperature, single pass, multiple-effect evaporators; (2) freeze concentration with mechanical separation; and (3) low-temperature, recirculatory, high-vacuum evaporators.

Thermally Accelerated Short-Time Evaporator (TASTE)

At present, TASTE is the standard evaporator of the citrus industry. This unit has also been successfully applied to grape, apple, and other juices. TASTEs produce at least 90% of all juice concentrate in the western hemisphere. The first cost of this unit is among the lowest of all alternatives, and it has excellent thermal efficiency. Flavor quality and storage stability of juice processed in this unit compare favorably with those of juice from alternative methods of concentration.

The TASTE includes all standard methods of heat conservation. It is multiple-effect, uses concurrent vapor for staged preheating, and flashes condensate to discharge condensed vapors at the lowest possible temperature. Residence time is minimal because all stages are single-pass. Total residence time is 2 to 8 min, varying directly with evaporator capacity.

Figure 2 shows a schematic diagram of a typical TASTE. A triple-effect unit illustrates the basic design. Typical units are offered in four to seven effects (five to nine stages), and capacity ranges from about 10,000 to 90,000 lb/h water removal. The vapor-to-steam ratio is approximately the number of effects times 0.85. In Figure 2, note that juice enters the spray nozzles atop each stage at a temperature well above the saturation temperature at which that body operates. The resultant flashing of vapor, combined with the spray effect of the nozzle, distributes the juice on the vertical tube walls. Thus, the juice film and the water vapor flow concurrently down the tube; the juice flows down the wall as the vapor flows in the central area of the tube. The centrifugal pumps that transfer the juice to the succeeding stage operate continuously in a cavitating, or almost cavitating, condition.

As juice is transferred from stage 3 to stage 4, it must be reheated to a higher temperature (140°F) to effect sufficient flashing. Reheating is typical for any effect that has more than one stage. These stage divisions are necessary when water removal renders the remaining juice insufficient to wet the total required surface for that effect. Then flash cooling quickly reduces the concentrate from 117°F to the pump-out temperature of 60°F.

On a TASTE, vapor flow is fixed and virtually unalterable. Although the juice flow shown in Figure 2 is concurrent with the vapor, in many cases it is mixed flow, in which the juice may first be introduced to an intermediate effect before being delivered to the first effect. In all cases, however, the essence-bearing material must be recovered from the vapor stream that is produced where the juice first enters the evaporator.

Figure 2 also shows an essence recovery method. Several different systems for essence recovery are used, some of which are proprietary. The refrigeration load for essence recovery is relatively small, usually 3 to 5 tons, according to evaporator capacity.

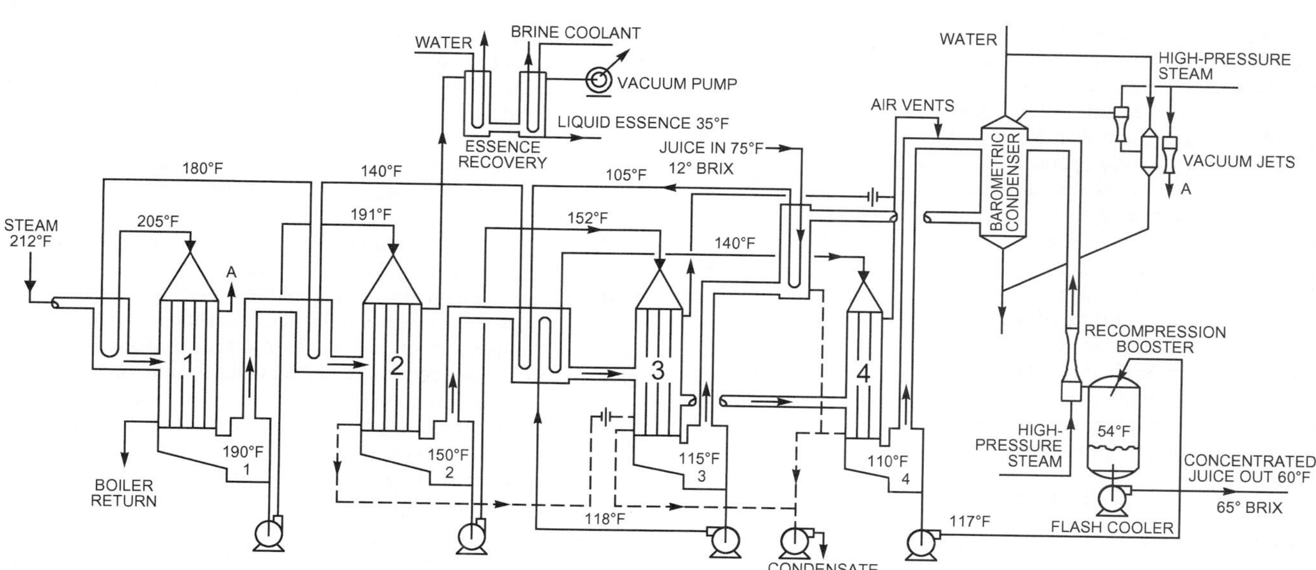

Fig. 2 Thermally Accelerated Short-Time Evaporator (TASTE) Schematic

Freeze Concentration

In the freeze concentration system, juice is introduced and pumped rapidly through a swept-surface heat exchanger in which ice nuclei are formed at about 28°F. This slurry is delivered to a recrystallizer, in which small crystals are melted to form larger crystals. The slurry of larger crystals, together with the resultant concentrate, is delivered to the wash column, where the ice rises. As it does, chilled water (melted ice) is introduced at the top to wash the ice crystals; these continue to rise and melt as they reach the top of the column. The concentrate (now ice-free) is drawn off the bottom of the wash column.

At first, centrifuges were used to separate ice crystals from concentrate, but because of poor separation and attendant problems with crystal washing, the losses of orange soluble solids were unacceptable. The proprietary wash-column process drastically reduces these losses. The output concentration of commercial freeze concentration equipment is presently limited to under 50° Brix.

When first-quality oranges are used to prepare the juice, freeze concentration makes a concentrate that is indistinguishable from fresh juice. However, first costs for equipment and installation are relatively high, and operating costs at best are the same as for modern evaporators. The cost is high despite the 144 Btu/lb required to freeze water, as opposed to about 1000 Btu/lb for evaporation. A contributing factor is the relative cost of electricity (compressors) versus steam from fossil fuel. Another factor is that, by using multiple-effect evaporators, the 1000 Btu/lb can be divided by the number of effects in the evaporator, thus reducing the energy requirement to the range of 150 to 250 Btu/lb, depending on the number of effects.

QUALITY CONTROL

The most important factor in quality control is using good, sound fruit, but quality must also be checked during processing and again in the final product. Standards for the various concentrates describe the quality and serve as a basis for grading. Concentrates are checked for Brix value, Brix-acid ratio, peel oil content, and other factors. Additionally, tests may be run to ensure that other requirements of a particular brand are met. At periodic intervals, bacteriological samples are taken during various stages in the plant, plated on orange serum agar, and examined after incubation. Although the necessary sanitation measures are known, bacteriological samples serve as a check and indicate whether the procedures have been effective.

In warm weather, cleanup must be more stringent and more frequent than during cold weather. TASTEs can run for only 6 to 12 h before substantial loss of evaporation or the release of hesperidin crystals forces a cleanup cycle. These cycles are normally spaced to coincide with other required cleaning in the juice extraction system and often take no more than 30 min of production time. Thermal conditions in a properly operated TASTE are so adverse to microorganisms and enzymes that these do not influence the cleaning cycle. Normally, low-temperature evaporators are cleaned at least every 24 h; however, in some instances when arrangements are favorable, they may run as long as 7 days between cleanups. Total plate counts in final product are generally maintained well below 10^6 organisms per millilitre of reconstituted juice. Counts of a few thousand per millilitre are common and indicate a high degree of plant cleanliness. Sanitation is based on asepsis rather than antisepsis. The natural acidity and high sugar content of citrus juice concentrate normally inhibit rapid growth of organisms. Generally, counts tend to decrease in storage.

CHILLED JUICE

Chilled juice is usually packed either in fiber cartons or in plastic bottles or jugs. The ideal storage temperature is about 30°F, but it is frequently handled in the retail trade at 40 to 45°F and is typically stored in household refrigerators. Normal shelf life is 3 to 4 weeks.

Chilled juice is marketed in two basic forms: "from-concentrate" and "not-from-concentrate." Because of higher overall costs, the not-from-concentrate product is higher-priced. Figure 1 shows the processing steps for each of the two products. The sole difference in the two products is the source of the juice.

From-Concentrate Juice. Bulk concentrate, taken either from partially thawed drums or from bulk storage, is mixed in a blending tank with water, essence, and cold-pressed oil so that it is reconstituted to about 12° Brix. This juice is then processed in a three-stage pasteurizer. First, the juice is preheated in a regenerative section that recovers heat from juice leaving the pasteurizing section. It then flows to the pasteurizer, where it is steam-heated to 180 to 190°F. Then it flows back through the regenerative section, where it is partially cooled by incoming juice. Finally, it passes through the cooling section, where it is cooled to about 30°F. A refrigerated glycol/water solution usually serves as the chilled brine for this purpose. These pasteurizer units are usually plate units, but tubular equipment is also used. From this point, every effort is made to maintain the juice at 30°F as it is packaged and placed in 30°F storage.

Not-from-Concentrate Juice. When fruit is mature and in season, fresh juice may be taken directly from extraction to pasteurizing and packaging. No blending is required, and the remaining process duplicates that for the from-concentrate product. However, unless special means are used during extraction, the peel-oil content of the juice may be excessive. A low-oil extraction method or heating and flashing the juice in a special deoiler before pasteurizing can remove excess oil.

Because not-from-concentrate chilled juice is in demand year-round and mature fruit is only available a maximum of 8 months per year, two storage methods are commonly used to ensure year-round supply: (1) store the juice in large bulk tanks in an aseptic environment at a temperature just above the freezing point or (2) freeze the fresh juice and store in a solid form at about 0°F.

The aseptic system usually involves a number of large storage tanks in a cold room. These tanks are commonly built to hold one million gallons or more of juice. Most of these tanks are constructed of carbon steel with a special interior lining; however, some are constructed entirely of stainless steel. The entire system must be carefully sanitized before use and must be so maintained during use. Juice is pasteurized and cooled before introduction to the tanks.

Freezing fresh juice predates the aseptic liquid method, and involves cooling the juice to about 30°F before placing it in a container for freezing. The choice of container is related to (1) the rate at which the juice will be frozen and (2) the method used to thaw the product and remove it from the container for later processing. In some instances, the advantage for freezing may be counterbalanced by disadvantages for thawing and removal.

For example, juice may be cooled, placed in an open-top drum, and immediately transferred to –10°F storage, where it will slowly freeze. The product quality will be satisfactory, but thawing and dumping the product from the large drum is difficult. In another method, the juice is frozen and stored in a specially constructed container. Later, it is fully thawed and pumped from the container. In still another method, the juice is encased in a plastic bag and then quick-frozen in a blast freezer room. Thawing and removal methods are similar to those for drums. Some juice is also stored in open ice blocks, which make recovery easier but make the juice more susceptible to contamination or losses.

A third distinct method of providing fresh juice in the off-season is to store fresh oranges under refrigeration for extraction of juice as required. This method requires large cold rooms held at about 30 to 32°F. In many cold storages, moisture must be added to keep the air at saturation. Added moisture, however, increases the refrigeration load because all condensate from the cooling coils must be recycled to the conditioned space. Because of cost and other limitations, fresh fruit is seldom stored for juice.

REFRIGERATION

Refrigeration Equipment

The choice of refrigerant for juice processing is almost universally R-717 (ammonia), although a few small systems use R-22. R-22 is sometimes used in freeze concentration systems, although ammonia has also been used for this purpose.

High-stage compressors have been of the reciprocating type, but to meet greater demands, there is a trend toward rotary screw compressors.

Most condensers are evaporatively cooled. Exposure to subtropical climatic conditions makes construction materials an important consideration. In Florida, water conditioning is critical because the available makeup water has a high mineral content.

The most common refrigerant evaporators in cold rooms and blast freezing tunnels are finned or plain coils. In larger installations, a single, low-pressure receiver serves a multitude of coils. Liquid refrigerant is pumped through the coils at two to three times the evaporation rate, and the liquid/gas mixture is returned to the low-pressure receiver, from which the compressors take their suction. At refrigerant temperatures below 32°F, coils must be defrosted on a regular basis, generally by hot gas from the compressor discharge manifold. Some smaller ammonia installations use air units, each with its own surge drum and controls for flooded operation. Small installations using R-22 are usually direct-expansion, and some use electric or water defrost.

To solve coil defrosting problems, some cold rooms and freezing tunnels operate with continuous defrost, in which a strong solution of propylene glycol is sprayed continuously over the coils. The weaker glycol solution is removed, and acquired water is driven off in an external concentrator. Unless extreme care is used in designing the eliminators, the glycol solution may be entrained and deposited on the containers. The resultant appearance of the containers is unacceptable. More information on coil defrost can be found in Chapter 3.

Another common evaporator is a shell-and-tube unit that chills a secondary coolant, typically propylene glycol. Ethylene glycol should not be used for food processing because it is toxic. Plate juice processors require large volumes of chilled polypropylene glycol. Although the refrigerant can be supplied directly to a gasketed plate heat exchanger, most heat exchangers do not satisfy the pressure requirements of the refrigerant.

Freezing tunnels using an alcohol-water solution chilled in shell-and-tube equipment have been constructed. The chilled solution was sprayed over the containers for quick freezing. These tunnels are now largely obsolete and have been replaced with blast units.

Swept-surface heat exchangers for concentrate precooling preparatory to can filling were also used. These units operated as flooded refrigerant systems, but most have been replaced by plate units, which also require glycol coolant. Today, application of swept-surface units is generally limited to freeze concentration systems.

A typical installation has a number of small loads, including

- Essence recovery (condensing of water vapor laden with essence). This load is typically handled with shell-and-tube equipment either by direct expansion or with a secondary coolant (glycol).
- Winterizing of cold-pressed oil in order to form precipitates of undesired materials in a quiescent storage vessel. The process goes forward at about 0 to –40°F and requires months of storage. Winterizing may be done in 55 gal drums in a cold room or, preferably, in a jacketed tank using direct refrigerant expansion or very cold glycol.
- Storage of essence. Tanks of essence may be stored in a 30 to 33°F cold room or in an outdoor insulated storage tank in which heat gain is removed by continuous recirculation of the essence through an external plate heat exchanger cooled by glycol brine.

Refrigeration Loads

Refrigeration loads vary not only with plant throughput but also according to the relative rates at which different products are processed. For a given amount of consumable product, the volumes of chilled juice are higher than those for concentrates by a factor of 4:1 to 6:1. Thus, the refrigeration load depends largely on the proportions of concentrates and chilled juices being processed.

To illustrate the magnitude of the refrigeration loads, the following fixed set of processing conditions is assumed. All percentages are percentage of total throughput of **soluble solids**.

- The plant has an operating evaporator capacity of 100,000 lb/h of water removal.
- 10% is packed as not-from-concentrate chilled juice. Of this, half is packaged directly for immediate sale; the remainder is frozen for storage and use during the off-season.
- 90% is processed into concentrate, which is made up of 85.5% fed to the evaporator and 4.5% reserved for use as cutback to make 42° Brix retail packs.
- Of the concentrate from the evaporator, 30% is packed in bulk for later use as from-concentrate chilled juice; an additional 24% is similarly packed for later use as retail frozen concentrate.
- The remaining 31.5% from the evaporator is blended with the 4.5% cutback (to total 36%). This is packed into retail containers at 42° Brix and immediately frozen.
- The chilled juice line processes 30% of the concentrate (i.e., same as quantity from the evaporator stored for this purpose).

Given these conditions, plant refrigeration loads would be approximately as follows:

	Refrigeration, Tons	
	Low-Stage	High-Stage
Process	250	640
Cold storage	150	280
Miscellaneous loads	—	100
Total	400	1020

Compressor Manifolding

In most cases, low-stage compressors have their own isolated high-stage machines, with an intermediate pressure of 30 to 35 psig. The remaining high-stage compressors are usually paralleled to handle all other high-stage loads at a somewhat lower suction pressure (15 to 20 psig) to optimize power requirements.

Frequently, these systems operate as central systems, with all condensers manifolded using a common receiver.

PURE FRUIT JUICE POWDERS

One manufacturer successfully produced a vacuum-dried orange juice powder. In this **puff-drying** process, concentrate of about 58° Brix is introduced into a vacuum chamber, where it is dried on a moving stainless steel belt. The dried powder is flavored with a locked-in orange oil prepared by dispersing orange oil in a mixture of molten sugars, extruding, and cooling rapidly. The oil is thus kept out of contact with the powdered orange juice until water is added to reconstitute the juice.

Another process, known as **foam-mat drying**, has been under investigation for citrus juices. In this process, a small amount of foam stabilizer is added (0.5% of dry solids content), and the chilled concentrate of about 50° Brix is beaten into a foam. This foam is laid out in a sheet about 1/8 in. thick on perforated trays. An air blast clears the foam from the holes, and the stacked trays are conveyed up a column while hot air passes up through them and reduces the moisture to about 1.25% in 12 min. The product is then chilled until it hardens and is scraped from the trays. In a variation, a thin layer of foam is placed on a polished stainless steel belt, dried in a stream of hot air, and finally stripped from the belt in dried form with a doctor blade.

Both puff-drying and foam-mat drying were developed through U.S. Department of Agriculture (USDA) research.

OTHER CITRUS JUICES

Grapefruit Juice

Production of frozen concentrated grapefruit juice uses essentially the same equipment as production of frozen concentrated orange juice does. Some adjustments at the extractors are necessary to accommodate grapefruit. Because bittering is considered a defect, debittering systems may also be used to improve the flavor.

Both sweetened and unsweetened frozen concentrate grapefruit juices are prepared (the sweetened product in greater quantities). The final unsweetened product may vary from 38 to 42° Brix. The sweetened product must contain at least 3.47 lb of soluble grapefruit solids per gallon exclusive of added sweetening ingredients. The final Brix may vary from 38 to 48°. In Grade A unsweetened concentrate, the Brix-acid ratio may vary from 9:1 to 14:1; in the sweetened product, the ratio may vary from 10:1 to 13:1. All the types mentioned are 3-plus-1 concentrate. Either seedless or seeded grapefruit can be used, but seeded varieties, such as Duncan, are generally preferred.

Blended Grapefruit and Orange Juice

The same procedures are used in producing a grapefruit and orange juice blend as are used in preparing the separate products. USDA grade standards recommend no less than 50% orange juice in the mixture and as much as 75% orange juice when it is light in color. Military specifications require 60 to 75% orange juice. USDA grades require 40 to 44° Brix in unsweetened concentrates. In sweetened concentrates, the Brix must be at least 38° before sweetening and 40 to 48° after sweetening. For Grade A, the Brix-acid ratios in the packed concentrate may vary from 10:1 to 16:1 if unsweetened, and from 11:1 to 13:1 if sweetened.

Tangerine Juice

Tangerines require different methods of handling during picking, hauling, and storage at the plant. Whereas the grapefruit and orange are generally round, quite firm, and able to withstand considerable rough handling, the tangerine is somewhat flat and irregular in shape and has a loose, tender skin that is easily broken. If the skin is broken and the fruit bruised, bacteria and yeasts readily attack the fruit, and undesirable enzyme actions occur. Thus, tangerines cannot be handled in orange bins but must be handled in boxes or loose in trucks to a depth of no more than 2 ft.

The processes and equipment used in manufacturing concentrated tangerine juice are practically the same as those used with oranges. Because the fruit is smaller, the yield of juice from a given number of extractors is smaller, and about twice the extracting equipment is required to furnish enough juice to keep the evaporators operating at full capacity. Generally, values for Brix-acid ratio, peel oil content, and concentration have followed those prescribed for the orange product. Recently, more of the pack has been sweetened, and there has been a trend toward packing at a higher concentration. A Brix of 44° is common for a 3-plus-1 concentrate.

NONCITRUS JUICES

PINEAPPLE JUICE

Pineapple juice is prepared from small fruit and the parts of larger pineapples that are unsuitable for packing as fruit pieces. The main sources are the cores, the layer of flesh between the shell and cylinder that is cut for the preparation of pineapple slices, and the juice that drains from crushed pineapple: altogether, about one-third of the weight of the fresh fruit. Pieces of shell and spoiled flesh are removed during inspection. Juice is extracted by passing through disintegrators and screw presses. It is then centrifuged to remove heavy foreign material and excessive insoluble solids. Processing up to this point is the same for both single-strength and concentrate.

Pineapple concentrate is produced from single-strength juice in equipment similar to that used to produce orange and other fruit juice concentrates. The first step in the concentrating operation is to strip out the volatile flavoring materials. These are separated as about a 100-fold concentrate and added back to the final concentrate. Concentration occurs in multiple-effect evaporators.

Pineapple concentrate is produced either as a 3:1 product with a Brix of about 46.5° or as a 4 1/2:1 product with a Brix of about 61°. The 3:1 concentrate is produced in both a sterile form and a frozen form. However, even the sterile product is stored and sold under refrigeration in order to preserve quality. The 4 1/2:1 concentrate is also produced in both a sterile form and a frozen form. It may be held for short periods without refrigeration, but it should be stored at 40°F or lower. The frozen 61° Brix concentrate is packaged in polyethylene bags and held in 7 gal fiber containers. This product is stored under refrigeration.

Bulk pineapple concentrate is used principally for mixing with citrus concentrate to produce frozen juice blends. Pineapple concentrate is also used as an ingredient in many types of canned fruit drinks.

The composition of pineapple juice varies greatly. Brix varies between 12 and 18°, with an average of about 13.5 to 14°. Brix-acid ratio ranges from 12:1 to over 20:1 and usually averages between 16:1 and 17:1. Because pineapple concentrate is produced at a standard Brix, variation in composition shows up only in the acidity and the Brix-acid ratio.

APPLE JUICE

Evaporative procedures include some method of essence (ester) recovery for incorporating the volatile components of apple flavor into the final concentrate. These procedures are designed to take advantage of the fact that in the distillation of apple juice, most of the volatile flavors are found in the first 10% of the distillate. This portion of distillate is passed through a fractionating column to obtain the volatile flavors in concentrated form, usually about 100-fold compared to the fresh juice. The remaining 90% of the original juice, now stripped of its volatile flavors, is then concentrated under vacuum to somewhat more than the concentration desired for the final product.

For example, 100 gal of apple juice prepared in a conventional manner may be treated to yield 1 gal of 100-fold essence and 24 gal of concentrated stripped juice. The combination of these two fractions yields full-flavored fourfold concentrate. If a higher concentrate is desired, the stripped juice is concentrated to a greater extent. In such instances, however, the juice must be depectinized to avoid excessive viscosity and gelation of the highly concentrated juice.

One report shows that fourfold apple juice (depectinized) concentrate was essentially unchanged after one year of storage at 0°F. Similar information on concentrate prepared without depectinization is not available.

GRAPE JUICE

Concord Grapes

Most of the concentrated grape juice marketed in North America is prepared from Concord grapes (*Vitis labrusca*) grown in New York, Michigan, Washington, Pennsylvania, Ohio, Arkansas, and Ontario. The grapes are harvested when the soluble solids reach a concentration of 15 to 16%. This varies with maturity and is influenced by cultural and climatic factors.

After washing, Concord grapes are conveyed to the stemmer, which consists of a perforated, slowly revolving (20 rpm) horizontal drum with several beaters inside, revolving at a much faster speed (200 rpm), to knock the berries off the cluster and partially crush them before they are discharged through the drum perforations. Cluster stems are expelled from the open end of the drum. The crushed fruit is then pumped through a tubular heat exchanger, where it is heated to 140 to 145°F for good extraction of pigments and juice. The hot pulp then goes to hydraulic presses, where juice is removed in the same manner as apple juice. Expressed juice may be clarified in a centrifuge or filter press; the latter uses 1 to 2% diatomaceous earth to maintain a high filtering rate and remove a substantial amount of suspended matter. Under normal operating conditions, 190 to 195 gal of juice are obtained from a ton of grapes. In some plants, screw presses are used to remove juice from all or part of the crushed grapes, but this increases the amount of suspended matter to be removed later.

Clarified juice is pasteurized in tubular or plate heat exchangers to a temperature of 180 to 190°F and cooled immediately to 30°F before storage in tanks in refrigerated rooms maintained at 28°F. The juice is usually cooled in two or more steps. Some heat exchange systems begin with a regeneration cycle, in which hot juice leaving the pasteurizer preheats entering juice. Cooling water discharged from the heat exchangers may be piped to the washers to heat the water applied to the incoming grapes.

The method of handling the cooled juice depends on its intended use. If it is to be used in jelly manufacture, the juice is stored at 28°F for 1 to 6 months to permit settling of the **argols**; these consist of potassium acid tartrate, tannins, and some colored materials that would give a gritty texture to the jelly or detract from its clarity. Clear juice is siphoned off the precipitate in the storage tanks and may be refiltered. If the concentrate is to be sold as a blend formed by mixing in sugar and ascorbic acid before canning and freezing, the cold storage tank merely serves as a surge tank, because juice is pumped out to the concentrator within a few hours. A polishing filter is used before the concentrator to minimize fouling of the evaporator tubes. Concentrates for both jelly manufacture and blended juice may be stored in tanks at 27°F before processing. Whenever single-strength juice is bulk-stored at 27°F before concentration, spoilage by fermentation is a danger. To minimize yeast growth during storage, all pipelines and equipment from the pasteurizer to the cold room should be designed for ready and frequent cleaning. Interior surfaces of storage tanks must be relatively smooth and free from crevices, and the tank should be thoroughly cleaned before use.

Concentration involves two steps. First, volatile flavoring materials are stripped from the juice, and the stripped juice is then concentrated to the desired density. Volatile components are removed by heating the single-strength juice to 220 to 230°F for a few seconds in a heat exchanger, flashing a percentage of the liquid into a vapor in a jacketed tube bundle, and discharging the liquid and vapor tangentially through an orifice into a separator. The separator should be large enough to reduce the vapor velocity to 10 fps or less for minimal entrainment. Twenty to 30% by mass of the original juice flashes off as vapor, which is led into the base of a fractionating column filled with ceramic saddles or rings. A reflux condenser on the vapor line from the column and a reboiler section at the base of the column provide the necessary reflux ratio. Vent gases from the reflux condenser are then chilled in a heat exchanger, and the condensate containing the essence is collected at a rate equivalent to 1/150 of the volume of entering flavoring material.

Methyl anthranilate, which has a boiling point of 512°F and is only slightly soluble in water, is an important flavor component of Concord grape juice. Its high boiling point and low solubility cause losses when the stripping column's efficiency is low. These losses may be reduced by increasing the vaporizing temperature.

In a typical formulation, grape juice is concentrated to a little over 34° Brix, and essence or fresh cutback juice is added to reduce it to 34° Brix. Sucrose is added to achieve 47° Brix, and citric acid is added until the total acidity is 1.8% calculated at tartaric acid. The concentrate may be cooled to 20 to 30°F in a heat exchanger or cold wall tank, filled into cans, sealed, cased, and allowed to freeze in subzero storage. When diluted with an equal quantity of water, this concentrate yields the equivalent of sweetened single-strength juice; however, for a more palatable beverage, three parts of water are added. The product is labeled as concentrated, sweetened grape juice.

Muscadines

Muscadines (*Vitis rotundifolia*), typical of the southeastern United States, differ from *V. labrusca* in that they grow not in clusters but as individual berries, so there is no need to destem them. These grapes are processed into juice as follows: grapes are harvested, transported in trucks to the receiving station, dumped into hoppers, and crushed with potassium persulfate ($K_2S_2O_8$) to give 50 ppm free sulfite.

After crushing, the grapes are conveyed into a pneumatic press (without heating), and pressure up to 75 psi is exerted. Once the juice is extracted, it is quickly pasteurized as it passes through a plate heat exchanger that heats it to about 185°F. The juice is partially cooled, and a mixture of pectinases and cellulases is added to clarify it. After 1 to 2 h, the juice is passed through an ultrafiltration (UF) tubular unit with 2.0 μm pores for filtering. The juice is then repasteurized and cooled to 45°F in plate heat exchangers and sent to large refrigerated storage tanks for removal of tartrate and sediment at 28°F. Some of the juice is concentrated to about 65° Brix in a triple-effect evaporator with an essence recovery system. This concentrate is field-frozen and used later for bottle juice. When ready to bottle, juice is pumped through the UF unit, bottled and capped, and passed through a pasteurizer/cooler in a conveyor belt.

STRAWBERRY AND OTHER BERRY JUICES

Frozen strawberry juice concentrate, a sevenfold concentrate with separately packed concentrated (100-fold) essence, is used for manufacturing, especially jellies. Availability of sevenfold frozen concentrate also allows marketing of high-quality strawberry juice solids. Concentrates of red raspberry, black raspberry, and blackberry juices are also available in limited quantities.

Preparation of strawberry and other berry juice concentrates involves essence recovery in which 12 to 20% of the juice is separated by a stripping process using a steam injection heater. Vapors containing volatile flavors are concentrated in a fractionating column to the desired degree. Juice remaining after the essence recovery step is concentrated under vacuum three- to sevenfold by volume. For strawberry juice, a maximum temperature of 100°F for 2.5 h should not be exceeded, whereas temperatures up to 130°F may be used in preparing batches of boysenberry.

Preparation of juice for concentration involves chopping or coarse milling of cold, sound berries and mixing with pectic enzymes and filter aid. After several hours (4 to 5 h at room temperature), juice is expressed with a bag press or rack and clothes press. The cloudy juice is clarified in a filter press. Recovered essences are concentrated and packaged separately so that the jelly manufacturer can incorporate the essence in the jelly just before filling. This procedure greatly reduces the amount of essence lost by volatilization. The essence can also be incorporated in the concentrate to make a full-flavored product for shipping as a single unit.

Concentrated juice without essence can be packed in enamel-lined containers that need only be liquid-tight. Concentrated essence should be kept in carefully sealed cans to avoid loss of the highly volatile flavor. Both juice concentrate and essence are kept frozen for proper quality retention.

CHAPTER 26

BEVERAGES

THIS chapter discusses the processes and use of refrigeration in breweries, wineries, and carbonated beverage plants.

BREWERIES

MALTING

Malt is the primary raw ingredient in brewing beer. Although adjuncts such as corn grits and rice contribute considerably to the composition of the extract, they do not possess the necessary enzymatic components required for preparing the wort. They lack nutrients (amino acids) required for yeast growth, and contribute little to the flavor of beer. Malting is the initial stage in preparing raw grain to make it suitable for mashing. Traditionally, this operation was carried out in the brewery, but in the past century, this phase has become so highly specialized that it is now almost entirely the function of a separate industry.

Various grains such as wheat, oats, rye, and barley can be malted; however, barley is the predominate grain used in preparing malt because it has a favorable protein-to-starch ratio. It has the proper enzyme systems required for conversion, and the barley hull provides an important filter bed during lautering. Also, barley is readily available in most of the world.

There are three steps to malting barley. In **steeping**, the raw grain is soaked in 40 to 65°F water for 2 to 3 days. The moisture content of the barley kernel increases from 12% to approximately 45%. The water is changed frequently and the grain is aerated. After two or three days, the kernels start to germinate and the white tips of rootlets appear at the end of the kernels. At this time, the water is drained and the barley is transferred to where it is germinated.

During 4 to 5 days of **germination**, the kernel continues to grow. The green malt is constantly turned over to ensure uniform growth of the kernels. Slowly revolving drums can be used to turn over the growing malt. In a compartment system, slowly moving, mechanically driven plowlike agitators are used for mixing. Cool (50 to 65°F) saturated moist air is used to maintain temperature and green malt moisture levels. At the desired stage in its growth, the green malt is transferred to a kiln.

Kilning, the final step, stops the growth of the barley kernel by reducing its moisture level. Warm (120 to 150°F) dry air is used to remove moisture from the green malt. Kilning is usually done in two stages. First, the malt's moisture content is reduced to approximately 8 to 14%; then, the heat is increased until the moisture is further reduced to about 4%. Using this heating procedure reduces excessive destruction of enzymes. The desired color and aroma are obtained by controlling the final degree of heat.

After kilning, the malt is cleaned to separate dried rootlets from the grain, which is then stored for future use. The finished malt differs from the original grain in several significant ways. The hard endosperm was modified and is now chalky and friable. The enzymatic activity has been greatly increased, especially alpha amylase, which is not present in unmalted barley. The moisture content is reduced, making it more suitable for storing and subsequent crushing. It now has a distinctive flavor and aroma, and the starches and enzymes are readily extractable in the brewhouse.

PROCESS ASPECTS

Two distinct types of chemical reactions are used in brewing beer. **Mashing** is carried out in the brewhouse. Starches in the malted grain are hydrolyzed into sugars and complex proteins are broken down into simpler proteins, polypeptides, and amino acids. These reactions are brought about by crushing the malt and suspending it in warm (100 to 122°F) water by means of agitation in the mash tun. When adjuncts (usually corn grits or rice) are used, a portion of the malt is cooked separately with the adjunct. After boiling, this mixture is combined with the main mash, which has been proportioned so that a combining temperature generally in the range of 145 to 162°F results. Within this temperature range, the alpha and beta amylases degrade the starch to mono-, di-, tri-, and higher saccharides. By suitably choosing a time and temperature regimen, the brewer controls the amount of fermentable sugars produced. The enzyme diastase (essentially a mixture of alpha and beta amylase), which induces this chemical reaction, is not consumed but acts merely as a catalyst. Some of the maltose is subsequently changed by another enzyme, maltase, into a fermentable monosaccharide, glucose.

Mashing is complete when the starches are converted to iodine-negative sugars and dextrins. At this point, the temperature of the mash is raised to a range of 167 to 172°F, which is the "mashing-off" temperature. This stops the amylolytic action and fixes the ratio of fermentable to nonfermentable sugars. The **wort** is separated from the mash solids using a lauter tub, a mash filter, or other proprietary equipment (MBAA 1999). Hot water (168 to 170°F) is then "sparged" through the grain bed to recover additional extract. Wort and sparge water are added to the brew kettle and boiled with hops, which may be in the form of pellets, extract, or whole cones. After boiling, the brew is quickly cooled and transferred to the fermentation cellar, where yeast is added to induce fermentation. Figure 1 shows a double-gravity system with grains stored at the

The preparation of this chapter is assigned to TC 10.9, Refrigeration Application for Foods and Beverages.

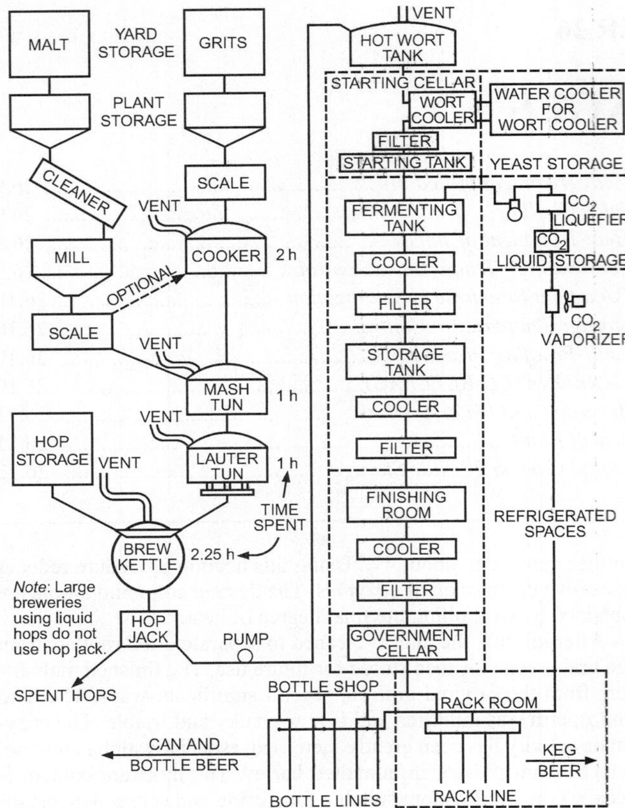

Fig. 1 Brewery Flow Diagram

top of the brewhouse. As processing continues, gravity creates a downward flow. Hot wort from the bottom of the brewhouse is then pumped to the top of the stockhouse, where it is cooled and again proceeds by gravity through fermentation and lagering.

After the wort cools, yeast and sterile air are injected into it. The yeast is pumped in as a slurry at a rate of 1 to 3 lb of slurry per barrel of wort. Normally, oil-free compressed air is filtered and treated with ultraviolet light and then added to the wort, which is nearly saturated with approximately 8 ppm of oxygen. However, the wort may also be oxygenated with pure oxygen.

Fermentation takes place in two phases. During the first phase, called the respiratory or aerobic phase, the yeast consumes the oxygen present. It uses a metabolic pathway, preparing it for the anaerobic fermentation to follow. The process typically lasts 6 to 8 h.

Oxygen depletion causes the yeast to start anaerobically metabolizing the sugars in the extract, releasing heat and producing CO_2 and ethanol as metabolic by-products.

During early fermentation, the yeast multiplies rapidly then more slowly as it consumes the available sugars. Normal multiplication for the yeast is approximately 3 times. A representative value for the heat released during fermentation is 280 Btu/lb of extract (sugar) fermented.

Wort is measured by the saccharometer (measures sugar content), which is a hydrometer calibrated to read the percentage of maltose solids in solution with water. The standard instrument is the Plato saccharometer, and the reading is referred to as percentage of solids by saccharometer, or degrees Plato (°P). Table 1 illustrates the various data deducible from reading the saccharometer.

The same instrument is used to check fermentation progress. Although it still gives an accurate measure of density of the fermenting liquid, it is no longer a direct indicator of dissolved solids because the solution now contains alcohol, which is less dense than

Table 1 Total Solids in Wort

% Solids*	Specific Gravity	Weight per Barrel, lb		Specific Heat, Btu/lb·°F
		Total	Solids	
0	1.0000	258.7	0.00	1.000
1	1.0039	259.7	2.60	0.993
2	1.0078	260.7	5.21	0.986
3	1.0118	261.8	7.85	0.979
4	1.0157	262.8	10.51	0.972
5	1.0197	263.8	13.19	0.965
6	1.0238	264.9	15.89	0.958
7	1.0278	265.9	18.61	0.951
8	1.0319	267.0	21.36	0.944
9	1.0360	268.0	24.12	0.937
10	1.0402	269.1	26.91	0.930
11	1.0443	270.2	29.72	0.923
12	1.0485	271.2	32.55	0.916
13	1.0528	272.4	35.41	0.909
14	1.0570	273.4	38.28	0.902
15	1.0613	274.6	41.18	0.895
16	1.0657	275.7	44.11	0.888
17	1.0700	276.7	47.06	0.881
18	1.0744	277.9	50.03	0.874
19	1.0788	279.1	53.03	0.867
20	1.0833	280.2	56.05	0.860

*Saccharometer readings.

water. This saccharometer reading is called the apparent extract, which is always less than the real extract (apparent attenuation is calculated from the hydrometer reading of apparent extract and the original extract). In engineering computations, 81% of the change in apparent extract is considered a close approximation of the change in real extract. Thus, 81% of the difference between the solids shown in Table 1 for saccharometer readings before and after fermentation represents the mass of maltose fermented. This weight in pounds per barrel times 280 Btu/lb gives the heat of fermentation in Btu per barrel. (Each barrel has a capacity of 31 gal.) The difference between the original solids and mass of fermented solids gives the residual solids per barrel. It is assumed that there is no change in the volume because of fermentation. The specific heat of beer is assumed to be the same as that of the original wort, but the mass per barrel decreases according to the apparent attenuation.

Bottom-fermentation yeast (e.g., *Saccharomyces uvarium*, formerly *carlsbergensis*) is used in fermenting lager beer. Top-fermentation yeast (e.g., *Saccharomyces cerevisiae*) is used in making ale. They are so called because, after fermentation, one settles to the bottom and the other rises to the top. A more significant difference between the two types is that in the top-fermentation type, the fermenting liquid is allowed to attain a higher temperature before a continued rise is checked. The following characteristics of brewing ale make it different from brewing lager:

- A more highly kilned darker malt is used.
- Malt forms a greater proportion of the total grist (less adjunct).
- Infusion mashing is used and a wort of higher original specific gravity is generally produced.
- More hops are added during the kettle boil.
- A different yeast and temperature of fermentation are used.

Therefore, ale may have a somewhat higher alcohol content and a fuller, more bitter flavor than lager beer. With bottom-fermentation yeasts, fermentation is generally carried out between 45 and 65°F and most commonly between 50 and 60°F. Ale fermentations are generally carried out at somewhat higher temperatures, often peaking in the range of 70 to 75°F. In either type, the temperature during fermentation would continue to rise above that desired if not checked by cooling coils or attemperators, through which a cooling medium such as propylene glycol, ice water, brine, or ammonia is circulated. In the past, these attemperators were manually controlled, but more recent installations are automatic.

PROCESSING

Wort Cooling

To prepare boiling wort from the kettle for fermentation, it must first be cooled to a temperature of 45 to 55°F. To avoid contamination with foreign organisms that would adversely affect subsequent fermentation, cooling must be done as quickly as possible, especially through the temperatures around 100°F. Besides the primary function of wort cooling, other beneficial effects accrue that are essential to good fermentation, including precipitation, coagulation of proteins, and aeration (natural or induced, depending on the type of cooler used).

In the past, the Baudelot cooler was almost universally used because it is easy to clean and provides the necessary wort aeration. However, the traditional open Baudelot cooler was replaced by one consisting of a series of swinging leaves encased within a removable enclosure into which sterilized air was introduced for aeration. This modified form, in turn, has virtually been replaced by the totally enclosed heat exchanger. Air for aeration is admitted under pressure into the wort steam, usually at the discharge end of the cooler. The air is first filtered and then irradiated to kill bacteria, or it can be sterilized by heating in a double-pipe heat exchanger with steam. By injecting 0.167 ft^3 of air per barrel (bbl) of wort, which is the amount necessary to saturate the wort, normal fermentation should result. The quantity can be accurately increased or diminished as the subsequent fermentation indicates.

The coolant section of wort coolers is usually divided into two or three sections. For the first section, a potable source of water is used. The heated effluent goes to hot-water tanks where, after additional heating, it is used for subsequent mashing and sparging in the brew house. Final cooling is done in the last section, either by direct expansion of the refrigerant or by means of an intermediate coolant such as chilled water or propylene glycol. Between these two, a third section may be used from which warm water can be recovered and stored in a wash-water tank for later use in various washing and cleaning operations around the plant.

Closed coolers save on space and money for expensive cooler room air-conditioning equipment. They also allow a faster cooling rate and provide accurate control of the degree of aeration. To maintain good heat transfer, closed coolers may be flushed with hot water between brews or circulated for a few minutes with cleaning solutions. More thorough cleaning, perhaps done weekly, is accomplished by much longer periods of circulation with cleaning solutions, such as 2 to 4% caustic at 175°F. Reverse flow of the cleaning solution may also be used to help dislodge deposits of protein, hops, and other materials.

In selecting a wort cooler, consider the following:

- The cooling rate should allow the contents of the kettle to be cooled in 1 or, at most, 2 h.
- The heat transfer surfaces to be apportioned between the first section, using an available water supply, and the second section, using refrigeration, should make the most economical use of each of these resources. Cost of water, its temperature, and its availability should be balanced against the cost of refrigeration. Usual design practice is to cool the wort in the first section to within 10°F of the available water.
- Usable heat should be recovered (effluent from the first section is a good source of preheated water). After additional heating, it can be used for succeeding brews and as wash water in other parts of the plant. At all times, the amount of heat recovered should be consistent with the overall plant heat balance.
- Meticulous sanitation and maintenance costs are important.

Wort cooler size is determined by the rate of cooling desired, rate of water flow, and temperature differences used. A brew, which may vary in size from 50 to 1000 bbl and over, is ordinarily cooled in 1 or, at most, 2 h. Open coolers are made in stands up to about 20 ft long. Where more length is needed, two or more stands are operated in parallel.

Open coolers are best operated with a wort flow of 10 to 11.5 bbl/h per foot of stand. As flow increases beyond this rate, an increasingly larger part of the wort splashes from the top tube of the cooler and drops directly into the collecting pan below without contacting the cooler surfaces. An increased amount of wort flowing over the surfaces must be subcooled to offset what has been bypassed.

In plate coolers, this bypassing does not occur, and wort velocities can be increased to a point where friction pressure through the cooler approaches the maximum design pressure of the press and gasketing. The number of passes and streams per pass afford the designer much latitude in selecting the most favorable parameters for optimum performance and economical design. This design is based on (1) the specific heat of wort, (2) its initial temperature and range through which it is to be cooled, (3) temperature of the available water supply, and (4) ratio of the quantity of cooling water to wort that is to be used. Design and operating features of a typical plate cooler are as follows:

Specifications

Quantity of wort to be cooled	17,000 lb/h
Temperature of hot wort	210°F
Temperature of cooled wort	40°F
Temperature of available water (maximum)	70°F
Water used, not to exceed	34,000 lb/h
Temperature of water leaving cooler	140°F
Temperature of wort leaving first section of cooler	80°F
Temperature of incoming recirculated chilled water	34°F

Plate cooler (first section)

Number of plates	40
Heat transfer surface per plate	4.3 ft^2
Heat transfer surface in first section	172 ft^2
Number of passes	5
Number of streams per pass	4
Water flow rate	34,000 lb/h
Wort flow rate	17,000 lb/h

Plate cooler (second section)

Number of plates	24
Heat transfer surface per plate	4.3 ft^2
Heat transfer surface in second section	103 ft^2
Number of passes	3
Number of streams per pass	4
Chilled-water flow rate	52,000 lb/h

A shell-and-tube or plate cooler with two stages of cooling can cool the wort efficiently. In the first (hot) stage, potable water is used counterflow to the wort, and the usual discharge temperature is about 169 to 171°F. This hot water is then used in the following brews at various blended temperatures. Excess is used in the brewery's general operations.

The second stage of wort cooling is accomplished at about 36°F by a closed system of refrigerated water through a closed cooler, which cools the wort to 50°F or lower, depending on the brewer. Lower-temperature water (33°F) may be used in open units where no danger of freezing exists.

Wort cooling may be accomplished in one stage, depending on the potable water temperature available and plant refrigeration capacity. If chilled water (33 to 36°F) is available, water use is typically 1.1 to 1.4 times the volume of wort and exits the cooler at temperatures suitable for immediate use in brewing. If ambient water is used, much larger volumes are required and water costs must be considered. Also, excess hot water may be sewered, leading to increased waste effluent charges.

Fermenting Cellar

After cooling, the wort is pitched with yeast and collected in a fermenting tank, where respiration and fermentation occur according to the chemical reaction previously discussed. The daily rate of fermentation varies depending on the operating procedure adopted in each plant. On the first day, a representative rate might be 2 lb of converted maltose per barrel of wort. The rise in temperature caused by fermentation and by the growth and changing physiology of the yeast increases this rate to 7 lb/bbl on the second day. By now, the maximum desired temperature has been attained, and a further rise is checked by an attemperator, so that on the third day another 7 lb is converted. This rate continues through the fourth day. Two examples of the fermentation rate follow; one is for normal-gravity brewing, and the other is for high- (heavy) gravity brewing.

Example 1 Normal-Gravity Brewing

Fermentation Day	°Plato	Real Extract	Extract per bbl, lb	Extract Fermented per bbl, lb
0	11	11.0	29.63	—
1	10	10.2	27.38	2.25
2	8	8.6	22.95	4.43
3	5	6.1	16.12	6.83
4	3	4.5	11.81	4.31
5	2.5	4.1	10.75	1.06
				18.88

Example 2 High-Gravity Brewing

Fermentation Day	°Plato	Real Extract	Extract per bbl, lb	Extract Fermented per bbl, lb
0	16	16.0	43.97	—
1	15	15.2	41.64	2.33
2	12	12.8	34.73	6.91
3	7	8.7	23.11	11.51
4	4	6.3	16.66	6.56
5	3.5	5.9	15.58	1.08
				28.39

By now, the amount of unconverted maltose remaining in the beer is greatly diminished. Because alcohol, carbon dioxide, and other products of fermentation inhibit further yeast propagation, the action nearly stops on the fifth day, when only about 3 lb is converted per barrel. At this stage the yeast begins to flocculate (clump together) and either settles to the bottom of the fermenter (bottom yeast) or rises to the top (top yeast). Because of the reduced fermentation rate, the temperature of the beer begins to fall, either as the result of increased attemperation applied to the tank itself, heat loss from the tank to the surrounding area, or both. Many fermentation programs call for the beer to be cooled to 35 to 45°F at this time. This period of more rapid cooling helps settle the yeast. At the completion of this cooling period, the fermentation rate is essentially zero, and the beer is ready to be transferred off the settled yeast. Complete fermentation generally occurs in about 7 days. The introduction of new types of beers (e.g., reduced calorie, reduced alcohol) and the more general use of high-gravity brewing have led to the use of a variety of fermentation programs both between brewers making the same product and within the same brewery for different products.

Complete fermentation can be accomplished in less than 7 days, but most modern brewers take 7 to 10 days for the fermentation and subsequent cooling. The time depends on original gravity, whether a secondary fermentation is used, and available cooling capacity. Most brewers cool beer to between 42 and 38°F after ending fermentation or after the final days of quick cooldown in the fermenting tank. In addition, the long rest allows time for the yeast to settle. Some brewers agitate the beer in cylindrical fermenters, which enables them to ferment the beer faster and then to separate out the yeast by centrifuge. Most brewers cool the beer to the desired 29 to 45°F temperature before it goes into storage for resting and settling between fermentation and final aging.

Fermenting Cellar Refrigeration

The agitation necessary for heat exchange between the attemperator and the beer is provided partly by convection resulting from temperature gradients in the beer. Agitation is principally by the ebullition caused by the carbon dioxide (CO_2) bubbles rising to the surface of the liquid. In estimating the heat transfer surface required, a heat transfer rate range from 15 to 30 Btu/h·ft^2·°F is reasonable. Heat loss from tank walls and the surface of the liquid may be disregarded when calculating attemperator coil surface requirements. However, if the room temperature drops appreciably below 50°F, heat dissipated through the metal tank walls becomes important. Depending on the degree of heat dissipation, fermentation may be retarded or even inhibited. In such instances, insulating the fermenter walls and bottom is required so that control over heat removal remains in the attemperator.

Refrigeration requirements are based on the maximum volume of wort being fermented, as illustrated by Example 3.

Example 3. Figure 2 illustrates the volume of wort production based on a 500 bbl/day production rate. Days are represented by the abscissa, and the pounds of solids converted per day by the ordinate. The individual brews in fermentation on any particular day are additive. For example, on the fifth day, Brew No. 1 is finishing with a conversion rate of 3 lb/bbl for that day; Brew No. 5, which is just beginning the fermentation cycle, is fermenting at the rate of 2 lb/bbl; and Brews No. 2, 3, and 4 are each at the maximum rate of 7 lb/bbl per day. The total solids fermented on this day are 26 lb/bbl for the 2500 bbl in fermentation, totaling 13,000 lb of solids converted per day. Because the heat of fermentation is 280 Btu/lb, the refrigeration load is

$$(13{,}000 \times 280)/(24 \times 12{,}000) = 12.6 \text{ tons}$$

Calculations for sizing attemperators must consider the (1) internal dimensions of the fermenting tank and its capacity; (2) temperature

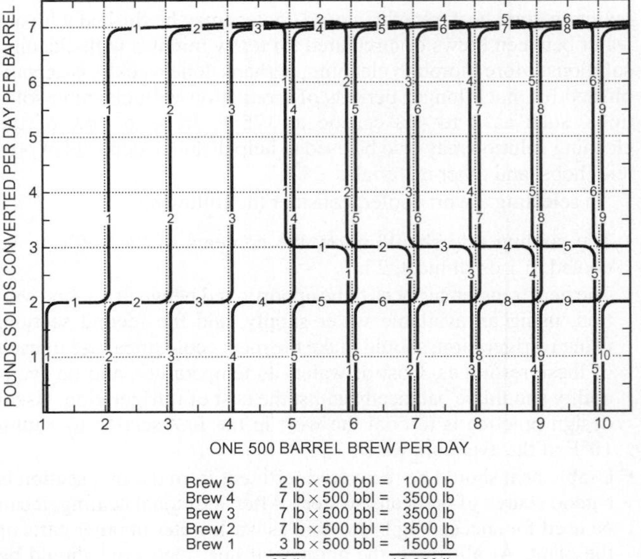

Fig. 2 Solids Conversion Rate

difference between the coolant and fermenting beer; (3) maximum daily sugar conversion rate; and (4) heat evolved, which is at the rate of 280 Btu/lb of fermentable sugar converted.

Assuming a square fermenting tank 13 ft per side to hold a brew of 500 bbl and allowing 1 ft between the tank wall and the attemperator for easy cleaning, an 11 ft^2 attemperator can be used, giving 44 feet of tubing.

From Figure 2, the maximum daily conversion rate is 7 lb/bbl. Calculating for 500 bbl per day at 280 Btu/lb of sugar converted,

$$7 \times 500 \times 280 = 980{,}000 \text{ Btu/day or } 40{,}833 \text{ Btu/h}$$

Assuming a 50°F fermenting beer temperature and 20°F brine (a temperature difference of 30°F) and a heat transfer rate of 15 Btu/h·ft^2·°F for the attemperator, the surface area required is

$$40{,}833/(15 \times 30) = 90.7 \text{ ft}^2$$

Considering 4 in. OD tubing with an external area of 1.05 ft^2/ft, the length required is

$$90.7/1.05 = 86.4 \text{ ft}$$

Two attemperators (each 11 ft square) give 88 ft of tubing, which is adequate for the conditions outlined.

Old attemperators usually consisted of one or more rings of 3 or 4 in. copper or stainless tubing, concentric with the walls of the tank and supported at about two-thirds of the height of the liquid. Almost all modern fermenter designs use exterior jackets for temperature control. The side wall of the fermenter may have two or three individual jackets and the cone or bottom of vertical fermenters may also have one or two small jackets. Each jacket uses a baffled flow or dimple plate design that maintains good flow and good heat transfer across all areas of the jacket. A glycol solution or liquid ammonia may be circulated through the cooling jackets. These tank changes, dictated by automation and economics, allow easier in-place tank cleaning and provide more cooling effect in fermenting.

Stock Cellar

The stock cellar may be a refrigerated room containing storage tanks that do not have any cooling capacity, or it may be an ambient room containing storage tanks with exterior jackets or interior cooling surfaces. Cooled beer from fermentation is transferred into these tanks for aging or maturing, as the process is sometimes called. Some brewers prefer some yeast carry-over into aging, so they simply transfer the fermented beer into the stock cellar tanks. Other brewers do not want as much (or perhaps no) yeast carry-over. Storage residence time varies, depending on the wishes of the brewer. Typical residence times for modern breweries range from 5 to 15 days, but much longer times may be used. Under cold-storage conditions, slow, subtle chemical changes take place that are very important to the final flavor and aroma profile of the beer or ale. Physical changes, such as precipitation of insoluble proteins, also occur. These changes are important for preventing haze formation in the finished product.

Modern aging tanks are normally pressurized with carbon dioxide to prevent air from coming into contact with the aging beer. For stock cellars that use storage tanks that are vented to the atmosphere, adequate provision must be made to supply fresh air in sufficient amounts to keep the CO_2 concentration below 0.5%. Air-conditioning equipment, using chemical dehumidification and refrigeration, is generally used to maintain dry conditions such as 32°F and 50% rh in storage areas. This decreases mold growth and rusting of steel girders and other steel structures. To maintain lower CO_2 concentration in tightly closed cellars and to reduce operational cost, heat exchange sinks and thermal wheels are used to cool incoming fresh air and to exhaust cold stale air.

Air compressor systems commonly use air driers with refrigerated aftercoolers, 32°F glycol coolers, desiccant drying, or a combination. This is necessary if lines pass through areas below 32°F.

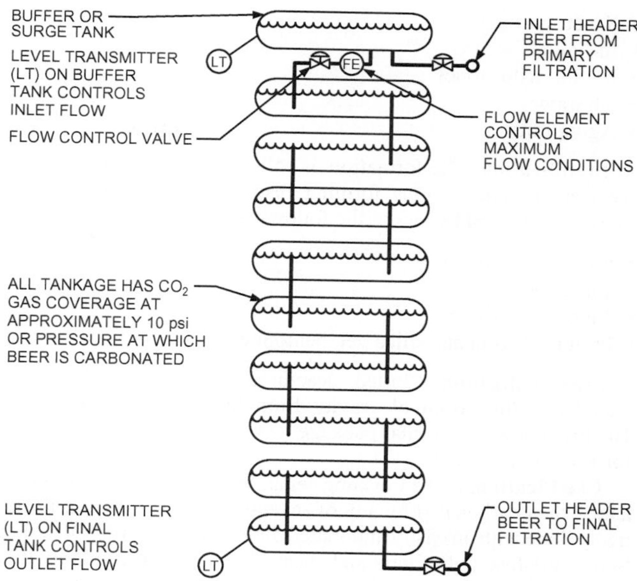

Fig. 3 Continuous Aging Gravity Flow

A continuous aging process used in multistory buildings, all gravity flow, is shown in Figure 3. The process is better for larger operations that principally produce one brand of beer.

Kraeusen Cellar

Instead of carbonating the beer during the finishing step, some brewers prefer to carbonate by the Kraeusen method. In this procedure, fully fermented beer is moved from the fermenting tank to a tank capable of holding about 20 psig. A small percentage of actively fermenting beer is added. The tank is allowed to vent freely for 24 to 48 h, then is closed and the CO_2 pressure allowed to build. Because the amount of CO_2 retained in the beer is a function of temperature and pressure, the brewer can achieve the desired carbonation level by controlling either or both pressure and temperature. After Kraeusen fermentation, generally a week or more, the beer may be moved to another storage tank. However, the brewer can accomplish the same effect by leaving the beer in the Kraeusen tank and cooling the beer by space cooling, tank coils, or both.

Heat is generated by this secondary fermentation, but the temperature of the liquid does not rise as high as it did in the fermenter because fermentable sugars are only available from the small percentage of actively fermenting beer, added as Kraeusen. Furthermore, the bulk of the liquid may have a lower starting temperature than in primary fermentation. Typically, a temperature of 40 to 50°F may be reached at the peak, after which the liquid cools to the ambient temperature of the room. This cooling can be accelerated in the tanks by circulating a cooling liquid, such as propylene glycol, through attemperators. Because heat is generated during Kraeusen fermentation, refrigeration load calculations must include removal of this heat by transfer to air in the cellar, by tank coils, or by a combination of both. Furthermore, if the tank is to be used as a storage tank, the calculation must include the necessary heat removal to reduce the beer temperature to the desired level.

Finishing Operations

After flavor maturation and clarification in the storage tanks, the beer is ready for finishing. Finishing includes carbonation, stabilization, standardization, and clarification.

Carbonation. Any of the following processes are used to raise the CO_2 concentration from 1.2 to 1.7 volumes/volume to about 2.7 volumes/volume:

- Kraeusen
- In-line
- In-tank with stones
- Saturator
- Aging train

Stabilization. The formation of colloidal haze, caused by soluble proteins and tannins forming insoluble protein/tannin complexes, is reduced by any of the following materials:

- Enzymes (papain)
- Tannic acid
- Tannin absorbents
- Protein absorbents, silica gel, bentonite

Standardization. Chilled, deaerated, and carbonated water is added to adjust original gravity from high-gravity level (14 to 16° Brix) down to normal package levels (10 to 12° Brix or lower for low-calorie beer).

Clarification. In the finishing cellar, beer is polished by filtration and is then carbonated by any of several methods. Filtering is normally done with an easily automated diatomaceous earth (DE) filter. Some cellulose pulp filters and sheet filters are still used, but they require more labor. After the filtration step, frequently a cartridge-type filter will be used to trap any particles that may still be present. Recently, various types of cartridges and membranes have been used to produce products that are essentially sterile. The number of filters used depends on the brilliance desired in the finished product. After this final processing, beer is transferred to the government cellar and held until it is needed for filling kegs in the racking room or bottles or cans in the packaging plant. In some breweries, initial clarification is accomplished using centrifuges. This reduces the load on the filtration system, allowing higher flow rates and longer filter runs.

Outdoor Storage Tanks

Some breweries use vertical outdoor fermenting and holding tanks (similar to those popular with dairies). These tanks have working capacities of 2000 to 10,000 barrels. The geometry of these tanks includes a conical bottom and height-to-diameter ratios from 1:1 to 5:1. The tanks are jacketed and use propylene glycol or the direct expansion of ammonia for cooling. Insulation is usually 4 to 6 in. thick polyurethane foam with a stainless steel cladding. They may be built as fermenters or as aging tanks, or in many cases, the same tank may serve for both fermenting and aging, with no beer transfer.

Hop Storage

If raw hop cones are used, they should be stored at a temperature of 32 to 34°F with 55 to 65% rh and very little air motion to prevent excessive drying. Sweating of the bales should not be permitted because this would carry off the light aromatic esters and deteriorate the fine hop character. Nothing else should be stored in the hops cellar because foreign odors may be absorbed by the hops, which would result in off-flavors in the beer. Hops pellets are packed in airtight, sealed containers, but should be stored near or below 32°F to prevent flavor and aroma deterioration. Hop extracts are generally very stable and may be stored at ambient temperatures.

Yeast Culture Room

In the yeast culture room, yeast is propagated to be used in reseeding and replacing yeast that has lost its viability. Normal fermentation of aerated wort also propagates yeast. The amount of yeast roughly triples during fermentation, depending on the degree of aeration. A portion of this yeast is repitched (reused) in later fermentation, and the balance is discarded as waste yeast, which is sometimes sold for other purposes. Clean yeast, usually the middle layer of the yeast deposit that remains in a fermenting tank after removal of the beer, is selected for repitching.

Repitched yeast is carefully handled to avoid contamination with bacteria and is stored in the yeast room as a liquid slurry (yeast balm) in suitable vats. If open vats are used, 80% rh is required to prevent the yeast from hardening on the vat walls. The CO_2 blanket on top of the vats should not be disturbed by excessive air motion. There is considerable variation in yeast handling and recycling practices.

PASTEURIZATION

Plate pasteurizers heat beer to a temperature sufficient for proper pasteurization (15 s at 160°F or 10 s at 165°F) and then cool the pasteurized product with incoming cold beer. Plate pasteurizers and microfiltration are used to produce a beer that is similar to draft beer but does not require refrigeration to prevent spoilage. It is distributed in bottles and cans that can be of slightly lighter construction because they do not have to withstand the high pressure created in tunnel pasteurizers.

CARBON DIOXIDE

The amount of CO_2 produced per barrel of wort depends on the original gravity (starting sugar concentration) and the final gravity of the beer (ending sugar concentration). Depending on the type of fermenter and amount of free head space at the start of fermentation, over 75% of this gas may be collected, purified, and liquefied for later use in the brewery. In addition to carbonating the beer to the proper level, CO_2 is also used to purge air from tanks, to push beer from one tank to another, for pressure in tanks of beer, and for operating bottle, can, and keg lines.

Decades ago, open-top fermenters were common. Carbon dioxide from fermentation filled the fermenting room and was a serious health hazard. Concentrations below 0.5% were generally considered to be safe for the operators, but higher concentrations reduced a worker's efficiency. Concentrations between 4 and 5% were considered too dangerous to work in for more than a minute or two. Because carbon dioxide is heavier than air, it tends to settle to the floor. Fermenting rooms were constructed with outlets near the floor where the elevated concentrations of carbon dioxide could be withdrawn. Fresh air inlets were located at the upper levels of the room.

Almost all modern breweries now use fully closed fermentation tanks. Not only do these closed tanks protect the process from accidental contamination, but all carbon dioxide can be either vented outside of the room, or directed to a collection system. As federal and state regulation became more common, very strict safety standards were adopted for exposure to carbon dioxide in the working areas. To comply with these standards and avoid potential penalties from agencies such as the Occupational Safety and Health Administration (OSHA), many modern breweries now use monitoring systems to detect elevated concentrations of carbon dioxide in the various enclosed work areas in the brewery.

Collection

Carbon dioxide gas, produced as a by-product of fermentation, can be collected from closed fermenters, compressed, and stored in pressure tanks for later use. It may be used for final carbonation, counterpressure in storage and finishing tanks, transfer, and bottling and canning. In the past, the CO_2 was stored in the gaseous state at about 250 psig. Today, however, in most medium and large breweries, the gas is collected and, after thorough washing and purification, it is liquefied and stored. Carbon dioxide stored in the liquid state occupies about 2% of the volume of an equal mass of gas at the same pressure at room temperature.

As an example, from each barrel of wort fermented, about 13 lb of CO_2 is generated over a period of five days, though not at a constant daily rate. Therefore, brews must be carefully scheduled to provide the necessary CO_2 gas, thereby minimizing storage requirements. As a general rule, only about 50 to 60% of the total gas generated is collected. Gas generated at the beginning and end of the fermentation cycle is discarded because of excessive air content and other impurities.

From the fermenting tank, the gas is piped through a foam trap to a gas pressure booster. Surplus gas is discharged to the outside from a water-column safety relief tank, which also protects the fermenting tank from excessive gas pressure. To compensate for friction pressure loss in the long lines to the compressor and to increase its capacity, the booster raises the pressure from as low as 1 in. of water to 5 or 6 psig.

Compressors, which in the past were two-stage with water injection, are being replaced by nonlubricated compressors that use carbon or nonstick fluorocarbon rings. Today, lubricated screw and reciprocating compressors are used for food and beverage-grade CO_2 production in commercial CO_2 plants and in some large breweries. These may be single two-stage compressors or two compressors comprising individual high and low stages. A complete collection system consists of suction and foam trap; rotary boosters, where required; scrubber; deodorizer; compressor (or compressors); intercoolers and aftercoolers; dehumidifying tower; condenser (with refrigeration from a separate system to ensure that no CO_2 enters the main system through leaks); liquid storage tanks; and vaporizers, all interconnected and automatically controlled.

Liquefaction

The condensing pressures of carbon dioxide at several temperatures are

–20°F	200 psig
–14°F	225 psig
–8°F	252 psig

The latent heat at saturation temperature is about 120 Btu/lb. The refrigerant for liquefying the compressed gas should be about –21°F to condense the CO_2 effectively. Most of the moisture must be removed from the compressed gas; this may be done by passing the gas through a horizontal-flow finned coil (located in a 36°F cellar), which condenses out about 80% of the moisture (i.e., the condensate is drained from the system). Also, this is done effectively with refrigerant-cooled precoolers, intercoolers, and aftercoolers. Sending the gas through desiccant driers removes additional moisture. The emerging gas has a slightly higher temperature, but has a dew point around –70°F or lower.

Under these conditions, gas is liquefied when it comes in contact with the liquefying surfaces, which stay ice-free because of the low moisture content (–40°F dew point) of the gas, thus ensuring continuous service. Dryers are installed in duplicate with automatic timing for regeneration of the desiccant material. Desiccant dryers usually rely on heated CO_2 as the regeneration gas. An earlier method used dual sets of double-pipe dryers, which froze out moisture and retained it in the heat exchanger.

Liquefiers are vertical shell-and-tube, inclined double-pipe, or shell-and-tube types. The refrigerant side is operated fully flooded, with refrigerant supplied from a system separate from the main system. Carbonating systems have changed with all-closed fermenters, refrigerated condensing systems, and large liquid CO_2 holding tanks. See Figure 4 for collecting and liquefaction system flow diagrams.

CO_2 Storage and Reevaporation

Condensed CO_2 drains into a storage tank, which is usually designed for a working pressure of 300 psig and varying storage capacities of 10,000 to 120,000 lb each. The vessel is insulated and is equipped with equalizing connections, safety valves, liquid-level indicators, and electric heating units. Gas purity tests are regularly conducted from samples withdrawn from above the liquid level.

As liquid is withdrawn from the tank, it is introduced to a steam-heated liquid vaporizer, which is automatically controlled to give the desired superheat to the vaporized gas. This type of vaporizer is now replacing other types because of its ability to control the temperature of CO_2 gas. Vaporized gas is directed to large, high-pressure surge tanks, where the pressure may be over 200 psig. Carbon dioxide from these tanks passes through pressure reduction valves to supply the brewery with gas that normally ranges from 80 to 100 psig.

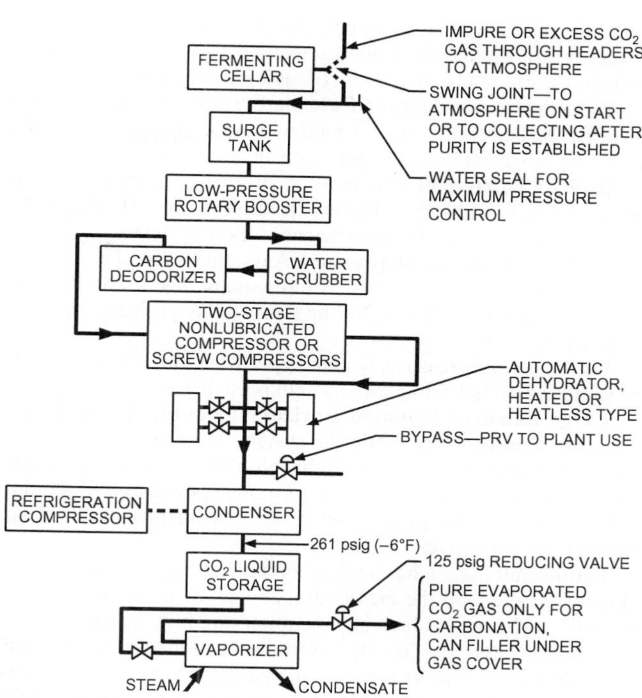

Fig. 4 Typical Arrangement of CO_2 Collecting System

HEAT BALANCE

Most of the steam required for processing, heating water, and general plant heating can be obtained as a by-product. Because the manufacture of beer is a batch process with various peaks occurring at different times, the study of the best heat balance possible is difficult. In a given plant, it depends on many variables, and a comprehensive study of all factors is necessary.

In brewery plant locations where electric energy costs are high, installation of cogeneration facilities can be favorable. However, in plants that produce in excess of 3 million bbl annually, the steam turbine as a prime mover often comes into prominence. A bleeder steam turbine operating at 400 psig can be used to drive a refrigeration compressor, electrical generator, or both; steam bled from it can be used for process and other needs requiring low-pressure steam. In smaller plants, less favorable heat balances must be accepted in line with more economical plant investment programs. Each brewery requires individual study to develop the most economical program.

Many process steam loads are highly variable; for example, the brew kettle's warm-up cycle (both equipment mass and the content's mass) requires three to five times as much steam as brewing does. Steam plant (boiler) sizing is significantly affected by this changeability. Initial boiler size/capacity can be reduced by installing a heat storage tank (e.g., for brewing water) that can be heated over a longer period of time, thus reducing the dynamic peak load required at batch start-up.

COMMON REFRIGERATION SYSTEMS

Absorption Machine for Heat Balance (especially for air conditioning and water cooling for wort cooling). The unit requires a careful heat balance study to determine whether it is economical. These machines may be a good selection if excess 20 psig steam (single-effect) or 110 psig steam (double-effect) is available.

Halocarbon Refrigerant Cascade System. Oil-free ammonia as brine can be pumped at a 1:1 ratio with water. The ratio of motive power to refrigeration at 25 and 185 psig is 1.4 bhp/ton.

Direct Centrifugal. This is often an oil-free ammonia system with a 1:1 ratio with water. The unit requires pumps. This is usually the most expensive and least efficient of these systems.

Oil-Sealed Screw Compressors. Ammonia is circulated at a 1:1 ratio in this system. The units require pumps. The ratio of motive power to refrigeration is 1.26 bhp/ton. This is probably the least expensive system.

Oil-Free Compressors, Screw Type. The ratio of motive power to refrigeration is 1.5 bhp/ton. System noise levels can be high.

Large Balanced Opposed Horizontal Double-Acting Reciprocating Compressor. While the system is not oil-free, good oil separation equipment minimizes this problem. It can use recirculation or direct expansion. The ratio of motive power to refrigeration is 1.25 bhp/ton. This system requires the most maintenance, and can be the most expensive.

Further automation has been accomplished by programming the flow of materials in the brewhouse, as well as the entire brewing operation. The newest brewing operations are fully automated.

Where necessary, a cooling tower may be used to reduce thermal pollution or to conserve water in the pasteurizing phase. Ecology plays an important part in the brewery; stacks are monitored for particulates, effluent is checked, and heat from kettle vents and others is recovered. Water use is more closely regulated, and refrigeration systems use water-saving equipment, including evaporative condensers.

Food-Grade Brines. Some of the cooling temperatures (e.g., ingredient water required for finished product mixing from heavy-gravity brewing) are near freezing, thus requiring coolant temperatures below freezing and consequently the necessary brine in the coolant. In food and beverage facilities, this brine should be food-grade propylene glycol so that minor leaks (e.g., heat exchanger pinholes) into the ingredient will not render the finished product nonsalable. If other plant systems require brine solutions, they should also use food-grade propylene glycol to eliminate the chance of injecting a non-food-grade brine into a food-grade system.

VINEGAR PRODUCTION

Vinegar is produced from any liquid capable of first being converted to alcohol (e.g., wine, cider, malt) and syrups, glucose, molasses, and the like.

First Stage: Conversion of sugar to alcohol by yeast (anaerobic)

$$C_6H_{12}O_6 \rightarrow 2CH_3CH_2OH + 2CO_2$$

Second Stage: Conversion of alcohol to acetic acid by bacterial action (aerobic)

$$CH_3CH_2HO + O_2 \rightarrow CH_3COOH + H_2O$$

Bacteria are active only at the surface of the liquid where air is available. Two methods are used to increase the air-to-vinegar surface:

The **packed** (or **Frings**) **generator** is a vertical cylinder with a perforated plate and is filled with oak shavings or other inert support material intended to increase column surface area. The weak alcohol and vinegar culture are introduced, and the solution is continuously circulated through a sparger arm, with air introduced through drilled holes in the top of the tank. A heat exchanger is used to remove the heat generated and to maintain the solution at 86°F. This is a batch process requiring 72 h.

In **submerged fermentation**, air is distributed to the bacteria by continuously disbursing air bubbles through the mash in a tank filled with cooling coils to maintain the 86°F temperature. This also is a batch process, requiring 39 h.

Concentration is best accomplished by removing some of the water in the form of ice, which increases the acid concentration by 12 to 40%. In freezing out water, a rotator is often used. About 0 to 10°F is required on the evaporating surface to produce the best crystals; the ice is separated in a centrifuge. The vinegar is then stored 30 days before filtering. Effective concentration can also be achieved by distillation (as is done for distilled white vinegar).

WINE MAKING

The use of refrigeration to control the rates of various physical, chemical, enzymatic, and microbiological reactions in commercial wine making is well established. Periods at elevated temperatures, followed by rapid cooling, can be used to denature oxidative enzymes and proteins in grape juices, to retain desirable volatile constituents of grapes, to enhance the extraction of color pigments from skins of red grapes, to modify the aroma of juices from certain white grape cultivars, and to inactivate the fungal populations of mold-infected grapes. Reduced temperatures can slow the growth rate of natural yeast and of the enzymatic oxidation of certain phenolic compounds, assist in the natural settling of grape solids in juices, and favor the formation of certain by-products during fermentation. Also, reduced temperatures can be used to enhance the nucleation and crystallization of potassium bitartrate from wines, to slow the rate of aging reactions during storage, and to promote the precipitation of wood extractives of limited solubility from aged brandies.

The extent to which refrigeration is used in these applications depends on such factors as the climatic region in which the grapes have been grown, the grape cultivars used, the physical condition of the fruit at harvest, the styles and types of wines being produced, and the discretion of the winemaker.

Presently, the wine industry in the United States is heavily committed to the production of table wines (ethanol content less than 14% by volume). Considerably less emphasis is being placed on the production of dessert wines and brandies than in the past. Additionally, the recent growth in wine cooler popularity has significantly altered winery operations where they are produced. A variety of enological practices and winery equipment can be found between the batch emphasis of small wineries (crushing tens of tons per season) and the continuous emphasis of large wineries (crushing hundreds or thousands of tons per season).

The applications of refrigeration will be classified and considered in the following order:

1. Must cooling
2. Heat treatment of red musts
3. Juice cooling
4. Heat treatment of juices
5. Control of fermentation temperature
6. Potassium bitartrate crystallization
7. Control of storage temperatures
8. Chill-proofing of brandies

MUST COOLING

Must cooling is the cooling of crushed grapes before separating the juice from the skins and seeds. White wine grape musts will often be cooled before being introduced to a juice-draining system or a skin-contacting tank; this is done to reduce the rate of oxidation of certain juice components, as well as to prevent the onset of spontaneous fermentation by wild and potentially undesirable organisms. Must cooling can be used when grapes are delivered to the winery at excessively high temperatures or when they have been heated to aid in pressing or extracting red color pigments.

In general, tube-in-tube or spiral heat exchangers are used. Tubes of at least 4 in. internal diameter with detachable end sections of large-radius return bends are necessary to reduce the possibility of blockage by any stems that might be left in the must after the crushing-destemming operation. The cooling medium can be chilled water, a glycol solution, or a directly expanding refrigerant. Overall heat transfer coefficients for must cooling range between 70 and 125 Btu/h·ft^2·°F, depending on the proportions of juice and skins, with the must side providing the controlling resistance. In small wineries, jacketed draining tanks and fermenters are often used to cool musts in a relatively inefficient batch procedure in which the overall coefficients are on the order of 1.70 to 5.30 Btu/h·ft^2·°F because the must is stationary and, therefore, rate-controlling.

HEAT TREATMENT OF RED MUSTS

Most red grapes have white or greenish-white flesh (pulp) and juice. The coloring matter or pigments (anthocyanins) reside in the skins. Color can be rapidly extracted from these varieties by heat treating musts so that the pigment-containing cells are disrupted before actual fermentation. This process is done in several countries throughout the world when grape skins are low in color or when color extraction during fermentation is poor. Must heating is necessary to produce the desirable flavor profile of some varieties, most notably Concord, when manufacturing juices, jellies, and wines. The series of operations is referred to as thermovinification. The must is heated to temperatures in the range of 135 to 167°F, generally by draining off some of the juice, condensing steam, and returning the hot juice to the skins for a given contact time, often 30 min. The complete must can be cooled prior to separation and pressing, or the colored juice can be drawn off and cooled prior to the fermentation.

Heat treatment can also be used to inactivate the more active oxidative enzymes found in red grapes infected by the mold *Botrytis cinerea*. It can further aid the action of pectic enzyme preparations added to facilitate pressing of some cultivars. In all cases, the temperature/time pattern used is a compromise between desirable and undesirable reactions. The two most undesirable reactions are caramelization and accelerated oxidation of the juice. Condensing steam and tube-in-tube exchangers are generally used for these applications, with design coefficients similar to those given previously for must cooling.

JUICE COOLING

Juices separated from the skins of white grapes are usually cooled to between 35 and 70°F to aid natural settling of suspended grape solids, to retain volatile components in the juice, and to prepare it for cool fermentation. Tube-in-tube, shell-and-tube, and spiral exchangers and small jacketed tanks are used with either direct expansion refrigerant, propylene glycol solution, or chilled water as the cooling medium. Overall coefficients for juice cooling range between 95 and 150 Btu/h·ft^2·°F for the exchanger and 4.40 to 8.80 Btu/h·ft^2·°F for small jacketed tanks.

The small jacketed tank values can be improved significantly by juice agitation. Transport and thermal properties of 24% (by mass) sucrose solutions can be used for grape juices. There is a general tendency for medium and large wineries to use continuous-flow juice cooling arrangements of jacketed tanks.

HEAT TREATMENT OF JUICES

Juices from sound grapes can be exposed to a high-temperature, short-time (HTST) treatment to denature grape proteins, reduce the number of unwanted microorganisms, and, some winemakers believe, enhance the varietal aroma of certain juices. Denaturation of proteins reduces the need for their removal (by absorptive clays such as bentonite) from the finished wine. However, turbid juices and wines can result from this treatment, presumably because of modification of the pectin and polysaccharide fraction. Clarified juices from mold-infected grapes can be treated in a similar way to denature oxidative enzymes and to inactivate the molds. Most wineries in the United States rely on "pure culture" fermentation to achieve consistently desirable results; hence the value of HTST treatment in the control of microbial populations.

A typical program includes rapidly raising the juice temperature to 194°F, holding it for 2 s, and rapidly cooling it to 60°F. Plate heat exchangers are used because of their thin film paths and high overall coefficients. Grape pulp and seeds can cause problems in this equipment if they are not completely removed beforehand.

FERMENTATION TEMPERATURE CONTROL

Anaerobic conversion of grape sugars to ethanol and carbon dioxide by yeast cells is exothermic, although the yeast is capturing a significant quantity of the overall energy change in the form of high-energy phosphate bonds. Experimental values of the heat of reaction range between 79.3 and 95.3 Btu/mole, with 94.4 Btu/mole being generally accepted for fermentation calculations (Bouffard 1985).

One gallon of juice at 16.37 lb/ft^3 sucrose (24° Brix) will produce approximately 60 gal of carbon dioxide during fermentation. Allowing for the enthalpy lost by this gas, with its saturation levels of water and ethanol vapors, the corrected heat release value is 90.9 Btu/mole at 59°F and 83.6 Btu/mole at 77°F. The adiabatic temperature rise of the 16.37 lb/ft^3 juice would then be 87°F, based on the 59°F value, and 82°F, based on the 77°F value. Whether a fermentation approaches these adiabatic conditions depends on the difference between the rate of heat generation by fermentation and the rate of heat removal by the cooling system. For constant-temperature fermentations, which are the most common type of temperature control practiced, these rates must be equal. Red wine fermentations are generally controlled at temperatures between 75 and 90°F, whereas white wines are fermented at 50, 60, or 68°F, depending on the cultivar and type of wine being produced. The more rapid fermentations of red wines are used in the cooling load calculations of individual fermenters; a more involved composite calculation, allowing for both red and white fermentations staggered in time, is necessary for the overall daily fermentation loads.

At 68°F, red wines have average fermentation rates in the range of 2.5 to 3.1 lb/ft^3 per day, which correspond to heat release rates of approximately 21 to 26 Btu/ft^3 per hour. The peak fermentation rate is generally 1.5 times the average, leading to values of 31.5 to 39.0 Btu/ft^3 per hour. This value, multiplied by the volume of must fermenting, provides the maximum rate of heat generation. The heat transfer area of the jacket or external exchanger can then be calculated from the average coolant temperature and overall heat transfer coefficient.

The largest volume of a fermenter of given proportions, with fermentation that can be controlled by jacket cooling alone, is a function of the maximum fermentation rate and the coolant temperature. The limitation occurs because the volume (and hence the heat generation rate) increases with the diameter cubed, whereas the jacket area (and hence the cooling rate) only increases with the diameter squared. Similarly, the temperature rise in small fermenters, cooled only by ambient air, depends on the fermenter's volume and shape and the ambient air temperature (Boulton 1979a).

Development of a kinetic model for wine fermentations (Boulton 1979b) has made it possible to predict the daily or hourly cooling requirements of a winery. The many different fermentation temperatures, volumes, and starting times can now be incorporated into algorithms that predict future demands and schedule off-peak electricity usage, allowing for optimal control of refrigeration compressors.

POTASSIUM BITARTRATE CRYSTALLIZATION

Freshly pressed grape juices are usually saturated solutions of potassium bitartrate. The solubility of this salt decreases as alcohol accumulates, so newly fermented grape wines are generally supersaturated. The extent of supersaturation and even the solubility of this salt depend on the type of wine. Young red wines can hold almost twice the potassium content at the same tartaric acid level as young white wines, with other effects caused by pH and ethanol concentration. Because salt solubility also decreases with temperature, wines will usually be cold-stabilized so that crystallization occurs in the tank rather than in the bottle if the finished wine is chilled. In the past, this was done by holding the wine close to its congealing temperature (usually 25 to 21°F for table wines) for two to three weeks. Crystallization at these temperatures can be increased dramatically by introduction of nuclei, either potassium bitartrate powder or other neutral particles, and subsequent agitation. In modern wineries, it is particularly important to supply nuclei for crystallization, because unlike older wooden cooperage, stainless steel tanks do not offer convenient sites for rapid growth. Holding times can be reduced to 1 to 4 h by these methods. Several continuous and semicontinuous processes have been developed, most incorporating an interchange of the cold exit stream with warmer incoming wine (Riese and Boulton 1980). Dessert wines can be stabilized in the same manner, except that the congealing temperatures are usually in the range of 12 to 7°F. In all wines, it is usual for stabilization to occur sometime after grape harvest and for suction temperatures of the refrigeration compressors to be adjusted in favor of low coolant temperatures rather than refrigeration capacity.

STORAGE TEMPERATURE CONTROL

Control of storage temperature is perhaps the most important aspect of postfermentation handling of wines, particularly generic white wines. Transferring wine from a fermenter to a storage vessel generally results in at least a partial saturation with oxygen. The rates at which oxidative browning reactions (and the associated development of acetaldehyde) advance depend on the wine, its pH and free sulfur dioxide level, and its storage temperature. Berg and Akiyoshi (1956) indicate that, in oxidation of white wines, for temperatures below ambient, the rate was reduced to one-fifth its value for each 18°F reduction in temperature. Similar studies of hydrolysis of carboxylic esters (Ramey and Ough 1980) produced during low-temperature fermentations indicate that the rate was more than halved for each 18°F reduction in temperature. These latter data suggest that, on average, the esters have half-lives of 380, 600, and 940 days when wine is stored at 60, 50, and 40°F, respectively. As a result, wines should generally be stored at temperatures between 40 and 50°F if oxidation and ester hydrolysis are to be reduced to acceptable levels. The importance of cold bulk wine storage will likely increase as vintners strive to reduce the amount of sulfur dioxide used to control undesirable yeasts.

Cooling requirements during storage are easily calculated using the vessels' dimensions and construction materials as well as the thickness of the insulation used.

CHILL-PROOFING BRANDIES

In brandy production, refrigeration is used in the chill-proofing step, just before bottling. When the proof is in the range of 100 to 120 (50 to 60% v/v ethanol), aged brandies contain polysaccharide fractions extracted from the wood of the barrels. When the proof is reduced to 80 for bottling, some of these components with limited solubility become unstable and precipitate, often as a dispersed haze. These components are removed by rapidly chilling the diluted brandy with a plate heat exchanger to a temperature in the 0 to 32°F range and filtering while cold with a pad filter. The outgoing filtered brandy is then used to precool the incoming stream, thus reducing the cooling load. Calculations can be made by using the properties of equivalent ethanol/water mixtures, with particular attention to the viscosity effects on the heat transfer coefficient.

CARBONATED BEVERAGES

Refrigeration equipment is used in many carbonated beverage plants. The refrigeration load varies with plant and production conditions; small plants may use 150 tons of refrigeration and large plants may require over 1500 tons.

Dependency on refrigeration equipment has diminished in carbonated beverage plants using modern deaerating, carbonating, and high-speed beverage container-filling equipment. In facilities that use refrigeration, product water is often deaerated before cooling to aid carbonation. In addition, cooling the product at this stage of production (1) facilitates carbonation to obtain maximum stability of the carbonated beverage during filling (reduces foaming), (2) permits reducing the pressure at which the beverage is filled into the container (minimizing glass bottle breakage at filler), and (3) reduces overall filling equipment size and investment.

Immediately before filling, beverage product preparation requires the use of equipment for proportioning, mixing, and carbonating so that the finished beverage has the proper release of carbon dioxide gas when it is served. The equipment for these functions is frequently found as an integrated apparatus, often called a mixer-carbonator or a proportionator.

Table 2 lists the volume of carbon dioxide dissolved per volume of water at various temperatures. At 60°F and atmospheric pressure, a given volume of product water will absorb an equal volume of carbon dioxide gas. If the carbon dioxide gas is supplied to the product water under a pressure of approximately 15 psig (interpolating between 10 and 20 psig), it will absorb two volumes. For each additional 15 psig, one additional volume of gas is absorbed by the water. Reducing the temperature of the product water to 32°F increases the absorption rate to 1.7 volumes. Therefore, at 32°F product temperature, each increase of 15 psig in CO_2 pressure results in the absorption of an additional 1.7 volumes instead of one volume as when the product water temperature is 60°F. Carbonated levels for different products vary from less than 2.0 volumes to around 5.0 volumes.

Table 2 Volume of CO_2 Gas Absorbed in One Volume of Water

Temperature °F	Pressure in Bottle, psig										
	0	10	20	30	40	50	60	70	80	90	100
32	1.71	2.9	4.0	5.2	6.3	7.4	8.6	9.7	10.9	12.2	13.4
40	1.45	2.4	3.4	4.3	5.3	6.3	7.3	8.3	9.2	10.3	11.3
50	1.19	2.0	2.8	3.6	4.4	5.2	6.0	6.8	7.6	8.5	9.5
60	1.00	1.7	2.3	3.0	3.7	4.3	5.0	5.7	6.3	7.1	7.8
70	0.85	1.4	2.0	2.5	3.1	3.7	4.2	4.8	5.4	6.1	6.6
80	0.73	1.2	1.7	2.2	2.7	3.2	3.6	4.1	4.6	5.2	5.7
90	0.63	1.0	1.5	1.9	2.3	2.7	3.2	3.6	4.0	4.5	4.9
100	0.56	0.9	1.3	1.7	2.0	2.4	2.8	3.2	3.5	3.9	4.3

BEVERAGE AND WATER COOLERS

The main sanitation requirements for beverage and/or water coolers are hygiene and ease of cleaning, particularly if the beverage is cooled rather than the water. The key point is that water freezes easily, and cooler equipment design needs to avoid this. The early Baudelot tank solved the problem by not forming ice; however, because sanitation of such systems is a problem, their use is not recommended.

If a cooler is needed, most plants choose plate heat exchangers and careful control of temperature. Plate heat exchangers reduce ice formation through high turbulence, which reduces thermal gradients. Furthermore, they are hygienic and easy to clean. These heat exchange devices are normally fed by a brine (not direct refrigeration) and are protected against brine leakage, for example, by ensuring that the brine pressure is lower than the beverage product/water pressure.

Many beverage plants use coolers with patented direct-expansion refrigeration equipment to achieve system security, hygiene, ease of cleaning, etc., using a Baudelot-type system. However, this equipment is only used for cooling water (not product), making it easier to clean. This is achieved by the equipment manufacturer's long experience with a proprietary system and its attention to detailed equipment design.

When coolers are necessary, it is recommended that this component of the refrigeration system be located adjacent to, or integrated with, the proportioner-mixer-carbonator. Usually these devices are physically positioned next to the beverage container filler. Normally, the refrigeration plant itself should be housed separately from product processing and filling areas, preferably located together with the other plant utilities (boilers, hot water heater, air compressors, etc.).

It is most important to keep the beverage free from contamination by foreign substances or organisms picked up from the atmosphere or from metals dissolved in transit. Consequently, coolers are designed for easy cleaning and freedom from water stagnation. The coolers and product water piping are fabricated of corrosion-resistant nontoxic metal (preferably stainless steel); however, certain plastics are usable. For example, acrylonitrile-butadiene-styrene (ABS) is used in the beverage industry for raw water piping.

Refrigeration Plant

Halogenated hydrocarbons or ammonia refrigerants are commonly used for plants requiring beverage product and/or water coolers. Refrigeration compressors vary from two-cylinder vertical units to larger, multicylinder V-style compressors.

The refrigeration plant should be centralized in larger production facilities. With the multiplicity of product sizes, production speeds, and other factors affecting refrigeration load, an automatically controlled central plant conserves energy, reduces electrical energy costs, and improves opportunities for a preventive maintenance program.

Makeup water and electrical energy costs encourage careful selection and use of compressors, air-cooled condensers, evaporative condensers, and cooling towers. Some plants have economized by using spent water from empty can and bottle rinsers as makeup water for evaporative condensers and cooling towers. Also, thermal storage (e.g., cold glycol storage tanks) can be used to reduce refrigeration equipment size.

As indicated earlier, the temperature to which the product must be cooled depends on the type of filling machinery used, as well as the deaerating-mixing-proportioning-carbonating equipment used. Cooling needs may be divided into three general categories: those that use water (1) at supply temperature or less, (2) at 45 to 55°F, and (3) at 40°F or lower. The exact temperature to which the product should be cooled depends on the specific requirements of the beverage product and the needs of the particular plant. These requirements are primarily based on product preparation, production equipment availability, and capital costs versus operating costs.

The refrigeration load per case has been reduced by improved filling technology. Fill temperatures between 58 and 60°F have been achieved, which raises the required coolant temperature (thus raising refrigerant suction temperatures and lowering compressor power input).

Refrigeration Load

Refrigeration load is determined by the amount of water being cooled per unit of time. This is derived from the maximum fluid output of the beverage filler. Most cooling units are of the instantaneous type; they must furnish the desired output of cold water continuously, without relying on storage reserve.

Knowing the water temperature from the supply source, temperature to which the water is to be cooled, and water demand, refrigeration load can be determined by

$$q_R = Qc_p(t_s - t_c)/24$$

where

q_R = cooling load, tons of refrigeration
Q = water flow rate, gpm
t_s = supply water temperature, °F
t_c = cold-water temperature, °F
c_p = specific heat of water = 1 Btu/lb·°F

In computing the refrigeration load, one of the most troublesome values to determine is the highest temperature the incoming supply water can be expected to reach. This temperature usually occurs during the hottest summer period. Allow for additional supply water warming from flowing through piping and water treatment equipment in the beverage plant.

SIZE OF PLANT

Output of each plant depends on the beverage-filling capacity of the plant production equipment. Small, individual filling units turn out approximately 600 cases of 24 beverage containers (approximately half-pint capacity) per hour, or 240 containers per minute (cpm); intermediate units turn out up to 1200 cases per hour (480 cpm); and high-speed, fully automatic machines begin at approximately 1200 cases per hour and go through several increases in size up to the largest units, which approach 5000 cases per hour (2000 cpm).

Operation of these filling machines, which also determines demand on refrigeration machinery, usually exceeds 8 h per day, especially during summer, when market demands are highest.

An arbitrary classification of beverage plants may be (1) small plants that produce under 1.25 million cases per year, (2) intermediate plants that produce about 2.5 million cases per year, and (3) large plants that produce 15 million or more cases per year. The latter require installation of multiple-filling lines.

The usual distribution area of finished beverages is within the metropolitan area of the city in which the plant is located. Some plants have built such a reputation for their goods that they ship to warehouses several hundred miles away. Local distribution is made from there. A few nationally known products are shipped long distances from producing plants to specialized markets.

In the warehouse, cans and nonreturnable bottles filled with precooled beverage are commonly warmed to a temperature exceeding the dew-point temperature to prevent condensation and resulting package damage. Bottled goods should be protected against excessive temperature and direct sunlight while in storage and transportation. At the point of consumption, the carbonated beverage is often cooled to temperatures close to 32°F.

LIQUID CARBON DIOXIDE STORAGE

Liquefied carbon dioxide, used to carbonate water and beverages, is truck-delivered in bulk to the beverage plant. The liquid is then piped to large outdoor storage-converter tanks equipped with mechanical refrigeration and electrical heating. The typical tank unit is maintained at internal temperatures not exceeding 0°F, so that the equilibrium pressure of the carbon dioxide does not exceed 300 psig and the storage tanks need not be built for excessively high pressures. Full-controlled equipment heats or refrigerates, and safety relief valves discharge sufficient carbon dioxide to relieve excess pressure.

REFERENCES

Berg, H.W. and M. Akiyoshi. 1956. Some factors involved in the browning of white wines. *American Journal of Enology* 7:1.

Bouffard, A. 1985. Determination de la chaleur degagée dans la fermentation alcoölique. Comptes rendus hebdomadaires des séances, Academie des Sciences, Paris 121:357. *Progres agricole et viticole* 24:345.

Boulton, R. 1979a. The heat transfer characteristics of wine fermentors. *American Journal of Enology and Viticulture* 30:152.

Boulton, R. 1979b. A kinetic model for the control of wine fermentations. *Biotechnology and Bioengineering Symposium* 9:167.

MBAA. 1999. *The practical brewer*, 3rd ed. John T. McCabe, ed. Master Brewers Association of the Americas, Madison, WI.

Ramey, D.R. and C.S. Ough. 1980. Volatile ester hydrolysis or formation during storage of model solutions and wine. *Journal of Agriculture and Food Chemistry* 28:928.

Riese, H. and R. Boulton. 1980. Speeding up cold stabilization. *Wines and Vines* 61(11):68.

BIBLIOGRAPHY

Kunze, W. 1999. *Technology brewing and malting*, 2nd ed. Westkreuz-Druckerei Ahrens KG, Berlin.

Priest, F.G. and I. Campbell. 2003. *Brewing microbiology*, 3rd ed. Kluwer, Amsterdam

CHAPTER 27

PROCESSED, PRECOOKED, AND PREPARED FOODS

THERE are many categories of prepared foods. This chapter covers prepared meals, fruits, vegetables, and potato products and gives an overview of the HVAC&R requirements of facilities that process these products.

Processed, precooked, and prepared foods all come under the regulations of the Food and Drug Administration (FDA) 21CFR, and facilities using meat products also come under regulation by the U.S. Department of Agriculture (USDA).

The FDA has also implemented bioterrorism regulations for processors of fruits, vegetables, and potato products. Records must identify previous sources and initial recipients of food products, including its packaging.

To ensure the safety of food products, the Hazard Analysis and Critical Control Point (HACCP) system is now mandated by the FDA. It is a preventive system that builds safety control features into the food product's design and the process by which it is produced. HACCP is used to manage physical, chemical, and biological hazards.

Each food manufacturing site should develop, implement, and maintain its HACCP plan. Areas of concern include HVAC&R equipment and the associated air distribution system, sanitation equipment and procedures, and processing (automation, heating, cooling, temperature control). Chapter 12 includes additional information on HACCP.

Each plant should be equipped with HVAC systems that provide clean, filtered fresh air in quantities to offset any plant process exhaust systems. Negative plant air pressure should be avoided. In addition, HVAC systems should direct airflows and pressures from finished product areas toward initial processing areas to minimize aerosol bacteriological contamination.

MAIN DISHES, MEALS

Main dishes constitute the largest category of processed, precooked and prepared foods. They are primarily frozen products, but many are also refrigerated. Most can be prepared in a conventional or microwave oven. Many contain sauces and/or gravies, so sauce and/or gravy kitchens may be an integral part of production facilities. A principal characteristic of main dishes is the requirement for a substantial number of ingredients, several unit operations, an assembly-type packaging line, and subsequent cooling or freezing in individual cartons or cases. Examples of these products include

- Soups and chowders
- Main dishes of meat, poultry, fish, or pasta
- Complete dinners, each with a main dish (usually with sauce and/or gravies), a vegetable, and dessert
- Lunches and breakfasts
- Ethnic main dishes and dinners, particularly Italian, Mexican, and Asian styles
- Low-calorie or diet versions of many of the above
- Snack foods such as pizza, fish sticks, and breaded items

General Plant Characteristics

Plant facilities for preparing, processing, packaging, cooling, freezing, casing, and storing products vary widely. The variety of products is diverse, and it is beyond the scope of this section to cover all the operations involved in all product formulations, food chemistry, and process details of prepared foods. This chapter also does not cover the basic process details for preparing meat, poultry, and fishery products. For information on these subjects, refer to Chapters 17, 18, and 19, respectively.

A prepared foods plant has production areas and/or rooms for the following: receiving; storage for packaging materials and supplies; storage for ambient, refrigerated, and frozen ingredients; thawing and defrosting; refrigerated in-process storage; mixing, cutting, chopping, and assembly; sauce and/or gravy kitchens; cooking and cooling rice, pasta, and/or other starches; unit operations for preparing main dishes such as meat patties, ethnic foods, and poultry items; dough manufacturing for pies and pizzas; assembly, filling, and packaging; cooling and freezing facilities; casing and palletizing operations; finished goods storage for refrigerated and frozen products; and shipping of outbound finished goods by truck or rail.

In addition, these operations support equipment and utensil sanitation; personnel facilities; and areas for utilities such as refrigeration, steam, water, wastewater disposal, electric power, natural gas, air, and vacuum.

Plants and equipment for producing prepared foods should be constructed and operated to provide for minimum bacteriological contamination and easy cleanup and sanitation. Sound sanitary practices should be followed at all stages of production. This is of particular concern in prepared food plants where the finished product may not receive a kill step (in which harmful microorganisms are inactivated by high temperature, high pressure, electrical fields, etc.) before or after packaging. (See the section on Destruction of Organisms for more information.) All U.S. meat and poultry plants, and hence many prepared food plants, operate under USDA regulations. Finished and raw product paths are controlled to prevent cross contamination, and sanitary standards are strictly followed.

Preparation, Processing, Unit Operations

Initial steps in production of prepared foods involve preparation, processing, and unit manufacture of items for assembly and filling on the packaging line. These generally include scheduling ingredients; thawing or defrosting frozen ingredients, where applicable; manufacturing sauces and gravies; cooking and cooling rice, pasta, and/or other starches; unit operations for manufacturing meat patties and ethnic foods, such as burritos; mixing vegetables and/or vegetables and rice, pasta, or other starches; manufacturing dough; and cooling, storage, and transport operations before packaging.

The preparation of this chapter is assigned to TC 10.9, Refrigeration Application for Foods and Beverages.

These processes require refrigeration for controlled tempering in refrigerated rooms; plate heat exchangers or swept-surface heat exchangers using chilled water or propylene glycol for preparation of sauces and gravies in processors; chilled water for cooling rice, pasta, and/or other starches; refrigerated preparation rooms for meat and poultry products; in-line coolers or freezers for meat patties, burritos, etc.; ice for dough manufacture; and cooler rooms for in-process storage and inventory control.

Refrigeration loads for each of these categories should be calculated individually using the methods described in Chapter 13. In addition, loads should be tabulated by time of day and classified by evaporator temperature, chilled-water, ice, and propylene glycol requirements. An assessment should be made whether an ice builder is appropriate for chilled-water requirements to reduce refrigeration compressor capacity.

Refrigerated rooms should be amply sized, be able to maintain temperatures from 31 to 49°F as required for the specific application, have power-operated doors equipped with infiltration reduction devices, and have evaporators that are easily sanitized. Evaporators for rooms where personnel are working should be equipped with gentle airflow to minimize drafts. Rooms with temperature maintained at 38°F or below should have evaporators equipped for automatic coil defrost. Temperature controls for such rooms should be tamperproof.

Proper safeguards and good manufacturing practices during preparation and processing minimize bacteriological contamination and growth. This involves using clean raw materials, clean water and air, sanitary handling of product throughout, proper temperature control, and thorough sanitizing of all product contact surfaces during cleanup. Sauces, gravies, and cooked products must be cooled quickly to prevent conditions favorable for microbial growth. Refer to Chapter 12 for further information on control of microorganisms, cleaning, and sanitation.

Assembly, Filling, and Packaging

These activities include transporting components to packaging lines; preparing and depositing doughs for pies, ethnic dishes, and pizzas; filling or placing components into containers; placing containers into packages; coding, closing, and checking the packages; and transporting packages to cooling and freezing equipment.

Items that are pumpable, such as gravies and sauces, are usually pumped to a hopper or tank adjacent to the packaging lines. Items such as free-flowing individually quick frozen (IQF) vegetables may be transported to the lines in bulk boxes by lift truck directly from cold storage. For example, mixes of vegetables, rice, pasta, and/or other starches that have been prepared at the plant may be wheeled to the packaging lines in portable tanks or vats. Tempered, prefrozen meat or poultry rolls may be placed on special carts and wheeled to the packaging lines.

A typical packaging line for meals consists of a timed conveyor system with equipment for dispensing containers; a filler or fillers for components that can be filled volumetrically; volumetric timed dispensers for sauces, gravies, and desserts; net weight filler systems for components and mixes that cannot be placed volumetrically; slicing and dispensing apparatus for placing components such as tempered meat or poultry rolls and similarly configured items; line space for personnel to place components that must be placed manually; liquid dispensers for placing spices and flavorings; and a sealing mechanism for placing a sealed plastic sheet over the containers. This system may be two or three compartments wide for high-volume production.

The timed conveyor system is followed by a single filer to align containers in single file before indexing into a cartoner. The containers are automatically inserted into cartons, after which they are coded and sealed. After leaving the cartoner, filled cartons are automatically checked for underweights and tramp metal, and then conveyed to cooling or freezing equipment. Other types of packaging lines may include some or all of the mentioned apparatus; they are usually specific for the prepared food items being filled and packaged.

Comments regarding product and equipment safeguards, good manufacturing practices, and actions to minimize microbial growth apply to packaging activities as well.

It is good practice to air-condition filling and packaging areas, particularly when these areas are subject to ambient temperatures and humidities that can affect product quality and significantly increase potential bacteriological exposure. Also, workers are more productive in air-conditioned areas. Packaging-area air conditioning is usually supplied through air-handling units that also provide ventilation, heat, and positive pressure to the area. Some applications use separate refrigeration systems; others use chilled water or propylene glycol from systems installed for product chilling. Generally, no other significant refrigeration is required for filling and packaging areas.

Cooling, Freezing, Casing

Cartons from packaging are cooled or frozen in different ways, depending on package sizes and shapes, speed of production, cooling or freezing time, inlet temperatures, plant configuration, available refrigeration systems, and labor costs relative to production requirements. Refer to Chapter 16 for additional information.

In small plants, stationary air-blast or push-through trolley freezers are used when flexibility is required for a variety of products. In these cases, fully mechanized in-line freezers are not economically justified. In larger plants with high production rates, mechanized freezers (e.g., automatic plate, belt, spiral belt) are used extensively. These freezers significantly reduce labor costs and provide for in-line freezing.

The prepared foods business has many line extension additions and changes. Each product change results in component differences that may change the freezing load and/or time because of inlet temperatures, latent heat of freezing, and/or package size (particularly depth). Each product should be checked to ensure that production rates for the line extension or different products match the current freezing capacity.

Over time, experience and advances in packing line technology tend to make packaging lines more efficient and capable of higher production rates. These advances often are implemented with existing cooling or freezing capacity, which is relatively constant. This may result in freezer exit temperatures above 0°F if reserve freezing capacity is not available or cannot be physically added because of space limitations. For new or expanded plants, allow space and/or reserve freezer capacity. Increases in packaging line speed and efficiency of 25 to 50% are reasonable to expect. Each instance should be individually evaluated.

Casing the cartoned product follows freezing with in-line production of meals and main dishes. Manual casing is used in small plants and/or with low production rates. High-speed production lines use semiautomatic and automatic casing methods to increase productivity and lower labor costs. Inspection at this point is necessary to ensure that freezing has been satisfactory and that the cartons are properly sealed and not disfigured. It is also important to ensure that cartons do not defrost during conveyor hangups or line stoppages.

Product palletizing follows product casing. Manual palletizing is used for small plants and/or low production rates, and automatic palletizers are used for higher production rates. Most palletizing is done adjacent to or near cold-storage facilities, and the pallets are transported to the cold-storage rooms with a lift truck. Some manual palletizing occurs inside cold-storage rooms, particularly for slow production lines, to prevent product warm-up, but it is more costly because labor rates are higher for workers in cold-storage rooms.

Some plants are equipped with air-blast freezing cells to augment in-line freezing capacity. These are used primarily for cased products when existing in-line freezers are overloaded and to reduce temperatures of products that have been frozen through the latent zone but have not been sufficiently pulled down for placement in cold storage. These products are usually loosely stacked on pallets with enough air space around the sides of the cases to achieve rapid pulldown to 0°F or lower.

Finished Goods Storage and Shipping

Larger prepared foods operations usually have enough cold storage space to store the necessary refrigerated and frozen ingredients and at least 72 h of finished goods production. This volume of space allows proper inventory control, adequate scheduling of ingredients for production, and sufficient control of finished goods to ensure that the product is 0°F or lower before shipment and that it meets the criteria established for product quality and bacteriological counts.

Refrigeration loads for these production warehouses are calculated as suggested in Chapters 13 and 14. Special attention, however, should be given to product pulldown loads and infiltration. Liberal allowances should be made for product pulldown, because freezing problems do occur, and some plants or sections of plants are under negative pressure at times because more air is exhausted than supplied through ventilation, almost always caused by the need to remove an undesirable component (e.g., steam, heat, dust) without commensurate mechanical air supply to offset the exhaust. This can result in infiltration by direct inflow, which is a serious refrigeration load (see Chapter 13) that should be corrected. This problem is not only costly in energy use, but it also makes it difficult to maintain proper storage temperatures.

Refrigerated trucks and/or rail cars are generally used for shipping. Shipping areas are usually refrigerated to 35 to 45°F, and the truck loading doors are equipped with cushion-closure seals to reduce infiltration of outside air. See Chapter 14 for additional information on refrigerated docks. Where refrigerated docks are not provided, take special care to ensure that the frozen product is rapidly handled to prevent undue warming.

Refrigeration Loads

Refrigeration loads cover a wide range of evaporator temperatures and different types of equipment. Most plants cover these with two or three basic saturated suction temperatures. Where two suction temperatures are provided, they are usually at –35 to –45°F for freezing and cold storage and 10 to 20°F for cooling loads. Where provided, a third suction temperature is usually from –20 to –10°F for frozen product storage rooms and other medium- to low-temperature loads. The third suction temperature is advantageous with relatively large frozen product storage loads to reduce energy costs.

Refrigeration loads should be tabulated by time of day, season, evaporator temperature, and equipment type or function. This record should be made periodically for existing operations; it is essential for new or expanded plants. These tabulations reveal the loading diversity and provide guidance for existing operations and for equipment sizing for added capacity or new plants.

In addition, loads should be tabulated for off-shift production, weekends, and holidays to provide proper equipment sizing and for economic operation for these relatively small loads. See Chapter 13 for information on load calculation procedures.

Refrigeration Systems

Most refrigeration systems for prepared foods use ammonia as the refrigerant. Two-stage compression systems are dominant, because compression ratios are high when freezing is involved and energy savings warrant the added expense and complexity. Evaporative condensers are extensively used for condensing the refrigerant. Evaporators are designed for full-flooded or liquid overfeed operation, depending on the equipment or application. Direct-expansion evaporators are not used extensively.

New plant designs limit plant employees' exposure to large quantities of ammonia. This can be accomplished by locating evaporators, such as propylene glycol chillers, water chillers, and ice builders, in or near machine rooms and away from production employees. Freezers can be located in isolated clusters near machine rooms so that the low-pressure receivers are also in or near the machine room. These practices limit exposure and provide a means for close supervision by trained, competent operators.

Some plants use glycol chillers to circulate propylene glycol to evaporators located in production areas. This results in an energy penalty because of the secondary heat transfer, but it is deemed affordable because of less potential exposure to employees and product in the event of an ammonia spill.

Other plants install most of the ammonia main lines on the roof to reduce exposure to personnel and product and to provide accessibility. These mains require close monitoring and inspection, because an ammonia spill can still injure personnel both inside and outside the plant, as well as damage products.

Machine room equipment not only should be sized for the full refrigeration loads imposed, but also should be able to handle relatively small off-shift and weekend loads without using large compressors and components at low capacity levels.

Many plants use energy-saving measures, such as floating head pressure controls with oversized evaporative condensers coupled with two-speed fans, single-stage refrigeration for small areas and loads; variable-speed pumps for glycol chiller systems; ice builders to compensate for peak loads; door infiltration protection devices; added insulation; and computerized control systems for monitoring and controlling the system. These measures should be considered for existing plants that are not so equipped and should be included in the design of plant expansions and new plants.

All refrigeration systems for prepared food plants should comply with applicable codes and standards (see Chapter 49).

Plant Internal Environment

HVAC systems that provide proper ventilation and positive plant pressure relationships are very important in processed and prepared food plants to prevent condensation on building and equipment components and to minimize potential bacteriological cross-contamination between raw and cooked products. New plants and expansions should be so equipped initially, and existing plants should correct conditions that result in condensation and negative plant air pressure.

VEGETABLES

Frozen vegetables are prepared foods in that they are essentially precooked and require minimal preparation. Chapters 15 and 24 have further information on handling, cooling, and storing fresh vegetables. Most vegetables to be frozen are received directly from harvest. Some are cooled and stored to smooth out production, and others are processed directly. They are cleaned, washed, and graded; cut, trimmed, or chopped (if necessary); and then blanched, cooled, and inspected before freezing. At this point, some vegetables are packed into cartons before freezing, whereas others are frozen and then filled into packages, bags, cases, or bulk bins.

Products cartoned before freezing generally can only be manually packed and/or have to be check-weighed before closing the cartons. Examples include broccoli and asparagus spears, leaf spinach, French cut green beans, okra, cauliflower florets, and some volumetrically machine-filled vegetables such as peas, cut corn, and cut green beans. These products are usually frozen in manual plate freezers, automatic plate freezers, stationary airblast tunnels, and push-through trolley freezers.

Products that are frozen before packaging are included in the free-flowing or individually quick frozen (IQF) categories. These include true IQF products such as peas, cut corn, cut green beans, diced carrots, and lima beans, as well as products that are more difficult to IQF such as broccoli and cauliflower florets, sliced carrots or squash, and chopped onions.

Products that are easy to IQF are usually frozen in straight belt freezers, fluidized bed freezers, or fluidized belt freezers. More difficult products to IQF are usually frozen in fluidized bed freezers and fluidized belt freezers. Cryomechanical freezers are sometimes used where high-value, sticky products are frozen. Hydrocooling these products to 45 to 60°F before freezing reduces the freezing load and overall energy requirements if the product can be adequately drained before freezing. Sliced vegetables with large, flat surfaces are particularly difficult to drain.

Products from IQF freezers are packed either directly into cartons or polyethylene bags, into cases for bulk shipment, or into tote bins for repacking at a later date or shipment to other customers for repack or use in prepared foods. The products packed into tote bins for repacking into cartons or polyethylene bags are used for single products, products of various vegetable mixes, and products with butter or cheese sauces. Repacking rather than direct in-season packing is preferred by most companies, because it allows the packer to produce directly for orders, which saves buying finished goods packaging material until required.

Products in tote bins for repacking are placed in dumpers in a cold-storage room adjacent to the packaging lines. Products are metered onto a conveyor in proportion to the end mix required and pneumatically or mechanically conveyed to the filler hopper for volumetrically filled mixes or single products. Products and product mixes that require weighing are generally conveyed to net weight filler systems. After packaging, products are usually cased semiautomatically or automatically and returned to cold storage immediately. The products are rapidly handled to minimize warm-up and clustering.

Corn on the cob is prepared in the same manner as other vegetables, but because of its bulk and to retain quality, it is usually cooled with refrigerated water. After cooling, it is either packaged in polyethylene bags and frozen in stationary blast cells, push-through trolleys, or manual plate freezers; or frozen bare in this same equipment or in straight or multipass straight belt freezers, and packed into cases for institutional use or tote bins for repacking throughout the year.

Raw, whole onions for French fried onion rings are cleaned, sliced, and prepared before being coated in breading machines; then, they are fried in an oil fryer, cooled, and frozen. Almost all production is IQF for restaurants and food service. The product is frozen in various types of belt freezers. Refrigerated precoolers or precooler sections coupled with IQF freezers are common. Handling must be gentle, and product should not be cooled below 5°F before discharge from the freezer because it can become brittle and fractured, resulting in some product downgrading.

International Production

U.S. production of some frozen vegetables, mainly those that require large amounts of hand labor for harvesting and processing or those that have a short production season in the United States, has largely moved to Mexico and Central and South America. This is coupled with an increased demand for these products and other prepared foods for both retail and food service markets.

Products requiring large amounts of hand labor include asparagus, broccoli, cauliflower, Brussels sprouts, okra, and strawberries. These products are packaged for retail distribution in the United States, or packaged in bulk containers for further processing elsewhere as single items or as part of prepared foods. Freezing equipment for these products is primarily manually operated, because automation cannot be financially justified (except for some types of belt freezers).

Short-season products include peas, Lima beans, green beans, and sweet corn. The incentive for producing these products outside the United States is savings in inventory, cold-storage costs, and ability to determine yearly supply more accurately. If an entire year's estimated requirement is based on one short processing season, it must be stored in freezer warehouses at considerable expense. On the other hand, if some of the estimated requirement is produced approximately six months later in another location, storage requirements can be reduced substantially, and the requirements can be estimated more accurately.

Vegetables in Other Prepared Foods

Inclusion of frozen vegetables in other prepared foods has increased as the variety and type of frozen dishes and meals have proliferated for both retail and food service markets. This is driven not only by increased prepared foods sales volume but also by the importance health authorities place on increased vegetable consumption as part of a healthy diet. Vegetable production is in bulk, either in cases or tote bins. Some products are used frozen as a main vegetable or as part of a mix. In mixes, some are used frozen; others are defrosted to mix with additional items before becoming a portion of a meal.

The emphasis on vegetables as part of a healthy diet has also greatly increased production of prepackaged fresh, refrigerated vegetables, either as single items or as combinations of products and salad items. These products have substantially displaced frozen vegetables as a product in areas of the United States where vegetables can be grown for most of the year. Thus, the movement of frozen vegetable production to countries outside the United States has not had a negative economic effect.

Unit operations need to be sanitary and efficient, to preserve product quality. This is especially important for products that have not had a process step to inactivate microorganisms (kill step) before packaging.

Refrigeration Loads and Systems

The principal refrigeration loads for vegetable operations are raw product cooling and storage, product cooling after blanching, freezing, process equipment located in freezer storage facilities, and freezer storage warehouses. Not all vegetable operations have all of these loads; each plant has unique conditions. Raw product cooling and storage are covered in Chapters 15 and 24.

Cooling after blanching is done with fresh water, with refrigerated water, or with evaporation cooling. Fresh water is usually well water at 55 to 60°F that is used once or twice before blanching for product washing, cleaning, or waste product transfer. Refrigerated water is often used in combination with well or municipal water. Refrigerated water at 35 to 40°F reduces freezing loads and, for some products (e.g., cut green beans), enhances quality.

Freezing loads and corresponding freezing capacity vary widely between different vegetables, depending on the initial product temperature and the latent heat of fusion. Special attention must be paid to the variety and size of the particular vegetables to be frozen. A belt freezer designed to freeze 10,000 lb/h of lima beans may only freeze 7500 lb/h of 1 in. cut green beans. In addition, freezing time is approximately twice as long for cut green beans because of differences in shape and bulk. Specific and latent heats of fusion for vegetables are listed in Chapter 9.

Freezer warehouse loads are calculated as suggested in Chapters 13 and 14. Freezer storages in vegetable processing plants have three additional potential loads to consider: (1) the extra capacity reserve needed for product pulldown during peak processing; (2) the negative pressure almost all vegetable facilities are under, which can substantially increase infiltration by direct flow-through; and (3) the process machinery load (particularly pneumatic conveyors) associated with repack operations.

Almost all vegetable freezing operations use ammonia as the refrigerant. Two-stage compression systems are in general use, even in those with short peak seasons. The product cooling load is done at the intermediate suction pressure. Freezing and freezer warehouse loads use the first-stage suction. Design saturated suction temperatures vary from −32 to −40°F in the first stage to 10 to 20°F in the second stage. Design saturated condensing temperatures vary from 85 to 95°F.

A unique feature of vegetable facility refrigeration is a lack of spare equipment and redundancy for those that operate for short periods at peak capacity (usually 1500 to 2500 h/year). In these applications, spare capacity cannot be justified financially. Extra care and maintenance are usually provided before peak season to ensure that full capacity is available and to minimize downtime.

Flooded evaporators are often used for product cooling and cold-storage facilities. Liquid overfeed systems are used for freezing apparatus and for cold-storage facilities and product cooling. Direct-expansion evaporators are rarely used.

Condensing is usually done with evaporative condensers. Some older plants use shell-and-tube condensers with once-through water usage. The warmed water is reused for product washing and cleaning before blanching. This method provides some risk if ammonia leaks into the water stream and should not be used in new installations.

FRUITS

Frozen fruits are processed foods that are thawed before serving, except for specialties such as fruit pies, which are cooked. This section covers fruits that can be successfully frozen. Chapters 15, 22, and 23 have further information on handling, cooling, and storing fruits before processing.

Most fruits to be frozen are received directly from harvest. Some are cooled and stored to even out production, and others are processed directly. Fruits are typically cleaned, washed, and graded; cut, trimmed, or sliced (if required); and then inspected before freezing. At this point, some fruits are packaged with sugar or syrup before freezing, whereas others are frozen and then filled into polyethylene bags, cases, and bulk bins.

Usually, no special step is taken to kill pathogens in processed fruit. Hence, it is imperative that strict sanitary practices and standards be imposed to minimize the presence of pathogenic organisms. The acidity of most fruits is a bacteriological deterrent, but it is not a guarantee of bacteriological safety.

Products cartoned before freezing include sliced strawberries mixed with sugar in a 4 to 1 ratio and whole strawberries, other berries, mixed fruits, and melon balls in a sugar syrup. The cartons and containers are liquidtight to prevent spills. These products are usually frozen in manual or automatic plate freezers, stationary air-blast tunnels, and push-through trolley freezers. These types of products are losing market share to other forms such as fresh fruit.

Products that are frozen before packaging are usually in the free-flowing or IQF categories. These include whole fruits such as strawberries, cherries and grapes; and pieces (slices, dices, halves, balls) of fruits such as apples, peaches, melons, pineapples, and citrus.

IQF products are usually frozen in straight belt freezers, fluidized belt freezers, fluidized bed freezers, and cryomechanical freezers. Many fruits and fruit pieces are sticky, fragile, and have a relatively high latent heat value. There is a trend toward using cryomechanical freezers for those applications both in new installations and in retrofits, because they often provide a superior IQF product at a reasonable cost. Chapter 16 has further information on freezers. Products from IQF freezers are packed and handled as described under the Vegetables section.

Fruit pies are a specialty pack. The IQF or fresh fruits are deposited in a dough shell and a top dough sheet is added on an assembly, filling, and packaging line as described in the section on Main Dishes, Meals. Fruit pies are usually frozen in automatic plate freezers or spiral belt freezers.

Refrigeration Loads and Systems

The principal refrigeration load calculations for fruits are similar to those described under the Vegetables section, except there is no cooling after blanching. The Vegetables section discussion of freezing, freezer warehouse loads, and refrigeration systems applies to fruits, as well. Fruits do, however, respond better to a lower storage temperature (−10°F versus 0°F) because of their higher sugar content.

POTATO PRODUCTS

The primary frozen potato products include various types of French fried potatoes for fast food restaurant, regular restaurant, and institutional uses. They include regular, shoestring, crinkle cut, and curly fries. Potato sales for retail consumption are far less than those for institutional use. Other potato products include potato puffs, tots, and wedges, which are a formed product made from waste-stream potatoes and rejected raw strips. Specialty potato products include hash browns, twice-baked potatoes, refrigerated French fries, potato skins, and boiled potatoes (with or without skins). The refrigerated product has a relatively short shelf life and requires close control of handling, shipping, and storage.

French Fries

French fried potatoes and formed product are processed year-round. In the northern United States, product is received in bulk directly from the field for 1 1/2 to 2 months in the fall and thereafter from storage. Storing fresh potatoes for processing is discussed in Chapter 24.

French fried potato lines are usually designed to produce large volumes, often more than 50,000 to 100,000 lb of raw product per hour. This results in a finished rate of 25,000 to 50,000 lb/h of frozen French fries, depending upon final product solids and oil content. Many plants also produce an additional 5000 to 10,000 lb/h of formed or dehydrated products from the small potatoes and the shorts and slivers produced in the process.

Bulk potatoes are metered into the processing lines, after which they are conveyed through a destoner to provide initial washing and to remove stones, vines, and other debris. They are then graded, and small potatoes are pulled for animal feed, starch production, manufacturing into other products, or the formed product line. Some processors also grade out large potatoes to be sold fresh as baking potatoes. The potatoes for French fries are then washed, steam-peeled (unless the end product should have its skin), inspected for foreign material, and sometimes preheated whole to minimize fracturing during cutting. Whole peeled potatoes are cut into the desired fry shapes, and the slivers and nubbins (short strips) are automatically graded out and diverted to the formed product line. The remaining strips are then electronically scanned, and those with defects are sorted out and the imperfections automatically removed.

The fry shapes are then processed in one of two ways. The first method is to blanch, cool, blanch, minidry, and fry the potatoes, producing a product of approximately 28 to 34% solids. This product has a high moisture content and is more difficult to freeze because of its higher latent heat of fusion. The second method is to blanch, process the product through a two- or three-stage drier, and then fry the potatoes, which produces a product of approximately 34 to 36% solids, a "high-solids fry." Many French fries are now battered before frying. Battering improves the shelf life of fried product at the fast food restaurant and provides opportunities for flavor addition to the strip surface.

Just before freezing, excess oil is removed from the par-fried strips, which are then cooled from 200°F to 70 to 80°F. Filtered ambient air, direct-flooded ammonia evaporator coils, ammonia thermosiphon system, or water coils (where plant water is heated for reuse in processing) provide the necessary cooling air. The freeze tunnel is a straight belt freezer system using ammonia recirculation evaporator coils and operating at 0 psig or less suction pressure. It is usually composed of two belts, so that product just starting to surface-freeze discharges from the first belt onto the second and breaks up any product clusters that have formed. Final product temperature at the tunnel discharge is kept at 5 to 10°F, because the strips are very fragile below these temperatures and may be broken in subsequent handling. Once packaged and in cold storage, product is cooled to 0 to –10°F.

After freezing, the fries are size-graded, with the longest lengths slated for institutional markets, the shorter lengths for retail, and the slivers and pieces relegated for use as cattle feed. The grading occurs in an area usually maintained at 15°F. Product from the graders goes directly to packaging and off-grade product to totes and storage for packaging under a different label.

Fries are packaged using net-weight fillers. Institutional product is packaged into 5 to 6 lb kraft/poly bags, and retail product into poly bags. The packaged product is automatically loaded into cases, palletized and on to 0 to –10°F cold storage.

Formed Potato Products

Puffs, tots, and wedges are manufactured from small whole potatoes that were graded out before peeling, and slivers and nubbins that were graded out after cutting the whole potatoes into shapes from the French fry line. The graded-out product is about 7% of the French fried potato production. If more formed products are desired, small whole potatoes are used.

The graded-out product and/or whole potatoes that have been steam-peeled or diced are inspected to remove blemishes and are then blanched. The product is then conveyed to a retrograde cooler, where the product is reduced to approximately 35°F and partially dehydrated. From the cooler, the potatoes are conveyed to other equipment where they are chopped into small pieces, mixed with flavorings and condiments, and formed into the desired shapes. From the formers, the product is conveyed to the oil fryers and then to the freezer.

These products are usually frozen on spiral belt freezers, because the freezing time is relatively long because of the bulky shape and high product inlet temperature of approximately 180°F or more, depending upon the amount of ambient cooling available. These products are usually not cooled in a refrigerated cooler before entering the spiral freezer.

Products can be packaged directly from the freezer or stored in bins for later packaging. They are distributed in both retail and institutional markets and primarily packaged in polyethylene bags.

Hash Brown Potatoes

Hash browns are manufactured from steam-peeled whole potatoes that are too small for French fries. They are then blanched and cooled conventionally or by retrograde cooling in bins, after which they are conveyed to a slicer and sliced into very thin strips.

They are placed on a conveyor belt and formed into shapes that are scored and pulled apart to make discrete groupings of patties. They are conveyed to a straight line belt freezer and frozen to 0°F or below. The product is generally packed into polyethylene bags for either retail or institutional distribution.

Refrigeration Loads and Systems

The refrigeration loads result primarily from cooling and freezing products and the associated freezer storage for in-process and finished goods storage. As noted previously, the production rates and total capacity of these plants are high, to take advantage of the economics of scale.

Belt freezer refrigeration loads are the major component, and a careful analysis should be made to ensure that performance meets the capacity requirements of the various product forms to be frozen. Refrigeration loads and capacity levels for the same freezing apparatus change significantly because of differences in latent heats of fusion, inlet and outlet temperatures, specific heats, and the size and shape of individual product pieces. Additional information on freezing times and refrigeration loads for specific foods may be found in Chapters 9 and 10.

Freezer warehouse loads are calculated as suggested in Chapters 13 and 14. Freezer storage in potato processing plants has additional refrigeration requirements because of product pull-down loads. French fries are often discharged from belt freezers at 5 to 10°F to reduce or eliminate product breakage from brittleness at lower temperatures. This discharge temperature, coupled with subsequent packaging, can result in product inlet temperature to the freezer warehouse of 15°F. Product temperature should be lowered promptly in the freezer to 0°F or below before shipping. Product quantities for a typical plant can be several million pounds per day. Also, some freezer warehouses may be attached to plants under negative pressure, which can substantially increase infiltration. This should be corrected, because it is very difficult and costly to offset with refrigeration and infiltration reduction devices.

Potato freezing plants primarily use ammonia as a refrigerant. Two-stage compression systems are in general use, but some single-stage plants are in operation, particularly in the western United States. In two-stage systems, product cooling is done at the intermediate pressure. Freezing and freezer warehouse loads use the first stage. Design saturated suction temperatures vary from –32 to –40°F in the first stage and 10 to 20°F in the second stage. Design saturated condensing temperatures vary from 85 to 95°F.

In single-stage systems, separate compressors are used for the 10 to 20°F product cooling and liquid refrigerant precooling. Separate compressors are used for freezing and freezer storage, with design saturated suction temperatures of –28 to –32°F. The higher design suction temperature is achieved with more evaporator surface. Design saturated condensing temperatures are usually 85°F in the low-wet-bulb design temperature areas where these plants are located. Because wet-bulb temperatures are even lower during the early production months in the fall, winter, and spring, the compression ratios are tolerable. Some firms find inadequate or borderline financial justification in electric power savings for the extra capital associated with two-stage systems.

Single-stage systems are simple and have a low first cost. Most of these plants are in rural areas, and it is easier and usually more satisfactory to train operators and mechanics for single-stage systems.

Condensing is usually done with evaporative condensers. New plants are often designed for 85°F condensing temperature with floating condensing pressures for lower wet-bulb temperatures and partial loads.

Potato processing plants have heavy refrigeration loads and operate for 6000 to 7000 h per year. Some spare machine room equipment may be needed, but it is not generally provided. Major maintenance is performed during periods of lower production as well as during downtime periods totaling 4 to 6 weeks per year.

French fried potato freezers function under heavy loads and severe duty. Features include modular, rugged construction for easy installation, evaporators of aluminum or hot-dipped galvanized tubing with variable fin spacing, axial or centrifugal fans with updraft airflow, noncorrosive materials for product contact parts, regular or sequential water defrost, belt washing apparatus, catwalks for access, and insulated panel housings with interiors constructed to withstand periodic washdown. A few of these freezers are designed to operate continuously to provide full capacity at all times. Most are designed to be defrosted every 7 to 7.5 h, between shifts, to maintain capacity. In the latter case, the freezer should provide full capacity at the end of the shift. Spiral freezers for formed product and straight belt freezers for hash browns are similar to those described in Chapter 16.

OTHER PREPARED FOODS

Several other types of prepared foods, including appetizers, sandwiches, breads, rolls, cakes, cookies, fruit pies, toppings, ice creams, sherbets, yogurts, and frozen novelties, are covered in other chapters. Bakery products are covered in Chapter 28, and ice cream products are covered in Chapter 20.

LONG-TERM STORAGE

Most prepared foods are not produced with long-term frozen storage as an objective. Inventories are closely supervised, and production and sales are closely linked to minimize inventory. Profitability is reduced by having finished goods in storage and in the distribution chain.

One exception is some vegetables and fruits that can be processed and frozen only during the harvest season. Even here, steps are usually taken to maximize in-process storage of bulk products and to package them as required by sale orders and projections.

Some components and ingredients for prepared foods, however, must withstand long-term frozen storage if they are only produced annually or infrequently. These products require close monitoring to ensure that the quality still meets standards when used.

Regardless of the length of storage, it is important that ingredients, components, and finished goods are stored at 0°F or below with minimal temperature fluctuations.

BIBLIOGRAPHY

FDA. 2001. HACCP: A state-of-the-art approach to food safety. *FDA Backgrounder.* http://www.cfsan.fda.gov/~lrd/bghaccp.html.

Food Safety and Inspection Service, Department of Agriculture, *Code of Federal Regulations*, U.S. Government Printing Office, Washington, D.C.:

2005. 9CFR304. Application for inspection; Grant of inspection.

2005. 9CFR416. Sanitation.

Food and Drug Administration, Department of Health and Human Services, *Code of Federal Regulations*, U.S. Government Printing Office, Washington, D.C.:

2005. 21CFR1. General enforcement regulations.

2005. 21CFR11. Electronic records; Electronic signatures.

2005. 21CFR110. Current good manufacturing practice in manufacturing, packing, or holding human food.

CHAPTER 28

BAKERY PRODUCTS

THIS chapter addresses refrigeration and air conditioning as applied to bakery products, including items distributed (1) at ambient temperature, (2) refrigerated but unfrozen, and (3) frozen. Refrigeration plays an important part in modern bakery production.

Some of the major uses of refrigeration in the baking industry are

- Ingredient cooling
- Dough and batter temperature control during mixing
- Refrigerating dough products
- Freezing dough for the food service industry and supermarkets
- Freezing bread for later holding, thawing, and sale
- Freezing fried and baked products for sale to consumers

Refrigeration methods and equipment needed to accomplish these uses include

- Normal air conditioning
- Dough mixers with jackets through which chilled water, low-temperature antifreeze, or direct-expansion refrigerant passes
- CO_2 chips or CO_2 fog placed directly into dough mixer bowls
- Cooling tunnels with refrigerated air flowing counterflow to the product
- Medium-temperature cool rooms for storage of refrigerated dough products
- Freezing tunnels for dough
- Kettles for fillings
- Spiral freezer chambers in which the product remains for about 20 min to 1 h and is subjected to air at about −30°F for freezing
- Holding freezers: 9 to −20°F

Total plant air conditioning is increasingly used in new plant construction, except in areas immediately surrounding ovens, in final proofers, and in areas where cooking vessels prepare fruit fillings and hot icings. Total plant cooling was first used in plants producing Danish pastry, croissants, puff pastry, and pies, and has expanded to new-construction facilities for frozen dough operations and general production. Flour dust in the air should be filtered out because it fouls air passages in air-conditioning equipment, seriously reduces heat transfer rates, and is a potential respiratory health hazard.

INGREDIENT STORAGE

Raw materials are generally purchased in bulk, except in small operations. Deliveries are made by truck or railcar and stored in bins or tanks with required temperature protection while in transit and storage.

Flour. Flour is stored in bins at ambient temperature. Some bakeries locate these bins outside their buildings; however, inside storage is recommended where outside temperatures vary greatly. This improves control of product temperature and decreases the risk of moisture condensation inside the storage bins. Pneumatic conveyance and subsequent sifting before use generally increase flour temperature a few degrees. Smaller quantities of other flours, such as clear, rye, and whole wheat, are usually received in bags and stored on pallets.

Sugars and Syrups. Sugar is handled in both dry and liquid bulk forms by many large production bakeries. Although most prefer liquid, many cake and sweet goods plants produce their own powdered sugar for icings by passing granulated sugar through a pulverizer. Refrigeration dehumidifiers are sometimes used to minimize caking or sticking of the powdered sugar for proper pneumatic handling. Liquid sucrose (cane or beet sugar), generally with a solids content of 66 to 67%, is stored at ambient temperature; however, it can be cooled to as low as 45°F without crystallizing out of solution. Corn syrups and various blends of sucrose and corn syrups should be stored at 90 to 100°F to improve fluidity and pumpability. Unlike sucrose, corn syrups become more viscous when cooled. High-fructose corn syrups are best handled at 80 to 90°F. Lower storage temperatures cause sugars to crystallize, and higher temperatures accelerate caramelizing. Dextrose (corn sugar) solutions containing 65 to 67% solids must be stored in heated tanks at 130°F to prevent crystallization. Many bakeries use high-fructose corn syrups. Because these syrups are stored at a lower temperature than conventional syrup, less thermal input is required during storage, and the refrigeration load during mixing is significantly reduced. Smaller-volume and specialized sugars are received in poly-lined bags and stored at ambient temperature.

Shortenings. Shortenings are stored in heated tanks or a "hot room," where the temperature is maintained at 10°F above the American Oil Chemists' Society (AOCS) capillary closed-tube melting point of the fat (AOCS 1999, 2004). Lard, for example, should be stored at 120°F to be totally liquid. Other shortenings need slightly higher temperatures. Fluid shortenings and oils are stored at room temperature, but fluid shortenings need constant slow-speed agitation to prevent hard fats from separating to the bottom of the tanks.

Yeast. Fresh yeast comes in 1 lb blocks packaged in cartons of various sizes, in crumbled form in 50 lb bags, and in liquid cream form handled in bulk tanks. Refrigerated storage temperatures ranging from 45°F to the freezing point of the product are required. For maximum storage life, 34 to 36°F is considered best. Active dry and instant dry forms of yeast are available that do not need refrigeration.

Egg Products. Liquid egg products (whole, whites, yolks, and fortified) are commonly used in small retail and large cake and sweet goods bakeries. They generally come frozen in 30 lb containers that must be thawed under refrigeration or cold-water baths. Where large quantities are needed, liquid bulk refrigerated handling can be an economic advantage. Storage temperatures for liquid egg products should be less than 40°F, with 35 to 38°F being the ideal storage temperature range. Dried egg solids, which need no refrigeration, are also used. A shelf-stable whole egg that requires no

The preparation of this chapter is assigned to TC 10.9, Refrigeration Application for Foods and Beverages.

refrigeration has been introduced. This stability was achieved by removing two-thirds of the water from the eggs and replacing it with sugar, thereby lowering the water activity to the point that most organisms cannot grow.

Other. Dried milk products, cocoa, spices, and other raw ingredients in baking are usually put into dry storage, ideally at 70°F. Ideal storage is rarely achieved under normal bakery conditions. Refrigerated storage is sometimes used where longer shelf lives are desired or high storage temperatures are the norm. This decreases flavor loss and change, microbial growth, and insect infestation.

MIXING

Bread, buns, sweet rolls/Danish, yeast-raised doughnuts, and honey buns are the most important yeast-leavened baked products in terms of production volume. After scaling the ingredients, mixing is the next active step in production. Proper development of the flour's gluten proteins is what gives doughs their gas-retaining properties, which affect the volume and texture of the baked products. Temperature control during mixing is essential. Refrigeration is generally required because of the heat generation and the necessity of controlling dough temperature at the end of mixing. However, ingredient temperatures combined with room temperature may require addition of warm water to produce the desired finished dough temperature.

Yeast metabolism is materially affected by the temperatures to which the yeast is exposed. During dough mixing, the following heat factors are encountered: (1) **heat of friction**, by which the electrical energy input of the mixer motor is converted to heat; (2) **specific heat** of each ingredient; and (3) **heat of hydration**, generated when a dry material absorbs water. If ice is used for temperature control, **heat of fusion** is involved. Finally, the temperature of the dough ingredients must be considered. Yeast acts very slowly below 45°F. It is extremely active in the presence of water and fermentable sugars at 80 to 100°F, but all yeast cells are killed at 140°F and at a lower but sustained pace below its freezing point of 26°F. Precise temperature control is essential at all stages of storage and production, especially during mixing, because of its effect on downline processing.

Mixers

The three most common styles of mixers are the horizontal, vertical or planetary, and spiral. **Horizontal mixers** are primarily used by wholesale bakeries and are designed with horizontal agitator bars. They range in capacity from 200 to 3000 lb. Because of the large dough sizes, these mixers are generally jacketed with some form of cooling. **Vertical mixers** are more common in retail bakeries and are categorized by their largest bowl capacity. Bowls range in size from 12 to 20 qt for tabletop models to as high as 340 qt in some large wholesale plants. The bowls have no refrigeration jacket and can be removed from the mixers. The hook or agitator revolves as it travels around the inside of the bowl. **Spiral mixers** are somewhat newer and are gaining in popularity with retail and specialty bread bakers. Here the bowl revolves, bringing the ingredients to the off-center spiral agitator. Bowls for smaller models (50 to 400 lb) are not removable, but those for the larger models (up to 1000 lb) can be removed. Like vertical mixers, spiral mixers are not jacketed.

Where flour is pneumatically transferred to the mixer, liquid CO_2 can be injected directly into the flour stream. This technique has been used for mixers, such as vertical and spiral mixers, that are not jacketed for temperature control. Dry ice (CO_2) chips have also been used in frozen dough production, where dough temperatures below 70°F are required. Dry ice is used as an aid to other forms of refrigeration. Because of expansion of CO_2 gas, horizontal mixers should be left open slightly.

Dough Systems

The four principal types of batch dough mixes are straight, no-time, sponge, and liquid ferment. These methods are called **dough systems** or **dough process** in the baking industry. The type of dough system determines the stages in the process, the equipment needed, and the general processing parameters (times and temperatures). **Straight dough** and **no-time dough** systems are the two most common in retail bakeries. All of the ingredients are mixed at one time. Straight doughs require fermentation, whereas no-time doughs do not. The no-time system is also used in wholesale or large plant bakeries that produce hearth and specialty breads, in which fermentation flavor is not as important as compared to white pan bread and buns.

The more common systems used in wholesale bakeries are the sponge dough and liquid ferment or liquid sponge. The **sponge dough** process requires more equipment and longer fermentation times than the straight and no-time dough systems. Here, only a part of the total amount of flour and water required are mixed with all of the yeast and yeast food. The resulting mixture, or sponge, is then fermented before it is given the final mix or remix with the remaining ingredients. This is done just before makeup, improving both tolerance of schedule disruptions and dough machinability.

The principal heat generated during mixing comes from hydration, as the flour absorbs water, and from the friction of the mixer. To absorb this excess heat and maintain the dough at 78 to 82°F, the dough-side water is usually supplied to the mix at 35 to 39°F, and horizontal mixers are generally jacketed to circulate a cooling medium around the bowl.

The **liquid ferment** or **liquid sponge** process has gained popularity with wholesale bakeries because of its excellent temperature control and acceptable product quality. Liquid sponge ingredients can be incorporated and fermented in special equipment either in batches or on an uninterrupted basis. To render the mixture pumpable, more water is incorporated than the actual amount used in a sponge to be fermented in a trough. After fermentation, the liquid sponge (at required pH and titratable acid levels) is chilled through heat exchange equipment from about 79 to 88°F to about 45 to 55°F. The cold liquid sponge is then maintained at the required temperature in a storage or feed tank until it is weighed or metered and pumped to the mixer, where it is combined with the remaining ingredients before being remixed into a dough. Regular sponges come back for remixing at about 84°F. Often, it is necessary to use the refrigerated surface on the mixer jacket in conjunction with ice water or ice to achieve the required dough temperature after mixing.

Positive dough temperature control is achieved using a cold liquid sponge at 45 to 55°F, which limits the need for the refrigeration jacket and eliminates using ice in the doughs. In cold weather, remaining dough water temperatures of 100°F or higher are often required.

A fifth dough system, the **continuous mix** method, was developed in the 1950s and requires specially designed continuous mix equipment for making bread and buns. Originally, a liquid ferment, then called a brew or broth, was formed, using 0 to 10% of the flour, 10 to 15% of the sugar, 25 to 50% of the salt, all of the yeast, and yeast food in about 85% of the total water. Through the years, the amount of flour has increased to as high as 50% and the needed sugar decreased. Using mass flow meters, the liquid ferment is metered into an incorporator or premixer, where the remaining ingredients are added. The resultant thicker batter finally passes through a developer head/mixer, from which the finished dough is extruded and deposited directly into a greased baking pan. Properly formulated dough can also be deposited onto floured belts that then pass through conventional makeup equipment. Final dough temperatures could reach as high as 118°F, though today, with higher levels of flour in the ferment, they could be as low as 90°F. It is estimated that less than 5% of the U.S. bread and bun market uses this system.

Table 1 Size of Condensing Units for Various Mixers

Dough Capacity, lb	Condensing Unit Size, hp
800	5
1000	7.5 to 10
1300	10 to 15
1600	15 to 20
2000	20

Hot dog and hamburger rolls are also produced in quantity by pumping bun dough from the continuous mix developer directly to the hopper of the makeup equipment. A coating of flour on outside dough surfaces affects gas development and retention, which in turn leads to a grain/texture more closely resembling that of sponge dough products. External symmetry and crust characteristics are similarly changed.

Dough Cooling

Some dough mixers are cooled by direct-expansion refrigerant, but the most common means of cooling is with chilled water or an antifreeze such as propylene glycol. The temperature of the evaporating refrigerant or antifreeze supplied may often be as low as 30°F to maintain the dough at the desired temperature. When the dough mixers are cooled by evaporating refrigerant, the condensing unit is usually located close to the mixers. Table 1 lists the sizes of condensing units commonly selected for mixers. The ingredient water is cooled in separate liquid chillers, and when the mixer bowl is cooled by antifreeze, the liquid chiller may be remotely located.

When large batches of dough are handled in the mixers, the required cooling is sometimes greater than the available heat transfer surface can produce at 30°F, and the refrigerant temperature must be lowered. Refrigerant temperatures below 30°F can, however, cause a thin film of frozen dough to form on the jacketed surface of the mixer, which effectively insulates the surface and impairs heat transfer from the dough to the refrigerant.

Formulas, batch sizes, mixing times, and almost every other part of the mixing process vary considerably from bakery to bakery and must be determined for each application. These variations usually fall within the following limits:

Final batch dough weight	Up to 3000 lb
Ratio of sponge to final dough weight	50 to 75%
Ratio of flour to final dough weight	50 to 65%
Ratio of water to flour	50 to 65%
Sponge mixing time	240 to 360 s
Final dough mixing time	480 to 720 s
Number of mixes per hour	2 to 5
Continuous mix production rates	Up to 7000 lb/h

The total cooling load is the sum of the heat that must be removed from each ingredient to bring the homogeneous mass to the desired final temperature plus the generated heat of hydration and friction. In large batch operations, the sponge and final dough are mixed in different mixers, and refrigeration requirements must be determined separately for each process. In small operations, a single mixer is used for both sponge and final dough. The cooling load for final dough mix is greater than that for sponge mix and is used to establish refrigeration requirements.

Some bakery manufacturers inject CO_2 into a mixer to chill ingredients before mixing; this helps obtain lower dough temperatures when mechanical refrigeration is not adequate for desired mixing times. This technique is principally applied to laminated and frozen doughs.

FERMENTATION

After completion of the sponge mix, the sponge is placed in large troughs that are rolled into an enclosed conditioned space for a fermentation period of 3 to 5 h, depending on the dough formula. The sponge comes out of the mixer at 72 to 76°F. During fermentation, the sponge temperature rises 6 to 10°F as a result of the heat produced by the yeast and fermentation, or about 1.8°F per hour.

To equalize the temperature substantially throughout the dough mass, room temperature is maintained at the approximate mean sponge temperature of 80°F. Even temperature throughout the batch produces even fermentation action and a uniform product.

Water makes up a large part of the sponge, and uncontrolled evaporation causes significant variations in the quality and weight of the bread. The rate of evaporation from the sponge surface varies with the ambient air's relative humidity and airflow rate over the surface. The rate of moisture movement from the inside of the sponge to the surface does not react similarly to external conditions, and surface drying and crust formation can result. Crusted dough is inactive and does not develop. When folded into the dough mass later, undeveloped portions produce hard, dark streaks in the finished bread.

To control the evaporation rate from the sponge surface, air is maintained at 75% rh, and the conditioned air is moved into, through, and out of the process room without producing crust-forming drafts.

In calculating the room cooling load, the product is not considered because the air temperature is maintained at approximately the mean of the various dough temperatures in the room. Transmission heat loss through walls, ceiling, and floor is the principal load source. Infiltration is estimated as 1.5 times the room volume per hour. The lighting requirement for a fermentation room is usually about 75 W per 400 ft^2 of floor area. The conditioning units are often placed within the conditioned space, and the full motor heat load must be considered.

The only source of latent heat is the approximate 0.5% weight loss in the sponge. Under full operating load, this could account for a 1.5°F increase in dew-point temperature. For conditions of 80°F db and 75% rh, the dew-point temperature would be 71.5°F, and the supply air would be introduced into the conditioned space at 72°F db and 70°F dew point.

In large rooms, sufficient air volume can be introduced to pick up the sensible heat load with an 8°F rise in air temperature. In smaller rooms, a latent heat load may need to be added by spraying water directly in the room through compressed-air atomizing nozzles. Water sprays have been most successful when a relatively large number of nozzles is spaced around the periphery of the room.

Because a high removal ratio of sensible to latent heat is desirable, the condensing unit is usually specified for operation as close to 60°F evaporator temperature as is practicable with the temperature and quantity of condensing water available.

BREAD MAKEUP

After the dough is mixed using conventional mixers, and perhaps given floor time to become more elastic and less tacky, it is placed into the hopper of the divider. There are two types of dividers used in wholesale bakeries. With the **ram and piston divider**, the dough is forced into cylinders; the pistons adjust the cylinder opening, which controls the unit weight. The **rotary (extrusion) divider** extrudes the dough through an opening using a metering pump; a rotating knife then cuts off the dough. The unit weight is adjusted by the speed of the metering pump and/or the speed of the knife. Because dough density changes with time and the dividers work on the principle of volume, the baker must routinely check scaling weights and adjust the dividers.

Next, the irregularly shaped units are rounded into dough balls for easier handling in subsequent processing. This **rounding** is done on drum, cone, or bar/belt rounders. The dough pieces are well floured, and a smooth skin is established on the outer surface to reduce sticking.

A short resting or relaxation period on an intermediate proofer follows rounding. Flour-dusted trays or belts hold the dough pieces and carry them to the next pieces of equipment via a transfer shoot. Residency time is 1 to 8 min.

The dough is next formed into a loaf by the **sheeter** and the **moulder**. The relaxed dough ball is sheeted (e.g., to about 1/8 in. thick for white pan bread) by passing through sets of rollers. This reduces the size of gas cells and multiplies them, producing a finer grain in the baked bread. Next the dough goes to the moulder, where it receives its final length and shape before it is placed into the greased baking pan. There are three styles of moulders: (1) straight grain, (2) cross grain, and (3) tender-curl.

FINAL PROOF

After the loaves are formed, they are placed in baking pans, and set in the **proofer** for 50 to 75 min. The proofer is an insulated enclosure with a controlled atmosphere in which the dough receives its final fermentation or proof before it is baked. To stimulate the yeast's fermentative ability, the temperature is maintained at 95 to 110°F, depending on the exact formula, prior intensity of dough handling, and desired character of the baked loaf.

For proper crust development during baking, the exposed surface of the dough must be kept pliable by maintaining the relative humidity of the air in the range of 75 to 95%. Some bakeries find it necessary to compromise on a lower humidity range because of the effect on dough flow.

With dew points inside the proofer at about 92 to 106°F and ambient conditions as low as 65°F at times, the proofer must be adequately insulated to keep the inside surface warm enough to prevent condensation from forming because mold growth can be a serious problem on warm, moist surfaces. A thermal conductance of $C = 0.12$ Btu/h·ft^2·°F is adequate for most conditions.

In small and midsized bakeries, one of the more commonly used proofers for this process consists of a series of aisles with doors at both ends. Frequent opening of the doors to move racks of panned dough in and out makes control of the conditioned air circulation an unusually important engineering consideration. The air is recirculated at 90 to 120 changes per hour and is introduced into the room to temper the infiltration air and cause a rolling turbulence throughout the enclosure. This brings the newly arrived racks and pans up to room temperature as soon as possible.

The problem of air circulation is somewhat simpler when large automatically loaded and unloaded tray and conveyer or spiral proofers are used. These proofers have only minimal openings for the entrance and exit of the pans, and the thermal load is reduced by elimination of the racks moving in and out.

BAKING

Most 16 to 24 oz bread loaves are baked in ovens at around 400 to 450°F for 18 to 30 min. High temperatures are used for hearth breads, and lower temperatures for denser styles of breads. Buns and rolls are baked at 420 to 450°F for 10 to 12 min. Because of their small size, a quick bake is desired so as not to dry out the product during the bake.

Deck ovens are found in some smaller retail establishments, specialty cookie and hearth bread bakeries, pizzerias, and restaurants. The baking surfaces are stationary, requiring manual loading and unloading. Uneven heat distribution, little air movement, and low clearance between decks make the deck oven unsuitable for some products. Newer designs have made this oven the choice for retail baking of the denser, heavy, European-type hearth breads.

Rack ovens are very popular in in-store supermarket bakeries, freestanding retail bakeries, and some wholesale bakeries. These ovens have a small chamber into which one or two racks of panned product can be rolled at one time. Using forced-air convection, baking can be accomplished at lower temperatures for shorter baking times. These ovens are highly efficient in heat transfer and easy to operate. They generate their own steam without boilers and are well suited for crusty breads.

There are two popular styles of tray ovens, which are loaded and unloaded at the same end. The **reel-type tray oven**, which looks like a Ferris wheel inside a heated baking chamber, is one of the most popular ovens in retail bakeries and is used by some large wholesalers. This design, with its manual loading and unloading, allows different products to be baked simultaneously. Because of the revolving action, hot spots and uneven baking are less likely than in deck ovens.

The **single-lap traveling tray oven** is used principally in large production bakeries and is usually equipped with pan loaders and unloaders for efficient product handling and continuous baking. There is also a double-lap design, which travels back and forth twice, but it is not currently popular.

Wholesale bakeries are equipped with either tunnel (traveling hearth) ovens or conveyor-style ovens. **Tunnel ovens**, as the name suggests, have the product loaded in at one end and unloaded at the other. This permits ideal temperature control for the entire baking time. **Conveyor-style ovens** have a continuous spiral climbing to the top of a heated chamber and then back down again, with each pan carried on its own section of the conveyor and every product passing through the same spot within the baking chamber. Another style of tunnel oven is the **impingement oven**. Its directional flow of heated air to the top and bottom of the product provides excellent baking speed and efficiency for pizza and other low-profile products.

Ovens are usually gas fired, but electric or oil firing may be more economical in some areas. Where gas interruption is a problem, combination oil and gas burners are an alternative. Gas firing can be direct or indirect, but indirect firing is generally more popular because of its greater flexibility. Ribbon burners running across the width of the oven are operated as an atmospheric system at pressures of 6 to 8 in. of water.

Burners are located directly beneath the path of the hearth or trays, and the flames may be adjusted along the length of the burner to equalize heat distribution across the oven. The oven is divided into zones, with the burners zone-controlled so that heat can be varied for different periods of the bake.

Controlled air circulation inside the oven ensures uniform heat distribution around the product, which produces desired crust color and thickness. When baking hearth breads, ovens are equipped with steam ejection. Low-pressure (2 to 5 psi), high-volume wet steam is used for the first few minutes of baking to control loaf expansion and crust crispness and shine. This is accomplished with a series of perforated steam tubes located in the first zone of baking in place of the air-recirculating tubes. Heating calculations are based on 450 Btu per pound of bread baked. Steam requirements run approximately 1 boiler hp per 125 lb of bread baked per hour.

Many wholesale and plant bakeries use the internal temperature of product exiting the oven as a guide to judge the proper bake. A judgment to minimize baking time is based on establishing product structure and, at the same time, minimizing bake loss. Desired crust color is essential; however, a target internal finished product temperature of 196 to 205°F can help control product characteristics and maximize oven throughput by minimizing baking time.

BREAD COOLING

Baked loaves come out of the oven with an internal temperature of 196 to 205°F because of the evaporative cooling effect of moisture that is driven off during baking. The crust temperature is closer to the oven baking temperature, 450°F.

Loaves are then removed from the pans and allowed to cool to an internal temperature of 95 to 106°F. A hygroscopic material cools in two phases, which are not distinct periods. When the bread first

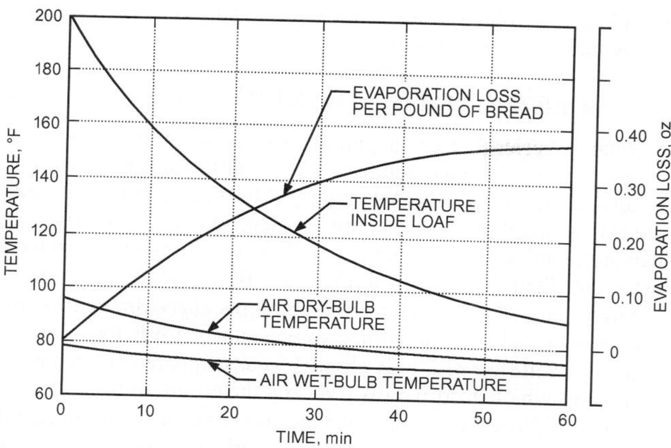

Fig. 1 Moisture Loss and Air Temperature Rise in Counterflow Bread-Cooling Tunnel

comes out of the oven, the vapor pressure of moisture in the loaf is high compared to that of moisture in the surrounding air. Moisture is rapidly evaporated, with a resultant cooling effect. This **evaporative cooling**, combined with **heat transmission** from bread to air caused by a relatively high temperature differential, causes rapid cooling in the early cooling stage, as indicated by the steepness of the temperature versus time curve (Figure 1). As vapor pressure approaches equilibrium, heat transfer is mainly by transmission, and the cooling curve flattens rapidly.

Small bakeries still cool bread on racks standing on the open floor for 1 to 3 h, depending on air conditions, spacing, and size of product. Many large operations cool bread while it is moving continuously on belt, spiral, or tray conveyors. Cooling is mostly atmospheric, even on these conveyors. However, to ensure a uniform final product, cooling is often handled in air-conditioned enclosures with a conventional counterflow movement of air in relation to the product.

An internal temperature of 95 to 106°F stabilizes moisture in the bread enough to accomplish proper slicing and reduce excessive condensation inside the wrapper, thus discouraging mold development. Approximately 50 to 75 min of cooling is required to bring the internal loaf temperature to 95°F, as shown in Figure 1.

In counterflow cooling, the optimum temperature for air introduced into the cooling tunnel is about 75°F. Air at 80 to 85% rh controls moisture loss from the bread during the latter stage of cooling and improves the bread's keeping quality.

Attempting to shorten cooling time by forced-air cooling with air temperatures below 75°F is not satisfactory. The rate of heat and moisture loss from the surface becomes much greater than that from the interior to the surface, causing the crust to shrivel and crack.

Product heat load is calculated assuming sensible heat transfer based on a specific heat of 0.70 Btu/lb·°F for bread and a temperature reduction from 180°F to 95°F. For calculating purposes, moisture evaporation from the bread may be considered responsible for reducing the bread temperature to 95°F.

Under normal conditions, air conditioning can best be accomplished by evaporative cooling in an air washer. When the outdoor wet-bulb temperature exceeds 72°F, which is the maximum allowable based on 75°F db and 85% saturation, refrigeration is required.

The amount of air required is rather high because the maximum temperature rise of air passing over the bread is usually about 20°F; consequently, the refrigeration load is comparatively high. Where wet-bulb temperatures above 72°F are rare or of short duration, the expense of refrigeration is not considered justified, and the bread is sliced at a higher than normal temperature.

SLICING AND WRAPPING

Bread from the cooler goes through the slicer. The slicer's high-speed cutting blades (similar to band saw blades) cut cleanly through properly cooled bread. If the moisture evaporation rate from the surface and replacement rate of surface moisture from the interior are not kept in balance, the bread develops a soggy undercrust that fouls the blades, causing the loaf to crush during slicing. A brittle crust may also develop, which leads to excessive crumbing during slicing. From the slicer, the bread moves automatically into the bagger.

BREAD FREEZING

Bagged bread normally moves into the shipping area for delivery to various markets by local route trucks or by long-distance haulers. Part of the production may go into a quick-freezing room and then into cold storage.

Bakers face two important problems in freezing bread and other bakery products. The first is connected with the short work week. Most bread and roll production bakeries are inoperative on Saturday; thus, production near the end of the week is much larger than for the earlier part of the week. The problem increases for bakeries on a 5 day week, with Tuesday usually being the second day off. Freezing a portion of each day's production for distribution on the days off can enable a more even daily production schedule.

A second problem is increased demand for variety breads and other products. The daily production run of each variety is comparatively small, so the constant setup change is expensive and time-consuming. Running a week's supply of each variety at one time and freezing it to fill daily requirements can reduce operating cost.

Both problems concern staleness, because bread is a perishable commodity. After baking, starch from the loaf progressively crystallizes and loses moisture until a critical point of moisture loss is reached. A tight wrap helps keep the moisture content high over a reasonable time. Starch crystallization, when complete, produces the crumbly texture of stale bread. The rate of this spontaneous action increases as either moisture or temperature decreases. Starch crystallization accelerates as the product passes through a critical temperature zone of 50°F to the freezing point of the product. The rate then decreases until the temperature reaches 0°F, where moisture loss seems to be somewhat arrested.

Bread freezes at 16 to 20°F (Figure 2). Bread should be cooled through the freezing or latent heat removal phase as fast as possible to preserve the cell structure. Because moisture loss rate increases with reduced temperature, bread should be cooled rapidly through the entire range from the initial temperature down to, and through, the freezing points. Successful freezing has been reported in room temperatures of 0, –10, –20, and –30°F.

In U.S. Department of Agriculture (USDA) laboratory tests, core temperatures of loaves of wrapped bread placed in 700 fpm cold air blasts were brought down from 70°F to 15°F in the times given:

Freezer Air Temperature, °F	Cooling Time, 70°F to 15°F, h	Bread Core Temperature at End of 2 h, °F
–40	2	15
–30	2.25	16
–20	3	19
–10	3.75	21
0	5	22

Changes in air velocity from 200 to 1300 fpm had relatively little effect on cooling of wrapped bread.

Cooling from 0 to –20°F is 10 to 30 min faster for unwrapped bread than for wrapped bread. However, the wrapper's value in retaining moisture during freezing and thawing makes freezing wrapped product advisable.

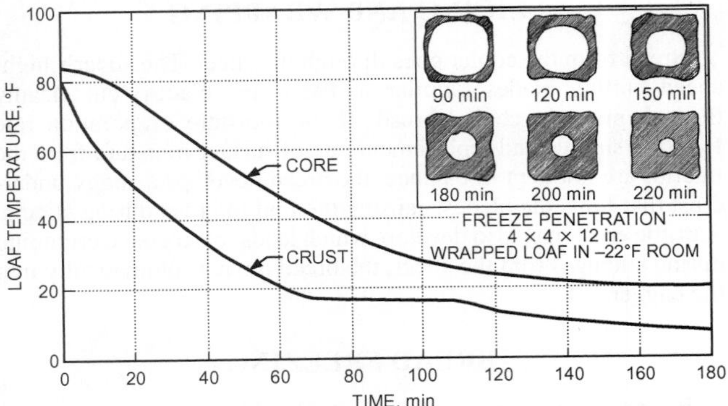

Fig. 2 Core and Crust Temperatures in Freezing Bread

Table 2 Important Heat Data for Baking Applications

Specific heat	
Baked bread (above freezing)	0.70 Btu/lb·°F
Baked bread (below freezing)	0.34 Btu/lb·°F
Butter	0.57 Btu/lb·°F
Dough	0.60 Btu/lb·°F
Flour	0.42 Btu/lb·°F
Ingredient mixture	0.40 Btu/lb·°F
Lard	0.45 Btu/lb·°F
Milk (liquid whole)	0.95 Btu/lb·°F
Liquid sponge (50% flour)	0.70 Btu/lb·°F
Heat of friction per horsepower of mixer motor	42.4 Btu/min
Heat of hydration of dough or sponge	6.49 Btu/lb
Latent heat of baked bread	46.90 Btu/lb
Specific heat of steel	0.12 Btu/lb·°F

Some commercial installations freeze wrapped bread in corrugated shipping cartons. The additional insulation provided by the carton increases freezing time considerably and causes a wide variation in freezing time between variously located loaves. One test found that a corner loaf reached 15°F in 5.5 h, whereas the center loaf required 9 h.

Most freezers are batch-loaded rooms in which bread is placed on wire shelves of steel racks. One of the principal disadvantages of this arrangement is that the racks must be moved in and out of the room manually. Continuous freezers with wire belt or tray conveyors allow steady flow and do not expose personnel to the freezer temperature. See Chapter 16 for freezer descriptions.

For air freezing temperatures of −20°F or below, consider using two-stage compression systems for overall economical operation.

In addition to primary air blowers designed for about 10 cfm per pound of bread frozen per hour, a series of fans is used to ensure good air turbulence in all parts of the room. Heat load calculations are based on the specific and latent heat values in Table 2.

After the quick freeze, bread is moved into a −10°F holding room where the temperature throughout the loaf equalizes. Bread is often placed in shipping cartons after freezing and stacked tightly on pallets.

Defrosting. Frozen bread must be thawed or defrosted for final use. Slow, uncontrolled defrosting requires only that the frozen item be left to stand, usually in normal atmospheric conditions. For quality control, the defrosting rate is just as critical as the freezing rate. Passing the product rapidly through the critical temperature range of 50°F to the product's freezing point yields maximum crumb softness. Too high a relative humidity causes excessive condensation on the wrapper, with some resultant susceptibility to handling damage. The product defrosts in about 1.75 h when placed in air at 120°F and 50% rh or less. Good air movement over the entire product surface at 200 fpm or higher helps minimize condensation and make the defrosting rate more uniform.

FREEZING OTHER BAKERY PRODUCTS

Retail bakeries freeze many products to meet fluctuating demand. Cakes, pies, sweet yeast dough products, soft rolls, and doughnuts are all successfully frozen. A summary of tests and commercial practice shows that these products are less sensitive to the freezing rate than are bread and rolls. Freezing at 0 to 10°F apparently produces just as satisfactory results as freezing at −10 to −20°F.

Storage at −10 to −20°F or below keeps packaged dinner rolls and yeast-raised and cake doughnuts satisfactorily fresh for 8 weeks. Cinnamon rolls keep for only about 3 weeks, apparently because of the raisins, which absorb moisture from the crumb of the roll.

Pound, yellow layer, and chocolate layer cakes can be frozen and held at 10°F for 3 weeks without significantly affecting their quality. Sponge and angel food cakes tend to be much softer as the freezing temperature is reduced to 0°F. Layer cakes with icing freeze well, but condensation on exposed icing during thawing ruins the gloss; therefore, these cakes are wrapped before freezing.

Unlike cakes, which can be satisfactorily frozen after baking, pies frozen after baking have an unsatisfactory crust color, and the bottom crust of fruit pies becomes soggy when the pie is thawed. Freezing unbaked fruit pies is highly successful. Freezing time has little, if any, effect on product quality, but storage temperature does have an effect. Frozen pies stored at temperatures above 0°F develop badly soaked bottom crusts after 2 weeks, and fillings tend to boil out during baking, possibly because of moisture migration from starch-based syrups. Freezing baked or fried products is generally a high-production operation carried out in freezing tunnels or spiral freezers.

One of the fastest growing applications of refrigeration is freezing dough for in-store bakeries. A few frozen dough products are sold in supermarkets directly to consumers, but most is destined either for food service or for supermarket bakeries.

Danish and sweet dough products are frozen baked or unbaked, depending on how quickly they will be required for sale after they are removed from the freezer. Custard and chiffon pie fillings have not had uniformly good results, but some retail bakeries have achieved satisfactory results by carefully selecting starch ingredients. Meringue toppings made with proper stabilizers stand up very well, and whipped cream seems to improve with freezing. Cheesecake, pizza, and cookies also freeze well.

Although some products are of better quality if frozen at −10°F and others at −20°F, variety shops must compromise on a single freezer temperature so that all products can be placed in one freezer. The freezer is usually maintained between −10 and −20°F. Freezing time is not a factor because the products are kept in storage in the freezer. Freezers range from large reach-in refrigerators in retail shops to walk-in boxes in wholesale shops.

FROZEN PRE-PROOFED BAKERY PRODUCTS

Unlike yeast-leavened frozen dough products in today's market, which require thawing, proofing, and baking, frozen "pre-proofed" dough products are partially proofed (≈80% of full proof) before freezing at −4°F to −22°F and do not need to be thawed and proofed before baking. Pre-proofed products eliminate the need for expert thawing and proofing, while still providing optimum quality for the end user. The products can go directly from the freezer onto baking trays and into the oven. Baking temperatures for various products are very important: sweet rolls and Danish pastries require 302 to 320°F, whereas croissants require 320 to 338°F. The product thaws and expands some during the first part of baking. Advantages include the following: (1) no proof box is needed, (2) no thawing or proofing time is required, and (3) there is no chance of over- or underproofing the product. This allows the freshest baked product delivered on demand in the shortest possible time. Some breads, bread rolls, sweet rolls, croissants, and Danish pastry are produced by this method.

Some companies use normal yeast levels, as for fresh products; others use higher levels (≈2% based on flour weight), as for frozen doughs. Other adjustments to formulation are a combination of gums to improve moisture retention and the use of oxidants for increased dough strength. Processing changes seem to be of great importance. Laminating dough adds strength to the gluten structure for optimum gas retention and product height, and partially proofing moulded dough pieces at lower temperatures than normal (≤80°F) reduces weakness produced by yeast. Some operations temper/refrigerate proofed product to an internal temperature of 60°F before freezing and packaging. Depending on the type and amount of product to be baked, a baking temperature 27 to 40°F lower than for conventional products is used, with a steam injection during the first one-third to one-half of the bake. Using lower temperatures and steam keeps the crust from setting too quickly, allowing proper product expansion. This new method of production ensures top-quality product for restaurant, food commissary, and in-store bakery markets. Product shelf life is said to be 9 months to 1 year. The primary disadvantages are that (1) frozen product is easily thawed during transport, and (2) more freezer space is required.

RETARDING DOUGHS AND BATTERS

Freezing is usually used if the products are to be held for 3 days to 3 weeks. For shorter holding periods, such as might be required to have freshly baked products all day from one batch mix, a temperature just cold enough to retard fermentation action in the dough is applied. Retarder temperatures of 32 to 40°F slow the yeast action sufficiently to allow holding for 3 h to 3 days.

Doughs to be retarded are sometimes made up into final shaped units ready for proofing and baking. Cold slabs of dough can also be stored, with baking units made up after thawing. This method is especially satisfactory for Danish pastry dough and other doughs with rolled-in shortening, such as croissants and puff pastries. Chilling to retarded temperatures improves flakiness of these products. Refrigeration load calculations should be made following the recommendations included in Chapter 13.

In the retarding refrigerator, about 85% rh is required to prevent the product from drying out. Condensation on the product is undesirable. Complete batch loading is usually used, so refrigeration calculations must be based on introduction of the batch over a short period of time. Refrigeration equipment be able to absorb product and carrier heat loads in 0.75 to 3 h, depending on cabinet size and handling technique. Products most commonly handled in this manner are Danish pastry, dough for sweet rolls and coffee cake, cookies, layer cake mixes, pie crust mixes, and bun doughs.

Temperatures required for retarding are very similar to those required for storing ingredients, and refrigerators are usually designed to handle both ingredient storage and dough retarding.

CHOICE OF REFRIGERANTS

The most popular refrigerants in the baking industry have been R-12, R-22, R-502, and ammonia (R-717). R-12 is a chlorofluorocarbon (CFC) and can no longer be used; its most successful replacement has been R-134a. In general, any R-12 system can be refitted with R-134a along with the substitution of the proper lubricant. R-22 is a hydrochlorofluorocarbon (HCFC), so it is destined for phaseout over several decades. R-502 is an azeotrope containing the CFC R-115, so no new R-502 systems are being installed. Hydrofluorocarbon (HFC) substitutes for R-502 and R-22 are available, but they cost much more. Therefore, the preferred configuration of the system may be different than for traditional refrigerants.

For water chillers, R-134a is becoming popular. For chilling antifreeze to lower temperatures, the HFC replacements for R-22 and R-502 are possibilities. For large freezing facilities (e.g., spiral freezers), ammonia dominates because of its excellent low-temperature performance. Some producers also choose expendable refrigerants, such as carbon dioxide (CO_2) or nitrogen (N_2), which many believe provides a superior frozen product because of the freezing medium's low temperature. Hybrid systems can be used in which CO_2 and N_2 rapidly freeze a crust on the product surface; vapor compression, with its lower operating cost, completes the freezing process.

REFERENCES

AOCS. 1999. *Physical and chemical characteristics of oils, fats and waxes.* D. Firestone, ed. American Oil Chemists' Society, Champaign, IL.

AOCS. 2004. *Official methods and recommended practices of the AOCS*, 5th ed. American Oil Chemists' Society, Champaign, IL.

BIBLIOGRAPHY

AIB. 1995. *Baking science and technology bread lecture book.* American Institute of Baking, Manhattan, KS.

Kulp, K., K. Lorenz, and J. Brümmer. 1995. *Frozen and refrigerated doughs and batters*. American Association of Cereal Chemists, St. Paul, MN.

Matz, S.A. 1987. *Formulas and processes for bakers.* Pan-Tech International, McAllen, TX.

Matz, S.A. 1988. *Equipment for bakers.* Pan-Tech International, McAllen, TX.

Poulos, G. 2000. Par-baked for recapturing market shares. *Proceedings of the American Society of Baking*, pp. 259-263.

Pyler, E.J. 1988. *Baking science and technology,* 3rd ed. Sosland Publishing, Merriam, KS.

Sullivan, R.C. 1995. Frozen par-baked products. *Proceedings of the American Society of Bakery Engineers*, pp. 49-56.

CHAPTER 29

CHOCOLATES, CANDIES, NUTS, DRIED FRUITS, AND DRIED VEGETABLES

CANDY MANUFACTURE

Air conditioning and refrigeration are essential for successful candy manufacturing. Proper atmospheric control increases production, lowers production costs, and improves product quality.

Every plant has one or more of several standardized spaces or operations, including hot rooms; cold rooms; cooling tunnels; coating kettles; packing, enrobing, or dipping rooms; and storage.

Sensible heat must be absorbed by air-conditioning and refrigeration equipment, which includes the air distribution system, plates, tables, cold slabs, and cooling coils in tunnels or similar coolers. In calculating loads, sensible heat sources such as people, power, lights, sun effect, transmission losses, infiltration, steam and electric heating apparatus, and the heat of the entering product must be considered. See Chapter 13 for more information. Table 1 summarizes the optimum design conditions for refrigeration and air conditioning.

Two of the basic ingredients in candy are sucrose and corn syrup. These change easily from a crystalline form to a fluid, depending on temperature, moisture content, or both. The surrounding temperature and humidity must be controlled to prevent moisture gain or loss, which affects the product's texture and storage life. Temperature should be relatively low, generally below 70°F. The relative humidity should be 50% or less, depending on the type of sugar used. For chocolate coatings, temperatures of 65°F or less are desirable, with 50% rh or less.

In processing areas where lower relative humidity and temperature are required and production demands are high, serious consideration should be given to using ASHRAE extreme conditions as the design criteria for the air-handling equipment.

Table 1 Optimum Design Air Conditions[a]

	Dry-Bulb Temperature, °F	Relative Humidity, %
Department or Process		
Chocolate pan supply air	55 to 62	55 to 45
Enrober room	80 to 85	30 to 25
Single cooling tunnel	36 to 45	85 to 70
Double cooling tunnel		
entering	50 to 55	
leaving	38 to 45	
Hand dipper	62	45
Molded goods cooling	36 to 45	85 to 70
Chocolate packing room	65	50
Chocolate finished stock storage	65	50
Centers tempering room	75 to 80	35 to 30
Marshmallow setting room	75 to 78	45 to 40
Grained marshmallow (deposited in starch) drying	110	40
Gum (deposited in starch) drying	125 to 150	25 to 15
Sanded gum drying	100	25 to 40
Gum finished stock storage	50 to 65	65
Sugar pan supply air (engrossing)	85 to 105	30 to 20
Polishing pan supply air	70 to 80	50 to 40
Pan rooms	75 to 80	35 to 30
Nonpareil pan supply air	100 to 120	20
Hard candy cooling tunnel supply air	60 to 70	55 to 40
Hard candy packing	70 to 75	40 to 35
Hard candy storage	50 to 70	40
Caramel rooms	70 to 80	40
Raw Material Storage		
Nuts (insect)	45	60 to 65
Nuts (rancidity)	34 to 38	85 to 80
Eggs	30	85 to 90
Chocolate (flats)	65	50
Butter	20	
Dates, figs, etc.	40 to 45	75 to 65
Corn syrup[b]	90 to 100	
Liquid sugar	75 to 80	40 to 30
Comfort air conditions	75 to 80	60 to 50

Note: Conditions given are intended as a guide and represent values found to be satisfactory for many installations. However, specific cases may vary widely from these values because of factors such as type of product, formulas, cooking process, method of handling, and time. Acceleration or deceleration of any of the foregoing changes temperature, humidity, or both to some degree.

[a]Temperature and humidity ranges are given in respective order (i.e., first temperature corresponds to first humidity).

[b]Depends on removal system. With higher temperatures, coloration and fluidity are greater.

MILK AND DARK CHOCOLATE

Cocoa butter is either the only fat or the principal fat in chocolate, constituting 25 to 40% or more of various types. Cocoa butter is a complex mixture of triglycerides of high molecular weight fatty acids, mostly stearic, oleic, and palmitic. Because cocoa butter is present in such large amounts in chocolate, anything affecting cocoa butter affects the chocolate product as well.

Because cocoa butter is a mixture of triglycerides, it does not act as a pure compound. Its physical properties, melting point, solidification point, latent heat, and specific heat affect the mixture. Cocoa butter softens over a wide temperature range, starting at about 80°F and melting at about 94°F. It has no definite solidification point; this varies from just below its melting point to 80°F or lower,

The preparation of this chapter is assigned to TC 10.9, Refrigeration Application for Foods and Beverages.

depending on the quantity and hardness of cocoa butter and the time it is held at various temperatures. The presence of milkfat in milk chocolate lowers both the melting point and the solidification point of the cocoa butter. High-quality milk chocolate remains fluid for easy handling at temperatures as low as 86 to 88°F. Sweet chocolate remains fluid as low as 90 to 92°F.

Chocolate can be subcooled below its melting point without crystallization. In fact, it does not crystallize en masse but rather in successive stages, as solid solutions of a very unstable crystalline state are formed under certain conditions. The latent heat of crystallization (or fusion) is a direct function of the manner in which the chocolate has been cooled and solidified. Once crystallization has started, it continues until completion, taking from several hours to several days, depending on exposure to cooling, particularly to low temperatures (subcooling).

The latent heat of solidification of the grades of chocolate commonly used in candy manufacture varies from approximately 36 to 40 Btu/lb. Average values for the specific heat of chocolate may be taken as 0.56 Btu/lb·°F before solidification and 0.30 Btu/lb·°F after solidification. The average value for the specific heat of cocoa butter is 0.5 Btu/lb·°F; for milk chocolate, 0.484 Btu/lb·°F; and for roasted cocoa bean, 0.44 Btu/lb·°F. In calculating the cooling load, a margin of safety should be added to these figures.

Cocoa butter's cooling and solidification properties exist in five polymorphic forms: one stable form and four metastable or labile ones. Cocoa butter usually solidifies first in one of its metastable forms, depending on the rate and temperature at which it solidifies. In solidified cocoa butter, the lower-melting labile forms change rapidly to the higher-melting forms. The higher-melting labile forms change slowly, and seldom completely, to the stable form.

Commercial chocolate blocks are cast in metal or plastic polycarbonate molds after tempering. During this process, it is desirable to cool the chocolate in the molds as quickly as possible, thus requiring the shortest possible cooling tunnel. However, cooling blocks too quickly (particularly large commercial blocks, which can range from 10 to 50 lbs) may cause checking or cracking, which, though not injurious to quality, adversely affects the block's appearance and strength. Depositing chocolate into molds at 85 to 90°F is common.

Dark chocolate should be cooled very slowly at 90 to 92°F; milk chocolate, at 86 to 88°F. Air entering the cooling tunnel, where the goods are unmolded, may be 40°F. The air may be 62°F where the goods enter the tunnel. After the chocolate is deposited in the mold, it can be moved into a cooling tunnel for a continuous cooling process, or the molds can be stacked up and placed in a cooling room with forced-air circulation. In either case, temperatures of 40 to 50°F are satisfactory. The discharge room from the cooling tunnel or the room to which molds are transferred for packing should be maintained at a dew point low enough to prevent condensation on the cooled chocolate. In load calculations for the cooling or cold room, it is necessary to account for transmission and infiltration losses, any load derived from further cooling of the molds, and the sensible and latent heat cooling loads of the chocolate itself.

The tunnel is designed to introduce 40°F air countercurrent to the flow of chocolate; the coldest air enters the tunnel where the cooled chocolate leaves the tunnel. Because the tunnel air warms on its way out, the warmest air leaves the tunnel at the point where the warmest molten chocolate enters. The leaving chocolate is markedly cooler than the entering chocolate, and the subcooling is greatly reduced. This in turn reduces the large temperature difference between the chocolate and the cooling air along the entire tunnel length.

For any particular application, only testing will determine the length of time the chocolate should remain in the tunnel and the subsequent temperature requirements. Good cooling is generally a function of tunnel length, belt speed, and the actual time the product contacts the cooling medium.

HAND DIPPING AND ENROBING

The candy centers of chocolate-coated candies are either formed by hand or cast in starch or rubber molds. They are then dipped by hand or enrobed mechanically. The chocolate supply for hand dipping is normally kept in a pan maintained at the lowest temperature that still ensures sufficient fluidity for the process. Because this temperature is higher than the dipping room temperature, a heat source, such as electrically heated **dipping pans** with thermostatic controls, is required. Dipped candy is placed either on trays or on belts while the chocolate coating sets.

Setting is controlled by conditioning the dipping room air. A dry-bulb temperature of 35 to 40°F best promotes rapid setting and provides a high gloss on the finished goods. However, the temperature in the dipping room is raised for human comfort. Suggested conditions for hand-dipping rooms are 64°F db and a relative humidity not exceeding 50 to 55%. The principal aim is to achieve uniform air distribution without objectionable drafts. Loads for this room include transmission, lights, and people, as well as heat load from the chocolate and heat used to warm dipping pans.

In high-speed production of bar candy, the chocolate coating is applied in an **enrober machine**, which consists mainly of a heated and thermostatically controlled reservoir for the fluid chocolate. This chocolate is pumped to an upper flow pan that allows it to flow in a curtain down to the main reservoir. An open chain-type belt carries the centers through the flowing chocolate curtain, where they pick up the coating. At the same time, grooved rolls pick up some chocolate and apply it to the bottom of the centers. Centers should be cooled to 75 to 80°F to help solidification and retention of the proper amount of coating.

The coated pieces are transferred from the enrober to the **bottomer slab** and then pass into the enrober cooling tunnel. The function of the bottomer slab is to set the bottom coating as rapidly as possible, to form a firm base for the pieces as they pass through the enrober tunnel. The bottomer slab is often a flat-plate heat exchanger fed with chilled water or propylene glycol or directly supplied with refrigerant. The belt carrying the candy passes directly over this plate, and heat transfer must take place from the candy through the belt to the surfaces of the bottomer slab. The bottomer slab is sometimes located before the enrober to create a good bottom before full coverage.

The **enrober cooling tunnel** sets the balance of the chocolate coating as rapidly as is consistent with high quality and good appearance of the finished pieces. Typical enrober cooling tunnel times are approximately 7 to 8 min for milk chocolate, and can be as low as 3 to 6 min for vegetable-fat-based coatings. The discharge end of the enrober tunnel is normally in the packing room, where the finished candy is wrapped and packed.

Although not absolutely necessary, air conditioning the enrobing room is desirable. Because the coating is exposed to the room atmosphere, the atmosphere should be clean to prevent contamination of the coating with foreign material. It is advisable to maintain conditions of 75 to 80°F dry bulb and 50 to 55% rh; that is, low enough to prevent centers from warming and to help set the chocolate after it is applied.

BAR CANDY

Production of bar candy calls for high-speed semiautomatic operations to minimize production costs. From the kitchen, the center material is either delivered to spreaders, which form layers on tables, or cast in starch molds. Depending on the composition of the center, the material may be delivered at temperatures as high as 160 to 180°F. Successive layers of different color or flavor may be deposited to build up the entire center. These layers usually consist of nougat, caramel, marshmallow whip, or similar ingredients, to which peanuts, almonds, or other nuts may be added. Because each ingredient requires a different cooking process, each separate ingredient is deposited in a separate operation. Thus, a 1/8 in. layer of caramel may be deposited first, then a layer of peanuts, followed by a 3/4 in. layer of nougat. Except for nuts, it is necessary to allow time for each successive layer to set before the next is applied. If the candy is spread in slabs, the slab must be cooled and then cut with rotary knives into pieces the size of the finished center.

HARD CANDY

Manufacturing hard candy with high-speed machinery requires air conditioning to maintain temperature and humidity. Candy made of cane sugar has somewhat different requirements from that made partly with corn syrup. For example, a dry-bulb temperature of 75 to 80°F with 40% rh is satisfactory for corn syrup (as the corn syrup percentage increases, the relative humidity must decrease), whereas the same temperature with 50% rh is satisfactory for cane sugar.

Where relative humidity is to be maintained at 40% or less, standard dehydrating systems using chemicals such as lithium chloride, silica gel, or activated alumina should be used. A combination of refrigeration and dehydration is also used.

The amount of air required is a direct function of the sensible heat of the room. Approximate rules indicate that the quantity should be between 1.5 and 2.5 cfm per square foot of floor area, with a minimum of 15% outdoor air, or 30 cfm per person. The sensible heat in hard candy, which is at a high temperature to keep it pliable during forming, must also be taken into account.

If concentrations of the finished product in containers or tubs are located in the general conditioned area, the quantity of air must be increased to prevent the product from sticking to the container.

Unitary air conditioners using dry coils are satisfactory if they have a sufficient number of rows and adequate surface. A central station apparatus using cooling and dehumidifying coils of similar design may also be used. Good filtration is essential for air purity as well as for preventing dirt accumulation on cooling coils. Reheat is required for some temperature and humidity conditions. Air distribution should be designed to provide uniform conditions and to minimize drafts.

HOT ROOMS

Jellies and gums are best dried in air-conditioned hot rooms. These products are normally cast into starch molds. The molds are contained in a tray approximately 35 by 15 by 1.5 in. with an extra 0.50 to 0.75 in. blocking at the bottom for air circulation. These trays are racked on trucks, with the number of trays per truck (usually 25 to 30) determined by the method of loading. Trucks are loaded into the hot room where the actual drying is accomplished.

When starch drying is used, careful consideration must be given to proper design of all process and utility equipment in the immediate area because of the explosive nature of dry starch. Refer to the National Electrical Codes and most recent NFPA guidelines for design criteria.

Normal drying conditions average between 120 and 150°F db with 15 to 20% rh. Although humidity is important, close humidity control is not necessary.

Some operators prefer manual humidity control, which requires frequent inspection of actual conditions; others prefer automatic humidity control by instruments calibrated to maintain desired dry-bulb temperature and relative humidity in the hot room regardless of the moisture from the candy. With full automation, the supply air system should provide dry air to the unit and also cool the air usually needed in hot weather to purge the hot room after the drying cycle.

For proper air distribution in the hot room, the maximum amount of air must be in contact with the product. Providing space between trays is one means of accomplishing this. Trucks in the hot room must be placed to ensure continuous airflow from truck to truck with the shortest airflow path. Space must also be maintained at the entering and leaving air sides to ensure flow from the top to the bottom tray for each truck. A large air quantity is required to secure uniformity over the entire product zone.

One device for achieving uniformity is the **ejector nozzle system**, which consists of a supply header fitted with conical ejector nozzles designed for a tip velocity of 2000 to 5000 fpm with a static pressure behind the nozzle as high as 12 in. of water. Nozzles arranged in this way induce an airflow about three times that actually supplied by the nozzles. This ratio gives the most economical balance between air quantity supplied and fan power. The ejector system causes the primary and induced airstreams to mix over the product and the space between the top tray and the ceiling. Ceiling height must be sufficient for this mixing, which rapidly decreases the differential between air supply and actual room temperatures. The high airflow thus created decreases temperature drop across the product. Because temperature drop is proportional to heat pickup, a greater airflow has a lower temperature drop. Thus, the spread between air temperatures entering and leaving the product zone is reduced, promoting uniform drying.

When drying is completed, the product must be cooled rapidly to facilitate unloading. This quick cooling is provided by a second outdoor air intake, which bypasses the heating coil or air-conditioning unit. When this bypass intake is activated, drop dampers open in the bottom of the ejector header, so the air also bypasses the ejector nozzles and removes heat from the room. A ceiling exhaust fan is also recommended for removing rising heated air.

Equipment for this operation consists of a fan and heating coil located outside the hot room. No electric motors should be in the room because of the hazard of sparking. This unit has outdoor air intakes, ejector headers, return air dampers, and dampers for the outdoor air intake. A recording controller to maintain an accurate record of each batch is recommended. The controller simply regulates the flow of steam to the heating coil to maintain the desired room temperature. Control switches should be provided to position outdoor and return air dampers, because a rise in humidity requires more outdoor air, and a drop in humidity requires more return air. In some cases, this function can be achieved automatically with a humidity control. An end position can be included on the control switches for the cooling-down period to start the exhaust fan when the outdoor air damper is opened wide.

COLD ROOMS

Many confectionary products (e.g., marshmallows, certain types of bar centers, and cast cream centers) require chilling and drying but cannot withstand high temperatures. Drying conditions of about 75°F and 45% rh are required in the cold room, and the drying period varies from 24 to 48 h. An ejector-type system similar to that for hot rooms is used, except that cooling coils are provided in the unit. The sensible and latent heat components of the load must be carefully determined so the actual air quantity, as well as the air supply and refrigerant temperatures, can be calculated.

Controlling relative humidity is an inherent part of the system design. The control system is similar to the one used for the hot room, except that its recording regulator must control the flow of steam to the heating coil in winter and regulate the flow of refrigerant to the cooling coil in summer. Flushing dampers and a cooldown cycle are necessary. One precaution in connection with starch or sugar dust picked up in the return airstream must be observed. During the cooling cycle, condensate forms on the cooling coils, turning any starch or sugar dust deposits on the coils into a paste. This reduces capacity and necessitates frequent equipment maintenance and cleaning. Thus, air filters should always be used for outdoor and return air entering the cooling coil. In addition, a coil wash system should be provided to aid in cleaning the coils.

COOLING TUNNELS

Various candy plant cooling requirements can best be handled in a cooling tunnel, including the cooling of (1) coated centers after they leave the enrober, (2) cast chocolate bars, and (3) hard candy. These operations are usually set up for a continuous flow of high-rate production. Product is normally conveyed on belts through the enrober or casting machine and then through a cooling chamber.

A cooling tunnel is an insulated box placed around the conveyor so that product may travel through it in a continuous flow. Refrigerated air is supplied to this enclosure to cool the product. To achieve maximum heat transfer between the air and product, air should flow counter to the material flow. In general, air supply temperatures of 35 to 45°F with air speeds up to 500 fpm have been found satisfactory.

Cooling tunnel air speed can significantly affect the degree of chocolate temper found in the finished candy. **Temper** is a measure of the percentage of stable fat (cocoa butter) crystals that form during chocolate solidification. Too rapid a cooling often results in poor final temper, causing the finished candy to have a dull matte finish and poor snap (soft chocolate), and over time results in cocoa butter migrating to the surface of the piece, giving it a grayish cast. It is generally recommended to expose the chocolate candy to higher initial cooling temperatures (55 to 60°F) and lower initial air speeds to avoid cooling the candy too rapidly. Product temperature at the exit of the cooling tunnel is also critical. Finished product must exit the cooling tunnel at a temperature well above the dew point of the packaging area, to avoid condensation on the surface of the candy. On larger, higher-speed cooling tunnels, this is often accomplished by dividing the cooling tunnel into several zones, where air temperature and velocity can be adjusted independently.

The actual size of the tunnel is determined by the size of the conveyor belt and the air quantity, which depends on heat load and desired rate of cooling. The rise in air temperature through the tunnel should be limited to 15 to 20°F maximum. Each tunnel generally has one refrigerated air handler, which normally consists of a fan and coil with the necessary duct connections to and from the unit. An outdoor air intake is advisable in appropriate climates because cooling can at times be accomplished with outdoor air (without operating the refrigeration plant). The outside air intake should have suitable air filters to minimize contamination. The tunnel should be made as tight as possible, and the entrance and exit openings for the candy should be as small as possible to limit air loss from or infiltration into the tunnel; in some cases, it is advisable to use a flexible canvas curtain to control airflow. Because some loss is unavoidable, it is practical in some applications to take a small amount of outdoor air or air from adjoining spaces to provide a slight excess pressure in the tunnel.

For chocolate enrobing, the condition of the air is the paramount factor in securing the best possible luster and most even coating. The best results are obtained with rather slow cooling, but this requires either a low production rate or excessively long tunnels; the final design is a compromise. The coating must be in the proper condition when it is poured over the centers, because improper temperature at this point causes blushing or loss of luster. Proper temperature, however, is a function of the enrober machine and its operation; no amount of correction in the tunnel can compensate.

A variation of the standard single-pass counterflow tunnel has been used for enrobing. The tunnel is divided horizontally by an uninsulated sheet metal partition. The belt carrying the candy rides directly on this partition, and the return belt is brought back through the space below the partition. Cold air is supplied to the lower chamber near the enrober, progresses to the opposite end of the tunnel, is transferred to the top chamber, moves back to a point near the enrober, and is then returned to the cooling equipment. This tunnel has two important advantages: (1) it chills the return belt so that the belt can act as a bottomer slab to quickly set the base of the coated piece, and (2) the uninsulated partition, with the coldest air below it, assists in this bottoming operation. Thus, the air supply has already absorbed some of its heat load by the time it is actually introduced to the product-cooling zone. The method approaches the advantage of slow cooling but keeps the tunnel at a minimum length.

For some applications, a spiral belt cooling tunnel can be used to save floor space. Products that are sensitive to injury or abrasion from lateral movement of the belt underneath are generally unsuitable for a spiral conveyor. Typical cooling tunnel configurations are described and illustrated in Chapter 16.

COATING KETTLES OR PANS

Originally, revolving coating kettles or pans were merely supplied with warm air, ranging from 80 to 125°F. This air was then exhausted from the kettle to the room, creating a severe nuisance from sugar dust blowing out of the kettle. Another difficulty was that some of the energy required to rotate the kettle was converted to heat in the centers being coated, causing enough expansion to produce cracking or checking. To mitigate this problem, the following practice evolved: a portion of the coating is applied, the product is withdrawn for a seasoning period of up to 24 h, and the material is then returned to the kettles for additional coating.

Some installations overcome most of the sugar dust problem by providing a conditioned air supply to the kettles and positive exhaust from the kettles. The wet- and dry-bulb temperatures of air supplied to the kettles is controlled so that the rate of coating evaporation and drying are uniform, at a high production rate and reduced labor cost. Evaporation of moisture in the coating material tends to occur at the wet-bulb temperature of the air supplied to the kettle. Much of the heat of crystallization entering the product is absorbed, and the centers are not overheated. This eliminates the need for a seasoning period and allows continuous operation.

To minimize drying times, a relative humidity of less than 40% is often used for the panning supply air. This requires using a desiccant system with chemicals such as lithium chloride or activated alumina. The design of any desiccant system for year-round candy production should be based on ASHRAE extreme conditions to avoid quality problems as seasonal changes occur.

With air conditioning applied to coating kettles, the number of rejects caused by splitting, cracking, uneven coating, or doubles can be reduced considerably. Sugar dust recovery is accomplished with cyclone-type dust-collecting devices.

PACKING ROOMS

Manufacturers spend considerable time and effort in the design and application of packaging materials because of their effects on product keeping quality. Important packaging considerations are moisture-proof containers, vapor retarders for abnormally high humidity, and protection against freezing or extreme heat. Controlling air in the packing room is essential for proper packing.

For example, air surrounding products packed in a room at 85°F db and 60% rh, which is not unusual in a normal summer, has a 69°F dew point. If this sealed package were subjected to temperatures below 69°F, air in the package would become supersaturated and moisture would condense on container and product surfaces. If the product were then subjected to a higher temperature, the moisture would reevaporate. In the process, chocolates would lose their luster or show sugar bloom, and marshmallows would develop either a sticky or a grained surface, depending on the formula used.

In practice, packing room conditions of 65°F db and 50% rh have been found most practicable. Results improve if the relative humidity is reduced several points. For hard candy, which is intensely hygroscopic, the relative humidity should be reduced to 35 to 40% at 70 to 75°F.

REFRIGERATION PLANT

Large candy manufacturers often use a central refrigeration plant for cooling water and/or a secondary coolant for circulation throughout the plant to meet the various load requirements. Propylene glycol is the secondary coolant of choice for food plants. See Chapter 4 for design and application information on secondary coolants.

The central refrigeration plant for cooling water and/or propylene glycol may use ammonia or one of the environmentally suitable halocarbon refrigerants. Heat transfer equipment both for new plants and for retrofits for cooling water or propylene glycol may be welded-plate heat exchangers or extended-surface shell-and-tube heat exchangers to increase plant efficiency and to reduce the refrigerant charge. They are usually piped for gravity-flooded or liquid overfeed operation. Compressors vary with refrigerant used, plant size, initial and operating costs, and plans for future expansion. All installations should adhere to applicable refrigeration codes and regulations.

Secondary coolant temperatures usually vary between 28 and 32°F for central refrigeration plants. These temperatures seldom require an artificial defrost system for the evaporator coils. Some cooling tunnel applications may require lower coolant temperatures and associated defrost cycles.

A secondary coolant distribution system makes it feasible to connect all service points to one source of refrigeration and, with control systems, to maintain dry-bulb and dew-point temperatures precisely. This system also can be extended to comfort cooling in offices and other nonproduction areas.

Smaller manufacturing plants may be better served by multiple condensing units and direct-expansion evaporators using an environmentally suitable refrigerant such as R-134a. These condensing unit/evaporator combinations may be grouped at appropriate refrigeration or air-conditioning load points as dictated by the plant layout. This arrangement provides flexibility to expand or retrofit a plant a portion at a time at a reasonable investment and operating cost.

STORAGE

CANDY

Most candies are held for 1 week to over a year between manufacture and consumption. Storage may be in the factory, in warehouses during shipping, or in retail outlets. It is important that the candy maintain quality during that time.

Low-temperature storage does not produce undesirable results if the following conditions are met:

- Candy ingredients are appropriate for refrigerated storage.
- Packages have a moisture barrier.
- Storage room humidity is held at equilibrium with desirable moisture conditions for preserving the candy.
- Candy is brought to room temperature before packages are opened.

The storage period depends on (1) marketing season of the candy; (2) stability of the candy; and (3) storage temperature and humidity (see Table 2).

A candy's shelf life is determined by the stability of its individual ingredients. Common ingredients are sugar (including sucrose, dextrose, corn syrups, corn solids, and invert syrups), dark and milk chocolates, nuts (including coconut, peanuts, pecans, almonds, walnuts, and others), fruits (including cherries, dates, raisins, figs, apricots, and strawberries), dried milk and milk products, butter, dried eggs, cream of tartar, gelatin, soybean flour, wheat flour, starch, and artificial colors.

Refrigerated storage of candy ingredients is especially advantageous for seasonal products, such as peanuts, pecans, almonds, cherries, coconut, and chocolate. Ingredients with delicate flavors and colors, such as butter, dried eggs, and dried milk, retain quality more evenly year-round if kept properly refrigerated. Otherwise, ingredients containing fats or proteins may lose considerable flavor or develop off-flavors before being used.

Table 2 Expected Storage Life for Candy

Candy	Moisture Content, %	Relative Humidity, %	Storage Life, Months — Storage Temperature, °F: 68	48	32	0
Sweet chocolate	0.36	40	3	6	9	12
Milk chocolate	0.52	40	2	2	4	8
Lemon drops	0.76	40	2	4	9	12
Chocolate-covered peanuts	0.91	40 to 45	2	4	6	8
Peanut brittle	1.58	40	1	1.5	3	6
Coated nut roll	5.16	45 to 50	1.5	3	6	9
Uncoated peanut roll	5.89	45 to 50	1	2	3	6
Nougat bar	6.14	50	1.5	3	6	9
Hard creams	6.56	50	3	6	12	12
Sugar bonbons	7.53	50	3	6	12	12
Coconut squares	7.70	50	2	3	6	9
Peanut butter taffy kisses	8.20	40	2	3	5	10
Chocolate-covered						
creams	8.09	50	1	3	6	9
soft	8.22	50	1.5	3	5	9
Plain caramels	9.04	50	3	6	9	12
Fudge	10.21	65	2.5	5	12	12
Gumdrops	15.11	65	3	6	12	12
Marshmallows	16.00	65	2	3	6	9

Candies are semiperishable: the finest candies or candy ingredients may be ruined by a few weeks of improper storage. This includes many candy bars and packaged candies and some choice bulk candies. Unless refrigeration is provided from time of manufacture through the retail outlet, the types of candies offered for sale must be greatly reduced in the summer.

Benefits of refrigerated storage of candies, especially during summer, are the following:

- Insects are rendered inactive below 48°F.
- The tendency to become stale or rancid is reduced.
- Candies remain firm, protecting against sticking to the wrapper or being smashed.
- Loss of color, aroma, and flavor is reduced.
- Candies can be manufactured year-round and accumulated for periods of heavy sales.

Color

Many colors used in hard candies, hard creams, and bonbons gradually fade during storage at room temperature, especially in the light. However, the most marked effect of storage temperature on color occurs with chocolates. In candies high in protein and nuts, there is a gradual darkening of color, especially at higher storage temperatures.

Temperatures of 85 to 95°F cause graying of chocolates in only a few hours and darkening of nuts within a month. **Graying** or **fat blooming** is caused by crystals of fat on the surface of the chocolate coating. This condition is usually associated with old candies, but new candies can become gray after one day under adverse storage conditions. Chilling chocolates after exposure to high temperatures produces graying very quickly, but chilling without previous heat exposure does not.

Sugar blooming of chocolate looks similar to graying or fat blooming and is caused by crystallized sugar deposited on the surface by moisture condensation when candy is removed from refrigerated storage without proper tempering. Chocolate tempered by gradually raising the temperature to normal without opening the package does not incur sugar blooming, even after storage at 0°F or lower. Sugar bloom may also occur because of storage in overhumid air and migration of moisture from the centers to the surface.

Flavor

Keeping candy fresh is one of the chief reasons for refrigerated storage. Most flavors added to candies, including peppermint, lemon, orange, cherry, and grape, are distinctive and stable during storage. Proper design of flavor and color storage areas is critical. Many flavors and colors have very low flash points because of their alcohol bases and high levels of organic volatiles. Storage areas for these types of compounds should be adequately ventilated and comply with recent NFPA guidelines. Less-pronounced flavors such as those of butter, milk, eggs, nuts, and fruits are more sensitive to high temperatures.

Low temperatures retard development of staleness and rancidity in fats, preserve flavors in fruit ingredients, and prevent staleness and other off-flavors in candies containing semiperishable ingredients such as milk, eggs, gelatin, nuts, and coconut. Candies containing fruit become strong in flavor when stored at room temperature or higher for more than a few weeks; those containing nuts become rancid. There are no specific critical temperatures at which undesirable changes occur, but the lower the temperature, the more slowly they take place.

Texture

Candy becomes increasingly soft at high temperatures and increasingly hard and brittle at low temperatures, reaching an optimum for eating at about 70°F. Changes in texture are reversible from below 0°F to 80°F, enabling refrigerated candies to be returned satisfactorily to any desired temperature for eating. This is extremely important because the texture of most candies subjected to very low temperatures (or even shipped in contact with dry ice at about –110°F) is not permanently changed.

Most candies are manufactured at controlled temperatures. Their texture (except in hard candies and hard creams) is maintained best below 68°F.

Insects

Candies containing fruit, chocolate, nuts, or coconut are favorite hosts for insects. Because fumigation and insect repellants are seldom permissible, refrigeration is used to inactivate insects in candy and candy ingredients.

Common insects become active at about 50°F, and activity increases as the temperature is raised to 100°F. Although common cold storage temperatures do not kill many insects, temperatures below 50°F do inactivate them. Both adults and eggs may exist for months at above-freezing temperatures without feeding or propagating. Candies with insect eggs on either the product or wrappers may be refrigerated for long periods with no apparent damage, but when they are warmed up, a serious insect infestation may develop.

Both adults and eggs are killed at storage temperatures of about 0°F. Storing candies at 0°F for a few weeks usually destroys all forms of insect life. Lower temperatures and long storage periods are lethal to insects.

Storage Temperature

The effect of storage temperature is difficult to separate from that of humidity, but the latter is more important. There are no specific critical refrigerated temperatures at which certain types of candy must be held. In general, the lower the temperature, the longer the storage life, but the greater the problem of moisture condensation on removal.

Air Conditioning (68 to 70°F). Because candy storage begins in the tempering room of the manufacturing plant, 68°F and 50% rh is desirable to prevent pieces from being packaged when they are soft and sticky. Under these conditions, all candies remain firm and there is little or no graying of chocolates; their original luster can be held.

Hard candies and candies containing only sugar ingredients keep in good condition for more than 6 months at 68 to 70°F. Other types become stale, lose flavor and luster, and darken in color.

Under prolonged storage, candy containing nuts or chocolate becomes musty or rancid, and even the colors and flavors of some hard candies may fade. Only **summer candies** should be held for more than a few weeks at 68 to 70°F or higher. Unless precautions are taken, the temperature of truck or rail shipments may rise to the melting point of semisoft candies or to the graying point of chocolates. Candies temporarily stored in the sunshine or in warm places in buildings may suffer severe loss of shape, luster, and color. Some companies use refrigerated trucks and railroad cars for hauling candies. Portable refrigerated containers provide a viable means of ensuring that candies maintain their quality from manufacturer to retail outlet.

Cool Storage (48 to 50°F). Candies stored at this temperature remain firm and retain good texture and color; only those containing nuts, butter, cream, or other fats become stale or rancid within 4 months. Candies that remain practically fresh for 4 months are fudge, caramels, sugar bonbons, gumdrops, marshmallows, lemon drops, hard creams, and semisweet chocolates; they are wrapped in aluminum foil to give added protection. Peanut butter taffy kisses, peanut brittle, uncovered peanut rolls, chocolate-covered peanut rolls, and nougat bars become stale at this temperature.

Cold Storage (32 to 34°F). Most candies can be successfully held in cold storage for at least one year, and many for much longer. Only those containing nuts, coconut, chocolate, or other fatty materials become stale or rancid.

Freezer Storage (0°F). The economic justification for freezing candy is the same as for any other food: better preservation for a longer time. This method of preservation is suitable for candies (1) in which high quality standards must be maintained; (2) in which a longer shelf life is desired than is accomplished from other methods of storage; (3) that are normally manufactured 6 to 9 months in advance of consumption; and (4) that are especially suitable for retailing as frozen items. Because of their high sugar content and low moisture content, little ice forms in candies at 0°F.

One of the chief reasons for freezing candies is to hold them in an unchanged condition for as long as 9 months, then thaw and sell them as fresh candies. Experience shows that this is not only possible but also practical if the manufacturer (1) freezes only those candies that would lose quality when held at a higher temperature; (2) eliminates the few kinds that crack during freezing; (3) packages the candies in moistureproof containers similar to those used for other frozen foods; and (4) thaws the packages unopened to avoid condensation of moisture on the surface.

Moistureproof Packaging. Experiments show that candies for freezing require more protection than those for common storage because the storage period is usually longer and condensation on removal is more likely. A single layer of moistureproof material (e.g., aluminum foil, plastic film, or glassine) affords adequate protection. Candies not fully protected from desiccation become hard and grainy and lose flavor.

Adequate protection is provided when the moisture barrier is in contact with the candy in the form of a sealed, individual wrapper. Inner liners for the boxes protect candy, provided they are sealed (which is difficult and seldom accomplished). Moisture barriers are usually applied as overwraps for the boxes, chiefly because they are easiest to apply and seal by machines. Overwraps provide less protection than wraps for individual pieces of candy because of the larger amount of air enclosed in the box. Also, boxes with extended edges offer less protection.

Thawing. Frozen candies should be thawed in unopened packages. Freezing itself affects only a few types of candies, but the manner of removal from storage affects all candies, especially those unprotected by special coatings or individual wraps.

Improvement. Candies that are improved in freshness, mellowness, or textural smoothness by freezing include those with high moisture content and without protective coatings or individual wrapping. Usually, these are candies ordinarily subject to surface drying. Marshmallows, jellies, caramels, fudges, divinities, coconut macaroons, fruit loaves, coconut bonbons, panned Easter eggs, malted milk balls, and chocolate puffs are in this group.

Stability. The stability of candies after freezing is good. Candies may be held frozen for 6 months or more, carefully thawed, and then sold as fresh. Some candies are prepared especially for freezing. These are made of low-melting-point fat, have more flavor and softer texture than most candies, and should be eaten while they are cold.

Humidity Requirements

Sugar ingredients are stable over a wide range of storage temperatures, but they are sensitive to high or low humidity. The initial moisture content of candies largely determines the optimum relative humidity of the storage atmosphere. Candies with a moisture content of 12 to 16% (marshmallows, gumdrops, coconut sticks, jelly beans, and fudge) should be stored at 60 to 65% rh to avoid becoming (1) sticky, runny, or moldy or (2) hard and crusty. Candies with a moisture content of 5 to 9% (most fine candies, nougat bars, nut bars, hard and soft creams, bonbons, and caramels) should be stored at 50 to 55% rh to retain their original weight, finish, and texture. Candies with a moisture content below 2% (milk chocolate bars, chocolate-covered nuts, and all kinds of hard candies) should be stored at 45% rh or lower.

The hygroscopicity of the ingredients also determines the relative humidity at which candies must be held to retain their original firmness and finish. Candies with a high proportion of invert syrups, such as taffy kisses, must be kept very dry, even though their moisture content is not extremely low. Other candies containing high proportions of invert syrup, honey, or corn syrup must be held in a drier atmosphere than the moisture content indicates.

Candies stored at low temperatures have a wider range of critical relative humidities than those stored at high temperatures. For example, nougat bars stored at 65% rh (10% too high) become sticky within a few days at room temperature, but at 40°F or lower, stickiness might not develop for many weeks. Similarly, marshmallows stored in a room with 55% rh (10% too low) become dry and crusty within a few days at room temperature, but at 40°F or lower, they show little change for several weeks. Refrigeration retards the ill effects of storage under improper humidity conditions. Humectants, such as sorbitol, glycerine, and high conversion corn syrup, are advantageous for maintaining original moisture content of certain candies.

NUTS

Commonly refrigerated nuts include peanuts, walnuts, pecans, almonds, filberts, chestnuts, and imported cashews and Brazil nuts. The advantages of refrigerated storage of nuts are the following:

- Marketable life is increased as much as 10 times.
- Natural texture, color, and flavor are retained almost perfectly from one season to the next.
- Staleness, rancidity, and molding are retarded for more than 2 years, depending on the temperature.
- Insect activity is arrested at temperatures below 48°F.

With optimum temperature, humidity, atmospheric conditions, and packaging, good-quality nuts may be successfully stored for up to 5 years.

Temperature

Other conditions being equal, the lower the temperature, the longer the storage life of nuts. Storage life may be doubled or tripled with each 20°F drop in temperature. The freezing points of nuts, depending on the moisture content, are about 23°F for chestnuts; 14°F for walnuts, pecans, and filberts; and 13°F for peanuts. Normal moisture content for stored nuts is as follows: chestnuts, 30%; peanuts, 6%; walnuts, 4.5%; pecans, 4%; and filberts, 3.5%.

Shelled nuts to be stored from one harvest season to the next without appreciable loss in quality must be held at 36°F or lower; those to be stored for 6 to 9 months must be kept at 48°F or lower; and all nuts stored for 4 to 6 months should be held below 68°F. Storage life at a given temperature doubles if the nuts are unshelled.

Relative Humidity

Although the storage temperature of nuts may range from 68 to –20°F or lower, relative humidity must remain between 65 and 75% to maintain the optimum moisture content for desired texture, color, flavor, and stability. If the moisture content rises as much as 2% above normal, the nuts (except chestnuts) darken, become stale, and may become moldy. If the moisture drops more than 2% below normal, the nuts become objectionably hard and brittle.

When the relative humidity is suitable, nuts that are too high or too low in moisture may be safely stored with the assurance that rapid air circulation will bring the moisture content to a safe level. In this sense, the storage room acts as a conditioning room.

Atmosphere

All nuts (again, except chestnuts) contain 45% or more oil and readily absorb odors and flavors from the atmosphere and surrounding products. Certain gases, particularly ammonia, react with tannin in the seed coats of nuts, causing them to turn black. Therefore, the atmosphere in the nut storage room must be free of all odors. This includes the containers, walls of the room, pallets, and other stored products.

Products that can be safely stored with nuts include dried fruits, candies, rice, and goods packaged in cans, bottles, or barrels. Onions, meats, cheese, chocolate, fresh fruits, and other products having an odor or a high moisture content should not be stored with nuts.

Packaging

The storage life of nuts may be greatly influenced by the choice of package. Nuts become bruised when they are shelled, and oil crawls over the surface and onto the package in a very thin film. Unless this crawling is retarded by a package that acts as a barrier, contains an antioxidant, or removes air by vacuum, the nuts become stale and rancid. Furthermore, some packaging materials (e.g., polyethylene) should be avoided because they impart an undesirable odor to the nuts.

DRIED FRUITS AND VEGETABLES

Dried fruits and vegetables differ in the following ways:

1. Fruits contain sugars that render them more hygroscopic, harder to dry, and greater absorbers of moisture during storage.
2. Moisture in dried fruits may range from 32% with sorbate treatment to as low as 2%, whereas that in vegetables ranges from 7% to a low of 0.3%.
3. Dried fruits are acid, more highly colored, and more stable during storage than vegetables, which are nonacid.
4. Fruits are generally dried raw with active enzyme and respiration systems, whereas vegetables are blanched or precooked, with no active enzymes (thus, dried fruits are more responsive to storage temperature and humidity conditions than are vegetables).
5. The high sugar and acid content of dried fruits provides an adverse physical environment for bacteria, thus making their growth almost impossible even though the moisture level is higher than that in dried vegetables.

Dried fruits and vegetables maintain quality longer when stored at low temperatures. The storage life of dehydrated vegetables has been extended by packing them in a nitrogen or carbon dioxide atmosphere in the presence of a desiccant to achieve further reduction in moisture content. Staleness (charred flavor or off-flavor) and other deterioration in dehydrated vegetables are inhibited by packing in nitrogen. In air-packed samples, the staling rate is reduced at low temperature. Cut fruits and dried vegetables are widely treated with sulfite to retain color and extend storage life. A light coating of laundry starch applied to diced carrots prior to dehydration has achieved excellent results in retention of color and other quality factors during storage in cellophane at 84°F without sulfite.

Refrigerated storage at 40 to 50°F or lower retards and controls insect infestation. Substantial killing occurs with exposure at 32°F for 6 months or longer, and a temperature of 0°F kills insects within a few hours. An alternative method is fumigation, which is generally used in commercial practice.

Nonenzymatic browning (browning in products that have been scaled or blanched adequately to inactivate enzymes) is reduced in dehydrated vegetables at low temperature. Cold storage offers protection for several years.

Increasing the temperature of dried apricots accelerates oxygen consumption, carbon dioxide production, disappearance of sulfur dioxide, and darkening.

Molds and yeasts do not grow in dried fruits that have an adequate sulfur dioxide content or less than 25% moisture. At 32°F, relative humidity is less important than at higher temperatures.

Methods of dehydration include the following:

- **Dehydrofreezing.** Raw, prepared product is dried to about 50% in weight, followed by freezing. This method yields excellent fruit and vegetable products with storage at 0°F or lower. Concentration of juices by low temperature and high vacuum followed by preservation of the concentrate by freezing (Chapter 25) is another application of the process.
- **Freeze Drying.** Products to be freeze dried are usually frozen slowly to form larger ice crystals to increase process efficiency while retaining quality. Frozen product is placed in freeze-dry chambers, where moisture is removed by sublimation through the application of vacuum and low heat such that product porosity is preserved for subsequent reconstitution. This method is successfully used with products of high value, high protein, low fat, and low sugar content. Although refrigerated storage is not necessary to prevent freeze-dried food from spoiling, it is necessary to preserve maximum flavor and natural color.

Dried Fruit Storage

Refrigeration augments drying as a means of preserving fruits. The optimum conditions for holding most dried fruits are about 55% rh and just above the individual fruit's freezing temperature. Because the sugar content of these products is high, the freezing point varies from about 22°F to 26°F. Refrigerated storage helps retain natural flavor, ascorbic acid, carotene, and sulfur dioxide, and helps control browning, insects, rancidity, and molding. Other than for insect control, low humidity is more important than low temperature for storing dried fruit. Packaged, sulfured cut fruits keep adequately at higher humidities.

Although most dried fruits are adversely affected by softening and injured by freezing, dates are held best by freezing. Before storage, dried fruits should be brought to the desired moisture content.

When practical, dried fruits should be packed in moistureproof containers made of metal or foil that not only ensure a constant moisture content in storage, but also prevent injury from moisture condensation on removal from storage. Packaging permeability is extremely important because of the adverse effects of storage humidity. Recommendations are as follows:

Raisins. At 32 to 40°F and 50 to 60% rh, sugaring is prevented for one year, if the moisture content of the dried fruit is not unusually high. Raisins contain 15 to 18% moisture; for extremely long storage, the lowest possible moisture content should be maintained.

Figs. These may be held for a year at 32 to 40°F and 50 to 60% rh. A temperature of 55°F or lower prevents darkening for more than 5 months, and low humidity controls sugaring.

Prunes. These may be held for a year at 32 to 40°F and 50 to 60% rh. For storage of 4 to 5 months, 75 to 80% rh is not detrimental.

Apples. At 32 to 40°F and 55 to 65% rh, dried apples retain excellent color and texture for more than a year. A relative humidity of 70 to 80% is not objectionable at 32°F, but at 40°F enough moisture may be gained to cause molding within 8 months. Browning develops gradually at 40°F and above.

Pears. Same as for apples.

Peaches. Sun-dried freestone peaches are harder to store than most dried fruits. Therefore, the temperature should be held close to 32°F with 55 to 65% rh. At 40°F and moderate humidity, moisture pickup causes rapid molding and browning.

Clingstone peaches (dehydrated after steam scalding) should be stored at 32 to 40°F and 55 to 75% rh. Sun-dried peaches tolerate a slightly higher humidity than dehydrated peaches.

Apricots. Dried apricots are easy to keep in refrigerated storage at 32 to 40°F and 55 to 65% rh. They remain in excellent condition for more than a year. At 40°F and moderate humidity, there is enough gain in moisture content to cause molding.

Dates (sucrose or hard type). For storage of 6 months or less, dates may be held at 32°F and 70 to 75% rh, but for longer storage they should be stored at 24 to 26°F. Usually it is more convenient to store them at 32°F, at which temperature they can be stored for over a year.

Soft or invert sugar-type dates may be held for 6 months at 28 to 32°F, but if storage is for 9 to 12 months, the temperature should be 0 to 10°F. Uncured dates should be stored at 0 to 10°F.

Dried Vegetable Storage

Few specific recommendations for dehydrated vegetable storage temperatures are available. Low storage temperatures retard deterioration, but cold storage is considered necessary only for long storage periods. Advantages include (1) control of insects at 45°F or lower, (2) preservation of natural colors, and (3) retention of initial flavors and vitamins.

Because most dried vegetables have very low moisture content, are well packaged, and are usually surrounded by an atmosphere of nitrogen or rarefied air, refrigeration is less essential than it is for fresh vegetables.

CONTROLLED ATMOSPHERE

Low oxygen in the storage atmosphere suppresses growth of insects and molds, retards rancidity and staleness, and reduces oxidative changes in flavors, odors, and colors. In small packages, oxygen can be reduced by vacuum; in storage and large shipping containers, oxygen can best be flushed out with nitrogen. Excellent results have been obtained by substituting up to 98% of the atmosphere with nitrogen. Nitrogen is preferred to carbon dioxide, ethylene, or other gases for storage of low-moisture products; it greatly extends shelf life even with refrigeration.

CHAPTER 30

CARGO CONTAINERS, RAIL CARS, TRAILERS, AND TRUCKS

TRANSPORT of commodities may be as simple as direct delivery of fresh vegetables from garden to market in a wagon. However, travel time, ambient temperature, and risk of spoilage often make temperature-controlled transport necessary. Because some commodities are sensitive to the relative humidity and chemical composition of their surrounding atmosphere, these conditions may also need to be controlled. Today many commodities travel to distant markets intermodally (i.e., by some combination of highway, ocean, and railroad). This chapter discusses the vehicles, equipment, and related factors that combine to preserve temperature-sensitive commodities as they travel.

Users are urged to regard the vehicle and its equipment as a system, particularly when making insulation and equipment sizing decisions.

VEHICLES

Vehicles used for temperature-controlled transport are similar in construction and outward appearance to those in general freight service, but have three fundamental differences: they have (1) insulation that is usually foamed in place, (2) provisions for conditioned air circulation through and around the cargo, and (3) machinery for cooling and/or heating. A brief description of the four main vehicle types follows.

Cargo containers are usually 8 ft wide, 8 to 9.5 ft high, and 20 or 40 ft long (Figure 1). They have hinged doors in one end for cargo loading and other access to the interior. The machinery comprises the opposite end, so it must also provide structural rigidity and insulation. As shown in Figure 1, containers have standardized corner fittings to secure them to vessels, railway cars, and highway vehicles. Standards also govern their exterior dimensions. (Refer to ANSI *Standard* MH5.1.1.5 and ISO *Standard* 668.)

Railway refrigerator cars are insulated boxcars, usually 50 to 70 ft long (Figure 2). As illustrated, they may have a machinery compartment at one end.

Trailers range in size from 8 to 8.5 ft wide, 12 to 13.5 ft high, and 24 to 55 ft long. Their doors are usually hinged, but they may have insulated roll-up doors if used for multistop delivery service. Some include a curbside door in addition to rear doors. Several interior compartments for different temperatures may be provided. For hanging uncut meat, overhead rails are used. Specially designed trailers riding on railway flat cars are quite common. Another, bimodal design can be mounted directly on specially configured railway bogies and pulled by a locomotive in a train of similar trailers.

As with ordinary trucks, those built for temperature-controlled duty come in a wide variety of designs and sizes. Their bodies may have insulated hinged or roll-up doors; these may be on the sides and rear. Smaller vehicles may include a refrigeration compressor as an engine-driven accessory (see Figure 7).

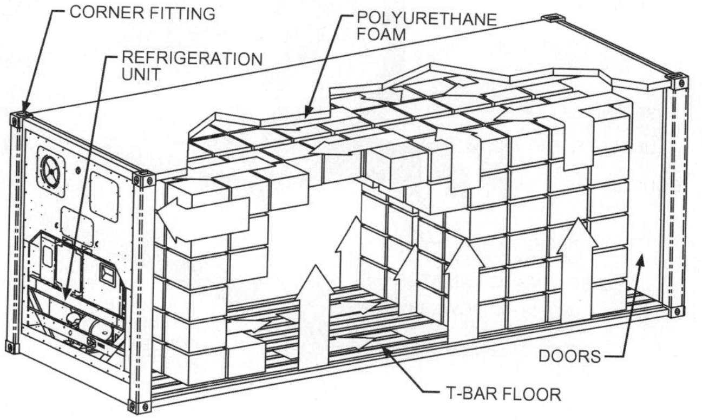

The preparation of this chapter is assigned to TC 10.6, Transport Refrigeration.

Fig. 1 Refrigerated Cargo Container

VEHICLE DESIGN CONSIDERATIONS

Insulation and Vapor Barrier

Envelope design factors to be considered are similar to those for stationary refrigerated facilities, and include the following:

- Extremes of exterior conditions: temperature, relative humidity, wind, and solar effect
- Desired interior conditions: temperature and relative humidity
- Insulation properties: thermal conductivity, moisture permeability and retention, chemical and physical stability, adhesion, uniformity of application, fire resistance, cost of material and application, and presence of structural members
- Infiltration of air and moisture
- Tradeoffs between construction cost and operating expense

When applied to refrigerated vehicles, these five factors are complicated by others unique to transportation. Exterior dimension constraints are imposed by domestic or international standards and regulations, and shippers want maximum cargo space (which limits insulation thickness) and minimum tare weight. The frequency and duration of door openings may be considerable. Long trips at highway or railway cruising speeds affect infiltration. Physical deterioration from the shock and vibration of travel and cargo shifting is likely. Also, there is potential for damage to insulation and vapor barriers from vehicle accidents and cargo handling mishaps.

Closed-cell foamed-in-place insulation, such as polyurethane, is generally recommended to achieve an approximate thermal conductivity k of 0.15 Btu·in/h·ft^2·°F. It also helps limit air and

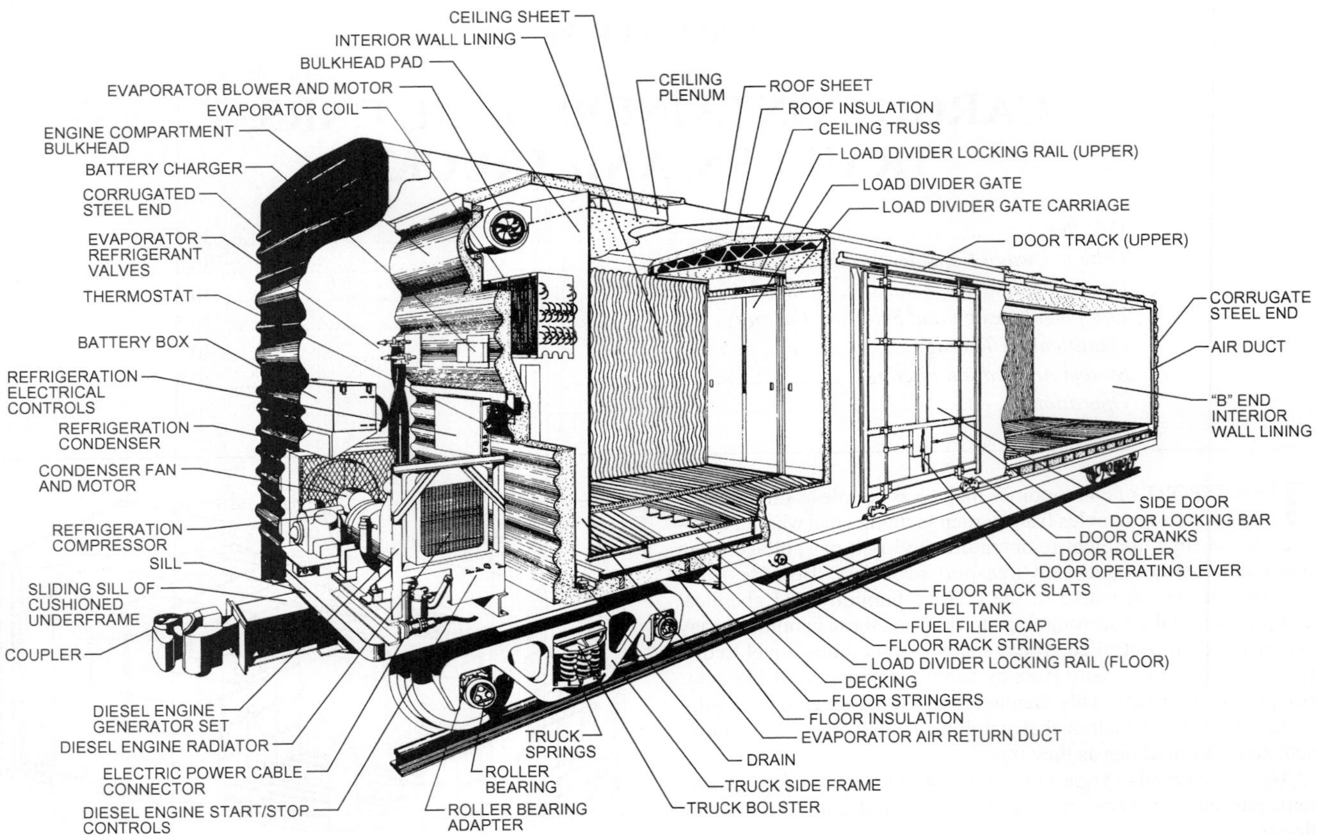

Fig. 2 Mechanical Railway Refrigerator Car

water vapor infiltration. Buyers often specify the *UA* or maximum heat transfer rate, usually at 100°F and 50% rh outside, and 0°F inside, expressed as Btu/h·°F for the entire vehicle.

Environmental considerations affect and are affected by vehicle insulation and vapor barrier choices. Mandated changes to insulation frothing agents with little or no adverse environmental impact may increase insulation *k* value, and moisture permeability and retention. Chemical and physical characteristics such as adhesion, durability, and stability may also be degraded. Because reduced insulation effectiveness increases energy use, it adds to air pollution and global warming concerns. Finally, the potential for materials recycling at the end of useful vehicle life must be considered.

Cargo containers usually have polyurethane insulation at 3 in. thickness in walls and floors, and 4 in. in ceilings. Rail cars often use 3 to 6 in. in walls, 5 to 8 in. in floors, and 5 to 8 in. in ceilings. Trailers and trucks generally use 1.5 to 4 in. in walls, floors, and ceilings for frozen loads, and 1 to 2.5 in. in walls, floors, and ceilings for nonfrozen loads. Vehicle front walls are sometimes thicker because of structure to resist cargo shifting and support equipment.

As mentioned previously, exterior dimensions are restricted and shippers want maximum cargo space. Increasing insulation thickness from 3 in. to 4 in. in a 40 ft long trailer decreases cargo space by 100 ft^3, or about 4%. However, the vehicle's *UA* will improve, affecting equipment selection and improving operating economy. This exemplifies the need to regard the vehicle and its equipment as a system.

Floors in all vehicles must support cargo and cargo-handling equipment. They frequently include rigid polystyrene or polyurethane foam to eliminate beams. Floors must be watertight and joined to walls to exclude water from insulation; a skirt bonded to the floor and extending at least 6 in. up walls may be needed to control water running down walls and collecting on the floor. Floor drains, if used, must be trapped or capped to prevent infiltration of outside air.

Infiltration of moisture and air is affected by the integrity of a vehicle's exterior surfaces (usually sheet metal with riveted joints). The molded fiberglass-reinforced plastic sometimes used for truck and trailer exteriors is quite effective. There is some experimentation with composite materials for cargo container bodies. Inside, it is common to use a vapor barrier, such as aluminum foil coated with plastic binder and sealed at joints. Integrity of foamed-in-place insulation (the absence of voids and breaks) is also important. Other physical contributors to limiting infiltration include the effectiveness of door gaskets and sealing around all exterior-to-interior penetrations. Operational factors that influence infiltration are vehicle travel speed and the frequency and duration of door openings.

Purchasers of new refrigerated vehicles may require air leakage tests. Purchasers' criteria for these tests vary, depending on vehicle size and intended use. A cargo container for modified-atmosphere service (see the discussion on Controlled and Modified Atmosphere in the Equipment section) must be especially tight. The purchaser may specify that air pressure in a 40 ft long container drop from 3 to 2 in. H_2O in not less than 8 minutes, for a leakage rate of approximately 48 ft^3/h. A 48 ft trailer for general refrigeration service may be tested at 0.5 in. H_2O with a leakage limit of 120 ft^3/h.

Infiltration into insulated vehicles occurs even when they are stationary, probably because of stack effect caused by the inside-to-outside temperature difference. The infiltration driving force for a vehicle 8 ft high with a 100°F difference is about 0.03 in. H_2O (Phillips et al. 1960). Openings with an aggregate area of 1 in^2 each at the top and bottom allow infiltration of about 120 ft^3/h if assumed to be thin-plate orifices.

Eby and Collister (1955) discuss the infiltration load from air entering through cracks in the front of a moving vehicle. The ram air pressure is 1.21 in. H_2O at 50 mph, and an exposed 1 in^2 opening can allow 1150 ft^3/h of air to enter. At ambient conditions of 100°F and 50% rh and a vehicle temperature of 0°F, the extra infiltration load is approximately 3800 Btu/h. Figure 3 illustrates heat gain into a 0°F vehicle resulting from infiltration of ambient air at various conditions.

Air Circulation

To avoid spoilage during transport,

- *Surround the cargo with a flow of conditioned air sufficient to remove heat that enters the vehicle by conduction and infiltration.* To do this, interior surfaces must have channels for flow of conditioned air. There may be space between the top of the cargo and the ceiling, or flexible duct(s) in that space, or a fixed duct (false ceiling). Walls may have batten strips, or channels formed into the wall surfaces, or fixed ducts (false walls). The floor may have fixed longitudinal T-bars or "hat" sections, or movable racks.
- *For commodities that respire or require in-transit cooling (e.g., fresh fruits, vegetables, and flowers), provide an adequate flow of conditioned air between and through packages.* This process relies on the air circulation ability of the equipment (see Equipment Selection in the section on System Application Factors), and commodity loading practices (see Vehicle Use Practices in the section on Operations).

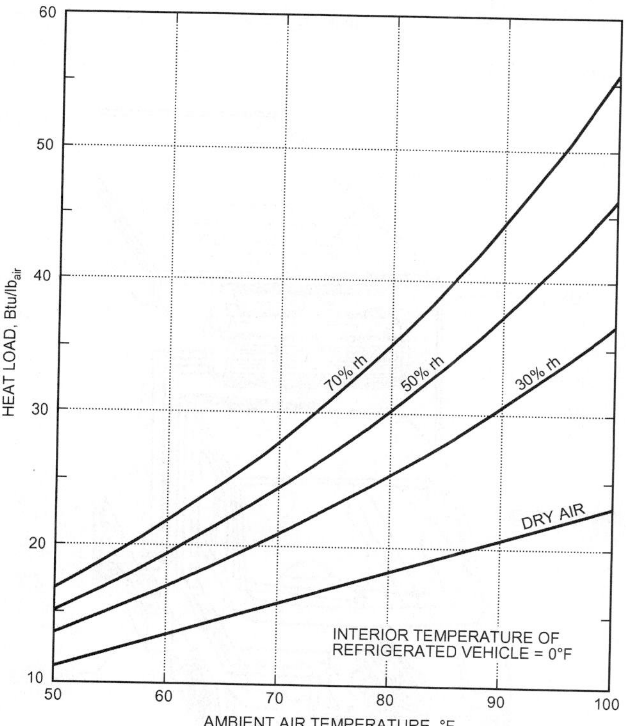

Fig. 3 Heat Load from Air Leakage

Conditioned air may enter the cargo space over the top of cargo (normally used in rail cars, trailers, and trucks), or under the cargo (normally used in cargo containers). Figure 4 shows both methods, using a trailer as an example.

Equipment Attachment Provisions

All components must be securely fastened to the vehicle to resist shock, vibration, and vandalism. Vehicle-to-equipment interfaces must have structure capable of secure support under all conditions of dynamic loading caused by vehicle travel and equipment operation (e.g., engine and compressor vibration). Suitable fastening provisions (mounting holes, studs, or captive nuts) are needed. Wall openings for equipment, which may be large (e.g., the entire front wall of cargo containers), must have provisions to limit infiltration, using gaskets or other sealing methods.

Sanitation

Vehicle internal cleanliness is enhanced by eliminating interior crevices where fungi and bacteria can grow, and by using surfaces that tolerate cleaning materials such as hot water, disinfectants, detergents (including harsh cleaning solutions), and metal brighteners. Vehicle interiors should enable access to equipment components exposed to conditioned air (e.g., fans, cooling coils, and condensate pans) for periodic cleaning.

EQUIPMENT

Mechanical Cooling and Heating

Refrigerated cargo containers typically have unitary equipment that comprises the entire front wall of the container. The refrigeration unit depth is approximately 16 in. and provides structure and insulation to the container front wall. Figure 5 illustrates a typical unit. The equipment has a vapor compression refrigeration system and uses an external source of electricity for its compressor and fan motors, resistance heaters, and operating controls. It usually uses bottom air delivery, as shown in Figure 4. The unit may have a detachable diesel engine-generator set (with integral fuel tank) accompany it while traveling by land.

Rail cars may have field-installed components. A three-phase ac diesel engine-generator set, condensing unit, and refrigerant and electrical operating controls are usually located in a machinery compartment at one end of the car. An evaporator fan-coil package, or separately mounted evaporator and fan, is typically adjacent to the machinery compartment but inside the insulated space. Electric heaters located in or under the evaporator are used for heating and defrost. This equipment usually uses top air delivery, as shown in Figure 4. Fuel tanks are generally located under the car. Newer rail cars may use end-mounted unitary equipment similar to trailers.

Trailers typically have unitary equipment that consists of a diesel engine with battery-charging alternator, compressor, condenser and engine radiator with fan, evaporator with fan, and refrigerant and

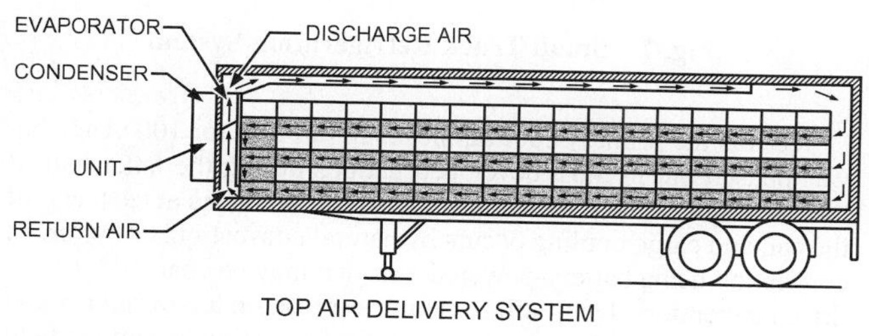

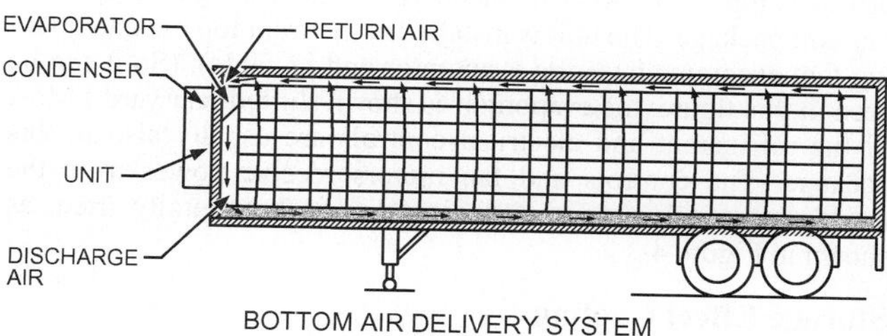

Fig. 4 Sections of Vehicle Showing Air Circulation

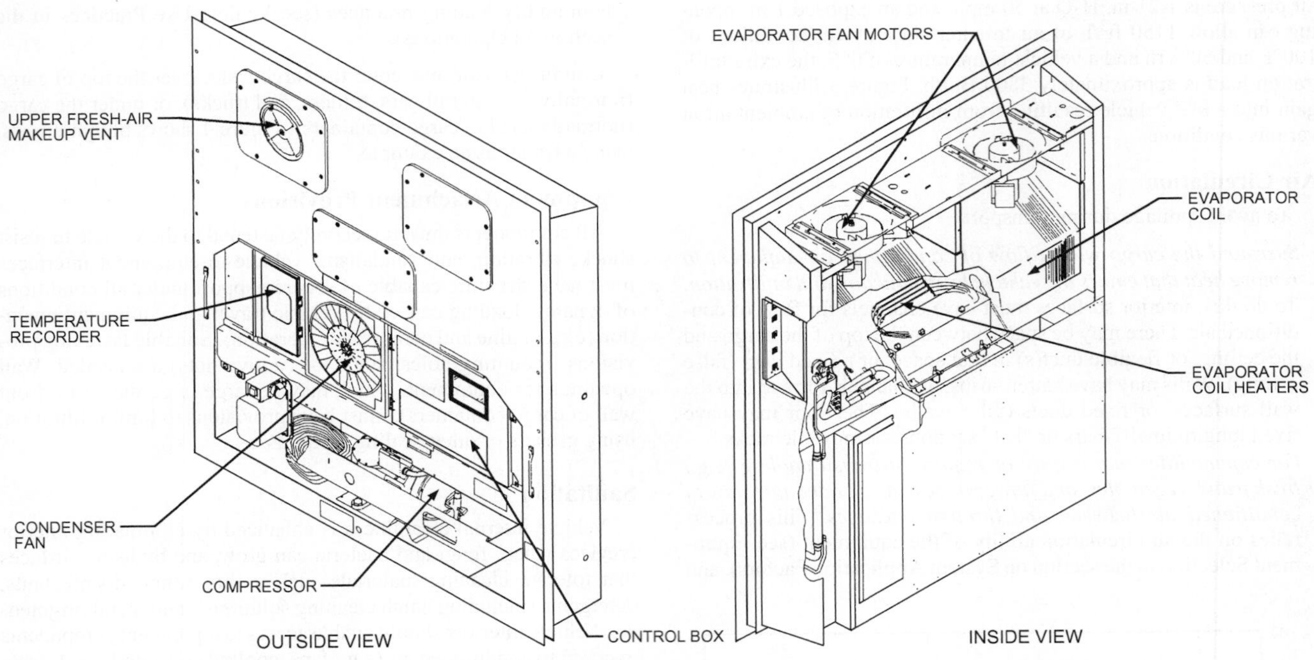

Fig. 5 Container Refrigeration Unit

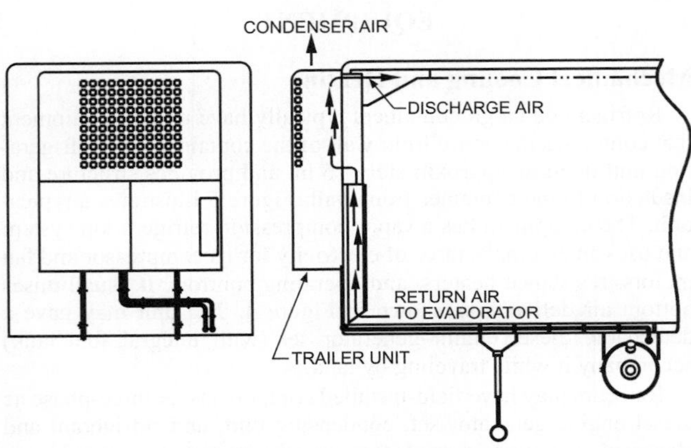

Fig. 6 Trailer Unit Installation

electrical controls. It is installed on the front of the vehicle near the top, over an opening that accommodates the evaporator and fan, as shown in Figure 6. Top air delivery is usually used as shown here and in Figure 4. The fuel tank is mounted under the trailer.

Large trucks typically have unitary equipment that is similar to trailer equipment, but more compact. Small trucks may have unitary equipment similar to that for large trucks, or field-installed components. The latter, as shown in Figure 7, have a truck-engine-driven compressor. Also included is a condenser and evaporator fan-coil package. The unit is installed at the front top over an opening that accommodates the evaporator and its fan(s). (So it can be seen in this figure, the evaporator is shown shifted rearward.) Most of the refrigerant and electrical controls are usually also in this package. The controls and fan motors receive power from the truck's electrical system. Top air delivery is generally used, as shown in Figure 4.

Storage Effect Cooling

Although mechanical cooling now dominates, stored thermal energy is still occasionally used for transport of commodities.

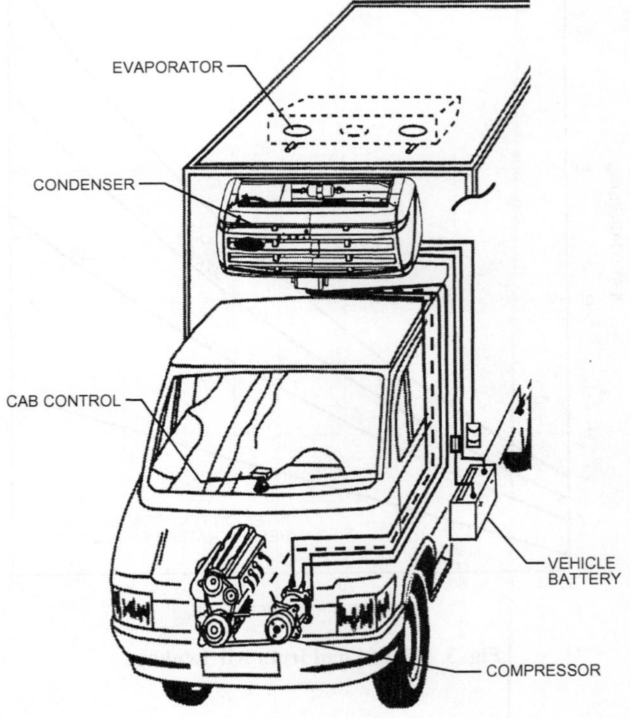

Fig. 7 Small Truck Refrigeration System

Ice was the primary cooling means in rail cars for 100 years, but was phased out in most developed nations during the latter half of the twentieth century. When used, it is put in bunkers at each end of the car, and cargo cooling occurs by natural convection, or by forced convection using battery-powered fans that may be charged by axle-driven generators. Ice is still sometimes spread on top of rail car and trailer loads to supplement mechanical refrigeration to remove field heat, or as an enhancement to mechanical refrigeration.

Solid carbon dioxide (dry ice) is sometimes used in small insulated boxes of very cold commodity. Liquid CO_2 or nitrogen, when used in vehicles, may be expanded directly into the cargo space. Alternatively, CO_2 is evaporated in a heat exchanger that cools cargo compartment air, and the gas is exhausted outside. The fluids are carried in storage vessels located in or outside the refrigerated space. *Note*: when these fluids are directly expanded into the cargo space for cooling, worker safety precautions must be observed to avoid asphyxiation hazards.

Eutectic plates are sometimes used in local delivery vehicles and are mounted on walls, ceilings, or both, or may be used as shelves or compartment dividers. Most have internal heat exchangers to allow recooling from a stationary or vehicle-mounted refrigeration system while the vehicle is not in motion.

The thermal storage effect of precooled milk or fruit juice, and some hot liquids, can hold the commodity within satisfactory temperature limits if carried in sufficiently insulated tanks.

Heating Only

Commodities sensitive to low temperature are sometimes hauled in insulated vehicles with heating capability only. Direct-combustion heaters may be fueled by alcohol, butane, charcoal, kerosene, or propane. These are hazardous and must be appropriately designed and carefully used.

Engine-driven units, consisting of a diesel engine with battery-charging alternator, engine-coolant heat exchanger, engine-driven fan, and operating controls, are available. They are installed on the front of the vehicle near the top, over an opening that accommodates the heat exchanger and fan. Top air delivery is typically used, as shown in Figure 4. The fuel tank is placed under the vehicle.

Ventilation

Ventilation for temperature control is usually accomplished by adjusting the opening of small doors at the front and rear to establish a flow of outside air through the vehicle. It is strictly limited by outdoor conditions. Ventilation is sometimes used with mechanical refrigeration to reduce the concentration of ripening gases in fresh commodities (see the section on Operations).

Controlled and Modified Atmosphere

The benefits of controlled-atmosphere (CA) and modified-atmosphere (MA) transport are proportional to the duration of commodity exposure, so these supplements to mechanical refrigeration tend to be used when shipboard travel is involved. For further information, see Control of Atmospheric Chemistry in the Operations section.

Seagoing cargo container equipment sometimes has CA capability incorporated. Typical systems sense CO_2 and O_2 and are able to boost N_2 and/or CO_2 concentration in the cargo space, depending on commodity needs. Nitrogen levels are usually raised by passing a small flow of outside air through a N_2 separator, and venting O_2 and trace gases outside the vehicle. A supplemental supply of CO_2 is required to increase its concentration.

For short trips, or when replenishment is practical, MA may be used. It involves injecting an appropriate gas into the vehicle or into special commodity packaging, so it does not affect equipment design.

Control Systems

Equipment control system functions normally include temperature control, defrost, and safety provisions. Cargo space return or supply (and sometimes both) air temperatures are monitored and thermostatically controlled. In addition to off/on cooling and heating, more sophisticated systems include gradual capacity modulation to achieve a commodity temperature closer to the set point. Evaporator coil defrost may be initiated by sensing system performance parameters (e.g., evaporator airway pressure differential) or at timed intervals, and is usually terminated by sensing temperature on some part of the evaporator. Safety provisions essential to avoiding equipment damage and hazards to people are also incorporated.

Equipment control system functions may also include the following:

- Monitoring and displaying important operating parameters such as return air temperature
- Logging a detailed record of equipment performance during trips
- Providing alarms if unacceptable conditions occur
- Monitoring probes in the cargo space (e.g., for commodity pulp temperature)
- Stopping and starting engine-driven equipment, depending on the need for cooling or heating, to improve fuel economy
- Adjusting system capacity to match engine power capability during cargo space temperature pulldown and at different ambient temperatures
- Monitoring and controlling cargo space atmospheric chemistry and relative humidity

Many of these control system functions are made practical by microprocessors. They enhance equipment response to varying operating conditions, such as ambient temperature. Their memory capability facilitates pretrip equipment operational checks, and enables tracking of equipment performance and analysis of possible malfunctions. Also, microprocessors can be used with radio telemetry to enable a central location to monitor thermostat set point, return and supply air temperature, operating mode, and alarm status.

EQUIPMENT DESIGN AND SELECTION FACTORS

This information is intended as a source of typical transportation duty guidelines for equipment design engineers, and for persons who select, specify, or apply this equipment. Specific applications may have more or less severe requirements.

Time

Time is a design factor for some components. Two examples illustrate how it affects the objective of producing a robust system: (1) an oversized refrigerant filter-drier is often used because there may not be enough time for thorough refrigeration system evacuation during emergency service operations; moisture remaining in the system can be removed by one or more changes of a large filter-drier; (2) a large engine oil reservoir reduces the frequency of oil level inspections and extends time between oil changes without sacrificing engine reliability.

Shock and Vibration

Shock and vibration are primary design concerns because equipment travels with the vehicle it serves, often includes an internal combustion engine, and may be subjected to occasional rough handling. Most system components are affected, but particular attention must be given to design of

- Structural frames
- Heavy component attachment
- Shock and vibration isolation devices
- Finned-tube heat exchangers
- Refrigerant piping (including capillaries) and its supports
- Refrigerant filter-driers
- Wiring supports and routing
- Electronic control devices
- Air impellers

Table 1 provides general guidance to the engineer in establishing design load factors for structural calculations. Impacts of these magnitudes may occasionally occur during vehicle travel. The resulting forces transmitted to the equipment may be amplified or

Table 1 Typical Peak Shock Levels

Equipment	Peak Shock, by Axis, *g*[a] Vertical	Longitudinal[b]	Lateral[c]
Containers			
Highway	9	2	4
Rail (on flat car)	7	7	3
Railway cars			
Standard draft gear	9	14	5
Cushioned	7	7	3
Trailers/trucks, highway	9	2	4
Trailers, railway			
On flat car	3	6	2
Bimodal[d]	9	6	5

[a]Acceleration of gravity, 32 ft/s^2.
[b]Parallel to direction of travel.
[c]Cross-wise of vehicle.
[d]See Vehicles section.

attenuated by the response of the intervening vehicle structure. Also, adjustment may be needed for local conditions (e.g., highways in very poor condition). Although *g*-levels under such conditions may be only slightly higher, the frequency of occurrence may be several times greater. Table 1 is based on data in four reports prepared by the Association of American Railroads (AAR 1987, 1991, 1992a, 1992b).

Vibration criteria are difficult to establish because of the many variables involved. For example, vehicle speed and road characteristics have an input that varies widely. Equipment manufacturers often have criteria based on their testing and experience.

It is good design practice to avoid equipment natural frequencies of 10 Hz or less (these are typical vehicle wheel rotation and suspension inputs). Also, equipment should not have natural frequencies close to the equipment engine's firing frequency, compressor's pumping frequency, or rpm of any of its rotating components (see the section on Qualification Testing).

Ambient Temperature Extremes

Ambient temperature is a primary design consideration because equipment must be able to start, run, and perform under conditions that may include summer desert, summer tropical, and winter near-arctic. Many items are affected, but particular attention must be given to design of

- Heat exchangers using ambient air as a heat sink
- Other components dependent on ambient air for cooling, especially motors, alternators and generators, and electronic devices
- Systems relying on heat of refrigerant compression for heating or evaporator defrost
- Components that include elastomers and plastics, whose physical properties may be degraded
- Engine cranking motors, to account for engine oil viscosity increases and battery voltage decreases during cold weather

Typical ambient temperature maximum and minimum values for several geographic areas are shown in Table 2 as a guide for equipment design. Some vehicles (e.g., cargo containers) may travel anywhere, and their equipment should be designed for global extremes. Others, such as delivery trucks, usually have equipment designed for local conditions only. Engine starting below –20°F may be difficult and require special procedures, but equipment operation can be required to heat the vehicle at very low ambient temperatures.

The values in Table 2 are very general for vehicle equipment operation in the areas considered. Consult Chapter 28 of the 2005 *ASHRAE Handbook—Fundamentals* for more specific data. Caution is required when using climatic design information because (1) more extreme temperatures may be encountered at locations in the region that are some distance from the station where data are taken,

Table 2 Ambient Temperatures for Equipment Design in Several Geographical Regions

	Asia	Europe	Mideast	North America	Tropics	Global
Maximum, °F	120	110	125	120	110	125
Minimum, °F	–40	–30	–30	–40	30	–40

and (2) higher transient values of 135°F or more may occur because of recirculation in sheltered areas or heat rejected from other nearby sources (see the section on Safety).

Other Ambient Design Factors

Precipitation, such as snow, hail, rain, and freezing rain, affects electric motors, electrical component enclosures, and cable connections.

Sea water, salt-laden sea air, and wintertime road salt tend to corrode metal parts, including electric motors, compressors, and electrical enclosures and cable connections. Salt also affects finned-tube heat exchangers, air impellers, fasteners, structural frames, and sheet metal parts.

Air pollutants (e.g., sulfur dioxide in diesel engine exhaust of the vehicle and equipment) combined with atmospheric moisture can also contribute to degradation of metals, especially aluminum or copper-finned heat exchangers.

Interior air quality is also important. Some materials, including plastics and elastomers, may emit chemical odors that adversely affect the taste and smell of certain foods. Their use in parts exposed to conditioned air should be avoided. British Standards Institute (BSI) *Standard* 3755, known as the butter taint test, may be used to check this aspect of a material's suitability.

Dirt and debris are common along highways and railways and inside cargo spaces. Also encountered are insects and fluff from vegetation. Equipment designs must exclude these materials from critical components where possible, and include provisions for essential cleaning in both vehicles and equipment.

Solar radiation is unavoidable. Some components may be affected by ultraviolet radiation, including paint, plastics, and elastomers; some may also be sensitive to heat. Solar radiation also contributes to the cooling load (see Load Calculations in the section on System Application Factors).

Vandalism and theft can occur, and are difficult to thwart. Among the concerns are storage batteries, electrical cables, and other components with obvious salvage value.

Equipment to be used internationally may have to meet anti-smuggling requirements of the *TIR Handbook* (UN 1988). For example, ventilation (outside-air) ports in cargo container units include sturdy screens.

Operating Economy

This is a complex consideration that includes reliability, serviceability, and fuel or power consumption.

Equipment should be reliable because

- It may be preserving a high-value commodity.
- It often operates unattended during trips.
- It frequently travels far from knowledgeable service technicians, repair parts, and supplies of refrigerant and other operating fluids.

Laboratory and field experience must be combined to demonstrate the durability of operating components; the ability of the system to start, run, and perform under all expected ambient conditions; and structural integrity.

Good equipment serviceability is important for the same three reasons. Components and assemblies should be designed for easy access. This is important for testing and trouble analysis, routine maintenance, and repair or replacement of components.

Another factor peculiar to transportation refrigeration service is the availability of knowledgeable service technicians, repair parts, and supplies (e.g., suitable fuel; the exact refrigerant for which the system is designed; engine coolant; and lubricants specified for the compressor, engine, and other components).

Fuel (or power) consumption has an obvious role; it is affected by the thermodynamic and mechanical efficiencies of engines, motors, compressors, and power transmission devices. Thermodynamic performance of condensers and evaporators has a major role, as do system provisions for part-load operation. Aerodynamic efficiency of air impellers and related flow paths has an effect, too.

Airborne Sound

Sound may be a concern when residential and commercial areas abut or government-imposed sound limits exist. Design and selection considerations include using the following:

- Engines, compressors, and air impellers that are inherently relatively quiet
- Acoustical treatments and sound attenuation features that are durable and likely to remain in use

Safety

Safety in transport equipment is of concern because cargo handlers, vehicle operators, and service technicians are often close to the equipment. Also, at times the general public may be near refrigerated vehicles. Although there are no known safety codes specifically for transport equipment, those mentioned in this section are sufficiently general in scope to apply, or good engineering practice suggests their use for guidance.

All potentially dangerous parts, including fans, belts, rotating parts, hot surfaces, and electrical items, require appropriate guards, enclosures, and/or warning labels. Labels should emphasize universally understandable graphics; suitable designs can be provided by the International Organization for Standardization (ISO) or similar groups.

Refrigerant pressure vessels must comply with applicable safety codes; some common codes are listed here:

- American Society of Mechanical Engineers *Boiler and Pressure Vessel Code*, Section VIII (ASME 2004), for vessels larger than 6 in. inside diameter
- European Committee for Standardization (CEN) Pressure Equipment Directive 97/23/EC
- CEN *Standard* EN 378-2:2000
- Underwriters Laboratories (UL) *Standard* 1995, for vessels of 6 in. or less inside diameter

Refrigeration system design pressures should account for worst-case temperature extremes for any intended application of equipment. Because the presence of some liquid is likely, either by design or from excess refrigerant in the system, saturation pressures are used.

Under nonoperating conditions, usually 150°F (saturation) is used for low and high sides, based on new product shipping experience. Vehicles standing still in the sun without wind should meet this criterion, as well. *Note*: shipping package design and storage practices must anticipate potential excessive temperature hazards. For example, equipment covered with plastic shrinkwrap may experience temperatures greater than 150°F when left in direct sunlight.

While operating, the low side is unlikely to exceed the nonoperating criterion. The high-side criterion of 150°F (saturation) is based on the nonoperating value because the worst-case high-side value, from ASHRAE *Standard* 15-2004, par. 9.2.1(c), is lower. For example,

- 130°F Europe worst case: Sevilla, Spain at 100°F 1% db
- 139°F North America worst case: Yuma, AZ at 109°F 1% db
- 145°F Mideast worst case: Kuwait, Kuwait at 115°F 1% db

High-side saturation temperature excursions above 150°F can result from the 135°F (or more) transients cited in the Ambient Temperature Extremes section. For example, 155°F is sometimes used for trailer equipment.

The choice of over-pressure protection in the high side, as for a liquid receiver, also affects design pressure.

Example 1. What design pressures are required if a pressure-relief device (valve or rupture disk) is used, or if a fusible plug is used? The highest normal pressure for a R-134a system high side, assuming 155°F transient exposure, is 281 psig. The critical pressure for R-134a is 574 psig.

Solutions:

Pressure-relief device: The recommended 25% margin between relief pressure and the highest normal pressure (to avoid accidental discharge) raises minimum relief pressure from 281 to 351 psig. Therefore, a design pressure of 355 psig is appropriate. Pressure vessel testing at 1.25 or 1.5 times design pressure is required by major safety codes.

Fusible plug: The design pressure must be 281 psig or greater. So, a design pressure of 285 psig is appropriate. If a fusible plug with a nominal setting of 173°F is chosen, the corresponding nominal release pressure is 354 psig. Some codes (e.g., UL *Standard* 1995) require pressure vessel testing at 2.5 times the nominal release pressure or the critical pressure, whichever is less; in this example, 2.5 times release pressure (in psia) would be used.

Safety codes (e.g., ANSI/ASHRAE *Standard* 15-2004) require overpressure protection such as a relief valve, rupture disk, or fusible plug. Also required is a pressure-limiting device (i.e., a high-pressure switch to stop compressor operation). Overpressure protection design decisions must consider possible regional or local regulations that reflect environmental concerns. Rupture-disk discharge to atmosphere may be prohibited in some areas unless in series with a relief valve. Fusible plugs may be prohibited in some areas unless connected to the low-pressure side.

QUALIFICATION TESTING

This section provides an overview of testing usually done by equipment manufacturers to determine whether equipment meets operating design criteria and performs satisfactorily in the transportation environment. Prospective users may wish to be guided by evidence of successful completion of appropriate tests, or may wish to consider special testing of their own.

Operational tests are done to establish the ability of equipment to provide satisfactory control of temperature in a typical vehicle, especially at set points between 30 and 60°F. Tests normally include cooling and heating, with ambient temperatures above, at, and below each set point. Operation over the entire range of ambient and internal temperatures expected for the intended vehicles is also tested.

Psychrometric testing is appropriate because cargo space relative humidity affects nonfrozen commodity desiccation. Desiccation is limited by keeping the evaporator surface temperature as high as possible by its thermodynamic design and by refrigeration capacity control methods. In some equipment, additional control is achieved by sensing relative humidity and atomizing water into the conditioned air stream as needed to maintain the desired level for commodity storage.

It is also necessary to verify operation of other equipment functions, such as defrost effectiveness and evaporator fan performance (static pressure versus flow). If controlled atmosphere is an equipment option, its effectiveness may need to be checked.

For equipment rating purposes, refrigeration capacity is normally determined at the following conditions:

- 100°F ambient in North America, and 86°F in Europe
- 35, 0, and –20°F cargo space (return air) in North America, and 32 and –4°F in Europe

Capacity may be certified to meet ARI *Standard* 1110 or other industry standards. Most qualification testing programs also include establishing capacity at other conditions.

Heating capacity (if not electric resistance) is usually, as a minimum, tested at 0°F ambient and 35°F cargo space (return air).

Shock and vibration qualification demonstrates that equipment meets guidelines presented in the Equipment Design and Selection Factors section. Search for natural frequencies of 10 Hz or less, and for any that are close to the firing frequency of the equipment's engine, pumping frequency of the compressor, or rpm of any of its rotating components. This testing may be done on laboratory apparatus that shakes equipment in each of its three primary axes. Endurance testing usually follows, using an amplitude and frequency spectrum based on field experience. A special test for cargo container equipment imposes typical lateral (racking) and end-loading forces. Field trials are essential; they usually take several months, and are followed by disassembly and thorough inspection. Some redesign may be needed; testing should be repeated as needed to confirm its success.

Testing at extremes of ambient temperature (and sometimes relative humidity) is usually done on major mechanical and electrical components by their suppliers (including engines, compressors, motors, alternators, generators, solenoids, and electronic devices). It is also done on complete equipment in test chambers capable of operation over the temperature ranges expected for both high and low sides. Of particular interest is the ability of equipment to start and run at ambient and cargo space extremes, and evaporator defrosting performance.

Qualification for other ambient design factors generally includes the following:

- Rain testing is usually done by component suppliers on items such as electric motors. It is also done on complete equipment to check integrity of electrical cabinets and cable entries.
- Components susceptible to ocean-environment corrosion, especially for cargo containers, are laboratory-tested in specially designed salt-spray apparatus.
- Visual inspection and experience-based judgment are usually used for concerns such as potential for clogging with dirt and debris, and susceptibility to vandalism and theft.
- Items exposed to sunlight, including elastomers, plastics, labels, and paints, are laboratory-tested for ultraviolet resistance.

SYSTEM APPLICATION FACTORS

As noted in the introduction, users of this information are urged to regard the vehicle and its equipment as a system. This section provides guidance for that design effort. Three steps, discussed in the following sections, are recommended.

Load Calculations

The objective of load calculations in this application is to determine average conduction and infiltration loads per hour. Familiarity with Chapter 13 is helpful.

The vehicle's heat transfer factor (*UA*) should be determined. If more than one insulation system is being considered (e.g., several thicknesses), there will be a *UA* factor for each. This information is normally available from the vehicle manufacturer. It may be calculated with guidance from Chapter 13 and confirmed (or determined) by test.

Typical vehicle travel time (hours) is needed. For long hauls, add the vehicle's stationary time at each end while loaded with the equipment operating. The time for short hauls is usually a typical working day.

Determine the range of average ambient temperatures the vehicle is likely to encounter. (See Ambient Temperature Extremes in the section on Equipment Design and Selection Factors) The range of commodity types and temperatures the vehicle is likely to encounter must also be known.

Solar radiation may be approximated using data in Table 3 of Chapter 13. For example, assume the vehicle roof and one side are exposed to the sun. Divide the sum of the products of areas and temperatures by the total area to get an adjusted ambient temperature. (If this were included in Example 2, the ambient temperature would be about 3°F higher, increasing conduction and infiltration about 6%, and total load about 4%.)

Infiltration through the vehicle body and closed doors is usually included in the *UA* value. Infiltration during door openings is not, however. This is a significant consideration for delivery vehicles; estimates may be made using information from vehicle or equipment manufacturers.

Vehicle aging effects should be considered, because the vehicle's ability to protect low-temperature commodities decreases with time. Insulation and door seal deterioration can increase *UA* by 25% or more in vehicles older than 3 years.

Cooling load calculations in Example 2 include commodity temperature pulldown and heat of respiration. Both of these vary widely with the type of commodity and its treatment before loading into the vehicle. Some commodities (e.g., frozen goods and meat) have no heat of respiration. Recommended practices include precooling or preheating the vehicle to bring its interior surfaces to the planned thermostat setting; that load is ignored. Also ignored are the minor loads associated with cooling (1) air not displaced by the commodity and (2) its packaging materials; both have relatively low mass and poor thermal conductivity.

Example 2. Determine the conduction and infiltration load, commodity temperature pulldown load, commodity heat of respiration, total load, and average load for the following shipment:

Vehicle *UA* factor: 155 Btu/h·°F
Vehicle travel time: 72 h (3 days)
Assumed average ambient temperature: 85°F
Initial average commodity temperature: 40°F
Final average commodity temperature: 34°F
Assumed average commodity temperature en route: 37°F
Thermostat setting temperature: 36°F
Commodity: Elberta peaches
Quantity: 38,000 lb (19 tons)
Specific heat above freezing: 0.9 Btu/lb·°F (from Table 3 of Chapter 9)
Heat of respiration at 37°F: 1169 Btu/24 h·ton (from Table 9 of Chapter 9, interpolated)

Solutions:

Conduction and infiltration load (155)(72)(85 – 36) = 546,840 Btu
Commodity temperature pulldown load (0.9)(38,000)(6) = 205,200 Btu
Commodity heat of respiration = (1169)(19)(3) = 66,633 Btu
Total load for trip = 818,673 Btu
Average load = 818,673/72 = 11,370 Btu/h

Note: One vehicle insulation system was assumed in this example; if several are to be compared, iterations with the different UA factors are required.

If cold-weather travel is likely, do a similar calculation for the heating load. In this calculation, average ambient temperature should ignore solar radiation. Commodity temperature pulldown load may be nil. Heat of respiration of the commodity reduces the total and average loads.

Equipment Selection

Selection of equipment from manufacturer's product data begins with matching its cooling and heating capacity to new vehicle needs, as determined by load calculations. It may include consideration of aging effects on vehicle *UA* (see the note following Example 2), and equipment cooling and heating capacity.

Also important is the equipment's ability to properly control cargo space conditions, especially temperature. If relative humidity and/or atmospheric chemistry control options are being considered, their effectiveness needs review.

Microprocessors, multiple temperature sensors, and sophisticated equipment capacity controls enable close control of air temperatures, some to ±0.5°F of desired temperature. Depending on the level of system sophistication, it is possible to control either return or supply temperature, or both. Maintaining suitable return air temperature is essential for all commodities. Control of supply air temperature helps prevent freezing damage to commodities like fruits and vegetables.

Evaporator fan performance also has a role in control of cargo space temperature, and to some extent, cargo space relative humidity. Airflow should be adequate to ensure that the commodity is surrounded by air at the proper temperature, and to minimize the supply-to-return air temperature difference. Fan static pressure must be sufficient to force air through the distribution system and cargo. Also, the evaporator may be partially frosted, which suggests the need to have fans with steep characteristic curves (static pressure versus flow).

Evaporator defrost effectiveness should be judged under frozen load conditions. It must be initiated when needed and must be thorough, to avoid ice build-up on the evaporator.

Details of vehicle selection are beyond the scope of this chapter, but it should be guided by the section on Vehicle Design Considerations. Three sections under Equipment Design and Selection Factors also affect vehicle selection: Shock and Vibration, Ambient Temperature Extremes, and Other Ambient Design Factors.

Operating economy data for life cycle cost analysis should be obtained during selection for both the vehicle and equipment. Some of these data (e.g., fuel or power consumption at various operating conditions) may be available from manufacturers. Some, such as annual cost of emergency service, preventive maintenance, repair parts, and supplies, will come from user records and experience. Other data, such as cost of fuel or power, may have to be forecasts.

Owning and Operating Costs

Chapter 36 of the 2003 *ASHRAE Handbook—HVAC Applications* discusses this topic in detail for HVAC systems, and its contents may be adapted for vehicles and equipment used for transport of commodities. Table 1, Owning and Operating Cost Data and Summary, in that chapter can be modified to suit. Item I is the total of vehicle and equipment cost. Item V requires deleting factors that do not apply. Other factors peculiar to the transportation business may need to be added to either owning or operating costs. Once established for a particular vehicle-equipment combination, calculations may be replicated for tradeoff comparisons of vehicle and equipment options, such as insulation thickness, fuel economy, and vehicle revenue.

Some cost comparison of vehicle and equipment choices is usual in procurement. Its level of sophistication may depend on one or more of the following:

- Availability of information
- Size of planned procurement
- Management requirements for decision making
- Familiarity with engineering economic analysis techniques

Two possible cost comparison choices follow.

Life-cycle cost analysis: See Chapter 36 of the 2003 *ASHRAE Handbook—HVAC Applications*. It would include each combination of vehicle and equipment options, and have a column similar to Table 1 in that chapter for each year of useful life. This method accounts for the time value of money.

First-year-only analysis: For each combination of vehicle and equipment options, comparisons would be made of initial cost, estimated fuel or power cost, and expected vehicle revenue. Organizing information as in Table 1 in Chapter 36 of the 2003 *ASHRAE Handbook—HVAC Applications* is useful.

OPERATIONS

Commodity Precooling

Ideally, every commodity should be brought to its optimum storage temperature quickly and held there until it is placed in the vehicle. Otherwise, effects on commodity quality can range from slight to significant. This topic is addressed in more detail in chapters on food refrigeration in this volume and other publications on refrigerated transport (see the References and Bibliography).

Should a commodity be loaded at a temperature above optimum, the vehicle's equipment will attempt to reduce it during transit (see Example 2). If the shipping carton design and loading pattern provide good air circulation around the commodity, and if the equipment has adequate cooling capacity, eventually the commodity temperature will approach the thermostat setting. However, any time spent away from optimum storage temperature makes some loss of product quality inevitable. Therefore, regular reliance on vehicle refrigeration equipment to compensate for bad practices in precooling, storage, and loading is not recommended.

Vehicle Use Practices

Cleanliness of vehicle interiors, including conditioned air paths, is essential to food commodity safety and quality. It avoids bacterial, chemical, and odor contamination, and in some cases (e.g., transport of fresh meat products) may be required by government food safety regulations. Regular cleaning and sanitizing are recommended; details appear in publications such as *Protecting Perishable Foods During Transport by Truck* (Ashby 2000) and *Guide to Refrigerated Transport* (IIR 1995).

Precool the closed vehicle to the desired commodity temperature before loading. In cold weather, preheat for fresh commodities. This may take several hours under extreme ambient temperature conditions. However, precooling or preheating helps avoid overwhelming the equipment's capacity and possible damage to portions of the cargo. *Note*: to avoid significant frost buildup on the evaporator, do not operate equipment in the cooling mode when loading at an open, unrefrigerated dock.

Prompt cargo transfer between refrigerated warehouses and vehicles is important to maintenance of product quality. Ideally, loading and unloading should be done using refrigerated dock areas. When this is not possible, commodity movement methods and packaging that help minimize exposure to warm air (or very cold air for fresh commodities) become very important.

Commodity loading practices are an important factor in helping equipment maintain good temperature distribution within the cargo. Main points to remember include the following:

- Use packages and stack heights that avoid crushing (which blocks airflow between packages)
- Provide spaces between packages for airflow through the load
- Leave adequate space between ceiling and cargo for airflow
- Support cargo away from walls and doors so that air, rather than the commodity, absorbs transmitted heat

For further information, see Ashby (2000) and IIR (1995).

Commodity arrangement is important in delivery vehicles that have frequent door openings to unload cargo. Items to be removed at the first scheduled stop should be located close to the door(s) to expedite unloading and minimize open door time. Next in the arrangement should be the items for the second stop, then the third, etc.

Temperature Settings

The cargo space must be held close to a temperature that helps maintain commodity quality and provides desired shelf life at its destination. This volume, as well as Ashby (2000) and IIR (1995), are good sources for the recommended storage temperatures for various commodities. Because these sources draw from several sets of research and field experience, there are minor differences in the recommendations, but using any of them (or other, similarly reliable sources) should yield good results. Users sometimes create their own tables of settings based on the work of others and experience with shipments.

Thermostat settings for nonfrozen commodities must be chosen carefully to avoid possible damage from lengthy exposure to subfreezing evaporator air discharge temperature. The following example is related to the Load Calculations example, and shows that portions of the commodity exposed to the supply air stream could suffer freezing damage.

Example 3. Determine the approximate supply air temperature (SAT), at both full and reduced refrigeration capacities, for the following shipment:

Return air temperature: 36°F
Refrigeration capacity: 48,000 Btu/h
Refrigeration capacity, reduced by equipment's capacity control: 13,000 Btu/h
Evaporator airflow: 3000 cfm
Approximate specific heat of air: 0.24 Btu/lb·°F
Approximate density of air: 0.075 lb/ft^3
Commodity: Elberta peaches
Initial freezing point: 30.4°F

Solution:

SAT (full capacity)
$36 - [48{,}000/(3000)(60)(0.24)(0.075)] = 21.2°F$
SAT (reduced capacity)
$36 - [13{,}000/(3000)(60)(0.24)(0.075)] = 32°F$

As discussed in the section on Control Systems and Equipment Selection, equipment with microprocessors and sophisticated capacity controls can achieve very close air (and commodity) temperature control. Some equipment can control both return and supply air temperatures.

Other Cargo Space Considerations

The number of days spent in a refrigerated vehicle determines whether fresh commodities benefit significantly from ventilation, or control of relative humidity and/or cargo space atmospheric chemistry. Because their itineraries often include lengthy sea voyages, cargo containers are more likely to need one or more of the three following provisions.

Ventilation. Cargo container equipment may include means to admit outside air to reduce concentration of gases, primarily ethylene and CO_2, that are produced by fresh product respiration. Their outside air openings are adjustable and usually calibrated for several air exchange rates. Some container users publish suggested rates for various fresh products, ranging from 15 to 150 cfm. Information may be available from local grower's cooperatives, agricultural universities, and similar organizations.

Other vehicles, such as trailers, sometimes have small doors in their front and rear to admit and exhaust air as discussed in the section on Equipment. If ambient temperatures are moderate, a few products (e.g., unripened honeydew melons) may be transported with ventilation only.

Relative Humidity Control. Equipment, particularly for cargo containers, may have relative humidity sensing and control capability. Lengthy trips may make humidification desirable to help limit commodity desiccation. To maintain sanitary conditions within the cargo space, it is essential that potable water be put in the humidifier reservoir. Some equipment may have provisions to automatically replenish the reservoir with condensate from evaporator defrost operation. For further information, see Chapters 17 to 29 in this volume, Ashby (2000), and Nichols (1985).

Control of Atmospheric Chemistry. The chemistry of the cargo space atmosphere may affect fresh product quality at its destination. Commodity respiration, the combining of natural sugars with oxygen, can be slowed by reducing the ambient O_2 level. Increasing the CO_2 level slows respiration further, and helps prevent premature aging. Specific combinations of gases will control, and in some cases eliminate, certain pathogens and insects. Also, texture and color can be maintained by limiting ethylene, a natural ripening agent. Each product has unique physiological characteristics that dictate specific O_2 and CO_2 levels. Depending on travel time, monitoring and control of these levels can be critical to maintenance of product quality. Ashby (2000) only provides recommendations for berries and cherries (perhaps because most truck and trailer trips are of a few days' duration); a 10 to 20% CO_2 atmosphere is normally used as a mold retardant.

For long hauls, especially in seagoing cargo containers, controlled-atmosphere capability may be needed. As discussed in the Equipment section, CO_2 and O_2 are sensed. N_2 and/or CO_2 concentrations in the cargo space are then adjusted, depending on commodity needs.

On short hauls, or when replenishment from a stationary source while en route is practical, modified atmosphere may be used. The entire vehicle may be treated after loading, but requires a seal of plastic film at the doorway(s), successive purging operations to drive out much of the air, and injection of the gas treatment. Sometimes pallet loads of commodity are sealed, evacuated, and injected with the appropriate gas.

For further information, including advice on control settings, see Hardenburg et. al. (1990), Kader et al. (1992), and Nichols (1985). Also see Chapter 11 in this volume.

Maintenance

Successful operations depend on regular pretrip inspections and scheduled maintenance. Prompt action to correct vehicle or equipment deficiencies is required. *Note*: all appropriate safety precautions must be taken during pretrip and scheduled maintenance work.

Proper vehicle maintenance helps ensure system effectiveness in preserving temperature-sensitive commodities. Periodic inspection, preferably before each trip, is essential. It should include the following:

- Attachment of equipment components: correct loose or damaged fasteners
- Insulation, vapor barrier, and door seals: repair or replace damaged areas
- Air distribution system (ducts): repair or replace damaged parts
- Evaporator condensate outlets: clear drain lines and replace or correct faulty air traps
- Floor drains: repair or correct faulty air traps
- Interior cleanliness (discussed in the section on Vehicle Use Practices)

Equipment maintenance items may be categorized as pretrip and scheduled. Manufacturer's operation and service manuals provide valuable guidance on both. All who use and maintain equipment should be thoroughly familiar with them and follow all of their instructions carefully. Highlights typical of these manuals follow:

Pretrip Inspection (usually daily)

- Physical appearance of equipment components: repair or replace as required
- Evaporator, condenser, and engine radiator (if used): clean if airflow or heat transfer are obstructed
- Evaporator condensate drain: clean if obstructed, check and service drain trap if faulty
- Refrigerant moisture indicator (if used): service if "wet" indication occurs
- Refrigerant charge level: if low, check for and repair leaks; add proper refrigerant as needed

- Compressor oil level: unlike engines, compressors do not consume oil, and addition should only be needed if a refrigerant leak or service procedure has resulted in loss
- Engine oil and coolant levels: if low, check for and repair leaks; add proper fluid as needed
- Check all equipment functions (microprocessor controls usually do this automatically): if any are faulty, troubleshoot and repair

Scheduled Maintenance (inspect and service, following equipment manual's instructions)

- Mechanical and electrical components, including
  - Fasteners, latches, hinges, and covers (and their gaskets, if any)
  - Gages, switches, and electrical connections
  - Belts and shaft couplings
  - Fans
- Refrigeration system components, including
  - Evaporator and condenser: airflow and heat transfer must not be obstructed
  - Evaporator condensate drain: must be clear and air traps in good working order
  - Filter-drier: must not be clogged (no detectable temperature drop across it)
  - Refrigerant moisture indicator (if used): must indicate "dry"
  - Refrigerant charge level: must be within normal limits
  - Compressor oil level: must be within normal limits
  - System operation in all modes: cooling, heating, and defrost must work properly
  - Operating pressures and temperatures: must be within normal limits
- Engine systems (if engine-driven), including
  - Lubrication: no leaks; change oil and filter as required
  - Cooling: no leaks; airflow and heat transfer must not be obstructed; change coolant as required
  - Fuel system: no leaks; change filter(s) as required
  - Exhaust: no leaks; sound level must be normal
  - Combustion air: service and change filter as required
  - Starting: cranking motor engagement must be normal
  - Battery charging: output must be normal

REFERENCES

AAR. 1987. *Environmental analysis—Yard handling of TOFC traffic.* File 204/400. Association of American Railroads, Washington, D.C.

AAR. 1991. A technical summary of the intermodal environment study. *Report* DP 3-91. Association of American Railroads, Washington, D.C.

AAR. 1992a. Study of the railroad shock and vibration environment for roadrailer equipment. *Report* DP 1-92. Association of American Railroads, Washington, D.C.

AAR. 1992b. Multi-level environment study with Ford Motor Company. *Report* DP 4-92. Association of American Railroads, Washington, D.C.

ANSI. 1990. Road/rail closed dry van containers. *Standard* MH5.1.1.5-1990 (R1997). American National Standards Institute, New York.

ARI. 2001. Mechanical transport refrigeration units. *Standard* 1110-01. Air-Conditioning and Refrigeration Institute, Arlington, VA.

Ashby, B.H. 2000. Protecting perishable foods during transport by truck. *Handbook* 669. U.S. Department of Agriculture, Washington, D.C.

ASHRAE. 2004. Safety code for mechanical refrigeration. ANSI/ASHRAE *Standard* 15-2004.

ASME. 2004. *Boiler and pressure vessel code*, Section VIII. American Society of Mechanical Engineers, New York.

BSI. 1964. Methods of test for the assessment of odour from packaging materials used for foodstuffs. *Standard* BS 3755:1964. British Standards Institution, London.

CEN. 1997. Pressure equipment directive 97/23/EC. *Document* 397L0023. European Committee for Standardization, Brussels.

CEN. 2000. Refrigerating systems and heat pumps—Safety and environmental requirements—Part 2: Design, construction, testing, marking and documentation. *Standard* EN 378-2:2000. European Committee for Standardization, Brussels.

Eby, C.W. and R.L. Collister. 1955. Insulation in refrigerated transportation body design. *Refrigerating Engineering* (July):51.

Hardenburg, R.E., A.E. Watada, and C.Y. Wang. 1990. The commercial storage of fruits, vegetables, and florist and nursery stocks. *Agriculture Handbook* 66. U.S. Department of Agriculture, Washington, D.C.

IIR. 1995. *Guide to refrigerated transport*. International Institute of Refrigeration, Paris.

ISO. 1995. Series 1 freight containers—Classifications, dimensions and ratings, 4th ed. *Standard* 668:1995. International Organization for Standardization, Geneva.

Kader, A.A., R.F. Kasmire, F.G. Mitchell, M.S. Reid, N.F. Sommer, and J.F. Thompson. 1992. Postharvest technology of horticultural crops. *Special Publication* 3311. Cooperative Extension, University of California, Division of Agriculture and Natural Resources.

Nichols, C.J. 1985. Export handbook for U.S. agricultural products. *Agriculture Handbook* 593. U.S. Department of Agriculture, Washington, D.C.

Phillips, C.W., W.F. Goddard, and P.R. Achenbach. 1960. A rating method for refrigerated trailer bodies hauling perishable foods. *Marketing Research Report 433*. Agricultural Marketing Service, U.S. Department of Agriculture, Washington, D.C.

UL. 1995. Heating and cooling equipment. *Standard* 1995 (2nd ed.). Underwriters Laboratories, Northbrook, IL.

UN. 1988. *TIR handbook.* ECE/TRANS/TIR/1. Economic Commission for Europe (Geneva), United Nations, New York.

BIBLIOGRAPHY

AFDO. *Guidelines for the transportation of food*. Association of Food and Drug Officials, York, PA.

Bioteknisk Institut and Technical University of Denmark. 1989. *Guide to food transport—Fruit and vegetables*. Mercantila Publishers, Copenhagen.

Danish Meat Products Laboratory and Danish Meat Research. 1990. *Guide to food transport—Fish, meat and dairy products*. Mercantila Publishers, Copenhagen.

IIR. 1986. *Recommendations for the processing and handling of frozen foods*. International Institute of Refrigeration, Paris.

IIR. 1992. *Compression cycles for environmentally acceptable refrigeration, air conditioning and heat pump system*. International Institute of Refrigeration, Paris.

IIR. 1993. *Cold store guide*. International Institute of Refrigeration, Paris.

IIR. 1994. *New applications of refrigeration to fruit and vegetables processing*. International Institute of Refrigeration, Paris.

IIR. 1995. *Refrigeration and the quality of fresh vegetables*. International Institute of Refrigeration, Paris.

IIR. 1996. *New developments in refrigeration for food safety and quality.* International Institute of Refrigeration, Paris.

IIR. 1996. *Refrigeration, climate control and energy conservation*. International Institute of Refrigeration, Paris.

IIR. 1996. *Research, design and construction of refrigeration and air conditioning equipment in Eastern European countries*. International Institute of Refrigeration, Paris.

Ryall, A.L. and W.J. Lipton. 1972. *Handling, transportation, and storage of fruits and vegetables,* vol. 1—*Vegetables and melons*. AVI Publishing, Westport, CT.

Ryall, A.L. and W.T. Pentzer. 1982. *Handling, transportation, and storage of fruits and vegetables,* vol. 2—*Fruits and tree nuts*, 2nd ed. AVI Publishing, Westport, CT.

Serek, M. and M.S. Reid. 1999. *Guide to food transport—Controlled atmosphere*. Mercantila Publishers, Copenhagen.

Sinclair, Joseph. 1999. *Refrigerated transportation*. Witherby & Co., London.

CHAPTER 31

MARINE REFRIGERATION

MARINE refrigeration systems are used aboard seagoing vessels and offshore facilities and generally include cargo hold refrigeration, domestic refrigeration services, and refrigerated containers. These systems differ from stationary systems not only in physical aspects but also in the fact that marine systems must be designed to handle frequent starting and stopping. Process freezing or chilling plants on vessels might run continuously for weeks, but under certain conditions may be started and stopped daily. Cold storages are usually shut down after the cargo is discharged, and are restarted before new cargo is loaded.

Personnel changes of engineers and refrigeration crew members require that those unfamiliar with the installation be able, on short notice, to trace well-labeled systems and place the plant in operation or maintain it without undue hazards to the machinery, cargo, or personnel.

Plant layout aboard ships should be as simple as possible without sacrificing reliability. The machinery plant should be close to the main power plant to provide short piping and power connections and facilitate close supervision by operating personnel. Machinery space should be uncrowded, even at the expense of revenue space, to give ample room for operation, maintenance, and repair of both the apparatus and the ship's structure.

All machinery must have sturdy foundations, and all components should be secured against vibration from either themselves or other machinery. High-speed machinery should be mounted fore and aft, and all feeds, drains, and vessels must be installed with full consideration of the effects of pitch, roll, trim, and list.

Refrigeration equipment should not, in general, be kept in the same enclosed space as internal combustion engines, because engine damage can occur in the event of a refrigerant leak. Locating refrigeration equipment close to the main engine space usually improves economy of space and provides easy connection to power and cooling.

REFRIGERATION LOAD

A detailed discussion of load calculations has been omitted from this chapter, because the loads that might be encountered in a marine refrigeration plant are so widely varied. However, the methods used to calculate them can be found in Chapter 13, and load calculation considerations are discussed in this chapter in the section on Specific Vessels.

REFRIGERATION SYSTEM

Refrigerants

Refrigerants for shipboard use must meet the same environmental regulations that apply to land-based systems. The choices are similar, but special attention should be given to the availability of refrigerants and compressor lubricants at all ports of call.

The preparation of this chapter is assigned to TC 10.6, Transport Refrigeration.

Compressors

Generally, all of the same types of compressors used in stationary refrigeration plants can also be applied on ships. Chapter 34 of the 2004 *ASHRAE Handbook—HVAC Systems and Equipment* describes compressors in detail and discusses their application and control.

Intermittent operation of compressors should be minimized to ease the starting load burden on the vessel's electrical generating plant. Automatic capacity control should be used to react to varying loads. Oversized compressors should be avoided.

The shafts of rotating equipment are usually oriented fore and aft to minimize the gyroscopic bearing loads that occur when a vessel rolls. Compressor lubrication systems must be able to function under all conditions of pitch, roll, trim, and list.

Reserve capacity and spare parts must be taken into account in the design. There must be redundancy built into the system, a complement of spare parts to ensure the ability to maintain temperature, or some combination of the two. ANSI/ASHRAE *Standard* 26 lists spare parts and tools to be provided on board. Table 1 suggests reserve capacities for various installations.

Condensers and Coolers

Shipboard condensers are most often of shell-and-tube design, using seawater as the condensing medium. Other types of condensers, such as plate-and-frame and double pipe, are sometimes used. Surfaces exposed to seawater must be resistant to corrosion. Cupronickel is the most common tube and tube sheet material for refrigerants other than ammonia. During selection of equipment, installation, and operation, special consideration must be given to preventing damage from galvanic corrosion, erosion, electrolysis, and anaerobic corrosion. Epoxy coatings and sacrificial anodes are often used as preventive measures.

Considerations for brine (including seawater) coolers and water-cooled oil coolers or subcoolers are similar to those for condensers. Materials of construction must be compatible with the medium being cooled or being used for cooling.

Shell-and-tube condensers are normally fitted with dual drains in order to drain freely under all conditions of pitch, roll, trim, or list. As an alternative, they may be installed on an angle great enough to compensate for the maximum angle of vessel trim or list that may be encountered.

Table 1 Operating and Reserve Capacities of Condensing Units

No. of Units, 100% Load	Additional or Reserve Unit, %	Total No. of Units
1	100	2
2	50	3
3	33 1/3	4
4	25	5
5 or more	20	6 or more

Receivers and Refrigerant Distribution

Receivers, either vertical or horizontal, must be installed so as to retain a liquid seal at their outlet under all conditions of pitch, roll, trim, or list. They should be fitted with an impact-resistant level glass, and may be additionally fitted with electronic level indication.

All of the same methods of refrigerant distribution that are used in stationary refrigeration plants are also used in shipboard refrigeration, including the use of secondary refrigerants. Generally, the same requirements must be met, in addition to those imposed by operating at sea.

Take care to ensure proper operation at any vessel angle that may be encountered. For direct-expansion systems, piping must ensure adequate oil return. Liquid level controls used for flooded and recirculating systems should be located in the middle of vessels rather than at either end. Provisions must be taken in vessel design to minimize liquid sloshing caused by sea conditions.

System piping must be able to withstand the stresses of operation at sea, including vibration, impact, and flexing of the ship's structure.

Controls

Recent technological developments have significantly changed how marine refrigeration plants are controlled. Electromechanical controls, which in earlier decades supplanted manual controls, are now increasingly being supplanted with solid-state controls in new and existing installations. The proliferation of electronics has influenced temperature and pressure controls, motor controls, level controls, data and trend logging, compressor sequencing, and leak detection. Microprocessors are becoming the common method of compressor control. As solid-state technology continues to advance, its advantages over prior methods of control are becoming increasingly pronounced. Automatic sequencing of multiple compressors has become as simple as entering parameters on a keypad. Electronic temperature controls are very precise, can easily be provided with multiple set points for varying duties, and can be located several hundred feet away from the space they are being used to control.

Pressures, temperatures, amperages, flow rates, liquid levels, events, and virtually any other information required can be delivered electronically to one or more central locations for monitoring and control. Computer technology allows logging these data for long- and short-term storage. The data can even be transmitted from a vessel at sea to a shore-based facility via satellite communication.

Electronic leak detection equipment can reduce the potential for accidental exposure to dangerous levels of refrigerant vapor in confined spaces.

The applications of solid-state technology are too numerous, and evolving too quickly, to list completely in this chapter. The changes in controls technology are probably the most significant in marine refrigeration over the last few years. It should be noted, though, that sophisticated controls are not a substitute for sound design and construction. Whatever control system is used, proper documentation must be provided; operating instructions must provide enough detail for users to operate the system with a minimum amount of training.

Thermometers and Thermostats

The thermometer (or thermostat display) is the principal indicator of how a refrigeration plant is functioning. Accurate control of space/room temperature depends on proper placement of the sensor(s) in that space. In spaces where the product is held just above freezing, placement is critical to avoid freezing the product. Sensors used with forced-air evaporators in this type of application should be placed in the delivery airstream, which is the point of critical temperature. Thus, if 32 to 33°F air is delivered, the product will be cooled as effectively as possible without risk of freezing. In spaces where a wider variance in temperature is tolerable, sensor placement is less critical. However, always attempt to place them in the most representative location.

Temperature recorders are essential to proper operation and control of cargo refrigeration systems. These can be paper charts, electronic media storage, or some combination of the two. The shipper or buyer of the product or cargo will often specify the type of recording device.

Electronic thermometers and thermostats have come into wide use in recent years. They allow for long distances between the sensor and the controller or display, are very precise, can often be tied in with microprocessor system controls and data logging systems, and many can be calibrated in the field. In a large refrigerated space, multiple sensors can be combined to give an average reading to the display or controller.

CARGO HOLDS

Arrangement

Arrangement and dimensions are determined by the ship's structure, compliance with compartmentalization of the hull as related to watertight integrity, vessel stability, and fire resistance. Cargo holds should not be designed exclusively for high-temperature service unless it is certain that the vessel will always remain in that limited trade.

Refrigeration controls must be located where they can be readily accessed by operating personnel regardless of whether holds are full of cargo. When controls, piping, or other equipment, such as evaporators, are located near hatches, adequate measures must be taken to prevent damage from impact by cargo, hatch covers, etc.

The greater the number of subdivisions in the refrigerated compartments, the greater the loss to the ship's revenue-generating spaces, because of the volumes occupied by insulated partitions, cooling apparatus, piping, and accesses. Thus, the all-refrigerated ship, with only the main structural boundaries insulated, makes the most efficient use of a ship's refrigerated enclosures. This efficiency comes with disadvantages, such as the difficulty in providing uniform temperatures throughout, and the inability to maintain different cargoes at different temperatures.

Space Cooling

Cargo is tightly packed in refrigerated holds of all types of vessels, with no aisles or clearances, presenting challenges to the designer. Cooling can be by extended-surface overhead-mounted coils, prime surface coils, or forced air.

For operation below freezing, the designer must consider whether defrosting will occur during operation, as with forced-air handlers, or after cargo is unloaded, as with prime surface coils. Draining defrost water from the space must be allowed for in the design of the hold, because it cannot usually be discharged to the outside of the space, as in stationary plants.

Insulation and Construction

Insulation. Moisture-, vapor-, and water-resistant insulation is of particular importance aboard ship because of frequent and extreme temperature cycles caused by intermittent refrigeration. On termination of refrigeration at discharging ports, insulation is at lower temperatures than the open room; often, the room surfaces are dripping wet with atmospheric moisture, which enters through the open door or hatch. Both warm and cold sides should be moisture-sealed equally; cold-side breather ports are not recommended. Other common sources of water in ships' cold storages are melting ice and defrosting cool surfaces.

Severe service conditions, which subject the insulation to injury or change by mechanical damage or vibration, and intermittent refrigeration place exacting requirements on insulation for ships' cold storages. The ideal shipboard composite insulation should have the following characteristics:

- High insulating value
- Imperviousness to moisture from any source

- Light in weight
- Flexibility and resilience to accommodate ships' stresses and loading
- Good structural strength
- Resistance to infiltrating air
- Resistance to disintegration or deterioration
- Fire resistance or fireproof self-extinguishing qualities
- Odorlessness
- Not conducive to harboring rodents or vermin
- Reasonable installation cost
- Workability in construction

In the United States, the properties of the insulation and the details of construction should meet approval by the U.S. Coast Guard and U.S. Public Health Service. For information on insulation materials and moisture barriers, see Chapters 23 to 26 of the 2005 *ASHRAE Handbook—Fundamentals*.

Construction. The three principal parts to the cold storage boundary are the envelope or basic structure, the insulating material, and the room lining.

The **envelope** is usually partly composed of the ship's hull, watertight decks, or watertight main bulkheads with members that resist entry of vapor from the warm side. Inboard boundaries outlining cold storages should have an equal ability to resist moisture. A continuous steel internal bulkhead with lap seams and welded stiffeners provides a boundary of adequate strength and tightness. Details of design may accommodate dimensioned insulators or facilitate means of fastening these materials. Doorway main bucks of steel channel provide good structure, but are usually a source of sweating on low-temperature rooms because of heat gain through the metal. Wooden door bucks minimize sweating but are a retreat from efforts to eliminate concealed wooden structure.

Partitioning bulkheads may be of similar detail, but airtight sealing is less important. Some installations are framed with angle-bar grids, between which the insulator is installed. In passenger vessels (over 12 passengers), Coast Guard regulations governing fire-resistant construction restrict wood assembly. Under no circumstances should wood be a part of the deck assembly, because it deteriorates rapidly under the prevailing conditions.

The assembled boundary of a ship's cold storage must withstand heavy deck loads and several bulkhead thrusts of cargo when the vessel rolls or pitches in a heavy sea; it must also be able to flex with the hull structure being stressed in any angle. The assembly must resist vibration caused by propelling machinery, the sea, and careless handling of cargo. The vapor seal of all surfaces must remain intact.

Only in extreme cases should voids in the **insulation** assembly be concealed. Filling such volumes with insulating material is cheaper and more effective than constructing internal framing. The exceptions to this rule are in the deep volumes formed by bilge brackets, deck brackets, and open box girders. Solid filling results in more insulation thickness than is needed for a heat barrier in the overhead and the ship's side, where beams and frames are deep.

The **room lining** must be sturdy enough to withstand the impact of frequent cargo loading and handling. On passenger vessels, U.S. Coast Guard regulations require that the lining be fire-resistant. Tongue-and-groove lumber is considered obsolete for any ship; on freight vessels where wood is permissible, exterior-grade plywood is sometimes applied. A few installations have been made either of laminated plastic sheets or wood fiber hardboard; both are satisfactory when properly supported. Steel linings are costly, impractical, and difficult to maintain and repair. The favored lining is the cement and fire-resistant fiber hardboard panel with aluminum sandwich lamination. When the insulator is secured with adhesives containing volatile solvents, the aluminum laminations should be of perforated metal or mesh.

The U.S. Public Health Service requires all linings to be ratproofed. Vulnerable linings should have an underlay of 1/4 in., 16 ga galvanized wire mesh for ratproofing.

Applying Insulation

When applying panels with adhesives over block insulators, butted joints should be separated sufficiently for the adhesive to extrude, provide a moisture-sealed joint, and accommodate movement of the panels by flexing of the ship's structure. Joints may be covered by cargo battens, which are also secured by adhesive and with brass screws. Bulkheads of all cold storages should be fitted with vertical cargo battens on 15 to 18 in. centers to hold cargo clear of the insulated bulkhead. This spacing allows circulation of air and prevents the contacting package from assuming the role of a room insulator.

Container vessels of cellular construction move containers along guiding columns; the boundaries of the hold, except for the hatch coamings, are not subject to mechanical damage. Here the insulation may be of the simplest, low-cost, ratproofed form without fitted hard panels or cargo battens.

Urethane foam deck insulation requires a restricting surface to confine and distribute the expanding material. In one method, plywood is secured over foam spacer blocks, and the material is injected into the space through properly spaced holes. The membrane and wearing surface covering are laid over the plywood. Wood in the deck is not good practice, however, and wood is not a suitable base for a heavy-duty wearing surface material. An alternative method calls for laying expanded board or blocks in an approved adhesive. Using the more resilient corkboard laid in adhesives may be good practice in some applications.

The greatest weakness in ship cold storage construction is the deck covering. Research and practice have not developed a totally satisfactory covering. Unlike a warehouse, a tightly packed cargo cold storage must have deck gratings both to ensure air circulation and to protect the bottom tier of products from heat leakage. Grating supports carry the cargo weight to concentrated load areas of the deck. The room may be filled with warm general cargo in alternative service, and a thermoplastic covering may be punctured by the supports.

The deck covering must be flexible enough to withstand the flexing of a ship or extreme temperature fluctuations, as well as maintain a moistureproof cover over the insulation. The most satisfactory material is a mastic composed of emulsified asphalt, sand, and cement. This material is applied cold; on setting, it has good load-bearing qualities, is impervious to water, has a small degree of ductility, and may be used in thinner layers than concrete. It should always be reinforced, and expansion joints should be included to accommodate shrinkage and adjustment to movements of the ship.

All rigid or semirigid deck coverings should have rubber-base composition expansion joints capable of bonding to the edges of the deck slabs or bulkhead and not subject to shrinking from age. The expansion joints should trace the periphery of the room, the line of all underdeck girder systems, bulkhead offsets to pillars, and similar lines of anticipated ship stress.

Water from ice-packed vegetables and from defrosting requires the deck covering to be impervious to moisture in the slab as well as the joints. If the deck insulation is wetted, it will deteriorate, lose its efficiency as an insulator, and give off odors. If wet deck insulation freezes, it will lift the deck covering and destroy it. If water penetrates to the steel, the ship's structure will corrode unnoticed.

All deck coverings will crack or become damaged during the ship's life. As a secondary security against water, a membrane should be laid between the insulator and the deck covering. The membrane need not be flashed to the bulkhead if suitable sealing expansion joints are fitted at the juncture of the bulkhead panel and the deck covering.

When an attempt is made to seal insulation with waterproof paper, it should be applied in double thickness, and the laps and perforations should be cement-sealed. Bulkhead and ceiling panels attached with suitable adhesives do not require waterproof paper inner linings.

Decks and Doors

The deck is the weakest point in the ship's refrigerated cargo hold. The weakest element in the deck is the deck drain, because of the difficulty in bonding the deck covering with the metal drain fitting. Water often finds its way between the covering and the insulation. The conventional deck drain is fitted with a perforated plate flush with the deck covering and hidden by the deck gratings. If the perforations become clogged by debris, water will accumulate at the drain. A drain fitting near a bulkhead or corner can create a weak section in the deck covering and develop cracks running to the bulkhead or across the corner. Figure 1 shows a satisfactory drain fitting. It is flush with the top of the gratings, has a lift-out cover for easy cleaning, and is bonded to the deck covering with expansion joint material. Drains between decks should be omitted wherever practical.

Refrigerated enclosure doors are generally a manufactured product. They should have generously designed steel hardware, and the door and frame should be metal-sheathed and have a flat sill and double gasket. Very large or double doors should have additional dogs to assure proper sealing when closed.

Sliding doors should be installed wherever possible to reduce interferences and conserve adjacent revenue space. When used, brackets should be installed to support portable horizontal spars inside the doors to prevent cargo from falling against the doors in a seaway. Molded glass fiber swinging doors insulated with urethane foams poured in place are available. They are strong and lightweight, are easily handled, and can be fitted with lightweight hardware.

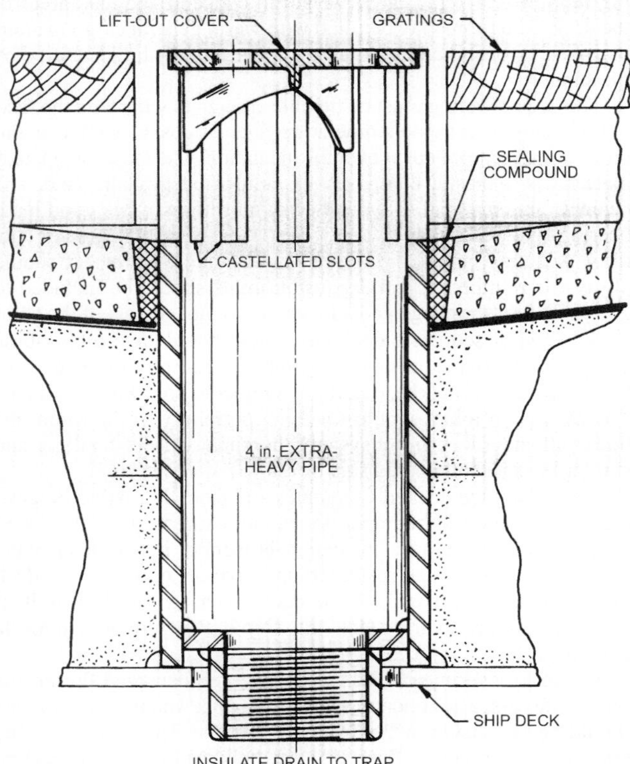

Fig. 1 Floor Drain Fitting

In finishing, wooden surfaces should be varnished rather than shellacked, because shellac has little protective penetration. Manufactured nonmetallic-surfaced materials may be painted or varnished, but if they are nonhygroscopic, their original surface will usually present a good appearance for longer than a painted coating.

Low-temperature apparatus or piping should be inspected carefully during installation. All joints and surfaces should be generously sealed to keep out atmospheric moisture. Special attention should be given to pipe covering ends, valves, and bulkhead penetrations. On below-freezing services, special composition adhesives should be used. The smallest omission or breach of a seal will allow progressive destruction of the covering.

On ships, where piping systems are relatively short, insulation functions more to prevent sweating or frosting of cold surfaces than to prevent heat gain.

SHIPS' REFRIGERATED STORES

Most vessels carry enough provisions for long voyages without replenishing en route. The refrigerating equipment must operate under extreme ambient conditions. Storage of frozen foods, packaging, humidity, air circulation, and space requirements are important factors.

Perishable foods can be fresh, dehydrated, canned, smoked, salted, and frozen. For most, refrigeration is necessary; for some, it may be omitted if the storage period is not too long. Space aboard ship is costly and limited; many rooms at different temperatures cannot be provided. Suitable product storage can be obtained by providing conditions outlined in Table 2 and described in the following sections.

Table 2 Classifications for Ships' Refrigeration Services

Service	Temp., °F	Passenger Vessels	Freight Vessels
Freezer rooms			
Meats/poultry	–20	X	X
Frozen foods	–20	X	X
Ice cream	–20	X	X
Fish	–20	X	X
Ice	28	X	
Bread	0	X	
Chill rooms			
Fresh fruit/vegetables	34	X	X
Dairy products/eggs	32	X	X
Thaw rooms	40 to 45	X	X
Wine rooms	48	X	
Bon voyage packages	40	X	
Service boxes in main galley			
Cooks' boxes	40	X	X
Butchers' boxes	40	X	
Bakers' boxes	40	X	
Salad pantry refrigerator	40	X	
Coffee pantry refrigerator	40	X	
Ice cream cabinet	10	X	
Mess rooms or pantries	40	X	X
Deck pantries	40	X	
Wine stewards' box	40	X	
Bars/fountains	Various	X	
Miscellaneous			
Ice cube freezers	See text	X	
Ice cream freezers	See text	X	
Biologicals	40	X	
Drinking water systems	See text	X	X
Ventilated stores			
Hardy root vegetables	See text	X	X
Flour/cereals	See text	X	X

COMMODITIES

Meats and Poultry

Substantial savings in space and preparation labor and better quality can be obtained with precut, boned, frozen meat and poultry packed in moisture- and vaporproof cartons and wrappers. For this reason, increased 0°F storage space should be anticipated. Fresh meats are less suitable because of their relatively short storage life. Also, the space required for fresh meats is two to four times more than that needed for prepackaged meats.

Fish, Ice Cream, and Bread

Good-quality fish, properly prepared and packaged, will remain odorless and palatable for a long time.

Commercially prepared ice cream is nearly always available and used to a great extent for both passenger and freight vessels. Ice cream for immediate use should be kept at a slightly higher temperature in an ice cream cabinet in the galley or pantry. If ice-cream-making equipment is used, provision must be made for hardening the ice cream, ices, and sherbets.

Excellent results can be obtained by purchasing freshly baked bread, sealing it in moistureproof wrappers, and storing it at 0°F. This supply may be supplemented by bread and other bakery items made on the ship. Frozen bread may be thawed in its wrapper in a few hours.

Fruits and Vegetables

Packaged frozen fruits, fruit juices or concentrates, and vegetables may be stored in any freezer room. All packaged frozen products may be held in a common 0°F storage space. However, improved accessibility, especially on large passenger vessels, may justify separate refrigerated spaces for some products.

In some cases, fresh-grown product is desired. These items may be stored in a common chill room, but some compromises with their optimum storage conditions must be expected.

Dairy Products, Ice, and Drinking Water

All dairy products may be stored in a single room, following customary shoreside practice. Strong cheeses with odors that might be adsorbed by other foods should be stored in a tightly enclosed chest or cabinet in the dairy refrigerator. Eggs may be processed by oil dipping or heat stabilization to make them less sensitive to unfavorable humidity conditions or odors. Large passenger vessels should be fitted with a separate egg storage room. Butter for reserve supply should come aboard frozen and be kept in a freezer. Frozen homogenized milk has been perfected to a degree that it can be carried for reasonably long periods. Aseptically canned whole milk may be stored without benefit of refrigeration, but this product has some limitations because of the detectable cooked flavor.

Flake ice machines and automatic ice cube makers are common on passenger and freight vessels. Chilled drinking water is piped to many parts of a ship. The water is cooled in closed-system scuttlebutts, and the necessary circulating lines serve living and machinery spaces where drinking fountains and carafe-filling taps are installed. Remote stations that would require unusually long insulated piping runs are better served by independent refrigerated drinking fountains.

STORAGE AREAS

Many borderline perishables, such as potatoes and onions, are satisfactorily stored with ventilation only. Hardy root vegetables carried on freight vessels not destined for winter zones are kept in slatted bins on a protected weather deck. Flour and cereals must be stored in cool, well-ventilated spaces to minimize conditions conducive to propagation of weevils and other insects.

Storage Space Requirements

Space requirements for refrigerated ships' stores can be approximated by formulas. However, catering officials and supervising stewards have specific ideas regarding the total volume and subdivisions, and these sometimes vary greatly. A freight ship in ordinary scheduled service seldom exceeds 45 days between replenishment of stores, and passenger vessels considerably less. In addition, deliveries en route are possible.

The space provided should allow suitable working floor areas for good storekeeping. When possible, stable piling 6 ft high is good practice; if the clear height of the room is less than 7 ft, allowances must be made for air circulation. A storage factor of 90 ft^3/ton of goods should suffice and allow floor working area. In the absence of a directing caterer or steward for consultation, Equation (1) may be used to estimate the total refrigerated storage space for merchant vessels. Ice storage is not included because of the various methods used in supplying it.

$$V = \left(\frac{NPD}{2000}\right)F \tag{1}$$

where

V = total volume of refrigerated storage (not including ice), ft^3
N = number of crew and passengers
D = number of days between re-storing
P = mass of refrigerated perishables per person per day, lb
= 10 lb for freighters
= 13.5 lb for passenger vessels
F = stowage factor (approximately 90 ft^3 per ton of goods)
2000 = lb/ton

For example, for a freighter on a 45-day voyage with a crew of 53 and 12 passengers,

$$V = \left(\frac{65 \times 10 \times 45}{2000}\right)90 = 1316 \text{ ft}^3$$

With 6 ft high stowage, the net floor area would be 219 ft^2. With 8 ft high ceilings, the gross volume would be 1752 ft^3.

Gross volume represents actual space available for storage of foods up to ceiling height and does not include the space occupied by cooling units, coils, gratings, or other equipment.

Stores' Arrangement and Location

Next to the arrangement of ships to meet their major purpose, the planning of ships' housekeeping facilities is most important. Efficient operation by culinary workers requires not only well-arranged working spaces, but also convenient supply stores. Storerooms are usually located in spaces least suitable for living quarters or revenue-earning volumes and in areas adjacent to the main galley and pantry. The arrangement should provide easy access, which generally places the reserve storage refrigerators on the deck below the galley.

Aboard freight vessels, refrigerators serve for daily issue as well as reserve storage. Aboard passenger vessels, reserve storerooms are less frequently entered, and greater use is made of the service or work boxes.

Passenger vessels carry a corps of steward's storekeepers, who should have an issue counter and office located within sight of the exits serving this area. The storerooms should extend to the ship's sides or have passageways reaching to sideport doors through which stores may be loaded directly into the ship. However, the arrangement of passenger ship stores will likely be compromised because of the interferences of structure, machinery or access hatches, and ventilation trunks.

In addition to the requirements for reserve storage of perishable foods, refrigerators (often referred to as working boxes) must be provided for the galley and pantry crew. On cargo vessels, a large domestic-type refrigerator will suffice. When more space is needed, a commercial walk-in box can be used. On passenger vessels, larger boxes are built-in like reserve refrigerators. Capacity of the passenger ship refrigerators is governed by the number of passengers carried, the variety of the menu, and the arrangement of the galley and pantries.

Ice cream stored in the reserve boxes is too hard for serving; hence, a dry or closed serving cabinet that will maintain temperatures from 5 to 10°F must be installed in the pantry. Passenger vessels may require ice-cream-making machines as well as bar and soda fountain equipment, the latter being fitted with commercial, independent refrigerating units.

SHIP REFRIGERATED ROOM DESIGN

Marine refrigeration equipment for offshore vessels should be designed, selected, and applied to function properly under extreme conditions with minimal dependence on expert servicing.

Refrigerated Room Construction

Free water that might enter the insulation through faulty floor or wall surfaces is the most harmful element to ships' refrigerated rooms. Room linings and floor coverings should be made of materials and have surface character that will give lifelong resistance to water absorption by the insulation and adherence of moisture on the room's interior surfaces. Construction of reserve and built-in refrigerated rooms should follow details similar to those of conventionally designed cargo holds.

Adequate floor drains of the type that may be cleaned without lifting floor gratings should be provided and located so that, with the probable stowage plan, the drains will be accessible for cleaning without moving shelving or excessive weights of stores.

All details should be in compliance with the regulations of the U.S. Public Health Service, which also emphasizes ratproofing. U.S. ships are also subject to strict fire-resistance regulations.

All doors and frames should be of sturdy construction to resist frequent slamming and should have metal sheathing or reinforced glass fiber doors. They should be large enough to facilitate loading of stores. The locking device must allow release of its fastenings from the inside by a person accidentally locked in.

All rooms should have galvanized or stainless steel racks or shelves to meet storekeeping needs, and they should be easily removable from their supports for rapid and thorough cleaning. The meat room and thaw room should have a single fore-and-aft meat rail for miscellaneous uses and thawing, respectively. Floor gratings or duckboards, fitted to each room, should be of a size and weight to facilitate removal and cleaning of gratings and the room.

The refrigerators should be fitted with waterproof electrical fixtures well guarded from damage by storing operations. Mount lighting switches inside each room at the door, with indicating lights in the outer passageway. Each room should have an audible alarm for use by any person inadvertently locked inside.

Remote reading thermometers, from which room temperatures can be read in the outside passageway, are essential to good operations. The sensor should be located in a representative location in the room, generally the geometric center at the ceiling. A large passenger installation justifies a duplicate electronic thermometer, with the instrument located in the refrigeration machinery room.

Service boxes in the galley and pantries should be constructed with a minimum amount of wood. The linings and shelving should be made of materials and have a surface character that facilitates thorough cleaning. Service refrigerators should not have raised door sills, and the floor should present a flush surface that is easily drained and cleaned. Cooling surfaces should be totally accessible for cleaning. Small units should be mounted without floor clearance on elevated bases, or be provided with at least an 8 in. clearance to facilitate scrubbing underneath.

SPECIFIC VESSELS

Cargo Vessels

Marine transport is an interim storage operation between preshipment storage of indeterminate duration, and distribution at destination ports. Good design requires the application of criteria that meet or even exceed those applied to shore-side cold storages.

The increased use of cargo containers has effected savings by providing faster loading and unloading of vessels. Containers also allow cargo to remain refrigerated during vessel loading and unloading. Even with these advantages, containers will never completely supplant built-in cargo refrigeration systems for many types of vessels, including passenger ships, logistical supply ships, refrigerated fruit carriers, refrigerated seafood carriers, and other types of special-service vessels.

Specifications

Refrigeration specifications should set forth the extreme operating conditions of loading, ambient and sea temperatures, and rates of pulldown. In the all-purpose installation, each compartment should be designed for refrigeration of warm, fresh products from the field or orchard; for the overall condition, a percentage division of chill and freezer cargo with simultaneous and total loading should be stated.

Typical conditions for a cargo vessel operating in all oceans with all cargoes include the following:

- Arrangement and net volume capacities of the refrigerated compartments
- Thicknesses and kinds of insulation
- Ambient temperatures

Weather surfaces	100°F
Adjacent machinery spaces	100°F
Other adjacent spaces	85°F
Sea temperatures	85°F
• Overall stowage factor	70 ft^3/ton
• Percentage total loading as chill	75%
Percentage total loading as freezer	25%
• Receiving temperature, chilled cargo	80°F
Receiving temperature, frozen cargo	25°F
• Carrying temperature, chilled cargo	34°F
Carrying temperature, frozen cargo	0°F
• Initial period of cargo heat removal (equivalent)	72 h
• Replacement air at 85°F db, 75°F wb	3%

The specification writer should describe the kind of refrigeration system to be installed and specify the number of compressors and other auxiliary parts or apparatus together with sources of emergency pumping and water facilities. All equipment and installations should be specified as complying with the rules and regulations of the Classification Societies (ABS, Lloyd's Register, and others), the U.S. Coast Guard, the U.S. Public Health Service, ASHRAE *Standard* 15, and ASHRAE *Standard* 26.

The owner's representative should obtain from the vendors full descriptions, details, capacities, and specifications of the equipment proposed, to allow comparative analysis.

Completion tests should be required to determine workmanship and functional performance. Performance guarantees should cover operations under loaded service conditions.

Calculations

The following discussion of refrigeration load calculation considerations for a general-service plant carrying heterogeneous chilled cargo may appear simplified, but it is justified by the great range of conditions common to marine installations. Refrigeration loads for freezer cargo may be calculated in a similar manner using the same stowage factor, a specific heat of 0.40 Btu/lb·°F, and an equivalent pulldown period of 72 h. The loads for respiration heat, replacement air, or latent heat of fusion will not be present.

Specialized service in known ambient conditions may be calculated more precisely, but an arbitrary 10% margin should still be added to the results to compensate for aging and unforeseen heat gains.

With general calculations, the following operating conditions should be assumed:

- *Weather ambient conditions:* Up to 100°F.
- *Ambient sea conditions*: Up to 85°F.
- *Conductivity of insulation*: According to standards given for the material, urethane foam with an installed thermal conductivity k of about 0.15 Btu·in/h·ft²·°F is suggested.
- *Resistivity of outer boundaries, inner linings, and surface films*: These factors should be ignored because boundaries and linings are usually dense and have high conductivity values.
- *Infiltration and open-door leakage*: For cargo refrigeration installations, such losses at sea are nil. Port exposures during loading and discharge reestablish pulldown conditions.
- *Ventilation or replacement air*: This factor is often omitted when carrying heterogeneous cargo for short to medium-length voyages. For specialized service, it may be as much as 300% of the gross room volume per hour.
- *Electrical energy conversion*: The energy load from fans and brine pumps will be on demand load rather than connected load. An arbitrary value of 3000 Btu/h per brake horsepower may be used.
- *Product load*: This factor ranges widely for heterogeneous cargo. Volume ranges from 40 to 120 ft³/ton; an average volume is 70 ft³/ton. Specific heat ranges from 0.22 to 0.95 Btu/lb·°F. The gross weight of the package and the specific heat of the product should be used.
- *Receiving temperature of cargo*: Chilled cargo ranges from carrying temperature to ambient. Frozen cargo ranges from –20 to 28°F.
- *Carrying temperature of cargo*: Ranges from 32 to 55°F for chilled cargo, and –20 to 0°F for frozen foods.
- *Respiratory heat of chilled cargo*: Meat products, eggs, and dairy products have no respiratory heat. Chapter 9 lists heat of respiration of many horticultural products at various storage temperatures.

FISHING VESSELS

Nearly all types of fishing vessels, from small, open gill-netters to large factory processing ships, use refrigeration in one form or another to preserve their catch. Methods range from ice taken aboard daily to sophisticated low-temperature production freezing systems and cold storages.

REFRIGERATION SYSTEM DESIGN

When designing a refrigeration system, the following issues should be considered:

- The vessel owner and design engineer must be aware of the monetary value of a fully loaded fish hold. Money saved by selecting and using substandard equipment may be a needless and expensive gamble.
- The vessel may be several hundred miles from a qualified service technician and have very limited resources on board for emergency repair. In the event of a system failure, effective initial design may maintain temperatures longer, thus preserving the product.
- Marine refrigeration systems are subjected to severe conditions, including high engine room temperatures, low ambient temperatures, electrolysis, corrosion, impacts, and vibrations. In some cases, these conditions are compounded by little or no maintenance, or worse, abusive maintenance.
- The system should be well laid out, and designed to allow new operators to adapt to the system quickly.
- All safety and operating controls should be used. In the event of a component failure, a back-up system should be available, or, ideally, built into the system. On vessels with production freezing systems, it is advisable to provide enough redundancy to enable the vessel to reach port with the already frozen product preserved in a frozen state, even if there is a failure of the production freezing plant.
- Upon completion, the vessel should be provided with all wiring and refrigerant flow diagrams, an operator's manual, and a supply of spare parts.

In the initial planning, the designer must know the following:

- For what fisheries the vessel is being equipped and in what area of the world the vessel will operate.
- In what future fisheries the vessel may be required to work. (At this point, such considerations will probably add little or no cost to the system.) Necessary alterations may be as small as increasing the spacing in the freezing racks.

Hold Preparation

On any vessel presently being refrigerated or being fitted for future refrigeration installations, 4 in. or more of insulating spray-on urethane is recommended. Special attention must be given to insulating areas of high heat, such as engine rooms, bulkheads, and the underside of the main decks. High-heat sources, such as the hatch coaming, shaft log, and fuel tanks with fuel returns from the engines, must also be insulated. The insulation must be protected to prevent moisture from destroying its insulating quality. Laid-up fiberglass is often used because of its strength, light weight, and versatility. Pen board guides, mounting brackets, and plate racks are at times fiberglassed into the liner, and thus become a very secure part of the vessel. Fiberglass has the advantage of being easily cleaned and sanitized.

REFRIGERATION WITH ICE

Ice is commonly used to preserve groundfish, shrimp, halibut, and most other commercial species. Bin or pen boards are installed to divide the hold as desired (Figure 2). Ice is usually stored in alternate bins so that it is handy for packing around the fish as the fish is loaded into the adjacent bin. The crushed ice varies in size up to 5 in. lumps. As the fish are stowed with crushed ice, each pen is generally divided horizontally by inserting boards so that the bottom fish will not be crushed. The compartmentalized sections should not be more than 30 in. high if undesirable crushing and bruising are to be avoided.

The approximate amount of ice required is 1 ton for each 2 tons of fish in summer, and 1 ton to each 3 tons of fish in winter, based

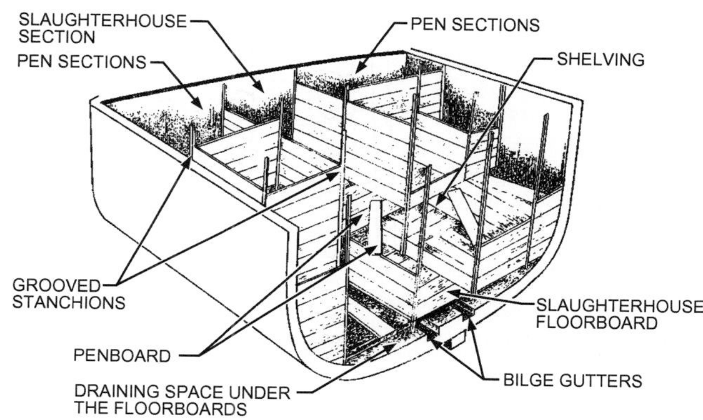

Fig. 2 Typical Layout of Pens in Hold

on a voyage of about 8 days. Less ice is needed if the ship has supplemental refrigeration.

The method of stowing the fish in ice is very important to the keeping quality. The depth of ice on the floor of the pen should be a minimum of 2 in. at the end of the voyage. This is obtained by having the initial bedding of ice 1 in. thick for each day of the voyage. In stowing, one or two layers of fish are laid on the bedding ice so that the ice is just completely covered. In no case should the layer of fish exceed 12 in. in thickness. The top covering layer of ice is about 9 in. thick, heaped up higher in the center than along the sides. This method of stowing permits the pile to adjust itself to melting and settling and results in good drainage of water and fish slime.

Many small fishing vessels are constructed of wood, with uninsulated holds. Larger vessels are usually of steel construction with insulated holds. Mechanical refrigeration is used on some vessels to keep ice from melting quickly and to maintain lower temperatures. The most common mechanical system uses direct-expansion cooling coils under the overhead in the hold and sometimes around the entire shell.

REFRIGERATION WITH SEAWATER

Refrigerated seawater is commonly used instead of ice for holding fish in satisfactory condition. The seawater is continuously pumped either around the fish and over cooling coils placed along the sides of the insulated tank, or through external chillers and then through, or over, the fish. Capacity requirements vary widely, and are primarily determined by how quickly the water needs to be chilled before taking on fish.

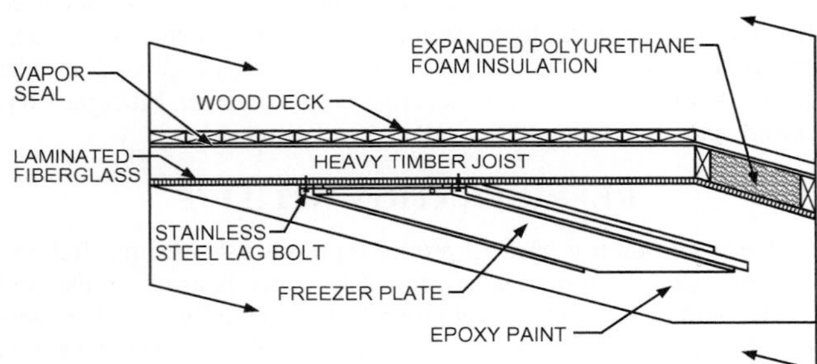

Fig. 3 Typical Underdeck Freezer Plate Installation

Design of refrigerated seawater systems is unusual, in that their primary function is to operate at full capacity during pulldown. Extra care must be taken by the designer to ensure that the compressor drive motors and condenser(s) are sized for the extremes, not just for the final temperature.

PROCESS FREEZING AND COLD STORAGE

Distant-water vessels are usually equipped for freezing and handling the catch at sea because they stay out for weeks or even months, making storage with ice or refrigerated seawater unfeasible. Although many different vessels and freezing systems are used, the general types can be classified as either those freezing large whole fish, such as tuna and halibut, and those freezing processed and semiprocessed fishery products, such as fish blocks or groundfish in bulk lots.

The method of freezing is determined by the physical and biochemical characteristics of the fish and the desired end product. For the most part, large fish such as tuna, which are eventually canned and somewhat resistant to salt intake, are conveniently frozen in brine wells where space savings and ease of handling offer convincing benefits. Cod, haddock, hake, pollack, and similar demersal or midwater species, which are more delicate than tuna, are usually frozen rapidly either in vertical or horizontal plate freezers (see Figures 3 and 4), or in air-blast freezers. Other products, such as crab, are frozen in flow-through sodium chloride brine tanks.

Virtually every type of equipment and method of freezing used in stationary installations is also used on vessels that freeze seafood at sea. Specialized products demand exacting methods of handling, freezing, and packaging. Additional design and application information can be found in other documents not specifically related to marine systems.

REFERENCES

ASHRAE. 2004. Safety standard for refrigeration systems. ANSI/ASHRAE *Standard* 15-2004.

ASHRAE. 1996. Mechanical refrigeration and air-conditioning installations aboard ship. ANSI/ASHRAE *Standard* 26-1996.

BIBLIOGRAPHY

ABS. 1998. *Rules for certification of cargo containers.* American Bureau of Shipping, Houston.

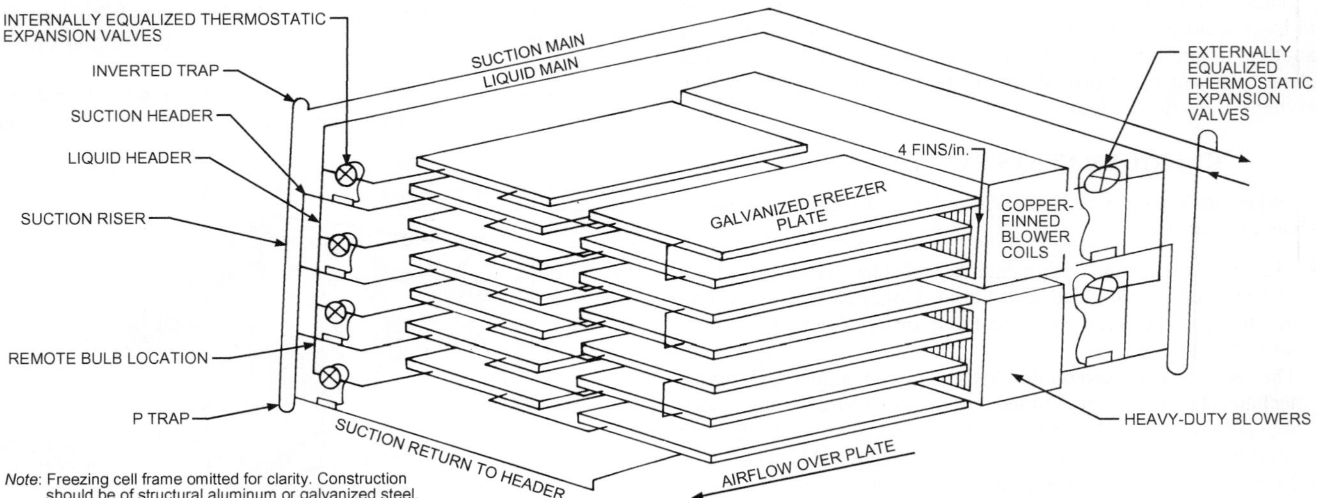

Fig. 4 Typical Marine Freezing Cell

CHAPTER 32

AIR TRANSPORT

AIR freight service is provided by all-cargo carriers and passenger airlines. The latter companies also have all-cargo aircraft. Wide-body aircraft have a passenger and cargo mix on the main deck, increasing cargo capacity (Figures 1 and 2). All lines maintain regularly scheduled flights so shippers may adequately plan delivery time. Special charter flights are also available from regular terminals and from airports located close to producing areas. Payload range comparisons of wide-body jets are shown in Figure 2.

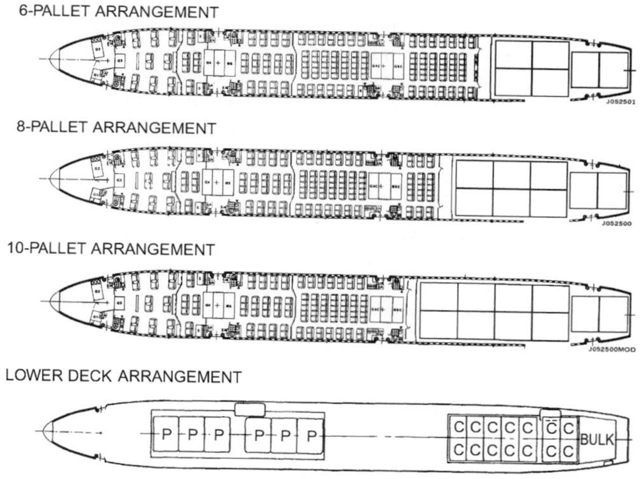

6-PALLET ARRANGEMENT

8-PALLET ARRANGEMENT

10-PALLET ARRANGEMENT

LOWER DECK ARRANGEMENT

Fig. 1 Flexible Passenger/Cargo Mix

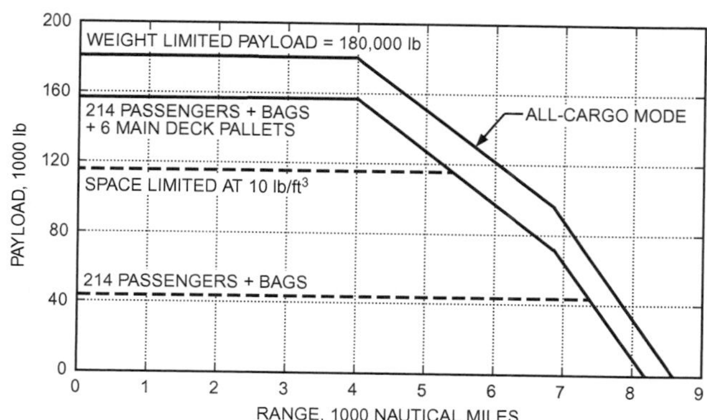

Fig. 2 Payload/Range Comparison for Wide-Body Jet

The preparation of this chapter is assigned to TC 10.6, Transport Refrigeration.

Prospective shippers should contact the airlines serving their locality to obtain specific details for handling perishable shipments.

PERISHABLE AIR CARGO

Some aircraft have cargo compartment temperature control with options ranging from just above freezing to normal room temperature. Most compartments have a single temperature control. The control is achieved by balancing skin heat loss with the supply of expended passenger cabin air and, when necessary, introduction of hot jet engine bleed air through eductors. Skin heat exchangers are used to help maintain the lower temperatures at high (cold) altitudes. This mode of refrigeration is not available at low altitudes or on the ground, where skin temperatures can exceed the compartment temperature significantly. Refrigeration techniques for aircraft rely primarily on precooling, insulated containers, dry-ice-charged containers, quick handling, and short-time exposure to adverse conditions. Airports seeking to expand cargo operations are adding refrigerated warehouses internationally. The availability of refrigerated warehouses is generally the result of specific market demands and competition.

Fruits and vegetables, flowers and nursery stock, poultry and baby chicks, hatching eggs, meats, seafoods, dairy products, live animals, whole blood, body organs, and drugs (biologicals) are transported by air. Items so shipped are generally so perishable that slower modes of transportation result in excessive deterioration in transit, making air movement the only possible means of delivery. Certain early-season and specialty fruits and vegetables can be flown to distant markets economically because of the high market prices when there is a short supply. Some items, such as cut flowers and papayas, arrive at distant markets in better condition than they would otherwise, so the extra transportation cost is justified. Flowers are shipped on a regular basis from Hawaii to the mainland United States and from California and Florida to large midwestern and eastern cities. Air movement of strawberries has increased tremendously, including direct shipments to global destinations. Papayas are shipped from Hawaii almost exclusively by air.

When carefully handled, ice cream is shipped successfully to overseas markets from the United States; however, some unsuccessful shipments have occurred because customs inspectors have opened containers for inspection and have taken too much time. Lowered trade barriers have reduced this risk.

Fruits and Vegetables

All fresh fruits, vegetables, and cut flowers remain living throughout their entire salable period. Being alive, they respond to their environment and have definite limitations on the conditions they can tolerate. They remain alive through respiration, which breaks down stored foods into energy, carbon dioxide, and water, with the uptake of atmospheric oxygen. Respiration, together with accompanying chemical changes, results in quality changes and the eventual death of the commodity. These internal changes associated with life cannot be stopped but should be retarded if high quality is to be retained for a prolonged period.

Seafood

Seafood and fish also benefit from the speed of air freight. The abundance of fresh fish at restaurants and markets throughout the United States is the result of air shipment.

Animals

Design of aircraft cargo compartments for animals is based on Society of Automotive Engineers (SAE) *Standard* AIR 1600 and the U.S. *Code of Federal Regulations* (CFR), Title 9. Temperature and ventilation regulations as well as recommendations for birds and animals of all sizes are included in these documents. Air transportation limits exposure to the extremes that would otherwise require special handling and additional cost for animal safety in accordance with the regulations.

PERISHABLE COMMODITY REQUIREMENTS

Justification for air transport of perishable commodities is based on (1) time and (2) the delivery of a higher-quality product than is possible by other modes of transportation. Better delivered quality increases returns to the shipper. This not only offsets the added transportation costs but also increases consumer demand and acceptance. The market quality of perishable items is definitely controlled by a time and temperature relationship. Temperature cannot be ignored even for the few hours now required for transcontinental air movement. Proper temperature and humidity must be maintained at all times.

Pentzer et al. (1958) lists desirable transit environments for most perishable horticultural commodities. Figure 3 shows the result of a test of air shipments of strawberries from California to Chicago in a refrigerated but uninsulated container. The shipments were exposed to high ambient temperatures during ground handling at origin, resulting in fruit temperatures ranging from 50 to 60°F instead of the desired 32 to 34°F. These berries were compared with those shipped by rail in 4.5 days with temperatures averaging 38°F for the transit period. Appearance and decay on delivery were about the same for both lots. Thus, the advantage of the short 22 h air movement was offset by a loss in quality caused by unfavorable temperature.

Top quality of many of the most perishable commodities can be significantly reduced by only a few hours' exposure to unfavorably high temperatures. Many drugs (biologicals) and other items, such as whole blood, can be rendered completely ineffective or toxic if not kept at the specified low temperature.

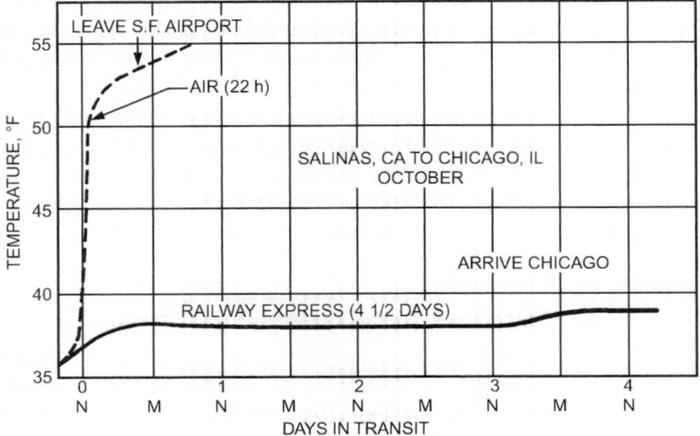

Fig. 3 Temperature of Strawberries Shipped by Air and Rail

Some flowers, fruits, and vegetables respond favorably to reduced oxygen levels, increased amounts of carbon dioxide, or both, which could be maintained by gastight packaging or containers.

Maintaining temperatures near freezing is not desirable for all products because some are subject to chilling injury, even at temperatures well above the freezing point. Chilling injury is most pronounced in tropical products, such as bananas, tomatoes, cucumbers, avocados, and orchids. Temperatures above 55°F are usually safe for cold-sensitive commodities. Other items, such as cut flowers, require temperatures between 32 and 55°F.

Certain fruits and vegetables require humidity control. Humidity should be kept between 85 and 95% to prevent wilting and general loss of water. The relative humidity in the cabin of an airplane flying at about 40,000 ft is generally less than 10%. However, respiration of fruits and vegetables, placed in closed containers with recirculated cooling air, should produce the required humidity level with no additional water added.

Certain vegetables, such as peas, broccoli, lettuce, and sweet corn, have high respiration rates and may produce heat equivalent to 250 lb or more of ice meltage per ton of vegetables per day at 60°F. In designing the refrigeration systems for aircraft containers, the additional evaporator capacity required to handle the heat of respiration should be considered.

DESIGN CONSIDERATIONS

A refrigeration or air-conditioning system for a cargo airplane or airborne cargo containers has conflicting design temperature requirements, depending on the type of cargo to be carried, which makes it difficult to use one optimum refrigeration system for all kinds of cargo. For example, frozen foods should have a temperature of 0°F or lower, fresh meat and produce 30 to 45°F, and live animals generally require temperatures in the same comfort range as for humans. Many commercial jet cargo planes operate with the main cabin divided between cargo compartments and passenger compartments, and they are supplied by a single air system controlled to the comfort of the human occupants. In this case, perishable cargo must be packed in containers, insulated, and iced or precooled.

The design ambient temperatures that an airplane experiences in flight are given in Chapter 10 of the 2003 *ASHRAE Handbook—HVAC Applications*. A cargo jet cruising close to Mach 0.9 has an increase in skin temperature over ambient of about 50°F. With an all-cargo load, the basic air-conditioning systems are capable of maintaining main cargo compartment temperatures on a design hot day from 40°F at 30,000 ft to 30°F at 40,000 ft.

Air-conditioning systems are equipped with controls to prevent freezing of moisture condensed from the air at low altitudes. With the extremely dry air prevailing at cruise altitudes, an override of these anti-icing controls would permit an even lower cabin temperature, although it is doubtful that storage temperatures of frozen goods could be met. Thus, some insulation would still be required in frozen food containers. Also, airplanes are often required to hold at relatively low altitudes of 20,000 ft or less (because of heavy traffic at busier airports) for 30 min or more.

Permanent attachment of a mechanical refrigeration system to a cargo container may not be desirable for several reasons: increased load, reduction in usable volume, and difficulty in rejecting the condensing unit heat load overboard. These objections are particularly applicable to containers carried in the main cargo hold. On the other hand, permanently attached units allow refrigeration of just part of the cargo load while the remainder is held at temperatures in the normal human or animal comfort zone. Temperature control of products requiring widely differing transit and storage temperatures would be more feasible with onboard refrigeration units.

SHIPPING CONTAINERS

Fruits and vegetables are generally shipped in the same containers used for surface transportation: wooden boxes, veneer crates of various types, or fiberboard cartons. Most flower containers are constructed of either plain or corrugated fiberboard materials. Wooden cleats are used for bracing, generally as dividers or corner braces inside the cargo box. Where lading may be exposed to very cold surfaces, external cleats may be used as spacers to prevent direct contact. Certain flowers, such as gardenias and orchids, may be packaged in individual cellophane-wrapped boxes or trays and placed in a master container. Any tightly sealed film wrap must be perforated by at least one small hole to allow release of air from the container during ascent to high altitudes.

Containers, built on pallets and shaped to make maximum use of the interior airplane volume, are in use. Airline containers presently in use are described in the International Air Transport Association's (IATA) *Unit Load Devices Technical Manual*. Containers for aircraft, except for belly cargo holds, are not shaped to make maximum use of the interior volume of the airplane. One reason for this is that the individual packages filling the containers are generally rectangular anyway. Another reason is to allow easier intermodal transport (e.g., from motor truck to the airplane and vice versa). Because of the size of aircraft loading doors and irregular aircraft cross sections compared to surface vehicles and vessels, containerization may require this compromise.

Containerization is a system of moving goods in sealed, reusable freight containers too large for manual handling and without permanently attached wheels. Advantages include far less cargo damage and pilferage, lower packaging costs, minimized handling, and lower shipping rates. Presently, these containers may be loaded at the air freight terminal or loaded at the shipper's facility and transported by flatbed truck-trailer or railroad, or both, to the air terminal.

The critical condition for design of insulation and refrigeration systems (detachable plug-in type or permanently installed) for cargo containers is the time that the container is on the dock in the hot sun waiting for shipment. For this condition, an ambient temperature of 100°F db is assumed. The average outside skin temperature of an unpainted metal container is about 115°F.

Under these conditions, the 8 by 8 by 10 ft container with 0.5 in. of high-efficiency insulation (recirculating the air and considering no latent load) requires about 18,000 Btu/h of refrigeration to maintain 35°F inside and about 24,000 Btu/h to maintain 0°F inside. For quick pulldown to these temperatures of the container and fresh perishable contents (assuming prefreezing of frozen products), capacities should increase by about 50%.

Fresh fish, shrimp, and oysters may be packed in boxes, barrels, or special containers. Proper precaution must be taken to prevent drippage from melted ice into the cargo space. Live lobsters are packed in insulated containers with saltwater seaweed. Frozen foods are always packed in insulated containers. Whole blood is shipped in specially developed containers. Insulated bags are also used.

The configurations and dimensions of two insulated containers are shown in Figure 4. Insulated with closed-cell, rigid plastic foam, the containers are a fabricated sandwich structure and are sized to fit conventional pallets and materials handling systems. The heat transfer rate for the entire standard container is 28 Btu/h·°F, and 32 Btu/h·°F for the commercial size. More recent aircraft, such as the MD-11 depicted in Figure 1, use pallets 125 in. wide by 64.4, 88, and 96 in., which may be loaded to 64 in. high and retained by straps. Load capacities are 6700, 10,000, and 11,000 lb, respectively. Containers are LD-3 (half width) and LD-6 (full width), the latter having twice the capacity and width. Both are 60.4 in. deep and 64 in. high. The LD-3 width is 79 in., the volume is 158 ft^3, and the capacity is 3300 lb. A plug-in portable mechanical refrigerating unit may be positioned in the doorway for standby operation. Tight construction permits controlled atmosphere application. A smaller shipping unit, insulated with a foamed plastic, has inside dimensions of 45 by 21 by 24 in. (i.e., a total area of 35.1 ft^2 and a capacity of 13.1 ft^3). The heat transfer rate of the entire container is 2.8 Btu/h·°F.

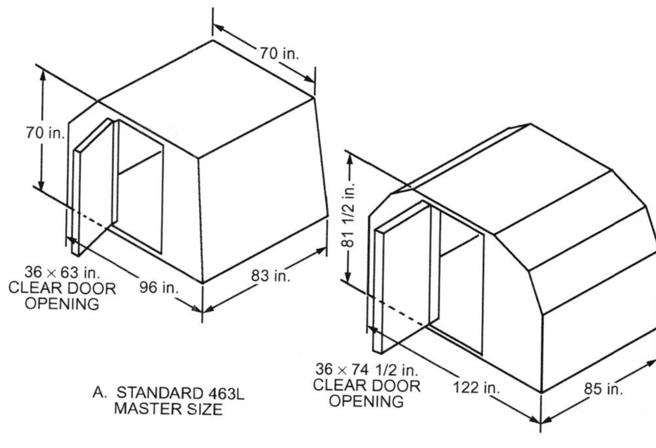

Fig. 4 Insulated Containers Designed to Fit Configuration of Cargo Aircraft

TRANSIT REFRIGERATION

Many commodities must receive refrigeration in transit. In most cases, this is accomplished by a refrigerant in the package. Water ice, dry ice, and certain proprietary refrigerants are used. Because no method of transit refrigeration can economically cool a warm commodity to its desired transit temperature, all perishable items must be cooled before shipment. Flowers are generally wrapped in several layers of paper or light insulating material and kept cool by water ice. The ice (solid, chopped, or flaked) is placed in a plastic bag or wrapped in many layers of paper and tied to one of the cleats of the container. In some instances, the ice may be chilled to 0°F before putting it in the package, thereby obtaining a slightly greater refrigeration capacity. Newspapers are sometimes wadded up, thoroughly wetted, and then frozen to 0°F or lower. In all cases, the paper helps absorb meltage and reduces the chances for leakage into the cargo space.

Some voids should be left in the containers to allow air circulation and uniform cooling. Boxes should be sealed to prevent air exchange. Placing ice or water for freezing in sealed plastic bags eliminates drippage, but melting ice in open containers increases humidity, which is particularly desirable for cut flowers. A packaged refrigerant with no escape of free liquid must be used with commodities that would be damaged by water. Water ice acts as refrigerant in special blood containers.

Dry ice is used extensively with frozen products and fresh strawberries, the amount depending on the type of container and length of journey. Sometimes it is used with water ice, not only for its own refrigerating value, but also to slow down meltage of the water ice and extend its value to the end of the transit period. Dry ice alone is seldom used for flowers because its very low temperature may cause freezing damage to adjacent blooms if not properly spaced or insulated. Unless proper ventilation of the cargo compartment is provided, the use of large amounts of dry ice may cause a build-up of carbon dioxide gas in concentrations dangerous to humans and animals.

When the heat transfer rate of insulated shipping containers is known, the amount of refrigeration required can be estimated with reasonable accuracy from

$$Q = HD\,\Delta t$$

where

Q = total heat transfer, Btu
H = heat transfer rate of entire container, Btu/h·°F

Δt = difference between ambient temperature and that at which product is to be carried, °F
D = duration of transit, h

For example, assume the small container previously described holds 15 standard strawberry trays, each holding 13 lb of berries (195 lb total), with the fruit cooled to, and carried at, 35°F at an ambient temperature of 75°F for a transit time of 24 h:

$$H = 2.8 \text{ Btu/h}\cdot°\text{F}$$

$$\Delta t = 75 - 35 = 40°\text{F}$$

$$D = 24 \text{ h}$$

$$Q = 2.8 \times 40 \times 24 = 2688 \text{ Btu}$$

The heat of respiration generated by the berries at 35°F is about 4000 Btu/ton·24 h (see Table 9 in Chapter 9). For 195 lb of berries 24 h in transit,

$$4000(195/2000) = 390 \text{ Btu}$$

The ice required to absorb this heat is

$$\frac{2688 + 390}{144} = 21.4 \text{ lb}$$

The amount of dry ice required would be about 11.9 lb (calculated using the heat of sublimation for dry ice instead of the heat of fusion for ice).

These simple calculations can be made only when the thermal conductance or heat transfer rate of the container is known. It would therefore be of considerable value to the shipper, carrier, and receiver to have this factor determined and clearly displayed for all insulated shipping containers, as on truck-trailer bodies (see Chapter 30).

When package refrigeration is not available, rapid warm-up can be retarded by insulated containers or blanket insulation over stacks or pallet loads. This method has been satisfactory with some less perishable fruits, such as peaches. Temperatures can be maintained for several in-flight hours with the proper use of these blankets. Care must be used in loading to ensure that insulating material is wrapped completely around the cargo and that containers are not in direct contact with hot or cold surfaces. Some cargo compartments on passenger aircraft may be cooled by the air-conditioning system, but the temperature will not be in the optimum range for most perishable commodities.

GROUND HANDLING

All the advantages of speed can be lost if the shipper, carrier, and receiver do not follow the good handling practices that keep deterioration to a minimum.

Ground handling can amount to over 70% of the total elapsed time from shipper to receiver. Reducing ground time involves using load palletization and special pallet carriers and loaders, in conjunction with improved load-handling systems aboard the aircraft. Air freight terminals are now designed and built to use these new handling techniques. Combination cargo/passenger jets present unique loading techniques. A typical ground service equipment arrangement is shown in Figure 5.

Fast pickup and delivery are also essential. Because of the generally high ground temperatures at shipping point terminals and intermediate points, most perishable agricultural commodities should be cooled as soon after harvesting as possible and delivered to the air terminal in properly refrigerated vehicles, particularly if they are shipped in uninsulated containers. At the terminal, these shipments must be held at proper temperatures if prompt loading from the pickup vehicle is not possible. Holding rooms, refrigerated mechanically or by ice, should be provided. During seasonal loading peaks, refrigerated trucks or trailers may be used as temporary holding rooms. During hot weather, cargo space must be cooled before loading. Portable air-conditioning equipment, such as that used for passenger aircraft, is used.

Airlines have rules for handling various perishable commodities. These include the temperatures desired in transit, the amount of seasonal protection needed, loading methods for various types of containers, and other factors involved in proper handling.

GALLEY REFRIGERATION

Requirements imposed on aircraft food service by the U.S. Public Health Service (USPHS) require that potentially hazardous

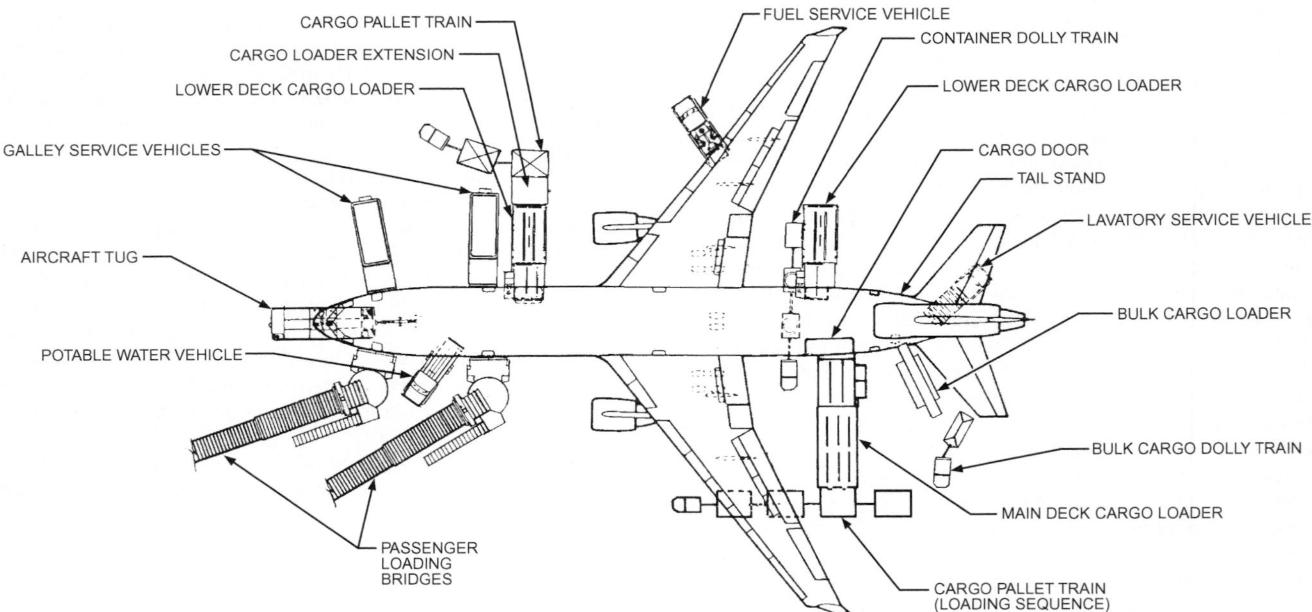

Fig. 5 Typical Ground Service Equipment Arrangement

(perishable) food carried in airplane galleys either is served within two hours of preparation or is stored at temperatures of 45°F or less (USPHS 1964). Some other countries, notably France and Britain, require temperatures as low as 41°F. There is some discussion, including within the United States, of requiring temperatures as low as 39.2°F, which would require significantly more food refrigeration capacity in airplanes; part of the capacity in the short term is likely to be dry ice.

Passenger meals are prepared in ground kitchens well in advance of actual catering Meals are refrigerated and held at the catering facility. When airplanes require catering, refrigeration is often provided by packing the meals with dry ice in the service carts, which are transferred by truck from the ground kitchen to the aircraft's galleys. Consequently, recent air quality surveys using CO_2 as a surrogate measurement have found that CO_2 concentrations are higher, but far from toxic, near the galleys. CO_2 levels are used as an indicator of the ventilation rate, considering passenger respiration release of CO_2. The indicator is therefore not valid in the galley areas where additional sources of CO_2 are located.

Most airlines flying shorter domestic routes use dry ice as a primary refrigeration method when perishable foods are to be served later in the flight. Any route of around 4 h or less can carry a perishable meal service in this way.

Most airlines flying longer domestic and international routes use vapor cycle refrigeration units installed in the galleys or in the airplane structure near the galleys that store perishable food. The current industry standard uses a self-contained refrigeration unit that uses fans to move air through heat exchangers both to provide chilled air (via the evaporator heat exchanger) for refrigeration and to remove waste heat (via the condenser heat exchanger) from the unit. Chilled air is ducted to the food storage areas in the galleys. Airflow carrying waste heat (often called exhaust air) is ducted to areas of the airplane where it can be dissipated without effect on the cabin environmental control system.

Chilled air delivered to galley units is generally handled in one of two ways. An **air-through** system delivers air to valve manifolds that interface with vent openings directly in the service carts. Its advantage is capability to provide refrigeration pulldown if perishable food temperatures temporarily exceed the limits. These systems are also more complex, and because of the many branches in the airflow, they are more prone to losses (such as cooling, airflow, pressure, and temperature) because of airflow restriction, leakage, and heat transfer through duct and cart walls.

An **air-over** system distributes air in enclosed cart compartments, blanketing the service carts with chilled air. These systems are simpler and more efficient in the volume of perishables that can be refrigerated per unit of refrigeration capacity. Because food in the service cart is separated from the chilled air, pulldown performance is very poor. These systems often require complex baffling to ensure air is evenly distributed in the compartments. An advantage to this type of system is that service carts intended for air-through systems can be used in galleys designed for air-over refrigeration.

Some galleys are equipped with fully contained refrigerators that are much like household units.

It is common practice to load service carts with dry ice even if the airplane galley system is equipped with chillers. Fundamentally, airplane designers do not consider food refrigeration a primary airplane function; thus, airplane systems are not well adapted to accommodate vapor-cycle refrigeration machines. Also, there is less rigor in ensuring that mechanical food refrigeration systems remain functional in flight. This has the following effects:

- Vapor-cycle systems in use today have difficulties in ground operations, especially in very hot environments. Galley chiller units are often located in the lower lobe or in the overhead crown areas, where air temperatures can exceed cabin and outside ambient air temperatures significantly. High air temperature for condensing refrigerant reduces chiller efficiency, often requiring operators to supplement mechanical refrigeration of perishables with dry ice.
- These systems are generally not robust. When not maintained diligently, they are prone to breakdowns. Dry ice is used as the backup refrigeration method.
- Flight kitchens and other catering points try to time the delivery of food to the airplane so that it arrives just in time for the food to be catered and flown away. This is critical because most caterers do not use refrigerated trucks to deliver the food to the airplane. Although caterers are provided with flight information (arrival and departure times, delays, etc.), occasionally catering trucks have to wait at the airport gate for the airplane that they are to service. To guard against having to condemn a truckload of food, airlines often require that service trolleys be loaded with dry ice.

For these reasons, despite the availability of mechanical refrigeration, some dependence remains on dry ice in airplane galley refrigeration. Systems are in development to decrease reliance on both dry ice and mechanical systems, but none are yet in wide production.

REFERENCES

CFR. 2005. Animals and animal products. 9CFR. *Code of Federal Regulations*, U.S. Government Printing Office, Washington, D.C.

IATA. Updated annually. *ULD technical manual.* International Air Transport Association, Montreal.

Pentzer, W.T., Jr., et al. 1958. Air transportation of fruits, vegetables and cut flowers: Temperature and humidity requirements and perishable nature. AMS *Report* 280. U.S. Department of Agriculture, Washington, D.C.

SAE. 1997. Animal environment in cargo compartments. ANSI/SAE *Standard* AIR1600. Society of Automotive Engineers, Warrendale, PA.

USPHS. 1964. Handbook on sanitation of airlines: Standards for sanitation for the construction and operation of commercial passenger aircraft and servicing and catering facilities. *Publication* 308. U.S. Public Health Service, Washington, D.C.

BIBLIOGRAPHY

Taylor, W.P. 1990. In *Proceedings of the 25th Intersociety Energy Conversion Engineering Conference* 4:285-87.

CHAPTER 33

INSULATION SYSTEMS FOR REFRIGERANT PIPING

THIS chapter is a guide to specifying insulation systems for refrigeration piping, fittings, and vessels operated at temperatures ranging from 35 to −100°F. It does not deal with HVAC systems or applications such as chilled-water systems. Refer to Chapters 23, 24, 25, and 26 in the 2005 *ASHRAE Handbook—Fundamentals* for information about insulation and vapor barriers for these systems.

The success of an insulation system for cold piping, such as refrigerant piping, depends on factors such as

- Correct refrigeration system design
- Correct specification of insulating system
- Correct specification of insulation thickness
- Correct installation of insulation and related materials (e.g., vapor retarders)
- Installation quality
- Adequate maintenance of the insulating system

Various insulation materials are used for HVAC, steam, and hot-water lines that either run hot or cycle between cold and hot. These lines are insulated for the following reasons:

- Energy conservation
- Economics (to minimize annualized costs of ownership and operation)
- External surface condensation control
- Prevention of gas condensation inside the pipe
- Process control (i.e., for freeze protection and to limit temperature change of process fluids)
- Personnel protection
- Fire protection
- Sound and vibration control

Design features for typical refrigeration insulation applications recommended in this chapter may be followed unless they conflict with applicable building codes. A qualified engineer may be consulted to specify both the insulation material and the insulation thickness (see Tables 2 through 11) based on specific design conditions. All fabricated pipe, valve, and fitting coverings should have dimensions and tolerances in accordance with ASTM *Standards* C450 and C585. The installation of all materials used for thermal insulation should be carried out in accordance with the Midwest Insulation Contractors Association's (MICA) National Commercial and Industrial Insulation Standards or the recommendations of the manufacturers for materials not presented in this standard.

DESIGN CONSIDERATIONS FOR BELOW-AMBIENT REFRIGERANT PIPING

Below-ambient refrigerant lines are insulated primarily to (1) minimize heat gain to the internal fluids, (2) control surface condensation, and (3) prevent ice accumulations. Noise reduction and personnel protection are also reasons for providing thermal insulation. For most installations, the thickness required to prevent surface condensation will control the design. Given appropriate design conditions and insulation properties, computer programs such as NAIMA 3E Plus may be helpful in calculating the required insulation thickness. Tables 3 through 12 give estimates for several typical design conditions for a variety of insulation materials.

In many refrigeration systems, operation is continuous; thus, the vapor drive is unidirectional. Water vapor that condenses on the pipe surface or in the insulation remains there (as liquid water or as ice) unless removed by other means. An insulation system must deal with this unidirectional vapor drive by providing a continuous and effective vapor retarder to limit the amount of vapor entering the insulation.

Various insulation and accessory materials are used in systems for refrigerant piping. Successful system design provides the best solution for material selection, installation procedures, operations, and maintenance to achieve long-term satisfactory performance, meeting all criteria imposed by the owner, the designer, and code officials.

INSULATION PROPERTIES AT BELOW-AMBIENT TEMPERATURES

Insulation properties important for the design of below-ambient systems include thermal conductivity, water vapor permeance, water absorption, coefficient of thermal expansion, and wicking of water. See Table 2 for material properties.

Thermal conductivity of insulation materials varies with temperature, generally decreasing as temperature is reduced. For pipe insulation, conductivity is determined by ASTM *Standard* C335. This method is generally run at above-ambient conditions and the results extrapolated for below-ambient applications. In some cases, conductivity is determined on flat specimens (using ASTM *Standard* C177 or C518). The designer should be aware of the method used and its inherent limitations.

Water vapor permeance is a measure of the time rate of water vapor transmission through a unit area of material or construction induced by a unit vapor pressure difference through two specific surfaces, under specified temperature and humidity conditions. The lower the permeance, the higher the resistance of the material or system to passing water vapor. The unit of water vapor permeance is the perm, and data are determined by ASTM *Standard* E96. As with thermal conductivity, permeance can vary with conditions. Data for most insulation materials are determined at room temperature using the desiccant method. Water vapor permeance can be critical in design because water vapor can penetrate materials or systems that are unaffected by water in the liquid form. Water vapor diffusion is a particular concern to insulation systems subjected to a thermal gradient. Pressure differences between ambient conditions and the colder operating conditions of the piping drive water vapor into the insulation. There it may be retained as water vapor, condense to liquid water, or condense and freeze to form ice, and can eventually cause the insulation to pop off the pipe. Thermal

This preparation of this chapter is assigned to TC 10.3, Refrigerant Piping.

properties of insulation materials are negatively affected as the moisture or vapor content of the insulation material increases.

The coefficient of thermal expansion is important both for insulation systems that operate continuously at below-ambient conditions and systems that cycle between below-ambient conditions and elevated temperatures. Thermal contraction of insulation materials may be substantially different from that of the metal pipe. A large difference in contraction between insulation and piping may open joints in the insulation, which not only create a thermal short circuit at that point, but may also affect the integrity of the entire system. Insulation materials that have large coefficients of thermal expansion and do not have a high enough tensile or compressive strength to compensate may experience shrinkage and subsequently crack. At the high-temperature end of the cycle, the reverse is a concern. High thermal expansion coefficients may cause permanent warping or buckling in some insulation material. In this instance, the possible stress on an external vapor retarder or weather barrier should be considered.

Water absorption is a material's ability to absorb and hold liquid water. Water absorption is important where systems are exposed to water. This water may come from a number of external sources such as rain, surface condensation, or washdown water. The property of water absorption is especially important on outdoor systems and when vapor or weather retarder systems fail. Collected water in an insulation system degrades thermal performance, enhances corrosion potential, and shortens the system's service life.

Wicking is the tendency of an insulation material to absorb liquid through capillary action. Wicking is measured by partially submerging a material and measuring both the mass of liquid that is absorbed and the volume that the liquid has filled within the insulation material.

Insulation System Water Resistance

Refrigeration systems are often insulated to conserve energy and prevent surface condensation. An insulation system's resistance to the intrusion of water is a critical consideration for many refrigerant piping installations. When the vapor retarder system fails, water vapor will move into the insulation material. This may lead to partial or complete failure of the insulation system. The problem becomes more severe at lower operating temperatures and when operating continuously at cold temperatures. The driving forces are greater in these cases and water vapor condenses and freezes on or within the insulation. As more water vapor is absorbed, the thermal conductivity of the insulation material increases, which leads to a lower surface temperature. This lower surface temperature leads to more condensation, which may cause the insulation material to pop off because of ice formation. With refrigeration equipment operating at 35°F or lower, the problem may be severe.

If a low-permeance vapor retarder is properly installed on the insulation system and is not damaged in any way, then the water resistance of the insulation material is not as important. In practice, it is very difficult to achieve and maintain perfect performance in a vapor retarder. Therefore, the water resistance of the insulation material is an important design consideration. The water absorption and water vapor permeability properties of an insulation material are good indicators of its resistance to water. Because intrusion of water into an insulation system has numerous detrimental effects, better long-term performance can be achieved by limiting this intrusion. For these reasons, insulation materials with high resistance to moisture (low absorption and permeability) should be used for refrigerant piping operating at temperatures below 35°F.

INSULATION SYSTEMS

The elements of a below-ambient temperature insulation system include

- Pipe preparation
- Insulation material
- Insulation joint sealant/adhesive
- Vapor retarders
- Weather barrier/jacketing

Pipe Preparation for Corrosion Control

Before any insulation is applied, all equipment and pipe surfaces to be insulated **must** be dry and clean of contaminants and rust. Corrosion of any metal under any thermal insulation can occur for a variety of reasons. The outer surface of the pipe should be properly prepared before installation of the insulation system. The pipe can be primed to minimize the potential for corrosion. Careful consideration during insulation system design is essential. The prime concern is to keep the piping surface dry throughout its service life. A dry, insulated pipe surface will not have a corrosion problem. Wet, insulated pipe surfaces are the problem.

Insulated carbon steel surfaces that operate continuously below 25°F do not present major corrosion problems. However, equipment or piping operating either steadily or cyclically at or above these temperatures may have significant corrosion problems if water or moisture is present. These problems are aggravated by inadequate insulation thickness, improper insulation material, improper insulation system design, and improper installation of insulation.

Common flaws include the following:

- Incorrect insulation materials, joint sealants/adhesives or vapor retarders used on below-ambient temperature systems
- Improper specification of insulation materials by generic type rather than by specific material properties required for the intended service
- Improper or unclear application methods

Carbon Steel. Carbon steel corrodes not because it is insulated, but because it is contacted by aerated water and/or a waterborne corrosive chemical. For corrosion to occur, water must be present. Under the right conditions, corrosion can occur under all types of insulation. Examples of insulation system flaws that create corrosion-promoting conditions include

- Annular space or crevice for water retention
- Insulation material that may wick or absorb water
- Insulation material that may contribute contaminants that can increase the corrosion rate

The corrosion rate of carbon steel depends on the temperature of the steel surface and the contaminants in the water. The two primary sources of water are infiltration of liquid water from external surfaces and condensation of water vapor on cold surfaces.

Infiltration occurs when water from external sources enters an insulated system through breaks in the vapor retarder or in the insulation itself. The breaks may result from inadequate design, incorrect installation, abuse, or poor maintenance practices. Infiltration of external water can be reduced or prevented.

Condensation results when the metal temperature or insulation surface temperature is lower than the dew point. Insulation systems cannot always be made completely vaportight, so condensation must be recognized in the system design.

The main contaminants found in insulation are chlorides and sulfates, introduced during manufacture of the insulation or from external sources. These contaminants may hydrolyze in water to produce free acids, which are highly corrosive.

Table 1 lists a few of many protective coating systems that can be used for carbon steel. For other systems or for more details, contact the coating manufacturer.

Copper. External stress corrosion cracking (ESCC) is a type of localized corrosion of various metals, notably copper. For ESCC to occur in a refrigeration system, the copper must undergo the combined effects of sustained stress and a specific corrosive species.

Table 1 Protective Coating Systems for Piping

Substrate	Temperature Range	Surface Prep.[d]	Surface Profile	Prime Coat[a]	Intermediate Coat[a]	Finish Coat[a]
Carbon Steel System No.1	–50 to 140°F	NACE *Standard* 2	0.002 to 0.003 in.	0.005 in. high-build (HB) epoxy	N/A	0.005 in. HB epoxy
Carbon Steel System No.2	–50 to 140°F	NACE *Standard* 2	0.002 to 0.004 in.	0.007 to 0.010 in. metallized aluminum	0.0005 to 0.00075 in. of MIL-P-24441/1[b] epoxy polyamide (EPA) followed by 0.003 in. of MIL-P-24441/1[c] EPA	0.003 in. of MIL-P-24441/2[c] EPA
Carbon Steel System No.3	200°F maximum	NACE *Standard* 2	0.002 to 0.003 in.	0.002 to 0.003 in. moisture-cured urethane aluminum primer	0.002 to 0.003 in. moisture-cured micaceous aluminum	Two 0.003 in. coats of acrylic urethane
Carbon Steel System No.3	–50 to 300°F	NACE *Standard* 2	0.002 to 0.003 in.	0.006 in. epoxy/phenolic or high-temperature rated amine-cured coal tar epoxy	N/A	0.006 in. epoxy/phenolic or high-temperature rated amine-cured coal tar epoxy

[a]Coating thicknesses are typical dry film values. [b]MIL-P-24441, Part 1. [c]MIL-P-24441, Part 2. [d]NACE *Standard* 2/SSPC-SP 10

During ESCC, copper degrades so that localized chemical reactions occur, often at the grain boundaries in the copper. The localized corrosion attack creates a small crack that advances under the influence of the tensile stress. The common form of ESCC (intergranular) in copper results from grain boundary attack. Once the advancing crack extends through the metal, the pressurized refrigerant leaks from the line.

ESCC occurs in the presence of

- Oxygen (air).
- Tensile stress, either residual or applied. In copper, stress can be put in the metal at the time of manufacture (residual) or during installation (applied) of a refrigeration system.
- A chemical corrosive.
- Water (or moisture) to allow copper corrosion to occur.

The following precautions reduce the risk of ESCC in refrigeration systems:

- Properly seal all seams and joints of the insulation to prevent condensation between insulation and copper tubing.
- Avoid introducing applied stress to copper during installation. Applied stress can be caused by any manipulation, direct or indirect, that stresses the copper tubing; for example, applying stress to align a copper tube with a fitting or physically damaging the copper before installation.
- Never use chlorinated solvents such as 1,1,1-trichloroethane to clean refrigeration equipment. These solvents have been linked to rapid corrosion.
- Use no acidic substances such as citric acid or acetic acid (vinegar) on copper. These acids are found in many cleaners.
- Make all soldered connections gastight because a leak could cause the section of insulated copper tubing to fail. A gastight connection prevents self-evaporating lubricating oil, and even refrigerants, from reacting with moisture to produce corrosive acidic materials such as acetic acid.
- Choose the appropriate thickness of insulation for the environment and operating condition to avoid condensation on tubing.
- Never mechanically constrict or adhere insulation to copper. An example of mechanical constriction is using wire ties to compress the insulation. This may result in water pooling between the insulation and copper tubing.
- Prevent extraneous chemicals or chemical-bearing materials such as corrosive cleaners containing ammonia and/or amine salts, wood smoke, nitrites, and ground or trench water, from contacting insulation or copper.
- Prevent water from entering between the insulation and the copper. Where system layout is such that condensation may form and run along uninsulated copper by gravity, completely adhere and seal the beginning run of insulation to the copper or install vapor stops.
- Use copper that complies with ASTM *Standard* B280. Buy copper from a reputable manufacturer.
- When pressure-testing copper tubing, take care not to exceed its specific yield point.
- When testing copper for leaks, use only a commercial refrigerant leak detector solution specifically designed for that purpose. Assume that all commercially available soap and detergent products contain ammonia or amine-based materials, all of which contribute to formation of stress cracks.
- Replace any insulation that has become wetted or saturated with refrigerant lubricating oils, which can react with moisture to form corrosive materials.

Stainless Steel. Certain grades of stainless steel piping are susceptible to ESCC. ESCC occurs in austenitic steel piping and equipment when chlorides in the environment or insulation material are transported in the presence of water to the hot stainless steel surface and are then concentrated by evaporation of the water. This situation occurs most commonly beneath thermal insulation, but the presence of insulation is not a requirement. Thermal insulation simply provides a medium to hold and transport water, with its chlorides, to the metal surface.

Most ESCC failures occur when metal temperature is in the hot-water range of 120 to 300°F. Below 120°F, the reaction rate is slow and the evaporative concentration mechanism is not significant. Equipment that cycles through the water dew-point temperature is particularly susceptible. Water present at the low temperature evaporates at the higher temperature. During the high-temperature cycle, chloride salts dissolved in the water concentrate on the surface.

As with copper, sufficient tensile stress must be present in the stainless steel for ESCC to develop. Most mill products, such as sheet, plate, pipe, and tubing, contain enough residual processing tensile stresses to develop cracks without additional applied stress. When stainless steel is used, coatings may be applied to prevent ESCC. A metallurgist should be consulted to avoid catastrophic piping system failures.

Insulation Materials

All insulation must be stored in a cool, dry location and be protected from the weather before and during application. Vapor retarders and weather barriers must be installed over dry insulation. The insulation system should have a low thermal conductivity with low water vapor permeability.

Cellular glass, closed-cell phenolic, flexible elastomeric, polyisocyanurate, and polystyrene are insulation materials commonly

Table 2 Properties of Insulation Materials

	Cellular Glass	Flexible Elastomeric	Closed-Cell Phenolic	Polyisocyanurate	Polystyrene
Standard that specifies material and temperature requirements	ASTM C552	ASTM C534	ASTM C1126	ASTM C591	ASTM C578
Suitable temp. range, °F	−450 to 800	−70 to 220	−290 to 250	−297 to 300	−65 to 165
Flame spread rating[a]	5	25	25	25	25
Smoke developed rating[a]	0	50	50	50	115
Water vapor permeability,[b] perm-inches	0.005	0.1	2.0	4.5	1.5
Thermal conductivity,[c] Btu·in/h·ft^2·°F					
At 0°F mean temperature	0.27	0.26	—	0.19	—
At +75°F mean temperature	0.31	0.28	0.13	0.19	0.24
At +120°F mean temperature	0.33	0.30	0.15	0.21	0.26

[a]Tested in accordance with ASTM *Standard* E84 for 1 in. thick insulation.
[b]Tested in accordance with ASTM *Standard* E96, Procedure A. Cellular glass tested with ASTM *Standard* E96, Procedure B.
[c]Tested at 180 days of age in accordance with ASTM *Standard* C177 or C518.

used in refrigerant applications. Designers should specify compliance with the material properties for each insulation in Table 2. Table 2 lists physical properties and Tables 3 through 12 list recommended thicknesses for pipe insulation based on condensation control or for limiting heat gain.

- **Cellular glass** has excellent compressive strength, but it is rigid. Density varies between 6.3 and 8.6 lb/ft^3, but does not greatly affect thermal performance. It is fabricated to be used on piping and vessels. When installed on applications that are subject to excessive vibration, the inner surface of the material may need to be coated. The coefficient of thermal expansion for this material is relatively close to that of carbon steel. When installed on refrigeration systems, provisions for expansion and contraction of the insulation are usually only recommended for applications that cycle from below-ambient to high temperatures.
- **Flexible elastomerics** are soft and flexible. This material is suitable for use on nonrigid tubing, and its density ranges from 3.0 to 8.5 lb/ft^3. It has a low vapor permeability and normally requires no supplemental vapor retarder protection.
- **Closed-cell phenolic foam insulation** has a very low thermal conductivity, and can provide the same thermal performance as other insulations at a reduced thickness. Its density is 2.0 to 3.0 lb/ft^3.
- **Polyisocyanurate insulation** has low thermal conductivity and excellent compressive strength. Density ranges from 1.8 to 6.0 lb/ft^3.
- **Polystyrene insulation** has good compressive strength. Typical density range is 1.5 to 2.5 lb/ft^3.

Insulation Joint Sealant/Adhesive

All insulation materials that operate in below-ambient conditions should be protected by a continuous vapor retarder system. Joint sealants contribute to the effectiveness of this system. The sealant should resist liquid water and water vapor, and should bond to the specific insulation surface. The sealant should be applied at all seams, joints, terminations, and penetrations to retard the transfer of water and water vapor into the system.

Vapor Retarders

Insulation materials should be protected by a continuous and effective vapor retarder, either integral to the insulation or a vapor retarder material applied to the exterior surface of the insulation.

Service life of the insulation and pipe depends primarily on the installed water vapor permeance of the system, comprised of the permeance of the insulation, vapor retarders on the insulation, and the sealing of all joints, seams, and penetrations. Therefore, the vapor retarder must be free of discontinuities and penetrations. It must be installed to allow expansion and contraction without compromising the vapor retarder's integrity. The manufacturer should have specific design and installation instructions for their products.

Vapor retarders may be of the following types:

- **Metallic foil** or **all-service jacket (ASJ) retarders** are applied to the insulation surface by the manufacturer or in the field. This type of jacket has a low water vapor permeance under ideal conditions (0.02 perms). These jackets have longitudinal joints and butt joints, so achieving low permeability depends on complete sealing of all joints and seams. Jackets may be sealed with a contact adhesive applied to both overlapping surfaces. Manufacturers' instructions must be strictly followed during the installation. Butt joints are sealed similarly using metallic-faced ASJ material and contact adhesive. ASJ jacketing, when used outdoors with metal jacketing, may be damaged by the metal jacketing, so extra care should be taken when installing it. Pressure-sensitive adhesive systems for lap and butt joints may be acceptable, but they must be properly sealed.
- **Coatings**, **mastics**, and **heavy, paint-type products** applied by trowel, brush, or spraying, are available for covering insulation. Material permeability is a function of the thickness applied. Some products are recommended for indoor use only, whereas others are available for indoor or outdoor use. These products may impart odors, and manufacturers' instructions should be meticulously followed. Ensure that mastics used are chemically compatible with the insulation system.

  Mastics should be applied in two coats (with an open-weave fiber reinforcing mesh) to obtain a total dry-film thickness as recommended by the manufacturer. The mastic should be applied as a continuous monolithic retarder and extend at least 2 in. over any membrane, where applicable. This is typically done only at valves and fittings. Mastics must be tied to the rest of the insulation or bare pipe at the termination of the insulation, preferably with a 2 in. overlap to maintain continuity of the retarder.
- A **laminated membrane retarder,** consisting of a rubber bitumen layer adhered to a plastic film, is also an acceptable and commonly used vapor retarder. This type of retarder has a very low permeance of 0.015 perms. Some solvent-based adhesives can attack this vapor retarder. All joints should have a 2 in. overlap to ensure adequate sealing. Other types of finishes may be appropriate, depending on environmental or other factors.
- **Homogeneous polyvinylidene chloride films** are another type of commonly and successfully used vapor retarder. This type of vapor retarder is available in thicknesses ranging from 0.002 to 0.006 in. Its permeance is very low, is dependent on thickness, and ranges from 0.01 to 0.02 perms. Some solvent-based adhesives can attack this vapor retarder. All joints should have a 1 to 2 in. overlap to ensure adequate sealing and can be sealed with tapes made from the same film or various adhesives.

Table 3 Flexible Elastomeric Insulation Thickness for Outdoor Design Conditions
(100°F Ambient Temperature, 90% Relative Humidity, 0.4 Emittance, 7.5 mph Wind Velocity)

Nominal Pipe Size, in.	Pipe Operating Temperature, °F							
	40	20	0	−20	−40	−60	−80	−100
0.50	1.5	2.0	2.5	2.5	2.5	3.0	3.0	3.0
0.75	2.0	2.5	2.5	2.5	3.0	3.0	3.5	3.5
1.00	2.0	2.5	2.5	3.0	3.0	3.5	3.5	4.0
1.50	2.0	2.5	3.0	3.0	3.0	3.5	4.0	4.0
2.00	2.0	3.0	3.0	3.0	3.5	4.0	4.0	4.5
2.50	2.5	3.0	3.0	3.0	3.5	4.0	4.0	4.5
3.00	2.5	3.0	3.5	3.5	4.0	4.5	4.5	5.0
4.00	2.5	3.0	3.5	4.0	4.5	4.5	5.0	5.0
5.00	2.5	3.5	4.0	4.0	4.5	5.0	5.0	5.5
6.00	2.5	3.5	4.0	4.5	4.5	5.0	5.5	6.0
8.00	3.0	3.5	4.5	4.5	5.0	5.5	6.0	6.5
10.00	3.0	4.0	4.5	5.0	5.5	6.0	6.5	7.0
12.00	3.0	4.0	4.5	5.5	5.5	6.0	6.5	7.0
14.00	3.5	4.0	5.0	5.5	6.0	6.5	6.5	7.0
16.00	3.5	4.5	5.0	5.5	6.0	6.5	7.0	7.5
18.00	3.5	4.5	5.0	5.5	6.0	6.5	7.0	7.5
20.00	3.5	4.5	5.0	5.5	6.0	6.5	7.0	7.5
24.00	3.5	4.5	5.0	5.5	6.5	7.0	7.5	8.0
28.00	3.5	4.5	5.5	6.0	6.5	7.0	7.5	8.0
30.00	3.5	4.5	5.5	6.0	6.5	7.0	7.5	8.0
36.00	3.5	4.5	5.5	6.0	7.0	7.0	7.5	8.0

Notes:
1. Insulation thickness is chosen either to prevent or minimize condensation on outside pipe surface or to limit heat gain to 8 Btu/h·ft^2, whichever thickness is greater.
2. All thicknesses are in inches.
3. Values do not include safety or aging factor. Actual operating conditions may vary. Consult a design engineer for appropriate recommendation for your specific system.
4. Data calculated using NAIMA 3E Plus program.

Table 4 Cellular Glass Insulation Thickness for Outdoor Design Conditions
(100°F Ambient Temperature, 90% Relative Humidity, 0.4 Emittance, 7.5 mph Wind Velocity)

Nominal Pipe Size, in.	Pipe Operating Temperature, °F							
	40	20	0	−20	−40	−60	−80	−100
0.50	1.5	2.0	2.5	3.0	3.5	3.5	4.0	4.0
0.75	2.0	2.5	3.0	3.5	3.5	3.5	3.5	4.0
1.00	2.0	2.5	2.5	3.0	3.5	4.0	4.0	4.5
1.50	2.5	3.0	3.0	3.5	4.0	4.5	4.5	5.0
2.00	2.0	2.5	3.0	3.5	4.0	4.5	4.5	5.0
2.50	2.5	3.0	3.5	4.0	4.5	5.0	5.0	5.5
3.00	2.5	3.0	3.5	4.0	4.5	5.0	5.0	5.5
4.00	2.5	3.0	3.5	4.0	4.5	5.0	5.5	6.0
5.00	2.5	3.5	4.0	4.5	5.0	5.5	6.0	6.5
6.00	2.5	3.5	4.0	4.5	5.0	5.5	6.0	6.5
8.00	3.0	3.5	4.5	5.0	5.5	6.0	6.5	7.0
10.00	3.0	4.0	4.5	5.5	6.0	7.0	7.0	7.5
12.00	3.0	4.0	4.5	5.5	6.0	7.0	7.5	8.0
14.00	3.5	4.0	5.0	5.5	6.5	7.0	7.5	8.0
16.00	3.5	4.5	5.0	6.0	6.5	7.0	7.5	8.5
18.00	3.5	4.5	5.0	6.0	6.5	7.5	8.0	8.5
20.00	3.5	4.5	5.0	6.0	7.0	7.5	8.0	8.5
24.00	3.5	4.5	5.0	6.0	7.0	7.5	8.0	9.0
28.00	3.5	4.5	5.5	6.5	7.0	8.0	8.5	9.0
30.00	3.5	4.5	5.5	6.5	7.0	8.0	8.5	9.0
36.00	3.5	4.5	5.5	6.5	7.5	8.0	9.0	9.5

Notes:
1. Insulation thickness is chosen either to prevent or minimize condensation on outside pipe surface or to limit heat gain to 8 Btu/h·ft^2, whichever thickness is greater.
2. All thicknesses are in inches.
3. Values do not include safety or aging factor. Actual operating conditions may vary. Consult a design engineer for appropriate recommendation for your specific system.
4. Data calculated using NAIMA 3E Plus program.

Table 5 Flexible Elastomeric Insulation Thickness for Indoor Design Conditions
(90°F Ambient Temperature, 80% Relative Humidity, 0.9 Emittance, 0 mph Wind Velocity)

Nominal Pipe Size, in.	Pipe Operating Temperature, °F							
	40	20	0	−20	−40	−60	−80	−100
0.50	1.0	1.0	1.5	1.5	2.0	2.0	2.0	2.0
0.75	1.0	1.0	1.5	2.0	2.0	2.0	2.5	2.5
1.00	1.0	1.0	1.5	2.0	2.0	2.0	2.5	2.5
1.50	1.0	1.0	1.5	2.0	2.0	2.5	2.5	3.0
2.00	1.0	1.0	2.0	2.0	2.0	2.5	3.0	3.0
2.50	1.0	1.5	2.0	2.0	2.5	2.5	3.0	3.0
3.00	1.0	1.5	2.0	2.0	2.5	2.5	3.0	3.0
4.00	1.0	1.5	2.0	2.5	2.5	3.0	3.0	3.0
5.00	1.5	1.5	2.0	2.5	2.5	3.0	3.5	3.5
6.00	1.5	2.0	2.0	2.5	3.0	3.0	3.5	3.5
8.00	1.5	2.0	2.0	2.5	3.0	3.0	3.5	3.5
10.00	1.5	2.0	2.0	2.5	3.0	3.5	3.5	3.5
12.00	1.5	2.0	2.0	2.5	3.0	3.5	4.0	4.0
14.00	1.5	2.0	2.5	2.5	3.0	3.5	4.0	4.0
16.00	1.5	2.0	2.5	2.5	3.5	3.5	4.0	4.0
18.00	1.5	2.0	2.5	2.5	3.5	3.5	4.0	4.5
20.00	1.5	2.0	2.5	3.0	3.5	3.5	4.0	4.5
24.00	1.5	2.0	2.5	3.0	3.5	4.0	4.0	4.5
28.00	1.5	2.0	2.5	3.0	3.5	4.0	4.0	4.5
30.00	1.5	2.0	2.5	3.0	3.5	4.0	4.0	4.5
36.00	1.5	2.0	2.5	3.0	3.5	4.0	4.5	4.5

Notes:
1. Insulation thickness is chosen either to prevent or minimize condensation on outside pipe surface or to limit heat gain to 8 Btu/h·ft^2, whichever thickness is greater.
2. All thicknesses are in inches.
3. Values do not include safety or aging factor. Actual operating conditions may vary. Consult a design engineer for appropriate recommendation for your specific system.
4. Data calculated using NAIMA 3E Plus program.

Table 6 Flexible Elastomeric Insulation Thickness for Outdoor Design Conditions
(100°F Ambient Temperature, 90% Relative Humidity, 0.4 Emittance, 7.5 mph Wind Velocity)

Nominal Pipe Size, in.	Pipe Operating Temperature, °F							
	40	20	0	−20	−40	−60	−80	−100
0.50	1.5	2.0	2.5	2.5	2.5	3.0	3.0	3.0
0.75	2.0	2.5	2.5	2.5	3.0	3.0	3.5	3.5
1.00	2.0	2.5	2.5	3.0	3.0	3.5	3.5	4.0
1.50	2.0	2.5	3.0	3.0	3.0	3.5	4.0	4.0
2.00	2.0	3.0	3.0	3.0	3.5	4.0	4.0	4.5
2.50	2.5	3.0	3.0	3.0	3.5	4.0	4.0	4.5
3.00	2.5	3.0	3.5	3.5	4.0	4.5	4.5	5.0
4.00	2.5	3.0	3.5	4.0	4.5	4.5	5.0	5.0
5.00	2.5	3.5	4.0	4.0	4.5	5.0	5.0	5.5
6.00	2.5	3.5	4.0	4.5	4.5	5.0	5.5	6.0
8.00	3.0	3.5	4.5	4.5	5.0	5.5	6.0	6.5
10.00	3.0	4.0	4.5	5.0	5.5	6.0	6.5	7.0
12.00	3.0	4.0	4.5	5.5	5.5	6.0	6.5	7.0
14.00	3.5	4.0	5.0	5.5	6.0	6.5	6.5	7.0
16.00	3.5	4.5	5.0	5.5	6.0	6.5	7.0	7.5
18.00	3.5	4.5	5.0	5.5	6.0	6.5	7.0	7.5
20.00	3.5	4.5	5.0	5.5	6.0	6.5	7.0	7.5
24.00	3.5	4.5	5.0	5.5	6.5	7.0	7.5	8.0
28.00	3.5	4.5	5.5	6.0	6.5	7.0	7.5	8.0
30.00	3.5	4.5	5.5	6.0	6.5	7.0	7.5	8.0
36.00	3.5	4.5	5.5	6.0	7.0	7.0	7.5	8.0

Notes:
1. Insulation thickness is chosen either to prevent or minimize condensation on outside pipe surface or to limit heat gain to 8 Btu/h·ft^2, whichever thickness is greater.
2. All thicknesses are in inches.
3. Values do not include safety or aging factor. Actual operating conditions may vary. Consult a design engineer for appropriate recommendation for your specific system.
4. Data calculated using NAIMA 3E Plus program.

Table 7 Closed-Cell Phenolic Foam Insulation Thickness for Indoor Design Conditions
(90°F Ambient Temperature, 80% Relative Humidity, 0.9 Emittance, 0 mph Wind Velocity)

Nominal Pipe Size, in.	Pipe Operating Temperature, °F 40	20	0	−20	−40	−60	−80	−100
0.50	1.0	1.0	1.0	1.0	1.5	1.5	1.5	1.5
0.75	1.0	1.0	1.0	1.5	1.5	1.5	1.5	1.5
1.00	1.0	1.0	1.0	1.5	1.5	1.5	1.5	1.5
1.50	1.0	1.0	1.0	1.5	1.5	1.5	1.5	1.5
2.00	1.0	1.0	1.0	1.5	1.5	1.5	1.5	1.5
2.50	1.0	1.0	1.0	1.5	1.5	1.5	1.5	1.5
3.00	1.0	1.0	1.0	1.5	1.5	2.0	2.0	2.0
4.00	1.0	1.0	1.5	1.5	1.5	2.0	2.0	2.0
5.00	1.0	1.0	1.5	1.5	1.5	2.0	2.0	2.5
6.00	1.0	1.0	1.5	1.5	2.0	2.0	2.0	2.5
8.00	1.0	1.0	1.5	1.5	2.0	2.0	2.0	2.5
10.00	1.0	1.0	1.5	1.5	2.0	2.0	2.0	2.5
12.00	1.0	1.0	1.5	1.5	2.0	2.0	2.0	2.5
14.00	1.0	1.0	1.5	1.5	2.0	2.0	2.0	2.5
16.00	1.0	1.0	1.5	1.5	2.0	2.0	2.5	2.5
18.00	1.0	1.0	1.5	1.5	2.0	2.0	2.5	2.5
20.00	1.0	1.0	1.5	1.5	2.0	2.0	2.5	2.5
24.00	1.0	1.0	1.5	1.5	2.0	2.0	2.5	2.5
28.00	1.0	1.0	1.5	1.5	2.0	2.0	2.5	2.5
30.00	1.0	1.0	1.5	1.5	2.0	2.0	2.5	2.5
36.00	1.0	1.0	1.5	2.0	2.0	2.0	2.5	2.5

Notes:
1. Insulation thickness is chosen either to prevent or minimize condensation on outside pipe surface or to limit heat gain to 8 Btu/h·ft^2, whichever thickness is greater.
2. All thicknesses are in inches.
3. Values do not include safety or aging factor. Actual operating conditions may vary. Consult a design engineer for appropriate recommendation for your specific system.
4. Data calculated using NAIMA 3E Plus program.

Table 8 Closed-Cell Phenolic Foam Insulation Thickness for Outdoor Design Conditions
(100°F Ambient Temperature, 90% Relative Humidity, 0.4 Emittance, 7.5 mph Wind Velocity)

Nominal Pipe Size, in.	Pipe Operating Temperature, °F 40	20	0	−20	−40	−60	−80	−100
0.50	1.0	1.0	1.5	1.5	1.5	2.0	2.0	2.0
0.75	1.0	1.5	1.5	1.5	2.0	2.0	2.0	2.5
1.00	1.0	1.5	1.5	1.5	2.0	2.0	2.0	2.5
1.50	1.0	1.5	1.5	1.5	2.0	2.0	2.0	2.5
2.00	1.0	1.5	1.5	1.5	2.0	2.0	2.5	2.5
2.50	1.0	1.5	1.5	1.5	2.0	2.0	2.5	2.5
3.00	1.0	1.5	2.0	2.0	2.5	2.5	3.0	3.0
4.00	1.5	1.5	2.0	2.5	2.5	3.0	3.0	3.0
5.00	1.5	2.0	2.0	2.5	2.5	3.0	3.0	3.5
6.00	1.5	2.0	2.0	2.5	3.0	3.0	3.5	3.5
8.00	1.5	2.0	2.5	2.5	3.0	3.0	3.5	4.0
10.00	1.5	2.0	2.5	2.5	3.0	3.5	3.5	4.0
12.00	1.5	2.0	2.5	3.0	3.0	3.5	4.0	4.0
14.00	1.5	2.0	2.5	3.0	3.5	3.5	4.0	4.5
16.00	1.5	2.0	2.5	3.0	3.5	4.0	4.0	4.5
18.00	1.5	2.5	2.5	3.0	3.5	4.0	4.0	4.5
20.00	2.0	2.5	2.5	3.0	3.5	4.0	4.0	4.5
24.00	2.0	2.5	3.0	3.0	3.5	4.0	4.5	5.0
28.00	2.0	2.5	3.0	3.0	3.5	4.0	4.5	5.0
30.00	2.0	2.5	3.0	3.5	3.5	4.0	4.5	5.0
36.00	2.0	2.5	3.0	3.5	3.5	4.0	4.5	5.0

Notes:
1. Insulation thickness is chosen either to prevent or minimize condensation on outside pipe surface or to limit heat gain to 8 Btu/h·ft^2, whichever thickness is greater.
2. All thicknesses are in inches.
3. Values do not include safety or aging factor. Actual operating conditions may vary. Consult a design engineer for appropriate recommendation for your specific system.
4. Data calculated using NAIMA 3E Plus program.

Table 9 Polyisocyanurate Foam Insulation Thickness for Indoor Design Conditions
(90°F Ambient Temperature, 80% Relative Humidity, 0.9 Emittance, 0 mph Wind Velocity)

Nominal Pipe Size, in.	Pipe Operating Temperature, °F 40	20	0	−20	−40	−60	−80	−100
0.50	1.0	1.0	1.5	1.5	1.5	1.5	2.0	2.0
0.75	1.0	1.0	1.5	1.5	1.5	2.0	2.0	2.0
1.00	1.0	1.0	1.5	1.5	1.5	2.0	2.0	2.0
1.50	1.0	1.0	1.5	1.5	1.5	2.0	2.0	2.0
2.00	1.0	1.0	1.5	1.5	1.5	2.0	2.0	2.5
2.50	1.0	1.0	1.5	1.5	1.5	2.0	2.0	2.5
3.00	1.0	1.0	1.5	1.5	2.0	2.5	2.5	2.5
4.00	1.0	1.0	1.5	1.5	2.0	2.5	2.5	3.0
5.00	1.0	1.5	1.5	2.0	2.0	2.5	2.5	3.0
6.00	1.0	1.5	1.5	2.0	2.0	2.5	2.5	3.0
8.00	1.0	1.5	1.5	2.0	2.0	2.5	2.5	3.0
10.00	1.0	1.5	1.5	2.0	2.0	3.0	3.0	3.5
12.00	1.0	1.5	1.5	2.0	2.5	3.0	3.0	3.5
14.00	1.0	1.5	1.5	2.0	2.5	3.0	3.0	3.5
16.00	1.0	1.5	2.0	2.0	2.5	3.0	3.0	3.5
18.00	1.0	1.5	2.0	2.0	2.5	3.0	3.5	3.5
20.00	1.0	1.5	2.0	2.0	2.5	3.0	3.5	3.5
24.00	1.0	1.5	2.0	2.0	2.5	3.0	3.5	3.5
28.00	1.0	1.5	2.0	2.0	2.5	3.0	3.5	4.0
30.00	1.0	1.5	2.0	2.0	2.5	3.0	3.5	4.0
36.00	1.0	1.5	2.0	2.0	2.5	3.0	3.5	4.0

Notes:
1. Insulation thickness is chosen either to prevent or minimize condensation on outside pipe surface or to limit heat gain to 8 Btu/h·ft^2, whichever thickness is greater.
2. All thicknesses are in inches.
3. Values do not include safety or aging factor. Actual operating conditions may vary. Consult a design engineer for appropriate recommendation for your specific system.
4. Data calculated using NAIMA 3E Plus program.

Table 10 Polyisocyanurate Foam Insulation Thickness for Outdoor Design Conditions
(100°F Ambient Temperature, 90% Relative Humidity, 0.4 Emittance, 7.5 mph Wind Velocity)

Nominal Pipe Size, in.	Pipe Operating Temperature, °F 40	20	0	−20	−40	−60	−80	−100
0.50	1.0	1.5	1.5	2.0	2.0	2.5	2.5	2.5
0.75	1.0	1.5	2.0	2.0	2.5	2.5	2.5	3.0
1.00	1.0	1.5	2.0	2.0	2.5	2.5	3.0	3.5
1.50	1.5	1.5	2.0	2.0	2.5	2.5	3.0	3.5
2.00	1.5	1.5	2.0	2.5	3.0	3.0	3.5	4.0
2.50	1.5	1.5	2.0	2.5	3.0	3.0	3.5	4.0
3.00	1.5	2.0	2.5	3.0	3.0	3.5	4.0	4.5
4.00	1.5	2.0	2.5	3.0	3.5	3.5	4.0	4.5
5.00	1.5	2.0	2.5	3.0	3.5	4.0	4.5	5.0
6.00	2.0	2.5	3.0	3.0	3.5	4.0	4.5	5.0
8.00	2.0	2.5	3.0	3.5	4.0	4.5	5.0	5.5
10.00	2.0	2.5	3.0	3.5	4.0	4.5	5.0	6.0
12.00	2.0	2.5	3.0	3.5	4.5	5.0	5.5	6.0
14.00	2.0	2.5	3.5	4.0	4.5	5.0	5.5	6.0
16.00	2.0	3.0	3.5	4.0	4.5	5.0	6.0	6.5
18.00	2.0	3.0	3.5	4.0	4.5	5.5	6.0	6.5
20.00	2.0	3.0	3.5	4.0	4.5	5.5	6.0	6.5
24.00	2.0	3.0	3.5	4.0	5.0	5.5	6.0	7.0
28.00	2.0	3.0	3.5	4.0	5.0	5.5	6.0	7.0
30.00	2.5	3.0	3.5	4.0	5.0	5.5	6.5	7.0
36.00	2.5	3.0	3.5	4.0	5.0	5.5	6.5	7.0

Notes:
1. Insulation thickness is chosen either to prevent or minimize condensation on outside pipe surface or to limit heat gain to 8 Btu/h·ft^2, whichever thickness is greater.
2. All thicknesses are in inches.
3. Values do not include safety or aging factor. Actual operating conditions may vary. Consult a design engineer for appropriate recommendation for your specific system.
4. Data calculated using NAIMA 3E Plus program.

Table 11 Polystyrene Foam Insulation Thickness for Indoor Design Conditions

(90°F Ambient Temperature, 80% Relative Humidity, 0.9 Emittance, 0 mph Wind Velocity)

Nominal Pipe Size, in.	Pipe Operating Temperature, °F							
	40	20	0	−20	−40	−60	−80	−100
0.50	1.0	1.5	1.5	2.0	2.0	2.0	2.5	2.5
0.75	1.5	1.5	1.5	2.0	2.0	2.5	2.5	2.5
1.00	1.5	1.5	1.5	2.0	2.0	2.5	2.5	2.5
1.50	1.5	1.5	2.0	2.0	2.0	2.5	2.5	2.5
2.00	1.5	1.5	2.0	2.0	2.5	2.5	2.5	3.0
2.50	1.5	1.5	2.0	2.0	2.5	2.5	2.5	3.0
3.00	1.5	2.0	2.0	2.5	2.5	3.0	3.0	3.5
4.00	1.5	2.0	2.0	2.5	3.0	3.0	3.0	3.5
5.00	1.5	2.0	2.5	2.5	3.0	3.0	3.5	3.5
6.00	1.5	2.0	2.5	2.5	3.0	3.5	3.5	3.5
8.00	1.5	2.0	2.5	2.5	3.0	3.5	3.5	4.0
10.00	1.5	2.0	2.5	3.0	3.0	3.5	4.0	4.0
12.00	1.5	2.0	2.5	3.0	3.5	3.5	4.0	4.0
14.00	1.5	2.0	2.5	3.0	3.5	4.0	4.0	4.0
16.00	1.5	2.0	2.5	3.0	3.5	4.0	4.0	4.5
18.00	1.5	2.0	2.5	3.0	3.5	4.0	4.0	4.5
20.00	1.5	2.5	3.0	3.0	3.5	4.0	4.0	4.5
24.00	1.5	2.5	3.0	3.5	3.5	4.0	4.0	4.5
28.00	1.5	2.5	3.0	3.5	3.5	4.0	4.5	4.5
30.00	1.5	2.5	3.0	3.5	3.5	4.0	4.5	4.5
36.00	2.0	2.5	3.0	3.5	4.0	4.0	4.5	5.0

Notes:
1. Insulation thickness is chosen either to prevent or minimize condensation on outside pipe surface or to limit heat gain to 8 Btu/h·ft^2, whichever thickness is greater.
2. All thicknesses are in inches.
3. Values do not include safety or aging factor. Actual operating conditions may vary. Consult a design engineer for appropriate recommendation for your specific system.
4. Data calculated using NAIMA 3E Plus program.

Table 12 Polystyrene Foam Insulation Thickness for Outdoor Design Conditions

(100°F Ambient Temperature, 90% Relative Humidity, 0.4 Emittance, 7.5 mph Wind Velocity)

Nominal Pipe Size, in.	Pipe Operating Temperature, °F							
	40	20	0	−20	−40	−60	−80	−100
0.50	1.5	2.0	2.5	2.5	2.5	3.0	3.0	3.0
0.75	1.5	2.0	2.5	2.5	3.0	3.0	3.5	3.5
1.00	1.5	2.0	2.5	3.0	3.0	3.5	3.5	4.0
1.50	2.0	2.0	2.5	3.0	3.0	3.5	4.0	4.0
2.00	2.0	2.5	3.0	3.0	3.5	4.0	4.0	4.5
2.50	2.0	2.5	3.0	3.0	3.5	4.0	4.0	4.5
3.00	2.5	3.0	3.5	3.5	4.0	4.5	4.5	5.0
4.00	2.5	3.0	3.5	4.0	4.5	4.5	5.0	5.0
5.00	2.5	3.0	3.5	4.0	4.5	5.0	5.0	5.5
6.00	2.5	3.5	3.5	4.5	4.5	5.0	5.5	6.0
8.00	2.5	3.0	4.5	4.5	5.0	5.5	6.0	6.5
10.00	3.0	3.5	4.5	5.0	5.5	6.0	6.5	7.0
12.00	3.0	3.5	4.5	5.0	5.5	6.0	6.5	7.0
14.00	3.0	4.0	4.5	5.5	6.0	6.5	6.5	7.0
16.00	3.0	4.0	5.0	5.5	6.0	6.5	7.0	7.5
18.00	3.5	4.0	5.0	5.5	6.0	6.5	7.0	7.5
20.00	3.5	4.0	5.0	5.5	6.0	6.5	7.0	7.5
24.00	3.5	4.0	5.0	5.5	6.5	7.0	7.5	8.0
28.00	3.5	4.0	5.0	6.0	6.5	7.0	7.5	8.0
30.00	3.5	4.0	5.0	6.0	6.5	7.0	7.5	8.0
36.00	3.5	4.5	5.0	6.0	6.5	7.0	7.5	8.0

Notes:
1. Insulation thickness is chosen either to prevent or minimize condensation on outside pipe surface or to limit heat gain to 8 Btu/h·ft^2, whichever thickness is greater.
2. All thicknesses are in inches.
3. Values do not include safety or aging factor. Actual operating conditions may vary. Consult a design engineer for appropriate recommendation for your specific system.
4. Data calculated using NAIMA 3E Plus program.

Weather Barrier Jacketing

Weather barrier jacketing on insulated pipes and vessels protects the vapor retarder system and insulation. Various plastic and metallic products are available for this purpose. Some specifications suggest that the jacketing should preserve and protect the sometimes fragile vapor retarder over the insulation. This being the case, bands must be used to secure the jacket. Pop rivets, sheet metal screws, staples, or any other items that puncture should not be used because they will compromise the vapor retarder system. Use of such materials may indicate that the installer does not understand the vapor retarder concept, and corrective education steps should be taken.

Protective jacketing is designed to be installed over the vapor retarder and insulation to prevent weather and abrasion damage. The protective jacketing must be installed independently and in addition to any factory- or field-applied vapor retarder. Ambient-temperature cycling causes the jacketing to expand and contract. The manufacturer's instructions should show how to install the jacketing to permit this expansion and contraction.

Metal jacketing may be smooth, textured, embossed, or corrugated aluminum or stainless steel with a continuous moisture retarder. Metallic jackets are recommended for exposed, roof-mounted piping.

Protective jacketing is required whenever piping is exposed to washing, physical abuse, or traffic. White PVC (0.03 in. thick) is popular inside buildings where degradation from sunlight is not a factor. Colors can be obtained at little, if any, additional cost. All longitudinal and circumferential laps should be seal-welded using a solvent welding adhesive. Laps should be located at the ten o'clock or two o'clock positions. A sliding lap (PVC) expansion/contraction joint should be located near each endpoint and at intermediate joints no more than 20 ft apart. Where very heavy abuse and/or hot, scalding washdowns are encountered, a CPVC material is required. These materials can withstand temperatures as high as 225°F, whereas standard PVC will warp and disfigure at 140°F.

Roof piping should be jacketed with a minimum 0.016 in. aluminum (embossed or smooth finish depending on aesthetic choice). On pitched lines, this jacketing should be installed with a minimum 2 in. overlap arranged to shed any water in the direction of the pitch. Only stainless steel bands should be used to install this jacketing (1/2 in. wide by 0.02 in. thick 304 stainless) and spaced every 12 in. Jacketing on valves and fittings should match that of the adjacent piping.

INSTALLATION GUIDELINES

Preliminary Preparation. Corrosion of any metal under any thermal insulation can occur for many reasons. With any insulation, the pipe can be primed to minimize the potential for corrosion. Before installing insulation,

- Complete all welding and other hot work.
- Complete hydrostatic and other performance testing.
- Remove oil, grease, loose scale, rust, and foreign matter from surfaces to be insulated. Surface must also be dry and free from frost.
- Complete site touch-up of all shop coating, including preparation and painting at field welds. (*Note*: Do not use varnish on welds of ammonia systems.)

Insulating Fittings and Joints. Insulation for fittings, flanges, and valves should be the same thickness as for the pipe and must be fully vapor-sealed. The following guidelines also apply:

- If valve design allows, valves should be insulated to the packing glands.
- Stiffener rings, where provided on vacuum equipment and/or piping, should be insulated with the same thickness and type of

insulation as specified for that piece of equipment or line. Rings should be fully independently insulated.

- Where multiple layers of insulation are used, all joints should be staggered or beveled where appropriate.
- Insulation should be applied with all joints fitted to eliminate voids. Large voids should not be filled with vapor sealant or fibrous insulation, but eliminated by refitting or replacing the insulation.
- All joints, except for contraction joints and the inner layer of a double-layer system, should be sealed with either the proper adhesive or a joint sealer during installation.
- Each line should be insulated as a single unit. Adjacent lines must not be enclosed within a common insulation cover.

Planning Work. Insulations require special protection during storage and installation to avoid physical abuse and to keep them clean and dry. All insulation applied in one day should also have the vapor barrier installed. When specified, at least one coat of vapor retarder mastic should be applied the same day. If applying the first coat is impractical, the insulation must be temporarily protected with a moisture retarder, such as an appropriate polyethylene film, and sealed to the pipe or equipment surface. All exposed insulation terminations should be protected before work ends for the day.

Vapor Stops. Vapor stops should be installed using either sealant or the appropriate adhesive at all directly attached pipe supports, guides, anchors, and at all locations requiring potential maintenance, such as valves, flanges, and instrumentation connections to piping or equipment. If valves or flanges must be left uninsulated until after plant start-up, temporary vapor stops should be installed using either sealant or the appropriate adhesive approximately every 10 ft on straight runs.

Securing Insulation. When applicable, the innermost layer of insulation should be applied in two half-sections and secured with 3/4 in. wide pressure-sensitive filament tape banding spaced a maximum of 9 in. apart and applied with a 50% overlap. Single and outer layers more than 18 in. in diameter and inner layers with radiused and beveled segments should be secured by 3/8 in. wide stainless steel bands spaced on 9 in. maximum centers. Bands must be firmly tensioned and sealed.

Applying Vapor Retarder Coating and Mastic. *First coat*: Irregular surfaces and fittings should be vapor-sealed by applying a thin coat of vapor retarder mastic or finish with a minimum wet-film thickness as recommended by the manufacturer. While the mastic or finish is still tacky, an open-weave glass fiber reinforcing mesh should be laid smoothly into the mastic or finish and should be thoroughly embedded in the coating. Care should be taken not to rupture the weave. The fabric should be overlapped a minimum of 2 in. at joints to provide strength equal to that maintained elsewhere.

Second coat: Before the first coat is completely dry, a second coat should be applied over the glass fiber reinforcing mesh with a smooth, unbroken surface. The total thickness of mastic or finish should be in accordance with the coating manufacturer's recommendation.

Pipe Supports and Hangers. When possible, the pipe hanger or support should be located outside of the insulation. Supporting the pipe outside of the protective jacketing eliminates the need to insulate over the pipe clamp, hanger rods, or other attached support components. This method minimizes the potential for vapor intrusion and thermal bridges because a continuous envelope surrounds the pipe.

ASME *Standard* B31 establishes basic stress allowances for piping material. Loading on the insulation material is a function of its compressive strength. Table 13 suggests spacing for pipe supports. Related information is also in Chapter 41 of the 2004 *ASHRAE Handbook—HVAC Systems and Equipment*.

Insulation material may or may not have the compressive strength to support loading at these distances. Therefore, force

Table 13 Suggested Pipe Support Spacing for Straight Horizontal Runs

Nominal Pipe OD, in.	Standard Steel Pipe[a, b]	Copper Tube
	Support Spacing, ft	
1/2	6	5
3/4	6	5
1	6	6
1 1/2	10	8
2	10	8
2 1/2	11	9
3	12	10
4	14	12
6	16	—
8	16	—
10	16	—
12	16	—
14	16	—
16	16	—
18	16	—
20	16	—
24	16	—

Source: Adapted from MSS *Standard* SP-69 and ASME *Standard* B31.1
[a] Spacing does not apply where span calculations are made or where concentrated loads are placed between supports such as flanges, valves, specialties, etc.
[b] Suggested maximum spacing between pipe supports for horizontal straight runs of standard and heavier pipe.

from the piping and contents on the bearing area of the insulation should be calculated. In refrigerant piping, bands or clevis hangers typically are used with rolled metal shields or cradles between the band or hanger and the insulation. Although the shields are typically rolled to wrap the outer diameter of the insulation in a 180° arc, the bearing area is calculated over a 120° arc of the outer circumference of the insulation multiplied by the shield length. If the insulated pipe is subjected to point loading, such as where it rests on a beam or a roller, the bearing area arc is reduced to 60° and multiplied by the shield length. In this case, rolled plate may be more suitable than sheet metal. Provisions should be made to secure the shield on both sides of the hanger (metal band), and the shield should be centered in the support. Table 14 lists widths and thicknesses for pipe shields.

Expansion Joints. Some installations require an expansion or contraction joint. These joints are normally required in the innermost layer of insulation, and may be constructed in the following manner:

1. Make a 1 in. break in insulation.
2. Tightly pack break with fibrous insulation material.
3. Secure insulation on either side of joint with stainless steel bands that have been hand-tightened.
4. Cover joint with appropriate vapor retarder and seal properly.

The presence and spacing of expansion/contraction joints is an important design issue in insulation systems used on refrigerant piping. Spacing may be calculated using the following equation:

$$S = \frac{L}{\left[\left(|T_i - T_o| \times |\alpha_i - \alpha_p| \times \frac{L}{d}\right) + 1\right]}$$

where

S = worst-case maximum spacing of contraction joints, ft
T_i = temperature during insulation installation, °F
T_o = coldest service temperature of pipe, °F
α_i = coefficient of linear thermal expansion (COLTE) of insulation material, in/ft·°F
α_p = COLTE of the pipe material, in/ft·°F

Table 14 Shield Dimensions for Insulated Pipe and Tubing

Insulation Diameter, in.	Shield Thickness, gage (in.)	Shield Arc Length, in.	Shield Length, in.	Shield Radius, in.
2.5	20 (0.036)	2.5	12	1.25
3	20 (0.036)	3	12	1.5
3.5	18 (0.048)	3.5	12	1.75
4	18 (0.048)	4	12	2
4.5	18 (0.048)	5	12	2.25
5	16 (0.060)	5.5	12	2.5
6	16 (0.060)	6.5	12	3
8	16 (0.060)	8.5	18	4
10	14 (0.075)	10.5	18	5
12	14 (0.075)	12.5	18	6
14	14 (0.075)	14.5	18	7
16	12 (0.105)	19	18	9
20	12 (0.105)	21	18	10
22	12 (0.105)	23	18	11
24	12 (0.105)	25	18	12
26	12 (0.105)	27	18	13
28	12 (0.105)	29.5	18	14
30	12 (0.105)	31.5	18	15

Source: Adapted from IIAR *Ammonia Refrigeration Handbook*.
Note: Protection shield gages listed are for use with band-type hangers only. For point loading, increase shield thickness and length.

Table 15 COLTE Values for Various Materials

Material	COLTE,[a] in/ft·°F
Pipe	
Carbon steel	6.78×10^{-5}
Stainless steel	10.5×10^{-5}
Aluminum	13.5×10^{-5}
Ductile iron	6.1×10^{-5}
Copper[b]	11.3×10^{-5}
Insulation	
Cellular glass	4.0×10^{-5}
Flexible elastomeric	N/A
Closed-cell phenolic	34×10^{-5}
Polyisocyanurate	60×10^{-5}
Polystyrene	42×10^{-5}

[a]Mean COLTE between 70 and –100°F from *Perry's Chemical Engineer's Handbook*, 7th ed., Table 10-52.
[b]COLTE between 68 and 212°F from *Perry's Chemical Engineer's Handbook,* 7th ed., Table 28-4.

L = pipe length, ft
d = amount of expansion or contraction that can be absorbed by each insulation contraction joint, in.

Table 15 provides COLTEs for various pipe and insulation materials. The values can be used in this equation as α_i and α_p.

MAINTENANCE OF INSULATION SYSTEMS

Periodic inspections of refrigerant piping systems are needed to determine the presence of moisture, which degrades an insulation system's thermal efficiency and shortens its service life. The frequency of inspection should be determined by the critical nature of the process, external environment, and age of the insulation. A *routine* inspection should include the following checks:

- Look for signs of moisture or ice on lower part of horizontal pipe, at bottom elbow of a vertical pipe, and around pipe hangers and saddles (moisture may migrate to low areas).
- Look for mechanical damage and jacketing penetrations, openings, or separations.
- Check jacketing to determine whether banding is loose.
- Look for bead caulking failure, especially around flange and valve covers.
- Look for loss of jacketing integrity and for open seams around all intersecting points, such as pipe transitions, branches, and tees.
- Look for cloth visible through mastic or finish if pipe is protected by a reinforced mastic weather barrier.

An *extensive* inspection should also include the following:

- Use thermographic equipment to isolate areas of concern.
- Design a method to repair, close, and seal any cut in insulation or vapor retarder to maintain a positive seal throughout the entire system.
- Examine pipe surface for corrosion if insulation is wet.

The extent of moisture present in the insulation system and/or the corrosion of the pipe determines the need to replace the insulation. All wet parts of the insulation must be replaced.

REFERENCES

ASME. 2001. Power piping. *Standard* B31.1-2001. American Society of Mechanical Engineers, New York.

ASTM. 2003. Specification for seamless copper tube for air conditioning and refrigeration field service. *Standard* B280-03. American Society for Testing and Materials, West Conshohocken, PA.

ASTM. 2004. Test method for steady-state heat flux measurements and thermal transmission properties by means of the guarded hot-plate apparatus. *Standard* C177-04. American Society for Testing and Materials, West Conshohocken, PA.

ASTM. 2005. Test method for steady-state heat transfer properties of pipe insulation. *Standard* C335. American Society for Testing and Materials, West Conshohocken, PA.

ASTM. 2002. Practice for fabrication of thermal insulating fitting covers for NPS piping, and vessel lagging. *Standard* C450-02. American Society for Testing and Materials, West Conshohocken, PA.

ASTM. 2004. Test method for steady-state thermal transmission properties by means of the heat flow meter apparatus. *Standard* C518-04. American Society for Testing and Materials, West Conshohocken, PA.

ASTM. 2005. Specification for preformed flexible elastomeric cellular thermal insulation in sheet and tubular form. *Standard* C534-05. American Society for Testing and Materials, West Conshohocken, PA.

ASTM. 2003. Specification for cellular glass thermal insulation. *Standard* C552-03. American Society for Testing and Materials, West Conshohocken, PA.

ASTM. 2005. Specification for rigid, cellular polystyrene thermal insulation. *Standard* C578-05a. American Society for Testing and Materials, West Conshohocken, PA.

ASTM. 1990. Practice for inner and outer diameters of rigid thermal insulation for nominal sizes of pipe and tubing (NPS System). *Standard* C585-90(2004). American Society for Testing and Materials, West Conshohocken, PA.

ASTM. 2005. Specification for unfaced preformed rigid cellular polyisocyanurate thermal insulation. *Standard* C591-05. American Society for Testing and Materials, West Conshohocken, PA.

ASTM. 2004. Specification for faced or unfaced rigid cellular phenolic thermal insulation. *Standard* C1126-04. American Society for Testing and Materials, West Conshohocken, PA.

ASTM. 2005. Test method for surface burning characteristics of building materials. *Standard* E84-05e1. American Society for Testing and Materials, West Conshohocken, PA.

ASTM. 2005. Test methods for water vapor transmission of materials. *Standard* E96/E96M-05. American Society for Testing and Materials, West Conshohocken, PA.

IIAR. 2000. *Ammonia refrigeration piping handbook*. International Institute of Ammonia Refrigeration, Arlington, VA.

MIL-P-24441. *General specification for paint, epoxy-polyamide*. Naval Publications and Forms Center, Philadelphia, PA.

MSS. 2003. Pipe hangers and supports—Selection and application. *Standard* SP-69-2003. Manufacturers Standardization Society of the Valve and Fittings Industry, Inc., Vienna, VA.

NACE. 1999. Near-white metal blast cleaning. *Standard* 2/SSPC-SP10. National Association of Corrosion Engineers International, Houston, and Steel Structures Painting Council, Pittsburgh.

Perry, R.H. and D.W. Green. 1997. *Perry's chemical engineer's handbook*, 7th ed. McGraw-Hill.

SofTech[2]. 1996. NAIMA 3E Plus. Grand Junction, CO.

BIBLIOGRAPHY

Hedlin, C.P. 1977. Moisture gains by foam plastic roof insulations under controlled temperature gradients. *Journal of Cellular Plastics* (Sept./ Oct.):313-326.

Lenox, R.S. and P.A. Hough. 1995. Minimizing corrosion of copper tubing used in refrigeration systems. *ASHRAE Journal* 37:11.

Kumaran, M.K. 1989. Vapor transport characteristics of mineral fiber insulation from heat flow meter measurements. In ASTM STP 1039, *Water vapor transmission through building materials and systems: Mechanisms and measurement*, pp. 19-27. American Society for Testing and Materials, West Conshohocken, PA.

Kumaran, M.K., M. Bomberg, N.V. Schwartz. 1989. Water vapor transmission and moisture accumulation in polyurethane and polyisocyanurate foams. In ASTM STP 1039, *Water vapor transmission through building materials and systems: Mechanisms and measurement*, pp. 63-72. American Society for Testing and Materials, West Conshohocken, PA.

Malloy, J.F. 1969. *Thermal insulation*. Van Nostrand Reinhold, New York.

NACE. 1997. *Corrosion under insulation*. National Association of Corrosion Engineers International, Houston.

CHAPTER 34

ICE MANUFACTURE

MOST commercial ice production is done with ice makers that produce three basic types of fragmentary ice (flake, tubular, and plate), which vary according to the type and size required for a particular application. Among the many applications for manufactured ice are

- Processing: fish, meat, poultry, dairy, bakery products, and hydrocooling
- Storage and transportation: fish, meat, poultry, and dairy products
- Manufacturing: chemicals and pharmaceuticals
- Others: retail consumer ice, concrete mixing and curing, and off-peak thermal storage

ICE MAKERS

Flake Ice

Flake ice is produced by applying water to the inside or outside of a refrigerated drum or to the outside of a refrigerated disk. The drum is either vertical or horizontal and may be stationary or fixed. The disk is vertical and rotates about a horizontal axis.

Ice removal devices fracture the thin layer of ice produced on the freezing surface of the ice maker, breaking it free from the freezing surface and allowing it to fall into an ice bin, which is generally located below the ice maker.

Thickness of ice produced by flake ice machines can be varied by adjusting the speed of the rotating part of the machine, varying evaporator temperature, or regulating water flow on the freezing surface. Flake ice is produced continuously, unlike tubular and plate ice, which are produced in an intermittent cycle or harvest operation. The resulting thickness ranges from 0.04 to 0.18 in. Continuous operation (without a harvest cycle) requires less refrigeration capacity to produce a ton of ice than any other type of ice manufacture with similar makeup water and evaporating temperatures. The exact amount of refrigeration required varies by machine type and design. Typical flake ice machines are shown in Figures 1 and 2.

All water used by flake ice machines is converted into ice; therefore, there is no waste or spillage. Flake ice makers usually operate at a lower evaporating temperature than tube or plate ice makers, and the ice is colder when it is removed from the ice-making surface. The surface of flake ice is not wetted by thawing during removal from the freezing surface, as is common with other types of ice. Because it is produced at a colder temperature, flake ice is most adaptable to automated storage, particularly when low-temperature ice is desired.

Rapid freezing of water on the freezing surface entrains air in the flake ice, giving it an opaque appearance. For this reason, flake ice is not commonly used for applications where clear ice is important. Where rapid cooling is important, such as in chemical processing and concrete cooling, flake ice is ideal because the flakes present the maximum amount of cooling surface for a given amount of ice.

When used as ingredient ice in sausage making or other food grinding and mixing, flake ice provides rapid cooling while minimizing mechanical damage to other ingredients and wear on mixing/cutting blades.

Some flake ice machines can produce salty ice from seawater. These are particularly useful in shipboard applications. Other flake ice machines require adding trace amounts of salt to the makeup water to enhance the release of ice from the refrigerated surface. In

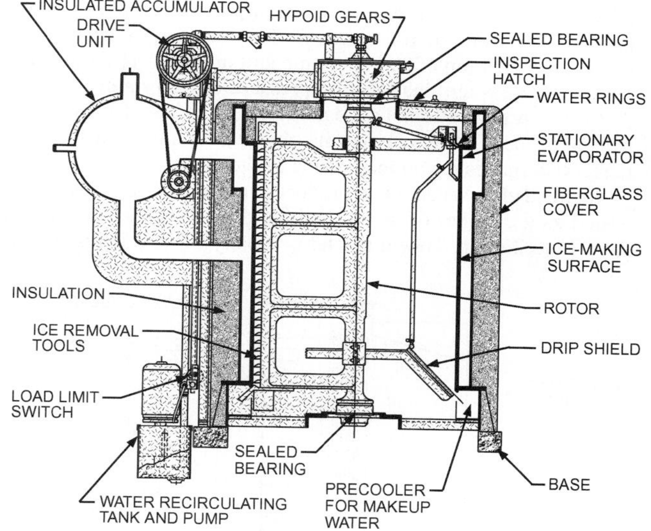

Fig. 1 Flake Ice Maker

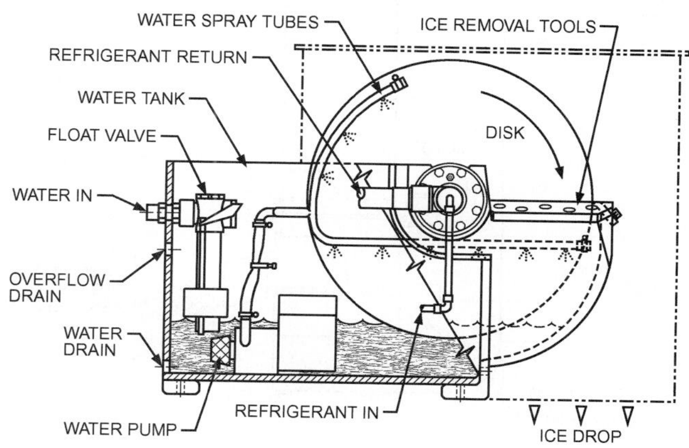

Fig. 2 Disk Flake Ice Maker

The preparation of this chapter is assigned to TC 10.2, Automatic Icemaking Plants and Skating Rinks.

rare cases, the presence of salt in the finished product may be objectionable.

Tubular Ice

Tubular ice is produced by freezing a falling film of water either on the outside of a tube with evaporating refrigerant on the inside, or on the inside of tubes surrounded by evaporating refrigerant on the outside.

Outside Tube. When ice is produced on the outside of a tube, the freezing cycle is normally 8 to 15 min, with the final ice thickness 0.2 to over 0.5 in., following the tube's curvature. The refrigerant temperature inside the tube continually drops from an initial suction temperature of about 25°F to the terminal suction temperature in the range of 10 to –15°F. At the end of the freezing cycle, the circulating water is shut off, and hot discharge gas is introduced to harvest the ice. To maintain proper harvest temperatures, typical discharge gas pressure is 160 psia. This drives the liquid refrigerant in the tube up into an accumulator and melts the inside of the tube of ice, which slides down through a sizer and mechanical breaker, and finally down into storage. The defrost cycle is normally about 30 s. The unit returns to the freezing cycle by returning the liquid refrigerant to the tube from the accumulator.

This type of ice maker operates with R-717, R-404A, R-507, and R-22. R-12 may be found in some older units. Higher-capacity units of 10 tons per 24 h and larger usually use R-717. Unit capacity increases as terminal suction pressure decreases. A typical unit with 70°F makeup water and R-717 as the refrigerant produces 19.3 tons of ice per 24 h with a terminal suction pressure of 38.5 psia and requires 35.7 tons of refrigeration. This equates to 1.85 tons of refrigeration per ton of ice. The same unit produces 41.6 tons of ice per 24 h with a terminal suction pressure of 21 psia and requires 80 tons of refrigeration. This equates to 1.92 tons of refrigeration per ton of ice. Figure 3 shows the physical arrangement for an ice maker that makes ice on the outside of the tubes.

Inside Tube. When ice is produced inside a tube, it can be harvested as a cylinder or as crushed ice. The freezing cycle ranges from 13 to 26 min. Tube diameter is usually 0.9 to 2 in., producing a cylinder that can be cut to desired lengths. The refrigerant temperature outside the tube is continually dropping, with an initial temperature of 25°F and a terminal suction temperature ranging from 20 to –5°F. At the end of the freezing cycle, the circulating water is shut off and ice is harvested by introducing hot discharge gas into the refrigerant in the freezing section. To maintain gas temperature, typical discharge gas pressure is 180 psia. This releases the ice from the tube; the ice descends to a motor-driven cutter plate that can be adjusted to cut ice cylinders to the length desired (up to 1.5 in.). At the end of the defrost cycle, the discharge gas valve is closed and water circulation resumes.

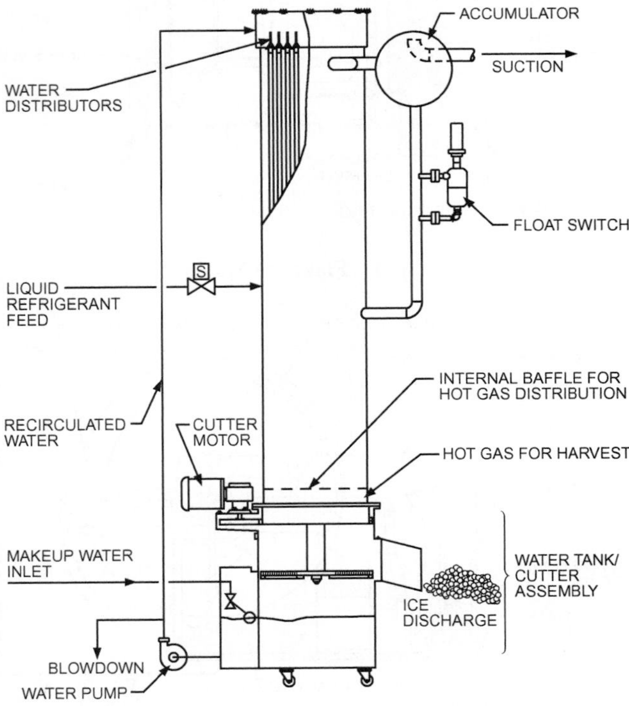

Fig. 3 Tubular Ice Maker

These units can use refrigerants R-717 and R-22; R-12 may be found in older units. Again, capacity increases as terminal suction pressure decreases. A typical unit with 70°F makeup water and R-717 as the refrigerant produces 43 tons of ice per 24 h with a terminal suction pressure of 40 psia and requires 74.5 tons of refrigeration. This equates to 1.73 tons of refrigeration per ton of ice. The same unit produces 66 tons of ice per 24 h with a terminal suction pressure of 30 psia and requires 135 tons of refrigeration. This equates to 2.04 tons of refrigeration per ton of ice.

Tubular ice makers are advantageous because they produce ice at higher suction pressures than other types of ice makers. They can make relatively thick and clear ice, with curvatures that help prevent bridging in storage. Tubular ice makers have a greater height requirement for installation than do plate or flake ice makers, but a smaller footprint. Provision must be made in the refrigeration system high side to accommodate the volume of refrigerant required for the proper amount of harvest discharge gas. Ice temperatures are generally higher than with flake ice makers.

Supply Water. Supply water temperature greatly affects capacity of either type of tubular ice maker. If supply water temperature is reduced from 70 to 40°F, ice production of the unit increases approximately 18%. In larger systems, the economics of precooling water in a separate system with higher suction pressures should be considered.

Plate Ice

Plate ice makers are commonly defined as those that build ice on a flat vertical surface. Water is applied above freezing plates and flows by gravity over the freezing plates during the freeze cycle. Liquid refrigerant at a temperature between –5 and 20°F is contained in circuiting inside the plate. The length of the freezing cycle governs the thickness of ice produced. Ice thicknesses in the range of 0.25 to 0.75 in. are quite common, with freeze cycles varying from 12 to 45 min. Figure 4 shows a flow diagram of a plate ice maker using water for harvest. All plate ice makers use a sump and recirculating pump concept, whereby an excess of water is applied to the freezing surface. Water not converted to ice on the plate is collected in the sump and recirculated as precooled water for ice making.

Ice is harvested by one of two methods. One method involves applying hot gas to the refrigerant circuit to warm the plates to 40 to 50°F, causing the ice surface touching the plate to reach its melting point and thereby release the ice from the plate. The ice falls by gravity to the storage bin below or to a cutter bar or crusher that further reduces the ice to a more uniform size. Plate ice makers using the hot-gas method of harvesting can produce ice on one or two sides of the plate, depending on the design.

In the second method of harvesting ice, warm water flows on the back side of the plate. This heats the refrigerant inside the plate above the ice melting point, and the ice is released. Ice makers using this approach manufacture ice on one side of a plate only. Harvest water is chilled by passing over the plates. It is then collected in the sump and recirculated to become precooled water for the next batch of ice.

For plate ice makers, freezing time, harvest time, water, pump, and refrigeration are controlled by adjustable electromechanical or electronic devices. Using the wide variety of thicknesses and freezing

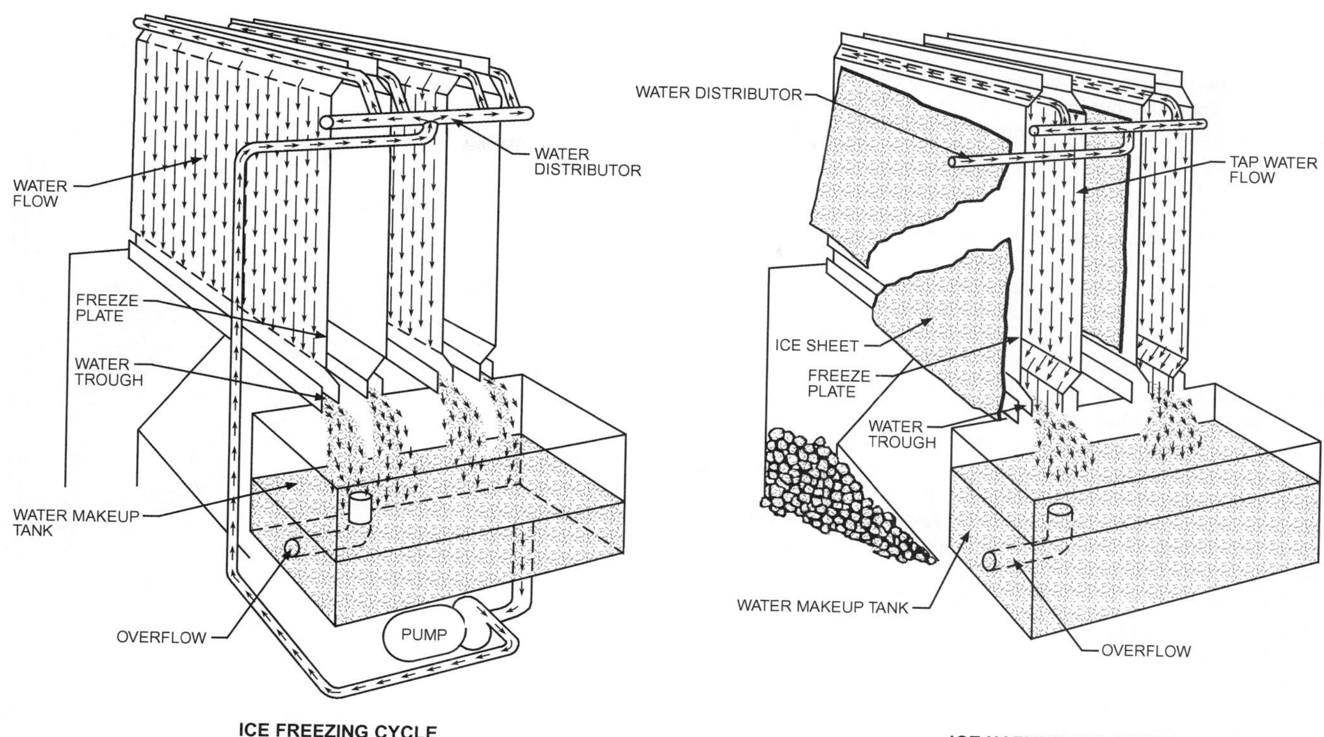

Fig. 4 Plate Ice Maker

times available, plate ice makers can produce clear ice. Thus, the plate ice maker is commonly used in applications requiring clear ice.

Because of the harvest cycle involved, plate ice makers require more refrigeration per unit mass of ice produced than flake ice makers. This disadvantage is offset by the ability of plate ice makers to operate at higher evaporating temperatures; thus, connected motor power per ton of refrigeration is usually less than that of flake ice makers. During the harvest cycle, the suction pressure rises considerably, depending on the design of the ice maker. When a common refrigeration system is used for multiple refrigerated requirements, a stable suction pressure can be maintained for all the refrigeration loads by using a dedicated compressor for the ice machine, or by using a dual pressure suction regulator at each ice machine to minimize the load placed on the suction main during harvest. This may occur in large processing plants, refrigerated warehouses, and so forth. Large plate ice makers can be arranged such that only sections or groups of plates are harvested at one time. Properly adjusting the timing of harvesting each section can reduce the fluctuation in suction pressure.

Plate ice makers using the water harvest principle rely on the temperature of the water for harvesting. A minimum of 65°F is usually recommended to minimize both harvest cycle time and harvest water consumption. For installations in cold-water areas, or where wintertime inlet water temperatures are low, it is advisable to provide auxiliary means of warming the inlet water to 65°F.

Ice Builders

Ice builders comprise various types of apparatus that produce ice on the refrigerated surfaces of coils or plates submerged in insulated tanks of water. This equipment is commonly known as an ice bank water chiller. Ice built on the freezing coils is not used as a manufactured ice product but rather as a means of cooling water circulating through the tank as the ice melts from the coils. The ice builder is most often used for thermal storage applications with high peak and intermittent cooling loads that require chilled water. See Chapter 34 of the 2003 *ASHRAE Handbook—HVAC Applications* for more information.

Scale Formation

Performance of all ice makers is affected by the characteristics of the inlet water used. Impurities and excessive hardness can cause scale to be deposited on the freezing surface of the ice maker. The deposit reduces the heat transfer capability of the freezing surface, thereby reducing ice-making capacity. Deposited scale may further reduce ice-making capacity by causing the ice to stick on the freezing surface during the harvest process. The rated capacity of all ice makers is based on the substantial release of all the ice from the freezing surface during the removal period. Because the process of freezing water tends to freeze a greater proportion of pure water on the ice maker's freezing surface, impurities tend to remain in the excess or recirculated water. A blowdown, or bleedoff, whereby a portion of the recirculated water is bled off and discharged, can be installed. The bleedoff system can control the concentration of chemicals and impurities in the recirculated water. The necessity of a bleedoff system and the effectiveness of this concept for controlling scale deposits depend on local water conditions. Some refrigeration system loss occurs because recirculated water that is bled off to drain is precooled. Water that is bled off may be passed through a heat exchanger to precool incoming makeup water. Water conditions, treatment, and related problems in ice making are covered in Chapter 48 of the 2003 *ASHRAE Handbook—HVAC Applications*.

THERMAL STORAGE

Interest in energy conservation has renewed interest in using ice to provide thermal storage of cooling capacity for air-conditioning or process applications. The ice is produced and stored using lower off-peak and weekend power rates. During the day, stored ice provides refrigeration for the chilled water system. The design and features of thermal storage equipment are covered in Chapter 34 of the 2003 *ASHRAE Handbook—HVAC Applications*.

ICE STORAGE

Fragmentary ice makers can produce ice either on a continuous basis or in a constant number of harvest cycles per hour. The use of the ice is generally not at a constant rate but on a batch basis. Batches vary greatly, based on user requirements. The ice must be stored and recovered from storage on demand. Labor savings, economics, quantity of ice to be stored, amount of automation desired, and user delivery requirements must all be considered in ice storage and storage bin design.

Ice makers can produce ice 24 h a day. Making ice during off-shifts and weekends, as well as during work shifts, can lead to considerable savings in total ice-making and refrigeration system requirements. In addition, by using electrical power during off-peak hours, peak loads on the power system are reduced during the day. Many power companies offer reduced rates during off-peak hours.

Ice storages vary in type from short- to prolonged-term. Degree of automation for filling and discharge ranges from manual shoveling to a completely automatic rake system.

Short-term storage generally requires provision for one day's ice production. The ice maker is mounted over a bin, and ice falls by gravity into the bin. The bin is an insulated, airtight enclosure with one or more insulated doors for access. Ice is removed by shoveling or scooping. In such storage, the subcooling effect of the ice generally offsets the heat loss through the insulated bin walls without excessive melting. In most situations, it is not necessary to provide refrigeration units in an ice storage bin where ice production is being used on a daily basis and ambient temperatures are reasonable.

Prolonged ice storage requires a refrigerated, airtight, insulated storage bin. Some designs provide for false walls and floor, which produce an envelope effect that allows cold air to circulate completely around the mass of ice in storage. If wet ice is placed in a bin refrigerated to a temperature below 32°F, it will freeze together and may be difficult to remove.

Time and pressure affect the storage quality of fragmentary ice. Even though a bin is refrigerated to a temperature well below 32°F, pressure can cause local melting near the bottom. Thus, there are limits to the size and configuration of a gravity-filled storage bin. Ice falling from an ice maker forms a cone directly underneath the drop in the bin. With slight variation because of the type of ice, the angle of repose is approximately 30°. Fusion of ice under pressure limits the practical ice storage depth to 10 to 12 ft. To use the bin's volume more efficiently, a leveling screw mounted in the overhead can be used to carry the ice away from the top of the ice cone.

There is also a practical limit to the size of a storage bin from which ice can be manually removed through refrigerator doors. The simplest device used to remove ice from a bin is a screw conveyor with a trough at floor level, which is equipped with gratings and removable sectional covers. The removable covers protect the screw from ice blockage when the conveyor is not running. The gratings are for personnel protection.

Ice Rake and Live Bottom Bins

The ice rake system is used for larger and fully automated storages. These storages generally have a 10 to 300 ton capacity for a single rake system. Depending on plant demands, combinations of rake systems can be developed into an integrated production, storage, and delivery system. Such a mechanism could have capabilities of up to 1000 tons of storage with multiple screw or pneumatic conveying delivery systems. A range of delivery rates, up to 60 tons of ice per hour, can be achieved. Advantages include elimination of labor for storing and transporting ice, longer and more effective distribution systems, faster delivery and termination of ice flow, less waste of ice, and the elimination of physical contamination.

Storages that incorporate rake systems are of two basic types. One type encloses the ice storage and rake system in an arrangement of steel framework and panels, with the complete unit installed in a refrigerated room. The second type constructs an insulated enclosure around the ice bin and rake system. This type can be installed inside a building or outside, depending on the weather protection provided. For either type, the ice makers are mounted outside of the refrigerated space. Figure 5 shows the arrangement of components for a typical rake system.

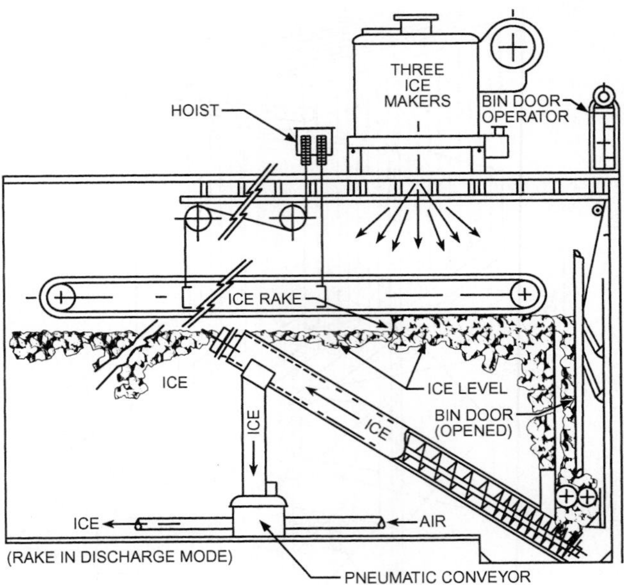

Fig. 5 Ice Rake System

Once fragmentary ice comes to rest in the storage bin, it will not flow freely, and a mechanical force is necessary to start it moving. The deeper the ice is stored, the greater the pressure on ice near the bottom. Ice near the bottom tends to fuse together faster. Rake systems work from the top of the ice; they continuously level and fill the storage bin, as well as automatically remove the ice on demand. The systems operate in nonrefrigerated or refrigerated bins; because most users of large storages also want ice to be dry for ease of handling, large automated storages are usually refrigerated.

The ice rake itself consists of a structural steel mechanism with drive. It operates similarly to the tracks of a crawler tractor. By means of a hoist and timer, the rake is raised or lowered to automatically maintain its position suspended over, and in close contact with, the ice level. Wide scraper conveyors, mounted across the tracks along the full width of the bin, spread the ice out and drag it toward the back of the bin during the filling mode. To dispense the ice at delivery, the scraper conveyors reverse direction and drag the ice to the opposite end of the bin, dropping it into a screw conveyor mechanism. From this point, the ice is transferred to the external delivery system of screw conveyors.

Some rake systems have features that allow ice deliveries from the bin to be remotely controlled and volumetrically metered. Ice deliveries can be recorded on digital counters at the storage bin, remote stations, and control centers. Accuracy is in the range of ±2%. Another method of metering ice from a rake system storage involves the screw conveyor delivering the ice to a weigh belt. As the ice passes along the moving belt, it is electronically weighed and the weight is recorded. Selection of the belt material carrying the ice is critical in preventing ice from sticking to the belt. The weigh belt is often installed in a refrigerated area adjacent to the ice storage.

Another type of ice storage with delivery system capabilities is the **live bottom** type, with multiple screws arranged in various configurations on the bottom of the storage bin. Because of ice fusion, these bins are limited to short-term storage. Success of this type of bin depends on the type and quality of fragmentary ice being stored and the ability of the design to overcome particle fusion. Particle fusion may result in ice bridges forming over the top of the screws; then the screws will bore holes in the ice rather than empty the bin.

Primarily for the consumer bagged-ice industry, a bin and automatic storage system is used in which the entire floor moves, carrying the ice load into slowly rotating beaters. As the ice breaks loose, it drops to a screw conveyor, which feeds an ice bagger. This type of bin is located in a refrigerated room, and the ice makers are located away from the bin. The ice makers must be shut off so that no ice can flow into the storage during the discharge and bagging process.

The **ice silo** is used for long- or short-term storage with capacities in the range of 20 to 100 tons. The silo tank comprises a cylindrical part and a tapered, conical part leading the ice to the outlet at the bottom of the tank. From this point, the ice is transported by a screw delivery system. A rotating flexible chain arrangement is provided in the silo to assist in ice removal and to partially overcome the fusion problem. The ice maker is mounted over the top of the silo, and the ice falls into the storage. No leveling of ice is required in the bin because the silo's diameter is sized to be compatible with the ice maker. The larger the ice storage, the higher the silo. As a result, proper ice discharge becomes more critical in the design when considering the fusion of ice and the fact that the ice must finally pass through a relatively small opening at the bottom of a tapered zone.

DELIVERY SYSTEMS

The location of ice manufacture is rarely the location of ice usage. Usually it is necessary to move ice from the ice machine or storage bin to some other area where it will be used; thus, a conveying system is required. Most conveyor applications use screws, belts, or pneumatic systems. Great care must be taken in selecting the size and type of conveyor to be used because no matter what type of fragmentary ice is being handled, problems such as fines, freeze-up, and ice jams can be encountered with an improperly designed system. Fines, or snow, are small particles of ice that chip off the larger pieces during harvesting, crushing, or conveying operations.

Screw and Belt Conveyors

Screw conveyors are the most popular of all the conveyances used for transporting ice. They are manufactured in sizes of 4 in. diameter and up, as well as in various screw pitches. Most ice-conveying operations use 6, 9, and 12 in. diameter screws.

The sizing and drive power requirements of screw conveyors are determined by the ice delivery rate, inclination of the conveyor, and conveyor screw pitch. With fragmentary ice, selection of an undersized conveyor results in excessive conveyor speed or requires that the conveyor run too full of ice. These conditions can produce excessive fines.

When screw conveyors transport ice through high-ambient inside areas or outside in the weather (e.g., in icing fishing vessels), it is advisable to insulate the screw conveyor trough and provide the conveyor with insulated covers. Rain is as problematic as sunshine for contributing to ice meltage and delivery difficulties. For this reason, most screw conveyors operating in the weather are provided with sectional and removable covers.

Belt conveyors are often used when excess moisture has to be removed from the ice or to minimize the fines. The mesh belts allow snow and excess water to fall through. Stainless steel, galvanized steel, or high-density polyethylene are commonly used for belting.

Pneumatic Ice Conveying

Pneumatic ice conveying systems have proved desirable, economical, and practical for transporting fragmentary ice distances of 100 ft or more and when multiple delivery stations must be served. A pneumatic system is advantageous when delivery stations are in different directions or at different elevations, when delivery through a pressure hose is needed, or when flexibility is required for future changes or addition of delivery stations.

The basic principle of conveying ice by a pneumatic system involves a rotary blower, which delivers air to a rotary air-lock valve or conveying valve. Ice is fed into the conveying valve, and compressed air conveys the mixture of air and ice at high velocity through thin-walled tubing (aluminum, stainless steel, or plastic). Figure 6 shows the diagrammatic arrangement of pneumatic system components.

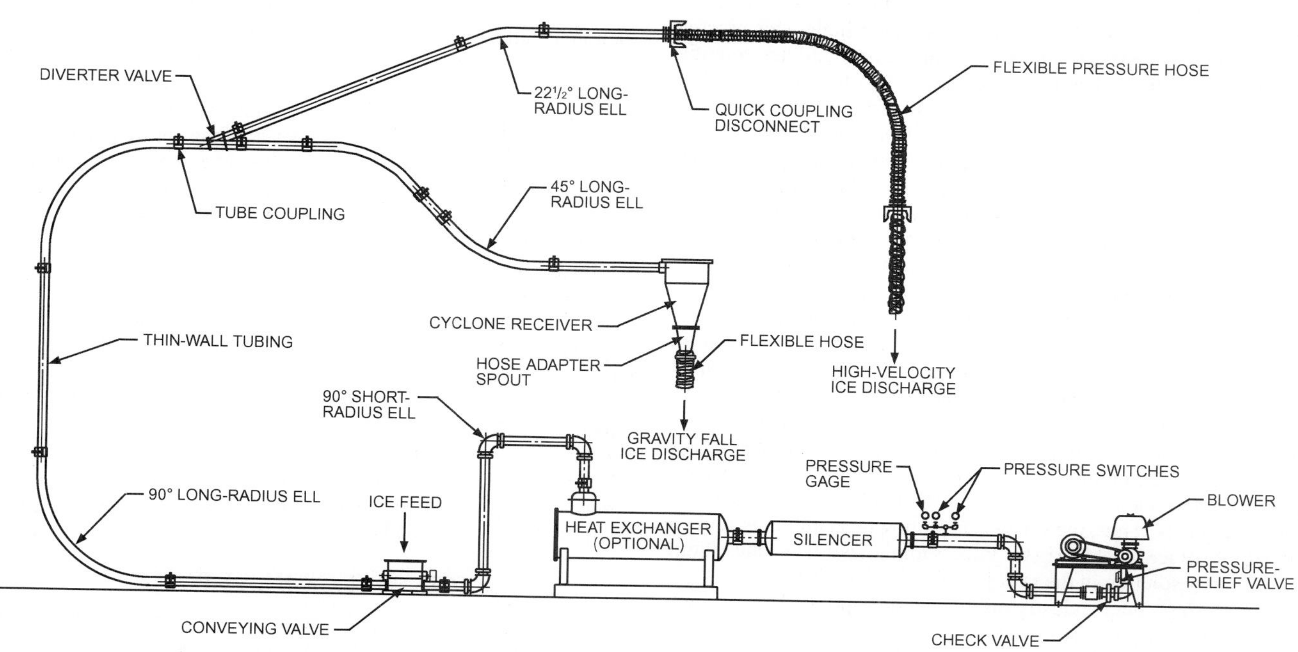

Fig. 6 Typical Flake Ice Pneumatic Conveying System

Delivery rates between 10 and 40 ton/h and conveying distances up to 600 ft are common. Delivery distances exceeding 600 ft can be achieved at reduced delivery rates, with the maximum practical distance being approximately 1000 ft. Conveying pressures range from 4 to 10 psig, depending on the delivery rate and the maximum distance the ice is to be conveyed. The air velocity required to keep ice in suspension in the conveying line will vary among the different types of fragmentary ice. A pneumatic system cannot satisfactorily convey all sizes of fragmentary ice. The equipment manufacturer should be consulted for recommended line velocities. Sometimes, storage bins for ice plants using a pneumatic delivery system are refrigerated to ensure cold, free-flowing ice with minimum moisture. Because the ice remains in the tubing a very short time, tubing insulation is seldom needed or used. However, in warm climates, shading the conveying line reduces the solar load. The tubing typically used has a diameter of 4 to 8 in. and requires minimum support, making installation easy and economical.

Multiple delivery points are served by automatic Y-type diverter valves or multiple-way slide valves, either air or electrically operated. Pneumatically blown ice can be delivered under pressure out of the end of a hose. An alternative method of delivery is a cyclone receiver, which takes the ice at high velocity, dissipates the air, and drops the ice by gravity. Combinations of hose stations and cyclone delivery stations in the same system are common.

When a pneumatic conveying system is used in areas of high ambient and wet-bulb temperatures on an application requiring higher conveying pressures, a heat exchanger is often used to cool and dehumidify the pneumatic air before it enters the conveying valve. The heat exchanger is provided with a cooling coil, either refrigerant- or chilled-water-cooled, a demister or other means of separating moisture from the air, and a condensate trap to expel entrapped moisture. Geographical location, system pressure, and quality and use of ice at the delivery point must be considered when determining whether to use a heat exchanger.

Slurry Pumping

A mixture of particle ice and water can be pumped as a slurry, which has some advantages for transporting ice. Generally, the slurry is approximately 50% water and 50% ice, but for specialized application, mixtures of up to 80% ice and 20% water can be successfully pumped. Delivery distances of 800 ft have been achieved with delivery rates of 60 tons of slurry mix per hour. This practice has been extensively used in the produce industry and has potential in concrete cooling, chemical processing, and other ice or chilled-water applications. Ice slurry mixes are of particular interest where there is a need for low-temperature chilled water at or near 32°F. In converting ice to water, the absorption of latent heat at the usage point enables more cooling to be done with a slurry mix than with straight chilled-water cooling. Pumping volumes and line sizes are minimized, and ice meltage during mixing and pumping does not normally exceed 1 to 3%. The system is capable of automation for continuous operation.

The basic system for slurry pumping includes a mixing tank, in which the ice and water are mixed. The ice is carried by any of the conventional conveying methods from the ice storage bin to the slurry mix tank. Agitators in the tank operate continuously to maintain a mixture with uniform consistency. Pumps discharge the slurry mix through pipelines to the point of use. The pumps are of the centrifugal type, modified for pumping slurry. When the icing cycle at the usage points is intermittent, a recirculating system returns unused slurry to the mixing tank. Thus, the slurry is kept moving at all times, and the possibility of ice blockage in the lines is minimized. The temperature of the slurry solution in the tank is maintained at 32°F.

The fresh produce industry offers a unique application for slurry mixes. Body icing of the fresh produce, which is generally of non-uniform size and configuration, is achieved by applying slurry to the dry packed product. Drain holes are provided in the shipping container to remove water. Because of its suspension in water, ice is carried to all parts of the container. The ice solidifies as water drains from the container. The product is then completely surrounded with ice, the voids are filled, and potential hot spots are eliminated. Drained water can be collected and returned to the mixing tank.

COMMERCIAL ICE

Commercial ice is primarily used for human consumption. It is also called **packaged** or **consumer** ice and is used for cooling beverages and for other applications in restaurants, hotels, and similar institutions. This use requires packaging at the ice plant for storage and eventual distribution. In bagged form, commercial ice is also available for sale in grocery stores and in automatic, coin-operated vending machines. When the ice is to be used in beverages, ice produced by plate or tube ice makers is preferred because of its clear appearance and the fact that it can be made in greater thicknesses. Rake systems are often used to store packaged ice and convey it into the packaging system.

Packaging. A packaging system normally comprises an ice bagger and a bag closer. These components are available from ice packaging equipment manufacturers in various types and sizes. In the bagging process, the ice is fed from the ice storage bin into the bagging machine by a screw or belt conveyor. The bagging machine meters ice into a bag placed below its discharge chute. The amount of ice measured into the bag can be determined by weight, volume, or sight approximation, depending on the equipment used.

When packaging by weight, the ice bag is placed on a weighing table on the bagging machine. Ice is then dispensed into the ice bag until the desired weight is in the bag. At this point, a switch mounted on the scale stops the flow of ice.

The volumetric bagging machine deposits the ice in a rotating chamber, which is adjustable in volume. After a predetermined volume of ice enters the chamber, the ice is discharged into the bag below. Because the shape and size of the ice is not constant, the volumetric chamber is usually set to produce a 3 to 5% overage by volume. Therefore, the proper minimum weight of ice in the bag is ensured.

The bag closer consists of a mechanical unit that ties and seals the top of the bag with a wire ring, wire twist tie, or plastic clip. Smaller bagging operations do not use a bag closer, and the bags are manually closed with plastic ties, wire rings, staples, and so forth. In more elaborate systems, the bag is formed from roll stock, filled with ice, automatically removed from the bagging machine, and automatically closed before it is dropped onto a conveyor, which carries the bagged ice to the refrigerated storage room. The degree of automation for the bagging and closing operation is determined by the number of bags of ice to be produced per day, bag size, and cost-benefit relationship between automated equipment and reduced labor costs.

Ice bags are made of plastic, most often polyethylene, or, very rarely, heavy moisture-resistant paper. Plastic bags are used in most modern plants.

Storage. Packaged ice must be stored in a refrigerated warehouse or room before distribution. The ice storage area is sized to meet the daily production of the plant and distribution requirements. Generally, a bag ice storage facility can store 3 to 7 days' production. Although the ice will not melt at a storage temperature below 32°F, it is important that storage be maintained at a temperature between 10 and 25°F. The lower temperature subcools the ice and avoids meltage during distribution. Depending on the type and quality of the ice, bagged ice can contain some water (from 0 to 5%). For this reason, provision is made in the storage room refrigeration system for the product load of refreezing the water.

ICE-SOURCE HEAT PUMPS

Ice-making systems can be configured to provide heating alone, or heating and cooling, for a building or process. Conversion of water to ice occurs at a relatively high evaporator temperature and coefficient of performance compared to air-source heat pumps operating at low ambient temperatures. Systems can provide necessary heating, with the resulting ice disposed of by melting with low-grade heat, such as solar. In addition, ice can be used for useful cooling through daily, weekly, or seasonal storage.

The concept was originally considered mainly for residential heating and cooling, but installations are proving feasible for larger structures, such as office buildings. Energy consumption savings from the coefficient of performance of a conventional heat pump system are achieved. Using off-peak night and weekend rates can reduce power costs, and ice produced is used for building cooling requirements. As a result, a system can be developed that consumes less energy, at a lower utility rate.

Ice-source heat pumps follow two basic approaches. The first involves using the ice builder principle, with coils in a large tank, as the evaporator component of the heat pump. The second approach uses a fragmentary type of ice maker as the evaporator, with a mixture of ice and water stored in a tank. Many variations, combinations, and adaptations may be developed from these basic systems. The requirement for thermal storage of a large quantity of ice dictates new planning in architectural building design. Heating and cooling system designs for the building are also influenced.

BIBLIOGRAPHY

Dorgan, C.E. 1985. Ice-maker heat pumps operation and design. *ASHRAE Transactions* 91(1B):856-862.

Dorgan, C.E., G.C. Nelson, and W.F. Sharp. 1982. Ice-maker heat pump performance—Reedsburg Center. *ASHRAE Transactions* 88(1):1271-1278.

CHAPTER 35

ICE RINKS

ANY level sheet of ice made by refrigeration (the term **artificial ice** is sometimes used) is referred to in this chapter as an ice rink regardless of use and whether it is located indoors or outdoors. Bobsled-luge tracks are not referred to as rinks but are referenced under this chapter.

An ice sheet is usually frozen by circulation of a heat transfer fluid through a network of pipes or tubes located below the surface of the ice. The heat transfer fluid is predominantly a secondary coolant such as glycol, methanol, or calcium chloride (see Chapter 21 of the 2005 *ASHRAE Handbook—Fundamentals*).

R-22 and R-717 are most frequently used for chilling secondary coolants for ice rinks. R-12 and R-502 have also been used; however, because of the phaseout of the CFC refrigerants, they should no longer be considered for use. R-22 will also be phased out, so for new rink equipment selection, R-22 and CFC replacements should be evaluated according to status and availability.

In some rinks, R-22, and R-717 to a lesser degree, have been applied as a direct coolant for freezing ice. The direct-refrigerant rinks operate at higher compressor suction pressures and temperatures, thus achieving an increased COP, compared to secondary coolants. The primary refrigerant charge is greatly increased with this method of freezing. Because of emissions regulations, the projected R-22 phaseout, building codes, and fire regulations, R-22 and R-717 should not be used to freeze ice *directly* in indoor rinks.

APPLICATIONS

Most ice surfaces are used for a variety of sports, although some are constructed for specific purposes and are of specific dimensions. Usual rink sizes include the following:

Hockey. The accepted North American hockey rink size is 85 by 200 ft. Radius corners of 28 ft are recommended by professional and amateur rules. The Olympic and international hockey rink size is 96 by 196 ft, with 20 ft radius corners. Many rinks are considered adequate with dimensions of 85 by 185 ft, 80 by 180 ft, and 70 by 170 ft. In substandard size rinks, a corner radius of not less than 20 ft should be provided to allow use of mechanical resurfacing equipment.

Curling. Regulation surface for this sport is 14 by 146 ft; however, the width of the ice sheet is often increased to allow space for installation of dividers between the sheets, particularly at the circles. Most curling rinks are laid out with ice sheets measuring 15 by 150 ft.

Figure Skating. School or compulsory figures are generally done on a "patch" measuring approximately 16 by 40 ft. Freestyle and dance routines generally require an area of 60 by 120 ft or more.

Speed Skating. Indoor speed skating has traditionally been performed on hockey-sized rinks. The Olympic-sized outdoor speed skating track is a 1400 ft oval, 35 ft wide with 392 ft straightaways and curves with an inner radius of 87.5 ft. Most speed skating ovals were originally constructed outdoors; however, some have now been constructed indoors.

Recreational Skating. Recreational skating can be done on any size or shape rink, as long as it can be efficiently resurfaced. Generally, 25 to 30 ft^2 is allowed for each person actually skating. This ratio may vary for large numbers of beginner skaters. An 85 by 200 ft hockey rink with 28 ft radius corners has an area of 16,327 ft^2 and will accommodate a mixed group of about 650 skaters.

Public Arenas, Auditoriums, and Coliseums. Public arenas, auditoriums, field houses, etc., are designed primarily for spectator events. They are used for ice sports, ice shows, and recreational skating, as well as for non-ice events, such as basketball, boxing, tennis, conventions, exhibits, circuses, rodeos, tractor events, and stock shows. The refrigeration system can be designed so that, with adequate personnel, the ice surface can be produced within 12 to 16 h. However, general practice is to leave the ice sheet in place and to hold other events on an insulated floor placed on the ice. This approach saves significant time, labor, and energy.

Bobsled-Luge Tracks. The bobsled-luge track usually incorporates steel piping embedded in the track and fed by an ammonia liquid recirculation system. Approximately 280,000 to 300,000 ft of piping is required for an Olympic-sized track. The total refrigerated surface is 90,000 to 100,000 ft^2. Refrigeration plant capacities in the range of 1100 to 1400 tons are required, depending on ambient design conditions, wind, and sun loads. The ammonia charge can exceed 200,000 lb. Because elevation changes are significant, care must be used in placing liquid recirculators, selecting ammonia pumps, and circuiting floor piping.

REFRIGERATION REQUIREMENTS

The heat load factors considered in the following section include type of service, length of season, use, type of enclosure, radiant load from roof and lights, and geographic location of the rink with associated wet- and dry-bulb temperatures. For outdoor rinks, the sun effect and weather conditions (wind velocity and rain) must also be considered.

Refrigeration requirements can be estimated fairly accurately based on data from a number of rink installations with the pipes covered by not more than 1 in. of sand or concrete and not more than 1.5 in. of ice (a total of 2.5 in. sand or concrete and ice).

Refrigeration load may be estimated by considering the larger of (1) the refrigeration necessary to freeze the ice to required conditions in a specified time, or (2) the refrigeration necessary to maintain the ice surface and temperature during the most severe usage and operating conditions that coincide with the maximum ambient environmental conditions.

In the time-to-freeze method, determine the (1) quantity of ice required (rink surface area multiplied by thickness); (2) heat load to reduce the water from application temperature to 32°F, freeze the water to ice, and reduce the ice to the required temperature; and (3) heat loads and system losses during the freezing period. The total requirement is divided by system efficiency and freezing period to determine the required refrigeration load or rate of heat removal.

The preparation of this chapter is assigned to TC 10.2, Automatic Ice-Making Plants and Skating Rinks.

Example 1. Calculate the refrigeration required to build 1 in. thick ice on a 16,300 ft^2 rink in 24 hours.

Assume the following material properties and conditions:

Material	Specific Heat, Btu/lb·°F	Temperature, °F Initial	Temperature, °F Final	Density or Weight
6 in. concrete slab	0.16	35	20	150 lb/ft^3
Supply water	1.0	52	32	62.5 lb/ft^3
Ice	0.49	32	24	—
Ethylene glycol, 35%	0.83	40	15	32,000 lb

Latent heat of freezing water = 144 Btu/lb
Building and pumping heat load = 50 tons of refrigeration
System losses = 15%
Mass of water = 16,300 ft^2 × 1/12 ft × 62.5 lb/ft^3 = 85,000 lb
Mass of concrete = 16,300 ft^2 × 1/2 ft × 150 lb/ft^3 = 1,230,000 lb

Then,

$$q_R = (\text{Sys. losses})(q_F + q_C + q_{SR} + q_{HL})$$

where

q_R = refrigeration requirement
q_F = water chilling and freezing
q_C = concrete chilling load
q_{SR} = refrigeration to cool secondary coolant
q_{HL} = building and pumping heat load

$$q_F = \frac{85{,}000\text{ lb}[1(52-32) + 144\text{ Btu/lb} + 0.49(32-24)]}{24\text{ h} \times 12{,}000\text{ Btu/h}\cdot\text{ton}}$$

$$= 49.6\text{ tons}$$

$$q_C = \frac{1{,}230{,}000 \times 0.16(35-20)}{24 \times 12{,}000} = 10.3\text{ tons}$$

$$q_{SR} = \frac{32{,}000 \times 0.83(40-15)}{24 \times 12{,}000} = 2.3\text{ tons}$$

$$q_R = 1.15(49.6 + 10.3 + 2.3 + 50) = 129\text{ tons}$$

When no time restrictions for making ice apply, the estimated refrigeration load is the amount of heat removal needed to offset the usage loads plus the coincidental heat loads during the most severe operating conditions. Table 1 lists approximate refrigeration requirements for various rinks with controlled and uncontrolled atmospheric conditions. Table 1 should only be used to check the calculated refrigeration requirements. Table 2 shows the distribution of various load components for basic construction and the estimated potential load reductions that may be obtained when energy-conserving design and operating techniques are used.

Heat Loads

Energy and operating costs for ice rinks are very significant, and these costs should be analyzed during design. A good estimate of required refrigeration can be calculated by summing the heat load components at design operating conditions. Heat loads for ice rinks consist of conductive, convective, and radiant components. Connelly (1976) collected the performance data summarized in Tables 2 and 3. The amount of control over each load source is indicated as an approximate percentage of the maximum reduction possible through effective design and operation.

Table 1 Range of Refrigeration Capacities for Ice Rinks

Type of Facility	Up to 7 months (spring, fall, winter), ft^2/ton	8 months to year-round, ft^2/ton
Outdoors, unshaded	80 to 140	—
Outdoors, shaded	100 to 190	—
Sports arena	110 to 160	100 to 140
Sports arena, accelerated ice making	80 to 135	75 to 120
Ice recreation center	170 to 240	140 to 190
Curling rinks	200 to 380	150 to 200
Ice shows	80 to 160	75 to 130

Conductive Loads. If a rink is uninsulated, **heat gain from the ground** below the rink and at the edges averages 2 to 4% of the total heat load. Permafrost may accumulate and frost heaving, which is detrimental to both the rink and the piping, may result. Heaving also makes it more difficult to maintain a usable ice surface, can affect the structural integrity of the building, and is dangerous to the users.

Heat gain from the ground and perimeter is highest when the system is first placed in operation; however, it decreases as the temperature of the mass beneath the rink decreases and permafrost accumulates. Ground heat gain is reduced substantially with insulation. Chapter 25 of the 2005 *ASHRAE Handbook—Fundamentals* gives details on computing heat gain with insulation.

Heat gain to the piping is normally about 2 to 4% of the total refrigeration load, depending on length of piping, surface area, and ambient temperatures. The ice and frost that naturally accumulate on headers reduce the heat gain. Insulation can be applied to reduce heat gain to the piping and keep ice from accumulating. However, insulating headers while maintaining visual inspection of joints (floor piping to the headers) is usually impractical. Headers may, with precautions and the use of steel headers and piping, be embedded in the rink floor. Embedded headers contribute to ice freezing and eliminate the trench to rink floor piping penetrations. When headers are embedded in concrete, all joints from the steel floor piping to the headers should, ideally, be welded. It may be difficult to remove air from this type of floor system.

A circuit loop should be placed around the rink perimeter to prevent soft ice from developing at the edges (see the section on Rink

Table 2 Ice Rink Heat Loads, Indoor Rinks

Load Sources Category	Approximate Maximum of Total Load,* %	Maximum Reduction of Load Category Through Design and Operation, %
Conductive loads:		
Ice resurfacing	12	60
System pump work	15	80
Ground heat	4	80
Header heat gain	2	40
Skaters	4	0
Convective loads:		
Rink air temperature	13	50
Rink humidity	15	80
Radiant loads:		
Ceiling radiation	28	90
Lighting radiation	7	40
Total	100	

*Load distribution for basic rink without insulation below rink floor.

Table 3 Ice Rink Heat Loads, Outdoor Rinks

Load Sources Category	Approximate Maximum of Total Load,* %	Maximum Reduction of Load Category Through Design and Operation, %
Conductive loads:		
Ice resurfacing	9	50
System pump work	12	80
Ground heat	2	40
Header heat gain	1	30
Skaters	1	0
Convective loads:		
Air velocity	0 to 15	10
Air temperature	0 to 15	0
Humidity	0 to 15	0
Radiant loads:		
Solar load	10 to 30	60
Total	100	

*Load distribution for basic rink without insulation below rink floor.

Piping and Pipe Supports). A circuit loop is especially important if return bends are used and embedded in the concrete. If return bends are embedded in the concrete, the pipe and the return bend should be steel with welded joints.

Heat gain from coolant circulating pumps can represent up to 15% of the refrigeration load. Some facilities operate continuously. Energy consumption from pump operation can be reduced by using pump cycling, two-speed motors, multiple pumps, multiple motors driving a single pump, or variable-speed motors with the appropriate controls. High-efficiency pumps and motors should be used. Proprietary variable motor speed controls are also available. Coolant flow should be sufficient at all times for acceptable chiller operation and to maintain a balanced flow through the piping grid.

Equipment components should be selected for low energy consumption; they may be selected to operate at or feature low discharge pressure (oversized condenser), high suction pressure (oversized chiller), multiple compressors, and an intelligent control system. Computer control of the refrigeration system is recommended.

Ice resurfacing represents a significant operating heat load. Water is flooded onto the ice surface, normally at temperatures between 130 and 180°F, to restore the ice surface condition. The heat load resulting from the flood water application may be calculated as follows:

$$Q_f = 8.33V_f[1.0(t_f - 32) + 144 + 0.49(32 - t_i)]$$

where

Q_f = heat load per flood, Btu
V_f = flood water volume (typically 120 to 180 gal. for a 100 by 200 ft rink), gal
t_f = flood water temperature, °F
t_i = ice temperature, °F

The resurfacing water temperature affects the load and time required to freeze the flood water. Maintaining good water quality through proper treatment may permit the use of lower flood water temperature and less volume.

Convective Loads. Convective load from air to ice may be as much as 28% or more of the total heat load to the ice (see Tables 2 and 3). The convective heat load is affected by air temperature, relative humidity, and air velocity near the ice surface. Precautions should be taken to minimize the influence of air movement across the ice surface in the design of the rink heating and dehumidification air distribution system. The convection heat load may be estimated using the procedure from Appendix 5 in "Energy Conservation in Ice Skating Rinks" (DOE 1980). The estimated convective heat transfer coefficient can be calculated as follows:

$$h = 0.6 + 0.00318V$$

where

h = convective heat transfer coefficient, Btu/h·ft²·°F
V = air velocity over the ice, ft/min

The effective heat load (including the latent heat effect of convective mass transfer) is given by the following equation:

$$Q_{cv} = h(t_a - t_i) + [K(X_a - X_i)(1226 \text{ Btu/lb})(18 \text{ lb/mol})]$$

where

Q_{cv} = convective heat load, Btu/h·ft²
K = mass heat transfer coefficient
t_a = air temperature, °F
t_i = ice temperature, °F
X_a = mole fraction of water vapor in air, lb mol/lb mol
X_i = mole fraction of water in saturated ice, lb mol/lb mol

When the mole fraction of air is calculated using a relative humidity of 80% and a dry bulb of 38°F, X_a is approximately 6.6×10^{-3}, and X_i for saturated ice at 100% and a temperature of 21°F is 3.6×10^{-3}. On the basis of the Chilton/Colburn analogy, $K \approx 0.17$ lb/h·ft² (DOE 1980).

In locations with high ambient wet-bulb temperatures, dehumidification of the building interior should be considered. This process lowers the load on the icemaking plant and reduces condensation and fog formation. Traditional air conditioners are inappropriate because the large ice slab tends to maintain a lower than normal dry-bulb temperature.

Radiant Loads. Indoor ice rinks create a unique condition where a large, relatively cold plane (the ice sheet) is maintained beneath an equally warm plane (the ceiling). The ceiling is warmed by conductive heat flow from the outside and by normal stratification of arena air. Up to 35% of the heat load on the ice sheet comes from radiant sources. On outdoor rinks radiant sources are the sun or a warm cloud cover. Vertical hanging cloth suspended from east-west horizontal overhead wires has been used to reduce the winter sun load.

In indoor and covered rinks, lighting is a major source of radiant heat to the ice sheet. The actual quantity depends on the type of lighting and how the lighting is applied. The direct radiant heat component of the lighting can be as much as 60% of the kilowatt rating of the luminaires. A radiant heating system can be another source of radiant heat gain to the ice. If radiant heat is used to maintain the comfort level in the promenade or spectator area, the radiant heaters should be located and directed to avoid direct radiation to the ice surface. The infrared components of the lighting can be estimated from manufacturers' data.

The infrared heat gain component from the ceiling and building structure, which is warmer than the ice surface, can be calculated by applying the Stefan-Boltzmann equation as follows:

$$q_r = A_c f_{ci} \sigma\left(T_c^4 - T_i^4\right)$$

$$f_{ci} = \left[\frac{1}{F_{ci}} + \left(\frac{1}{\varepsilon_c} - 1\right) + \frac{A_c}{A_i}\left(\frac{1}{\varepsilon_i} - 1\right)\right]^{-1}$$

where

q_r = radiant heat load, Btu/h
A_c = ceiling area, ft²
A_i = ice area, ft²
ε = emissivity
f_{ci} = gray body configuration factor, ceiling to ice surface
F_{ci} = angle factor, ceiling to ice interface (from Figure 1)
T = temperature, °R
σ = Stefan-Boltzmann constant = 0.1714×10^{-8} Btu/h·ft²·°R⁴

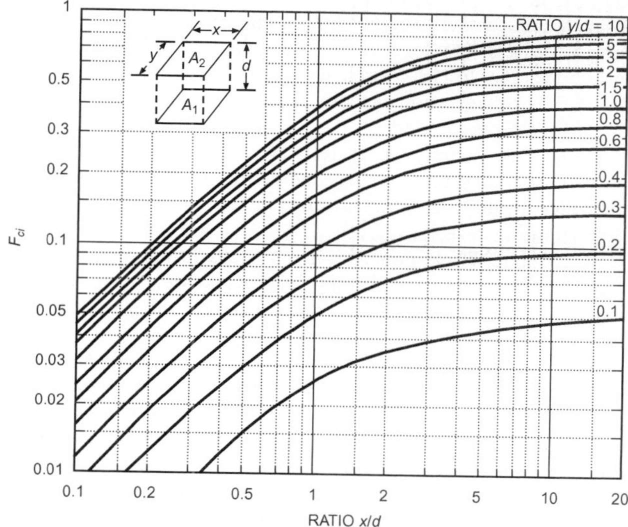

Fig. 1 Angle Factor for Radiation Between Parallel Rectangles F_{ci}

Example 2. An ice rink has the following conditions:

Ice dimension: 85 × 200 ft = 17,000 ft^2
Ice temperature: 24°F (484°R), $\varepsilon_i = 0.95$
Ceiling radiating area: 90 × 200 ft = 18,000 ft^2
Ceiling mid-height: 25 ft
Ceiling temperature: 60°F (520°R), $\varepsilon_c = 0.90$

$$x/d = 90/25 = 3.6$$

$$y/d = 200/25 = 8.0$$

From Figure 1, $F_{ci} = 0.68$

$$f_{ci} = \left[\frac{1}{0.68} + \left(\frac{1}{0.90} - 1\right) + \frac{18{,}000}{17{,}000}\left(\frac{1}{0.95} - 1\right)\right]^{-1} = 0.611$$

Then

$$q_r = 18{,}000 \times 0.611 \times 0.1714 \times 10^{-8}(520^4 - 484^4)$$
$$= 343{,}840 \text{ Btu/h}$$

The ceiling radiant heat load can be reduced by lowering the temperature of the ceiling, keeping warm air away from the ceiling, increasing the roof insulation, and, more significantly, lowering the ceiling material's emissivity to shield the ice from the building structure.

Ceiling and roof materials and exposed structural members may have an emissivity as high as 0.9. Special aluminum paint can lower the emissivity to between 0.5 and 0.2. Polished metal such as polished aluminum or aluminum foil have an emissivity of 0.05.

Because a low-emissivity ceiling is cooled very little by radiant loss, most of the time its temperature remains above the dew point of the rink air. Thus, condensation and dripping is substantially reduced or eliminated.

Low-emissivity fabric or tiled ceilings are frequently incorporated into new and existing rinks to reduce radiation loads, decrease condensation problems, and reduce the overall lighting required.

Radiant heat gain to the ice, especially in outdoor rinks, can be further controlled by painting the ice about 1 in. below the surface with whitewash or slaked lime. Commercial paints, generally water based, with a low solar absorptivity are also available.

ICE RINK CONDITIONS

Properly designed indoor rinks, as well as properly designed renovated rinks, can be operated year-round without shutdown. However, some indoor rinks operate from 6 to 11 months and shut down for various reasons, including maintenance, rink construction, inability to control indoor conditions, or unprofitable operation during part of the year. Outdoor, uncovered rinks generally operate from early November to mid-March above 40° North latitude. However, if sufficient refrigeration capacity is provided, the ice can be maintained for a longer period.

Indoor rinks operate successfully even in warm tropical climates. Relative humidity, temperature, and ceiling radiant losses must be controlled in these climates to prevent fog, ceiling dripping, and high operating cost.

Steel frame, brick, concrete, and various forms of plastic have been used to enclose ice skating rinks. Rinks have also been built under air-supported structures for seasonal use and are usually over a multipurpose surface.

Arena heating is frequently provided for skater and/or spectator comfort. Where airflow may be directed to the ice surface, space heating should not be combined with a dehumidification system. Space heating and dehumidification have different objectives: the dehumidification system removes moisture from the ice surface, whereas the space heating system provides comfort conditions for spectators. Warm air movement over the ice surface is not desirable; any air movement over the ice surface is detrimental to the control of ice and space temperature. Heat recovery from the refrigeration system may be used for limited heating, supplementing the heating system. Infrared heating has been used successfully for spectator areas. Ice rink temperatures are usually maintained between 40 and 60°F; however, for skater or spectator comfort, higher temperatures are sometimes preferred. Relative humidity in the arena depends on factors such as building construction, indoor temperature, and outdoor wet bulb temperature.

The system should be designed to prevent fogging and surface condensation. A maximum dew-point temperature of 45°F is usually sufficient to eliminate fogging; however, condensation can occur on the ceiling or roof structure because of radiation from the building structure to the ice. Low relative humidity is needed to reduce this condition when a high-emissivity ceiling is exposed to the ice surface. Ceiling emissivity and height are critical factors in controlling roof and ceiling condensation. Low ceilings and dark-colored structures promote condensation because these features favor radiant heat flow toward the ice surface. The result is a low structure temperature that could be near the dew-point temperature of the space. Wire-suspended, low-emissivity ceiling curtains are known to raise the inside surface temperature of the roof structure, thus eliminating the condition where condensation could occur. Low-emissivity ceilings not only reduce heat flow between the roof and the ice surface, but also reflect light. This reduces the lighting requirement and therefore reduces the cooling load imposed on the refrigeration plant. The low-emissivity ceiling must resist damage from hockey pucks and allow free air circulation around its perimeter. Providing too much roof insulation can promote condensation by reducing the inside ceiling temperature.

Ventilation should be the minimum required for the building occupancy so that humidity introduced with outdoor air is kept as low as is feasible, but enough outdoor air must enter to maintain acceptable indoor air quality (see ASHRAE *Standard* 62.1). Makeup air should always be conditioned before being introduced to the arena space. Because of the high enthalpy difference between indoor and outdoor air, exhaust air energy recovery using enthalpy wheels improves efficiency. Mechanical makeup air dehumidification systems may be downsized up to 50% when using enthalpy wheels. It is more energy-efficient to dehumidify makeup air separately from the recirculated air, because makeup air usually has a much higher dew-point temperature. Self-contained, air-cooled, compressor-equipped dehumidifying units, as well as desiccant drier types with gas or electric regeneration, are available to control humidity. The owning and operating costs of various dehumidification and defogging systems should be evaluated.

Carbon monoxide and nitrogen dioxide are pollutant emissions from gasoline- or propane-fueled ice resurfacers. The concentration of these chemicals can reach dangerously high levels if they are not controlled or eliminated. In some areas, regulations require sensors to detect and alarm at unsafe chemical concentrations. Check health regulations for local requirements. Air circulation is conducive to removing carbon monoxide produced by ice-resurfacing equipment. Carbon monoxide is usually in the highest concentration below the top of the boards and near the ice surface. Gas-engine resurfacing machines should be equipped with catalytic exhaust convertors to reduce carbon monoxide emissions. Electric-powered resurfacing machines eliminate the need for additional makeup air otherwise required to dilute and ventilate the combustion products generated by internal combustion engines.

Each rink user group has its own preference for the type of ice used. Hockey players and curlers prefer hard ice; figure skaters prefer softer (i.e., warmer) ice so they can clearly see the tracings of their skates; and recreational skaters prefer even softer ice, which minimizes the buildup of shavings and scrapings.

With approximately 45°F air temperature and 1 in. ice thickness, ice at 20 to 22°F is satisfactory for hockey, 24 to 26°F for figure skating, and 26 to 28°F for recreational skating. A 1°F higher ice temperature may be feasible when water with a low mineral content is used for resurfacing. To achieve these ice temperatures, the coolant temperature is maintained about 6 to 10°F lower than the ice temperature. The temperature of the coolant must be lowered to maintain the same ice conditions when there are higher wet-bulb temperatures or abnormally high loads, such as when television lighting is used.

EQUIPMENT SELECTION

Compressors

Two or more refrigeration compressors should be used in an ice rink system. When two compressors are used, one compressor should be specified with ample capacity to maintain the ice sheet under normal load and operating conditions. When greater capacity is required during initial ice freezing or under high heat loads, the second compressor picks up the load. In multiple-compressor installations, a multistage thermostat microprocessor control, programmable logic controller (PLC), or computerized control system may be used to control the operation of the compressors. The multiple compressors serve as backups; they maintain the ice in the event of compressor failure or a service requirement.

Compressors and evaporators normally operate at a suction pressure corresponding to a mean temperature difference of 8 to 10°F between the coolant and primary refrigerant in systems operating with secondary coolants, or between the ice and the refrigerant in direct-refrigerant rinks.

Most arenas with a single sheet of ice use two or three reciprocating compressors. With trends toward multiple ice sheets served from a central plant, screw compressors are widely used. Development of smaller, more economical screw compressors has led to the use of screw compressors in applications that traditionally used reciprocating compressors.

Evaporators

Ideally, there should be one chiller (evaporator) for each ice surface. However, economics sometimes dictates one chiller serving multiple ice surfaces.

Chillers for indirect systems normally are shell-and-tube (with and without surface enhancement), immersed-tube, or plate-and-frame. Gravity-flooded or direct-expansion feed is usually used for the primary refrigerant.

Flooded shell-and-tube chillers used for cooling glycol or calcium chloride may be manufactured from carbon steel. Stainless steel is recommended for constructing plate-and-frame chillers used for cooling glycol. Titanium is recommended for plate-and-frame chillers used for cooling calcium chloride. Aluminum is not recommended for chillers cooling calcium chloride with ammonia.

The chiller for direct-cooled systems is composed of piping installed in the floor under the ice surface. Typically, direct halocarbon systems are fed by direct expansion or liquid recirculation. The operating charge should be considered for this type of system. Local codes may impose restrictions on the size and use of direct systems.

Care should be taken in designing common evaporators for multiple-rink facilities; the high load from one rink should not affect ice temperature on the other rinks.

Condensers and Heat Recovery

Ice arenas and curling rinks typically reject heat to a water source or the atmosphere.

Wells, lakes, or rivers can be good sources of condenser cooling water, if they are available. Capacity is easy to regulate and the low coolant temperature maintains low condensing pressures, which saves energy. Condensers require high-quality water, though, which may need treatment to prevent scale formation, fouling, or corrosion in the condenser tubes. Water and sewage costs usually prohibit the use of water for condensing on a "once-through" basis.

Cooling towers used in conjunction with water-cooled condensers, evaporative condensers, or air-cooled condensers are alternatives to once-through water-cooled condensers. When selecting a cooling tower or evaporative condenser, not only the maximum expected wet-bulb temperature during the skating season should be considered, but also suitable controls to cover the wide range in capacities and protection against freeze-up needed in cold weather. A water treatment specialist should also be consulted.

Air-cooled condensers can be designed to produce reasonable discharge pressures in northern climates, particularly where the rink is used mostly in the spring, fall, and winter months. They can be economically sized and require no water, so the possibility of freezing is eliminated. This type of condenser, however, is not economical for year-round operation, and for seasonal operation it must have wide-range capacity control.

One alternative to rejecting heat to a water source or to the atmosphere is to recover waste heat and put it to use. This process harnesses all or a portion of this wasted heat and uses it to preheat a secondary fluid before the primary refrigerant enters the condenser. Both reciprocating and screw compressors have potential for heat reclaim. Abundant heating energy is available from discharge gas, by either desuperheating or condensing.

In addition to the power saved by heating a secondary fluid with the refrigerant, the load to the condenser is reduced and reduction in condenser fan and pump motor operation results in further electrical savings. A lower operating head pressure reduces compressor motor power requirements and increases the operating life of the refrigeration equipment. Condenser fouling is also reduced because of the lower discharge temperature.

Superheat from discharge gas is reclaimed via a thermal contact surface in which superheated refrigerant transfers its energy to the cooler fluid on the opposing side. When water is used as the fluid to be heated, it is often recirculated through a storage tank and used on demand.

Heat of condensation is typically reclaimed through a designated condenser piped in parallel with the main condenser.

Whether to reclaim superheat or heat of condensation depends on the amount of heat that may be used and the temperature level the designer wishes to attain.

To minimize the possibility of contamination of potable fluids by a rupture in the exchanger, a double wall of heat transfer surface between the fluid and the refrigerant is required. Any leakage of primary refrigerant into the space between the inner and outer walls is vented to the atmosphere. Double wall-vented construction has become the accepted standard for heat-recovery units heating potable fluids and is required by law when used for this purpose.

The exchanger material must be suitable for use with both the refrigerant and the process fluid and must meet code requirements. Standard materials that usually meet these requirements are 304 and 316 stainless steel and titanium. Some applications for using waste heat are space heating, ice resurfacing, underfloor heating, building makeup air, boiler water makeup and domestic hot water use.

Most installations in large plants show a payback period of 18 to 30 months. Ice facilities with an 8 month operating season typically have a 3 to 5 year payback. Paybacks depend on the degree of heat reclaim, hours of operation, and cost of the fuel source, and must be analyzed on a project-by-project basis.

Figure 2 illustrates a desuperheater and a condenser piped for waste heat recovery.

Ice Temperature Control

Ice temperature may be controlled by (1) sensing the average temperature of the secondary coolant, (2) infrared sensor(s) hung over the ice surface, (3) thermocouples or thermistors embedded underneath or in the ice, or (4) thermocouples or thermistors placed in a well in the concrete floor under the ice surface.

Thermostats that sense return coolant temperature or the differential temperature between the supply and return coolant can be

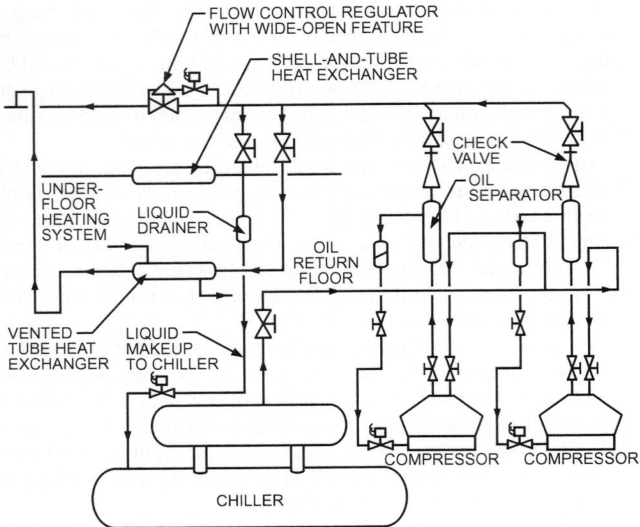

Fig. 2 Typical Waste Heat Recovery Piping

used to control the refrigeration system. They may also be used in controlling operation of the coolant pump. To be effective, a differential sensor should sense a small temperature difference. The return coolant temperature can be sensed by multistage sensors that sense a larger temperature difference. Another strategy varies coolant flow by controlling the pump speed on a signal from an ice temperature sensor. Direct-refrigerant systems can be controlled by regulating compressor operation from an ice temperature sensor. This method has been used with a direct refrigerant impulse pumping system. Compressor capacity and pump operation may be controlled from the low-pressure receiver when refrigerant pumps are used to circulate refrigerant.

Rink Piping and Pipe Supports

High-flow-rate secondary systems use standard mild steel pipe 0.75, 1, or 1.25 in. in diameter; thin-walled polyethylene plastic pipe 1 in. in diameter; or UHMW (ultrahigh molecular weight) polyethylene plastic pipe 1 in. in diameter. These are placed at 3.5 or 4 in. centers on the rink floor. One proprietary low-flow-rate secondary coolant system uses 0.25 in. tubing made of flexible plastic with tube spacing averaging 0.75 in. or one dual tube every 1.5 in. Direct-refrigerant rinks generally use 0.63 to 0.88 in. steel tubing, which is placed on 3 in. centers for outdoor rinks and 4 in. centers for indoor rinks.

The pipe grid must be kept as close to level as possible, regardless of the rink piping system used. When a pipe rink surface is open with sand fill around and over the pipes, the pipe usually rests on pressure-treated sleepers set level with the subbase; however, the sleepers can be omitted in a rink that is to be operated year-round. The piping is then spaced with clips, plastic stripping, or punched metal spacers.

In permanent concrete floors, the pipe or steel tubes are supported on notched iron supports or welded chair supports. The latter must be used in the case of plastic pipe.

Headers and Expansion Tanks

Secondary coolant rinks using large-diameter pipe generally run the piping lengthwise, with the supply and return headers across one end and the return bends located on the opposite end. Supply and return headers may be positioned in a header trench and the return bends may be positioned in a return bend trench. This allows regular inspection of the clamps and joints. However, to avoid having header and return bend trenches, some facilities have successfully used straight headers running across the rink between the blue line

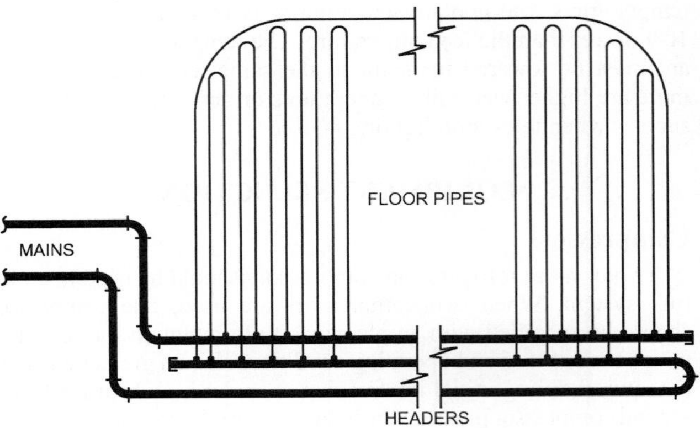

Fig. 3 Reverse-Return System of Distribution

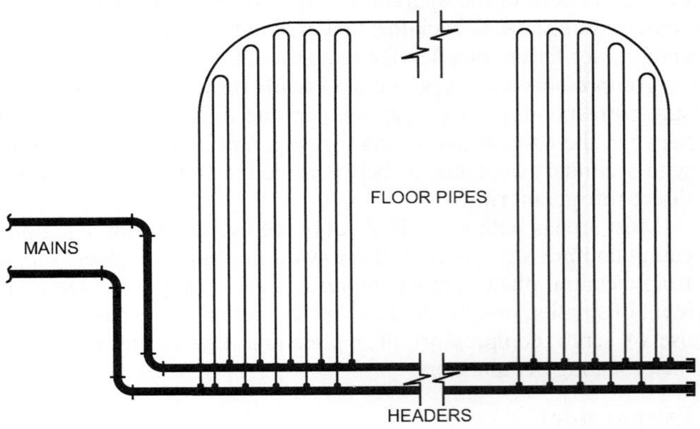

Fig. 4 Two-Pipe Header and Distribution

and center ice, buried below the floor piping grid. The cooling grid piping then crosses over the top of the headers to ensure consistent slab temperatures. Curved headers buried at one end of the rink have also been used. Rinks using small-diameter tubing generally run crosswise, with the supply and return headers along one side. Direct-refrigerant rinks generally run lengthwise, with the supply header at one end and the return header at the opposite end in a balanced system. The header must be sized to ensure an even distribution of coolant through every pipe. The systems are generally designed with low coolant velocities, which do not need balancing valves. If at all possible, the return header should be placed at the same elevation as the rink piping, with a minimum of two air vents to eliminate trapping of air.

The three-pipe reverse-return header and distribution arrangement (Figure 3) is used occasionally. A properly sized two-pipe header system (Figure 4) is frequently applied and gives nearly uniform circuit flow with no discernible differences in the ice surface temperature or conditions. To allow for thermal contraction and expansion, headers and main piping should be free to move without producing excessive stress.

Polyvinyl chloride (PVC) distribution headers are becoming popular because they have a very low maintenance requirement and accumulate less frost than steel headers. These headers should be used with proper allowances for expansion and contraction. The coefficient of thermal expansion for steel is relatively low and very close to that of concrete; the PVC pipe expansion coefficient is much higher. Schedule 80 wall thickness is used to provide a solid connection to the pipe nipples leading to the floor piping. PVC pipe can become brittle at low temperatures, so the pipe should not be used to support equipment weights, and the nipples should not be placed where they could be knocked or stood on. Pipe clamp connections are not considered permanent joints and should ideally remain accessible for inspection and tightening. However, use of clamped return bends and header connections cast in the concrete floor slab has been successful.

A closed secondary coolant system requires an expansion tank to safely accommodate the expansion and contraction of the coolant resulting from fluid temperature changes. The expansion tank must be installed so that it cannot be isolated from the system.

Coolant Equipment

The coolant circulating pumps must be sized for the particular type of rink and system involved. Large-diameter pipe rinks require 10 to 15 gpm per ton of refrigeration to maintain the required 2 to 3°F temperature differential between incoming and outgoing coolant. These operate at approximately 20 to 25 psig. Low-flow-rate dual tubing or mat rinks use about 2.4 gpm per ton. Differentials of 4 to 6°F are normal, but 10 to 12°F differentials can be experienced in high-load conditions. Temperature averaging is achieved in mat rinks by closely spacing adjoining counterflow tubes operating at approximately 40 to 50 psig pressure.

Ice Removal

For auditoriums and sports arenas, the rink surface should have provision for deicing in less than 4 h. In deicing, the floor is heated to about 50°F so that the bond between the floor and ice is broken; the ice is then removed with power tractors.

A standard heat exchanger can be used, with piping arranged so that all the coolant can be pumped through the heater, with coolant flowing in the tubes and steam or hot water in the shell. Approximately 200 to 350 Btu/h per square foot of rink surface is needed to heat the coolant in the system enough to warm the floor and break the ice bond.

Storage Accumulators

To reduce large cooling demands associated with frequently producing ice in short time periods, some older, large-event facilities incorporate storage accumulators that act as a source of low-temperature, large-volume secondary coolant.

When refrigeration on the ice surface is not required, a large volume of coolant in the accumulator may be cooled to approximately –25°F and be ready to be pumped into the rink piping when needed. The cold-coolant accumulator should store enough to cool the entire system coolant volume from 65 to 0°F. This cold-coolant tank usually holds more than three times the volume of the cooling system's charge.

Use of accumulators has been declining. Instead, ice-making equipment is sized to handle the demand loads.

Energy Consumption

Energy consumption for an ice rink facility is unique. Maintenance of internal conditions is affected by the cold ice sheet. Lighting, ventilation, heating, and dehumidification systems depend on the facility's use and occupancy. Energy consumed by refrigeration equipment is affected by construction, operation, water quality, and various use factors. To reduce heat load and energy consumption,

- Install low-emissivity ceilings to reduce refrigeration and lighting loads and to allow compressors to operate at a higher saturated suction temperature.
- Reclaim refrigerant superheat to preheat shower water, heat ice resurfacing water, melt ice shavings, heat the subfloor, etc.
- Select a pumping system and controls to reduce or stop coolant flow during part load conditions.
- Install an energy management system.
- Insulate the subfloor and header piping.
- Control temperature and humidity in the arena to reduce sensible and latent heat gain to the ice.
- Install high-efficiency luminaires.
- Use demineralized or very-low-mineral-content water for ice and resurfacing.
- Do not operate the underfloor heating system more than necessary to prevent frost formation.
- Maintain the secondary coolant temperature no lower than necessary to maintain the desired ice quality.
- Maintain high suction pressure and low discharge pressure.

Dehumidifiers

To minimize the potential for rink fog or ceiling dripping, a properly designed dehumidification system should be installed. Operating season, arena location, and utility costs should all be considered when selecting a system.

Desiccant systems use adsorption or absorption to remove moisture from rink air. These systems can provide arena ventilation while delivering dew-point temperatures below freezing. Desiccant systems come in many forms, from stand-alone dehumidifiers to total environmental control systems incorporating air conditioning, heating, and energy recovery. Single desiccant units, centrally located, have been used successfully to dehumidify multiple rinks.

Self-contained mechanical refrigeration units rely on the moisture removal capabilities of an evaporator to reduce the moisture content of air inside a rink. Either hot-gas or electric defrost is provided to remove ice that forms on the evaporator surface. These units are typically manufactured with hermetic or semihermetic compressors in the range of 5 to 7 1/2 hp. Multiple units are selected to suit the size of the facility and moisture removal needs. Typically, a standard hockey rink requires two self-contained units.

Some older facilities use secondary coolant from the arena to dehumidify the air. The temperature of the secondary coolant is too low for dehumidification purposes and should be mixed by recirculation to suit the application. This method of dehumidification is not energy efficient.

RINK FLOOR DESIGN

Generally, five types of rink surface floors are used (Figure 5):

- Open or sand fill, for plastic or metal piping or tubing
- Permanent, general-purpose, with piping or tubing embedded in concrete on grade
- All-purpose, with piping or tubing embedded in concrete with floor slab insulated on grade
- All-purpose floors, supported on piers or walls
- All-purpose floor with reheat; for use when the water table and moisture are severe problems or when the rink is to operate for more than 6 months

The open sand fill floor is the least expensive type of rink floor. The cooling pipes rest on wood sleepers over a bed of crushed stone or other fill. Clean, washed sand is filled in around the pipes. Curling rink floors, as well as hockey and skating rinks, where first cost is a factor and the building is not intended for other uses, are usually constructed in this manner. Clay or cinders should never be used in the bed or for fill around the pipes. Tubing rinks do not need supports or sleepers; the tubes are laid on accurately leveled sand.

Rinks using 1 in. plastic pipe or the mat type are usually covered with sand to a depth of 0.5 to 1 in. to provide additional strength to the ice surface and to reduce cracking. Many portable outdoor rinks have used this arrangement for laying plastic pipes or tubing mats on top of existing sodded areas, black top, or concrete. More permanent installations of outdoor semiportable rinks have used this same arrangement where recreational space is at a premium. Such an installation consists of steel pipes supported on notched steel sleepers, which in turn are supported on concrete piers down to solid ground.

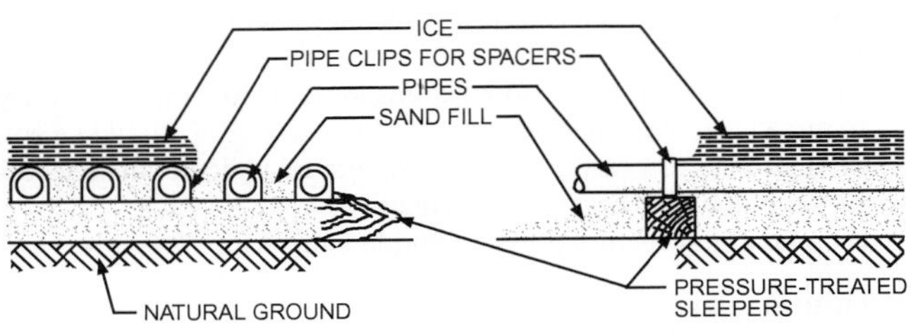

OPEN- OR SAND-FILL FLOOR WITH NO INSULATION

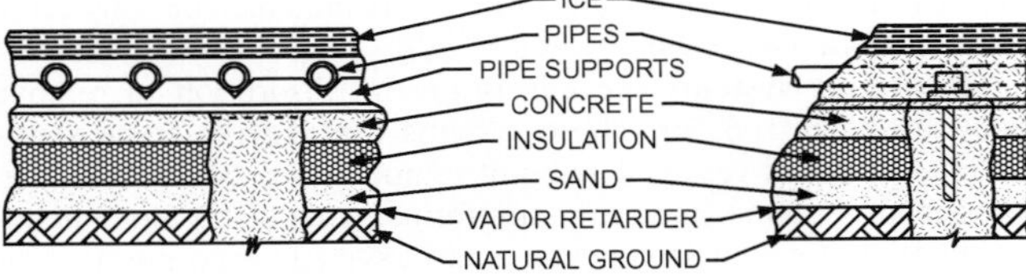

PERMANENT OR GENERAL-PURPOSE FLOOR

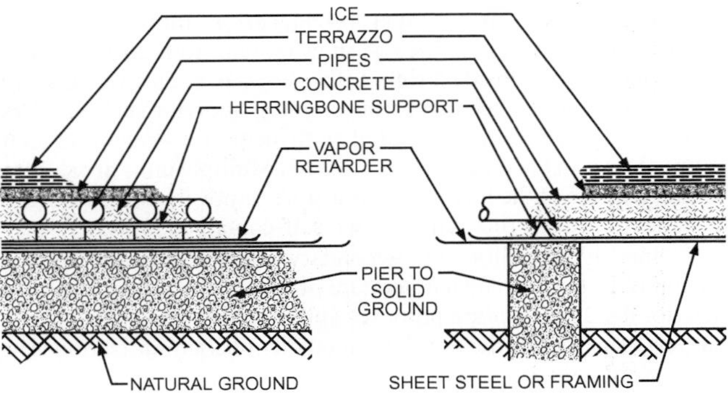

PERMANENT OR ALL-PURPOSE FLOOR

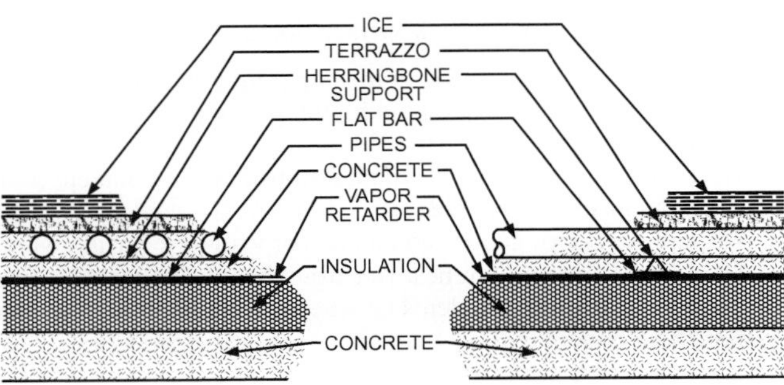

ALL-PURPOSE FLOOR

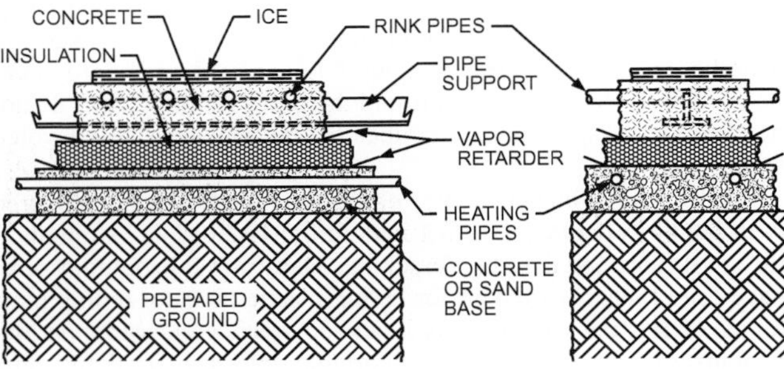

ALL-PURPOSE FLOOR WITH REHEAT

Fig. 5 Ice Rink Floors

To obtain a better return on investment, most indoor rinks that operate with an ice surface for only a portion of the year have a permanent general-purpose concrete floor with subfloor insulation and heat pipes so that the floor may be used for other purposes when the skating season is over. The floor should withstand the average street load and is usually designed with 1 or 1.25 in. steel or plastic pipe embedded in a steel-reinforced concrete slab 4 to 6 in. thick, depending on the anticipated loading and coolant pipe diameter.

In sports arenas, where the ice is removed and the floor made ready for other sports and entertainment, the ice floor must be constructed to withstand the frequent change from hot to cold. The refrigerating machinery must be of sufficient capacity to freeze a sheet of ice 0.63 in. thick in 12 h. This type of floor is always insulated.

Subfloor insulation must be installed when quick changeovers are desired, when the subsoil has a high moisture content, when the floor is elevated, or when the rink is in continuous use for more than 9 months. Subfloor insulation reduces refrigeration load on ice-making equipment and slows down, but does not eliminate, cooling of the subsoil on surfaces installed on grade.

Drainage

The suitability of an ice rink's subsoil greatly influences the rink's success. Complete ice surfaces have had to be rebuilt because of poor drainage and the ultimate heaving of the ice surface. Thus, skating rinks should not be built on swampy or low-lying land unless adequate drainage is provided.

Moist subsoil freezes in the ground to a depth of 4 ft or more. The frozen water will heave the ice surface when freezing occurs 6 in. deep or more. Heaving creates an uneven skating surface; moves and raises walls, piers, and header trenches; cracks walls and piping; and eventually necessitates drainage and rebuilding of the rink floor.

Not only should there be a complete drainage system around the footings of the rink to prevent seepage, but there should also be one under the rink surface itself. This is particularly important when a sand fill floor is used; a good system ensures that ice melted after the skating season completely drains away and the sand dries out as quickly as possible.

Subfloor Heating for Freeze Protection

Subfloor heating, by electrical heating cables or a pipe or tubing recirculating system using a warm antifreeze solution, is found in most new rinks to prevent below-floor permafrost development and heaving. Pipes or tubing are on 12 to 24 in. centers located under 2 to 4 in. of insulation. They are generally installed in sand rinks, which are used year-round, although they may be poured into a concrete base slab with insulation between the base slab and the rink slab (see Figure 5).

Alternatively, heating pipes may be laid directly in the subfoundation below the rink pipe or insulation. However, an uninsulated installation requires a greater depth between heating and ice-making pipes to prevent an increased load.

Neither water nor warm air should be used for subfloor heating. Water, if inadvertently allowed to freeze, cannot be readily melted out. In time, warm air ducts become filled with frost and ice because of high rink humidity and air duct leakage.

Usually, the same fluid used for the coolant in the ice-making system is used for subfloor heating; it can be heated to the necessary temperature (40 to 42°F) in a heat exchanger warmed by compressor waste heat. Subfloor insulation should be of a rigid moisture-proof board, such as high-density polystyrene foam, and be completely enveloped in a polyethylene vapor retarder.

Preparation of Rink Floor

When building on natural ground, regardless of whether a sand fill or a permanent general-purpose floor is intended, proper preparation of the bed is important unless the rink is built on elevated sand

and gravel subsoil. If the rink is to be built on clay, part clay, or rock subsoil, water should be prevented from collecting in low areas. Either the clay or rock should be excavated or the rink level should be built up with crushed stone and gravel to a height of about 4 ft, after which it should be well rolled. Water should not be used for settling the fill.

For sand-fill rinks, quickly draining melted ice at the end of the skating season ensures rapid drying of the sand and rink piping and results in a longer life for the steel piping. Cinders should never be used as fill in open sand-fill rinks because of the possibility of sulfur in the cinders, which, when damp, accelerates corrosion of steel piping.

Ensure a level surface over the entire rink, with no more than ±0.13 in. in any 10 ft^2 area and ±0.25 in. overall.

Permanent General-Purpose Rink Floor

When constructing a permanent general-purpose floor, the same subsoil precaution must be taken as for a sand-fill rink. The concrete floor should withstand, at a minimum, the average road pavement load.

When local conditions make it advisable, the rink floor should be insulated. Insulation may be laid on a level concrete or sand base.

The concrete mixture should have a 28 day strength of 3000 to 5000 lb/ft^2 and be put in place properly (a concrete engineer is recommended to specify concrete, its placement, and curing). Suitable cross-reinforcing and pipe supports are necessary.

Concrete floors with mat tubing are poured in two courses. A first course is poured and leveled; the mats are then rolled out and positioned. A 6 by 6 in. wire mesh is laid on top of the mats; then a second course, with grouting between it and the first, is poured on top of the first course, mats, and wire. Water pressure should be kept in the tubing to spot any leaks or cuts that may develop. Once started, the pouring of each course of the concrete floor should be continuous, with interruptions not to exceed 15 min.

General-purpose rink floors should not be defrosted too frequently. When a rink constructed with a general-purpose floor is to be used during the ice season for purposes that require an ice-free floor, it is preferable to place an insulated portable-section wood floor over the ice for each occasion.

All-Purpose Floors

If a rink floor as used in sports arenas is to withstand both the expansion and contraction of frequent frosting and defrosting and thermal shock because of the circulation of very-low-temperature coolant, then extra precautions must be taken in its construction, such as provisions for the free movement of the freezing slab with respect to the subfloor.

Header Trench

A well-constructed header trench of sufficient size to house the headers and connections and the subfloor heating system, if applicable, is essential unless the steel distribution headers are cast into the concrete slab as part of the rink. Provisions for movement of pipes caused by thermal expansion and contraction should be incorporated into the design. This trench should be equipped with removable covers and be well drained to facilitate drying out. The headers and piping in the trench are not usually insulated, which allows for periodic inspection and painting of the piping. Provision must be made for purging air from the rink piping and header system.

Snow Melt Pit

When ice is resurfaced mechanically, a thin layer of ice is removed (as snow) and replaced by a thin layer of clean, warm water. The snow may be placed outdoors and allowed to melt, but in many jurisdictions, this is not allowed, and it must be disposed of by other means. A common method of disposal is to place the snow in a snow melting pit.

The snow melting pit should be provided in the ice-resurfacer holding area. For a single rink, the capacity of the pit should be sufficient to hold and melt double the quantity of snow removed in one resurfacing (approximately 3200 lb of snow or ice). In addition, space must be allowed for water spraying, water reservoir, and free-board. About 1000 to 1600 lb of snow can be generated every 45 min for each active rink in a facility.

Heat to melt this snow can be obtained from a number of sources. By maintaining sufficient standing water in the pit, the stored heat can be used to melt the snow as it is dumped. The heat retained in the standing water should be sufficient to melt the entire load of snow as it is dumped. The temperature of water in the pit can be then restored over the next 45 min by several heating sources:

- Cold domestic water supply
- Hot domestic water supply
- Waste heat from the refrigeration plant on a recirculated system
- Waste heat from the jacket-cooling water from ammonia compressors

To minimize water consumption, a supplementary heat source is recommended, such as the waste heat from the refrigeration plant. Approximately 300,000 Btu/h of waste heat at 70 to 80°F is recommended for one rink. The standing water in the snow melt pit should not be circulated directly through conventional heat exchangers.

One successful layout of a snow melt pit is shown in Figure 6. Typical pit dimensions for a single rink are 8 by 10 by 6 ft. A large standpipe drain is required to handle the overflow of water during the snow dump. This drain should be equipped with a removable screen (minimum 12 in. diameter) to handle the large volume of trash scraped off the ice with the snow. The standpipe drain also allows for a standing water level to be maintained. For cleaning purposes, the pit can be pumped out with a portable sump pump when required.

Spray headers are also recommended around the top of the pit 2 ft above the water level to assist in snow melting if the water temperature is not high enough. Spray nozzles on 2 ft centers with a cone spray pattern are recommended. The spray header should be located so snow being dumped into the pit does not hit the header. Spray headers can be supplied with warm or cold domestic water. The snow melt pit should have a closable lid to prevent moist air from infiltrating the refrigerated rink space.

The snow melt pit is not intended for disposal of ice paint, which clogs the drain and is not permitted in most sewer systems. The layer of painted ice is not affected when the ice resurfacer is used to maintain the ice surface. When ice is being removed at the end of the skating season, the entire ice surface is scraped up and should be disposed of in a temporary dump that allows the ice paint to be separated. The temporary dump can be constructed in a parking lot using a wood frame and plastic ground sheet. When the ice has melted, the ice paint can be rolled up in the ground sheet and disposed of in an environmentally friendly manner.

BUILDING, MAINTAINING, AND PLANING ICE SURFACES

Regardless of the type of rink floor used, when the plant is first placed in operation, the equipment should be operated long enough for a sharp frost to appear on the surface. Then the entire surface should be uniformly covered with a fine spray. This process should be repeated until a 0.5 in. thickness of ice is built, or until the surface is level. After applying a layer of water-based white paint, another 0.38 in. thick layer of ice is built before painting the red and blue lines. Red and blue lines are available in plasticized paper; however, they need to be covered with a minimum of 0.5 in. of ice to protect against damage. It is essential that sand floors be thoroughly wet

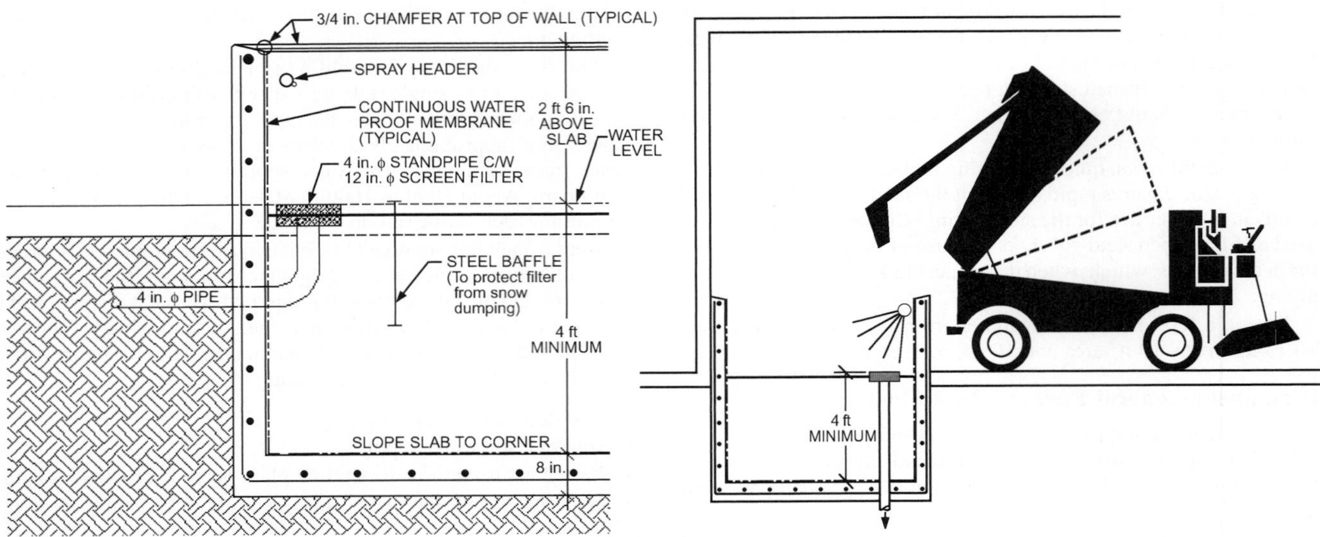

Fig. 6 Snow Melt Pit

before freezing because dry sand has poor conductivity. The surface should not be frozen any colder than required after this buildup, to allow the ice to temper before it is used for skating, and also to deter cracking.

To maintain an ice surface, it is customary to scrape off the snow after each skating session or hockey period. In all but the smallest rinks, this is done by a motorized resurfacer. On small rinks, scraping is done manually with a wide hardened-steel scraper blade. The most satisfactory method of resurfacing the ice between sessions is to wheel a sprinkler tank filled with hot water over the ice. The sprinkler has an adjustable valve to control the quantity of water, which is sprayed into a terry cloth bag that wipes the fine snow off the ice surface and fills the crevices cut by the skaters. In this manner, the least amount of water is added, reducing ice build-up and refrigeration load.

By far the most common method is the use of automatic resurfacing machines. Mounted on four-wheel drive chassis, the machines plane the ice, pick up the snow, and lay down a new ice surface using hot or cold water. Hot water generally gives harder ice, because air bubbles are removed, but energy costs have led many rinks to alternate hot- and cold-water resurfacings. Rink corners should be at least a 20 ft (preferably 28 ft) radius for effective use of this equipment. Smaller equipment is available for studio and small rinks.

Because of inattentive ice making, improper sprinkling equipment, or deep cutting of the ice during public skating, the ice may become uneven and excessively thick. There may be a fairly slight variation in ice thickness across the rink, but more serious is the resulting variation in ice condition. In any case, low spots on the ice must be built up, increasing the thickness and refrigeration requirements.

For example, under assumed conditions, where 18°F coolant would be cold enough to hold a 1.5 in. thickness of ice, calculations show that −5°F coolant would be required if the ice were allowed to build up to 6 in., with a corresponding decrease in effective refrigeration capacity and an increase in operating costs. In other words, every additional inch of ice thickness required from the refrigeration system increases costs 8 to 15%, depending on system heat load (DOE 1980).

Because ice 0.5 to 1 in. thick is satisfactory for skating and is the most economical to freeze and hold, the ice should be periodically planed to maintain this desired thickness.

Pebbling

Pebbling is a term used to describe the surface finish applied to curling ice. The pebbles, actually water droplets frozen to the ice surface, reduce friction between the bottom of the curling rock and the ice. This makes the rock glide more easily and promotes the "curl" of the rock when a turn is applied to the handle of the rock on release.

The temperature of water used for pebbling is critical and varies by facility, depending on ice surface temperature, water quality, humidity, and application techniques. If the pebbling water temperature is too warm, the pebbles will be too flat. If the pebbling water temperature is too cold, the pebbles can break off when the rock passes over. Pebbles are applied manually from a water can with a hose connected to a perforated sprinkler head. The water can is carried by a shoulder strap and the sprinkler head is held in one hand. The person applying the pebbles usually walks backward down the curling sheet, sprinkling the water in a rhythmic side-to-side motion.

Water Quality

Water quality affects energy consumption and ice quality. Water contaminants, such as minerals, organic matter, and dissolved air, can affect both the freezing temperature and the ice thickness necessary to provide satisfactory ice conditions. Proprietary treatment systems for arena flood water are available. When these treatments are properly applied, they reduce or eliminate the effects of contaminants and improve ice conditions.

IMITATION ICE-SKATING SURFACES

A number of different imitation ice-skating surfaces have been marketed; these use semiporous plastic panels dressed with a synthetic lubricant. The coefficient of friction of ice is approximately 0.03 at 26°F and is further reduced by the film of water produced by pressure under the skate. In considering the use of imitation surfaces, the actual friction coefficients of these surfaces, both when freshly lubricated and after a period of use, should be investigated.

REFERENCES

ASHRAE. 2004. Ventilation for acceptable indoor air quality. ANSI/ASHRAE *Standard* 62.1-2004.

Connelly, J.J. 1976. ASHRAE Seminar on Ice Rinks (February), Dallas.

DOE. 1980. *Energy conservation in ice skating rinks*. Prepared by B.K. Dietrich and T.J. McAvoy. U.S. Department of Energy, Washington, D.C.

BIBLIOGRAPHY

Albern, W.F. and J.J. Seals. 1983. Heat recovery in an ice rink? They did it at Cornell University. *ASHRAE Journal* 25(9):38-39.

ASHRAE. 1968. Ice skating rinks. Symposium at ASHRAE meeting in Columbus, OH.

Banks. N.J. 1990. Desiccant dehumidifiers in ice arenas. *ASHRAE Transactions* 96(1):1269-1272.

Blades, R.W. 1992. Modernizing and retrofitting ice skating rinks. *ASHRAE Journal* 34(4):34-42.

Brauer, M., J.D. Spengler, K. Lee, and Y. Yanagisana. 1992. Air pollutant exposures inside hockey rinks: Exposure assessment and reduction strategies. *Proceedings of the Second International Symposium on Safety in Ice Hockey*, Pittsburgh, PA.

Canadian Electricity Association. 1992. Potential electricity savings in ice arenas and curling rinks through improved refrigeration plant. CEA *Publication* 9129 U 858.

Matus, S.E., A.H. Stern, M. Hopkins, R. Martinez, B. Johnson, and W. Wallace. 1988. Carbon monoxide poisoning at an indoor ice skating facility. *Proceedings of the ASHRAE IAQ '88 Conference*, pp. 275-283.

Minnesota Department of Health. 1990. *Indoor air quality unit: Regulating air quality in ice arenas*. St. Paul.

Rein, R.G. and C.M. Burrows. 1981. Basic concepts of frost heaving. *ASHRAE Transactions* 87(2):1087-1097.

Strong, R.H. 1990. Refrigeration Theory and Safety Course for Arena Operators.

CHAPTER 36

CONCRETE DAMS AND SUBSURFACE SOILS

REFRIGERATION is one of the more important tools of the heavy construction industry, particularly in the temperature control of large concrete dams. It is also used to stabilize both water-bearing and permanently frozen soil. This chapter briefly describes some of the cooling practices that have been used for these purposes.

CONCRETE DAMS

Without the application of mechanical refrigeration during construction of massive concrete dams, much smaller construction blocks or monoliths would have to be used, which would slow construction. By removing unwanted heat, refrigeration can speed construction, improve the quality of the concrete, and lower the overall cost.

METHODS OF TEMPERATURE CONTROL

Temperature control of massive concrete structures can be achieved by (1) selecting the type of cement, (2) replacing part of the cement with pozzolanic materials, (3) using embedded cooling coils, or (4) precooling the materials. The measures used depend on the size and type of structure and on the time permitted for its construction.

Cement Selection and Pozzolanic Admixtures

The temperature rise that occurs after concrete is placed is due principally to the cementing materials' heat of hydration. This temperature rise varies directly with cement content per unit volume and, more significantly, with the type of cement. Ordinary Portland cement (Type I) releases about 180 Btu/lb, half of which is typically generated in the first day after the concrete is placed. Depending on specifications, Type II cement may generate slightly less heat. Type IV is a low-heat cement that generates less heat at a slower rate.

Pozzolanic admixtures, which include fly ash, calcined clays and shales, diatomaceous earths, and volcanic tuffs and pumicites, may be used in lieu of part of the cement. Heat-generating characteristics of pozzolans vary, but are generally about one-half that of cement.

When determining system refrigeration load, heat release data for the cement being used should be obtained from the manufacturer.

Cooling with Embedded Coils

In the early to mid-1900s, the heat of curing on large concrete structures was removed by embedded cooling coils for glycol or water recirculation. These coils also lowered the structure's temperature to its final state during construction. This is desirable where volumetric shrinkage of a large mass is necessary during construction (e.g., to allow the contraction joint grouting of intermediate abutting structures to be completed).

In an embedded coil system, thin-wall tubing is placed as a grid-like coil on top of each 5 or 7.5 ft lift of concrete in the monoliths. Chilled water is then pumped through the tubing, using a closed loop system to remove the heat. A typical system uses 1 in. OD tubing with a flow of about 4 gpm through each embedded coil. Although the number of coils in operation at any time varies with the size of the structure, 150 coils is not uncommon in larger dams. Initially, the temperature rise in each coil can be as much as 8 to 10°F, but it later becomes 3 to 4°F. An average temperature rise of 6°F is normal. When sizing refrigeration equipment, the heat gain of all circuits is added to the heat gain through the headers and connecting piping.

For a typical system with 150 coils based on a design temperature rise of 6°F in the embedded coils and a total heat loss of 6°F elsewhere, the size of the refrigeration plant is about 300 tons. Figure 1 shows a flow diagram of a typical embedded coil system.

Cooling with Chilled Water and Ice

The actual temperature of the mix at the time of placement has a greater effect on the overall temperature changes and subsequent contraction of the concrete than any change caused solely by varying the heat-generating characteristics of the cementing materials. Further, placing the concrete at a lower temperature normally results in a smaller overall temperature change than that obtained with embedded-coil cooling. Because of these inherent advantages, precooling measures have been applied to most concrete dams.

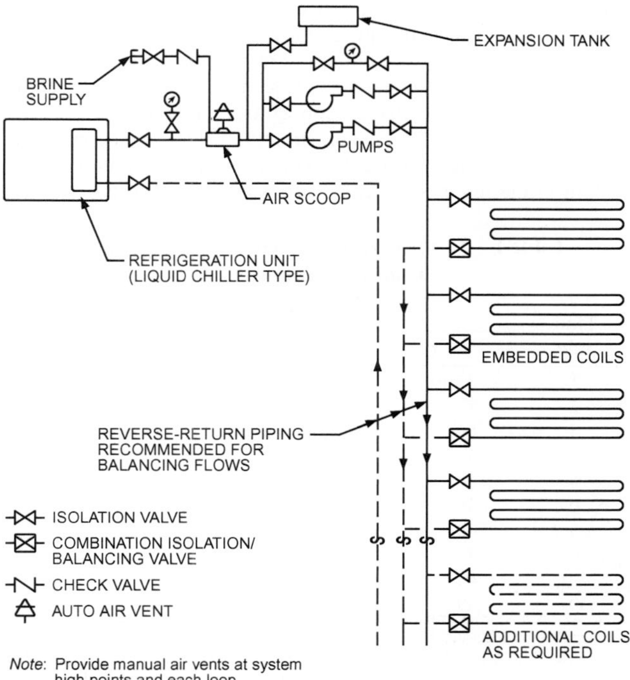

Fig. 1 Flow Diagram of Typical Embedded Coil System

The preparation of this chapter is assigned to TC 10.1, Custom-Engineered Refrigeration Systems.

Glen Canyon Dam illustrates the installation required. The concrete was placed at a maximum placing temperature of 50°F during summer, when the aggregate temperature was about 87°F, cement temperature was as high as 150°F, and the river water temperature was about 85°F. Maximum air temperatures averaged over 100°F during the summer months. The selected system included cooling aggregates with 35°F water jets on the way to the storage bins, adding refrigerated mix water at 35°F, and adding flaked ice for part of the cold-water mix. Subsequent cooling of the concrete to temperatures varying from 40°F at the base of the dam to 55°F at the top was also required. The total connected brake power of the ammonia compressors in the plant was 6200 hp, with a refrigeration capacity equal to making 6000 tons of ice per day.

The maximum amount of chilled water that may be added to the concrete mix is determined by subtracting the amount of surface water from the total mix water, which is free water. Frequently, if a chemical admixture is specified, some water (usually about 20% of the total free water) must be added to dissolve the admixture. This limits the amount of ice that can be added to the remaining 80% of free water available. After the amount of ice needed for cooling is determined, the size of the ice-making equipment can be fixed. When determining equipment capacity, allowances should also be made for cleaning, service time, and ice storage during nonproductive times.

When calculating heat removal, consider ice to be 32°F when introduced into the mixer. Chilled water is assumed to be 40°F entering the mixer, even though it may be supplied at a lower temperature.

Cooling by Inundation

The temperatures specified today cannot be achieved solely by adding ice to the mix. In fact, it is not possible on heavy construction of this type (in view of low cement content and low water/cement ratio specified) to put enough ice in the mix to obtain the specified temperatures. As a result, inundation (deluging or overflowing) of aggregates in refrigerated water was developed and was one of the first uses of refrigeration in dams.

When aggregates are cooled by inundation with water, generally the three largest sizes are placed in large cylindrical tanks. Normally, two tanks are used for each of the three aggregate sizes to provide back-up capacity and a constant flow of materials. Cooling tanks, loaders, unloaders, chutes, screens, and conveyor systems from the tanks into the concrete plant should be enclosed and cooled from 45 to 40°F by refrigeration units, with blowers placed at appropriate points in the housing around the tanks and conveyors.

Peugh and Tyler (House 1949) determined the inundation (or soaking) time required by calculations and actual tests. Pilot tests corroborated their computations of aggregates' cooling times as indicated in Table 1.

This study indicated that immersion for about 40 min brings the aggregate down to an average temperature of 40°F. Theoretically, smaller sizes can be brought to the desired temperature in less time. Considerations such as the rate at which cooling water can be pumped make it unlikely that a cooling period less than 30 min should be considered. Any excess cooling provides the needed safety factor. However, the limiting factor on the overall cycle is the cooling time for the largest aggregate, which is nearly 45 min, plus about 15 min for loading and unloading. Back-up capacity should be considered for maintaining a constant flow of materials.

Table 1 Temperature of Various Size Aggregates Cooled by Inundation

Time, min.	Aggregate Size, in.				
	6	3	1.5	0.75	0.375
1	85°F	69°F	49°F	39°F	38°F
2	77	59	41	38 (42)	
5	66	46 (45)	38 (42)		
10	56	40 (42)			
15	50	38			
20	46				
25	44				
30	42				
40	40				
50	39				

Source: Peugh, V.L. and I. Tyler. 1934. "Mathematical theory of cooling concrete aggregates." In House (1949). Numbers in parentheses are from tests by R. McShea.
Note: Temperatures listed are at cobble center with assumed thermal diffusivity of 0.07 ft²/h. Aggregate initial temperature is 90°F and cooling water temperature is 35°F.

Air-Blast Cooling

Air-blast cooling, a more recent development than inundation, does not require particular changes to material handling or additional tanks for inundation; instead, cold air is blown through the aggregate in batching bins above the concrete mixers. Also, the air cycle used in cooling can be used in heating aggregates during cold weather. The aggregate is cooled during the final stage of handling; this does not increase the moisture content.

The compartmented bins where air-blast cooling usually occurs are generally sized so that if any supply breakdowns occur, the mixing plant will not have to shut down before a particular pour can be completed. On this basis, the average concrete octagonal bin above the mixing plant on a large job holds at least 600 yd³, which is usually more than adequate to allow time for air cooling. However, certain minimum requirements must be considered. If possible, based on a 1 h loading and cooling schedule, at least 2 h of storage volume should be provided for each size aggregate that air will cool. The minimum volume should be 1.5 h plus cycling time, based on the tables shown for cooling 6 in. aggregate.

The bin compartment analysis shown in Table 2 may be used as a starting point in determining the air refrigeration loads and static pressures. This type of analysis should give approximately equal storage periods for each aggregate size used. In practice, after air-cycle cooling is calculated, more air volume is needed in the smaller aggregate compartments, and less in the sand and aggregate, to obtain maximum cooling. This is because of the higher air resistance in the smaller aggregate sections and the fact that air does not cool the sand compartment as effectively.

Computing Air-Blast Cooling Loads. To calculate the required cooling, several assumptions must be made:

Assumption 1. Normally, the lowest temperature of air leaving the cooling coils is 38 to 40°F. Lower air temperatures may be achieved, but should not be trusted: a temperature lower than 35°F usually causes rapid frosting of the coils.

Assumption 2. Heat transfer between the aggregate and air is only 80 to 90% effective. To allow for air temperature rise in the ducts, heat leakage, pressure drop, etc., an empirical factor of 85% may be used.

Table 2 Bin Compartment Analysis for Determining Refrigeration Loads and Static Pressures

Material Size, in.	lb/yd³	% of Total	No. of Bin Compartments	Bin %
Stones				
6 to 3	800	21.33	4	25.00
3 to 1.5	700	18.67	3	18.75
1.5 to 0.75	650	17.33	2	12.50
0.75 to 0.25	600	16.00	3	18.75
Sand	1000	26.67	4	25.00
Total	3750	100.00	16	100.00

Note: In practice, relative sizes vary; amounts shown are for an assumed design mix of principal classes of concrete. On any given job, several design mixes requiring different amounts for each size are needed.

Table 3 Resistance Pressure

Aggregate Size, in.	Velocity, fpm	Resistance Pressure, in. of water/ft of ht.
6 to 3	300	0.29
3 to 1.5	200	0.32
1.5 to 0.75	100	0.27
0.75 to 0.25	60	0.30

Thus, with 80°F aggregate and 40°F cooling air, the effective temperature differential would be 0.85(80 – 40) = 34°F.

Assumption 3. The rise in temperature of air passing through the aggregate compartment normally will not exceed 80% of the difference between the entering aggregate and the entering air temperatures. Thus with 80°F aggregate and 40°F air, the temperature rise of the air is 0.80(80 – 40) = 32°F. The maximum temperature of the return air is 40 + 32 = 72°F.

Assumption 4. An allowance should be included for heat leakage into air ducts on the aggregate bin sides, normally about 2% of a total air-blast cooling load.

Another important consideration is the static pressure against air flowing through a body of aggregates. The resistance to air flowing through a bin varies as the square of the air volume or velocity; this is summarized in Table 3. The resistance pressure listed is for a unit height of aggregate. To use values in Table 3, the cross-sectional area of the aggregate compartment and the height of the aggregate column must be known. Manufacturers of concrete mixing bins and mixing equipment can supply this information.

Other Cooling Methods

As specifications require lower and lower placing temperatures, direct methods of cooling sand and cement have been attempted to obtain or lower the heat removal required. No method has been proven to cool cement. Sand-cooling methods that have been tried and found to be unsuccessful are

- **Water inundation.** The increase in free moisture content of batched sand makes correctly proportioning the mix difficult.
- Moving sand through **screw conveyors with hollow flights**. Chilled water was pumped through the flights, but because components were cooled below the dew point, resulting in condensation, serious handling problems occurred.
- **Vacuum systems,** which evaporate surface moisture to reduce aggregate temperature. The unreliability of the equipment and the batch nature of the process precluded success.

An alternative method of cooling sand, **air cooling**, is being tried on some projects, but it is too early for final results of this method.

For small pours, **liquid nitrogen** is sometimes used to reduce the temperature of the mixture to the final pour temperature. Nitrogen is used because the initial capital costs are considerably lower than a mechanical system, but because of the high cost of manufacturing nitrogen, the operation cost is much greater than a mechanical system. Cost comparison must be done on a case-by-case basis.

SYSTEM SELECTION PARAMETERS

For larger installations, plant selection depends on a number of factors, including the following:

- Normal pouring rate, yd^3/h (contractor or contract specified).
- Maximum pouring rate, yd^3/h (contractor or contract specified).
- Total allowable mixing water, lb/yd^3 (usually contract specified).
- Required concrete placement temperature (usually contract specified).
- Concrete temperature when coming from the mixer (to be determined considering materials handling to placement site, time in transit, and storage at placement site).
- Average ambient temperature and aggregate temperature during period of maximum placement. The average ambient temperature of the aggregates is assumed to be the mean ambient temperature (including night and day) during the period of storage, which is determined by the amount of storage capacity provided by the contractor and the rate of concrete placement. If the minimum live storage is, for example, 100,000 tons of aggregates and the pouring rate is 2000 yd^3/day, consumption is approximately 3800 tons of aggregates per day. If the job is working a 6 day week, this provides 26 days of storage. Unless weather conditions are unusual, the temperature of rock delivered to the reclaiming tunnel is assumed to equal the average ambient temperature for the 26 days preceding the delivery.
- Specific heat of materials. The specific heat of the aggregates and sand may vary with project location. Typical values are sand = 0.106 Btu/lb·°F; aggregates = 0.12 Btu/lb·°F; water = 1.0 Btu/lb·°F; cement = 0.12 Btu/lb·°F.
- Heat release rate for cement (material specifications).

Where the aggregate cooling range from initial to final temperature of the mix is relatively small (15 to 20°F), or where the required pour temperature is relatively high (65°F or more), chilled water plus ice in the mix or chilled water plus air blast on larger aggregates may handle the entire cooling load. When the overall temperature reduction is greater than this, or when lower pour temperatures are specified, such as 50°F or less, a combination of all three types of cooling will probably be required because only a limited amount of heat removal can be obtained by one of these methods alone.

Cooling by air blast alone is limited by the entering air temperature, which must be maintained high enough to prevent coil frosting. Cooling by inundation, although it requires large inundation tanks, offers the most positive and sure method of cooling. Ice can also be added to the mix to remove the remainder of the heat. The result is a very satisfactory blending of the aggregates and exact control of the amount of water in the mix.

CONTROL OF SUBSURFACE WATER FLOW

Refrigeration has been used successfully since 1880 to freeze moisture in unstable and water-bearing soil and to stop underground flows of water in pervious material or gravelly stream beds. Other common methods of stopping water flow include sheet piling, cement grout, chemicals, and well points. In many cases, freezing has been the last resort after other methods are unsuccessful. In a number of cases, a combination of well points, grouting, and freezing has solved the problem.

Applications include

- Deep excavations for building foundations (to prevent water from seeping into the excavation before the foundation is poured)
- Deep ditches for laying pipe (to keep the banks from sliding in)
- Large dam excavations (to stop water seepage until the dam footings are poured)
- Mining (to temporarily stabilize areas affected by water seepage)

Using refrigeration, the common practice is to lay a series of concentric pipes in a line or arch pattern in the path of the subsurface water flow. A wall of earth is frozen by pumping cold brine down the inside pipe and letting it flow back through the annular space between the inside and outside pipe. The growth of frozen soil on the outside of the concentric pipe proceeds until it connects with the frozen cylinder formed on the adjacent pipe.

Spacing between pipes can vary, depending on the time available to complete the wall of ice, but spacings of 2 to 4 ft on center have been used successfully. Freezing pipes have been successfully used to control subsurface water in both dam and

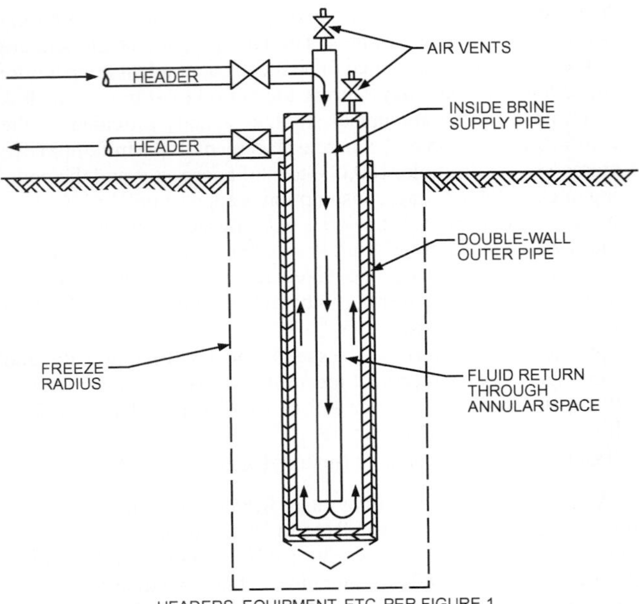

Fig. 2 Typical Freezing Point

mining projects. Figure 2 shows a typical freezing pipe and system diagram.

The brine or refrigerant and method of containment should be carefully selected in system design. If a leak develops in the system, groundwater contamination could result or the soil saturated with the refrigerant may have to be excavated and cleaned, depending on the type of refrigerant used. Double-wall piping, or an environmentally safe refrigerant that vaporizes when exposed to air, should be considered.

Ammonia is generally the basic refrigerating medium, and brine, chilled by the ammonia, is circulated through the freezing pipes. The brine is commonly calcium chloride ($CaCl_2$), but magnesium chloride ($MgCl_2$) is recommended because it is less likely to precipitate and clog piping at low temperatures. To monitor freezing progress, thermocouples are normally located throughout the area to be frozen. Brine temperatures may range from +15 to –20°F, depending on the state of freezing and amount of time available to complete the task.

Liquid nitrogen has been used for small projects.

SOIL STABILIZATION

In latitudes where areas of permanently frozen earth, or permafrost, prevail, methods of soil stabilization for building and equipment foundations are required. These methods range from providing a non-frost-susceptible gravel pad to rigid below-grade insulation plus an active or passive refrigeration system to freeze the soil and keep it frozen. Because many systems are in remote areas, simplicity and reliability are major factors in system design.

THERMAL DESIGN

Piling Design

Frost heave has long been a problem for designers of piles and buildings in the arctic or subarctic. Uplift forces as high as 13.5 tons/ft of perimeter must be taken into account when designing nonthermal piling (Long and Yarmak 1982). Designers have used sleeves, greases, waxes, and plastics to reduce the adfreeze bond in the active layer and to reduce frost heave forces, but Long and Yarmak found that almost all these methods were only temporary. Increased pile embedment remains the only sure way to prevent frost heaving of conventional piling. Thermopiles can effectively eliminate frost heaving forces. The thermopile freezes the active layer radially from the pile. As a result, active layer temperatures adjacent to the pile remain at the same temperature as the rest of the pile.

Slab-on-Grade Buildings, Outdoor Slabs, and Equipment Pads

When a slab-on-grade building, outdoor slab, or equipment pad is constructed on a permafrost area, the resulting soil thawing or thaw bulb must be considered. Thermopiles, thermoprobes, or an active refrigerated foundation system may be required to stabilize the structure, slab, or pad.

Design Considerations

Soil properties have a pronounced effect on the capacity requirements for passive or active refrigeration systems. Soil within the radius of influence is an integral part of the refrigeration system. Highly conductive soils increase the system's radius of influence and allow more heat to be pumped out of the subsoils. Conversely, poorly conductive soils reduce the radius of influence; thermal lag through the soils is high, and the heat transfer rate low. Soil moisture contents and soil classifications are valuable for estimating thermal conductivities. Back-up capabilities for active and passive systems should be considered to avoid foundation failure.

PASSIVE COOLING

The three processes used by passive systems for heat removal are air convection, liquid convection, and two-phase liquid/vapor convection. All passive refrigeration systems rely on the temperature differential between the soil and winter air to operate. When the temperature of soil in contact with the refrigeration system is lower than the air temperature, the system is dormant.

Air Convection Systems

Air convection has been used to provide subgrade cooling below on-grade and pile-supported structures. For a passive air convection system to work, two design criteria must be met: the air outlet must be higher than the inlet to promote convection, and the air distribution system must be designed so that the friction imposed by the distribution system is low enough to allow convection to begin. Air convection systems are usually designed to take advantage of the prevailing winds at a specific site. The wind can push air through a distribution system at greater velocities than convection would allow; however, wind in the arctic frequently carries large quantities of snow. Distribution ducts can be blocked by snow and ice, causing the system to fail (Long and Yarmak 1982). Ducting fans are built into many air convection systems for active refrigeration back-up.

Liquid Convection Systems

Liquid convection has been used to provide subgrade cooling below on-grade structures. Radiator and heat absorber portions of the system are normally connected by either a double or single pipe with a flow splitter to decrease frictional losses in the system and to provide maximum cooling of the working fluid. Examples of some working fluids are trichloroethylene, kerosene, and methanol and water. Frictional losses within the liquid system are high and limit heat transfer rates. Large-diameter pipes may be used to overcome frictional losses so that high heat transfer rates can be achieved. Some liquid convection systems allow an option for mechanical circulation of the working fluid as an active refrigeration back-up. Liquid systems must be sealed to avoid leakage into subsoils. Introducing the working fluid into permafrost subsoils could depress the soil freezing point and increase the probability of foundation failure.

Two-Phase Systems (Heat Pipes)

Two-phase liquid/vapor convection systems are the most widely used passive refrigeration systems for permafrost foundations and earth stabilization. A typical two-phase unit is constructed of pipe enclosed at both ends and charged with a passive refrigerant gas. The radiator (aboveground condenser) portion of the unit can have a bare or finned surface, depending on heat transfer requirements (Figure 3). The evaporator portion of the unit can have any configuration as long as a slope remains between the evaporator and the radiator (Figure 4).

Refrigeration of the subgrade occurs when the radiator has a lower temperature than the soil in contact with the bottom of the evaporator, where the liquid portion of the refrigerant pools (Yarmak and Long 1982). Condensation occurs in the radiator, initiating evaporation of the refrigerant in the evaporator. The condensate wets the walls of the unit and flows down to the evaporator. Reevaporation of refrigerant condensate with subsequent cooling occurs where the soil in contact with the evaporator is warmer than the soil adjacent to the liquid pool of refrigerant at the bottom of the unit. Then the entire evaporator unit is reduced in temperature, cooling the surrounding soil.

A two-phase system can start with a temperature differential as little as 0.01°F between the radiator and the evaporator. Liquid and air convection systems may require temperature differentials of 4 to 15°F before they start.

Propane, butane, halocarbons, anhydrous ammonia, and carbon dioxide have been used as the refrigerant gas in two-phase systems. Choice of refrigerant gas depends primarily on the allowable internal pressure capabilities of the vessel containing the gas, quality of available gases, molecular stability of the gas, and preference of either the customer or manufacturer of the system. Relatively low-pressure systems using propane, halocarbons, or anhydrous ammonia have been known to gas lock (i.e., gases other than the refrigerant accumulate in the radiator portion of the unit and prevent the refrigerant gas from condensing). Purging or venting of noncondensable gases may be required after start-up.

ACTIVE SYSTEMS

Active ground-freezing refrigeration systems are used to keep building or equipment foundations stable when a passive system will not maintain the required stability. The system is normally a network of underground ductwork or piping through which a cooling fluid flows. Heat is removed using a heat exchanger or cooling coil, with refrigeration provided by medium- or low-temperature refrigeration units. System design can include provisions to bypass the refrigeration units during cold weather. Outside air should

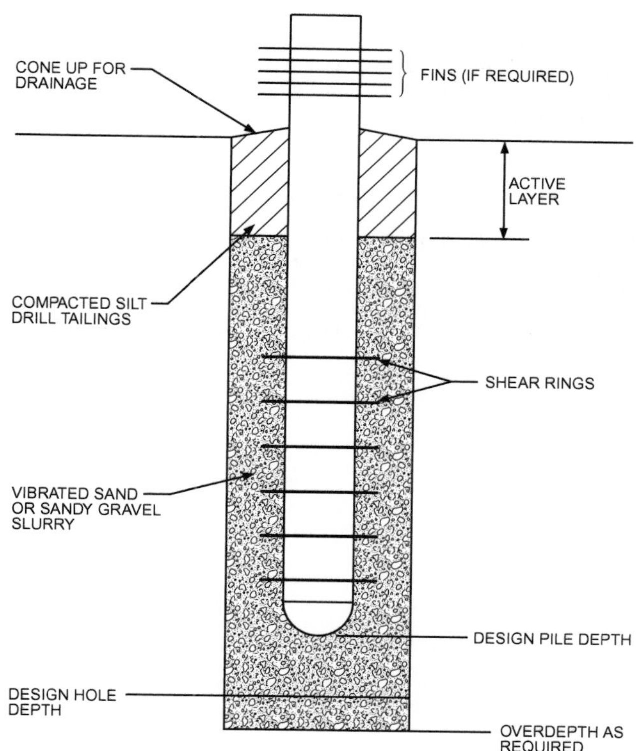

Fig. 3 Thermo Ring Pile Placement

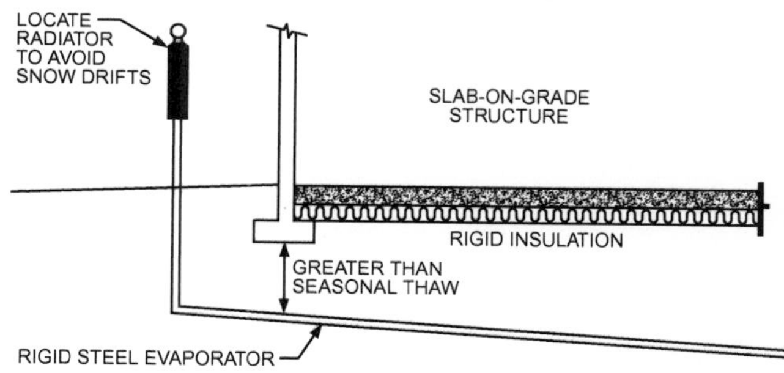

Fig. 4 Typical Thermo-Probe Installation

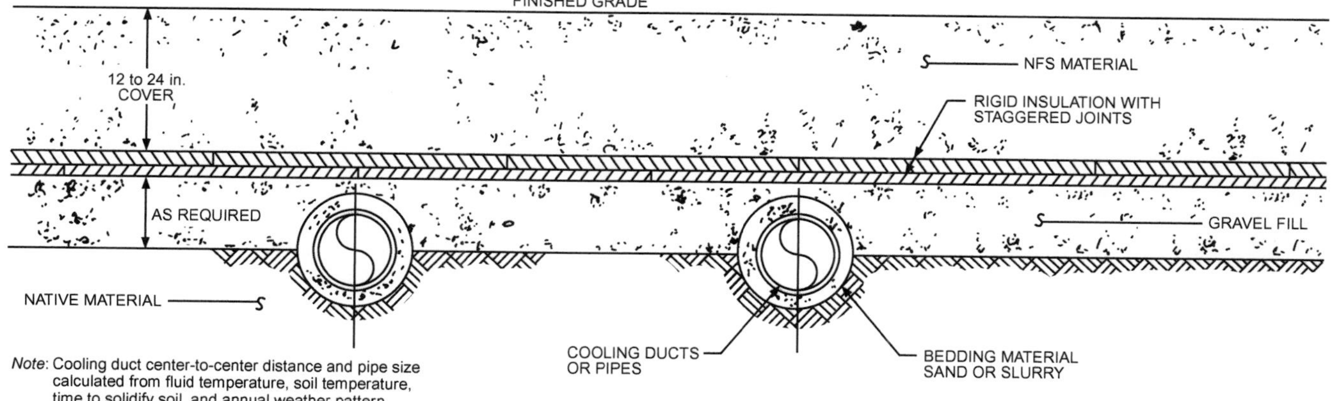

Fig. 5 Active Ground Stabilization System

not be used directly in systems that use air as a cooling fluid because moisture can be introduced into the system in the form of snow and humidity. A defrost system for the cooling coil in a circulating air system should be considered. Figure 5 shows a typical system installation.

REFERENCES

House, R.F. 1949. *The inundation method of cooling concrete aggregates at Bull Shoals Dam*. U.S. Corps of Engineers, Little Rock, AR.

Long, E.L. and E. Yarmak, Jr. 1982. *Permafrost foundations maintained by passive refrigeration*. Petroleum Division, Ocean Engineering Division, American Society of Mechanical Engineers, New York.

Yarmak, E., Jr. and E.L. Long. 1982. *Some considerations regarding the design of two-phase liquid/vapor convection type passive refrigeration systems*. Petroleum Division, Ocean Engineering Division, American Society of Mechanical Engineers, New York.

BIBLIOGRAPHY

Kinley, F.B. 1955. Refrigeration for cooling concrete mix. *Air Conditioning, Heating, and Ventilating* (March):94.

Long, E.L. 1963. The long thermopile, permafrost. *Proceedings of International Conference*, National Academy of Sciences, Washington, D.C., pp. 487-491.

Townsend, C.L. 1965. Control of cracking in mass concrete structures. *Engineering Monograph* 34. U.S. Department of Interior, Bureau of Reclamation, Washington, D.C.

CHAPTER 37

REFRIGERATION IN THE CHEMICAL INDUSTRY

CHEMICAL industry refrigeration systems range in capacity from less than one ton of refrigeration to thousands of tons. Temperature levels range from those associated with chilled water through the cryogenic range. The degree of sophistication and interrelation with the chemical process ranges from that associated with comfort air conditioning of laboratories or offices to that where reliable refrigeration is vital to product quality or to safety.

Two significant characteristics identify most chemical industry refrigeration systems: (1) almost exclusively, they are engineered one-of-a-kind systems, and (2) equipment used for normal commercial application may be unacceptable for chemical plant service.

This chapter gives guidance to refrigeration engineers working with chemical plant designers so they can design an optimum refrigeration system. Refrigeration engineers must be familiar with the chemical process for which the refrigeration facilities are being designed. Understanding the overall process is also desirable. Computer programs are also available that can calculate cooling loads based on the gas chromatographic analysis of a process fluid. These programs accurately define not only the thermodynamic performance of the fluid to be chilled, but also the required heat transfer characteristics of the chiller.

Occasionally, because the process is proprietary, refrigeration engineers may have limited access to process information. In such cases, chemical plant design engineers must be aware of the restrictions this may place on providing a satisfactory refrigeration system.

FLOW SHEETS AND SPECIFICATIONS

The starting point in attaining a sound knowledge of the chemical process is the flow sheet. Flow sheets serve as a road map to the unit being designed. They include information such as heat and material balances around major system components and pressures, temperatures, and composition of the various streams in the system. Flow sheets also include refrigeration loads, the temperature level at which refrigeration is to be provided, and the manner in which refrigeration is to be provided to the process (e.g., by a primary refrigerant or a secondary coolant). They indicate the nature of the chemicals and processes to be anticipated in the vicinity in which the refrigeration system is to be installed. This information should indicate the need for special safety considerations in system design or for construction materials that resist corrosion by process materials or process fumes.

Different portions of a process flow sheet may be developed by different process engineers; consequently, the temperature levels at which refrigeration is specified may vary by only a few degrees. Study may reveal that a single temperature is satisfactory for several or even all users of refrigeration, which could reduce project cost by eliminating multilevel refrigeration facilities.

Most process flow sheets indicate the design maximum refrigeration load required. The refrigeration engineer should also know the minimum design load. Process loads in the chemical industry tend to fluctuate through a wide range, creating potential operational problems.

Flow sheets also indicate the significance of the refrigeration system to the overall process and the desirability of providing redundant systems, interlocking systems, and so forth. In some cases, refrigeration is mandatory to ensure safe control of a process chemical reaction or to achieve satisfactory product quality control. In other cases, loss or malfunction of the refrigeration system has much less significance.

Other sources of information are also valuable. A properly prepared set of specifications and process data expands on flow sheet information. These generally cover the proposed process design in much more detail than the flow sheets and may also detail the mechanical systems. Information about the design principles, including continuity of operation, safety hazards, degree of automation, and special start-up requirements, is generally found in the specifications. Equipment capacities, design pressures and temperatures, and materials of construction may be included. Specifications for piping, insulation, instrumentation, electrical, pressure vessels and heat exchangers, painting, and so forth are normally issued as part of the design package available to a refrigeration engineer.

It is imperative that refrigeration engineers establish effective communication with chemical process engineers. The refrigeration engineer must know what information to request and what information to give to the chemical process engineer for design optimization. The following sections outline some of the significant characteristics of chemical industry refrigeration systems. A full understanding of these peculiarities is of value in achieving effective communication with chemical plant designers.

REFRIGERATION: SERVICE OR UTILITY

Refrigeration engineers unfamiliar with the chemical industry must understand that, unless the chemical process is cryogenic, chemical plant designers probably consider refrigeration merely as a service or utility of the same nature as steam, cooling water, compressed air, and the like. Chemical engineers expect the reliability of the refrigeration to be of the same quality as other services. When a steam valve is opened, chemical engineers expect steam to be available instantly, in whatever quantities demanded. When steam is no longer required, the engineer expects to be able to shut off the steam supply at any time without adversely affecting any other steam user or the source of steam. The same response from the refrigeration system will be expected. This high degree of reliability is usually so strongly implied that no specific mention of it may be made in specifications.

Because refrigeration is frequently considered a service, process designers spend insufficient time analyzing temperature levels, potential load combinations, energy recovery potentials, and the like. The potential for minimizing the size of the refrigeration system, the total plant investment, or both, by providing refrigeration at a minimum number of temperature levels, is frequently not investigated by

The preparation of this chapter is assigned to TC 10.1, Custom-Engineered Refrigeration Systems.

process engineers. Likewise, the potential for power recovery is frequently overlooked.

Part of the reason for this attitude is that refrigeration facilities represent only a minor part of the total plant investment. The entire utilities installation for the chemical industry usually falls in the range of 5 to 15% of the total plant investment, with the refrigeration system only a small portion of the utilities investment. Process requirements may be overruling, but process engineers must recognize legitimate process necessities and avoid unnecessary and costly restrictions on the refrigeration system design.

LOAD CHARACTERISTICS

Flow sheet values generally indicate direct process refrigeration requirements and do not include heat gains from equipment or piping. Flow sheet peaks and average loads generally do not allow for unusual start-up conditions or off-normal process operation that may impose unusual refrigeration loads. This information must be gained by a thorough understanding of the process and by discussing the potential effect on refrigeration system design for off-normal process conditions with the process engineer.

Once the true peak loads are established, the duration and frequency of the peaks must be considered. For some simple processes this information is fairly straightforward; however, if the plant is designed for both batch operation and multiproduct manufacture, this can be an enormous task. Computer simulation of such a combination of processes has led to optimization of not only the refrigeration equipment but also the process equipment. Computer simulation ensures that the refrigeration machine, secondary coolant storage tanks, and circulating pumps are properly sized to handle both average and peak loads with a minimum investment. Few applications require computer simulation, but a thorough understanding of the relationship of peak loads to average loads and their influence on refrigeration system component sizing is vital to good design.

Sometimes unusually light load conditions must be met. If the process is cyclic and on/off operation is undesirable, the refrigeration system may run for significant periods under a very light load or even no load. Light loads often require special design of the system controls, such as multistep unloading and hot-gas bypass with reciprocating compressors, a combination of suction throttling or prerotation vanes and hot-gas bypass with centrifugal compressors, and slide valve unloading and hot-gas bypass with screw compressors. A secondary coolant system might require a bypass arrangement.

Investment in refrigeration equipment can be minimized when only a few levels of refrigeration are required. Checking the specified temperatures to be sure they are based on some process requirement may show that fewer temperature levels than shown on the flow sheet are necessary. If multiple levels of refrigeration are required, a compound system should be evaluated. The evaluation must consider the limited ability of a compound system to provide the precise temperatures required of some processes.

Production Philosophy

The specifications generally indicate whether a chemical process is in continuous or batch operation, but further research may be required to understand the required continuity of service for the refrigeration facilities. The chemical industry frequently requires a high degree of continuity; plant production rates are often based on 8000 h or more per year. In general, the refrigeration equipment is worked extremely hard, with no off-peak period because of seasonal changes, and any unscheduled interruption of refrigeration service may create large production losses. In most cases, scheduled maintenance shutdowns are not only infrequent but also highly vulnerable to cancellation or delays because of production requirements.

As a result, reliability is key in the design of chemical industry systems. Equipment that is satisfactory in commercial or light industrial service is frequently unfit where high service rates and minimum availability for maintenance are the rule. In some cases, duplicate systems are justified. More often, multiple part-capacity units are installed so that a refrigeration system breakdown will not create a total process production loss. Major equipment and hardware items that require minimal maintenance or that allow maintenance with the refrigeration system in operation should be selected (e.g., dual lube oil pumps, oil filters, and in many cases dual oil coolers, compressors, and bypasses around control valves) to allow operation during service of control valve and other components; many systems are built to American Petroleum Institute (API) specification. Particular attention should be paid to equipment layout, so that adequate access, tube pullout space, and laydown space are available to minimize refrigeration system maintenance time. In some cases, overhead steel supports for rigging heavy equipment, or permanent monorails, are justifiable.

Flexibility Requirements

The chemical industry constantly develops new processes; consequently, the usual chemical plant undergoes constant modification. On occasion, total processes are rendered obsolete and scrapped before design production rates are ever reached. Thus, flexibility should be designed into the refrigeration system so it may be adaptable to some process modification. Designing for optimum flexibility is difficult; however, study of the potentials or probabilities of process modifications and of the expected life of the process facility helps in making design decisions.

SAFETY REQUIREMENTS

Most chemical processes require special design to ensure safe operation. Many raw materials, intermediates, or finished products are themselves corrosive or toxic or are potential fire or explosion hazards. Frequently, chemical reactions in the process generate extremes of pressure or temperature that must be properly contained for safe operation. Refrigeration engineers must be aware of these potential hazards as well as any abnormal hazards that may develop during start-up, unscheduled shutdown, or other upset to the chemical process. When designing modifications or expansions for an existing facility, the possibility that certain construction or maintenance techniques may be safety hazards must be considered.

Corrosion

The shell, tubes, tube sheets, gaskets, packing, O rings, seal materials, and components of instrument or control hardware must be properly specified. The potential hazards of leakage between the normal process side and the refrigeration side of heat exchange equipment must be investigated, because an undesirable chemical reaction may occur between the process material and the refrigerant.

An additional corrosive hazard may be leaks, spills, or upsets in the process area. Safe chemical plant design must anticipate unusual as well as usual hazards. For example, if refrigerant piping runs adjacent to a flanged piping system containing a highly corrosive material, special materials for the piping and insulation systems may be justified.

Toxicity

If the refrigeration system indirectly contacts a toxic material through heat exchange equipment, flanges and elements such as gaskets, packing, and seals in direct contact with the toxic material must be designed carefully. The possibility of a leak in equipment that might allow refrigerants or secondary coolants to mix with process chemicals and cause a toxic or otherwise dangerous reaction must also be considered. In some cases, the potential for toxic leaks may be so high that a special ventilation system may be required, double-tube sheets are used, or the tubes are welded and rolled into the tube sheets of shell-and-tube heat exchangers. Intermediate reboilers may also be used to isolate the process.

Even though containment and ventilation can handle toxic materials under normal process conditions, toxic material may need to be vented in abnormal situations to avoid the hazards of fire or explosion. In such cases, the toxic material is frequently vented or diluted through a flare stack so that ground- or operating-level concentrations do not reach toxic limits. Refrigeration engineers should evaluate the desirability of locating the refrigeration equipment itself or its controls outside the operating areas. An alternative is to ventilate either the refrigeration system or its controls with a system that has a remote air intake.

Toxicity hazards are not confined to the chemical process itself. Some common refrigeration chemicals are toxic in varying degrees. Because of the chemical industry's interest in safety, refrigeration engineers are required to treat some of the common refrigerants and secondary coolants with much more caution than required for the usual commercial or industrial system.

Fire and Explosion

Plant specifications normally define the area classifications, which determine the need for special enclosures for electrical equipment. For areas designated as being in an explosionproof environment, standard components are grouped in a single large explosionproof cabinet. Intrinsically safe control systems, which eliminate the need for explosionproof enclosures at all end devices, are also used. Intrinsically safe shutdown systems eliminate arcing at the end device by using low-voltage signals controlled by a microprocessor. Control equipment can also be mounted in standard or weatherproof cabinets and kept at a positive pressure with an inert gas purge system.

Items other than the electrical system may require special consideration in areas of high fire or explosion hazard. Use of flammable materials should be carefully reviewed. Stainless steel instrument tubing should be chosen rather than unprotected plastic tubing, for example. Both insulation materials and insulation finish systems should minimize flame spread in the event of a fire. In some extremely hazardous areas, nonflammable refrigerants or secondary coolants may be required, rather than flammable materials that would provide otherwise superior performance.

In areas of high explosion potential, usual refrigeration system design may need to be modified. Design pressures for refrigeration vessels or piping may be determined by process considerations as well as refrigeration system requirements. Special pressure-relief systems, such as rupture disks in series with relief valves, dual full-sized relief valves, or a parallel relief valve and rupture disk with transfer valves between them, are often required, using API relief valves.

Refrigeration System Malfunction

A malfunction can itself be a significant safety hazard, whether the upset is caused by an internal system failure or by fire, explosion, or some other catastrophe. Some process areas designated as being in a hazardous environment are protected by automatic gas-detection systems that shut down the refrigeration system when explosive mixtures are detected in an area. Loss of refrigeration may allow a process reaction to run out of control and cause loss of product, fire, explosion, or release of toxic material in an area remote from the original source of trouble. Thus, one or several degrees of redundancy may be needed to minimize the consequence of refrigeration system malfunction.

Storage of cold coolant, ice, or cold eutectic, or an alternative emergency supply of cooling water may be necessary to meet emergency peaks. Alternative sources of electric power to the refrigeration system may be desirable. Dual drive capability by either electric motor or steam turbine might even be justified. Uninterruptible power systems may also be used for control systems.

Frequently, special facilities are used to protect against the extreme consequences of refrigeration system malfunction. One example is quenching the process reaction by an inhibiting or neutralizing chemical introduced to the process in the event of an unusual pressure or temperature rise. Another simpler and more common example requires closing one or more process valves in the event of a loss of refrigeration.

Maintenance

In an operating chemical plant, because of hazards frequently encountered, normal maintenance procedures may not be allowed. Because welding, burning, or using an open flame is often prohibited throughout large areas of a chemical plant, maintenance flanges or screwed connections are used to permit replacement of piping and equipment. Sometimes extra access space or handling facilities (such as monorails) are provided to permit efficient removal of machinery to an area where welding, burning, and the like is permitted.

EQUIPMENT CHARACTERISTICS

Automation

In chemical plant operations, instrumentation represents a significant percentage of the plant investment. For this reason, most chemical plant designers insist on standardization of instrumentation throughout the plant. These requirements may not include familiar refrigeration components and may create problems in refrigeration design. Therefore, it is vital that refrigeration and chemical or instrumentation engineers agree on instrument requirements early in the design phase.

A second concept frequently adopted is control and monitoring of all plant operation from a single central control room. It is not unusual to operate a multimillion-dollar processing facility with two central control room operators and perhaps a single roving operator. This concept influences the refrigeration system in several ways. Refrigeration system controllers and alarm and shutdown lights are mounted in the distributive control system (DCS) panel, as are recorders or indicators that display refrigeration system temperatures, pressures, flows, and so forth.

Even with DCS operation, a local panel is required for starting, stopping, displaying, or recording additional information that can aid in troubleshooting an emergency shutdown. Frequently, start-up control is available only at this local panel, to ensure that start-up is not attempted unless an operator is present to witness the operation of major items of refrigeration equipment. DCS-mounted hardware should certainly conform to the standards of the process instrumentation to minimize operator confusion either in reading informative devices or in operating control devices. Most plants now use centralized computer or microprocessor controls. In most cases, the refrigeration unit is controlled from the local PLC or microprocessor.

When applying DCS, it is important to determine exactly how much information and control are to be provided at the control room and how much are to be provided locally. Transmission of unnecessary information to the DCS can be costly, but if enough information is not available, serious process upsets are inevitable. Most process operators are not trained to understand the intricacies of refrigeration machinery. The DCS must allow monitoring of the refrigeration system performance and control of that performance to suit process needs.

Operators require alarms to indicate abnormal conditions for which they can make corrections, either at the DCS or in the field, and to indicate system malfunction or shutdown. They also require a locally mounted manual shutdown station in the event of an emergency. Devices such as sequencing alarms and lube oil or bearing temperature recorders, which are troubleshooting aids, can be checked or logged by the local PLC.

Designing for a minimum of operator attention with personnel who are not refrigeration specialists is yet another reason that a high degree of reliability is required for a chemical industry refrigeration system.

Outdoor Construction

Another chemical industry characteristic is outdoor construction. The chemical industry installs sophisticated process and auxiliary facilities outdoors all over the world. Problems imposed by low temperatures, heavy snows and freezing rain, dust storms, baking heat, or hurricane-force winds with salt-laden rains are generally unfamiliar to the uninitiated refrigeration engineer. For example, explosionproof electrical construction is not necessarily weather resistant. Lube oil heaters and prestart circulation of heated lube oil may be required for compressors or other rotating machinery. In areas with high winds, special attention must be paid to the detailed installation instructions for insulation jacketing applied to pipe or vessels.

Winter operation of cooling towers may require multiple-cell construction, two-speed or reverse rotation cooling-tower fans, or even facilities for steam heating the cooling-tower basin. Instrument air for transmission of signals or power to pneumatic operators should be dried to a dew point (under pressure conditions) lower than anticipated ambient conditions to avoid condensate or ice forming in the instrument air lines or in the instruments themselves. In fact, the almost-standard expectation is that the instrument air provided is both oil-free and dried to a low dew point. The effect of ambient temperature, as well as radiation from the sun, should be considered in determining system design pressures, especially when equipment may be idle. Ambient temperatures up to 120°F and vessel skin temperatures of 165°F can be experienced in hot climates. To determine whether purchased equipment meets the requirements for outdoor installation, a detailed check of vendors' drawings, specifications, descriptive literature, and vendor-procured components is required.

Energy Recovery

Both installed and operating costs for the refrigeration system of some chemical processes can be reduced by intelligent use of energy recovery techniques. If the process requires large quantities of low-pressure steam, using back-pressure turbines to drive the refrigeration compressors could significantly reduce refrigeration system energy costs. On the other hand, if the process generates excess low-pressure steam, an absorption system may provide an overall saving. For processes with excess low-pressure steam only during the summer months when heating requirements are at a minimum, a condensing turbine may be economical if it is sized to operate at the low steam pressure when the excess is available and at a higher pressure during the heating season.

Other energy imbalances occurring within a chemical process can be advantageous. A waste heat boiler installed in a high-temperature gas stream may provide a source for low-cost steam. If the gas stream is at a moderate pressure as well as a high temperature, the possibility of a gas-driven power recovery turbine should be considered.

Another way to reduce operating costs is reusing once-through cooling water. Frequently, turbine-driven centrifugal refrigeration machines can use cooling water from the refrigerant condenser to condense the turbine exhaust steam, either in a shell-and-tube or a low-level jet condenser. Another possibility is the reuse of refrigeration system cooling water in process heat exchangers. Again, a thorough understanding of the process is a prerequisite for understanding the energy recovery potential, and a flow sheet should be helpful.

Performance Testing

Frequently, a requirement for performance testing of the refrigeration facilities is included within the contract for a chemical industry processing unit. Agreement should be reached as early as possible between the owner and the contractor regarding the exact procedure to be used for testing. If the test is to be run at some condition other than design conditions, both parties must agree on the methods of converting the test results to design conditions. Approximation techniques, such as those outlined in Air-Conditioning and Refrigeration Institute (ARI) *Standard* 550/590, are usually unacceptable in the chemical industry. The refrigeration engineer must be ensured that adequate facilities for an equitable test are designed into the refrigeration system. This may require additional flow-metering devices or more accurate temperature-measuring devices than are required for normal plant operation.

Insulation Requirements

The service conditions imposed by the chemical industry on both piping and equipment insulation are frequently more exacting than those in usual commercial or industrial installation. Not only must the initial integrity be as near perfect as possible, but it must also resist the high degree of both physical and chemical abuse that it is likely to incur during its lifetime. To achieve both a minimum permeability and maximum resistance to abuse, multicomponent finish systems may be required. In some cases, a vapor barrier mastic coating system (which usually includes reinforcing cloth) is covered with aluminum, stainless steel, or an epoxy-coated carbon steel jacket to protect against physical and chemical abuse. Piping and equipment insulated under ideal shop working conditions must be designed to withstand loading and unloading and erecting into position on the job site without damaging the vapor barrier.

Because the fire hazard is generally high and the potential loss of personnel and investment resulting from a fire is prohibitive, a strict limit is usually placed on insulation systems having a high flame-spread rating, particularly in indoor construction.

The frequent use of stainless steel in the chemical industry for piping and equipment creates another problem with regard to insulation system design. Many stainless steels fail when they are exposed to chlorides. Stress corrosion cracking can occur in a matter of hours. Consequently, chloride-bearing insulation materials must not contact stainless steels even in minute quantities and should not be used anywhere in the system unless a valid vapor retarder is interposed. Chapter 24 of the 2005 *ASHRAE Handbook—Fundamentals* and Chapter 33 of this volume cover the general subject of thermal insulation and water vapor retarders.

Nonflammable insulation material is usually used, such as foam glass (even though it has a much lower resistance to heat flow).

Design Standards and Codes

Relatively few suppliers of refrigeration equipment regularly manufacture to meet codes or standards that apply to the chemical industry. Another variation from commercial or industrial design practice is the use of company standards. For the usual commercial or industrial plant, the client at best provides performance specifications and a statement of what is to be done, leaving preparation of detailed specifications for equipment, piping, ducting, insulation, and painting to the designer. However, many such items are covered by company standards, which, though established primarily for use in the process cycles, can yield corporate benefits if they are also used in design of the refrigeration systems. A request for all applicable company standards at the start of the design and an effort to use them will avoid costly rework of the design following a review by the client.

START-UP AND SHUTDOWN

Processes are most hazardous during start-up and shutdown. Although present in batch or discontinuous processing units, the problem is usually more severe in continuously operating units for the following reasons:

- Instrumentation and control must be designed for the normal condition, and the cost of features intended for use only during start-up or shutdown often cannot be justified. Frequently, conditions at start-up or shutdown fall outside the range of the operating instruments and control, so that manual control is necessary.
- The same argument holds for much of the process equipment, so that extraordinary measures, such as severely throttled flow, minimum-flow bypasses, and recycling, may be needed.
- Operators go through these conditions only infrequently and may have forgotten the techniques of operation at the time they are most needed.
- Process conditions at start-up and shutdown are usually not recorded on flow sheets or in descriptions because they occur so infrequently and usually vary continually as units are brought on and off stream. Consequently, designers tend to overlook them and concentrate on conditions in the operating range.

Refrigeration engineers must inquire whether start-up or shutdown is likely to impose any special conditions on the refrigeration system. Start-up and shutdown are a special burden in this respect, because operators are particularly busy with the processing equipment and cycle during these times and generally cannot monitor or adjust operation of what they regard as a service system. Therefore, process engineers must be made thoroughly aware of the precise limitations that start-up or shutdown of the refrigeration system may impose on process operation.

REFRIGERANTS

Factors such as flammability, toxicity, and compatibility with proposed construction materials may influence final selection of a refrigerant more than in other applications. Special attention should be paid to the consequences of leakage between the process materials and the refrigerant or the secondary coolant. Chapters 19 and 21 of the 2005 *ASHRAE Handbook—Fundamentals* discuss refrigerants and secondary coolants in detail.

In addition to traditional refrigerants, other refrigerants used in the petrochemical industry are hydrocarbons. Produced in many refineries and readily available, they are very good refrigerants, but are flammable. When hydrocarbons are used as a refrigerant in a plant, the area around the system is process-classified as flammable. The most common hydrocarbon refrigerants are propane, propylene, ethane, and ethylene; they are used in many cases where the process stream involves them as constituents. Some cryogenic processes use a mixed hydrocarbon cycle.

Because of their nontoxic, nonflammable properties, halogenated hydrocarbons have been used predominantly, but recent environmental concerns have reduced their application throughout the chemical industry.

Of the secondary coolants, calcium and sodium chloride brines have been used most often, although glycols and such halocarbons as methylene chloride, trichloroethylene, R-11, and R-12 also have been frequent choices. Again, environmental concerns predicate against using R-11 and R-12 in new facilities.

Many of the same factors that influence refrigerant selection also affect secondary coolant choice. Corrosivity, toxicity, and stability are of special significance in determining suitability for chemical plant service.

REFRIGERATION SYSTEMS

An indirect system, in which brine or chilled water is circulated to air washers, cooling coils, and process heat exchangers from a central refrigeration plant, is much more prevalent in the chemical industry than in the food industry or in residential or light commercial comfort air conditioning. This is particularly true where large capacities or low temperature levels are involved. An indirect system allows centralization of the refrigeration equipment and associated auxiliaries, which may offer significant advantages in operation and maintenance, particularly if remote location of the refrigeration equipment permits design, operation, and maintenance in a nonhazardous location. It also may allow the installation of a minimum number of large units rather than many small units located in remote areas. For low-temperature systems of significant capacity, an indirect brine cooling system installed in the process area close to the process users is common.

Where the number of process heat exchangers requiring cooling and the length of piping can be kept to a minimum, a direct system, which uses the refrigerant in the process heat exchange equipment, often is the optimum design, particularly for small or medium loads. Because an indirect heat exchanger is not required in this case, a higher operating suction pressure and consequent lower operating and investment costs may be possible. Direct systems are also used when a refrigerant is involved in the manufacturing process stream, as in the production of ammonia or many petrochemicals. Here, the length of refrigerant lines, with possible high refrigerant losses because of leakage, is a less significant factor in system selection. Direct systems are usually of the flooded evaporator design; flooded coil systems of the gravity feed or pumped liquid overfeed design are relatively uncommon.

Direct systems for larger-capacity, low-temperature, multiple-user service have the following disadvantages that must be considered for proper system selection:

- Keeping an extensive refrigeration piping system leak-free is difficult. Leakage from piping for secondary coolants is frequently less objectionable than the refrigerant gases. Checking for refrigerant leaks or repairing them in some high-explosion-hazard process areas can be a problem because electronic leak detectors may not be permitted and burning or welding may not be possible without plant shutdown. If air or moisture leaks into a system operating at vacuum conditions, extensive icing and corrosion problems can result.
- Higher piping costs are often involved when all items are considered, including large and expensive vapor and liquid-control valves at individual heat exchangers. Generous refrigerant knockout separators are necessary at each stage. Both refrigerant and secondary coolant lines require insulation to prevent capacity losses, sweating, and icing. Refrigerant lines are about the same size as vapor return to the compressor.
- No system reserve capacity is available as is the case with a secondary coolant, particularly if the latter is designed as a storage system. Process upsets can directly and suddenly increase the load on the refrigeration unit, causing rapid cycling of the equipment. For some processes, meeting short, sharp load peaks is of paramount importance to avoid off-standard quality and unsafe operating conditions.
- Constant temperature control is sometimes more difficult or costly to maintain with direct refrigeration than with a secondary coolant.
- Initial pressure testing of an extensive direct refrigeration system may be a significant problem. Testing must be done pneumatically rather than hydrostatically to prevent problems associated with water left in the refrigerant system. Pneumatic testing is considered a hazardous operation and is avoided where possible in some chemical plants. The alternative, extensive posttesting dehydration, is usually expensive and time consuming.
- Initial cost for refrigerants is usually much lower in an extensive direct refrigeration system using hydrocarbons than in a secondary coolant system operating at temperatures most frequently encountered in the chemical industry. In the case of system leaks, the costs of makeup coolant are generally about the same as those for makeup primary refrigerant.

Advantages offered by direct refrigeration systems include the following:

- No careful control of corrosion inhibitors is needed, such as may be necessary to keep secondary coolants stable so that they do not cause extensive equipment damage.
- Less equipment and maintenance may be required; secondary coolant circulation and control or coolant mixing and makeup facilities are not needed.
- Power costs are generally lower because of higher suction pressures and, in some designs, because pumps are not required.
- Damage because of equipment freezing is not likely, though it can occur in a secondary coolant system if the coolant condition or the refrigeration plant is not properly operated.

Thus, the broad scope of refrigeration applications in the chemical industry allows the use of virtually any refrigeration system under the proper process conditions.

REFRIGERATION EQUIPMENT

For the most part, the refrigeration equipment used in the chemical industry is identical to, or closely parallels, equipment used in other industries. The chemical industry is unique, however, in the wide variety of applications, large temperature ranges covered by these applications, diversity of equipment usage, and variation of mechanical specifications required. Where possible, the chemical industry uses standard equipment, but this is frequently impossible because of the particularly rigorous demands of chemical plant service. Therefore, this section only briefly describes the application and modification of refrigeration equipment for chemical plant service.

Compressors

Refrigeration engineers may have difficulty in applying conventional refrigeration compressors to chemical plant service. Most process engineers are familiar with heavy duty, forged steel, high-pressure, single- or double-throw reciprocating gas compressors; they are uncomfortable with the high-speed, cast iron or steel compressors that are standard to the refrigeration industry. Another difference between commercial and chemical plant usage is the greater use of open-drive equipment in the chemical plant.

The large capacities and low temperatures frequently encountered in chemical plant duty have led to wide use of either centrifugal compressors or high-capacity rotary or screw compressors. These large machines vary from standard commercial equipment principally in the amount and complexity of controls or other auxiliaries provided. Load control devices such as multistep unloaders or hot-gas bypass systems are often required to allow a compressor turndown to 10% of full load or, in some cases, to permit no-load operation without either compressor surge problems or on-off operation. Most systems with large multistage centrifugal compressors use economizers to minimize power and suction volume requirements. Compressor lube oil systems are often provided with auxiliary oil pumps, dual oil filters, dual oil coolers, and the like to allow routine maintenance without shutdown and to minimize shutdown frequency. Compressor control and alarm systems are frequently tied into central control room panel boards for monitoring and/or control of compressors.

Compressors for hydrocarbon gas refrigerants find their greatest use in the chemical industry, particularly in the field of petrochemicals. The relatively low cost and ready availability of pure hydrocarbons and hydrocarbon mixtures frequently dictate their use. Many offer the additional advantage of positive-pressure operation throughout the entire refrigeration cycle.

Because refrigeration systems in the chemical industry are often required to operate for a year or more without shutdown, standby compression equipment is frequently installed. Even the larger refrigeration loads sometimes require 100% standby protection. Special controls may be required to provide rapid and automatic start-up of the standby equipment. The main drive is commonly an electric motor; the standby drive may be either a steam turbine or an internal combustion engine. Provisions must be made via nonelectric drivers or emergency generating equipment to keep all necessary auxiliaries and controls operative during an electrical outage. Oversized crankcase heaters may be required, as well as electric or steam tracing of various lubricant system components.

High in-service requirements, plant standardization, explosion hazards, and corrosive atmospheres all require special controls. Often, copper instrument tubing normally used on commercial equipment must be replaced with steel or stainless steel tubing more suitable to the proposed plant atmosphere. Lubricant piping must be stainless steel with nonferrous valves, coolers, and filters. This requirement is primarily to minimize expensive delays in initial plant start-up that may result when rust or scale in lubricant systems causes damage to bearings or seals in high-speed centrifugal or screw compressors.

Absorption Equipment

Chapter 1 of the 2005 *ASHRAE Handbook—Fundamentals* and Chapter 41 of this volume discuss absorption equipment in detail. Absorption equipment has seen little recent use in chemical plants, even though plant waste heat may be available to operate it, because of the proximity of the heat source to the refrigeration requirement.

By special design and reselection of materials, hot streams of many fluids can be used as the energy source instead of using hot water or steam. Direct-fired units are available. Hot condensable vapors can also be used as the energy source.

Lithium bromide absorption equipment must be modified for outdoor operation. Follow manufacturers' recommendations regarding changes necessary to prevent freezing on the water side and solution crystallization on the absorption side of the equipment, particularly during shutdown.

Condensers

Water-Cooled Condensers. Units for chemical plant service require relatively minor design changes from those provided for industrial installations. Because cooling water is frequently of low quality, special construction materials may be required throughout the tube side. An example is the necessity to switch from copper to a cupronickel when cooling water comes from a brackish source that is high in chlorides. If the cooling water is high in mud or silt content, it is sometimes justifiable to install piping and valving that will allow backflushing the condenser without requiring a refrigeration machine shutdown. Chemical plant requirements normally dictate shell-and-tube condensers of the replaceable tube type. Process engineers may insist on conservative tube-side velocities (8 fps or less as a maximum for copper tubes) and removable bundles. The long hours of required operation without opportunity for cleaning and the types of cooling water used frequently require that a higher water-side fouling factor be assumed than on industrial installations.

Air-Cooled Condensers. With increasing restrictions on using water for condensing, air-cooled condensing systems have been used in many instances, even in larger centrifugal-type plants. These have usually been installed in humid locations where the increase in condensing pressure (temperature) over that from the use of cooling tower water or once-through water systems is minimal.

Air-cooled condensers for chemical plant service are normally fabricated to one of the API standards for forced-convection coolers. Care must be taken when specifying these coolers so that the manufacturer understands the type of duty associated with a condensing refrigerant. The service required of an air-cooled condenser in a chemical plant atmosphere dictates either the use of more expensive alloys in the tube construction or conventional materials of greater wall thickness to give acceptable service life. Air coolers may be more difficult to locate because recirculation of hot discharge air or fouling by hot process exhaust gases must be avoided.

Evaporative Condensers. Evaporative condensers, particularly for smaller refrigerating loads, are used extensively in the chemical industry, and they should become more prevalent as emphasis on reducing thermal contamination of rivers, lakes, and streams increases. In a few larger installations, the combination of an air cooler and an evaporative condenser operating in series satisfies the condensing requirements.

In most cases, the commercial evaporative condenser is totally unsuitable for chemical plant service, but satisfactory results can be obtained if this equipment is carefully specified. The major items of concern are the atmospheric conditions to which equipment may be exposed and the long in-service requirements of the chemical plant. The chemical plant atmosphere, which may abound in vapors or dusts that are corrosive in themselves, can be an even more serious problem when these vapors and dusts pass over surfaces that are constantly being wetted. Another problem is that dusts from nearby raw material storages or grinding operations may infiltrate the water recirculating system and plug the spray nozzles. The problems of water treatment and winter freeze protection are usually much more severe in chemical plant service because of the lower-quality water that is frequently available and the demand for both year-round operation and a high turndown ratio. Light load operation in freezing weather calls for extreme care in design to avoid freezeup.

Two other areas of commercial evaporative condenser design that frequently must be strengthened for chemical plant duty are the electrical equipment, which must be satisfactory for the plant environment, and the fans, dampers, and recirculating pumps, which must be suitable for long-life, low-maintenance service.

Evaporators

The general familiarity of chemical plant design personnel with heat exchanger design and application may sometimes lead them to suggest that refrigeration evaporators for the chemical plant should be designed similarly to evaporators in nonrefrigeration service. Although the general laws of heat transfer apply in either case, there are special requirements for evaporators in refrigeration service that are not always present in other types of heat exchanger design. Refrigeration engineers must coordinate the process engineers' experience with the special requirements of a refrigeration evaporator. To do so, Tubular Exchanger Manufacturers Association (TEMA) standards should be consulted to ensure that the end product is familiar to the plant engineer while still performing efficiently as a refrigeration chiller.

Paramount in these special requirements are the proper treatment of oil circulation in the refrigeration evaporator and proper evaluation of liquid submergence as it may affect low-temperature evaporator performance. When the evaporator in chemical plant service is being used with reciprocating and rotary screw compression equipment, continuous oil return from the evaporator must normally be provided. If continuous oil return is not possible, an adequate oil reservoir for the compression equipment, with periodic transfer of oil from the low side of the system, may be needed. On evaporators used with centrifugal compression equipment, continuous oil return from the evaporators is not necessary. In general, centrifugal compressors pump very little oil, so oil contamination of the low side of the system is not as serious as with positive-displacement equipment. However, even with centrifugal equipment, the low-side evaporators eventually become contaminated with oil, which must be removed. Most centrifugal systems operate for several years before oil accumulation in the evaporator adversely affects evaporator performance. Newer tube surfaces with porous coatings may be more sensitive to the presence of oil in the refrigerant than would be conventional finned surfaces. For newer surfaces, a continuous oil return system may be essential for centrifugal systems.

Flooded shell refrigeration evaporators operating at extremely low temperatures and low suction pressures may build up an excessive liquid head, which can create higher evaporating pressures and temperatures at the bottom of the evaporator than at the top. Spray-type evaporators with pump recirculation of refrigerant eliminate this static head penalty.

Special materials for evaporator tubes and shells of particularly heavy wall thickness are frequently dictated to cool highly corrosive process streams. Corrosion allowances in evaporator design, which are seldom a factor in the commercial refrigeration field, are often required in chemical service. Ranges of permissible velocities are frequently specified to prevent sludge deposits or erosion at tube ends.

Process-side construction suitable for high pressures seldom encountered in usual refrigeration applications is frequently necessary. Choice of process-side scale factors must also be made carefully without overstatement.

Differences between process inlet and outlet temperatures of 100°F or more are not uncommon. For this reason, special consideration must be given to thermal stresses within the refrigerant evaporator. U-tube or floating-tube sheet construction is frequently specified in chemical plant service, but minor process-side modifications may allow use of less expensive standard fixed-tube sheet design. The refrigerant side of the evaporator may be required to withstand pressures resulting from maximum process temperature or the evaporator must be able to bypass the process stream under certain high-temperature conditions (e.g., in a refrigeration system failure).

Relief devices and safety precautions common to the refrigeration field normally meet chemical plant needs but should be reviewed against individual plant standards and local statutory requirements. Forged steel relief valves are becoming more common as they meet the applicable refinery piping codes. In hazardous service, relief valves are sized for emergency discharge in the event of fire. Effects of chemical vapors on downstream (outlet) internal parts of relief valves may call for special materials or trapped outlet piping with isolating liquid seals.

Process requirements frequently call for sudden or unexpected load changes on the refrigeration evaporator. Possible thermal shocks, with attendant stresses, must be evaluated, and the evaporator must be designed to meet any such conditions.

Evaporators in chemical plant duty normally require annual inspection and cleaning. For this reason, they should be located for accessibility and ease of tube replacement. Possible contamination of the process stream or refrigerant side by leakage should be evaluated. Special means of leak detection from one side of the evaporator to the other may be justified on occasion.

Low-temperature refrigeration in the chemical industry often creates extremely high viscosities on the process side of the equipment. Special evaporator designs may be needed to minimize pressure drops on the process side and to maintain optimum heat transfer performance. Small tube diameters may not be compatible with the process stream because some processes may call for extra-large tubes. For extremely low-temperature and high-viscosity duties, evaporators are sometimes provided with rotating internal wall scrapers to ensure flow of high-viscosity fluids through the evaporators. Similarly, jacketed process vessels are used to cool highly viscous materials, while rotary scrapers keep the vessel walls clean.

For proper process flow, evaporators usually are remote from other refrigeration equipment to minimize piping and pumping costs. Because remotely located evaporators place special emphasis on proper refrigerant piping practices, secondary coolant systems may be used. Chemical plants frequently use flooded refrigeration systems, which pump refrigerant from the central compressor station to remote evaporators. These systems often reduce design difficulties in ensuring adequate oil return, and special provisions must be made at the central refrigeration station to protect the compressor against liquid carryover in the suction gas. The system must have an adequate accumulator to ensure dry gas to the compressor.

Standard air-side evaporators may require modification, mainly to solve special corrosion problems in handling air or process gases that attack standard coil materials. Occasionally, process requirements demand coil designs that do not match standard commercial air-side pressure drops, design temperature range, or both. Coils of special depth and finning may be required, and coil casings and fan casings of alloy steel are common.

Instrumentation and Controls

Because the heart of the chemical plant is its instrument control system, it follows that instrumentation and control are much more advanced in the chemical industry than in commercial or usual industrial refrigeration applications. As previously discussed, chemical industry refrigeration instrumentation hardware is much more sophisticated in design, particularly in regard to providing increased safety, reliability, and compatibility with process instrumentation devices. This sophistication extends to the application and design of individual hardware items. The chemical industry seldom settles for integral control devices such as self-contained pressure regulators or capillary-actuated thermal control valves. The usual chemical industry control loop consists of a sensing device, a transmitter, a recorder/controller, a positioner, and an operator, all pneumatically or electrically interconnected. Many plants use central computer and microprocessor controls. Interfacing between the refrigeration system and control system may be necessary.

Cooling Towers and Spray Ponds

A refrigeration system using water-cooled rather than air-cooled or evaporative condensers may reject heat to once-through cooling water, spray ponds, or cooling towers. The chemical industry uses mechanical draft towers almost exclusively. These are generally of the induced draft design and are about evenly divided between crossflow and counterflow operation. Although familiarity with these items is necessary, chemical plant engineers are usually responsible for their design.

Miscellaneous Equipment

Pumps. Refrigeration system pumps are usually of a high-quality centrifugal design, the primary exception being small positive-displacement pumps for compressor lube oil systems. In the past, heavy-duty design was the rule rather than the exception, and secondary coolant and chilled-water units were usually of a horizontal split-case design, patterned after boiler-house or water plant construction. Chemical process designers have advocated standard chemical plant pump designs, which usually have a vertically split case and end suction. If selection is made carefully, this design is successful in many applications, and the resultant savings in pump costs, space requirements, and spare parts stocking requirements make it economically attractive.

For pumped materials difficult to contain, such as most refrigerants and many secondary coolants, mechanical shaft seals of various designs are frequently used. As a result of their highly successful use in pumping difficult process fluids, canned or sealless pumps are used in applications such as liquid overfeed systems using halocarbons. Because pumping difficult fluids is a common problem, chemical process designers can be of invaluable assistance.

Piping. As a consequence of several factors, including low fluid temperatures, large pipe sizes, congested pipe alley space, and the industry's reluctance to use expansion joints for high-duty service and in corrosive atmospheres, piping flexibility problems are much more complex. Expansion joints are often prohibited, which increases space requirements dramatically. Secondly, piping and valve standards that apply to both process and service facilities are frequently established by the process designer. The engineer who is accustomed to using carbon steel piping systems with tongue and groove flanging and valves may find that plant standards call for a welded nickel steel system with raised face flanges, spirally wound stainless steel gaskets, and cast steel valving.

Most piping construction problems resulting from the difference between expectations of the process engineer and the experience of the refrigeration engineer can be resolved by constructing the system to meet ASME *Standard* B31.3. The ASME B16 series of standards that defines the flanges and fittings of the process industry should also be followed. Chapters 1 to 4 also discuss piping sizing for various refrigeration systems.

Tanks. Chemical plants use storage tanks for both refrigerants and secondary coolants more frequently than most commercial or industrial plants. In chilled-water or brine circulation systems, storage tanks often serve a dual purpose: to store secondary coolants (1) during operation to provide reserve capacity and thus smooth out short-term peak requirements and (2) during maintenance shutdown of process evaporators. In some cases, brine mix and storage facilities are provided, so that any brine lost from leakage or unusual maintenance demands can be quickly replaced, thus minimizing unscheduled process outages. In many cases, refrigerant pumpout compressors and storage receivers can minimize loss of the refrigerant and unscheduled outage time because of refrigeration system failures on the refrigerant side.

The chemical industry designs all pressure vessels in accordance with the ASME *Boiler and Pressure Vessel Code*, in particular Section VIII, Division 1, for unfired pressure vessels, regardless of local government regulations requiring such design. Most plants establish standards for items such as pressure-relief devices, manhole design, insulation supports, and tank supports. A thorough knowledge of the plant standards to be applied should be gained before specifications and design details are established for refrigeration system tankage.

REFERENCES

ARI. 2003. Centrifugal and rotary screw water-chilling packages. ANSI/ARI *Standard* 550/590-2003. Air-Conditioning and Refrigeration Institute, Arlington, VA.

ASME. 2004. Rules for construction of pressure vessels. ANSI/ASME *Boiler and pressure vessel code*, Section VIII, Division 1. American Society of Mechanical Engineers, New York.

ASME. 2002. Process piping. ANSI/ASME *Standard* B31.3-2002.American Society of Mechanical Engineers, New York.

CHAPTER 38

CRYOGENICS

CRYOGENICS is a term normally associated with low temperatures. However, the location on the temperature scale at which refrigeration generally ends and cryogenics begins has never been well defined. Most scientists and engineers working in this field restrict cryogenics to a temperature below −235°F (225°R), because the normal boiling points of most permanent gases (e.g., helium, hydrogen, neon, nitrogen, argon, oxygen, and air) occur below this temperature. In contrast, most common refrigerants have boiling points above this temperature.

Cryogenic engineering therefore is involved with the design and development of low-temperature systems and components. In such activities the designer must be familiar with the properties of fluids used to achieve these low temperatures as well as the physical properties of components used to produce, maintain, and apply such temperatures.

GENERAL APPLICATIONS

The application of cryogenic engineering has become extensive. In the United States, for example, nearly 30% of the oxygen produced by cryogenic separation is used by the steel industry to reduce the cost of high-grade steel, and another 20% is used in the chemical process industry to produce a variety of oxygenated compounds. Liquid hydrogen production has risen from laboratory quantities to over 200 tons/day. Similarly, liquid helium demand has required the construction of large plants to separate helium from natural gas cryogenically. Energy demand likewise has accelerated construction of large base-load liquefied natural gas (LNG) plants. Applications include high-field magnets and sophisticated electronic devices that use the superconductivity of materials at low temperatures. Space simulation requires cryopumping (freezing residual gases in a chamber on a cold surface) to provide the ultrahigh vacuum representative of conditions in space. This concept has also been used in commercial high-vacuum pumps.

The food industry uses large amounts of liquid nitrogen to freeze expensive foods such as shrimp and to maintain frozen food during transport. Liquid-nitrogen-cooled containers are used to preserve whole blood, bone marrow, and animal semen for extended periods. Cryogenic surgery is performed to treat disorders such as Parkinson's disease. Medical diagnosis uses magnetic resonance imaging (MRI), which requires cryogenically cooled superconducting magnets. Superconducting magnets are now an essential component in high-energy accelerators and target chambers. Finally, the chemical processing industry relies on cryogenic temperatures to recover valuable heavy components or upgrade the heat content of fuel gas from natural gas, recover useful components such as argon and neon from air, purify various process and waste streams, and produce ethylene from a mixture of olefin compounds.

This preparation of this chapter is assigned to TC 10.4, Ultralow-Temperature Systems and Cryogenics.

LOW-TEMPERATURE PROPERTIES

Test data are necessary because properties at low temperatures are often significantly different from those at ambient temperatures. For example, the onset of ductile-to-brittle transitions in carbon steel, the phenomenon of superconductivity, and the vanishing of specific heats cannot be inferred from property measurements obtained at ambient temperatures.

Fluid Properties

Some thermodynamic data for cryogenic fluids are given in Chapter 20 of the 2005 *ASHRAE Handbook—Fundamentals*. Computer-compiled tabulations include those of MIPROPS prepared by NIST; GASPAK, HEPAK, and PROMIX developed by Cryodata (Arp 1998); and EES [Klein (continuously updated)]. Some key properties for selected cryogens are summarized in Table 1, including the normal boiling point (i.e., boiling point at atmospheric pressure), critical point, and triple point (nominally equal to the freezing point at atmospheric pressure). Table 1 also presents the volumetric enthalpy change associated with evaporation at atmospheric pressure, and the volumetric enthalpy change associated with heating saturated vapor at atmospheric pressure to room temperature. These quantities reflect the value of the cryogen in the conventional situation (where only the latent heat of evaporation is used) and the less typical situation where the sensible heat is also recovered.

Several cryogens have unique properties, discussed in the following sections.

Helium. Helium exists in two isotopic forms, the more common being helium 4. The rarer form, helium 3, exhibits a much lower vapor pressure, which has been exploited in the development of the helium dilution refrigerator to attain temperatures as low as 0.03 to 0.09°R. Whenever helium is referenced without isotopic designation, it can be assumed to be helium 4.

As a liquid, helium exhibits two unique phases: liquid helium I and liquid helium II (Figure 1). Helium I is labeled as the normal fluid and helium II as the superfluid because, under certain conditions, the fluid exhibits no viscosity. The phase transition between these two liquids is identified as the lambda (λ) line. Intersection of helium II with the vapor pressure curve is known as the λ point. Immediately to the right of the λ line, the specific heat of helium I increases to a large but finite value as the temperature approaches this line; therefore, although there is no specific volume change or latent heat associated with the helium I to II transition, a significant energy change is required. Once below the λ line, the specific heat of helium II rapidly decreases to zero. Figure 2 illustrates the specific heat capacity of helium at low temperatures, both above and below the λ line, and various pressures (data from HEPAK). Notice the sharp rise in specific heat capacity near –455.76°F (the λ line) at all pressures (essentially independent of pressure). Also note the specific heat fluctuations at higher temperatures, related to the normal two-phase behavior of a substance near its critical point.

The thermal conductivity of helium I decreases with decreasing temperature. However, once the transition to helium II has been

Table 1 Key Properties of Selected Cryogens

Cryogen	Normal Boiling Temperature, °R	Critical Temperature, °R	Triple-Point Temperature, °R	Density of Saturated Liquid at 1 atm, lb_m/ft^3	Density of Saturated Vapor at 1 atm, lb_m/ft^3	Volumetric Enthalpy of Vaporization at 1 atm, Btu/ft^3*	Volumetric Enthalpy to Warm Vapor to 537°R at 1 atm, Btu/ft^3*
Helium	7.61	9.35	—	7.787	1.046	69.4	5212
Hydrogen	36.70	59.74	25.12	4.420	0.0836	845.5	4005
Neon	48.79	80.09	44.21	75.35	0.5979	2774	7476
Oxygen	162.34	278.25	97.85	71.24	0.2789	6535	8707
Nitrogen	139.24	227.15	113.67	50.32	0.2879	4303	6370
Argon	157.14	271.24	150.85	87.11	0.3604	6035	7629
Methane	201.00	343.02	163.25	26.37	0.1134	5790	7125

*Per cubic foot of saturated liquid cryogen at 1 atm.

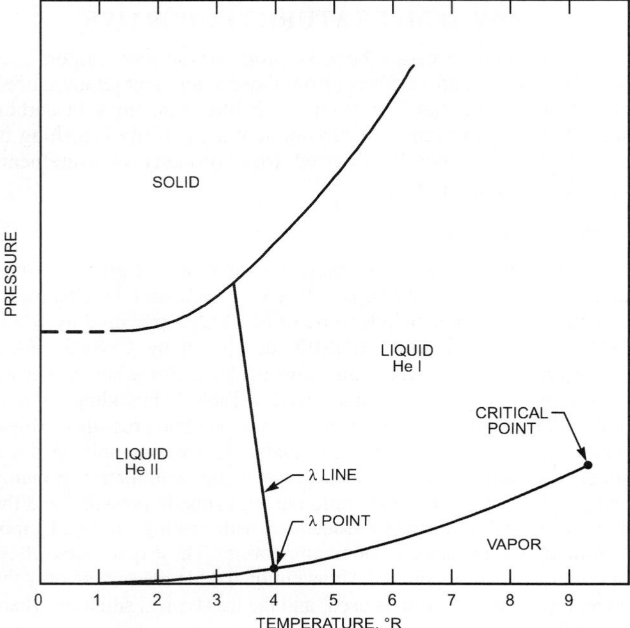

Fig. 1 Phase Diagram for Helium 4

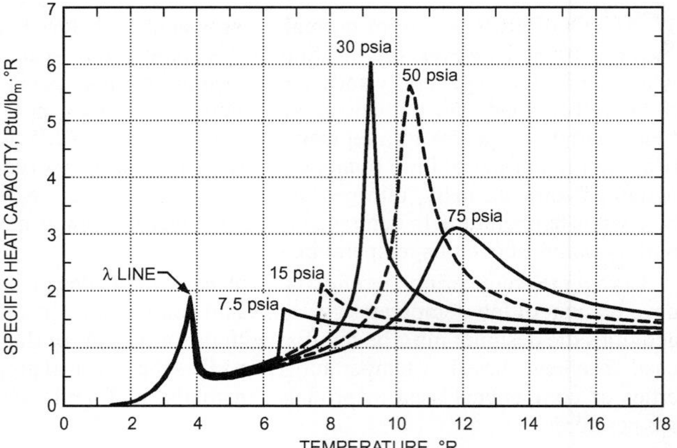

Fig. 2 Specific Heat for Helium 4 as Function of Temperature for Various Pressures

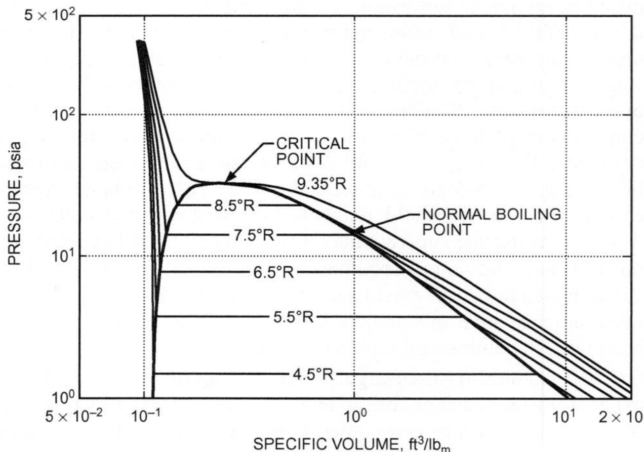

Fig. 3 Pressure/Volume Diagram for Helium 4 near Its Vapor Dome

made, the thermal conductivity of the liquid has no real physical meaning, yet the heat transfer characteristics of helium II are spectacular. As the vapor pressure above helium I is reduced, the fluid boils vigorously. As the liquid pressure decreases, its temperature also decreases as the liquid boils away. When the temperature reaches the λ point and the helium transitions to helium II, all bubbling suddenly stops. The liquid becomes clear and quiet, although it is still vaporizing quite rapidly at the surface. The apparent thermal conductivity of helium II is so large that vapor bubbles do not have time to form within the body of the fluid before the heat is conducted to the surface of the liquid. Liquid helium I has a thermal conductivity of approximately 0.0139 Btu/h·ft·°R, whereas liquid helium II can have an apparent thermal conductivity as large as 49,112 Btu/h·ft·°R, approximately six orders of magnitude larger. This characteristic makes He II the ideal coolant for low-temperature applications, including superconducting magnets (Barron 1985). Also, helium 4 has no triple point and requires a pressure of 360 psia or more to exist as a solid below a temperature of –455°F (5°R).

Figure 3 illustrates the pressure/volume diagram of helium 4 near its vapor dome, and clearly shows the critical point and normal boiling point. Note that the density of the liquid and vapor phases of liquid helium far from the critical point differ only by a factor of about 7.5, compared to 1000 for many substances. Also, the latent heat of vaporization for helium is only 9 Btu/lb (or 70 Btu/ft^3 of liquid helium), which is very small; thus, the amount of heat that can be absorbed by a bath of liquid helium is limited, so liquid nitrogen shielding is needed, as well as stringent requirements for thermal isolation. Notice that the amount of energy that can be absorbed by evaporation of liquid helium if the sensible heat capacity of the vapor is included is far larger: 5260 Btu/lb^3 (Table 1). Therefore, the flow of the helium vapor as it warms to room temperature is often controlled to ensure that this sensible heat is properly used.

Hydrogen. A distinctive property of hydrogen is that it can exist in two molecular forms: orthohydrogen and parahydrogen. These forms differ by having parallel (orthohydrogen) or opposed (parahydrogen) nuclear spins associated with the two atoms forming the hydrogen molecule. At ambient temperatures, the equilibrium mixture of 75% orthohydrogen and 25% parahydrogen is designated as normal hydrogen. With decreasing temperatures, the thermodynamics shift to 99.79% parahydrogen at –423°F (36.7°R), the normal boiling point of hydrogen. Conversion from normal hydrogen to parahydrogen is exothermic and evolves sufficient energy to vaporize ~1% of the stored liquid per hour, assuming negligible heat leak into the storage container. The fractional rate of conversion is given by

$$\frac{dx}{d\theta} = -kx^2 \tag{1}$$

where x is the orthohydrogen fraction at time θ in hours and k is the reaction rate constant, 0.0114/h. The fraction of liquid remaining in a storage dewar at time θ is then

$$\ln\frac{m}{m_o} = \frac{1.57}{1.33 + 0.0114\theta} - 1.18 \tag{2}$$

Here m_o is the mass of normal hydrogen at $\theta = 0$ and m is the mass of remaining liquid at time θ. If the original composition of the liquid is not normal hydrogen at $\theta = 0$, a new constant of integration based on the initial orthohydrogen concentration can be evaluated from Equation (1). Figure 4 summaries the calculations.

To minimize such losses in commercial production of liquid hydrogen, a catalyst is used to hasten the conversion from normal hydrogen to the thermodynamic equilibrium concentration during liquefaction. Hydrous iron oxide, Cr_2O_3 on an Al_2O_3 gel carrier, or NiO on an Al_2O_3 gel are used as catalysts. The latter combination is about 90 times as rapid as the others and is therefore the preferred choice.

Figure 5 shows a pressure/volume diagram for hydrogen.

Oxygen. Unlike other cryogenic fluids, liquid oxygen (LOX) is slightly magnetic. Its paramagnetic susceptibility is 1.003 at its normal boiling point. This characteristic has prompted the use of a magnetic field in a liquid oxygen dewar to separate the liquid and gaseous phases under zero-gravity conditions.

Both gaseous and liquid oxygen are chemically reactive, particularly with hydrocarbon materials. Because oxygen presents a serious safety problem, systems using liquid oxygen must be maintained scrupulously clean of any foreign matter. Liquid oxygen cleanliness in the space industry has come to be associated with a set of elaborate cleaning and inspection specifications representing a near ultimate in large-scale equipment cleanliness.

Nitrogen. Liquid nitrogen (LIN) is of considerable importance as a cryogen because it is a safe refrigerant. Because it is rather inactive chemically and is neither explosive nor toxic, liquid nitrogen is commonly used in hydrogen and helium liquefaction cycles as a precoolant. Figure 6 illustrates the pressure/volume diagram for nitrogen near its vapor dome.

Liquefied Natural Gas (LNG). Liquefied natural gas is the liquid form of natural gas, consisting primarily of methane, a mixture of heavier hydrocarbons, and other impurities such as nitrogen and hydrogen sulfide. Liquefying natural gas reduces its specific volume by a factor of approximately 600 to 1, which makes handling and storage economically possible despite the added cost of liquefaction and the need for insulated transport and storage equipment.

Thermal Properties

Specific heat, thermal conductivity, and thermal expansivity are of major interest at low temperatures.

Specific Heat. Specific heat can be predicted fairly accurately by mathematical models through statistical mechanics and quantum theory. For solids, the Debye model gives a satisfactory representation of

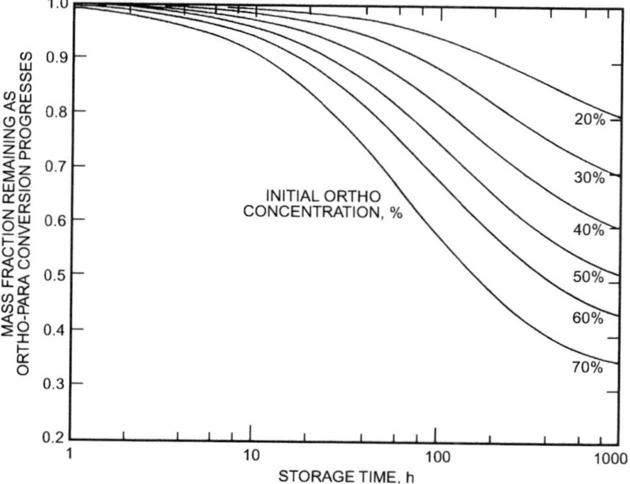

Fig. 4 Fraction of Liquid Hydrogen Evaporated due to Ortho-Parahydrogen Conversion as Function of Storage Time

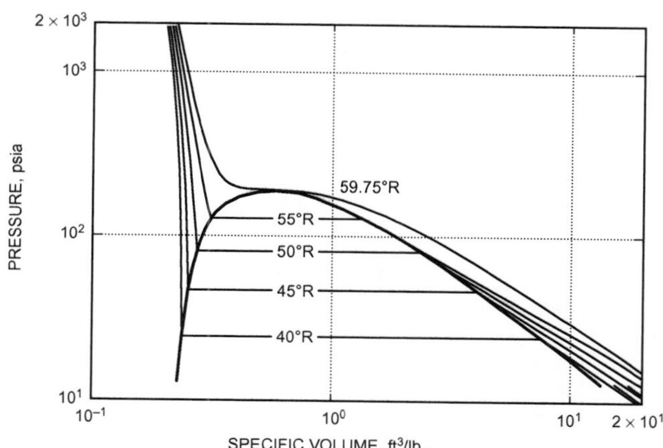

Fig. 5 Pressure/Volume Diagram for Hydrogen near Its Vapor Dome

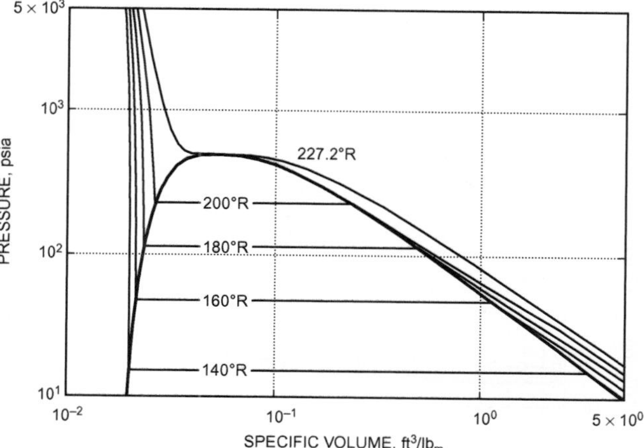

Fig. 6 Pressure/Volume Diagram for Nitrogen near Its Vapor Dome

specific heat with changes in temperature. However, difficulties are encountered when applying the Debye theory to alloys and compounds.

Several computer programs provide thermal data for many metals used in low-temperature equipment. METALPAK (Arp 1997), for example, is a reference program for the thermal properties of 13 metals used in low-temperature systems.

The specific heat of cryogenic liquids generally decreases in a manner similar to that observed for crystalline solids as the temperature is lowered. At low pressures, specific heat decreases with a decrease in temperature. However, at high pressures near the critical point, humps appear in the specific heat curve for all cryogenic fluids.

Figure 7 illustrates the specific heat capacity of several commonly used solid materials as a function of temperature. In general, specific heat decreases with decreasing temperature.

Often, the specific heat must be known to determine the amount of energy to remove from a material to cool it from room temperature (80.3°F) to cryogenic temperatures. The integrated average specific heat $\bar{c}$ is useful for this type of calculation:

$$\bar{c}(T) = \frac{1}{(T_{room} - T)} \int_T^{T_{room}} c(T)dT \qquad (3)$$

Figure 8 illustrates the integrated average (from room temperature) specific heat for various solid materials.

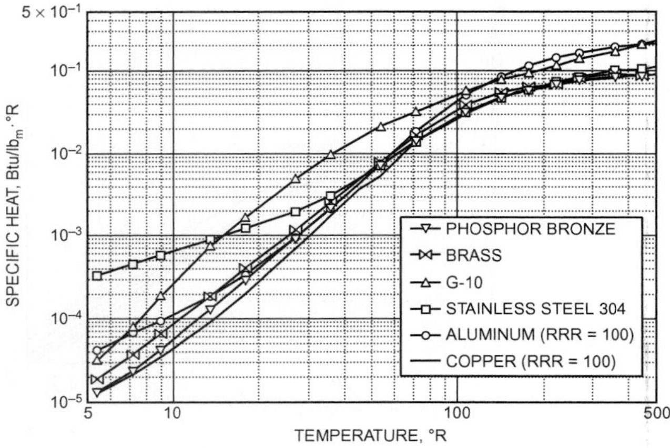

Fig. 7 Specific Heat of Common Cryogenic Materials

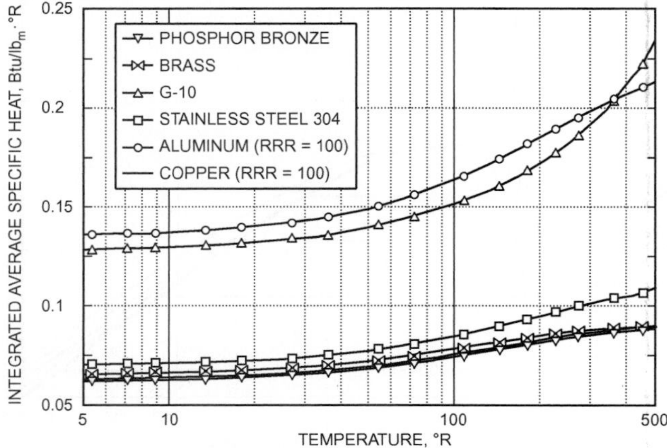

Fig. 8 Integrated Average Specific Heat (from 540°R) for Common Cryogenic Materials

Table 2 summarizes the integrated average specific heat for various materials and for several temperature ranges.

Thermal Conductivity. Thermal conductivity of pure metals at low temperatures can be accurately predicted with the Wiedeman-Franz law, which states that the ratio of thermal conductivity to the product of the electrical conductivity and absolute temperature is a constant. This ratio for high-conductivity metals extrapolates close to the Sommerfeld value of 2.58×10^{-8} Btu·Ω/(h·°R^2) at absolute zero, but is considerably below this value at higher temperatures. However, high-purity aluminum and copper peak in thermal conductivity between 423 and –370°F (36 and 90°R), but these peaks are rapidly suppressed with increased impurity levels or cold work of the metal. Aluminum alloys, for example, steadily decrease in thermal conductivity as temperature decreases. Other structural alloys, such as those of nickel-copper (67% Ni/30% Cu), nickel-chromium-iron (78% Ni/16% Cr/6% Fe), and stainless steel, exhibit similar thermal conductivity properties and thus are helpful in reducing heat leak into a cryogenic system.

Figure 9 illustrates the thermal conductivity of some common cryogenic materials. Note that the thermal conductivity of copper and other metals is extremely sensitive to the purity of the material below about –350°F. The purity of copper is often reported using the residual resistance ratio (RRR) value, which is the ratio of the electrical resistivity of the material at room temperature to its resistivity at –452°F. An RRR value of 100 is typical for commercial-grade copper.

Often, the thermal conductivity is needed to determine the conduction heat leak associated with a structure made of the material between room temperature (80.3°F) and some cryogenic temperatures. The integrated average thermal conductivity $\bar{k}$ is useful for this type of calculation:

$$\bar{k}(T) = \frac{1}{(T_{room} - T)} \int_T^{T_{room}} k(T)dT \qquad (4)$$

Figure 10 illustrates the integrated average (from room temperature) thermal conductivity for various solid materials. Table 3

Table 2 Integrated Average Specific Heat for Cryogenic Materials, in Btu/lb$_m$·°R

	540 to 270°R	540 to 150°R	540 to 35°R	540 to 7°R
Copper (RRR = 100)	0.0866	0.0800	0.0673	0.0639
Aluminum (RRR = 100)	0.196	0.176	0.145	0.137
304 stainless steel	0.100	0.0906	0.0755	0.0716
G-10	0.186	0.1620	0.135	0.129
Brass	0.0882	0.0825	0.0702	0.0666
Phosphor bronze	0.0850	0.0785	0.0663	0.0629

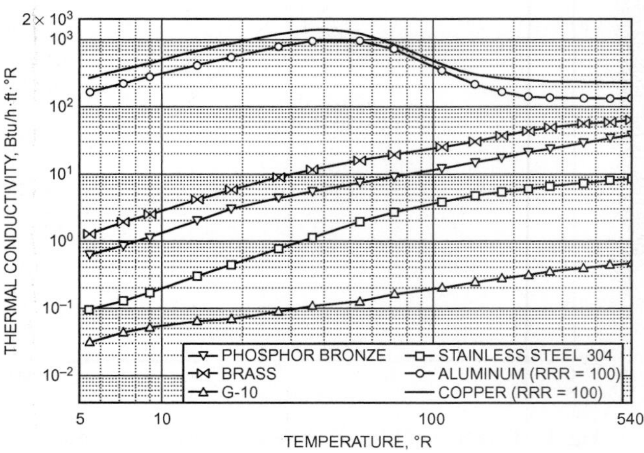

Fig. 9 Thermal Conductivity of Common Cryogenic Materials

summarizes the integrated average thermal conductivity for various materials and for several temperature ranges.

Thermal Expansion Coefficient. The expansion coefficient of a solid can be estimated with an approximate thermodynamic equation of state for solids that equates the volumetric thermal expansion coefficient with the quantity $\gamma c_v \rho / B$, where γ is the Grüneisen dimensionless ratio, c_v the specific heat of the solid, ρ the density of the material, and B its bulk modulus. For face-centered cubic metals, the average value of the Grüneisen constant is approximately 2.3. However, this constant tends to increase with atomic number. Figure 11 illustrates the integrated average coefficient of thermal expansion $\bar{\alpha}$ from room temperature, defined as

$$\bar{\alpha}(T) = \frac{1}{(T_{room} - T)} \int_{T}^{T_{room}} \alpha(T) dT \quad (5)$$

Electrical and Magnetic Properties

The ratio of the electrical resistivity of most pure metallic elements at ambient temperature to that at moderately low temperatures is approximately proportional to the ratio of the absolute temperatures. However, at very low temperatures, the electrical resistivity (except of superconductors) approaches a residual value almost independent of temperature. Alloys, on the other hand, have resistivities much higher than those of their constituent elements and exhibit very low resistance temperature coefficients. Electrical resistivity is thus largely independent of temperature and may often be of the same magnitude as the ambient temperature value.

The insulating quality of solid electrical conductors usually improves with decreased temperature. However, from 2 to 10°R, the electrical resistivity of many semiconductors increases quite rapidly with a small decrease in temperature. This has formed the basis for the development of numerous sensitive semiconductor resistance thermometers for very-low-temperature measurements. Figure 12 illustrates the electrical resistivity of several common cryogenic materials.

Table 3 Integrated Average Thermal Conductivity for Cryogenic Materials, in Btu/h·ft·°F

Material	80.3 to −190°F	80.3 to −316°F	80.3 to −424°F	80.3 to −452°F
Copper (RRR = 100)	2806	2902	4135	4544
Aluminum (RRR = 100)	1627	1711	2798	3045
304 Stainless steel	93.60	85.98	75.57	72.11
G-10	5.061	4.645	4.091	3.952
Brass	693.3	630.3	551.9	526.9
Phosphor bronze	380.6	334.2	291.9	276.6

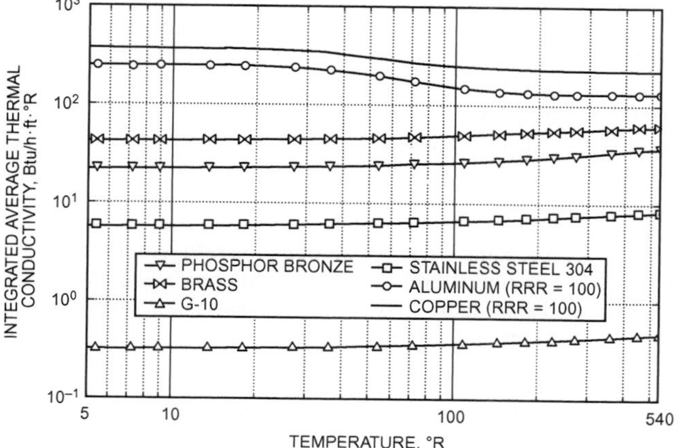

Fig. 10 Integrated Average Thermal Conductivity (from 540°R) for Common Cryogenic Materials

Superconductivity. Superconductivity is described as the disappearance of electrical resistance in certain materials that are maintained below a characteristic temperature, electrical current, and magnetic field, and the appearance of perfect diamagnetism, which is the most distinguishing characteristic of superconductors.

More than 26 elements have been shown to be superconductors at low temperatures at ambient pressure, and at least 10 more at higher pressure. In fact, the number of materials with identified superconducting properties extends into the thousands. Bednorz and Müller's (1986) discovery of high-temperature superconductors and the intensive research to extend the upper temperature limit above 243°R have further increased this list of superconductors.

For all superconductors, the superconducting state is defined by the region below three interdependent critical parameters: temperature, current density, and magnetic field. At zero field and current density, the critical temperature for all elemental superconductors is below 18°R; for low-temperature alloys, it is below 45°R; and for the high-temperature compounds, it extends up to 243°R. Critical current densities of practical interest for magnet and electronics applications range from 1000 to 50,000 A/mm^2 for fields less than 50,000 gauss, and fall to zero, depending on the superconductor, between 90,000 and 280,000 gauss. The BSCCO (bismuth, strontium,

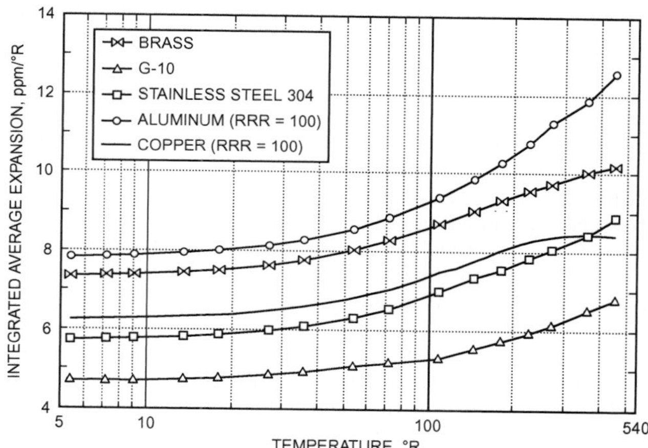

Fig. 11 Integrated Average Thermal Coefficient of Expansion (from 540°R) for Common Cryogenic Materials

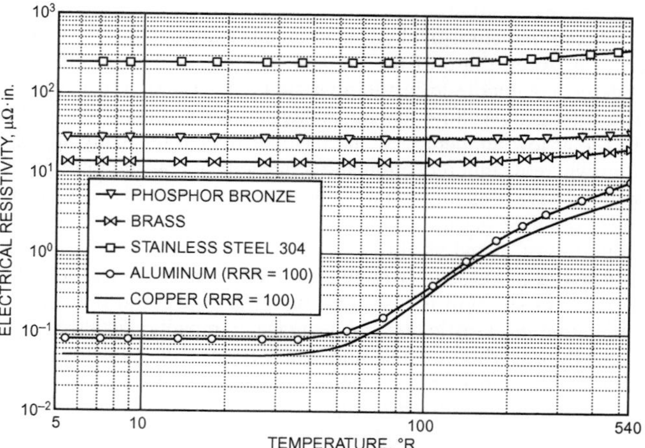

Fig. 12 Electrical Resistivity of Some Common Cryogenic Materials
(Data from Eckels Engineering)

copper, carbon, and oxygen) and YBCO (yttrium, barium, copper, and oxygen) superconductors display the highest critical magnetic fields (above 250,000 gauss).

Properties affected when a material becomes superconducting include specific heat, thermal conductivity, electrical resistance, magnetic permeability, and thermoelectric effect. Consequently, use of superconducting materials in construction of equipment subjected to operating temperatures below their critical temperatures needs to be evaluated carefully.

Thermal conductivities of superconducting materials are significantly lower than the same materials in their normal (or nonsuperconducting) state. For example, increasing the magnetic field around a sample of lead (Pb) from 0 to 1000 gauss at a constant temperature of 4.5°R, and thereby causing it to transition from the superconducting to normal state, increases thermal conductivity tenfold. This property is used for thermal switches in low-temperature technologies. The relatively low thermal conductivity of high-temperature superconductors (0.58 to 2.33 Btu/h·ft·°R) provide a convenient means to transfer current with minimal accompanying heat from −316°F (144°R) stages down to superconducting systems at −452.47°F (7.2°R).

Mechanical Properties

Mechanical properties of interest for the design of a facility subjected to low temperatures include ultimate strength, yield strength, fatigue strength, impact strength, hardness, ductility, and elastic moduli. Chapter 40 and Wigley (1971) include information on some of these properties.

Mechanical properties of metals at low temperature are classified most conveniently by their lattice symmetry. **Face-centered cubic (fcc)** metals and their alloys are most often used in constructing cryogenic equipment. Al, Cu, Ni, their alloys, and the austenitic stainless steels of the 18-8 type are fcc and do not exhibit an impact ductile-to-brittle transition at low temperatures. As a general rule, the mechanical properties of these metals improve as the temperature is reduced. The yield strength at −423°F (36°R) is considerably larger than at ambient temperature; Young's modulus is 5 to 20% larger at the lower temperature, and fatigue properties, except those of 2024 T4 aluminum, are improved at the lower temperature. Because annealing can affect both the ultimate and yield strengths, care must be exercised under these conditions.

Body-centered cubic (bcc) metals and alloys are normally undesirable for low-temperature construction. This class includes Fe, the martensitic steels (low-carbon and the 400 series of stainless steels), Mo, and Nb. If not brittle at room temperature, these materials exhibit a ductile-to-brittle transition at low temperatures. Hard working of some steels, in particular, can induce the austenite-to-martensite transition.

Hexagonal close-packed (hcp) metals exhibit mechanical properties between those of the fcc and bcc metals. For example, Zn undergoes a ductile-to-brittle transition, but Zr and pure Ti do not. The latter and its alloys, having an hcp structure, remain reasonably ductile at low temperatures and have been used for many applications where weight reduction and reduced heat leakage through the material have been important. However, small impurities of O, N, H, and C can have detrimental effects on the low-temperature ductility properties of Ti and its alloys.

Plastics increase in strength as temperature decreases, but this increase in strength is also accompanied by a rapid decrease in elongation in a tensile test and decrease in impact resistance. Nonstick fluorocarbon resins and glass-reinforced plastics retain appreciable impact resistance as temperature decreases. Glass-reinforced plastics also have high strength-to-weight and strength-to-thermal-conductivity ratios. All elastomers, on the other hand, become brittle at low temperatures. Nevertheless, many of these materials, including rubber, polyester film, and nylon, can be used for static seal gaskets if they are highly compressed at room temperature before cooling.

The strength of **glass** under constant loading also increases with decreased temperature. Because failure occurs at a lower stress when the glass surface contains surface defects, strength can be improved by tempering the surface.

REFRIGERATION AND LIQUEFACTION

A refrigeration or liquefaction process at cryogenic temperatures usually involves ambient compression of a suitable fluid with heat rejected to a coolant. The fluid's enthalpy and entropy decrease during compression and cooling, but increase at the cryogenic temperature where heat is absorbed. The temperature of the process fluid is usually reduced by heat exchange with a returning colder fluid and then followed with an expansion of the process fluid. This expansion may take place using either a throttling device approximating an isenthalpic expansion, with only a reduction in temperature, or a work-producing device approximating an isentropic expansion, in which both temperature and enthalpy decrease.

Normal commercial refrigeration generally uses a vapor compression process. Temperatures down to about −100°F (360°R) can be obtained by cascading vapor compression, in which refrigeration is obtained by liquid evaporation in each stage. Below this temperature, isenthalpic and isentropic expansion are generally used, either singly or in combination. With few exceptions, refrigerators using these methods also absorb heat by vaporization of liquid. If no suitable liquid exists to absorb the heat by evaporation over a temperature range, a cold gas must be available to absorb the heat. This is generally accomplished by using a work-producing expansion engine.

In a continuous refrigeration process, no refrigerant accumulates in any part of the system. This is in contrast with a liquefaction system, where liquid accumulates and is continuously withdrawn. Thus, in a liquefaction system, the total mass of returning cold gas that is warmed before compression is less than the mass of gas that is to be cooled, creating an imbalance of flow in the heat exchangers. In a refrigerator, the mass flow rates of the warm- and cold-gas streams in the heat exchangers are usually equal, unless a portion of the warm-gas flow is diverted through a work-producing expander. This condition of equal flow is usually called a **balanced flow condition** in the heat exchangers.

Isenthalpic Expansion

The thermodynamic process identified either as the simple **Linde-Hampson cycle** or the **Joule-Thomson cycle (J-T cycle)** is shown schematically in Figure 13A. In this cycle, the gaseous refrigerant is compressed isothermally at ambient temperature by rejecting heat to a coolant. The compressed refrigerant is then cooled in a heat exchanger, with the cold-gas stream returning to the compressor intake. Joule-Thomson cooling accompanying the expansion of gas exiting the heat exchanger further reduces the temperature of the refrigerant until, under steady-state conditions, a small fraction of the refrigerant is liquefied. For a refrigerator, the liquid fraction is vaporized by absorbing the heat Q that is to be removed, combined with the unliquefied fraction, and, after warming in the heat exchanger, returned to the compressor. Figure 13B shows the ideal process on a temperature-entropy diagram.

Applying the first law to this refrigeration cycle, assuming no heat leaks to the system as well as negligible kinetic and potential energy changes in the refrigeration fluid, the refrigeration effect per unit mass of refrigeration is simply the difference in enthalpies of streams 1 and 2 in Figure 13A. Thus, the coefficient of performance (COP) of the ideal J-T cycle is given by

$$\text{COP} = \frac{Q}{W} = \frac{h_1 - h_2}{T_1(s_1 - s_2) - (h_1 - h_2)} \tag{6}$$

where Q is the refrigeration effect, W the work of compression, h_1 and s_1 are the specific enthalpy and entropy at point 1, and h_2 and s_2 the specific enthalpy and entropy at point 2 of Figure 13A.

For a liquefier, the liquefied portion is continuously withdrawn from the liquid reservoir and only the unliquefied portion of the fluid is warmed in the heat exchanger and returned to the compressor. The fraction y that is liquefied is determined by applying the first law to the heat exchanger, J-T valve, and liquid reservoir. This results in

$$y = \frac{h_1 - h_2}{h_1 - h_f} \tag{7}$$

where h_f is the specific enthalpy of the saturated liquid being withdrawn. The maximum liquefaction occurs when the difference between h_1 and h_2 is maximized.

To account for any heat leak Q_L into the system, Equation (7) needs to be modified to

$$y = \frac{h_1 - h_2 - Q_L}{h_1 - h_f} \tag{8}$$

resulting in a decrease in the fraction liquefied. The work of compression is identical to that determined for the J-T refrigerator. The figure of merit (FOM) is defined as $(W/m_f)_i/(W/m_f)$, where $(W/m_f)_i$ is the work of compression per unit mass liquefied for the ideal liquefier and (W/m_f) is the work of compression per unit mass liquefied for the J-T liquefier. The FOM reduces to

$$\text{FOM} = \left[\frac{T_1(s_1 - s_f) - (h_1 - h_f)}{T_1(s_1 - s_2) - (h_1 - h_2)}\right]\left(\frac{h_1 - h_2}{h_1 - h_f}\right) \tag{9}$$

Liquefaction by this cycle requires that the inversion temperature of the refrigerant be above ambient temperature. Auxiliary refrigeration is required if the J-T cycle is to be used to liquefy fluids with an inversion temperature below ambient (e.g., helium, hydrogen, neon). Liquid nitrogen is the optimum auxiliary refrigerant for hydrogen and neon liquefaction systems, and liquid hydrogen accompanied by liquid nitrogen are the normal auxiliary refrigerants for helium liquefaction systems. Upper operating pressures for the J-T cycle are often as high as 3000 psia.

To reduce the work of compression in the previous cycle, a two-stage or dual-pressure cycle may be used in which the pressure is reduced by two successive isenthalpic expansions, as shown in Figure 14. Because the work of compression is approximately proportional to the logarithm of the pressure ratio, and the Joule-Thomson cooling is roughly proportional to the pressure difference, compressor work is reduced by more than the refrigeration performance. Hence, the dual-pressure process produces the same quantity of refrigeration with less energy input than the simple J-T process.

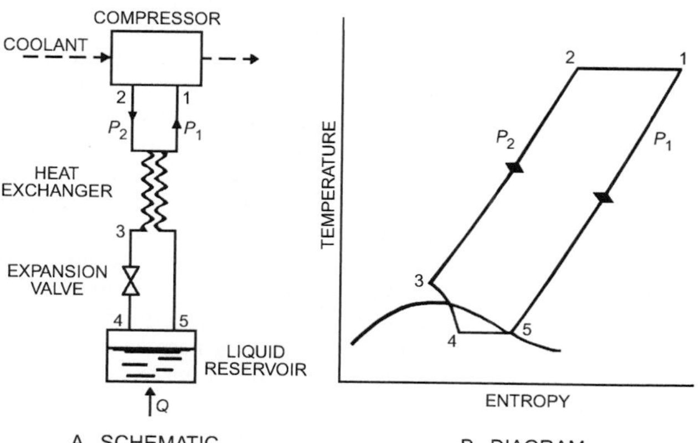

Fig. 13 Schematic and Temperature-Entropy Diagram for Simple Joule-Thomson Cycle Refrigerator

The theoretical liquid yield for the dual-pressure J-T cycle is obtained by making an energy balance around the cold heat exchanger, lower J-T valve, and liquid reservoir as follows:

$$(m - m_f - m_i)h_1 + m_i h_2 + m_f h_f - m h_3 = 0 \tag{10}$$

where m, m_f, and m_i are the mass flow rates of the streams designated in Figure 14. Solving for the liquid yield y for the ideal dual-pressure cycle gives

$$y = \frac{h_1 - h_3}{h_1 - h_f} - \frac{m_i}{m}\left(\frac{h_1 - h_2}{h_1 - h_f}\right) \tag{11}$$

The intermediate pressure in this cycle must be optimized. For a cycle with an upper pressure of 3000 psia, the intermediate optimum pressure generally occurs between 580 and 1015 psia.

Isentropic Expansion

Because temperature always decreases in a work-producing expansion, cooling does not depend on being below the inversion temperature before expansion. In large industrial refrigerators, work produced during the expansion is conserved with an expansion turbine. In small refrigerators, energy from expansion is usually dissipated in a gas or hydraulic pump or other suitable device.

A schematic of a refrigerator using the work-producing expansion principle and the corresponding temperature-entropy diagram are shown in Figure 15. Gas compressed isothermally at ambient temperature is cooled in a heat exchanger by gas being warmed on its return to the compressor intake. Further cooling takes place in the expansion engine. In practice, this expansion is never truly isentropic, as shown by path 3-4 on the temperature-entropy diagram. The refrigerator shown produces a cold gas that absorbs heat during path 4-5 and provides a method of refrigeration that can be used to obtain temperatures between those of the boiling points of the lower-boiling cryogenic fluids.

The coefficient of performance of an ideal gas refrigerator with varying refrigerator temperature can be obtained from the relation

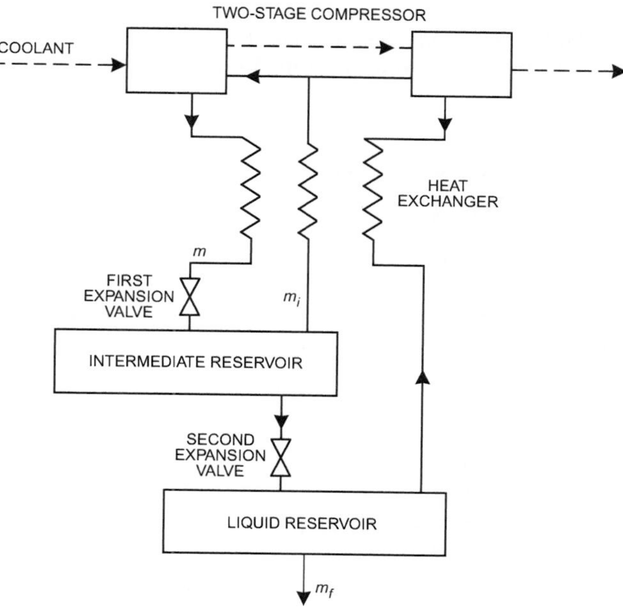

Fig. 14 Dual-Pressure Joule-Thomson Cycle Used as Liquefier

$$\text{COP} = \frac{T_5 - T_4}{T_1 \ln(T_5/T_4) - (T_5 - T_4)} \quad (12)$$

where the absolute temperatures refer to the designated points shown in Figure 15 and T_1 is the ambient temperature.

Combined Isenthalpic and Isentropic Expansion

To take advantage of the increased reversibility gained by using a work-producing expansion engine while minimizing the problems associated with the formation of liquid in such a device, Claude developed a process that combined both expansion processes in the same cycle. Figure 16 shows a schematic of this cycle and the corresponding temperature-entropy diagram.

An energy balance for this system, which incorporates the heat exchangers, expansion engine, J-T expansion valve, and liquid reservoir, allows evaluation of the refrigeration effect or, in the case of a liquefier, of the liquid yield. For the refrigerator, the refrigeration effect is

$$Q = m(h_1 - h_2) + m_e(h_3 - h_e) \quad (13)$$

where m_e and h_e are the mass flow rate through the expander and the specific enthalpy of the stream leaving the expander, respectively.

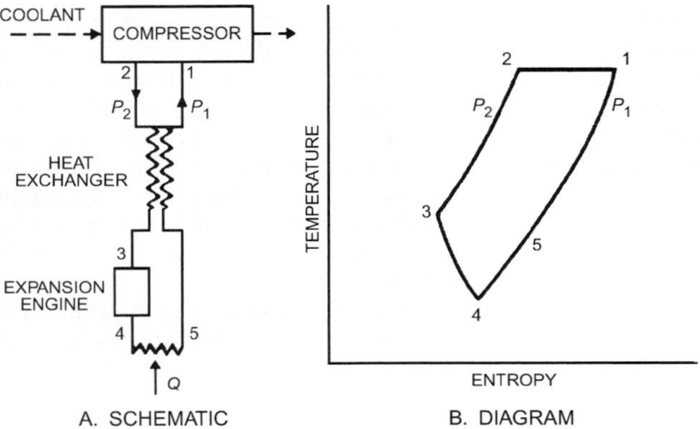

Fig. 15 Schematic for Cold-Gas Expansion Refrigerator and Temperature-Entropy Diagram for Cycle

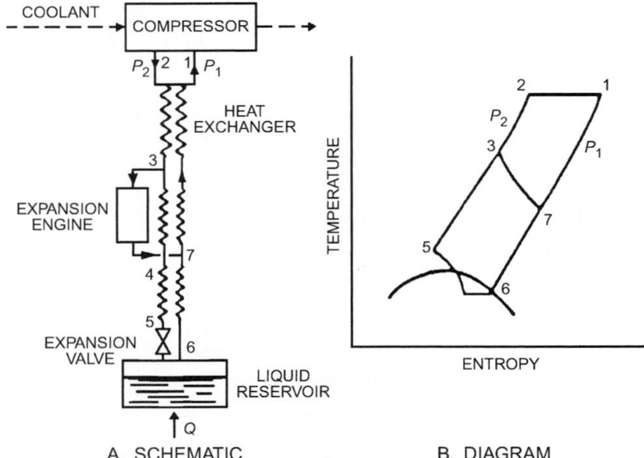

Fig. 16 Schematic for Claude Cycle Refrigerator and Temperature-Entropy Diagram for Cycle

For a liquefier, the energy balance yields

$$mh_2 - (m - m_f)h_1 - m_e(h_3 - h_e) - m_f h_f = 0 \quad (14)$$

Collecting terms and using previously adopted symbols gives the liquid fraction obtained as

$$y = \frac{h_1 - h_2}{h_1 - h_f} - \frac{m_e}{m}\left(\frac{h_3 - h_e}{h_1 - h_f}\right) \quad (15)$$

For a nonideal expander, the h_e term would be replaced with $h_{e'}$.

One modification of the **Claude cycle** that has been used extensively in high-pressure liquefaction plants for air is the **Heylandt cycle**. In this cycle the first warm heat exchanger in Figure 16 is eliminated, allowing the inlet of the expander to operate at ambient temperatures, which minimizes many of the lubrication problems that are often encountered at lower temperatures.

Other modifications of the basic Claude cycle are replacing the throttling valve with a "wet" expander operating in the two-phase region and adding a saturated vapor compressor after the liquid reservoir. The two-phase expander is used with systems involving helium as the working fluid, because the thermal capacity of the compressed gas is, in many cases, larger than the latent heat of the liquid phase. Experience has shown that operation of the expander with helium in the two-phase region has little effect on expander efficiency as experienced with expanders handling either nitrogen or air. Adding the saturated vapor compressor actually improves the thermodynamic performance of the system.

Another modification of the basic Claude cycle is the **dual-pressure Claude cycle**, similar in principle to the dual-pressure system shown in Figure 14. Only the gas that flows through the expansion valve is compressed to the high pressure; this modification reduces the work requirement per unit mass of gas liquefied. Barron (1985) compared the Claude dual-pressure cycle with the Linde dual-pressure cycle and showed that, in liquefaction of air, the liquid yield can be doubled and the work per unit mass liquefied can be halved when the Claude dual-pressure cycle is selected over the Linde dual-pressure cycle.

Still another extension of the Claude cycle is the **Collins helium liquefier**. Depending on the helium inlet pressure, two to five expansion engines are used in this system. The addition of a liquid nitrogen precooling bath results in a two- to threefold increase in liquefaction performance.

Mixed-Refrigerant Cycle

With the advent of large natural gas liquefaction plants, the mixed-refrigerant cycle (MRC) is the refrigeration process primarily used in LNG production. This cycle, in principle, resembles the cascade cycle occasionally used in LNG production, as shown in Figure 17. After purification, the natural gas stream is cooled by the successive vaporization of propane, ethylene, and methane. These gases have each been liquefied in a conventional refrigeration loop. Each refrigerant may be vaporized at two or three different pressure levels to increase the natural gas cooling efficiency, but at a cost of considerably increased process complexity.

Cooling curves for natural gas liquefaction by the cascade process are shown in Figures 18A and 18B. The cascade cycle efficiency can be improved considerably by increasing the number of refrigerants used. For the same flow rate, the actual work required for the nine-level cascade cycle is approximately 80% of that required by the three-level cascade cycle. This increase in efficiency is achieved by minimizing the temperature difference throughout the cooling curve.

The cascade system can be designed to approximate any cooling curve; that is, the quantity of refrigeration provided at the various temperature levels can be chosen so that the temperature differences between the warm and cold streams in the evaporators and heat exchangers approach a practical minimum (smaller temperature differences equate to lower irreversibilities and therefore lower power consumption).

The simplified version of the mixed-refrigerant cycle shown in Figure 19 is a variation of the cascade cycle. This version involves the circulation of a single mixed-refrigerant system, which is repeatedly condensed, vaporized, separated, and expanded. As a result, these processes require more sophisticated design approaches and more complete knowledge of the thermodynamic properties of gaseous mixtures than expander or cascade cycles. Nevertheless, simplifying compression and heat exchange in such cycles offers potential for reduced capital expenditure over conventional cascade cycles. Note the similarity of the temperature versus enthalpy diagram shown in Figure 20 for the mixed-refrigerant cycle with the corresponding diagram for the nine-level cascade cycle in Figure 18B.

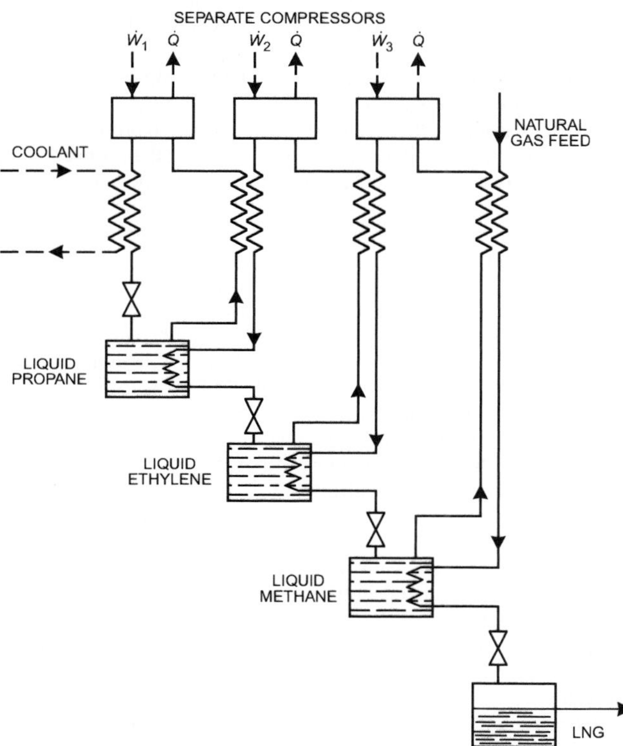

Fig. 17 Classical Cascade Compressed-Vapor Refrigerator

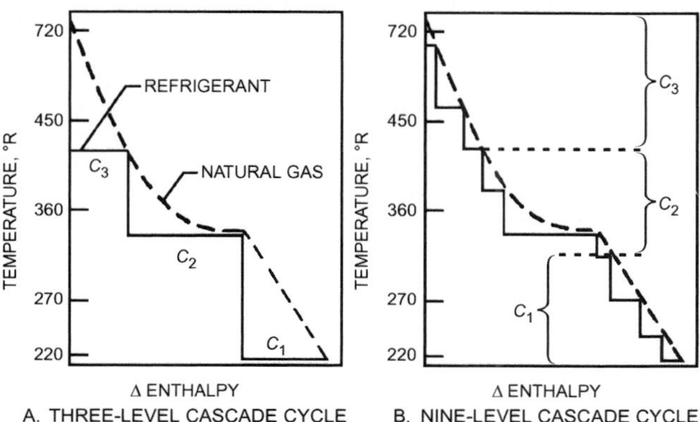

Fig. 18 Three-Level and Nine-Level Cascade Cycle Cooling Curves for Natural Gas

Proprietary variations of the mixed refrigerant have been developed. In one commercial process, for example, the gas mixture is obtained by condensing part of the natural gas feed. This has the advantage of requiring no fluid input other than the natural gas itself; however, this procedure can cause a slow start-up because of the refrigerant that must be collected and the time required to adjust its composition. Another version uses a multicomponent refrigerant that is circulated in a completely separate flow loop. The mixture is prepared from pure gaseous components to obtain the desired composition. Thereafter, only occasional makeup gases are added,

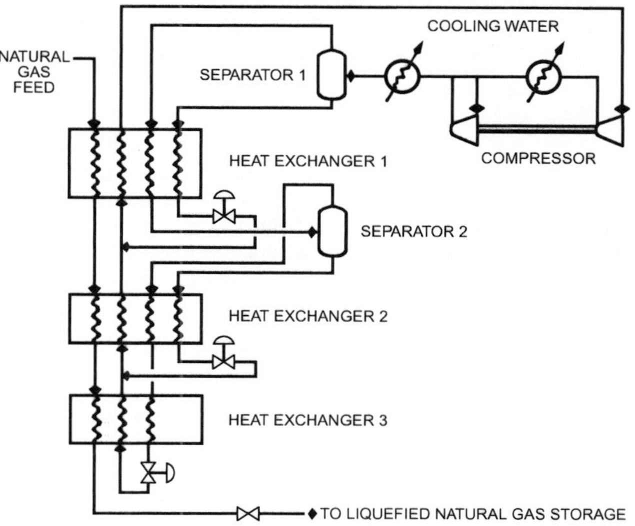

Fig. 19 Mixed-Refrigerant Cycle for Natural Gas Liquefaction

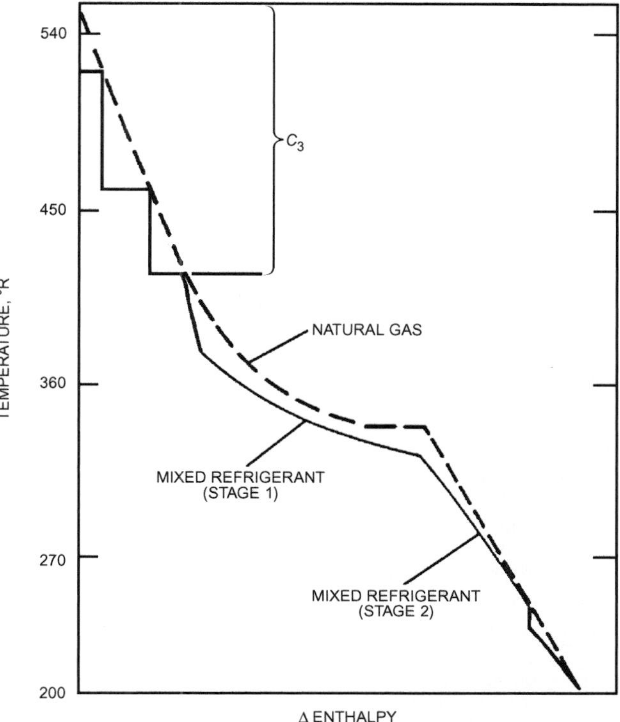

Fig. 20 Propane-Precooled Mixed-Refrigerant Cycle Cooling Curve for Natural Gas Liquefaction

usually from cylinders. This process is simpler and allows for rapid start-up; however, the refrigerant mixture must be stored when the process is shut down. Accordingly, suction and surge drums must be provided, which increases capital expenditure. Gaumer (1986) presents a good review of the mixed-refrigerant cycle.

Comparison of Refrigeration and Liquefaction Systems

The measure of the thermodynamic quality associated with either a piece of equipment or an entire process is its reversibility. The second law, or more precisely the entropy increase, is an effective guide to this degree of irreversibility. However, to obtain a clearer picture of what these entropy increases mean, it is convenient to relate such an analysis to the additional work required to overcome these irreversibilities. The fundamental equation for such an analysis is

$$W = W_{rev} + T_o \sum m \Delta s \tag{16}$$

where the total work is the sum of the reversible work W_{rev} plus a summation of the losses in availability for various steps in the analysis. Here T_o is the reference temperature (normally ambient), m the flow rate through each individual process step, and Δs the change in specific entropy across these same process steps.

Table 4 Comparison of Several Liquefaction Systems Using Air as Working Fluid

Air Liquefaction System (Inlet conditions of 70°F and 14.7 psia)	Liquid Yield $y = m_f/m$	Work per Unit Mass Liquefied, Btu/lb	Figure of Merit
Ideal reversible system	1.000	307	1.000
Simple Linde system p_2 = 2900 psia, η_c = 100%, ε = 1.0	0.086	2253	0.137
Simple Linde system p_2 = 2900 psia, η_c = 70%, ε = 0.95	0.061	4566	0.068
Simple Linde system observed	—	4437	0.070
Precooled simple Linde system p_2 = 2900 psia, T_3 = –49°F, η_c = 100%, ε = 1.00	0.179	963	0.320
Precooled simple Linde system p_2 = 2900 psia, T_3 = –49°F, η_c = 70%, ε = 0.95	0.158	1591	0.194
Precooled simple Linde system, observed	—	2399	0.129
Linde dual-pressure system, p_3 = 2900 psia, p_2 = 870 psia, i = 0.8, η_c = 100%, ε = 1.00	0.060	1180	0.261
Linde dual-pressure system, p_3 = 2900 psia, p_2 = 870 psia, i = 0.8, η_c = 70%, ε = 0.95	0.032	3439	0.090
Linde dual-pressure system, observed	—	2726	0.113
Linde dual-pressure system, precooled to –49°F, observed	—	1539	0.201
Claude system, p_2 = 580 psia, $x = m_e/m$ = 0.7, $\eta_c = \eta_e$ = 100%, ε = 1.00	0.260	383	0.808
Claude system, p_2 = 580 psia, $x = m_e/m$ = 0.7, η_c = 70%, $\eta_{e,ad}$ = 80%, $\eta_{e,m}$ = 90%, ε = 0.95	0.189	868	0.356
Claude system, observed	—	1539	0.201
Cascade system, observed	—	1399	0.221

η_c = compressor overall efficiency
η_e = expander overall efficiency
$\eta_{e,ad}$ = expander adiabatic efficiency
$\eta_{e,m}$ = expander mechanical efficiency
ε = heat exchanger effectiveness
$i = m_i/m$ = mass in intermediate stream divided by mass through compressor
$x = m_e/m$ = mass through expander divided by mass through compressor

Use caution when accepting comparisons made in the literature, because it is difficult to put all processes on a comparable basis. Many assumptions must be made in the course of the calculations, and these can have considerable effect on the conclusions. The most important assumptions generally include heat leak, temperature differences in heat exchangers, efficiencies of compressors and expanders, number of stages of compression, fraction of expander work recovered, and state of flow. For this reason, differences in power requirements of 10 to 20% can readily be due to differences in assumed variables and can negate the advantage of one cycle over another. Table 4 illustrates this point by comparing some of the more common liquefaction systems (described earlier) that use air as the working fluid with a compressor inlet gas temperature and pressure of 70°F and 1 atm (530°R and 14.7 psia), respectively.

To avoid these pitfalls, an analysis should compare the minimum power requirements to produce a unit of refrigeration. Table 5 provides such data for some of the more common cryogens, in addition to the minimum power requirements for liquefaction. The warm temperature in all cases is fixed at 80°F (540°R), and the cold temperature is assumed to be the normal boiling temperature T_{bp} of the fluid. The specific power requirements increase rapidly as the boiling points of the fluids decrease.

Table 5 also shows that more power is required to produce a given amount of cooling with the liquid from a liquefier than is needed for a continuously operating refrigerator. This greater power is needed because less refrigeration effect is available for cooling the feed gas stream when the liquid is evaporated at another location (other than the source of liquefaction itself).

An ideal helium liquefier requires a power input of 3048 Btu/h to produce liquid at the rate of 1 gal/h. Because the heat of vaporization of helium is low, 1 Btu/h will evaporate 0.107 gal/h. Thus, 326 Btu/h would be required to ideally power a liquefier with a liquid product used to absorb 1 Btu/h of refrigeration at –452.4°F (7.6°R). An ideal refrigerator, on the other hand, would only require 70.4 Btu/h of input power to produce the same quantity of refrigeration also at –452.4°F (7.6°R). This difference in power requirement does not mean that a refrigerator will always be chosen over a liquefier for all cooling applications; in some circumstances, a liquefier may be the better or the only choice. For example, a liquefier is required when a constant-temperature bath is used in a low-temperature region. Also, a single centralized liquefier can supply liquid helium to a large number of users.

Another comparison of low-temperature refrigeration examines the ratio of W/Q for a Carnot refrigerator with the ratio obtained for the actual refrigerator. This ratio indicates the extent to which an actual refrigerator approaches ideal performance. (The same ratio can be formed for liquefiers using the values from Table 5 and the actual power consumption per unit flow rate.) This comparison with

Table 5 Reversible Power Requirements

		Ideal Power Input for		
Fluid	T_{bp}, °R	1 Btu/h Refrigerant Capacity, (Btu/h)/(Btu/h)	1 gal Liquid, (Btu/h)/gal	1 Btu/h Refrig. Capacity from Liquid,* (Btu/h)/(Btu/h)
Helium	7.6	70.4	3048	326
Hydrogen	36.7	13.7	3590	31.7
Neon	48.8	10.1	5773	15.5
Nitrogen	139.3	2.88	2234	3.87
Fluorine	153.0	2.53	3074	3.26
Argon	157.1	2.44	2389	2.95
Oxygen	162.3	2.33	2518	2.89
Methane	201	1.69	1666	2.15

*Obtained by dividing ideal liquefaction power requirement by heat of vaporization of fluid.

ideal performance is plotted in Figure 21 as a function of refrigeration capacity for actual refrigerators and liquefiers. The capacity of the liquefiers included in this comparison was obtained by determining the percent of Carnot performance that these units achieved as liquefiers and then calculating the refrigeration output of a refrigerator operating at the same efficiency with the same power input.

The data for these low-temperature refrigerators cover a wide range of capacity and temperatures. The more efficient facilities are the larger ones that can more advantageously use complex thermodynamic cycles. The same performance potential may exist in the smaller units, but costs are prohibitive and the savings in electrical power have generally not justified the greater capital expenditure.

CRYOCOOLERS

Small, low-temperature refrigerators that provide no more than a few watts of cooling are generally called cryocoolers. Problems of efficiency, unreliability, size, weight, vibration, and cost have been major concerns for cryocooler developers. The seriousness of any one of these problems depends on the application of the cryocooler. The largest application of cryocoolers has been in cooling infrared sensors for night vision in satellites by the military, primarily using Stirling cryocoolers with a refrigeration capacity of about 0.85 Btu/h at −316°F (144°R). These devices have a mean time to failure of about 7500 h (M-CALC IV 2003), but this is inadequate to meet the requirements of space operations and has provided the impetus for major innovations for Stirling units and the rapid growth of research and development of pulse tube refrigerators.

The largest application of cryocoolers in the commercial area has been in cryopumping for the manufacture of semiconductors. Gifford-McMahon cryocoolers (see the section on these cryocoolers under Regenerative Systems) providing a few watts of refrigeration at a temperature of about −433°F (27°R) are the most popular devices for this area. This choice may change, however, because of the vibration characteristics of this cryocooler. Semiconductor manufacturers are forced to achieve narrow line widths in their computer chips to provide more compact packaging of semiconductor circuits.

Even though cryocoolers can be classified by the thermodynamic cycle that is followed, generally there are only two types: recuperative or regenerative units. **Recuperative** cryocoolers use only recuperative heat exchangers; a **regenerative** cryocooler must use at least one regenerative heat exchanger (regenerator) in the unit.

Because recuperative heat exchangers provide two separate flow channels for the refrigerant, refrigerant flow is always continuous

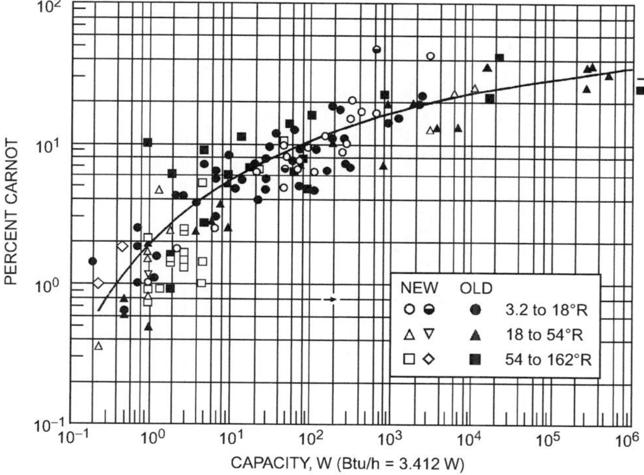

Fig. 21 Efficiency as Percent of Carnot Efficiency
(For low-temperature refrigerators and liquefiers as function of refrigeration capacity)
(Strobridge 1974)

and in one direction, analogous to a dc electrical system. This heat exchanger requires either valving with reciprocating compressors and expanders or rotary or turbine compressors and expanders. In regenerative cycles, the refrigerant flow oscillates, analogous to an ac electrical system. This oscillatory effect allows the regenerator matrix to store energy in the matrix for the first half of the cycle and release the energy during the next half. To be effective, the solid matrix material in the regenerator must have high heat capacity and good thermal conductivity. Some advances in both types of cryocoolers are reviewed in the following sections.

Recuperative Systems

The J-T and the Brayton cycles, compared schematically in Figure 22, are two recuperative systems.

Joule-Thomson Cryocoolers. The J-T effect of achieving cooling by throttling a nonideal gas is one of the oldest but least efficient methods for attaining cryogenic temperatures. However, J-T cryocoolers have been significantly improved by applying novel methods of fabrication, incorporating more complex cycles, and using special gas mixtures as refrigerants. These developments have made the J-T cryocooler competitive in many applications, even when compared with cryocoolers that generally exhibit more efficient cooling cycles. Their relative simplicity combined with their small size, low mass, and lack of mechanical noise or vibration have been additional advantages for these small refrigerators. In the past, J-T cryocoolers were fabricated by winding a finned capillary tube on a mandrel, attaching an expansion nozzle at the end of the capillary tube, and inserting the entire unit in a tightly fitting tube closed at one end with inlet and exit ports at the other end. Little (1984) introduced a method of fabricating J-T cryocoolers using a photolithographic manufacturing technique in which gas channels for the heat exchangers, expansion capillary, and liquid reservoir are etched on thin, planar glass substrates that are fused together to form a sealed unit. These miniature cryocoolers have been fabricated in a wide range of sizes and capacities. One cryocooler operating at −316°F (144°R) with a refrigeration capacity of 0.8 Btu/h uses a heat exchanger with channels 0.0079 in. wide and 0.0012 in. deep. The channels etched on the glass substrate must be controlled to a tolerance of ±0.00008 in., and the bond between the different substrates must withstand pressures on the order of 2000 to 3000 psia. This cryocooler, used in spot cooling of electronic systems, has an overall dimension of only 3 by 0.55 by 0.08 in (not including the compressor).

Fabrication of miniature J-T cryocoolers by this approach has made it much simpler to use more complex refrigeration cycles and multistage configurations. The dual-pressure J-T cycle, in which the refrigerant pressure is reduced by two isenthalpic expansions, provides either a lower spot-cooling temperature or a higher coefficient of performance for the same power input. Temperatures

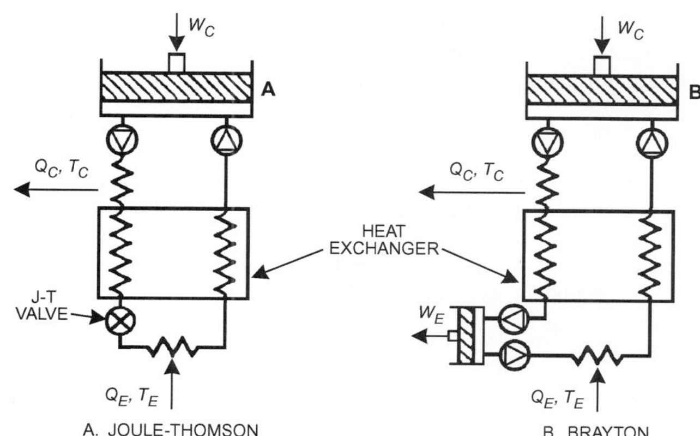

Fig. 22 Schematic of Joule-Thomson and Brayton Cycles
(Two recuperative types of cycles used in cryocoolers)

below –406°F (54°R) have been achieved with two-stage systems. A nitrogen refrigeration stage provides precooling to –321°F (139°R), and a neon J-T stage to achieve operation at –411°F (49°R). Operation at –424°F (36°R) is also possible using hydrogen in place of neon. However, use of multistage units in these miniature cryocoolers requires an order of magnitude better dimensional control of the etching process to match the desired flows and capacities specified for the heat exchangers, expansion capillaries, and liquid reservoirs.

To attain temperatures of –80°F (380°R), pure nitrogen has been used as the refrigerant in J-T cryocoolers. At 80°F (540°R), nitrogen must be compressed to a very high pressure (1500 to 3000 psia) to achieve any significant enthalpy change. The high pressure required leads to a low compression efficiency with high stresses on compressor components, and the small enthalpy change results in a low cycle efficiency. Alfiev et al. (1973), using a gaseous mixture of 30 mol% nitrogen, 30 mol% methane, 20 mol% ethane, and 20 mol% propane, achieved a temperature of –320°F (140°R) using a 50:1 pressure ratio. The system efficiency with this gas mixture was 10 to 12 times better than when pure nitrogen gas was used as the refrigerant. Temperatures below –334°F (126°R) were obtained by adding neon, hydrogen, or helium to the mixture.

Little (1990) established that the addition of the fire retardant CF_3Br (halon) to the nitrogen/hydrocarbon mixture was sufficient to render the mixture nonflammable and was retained in the resulting liquid solution down to –321°F (139°R) or lower without precipitation because of the excellent solvent properties of the mixture. As a result, a series of nitrogen/hydrocarbon gas mixtures that are reasonably safe to use and provide high refrigeration efficiencies were available. However, halon production was discontinued in the United States in 1992 because of its high ozone-depletion potential, and the EPA has additional restrictions on handling recycled halon.

The high cooling capacity of nitrogen/hydrocarbon mixtures is illustrated in the following example. Consider the temperature-entropy diagram shown in Figure 23 for a hydrocarbon mixture of 27% methane, 50% ethane, 13% propane, and 10% butane on a volumetric basis. A throttling process for this gas mixture, initially at 80°F (540°R) and 45 atm (660 psia), can ideally (constant enthalpy) achieve an exit temperature of –100°F (360°R) at a final exit pressure of 1 atm (14.7 psia). Pure nitrogen gas undergoing a similar throttling process from the same inlet conditions to the same final exit pressure only achieves an exit gas temperature of 64°F (524°R). Thus, there can be as much as an elevenfold increase in refrigerant temperature drop after the throttling process by using the gas mixture instead of pure nitrogen gas over these pressure and temperature conditions. That is, refrigeration performance comparable to that using pure nitrogen gas at 1700 to 2000 psia inlet pressures can be achieved with specific gas mixtures at pressure as low as 400 to 700 psia.

The effect of various gas mixture concentrations on the efficiency of any cycle can be analyzed by evaluating the coefficient of performance (COP) of the cycle. The refrigeration effect Q of a J-T refrigerator using such gas mixtures is given by

$$Q = n(h_{lp} - h_{hp})_{min} = n\Delta h_{min} \quad (17)$$

where n is the molar flow rate of the gas mixture, h_{lp} is the molar enthalpy of the low-pressure stream, and h_{hp} is the molar enthalpy of the high-pressure stream at the location in the recuperative heat exchanger that provides a minimum difference in molar enthalpies Δh_{min} between the two streams. The ideal work of compression is evaluated from

$$W_{ideal} = n[(h_2 - h_1) - T_o(s_2 - s_1)] = n\Delta g_o \quad (18)$$

where s_1 and s_2 are the molar entropies at the inlet and outlet from the compressor, respectively, at a constant compression temperature of T_o. The Δg_o is the change in the molar Gibbs free energy also at T_o. The ideal COP of the refrigerant cycle is then

$$COP = Q/W_{ideal} = \Delta h_{min}/\Delta g_o \quad (19)$$

This indicates that a maximum efficiency in the J-T cycle is achieved when the value of $\Delta h_{min}/\Delta g_o$ is maximized for the refrigerant mixture in the temperature range of interest.

Brayton Cryocoolers. Expanding refrigerant with an expansion engine in the Brayton cycle leads to higher cycle efficiencies than are attainable with J-T cryocoolers. The Brayton cycle is commonly used in large liquefaction systems accompanied by a final J-T expansion. The units are highly reliable because they use turboexpanders operating with gas bearings. For small cryocoolers, the challenge has been in fabricating the miniature turboexpanders while maintaining a high expansion efficiency and minimizing heat leakage. Swift and Sixsmith (1993) addressed this challenge by developing a single-stage Brayton cryocooler with a small turboexpander (rotor diameter of 1/8 in.) providing 17 Btu/h of refrigeration at –343°F (117°R) with neon as the working fluid. The compressor also uses gas bearings with an inlet pressure of 16.2 psia and a pressure ratio of 1.6. The unit operates between –343 and 44°F (117 and 504°R) with a Carnot efficiency of 7.7%. However, the present cost of such cryocoolers limits their service to space applications, which require high reliability, high thermodynamic efficiency, and low vibration.

Mixed-Refrigerant Systems. The possibility of achieving high refrigeration efficiency at cryogenic temperatures with simple closed-cycle J-T (throttle expansion) systems has sparked interest in commercial development of such cryocoolers in the United States. Missimer (1973) described a multistage system in which liquid condensate was withdrawn from the compressed vapor/liquid refrigerant mixture after each of a series of heat exchangers, throttling the withdrawn condensate to a lower pressure, and returning the cold refrigerant to the compressor via counterflow exchangers. Temperatures to 280°R were achieved with this system. However, the system was complicated and has found limited application.

Single-Stage Throttle Expansion Cryocoolers. Longsworth (1997a) described a simple, single-stage throttle expansion refrigerator (see Figure 13) using a single-stage compressor of the type used

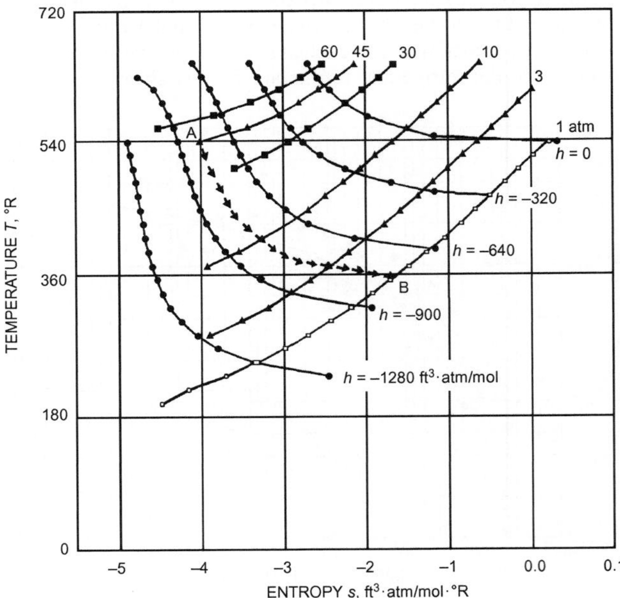

Fig. 23 Isenthalpic Expansion of Multicomponent Gaseous Mixture from A to B

(27% methane, 50% ethane, 13% propane, and 10% butane; h is enthalpy)

in domestic refrigerators and air-conditioning units. Temperatures to 126°R have been achieved with this system, which evolved out of a Gifford-McMahon (GM) compressor system combined with a J-T heat exchanger. The compressor is oil lubricated, and requires an efficient oil separator and special pretreatment of the oil and system. These additional elements add to the cost of an otherwise simple system (Longsworth 1997b). Over a thousand of these units have been manufactured and are used to cool gamma ray detectors, vacuum system cold traps, infrared (IR) detectors, and laboratory instrumentation.

Alexeev et al. (1999) achieved a refrigeration capacity of 0.03 ton at 180°R with power input of 1100 W using a mixed-refrigerant throttle expansion refrigerator with a precooling stage. This is 18% of Carnot efficiency, and 1.5 times more efficient than a comparable GM refrigerator at the same temperature.

Mixed-Refrigerant Cascade (Kleemenko Cycle) Cryocoolers. Kleemenko (1959) described a one-flow cascade refrigeration cycle using multicomponent refrigerant mixtures. The cycle was based on two key features that promised to give high refrigeration efficiency. As is well known, much of the inefficiency of a throttle expansion system lies in irreversibility of heat transfer in the heat exchanger and in the expansion process. Kleemenko pointed out that heat exchanger inefficiency is exacerbated by the fact that heat capacity of the fluid in the high-pressure stream and of that of the low-pressure stream generally differ, and, as a result, the temperature difference between the two streams diverges along the length of the heat exchanger. If, however, a suitably designed mixture of refrigerants is used instead of a single component, it is possible to keep these two capacities similar and thus minimize the temperature difference between them, reducing the irreversibility. Secondly, he noted that the thermodynamic reversibility of throttling is much greater for a fluid in the liquid state than for one in the gaseous state. He demonstrated the improvement that could be achieved by applying these two factors in a large, liquid natural gas plant.

These factors are the basis for development of a new class of small, low-cost cryocoolers with good efficiency and exceptionally high reliability. The history of this development is given by Little (1998). These coolers use the refrigerant cycle shown in Figure 24. The compressor (1) is an oil-lubricated, hermetically sealed, home refrigerator compressor. Some oil is entrained in the high-pressure refrigerant stream, and most of it is removed in a small cyclone oil separator (2) and returned to the compressor via a small capillary return line (11). The high-pressure vapor is then cooled in the air-cooled condenser (3), where the highest-boiling components of the mixture condense. This two-phase mixture is injected tangentially into the second cyclone separator (4), where the liquid condensate and any remaining oil collect at the bottom, are fed through a short heat exchanger to the throttle expander (6), and ultimately return to the compressor via the line (5) that cools the upper end of the separator. This upper part of the separator is filled with platelets that act as a fractionating column; its large surface area ensures that the vapor and liquid fractions are in equilibrium with one another. Additional refrigerant condenses on these platelets and drips down to the liquid outlet (4), while the remaining vapor exits from the top of the column.

This cleansed vapor then passes to the heat exchanger and is precooled by evaporating liquid from the throttle restrictor (6) in the heat exchanger. It flows through the remainder of the exchanger (7), condensing as it goes, and passes through the second throttle (8) as a liquid, which, upon evaporating, cools the load (9).

An important difference between the Kleemenko cycle cooler and the single-stage throttle expansion cooler is the former's use of the fractionator column to remove residual oil and any other impurities from the vapor stream automatically, and return them to the compressor. In the single-stage cooler, oil is trapped in a zeolite or charcoal absorber, which must be replaced when it is saturated. No such maintenance is needed for the Kleemenko system. This self-cleaning feature (Little 1997a; Little and Sapozhnikov 1998) accounts for the cooler's long life and maintenance-free operation. Continuous operation at 216°R for over 67,000 h has been achieved for early prototypes of these coolers, and over 35,000 h to date for units operating at 144°R (Little 2003). The system's simplicity (having only one liquid/vapor separator, which is at ambient temperature) and the use of common commercial refrigeration components (i.e., compressors, condensers, capillaries, and dewars) has reduced the cost of these cryocoolers to little more than the cost of vapor-compression systems.

Tests at the Naval Research Laboratory of low-cost, long-life cryocoolers developed under a DARPA program found they were exceptionally reliable. Vendors supplied demonstration systems for each of the following cycles: Stirling, Gifford-McMahon, single-stage throttle, Kleemenko cycle, and pulse tube. The only coolers to operate continuously within their specification for the 5000 h test were the Kleemenko coolers (Kawecki and James 1999). These coolers have now logged almost 70,000 h at 216°R and 30,000 h at 144°R to date (Little 2003).

A major advantage of the J-T or throttle expansion cycle coolers over other cryocoolers is the absence of moving parts at the cold end. This is important because cooling is frequently used to reduce noise in sensors or detectors, the performance of which can be degraded by any vibration or noise. For this reason, gamma and x-ray detectors had always been cooled with liquid nitrogen. Now, though, the vibration levels of the Kleemenko cycle cold stage are low enough to be comparable to that of boiling liquid nitrogen (Broerman et al. 2001), enabling their use for these detectors.

Refrigerants in these coolers typically are mixtures of 5 to 10 components. Their thermodynamic properties can be calculated with programs such as SUPERTRAPP and DDMIX (NIST 2002a, 2002b), available from NIST, and other commercial programs. Methods for optimizing their refrigerant properties are also available (Dobak 1998; Keppler et al. 2004; Little 1997). However, in practice, the problem is more complicated. Low-cost compressors are oil-lubricated and contain about 1 lb of oil. Each component of the mixture dissolves to a different extent in the oil, and these solubilities are generally a strong function of temperature. Consequently, the composition of the mixture with which the unit is charged differs from that which circulates during operation, and this composition varies as compressor temperature changes. The problem is exacerbated by the fact that some of the higher-boiling-point components condense in the heat exchanger during cooldown, reducing the fraction of these in the refrigerant that circulates. These factors have to be taken into account in designing the appropriate mixture to charge the units.

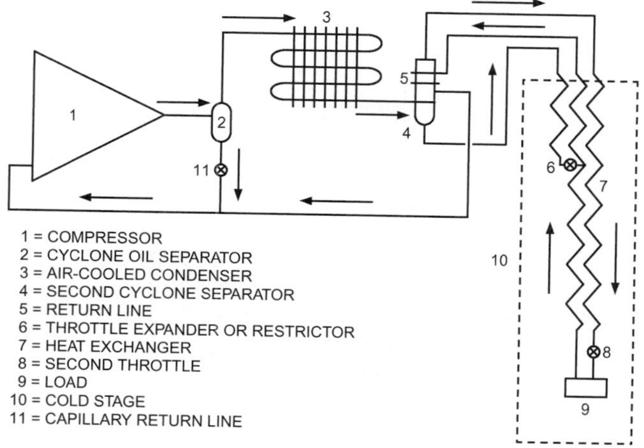

Fig. 24 Kleemenko Cycle Cooler

The low cost, low noise, high reliability, and good efficiency of these new coolers have changed the landscape for cooling in the temperature range from –40 to –330°F. Applications have increased dramatically and now include cryosurgical devices, chip handlers for automated test equipment, low-noise microwave amplifiers, and x-ray and gamma detectors; they have also become the enabling technology for low-cost nitrogen liquefiers for the dermatology market, and oxygen liquefiers for the home care market.

Regenerative Systems

Stirling Cryocoolers. The Stirling refrigerator, which boasts the highest theoretical efficiency of all cryocoolers, is the oldest and most common of the regenerative systems. The elements of the Stirling refrigerator normally include two variable volumes at different temperatures, coupled together through a regenerative heat exchanger, a heat exchanger rejecting the heat of compression, and a refrigerator absorbing the refrigeration effect. These elements can be arranged in a wide variety of configurations and operate as either single- or double-acting systems. The single-acting units are either two-piston or piston-displacer systems, as shown in Figure 25.

The ideal Stirling cycle consists of four processes:

1. Isothermal compression of the refrigerant in the compression space at ambient temperature by rejecting heat Q_c to the surroundings.
2. Constant-volume regenerative cooling, transferring heat from the working fluid to the regenerator matrix. The reduction in temperature at constant volume causes a reduction in pressure.
3. Isothermal expansion in the expansion space at refrigeration temperature T_E. Heat Q_E is absorbed from the surroundings of the expansion space.
4. Constant-volume regenerative heating in which heat is transferred from the regenerator matrix to the working fluid. The increase in temperature at constant volume increases pressure back to the initial conditions.

Successful operation of the cycle requires that volume variations in the expansion space lead those in the compression space.

Thousands of small, single-stage Stirling cryocoolers have been manufactured. Capacities range from about 0.03 to 3.2 Btu/h at –316°F (144°R). The largest units (not including the compressor) are generally no larger than 6 in. in any dimension, and weigh less than 6.5 lb. Power inputs range from 40 to 50 Btu/h per Btu/h of refrigeration, equivalent to an efficiency of 6 to 7% of Carnot limit.

Many recent developments in Stirling cryocoolers have been directed toward improved reliability. In most applications, for example, linear motor drives have replaced rotary drives to reduce moving parts as well as reduce the side forces between the piston and cylinder. Lifetimes of about 4000 h are the norm with linear compressors; however, lifetimes greater than 15,000 h have been achieved by using improved materials for the rubbing contact. Longer lifetimes have been achieved with piston devices by using flexure, gas, or magnetic bearings to center the piston and displacer in the cylinder housing. Davey (1990) reviews the development of these cryocoolers.

Orifice Pulse Tube Refrigerators. Spaceflight applications require lifetimes of 10 to 15 years, low mass, and low energy consumption. These considerations have directed research on the orifice pulse tube refrigerator (OPTR) shown schematically in Figure 26. This unit is a variation of the Stirling cryocooler in which the moving displacer is replaced by a pulse tube, orifice, and reservoir volume. Radebaugh (1990) gives a detailed review of pulse tube refrigerators.

The orifice pulse tube refrigerator operates on a cycle similar to the Stirling cycle, except that proper phasing between mass flow and pressure is established by the passive orifice rather than by the moving displacer. In this cycle, a low-frequency compressor raises the pressure of the helium refrigerant gas to between 70 and 350 psia during the first half of a sinusoidal compression cycle. The oscillating pressure for the OPTR can be provided either by a compressor similar to that used in the Stirling refrigerator, or by a Gifford-McMahon compressor that has been modified with appropriate valving to achieve the required oscillating pressure. The high-pressure gas, after being cooled in the regenerator, adiabatically compresses the gas in the pulse tube.

Approximately one-third of the compressed gas originally in the pulse tube flows through the orifice to the reservoir volume, with the heat of compression being removed in the hot-exchanger. During the latter half of the sinusoidal cycle, the gas in the pulse tube expands adiabatically, which causes a cooling effect. The cold, expanded gas is forced past the cold heat exchanger and a buffer volume of gas that allows a temperature gradient to exist between the hot and cold ends of the pulse tube. Some mixing or turbulence occurs in the buffer volume because the time-averaged enthalpy flow that represents the gross refrigeration capacity is only 55 to 85% of the ideal enthalpy flow.

By assuming simple harmonic pressure, mass flow, and temperature oscillations in the entire pulse tube refrigerator as well as adiabatic operation in the pulse tube itself, researchers at NIST developed an analytical model that reasonably predicts refrigeration performance (Storch and Radebaugh 1988). Thermoacoustic theories include a linear approximation with higher harmonics and realistic heat transfer and viscous effects between the gas and pulse tube wall. Losses accounted for in these models result in a time-averaged enthalpy flow that agrees closely with experimental values.

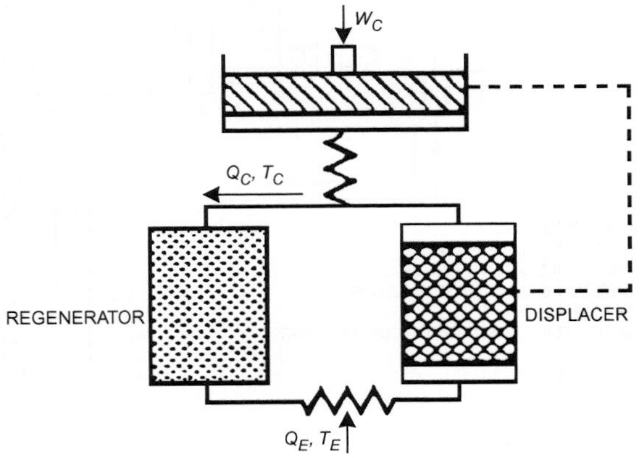

Fig. 25 Schematic of Stirling Cryocooler

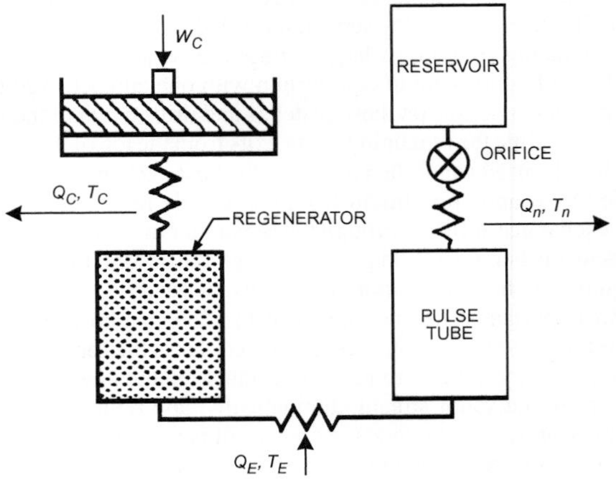

Fig. 26 Schematic for Orifice Pulse Tube Cryocooler

The orifice concept for pulse tube refrigerators achieves a refrigeration temperature of –352°F (108°R) with only one stage. Further refinement can achieve temperatures below –370°F (90°R). However, the improved Stirling refrigerator remains the choice for most spaceflight applications because its Carnot efficiency is typically higher than those obtained from the OPTR.

Zhu et al. (1990) improved efficiencies for pulse tube refrigerators with higher operating frequencies by adding a second orifice, as shown in Figure 27. This double-inlet concept allows gas flow needed to compress and expand the gas at the warm end of the pulse tube to bypass the regenerator and pulse tube. The reduced mass flow through the regenerator reduces regenerator losses, particularly at high frequencies where these losses become large. The second orifice can reduce refrigerator temperature by at least 27 to 36°F in a well-designed pulse tube operating at frequencies of 40 to 60 Hz. This was substantiated in 1994 with a temperature of –425°F (34°R), the lowest temperature achieved to date with a single-stage, double-inlet arrangement.

Figure 28 compares the percent of Carnot efficiency obtained for the improved pulse tube refrigerators with Stirling refrigerators. The shaded area represents the efficiency range obtained for most of the recent Stirling refrigerators, and the circles represent the individual efficiencies obtained from recent pulse tube refrigerators. The highest-power and highest-efficiency unit is the pulse tube refrigerator described by Radebaugh (1995). This unit provided 106 Btu/h of refrigeration at –316°F (144°R) with a rejection temperature of 109°F (569°R), equivalent to a relative Carnot efficiency of 13%. The average operating pressure was 360 psia, and the operating frequency was maintained at 4.5 Hz. The other two circles with lower efficiencies in the –343 to –316°F (117 to 144°R) range represent the efficiencies obtained for the same unit but with different input powers and different cold-end temperatures. The efficiency shown in the –406 to –397°F (54 to 63°R) range is for a small unit developed by Burt and Chan (1995). Even though the data for pulse tube refrigerators are limited, the efficiencies of the most recent pulse tube refrigerators are becoming quite competitive with the best Stirling refrigerators of comparable size.

Two or more pulse tube refrigerator stages are normally used to maintain high efficiency when temperatures below about –370°F (90°R) are desired. Purposes of the staging are to provide net cooling at an intermediate temperature and to intercept regenerator and pulse tube losses at a higher temperature. Three methods exist for the staging arrangement. The first uses a parallel arrangement of a separate regenerator and pulse tube for each stage, with the warm end of each pulse tube at ambient temperature. In the second method, shown in Figure 29, the warm end of the lower-stage pulse tube is thermally anchored to the cold end of the next higher stage in a series configuration. The third method uses a third orifice to permit a fraction of the gas removed from an optimized location in the regenerator to enter the pulse tube at an intermediate temperature. This staging configuration, the multi-inlet arrangement, maintains the simple geometrical arrangement of a single pulse tube, although it would normally require a change in diameter at the tube

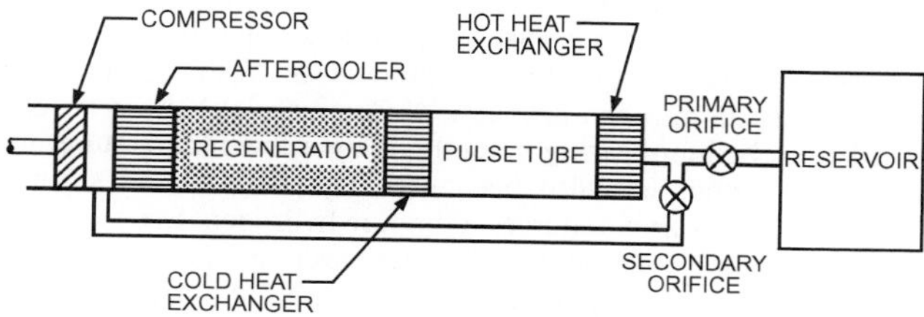

Fig. 27 Schematic of Double-Inlet Pulse Tube Refrigerator Using Secondary Orifice

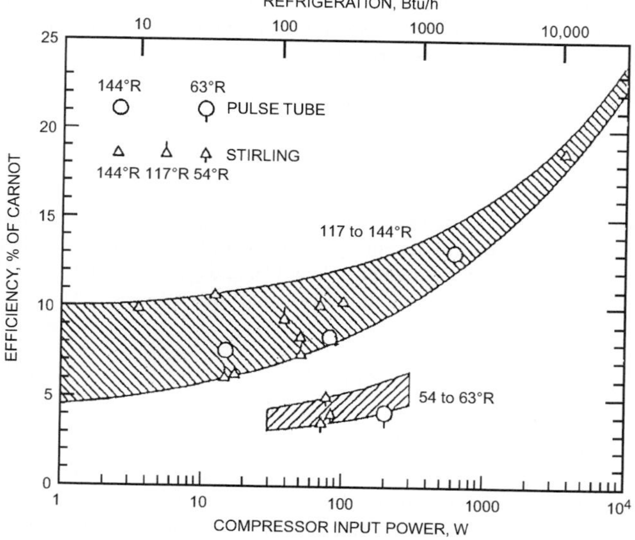

Fig. 28 Comparison of Carnot Efficiency for Several Recent Pulse Tube Cryocoolers with Similarly Powered Stirling Cryocoolers
(Strobridge 1974)

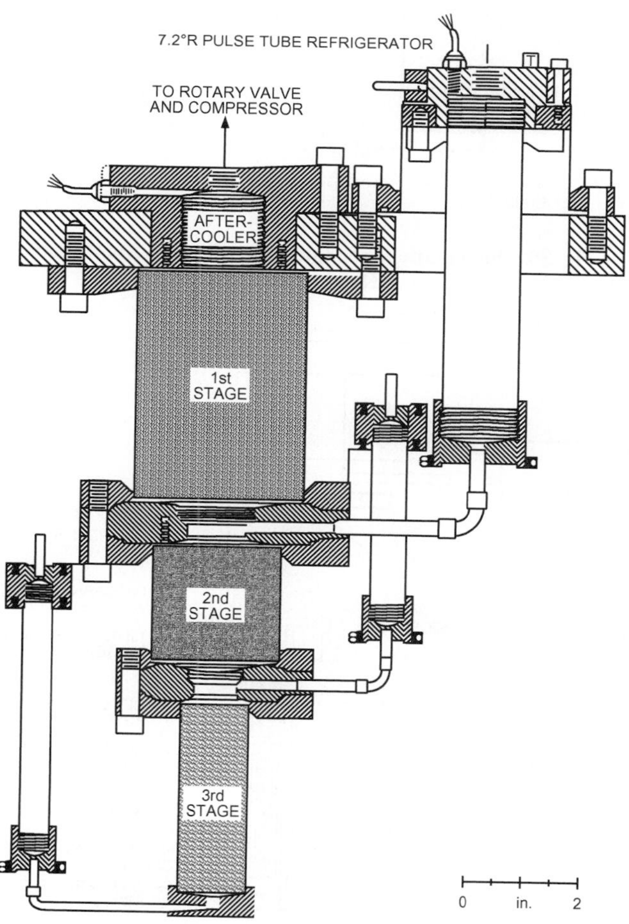

Fig. 29 Three-Stage Series Orifice Pulse Tube Cryocooler for Liquefying Helium

junction with the pulse tube to maintain constant gas velocity in the pulse tube. The lowest temperature attained with a two-stage parallel arrangement of pulse tube refrigerators is −455.7°F (4.0°R).

Gifford-McMahon Refrigerator. J-T cryocoolers using pure gas refrigerants require very high operating pressures; therefore, most commercial closed-cycle cryocoolers use one or more expanders to achieve part of the cooling effect. One of the most widely used regenerative cryocoolers is the Gifford-McMahon refrigerator, schematically shown in Figure 30. These units can achieve temperatures of −343 to −316°F (117 to 144°R) with one stage of expansion and −433 to −453°F (27 to 6.7°R) with two stages of expansion. Precooling the expansion stage is accomplished with regenerators using carefully selected matrix materials. Because regenerators essentially store energy, the matrix materials must possess a high heat capacity as well as a good thermal conductivity. Lead shot has been the regenerator material selected for most regenerative cryocooler operation between −442 to −343°F (18 to 117°R). However, its heat capacity becomes too low to be effective below this temperature range.

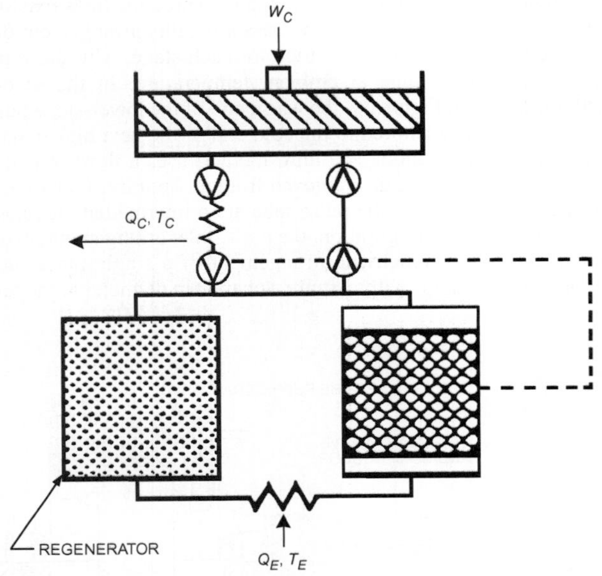

Fig. 30 Schematic for Single-Stage Gifford-McMahon Refrigerator

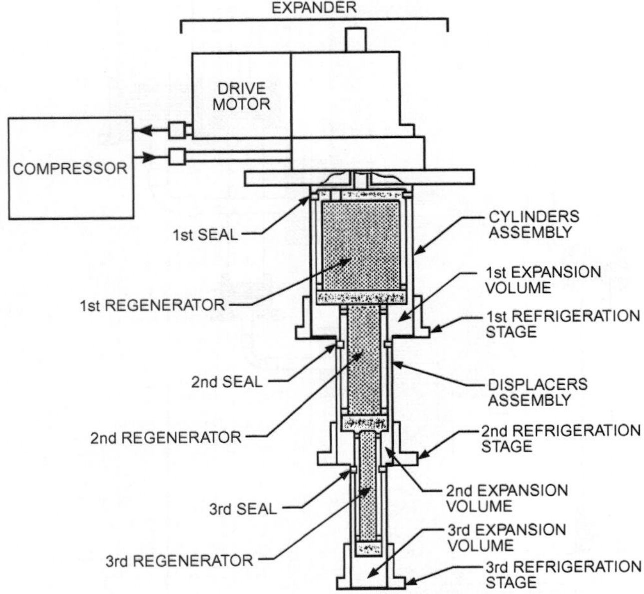

Fig. 31 Cross Section of Three-Stage Gifford-McMahon Refrigerator

Without a suitable matrix material, addition of a third stage with its accompanying regenerator has made it impossible for any regenerative cryocooler to achieve a temperature below −442 to −438°F (18 to 22°R). Attaining a temperature of −452.4°F (7.3°R) to reliquefy helium boil-off has required a two-stage Gifford-McMahon refrigerator equipped with a J-T loop using a compact countercurrent heat exchanger. However, the availability of new matrix materials (rare earth compounds) has made it possible to use a three-stage Gifford-McMahon refrigerator to provide sufficient cooling to reliquefy helium boil-off from superconducting magnets serving MRI units. Nagao et al. (1994) described such a device, shown in Figure 31. It uses $Er_{1.5}Ho_{1.5}Ru$ as the matrix material in the third-stage regenerator and provides a refrigeration capacity of more than 0.5 Btu/h at −452.4°F (7.3°R). This unit, which is smaller than the conventional −453°F (6.7°R) Gifford-McMahon refrigerator, also has greater reliability as well as lower operating costs.

SEPARATION AND PURIFICATION OF GASES

The major application of low-temperature processes in industry involves separation and purification of gases. Much commercial oxygen and nitrogen and all of the neon, argon, krypton, and xenon are separated from air. Pressure-swing adsorption processes account for the oxygen and nitrogen production that is not obtained by cryogenic separation of air. Membranes are also used for small applications in nitrogen production. Commercial helium is separated from helium-bearing natural gas by a low-temperature process. Cryogenics has also been used commercially to separate hydrogen from various sources of impure hydrogen. Even the valuable low-boiling components of natural gas (e.g., methane, ethane, ethylene, propane, propylene) are recovered and purified by various low-temperature schemes. Separation of these gases is dictated by the thermodynamic principles of phase equilibria. The degree to which they separate is based on the physical behavior of the liquid and vapor phases. This behavior is governed, for ideal gas conditions, by the laws of Raoult and Dalton.

The energy required to reversibly separate gas mixtures is the same as the work needed to isothermally compress each component in the mixture from its own partial pressure in the mixture to the final pressure of the mixture. This reversible isothermal work per unit mass is given by the relation

$$(W/m)_i = T_1(s_1 - s_2) - (h_1 - h_2) \qquad (20)$$

where s_1 and h_1 refer to the entropy and enthalpy before separation and s_2 and h_2 refer to the entropy and enthalpy after separation. For a binary system of components A and B, and assuming an ideal gas for both components, Equation (20) simplifies to

$$(W/n_T)_i = -RT\left(n_A \ln\frac{p_T}{p_A} + n_B \ln\frac{p_T}{p_B}\right) \qquad (21)$$

in which n_A and n_B are the moles of components A and B in the mixture, p_A and p_B are the partial pressures of these two components in the mixture, and p_T is the total pressure of the mixture.

The figure of merit for a separation system is defined in a manner similar to that for a liquefaction system, namely

$$\text{FOM} = \frac{(W/m)_i}{(W/m)_{act}} \qquad (22)$$

The number of stages or plates to effect a low-temperature separation is determined by the same procedures as developed for normal separations. A computer is programmed to make interactive mass and energy balances around each plate in a separation column to determine the number of plates required to effect a desired separation. Meaningful computations require accurate thermodynamic data for mixtures and an understanding of the efficiency of separation that can be expected on each plate. Efficiency factors can vary from 65 to 100%.

Air Separation

Figure 32 provides a simplified schematic of the Linde single column originally used for air separation. This cycle produces a high-impurity nitrogen as a by-product. The separation scheme shown uses the simple J-T liquefaction cycle considered earlier but with a rectification column substituted for the liquid reservoir. (Any other liquefaction cycle could have been used in place of the J-T cycle; it is immaterial as to how the liquefied air is furnished to the column). As shown here, purified compressed air is precooled in a three-channel heat exchanger if gaseous oxygen is the desired product. (If liquid oxygen is recovered from the bottom of the column, a two-channel heat exchanger is used for the compressed air and waste nitrogen streams.) The precooled air then flows through a coil in the boiler of the rectifying column, where it is further cooled to saturation while serving as the heat source to vaporize the liquid in the boiler. After leaving the boiler, the compressed fluid expands essentially to atmospheric pressure through a throttling valve and enters the top of the column as reflux for the separation process. Rectification in the column occurs in a manner similar to that observed in ambient-temperature columns. If oxygen gas is to be the product, the air must be compressed to 450 to 900 psia; if it is to be liquid oxygen, pressures of 1500 to 3000 psia are necessary.

Although the oxygen product purity is high from a simple single-column separation scheme, the nitrogen effluent stream always contains about 6 to 7 mol% oxygen. This means that approximately one-third of the oxygen liquefied as feed to the column appears in the nitrogen effluent. This loss is not only undesirable but wasteful in terms of compression requirements. This problem was solved by the introduction of the Linde double-column gas-separation system, in which two columns are placed one on top of the other (Figure 33).

In this system, liquid air is introduced at an intermediate point in the lower column. A condenser-evaporator at the top of the lower column provides the reflux needed for both columns. Because the condenser must condense nitrogen vapor in the lower column by evaporating liquid oxygen in the upper column, the lower column must operate at a higher pressure (between 75 and 90 psia), while the upper column operates just above 14.7 psia. This requires throttling the overhead nitrogen and the ~45 mol% oxygen products from the lower column as they are transferred to the upper column. The reflux and rectification process in the upper column produce high-purity oxygen at the bottom and high-purity nitrogen at the top of the column, provided that argon and the rare gases have previously been removed.

Figure 34 illustrates the scheme for removing and concentrating the argon. The upper column is tapped at a level where the argon concentration is highest in the column. This gas is fed to an auxiliary column where a large fraction of the argon is separated from the oxygen and nitrogen mixture, which is returned to the appropriate level in the upper column. In modern separation plants, argon is recovered at two purities: either 0.5 or 4% oxygen (by mole). This is called crude argon. Oxygen is readily removed by chemical reduction or adsorption. Nitrogen content is variable, but may be maintained at low levels by proper operation of the upper column. For high-purity argon, the nitrogen must be removed with the aid of another separation column.

Because helium and neon have boiling points considerably below that of nitrogen, these components from the air feed stream collect on the nitrogen side of the condenser-reboiler unit. These gases are recovered by periodically removing a small portion of the gas in the dome of the condenser and sending the gas to a small nitrogen-refrigerated condenser-rectifier. The resulting crude helium and neon are further purified to provide high purity.

Atmospheric air contains only very small concentrations of krypton and xenon. As a consequence, very large amounts of air must be processed to obtain appreciable amounts of these rare gases. Because krypton and xenon tend to collect in the oxygen product, liquid oxygen from the reboiler of the upper column is first sent to an auxiliary condenser-reboiler to increase the concentration of these

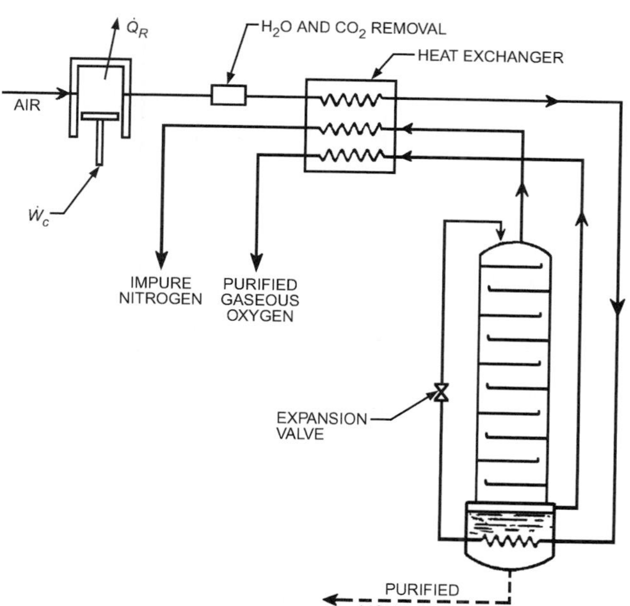

Fig. 32 Linde Single-Column Gas Separator

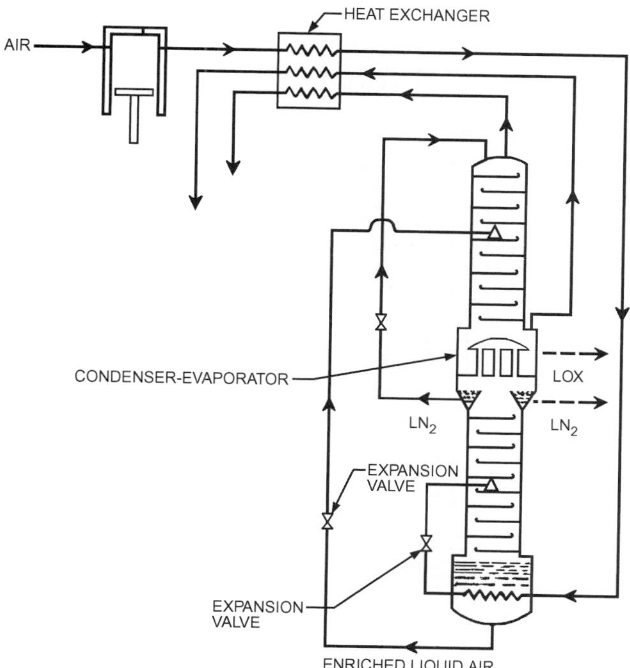

Fig. 33 Traditional Linde Double-Column Gas Separator

two components. The enriched product is further concentrated in another separation column before being vaporized and passed through a catalytic furnace to remove any remaining hydrocarbons with oxygen. The resulting water vapor and carbon dioxide are removed by a caustic trap and the krypton and xenon absorbed in a silica gel trap. The krypton and xenon are finally separated either with another separation column or by a series of adsorptions and desorptions on activated charcoal.

Figure 35 shows a schematic of the double-column gas-separation system presently used to produce gaseous oxygen. Such a column has both theoretical and practical advantages over the Linde double column. A second-law analysis for the two columns shows that the modern double column has fewer irreversibilities than are present in the Linde double column, which results in lower power requirements. From a practical standpoint, only two pressure levels are needed in the modern column, instead of the three required in the Linde double column. A further advantage of the double column in Figure 35 is that it does not require a reboiler in the bottom of the lower column, thereby simplifying the heat transfer process for providing the needed vapor flow in this column. The cooled gaseous air from an expansion turbine provides additional feed to the upper column. High-purity oxygen vapor is available from the vapor space above the liquid in the reboiler of the upper column if impurities in the air stream are removed at the appropriate locations in the column, as with the traditional Linde double-column gas separation system. Further details on modifications made to modern cryogenic air separation plants are given by Grenier and Petit (1986).

Helium Recovery

The major source of helium in the United States is natural gas. Because the major constituents of natural gas have boiling points considerably higher than that of helium, the separation can be accomplished with condenser-evaporators rather than with the more expensive separation columns.

A typical scheme pioneered by the U.S. Bureau of Mines for separating helium from natural gas is shown in Figure 36. In this scheme, the natural gas is treated to remove impurities and compressed to approximately 600 psia. The purified and compressed natural gas stream is then partially condensed by the returning cold low-pressure natural gas stream, throttled to a pressure of 250 psia, and further cooled with cold nitrogen vapor in a heat exchanger separator, where 98% of the gas is liquefied. The cold nitrogen vapor, supplied by an auxiliary refrigeration system, not only provides necessary cooling but also results in some rectification of the gas phase in the heat exchanger, thereby increasing the helium concentration. The remaining vapor phase, consisting of about 60 mol% helium and 40 mol% nitrogen with a very small amount of methane, is warmed to ambient temperature for further purification. The liquid phase, now depleted of helium, furnishes the refrigeration required to cool and partially condense the incoming high-pressure gas. The process is completed by recompressing the stripped natural gas and returning it to the natural gas pipeline with a higher heating value.

The crude helium is purified by compressing the gas to 2700 psia and cooling it first in a heat exchanger and then in a separator that is immersed in a bath of liquid nitrogen. Nearly all of the nitrogen in the crude helium gas mixture is condensed in the separator and removed as a liquid. The latter contains some dissolved helium, which is released and recovered when the pressure is reduced to 250 psia. Helium gas from the separator has a purity of about 98.5 mol%. Final purification to 99.995% is accomplished by sending the cold helium through charcoal adsorption purifiers to remove the nitrogen impurity.

Natural Gas Processing

The need for greater recoveries of the light hydrocarbons in natural gas has led to expanded use of low-temperature processing of these streams. Cryogenic processing of natural gas brings about a phase change and involves physical separation of the newly formed phase from the main stream. The lower the temperature for a given pressure, the greater the selectivity of the phase separation for a particular component.

Most low-temperature natural gas processing uses the turbo-expander cycle to recover light hydrocarbons. Feed gas is normally

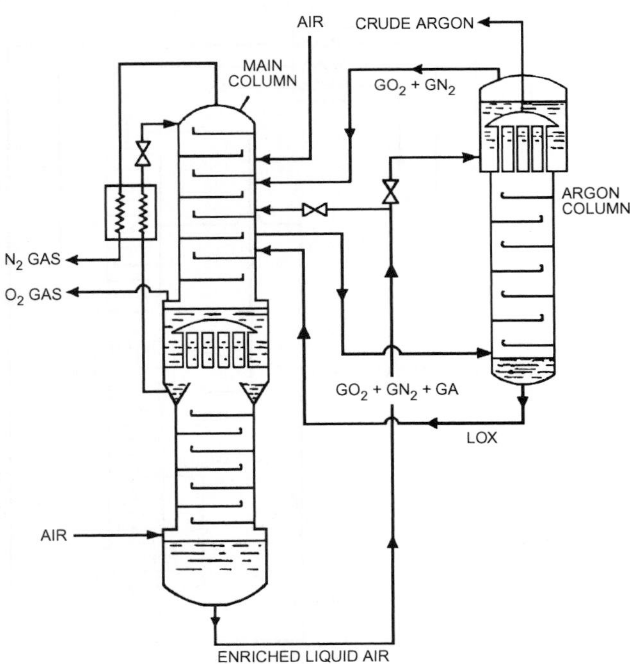

Fig. 34 Argon Recovery Subsystem

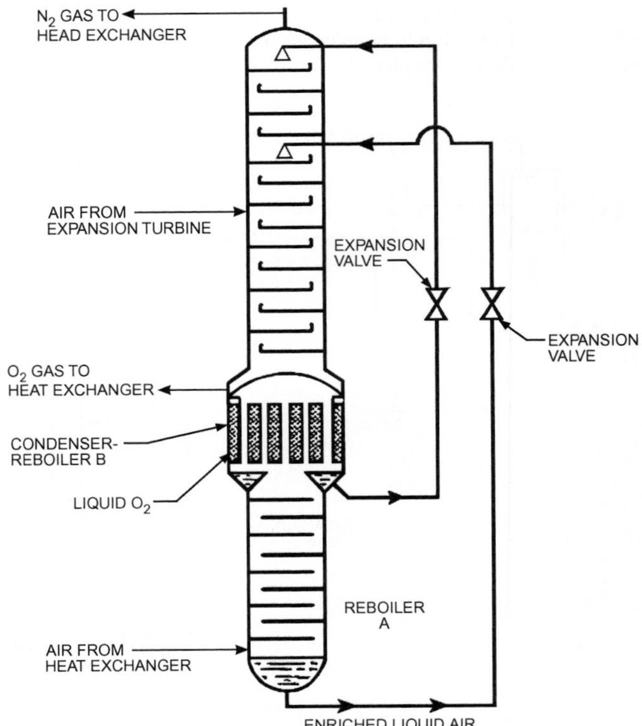

Fig. 35 Contemporary Double-Column Gas Separator

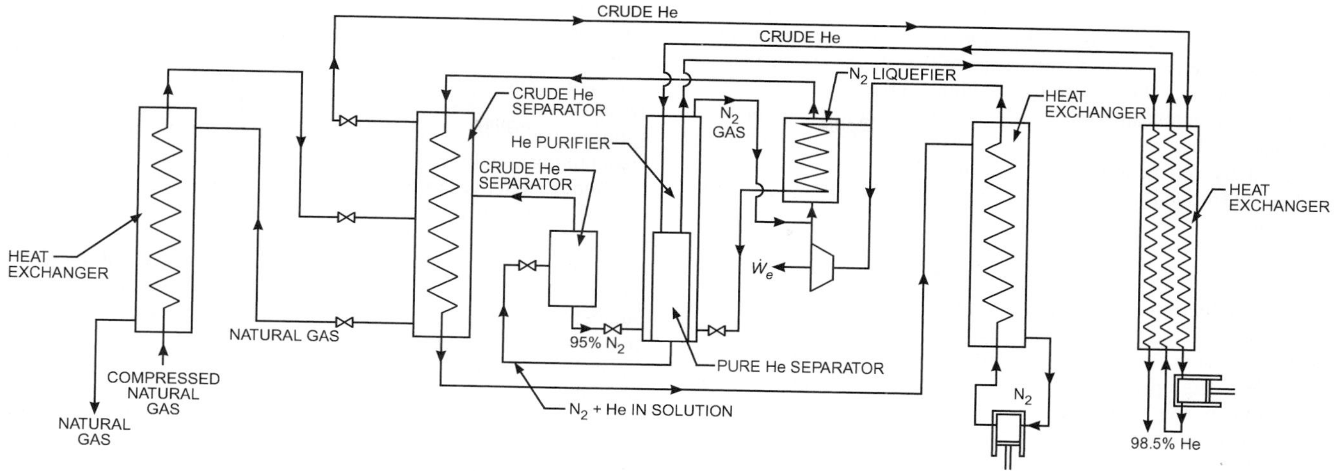

Fig. 36 Schematic of U.S. Bureau of Mines Helium Separation Plant

available from 150 to 1500 psia. The gas is first dehydrated to dew points of −100°F (360°R) and lower. After dehydration, the feed is cooled with cold residue gas. Liquid produced at this point is separated before entering the expander and sent to the condensate stabilizer. Gas from the separator flows to the expander. The expander exhaust stream can contain as much as 20 mass % liquid. This two-phase mixture is sent to the top section of the stabilizer, which separates the two phases. The liquid is used as reflux in this unit, and the cold gas exchanges heat with fresh feed and is recompressed by the expander-driven compressor. Many variations to this cycle are possible and have found practical applications.

Purification Procedures

The nature and concentration of impurities to be removed depend entirely on the process involved. For example, in the production of large amounts of oxygen, impurities such as water and carbon dioxide must be removed to avoid plugging the cold process lines or to avoid build-up of hazardous contaminants. Helium, hydrogen, and neon accumulate on the condensing side of the oxygen reboiler and reduce the rate of heat transfer unless removed by intermittent purging. Acetylene build-up can be dangerous even if the feed concentration of the air is no greater than 0.04 ppm or 4×10^{-8} lb/lb.

Refrigeration purification is a relatively simple method for removing water, carbon dioxide, and other contaminants from a process stream by condensation or freezing. (Either regenerators or reversing heat exchangers may be used for this purpose, because flow reversal is periodically necessary to reevaporate and remove the solid deposits.) Effectiveness depends on the vapor pressure of impurities relative to that of the major process stream components at the refrigeration temperature. Thus, assuming ideal gas behavior, the maximum impurity content in a gas stream after refrigeration would be inversely proportional to its vapor pressure. However, at higher pressures, the impurity content can be significantly greater than that predicted for the ideal situation. Data on this behavior are available as **enhancement factors**, defined as the ratio of the actual molar concentration to the ideal molar concentration of a specific impurity in a given gas.

Purification by a solid adsorbent is one of the most common low-temperature methods for removing impurities. Materials such as silica gel, carbon, and synthetic zeolites (molecular sieves) are widely used as adsorbents because of their extremely large effective surface areas. Carbon and most of the gels have pores of varying sizes in a given sample, but the synthetic zeolites are manufactured with closely controlled pore size openings ranging from 0.4 to about 1.3 nm. This pore size makes them even more selective than other adsorbents because it permits separation of gases on the basis of molecular size.

The equilibrium adsorption capacity of the gels and carbon is a function of temperature, the partial pressure of the gas to be adsorbed, and the properties of the gas. An approximation generally exists between the amount adsorbed per unit of adsorbent and the volatility of the gas being adsorbed. Thus, carbon dioxide would be adsorbed to a greater extent than nitrogen under comparable conditions. In general, the greater the difference in volatility of the gases, the greater the selectivity for the more volatile component.

The design of low-temperature adsorbers requires knowledge of the equilibrium between the solid and the gas and the rate of adsorption. Equilibrium data for the common systems generally are available from the suppliers of such material. The rate of adsorption is usually very rapid and the adsorption is essentially complete in a relatively narrow zone of the adsorber. If the concentration of adsorbed gas is more than a trace, then the heat of adsorption may also be a factor of importance in the design. (The heat of adsorption is usually of the same order as or larger than the normal heat associated with a phase change.) Under such situations, it is generally advisable to design the purification process in two steps: first removing a significant portion of the impurity, either by condensation or chemical reaction, and then completing the purification with a low-temperature adsorption system.

In normal plant operation, at least two adsorption units are used: one is in service while the other is being desorbed of its impurities. In some cases, a third adsorbent unit offers some advantage: one adsorbs, one desorbs, and one is cooled to replace the adsorbing unit as it becomes saturated. Adsorption units are generally cooled by using some of the purified gas, to avoid adsorption of additional impurities during the cooling period.

Low-temperature adsorption systems are used for many applications. For example, such systems are used to remove the last traces of carbon dioxide and hydrocarbons in air separation plants. Adsorbents are also used in hydrogen liquefaction to remove oxygen, nitrogen, methane, and other trace impurities. They are also used in the purification of helium suitable for liquefaction (Grade A) and for ultrapure helium (Grade AAA, 99.999% purity). Adsorption at −397°F (63°R), in fact, yields a helium with less than 2×10^{-9} lb/lb (2 ppb) of neon, which is the only detectable impurity in the helium after this treatment.

Even though most chemical purification methods are not carried out at low temperatures, they are useful in several cryogenic gas separation systems. Ordinarily, water vapor is removed by refrigeration and adsorption methods. However, for small-scale purification, the gas can be passed over a desiccant, which removes water vapor as water of crystallization. In the krypton-xenon purification system, carbon dioxide is removed by passage of the gas through a caustic, such as sodium hydroxide, to form sodium carbonate.

When oxygen is an impurity, it can be removed by reacting with hydrogen in the presence of a catalyst to form water, which is then removed by refrigeration or adsorption. Palladium and metallic nickel have proved to be effective catalysts for the hydrogen/oxygen reaction.

EQUIPMENT

The production and use of low temperatures require the use of highly specialized equipment, including compressors, expanders, heat exchangers, pumps, transfer lines, and storage tanks. As a general rule, design principles applicable at ambient temperature are also valid for low-temperature design. However, underlying each aspect of design must be a thorough understanding of temperature's effects on the properties of the fluids being handled and the materials of construction being selected.

Compression Systems

Compression power accounts for more than 80% of the total energy required to produce industrial gases and liquefy natural gas. The three major types of compressors used today are reciprocating, centrifugal, and screw. No particular type of compressor is generally preferred for all applications. The final selection ultimately depends on the specific application, the effect of plant site, available fuel source and its reliability, existing facilities, and power structure.

The key feature of reciprocating compressors is their adaptability to a wide range of volumes and pressures with high efficiency. Some of the largest units for cryogenic gas production range up to 15,000 bhp. They use the balanced-opposed machine concept in multistage designs with synchronous motor drive. When designed for multistage, multiservice operation, these units incorporate manual or automatic, fixed- or variable-volume clearance packets, and externally actuated unloading devices where required. Balanced-opposed units not only minimize vibrations, resulting in smaller foundations, but also allow compact installation of coolers and piping, further increasing the savings.

Air compressors for constant-speed service normally use piston suction valve loaders for low-pressure lubricated machines. Nonlubricated units require diaphragm-operated unloaders. Medium-pressure compressors for argon and hydrogen often use this type of unloader as well. The trend towards nonlubricating machines has led to piston designs using glass-filled PTFE (polytetrafluoroethylene) rider rings and piston rings, with cooled packing for the piston rods.

Larger units operate as high as 277 rpm with piston speeds for air service up to 850 fpm. Larger compressors with provision for multiple services reduce the number of motors or drivers and minimize the accessory equipment, resulting in lower maintenance cost.

Nonlubricated compressors used in oxygen compression have carbon- or bronze-filled PTFE piston rings and piston rod packing. The suction and discharge valves are specially constructed for oxygen service. The distance pieces that separate the cylinders from the crankcase are purged with an inert gas such as nitrogen, to preclude the possibility of high concentrations of oxygen in the area in the event of excessive rod packing leakage. Compressors for oxygen service are characteristically operated at lower piston speeds of the order of 650 fpm. Maintaining these machines requires rigid control of cleaning procedures and inspection of parts to ensure the absence of oil in the working cylinder and valve assemblies.

Variable-speed engine drives can generally operate over a 10 to 100% range in the design speed with little loss in operating efficiency because compressor fluid friction losses decrease with lower revolutions per minute.

Technological advances achieved in centrifugal compressor design have resulted in improved high-speed compression equipment with capacities exceeding 600,000 cfm in a single unit. Discharge pressure of such units is usually between 60 to 105 psia. Large centrifugal compressors are generally provided with adjustable inlet guide vanes to facilitate capacity reductions of up to 30% while maintaining economical power requirements. Because of their high efficiency, better reliability, and design upgrading, centrifugal compressors have become accepted for low-pressure cryogenic processes such as air separation and base load LNG plants.

Separately driven centrifugal compressors are adaptable to low-pressure cryogenic systems because they can be coupled directly to steam turbine drives, are less critical from the standpoint of foundation design criteria, and lend themselves to gas turbine or combined cycle applications. Isentropic efficiencies of 80 to 85% are usually obtained.

Most screw compressors are oil-lubricated. They either are semihermetic (the motor is located in the same housing as the compressor) or have an open-drive (the motor is located outside of the compressor housing and thus requires a shaft seal). The only moving parts in screw compressors are two intermeshing helical rotors. Because rotary screw compression is a continuous positive-displacement process, no surges are created in the system.

Screw compressors require very little maintenance because the rotors turn at conservative speeds and they are well lubricated with a cooling lubricant. Fortunately, most of the lubricant can easily be separated from the gas in screw compressors. Typically, only small levels of impurities (1 to 2 ppm by weight) remain in the gas after separation. Charcoal filters can be used to reduce the impurities further.

A major advantage of screw compressors is that they can attain high pressure ratios in a single mode. To handle these same large volumes with a reciprocating compressor requires a double-stage unit. Because of this and other advantages, screw compressors are now preferred over reciprocating compressors for helium refrigeration and liquefaction applications. They are competitive with centrifugal compressors in other applications as well.

Expansion Devices

The primary function of a cryogenic expansion device is to reduce gas temperature to provide useful refrigeration for the process. In expansion engines, the temperature is reduced by converting part of the energy of the high-pressure gas stream into mechanical work. In large cryogenic facilities, this work is recovered and used to reduce the overall compression requirements of the process. A gas can also be cooled by expanding it through an expansion valve (provided that its initial temperature is below the inversion temperature of the gas), converting part of the energy of the high-pressure gas stream into kinetic energy. No mechanical work is obtained from such an expansion.

Expanders are of either the reciprocating or the centrifugal type. **Centrifugal expanders** have gradually displaced the reciprocating type in large plants. However, the **reciprocating expander** is still popular for those processes where the inlet temperature is very low, such as for hydrogen or helium gas. Units up to 3600 hp are in service for nitrogen expansion in liquid hydrogen plants, whereas nonlubricated expanders with exhausts well below –400°F (60°R) are used in liquid hydrogen plants developed for the space program.

For reciprocating expanders, efficiencies of 80% are normally quoted; values of 85% are quoted for high-capacity centrifugal types (generally identified as turboexpanders). Usually, reciprocating expanders are selected when the inlet pressure and pressure ratio are high and the volume of gas handled is low. The inlet pressure to expansion engines used in air separation plants varies from 600 to 3000 psia, and capacities range from 200 to 6500 cfm.

The design features of reciprocating expanders used in low-temperature processes include rigid, guided cam-actuated valve gears; renewable hardened valve seats; helical steel or air springs;

and special valve packing that eliminates leakage. Cylinders are normally steel forgings effectively insulated from the rest of the structure. Removable nonmetallic cylinder liners and floating piston design offer wear resistance and good alignment in operation. Piston rider rings serve as guides for the piston. Nonmetallic rings are used for nonlubricated service. Both horizontal and vertical design, and one- and two-cylinder versions, have been used successfully.

Nonlubricated reciprocating expansion engines are generally used whenever possible oil contamination is unacceptable or where extremely low operating temperatures preclude using cylinder lubricants. This type of expansion engine is found in hydrogen and helium liquefaction plants and in helium refrigerators.

Reciprocating expanders in normal operation should not accept liquid in any form during the expansion cycle. However, the reciprocating device can tolerate some liquid for short periods if none of the constituents freeze in the expander cylinder and cause serious mechanical problems. Inlet pressure and temperature must be changed to eliminate any possibility of entering the liquid phase and especially the triple point range on expansion during normal operation.

Turboexpanders are classified as either axial or radial. Most turboexpanders built today are radial, because of their generally lower cost and reduced stresses for a given tip speed. This design allows them to run at higher speeds with higher efficiencies and lower operating costs. On the other hand, axial flow expanders are more suitable for multistage expanders because these units provide an easier flow path from one stage to the next. Where low flow rates and high enthalpy reductions are required, an axial-flow two-stage expander is generally used, with nozzle valves controlling the flow. For example, ethylene gas leaving the demethanizer is normally saturated, and processing expansion conditions cause a liquid product to exit from the expander. Up to 15 to 20% liquid at the isentropic end point can be handled in axial-flow impulse-turbine expanders, so recovery of ethylene is feasible. Depending on the initial temperature and pressure entering the expander and the final exit pressure, good flow expanders can reduce the enthalpy of an expanded fluid by between 75 to 150 Btu/lb, and this may be multistaged. The change in enthalpy drop can be regulated by turbine speed.

Highly reliable and efficient turboexpanders have made large-capacity air separation plants and base-load LNG facilities a reality. Notable advances in turboexpander design center on improved bearings, lubrication, and wheel and rotor design to allow nearly ideal rotor assembly speeds with good reliability. Pressurized labyrinth sealing systems use dry seal gas under pressure mixed with cold gas from the process to provide seal output temperatures above the frost point. Seal systems for oxygen compressors are more complex than those for air or nitrogen and prevent lubricant carryover to the processed gas. By combining variable-area nozzle grouping or partial admission of multiple nozzle grouping, efficiencies up to 85% have been obtained with radial turboexpanders.

Turboalternators were developed to improve the efficiency of small cryogenic refrigeration systems. This is accomplished by converting the kinetic energy in the expanding fluid to electrical energy, which in turn is transferred outside the system where it can be converted to heat and dissipated to an ambient heat sink.

The **expansion valve** (often called the J-T valve) is an important component in any liquefaction system, although not as critical as the others mentioned in this section. This valve resembles a normal valve that has been modified (e.g., exposing the high-pressure stream to the lower part of the valve seat to reduce sealing problems, lengthening the valve stem and surrounding it with a thin-walled tube to reduce heat transfer) to handle the flow of cryogenic fluids.

Heat Exchangers

One of the more critical components of any low-temperature liquefaction and refrigeration system is the heat exchanger. This point is demonstrated by considering the effect of heat exchanger effectiveness on the liquid yield of nitrogen in a simple J-T liquefaction process operating between 15 to 300 psia. The liquid yield under these conditions is zero if the effectiveness of the heat exchanger is less than 85%. (Heat exchanger effectiveness is defined as the ratio of actual heat transfer to the maximum possible heat transfer in the heat exchanger.)

Except for helium II, the behavior of most cryogens may be predicted by using the principles of mechanics and thermodynamics that apply to many fluids at room temperature. This behavior has allowed the formulation of convective heat transfer correlations for low-temperature designs of heat exchangers similar to those used at ambient conditions and ones that use Nusselt, Reynolds, Prandtl, and Grashof numbers.

However, the need to operate more efficiently at low temperatures has made the use of simple exchangers impractical in many cryogenic applications. One of the important advances in cryogenic technology is the development of complex but very efficient heat exchangers. Some of the criteria that have guided the development of these units for low-temperature service are (1) small temperature differences at the cold end of the exchanger to enhance efficiency, (2) large heat exchange surface area to heat exchanger volume ratios to minimize heat leak, (3) high heat transfer rates to reduce surface area, (4) low mass to minimize start-up time, (5) multichannel capability to minimize the number of exchangers, (6) high pressure capability to provide design flexibility, (7) low or reasonable pressure drops in the exchanger to minimize compression requirements, and (8) minimum maintenance to minimize shutdowns.

Minimizing the temperature difference at the cold end of the exchanger has some problems, particularly if the specific heat of the cold fluid increases with increasing temperature, as with hydrogen. In such cases, a temperature pinch, or a minimum temperature difference between the two streams in the heat exchanger, can occur between the warm and cold ends of the heat exchanger. This problem is generally alleviated by adjusting the mass flow of the key stream into the heat exchanger. In other words, the capacity rate is adjusted by controlling the mass flow to offset the change in specific heats. Problems of this nature can be avoided by balancing enthalpy in incremental steps from one end of the exchanger to the other.

Selection of an exchanger for low-temperature operation is normally determined by process design requirements, mechanical design limitations, and economic considerations. The principal industrial exchangers used in cryogenic applications are coiled-tube, plate-fin, reversing, and regenerator units.

Construction. A large number of aluminum tubes are wound around a central core mandrel of a **coiled-tube exchanger**. Each exchanger contains many layers of tubes, along both the principal and radial axes. Pressure drops in the coiled tubes are equalized for each specific stream by using tubes of equal length and carefully varying their spacing in the different layers. A shell over the outer tube layer together with the outside surface of the core mandrel form the annular space in which the tubes are nested. Coiled-tube heat exchangers offer unique advantages, especially for low-temperature conditions where simultaneous heat transfer between more than two streams is desired, a large number of heat transfer units is required, and high operating pressures in various streams are encountered. The geometry of these exchangers can be varied to obtain optimum flow conditions for all streams and still meet heat transfer and pressure drop requirements.

Optimizing a coiled-tube heat exchanger involves variables such as tube and shell flow velocities, tube diameter, tube pitch, and layer spacing. Other considerations include single- and two-phase flow, condensation on either the tube or shell side, and boiling or evaporation on either the tube or shell side. Additional complications occur when multicomponent streams are present, as in natural gas liquefaction, because mass transfer accompanies the heat transfer in the two-phase region.

The largest coiled-tube exchangers contained in one shell have been constructed for LNG base-load service. These exchangers handle liquefaction rates in excess of 60,000 cfm with a heat transfer surface of 250,000 ft^2, an overall length of 200 ft, a maximum diameter of 15 ft, and a mass of over 400,000 lb.

Plate-and-fin heat exchangers are fabricated by stacking layers of corrugated, high-uniformity, die-formed aluminum sheets (fins) between flat aluminum separator plates to form individual flow passages. Each layer is closed at the edge with aluminum bars of appropriate shape and size. Figure 37 illustrates the elements of one layer and the relative position of the components before being joined by brazing to form an integral structure with a series of fluid flow passages. These flow passages are combined at the inlet and exit of the exchanger with common headers. Several sections can be connected to form one large exchanger. The main advantage is that it is compact (about nine times as much surface area per unit volume as conventional shell-and-tube exchangers), yet permits wide design flexibility, involves minimum weight, and allows design pressures to 1000 psia from –453 to 152°F (7 to 612°R).

Fins for these heat exchangers are typically 0.4 in. high and can be manufactured in a variety of configurations that can significantly alter the exchanger's heat transfer and pressure drop characteristics. Various flow patterns can be developed to provide multipass or multistream arrangements by incorporating suitable internal seals, distributors, and external headers. The type of headers used depends on the operating pressures, the number of separate streams involved, and, in the case of counterflow exchangers, whether reversing duty is required.

Plate-and-fin exchangers can be supplied as single units or as manifolded assemblies that consist of multiple units connected in parallel or in series. Sizes of single units are presently limited by manufacturing capabilities and assembly tolerances. Nevertheless, the compact design of brazed aluminum plate-and-fin exchangers makes it possible to furnish more than 350,000 ft^2 of heat transfer surface in one manifolded assembly. These exchangers are used in helium liquefaction, helium extraction from natural gas, hydrogen purification and liquefaction, air separation, and low-temperature hydrocarbon processing. Design details for plate fin exchangers are available in most heat exchanger texts.

Removal of Impurities. Continuous operation of low-temperature processes requires that impurities in feed streams be removed almost completely before cooling the streams to very low temperatures. Removing impurities is necessary because their accumulation in certain parts of the system creates operational difficulties or constitutes potential hazards. Under certain conditions, the necessary purification steps can be accomplished by using reversing heat exchangers.

A typical arrangement of a reversing heat exchanger for an air separation plant is shown in Figure 38. Channels A and B constitute the two main reversing streams. During operation, one of these streams is cyclically changed from one channel to the other. The reversal normally is accomplished by pneumatically operated valves on the warm end and by check valves on the cold end of the exchanger. The warm-end valves are actuated by a timing device, which is set to a period such that the pressure drop in the feed channel is prevented from increasing beyond a certain value because of the accumulation of impurities. Feed enters the warm end of the exchanger and as it is progressively cooled, impurities are deposited on the cold surface of the exchanger. When the flows are reversed, the return stream reevaporates deposited impurities and removes them from the system.

Proper functioning of the reversing exchanger depends on the relationship between the pressures and temperatures of the two streams. Because pressures are normally fixed by other considerations, the purification function of the exchanger is usually controlled by proper selection of temperature differences throughout the exchanger. These differences must be such that, at every point in the exchanger where reevaporation takes place, the vapor pressure of the impurity must be greater than the partial pressure of the impurity in the scavenging stream. Thus, a set of critical values for the temperature differences exists, depending on the pressures and temperatures of the two streams. Because ideal equilibrium concentrations can never be attained in an exchanger of finite length, allowances must be made for an exit concentration in the scavenging stream sufficiently below the equilibrium one. Generally, a value close to 85% of equilibrium is selected.

The use of regenerators was proposed by Frankl in the 1920s for simultaneous cooling and purification of gases in low-temperature processes. In contrast to reversing heat exchangers, in which the flows of the two fluids are continuous and countercurrent during any period, the regenerator operates periodically by storing heat in a high-heat-capacity packing in one half of the cycle and then releasing this stored heat to the fluid in the other half of the cycle. Such an

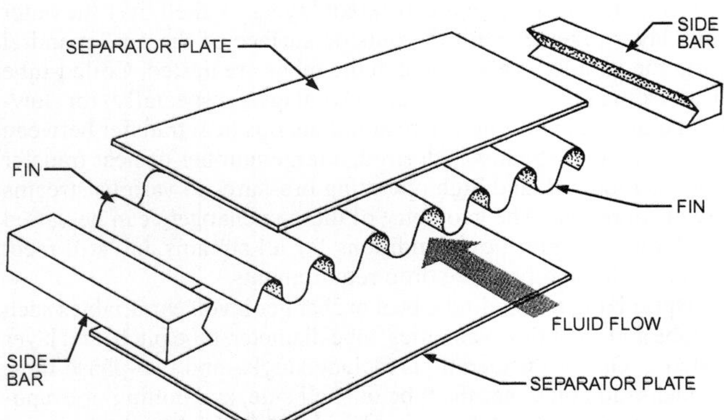

Fig. 37 Enlarged View of One Layer of Plate-and-Fin Heat Exchanger Before Assembly

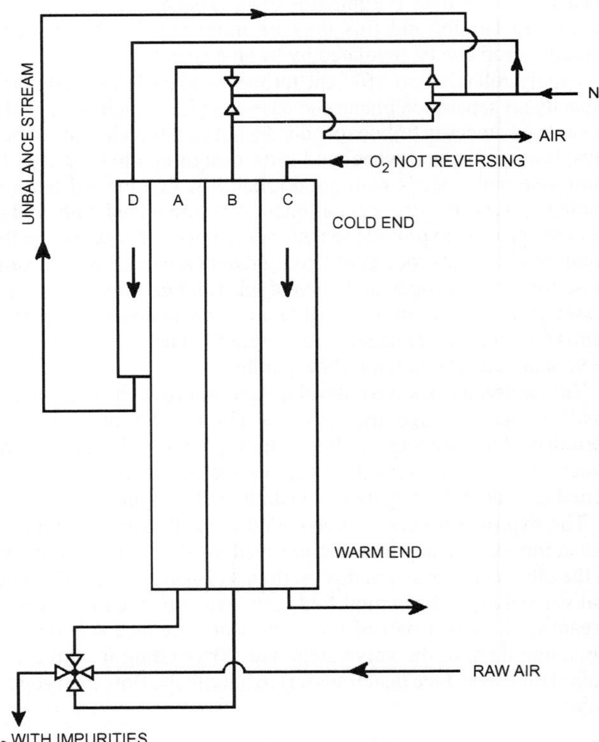

Fig. 38 Typical Flow Arrangement for Reversing Heat Exchanger in Air Separation Plant

exchanger, shown in Figure 39, consists of two identical columns packed with typical matrix materials such as metal screens or lead shot, through which a cyclical flow of gases is maintained. In cooldown, the warm feed stream deposits impurities on the cold surface of the packing. When the streams are switched, the impurities reevaporate as the cold stream is warmed while cooling the packing. Thus, the purifying action of the regenerator is based on

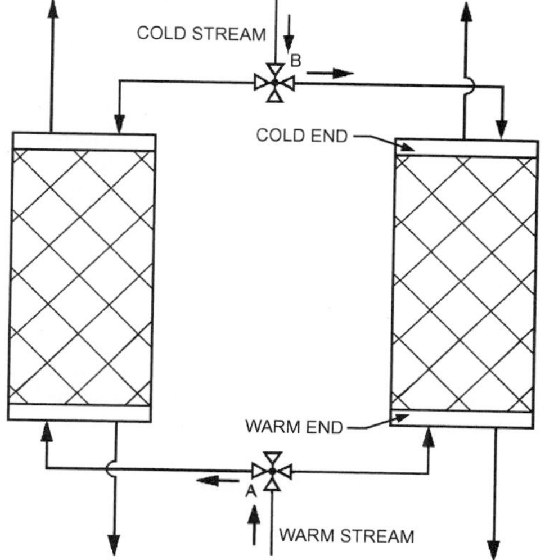

Fig. 39 Flow Arrangement in Regenerator Operation

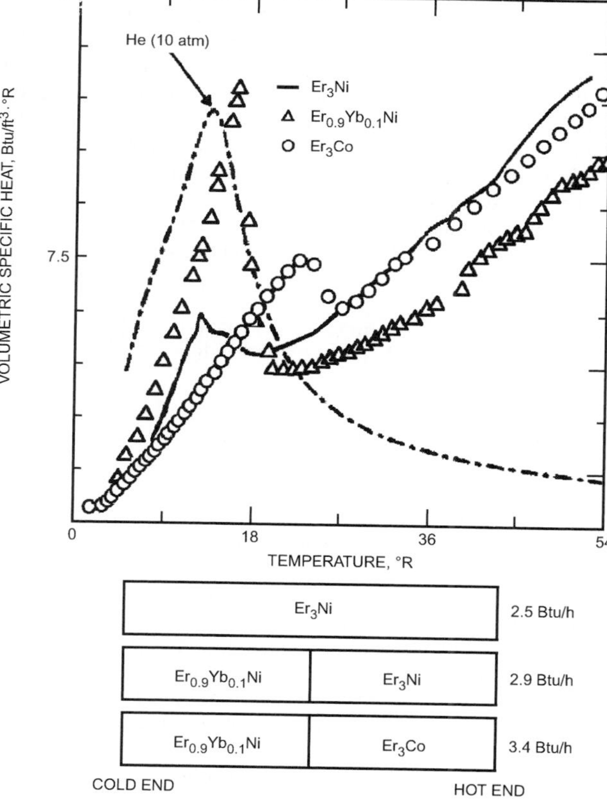

Fig. 40 Specific Heat of Several Rare Earth Matrix Materials
(Kuriyama et al. 1994; reprinted by permission of Springer Science and Business Media))

the same principles as the reversing exchanger, and the same limiting critical temperature differences must be observed if complete reevaporation of the impurities is to take place.

Regenerators frequently are selected for applications in which the heat transfer effectiveness, defined as Q_{actual}/Q_{ideal}, must be greater than 0.98. A high regenerator effectiveness requires a matrix material with a high heat capacity per unit volume and also a large surface area per unit volume. Until the early 1990s, recuperative heat exchangers rather than regenerators were used in cryocoolers, because the heat capacity of typical matrix materials rapidly decreases to a negligible value below −442°F (18°R). Because an increase in specific heat of a material can only occur when a physical transition occurs in the material, studies have been directed to heavy rare earth compounds that exhibit a magnetic phase transition at these low temperatures. Some of the experimental results are shown in Figure 40. Hashimoto et al. (1992) determined that specific heats of the $ErNi_{i-x}Co_x$ system are more than twice the values obtained for Er_3Ni at −448°F (12°R). Kuriyama et al. (1994) used layered rare earth matrix materials with higher heat capacities than Er_3Ni by itself in the cold end of the second stage of a Gifford-McMahon refrigerator and increased the refrigeration power of the refrigerator by as much as 40% at −453°F (7°R).

The low cost of the heat transfer surface along with the low pressure drop are the principal advantages of regenerators. However, contamination of fluid streams by mixing caused by periodic flow reversals and the difficulty of designing a regenerator to handle three or more fluids has restricted its use and favored the adoption of the plate-and-fin exchangers for air separation plants.

LOW-TEMPERATURE INSULATIONS

The effectiveness of a liquefier or refrigerator depends largely on the amount of heat leaking into the system. Because heat removal becomes more costly as temperature is reduced (the Carnot limitation), most cryogenic systems include some form of insulation to minimize this effect. Cryogenic insulations can be divided into five general categories: high-vacuum, multilayer evacuated insulation, evacuated powder, homogeneous material insulation (cellular glass, polyisocyanurate foam), and composite material systems insulations. The type of insulation chosen for a given cryogenic use depends on the specific application. Homogeneous or composite insulation material itself is only part of a system; other components (e.g., joint sealant, vapor retarder jacketing) need equal consideration to achieve the design goal. Generally, insulation performance is determined by material properties such as thermal conductivity, emissivity, percent moisture content by volume, evacuability, porosity, water vapor permeability, and flammability. In cryogenic service, the dimensional stability and coefficient of linear thermal expansion/contraction of a material are also of particular importance.

Heat flows through an insulation by solid conduction, gas conduction (convection), and radiation. Because these heat transfer mechanisms operate simultaneously and interact with each other, an apparent thermal conductivity k is used to characterize the insulation. The value of k is measured experimentally during steady-state heat transfer and evaluated from the basic one-dimensional Fourier equation. An insulation system is exposed to cold temperatures on the process side and warm temperatures on the ambient side. Consequently, thermal conductivity at the mean temperature of the application is used in calculating the insulation thickness. The mean temperature is determined by adding the process temperature to the ambient temperature, then dividing by two. Each homogeneous or composite material insulation has an associated polynomial equation to generate its thermal conductivity curve. Basically, thermal conductivity is a data point on a particular material curve at a certain mean temperature. ASTM *Standard* C1045 is the standard for this curve, and calculation methods are based on the ASTM *Standard* C680 methodology. Commercially available software packages can

Table 6 Apparent Thermal Conductivity of Selected Insulations

Type of Insulation	Apparent Thermal Conductivity k_a (between −200 and 75°F mean temperature), Btu/h·ft·°F	Bulk Density ρ, lb/ft³
Pure gas (at 1 atm, 324°R)		
n-H_2	0.07	0.0082
N_2	0.00042	0.119
Pure vacuum ($p = 10^{-6}$ torr or less)	0.00983	Nil
Foam insulation		
Polystyrene foam	0.0104 to 0.0216	1.6 to 2.2
Polyisocyanurate	0.0117 to 0.0158	1.6 to 8
Cellular glass	0.0242 to 0.0267	7 to 9
Evacuated powder		
Perlite ($p = 10^{-3}$ torr)	0.0006 to 0.0016	4 to 9
Silica ($p = 10^{-3}$ torr)	0.00098 to 0.00121	4 to 6
Multilayer evacuated insulation		
Aluminum foil and fiberglass		
(30–71 layers/in., $p = 10^{-6}$ torr)	0.00002 to 0.00004	4 to 7
(76–152 layers/in., $p = 10^{-6}$ torr)	0.00001	7.5
Aluminum foil and nylon net		
(81 layers/in., $p = 10^{-6}$ torr)	0.00002	5.6

do the calculations on this basis. Insulation manufacturers characterize materials at a thermal conductivity reported at a 75°F mean temperature. This value can be used for comparative purposes between materials at that temperature, but should not be used in cryogenic design. Material manufacturers should be able to provide thermal and property data for performance at cryogenic temperatures. Typical k values for a variety of insulations used in cryogenic service are presented in Table 6.

High-Vacuum Insulation

The mechanism of heat transfer prevailing across an evacuated space (10^{-5} torr or less) is by radiation and conduction through the residual gas. Radiation is generally the more predominant mechanism and can be approximated by

$$\frac{q_r}{A_1} = \sigma\left(T_2^4 - T_1^4\right)\left[\frac{1}{\varepsilon_1} + \frac{A_1}{A_2}\left(\frac{1}{\varepsilon_2} - 1\right)\right]^{-1} \tag{23}$$

where q_r/A_1 is the radiant heat flux, σ the Stefan-Boltzmann constant, and ε the emissivity of the surface. The subscripts 1 and 2 refer to the cold and warm surfaces, respectively. The bracketed term on the right is the emissivity for an evacuated space with diffuse radiation between spheres or cylinders (with length much greater than diameter). At pressures below 10^{-5} torr, the heat transferred by a gas is directly proportional to the gas pressure and temperature difference. When the molecular mean-free path is much larger than characteristic dimension of the body, molecules that impinge on the body and are then reemitted will, on average, travel a long distance before colliding with other molecules.

The number of reemitted molecules depends on the interaction between the impinging particles and the surface. Gaseous heat conduction under free molecular conditions for most cryogenic applications is given by

$$\frac{q_{gc}}{A_1} = \frac{\gamma + 1}{\gamma - 1}\left(\frac{R}{8\pi MT}\right)^{1/2} \alpha p(T_2 - T_1) \tag{24}$$

where α, the overall accommodation coefficient, is defined by

$$\alpha = \frac{\alpha_1\alpha_2}{\alpha_2 + \alpha_1(1 - \alpha_2)(A_1/A_2)} \tag{25}$$

Table 7 Accommodation Coefficients for Several Gases

Temperature, °R	Helium	Hydrogen	Air
540	0.29	0.29	0.8 to 0.9
140	0.42	0.53	1
36	0.59	0.97	1

and γ is the ratio of the heat capacities, R the molar gas constant, M the relative molecular mass of the gas, and T the temperature of the gas at the point where the pressure p is measured. A_1 and A_2, T_1 and T_2, and α_1 and α_2 are the areas, temperatures, and accommodation coefficients of the cold and warm surfaces, respectively. The accommodation coefficient depends on the specific gas surface combination and the surface temperature. It is defined as representing the fractional extent to which those molecules that fall on the surface and are reflected or reemitted from it have their mean energy adjusted or "accommodated" toward what it would be if the returning molecules were issuing as a stream out of a mass of gas at the temperature of the wall. Table 7 presents accommodation coefficients of three gases at several temperatures.

Heat transfer across an evacuated space by radiation can be reduced significantly by inserting one or more low-emissivity floating shields within the evacuated space. These shields reduce the emissivity factor. The only limitations on the number of floating shields used are system complexity and cost.

Evacuated Multilayer Insulations

Multilayer insulation provides the most effective thermal protection available for cryogenic storage and transfer systems. It consists of alternating layers of highly reflective material, such as aluminum foil or aluminized polyester film, and a low-conductivity spacer material or insulator, such as submicron-diameter glass fibers, paper, glass fabric, or nylon net, all under high vacuum. (The desired vacuum of 10^{-6} torr or less is maintained by using a getter such as activated charcoal to adsorb gases that desorb from the surfaces within the evacuated space.) When properly applied at the optimum density, this type of insulation can have an apparent thermal conductivity as low as 0.00001 to 0.00004 Btu/h·ft·°R between −424 and 80°F (36 and 540°R). The very low thermal conductivity of multilayer insulations can be attributed to the fact that all modes of heat transfer are reduced to a minimum.

The apparent thermal conductivity of a highly evacuated (pressures on the order of 10^{-6} in. Hg or less) multilayer insulation can be determined from

$$k_a = \frac{1}{N/\Delta x}\left\{h_s + \frac{\sigma e T_2^3}{2 - \varepsilon}\left[1 + \left(\frac{T_1}{T_2}\right)^2\right]\left(1 + \frac{T_1}{T_2}\right)\right\} \tag{26}$$

where $N/\Delta x$ is the number of complete layers (reflecting shield plus spacer) of insulation per unit thickness, h_s the solid conductance for the spacer material, σ the Stefan-Boltzmann constant, ε the effective emissivity of the reflecting shield, and T_2 and T_l the absolute temperatures of the warm and cold surfaces of the insulation, respectively. Equation (26) indicates that apparent thermal conductivity is inversely proportional to the number of complete layers used in the evacuated space. However, as the multilayer insulation is compressed, the increase in solid conductivity outweighs the decrease in radiative heat, thereby establishing an optimum layer density.

The effective thermal conductivity values generally obtained with actual cryogenic storage and transfer systems often are at least a factor of two greater than the thermal conductivity values shown in Figure 41, which were obtained under carefully controlled conditions. This degradation in insulation thermal performance is caused by the combined presence of edge exposure to isothermal boundaries, gaps, joints, or penetrations in the insulation blanket required for structural supports, fill and vent lines, and the high lateral thermal conductivity of these insulation systems.

Evacuated Powder and Fibrous Insulations

The difficulties encountered with applying multilayer insulation to complex structural storage and transfer systems can be minimized by using evacuated powder or fibrous insulation. This substitution in insulation materials, however, decreases the overall thermal effectiveness of the insulation system by tenfold.

A powder insulation system consists of a finely divided particulate material such as perlite, colloidal silica, calcium silicate, diatomaceous earth, or carbon black inserted between the surfaces to be insulated. When used at 14.7 psia gas pressure (generally with an inert gas), the powder reduces both convection and radiation and, if the particle size is sufficiently small, can also reduce the mean free path of the gas molecules. The apparent thermal conductivity of gas-filled powders is given by the expression

$$k_a = \left[\frac{V_r}{k_s} + \frac{1}{k_g/(1 - V_r) + 4\sigma T^3 d/V_r}\right] \tag{27}$$

where V_r is the ratio of the solid powder volume to the total volume, k_s the thermal conductivity of the powder, k_g the thermal conductivity of the residual gas, σ the Stefan-Boltzmann constant, T the mean temperature of the insulation, and d the mean diameter of the individual powder particles.

The insulating value of powders is increased considerably by removing the interstitial gas. Thus, when powders are used at pressures of 10^{-3} torr or less, gas conduction is negligible and heat transport is mainly by radiation and solid conduction. Figure 42 shows the apparent thermal conductivity of several powders as a function of interstitial gas pressure.

The radiation contribution for evacuated powders near room temperature is larger than the solid conduction contribution to the total heat transfer rate. On the other hand, the radiant contribution is smaller than the solid conduction contribution for temperatures between −321 and −424 or −453°F (139 and 36 or 7°R). Thus, evacuated powders can be superior to vacuum alone (for insulation thicknesses greater than 4 in.) for heat transfer between ambient and liquid nitrogen temperatures. Conversely, because solid conduction becomes predominant at lower temperatures, vacuum alone is usually better for reducing heat transfer between two cryogenic temperatures.

Very similar considerations govern the use of both powder and fiber insulations. When fibers are used as spacers for multilayer insulation, they are prepared without any binders or lubricants to reduce the possibility of outgassing.

Homogeneous Material Insulations

This category describes materials used in their manufactured state. The most common types of materials used are glass and plastic. Both materials are closed-cell and cellular in structure. Density does not generally affect the performance of homogeneous material insulations in the lower ranges. This does not hold true for high-density inserts used in engineered supports, which typically have densities in the 10 to 20 lb/ft³ range. These materials have higher thermal conductivities than their lower-density counterparts do. The **coefficient of linear thermal expansion (COLTE)** of the insulation material is one of several important factors to consider when designing the insulation system. The COLTE of each material is a property specific to that material. The design of the insulated system must compensate for differences in expansion/contraction of the insulation versus the pipe metal. Homogeneous material insulations can be used alone or combined with other materials to form a thermal insulation system. The apparent thermal conductivity of homogeneous material insulations depends on the bulk density of the foamed material, the gas contained in the cells (which is a function of age and service temperature of the materials used as the foaming agent), the size of the cells, and the temperature levels to which the insulation is exposed. Heat transport across a foam is determined by convection and radiation in the cells of the foam and by conduction in the solid structure. Evacuation of a foam can effectively reduce its thermal conductivity, provided there is at least a partially open cellular structure, but the resulting values are still considerably higher than either evacuated multilayer, evacuated powder or evacuated fibrous insulations. The opposite effect,

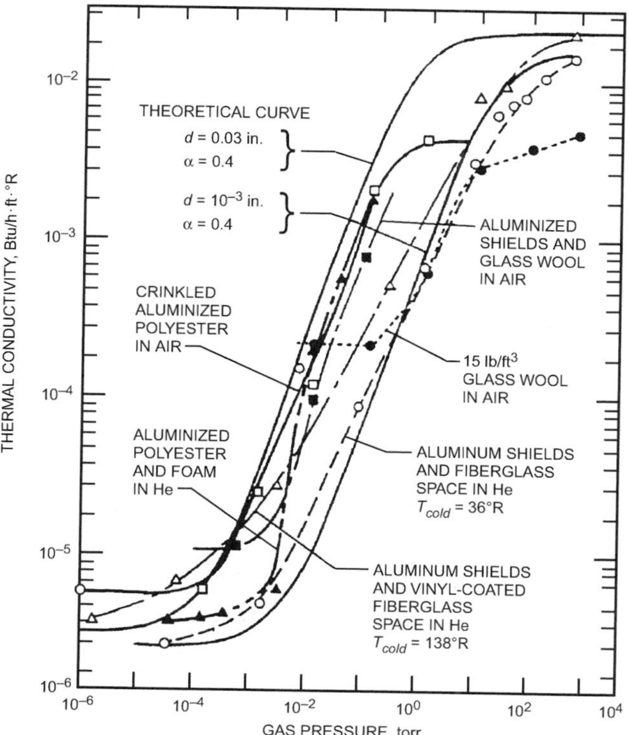

Fig. 41 Effect of Residual Gas Pressure on Apparent Thermal Conductivity of Multilayer Insulation

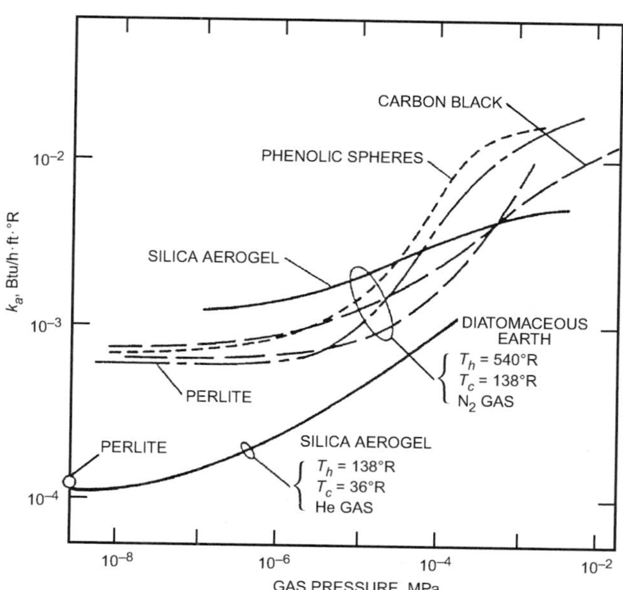

Fig. 42 Apparent Thermal Conductivity of Several Powder Insulations as Function of Residual Gas Pressure

diffusion of atmospheric gases into the cells, can increase the apparent thermal conductivity. In particular, the diffusion of hydrogen and helium into the cells can increase the foam's apparent thermal conductivity by a factor of three or four. Of all the foams, polyurethane-modified polyisocyanurates and polystyrene have received the widest use at low temperatures.

When applied to cryogenic systems, some foams tend to crack on repeated thermal cycling and lose their insulation value. The foam material selected must be suitable for the temperature range and cycling conditions expected.

Composite Material Insulations Systems

The optimum insulation system should combine maximum insulation effectiveness, minimum weight, ease of fabrication, and reasonable cost. It would be desirable to use only one insulation material, but because no single insulation has all the desirable physical and strength characteristics required in many applications, composite insulations have been developed. One such insulation consists of a polyurethane foam, reinforcement materials to provide adequate compressive strength, adhesives for sealing and securing the foam to the container, enclosures to prevent damage to the foam from external sources, and vapor barriers to maintain a separation between the foam and atmospheric gases. Another external insulation system for space applications uses honeycomb structures. Phenolic-resin-reinforced fiberglass cloth honeycomb is most commonly used. Filling the cells with a low-density polyurethane foam further improves the thermal effectiveness of the insulation.

STORAGE AND TRANSFER SYSTEMS

Storage vessels range from low-performance containers where the liquid in the container boils away in a few hours to high-performance containers and dewars where less than 0.1% of the fluid contents is evaporated per day.

Storage Systems

The essential elements of a storage vessel consist of an inner vessel, which encloses the cryogenic fluid to be stored, and an outer vessel, which contains the appropriate insulation and serves as a vapor barrier to prevent water and other condensables from reaching the cold inner vessel. The value of the cryogenic liquid stored dictates whether the insulation space is evacuated. In small laboratory dewars designed for liquid nitrogen and oxygen, as shown in Figure 43, insulation is obtained by coating the two surfaces facing the insulation space with low-emissivity materials and then evacuating the space to a pressure of 10^{-6} in. Hg or lower. Laboratory storage of liquid hydrogen and liquid helium, on the other hand, requires evacuated multilayer insulation. In larger vessels, insulations such as powders, evacuated powders, or evacuated multilayer insulations are used, depending on the type of fluid and storage applications. Table 8 provides a tabulation of storage systems based on the preceding criteria.

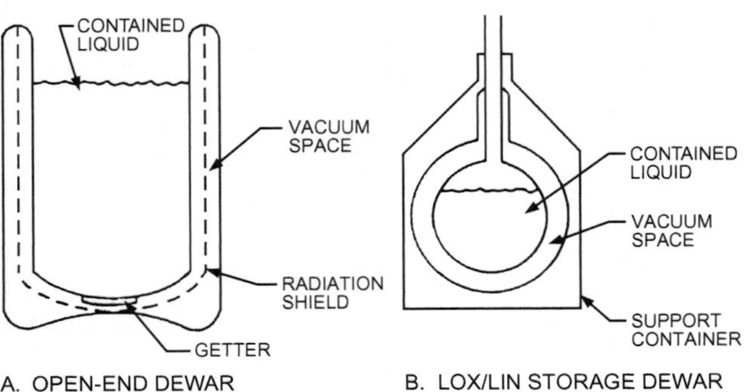

Fig. 43 Laboratory Storage Dewars for Liquid Oxygen and Nitrogen

The transport of cryogens for more than a few hundred feet generally requires specially built transport systems for truck, railroad, or airline delivery. Volumes from 0.7 to more than 3500 ft³ have been transported successfully by these carriers. Large barges and ships built specifically for shipment of cryogens, particularly LNG, have increased the volume transported manyfold.

Several requirements must be met in the design of the inner vessel. The material of construction must be compatible with the stored cryogen. Nine percent nickel steels are acceptable for high-boiling cryogens ($T > -325$ °F); many aluminum alloys and austenitic steels are usually structurally acceptable throughout the entire temperature range. Economic and cooldown considerations dictate that the inner shell be as thin as possible. Accordingly, the inner container is designed to withstand only the internal pressure and bending forces, and stiffening rings are used to support the weight of the fluid. The minimum thickness of the inner shell for a cylindrical vessel under such a design arrangement is given in Section VIII of the ASME (2004) *Boiler and Pressure Vessel Code*.

The outer shell of the storage vessel, on the other hand, is subjected to atmospheric pressure on the outside and evacuated conditions on the inside. Such a pressure difference requires an outer shell of sufficient material thickness with appropriately placed stiffening rings to withstand collapsing or buckling. Again, consult the ASME code for specific design charts.

Heat leaks into cryogen storage vessels by radiation and conduction through the insulation and by conduction through the inner shell supports, piping, instrumentation leads, and access ports. Conduction losses are reduced by making long heat-leak paths, by making the cross-sectional areas for heat flow small, and by using materials with low thermal conductivity. Radiation losses, a major factor in the heat leak through insulations, are reduced by using radiation shields (such as multilayer insulation), boil-off vapor-cooled shields, and opacifiers in powder insulation.

Most storage vessels for cryogens are designed for a 90% liquid volume and a 10% vapor or ullage volume. The latter permits reasonable vaporization of the liquid contents by heat leakage without incurring too rapid a build-up of pressure in the vessel. This, in turn, allows closing the container for short periods either to avoid

Table 8 Insulation Selection for Various Cryogenic Storage Vessels

Fluid	Application	Volume, ft³	Insulation
Liquid natural gas	Liquefier storage	Up to 57,000	Powder or homogeneous material
	Sea transport	Up to 1.4×10^6	Powder and foam with N_2 purge
Liquid oxygen or nitrogen	Laboratory use	Up to 3.5	Vacuum
	Transport	Up to 7	Multilayer
	Liquefier storage	Up to 1800	Evacuated powder or homogeneous material
	Liquefier storage	Above 1800	Powder or homogeneous material
Liquid hydrogen or helium	Laboratory use	Up to 5.3	Multilayer
Liquid hydrogen	Truck transport	Up to 3500	Multilayer
	Liquefier storage	Up to 134,000	Evacuated powder or homogeneous material
Liquid helium	Truck transport	Up to 1200	Multilayer
	Liquefier storage	Up to 4000	Multilayer

Note: This tabulation should only serve as a guideline; economics will determine the final selection.

partial loss of the contents or to allow the safe transport of flammable or hazardous cryogens.

Transfer Systems

A cryogen is transferred from the storage vessel by one of three methods: self-pressurization of the container, external gas pressurization, or mechanical pumping. Self-pressurization involves removing some of the fluid from the container, vaporizing the extracted fluid, and then reintroducing the vapor into the ullage space to displace the contents of the container. External gas pressurization uses an external gas to displace the container contents. In the mechanical pumping method, the contents of the storage vessel are removed by a cryogenic pump located in the liquid drain line.

Several types of pumps have been used with cryogenic fluids. In general, positive-displacement pumps are best suited for low flow rates at high pressures. Centrifugal or axial-flow pumps are generally best for high-flow applications. Centrifugal or axial-flow pumps have been built and used for liquid hydrogen with flow rates of up to 8000 cfm and pressures of more than 1000 psia. Cryogen subcooling, thermal contraction, lubrication, and compatibility of materials must be considered carefully for successful operation.

Cryogenic fluid transfer lines are generally classified as one of three types: uninsulated, foam-insulated lines, and vacuum-insulated lines. The latter may entail vacuum insulation alone, evacuated powder insulation, or multilayer insulation. A vapor retarder must be applied to the outer surface of foam-insulated transfer lines to minimize insulation degradation that occurs when water vapor and other condensables diffuse through the insulation to the cold surface of the lines.

Cooldown of a transfer line always involves two-phase flow. Severe pressure and flow oscillations occur as the cold liquid comes in contact with the successive warm sections of the line. Such instability continues until the entire transfer line is cooled down and filled with liquid cryogen.

INSTRUMENTATION

Cryogenic instrumentation is used primarily to determine the condition or state of cryogenic fluids, such as pressure and temperature. Such information is typically required for process optimization and control. In addition, the question of quantity and quality transferred or delivered has become commercially important. Accordingly, the instrumentation system must also be able to indicate liquid level, density, and flow rate accurately.

Pressure Measurements

Pressure in cryogenic systems has been measured by simply attaching gage lines from the points where the pressure is to be measured to some convenient location at ambient temperature where a suitable pressure-measuring device is available. This method works well for many applications, but it can have problems of independent frequency response and thermal oscillations. In addition, heat leak, uncertainties in hydrostatic pressure in gage lines, and fatigue failure of gage lines can be significant in some applications. Such problems can be eliminated by installing pressure transducers at the point of measurement. Many pressure transducers used in the cryogenic environment are similar to those used at ambient conditions, such as strain gage and capacitance transducers. However, because pressure-sensing devices often behave quite differently under cryogenic conditions, a systematic testing program can identify the device most suitable for the specific application.

Thermometry

Most low-temperature temperature measurements are made with metallic resistance thermometers, nonmetallic resistance thermometers, or thermocouples. Vapor-pressure thermometers find limited application, but they can provide convenient temperature check points. Important factors in selecting a thermometer include absolute accuracy, reproducibility, sensitivity, heat capacity, self heating, heat conduction, stability, simplicity, response to magnetic field, strain sensitivity, convenience of operation, ruggedness, and cost.

The resistivity of a metallic element or compound varies with a change in temperature. Although many metals are suitable for resistance thermometry, platinum is predominant, mainly because it is chemically inert, easy to fabricate, sensitive down to −424°F (36°R), and very stable.

Many semiconductors, such as germanium, silicon, and carbon, also have useful thermometric properties at low temperatures. Carbon, though not strictly a semiconductor, is included in this group because of its similarity in behavior to semiconductors.

Thermocouples can be very small, so that the disturbance to the object being sensed is very slight and the response time very fast. However, such devices generate rather small voltages, which become smaller as the temperature drops. Copper and constantan is the most commonly used thermocouple pair for low-temperature thermometry, giving an accuracy of ±0.2°F and a sensitivity of 20 μV/°R at room temperature, 9.5 μV/°R at liquid oxygen temperatures, and 3 μV/°R at liquid hydrogen temperatures. Even though other thermocouple combinations such as gold (2.1% atomic cobalt) and copper have larger thermoelectric powers at low temperatures, copper-constantan is still favored because it suffers less from the inhomogeneities common with other thermocouples.

Liquid-Level Measurements

Liquid level is one of several measurements needed to establish the contents of a cryogenic container. Other measurements may include volume as a function of depth, density as a function of physical storage conditions, and, sometimes, useful contents from total contents. Of these measurements, liquid level is presently the most advanced; it can be as accurate and precise as thermometry, and often with greater simplicity.

The operation of cryogenic liquid-level sensors generally depends on a large property change that occurs at the liquid/vapor interface (e.g., a significant change in density). This change may not occur if the fluid is stored near its critical point or if it has stratified after prolonged storage. A convenient way to classify such liquid-level sensors is according to whether the output is discrete (point sensors) or continuous. Point level sensors are tuned to detect sharp property differences and give an on/off signal.

The **capacitance liquid-level sensor** recognizes the differences in dielectric constant between the liquid and the vapor, which is closely related to the fact that the liquid is more dense than the vapor. The **thermal** or **hot wire sensor** detects the large difference in heat transfer between the liquid and vapor phases. The **optical sensor**, on the other hand, detects the change in refractive index between the liquid and the vapor, which is also related to the dielectric constant and density. Several **acoustic** and **ultrasonic devices** operate on the principle that the damping of a vibrating member is greater in liquid than in a vapor.

Continuous liquid-level sensors take many forms. Some sensors determine mass, whereas others are just continuous analogs of some point sensor and merely follow the liquid/vapor interface. Mass sensors include direct weighing, nuclear radiation attenuation, and radio frequency techniques to detect mass. Differential pressure sensors, capacitance, and acoustic devices are used to follow the vapor/liquid interface.

Density Measurements

Measurements of liquid density are closely related to quantity and liquid-level measurements because both are often required simultaneously to establish the mass contents of a tank. The same physical principle may often be used for either measurement, because liquid-level detectors sense the steep density gradient at the liquid/vapor interface. Density may be detected by direct weighing, differential pressure, capacitance, optical or acoustic means, or nuclear radiation attenuation. In general, the various liquid-level principles also apply to density measurement techniques.

Two exceptions are noteworthy. In the case of homogeneous pure fluids, density can usually be determined more accurately by an indirect measurement. That is, density can be calculated from accurate thermophysical properties data by measuring pressure and temperature.

Nonhomogeneous fluids are quite different. LNG, for example, is often a mixture of five or more components with varying composition and, hence, varying density. Accordingly, temperature and pressure measurements alone will not suffice to determine density. In that case, a dynamic, direct measurement, using liquid-level measurement, is required.

Flow Measurements

Three types of flow meters are useful for liquid cryogens: pressure drop or head type, turbine, and momentum.

The **pressure drop** or **head meter** is the oldest method of measuring flowing fluids. Its distinctive feature is a restriction that is used to reduce the static pressure of the flowing fluid. This static pressure difference between the pressures upstream and downstream of the restriction is measured. These meters are simple, and they do not need calibration if proper design, application theory, and practices are followed. Orifice accuracy is generally within ±3% of full scale and repeatability is ±1%. However, transient fluctuation in flow can cause erratic readout, particularly during cooldown and warm-up. Also, static liquid pressures within the meter should be well above the fluid saturation pressure to avoid an erroneous flow measurement from possible cavitation in the flowing stream. In spite of these shortcomings, head meters are widely used and quite reliable.

The **turbine volumetric flow meter** is probably the most popular of the various flow-measuring instruments because of its simple mechanical design and demonstrated repeatability. This meter consists of a freely spinning rotor with multiple blades. The rotor is supported in guides or bearings mounted in a housing that forms a section of the pipeline. The primary requirement is that the angular velocity of the rotor be directly proportional to the volumetric flow rate or, more correctly, to some average velocity of the fluid in the pipe.

Mass-reaction or **momentum-flow meters** are primarily of three types. In one type, an impeller imparts a constant angular momentum to the fluid stream, which is measured as a variable torque on a turbine. In another type, an impeller is driven at a constant torque and the variable angular velocity of the impeller is measured. The third type drives a loop of fluid at either a constant angular speed or a constant oscillatory motion, and the mass reaction is measured. Several of these momentum mass flow meters provide liquid hydrogen mass flow measurement accuracies on the order of ±0.5%. However, two-phase flow degrades mass flow measurement accuracy.

HAZARDS OF CRYOGENIC SYSTEMS

Hazards can best be classified as those associated with the response of the human body and the surroundings to cryogenic fluids and their vapors, and those associated with reactions between certain of the cryogenic fluids and their surroundings.

Physiological Hazards

Exposing the human body to cryogenic fluids or to surfaces cooled by cryogenic fluids can cause severe cold burns (i.e., damage to the skin or tissue similar to that caused by an ordinary burn). Severity of the burn depends on the contact area and contact time: prolonged contact results in deeper burns. Severe burns are seldom sustained if withdrawal is rapid. Cold gases may not be damaging if turbulence in the gas is low, particularly because the body can normally withstand a heat loss of nearly 30 Btu/h·ft^2 for an area of limited exposure. If a burn is inflicted, the only first aid treatment is to liberally flood the affected area with lukewarm water. Massaging the affected area could cause additional damage.

Protective clothing, including safety goggles, gloves, and boots, is imperative for personnel who handle liquid cryogens. Such operations should only be attempted when sufficient personnel are available to monitor the activity.

Whenever the oxygen content of the atmosphere is diluted by nitrogen spills or leaks, there is the danger of nitrogen asphyxiation. Because nitrogen is a colorless, odorless, inert gas, personnel must be aware of associated respiratory and asphyxiation hazards. In general, the oxygen content of air for breathing purposes should be kept between 16 and 25%.

In contrast, an oxygen-enriched atmosphere produces exhilarating effects when breathed. However, lung damage can occur if the oxygen concentration in the air exceeds 60%, and prolonged exposure to an atmosphere of pure oxygen may initiate bronchitis, pneumonia, or lung collapse. Additional threats from oxygen-enriched air are increased flammability and explosion hazards. In this application, the presence of organic materials is typically limited or prohibited altogether.

Construction and Operations Hazards

Most failures of cryogenic equipment can be traced to improper selection of construction materials or disregard for the change of some material property at low temperatures. For example, low temperatures make some construction materials brittle or less ductile. This behavior is further complicated because some materials become brittle at low temperatures but still can absorb considerable impact strength (see Chapter 40 for additional details). Brittle fracture can cause almost instantaneous failure, which can cause shrapnel damage if the system is under pressure, and release of a fluid such as oxygen can cause a fire or an explosion.

Low-temperature equipment can also fail because thermal stresses cause thermal contraction of the materials. Figure 44 provides information for evaluating the thermal contraction exhibited by several metals used widely in low-temperature construction. Solder in joints must be able to withstand stresses caused by differential contraction where two dissimilar metals are joined. Contraction in long pipes is also a serious problem; for example, a stainless steel pipeline 100 ft long will contract approximately 2.3 in. when filled with liquid oxygen or nitrogen. Provisions must be made for this change in length

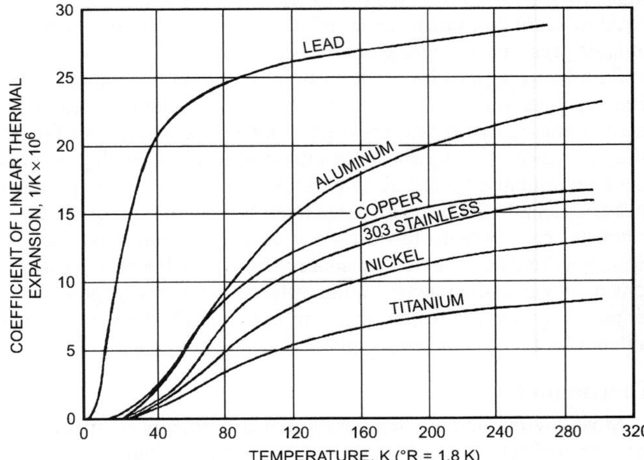

Fig. 44 Coefficient of Linear Expansion for Several Metals as Function of Temperature

during both cooling and warming of the pipeline by using bellows, expansion joints, or flexible hose. Pipe anchors, supports, and so on must also be carefully designed to allow contraction and expansion. The primary hazard of failure caused by thermal contraction is cryogen spillage and the possibility of fire or explosion.

Overpressure. All cryogenic systems should be protected against overpressure caused by phase change from liquid to gas. Systems containing liquid cryogens can reach bursting pressures, if not relieved, simply by trapping the liquid in an enclosure. Figure 45 shows the pressure build-up in a closed vessel of liquid nitrogen. The rate of pressure rise in storage containers for cryogens depends on the rate of heat transfer into the liquid. In uninsulated systems, liquid vaporizes rapidly and pressure in the closed system can rise very rapidly. The more liquid that is originally in the tank before it is sealed off, the greater will be the resulting final pressure.

Heat leakage and vaporization of cryogenic fluid trapped in valves, fittings, and sections of piping can cause excessive pressure buildup and possible rupture of the equipment. For instance, liquid hydrogen expands about 850 times its volume when warmed to ambient temperature. Relief valves and burst disks are normally used to relieve piping systems at a pressure slightly above the operating design pressure of the equipment. Relief should be provided between valves, on tanks, and at all locations of possible (though perhaps unintentional) pressure rise in a piping system.

Overpressure in cryogenic systems can also occur more subtly. Vent lines without appropriate rain traps can collect rainwater and freeze closed, as can exhaust tubes on relief valves and burst disks. Small-necked, open-mouth dewars can collect moisture from the air and freeze closed. Cold liquids or gases can be trapped by freezing water or other condensables in some portion of the cold system. If this occurs in an unanticipated location, the relief valve or burst disk may be isolated and afford no protection.

Another source of system overpressure that is frequently overlooked is cooldown surges. If a liquid cryogen is admitted to a warm line with the intention of transferring the liquid from one point to another, severe pressure surges will occur. These pressure surges can be 10 times the operating or transfer pressure and can even cause backflow into the storage container. The overall design and operating procedures for the transfer systems must protect against such overpressure.

Release of Cryogens. Although several cryogenic liquids produce vapors of lower molecular mass than air, the lower temperatures result in a denser vapor. Released vapors travel along the ground and collect in low places. Exposure to these vapors is hazardous. Exposure to oxygen vapors has caused clothing or any equipment with oil-lubricated parts to become oxygen-enriched and cause fires.

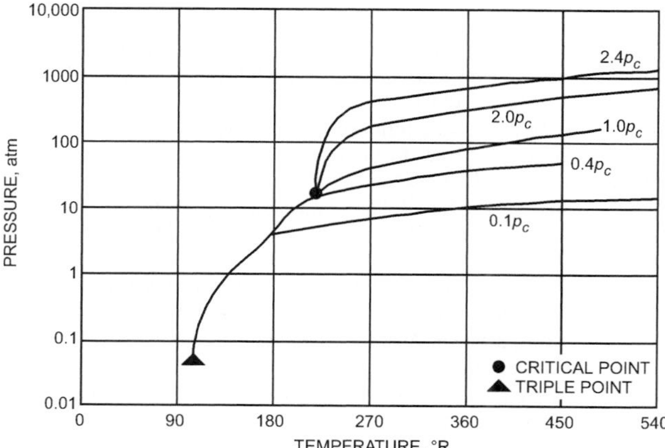

Fig. 45 Pressure Developed During Warming of Liquid Nitrogen in Closed Container
(Courtesy of Air Products and Chemicals, Inc., Allentown, PA)

In making an accident or safety analysis, the possibility of encountering even more serious secondary effects from any cryogenic accident should be considered. For example, any one of the failures discussed previously (brittle fracture, contraction, overpressure, etc.) may release sizable quantities of cryogenic liquids or cold gases, causing severe fire or explosion hazards, asphyxiation dangers, further brittle fracture problems, or shrapnel damage to other flammable or explosive materials. In this way the accident can rapidly and progressively become much more serious.

Flammability and Detonability Hazards

Almost any flammable mixture will, under favorable conditions of confinement, support an explosive flame propagation or even a detonation. A fuel/oxidant composition favorable for high-speed combustion first loses its capacity to detonate when it is diluted with an oxidant, fuel, or inert substance. Further dilution causes it to lose its capacity to burn explosively. Eventually, the lower or upper flammability limits will be reached and the mixture will not maintain its combustion temperature and will extinguish itself. These principles apply to the combustible cryogens, hydrogen and methane. The flammability and detonability limits for these two cryogens with either air or oxygen are presented in Table 9. Because the flammability limits are rather broad, great care must be exercised to exclude oxygen from these cryogens, particularly from hydrogen: even trace amounts of oxygen will condense, solidify, and build up with time in the bottom of the liquid hydrogen storage container, and eventually the upper flammability limits will be reached. Then some ignition source, such as a mechanical or electrostatic spark, may initiate a fire or possibly an explosion.

Liquid hydrogen and LNG spills form a vapor blanket that includes zones of combustible mixtures that could ignite the entire spilled fuel. Both these fluids burn clean; hydrogen produces a nearly invisible flame. Compared to a flammability limit of 2 to 9% (by volume) for jet fuel in air, hydrogen has a flammability limit of 4 to 75%, and LNG from 5 to 15%. The ignition of explosive mixtures of hydrogen with oxygen or air occurs with low energy input, about one-tenth that of a gasoline/air mixture. All ignition sources should be eliminated, and all equipment and connections should be grounded. Lightning protection (lightning rods, aerial cables, and ground rods suitably connected) should be provided at all preparation, storage, and use areas for these flammable cryogenic fluids.

The methane/oxygen/nitrogen flammability limits diagram of Figure 46 can be used to analyze a fuel/oxidant/diluent. The diagram is typical of any system where methane represents any fuel, oxygen represents any oxidant, and nitrogen represents any diluent. Gas mixtures that lie outside the flammability envelope will not burn or detonate because insufficient heat is released upon combustion to attain a temperature at which combustion is supported. Flame temperatures are at a minimum around the edges of the envelope but increase toward the center of the envelope. The line drawn from a 21% O_2/79% N_2 binary to pure CH_4 represents the compositions that can result from mixing methane with air. Flammability occurs from about 5 to 15% methane in air. If an LNG tank is

Table 9 Flammability and Detonability Limits of Hydrogen and Methane Gas

Mixture	Flammability Limits, mol %	Detonability Limits of Mixture, mol %
H_2/Air	4 to 75	20 to 65
H_2/O_2	4 to 95	15 to 90
CH_4/Air	5 to 15	6 to 14
CH_4/O_2	5 to 61	10 to 50

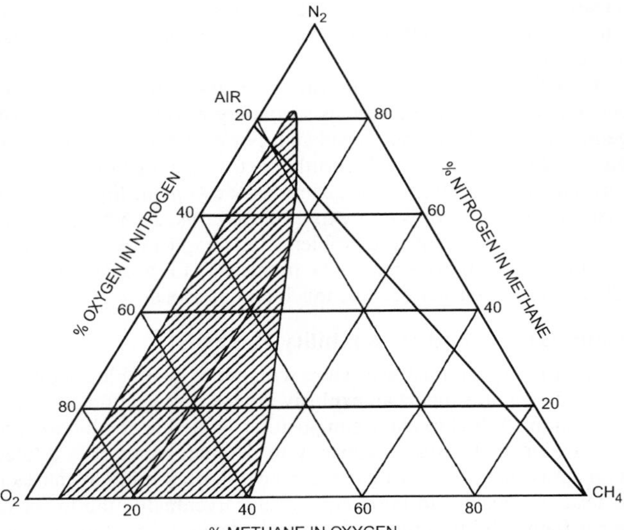

Fig. 46 Flammable Limits for $O_2/N_2/CH_4$ System
(Courtesy of Air Products and Chemicals, Inc., Allentown, PA)

purged with air, the resulting gas mixture would at first be too rich in methane for combustion. However, as purging continues, flammable mixtures would develop, resulting in a hazardous condition.

Such predictable behavior, however, depends on the homogeneous composition of the fuel/oxidant/diluent mixture achievable in gases. Where local concentration gradients exist, combustion behavior is less certain. In such cases, flame velocity depends on such factors as particle size, droplet size, and heat capacities of components in the system. Flame arresters, which are inserted in a line or vessel, prevent propagation of a flame front by absorbing energy from the combustion process and lowering the flame temperature below a level that supports combustion. The minimum ignition energy, however, varies considerably, being as much as an order of magnitude lower for a hydrogen/air mixture than for a methane/air mixture.

Flammable cryogens are best disposed of in a burnoff system in which the liquid or gas is piped to a remote area and burned with air in multiple burner arrangements. Such a system should include pilot ignition methods, flameout warning systems, and means for purging the vent line.

Because of its chemical activity, oxygen also presents a safety problem in handling. Liquid oxygen is chemically reactive with hydrocarbon materials. Ordinary hydrocarbon lubricants are even dangerous to use in oxygen compressors and vacuum pumps exhausting gaseous oxygen. In fact, valves, fittings, and lines used with oil-pumped gases should never be used with oxygen. Serious explosions have resulted from combining oxygen and hydrocarbon lubricants. To ensure against such unwanted chemical reactions, systems using liquid oxygen must be kept scrupulously clean of any foreign matter.

Liquid oxygen equipment must also be constructed of materials incapable of initiating or sustaining a reaction. Only a few polymeric materials can be used in such equipment, because most will react violently with oxygen under mechanical impact. Also, reactive metals such as titanium and aluminum should be used cautiously, because they are potentially hazardous. Once the reaction is started, an aluminum pipe containing oxygen burns rapidly and intensely. With proper design and care, however, liquid oxygen systems can be operated safely.

Although nitrogen is an inert gas and will not support combustion, in some subtle ways a flammable or explosive hazard may develop. Cold traps or open-mouth dewars containing liquid nitrogen can condense air and cause oxygen enrichment of the liquid nitrogen. The composition of air as it condenses into the liquid nitrogen container is about 50% oxygen and 50% nitrogen. As the liquid nitrogen evaporates, the liquid oxygen content steadily increases so that the last portion of liquid to evaporate will have a relatively high oxygen concentration. The nitrogen container must then be handled as if it contained liquid oxygen; fire, compatibility, and explosive hazards all apply to this oxygen-enriched liquid nitrogen.

Because air condenses at a temperature above the normal boiling point of liquid nitrogen, uninsulated pipelines transferring liquid nitrogen will condense air. This oxygen-enriched condensate can drip on combustible materials, causing an extreme fire hazard or explosive situation. The oxygen-rich condensate can saturate clothing, rags, wood, asphalt pavement, etc., and cause the same problems associated with the handling and spillage of liquid oxygen.

Hazard Evaluation Summary

The best-designed facility is no better than the detailed attention paid to every aspect of safety. Such attention cannot be considered once and forgotten. Rather, it is an ongoing activity that requires constant review of every conceivable hazard that might be encountered.

REFERENCES

Alexeev, A., C. Haberstroh, and H. Quack. 1999. Mixed gas J-T cryocooler with precooling stage. *Cryocoolers*, vol. 10, R.G. Ross, Jr., ed. Kluwer Academic/Plenum Publishers.

Alfiev, V.N., V.M. Brodyansky, V.M. Yagodin, V.A. Nikolsky, and A.V. Invantso. 1973. *Refrigerant for a cryogenic throttling unit*. U.K. Patent 1,336,892.

ASME. 2004. *Boiler and pressure vessel code*. American Society of Mechanical Engineers, New York.

ASTM. 2004. Standard practice for estimate of the heat gain or loss and the surface temperatures of insulated flat, cylindrical, and spherical systems by use of computer programs. *Standard* C680-04e1. American Society for Testing and Materials, West Conshohocken, PA.

ASTM. 2001. Standard practice for calculating thermal transmission properties under steady-state conditions. *Standard* C1045-01. American Society for Testing and Materials, West Conshohocken, PA.

Arp, V. 1997. Electrical and thermal conductivities of elemental metals below 300 K. *Proceedings of the 13th Symposium on Thermophysical Properties*, Boulder, CO.

Arp, V. 1998. A summary of fluid properties including near-critical behavior. *Proceedings of the International Conference on Cryogenics and Refrigeration*, Hangzhou, China.

Barron, R.F. 1985. *Cryogenic systems*, 2nd ed. Oxford University Press, London.

Bednorz, J.G. and K.A. Müller. 1986. Possible high superconductivity in the Ba-La-Cu-O system. *Zeitschrift für Physik B* (now *European Physical Journal*) 64(2):189-193.

Broerman, E., et al. 2001. Performance of a new type of electrical cooler for HPGe detector systems. Institute of Nuclear Materials Management Conference, India Head, CA.

Burt, W.W. and C.K. Chan. 1995. Demonstration of a high performance 35 K pulse-tube cryocooler. *Cryocoolers 8*, p. 313. Plenum Press, New York.

Davey, G. 1990. Review of the Oxford cryocooler. *Advances in Cryogenic Engineering*, vol. 35, p. 1423. Plenum Press, New York.

Dobak, J.D., R. Radebaugh, M.L. Huber, and E.D. Marquardt. 1998. *Mixed gas refrigerant method*. U.S. Patent 5,787,715.

Eckels Engineering. 2003. *CRYOCOMP*, v. 2.0. Available at http://www.htess.com/cryocomp.htm.

Gaumer, L.S. 1986. LNG processes. *Advances in Cryogenic Engineering*, vol. 31, p. 1095. Plenum Press, New York.

Grenier, M. and P. Petit. 1986. Cryogenic air separation: The last twenty years. *Advances in Cryogenic Engineering*, vol. 31, p. 1063. Plenum Press, New York.

Hashimoto, T., M. Ogawa, A. Hayaski, M. Makino, R. Li, and K. Aoki. 1992. Recent progress on rare earth magnetic regenerator materials. *Advances in Cryogenic Engineering*, vol. 37, p. 859. Plenum Press, New York.

Keppler, F., G. Nellis, and S.A. Klein. 2004. Optimization of the composition of a gas mixture in a Joule-Thomson cycle. *International Journal of HVAC&R Research* 10(2):213-230.

Kleemenko, A.P. 1959. One flow cascade cycle. *Proceedings of the Xth International Congress of Refrigeration*, Copenhagen, vol. 1, pp. 34-39. Pergamon Press, London.

Klein, S.A. Continuously updated. *EES—Engineering equation solver.* F-Chart Software, Madison, WI. (fchart.com)

Kuriyama, T., M. Takahashi, H. Nakagoma, T. Hashimoto, H. Nitta, and M. Yabuki. 1994. Development of 1 watt class 4K GM refrigerator with magnetic regenerator materials. *Advances in Cryogenic Engineering*, vol. 39, p. 1335. Plenum Press, New York.

Little, W.A. 1984. Microminiature refrigeration. *Review of Scientific Instruments* 55:661-680.

Little, W.A. 1990. Advances in Joule Thomson cooling. *Advances in Cryogenic Engineering*, vol. 35, p. 1305. Plenum Press, New York.

Little, W.A. 1997a. *Self-cleaning low-temperature refrigeration system.* U.S. Patent 5,617,739.

Little, W.A. 1997b. *Method for efficient counter current heat exchange using optimized mixtures.* U.S. Patent 5,644,502.

Little, W.A. 1998. Kleemenko cycle coolers: Low cost refrigeration at cryogenic temperatures. ICEC 17, Bournemouth, U.K. Institute of Physics Publishing, Bristol and Philadelphia.

Little, W.A. 2003. Seven year history of continuous operation of Kleemenko cryocoolers. M-CALC IV, Strategic Analysis, Inc.

Little, W.A. and I. Sapozhnikov. 1998. *Self-cleaning cryogenic refrigeration system.* U.S. Patent 5,724,832.

Longsworth, R.C. 1997a. *Cryogenic refrigerator with single stage compressor.* U.S. Patent 5,337,572.

Longsworth, R.C. 1997b. 80 K throttle-cycle refrigerator cost reduction. *Cryocoolers 9*, R.G. Ross, Jr., ed. Plenum Press, New York.

M-CALC-IV. 2003. Fourth workshop on military and commercial applications of low-cost cryocoolers, November 20-21, 2003, Hyatt Islandia, San Diego, CA. Strategic Analysis, Inc. (www.sainc.com/MCALC4)

Missimer, D. J. 1973. *Self-balancing low temperature refrigeration system.* U.S. Patent 3,768,273.

Nagao, M., T. Anaguchi, H. Yoshimura, S. Nakamura, T. Yamada, T. Matsumoto, S. Nakagawa, K. Moutsu, and T. Watanabe. 1994. 4 K three stage Gifford-McMahon cycle refrigerator for MRI magnet. *Advances in Cryogenic Engineering*, vol. 39, p. 1327. Plenum Press, New York.

NIST. 2002a. *NIST standard reference database* 4, *SUPERTRAPP.* National Institute of Standards and Technology, Gaithersburg, MD.

NIST. 2002b. *NIST standard reference database* 14, *DDMAX.* National Institute of Standards and Technology, Gaithersburg, MD.

Radebaugh, R. 1990. A review of pulse tube refrigeration. *Advances in Cryogenic Engineering*, vol. 35, p. 1191. Plenum Press, New York.

Radebaugh, R. 1995. Recent developments in cryocoolers. *Proceedings of the XIX International Congress of Refrigeration* IIIb:973. IIR, Paris.

Strobridge, T.R. 1974. Cryogenic refrigerators—An updated survey. NBS *Technical Note* 655. National Bureau of Standards, U.S. Government Printing Office, Washington, D.C.

Swift, W.L. and H. Sixsmith. 1993. Performance of a long life reverse Brayton cryocooler. *Proceedings of the 7th International Cryocooler Conference*, vol. 84. Air Force *Report* PL CP 93 1001.

Wigley, D.A. 1971. *Mechanical properties of materials at low temperatures.* Plenum Press, New York.

Zhu, S., P. Wu, and Z. Chen. 1990. Double inlet pulse tube refrigerators: An important improvement. *Cryogenics* 30:514.

BIBLIOGRAPHY

Anderson, J.E., D.A. Fester, and A.M. Czysz. 1990. Evaluation of long term cryogenic storage system requirements. *Advances in Cryogenic Engineering*, vol. 35, p. 1725. Plenum Press, New York.

Augustynowicz, S.D., J.A. Demko, and V.I. Datskov. 1994. Analysis of multilayer insulation between 80 K and 300 K. *Advances in Cryogenic Engineering*, vol. 39, p. 1675. Plenum Press, New York.

Edeskuty, F.J. and W.F. Stewart. 1986. *Safety in the handling of cryogenic fluids.* Plenum Press, New York.

Edeskuty, F.J. and K.D. Williamson. 1983. *Liquid cryogens*, vols. 1 and 2. CRC Press, Boca Raton, FL.

Gistan, G.M., J.C. Villard, and F. Turcat. 1990. Application range of centrifugal compressors. *Advances in Cryogenic Engineering*, vol. 35, p. 1031. Plenum Press, New York.

Hashimoto, T., K. Gang, H. Makuuchi, R. Li, A. Oniski, and Y. Kanazawa. 1994. Improvement of refrigeration characteristics of the G-M refrigerator using new magnetic regenerator material $ErNi_{1-x}CO_x$ system. *Advances in Cryogenic Engineering*, vol. 37, p. 859. Plenum Press, New York.

Haynes, W.M. and D.G. Friend. 1994. Reference data for thermophysical properties of cryogenic fluids. *Advances in Cryogenic Engineering*, vol. 39, p. 1865. Plenum Press, New York.

Jacobsen, R.T. and S.G. Penoncello. 1997. *Thermodynamic properties of cryogenic fluids.* Plenum Press, New York.

Johnson, P.C. 1986. Updating LNG plants. *Advances in Cryogenic Engineering*, vol. 31, p. 1101. Plenum Press, New York.

Kittel, P. 1992. Ideal orifice pulse tube refrigerator performance. *Cryogenics* 32:843.

Kun, L.C. 1988. Expansion turbines and refrigeration for gas separation and liquefaction. *Advances in Cryogenic Engineering*, vol. 33, p. 963. Plenum Press, New York.

Lee, J.M., P. Kittel, K.D. Timmerhaus, and R. Radebaugh. 1993. Flow patterns intrinsic to the pulse tube refrigerator. *Proceedings of the 7th International Cryocooler Conference*, p. 125. Air Force *Report* PL CP 93 1001.

Lemmon, E.W., R.T. Jacobsen, S.G. Penoncello, and S.W. Beyerlein. 1994. Computer programs for the calculation of thermodynamic properties of cryogens and other fluids. *Advances in Cryogenic Engineering*, vol. 39, p. 1891. Plenum Press, New York.

Longsworth, R.C. 1988. 4 K Gifford-McMahon/Joule-Thomson cycle refrigerators. *Advances in Cryogenic Engineering*, vol. 33, p. 689. Plenum Press, New York.

Pavese, F. and G. Molinas. 1992. *Modern gas-based temperature and pressure measurements.* Plenum Press, New York.

Rao, M.G. and R.G. Scurlock. 1986. Cryogenic instrumentation with cold electronics: A review. *Advances in Cryogenic Engineering*, vol. 31, p. 1211. Plenum Press, New York.

Sarwinski, R.E. 1988. Cryogenic requirements for medical instrumentation. *Advances in Cryogenic Engineering*, vol. 33, p. 87. Plenum Press, New York.

Scurlock, R.G. 1992. A brief history of cryogenics. *Advances in Cryogenic Engineering*, vol. 37, p. 1. Plenum Press, New York.

Shuh, Q.S., R.W. Fast, and H.L. Hart. 1988. Theory and technique for reducing the effect of cracks in multilayer insulation from room temperature to 77 K. *Advances in Cryogenic Engineering*, vol. 33, p. 291. Plenum Press, New York.

Sixsmith, H., R. Hasenbein, J.A. Valenzuela, J.C. Theilacker, and J. Fuerst. 1990. A miniature wet turboexpander. *Advances in Cryogenic Engineering*, vol. 35, p. 989. Plenum Press, New York.

Spradley, I.E., T.C. Nast, and D.J. Frank. 1990. Experimental studies of MLI systems at very low boundary temperatures. *Advances in Cryogenic Engineering*, vol. 35, p. 477. Plenum Press, New York.

Storch, P.J. and R. Radebaugh. 1988. Development and experimental test of an analytical model of the orifice pulse tube refrigerator. *Advances in Cryogenic Refrigeration* 33:851-859.

Timmerhaus, K.D. and T.M. Flynn. 1989. *Cryogenic process engineering.* Plenum Press, New York.

Van Sciver, S.W. 1991. *Helium cryogenics.* Plenum Press, New York.

Vinen, W.F. 1990. Fifty years of superfluid helium. *Advances in Cryogenic Engineering*, vol. 35, p. 1. Plenum Press, New York.

Walker, G. 1983. *Cryocoolers.* Plenum Press, New York.

CHAPTER 39

ULTRALOW-TEMPERATURE REFRIGERATION

ULTRALOW-TEMPERATURE refrigeration is defined here as refrigeration in the temperature range of –58 to –148°F. What is considered low temperature for an application depends on the temperature range for that specific application. Low temperatures for air conditioning are around 32°F; for industrial refrigeration, –31 to –58°F; and for cryogenics, approaching 0°R. Applications such as freeze-drying, as well as the pharmaceutical, chemical, and petroleum industries, use refrigeration in the temperature range designated low in this chapter.

The –58 to –148°F temperature range is treated separately because design and construction considerations for systems that operate in this range differ from those encountered in industrial refrigeration and cryogenics, which bracket it. Designers and builders of cryogenic facilities are rarely active in the low-temperature refrigeration field. One major type of low-temperature system is the **packaged type**, which often serves such applications as environment chambers. The other major category is **custom-designed and field-erected** systems. Industrial refrigeration practitioners are the group most likely to be responsible for these systems, but they may deal with low-temperature systems only occasionally; the experience of a single organization does not accumulate rapidly. The objective of this chapter is to bring together available experience for those whose work does not require daily contact with low-temperature systems.

The refrigeration cycles presented in this chapter may be used in both standard packaged and custom-designed systems. Cascade systems are emphasized, both autocascade (typical of packaged units) and two-refrigerant cascade (found in custom-engineered low-temperature systems).

AUTOCASCADE SYSTEMS

An autocascade refrigeration system is a complete, self-contained refrigeration system in which multiple stages of cascade cooling effect occur simultaneously by means of vapor/liquid separation and adiabatic expansion of various refrigerants. Physical and thermodynamic features, along with a series of counterflow heat exchangers and an appropriate mixture of refrigerants, allow the system to reach low temperature.

Autocascade refrigeration systems offer many benefits, such as a low compression ratio and relatively high volumetric efficiency. However, system chemistry and heat exchangers are complex, refrigerant compositions are sensitive, and compressor displacement is large.

Operational Characteristics

Components of an autocascade refrigeration system typically include a vapor compressor, an external air- or water-cooled condenser, a mixture of refrigerants with descending boiling points, and a series of insulated heat exchangers. Figure 1 is a schematic of a simple system illustrating a single-stage of autocascade.

The preparation of this chapter is assigned to TC 10.4, Ultralow-Temperature Systems and Cryogenics.

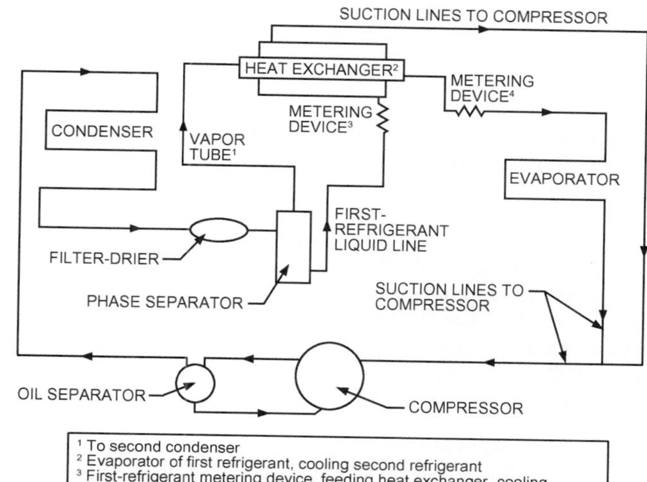

Fig. 1 Simple Autocascade Refrigeration System

In this system, two refrigerants with significantly different boiling points are compressed and circulated by one vapor compressor. Assume that one refrigerant is R-23 (normal boiling point, –115.6°F) and the second refrigerant is R-404a (normal boiling point, –52°F). Assume that ambient temperature is 77°F and that the condenser is 100% efficient.

With properly sized components, this system should be able to achieve –76°F in the absorber while the compression ratio is maintained at 5.1 to 1. As the refrigerant mixture is pumped through the main condenser and cooled to 77°F at the exit, compressor discharge pressure is maintained at 221 psig. At this condition, virtually all R-404a is condensed at 95°F and then further chilled to subcooled liquid. Although R-23 molecules are present in both liquid and vapor phases, the R-23 is primarily vapor because of the large difference in the boiling points of the two refrigerants. A phase separator at the outlet of the condenser collects the liquid by gravitational effect, and the R-23-rich vapor is removed from the outlet of the phase separator to the heat exchanger.

At the bottom of the phase separator, an expansion device adiabatically expands the collected R-404a-rich liquid such that the outlet of the device produces a low temperature of –2.2°F at 32 psig (Weng 1995). This cold stream is immediately sent back to the heat exchanger in a counterflow pattern to condense the R-23-rich vapor to liquid at –1.4°F and 221 psig. The R-23-rich liquid is then adiabatically expanded by a second expansion device to –76°F. As it absorbs an appropriate amount of heat in the absorber, the R-23 mixes with the expanded R-404a and evaporates in the heat exchanger, providing a cold source for condensing R-23 on the high-pressure side of the heat exchanger. Leaving the heat exchanger at superheated conditions, the vapor mixture then returns to the suction of the compressor for the next cycle.

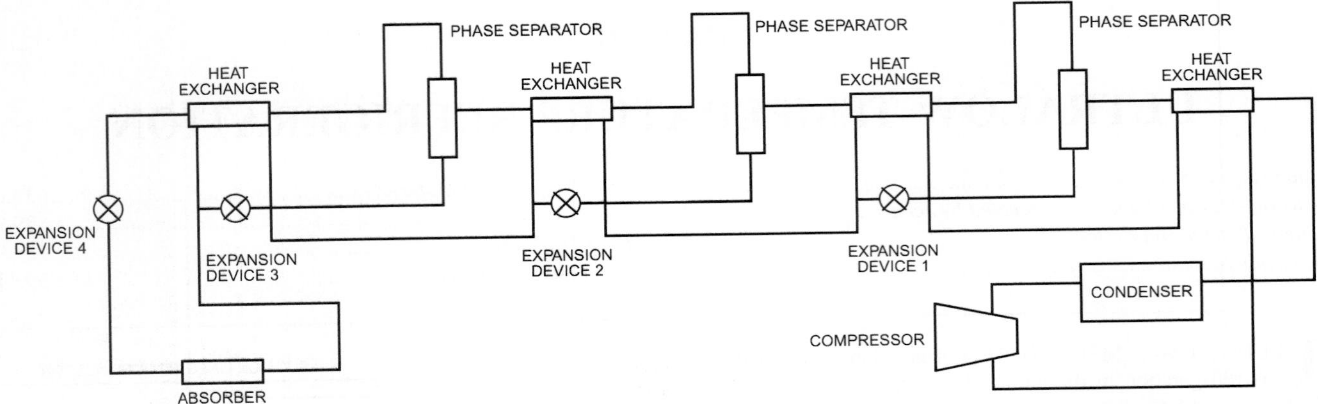

Fig. 2 Four-Stage Autocascade System

As can be seen from this simple example, the autocascade effect derives from a **short cycle** of the refrigerant circuit within the system that performs only internal work to condense the lower boiling point refrigerant.

The concept of the single-stage cycle can be extended to multiple stages. Figure 2 shows the flow diagram of a four-stage system. The condensation and subsequent expansion of one refrigerant provides the cooling necessary to condense the next refrigerant in the heat exchanger downstream. This process continues until the last refrigerant with the lowest boiling point is expanded to achieve extremely low temperature.

Design Considerations

Compressor Capacity. As can be seen from Figures 1 and 2, a significant amount of compressor work is used for internal evaporating and condensing of refrigerants. The final gain of the system is therefore relatively small. Compressor capacity must be enough to produce an appropriate amount of final refrigerating effect.

Heat Exchanger Sizing. Because there is a significant amount of refrigerant vapor in each stage of the heat exchanger, the overall heat transfer coefficients on both the evaporating and condensing sides are rather small compared to those of pure components at phase-changing conditions. Therefore, generous heat-transfer area should be provided for energy exchange between refrigerants on the high- and low-pressure sides.

Expansion Devices. Each expansion device is sized to provide sufficient refrigerating effect for the adjacent downstream heat exchanger.

Compressor Lubrication. General guidelines for lubrication of refrigeration systems should be adopted.

CUSTOM-DESIGNED AND FIELD-ERECTED SYSTEMS

If refrigeration is to maintain a space at a low temperature to store a modest quantity of product in a chest or cabinet, the packaged low-temperature system is probably the best choice. Prefabricated walk-in environmental chambers are also practical solutions when they can accommodate space needs. When the required refrigeration capacity exceeds that of packaged systems, or when a fluid must be chilled, a custom-engineered system should be considered.

The refrigeration requirement may be to chill a certain flow rate of a given fluid from one temperature to another. Part of the design process is to choose the type of system, which may be a multistage plant using a single refrigerant or a two-circuit cascade system using a high-pressure refrigerant for the low-temperature circuit. The compressor(s) and condenser(s) must be selected, and the evaporator and interstage heat exchanger (in the case of the cascade system) must be either selected or custom-designed.

The design process includes selection of (1) metal for piping and vessels and (2) insulating material and method of application. The product to be refrigerated may actually pass through the evaporator, but in many cases a secondary coolant transfers heat from the final product to the evaporator. Brines and antifreezes that perform satisfactorily at higher temperatures may not be suitable at low temperatures. Compressors are subjected to unusual demands when operating at low temperatures, and, because they must be lubricated, oil selection and handling must be addressed.

SINGLE-REFRIGERANT SYSTEMS

Single-refrigerant systems are contrasted with the cascade system, which consists of two separate but thermally connected refrigerant circuits, each with a different refrigerant (Stoecker and Jones 1982).

In the industrial refrigeration sector, the traditional refrigerants have been R-22 and ammonia (R-717). Because R-22 will ultimately be phased out, various hydrofluorocarbon (HFC) refrigerants and blends are proposed as replacements. Two that might be considered are R-507 and R-404a.

Two-Stage Systems

In systems where the evaporator operates below about −4°F, two-stage or compound systems are widely used. These systems are explained in Chapter 3 of this volume and in Chapter 1 of the 2005 *ASHRAE Handbook—Fundamentals*. Advantages of two-stage compound systems that become particularly prominent when the evaporator operates at low temperature include

- Improved energy efficiency because of removal of flash gas at the intermediate pressure and desuperheating of discharge gas from the low-stage compressor before it enters the high-stage compressor.
- Improved energy efficiency because two-stage compressors are more efficient operating against discharge-to-suction pressure ratios that are lower than for a single-stage compressor.
- Avoidance of high discharge temperatures typical of single-stage compression. This is important in reciprocating compressors but of less concern with oil-injected screw compressors.
- Possibility of a lower flow rate of liquid refrigerant to the evaporator because the liquid is at the saturation temperature of the intermediate pressure rather than the condensing pressure, as is true of single-stage operation.

Refrigerant and Compressor Selection

The compound, two-stage (or even three-stage) system is an obvious possibility for low-temperature applications. However, at very low temperatures, limitations of the refrigerant itself appear: freezing point, pressure ratios required of the compressors, and volumetric flow at the suction of the low-stage compressor per unit

Table 1 Low-Temperature Characteristics of Several Refrigerants at Three Evaporating Temperatures

Refrig-erant	Freezing Point	Pressure Ratio with Two-Stage System			Volumetric Flow of Refrigerant, cfm per ton		
		Evaporating Temp.			Evaporating Temp.		
		−58°F	−94°F	−130°F	−58°F	−94°F	−130°F
R-22	−256°F	4.6	8.14	17.9	11.7	35.9	147.6
R-507	≤148°F	4.4	7.8	16.1	9.62	30.4	130.5
R-717	−108°F	5.75	11.1	25.4	15.3	54.0	—
R-404a	≤148°F	4.36	7.58	15.2	9.99	30.8	124.9

refrigeration capacity. Table 1 shows some key values for four candidate refrigerants, illustrating some of the concerns that arise when considering refrigerants that are widely applied in industrial refrigeration systems. Hydrocarbons (HCs), which are candidates particularly in the petroleum and petrochemical industry, where the entire plant is geared toward working with flammable gases, are not included in Table 1.

The **freezing point** is not a limitation for the halocarbon refrigerants, but ammonia freezes at −108°F, so its use must be restricted to temperatures safely above that temperature.

The **pressure ratios** the compressors must operate against in two-stage systems are also important. A condensing temperature of 95°F is assumed, with the intermediate pressure being the geometric mean of the condensing and evaporating pressures. Many low-temperature systems may be small enough that a reciprocating compressor would be favorable, but the limiting pressure ratio with reciprocating compressors is usually about 8, a value chosen to limit the discharge temperature. An evaporating temperature of −94°F is about the lowest permissible for systems using reciprocating compressors. For evaporating temperatures lower than −94°F, consider using a three-stage system. An alternative to the reciprocating compressor is the screw compressor, which operates with lower discharge temperatures because it is oil flooded. The screw compressor can therefore operate against larger pressure ratios than the reciprocating compressor, and is favored in larger systems.

The required **volumetric pumping capacity** of the compressor is measured at the compressor suction. This value is an indicator of the physical size of the compressor; the values become huge at the −130°F evaporating temperature.

Some conclusions from Table 1 are

- A single-refrigerant, two-stage system can adequately serve a plant in the higher-temperature portion of the range considered here, but it becomes impractical in the lower-temperature portion.
- Ammonia, which has many favorable properties for industrial refrigeration, has little appeal for low-temperature refrigeration because of its relatively high freezing point and pressure ratios.

Special Multistage Systems

Special high-efficiency operations to recover volatile compounds such as hydrocarbons use the **reverse Brayton cycle**. This consists of one or two conventional compressor refrigeration cycles with the lowest stage ranging from −76 to −148°F. This final stage is achieved by using a turbo compressor/expander and enables the collection of liquefied hydrocarbons (Emhö 1997; Enneking and Priebe 1993; Jain and Enneking 1995).

CASCADE SYSTEMS

The cascade system (Figure 3) confronts some of the problems of single-refrigerant systems. It consists of two separate circuits, each using a refrigerant appropriate for its temperature range. The two circuits are thermally connected by the cascade condenser, which is the condenser of the low-temperature circuit and the evaporator of the high-temperature circuit. Typical refrigerants for the

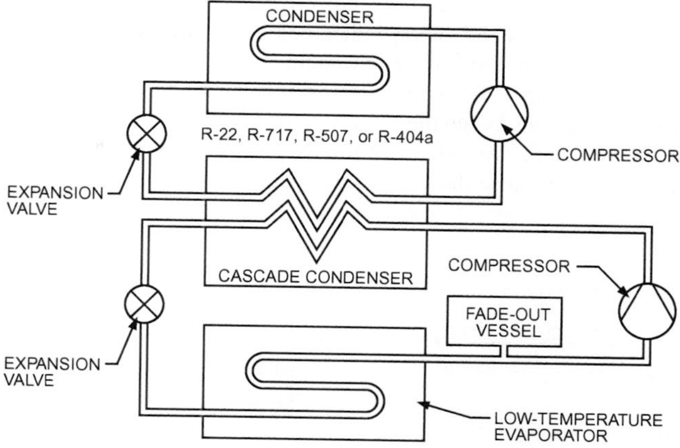

Fig. 3 Simple Cascade System

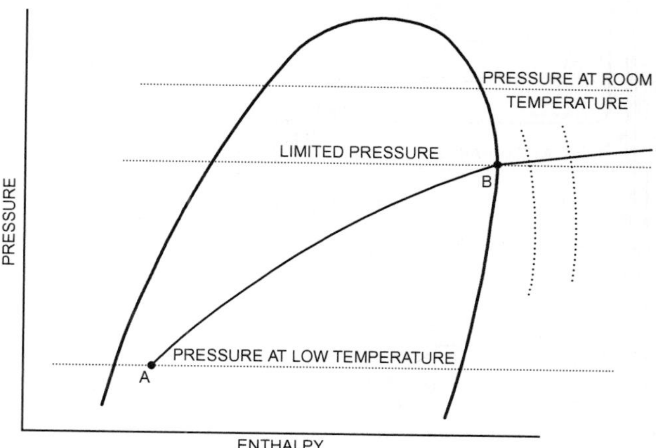

Fig. 4 Simple Cascade Pressure-Enthalpy Diagram

high-temperature circuit include R-22, ammonia, R-507, and R-404a. For the low-temperature circuit, a high-pressure refrigerant with a high vapor density (even at low temperatures) is chosen. For many years, R-503, an azeotropic mixture of R-13 and R-23, was a popular choice, but R-503 is no longer available because R-13 is an ozone-depleting chlorofluorocarbon (CFC). R-23 could be and has been used alone, but R-508b, an azeotrope of R-23 and R-116, has superior properties, as discussed in the section on Refrigerants for Low-Temperature Circuit.

The cascade system has some of the thermal advantages of two-stage, single-refrigerant systems: it approximates flash gas removal and allows each compressor to take a share of the total pressure ratio between the low-temperature evaporator and the condenser. The cascade system has the thermal disadvantage of needing to provide an additional temperature lift in the cascade condenser because the condensing temperature of the low-temperature refrigerant is higher than the evaporating temperature of the high-temperature refrigerant. There is an optimum operating temperature of the cascade condenser for minimum total power requirement, just as there is an optimum intermediate pressure in two-stage, single-refrigerant systems.

Figure 3 shows a fade-out vessel, which limits pressure in the low-temperature circuit when the system shuts down. At room temperature, the pressure of R-23 or R-508b in the system would exceed 600 psia if liquid were present. The entire low-temperature system must be able to accommodate this pressure. The process occurring in the fade-out vessel is at constant volume, as shown on the pressure-enthalpy diagram of Figure 4. When the system operates at low temperature, the refrigerant in the system is a mixture of liquid and vapor, indicated by point A. When the system shuts down, the refrigerant begins to warm and follows the constant-volume line, with pressure increasing according to the saturation curve. When the saturated vapor line is reached at point B, further increases in temperature result in only slight increases in pressure because the refrigerant is superheated vapor.

Larger cascade systems are field engineered, but packaged systems are also available. Figure 5 shows a two-stage system with the necessary auxiliary equipment. Figure 6 shows a three-stage cascade system.

Refrigerants for Low-Temperature Circuit

R-503 is no longer available for general use. R-23 can be used, but R-508b, an azeotropic mixture of non-ozone-depleting R-23 and

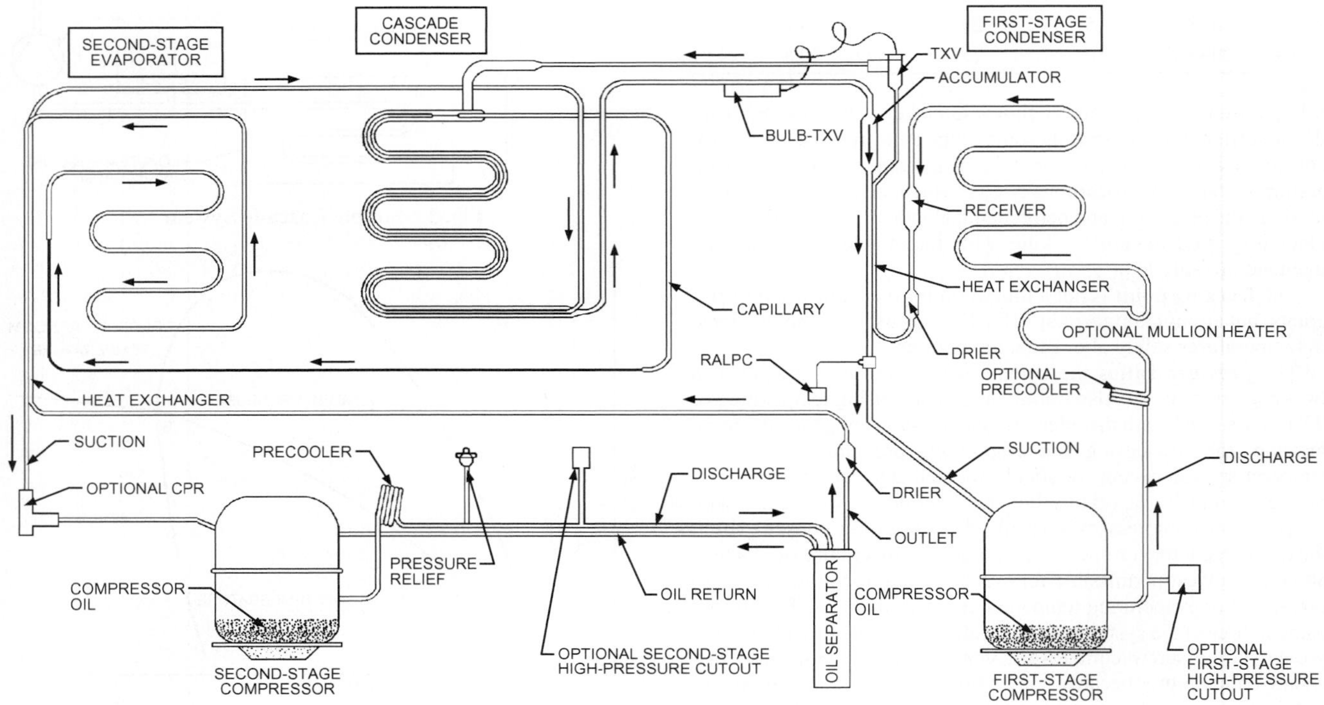

Fig. 5 Two-Stage Cascade System

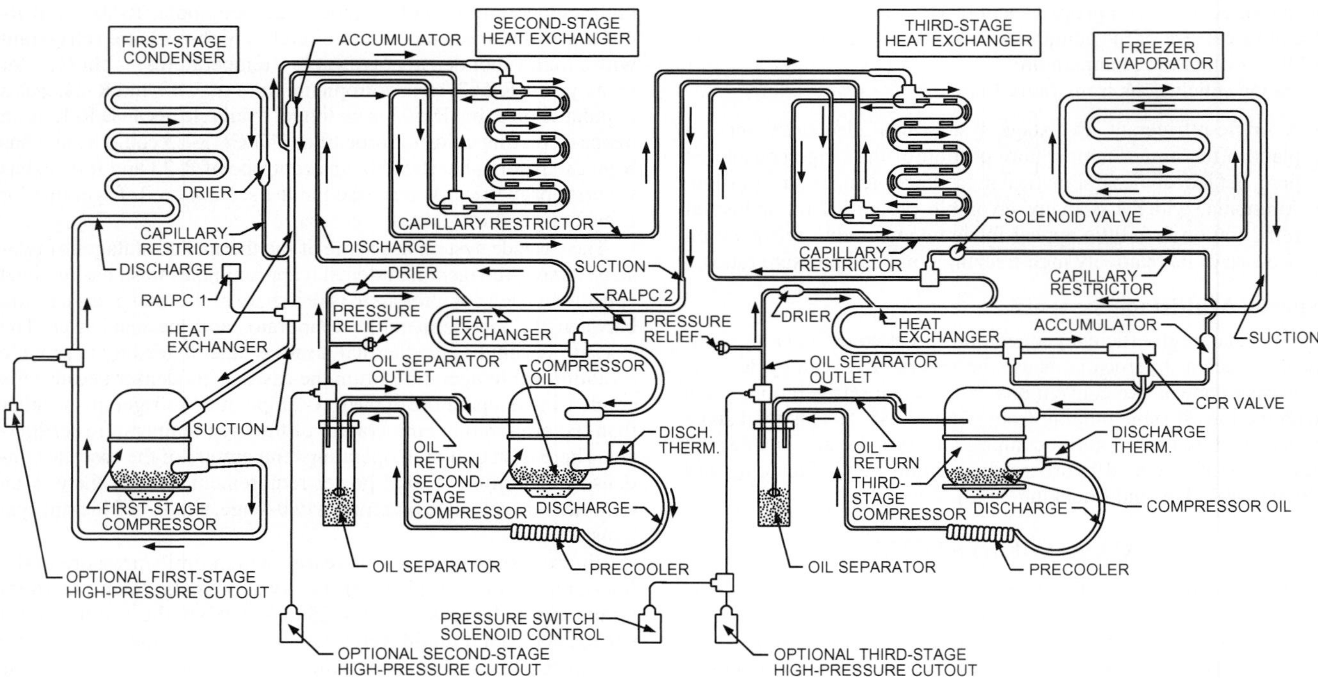

Fig. 6 Three-Stage Cascade System

Table 2 Properties of R-508b

Boiling point (1 atm)	–126.5°F
Critical temperature	56.7°F
Critical pressure	571 psia
Latent heat of vaporization at boiling point	72.4 Btu/lb
Ozone depletion potential (R-12 = 1)	0
Flammability	Nonflammable
Exposure limit (8 and 12 h)*	1000 ppm

*The exposure limit is a calculated limit determined from the DuPont airborne exposure limit (AEL) of the individual components. The AEL is the maximum amount to which nearly all workers can be repeatedly exposed during a working lifetime without adverse effects.

Table 3 Theoretical Performance of Cascade System Using R-13, R-503, R-23, or R-508b

	R-503	R-13	R-23	R-508b
Capacity (R-503 = 100)	100	71	74	98
Efficiency (R-503 = 100)	100	105	95	103
Discharge pressure, psia	145	104	123	147
Suction pressure, psia	16	12	13	16
Discharge temperature, °F	224	198	280	188

Note: Operating conditions are –120°F evaporator, –31°F condenser; 10°F subcooling; 0°F suction temperature; 70% isentropic compression efficiency, 4% volumetric clearance.

Table 4 Theoretical Compressor Performance Data for Two Different Evaporating Temperatures

Evaporating Temperature, °F	Refrigerant	Pressure Ratio	Discharge Temperature, °F	Volumetric Flow, cfm per ton
–112	R-23	7.49	136	8.19
	R-508b	6.51	90	6.46
–148	R-23	26.88	161	28.76
	R-508b	21.88	102	21.89

Basis: –31°F condensing temperature; compressor efficiency of 70%; volumetric efficiency of 100%; 18°F subcooling, and 90°F suction superheat.

R-116, is superior. It is nonflammable and has zero ozone depletion potential (ODP). Table 2 lists some properties of R-508b.

R-508b offers excellent operating characteristics compared to R-503 and R-13. Capacity and efficiency values are nearly equivalent to R-503's and superior to R-13's. The compressor discharge temperature is lower than for R-23; lower discharge temperatures may equate to longer compressor life and better lubricant stability. The estimated operating values of a cascade system running with R-508b are shown in Table 3. Performance parameters of R-503, R-13, and R-23 are shown for comparison.

Table 4 shows calculated data for R-23 and R-508b for two operating ranges. The volumetric efficiency is 100%. Actual compressor performance varies with pressure ratios and yields lower capacity and efficiencies and higher discharge temperatures and flow requirements than shown.

Compressor Lubrication

When selecting a lubricant to use with R-508b in an existing low-temperature system, consider (1) refrigerant/lubricant miscibility, (2) chemical stability, (3) materials compatibility, and (4) refrigeration system design. Original equipment manufacturers and compressor suppliers should be consulted.

Using additives to enhance system performance is well established in the low-temperature industry, and may be applied to R-508b. The miscibility of R-508b with certain polyol esters (POEs) is slightly better than the limited miscibility of R-13 and R-503 with mineral oil and alkylbenzene, which helps oil circulation at the low evaporator temperatures. Even with increased miscibility, additives may enhance performance. Some POE oils designed for use in very-low-temperature systems have been used successfully with R-508b in equipment retrofits. Consult compressor manufacturers and suppliers before a final decision on lubricants and any additives.

Compressors

Larger cascade systems typically use standard, positive-displacement compressors in the dual refrigeration system. The evaporator of the higher-temperature refrigerant system serves as the condenser for the lower-temperature one. This allows rather normal application of the compressors on both systems in relation to the pressures, compression ratios, and oil and discharge temperatures within the compressors. However, several very important items must be considered for both the high and low sides of the cascade system to avoid operational problems. Commercially available compressors must be analyzed for both sides of the cascade to determine the best combination for a suitable intermediate high-side evaporator/low-side condenser and minimum (or economical) system power usage.

High-Temperature Circuit. The higher-temperature system is generally a single- or two-stage system using a commercial refrigerant (R-134a, R-22, R-404a, or R-717); evaporating temperature is approximately –10 to –50°F, and condensing is at normal ambient conditions. Commercially available reciprocating and screw compressors are suitable. If compressor evaporating temperature is below –50°F, a suction-line heat exchanger is needed to superheat compressor suction gas to at least –45°F to avoid the metal brittleness associated with lower temperatures at the compressor suction valve and body. Suction piping materials and the evaporator/condenser (evaporator side) must also be suitable for these low temperatures per American Society of Mechanical Engineers (ASME) code requirements.

Lubricant for the higher-temperature system must be compatible with the refrigerant and suitable for the type of system, considering oil carryover and return from the evaporator and the low-temperature conditions within the evaporator.

Low-Temperature Circuit. The compressor may also be a standard refrigeration compressor, if the compressor suction gas is superheated to at least –45°F to avoid low-temperature metal brittleness. Operation is typically well within standard pressure, oil temperature, and discharge temperature limits. The refrigerant is usually R-23 or R-508b. Because temperatures in the low side are below –50°F, all piping, valves, and vessels must be of materials that comply with ASME codes for these temperatures.

It may be difficult to obtain compressor rating data for low-temperature applications with these refrigerants because few actual test data are available, and the manufacturer may be reluctant to be specific. Therefore, the low side should not be designed too close to the required specification. Good practice is to calculate the actual volumetric flow rate to be handled by the compressor (at the expected superheat) to be certain that it can perform as required.

Capacity loss from high superheat is more than recouped (for a net capacity gain) from the liquid subcooling obtained by the suction-line heat exchanger. In rare cases, it may be necessary to inject a small quantity of hot gas into the suction to ensure maximum suction temperature.

The lubricant selected for the low side must be compatible with the specific refrigerant used and also suitable for the low temperatures expected in the evaporator. It is important that adequate coalescing oil separation (5 ppm) is provided to minimize oil carryover from compressor to evaporator.

In direct-expansion evaporators, any oil is forced through the tubes, but speed in the return lines must be high enough to keep this small amount of oil moving back to the compressor. If the system has capacity control, then multiple suction risers or alternative design procedures may be required to prevent oil logging in the evaporator and ensure oil return. At these ultralow temperatures, it is imperative to select a lubricant that remains fluid and does not plate out on the evaporator surfaces, where it can foul heat transfer.

Choice of Metal for Piping and Vessels

The usual construction metal for use with thermal fluids is carbon steel. However, carbon steel should not be used below –20°F because of its loss of ductility. Consider using 304 or 316 stainless steel because of their good low-temperature ductility. Another alternative is to use carbon steel that has been manufactured specifically to retain good ductility at low temperatures (Dow Corning USA 1993). For example,

Carbon steel	Down to –20°F
SA - 333 - GR1	–20 to –50°F
SA - 333 - GR7	–50 to –100°F
SA - 333 - GR3	–75 to –150°F
SA - 333 - GR6	To –50°F

LOW-TEMPERATURE MATERIALS

Choosing material for a specific low-temperature use is often a compromise involving several factors:

- Cost
- Stress level at which the product will operate
- Manufacturing alternatives
- Operating temperature
- Ability to weld and stress-relieve welded joints
- Possibility of excessive moisture and corrosion
- Thermal expansion and modulus of elasticity characteristics for bolting and connection of dissimilar materials
- Thermal conductivity and resistance to thermal shock

Effect of Low Temperature on Materials. When the piping and vessels are to contain refrigerant at low temperature, special materials must usually be chosen because of the effect of low temperature on material properties. Chemical interactions between the refrigerant and containment material must also be considered. Mechanical and physical properties, fabricability, and availability are some of the important factors to consider. Few generalizations can be made, except that decreased temperature increases hardness, strength, and modulus of elasticity. The effect of low temperatures on ductility and toughness can vary considerably between materials. With a decrease in temperature, some metals show increased ductility; others show increased some limiting low temperature, followed by a decrease at lower temperatures. Still other metals decrease in toughness and ductility as temperature decreased below room temperature.

The effect of temperature reduction on polymers depends on the type of polymer. Thermoplastic polymers, which soften when heated above their glass transition temperature T_g, become progressively stiffer and finally brittle at low temperatures. Thermosetting plastics, which are highly cross-linked and do not soften when heated, are brittle at both ambient and lower temperatures. Elastomers (rubbers) are lightly cross-linked and stiffen like thermoplastics as the temperature is lowered, becoming fully brittle at very low temperatures.

Although polymers become brittle and may crack at low temperatures, their unique combination of properties (excellent thermal and electrical insulation capability, low density, low heat capacity, and nonmagnetic character) make them attractive for a variety of lower-temperature applications. At extremely low temperatures, all plastic materials are very brittle and have low thermal conductivity and low strength relative to metals and composites, so selection and use must be carefully evaluated.

Fiber composite materials have gained widespread use at low temperatures, despite their incorporation of components that are often by themselves brittle. A factor that must be considered in the use of composites is the possibility of **anisotropic** behavior, in which they exhibit properties with different values when measured along axes in different directions. Composites with aligned fibers are highly anisotropic.

Metals

The relation of tensile strength to temperature for common structural metals at low temperatures is shown in Figure 7 (Askeland 1994). The slopes of the curves indicate that the increase in strength with decrease in temperature varies. However, tensile strength is not the best criterion for determining the suitability of a material for low-temperature service, because most failures result from a loss of ductility.

Lower temperatures can have a dramatic effect on the **ductility** of metal; the effect depends to a large extent on crystal structure. Metals and alloys that are face-centered cubic (FCC) and ductile at ambient temperatures remain ductile at low temperatures; this category includes aluminum, copper, copper-nickel alloys, nickel, and austenitic stainless steels. Metals and alloys that are body-centered cubic (BCC), such as pure iron, carbon steel, and many alloy steels, become brittle at low temperatures. Many BCC metals and alloys exhibit a ductile-to-brittle transition at lower temperatures (see 1020 steel in Figure 8). This loss of ductility comes from a decrease in the number of operating slip systems, which accommodate dislocation motion. Hexagonal close-packed (HCP) metals and alloys occupy an intermediate place between FCC and BCC materials and may

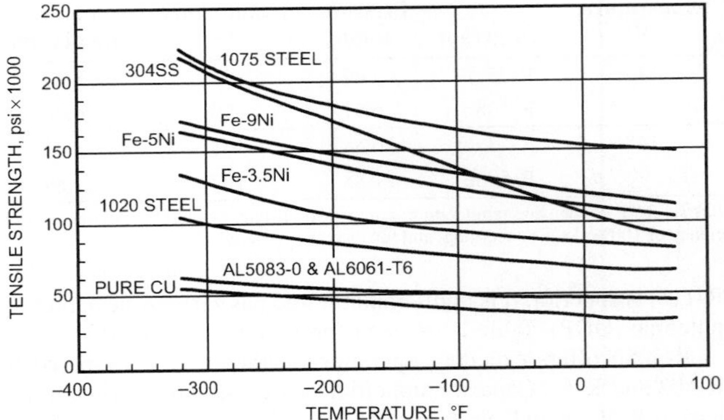

Fig. 7 Tensile Strength Versus Temperature of Several Metals

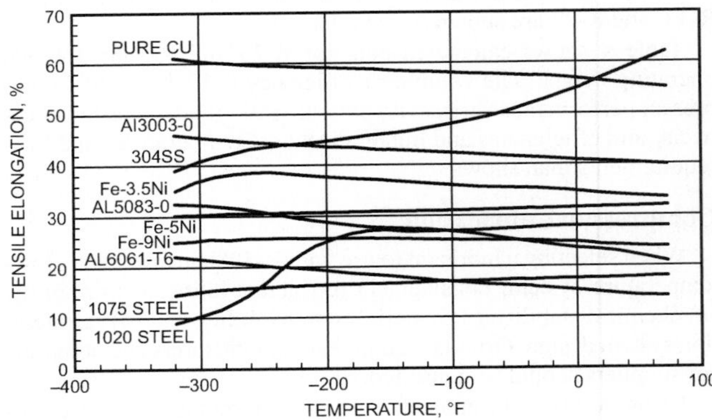

Fig. 8 Tensile Elongation Versus Temperature of Several Metals

Table 5 Several Mechanical Properties of Aluminum Alloys at –321°F

Aluminum Alloy	Modulus of Elasticity, psi × 10^6	Yield Strength, psi × 10^3	Ultimate Tensile Strength, psi × 10^6	Percent Elongation at Failure, %	Plane Strain Fracture Toughness, psi × $10^3 \cdot in^{1/2}$
1100-0	11.3	7.2	27.6	—	—
2219-T851	12.3	63.8	82.4	14	41
5083-0	11.6	22.9	62.9	32	56
6061-T651	11.2	48.9	58.3	23	38

remain ductile or become brittle at low temperatures. Zinc becomes brittle, whereas pure titanium (Ti) and many Ti alloys remain ductile.

Ductility values obtained from the **static tensile test** may give some clue to ductility loss, but the **notched-bar impact test** gives a better indication of how the material performs under dynamic loading and how it reacts to complex multidirectional stress. Figure 8 shows ductility, as measured by percent elongation in the tensile test, in relation to temperature for several metals. As temperature drops, the curves for copper and aluminum show an increase in ductility, while AISI 304 stainless steel and Ti-6%Al-4%V show a decrease.

Aluminum alloys are used extensively for low-temperature structural applications because of cost, weldability, and toughness. Although their strength is considered modest to intermediate, they remain ductile at lower temperatures. Typical mechanical properties at –321°F are listed in Table 5. Property values between ambient (72°F) and –321°F are intermediate between those at these temperatures.

Aluminum 1100 (relatively pure at 99% Al) has a low yield strength but is highly ductile and has a high thermal conductivity. It is used in nonstructural applications such as thermal radiation shields. For structural purposes, alloys 5083, 5086, 5454, and 5456 are often used. Alloys such as 5083 have a comparatively high strength when annealed (0) and can be readily welded with little loss of strength in the heat-affected zone; post-welding heat treatments are not necessary. These alloys are used in the storage and transportation areas. Alloy 3003 is widely used for plate-fin heat exchangers because it is easily brazed with an Al-7%Si filler metal. Aluminum-magnesium alloys (6000 series) are used as extrusions and forgings for such components as pipes, tubes, fittings, and valve bodies.

Copper alloys are rarely used for structural applications because of joining difficulties. Copper and its alloys behave similarly to aluminum alloys as temperature decreases. Strength is typically inversely proportional to impact resistance; high-strength alloys have low impact resistance. Silver soldering and vacuum brazing are the most successful methods for joining copper. Brass is useful for small components and is easily machined.

Nickel and **nickel alloys** do not exhibit a ductile-to-brittle transition as temperature decreases and can be welded successfully, but their high cost limits use. High-strength alloys can be used at very low temperatures.

Iron-based alloys that are body-centered cubic usually exhibit a ductile-to-brittle transition as the temperature decreases. The BCC phase of iron is ferromagnetic and easily identified because it is attracted to a magnet. Extreme brittleness is often observed at lower temperatures. Thus, BCC metals and alloys are not normally used for structural applications at lower temperatures. Notable exceptions are iron alloys with a high nickel content.

Nickel and manganese are added to iron to stabilize the austenitic phase (FCC), promoting low-temperature ductility. Depending on the amount of Ni or Mn added, a great deal of low-temperature toughness can be developed. Two notable high-nickel alloys for use below ambient temperature include 9% nickel steel and austenitic 36% Ni iron alloy. The 9% alloy retains good ductility down to 180°R (–280°F). Below 180°R, ductility decreases slightly, but a clear ductile-to-brittle transition does not occur. Iron containing 36% Ni has the unusual feature of nearly zero thermal contraction during cooling from room temperature to near absolute zero. It is therefore an attractive metal for subambient use where the thermal stress associated with differential thermal contraction is to be avoided. Unfortunately, this alloy is quite expensive and therefore sees limited use.

Lesser amounts of Ni can be added to Fe to lower cost and depress the ductile-to-brittle transition temperature. Iron with 5% Ni can be used down to 270°R (–190°F), and Fe with 3.5% Ni remains ductile to 307°R (–153°F). High-nickel steels are usually heat treated before use by water quenching from 1471°F, followed by tempering at 1076°F. The 1076°F heat treatment tempers martensite formed during quenching and produces 10 to 15% stable austenite, which is responsible for the improved toughness of the product.

The austenitic stainless steels (300 series) are widely used for low-temperature applications. Many retain high ductility down to 7°R (–452°F) and below. Their attractiveness is based on good strength, stiffness, toughness, and corrosion resistance, but cost is high compared to that of Fe-C alloys. A stress relief heat treatment is generally not required after welding, and impact strengths vary only slightly with decreasing temperature. A popular, readily available steel with moderate strength for low-temperature service is AISI type 304, with the low-carbon grade preferred. Where higher strengths are needed and welding can be avoided, strain-hardened or high-nitrogen grades are available. Castable austenitic steels are also available; a well-known example (14-17%Cr, 18-22%Ni, 1.75-2.75%Mo, 0.5%Si max, and 0.05%C max) retains excellent ductility and strength to extremely low temperatures.

Titanium alloys have high strength, low density, and poor thermal conductivity. Two alloys often used at low temperatures are Ti-5%Al-2.5%Sn and Ti-6%Al-4%V. The Ti-6-4 alloy has the higher yield strength, but loses ductility below about 144°R (–315°F). The low-temperature properties are dramatically affected by oxygen, carbon, and nitrogen content. Higher levels of these interstitial elements increase strength but decrease ductility. Extra-low interstitial (ELI) grades containing about half the normal levels are usually specified for low-temperature applications. Both Ti-6-4 and Ti-5-2.5 are easily welded but expensive and difficult to form. They are used where a high strength-to-weight or strength-to-thermal conductivity ratio is attractive. Titanium alloys are not recommended for applications where an oxidation hazard exists.

Thermoplastic Polymers

Reducing the temperature of thermoplastic polymers restricts molecular motion (bond rotations and molecules sliding past one another), so that the material becomes less deformable. Behavior normally changes rapidly over a narrow temperature range, beginning at the material's glass transition temperature T_g.

Figure 9 shows the general mechanical response of linear amorphous thermoplastics to temperature. At or above the melting temperature T_m, bonding between polymer chains is weak, the material flows easily, and the modulus of elasticity is nearly zero. Just below T_m, the polymer becomes rubbery; with applied stress, the material deforms by elastic and plastic strain. The combination of these deformations is related to the applied stress by the shear modulus. At still lower temperatures, the polymer becomes stiffer, exhibiting "leathery" behavior and a higher stress at failure. Many commercially available polymers (e.g., polyethylene) are used in this condition. T_g is at the transition between the leathery and

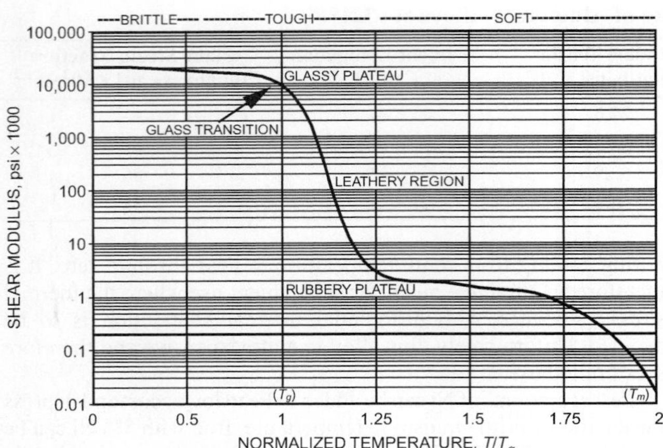

Fig. 9 Shear Modulus Versus Normalized Temperature (T/T_g) for Thermoplastic Polymers

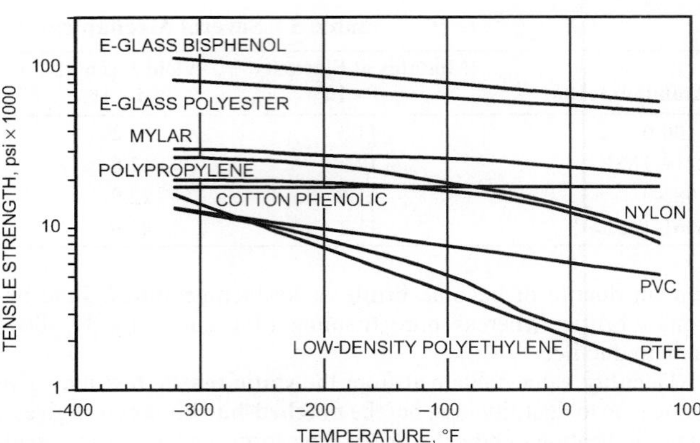

Fig. 10 Tensile Strength Versus Temperature of Plastics and Polymer Matrix Laminates

Table 6 Approximate Melting and Glass Transition Temperatures for Common Polymers

Polymer	Melting Temperature T_m °R	Melting Temperature T_m °F	Glass Transition Temperature T_g °R	Glass Transition Temperature T_g °F
Addition polymers				
Low-density polyethylene	700	240	275	–185
High-density polyethylene	745	285	275	–185
Polyvinyl chloride	—	—	655	195
Polypropylene	800	340	465	5
Polystyrene	—	—	680	220
Polytetrafluoroethylene	1085	625	—	—
Polymethyl methacrylate (acrylic)	—	—	665	205
Condensation polymers				
6-6 Nylon	970	510	580	120
Polycarbonate	—	—	755	295
Polyester	950	490	630	170
Elastomers				
Silicone	—	—	265	–195
Polybutadiene	710	250	330	–130
Polychloroprene	635	175	400	–60
Polyisoprene	545	85	365	–95

Source: Askeland (1994). Derived from Table 15-2, p. 482.

glassy regions, and is usually 0.5 to 0.75 times the absolute melting temperature T_m. Table 6 lists T_g and T_m values for common polymers. In the glassy state, at temperatures below T_g, the polymer is hard, brittle, and glass-like. Although polymers in the glassy region have poor ductility and formability, they are strong, stiff, and creep-resistant.

For thermoplastic polymers, the temperature at which stress is applied and the rate of stress application are interdependent based on time/temperature superposition. This relationship allows different types of tests, such as creep or stress relaxation, to be related through a single curve that describes the viscoelastic response of the material to time and temperature. Applying stress more rapidly has an effect equivalent to applying stress at lower temperatures. Figure 10 shows tensile strength versus temperature for plastic and polymer composites.

Thermoplastic polymers such as polyethylene and polyvinyl chloride (PVC) may be used for plastic films and wire insulation but are not generally suitable for structural applications because of their brittleness at temperatures below T_g. The only known polymer that exhibits appreciable ductility at temperatures substantially below T_g is polytetrafluoroethylene (PTFE). Because of their large thermal contraction coefficients, thermoplastic polymers should not be restrained during cooldown. Large masses should be cooled slowly to ensure uniform thermal contraction; the coefficient of thermal contraction decreases with temperature. Contractions of 1 to 2% in cooling from ambient to –321°F are common. For instance, nylon, PTFE, and polyethylene contract 1.3, 1.9, and 2.3%, respectively. These values are large compared to those for metals, which contract 0.2 to 0.3% over the same temperature range.

Thermosetting Plastics

Thermosetting plastics such as epoxy are relatively unaffected by changes in temperature. As a class, they are brittle and generally used in compression and not tension. Care must be taken in changing their temperature to avoid thermally induced stress, which could lead to cracking. Adding particulate fillers such as silica (SiO_2) to thermosetting resins can increase elastic modulus and decrease strength. The main reasons for adding fillers are to reduce the coefficient of thermal expansion and to improve thermal conductivity. Filled thermosetting resins such as epoxy and polyester can be made to have coefficients of thermal expansion that closely match those of metal; they may be used as insulation and spacers but are not generally used for load-bearing structural applications.

Fiber Composites

Nonmetallic filamentary reinforced composites have gained wide acceptance for low-temperature structural applications because they have good strength, low density, and low thermal conductivity. Nonmetallic insulating composites are usually formed by laminating together layers of fibrous materials in a liquid thermosetting resin such as polyester or epoxy. The fibers are often but not necessarily continuous; they can take the form of bundles, mats, yarns, or woven fabrics. The most frequently used fiber materials include glass, aramid, and carbon. Reinforcing fibers add considerable mechanical strength to otherwise brittle matrix material and can lower the thermal expansion coefficient to a value comparable to that of metals. High figures of merit (ratio of thermal conductivity to elastic modulus or strength) can reduce refrigeration costs substantially from those obtained with fully metallic configurations. Nonmetallic composites can be used for tanks, tubes, struts, straps, and overlays in low-temperature refrigeration systems. These materials perform well in high-loading environments and under cyclic stress; they do not degrade chemically at low temperatures.

Different combinations of fiber materials, matrices, loading fractions, and orientations yield a range of properties. Material properties are often anisotropic, with maximum properties in the fiber direction. Composites fail because of cracking in the matrix

Table 7 Tensile Properties of Unidirectional Fiber-Reinforced Composites

Composite	Test Temperature, °F	Tensile Strength, psi × 10^3	Tensile Modulus, psi × 10^6
E glass (50%)			
Longitudinal	72	152.2	5.94
	–321	194.3	6.52
Transverse	72	1.30	1.59
	–321	1.16	1.74
Aramid fibers (63%)			
Longitudinal	77	163.8	10.29
	–321	166.7	14.35
Transverse	77	0.609	0.362
	–321	0.522	0.522

Source: Hands (1986). Table 11.3.

layer perpendicular to the direction of stress. Cracking may propagate along the fibers but does not generally lead to debonding. Maximum elongations at failure for glass-reinforced composites are usually 2 to 5%; the material is generally elastic all the way to failure.

A major advantage of using glass fibers with a thermosetting binder matrix is the ability to match thermal contraction of the composite to that of most metals. Aramid fibers produce laminates with lower density but higher cost. With carbon fibers, it is possible to produce components that show virtually zero contraction on cooling.

Typical tensile mechanical property data for glass-reinforced laminates are given in Table 7. Under compressive loading, strength and modulus values are generally 60 to 70% of those for tensile loading because of matrix shrinkage away from fibers and microbuckling of fibers.

Adhesives

Adhesives for bonding composite materials to themselves or to other materials include epoxy resins, polyurethanes, polyimides, and polyheterocyclic resins. Epoxy resins, modified epoxy resins (with nylon or polyamide), and polyurethanes apparently give the best overall low-temperature performance. The joint must be properly designed to account for the different thermal contractions of the components. It is best to have adhesives operate under compressive loads. Before bonding, surfaces to be joined should be free of contamination, have uniform fine-scale roughness, and preferably be chemically cleaned and etched. An even bond gap thickness of 0.004 to 0.007 in. is usually best.

INSULATION

Refrigerated pipe insulation, by necessity, has become an engineered element of the refrigeration system. The complexity and cost of this element now rival that of the piping system, particularly for ultralow-temperature systems.

Some factory-assembled, close-coupled systems that operate intermittently can function with a relatively simple installation of flexible sponge/foam rubber pipe insulation. Larger systems that operate continuously require much more investment in design and installation. Higher-technology materials and techniques, which are sometimes waived (at risk of invested capital) for systems operating at warmer temperatures, are critical for low-temperature operation. Also, the nature of the application does not usually allow shutdown for repair.

Pipe insulation systems are distinctly different from cold-room construction. Cold-room construction vapor leaks can be neutral if they reach equilibrium with the dehumidification effect of the refrigeration unit. Moisture entering the pipe insulation can only accumulate and form ice, destroying the insulation system. At these low temperatures, it is proper to have redundant vapor retarders

Table 8 Components of a Low-Temperature Refrigerated Pipe Insulation System

Insulation System Component	Primary Roles	Secondary Roles	Typical Materials
Insulation	Efficiently insulate pipe Provide external hanger support	Limit water movement toward pipe Reduce rate of moisture/vapor transfer toward pipe Protect vapor retarder from external damage	Polyurethane-modified polyisocyanurate foams Extruded polystyrene foams Cellular glass
Elastomeric joint sealant	Limit liquid water movement through insulation cracks Reduce rate of moisture/vapor transfer toward pipe		Synthetic rubbers Resins
Vapor retarder	Severely limit moisture transfer toward pipe Eliminate liquid water movement toward pipe		Mastic/fabric/mastic Laminated membranes and very-low-permeance plastic films
Protective jacket	Protect vapor retarder from external damage Limit water movement toward pipe	Reduce moisture/vapor transfer toward pipe	Aluminum Stainless steel PVC
Protective jacket joint sealant	Prevent liquid water movement through gaps in protective jacket	Limit rate of moisture/vapor transfer toward pipe	
Vapor stops	Isolate damage caused by moisture penetration		Mastic/fabric/mastic

(e.g., reinforced mastic plus membrane plus sealed jacket). Insulation should be multilayer to allow expansion and contraction, with inner plies allowed to slide and the outer ply joint sealed. Sealants are placed in the warmest location because they may not function properly at the lower temperature of inner plies. Insulation should be thick enough to prevent condensation (above dew point) at the outside surface.

The main components of a low-temperature refrigerated pipe insulation system are shown in Table 8. See Chapter 33 for more information on insulation systems for refrigerant piping.

HEAT TRANSFER

The heat transfer coefficients of boiling and condensing refrigerant and the convection heat transfer coefficients of secondary coolants are the most critical heat transfer issues in low-temperature refrigeration. In a cascade system, for example, the heat transfer coefficients in the high-temperature circuit are typical of other refrigeration applications at those temperatures. In the low-temperature circuit, however, the lower temperatures appreciably alter the refrigerant properties and therefore the boiling and condensing coefficients.

The expected changes in properties with a decrease in temperature are as follows. As temperature drops,

- Density of liquid increases
- Specific volume of vapor increases
- Enthalpy of evaporation increases
- Specific heat of liquid decreases
- Specific heat of vapor decreases
- Viscosity of liquid increases
- Viscosity of vapor decreases
- Thermal conductivity of liquid increases
- Thermal conductivity of vapor decreases

In general, increases in liquid density, enthalpy of evaporation, specific heats of liquid and vapor, and thermal conductivity of liquid and vapor cause an increase in the boiling and condensing heat transfer coefficients. Increases in specific volume of vapor and viscosities of liquid and vapor decrease these heat transfer coefficients.

Data from laboratory tests or even field observations are scarce for low-temperature heat transfer coefficients. However, heat transfer principles indicate that, in most cases, lowering the temperature level at which heat transfer occurs reduces the coefficient. The low-temperature circuit in a custom-engineered cascade system encounters lower-temperature boiling and condensation than are typical of industrial refrigeration. In some installations, refrigerant boiling is within the tubes; in others, it is outside the tubes. Similarly, the designer must decide whether condensation at the cascade condenser occurs inside or outside the tubes.

Some relative values based on correlations in Chapter 4 of the 2005 *ASHRAE Handbook—Fundamentals* may help the designer determine which situations call for conservative sizing of heat exchangers. The values in the following subsections are based on changes in properties of R-22 because data for this refrigerant are available down to very low temperatures. Other halocarbon refrigerants used in the low-temperature circuit of the cascade system are likely to behave similarly. Predictions are complicated by the fact that, in a process inside tubes, the coefficient changes constantly as the refrigerant passes through the circuit. For both boiling and condensing, temperature has a more moderate effect when the process occurs outside the tubes than when it occurs inside the tubes.

A critical factor in the correlations for boiling or condensing inside the tubes is the mass velocity G in lb/h·ft^2. The relative values given in the following subsections are based on keeping G in the tubes constant. The result is that G drops significantly because the specific volume of vapor experiences the greatest relative change of all the properties. As the vapor becomes less dense, the linear velocity can be increased and still maintain a tolerable pressure drop of the refrigerant through the tubes. So G would not drop to the extent used in the comparison below, and the reductions shown for tube-side boiling and condensing would not be as severe as shown.

Table 9 Overview of Some Secondary Coolants

Coolant	Flash Point, °F	Freezing Point, °F	Boiling Point, °F	Temperature at Which Viscosity > 10 cSt
Polydimethyl-siloxane	116	−168	347	−76
d-Limonene	115	−142	310	−112
Diethylbenzene*	136.4	≤119.2	357.8	−112
	134.6	−103	357.8	−94
Hydrofluoroether	not flamm.	−202	140	−22
Ethanol	53.6	−178.6	172.4	−76
Methanol	51.8	−144.4	147.2	−130

*Two proprietary versions containing different additives.

Table 10 Refrigerant Properties of Some Low-Temperature Secondary Coolants

Temperature, °F	Viscosity, lb$_m$/ft·s	Density, lb/ft^3	Heat Capacity, Btu/lb·°F	Thermal Conductivity, Btu·ft/h·ft^2·°F
Polydimethylsiloxane[a]				
−148	0.0528	61.12	0.363	0.0774
−130	0.0226	60.50	0.368	0.0764
−112	0.0135	59.87	0.372	0.0754
−94	0.0089	59.25	0.377	0.0744
−76	0.0063	58.56	0.382	0.0733
−58	0.0043	57.93	0.913	0.0722
d-Limonene[b]				
−112	0.0012	58.10	—	0.0803
−94	0.0011	57.56	—	0.0791
−76	0.0010	57.01	—	0.0779
−58	0.0010	56.47	0.332	0.0768
Diethylbenzene[a, c]				
−130	Below Freezing Point			
−112	0.00672	58.35	0.375	0.0865
−94	0.00478	57.92	0.381	0.0852
−76	0.00344	57.50	0.386	0.0840
−58	0.00254	57.06	0.391	0.0829
Hydrofluoroether[d]				
−148	0.01430	113.37	0.223	0.0537
−130	0.00726	111.75	0.228	0.0526
−112	0.00431	110.12	0.233	0.0514
−94	0.00285	108.56	0.237	0.0503
−76	0.00203	106.93	0.242	0.0491
−58	0.00153	105.37	0.247	0.0479
Ethanol[e]				
−148	0.0316	44.81	0.450	0.1150
−130	0.0190	45.37	0.458	0.1144
−112	0.0122	45.94	0.464	0.1138
−94	0.0084	46.50	0.469	0.1127
−76	0.0058	47.06	0.474	0.1121
−58	0.0043	47.63	0.481	0.1109
Methanol[e]				
−148	0.0108	45.00	0.521	0.1294
−130	0.0059	45.56	0.527	0.1288
−112	0.0038	46.12	0.533	0.1282
−94	0.0270	46.68	0.538	0.1277
−76	0.0020	47.25	0.544	0.1271
−58	0.0152	47.81	0.550	0.1265
Acetone				
−137	Freezing Point			
−130	—	—	0.477	0.0867
−112	0.0008	—	0.488	0.0855
−94	0.0006	—	0.480	0.0843
−76	0.0005	—	0.483	0.0838
−58	0.0005	—	0.485	0.0826
68	—	49.43	—	—

Sources:
[a]Dow Corning USA (1993)
[b]Florida Chemical Co. (1994)
[c]Therminol LT (1992)
[d]3M Company (1996)
[e]Raznjevic (1997)

Condensation Outside Tubes. Based on Nusselt's film condensation theory, the condensing coefficient at –4°F, a temperature that could be encountered in a cascade condenser, would actually be 17% higher than the condensing coefficient in a typical condenser at 86°F because of higher latent heat, liquid density, and thermal conductivity. The penalizing influence of the increase in specific volume of vapor is not present because this term does not appear in the Nusselt equation.

Condensation Inside Tubes. Using the correlation of Ackers and Rosson (Table 3, Chapter 4 of the 2001 *ASHRAE Handbook—Fundamentals*) with a constant velocity and thus decreasing the value of G by one-fifth, the condensation coefficient at –4°F is one-fourth that at 86°F.

Boiling Inside Tubes. Using the correlation of Pierre [Equation (1) in Table 2, Chapter 4 of the 2001 *ASHRAE Handbook—Fundamentals*] and maintaining a constant velocity, when the temperature drops to –94°F, the boiling coefficient drops to 46% of the value at –4°F.

Boiling Outside Tubes. In a flooded evaporator with refrigerant boiling outside the tubes, the heat-transfer coefficient also drops as the temperature drops. Once again, the high specific volume of vapor is a major factor, restricting the ability of liquid to be in contact with the tube, which is essential for good boiling. Figure 4 in Chapter 4 (Perry 1950; Stephan 1963a, 1963b, 1963c) of the 2005 *ASHRAE Handbook—Fundamentals* shows that the heat flux has a dominant influence on the coefficient. For the range of temperatures presented for R-22, the boiling coefficient drops by 12% as the boiling temperature drops from 5°F to –42°F.

SECONDARY COOLANTS

Secondary coolant selection, system design considerations, and applications are discussed in Chapter 4; properties of brines, inhibited glycols, halocarbons, and nonaqueous fluids are given in Chapter 21 of the 2005 *ASHRAE Handbook—Fundamentals*. The focus here is on secondary coolants for low-temperature applications in the range of –58 to –148°F.

An ideal secondary coolant should

- Have favorable thermophysical properties (high specific heat, low viscosity, high density, and high thermal conductivity)
- Be nonflammable, nontoxic, environmentally acceptable, stable, noncorrosive, and compatible with most engineering materials
- Possess a low vapor pressure

Only a few fluids meet these criteria, especially in the entire –58 to –148°F range. Some of these fluids are hydrofluoroether (HFE), diethylbenzene, d-limonene, polydimethylsiloxane, trichloroethylene, and methylene chloride. Table 9 provides an overview of these coolants. Table 10 gives refrigerant properties for the coolants at various low temperatures.

Polydimethylsiloxane, known as silicone oil, is environmentally friendly, nontoxic, and combustible and can operate in the whole range. Because of its high viscosity (greater than 0.00672 $lb_m/ft \cdot s$), its flow pattern is laminar at lower temperatures, which limits heat transfer.

d-Limonene is optically active terpene ($C_{10}H_{16}$) extracted from orange and lemon oils. This fluid can be corrosive and is not recommended for contact with some important materials (polyethylene, polypropylene, natural rubber, neoprene, nitrile, silicone, and PVC). Some problems with stability, such as increased viscosity with time, are also reported. Contact with oxidizing agents should be avoided. The values listed are based on data provided by the manufacturer in a limited temperature range. d-Limonene is a combustible liquid with a flash point of 115°F.

The synthetic aromatic heat transfer fluid group includes **diethylbenzene**. Different proprietary versions of this coolant contain different additives. In these fluids, the viscosity is not as strong a function of temperature. Freezing takes place by crystallization, similar to water.

Hydrofluoroether (1-methoxy-nonafluorobutane, $C_4F_9CH_3$), is a new fluid, so there is limited experience with its use. It is nonflammable, nontoxic, and appropriate for the whole temperature range. No ozone depletion is associated with its use, but its global warming potential is 500 and its atmospheric lifetime is 4.1 years.

The **alcohols (methanol and ethanol)** have suitable low-temperature physical properties, but they are flammable and methanol is toxic, so their application is limited to industrial situations where these characteristics can be accommodated.

Another possibility for a secondary coolant is **acetone** (C_3H_6O).

REFERENCES

Askeland, D.R. 1994. *The science and engineering of materials*, 3rd ed. PWS Publishing, Boston.

Dow Corning USA. 1993. *Syltherm heat transfer fluids*. Dow Corning Corporation, Midland, MI.

Emhö, L.J. 1997. HC-recovery with low temperature refrigeration. Presented at ASHRAE Annual Meeting, Boston.

Enneking, J.C. and S. Priebe. 1993. Environmental application of Brayton cycle heat pump at Savannah River Project. Meeting Customer Needs with Heat Pumps, Conference/Equipment Show.

Florida Chemical Co. 1994. *d-Limonene product and material safety data sheets*. Winter Haven, FL.

Hands, B.A. 1986. *Cryogenic engineering*. Academic Press, New York.

Jain, N.K. and Enneking, J.C. 1995. Optimization and operating experience of an inert gas solvent recovery system. Air and Waste Management Association Annual Meeting and Exhibition, San Antonio, June 18-23.

Perry, J.H. 1950. *Chemical engineers handbook*, 3rd ed. McGraw-Hill, New York.

Raznjevic, K. 1997. *Heat transfer*. McGraw-Hill, New York.

Stephan, K. 1963a. The computation of heat transfer to boiling refrigerants. *Kältetechnik* 15:231.

Stephan, K. 1963b. Influence of oil on heat transfer of boiling Freon-12 and Freon-22. Eleventh International Congress of Refrigeration, I.I.R. *Bulletin* No. 3.

Stephan, K. 1963c. A mechanism and picture of the processes involved in heat transfer during bubble evaporation. *Chemic. Ingenieur Technik* 35:775.

Stoecker, W.F. and J.W. Jones. 1982. *Refrigeration and air conditioning*, 2nd ed. McGraw-Hill, New York.

Therminol LT. 1992. *Technical Bulletin No. 9175*. Monsanto, St. Louis.

3M Company. 1996. Performance Chemicals and Fluids Laboratory, St. Paul, MN.

Weng, C. 1995. *Non-CFC autocascade refrigeration system*. U.S. Patent 5,408,848 (April).

BIBLIOGRAPHY

Wark, K. 1982. *Thermodynamics*, 4th ed. McGraw-Hill, New York.

Weng, C. 1990. *Experimental study of evaporative heat transfer for a non-azeotropic refrigerant blend at low temperature*. M.A. thesis, Ohio University.

CHAPTER 40

BIOMEDICAL APPLICATIONS OF CRYOGENIC REFRIGERATION

THE controlled exposure of biological materials to subfreezing states has multiple practical applications, which have been rapidly multiplying in recent times. Primary among these applications are long-term preservation of cells and tissues, the selective surgical destruction of tissue by freezing, the preparation of aqueous specimens for electron microscopy imaging, and the study of biochemical mechanisms used by a multitude of living species to withstand the rigors of extreme environmental cold. Some of the applications are restricted to the research laboratory, but clinical and commercial environments are increasingly frequent venues for activities in low-temperature biology. The success of much of this work depends on the design and availability of an apparatus that can control temperatures and thermal histories. This apparatus can be adapted and programmed to meet the specific needs of particular applications.

This chapter briefly describes many of the principles driving the present growth and development of low-temperature biological applications. An understanding of these principles is required to optimize design of practical apparatus for low-temperature biological processes. Although this field is growing in both breadth and sophistication, this chapter is restricted to processes that involve temperatures below which ice formation is normally encountered (i.e., 32°F), and to an overview of the state of the art.

PRESERVATION APPLICATIONS

Principles of Biological Preservation

Successful cryopreservation of living cells and tissues is coupled to control of the thermal history during exposure to subfreezing temperatures. The objective of cryopreservation is to reduce the specimen's temperature to such an extent that the rates of chemical reactions that control processes of degeneration become very small, creating a state of effective suspended animation. An Arrhenius analysis (Benson 1982) shows that temperatures must be maintained well below freezing to reduce reaction kinetics enough to store specimens injury-free for an acceptable time (usually measured in years). Consequently, one of two types of processes is typically encountered: either the specimen freezes or it undergoes a transition to a glassy state (**vitrification**). Although both of these phenomena may lead to irreversible injury, most of the destructive consequences of cryopreservation can be avoided.

A change in chemical composition occurs with freezing as water segregates in the solid ice phase, leaving a residual solution that is rich in electrolytes. This process occurs progressively as the solidification process proceeds through a temperature range that defines a "mushy zone" between the ice nucleation and eutectic states (Körber 1988). If this process follows a series of equilibrium states, the liquidus line on the solid/liquid phase diagram for a system of the chemical composition of the specimen defines the relationship between the system temperature and the solute concentration. The fraction of total water that is solidified increases as the temperature is reduced, according to the function defined by applying the lever rule to the phase diagram liquidus line for the initial composition of the specimen (Prince 1966). This relationship has been worked out for a simple binary model system of water and sodium chloride and has been used to calculate the thermal history of a specimen of defined geometry during cryopreservation (Diller et al. 1985). As explained later, the osmotic stress on the cells with a concurrent efflux of intracellular water results from chemical changes. The critical range of states over which this process occurs corresponds closely to the temperature extremes defined by the mushy zone. At higher temperatures there is no phase change, so osmotic stress does not exist. At lower temperatures, the permeability of the cell plasma membrane is reduced significantly (as described via an Arrhenius function), and the membrane transport impedance is so high that no significant efflux can occur. Thus, the specimen's chemical history and osmotic response are coupled to its thermal history as defined by the phase diagram properties.

The property of a cell that dictates response to freezing is the permeability of the plasma membrane to water and permeable solutes. The permeability determines the mass exchange between a cell and its environment when osmotic stress develops during cryopreservation. The magnitude of permeability decreases exponentially with absolute temperature. Thus, resistance to the movement of chemical species in and out of the cell becomes much larger as the temperature is reduced during freezing. Because the osmotic driving force also increases as temperature decreases, in general, the balance between the osmotic force and resistance determines the extent of mass transfer that occurs during freezing. At high subfreezing temperatures (generally defined by the mushy zone), the osmotic force dominates and extensive transport occurs. At low subfreezing temperatures, the resistance dominates and the chemical species are immobilized either inside or outside the cells. The amount of mass exchanged across the membrane is a direct function of the amount of time spent in states for which the osmotic force dominates the resistance. Thus, at slow cooling rates, the cells of a sample dehydrate extensively, and at rapid cooling rates, very little net transport occurs. The absolute magnitude of the cooling rate that defines the slow and rapid regimes for a specific cell depends on the plasma membrane permeability. A cell with high permeability requires a rapid cooling rate to prevent extreme transport. The converse holds for cells with low membrane permeability: they require prolonged high-temperature exposure to effect significant accumulated transport.

When very little transport occurs before low temperatures are reached, water becomes trapped within the cell in a subcooled state. Chemical equilibration is achieved with extracellular ice by the intracellular nucleation of ice. This phenomenon is referred to as **intracellular freezing**. In this process, a substantial degree of liquid subcooling occurs before nucleation, so the resulting ice structure is dominated by numerous, very small crystals. Further, at low temperatures, the extent of subsequent recrystallization is minimal and the intracellular solid-state surface energy is high.

The preparation of this chapter is assigned to TC 10.4, Ultralow-Temperature Systems and Cryogenics.

At slow cooling rates and at high subfreezing temperatures, both extensive dehydration of cells and an extended period of exposure to concentrated electrolyte solutions occur. There is clear evidence that some combination of dehydration and exposure to concentrated solutes leads to irreversible injury (Mazur 1970; Meryman et al. 1977). Recently, Han and Bischof (2004a) showed that eutectic solidification during freezing can also contribute to cellular injury. Mazur (1977) also demonstrated that freezing at cooling rates that are rapid enough to cause intracellular ice formation causes a second mechanism of irreversible cell injury. These processes are illustrated in Figure 1, which shows that each extreme of the cooling process during freezing produces a potential for damaging cells. Figure 1 also implies that an intermediate cooling rate should minimize the aggregate effects of these injury processes and define the conditions at which optimum recovery from cryopreservation can be achieved.

Experimental data have been obtained for the survival of a large number of cell types for freezing and thawing as a function of the cooling rate. Nearly without exception, the survival function follows an inverted V profile when plotted against cooling rate (Figure 2). This plot has been described as the **survival signature** of a cell; it illustrates the tradeoff between competing heat and mass transfer processes that govern the cryopreservation process. Solution concentration/osmotic effects lead to slow cooling rate injury. In this state, there is adequate time for transport of water out of the cell before sufficient heat transport occurs to lower the temperature enough to drive the membrane permeability to nearly zero. Conversely, at rapid cooling rates, the cell temperature is lowered so quickly that there is insufficient time for dehydration, and injury is caused by formation of intracellular ice. The magnitude of the optimum intermediate cooling rate is a function of the magnitude of the membrane transport permeability. Higher permeabilities result in higher optimum cooling rates. Thus, the optimum thermal history for any cell type must be tailored for its unique constitutive properties.

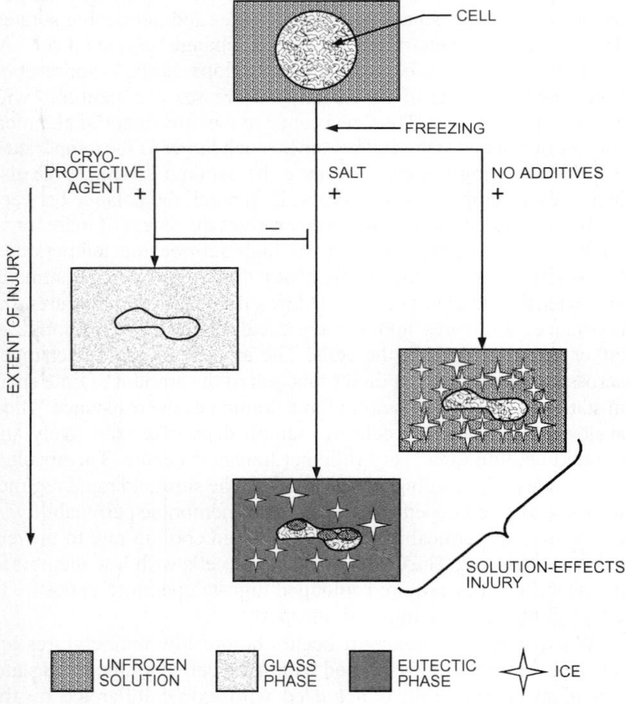

At rapid cooling rates, water is trapped in the cell and ice crystals form intracellularly. At slow cooling rates, solution effects, including dehydration, high electrolyte concentration, and eutectic solidification, become dominant injury mechanisms.

Fig. 1 Schematic of Response of Single Cell During Freezing as Function of Cooling Rate

For most cell types, the bandwidth of cooling rates for optimum cryopreservation survival is small, and the highest achievable survival is unacceptably low. Fortunately, for practical clinical applications, the spectrum of working cooling rates can be broadened and the maximum survival increased by adding a **cryoprotective agent (CPA)** to the sample before freezing. Although a wide range of chemicals exhibit cryoprotective properties, as summarized in Table 1, the most commonly used include glycerol, dimethyl sulfoxide (DMSO), and polyethylene glycol. Numerous theories have been postulated to explain the action of CPAs. In simplest terms, they modify the processes of solute concentration and/or intracellular freezing (e.g., Lovelock 1954; Mazur 1970). Introducing CPAs to cell systems results in a major modification of the phase diagram for the system (Fahy 1980). In particular, the rate of electrolyte concentration with decreasing temperature may be reduced by nearly ten times, and the eutectic state depressed by as much as 108 to 144°F. These consequences greatly extend the regime of the mushy zone during solidification (Cocks et al. 1975; Jochem and Körber 1987).

Although phase diagrams provide much information for understanding the possible states that may occur during cryopreservation of living tissues, their interpretation is limited by two major factors. First, the chemical complexity of living systems is far greater than the simple binary, ternary, or quaternary mixtures that are used to model their behavior. Second, and more importantly, the thermal data used to generate phase diagrams are usually obtained for near-equilibrium conditions. In contrast, most cryopreservation is executed under conditions far from the equilibrium state. Han and Bischof (2004b) describe the significance of nonequilibrium phase change in the presence of CPAs. For some situations, the goal is to maintain a state of disequilibrium; this includes vitrification methods

Table 1 Summary of Cryoprotective Agents (CPAs)

Category	CPAs	Comments
Permeable	Glycerol, ethylene glycol, DMSO	Low molecular mass Osmotically transportable across cellular membrane Shrink/swell cellular response
Impermeable	Sugar group: sucrose, raffinose, trehalose	High molecular mass
	Polymer group: polyvinyl pyrrolidone (PVP), polyethylene glycol (PEG), hydryoxyethyl starch (HES)	Osmotically untransportable across cellular membrane Shrink-only cellular response

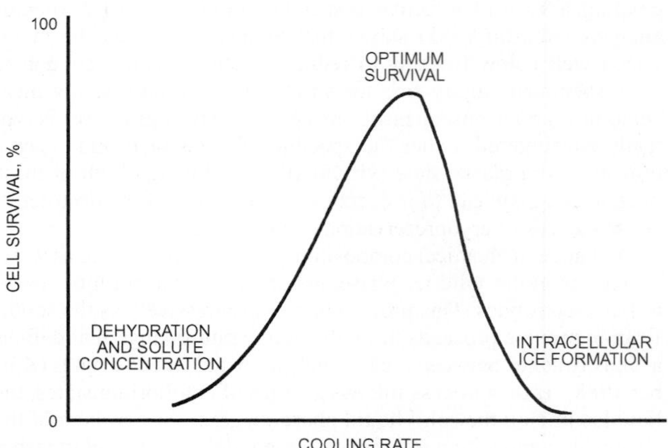

Fig. 2 Generic Survival Signature Indicating Independent Injury Mechanisms Associated with Extremes of Slow and Rapid Cooling Rates During Cell Freezing

Table 2 Spectrum of Various Types of Living Cells and Tissues Commonly Stored by Freezing (as of 1993)

Tissue	Comments	References
Blood vessels	DMSO used for CPA; cooling rate < 2 °F/min.	Gottlob et al. (1982)
Bone marrow stem cells	DMSO is usual CPA; widely used in cancer therapy.	McGann et al. (1981)
Cornea	DMSO is usual CPA.	Armitage (1991)
Erythrocytes	Usual CPA is glycerol; high concentrations for slow cooling and low concentrations for rapid cooling; wide spread clinical use.	Turner (1970), Valeri (1976), AABB (1985), Huggins (1985)
Embryos		
Mouse	Many species of mammalian embryos have been cryopreserved successfully. Common CPAs for these applications are glycerol and DMSO. 1,2-Propanediol is used for humans. Variations in the required thermal protocol and processing steps exist among different species. Ice crystal nucleation is usually controlled by seeding.	CIBA (1977), Zeilmaker (1981)
Rat		Whittingham et al. (1972), Wilmut (1972)
Goat		Whittingham (1975)
Sheep		Bilton and Moore (1976)
Rabbit		Willadsen et al. (1976)
Bovine		Bank and Maurer (1974)
Drosophila		Wilmut and Rowson (1973)
Human		Mazur et al. (1992)
		Troundson and Mohr (1983)
Heart valves	DMSO is usual CPA; cooling rate of 2 to 3.5 °F/min.	Angell et al. (1987)
Hepatocytes	DMSO is usual CPA; cooling rate of 4.5 °F/min.	Fuller and Woods (1987)
Islets	DMSO used for CPA; cooling rate < 2 °F/min.	Rajotte et al. (1983), Taylor and Benton (1987)
Lymphocytes	DMSO is usual CPA; primary application is in clinical testing.	Knight (1980), Scheiwe et al. (1981)
Microorganisms	DMSO and glycerol used for CPAs.	James (1987)
Oocytes		
Hamster	Many species of mammalian oocytes have been cryopreserved successfully. The most common CPAs are glycerol and DMSO. Variations in the required thermal protocol and processing steps exist among the different species. Clinical applications in humans have been difficult to achieve.	Bernard (1991)
Mouse		Critser et al. (1986)
Primate		Whittingham (1977)
Rabbit		DeMayo et al. (1985)
Rat		Diedrick et al. (1986)
Human		Kasai et al. (1979)
		Van Uem et al. (1987)
Parathyroid	DMSO used for CPA; cooling rate of 2 °F/min.	Wells et al. (1977)
Periosteum	DMSO used for CPA; cooling rate of 2 °F/min.	Kreder et al. (1993)
Plants	Selected plants are cold hardy; some germplasm is cryopreserved.	Grout (1987), Withers (1987)
Platelets	Best success is with DMSO as CPA; high sensitivity to freezing and osmotic injury.	Schiffer et al. (1985) Sputtek and Körber (1991)
Skin	Both glycerol and DMSO used as CPA.	Aggarwal et al. (1985)
Sperm		
Animal	First mammalian cells frozen successfully. Broad applications for animals and humans using glycerol as CPA.	Polge (1980)
Human		Sherman (1973)

that are applied to reach a solid glassy state that avoids ice crystal formation, latent heat effects, and solute concentration effects. In many cases, the degree of thermodynamic equilibrium reached for the low-temperature storage state may differ significantly between the intracellular and extracellular volumes (Mazur 1990). The equilibration can be controlled by manipulating the thermal boundary conditions of the cryopreservation protocol and by altering the system's chemical composition prior to initiating cooling. Many of the same chemicals used for cryoprotection may be added at higher concentrations to decrease the probability of ice crystal formation at subzero temperatures and elicit vitrification (Fahy 1988).

In addition to the thermal history of the interior of a specimen, the thermodynamic relations determining the release of the latent heat of fusion as a function of temperature in the mushy zone (Hayes et al. 1988) must be considered. A specimen of finite dimension has a distribution of thermal histories within it during the freezing process (Meryman 1966). The pattern assumed for modeling the evolution of latent heat during freezing has a large effect on the cooling rates predicted as a function of local position in a specimen. Consequently, the anticipated spatial distribution of cell survival as a consequence of the preservation process may depend strongly on the model chosen for the thermodynamic coupling between the system's thermal and osmotic properties. Hartman et al. (1991) applied this principle to evaluate how to choose the optimum location for a thermal sensor to record the most representative thermal history during the freezing of a specimen of finite dimensions. Hartman et al.'s analysis indicated that the geometric center of a sample is a poor selection for positioning the sensor. A position approximately one-third of the distance from the center to the periphery more accurately represents the integrated thermal history experienced by the mass during freezing.

Preservation of Biological Materials by Freezing

Biological materials are primarily cryopreserved by freezing them to deep below freezing temperatures. Among clinical and commercial tissue banks, freezing is the predominant method for preservation. Following the discovery of the cryoprotective properties of glycerol (Polge et al. 1949) and other CPAs, procedures for cryopreservation were developed for storing a variety of cells and tissues. Table 2 summarizes representative research efforts to preserve various cells and tissues by freezing.

A typical protocol for cryopreservation consists of the following steps:

1. Place specimen in an appropriate container.
2. Add CPA by sequential increments at reduced temperatures.

3. Cool to below 0°F.
4. Possibly induce extracellular ice nucleation followed by a controlled period of thermal and osmotic equilibration.
5. Cool through high subfreezing temperatures with the greatest degree of thermal control invoked for the entire process, then quench to storage temperature, usually in liquid nitrogen.
6. Store for extended periods.
7. Warm relatively rapidly by immersion in a heated water bath.
8. Serially dilute to remove CPA using nonpenetrating solutes to control the intracellular/extracellular osmotic balance.
9. Harvest the specimen for its intended application.

Details vary among individual tissue types; the references in Table 2 give sources of specific parameter values for individual tissues. Refrigeration requirements may vary considerably among different tissues, but basic principles and processes of the cryopreservation processes are generally consistent. The Bibliography, and the references in Table 2, identify appropriate introductory references, and Han and Bischof (2004c) reviewed engineering challenges in cryopreservation.

Most initial applications of cryobiology were in clinical, research, and nonprofit (e.g., the Red Cross) venues. More recently, however, the commercial sector has adopted cryopreservation methods (McNally and McCaa 1988). As the arsenal of practical cryopreservation methods has grown, the profit potential of freezing tissues for prolonged storage is being recognized and exploited. Thus, an added set of incentives and motives is driving the development of techniques that make use of challenging refrigeration schemes.

Preservation of Biological Materials by Freeze Drying

Freeze drying extends long-term storage at ambient temperatures without the threat of product deterioration. This process removes water from the specimen by sublimation while it is frozen. As a result, no thawing occurs during rewarming. Thus, none of the decay processes associated with the presence of water in the liquid state are active. Freeze drying has been applied widely in the food processing industry, where the product need not be rehydrated in the living state. Other applications, such as taxidermy, also avoid this stringent requirement. The list of biological materials frequently processed by freeze drying is extensive and encompasses various microorganisms, protein solutions, pharmaceuticals, and bone. Rowe (1970) reviewed the early state of the art of the physical and engineering aspects of freeze drying, and Franks (1990) reviewed the physical and chemical principles that govern the freeze-drying process from the perspective of achieving an optimal process design.

Figure 3 summarizes the processing steps for freeze drying (Franks 1990). In the figure, note the alternative process pathways from the native to the stored state. The path can be controlled by equipment design and operator intervention. The material initially is in the native state from which cooling is initiated. As subfreezing temperatures are reached, either the material remains subcooled in the liquid state or ice crystals form (either by spontaneous nucleation or by active seeding a substrate on which a solid phase may form). The material will vitrify if sufficiently subcooled. Ice crystals will grow in a nucleated material, with the rate of temperature change determining structure and size distribution. Simultaneously, the solute becomes concentrated until the eutectic state is reached. At this state, an additional solid phase may form, or the liquid solution may become supersaturated as the temperature reduces further.

The material is dried, in either the crystalline or vitreous state, by drawing a vacuum on the system at low temperature. Finally, the dried material may be stored at ambient temperatures, although the material is often stored at high subfreezing temperatures to minimize the probability of product deterioration by the activity of residual water. Production methods have been developed mainly by empirical experience and art. However, Franks (1990) argues for the need of a stronger scientific base to increase the process productivity and quality.

During freeze drying, many phase transitions either never occur or are precipitated at states far from equilibrium, and the slow kinetics of subsequent diffusion processes at low temperatures limit the system from moving toward equilibrium. Even if water crystallizes, the residual solutes are likely never to crystallize fully, if at all. As a supersaturated solution is cooled, the viscosity becomes so large that crystallization processes become undetectable. The so-called **glass transition temperature** is the intersection of the liquidus curve on the phase diagram and an isoviscosity curve for which the mechanical properties of the material are glasslike. These states are illustrated on a state diagram (Figure 4A) and compared with a simple binary mixture phase diagram (Figure 4B). By definition, the state diagram does not represent a locus of the system's equilibrium states, but it provides a map of the temperature and composition combinations of defined kinetic properties (Franks 1985).

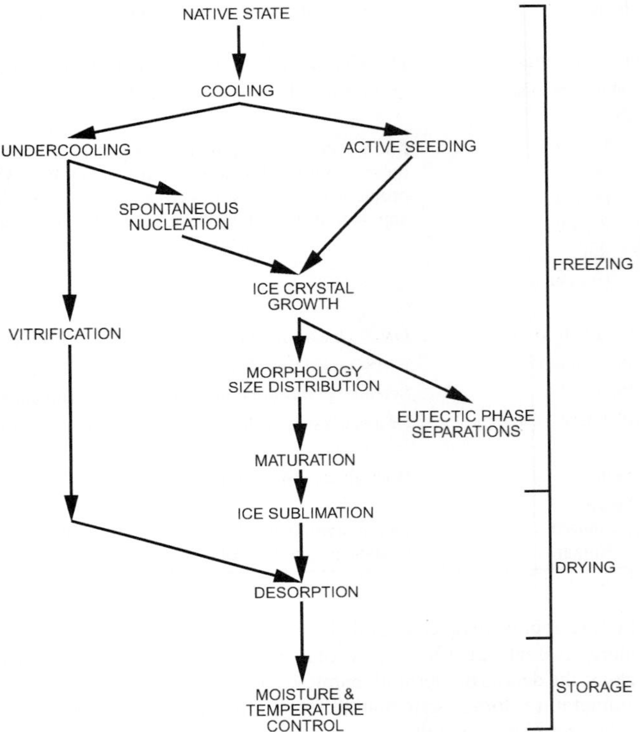

Fig. 3 Key Steps in Freeze-Drying Process

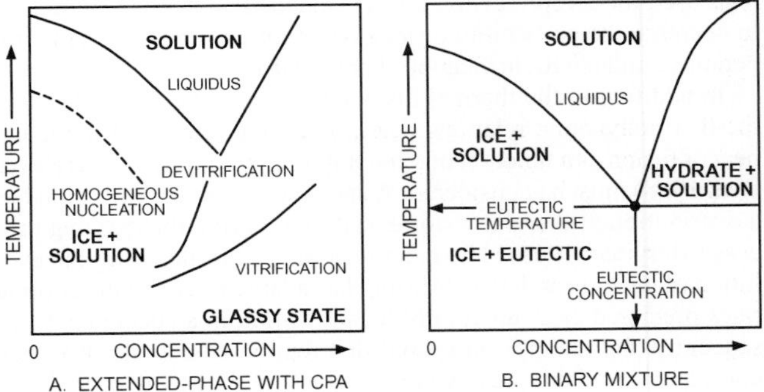

Fig. 4 Phase Diagrams of Aqueous Solutions

The glass transition curve (i.e., vitrification) defines a specific glass transition temperature T_g, which depends on the combination of solute concentration and composition. At states above the glass transition threshold, the material behaves as a viscoelastic medium, which is unacceptable for long-term storage. An important aspect of the state diagram is that the slope of the glass transition curve is very steep at high concentrations of solute (not shown in Figure 4A). Consequently, the glass transition temperature T_g is well above 32°F for a pure solute, thus providing for stable storage conditions. However, because small amounts of residual water in the system can significantly lower the glass transition temperature, it is important to check and control the moisture content of a freeze-dried product. To this end, Levine and Slade (1988) provided extensive data on the glass transition temperature and unfreezable water fraction of many molecular solutions of interest in the design of freeze-drying processes. It is most important that the water is removed from the material at a state temperature lower than the glass transition value at the local solute concentration value. If salts do not precipitate into a solid phase, then unfrozen water remains, which can affect the freeze-drying kinetics (Murase et al. 1991).

Preservation of Biological Materials by Vitrification

The vitrified state plays an integral, albeit partial and secondary, role in preservation of tissues by freezing and by freeze drying. Vitrification may also be used as a storage technique in its own right. Because solidification is avoided, problems associated with the freezing concentration of solutes are avoided. In addition, there are no complications caused by latent heat removal from the specimen at a moving phase front. However, water cannot achieve a vitreous state simply by cooling a solution of physiological composition. Therefore, vitrification is achieved only with the prior addition of high concentrations of solutes (i.e., CPAs) to alter the kinetics of the crystallization process and the locus of the liquidus and glass transition curves on the state diagram. Although the CPA concentration that must be achieved before cooling is higher for vitrification than for freezing, vitrification produces no subsequent solution concentration phenomenon as does freezing. Therefore, if the higher initial CPA concentrations can be tolerated without injury above 32°F where the addition occurs, vitrification may present a distinct benefit as an approach to long-term cryopreservation.

Fahy (1988) summarizes various constitutive properties of candidate CPAs for vitrification, as well as empirical data for the crystallization properties of solutes in aqueous solutions. Another important source of data for the design of vitrification processes is Boutron's research on thermal and glass-forming properties of solutions particularly relevant to cryopreservation. For example, Boutron (1993) deals with the glass forming tendency and stability of the amorphous (glassy) state of 2,3-butanediol in physiological solutions of varying chemical complexity.

In addition to modifying a system to be cryopreserved by adding a CPA, Fahy et al. (1984) explored modifying the state behavior of tissues by cooling under high pressures. Pressures of up to 15,000 psi were used during cooling to reduce the melting temperature of water to about 16°F and the homogeneous nucleation temperature to –65°F, which is equivalent to the reduction in phase change state achieved by introducing a 3 molar concentration of a common CPA. However, limiting factors associated with thermodynamic properties and design of apparatus must be solved before this technique can be considered for practical applications.

The growth of submicroscopic (light) ice crystals, primarily during warming, has been hypothesized to be injurious to vitrified cells and tissues. Several approaches have been pursued to control this process. Rapid warming through the region of sensitive temperatures where crystal nucleation and growth are most probable is used to reduce the time of exposure to these processes (e.g., Marsland 1987). Problems with this technique have included ensuring a homogeneous temperature throughout the tissue and matching the hardware to the impedance properties of the specimen, especially for large organs composed of heterogeneous tissues. Alternatively, Rubinsky et al. (1992) used biological antifreezes from polar fishes (which adsorb to specific faces of ice crystals to inhibit crystal growth) as a CPA constituent to reduce the susceptibility of mammalian tissues to injury. Accordingly, antifreeze glycopeptides have been added to the vitrifying solution to increase the post-thaw viability of vitrified porcine oocytes and embryos. More recently, Wowk et al. (2000) used a low concentration of synthetic polymer polyvinyl alcohol, which inhibited formation of ice in vitrified samples.

Vitrification of tissues and freezing cryopreservation have been most successful with small specimens (e.g., suspensions of isolated cells and small multicellular tissues). One major anticipated advantage of vitrification is in processing whole organs for cryopreservation. To date, this potential has not been realized, in part because of difficulty in solving engineering problems associated with processing. The specimen must cool rapidly throughout to prevent significant numbers of ice crystals from forming in any portion of the tissue volume, which could then later propagate into other areas. Unfortunately, boundary conditions and heat transfer characteristics of relatively large organs do not allow such rapid cooling. The threshold cooling rates can be altered as a function of the tissue's chemical composition by adding a CPA: the most promising approach to resolving this limitation is likely to be chemical rather than thermal. Nonetheless, more effective control of the thermal boundary conditions could be beneficial. Fahy et al. (2004) summarize current challenges for preserving tissues and organs by vitrification.

The cooling process also produces a second problem that is in direct conflict with satisfying the threshold cooling rate requirement. As progressively larger temperature gradients are created within the specimen to boost the cooling rate, corresponding internal thermal stresses are generated. In the glassy state the elastic strength of the vitrified tissue can easily be exceeded, causing mechanical fracture of the tissue (Fahy et al. 1990). This phenomenon is obviously irreversible and totally unacceptable. Thus, cooling must be designed to reduce the temperature fast enough to avoid ice nucleation but slow enough to avoid mechanical fracture. Fortunately, some possible solutions to this quandary have been tested (e.g., annealing stages at appropriate thermal states) and hold promise for vitrification of large organs.

Preservation of Biological Materials by Undercooling

One option for cryopreservation in the undercooled state has found a limited range of applications. This technique avoids heterogeneous nucleation of ice crystals in subcooled water and maintains the storage temperature above the value at which homogeneous nucleation occurs (Franks 1988).

Undercooling is based on the fact that aqueous solutions can be cooled to temperatures substantially below the equilibrium phase change state without nucleation of ice crystals. The temperature of spontaneous homogeneous nucleation for pure water is approximately –40°F. Thus, if externally induced heterogeneous nucleation can be blocked, a substantial window of subzero temperatures can be used for storage of biological materials. This approach avoids the injurious effects of ice formation and the freeze concentration of solutes as well as the need to add and remove chemical CPAs from the specimen, although the temperature range available is not low enough to ensure long-term storage without product deterioration. Because the physical basis of undercooling is much different from the alternative methods described previously, the strategy for developing effective storage is also substantially different.

The key to undercooled storage is the ability to control (prohibit) the nucleation of ice in the specimen. Although the homogeneous nucleation temperature is about 72°F below the equilibrium freezing state, in practice it is difficult to reach even –4°F due to heterogeneous nucleation by particulate matter in the specimen. Further, the presence of just a single ice crystal nucleus is adequate to feed the growth of ice throughout a large volume of aqueous medium. However, because heterogeneous nucleation occurs in the extracellular subvolume of a cell suspension, Franks et al. (1983) suspended the biological material in a medium of innocuous oil formed into microdroplets, thereby dispersing the bulk aqueous suspending solution. In effect, the material, such as cells, was suspended in a very thin film of aqueous solution, which dramatically depresses the ice nucleating tendency of the extracellular matrix. By this method, living cells may be undercooled to nearly the homogeneous nucleation temperature (Franks et al. 1983). Subsequently, many different types of cells have been undercooled in water-in-oil dispersions to –4°F or lower without injury (Mathias et al. 1985).

A similar approach has been developed for storing biochemicals. For example, an aqueous protein solution can be dispersed in an oil carrier formulated to form a gel, thereby trapping the biological material in very small isolated droplets in the inert matrix. Each of the microdroplets is unable to communicate with any neighboring droplets, thus preventing local ice nuclei from providing a substrate for ice growth in the material. Challenges of this process involve creating microdroplet dispersions for effective storage that recover when returned to ambient temperatures. The temperature must be precisely controlled to avoid both homogeneous nucleation by becoming too cold and accelerated product deterioration by becoming too warm. Typical storage temperatures are around –4°F.

RESEARCH APPLICATIONS

Electron Microscopy Specimen Preparation

Freezing is a widely adopted method of preparing specimens for electron microscopy. The advantages of freezing are that it need not involve chemical modification of the specimen in the active liquid state and that the physical substructure of components may be preserved. Conversely, the cooling process may cause ice crystals to form, which would alter or mask the structure to be imaged and which could concentrate the solute locally and cause internal osmotic flows that would produce image artifacts. Thus, control of the thermal history during cooling is critical in obtaining a high-quality preparation for viewing on the microscope. Cooling rates of 2×10^5 to 2×10^6°F/s or higher are desirable to minimize osmotic dehydration of cells and to avoid ice crystal nucleation and growth. Cooling removes heat from the surface of the specimen, and in most cases, the highest cooling rates occur at the boundary of the specimen. Thus, the quality of preparation may vary significantly as a function of position, so the specimen should be mounted so that the dimension normal to the primary direction of heat transfer is as small as possible. Echlin (1992) comprehensively summarizes cryoprocessing of materials for electron microscopy.

Bald (1987) analyzed factors that govern the cooling process during specimen cryopreparation. In each case, the objective is to cool the specimen as rapidly as possible. Three different approaches have been developed for cryofixation: slamming, plunging, and spraying. Cooling by **slamming** is effected by mechanically driving the specimen and its mounting holder onto the surface of a cryogenically refrigerated solid block, which has a large thermal inertia in comparison with the specimen. The impact velocity of the specimen against the cold block is high, to achieve as rapid a change in the thermal boundary conditions as possible. The drive mechanism is spring-loaded to maintain continuous contact with the block after impact so that thermal resistance to the specimen is minimized.

Plunging uses a liquid rather than a solid refrigeration sink. As in slamming, the specimen is driven into a relatively large volume of cryogenic liquid. In common practice, the liquid is prepared in a subcooled or supercritical state so that heat transport from the specimen is not limited by a boiling boundary layer at the interface (Bald 1984). It is also important to eliminate a stratified layer of chilled vapor above the liquid through which the specimen would pass during plunging. Such a vapor layer would cool the specimen somewhat before contact with the liquid cryogen in the vapor medium, but because it has a relatively low convective coefficient, the effective cooling rate is substantially reduced.

For **spraying**, the specimen is held in a stationary mount, and a jet of liquid cryogen is directed onto the specimen. Heat is removed by a combination of evaporation and convection of the cryogen.

Analysis by Bald (1987) indicated that slamming is potentially the most effective method of rapid cooling for cryofixation. The velocity of the specimen during plunging must be 66 fps or greater to reach thermal performance levels characteristic of slamming. In general, it is easier to design apparatus to achieve the velocities required for satisfactory performance by spraying than plunging. Further, high plunge velocities are more likely to damage the specimen than are equivalent spray velocities. The most effective cryogen for both plunging and spraying is subcritical ethane.

After the temperature is reduced, further preparation for viewing on the electron microscope may involve mechanical fracture of the specimen, chemical substitution of one constituent such as water (Hunt 1984), or removal of a chemical constituent such as by vacuum sublimation of water (Echlin 1992; Linner and Livesey 1988; Livesey and Linner 1988). Sectioning and fracturing techniques are used to expose internal structure and constituents of a specimen. This approach to preparation is particularly appropriate at cryogenic temperatures, because biological materials become quite brittle and very little plastic deformation occurs that would alter the morphology. The exposed internal surfaces may be either imaged directly or modified mechanically or chemically.

Cryomicroscopy

Initial investigations using cryomicroscopy were conducted in the early 1800s and have been pursued ever since. From its earliest adoption, cryomicroscopy made it possible to obtain useful information about the behavior of living tissues at subfreezing temperatures, but application has been limited primarily by the difficulty in controlling the refrigeration applied to the specimen.

Diller and Cravalho (1970) designed a cryomicroscope in which independently regulated refrigerating and heating sources controlled the specimen temperature and its time rate of change during both cooling and heating. Heating was produced by applying a variable voltage across a transparent, electrically resistive thick film coating deposited on the underside of a glass plate, on which the biological specimen was mounted. The local temperature was monitored via a microthermocouple positioned in direct contact with the specimen, and this signal was applied as the input to the electronic control system. By miniaturizing the thermal masses of all components of the system, much higher rates of temperature change were achieved with this system than were previously possible (cooling rates approaching 2×10^5°F/min). This system was cooled by circulating a chilled refrigerant fluid through a closed chamber directly beneath the plate on which the specimen was mounted.

This design was modified by McGrath et al. (1975) to eliminate the flow of refrigerant fluid passing through the optical path of the microscope. Rather, heat was conducted away from the specimen via a thin radial plate that was chilled on its periphery by a refrigerant. This design is mechanically more satisfactory and offers a thinner working cross section through the optical path, but the lateral temperature gradients are much higher. These two designs are known as **convection** and **conduction cryomicroscopes**, respectively (Diller 1988). The former has been adapted to allow for simultaneous alteration of the specimen's chemical and thermal environments (Walcerz and Diller 1991), and the latter has been commercially marketed with a computer control system (McGrath 1987).

These cryomicroscopes modulate the temperature where the specimen is mounted on the microscope to create the desired thermal history for an experimental trial. The dimensions of the specimen are limited by the field of view of the microscope optics, because the specimen is stationary during a trial. An alternative approach has been adapted to study the control of a different set of variables. In this system, a steady-state temperature gradient is established across the viewing area of the microscope, and the specimen is moved in time through the gradient to produce the desired temperature history (Körber 1988; Rubinsky and Ikeda 1985). Advantages of this system are that macroscopic specimens may be frozen, because it has been adapted to controlled thermal preparation of specimens for electron microscopy (Bischof et al. 1990), and the cooling rate applied to a specimen can be investigated as defined by the product of the spatial temperature gradient and the velocity of advance of the phase interface (Beckmann et al. 1990). A similar gradient stage was built by Koroush and Diller (1984) for analysis of solidification processes. This system included feedback control of the temperatures at the ends of the gradient to view a stationary specimen.

Innovations in cryomicroscope design continue. A new optical axis freezing stage for laser scanning confocal microscopy provides an end-on view of a growing ice interface (Neils and Diller 2004). This system has the advantage of imaging the freezing of a truly three-dimensional specimen in which the dimensions of the phase interface are not physically constrained within a narrow capillary tube or microscope slide typical of other cryostages. The resulting images can be processed to quantify the lamellar structure of the ice interface. A second system incorporates into a single device the capability for simultaneous optical cryomicroscopy and differential scanning calorimetry (DSC) (Yuan and Diller 2005). This instrument can be used to obtain both visual and thermal data for an individual specimen subjected to a defined freezing and thawing protocol, with very little compromise in quality or range of data available in comparison with dedicated single instruments.

Cryomicrotome

The refrigerated microtome maintains tissue specimens at a subfreezing temperature in a mechanically rigid state, so that very thin sections may be cut for viewing by electron microscopy. The degree of rigidity required is a function of the thickness of the specimen to be cut; thinner sections require greater rigidity, which is achieved by lower temperatures. Stumpf and Roth (1965) have determined that temperatures above −22°F are adequate to obtain sections 1 μm thick, and temperatures below −94°F facilitate cutting of sections thinner than 1 μm. Thus, the apparatus must produce both a wide range of temperatures and accurate thermal control during processing. The apparatus must also be designed to exclude environmental moisture that could contaminate the specimen, and to isolate the refrigeration apparatus from the sectioning chamber to minimize mechanical vibrations that could compromise the dimensional integrity of the delicate cutting process.

CLINICAL APPLICATIONS

Hypothermia

Although accidental hypothermia is the most widely encountered clinical condition of lowered body core temperature, induced hypothermia has been developed as a method of reducing the metabolic rate of selected organs, such as the heart and brain, during surgical procedures. This procedure is of particular benefit in neonatal patients, whose blood vessels and surgical field are too small to effectively apply standard cardiac bypass procedures for maintaining peripheral circulation during surgery. If the temperature can be reduced to a suitably low level (54 to 68°F), then it is possible to stop the heart and to pursue surgical procedures (in the absence of blood perfusion) without incurring irreversible injury. The period for which the body can be subjected to the absence of perfused oxygenated blood is a function of the hypothermic temperature, and may last as long as an hour. These procedures require (1) the temperature of the organ to be within tolerances that limit tissue damage, and (2) the ability to lower and raise the temperature quickly to provide the maximum fraction of the low-temperature period for the surgical procedure. For example, Eberhart addressed the challenge of achieving a suitably rapid rate of cooling for the brain by perfusion through the vascular network with a chilled solution (Dennis et al. 2003; Olson et al. 1985).

The most effective approach to cooling an internal organ is to circulate the blood through a heat exchanger outside the body. The blood is then perfused through the vascular system of the organ, which acts as a physiological heat exchanger. Weinbaum and Jiji (1989) demonstrated the efficacy of thermal equilibration between various components of the vascular tree and the local embedding tissue. Earlier procedures relied primarily on surface cooling to chill internal organs, which is significantly less effective than perfusion in most applications. The results of Olson et al. (1985) indicate that the brain can be very rapidly cooled to a hypothermic state by infusion of cold arterial blood. However, when blood circulation was stopped for cardiac surgical procedures, a gradual but significant rewarming of the brain occurred because of parasitic heat flow from surrounding structures that had not been cooled. Thus, a combination of cold perfusion through the vascular system and surface cooling seems to provide the best control of the body core temperature during hypothermic surgery.

Cryosurgery

In contrast with the previous applications, in which the objective is to maximize the survival of tissues exposed to freezing and thawing, cryosurgery has the goal of selective total destruction of a targeted area of tissue within the body. Cryosurgery is applied to destroy and/or excise tissue that is either dead or diseased. It is usually one of several treatment alternatives and has risen and fallen in favor as a method of treating various types of lesions. In general, it has been most effective in treating lesions for which there is direct or easy external access to allow mechanical placement of a cryoprobe or the spray of a cryogenic fluid. The most commonly accepted uses of cryosurgery include the treatment of skin, mucosal, and gynecological lesions; liver cancer; and in cardiac surgery for treatment of tachyarrhythmias (Gage 1992). Other uses that have demonstrated efficacy but not such broad adoption are the treatment of hemorrhoids; oral, prostate, and anorectal cancer; bone tumors; vertigo; retinal detachment; and visceral tumors.

Primary advantages of cryosurgery are that (1) it provides a bloodless approach to surgery, (2) in some applications it reduces the rate of death, and (3) the extent of destruction inside the affected area can be imaged with noninvasive methods (Gilbert et al. 1985). This latter process makes use of a continuous ultrasonic scan of the freezing zone to monitor the interface between the solid and liquid phases as it grows into the targeted tissue. Experimental evidence indicates that a close correlation exists between the extent of phase interface propagation and the boundary of the zone of tissue destruction (Rubinsky et al. 1990), and these results may be explained in large part by a model for the mechanism of destruction of the freezing process (Rubinsky and Pegg 1988). The model asserts that, during tissue freezing, ice forms preferentially in the vascular network. The ice also propagates through the vessels as the solidification front advances. Cells near the vascular network dehydrate from osmotic stress, and this water then freezes in the vascular lumina. As a result, vessels may expand by as much as a factor of two [for electron micrographs, see Rubinsky et al. (1990)], causing irreversible injury. Thus, the primary action of freezing in destroying tissue during cryosurgery may be by rendering the vascular system nonfunctional rather than by causing direct cryoinjury. Without an active microcirculatory blood flow, the thawed tissue will die rapidly. Hoffman and Bischof (2001a, 2001b) correlated thermal conditions during a cryosurgery to in vivo injury characteristics, and found that vascular injury is the primary in vivo tissue injury mechanism.

Cryogens are usually liquid nitrogen at −320°F or pressurized argon gas, which can reach −303°F via the Joule-Thompson effect.

Table 3 Adjuvants for Cryosurgical Application

Adjuvants	Comments	References
Antifreeze proteins (AFPs)	Alters ice crystal morphology	Pham and Rubinsky (1998) Koushafar et al. (1997)
Eutectic inducer	Inducing secondary (i.e., eutectic) freezing within cryolesion	Han and Bischof (2004a)
Cytokines	TNF-α: Enhancing vascular injury	Chao et al. (2004)
Chemotherapeutic agents	Increased membrane permeability by freezing	Ikekawa et al. (1985) Clarke et al. (2001) Mir and Rubinsky (2002)

The size of the probe and flow rate of cryogen through it determine the volume of tissue that may be frozen. For example, a 0.4 in. diameter probe will produce in tissue an ice ball with a diameter as large as 1 in. (Dilley et al. 1993). Frequently, tumors exceed the capacity of a single probe, but at present, commercial multiprobe cryosurgery systems do not exist. As a result, multiple systems are used, which are hardware-intensive and compromise control over the freezing process (Onik and Rubinsky 1988). Thus, opportunities exist to improve cryosurgical apparatus.

Recent innovations have included operating the refrigerant system under vacuum, thereby creating liquid-phase heat transfer with the active heat transfer surface of the probe, which has a considerably lower thermal resistance than a boiling interface (Baust et al. 1992). This approach to enhancing thermal performance is similar to that used to cool specimens for electron microscopy rapidly (Bald 1987).

Other problems in the design of cryosurgical equipment remain to be solved. For example, parasitic heat leakage along the probe stem to the cold tip extends the active surface capable of causing tissue damage away from the area designed for destruction. This leakage is particularly compromising to the surgical procedure for treating malignant diseases in locations other than on the body surface (Onik and Rubinsky 1988). The simple and convenient interchangeability of probe tips with various geometries and thermal capacities would enhance the flexibility of cryosurgical apparatus. Further, the increasing incidence of sexually transmitted diseases dictates the need for a cryosurgical probe that may either be effectively sterilized (Evans 1992) or be disposable (Baust 1993). Recent research efforts in cryosurgery have focused on the enhancement of cell/tissue injury within cryolesion by use of various adjuvants, as summarized in Table 3.

REFRIGERATION HARDWARE FOR CRYOBIOLOGICAL APPLICATIONS

In general, two classes of refrigeration sources have been adapted successfully to biological applications: vapor compression cycle cooling and boiling of liquid cryogens. Also, two types of thermal performance standards may be required of these refrigeration sources. As indicated in the previous sections, the thermal history during cooling is very often a critical factor in determining the success of a cryobiological procedure. The refrigerating apparatus must achieve a critical cooling rate within a specimen and regulate the cooling rate within specified tolerances over a designated range of temperatures. If the refrigeration apparatus is designed for general applications, these criteria will be demanded for a large variety of procedures.

A second important performance standard is the minimum specimen temperature that can be maintained in the system. Many biological applications depend on continuously holding the specimen at a temperature below a value at which significant process kinetics may occur. Of most importance are (1) control of the nucleation of ice or other solid phases in vitrified materials, and (2) limitation of recrystallization of small ice crystals that form during cooling. Many cryopreservation procedures require that the specimen be

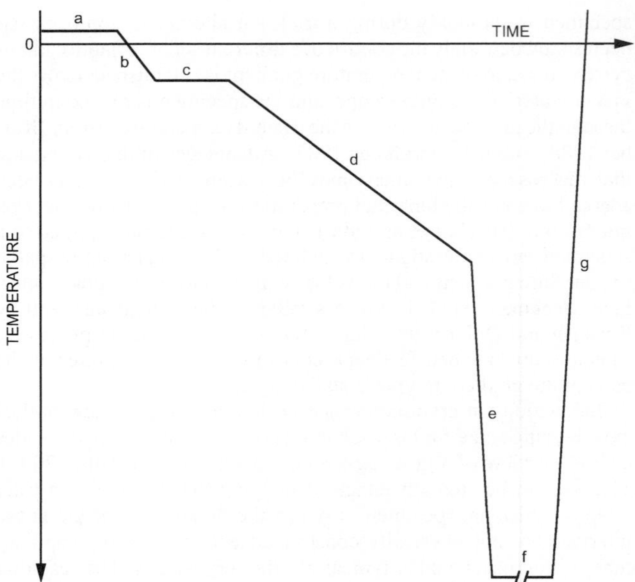

Fig. 5 Generic Thermal History for Example Cryopreservation Procedure

warmed from the stored state as rapidly as possible, to avoid these phenomena, for which the kinetics are most favorable at higher subfreezing temperatures. For long-term storage of biological materials, temperatures below −184°F are generally considered to be safe from the effects of devitrification and crystal growth. This state pushes the limits of refrigeration that can be produced by mechanical means.

An example of a generic cooling, storage, and warming protocol for cryopreservation is shown in Figure 5. The protocol is divided into seven steps. The first (a) consists of adding a cryoprotective agent at a temperature slightly above freezing. This operation is usually executed with the specimen held in a constant-temperature circulating bath. The mixing and osmotic equilibration process may occur in several serial steps and last for half an hour or longer. The specimen is then immersed into a second constant-temperature bath held at a high subfreezing temperature (such as 14°F). The cooling rate during this process (b) is uncontrolled, governed by the inherent heat transfer characteristics of the container and the refrigerant fluid. This constant-temperature holding period (c) enables nucleation of ice in the specimen at a predetermined thermodynamic state and provides time for release of the latent heat of fusion and for osmotic equilibration between the intracellular and extracellular volumes. Subsequently, the specimen is placed into a controlled-rate refrigerator (d) and the temperature is reduced at a rate that maintains a balance between an acceptable osmotic state of the cells and avoids intracellular ice formation. The absolute magnitude of this cooling rate depends on the properties of the subject cell, and it may vary over several orders of magnitude for different specimen types. When the specimen reaches a temperature where kinetic rate processes approach zero (e.g., −112°F), the specimen may be plunged (e) into a liquid nitrogen bath for long-term storage (f). Finally, the specimen is warmed and thawed by removing it from the refrigerator and immersing it directly in a water bath (g).

In practice, many variations exist on the cryopreservation scheme shown in Figure 5. One of the most frequent simplifications is to eliminate one or more of the steps (b through d). Whether this simplification is acceptable depends on the specimen's sensitivity to variations in thermal history, which is determined by the properties of the cells, the physical geometry of the specimen and its packaging for cryopreservation, and chemical modifications performed during step (a).

As the scientific basis for understanding and designing optimal protocols for processes in cryobiology has been strengthened, the specificity and sophistication of the associated refrigeration apparatus has likewise progressed. Therefore, considerable opportunity for improvements in cryobiology hardware remains. The 1980s and 1990s witnessed the founding of many new commercial ventures with the objective of exploiting this potential. A common theme was an effective link to the scientific and/or medical community to ensure that equipment was designed to address the needs of the customers.

REFERENCES

AABB. 1985. *Technical manual of the American Association of Blood Banks*. American Association of Blood Banks, Arlington, VA.

Aggarwal, S.J., C.R. Baxter, and K.R. Diller. 1985. Cryopreservation of skin: An assessment of current clinical applicability. *Journal of Burn Care & Rehabilitation* 6:469-476.

Angell, W.W., J.D. Angell, J.H. Oury, J.J. Lamberti, and T.M. Greld. 1987. Long-term follow-up of viable frozen aortic homografts. A viable homograft valve bank. *Journal of Thoracic and Cardiovascular Surgery* 93:815-822.

Armitage, W.J. 1991. Preservation of viable tissues for transplantation. In *Clinical applications of cryobiology*, pp. 170-189. B.J. Fuller and B.W.W. Grout, eds. CRC Press, Boca Raton, FL.

Bald, W.B. 1984. The relative efficiency of cryogenic fluids used in the rapid quench cooling of biological samples. *Journal of Microsurgery* 134:261-270.

Bald, W.B. 1987. *Quantitative cryofixation*. Adam Hilger, Bristol, U.K.

Bank, H. and R.R. Maurer. 1974. Survival of frozen rabbit embryos. *Experimental Cell Research* 89:188-196.

Baust, J.G. 1993. Cautions in cryosurgery. *Cryo-Letters* 14:1-2.

Baust, J.G., Z. Chang, and T.C. Hua. 1992. Emerging technology in cryosurgery. *Cryobiology* 29:777.

Beckmann, J., C. Körber, G. Rau, A. Hubel, and E.G. Cravalho. 1990. Redefining cooling rate in terms of ice front velocity and thermal gradient: First evidence of relevance to freezing injury of lymphocytes. *Cryobiology* 27:279-287.

Benson, S.W. 1982. *The foundation of chemical kinetics*. Robert E. Kreiger, Malabar, FL.

Bernard, A.G. 1991. Freeze preservation of mammalian reproductive cells. In *Clinical applications of cryobiology*, pp. 149-168. B.J. Fuller and B.W.W. Grout, eds. CRC Press, Boca Raton, FL.

Bilton, F.J. and N.M. Moore. 1976. In vitro culture, storage and transfer of goat embryos. *Australian Journal of Biological Science* 29:125-129.

Bischof, J., C.J. Hunt., B. Rubinsky, A. Burgess, and D.E. Pegg. 1990. Effects of cooling rate and glycerol concentration on the structure of the frozen kidney: Assessment by cryo-scanning electron microscopy. *Cryobiology* 27:301-310.

Boutron, P. 1993. Glass-forming tendency and stability of the amorphous state in solutions of a 2,3-butanediol containing mainly the levo and dextro isomers in water, buffer, and Euro-Collins. *Cryobiology* 30:86-97.

Chao, B.H., X. He, and J.C. Bischof. 2004. Pre-treatment inflammation induced by TNF-α augments cryosurgical injury on human prostate cancer. *Cryosurgery* 49(1):10-27.

CIBA Foundation. 1977. *The freezing of mammalian embryos*. North Holland/Elsevier, Amsterdam.

Cocks, F.H., W.H. Hildebrandt, and M.L. Shepard. 1975. Comparison of the low-temperature crystallization of glasses in the ternary systems H_2O-NaCl-dimethyl sulfoxide and H_2O-NaCl-glycerol. *Journal of Applied Physics* 46(8):3444-3448.

Clarke, D.M., J.M. Baust, R.G. Van Buskirk, and J.G. Baust. 2001. Chemo-cryo combination therapy: An adjunctive model for the treatment of prostate cancer. *Cryobiology* 42:274-285.

Critser, J.K., B.W. Arneson, D.V. Aaker, and G.D. Ball. 1986. Cryopreservation of hamster oocytes: Effects of vitrification or freezing on human sperm penetration of zona-free hamster oocytes. *Fertility and Sterility* 46:277-284.

Dennis, B.H., R.C. Eberhart, G.S. Dulikravich, and S.W. Radons. 2003. Finite-element simulation of cooling of a realistic human head and neck. *Journal of Biomechanical Engineering* 125:832-840.

DeMayo, F.J., R.G. Rawlins, and W.R. Dukelow. 1985. Xenogenous and in vitro fertilisation of frozen/thawed primate oocytes and blastomere separation of embryos. *Fertility and Sterility* 43:295-300.

Diedrick, K., S. al-Hasani, H. Van der Ven, and D. Krebs. 1986. Successful in vitro fertilisation of frozen thawed rabbit and human oocytes. *Journal of In Vitro Fertilization and Embryo Transplantation* 3:65.

Diller, K.R. 1988. Cryomicroscopy. In *Low temperature biotechnology: Emerging applications and engineering contributions*, pp. 347-362. J.J. McGrath and K.R. Diller, eds. American Society of Mechanical Engineers, New York.

Diller, K.R. and E.G. Cravalho. 1970. A cryomicroscope for the study of freezing and thawing processes in biological cells. *Cryobiology* 7:191-199.

Diller, K.R., L.J. Hayes, and M.E. Crawford. 1985. Variation in thermal history during freezing with the pattern of latent heat evolution. *AIChE Symposium Series* 81:234-239.

Dilley, A.V., D.Y. Dy, A. Warlters, S. Copeland, A.E. Gillies, R.W. Morris, D.B. Gibb, T.A. Cook, and D.L. Morris. 1993. Laboratory and animal model evaluation of the Cryotech LCS 2000 in hepatic cryotherapy. *Cryobiology* 30:74-85.

Echlin, P. 1992. *Low-temperature microscopy and analysis*. Plenum Press, New York.

Evans, D.T.P. 1992. In search of an optimum method for the sterilization of a cryoprobe in a sexually transmittable diseases clinic. *Genitourinary Medicine* 68:275-276.

Fahy, G.M. 1980. Analysis of "solution effects" injury: Equations for calculating phase diagram information for the ternary systems NaCl-dimethylsulfoxide-water and NaCl-glycerol-water. *Biophysical Journal* 32:837-850.

Fahy, G.M. 1988. Vitrification. In *Low temperature biotechnology: Emerging applications and engineering contributions*, pp. 113-146. J.J. McGrath and K.R. Diller, eds. American Society of Mechanical Engineers, New York.

Fahy, G.M., D.R. MacFarlane, C.A. Angell, and H.T. Meryman. 1984. Vitrification as an approach to cryopreservation. *Cryobiology* 21:407-426.

Fahy, G.M., J. Saur, and R.J. Williams. 1990. Physical problems with the vitrification of large biological systems. *Cryobiology* 27:465-471.

Fahy, G.M., B. Wowk, J. Wu, J. Phan, C. Rasch, A. Chang, and E. Zendejas. 2004. Cryopreservation of organs by vitrification: Perspectives and recent advances. *Cryobiology* 48:157-178.

Franks, F. 1985. *Biophysics and biochemistry at low temperatures*. Cambridge University Press, U.K.

Franks, F. 1988. Storage in the undercooled state. In *Low temperature biotechnology: Emerging applications and engineering contributions*, pp. 107-112. J.J. McGrath and K.R. Diller, eds. American Society of Mechanical Engineers, New York.

Franks, F. 1990. Freeze drying: From empiricism to predictability. *Cryo-Letters* 11:93-110.

Franks, F., S.F. Mathias, P. Galfre, S.D. Webster, and D. Brown. 1983. Ice nucleation and freezing in undercooled cells. *Cryobiology* 20:298-309.

Fuller, B.J. and R.J. Woods. 1987. Influence of cryopreservation on uptake of 99m Tc Hida by isolated rat hepatocytes. *Cryo-Letters* 8:232-237.

Gage, A. 1992. Progress in cryosurgery. *Cryobiology* 29:300-304.

Gilbert, J.C., G.M. Onik, W.K. Hoddick, and B. Rubinsky. 1985. Real time ultrasonic monitoring of hepatic cryosurgery. *Cryobiology* 22:319-330.

Gottlob, R., L. Stockinger, and G.F. Gestring. 1982. Conservation of veins with preservation of viable endothelium. *Journal of Cardiovascular Surgery* 23:109-116.

Grout, B.W.W. 1987. Higher plants at freezing temperatures. In *The effects of low temperatures on biological systems*, pp. 293-314. B.W.W. Grout and G.J. Morris, eds. Edward Arnold, London.

Han, B. and J.C. Bischof. 2004a. Direct cell injury associated with eutectic crystallization during freezing. *Cryobiology* 48:8-21.

Han, B. and J.C. Bischof. 2004b. Thermodynamic non-equilibrium phase change behavior and thermal properties of biological solutions for cryobiology applications. *Journal of Biomechanical Engineering* 126:196-203.

Han, B. and J.C. Bischof. 2004c. Engineering challenges in tissue preservation. *Cell Preservation Technology* 2:91-112.

Hartman, U., B. Nunner, C. Körber, and G. Rau. 1991. Where should the cooling rate be determined in an extended freezing sample? *Cryobiology* 28:115-130.

Hayes, L.J., K.R. Diller, H.J. Chang, and H.S. Lee. 1988. Prediction of local cooling rates and cell survival during the freezing of cylindrical specimens. *Cryobiology* 25:67-82.

Hoffman, N.E. and J.C. Bischof. 2001a. Cryosurgery of normal and tumor tissue in the dorsal skin flap chamber, I—Thermal response. *Journal of Biomechanical Engineering* 123:301-309.

Hoffman, N.E. and J.C. Bischof. 2001b. Cryosurgery of normal and tumor tissue in the dorsal skin flap chamber, II—Injury response. *Journal of Biomechanical Engineering* 123:310-316.

Huggins, C.E. 1985. Preparation and usefulness of frozen blood. *Annual Review of Medicine* 36:499-503.

Hunt, C.J. 1984. Studies on cellular structure and ice location in frozen organs and tissues: The use of freeze-substitution and related techniques. *Cryobiology* 21:385-402.

Ikekawa, S., K. Ishihara, S. Tanaka, and S. Ikeda. 1985. Basic studies of cryochemotherapy in a murine tumor system. *Cryobiology* 22:477-483

James, E. 1987. The preservation of organisms responsible for parasitic diseases. In *The effects of low temperatures on biological systems*, pp. 410-431. B.W.W. Grout and G.J. Morris, eds. Edward Arnold, London.

Jochem, M. and C. Körber. 1987. Extended phase diagrams for the ternary solutions H_2O-NaCl-glycerol and H_2O-NaCl-hydroxyethylstarch (HES) determined by DSC. *Cryobiology* 24:513-536.

Kasai, M., A. Iritani, and B.C. Chang. 1979. Fertilisation in vitro of rat ovarian oocytes after freezing and thawing. *Biology of Reproduction* 21: 839-844.

Knight, S.C. 1980. Preservation of leukocytes. In *Low temperature preservation in medicine and biology*, pp. 121-128. M.J. Ashwood-Smith and J. Farrant, eds. University Park Press, Baltimore.

Körber, C. 1988. Phenomena at the advancing ice-liquid interface: Solutes, particles and biological cells. *Quarterly Reviews of Biophysics* 21:229-298.

Kourosh, S. and K.R. Diller. 1984. A unidirectional temperature gradient stage for solidification studies in aqueous solutions. *Journal of Microscopy* 135(1):39-48.

Koushafar, H, L.D. Pham, C. Lee, and B. Rubinsky. 1997. Chemical adjuvant cryosurgery with antifreeze proteins. *Journal of Surgical Oncology* 66:114-121.

Kreder, H.J., F.W. Keeley, and R. Salter. 1993. Cryopreservation of periosteum for transplantation. *Cryobiology* 30:107-112.

Levine, H. and L. Slade. 1988. Principles of "cryostabilization" technology from structure/property relationships of carbohydrate/water systems. *Cryo-Letters* 9:21-63.

Linner, J.G. and S.A. Livesey. 1988. Low temperature molecular distillation drying of cryofixed biological samples. In *Low temperature biotechnology: Emerging applications and engineering contributions*, pp. 117-158. J.J. McGrath and K.R. Diller, eds. American Society of Mechanical Engineers, New York.

Livesey, S.A. and J.G. Linner. 1988. Cryofixation methods for electron microscopy. In *Low temperature biotechnology: Emerging applications and engineering contributions*, pp. 159-174. J.J. McGrath and K.R. Diller, eds. American Society of Mechanical Engineers, New York.

Lovelock, J.E. 1954. The protective action by neutral solutes against haemolysis by freezing and thawing. *Biochemical Journal* 56:265-270.

Marsland, T.P. 1987. The design of an electromagnetic rewarming system for cryopreserved tissue. In *The biophysics of organ cryopreservation*, pp. 367-385. D.E. Pegg and A.M. Karow, Jr., eds. Plenum Press, New York.

Mathias, S.F., F. Franks, R.H.M. Hatley. 1985. Preservation of viable cells in the undercooled state. *Cryobiology* 22:537-546.

Mazur, P. 1970. Cryobiology: The freezing of biological systems. *Science* 168:939-949.

Mazur, P. 1977. The role of intracellular freezing in the death of cells cooled at supraoptimal rates. *Cryobiology* 14:251-272.

Mazur, P. 1990. Equilibrium, quasi-equilibrium and nonequilibrium freezing of mammalian embryos. *Cell Biophysics* 17:53-92.

Mazur, P., K.W. Cole, J.W. Hall, P.D. Schreuders, and A.P. Mahowald. 1992. Cryobiological preservation of *Drosophila* embryos. *Science* 258:1932-1935.

McGann, L.E., A.R. Turner, M.J. Allalunis, and J.M. Turc. 1981. Cryopreservation of human peripheral blood stem cells: Optimal cooling and warming conditions. *Cryobiology* 18:469-472.

McGrath, J.J. 1987. Temperature-controlled cryogenic light microscopy—An introduction to cryomicroscopy. In *The effects of low temperatures on biological systems*, pp. 234-267. B.W.W. Grout and G.J. Morris, eds. Edward Arnold, London.

McGrath, J.J., E.G. Cravalho, and C.E. Huggins. 1975. An experimental comparison of intracellular ice formation and freeze-thaw survival of hela S-3 cells. *Cryobiology* 12:540-550.

McNally, R.T. and C. McCaa. 1988. Cryopreserved tissues for transplant. In *Low temperature biotechnology: Emerging applications and engineering contributions*, pp. 91-106. J.J. McGrath and K.R. Diller, eds. American Society of Mechanical Engineers, New York.

Meryman, H.T. 1966. The interpretation of freezing rates in biological materials. *Cryobiology* 2:165-170.

Meryman, H.T., R.J. Williams, and M. St J. Douglas. 1977. Freezing injury from "solution" effects and its prevention by natural or artificial cryoprotection. *Cryobiology* 14:287-302.

Mir, L.M. and B. Rubinsky. 2002. Treatment of cancer with cryochemotherapy. *British Journal of Cancer* 86:1658-1660.

Murase, N., P. Echlin, and F. Franks. 1991. The structural states of freeze-concentrated and freeze-dried phosphates studied by scanning electron microscopy and differential scanning calorimetry. *Cryobiology* 28: 364-375.

Neils, C. M. and K.R. Diller. 2004. An optical-axis freezing stage for laser-scanning microscopy of broad ice-water interfaces. *Journal of Microscopy* 216(3):249-262.

Olson, R.W., L.J. Hayes, E.H. Wissler, H. Nikaidoh, and R.C. Eberhart. 1985. Influence of hypothermia and circulatory arrest on cerebral temperature distributions. *ASME Transactions, Journal of Biomechanical Engineering* 107:354-360.

Onik, G. and B. Rubinsky. 1988. Cryosurgery: New developments in understanding and technique. In *Low temperature biotechnology: Emerging applications and engineering contributions*, pp. 57-80. J.J. McGrath and K.R. Diller, eds. American Society of Mechanical Engineers, New York.

Pham, L.D. and B. Rubinsky. 1998. Breast tissue cryosurgery with antifreeze proteins. *ASME Advances in Heat and Mass Transfer in Biotechnology*, HTD 362/BED 49, pp. 171-175.

Polge, C. 1980. Freezing of spermatozoa. In *Low temperature preservation in medicine and biology*, pp. 45-64. M.J. Ashwood-Smith and J. Farrant, eds. University Park Press, Baltimore.

Polge, C., A.U. Smith, and A.S. Parkes. 1949. Revival of spermatazoa after vitrification and dehydration at low temperatures. *Nature* (London) 49:666.

Prince, A. 1966. *Alloy phase equilibria*. Elsevier, Amsterdam.

Rajotte, R.V., G.L. Warnock, L.C. Bruch, and A.W. Procyshyn. 1983. Transplantation of cryopreserved and fresh rat islets and canine pancreatic fragments: Comparison of cryopreservation protocols. *Cryobiology* 20:169-184.

Rowe, T.W.G. 1970. Freeze-drying of biological materials: Some physical and engineering aspects. In *Current trends in cryobiology*, pp. 61-138. A.U. Smith, ed. Plenum Press, New York.

Rubinsky, B., A. Arav, and A.L. DeVries. 1992. The cryoprotective effect of antifreeze glycopeptides from antarctic fishes. *Cryobiology* 29:69-79.

Rubinsky, B. and M. Ikeda. 1985. A cryomicroscope using directional solidification for the controlled freezing of biological material. *Cryobiology* 22:55-68.

Rubinsky, B., C.Y. Lee, L.C. Bastacky, and G. Onik. 1990. The process of freezing and the mechanism of damage during hepatic cryosurgery. *Cryobiology* 27:85-97.

Rubinsky, B. and D.E. Pegg. 1988. A mathematical model for the freezing process in biological tissue. *Proceedings of the Royal Society, London* B 234:343-358.

Scheiwe, M.W., Z. Pusztal-Markos, U. Essers, R. Seelis, G. Rau, C. Körber, K.H. Stürner, H. Jung, and B. Liedtke. 1981. Cryopreservation of human lymphocytes and stem cells (CFU-c) in large units for cancer therapy—A report based on the data of more than 400 frozen units. *Cryobiology* 18:344-356.

Schiffer, C.A., J. Aisner, and J.P. Dutcher. 1985. Platelet cryopreservation using dimethyl sulfoxide. *Annals of the New York Academy of Science* 459:353-361.

Sherman, J.K. 1973. Synopsis of the use of frozen human semen since 1964: State of the art of human semen banking. *Fertility and Sterility* 24: 397-412.

Sputtek, A. and C. Körber. 1991. Cryopreservation of red blood cells, platelets, lymphocytes, and stem cells. In *Clinical applications of cryobiology*, pp. 95-147. B.J. Fuller and B.W.W. Grout, eds. CRC Press, Boca Raton, FL.

Stumpf, W.F. and L.J. Roth. 1965. Frozen sectioning below –60°C with a refrigerated microtome. *Cryobiology* 1:227-232.

Taylor, M.J. and M.J. Benton. 1987. Interaction of cooling rate, warming rate and extent of permeation of cryoprotectant in determining survival of isolated rat islets of langerhans during cryopreservation. *Diabetes* 36:59-65.

Troundson, A. and L. Mohr. 1983. Human pregnancy following cryopreservation, thawing and transfer of an 8-cell embryo. *Nature* 305:707-709.

Turner, A.R. 1970. *Frozen blood—A review of the literature 1949-1968*. Gordon and Breach, London.

Valeri, C.R. 1976. *Blood banking and the use of frozen blood products*. CRC Press, Boca Raton, FL.

Van Uem, J.F.H.M., D.R. Siebzehnrueble, B. Schuh, R. Koch, S. Trotnow, and N. Lang. 1987. Birth after cryopreservation of unfertilized oocytes. *Lancet* 1:752-753.

Walcerz, D.B. and K.R. Diller. 1991. Quantitative light microscopy of combined perfusion and freezing processes. *Journal of Microscopy* 161:297-311.

Weinbaum, S. and L.M. Jiji. 1989. The matching of thermal fields surrounding countercurrent microvessels and the closure approximation in the Weinbaum–Jiji Equation. *ASME Transactions, Journal of Biomechanical Engineering* 111:234-237.

Wells, S.A., J.C. Gunnells, R.A. Gutman, J.D. Shelburne, S.G. Schneider, and L.M. Sherwood. 1977. The successful transplantation of frozen parathyroid tissue in man. *Surgery* 81:86-91.

Whittingham, D.G. 1975. Survival of rat embryos after freezing and thawing. *Journal of Reproduction and Fertility* 43:575-778.

Whittingham, D.G. 1977. Fertilisation in vitro and development to term of unfertilised mouse oocytes previously stored at –196°C. *Journal of Reproduction and Fertility* 49:89-94.

Whittingham, D.G., P. Mazur, and S.P. Leibo. 1972. Survival of mouse embryos frozen to –196°C and –269°C. *Science* 178:411-414.

Willadsen, S.M., C. Polge, L.E.A. Rowson, and R.M. Moor. 1976. Deep freezing of sheep embryos. *Journal of Reproduction and Fertility* 46: 151-154.

Wilmut, I. 1972. The effect of cooling rate, cryoprotective agent and stage of development on survival of mouse embryos during freezing and thawing. *Life Science* 11:1071-1079.

Wilmut, I. and L.E.A. Rowson. 1973. Experiments on the low-temperature preservation of cow embryos. *Veterinary Record* 93:686-690.

Withers, L.A. 1987. The low temperature preservation of plant cell, tissue and organ cultures and seed for genetic conservation and improved agricultural practice. In *The effects of low temperatures on biological systems*, pp. 389-409. B.W.W. Grout and G.J. Morris, eds. Edward Arnold, London.

Wowk, B., E. Leitl, C.M. Rasch, N. Mesbah-Karimi, S.B. Harris, and G.M. Fahy. 2000. Vitrification enhancement by synthetic ice blocking agents. *Cryobiology* 40(3):228-236.

Yuan, S. and K.R. Diller. 2005. An optical differential scanning calorimeter cryomicroscope. *Journal of Microscopy* 218(2):85-93.

Zeilmaker, G., ed. 1981. *Frozen storage of laboratory animals*. Gustav Fischer, Stuttgart.

BIBLIOGRAPHY

Primary literature: The main English-language sources for general literature on cryobiology are two archival journals: *Cryobiology* (founded 1964) and *Cryo-Letters* (founded 1979). In addition, *Cell Preservation Technology* was founded in 2002, and the *Bulletin of the International Institute of Refrigeration* provides a timely listing of world literature in low-temperature biology. Other references are distributed among a large number of journals that are either more general or are oriented toward specific physiological or applications areas.

Monographs: Numerous monographs have been written on the principles and applications of low-temperature biology. In general, these have been edited works in which contributing authors provide a series of expositions in focused areas of expertise.

Ashwood-Smith, M.J. and J. Farrant, eds. 1980. *Low temperature preservation in medicine and biology*. University Park Press, Baltimore, MD.

Bald, W.B. 1987. *Quantitative cryofixation*. Adam Hilger, Bristol.

Davenport, J. 1992. *Animal life at low temperature*. Chapman & Hall, London.

Diller, K.R. 1992. Modeling of bioheat transfer processes at high and low temperatures. In *Advances in heat transfer: Bioengineering heat transfer* 22, pp. 157-357. Y.I. Cho, ed. Academic Press, Boston.

Echlin, P. 1992. *Low-temperature microscopy and analysis*. Plenum Press, New York.

Fennema, O.R., W.D. Powrie, and E.H. Marth, eds. 1973. *Low-temperature preservation of foods and living matter*. Dekker, New York.

Franks, F. 1985. *Biophysics and biochemistry at low temperatures*. Cambridge University Press, U.K.

Franks, F., ed. 1972-1982. *Water: A comprehensive treatise*. Plenum Press, New York.

Fuller, B.J. and B.W.W. Grout, eds. 1991. *Clinical applications of cryobiology*. CRC Press, Boca Raton, FL.

Fuller, B.J., N. Lane, and E.E. Benson. 2004. *Life in the frozen state*. CRC Press, Boca Raton, FL.

Grout, B.W.W. and G.J. Morris, eds. 1987. *The effects of low temperatures on biological systems*. Edward Arnold, Ltd., London.

Hobbs, P.V. 1974. *Ice physics*. Clarendon Press, Oxford.

Karow, A. and D.E. Pegg. 1981. *Organ preservation for transplantation*. Marcel Dekker, New York.

Kavaler, L. 1970. *Freezing point: Cold as a matter of life and death*. John Day, New York.

Lozina-Lozinskii, L.K. 1974. *Studies in cryobiology: Adaptation and resistance of organisms and cells to low and ultralow temperatures* (in Russian). P. Harry, translator. Wiley, New York.

McGrath, J.J. and K.R. Diller, eds. 1988. *Low temperature biotechnology: Emerging applications and engineering contributions*. American Society of Mechanical Engineers, New York.

Morris, G. 1983. *Effects of low temperatures on biological membranes*. Academic Press, Boston.

Onik, G.M., B. Rubinsky, G. Watson, and R.J. Ablin. 1995. *Percutaneous prostate cryoablation*. Quality Medical Publishing, St. Louis.

Pegg, D.E. and A.M. Karow, eds. 1987. *The biophysics of organ cryopreservation*. Plenum Press, New York.

Robards, A.W., U.B. Sleytr, and A.M. Glauert, eds. 1985. *Low temperature methods in biological electron microscopy*, vol. 10: *Practical methods in electron microscopy*. Elsevier, Amsterdam.

Roos, A., J. Morgan, and N. Roos. 1990. *Cryopreparation of thin biological specimens for electron microscopy: Methods and applications*. BIOS Scientific.

Shitzer, A. and R.C. Eberhart, eds. 1985. *Heat transfer in medicine and biology: Analysis and applications*, vols. I and II. Plenum Press, New York.

Sibinga, C.T.S. and P.C. Das. 1990. Cryopreservation and low temperature biology in blood transfusion. In *Developments in hematology and immunology* 24. Kluwer Academic, Boston.

Smith, A.U., ed. 1970. *Current trends in cryobiology*. Plenum Press, New York.

Steponkus, P.L., ed. 1992, 1993, 1996. *Advances in low temperature biology*, vols. 1, 2 and 3. JAI Press, London.

Wolsternolme, G.E.W. and M. O'Connor, eds. 1970. *The frozen cell: A Ciba Foundation symposium*. Churchill, London.

CHAPTER 41

ABSORPTION COOLING, HEATING, AND REFRIGERATION EQUIPMENT

THIS chapter surveys and summarizes the types of absorption equipment that are currently manufactured and/or commonly encountered. The equipment can be broadly categorized by whether it uses water or ammonia as refrigerant. The primary products in the water refrigerant category are large commercial chillers, which use lithium bromide (LiBr) as absorbent. There are three primary products in the ammonia refrigerant category: (1) domestic refrigerators, (2) residential chillers, and (3) large industrial refrigeration units.

This chapter focuses on hardware (i.e., cycle implementation), not on cycle thermodynamics. Cycle thermodynamic descriptions and calculation procedures, along with a tabulation of the types of absorption working pairs and a glossary, are presented in Chapter 1 of the 2005 *ASHRAE Handbook—Fundamentals*.

Absorption units provide two major advantages: (1) they are activated by heat, and (2) no mechanical vapor compression is required. They also do not use atmosphere-harming halogenated refrigerants, and reduce summer electric peak demand. No lubricants, which are known to degrade heat and mass transfer, are required. The various equipment can be direct-fired by combustion of fuel, directly heated by various waste fluids, or heated by steam or hot water (from either direct combustion or from hot waste fluids). Figure 1 illustrates the similarities between absorption and vapor compression systems.

With natural gas firing, absorption chilling units level the year-round demand for natural gas. From an energy conservation perspective, the combination of a prime mover plus a waste-heat-powered absorption unit provides unparalleled overall efficiency.

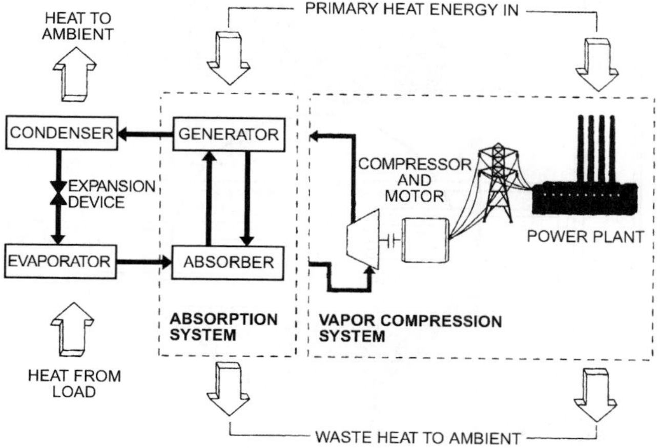

Fig. 1 Similarities Between Absorption and Vapor Compression Systems

The preparation of this chapter is assigned to TC 8.3, Absorption and Heat Operated Machines.

WATER/LITHIUM BROMIDE ABSORPTION TECHNOLOGY

Components and Terminology

Absorption equipment using water as the refrigerant and lithium bromide as the absorbent is classified by the method of heat input to the primary generator (firing method) and whether the absorption cycle is single- or multiple-effect.

Machines using steam or hot liquids as a heat source are **indirect-fired**, and those using direct combustion of fossil fuels as a heat source are **direct-fired**. Machines using hot waste gases as a heat source are also classified as indirect-fired, but are often referred to as **heat recovery chillers**.

Solution recuperative heat exchangers, also referred to as **economizers**, are typically shell-and-tube or plate heat exchangers. They transfer heat between hot and cold absorbent solution streams, thus recycling energy. The material of construction is mild steel or stainless steel.

Condensate subcooling heat exchangers, a variation of solution heat exchangers, are used on steam-fired, double-effect machines and on some single-effect, steam-fired machines. These heat recovery exchangers use the condensed steam to add heat to the solution entering the generator.

Indirect-fired generators are usually shell-and-tube, with the absorbent solution either flooded or sprayed outside the tubes, and the heat source (steam or hot fluid) inside the tubes. The absorbent solution boils outside the tubes, and the resulting intermediate- or strong-concentration absorbent solution flows from the generator through an outlet pipe. The refrigerant vapor evolved passes through a vapor/liquid separator consisting of baffles, eliminators, and low-velocity regions and then flows to the condenser section. Ferrous materials are used for absorbent containment; copper, copper-nickel alloys, stainless steel, or titanium are used for the tube bundle.

Direct-fired generators consist of a fire-tube section, a flue-tube section, and a vapor/liquid separation section. The fire tube is typically a double-walled vessel with an inner cavity large enough to accommodate a radiant or open-flame fuel oil or natural gas burner. Dilute solution flows in the annulus between the inner and outer vessel walls and is heated by contact with the inner vessel wall. The flue tube is typically a tube or plate heat exchanger connected directly to the fire tube.

Heated solution from the fire-tube section flows on one side of the heat exchanger, and flue gases flow on the other side. Hot flue gases further heat the absorbent solution and cause it to boil. Flue gases leave the generator, and the partially concentrated absorbent solution and refrigerant vapor mixture pass to a vapor/liquid separator chamber. This chamber separates the absorbent solution from the refrigerant vapor. Materials of construction are mild steel for the absorbent containment parts and mild steel or stainless steel for the flue gas heat exchanger.

Secondary or **second-stage generators** are used only in double- or multistage machines. They are both a generator on the low-pressure side and a condenser on the high-pressure side. They are usually of the shell-and-tube type and operate similarly to indirect-fired generators of single-effect machines. The heat source, which is inside the tubes, is high-temperature refrigerant vapor from the primary generator shell. Materials of construction are mild steel for absorbent containment and usually copper-nickel alloys or stainless steel for the tubes. Droplet eliminators are typically stainless steel.

Evaporators are heat exchangers, usually shell-and-tube, over which liquid refrigerant is dripped or sprayed and evaporated. Liquid to be cooled passes through the inside of the tubes. Evaporator tube bundles are usually copper or a copper-nickel alloy. Refrigerant containment parts are mild steel. Mist eliminators and drain pans are typically stainless steel.

Absorbers are tube bundles over which strong absorbent solution is sprayed or dripped in the presence of refrigerant vapor. The refrigerant vapor is absorbed into the absorbent solution, thus releasing heat of dilution and heat of condensation. This heat is removed by cooling water that flows through the tubes. Weak absorbent solution leaves the bottom of the absorber tube bundle. Materials of construction are mild steel for the absorbent containment parts and copper or copper-nickel alloys for the tube bundle.

Condensers are tube bundles located in the refrigerant vapor space near the generator of a single-effect machine or the second-stage generator of a double-effect machine. The water-cooled tube bundle condenses refrigerant from the generator onto tube surfaces. Materials of construction are mild steel, stainless steel, or other corrosion-resistant materials for the refrigerant containment parts and copper for the tube bundle. For special waters, the condenser tubes can be copper-nickel, which derates the performance of the unit.

High-stage condensers are found only in double-effect machines. This type of condenser is typically the inside of the tubes of the second-stage generator. Refrigerant vapor from the first-stage generator condenses inside the tubes, and the resulting heat is used to concentrate absorbent solution in the shell of the second-stage generator when heated by the outside surface of the tubes.

Pumps move absorbent solution and liquid refrigerant in the absorption machine. Pumps can be configured as individual (one motor, one impeller, one fluid stream) or combined (one motor, multiple impellers, multiple fluid streams). The motors and pumps are hermetic or semihermetic. Motors are cooled and bearings lubricated either by the fluid being pumped or by a filtered supply of liquid refrigerant. Impellers are typically brass, cast iron, or stainless steel; volutes are steel or impregnated cast iron, and bearings are babbitt-impregnated carbon journal bearings.

Refrigerant pumps (when used) recirculate liquid refrigerant from the refrigerant sump at the bottom of the evaporator to the evaporator tube bundle in order to effectively wet the outside surface and enhance heat transfer.

Dilute solution pumps take dilute solution from the absorber sump and pump it to the generator.

Absorber spray pumps recirculate absorbent solution over the absorber tube bundle to ensure adequate wetting of the absorber surfaces. These pumps are not found in all equipment designs. Some designs use a jet eductor for inducing concentrated solution flow to the absorber sprays. Another design uses drip distributors fed by gravity and the pressure difference between the generator and absorber.

Purge systems are required on lithium bromide absorption equipment to remove noncondensables (air) that leak into the machine or hydrogen (a product of corrosion) that is produced during equipment operation. Even in small amounts, noncondensable gases can reduce chilling capacity and even lead to solution crystallization. Purge systems for larger sizes above 100 tons of refrigeration typically consist of these components:

- Vapor pickup tube(s), usually located at the bottom of large absorber tube bundles
- Noncondensable separation and storage tank(s), located in the absorber tube bundle or external to the absorber/evaporator vessel
- A vacuum pump or valving system using solution pump pressure to periodically remove noncondensables collected in the storage tank

Some variations include jet pumps (eductors), powered by pumped absorbent solution and placed downstream of the vapor pickup tubes to increase the volume of sampled vapor, and water-cooled absorbent chambers to remove water vapor from the purged gas stream.

Because of their size, smaller units have fewer leaks, which can be more easily detected during manufacture. As a result, small units may use variations of solution drip and entrapped vapor bubble pumps plus purge gas accumulator chambers.

Palladium cells, found in large direct-fired and small indirect-fired machines, continuously remove the small amount of hydrogen gas that is produced by corrosion. These devices operate on the principle that thin membranes of heated palladium are permeable to hydrogen gas only.

Corrosion inhibitors, typically lithium chromate, lithium nitrate, or lithium molybdate, protect machine internal parts from the corrosive effects of the absorbent solution in the presence of air. Each of these chemicals is used as a part of a corrosion control system. Acceptable levels of contaminants and the correct solution pH range must be present for these inhibitors to work properly. Solution pH is controlled by adding lithium hydroxide or hydrobromic acid.

Performance additives are used in most lithium bromide equipment to achieve design performance. The heat and mass transfer coefficients for the simultaneous absorption of water vapor and cooling of lithium bromide solution have relatively low values that must be enhanced. A typical additive is one of the octyl alcohols.

Single-Effect Lithium Bromide Chillers

Figure 2 is a schematic diagram of a commercially available single-effect, indirect-fired liquid chiller, showing one of several configurations of the major components. Table 1 lists typical characteristics of this chiller. During operation, heat is supplied to tubes

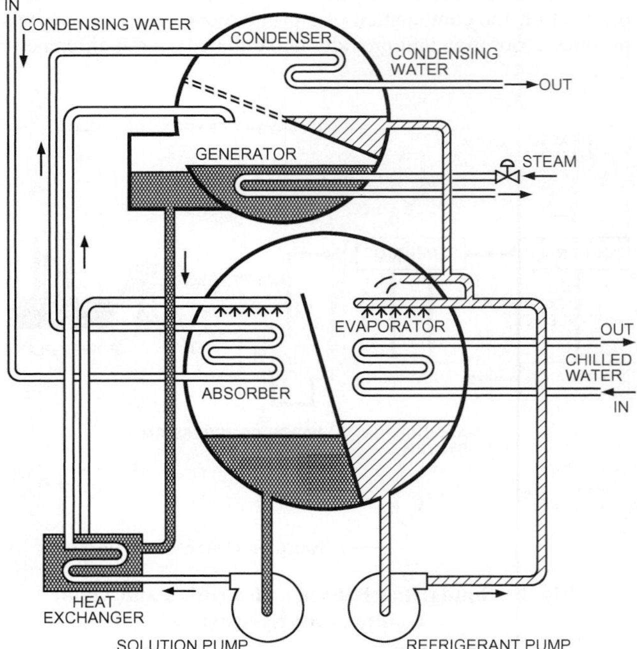

Fig. 2 Two-Shell Lithium Bromide Cycle Water Chiller

Table 1 Characteristics of Typical Single-Effect, Indirect-Fired, Water/Lithium Bromide Absorption Chiller

Performance Characteristics	
Steam input pressure	9 to 12 psig
Steam consumption	18.3 to 18.7 lb/ton·h
Hot-fluid input temp.	240 to 270°F, with as low as 190°F for some smaller machines for waste heat applications
Heat input rate	18,100 to 18,500 Btu/ton·h, with as low as 17,100 Btu/ton·h for some smaller machines
Cooling water temp. in	85°F
Cooling water flow	3.6 gpm/ton, with up to 6.4 gpm/ton for some smaller machines
Chilled-water temp. off	44°F
Chilled-water flow	2.4 gpm/ton, with 2.6 gpm/ton for some smaller international machines
Electric power	0.01 to 0.04 kW/ton with a minimum of 0.004 kW/ton for some smaller machines
Physical Characteristics	
Nominal capacities	50 to 1660 tons, with 5 to 10 tons for some smaller machines
Length	11 to 33 ft, with as low as 3 ft for some smaller machines
Width	5 to 10 ft, with 3 ft minimum for some smaller machines
Height	7 to 14 ft, with 6 ft for some smaller machines
Operating weight	11,000 to 115,000 lb, with 715 lb for some smaller machines

of the **generator** in the form of a hot fluid or steam, causing dilute absorbent solution on the outside of the tubes to boil. This desorbed refrigerant vapor (water vapor) flows through eliminators to the **condenser**, where it is condensed on the outside of tubes that are cooled by a flow of water from a heat sink (usually a cooling tower). Both boiling and condensing occur in a vessel that has a common vapor space at a pressure of about 0.9 psia.

The condensed refrigerant passes through an orifice or liquid trap in the bottom of the condenser and enters the **evaporator**, in which liquid refrigerant boils as it contacts the outside surface of tubes that contain a flow of water from the heat load. In this process, water in the tubes cools as it releases the heat required to boil the refrigerant. Refrigerant that does not boil collects at the bottom of the evaporator, flows to a **refrigerant pump**, is pumped to a distribution system located above the evaporator tube bundle, and is sprayed over the evaporator tubes again.

The dilute (weak in absorbing power) absorbent solution that enters the generator increases in concentration (percentage of sorbent in the water) as it boils and releases water vapor. The resulting strong absorbent solution leaves the generator and flows through one side of a **solution heat exchanger**, where it cools as it heats a stream of weak absorbent solution passing through the other side of the solution heat exchanger on its way to the generator. This increases the machine's efficiency by reducing the amount of heat from the primary heat source that must be added to the weak solution before it begins to boil in the generator.

The cooled, strong absorbent solution then flows (in some designs through a jet eductor or solution spray pumps) to a solution distribution system located above the **absorber tubes** and drips or is sprayed over the outside surface of the absorber tubes. The absorber and evaporator share a common vapor space at a pressure of about 0.1 psia. This allows refrigerant vapor, which is evaporated in the evaporator, to be readily absorbed into the absorbent solution flowing over the absorber tubes. This absorption process releases heat of condensation and heat of dilution, which are removed by cooling water flowing through the absorber tubes. The resulting weak absorbent solution flows off the absorber tubes and then to the absorber sump and **solution pump**. The pump and piping convey the weak absorbent solution to the heat exchanger, where it accepts heat from the strong absorbent solution returning from the generator. From there, the weak solution flows into the generator, thus completing the cycle.

These machines are typically fired with low-pressure steam or medium-temperature liquids. Several manufacturers have machines with capacities ranging from 50 to 1660 tons of refrigeration. Machines of 5 to 10 ton capacities are also available from international sources.

Typical coefficients of performance (COPs) for large single-effect machines at Air Conditioning and Refrigeration Institute (ARI) rating conditions are 0.7 to 0.8.

Single-Effect Heat Transformers

Figure 3 shows a schematic of a single-effect heat transformer (or Type 2 heat pump). All major components are similar to the single-effect, indirect-fired liquid chiller. However, the absorber/evaporator is located above the desorber (generator)/condenser because of the higher pressure level of the absorber and evaporator compared to the desorber/condenser pair, which is the opposite of a chiller.

High-pressure refrigerant liquid enters the top of the evaporator, and heat released from a waste hot-water stream converts it to a vapor. The vapor travels to the absorber section, where it is absorbed by the incoming rich solution. Heat released during this process is used to raise the temperature of a secondary fluid stream to a useful level.

The diluted solution leaves the bottom of the absorber shell and flows through a solution heat exchanger. There it releases heat in counterflow to the rich solution. After the solution heat exchanger, the dilute solution flows through a throttling device, where its pressure is reduced before it enters the generator unit. In the generator, heat from a waste hot-water system generates low-pressure refrigerant vapor. The rich solution leaves the bottom of the generator shell and a solution pump sends it to the absorber.

Low-pressure refrigerant vapor flows from the generator to the condenser coil, where it releases heat to a secondary cooling fluid and condenses. The condensate flows by gravity to a liquid storage sump and is pumped into the evaporator. Unevaporated refrigerant collects at the bottom of the evaporator and flows back into the storage sump below the condenser. Measures must be taken to control the refrigerant pump discharge flow and to prevent vapor from blowing back from the higher-pressure evaporator into the condenser during start-up or during any other operational event that causes low condensate flow. Typically, a column of liquid refrigerant is used to seal the unit to prevent blowback, and a float-operated valve controls the refrigerant flow to the evaporator. Excess refrigerant flow is maintained to adequately distribute the liquid with only fractional evaporation.

Double-Effect Chillers

Figure 4 is a schematic of a commercially available, double-effect indirect-fired liquid chiller. Table 2 lists typical characteristics of this chiller. All major components are similar to the single-effect chiller except for an added generator (first-stage or primary generator), condenser, heat exchanger, and optional condensate subcooling heat exchanger.

Operation of the double-effect absorption machine is similar to that for the single-effect machine. The primary generator receives heat from the external heat source, which boils dilute absorbent solution. Pressure in the primary generator's vapor space is about 15 psia. This vapor flows to the inside of tubes in the second-effect generator. At this pressure, the refrigerant vapor has a condensing temperature high enough to boil and concentrate absorbent solution on the outside of these tubes, thus creating additional refrigerant vapor with no additional primary heat input.

The extra solution heat exchanger (high-temperature heat exchanger) is placed in the intermediate and dilute solution streams flowing to and from the primary generator to preheat the dilute

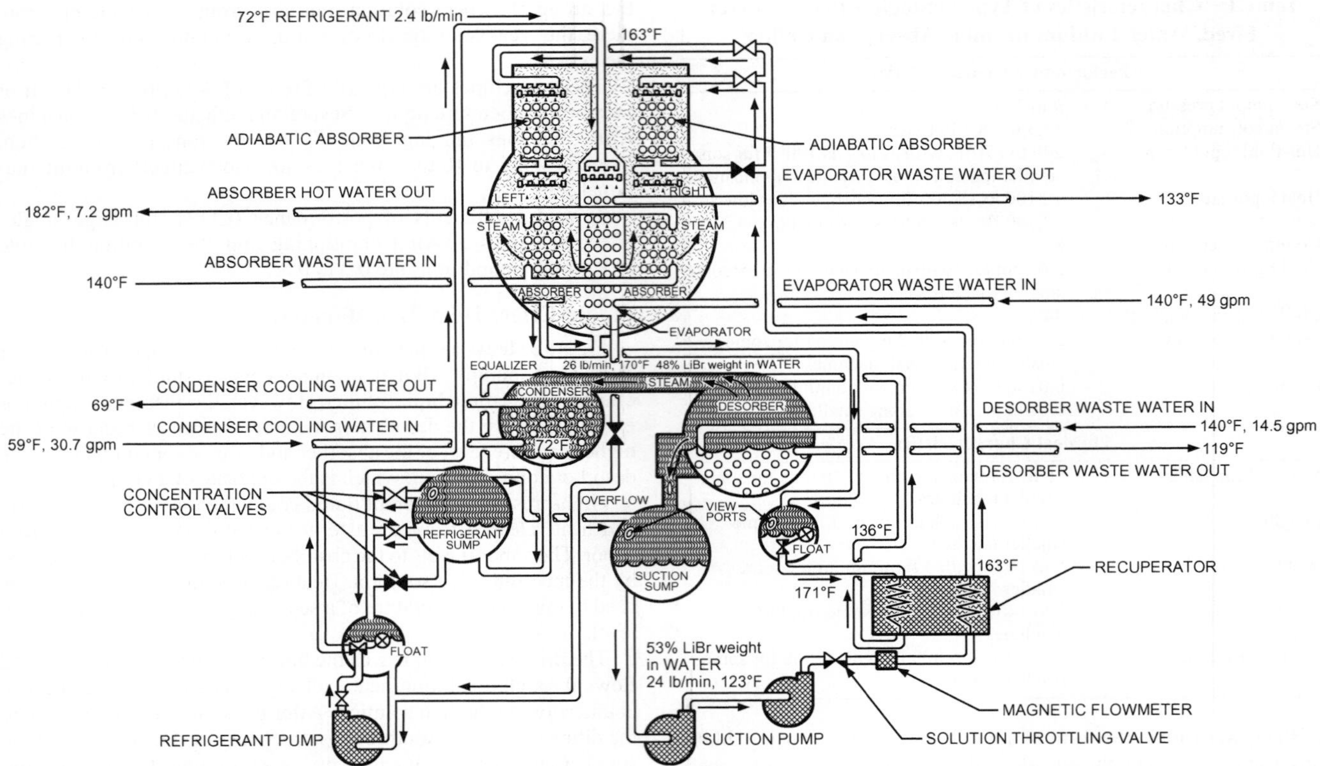

Fig. 3 Single-Effect Heat Transformer

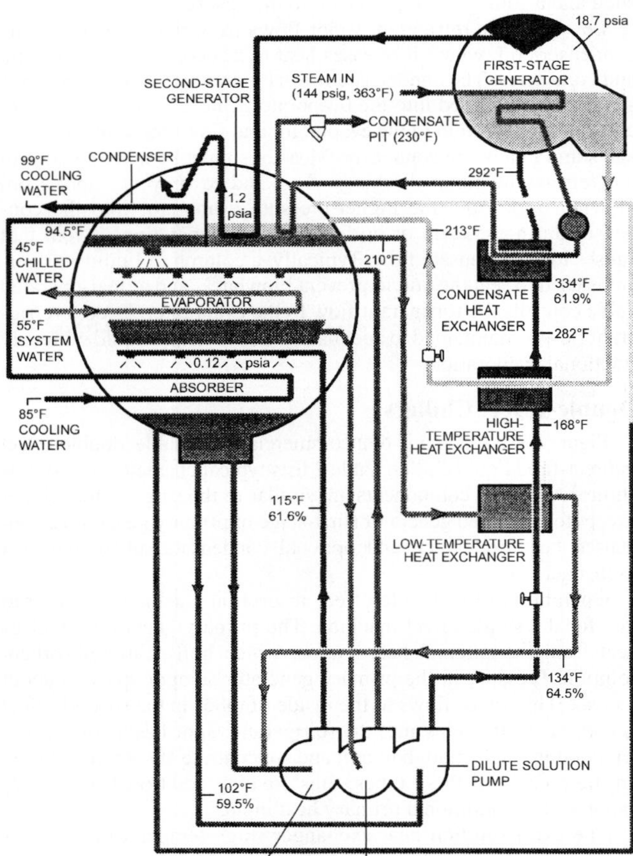

Fig. 4 Double-Effect Indirect-Fired Chiller

Table 2 Characteristics of Typical Double-Effect, Indirect-Fired, Water/Lithium Bromide Absorption Chiller

Performance Characteristics	
Steam input pressure	115 psig
Steam consumption (condensate saturated conditions)	9.7 to 10 lb/ton·h
Hot-fluid input temperature	370°F
Heat input rate	10,000 Btu/ton·h
Cooling water temperature in	85°F
Cooling water flow	3.6 to 4.5 gpm/ton
Chilled water temperature off	44°F
Chilled water flow	2.4 gpm/ton
Electric power	0.01 to 0.04 kW/ton
Physical Characteristics	
Nominal capacities	100 to 1700 tons
Length	10 to 31 ft
Width	6 to 12 ft
Height	8 to 14 ft
Operating weight	15,000 to 132,000 lb

solution. Because of the relatively large pressure difference between the vapor spaces of the primary and secondary generators, a mechanical solution flow control device is required at the outlet of the high-temperature heat exchanger to maintain a liquid seal between the two generators. A valve at the heat exchanger outlet that is controlled by the liquid level leaving the primary generator can maintain this seal.

One or more condensate heat exchangers may be used to remove additional heat from the primary heat source steam by subcooling the steam condensate. This heat is added to the dilute or intermediate solution flowing to one of the generators. The result is a reduction in the quantity of steam required to produce a given refrigeration effect; however, the required heat input remains the same. The COP is not improved by condensate exchange.

As with the single-effect machine, the strong absorbent solution flowing to the absorber can be mixed with dilute solution and pumped over the absorber tubes or can flow directly from the low-temperature heat exchanger to the absorber. Also, as with the single-effect machines, the four major components can be contained in one or two vessels.

The following solution flow cycles may be used:

Series flow. All solution leaving the absorber runs through a pump and then flows sequentially through the low-temperature heat exchanger, high-temperature heat exchanger, first-stage generator, high-temperature heat exchanger, second-stage generator, low-temperature heat exchanger, and absorber, as show in Figure 4.

Parallel flow. Solution leaving the absorber is pumped through appropriate portions of the combined low- and high-temperature solution heat exchanger and is then split between the first- and second-stage generators. Both solution flow streams then return to appropriate portions of the combined solution heat exchanger, are mixed together, and flow to the absorber.

Reverse parallel flow. All solution leaving the absorber is pumped through the low-temperature heat exchanger and then to the second-stage generator. Upon leaving this generator, the solution flow is split, with a portion going to the low-temperature heat exchanger and on to the absorber. The remainder goes sequentially through a pump, high-temperature heat exchanger, first-stage generator, and high-temperature heat exchanger. This stream then rejoins the solution from the second-stage generator; both streams flow through the low-temperature heat exchanger and to the absorber, as shown in Figure 5.

These machines are typically fired with medium-pressure steam of 80 to 144 psig or hot liquids of 300 to 400°F. Typical operating COPs are 1.1 to 1.2. These machines are available commercially from several manufacturers and have capacities ranging from 100 to 1700 tons of refrigeration.

Figure 5 is a schematic of a commercially available double-effect, direct-fired chiller with a reverse parallel flow cycle. Table 3 lists typical characteristics of this chiller. All major components are

Table 3 Characteristics of Typical Double-Effect, Direct-Fired, Water/Lithium Bromide Absorption Chiller

Performance Characteristics	
Fuel consumption (high heating value of fuel)	12,000 to 13,044 Btu/ton·h
COP (high heating value)	0.92 to 1.0
Cooling water temperature in	85°F
Cooling water flow	4.4 to 4.5 gpm/ton
Chilled water temperature off	44°F
Chilled water flow	2.4 gpm/ton
Electric power	0.01 to 0.04 kW/ton
Physical Characteristics	
Nominal capacities	100 to 1500 tons
Length	10 to 34 ft, with minimum of 5 ft for some machines
Width	5 to 21.3 ft, with minimum of 4 ft for some machines
Height	7 to 12 ft
Operating weight	11,000 to 174,600 lb, with a minimum of 3300 lb for some machines

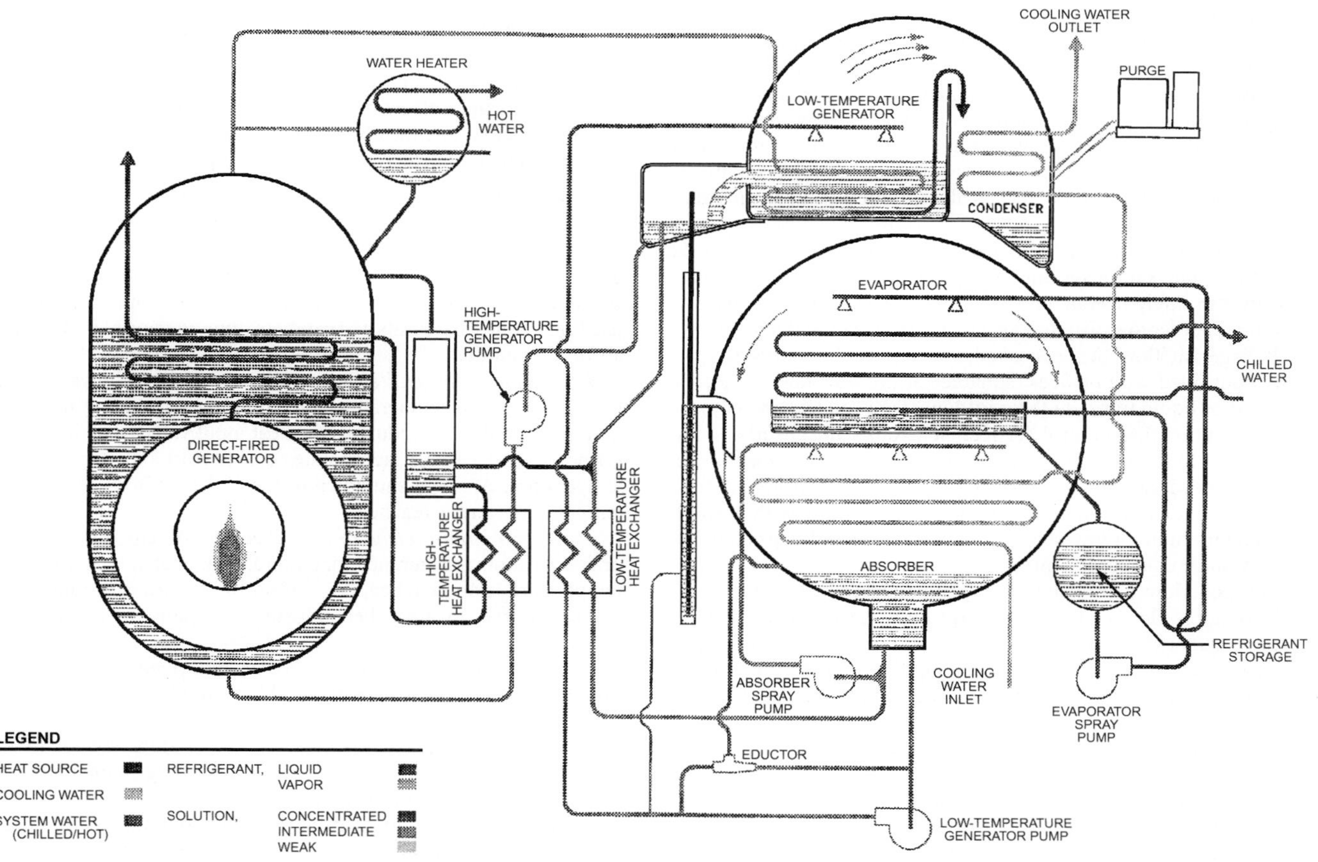

Fig. 5 Double-Effect, Direct-Fired Chiller

similar to the double-effect indirect-fired chiller except for substitution of the direct-fired primary generator for the indirect-fired primary generator and elimination of the steam condensate subcooling heat exchanger. Operation of these machines is identical to that of the double-effect indirect-fired machines. The typical direct-fired, double-effect machines can be ordered with a heating cycle. Some units also offer a simultaneous cycle, which provides about 180°F water, via a heat exchanger, and chilled water, simultaneously. The combined load is limited by the maximum burner input.

These machines are typically fired with natural gas or fuel oil (most have dual fuel capabilities). Typical operating COPs are 0.92 to 1.0 on a fuel input basis. These machines are available commercially from several manufacturers and have capacities ranging from 100 to 1500 tons. Machine capacities of 20 to 100 tons are also available from international sources.

Operation

Modern water/lithium bromide chillers are trouble-free and easy to operate. As with any equipment, careful attention should be paid to operational and maintenance procedures recommended by the manufacturer. The following characteristics are common to all types of lithium bromide absorption equipment.

Operational Limits. Chilled-water temperature leaving the evaporator should normally be between 40 and 60°F. The upper limit is set by the pump lubricant and is somewhat flexible. The lower limit exists because the refrigerant (water) freezes at 32°F.

Cooling water temperature entering the absorber tubes is generally limited to between 55 and 110°F, although some machines limit it to between 70 and 95°F. The upper limit exists because of hydraulic and differential pressure limitations between the generator-absorber, the condenser-evaporator, or both, and to reduce absorbent concentrations and corrosion effects. The lower temperature limit exists because, at excessively low cooling water temperature, the condensing pressure drops too low and excessive vapor velocities carry over solution to the refrigerant in the condenser. Sudden lowering of cooling water temperature at high loads also promotes crystallization; therefore, some manufacturers dilute the solution with refrigerant liquid to help prevent crystallization. The supply of refrigerant is limited, however, so this dilution is done in small steps.

Operational Controls. Modern absorption machines are equipped with electronic control systems. The primary function of the control system is to safely operate the absorption machine and modulate its capacity in order to satisfy the load requirements placed upon it.

Refrigerant flow between condensers and evaporators is typically controlled with orifices (suitable for high- or low-stage condensers) or liquid traps (suitable for low-stage condensers only).

For solution flow control between generators and absorbers, use flow control valves (primary generator of double-effect machines), variable-speed solution pumps, or liquid traps. Refrigerant flow between condensers and evaporators is controlled with orifices (suitable for high- or low-stage condensers) or liquid traps (suitable for low-stage condensers only).

Solution flow control between generators and absorbers typically requires flow control valves (primary generator of double-effect machines), variable-speed solution pumps, or liquid traps. The temperature of chilled water leaving the evaporator is set at a desired value. Deviations from this set point indicate that the machine capacity and the load applied to it are not matched. Machine capacity is then adjusted as required by modulation of the heat input control device. Modulation of heat input results in changes to the concentration of absorbent solution supplied to the absorber if the pumped solution flow remains constant.

Some equipment uses solution flow control to the generator(s) in combination with capacity control. The solution flow may be reduced with modulating valves or solution pump speed controls as the load decreases (which reduces the required sensible heating of solution in the generator to produce a given refrigeration effect), thereby improving part-load efficiency.

Operation of lithium bromide machines with low entering cooling water temperatures or a rapid decrease in cooling water temperature during operation can cause liquid carryover from the generator to the condenser and possible crystallization of absorbent solution in the low-temperature heat exchanger. For these reasons, most machines have a control that limits heat input to the machine based on entering cooling water temperature. Because colder cooling water enhances machine efficiency, the ability of machines to use colder water, when available, is important.

Use of electronic controls with advanced control algorithms has improved part-load and variable cooling water temperature operation significantly, compared to older pneumatic or electric controls. Electronic controls have also made chiller setup and operation simpler and more reliable.

The following steps are involved in a typical start-run-stop sequence of an absorption chiller with chilled and cooling water flows preestablished (this sequence may vary from one product to another):

1. Cooling required signal is initiated by building control device or in response to rising chilled water temperature.
2. All chiller unit and system safeties are checked.
3. Solution and refrigerant pumps are started.
4. Heat input valve is opened or burner is started.
5. Chiller begins to meet the load and controls chilled-water temperature to desired set point by modulation of heat input control device.
6. During operation, all limits and safeties are continually checked. Appropriate action is taken, as required, to maintain safe chiller operation.
7. Load on chiller decreases below minimum load capabilities of chiller.
8. Heat input device is closed.
9. Solution and refrigerant pumps continue to operate for several minutes to dilute the absorbent solution.
10. Solution and refrigerant pumps are stopped.

Limit and Safety Controls. In addition to capacity controls, these chillers require several protective devices. Some controls keep units operating within safe limits, and others stop the unit before damage occurs from a malfunction. Each limit and safety cutout function usually uses a single sensor when electronic controls are used. The following limits and safety features are normally found on absorption chillers:

Low-temperature chilled water control/cutout. Allows the user to set the desired temperature for chilled water leaving the evaporator. Control then modulates the heat input valve to maintain this set point. This control incorporates chiller start and stop by water temperature. A safety shutdown of the chiller is invoked if a low-temperature limit is reached.

Low-temperature refrigerant limit/cutout. A sensor in the evaporator monitors refrigerant temperature. As the refrigerant low-limit temperature is approached, the control limits further loading, then prevents further loading, then unloads, and finally invokes a chiller shutdown.

Chilled water, chiller cooling water, and pump motor coolant flow. Flow switches trip and invoke chiller shutdown if flow stops in any of these circuits.

Pump motor over-temperature. A temperature switch in the pump motor windings trips if safe operating temperature is exceeded and shuts down the chiller.

Pump motor overload. Current to the pump motor is monitored, and the chiller shuts down if the current limit is exceeded.

Absorbent concentration limit. Key solution and refrigerant temperatures are sensed during chiller operation and used to determine the temperature safety margin between solution temperature and solution crystallization temperature. As this safety margin is reduced, the control first limits further chiller loading, then prevents further chiller loading, then unloads the chiller, and finally invokes a chiller shutdown.

In addition to this type of control, most chiller designs incorporate a built-in overflow system between the evaporator liquid storage pan and the absorber sump. As the absorbent solution concentration increases in the generator/absorber flow loop, the refrigerant liquid level in the evaporator storage pan increases. The initial charge quantities of solution and refrigerant are set such that liquid refrigerant will begin to overflow the evaporator pan when maximum safe absorbent solution concentration has been reached in the generator/absorber flow loop. The liquid refrigerant overflow goes to the absorber sump and prevents further concentration of the absorbent solution.

Burner fault. Operation of the burner on direct-fired chillers is typically monitored by its own control system. A burner fault indication is passed on to the chiller control and generally invokes a chiller shutdown.

High-temperature limit. Direct-fired chillers typically have a temperature sensor in the liquid absorbent solution near the burner fire tube. As this temperature approaches its high limit, the control first limits further loading, then prevents further loading, then unloads, and finally invokes a chiller shutdown.

High-pressure limit. Double-effect machines typically have a pressure sensor in the vapor space above the first-stage generator. As this pressure approaches its high limit, the control first limits further loading, then prevents further loading, then unloads, and finally invokes a chiller shutdown.

The performance of lithium bromide absorption machines is affected by operating conditions and the heat transfer surface chosen by the manufacturer. Manufacturers can provide detailed performance information for their equipment at specific alternative operating conditions.

Machine Setup and Maintenance

Large-capacity lithium bromide absorption water chillers are generally put into operation by factory-trained technicians. Proper procedures must be followed to ensure that machines function as designed and in a trouble-free manner for their intended design life (20+ years). Steps required to set up and start a lithium bromide absorption machine include the following:

1. Level unit so internal pans and distributors function properly.
2. Isolate unit from foundations with pads if it is located near noise-sensitive areas.
3. Confirm that factory leaktightness has not been compromised.
4. Charge unit with refrigerant water (distilled or deionized water is required) and lithium bromide solution.
5. Add corrosion inhibitor to absorbent solution if required.
6. Calibrate all control sensors and check all controls for proper function.
7. Start unit and bring it slowly to design operating condition while adding performance additive (usually one of the octyl alcohols).
8. If necessary to obtain design conditions, adjust absorbent and/or refrigerant charge levels. If done correctly, this procedure, known as trimming the chiller, allows the chiller to operate safely and efficiently over its entire operating range.
9. Fine-tune control settings.
10. Check purge operation.

Recommended periodic operational checks and maintenance procedures typically include the following:

- Purge operation and air leaks. Confirm that the purge system operates correctly and that the unit does not have chronic air leaks. Continued air leakage into an absorption chiller depletes the corrosion inhibitor, causes corrosion of internal parts, contaminates the absorbent solution, reduces chiller capacity and efficiency, and may cause crystallization of the absorbent solution.
- Sample absorbent and refrigerant periodically and check for contamination, pH, corrosion-inhibitor level, and performance additive level. Use these checks to adjust the levels of additives in the solution and as an indicator of internal machine malfunctions.

Mechanical systems such as the purge, solution pumps, controls, and burners all have periodic maintenance requirements recommended by the manufacturer.

AMMONIA/WATER ABSORPTION EQUIPMENT

Residential Chillers and Components

In the 1950s, under sponsorship from natural gas utilities, three companies developed a gas-fired, air-cooled residential chiller. Manufacturing volume reached 150,000 units per year in the 1960s, but only a single manufacturer remains at the start of the twenty-first century; the product line is now being changed over to the GAX cycle, described in the section on Special Applications and Emerging Products.

Figure 6 shows a typical schematic of an ammonia/water machine, which is available as a direct-fired, air-cooled liquid chiller in capacities of 3 to 5 tons. Table 4 lists physical characteristics of this chiller. Ammonia/water equipment varies from water/lithium bromide equipment in three main ways:

- Water (the absorbent) is also volatile, so the regeneration of weak absorbent to strong absorbent is a fractional distillation process.
- Ammonia (the refrigerant) causes the cycle to operate at condenser pressures of about 280 psia and at evaporator pressures of approximately 70 psia. As a result, vessel sizes are held to a diameter of 6 in. or less to avoid construction code requirements on small systems, and positive-displacement solution pumps are used.
- Air cooling requires condensation and absorption to occur inside the tubes so that the outside can be finned for greater air contact.

The vertical vessel is finned on the outside to extract heat from the combustion products. Internally, a system of analyzer plates creates

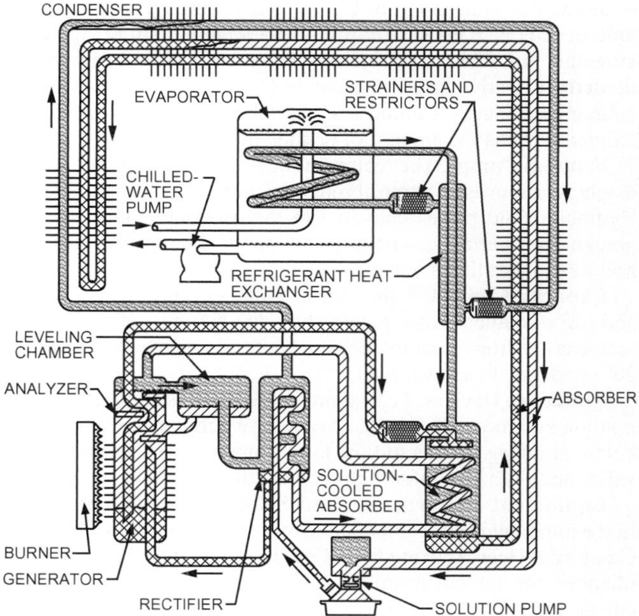

Fig. 6 Ammonia/Water Direct-Fired Air-Cooled Chiller

Table 4 Physical Characteristics of Typical Ammonia/Water Absorption Chiller

Cooling capacities	36,000 to 60,000 Btu/h
Length	40 to 48.5 in.
Width	29.1 to 33.5 in.
Height	37.6 to 46 in.
Weight	550 to 775 lb

intimate counterflow contact between the vapor generated, which rises, and the absorbent, which descends. Atmospheric gas burners depend on the draft of the condenser air fan to sustain adequate combustion airflow to fire the generator. Exiting flue products mix with the air that has passed over the condenser and absorber.

Heat exchange between strong and weak absorbents takes place partially within the generator-analyzer. A tube bearing strong absorbent (nearly pure water) spirals through the analyzer plates, releasing heat to the generation process. Strong absorbent, metered from the generator through the solution capillary, passes over a helical coil bearing weak absorbent, called the solution-cooled absorber. The strong absorbent absorbs some of the vapor from the evaporator, thus releasing the heat of absorption within the cycle to improve the COP. The strong absorbent and unabsorbed vapor continue from the solution-cooled absorber into the air-cooled absorber, where absorption is completed and the heat of absorption is rejected to the air.

The **solution-cooled rectifier** is a spiral coil through which weak absorbent from the solution pump passes on its way to the absorber and generator. Some type of packing is included to assist counterflow contact between condensate from the coil (which is refluxed to the generator) and the vapor (which continues on to the air-cooled condenser). The function of the rectifier is to concentrate the ammonia in the vapor from the generator by cooling and stripping out some of the water vapor.

Absorber and Condenser. These finned-tube air exchangers are arranged so that most of the incoming air flows over the condenser tubes and most of the exit air flows over the absorber tubes.

Evaporator. Liquid to be chilled drips over a coil bearing evaporating ammonia, which absorbs the refrigeration load. On the chilled-water side, which is at atmospheric pressure, a pump circulates the chilled liquid to the load source. Refrigerant to the evaporator is metered from the condenser through restrictors. A tube-in-tube heat exchanger provides the maximum refrigeration effect per unit mass of refrigerant. The tube-in-tube design is particularly effective in this cycle because water present in the ammonia produces a liquid residue that evaporates at increasing temperatures as the amount of residue decreases.

Solution Pumps. The reciprocating motion of a flexible sealing diaphragm moves solution through suction and discharge valves. Hydraulic fluid pulses delivered to the opposite side of the diaphragm by a hermetic vane or piston pump at atmospheric suction pressure impart this motion.

Capacity Control. A thermostat usually cycles the machine on and off. A chilled-water switch shuts the burners off if the water temperature drops close to freezing. Units may also be underfired by 20% to derate to a lower load.

Protective Devices. Typical protective devices include (1) flame ignition and monitor control, (2) a sail switch that verifies airflow before allowing the gas to flow to the burners, (3) a pressure relief valve, and (4) a generator high-temperature switch.

Equipment Performance and Selection. Ammonia absorption equipment is built and rated to meet ANSI *Standard* Z21.40.1, Gas-Fired, Heat-Activated Air Conditioning and Heat Pump Appliances, for outdoor installation. The rating conditions are ambient air at 95°F dry bulb and 75°F wet bulb and chilled water delivered at the manufacturer's specified flow at 45°F. A COP of about 0.5 is realized, based on the higher heating value of the gas.

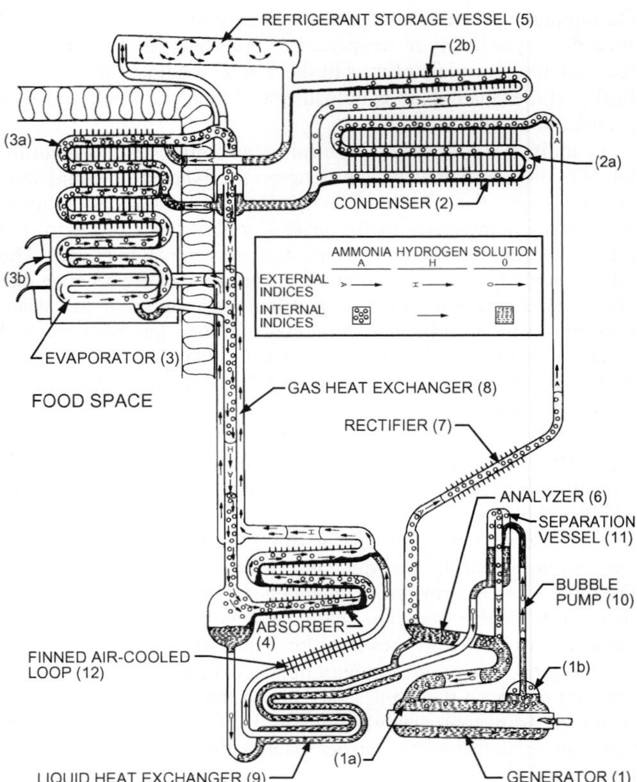

Fig. 7 Domestic Absorption Refrigeration Cycle

Although most units are piped to a single furnace, duct, or fan coil and operated as air conditioners, multiple units supplying a multicoil system for process cooling and air conditioning are also encountered. Also, chillers can be packaged with an outdoor boiler and can supply chilled or hot water as the cooling or heating load requires.

Domestic Absorption Refrigerators and Controls

Domestic absorption refrigerators use a modified absorption cycle with ammonia, water, and hydrogen as working fluids. Wang and Herold (1992) reviewed the literature on this cycle. These units are popular for recreational vehicles because they can be dual-fired by gas or electric heaters. They are also popular for hotel rooms because they are silent. The refrigeration unit is hermetically sealed. All spaces in the system are open to each other and, hence, are at the same total pressure, except for minor variations caused by fluid columns used to circulate the fluids.

The key elements of the system shown in Figure 7 include a generator (1), a condenser (2), an evaporator (3), an absorber (4), a rectifier (7), a gas heat exchanger (8), a liquid heat exchanger (9), and a bubble pump (10). The following three distinct fluid circuits exist in the system: (I) an ammonia circuit, which includes the generator, condenser, evaporator, and absorber; (II) a hydrogen circuit, which includes the evaporator, absorber, and gas heat exchanger; and (III) a solution circuit, which includes the generator, absorber, and liquid heat exchanger.

Starting with the generator, a gas burner or other heat source applies heat to expel ammonia from the solution. The ammonia vapor generated then flows through an analyzer (6) and a rectifier (7) to the condenser (2). The small amount of residual water vapor in the ammonia is separated by atmospheric cooling in the rectifier and drains to the generator (1) through the analyzer (6).

The ammonia vapor passes into section (2a) of the condenser (2), where it is liquified by air cooling. Fins on the condenser increase the cooling surface. Liquified ammonia then flows into an intermediate point of the evaporator (3). A liquid trap between the condenser section (2a) and the evaporator prevents hydrogen from entering the condenser. Ammonia vapor that does not condense in the condenser section (2a) passes to the other section (2b) of the condenser and is liquified. It then flows through another trap into the top of the evaporator.

The evaporator has two sections. The upper section (3a) has fins and cools the freezer compartment directly. The lower section (3b) cools the refrigerated food section.

Hydrogen gas, carrying a small partial pressure of ammonia, enters the lower evaporator section (3) and, after passing through a precooler, flows upward and counterflow to the downward-flowing liquid ammonia, increasing the partial pressure of the ammonia in the vapor as the liquid ammonia evaporates. Although the total pressures in the evaporator and the condenser are the same, typically 20 bar, substantially pure ammonia is in the space where condensation takes place, and the vapor pressure of the ammonia essentially equals the total pressure. In contrast, the ammonia partial pressures entering and leaving the evaporator are typically 1 and 3 bars, respectively.

The gas mixture of hydrogen and ammonia leaves the top of the evaporator and passes down through the center of the gas heat exchanger (8) to the absorber (4). Here, ammonia is absorbed by liquid ammonia/water solution, and hydrogen, which is almost insoluble, passes up from the top of the absorber, through the external chamber of the gas heat exchanger (8), and into the evaporator. Some ammonia vapor passes with the hydrogen from absorber to evaporator. Because of the difference in molecular mass of ammonia and hydrogen, gas circulation is maintained between the evaporator and absorber by natural convection.

Countercurrent flow in the evaporator allows placing the box cooling section of the evaporator at the top of the food space (the most effective location). Gas leaving the lower-temperature evaporator section (3b) also can pick up more ammonia at the higher temperature in the box cooling evaporator section (3a), thus increasing capacity and efficiency. In addition, liquid ammonia flowing to the lower-temperature evaporator section is precooled in the upper evaporator section. The dual liquid connection between condenser and evaporator allows extending the condenser below the top of the evaporator to provide more surface, while maintaining gravity flow of liquid ammonia to the evaporator. The two-temperature evaporator partially segregates the freezing function from the box cooling function, thus giving better humidity control.

In the absorber, strong absorbent flows counter to and is diluted by direct contact with the gas. From the absorber, the weak absorbent flows through the liquid heat exchanger (9) to the analyzer (6) and then to the weak absorbent chamber (1a) of the generator (1). Heat applied to this chamber causes vapor to pass up through the analyzer (6) and to the condenser. Solution passes through an aperture in the generator partition into the strong absorbent chamber (1b). Heat applied to this chamber causes vapor and liquid to pass up through the small-diameter bubble pump (10) to the separation vessel (11). While liberated ammonia vapor passes through the analyzer (6) to the condenser, the strong absorbent flows through the liquid heat exchanger (9) to the absorber. The finned air-cooled loop (12) between the liquid heat exchanger and the absorber precools the solution further. The heat of absorption is rejected to the surrounding air.

The refrigerant storage vessel (5), which is connected between the condenser outlet and the evaporator circuit, compensates for changes in load and the heat rejection air supply temperature.

The following controls are normally present on the refrigerator:

Burner Ignition and Monitoring Control. These controls are either electronic or thermomechanical. Electronic controls ignite, monitor, and shut off the main burner as required by the thermostat. For thermomechanical control, a thermocouple monitors the main flame. The low-temperature thermostat then changes the input to the main burner in a two-step mode. A pilot is not required because the main burner acts as the pilot on low fire.

Low-Temperature Thermostat. This thermostat monitors temperature in the cabinet and controls gas input.

Safety Device. Each unit has a fuse plug to relieve pressure in the event of fire. Gas-fired installations require a flue exhausting to outside air. Nominal operating conditions are as follows:

Ambient temperature	95°F
COP	0.22
Freezer temperature	10°F
Heat input	100 Btu/h·ft^3 of cabinet interior

Industrial Absorption Refrigeration Units

Industrial absorption refrigeration units (ARUs) were pioneered by the Carre brothers in France in the late 1850s. They were first used in the United States for gunpowder production during the Civil War. The technology was placed on a firm footing some 20 years later, when the principles of rectification became known and applied. Rectification is necessary in ammonia/water cycles because the absorbent (water) is volatile.

Industrial ARUs are essentially custom units, because each application varies in capacity, chilling temperature, driving heat, heat rejection mode, or other key parameters. They are almost invariably waste-heat-fired, using steam, hot water, or process fluids. The economics improve relative to mechanical vapor compression at lower refrigeration temperatures and at higher utility rates. These units can produce refrigeration temperatures as low as –70°F, but are more commonly rated for –20 to –50°F.

Industrial ARUs are rugged, reliable, and suitable for demanding applications. For example, they have been directly integrated into petroleum refinery operations. In one early example, the desorber contained hot gasoline, and the evaporator directly cooled lean oil for the oil refinery sponge absorbers. In a recent example, 280°F reformate heated the shell side of the desorber, and the evaporator directly chilled trat gas to –20°F to recover liquified petroleum gas (Erickson and Kelly 1998).

SPECIAL APPLICATIONS AND EMERGING PRODUCTS

Systems Combining Power Production with Waste-Heat-Activated Absorption Cooling

Most prime movers require relatively high-temperature heat to operate efficiently, and reject large amounts of low-temperature heat. In contrast, absorption cycles are uniquely capable of operating at high second-law efficiency with low-temperature heat input. Thus, it is not surprising that many combination systems comprised of fuel-fired prime mover and waste-heat-powered absorption unit have been demonstrated.

These systems come in many forms, usually in ad hoc, one-of-a-kind custom systems. Examples include (1) engine rejects heat to a heat recovery steam generator, and steam powers the absorption cycle; (2) steam boiler powers a steam turbine, and turbine extraction steam powers the absorption cycle; (3) hot engine exhaust directly heats the absorption unit generator; and (4) engine jacket cooling water powers the absorption unit.

Recent programs are under way to better integrate and standardize these combined systems to make them more economical and replicable.

A related technology is derived from the effect of cooling on the inlet air to a compressor. When the compressor supplies a prime mover, the power output is similarly benefitted. Hence, applications are found where combustion turbine waste heat supplies an absorption refrigeration unit, and the cooling in turn chills the inlet air.

Triple-Effect Cycles

Triple-effect absorption cooling can be classified as single-loop or dual-loop cycles. Single-loop triple-effect cycles are basically double-effect cycles with an additional generator and condenser. The resulting system with three generators and three condensers operates similarly to the double-effect system. Primary heat (from a natural gas or fuel oil burner) concentrates absorbent solution in a first-stage generator at about 400 to 450°F. A fluid pair other than water/lithium bromide must be used for the high-temperature cycle. The refrigerant vapor produced is then used to concentrate additional absorbent solution in a second-stage generator at about 300°F. Finally, the refrigerant vapor produced in the second-stage generator concentrates additional absorbent solution in a third-stage generator at about 200°F. The usual internal heat recovery devices (solution heat exchangers) can be used to improve cycle efficiency. As with the double-effect cycles, several variations of solution flow paths through the generators are possible.

Theoretically, these triple-effect cycles can obtain COPs of about 1.7 (not taking into account burner efficiency). Difficulties with these cycles include the following:

- High solution temperatures pose problems to solution stability, performance additive stability, and material corrosion.
- High pressure in the first-stage generator vapor space requires costly pressure vessel design and high-pressure solution pump(s).

A double-loop triple-effect cycle consists of two cascaded single-effect cycles. One cycle operates at normal single-effect operating temperatures and the other at higher temperatures. The smaller high-temperature topping cycle is direct-fired with natural gas or fuel oil and has a generator temperature of about 400 to 450°F. A fluid pair other than water/lithium bromide must be used for the high-temperature cycle. Heat is rejected from the high-temperature cycle at 200°F and is used as the energy input for the conventional single-effect bottoming cycle. Both the high- and low-temperature cycles remove heat from the cooling load at about 44°F.

Theoretically, this triple-effect cycle can obtain an overall COP of about 1.8 (not taking into account burner efficiency).

As with the single-loop triple-effect cycle, high temperatures create problems with solution and additive stability and material corrosion. Also, using a second loop requires additional heat exchange vessels and additional pumps. However, both loops operate below atmospheric pressure and, therefore, do not require costly pressure vessel designs.

GAX (Generator-Absorber Heat Exchange) Cycle

Current air-cooled absorption air-conditioning equipment operates at gas-fired cooling COPs of just under 0.5 at ARI rating conditions. The absorber heat exchange cycle of past air conditioners had a COP of about 0.67 at the rating conditions. In recent years, several projects have been initiated around the world to develop generator-absorber heat exchange (GAX) cycle systems. The best-known programs have been directed toward cycle COPs of about 0.9.

The GAX cycle is a heat-recovering cycle in which absorber heat is used to heat the lower-temperature section of the generator as well as the rich ammonia solution being pumped to the generator. This cycle, like others capable of higher COPs, is more difficult to develop than ammonia single-stage and absorber heat exchange cycles, but its potential gas-fired COPs of 0.7 in cooling mode and 1.5 in heating mode make it capable of significant annual energy savings. In addition to providing a more effective use of heat energy than the most efficient furnaces, the GAX heat pump is able to supply all the heat a house requires to outdoor temperatures below 0°F without the use of supplemental heat.

Solid-Vapor Sorption Systems

Solid-vapor heat pump technology is being developed for zeolite, silica-gel, activated-carbon, and coordinated complex adsorbents. The cycles are periodic in that the refrigerant is transferred periodically between two or more primary vessels. Several concepts providing quasi-continuous refrigeration have been developed. One advantage of solid-vapor systems is that no solution pump is needed. The main challenge in designing a competitive solid-vapor heat pump is to package the adsorbent in such a way that good heat and mass transfer are obtained in a small volume. A related constraint is that good thermal performance of periodic systems requires that the thermal mass of the vessels be small to minimize cyclic heat transfer losses.

Liquid Desiccant/Absorption Systems

In efforts to reduce a building's energy consumption, designers have successfully integrated liquid desiccant equipment with standard absorption chillers. These applications have been building-specific and are sometimes referred to as application hybrids. In a more general approach, the absorption chiller is modified so that rejected heat from its absorber can be used to help regenerate liquid desiccant. Only liquid desiccants are appropriate for this integration because they can be regenerated at lower temperatures than solid desiccants.

The desiccant dehumidifier dries ventilation air sufficiently that, when it is mixed with return air, the building's latent load is satisfied. The desiccant drier is cooled by cooling tower water so that a significant amount of the cooling load is transferred directly to the cooling water. Consequently, absorption chiller size is significantly reduced, potentially to as little as 60% of the size of the chiller in a conventional installation.

Because the air handler is restricted to sensible load, the evaporator in the absorption machine will run at higher temperatures than normal. Consequently, a machine operating at normal concentrations in its absorber rejects heat at higher temperatures. For convenient regeneration of liquid desiccant, only moderate increases in solution concentration are required. These are subtle but significant modifications to a standard absorption chiller.

Combined systems seem to work best when about one-third of the supply air comes from outside the conditioned space. These systems do not require 100% outside air for ventilation, so they should be applicable to conventional buildings as newly mandated ventilation standards are accommodated. Because they always operate in a form of economizer cycle, they are particularly effective during shoulder seasons (spring and fall). As lower-cost liquid desiccant systems become available, reduced first costs may join the advantages of decreased energy use, better ventilation, and improved humidity control.

INFORMATION SOURCES

The are four modern textbooks on absorption: Alefeld and Radermacher (1994), Bogart (1981), Herold et al. (1995), and Niebergall (1981). Other sources of information include conference proceedings, journal articles, newsletters, trade association publications, and manufacturers' literature.

The only recurring conference that focuses exclusively on absorption technology is the triennial Absorption Experts conference, most recently identified as the "International Sorption Heat Pump Conference." Proceedings from these conferences are available from Berlin (1982), Paris (1985), Dallas (1988), Tokyo (1991), New Orleans (1994), Montreal (1996), and Munich (1999).

Technical Committee 8.3 of ASHRAE sponsors symposia on absorption technology at least annually, and the papers appear in *ASHRAE Transactions*.

The Advanced Energy Systems Division of ASME sponsors heat pump symposia approximately annually, with attendant proceedings.

The International Congress of Refrigeration is held quadrennially, under auspices of the International Institute of Refrigeration (IIR). IIR publishes the conference proceedings, and also the

International Journal of Refrigeration, both of which include articles on absorption.

A newsletter covering absorption topics is the *International Energy Agency Heat Pump Center Newsletter.*

The American Gas Cooling Center publishes a comprehensive *Natural Gas Cooling Equipment and Services Guide* plus a periodic journal, *Cool Times*.

REFERENCES

Absorption Experts. Various years. *Proceedings of the International Sorption Heat Pump Conference*.

Alefield, G. and R. Radermacher. 1994. *Heat conversion systems.* CRC, Boca Raton, FL.

ANSI. 1996. Gas-fired, heat-activated air conditioning and heat pump appliances. ANSI *Standard* Z21.40.1-1996/CGA 2.91-M96. American National Standards Institute, Washington, D.C.

Bogart, M. 1981. *Ammonia absorption refrigeration in industrial processes.* Gulf Publishing, Houston.

Erickson, D.C. and F. Kelly. 1998. LPG recovery from refinery flare by waste heat-powered absorption refrigeration. Intersociety Engineering Conference on Energy Conversion, Colorado Springs.

Herold, K.E., R. Radermacher, and S.A. Klein. 1995. *Absorption chillers and heat pump.* CRC, Boca Raton, FL.

Niebergall, W. 1981. *Handbuch der Kältetechnik*, vol. 7: *Sorptionsmaschinen*. R. Plank, ed. Springer Verlag, Berlin.

Wang, L. and K.E. Herold. 1992. *Diffusion-absorption heat pump.* Annual Report to Gas Research Institute, GRI-92/0262.

BIBLIOGRAPHY

Alefeld, G. 1985. *Multi-stage apparatus having working-fluid and absorption cycles, and method of operation thereof.* U.S. Patent No. 4,531,374.

Eisa, M.A.R., S.K. Choudhari, D.V. Paranjape, and F.A. Holland. 1986. Classified references for absorption heat pump systems from 1975 to May 1985. *Heat Recovery Systems* 6:47-61. Pergamon, U.K.

Hanna, W.T. and W.H. Wilkinson. 1982. Absorption heat pumps and working pair developments in the U.S. since 1974: New working pairs for absorption processes, pp. 78-80. *Proceedings of Berlin Workshop* by the Swedish Council for Building Research, Stockholm.

Huntley, W.R. 1984. Performance test results of a lithium bromide-water absorption heat pump that uses low temperature waste heat. Oak Ridge National Laboratory *Report* ORNL/TM9702, Oak Ridge, TN.

IIR. 1991. *Proceedings of the XVIIIth International Congress of Refrigeration*, Montreal, Canada, vol. III. International Institute of Refrigeration, Paris.

IIR. 1992. *Proceedings of Solid Sorption Refrigeration Meetings of Commission B1*, Paris, France. International Institute of Refrigeration, Paris.

Phillips, B.A. 1990. *Development of a high-efficiency, gas-fired, absorption heat pump for residential and small-commercial applications: Phase I Final Report: Analysis of advanced cycles and selection of the preferred cycle*. ORNL/Sub/86-24610/1, September.

Scharfe, J., F. Ziegler, and R. Radermacher. 1986. Analysis of advantages and limitations of absorber-generator heat exchange. *International Journal of Refrigeration* 9:326-333.

Vliet, G.C., M.B. Lawson, and R.A. Lithgow. 1982. Water-lithium bromide double-effect cooling cycle analysis. *ASHRAE Transactions* 88(1):811-823.

Wilkinson, W.H. 1991. A simplified high efficiency DUBLSORB system. *ASHRAE Transactions* 97(1).

CHAPTER 42

FORCED-CIRCULATION AIR COOLERS

FORCED-CIRCULATION unit coolers and product coolers are designed to operate continuously in refrigerated enclosures; a cooling coil and motor-driven fan are their basic components, and provide cooling or freezing temperatures and proper airflow to the room. Coil defrost equipment is added for low-temperature operations when coil frosting might impede performance.

Any unit (e.g., blower coil, unit cooler, product cooler, cold diffuser unit, air-conditioning air handler) is considered a forced-air cooler when operated under refrigeration conditions. Many design and construction choices are available, including (1) various coil types and fin spacing; (2) electric, gas, air, water, or hot-brine defrosting; (3) discharge air velocity and direction; (4) centrifugal or propeller fans, either belt- or direct-driven; (5) ducted or nonducted; and/or (6) freestanding or ceiling-suspended, or penthouse (roof-mounted).

Fans in these units direct air over a refrigerated coil contained in an enclosure. For nearly all applications of these units, the coil lowers airflow temperature below its dew point, which causes condensate or frost to form on the coil surface. However, the normal refrigeration load is a sensible heat load; therefore, the coil surface is considered dry. Rapid and frequent defrosting on a timed cycle can maintain this dry-surface condition, or the coil and airflow can be designed to reduce frost accumulation and its effect on refrigeration capacity.

TYPES OF FORCED-CIRCULATION AIR COOLERS

Figures 1 to 4 illustrate features of some types of air coolers.

Sloped-front unit coolers, often called reach-in unit coolers, range from 5 to 10 in. high (Figure 1). Their distinctive sloped fronts are designed for horizontal top mounting as a single unit, or for installation as a group of parallel connected units. Direct-drive fans are sloped to fit in the restricted return airstream, which rises past the access doors and across the ceiling of the enclosure. Airflows are usually less than 150 cfm per fan. Commonly, these units are installed in back-bar and under-the-counter fixtures, as well as in vertical, self-serve, glass door reach-in enclosures.

Low-air-velocity units feature a long, narrow profile (Figure 2). They have a dual-coil arrangement, and usually two or more fans. These units are used in above-freezing meat-cutting rooms and in carcass and floral walk-in enclosures, as well as 28°F meat carcass holding rooms. They are designed to maintain as high a humidity as possible in the enclosure. The unit's airflow velocity is low and fins on the coil are amply spaced, which reduces the coil's wetted surface area and thus the amount of dew-point contact area for the air stream. Discharge air velocities at the coil face range from 85 to 200 fpm.

Medium-air-velocity unit coolers originally had a half-round appearance, although the more common version (often called **low-profile units**) features a long, narrow, dual-coil unit design (Figure 3). Both types of units are equipped with higher-volume fans. They are used in vegetable preparation rooms, walk-in rooms for wrapped fresh meat, and dairy coolers. These units normally extract more moisture from ambient air than low-velocity units do. Discharge air velocities at the coil face range from 200 to 400 fpm.

Low-silhouette units are 12 to 15 in. high. **Medium- or mid-height units** are 18 to 30 in. high. Those over 30 in. high are

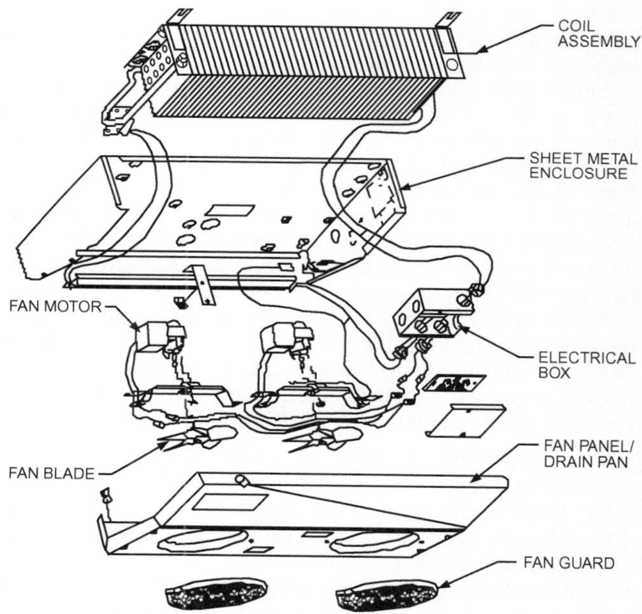

Fig. 1 Sloped-Front Unit Cooler for Reach-In Cabinets

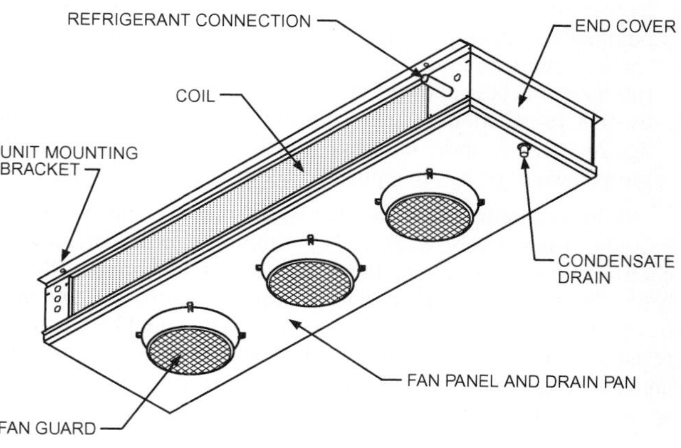

Fig. 2 Low-Air-Velocity Unit

The preparation of this chapter is assigned to TC 8.4, Air-to-Refrigerant Heat Transfer Equipment.

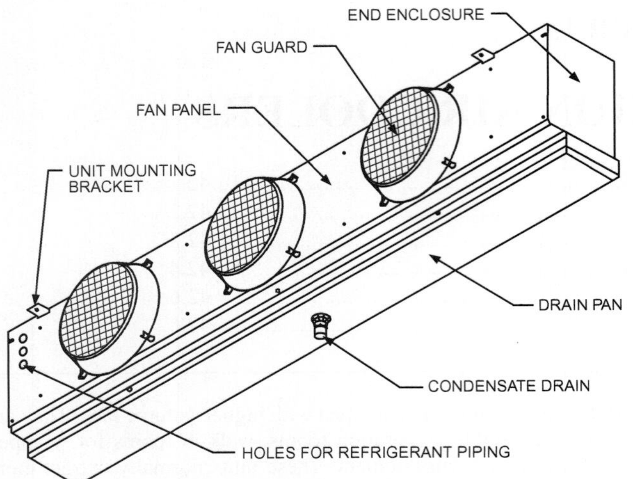

Fig. 3 Low-Profile Cooler

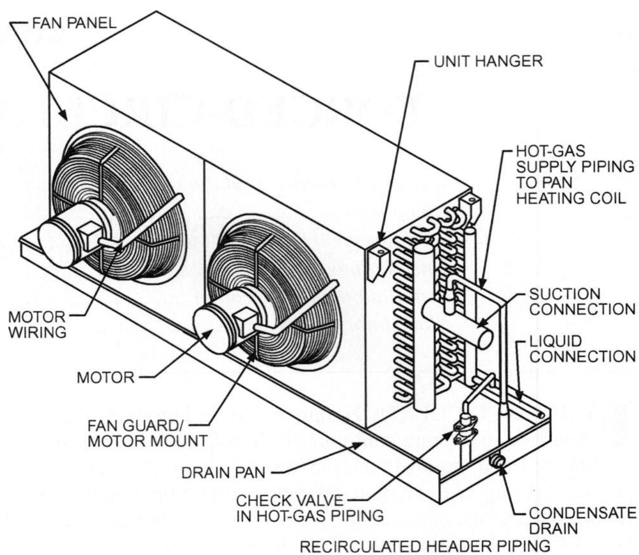

Fig. 4 Liquid Overfeed Unit Cooler

classified as high-silhouette unit coolers, which are used in warehouse-sized coolers and freezers. Air velocity at the coil face can be over 600 fpm. Outlet air velocities range from 1000 to 2000 fpm when the unit is equipped with cone-shaped fan discharge venturis for extended air throw.

Spray coils feature a saturated coil surface that can cool processed air closer to the coil surface temperature than can a regular (nonsprayed) coil. In addition, the spray continuously defrosts the low-temperature coil. Unlike unit coolers, spray coolers are usually floor-mounted and discharge air vertically. Unit sections include a drain pan/sump, coil with spray section, moisture eliminators, and fan with drive. The eliminators prevent airborne spray droplets from discharging into the refrigerated area. Typically, belt-driven centrifugal fans draw air through the coil at 600 fpm or less.

Water can be used as the spray medium for coil surfaces with temperatures above freezing. For coil surfaces with temperatures below freezing, a suitable chemical must be added to the water to lower the freezing point to 12°F, or below the coil surface temperature. Some suitable recirculating solutions include the following:

- **Sodium chloride** solution is limited to a room temperature of 10°F or higher. Its minimum freezing point is –6°F.
- **Calcium chloride** solution can be used for enclosure temperatures down to about –10°F, but its use may be prohibited in enclosures containing food products.
- **Aqueous glycol** solutions are commonly used in water and/or sprayed-coil coolers operating below freezing. Food-grade propylene glycol solutions are commonly used because of their low oral toxicity, but they generally become too viscous to pump at temperatures below –13°F. Ethylene glycol solutions may be pumped at temperatures as low as –40°F. Because of its toxicity, sprayed ethylene glycol in other than sealed tunnels or freezers (no human access allowed during process) is usually prohibited by most jurisdictions. When a glycol mix is sprayed in food storage rooms, any spray carryover must be maintained within the limits prescribed by all applicable regulations.

All brines are hygroscopic; that is, they absorb condensate and become progressively weaker. This dilution can be corrected by continually adding salt to the solution to maintain a sufficient below-freezing temperature. Salt is extremely corrosive, and must be contained in the sprayed-coil unit with suitable corrosive-resistant materials or coatings, which must be periodically inspected and maintained. All untreated brines are corrosive: neutralizing the spray solution relative to its contact material is required.

Sprayed-coil units are usually installed in refrigerated enclosures requiring high humidity (e.g., chill coolers). Paradoxically, the same sprayed-coil units can be used in special applications requiring low relative humidity. For these applications, both a high brine concentrate (near its eutectic point) and a large difference between the process air and the refrigerant temperature are maintained. Process air is reheated downstream from the sprayed coil to correct the dry-bulb temperature.

COMPONENTS

Draw-Through and Blow-Through Airflow

Unit fans may draw air through the cooling coil and discharge it through the fan outlet into the enclosure, or they may blow air through the cooling coil and discharge it from the coil face into the enclosure. Blow-through units have a slightly higher thermal efficiency because heat from the fan is removed from the forced airstream by the coil, but their air distribution pattern is less effective than the draw-through design. Draw-through fan energy adds to the heat load of the refrigerated enclosure, but heat gain from fractional horsepower or small three-phase integral fan motors is not significant. Selection of draw-through or blow-through depends more on a manufacturer's design features for the unit size required, air throw required for the particular enclosure, and accessibility of the coil for periodic surface cleaning.

The blow-through design has a lower discharge air velocity because the entire coil face area is usually the discharge opening (grilles and diffusers not considered). Throw of 33 ft or less is common for the average standard air velocity from a blow-through unit. Greater throw, in excess of 100 ft, is normal for draw-through centrifugal units. The propeller fan in the high-silhouette draw-through unit cooler is popular for intermediate ranges of air throw.

Fan Assemblies

Direct-drive propeller fans (motor plus blade) are popular because they are simple, economical, and can be installed in multiple assemblies in a unit cooler housing. Additionally, they require less motor power for a given airflow capacity.

The centrifugal fan assembly usually includes belts, bearings, sheaves, and coupler drives, each with inherent maintenance problems. This design is necessary, however, for applications having high air distribution static pressure losses (e.g., enclosures with ductwork runs, tunnel conveyors, and densely stacked products). Centrifugal-fan-equipped units are also used in produce-ripening rooms, where a large air blast and 0.5 to 0.75 in. discharge air static is needed for proper air circulation around all the product in the enclosure, to ensure uniform batch ripening.

Casing

Casing materials are selected for compatibility with the enclosure environment. Aluminum (coated or uncoated) or steel (galvanized or suitably coated) are typical casing materials. Stainless steel is also used in food storage or preparation enclosures where sanitation must be maintained. On larger cooler units, internal framing is fabricated of sufficiently substantial material, such as galvanized steel, and casings are usually made with similar material. Some plastic casings are used in small unit coolers, whereas some large, ceiling-suspended units may have all-aluminum construction to reduce weight.

Coil Construction

Coil construction varies from uncoated (all) aluminum tube and fin to hot-dipped galvanized (all) steel tube and fin, depending on the type of refrigerant used and the environmental exposure of the coil. The most popular unit coolers have coils with copper tubes and aluminum fins. Ammonia refrigerant evaporators never use copper tubes because ammonia corrodes copper. Also, sprayed coils are not constructed with aluminum fins unless they are completely protected with a baked-on phenolic dip coating or similar protection applied after fabrication. Coils constructed with stainless steel tubes and fins are preferred in corrosive environments, and all-stainless construction, or with aluminum fins, is preferred in environments where high standards of sanitation are maintained.

Fin spacings vary from 6 to 8 fins per inch for coils with surfaces above 32°F when latent loads are insignificant. Otherwise, 3 to 6 fins per inch is the accepted spacing for coil surfaces below 32°F, with a spacing of 4 fins per inch when latent loads exceed 15% of the total load. One and two fin(s) per inch are used when defrosting is set for once a day, such as in low-temperature supermarket display cases. Staged fins in a row of coils, such as a 1-2-4 fins per inch spacing combination, greatly reduce fin blockage by frost accumulation (Ogawa et al. 1993).

Even distribution of the refrigerant flow to each circuit of the coil is vital for maximizing cooler coil performance. Distributor assemblies are used for direct-expansion halocarbon refrigerants and occasionally for large, medium-temperature ammonia units. Application requires that they be precisely sized. Distributor design and construction material may vary by refrigerant type and application. Application information from the distributor manufacturer should be closely followed, particularly regarding orifice sizing and assembly mounting orientation on the coil.

For liquid pumped recirculating systems, orifice disks are usually used in lieu of a distributor assembly. These disks are sized and installed by the coil manufacturer. They fit in the inlet (supply) header, at the connection spuds of each coil circuit. The specifying engineer may require a down-feed distributor assembly, less any orifice, if significant flash gas is anticipated.

Headers and their piping connections are part of the coil assembly. Usually, header lengths equal the coil height dimension; therefore each header is sized to the coil capacity for the application, based on refrigerant flow velocities and not on the temperature equivalent of the saturated suction temperature drop. Velocities of approximately 1500 fpm are used to compute the size of the return gas header and its connection size. In the field, connection size is often mistaken to be the recommended return line size, but the size of lines installed in the field should be based on the suction drop calculation method (see Chapters 1, 2, and 3).

Frost Control

Coils must be defrosted when frost accumulates on their surfaces. The frost (or ice) is usually greatest at the coil's air entry side; therefore, the required defrost cycle is determined by the inlet surface condition. In contrast, a reduced secondary-surface-to-primary-surface ratio produces greater frost accumulations at the coil outlet face. A long-held theory is that accumulation of relatively more frost at the coil entry air surface somewhat improves the heat transfer capacity of the coil. However, overall accumulated coil frost usually has two negative effects: it (1) impedes heat transfer because of its insulating effect, and (2) reduces airflow because it restricts the free airflow area within the coil. Both effects, to different degrees, result from combinations of airflow, fin spacing, frost density, and ambient air conditions.

Depending on the defrost method, as much as 80% of the defrost head load of the unit could be transferred into the enclosure. This heat load is not normally included as part of the enclosure heat gain calculation. The unit's refrigeration capacity rating is averaged over a 24 h period, by a factor that estimates the typical hours per day of refrigeration running time, including the defrost cycles.

As previously mentioned, a longer time between defrost cycles can be achieved by using more coil tube rows and wider fin spacing. Ice accumulation, which interferes with airflow, should be avoided to reduce both the frequency and duration of the defrost cycles. For example, in low-temperature applications having high latent loads, unit coolers should not be located above freezer entry or exit doors.

Operational Controls

In the simplest form, electromechanical controls cycle the refrigeration system components to maintain the desired enclosure temperature and defrost cycle. Pressure-responsive modulating control valves, such as evaporator-pressure regulators and head-pressure controls, are also used. A temperature control could be a thermostat mounted in the enclosure, used to cycle the compressor on and off, or a liquid-line solenoid valve that allows liquid refrigerant to flow to the evaporator coil. A suction-pressure switch at the compressor can substitute for the wall-mounted thermostat.

Electronic controls have made electromechanical controls obsolete, except on very small unit installations. Microprocessor controllers mounted at the compressor receive and process signals from one or more temperature diode sensors and/or pressure transducers. These signals are converted to coordinate precise control of the compressor and the suction, discharge, and liquid-line flow-control valves. Defrost cycling, automatic callout for service, and remote site operation checks are standard options on the typical type of microprocessor controller used in refrigeration. For large warehouses and supermarkets, an electronically based energy management system (EMS) can easily incorporate multicompressor systems into virtually any type of control system.

AIR MOVEMENT AND DISTRIBUTION

Air distribution is an important concern in refrigerated enclosure design and location of unit coolers. The direction of the air and air throw should be such that air moves where there is a heat gain. This principle implies that the air sweeps the enclosure walls and ceiling as well as to the product. Nearly all unit coolers are ceiling-mounted and should be placed (1) so they do not discharge air at any doors or openings, (2) away from doors that do not incorporate an entrance vestibule or pass to another refrigerated enclosure to keep from inducing additional infiltration into the enclosure, and (3) away from the airstream of another unit to avoid defrosting difficulties.

The velocity and relative humidity of air passing over an exposed product affect that product's surface drying and weight loss. Air velocities up to 500 fpm over the product are typical for most freezer applications. Higher velocities require additional fan power and, in many cases, only slightly decrease cooling time. For example, air velocities over 500 fpm for freezing plastic-wrapped bread reduce freezing time very little. However, increasing air velocity from 500 to 1000 fpm over unwrapped pizza reduces freezing time and product exposure by almost half. This variation shows that product testing is necessary to design the special enclosures intended for blast freezing and/or automated food processing. Sample tests should yield the following information: ideal air temperature, air velocity, product weight loss, and dwell time. With this information, the proper unit or product coolers, as well as supporting refrigeration equipment and controls, can be selected.

UNIT RATINGS

No industry standard exists for rating unit and product coolers. Part of the difficulty in developing a workable standard is the many variables encountered. Cooler coil performance and capacities should be based on a fixed set of conditions, and they greatly depend on (1) air velocity, (2) refrigerant velocity, (3) circuit configuration, (4) refrigerant blend glide, (5) temperature difference, (6) frost condition, and (7) superheating adjustment. The most significant items are refrigerant flow rate, as related to refrigerant feed through the coil, and frost condition defrosting in low-temperature applications. The following sections discuss a number of performance differences relative to some of the available unit cooler variations.

Refrigerant Velocity

Depending on the commercially available refrigerant feed method used, both the cooler's capacity ratings and its refrigerant flow rates vary. The following feed methods are used:

Dry Expansion. In this system, a thermostatically controlled, direct-expansion valve allows just enough liquid refrigerant into the cooling coil to ensure that it vaporizes at the outlet. In addition, 5 to 15% of the coil surface is used to superheat the vapor. Direct-expansion (DX) coil flow rates are usually the lowest of all the feed methods.

Recirculated Refrigerant. This system is similar to a dry expansion feed except it includes a recirculated refrigerant drum (i.e., a low-pressure receiver) and a liquid refrigerant pump connected to the coil. It also has a hand expansion valve, which is the metering device used to control the flow of the entering liquid refrigerant. The coil is intentionally overfed liquid refrigerant by the pump, such that complete coil flooding eliminates superheating of the refrigerant in the evaporator. The amount of liquid refrigerant pumped through the coil may be two to six times greater (overfeed: 1 to 5) than that passed through a dry DX coil. As a result, this coil's capacity is higher than that of a dry expansion feed. To accurately calculate rated capacity, supply refrigerant temperature and pressure for the operating evaporator temperature should be provided by the air cooler's manufacturer (see Chapter 1 for further information).

Flooded. This system has a liquid reservoir (surge drum or accumulator) located next to each unit or set of units. The surge drum is filled with subcooled refrigerant and connected to the cooler coil. To ensure gravity flow of the refrigerant and a completely wet internal coil surface, the liquid level in the surge drum must be equal to the top of the coil. Gravity-recirculated feed capacity is usually the highest attainable, in part because large coil tubes (≥1 in. OD) are required so that virtually no evaporator pressure drop exists. In flooded gravity systems, the relative position of the surge drum to the air cooler, as well as their interconnecting piping and valves, are all important for proper operation. The intended location of these components and valves should be provided by the manufacturer.

Brine. In this chapter, "brine" encompasses any liquid or solution that absorbs heat in the coil without a change in state; these fluids are also called secondary refrigerants. Aqueous glycols, ethylene, and propylene are well accepted and thus most often used. Food-grade propylene glycol should be used in food-processing applications. Calcium chloride or sodium chloride in water (for extra-low-temperature applications) and R-30 can be used only under tightly controlled and monitored conditions. For corrosion protection, most of these solutions must be neutralized or inhibited (preferably by the chemical manufacturer) before being introduced into the system.

The capacity rating for a brine coil depends on the thermal properties of the brine (freeze point, thermal conductivity, viscosity, specific heat, density) and its flow rate in the coil. This rating is usually obtained by special request from the coil manufacturer. Generally, coils handling a commercial inhibited glycol solution have about 11% less capacity at low temperatures and 14% less capacity at medium temperatures than comparable direct-expansion halocarbon refrigerants. The glycol temperature must run 8 to 10°F lower than the comparable saturated suction temperature of a comparable DX coil to obtain the same capacity.

Frost Condition

Frost accumulation on the coil and its defrosting are perhaps the most indeterminate variables that affect the capacity of forced-air coolers. Ogawa et al. (1993) showed that a light frost accumulation slightly improves the heat transfer of the coil. Continuous accumulation has a varying result, depending on the airflow. Performance suffers when airflow through the coil is reduced because coil surface frosting increases air-side static pressure (e.g., as in prop-fan unit coolers). But if airflow through a frosting coil is maintained (e.g., a variable-speed fan arrangement), frost reduces capacity somewhere between 2 to 10% (Kondepudi and O'Neal 1990; Rite and Crawford 1991). Thermal resistance of the frost (ice) varies with time and temperature, and ice pack growth is a product of operating at a surface temperature below the air dew point. Ultimately, defrosting is the only way to return to rated performance. This is usually initiated when unit performance drops to 75 to 80% of rated.

Controlled lab tests also showed that frost growth on a finned surface is not uniform with coil depth. Fin spacing is by far the biggest factor in restricting airflow through the coil. For DX coils, the location of the superheat region in the coil had the most effect on uniformity. Oskarsson et al. (1990) discussed the effect of the length of time of frosting on uniformity. The industry generally considers that ice formation is uniform through a coil with a wide fin spacing (i.e., <5 fins/in.). This spacing is used to determine an interfin free-air area to estimate the air static pressure drop through a coil operating under frosting conditions.

Defrosting

The defrost cycle may be initiated and terminated in a number of ways. Microprocessor control, which has largely replaced the mechanical time clock, has reduced energy use and helped to maintain product quality (by reducing temperature rise in the enclosure during defrost). Accurate, short-time defrosting is now a health-safety concern. Too long a defrost cycle can result in an unacceptable product core temperature rise. These conditions are vigorously monitored by most local and state health departments. In addition, proper defrost initiation and termination are needed. Accurate defrosting also provides better protection for the refrigeration equipment. Improper and/or incomplete defrosting can damage the compressor and evaporator coil, to the extent that irreparable refrigerant leaks develop when ice is allowed to build up and crush one or more of the coil tubes. The following defrosting methods are in use.

For Enclosure Air Temperature Above 35°F. Enclosure air that is 35°F or slightly warmer can be used to defrost a cooler coil. Fans are left on and defrosting occurs during compressor off cycles. However, some moisture on the coil surface evaporates, which is undesirable for a low-humidity application. The following methods of control are commonly used:

- If the refrigeration cycle is interrupted by a **defrost timer**, the continually circulating air melts the coil frost and ice. The timer can operate either the compressor or a liquid-line solenoid valve.

- An oversized unit cooler controlled by a **wall thermostat** defrosts during its normal off/on cycling. The thermostat can control a refrigeration solenoid in a multiple-coil system or the compressor in a unitary installation. *Note*: An oversized unit is sized to handle a 24 h cooling load in 16 h.
- A **pressure control** can be used for slightly oversized unitary equipment. A low-pressure switch connected to the compressor suction line is set at a cut-out point such that the design suction pressure corresponds to the saturated temperature required to handle the maximum enclosure load. The suction pressure at the compressor drops and causes the compressor motor to stop as the enclosure load fluctuates, or as the oversized compressor overcomes the maximum loading.

The thermostatic expansion valve on the unit cooler controls evaporator temperature by regulating its liquid refrigerant flow, which varies with the load. The cut-in point, which restarts the compressor motor, should be set at the suction pressure that corresponds to the equivalent saturated temperature of the desired refrigerated enclosure air temperature. The pressure differential between the cut-in and cutout points corresponds to the temperature difference between the enclosure air and coil. Pressure settings should allow for the pressure drop in the suction line.

For Enclosure Air Temperature Below 35°F. When enclosure air is below 35°F, supplementary heat must be introduced into the enclosure to defrost the coil surface and drain pan. Unfortunately, some of this defrost heat remains in the enclosure until the unit starts operation after completion of the defrost cycle. The following supplemental heat sources are used for defrosting:

- **Gas defrosting** can be the fastest and most efficient method if an adequate supply of hot gas is available. Besides performing the defrost function, hot refrigerant discharge gas internally clears the coil and drain pan tube assembly of accumulated compressor oil. This aids in returning the oil to the compressor. Gas defrosting is used for small, commercial single and multiplex units, as well as for large, industrial central plants; it is broadly used on most low-temperature applications. Hot-gas defrosting also increases the capacity of a large, continuously operating compressor system because it removes some of the load from the condenser as it alternately defrosts the multiple evaporators. This method of defrost puts the least amount of heat into the enclosure ambient. A further improvement on hot-gas defrosting is using latent gas (sometimes called cool gas) from the top part of the receiver.
- **Electric defrost** effectiveness depends on the location of the electric heating elements. The elements can be either attached to the finned coil surface or inserted inside special fin holes or dummy tubes in the coil element. Electric defrost can be efficient and rapid. It is simple to operate and maintain, but it dissipates the most heat into the enclosure, and, depending on energy costs, may not be as economical to operate as gas defrosting.
- **Heated air** may be circulated in a loop within freezer units that are constructed so as to isolate the frosted coil from the cold enclosure air. This is mostly done in packaged units, with dampers used to isolate the cooling coil. Once the coil is isolated, the unit's airflow is heated by a hot-gas reheat coil or electric heating elements. Heated air circulates in the unit to perform the defrost, and also must heat a drain pan, which is needed in all enclosures at temperatures of 34°F or less. Some units have specially constructed housings and ducting to draw warm air from adjoining areas.
- **Water defrost** is the quickest method of defrosting a unit. It is efficient and effective for rapid cleaning of the complete coil surface. Water defrost can be performed manually or on an automatic timed cycle. This method becomes less desirable as the enclosure temperature decreases much below freezing, but it has been successfully used in applications as low as –40°F. Water defrost is used more for large units used for cooling industrial products. This application typically has a large reservoir of warm condenser water provided by heat reclaim from the water-cooled condenser.
- **Hot brine** can be used to defrost brine-cooled coils by remotely heating the brine for the defrost cycle. This system heats from within the coil and is as rapid as hot-gas defrost. The heat source can be steam, electric resistance elements, or condenser water.

Defrost Control. For the most part, defrosting is done with the fan turned off. Inadequate defrost time and over-defrosting both can degrade overall performance; thus, a defrost cycle is best ended by monitoring temperature. A thermostat may be mounted in the cooler coil to sense a rise in the temperature of the finned or tube surface. A temperature of at least 45°F indicates frost removal and automatically returns the unit to the cooling cycle.

Fan operation is delayed, usually by the same thermostat, until coil surface temperature approaches its normal operating level. This practice prevents unnecessary heating of the enclosure after defrost. It also prevents drops of defrost water from being blown off the coil surface, which avoids icing of the fan blade, guard, and orifice ring. In some applications, fan delay after defrost is essential to prevent a rapid buildup of ambient air pressure, which could structurally damage the enclosure.

Defrost initiation can be automated by time clocks, running time monitors, or air-pressure-differential controls, or by monitoring the air temperature difference through the coil (which increases as frost accumulation reduces the airflow). Adequate supplementary heat for the drain pan and condensate drain lines should be considered. It is not uncommon for two methods to run simultaneously (e.g., hot-gas and electric) to simplify drain pan defrosting and shorten the defrost cycle. Drain lines should be properly pitched, insulated, and trapped outside the freezer, preferably when traversing a warm area.

Basic Cooling Capacity

Most rating tables state gross capacity and assume fan assembly or defrost heat is included in enclosure load calculation. Some manufacturers' cooler coil ratings may appear as sensible capacity; others may be listed as total capacity, which includes both sensible and latent capacities. Some ratings include reduction factors to account for frost accumulation in low-temperature applications or for some unusual condition. Others include capacity multiplier factors for various refrigerants.

The published rating, defined as the basic cooling capacity, is based on the temperature difference (TD) between inlet air and refrigerant in the coil (Btu/h per degree TD). The coil inlet air temperature is considered to be the same as the enclosure air temperature, and the refrigerant temperature is assumed to be the temperature equivalent to the saturated pressure at the coil outlet. This practice is common for both cooler and freezer enclosure (unit coolers) applications. For heavy-duty use (e.g., for a blast freezer or process conveyor work), manufacturers' ratings may be based somewhat differently, such as on the average of the coil inlet-to-outlet air temperatures considered as the enclosure temperature.

The TD necessary to obtain the unit cooler capacity varies with the application. It may be as low as 8°F for wet storage coolers and as high as 25°F for gut storage and workrooms. TD can be related to the desired humidity requirements. The smaller the TD, the less dehumidification from coil operation. The following is general guidance for selecting a proper TD for medium-temperature applications above 25°F saturated suction:

- For *very high* relative humidity (about 90%), a TD of 8 to 10°F is common.
- For *high* relative humidity (approximately 80%), a TD of 10 to 12°F is recommended.
- For *medium* relative humidity (approximately 75%), a TD of 12 to 16°F is recommended.

Temperature differences above these limits usually result in low enclosure humidities, which dry the product. However, for packaged products and workrooms, a TD of 25 to 30°F is not unusual. Paper storage or similar products also require a low humidity level, and a TD of 20 to 30°F may be necessary.

For low-temperature applications below 25°F saturated suction, the TD is generally kept below 15°F because of system economics and frequency of defrosting rather than for humidity control.

Refer to ASHRAE *Standard* 25 for unit cooler testing methods and to ARI *Standard* 420 for unit cooler rating procedures. It is advisable that the specifying engineer check the individual manufacturer's literature for all such rating factors.

INSTALLATION AND OPERATION

Whenever possible, refrigerating air-cooling units should be located away from enclosure entrance doors and passageways. This practice helps reduce coil frost accumulation and fan blade icing. The cooler manufacturer's installation, start-up, and operation instructions generally give the best information. On installation, the unit nameplate data (model, refrigerant type, electrical data, warning notices, certification emblems, etc.) should be recorded and compared to the job specifications and to the manufacturer's instructions for correctness.

MORE INFORMATION

Additional information on the selection, ratings, installation, and maintenance of cooler units is available from the manufacturers of that type of equipment. Chapters 9 to 29 of this volume have specific product cooling information.

REFERENCES

ARI. 2000. Unit coolers for refrigeration. *Standard* 420. Air Conditioning and Refrigeration Institute, Arlington, VA.

ASHRAE. 2001. Method of testing forced and natural convection air coolers for refrigeration. ANSI/ASHRAE *Standard* 25-2001.

Kondepudi, S.N. and D.L. O'Neal. 1990. The effect of different fin configurations on the performance of finned-tube heat exchangers under frosting conditions. *ASHRAE Transactions* 96(2):439-444.

Ogawa, K., N. Tanaka, and M. Takashita. 1993. Performance improvement of plate fin-and-tube heat exchangers under frosting conditions. *ASHRAE Transactions* 99(1):762-771.

Oskarsson, S.P., K.I. Krakow, and S. Lin. 1990. Evaporator models for operation with dry, wet, and frosted finned surfaces—Part II: Evaporator models and verification. *ASHRAE Transactions* 96(1):381-392.

Rite, R.W. and R.R. Crawford. 1991. The effect of frost accumulation on the performance of domestic refrigerator freezer finned-tube evaporator coils. *ASHRAE Transactions* 97(2):428-437.

CHAPTER 43

COMPONENT BALANCING IN REFRIGERATION SYSTEMS

THIS chapter describes methods and components used in balancing a primary refrigeration system. A refrigerant is a fluid used for heat transfer in a refrigeration system. The fluid absorbs heat at a low temperature and pressure and transfers heat at a higher temperature and pressure. Heat transfer can involve either a complete or partial change of state in the case of a primary refrigerant. Energy transfer is a function of the heat transfer coefficients; temperature differences; and amount, type, and configuration of the heat transfer surface and, hence, the heat flux on either side of the heat transfer device.

REFRIGERATION SYSTEM

A typical basic direct-expansion refrigeration system includes an evaporator, which vaporizes incoming refrigerant as it absorbs heat, increasing the refrigerant's heat content or enthalpy. A compressor pulls vapor from the evaporator through suction piping and compresses the refrigerant gas to a higher pressure and temperature. The refrigerant gas then flows through the discharge piping to a condenser, where it is condensed by rejecting its heat to a coolant (e.g., other refrigerants, air, water, or air/water spray). The condensed liquid is supplied to a device that reduces pressure, cools the liquid by flashing vapor, and meters the flow. The cooled liquid is returned to the evaporator. For more information on the basic refrigeration cycle, see Chapter 1 of the 2005 *ASHRAE Handbook—Fundamentals*.

Gas compression theoretically follows a line of constant entropy. In practice, adiabatic compression cannot occur because of friction and other inefficiencies of the compressor. Therefore, the actual compression line deviates slightly from the theoretical. Power to the compressor shaft is added to the refrigerant, and compression increases the refrigerant's pressure, temperature, and enthalpy.

In applications with a large compression ratio (e.g., low-temperature freezing, multitemperature applications), multiple compressors in series are used to completely compress the refrigerant gas. In multistage systems, interstage desuperheating of the lower-stage compressor's discharge gas protects the high-stage compressor. Liquid refrigerant can also be subcooled at this interstage condition and delivered to the evaporator for improved efficiencies.

An intermediate-temperature condenser can serve as a cascading device. A low-temperature, high-pressure refrigerant condenses on one side of the cascade condenser surface by giving up heat to a low-pressure refrigerant that is boiling on the other side of the surface. The vapor produced transfers energy to the next compressor (or compressors); heat of compression is added and, at a higher pressure, the last refrigerant is condensed on the final condenser surface.

Heat is rejected to air, water, or water spray. Saturation temperatures of evaporation and condensation throughout the system fix the terminal pressures against which the single or multiple compressors must operate.

Generally, the smallest differential between saturated evaporator and saturated condensing temperatures results in the lowest energy requirement for compression. Liquid refrigerant cooling or subcooling should be used where possible to improve efficiencies and minimize energy consumption.

Where intermediate pressures have not been specifically set for system operation, the compressors automatically balance at their respective suction and discharge pressures as a function of their relative displacements and compression efficiencies, depending on load and temperature requirements. This chapter covers the technique used to determine the balance points for a typical brine chiller, but the theory can be expanded to apply to single- and two-stage systems with different types of evaporators, compressors, and condensers.

COMPONENTS

Evaporators may have flooded, direct-expansion, or liquid overfeed cooling coils with or without fins. Evaporators are used to cool air, gases, liquids, and solids; condense volatile substances; and freeze products.

Ice-builder evaporators accumulate ice to store cooling energy for later use. Embossed-plate evaporators are available (1) to cool a falling film of liquid; (2) to cool, condense, and/or freeze out volatile substances from a fluid stream; or (3) to cool or freeze a product by direct contact. Brazed- and welded-plate fluid chillers can be used to improve efficiencies and reduce refrigerant charge.

Ice, wax, or food products are frozen and scraped from some freezer surfaces. Electronic circuit boards, mechanical products, or food products (where permitted) are flash-cooled by direct immersion in boiling refrigerants. These are some of the diverse applications demanding innovative configurations and materials that perform the function of an evaporator.

Compressors can be positive-displacement, reciprocating-piston, rotary-vane, scroll, single and double dry and lubricant-flooded screw devices, and single-stage or multistage centrifugals. They can be operated in series or in parallel with each other, in which case special controls may be required.

Drivers for compressors can be direct hermetic, semihermetic, or open with mechanical seals on the compressor. In hermetic and semihermetic drives, motor inefficiencies are added to the refrigerant as heat. Open compressors are driven with electric motors, fuel-powered reciprocating engines, or steam or gas turbines. Intermediate gears, belts, and clutch drives may be included in the drive.

Cascade condensers are used with high-pressure, low-temperature refrigerants (such as R-23) on the bottom cycle, and high-temperature refrigerants (such as R-22, azeotropes, and refrigerant blends or zeotropes) on the upper cycle. Cascade condensers are manufactured in many forms, including shell-and-tube, embossed plate, submerged, direct-expansion double coils, and brazed or welded plate heat exchangers. The high-pressure refrigerant from the compressor(s) on the lower cycle condenses at a given intermediate temperature. A separate, lower-pressure refrigerant evaporates on the other side of the surface at a somewhat lower

The preparation of this chapter is assigned to TC 10.1, Custom Engineered Refrigeration Systems.

temperature. Vapor formed from the second refrigerant is compressed by the higher-cycle compressor(s) until it can be condensed at an elevated temperature.

Desuperheating suction gas at intermediate pressures where multistage compressors balance is essential to reduce discharge temperatures of the upper-stage compressor. Desuperheating also helps reduce oil carryover and reduces energy requirements. Subcooling improves the net refrigeration effect of the refrigerant supplied to the next-lower-temperature evaporator and reduces system energy requirements. The total heat is then rejected to a condenser.

Subcoolers can be of shell-and-tube, shell-and-coil, welded-plate, or tube-in-tube construction. Friction losses reduce the liquid pressure that feeds refrigerant to an evaporator. Subcoolers are used to improve system efficiency and to prevent refrigerant liquid from flashing because of pressure loss caused by friction and the vertical rise in lines. Refrigerant blends (zeotropes) can take advantage of temperature glide on the evaporator side with a direct-expansion-in-tube serpentine or coil configuration. In this case, temperature glide from the bubble point to the dew point promotes efficiency and lower surface requirements for the subcooler. A flooded shell for the evaporating refrigerant requires use of only the higher dew-point temperature.

Lubricant coolers remove friction heat and some of the superheat of compression. Heat is usually removed by water, air, or a direct-expansion refrigerant.

Condensers that reject heat from the refrigeration system are available in many standard forms, such as water- or brine-cooled shell-and-tube, shell-and-coil, plate-and-frame, or tube-in-tube condensers; water cascading or sprayed over plate or coil serpentine models; and air-cooled, fin-coil condensers. Special heat pump condensers are available in other forms, such as tube-in-earth and submerged tube bundle, or as serpentine and cylindrical coil condensers that heat baths of boiling or single-phase fluids.

SELECTING DESIGN BALANCE POINTS

Refrigeration load at each designated evaporator pressure, refrigerant properties, liquid refrigerant temperature feeding each evaporator, and evaporator design determine the required flow rate of refrigerant in a system. The additional flow rates of refrigerant that provide refrigerant liquid cooling, desuperheating, and compressor lubricant cooling, where used, depend on the established liquid refrigerant temperatures and intermediate pressures.

For a given refrigerant and flow rate, the suction line pressure drop, suction gas temperature, pressure ratio and displacement, and volumetric efficiency determine the required size and speed of rotation for a positive displacement compressor. At low flow rates, particularly at very low temperatures and in long suction lines, heat gain through insulation can significantly raise the suction temperature. Also, at low flow rates, a large, warm compressor casing and suction plenum can further heat the refrigerant before it is compressed. These heat gains increase the required displacement of a compressor. The compressor manufacturer must recommend the superheating factors to apply. The final suction gas temperature from suction line heating is calculated by iteration.

Another concern is that more energy is required to compress refrigerant to a given condenser pressure as the suction gas gains more superheat. This can be seen by examining a pressure-enthalpy diagram for a given refrigerant such as R-22, which is shown in Figure 2 in Chapter 20 of the 2005 *ASHRAE Handbook—Fundamentals*. As suction superheat increases along the horizontal axis, the slopes of the constant entropy lines of compression decrease. This means that a greater enthalpy change must occur to produce a given pressure rise. For a given flow, then, the power required for compression is increased. With centrifugal compressors, pumping capacity is related to wheel diameter and speed, as well as to volumetric flow and acoustic velocity of the refrigerant at the suction entrance. If the thermodynamic pressure requirement becomes too great for a given speed and volumetric flow, the centrifugal compressor experiences periodic backflow and surging.

Figure 1 shows an example system of curves representing the maximum refrigeration capacities for a brine chilling plant. The example shows only one type of positive-displacement compressor using a water-cooled condenser in a single-stage system operating at a steady-state condition. The figure is a graphical method of expressing the first law of thermodynamics with an energy balance applied to a refrigeration system.

One set of nearly parallel curves (A) represents cooler capacity at various brine temperatures versus saturated suction temperature (a pressure condition) at the compressor, allowing for suction line pressure drops. The (B) curves represent compressor capacities as the saturated suction temperature varies and the saturated condenser temperature (a pressure condition) varies. The (C) curves represent heat transferred to the condenser by the compressor. It is calculated by adding the heat input at the evaporator to the energy imparted to the refrigerant by the compressor. The (D) curves represent condenser performance at various saturated condenser temperatures as the inlet temperature of a fixed quantity of cooling water is varied.

The (E) curves represent the combined compressor and condenser performance as a "condensing unit" at various saturated suction temperatures for various cooling water temperatures. These curves were cross plotted from the (C) and (D) curves back to the set of brine cooler curves as indicated by the dashed construction lines for the 80 and 92°F cooling water temperatures. Another set of construction lines (not shown) would be used for the 86°F cooling water. The number of construction lines used can be increased as necessary to adequately define curvature (usually no more than three per condensing-unit performance line).

The intersections of curves (A) and (E) represent the maximum capacities for the entire system at those conditions. For example, these curves show that the system develops 150 tons of refrigeration when cooling the brine to 44°F at 36.4°F (saturated) suction and using 80°F cooling water. At 92°F cooling water, capacity drops to 134.5 tons if the required brine temperature is 42°F and the required saturated suction temperature is 35°F. The corresponding saturated condensing temperature for 42°F brine with an accompanying suction temperature of 36.4°F and using 80°F water is graphically projected on the brine cooler line with a capacity of 150 tons of refrigeration to meet a newly constructed 36.4°F saturated suction temperature line (parallel to the 34°F and 37°F lines). At this junction, draw a horizontal line to intersect the vertical saturated condensing temperature scale at 93.5°F. The condenser heat rejection is apparent from the (C) curves at a given balance point.

The equation at the bottom of Figure 1 may be used to determine the shaft horsepower (BHP) required at the compressor for any given balance point. A sixth set of curves could be drawn to indicate the power requirement as a function of capacity versus saturated suction and saturated condensing temperatures.

The same procedure can be repeated to calculate cascade system performance. Rejected heat at the cascade condenser would be treated as the chiller load in making a cross plot of the upper-cycle, high-temperature refrigeration system.

For cooling air at the evaporator(s) and for condenser heat rejection to ambient air or evaporative condensers, use the same procedures. Performance of coils and expansion devices such as thermostatic expansion valves may also be graphed, once the basic concept of heat and mechanical energy input equivalent combinations is recognized. Chapter 1 of the 2005 *ASHRAE Handbook—Fundamentals* has further information.

This method finds the natural balance points of compressors operating at their maximum capacities. For multiple-stage loads at several specific operating temperatures, the usual way of controlling compressor capacities is with a suction pressure control and

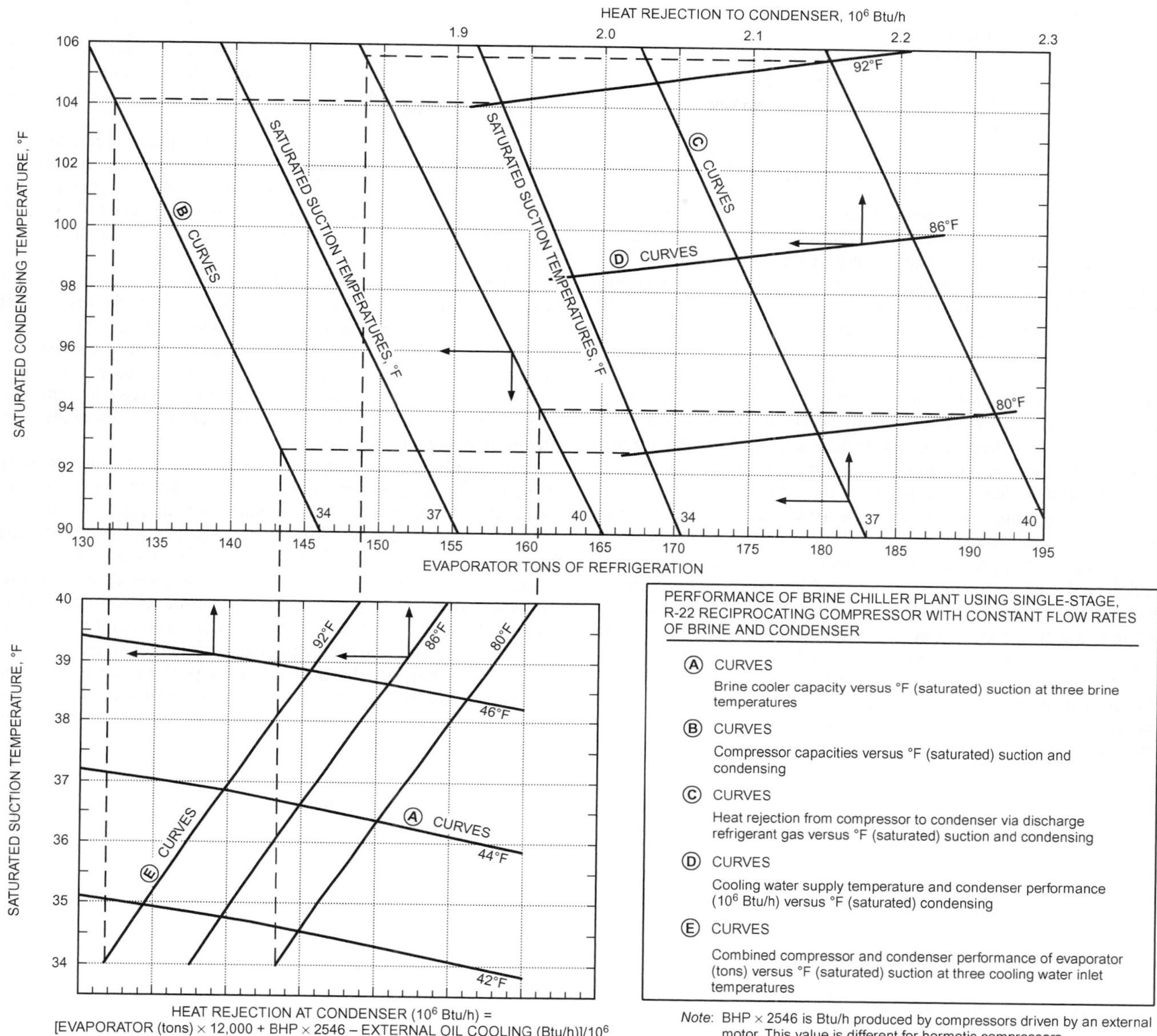

Fig. 1 Brine Chiller Balance Curve

compressor capacity control device. This control accommodates any mismatch in pumping capabilities of multistage compressors, instead of allowing each compressor to find its natural balance point.

Computer programs could be developed to determine balance points of complex systems. However, because applications, components, and piping arrangements are so diverse, many designers use available capacity performance data from vendors and plot balance points for chosen components. Individual computer programs may be available for specific components, which speeds the process.

ENERGY AND MASS BALANCES

A systematic, point-to-point flow analysis of the system (including piping) is essential in accounting for pressure drops and heat gains, particularly in long suction lines. Air-cooled condensers, in particular, can have large pressure drops, which must be included in the analysis to estimate a realistic balance. Making a flow diagram of the system with designated pressures and temperatures, loads, enthalpies, flow rates, and energy requirements helps identify all important factors and components.

An overall energy and mass balance for the system is also essential to avoid mistakes. The overall system represented by the complete flow diagram should be enclosed by a dotted line envelope. Any energy inputs to or outputs from the system that directly affect the heat content of the refrigerant itself should cross the dotted envelope line and must enter the energy balance equations. Accurate estimates of the ambient heat gains through insulation and heat losses from discharge lines where they are significant improve the comprehensiveness of the energy balance and accuracy of equipment selections.

Cascade condenser loads and subcooler or desuperheating loads carried by a refrigerant are internal to the system and thus do not enter into the overall energy balance. The total energy entering the system equals the total energy leaving the system. If calculations do not show an energy balance within reasonable tolerances for the accuracy of data used, then an omission occurred or a mistake was made and should be corrected.

The dotted envelope technique can be applied to any section of the system, but all energy transmissions must be included in the equations, including the enthalpies and mass flow rates of streams that cross the dotted line.

SYSTEM PERFORMANCE

Rarely are sufficient sensors and instrumentation devices available, nor are conditions proper at a given job site to allow calculation of a comprehensive, accurate energy balance for an operating system. Water-cooled condensers and oil coolers for heat rejection and the use of electric motor drives, where motor efficiency and power factor curves are available, offer the best hope for estimating the actual performance of the individual components in a system. Evaporator heat loads can be derived from the measured heat rejection and derived mechanical or measured electrical energy inputs. A comprehensive flow diagram assists in a field survey.

Various coolant flow detection devices are available for direct measurement inside a pipe and for measurement from outside the pipe with variable degrees of accuracy. Sometimes flow rates may be estimated by simply weighing or measuring an accumulation of coolant over a brief time interval.

Temperature and pressure measurement devices should be calibrated and be of sufficient accuracy. Calibrated digital scanning devices for comprehensive simultaneous readings are best. Electrical power meters are not always available, so voltage and current at each leg of a motor power connection must be measured. Voltage drops for long power leads must be calculated when the voltage measurement points are far removed from the motor. Motor load versus efficiency and power factor curves must be used to determine motor output to the system.

Gears and belt or chain drives have friction and windage power losses that must be included in any meaningful analysis.

Stack gas flows and enthalpies for engine or gas turbine exhausts as well as air inputs and speeds must be included. In this case, performance curves issued by the vendor must be heavily relied on to estimate the energy input to the system.

Calculating steam turbine performance requires measurements of turbine speed, steam pressures and temperatures, and condensate mass flow coupled with confidence that the vendor's performance curves truly represent the current mechanical condition. Plant personnel normally have difficulty in obtaining operating data at specified performance values.

Heat rejection from air-cooled and evaporative condensers or coolers is extremely difficult to measure accurately because of changing ambient temperatures and the extent and scope of airflow measurements required. Often, one of the most important issues is the wide variation or cycling of process flows, process temperatures, and product refrigeration loads. Hot-gas false loading and compressor continuous capacity modulations complicate any attempt to make a meaningful analysis.

Prediction and measurement of performance of systems using refrigerant blends (zeotropes) are especially challenging because of temperature variations between bubble points and dew points.

Nevertheless, ideal conditions of nearly steady-state loads and flows with a minimum of cycling sometimes occur frequently enough to permit a reasonable analysis. Computer-controlled systems can provide the necessary data for a more accurate system analysis. Several sets of nearly simultaneous data at all points over a short time enhance the accuracy of any calculation of performance of a given system. In all cases, properly purging condensers and eliminating excessive lubricant contamination of the refrigerant at the evaporators are essential to determine system capabilities accurately.

CHAPTER 44

REFRIGERANT-CONTROL DEVICES

CONTROL of refrigerant flow, temperatures, pressures, and liquid levels is essential in any refrigeration system. This chapter describes a variety of devices and their application to accomplish these important control functions.

Most examples, references, and capacity data in this chapter refer to the more common refrigerants. For further information on control fundamentals, see Chapter 15 of the 2005 *ASHRAE Handbook—Fundamentals* and Chapter 46 of the 2003 *ASHRAE Handbook—HVAC Applications*.

CONTROL SWITCHES

A control switch includes both a sensor and a switch mechanism capable of opening and/or closing an electrical circuit in response to changes in the monitored parameter. The control switch operates one or more sets of electrical contacts, which are used to open or close water or refrigerant solenoid valves; engage and disengage automotive compressor clutches; activate and deactivate relays, contactors, magnetic starters, and timers; etc. Control switches respond to a variety of physical changes, such as pressure, temperature, and liquid level.

Liquid-level-responsive controls use floats, mercury balance tubes, or electronic probes to operate (directly or indirectly) one or more sets of electrical contacts.

Refrigeration control switches may be categorized into three basic groups:

- **Operating** controls (e.g., thermostats) turn systems on and off.
- **Primary** controls provide safe continuous operation (e.g., compressor or condenser fan cycling).
- **Limit** controls (e.g., high-pressure cutout switch) protect a system from unsafe operation.

PRESSURE SWITCHES

Pressure-responsive switches have one or more power elements (e.g., bellows, diaphragms, bourdon tubes) to produce the force needed to operate the mechanism. Typically, pressure-switch power

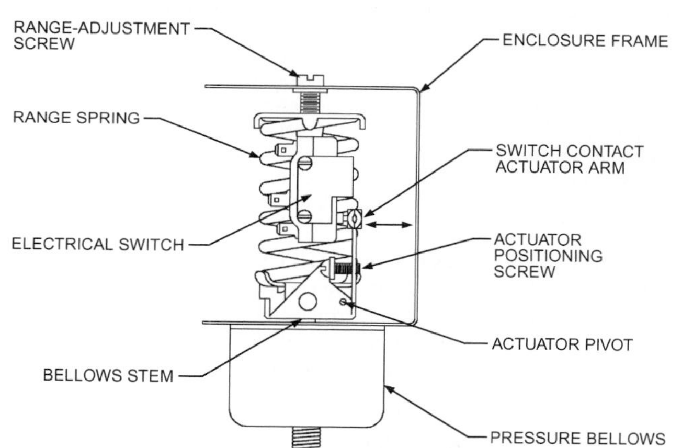

Fig. 1 Typical Pressure Switch

elements are all metal, although some miniaturized devices for specific applications, such as automotive air conditioning, may use synthetic diaphragms. Refrigerant pressure is applied directly to the element, which moves against a spring that can be adjusted to control an operation at the desired pressure (Figure 1). If the control is to operate in the subatmospheric (or vacuum) range, the bellows or diaphragm force is sometimes reversed to act in the same direction as the adjusting spring.

The force available for doing work (i.e., operating the switch mechanism) in this control depends on the pressure in the system and on the area of the bellows or diaphragm. With proper area, enough force can be produced to operate heavy-duty switches. In switches for high-pressure service, the minimum differential is relatively large because of the high-gradient-range spring required.

Miniaturized pressure switches may incorporate one or more snap disks, which provide positive snap action of the electrical contacts. Snap-disk construction ensures consistent differential pressure between on and off settings (Figure 2). Another important benefit of snap-disk construction is the substantial reduction in electrical contact bounce or flutter, which can damage compressor clutch assemblies, relays, and electronic control modules. Some snap-disk switches are built to provide multiple functions in a single

The preparation of this chapter is assigned to TC 8.8, Refrigerant System Controls and Accessories.

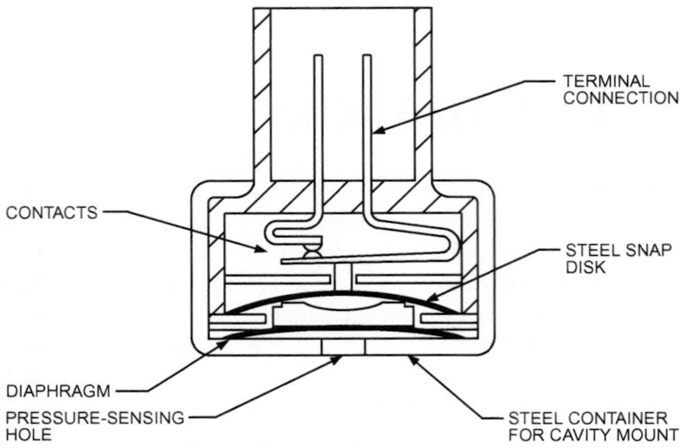

Fig. 2 Miniaturized Pressure Switch

Table 1 Various Types of Pressure Switches

Type	Function
High-pressure cutout (HPCO)	Stops compressor when excessive pressure occurs
High-side low-pressure (HSLP)	Prevents compressor operation under low ambient or loss of refrigerant conditions
High-side fan-cycling (HSFC)	Cycles condenser fan on and off to provide proper condenser pressure
Low-side low-pressure (LSLP)	Initiates defrost cycle; stops compressor when low charge or system blockage occurs
Low-side compressor cycling (LSCC)	Cycles compressor on and off to provide proper evaporator pressure and load temperature
Lubricant pressure differential failure (LPDF)	Stops compressor when difference between oil pressure and crankcase pressure is too low for adequate lubrication

unit, such as high-pressure cutout (HPCO), high-side low-pressure (HSLP), and high-side fan-cycling (HSFC) switches.

Pressure switches in most refrigeration systems are used primarily to start and stop the compressor, cycle condenser fans, and initiate and terminate defrost cycles. Table 1 shows various types of pressure switches with their corresponding functions.

TEMPERATURE SWITCHES (THERMOSTATS)

Temperature-responsive switches have one or more metal power elements (e.g., bellows, diaphragms, bourdon tubes, bimetallic snap disks, bimetallic strips) that produce the force needed to operate the switch.

An **indirect temperature switch** is a pressure switch with the pressure-responsive element replaced by a temperature-responsive element. The temperature-responsive element is a hermetically sealed system comprised of a flexible member (diaphragm or bellows) and a temperature-sensing element (bulb or tube) that are in pressure communication with each other (Figure 3). The closed system contains a temperature-responsive fluid.

The exact temperature/pressure or temperature/volume relationship of the fluid used in the element allows the bulb temperature to control the switch accurately. The switch is operated by changes in pressure or volume that are proportional to changes in sensor temperature.

A **direct temperature switch** typically contains a bimetallic disk or strip that activates electrical contacts when the temperature increases or decreases. As its temperature increases or decreases, the bimetallic element bends or strains because of the two metals' different coefficients of thermal expansion, and the linked electrical contacts engage or disengage. The disk bimetallic element provides snap action, which results in rapid and positive opening or closing of the electrical contacts, minimizing arcing and bounce. A bimetallic

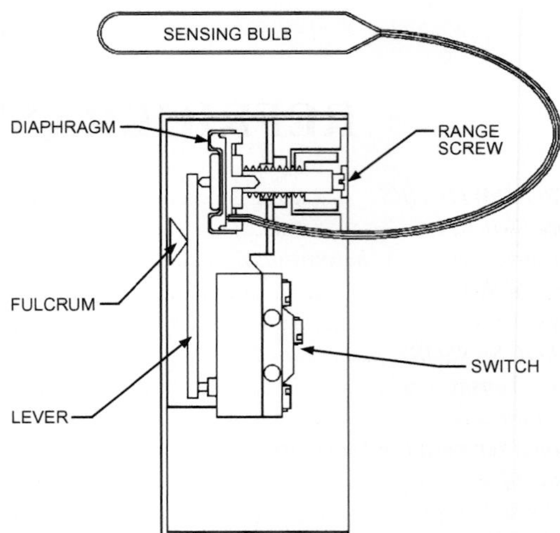

Fig. 3 Indirect Temperature Switch

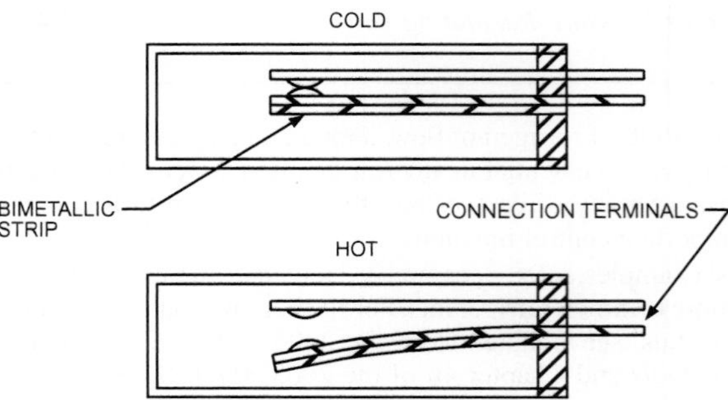

Fig. 4 Direct Temperature Switch

strip (Figure 4) produces very slow contact action and is only suitable for use in very-low-energy electrical circuits. This type of switch is typically used for thermal limit control because the switch differentials and precision may be inadequate for many primary refrigerant control requirements.

DIFFERENTIAL SWITCHES

Differential control switches typically maintain a given difference in pressure or temperature between two pipelines, spaces, or loads. An example is the lubricant pressure differential failure switch used with reciprocating compressors that use forced-feed lubrication.

Figure 5 is a schematic of a differential switch that uses bellows as power elements. Figure 6 shows a differential pressure switch used to protect compressors against low oil pressure. These controls have two elements (either pressure- or temperature-sensitive) simultaneously sensing conditions at two locations. As shown, the two elements are rigidly connected by a rod, so that motion of one causes motion of the other. The connecting rod operates contacts (as shown). The scale spring is used to set the differential pressure at which the device operates. At the control point, the sum of forces developed by the low-pressure bellows and spring balances the force developed by the high-pressure bellows.

Instrument differential is the difference in pressure or temperature between the low- and the high-pressure elements for which the instrument is adjusted. **Operating differential** is the change in differential pressure or temperature required to open or close the switch contacts. It is actually the change in instrument differential

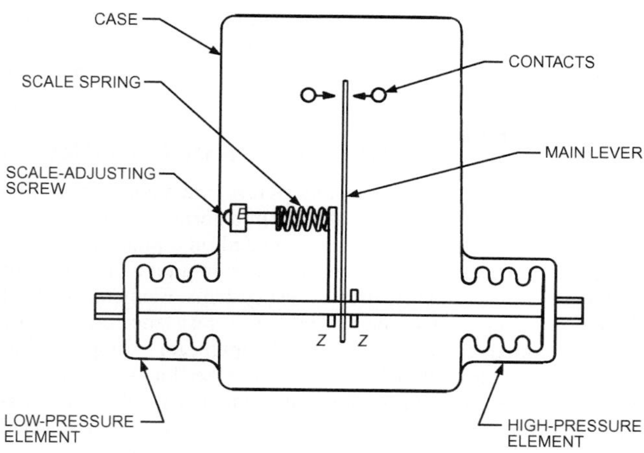

Fig. 5 Differential Switch Schematic

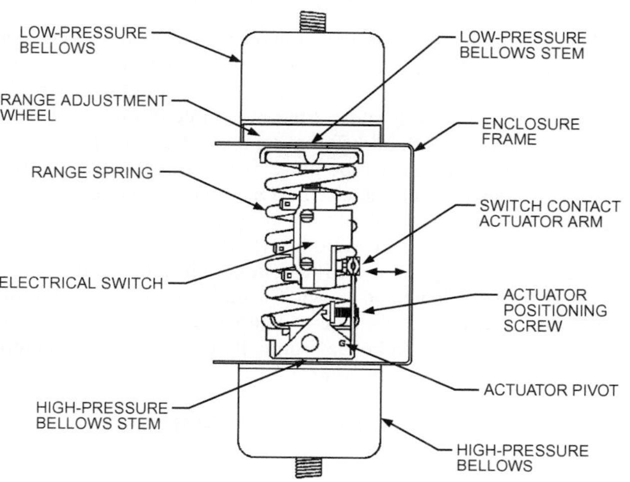

Fig. 6 Differential Pressure Switch

from cut-in to cutout for any setting. Operating differential can be varied by a second spring that acts in the same direction as the first and takes effect only at the cut-in or cutout point without affecting the other spring. A second method is adjusting the distance between collars *Z-Z* on the connecting rod. The greater the distance between them, the greater the operating differential.

If a constant instrument differential is required on a temperature-sensitive differential control switch throughout a large temperature range, one element may contain a different temperature-responsive fluid than the other.

A second type of differential-temperature control uses two sensing bulbs and capillaries connected to one bellows with a liquid fill. This is known as a constant-volume fill, because the operating point depends on a constant volume of the two bulbs, capillaries, and bellows. If the two bulbs have equal volume, a temperature rise in one bulb requires an equivalent fall in the other's temperature to maintain the operating point.

FLOAT SWITCHES

A float switch has a float ball, the movement of which operates one or more sets of electrical contacts as the level of a liquid changes. Float switches are connected by equalizing lines to the vessel or an external column in which the liquid level is to be maintained or monitored. The switch mechanism is generally hermetically sealed. Small heaters can be incorporated to prevent moisture from permeating the polycarbonate housing in cold operating conditions. Other nonmechanical devices, such as capacitance probes, use other methods to monitor the change in liquid level.

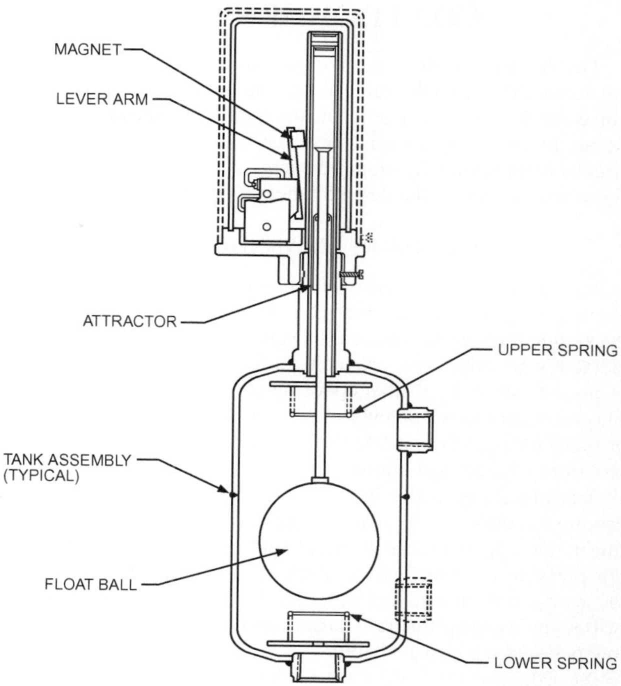

Fig. 7 Magnetic Float Switch

Operation and Selection

Some float switches (Figure 7) operate from movement of a magnetic armature located in the field of a permanent magnet. Others use solid-state circuits in which a variable signal is generated by liquid contact with a probe that replaces the float; this method is adapted to remote-controlled applications and is preferred for ultralow-temperature applications.

Application

The float switch can maintain or indicate the level of a liquid, operate an alarm, control pump operation, or perform other functions. A float switch, solenoid liquid valve, and hand expansion valve combination can control refrigerant level on the high- or low-pressure side of the refrigeration system in the same way that high- or low-side float valves are used. The hand expansion valve, located in the refrigerant liquid line immediately downstream of the solenoid valve, is initially adjusted to provide a refrigerant flow rate at maximum load to keep the solenoid liquid valve in the open position 80 to 90% of the time; it need not be adjusted thereafter. From the outlet side of the hand expansion valve, refrigerant passes through a line and enters either the evaporator or the surge drum.

When the float switch is used for low-pressure level control, precautions must be taken to provide a calm liquid level that falls in response to increased evaporator load and rises with decreased evaporator load. The same recommendations for insulation of the body and liquid leg of the low-pressure float valve apply to the float switch when it is used for refrigerant-level control on the low-pressure side of the refrigeration system. To avoid floodback, controls should be wired to prevent the solenoid liquid valve from opening when the solenoid suction valve closes or the compressor stops.

CONTROL SENSORS

The control sensor is the component in a control system that measures and signals the value of a parameter but has no direct function control. Control sensors typically require an auxiliary source of energy for proper operation. They may be integrated into electronic circuits that provide the required energy and condition the sensor's signal to accomplish the desired function control.

PRESSURE TRANSDUCERS

Pressure transducers sense refrigerant pressure through a flexible element (diaphragm, bourdon tube, or bellows) that is exposed to the system refrigerant pressure. The pressure acts across the flexible element's effective area, producing a force that causes the flexible element to strain against an opposing spring within the transducer. The transducer uses a potentiometer, variable capacitor, strain gage, or piezo element to translate the flexible element's movement to a proportional electrical output.

Transducers typically include additional electronic signal processing circuitry to temperature-compensate, modify, amplify, and linearize the final analog electrical output. Typically, the outside of the pressure-sensing flexible element is exposed to atmospheric pressure and the transducer's electrical output is proportional to the refrigerant's gage pressure. Transducers capable of measuring absolute pressure are also available.

Transducers are usually used as control sensors in electronic control systems, where the continuous analog pressure signal provides data to comprehensive algorithm-based control strategies. For example, in automotive air-conditioning systems, engine load management can be significantly enhanced. Based on a correlation between refrigerant pressure and compressor torque requirements, the vehicle electronic engine controller uses the transducer signal to regulate the engine air and fuel flow compensating for compressor load variations. This improves fuel economy and eliminates the power drain experienced when the compressor starts.

THERMISTORS

Thermistors are cost-effective and reliable temperature sensors. They are typically small and are available in a variety of configurations and sheath materials. Thermistors are beads of semiconductor materials with electrical resistances that change with temperature. Materials with negative temperature coefficients (NTC) (i.e., resistance decreases as temperature increases) are frequently used. NTC thermistors typically produce large changes in resistance with relatively small changes of temperature, and their characteristic curve is nonlinear (Figure 8).

Thermistors are used in electronic control systems that linearize and otherwise process their resistance change into function control actions such as driving step motors or bimetallic heat motors for function modulation. Their analog signal can also be conditioned to perform start/stop functions such as energizing relays, contactors, or solenoid valves.

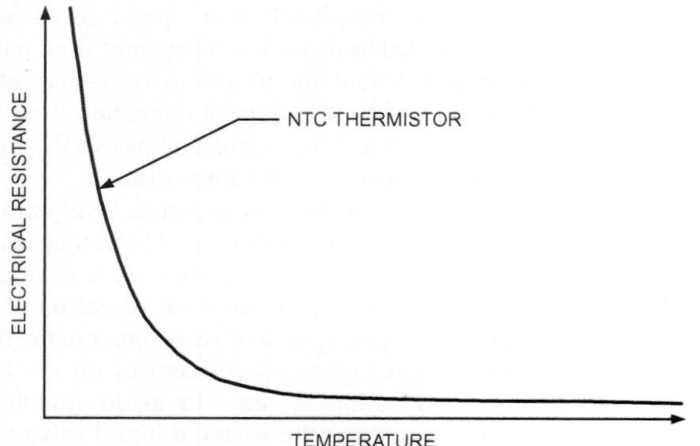

Fig. 8 Typical NTC Thermistor Characteristic

RESISTANCE TEMPERATURE DETECTORS

Resistance temperature detectors (RTDs) are made of very fine metal wire or films coiled or shaped into forms suitable for the application. The elements may be mounted on a plate for surface temperature measurements or encapsulated in a tubular sheath for immersion or insertion into pressurized systems. Elements made of platinum or copper have linear temperature-resistance characteristics over limited temperature ranges. Platinum, for example, is linear within 0.3% from 0 to 300°F and minimizes long-term changes caused by corrosion. RTDs are often mated with electronic circuitry that produces a 4 to 20 mA current signal over a selected temperature range. This arrangement eliminates error associated with connecting line electrical resistance.

THERMOCOUPLES

Thermocouples are formed by the junction of two wires of dissimilar metals. The electromotive force between the wires depends on the wire material and the junction temperature. When the wires are joined at both ends, a thermocouple circuit is formed. When the junctions are at different temperatures, an electric current proportional to the temperature difference between the two junctions flows through the circuit. One junction, called the cold junction, is kept at a constant known temperature (e.g., in an ice bath). The temperature of the other (hot) junction is then determined by measuring the net voltage in the circuit. Electronic circuitry is often arranged to provide a built-in cold junction and linearization of the net voltage-to-temperature relationship. The resulting signal can then be electronically conditioned and amplified to implement function control.

LIQUID LEVEL SENSORS

Capacitance probes (Figure 9) can provide a continuous range of liquid-level monitoring. They compare the impedance value of

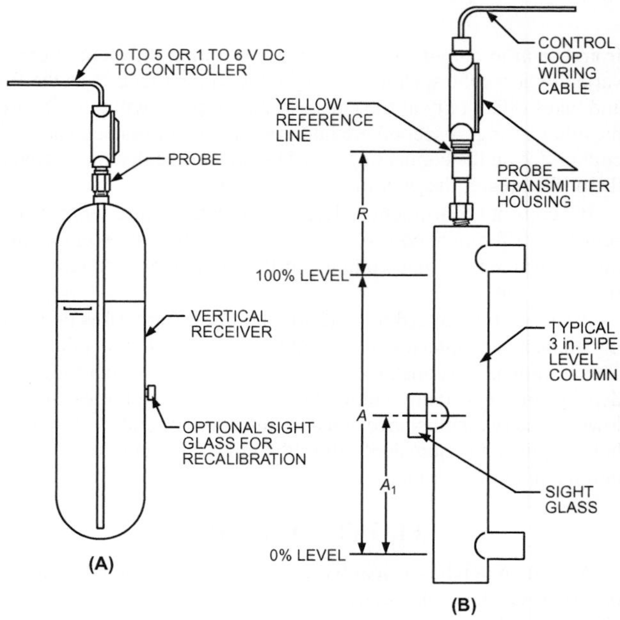

Fig. 9 Capacitance Probe in (A) Vertical Receiver and (B) Auxiliary Level Column

the amount of probe wetted with liquid refrigerant to that in the vapor space. The output can be converted to a variable signal and sent to a dedicated control device with multiple switch points or a computer/programmable logic controller (PLC) for programming or monitoring the refrigerant level. These probes can replace multiple float switches and provide easy level adjustability.

Operation and Selection

The basic principle is that the electrical capacitance of a vertical conducting rod, centered within a vertical conducting cylinder, varies approximately in proportion to the liquid level in the enclosure. The capability to accomplish this depends on the significant difference between the dielectric constants of the liquid and the vapor above the liquid surface.

Capacitance probes are available in a variety of configurations, using a full range of refrigerants. Active lengths vary from 6 in. to 13 ft; output signals vary from 0 to 5 or 1 to 6 V, 4 to 20 mA, or digital readout. Operating temperatures range from −100 to 150°F. Both internal and external vessel mountings are available.

CONTROL VALVES

Control valves are used to start, stop, direct, and modulate refrigerant flow to satisfy system requirements in accordance with load requirements. To ensure satisfactory performance, valves should be protected from foreign material, excessive moisture, and corrosion by properly sized strainers, filters, and/or filter-driers.

THERMOSTATIC EXPANSION VALVES

The thermostatic expansion valve controls the flow of liquid refrigerant entering the evaporator in response to the superheat of gas leaving the evaporator. It keeps the evaporator active without allowing liquid to return through the suction line to the compressor. This is done by controlling the mass flow of refrigerant entering the evaporator so it equals the rate at which it can be completely vaporized in the evaporator by heat absorption. Because this valve is operated by superheat and responds to changes in superheat, a portion of the evaporator must be used to superheat refrigerant gas.

Unlike the constant-pressure valve, the thermostatic expansion valve is not limited to constant-load applications. It is used for controlling refrigerant flow to all types of direct-expansion evaporators in air-conditioning and in commercial (medium-temperature), low-temperature, and ultralow-temperature refrigeration applications.

Operation

Figure 10 shows a schematic cross section of a typical thermostatic expansion valve, with the principal components identified. The following pressures and their equivalent forces govern thermostatic expansion valve operation:

P_1 = pressure of thermostatic element (a function of bulb's charge and temperature), which is applied to top of diaphragm and acts to open valve

P_2 = evaporator pressure, which is applied under diaphragm through equalizer passage and acts in closing direction

P_3 = pressure equivalent of superheat spring force, which is applied underneath diaphragm and is also a closing force

At any constant operating condition, these pressures (forces) are balanced and $P_1 = P_2 + P_3$.

An additional force, which is small and not considered fundamental, arises from the pressure differential across the valve port. To a degree, it can affect thermostatic expansion valve operation. For the configuration shown in Figure 11, the force resulting from port imbalance is the product of pressure drop across the port and the area of the port; it is an opening force in this configuration. In other designs, depending on the direction of flow through the valve, port imbalance may result in a closing force.

The principal effect of port imbalance is on the stability of valve control. As with any modulating control, if the ratio of the diaphragm area to the port is kept large, the unbalanced port effect is minor. However, if this ratio is small or if system operating conditions

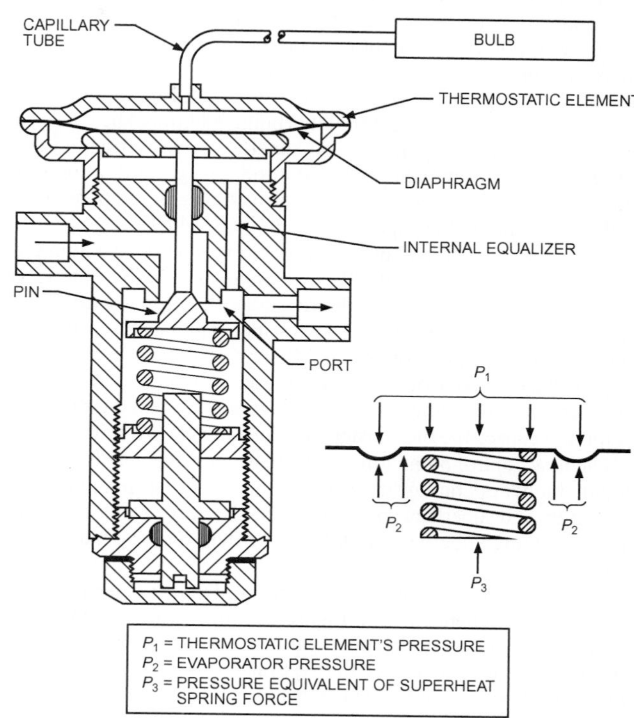

Fig. 10 Typical Thermostatic Expansion Valve

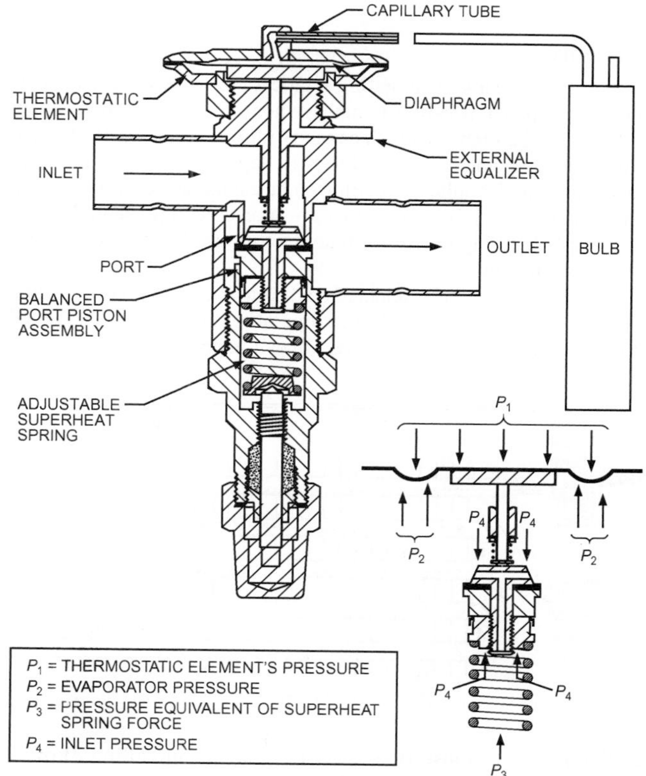

Fig. 11 Typical Balanced Port Thermostatic Expansion Valve

require, a balanced port valve can be used. Figure 11 shows a typical balanced port design.

Figure 12 shows an evaporator operating with R-22 at a saturation temperature of 40°F (68.5 psig). Liquid refrigerant enters the expansion valve, is reduced in pressure and temperature at the valve port, and enters the evaporator at point A as a mixture of saturated liquid and vapor. As flow continues through the evaporator, more of the refrigerant is evaporated. Assuming there is no pressure drop, the refrigerant temperature remains at 40°F until the liquid is entirely evaporated at point B. From this point, additional heat absorption increases the temperature and superheats the refrigerant gas, while the pressure remains constant at 68.5 psig, until, at point C (the outlet of the evaporator), the refrigerant gas temperature is 50°F. At this point, the superheat is 10°F (50 – 40°F).

An increased heat load on the evaporator increases the temperature of refrigerant gas leaving the evaporator. The bulb of the valve senses this increase, and the thermostatic charge pressure P_1 increases and causes the valve to open wider. The increased flow results in a higher evaporator pressure P_2, and a balanced control point is again established. Conversely, decreased heat load on the evaporator decreases the temperature of refrigerant gas leaving the evaporator and causes the thermostatic expansion valve to start closing.

The new control point, after an increase in valve opening, is at a slightly higher operating superheat because of the spring rate of the diaphragm and superheat spring. Conversely, decreased load results in an operating superheat slightly lower than the original control point.

These superheat changes in response to load changes are illustrated by the gradient curve of Figure 13. Superheat at no load, distance 0-A, is called static superheat and ensures sufficient spring force to keep the valve closed during system shutdown. An increase in valve capacity or load is approximately proportional to superheat until the valve is fully open. Opening superheat, represented by the distance A-B, is the superheat increase required to open the valve to match the load; operating superheat is the sum of static and opening superheats.

Capacity

The factory superheat setting (static superheat setting) of thermostatic expansion valves is made when the valve starts to open. Valve manufacturers establish capacity ratings on the basis of opening superheat, typically from 4 to 8°F, depending on valve design, valve size, and application. Full-open capacities usually exceed rated capacities by 10 to 40% to allow a reserve, represented by the distance B-C in Figure 13, for manufacturing tolerances and application contingencies.

A valve should not be selected on the basis of its reserve capacity, which is available only at higher operating superheat. The added superheat may have an adverse effect on performance. Because valve gradients used for rating purposes are selected to produce optimum modulation for a given valve design, manufacturers' recommendations should be followed.

Thermostatic expansion valve capacities are normally published for various evaporator temperatures and valve pressure drops. (See ASHRAE *Standard* 17 and ARI *Standard* 750 for testing and rating methods.) Nominal capacities apply at 40°F evaporator temperature. Capacities are reduced at lower evaporator temperatures. These capacity reductions result from the changed refrigerant pressure/temperature relationship at lower temperatures. For example, if R-22 is used, the change in saturated pressure between 40 and 45°F is 7.5 psi, whereas between –20 and –15°F the change is 3.0 psi. Although the valve responds to pressure changes, published capacities are based on superheat change. Thus, the valve opening and, consequently, valve capacity are less for a given superheat change at lower evaporator temperatures.

Pressure drop across the valve port is always the net pressure drop available at the valve, rather than the difference between compressor discharge and compressor suction pressures. Allowances must be made for the following:

- Pressure drop through condenser, receiver, liquid lines, fittings, and liquid line accessories (filters, driers, solenoid valves, etc.).
- Static pressure in a vertical liquid line. If the thermostatic expansion valve is at a higher level than the receiver, there will be a pressure loss in the liquid line because of the static pressure of liquid.
- Distributor pressure drop.
- Evaporator pressure drop.
- Pressure drop through suction line and accessories, such as evaporator-pressure regulators, solenoid valves, accumulators, etc.

Variations in valve capacity related to changes in system conditions are approximately proportional to the following relationship:

$$q = C\sqrt{\rho \Delta p}\,(h_g - h_f) \tag{1}$$

where

q = refrigerating effect
C = thermostatic expansion valve flow constant
ρ = entering liquid density
Δp = valve pressure difference
h_g = enthalpy of vapor exiting evaporator
h_f = enthalpy of liquid entering thermostatic expansion valve

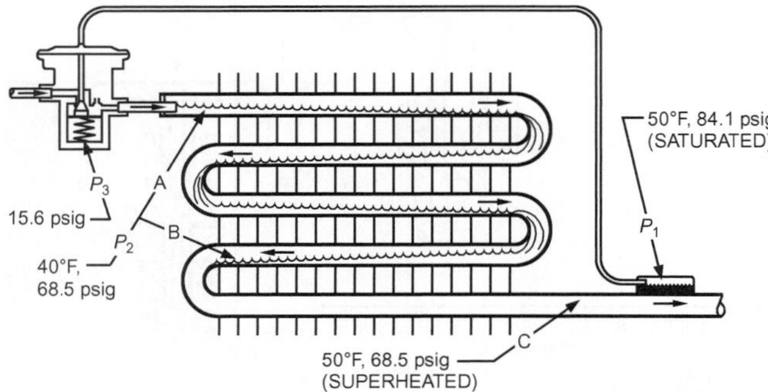

Fig. 12 Thermostatic Expansion Valve Controlling Flow of Liquid R-22 Entering Evaporator, Assuming R-22 Charge in Bulb

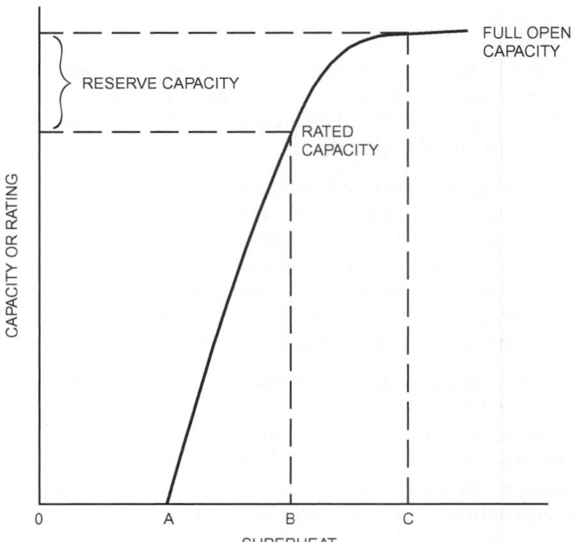

Fig. 13 Typical Gradient Curve for Thermostatic Expansion Valves

Thermostatic expansion valve capacity is dependent on vapor-free liquid entering the valve. If there is flash gas in the entering liquid, valve capacity is reduced substantially because

- Refrigerant mass flow passing through the valve is significantly diminished because the two-phase flow has a lower density
- Flow of the compressible vapor fraction chokes at pressure ratios that typically exist across expansion valves and further restricts liquid-phase flow rate
- Vapor passing through the valve provides no refrigerating effect

Flashing of liquid refrigerant may be caused by pressure drop in the liquid line, filter-drier, vertical lift, or a combination of these. If refrigerant subcooling at the valve inlet is not adequate to prevent flash gas from forming, additional subcooling means must be provided.

Thermostatic Charges

There are several principal types of thermostatic charges, each with certain advantages and limitations.

Gas Charge. Conventional gas charges are limited liquid charges that use the same refrigerant in the thermostatic element that is used in the refrigeration system. The amount of charge is such that, at a predetermined temperature, all of the liquid has vaporized and any temperature increase above this point results in practically no increase in element pressure. Figure 14 shows the pressure-temperature relationship of the R-22 gas charge in the thermostatic element. Because of the characteristic pressure-limiting feature of its thermostatic element, the gas-charged valve can provide compressor motor overload protection on some systems by limiting the maximum operating suction pressure (MOP). It also helps prevent floodback (return of refrigerant liquid to the compressor through the suction line) on starting. Increasing the superheat setting lowers the maximum operating suction pressure; decreasing the superheat setting raises the MOP because the superheat spring and evaporator pressure balance the element pressure through the diaphragm.

Gas-charged valves must be carefully applied to avoid loss of control from the bulb. If the diaphragm chamber or capillary tube becomes colder than the bulb, the small amount of charge in the bulb condenses at the coldest point. This results in the valve throttling or closing, as detailed in the section on Application.

Liquid Charge. Straight liquid charges use the same refrigerant in the thermostatic element and refrigeration system. The volumes of the bulb, bulb tubing, and diaphragm chamber are proportioned so that the bulb contains some liquid under all temperatures. Therefore, the bulb always controls valve operation, even with a colder diaphragm chamber or bulb tubing.

The straight liquid charge (Figure 15) results in increased operating superheat as evaporator temperature decreases. This usually limits use of the straight liquid charge to moderately high evaporator temperatures. The valve setting required for a reasonable operating superheat at a low evaporator temperature may cause floodback during cooling from normal ambient temperatures.

Liquid Cross Charge. Liquid cross charges, unlike conventional liquid charges, use a liquid in the thermostatic element that is different from the refrigerant in the system. Cross charges have flatter pressure/temperature curves than the system refrigerants with which they are used. Consequently, their superheat characteristics differ considerably from those of straight liquid or gas charges.

Cross charges in the commercial temperature range generally have superheat characteristics that are nearly constant or that deviate only moderately through the evaporator temperature range. This charge, also illustrated in Figure 15, is generally used in the evaporator temperature range of 40 to 0°F or slightly below.

For evaporator temperatures substantially below 0°F, a more extreme cross charge may be used. At high evaporator temperatures, the valve controls at a high superheat. As the evaporator temperature falls to the normal operating range, the operating superheat also falls to normal. This prevents floodback on starting, reduces load on the compressor motor at start-up, and allows rapid pulldown of suction pressure. To avoid floodback, valves with this type of charge must be set for the optimum operating superheat at the lowest evaporator temperature expected.

Gas Cross Charge. Gas cross charges combine features of the gas charge and liquid cross charge. They use a limited amount of liquid, thereby providing a maximum operating pressure. The liquid used in the charge is different from the refrigerant in the system and is chosen to provide superheat characteristics similar to those of the liquid cross charges (low temperature). Consequently, they provide both the superheat characteristics of a cross charge and the maximum operating pressure of a gas charge (Figure 15). A commercial (medium-temperature) gas cross charge is possible, but its uses are limited.

Adsorption Charge. Typical adsorption charges depend on the property of an adsorbent, such as silica gel or activated carbon, that is used in an element bulb to adsorb and desorb a gas such as carbon dioxide, with accompanying changes in temperature. The amount of

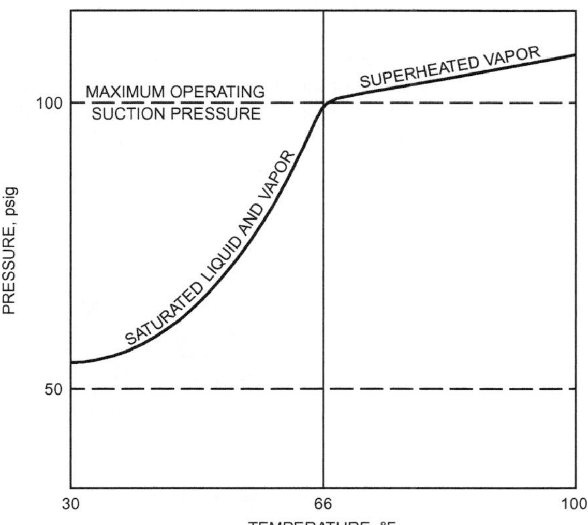

Fig. 14 Pressure-Temperature Relationship of R-22 Gas Charge in Thermostatic Element

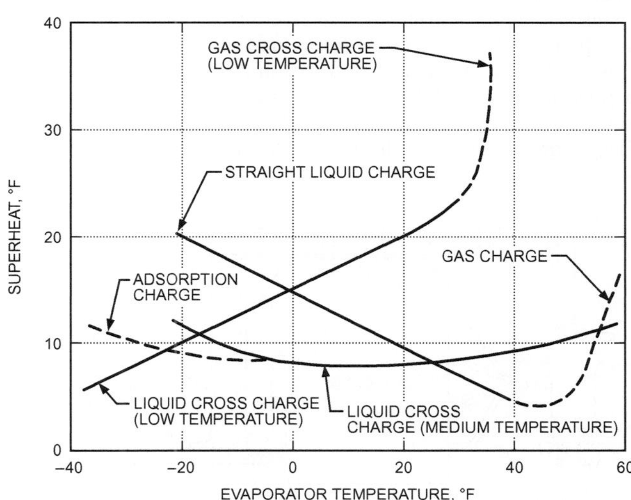

Fig. 15 Typical Superheat Characteristics of Common Thermostatic Charges

adsorption or desorption changes the pressure in the thermostatic element. Because adsorption charges respond primarily to the temperature of the adsorbent material, they are the charges least affected by the ambient temperature surrounding the bulb, bulb tubing, and diaphragm chamber. The comparatively slow thermal response of the adsorbent results in a charge characterized by its stability. Superheat characteristics can be varied by using different charge fluids, adsorbents, and/or charge pressures. The pressure-limiting feature of the gas or gas cross charges is not available with the adsorption element.

Type of Equalization

Internal Equalizer. When the refrigerant pressure drop through an evaporator is relatively low (e.g., equivalent to 2°F change in saturation temperature), a thermostatic expansion valve that has an internal equalizer may be used. Internal equalization describes valve outlet pressure transmitted through an internal passage to the underside of the diaphragm (see Figure 10).

Pressure drop in many evaporators is greater than the 2°F equivalent. When a refrigerant distributor is used, pressure drop across the distributor causes pressure at the expansion valve outlet to be considerably higher than that at the evaporator outlet. As a result, an internally equalized valve controls at an abnormally high superheat. Under these conditions, the evaporator does not perform efficiently because it is starved for refrigerant. Furthermore, the distributor pressure drop is not constant but varies with refrigerant flow rate and therefore cannot be compensated for by adjusting the superheat setting of the valve.

External Equalizer. Because evaporator and/or refrigerant distributor pressure drop causes poor system performance with an internally equalized valve, a valve that has an external equalizer is used. Instead of the internal communicating passage shown in Figure 10, an external connection to the underside of the diaphragm is provided. The external equalizer line is connected either to the suction line, as shown in Figure 16, or into the evaporator at a point downstream from the major pressure drop.

Alternative Construction Types

Pilot-operated thermostatic expansion valves are used on large systems in which the required capacity per valve exceeds the range of direct-operated valves. The pilot-operated valve consists of a piston-type pilot-operated regulator, which is used as the main expansion valve, and a low-capacity thermostatic expansion valve, which serves as an external pilot valve. The small pilot thermostatic expansion valve supplies pressure to the piston chamber or, depending on the regulator design, bleeds pressure from the piston chamber in response to a change in the operating superheat. Pilot operation allows the use of a characterized port in the main expansion valve to provide good modulation over a wide load range. Therefore, a very carefully applied pilot-operated valve can perform well on refrigerating systems that have some form of compressor capacity reduction, such as cylinder unloading. Figure 17 illustrates such a valve applied to a large-capacity direct-expansion chiller.

The auxiliary pilot controls should be sized to handle only the pilot circuit flow. For example, in Figure 17 a small solenoid valve in the pilot circuit, installed ahead of the thermostatic expansion valve, converts the pilot-operated valve into a stop valve when the solenoid valve is closed.

Equalization Features. When the compressor stops, a thermostatic expansion valve usually moves to the closed position. This movement sustains the difference in refrigerant pressures in the evaporator and the condenser. Low-starting-torque motors require that these pressures be equalized to reduce the torque needed to restart the compressor. One way to provide pressure equalization is to add, parallel to the main valve port, a small fixed auxiliary passageway, such as a slot or drilled hole in the valve seat or valve pin. This opening permits limited fluid flow through the control, even when the valve is closed, and allows the system pressures to equalize on the off cycle. The size of such a fixed auxiliary passageway must be limited so its flow capacity is not greater than the smallest flow that must be controlled in normal system operation.

Another, more complex control is available for systems requiring shorter equalizing times than can be achieved with the fixed auxiliary passageway. This control incorporates an auxiliary valve port, which bypasses the primary port and is opened by the element diaphragm as it moves toward and beyond the position at which the primary valve port is closed. Flow capacity of an auxiliary valve port can be considerably larger than that of the fixed auxiliary passageway, so pressures can equalize more rapidly.

Flooded System. Thermostatic expansion valves are seldom applied to flooded evaporators because superheat is necessary for proper valve control; only a few degrees of suction vapor superheat in a flooded evaporator incurs a substantial loss in system capacity. If the bulb is installed downstream from a liquid-to-suction heat exchanger, a thermostatic expansion valve can be made to operate at this point on a higher superheat. Valve control is likely to be poor because of the variable rate of heat exchange as flow rates change (see the section on Application).

Expansion valves with modified thermostatic elements in which electric heat is supplied to the bulb are available. The bulb is inserted in direct contact with refrigerant liquid in a low-side accumulator. The contact of cold refrigerant liquid with the bulb overrides the artificial heat source and throttles the expansion valve. As liquid falls away from the bulb, the valve feed increases again. Although similar in construction to a thermostatic expansion valve, it is essentially a modulating liquid-level control valve.

Desuperheating Valves. Thermostatic expansion valves with special thermostatic charges are used to reduce gas temperatures (superheat) on various air-conditioning and refrigeration systems. Suction gas in a single-stage system can be desuperheated by injecting liquid directly into the suction line. This cooling may be required with or without discharge gas bypass used for compressor

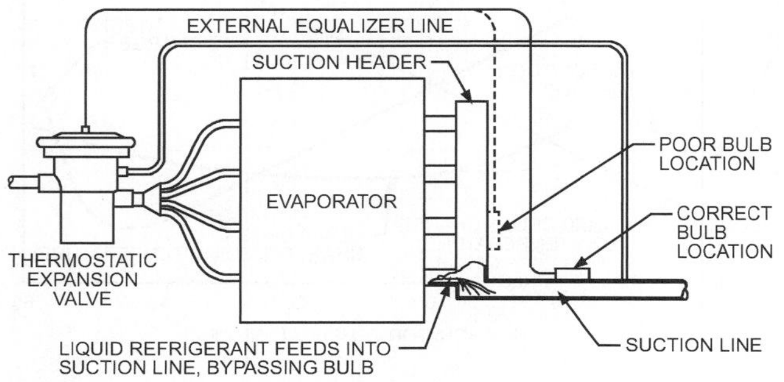

Fig. 16 Bulb Location for Thermostatic Expansion Valve

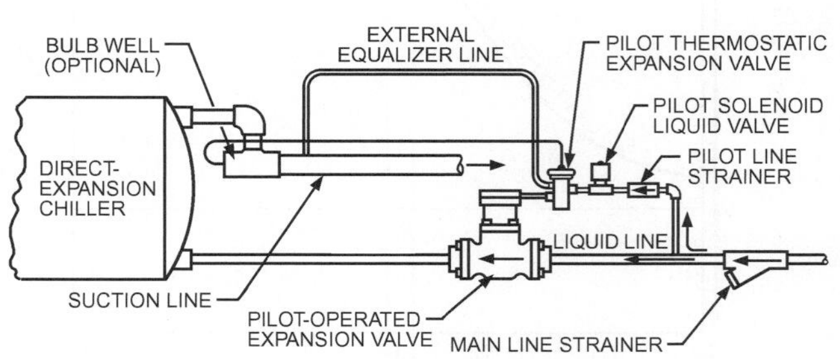

Fig. 17 Pilot-Operated Thermostatic Expansion Valve Controlling Liquid Refrigerant Flow to Direct-Expansion Chiller

capacity control. The line upstream of the valve bulb must be long enough so the injected liquid refrigerant can mix adequately with the gas being desuperheated. On compound compression systems, a specially selected expansion valve may be used to inject liquid directly into the interstage line upstream of the valve bulb to provide intercooling.

Application

Hunting is alternate overfeeding and starving of the refrigerant feed to the evaporator. It produces sustained cyclic changes in the pressure and temperature of the refrigerant gas leaving the evaporator. Extreme hunting reduces refrigeration system capacity because mean evaporator pressure and temperature are lowered and compressor capacity is reduced. If overfeeding of the expansion valve causes intermittent flooding of liquid into the suction line, the compressor may be damaged.

Although hunting is commonly attributed to the thermostatic expansion valve, it is seldom solely responsible. One reason for hunting is that all evaporators have a time lag. When the bulb signals for a change in refrigerant flow, the refrigerant must traverse the entire evaporator before a new signal reaches the bulb. This lag or time lapse may cause continuous overshooting of the valve both opening and closing. In addition, the thermostatic element, because of its mass, has a time lag that may be in phase with the evaporator lag and amplify the original overshooting.

It is possible to alter the thermostatic element's response rate by either using thermal ballast or changing the mass or heat capacity of the bulb, thereby damping or even eliminating hunting. A change in valve gradient may produce similar result.

Slug flow or percolation in the evaporator can also cause hunting. Under these conditions, liquid refrigerant moves in waves (slugs) that fill a portion of the evaporator tube and erupt into the suction line. These unevaporated slugs chill the bulb and temporarily reduce the feed of the valve, resulting in intermittent starving of the evaporator.

On multiple-circuit evaporators, a lightly loaded or overfed circuit also floods into the suction line, chills the bulb, and throttles the valve. Again, the effect is intermittent; when the valve feed is reduced, flooding ceases and the valve reopens.

Hunting can be minimized or avoided in the following ways:

- Select the proper valve size from the valve capacity ratings rather than nominal valve capacity; oversized valves aggravate hunting.
- Change the valve adjustment. A lower superheat setting usually (but not always) increases hunting.
- Select the correct thermostatic element charge. Cross-charged elements are less susceptible to hunting.
- Design the evaporator section for even flow of refrigerant and air. Uniform heat transfer from the evaporator is only possible if refrigerant is distributed by a properly selected and applied refrigerant distributor and air distribution is controlled by a properly designed housing. (Air-cooling and dehumidifying coils, including refrigerant distributors, are detailed in Chapter 21 of the 2004 *ASHRAE Handbook—HVAC Systems and Equipment.*)
- Size and arrange suction piping correctly.
- Locate and apply the bulb correctly.
- Select the best location for the external equalizer line connection.

Bulb Location. Most installation requirements are met by strapping the bulb to the suction line to obtain good thermal contact between them. Normally, the bulb is attached to a horizontal line upstream of the external equalizer connection (if used) at a 3 or 9 o'clock position as close to the evaporator as possible. The bulb is not normally placed near or after suction-line traps, but some designers test and prove locations that differ from these recommendations. A good moisture-resistant insulation over the bulb and suction line diminishes the adverse effect of varying ambient temperatures at the bulb location.

Occasionally, the bulb of the thermostatic expansion valve is installed downstream from a liquid-suction heat exchanger to compensate for a capacity shortage caused by an undersized evaporator. Although this procedure seems to be a simple method of maximizing evaporator capacity, installing the bulb downstream of the heat exchanger is undesirable from a control standpoint. As the valve modulates, the liquid flow rate through the heat exchanger changes, causing the rate of heat transfer to the suction vapor to change. An exaggerated valve response follows, resulting in hunting. There may be a bulb location downstream from the heat exchanger that reduces the hunt considerably. However, the danger of floodback to the compressor normally overshadows the need to attempt this method.

Certain installations require increased bulb sensitivity as a protection against floodback. The bulb, if located properly in a well in the suction line, has a rapid response because of its direct contact with the refrigerant stream. Bulb sensitivity can be increased by using a bulb smaller than is normally supplied. However, use of the smaller bulb is limited to gas-charged valves. Good piping practice also affects expansion valve performance.

Figure 18 illustrates the proper piping arrangement when the suction line runs above the evaporator. A lubricant trap that is as short as possible is located downstream from the bulb. The vertical riser(s) must be sized to produce a refrigerant velocity that ensures continuous return of lubricant to the compressor. The terminal end of the riser(s) enters the horizontal run at the top of the suction line; this avoids interference from overfeeding any other expansion valve or any drainback during the off cycle.

If circulated with lubricant-miscible refrigerant, a heavy concentration of lubricant elevates the refrigerant's boiling temperature. The response of the thermostatic charge of the expansion valve is related to the saturation pressure and temperature of pure refrigerant. In an operating system, the false pressure/temperature signals of lubricant-rich refrigerants cause floodback or operating superheats considerably lower than indicated, and quite often cause erratic valve operation. To keep lubricant concentration at an acceptable level, either the lubricant pumping rate of the compressor must be reduced or an effective lubricant separator must be used.

The **external equalizer** line is ordinarily connected at the evaporator outlet, as shown in Figure 18. It may also be connected at the evaporator inlet or at any other point in the evaporator downstream of the major pressure drop. On evaporators with long refrigerant circuits that have inherent lag, hunting may be minimized by changing the connection point of the external equalizer line.

In application, the various parts of the valve's thermostatic element are simultaneously exposed to different thermal influences from the surrounding ambient air and the refrigerant system. In some situations, cold refrigerant exiting the valve dominates and cools the thermostatic element to below the bulb temperature. When

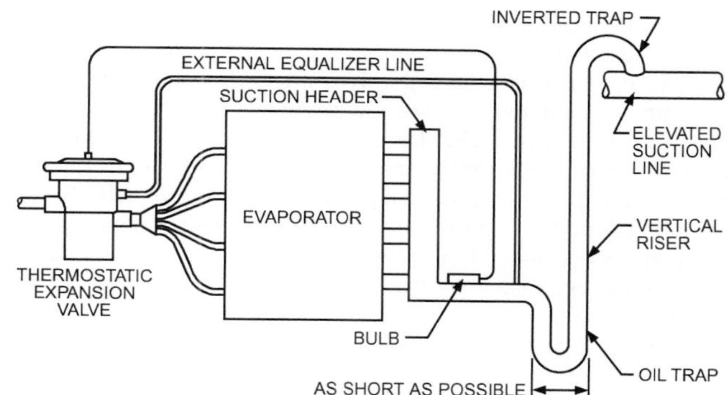

Fig. 18 Bulb Location When Suction Main Is Above Evaporator

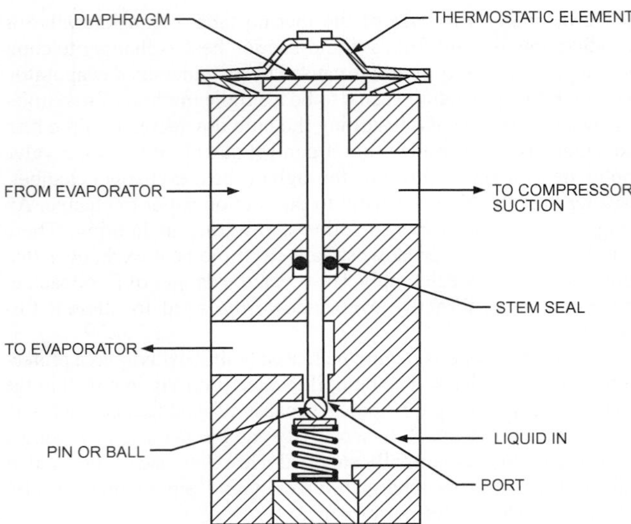

Fig. 19 Typical Block Valve

this occurs with a gas-charged or gas cross-charged valve, the charge condenses at the coldest point in the element and control of refrigerant feed moves from the bulb to the thermostatic element (diaphragm chamber). Pressure applied to the top of the diaphragm diminishes to saturation pressure at the cold point. Extreme starving of the evaporator, progressing to complete cessation of refrigerant flow, is characteristic. For this reason, gas-charged or gas cross-charged valves should be applied only to multicircuited evaporators that use refrigerant distributors. The distributor typically provides sufficient pressure drop to maintain a saturation temperature at the valve outlet well above the temperature at the bulb location.

Internally equalized gas-charged or gas cross-charged valves should only be considered in very carefully selected applications where the risk of loss of control can be minimized. Some gas cross-charge formulations may be slightly less susceptible to the described loss of control than are straight gas charges, but they are far from immune. Gas-charged and gas cross-charged valves with specially constructed thermostatic power elements that positively isolate the charge fluids in the temperature-sensing element (bulb) have been applied in situations where there was high risk of control loss and the pressure-limiting feature of a gas-charged valve was required.

Gas-charged bulbless valves, frequently called block valves (Figure 19), are practically immune to loss of control because the thermostatic element (diaphragm chamber) is located at the evaporator outlet. The valve is constructed so that the temperature-sensing function of the remote bulb is integrated into the thermostatic element by purposely confining all of the charge fluid to this chamber.

Liquid, liquid cross-charged, and adsorption-charged valves are not susceptible to the same type of loss of control that gas-charged or gas cross-charged valves are. However, exposure to extreme ambient temperature environments causes shifting of operating superheats. The degree of superheat shift depends on the severity of the thermal exposure. High ambient temperatures surrounding thermally sensitive parts of the valve typically lower operating superheats, and vice versa. Gas-charged and gas cross-charged valves, including bulbless or block valves, respond to high ambient exposure similarly but starve the evaporator when exposed to ambient temperatures below evaporator outlet refrigerant temperatures.

ELECTRIC EXPANSION VALVES

Application of an electric expansion valve requires a valve, controller, and control sensors. The control sensors may include pressure transducers, thermistors, resistance temperature devices (RTDs), or other pressure and temperature sensors. See Chapter 14 in the 2005 *ASHRAE Handbook—Fundamentals* for a discussion of instrumentation. Specific types should be discussed with the electric valve and electronic controller manufacturers to ensure compatibility of all components.

Electric valves typically have four basic types of actuation:

- Heat-motor operated
- Magnetically modulated
- Pulse-width-modulated (on/off type)
- Step-motor-driven

Heat-motor valves may be either of two types. In one type, one or more bimetallic elements are heated electrically, causing them to deflect. The bimetallic elements are linked mechanically to a valve pin or poppet; as the bimetallic element deflects, the valve pin or poppet follows the element movement. In the second type, a volatile fluid is contained within an electrically heated chamber so that the regulated temperature (and pressure) is controlled by electrical power input to the heater. The regulated pressure acts on a diaphragm or bellows, which is balanced against atmospheric air pressure or refrigerant pressure. The diaphragm is linked to a pin or poppet.

A **magnetically modulated** (analog) valve functions by modulation of an electromagnet; a solenoid armature compresses a spring progressively as a function of magnetic force. The modulating armature may be connected to a valve pin or poppet directly or may be used as the pilot element to operate a much larger valve. When the modulating armature operates a pin or poppet directly, the valve may be of a pressure-balanced port design so that pressure differential has little or no influence on valve opening.

The **pulse-width-modulated valve** is an on/off solenoid valve with special features that allow it to function as an expansion valve through a life of millions of cycles. Although the valve is either fully opened or closed, it operates as a variable metering device by rapidly pulsing the valve open and closed. For example, if 50% flow is needed, the valve will be open 50% of the time and closed 50% of the time. The duration of each opening, or pulse, is regulated by the electronics.

A **step motor** is a multiphase motor designed to rotate in discrete fractions of a revolution, based on the number of signals or "steps" sent by the controller. The controller tracks the number of steps and can offer fine control of the valve position with almost absolute repeatability. Step motors are used in instrument drives, plotters, and other applications where accurate positioning is required. When used to drive expansion valves, a lead screw changes the rotary motion of the rotor to a linear motion suitable for moving a valve pin or poppet. The lead screw may be driven directly from the rotor, or a reduction gearbox may be placed between the motor and lead screw. The motor may be hermetically sealed within the refrigerant environment, or the motor and gearbox can operate outside the refrigerant system with an appropriate stem seal.

Electric expansion valves may be controlled by either digital or analog electronic circuits. Electronic control gives additional flexibility over traditional mechanical valves to consider control schemes that would otherwise be impossible, including stopped or full flow when required.

The electric expansion valve, with properly designed electronic controllers and sensors, offers a refrigerant flow control means that is not refrigerant specific, has a very wide load range, can be set remotely, and can respond to a variety of input parameters.

REGULATING AND THROTTLING VALVES

Regulating and throttling valves are used in refrigeration systems to perform a variety of functions. Valves that respond to and control their own inlet pressure are called **upstream pressure regulators**.

This type of regulator, when located in an evaporator vapor outlet line, responds to evaporator outlet pressure and is commonly called an **evaporator-pressure regulator**. A special three-way version of an upstream pressure regulator is designed specifically for air-cooled condenser pressure regulation during cold-weather operation. Valves that respond to and control their own outlet pressure are called **downstream pressure regulators**. Downstream pressure regulators located in a compressor suction line regulate compressor suction pressure and may also be called suction-pressure regulators, crankcase pressure regulators, or holdback valves. A downstream pressure regulator located at an evaporator inlet to feed liquid refrigerant into the evaporator at a constant evaporator pressure is known as a **constant-pressure** or **automatic expansion valve**.

A third category of pressure regulator, a **differential pressure regulator**, responds to the difference between its own inlet and outlet pressures.

Electronically controlled, electrically operated suction throttling valves have been developed to control temperature in food merchandising refrigerators and other refrigerated spaces (Figure 20). This type of valve regulates evaporator pressure, although it responds only to temperature in the space or load, rather than pressure in the evaporator or suction line. The system consists of a temperature sensor, an electronic control circuit that has been programmed by the manufacturer with a control strategy or algorithm, and an electrically driven suction throttling valve. The set point may be set or changed on site or at a remote location through communication software. The valve responds to the difference between set-point temperature and the sensed temperature in the space or load. A sensed temperature above the set point drives the valve further open, thereby reducing evaporator pressure and saturation temperature; a sensed temperature below set point modulates the valve in the closing direction, which increases evaporator pressure. During defrost, the control circuit usually closes the valve. Additional information on the drive and sensing mechanisms used with this valve type is given in the sections on Electric Expansion Valves and on Control Sensors.

Electronically controlled throttling valves may also be used in various other applications, such as discharge gas bypass capacity reduction, compressor suction throttling, condenser pressure regulation, gas defrost systems, and heat reclaim schemes.

EVAPORATOR-PRESSURE-REGULATING VALVES

The evaporator-pressure regulator is a regulating valve designed to control its own inlet pressure. Typically installed in the suction line exiting an evaporator, it regulates that evaporator's outlet pressure, which is the regulator's upstream or inlet pressure. For this reason evaporator-pressure regulators are also called **upstream pressure regulators**. They are most frequently used to prevent evaporator pressure (and saturation temperature) from dropping below a desired minimum. As declining regulator inlet pressure approaches the regulator set point, the valve throttles, thereby maintaining the desired minimum evaporator pressure (and temperature). Evaporator-pressure regulators are often used to balance evaporator capacity with varying load conditions and to protect against freezing at low loads, such as in water chillers.

The work required to drive pilot-operated valves is most commonly produced by harnessing the pressure loss caused by flow through the valve. Direct-operated regulating valves are powered by relatively large changes in the controlled variable (in this case, inlet pressure). Pilot- and direct-operated evaporator-pressure regulators may be classified as self-powered. Evaporator-pressure regulators are sometimes driven by a high-pressure refrigerant liquid or gas flowing from the system's high-pressure side, as well as electrically. These types are usually considered to be externally powered.

Operation

Direct-operated evaporator-pressure-regulating valves are relatively simple, as shown in Figure 21. The inlet pressure acts on the bottom of the seat disk and opposes the spring. The outlet pressure acts on the bottom of the bellows and the top of the seat disc. Because the effective areas of the bellows and the port are equal, these two forces cancel each other, and the valve responds to inlet pressure only. When the inlet pressure rises above the equivalent pressure exerted by the spring force, the valve begins to open. When inlet pressure falls, the spring moves the valve in the closing direction. In operation, the valve assumes an intermediate throttling position that balances the refrigerant flow rate with evaporator load.

Because both spring and bellows must be compressed through the entire opening valve stroke, a significant change in inlet pressure is required to open the valve to its rated capacity. Inlet pressure changes of 5 to 10 psi or more, depending on design, are typically required to move direct-operated evaporator-pressure regulators from closed position to their rated flow capacity. Therefore, these valves have relatively high gradients and the system may experience significant changes in regulated evaporator pressure when large load changes occur.

Pilot-operated evaporator-pressure-regulating valves are either self-powered or high-pressure-driven. The self-powered regulator

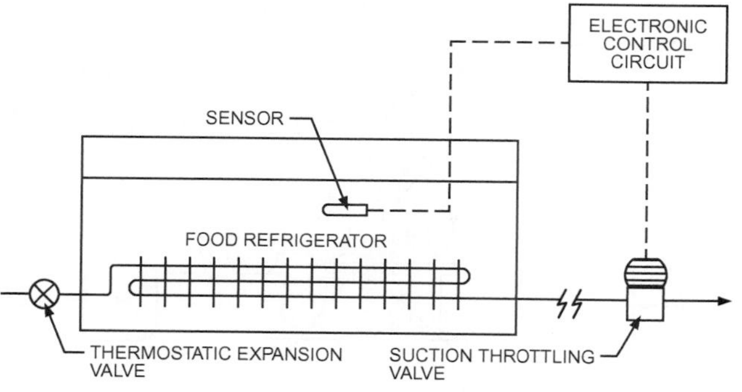

Fig. 20 Electronically Controlled, Electrically Operated Suction-Throttling Regulator

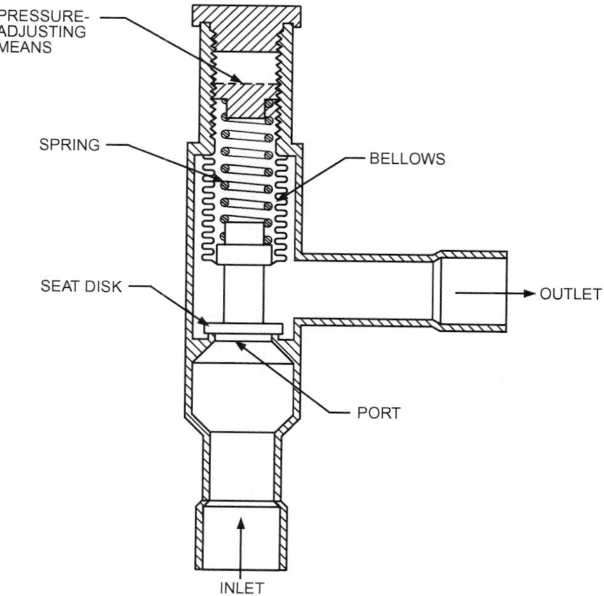

Fig. 21 Direct-Operated Evaporator-Pressure Regulator

(Figure 22) starts to open when the inlet pressure approaches the equivalent pressure setting of the diaphragm spring. The diaphragm lifts to allow inlet pressure to flow through the pilot port, which increases the pressure above the piston. This increase moves the piston down, causing the main valve to open. Flow through the opening valve relieves evaporator pressure into the suction line. As evaporator pressure diminishes, the diaphragm throttles flow through the pilot port, a bleed hole in the piston relieves pressure above the piston to the low-pressure outlet side of the main valve, and the main spring moves the valve in the closing direction. Balanced flow rates through the pilot port and piston bleed hole establish a stable piston pressure that balances against the main spring. The main valve assumes an intermediate throttling position that allows the refrigerant flow rate required to satisfy the evaporator load. Pilot-operated regulators have relatively low gradients and are capable of precise pressure regulation in evaporators that experience large load changes. Typically, pressure loss of up to 2 psi is needed to move the valve to its full open position.

Suction stop service can be provided with this style regulator by adding a pilot solenoid valve in the equalizer flow passage to prevent inlet pressure from reaching the underside of the pressure pilot diaphragm regardless of inlet pressure. Suction stop service is often required to facilitate and control evaporator defrost.

High-pressure-driven pilot-operated regulating valves are of a normally open design and require high-pressure liquid or gas to provide a closing force. One advantage of this design over self-powered regulators is that it does not require any suction-pressure drop across the valve or large inlet pressure change to operate. When valve inlet pressure increases above set point (Figure 23), the diaphragm moves up against the spring, allowing the pilot valve pin spring to move the pilot valve pin, pin carrier, and push rods (not shown) up toward closing the pilot valve port. The gas or liquid from the high-pressure side of the system is throttled by the pilot valve, and pressure in the top of the piston chamber bleeds to the valve's downstream side through a bleed orifice. As pressure on top of the piston diminishes, the main body spring moves the valve piston in the opening direction.

As inlet pressure diminishes, increased flow of high-pressure liquid or gas through the pilot valve drives the piston down toward a closed position.

A solenoid valve may be used to the drive the piston to the closed position for suction stop service, either by closing the bleed orifice or by supplying high pressure directly to the top of the piston chamber.

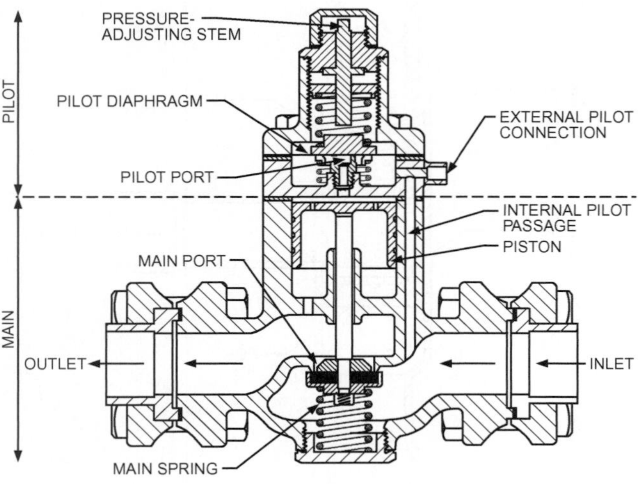

Fig. 22 Pilot-Operated Evaporator-Pressure Regulator (Self-Powered)

Note that, in the latter arrangement, a continuous but very small flow of liquid or gas from the system high side is discharged into the suction line downstream of the regulator while the valve is closed. In some applications, this bleed may enhance compressor cooling and lubricant return.

Selection

Selection of evaporator-pressure-regulating valves is based on the flow capacity required to satisfy the load imposed on the evaporator being regulated and the pressure drop available across the regulator. For example, if an evaporator is to be regulated to a pressure of 30 psig and the regulator discharges into a suction line that normally operates or is maintained at 20 psig, the regulator should be selected to satisfy the evaporator load at a 10 psi pressure loss across the valve. To select direct-operated regulators, consider the high gradient of this design; ensure that the variation of inlet pressure that occurs with load changes is acceptable for the application. For example, a direct-operated regulator set at high-load operating conditions to protect against chiller freeze-up may allow evaporator pressure to drop into the freeze-up danger zone at low loads because of the large reduction in inlet pressure needed to throttle the valve to near-closed stroke. Externally powered regulators should be selected to satisfy the flow requirements imposed by the evaporator load at pressure drops compatible with the application.

Grossly oversized regulating valves are very susceptible to unstable operation, which may in turn upset the stability of other controls in the system, significantly degrade system performance, and risk damage to other system components.

Application

Evaporator-pressure regulators are used on air-cooling evaporators to control frosting or prevent excessive dehumidification. They are also used on water and brine chillers to protect against freezing under low-load conditions.

When multiple evaporators are connected to a common suction line, as shown in Figure 24, evaporator-pressure regulators may be installed to control evaporator pressure in each individual unit or in a group of units operating at the same pressure. The regulators maintain the desired saturation temperature in evaporators serving the high- and medium-temperature loads; those for low-temperature loads may be directly connected to the suction main. In these systems, the compressor(s) are loaded, unloaded, and cycled to maintain suction main pressure as the combined evaporator loads vary.

The pilot-operated self-powered evaporator-pressure regulator, with internal pilot passage, receives its source of pressure to both

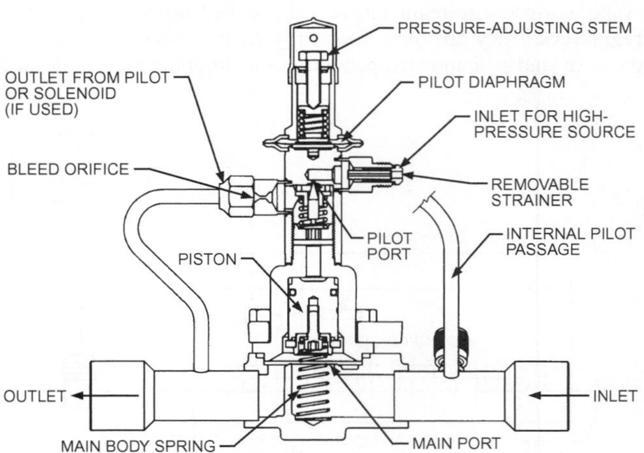

Fig. 23 Pilot-Operated Evaporator-Pressure Regulator (High-Pressure-Driven)

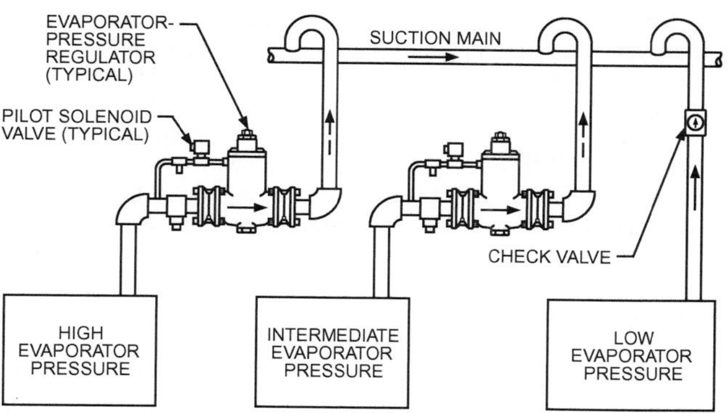

Fig. 24 Evaporator-Pressure Regulators in Multiple System

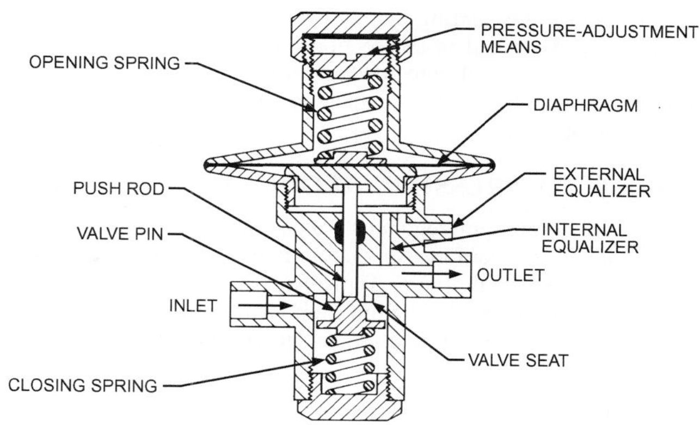

Valve is used with either internal or external equalizer, but not with both.

Fig. 25 Constant-Pressure Expansion Valve

power the valve and sense the controlled pressure at the regulator inlet connection. A regulator with an external pilot connection allows a choice of remote pressure source for both controlled pressure sensing and driving the valve. The external pilot connection can also facilitate use of remote pressure and solenoid pilot valves. Figure 23 shows a pilot solenoid valve installed in the external pilot line. This arrangement allows the regulator to function as a suction stop valve as well as an evaporator-pressure regulator. The suction stop feature is particularly useful on a flooded evaporator to prevent all of the refrigerant from leaving the evaporator when the load is satisfied and the evaporator is deactivated. The stop feature is also effective during evaporator defrost cycles, especially with gas defrost systems. When regulator inlet pressure is unstable to the point of upsetting regulator performance, the external pilot connection may be used to facilitate use of volume chambers and other non-flow-restricting damping means to smooth the pilot pressure source before it enters the regulator pilot connection.

A remote pressure pilot installed in the external pilot line can be located to facilitate manual adjustment of the pressure setting when the main regulator must be installed in an inaccessible location.

Multiple pilots, including temperature-actuated pilots, pressure pilots, and solenoid pilot stop valves, may be connected in various parallel-series arrangements to the external pilot connection, thus allowing the main valve to function in different modes and pressure settings, depending on which pilot is selected to control. The controlling pilot is then selected by activating the appropriate solenoid stop valve. Pressure pilots may also be adapted to accept connection to pneumatic control systems, allowing automatic resetting of the pressure pilot as part of a much more comprehensive control strategy.

Although evaporator-pressure regulation is the most common use of upstream pressure regulators, they are also used in a variety of other refrigeration system applications. Upstream pressure regulators may be adapted for internal pressure relief, air-cooled condenser pressure regulation during low ambient operation, and liquid receiver pressure regulation.

CONSTANT-PRESSURE EXPANSION VALVES

The constant-pressure expansion valve is a downstream pressure regulator that is positioned to respond to evaporator-pressure changes and meter the mass flow of liquid refrigerant entering the evaporator to maintain a constant evaporator pressure.

Operation

Figure 25 shows a cross section of a typical constant-pressure expansion valve. The valve has both an adjustable opening spring, which exerts its force on top of the diaphragm in an opening direction, and a spring beneath the diaphragm, which exerts its force in a closing direction. Evaporator pressure is admitted beneath the diaphragm, through either the internal or external equalizer passage, and the combined forces of evaporator pressure and closing spring counterbalance the adjustable opening spring force.

During normal system operation, a small increase in evaporator pressure pushes the diaphragm up against the adjustable opening spring, allowing the closing spring to move the pin in a closing direction. This restricts refrigerant flow and limits evaporator pressure. When evaporator pressure drops below the valve setting (a decrease in load), the opening spring moves the valve pin in an opening direction. As a result, refrigerant flow increases and raises the evaporator pressure, bringing the three primary forces in the valve back into balance.

Because constant-pressure expansion valves respond to evaporator load changes inversely, their primary application is in systems that have nearly constant evaporator loading.

Selection

The constant-pressure expansion valve should be selected to provide the required liquid refrigerant flow capacity at the expected pressure drop across the valve, and should have an adjustable pressure range that includes the required design evaporator (valve outlet) pressure. The system designer should decide whether off-cycle pressure equalization is required.

Application

Constant-pressure expansion valves overfeed the evaporator as load diminishes, and underfeed as load increases. Their primary function is to balance liquid flow rate with compressor capacity at constant evaporator pressure/temperature as loading varies, protecting against product freezing at low loads and compressor motor overload when evaporator loading increases. Because the valve responds inversely to evaporator load variations, other means to protect the compressor against liquid floodback at low loads and overheating at high loads (e.g., suction-line accumulators, enhanced compressor cooling or liquid injection devices) are needed in systems that experience significant load variation.

Constant-pressure expansion valves are best suited to simple single-compressor/single-evaporator systems when constant evaporator temperature is important and significant load variation does not occur. They are commonly used in drink dispensers, food dispensers, drinking fountains, ice cream freezers, and self-contained room air conditioners. They are typically direct-operated devices; however, they may be pilot-operated for applications requiring very large capacity. They are also used to regulate hot gas in discharge bypass capacity reduction arrangements, as described in the section on Discharge Bypass Valves.

Constant-pressure expansion valves close in response to the abrupt increase in evaporator pressure when the compressor cycles off, preventing flow during the off-cycle. A small, fixed auxiliary passageway, described in the section on Thermostatic Expansion Valves, can also be built into constant-pressure expansion valves to provide off-cycle pressure equalization for use with low-starting-torque compressor motors.

SUCTION-PRESSURE-REGULATING VALVES

The suction-pressure regulator is a downstream pressure regulator positioned in a compressor suction line to respond to and limit compressor suction pressure. Typically, they are used to prevent compressor motor overload from high suction pressure related to warm start-up, defrost termination, and intermittent high evaporator loading.

Operation

Direct-acting suction-pressure regulators respond to their own outlet or downstream pressure. They are relatively simple devices, as illustrated in Figure 26. The valve outlet pressure acts on the bottom of the disk and exerts a closing force, which is opposed by the adjustable spring force. The inlet pressure acts on the underside of the bellows and the top of the seat disk. Because the effective areas of the bellows and port are equal, these two forces cancel each other and the valve responds to outlet pressure only. When outlet pressure falls below the equivalent force exerted by the spring, the seat disk moves in an opening direction to maintain outlet pressure. If outlet pressure rises, the seat disk moves in a closing direction and throttles the refrigerant flow to limit the downstream pressure. The proper pressure setting for a specific system is one that is low enough to protect the compressor from overload without unnecessarily compromising system capacity. Because both spring and bellows must be compressed through the entire closing valve stroke, a significant change in outlet pressure is required to close the valve to its minimum capacity. Outlet pressure changes of 5 to 10 psi or more, depending on design, are typically required to move direct-operated downstream pressure regulators from open position to near closed position. Therefore, these valves have relatively high gradients, and regulated suction pressure may change significantly when large changes in load occur.

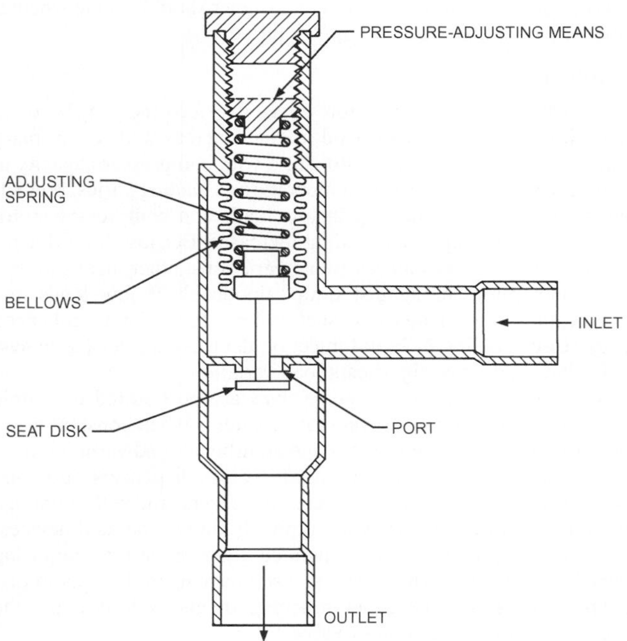

Fig. 26 Direct-Acting Suction-Pressure Regulator

Pilot-operated suction-pressure regulators are available for larger systems and applications requiring more precise pressure regulation over wide load and inlet pressure variations. Their design is significantly more complex, because of pilot operation. The method of operation is similar to that described in the discussion of pilot-operated evaporator-pressure regulators. However, in downstream pressure regulators, the pilot is reverse-acting and functions similarly to the constant-pressure expansion valve. Suction stop service can also be provided with this type of regulator.

Selection

The suction-pressure regulator should be selected to provide the required flow capacity at a low pressure loss to minimize system capacity penalty. However, take care to avoid oversizing, which can lead to unstable regulator operation. The significant change in outlet pressure required to stroke direct-operated regulators should also be considered during selection.

Application

Suction-pressure-regulating valves are primarily used to prevent compressor motor overload caused by excessive suction pressure related to warm evaporator start-up, defrost termination, or intermittent high-load operation. These regulating valves are typically designed to operate at normal refrigeration system low-side pressures and temperatures. However, similar-type downstream pressure regulators may be modified to include suitable seat materials and high-gradient springs for application in system high-side pressure and temperature conditions. For example, they may be used in a variety of schemes to maintain necessary operating pressures in air-cooled condensers during cold-weather operation. Additionally, modified regulators are used to bypass compressor discharge gas in refrigeration system capacity reduction schemes, as mentioned in the application sections in the Constant-Pressure Expansion Valves and Discharge Bypass Valves sections.

CONDENSER-PRESSURE-REGULATING VALVES

Various pressure-regulating valves are used to maintain sufficient pressure in air-cooled condensers during cold-weather operation. Both single- and two-valve arrangements have been used for this purpose. See Chapter 2 in this volume and Chapter 35 of the 2004 *ASHRAE Handbook—HVAC Systems and Equipment* for more information.

Operation

The first valve in the two-valve arrangement shown in Figure 27 is typically an upstream pressure regulator that is constructed and operates similarly to the evaporator-pressure-regulating valves shown in Figures 21 to 23. Pilot-operated regulators are typically used to meet the flow capacity requirements of large systems. It may have special features that make it suitable for high-pressure and high-temperature operating conditions. This control may be installed at either the condenser inlet or outlet; the outlet is usually preferred because a smaller valve can satisfy the system's flow capacity requirements. It throttles when the condenser outlet or compressor discharge pressure falls as a result of cold-weather operating conditions.

The second valve in the two-valve arrangement is installed in a condenser bypass line. It may be a downstream pressure regulator similar to the suction-pressure regulator in Figure 26, or a differential pressure regulator as in Figure 27. The differential regulator is often preferred for simplicity; however, the minimum opening differential pressure must be greater than the pressure drop across the condenser at full-load summer operating conditions. As the first valve throttles in response to falling compressor discharge or condenser outlet pressure, the second valve opens and allows hot gas to bypass the condenser to mix with and warm the cold liquid entering

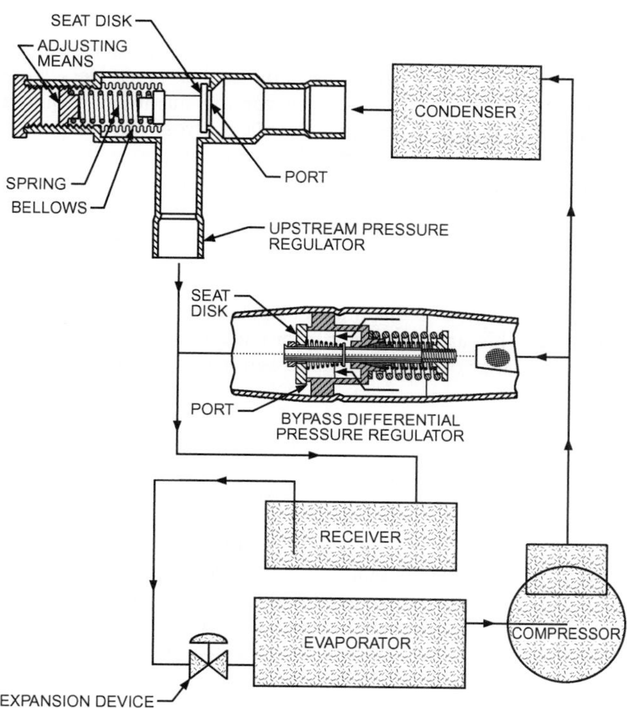

Fig. 27 Condenser Pressure Regulation (Two-Valve Arrangement)

the receiver, thereby maintaining adequate high-side saturation pressure.

Special-purpose three-way pressure-regulating valves similar to that shown in Figure 28 are also used for condenser pressure regulation. The valve in Figure 28 is a direct-acting pressure regulator with a second inlet that performs the hot-gas bypass function, eliminating the need for the second valve in the two-valve system described previously. The three-way valve simultaneously throttles liquid flow from the condenser and bypasses hot compressor discharge gas to the valve outlet, where it mixes with and warms liquid entering the receiver, thereby maintaining adequate high-side saturated liquid pressure.

In the three-way valve, the lower side of a flexible metal diaphragm is exposed to system high-side pressure, while the upper side is exposed to a noncondensable gas charge (usually dry nitrogen or air). A pushrod links the diaphragm to the valve poppet, which seats on either the upper or lower port and throttles either discharge gas or liquid from the condenser, respectively. Valves that respond to condenser pressure are frequently used; however, valve designs that respond to receiver pressure are available.

During system start-up in extremely cold weather, the poppet may be tight against the lower seat, stopping all liquid flow from the condenser and bypassing discharge gas into the receiver until adequate system high-side pressure is generated. During stable operation in cold weather, the poppet modulates at an intermediate position, with liquid flow from the condensing coil mixing with compressor discharge gas in the valve outlet and flowing to the receiver. With higher condensing pressure during warm weather, the poppet seats tightly against the upper port, allowing free flow of liquid from condenser to receiver and preventing compressor discharge gas from bypassing the condenser.

The three-way condenser-pressure-regulating valve set point is usually not field-adjustable. The pressure setting is established by the pressure of the gas charge placed in the dome above the diaphragm during manufacture. Some designs allow field selection between two factory-predetermined set points.

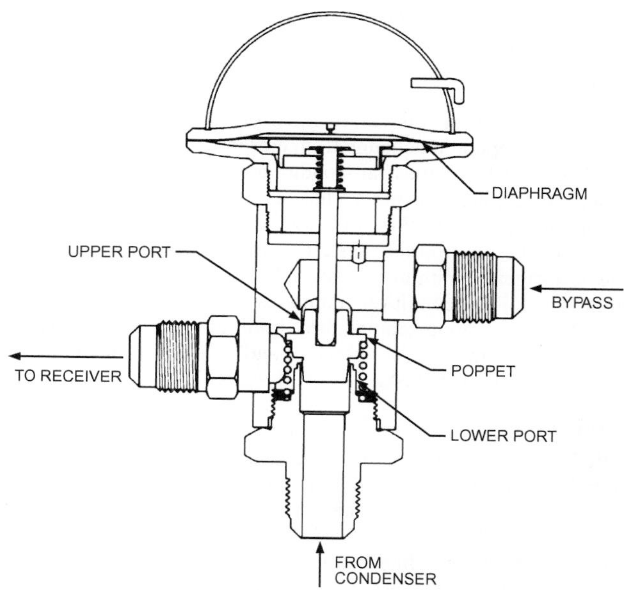

Fig. 28 Three-Way Condenser-Pressure-Regulating Valve

Application

Systems using pressure regulators for air-cooled condenser pressure control during cold-weather operation require careful design. Condenser pressure is maintained by partially filling the condenser with liquid refrigerant, reducing the effective condensing surface available. The condenser is flooded with liquid refrigerant to the point of balancing condenser capacity at low ambient with condenser loading. The system must have adequate refrigerant charge and receiver capacity to maintain a liquid seal at the expansion valve inlet while allowing sufficient liquid for head pressure control to accumulate in the condenser. If the system cycles off or otherwise becomes idle during cold weather, the receiver must be kept warm during the off time so that adequate liquid pressure is available at start-up. When receivers are exposed to low temperatures, it may be necessary to provide receiver heaters and insulation to ensure start-up capability. A check valve at the receiver inlet may be advisable.

Because these systems necessarily contain abnormally large refrigerant charges, careful consideration must be given to controlling refrigerant migration during system idle times under adverse ambient temperatures.

DISCHARGE BYPASS VALVES

The discharge bypass valve (or hot-gas bypass valve) is a downstream pressure regulator located between the compressor discharge line and the system low-pressure side, and responds to evaporator pressure changes to maintain a desired minimum evaporator pressure. Typically, they are used to limit the minimum evaporator pressure during periods of reduced load to prevent coil icing or to avoid operating the compressor at a lower suction pressure than recommended. See Chapter 2 for more information.

Operation

A typical mechanical discharge bypass valve has the same basic configuration as the constant-pressure expansion valve in Figure 25. The construction materials for the discharge bypass valves are suitable for application at high pressure and temperature.

The equivalent pressure from the adjustable spring is balanced across the diaphragm against system evaporator pressure. When the evaporator pressure falls below the valve setting, the spring strokes the valve member in the opening direction.

Selection

Discharge bypass valves are rated on the basis of allowable evaporator temperature change from closed position to rated opening. This is 6°F for most air-conditioning applications, although capacity multipliers for other changes are available to make the appropriate valve selection. Because several basic system factors are involved in appropriate selection, it is important to know the type of refrigerant, minimum allowable evaporating temperature at the reduced load condition, compressor capacity at minimum allowable evaporating temperature, minimum evaporator load at which the system is to be operated, and condensing temperature when minimum load exists.

Application

Refrigeration systems experience load variations to some degree throughout the year and may use discharge bypass valves to balance compressor capacity with system load. Depending on the system size and type of compressor used, this valve type may be used in place of compressor cylinder unloaders or to handle the unloading requirements below the last step of cylinder unloading. Depending on the system components and configuration, hot gas may be introduced (1) directly into the suction line, in which case a desuperheating thermostatic expansion valve is required to control suction gas temperature, or (2) into the evaporator inlet through a specially designed refrigerant distributor with an auxiliary side connection, where it mixes with cold refrigerant from the system thermostatic expansion valve to control suction gas temperature before the gas reaches the compressor.

The location of the discharge bypass valve and necessary accessories depend on the type of system. Valve manufacturers' literature gives proper locations and piping recommendations for their products.

HIGH-SIDE FLOAT VALVES

Operation

A high-side float valve controls the mass flow of refrigerant liquid entering the evaporator so it equals the rate at which the refrigerant gas is pumped from the evaporator by the compressor. Figure 29 shows a cross section of a typical valve. The refrigerant liquid flows from the condenser into the high-side float valve body, where it raises the float and moves the valve pin in an opening direction, permitting the liquid to pass through the valve port, expand, and flow into the evaporator or low-pressure receiver. Most of the system refrigerant charge is contained in the evaporator or low-pressure receiver at all times.

Selection

For acceptable performance, the high-side float valve is selected for the refrigerant and a rated capacity neither excessively large nor too small. The orifice is sized for the maximum required capacity with the minimum pressure drop across the valve. The valve operated by the float may be a pin-and-port construction (Figure 29), a butterfly valve, a balanced double-ported valve, or a sliding gate or spool valve. The internal bypass vent tube allows installation of the high-side float valve near the evaporator and above the condenser without danger of the float valve becoming gas-bound. Some large-capacity valves use a high-side float valve for pilot operation of a diaphragm or piston spring-loaded expansion valve. This arrangement can provide improved modulation over a wide range of load and pressure-drop conditions.

Application

A refrigeration system in which a high-side float valve is typically used may be a simple single evaporator/compressor/condenser system or have a low-pressure liquid receiver with multiple evaporators and compressors. The high-pressure receiver or liquid sump at the condenser outlet can be quite small. A full-sized high-pressure receiver may be required for pumping out flooded evaporator(s) and/or low-pressure receivers. The amount of refrigerant charge is critical with a high-side float valve in simple single evaporator/compressor/condenser systems. An excessive charge causes floodback, whereas insufficient charge reduces system capacity.

LOW-SIDE FLOAT VALVES

Operation

The low-side float valve performs the same function as the high-side float valve, but it is connected to the low-pressure side of the system. When the evaporator or low-pressure receiver liquid level drops, the float opens the valve. Liquid refrigerant then flows from the liquid line through the valve port and directly into the evaporator or surge drum. In another design, the refrigerant flows through the valve port, passes through a remote feed line, and enters the evaporator through a separate connection. (A typical direct-feed valve construction is shown in Figure 30.) The low-side float system is a flooded system.

Selection

Low-side float valves are selected in the same way as the high-side float valves discussed previously.

Application

In the low-side float valve system, the refrigerant charge is not critical. The low-side float valve can be used with multiple evaporators such that some evaporators may be controlled by other low-side float valves and some by thermostatic expansion valves.

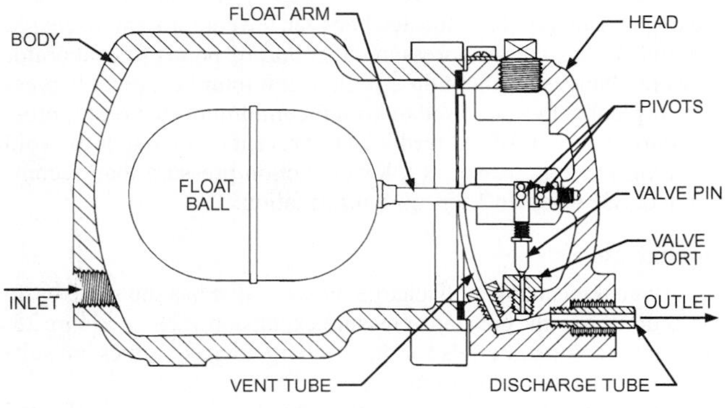

Fig. 29 High-Side Float Valve

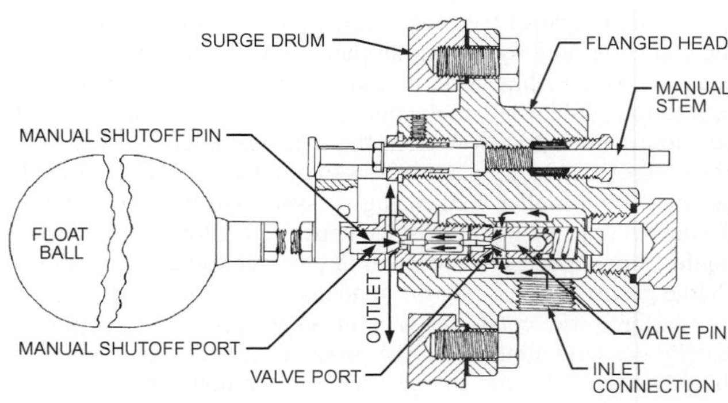

Fig. 30 Low-Side Float Valve

The float valve is mounted either directly in the evaporator or surge drum or in an external chamber connected to the evaporator or surge chamber by equalizing lines (i.e., a gas line at the top and a liquid line at the bottom). In the externally mounted type, the float valve is separated from the float chamber by a gland that maintains a calm level of liquid in the float chamber for steady actuation of the valve.

In evaporators with high boiling rates or restricted liquid and gas passages, the boiling action of the liquid raises the refrigerant level during operation. When the compressor stops or the solenoid suction valve closes, boiling of the liquid refrigerant ceases, and the refrigerant level in the evaporator drops. Under these conditions, the high-pressure liquid line supplying the low-side float valve should be shut off by a solenoid liquid valve to prevent overfilling the evaporator. Otherwise, excess refrigerant will enter the evaporator on the off-cycle, which can cause floodback when the compressor starts or the solenoid suction valve opens.

When a low-side float valve is used, ensure that the float is in a calm liquid level that falls properly in response to increased evaporator load and rises with decreased evaporator load. In low-temperature systems particularly, it is important that the equalizer lines between the evaporator and either the float chamber or surge drum be generously sized to eliminate any reverse response of the refrigerant liquid level near the float. Where the low-side float valve is located in a nonrefrigerated room, the equalizing liquid and gas lines and float chamber must be well insulated to provide a calm liquid level for the float.

SOLENOID VALVES

Solenoid valves, also called **solenoid-operated valves**, are comprised of a soft-iron armature positioned in the central axis of a copper wire coil. When electric current flows through the coil, a magnetic field is created that draws the movable armature to it. The armature is adapted to open or close a valve port as it is moved by the magnetic field. This basic operating mechanism is adapted to a wide variety of valve designs and sizes for refrigerant service.

Solenoid valves for refrigerant service are typically two-position devices (i.e., solenoid energized or deenergized). In energized mode, the armature is drawn into the coil by the magnetic field. The electromagnetic coil must provide the work required to overcome the spring or gravity plus the work necessary to open or close the valve. In deenergized mode, the armature, sometimes called a plunger or core, is moved to the extended end of its stroke by a spring, or in some designs by gravity.

Refrigerant-service solenoid valves have the plunger enclosed in a thin-walled nonmagnetic metal tube (usually nonmagnetic stainless steel). One end of the tube is closed pressuretight by welding or brazing a soft-iron-bearing magnetic metal plug into the tube. The open end is adapted to the valve body, usually in axial alignment with the valve port. This construction eliminates the need for a dynamic stem seal between the solenoid operator and the valve. Figure 31 shows a semihermetic construction in which the tube, top plug, and lower nut are welded or brazed together. The tube assembly is threaded to the valve body using a metal-to-metal pressuretight seal that eliminates the need for synthetic materials. Some small valves are made completely hermetic by welding or brazing the valve body to the enclosing tube containing the magnetically movable plunger assembly. Most often, the connection between magnetic assembly and valve body is made pressuretight with synthetic gaskets or O rings, as shown in Figure 32.

The copper wire electromagnetic coil, with its associated electrical insulation system, is closely fitted to the outer diameter of the enclosing tube. This arrangement places the coil and insulation outside the pressurized refrigerant environment. The electromagnetic circuit includes the plunger, top plug, metallic housing surrounding the copper wire coil assembly, and any metal spacers or sleeves used to properly position the coil on the tube, as shown in Figure 31. These components are fabricated of soft-iron-bearing magnetic materials and are absolutely essential to acceptable solenoid valve performance. Combined, they form a well-defined and complete magnetic circuit. Many contemporary designs use molded resin coil assemblies that encapsulate the coil and outer soft-iron housing within the resin. This construction enhances heat dissipation, protects electrical parts from moisture and mechanical damage, and simplifies field assembly.

Operation

A number of valve designs have been developed that can be reliably operated with relatively low-powered solenoids, which helps to minimize energy consumption. The major force that must be overcome by the solenoid operator to open a normally closed valve or close a normally open valve is related to the port area multiplied by the pressure differential across the valve when it is closed. The maximum pressure differential that a specific solenoid valve will reliably

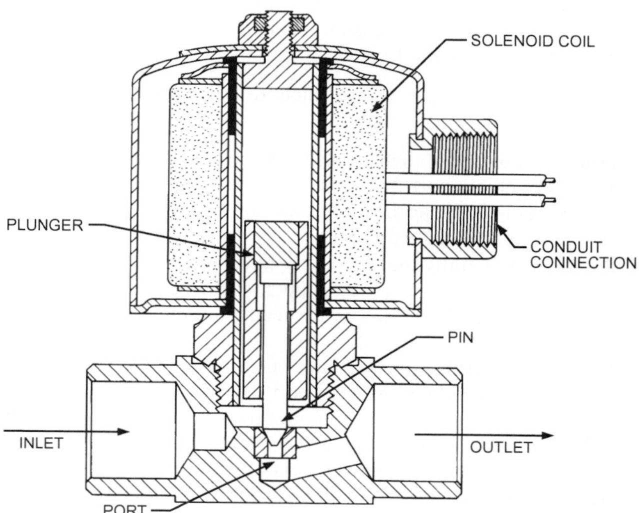

Fig. 31 Normally Closed Direct-Acting Solenoid Valve with Hammer-Blow Feature

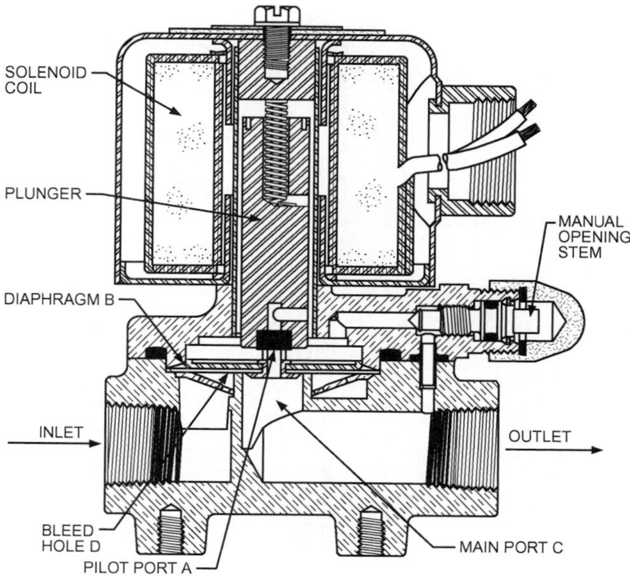

Fig. 32 Normally Closed Pilot-Operated Solenoid Valve with Direct-Lift Feature

open against is called the **maximum operating pressure differential (MOPD)**. To provide acceptable MOPD for refrigeration applications with reasonably powered solenoids, it is necessary to design valves with small ports. Small-ported valves can satisfy low-flow-capacity requirements and pilot duty applications. Thus, direct-acting solenoid valves are limited to low-capacity or low-MOPD applications. Large-capacity and high-MOPD applications use pilot-operated valves similar to those in Figures 32 and 33.

Solenoid valves are divided into two basic categories related to flow capacity and MOPD:

- **Direct-acting valves** (see Figure 31) are capable of relatively small flow capacities with MOPDs suitable for most refrigerant applications, are useful as pilots on larger valves, and can be designed for medium flow capacities with low MOPD.
- **Pilot-operated valves** (see Figures 32 and 33) are capable of large flow capacities with a full range of MOPDs suitable for refrigerant applications. They have small direct-acting valves embedded in the main valve body that pilot the main valve. Opening the pilot port creates a pressure imbalance within the main valve mechanism, causing it to open.

These basic categories are each divided into several subcategories:

- **Two-way normally closed valves** have one inlet, one outlet, and one intermediate port that is closed when the solenoid is deenergized. This is by far the most common configuration in a wide variety of refrigerant stop services.
- **Two-way normally open valves** have one inlet, one outlet, and one intermediate port that is open when the solenoid is deenergized. This configuration is particularly useful in applications requiring a valve that will "fail" to a normally open position.
- **Three-way diverting valves** have one inlet and two outlets; each outlet is associated with its own port so that when the solenoid is deenergized, flow from the common inlet exits through outlet 1, and when the solenoid is energized, outlet 1 is stopped and flow exits through outlet 2. This type may be used to divert hot compressor discharge gas from the normal condenser to the heat reclaim heat exchanger.
- **Three-way mixing valves** have two inlets and one outlet, each inlet being associated with its own port so that when the solenoid is deenergized, refrigerant flows from inlet 1 to the common outlet, and when the solenoid is energized, refrigerant flows from inlet 2 to the common outlet. This style, in the direct-acting version, may be used to activate compressor cylinder unloading mechanisms or as a pilot for large gas-powered valves that use system high-side pressure to close normally open stop valves.
- **Four-way reversing valves** have two inlets and two outlets. One connection always functions as an inlet. Another one of the four always functions as an outlet. Flow in the remaining two connections is reversed when the solenoid is energized or deenergized. This configuration is used almost exclusively to switch heat pumps between cooling and heating modes. The direct-acting version is used to pilot the main valve, and both main and pilot valves are most often hybrid spool valves. Figures 34 and 35 show these valves schematically in both energized and deenergized modes.

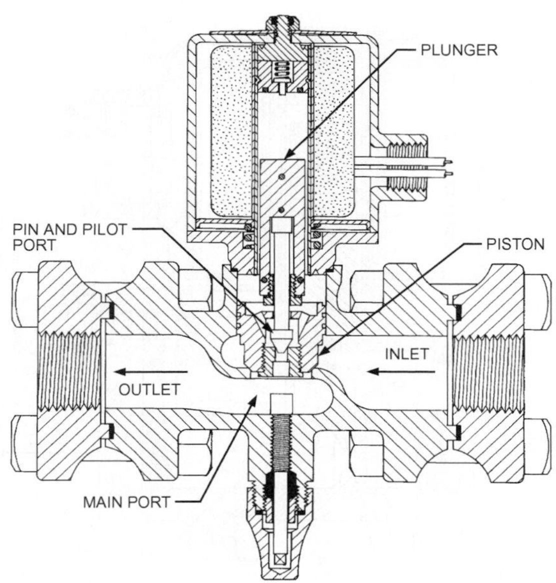

Fig. 33 Normally Closed Pilot-Operated Solenoid Valve with Hammer-Blow and Mechanically Linked Piston-Pin Plunger

Figure 31 shows a direct-acting normally closed two-way valve with a long-stroke solenoid that uses a "lost-motion hammer-blow" mechanism to hit the valve pin and overcome the pressure differential force holding the valve closed. This design, though providing some additional opening capability, is less reliable with respect to MOPD repeatability and has a shorter cyclic life because of the high-momentum impacts between plunger, pin, and top plug. It also relies entirely on gravity to move the valve in the closing direction when the solenoid is deenergized. This reliance limits the installed position. This type of valve must be installed in an upright position with the magnetic assembly on top.

A common design of direct-acting solenoid is the direct-lift type such as the solenoid operator portion of the pilot-operated valve in Figure 32. The short stroke minimizes mechanical damage related to impact of the plunger with the top plug and minimizes inrush current in alternating current systems. The closing spring allows installation in any position. Some solenoids of this short-stroke design can survive extended periods of time in failure mode (i.e., when the coil is energized but the valve fails to open because of excessive pressure, low voltage, or other reasons) without significant damage to the coil.

Figure 32 illustrates a pilot-operated valve that uses a semi-flexible diaphragm to operate the main valve. When the solenoid is energized, the pilot port (shown centered over the main port) opens and allows pressure above the diaphragm to diminish as it discharges to low pressure at the valve outlet. The higher valve inlet pressure on the underside of the diaphragm surrounding the main port causes the diaphragm to move up, opening the main port to full flow. The ratio of diaphragm area to port area as well as the relative flow rates of bleed hole D and the pilot port are carefully balanced in the design. This ensures that adequate opening force develops to meet the MOPD requirements of the valve specification. When the solenoid is deenergized, the plunger moves down, closing the pilot port, and inlet pressure flowing through bleed hole D allows pressure above the diaphragm to equalize with valve inlet pressure. The entire top of the diaphragm is exposed to inlet pressure; on the bottom side, only the annular area surrounding the port is exposed to inlet pressure; the center portion is exposed to outlet pressure. The net effect of the pressure and area differences is a force that pushes the diaphragm down to close the main port. The spring in the top of the plunger causes the plunger to follow the falling pilot port, keeping it sealed and providing additional downward push on the diaphragm to help close the main port. In the valve of Figure 32, the pilot operator is centrally located directly over the main port. In this configuration, the diaphragm stroke is limited. When larger diaphragm strokes are required for greater flow capacity, the pilot operator and pilot port are relocated to a point beyond the perimeter of the diaphragm over the main valve outlet connection. Separate flow passages in the valve body are provided to conduct pilot port flow from the top of the diaphragm to the valve outlet. This arrangement is usually called an "offset pilot" and retains the benefits of a short-stroke solenoid operator while allowing a diaphragm stroke commensurate with full flow through the main port.

The pilot-operated valve shown in Figure 33 operates according to the same principles as the diaphragm valve, but uses a piston instead of a diaphragm. The carefully controlled annular clearance between piston and inside bore of the valve body is often used to perform the function of bleed hole D (shown in Figure 32), eliminating the need for a separate bleed hole in the piston. The valve in Figure 33 uses a long-stroke hammer-blow pilot operator, which allows long main valve strokes to accommodate large flow capacity requirements without offsetting the pilot. The centered pilot allows mechanically linking the main piston to the plunger assembly to help hold the valve wide open at near-zero pressure drops in low-temperature suction-line applications.

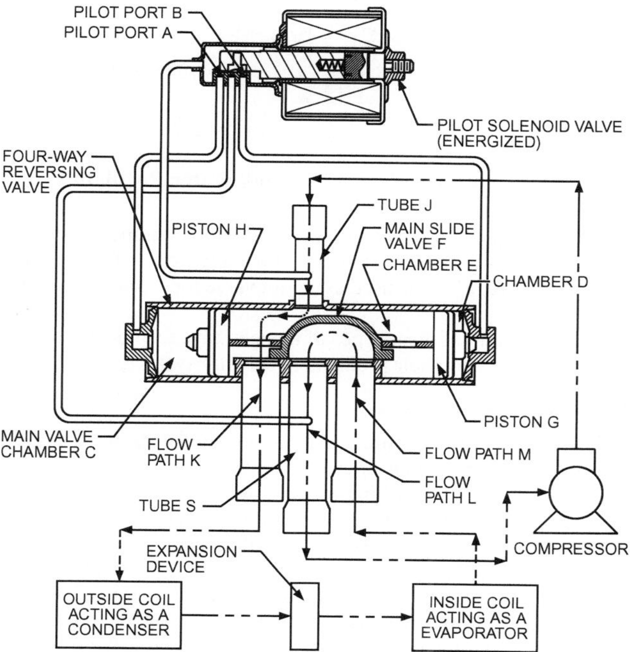

Fig. 34 Four-Way Slide Refrigerant-Reversing Valve Used in Cooling (or Defrosting) Cycle of Refrigeration System

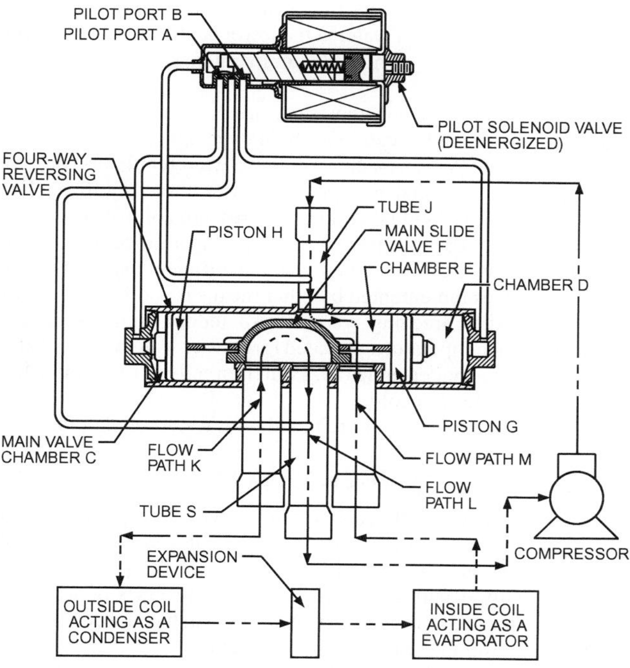

Fig. 35 Four-Way Slide Refrigerant-Reversing Valve Used in Heating Cycle of Refrigeration System

Figure 34 shows a four-way valve piloted by a four-way direct-acting valve shown in the energized position. The main valve slide F is positioned to connect flow path M, coming from the evaporator (inside coil), to flow path L going to compressor suction through tube S. At the same time, high-pressure hot gas flows from the compressor discharge through tube J, around slide F and through flow path K to the condenser (outside coil). High pressure from tube J passes through pilot port A to main valve chamber C. The main valve slide is held in this position by the high pressure in chamber C pushing piston H to the right. When the pilot solenoid is deenergized, the pilot valve plunger is moved to the left by the spring (as shown in Figure 35), allowing high pressure from tube J to flow through pilot port B to chamber D at the right-hand end of the main valve body. Simultaneously, chamber C is connected through pilot port A to low pressure in tube S. The pressure in chamber D rises as the pressure in chamber C falls, driving slide F to the left, as shown in Figure 35. Flows in paths K and M reverse and the outside coil becomes the evaporator and the inside coil becomes the condenser. The system has been transferred into heating mode. When the pilot solenoid is reenergized, the processes reverses and the system reverts to cooling mode.

Application

Solenoid valves are generally vulnerable to particles in the refrigerant stream and should be protected by a filter-drier.

Valves that are attitude-sensitive must be carefully oriented and properly supported to ensure reliable operation. Take care to avoid overheating sensitive valve parts when installation involves soldering, brazing, or welding.

Electrical service provided to solenoid operators deserves careful attention. Most solenoid valve performance failures are related to improper or inadequate provision of electric power to the solenoid. Undervoltage when attempting to open seriously compromises MOPD and causes failure to open. Continued application of power to an alternating current (ac) solenoid coil installed on a valve that is unable to open overheats the coil and may lead to premature coil failure, even at undervoltage. Although direct current (dc) solenoids may tolerate a little more voltage variation, overvoltage leads to overheating, even when the valve successfully opens, and shortens coil life; undervoltage reduces the MOPD.

The probability of experiencing undervoltage at the moment of opening with ac systems increases when a control transformer of limited capacity supplies power to the solenoid. This type of transformer is commonly used to supply power to low-voltage control systems using class 2 wiring. The situation is aggravated when more than one device served by the same transformer is energized simultaneously.

CONDENSING WATER REGULATORS

Condensing water regulators are used for head pressure control during year-round operation of refrigeration systems. Additional information can be found in Chapter 13 of the 2004 *ASHRAE Handbook—HVAC Systems and Equipment*, in the section on Operation Optimization in Chapter 46 of the 2003 *ASHRAE Handbook—HVAC Applications*, and in Chapter 2 of this volume, under Head Pressure Control for Refrigerant Condensers.

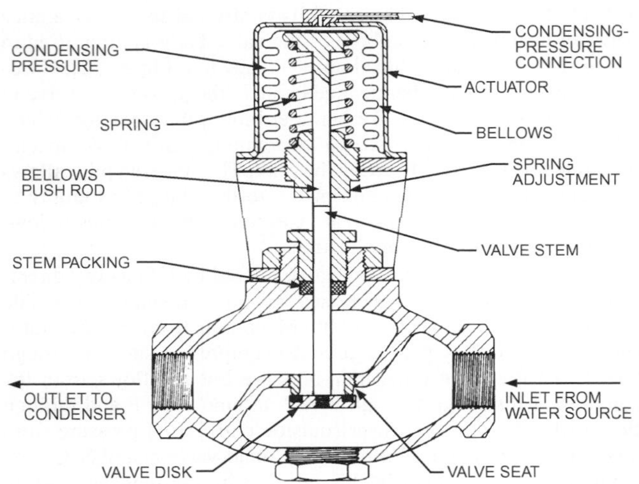

Fig. 36 Two-Way Condensing Water Regulator

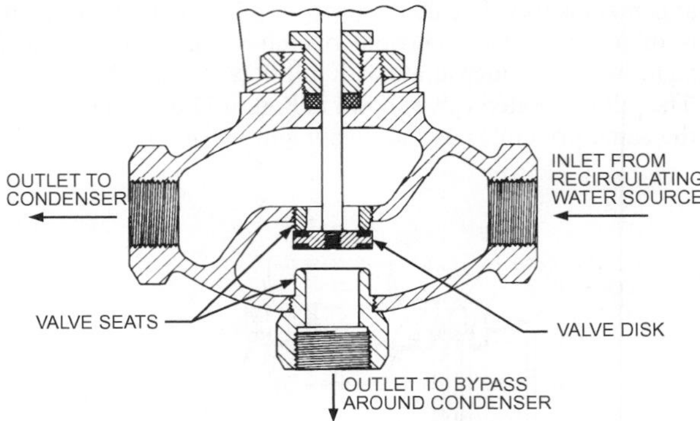

Fig. 37 Three-Way Condensing Water Regulator

Two-Way Regulators

Condensing water regulators modulate the quantity of water passing through a water-cooled refrigerant condenser in response to the condensing pressure. They are available for use with most refrigerants, including ammonia (R-717). Most manufacturers stress that these valves are designed for use only as operating devices. Where system closure, improper flow, or loss of pressure caused by valve failure can result in personal injury, damage, or loss of property, separate safety devices must be added to the system.

These devices are used on vapor-cycle refrigeration systems to maintain satisfactory condensing pressure. The regulator automatically modulates to correct for both variations in temperature or pressure of the water supply and variations in the quantity of refrigerant gas being pumped into the condenser.

Operation. The condensing water regulator consists of a valve and an actuator linked together, as shown in Figure 36. The actuator consists of a metallic bellows and adjustable spring combination connected to the system high side.

After a compressor starts, the compressor discharge pressure begins to rise. When the opening pressure setting of the regulator spring is reached, the bellows moves to open the valve disk gradually. The regulator continues to open as condenser pressure rises, until water flow balances the required heat rejection. At this point, the condenser pressure is stabilized. When the compressor stops, the continuing water flow through the regulator causes the condenser pressure to drop gradually, and the regulator becomes fully closed when the opening pressure setting of the regulator is reached.

Selection. To avoid hunting or internal erosion caused by high pressure drops through an oversized valve because it operates only partially open for most of its duty cycle, the regulator should be selected from the manufacturer's data on the basis of maximum required flow, minimum available pressure drop, water temperature, and system operating conditions. Also, depending on the specific refrigerant being used, special components may be required (e.g., stainless steel rather than brass bellows for ammonia).

The water flow required depends on condenser performance, temperature of available water, quantity of heat that must be rejected to the water, and desired operating condenser pressure. For a given opening of the valve seat, which corresponds to a given pressure rise above the regulator opening point, the flow rate handled by a given size of water regulator is a function of the available water-pressure drop across the valve seat.

Application. Because there are two types of control action available (direct- or reverse-acting), these regulators can be used for various applications (e.g., water-cooled condensers, bypass service on refrigeration systems, ice machines, heat pump systems that control water temperature). For equipment with a large water flow requirement, a small regulator is used as a pilot valve for a diaphragm main valve.

Manufacturers of these types of devices have technical literature to assist in applying their products to specific systems.

Three-Way Regulators

These regulators are similar to two-way regulators, but they have an additional port, which opens to bypass water around the condenser as the port controlling water flow to the condenser closes (Figure 37). Thus, flow through the cooling tower decking or sprays and the circulating pump is maintained, although the water supply to individual or multiple condensers is modulated for control.

Operation. The three-way regulator operates akin to the two-way regulator. Low refrigerant head pressure, which may result from low cooling-tower water temperature, decreases the refrigeration system's cooling ability rapidly. The three-way regulator senses the compressor head pressure and allows cooling water to flow to the condenser, bypass the condenser, or flow to both the condenser and bypass line to provide correct refrigerant head pressures.

The regulator allows water flow to the tower through the bypass line, even though the condenser does not require cooling. This provides an adequate head of water at the tower at all times so the tower can operate efficiently with minimum maintenance on nozzles and wetting surfaces.

Selection. Selection considerations are the same as for two-way regulators, including the cautions about oversizing.

Application. Pressure-actuated three-way regulators are for condensing units cooled by atmospheric or forced-draft cooling towers requiring individual condenser-pressure control. They may be used on single or multiple condenser piping arrangements to the tower to provide the most economical and efficient use. These regulators must be supplemented by other means if cooling towers are to be operated in freezing weather. An indoor sump is usually required, and a temperature-actuated three-way water control valve diverts all of the condenser leaving water directly to the sump when the water becomes too cold.

Strainers are not generally required with water regulators.

CHECK VALVES

Refrigerant check valves are normally used in refrigerant lines in which pressure reversals can cause undesirable reverse flows. A check valve is usually opened by a portion of the pressure drop. Closing usually occurs either when pressure reverses or when the pressure drop across the check valve is less than the minimum opening pressure drop in the normal flow direction.

The conventional large check valve uses piston construction in a globe-pattern valve body, whereas in-line designs are common for 2 in. or smaller valves. Either design may include a closing spring;

a heavier spring gives more reliable and tighter closing but requires a greater pressure differential to open. Although conventional check valves may be designed to open at less than 1 psi, they may not be reliable below –25°F because the light closing springs may not overcome viscous lubricants.

Seat Materials

Although precision metal seats may be manufactured nearly bubbletight, they are not economical for refrigerant check valves. Seats made of synthetic elastomers provide excellent sealing at medium and high temperatures, but may leak at low temperatures because of their lack of resilience. Because high temperatures deteriorate most elastomers suitable for refrigerants, plastic materials have become more widely used, despite being susceptible to damage by large pieces of foreign matter.

Applications

In compressor discharge lines, check valves are used to prevent flow from the condenser to the compressor during the off cycle or to prevent flow from an operating compressor to an idle compressor. Although a 2 to 6 psi pressure drop is tolerable, the check valve must resist pulsations caused by the compressor and the temperature of discharge gas. Also, the valve must be bubbletight to prevent liquid refrigerant from accumulating at the compressor discharge valves or in the crankcase.

In liquid lines, a check valve prevents reverse flow through the unused expansion device on a heat pump or prevents backup into the low-pressure liquid line of a recirculating system during defrosting. Although a 2 to 6 psi pressure drop is usually acceptable, the check-valve seat must be bubbletight.

In the suction line of a low-temperature evaporator, a check valve may be used to prevent transfer of refrigerant vapor to a lower-temperature evaporator on the same suction main. In this case, the pressure drop must be less than 2 psi, the valve seating must be reasonably tight, and the check valve must be reliable at low temperatures.

In hot-gas defrost lines, check valves may be used in the branch hot-gas lines connecting the individual evaporators to prevent crossfeed of refrigerant during the cooling cycle when defrost is not taking place. In addition, check valves are used in the hot-gas line between the hot-gas heating coil in the drain pan and the evaporator to prevent pan coil sweating during the refrigeration cycle. Tolerable pressure drop is typically 2 to 6 psi, seating must be nearly bubbletight, and seat materials must withstand high temperatures.

Oversized check valves may chatter or pulsate.

RELIEF DEVICES

A refrigerant relief device has either a safety or functional use. A safety relief device is designed to relieve positively at its set pressure for one crucial occasion without prior leakage. Relief may be to the atmosphere or to the low-pressure side.

A functional relief device is a control valve that may be required to open, modulate, and close with repeatedly accurate performance. Relief is usually from a portion of the system at higher pressure to a portion at lower pressure. Design refinements of the functional relief valve usually make it unsuitable or uneconomical as a safety relief device.

Safety Relief Valves

These valves are most commonly a pop-type design, which open abruptly when the inlet pressure exceeds the outlet pressure by the valve setting pressure (Figure 38). Seat configuration is such that once lift begins, the resulting increased active seat area causes the valve seat to pop wide open against the force of the setting spring. Because the flow rate is measured at a pressure of 10% above the setting, the valve must open within this 10% increase in pressure.

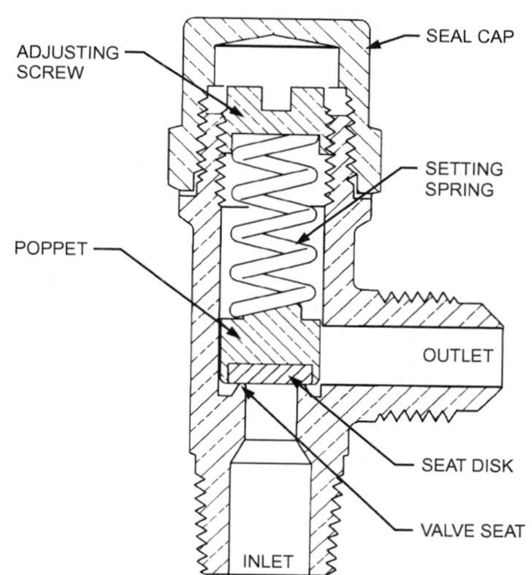

Fig. 38 Pop-Type Safety Relief Valves

This relief valve operates on a fixed pressure differential from inlet to outlet. Because the valve is affected by back pressure, a rupture disk must not be installed at the valve outlet.

Relief valve seats are made of metal, plastic, lead alloy, or synthetic elastomers. Elastomers are commonly used because they have greater resilience and, consequently, reseat more tightly than other materials. Some valves that have lead-alloy seats have an emergency manual reseating stem that allows reforming the seating surface by tapping the stem lightly with a hammer. Advantages of the pop-type relief valve are simplicity of design, low initial cost, and high discharge capacity.

Capacities of pressure-relief valves are determined by test in accordance with the provisions of the ASME *Boiler and Pressure Vessel Code*. Relief valves approved by the National Board of Boiler and Pressure Vessel Inspectors are stamped with the applicable code symbol(s). (Consult the *Boiler and Pressure Vessel Code* for specific text and marking details.) In addition, the pressure setting and capacity are stamped on the valve.

When relief valves are used on pressure vessels of 10 ft^3 internal gross volume or more, a relief system consisting of a three-way valve and two relief valves in parallel is required.

Pressure Setting. The maximum pressure setting for a relief device is limited by the design working pressure of the vessel to be protected. Pressure vessels normally have a safety factor of 5. Therefore, the minimum bursting pressure is five times the rated design working pressure. The relief device must have enough discharge capacity to prevent pressure in the vessel from rising more than 10% above its design pressure. Because the capacity of a relief device is measured at 10% above its stamped setting, the setting cannot exceed the design pressure of the vessel.

To prevent loss of refrigerant through pressure-relief devices during normal operating conditions, the relief setting must be substantially higher than the system operating pressure. For rupture members, the setting should be 50% above a static system pressure and 100% above a maximum pulsating pressure. Failure to provide this margin of safety causes fatigue of the frangible member and rupture well below the stamped setting.

For relief valves, the setting should be 25% above maximum system pressure. This provides sufficient spring force on the valve seat to maintain a tight seal and still allow for setting tolerances and other factors that cause settings to vary. Although relief valves are set at the factory to be close to the stamped setting, the variation may be as much as 10% after the valves have been stored or placed in service.

Discharge Piping. The size of the discharge pipe from the pressure-relief device or fusible plug must not be less than the size of the pressure-relief device or fusible plug outlet. The maximum length of discharge piping is provided in a table or may be calculated from the formula provided in ASHRAE *Standard* 15.

Selection and Installation. When selecting and installing a relief device,

- Select a relief device with sufficient capacity for code requirements and suitable for the type of refrigerant used.
- Use the proper size and length of discharge tube or pipe.
- Do not discharge the relief device before installation or when pressure-testing the system.
- For systems containing large quantities of refrigerant, use a three-way valve and two relief valves.
- Install a pressure vessel that allows the relief valve to be set at least 25% above maximum system pressure.

Functional Relief Valves

Functional relief valves are usually diaphragm types; system pressure acts on a diaphragm that lifts the valve disk from the seat (Figure 39). The other side of the diaphragm is exposed to both the adjusting spring and atmospheric pressure. The ratio of effective diaphragm area to seat area is high, so outlet pressure has little effect on the operating point of the valve.

Because the diaphragm's lift is not great, the diaphragm valve is frequently built as the pilot or servo of a larger piston-operated main valve to provide both sensitivity and high flow capacity. Construction and performance are similar to the previously described pilot-operated evaporator-pressure regulator, except that diaphragm valves are constructed for higher pressures. Thus, they are suitable for use as defrost relief from evaporator to suction pressure, as large-capacity relief from a pressure vessel to the low side, or as a liquid refrigerant pump relief from pump discharge to the accumulator to prevent excessive pump pressures when some evaporators are valved closed.

Other Safety Relief Devices

Fusible plugs and **rupture disks** (Figure 40) provide similar safety relief. The former contains a fusible member that melts at a predetermined temperature corresponding to the safe saturation pressure of the refrigerant, but is limited in application to pressure vessels with internal gross volumes of 3 ft^3 or less and internal diameters of 6 in. or less. The rupture member contains a preformed disk designed to rupture at a predetermined pressure. These devices may be used as stand-alone devices or installed at the inlet to a safety relief valve.

When these devices are installed in series with a safety relief valve, the chamber created by the two valves must have a pressure gage or other suitable indicator. A rupture disk will not burst at its design pressure if back pressure builds up in the chamber.

The rated relieving capacity of a relief valve alone must be multiplied by 0.9 when it is installed in series with a rupture disk (unless the relief valve has been rated in combination with the rupture disk).

Discharge Capacity. The minimum required discharge capacity of the pressure-relief device or fusible plug for each pressure vessel is determined by the following formula, specified by ASHRAE *Standard* 15:

$$C = fDL \tag{2}$$

where

C = minimum required air discharge capacity of relief device, lb/min
D = outside diameter of vessel, ft
L = length of vessel, ft
f = factor dependent on refrigerant, as shown in Table 2

Equation (2) determines the required relief capacity for a pressure vessel containing liquid refrigerant. See ASHRAE *Standard* 15 for other relief device requirements, including relief of overpressure caused by compressor flow rate capacity.

Table 2 Values of f for Discharge Capacity of Pressure-Relief Devices

Refrigerant	f
On the low side of a limited-charge cascade system:	
R-13, R-13B1, R-503	0.163
R-14	0.203
R-23, R-170, R-508A, R-508B, R-744, R-1150	0.082
Other applications:	
R-11, R-32, R-113, R-123, R-142b, R-152a, R-290, R-600, R-600a, R-764	0.082
R-12, R-22, R-114, R-124, R-134a, R-401A, R-401B, R-401C, R-405A, R-406A, R-407C, R-407D, R-407E, R-409A, R-409B, R-411A, R-411B, R-411C, R-412A, R-414A, R-414B, R-500, R-1270	0.131
R-115, R-402A, R-403B, R-404A, R-407B, R-410A, R-410B, R-502, R-507A, R-509A	0.203
R-143a, R-402B, R-403A, R-407A, R-408A, R-413A	0.163
R-717	0.041
R-718	0.016

Notes:
1. Listed values of f do not apply if fuels are used within 20 ft of pressure vessel. In this case, use methods in API (2000, 2003) to size pressure-relief device.
2. When one pressure-relief device or fusible plug is used to protect more than one pressure vessel, required capacity is the sum of capacities required for each pressure vessel.
3. For refrigerants not listed, consult ASHRAE *Standard* 15.

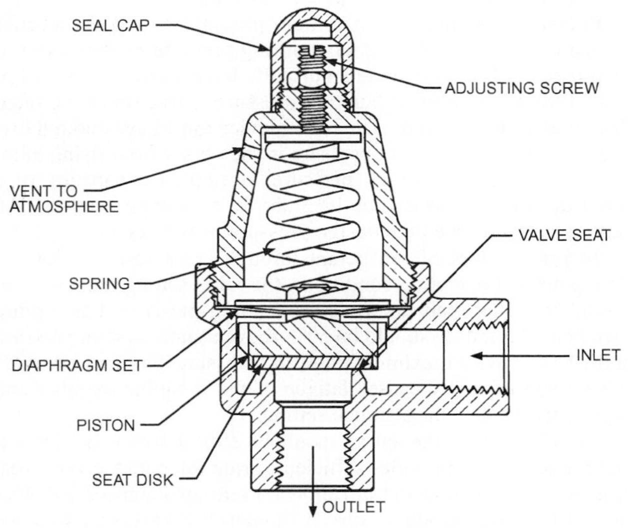

Fig. 39 Diaphragm Relief Valve

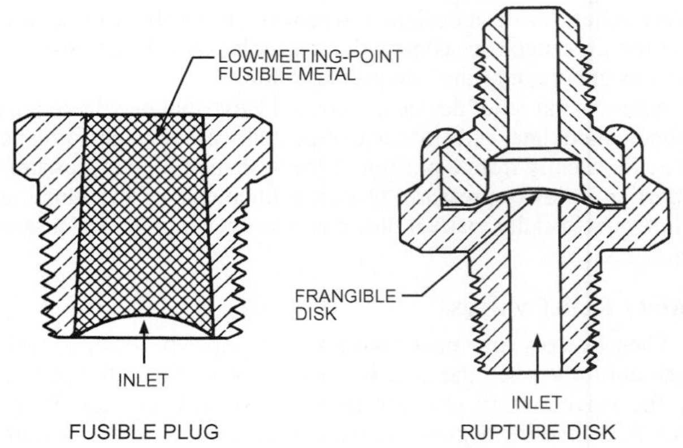

Fig. 40 Safety Relief Devices

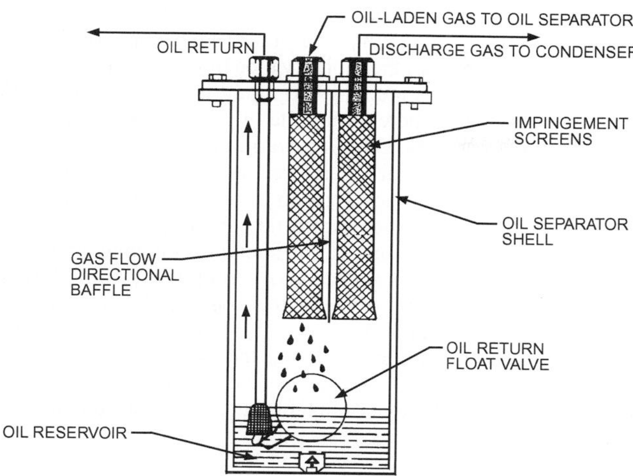

Fig. 41 Discharge-Line Lubricant Separator

DISCHARGE-LINE LUBRICANT SEPARATORS

The **discharge-line lubricant separator** removes lubricant from the discharge gas of helical rotary (screw) and reciprocating compressors. Lubricant is separated by (1) reducing gas velocity, (2) changing direction of flow, (3) impingement on baffles, (4) mesh pads or screens, (5) centrifugal force, or (6) coalescent filters. The separator reduces the amount of lubricant reaching the low-pressure side, helps maintain the lubricant charge in the compressor sump, and muffles the sound of gas flow.

Figure 41 shows one type of separator incorporating inlet and outlet screens and a high-side float valve. A space below the float valve allows for dirt or carbon sludge. When lubricant accumulates it raises the float ball, then passes through a needle valve and returns to the low-pressure crankcase. When the level falls, the needle valve closes, preventing release of hot gas into the crankcase. Insulation and electric heaters may be added to prevent refrigerant from condensing when the separator is exposed to low temperatures. A wide variety of horizontal and vertical flow separators is manufactured with centrifuges, baffles, wire mesh pads, and/or cylindrical filters.

Selection

Separators are usually given capacity ratings for several refrigerants at several suction and condensing temperatures. Another method rates capacity in terms of compressor displacement volume. Some separators also show a marked reduction in separation efficiency at some stated minimum capacity. Because compressor capacity increases when suction pressure rises or condensing pressure drops, system capacity at its lowest compression ratio should be the criterion for selecting the separator.

Application

Discharge-line lubricant separators are commonly used for ammonia or hydrocarbon refrigerant systems to reduce evaporator fouling. With lubricant-soluble halocarbon refrigerants, only certain flooded systems, low-temperature systems, or systems with long suction lines or other lubricant return problems may need lubricant separators. (See Chapters 2 and 3 for more information about lubricant separators.)

CAPILLARY TUBES

Every refrigerating system requires a pressure-reducing device to meter refrigerant flow from the high-pressure side to the low-pressure side according to load demand. The capillary tube is especially popular for smaller single-compressor/single-evaporator systems such as household refrigerators and freezers, dehumidifiers, and room air conditioners. Capillary tube use may extend to larger single-compressor/single-evaporator systems, such as unitary air conditioners up to 10 tons capacity.

The capillary operates on the principle that liquid passes through it much more readily than vapor. It is a length of drawn copper tubing with a small inner diameter. When used for controlling refrigerant flow, it connects the outlet of the condenser to the inlet of the evaporator. The term “capillary tube” is a misnomer because the inner bore, though narrow, is much too large to allow capillary action. In some applications, the capillary tube is soldered to the suction line and the combination is called a **capillary-tube/suction-line heat exchanger system**. Refrigeration systems that use a capillary tube without the heat exchanger relationship are often referred to as **adiabatic capillary tube systems**.

A high-pressure liquid receiver is not normally used with a capillary tube; consequently, less refrigerant charge is needed. In a few applications, such as household refrigerators, freezers, room air conditioners, and heat pumps, a suction-line accumulator may be used. Because the capillary tube allows pressure to equalize when the refrigerator is off, a compressor motor with a low starting torque may be used. A capillary tube system does not control as well over as wide a range of conditions as does a thermostatic expansion valve; however, a capillary tube may be less expensive and may provide adequate control for some systems.

Theory

A capillary tube passes liquid much more readily than vapor because of the latter’s increased volume; as a result, it is a practical metering device. When a capillary tube is sized to permit the desired flow of refrigerant, the liquid seals its inlet. If the system becomes unbalanced, some vapor (uncondensed refrigerant) enters the capillary tube. This vapor reduces the mass flow of refrigerant considerably, which increases condenser pressure and causes subcooling at the condenser exit and capillary tube inlet. The result is increased mass flow of refrigerant through the capillary tube. If properly sized for the application, the capillary tube compensates automatically for load and system variations and gives acceptable performance over a limited range of operating conditions.

A common flow condition is to have subcooled liquid at the entrance to the capillary tube. Bolstad and Jordan (1948) described the flow behavior from temperature and pressure measurements along the tube (Figure 42) as follows:

> With subcooled liquid entering the capillary tube, the pressure distribution along the tube is similar to that shown in the graph. At the entrance to the tube, section 0-1, a slight pressure drop occurs, usually unreadable on the gauges. From point 1 to point 2, the pressure drop is linear. In the portion of the tube 0-1-2, the refrigerant is entirely in the liquid state, and at point 2, the first bubble of vapor forms. From point 2 to the end of the tube, the pressure drop is not linear, and the pressure drop per unit length increases as the end of the tube is approached. For this portion of the tube, both the saturated liquid and saturated vapor phases are present, with the percent and volume of vapor increasing in the direction of flow. In most of the runs, a significant pressure drop occurred from the end of the tube into the evaporator space.
>
> With a saturation temperature scale corresponding to the pressure scale superimposed along the vertical axis, the observed temperatures may be plotted in a more efficient way than if a uniform temperature scale were used. The temperature is constant for the first portion of the tube 0-1-2. At point 2, the pressure has dropped to the saturation pressure corresponding

to this temperature. Further pressure drop beyond point 2 is accompanied by a corresponding drop in temperature, the temperature being the saturation temperature corresponding to the pressure. As a consequence, the pressure and temperature lines coincide from point 2 to the end of the tube.

Li et al. (1990) and Mikol (1963) showed that the first vapor bubble is not generated at the point where the liquid pressure reaches the saturation pressure (point 2 on Figure 42), but rather the refrigerant remains in the liquid phase for some limited length past point 2, reaching a pressure below the saturation pressure. This delayed evaporation, often referred to as a metastable or superheated liquid condition, must be accounted for in analytical modeling of the capillary tube, or the mass flow rate of refrigerant will be underestimated (Kuehl and Goldschmidt 1991; Wolf et al. 1995).

The rate of refrigerant flow through a capillary tube always increases with an increase in inlet pressure. Flow rate also increases with a decrease in external outlet pressure down to a certain critical value, below which flow does not change (choked flow). Figure 42 illustrates a case in which outlet pressure inside the capillary tube has reached the critical value (point 3), which is higher than the external pressure (point 4). This condition is typical for normal operation. The point at which the first gas bubble appears is called the **bubble point**. The preceding portion of capillary tube is called the **liquid length**, and that following is called the **two-phase length**.

System Design Factors

A capillary tube must be compatible with other components. In general, once the compressor and heat exchangers have been selected to meet the required design conditions, capillary tube size and system charge are determined. However, detailed design considerations may be different for different applications (e.g., domestic refrigerator, window air conditioner, residential heat pump).

Capillary tube size and system charge together are used to determine subcooling and superheat for a given design. Performance at off-design conditions should also be checked. Capillary tube systems are generally much more sensitive to the amount of refrigerant charge than expansion valve systems.

The high-pressure side must be designed for use with a capillary tube. To prevent rupture in case the capillary tube becomes blocked, the high-side volume should be sufficient to contain the entire refrigerant charge. A sufficient refrigerant storage volume (such as additional condenser tubes) may also be needed to protect against excessive discharge pressures during high-load conditions.

Pressure equalization during the off-period is another concern in designing the high side. When the compressor stops, refrigerant continues to pass through the capillary tube from the high side to the low side until pressures are equal. If liquid is trapped in the high side, it will evaporate there during the off cycle, pass through the capillary tube to the low side as a warm gas, condense, and add latent heat to the evaporator. Therefore, good liquid drainage to the capillary tube during this equalization interval should be provided. Liquid trapping may also increase the time for the pressure to equalize after the compressor stops. If this interval is too long, pressures may not be sufficiently equalized to permit low-starting-torque motor compressors to start when the thermostat calls for cooling.

The maximum quantity of refrigerant is in the evaporator during the off-cycle and the minimum during the running cycle. Suction piping should be arranged to reduce the adverse effects of the variable-charge distribution. A suitable suction-line accumulator is sometimes needed.

Capacity Balance Characteristic

Selection of a capillary tube depends on the application and anticipated range of operating conditions. One approach to the problem involves the concept of **capacity balance**. A refrigeration system operates at capacity balance when the capillary tube's resistance is sufficient to maintain a liquid seal at its entrance without excess liquid accumulating in the high side (Figure 43). Only one such capacity balance point exists for any given compressor discharge pressure. A curve through the capacity balance points for a range of compressor discharge and suction pressures (as in Figure 44) is called the capacity balance characteristic of the system. Ambient temperatures for a typical air-cooled system are shown in Figure 44. A given set of compressor discharge and suction pressures associated with condenser and evaporator pressure drops establish the capillary tube inlet and outlet pressures.

The capacity balance characteristic curve for any combination of compressor and capillary tube may be determined experimentally by the arrangement shown in Figure 45. Although Figure 45 shows the capillary tube suction-line heat exchanger application, a similar test setup without heat exchange would be used for adiabatic capillary tube systems. This test arrangement makes it possible to vary suction and discharge pressures independently until capacity balance is obtained. The desired suction pressure may be obtained by regulating heat input to the low side, usually by electric heaters. The desired discharge pressure may be obtained by a

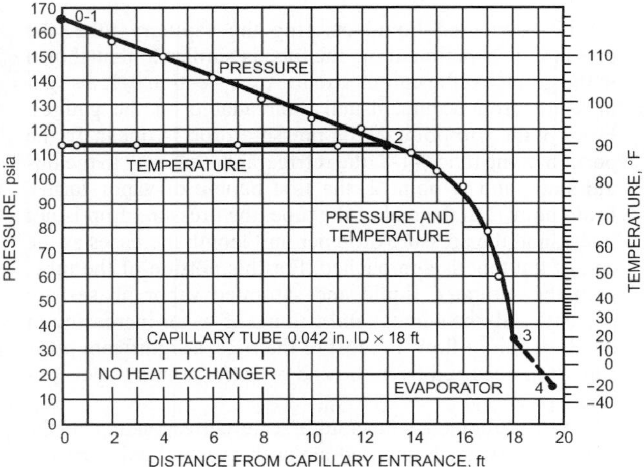

Fig. 42 Pressure and Temperature Distribution along Typical Capillary Tube
(Bolstad and Jordan 1948)

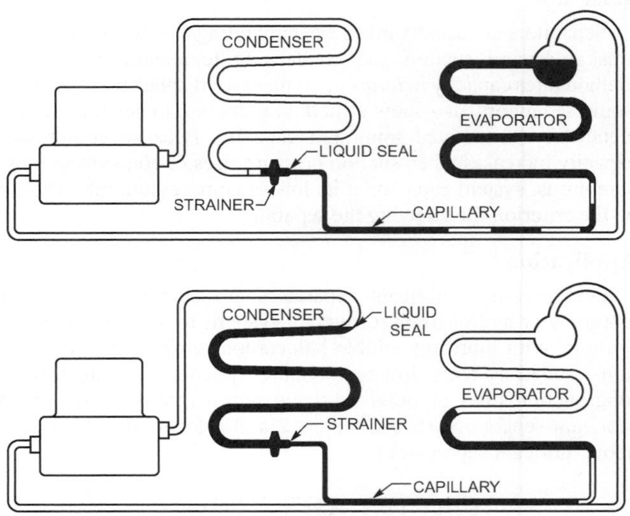

Fig. 43 Effect of Capillary Tube Selection on Refrigerant Distribution

suitably controlled water-cooled condenser. A liquid indicator is located at the entrance to the capillary tube. The usual test procedure is to hold high-side pressure constant and, with gas bubbling through the sight glass, slowly increase suction pressure until a liquid seal forms at the capillary tube entrance. Repeating this procedure at various discharge pressures determines the capacity balance characteristic curve similar to that shown in Figure 44. This equipment may also be used as a calorimeter to determine refrigerating system capacity.

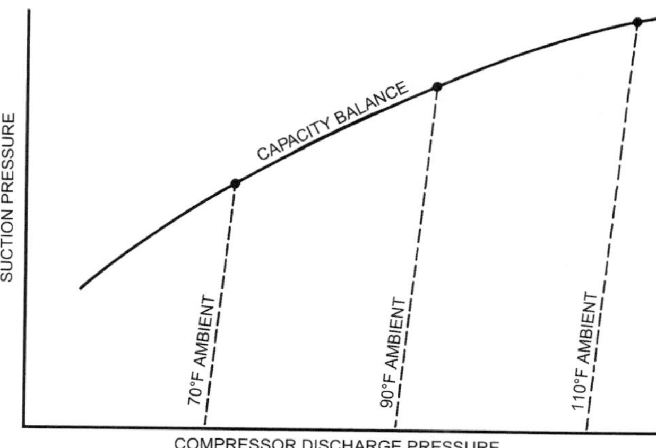

Note: Operation below this curve results in a mixture of liquid and vapor entering the capillary. Operation above capacity balance points causes liquid to back up in the condenser and elevate its pressure.

Fig. 44 Capacity Balance Characteristic of Capillary System

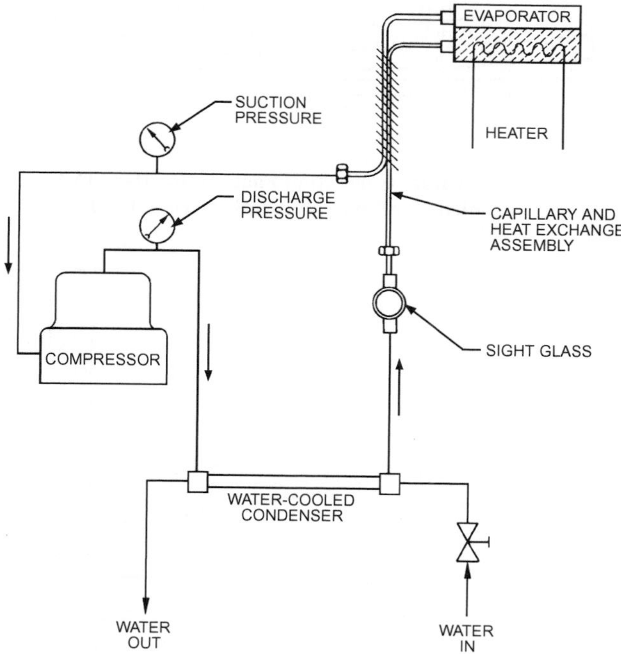

Fig. 45 Test Setup for Determining Capacity Balance Characteristic of Compressor, Capillary, and Heat Exchanger

Optimum Selection and Refrigerant Charge

Whether initial capillary tube selection and charge are optimum for the unit is always questioned, even in simple applications such as a room air-conditioning unit. The refrigerant charge in the unit can be varied using a small refrigerant bottle (valved off and sitting on a scale) connected to the circuit. The interconnecting line must be flexible and arranged so that it is filled with vapor instead of liquid. The charge is brought in or removed from the unit by heating or cooling the bottle.

The only test for varying capillary tube restriction is to remove the tube, install a new capillary tube, and determine the optimum charge, as outlined previously. Occasionally, pinching the capillary tube is used to determine whether increased resistance is needed.

The refrigeration system should be operated through its expected range of conditions to determine power requirements and cooling capacity for any given selection and charge combination.

Application

Processing and Inspection. To prevent mechanical clogging by foreign particles, a strainer should be installed ahead of the capillary tube. Also, all parts of the system must be evacuated adequately to eliminate water vapor and noncondensable gases, which may cause clogging by freezing and/or corrosion. The lubricant should be free from wax separation at the minimum operating temperature.

The interior surface of the capillary tube should be smooth and uniform in diameter. Although plug-drawn copper is more common, wire-drawn or sunk tubes are also available. Life tests should be conducted at low evaporator temperatures and high condensing temperatures to check the possibility of corrosion and plugging. Material specifications for seamless copper tube are given in ASTM *Standard* B75, and for hard-drawn copper tubes in ASTM *Standard* B360.

Establish a procedure to ensure uniform flow capacities, within reasonable tolerances, for all capillary tubes used in product manufacture. The following procedure is a good example:

1. Remove the final capillary tube, determined from tests, from the unit and rate its airflow capacity using the wet-test meter method described in ASHRAE *Standard* 28.
2. Produce master capillary tubes using the wet-test meter airflow equipment, to provide maximum and minimum flow capacities for the particular unit. The maximum-flow capillary tube has a flow capacity equal to that of the test capillary tube, plus a specified tolerance. The minimum-flow capillary tube has a flow capacity equal to that of the test capillary tube, less a specified tolerance.
3. Send one sample of the maximum and minimum capillary tubes to the manufacturer, to be used as tolerance guides for elements supplied for a particular unit. Also send samples to the inspection group for quality control.

Considerations. In selecting a capillary tube for a specific application, practical considerations influence the length. For example, the minimum length is determined by geometric considerations such as the physical distance between the high and low sides and the length of capillary tube required for optimum heat exchange. It may also be dictated by exit velocity, noise, and the possibility of plugging with foreign materials. Maximum length may be determined primarily by cost. It is fortunate, therefore, that flow characteristics of a capillary tube can be adjusted independently by varying either its bore or its length. Thus, it is feasible to select the most convenient length independently and then (within certain limits) select a bore to give the desired flow. An alternative procedure is to select a standard bore and then adjust the length, as required.

ASTM *Standard* B360 lists standard diameters and wall thicknesses for capillary tubes. Many nonstandard tubes are also used, resulting in nonuniform interior surfaces and variations in flow.

ADIABATIC CAPILLARY TUBE SELECTION PROCEDURE

Wolf et al. (1995) developed refrigerant-specific rating charts to predict refrigerant flow rates through adiabatic capillary tubes. The methodology involves determining two quantities from a series of curves, similar to rating charts for R-12 and R-22 developed by Hopkins (1950). The two quantities necessary to predict refrigerant flow rate through the adiabatic capillary tube are a flow rate through a reference capillary tube and a flow factor ϕ, which is a geometric correction factor. These two quantities are multiplied together to calculate the flow rate.

Figures 46 and 47 are rating charts for pure R-134a through adiabatic capillary tubes. Figure 46 plots the capillary tube flow rate as a function of inlet condition and inlet pressure for a reference capillary tube geometry of 0.034 in. ID and 130 in. long. Figure 47 is a geometric correction factor. Using the desired capillary tube geometry, ϕ may be determined and then multiplied with the flow rate from Figure 46 to determine the predicted capillary tube flow rate. *Note*: For quality inlet conditions for R-134a, an additional correction factor of 0.95 is necessary to obtain the proper results.

Figures 48 to 50 present the rating charts for pure R-410A through adiabatic capillary tubes. The method of selection from these charts is identical to that previously presented for R-134a, except that an additional correction factor for quality inlet conditions is not used. A separate flow factor chart (Figure 50), however, is provided to determine the value of the geometric correction factor for quality inlet conditions. The same methodology using Figures 51, 52, and 53 can be used to determine the mass flow rate of R-22 through adiabatic capillary tubes. Wolf et al. (1995) developed additional rating charts for R-152a.

Wolf et al. (1995) also presented limited performance results for refrigerants R-134a, R-22, and R-410A with 1.5% lubricant.

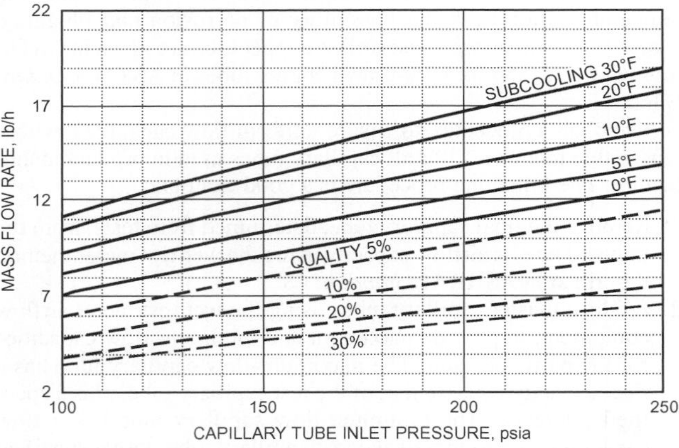

Fig. 46 Mass Flow Rate of R-134a Through Capillary Tube
(Capillary tube diameter is 0.034 in. ID and length is 130 in.)

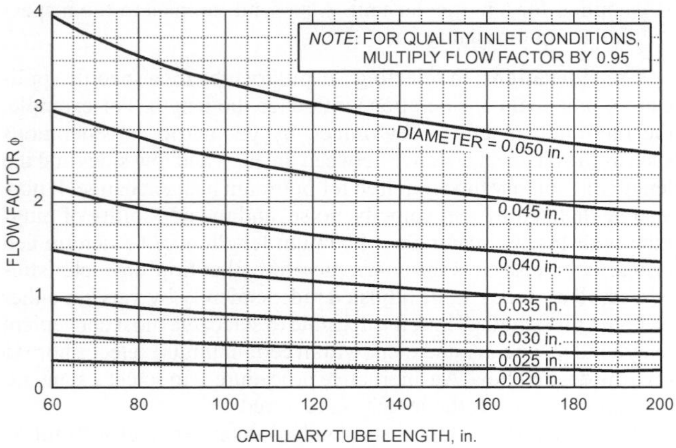

Fig. 47 Flow Rate Correction Factor ϕ for R-134a

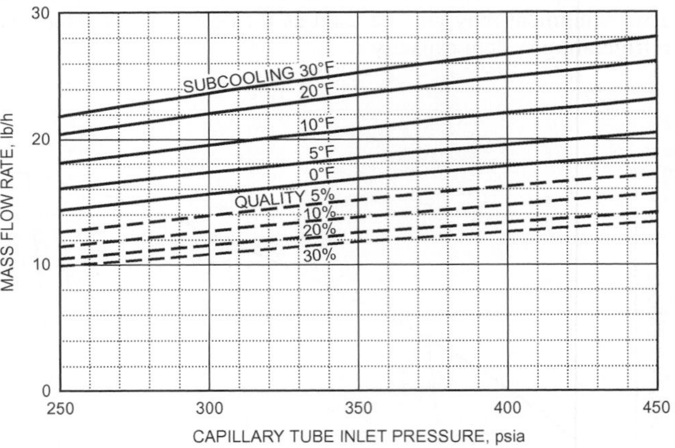

Fig. 48 Mass Flow Rate of R-410A Through Capillary Tube
(Capillary tube diameter is 0.034 in. ID and length is 130 in.)

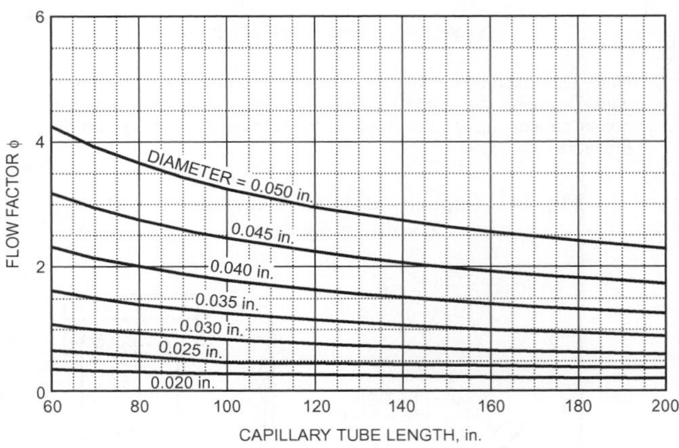

Fig. 49 Flow Rate Correction Factor ϕ for R-410A for Subcooled Condition at Capillary Tube Inlet

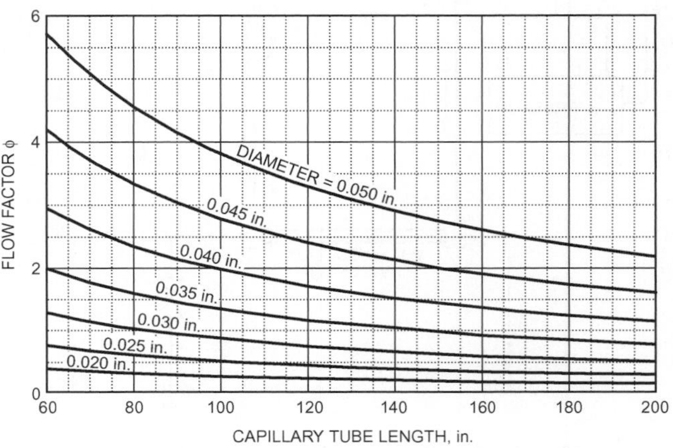

Fig. 50 Flow Rate Correction Factor ϕ for R-410A for Two-Phase Condition at Capillary Tube Inlet

Table 3 Capillary Tube Dimensionless Parameters

π Term	Definition	Description
π_1	L_c/d_c	Geometry effect
π_2	$d_c^2 h_{fg}/v_f^2\mu_f^2$	Vaporization effect
π_3	$d_c\sigma/v_f\mu_f^2$	Bubble formation
π_4	$d_c^2 p_{in}/v_f\mu_f^2$	Inlet pressure
π_5 (subcooled)	$d_c^2 c_p \Delta t_{sc}/v_f^2\mu_f^2$	Inlet condition
π_5 (quality)	x	Inlet condition
π_6	v_g/v_f	Density effect
π_7	$(\mu_f - \mu_g)/\mu_g$	Viscous effect
π_8	$\dot{m}/d_c\mu_f$	Flow rate

where

c_p = liquid specific heat
d_c = capillary tube diameter
h_{fg} = latent heat or vaporization
L_c = capillary tube length
$\dot{m}$ = mass flow rate
x = quality (decimal)
Δt_{sc} = degree of subcooling
μ_f = liquid viscosity
μ_g = vapor viscosity
v_f = liquid specific volume
v_g = vapor specific volume
ρ_{in} = capillary tube inlet pressure
σ = surface tension

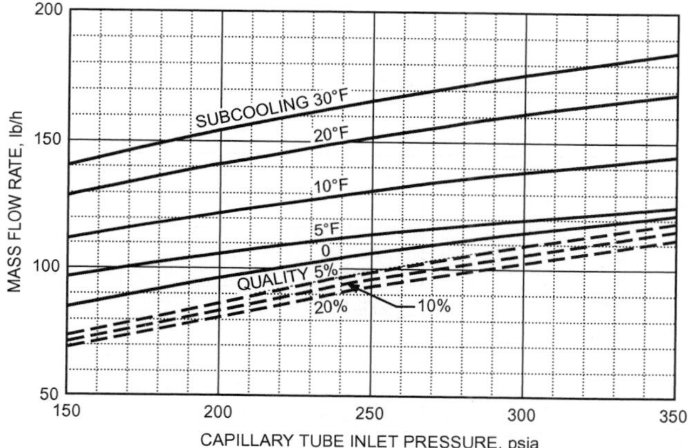

Fig. 51 Mass Flow Rate of R-22 Through Capillary Tube
(Capillary tube diameter is 0.066 in. ID and length is 60 in.)

Though these results did indicate a 1 to 2% increase in refrigerant mass flow through the capillary tubes, this was considered insignificant.

Generalized Prediction Equations. Based on tests with R-134a, R-22, and R-410A, Wolf et al. (1995) developed a general method for predicting refrigerant mass flow rate through a capillary tube. In this method the Buckingham π theorem was applied to the physical factors and fluid properties that affect capillary tube flow rate. The physical factors include capillary tube diameter and length, capillary tube inlet pressure, and refrigerant inlet condition; the fluid properties included specific volume, viscosity, surface tension, specific heat, and latent heat of vaporization. The result of this analysis was a group of eight dimensionless π terms shown in Table 3.

All fluid properties for respective π terms, both liquid and vapor, are evaluated at the saturation state using the capillary tube inlet temperature. Separate π_5 terms are presented for subcooled and two-phase refrigerant conditions at the capillary tube entrance. *Note*: The bubble formation term π_3 is statistically insignificant and, therefore, does not appear in prediction Equations (3) and (4).

Regression analysis of the refrigerant flow rate data for the subcooled inlet results produced the following equation for subcooled inlet conditions:

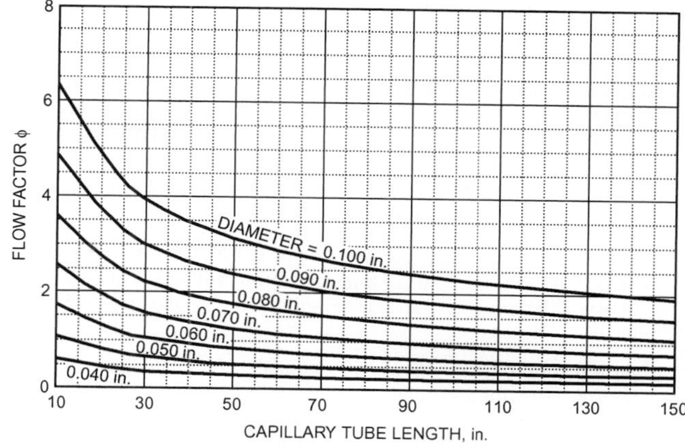

Fig. 52 Flow Rate Correction Factor ϕ for R-22 for Subcooled Condition at Capillary Tube Inlet

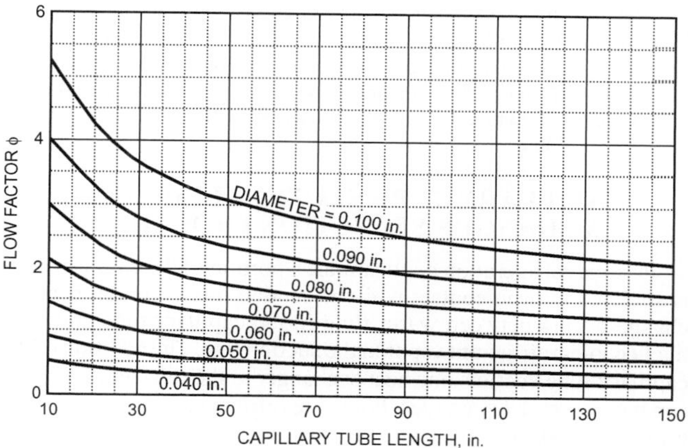

Fig. 53 Flow Rate Correction Factor ϕ for R-22 for Two-Phase Condition at Capillary Tube Inlet

$$\pi_8 = 1.8925\pi_1^{-0.484}\pi_2^{-0.824}\pi_4^{1.369}\pi_5^{0.0187}\pi_6^{0.773}\pi_7^{0.265} \tag{3}$$

where $\pi_5 = d_c^2 c_p \Delta t_{sc}/v_f^2\mu_f^2$ and $2°F < \Delta t_{sc} < 30°F$.

The following includes the π_5 term that only includes the quality x.

$$\pi_8 = 187.27\pi_1^{-0.635}\pi_2^{-0.189}\pi_4^{0.645}\pi_5^{-0.163}\pi_6^{-0.213}\pi_7^{-0.483} \tag{4}$$

where π_5 = quality x from $0.03 < x < 0.25$.

Wolf et al. (1995) compared these equations to R-134a results by Dirik et al. (1994) and R-22 results by Kuehl and Goldschmidt (1990). Wolf et al. also compared experimental results with R-152a and predictions by Equations (3) and (4) and found agreement within 5%. In addition, less than 1% of the R-134a, R-22, and R-410A experimental results used by Wolf et al. to develop correlation Equations (3) and (4) were outside ±5% of the flow rates predicted by the equations.

Sample Calculations

Example 1. Determine mass flow rate of R-134a through a capillary tube of 0.030 in. ID, 140 in. long, operating without heat exchange at 180 psia inlet pressure and 10°F subcooling.

Solution: From Figure 46 at 180 psia and 10°F subcooling, the flow rate of HFC-134a for a capillary tube 0.034 in. ID and 130 in. long is 13 lb/h. The flow factor ϕ from Figure 47 for a capillary tube of 0.030 in. ID and 140 in. long is 0.73. The predicted R-134a flow rate is then 13 × 0.73 = 9.49 lb/h.

Example 2. Determine the mass flow rate of R-134a through a capillary tube of 0.042 in. ID, 100 in. long, operating without heat exchange at 200 psia inlet pressure and 10% vapor content at the capillary tube inlet.

Solution: From Figure 46, at 200 psia and 10% quality, the flow rate of R-134a for a capillary tube 0.034 in. ID and 130 in. long is 7.9 lb/h. The flow factor ϕ from Figure 47 for a capillary tube of 0.042 in. ID and 100 in. long is 2.0. The predicted R-134a flow rate is then 0.95 × 7.9 × 2.0 = 15.0 lb/h, where 0.95 is the additional correction factor for R-134a quality inlet conditions.

Example 3. Determine the mass flow rate of R-410A through a capillary tube of 0.040 in. ID, 140 in. long, operating without heat exchange, 350 psia inlet pressure, and 15°F subcooling at the capillary tube inlet.

Solution: From Figure 48, at 350 psia and 15°F subcooling, the flow rate of R-410A for a capillary tube of 0.034 in. ID and 130 in. long is 22.4 lb/h. The flow factor ϕ from Figure 49 for a capillary tube of 0.040 in. ID and 140 in. long is 1.5. The predicted R-410A flow rate is then 22.4 × 1.5 = 33.6 lb/h.

Example 4. Determine the mass flow rate of R-22 through a capillary tube of 0.080 in. ID, 50 in. long, operating without heat exchange, 250 psia inlet pressure, and 5% vapor content at the capillary tube inlet.

Solution: From Figure 51, at 250 psia and 5% quality, the flow rate of R-22 for a capillary tube of 0.066 in. ID and 60 in. long is 99.5 lb/h. The flow factor ϕ from Figure 53 for a capillary tube of 0.080 in. ID and 50 in. long is 1.75. The predicted R-22 flow rate is then 99.5 × 1.75 = 174 lb/h.

CAPILLARY-TUBE/SUCTION-LINE HEAT EXCHANGER SELECTION PROCEDURE

In some refrigeration applications, a portion of the capillary tube is soldered to or in contact with the suction line such that heat exchange occurs between warm fluid in the capillary tube and relatively cooler refrigerant vapor in the suction line. The addition of the capillary-tube/suction-line heat exchanger to the refrigeration cycle can have several advantages. For some refrigerants, the addition increases the system's COP and volumetric capacity. Domanski et al. (1994) reported that this is not the case for all refrigerants and that some refrigeration systems may decrease in COP, depending on the particular refrigerant in the system. In typical household refrigerators, the suction line is soldered to the capillary tube so that heat exchanged with the suction line prevents condensation on the surface of the suction line during operation. The capillary-tube/suction-line heat exchanger also ensures that superheated refrigerant vapor exists at the compressor inlet and eliminates the possibility of liquid refrigerant returning to the compressor.

When a suction-line heat exchanger is used, the excess capillary length may be coiled and located at either end of the heat exchanger. Although more heat is exchanged with the excess coiled at the evaporator, system stability may be enhanced with the excess coiled at the condenser. Coils and bends should be formed carefully to avoid local restrictions. The effect of forming on the restriction should be considered when specifying the capillary.

Capillary Tube Selection

Wolf and Pate (2002) developed refrigerant-specific rating charts to predict refrigerant flow rates through capillary-tube/suction-line heat exchangers. The methodology involves determining four quantities (three for quality inlet conditions) from a series of curves and multiplying the results together to obtain the flow rate prediction. The calculation procedure is very similar to the adiabatic capillary tube selection process, with the exception of the additional flow rate factors. Figures 54 to 58 are the rating charts for pure R-134a through capillary-tube/suction-line heat exchangers. Figure 54 plots refrigerant flow rate as a function of capillary tube inlet pressure and inlet condition for a reference heat exchanger configuration (D_c = 0.034 in., L_c = 130 in., and L_{hx} = 60 in.). The thermodynamic state of the refrigerant vapor at the suction-line inlet to the heat exchanger has also been fixed (24 psia and 20°F superheat). Figure 55 plots flow rate correction factor ϕ_1 as a function of capillary tube diameter and length for subcooled inlet conditions at the capillary tube inlet. Figure 56 plots flow rate correction factor ϕ_2 as a function of suction-line inlet pressure and superheat level at the inlet to the suction line for subcooled inlet conditions at the capillary tube inlet. Figure 57 plots flow rate correction factor ϕ_3 as a function of heat exchange

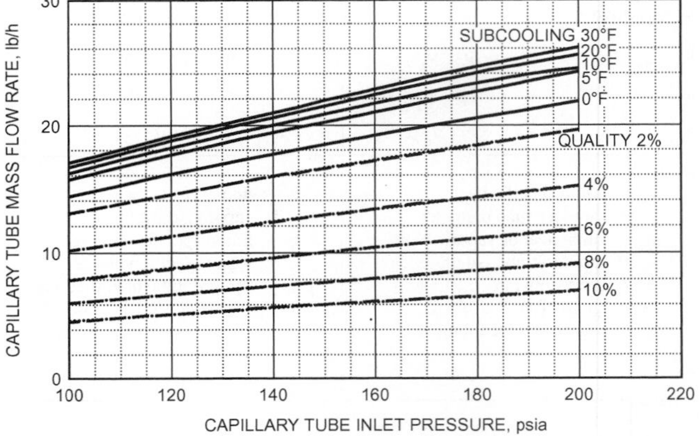

Fig. 54 Inlet Condition Rating Chart for R-134a

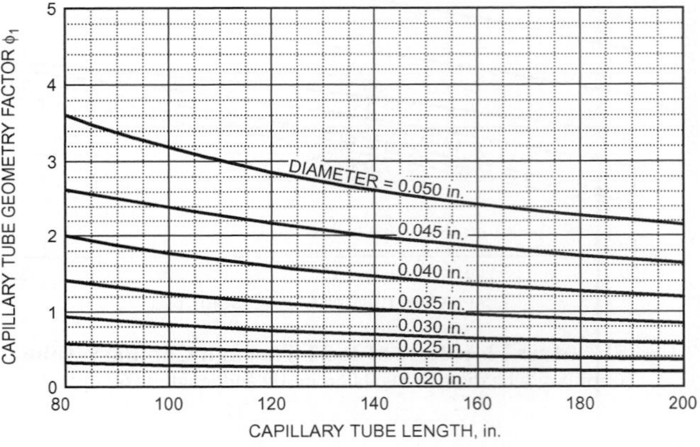

Fig. 55 Capillary Tube Geometry Correction Factor for Subcooled R-134a Inlet Conditions

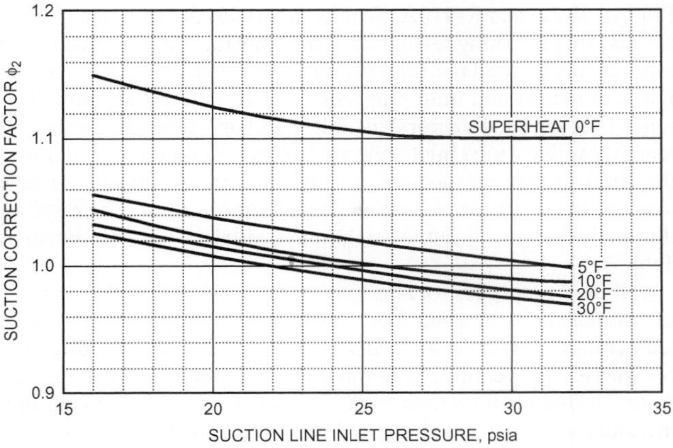

Fig. 56 Suction-Line Condition Correction Factor for R-134a Subcooled Inlet Conditions

length for subcooled inlet conditions at the capillary tube inlet. Figure 58 is the capillary tube geometry correction factor for quality inlet conditions at the capillary tube inlet, and Figure 59 is the suction-line condition correction factor for quality inlet conditions at the capillary tube inlet. There is no ϕ_3 correction factor for heat exchange length for quality inlet conditions. In general, heat exchange length delays the onset of vaporization of the refrigerant inside the capillary tube. However, without recondensation in the capillary tube as a result of the heat exchange effect, there is no vaporization for two-phase conditions at the capillary tube inlet.

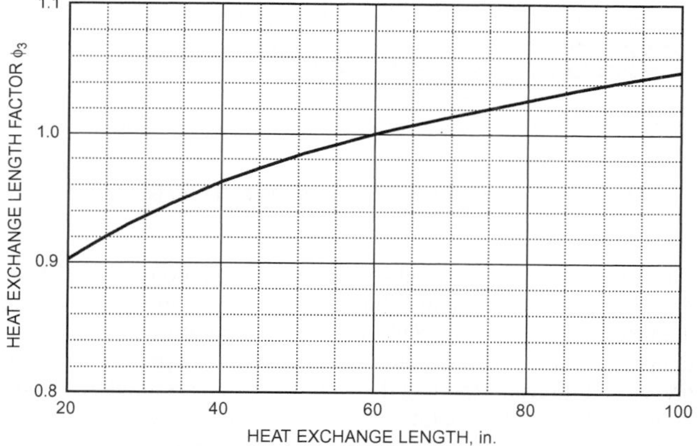

Fig. 57 Heat Exchange Length Correction Factor for R-134a Subcooled Inlet Conditions

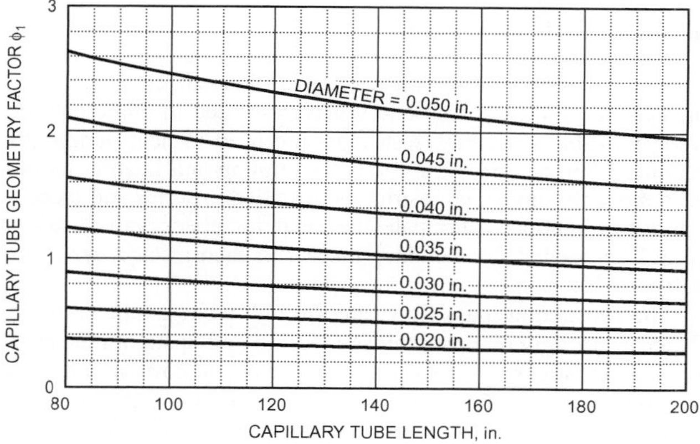

Fig. 58 Capillary Tube Geometry Correction Factor for R-134a Quality Inlet Conditions

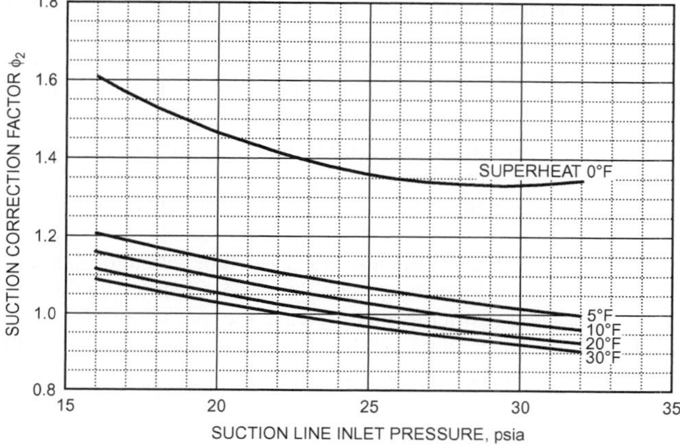

Fig. 59 Suction-Line Condition Correction Factor for R-134a Quality Inlet Conditions

To predict refrigerant mass flow rate for other conditions and geometries, multiply the necessary correction factors by the reference flow rate from Figure 54. Additional design variables examined by Wolf and Pate (2002) were the adiabatic upstream entrance length and inner diameter of the suction-line tubing. For adiabatic entrance lengths from 6 to 24 in., there was no observed effect. There was also no effect observed that could be attributed to the inner diameter of the suction line. Because of their lack of effect, these two design variables are not included when predicting refrigerant flow rate.

Generalized Prediction Equations

Based on tests with R-134a, R-22, R-410A, and R-600a, Wolf and Pate (2002) developed a general method for predicting refrigerant mass flow rate through a capillary-tube/suction-line heat exchanger. In this method, the Buckingham π theorem was applied to the physical factors and fluid properties that affect capillary tube flow rate through a capillary-tube/suction-line heat exchanger. Physical factors included capillary tube diameter, capillary tube length, suction-line diameter, heat exchange length, adiabatic entrance length of the capillary tube, capillary tube inlet pressure and condition, and suction-line inlet pressure and condition. Fluid properties included specific volume, viscosity, specific heat, and latent heat of vaporization. The result of this analysis was a group of 15 dimensionless π parameters, shown in Table 4.

All fluid properties for the respective π terms are evaluated at the capillary tube inlet temperature. Separate π_7 terms are presented for subcooled and quality inlet conditions to the capillary tube. The π terms in Table 4 that do not appear in Equations (5) and (6) were statistically insignificant and were not included in the final correlations.

Regression analysis of refrigerant flow rate data for subcooled inlet conditions produced the following equation:

$$\pi_9 = 0.07602\pi_1^{-0.4583}\pi_3^{0.07751}\pi_5^{0.7342}\pi_6^{-0.1204}\pi_7^{0.03774}\pi_8^{-0.04085}\pi_{11}^{0.1768} \quad (5)$$

where $\pi_7 = \Delta T_{sc} C_{pfc} D_c^2/\mu_{fc}^2 v_{fc}^2$ and $2°F < \Delta T_{sc} < 30°F$.

Equation (6) is valid for quality inlet conditions and contains the quality inlet term $\pi_7 = 1 - x$.

$$\pi_9 = 0.01960\pi_1^{-0.3127}\pi_5^{1.059}\pi_6^{-0.3662}\pi_7^{4.759}\pi_8^{-0.04965} \quad (6)$$

where x = quality (decimal) from $0.02 < x < 0.10$.

Wolf and Pate (2002) compared these equations to R-152a test results as well as published R-134a, R-152a, and R-12 data. Excellent agreement (±10%) was observed in nearly all cases. The only exception was the R-134a results by Dirik et al. (1994). Equation (5) overpredicted these results by an average of 45%. However, the adiabatic entrance length used was approximately 134 in., indicating that very long adiabatic entrance lengths may decrease refrigerant flow rates.

For further information on other refrigerants, please see the Bibliography.

Sample Calculations

Example 5. Determine the mass flow rate of R-134a through a capillary tube of 0.040 in. ID and 160 in. long operating with heat exchange at 160 psia capillary inlet pressure and 20°F subcooling. The heat exchange length is 70 in., and suction-line inlet conditions are 16 psia and 20°F superheat.

Solution: From Figure 54, at 160 psia and 20°F subcooling, the flow rate for a capillary tube 0.034 in. ID and 130 in. long is 22.5 lb/h. The capillary tube geometry correction factor from Figure 55 is 1.3. The

Table 4 Capillary-Tube/Suction-Line Heat Exchanger Dimensionless Parameters

π Parameter	Definition	Description
π_1	L_c/D_c	Geometry effect
π_2	L_i/D_c	Geometry effect
π_3	L_{hx}/D_c	Geometry effect
π_4	D_s/D_c	Geometry effect
π_5	$P_{capin}D_c^2/\mu_{fc}^2 v_{fcT}$	Inlet pressure
π_6	$P_{suctin}D_c^2/\mu_{fc}^2 v_{fc}$	Inlet pressure
π_7 (subcooled)	$\Delta T_{sc}C_{pfc}D_c^2/\mu_{fc}^2 v_{fc}^2$	Inlet condition
π_7 (quality)	$1 - x$	Inlet condition
π_8	$\Delta T_{sh}C_{pfc}D_c^2/\mu_{fc}^2 v_{fc}^2$	Inlet condition
π_9	$\dot{m}/D_c\mu_{fc}$	Flow rate
π_{10}	v_{gc}/v_{fc}	Density effect
π_{11}	$(\mu_{fc} - \mu_{gc})/\mu_{fc}$	Viscous effect
π_{12}	$h_{fgc}D_c^2/\mu_{fc}^2 v_{fc}^2$	Vaporization effect
π_{13}	μ_{gs}/μ_{fc}	Viscous effect
π_{14}	v_{gs}/v_{fc}	Density effect
π_{15}	C_{pgs}/C_{pfc}	Specific heat effect

where

C_{pfc} = capillary inlet liquid specific heat
C_{pgs} = suction inlet vapor specific heat
D_c = capillary tube inside diameter (0.026 to 0.042 in)
D_s = suction line inside diameter (0.194 to 0.319 in.)
h_{fgc} = capillary inlet latent heat of vaporization
L_c = capillary tube length (80 to 180 in.)
L_{hx} = heat exchange length (20 to 100 in.)
L_i = adiabatic entrance length (6 to 24 in.)
$\dot{m}$ = mass flow rate
P_{capin} = capillary tube inlet pressure
P_{suctin} = suction line inlet pressure
x = quality (2 to 10%)
ΔT_{sc} = capillary tube inlet subcool level (2 to 30°F)
ΔT_{sh} = suction line inlet superheat (6 to 40°F)
μ_{fc} = capillary inlet liquid viscosity
μ_{gc} = capillary inlet vapor viscosity
μ_{gs} = suction inlet vapor viscosity
v_{fc} = capillary inlet liquid specific volume
v_{gc} = capillary inlet vapor specific volume
v_{gs} = suction inlet vapor specific volume

suction-line inlet correction factor from Figure 56 is 1.033, and the heat exchange length factor from Figure 57 is 1.01. The predicted mass flow rate is then 22.5 × 1.37 × 1.033 × 1.01 = 32.2 lb/h.

Example 6. Determine the mass flow rate of R-134a through a capillary tube of 0.030 in. ID and 80 in. long operating with heat exchange at 140 psia capillary inlet pressure and 6% quality. The heat exchange length is 55 in., and suction-line inlet conditions are 20 psia and 10°F superheat.

Solution: From Figure 54, at 140 psia and 6% quality, the flow rate for a capillary tube 0.034 in. ID and 130 in. long is 9.5 lb/h. The capillary tube geometry correction factor from Figure 58 is 0.90. The suction-line inlet correction factor from Figure 59 is 1.09. There is no heat exchange length correction factor for quality inlet conditions, so the predicted mass flow rate is then 9.5 × 0.90 × 1.09 = 9.3 lb/h.

SHORT-TUBE RESTRICTORS

Application

Short-tube restrictors are widely used in residential air conditioners and heat pumps. They offer low cost, high reliability, ease of inspection and replacement, and potential elimination of check valves in the design of a heat pump. Because of their pressure-equalizing characteristics, short-tube restrictors allow the use of a low-starting-torque compressor motor.

Short-tube restrictors, as used in residential systems, are typically 3/8 to 1/2 in. in length, with a length-to-diameter (*L*/*D*) ratio greater than 3 and less than 20. Short-tube restrictors are also called **plug orifices** or **orifices**, although the latter is reserved for restrictors with an *L*/*D* ratio less than 3. Capillary tubes have an *L*/*D* ratio much greater than 20.

An **orifice tube**, a type of short-tube restrictor, is commonly used in automotive air conditioners. Its *L*/*D* ratio falls between that of a capillary tube and a short-tube restrictor. Most automotive applications use orifice tubes with *L*/*D* ratios between 21 and 35 and inside diameters from 0.04 to 0.08 in. An orifice tube allows the evaporator to operate in a flooded condition, which improves performance. To prevent liquid from flooding the compressor, an accumulator/dehydrator is installed to separate liquid from vapor and to meter a small amount of lubricant-rich refrigerant to the compressor. However, the accumulator/dehydrator does cause a pressure drop penalty on the suction side.

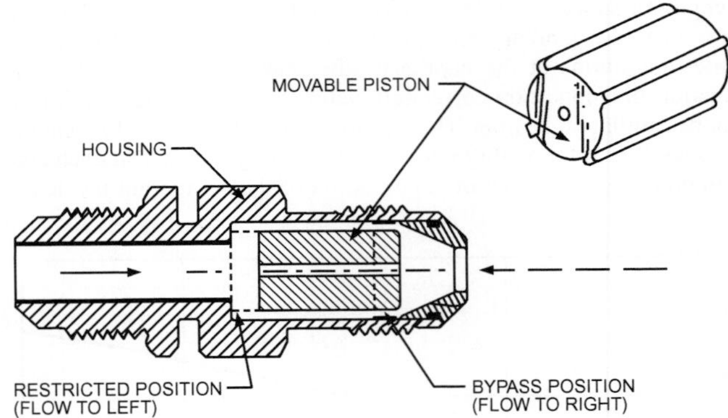

Fig. 60 Schematic of Movable Short-Tube Restrictor

There are two basic designs for short-tube restrictors: stationary and movable. Movable short-tube restrictors consist of a piston that moves within its housing (Figure 60). A movable short-tube restrictor restricts refrigerant flow in one direction. In the opposite direction, the refrigerant pushes the restrictor off its seat, opening a larger area for the flow. The stationary design is used in units that cool only; movable short-tube restrictors are used in heat pumps that require different flow restrictions for cooling and heating modes. Two movable short-tube restrictors, installed in series and faced in opposite directions, eliminate the need for check valves, which would be needed for capillary tubes and thermostatic expansion valves.

The refrigerant mass flow rate through a short-tube restrictor depends strongly on upstream subcooling and upstream pressure. For a given inlet pressure, inlet subcooling, and downstream pressure below the saturation pressure corresponding to the inlet temperature, flow has a very weak dependence on the downstream pressure, indicating a nearly choked flow. This flow dependence is shown in Figure 61, which presents R-22 test data obtained on a 0.5 in. long, laboratory-made short-tube restrictor at three different downstream pressures and the same upstream pressure.

A significant drop in downstream pressure from approximately 170 to 70 psia produces a smaller increase in the mass flow rate than does a modest change of downstream pressure from 190 to 170 psia. Pressure drops only slightly along the length of the short-tube restrictor. The large pressure drop at the entrance is caused by rapid fluid acceleration and inlet losses. The large pressure drop in the exit plane, typical for heat pump operating conditions and represented in Figure 61 by the bottom pressure line, indicates that choked flow nearly occurred.

Among geometric parameters, the short-tube restrictor diameter has the strongest influence on mass flow rate. Chamfering the inlet of the short-tube restrictor may increase the mass flow rate by as much as 25%, depending on the *L*/*D* ratio and chamfer depth. Chamfering the exit causes no appreciable change in mass flow rate.

Although refrigerant flow inside a short tube is different from flow inside a capillary tube, choked flow is common for both, making both types of tubes suitable as metering devices. Systems equipped with short-tube restrictors, as with capillary tubes, must be precisely charged with the proper amount of refrigerant. Inherently,

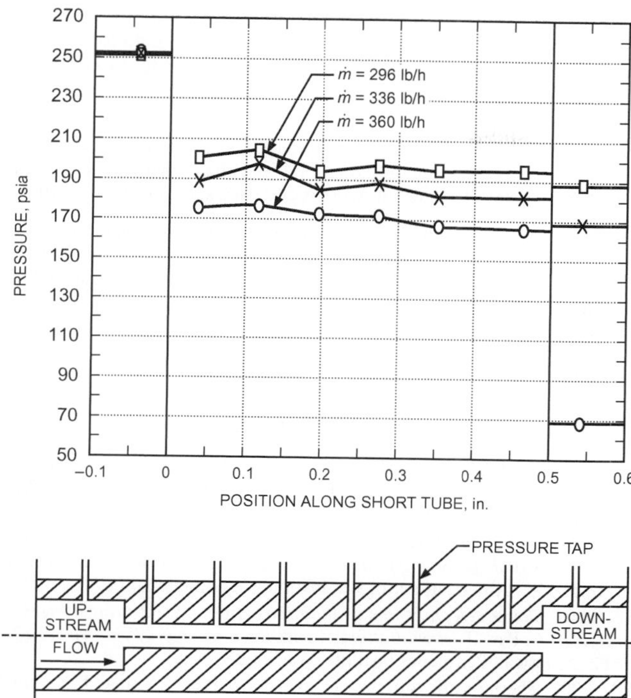

Fig. 61 R-22 Pressure Profile at Various Downstream Pressures with Constant Upstream Conditions: L = 0.5 in., D = 0.053 in., Subcooling 25°F
(Adapted from Aaron and Domanski 1990)

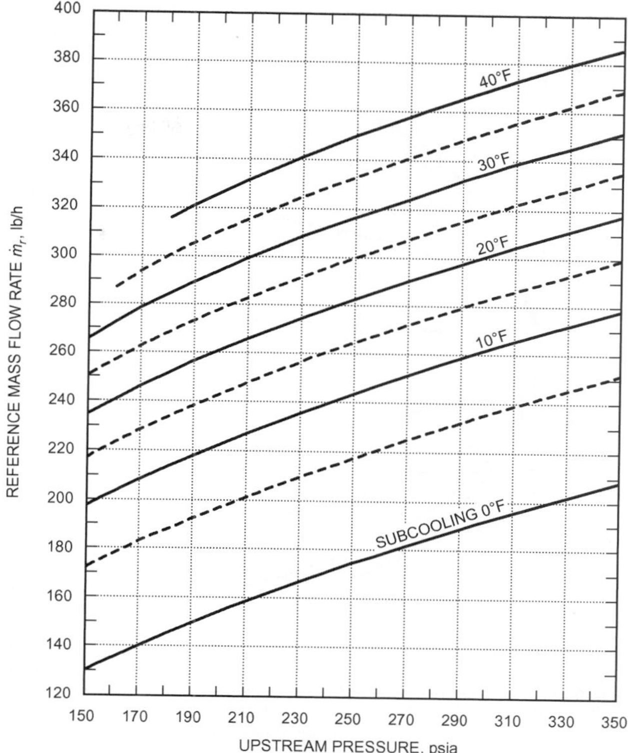

Fig. 62 R-22 Mass Flow Rate Versus Condenser Pressure for Reference Short Tube: L = 0.5 in., D = 0.053 in., Sharp-Edged
(Adapted from Aaron and Domanski 1990)

a short-tube restrictor does not control as well over a wide range of operating conditions as does a thermostatic expansion valve. However, its performance is generally good in a properly charged system.

Selection

Figures 62 to 65, from Aaron and Domanski (1990), can be used for preliminary evaluation of the mass flow rate of R-22 at a given inlet pressure and subcooling in air-conditioning and heat pump applications (i.e., applications in which downstream pressure is below the saturation pressure of refrigerant at the inlet). The method requires determination of mass flow rate for the reference short tube from Figure 62 and modifying the reference flow rate with multipliers that account for the short-tube geometry according to the equation

$$\dot{m} = \dot{m}_r \phi_1 \phi_2 \phi_3 \tag{7}$$

where

$\dot{m}$ = mass flow rate for short tube
$\dot{m}_r$ = mass flow rate for reference short tube from Figure 62
ϕ_1 = correction factor for tube geometry from Figure 63
ϕ_2 = correction factor for L/D versus subcooling from Figure 64
ϕ_3 = correction factor for chamfered inlet from Figure 65

Aaron and Domanski (1990) also provide a more accurate correlation than the graphical method. Neglecting downstream pressure on the graphs may introduce an error in the prediction as compared to the correlation results; however, this discrepancy should not exceed 3% because of the choked-flow condition at the tube exit. *Note*: The lines for 0 and 40°F subcooling in Figure 62 were obtained by extrapolation beyond the test data and may carry a large error.

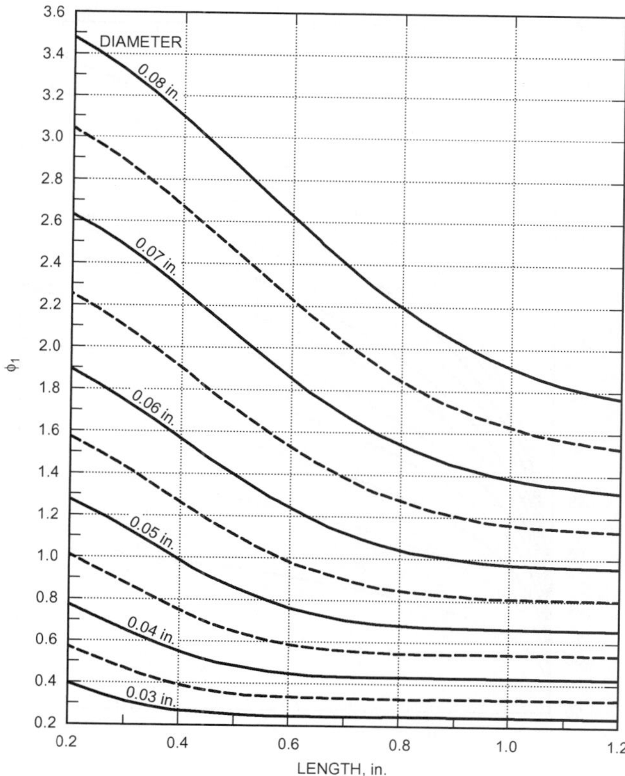

Fig. 63 Correction Factor for Short-Tube Geometry (R-22)
(Adapted from Aaron and Domanski 1990)

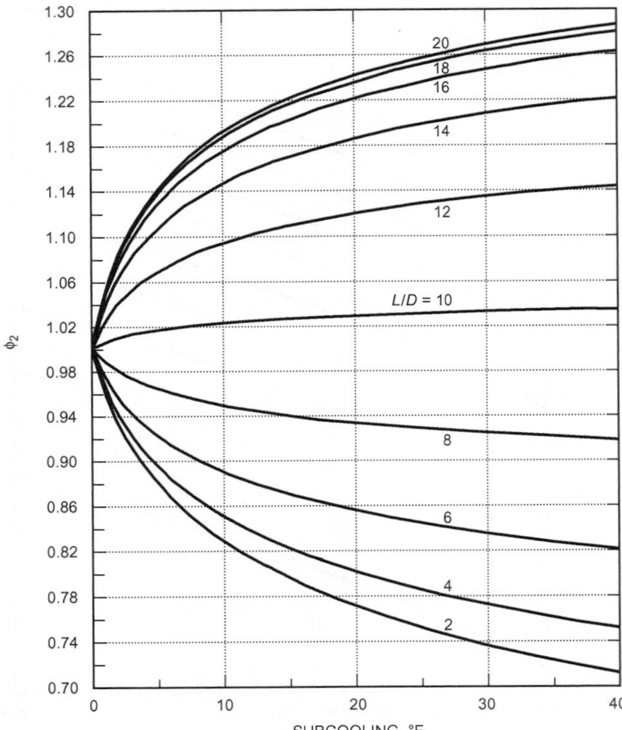

Fig. 64 Correction Factor for *L*/*D* Versus Subcooling (R-22)
(Adapted from Aaron and Domanski 1990)

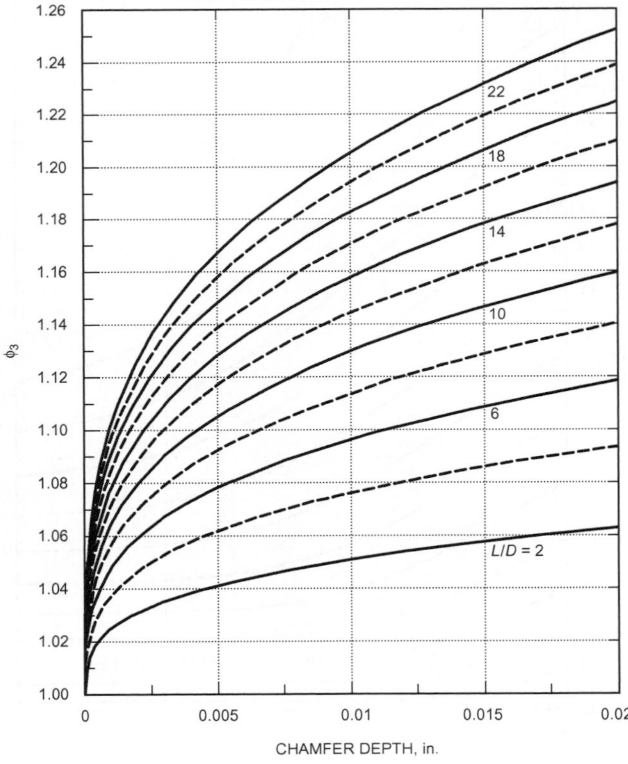

Fig. 65 Correction Factor for Inlet Chamfering (R-22)
(Adapted from Aaron and Domanski 1990)

Additional research has been done to add several other refrigerants' capacity ratings to this section. However, only data for R-410A and R-407c have been published in *ASHRAE Transactions* (Payne and O'Neal 1998, 1999).

Example 7. Determine the mass flow rate of R-22 through a short-tube restrictor 0.375 in. long, of 0.06 in. ID, and chamfered 0.01 in. deep at an angle of 45°. The inlet pressure is 240 psia, and subcooling is 10°F.

Solution: From Figure 62, for 240 psia and 10°F subcooling, the flow rate for the reference short tube is 240 lb/h. The value for ϕ_1 from Figure 63 is 1.62. The value for ϕ_2 from Figure 64 for 10°F subcooling and $L/D = 0.375/0.06 = 6.25$ is 0.895. The value of ϕ_3 from Figure 65 for $L/D = 6.25$ and a chamfer depth of 0.01 in. is 1.098. Thus, the predicted mass flow rate through the restrictor is $240 \times 1.62 \times 0.895 \times 1.098 = 382$ lb/h.

REFERENCES

Aaron, D.A. and P.A. Domanski. 1990. Experimentation, analysis, and correlation of Refrigerant-22 flow through short tube restrictors. *ASHRAE Transactions* 96(1):729-742.

API. 2000. *Sizing, selection, and installation of pressure-relieving devices in refineries*, Part I—*Sizing and selection*, 7th ed. *Recommended Practice* RP 520 P1. American Petroleum Institute, Washington, D.C.

API. 2003. *Sizing, selection, and installation of pressure-relieving devices in refineries*, Part II—*Installation*, 5th ed. *Recommended Practice* RP 520 P2. American Petroleum Institute, Washington, D.C.

ARI. 2001. Thermostatic refrigerant expansion valves. ANSI/ARI *Standard* 750. Air-Conditioning and Refrigeration Institute, Arlington, VA.

ASHRAE. 2004. Safety standard for refrigeration systems. ANSI/ASHRAE *Standard* 15-2004.

ASHRAE. 1998. Method of testing capacity of thermostatic refrigerant expansion valves. ANSI/ASHRAE *Standard* 17-1998 (RA2003).

ASHRAE. 1996. Method of testing flow capacity of refrigerant capillary tubes. ANSI/ASHRAE *Standard* 28-1996 (RA2002).

ASTM. 2002. Standard specification for seamless copper tube. *Standard* B75-02. American Society for Testing and Materials, West Conshohocken, PA.

ASTM. 2001. Standard specification for hard-drawn copper capillary tube for restrictor applications. *Standard* B360-01. American Society for Testing and Materials, West Conshohocken, PA.

Bolstad, M.M. and R.C. Jordan. 1948. Theory and use of the capillary tube expansion device. *Refrigerating Engineering* (December):519.

Dirik, E., C. Inan, and M.Y. Tanes. 1994. Numerical and experimental studies on adiabatic and non-adiabatic capillary tubes with HFC-134a. *Proceedings of the International Refrigeration Conference*, Purdue University, West Lafayette, IN.

Domanski, P.A., D.A. Didion, and J.P. Doyle. 1994. Evaluation of suction line-liquid line heat exchange in the refrigeration cycle. *International Journal of Refrigeration* 17(7):487-493.

Hopkins, N.E. 1950. Rating the restrictor tube. *Refrigerating Engineering* (November):1087.

Kuehl, S.J. and V.W. Goldschmidt. 1990. Steady flow of R-22 through capillary tubes: Test data. *ASHRAE Transactions* 96(1):719-728.

Kuehl, S.J. and V.W. Goldschmidt. 1991. Modeling of steady flows of R-22 through capillary tubes. *ASHRAE Transactions* 97(1):139-148.

Li, R.Y., S. Lin, Z.Y. Chen, and Z.H. Chen. 1990. Metastable flow of R-12 through capillary tubes. *International Journal of Refrigeration* 13(3):181-186.

Mikol, E.P. 1963. Adiabatic single and two-phase flow in small bore tubes. *ASHRAE Journal* 5(11):75-86.

Payne, W.V. and D.L. O'Neal. 1998. Mass flow characteristics of R-407c through short tube orifices. *ASHRAE Transactions* 104(1).

Payne, W.V. and D.L. O'Neal. 1999. Multiphase flow of Refrigerant 410A through short tube orifices. *ASHRAE Transactions* 105(2).

Wolf, D.A., R.R. Bittle, and M.B. Pate. 1995. Adiabatic capillary tube performance with alternative refrigerants. ASHRAE Research Project RP-762, *Final Report*.

Wolf, D.A. and M.B. Pate. 2002. Performance of a suction line/capillary-tube heat exchanger with alternative refrigerants. ASHRAE Research Project RP-948, *Final Report*.

BIBLIOGRAPHY

ASME. 2004. *Boiler and pressure vessel code*. American Society of Mechanical Engineers, New York.

Bittle, R.R. and M.B. Pate. 1996. A theoretical model for predicting adiabatic capillary tube performance with alternative refrigerants. *ASHRAE Transactions* 102(2):52-64.

Bittle, R.R., D.A. Wolf, and M.B. Pate. 1998. A generalized performance prediction method for adiabatic capillary tubes. *International Journal of HVAC&R Research* (now *HVAC&R Research*) 4(1):27-43.

Kuehl, S.J. and V.W. Goldschmidt. 1990. Transient response of fixed area expansion devices. *ASHRAE Transactions* 96(1):743-750.

Marcy, G.P. 1949. Pressure drop with change of phase in a capillary tube. *Refrigerating Engineering* (January):53.

Pate, M.B. and D.R. Tree. 1987. An analysis of choked flow conditions in a capillary tube–suction line heat exchanger. *ASHRAE Transactions* 93(1): 368-380.

Schulz, U. 1985. State of the art: The capillary tube for, and in, vapor compression systems. *ASHRAE Transactions* 91(1):92-105.

Whitesel, H.A. 1957a. Capillary two-phase flow. *Refrigerating Engineering* (April):42.

Whitesel, H.A. 1957b. Capillary two-phase flow—Part II. *Refrigerating Engineering* (September):35.

Wijaya, H. 1991. An experimental evaluation of adiabatic capillary tube performance for HFC-134a and CFC-12, pp. 474-483. *Proceedings of the International CFC and Halon Alternatives Conference*. Alliance of Responsible CFC Policy, Arlington, VA.

CHAPTER 45

FACTORY DEHYDRATING, CHARGING, AND TESTING

PROPER dehydration, charging, and testing of packaged refrigeration systems and components (compressors, evaporators, and condensing coils) help ensure proper performance and extend the life of refrigeration systems. This chapter covers the methods used to perform these functions. It does not address criteria such as allowable moisture content, refrigerant quantity, and performance, which are specific to each machine.

DEHYDRATION (MOISTURE REMOVAL)

Factory dehydration may be feasible only for certain sizes of equipment. On large equipment, which is open to the atmosphere when connected in the field, factory treatment is usually limited to purge and backfill, with an inert holding charge of nitrogen. In most instances, this equipment is stored for short periods only, so this method suffices until total system evacuation and charging can be done at the time of installation.

Excess moisture in refrigeration systems may lead to freeze-up of the capillary tube or expansion valve. It also has a negative effect on thermal stability of certain refrigeration oils [e.g., polyol ester (POE)]. Contaminants can cause valve breakage, motor burnout, and bearing and seal failure. Chapter 6 has more information on moisture and other contaminants in refrigerant systems.

Except for freeze-up, these effects are not normally detected by a standard factory test. Therefore, it is important to use a dehydration technique that yields a safe moisture level without adding foreign elements or solvents. In conjunction with dehydration, an accurate method of moisture measurement must be established. Many factors, such as the size of the unit, its application, and type of refrigerant, determine acceptable moisture content. Table 1 shows moisture limits recommended by various manufacturers for particular refrigeration system components.

Sources of Moisture

Moisture in refrigerant systems can be (1) retained on the surfaces of metals; (2) produced by combustion of a gas flame; (3) contained in liquid fluxes, oil, and refrigerant; (4) absorbed in the hermetic motor insulating materials; (5) derived from the factory ambient at the point of unit assembly; and (6) provided by free water. Moisture contained in the refrigerant has no effect on dehydration of the component or unit at the factory. However, because the refrigerant is added after dehydration, it must be considered in determining the overall moisture content of the completed unit. Moisture in oil may or may not be removed during dehydration, depending on when the oil is added to the component or system.

Bulk mineral oils, as received, have 20 to 30 ppm of moisture. Synthetic POE lubricants have 50 to 85 ppm; they are highly hygroscopic, so they must be handled appropriately to prevent moisture contamination. Refrigerants have an accepted commercial tolerance of 10 to 15 ppm on bulk shipments. Controls at the factory are needed to ensure these moisture levels in the oils and refrigerant are maintained.

Newer insulating materials in hermetic motors retain much less moisture compared to the old rag paper and cotton-insulated motors. However, tests by several manufacturers have shown that the stator, with its insulation, is still the major source of moisture in compressors.

Dehydration by Heat, Vacuum, or Dry Air

Heat may be applied by placing components in an oven or by using infrared heaters. Oven temperatures of 180 to 340°F are usually maintained. The oven temperature should be selected carefully to prevent damage to the synthetics used and to avoid breakdown of any residual run-in oil that may be present in compressors. Air in the oven must be maintained at low humidity. When dehydrating by heat alone, the time and escape area are critical; therefore, the size of parts that can be economically dehydrated by this method is restricted.

The **vacuum** method reduces the boiling point of water below the ambient temperature. The moisture then changes to vapor, which is pumped out by the vacuum pump. Table 3 in Chapter 6 of the 2005 *ASHRAE Handbook—Fundamentals* shows the relationship of temperature and pressure for water at saturation.

Vacuum is classified according to the following absolute pressure ranges:

Low Vacuum	29.92 to 1.0 in. Hg
Medium Vacuum	1.0 in. Hg to 1 μm Hg
High Vacuum	1 to 10^{-3} μm Hg
Very High Vacuum	10^{-3} to 10^{-6} μm Hg
Ultrahigh Vacuum	10^{-6} μm Hg and below

The degree of vacuum achieved and the time required to obtain the specified moisture level are a function of the (1) type and size of vacuum pump used, (2) internal volume of the component or system, (3) size and composition of water-holding materials in the system, (4) initial amount of moisture in the volume, (5) piping and fitting sizes, (6) shape of the gas passages, and (7) external temperatures maintained. The pumping rate of the vacuum pump is critical only if the unit is not evacuated through a conductance-limiting orifice such as a purge valve. Excessive moisture content, such as a pocket of puddled water, takes a long time to remove because of the volume expansion to vapor.

Vacuum measurements should be taken directly at the equipment (or as close to it as possible) rather than at the vacuum pump. Small tubing diameters or long tubing runs between the pump and the equipment should be avoided because line/orifice pressure drops reduce the actual evacuation level at the equipment.

If **dry air** or **nitrogen** is drawn or blown through the equipment for dehydration, it removes moisture by becoming totally or partially saturated. In systems with several passages or blind passages, flow may not be sufficient to dehydrate. The flow rate should obtain

---

The preparation of this chapter is assigned to TC 8.1, Positive Displacement Compressors.

Table 1 Typical Factory Dehydration and Moisture-Measuring Methods for Refrigeration Systems

Component	Dehydration Method	Moisture Audit	Moisture Limit
Coils and tubing	250°F oven, –70°F dry-air sweep	Dew point recorder	10 mg
Evaporator coils			
Small	–70°F dew point dry-air sweep, 240 s	P_2O_5	25 mg
Large	–70°F dew point dry-air sweep, 240 s	P_2O_5	65 mg
Evaporators/condensers	300°F oven, 1 h, dry-air sweep	Cold trap	200 mg
	Dry-air sweep	Nesbitt tube	8.5 mg/ft^2 surf. area
Condensing unit (0.25 to 7.5 ton)	Purchase dry	P_2O_5	25 to 85 mg
	Dry-air sweep	Nesbitt tube	8.5 mg/ft^2 surf. area
Air-conditioning unit	Evacuate to 240 μm Hg	P_2O_5	35 ppm
	3 h winding heat, 0.5 h vacuum	Refrigerant moisture check	25 ppm
Refrigerator	250°F oven, dc winding heat, vacuum	Cold trap	200 mg
Freezer	–70°F dew point dry-air ambient, –40°F dew point air sweep	P_2O_5	10 ppm
Compressors			
	dc Winding Heat		
	0.5 h dc winding heat 350°F, 0.25 h vacuum/repeat	Cold trap	200 mg
	dc winding heat 190°F, 0.5 h vacuum	Cold trap	1200 mg
2 to 60 ton semihermetic	dc winding heat, 30 min, evacuation, N_2 charge	Cold trap	1000 to 3500 mg
	Oven Heat		
	250°F oven, 4 h vacuum	Cold trap	180 mg
	250°F oven, 5.5 h at –60°F dew point air	Cold trap	200 mg
0.5 to 12 ton hermetic	300°F oven 4 h, –70°F dew point air 3.5 min	Cold trap	150 to 400 mg
50 to 100 ton	Oven at 270°F, 4 h evacuate to 1000 μm Hg	Cold trap	750 mg
1.5 to 5 ton hermetic	340°F oven, –100°F dew point dry air, 1.5 h	Cold trap	100 to 500 mg
2 to 40 ton semihermetic	250°F oven, –100°F dew point dry air, 3.5 h	Cold trap	100 to 1100 mg
5 to 150 ton open	175°F oven, evacuate to 1 mm Hg	Cold trap	400 to 2700 mg
Scroll 2 to 10 ton hermetic	300°F oven 4 h, 50 s evacuation and 10 s –70°F dew point air charge/repeat 7 times	Cold trap	300 to 475 mg
	Hot Dry Air, N_2		
3 to 5 ton	Dry air at 275°F, 3 h	Cold trap	250 mg
7.5 to 15 ton	Dry air at 275°F, 0.5 h vacuum	Cold trap	750 mg
20 to 40 ton	Dry N_2 sweep at 275°F, 3.5 h evacuate to 200 μm Hg	Cold trap	750 mg
	Dry N_2 Flush		
Reciprocating, semihermetic	N_2 run, dry N_2 flush, N_2 charge	—	—
Screw, hermetic/semihermetic	R-22 run, dry N_2 flush, N_2 charge	—	—
Screw, open	N_2 run, dry N_2 flush, N_2 charge	—	—
	Evacuation Only		
Screw, open, 50 to 1500 ton	Evacuate <1500 μm Hg, N_2 charge	—	—
Refrigerants	As purchased	Electronic analyzer	Typically 10 ppm
Lubricants			
Mineral oil	As purchased	Karl Fischer method	25 to 35 ppm
	As purchased and evacuation	Hygrometer	10 ppm
Synthetic polyol ester	As purchased	Karl Fischer method	50 to 85 ppm

optimum moisture removal, and its success depends on the overall system design and temperature.

Combination Methods

Each of the following methods can be effective if controlled carefully, but a combination of methods is preferred because of the shorter drying time and more uniform dryness of the treated system.

Heat and Vacuum Method. Heat drives deeply sorbed moisture to the surfaces of materials and removes it from walls; the vacuum lowers the boiling point, making the pumping rate more effective. The heat source can be an oven, infrared lamps, or an ac or dc current circulating through the internal motor windings of semihermetic and hermetic compressors. Combinations of vacuum, heat, and then vacuum again can also be used.

Heat and Dry-Air Method. Heat drives moisture from the materials. The dry air picks up this moisture and removes it from the system or component. The dry air used should have a dew point between –40 and –100°F. Heat sources are the same as those mentioned previously. Heat can be combined with a vacuum to accelerate the process. The heat and dry-air method is effective with open, hermetic, and semihermetic compressors. The heating temperature should be selected carefully to prevent damage to compressor parts or breakdown of any residual oil that may be present.

Advantages and limitations of the various methods depend greatly on the system or component design and the results expected. Goddard (1945) considers double evacuation with an air sweep between vacuum applications the most effective method, whereas Larsen and Elliot (1953) believe the dry-air method, if controlled carefully, is just as effective as the vacuum method and much less expensive, although it incorporates a 1.5 h evacuation after the hot-air purge. Tests by manufacturers show that a 280°F oven bake for 1.5 h, followed by a 20 min evacuation, effectively dehydrates compressors that use newer insulating materials.

MOISTURE MEASUREMENT

Measuring the correct moisture level in a dehydrated system or part is important but not always easy. Table 1 lists measuring methods used by various manufacturers, and others are described in the literature. Few standards are available, however, and acceptable moisture limits vary by manufacturer.

Cold-Trap Method. This common method of determining residual moisture monitors the production dehydration system to ensure that it produces equipment that meets the required moisture specifications. An equipment sample is selected after completion of the dehydration process, placed in an oven, and heated at 150 to 275°F (depending on the limitations of the sample) for 4 to 6 h. During this time, a vacuum is drawn through a cold-trap bottle immersed in an acetone and dry-ice solution (or an equivalent), which is generally held at about –100°F. Vacuum levels are between 10 and 100 μm Hg, with lower levels preferred. Important factors are leaktightness of the vacuum system and cleanliness and dryness of the cold-trap bottle.

Vacuum Leakback. Measuring the rate of vacuum leakback is another means of checking components or systems to ensure that no water vapor is present. This method is used primarily in conjunction with a unit or system evacuation that removes the noncondensables before final charging. This test allows a check of each unit, but too rapid a pressure build-up may signify a leak, as well as incomplete dehydration. The time factor may be critical in this method and must be examined carefully. Blair and Calhoun (1946) show that a small surface area in connection with a relatively large volume of water may only build up vapor pressure slowly. This method also does not give the actual condition of the charged system.

Dew Point. When dry air is used, a reasonably satisfactory check for dryness is a dew-point reading of the air as it leaves the part being dried. If airflow is relatively slow, there should be a marked difference in dew point between air entering and leaving the part, followed by a decrease in dew point of the leaving air until it eventually equals the dew point of the entering air. As is the case with all systems and methods described in this chapter, acceptable values depend on the size, usage, and moisture limits desired. Different manufacturers use different limits.

Gravimetric Method. In this method, described by ASHRAE *Standard* 35, a controlled amount of refrigerant is passed through a train of flasks containing phosphorous pentoxide (P_2O_5), and the weight increase of the chemical (caused by the addition of moisture) is measured. Although this method is satisfactory when the refrigerant is pure, any oil contamination produces inaccurate results. This method must be used only in a laboratory or under carefully controlled conditions. Also, it is time-consuming and cannot be used when production quantities are high. Furthermore, the method is not effective in systems containing only small charges of refrigerant because it requires 200 to 300 g of refrigerant for accurate results. If it is used on systems where withdrawal of any amount of refrigerant changes the performance, recharging is required.

Aluminum Oxide Hygrometer. This sensor consists of an aluminum strip that is anodized by a special process to provide a porous oxide layer. A very thin coating of gold is evaporated over this structure. The aluminum base and gold layer form two electrodes that essentially form an aluminum oxide capacitor.

In the sensor, water vapor passes through the gold layer and comes to equilibrium on the pore walls of the aluminum oxide in direct relation to the vapor pressure of water in the ambient surrounding the sensor. The number of water molecules absorbed in the oxide structure determines the sensor's electrical impedance, which modulates an electrical current output that is directly proportional to the water vapor pressure. This device is suitable for both gases and liquids over a temperature range of 158 to –166°F and a pressure range of about 10 μm Hg to 5000 psig. The **Henry's Law constant** (saturation parts per million by mass of water for the fluid divided by the saturated vapor pressure of water at a constant temperature) for each fluid must be determined. For many fluids, this constant must be corrected for the operating temperature at the sensor.

Christensen Moisture Detector. The Christensen moisture detector is used for a quick check of uncharged components or units on the production line. In this method, dry air is blown first through the dehydrated part and then over a measured amount of calcium sulfate ($CaSO_4$). The temperature of the $CaSO_4$ rises in proportion to the quantity of water it absorbs, and desired limits can be set and monitored. One manufacturer reports that coils were checked in 10 s with this method. Moisture limits for this detector are 2 to 60 mg. Corrections must be made for variations in desiccant grain size, the quantity of air passed through the desiccant, and the difference in instrument and component temperatures.

Karl Fischer Method. In systems containing refrigerant and oil, moisture may be determined by (1) measurement of the dielectric strength or (2) the Karl Fischer method (Reed 1954). In this method, a sample is condensed and cooled in a mixture of chloroform, methyl alcohol, and Karl Fischer reagent. The refrigerant is then allowed to evaporate as the solution warms to room temperature. When the refrigerant has evaporated, the remaining solution is titrated immediately to a dead stop electrometric end point, and the amount of moisture is determined. This method requires a 15 g sample of refrigerant and takes about 20 min. Multiple checks are run to confirm results. This method is generally considered inaccurate below 15 ppm; however, it can be used for checking complete systems because this method does not require that oil be boiled off the refrigerant. Reed points out that additives in the oil, if any, must be checked to ensure that they do not interfere with the reactions of the method. The Karl Fischer method may also be used for determining moisture in oil alone (ASTM *Standard* D117; Morton and Fuchs 1960; Reed 1954).

An alternative method is available. A 5 to 10 g refrigerant sample is injected directly into Karl Fischer reagents at a constant flow rate using a pressure-reducing device such as a capillary tube. After the refrigerant is completely passed through the reagent, the moisture content is determined by automatic titration of a dead stop electrometric end point. This alternative method takes about 1 h to perform and is typically considered to be accurate to 5 ppm.

Electrolytic Water Analyzer. Taylor (1956) describes an electrolytic water analyzer designed specifically to analyze moisture levels in a continuous process, as well as in discrete samples. The device passes the refrigerant sample, in vapor form, through a sensitive element consisting of a phosphoric acid film surrounding two platinum electrodes; the acid film absorbs moisture. When a dc voltage is applied across the electrodes, water absorbed in the film is electrolyzed into hydrogen and oxygen, and the resulting dc current, in accordance with Faraday's first law of electrolysis, flows in proportion to the weight of the products electrolyzed. Liquids and vapor may be analyzed because the device has an internal vaporizer. This device handles the popular halocarbon refrigerants, but samples must be free of oils and other contaminants. In tests on desiccants, this method is quick and accurate with R-22.

Sight-Glass Indicator. In fully charged halocarbon systems, a sight-glass indicator can be used in the refrigerant lines. This device consists of a colored chemical button, visible through the sight glass, that indicates excessive moisture by a change in color. This method requires that the system be run for a reasonable length of time to allow moisture to circulate over the button. This method compares moisture only qualitatively to a fixed standard. Sight-glass indicators have been used on factory-dehydrated split systems to ensure that they are dry after field installation and charging, and are commonly used in conjunction with filter driers to monitor moisture in operating systems.

Special Considerations. Although all methods described in this section can effectively measure moisture, their use in the factory requires certain precautions. Operators must be trained in the use of the equipment or, if the analysis is made in the laboratory, the proper method of securing samples must be understood. Sample flasks must be dry and free of contaminants; lines must be clean, dry, and properly purged. Procedures for weighing the sample, time during the cycle, and location of the sample part should be clearly defined and followed carefully. Checks and calibrations of the equipment must be made on a regular basis if consistent readings are to be obtained.

CHARGING

The accuracy required when charging refrigerant or oil into a unit depends on the size and application of the unit. Charging equipment must also be adapted to the particular conditions of the plant; equipment may be manual or automatic. Standard charging is used where extreme accuracy is not necessary or the production rate is not high. Fully automatic charging boards check the vacuum in the units, evacuate the charging line, and meter the desired amount of oil and refrigerant into the system. These devices are accurate and suitable for high production.

Refrigerant and oil must be handled carefully during charging; the place and time of oil and refrigerant charging greatly affect the life of a system. To avoid unnecessary complications (foaming, oil slugging, improper oil distribution, etc.), the unit should be charged with oil before the refrigerant. Charging with refrigerant should avoid liquid slugging during initial start-up; the best way to do this is to charge the unit at the high-pressure side. Refrigerant lines must be dry and clean, and all charging lines must be kept free of moisture and noncondensable gases. Also, new containers must be connected with proper purging devices. Carelessness in observing these precautions may lead to excess moisture and noncondensables in the refrigeration system.

Oil storage and charging systems should be designed and maintained to avoid contamination and direct contact between oil and air. Regular checks for moisture or contamination must be made at the charging station to ensure that oil and refrigerant delivered to the unit meet specifications. Compressors charged with oil for storage or shipment must be charged with dry nitrogen. Compressors without oil may be charged with dry air.

TESTING FOR LEAKS

Extended warranties and critical refrigerant charges add to the importance of proper leak detection before charging.

The U.S. Environmental Protection Agency (EPA) established an allowable leakage rate for certain refrigerants (e.g., no more than 0.1 oz per year of R-22 at 150 psig). A system that has 4 to 6 oz of refrigerant and a 5 year warranty must have virtually no leak, whereas in a system that has 10 to 20 lb of refrigerant, the loss of 1 oz of refrigerant in 1 year would not have much affect on system performance. Any leak on the low-pressure side of a system operating below atmospheric pressure is dangerous regardless of the size of the refrigerant charge.

Before any leak testing is done, the component or system should be strength tested at a pressure considerably higher than the leak test pressure. This test ensures safety when the unit is being tested under pressure in an exposed condition. Applicable design test pressures for high- and low-side components have been established by Underwriters Laboratories (UL), the American Society of Mechanical Engineers (ASME), the American National Standards Institute (ANSI), and ASHRAE. Units or components using composition gaskets as joint seals should have the final leak test after dehydration. Retorquing bolts after dehydration helps to reduce leaks past gaskets.

Leak Detection Methods

Water Submersion Testing. A water submersion test is a method of leak and strength testing. The test article is pressurized to the specified positive pressure and submerged in a well-lighted tank filled with clean water. It may take a few minutes for development of a small bubbles trace to visualize a small leak. Note that bubbles can develop on the surface as a result of outgassing, and development of a trace is a key factor. This method of leak testing is not as sensitive as the mass spectrometer or electronic leak detection methods, but is suitable for high-volume production.

Soap Bubble Leak Detection. High-rate leaks from a pressurized system can be found by applying a soapy liquid solution to the suspected leak areas. Bubbles that form in the solution indicate refrigerant leakage.

Fluorescent Leak Detection. This system involves infusing a small quantity of a fluorescent additive into the oil/refrigerant charge of an operating system. Leakage is observed as a yellow-green glow under an ultraviolet (UV) lamp. This method is suitable for halocarbon systems. Because the additive is in the oil, thorough cleanup is needed after the leak is fixed to avoid a false positive caused by leftover oil residue. It may also be a problem to identify fluorescent glow in daylight.

Pressure Testing. The test article is sealed off under pressure or vacuum, and any decrease or rise in pressure noted over time indicates leakage. Dry nitrogen is often used as the medium for pressure testing. The limitations of this method are the time required to conduct the test, the lack of sensitivity, and the inability to determine the location of any leak that may exist.

Electronic Leak Testing. The electronic leak detector consists of a probe that draws air over a platinum diode, the positive ion emission of which is greatly increased in the presence of a halogen gas. This increased emission is translated into a visible or audible signal. Electronic leak testing shares with halide torches the disadvantages that every suspect area must be explored and that contamination makes the instrument less sensitive; however, it does have some advantages: mainly, increased sensitivity. With a well-maintained detector, it is possible to identify leakage at a rate of 10^{-3} mm^3/s (standard), which is roughly equivalent to the loss of 1 oz of refrigerant in 100 years. The instrument also can be desensitized to the point that leaks below a predetermined rate are not found. Some models have an automatic compensating feature to accomplish this.

The problem of contamination is more critical with improved sensitivity, so the unit under test is placed in a chamber slightly pressurized with outside air, which keeps contaminants out of the production area and carries contaminating gas from leaky units. An audible signal allows the probe operator to concentrate on probing, without having to watch a flame or dial. Equipment maintenance presents a problem because the sensitivity of the probe must be checked at short intervals. Any exposure to a large amount of refrigerant causes loss of probe sensitivity. A rough check (e.g., air underwater testing) is frequently used to find large leaks prior to use of the electronic device.

Mass Spectrometer. In this method, the unit to be tested is evacuated and then surrounded by a helium-and-air mixture. The vacuum is sampled through a mass spectrometer; any trace of helium indicates one or more leaks. Many equipment manufacturers use the mass spectrometer leak detection method because of its high sensitivities: mass spectrometers can detect leaks of 10^{-7} mm^3/s. Test levels for production equipment are typically set near 10^{-2} mm^3/s. This method is normally used to measure the total leakage rate from all joints simultaneously. The main limitation for this method is that the costs for test equipment and consumables are higher than for other leak detection methods.

The required concentration of helium depends on the maximum leak permissible, the configuration of the system under test, the time the system can be left in the helium atmosphere, and the vacuum level in the system; the lower the vacuum level, the higher the helium readings. The longer a unit is exposed to the helium atmosphere, the lower the concentration necessary to maintain the required sensitivity. If, because of the shape of the test unit, a leak is distant from the point of sampling, a good vacuum must be drawn, and sufficient time must be allowed for traces of helium to appear on the mass spectrometer.

As with other methods described in this chapter, the best testing procedure in using the spectrometer is to locate and characterize calibrated leaks at extreme points of the test unit and then to adjust exposure time and helium concentration to allow cost-effective testing. One manufacturer reportedly found leaks of 0.05 oz of refrigerant per year by using a 10% concentration of helium and exposing the tested system for 10 min.

The sensitivity of the mass spectrometer method can be limited by the characteristics of the tested system. Because only the total leakage rate is found, it is impossible to tell whether a leakage rate of, for example, 1 oz per year is caused by one fairly large leak or several small leaks. If the desired sensitivity rejects units outside the sensitivity range of tests listed earlier in this chapter, it is necessary to use a **helium probe** to locate leaks. In this method, the component or system to be probed is fully evacuated to clear it of helium; then, while the system is connected to the mass spectrometer, a fine jet of helium is sprayed over each joint or suspect area. With large systems, a waiting period is necessary because some time is required for the helium to pass from the leak point to the mass spectrometer. To save time, isolated areas (e.g., return bends on one end of a coil) may be hooded and sprayed with helium to determine whether the leak is in the region.

Special Considerations

There are two general categories of leak detection: those that allow a leak check before refrigerant is introduced into the system, and those that require refrigerant. Methods that do not use refrigerant have the advantage that heat applied to repair a joint has no harmful effects. On units containing refrigerant, the refrigerant must be removed and the unit vented before any welding, brazing, or soldering is attempted. This practice avoids refrigerant breakdown and pressure build-up, which would prevent the successful completion of a sound joint.

All leak-testing equipment must be calibrated frequently to ensure maximum sensitivity. The electronic leak detector and the mass spectrometer are usually calibrated with equipment furnished by the manufacturer. Mass spectrometers are usually checked using a flask containing helium. A glass orifice in the flask allows helium to escape at a known rate; the operator calibrates the spectrometer by comparing the measured escape rate with the standard.

The effectiveness of the detection system can best be checked with calibrated leaks made of glass, which can be bought commercially. These leaks can be built into a test unit and sent through the normal leak detection cycles to evaluate the detection method's effectiveness. Ensure that the test leak site does not become closed; the leakage rate of the test leak must be determined before and after each system audit.

From a manufacturing standpoint, use of any leak detection method should be secondary to leak prevention. Improper brazing and welding techniques, unclean parts, untested sealing compounds or improper fluxes and brazing materials, and poor workmanship result in leaks that occur in transit or later. Careful control and analysis of each joint or leak point make it possible to concentrate tests on areas where leaks are most likely to occur. If operators must scan hundreds of joints on each unit, the probability of finding all leaks is rather small, whereas concentration on a few suspect areas reduces field failures considerably.

PERFORMANCE TESTING

Because there are many types and designs of refrigeration systems, this section only presents specific information on reciprocating compressor testing and covers some important aspects of performance testing of other components and complete systems.

Compressor Testing

The two prime considerations in compressor testing are power and capacity. Secondary considerations are leakback rate, low-voltage starting, noise, and vibration.

Testing Without Refrigerant. A number of tests measure compressor power and capacity before the unit is exposed to refrigerant. In cases where excessive power is caused by friction of running gear, **low-voltage tests** spot defective units early in assembly. In these tests, voltage is increased from a low or zero value to the value that causes the compressor to break away, and this value is compared with an established standard. When valve plates are accessible, performance can be tested by using an air pump for **leakback tests**. Air at fixed pressure is put through the unit to determine the flow rate at which valves open properly. The air pressure exerted against the closing side of the valve indicates its efficiency. This method is effective only when the valves are reasonably tight, and is difficult to use on valves that must be run in before seating properly.

Extreme care should be taken when a compressor is used to pump air because the combination of oil, air, and high temperatures caused by compression can result in a diesel effect or an explosion.

In a common test using the compressor as an air pump, the discharge airflow is measured through a flowmeter, orifice, or other flow-measuring device. When the volumetric efficiency of the compressor with refrigerant is known, the flow rate that can be expected with air at a given pressure may be calculated. Because this test adiabatically compresses the air, the discharge pressure must be low to prevent overheating of discharge lines and oil oxidation if the test lasts longer than a few minutes. (The temperature of adiabatic compression is 280°F at 35 psig, but 540°F at 125 psig.) When the compressor is run long enough to stabilize temperatures, both power and flow can be compared with established limits. Discharge temperature readings and speed measurements aid in analyzing defective units. If a considerable amount of air is discharged or trapped, the air used in the test must be dry enough to prevent condensation from causing rust or corrosion on the discharge side.

Another method of determining compressor performance requires the compressor to pump from a free air inlet into a fixed volume. The time required to reach a given pressure is compared against a maximum standard acceptable value. The pressure used in this test is approximately 125 psig, so that a reasonable time spread can be obtained. The time needed for measuring the capacity of the compressor must be sufficient for accurate readings but short enough to prevent overheating. Power readings can be recorded at any time in the cycle. By shutting off the compressor, the leakback rate can be measured as an additional check. In addition to the pump-up and leakback tests noted above, a vacuum test should also be performed.

The **vacuum test** should be performed by closing off the suction side with the discharge open to the atmosphere. The normal vacuum obtained under these conditions is 1 to 1.5 psia. Abrupt closing of the suction side also allows the oil to serve as a check on the priming capabilities of the pump because of the suppression of the oil and attempt to deaerate. This test also checks for porosity and leaking gaskets. To establish reasonable pump-up times, leakback rates, and suctions, a large number of production units must be tested to determine the range of production variation.

In any capacity test using air, only clean, dry air should be used in order to prevent compressor contamination.

Observing performance while testing compressors of known capacity and power best establishes the acceptance test limits described. Take precautions to prevent oil that has been used repeatedly to lubricate many compressors from becoming acidic or contaminated.

Testing with Refrigerant. Calorimeter and flow meter testing methods for rating positive-displacement compressors are described in ASHRAE *Standard* 23. This type of testing is typically conducted on an audit basis. If the purpose of the testing is not an accurate determination of the unit's capacity and efficiency, alternative methods can be used, such as testing on vapor or desuperheating stands. The vapor stand requires an expansion device (TXV) and a heat exchanger (or condenser) large enough to handle the heat equivalent to the motor power. The gas compressed by the compressor is cooled until its enthalpy is the same as that at suction conditions. It is then adiabatically expanded back to the suction state. This method eliminates the need for an evaporator and uses a smaller heat exchanger (condenser). On small-capacity compressors, a piece of tubing that connects discharge to suction and has a hand expansion valve can be used effectively. The measure of performance is usually the relationship of suction and discharge pressures to power. When a water-cooled heat exchanger (condenser) is used, the discharge pressure is usually known, and the water temperature rise and flow are used as capacity indicators. Operation of the desuperheating stand is similar, but in addition to a condenser and TXV, it also requires a hot-gas bypass valve (HGBV). Liquid refrigerant from a condenser and hot discharge gas are mixed by the HGBV to provide adequate suction pressure and temperature to the compressor: the HGBV controls suction pressure and the TXV, acting as a quench valve, controls superheating. Note that higher range and stability during operation are achieved by using a desuperheating stand instead of a vapor stand.

As a further refinement, flow-measuring devices can be installed in the refrigerant lines. This system is charge-sensitive if predetermined discharge and suction pressures and temperatures are to be obtained. This is satisfactory when all units have the same capacity and one test point is acceptable, because the charge desired can be determined with little experimentation. When various sizes are to be tested, however, or more than one test point is desired, a liquid receiver after the condenser can be used for full-liquid expansion.

The refrigerant must be free of contamination, inert gases, and moisture; the tubing and all other components should be clean and sealed when they are not in use. In the case of hermetic and semi-hermetic systems, a motor burnout on the test stand makes it imperative not to use the stand until it has been thoroughly flushed and is absolutely acid-free. In all tests, oil migration must be observed carefully, and the oil must be returned to the crankcase.

The length of a compressor performance test depends on various factors. Stable conditions are required for accuracy. If oil pump or oil charging problems are inherent, the compressor should be run long enough to ensure that all defects are detected.

Testing Complete Systems

In a factory, testing of any system may be done at a controlled ambient temperature or at an existing shop ambient temperature. In both cases, tests must be run carefully, and any necessary corrections must be made. Because measuring air temperature and flow is difficult, production-line tests are usually more reliable when secondary conditions are used as capacity indicators. Measurements of water temperature and flow, power, cycle time, refrigerant pressures, and refrigerant temperatures are reliable capacity indicators.

When testing self-contained air conditioners, for example, a fixed load may be applied to the evaporator using any air source and either a controlled ambient or shop ambient temperature. As long as the load is relatively constant, its absolute value is not important. For water-cooled units, in which water flow can be absolutely controlled, capacity is best measured by the heat rejected from the condenser. Suction and discharge pressures can be measured for the analysis.

Suction and discharge pressures and temperatures can be used as an indirect measure of capacity in units with air-cooled condensers. As long as the load is relatively constant, the absolute value is not important. Air distribution, velocity, or temperature over the test unit's coil must be kept constant during the test, and the performance of the test unit must then be correlated with the performance of a standard unit. Power measurements supplement the suction and discharge parameter readings.

The primary function of the factory performance test is to ensure that a unit is constructed and assembled properly. Therefore, all equipment must be compared to a standard unit, which should be typical of the unit used to pass the Air-Conditioning and Refrigeration Institute (ARI) and Association of Home Appliance Manufacturers (AHAM) certification programs for compressors and other units. ARI and AHAM provide rating standards with applicable maximum and minimum tolerances. Several ASHRAE and International Organization for Standardization (ISO) standards specify applicable rating tests.

Normal causes of malfunction in a complete refrigeration system are overcharging, undercharging, presence of noncondensable gases in the system, blocked capillaries or tubes, and low compressor efficiency. To determine the validity and sensitivity of any test procedure, it is best to use a unit with known characteristics and then establish limits for deviations from the test standard. If the established limits for charging are ±1 oz of refrigerant, for example, the test unit is charged first with the correct amount of refrigerant and then with 1 oz more and 1 oz less. If this procedure does not establish clearly defined limits, it cannot be considered satisfactory and new values must be established. This same procedure should be followed regarding all variables that influence performance and cause deviations from established limits. All equipment must be maintained carefully and calibrated if tests are to have any significance. Gages must be checked at regular intervals and protected from vibration. Capillary test lines must be kept clean and free of contamination. Power leads must be kept in good repair to eliminate high-resistance connection, and electrical meters must be calibrated and protected to yield consistent data.

In plants where component testing and manufacturing control have been so well managed that the average unit performs satisfactorily, units are tested only long enough to find major flaws. Sample lot testing is sufficient to ensure product reliability. This approach is sound and economical because complete testing taxes power and plant capacity and is not necessary.

When the evaporator load is static (e.g., for refrigerators or freezers), time, temperature, and power measurements are used to measure performance. Performance is determined by the time elapsed between start and first compressor shutoff or by the average on-and-off period during a predetermined number of cycles in a controlled or known ambient temperature. Also, concurrent suction and discharge temperatures in connection with power readings are used to establish conformity to standards. On units where the necessary connections are available, pressure readings may be taken. Such readings are usually possible only on units where refrigerant loss is not critical because some loss is caused by gages.

Units with complicated control circuits usually undergo an operational test to ensure that controls function within design specifications and operate in the proper sequence.

Testing of Components

Component testing must be based on a thorough understanding of the use and purpose of the component. Pressure switches may be calibrated and adjusted with air in a bench test and need not be checked again if there is no danger of blocked passages or pulldown tripout during the operation of the switch. However, if the switch is brazed into the final assembly, precautions are needed to prevent blocking the switch capillary.

Capillaries for refrigeration systems are checked by air testing. When the capillary limits are known, it is relatively easy to establish a flow rate and pressure drop test for eliminating crimped or improperly sized tubing. When several capillaries are used in a distributor, a series of water manometers check for unbalanced flow and can find damaged or incorrectly sized tubes.

In plants with good manufacturing control, only sample testing of evaporators and condensers is necessary. Close control of coils during manufacture leads to the detection of improper expansion, poor bonding, split fins, or uneven spacing. Proper inspection eliminates the need for costly test equipment. In testing the sample, either a complete evaporator or condenser or a section of the heat transfer surface is tested. Because liquid-to-liquid is the most easily and accurately measurable method of heat transfer, a tube or coil can be tested by flowing water through it while it is immersed in a bath of water. The temperature of the bath is kept constant, and the capacity is calculated by measuring the coil flow rate and the temperature differential between water entering and leaving the coil.

REFERENCES

ASHRAE. 2005. Methods of testing for rating positive displacement refrigerant compressors and condensing units. ANSI/ASHRAE *Standard* 23-2005.

ASHRAE. 2004. Designation and safety classification of refrigerants. ANSI/ASHRAE *Standard* 34-2004.

ASHRAE. 1992. Method of testing desiccants for refrigerant drying. ANSI/ASHRAE *Standard* 35-1992.

ASTM. 2002. Standard guide for sampling, test methods, and specifications for electrical insulating oils of petroleum origin. *Standard* D117-02. American Society for Testing and Materials, West Conshohocken, PA.

Blair, H.A. and J. Calhoun. 1946. Evacuation and dehydration of field installations. *Refrigerating Engineering* (August):125.

Goddard, M.B. 1945. Moisture in Freon refrigerating systems. *Refrigerating Engineering* (September):215.

Larsen, L.W. and J. Elliot. 1953. Factory methods for dehydrating refrigeration compressors. *Refrigerating Engineering* (December):1325.

Morton, J.D. and L.K. Fuchs. 1960. Determination of moisture in fluorocarbons. *ASHRAE Transactions* 66:434.

Reed, F.T. 1954. Moisture determination in refrigerant oil solutions by the Karl Fischer method. *Refrigerating Engineering* (July):65.

Taylor, E.S. 1956. New instrument for moisture analysis of "Freon" fluorinated hydrocarbons. *Refrigerating Engineering* (July):41.

CHAPTER 46

RETAIL FOOD STORE REFRIGERATION AND EQUIPMENT

IN the United States, almost 200,000 retail food stores operate their refrigeration systems around the clock to ensure proper merchandising and safety of their food products. Figure 1 shows that supermarkets and convenience stores make the largest contribution to this total (Food Marketing Institute 2004). In U.S. retail food stores, refrigeration consumes about 2.3% of the total electricity consumed by all commercial buildings (EIA 2003). As shown in Figure 2, refrigeration accounts for roughly 50% of the electric energy consumption of a typical supermarket (A.D. Little 1996). Supermarkets and grocery stores have one of the highest electric usage intensities in commercial buildings, at 43 kWh/ft^2 per year. Use for larger supermarkets with long operating hours has been measured at 70 kWh/ft^2 per year (Komor et al. 1998).

The modern retail food store is a high-volume sales outlet with maximum inventory turnover. The Food Marketing Institute defines a **supermarket** as any full-line self-service grocery store with an annual sales volume of at least $2 million (Food Marketing Institute 2004). These stores typically occupy approximately 50,000 ft^2 and offer a variety of meat, produce, and groceries. A new category of supermarkets, called **supercenters**, incorporates a supermarket section and a general merchandise/dry goods section in one building. Almost half of retail food sales are of perishable or semiperishable foods requiring refrigeration, including fresh meats, dairy products, perishable produce, frozen foods, ice cream and frozen desserts, and various specialty items such as bakery and deli products and prepared meals. These foods are displayed in highly specialized and flexible storage, handling, and display apparatus. Many supermarkets also incorporate food service operations that prepare the food.

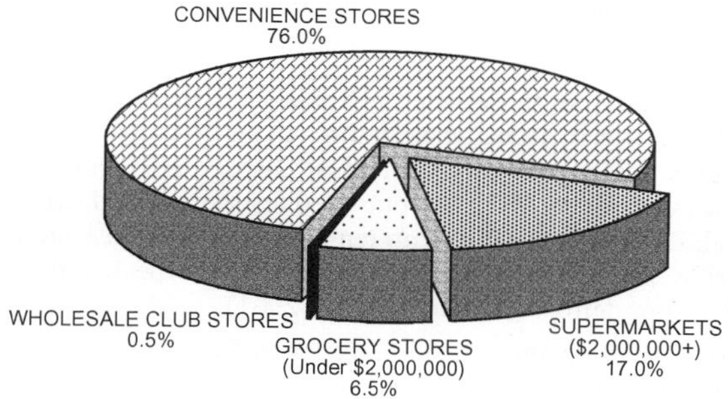

Fig. 1 Distribution of Stores in Retail Food Sector

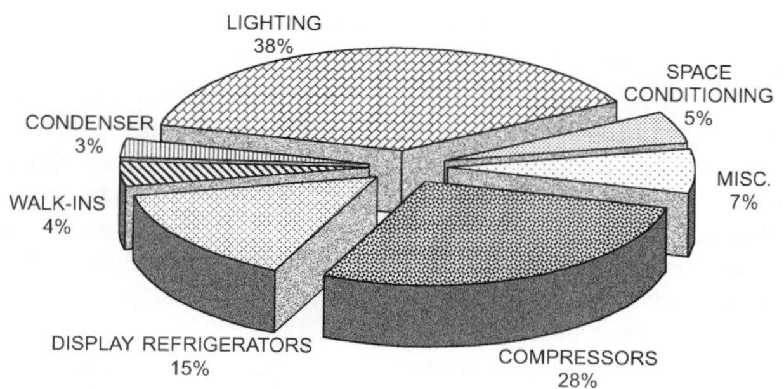

Fig. 2 Percentage of Electric Energy Consumption, by Use Category, of a Typical Large Supermarket

These food products must be kept at safe temperatures during transportation, storage, and processing, as well as during display. The back room of a food store is both a processing plant and a warehouse distribution point that includes specialized refrigerated rooms. All refrigeration-related areas must be coordinated during construction planning because of the interaction between the store's environment and its refrigeration equipment. Chapter 2 of the 2003 *ASHRAE Handbook—HVAC Applications* also covers the importance of coordination.

Refrigeration equipment used in retail food stores may be broadly grouped into display refrigerators, storage refrigerators, processing refrigerators, and mechanical refrigeration machines. Chapter 47 presents food service and general commercial refrigeration equipment. Equipment may also be categorized by temperature: **medium-temperature** refrigeration equipment maintains an evaporator temperature between 0 and 40°F and product temperatures above freezing; **low-temperature** refrigeration equipment maintains an evaporator temperature between –40 and 0°F and product temperatures below freezing.

DISPLAY REFRIGERATORS

Each category of perishable food has its own physical characteristics, handling logistics, and display requirements that dictate specialized display shapes and flexibility required for merchandising. Also, the same food product requires different display treatment in different locations, depending on local preferences, local income level, store size, sales volume, and local availability of food items by type. Display refrigerators provide easy product access and viewing, and typically include additional lighting to highlight the product for sale.

Open display refrigerators for medium and low temperatures are widely used in food markets. However, glass door multideck models have also gained popularity. Decks are shelves, pans, or racks that support the displayed product.

Medium- and low-temperature display refrigerator lineups account for roughly 68 and 32%, respectively, of a typical supermarket's total display refrigerators (Figure 3). In addition, open vertical meat, deli, and dairy refrigerators comprise about 46% of the total display refrigerators (Faramarzi 2000).

Many operators combine single- and multideck models in most departments where perishables are displayed and sold. Closed-service refrigerators are used to display unwrapped fresh meat,

The preparation of this chapter is assigned to TC 10.7, Commercial Food and Beverage Cooling, Display, and Storage.

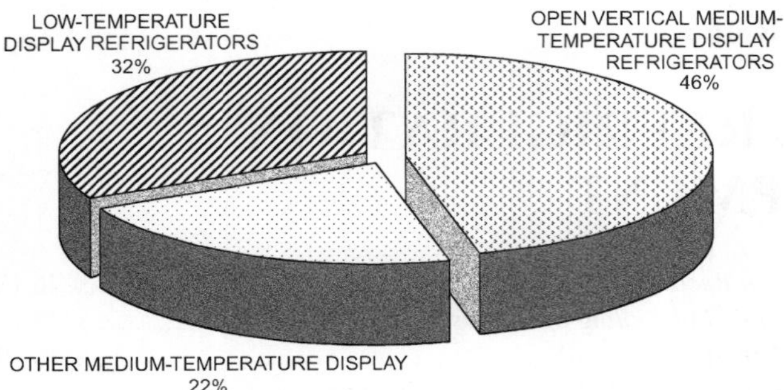

Fig. 3 Percentage Distribution of Display Refrigerators, by Type, in a Typical Supermarket

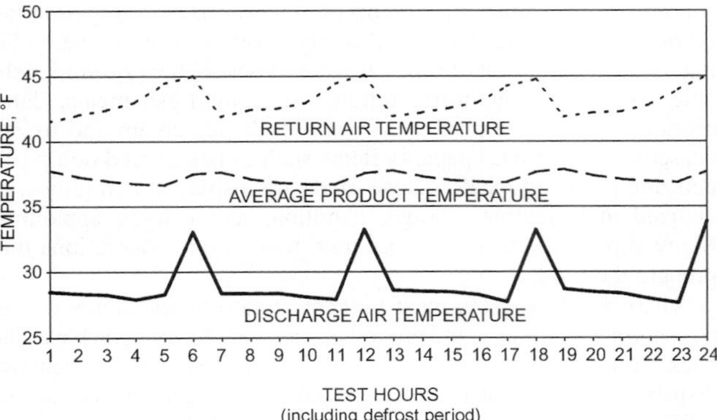

Fig. 4 Selected Temperatures in an Open Vertical Meat Display Refrigerator

delicatessen food, and, frequently, fish on crushed ice supplemented by mechanical refrigeration. A store employee assists the customer by obtaining product out of the service-type refrigerator. More complex layouts of display refrigerators have been developed as new or remodeled stores strive to be distinctive and more attractive. Refrigerators are allocated in relation to expected sales volume in each department. Thus, floor space is allocated to provide balanced stocking of merchandise and smooth flow of traffic in relation to expected peak volume periods.

Small stores accommodate a wide variety of merchandise in limited floor space. Thus, managers of these stores want to display more quantity and variety of merchandise in the available floor space. The concentration of large refrigeration loads in a small space makes year-round space temperature and humidity control essential.

Product Temperatures

Display refrigerators are designed to merchandise food to maximum advantage while providing short-term storage. Proper maintenance of product temperature plays a critical role in food safety. An estimated 24 to 81 million people annually become ill from microorganisms in food, resulting in an estimated 10,000 needless deaths every year. As a result, in 1995 the Food and Drug Administration (FDA) *Food Code* recommended a lower storage temperature for certain refrigerated food products for further prevention of foodborne diseases. The FDA 2001 *Food Code* requires that the core temperature of meat, poultry, fish, dairy, deli, and cut produce not exceed 41°F throughout packaging, shipping, receiving, loading, and storing (FDA 2001).

Proper maintenance of product temperature relies heavily on the temperature of air discharged into the refrigerator. Table 1 lists discharge air temperatures in various display refrigerators. Compliance with FDA requirements may require different refrigerator air

Table 1 Air Temperatures in Display Refrigerators

	Air Discharge Temperatures, °F [a]	
Type of Fixture	Minimum	Maximum
Dairy		
Multideck	34	38
Produce, packaged		
Single-deck	35	38
Multideck	35	38
Meat, unwrapped (closed display)		
Display area	36[b]	38[b]
Deli smoked meat		
Multideck	32	36
Meat, wrapped (open display)		
Single-deck	24	26
Multideck	24	26
Frozen food		
Single-deck	c	−13[c]
Multideck, open	c	−10[c]
Glass door reach-in	c	−5[c]
Ice cream		
Single-deck	c	−24[c]
Glass door reach-in	c	−13[c]

[a]Air temperatures measured with thermometer in outlet of refrigerated airstream and not in contact with displayed product.

[b]Unwrapped fresh meat should only be displayed in a closed, service-type display refrigerator. Meat should be cooled to 36°F internal temperature before placing on display. Refrigerator air temperature should be adjusted to keep internal meat temperature at 36°F or lower for minimum dehydration and optimum display life. Display refrigerator air temperature varies with manufacturer.

[c]Minimum temperatures for frozen foods and ice cream are not critical (except for energy conservation); maximum temperature is important for proper preservation of product quality. Differences in display temperatures among the three different styles of frozen food and ice cream display refrigerators are caused by orientation of refrigeration air curtain and size and style of opening. Single-deck refrigerators have a horizontal air curtain and opening of approximately 30 to 42 in. Multideck, open refrigerators have a vertical air curtain and an opening of about 42 to 50 in. Glass door reach-in refrigerators have a vertical air curtain protected by a multiple-pane insulated glass door.

temperatures from those listed in Table 1. Figure 4 depicts a relationship between discharge air, return air, and average product temperatures for an open vertical meat display refrigerator. These profiles were obtained from controlled tests conducted over a 24 h period. Discharge and return air temperatures were measured at the air grille. As shown, all temperatures reach their peak at the end of each of four defrosts (Faramarzi et al. 2001).

Product temperatures inside a display refrigerator may also vary, depending on the location of the product. Figure 5 depicts product temperature profiles and variations for an open vertical meat display refrigerator over a period of 24 h. As shown, the lowest product temperatures are observed at the top shelf near the discharge air grille, and the highest product temperatures are at the bottom shelf near the return air grille (Gas Research Institute 2000).

Display refrigerators are not designed to cool the product; they are designed to maintain product temperature. When put into the refrigerator, merchandise should be at or near the proper temperature. Food placed directly into the refrigerator or into another adequately refrigerated storage space on delivery to the store should come from properly refrigerated trucks. Little or no delay in transferring perishables from storage or trucks to the display refrigerator or storage space should be permitted.

Display refrigerators should be loaded properly. Most manufacturers provide indicators of physical load limits that define the refrigerated zone. The product on display should never be loaded so that it is out of the load limit zone or be stacked so that circulation of refrigerated air is blocked. The load line recommendations of the manufacturer must be followed to obtain good refrigeration performance. Proper refrigerator design and loading minimize energy use,

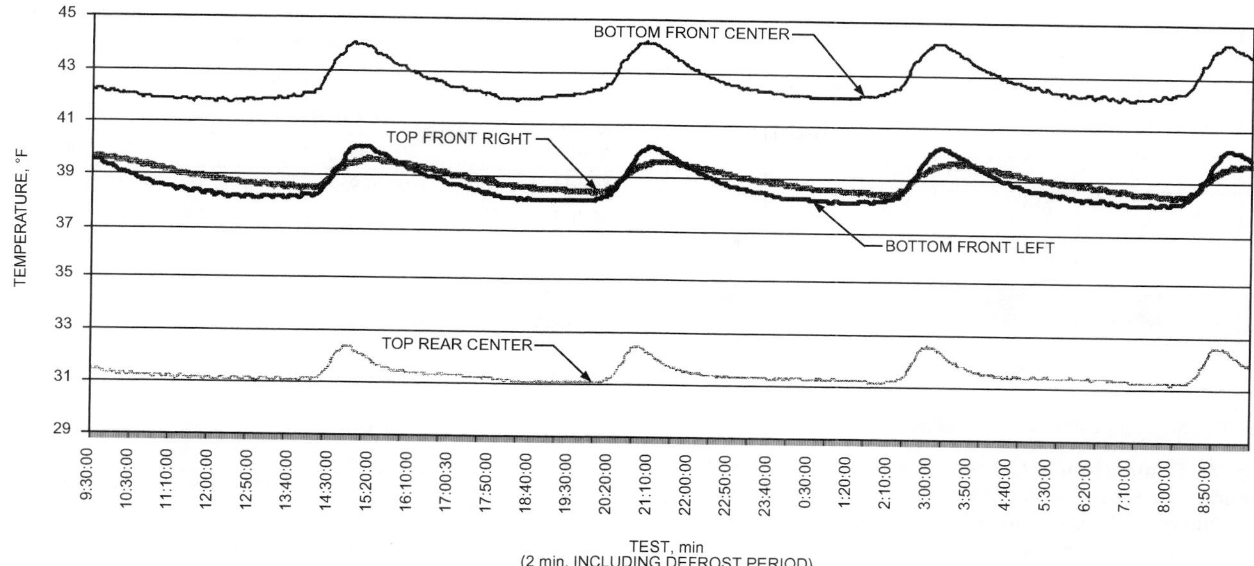

Fig. 5 Product Temperature Profiles at Four Different Locations Inside a Multideck Meat Refrigerator (Average Discharge Air Temperature of 29°F)

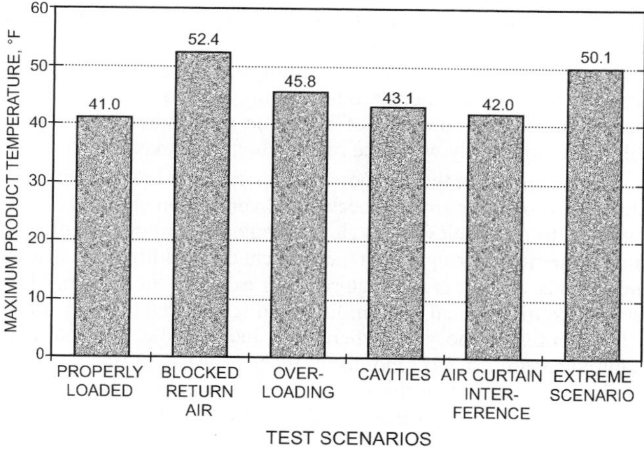

Fig. 6 Comparison of Maximum Product Temperature Variations Under Different Improper Product Loading Scenarios in an Open Vertical Meat Display Refrigerator

maximize efficiency of the refrigeration equipment, maximize food safety, and minimize product loss.

In actual applications, however, products may not always be loaded properly. Survey results (Faramarzi 2003) reveal that improper loading of products inside display refrigerators may fall into the following categories:

- Blocked return air (products block the return air grille)
- Overloading (products loaded beyond the load limit zones)
- Cavities (products loaded nonuniformly, leaving empty spots or voids on the shelves)
- Blocked air curtain (products suspended in the path of air curtain)
- Extreme (combination of blocked return air, blocked air curtain, and overloading)

Improper loading of the products can significantly affect maximum product temperatures, which adversely affects food safety and product loss. Figure 6 depicts the consequences of various improper product-loading scenarios on maximum product temperature of an open vertical meat display refrigerator (Faramarzi 2003).

Additionally, packaging may also affect food temperatures. The surface temperature of a loosely wrapped package of meat with an air space between the film and surface may be 2 to 4°F higher than the surrounding air inside the display refrigerator.

Table 2 Average Store Conditions in the United States

Season	Dry-Bulb Temperature, °F	Wet-Bulb Temperature, °F	Pounds Moisture per Pound Dry Air	rh, %
Winter	69	54	0.0054	36
Spring	70	58	0.0079	50
Summer	71	61	0.0091	56
Fall	70	58	0.0079	50

Store Conditions Survey conducted by Commercial Refrigerator Manufacturers' Association from December 1965 to March 1967. About 2000 store readings in all parts of the country, in all types of stores, during all months of the year reflected the above ambient store conditions.

Store Ambient Effect

Display fixture performance is affected significantly by the temperature, humidity, and movement of surrounding air. Display refrigerators are designed primarily for supermarkets, virtually all of which are air conditioned.

Table 2 summarizes a study of ambient conditions in retail food stores. Individual store ambient readings showed that only 5% of all readings (including those when the air conditioning was not operating) exceeded 75°F db or 0.0102 lb of moisture per pound of dry air. Based on these data, the industry chose 75°F db and 64°F wb (55% rh, 57.5°F dew point) as summer design conditions. This is the ambient condition at which refrigeration load for food store display refrigerators is normally rated.

Store humidity is one of the most critical variables that can affect performance of display refrigerators and refrigeration systems. Store relative humidity may depend on climatic location, seasonal changes, and, most importantly, on the store dehumidification or HVAC system.

Figure 7 shows an example of the relationship between refrigerator condensate and relative humidity. The increase in frost accumulation on the evaporator coils, and consequent increase in

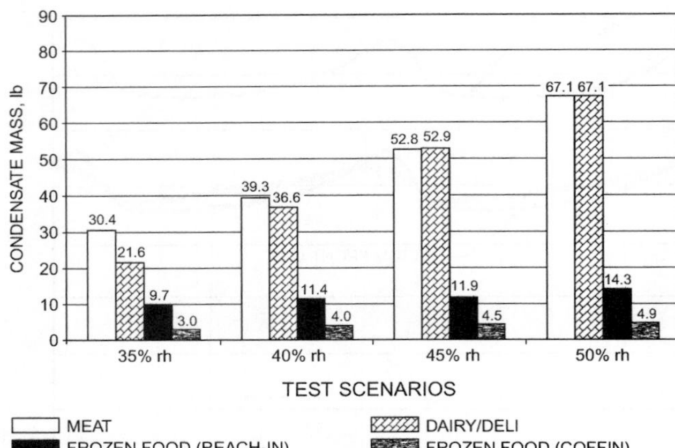

Fig. 7 Comparison of Collected Condensate vs. Relative Humidity for Open Vertical Meat, Open Vertical Dairy/Deli, Narrow Island Coffin, and Glass Door Reach-In Display Refrigerators
(Gas Research Institute 2000)

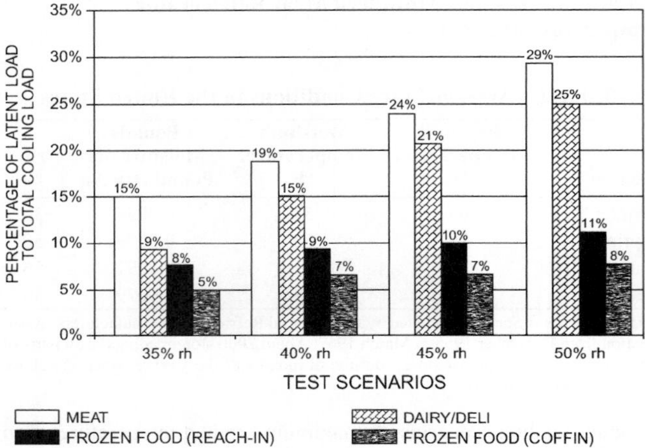

Fig. 8 Percentage of Latent Load to Total Cooling Load at Different Indoor Relative Humidities
(Gas Research Institute 2000)

condensate weight, is more drastic for open vertical display refrigerators. In other words, open vertical fixtures demonstrate more vulnerability to humidity variations and remove more moisture from the ambient (or store) air than other types of display refrigerators (Gas Research Institute 2000).

Increased frost formation from higher relative humidities increases latent load, which the refrigeration system must remove (Figure 8). Additional defrosts may be needed to maintain the product at its desired temperature.

When store ambient relative humidity is different from that at which the refrigerators were rated, the energy requirements for refrigerator operation will vary. Howell (1993a, 1993b) concludes that, compared to operation at 55% store rh, display refrigerator energy savings at 35% rh range from 5% for glass door reach-in refrigerators to 29% for multideck deli refrigerators. Table 3 lists correction factors for the effect of store relative humidity on display refrigerator refrigeration requirements when the dry-bulb temperature is 70 and 78°F.

Manufacturers sometimes publish ratings for open refrigerators at lower ambient conditions than the standard because the milder conditions may significantly reduce the cooling load on the refrigerators. In addition, lower ambient conditions may permit both reductions in antisweat heaters and fewer defrosts, allowing substantial energy savings on a storewide basis.

Table 3 Relative Refrigeration Requirements with Varying Store Ambient Conditions

Refrigerator Model	70°F db					78°F db		
	Relative Humidity, %					Relative Humidity, %		
	30	40	55	60	70	50	55	65
Multideck dairy	0.90	0.95	1.00	1.08[a]	1.18[b]	0.99	1.08[a]	1.18[b]
Multideck low-temperature	0.90	0.95	1.00	1.08[a]	1.18[b]	0.99	1.08[a]	1.18[b]
Single-deck low-temperature	0.90	0.95	1.00	1.08[a]	1.15	0.99	1.05	1.15
Single-deck red meat	0.90	0.95	1.00	1.08[a]	1.15	0.99	1.05	1.15
Multideck red meat	0.90	0.95	1.00	1.08[a]	1.18[b]	0.99	1.08[a]	1.18[b]
Low-temperature reach-in	0.90	0.95	1.00	1.05[a]	1.10	0.99	1.05[a]	1.10

Note: Package warm-up may be more than indicated. Standard flood lamps are clear PAR 38 and R-40 types.
[a]More frequent defrosts required.
[b]More frequent defrosts required plus internal condensation (not recommended).

The application engineer needs to verify that the year-round store ambient conditions are within the performance ratings of the various refrigerators selected for the store. Because relative humidity varies throughout the year, the dew point for each period should be analyzed. The sum of these refrigerator energy requirements provides the total annual energy consumption. In a store designed for a maximum relative humidity of 55%, the air-conditioning system will dehumidify only when the relative humidity exceeds 55%.

In climates where the outdoor air temperature is low in winter, infiltration of outdoor air and mechanical ventilation can cause store humidity to drop below 55% rh. Separate calculations need to be done for periods during which mechanical dehumidification is used and periods when it is not required. For example, in Boston, Massachusetts, mechanical dehumidification is required for only about 3 1/2 months of the year, whereas in Jacksonville, Florida, it is required for almost 7 1/2 months of the year. Also, in Boston, there are 8 1/2 months when the store relative humidity is below 40%, whereas Jacksonville has these conditions for only 4 1/2 months. The engineer must weigh the savings at lower relative humidity against the cost of the mechanical equipment required to maintain relative store humidity levels at, for example, below 40% instead of 55%.

Additional savings can be achieved by controlling antisweat heaters and reducing defrost frequency at ambient relative humidities below 55%. Energy savings credit for reduced use of display refrigerator antisweat heaters can only be taken if the display refrigerators are equipped with humidity-sensing controls that reduce the amount of power supplied to the heaters as the store dew point decreases. Also, defrost savings can be considered when defrost frequency or duration is reduced. Controls can reduce the frequency of defrost as store relative humidity decreases (**demand defrost**). Individual manufacturers give specific antisweat and defrost values for their equipment at stated store conditions. Less defrosting is needed as store dew point temperature or humidity decreases from the design conditions.

Attention should also be given to the condition in which store dry-bulb temperatures are higher than the industry standard, because this raises the refrigeration requirements and consequently the energy demand.

Display Refrigerator Cooling Load and Heat Sources

Heat transfer in a display refrigerator involves interactions between the product and the internal environment of the refrigerator, as well as heat from the surroundings that enters the refrigerator. Heat components from the surrounding environment include transmission (or conduction), radiation, and infiltration, whereas heat components from the internal environment include lights and evaporator fan motor(s). In addition, defrost and antisweat heaters also increase the cooling load of a display refrigerator. Conduction, radiation, and infiltration loads from the surroundings into the refrigerator, as well as heat exchanges between the product and parts of the refrigerator, depend on the temperatures of ambient air and air within the refrigerator. Open vertical display refrigerators rely on their air curtains to keep warm ambient air from penetrating into the cold environment inside the refrigerator. An air curtain consists of a stream of air discharged from a series of small nozzles through a honeycombed baffle at the top of the display refrigerator. Air curtains play a significant role in the thermal interaction of the display refrigerator with the surrounding air (see Figure 10).

The cooling load of a typical display refrigerator has both sensible and latent components. In general, the sensible portion consists of heat gain from lights, fan motor(s), defrost (electric and hot gas), antisweat heater, conduction, radiation, infiltration, and product pulldown load. The latent portion consists of infiltration and product latent heat of respiration.

Conduction Load. The conduction load refers to the heat transmission through the physical envelope of the display refrigerator. The temperature difference between air in the room and air inside the refrigerator is the main driving force for this heat transfer.

Radiation Load. The heat gain of the display refrigerator through radiation is a function of conditions inside the refrigerator, including surface temperature, surface emissivity, surface area, view factor with respect to the surrounding (store) walls/objects, floor, ceiling, and their corresponding emissivities and areas.

Infiltration Load. The infiltration load of the display refrigerator refers to the net entrainment of warm, moist air through the air curtain into the refrigerated space. The infiltration load has two components: sensible and latent. The total performance of the air curtain and the amount of heat transferred across it may depend on several factors, including

- Air curtain velocity and temperature profile
- Number of jets
- Air jet width and thickness
- Dimensional characteristics of the discharge air honeycomb
- Store and display refrigerator temperatures and humidity ratios
- Rate of air curtain agitation caused by shoppers passing
- Thermo-fluid boundary condition in the initial region of the jet

Sensible Infiltration. The sensible portion of infiltration refers to the direct heat added by the temperature difference between cold air in the refrigerator and warm room air drawn into the refrigerator.

Latent Infiltration. The latent portion of infiltration refers to the heat content of the moisture added to the refrigerator by the room air drawn into the refrigerator.

Internal Loads. The internal load includes heat from refrigerator lights and evaporator fan motors. The lamps, ballasts, and fan motors are typically located within the thermodynamic boundary of the display refrigerator; therefore, their total heat dissipation should be considered part of the refrigerator load. High-intensity lighting raises product temperatures and can discolor meats. Refrigerator shelf ballasts are sometimes located out of the refrigerated space to reduce refrigerator cooling load. Standard lighting equipment, which typically consists of T12 fluorescent lamps with magnetic ballast, draws approximately 0.73 A at 120 V.

Defrost Load. Refrigeration equipment in applications where frost can accumulate on the evaporator coils have some type of defrost mechanism. During defrost, refrigeration is stopped on the defrosting circuits and heat is introduced into the refrigerator. Defrost methods vary, depending on the refrigeration application and storage temperatures, as discussed in the section on Methods of Defrost. Some defrost methods deliver more heat than is needed to melt the ice. A large portion of the extra heat warms the coil metal, product (see Figures 4 and 5), and refrigerator. This extra heat adds to the refrigeration load and is called the postdefrost pulldown load (Faramarzi 1999).

Antisweat Heaters (ASH) Load. The antisweat heater load refers to the portion of the electrical load of the ASH that ends up as sensible heat inside the refrigerator. Antisweat heaters are used on most low-temperature open display refrigerators, as well as reach-in refrigerators with glass doors. These electric resistance heaters are located around the handrails of tub refrigerators and door frame/mullions of reach-in refrigerators to prevent condensation on metal surfaces. They also reduce fogging of the glass doors of reach-in refrigerators, a phenomenon that can hurt product merchandising. Without appropriate control systems, ASH units stay on round the clock. The cooling load contribution of ASH in a typical reach-in display refrigerator can reach 35% of their connected electric load (Faramarzi et al. 2001).

Pulldown Load. The pulldown load has two components (Faramarzi 1999):

- **Case product load.** This pulldown load is caused by product delivery into the refrigerator at a temperature higher than the designated storage temperature. It is the amount of cooling required to lower the product temperature to a desired target temperature.
- **Postdefrost load.** During the defrost cycle, product temperature inside the refrigerator rises. Once defrost is complete, the refrigeration system turns on and must remove the accumulated defrost heat and lower the product temperature to a desirable set point.

According to a test report by Gas Research Institute (2000), the major contributor to the total cooling load of open display refrigerators are infiltration and radiation (Figure 9). Infiltration constitutes approximately 80% of the cooling load of a typical medium-temperature open vertical display refrigerator. The relative role of infiltration diminishes for low-temperature open coffin (or tub) refrigerators, and is supplanted by radiation. Infiltration and radiation constitute roughly 24 and 43%, respectively, of the cooling load of a typical open coffin refrigerator.

Multideck open refrigerator shelves are an integral part of the air curtain and airstream. Without shelves, there will be substantial air distribution problems. An air deflector may be required when shelves are removed. As shown in Figure 9, infiltration through the air curtain plays a significant role in the cooling load of open vertical display refrigerators (Faramarzi 1999). Figure 10 depicts the air curtain velocity streamlines of an 8 ft open vertical meat display refrigerator. These velocity streamlines represent the actual airflow patterns using

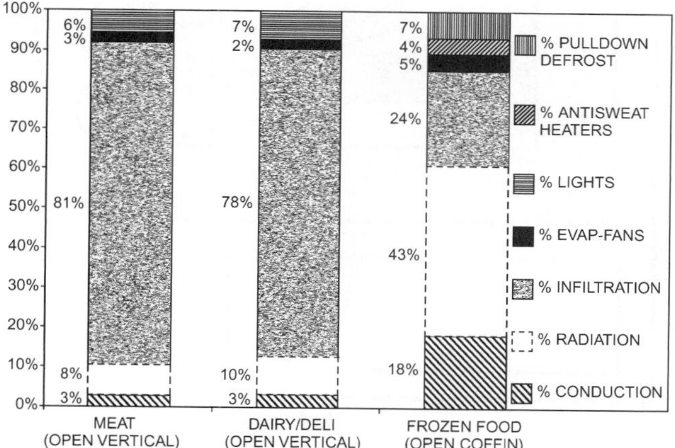

Fig. 9 Components of Refrigeration Load for Several Display Refrigerator Designs at 75°F db and 55% rh

digital particle image velocimetry. As shown, warm air is entrained into the display refrigerator at several locations along the plane of the air curtain. Based on the law of conservation of mass, an equal (and substantial) amount of cold air from the display refrigerator spills into the room near the return air grille of the fixture.

Refrigerator Construction

Commercial refrigerators for market installations are usually of the endless construction type, which allows a continuous display as refrigerators are joined. Clear plastic panels are often used to separate refrigerator interiors when adjacent refrigerators are connected to different refrigeration circuits. Separate end sections are provided for the first and last units in a continuous display. Methods of joining self-service refrigerators vary, but they are usually bolted or cam-locked together.

All refrigerators are constructed with surface zones of transition between the refrigerated area and the room atmosphere. Thermal breaks of various designs separate the zones to minimize the amount of refrigerator surface that is below the dew point. Surfaces that may be below the dew point include (1) in front of discharge air nozzles, (2) the nose of the shelving, and (3) front rails or center flue of the refrigerator. In glass door reach-in freezers or medium-temperature refrigerators, the frame jambs and glass can be below the dew point. In these locations, resistance heat is used effectively to raise the exterior surface temperature above the dew point to prevent accumulation of condensation.

With the current emphasis on energy efficiency, designers have developed means other than resistance heat to raise the surface temperatures above the dew point. However, when no other technique is known, resistance heating becomes necessary. Control by cycling and/or proportional controllers to vary heat with store ambient changes can reduce energy consumption.

Store designers can do a great deal to promote energy efficiency. Not only does controlling the atmosphere within a store reduce refrigeration requirements, it also reduces the need to heat the surfaces of refrigerators. This heat not only consumes energy, but also places added demand on the refrigeration load.

Evaporators and air distribution systems for display refrigerators are highly specialized and are usually fitted precisely into the particular display refrigerator. As a result, they are inherent in the fixture and are not standard independent evaporators. The design of the air circuit system, the evaporator, and the means of defrosting are the result of extensive testing to produce the particular display results desired.

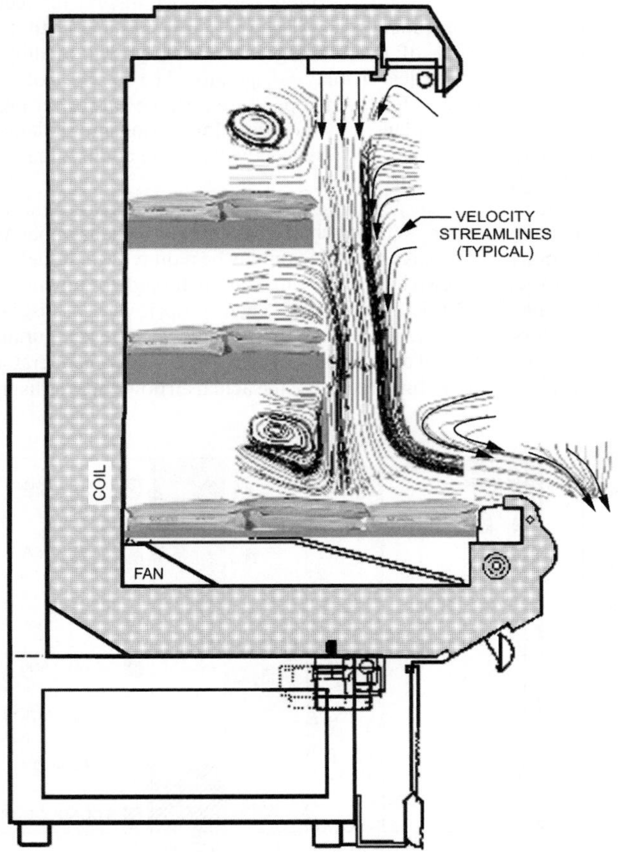

Fig. 10 Velocity Streamlines of a Single-Band Air Curtain in an Open Vertical Meat Display Refrigerator, Captured Using Digital Particle Image Velocimetry Technique

Cleaning and Sanitizing Equipment

Because the evaporator coil is the most difficult part to clean, consider the judicious use of high-pressure, low-liquid-volume sanitizing equipment. This type of equipment enables personnel to spray cleaning and sanitizing solutions into the duct, grille, coil, and waste outlet areas with minimum disassembly and maximum effectiveness. However, this equipment must be used carefully because the high-pressure stream can easily displace sealing and caulking materials. High-pressure streams should not be directed toward electrical devices. Hot liquid can also break the glass on models with glass fronts and on closed-service fixtures.

Refrigeration Systems for Display Refrigerators

Self-Contained. Self-contained systems, in which the condensing unit and controls are built into the refrigerator structure, are usually air-cooled and are of two general types. The first type has the condensing unit beneath the cabinet; in some designs, it takes up the entire lower part of the refrigerator, but in others it occupies only one lower corner. The second type has the condensing unit on top.

Remote. Remote refrigeration systems are often used if cabinets are installed in a hot or otherwise unfavorable location where the noise or heat of the condensing units would be objectionable. Remote systems can take advantage of cool ambient air and provide lower condensing temperatures, which allows more efficient operation of the refrigeration system.

Merchandising Applications

Dairy Display. Dairy products include items with significant sales volume, such as fresh milk, butter, eggs, and margarine. They also include a myriad of small items such as fresh (and sometimes processed) cheeses, special above-freezing pastries, and other perishables. Available display equipment includes the following:

1. Full-height, fully adjustable shelved display units without doors in back for use against a wall (Figure 11); or with doors in back for rear service or for service from the rear through a dairy cooler. The effect of rear service openings on the surrounding refrigeration must be considered. The front of the refrigerator may be open or have glass doors.
2. Closed-door displays built in the wall of a walk-in cooler with adjustable shelving behind doors. Shelves are located and stocked in the cooler (Figure 12).
3. A variety of other special display units, including single-deck and island-type display units, some of which are self-contained and reasonably portable for seasonal, perishable specialties.
4. A refrigerator, similar to Item 1, but able to receive either conventional shelves and a base shelf and front or premade displays on pallets or carts. This version comes with either front-load

capability only or rear-load capability only (Figure 13). These are called front roll-in or rear roll-in display refrigerators.

Meat Display. Most meat is sold prepackaged. Some of this product is cut and packaged on the store premises. Control of temperature, time, and sanitation from the truck to the checkout counter is important. Meat surface temperatures over 40°F shorten its salable life significantly and increase the rate of discoloration.

The design of open fresh meat display refrigerators, either tub-type single-deck or vertical multideck, is limited by the freezing point of meat. Ideally, refrigerators are set to operate as cold as possible without freezing the meat. Temperatures are maintained with minimal fluctuations (with the exception of defrost) to ensure the coldest possible stable internal and surface meat temperatures.

Sanitation is also important. If all else is kept equal, good sanitation can increase the salable life of meat in a display refrigerator. In this chapter, sanitation includes limiting the amount of time meat is exposed to temperatures above 40°F. If meat has been handled in a sanitary manner before being placed in the display refrigerator, elevated temperatures can be more tolerable. When meat surfaces are contaminated by dirty knives, meat saws, table tops, etc., even optimum display temperatures will not prevent premature discoloration and subsequent deterioration of the meat. See the section on Meat Processing Rooms for information about the refrigeration requirements of the meat-wrapping area.

Along with molds and natural chemical changes, bacteria discolor meat. With good control of sanitation and refrigeration, experiments in stores have produced meat shelf life of one week and more. Bacterial population is greatest on the exposed surface of displayed meat because the surface is warmer than the interior. Although cold airflow refrigerates each package, the surface temperature (and thus bacterial growth) is cumulatively increased by

- Infrared rays from lights
- Infrared rays from the ceiling surface
- High stacking of meat products
- Voids in display
- Store drafts that disturb refrigerator air

Improper control of these factors may cause meat surface temperatures to rise above values allowed by food-handling codes. It takes great care in every building and equipment detail, as well as in refrigerator loading, to maintain meat surface temperature below 40°F. However, the required diligence is rewarded by excellent shelf life, improved product integrity, higher sales volume, and less scrap or spoilage.

Surface temperatures rise during defrost. Tests have compared matched samples of meat: one goes through normal defrost, and the other is removed from the refrigerator during its defrosting cycles.

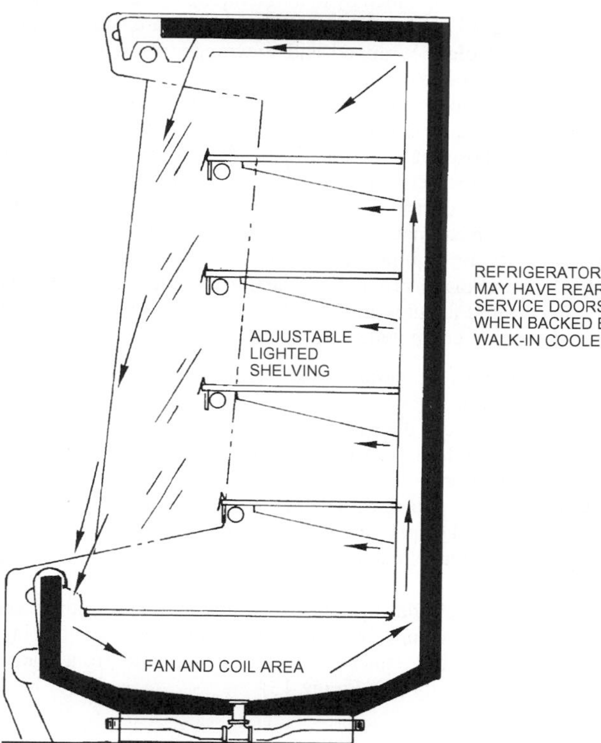

Fig. 11 Multideck Dairy Display Refrigerator

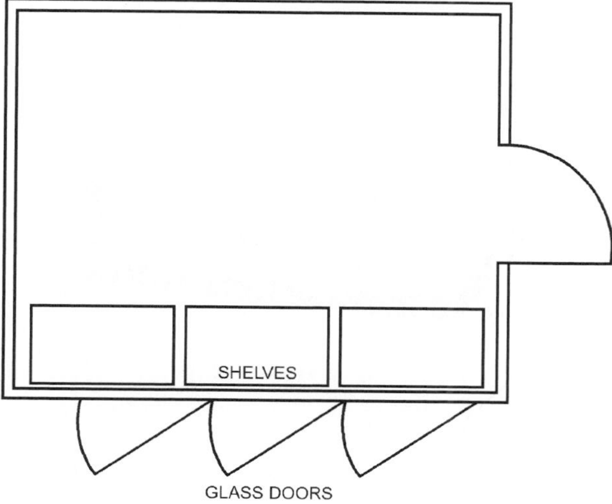

Fig. 12 Typical Walk-In Cooler Installation

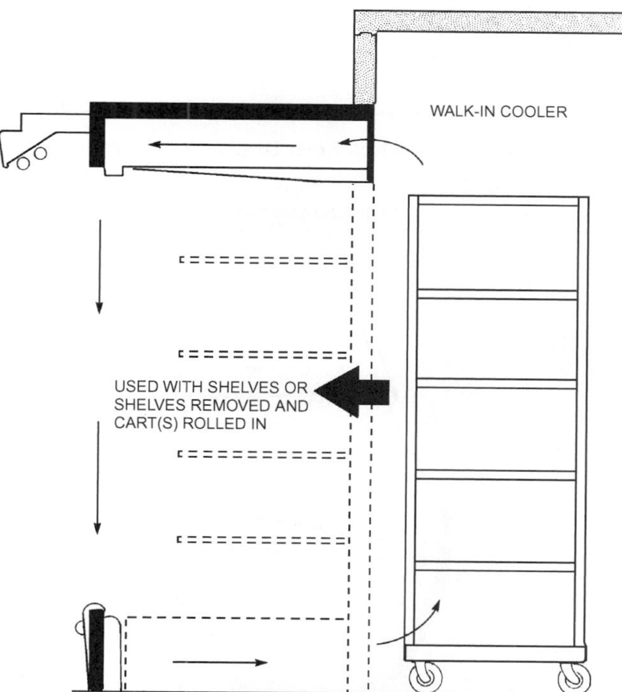

Fig. 13 Vertical Rear-Load Dairy (or Produce) Refrigerator with Roll-In Capability

Although defrosting characteristics of refrigerators vary, such tests have shown that the effects on shelf life of properly handled defrosts are negligible. Tests for a given installation can easily be run to prove the effects of defrosting on shelf life for that specific set of conditions.

Self-Service Meat Refrigerators. Self-service meat products are displayed in packaged form. Processed meat can be displayed in similar refrigerators as fresh packaged meat, but at slightly higher temperatures. The meat department planner can select from a wide variety of available meat display possibilities:

- Single-deck refrigerators, with optional rear or front access storage doors (Figure 14)
- Multideck refrigerators, with optional rear access (Figure 15)
- Either of the preceding, with optional glass fronts

All these refrigerators are available with a variety of lighting, superstructures, shelving, and other accessories tailored to special merchandising needs. Storage compartments are rarely used in self-service meat refrigerators.

Closed-Service Meat or Deli Refrigerators. Service meat products are generally displayed in bulk, unwrapped. Generally, closed refrigerators can be grouped in one of the following categories:

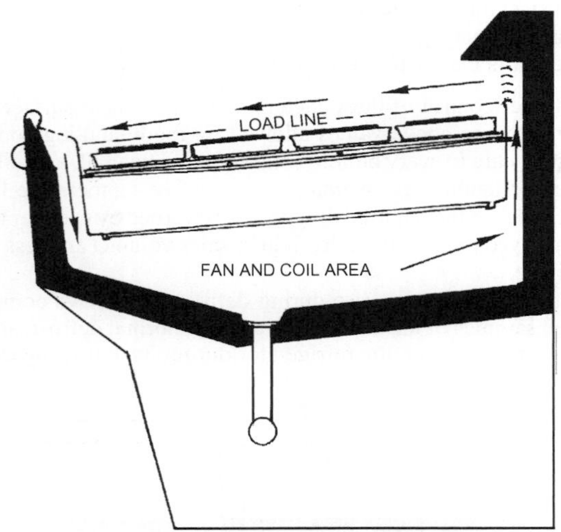

Fig. 14 Single-Deck Meat Display Refrigerator

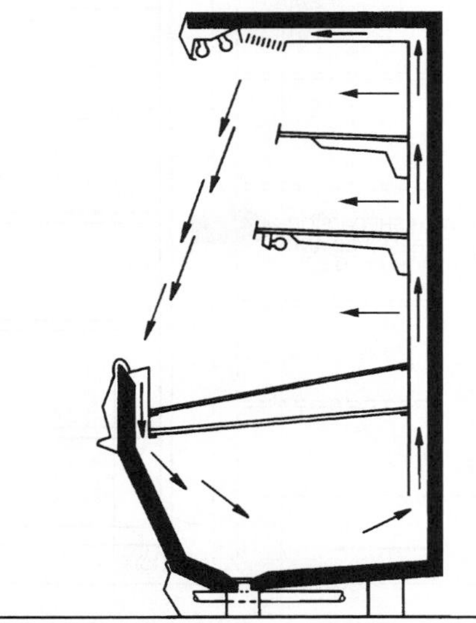

Fig. 15 Multideck Meat Refrigerator

- Fresh red meat, with optional storage compartment (Figure 16)
- Deli and smoked or processed meats, with optional storage
- Fresh fish and poultry, usually without storage but designed to display products on a bed of cracked ice

Closed-service meat display refrigerators are offered in a variety of configurations. Their fronts may be nearly vertical or angled up to 20° from vertical in flat or curved glass panels, either fixed or hinged, and they are available with gravity or forced-convection coils. Gravity coils are usually preferred for more critical products, but forced-air coil models using various forms of humidification systems are also common.

These service refrigerators typically have sliding rear access doors, which are sometimes removed during busy periods. This practice is not recommended by manufacturers, however, because it affects the internal product display zone temperature and humidity.

Produce Display. Wrapped and unwrapped produce is often intermixed in the same display refrigerator. Ideally, unwrapped produce should have low-velocity refrigerated air forced up through the loose product. Water is usually also sprayed, either by manually operated spray hoses or by automatic misting systems, on leafy vegetables to retain their crispness and freshness. Produce is often displayed on a bed of ice for visual appeal. However, packaging prevents air from circulating through wrapped produce and requires higher-velocity air. Equipment available for displaying both packaged and unpackaged produce is usually a compromise between these two desired features and is suitable for both types of product. Available equipment includes the following:

1. Wide or narrow single-deck display units with or without mirrored superstructures.
2. Two- or three-deck display units, similar to the one in Figure 17, usually for multiple-refrigerator lineups near single-deck display refrigerators.
3. Because of the nature of produce merchandising, a variety of nonrefrigerated display units of the same family design are usually designed for connection in continuous lineup with the refrigerators.
4. A refrigerator, similar to Item 2, but able to receive either conventional shelves and a base shelf and front or premade displays on pallets and carts. This version comes with either front-load or rear-load capability (see Figure 13).

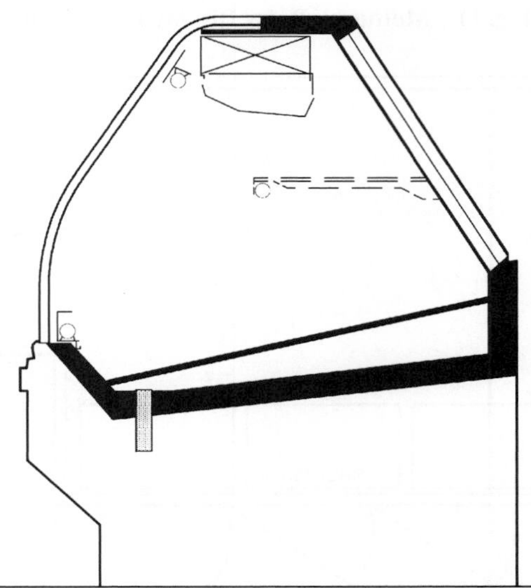

Fig. 16 Closed-Service Display Refrigerator (Gravity Coil Model with Curved Front Glass)

Produce equipment is generally available with a variety of merchandising and other accessories, including bag compartments, sprayers for wetting the produce, night covers, scale racks, sliding mirrors, and other display shelving and apparatus.

Frozen Food and Ice Cream Display

To display frozen foods most effectively (depending on varied need), many types of display refrigerators have been designed and are available. These include the following:

1. Single-deck tub-type refrigerators for one-side shopping (Figure 18). Many types of merchandising superstructures for related nonrefrigerated foods are available. Configurations are designed for matching lineup with fresh meat refrigerators, and there are similar refrigerators for matching lineup of ice cream refrigerators with their frozen food counterparts. These refrigerators are offered with or without glass fronts.
2. Single-deck island for shop-around (Figure 19). These are available in widths ranging from the single-deck refrigerators in Item 1 to refrigerators of double width, with various sizes in between. Some across-the-end increments are available with or without various merchandising superstructures for selling related nonrefrigerated food items to complete the shop-around configuration.
3. Freezer shelving in two to six levels with many refrigeration system configurations (Figure 20). Multideck self-service frozen food and ice cream fixtures are generally more complex in design and construction than single-deck models. Because they have wide, vertical display compartments, they are more affected by ambient conditions in the store. Generally, open multideck models have two or three air curtains to maintain product temperature and shelf life requirements.
4. Glass door, front reach-in refrigerators (Figure 21), usually of a continuous lineup design. This style allows for maximum inventory volume and variety in minimum floor space. The front-to-back interior dimension of these cabinets is usually about 24 in.

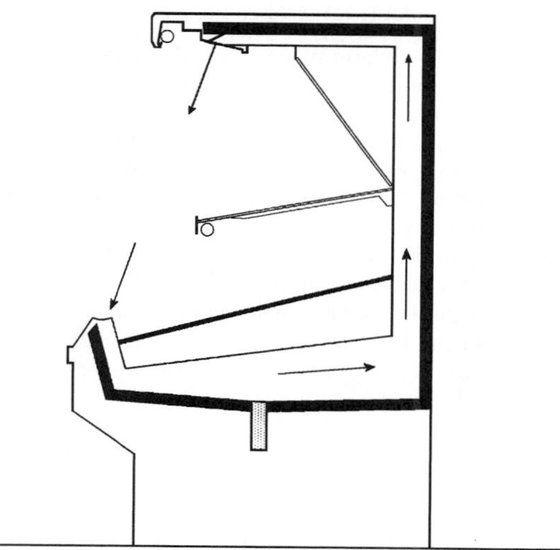

Fig. 17 Multideck Produce Refrigerator

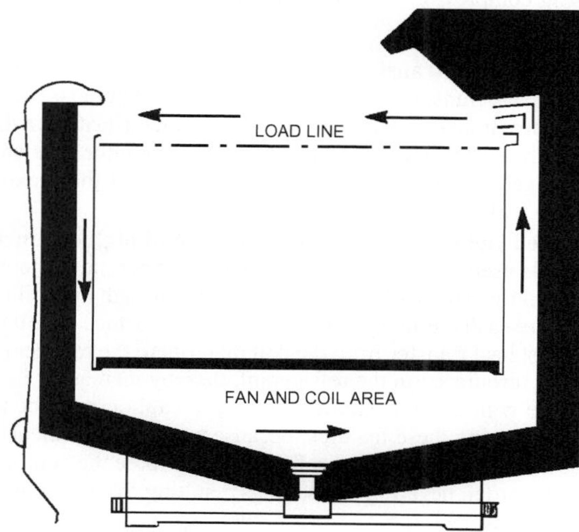

Fig. 18 Single-Deck Tub-Type Frozen Food Refrigerator

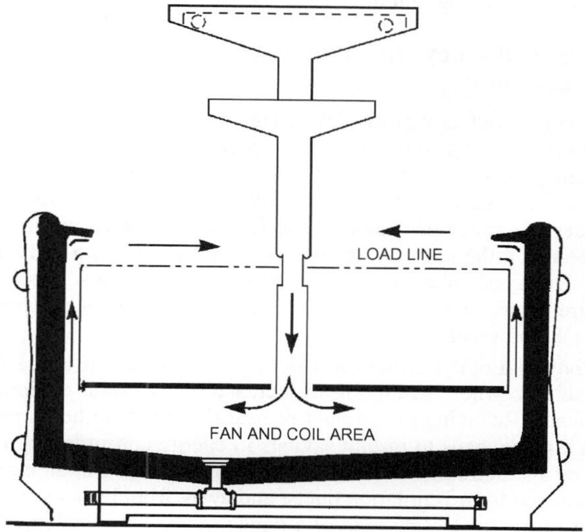

Fig. 19 Single-Deck Island Frozen Food Refrigerator

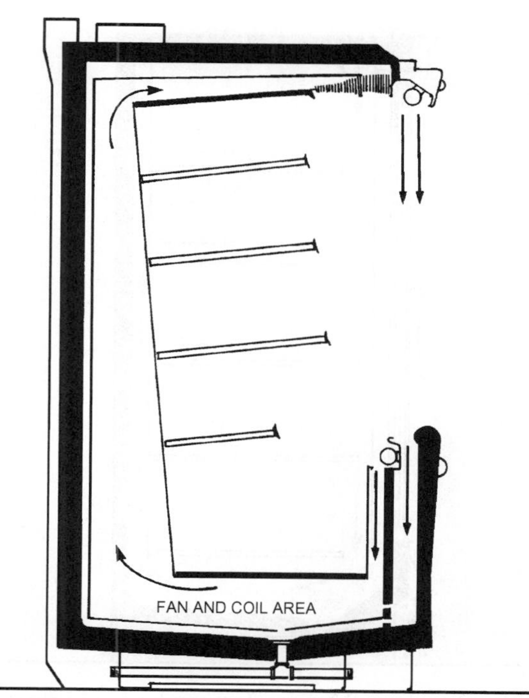

Fig. 20 Multideck Frozen Food Refrigerator

Greater attention must be given to the back product to provide the desired rotation. Although these refrigerators generally consume less energy than open multideck low-temperature refrigerators, specific comparisons by model should be made to determine capital and operating costs.

5. Spot merchandising refrigerators, usually self-contained and sometimes arranged for quick change from nonfreezing to freezing temperature to allow for promotional items of either type (e.g., fresh asparagus or ice cream).
6. Versions of most of the above items for ice cream, usually with modified defrost heaters and other changes necessary for the approximately 10°F colder required temperature. As display temperature decreases to below 0°F (product temperature), the problem of frost and ice accumulation in flues and in the product zone increases dramatically. Proper product rotation and frequent restocking minimize frost accumulation.

Energy Efficiency Opportunities in Display Refrigerators

Energy efficiency of display refrigerators can be improved by carefully selecting components and operating practices. Typically, efficiency is increased through one or more of the methods discussed in this section. Different products use different components and design strategies. Some of the following options are mature and tested in the industry, whereas others are emerging technologies. Designers must balance energy savings against customer requirements, manufacturing cost, system performance, reliability, and maintenance costs.

Cooling Load Reduction. Cooling load reduction is the first step to take when attempting to increase refrigeration equipment efficiency. Reducing the amount of heat that needs to be removed from a space leads to instant savings in energy consumption. Display refrigerators should be located to minimize drafts or air curtain disturbance from ventilation ducts, and away from heat sources or direct sunlight. Cooling load of a typical refrigerator is dependent on infiltration, conduction, and radiation from surroundings, as well as heat dissipation from internal components.

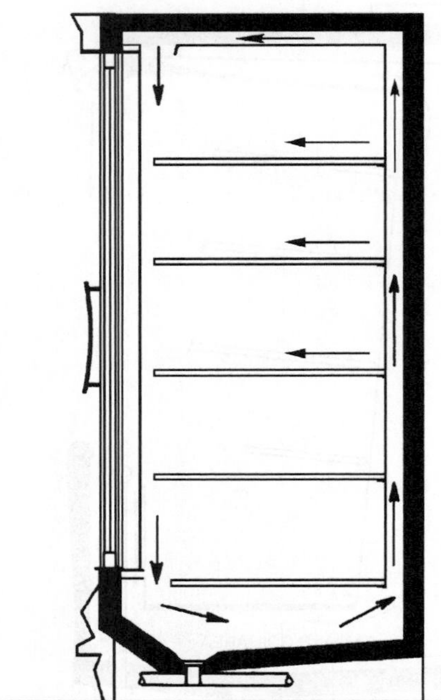

Fig. 21 Glass Door, Medium-Temperature and Frozen Food Reach-In Refrigerator

Infiltration. Research indicates that infiltration of warm and moist air from the sales area into an open vertical display refrigerator accounts for 70 to 80% of the display refrigerator total cooling load (Faramarzi 1999). Infiltrated air not only raises product temperatures, but moisture in the air also becomes frost on the evaporator coil, reducing its heat transfer abilities and forcing the fan to work harder to circulate air through the refrigerator. There are several ways to reduce the amount of infiltration into refrigerators:

- Installing **glass doors on open vertical display refrigerators** provides a permanent barrier against infiltration. Similarly, vertical refrigerators with factory-installed doors eliminate most infiltration and significantly reduce cooling load.
- **Optimizing the air curtain** can drastically reduce its entrainment of ambient air. This ensures that a larger portion of cold air supplied by the refrigerator makes it back to the evaporator through the return air duct.
- In stores that do not operate 24 h per day, installing **night covers** can provide an infiltration barrier during unoccupied hours. Faramarzi (1997) found that 6 h of night cover use can reduce the cooling load by 8% and the compressor power requirement by 9%. Select night curtains that do not condense water on the outside, creating potential for slippery floors. Local health inspectors should also be consulted to ensure that the curtain is considered cleanable and acceptable for use in a grocery store.

Thermal Radiation. Warm objects near the display refrigerator radiate heat into the refrigerated space. Night covers protect against radiation heat transfer.

Thermal Conduction. Improving the R-value of insulation, whether by using materials with low thermal conductivity or simply increasing insulation thickness, reduces conduction heat transfer through walls of the refrigerated space. Conduction accounts for less than 5% of cooling load of medium-temperature refrigerators but almost 20% for low-temperature refrigerators (see Figure 9).

Display Refrigerator Component Improvements. Careful selection of components based on proper application, energy efficiency attributes, and correct sizing can play a significant role in increasing overall system efficiency.

Evaporator. Evaporator coil design can significantly affect refrigerator performance. Efficient evaporator coils allow the refrigerator to maintain its target discharge air temperature while operating at a higher evaporator temperature. Higher evaporator temperature (or suction pressure) has the benefit of increasing its refrigeration effect; however, it also hampers refrigeration system performance by increasing the density of refrigerant entering the compressor, thus increasing compressor work. Evaporator coil characteristics can be improved in four ways:

- **Increased heat transfer effectiveness.** Efficient coils have a greater heat transfer surface area made of materials with improved heat transfer properties to absorb as much heat from the air as possible using optimized fin design. Evaporator fans should also be selected to evenly distribute air through the maximum possible coil face area.
- **Improved coil tube design: low friction and high conduction.** Materials used to construct coils, such as copper, have increased conductivity, which allows heat to transfer through the coil materials more easily. Enhancements to the inside surface of coil tubes can assist heat transfer from the coil material to the refrigerant by creating turbulence in the refrigerant, thereby increasing its contact time with the tube surface. However, caution must be taken when designing these features, because excessive turbulence can cause a pressure drop in the refrigerant and force the compressor to work harder, negating any savings resulting from the enhancement (Dossat 1997).
- **Improved refrigerant distribution.** Coil performance depends on the refrigerant's path through the evaporator coil. For optimal

coil design, the coldest refrigerant should come into contact with the coldest air to ensure maximum heat transfer capability.

- **Frost-tolerant surface.** Typically, the leading edge of the coil shows the worst frosting because moisture in return air condenses as soon as it hits the cold surface. This frost can grow to the point that it severely restricts airflow through the coil. Coils can be manufactured from modules with different fin spacing so that the frost formation is controlled. Larger fin spacing on the leading edge allows moisture to be removed and frost to build, but prevents the coil from becoming totally clogged. Smaller fin spacing can be used toward the trailing edge to maximize heat transfer to lower the air temperature to required levels.

Defrost. Heat added while the refrigeration system is in defrost can raise product temperatures and must be removed later. Defrost methods should be chosen so that the minimum amount of heat is added to the refrigerator. For example, hot-gas defrost can be considered an improved technique.

Demand defrost technologies can sense frost formation on the coil, enabling a controller to determine exactly when the refrigerator should begin its defrost cycle. Unnecessary defrosts and excessive frost formation leading to coil blockage can be eliminated. Care must be used when selecting a demand defrost system: if the system malfunctions, the refrigerators will require service, and there is the potential for product loss.

Sensors may also be used to verify the end of defrost cycles (**intelligent defrost termination**). Typically, the refrigerator is allowed to defrost for a set amount of time or until the air temperature leaving the coil reaches a specified level. This usually means that the defrost cycle is running for a longer period of time than necessary, allowing more heat to enter the refrigerator and raise product temperatures. Intelligent defrost termination sensors can determine exactly when the coil is free of frost and immediately restart the refrigeration system. An intelligent defrost termination sensor can be a simple electromechanical thermostat, a solid-state sensor, or other device.

Antisweat. Antisweat heaters (ASHs) with a low watt-per-door rating should be used whenever possible. In addition to using less energy at the antisweat heater level, less heat will be introduced into the refrigerated space, thus indirectly reducing the cooling load.

Some controllers can recognize the antisweat heat needs of the door and ensure that the heaters only operate when needed. They adjust their operation accordingly, through pulsation or other mechanisms. **Condensate sensors** on reach-in glass doors activate ASHs when droplets are detected; **RH-based controllers** sense the psychrometric properties of air and activate ASHs when needed.

New methods of glass **door construction** have brought products that require little or no antisweat heat to maintain customer-friendly fog-free panes. This performance is achieved by either using advanced glass types or special door frames, both of which greatly reduce or eliminate the amount of glass heating necessary to resist condensation.

Alternative Expansion Valves. **Dual-port thermostatic expansion valves (TXVs)** have capacity modulation capabilities not seen in other expansion valves. When the refrigerator emerges from defrost, there is typically a much higher load because of increased product temperatures. In this case, the large port of the expansion valve opens, allowing the system to operate at a higher capacity to account for the increased pulldown load.

Superheat can be most easily controlled by **electronic expansion valves**, which have a much faster response time than bulb-sensing TXVs. Manufacturers should test the valve and controller to ensure it maintains stable control at targeted superheats.

Liquid-to-Suction Heat Exchanger. Liquid-to-suction heat exchangers allow suction gas exiting the display refrigerator to absorb heat from liquid refrigerant entering the display refrigerator, increasing the cooling capacity of the refrigerant (Figure 22). These devices are most effective for low- and very-low-temperature applications (EPRI 1992). The effectiveness of liquid-to-suction heat exchangers also depends on which refrigerant is chosen. The system designer must be cautious in choosing when to use a liquid-to-suction heat exchanger (Klein et al. 2000).

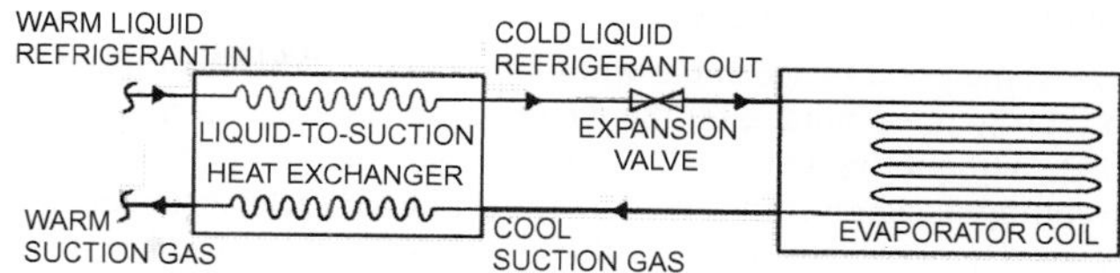

Fig. 22 External Liquid-to-Suction Heat Exchanger
(EPRI 1992)

Sophisticated Refrigerator Controls. All components of a display refrigerator should be linked to one master control system, which can optimally control the operation of individual components.

Power-Reducing Measures. Reducing power use of individual components will result in energy savings over time, and can also reduce the cooling load for components located inside the refrigerated space.

Energy-efficient **evaporator fan motors** such as electronically commutated motors (ECMs) and permanent split capacitor (PSC) motors consume about half the power of standard shaded-pole motors (Faramarzi and Kemp 1999). These motors, located inside the refrigerated space, produce less heat, thereby reducing the load on the refrigeration equipment. These motors also can incorporate variable-speed controls to slow fans as the cooling load is satisfied.

Standard **lighting** equipment, which typically consists of T12 fluorescent lamps with magnetic ballast, draws about 0.73 A at 120 V. More efficient lamps (T8 fluorescent lamps with electronic ballast) draw only 0.49 A at 120 V. As a result, they introduce less heat into the refrigerated space, which in turn reduces the refrigerator cooling load and improves maintenance of target product temperature without sacrificing light quality.

REFRIGERATED STORAGE ROOMS

Meat Processing Rooms

In a self-service meat market, cutting, wrapping, sealing, weighing, and labeling operations involve precise production control and scheduling to meet varying sales demands. The faster the processing, the less critical the temperature and corresponding refrigeration demand.

The wrapping room should not be too dry, but condensation on the meat, which provides a medium for bacterial growth, should be avoided by maintaining a dew-point temperature within a few degrees of the sensible temperature. Fan-coil units should be selected with a maximum of 10°F temperature difference (TD) between the entering air and the evaporator temperature. Low-velocity fan-coil units are generally used to reduce the drying effect on exposed meat. Gravity coils are also available and have the advantage of lower room air velocities.

The meat wrapping area is generally cooled to about 45 to 55°F, which is desirable for workers but not low enough for meat storage. Thus, meat should be held in that room only for cutting and packaging; then, as soon as possible, it should be moved to a packaged product storage cooler held at 28 to 32°F. The meat wrapping room may be a refrigerated room adjacent to the meat storage cooler or one compartment of a two-compartment cooler. In such a cooler, one compartment is refrigerated at about 28 to 32°F and used as a meat storage cooler, and the second compartment is refrigerated at 45 to 55°F and used as a cutting and packaging room. Best results are attained when meat is cut and wrapped to minimize exposure to temperatures above 28 to 32°F.

Wrapped Meat Storage

At some point between the wrapping room and display refrigerator, refrigerated storage for the wrapped cuts of meat must be provided. Without this space, a balance cannot be maintained between the cutting/packaging rate and the selling rate for each particular cut of meat. Display refrigerators with refrigerated bottom storage compartments, equipped with racks for holding trays of meats, offer one solution to this problem. However, the amount of stored meat is not visible, and the inventory cannot be controlled at a glance.

A second option is a pass-through, reach-in cabinet. This cabinet has both front and rear insulated glass doors and is located between the wrapping room and the display refrigerators. After wrapping, the meats are passed into the cabinet for temporary storage at 28 to 32°F and then are withdrawn from the other side for restocking the display refrigerator. Because these pass-through cabinets have glass doors, the inventory of wrapped meats is visible and therefore controllable.

The third and most common option involves a section of the back room walk-in meat storage cooler or a completely separate packaged meat storage cooler. The cooler is usually equipped with rolling racks holding slide-in trays of meat. This method also offers visible inventory control and provides convenient access to both the wrapping room and the display refrigerators.

The overriding philosophy in successful meat wrapping and merchandising can be summarized thus: keep it clean, keep it cold, and keep it moving.

Walk-In Coolers and Freezers

Each category of displayed food product that requires refrigeration for preservation is usually backed up by storage in the back room. This storage usually consists of refrigerated rooms with sectional walls and ceilings equipped with the necessary storage racks for a particular food product. Walk-in coolers are required for storage of meat, some fresh produce, dairy products, frozen food, and ice cream. Medium and large stores have separate produce and dairy coolers, usually in the 35 to 40°F range. Meat coolers are used in all food stores, with storage conditions between 28 and 32°F. Unwrapped meat, fish, and poultry should each be stored in separate coolers to prevent odor transfer. Walk-in coolers, which serve the dual purpose of storage and display, are equipped with either sliding or hinged glass doors on the front. These door sections are often prefabricated and set into an opening in the front of the cooler. In computing refrigeration load, allow for the extra service load.

Moisture conditions must be confined to a relatively narrow range because excessive humidity encourages bacteria and mold growth, which leads to sliming. Too little moisture leads to excessive dehydration.

Air circulation must be maintained at all times to prevent stagnation, but it should not be so rapid as to cause drying of an unwrapped product. Forced-air blasts must not be permitted to strike products; therefore, low-velocity coils are recommended.

For optimum humidity control, unit coolers should be selected at about a 10°F TD between entering air temperature and evaporator temperature. Note that the published ratings of commercial unit coolers do not reflect the effect of frost accumulation on the evaporator. The unit cooler manufacturer can determine the correct frost derating factor for its published capacity ratings. From experience, a minimum correction multiplier of 0.80 is typical.

A low-temperature storage capacity equivalent to the total volume of the low-temperature display equipment in the store is satisfactory. Storage capacity requirements can be reduced by frequent deliveries.

Generally, forced-air coils are selected for low-temperature coolers where humidity is not critical for packaged products. For low-temperature coolers, gas or electric defrost is required. Off-cycle defrosts are used in produce and dairy coolers. Straight time or time-initiated, time- or temperature-terminated gas or electric defrosts are generally used for meat coolers. For more details, see the section on Walk-In Coolers/Freezers in Chapter 47.

REFRIGERATION SYSTEMS

Food stores sell all types of perishable foods that require a variety of refrigeration systems to best preserve and most effectively display each product. Moreover, the refrigerating system must be highly reliable because it must operate 24 h per day for 10 or more years, to protect the large investment in highly perishable foods. Temperature controls vary greatly, from a produce preparation room (which may operate with a wet coil) requiring no defrost to the ice cream refrigerator requiring induced heat to defrost the coil periodically.

Design Considerations

When selecting refrigeration equipment to operate display refrigerators and storage rooms for food stores, consider (1) cost/space limitations, (2) reliability, (3) maintainability and complexity, and (4) operating efficiency. Solutions span the very simple (one compressor and associated controls on one refrigerator) to the complex (central refrigeration plant operating all refrigerators in a store).

Suction Groups. Various refrigerators have different evaporator pressure/temperature requirements. Produce and meat wrapping rooms, which have the highest requirements, may approach the suction pressures used in air-conditioning applications. Open ice cream display refrigerators, which have the lowest, may have suction pressures corresponding to temperatures as low as –40°F. All other refrigerators and coolers fall between these extremes.

Refrigeration Loads. Refrigerator requirements are often given as refrigeration load per unit length, with a lower value sometimes allowed for more complex parallel systems. The rationale for this lower value is that peak loads are smaller with programmed defrost, making refrigerator temperature recovery after defrost less of a strain than on a single-compressor system.

Published refrigerator load requirements allow for extra capacity for temperature pulldown after defrost, per ASHRAE *Standards* 72 and 117. The industry considers a standard store ambient condition to be 75°F and 55% rh, which should be maintained with air conditioning. A portion of this air-conditioning load is carried by the open refrigerators, and credit for heat removed by them should be considered in sizing the air-conditioning system.

Equipment Selection. The designer matches the load requirements of the refrigerator lineups to the capacity of the chosen refrigeration system. Manufacturers publish load ratings to help match the proper refrigeration system with the fixture loads. For single-compressor applications only, the ratings can be stated (for selection convenience) as the capacity the condensing unit must deliver at an arbitrary suction pressure (evaporator temperature). In general, manufacturers of display refrigerators use ASHRAE *Standards* 72 and 117, which specify standard methods of testing open and closed refrigerators for food stores. These standards establish refrigeration load requirements at rated ambient conditions of 75°F and 55% rh in the sales area with specific door-opening patterns. Display refrigerators for similar applications are commercially available from many manufacturers. Manufacturers' recommendations must be followed to achieve proper results in both efficiency and product integrity. Appropriate equipment selection depends on a number of factors.

Life-Cycle Cost. The total cost elements of the refrigeration system include not only the purchase price but also the operating cost (energy), cost of installation and commissioning, cost of maintenance and service, and the environmental cost.

Space Limitations. Store size, location, and price per square foot play a role in determining the type and location of equipment. Locations can include an equipment room at the back of the store, on a mezzanine, in a machine house on the roof, or distributed throughout or on top of the store.

Refrigerant Selection. Selection of a suitable refrigerant for food stores has been affected by international concern about the ozone-depleting effect of chlorine-containing refrigerants. International treaties no longer allow developed nations to manufacture equipment that uses chlorofluorocarbon refrigerants.

Hydrochlorofluorocarbon refrigerants, such as R-22, are still popular while their prices remain low and availability is assured for a reasonable time, although their consumption and production are scheduled to be phased out entirely by 2030. Current hydrofluorocarbon alternatives in the United States include R-404A, R-134a, and R-507. Other refrigerants are listed in ASHRAE *Standard* 34. Secondary loop systems are covered in the section on Low-Charge Systems in this chapter.

Compressor performance and material compatibility are two major concerns in selecting new refrigerants. Research has found good equipment reliability. Retrofit recommendations have also been developed by equipment and refrigerant manufacturers to guide stores in converting from ozone-depleting substances to alternatives; close consultation with equipment manufacturers is necessary to stay current on this issue.

Concern about ozone depletion has led to U.S. Environmental Protection Agency regulations to minimize refrigerant emissions. Intentional venting of all refrigerants, including the substitutes, is prohibited. Additional regulations apply to chlorine-containing refrigerants such as R-22. If systems that contain more than 50 lb of refrigerant leak at an annual rate exceeding 35%, equipment repairs are required. Certain servicing and record-keeping practices are also required (EPA 1990). Proposed regulations extend these regulations to the hydrofluorocarbon substitutes and tighten the leak repair requirements. These developments should be monitored. Chapters 19 and 20 of the 2005 *ASHRAE Handbook—Fundamentals* have more information on refrigerants and their properties.

Refrigerant Lines. Sizing liquid and suction refrigerant lines is critical in the average refrigeration installation, because of the typically long horizontal runs and frequent use of vertical risers. Correct liquid-line sizes are essential to ensure a full feed of liquid to the expansion valve; oversizing must be avoided to prevent system pumpdown or defrost cycles from operating improperly in single-compressor systems.

Proper suction-line sizing is required to ensure adequate oil return to the compressor without excessive pressure drop. Oil separates in the evaporator and moves toward the compressor more slowly than the refrigerant. Unless the suction line is properly installed, oil can accumulate in low places, causing problems such as compressor damage from liquid slugging or insufficient lubrication, excessive pressure drop, and reduced system capacity. To prevent these problems, horizontal suction lines must pitch down as gas flows toward the compressor, the bottoms of all suction risers must be trapped, and refrigerant speed in suction risers must be maintained according to piping practices described in Chapters 2 and 3. To overcome the larger pressure drop necessary in suction risers, suction lines may be oversized on long horizontal runs; however, they still must pitch down toward the compressor for good oil return.

Manufacturers' recommendations and appropriate line sizing charts should be followed to avoid adding heat to either suction or liquid lines. In large stores, both types of lines can be insulated profitably, particularly if subcooling is used.

Typical Systems

Refrigeration systems in use today can generally be categorized into one of the following types: single (a single compressor connected to one or more evaporator loads), multiplex (or parallel compressor) rack, loop, distributed, and secondary refrigerant. Each type has distinct advantages and disadvantages, and may be chosen based on the weight a designer assigns to the different components of equipment life-cycle cost.

The most common compressors used in a typical supermarket refrigeration system include reciprocating, scroll, and screw compressors, which are discussed in Chapter 34 of the 2004 *ASHRAE Handbook—HVAC Systems and Equipment.* Planning load management and sizing the compressors are very important to a successful refrigeration installation.

Single System. A single-compressor/single-evaporator system is sometimes referred to as a *conventional system*. Each compressor may be piped to an individual condenser, or several single compressors may be piped to a larger condenser with multiple circuits. Some single-compressor systems are connected to two or more evaporator systems, in which case each evaporator system uses its own liquid and suction lines and is controlled independently.

A solid-state pressure control for single systems can help control excess capacity when ambient temperature drops. The control senses the pressure and adjusts the cutout point to eliminate short-cycling, which ruins many compressors in low-load conditions. This control also saves energy by maintaining a higher suction pressure than would otherwise be possible and by reducing overall running time.

Multiplex System. Another common refrigeration technique couples two or more compressors in parallel, piped together with common suction and discharge lines. The compressors share a common oil management system and usually operate connected to one or more large condensers. The condensers are usually remotely air- or evaporatively cooled, but they can also be built as part of the compressor rack assembly. The multiplex rack system has several evaporator systems, individually controlled and connected to the compressor rack's common suction line.

Multiple-evaporator systems are usually designed such that each evaporator system operates at a different saturated suction pressure (temperature). Because they are connected to one common suction pressure, the compressors are forced to operate at the lowest evaporator pressure to achieve the coldest evaporator system temperature. The obvious result is a sacrifice in efficiency. Running all the equipment at the low suction pressure required for ice cream (on low-temperature systems) or for meat (on medium-temperature systems) causes all the compressors to operate at lower suction pressures than are necessary. To overcome this inefficiency, large parallel systems frequently isolate ice cream and meat refrigeration. Satellite compressors may be used for extreme loads. The satellite compressor has its own independent suction but shares the rack system's common discharge piping and oil management system. Split-suction manifolds are often used for larger loads: different suction pressures are obtained, but all compressors discharge into a common header and share the oil management system.

Manufacturers should be consulted to determine the appropriate suction pressure (temperature) at the fixture and the load that each system adds to the total. The multiplex rack system must then be designed to deliver the total of all the loads at a common suction pressure no higher than the lowest system pressure requirement less the suction line pressure drop. Systems designed to operate at suction pressures higher than the common must use some means of suction line regulation to prevent higher-temperature evaporators from operating at temperatures below what is necessary to maintain product temperatures.

Suction pressure can be regulated with either electronically [**electric evaporator pressure regulating (EEPR)**] or mechanically actuated [**evaporator pressure regulating (EPR)**] valves. When sized according to manufacturers' recommendations, these valves cause little or no pressure drop in the full-open position. When regulating, they create pressure drop to maintain the fixtures using them at their design condition above the common rack suction pressure. Larger pilot-operated EPR valves may use discharge pressure to open and close the valves, or may be internally piloted, with upstream pressure used to open and close. Although each type has advantages and disadvantages, electric valves are being used more frequently because of their ability to communicate with the rack's energy management system.

The suction gas temperature leaving display fixtures should be superheated to ensure that only vapor enters the compressor suction intake. Particularly on low-temperature fixtures, the suction line gas temperature increase from heat gained from the store ambient can be substantial and adversely affect both refrigeration system capacity and compressor discharge gas temperature. This must be considered for system design. One solution to reduce excessively high superheat is to run the suction and liquid lines tightly together between the fixture and compressor system if the liquid is subcooled, with the pair insulated together for a distance of 30 to 60 ft from the fixture outlet. This technique cannot be used with gas defrost or refrigerants requiring low suction superheat at the compressor suction (for example, low-temperature single-stage R-22 systems). Suction-to-liquid line heat exchangers can be installed in the display fixture. This technique allows the suction gas to pick up heat from the liquid instead of the store ambient. Under all conditions, the suction line should be insulated from the point where it leaves the display refrigerator to the suction service valve on the compressor. The insulation and its installation must be vapor resistant.

To ensure proper thermostatic expansion valve operation, the engineer should verify that the liquid entering the fixture is subcooled. Some refrigerator and/or system designs require liquid-line insulation, which is very important when ambient outdoor air or mechanical subcooling is used to improve system efficiency.

Parallel operation is also applied in two-stage or compound systems for low-temperature applications. Two-stage compression includes interstage gas cooling before the second stage of compression to avoid excessive discharge temperatures. A multiplex rack system with multiple compressors of equivalent capacity is called an *even parallel system*; with compressors of different capacities, it is called an *uneven parallel system*.

Parallel compressor systems must be designed to maintain proper refrigerator temperatures under peak summer load. During the rest of the year, store conditions can be easily maintained at a more ideal condition, and refrigeration load will be lower. In the past, refrigeration systems were operated at 90°F condensing conditions or above to maintain enough high-side pressure to feed the refrigerated display fixture expansion valves properly. When outdoor ambient conditions allow, current technology permits the condensing temperature to follow the ambient down to about 70°F or less. When proper liquid-line piping practices and valve selection guidelines are observed, the expansion valves will feed the evaporators properly under these low condensing pressures (temperatures). Therefore, at partial load, the system has excessive capacity to perform adequately.

Multiple compressors may be controlled or staged based on a drop in system suction pressure. If the compressors are equal in size, a mechanical device can turn off one compressor at a time until only one is running. The suction pressure will be perhaps 5 psi or more below optimum. Microprocessors offer the option of remote control and system operation for all types of compressors, managing compressor cycling and run time for each compressor, and ensuring the common suction pressure is optimized. Satellite compressors can be controlled accurately with one control that also monitors other components, such as oil pressure and alarm functions. To match changing evaporator loads, rack capacity can be varied by cycling compressors, varying the speed of one or more compressors, and/or unloading compressor cylinders by closing valves or moving ports on screw compressors.

Unequally sized compressors can be staged to obtain more steps of capacity than the same number of equally sized compressors. Figure 23 shows seven stages of capacity from a 5, 7, and 10 hp compressor parallel arrangement.

Loop Systems. A loop system is simply a variation of the multiplex rack system. Rather than piping the different evaporator systems (or circuits) back to the machine room, the loop system is designed so that a single suction and liquid "loop" is piped out to the store for each common suction pressure. The individual circuits are then connected to the loop near the fixtures. If EPRs and solenoid valves are used, they will typically be installed nearer the refrigerator lineups.

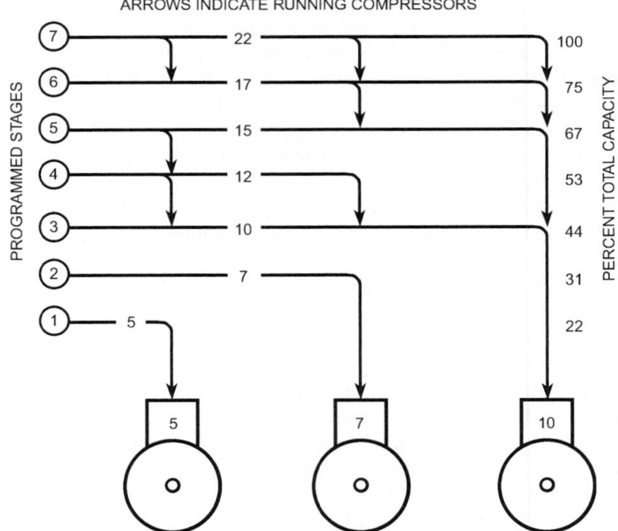

Fig. 23 Stages with Mixed Compressors

Factory-Assembled Equipment. Factory assembly of the necessary compressor systems with either a direct air-cooled condenser or any style of remote condenser is common practice. Both single and parallel systems can be housed, prepiped, and prewired at the factory. The complete unit is then delivered to the job site for placement on the roof or beside the store.

Prefabricated Equipment Rooms. Many supermarket designers choose to have compressor equipment installed in factory-prefabricated housing, commonly called a **mechanical center**, to reduce real estate costs for the building. The time requirements for installation of piping and wiring may also be reduced with prefabrication. Most of the rooms are modular and prewired and include some refrigeration piping. Their fabrication in a factory setting should offer good quality control of the assembly. They are usually put into operation quickly upon arrival at the site.

Energy Efficiency. A typical supermarket includes one or more medium-temperature parallel compressor systems for meat, deli, dairy, and produce refrigerators and medium-temperature walk-in coolers. The system may have a satellite compressor for the meat or deli refrigerators, or all units may have a single compressor. Energy efficiency ratios (EERs) typically range from 8 to 9 Btu/h per watt for the main load. Low-temperature refrigerators and coolers are grouped on one or more parallel systems, with ice cream refrigerators on a satellite or on a single compressor. EERs range from 4 to 5 Btu/h per watt for frozen-food units to as low as 3.5 to 4.0 Btu/h per watt for ice cream units. Cutting and preparation rooms are most economically placed on a single unit because the refrigeration EER is nearly 10 Btu/h per watt. Air-conditioning compressors are also separate because their EERs can range up to 11 Btu/h per watt (Figure 24).

Low-Charge Systems

Over the last decade, different supermarket refrigeration system configurations with lower refrigerant charges have been considered in attempts to mitigate the environmental issues of ozone depletion and global warming. The Montreal Protocol established due dates to

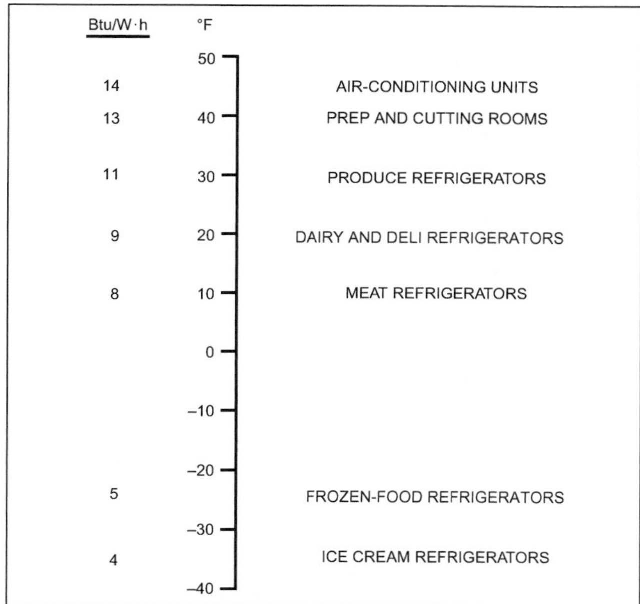

Fig. 24 Typical Single-Stage Compressor Efficiency

phase out different refrigerants worldwide. The production and use of hydrochlorofluorocarbons (HCFCs), such as R-22, in refrigeration systems will be totally phased out by the year 2030. Commercial refrigeration is one of the largest consumers of refrigerant worldwide, and special attention has been devoted to minimizing use of refrigerant in existing and new sites. This section discusses the following three types of low-charge systems: secondary loop, distributed, and liquid-cooled self-contained.

Secondary Loop. In secondary coolant systems, heat is removed from refrigerated spaces and display cabinets by circulating a chilled fluid in a secondary loop cooled by a primary refrigeration system. Fluid circulation is typically provided by a centrifugal pump(s) designed for the flow rate and pressure drop required by the system load and piping arrangement.

Selection of secondary fluid is critical to system efficiency because viscosity and heat transfer properties directly affect system performance. In most cases, the secondary fluid is in a single-phase state, removing heat through a sensible temperature change. Inhibited propylene glycol solutions are most often used for medium-temperature systems, typically at fluid temperatures not lower than 15°F. Low-temperature fluids are commonly composed of solutions of various potassium-based organic salts and inhibitors, though several alternatives are available and corrosion remains a concern. Fluids involving a phase change, including carbon dioxide and water-based ice slurries, are also possible. For an explanation of various options for secondary fluids, including safety considerations, see Chapter 4.

Heat can be removed from the fluid using a chiller of any design, but commonly a plate type is used for highest efficiency. Coils engineered to remove heat effectively from refrigerated spaces are generally designed differently from those used for volatile refrigerants. Liquid should enter the bottom of the coil, leave at the top, and avoid trapping air. Drain and vent valves must also be equipped to assist air removal and service.

Typically, the entire refrigeration system for supermarkets is divided into two temperature groups: one low-temperature (frozen food, ice cream) and one medium-temperature (meat, dairy, produce, preparation rooms). To increase efficiency, the systems may be further divided into additional temperature groups, although often at a higher capital cost. Temperature is controlled by regulating flow using a balance valve, or cycling flow around a set point using a solenoid valve. Piping may be in circuited or loop arrangement, or a combination of the two. Circuited systems have the advantage of containing most of the control valves in a central location, but at the cost of a greater amount of installed piping.

Performance Characteristics. Secondary coolant systems have several advantages. Because primary refrigeration piping is located almost wholly within the machine room, the amount of piping and refrigerant required can be reduced by as much as 80 to 90%. Because field piping of the primary system is typically limited to only a few joints, the majority of the primary system piping joints are factory-installed. Factory-installed joints are generally higher-quality than field-installed joints, because they are formed in controlled conditions by skilled labor, using nitrogen and a variety of pressure-testing and leak-identification methods. Higher-quality joints combined with a lower refrigerant charge can significantly lower refrigerant leakage rates, which reduce the environmental effects associated with the primary refrigerant. The compressors and evaporator are close-coupled, so suction line pressure losses and heat gains are minimized, enhancing system performance. Secondary coolant systems are inherently less complex than direct-expansion types, requiring fewer and less complicated valves and control devices. Less expensive nonmetallic piping systems and components can also be used, because the system operating pressure is low, typically less than 60 psig. Service of the refrigeration system is basically limited to the machine room area, and maintenance costs can be reduced. Because a fluid loop is used, thermal storage may be applied to reduce peak power demands and take advantage of lower off-peak utility rates. Ambient or free cooling may be applied in areas with colder climates. Secondary systems also can use primary refrigerants not typically suitable for direct expansion systems, including ammonia and hydrocarbons.

Disadvantages of secondary systems include thermodynamic loss inherent in the additional step of heat transfer in the chiller, as well as the energy consumed by the fluid pump and the heat it transfers to the circulating fluid. Insulation must also be applied to both coolant supply and return lines to minimize heat gain.

Distributed Systems. Distributed systems eliminate the long lengths of piping needed to connect display fixtures with compressor racks in back-room parallel compressor systems. The compressors are located in cabinets, close-coupled to the display refrigerator lineups, placed either at the end of the refrigerator lineup or, more often, behind the refrigerators around the store's perimeter.

Distributed systems are typically located in the store to provide refrigeration to a particular food department, such as meat, dairy, or frozen food. With this arrangement, the saturated suction temperature (SST) for each rack closely matches the evaporator temperature of the display refrigerators and walk-in coolers. This is not always the case for parallel-rack DX systems, because a single rack often serves display refrigerators with three or four different evaporator temperatures, and the parallel-rack DX system must operate at an SST that will satisfy the requirements of the lowest-temperature one connected. Better evaporator temperature matching with distributed systems can decrease the system's overall energy consumption.

Distributed systems typically require a much lower refrigerant charge than parallel-rack DX systems, because of the former's shorter suction and liquid lines to display refrigerators. Refrigerant piping to remote condensers can be eliminated by using a closed-loop water-cooled system.

Close-coupling display refrigerators to distributed systems has other ramifications for energy consumption. Shorter suction lines mean that pressure drop between evaporators and the compressor suction manifold is less than with parallel-rack DX systems, so the SST of distributed systems will be closer to the display refrigerator evaporator temperature: about 1 to 2°F less than refrigerator evaporator temperature, compared to 2 to 4°F difference with a parallel-rack DX. Shorter suction lines also mean less heat gain.

For a closed-loop water-cooled system, a central pump station contains the circulation pump and all valves needed to control fluid flow between the parallel compressor cabinets and fluid cooler. Inlet and outlet pipes sized for the entire system flow are provided to and from the fluid cooler and pump station. Flow to each distributed system is branched from these central supply and return pipes at a continuous rate; flow to each distributed compressor system is controlled by manual balancing valves set at installation to ensure proper flow to each cabinet.

Liquid-Cooled, Self-Contained Systems. In these systems, refrigeration condensing units connected (underneath, behind, above, or in a nearby enclosure) to one or more refrigerators are located in the display area of the supermarket. A low-temperature fluid or coolant, typically a brine or glycol solution, is pumped through a refrigerant-to-liquid heat exchanger, which serves as the condenser. The heated coolant then flows to a remote refrigeration system or chiller, which removes the heat and then pumps it back out to the refrigerator.

As with other systems, there are advantages and disadvantages. As much as 80% less refrigerant charge is needed, there is less potential for refrigerant loss by leakage, and initial equipment costs may be lower. In addition, refrigerators can be performance-tested before they are shipped from the factory, and installations may be less labor-intensive.

As with secondary cooling systems, the biggest disadvantage is the increased energy requirement from the additional step of heat transfer and the secondary fluid pumps. Noise levels can also be higher, and compressor service must be done in the display area of the supermarket. Advances in compressor technology leading to quieter, more compact, and energy-efficient systems would allow liquid-cooled, self-contained systems to become more feasible low-charge alternatives for widespread applications.

Environmental Considerations: Total Equivalent Warming Impact (TEWI). The environmental benefit of advanced low-charge refrigeration systems is a significant reduction in the amount of halogenated refrigerants now used in supermarkets. Present supermarkets use as much as 3000 lb of refrigerant, most of which is HCFC-22, which has an ozone depletion potential (ODP) of 0.055 and a global warming potential (GWP) of 1700. The latest replacement refrigerants are HFCs, such as R-134a, R-404A, and R-507, which have ODPs of 0, but have high GWP values (1300, 3260, and 3300, respectively).

All refrigeration systems considered here offer better approaches in terms of reduction and containment of refrigerant. There is some variation in charge requirement, depending on the type of heat rejection. The lowest charge is required by systems using a fluid loop for heat rejection. The charge requirement for close-coupled distributed and secondary loop systems is less because of reduced suction-side piping.

The total equivalent warming impact accounts for both direct and indirect effects of refrigeration systems on global warming potential:

$$\underset{\text{(refrigerant leakage and recovery losses)}}{\text{Direct effect}} + \underset{\text{(greenhouse gas emissions from power generation)}}{\text{Indirect effect}} = \text{TEWI}$$

CONDENSING METHODS

Many commercial refrigeration installations use air-cooled condensers, although evaporative or water-cooled condensers with cooling towers may be specified. To obtain the lowest operating costs, equipment should operate at the lowest condensing pressure allowed by ambient temperatures, determined by other design and component considerations; the equipment manufacturer should be consulted for recommendations. Techniques that allow a system to operate satisfactorily with lower condensing temperatures include (1) insulating liquid lines and/or receiver tank, (2) optimum subcooling of liquid refrigerant by design, and (3) connecting the receiver as a surge tank with appropriate valving. Condensing pressure must still be controlled, at least to the lower limit required by the expansion valve, gas defrosting, and heat reclaim. Expansion valve capacity is affected by entering liquid temperature and pressure drop across its port. If selected properly, the thermostatic expansion valve can feed the evaporator at lower pressures, assuming that liquid refrigerant is always supplied to the expansion valve inlet.

To minimize energy consumption, refrigeration condensers should be sized more generously and based on lower TDs than for typical air-conditioning applications. Condenser selection is usually based on the TD between the cooling medium entering the condenser and the saturated condensing temperature.

Condenser Types

Air-Cooled. The remote condenser may be placed outdoors or indoors (to heat portions of the building in winter). Regardless of the arrangement, the following design points are relevant. The air-cooled condenser may be either a single-circuit or a multiple-circuit condenser. The manufacturer's heat rejection factors should be followed to ensure that the desired TD is accommodated.

Pressure must be controlled on most outdoor condensers. Fan-cycle controls work well down to 50°F on condensers with single or parallel groups of compressors. Below 50°F, condenser flooding (using system refrigerant) can be used alone or with fan-cycle controls. Flooding requires a larger refrigerant charge and liquid receiver. In conjunction, splitting condensers with solenoid valves in the hot-gas lines can reduce the condenser surface during cold weather, thereby minimizing the additional refrigerant charge. Natural subcooling can be integrated into the design to save energy.

Fans are controlled by pressure controls, liquid-line thermostats, or a combination of both. Ambient control of condenser fans is common; however, it may not give the degree of condensing temperature control required in systems designed for high-efficiency gain. Thus, it is not recommended except in mild climates down to 50°F. Sometimes, pressure switches, in conjunction with gravity louvers, cycle the condenser fans. This system requires no refrigerant flooding charge.

The receiver tank on the high-pressure side, especially for remote condensers, must be sized carefully. Remote condenser installations, particularly when associated with heat recovery, have substantially higher internal high-side volume than other types of systems. Much of the high side is capable of holding liquid refrigerant, particularly if runs are long and lines are large.

Roof-mounted condensers should have at least 3 ft of space between the roof deck and bottom of the condenser slab to minimize the radiant heat load from the roof deck to the condenser surface. Also, free airflow to the condenser should not be restricted. Remote condensers should be placed at least 3 ft from any wall, parapet, or other airflow restriction. Two side-by-side condensers should be placed at least 6 ft from each other. Chapter 16 of the 2005 *ASHRAE Handbook—Fundamentals* discusses the problems of locating equipment for proper airflow.

Single-unit compressors with air-cooled condenser systems can be mounted in racks up to three high to save space. These units may have condensers sized so that the TD is in the 10 to 25°F range. Optionally available next-larger-size condensers are often used to achieve lower TDs and higher energy efficiency ratios (EERs) in some supermarkets, convenience stores, and other applications. Single compressors with heated crankcases and heated insulated receivers and other suitable outdoor controls are assembled into weatherproof racks for outdoor installations. Sizes range from 0.5 to 30 hp.

Generally accepted TDs for remote air-cooled refrigeration condenser sizing are 15°F for medium-temperature systems and 10°F for low-temperature systems.

Remote air-cooled condensers are popular for use with parallel compressors. Figure 25 illustrates a basic parallel system with an air-cooled condenser and heat recovery coil.

Air-Cooled Machine Room. Standard air-cooled condensing units in a separate air-cooled machine room are still used in some supermarkets. Dampers, which may be powered or gravity-operated, supply air to the room; fans or blowers controlled by room temperature at a thermostat exhaust the air.

A complete indoor air-cooled condensing unit requires ample, well-distributed ventilation. Ventilation requirements vary, depending on maximum summer conditions and evaporator temperature, but 750 to 1000 cfm per condensing unit horsepower has given proper results. Exhaust fans should be spaced for an even distribution of air (Figure 25).

Rooftop air intake units should be sized for 750 fpm velocity or less to keep airborne moisture from entering the room. When condensing units are stacked (as shown in Figure 25), the ambient air design should provide upper units with adequate ventilation. Rooftop intakes are preferred because they are not as sensitive to wind as side wall intakes, especially in winter in cold climates. Butterfly dampers installed in upblast exhaust fans, which are controlled by a thermostat in the compressor room, exhaust warm air from the space.

The air baffle helps prevent intake air from short-circuiting to the exhaust fans (Figure 25). Because air recirculation is needed around the condensers for proper winter control, intake air should not be baffled to flow only through the condensers.

Ventilation fans for air-cooled machine rooms normally do not have a capacity equal to the total of all the individual condenser fans. Therefore, if air is baffled to flow only through the condenser during maximum ambient temperatures, the condensers will not receive full free air volume when all or nearly all condensing units are in operation. Also, during winter operation, tight baffling of the air-cooled condenser prevents recirculation of condenser air, which is essential to maintaining sufficiently high room temperature for proper refrigeration system performance.

Machine rooms that are part of the building need to be airtight so that air from the store is not drawn by the exhaust fans into the machine room. Additional load is placed on the store air-conditioning system if the compressor machine room, with its large circulation of outside air, is not isolated from the rest of the store.

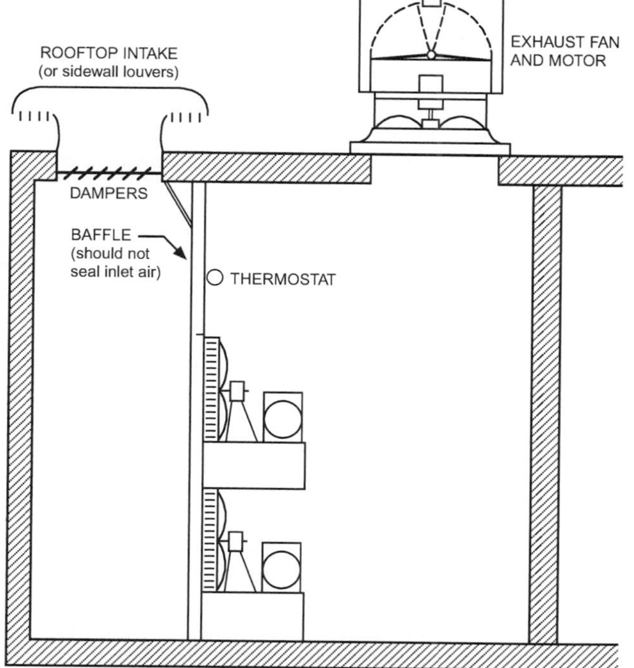

Fig. 25 Typical Air-Cooled Machine Room Layout

Evaporative. Evaporative condensers are equipped with a fan, circulating water-spray pump, and a coil. The circulating pump takes water from the condenser sump and sprays it over the surface of the coil, while the fan introduces an ambient airstream that comes into contact with the wet coil surface. Heat is transferred from condensing refrigerant inside the coil to the external wet surface and then into the moving airstream, principally by evaporation. Where the wet-bulb temperature is about 30°F below the dry-bulb, the condensing temperature can be 10 to 30°F above the wet-bulb temperature. This lower condensing temperature saves energy, and one evaporative condenser can be installed for the entire store. Chapter 35 of the 2004 *ASHRAE Handbook—HVAC Systems and Equipment* gives more details.

Evaporative condensers are also available as single- or multiple-circuit condensers. Manufacturer conversion factors for operating at a given condensing and wet-bulb temperature must be applied to determine the required size of the evaporative condenser.

In cold climates, the condenser must be installed to guard against freezing during winter. Evaporative condensers demand a regular program of maintenance and water treatment to ensure uninterrupted operation. The receiver tank should be capable of storing the extra liquid refrigerant during warm months. Line sizing must be considered to help minimize tank size.

The extremely high temperature of the entering discharge gas is the prime cause of evaporative condenser deterioration. The severity of deterioration can be substantially reduced by using the closed water condensing arrangement. The extent deterioration is reduced depends on how much the difference is reduced between the high discharge gas temperatures experienced even with generously sized evaporative condensers and the design entering water temperature for the closed water circuit.

Water-Cooled. Water-cooled conventional compressor units range in size from 0.5 to 30 hp and are best for hot, dry climates where air-cooled condensers will not operate properly or evaporative condensers are not economically feasible. Water-cooled condensers can also be applied to parallel-compressor systems. A city-water-cooled condensing unit that dumps hot water to a drain is usually no longer economical because of the high cost of water and sewer fees. Cooling towers or evaporative fluid coolers, which cool water for all compressor systems in a single loop, are used instead. If open cooling towers are used to remove heat from condensing water, shell-and-tube heat exchangers must be used, and brazed-plate heat exchangers avoided.

Water flow in the closed water circuit can be balanced between multiple condensers on the same evaporative fluid cooler circuit with water-regulating valves. Usually, low condensing temperatures are prevented by temperature control of the closed water circuit. Three-way valves provide satisfactory water distribution control between condensers.

Fluid Cooler. In a closed-loop water condenser/evaporative cooler arrangement, an evaporative fluid cooler removes heat from water instead of refrigerant. This water flows in a closed, chemically stabilized circuit through a regular water-cooled condenser (a two-stage heat transfer system). Heat from condensing refrigerant transfers to the closed water loop in the regular water-cooled condenser. The warmed water then passes to the evaporative cooler.

The water-cooled condenser and evaporative cooler must be selected considering the temperature differences between the (1) refrigerant and circulating water and (2) circulating water and available wet-bulb temperature. The double temperature difference results in higher condensing temperature than when the refrigerant is condensed in the evaporative condenser. On the other hand, this arrangement causes no corrosion inside the refrigerant condenser itself because the water flows in a closed circuit and is chemically stabilized.

Cooling Tower Arrangements. Few supermarkets use water-cooled condensing units; the trend is instead toward air cooling. Nearly all water-cooled condensing units are installed with a water-saving cooling tower because of the high cost of water and sewage disposal.

Designing water cooling towers for perishable foods is different than for air conditioning because (1) the hours of operation are much greater than for space conditioning; (2) refrigeration is required year-round; and (3) in some applications, cooling towers must survive severe winters. A thermostat must control the tower fan for year-round control of the condensing pressure. The control is set to turn off the fan when the water temperature drops to a point that produces the lowest desired condensing pressure. Water-regulating valves are sometimes used in a conventional manner. Dual-speed fan control is also used.

Some engineers use balancing valves for water flow control between condensers and rely on water temperature control to avoid low condensing temperature. Proper bleed-off is required to ensure satisfactory performance and full life of the cooling towers, condensers, water pumps, and piping. Water treatment specialists should be consulted because each locality has different water and atmospheric conditions. A regular program of water treatment is mandatory.

Energy Efficiency of Condensers

Hardware. Condenser design can significantly affect refrigeration equipment performance. The characteristics of condensers can be improved in three ways:

- **Increased heat transfer effectiveness.** Efficient coils are designed with an increased heat transfer surface area using materials with improved heat transfer properties to reject as much heat to the air as possible, using an optimized fin design.
- **Improved coil tube design: low friction and high conduction.** Materials (e.g., copper) used to construct the coils have increased conductivity, which allows heat to transfer through the coil materials more easily. The inside surface of tubes in the coil can also be enhanced to assist heat transfer from the coil material into the refrigerant: the enhancements create turbulence in the refrigerant, thus increasing its contact time with the tube surface. However, use caution when designing these features because excessive turbulence can cause a pressure drop in the refrigerant and force the compressor to work harder, negating any savings resulting from the enhancement.
- **Downsized fan motor.** Condenser fan motors are can be downsized if coils are efficient. Downsizing the fan motor decreases motor energy use but still allows sufficient heat transfer with the ambient air.

Controls. Allowing discharge pressure to float lower during low-ambient periods of can save considerable energy compared to fixed-pressure systems. Careful system design consideration is needed to ensure proper operation of the expansion valve and refrigerant feed to the evaporator coil during lower ambient conditions. Balanced-port thermostatic expansion valves and electronic expansion valves enhance the opportunity for floating pressures down with varying ambient temperatures.

Noise

Air-cooled condensing units located outdoors, either as single units with weather covers or grouped in prefabricated machine rooms, produce sounds that must be evaluated. The largest source of noise is usually propeller-type condenser fans. Other sources are compressor and fan motors, high-velocity refrigerant gas, general vibration, and amplification of sound where vibration is transmitted to mounting structures (most critical in roof-mounted units).

A fan-speed or fan-cycle control helps control fan air noise by ensuring that only the amount of air necessary to maintain proper condensing temperature is generated. Take care not to restrict discharge air; when possible, it should be discharged vertically upward.

Resilient mountings for fan motors and small compressors and isolation pads for larger motors and compressors help to reduce noise transmission. Proper discharge line sizing and mufflers are the best solution for high-velocity gas noise. Lining enclosures with sound-absorbing material is of minimal value. Isolation pads can help on roof-mounted units, but even more important is choosing the right location with respect to the supporting structure, so that structural vibration does not amplify the noise.

If sound levels are still excessive after these controls have been implemented, location becomes the greatest single factor. Distance from a sensitive area is most important in choosing a location; each time the distance from the source is doubled, the noise level is halved. Direction is also important. Condenser air intakes should face parking lots, open fields, or streets zoned for commercial use. In sensitive areas, ground-level installation close to building walls should be avoided because walls reflect sound.

When it is impossible to meet requirements by adjusting location and direction, barriers can be used. Although a masonry wall is effective, it may be objectionable because of cost and weight. If a barrier is used, it must be sealed at the bottom because any opening allows sound to escape. Barriers also must not restrict condenser entering air. Keep the open area at the top and sides at least equal to the condenser face area.

When noise is a consideration, (1) purchase equipment designed to operate as quietly as possible (e.g., 850 rpm condenser fan motors instead of higher-speed motors), (2) choose the location carefully, and (3) use barriers when the first two steps do not meet requirements.

See Chapter 47 of the 2003 *ASHRAE Handbook—HVAC Applications* for information on outdoor sound criteria, equipment sound levels, sound control for outdoor equipment, and vibration isolation.

HEAT RECOVERY STRATEGIES

Heat recovery may be important in refrigeration system design, parallel or single. Heat recovered from the refrigeration system can be used to heat a store or to heat water used in daily operations. The section on supermarkets in Chapter 2 of the 2003 *ASHRAE Handbook—HVAC Applications* has more information on the interrelation of the store environment and the refrigeration equipment.

Space Heating

Heat reclaim condensers and related controls operate as alternatives to or in series with the normal refrigeration condensers. They can be used in winter to return most of the refrigeration and compressor heat to the store. They may also be used in mild spring and fall weather when some heating is needed to overcome the cooling effect of the refrigeration system itself. Another use is for cooling coil reheat for humidity control in spring, summer, and fall. Excess humidity in the store can increase the display refrigerator refrigeration load as much as 20% at the same dry-bulb temperature, so it must be avoided.

In this application, a heat recovery coil is placed in the air handler for store heat. If the store needs heat, this coil is energized and usually run in series with the regular condenser (see Figure 26). The heat recovery coil can be sized for a 30 to 50°F TD, depending on the capacity in cool weather. Lower condensing temperature in parallel systems allow little heat recovery unless designed properly. When heat is required in the store, simple controls can create the higher condensing temperature needed during heat recovery. Compared with the cost of auxiliary gas or electric heat, the higher

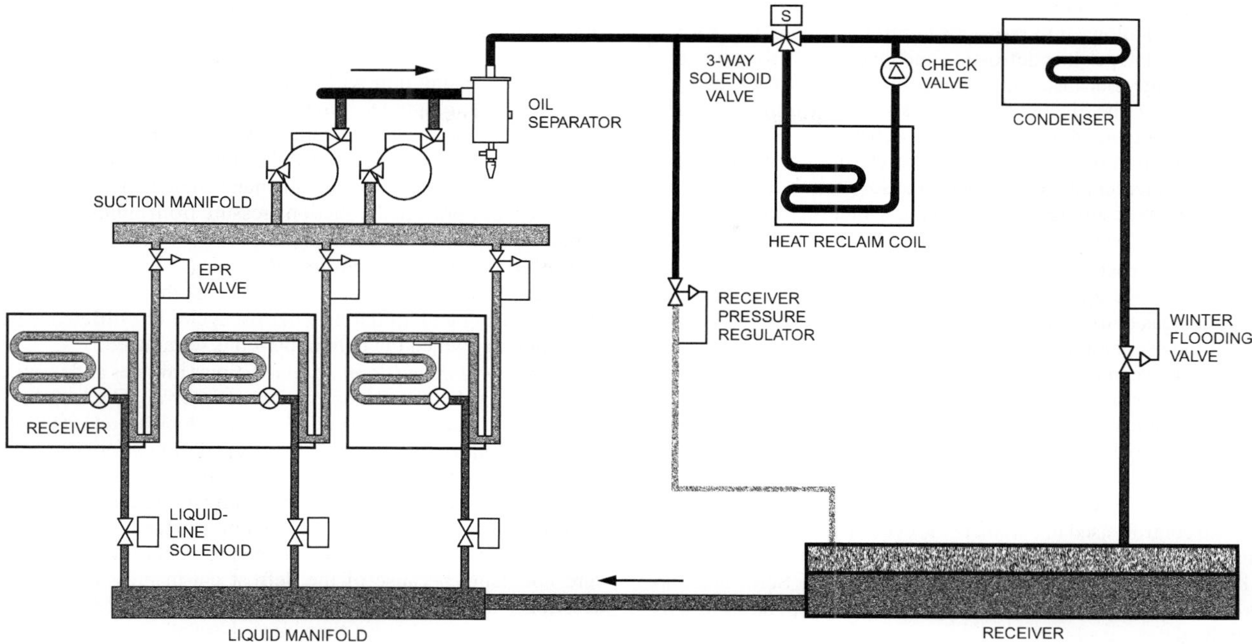

Fig. 26 Basic Parallel System with Remote Air-Cooled Condenser and Heat Recovery

energy consumption of the compressor system may be offset by the value of the heat gained.

Water Heating

Heat reclamation can also be used to heat water for store use. Recovery tanks are typically piped in series with the normal condenser and sized based on the refrigerant pressure drop through the tank and on the water temperature requirements.

On a large, single unit, water can be heated by a desuperheater; on two-stage or compound R-22 parallel systems, water is commonly heated by the interstage desuperheater.

LIQUID SUBCOOLING STRATEGIES

Allowing refrigerant to subcool in cool weather as it returns from the remote condenser can save energy if the system is designed properly. One method is to flood the condenser and allow the liquid refrigerant to cool close to the ambient temperature. The cooler liquid can then reduce the total mass flow requirements if used properly to feed the expansion valves. This may require a diverting valve around the warmer receiver or a special surge-type receiver design.

Mechanical subcooling may also be economical in many areas. This method uses a direct-expansion heat exchanger to cool the main liquid line feeding the evaporator systems. A subcooling satellite compressor can be used on one parallel system, or the medium-temperature rack can be designed with a circuit to handle the subcooling requirement of the low-temperature system. The advantage is that mechanical subcooling is accomplished at higher efficiency than the main system, thus saving energy through year-round liquid temperature control. The mechanical subcooling would be set to operate when the exiting liquid temperature is above the desired setpoint.

Given the wide range of loads on a mechanical subcooler, temperature control can be accomplished in various ways. Two solenoid valves may feed two different-sized thermostatic expansion valves, allowing for multiple stages. This method usually controls subcooling temperature by maintaining the evaporator pressure of the subcooler using a suction regulator. An electric expansion valve with a controller can be used to simplify the piping arrangement and to eliminate the need for suction regulation.

METHODS OF DEFROST

Defrosting is accomplished by latent heat reverse-cycle gas defrosting, selective ingestion of store air, electric heaters, or cycling the compressor. In defrost, particularly for low-temperature equipment, frost in the air flues and around the fan blades must be melted and completely drained.

Defrost methods use (1) off-time, (2) gas, (3) electric, and (4) ambient air induced into the refrigerator.

Parallel systems adapt easily to gas defrost. Compressor discharge gas, or gas from the top of the warm receiver at saturated conditions, flows through a manifold to the circuit requiring defrost. Electric, reverse-air, and off-cycle defrost can be used on both parallel and single-unit systems.

Conventional Refrigeration Systems

Gas Defrost. Gas defrost requires careful design consideration and the use of additional differential valves to keep liquid refrigerant from accumulating in the defrosting evaporator coils. One rule of thumb for gas defrost is that no more than 25% of the circuits can call for defrost at one time, to ensure that enough heat is available from circuits still in refrigeration mode to supply the gas necessary for those in defrost. Given the size of many modern supermarket refrigerator lineups, it is often practical to sequence the gas defrosts such that no two circuits are in defrost at the same time.

Hot-gas defrost uses heat from the compressor's discharge gas to defrost the evaporators. To remove the coil frost, discharge gas is introduced upstream of the suction stop control and directed to the evaporator system calling for defrost. Occasionally, supplemental electric refrigerator heaters are added to ensure rapid and reliable defrosting. Temperature generally terminates the defrost cycle, although a timers are used as a backup.

Saturated-gas defrost is similar to hot-gas defrost but is piped a little differently and uses saturated gas from the top of the liquid in the receiver for defrost purposes.

Off-Cycle or Off-Time Air Defrost. This method simply shuts off the unit and allows it to remain off until the evaporator reaches a temperature that permits defrosting and gives ample time for condensate drainage. Because this method obtains its defrost heat from air circulating in the display fixture, it is slow and limited to open fixtures operating at 34°F or above. Air defrost moves ambient air from the store into the refrigerator. A variety of systems are used; some use supplemental electric heat to ensure reliability. The heat content of the store ambient air during the winter is critical for good results from this method.

Electric Defrost. Electric defrost methods usually apply heat externally to the evaporator and require up to 1.5 times longer to defrost than gas defrost. The heating element may be in direct contact with the evaporator, relying on conduction for defrost, or may be located between the evaporator fans and the evaporator, relying on convection or a combination of conduction and convection for defrost. In both instances, the manufacturer generally installs a temperature-limiting device on or near the evaporator to prevent excessive temperature rise if any controlling device fails to operate.

Electric defrost simplifies installation of low-temperature fixtures. The controls used to automate the cycle usually include one or more of these devices: (1) defrost timer, (2) solenoid valve, (3) electrical contactor, and (4) evaporator fan delay switch. Some applications of open low-temperature refrigerators may operate the fans during the defrost cycle.

Low-Charge Systems

Two defrost methods, time-off and warm fluid, are most commonly applied to secondary systems. Time-off defrost can be used in some medium-temperature applications. However, the most effective method is warm-fluid defrost, which is used for all low-temperature applications and in selected medium-temperature refrigerators where product temperatures are critical or time-off defrost is not practical. Fluid for defrost is typically heated using refrigerant discharge gas, but system efficiency can be increased by heat exchange with liquid refrigerant. Warm fluid temperatures vary and must be optimized for the coil application; however, typical values are 50 to 60°F for medium-temperature systems and 70 to 80°F for low-temperature systems. Warm-fluid defrost is most often terminated by the fluid temperature exiting the coil and is preferable compared to time-off because of the small change in temperature imparted on the products, resulting in lower postdefrost pulldown loads.

Defrost Control Strategies

Defrost control methods include (1) suction pressure control (no time clock required), (2) time clock initiation and termination, (3) time clock initiation and suction pressure termination, (4) time clock initiation and temperature termination, and (5) demand defrost or proportional defrost.

Defrosting is usually controlled by a variety of clocks, which are often part of a compressor controller system. Electronic sensor control is the most accurate and can also provide a temperature alarm to prevent food loss. Electronic systems often have communication capabilities outside the store.

Liquid and/or suction line solenoid valves can be used to control the circuits for defrosting. Often, a suction-stop EPR is used to allow a single valve to isolate the defrosting circuit from the suction manifold and allow introduction of defrost gas upstream of the valve. Individual circuit defrosts are typically controlled by the rack's energy management system, or rack controller.

Suction Pressure Control. This control is adjusted for a cut-in pressure high enough to allow defrosting during the off cycle. This method is usually used in fixtures maintaining temperatures from 36 to 43°F. When the evaporator pressure is lowered to the cutout point of the control, the control initiates a defrost cycle to clear the evaporator.

However, condensing units and/or suction lines may, at times, be subjected to ambient temperatures below the evaporator's temperature. This prevents build-up of suction pressure to the cut-in point, and the condensing unit will remain off for prolonged periods. In such instances, fixture temperatures may become excessively high, and displayed product temperatures will increase.

A similar situation can exist if the suction line from a fixture is installed in a trench or conduit with many other cold lines. The other cold lines may prevent the suction pressure from building to the cut-in point of the control.

Initiation and Termination. Methods (2), (3), and (4) control defrosting using defrost time clocks to break the electrical circuit to the condensing unit, initiating a defrost cycle. The difference lies in the manner in which the defrost period is terminated.

Time Initiation and Termination. A timer initiates and terminates the defrost cycle after the selected time interval. The length of the defrost cycle must be determined and the clock set accordingly.

Time Initiation and Suction Pressure Termination. This method is similar to the first method, except that suction pressure terminates the defrost cycle. The length of the defrost cycle is automatically adjusted to the condition of the evaporator, as far as frost and ice are concerned. However, to overcome the problem of the suction pressure not rising because of the defrost cut-in pressure previously described, the timer has a fail-safe time interval to terminate the defrost cycle after a preset time, regardless of suction pressure.

Time Initiation and Temperature Termination. This method is also similar to the time initiation and termination method, except that temperature terminates the defrost cycle. The length of the defrost cycle varies depending on the amount of frost on the evaporator or in the airstream leaving the evaporator, as detected by a temperature sensor in either location. The timer also has a fail-safe setting in its circuit to terminate the defrost cycle after a preset time, regardless of the temperature.

Demand Defrost or Proportional Defrost. This system initiates defrost based on demand (need) or in proportion to humidity or dew point. Techniques vary from measuring change in the temperature spread between the air entering and leaving the coil, to changing the defrost frequency based on store relative humidity. Other systems use a device that senses the frost level on the coil.

SUPERMARKET AIR-CONDITIONING SYSTEMS

Major components of common store environmental equipment include rooftop packaged units or central air handler with (1) fresh makeup air mixing box, (2) air-cooling coils, (3) heat recovery coils, and (4) supplemental heat equipment. Additional items include (5) connecting ducts, and (6) termination units such as air diffusers and return grilles. Exhaust hoods, used for cooking, can dramatically affect store ventilation rates.

System Types

Constant Volume with Heat Reclaim Coils. This is typically done with one or two large HVAC units. The conditioned air must then be ducted throughout the store.

Multiple Zone. This is typically done with many smaller packaged rooftop units (RTUs), which reduces ductwork but increases electrical and gas infrastructure. Off-the-shelf RTUs do not typically accommodate heat reclaim coils, which is an energy disadvantage in both heating and dehumidification modes.

Comfort Considerations

Open display equipment often extracts enough heat from the store's ambient air to reduce the air temperature in customer aisles to as much as 16°F below the desired level. The air-conditioning return duct system or fans can be used to move chilled air from the floor in front of the refrigerators back to the store air handler. Lack of attention to this element can substantially reduce sales in these areas. This free cooling spilling out of refrigerated cases is commonly referred to as **case credits**. More information on display case effects can be found in the section on Supermarkets in Chapter 2 of the 2003 *ASHRAE Handbook—HVAC Applications* or in Pitzer and Malone (2005).

Interaction with Refrigeration

Rules for good air distribution in food stores are as follows.

Air Circulation. Supply fans operate 100% of the time the store is open, at a volumetric flow of 0.6 to 1 cfm per square foot of sales area. Some chains may have multiple-speed fans, or operate the fan with variable-speed drives (VSDs). Fan speed variation can be based on a number of variables (e.g., store temperature, hood operation, building pressurization, CO_2 level), with the primary objective of minimizing fan energy usage.

Air supply and return grilles must be located so they do not disturb the air in open display refrigerators and negatively affect refrigerator performance. Directional diffusers are helpful in directing air away from cases. Return air can also be positioned to pull treated air into areas with many open refrigerated cases, thus avoiding the higher air speeds created by diffusers.

Outside Air. Introduce outside air whenever the air handler is operating. Supply should meet the required indoor air quality or equal the total for all exhaust fans, whichever is greater, maintaining a positive store pressure. See ASHRAE *Standard* 62.1-2004 for more information on indoor air quality.

Supply Air. Discharge most or all of the air in areas where heat loss or gain occurs. This load is normally at the front of the store and around glass areas and doors.

Return Air. Locate return air registers as low as possible. With low registers, return air temperature may be 50 to 55°F. Low returns reduce heating and cooling requirements and temperature stratification. A popular practice, where store construction allows, is to return air under refrigerator ventilated bases and through floor trenches, or shafts built into walls behind the cases.

Environmental Control

Environmental control is the heart of energy management. Control panels designed for the unique heating, cooling, and humidity control requirements of food stores provide several stages of heating and cooling, plus a dehumidification stage. When high humidity exists in the store, cooling is activated to remove moisture, and the heat reclaim coil may be activated to prevent the store from overcooling. The controller receives input from temperature and dew-point sensors in the sales area. If the store does not need sensible cooling during dehumidification, then a heat reclaim coil is activated to temper the cold, dry air with waste heat from the refrigeration system.

Some controllers include night setback for cool climates and night setup for warm climates. This feature may save energy by modifying the nighttime store temperature, allowing the store temperature to fluctuate several degrees above or below the daytime setpoint temperature. However, store warm-up practices impose an energy use penalty to the display refrigeration systems and affect display case performance, particularly open models.

Energy Efficiency

Energy efficiency must be approached from a total-store perspective. Building envelope, lighting, HVAC, refrigeration, antisweat circuits, indoor air quality (IAQ), human comfort, and local utility cost all must be considered in the store design. Once the store is built and operational, effective commissioning and maintenance practices are critical to keeping energy cost at a minimum.

REFERENCES

ASHRAE. 2004. Designation and safety classification of refrigerants. ANSI/ASHRAE *Standard* 34-2004.

ASHRAE. 2004. Ventilation for acceptable indoor air quality. ANSI/ASHRAE *Standard* 62.1-2004.

ASHRAE. 1998. Method of testing open refrigerators. ASHRAE *Standard* 72-1998.

ASHRAE. 2002. Methods of testing closed refrigerators. ANSI/ASHRAE *Standard* 117-2002.

Arthur D. Little, Inc. 1996. *Energy savings potential for commercial refrigeration equipment—Final report.* Prepared for the U.S. Department of Energy.

CEC. 2004. Final report—Investigation of secondary loop supermarket refrigeration systems. *Report* 500-04-013. California Energy Commission.

Dossat, R. 1997. *Principles of refrigeration*, 4th ed. Prentice Hall, Upper Saddle River, NJ.

EIA. 2003. *1999 commercial building energy consumption survey.* U.S. Department of Energy, Energy Information Administration, Washington, D.C.

EPA. 1990. *Clean Air Act of 1990.* U.S. Environmental Protection Agency, Washington, D.C.

Faramarzi, R. 1999. Efficient display case refrigeration. *ASHRAE Journal* (November):46.

Faramarzi, R. 2000. Analyzing air curtain performance in a refrigerated display case. Seminar, ASHRAE Annual Meeting (June). Minneapolis.

Faramarzi, R. 2003. Effects of improper product loading on the performance of an open vertical meat display case. *ASHRAE Transactions* 109(1): 267-272.

Faramarzi, R., B. Coburn, and R. Sarhadian. 2001. Anti-sweat heaters in refrigerated display cases. *ASHRAE Journal* (June):64.

Faramarzi, R. and K. Kemp. 1999. Testing the old with the new. *Engineered Systems* (May):52.

Faramarzi, R. and M. Woodworth. 1999. Effects of the low-e shields on performance and power use of a refrigerated display case. *ASHRAE Transactions* 105(1):533-540.

FDA. 2001. *Food Code.* Food and Drug Administration, U.S. Department of Health and Human Services, Washington, D.C.

Food Marketing Institute, Inc. 2004. *Key industry facts.* http://www.fmi.org/facts_figs/.

Gas Research Institute. 2000. *Investigation of relative humidity impacts on the performance and energy use of refrigerated display cases.* Chicago.

Howell, R.H. 1993a. Effects of store relative humidity on refrigerated display case performance. *ASHRAE Transactions* 99(1):667-678.

Howell, R.H. 1993b. Calculation of humidity effects on energy requirements of refrigerated display cases. *ASHRAE Transactions* 99(1):679-693.

Komor, P., C. Fong, and J. Nelson. 1998. *Delivering energy services to supermarkets and grocery stores.* E Source, Boulder, CO.

Pitzer, R.S. and M.M. Malone. 2005. Case credits & return air paths for supermarkets. *ASHRAE Journal* 47(2):42-48.

CHAPTER 47

FOOD SERVICE AND GENERAL COMMERCIAL REFRIGERATION EQUIPMENT

FOOD service requires refrigerators that meet a variety of needs. This chapter covers refrigerators available for restaurants, fast-food restaurants, cafeterias, commissaries, hospitals, schools, convenience stores, grocery stores, and other specialized applications.

Many refrigeration products used in food service applications are self-contained, and the corresponding refrigeration systems are conventional. Some systems, however, do use ice for fish, salad pans, or specialized preservation and/or display. Chapters 46 and 48 have further information on some of these products.

Generally, electrical and sanitary requirements of refrigerators are covered by criteria, standards, and inspections of Underwriters Laboratories (UL), NSF International, and the U.S. Public Health Service.

REFRIGERATED CABINETS

Reach-In Cabinets

The **reach-in refrigerator** or freezer is an upright, box-shaped cabinet with straight vertical front(s) and hinged or sliding doors (Figure 1). It is usually about 2.5 to 3 ft deep and 6 ft high and ranges in width from about 3 to 10 ft. Capacities range from about 20 to 90 ft^3. Undercounter models 3 ft high with the same dimensions are also available. These capacities and dimensions are standard from most manufacturers.

The typical reach-in cabinet (Figure 1) is available in many styles and combinations, depending on its intended application. Other shapes, sizes, and capacities are available on a custom basis from some manufacturers. Chapter 46 discusses display cabinets in great detail.

There are many varied adaptations of refrigerated spaces for storing perishable food items. Reach-ins, by definition, are medium- or low-temperature refrigerators small enough to be moved into a building. This definition also includes refrigerators and freezers built for special purposes, such as mobile cabinets or refrigerators on wheels and display refrigerators for such products as beverages, pies, cakes, and bakery goods. The latter cabinets usually have glass doors and additional lighting to illuminate the product. Candy refrigerators are also specialized in size, shape, and temperature.

Refrigerated vending machines satisfy the general definition of reach-ins; however, because they also receive coins and dispense products individually, they are classified separately. Generally, the full product load of a vending machine is not accessible to the customer as in normal reach-in cabinets. Beverage-dispensing units dispense a measured portion into a cup rather than in a bottle or can.

Reach-in refrigerators have doors on the front. Refrigerators that have doors on both front and rear are called **pass-through** or **reach-through refrigerators** (Figure 2). Doors are either full height (one per section) or half height (two per section). Doors may have windows or be solid, hinged, or sliding.

Roll-In Cabinets

Roll-in cabinets are very similar in style and appearance to reach-in cabinets, but vary slightly in construction and functionality. Roll-ins (Figure 3) are usually part of a food-handling or other

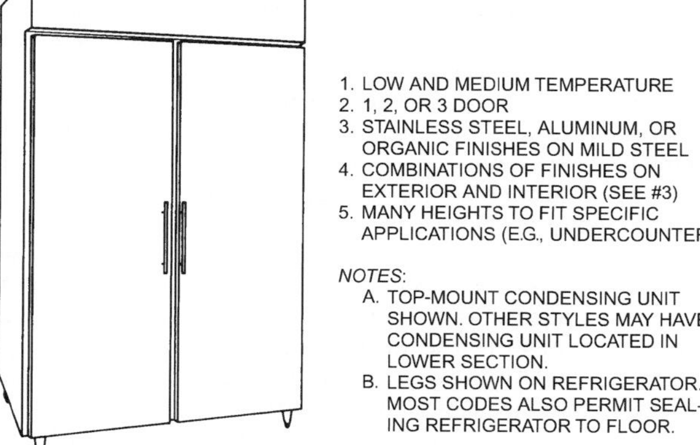

1. LOW AND MEDIUM TEMPERATURE
2. 1, 2, OR 3 DOOR
3. STAINLESS STEEL, ALUMINUM, OR ORGANIC FINISHES ON MILD STEEL
4. COMBINATIONS OF FINISHES ON EXTERIOR AND INTERIOR (SEE #3)
5. MANY HEIGHTS TO FIT SPECIFIC APPLICATIONS (E.G., UNDERCOUNTER)
NOTES:
A. TOP-MOUNT CONDENSING UNIT SHOWN. OTHER STYLES MAY HAVE CONDENSING UNIT LOCATED IN LOWER SECTION.
B. LEGS SHOWN ON REFRIGERATOR. MOST CODES ALSO PERMIT SEALING REFRIGERATOR TO FLOOR.

Fig. 1 Reach-In Food Storage Cabinet Features

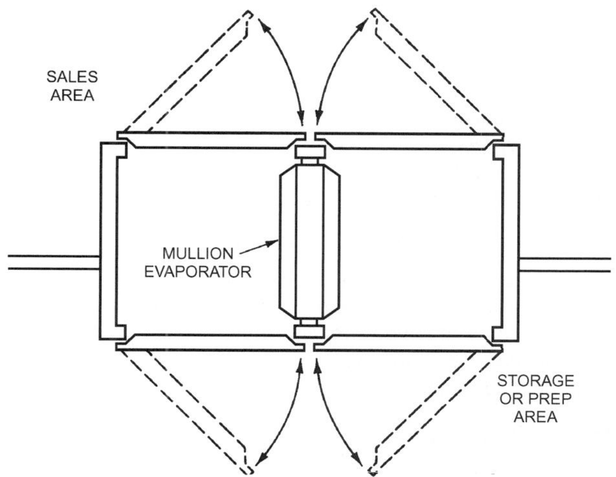

Fig. 2 Pass-Through (Reach-Through) Refrigerator

The preparation of this chapter is assigned to TC 10.7, Commercial Food and Beverage Cooling, Display and Storage.

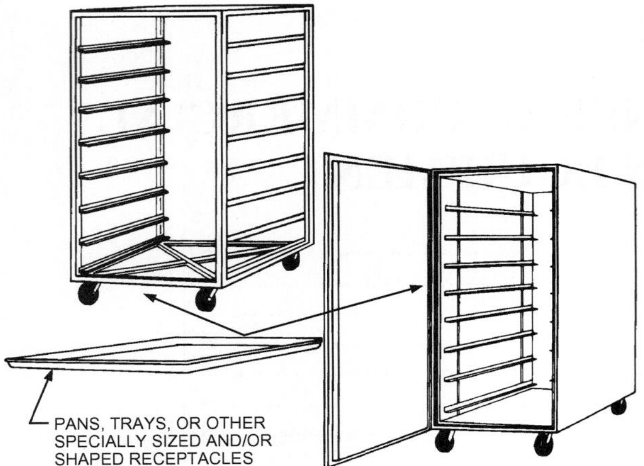

Fig. 3 Open and Enclosed Roll-In Racks

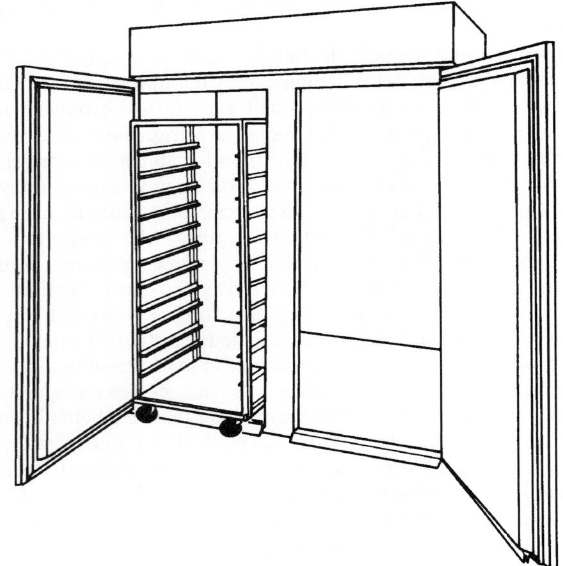

Fig. 4 Roll-In Cabinet, Usually Part of Food-Handling or Other Special-Purpose System

special-purpose system (Figure 4). Pans, trays, or other specially sized/shaped receptacles are used to serve a specific system need, such as the following:

- Food handling for schools, hospitals, cafeterias, and other institutional facilities
- Meal manufacturing
- Bakery processing
- Pharmaceutical products
- Body parts preservation (e.g., blood)

The roll-in differs from the reach-in in the following ways:

- The inside floor is at about the same level as the surrounding room floor, so wheeled racks of product can be rolled directly from the surrounding room into the cabinet interior.
- Cabinet doors are full height, with drag gaskets at the bottom.
- Cabinet interiors have no shelves or other similar accessories.

Product Temperatures

Refrigerators are available for medium- or low-temperature ranges. The medium-temperature range has a maximum of 41°F and a minimum of 33°F core product temperature, with the most desirable average temperature close to 38°F. Low-temperature refrigerators cover a range of core product temperatures between –10 and +10°F. The desirable average core product temperature is 0°F for frozen foods and –5°F for ice cream. Both temperature ranges are available in cabinets of many sizes, and some cabinets combine both ranges.

Typical Construction

Refrigerators are available in two basic types of construction. The older style is a wood frame substructure clad with a metal interior and exterior. The newer style is a welded assembly of exterior panels with insulation and liner inserts.

Exterior. Materials used on exteriors (and interiors) are stainless steel, painted steel, aluminum-coated steel, aluminum, and vinyl-clad steel with wood grain or other patterns. The requirements are for a material that (1) matches or blends with that used on nearby equipment; (2) is easy to keep clean; (3) is not discolored or etched by commonly used cleaning materials; (4) is strong enough to resist denting, scratching, and abrasion; and (5) provides the necessary frame strength. The material chosen by an individual purchaser depends a great deal on layout and budget.

Interior. Shelves, usually three or four per full-height section, are standard interior accessories. Generally, various types of shelf standards are used to provide vertical shelf adjustment.

Racks for roll-in cabinets are generally fitted with slides to handle 18 by 26 in. pans, although some newer systems call for either 12 by 20 in. or 12 by 18 in. steam table pans. Racks designed for special applications are available but usually custom designed.

Manufacturers and contractors offer various methods of floor insulation. This is important if the roll-in holds frozen food.

Specialty Applications

Reach-in and roll-in cabinets are regularly modified and adapted to fit the needs of many specialty applications. Variations from standard construction practices are needed to meet the different temperature, humidity, product volume, cleanliness, and other specifications of various refrigeration applications.

Food Service. These applications often require extra shelves or tray slides, pan slides, or other interior accessories to increase food-holding capacity or make operation more efficient. Because certain stored foods create a corrosive atmosphere in the enclosure, the evaporator coil may have special coatings or fin materials to prevent oxidation. As use of foods prepared off-premises increases, on-site storage cabinets are becoming more specialized; there is growing pressure for designs that consider new food shapes, as well as in-and-out handling and storage.

Beverage Service. If reach-ins are required, standard cabinets, are used, except when glass doors and special interior racks are needed for chilled product display. These cabinets generally have oversized refrigeration systems to allow for product pulldown cooling.

Meal Factories. These applications, which include airline or central feeding commissaries, require rugged, heavy-duty equipment, often fitted for bulk in-and-out handling.

Retail Bakeries. Special requirements of bakeries are the dough retarder refrigerator and the bakery freezer, which permit the baker to spread the work load over the entire week and to offer a greater variety of products. The recommended temperature for a dough retarder is 36 to 40°F. The relative humidity should be in excess of 80% to prevent crusting or other undesirable effects. In the freezer, the temperature should be held at 0°F. All cabinets or wheeled racks should be equipped with racks to hold 18 by 26 in. bun pans, which are standard throughout the baking industry.

Retail Stores. Stores use reach-ins for many different nonfood items. Drugstores often have refrigerators with special drawers for storing biological compounds. (See the section on Nonfood Installations.)

Retail Florists. Florists use reach-in refrigerators for displaying and storing flowers. Although a few floral refrigerator designs are considered conventional in the trade, the majority are custom built. The display refrigerator in the sales area at the front of the shop may include a picture window display front and have one or more display access doors, either swinging or sliding. A variety of open refrigerators may also be used.

For the general assortment of flowers in a refrigerator, most retail florists have found best results at temperatures from 40 to 45°F. The refrigeration coil and condensing unit should be selected to maintain high relative humidity. Some florists favor a gravity cooling coil because the circulating air velocity is low. Others, however, choose forced-air cooling coils, which develop a positive but gentle airflow through the refrigerator. The forced-air coil has an advantage when in-and-out service is especially heavy because it provides quick temperature recovery during these peak conditions.

Nonfood Installations. Various applications use a wide range of reach-ins, some standard except for accessory or temperature modifications and some completely special. Examples include (1) biological and pharmaceutical cabinets; (2) blood bank refrigerators; (3) low- and ultralow-temperature cabinets for bone, tissue, and red-cell storage; and (4) specially shaped refrigerators to hold column chromatography and other test apparatus.

Blood bank refrigerators for whole blood storage are usually standard models, ranging in size from under 20 to 45 ft^3, with the following modifications:

- Temperature is controlled at 37 to 41°F.
- Special shelves and/or racks are sometimes used.
- A temperature recorder with a 24 h or 7 day chart is furnished.
- An audible and/or visual alarm system is supplied to warn of unsafe blood temperature variation.
- An additional alarm system may be provided to warn of power failure.

Biological, laboratory, and mortuary refrigerators involve the same technology as refrigerators for food preservation. Most biological serums and vaccines require refrigeration for proper preservation and to retain highest potency. In hospitals and laboratories, refrigerator temperatures should be 34 to 38°F. The refrigerator should provide low humidity and should not freeze. Storage in mortuary refrigerators is usually short-term, normally 12 to 24 h at 34 to 38°F. Refrigeration is provided by a standard air- or water-cooled condensing unit with a forced-air cooling coil.

Items in biological and laboratory refrigerators are kept in specially designed stainless steel drawers sized for convenient storage, labeled for quick and safe identification, and perforated for proper air circulation.

Mortuary refrigerators are built in various sizes and arrangements, the most common being two- and four-cadaver self-contained models. The two-cadaver cabinet has two individual storage compartments, one above the other. The condensing unit compartment is above and indented into the upper front of the cabinet; also, ventilation grills are on the front and top of this section. The four-cadaver cabinet is equivalent to two two-cadaver cabinets set together; the storage compartments are two cabinets wide by two cabinets high, with the compressor compartment above. Six- and eight-cadaver cabinets are built along the same lines. The two-cadaver refrigerator is approximately 38 in. wide by 94 in. deep by 77 in. high and is shipped completely assembled.

Each compartment contains a mortuary rack consisting of a carriage supporting a stainless steel tray. The carriage is telescoping, equipped with roller bearings so that it slides out through the door opening, and is self-supporting even when extended. The tray is removable. Some specifications call for a thermometer to be mounted on the exterior front of the cabinet to show the inside temperature.

Refrigeration Systems

Reach-in cabinets can be supported by either remote or self-contained refrigeration systems. The following two types of systems apply to all types of refrigeration equipment.

Self-contained systems, in which the condensing unit and controls are built into the refrigerator structure, are usually air-cooled and are of two general types. The first type has the condensing unit beneath the cabinet; in some designs it takes up the entire lower part of the refrigerator, whereas in others it occupies only a corner at one lower end. The second type has the condensing unit on top.

Remote refrigeration systems are often used if cabinets are installed in a hot or otherwise unfavorable location where noise or heat of the condensing units would be objectionable. Other special circumstances may also make remote refrigeration desirable.

There are tradeoffs associated with locating a self-contained condensing unit beneath the refrigerator; although the air near the floor is generally cooler, and thus beneficial to the condensing unit, it is usually dirtier. Putting the condensing unit on top of the cabinet allows full use of cabinet space, and, although air passing over the condenser may be warmer, it is cleaner and less obstructed. Having the condensing unit and evaporator coil in the same location provides a refrigeration unit that can be removed, serviced, and replaced in the field as a whole. Servicing can then be done at an off-site repair facility.

FOOD FREEZERS

Some hospitals, schools, commissaries, and other mass-feeding operations use on-premises freezing to level work loads and operate kitchens efficiently on normal schedules. Industrial freezing equipment is usually too large for these applications, so operators use either regular frozen food storage cabinets for limited amounts of freezing or special reach-ins that are designed and refrigerated to operate as batch-type blast freezers.

Chapter 16 covers industrial freezing of food products.

BLAST CHILLERS AND BLAST FREEZERS

These types of units are designed to rapidly chill or freeze food immediately after it has been cooked. Blast chillers and freezers are used by food-service establishments, such as restaurants, hotels, and cafeterias, that cook large quantities of food items, chill or freeze them, and later reheat portions to be served. Blast chillers are designed to allow operators to comply with food preparation, handling, and storage guidelines on preventing the growth of dangerous bacteria. These guidelines mandate that food be cooked to a minimum core temperature of 160°F and held there for at least 2 minutes. The food is then immediately cooled to between 33 and 38°F within 2 to 4 hours. This not only prevents bacterial growth, but also helps preserve the appearance, flavor, texture, and nutritional value of the food. Once cooled, refrigerated food must be stored at a temperature range of 33 to 38°F for a period not to exceed 5 days. Frozen food must be maintained below 0°F and can be kept for 8 weeks or longer.

Blast chillers for refrigerated food, and blast freezers for frozen food, are available in reach-in and roll-in models in a variety of sizes and capacities. They are designed to operate both as blast chillers and as storage refrigerators or freezers. Most units automatically change over to storage mode when the blast-chill cycle is completed. Many models are equipped with sophisticated microprocessor control systems that allow the operator not only to program the chill cycle, but also to obtain readouts, printouts, and alarms that document and monitor the entire process. Built-in food probes are commonly used to take readings and allow the control system to make adjustments if necessary.

WALK-IN COOLERS/FREEZERS

Walk-in coolers/freezers are used in a wide variety of applications, but food sales and service facilities dominate all other uses.

This type of commercial refrigerator is a factory-made, prefabricated, modular version of the built-in, large-capacity cooling room.

A walk-in cooler's function is to store foods and other perishable products in larger quantities and for longer periods than reach-in refrigerators/freezers. Good refrigeration practice requires storing dissimilar unpackaged foods in separate rooms because they require different temperatures and humidity and to prevent odors from some foods from being absorbed by others. The food cooler/freezer is likely to be equipped with sturdy, adjustable shelving about 18 in. deep and arranged in tiers, three or four high, around the inside walls; another common option is rolling racks (basically shelving on wheels), which are rolled directly into and out of the cooler. Large food operations may have three rooms: one for fruits and vegetables, one for meats and poultry, and one for dairy products. A fourth room, at 0°F, may be added for frozen foods. Smaller food operations that use appropriate food packaging may require only two rooms: one for medium-temperature refrigeration and one for frozen storage.

Operating Temperatures

There are two major temperature classes of walk-ins: low (–20 to –10°F) and medium (–10 to 30°F). Coolers may be used to hold sides or quarters of beef, lamb carcasses, crates of vegetables, and other bulky items. Food operations now rarely use such items. If they do, the items are broken down, trimmed, or otherwise processed before entering refrigerated storage. The modern cooler is not a storage room for large items, but a temporary place for quantities of small, partially or totally processed products.

Typical Construction

Walk-in coolers/freezers are composed of prefabricated panels, which come in a variety of sizes that are shipped to the operator and assembled on site. The edges of the panels are usually of tongue-and-groove construction and either fitted with a gasket material or provided with suitable caulking material to ensure a tight vapor seal when assembled. In most cases, the panels, refrigeration components, and controls are ordered separately and assembled on site, although some smaller units are supplied fully assembled. Sizes are typically 80 to 750 ft^2 and 8 to 10 ft high, or about 640 to 7500 ft^3; the average size appears to be in the range of 2000 ft^3. The modular walk-in cooler/freezer offers flexibility over the built-in type. It can be easily erected and moved, and readily altered to meet changing requirements, uses, or layouts by adding standard sections. Modular walk-in coolers/freezers can be erected outside a building, providing more refrigerated storage with no building costs except for footings and an inexpensive roof supported by the cooler. Exterior and interior surfaces may be painted and can be made of galvanized steel, aluminum, aluminum-coated steel, stainless steel, or vinyl-clad steel.

The frames are filled with insulation and are covered with metal on both sides. Polyurethane and polystyrene are two common types of insulation. These foam plastic materials are both light and water-resistant, and have improved thermal insulation in both self-contained and remotely refrigerated sectional coolers.

Door Construction

A variety of optional accessories simplify opening and closing doors. Triple-pane windows (heated on freezers), digital thermometers, and light switches allow an operator to locate or inspect the contents of the cooler/freezer without entering. An interior/exterior kick plate provides protection when using ones foot to kick open or close the door. Cam-lift hinges allow the door to swing open easily, and should be coupled with an automatic door closer.

Walk-In Floors

Modular (insulated) floors can be purchased and integrated into walk-in coolers/freezers, but the raised entry level requires a step or ramp entry. Walk-in coolers (medium-temperature) called *floorless* by the supplier are furnished with floor splines to fasten to the existing floor to form a base for the wall sections. Generally, floor refrigeration losses are considered small in a floorless configuration.

Level entry is becoming more important as the use of hand and electric trucks increases. The advantage and convenience of level entry afforded by a floorless cooler can also be obtained by recessing a sectional insulated floor. Walk-in coolers (medium-temperature) are the only applications that can be floorless or mounted on slab concrete. Walk-in freezers must feature an insulated floor.

Refrigeration Systems

Walk-ins can be served by remote or self-contained refrigeration equipment (see the section on Refrigerated Cabinets for more details). There are various methods of application for self-contained refrigeration units (Figure 5). These self-contained units use complete refrigeration systems, usually air-cooled, in a single compact package. The units are installed in the sectional cooler/freezer wall or ceiling panels.

Compressors

Walk-in coolers/freezers typically use either hermetic or semihermetic compressors. **Hermetic compressors** are typically rated between 1/2 and 5 hp. The compressor and motor are sealed in a gastight shell; when repairs are needed, the shell must be cut open and then resealed by welding it closed. Typical applications for the hermetic compressor are smaller food service, ice machines, and beverage dispensers. **Semihermetic compressors** can have ratings between 1/4 and 15 hp. They need less maintenance and have longer life cycles (up to three times the life of hermetics), and provide a greater cooling capacity. Furthermore, the unit not being housed in a shell allows for easy serviceability. Semihermetic compressors are typically used in larger cooling applications. See Chapter 34 of the 2004 *ASHRAE Handbook—HVAC Systems and Equipment* for more compressor information.

Evaporators

Evaporator coils are constructed of rust-free aluminum housing containing staggered copper tubes expanded into corrugated aluminum fins. The evaporator fan motors are either 115 or 230 V. Walk-in coolers (medium-temperature) typically feature an off-cycle

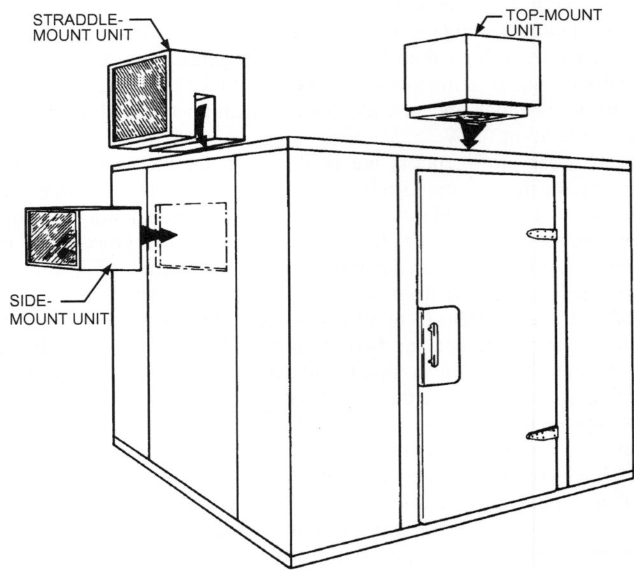

Fig. 5 Refrigeration Equipment Added to Make a Walk-In Cooler Self-Contained

defrost, whereas freezers (low-temperature) use electric defrost or hot-gas methods.

Refrigeration Sizing

When sizing the refrigeration system for a walk-in cooler or freezer, consider the following factors:

- Heat transmission (heat gained through the walls, floors, and ceiling)
- Air infiltration
- Product load (heat removed from a product to cool it, and heat of respiration from some fruits and vegetables)
- Supplemental loads (heat dissipated by people or mechanical equipment inside the cooler/freezer)
- Post-defrost pulldown load

Maintenance and Operation

- Clean the condenser coil quarterly. This will increase the refrigerator's efficiency and extend the life of the equipment by reducing compressor run time. Also, make sure air moves freely around the compressor and condenser to help the system disperse heat.
- To prevent warm, moist air infiltration into the cooler, ensure that the gaskets are in good condition, the door's hinges are lubricated, the auto-door closers are working, and, for walk-in coolers, that strip curtains have been installed.
- Keep the refrigerator's evaporator coils clean, and check them on a regular schedule. Also, check for plastic bags, which can get caught against the evaporator, on the back side of walk-in cooler units.
- Maintain proper loading of products inside the cooler to avoid blocking air circulation.
- Keep the refrigerator's evaporator drain line clean and open. Often, pooled water on the bottom a cooler results from a plugged drain line. This makes the unit work harder and longer, thus consuming more energy.
- In walk-in freezers, place heat tape in the evaporator's drain line to ensure proper drainage during defrost. In addition, make sure insulation is installed over the evaporator's drain line with heat tape, to reduce the amount of heat load inside the freezer.
- Make sure defrost time clocks are set properly and do not defrost more than necessary. Avoid defrosting during peak demand rate times, (typically between noon to 6:00 PM), when scheduled deliveries arrive, or when employees place a heavy use load on the freezer. When a freezer goes into its defrost cycle, heat is added to the evaporator (to defrost) and the compressor will not cycle on regardless of the cooler's temperature. Thus, avoiding adding heat to the freezer during the defrost times reduces a freezer's energy consumption.
- Turn off lights inside glass-door coolers at night, when customers are not present, and leave them off all of the time in noncustomer areas. This saves energy and add less heat load to the cooler.
- Ensure high integrity of vapor barriers and insulated panels. Check regularly for punctured or broken panels and breaches around pipe penetrations through the panels.

VENDING MACHINES

Refrigerated vending machines are designed to store food and beverages at a prescribed temperature and dispense product in exchange for currency. The U.S. refrigerated vending machine population is approximately 4,100,000, of which about 87% dispense canned and bottled beverages (Figure 6). The average energy usage for a typical vending machine is 3000 kWh/year and is affected by many factors (A.D. Little 1996). Generally, site owners are responsible for energy costs.

Types of Refrigerated Vending Machines

Vending machines can be divided into two main categories: closed-front and glass-front. **Closed-front units** house products inside a completely opaque insulated compartment. Some models have a display window where sample products are placed in view, but the products to be vended are contained behind an insulated door and cannot be seen by the consumer. These machines typically have a full-sized illuminated advertisement panel on the front. **Glass-front units** have a translucent panel that enables the purchaser to see the product as it is vended. In this type of machine, the product itself is illuminated and used to attract the purchaser's attention. There are various vending configurations but all machines fit into one of these categories.

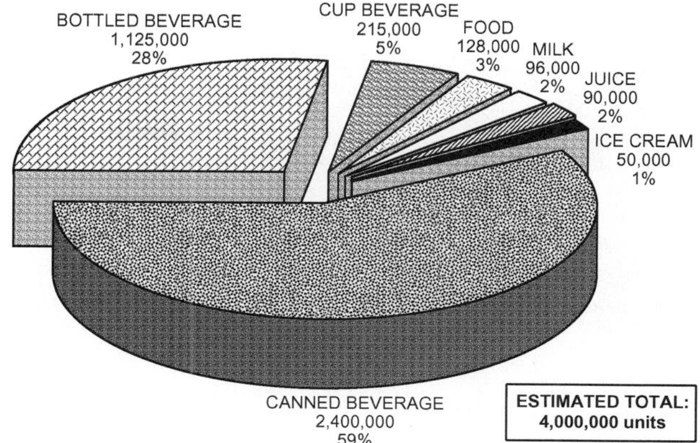

Fig. 6 Estimated 1994 Breakdown of Beverage Vending Machines by Type
(*Source*: A.D. Little 1996)

Refrigeration Systems

Vending machines use simple, self-contained refrigeration systems consisting of a compressor, evaporator coil, condenser, and capillary tube. The compressor and condenser are typically located at the bottom of the unit, between the ground and the refrigerated cabinet. Refrigerant is piped through the capillary tube to an evaporator inside the refrigerated cabinet.

The compressor consumes by far the most energy in the vending machine. Laboratory test results (Faramarzi 2005) indicate the compressor accounts for 65 to 75% of the total energy consumed (Figure 7). Oversized single-speed compressors are commonly used to provide excess cooling capacity to pull down product temperatures quickly after the machine is restocked. A condenser fan operates concurrently with compressor cycling and uses less than 10% of the total energy. The evaporator fan runs continuously to distribute air through the refrigerated cabinet and consumes about 8% of total machine energy. Lighting systems usually remain on the entire time a vending machine is plugged in. Different lighting systems are used, depending on the machine, but usually consume 5 to 20% of the total energy. Dispensing mechanisms operate intermittently, whenever a product is purchased, and therefore do not use a significant amount of energy (1 to 3%) compared to other components.

Cooling Load Components

Heat enters the refrigerated section of the cabinet by the following methods:

- Released by the evaporator fan motor
- Released by the lighting (glass-front units only)
- Conduction through the insulated envelope
- Infiltration through imperfections in construction such as cracks in the box or improperly sealed delivery doors
- Radiation through the glass front (glass-front units only)
- Product restocking (warm product placed into the machine must be cooled to acceptable temperature)

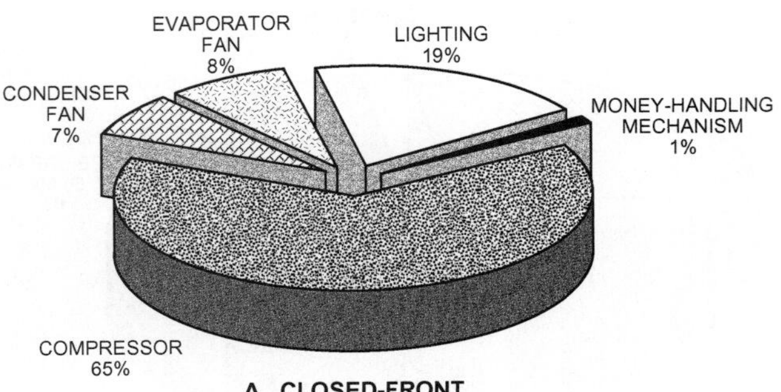

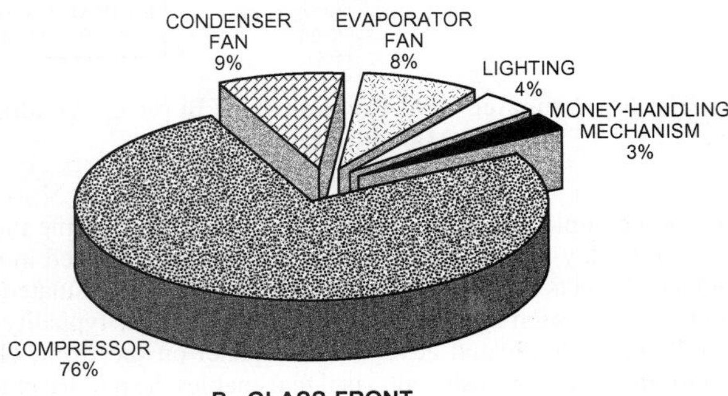

Fig. 7 Energy Use by Component For Typical Vending Machines

Data are for 24 h test with ambient conditions 95°F and 65% rh. (*Source*: Faramarzi 2005)

Sensitivity to Surroundings

A vending machine's refrigeration system is highly vulnerable to changes in its surroundings. Cooling load can be increased by high ambient temperatures, which raise head pressure, resulting in loss of refrigeration effect and increased compression ratio. It also increases conduction, radiation, and sensible infiltration loads. High humidity increases latent infiltration load, and can cause extreme ice build-up on the evaporator coil. This may severely restrict airflow through the unit and minimize the amount of heat absorbed by the refrigerant. Exposure to sunlight greatly increases heat gain into the refrigerated space through direct solar radiation, especially in glass-front machines, raises the sol-air temperature of the cabinet exterior, and increases conduction load. Exposure to high-intensity lighting fixtures can have a similar (but smaller) effect.

Condenser coils are usually located in the bottom of the machine, with air intakes at ground level. The coils often get dirty and clogged. Dirty condenser coils lose their heat rejection effectiveness, which reduces refrigeration capacity and increases compressor power use.

Steel security cages are often placed around machines, especially when they are outdoors. These cages trap heat rejected by the condenser as it exits the rear of the machine. Trapped heat increases the inlet air temperature at the condenser, causing operations at higher condensing temperatures.

Maintenance and Operation

The site owner can improve vending machine efficiency by taking these measures:

- Keep machines in cool/shaded locations to minimize sun exposure
- Keep condenser coils clean.
- Install a controller that will shift the unit to low power based on low traffic or low sales rate.

ICE MACHINES

Ice machines are used in many commercial applications, including bars, restaurants, gas stations, minimarts, delis, hotels, motels, hospitals, and other institutional facilities. Ice machines can harvest large quantities of ice and store it in holding bins, similar to those found in hotel hallways next to vending machines or in restaurant kitchens, where it is available for later use. Restaurants and minimarts, as well as other operations, require self-service ice production, typically integrated into fountain beverage dispensers.

There are three main categories of ice machines: (1) cubers, (2) flakers, and (3) nugget (chewable ice) makers. Ice machines come in various sizes, with harvest rates between 250 and 1400 lb per 24 h. The Air-Conditioning and Refrigeration Institute (ARI) maintains a database on certified commercial ice machines and storage bins (ARI 2005), which tabulates ice harvest capacity, potable water use, and energy consumption rate for all models. In addition, condenser water use is tabulated for water-cooled models.

Typical Operation and Construction

Ice is created and then harvested, dropping into the storage bin until the bin fills up, at which point ice production stops. When the level of ice in the storage bin falls below a threshold amount, the head unit cycles back on and refills the bin.

Similar to all mechanical refrigeration, ice machines are comprised of a compressor, evaporator, expansion valve, and condenser. Cubed ice forms on the evaporator plate. When the ice is fully formed, the refrigeration system cycles in reverse, heating the evaporator plate and melting a small layer of ice to release the sheet into the storage bin. Various methods are used to release the sheet of ice from the evaporator plate. Flake and nugget machines form ice on the inside of a cylinder, which features a screw (auger) that shaves off the ice. Shaved ice is forced out the top, where it falls into a storage bin. The nugget machine adds a cone for the ice to pass through, allowing the shavings to clump together and form nuggets. Flake and nugget machines use all the water that enters the ice machine to produce ice. Cube machines use a small amount of water to purge scale and mineral deposits, to prevent damage to the evaporator plate. Head units and storage bins can be mixed and matched to meet specific production and demand characteristics of the installation.

Storage bins come in multiple sizes, and are insulated with foamed-in-place polyurethane, with doors that are hinged to stay open during ice removal. No mechanical refrigeration is used to maintain the frozen ice in the storage bin (i.e., ice machines only use energy while making ice).

Refrigeration Systems

Each category of ice machine can have a remote or self-contained air- or water-cooled condenser. See the section on Refrigerated Cabinets for a more detailed description of refrigeration systems.

Maintenance and Operations

The site owner should ensure proper ice machine performance by adhering to the manufacturer's recommended maintenance schedule and monitoring the following parameters (Moore 2000):

- Perform regular cleaning and sanitization to discourage bacterial growth.
- Check water system and evaporator for scale build-up.
- Check for talc or mineral build-up in reservoir.
- Check pump motor operation (e.g., broken impeller, slow pumping).
- Check water flow through external water filter.
- Check strainer, inlet water valve screen, or float valve for obstruction.
- Inspect float valve assembly, adjustment, and operation.
- Check air filter, condenser fan blade, and coil for dust and grime.

- Check for proper drainage or water back-up in the bin that can melt ice away.
- Inspect the water overflow of the reservoir that washes ice away.
- Check the bin control for proper location and operation.

PREPARATION TABLES

The preparation (or "prep") table is a box-shaped cabinet with an open top section and sliding drawers or hinged doors on the bottom storage compartment (Figure 8). It is usually about 2 1/2 to 3 ft deep and 3 ft high, and ranges in width from about 3 to 10 ft. The prep table is designed to hold and provide easy access to pans of food or condiments. Cabinet capacities range from about 10 to 40 ft^3. A unit may or may not be equipped with a lower refrigerated compartment. These capacities and dimensions are standard from most manufacturers.

Typical types of prep tables include refrigerated sandwich units, pizza preparation tables, and buffet tables. Prep tables, by definition, are medium-temperature refrigerators small enough to be moved into a building.

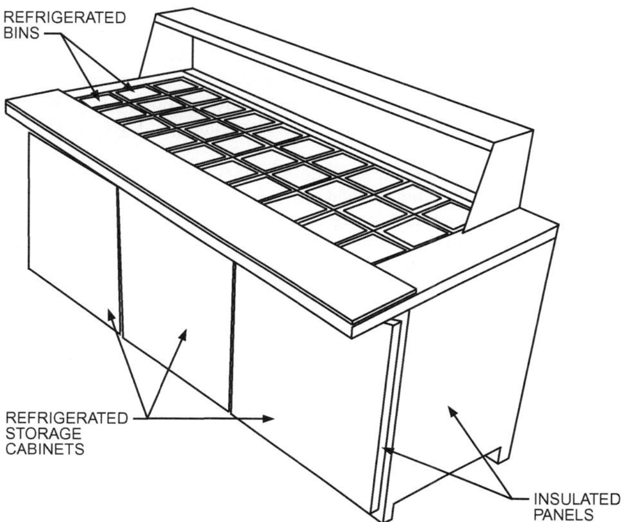

Fig. 8 Refrigerated Preparation Table

Product Temperatures

Refrigerated prep tables are available for medium-temperature ranges: a maximum of 41°F (for potentially hazardous foods) and a minimum of 33°F core product temperature, with the most desirable average temperature close to 38°F. The refrigerated rail is required to maintain food product between 33 and 41°F. Maintaining uniform bin product temperatures in prep tables is always a challenge. Cooling is provided by circulation of cold air under bins. Air does not reach each bin evenly, causing exposure of some bins to larger air volumes and, consequently, varying air temperature.

Typical Construction

Refrigerated prep tables are available in two basic types of construction. Cold wells may be designed to drop into existing countertops, and do not have a refrigerated cabinet. More common designs use a welded assembly of exterior panels with insulation and liner inserts.

Materials used on exteriors and interiors are stainless steel, painted steel, aluminum-coated steel, aluminum, and vinyl-clad steel with wood grain or other patterns. Materials must (1) match or blend with that used on nearby equipment; (2) be easy to keep clean; (3) not be discolored or etched by common cleaning materials; (4) be strong enough to resist denting, scratching, and abrasion; and (5) provide the necessary frame strength. The material chosen by an individual purchaser depends a great deal on layout and budget.

Shelves are standard interior accessories, and are usually adjustable and furnished three or four per full-height section.

ENERGY EFFICIENCY OPPORTUNITIES

Products discussed in this section use different components and design strategies. The following is a list of options that are or can be used in these products. Some of the options are mature and tested in the industry, whereas others are emerging technologies. Designers must balance energy savings against customer requirements, manufacturing cost, system performance, reliability, and maintenance costs. Several of these options can be applied to all refrigeration systems; others apply only to specific types of equipment. The following list defines some of the energy efficiency measures that may be implemented. Table 1 shows applicability to the equipment discussed in this chapter.

Table 1 Applicability of Energy-Efficiency Opportunities to Refrigeration Equipment

	Roll-Ins/ Reach-Ins	Walk-Ins	Vending Machines	Ice Machines	Preparation Tables
High-efficiency compressors with capacity modulation capability	X	X	X	X	X
High-efficiency evaporator and condenser coils	X	X	X	X	X
Condenser fan and evaporator fan ECM	X	X	X	X	X
Auto door-closers		X			
Compact fluorescent lights		X			
Liquid-to-suction heat exchangers	X	X	X	X	
Insulation for bare suction lines	X	X			
Expansion valve with superheat control	X	X			
Efficient defrost	X	X			
Strip curtains or plastic doors		X			
Occupancy sensors		X			
Evaporator fan controller		X (Coolers only)			
High-efficiency lighting system			X		
Improved airflow through refrigerated cabinet			X		X
Airtight cabinet construction			X		X
Improved insulation		X	X		X
Lids					X

- **High-efficiency compressors with capacity modulation capability** enable the variable-speed-driven compressor to match capacity with the varying cooling load.
- **High-efficiency evaporator and condenser coils** with increased surface area and conductivity can transfer heat with the air more effectively.
- **Condenser fan and evaporator fan electronically commutated motors (ECMs)** use less power than shaded-pole motors. They also can incorporate variable-speed controllers.
- **Auto door-closers** ensure that the door is pulled securely shut when it is within 1 in. of full closure, thereby reducing air infiltration.
- **Compact fluorescent lights (CFLs)** can replace the incandescent bulb, reducing energy consumption and heat production. CFLs can now operate in cooler/freezer temperature ranges.
- **Liquid-to-suction heat exchangers** allow suction gas leaving the evaporator to absorb heat from liquid refrigerant entering the evaporator, increasing the subcooling and cooling capacity of the system.
- **Insulation for bare suction lines** reduce system losses.
- **Expansion valve with superheat control** reduces the mass flow rate of refrigerant as a function of superheat, thereby reducing compressor power under low load.
- **Efficient defrost** methods, such as hot gas or cool gas, remove ice from the coil in a comparatively short period of time while adding little heat to the refrigerated space. Defrost should be controlled with temperature termination so that compressor off-time lasts only as long as necessary to melt the ice.
- **Strip curtains** or **plastic doors** provide a barrier against ambient air infiltration when the walk-in cooler/freezer door is open.
- **Occupancy sensors** can be installed to turn interior lights on and off as a function of occupancy and traffic. This also reduces a source of heat inside the cooler/freezer.
- An **evaporator fan controller for walk-in coolers** reduces airflow of evaporator fans in medium-temperature walk-in coolers when compressor(s) cycle off and there is no refrigerant flow through the evaporator. They should not be used if (1) the compressor runs all the time with high duty cycle, (2) the evaporator fan does not run at full speed all the time, (3) the evaporator fan motor runs on polyphase power, (4) the evaporator fan motor is not shaded pole, or (5) the evaporator does not use off-cycle or time-off defrost.
- **High-efficiency lighting systems** include electronic ballast and efficient light bulbs.
- **Improved airflow through refrigerated cabinets** can reduce the input energy required by the evaporator fan and promote uniform cooling of products inside the cabinet.
- **Airtight cabinet construction** eliminates infiltration of warm and moist air.
- **Improved insulation** reduces conductive heat transfer through the walls of the refrigerated cabinet.
- **Closing lids** on refrigerated prep tables during non-use periods substantially reduces energy use.

REFERENCES

ARI. 2005. *Directory of certified automatic ice-cube machines and ice storage bins 810/820.* Air-Conditioning and Refrigeration Institute, Arlington, VA. http://www.ari.org/cert/directories/acim/acim0001.pdf.

A.D. Little, Inc. 1996. *Energy savings potential for commercial refrigeration equipment.* Report prepared for Building Equipment Division, Office of Building Technologies, U.S. Department of Energy, Washington, D.C.

Davis Energy Group. 2002. *Codes and standards enhancement initiative for PY 2001: Title 20 standards development—Draft analysis of standards options for refrigerated beverage vending machines.* Report prepared for Pacific Gas and Electric Company, San Francisco.

Faramarzi, R. and S. Mitchell. 2005. *Performance evaluation of typical glass-front refrigerated beverage vending machines under various ambient conditions.* Southern California Edison Refrigeration and Thermal Test Center, Irwindale.

Moore, D. 2000. *Low ice production: What to do if your unit just won't keep up.* http://www.hoshizakiamerica.com.

CHAPTER 48

HOUSEHOLD REFRIGERATORS AND FREEZERS

THIS chapter covers design and construction of full-sized household refrigerators and freezers, the most common of which are illustrated in Figure 1. Some of these small refrigerators (Bansal and Martin 2000) use absorption systems, special forms of compressors, and, in some cases, **thermoelectric** (**Peltier effect**) refrigeration. Applications for water/ammonia **absorption** systems have developed for recreational vehicles, picnic coolers, and hotel room refrigerators, where noise is an issue.

Although there are various technologies for household refrigerators and freezers, this chapter only covers the **vapor-compression** cycle in detail, because it is a universally used system. In these applications, heat (usually ranging from 200 to 1400 Btu/h) is pumped through temperature differentials (from less than 50 to over 160°F) from evaporator to condenser. Other **electrically powered** systems compare unfavorably to vapor-compression systems in terms of manufacturing and operating costs. Typical coefficients of performance of the three most practical refrigeration systems are as follows for a 0°F freezer and 90°F ambient:

Thermoelectric	Approximately 0.3 Btu/watt-hour
Absorption	Approximately 1.5 Btu/watt-hour
Vapor-compression	Approximately 5.63 Btu/watt-hour

An absorption system may operate from gas at a lower cost per unit of energy, but the initial cost, size, and weight have made it unattractive to use gas systems for major appliances where electric power is available. Note, however, that if the electricity comes from an inefficient thermal power station, emissions from an absorption refrigerator (burning fossil fuel directly) will be of the same order as those from the power station running an equivalent vapor-compression unit. Because of its simplicity, thermoelectric refrigeration could replace other systems if (1) an economical thermoelectric material were developed and (2) design issues such as the need for a direct current (dc) power supply and an effective means for transferring heat from the module were addressed.

The preparation of this chapter is assigned to TC 8.9, Residential Refrigerators and Food Freezers.

PRIMARY FUNCTIONS

Providing food storage space at reduced temperature is the primary function of a refrigerator or freezer, with ice making an essential secondary function in some markets. To preserve fresh food, a general storage temperature between 32 and 39°F is desirable. Higher or lower temperatures or a humid atmosphere are more suitable for storing certain foods; the section on Cabinets discusses special-purpose storage compartments designed to provide these conditions. Food freezers and combination refrigerator-freezers for long-term storage are designed to hold temperatures near 0°F and always below 8°F during steady-state operation. In single-door refrigerators, the frozen food space is usually warmer than this and is not intended for long-term storage. Optimum conditions for food preservation are detailed in Chapters 9 through 29.

PERFORMANCE CHARACTERISTICS

A refrigerator or freezer must maintain desired temperatures and have reserve capacity to cool to these temperatures when started on a hot summer day. Most models cool down within hours in a 110°F ambient at rated voltage.

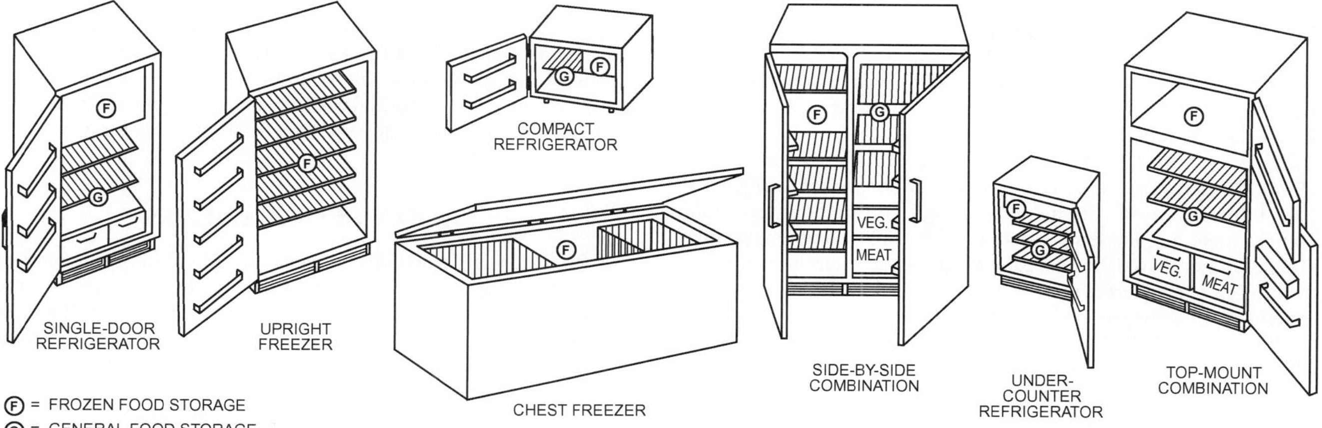

Note: A special cabinet (not shown) called an **all-refrigerator** has no frozen food storage or even ice making capacity. Also not shown is a **bottom freezer** cabinet, which is like the top-mount freezer except that the freezer compartment is on the bottom.

Fig. 1 Common Configurations of Contemporary Household Refrigerators and Freezers

Table 1 General Requirements for Various Test Standards

Cabinet Type or Parameters	Requirement	AS/NZS	ISO	U.S. DOE[a]	JIS Method C[b]	CNS/KS	GOST
Testing Parameters	Ambient T_A Relative Humidity	89.6 ± 0.9°F N/A	77/89.6 ± 0.9°F 45 to 75%	90 ± 1°F N/A	77°F ± 1.8°F 75 ± 5%	86 ± 1.8°F 75 ± 5%	77/89.6 ± 0.9°F N/A
All-Refrigerator	Fresh Food	37.4 ± 0.9°F	41°F	38°F	37.4 ± 0.9°F	37.4 ± 0.9°F	41°F
Refrigerator-Freezers[c]	Fresh Food	37.4 ± 0.9°F	41°F	45°F	37.4 ± 0.9°F	37.4 ± 0.9°F	41°F
	Freezer	5 ± 0.9°F	★ 21°F ★★ 10.4°F ★★★ –0.4°F	5°F	★ 21°F ★★ 10.4°F ★★★ –0.4°F	10.4/5°F	★ 21°F ★★ 10.4°F ★★★ –0.4°F
Freezers	Freezer	5 ± 0.9°F	–0.4°F	0°F	–0.4 ± 0.9°F	–0.4 ± 0.9°F	–0.4°F
Freezer Compartment	Loading of Test Packages	Unloaded	Loaded[d]	Sometimes[e]	Sometimes[e]	Unloaded	Loaded[d]
All Compartments	Door Openings Antisweat heaters Volume for labels/MEPS[f]	No Always on Gross	No When needed Storage (for EU)	No Average on and off Storage	Yes Always on Storage	No Always on Storage	No — Storage[g]
Energy Measurement Period		Lesser of 1 kWh or 16 h operation[h]	≥24 h	3 < t < 24 h, 2 or more cycles	24 h of testing	24 h of testing	24 < t < 48 h[h]

Source: Bansal 2003.
[a]Mexican and Canadian requirements are equivalent to U.S. DOE/AHAM.
[b]In previous Method A, 73% of consumption was weighted at an ambient of 59°F and 27% at 86°F.
[c]Per ISO, one-, two-, and three-star compartments are defined by their respective storage temperature being not higher than 21, 10.4, and –0.4°F. However, star ratings do not apply to AS/NZS, CNS, and U.S. DOE.
[d]Freezer temperature defined by warmest test package temperature that is below –0.4°F.
[e]Freezer temperature taken to be air temperature (contrary to ISO). Frost-free (forced-air) freezer compartments are generally unloaded. However, separate freezers in U.S. DOE are always loaded (to 75% of the available space) regardless of defrost type.
[f]Minimum energy performance standards.
[g]Freezer and fresh food compartment volumes are stated separately on the energy label.
[h]Note that test period for cyclic and frost-free models consists of a whole number of compressor and defrost cycles, respectively. Test must have at least one defrost cycle.
Abbreviations: **AS/NZS**: Australian/New Zealand Standard, **ISO**: International Organization for Standardization, **U.S. DOE**: American National Standards Institute, **JIS C**: Japanese International Standard (Method C), **CNS/KS**: Chinese National Standard/Korean Standard, **GOST**: Russian Committee of Standardization.

Overall system efficiency is important both because rising energy costs drive operating costs upward and because government energy standards dictate consumption limits. The challenge for the designer to control noise and vibration has been complicated by the need for fans for forced-air circulation and compressors with higher efficiencies and capacities. Vibrations from running or stopping the compressor must be isolated to prevent mechanical transmission to the cabinet or to the floor and walls, where it may cause additional vibration and noise.

Energy Efficiency Standards and Test Procedures

In many countries (see, e.g., the Collaborative Labeling and Appliance Standards Program at www.CLASPonline.org), regulators set efficiency standards for residential appliances. Periodically, these standards are reviewed and revised to promote the incorporation of emerging energy-saving technologies. For refrigerators and freezers, these standards are set in terms of the maximum annual electric energy consumption, which is measured according to a prescribed test procedure. In the United States, this is done under the U.S. Department of Energy's (DOE) National Appliance Energy Conservation Act (NAECA).

Different test procedures are used around the world (Bansal 2003) to determine energy consumption of household refrigerators (see Table 1). Most tests measure energy consumption at a food compartment internal temperature of 37°F and an ambient temperature of either 89.6 or 86°F. Exceptions are the International Organization for Standardization (ISO) and U.S. DOE tests, which specify food compartment temperatures of 41 and 45°F, respectively. In addition, ISO specifies two different ambient temperatures (77 and 89.6°F), depending on climate classification. However, the quoted energy consumption figures in ISO are usually based on the temperate climate classification of 77°F. Also, ISO is the only test procedure that specifies food loading in the freezer compartment of frost-free refrigerator-freezers, but with closed doors. The Chinese National Standard (CNS) requires the relative humidity of the ambient air to be 75 ± 5%, whereas the ISO specifies between 45 and 75%. Australian/New Zealand Standard (AS/NZS), Japanese Institute of Standards (JIS) and the U.S. DOE do not prescribe any humidity requirements. JIS Method A, the predecessor of JIS Method C (Banse 2000) shown in Table 1, was the only procedure that prescribed door openings of both compartments, but without loading of any food packs in either of the compartments. It required tests at a second test ambient of 59°F and weighted the two results, assuming 100 days at 86°F (27%) and 265 days at 59°F (73%), to evaluate the annual energy consumption. JIS Method C (Banse 2000), however, is compatible with ISO. The new procedure prescribes ambient and food compartment temperatures of 77 and 41°F, respectively, and the door opening frequency to be 25 times (25 minutes/day) and 8 times (25 minutes/day) for the food compartment and the freezer doors, respectively.

Maximum energy consumption varies with cabinet volume and by product class. The latest U.S. minimum energy performance standard (MEPS) level, introduced in 2001, set energy reductions at 30% below the 1993 MEPS levels, resulting in over six and a half quads of energy savings. In Australia and New Zealand, energy reductions from 1999 to 2005 MEPS levels vary from 25 to 50%, depending on product category. Other countries have other reductions on other timetables.

SAFETY REQUIREMENTS

Product safety standards are mandated in virtually all countries. These standards are designed to protect users from electrical shock, fire dangers, and other hazards under normal and some abnormal conditions. Product safety areas typically include motors, hazardous moving parts, earthing and bonding, stability (cabinet tipping), door-opening force, door-hinge strength, shelf strength, component restraint (shelves and pans), glass strength, cabinet and unit leakage current, leakage current from surfaces wetted by normal cleaning, high-voltage breakdown, ground continuity, testing and inspection of polymeric parts, and uninsulated live electrical parts accessible with an articulated probe. Flammability of refrigerants and foam-blowing agents are additional safety concerns that need to be considered. Most countries use IEC *Standard* 60335-2-24 or local variations. In the United States and Canada, however, products must comply with the joint Underwriters Laboratories/Canadian Standards UL *Standard* 250 CAN/CSA *Standard* C22.2. The United States, Canada, and Mexico are working to harmonize safety requirements for North America, based on IEC *Standard* 60335-2-24, with national differences as necessary.

DURABILITY AND SERVICE

Refrigerators and freezers are expected to last 15 to 20 years. The appliance therefore incorporates several design features that allow it to protect itself over this period. Motor overload protectors are normally incorporated, and an attempt is made to design fail-safe circuits so that the compressor's hermetic motor will not be damaged by failure of a minor external component, unusual voltage extremes, or voltage interruptions.

CABINETS

Good cabinet design achieves the optimum balance of

- Maximum food storage volume for floor area occupied by cabinet
- Maximum utility, performance, convenience, and reliability
- Minimum heat gain
- Minimum cost to consumer

Use of Space

The fundamental factors in cabinet design are usable food storage capacity and external dimensions. Food storage volume has increased considerably without a corresponding increase in external cabinet dimensions, by using thinner but more effective insulation and reducing the space occupied by the compressor and condensing unit.

Methods of computing storage volume and shelf area are described in various countries' standards [e.g., Association of Home Appliance Manufacturers (AHAM) *Standard* HRF-1 for the United States].

Special-Purpose Compartments

Special-purpose compartments provide a more suitable environment for storing specific foods. For example, some refrigerators have a meat storage compartment that can maintain storage temperatures just above freezing and may include independent temperature adjustment. Some models have a special compartment for fish, which is maintained at approximately 30°F. High-humidity compartments for storage of leafy vegetables and fresh fruit are found in practically all refrigerators. These drawers or bins, located in the food compartment, are generally tight-fitting to protect vulnerable foods from the desiccating effects of dry air circulating in the general storage compartment. The desired conditions are maintained in the special storage compartments and drawers by (1) enclosing them to prevent air exchange with the general storage area and (2) surrounding them with cold air to maintain the desired temperature.

Ice and Water Service

Through a variety of manual or automatic means, most units other than "all refrigerators" provide ice. For **manual** operation, ice trays are usually placed in the freezing compartment in a stream of air that is substantially below 32°F or placed in contact with a directly refrigerated evaporator surface.

Automatic Ice Makers. Automatic ice-making equipment in household refrigerators is increasingly common in the United States. Almost all U.S. automatic defrost refrigerators either include factory-installed automatic ice makers or can accept field-installable ice makers.

The ice maker mechanism is located in the freezer section of the refrigerator and requires attachment to a water line. Freezing rate is primarily a function of system design. Most ice makers are in no-frost refrigerators, and water is frozen by refrigerated air passing over the ice mold. Because the ice maker must share the available refrigeration capacity with the freezer and food compartments, ice production is usually limited by design to 4 to 6 lb per 24 h. A rate of about 4 lb per 24 h, coupled with an ice storage container capacity of 7 to 10 lb, is adequate for most users.

In the design of an ice maker, the various methods of accomplishing the basic functions must be evaluated to determine whether they meet the design objectives. The basic functions are as follows:

1. **Initiating** ejection of ice as soon as the water is frozen is necessary to obtain a satisfactory production rate. Ejection before complete freezing causes wet cubes to freeze together in the storage container and may cause the ice mold to overfill. One method is to initiate ejection in response to the temperature of a selected location in the mold that indicates complete freezing. Another successful method is to initiate ejection based on the time required to freeze the water under normal freezer temperatures. In either method, the temperature or time required may vary in different applications, depending on cooling air temperature and rate and direction of airflow.
2. **Ejecting** ice from the mold must be reliable. In several designs, ejection is accomplished by freeing the ice from the mold with an electric heater and pushing it from the tray into an ice storage container. In other designs, water is frozen in a plastic tray by passing refrigerated air over the top so that the water freezes from the top down. The natural expansion that occurs during freezing causes the ice to partially freeze free from the tray. Through twisting and rotation of the tray, the ice can be completely freed and ejected into a container.
3. **Driving** the ice maker is done in most designs by a gear motor, which operates the ice ejection mechanism and may also be used to time the freezing cycle and the water-filling cycle and to operate the stopping means.
4. **Filling** the ice mold with a constant volume of water, regardless of the variation in line water pressure, is necessary to ensure uniform-sized ice cubes and prevent overfilling. This is done by timing a solenoid flow-control valve or by using a solenoid-operated, fixed-volume slug valve.
5. **Stopping** is necessary after the ice storage container is filled until some ice is used. This is accomplished by using a feeler-type ice level control or a weight control.

Thermal Considerations

The total heat load imposed on the refrigerating system comes from both external and internal heat sources. The relative values of the basic or predictable components of the heat load (those independent of use) are shown in Figure 2. A large portion of the peak heat load may result from door openings, food loading, and ice making, which are variable and unpredictable quantities dependent on customer use. As the beginning point for the thermal design of the cabinet, the significant portions of the heat load are normally calculated and then confirmed by test.

The major predictable heat load is heat passing through the cabinet walls.

Foam Insulation. Polyurethane foam insulation has been used in refrigerator-freezer applications for over 40 years, originally using CFC-11 [an ozone-depleting substance (ODS)] as the blowing agent. Because of this ozone damage, the Montreal Protocol began curtailing its use in 1994. Most U.S. manufacturers of refrigerators and freezers then converted to HCFC-141b as a blowing agent as an interim solution, while those in many other parts of the world moved straight to cyclopentane as a blowing agent. Use of HCFC-141b was

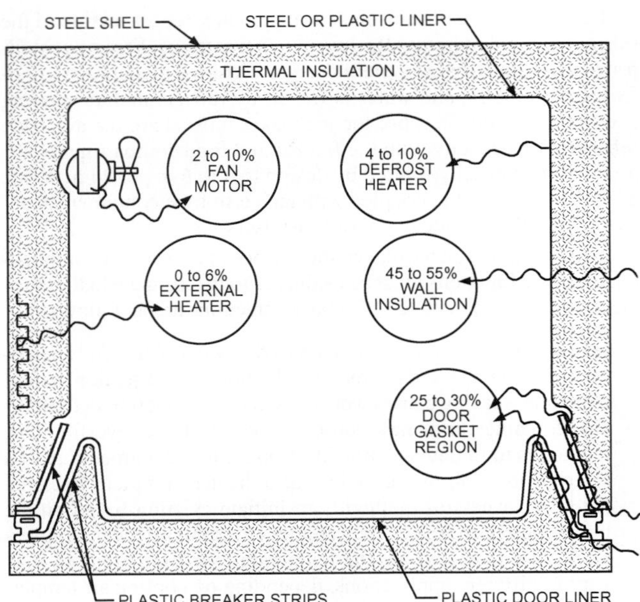

Fig. 2 Cabinet Cross Section Showing Typical Contributions to Total Basic Heat Load

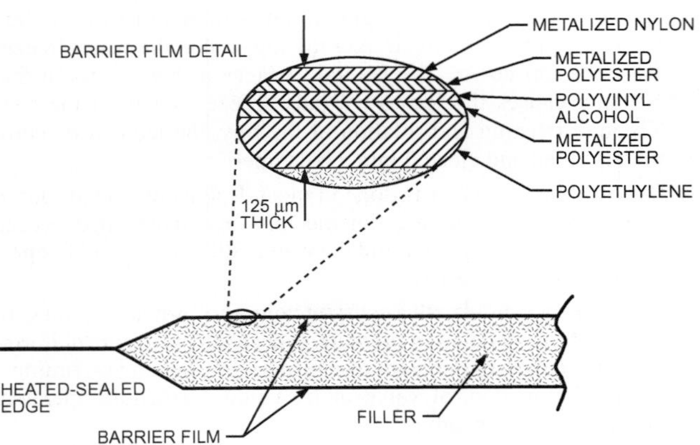

Fig. 3 Example Cross Section of Vacuum Insulation Panel

phased out in 2003 in the United States, and in most of the world. There are three alternatives to HCFC-141b:

- **Cyclopentane-blown foams** have the lowest foam material cost impact, require high capital cost for safety in foam process equipment, increase refrigerator energy use by about 4% compared to HCFC-141b, and can be difficult and expensive to implement in locations with very tight volatile organic compound restrictions.
- **HFC-134a-blown foams** have the next lowest foam material cost impact, require high-pressure-rated metering and mixing equipment, and increase refrigerator energy use by 8 to 10% compared to HCFC-141b.
- **HFC-245fa-blown foams** have the highest foam material cost impact, increase refrigerator energy use by 0 to 2% compared to HCFC-141b, require some revision to existing foam equipment, and retain insulating characteristics best over time.

Vacuum Insulation. Recently, flat vacuum-insulated panels have been developed to (Figure 3) inhibit heat conduction and provide highly effective insulation values down to 0.0277 Btu·in/h·ft^2·°F. A vacuum-insulated panel consists of a low-thermal-conductance fill and an impermeable skin. Fine mineral powders such as silicas, fiberglass, open-cell foam, and silica aerogel have all been used as fillers. The fill has sufficient compressive strength to support atmospheric pressure and can act as a radiation barrier. The skin must be highly impermeable, to maintain the necessary vacuum level over a long period of time. Getter materials are sometimes included to absorb small amounts of cumulative vapor leakage. The barrier skin provides a heat conduction path from the warm to the cool side of the panel, commonly referred to as the **edge effect**, which must be minimized if a high overall insulation value is to be maintained. Metalized plastic films are sufficiently impermeable while causing minimal edge effect. They have a finite permeability, so air gradually diffuses into the panel, degrading performance over time and limiting the useful life. There is also a risk of puncture and immediate loss of vacuum. Depending on how the vacuum panel is applied, the drastic reduction in insulation value from loss of vacuum may result in condensation on the outside wall of the cabinet, in addition to reduced energy efficiency. In commercial practice, vacuum-panel insulation is one of the least cost-effective options for improving efficiency, but, where thicker walls cannot be tolerated, they are a useful option for reaching specified minimum efficiency levels.

External sweating can be avoided by keeping exterior surfaces warmer than the ambient dew point. Condensation is most likely to occur around the hardware, on door mullions, along the edge of door openings, and on any cold refrigerant tubing that may be exposed outside the cabinet. In a 90°F room, no external surface temperature on the cabinet should be more than 5 or 6°F below the room temperature. If it is necessary to raise the exterior surface temperature to avoid sweating, this can be done either by routing a loop of condenser tubing under the front flange of the cabinet outer shell or by locating low-wattage wires or ribbon heaters behind the critical surfaces. Most refrigerators that incorporate electric heaters have power-saving electrical switches that allow the user to deenergize these electrical heaters when environmental conditions do not require their use.

Temporary condensation on internal surfaces may occur with frequent door openings, so the interior of the general storage compartment must be designed to avoid objectionable accumulation or drippage.

Figure 2 shows the design features of the throat section where the door meets the face of the cabinet. On products with metal liners, thermal breaker strips prevent metal-to-metal contact between inner and outer panels. Because the air gap between the breaker strip and the door panel provides a low-resistance heat path to the door gasket, the clearance should be kept as small as possible and the breaker strip as wide as practicable. When the inner liner is made of plastic rather than steel, there is no need for separate plastic breaker strips because they are an integral part of the liner.

Cabinet heat leakage can be reduced by using door gaskets with more air cavities to reduce conduction or by using internal secondary gaskets. Care must be taken so that the maximum door opening force as specified in safety standards is not exceeded; in the United States, this is specified in 16CFR1750.

Structural supports, if necessary to support and position the food compartment liner from the outer shell of the cabinet, are usually constructed of a combination of steel and plastics to provide adequate strength with maximum thermal insulation.

Internal heat loads that must be overcome by the system's refrigerating capacity are generated by periodic automatic defrosting, ice makers, lights, timers, fan motors used for air circulation, and heaters used to prevent undesirable internal cabinet sweating or frost build-up or to maintain the required temperature in a compartment.

Structure and Materials

The external shell of the cabinet is usually a single fabricated steel structure that supports the inner food compartment liner, door, and refrigeration system. Space between the inner and outer cabinet walls is usually filled with foam-in-place insulation. In general, the door and breaker strip construction is similar to that shown in Figure 2, although breaker strips and food liners formed of a single plastic sheet are also common. The doors cover the whole front of the cabinet, and plastic sheets become the inner surface for the doors, so no separate door breaker strips are required. Door liners are usually formed to provide an array of small door shelves and racks. Cracks and crevices are avoided, and edges are rounded and smooth to facilitate cleaning. Interior lighting, when provided, is usually from incandescent lamps controlled by mechanically operated switches actuated by opening the refrigerator door(s) or chest freezer lid.

Cabinet design must provide for the special requirements of the refrigerating system. For example, it may be desirable to refrigerate the freezer section by attaching evaporator tubing directly to the food compartment liner. Also, it may be desirable, particularly with food freezers, to attach condenser tubing directly to the shell of the cabinet to prevent external sweating. Both designs influence cabinet heat leakage and the amount of insulation required.

The method of installing the refrigerating system into the cabinet is also important. Frequently, the system is installed in two or more component pieces and then assembled and processed in the cabinet. Unitary installation of a completed system directly into the cabinet allows the system to be tested and charged beforehand. Cabinet design must be compatible with the method of installation chosen. In addition, forced-air systems frequently require ductwork in the cabinet or insulation spaces.

The overall structure of the cabinet must be strong enough to withstand shipping (and thus strong enough to withstand daily usage). However, additional support is typically provided in packaging material. Plastic food liners must withstand the thermal stresses they are exposed to during shipping and usage, and they must be unaffected by common contaminants encountered in kitchens. Shelves must be designed not to deflect excessively under the heaviest anticipated load. Standards typically require that refrigerator doors and associated hardware withstand a minimum of 300,000 door openings.

Foam-in-place insulation has had an important influence on cabinet design and assembly procedures. Not only does the foam's superior thermal conductivity allow wall thickness to be reduced, but its rigidity and bonding action usually eliminate the need for structural supports. The foam is normally expanded directly into the insulation space, adhering to the food compartment liner and the outer shell. Unfortunately, this precludes simple disassembly of the cabinet for service or repairs.

Outer shells of refrigerator and freezer cabinets are now typically of prepainted steel, thus reducing the volatile emissions that accompany the finishing process and providing a consistently durable finish to enhance product appearance and avoid corrosion.

Use of Plastics. As much as 15 or 20 lb of plastic is incorporated in a typical refrigerator or freezer. Use of plastic is increasing because of its

- Wide range of physical properties
- Good bearing qualities
- Electrical insulating ability
- Moisture and chemical resistance
- Low thermal conductivity
- Ease of cleaning
- Pleasing appearance with or without an applied finish
- Potential of multifunctional design in a single part
- Transparency, opacity, and colorability
- Ease of forming and molding
- Lower cost

A few examples illustrate the versatility of plastics. High-impact polystyrene (HIPs) and acrylonitrile butadiene styrene (ABS) plastics are used for inner door liners and food compartment liners. In these applications, no applied finish is necessary. These and similar thermoplastics such as polypropylene and polyethylene are also selected for evaporator doors, baffles, breaker strips, drawers, pans, and many small items. The good bearing qualities of nylon and acetal are used to advantage in applications such as hinges, latches, and rollers for sliding shelves. Gaskets, both for the refrigerator and for the evaporator doors, are generally made of vinyl.

Many items (e.g., ice cubes, butter) readily absorb odors and tastes from materials to which they are exposed. Accordingly, manufacturers take particular care to avoid using any plastics or other materials that impart an odor or taste in the interior of the cabinet.

Moisture Sealing

For the cabinet to retain its original insulating qualities, the insulation must be kept dry. Moisture may get into the insulation through leakage of water from the food compartment liner, through the defrost water disposal system, or, most commonly, through vapor leaks in the outer shell.

The outer shell is generally crimped, seam welded, or spot welded and carefully sealed against vapor transmission with mastics and/or hot-melt asphaltic or wax compounds at all joints and seams. In addition, door gaskets, breaker strips, and other parts should provide maximum barriers to vapor flow from the room air to the insulation. When refrigerant evaporator tubing is attached directly to the food compartment liner, as is generally done in chest freezers, moisture does not migrate from the insulation space, and special efforts must be made to vapor-seal this space.

Although urethane foam insulation tends to inhibit moisture migration, it tends to trap water when migrating vapor reaches a temperature below its dew point. The foam then becomes permanently wet, and its insulation value is decreased. For this reason, a vapor-tight exterior cabinet is equally important with foam insulation.

Door Latching and Entrapment

Door latching is accomplished by mechanical or magnetic latches that compress relatively soft compression gaskets made of vinyl compounds. Gaskets with embedded magnetic materials are generally used. Chest freezers are sometimes designed so that the weight of the lid acts to compress the gasket, although most of the weight is counterbalanced by springs in the hinge mechanism.

Safety standards mandate that appliances with any space large enough for a child to get into must be able to be opened from the inside. Doors or lids often must be removed when an appliance is discarded, as well.

Standards also typically mandate that any key-operated lock require two independent movements to actuate the lock, or be of a type that automatically ejects the key when unlocked. Some standards (e.g., IEC *Standard* 60335-2-24, UL *Standard* 250) also mandate safety warning markings.

Cabinet Testing

Specific tests necessary to establish the adequacy of the cabinet as a separate entity include (1) structural tests, such as repeated twisting of the cabinet and door; (2) door slamming test; (3) tests for vapor-sealing of the cabinet insulation space; (4) odor and taste transfer tests; (5) physical and chemical tests of plastic materials; and (6) heat leakage tests. Cabinet testing is also discussed later in the section on Evaluation.

REFRIGERATING SYSTEMS

The vapor-compression refrigerating systems used with modern refrigerators vary considerably in capacity and complexity, depending on the refrigerating application. They are hermetically sealed and normally require no replenishment of refrigerant or oil during the appliance's useful life. The components of the system must provide optimum overall performance and reliability at minimum cost. In addition, all safety requirements of the appropriate

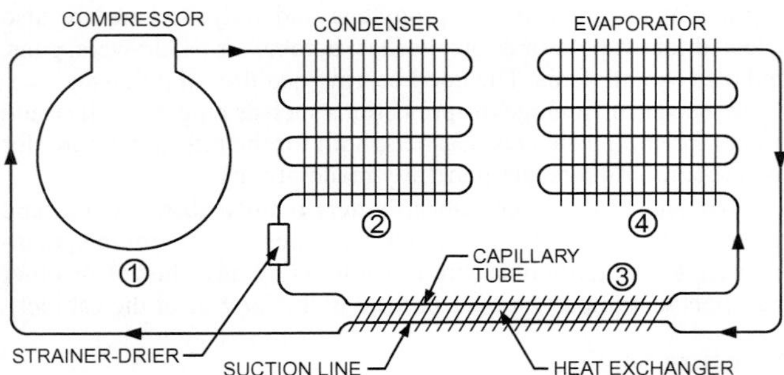

Fig. 4 Refrigeration Circuit

safety standard (e.g., IEC *Standard* 60335-2-24, UL *Standard* 250) must be met. The fully halogenated refrigerant R-12 was used in household refrigerators for many years. However, because of its strong ozone depletion property, appliance manufacturers have replaced R-12 with environmentally acceptable R-134a or isobutane.

Design of refrigerating systems for refrigerators and freezers has improved because of new refrigerants and oils, wider use of aluminum, and smaller and more efficient motors, fans, and compressors. These refinements have kept the vapor-compression system in the best competitive position for household application.

Refrigerating Circuit

Figure 4 shows a common refrigerant circuit for a vapor-compression refrigerating system. In the refrigeration cycle,

1. Electrical energy supplied to the motor drives a positive-displacement compressor, which draws cold, low-pressure refrigerant vapor from the evaporator and compresses it.
2. The resulting high-pressure, high-temperature discharge gas then passes through the condenser, where it is condensed to a liquid while heat is rejected to the ambient air.
3. Liquid refrigerant passes through a metering (pressure-reducing) capillary tube to the evaporator, which is at low pressure.
4. The low-pressure, low-temperature liquid in the evaporator absorbs heat from its surroundings, evaporating to a gas, which is again withdrawn by the compressor.

Note that energy enters the system through the evaporator (heat load) and through the compressor (electrical input). Thermal energy is rejected to the ambient by the condenser and compressor shell. A portion of the capillary tube is usually soldered to the suction line to form a heat exchange. Cooling refrigerant in the capillary tube with the suction gas increases capacity and efficiency.

A strainer-drier is usually placed ahead of the capillary tube to remove foreign material and moisture. Refrigerant charges of 0.3 lb or less are common. A thermostat (or cold control) cycles the compressor to provide the desired temperatures in the refrigerator. During the off cycle, the capillary tube allows pressures to equalize throughout the system.

Materials used in refrigeration circuits are selected for their (1) mechanical properties, (2) compatibility with the refrigerant and oil on the inside, and (3) resistance to oxidation and galvanic corrosion on the outside. Evaporators are usually made of bonded aluminum sheets or aluminum tubing, either with integral extruded fins or with extended surfaces mechanically attached to the tubing. Evaporators in cold-wall appliances are typically steel, copper, or aluminum. Condensers are usually made of steel tubing with an extended surface of steel sheet or wire. Steel tubing is used on the high-pressure side of the system, which is normally dry, and copper is used for suction tubing, where condensation can occur. Because of its ductility, corrosion resistance, and ease of brazing, copper is used for capillary tubes and often for small connecting tubing. Wherever aluminum tubing comes in contact with copper or iron, it must be protected against moisture to avoid electrolytic corrosion.

Defrosting

Manual Defrost. Manufacturers still make a few models that use manual defrost, in which the cooling effect is generated by natural convection of air over a refrigerated surface (evaporator) located at the top of the food compartment. The refrigerated surface forms some of the walls of a frozen food space, which usually extends across the width of the food compartment. Defrosting is typically accomplished by manually turning off the temperature control switch.

Cycle Defrosting (Partial Automatic Defrost). Combination refrigerator-freezers sometimes use two separate evaporators for the fresh food and freezer compartments. The fresh food compartment evaporator defrosts during each off cycle of the compressor, with energy for defrosting provided mainly by heat leakage (typically 10 to 20 W) into the fresh food compartment, though usually assisted by an electric heater, which is turned on when the compressor is turned off. The cold control senses the temperature of the fresh food compartment evaporator and cycles the compressor on when the evaporator surface is about 37°F. The freezer evaporator requires infrequent manual defrosting. This system is also commonly used in all-refrigerator units (see Figure 1 note).

Frost-Free Systems (Automatic Defrost). Most combination refrigerator-freezers and upright food freezers are refrigerated by air that is fan-blown over a single evaporator concealed from view. Because the evaporator is colder than the freezer compartment, it collects practically all of the frost, and there is little or no permanent frost accumulation on frozen food or on exposed portions of the freezer compartment. The evaporator is defrosted automatically by an electric heater located under the heat exchanger or by hot refrigerant gas, and the defrosting period is short, to limit food temperature rise. The resulting water is disposed of automatically by draining to the exterior, where it is evaporated in a pan located in the warm condenser compartment. A timer usually initiates defrosting at intervals of up to 24 h. If the timer operates only when the compressor runs, the accumulated time tends to reflect the probable frost load.

Adaptive Defrost. Developments in electronics have allowed the introduction of microprocessor-based control systems to some household refrigerators. An adaptive defrost function is usually included in the software. Various parameters are monitored so that the period between defrosts varies according to actual conditions of use. Adaptive defrost tends to reduce energy consumption and improve food preservation.

Forced Heat for Defrosting. All no-frost systems add heat to the evaporator to accelerate melting during the short defrosting cycle. The most common method uses a 300 to 1000 W electric heater. The traditional defrost cycle is initiated by a timer, which stops the compressor and energizes the heater.

When the evaporator has melted all the frost, a defrost termination thermostat opens the heater circuit. In most cases, the compressor is not restarted until the evaporator has drained for a few minutes and the system pressures have stabilized; this reduces the applied load for restarting the compressor. Commonly used defrost heaters include metal-sheathed heating elements in thermal contact with evaporator fins and radiant heating elements positioned to heat the evaporator.

Evaporator

The **manual defrost** evaporator is usually a box with three or four sides refrigerated. Refrigerant may be carried in tubing brazed to the walls of the box, or the walls may be constructed from double sheets of metal that are brazed or metallurgically bonded together with integral passages for the refrigerant. In this construction, the walls are usually aluminum, and special attention is required to avoid (1) contamination of the surface with other metals that would promote galvanic corrosion and (2) configurations that may be easily punctured during use.

The **cycle defrost** evaporator for the fresh food compartment is designed for natural defrost operation and is characterized by its low thermal capacity. It may be either a vertical plate, usually made from bonded sheet metal with integral refrigerant passages, or a serpentine coil with or without fins. In either case, the evaporator should be located near the top of the compartment and be arranged for good water drainage during the defrost cycle. In some designs, this cooling surface is located in an air duct remote from the fresh food space, with air circulated continuously by a small fan.

The **frost-free** evaporator is usually a forced-air fin-and-tube arrangement designed to minimize frost accumulation, which tends to be relatively rapid in a single-evaporator system. The coil is usually arranged for airflow parallel to the fins' long dimension.

Fins may be more widely spaced at the air inlet to provide for preferential frost collection and to minimize its air restriction effects. All surfaces must be heated adequately during defrost to ensure complete defrosting, and provision must be made for draining and evaporating the defrost water outside the food storage spaces. Some more efficient designs of new evaporators types (Bansal et al. 2001) are now commonly used in the industry. They are made of aluminum with continuous rectangular fins; fin layers are press-fitted onto the serpentine bent evaporator tube. These evaporators work in counter/parallel/cross flow configuration.

Freezers. Evaporators for chest freezers usually consist of tubing that is in good thermal contact with the exterior of the food compartment liner. Tubing is generally concentrated near the top of the liner, with wider spacing near the bottom to take advantage of natural convection of air inside. Most non-frost-free upright food freezers have refrigerated shelves and/or surfaces, sometimes concentrated at the top of the food compartment. These may be connected in series with an accumulator at the exit end. Frost-free freezers and refrigerator-freezers usually use a fin-and-tube evaporator and an air-circulating fan.

Condenser

The condenser is the main heat-rejecting component in the refrigerating system. It may be cooled by natural draft on free-standing refrigerators and freezers or fan-cooled on larger models and on models designed for built-in applications.

The **natural-draft condenser** is located on the back wall of the cabinet and is cooled by natural air convection under the cabinet and up the back. The most common form consists of a flat serpentine of steel tubing with steel cross wires welded on 1/4 in. centers on one or both sides perpendicular to the tubing. Tube-on-sheet construction may also be used.

The **hot-wall condenser**, another common natural-draft arrangement, consists of condenser tubing attached to the inside surface of the cabinet shell. The shell thus acts as an extended surface for heat dissipation. With this construction, external sweating is seldom a problem. Bansal and Chin (2003) provide an in-depth analysis of both these types of condensers.

The **forced-draft condenser** may be of fin-and-tube, folded banks of tube-and-wire, or tube-and-sheet construction. Various forms of condenser construction are used to minimize clogging caused by household dust and lint. The compact, fan-cooled condensers are usually designed for low airflow rates because of noise limitations. Air ducting is often arranged to use the front of the machine compartment for entrance and exit of air. This makes the cooling air system largely independent of the location of the refrigerator and allows built-in applications.

In hot and humid climates, defrosted water may not evaporate easily (Bansal and Xie 1999). Part of the condenser may be located under the defrost water evaporating pan to promote water evaporation.

For **compressor cooling**, the condenser may also incorporate a section where partially condensed refrigerant is routed to an oil-cooling loop in the compressor. Here, liquid refrigerant, still at high pressure, absorbs heat and is reevaporated. The vapor is then routed through the balance of the condenser, to be condensed in the normal manner.

Condenser performance may be evaluated directly on calorimeter test equipment similar to that used for compressors. However, final condenser design must be determined by performance tests on the refrigerator under a variety of operating conditions.

Generally, the most important design requirements for a condenser include (1) sufficient heat dissipation at peak-load conditions; (2) refrigerant holding capacity that prevents excessive pressures during pulldown or in the event of a restricted or plugged capillary tube; (3) good refrigerant drainage to minimize refrigerant trapping in the bottom of loops in low ambients, off-cycle losses, and the time required to equalize system pressures; (4) an external surface that is easily cleaned or designed to avoid dust and lint accumulation; (5) a configuration that provides adequate evaporation of defrost water; and (6) an adequate safety factor against bursting.

Fans

Advancements in small motor technology and electronic controls make high-efficiency fans advantageous. High-efficiency fan motors are typically dc and variable speed over a broad speed range. Energy improvements are approximately two times that of conventional ac shaded-pole fan motors.

Capillary Tube

The most commonly used refrigerant metering device is the capillary tube, a small-bore tube connecting the outlet of the condenser to the inlet of the evaporator. The regulating effect of this simple control device is based on the principle that a given weight of liquid passes through a capillary more readily than the same weight of gas at the same pressure. Thus, if uncondensed refrigerant vapor enters the capillary, mass flow is reduced, giving the refrigerant more cooling time in the condenser. On the other hand, if liquid refrigerant tends to back up in the condenser, the condensing temperature and pressure rise, resulting in an increased mass flow of refrigerant. Under normal operating conditions, a capillary tube gives good performance and efficiency. Under extreme conditions, the capillary either passes considerable uncondensed gas or backs liquid refrigerant well up into the condenser. Figure 5 shows the typical effect of capillary refrigerant flow rate on system performance. Because of

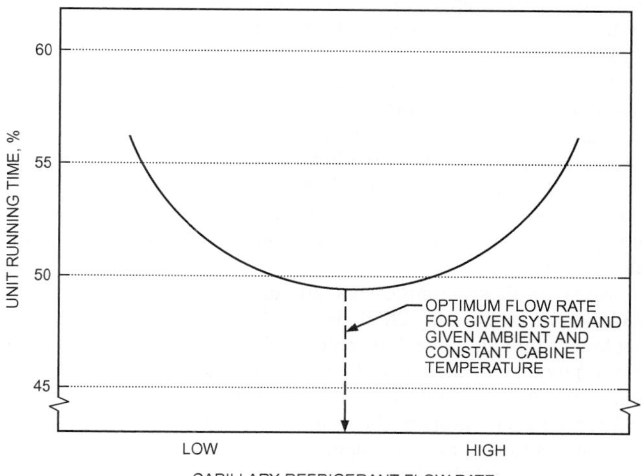

Fig. 5 Typical Effect of Capillary Tube Selection on Unit Running Time

these shortcomings and the difficulty of maintaining a match between the capillary restriction and the output of variable-pump-rate compressors, electronically controlled expansion valves are now used.

A capillary tube has the advantage of extreme simplicity and no moving parts. It also lends itself well to being soldered to the suction line for heat exchange purposes. This positioning prevents sweating of the otherwise cold suction line and increases refrigerating capacity and efficiency. Another advantage is that pressure equalizes throughout the system during the off cycle and reduces the starting torque required of the compressor motor. The capillary is the narrowest passage in the refrigerant system and the place where low temperature first occurs. For that reason, a combination strainer-drier is usually located directly ahead of the capillary to prevent it from being plugged by ice or any foreign material circulating through the system (see Figure 4). There are a number of studies available in the literature (e.g., Bansal and Xu 2002; Dirik et al. 1994; Mezavila and Melo 1996; Wolf and Pate 2002) on design and modeling of capillary tubes.

Selection. Optimum metering action can be obtained by variations in either the diameter or the length of the tube. Factors such as the physical location of system components and heat exchanger length (36 in. or more is desirable) may help determine the optimum length and bore of the capillary tube for any given application. Capillary tube selection is covered in detail in Chapter 44.

Once a preliminary selection is made, an experimental unit can be equipped with three or more different capillaries that can be activated independently. System performance can then be evaluated by using in turn capillaries with slightly different flow characteristics.

Final capillary selection requires optimizing performance under both no-load and pulldown conditions, with maximum and minimum ambient and load conditions. The optimum refrigerant charge can also be determined during this process.

Compressor

Although a more detailed description of compressors can be found in Chapter 34 of the 2004 *ASHRAE Handbook—HVAC Systems and Equipment*, a brief discussion of the small compressors used in household refrigerators and freezers is included here.

These products use positive-displacement compressors in which the entire motor-compressor is hermetically sealed in a welded steel shell. Capacities range from about 300 to about 2000 Btu/h measured at the ASHRAE rating conditions of –10°F evaporator, 130°F condenser, and 90°F ambient, with suction gas superheated to 90°F and liquid subcooled to 90°F, or CECOMAF rating conditions of –10°F evaporator, 131°F condenser, and 89.6°F ambient, with suction gas superheated to 89.6°F and liquid subcooled to 131°F.

Design emphasizes ease of manufacturing, reliability, low cost, quiet operation, and efficiency. Figure 6 illustrates the two reciprocating piston compressor mechanisms and two types of rotary compressors that are used in virtually all conventional refrigerators and freezers; no one type is much less costly than the others. Rotary compressors are somewhat more compact than reciprocating compressors, but a greater number of close tolerances is involved in their manufacture.

Most of these compressors are directly driven by two-pole (3450 rpm on 60 Hz, 2850 on 50 Hz) squirrel cage induction motors, although some four-pole (1750 rpm on 60 Hz, 1450 on 50 Hz) motors are also used. Field windings are insulated with special wire enamels and plastic slot and wedge insulation; all are chosen for their compatibility with the refrigerant and oil. During continuous runs at rated voltage, motor winding temperatures may be as high as 250°F when tested in a 110°F ambient temperature. In addition to maximum operating efficiency at normal running conditions, the motor must provide sufficient torque at the anticipated extremes of line voltage for starting and temporary peak loads from start-up and pulldown of a warm refrigerator and for loads associated with defrosting.

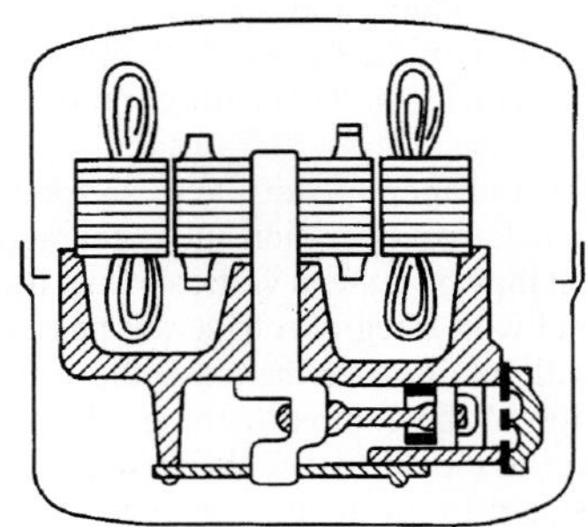

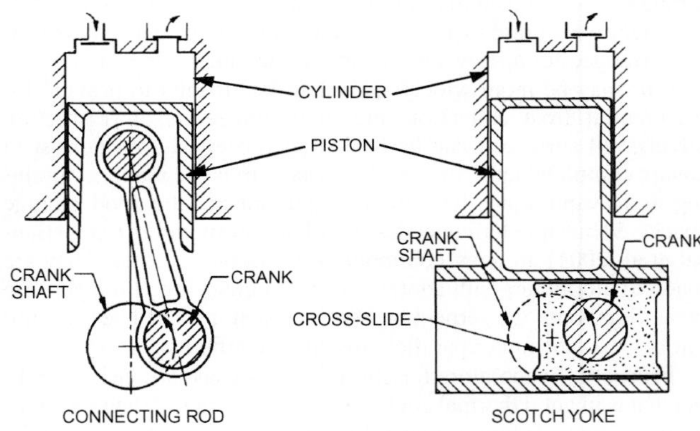

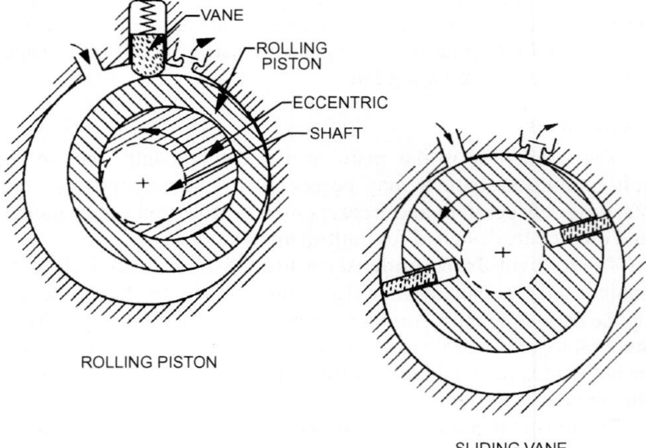

Fig. 6 Refrigerator Compressors

Starting torque is provided by a split-phase winding circuit, which in the larger motors may include a starting capacitor. When the motor comes up to speed, an external electromagnetic relay, positive temperature coefficient (PTC) device, or electronic switching device disconnects the start winding. A run capacitor is often used for greater motor efficiency. Motor overload protection is provided by an automatically resetting switch, which is sensitive to a combination of motor current and compressor case temperature or to internal winding temperature.

The compressor is cooled by rejecting heat to the surroundings. This is easily accomplished with a fan-cooled system. However, an oil-cooling loop carrying partially condensed refrigerant may be necessary when the compressor is used with a natural-draft condenser and in some forced-draft systems above 1000 Btu/h.

Temperature Control System

Temperature is often controlled by a thermostat consisting of an electromechanical switch actuated by a temperature-sensitive power element that has a condensable gas charge, which operates a bellow or diaphragm. At operating temperature, this charge is in a two-phase state, and the temperature at the gas/liquid interface determines the pressure on the bellows. To maintain temperature control at the bulb end of the power element, the bulb must be the coldest point at all times.

The thermostat must have an electrical switch rating for the inductive load of the compressor and other electrical components carried through the switch. The thermostat is usually equipped with a shaft and knob for adjusting the operating temperature. **Electronic temperature controls**, some using microprocessors, are becoming more common. They allow better temperature performance by reacting faster to temperature and load changes in the appliance, and do not have the constraint of requiring the sensor to be colder than the thermostat body or the phial tube connecting them. In some cases, both compartment controls use thermistor sensing devices that relay electronic signals to the microprocessor. Electronic temperature sensors provide real-time information to the control system that can be customized to optimize energy performance and temperature management. Electronic control systems provide a higher degree of independence in temperature adjustments for the two main compartments. Electronics also enable the engineer to use variable-speed fans and motorized dampers to further optimize temperature and energy performance.

In the simple gravity-cooled system, the controller's sensor is normally in close thermal contact with the evaporator. The location of the sensor and degree of thermal contact are selected to produce both a suitable cycling frequency for the compressor and the desired refrigerator temperature. For push-button defrosting, small refrigerators sold in Europe are sometimes equipped with a manually operated push button control to prevent the compressor from coming on until defrost temperatures are reached; afterward, normal cycling is resumed.

In a combination refrigerator-freezer with a split air system, location of the sensor(s) depends on whether an automatic damper control is used to regulate airflow to the fresh food compartment. When an auxiliary control is used, the sensor is usually located where it can sense the temperature of air leaving the evaporator. In manual-damper-controlled systems, the sensor is usually placed in the cold airstream to the fresh food compartment. Sensor location is frequently related to the damper effect on the airstream. Depending on the design of this relationship, the damper may become the freezer temperature adjustment or it may serve the fresh food compartment, with the thermostat being the adjustment for the other compartment. The temperature sensor should be located to provide a large enough temperature differential to drive the switch mechanism, while avoiding (1) excessive cycle length; (2) short cycling time, which can cause compressor starting problems; and (3) annoyance to the user from frequent noise level changes. Some combination refrigerator-freezers manage the temperature with a sensor for each compartment. These may manage the compressor, an automatic damper, variable-speed fans, or a combination of these. Such controls are almost certainly microprocessor-based.

System Design and Balance

A principal design consideration is selecting components that will operate together to give the optimum system performance and efficiency when total cost is considered. Normally, a range of combinations of values for these components meets the performance requirements, and the lowest cost for the required efficiency is only obtained through careful analysis or a series of tests (usually both). For instance, for a given cabinet configuration, food storage volume, and temperature, the following can be traded off against one another: (1) insulation thickness and overall shell dimensions, (2) insulation material, (3) system capacity, and (4) individual component performance (e.g., fan, compressor, and evaporator). Each of these variables affects total cost and efficiency, and most can be varied only in discrete steps.

The experimental procedure involves a series of tests. Calorimeter tests may be made on the compressor and condenser, separately or together, and on the compressor and condenser operating with the capillary tube and heat exchanger. Final component selection requires performance testing of the system installed in the cabinet. These tests also determine refrigerant charge, airflows for the forced-draft condenser and evaporator, temperature control means and calibration, necessary motor protection, and so forth. The section on Evaluation covers the final evaluation tests made on the complete refrigerator. The interaction between components is further addressed in Chapter 43. This experimental procedure assumes knowledge (equations or graphs) of the performance characteristics of the various components, including cabinet heat leakage and the heat load imposed by the customer. The analysis may be performed manually point by point. If enough component information exists, it can be entered into a computer simulation program capable of responding to various design conditions or statistical situations. Although the available information may not always be adequate for an accurate analysis, this procedure is often useful, although confirming tests must follow.

Processing and Assembly Procedures

All parts and assemblies that are to contain refrigerant are processed to avoid unwanted substances or remove them from the final sealed system and to charge the system with refrigerant and oil (unless the latter is already in the compressor as supplied). Each component should be thoroughly cleaned and then stored in a clean, dry condition until assembly. The presence of free water in stored parts produces harmful compounds such as rust and aluminum hydroxide, which are not removed by the normal final assembly process. Procedures for dehydration, charging, and testing may be found in Chapter 45.

Assembly procedures are somewhat different, depending on whether the sealed refrigerant system is completed as a unit before being assembled to the cabinet, or components of the system are first brought together on the cabinet assembly line. With the unitary installation procedure, the system may be tested for its ability to refrigerate and then be stored or delivered to the cabinet assembly line.

EVALUATION

Once the unit is assembled, laboratory testing, supplemented by field-testing, is necessary to determine actual performance.

Environmental Test Rooms

Controlled-temperature and -humidity test rooms are essential for performance-testing refrigerators and freezers. AHAM *Standard* HRF-1 describes the environmental conditions to be maintained for the U.S. and Canadian market, and ISO *Standard* 7371 describes conditions for many other markets. For some markets, ISO requires rooms to operate down to 50°F and up to 109°F. In-house testing may require a wider window than this. AHAM *Standard* HRF-1 requires test room temperatures from 70 to 109°F for North American markets accurate to within 0.9°F of the desired value. The temperature gradient and air circulation in the room should also be maintained closely. To provide more flexibility in testing, it may be desirable to have an additional test room that can cover the range down to 0°F for things such as plastic liner stress-crack testing. At least one test room should be able to maintain a desired relative humidity within a tolerance of ±2% up to 85% rh.

All instruments should be calibrated at regular intervals. Instrumentation should have accuracy and response capabilities of sufficient quality to measure the dynamics of the systems tested.

Computerized data acquisition systems that record power, current, voltage, temperature, and pressure are used in testing refrigerators and freezers. Refrigerator test laboratories have developed automated means of control and data acquisition (with computerized data reduction output) and automated test programming.

Standard Performance Test Procedures

AHAM *Standard* HRF-1 describes tests for determining the performance of refrigerators and freezers in the United States (see Table 1 for other standards). It specifies standard ambient conditions, power supply, and means for selecting samples and measuring temperatures. Test procedures include the following:

No-Load Pulldown Test. This tests the ability of the refrigerator or freezer in a 110°F ambient temperature to pull down from a stabilized warm condition to design temperatures within an acceptable period. This test is also a part of AS/NZS *Standard* 4474.1, but not of ISO regulations.

Simulated Load Test (*Refrigerators*) or Storage Load Test (*Freezers*). This test determines thermal performance under varying ambient conditions, as well as the percent operating time of the compressor motor, and temperatures at various locations within the cabinet at 70, 90, and 110°F ambient for a range of temperature control settings. Cabinet doors remain closed during the test. The freezer compartment is loaded with filled frozen packages. Heavy usage testing, although not generally in standards, is also almost always done by manufacturers (to their own procedures). This is typically testing with lots of door openings in high temperature and high humidity to see whether the defrosting system copes and recovers. AS/NZS *Standard* 4474.1 has neither a load test nor the compressor operating time test, and it tests at 50, 90, and 109°F (not 70, 90, and 110°F). Each test point may take 8 h or more to ensure steady-state condition and accuracy of data.

Freezers are tested similarly, but in a 90°F ambient. Under actual operating conditions in the home, with frequent door openings and ice making, performance may not be as favorable as that shown by this test. However, the test indicates general performance, which can serve as a basis for comparison.

Ice Making Test. This test, performed in a 90°F ambient, determines the rate of making ice with the ice trays or other ice-making equipment furnished with the refrigerator.

External Surface Condensation Test. This test determines the extent of moisture condensation on the external surfaces of the cabinet in a 90°F, high-humidity ambient when the refrigerator or freezer is operated at normal cabinet temperatures. Although AHAM *Standard* HRF-1 calls for this test to be made at a relative humidity of 75 ± 2%, it is customary to determine the sweating characteristics through a wide range of relative humidity up to 85%. This test also determines the need for, and the effectiveness of, anticondensation heaters in the cabinet shell and door mullions.

Internal Moisture Accumulation Test. This dual-purpose test is also run under high-temperature, high-humidity conditions. First, it determines the effectiveness of moisture sealing of the cabinet in preventing moisture from getting into the insulation space and degrading refrigerator performance and life. Secondly, it determines the rate of frost build-up on refrigerated surfaces, expected frequency of defrosting, and effectiveness of any automatic defrosting features, including defrost water disposal.

This test is performed in ambient conditions of 90°F and 75% rh with the cabinet temperature control set for normal temperatures. The test extends over 21 days with a rigid schedule of door openings over the first 16 h of each day: 96 openings per day for a general refrigerated compartment, and 24 per day for a freezer compartment and for food freezers.

Current Leakage Test. IEC *Standard* 60335-1 (not available in AHAM *Standard* HRF-1) allows testing on a component-by-component basis, determining the electrical current leakage through the entire electrical insulating system under severe operating conditions to eliminate the possibility of a shock hazard.

Handling and Storage Test. As with most other major appliances, it is during shipping and storage that a refrigerator is exposed to the most severe impact forces, vibration, and extremes of temperature. When packaged, it should withstand without damage a drop of several inches onto a concrete floor, the impact experienced in a freight car coupling at 10 mph, and jiggling equivalent to a trip of several thousand miles by rail or truck.

The widespread use of plastic parts makes it important to select materials that also withstand high and low temperature extremes that may be experienced. This test determines the cabinet's ability, when packaged for shipment, to withstand handling and storage conditions in extreme temperatures. It involves raising the crated cabinet 6 in. off the floor and suddenly releasing it on one corner. This is done for each of the four corners. This procedure is carried out at stabilized temperature conditions, first in a 140°F ambient temperature, and then in a 0°F ambient. At the conclusion of the test, the cabinet is uncrated and operated, and all accessible parts are examined for damage.

Special Performance Testing

To ensure customer acceptance, several additional performance tests are customarily performed.

Usage Test. This is similar to the internal moisture accumulation test, except that additional performance data are taken during the test period, including (1) electrical energy consumption per 24 h period, (2) percent running time of the compressor motor, and (3) cabinet temperatures. These data give an indication of the reserve capacity of the refrigerating system and the temperature recovery characteristics of the cabinet.

Low-Ambient-Temperature Operation. It is customary to conduct a simulated load test and an ice making test at ambient temperatures of 55°F and below, to determine performance under unusually low temperatures.

Food Preservation Tests. This test determines the food-keeping characteristics of the general refrigerated compartment and is useful for evaluating the utility of special compartments such as vegetable crispers, meat keepers, high-humidity compartments, and butter keepers. This test is made by loading the various compartments with food, as recommended by the manufacturer, and periodically observing the condition of the food.

Noise Tests. The complexity and increased size of refrigerators have made it difficult to keep the sound level within acceptable limits. Thus, sound testing is important to ensure customer acceptance.

A meaningful evaluation of the sound characteristics may require a specially constructed room with a background sound level of 30 dB or less. The wall treatment may be reverberant, semireverberant, or anechoic; reverberant construction is usually favored in making an instrument analysis. A listening panel is most commonly used for the final evaluation, and most manufacturers strive to correlate instrument readings with the panel's judgment.

High- and Low-Voltage Tests. The ability of the compressor to start and pull down the system after an ambient soak is tested with applied voltages at least 10% above and below the rated voltage. The starting torque is reduced at low voltage; the motor tends to overheat at high voltage.

Special Functions Tests. Refrigerators and freezers with special features and functions may require additional testing. Without formal procedures for this purpose, test procedures are usually improvised.

Energy Consumption Tests. Many countries use procedures relevant to their local conditions (see Table 1 for reference) to determine a refrigerator's energy consumption.

Materials Testing

The materials used in a refrigerator or freezer should meet certain test specifications [e.g., U.S. Food and Drug Administration (FDA) requirements]. Metals, paints, and surface finishes may be tested according to procedures specified by the American Society for Testing and Materials (ASTM) and others. Plastics may be tested according to procedures formulated by the Society of the Plastics Industry (SPI) appliance committee. In addition, the following tests on materials, as applied in the final product, are assuming importance in the refrigeration industry (*Federal Specification* A-A-2011).

Odor and Taste Contamination. This test determines the intensity of odors and tastes imparted by the cabinet air to uncovered, unsalted butter stored in the cabinet at operating temperatures.

Stain Resistance. The degree of staining is determined by coating cabinet exterior surfaces and plastic interior parts with a typical staining food (e.g., prepared cream salad mustard).

Environmental Cracking Resistance Test. This tests the cracking resistance of the plastic inner door liners and breaker strips at operating temperatures when coated with a 50/50 mixture of oleic acid and cottonseed oil. The cabinet door shelves are loaded with weights, and the doors are slammed on a prescribed schedule over 8 days. The parts are then examined for cracks and crazing.

Breaker Strip Impact Test. This test determines the impact resistance of the breaker strips at operating temperatures when coated with a 50/50 mixture of oleic acid and cottonseed oil. The breaker strip is hit by a 2 lb dart dropped from a prescribed height. The strip is then examined for cracks and crazing.

Component Life Testing

Various components of a refrigerator and freezer cabinet are subject to continual use by the consumer throughout the product's life; they must be adequately tested to ensure their durability for at least a 10 year life. Some of these items are (1) hinges, (2) latch mechanism, (3) door gasket, (4) light and fan switches, and (5) door shelves. These components may be checked by an automatic mechanism, which opens and closes the door in a prescribed manner. A total of 300,000 cycles is generally accepted as the standard for design purposes. Door shelves should be loaded as they would be for normal home usage. Several other important characteristics may be checked during the same test: (1) retention of door seal, (2) rigidity of door assembly, (3) rigidity of cabinet shell, and (4) durability of inner door panels.

Life tests on the electrical and mechanical components of the refrigerating system may be made as required.

Field Testing

Additional information may be obtained from a program of field testing in which test models are placed in selected homes for observation. Because high temperature and high humidity are the most severe conditions encountered, the Gulf Coast of the United States is a popular field test area. Laboratory testing has limitations in the complete evaluation of a refrigerator design, and field testing can provide the final assurance of customer satisfaction.

Field testing is only as good as the degree of policing and the completeness and accuracy of reporting. However, if testing is done properly, the data collected are important, not only in product evaluation, but also in providing criteria for more realistic and timely laboratory test procedures and acceptance standards.

REFERENCES

AHAM. 2004. Household refrigerators/household freezers. ANSI/AHAM *Standard* HRF-1, Washington, D.C.

AS/NZS. 1997. Performance of household electrical appliances—Refrigerating appliances, Part 1: Energy consumption and performance; Part 2: Energy labeling and minimum energy performance standard requirements. AS/NZS *Standard* 4474-1997. Standards Association of New Zealand, Wellington.

Bansal, P.K. 2003. Developing new test procedures for domestic refrigerators: Harmonization issues and future R&D needs—A review. *International Journal of Refrigeration* 26(7):735-748.

Bansal, P.K. and T. Chin. 2003. Heat transfer characteristics of wire-and-tube and hot-wall condensers. *International Journal of HVAC&R Research* (now *HVAC&R Research*) 9(3):277-290.

Bansal, P.K. and A. Martin. 2000. Comparative study of vapour compression, thermoelectric and absorption refrigerators. *International Journal of Energy Research* 24(2):93-107.

Bansal, P.K. and G. Xie. 1999. A simulation model for evaporation of defrosted water in domestic refrigerators. *International Journal of Refrigeration* 22(4):319-333.

Bansal, P.K. and B. Xu. 2002. Non-adiabatic capillary tube flow: A homogeneous model and process description. *Applied Thermal Engineering* 22(16):1801-1819.

Bansal, P.K., T. Wich, M.W. Browne, and J. Chen. 2001. Design and modeling of new egg-crate-type forced flow evaporators in domestic refrigerators. *ASHRAE Transactions* 107(2):204-213.

Banse, T. 2000. The promotion situation of energy saving in Japanese electric refrigerators. APEC Symposium on Domestic Refrigerator/Freezers, Wellington, New Zealand.

CFR. 2005. Standard for devices to permit the opening of household refrigerator doors from the inside. 16CFR1750. *Code of Federal Regulations*, U.S. Government Printing Office, Washington, D.C. http://www.gpoaccess.gov/ecfr/.

CNS. 1989. Electric refrigerators and freezers. *Chinese National Standard* CNS2062/C4048. National Bureau of Standards (Chinese), Taipei.

CNS. 1989. Method of test for electric refrigerators and freezers. *Chinese National Standard* CNS9577/C3164. National Bureau of Standards (Chinese), Taipei.

Consumer Product Safety Commission. 1956. Refrigeration safety act. *Public Law* 84-930.

Dirik, E., C. Inan, and M.Y. Tanes. 1994. Numerical and experimental studies on non-adiabatic capillary tubes. *Proceedings of the 1994 International Refrigeration Conference*, Purdue, IN, pp. 365-370.

GOST. 1991. Electric domestic refrigerating appliances. *Standard* 16317-87. Russian Committee for Standardization.

GSA. 1998. Refrigerators, mechanical, household (electrical, self-contained). *Federal Specification* A-A-2011. U.S. General Services Administration, Washington, D.C.

IEC. 2005. Household and similar electrical appliances—Safety: Particular requirements for refrigerating appliances, ice-cream appliances and ice-makers. *Standard* 60335-2-24. International Electrotechnical Commission, Geneva.

ISO. 1995. For the preservation of highly perishable foodstuffs. *Standard* 7371:1995. International Organization for Standardization.

JSA. 1999. Household electric refrigerators, refrigerator-freezers and freezers. *Standard* C 9607-1999. Japanese Standards Association, Akasaka.

Mezavila, M.M. and C. Melo. 1996. CAPHEAT: A homogeneous model to simulate refrigerant flow through non-adiabatic capillary tubes. *Proceedings of the International Refrigeration Conference*, Purdue, IN, pp. 95-100.

Wolf, D.A. and M.B. Pate. 2002. Performance of a suction-line/capillary-tube heat exchanger with alternative refrigerants. ASHRAE Research Project RP-948, *Final Report*.

UL. 1993. Household refrigerators and freezers. ANSI/UL *Standard* 250, CAN/CSA *Standard* C22.2. Underwriters Laboratories, Northbrook, IL.

BIBLIOGRAPHY

CFR. 2005. Energy conservation program for consumer products. 10CFR430. *Code of Federal Regulations*, U.S. Government Printing Office, Washington, D.C. http://www.gpoaccess.gov/ecfr/.

ISO. 1991. Compartments for the preservation of highly perishable foodstuffs. *Standard* 8187:1991. International Organization for Standardization.

CHAPTER 49

CODES AND STANDARDS

THE Codes and Standards listed here represent practices, methods, or standards published by the organizations indicated. They are useful guides for the practicing engineer in determining test methods, ratings, performance requirements, and limits of HVAC&R equipment. Copies of the standards can be obtained from most of the organizations listed in the Publisher column, from Global Engineering Documents at **global.ihs.com**, or from CSSINFO at **cssinfo.com**. Addresses of the organizations are given at the end of the chapter. A comprehensive database with over 250,000 industry, government, and international standards is at **www.nssn.org**.

Selected Codes and Standards Published by Various Societies and Associations

Subject	Title	Publisher	Reference
Air Conditioners	Commercial Applications, Systems, and Equipment, 1st ed.	ACCA	ACCA Manual CS
	Residential Equipment Selection, 2nd ed.	ACCA	ANSI/ACCA Manual S
	Methods of Testing for Rating Ducted Air Terminal Units	ASHRAE	ANSI/ASHRAE 130-1996
	Non-Ducted Air Conditioners and Heat Pumps—Testing and Rating for Performance	ISO	ISO 5151:1994
	Ducted Air-Conditioners and Air-to-Air Heat Pumps—Testing and Rating for Performance	ISO	ISO 13253:1995
	Guidelines for Roof Mounted Outdoor Air-Conditioner Installations	SMACNA	SMACNA 1998
	Heating and Cooling Equipment (1995)	UL/CSA	ANSI/UL 1995/C22.2 No. 236-95
Central	Performance Standard for Single Package Central Air-Conditioners and Heat Pumps	CSA	CAN/CSA-C656-05
	Performance Standard for Rating Large Air Conditioners and Heat Pumps	CSA	CAN/CSA-C746-98 (R2004)
	Performance Standard for Split-System and Single-Package Central Air Conditioners and Heat Pumps	CSA	CAN/CSA-C2656-05
	Heating and Cooling Equipment (1995)	UL/CSA	ANSI/UL 1995/C22.2 No. 236-95
Gas-Fired	Gas-Fired, Heat Activated Air Conditioning and Heat Pump Appliances	CSA	ANSI Z21.40.1-1996 (R2002)/CGA 2.91-M96
	Gas-Fired Work Activated Air Conditioning and Heat Pump Appliances (Internal Combustion)	CSA	ANSI Z21.40.2-1996 (R2002)/CGA 2.92-M96
	Performance Testing and Rating of Gas-Fired Air Conditioning and Heat Pump Appliances	CSA	ANSI Z21.40.4-1996 (R2002)/CGA 2.94-M96
Packaged Terminal	Packaged Terminal Air-Conditioners and Heat Pumps	ARI/CSA	ARI 310-380-04/CSA C744-04
Room	Room Air Conditioners	AHAM	ANSI/AHAM RAC-1-2003
	Method of Testing for Rating Room Air Conditioners and Packaged Terminal Air Conditioners	ASHRAE	ANSI/ASHRAE 16-1983 (RA99)
	Method of Testing for Rating Room Air Conditioner and Packaged Terminal Air Conditioner Heating Capacity	ASHRAE	ANSI/ASHRAE 58-1986 (RA99)
	Method of Testing for Rating Fan-Coil Conditioners	ASHRAE	ANSI/ASHRAE 79-2002
	Performance Standard for Room Air Conditioners	CSA	CAN/CSA-C368.1-M90 (R2001)
	Room Air Conditioners	CSA	C22.2 No. 117-1970 (R2002)
	Room Air Conditioners (1993)	UL	ANSI/UL 484
Unitary	Unitary Air-Conditioning and Air-Source Heat Pump Equipment	ARI	ANSI/ARI 210/240-2005
	Sound Rating of Outdoor Unitary Equipment	ARI	ARI 270-95
	Application of Sound Rating Levels of Outdoor Unitary Equipment	ARI	ARI 275-97
	Commercial and Industrial Unitary Air-Conditioning and Heat Pump Equipment	ARI	ARI 340/360-2004
	Methods of Testing for Rating Electrically Driven Unitary Air-Conditioning and Heat Pump Equipment	ASHRAE	ANSI/ASHRAE 37-2005
	Methods of Testing for Rating Heat-Operated Unitary Air-Conditioning and Heat Pump Equipment	ASHRAE	ANSI/ASHRAE 40-2002
	Methods of Testing for Rating Seasonal Efficiency of Unitary Air Conditioners and Heat Pumps	ASHRAE	ANSI/ASHRAE 116-1995 (RA05)
	Method of Testing for Rating Computer and Data Processing Room Unitary Air Conditioners	ASHRAE	ANSI/ASHRAE 127-2001
	Method of Rating Unitary Spot Air Conditioners	ASHRAE	ANSI/ASHRAE 128-2001
Ships	Specification for Mechanically Refrigerated Shipboard Air Conditioner	ASTM	ASTM F1433-97 (2004)
Accessories	Flashing and Stand Combination for Air Conditioning Units (Unit Curb)	IAPMO	IAPMO PS 120-2004
Air Conditioning	Commercial Applications Systems and Equipment, 1st ed.	ACCA	ACCA Manual CS
	Heat Pump Systems: Principles and Applications, 2nd ed.	ACCA	ACCA Manual H
	Residential Load Calculation, 8th ed.	ACCA	ANSI/ACCA Manual J
	Commercial Load Calculation, 4th ed.	ACCA	ACCA Manual N
	Comfort, Air Quality and Efficiency by Design	ACCA	ACCA Manual RS
	Environmental Systems Technology, 2nd ed. (1999)	NEBB	NEBB
	Installation of Air Conditioning and Ventilating Systems	NFPA	NFPA 90A-02

Selected Codes and Standards Published by Various Societies and Associations (*Continued*)

Subject	Title	Publisher	Reference
	Standard of Purity for Use in Mobile Air-Conditioning Systems	SAE	SAE J1991-1999
	HVAC Systems Applications, 1st ed.	SMACNA	SMACNA 1987
	HVAC Systems—Duct Design, 3rd ed.	SMACNA	SMACNA 1990
	Heating and Cooling Equipment (1995)	UL/CSA	ANSI/UL 1995/C22.2 No. 236-95
Aircraft	Air Conditioning of Aircraft Cargo	SAE	SAE AIR806B-1997
	Aircraft Fuel Weight Penalty Due to Air Conditioning	SAE	SAE AIR1168/8-1989
	Air Conditioning Systems for Subsonic Airplanes	SAE	SAE ARP85E-1991
	Environmental Control Systems Terminology	SAE	SAE ARP147E-2001
	Testing of Airplane Installed Environmental Control Systems (ECS)	SAE	SAE ARP217D-1999
	Guide for Qualification Testing of Aircraft Air Valves	SAE	SAE ARP986C-1997
	Control of Excess Humidity in Avionics Cooling	SAE	SAE ARP987A-1997
	Engine Bleed Air Systems for Aircraft	SAE	SAE ARP1796-1987
	Aircraft Ground Air Conditioning Service Connection	SAE	SAE AS4262A-1997
	Air Cycle Air Conditioning Systems for Military Air Vehicles	SAE	SAE AS4073-2000
Automotive	Refrigerant 12 Automotive Air-Conditioning Hose	SAE	SAE J51-2004
	Design Guidelines for Air Conditioning Systems for Off-Road Operator Enclosures	SAE	SAE J169-1985
	Test Method for Measuring Power Consumption of Air Conditioning and Brake Compressors for Trucks and Buses	SAE	SAE J1340-2003
	Information Relating to Duty Cycles and Average Power Requirements of Truck and Bus Engine Accessories	SAE	SAE J1343-2000
	Rating Air-Conditioner Evaporator Air Delivery and Cooling Capacities	SAE	SAE J1487-2004
	Recovery and Recycle Equipment for Mobile Automotive Air-Conditioning Systems	SAE	SAE J1990-1999
	R134a Refrigerant Automotive Air-Conditioning Hose	SAE	SAE J2064-2005
	Service Hose for Automotive Air Conditioning	SAE	SAE J2196-1997
Ships	Mechanical Refrigeration and Air-Conditioning Installations Aboard Ship	ASHRAE	ANSI/ASHRAE 26-1996
	Practice for Mechanical Symbols, Shipboard Heating, Ventilation, and Air Conditioning (HVAC)	ASTM	ASTM F856-97 (2004)
Air Curtains	Laboratory Methods of Testing Air Curtains for Aerodynamic Performance	AMCA	AMCA 220-05
	Air Terminals	ARI	ARI 880-98
	Standard Methods for Laboratory Airflow Measurement	ASHRAE	ANSI/ASHRAE 41.2-1987 (RA92)
	Method of Testing for Rating the Performance of Air Outlets and Inlets	ASHRAE	ANSI/ASHRAE 70-1991
	Rating the Performance of Residential Mechanical Ventilating Equipment	CSA	CAN/CSA C260-M90 (R2002)
	Air Curtains for Entranceways in Food and Food Service Establishments	NSF	NSF/ANSI 37-2005
Air Diffusion	Air Distribution Basics for Residential and Small Commercial Buildings, 1st ed.	ACCA	ACCA Manual T
	Test Code for Grilles, Registers and Diffusers	ADC	ADC 1062:GRD-84
	Method of Testing for Rating the Performance of Air Outlets and Inlets	ASHRAE	ANSI/ASHRAE 70-1991
	Method of Testing for Room Air Diffusion	ASHRAE	ANSI/ASHRAE 113-2005
Air Filters	Comfort, Air Quality, and Efficiency by Design	ACCA	ACCA Manual RS
	Industrial Ventilation: A Manual of Recommended Practice, 25th ed. (2004)	ACGIH	ACGIH
	Air Cleaners	AHAM	ANSI/AHAM AC-1-2006
	Residential Air Filter Equipment	ARI	ARI 680-2004
	Commercial and Industrial Air Filter Equipment	ARI	ARI 850-2004
	Agricultural Cabs—Engineering Control of Environmental Air Quality—Part 1: Definitions, Test Methods, and Safety Procedures	ASABE	ANSI/ASAE S525-1.2-2003
	Part 2: Pesticide Vapor Filters—Test Procedure and Performance Criteria	ASABE	ANSI/ASAE S525-2-2003
	Gravimetric and Dust-Spot Procedures for Testing Air-Cleaning Devices Used in General Ventilation for Removing Particulate Matter	ASHRAE	ANSI/ASHRAE 52.1-1992
	Method of Testing General Ventilation Air-Cleaning Devices for Removal Efficiency by Particle Size	ASHRAE	ANSI/ASHRAE 52.2-1999
	Code on Nuclear Air and Gas Treatment	ASME	ASME AG-1-2003
	Nuclear Power Plant Air-Cleaning Units and Components	ASME	ASME N509-2002
	Testing of Nuclear Air-Treatment Systems	ASME	ASME N510-1989 (RA95)
	Specification for Filter Units, Air Conditioning: Viscous-Impingement and Dry Types, Replaceable	ASTM	ASTM F1040-87 (2001)
	Test Method for Air Cleaning Performance of a High-Efficiency Particulate Air Filter System	ASTM	ASTM F1471-93 (2001)
	Specification for Filters Used in Air or Nitrogen Systems	ASTM	ASTM F1791-00
	Method for Sodium Flame Test for Air Filters	BSI	BS 3928:1969
	Particulate Air Filters for General Ventilation: Determination of Filtration Performance	BSI	BS EN 779:2002
	Electrostatic Air Cleaners (2000)	UL	ANSI/UL 867
	High-Efficiency, Particulate, Air Filter Units (1996)	UL	ANSI/UL 586
	Air Filter Units (1994)	UL	ANSI/UL 900
	Exhaust Hoods for Commercial Cooking Equipment (1995)	UL	UL 710
	Grease Filters for Exhaust Ducts (2000)	UL	UL 1046

Selected Codes and Standards Published by Various Societies and Associations (*Continued*)

Subject	Title	Publisher	Reference
Air-Handling Units	Commercial Applications Systems and Equipment, 1st ed.	ACCA	ACCA Manual CS
	Central Station Air-Handling Units	ARI	ANSI/ARI 430-99
	Non-Recirculating Direct Gas-Fired Industrial Air Heaters	CSA	ANSI Z83.4-2003/CSA 3.7-2003
Air Leakage	Residential Duct Diagnostics and Repair (2003)	ACCA	ACCA
	Air Leakage Performance for Detached Single-Family Residential Buildings	ASHRAE	ANSI/ASHRAE 119-1988 (RA04)
	Method of Determining Air Change Rates in Detached Dwellings	ASHRAE	ANSI/ASHRAE 136-1993 (RA01)
	Test Method for Determining Air Change in a Single Zone by Means of a Tracer Gas Dilution	ASTM	ASTM E741-00
	Test Method for Field Measurement of Air Leakage Through Installed Exterior Window and Doors	ASTM	ASTM E783-93 (2002)
	Test Method for Determining Air Leakage Rate by Fan Pressurization	ASTM	ASTM E779-03
	Practices for Air Leakage Site Detection in Building Envelopes and Air Retarder Systems	ASTM	ASTM E1186-03
	Test Method for Determining the Rate of Air Leakage Through Exterior Windows, Curtain Walls, and Doors Under Specified Pressure and Temperature Differences Across the Specimen	ASTM	ASTM E1424-91 (2000)
	Test Methods for Determining Airtightness of Buildings Using an Orifice Blower Door	ASTM	ASTM E1827-96 (2002)
	Practice for Determining the Effects of Temperature Cycling on Fenestration Products	ASTM	ASTM E2264-05
	Test Method for Determining Air Flow Through the Face and Sides of Exterior Windows, Curtain Walls, and Doors Under Specified Pressure Differences Across the Specimen	ASTM	ASTM E2319-04
	Test Method for Determining Air Leakage of Air Barrier Assemblies	ASTM	ASTM E2357-05
Boilers	Packaged Boiler Engineering Manual (1999)	ABMA	ABMA 100
	Selected Codes and Standards of the Boiler Industry (2001)	ABMA	ABMA 103
	Operation and Maintenance Safety Manual (1995)	ABMA	ABMA 106
	Fluidized Bed Combustion Guidelines (1995)	ABMA	ABMA 200
	Guide to Clean and Efficient Operation of Coal Stoker-Fired Boilers (2002)	ABMA	ABMA 203
	Guideline for Performance Evaluation of Heat Recovery Steam Generating Equipment (1995)	ABMA	ABMA 300
	Guidelines for Industrial Boiler Performance Improvement (1999)	ABMA	ABMA 302
	Measurement of Sound from Steam Generators (1995)	ABMA	ABMA 304
	Guideline for Gas and Oil Emission Factors for Industrial, Commercial, and Institutional Boilers (1997)	ABMA	ABMA 305
	Combustion Control Guidelines for Single Burner Firetube and Watertube Industrial/Commercial/Institutional Boilers (1999)	ABMA	ABMA 307
	Combustion Control Guidelines for Multiple-Burner Boilers (2001)	ABMA	ABMA 308
	Boiler Water Quality Requirements and Associated Steam Quality for Industrial/ Commercial and Institutional Boilers (2005)	ABMA	ABMA 402
	Commercial Applications Systems and Equipment, 1st ed.	ACCA	ACCA Manual CS
	Methods of Testing for Annual Fuel Utilization Efficiency of Residential Central Furnaces and Boilers	ASHRAE	ANSI/ASHRAE 103-1993
	Boiler and Pressure Vessel Code—Section I: Power Boilers; Section IV: Heating Boilers	ASME	BPVC-2004
	Fired Steam Generators	ASME	ASME PTC 4-1998
	Boiler, Pressure Vessel, and Pressure Piping Code	CSA	CSA B51-2003
	Testing Standard for Commercial Boilers	HYDI	HYDI BTS-2000
	Rating Procedure for Heating Boilers (2002)	HYDI	IBR
	Prevention of Furnace Explosions/Implosions in Multiple Burner Boilers	NFPA	ANSI /NFPA 8502-99
	Heating, Water Supply, and Power Boilers—Electric (2004)	UL	ANSI/UL 834
	Boiler and Combustion Systems Hazards Code	NFPA	NFPA 85-04
Gas or Oil	Gas-Fired Low-Pressure Steam and Hot Water Boilers	CSA	ANSI Z21.13-2004/CSA 4.9-2004
	Controls and Safety Devices for Automatically Fired Boilers	ASME	ASME CSD-1-2004
	Industrial and Commercial Gas-Fired Package Boilers	CSA	CAN 1-3.1-77 (R2001)
	Oil-Burning Equipment: Steam and Hot-Water Boilers	CSA	B140.7-2005
	Single Burner Boiler Operations	NFPA	ANSI/NFPA 8501-97
	Prevention of Furnace Explosions/Implosions in Multiple Burner Boilers	NFPA	ANSI/NFPA 8502-99
	Oil-Fired Boiler Assemblies (1995)	UL	UL 726
	Commercial-Industrial Gas Heating Equipment (1999)	UL	UL 795
	Standards and Typical Specifications for Tray Type Deaerators, 7th ed.	HEI	HEI 2003
Terminology	Ultimate Boiler Industry Lexicon: Handbook of Power Utility and Boiler Terms and Phrases, 6th ed. (2001)	ABMA	ABMA 101
Building Codes	ASTM Standards Used in Building Codes	ASTM	ASTM
	Practice for Conducting Visual Assessments for Lead Hazards in Buildings	ASTM	ASTM E2255-04
	Standard Practice for Periodic Inspection of Building Facades for Unsafe Conditions	ASTM	ASTM E2270-05
	Structural Welding Code—Steel	AWS	AWS D1.1/D1.1M:2006
	BOCA National Building Code, 14th ed. (1999)	BOCA	BNBC
	Uniform Building Code, vol. 1, 2, and 3 (1997)	ICBO	UBC V1, V2, V3

Selected Codes and Standards Published by Various Societies and Associations (*Continued*)

Subject	Title	Publisher	Reference
	International Building Code (2006)	ICC	IBC
	International Code Council Performance Code (2006)	ICC	ICC PC
	International Existing Building Code (2006)	ICC	IEBC
	International Energy Conservation Code (2006)	ICC	IECC
	International Property Maintenance Code (2006)	ICC	IPMC
	International Residential Code (2006)	ICC	IRC
	Directory of Building Codes and Regulations, State and City Volumes (annual)	NCSBCS	NCSBCS (electronic only)
	Building Construction and Safety Code	NFPA	ANSI/NFPA 5000-2006
	National Building Code of Canada (2005)	NRCC	NRCC
	Standard Building Code (1999)	SBCCI	SBC
Mechanical	Safety Code for Elevators and Escalators	ASME	ASME A17.1-2004
	Natural Gas and Propane Installation Code	CSA	CAN/CSA-B149.1-2005
	Propane Storage and Handling Code	CSA	CAN/CSA-B149.2-2005
	Uniform Mechanical Code (2006)	IAPMO	IAPMO
	International Mechanical Code (2006)	ICC	IMC
	International Fuel Gas Code (2006)	ICC	IFGC
	Standard Gas Code (1999)	SBCCI	SBC
	Standard Mechanical Code (1997)	SBCCI	SMC
Burners	Guidelines for Burner Adjustments of Commercial Oil-Fired Boilers (1996)	ABMA	ABMA 303
	Domestic Gas Conversion Burners	CSA	ANSI Z21.17-1998 (R2004)/ CSA 2.7-M98
	Installation of Domestic Gas Conversion Burners	CSA	ANSI Z21.8-1994 (R2002)
	Installation Code for Oil Burning Equipment	CSA	CAN/CSA-B139-04
	Oil-Burning Equipment: General Requirements	CSA	CAN/CSA-B140.0-03
	Vapourizing-Type Oil Burners	CSA	B140.1-1966 (R2001)
	Oil Burners: Atomizing-Type	CSA	CAN/CSA-B140.2.1-M90 (R2005)
	Pressure Atomizing Oil Burner Nozzles	CSA	B140.2.2-1971 (R2001)
	Oil Burners (2003)	UL	ANSI/UL 296
	Waste Oil-Burning Air-Heating Appliances (1995)	UL	ANSI/UL 296A
	Commercial-Industrial Gas Heating Equipment (1999)	UL	UL 795
	Commercial/Industrial Gas and/or Oil-Burning Assemblies with Emission Reduction Equipment (1999)	UL	UL 2096
Chillers	Commercial Applications Systems and Equipment, 1st ed.	ACCA	ACCA Manual CS
	Absorption Water Chilling and Water Heating Packages	ARI	ARI 560-2000
	Water Chilling Packages Using the Vapor Compression Cycle	ARI	ARI 550/590-2003
	Method of Testing Liquid-Chilling Packages	ASHRAE	ANSI/ASHRAE 30-1995
	Performance Standard for Rating Packaged Water Chillers	CSA	CAN/CSA C743-2002
Chimneys	Specification for Clay Flue Liners	ASTM	ASTM C315-02
	Specification for Industrial Chimney Lining Brick	ASTM	ASTM C980-88 (2001)
	Practice for Installing Clay Flue Lining	ASTM	ASTM C1283-03E01
	Guide for Design and Construction of Brick Liners for Industrial Chimneys	ASTM	ASTM C1298-95 (2001)
	Guide for Design, Fabrication, and Erection of Fiberglass Reinforced Plastic Chimney Liners with Coal-Fired Units	ASTM	ASTM D5364-93 (2002)
	Chimneys, Fireplaces, Vents, and Solid Fuel-Burning Appliances	NFPA	ANSI/NFPA 211-06
	Medium Heat Appliance Factory-Built Chimneys (2001)	UL	ANSI/UL 959
	Factory-Built Chimneys for Residential Type and Building Heating Appliance (2001)	UL	ANSI/UL 103
Cleanrooms	Practice for Cleaning and Maintaining Controlled Areas and Clean Rooms	ASTM	ASTM E2042-04
	Practice for Design and Construction of Aerospace Cleanrooms and Contamination Controlled Areas	ASTM	ASTM E2217-02
	Practice for Tests of Cleanroom Materials	ASTM	ASTM E2312-04
	Practice for Aerospace Cleanrooms and Associated Controlled Environments—Cleanroom Operations	ASTM	ASTM E2352-04
	Practice for Continuous Sizing and Counting of Airborne Particles in Dust-Controlled Areas and Clean Rooms Using Instruments Capable of Detecting Single Sub-Micrometre and Larger Particles	ASTM	ASTM F50-92 (2001)E01
	Test Method for Sizing and Counting Airborne Particulate Contamination in Clean Rooms and Other Dust-Controlled Areas Designed for Electronic and Similar Applications	ASTM	ASTM F25-04
	Procedural Standards for Certified Testing of Cleanrooms, 2nd ed. (1996)	NEBB	NEBB
Coils	Forced-Circulation Air-Cooling and Air-Heating Coils	ARI	ARI 410-2001
	Methods of Testing Forced Circulation Air Cooling and Air Heating Coils	ASHRAE	ANSI/ASHRAE 33-2000
Comfort Conditions	Threshold Limit Values for Physical Agents (updated annually)	ACGIH	ACGIH
	Good HVAC Practices for Residential and Commercial Buildings (2003)	ACCA	ACCA
	Comfort, Air Quality and Efficiency by Design (1997)	ACCA	ACCA Manual RS

Selected Codes and Standards Published by Various Societies and Associations (*Continued*)

Subject	Title	Publisher	Reference
	Thermal Environmental Conditions for Human Occupancy	ASHRAE	ANSI/ASHRAE 55-2004
	Classification for Serviceability of an Office Facility for Thermal Environment and Indoor Air Conditions	ASTM	ASTM E2320-04
	Hot Environments—Estimation of the Heat Stress on Working Man, Based on the WBGT Index (Wet Bulb Globe Temperature)	ISO	ISO 7243:1989
	Ergonomics of the Thermal Environment—Analytical Determination and Interpretation of Thermal Comfort Using Calculation of the PMV and PPD Indices and Local Thermal Comfort Criteria	ISO	ISO 7730:2005
	Ergonomics of the Thermal Environment—Determination of Metabolic Rate	ISO	ISO 8996:2004
	Ergonomics of the Thermal Environment—Estimation of the Thermal Insulation and Evaporative Resistance of a Clothing Ensemble	ISO	ISO 9920:1995
Compressors	Safety Standard for Air Compressor Systems	ASME	ASME B19.1-1995
	Safety Standard for Compressors for Process Industries	ASME	ASME B19.3-1991
	Displacement Compressors, Vacuum Pumps and Blowers	ASME	ASME PTC 9-1970 (RA97)
	Performance Test Code on Compressors and Exhausters	ASME	ASME PTC 10-1997 (RA03)
	Compressed Air and Gas Handbook, 6th ed. (2003)	CAGI	CAGI
Refrigerant	Variable Capacity Positive Displacement Refrigerant Compressors and Compressor Units for Air-Conditioning and Heat Pump Applications	ARI	ANSI/ARI 500-2000
	Positive Displacement Condensing Units	ARI	ARI 520-2004
	Positive Displacement Refrigerant Compressors and Compressor Units	ARI	ARI 540-2004
	Safety Standard for Refrigeration Systems	ASHRAE	ANSI/ASHRAE 15-2004
	Methods of Testing for Rating Positive Displacement Refrigerant Compressors and Condensing Units	ASHRAE	ANSI/ASHRAE 23-2005
	Testing of Refrigerant Compressors	ISO	ISO 917:1989
	Refrigerant Compressors—Presentation of Performance Data	ISO	ISO 9309:1989
	Hermetic Refrigerant Motor-Compressors (1996)	UL/CSA	UL 984/C22.2 No.140.2-96 (R2001)
Computers	Method of Rating Computer and Data Processing Room Unitary Air Conditioners	ASHRAE	ANSI/ASHRAE 127-2001
	Method of Test for the Evaluation of Building Energy Analysis Computer Programs	ASHRAE	ANSI/ASHRAE 140-2004
	Protection of Electronic Computer/Data Processing Equipment	NFPA	NFPA 75-03
Condensers	Commercial Applications Systems and Equipment, 1st ed.	ACCA	ACCA Manual CS
	Water-Cooled Refrigerant Condensers, Remote Type	ARI	ARI 450-99
	Remote Mechanical-Draft Air-Cooled Refrigerant Condensers	ARI	ARI 460-2005
	Remote Mechanical Draft Evaporative Refrigerant Condensers	ARI	ARI 490-2003
	Safety Standard for Refrigeration Systems	ASHRAE	ANSI/ASHRAE 15-2004
	Method of Testing for Rating Remote Mechanical-Draft Air-Cooled Refrigerant Condensers	ASHRAE	ANSI/ASHRAE 20-1997
	Methods of Testing for Rating Water-Cooled Refrigerant Condensers	ASHRAE	ANSI/ASHRAE 22-2003
	Methods of Laboratory Testing Remote Mechanical-Draft Evaporative Refrigerant Condensers	ASHRAE	ANSI/ASHRAE 64-2005
	Steam Surface Condensers	ASME	ASME PTC 12.2-1998
	Standards for Steam Surface Condensers, 9th ed., Addendum 1	HEI	HEI 2002
	Standards for Direct Contact Barometric and Low Level Condensers, 7th ed.	HEI	HEI 2004
	Refrigerant-Containing Components and Accessories, Nonelectrical (2001)	UL	ANSI/UL 207
Condensing Units	Commercial Applications Systems and Equipment	ACCA	ACCA Manual CS
	Commercial and Industrial Unitary Air-Conditioning Condensing Units	ARI	ARI 365-2002
	Methods of Testing for Rating Positive Displacement Refrigerant Compressors and Condensing Units	ASHRAE	ANSI/ASHRAE 23-2005
	Heating and Cooling Equipment (2005)	UL/CSA	ANSI/UL 1995/C22.2 No. 236-95
Containers	Series 1 Freight Containers—Classifications, Dimensions, and Ratings	ISO	ISO 668:1995
	Series 1 Freight Containers—Specifications and Testing; Part 2: Thermal Containers	ISO	ISO 1496-2:1996
	Animal Environment in Cargo Compartments	SAE	SAE AIR1600A-1997
Controls	Temperature Control Systems (2002)	AABC	National Standards, Ch. 12
	Energy Management Control Systems Instrumentation	ASHRAE	ASHRAE 114-1986
	BACnet®—A Data Communication Protocol for Building Automation and Control Networks	ASHRAE	ANSI/ASHRAE 135-2004
	Method of Test for Conformance to BACnet®	ASHRAE	ANSI/ASHRAE 135.1-2003
	Temperature-Indicating and Regulating Equipment	CSA	C22.2 No. 24-93 (R2003)
	Performance Requirements for Electric Heating Line-Voltage Wall Thermostats	CSA	C273.4-M1978 (R2003)
	Performance Requirements for Thermostats Used with Individual Room Electric Space Heating Devices	CSA	CAN/CSA C828-99 (R2005)
	Solid-State Controls for Appliances (2003)	UL	UL 244A
	Limit Controls (1994)	UL	ANSI/UL 353
	Primary Safety Controls for Gas- and Oil-Fired Appliances (1994)	UL	ANSI/UL 372
	Temperature-Indicating and -Regulating Equipment (1994)	UL	UL 873
	Tests for Safety-Related Controls Employing Solid-State Devices (2004)	UL	UL 991

Selected Codes and Standards Published by Various Societies and Associations (*Continued*)

Subject	Title	Publisher	Reference
	Control Centers for Changing Message Type Electric Signals (2003)	UL	UL 1433
	Automatic Electrical Controls for Household and Similar Use; Part 1: General Requirements (2002)	UL	UL 60730-1A
	Process Control Equipment (2002)	UL	UL 61010C-1
Commercial and Industrial	Guidelines for Boiler Control Systems (Gas/Oil Fired Boilers) (1998)	ABMA	ABMA 301
	Guideline for the Integration of Boilers and Automated Control Systems in Heating Applications (1998)	ABMA	ABMA 306
	Industrial Control and Systems: General Requirements	NEMA	NEMA ICS 1-2000 (R2005)
	Preventive Maintenance of Industrial Control and Systems Equipment	NEMA	NEMA ICS 1.3-1986 (R2005)
	Industrial Control and Systems, Controllers, Contactors, and Overload Relays Rated Not More than 2000 Volts AC or 750 Volts DC	NEMA	NEMA ICS 2-2000 (R2005)
	Industrial Control and Systems: Instructions for the Handling, Installation, Operation and Maintenance of Motor Control Centers Rated Not More than 600 Volts	NEMA	NEMA ICS 2.3-1995 (R2002)
	Industrial Control Equipment (1999)	UL	ANSI/UL 508
Residential	Manually Operated Gas Valves for Appliances, Appliance Connector Valves and Hose End Valves	CSA	ANSI Z21.15-1997 (R03)/CGA 9.1-1997
	Gas Appliance Pressure Regulators	CSA	ANSI Z21.18-2000/CSA 6.3-2000
	Automatic Gas Ignition Systems and Components	CSA	ANSI Z21.20-2005
	Gas Appliance Thermostats	CSA	ANSI Z21.23-2000
	Manually-Operated Piezo-Electric Spark Gas Ignition Systems and Components	CSA	ANSI Z21.77-2005/CGA 6.23-2005
	Manually Operated Electric Gas Ignition Systems and Components	CSA	ANSI Z21.92-2001/CSA 6.29-2001
	Residential Controls—Electrical Wall-Mounted Room Thermostats	NEMA	NEMA DC 3-2003
	Residential Controls—Surface Type Controls for Electric Storage Water Heaters	NEMA	NEMA DC 5-2002
	Residential Controls—Temperature Limit Controls for Electric Baseboard Heaters	NEMA	NEMA DC 10-1983 (R2003)
	Hot-Water Immersion Controls	NEMA	NEMA DC 12-1985 (R2002))
	Line-Voltage Integrally Mounted Thermostats for Electric Heaters	NEMA	NEMA DC 13-1979 (R2002)
	Residential Controls—Class 2 Transformers	NEMA	NEMA DC 20-1992 (R2003)
	Safety Guidelines for the Application, Installation, and Maintenance of Solid State Controls	NEMA	NEMA ICS 1.1-1984 (R2003)
	Electrical Quick-Connect Terminals (2003)	UL	ANSI/UL 310
Coolers	Refrigeration Equipment	CSA	CAN/CSA-C22.2 No. 120-M91 (R2004)
	Unit Coolers for Refrigeration	ARI	ARI 420-2000
	Refrigeration Unit Coolers (2004)	UL	ANSI/UL 412
Air	Methods of Testing Forced Convection and Natural Convection Air Coolers for Refrigeration	ASHRAE	ANSI/ASHRAE 25-2001
Drinking Water	Self-Contained, Mechanically Refrigerated Drinking-Water Coolers	ARI	ARI 1010-2002
	Methods of Testing for Rating Drinking-Water Coolers with Self-Contained Mechanical Refrigeration Systems	ASHRAE	ANSI/ASHRAE 18-1987 (RA97)
	Drinking-Water Coolers (1993)	UL	ANSI/UL 399
	Drinking Water System Components—Health Effects	NSF	NSF/ANSI 61-2005
Evaporative	Method of Testing Direct Evaporative Air Coolers	ASHRAE	ANSI/ASHRAE 133-2001
	Method of Test for Rating Indirect Evaporative Coolers	ASHRAE	ANSI/ASHRAE 143-2000
Food and Beverage	Terminology for Milking Machines, Milk Cooling, and Bulk Milk Handling Equipment	ASABE	ASAE S300.3-2003
	Methods of Testing for Rating Vending Machines for Bottled, Canned, and Other Sealed Beverages	ASHRAE	ANSI/ASHRAE 32.1-2004
	Methods of Testing for Rating Pre-Mix and Post-Mix Beverage Dispensing Equipment	ASHRAE	ANSI/ASHRAE 32.2-2003
	Manual Food and Beverage Dispensing Equipment	NSF	NSF/ANSI 18-2005
	Commercial Bulk Milk Dispensing Equipment	NSF	NSF/ANSI 20-2000
	Refrigerated Vending Machines (1995)	UL	ANSI/UL 541
Liquid	Refrigerant-Cooled Liquid Coolers, Remote Type	ARI	ARI 480-2001
	Methods of Testing for Rating Liquid Coolers	ASHRAE	ANSI/ASHRAE 24-2000 (RA05)
	Liquid Cooling Systems	SAE	SAE AIR1811A-1997
Cooling Towers	Cooling Tower Testing (2002)	AABC	National Standards, Ch 13
	Commercial Applications, Systems, and Equipment, 1st ed.	ACCA	ACCA Manual CS
	Bioaerosols: Assessment and Control (1999)	ACGIH	ACGIH
	Atmospheric Water Cooling Equipment	ASME	ASME PTC 23-2003
	Water-Cooling Towers	NFPA	NFPA 214-05
	Acceptance Test Code for Water Cooling Towers	CTI	CTI ATC-105 (00)
	Code for Measurement of Sound from Water Cooling Towers (2005)	CTI	CTI ATC-128 (05)
	Acceptance Test Code for Spray Cooling Systems (1985)	CTI	CTI ATC-133 (85)
	Nomenclature for Industrial Water Cooling Towers (1997)	CTI	CTI NCL-109 (97)
	Recommended Practice for Airflow Testing of Cooling Towers (1994)	CTI	CTI PFM-143 (94)
	Fiberglass-Reinforced Plastic Panels (2002)	CTI	CTI STD-131 (02)
	Certification of Water Cooling Tower Thermal Performance (R2004)	CTI	CTI STD-201 (02)
Crop Drying	Density, Specific Gravity, and Mass-Moisture Relationships of Grain for Storage	ASABE	ANSI/ASAE D241.4-2003
	Dielectric Properties of Grain and Seed	ASABE	ASAE D293.2-1989 (R2005)
	Shelled Corn Storage Time for 0.5% Dry Matter Loss	ASABE	ASAE D535-2005

Selected Codes and Standards Published by Various Societies and Associations (*Continued*)

Subject	Title	Publisher	Reference
	Thermal Properties of Grain and Grain Products	ASABE	ASAE D243.4-2003
	Moisture Relationships of Plant-Based Agricultural Products	ASABE	ASAE D245.5-19995 (R2001)
	Construction and Rating of Equipment for Drying Farm Crops	ASABE	ASAE S248.3-1976 (R2005)
	Cubes, Pellets, and Crumbles—Definitions and Methods for Determining Density, Durability, and Moisture Content	ASABE	ASAE S269.4-2001
	Resistance to Airflow of Grains, Seeds, Other Agricultural Products, and Perforated Metal Sheets	ASABE	ASAE D272.3-2001
	Moisture Measurement—Unground Grain and Seeds	ASABE	ASAE S352.2-2003
	Moisture Measurement—Meat and Meat Products	ASABE	ASAE S353-2003
	Moisture Measurement—Forages	ASABE	ASAE S358.2-2003
	Moisture Measurement—Peanuts	ASABE	ASAE S410.1-2003
	Energy Efficiency Test Procedure for Tobacco Curing Structures	ASABE	ASAE S416-2003
	Thin-Layer Drying of Agricultural Crops	ASABE	ANSI/ASAE S448.1-2001 (R2006)
	Moisture Measurement—Tobacco	ASABE	ASAE S487-2003
	Thin-Layer Drying of Agricultural Crops	ASABE	ASAE S488-1990 (R2005)
	Temperature Sensor Locations for Seed-Cotton Drying Systems	ASABE	ASAE S530-2001
Dehumidifiers	Commercial Applications, Systems, and Equipment, 1st ed.	ACCA	ACCA Manual CS
	Bioaerosols: Assessment and Control (1999)	ACGIH	ACGIH
	Dehumidifiers	AHAM	ANSI/AHAM DH-1-2003
	Testing for Rating Desiccant Dehumidifiers Utilizing Heat for the Regeneration Process	ASHRAE	ANSI/ASHRAE 139-1998
	Moisture Separator Reheaters	ASME	PTC 12.4-1992 (RA04)
	Dehumidifiers	CSA	C22.2 No. 92-1971 (R2004)
	Performance of Dehumidifiers	CSA	CAN/CSA C749-94 (R2005)
	Dehumidifiers (2004)	UL	ANSI/UL 474
Desiccants	Method of Testing Desiccants for Refrigerant Drying	ASHRAE	ANSI/ASHRAE 35-1992
Documentation	Preparation of Operating and Maintenance Documentation for Building Systems	ASHRAE	ASHRAE *Guideline* 4-1993
Driers	Liquid-Line Driers	ARI	ANSI/ARI 710-2004
	Method of Testing Liquid Line Refrigerant Driers	ASHRAE	ANSI/ASHRAE 63.1-1995 (RA01)
	Refrigerant-Containing Components and Accessories, Nonelectrical (2001)	UL	ANSI/UL 207
Ducts and Fittings	Hose, Air Duct, Flexible Nonmetallic, Aircraft	SAE	SAE AS1501C-1994
	Ducted Electric Heat Guide for Air Handling Systems, 2nd ed.	SMACNA	SMACNA 1994
	Factory-Made Air Ducts and Air Connectors (2005)	UL	ANSI/UL 181
Construction	Industrial Ventilation: A Manual of Recommended Practice, 25th ed. (2004)	ACGIH	ACGIH
	Preferred Metric Sizes for Flat, Round, Square, Rectangular, and Hexagonal Metal Products	ASME	ASME B32.100-2005
	Sheet Metal Welding Code	AWS	AWS D9.1:2000
	Fibrous Glass Duct Construction Standards, 5th ed.	NAIMA	NAIMA AH116
	Residential Fibrous Glass Duct Construction Standards, 3rd ed.	NAIMA	NAIMA AH119
	Fibrous Glass Duct Construction with 1-1/2″ Duct Boards	NAIMA	NAIMA AH120
	Thermoplastic Duct (PVC) Construction Manual, 2nd ed.	SMACNA	SMACNA 1995
	Accepted Industry Practices for Sheet Metal Lagging, 1st ed.	SMACNA	SMACNA 2002
	Fibrous Glass Duct Construction Standards, 7th ed.	SMACNA	SMACNA 2003
	HVAC Duct Construction Standards, Metal and Flexible, 3rd ed.	SMACNA	SMACNA 2005
	Rectangular Industrial Duct Construction Standards, 2nd ed.	SMACNA	SMACNA 2004
	Round Industrial Duct Construction Standards, 2nd ed.	SMACNA	ANSI/SMACNA 005-2003
Industrial	Round Industrial Duct Construction Standards, 2nd ed.	SMACNA	SMACNA 1999
	Rectangular Industrial Duct Construction Standards, 2nd ed.	SMACNA	SMACNA 2004
Installation	Flexible Duct Performance and Installation Standards, 4th ed.	ADC	ADC-91
	Installation of Air Conditioning and Ventilating Systems	NFPA	NFPA 90A-02
	Installation of Warm Air Heating and Air-Conditioning Systems	NFPA	NFPA 90B-06
Material Specifications	Specification for General Requirements for Flat-Rolled Stainless and Heat-Resisting Steel Plate, Sheet and Strip	ASTM	ASTM A480/A480M-05
	Specification for General Requirements for Steel, Sheet, Carbon, and High-Strength, Low-Alloy, Hot-Rolled and Cold-Rolled	ASTM	ASTM A568/A568M-05a
	Specification for Steel Sheet, Zinc-Coated (Galvanized) or Zinc-Iron Alloy-Coated (Galvannealed) by the Hot-Dipped Process	ASTM	ASTM A653/A653M-05a
	Specification for General Requirements for Steel Sheet, Metallic-Coated by the Hot-Dip Process	ASTM	ASTM A924/A924M-04
	Specification for Steel, Sheet and Strip, Cold-Rolled, Carbon, Structural, High-Strength Low-Alloy and High-Strength Low-Alloy with Improved Formability	ASTM	ASTM A1008-05b
	Specification for Steel, Sheet and Strip, Hot-Rolled, Carbon, Structural, High-Strength Low-Alloy and High-Strength Low-Alloy with Improved Formability	ASTM	ASTM A1011/A1011M-05a
	Practice for Measuring Flatness Characteristics of Coated Sheet Products	ASTM	ASTM A1030-05
System Design	Installation Techniques for Perimeter Heating and Cooling, 11th ed.	ACCA	ACCA Manual 4
	Residential Duct Systems	ACCA	ANSI/ACCA Manual D
	Commercial Low Pressure, Low Velocity Duct System Design, 1st ed.	ACCA	ACCA Manual Q

Selected Codes and Standards Published by Various Societies and Associations (*Continued*)

Subject	Title	Publisher	Reference
	Air Distribution Basics for Residential and Small Commercial Buildings, 1st ed.	ACCA	ACCA Manual T
	Method of Test for Determining the Design and Seasonal Efficiencies of Residential Thermal Distribution Systems	ASHRAE	ANSI/ASHRAE 152-2004
	Closure Systems for Use with Rigid Air Ducts (2005)	UL	ANSI/UL 181A
	Closure Systems for Use with Flexible Air Ducts and Air Connectors (2005)	UL	ANSI/UL 181B
Testing	Duct Leakage Testing (2002)	AABC	National Standards, Ch 5
	Residential Duct Diagnostics and Repair (2003)	ACCA	ACCA
	Flexible Air Duct Test Code	ADC	ADC FD-72 (R1979)
	Test Method for Measuring Acoustical and Airflow Performance of Duct Liner Materials and Prefabricated Silencers	ASTM	ASTM E477-05
	Method of Testing to Determine Flow Resistance of HVAC Ducts and Fittings	ASHRAE	ANSI/ASHRAE 120-1999
	Method of Testing HVAC Air Ducts	ASHRAE	ANSI/ASHRAE/SMACNA 126-2000
	HVAC Air Duct Leakage Test Manual, 1st ed.	SMACNA	SMACNA 1985
	HVAC Duct Systems Inspection Guide, 3rd ed.	SMACNA	SMACNA 2005
Electrical	Electrical Power Systems and Equipment—Voltage Ratings	ANSI	ANSI C84.1-2001
	Test Method for Bond Strength of Electrical Insulating Varnishes by the Helical Coil Test	ASTM	ASTM D2519-02
	Standard Specification for Shelter, Electrical Equipment, Lightweight	ASTM	ASTM E2377-04
	Canadian Electrical Code, Part I (20th ed.)	CSA	C22.1-2006
	Part II—General Requirements	CSA	CAN/CSA-C22.2 No.0-M91 (R2001)
	ICC Electrical Code, Administrative Provisions (2006)	ICC	ICCEC
	Enclosures for Electrical Equipment (1000 Volts Maximum)	NEMA	ANSI/NEMA 250-2003
	Molded Case Circuit Breakers and Molded Case Switches	NEMA	NEMA AB 1-2002
	Low Voltage Cartridge Fuses	NEMA	NEMA FU 1-2002
	Industrial Control and Systems: Terminal Blocks	NEMA	NEMA ICS 4-2005
	Industrial Control and Systems: Enclosures	NEMA	ANSI/NEMA ICS 6-1993 (R2001)
	Application Guide for Ground Fault Protective Devices for Equipment	NEMA	ANSI/NEMA PB 2.2-2004
	General Color Requirements for Wiring Devices	NEMA	NEMA WD 1-1999 (R2005)
	Wiring Devices—Dimensional	NEMA	ANSI/NEMA WD 6-2002
	National Electrical Code	NFPA	NFPA 70-05
	National Fire Alarm Code	NFPA	NFPA 72-02
	Compatibility of Electrical Connectors and Wiring	SAE	SAE AIR1329A-1988
Energy	Air-Conditioning and Refrigerating Equipment Nameplate Voltages	ARI	ARI 110-2002
	Comfort, Air Quality and Efficiency by Design	ACCA	ACCA Manual RS
	Energy Standard for Buildings Except Low-Rise Residential Buildings	ASHRAE	ANSI/ASHRAE/IESNA 90.1-2004
	Energy-Efficient Design of Low-Rise Residential Buildings	ASHRAE	ANSI/ASHRAE/IESNA 90.2-2004
	Energy Conservation in Existing Buildings	ASHRAE	ANSI/ASHRAE 100-1995
	Standard Methods of Measuring and Expressing Building Energy Performance	ASHRAE	ANSI/ASHRAE 105-1984 (RA99)
	Method of Test for the Evaluation of Building Energy Analysis Computer Programs	ASHRAE	ANSI/ASHRAE 140-2004
	Method of Test for Determining the Design and Seasonal Efficiencies of Residential Thermal Distribution Systems	ASHRAE	ANSI/ASHRAE 152-2004
	Fuel Cell Power Systems Performance	ASME	PTC 50-2002
	International Energy Conservation Code (2006)	ICC	IECC
	Uniform Solar Energy Code (2000)	IAPMO	IAPMO
	Model Energy Code, Thermal Envelope Compliance Guide for One and Two Family Dwellings	NAIMA	NAIMA BI407
	Energy Management Guide for Selection and Use of Fixed Frequency Medium AC Squirrel-Cage Polyphase Induction Motors	NEMA	NEMA MG 10-2001
	Energy Management Guide for Selection and Use of Single-Phase Motors	NEMA	NEMA MG 11-1977 (R2001)
	HVAC Systems Commissioning Manual, 1st ed.	SMACNA	SMACNA 1994
	Building Systems Analysis and Retrofit Manual, 1st ed.	SMACNA	SMACNA 1995
	Energy Systems Analysis and Management, 1st ed.	SMACNA	SMACNA 1997
	Energy Management Equipment (1998)	UL	UL 916
Exhaust Systems	Fan Systems: Supply/Return/Relief/Exhaust (2002)	AABC	National Standards, Ch 10
	Commercial Applications, Systems, and Equipment, 1st ed.	ACCA	ACCA Manual CS
	Industrial Ventilation: A Manual of Recommended Practice, 25th ed. (2004)	ACGIH	ACGIH
	Fundamentals Governing the Design and Operation of Local Exhaust Ventilation Systems	AIHA	ANSI/AIHA Z9.2-2001
	Safety Code for Design, Construction, and Ventilation of Spray Finishing Operations	AIHA	ANSI/AIHA Z9.3-1994
	Abrasive Blasting Operations—Ventilation and Safe Practices	AIHA	ANSI/AIHA Z9.4-1997
	Laboratory Ventilation	AIHA	ANSI/AIHA Z9.5-2003
	Exhaust Systems for Grinding, Polishing, and Buffing	AIHA	ANSI/AIHA Z9.6-1999
	Recirculation of Air from Industrial Process Exhaust Systems	AIHA	ANSI/AIHA Z9.7-1998
	Method of Testing Performance of Laboratory Fume Hoods	ASHRAE	ANSI/ASHRAE 110-1995
	Ventilation for Commercial Cooking Operations	ASHRAE	ANSI/ASHRAE 154-2003

Selected Codes and Standards Published by Various Societies and Associations (*Continued*)

Subject	Title	Publisher	Reference
	Performance Test Code on Compressors and Exhausters	ASME	PTC 10-1997 (RA03)
	Flue and Exhaust Gas Analyses	ASME	PTC 19.10-1981
	Mechanical Flue-Gas Exhausters	CSA	CAN 3-B255-M81 (R2005)
	Exhaust Systems for Air Conveying of Vapors, Gases, Mists, and Noncombustible Particulate Solids	NFPA	ANSI/NFPA 91-04
	Draft Equipment (1993)	UL	UL 378
Expansion Valves	Thermostatic Refrigerant Expansion Valves	ARI	ANSI/ARI 750-2001
	Method of Testing Capacity of Thermostatic Refrigerant Expansion Valves	ASHRAE	ANSI/ASHRAE 17-1998 (RA03)
Fan-Coil Units	Industrial Ventilation: A Manual of Recommended Practice, 25th ed. (2004)	ACGIH	ACGIH
	Room Fan-Coils	ARI	ARI 440-2005
	Methods of Testing for Rating Fan-Coil Conditioners	ASHRAE	ANSI/ASHRAE 79-2002
	Heating and Cooling Equipment (2005)	UL/CSA	ANSI/UL 1995/C22.2 No. 236-95
Fans	Residential Duct Systems	ACCA	ANSI/ACCA Manual D
	Commercial Low Pressure, Low Velocity Duct System Design, 1st ed.	ACCA	ACCA Manual Q
	Industrial Ventilation: A Manual of Recommended Practice, 25th ed. (2004)	ACGIH	ACGIH
	Standards Handbook	AMCA	AMCA 99-03
	Drive Arrangements for Centrifugal Fans	AMCA	ANSIAMCA 99-2404-03
	Inlet Box Positions for Centrifugal Fans	AMCA	ANSI/AMCA 99-2405-03
	Designation for Rotation and Discharge of Centrifugal Fans	AMCA	ANSI/AMCA 99-2406-03
	Motor Positions for Belt or Chain Drive Centrifugal Fans	AMCA	ANSI/AMCA 99-2407-03
	Operating Limits for Centrifugal Fans	AMCA	AMCA 99-2408-69
	Drive Arrangements for Tubular Centrifugal Fans	AMCA	ANSI/AMCA 99-2410-03
	Impeller Diameters and Outlet Areas for Centrifugal Fans	AMCA	ANSI/AMCA 99-2412-03
	Impeller Diameters and Outlet Areas for Industrial Centrifugal Fans	AMCA	ANSI/AMCA 99-2413-03
	Impeller Diameters and Outlet Areas for Tubular Centrifugal Fans	AMCA	ANSI/AMCA 99-2414-03
	Dimensions for Axial Fans	AMCA	ANSI/AMCA 99-3001-03
	Drive Arrangements for Axial Fans	AMCA	ANSI/AMCA 99-3404-03
	Air Systems	AMCA	AMCA 200-95
	Fans and Systems	AMCA	AMCA 201-02
	Troubleshooting	AMCA	AMCA 202-98
	Field Performance Measurement of Fan Systems	AMCA	AMCA 203-90
	Balance Quality and Vibration Levels for Fans	AMCA	ANSI/AMCA 204-05
	Laboratory Methods of Testing Air Circulator Fans for Rating	AMCA	ANSI/AMCA 230-99
	Laboratory Method of Testing Positive Pressure Ventilators for Rating	AMCA	ANSI/AMCA 240-96
	Reverberant Room Method for Sound Testing of Fans	AMCA	AMCA 300-05
	Methods for Calculating Fan Sound Ratings from Laboratory Test Data	AMCA	AMCA 301-05
	Application of Sone Ratings for Non-Ducted Air Moving Devices	AMCA	AMCA 302-73
	Application of Sound Power Level Ratings for Fans	AMCA	AMCA 303-79
	Recommended Safety Practices for Users and Installers of Industrial and Commercial Fans	AMCA	AMCA 410-96
	Industrial Process/Power Generation Fans: Site Performance Test Standard	AMCA	AMCA 803-02
	Mechanical Balance of Fans and Blowers	ARI	ARI *Guideline* G-2002
	Acoustics—Measurement of Noise and Vibration of Small Air-Moving Devices—Part 1: Airborne Noise Emission	ASA	ANSI S12.11-2003/Part 1/ISO 10302:1996 (MOD)
	Part 2: Structure-Borne Vibration	ASA	ANSI S12.11-2003/Part 2
	Laboratory Methods of Testing Fans for Aerodynamic Performance Rating	ASHRAE/ AMCA	ANSI/ASHRAE 51-1999 ANSI/AMCA 210-99
	Laboratory Method of Testing to Determine the Sound Power in a Duct	ASHRAE/ AMCA	ANSI/ASHRAE 68-1997 ANSI/AMCA 330-97
	Methods of Testing Fan Vibration—Blade Vibrations and Critical Speeds	ASHRAE	ANSI/ASHRAE 87.1-1992
	Laboratory Methods of Testing Fans Used to Exhaust Smoke in Smoke Management Systems	ASHRAE	ANSI/ASHRAE 149-2000 (RA05)
	Ventilation for Commercial Cooking Operations	ASHRAE	ANSI/ASHRAE 154-2003
	Fans	ASME	ANSI/ASME PTC 11-1984 (RA03)
	Fans and Ventilators	CSA	C22.2 No. 113-M1984 (R2004)
	Rating the Performance of Residential Mechanical Ventilating Equipment	CSA	CAN/CSA C260-M90 (R2002)
	Energy Performance of Ceiling Fans	CSA	CAN/CSA C814-96 (R2001)
	Electric Fans (1999)	UL	ANSI/UL 507
	Power Ventilators (2004)	UL	ANSI/UL 705
Fenestration	Test Method for Accelerated Weathering of Sealed Insulating Glass Units	ASTM	ASTM E773-01
	Specification for Classification of the Durability of Sealed Insulating Glass Units	ASTM	ASTM E774-97
	Practice for Calculation of Photometric Transmittance and Reflectance of Materials to Solar Radiation	ASTM	ASTM E971-88 (2003)
	Test Method for Solar Photometric Transmittance of Sheet Materials Using Sunlight	ASTM	ASTM E972-96 (2002)
	Test Method for Solar Transmittance (Terrestrial) of Sheet Materials Using Sunlight	ASTM	ASTM E1084-86 (2003)
	Practice for Determining the Load Resistance of Glass in Buildings	ASTM	ASTM E1300-04 E01

Selected Codes and Standards Published by Various Societies and Associations (*Continued*)

Subject	Title	Publisher	Reference
	Practice for Installation of Exterior Windows, Doors and Skylights	ASTM	ASTM E2112-01
	Test Method for Insulating Glass Unit Performance	ASTM	ASTM E2188-02
	Test Method for Testing Resistance to Fogging Insulating Glass Units	ASTM	ASTM E2189-02
	Specification for Insulating Glass Unit Performance and Evaluation	ASTM	ASTM E2190-02
	Guide for Assessing the Durability of Absorptive Electrochemical Coatings within Sealed Insulating Glass Units	ASTM	ASTM E2354-04
	Tables for Reference Solar Spectral Irradiance: Direct Normal and Hemispherical on 37° Tilted Surface	ASTM	ASTM G173-03
	Windows	CSA	A440-00
	Energy Performance Evaluation of Windows and Other Fenestration Systems	CSA	A440.2-04
	Window and Door Installation	CSA	A440.4-98
	Energy Performance Evaluation of Swinging Doors	CSA	A453-95 (R2000)
Filter-Driers	Flow-Capacity Rating of Suction-Line Filters and Suction-Line Filter-Driers	ARI	ARI 730-2005
	Method of Testing Liquid Line Filter-Drier Filtration Capability	ASHRAE	ANSI/ASHRAE 63.2-1996
	Method of Testing Flow Capacity of Suction Line Filters and Filter-Driers	ASHRAE	ANSI/ASHRAE 78-1985 (RA03)
Fireplaces	Factory-Built Fireplaces (1996)	UL	ANSI/UL 127
	Fireplace Stoves (1996)	UL	UL 737
Fire Protection	Test Method for Surface Burning Characteristics of Building Materials	ASTM/NFPA	ASTM E84-05e1/NFPA 255-06
	Test Methods for Fire Test of Building Construction and Materials	ASTM	ASTM E119-05a
	Test Method for Room Fire Test of Wall and Ceiling Materials and Assemblies	ASTM	ASTM E2257-03
	Test Method for Determining Fire Resistance of Perimeter Fire Barriers Using Intermediate-Scale Multi-Story Test Apparatus	ASTM	ASTM E2307-04e01
	Guide for Laboratory Monitors	ASTM	ASTM E2335-04
	Test Method for Fire Resistance Grease Duct Enclosure Systems	ASTM	ASTM E2336-04
	Practice for Specimen Preparation and Mounting of Paper or Vinyl Wall Coverings to Assess Surface Burning Characteristics	ASTM	ASTM E2404-05
	BOCA National Fire Prevention Code, 11th ed. (1999)	BOCA	BNFPC
	Uniform Fire Code	IFCI	UPC 1997
	International Fire Code (2006)	ICC	IFC
	International Mechanical Code (2006)	ICC	IMC
	International Urban-Wildland Interface Code (2006)	ICC	IUWIC
	Fire-Resistance Tests—Elements of Building Construction; Part 1: Gen. Requirements	ISO	ISO 834-1:1999
	Fire-Resistance Tests—Door and Shutter Assemblies	ISO	ISO 3008:1976
	Reaction to Fire Tests—Ignitability of Building Products Using a Radiant Heat Source	ISO	ISO 5657:1997
	Fire-Resistance Tests—Ventilating Ducts	ISO	ISO 6944:1985
	Fire Service Annunciator and Interface	NEMA	NEMA SB 30-2005
	Fire Protection Handbook (2003)	NFPA	NFPA
	National Fire Codes (issued annually)	NFPA	NFPA
	Fire Protection Guide to Hazardous Materials	NFPA	NFPA HAZ-01
	Uniform Fire Code	NFPA	NFPA 1-06
	Installation of Sprinkler Systems	NFPA	NFPA 13-2002
	Fire Protection for Laboratories Using Chemicals	NFPA	NFPA 45-04
	Flammable and Combustible Liquids Code	NFPA	NFPA 30-03
	National Fire Alarm Code	NFPA	NFPA 72-02
	Fire Doors and Fire Windows	NFPA	NFPA 80-99
	Health Care Facilities	NFPA	NFPA 99-05
	Life Safety Code	NFPA	NFPA 101-06
	Methods of Fire Tests of Door Assemblies	NFPA	NFPA 252-03
	Standard Fire Prevention Code (1999)	SBCCI	SFPC
	Fire, Smoke and Radiation Damper Installation Guide for HVAC Systems, 5th ed.	SMACNA	SMACNA 2002
	Fire Tests of Door Assemblies (1997)	UL	ANSI/UL 10B
	Heat Responsive Links for Fire-Protection Service (2003)	UL	ANSI/UL 33
	Fire Tests of Building Construction and Materials (2003)	UL	ANSI/UL 263
	Fire Dampers (1999)	UL	ANSI/UL 555
	Fire Tests of Through-Penetration Firestops (2003)	UL	ANSI/UL 1479
Smoke Management	Commissioning Smoke Management Systems	ASHRAE	ASHRAE *Guideline* 5-1994 (RA01)
	Laboratory Methods of Testing Fans Used to Exhaust Smoke in Smoke Management Systems	ASHRAE	ANSI/ASHRAE 149-2000 (RA05)
	Recommended Practice for Smoke-Control Systems	NFPA	NFPA 92A-06
	Smoke Management Systems in Malls, Atria, and Large Areas	NFPA	NFPA 92B-05
	Ceiling Dampers (1996)	UL	UL 555C
	Smoke Dampers (1999)	UL	UL 555S
Freezers	Energy Performance and Capacity of Household Refrigerators, Refrigerator-Freezers, and Freezers	CSA	C300-00 (R2005)

Selected Codes and Standards Published by Various Societies and Associations (*Continued*)

Subject	Title	Publisher	Reference
	Energy Performance Standard for Food Service Refrigerators and Freezers	CSA	C827-98 (R2003)
	Refrigeration Equipment	CSA	CAN/CSA-C22.2 No. 120-M91 (R2004)
Commercial	Dispensing Freezers	NSF	NSF/ANSI 6-2005
	Commercial Refrigerators and Freezers	NSF	NSF/ANSI 7-2001
	Commercial Refrigerators and Freezers (1995)	UL	ANSI/UL 471
	Ice Makers (1995)	UL	ANSI/UL 563
	Ice Cream Makers (2005)	UL	ANSI/UL 621
Household	Household Refrigerators, Refrigerator-Freezers and Freezers	AHAM	ANSI/AHAM HRF-1-2004
	Household Refrigerators and Freezers (1993)	UL/CSA	ANSI/UL 250/C22.2 No. 63-93 (R1999)
Fuels	Threshold Limit Values for Chemical Substances (updated annually)	ACGIH	ACGIH
	International Gas Fuel Code (2006)	AGA/NFPA	ANSI Z223.1/NFPA 54-2006
	Reporting of Fuel Properties when Testing Diesel Engines with Alternative Fuels Derived from Biological Materials	ASABE	ASAE EP552-2001
	Coal Pulverizers	ASME	PTC 4.2 1969 (RA03)
	Classification of Coals by Rank	ASTM	ASTM D388-05
	Specification for Fuel Oils	ASTM	ASTM D396-05
	Test Method for Determination of Homogeneity and Miscibility in Automotive Engine Oils	ASTM	ASTM D922-00a
	Specification for Diesel Fuel Oils	ASTM	ASTM D975-05
	Specification for Gas Turbine Fuel Oils	ASTM	ASTM D2880-03
	Specification for Kerosene	ASTM	ASTM D3699-05
	Practice for Receipt, Storage and Handling of Fuels	ASTM	ASTM D4418-00
	Test Method for Determination of Yield Stress and Apparent Viscosity of Used Engine Oils at Low Temperature	ASTM	ASTM D6896-03 E01
	Test Method for Total Sulfur in Naphthas, Distillates, Reformulated Gasolines, Diesels, Biodiesels, and Motor Fuels by Oxidative Combustion and Electrochemical Detection	ASTM	ASTM D6920-03
	Test Method for Measurement of Hindered Phenolic and Aromatic Amine Antioxidant Content in Non-Zinc Turbine Oils by Linear Sweep Voltammetry	ASTM	ASTM D6971-04
	Practice for Enumeration of Viable Bacteria and Fungi in Liquid Fuels—Filtration and Culture Procedures	ASTM	ASTM D6974-04a
	Test Method for Evaluation of Aeration Resistance of Engine Oils in Direct-Injected Turbocharged Automotive Diesel Engine	ASTM	ASTM D6984-05a
	Specification for Middle Distillate Fuel Oil-Military Marine Applications	ASTM	ASTM D6985-04a
	Test Method for Determination of Ignition Delay and Derived Cetane Number DCN of Diesel Fuel Oils by Combustion in a Constant Volume Chamber	ASTM	ASTM D6890-04
	Test Method for Determination of Total Sulfur in Light Hydrocarbon, Motor Fuels, and Oils by Online Gas Chromatography with Flame Photometric Detection	ASTM	ASTM D7041-04
	Test Method for Sulfur in Gasoline and Diesel Fuel by Monochromatic Wavelength Dispersive X-Ray Fluorescence Spectrometry	ASTM	ASTM D7044-04a
	New Draft Standard Test Method for Flash Point by Modified Continuously Closed Cup Flash Point Tester	ASTM	ASTM D7094-04
	Test Method for Determining the Viscosity-Temperature Relationship of Used and Soot-Containing Engine Oils at Low Temperatures	ASTM	ASTM D7110-05a
	Test Method for Determination of Trace Elements in Middle Distillate Fuels by Inductively Coupled Plasma Atomic Emission Spectrometry (ICPAES)	ASTM	ASTM D7111-05
	Test Method for Determining Stability and Compatibility of Heavy Fuel Oils and Crude Oils by Heavy Fuel Oil Stability Analyzer (Optical Detection)	ASTM	ASTM D7112-05a
	Test Method for Determination of Intrinsic Stability of Asphaltene-Containing Residues, Heavy Fuel Oils, and Crude Oils	ASTM	ASTM D7157-05
	Test Method for Hydrogen Content of Middle Distillate Petroleum Products by Low-Resolution Pulsed Nuclear Magnetic Resonance Spectroscopy	ASTM	ASTM D7171-05
	Gas-Fired Central Furnaces	CSA	ANSI Z21.47-2003/CSA 2.3-2003
	Gas Unit Heaters and Gas-Fired Duct Furnaces	CSA	ANSI Z83.8-2002/CSA-2.6-2002 (R2005)
	Industrial and Commercial Gas-Fired Package Furnaces	CSA	CGA 3.2-1976 (R2003)
	Uniform Mechanical Code (2006)	IAPMO	Chapter 13
	Uniform Plumbing Code (2006)	IAPMO	Chapter 12
	International Fuel Gas Code (2006)	ICC	IFGC
	Standard Gas Code (1999)	SBCCI	SGC
	Commercial-Industrial Gas Heating Equipment (1999)	UL	UL 795
Furnaces	Commercial Applications, Systems, and Equipment, 1st ed.	ACCA	ACCA Manual CS
	Residential Equipment Selection, 2nd ed.	ACCA	ANSI/ACCA Manual S
	Method of Testing for Annual Fuel Utilization Efficiency of Residential Central Furnaces and Boilers	ASHRAE	ANSI/ASHRAE 103-1993
	Prevention of Furnace Explosions/Implosions in Multiple Burner Boilers	NFPA	NFPA 8502-99

Selected Codes and Standards Published by Various Societies and Associations (*Continued*)

Subject	Title	Publisher	Reference
	Standard Mechanical Code (1997)	SBCCI	SMC
	Residential Gas Detectors (2000)	UL	ANSI/UL 1484
	Heating and Cooling Equipment (2005)	UL/CSA	ANSI/UL 1995/C22.2 No. 236-95
	Single and Multiple Station Carbon Monoxide Alarms (1996)	UL	ANSI/UL 2034
Gas	International Gas Fuel Code (2006)	AGA/NFPA	ANSI Z223.1/NFPA 54-2006
	Gas-Fired Central Furnaces	CSA	ANSI Z21.47-2003/CSA 2.3-2003
	Gas Unit Heaters and Gas-Fired Duct Furnaces	CSA	ANSI Z83.8-2002/CSA-2.6-2002
	Industrial and Commercial Gas-Fired Package Furnaces	CSA	CGA 3.2-1976 (R2003)
	International Fuel Gas Code (2006)	ICC	IFGC
	Standard Gas Code (1999)	SBCCI	SGC
	Commercial-Industrial Gas Heating Equipment (1999)	UL	UL 795
Oil	Specification for Fuel Oils	ASTM	ASTM D396-04a
	Specification for Diesel Fuel Oils	ASTM	ASTM D975-04b E01
	Test Method for Smoke Density in Flue Gases from Burning Distillate Fuels	ASTM	ASTM D2156-94 (2003)
	Standard Test Method for Vapor Pressure of Liquefied Petroleum Gases (LPG) (Expansion Method)	ASTM	ASTM D6897-2003a
	Oil Burning Stoves and Water Heaters	CSA	B140.3-1962 (R2001)
	Oil-Fired Warm Air Furnaces	CSA	B140.4-04
	Installation of Oil-Burning Equipment	NFPA	NFPA 31-06
	Oil-Fired Central Furnaces (1994)	UL	UL 727
	Oil-Fired Floor Furnaces (2003)	UL	ANSI/UL 729
	Oil-Fired Wall Furnaces (2003)	UL	ANSI/UL 730
Solid Fuel	Installation Code for Solid-Fuel-Burning Appliances and Equipment	CSA	B365-01
	Solid-Fuel-Fired Central Heating Appliances	CSA	CAN/CSA-B366.1-M91 (R2002)
	Solid-Fuel and Combination-Fuel Central and Supplementary Furnaces (1995)	UL	ANSI/UL 391
Heaters	Gas-Fired High-Intensity Infrared Heaters	CSA	ANSI Z83.19-2001/CSA 2.35-2001 (R2005)
	Gas-Fired Low-Intensity Infrared Heaters	CSA	ANSI Z83.20-2001/CSA 2.34-2001 (R2005)
	Threshold Limit Values for Chemical Substances (updated annually)	ACGIH	ACGIH
	Industrial Ventilation: A Manual of Recommended Practice, 25th ed. (2004)	ACGIH	ACGIH
	Thermal Performance Testing of Solar Ambient Air Heaters	ASABE	ANSI/ASAE S423-2001
	Air Heaters	ASME	ASME PTC 4.3-1968 (RA91)
	Guide for Construction of Solid Fuel Burning Masonry Heaters	ASTM	ASTM E1602-03
	Non-Recirculating Direct Gas-Fired Industrial Air Heaters	CSA	ANSI Z83.4-2003/CSA 3.7-2003
	Electric Duct Heaters	CSA	C22.2 No. 155-M1986 (R2004)
	Portable Kerosene-Fired Heaters	CSA	CAN3-B140.9.3 M86 (R2001)
	Standards for Closed Feedwater Heaters, 7th ed.	HEI	HEI 2004
	Electric Heating Appliances (2005)	UL	ANSI/UL 499
	Electric Oil Heaters (2003)	UL	ANSI/UL 574
	Oil-Fired Air Heaters and Direct-Fired Heaters (1993)	UL	UL 733
	Electric Dry Bath Heaters (2004)	UL	ANSI/UL 875
	Oil-Burning Stoves (1993)	UL	ANSI/UL 896
Engine	Electric Engine Preheaters and Battery Warmers for Diesel Engines	SAE	SAE J1310-1993
	Selection and Application Guidelines for Diesel, Gasoline, and Propane Fired Liquid Cooled Engine Pre-Heaters	SAE	SAE J1350-1988
	Fuel Warmer—Diesel Engines	SAE	SAE J1422-1996
Nonresidential	Installation of Electric Infrared Brooding Equipment	ASABE	ASAE EP258.3-2004
	Gas-Fired Construction Heaters	CSA	ANSI Z83.7-2000/CSA 2.14-2000
	Recirculating Direct Gas-Fired Industrial Air Heaters	CSA	ANSI Z83.18-2004
	Portable Industrial Oil-Fired Heaters	CSA	B140.8-1967 (R2001)
	Fuel-Fired Heaters—Air Heating—for Construction and Industrial Machinery	SAE	SAE J1024-1989
	Commercial-Industrial Gas Heating Equipment (1999)	UL	UL 795
	Electric Heaters for Use in Hazardous (Classified) Locations (1995)	UL	ANSI/UL 823
Pool	Method of Testing and Rating Pool Heaters	ASHRAE	ANSI/ASHRAE 146-1998
	Gas-Fired Pool Heaters	CSA	ANSI Z21.56A-2001/CSA 4.7-2001
	Oil-Fired Service Water Heaters and Swimming Pool Heaters	CSA	B140.12-03
Room	Specification for Room Heaters, Pellet Fuel Burning Type	ASTM	ASTM E1509-04
	Gas-Fired Room Heaters, Vol. II, Unvented Room Heaters	CSA	ANSI Z21.11.2-2002
	Gas-Fired Unvented Catalytic Room Heaters for Use with Liquefied Petroleum (LP) Gases	CSA	ANSI Z21.76-1994 (R2000)
	Vented Gas-Fired Space Heating Appliances	CSA	ANSI Z21.86-2004/CSA 2.32-2004
	Vented Gas Fireplace Heaters	CSA	ANSI Z21.88-2005/CSA 2.33-2005
	Unvented Kerosene-Fired Room Heaters and Portable Heaters (1993)	UL	UL 647
	Movable and Wall- or Ceiling-Hung Electric Room Heaters (2000)	UL	UL 1278

Selected Codes and Standards Published by Various Societies and Associations (*Continued*)

Subject	Title	Publisher	Reference
	Fixed and Location-Dedicated Electric Room Heaters (1997)	UL	UL 2021
	Solid Fuel-Type Room Heaters (1996)	UL	ANSI/UL 1482
Transport	Heater, Airplane, Engine Exhaust Gas to Air Heat Exchanger Type	SAE	SAE ARP86A-1996
	Installation, Heaters, Airplane, Internal Combustion Heater Exchange Type	SAE	SAE ARP266-1996
	Heater, Aircraft, Internal Combustion Heat Exchanger Type	SAE	SAE AS8040A-1996
	Motor Vehicle Heater Test Procedure	SAE	SAE J638-1998
	Heater, Aircraft, Internal Combustion Heat Exchanger Type	SAE	SAE AS8040A-1996
Unit	Gas Unit Heaters and Gas-Fired Duct Furnaces	CSA	ANSI Z83.8-2002/CSA-2.6-2002 (R2005)
	Oil-Fired Unit Heaters (1995)	UL	ANSI/UL 731
Heat Exchangers	Remote Mechanical-Draft Evaporative Refrigerant Condensers	ARI	ARI 490-2003
	Method of Testing Air-to-Air Heat Exchangers	ASHRAE	ANSI/ASHRAE 84-1991
	Boiler and Pressure Vessel Code—Section VIII, Division 1: Pressure Vessels	ASME	ASME BPVC-2004
	Single Phase Heat Exchangers	ASME	ASME PTC 12.5-2000 (RA05)
	Air Cooled Heat Exchangers	ASME	ASME PTC 30-1991 (RA05)
	Standard Methods of Test for Rating the Performance of Heat-Recovery Ventilators	CSA	C439-00 (R2005)
	Standards for Power Plant Heat Exchangers, 4th ed.	HEI	HEI 2004
	Standards of Tubular Exchanger Manufacturers Association, 8th ed. (1999)	TEMA	TEMA
	Refrigerant-Containing Components and Accessories, Nonelectrical (2001)	UL	ANSI/UL 207
Heating	Commercial Applications, Systems, and Equipment, 1st ed.	ACCA	ACCA Manual CS
	Comfort, Air Quality, and Efficiency by Design	ACCA	ACCA Manual RS
	Residential Equipment Selection, 2nd ed.	ACCA	ANSI/ACCA Manual S
	Heating, Ventilating and Cooling Greenhouses	ASABE	ANSI/ASAE EP406.4-2003
	Automatic Flue-Pipe Dampers for Use with Oil-Fired Appliances	CSA	B140.14-M1979 (R2001)
	Heater Elements	CSA	C22.2 No.72-M1984 (R2004)
	Determining the Required Capacity of Residential Space Heating and Cooling Appliances	CSA	CAN/CSA-F280-M90 (R2004)
	Heat Loss Calculation Guide (2001)	HYDI	HYDI H-22
	Residential Hydronic Heating Installation/Design	HYDI	IBR Guide
	Radiant Floor Heating (1995)	HYDI	HYDI 004
	Advanced Installation Guide (Commercial) for Hot Water Heating Systems (2001)	HYDI	HYDI 250
	Environmental Systems Technology, 2nd ed. (1999)	NEBB	NEBB
	Pulverized Fuel Systems	NFPA	NFPA 8503-97
	Aircraft Electrical Heating Systems	SAE	SAE AIR860A-2000
	Heating Value of Fuels	SAE	SAE J1498-2005
	Performance Test for Air-Conditioned, Heated, and Ventilated Off-Road Self-Propelled Work Machines	SAE	SAE J1503-2004
	HVAC Systems Applications, 1st ed.	SMACNA	SMACNA 1987
	Electric Baseboard Heating Equipment (1994)	UL	ANSI/UL 1042
	Electric Duct Heaters (2004)	UL	ANSI/UL 1996
	Heating and Cooling Equipment (2005)	UL/CSA	ANSI/UL 1995/C22.2 No. 236-95
Heat Pumps	Commercial Applications, Systems, and Equipment, 1st ed.	ACCA	ACCA Manual CS
	Geothermal Heat Pump Training Certification Program	ACCA	ACCA Training Manual
	Heat Pumps Systems, Principles and Applications, 2nd ed.	ACCA	ACCA Manual H
	Residential Equipment Selection, 2nd ed.	ACCA	ANSI/ACCA Manual S
	Industrial Ventilation: A Manual of Recommended Practice, 25th ed. (2004)	ACGIH	ACGIH
	Commercial and Industrial Unitary Air-Conditioning and Heat Pump Equipment	ARI	ARI 340/360-2004
	Ground Source Closed-Loop Heat Pumps	ARI	ARI 330-98
	Ground Water-Source Heat Pumps	ARI	ARI 325-98
	Water-Source Heat Pumps	ARI	ARI 320-98
	Methods of Testing for Rating Electrically Driven Unitary Air-Conditioning and Heat Pump Equipment	ASHRAE	ANSI/ASHRAE 37-2005
	Methods of Testing for Rating Seasonal Efficiency of Unitary Air-Conditioners and Heat Pumps	ASHRAE	ANSI/ASHRAE 116-1995
	Performance Standard for Split-System Central Air-Conditioners and Heat Pumps	CSA	CAN/CSA-C273.3-M94 (R2005)
	Installation Requirements for Air-to-Air Heat Pumps	CSA	C273.5-1980 (R2005)
	Performance of Direct-Expansion (DX) Ground-Source Heat Pumps	CSA	C748-94 (R1999)
	Water-Source Heat Pumps—Testing and Rating for Performance, Part 1: Water-to-Air and Brine-to-Air Heat Pumps	CSA	CAN/CSA C13256-1-01
	Part 2: Water-to-Water and Brine-to-Water Heat Pumps	CSA	CAN/CSA C13256-2-01 (R2005)
	Heating and Cooling Equipment (2005)	UL/CSA	ANSI/UL 1995/C22.2 No. 236-95
Gas-Fired	Gas-Fired, Heat Activated Air Conditioning and Heat Pump Appliances	CSA	ANSI Z21.40.1-1996 (R2002)/CGA 2.91-M96

Selected Codes and Standards Published by Various Societies and Associations (*Continued*)

Subject	Title	Publisher	Reference
	Gas-Fired, Work Activated Air Conditioning and Heat Pump Appliances (Internal Combustion)	CSA	ANSI Z21.40.2-1996 (R2002)/CGA 2.92-M96
	Performance Testing and Rating of Gas-Fired Air Conditioning and Heat Pump Appliances	CSA	ANSI Z21.40.4-1996 (R2002)/CGA 2.94-M96
Heat Recovery	Gas Turbine Heat Recovery Steam Generators	ASME	ANSI/ASME PTC 4.4-1981 (RA03)
	Water Heaters, Hot Water Supply Boilers, and Heat Recovery Equipment	NSF	NSF/ANSI 5-2005
Humidifiers	Commercial Applications, Systems, and Equipment, 1st ed.	ACCA	ACCA Manual CS
	Comfort, Air Quality, and Efficiency by Design	ACCA	ACCA Manual RS
	Bioaerosols: Assessment and Control (1999)	ACGIH	ACGIH
	Humidifiers	AHAM	ANSI/AHAM HU-1-2003
	Central System Humidifiers for Residential Applications	ARI	ARI 610-2004
	Self-Contained Humidifiers for Residential Applications	ARI	ARI 620-2004
	Commercial and Industrial Humidifiers	ARI	ANSI/ARI 640-2005
	Humidifiers (2001)	UL/CSA	ANSI/UL 998/C22.2 No. 104-93
Ice Makers	Performance Rating of Automatic Commercial Ice Makers	ARI	ARI 810-2003
	Ice Storage Bins	ARI	ARI 820-2000
	Methods of Testing Automatic Ice Makers	ASHRAE	ANSI/ASHRAE 29-1988 (RA05)
	Refrigeration Equipment	CSA	CAN/CSA-C22.2 No. 120-M91 (R2004)
	Performance of Automatic Ice-Makers and Ice Storage Bins	CSA	C742-98 (R2003)
	Automatic Ice Making Equipment	NSF	NSF/ANSI 12-2005
	Ice Makers (1995)	UL	ANSI/UL 563
Incinerators	Large Incinerators	ASME	ASME PTC 33-1978 (RA91)
	Incinerators and Waste and Linen Handling Systems and Equipment	NFPA	NFPA 82-04
	Residential Incinerators (1993)	UL	UL 791
Indoor Air Quality	Good HVAC Practices for Residential and Commercial Buildings (2003)	ACCA	ACCA
	Comfort, Air Quality, and Efficiency by Design (Residential) (1997)	ACCA	ACCA Manual RS
	Bioaerosols: Assessment and Control (1999)	ACGIH	ACGIH
	Ventilation for Acceptable Indoor Air Quality	ASHRAE	ANSI/ASHRAE 62.1-2004
	Ventilation and Acceptable Indoor Air Quality in Low-Rise Residential Buildings	ASHRAE	ANSI/ASHRAE 62.2-2004
	Test Method for Determination of Volatile Organic Chemicals in Atmospheres (Canister Sampling Methodology)	ASTM	ASTM D5466-01
	Guide for Using Probability Sampling Methods in Studies of Indoor Air Quality in Buildings	ASTM	ASTM D5791-95 (2001)
	Guide for Using Indoor Carbon Dioxide Concentrations to Evaluate Indoor Air Quality and Ventilation	ASTM	ASTM D6245-98 (2002)
	Guide for Placement and Use of Diffusion Controlled Passive Monitors for Gaseous Pollutants in Indoor Air	ASTM	ASTM D6306-98 (2003)
	Test Method for Determination of Metals and Metalloids Airborne Particulate Matter by Inductively Coupled Plasma Atomic Emissions Spectrometry (ICP-AES)	ASTM	ASTM D7035-04
	Test Method for Metal Removal Fluid Aerosol in Workplace Atmospheres	ASTM	ASTM D7049-04
	Practice for Emission Cells for the Determination of Volatile Organic Emissions from Materials/Products	ASTM	ASTM D7143-05
	Practice for Collection of Surface Dust by Micro-Vacuum Sampling for Subsequent Metals Determination	ASTM	ASTM D7144-05a
	Test Method for Determination of Beryllium in the Workplace Using Field-Based Extraction and Fluorescence Detection	ASTM	ASTM D7202-05
	Practice for Referencing Suprathreshold Odor Intensity	ASTM	ASTM E544-99 (2004)
	Guide for Specifying and Evaluating Performance of a Single Family Attached and Detached Dwelling—Indoor Air Quality	ASTM	ASTM E2267-04
	Classification for Serviceability of an Office Facility for Thermal Environment and Indoor Air Conditions	ASTM	ASTM E2320-04
	Practice for Continuous Sizing and Counting of Airborne Particles in Dust-Controlled Areas and Clean Rooms Using Instruments Capable of Detecting Single Sub-Micrometre and Larger Particles	ASTM	ASTM F50-92(2001)E01
	Ambient Air—Determination of Mass Concentration of Nitrogen Dioxide—Modified Griess-Saltzman Method	ISO	ISO 6768:1998
	Air Quality—Exchange of Data	ISO	ISO 7168:1999
	Environmental Tobacco Smoke—Estimation of Its Contribution to Respirable Suspended Particles—Determination of Particulate Matter by Ultraviolet Absorptance and by Fluorescence	ISO	ISO 15593:2001
	Indoor Air—Part 3: Determination of Formaldehyde and Other Carbonyl Compounds—Active Sampling Method	ISO	ISO 16000-3:2001
	Workplace Air Quality—Sampling and Analysis of Volatile Organic Compounds by Solvent Desorption/Gas Chromatography—Part 1: Pumped Sampling Method	ISO	ISO 16200-1:2001
	Part 2: Diffusive Sampling Method	ISO	ISO 16200-2:2000

Selected Codes and Standards Published by Various Societies and Associations (*Continued*)

Subject	Title	Publisher	Reference
	Workplace Air Quality—Determination of Total Isocyanate Group in Air Using 2-(1-Methoxyphenyl) Piperazine and Liquid Chromatography	ISO	ISO 16702-1:2001
	Installation of Household Carbon Monoxide (CO) Warning Equipment	NFPA	NFPA 720-2005
	Indoor Air Quality—A Systems Approach, 3rd ed.	SMACNA	SMACNA 1998
	IAQ Guidelines for Occupied Buildings Under Construction, 1st ed.	SMACNA	SMACNA 1995
	Single and Multiple Station Carbon Monoxide Alarms	UL	ANSI/UL 2034
Aircraft	Guide for Selecting Instruments and Methods for Measuring Air Quality in Aircraft Cabins	ASTM	ASTM D6399-04
	Guide for Deriving Acceptable Levels of Airborne Chemical Contaminants in Aircraft Cabins Based on Health and Comfort Considerations	ASTM	ASTM D7034-04
Insulation	Guidelines for Use of Thermal Insulation in Agricultural Buildings	ASABE	ANSI/ASAE S401.2-2003
	Terminology Relating to Thermal Insulating Materials	ASTM	ASTM C168-05a
	Test Method for Steady-State Heat Flux Measurements and Thermal Transmission Properties by Means of the Guarded-Hot-Plate Apparatus	ASTM	ASTM C177-04
	Test Method for Steady-State Heat Transfer Properties of Horizontal Pipe Insulations	ASTM	ASTM C335-05a
	Practice for Prefabrication and Field Fabrication of Thermal Insulating Fitting Covers for NPS Piping, Vessel Lagging, and Dished Head Segments	ASTM	ASTM C450-02
	Test Method for Steady-State and Thermal Transmission Properties by Means of the Heat Flow Meter Apparatus	ASTM	ASTM C518-04
	Specification for Preformed Flexible Elastometric Cellular Thermal Insulation in Sheet and Tubular Form	ASTM	ASTM C534-05
	Specification for Cellular Glass Thermal Insulation	ASTM	ASTM C552-03
	Specification for Rigid, Cellular Polystyrene Thermal Insulation	ASTM	ASTM C578-05a
	Practice for Inner and Outer Diameters of Rigid Thermal Insulation for Nominal Sizes of Pipe and Tubing (NPS System)	ASTM	ASTM C585-90 (2004)
	Specification for Unfaced Preformed Rigid Cellular Polyisocyanurate Thermal Insulation	ASTM	ASTM C591-05
	Practice for Determination of Heat Gain or Loss and the Surface Temperature of Insulated Pipe and Equipment Systems by the Use of a Computer Program	ASTM	ASTM C680-04e02
	Specification for Adhesives for Duct Thermal Insulation	ASTM	ASTM C916-8(2001)e1
	Classification of Potential Health and Safety Concerns Associated with Thermal Insulation Materials and Accessories	ASTM	ASTM C930-05
	Practice for Thermographic Inspection of Insulation Installations in Envelope Cavities of Frame Buildings	ASTM	ASTM C1060-90 (2003)
	Specification for Fibrous Glass Duct Lining Insulation (Thermal and Sound Absorbing Material)	ASTM	ASTM C1071-05
	Specification for Faced or Unfaced Rigid Cellular Phenolic Thermal Insulation	ASTM	ASTM C1126-04
	Practice for Installation and Use of Radiant Barrier Systems (RBS) in Building Construction	ASTM	ASTM C1158-05
	Test Method for Steady-State and Thermal Performance of Building Assemblies by Means of a Hot Box Apparatus	ASTM	ASTM C1363-05
	Specification for Perpendicularly Oriented Mineral Fiber Roll and Sheet Thermal Insulation for Pipes and Tanks	ASTM	ASTM C1393-00a
	Guide for Measuring and Estimating Quantities of Insulated Piping and Components	ASTM	ASTM C1409-98 (2003)
	Specification for Cellular Melamine Thermal and Sound Absorbing Insulation	ASTM	ASTM C1410-05a
	Guide for Selecting Jacketing Materials for Thermal Insulation	ASTM	ASTM C1423-98 (2003)
	Specification for Preformed Flexible Cellular Polyolefin Thermal Insulation in Sheet and Tubular Form	ASTM	ASTM C1427-04
	Specification for Polyimide Flexible Cellular Thermal and Sound Absorbing Insulation	ASTM	ASTM C1482-04
	Specification for Cellulosic Fiber Stabilized Thermal Insulation	ASTM	ASTM C1497-04
	Test Method for Characterizing the Effect of Exposure to Environmental Cycling on Thermal Performance of Insulation Products	ASTM	ASTM C1512-01
	Specification for Flexible Polymeric Foam Sheet Insulation Used as a Thermal and Sound Absorbing Liner for Duct Systems	ASTM	ASTM C1534-04
	Standard Guide for Development of Standard Data Records for Computerization of Thermal Transmission Test Data for Thermal Insulation	ASTM	ASTM C1558-03
	Guide for Determining Blown Density of Pneumatically Applied Loose Fill Mineral Fiber Thermal Insulation	ASTM	ASTM C1574-04
	Test Method for Determining the Moisture Content of Inorganic Insulation Materials by Weight	ASTM	ASTM C1616-05
	Specification for Cellular Polypropylene Thermal Insulation	ASTM	ASTM C1631-05
	Classification for Rating Sound Insulation	ASTM	ASTM E413-04
	Test Method for Determining the Drainage Efficiency of Exterior Insulation and Finish Systems (EIFS) Clad Wall Assemblies	ASTM	ASTM E2273-03
	Practice for Use of Test Methods E96 for Determining the Water Vapor Transmission (WVT) of Exterior Insulation and Finish Systems	ASTM	ASTM E2321-03
	Thermal Insulation—Materials, Products, and Systems—Vocabulary	ISO	ISO 9229:1991

Selected Codes and Standards Published by Various Societies and Associations (*Continued*)

Subject	Title	Publisher	Reference
	National Commercial and Industrial Insulation Standards, 5th ed.	MICA	MICA 1999
	Accepted Industry Practices for Sheet Metal Lagging, 1st ed.	SMACNA	SMACNA 2002
Louvers	Laboratory Methods of Testing Dampers for Rating	AMCA	AMCA 500-D-98
	Laboratory Methods of Testing Louvers for Rating	AMCA	AMCA 500-L-99
Lubricants	Methods of Testing the Floc Point of Refrigeration Grade Oils	ASHRAE	ANSI/ASHRAE 86-1994 (RA01)
	Test Method for Pour Point of Petroleum Products	ASTM	ASTM D97-05a
	Classification of Industrial Fluid Lubricants by Viscosity System	ASTM	ASTM D2422-97 (2002)
	Test Method for Relative Molecular Weight (Relative Molecular Mass) of Hydrocarbons by Thermoelectric Measurement of Vapor Pressure	ASTM	ASTM D2503-92 (2002) E01
	Test Method for Determination of Moderately High Temperature Piston Deposits by Thermo-Oxidation Engine Oil Simulation Test	ASTM	ASTM D7097-05
	Petroleum Products—Corrosiveness to Copper—Copper Strip Test	ISO	ISO 2160:1998
Measurement	Industrial Ventilation: A Manual of Recommended Practice, 25th ed. (2004)	ACGIH	ACGIH
	Engineering Analysis of Experimental Data	ASHRAE	ASHRAE *Guideline* 2-2005
	Standard Method for Measurement of Proportion of Lubricant in Liquid Refrigerant	ASHRAE	ANSI/ASHRAE 41.4-1996
	Standard Method for Measurement of Moist Air Properties	ASHRAE	ANSI/ASHRAE 41.6-1994 (RA01)
	Method of Measuring Solar-Optical Properties of Materials	ASHRAE	ANSI/ASHRAE 74-1988
	Standard Methods of Measuring and Expressing Building Energy Performance	ASHRAE	ANSI/ASHRAE 105-1984 (RA99)
	Method for Establishing Installation Effects on Flowmeters	ASME	ASME MFC-10M-2000
	Test Uncertainty	ASME	ASME PTC 19.1-1998 (RA04)
	Measurement of Industrial Sound	ASME	ANSI/ASME PTC 36-2004
	Test Methods for Water Vapor Transmission of Materials	ASTM	ASTM E96-05
	Specification for Temperature-Electromotive Force (EMF) Tables for Standardized Thermocouples	ASTM	ASTM E230-03
	Practice for Continuous Sizing and Counting of Airborne Particles in Dust-Controlled Areas and Clean Rooms Using Instruments Capable of Detecting Single Sub-Micrometre and Larger Particles	ASTM	ASTM F50-92(2001)E01
	Use of the International System of Units (SI): The Modern Metric System	IEEE/ASTM	IEEE/ASTM-SI10-2002
	Ergonomics of the Thermal Environment—Instruments for Measuring Physical Quantities	ISO	ISO 7726:1998
	Ergonomics of the Thermal Environment—Determination of Metabolic Rate	ISO	ISO 8996:2004
	Ergonomics of the Thermal Environment—Estimation of the Thermal Insulation and Evaporative Resistance of a Clothing Ensemble	ISO	ISO 9920:1995
Fluid Flow	Standard Methods of Measurement of Flow of Liquids in Pipes Using Orifice Flowmeters	ASHRAE	ANSI/ASHRAE 41.8-1989
	Calorimeter Test Methods for Mass Flow Measurements of Volatile Refrigerants	ASHRAE	ANSI/ASHRAE 41.9-2000
	Flow Measurement	ASME	ASME PTC 19.5-2004
	Glossary of Terms Used in the Measurement of Fluid Flow in Pipes	ASME	ASME MFC-1M-2003
	Measurement Uncertainty for Fluid Flow in Closed Conduits	ASME	ANSI/ASME MFC-2M-1983 (RA01)
	Measurement of Fluid Flow in Pipes Using Orifice, Nozzle, and Venturi	ASME	ASME MFC-3M-2004
	Measurement of Liquid Flow in Closed Conduits Using Transit-Time Ultrasonic Flowmeters	ASME	ASME MFC-5M-1985 (RA01)
	Measurement of Fluid Flow in Pipes Using Vortex Flowmeters	ASME	ASME MFC-6M-1998 (RA05)
	Fluid Flow in Closed Conduits: Connections for Pressure Signal Transmissions Between Primary and Secondary Devices	ASME	ASME MFC-8M-2001
	Measurement of Liquid Flow in Closed Conduits by Weighing Method	ASME	ASME MFC-9M-1988 (RA01)
	Measurement of Fluid Flow by Means of Coriolis Mass Flowmeters	ASME	ASME MFC-11M-2003
	Measurement of Fluid Flow Using Small Bore Precision Orifice Meters	ASME	ASME MFC-14M-2003
	Measurement of Fluid Flow in Closed Conduits by Means of Electromagnetic Flowmeters	ASME	ASME MFC-16M-1995 (R01)
	Measurement of Fluid Flow Using Variable Area Meters	ASME	ASME MFC-18M-2001
	Test Method for Determining the Moisture Content of Inorganic Insulation Materials by Weight	ASTM	ASTM C1616-05
	Test Method for Indicating Wear Characteristics of Petroleum Hydraulic Fluids in a High Pressure Constant Volume Vane Pump	ASTM	ASTM D6973-05
	Test Method for Dynamic Viscosity and Density of Liquids by Stabinger Viscometer and the Calculation of Kinematic Viscosity	ASTM	ASTM D7042-04
	Test Method for Indicating Wear Characteristics of Petroleum and Non-Petroleum Hydraulic Fluids in a Constant Volume Vane Pump	ASTM	ASTM D7043-04a
	Practice for Calculating Viscosity of a Blend of Petroleum Products	ASTM	ASTM D7152-05
	Test Method for Same-Different Test	ASTM	ASTM E2139-05
	Practice for Field Use of Pyranometers, Pyrheliometers, and UV Radiometers	ASTM	ASTM G183-05
Gas Flow	Standard Methods for Laboratory Airflow Measurement	ASHRAE	ANSI/ASHRAE 41.2-1987 (RA92)
	Method of Test for Measurement of Flow of Gas	ASHRAE	ANSI/ASHRAE 41.7-1984 (RA00)
	Measurement of Gas Flow by Means of Critical Flow Venturi Nozzles	ASME	ANSI/ASME MFC-7M-1987 (RA01)
	Measurement of Gas Flow by Turbine Meters	ASME	ANSI/ASME MFC-4M-1986 (RA03)
Pressure	Standard Method for Pressure Measurement	ASHRAE	ANSI/ASHRAE 41.3-1989

Selected Codes and Standards Published by Various Societies and Associations (*Continued*)

Subject	Title	Publisher	Reference
	Pressure Gauges and Gauge Attachments	ASME	ASME B40.100-1998
	Pressure Measurement	ASME	ANSI/ASME PTC 19.2-1987 (RA04)
Temperature	Standard Method for Temperature Measurement	ASHRAE	ANSI/ASHRAE 41.1-1986 (RA01)
	Thermometers, Direct Reading and Remote Reading	ASME	ASME B40.200-2001
	Temperature Measurement	ASME	ASME PTC 19.3-1974 (RA04)
	Total Temperature Measuring Instruments (Turbine Powered Subsonic Aircraft)	SAE	SAE AS793A-2001
Thermal	Method of Testing Thermal Energy Meters for Liquid Streams in HVAC Systems	ASHRAE	ANSI/ASHRAE 125-1992 (RA00)
	Test Method for Steady-State Heat Flux Measurements and Thermal Transmission Properties by Means of the Guarded-Hot-Plate Apparatus	ASTM	ASTM C177-04
	Test Method for Steady-State Heat Flux Measurements and Thermal Transmission Properties by Means of the Heat Flow Meter Apparatus	ASTM	ASTM C518-04
	Test Method for Thermal Performance of Building Assemblies by Means of a Calibrated Hot Box	ASTM	ASTM C976-90(1996)e1
	Practice for In-Situ Measurement of Heat Flux and Temperature on Building Envelope Components	ASTM	ASTM C1046-95(2001)
	Practice for Determining Thermal Resistance of Building Envelope Components from In-Situ Data	ASTM	ASTM C1155-95(2001)
Mobile Homes and Recreational Vehicles	Residential Load Calculation, 8th ed.	ACCA	ANSI/ACCA Manual J
	Recreational Vehicle Cooking Gas Appliances	CSA	ANSI Z21.57-2005
	Oil-Fired Warm Air Heating Appliances for Mobile Housing and Recreational Vehicles	CSA	B140.10-1974 (R2001)
	Mobile Homes	CSA	CAN/CSA-Z240 MH Series-92 (R2005)
	Recreational Vehicles	CSA	CAN/CSA-Z240 RV Series-99 (R2004)
	Gas Supply Connectors for Manufactured Homes	IAPMO	IAPMO TS 9-2003
	Fuel Supply: Manufactured/Mobile Home Parks & Recreational Vehicle Parks	IAPMO	Chapter 13, Part II
	Manufactured Housing Construction and Safety Standards	ICC/ANSI	ICC/ANSI 2.0-1998
	Manufactured Housing	NFPA	NFPA 501-05
	Recreational Vehicles	NFPA	NFPA 1192-05
	Plumbing System Components for Manufactured Homes and Recreational Vehicles	NSF	NSF/ANSI 24-1988
	Low Voltage Lighting Fixtures for Use in Recreational Vehicles (2005)	UL	ANSI/UL 234
	Liquid Fuel-Burning Heating Appliances for Manufactured Homes and Recreational Vehicles (1995)	UL	ANSI/UL 307A
	Gas-Burning Heating Appliances for Manufactured Homes and Recreational Vehicles (1995)	UL	UL 307B
	Gas-Fired Cooking Appliances for Recreational Vehicles (1993)	UL	UL 1075
Motors and Generators	Installation and Maintenance of Farm Standby Electric Power	ASABE	ANSI/ASAE EP364.2-2003
	Testing of Nuclear Air Treatment Systems	ASME	ASME N510-1989 (RA95)
	Nuclear Power Plant Air-Cleaning Units and Components	ASME	ASME N509-2002
	Fired Steam Generators	ASME	ASME PTC 4-1998
	Gas Turbine Heat Recovery Steam Generators	ASME	ASME PTC 4.4-1981 (RA03)
	Test Methods for Film-Insulated Magnet Wire	ASTM	ASTM D1676-03
	Test Method for Evaluation of Engine Oils in a High Speed, Single-Cylinder Diesel Engine—Caterpillar 1R Test Procedure	ASTM	ASTM D6923-05
	Test Method for Evaluation of Diesel Engine Oils in the T-11 Exhaust Gas Recirculation Diesel Engine	ASTM	ASTM D7156-05
	Energy Efficiency Test Methods for Three-Phase Induction Motors	CSA	C390-98 (R2005)
	Motors and Generators	CSA	C22.2 No. 100-04
	Emergency Electrical Power Supply for Buildings	CSA	CSA C282-05
	Energy Efficiency Test Methods for Single- and Three-Phase Small Motors	CSA	CAN/CSA C747-94 (R2005)
	Standard Test Procedure for Polyphase Induction Motors and Generators	IEEE	IEEE 112-1996
	Motors and Generators	NEMA	NEMA MG 1-2003
	Energy Management Guide for Selection and Use of Fixed Frequency Medium AC Squirrel-Cage Polyphase Industrial Motors	NEMA	NEMA MG 10-2001
	Energy Management Guide for Selection and Use of Single-Phase Motors	NEMA	NEMA MG 11-1977 (R2001)
	Magnet Wire	NEMA	NEMA MW 1000-2003
	Motion/Position Control Motors, Controls, and Feedback Devices	NEMA	NEMA ICS 16-2000
	Electric Motors (1994)	UL	UL 1004
	Electric Motors and Generators for Use in Division 1 Hazardous (Classified) Locations (2003)	UL	ANSI/UL 674
	Overheating Protection for Motors (1997)	UL	ANSI/UL 2111
Pipe, Tubing, and Fittings	Scheme for the Identification of Piping Systems	ASME	ASME A13.1-1996 (RA02)
	Pipe Threads, General Purpose (Inch)	ASME	ANSI/ASME B1.20.1-1983 (RA01)
	Wrought Copper and Copper Alloy Braze-Joint Pressure Fittings	ASME	ASME B16.50-2001
	Power Piping	ASME	ASME B31.1-2004

Selected Codes and Standards Published by Various Societies and Associations (*Continued*)

Subject	Title	Publisher	Reference
	Fuel Gas Piping	ASME	ASME B31.2-1968
	Process Piping	ASME	ASME B31.3-2004
	Refrigeration Piping and Heat Transfer Components	ASME	ASME B31.5-2001
	Building Services Piping	ASME	ASME B31.9-2004
	Practice for Obtaining Hydrostatic or Pressure Design Basis for "Fiberglass" (Glass-Fiber-Reinforced Thermosetting-Resin) Pipe and Fittings	ASTM	ASTM D2992-01
	Specification for Welding of Austenitic Stainless Steel Tube and Piping Systems in Sanitary Applications	AWS	AWS D18.1:1999
	Standards of the Expansion Joint Manufacturers Association, 8th ed. (2003)	EJMA	EJMA
	Pipe Hangers and Supports—Materials, Design and Manufacture	MSS	MSS SP-58-2002
	Pipe Hangers and Supports—Selection and Application	MSS	ANSI/MSS SP-69-2003
	Welding Procedure Specifications	NCPWB	NCPWB
	International Fuel Gas Code	AGA/NFPA	ANSI Z223.1/NFPA 54-2006
	Refrigeration Tube Fittings—General Specifications	SAE	SAE J513-1999
	Seismic Restraint Manual—Guidelines for Mechanical Systems, 2nd ed.	SMACNA	ANSI/SMACNA 001-2000
	Tube Fittings for Flammable and Combustible Fluids, Refrigeration Service, and Marine Use (1997)	UL	ANSI/UL 109
Plastic	Specification for Acrylonitrile-Butadiene-Styrene (ABS) Plastic Pipe, Schedules 40 and 80	ASTM	ASTM D1527-99 E01
	Specification for Poly (Vinyl Chloride) (PVC) Plastic Pipe, Schedules 40, 80, and 120	ASTM	ASTM D1785-05
	Specification for Polyethylene (PE) Plastic Pipe, Schedule 40	ASTM	ASTM D2104-03
	Specification for Polybutylene (PB) Plastic Tubing	ASTM	ASTM D2666-96a
	Test Method for Obtaining Hydrostatic or Pressure Design Basis for Thermoplastic Pipe Products	ASTM	ASTM D2837-04
	Specification for Polybutylene (PB) Plastic Hot- and Cold-Water Distribution Systems	ASTM	ASTM D3309-96a (2002)
	Specification for Perfluoroalkoxy (PFA)-Fluoropolymer Tubing	ASTM	ASTM D6867-03
	Specification for Polyethylene Stay in Place Form System for End Walls for Drainage Pipe	ASTM	ASTM D7082-04
	Specification for Chlorinated Poly (Vinyl Chloride) (CPVC) Plastic Pipe, Schedules 40 and 80	ASTM	ASTM F441/F441M-02
	Specification for Crosslinked Polyethylene/Aluminum/Crosslinked Polyethylene Tubing OD Controlled SDR9	ASTM	ASTM F2262-05
	Test Method for Evaluating the Oxidative Resistance of Polyethylene (PE) Pipe to Chlorinated Water	ASTM	ASTM F2263-05
	Specification for 12 to 60 in. Annular Corrugated Profile-Wall Polyethylene (PE) Pipe and Fittings for Gravity-Flow Storm Sewer and Subsurface Drainage Applications	ASTM	ASTM F2306/F2306M-05
	Specification for Series 10 Poly (Vinyl Chloride) (PVC) Closed Profile Gravity Pipe and Fittings Based on Controlled Inside Diameter	ASTM	ASTM F2307-03
	Standard Test Method for Evaluating the Oxidative Resistance of Multilayer Polyolefin Tubing to Hot Chlorinated Water	ASTM	ASTM F2330-04
	Test Method for Determining Chemical Compatibility of Thread Sealants with Thermoplastic Threaded Pipe and Fittings Materials	ASTM	ASTM F2331-04
	Test Method for Determining Thermoplastic Pipe Wall Stiffness	ASTM	ASTM F2433-05
	Specification for Steel Reinforced Polyethylene (PE) Corrugated Pipe	ASTM	ASTM F2435-05
	Electrical Polyvinyl Chloride (PVC) Tubing and Conduit	NEMA	NEMA TC 2-2003
	PVC Plastic Utilities Duct for Underground Installation	NEMA	NEMA TC 6 and 8-2004
	Smooth-Wall Coilable Polyethylene Electrical Plastic Duct	NEMA	NEMA TC 7-2005
	Fittings for PVC Plastic Utilities Duct for Underground Installation	NEMA	NEMA TC 9-2004
	Electrical Nonmetallic Tubing (ENT)	NEMA	NEMA TC 13-2005
	Plastics Piping System Components and Related Materials	NSF	NSF/ANSI 14-2004
	Rubber Gasketed Fittings for Fire-Protection Service (2004)	UL	ANSI/UL 213
Metal	Welded and Seamless Wrought Steel Pipe	ASME	ASME B36.10M-2004
	Stainless Steel Pipe	ASME	ASME B36.19-2004
	Specification for Pipe, Steel, Black and Hot-Dipped, Zinc-Coated, Welded and Seamless	ASTM	ASTM A53/53M-05
	Specification for Seamless Carbon Steel Pipe for High-Temperature Service	ASTM	ASTM A106/A106M-04b
	Specification for Pipe, Steel, Electric-Fusion Arc-Welded Sizes NPS 16 and Over	ASTM	ASTM A1034-05b
	Specification for Steel Line Pipe, Black, Furnace-Butt-Welded	ASTM	ASTM A1037/A1037M-05
	Specification for Composite Corrugated Steel Pipe for Sewers and Drains	ASTM	ASTM A1042-04
	Specification for Seamless Copper Pipe, Standard Sizes	ASTM	ASTM B42-02e1
	Specification for Seamless Copper Tube	ASTM	ASTM B75-02
	Specification for Seamless Copper Water Tube	ASTM	ASTM B88-03
	Specification for Seamless Copper Tube for Air Conditioning and Refrigeration Field Service	ASTM	ASTM B280-03
	Specification for Hand-Drawn Copper Capillary Tube for Restrictor Applications	ASTM	ASTM B360-01
	Specification for Welded Copper Tube for Air Conditioning and Refrigeration Service	ASTM	ASTM B640-00
	Specification for Copper-Beryllium Seamless Tube UNS Nos. C17500 and C17510	ASTM	ASTM B0937-04

Selected Codes and Standards Published by Various Societies and Associations (*Continued*)

Subject	Title	Publisher	Reference
	Test Method for Rapid Determination of Corrosiveness to Copper from Petroleum Products Using a Disposable Copper Foil Strip	ASTM	ASTM D7095-04
	Thickness Design of Ductile-Iron Pipe	AWWA	ANSI/AWWA C150/A21.50-02
	Fittings, Cast Metal Boxes, and Conduit Bodies for Conduit and Cable Assemblies	NEMA	NEMA FB 1-2003
	Polyvinyl-Chloride (PVC) Externally Coated Galvanized Rigid Steel Conduit and Intermediate Metal Conduit	NEMA	NEMA RN 1-2005
Plumbing	Backwater Valves	ASME	ASME A112.14.1-2003
	Plumbing Supply Fittings	ASME	ASME A112.18.1-2005
	Plumbing Waste Fittings	ASME	ASME A112.18.2-2005
	Performance Requirements for Backflow Protection Devices and Systems in Plumbing Fixture Fittings	ASME	ASME A112.18.3-2002
	Uniform Plumbing Code (2006) (with IAPMO Installation Standards)	IAPMO	IAPMO
	International Plumbing Code (2006)	ICC	IPC
	International Private Sewage Disposal Code (2006)	ICC	IPSDC
	2003 National Standard Plumbing Code (NSPC)	PHCC	NSPC 2003
	2003 National Standard Plumbing Code—Illustrated	PHCC	PHCC 2003
	Standard Plumbing Code (1997)	SBCCI	SPC
Pumps	Centrifugal Pumps	ASME	ASME PTC 8.2-1990
	Specification for Horizontal End Suction Centrifugal Pumps for Chemical Process	ASME	ASME B73.1-2001
	Specification for Vertical-in-Line Centrifugal Pumps for Chemical Process	ASME	ASME B73.2-2003
	Specification for Sealless Horizontal End Suction Metallic Centrifugal Pumps for Chemical Process	ASME	ASME B73.3-2003
	Specification for Thermoplastic and Thermoset Polymer Material Horizontal End Suction Centrifugal Pumps for Chemical Process	ASME	ASME B73.5M-1995 (RA01)
	Liquid Pumps	CSA	CAN/CSA-C.22.2 No. 108-01
	Energy Efficiency Test Methods for Small Pumps	CSA	CAN/CSA C820-02
	Performance Standard for Liquid Ring Vacuum Pumps, 3rd ed.	HEI	HEI 2005
	Centrifugal Pumps for Nomenclature and Definitions	HI	ANSI/HI 1.1-1.2 (2000)
	Centrifugal Pumps for Design and Applications	HI	ANSI/HI 1.3 (2000)
	Centrifugal Pumps for Installation, Operation, and Maintenance	HI	ANSI/HI 1.4 (2000)
	Vertical Pumps for Nomenclature and Definitions	HI	ANSI/HI 2.1-2.2 (2000)
	Vertical Pumps for Design and Application	HI	ANSI/HI 2.3 (2000)
	Vertical Pumps for Installation, Operation, and Maintenance	HI	ANSI/HI 2.4 (2000)
	Rotary Pumps for Nomenclature, Definitions, Application, and Operation	HI	ANSI/HI 3.1-3.5 (2000)
	Sealless Rotary Pumps for Nomenclature, Definitions, Application, Operation, and Test	HI	ANSI/HI 4.1-4.6 (2000)
	Sealless Centrifugal Pumps for Nomenclature, Definitions, Application, Operation, and Test	HI	ANSI/HI 5.1-5.6 (2000)
	Reciprocating Pumps for Nomenclature, Definitions, Application, and Operation	HI	ANSI/HI 6.1-6.5 (2000)
	Direct Acting (Steam) Pumps for Nomenclature, Definitions, Application, and Operation	HI	ANSI/HI 8.1-8.5 (2000)
	Pumps—General Guidelines for Types, Definitions, Application, Sound Measurement and Decontamination	HI	ANSI/HI 9.1-9.5 (2000)
	Centrifugal and Vertical Pumps for Allowable Nozzle Loads	HI	ANSI/HI 9.6.2 (2001)
	Centrifugal and Vertical Pumps for Allowable Operating Region	HI	ANSI/HI 9.6.3 (1997)
	Centrifugal and Vertical Pumps for Vibration Measurements and Allowable Values	HI	ANSI/HI 9.6.4 (2001)
	Centrifugal and Vertical Pumps for Condition Monitoring	HI	ANSI/HI 9.6.5 (2000)
	Pump Intake Design	HI	ANSI/HI 9.8 (1998)
	Engineering Data Book, 2nd ed.	HI	HI (1990)
	Circulation System Components and Related Materials for Swimming Pools, Spas/Hot Tubs	NSF	NSF/ANSI 50-2005
	Pumps for Oil-Burning Appliances (1997)	UL	UL 343
	Motor-Operated Water Pumps (2002)	UL	ANSI/UL 778
	Swimming Pool Pumps, Filters, and Chlorinators (1997)	UL	UL 1081
Radiators	Testing and Rating Standard for Baseboard Radiation, 8th ed. (2005)	HYDI	IBR
	Testing and Rating Standard for Finned Tube (Commercial) Radiation, 6th ed. (2005)	HYDI	IBR
Receivers	Refrigerant Liquid Receivers	ARI	ARI 495-2005
	Refrigerant-Containing Components and Accessories, Nonelectrical (2001)	UL	ANSI/UL 207
Refrigerants	Threshold Limit Values for Chemical Substances (updated annually)	ACGIH	ACGIH
	Specifications for Fluorocarbon Refrigerants	ARI	ARI 700-2004
	Refrigerant Recovery/Recycling Equipment	ARI	ARI 740-98
	Format for Information on Refrigerants	ASHRAE	ASHRAE *Guideline* 6-1996
	Method of Testing Flow Capacity of Refrigerant Capillary Tubes	ASHRAE	ANSI/ASHRAE 28-1996 (RA02)
	Designation and Safety Classification of Refrigerants	ASHRAE	ANSI/ASHRAE 34-2004
	Sealed Glass Tube Method to Test the Chemical Stability of Materials for Use Within Refrigerant Systems	ASHRAE	ANSI/ASHRAE 97-1999 (RA03)
	Refrigeration Oil Description	ASHRAE	ANSI/ASHRAE 99-1987

Selected Codes and Standards Published by Various Societies and Associations (*Continued*)

Subject	Title	Publisher	Reference
	Reducing the Release of Halogenated Refrigerants from Refrigerating and Air-Conditioning Equipment and Systems	ASHRAE	ANSI/ASHRAE 147-2002
	Test Method for Acid Number of Petroleum Products by Potentiometric Titration	ASTM	ASTM D664-06
	Test Method for Concentration Limits of Flammability of Chemical (Vapors and Gases)	ASTM	ASTM E681-04
	Refrigerant-Containing Components for Use in Electrical Equipment	CSA	C22.2 No. 140.3-M1987 (R2004)
	Refrigerants—Designation System	ISO	ISO 817:2005
	Procedure Retrofitting CFC-12 (R-12) Mobile Air-Conditioning Systems to HFC-134a (R-134a)	SAE	SAE J1661-1998
	Recommended Service Procedure for the Containment of CFC-12 (R-12)	SAE	SAE J1989-1998
	Standard of Purity for Recycled HFC-134a for Use in Mobile Air-Conditioning Systems	SAE	SAE J2099-1999
	HFC-134a (R-134a) Service Hose Fittings for Automotive Air-Conditioning Service Equipment	SAE	SAE J2197-1997
	Recommended Service Procedure for the Containment of HFC-134a	SAE	SAE J2211-1998
	HFC-134a (R-134a) Recovery/Recycling Equipment for Mobile Air-Conditioning Systems	SAE	SAE J2210-1999
	CFC-12 (R-12) Refrigerant Recovery Equipment for Mobile Automotive Air-Conditioning Systems	SAE	SAE J2209-1999
	Refrigerant-Containing Components and Accessories, Nonelectrical (2001)	UL	ANSI/UL 207
	Refrigerant Recovery/Recycling Equipment (2005)	UL	ANSI/UL 1963
	Field Conversion/Retrofit of Products to Change to an Alternative Refrigerant—Construction and Operation (1993)	UL	ANSI/UL 2170
	Field Conversion/Retrofit of Products to Change to an Alternative Refrigerant—Insulating Material and Refrigerant Compatibility (1993)	UL	ANSI/UL 2171
	Field Conversion/Retrofit of Products to Change to an Alternative Refrigerant—Procedures and Methods (1993)	UL	ANSI/UL 2172
	Refrigerants (1994)	UL	ANSI/UL 2182
Refrigeration	Safety Standard for Refrigeration Systems	ASHRAE	ANSI/ASHRAE 15-2004
	Mechanical Refrigeration Code	CSA	B52-05
	Refrigeration Equipment	CSA	CAN/CSA-C22.2 No. 120-M91 (R2004)
	Equipment, Design and Installation of Ammonia Mechanical Refrigerating Systems	IIAR	ANSI/IIAR 2-1999
	Refrigerated Medical Equipment (1993)	UL	ANSI/UL 416
Refrigeration Systems	Ejectors	ASME	ASME PTC 24-1976 (RA82)
	Reducing the Release of Halogenated Refrigerants from Refrigerating and Air-Conditioning Equipment and Systems	ASHRAE	ANSI/ASHRAE 147-2002
	Testing of Refrigerating Systems	ISO	ISO 916-1968
	Standards for Steam Jet Vacuum Systems, 5th ed.	HEI	HEI 2000
Transport	Mechanical Transport Refrigeration Units	ARI	ARI 1110-2001
	Mechanical Refrigeration and Air-Conditioning Installations Aboard Ship	ASHRAE	ANSI/ASHRAE 26-1996
	General Requirements for Application of Vapor Cycle Refrigeration Systems for Aircraft	SAE	SAE ARP731C-2003
	Safety Standard for Motor Vehicle Refrigerant Vapor Compression Systems	SAE	SAE J639-2005
Refrigerators	Method of Testing Commercial Refrigerators and Freezers	ASHRAE	ANSI/ASHRAE 72-2005
Commercial	Energy Performance Standard for Commercial Refrigerated Display Cabinets and Merchandise	CSA	C657-04
	Energy Performance Standard for Food Service Refrigerators and Freezers	CSA	C827-98 (R2003)
	Gas Food Service Equipment	CSA	ANSI Z83.11-2002/CSA 1.8A-2002
	Mobile Food Carts	NSF	NSF/ANSI 59-2002e
	Food Equipment	NSF	NSF/ANSI 2-2005a
	Commercial Refrigerators and Freezers	NSF	NSF/ANSI 7-2001
	Commercial Refrigerators and Freezers (1995)	UL	ANSI/UL 471
	Refrigerating Units (1994)	UL	ANSI/UL 427
	Refrigeration Unit Coolers (2004)	UL	ANSI/UL 412
Household	Household Refrigerators, Refrigerator-Freezers and Freezers	AHAM	ANSI/AHAM HRF-1-2004
	Refrigerators Using Gas Fuel	CSA	ANSI Z21.19-2002/CSA1.4-2002
	Energy Performance and Capacity of Household Refrigerators, Refrigerator-Freezers, and Freezers	CSA	CAN/CSA C300-00 (R2005)
	Household Refrigerators and Freezers (1993)	UL/CSA	ANSI/UL 250-1997/C22.2 No. 63-93
Retrofitting			
Building	Residential Duct Diagnostics and Repair (2003)	ACCA	ACCA
	Good HVAC Practices for Residential and Commercial Buildings (2003)	ACCA	ACCA
	Building Systems Analysis and Retrofit Manual, 1st ed.	SMACNA	SMACNA 1995
Refrigerant	Procedure for Retrofitting CFC-12 (R-12) Mobile Air Conditioning Systems to HFC-134a (R-134a)	SAE	SAE J1661-1998
	Field Conversion/Retrofit of Products to Change to an Alternative Refrigerant—Construction and Operation (1993)	UL	ANSI/UL 2170
	Field Conversion/Retrofit of Products to Change to an Alternative Refrigerant—Insulating Material and Refrigerant Compatibility (1993)	UL	ANSI/UL 2171

Selected Codes and Standards Published by Various Societies and Associations (*Continued*)

Subject	Title	Publisher	Reference
	Field Conversion/Retrofit of Products to Change to an Alternative Refrigerant—Procedures and Methods (1993)	UL	ANSI/UL 2172
Roof Ventilators	Commercial Low Pressure, Low Velocity Duct System Design, 1st ed.	ACCA	ACCA Manual Q
	Power Ventilators (2004)	UL	ANSI/UL 705
Safety Devices	Bursting Discs and Bursting Disc Devices	ISO	ISO 6718:1991
Solar Equipment	Thermal Performance Testing of Solar Ambient Air Heaters	ASABE	ANSI/ASAE S423-2001
	Testing and Reporting Solar Cooker Performance	ASABE	ASAE S580-2003
	Method of Measuring Solar-Optical Properties of Materials	ASHRAE	ASHRAE 74-1988
	Methods of Testing to Determine the Thermal Performance of Solar Collectors	ASHRAE	ANSI/ASHRAE 93-2003
	Methods of Testing to Determine the Thermal Performance of Solar Domestic Water Heating Systems	ASHRAE	ANSI/ASHRAE 95-1987
	Methods of Testing to Determine the Thermal Performance of Unglazed Flat-Plate Liquid-Type Solar Collectors	ASHRAE	ANSI/ASHRAE 96-1980 (RA89)
	Methods of Testing to Determine the Thermal Performance of Flat-Plate Solar Collectors Containing a Boiling Liquid	ASHRAE	ANSI/ASHRAE 109-1986 (RA03)
	Practice for Installation and Service of Solar Space Heating Systems for One and Two Family Dwellings	ASTM	ASTM E683-91 (2001)
	Practice for Evaluating Thermal Insulation Materials for Use in Solar Collectors	ASTM	ASTM E861-94 (2001)
	Practice for Installation and Service of Solar Domestic Water Heating Systems for One and Two Family Dwellings	ASTM	ASTM E1056-85 (2001)
	Reference Solar Spectral Irradiance at the Ground at Different Receiving Conditions—Part 1: Direct Normal and Hemispherical Solar Irradiance for Air Mass 1.5	ISO	ISO 9845-1:1992
	Solar Collectors	CSA	CAN/CSA F378-87 (R2004)
	Solar Domestic Hot Water Systems (Liquid to Liquid Heat Transfer)	CSA	CAN/CSA F379-1.88 (R1999)
	Seasonal Use Solar Domestic Hot Water Systems	CSA	CAN/CSA F379.2-M89 (R1999)
	Installation Code for Solar Domestic Hot Water Systems	CSA	CAN/CSA F383-87 (R2003)
	Solar Heating—Domestic Water Heating Systems—Part 2: Outdoor Test Methods for System Performance Characterization and Yearly Performance Prediction of Solar-Only Systems	ISO	ISO 9459-2:1995
	Test Methods for Solar Collectors—Part 1: Thermal Performance of Glazed Liquid Heating Collectors Including Pressure Drop	ISO	ISO 9806-1:1994
	Part 2: Qualification Test Procedures	ISO	ISO 9806-2:1995
	Part 3: Thermal Performance of Unglazed Liquid Heating Collectors (Sensible Heat Transfer Only) Including Pressure Drop	ISO	ISO 9806-3:1995
	Solar Water Heaters—Elastomeric Materials for Absorbers, Connecting Pipes and Fittings—Method of Assessment	ISO	ISO 9808:1990
	Solar Energy—Calibration of a Pyranometer Using a Pyrheliometer	ISO	ISO 9846:1993
	Solar Heating—Swimming Pool Heating Systems—Dimensions, Design and Installation Guidelines	ISO	ISO 12596:1995
Solenoid Valves	Solenoid Valves for Use with Volatile Refrigerants	ARI	ARI 760-2001
	Methods of Testing Capacity of Refrigerant Solenoid Valves	ASHRAE	ANSI/ASHRAE 158.1-2004
	Electrically Operated Valves (1999)	UL	UL 429
Sound Measurement	Threshold Limit Values for Physical Agents (updated annually)	ACGIH	ACGIH
	Specification for Sound Level Meters	ASA	ANSI S1.4-1983 (R2001)
	Specification for Octave-Band and Fractional-Octave-Band Analog and Digital Filters	ASA	ANSI S1.11-2004
	Microphones, Part 1: Specifications for Laboratory Standard Microphones	ASA	ANSI S1.15-1997/Part 1 (R2001)
	Part 2: Primary Method for Pressure Calibration of Laboratory Standard Microphones by the Reciprocity Technique	ASA	ANSI S1.15-1997/Part 2
	Specification for Acoustical Calibrators	ASA	ANSI S1.40-1984 (R2001)
	Measurement of Industrial Sound	ASME	ASME PTC 36-2004
	Test Method for Measuring Acoustical and Airflow Performance of Duct Liner Materials and Prefabricated Silencers	ASTM	ASTM E477-05
	Test Method for Determination of Decay Rates for Use in Sound Insulation Test Methods	ASTM	ASTM E2235-04e01
	Sound and Vibration Design and Analysis (1994)	NEBB	NEBB
Fans	Reverberant Room Method for Sound Testing of Fans	AMCA	AMCA 300-05
	Methods for Calculating Fan Sound Ratings from Laboratory Test Data	AMCA	AMCA 301-05
	Application of Sone Ratings for Non-Ducted Air Moving Devices	AMCA	AMCA 302-73
	Application of Sound Power Level Ratings for Fans	AMCA	AMCA 303-79
	Acoustics—Measurement of Noise and Vibration of Small Air-Moving Devices—Part 1: Airborne Noise Emission	ASA	ANSI S12.11-2003/Part 1/ISO 10302:1996 (MOD)
	Part 2: Structure-Borne Vibration	ASA	ANSI S12.11-2003/Part 2
	Laboratory Method of Testing to Determine the Sound Power in a Duct	ASHRAE/ AMCA	ANSI/ASHRAE 68-1997/ AMCA 330-97
Other Equipment	Sound Rating of Outdoor Unitary Equipment	ARI	ARI 270-95
	Application of Sound Rating Levels of Outdoor Unitary Equipment	ARI	ARI 275-97

Selected Codes and Standards Published by Various Societies and Associations (*Continued*)

Subject	Title	Publisher	Reference
	Sound Rating and Sound Transmission Loss of Packaged Terminal Equipment	ARI	ARI 300-2000
	Sound Rating of Non-Ducted Indoor Air-Conditioning Equipment	ARI	ARI 350-2000
	Sound Rating of Large Outdoor Refrigerating and Air-Conditioning Equipment	ARI	ARI 370-2001
	Method of Rating Sound and Vibration of Refrigerant Compressors	ARI	ARI 530-2005
	Method of Measuring Machinery Sound Within an Equipment Space	ARI	ARI 575-94
	Statistical Methods for Determining and Verifying Stated Noise Emission Values of Machinery and Equipment	ASA	ANSI S12.3-1985 (R2001)
	Sound Level Prediction for Installed Rotating Electrical Machines	NEMA	NEMA MG 3-1974 (R2003)
Techniques	Preferred Frequencies, Frequency Levels, and Band Numbers for Acoustical Measurements	ASA	ANSI S1.6-1984 (R2001)
	Reference Quantities for Acoustical Levels	ASA	ANSI S1.8-1989 (R2001)
	Measurement of Sound Pressure Levels in Air	ASA	ANSI S1.13-2005
	Procedure for the Computation of Loudness of Steady Sound	ASA	ANSI S3.4-2005
	Methods for Determining the Insertion Loss of Outdoor Noise Barriers	ASA	ANSI S12.8-1998 (R2003)
	Criteria for Evaluating Room Noise	ASA	ANSI S12.2-1995 (R1999)
	Engineering Method for the Determination of Sound Power Levels of Noise Sources Using Sound Intensity	ASA	ANSI S12.12-1992 (R2002)
	Procedures for Outdoor Measurement of Sound Pressure Level	ASA	ANSI S12.18-1994 (R2004)
	Guidelines for the Use of Sound Power Standards and for the Preparation of Noise Test Codes	ASA	ANSI S12.30-1990 (R2002)
	Precision Methods for the Determination of Sound Power Levels of Noise Sources in Anechoic and Hemi-Anechoic Rooms	ASA	ANSI S12.35-1990 (R2001)
	Methods for Measurement of Sound Emitted by Machinery and Equipment at Workstations and Other Specified Positions	ASA	ANSI S12.43-1997 (R2002)
	Methods for Calculation of Sound Emitted by Machinery and Equipment at Workstations and Other Specified Positions from Sound Power Level	ASA	ANSI S12.44-1997 (R2002)
	Acoustics—Determination of Sound Power Levels of Noise Sources Using Sound Pressure—Precision Method for Reverberation Rooms	ASA	ANSI S12.51-2002/ISO 3741:1999
	Acoustics—Determination of Sound Power Levels of Noise Sources—Engineering Methods for Small, Movable Sources in Reverberant Fields—Part 1: Comparison Method for Hard-Walled Test Rooms	ASA	ANSI S12.53/Part 1-1999 (R2004)/ISO 3743-1:1994
	Part 2: Methods for Special Reverberation Test Rooms	ASA	ANSI S12.53/Part 2-1999 (R2004)/ISO 3743-2:1994
	Acoustics—Determination of Sound Power Levels of Noise Sources Using Sound Pressure—Engineering Method in an Essentially Free Field over a Reflecting Plane	ASA	ANSI S12.54-1999 (R2004)/ISO 3744:1994
	Acoustics—Determination of Sound Power Levels of Noise Sources Using Sound Pressure—Survey Method Using an Enveloping Measurement Surface over a Reflecting Plane	ASA	ANSI S12.56-1999 (R2004)/ISO 3746:1995
	Test Method for Impedance and Absorption of Acoustical Materials by the Impedance Tube Method	ASTM	ASTM C384-04
	Test Method for Sound Absorption and Sound Absorption Coefficients by the Reverberation Room Method	ASTM	ASTM C423-02a
	Test Method for Measurement of Airborne Sound Insulation in Buildings	ASTM	ASTM E336-05
	Test Method for Impedance and Absorption of Acoustical Materials Using a Tube, Two Microphones and a Digital Frequency Analysis System	ASTM	ASTM E1050-98
	Test Method for Evaluating Masking Sound in Open Offices Using A-Weighted and One-Third Octave Band Sound Pressure Levels	ASTM	ASTM E1573-02
	Test Method for Measurement of Sound in Residential Spaces	ASTM	ASTM E1574-98
	Acoustics–Measurement of Sound Insulation in Buildings and of Building Elements; Part 1: Requirements for Laboratory Test Facilities with Suppressed Flanking Transmission	ISO	ISO 140-1:1997
	Part 4: Field Measurements of Airborne Sound Insulation Between Rooms	ISO	ISO 140-4:1998
	Part 5: Field Measurements of Airborne Sound Insulation of Facade Elements and Facades	ISO	ISO 140-5:1998
	Part 6: Laboratory Measurements of Impact Sound Insulation of Floors	ISO	ISO 140-6:1998
	Part 7: Field Measurements of Impact Sound Insulation of Floors	ISO	ISO 140-7:1998
	Part 8: Laboratory Measurements of the Reduction of Transmitted Impact Noise by Floor Coverings on a Heavyweight Standard Floor	ISO	ISO 140-8:1997
	Acoustics—Method for Calculating Loudness Level	ISO	ISO 532:1975
	Acoustics—Determination of Sound Power Levels of Noise Sources Using Sound Intensity; Part 1: Measurement at Discrete Points	ISO	ISO 9614-1:1993
	Part 2: Measurement by Scanning	ISO	ISO 9614-2:1996
	Procedural Standards for Measurement and Assessment of Sound and Vibration (1994)	NEBB	NEBB
Terminology	Acoustical Terminology	ASA	ANSI S1.1-1994 (R2004)
	Terminology Relating to Environmental Acoustics	ASTM	ASTM C 634-02
Space Heaters	Methods of Testing for Rating Combination Space-Heating and Water-Heating Appliances	ASHRAE	ANSI/ASHRAE 124-1991

Selected Codes and Standards Published by Various Societies and Associations (*Continued*)

Subject	Title	Publisher	Reference
	Gas-Fired Room Heaters, Vol. II, Unvented Room Heaters	CSA	ANSI Z21.11.2-2002
	Vented Gas-Fired Space Heating Appliances	CSA	ANSI Z21.86-2004/CSA 2.32-2004
	Movable and Wall- or Ceiling-Hung Electric Room Heaters (2000)	UL	UL 1278
	Fixed and Location-Dedicated Electric Room Heaters (1997)	UL	UL 2021
Symbols	Graphic Electrical/Electronic Symbols for Air-Conditioning and Refrigerating Equipment	ARI	ARI 130-88
	Graphic Symbols for Heating, Ventilating, Air-Conditioning, and Refrigerating Systems	ASHRAE	ANSI/ASHRAE 134-2005
	Graphical Symbols for Plumbing Fixtures for Diagrams Used in Architecture and Building Construction	ASME	ANSI/ASME Y32.4-1977 (RA04)
	Symbols for Mechanical and Acoustical Elements as Used in Schematic Diagrams	ASME	ANSI/ASME Y32.18-1972 (RA03)
	Practice for Mechanical Symbols, Shipboard Heating, Ventilation, and Air Conditioning (HVAC)	ASTM	ASTM F856-97 (2004)
	Standard Symbols for Welding, Brazing, and Nondestructive Examination	AWS	AWS A2.4:1998
	Recommended Practice for the Preparation and Use of Symbols	IEEE	IEEE 267-1966
	Standard Letter Symbols for Quantities Used in Electrical Science and Electrical Engineering	IEEE	IEEE 280-1985 (R2003)
	Graphic Symbols for Electrical and Electronics Diagrams	IEEE	ANSI 315-1975 (R1993)/IEEE 315A-1986
	Standard for Logic Circuit Diagrams	IEEE	IEEE 991-1986 (R1994)
	Use of the International System of Units (SI): The Modern Metric System	IEEE/ASTM	IEEE/ASTM-SI10-2002
	Abbreviations and Acronyms	ASME	ASME Y14.38-1999
	Engineering Drawing Practices	ASME	ASME Y14.100-2004
	Safety Color Code	NEMA	ANSI/NEMA Z535-2002
Terminals, Wiring	Electrical Quick-Connect Terminals (2003)	UL	ANSI/UL 310
	Wire Connectors (2003)	UL	ANSI/UL 486A-486B
	Splicing Wire Connectors (2004)	UL	ANSI/UL 486C
	Equipment Wiring Terminals for Use with Aluminum and/or Copper Conductors (1994)	UL	ANSI/UL 486E
Testing and Balancing	AABC Commissioning Guideline (2002)	AABC	AABC
	AABC National Standards for Total System Balance (2002)	AABC	AABC
	Industrial Process/Power Generation Fans: Site Performance Test Standard	AMCA	AMCA 803-02
	Guidelines for Measuring and Reporting Environmental Parameters for Plant Experiments in Growth Chambers	ASABE	ANSI/ASAE EP411.4-2002
	The HVAC Commissioning Process	ASHRAE	ASHRAE *Guideline* 1-1996
	Practices for Measurement, Testing, Adjusting, and Balancing of Building Heating, Ventilation, Air-Conditioning, and Refrigeration Systems	ASHRAE	ANSI/ASHRAE 111-1988
	Practices for Measuring, Testing, Adjusting, and Balancing Shipboard HVAR&R Systems	ASHRAE	ANSI/ASHRAE 151-2002
	Centrifugal Pump Tests	HI	ANSI/HI 1.6 (2000)
	Vertical Pump Tests	HI	ANSI/HI 2.6 (2000)
	Rotary Pump Tests	HI	ANSI/HI 3.6 (2000)
	Reciprocating Pump Tests	HI	ANSI/HI 6.6 (2000)
	Pumps—General Guidelines for Types, Definitions, Application, Sound Measurement and Decontamination	HI	HI 9.1-9.5 (2000)
	Submersible Pump Tests	HI	ANSI/HI 11.6 (2001)
	Procedural Standards for Certified Testing of Cleanrooms, 2nd ed. (1996)	NEBB	NEBB
	Procedural Standards for Testing, Adjusting, Balancing of Environmental Systems, 7th ed. (2005)	NEBB	NEBB
	HVAC Systems Testing, Adjusting and Balancing, 3rd ed.	SMACNA	SMACNA 2002
Thermal Storage	Thermal Energy Storage—A Contractor's Tool Book	ACCA	ACCA
	Method of Testing Active Latent-Heat Storage Devices Based on Thermal Performance	ASHRAE	ANSI/ASHRAE 94.1-2002
	Method of Testing Thermal Storage Devices with Electrical Input and Thermal Output Based on Thermal Performance	ASHRAE	ANSI/ASHRAE 94.2-1981 (RA02)
	Method of Testing Active Sensible Thermal Energy Devices Based on Thermal Performance	ASHRAE	ANSI/ASHRAE 94.3-1986 (RA02)
	Practices for Measurement, Testing, Adjusting, and Balancing of Building Heating, Ventilation, Air-Conditioning, and Refrigeration Systems	ASHRAE	ANSI/ASHRAE 111-1988
	Method of Testing the Performance of Cool Storage Systems	ASHRAE	ANSI/ASHRAE 150-2000 (RA04)
Transformers	Minimum Efficiency Values for Liquid-Filled Distribution Transformers	CSA	CAN/CSA C802.1-00
	Minimum Efficiency Values for Dry-Type Transformers	CSA	CAN/CSA C802.2-00
	Maximum Losses For Power Transformers	CSA	CAN/CSA C802.3-01
	Guide for Determining Energy Efficiency of Distribution Transformers	NEMA	NEMA TP-1-2002
Turbines	Steam Turbines	ASME	ASME PTC 6-2004
	Steam Turbines for Combined Cycle	ASME	ASME PTC 6.2-2004
	Hydraulic Turbines and Pump-Turbines	ASME	ASME PTC 18-2002
	Gas Turbines	ASME	ASME PTC 22-1997 (RA03)
	Wind Turbines	ASME	ASME PTC 42-1988 (RA04)

Selected Codes and Standards Published by Various Societies and Associations (*Continued*)

Subject	Title	Publisher	Reference
	Specification for Stainless Steel Bars for Compressor and Turbine Airfoils	ASTM	ASTM A1028-03
	Specification for Gas Turbine Fuel Oils	ASTM	ASTM D2880-03
	Land Based Steam Turbine Generator Sets	NEMA	NEMA SM 24-1991 (R2002)
	Steam Turbines for Mechanical Drive Service	NEMA	NEMA SM 23-1991 (R2002)
Valves	Face-to-Face and End-to-End Dimensions of Valves	ASME	ASME B16.10-2000 (R03)
	Pressure Relief Devices	ASME	ASME PTC 25-2001
	Valves—Flanged, Threaded, and Welding End	ASME	ASME B16.34-2004
	Manually Operated Metallic Gas Valves for Use in Aboveground Piping Systems up to 5 psi	ASME	ASME B16.44-2002
	Methods of Testing Capacity of Refrigerant Solenoid Valves	ASHRAE	ANSI/ASHRAE 158.1-2004
	Relief Valves for Hot Water Supply	CSA	ANSI Z21.22-1999 (R2003)/ CSA 4.4-M99 (R2004)
	Control Valve Capacity Test Procedures	ISA	ANSI/ISA-S75.02-1996
	Flow Equations for Sizing Control Valves	ISA	ANSI/ISA-S75.01.01-2002
	Industrial Valves—Part-Turn Actuator Attachments	ISO	ISO 5211-1:2001
	Metal Valves for Use in Flanged Pipe Systems—Face-to-Face and Centre-to-Face Dimensions	ISO	ISO 5752:1982
	Safety Valves for Protection Against Excessive Pressure, Part 1: Safety Valves	ISO	ISO 4126-1:1991
	Oxygen System Fill/Check Valve	SAE	SAE AS1225A-1997
	Valves for Anhydrous Ammonia and LP-Gas (Other Than Safety Relief) (1997)	UL	ANSI/UL 125
	Safety Relief Valves for Anhydrous Ammonia and LP-Gas (1997)	UL	ANSI/UL 132
	LP-Gas Regulators (1999)	UL	ANSI/UL 144
	Electrically Operated Valves (1999)	UL	UL 429
	Valves for Flammable Fluids (1997)	UL	ANSI/UL 842
Gas	Manually Operated Metallic Gas Valves for Use in Gas Piping Systems up to 125 psig (Sizes NPS 1/2 through 2)	ASME	ASME B16.33-2002
	Large Metallic Valves for Gas Distribution (Manually Operated, NPS-2 1/2 to 12, 125 psig Maximum)	ASME	ANSI/ASME B16.38-1985 (RA05)
	Manually Operated Thermoplastic Gas Shutoffs and Valves in Gas Distribution Systems	ASME	ASME B16.40-2002
	Manually Operated Gas Valves for Appliances, Appliance Connection Valves, and Hose End Valves	CSA	ANSI Z21.15-1997 (R2003)/CGA 9.1-M97
	Automatic Valves for Gas Appliances	CSA	ANSI Z21.21-2005/CGA 6.5-2005
	Combination Gas Controls for Gas Appliances	CSA	ANSI Z21.78-2005/CGA 6.20-2005
	Convenience Gas Outlets and Optional Enclosures	CSA	ANSI Z21.90-2001/CSA 6.24-2001
Refrigerant	Refrigerant Access Valves and Hose Connectors	ARI	ARI 720-2002
	Thermostatic Refrigerant Expansion Valves	ARI	ARI 750-2001
	Solenoid Valves for Use with Volatile Refrigerants	ARI	ARI 760-2001
	Refrigerant Pressure Regulating Valves	ARI	ARI 770-2001
Vapor Retarders	Practice for Selection of Vapor Retarders for Thermal Insulation	ASTM	ASTM C755-03
	Practice for Determining the Properties of Jacketing Materials for Thermal Insulation	ASTM	ASTM C921-03a
	Specification for Flexible, Low Permeance Vapor Retarders for Thermal Insulation	ASTM	ASTM C1136-03a
	Test Method for Water Vapor Transmission Rate of Flexible Barrier Materials Using an Infrared Detection Technique	ASTM	ASTM F372-99 (2003)
Vending Machines	Methods of Testing for Rating Vending Machines for Bottled, Canned, and Other Sealed Beverages	ASHRAE	ANSI/ASHRAE 32.1-2004
	Methods of Testing for Rating Pre-Mix and Post Mix Beverage Dispensing Equipment	ASHRAE	ANSI/ASHRAE 32.2-2003
	Vending Machines	CSA	C22.2 No.128-95 (R2004)
	Energy Performance of Vending Machines	CSA	CAN/CSA C804-96 (R2001)
	Vending Machines for Food and Beverages	NSF	NSF/ANSI 25-2005
	Vending Machines (1995)	UL	ANSI/UL 751
	Refrigerated Vending Machines (1995)	UL	ANSI/UL 541
Vent Dampers	Automatic Vent Damper Devices for Use with Gas-Fired Appliances	CSA	ANSI Z21.66-1996 (R2001)/CSA 6.14-M96
	Vent or Chimney Connector Dampers for Oil-Fired Appliances (1994)	UL	ANSI/UL 17
Ventilation	Commercial Applications, Systems, and Equipment, 1st ed.	ACCA	ACCA Manual CS
	Commercial Low Pressure, Low Velocity Duct System Design, 1st ed.	ACCA	ACCA Manual Q
	Comfort, Air Quality, and Efficiency by Design	ACCA	ACCA Manual RS
	Guide for Testing Ventilation Systems (1991)	ACGIH	ACGIH
	Industrial Ventilation: A Manual of Recommended Practice, 25th ed. (2004)	ACGIH	ACGIH
	Design of Ventilation Systems for Poultry and Livestock Shelters	ASABE	ASAE EP270.5-2003
	Design Values for Emergency Ventilation and Care of Livestock and Poultry	ASABE	ANSI/ASAE EP282.2-2004
	Heating, Ventilating and Cooling Greenhouses	ASABE	ANSI/ASAE EP406.4-2003
	Guidelines for Selection of Energy Efficient Agricultural Ventilation Fans	ASABE	ASAE EP566-2001
	Uniform Terminology for Livestock Production Facilities	ASABE	ASAE S501-2001

Selected Codes and Standards Published by Various Societies and Associations (*Continued*)

Subject	Title	Publisher	Reference
	Agricultural Ventilation Constant Speed Fan Test Standard	ASABE	ASABE S565-2005
	Ventilation for Acceptable Indoor Air Quality	ASHRAE	ANSI/ASHRAE 62.1-2004
	Ventilation and Acceptable Indoor Air Quality in Low-Rise Residential Buildings	ASHRAE	ANSI/ASHRAE 62.2-2004
	Method of Testing for Room Air Diffusion	ASHRAE	ANSI/ASHRAE 113-2005
	Measuring Air Change Effectiveness	ASHRAE	ANSI/ASHRAE 129-1997 (RA02)
	Method of Determining Air Change Rates in Detached Dwellings	ASHRAE	ANSI/ASHRAE 136-1993 (RA01)
	Ventilation for Commercial Cooking Operations	ASHRAE	ANSI/ASHRAE 154-2003
	Residential Mechanical Ventilation Systems	CSA	CAN/CSA F326-M91 (R2005)
	Parking Structures	NFPA	NFPA 88A-02
	Installation of Air Conditioning and Ventilating Systems	NFPA	NFPA 90A-02
	Ventilation Control and Fire Protection of Commercial Cooking Operations	NFPA	NFPA 96-04
	Food Equipment	NSF	NSF/ANSI 2-2005a
	Class II (Laminar Flow) Biosafety Cabinetry	NSF	NSF/ANSI 49-2004a
	Aerothermodynamic Systems Engineering and Design	SAE	SAE AIR1168/3-1989
	Heater, Airplane, Engine Exhaust Gas to Air Heat Exchanger Type	SAE	SAE ARP86A-1996
	Test Procedure for Battery Flame Retardant Venting Systems	SAE	SAE J1495-2005
Venting	Commercial Applications, Systems, and Equipment, 1st ed.	ACCA	ACCA Manual CS
	Draft Hoods	CSA	ANSI Z21.12-1990 (R2000)
	National Fuel Gas Code	AGA/NFPA	ANSI Z223.1/NFPA 54-2006
	Explosion Prevention Systems	NFPA	NFPA 69-02
	Smoke and Heat Venting	NFPA	NFPA 204-02
	Chimneys, Fireplaces, Vents and Solid Fuel-Burning Appliances	NFPA	NFPA 211-06
	Guide for Steel Stack Construction, 2nd ed.	SMACNA	SMACNA 1996
	Draft Equipment (1993)	UL	UL 378
	Gas Vents (1996)	UL	ANSI/UL 441
	Type L Low-Temperature Venting Systems (1995)	UL	ANSI/UL 641
Vibration	Balance Quality and Vibration Levels for Fans	AMCA	ANSI/AMCA 204-05
	Techniques of Machinery Vibration Measurement	ASA	ANSI S2.17-1980 (R2004)
	Vibrations of Buildings—Guidelines for the Measurement of Vibrations and Evaluation of Their Effects on Buildings	ASA	ANSI S2.47-1990 (R2001)
	Mechanical Vibration and Shock—Resilient Mounting Systems—Part 1: Technical Information to Be Exchanged for the Application of Isolation Systems	ASA	ISO 2017-1:2005
	Evaluation of Human Exposure to Whole-Body Vibration—Part 2: Vibration in Buildings (1 Hz to 80 Hz)	ISO	ISO 2631-2:2003
	Guidelines for the Evaluation of the Response of Occupants of Fixed Structures, Especially Buildings and Off-Shore Structures, to Low-Frequency Horizontal Motion (0.063 to 1 Hz)	ISO	ISO 6897:1984
	Procedural Standards for Measurement and Assessment of Sound and Vibration (1994)	NEBB	NEBB
	Sound and Vibration Design and Analysis (1994)	NEBB	NEBB
Water Heaters	Desuperheater/Water Heaters	ARI	ARI 470-2001
	Safety for Electrically Heated Livestock Waterers	ASABE	ASAE EP342.2-1995 (R2005)
	Methods of Testing to Determine the Thermal Performance of Solar Domestic Water Heating Systems	ASHRAE	ANSI/ASHRAE 95-1987
	Method of Testing for Rating Commercial Gas, Electric, and Oil Service Water Heating Equipment	ASHRAE	ANSI/ASHRAE 118.1-2003
	Method of Testing for Rating Residential Water Heaters	ASHRAE	ANSI/ASHRAE 118.2-1993
	Methods of Testing for Rating Combination Space-Heating and Water-Heating Appliances	ASHRAE	ANSI/ASHRAE 124-1991
	Methods of Testing for Efficiency of Space-Conditioning/Water-Heating Appliances That Include a Desuperheater Water Heater	ASHRAE	ANSI/ASHRAE 137-1995 (RA04)
	Boiler and Pressure Vessel Code—Section IV: Heating Boilers	ASME	BPVC-2004
	Section VI: Recommended Rules for the Care and Operation of Heating Boilers	ASME	BPVC-2004
	Gas Water Heaters—Vol. I: Storage Water Heaters with Input Ratings of 75,000 Btu per Hour or Less	CSA	ANSI Z21.10.1-2004/CSA 4.1-2004
	Vol. III: Storage, with Input Ratings Above 75,000 Btu per Hour, Circulating and Instantaneous Water Heaters	CSA	ANSI Z21.10.3-2004/CSA 4.3-2004
	Oil Burning Stoves and Water Heaters	CSA	B140.3-1962 (R2001)
	Oil-Fired Service Water Heaters and Swimming Pool Heaters	CSA	B140.12-03
	Construction and Test of Electric Storage-Tank Water Heaters	CSA	CAN/CSA-C22.2 No. 110-94 (R2004)
	Performance of Electric Storage Tank Water Heaters for Household Service	CSA	C191-04
	Energy Efficiency of Electric Storage Tank Water Heaters and Heat Pump Water Heaters	CSA	CSA C745-03
	One Time Use Water Heater Emergency Shut-Off	IAPMO	IGC 175-2003
	Water Heaters, Hot Water Supply Boilers, and Heat Recovery Equipment	NSF	NSF/ANSI 5-2005
	Oil-Fired Storage Tank Water Heaters (1995)	UL	ANSI/UL 732
	Commercial-Industrial Gas Heating Equipment (1999)	UL	UL 795

Selected Codes and Standards Published by Various Societies and Associations (*Continued*)

Subject	Title	Publisher	Reference
	Electric Booster and Commercial Storage Tank Water Heaters (2004)	UL	ANSI/UL 1453
	Household Electric Storage Tank Water Heaters (2004)	UL	ANSI/UL 174
Welding and Brazing	Boiler and Pressure Vessel Code—Section IX: Welding and Brazing Qualifications	ASME	BPVC-2004
	Structural Welding Code—Steel	AWS	AWS D1.1/D1.1M:2006
	Specification for Welding of Austenitic Stainless Steel Tube and Piping Systems in Sanitary Applications	AWS	AWS D18.1:1999
Wood-Burning Appliances	Threshold Limit Values for Chemical Substances (updated annually)	ACGIH	ACGIH
	Specification for Room Heaters, Pellet Fuel Burning Type	ASTM	ASTM E1509-04
	Guide for Construction of Solid Fuel Burning Masonry Heaters	ASTM	ASTM E1602-03
	Installation Code for Solid-Fuel-Burning Appliances and Equipment	CSA	CAN/CSA-B365-01
	Solid-Fuel-Fired Central Heating Appliances	CSA	CAN/CSA-B366.1-M91 (R2002)
	Chimneys, Fireplaces, Vents, and Solid Fuel-Burning Appliances	NFPA	ANSI/NFPA 211-06
	Commercial Cooking, Rethermalization and Powered Hot Food Holding and Transport Equipment	NSF	NSF/ANSI 4-2005e

ORGANIZATIONS

Abbrev.	Organization	Address	Telephone	http://www.
AABC	Associated Air Balance Council	1518 K Street NW, Suite 503 Washington, D.C. 20005	(202) 737-0202	aabchq.com
ABMA	American Boiler Manufacturers Association	4001 North 9th Street, Suite 226 Arlington, VA 22203-1900	(703) 522-7350	abma.com
ACCA	Air Conditioning Contractors of America	2800 Shirlington Road, Suite 300 Arlington, VA 22206	(703) 575-4477	acca.org
ACGIH	American Conference of Governmental Industrial Hygienists	1330 Kemper Meadow Drive Cincinnati, OH 45240	(513) 742-2020	acgih.org
ADC	Air Diffusion Council	1901 North Roselle Road, Suite 800 Schaumburg, IL 60195	(847) 706-6750	flexibleduct.org
AGA	American Gas Association	400 N. Capitol Street NW Washington, D.C. 20001	(202) 824-7000	aga.org
AHAM	Association of Home Appliance Manufacturers	111 19th Street NW, Suite 402 Washington, D.C. 20036	(202) 872-5955	aham.org
AIHA	American Industrial Hygiene Association	2700 Prosperity Avenue, Suite 250 Fairfax, VA 22031	(703) 849-8888	aiha.org
AMCA	Air Movement and Control Association International	30 West University Drive Arlington Heights, IL 60004-1893	(847) 394-0150	amca.org
ANSI	American National Standards Institute	1819 L Street NW, 6th Floor Washington, D.C. 20036	(202) 293-8020	ansi.org
ARI	Air-Conditioning and Refrigeration Institute	4100 North Fairfax Drive, Suite 200 Arlington, VA 22203	(703) 524-8800	ari.org
ASA	Acoustical Society of America	35 Pinelawn Road, Suite 114 E Melville, NY 11747-3177	(631) 390-0215	asa.aip.org
ASABE	American Society of Agricultural and Biological Engineers	2950 Niles Road St. Joseph, MI 49085-9659	(269) 429-0300	asabe.org
ASHRAE	American Society of Heating, Refrigerating and Air-Conditioning Engineers	1791 Tullie Circle, NE Atlanta, GA 30329	(404) 636-8400	ashrae.org
ASME	ASME	3 Park Avenue New York, NY 10016-5990	(973) 882-1167	asme.org
ASTM	ASTM International	100 Barr Harbor Drive, P.O. Box C700 West Conshohocken, PA 19428-2959	(610) 832-9585	astm.org
AWS	American Welding Society	550 N.W. LeJeune Road Miami, FL 33126	(305) 443-9353	aws.org
AWWA	American Water Works Association	6666 W. Quincy Avenue Denver, CO 80235	(303) 794-7711	awwa.org
BOCA	Building Officials and Code Administrators International	(*see* ICC)		
BSI	British Standards Institution	389 Chiswick High Road London W4 4AL, UK	44(0)20-8996-9000	bsi-global.com
CAGI	Compressed Air and Gas Institute	1300 Sumner Avenue Cleveland, OH 44115-2851	(216) 241-7333	cagi.org
CSA	Canadian Standards Association International	178 Rexdale Boulevard Toronto, ON M9W 1R3, Canada	(416) 747-4044	csa.ca
	Also available from CSA America	8501 East Pleasant Valley Road Cleveland, OH 44131-5575	(800) 463-6727	csa-america.org
CTI	Cooling Technology Institute	P.O. Box 73383 Houston, TX 77273-3383	(281) 583-4087	cti.org
EJMA	Expansion Joint Manufacturers Association	25 North Broadway Tarrytown, NY 10591	(914) 332-0040	ejma.org
HEI	Heat Exchange Institute	1300 Sumner Avenue Cleveland, OH 44115-2815	(216) 241-7333	heatexchange.org
HI	Hydraulic Institute	9 Sylvan Way Parsippany, NJ 07054-3802	(973) 267-9700	pumps.org
HYDI	Hydronics Institute Division of GAMA	35 Russo Place, P.O. Box 218 Berkeley Heights, NJ 07922-0218	(908) 464-8200	gamanet.org
IAPMO	International Association of Plumbing and Mechanical Officials	5001 E. Philadelphia Street Ontario, CA 91761-2816	(909) 472-4100	iapmo.org

ORGANIZATIONS (*Continued*)

Abbrev.	Organization	Address	Telephone	http://www.
ICBO	International Conference of Building Officials	(*see* ICC)		
ICC	International Code Council	5203 Leesburg Pike, Suite 600 Falls Church, VA 22041	(703) 931-4533	iccsafe.org
IEEE	Institute of Electrical and Electronics Engineers	45 Hoes Lane Piscataway, NJ 08854-1331	(732) 981-0060	ieee.org
IESNA	Illuminating Engineering Society of North America	120 Wall Street, Floor 17 New York, NY 10005-4001	(212) 248-5000	iesna.org
IFCI	International Fire Code Institute	(*see* ICC)		
IIAR	International Institute of Ammonia Refrigeration	1110 North Glebe Road, Suite 250 Arlington, VA 22201	(703) 312-4200	iiar.org
ISA	The Instrumentation, Systems, and Automation Society	67 Alexander Drive, P.O. Box 12777 Research Triangle Park, NC 27709	(919) 549-8411	isa.org
ISO	International Organization for Standardization	1, rue de Varembé, Case postale 56 CH-1211 Geneva 20, Switzerland	41-22-749-0111	iso.org
MCAA	Mechanical Contractors Association of America	1385 Piccard Drive Rockville, MD 20850	(301) 869-5800	mcaa.org
MICA	Midwest Insulation Contractors Association	2017 South 139th Circle Omaha, NE 68144-2149	(800) 747-6422	micainsulation.org
MSS	Manufacturers Standardization Society of the Valve and Fittings Industry	127 Park Street, N.E. Vienna, VA 22180-4602	(703) 281-6613	mss-hq.com
NAIMA	North American Insulation Manufacturers Association	44 Canal Center Plaza, Suite 310 Alexandria, VA 22314	(703) 684-0084	naima.org
NCPWB	National Certified Pipe Welding Bureau	1385 Piccard Drive Rockville, MD 20850-4340	(301) 869-5800	mcaa.org/ncpwb
NCSBCS	National Conference of States on Building Codes and Standards	505 Huntmar Park Drive, Suite 210 Herndon, VA 20170	(703) 437-0100	ncsbcs.org
NEBB	National Environmental Balancing Bureau	8575 Grovemont Circle Gaithersburg, MD 20877	(301) 977-3698	nebb.org
NEMA	National Electrical Manufacturers Association	1300 North 17th Street, Suite 1752 Rosslyn, VA 22209	(703) 841-3200	nema.org
NFPA	National Fire Protection Association	1 Batterymarch Park Quincy, MA 02169-7471	(617) 770-3000	nfpa.org
NRCC	National Research Council of Canada, Institute for Research in Construction	1200 Montreal Road Ottawa, ON K1A 0R6, Canada	(877) 672-2672	nrc-cnrc.ca
NSF	NSF International	P.O. Box 130140, 789 N. Dixboro Road Ann Arbor, MI 48113-0140	(734) 769-8010	nsf.org
PHCC	Plumbing-Heating-Cooling Contractors National Association	180 S. Washington Street, P.O. Box 6808 Falls Church, VA 22040	(703) 237-8100	phccweb.org
SAE	Society of Automotive Engineers International	400 Commonwealth Drive Warrendale, PA 15096-0001	(724) 776-4841	sae.org
SBCCI	Southern Building Code Congress International	(*see* ICC)		
SMACNA	Sheet Metal and Air Conditioning Contractors' National Association	4201 Lafayette Center Drive Chantilly, VA 20151-1209	(703) 803-2980	smacna.org
TEMA	Tubular Exchanger Manufacturers Association	25 North Broadway Tarrytown, NY 10591	(914) 332-0040	tema.org
UL	Underwriters Laboratories	333 Pfingsten Road Northbrook, IL 60062-2096	(847) 272-8800	ul.com

ASHRAE HANDBOOK

Additions and Corrections

This report includes additional information, and technical errors found between June 15, 2003, and April 21, 2006, in the inch-pound (I-P) editions of the 2003, 2004, and 2005 *ASHRAE Handbook* volumes. Occasional typographical errors and nonstandard symbol labels will be corrected in future volumes. The most current list of Handbook additions and corrections is on the ASHRAE Web site (www.ashrae.org).

The authors and editor encourage you to notify them if you find other technical errors. Please send corrections to: Handbook Editor, ASHRAE, 1791 Tullie Circle NE, Atlanta, GA 30329, or e-mail mowen@ashrae.org.

2003 HVAC Applications

p. 1.4, Evaporative Cooling. The reference to Chapter 50 should be to Chapter 51.

p. 8.2, Indoor Air Quality, 4th line. Delete "by a minimum of 3°F."

p. 13.6. Reverse the order of Figures 6 and 7.

p. 16.13, Temperature and Humidity, 3rd paragraph. If full-coverage smocks are not used, temperature set points can be higher, not lower.

Ch. 17. This chapter has been reassigned to TC 9.9, which recommends using ASHRAE's *Thermal Guidelines for Data Processing Environments* (2004) as a primary source of information for data processing environment design.

p. 22.9, Recommendations for Laying Houses with Birds in Cages. The space recommendation should be 50 to 65 in^2.

p. 23.7, Fig. 8. The bottom part of the figure was cut off; please refer to the full version here.

p. 27.5, 2nd col. In the bulleted list, the realistic water loading in a cooling tower should be 6 to 18 gpm per square foot.

p. 27.6. Step 6 should read, "The enthalpy of air h_{ai} at 83°F saturated and 15.226 psia is 45.95 Btu/lb. Σ_{ai} = 45.95 – (1)(0.0237)(83) = 43.98 Btu/lb.

p. 27.10, 2nd col, 5th full paragraph. Change "calorific" to "heat."

p. 27.11, Planning the Circuit section. In definitions for Equation (19), delete subscript t from R. Units for R should be in. of water·min/ft^6. Units for d and standard air density should be lb$_m$/ft^3. In definitions for Equation (20), units for k and the conversion factor should contain lb$_f$.

p. 27.12, Example 9 solution. For series operation, the solution should be 6.0×10^{-10} in. of water·min^2/ft^6. For parallel, the solution should be 1.92×10^{-11} in. of water·min^2/ft^6.

p. 31.5, Type II Hoods, Condensate hood list item. The second and third sentences should be, "The hood is designed to direct the condensate toward a perimeter gutter for collection and drainage, allowing none to drip onto the appliance below. Flow rates are typically based on 50 to 75 cfm per square foot of hood opening."

p. 31.11, Traditional Registers. Terminal velocities should be kept less than or equal to 50 fpm.

p. 32.4, 2nd col., last paragraph. In the second sentence, velocity should be limited to a maximum of 0.1 fps.

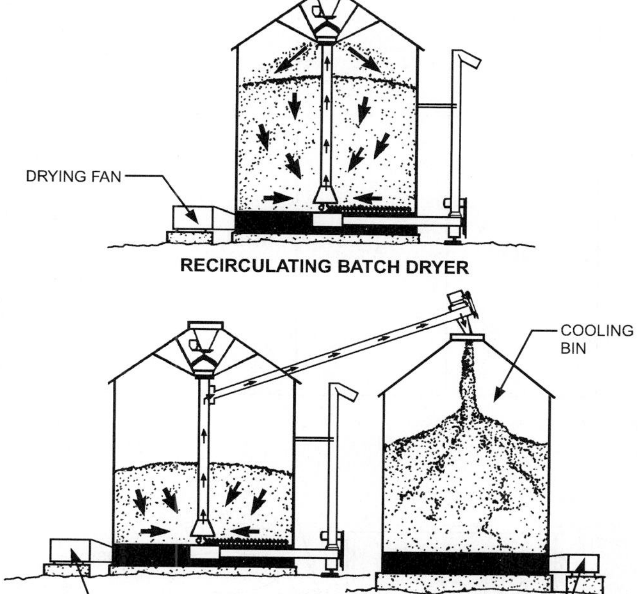

Fig. 8 Grain Recirculators Convert Bin Dryer to High-Speed Continuous-Flow Dryer
(2003 HVAC Applications, Chapter 23, p. 7)

p. 32.6, 2nd col., 4th full paragraph. Change "flashing" to "flushing." The units for free residual chlorine content should be ppm.

p. 32.17, Eqs. (5a) to (5c). Replace R_{gA}, R_{gB}, and R_{gC} with R_{ga}, R_{gm}, and R_{gd}.

p. 32.19, Fig. 18. In Nevada, change "15" to "59."

p. 36.3, Table 3. Add the following footnote:

3. For updated information on heat pump life, see Lovvorn and Hiller (2002).

p. 36.9, 1st col., 3rd line. The subscript for PWF should be "*ser.*"

p. 36.13, References. Add the following source after Lovvorn and Hiller (1985):

Lovvorn, N.C. and C.C. Hiller. 2002. Heat pump life revisited. *ASHRAE Transactions* 108(2):107-112.

p. 45.6, 1st col., end of last paragraph. The correct Research Project is RP-1134 (Tompkins 2002).

pp. 45.12-45.13, References section. Delete the Idem 2002 source, and insert the following after Sterling and Kabayashi 1981:

Tompkins, D. 2002. Evaluation of photocatalytic air cleaning capability. ASHRAE *Research Project* RP-1134.

p. 47.39, Equipment Vibration, 2nd paragraph. The second sentence should read, "Isolation efficiency is the percentage of vibratory force *not* transmitted to the support structure."

p. 47.50, Resources. The URL for the Institute of Noise Control Engineers (INCE) should be www.inceusa.org.

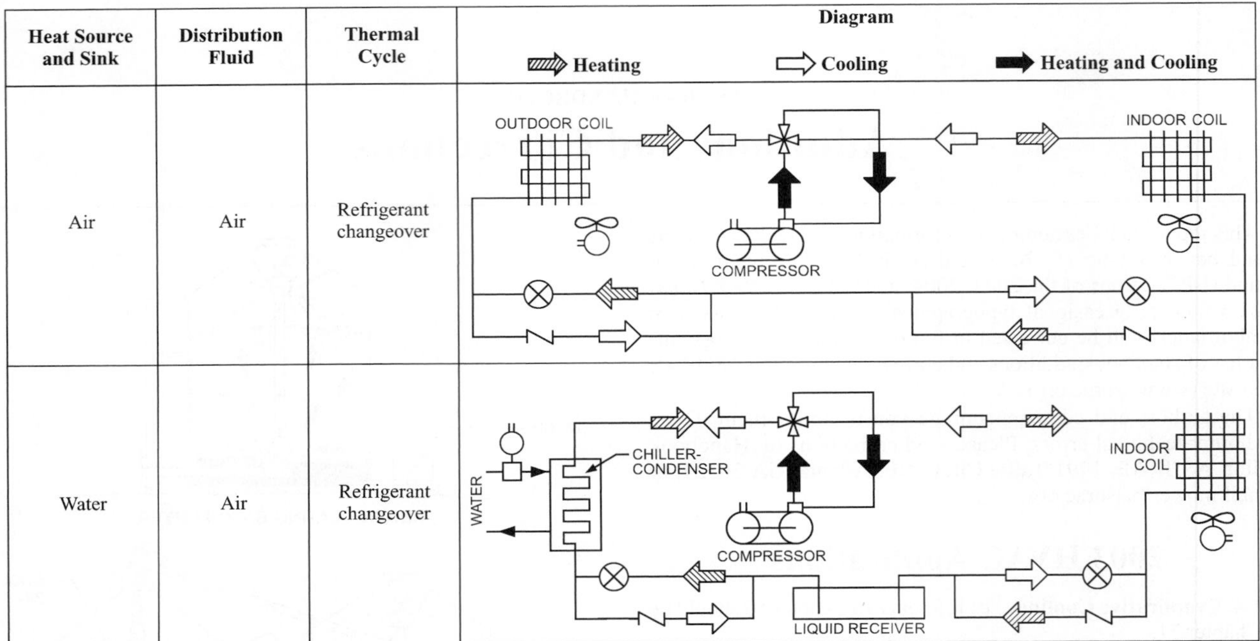

Heat Source and Sink
Distribution Fluid
Thermal Cycle
Diagram
Heating
Cooling
Heating and Cooling
Air
Air
Refrigerant changeover
OUTDOOR COIL
INDOOR COIL
COMPRESSOR
Water
Air
Refrigerant changeover
WATER
CHILLER-CONDENSER
INDOOR COIL
COMPRESSOR
LIQUID RECEIVER

Fig. 5 Heat Pump Types (first two rows)
(2004 HVAC Systems and Equipment, Ch. 8, p. 5)

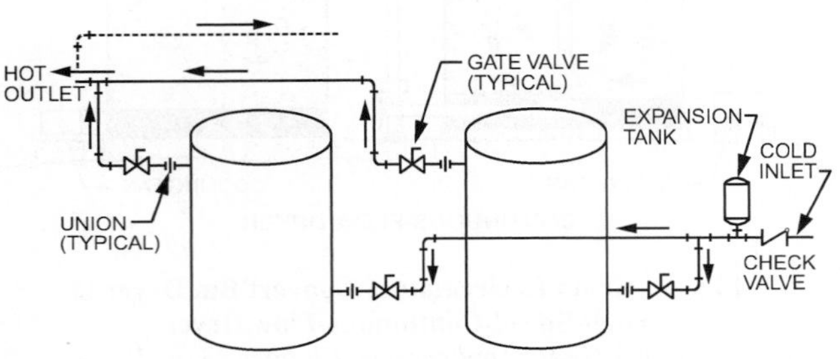

Fig. 7 Reverse/Return Manifold Systems
(2003 HVAC Applications, Ch. 49, p. 6)

p. 49.6, Fig. 7. The arrow for "Union (typical)" should point to the union symbol, not the gate valve symbol. The correct figure is presented here.

p. 50.1, Snow-Melting Heat Flux Requirement. The fourth list item should be "wind speed near the heated surface."

p. 50.24, Eq. (24). K_1 and K_2 should be k_1 and k_2.

p. 53.10, Symbols. The units for σ should be Btu/h·ft^2·R^4.

p. I.4, Index. Add "F15.12" to the entries for BAS and Building automation system.

2004 HVAC Systems and Equipment

p. 2.5, Fig. 8. The caption for Figure 9 should read "Chemical Dehumidification."

p. 8.5, Fig. 5. The top two figures, for air/air and water/air refrigerant changeover, were cut off and are provided here.

p. 12.9, Fig. 16. The figure for reverse-return two-pipe systems was incorrect; the corrected version is provided here.

p. 20.4. Under Additional Moisture Losses, the reference should be to Chapter 12, not Chapter 11.

p. 25.4, Eq. (5). There should be an equals symbol after D_{pc}.

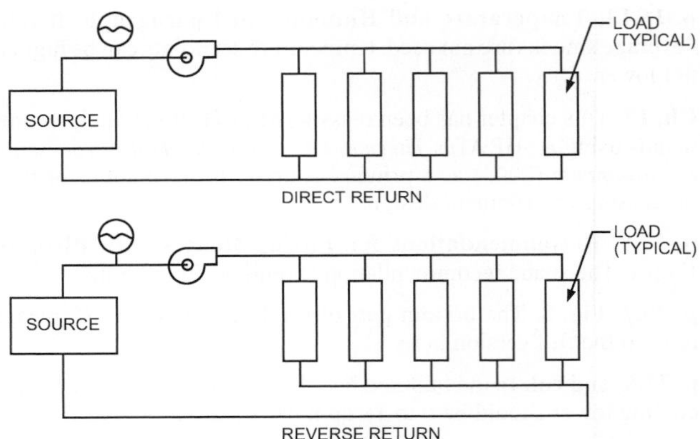

Fig. 16 Direct- and Reverse-Return Two-Pipe Systems
(2004 HVAC Systems and Equipment, Ch. 12, p. 9)

p. 35.19, 1st col., 4th line from bottom. Change "Chapters 35 and 38 of this volume" to "Chapter 38 of this volume."

p. 36.3, Fig. 4. The figure was partially cut off; it is supplied in complete form here.

p. 44.3-44.4, Example 1, Solution. Correct the following values: $h_1 = 30.6$, $h_3 = 28.15$, and $w_3 = 0.0093$. Thus, $w_2 = 0.0082$ lb/lb, $w_4 = 0.0082$ lb/lb, $q_L = -799$ Btu/min, and $q = 785$ Btu/min. In the last paragraph in the example, change 0.00795 to 0.0082, change 21.9 to 29.4, and change "to be" to "is."

p. 44.10, Fig. 5. Replace with the correct figure, provided here.

p. 44.19, Example 4, Step 7. The reference to Figure 17 should be to Figure 19.

p. 44.19, Example 5. Correct the following values: $h_3 = 23.7$ and $w_3 = 0.0052$. Thus, $h_4 = 13.43$ Btu/lb, $t_4 = 35.7$°F, $w_4 = 0.0044$ lb/lb, $m_w = 38.4$ lb/h, and $q_e = 505{,}900$ Btu/h.

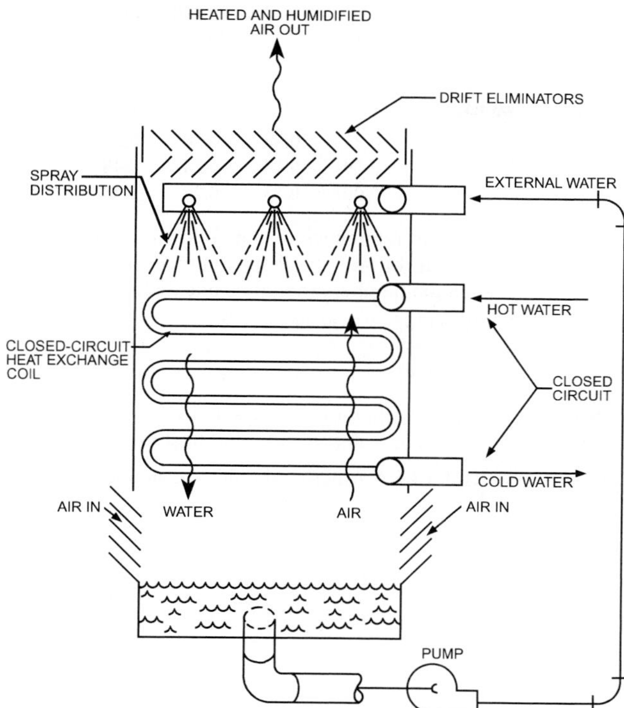

Fig. 4 Indirect-Contact or Closed-Circuit Evaporative Cooling Tower
(2004 HVAC Systems and Equipment, Chapter 36, p. 3)

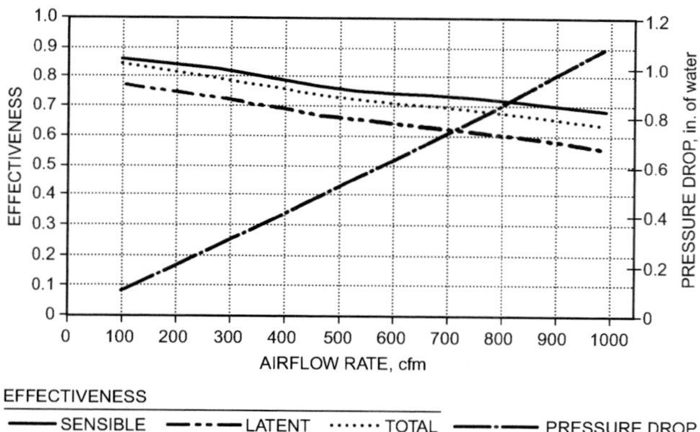

Fig. 5 Variation of Pressure Drop and Effectiveness with Air Flow Rates for a Membrane Plate Exchanger
(2004 HVAC Systems and Equipment, Ch. 44, p. 10)

pp. 44.19-44.20, Example 6. Correct the following values: h_1 = 44.6 Btu/lb, w_1 = 0.0198 lb/lb, h_3 = 28.5 Btu/lb, and w_3 = 0.0096 lb/lb. Thus, q_{max} (total) = 566,000 Btu/h, q_t = 321,000, q_{lat} = 203,000 Btu/h, h_2 = 35.5 Btu/lb, t_{w2} = 71.6°F, t_4 = 86.2°F, h_4 = 35.4 Btu/lb, w_4 = 0.0134, and t_{w4} = 71.6°F. In step 6, second equation, q_t = 319,000 Btu/h.

pp. 44.20-44.21, Example 7, Solution. Correct the following values: h_1 = 44.6 Btu/lb, w_1 = 0.0198 lb/lb, h_3 = 28.5 Btu/lb, and w_3 = 0.0096 lb/lb. Thus, h_2 = 33.16 Btu/lb, w_2 = 0.0129, t_{w2} = 69.2°F, h_4 = 39.5 Btu/lb, and t_{w4} = 76.1°F. In step 6, second equation, q_t = 41,460 Btu/h.

pp. 44.22-44.23, Example 9, Solution. In step 5a, entering enthalpy should be 44.6 Btu/lb. Thus, h_s = 38.6 Btu/lb; two lines below this equation, change the wet-bulb temperature for saturated air to 75°F. In step 5b, entering enthalpy is 21.3 Btu/lb; thus, h_4 = 27.3 Btu/lb.

p. 44.25, Bibliography. The correct entry for the ASHRAE (1974) source is as follows:

ASHRAE. 1974. Symposium on heat recovery. *ASHRAE Transactions* 80(1):302-332.

Index. The following chapter references are correct for these entries:

2005 Fundamentals

p. 1.17, Eq. (63). $\dot{Q}_{evap}$ should be $\dot{Q}_{cond}$.

p. 1.20, Symbols. Units for V should be ft/s.

p. 2.7, Table 2. Values for ε (right column) should be 60, 1800, 6000, and 10,200 μin.

p. 2.11, definitions for Eq. (38). In the definition for Δh, there should be parentheses around $p_1 - p_2$.

p. 3.2, Thermal Conduction, 2nd line from bottom. Change "steady" to "steady-state."

p. 3.5, Eq. (10). The equation should be as follows:

$$c_1 = \frac{2\,\text{Bi}}{(\mu_1^2 + \text{Bi}^2)\,J_0(\mu_1)}$$

p. 3.13, 1st col. Delete first repeated paragraph after Equation (30).

p. 3.21, 2nd col., last full sentence. Change to, "Depending on frequency and amplitude of vibration, forced convection from a wire to air is enhanced by up to 300% (Nesis et al. 1994)."

p. 3.28, Eq. (44). Delete second equals sign and second fraction.

p. 7.20, 1st col., last full paragraph. Change third sentence to read, "the resonance frequency of the system is maintained at 3.13 Hz, and the force transmitted to the structure remains at 12.5 lb_f." Change sixth sentence to read, "where a is acceleration, the maximum dynamic displacement of the mounted equipment is reduced by a factor of (M_1/M_2), where M_1 and M_2 are the masses before and after mass is added, respectively."

p. 15.3, Eq. (3) and following text. Change K_a to K_d in the equation, definitions, and following paragraph (three places total).

p. 16.11, Symbols. Add the following definitions:

h_s = exhaust stack height (typically above roof unless otherwise specified), ft (see Figure 3, and Chapter 44 in the 2003 *ASHRAE Handbook—HVAC Applications*)

S = stretched-string distance; shortest distance from exhaust to intake over obstacles and along building surface, ft [see Figure 3, and Equation (22) in the 2003 *ASHRAE Handbook—HVAC Applications*)

p. 23.6, Fig. 1. Change caption from "Adsorption Isotherms" to "Typical Adsorption."

p. 27.14, Figure 9. Air leakage should be at 0.2 in. of water.

p. 28.2, Table 1 (and all data tables), cols. 13a, c, and e. Because of a data processing error, the enthalpy values in these columns are systematically low by 7.687 Btu/lb. Thus, all enthalpy values in Table 1 and in all design climatic condition tables on the accompanying CD-ROM should be increased by that amount.

p. 29.8, Table 7. Units for OF_b should be °F.

p. 29.9, Table 9. The correct numbers for the last line of the table are as follows:

E_t	326	325	321	314	305	293	279	262	243

p. 29.9, Eq. (23). Units for CF_{slab} should be Btu/h·ft^2; 0.51 is a constant with units of Btu/h·ft^2; and 2.5 is a factor with units of °F.

p. 30.2, 2nd col., 6th paragraph, last line. Change "with" to "without."

p. 30.4, bottom of 2nd col. The reference to Equation (3) should be to Equation (4).

p. 30.27, Table 19. Footnote 7 should refer to Table 3 in Chapter 39.

p. 31.6, Table 2. In the footnote for Winter Conditions, add the following: $h_i = h_{ic} + h_{iR} = 0.30(\Delta T/L)^{0.25} + \varepsilon\Gamma(T_i^4 - T_g^4)/\Delta T$, where $\Delta T = T_i - T_g$, °R; L = glazing height, ft; T_g = glass temperature, °R.

p. 31.26, Table 13. For ID 1b, change the following values:

T	0.77	0.75	0.73	0.68	0.58	0.35	0.69
R^f	0.07	0.08	0.09	0.13	0.24	0.48	0.13
R^b	0.07	0.08	0.09	0.13	0.24	0.48	0.13

p. 32.5, 1st col. Cross references to the following equations in Chapter 30 should be as follows (the chapter number should stay the same; only the equation numbers should be updated):

Equation (36)	Equation (27)
Equation (35)	Equation (26)
Equation (34)	Equation (25)

p. 38.1, 1st col. The definition of an acre should be 43,560 ft^2. The conversion factor for ft of water to Pa should be 2989.

p. 40.26. The URL for CSA America should be csa-america.org.

COMPOSITE INDEX
ASHRAE HANDBOOK SERIES

This index covers the current Handbook series published by ASHRAE. The four volumes in the series are identified as follows:

A = 2003 HVAC Applications

S = 2004 HVAC Systems and Equipment

F = 2005 Fundamentals

R = 2006 Refrigeration

Alphabetization of the index is letter by letter; for example, **Heaters** precedes **Heat exchangers**, and **Floors** precedes **Floor slabs**.

The page reference for an index entry includes the book letter and the chapter number, which may be followed by a decimal point and the beginning page in the chapter. For example, the page number S31.4 means the information may be found in the 2004 HVAC Systems and Equipment volume, Chapter 31, beginning on page 4.

Each Handbook volume is revised and updated on a four-year cycle. Because technology and the interests of ASHRAE members change, some topics are not included in the current Handbook series but may be found in the earlier Handbook editions cited in the index.

COMMENT PAGE

ASHRAE publications strive to present the most current and useful information possible. If you would like to comment on chapters in this or any volume of the *ASHRAE Handbook*, please use one of the following methods:

- Fill out the comment form on the ASHRAE Web site (www.ashrae.org)
- E-mail the editor at mowen@ashrae.org
- Cut out this page and fax it to the editor at 678-539-2187, or mail it to

Handbook Editor
ASHRAE
1791 Tullie Circle
Atlanta, GA 30329 USA

Please provide your contact information if you would like a response. (Personal identification information will not be used for any purpose beyond responding to your comments.)

Name:

E-mail:

Address:

Phone:

Fax:

Preferred Contact Method(s):

Comment Page

ASHRAE publications strive to present the most current and useful information possible. If you would like to comment on chapters in this or any volume of the *ASHRAE Handbook*, please use one of the following methods:

- Fill out the comment form on the ASHRAE Web site (www.ashrae.org)
- E-mail the editor at mowen@ashrae.org
- Cut out this page and fax it to the editor at 678-539-2187, or mail it to

Handbook Editor
ASHRAE
1791 Tullie Circle
Atlanta, GA 30329 USA

Please provide your contact information if you would like a response. (Personal identification information will not be used for any purpose beyond responding to your comments.)

Name:

E-mail:

Address:

Phone:

Fax:

Preferred Contact Method(s):

COMMENT PAGE

ASHRAE publications strive to present the most current and useful information possible. If you would like to comment on chapters in this or any volume of the *ASHRAE Handbook*, please use one of the following methods:

- Fill out the comment form on the ASHRAE Web site (www.ashrae.org)
- E-mail the editor at mowen@ashrae.org
- Cut out this page and fax it to the editor at 678-539-2187, or mail it to

Handbook Editor
ASHRAE
1791 Tullie Circle
Atlanta, GA 30329 USA

Please provide your contact information if you would like a response. (Personal identification information will not be used for any purpose beyond responding to your comments.)

Name: ____________

E-mail: ____________

Address: ____________

Phone: ____________

Fax: ____________

Preferred Contact Method(s): ____________

COMMENT PAGE

ASHRAE publications strive to present the most current and useful information possible. If you would like to comment on chapters in this or any volume of the *ASHRAE Handbook*, please use one of the following methods:

- Fill out the comment form on the ASHRAE Web site (www.ashrae.org)
- E-mail the editor at mowen@ashrae.org
- Cut out this page and fax it to the editor at 678-539-2187, or mail it to

Handbook Editor
ASHRAE
1791 Tullie Circle
Atlanta, GA 30329 USA

Please provide your contact information if you would like a response. (Personal identification information will not be used for any purpose beyond responding to your comments.)

Name:

E-mail:

Address:

Phone:

Fax:

Preferred Contact Method(s):

Comment Page

ASHRAE publications strive to present the most current and useful information possible. If you would like to comment on chapters in this or any volume of the *ASHRAE Handbook*, please use one of the following methods:

- Fill out the comment form on the ASHRAE Web site (www.ashrae.org)
- E-mail the editor at mowen@ashrae.org
- Cut out this page and fax it to the editor at 678-539-2187, or mail it to

Handbook Editor
ASHRAE
1791 Tullie Circle
Atlanta, GA 30329 USA

Please provide your contact information if you would like a response. (Personal identification information will not be used for any purpose beyond responding to your comments.)

Name:

E-mail:

Address:

Phone:

Fax:

Preferred Contact Method(s):

Comment Page

ASHRAE publications strive to present the most current and useful information possible. If you would like to comment on chapters in this or any volume of the *ASHRAE Handbook*, please use one of the following methods:

- Fill out the comment form on the ASHRAE Web site (www.ashrae.org)
- E-mail the editor at mowen@ashrae.org
- Cut out this page and fax it to the editor at 678-539-2187, or mail it to

  Handbook Editor
  ASHRAE
  1791 Tullie Circle
  Atlanta, GA 30329 USA

Please provide your contact information if you would like a response. (Personal identification information will not be used for any purpose beyond responding to your comments.)

Name: ______

E-mail: ______

Address: ______

Phone: ______

Fax: ______

Preferred Contact Method(s): ______

COMMENT PAGE

ASHRAE publications strive to present the most current and useful information possible. If you would like to comment on chapters in this or any volume of the *ASHRAE Handbook*, please use one of the following methods:

- Fill out the comment form on the ASHRAE Web site (www.ashrae.org)
- E-mail the editor at mowen@ashrae.org
- Cut out this page and fax it to the editor at 678-539-2187, or mail it to

Handbook Editor
ASHRAE
1791 Tullie Circle
Atlanta, GA 30329 USA

Please provide your contact information if you would like a response. (Personal identification information will not be used for any purpose beyond responding to your comments.)

Name: ______________________ **Phone:** ______________________

E-mail: ______________________ **Fax:** ______________________

Address: ______________________ **Preferred Contact Method(s):** ______________________

Comment Page

ASHRAE publications strive to present the most current and useful information possible. If you would like to comment on chapters in this or any volume of the *ASHRAE Handbook*, please use one of the following methods:

- Fill out the comment form on the ASHRAE Web site (www.ashrae.org)
- E-mail the editor at mowen@ashrae.org
- Cut out this page and fax it to the editor at 678-539-2187, or mail it to

Handbook Editor
ASHRAE
1791 Tullie Circle
Atlanta, GA 30329 USA

Please provide your contact information if you would like a response. (Personal identification information will not be used for any purpose beyond responding to your comments.)

Name: ______________________ **Phone:** ______________________

E-mail: ______________________ **Fax:** ______________________

Address: ______________________ **Preferred Contact Method(s):** ______________________